CALCULUS

LAURA TAALMAN
James Madison University

PETER KOHN
James Madison University

W. H. Freeman and Company
New York

Senior Publisher: Ruth Baruth
Executive Editor: Terri Ward
Marketing Manager: Steve Thomas
Market Development Manager: Steven Rigolosi
Developmental Editors: Leslie Lahr, Katrina Wilhelm
Senior Media Editor: Laura Judge
Associate Editor: Jorge Amaral
Editorial Assistant: Liam Ferguson
Photo Editor: Ted Szczepanski
Cover Photo Researcher: Elyse Rieder
Cover Designer: Vicki Tomaselli
Text Designer: Marsha Cohen
Illustrations: Network Graphics
Illustration Coordinator: Bill Page
Production Coordinator: Susan Wein
Project Management and Composition: Aptara
Printing and Binding: RR Donnelley

Library of Congress Control Number: 2012947365

Complimentary Copy:
ISBN-13: 978-1-4641-2963-6
ISBN-10: 1-4641-2963-0

Student Edition Hardcover:
ISBN-13: 978-1-4292-4186-1
ISBN-10: 1-4292-4186-1

Student Edition Paperback:
ISBN-13: 978-1-4641-5108-8
ISBN-10: 1-4641-5108-3

Student Edition Loose-leaf:
ISBN-13: 978-1-4641-4005-1
ISBN-10: 1-4641-4005-7

Printed in the United States of America

First printing

W. H. Freeman and Company
41 Madison Avenue
New York, NY 10010
Houndmills, Basingstoke RG21 6XS, England
www.whfreeman.com

DEDICATION

To Leibniz and Newton

—Laura Taalman

To Newton and Leibniz

—Peter Kohn

CONTENTS

Contents iv
About the Authors vi
Preface vii
Media and Supplements ix
Features xi
Acknowledgments xiv
To the Student xvi

*Starred titles indicate optional material

I DIFFERENTIAL CALCULUS

Chapter 0
Functions and Precalculus* / 1

0.1 Functions and Graphs 2
0.2 Operations, Transformations, and Inverses 19
0.3 Algebraic Functions 35
0.4 Exponential and Trigonometric Functions 47
0.5 Logic and Mathematical Thinking* 63
Chapter Review, Self-Test, and Capstones 73

Chapter 1
Limits / 77

1.1 An Intuitive Introduction to Limits 78
1.2 Formal Definition of Limit 90
1.3 Delta–Epsilon Proofs* 100
1.4 Continuity and Its Consequences 109
1.5 Limit Rules and Calculating Basic Limits 123
1.6 Infinite Limits and Indeterminate Forms 138
Chapter Review, Self-Test, and Capstones 152

Chapter 2
Derivatives / 155

2.1 An Intuitive Introduction to Derivatives 156
2.2 Formal Definition of the Derivative 169
2.3 Rules for Calculating Basic Derivatives 187
2.4 The Chain Rule and Implicit Differentiation 201
2.5 Derivatives of Exponential and Logarithmic Functions 212
2.6 Derivatives of Trigonometric and Hyperbolic Functions 224
Chapter Review, Self-Test, and Capstones 236

Chapter 3
Applications of the Derivative / 239

3.1 The Mean Value Theorem 240
3.2 The First Derivative and Curve Sketching 250
3.3 The Second Derivative and Curve Sketching 264
3.4 Optimization 278
3.5 Related Rates 291
3.6 L'Hôpital's Rule 302
Chapter Review, Self-Test, and Capstones 313

II INTEGRAL CALCULUS

Chapter 4
Definite Integrals / 315

4.1 Addition and Accumulation 317
4.2 Riemann Sums 328
4.3 Definite Integrals 342
4.4 Indefinite Integrals 354
4.5 The Fundamental Theorem of Calculus 364
4.6 Areas and Average Values 375
4.7 Functions Defined by Integrals 388
Chapter Review, Self-Test, and Capstones 402

Chapter 5
Techniques of Integration / 407

5.1 Integration by Substitution 408
5.2 Integration by Parts 420
5.3 Partial Fractions and Other Algebraic Techniques 432
5.4 Trigonometric Integrals 444
5.5 Trigonometric Substitution 454
5.6 Improper Integrals 467
5.7 Numerical Integration* 480
Chapter Review, Self-Test, and Capstones 494

Chapter 6
Applications of Integration / 499

6.1 Volumes by Slicing 500
6.2 Volumes by Shells 514
6.3 Arc Length and Surface Area 526
6.4 Real-World Applications of Integration 542
6.5 Differential Equations* 559
Chapter Review, Self-Test, and Capstones 573

III SEQUENCES AND SERIES

Chapter 7
Sequences and Series / 577

7.1 Sequences 579
7.2 Limits of Sequences 594
7.3 Series 606
7.4 Introduction to Convergence Tests 617
7.5 Comparison Tests 626
7.6 The Ratio and Root Tests 633
7.7 Alternating Series 641
Chapter Review, Self-Test, and Capstones 655

Chapter 8
Power Series / 659

8.1 Power Series 660
8.2 Maclaurin Series and Taylor Series 672
8.3 Convergence of Power Series 681
8.4 Differentiating and Integrating Power Series 694
Chapter Review, Self-Test, and Capstones 703

IV VECTOR CALCULUS

Chapter 9
Parametric Equations, Polar Coordinates, and Conic Sections / 707

9.1 Parametric Equations 708
9.2 Polar Coordinates 724
9.3 Graphing Polar Equations 733
9.4 Computing Arc Length and Area with Polar Functions 749
9.5 Conic Sections* 758
Chapter Review, Self-Test, and Capstones 774

Chapter 10
Vectors / 777

10.1 Cartesian Coordinates 778
10.2 Vectors 792
10.3 Dot Product 803
10.4 Cross Product 814
10.5 Lines in Three-Dimensional Space 827
10.6 Planes 836
Chapter Review, Self-Test, and Capstones 847

Chapter 11
Vector Functions / 851

11.1 Vector-valued Functions 852
11.2 The Calculus of Vector Functions 862
11.3 Unit Tangent and Unit Normal Vectors 874
11.4 Arc Length Parametrizations and Curvature 881
11.5 Motion 891
Chapter Review, Self-Test, and Capstones 900

V MULTIVARIABLE CALCULUS

Chapter 12
Multivariable Functions / 903

12.1 Functions of Two and Three Variables 904
12.2 Open Sets, Closed Sets, Limits, and Continuity 919
12.3 Partial Derivatives 933
12.4 Directional Derivatives and Differentiability 946
12.5 The Chain Rule and the Gradient 955
12.6 Extreme Values 966
12.7 Lagrange Multipliers 977
Chapter Review, Self-Test, and Capstones 987

Chapter 13
Double and Triple Integrals / 991

13.1 Double Integrals over Rectangular Regions 992
13.2 Double Integrals over General Regions 1006
13.3 Double Integrals using Polar Coordinates 1017
13.4 Applications of Double Integrals 1029
13.5 Triple Integrals 1041
13.6 Integration using Cylindrical and Spherical Coordinates 1058
13.7 Jacobians and Change of Variables 1069
Chapter Review, Self-Test, and Capstones 1081

Chapter 14
Vector Analysis / 1085

14.1 Vector Fields 1086
14.2 Line Integrals 1097
14.3 Surfaces and Surface Integrals 1109
14.4 Green's Theorem 1122
14.5 Stokes' Theorem 1134
14.6 The Divergence Theorem 1143
Chapter Review, Self-Test, and Capstones 1152

Answers to Odd-Numbered Problems A-1
Index I-1

ABOUT THE AUTHORS

Laura Taalman and Peter Kohn are professors of mathematics at James Madison University, where they have taught calculus for a combined total of over 30 years.

Laura Taalman received her undergraduate degree from the University of Chicago and master's and Ph.D. degrees in mathematics from Duke University. Her research includes singular algebraic geometry, knot theory, and the mathematics of games and puzzles. She is a recipient of both the Alder Award and the Trevor Evans award from the Mathematical Association of America, and the author of five books on Sudoku and the mathematics of Sudoku. In her spare time, she enjoys being a geek.

Peter Kohn received his undergraduate degree from Antioch College, a master's degree from San Francisco State University, and a Ph.D. in mathematics from the University of Texas at Austin. His main areas of research are low-dimensional topology and knot theory. He has been a national judge for MathCounts since 2001. In his spare time, he enjoys hiking and riding his bicycle in the beautiful Shenandoah Valley.

Calculus books have become full of clutter, distracting margin notes, and unneeded features. This calculus book clears out that clutter so that students can focus on the important ideas of calculus. Our goal was to create a clean, streamlined calculus book that is accessible and readable for students while still upholding the standards required in science, mathematics, and engineering programs, and that is flexible enough to accommodate different teaching and learning styles.

Linear Flow with Clean Margins

One thing that is distinctive about this calculus book is that it follows a linear writing style. Figures and equations flow with the text as part of a clear, structured exposition instead of being scattered about in the margins. We feel that this approach greatly increases the clarity of the book and encourages focused reading.

Exposition Before Calculation

Another distinctive feature of this book is that in each section we have separated the exposition and illustrative examples from the longer, more complicated calculational examples. Including these longer examples separately from the exposition increases flexibility: Students who want to read and understand the development of the material can do so without being bogged down or distracted by large examples, while students who want to use the book as a reference for looking up examples that are similar to homework problems can also do that.

Examples to Learn From

Within the exposition of each section are short examples that quickly illustrate the concepts being developed. Following the exposition is a set of detailed, in-depth examples that explore both calculations and concepts. We took great pains to provide many steps and illustrations in each example in order to aid the student, including details about how to get started on a problem and choose an appropriate solution method. One of the elements of the book that we are most proud of is the "Checking the Answer" feature, which we have included after selected examples to encourage students to learn how to check their own answers.

Building Mathematics

We were very careful in this book to approach mathematics as a discipline that is developed logically, theorem by theorem. Whenever possible, theorems are followed by proofs that are written to be understood by students. We have included these proofs because they are part of the logical development of the material, but we have clearly labeled and indented each proof to indicate that it can be covered or skipped, according to instructor preference. Each exercise set contains an optional subsection of proofs, many of which are accessible even to beginning students. In addition, we have emphasized the interconnections among topics by providing "Thinking Back" and "Thinking Forward" exercises in each section and "Capstone" problems at the end of each chapter.

Consistency and Reliability

Another improvement in this book is that it has a consistent and predictable structure. For example, instructors can rely on every section concluding with a "Test Your Understanding" feature which includes five questions that students can use to self-test and that instructors can choose to use as pre-class questions. The exercises are always consistently split into subsections of different types of problems: "Thinking Back," "Concepts," "Skills," "Applications," "Proofs," and "Thinking Forward." In addition, the "Concepts" subsection always begins with a summary exercise, eight true/false questions, and three example construction exercises. Instructors and students alike can rely on this consistent structure when assigning exercises and choosing a path of study.

Flexibility

We recognize that instructors use calculus books in many different ways and that the real direction of a calculus course comes from the instructor, not any book. The streamlined, consistent structure of this book makes it easy to use with a wide variety of courses and pedagogical styles. In particular, instructors will find it easy to include or omit sections, proofs, examples, and exercises consistently according to their preferences and course requirements. Students can focus on mathematical development or on examples and calculations as they need to throughout the course. Later, they can use the book as a reliable reference.

We think it will be immediately clear to anyone opening this book that what we have written is substantially different from the other calculus books on the market today while still following the standard topics taught in most modern science, mathematics, and engineering calculus courses. Our hope is that faculty who use the book will find it flexible for different pedagogical approaches and that students will be able to read it on different levels as they learn to understand the beauty of calculus.

A Special Taalman/Kohn Option for Underprepared Calculus Students

Do some of your calculus students struggle with algebra and precalculus material? The Taalman/Kohn *Calculus* series has a ready-made option for such students, called *Calculus I with Integrated Precalculus*. This option includes all the material in Chapters 0–6 of Taalman/Kohn *Calculus*, but in a different order and with supplementary precalculus and algebra material.

- Chapters 0–3 of *Calculus I with Integrated Precalculus* cover the same development of differential calculus topics as Chapters 0–3 in Taalman/Kohn *Calculus*, but the more complicated calculational examples are deferred to later chapters.
- Chapters 4–6 of *Calculus I with Integrated Precalculus* revisit differential calculus through the lens of studying progressively more challenging types of functions. Any exercises or examples from Taalman/Kohn *Calculus* that were left out of Chapters 0–3 of *Calculus I with Integrated Precalculus* are included in Chapters 4–6. The requisite background precalculus and algebra material is built from the ground up.
- Chapters 7–9 of *Calculus I with Integrated Precalculus* are identical to Chapters 4–6 of Taalman/Kohn *Calculus* and cover all topics from integral calculus.

Students who learn Calculus I from *Calculus I with Integrated Precalculus* can continue with Calculus II using Taalman/Kohn *Calculus* or any other calculus textbook. Students who have weak algebra and precalculus skills can succeed in STEM-level calculus if given the right help along the way, and *Calculus I with Integrated Precalculus* is written specifically to address the needs of those students.

For an examination copy of *Calculus I with Integrated Precalculus*, please contact your local W. H. Freeman & Company representative.

For Instructors

Instructor's Solutions Manual

Single-variable ISBN: 1-4641-5017-6

Multivariable ISBN: 1-4641-5018-4

Contains worked-out solutions to all exercises in the text.

Test Bank

Computerized (CD-ROM), ISBN: 1-4641-2547-3

Includes multiple-choice and short-answer test items.

Instructor's Resource Manual

ISBN: 1-4641-2545-7

Provides suggested class time, key points, lecture material, discussion topics, class activities, worksheets, and group projects corresponding to each section of the text.

Instructor's Resource CD-ROM

ISBN: 1-4641-2548-1

Search and export all resources by key term or chapter. Includes text images, Instructor's Solutions Manual, Instructor's Resource Manual, and Test Bank.

For Students

Student Solutions Manual

Single-variable ISBN: 1-4641-2538-4

Multivariable ISBN: 1-4641-5019-2

Contains worked-out solutions to all odd-numbered exercises in the text.

Software Manuals

Maple™ and Mathematica® software manuals are available within CalcPortal. Printed versions of these manuals are available through custom publishing. They serve as basic introductions to popular mathematical software options and guides for their use with *Calculus*.

Book Companion Web Site at www.whfreeman.com/tkcalculus

For students, this site serves as a FREE 24–7 electronic study guide, and it includes such features as self-quizzes and interactive applets.

Online Homework Options

WebAssign Premium www.webassign.net/whfreeman

WebAssign Premium integrates the book's exercises into the world's most popular and trusted online homework system, making it easy to assign algorithmically generated homework and quizzes. Algorithmic exercises offer the instructor optional algorithmic solutions. WebAssign Premium also offers access to resources, including the new Dynamic Figures, CalcClips whiteboard videos, tutorials, and "Show My Work" feature. In addition, WebAssign Premium is available with a fully customizable e-Book option that includes links to interactive applets and projects.

calcportal www.yourcalcportal.com

CalcPortal combines a fully customizable e-Book, exceptional student and instructor resources, and a comprehensive online homework assignment center. Included are algorithmically generated exercises, as well as Precalculus diagnostic quizzes, Dynamic Figures, interactive applets, CalcClips whiteboard videos, student solutions, online quizzes, Mathematica and Maple manuals, and homework management tools, all in one affordable, easy-to-use, and fully customizable learning space.

WeBWorK webwork.maa.org

W. H. Freeman offers approximately 2,500 algorithmically generated questions (with full solutions) through this free, open-source online homework system at the University of Rochester. Adopters also have access to a shared national library test bank with thousands of additional questions, including 1,500 problem sets matched to the book's table of contents.

Additional Media

Freeman SolutionMaster

This easy-to-use Web-based version of the Instructor's Solutions Manual allows instructors to generate a solution file for any set of homework exercises. Solutions can be downloaded in PDF format for convenient printing and posting.

Interactive e-Book at ebooks.bfwpub.com/tkcalculus

The Interactive e-Book integrates a complete and customizable online version of the text with its media resources. Students can quickly search the text, and they can personalize the e-Book just as they would the print version, with highlighting, bookmarking, and note-taking features. Instructors can add, hide, and reorder content, integrate their own material, and highlight key text.

Course Management Systems

W. H. Freeman and Company provides courses for Blackboard, WebCT (Campus Edition and Vista), Angel, Desire2Learn, Moodle, and Sakai course management systems. These are completely integrated solutions that you can easily customize and adapt to meet your teaching goals and course objectives. Visit www.macmillanhighered.com/catalog/other/coursepack for more information.

i-clicker

This two-way radio frequency classroom response system was developed by educators for educators. University of Illinois physicists Tim Stelzer, Gary Gladding, Mats Selen, and Benny Brown created the i-clicker system after using competing classroom responses and discovering that they were neither appropriate for the classroom nor friendly to the student. Each step of i-clicker's development has been informed by teaching and learning. i-clicker is superior to other systems from both a pedagogical and a technical standpoint. To learn more about packaging i-clicker with this textbook, contact your local sales representative or visit **www.iclicker.com.**

Each section opens with a **list of the three main section topics**. The list provides a focus and highlights key concepts.

3.3 THE SECOND DERIVATIVE AND CURVE SKETCHING

- Using first and second derivatives to define and detect concavity
- The behavior of the first and second derivatives at inflection points
- Using the second-derivative test to determine whether critical points are maxima, minima, or neither

Definitions are clearly boxed, numbered, and labeled for easy reference. To reinforce their importance and meaning, definitions are followed by brief, often illustrated, examples.

DEFINITION 3.9 **Formally Defining Concavity**

Suppose f and f' are both differentiable on an interval I.

(a) f is ***concave up*** on I if f' is increasing on I.

(b) f is ***concave down*** on I if f' is decreasing on I.

How does this formal definition of concavity correspond with our intuitive notion of concavity? Consider the functions graphed next. On each graph four slopes are illustrated and estimated. Notice that when f is concave up, its slopes increase from left to right, and when f is concave down, its slopes decrease from left to right.

Slopes increase when f is concave up

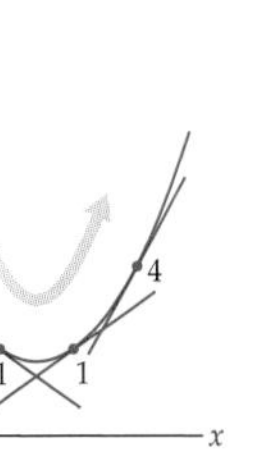

Slopes decrease when f is concave down

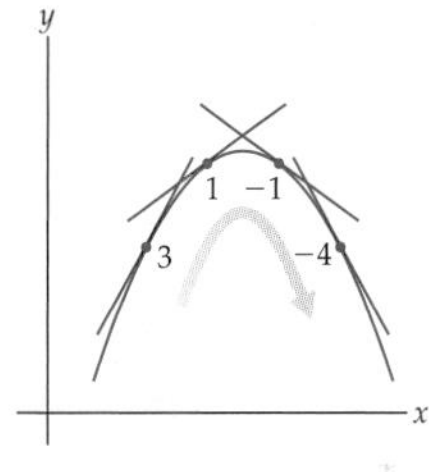

Theorems are developed intuitively before they are stated formally, and simple examples inform the discussion. **Proofs** follow most theorems, although they are optional, given instructor preference.

THEOREM 3.4 **Rolle's Theorem**

If f is continuous on $[a, b]$ and differentiable on (a, b), and if $f(a) = f(b) = 0$, then there exists at least one value $c \in (a, b)$ for which $f'(c) = 0$.

Actually, Rolle's Theorem also holds in the more general case where $f(a)$ and $f(b)$ are equal to each other (not necessarily both zero). For example, Rolle's Theorem is also true if $f(a) = f(b) = 5$, or if $f(a) = f(b) = -3$, and so on, because vertically shifting a function by adding a constant term does not change its derivative. However, the classic way to state Rolle's Theorem is with $f(a)$ and $f(b)$ both equal to zero.

Proof. Rolle's Theorem is an immediate consequence of the Extreme Value Theorem from Section 1.4 and the fact that every extremum is a critical point. Suppose f is continuous on the closed interval $[a, b]$ and differentiable on the open interval (a, b), with $f(a) = f(b) = 0$. By the Extreme Value Theorem, we know that f attains both a maximum and a minimum value on $[a, b]$. If one of these extreme values occurs at a point $x = c$ in the interior (a, b) of the interval, then $x = c$ is a local extremum of f. By the previous theorem, this means that $x = c$ is a critical point of f. Since f is assumed to be differentiable at $x = c$, it follows that $f'(c) = 0$ and we are done.

It remains to consider the special case where all of the maximum and minimum values of f on $[a, b]$ occur at the endpoints of the interval (i.e., at $x = a$ or at $x = b$). In this case, since $f(a) = f(b) = 0$, the maximum and minimum values of $f(x)$ must both equal zero. For all x in $[a, b]$ we would have $0 \le f(x) \le 0$, which means that f would have to be the constant function $f(x) = 0$ on $[a, b]$. Since the derivative of a constant function is always zero, in this special case we have $f'(x) = 0$ for *all* values of c in (a, b), and we are done. ■

Color is used consistently and pedagogically in **graphs and figures** to relate like concepts. For instance, the color used for rectangles in Riemann sum approximations is also quite purposefully used for linear approximations of arc length and rectangular solid approximations of volume.

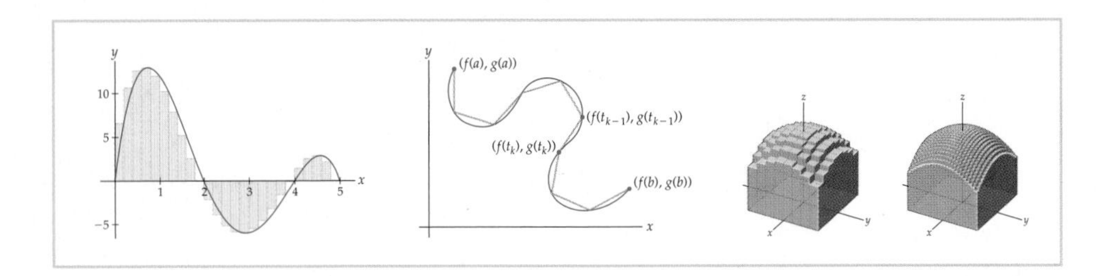

Cautions are appropriately placed at points in the exposition where students typically have questions about the nuances of mathematical thinking, processes, and notation.

It is important to note that although we use the notation x^{-1} to denote the reciprocal $\frac{1}{x}$, the notation f^{-1} does *not* stand for the reciprocal $\frac{1}{f}$ of f. The notation f^{-1} used in Definition 0.10 is pronounced "f inverse." We are now using the same notation for two very different things, but it should be clear from the context which one we mean.

Every section includes short illustrative examples as part of the discussion and development of the material. Once the groundwork has been laid, more complex **examples** and calculations are provided. Students find this approach easier to handle because the difficult calculations do not interfere with the development of why things work. Example **solutions** are explained in detail and include all the steps necessary for student comprehension.

EXAMPLE 4 Using critical points and Rolle's Theorem to find local extrema

The function $f(x) = x(x-1)(x-3)$ is a cubic polynomial with one local maximum and one local minimum. Use Rolle's Theorem to identify intervals on which these extrema exist. Then use derivatives to find the exact locations of these extrema.

SOLUTION

The roots of $f(x) = x(x-1)(x-3)$ are $x = 0$, $x = 1$, and $x = 3$. Since f is a polynomial, it is continuous and differentiable everywhere. Therefore Rolle's Theorem applies on the intervals $[0, 1]$ and $[1, 3]$, and it tells us that at least one critical point must exist inside each of these intervals.

The critical points of f are the possible locations of the local extrema that we seek. To find the critical points we must solve the equation $f'(x) = 0$. It is simpler to do some algebra before differentiating:

$$f'(x) = \frac{d}{dx}(x(x-1)(x-3)) = \frac{d}{dx}(x^3 - 4x^2 + 3x) = 3x^2 - 8x + 3.$$

By the quadratic formula, we have $f'(x) = 0$ at the points

$$x = \frac{-(-8) \pm \sqrt{8^2 - 4(3)(3)}}{2(3)} = \frac{8 \pm \sqrt{28}}{6} = \frac{4 \pm \sqrt{7}}{3}.$$

These x-values are approximately $x \approx 0.451$ and $x \approx 2.215$. If we look at the graph of f, then we can see that the smaller of these two x-values is the location of the local maximum and the larger is the location of the local minimum; see the figure that follows. □

Following many example solutions, **Checking the Answer** encourages students to learn to check their work, using technology such as a graphing calculator when appropriate.

The graph of $f(x) = x(x-1)(x-3)$ is shown next. Notice that the local extrema do seem to occur at the values we just found.

Extrema at $x \approx 0.451$ and $x \approx 2.215$

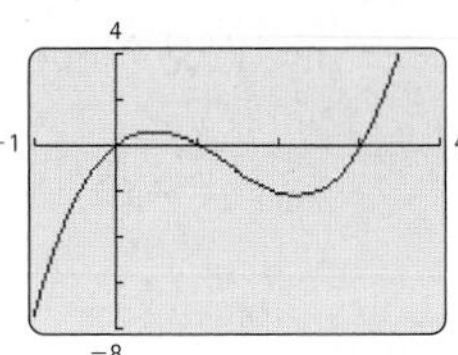

Each section closes with five **Test Your Understanding** questions that test students on the concepts and reading presented in the section. Because answers are not provided, instructors may choose to use these questions for discussion or assessment.

- Why could we not give a precise mathematical definition of concavity before this section?
- The domain points $x = c$ where $f''(c) = 0$ or where $f''(c)$ does not exist are the critical points of the function f'. Why?
- Why is it not clear to say a sentence such as "Because it is positive, it is concave up"? How could this information be conveyed more precisely?
- Why does it make sense that f' is increasing when f'' is positive?
- Suppose $x = c$ is a critical point with $f'(c) = 0$. Why does it make graphical sense that f has a local minimum at $x = c$ when f is concave up in a neighborhood around $x = c$?

Section Exercises are provided in a consistent format that offers the same types of exercises within each section. This approach allows instructors to tailor assignments to their course, goals, and student audience.

Thinking Back exercises ask students to review relevant concepts from previous sections and lessons.

Concepts exercises are consistently formatted to start with the following three problems:

- Problem 0 tests understanding.
- Problem 1 consists of eight true/false questions.
- Problem 2 asks the student to create examples based on their understanding of the reading.

Skills exercises offer ample practice, grouped into varying degrees of difficulty.

Applications exercises contain at least two in-depth real-world problems.

Proofs exercises can be completed by students in non-theoretical courses. Hints are often provided, and many exercises mimic work presented in the reading and examples. Often, these exercises are a continuation of a proof offered as a road map in the narrative.

Thinking Forward exercises plant seeds of concepts to come. In conjunction with the Thinking Back exercises, they offer a "tie together" of both past and future topics, thereby providing a seamless flow of concepts.

Chapter Review, Self-Test, and Capstones, found at the end of each chapter, present the following categories:

Definitions exercises prompt students to recall definitions and give an illustrative example.

Theorems exercises ask students to complete fill-in-the-blank theorem statements.

Formulas, Notation, and/or Rules exercises vary according to chapter content and ask students to show a working understanding of important formulas, equations, notation, and rules.

Skill Certification exercises provide practice with basic computations from the chapter.

Capstone Problems pull together the essential ideas of the chapter in more challenging mathematical and application problems.

ACKNOWLEDGMENTS

There are many people whose contributions to this project have made it immeasurably better. We are grateful to the many instructors from across the United States and Canada who have offered comments that assisted in the development of this book:

Jabir Abdulrahman, *Carleton University*
Jay Abramson, *Arizona State University*
Robert F. Allen, *University of Wisconsin–La Crosse*
Roger Alperin, *San Jose State University*
Matthew Ando, *University of Illinois*
Jorge Balbas, *California State University, Northridge*
Lynda Ballou, *New Mexico Institute of Mining and Technology*
E. N. Barron, *Loyola University Chicago*
Stavros Belbas, *University of Alabama*
Michael Berg, *Loyola Marymount University*
Geoffrey D. Birky, *Georgetown University*
Paul Blanchard, *Boston University*
Joseph E. Borzellino, *California Polytechnic State University, San Luis Obispo*
Eddie Boyd, Jr., *University of Maryland Eastern Shore*
James Brawner, *Armstrong Atlantic State University*
Jennifer Bready, *Mount Saint Mary College*
Mark Brittenham, *University of Nebraska*
Jim Brown, *Clemson University*
John Burghduff, *Lone Star College–CyFair*
Christopher Butler, *Case Western Reserve University*
Katherine S. Byler Kelm, *California State University, Fresno*
Weiming Cao, *The University of Texas at San Antonio*
Deb Carney, *Colorado School of Mines*
Lester Caudill, *University of Richmond*
Leonard Chastkofsky, *The University of Georgia*
Fengxin Chen, *University of Texas at San Antonio*
Dominic Clemence, *North Carolina A&T State University*
A. Coffman, *Indiana–Purdue Fort Wayne*
Nick Cogan, *Florida State University*
Daniel J. Curtin, *Northern Kentucky University*
Donatella Danielli-Garofalo, *Purdue University*
Shangrong Deng, *Southern Polytechnic State University*
Hamide Dogan-Dunlap, *The University of Texas at El Paso*
Alexander Engau, *University of Colorado, Denver*
Said Fariabi, *San Antonio College*
John C. Fay, *Chaffey College*
Tim Flaherty, *Carnegie Mellon University*
Stefanie Fitch, *Missouri University of Science & Technology*
Kseniya Fuhrman, *Milwaukee School of Engineering*
Robert Gardner, *East Tennessee State University*
Richard Green, *University of Colorado, Boulder*
Weiman Han, *University of Iowa*
Yuichi Handa, *California State University, Chico*
Liang (Jason) Hong, *Bradley University*
Steven Hughes, *Alabama A&M University*
Alexander Hulpke, *Colorado State University*
Colin Ingalls, *University of New Brunswick, Fredericton*
Lea Jenkins, *Clemson University*
Lenny Jones, *Shippensburg University*
Heather Jordan, *Illinois State University*
Mohammad Kazemi, *The University of North Carolina at Charlotte*
Dan Kemp, *South Dakota State University*
Boris L. Kheyfets, *Drexel University*
Alexander A. Kiselev, *University of Wisconsin–Madison*
Greg Klein, *Texas A&M University*
Evangelos Kobotis, *University of Illinois at Chicago*
Alex Kolesnik, *Ventura College*
Amy Ksir, *US Naval Academy*
Dan Kucerovsky, *University of New Brunswick*
Trent C. Kull, *Winthrop University*
Alexander Kurganov, *Tulane University*
Jacqueline La Vie, *SUNY College of Environmental Science and Forestry*
Melvin Lax, *California State University, Long Beach*
Dung Le, *The University of Texas at San Antonio*
Mary Margarita Legner, *Riverside City College*
Denise LeGrand, *University of Arkansas–Little Rock*
Mark L. Lewis, *Kent State University*
Xiezhang Li, *Georgia Southern University*
Antonio Mastroberardino, *Penn State Erie, The Behrend College*
Michael McAsey, *Bradley University*
Jamie McGill, *East Tennessee State University*
Gina Moran, *Milwaukee School of Engineering*
Abdessamad Mortabit, *Metropolitan State University*
Emilia Moore, *Wayland Baptist University*
Vivek Narayanan, *Rochester Institute of Technology*
Rick Norwood, *East Tennessee State University*
Gregor Michal Olsavsky, *Penn State Erie, The Behrend College*
Rosanna Pearlstein, *Michigan State University*
Kanishka Perera, *Florida Institute of Technology*
Cynthia Piez, *University of Idaho*
Jeffrey L. Poet, *Missouri Western State University*
Joseph P. Previte, *Penn State Erie, The Behrend College*
Jonathan Prewett, *University of Wyoming*
Elise Price, *Tarrant County College*
Stela Pudar-Hozo, *Indiana University of Northwest*
Don Redmond, *Southern Illinois University*
Dan Rinne, *California State University, San Bernardino*
Joe Rody, *Arizona State University*
John P. Roop, *North Carolina A&T State University*
Amber Rosin, *California State Polytechnic University, Pomona*
Nataliia Rossokhata, *Concordia University*
Dev K. Roy, *Florida International University*

Hassan Sedaghat, *Virginia Commonwealth University*
Asok Sen, *Indiana University–Purdue University*
Adam Sikora, *The State University of New York at Buffalo*
Mark A. Smith, *Miami University*
Shing Seung So, *University of Central Missouri*
David Stowell, *Brigham Young University–Idaho*
Jeff Stuart, *Pacific Lutheran University*
Howard Wainer, *Wharton School of the University of Pennsylvania*
Thomas P. Wakefield, *Youngstown State University*
Bingwu Wang, *Eastern Michigan University*
Lianwen Wang, *University of Central Missouri*
Antony Ware, *University of Calgary*
Talitha M. Washington, *Howard University*
Mary Wiest, *Minnesota State University, Mankato*
Mark E. Williams, *University of Maryland Eastern Shore*
G. Brock Williams, *Texas Tech University*
Dennis Wortman, *University of Massachusetts, Boston*
Hua Xu, *Southern Polytechnic State University*
Wen-Qing Xu, *California State University, Long Beach*
Yvonne Yaz, *Milwaukee School of Engineering*
Hong-Ming Yin, *Washington State University*
Mei-Qin Zhan, *University of North Florida*
Ruijun Zhao, *Minnesota State University, Mankato*
Yue Zhao, *University of Central Florida*
Jan Zijlstra, *Middle Tennessee State University*

We would also like to thank the Math Clubs at the following schools for their help in checking the accuracy of the exercises and their solutions:

CUNY Bronx Community College
Duquesne University
Fitchburg State College
Florida International University
Idaho State University
Jackson State University
Lander University
San Jose State University
Southern Connecticut State University
Texas A&M University
Texas State University–San Marcos
University of North Texas
University of South Carolina–Columbia
University of South Florida
University of Wisconsin–River Falls

Our students and colleagues at James Madison University have used preliminary versions of this text for the past two years and have helped to clarify the exposition and remove ambiguities. We would particularly like to thank our colleagues Chuck Cunningham, Rebecca Field, Bill Ingham, John Johnson, Brant Jones, Stephen Lucas, John Marafino, Kane Nashimoto, Edwin O'Shea, Ed Parker, Gary Peterson, Katie Quertermous, James Sochacki, Roger Thelwell, Leonard Van Wyk, Debra Warne, and Paul Warne for class-testing our book and for their helpful feedback. During the class-testing at JMU, hundreds of students provided feedback, made suggestions that improved the book, and, of course, showed us how they learned from the book! Thank you to all of our students, especially Lane O'Brien and Melissa Moxie for their meticulous review of an earlier draft of the text.

Chris Brazfield, now at Carroll Community College, helped with the initial development of the text and the ideas behind it. Kevin Cooper of Washington State University contributed many interesting and challenging real-world applications, and Elizabeth Brown and Dave Pruett of James Madison University contributed greatly to the development of the chapter on vector calculus. Roger Lipsett of Brandeis University wrote the excellent solution manual for the text, and at the same time eliminated any ambiguities in the exercises. We owe all of them great thanks for their expertise.

We also owe thanks to all of the people at W. H. Freeman who helped with the development of this text. Our developmental editor, Leslie Lahr, has been with this project from the beginning. Even under pressure, Leslie always maintains a positive attitude and finds a way for us to move forward. Without her support, we would not have made it through the rocky patches. Our executive editor, Terri Ward and developmental editor Katrina Wilhelm helped keep us on track while we wrote, rewrote, revised, revised, and revised some more, and we thank them for their support and patience. Brian Baker, our meticulous copy editor, made significant improvements to the text. Misplaced commas, dangling modifiers, and run-on sentences didn't stand a chance under his scrutiny. Ron Weickart and his team at Network Graphics took our graphs and sketches and turned them into the beautiful artwork in the text. Sherrill Redd and the compositing team at Aptara did a great job implementing all the design elements from our crazy LaTeX files.

Finally, we would like to thank our families and friends for putting up with us during the years of stress, turmoil, and tedium that inevitably come with any book project. Without their support, this book would not have been possible.

TO THE STUDENT

Learning something new can be both exciting and daunting. To gain a full understanding of the material in this text, you will have to read, you will have to think about the connections between the new topics and the topics that were previously presented, and you will have to work problems—many, many problems.

The structure of this text should help you understand the material. The material is laid out in a linear fashion that we think will facilitate your understanding. Each section is separated into two main parts: first, a presentation of new material and then second, a set of *Examples and Explorations*, where you will find problems that are carefully worked through. Working through these examples on your own, as you read the steps for guidance, will help prepare you for the exercises.

Reading a mathematics book isn't like reading a novel: You may have to read some parts more than once, and you may need to make notes or work things out on paper. Pay special attention to the "Checking Your Answer" features, so that you can learn how to check your own answers to many types of questions.

To succeed in calculus, you need to do homework exercises. The exercises in every section of this text are broken into six categories: "Thinking Back," "Concepts," "Skills," "Applications," "Proofs," and "Thinking Forward."

- As the title suggests, the *Thinking Back* problems are intended to tie the current material to material you've seen in previous sections or even previous courses.
- The *Concepts* problems are designed to help you understand the main ideas presented in the section without a lot of calculation. Every group of *Concepts* exercises begins by asking you to summarize the section, continues with eight true/false questions, and then asks for three examples illustrating ideas from the section.
- The bulk of the exercises in each section consists of *Skills* problems that may require more calculation.
- The *Applications* exercises use the concepts from the section in "real-world" problems.
- The *Proofs* exercises ask you to prove some basic theory from the section.
- Finally, the *Thinking Forward* questions use current ideas to introduce topics that you will see in subsequent sections.

We hope this structure allows you to tie together the material as you work through the book. We have supplied the answers to the odd-numbered exercises, but don't restrict yourself to those problems. You can check answers to even-numbered questions by hand or by using a calculator or an online tool such as wolframalpha.com. After all, on a quiz or test you won't have the answers, so you'll have to know how to decide for yourself whether or not your answers are reasonable.

Some students may like to work through each section "backwards," starting by attempting the exercises, then checking back to the examples as needed when they get stuck, and, finally, using the exposition as a reference when they want to see the big picture. That is fine; although we recommend that you at least try reading through the sections in order to see how things work for you. Either way, we hope that the separation of examples from exposition and the division of homework problems into subsections will help make the process of learning this beautiful subject easier. We have written this text with you, the student, in mind. We hope you enjoy using it!

CHAPTER 0

Functions and Precalculus

0.1 Functions and Graphs

What Is a Function?
Vertical and Horizontal Line Tests
Properties of Graphs
Examples and Explorations

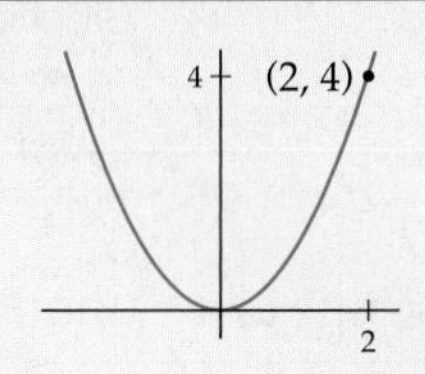

0.2 Operations, Transformations, and Inverses

Combinations of Functions
Transformations and Symmetry
Inverse Functions
Examples and Explorations

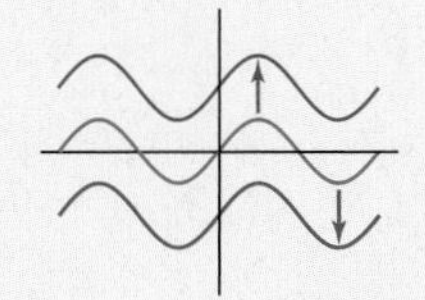

0.3 Algebraic Functions

Power Functions
Polynomial Functions
Rational Functions
Absolute Value Functions
Examples and Explorations

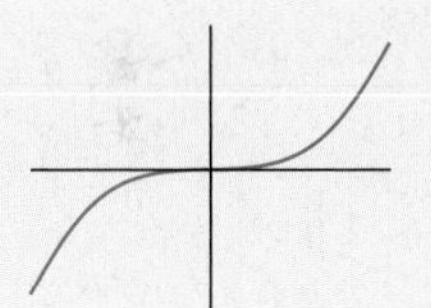

0.4 Exponential and Trigonometric Functions

Exponential Functions
Logarithmic Functions
Trigonometric Functions
Inverse Trigonometric Functions
Examples and Explorations

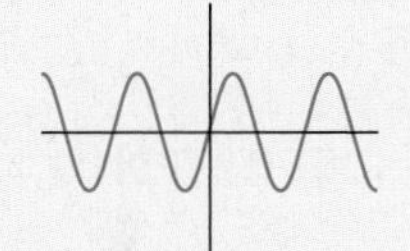

0.5 Logic and Mathematical Thinking*

From Definitions to Theorems
Quantifiers
Implications
Counterexamples
Simple Mathematical Proofs
Examples and Explorations

$A \Longrightarrow B$

Chapter Review, Self-Test, and Capstones

0.1 FUNCTIONS AND GRAPHS

- Definition of functions and their domain and range
- Graphs, horizontal and vertical line tests, and one-to-one-functions
- Graphical properties and features, asymptotes, and average rate of change

What Is a Function?

Mathematics is a language. In order to understand it, you have to learn how to read it and speak it with the correct vocabulary. Since calculus is at its heart the study of functions of real numbers, the universe we will spend most of our time exploring is the set of real numbers and the relationships between sets of real numbers. Therefore we must begin by setting out the mathematical language that describes these relationships we call "functions." Once we all speak the same language, we can start building the theory of calculus.

Functions and their properties will be at the core of everything we study in this text. In previous courses you likely encountered functions that were given in terms of formulas, such as

$$y(x) = x^2,$$

that relate two variables x and y. To set the stage for studying such functions, we must first be more precise about what functions are. Instead of thinking of functions merely as formulas, think of them as describing a certain kind of rule, relationship, or mapping from the elements of one set to the elements of another set.

DEFINITION 0.1

Functions

A ***function*** f from a set A to a set B is an assignment f that associates to each element x of the ***domain*** set A exactly one element $f(x)$ of the ***codomain***, or ***target***, set B.

We will use the notation

$$f : A \to B$$

to represent a function f together with its domain set A and target set B. This notation is pronounced "f from A to B." If x and y are variables that represent elements of the sets A and B, respectively, then we say that y is a function of x and write $y = f(x)$ or $y(x)$.

The variable x is called the ***independent variable*** and represents the "input" of the function. The function f sends each input x to one and only one "output," some value of the ***dependent variable*** y. Notice that y depends on x, according to the assignment defined by the function f.

For example, the assignment $f : \mathbb{R} \to \mathbb{R}$ that squares real numbers is a function, since each real number x is assigned to one and only one real-number square x^2. Here $\mathbb{R}$ denotes the set of all real numbers, and f assigns each real-number input to exactly one real-number output. Some real numbers (such as 3 and -3) get sent to the same square ($f(-3) = f(3) = 9$), but this does not violate the definition of function. You can think of a function as a machine that takes any given input value x and produces exactly one output value $f(x)$ (pronounced "f of x"), shown as follows:

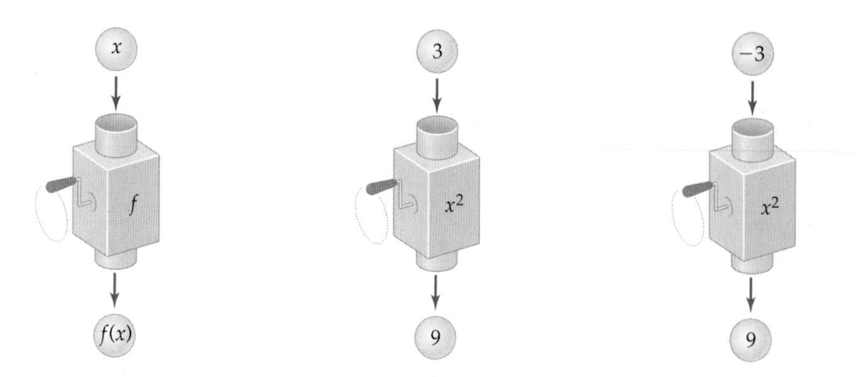

So, what *isn't* a function? If a rule assigns a real-number input to more than one output, then that rule is not a function. For example, consider the formula $y = \pm\sqrt{x}$. This assignment does not define y as a function of x, because the input $x = 4$ corresponds to two different y-values, both $y = -2$ and $y = 2$. In the "function machine" type of illustration just shown, the number 4 would go into the machine and two numbers, -2 and 2, would come out at once as outputs. This situation is not allowed for functions.

Returning to the squaring function $y = x^2$, notice that some real numbers can never serve as outputs, because squares of real numbers can never be negative. The ***range***, or set of possible outputs, of the squaring function is $[0, \infty)$. In this text we will usually work with functions whose domains and ranges are unspecified subsets of real numbers and whose rules are given by formulas such as $f(x) = x^2$.

DEFINITION 0.2

Domain and Range of a Function

If f is a function between unspecified subsets of $\mathbb{R}$, then we will take the ***domain*** of f to be the largest subset of $\mathbb{R}$ for which f is defined:

$$\text{Domain}(f) = \{x \in \mathbb{R} \mid f(x) \text{ is defined}\}.$$

The ***range*** of such a function is the set of all possible outputs that it can attain:

$$\text{Range}(f) = \{y \in \mathbb{R} \mid \text{there is some } x \in \text{Domain}(f) \text{ for which } f(x) = y\}.$$

For example, the function $f(x) = \sqrt{x-1}$ is defined only when $x \geq 1$, and therefore $f(x) = \sqrt{x-1}$ has domain $[1, \infty)$. When we write the square root symbol without the "$\pm$" before it, we always mean the positive square root. This means that $f(x) = \sqrt{x-1}$ can attain only nonnegative y-values. In fact, every nonnegative value can be expressed in the form $\sqrt{x-1}$ for some value of x, and therefore the function $f(x) = \sqrt{x-1}$ has range $[0, \infty)$.

A few notes about the notation we just used: The curly-brackets notation used in Definition 0.2 is called ***set notation***, and it is a way to describe a set of real numbers. In this case the set notation for the domain of f is pronounced "the set of all x contained in $\mathbb{R}$ such that $f(x)$ is defined." Notice in particular that the symbol "$\in$" means "contained in" and the vertical bar means "such that."

TECHNICAL POINT The name of a function is usually a single letter, such as "f." The name of the output of a function f evaluated at an input x is "$f(x)$." In this situation f is a function, or relationship, and $f(x)$ is a number that represents the output of the function at the input value x. However, it is sometimes convenient to write $f(x)$ (the name of the output of the function) instead of f (the name of the function itself). This allows us to indicate the name we are using for the independent variable when we reference the function. We may also write things like "consider the function $f(x) = x^2 + 1$," by which we mean "consider the function f whose output at a real number x is $f(x) = x^2 + 1$."

Vertical and Horizontal Line Tests

A function whose domain and range are sets of real numbers can be represented as a collection of pairs $(x, f(x))$ of real numbers. If we plot these pairs as points in the Cartesian plane, we obtain the ***graph*** of the function. In general we have the following definition:

DEFINITION 0.3 **The Graph of a Function**

The ***graph*** of a function f is the collection of ordered pairs $(x, f(x))$ for which x is in the domain of f. In set notation we can write

$$\text{Graph}(f) = \{\, (x, f(x)) \mid x \in \text{Domain}(f) \,\}.$$

For example, the graph of $f(x) = x^2$ is the collection of ordered pairs of the form (x, x^2), for $x \in \mathbb{R}$. Since $f(-1) = (-1)^2 = 1$ and $f(2) = 2^2 = 4$, the points $(-1, 1)$ and $(2, 4)$ are on the graph of $f(x) = x^2$. In contrast, the point $(1, 2)$ is *not* a part of the graph, because $f(1) \neq 2$, as shown in the following graph:

Graph of $f(x) = x^2$ and partial table of values

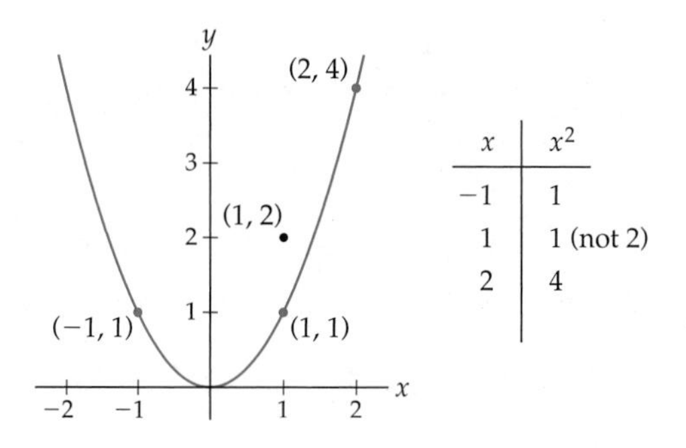

x	x^2
−1	1
1	1 (not 2)
2	4

A function always has exactly one output value for every input in the domain, which means that the graph of a function always passes the following test, which you will prove in Exercise 90:

THEOREM 0.4 **The Vertical Line Test**

A graph represents a function if and only if every vertical line intersects the graph in at most one point.

For example, consider the three graphs that follow this paragraph. The leftmost graph passes the vertical line test and thus is the graph of a function. The graph in the middle fails the vertical line test because the vertical line $x = 2$ intersects the graph in two points, $(2, 1)$ and $(2, 3)$; therefore the middle graph does not represent a function. The rightmost graph assigns the same output to two distinct inputs, but that is perfectly fine for a function. Because the graph on the right passes the vertical line test, it is the graph of a function.

A graph that is a function

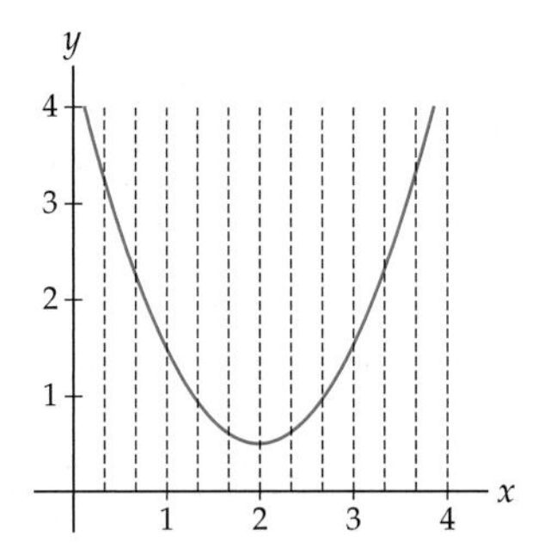

A graph that is not a function

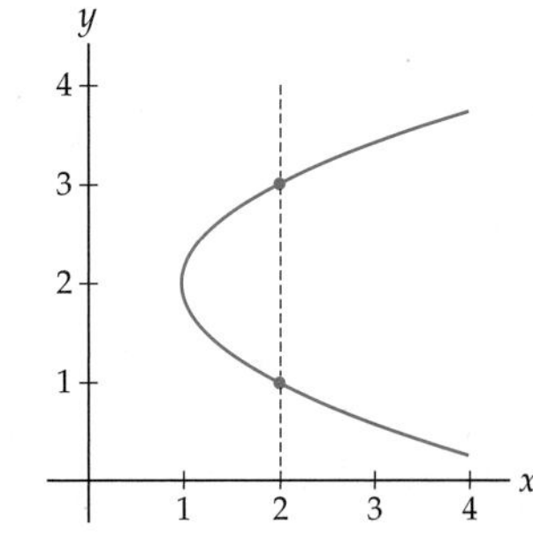

A function, but not one-to-one

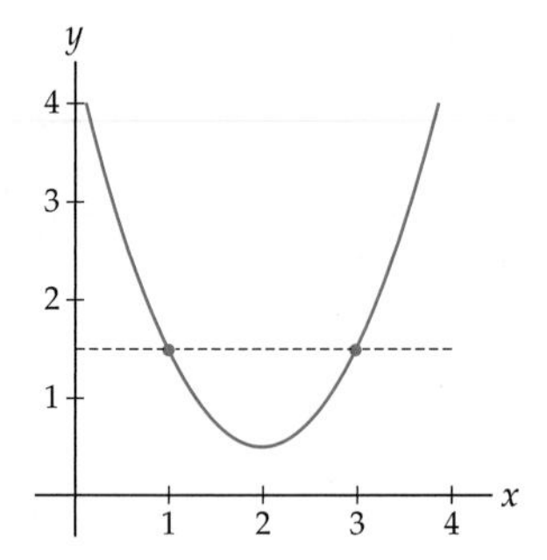

If each element of the range of a function is the output of exactly one element of the domain, then the function is said to be ***one-to-one***. Graphically, we can tell if a function f is one-to-one by checking to see if it passes the ***horizontal line test***: if f is one-to-one, then every horizontal line meets the graph of f at most once; see Exercise 91. Algebraically, this means that a function f is one-to-one if two distinct elements in the domain are always sent to different elements of the range:

DEFINITION 0.5

One-to-One Function

A function f is ***one-to-one*** if, for all a and b in the domain of f,

$$a \neq b \implies f(a) \neq f(b).$$

In this definition the notation $\Rightarrow$ is pronounced "implies," and it means that if the left-hand part of the expression is true, then the right-hand part of the expression is also true. In other words, the statement "$A \Rightarrow B$" is synonymous with the statement "if A, then B."

A logically equivalent form of Definition 0.5 is its so-called ***contrapositive***:

$$f(a) = f(b) \implies a = b.$$

As we will see in Section 0.5, the contrapositive of an implication $A \Rightarrow B$ is the equivalent statement (not B) $\Rightarrow$ (not A). The contrapositive form of Definition 0.5 is often easier to use, because it is an affirmative rather than a negative statement. For example, $f(x) = 3x$ is one-to-one because if $3a = 3b$, then we can guarantee that $a = b$. In contrast, the squaring function $f(x) = x^2$ is not one-to-one, because we cannot guarantee that if $a^2 = b^2$, then $a = b$ (for example, $(-3)^2 = 3^2$, but $-3 \neq 3$).

Properties of Graphs

The table that follows gives us vocabulary and precise mathematical definitions for various types of graphical behavior. Rows 1, 2, 5, and 6 describe behaviors that a function could exhibit at a specific point. The remaining rows describe graphical behaviors that occur over an interval I of real numbers. Much of the material in the early chapters of this book will be dedicated to developing techniques for properly defining and identifying these types of properties of functions. For now we present them just to set terminology and to familiarize ourselves with various types of function behavior.

Vocabulary	Definition	Behavior
f has a ***root*** at $x = c$	$f(c) = 0$	graph intersects the x-axis at $x = c$
f has a ***y-intercept*** at $y = b$	$f(0) = b$	graph intersects the y-axis at $y = b$
f is ***positive*** on I	$f(x) > 0$ for all $x \in I$	graph is above the x-axis on I
f is ***increasing*** on I	$f(b) > f(a)$ for all $b > a$ in I	graph moves up as we look from left to right on I
f has a ***local maximum*** at $x = c$	$f(c) \geq f(x)$ for all x near $x = c$	graph has a relative "hilltop" at $x = c$
f has a ***global maximum*** at $x = c$	$f(c) \geq f(x)$ for all $x \in \text{Domain}(f)$	graph is the highest at $x = c$
f is ***concave up*** on I	*will state precisely in Section 3.3*	graph curves upwards on I like part of a "U"
f has an ***inflection point*** at $x = c$	*will state precisely in Section 3.3*	graph of f changes concavity at $x = c$

Of course, there are similar definitions for local and global minima and for negative, decreasing, and concave-down behavior; see Exercises 20 and 21. Notice that we describe ***extrema*** (maxima and minima) by *where* on the x-axis they occur, since we can always find the corresponding y-values from these x-values. The concept of "near" in the description of a local maximum will be made more precise in Chapters 1 and 2. Inflection points and concavity cannot be precisely defined until we learn about ***derivatives*** in Chapters 2 and 3. In that chapter we will also learn ways for algebraically calculating the locations of extrema and inflection points. Until then, we will have to be content with examining such things graphically.

For example, the list that follows at the right describes some aspects of the graphical behavior of the graph $y = f(x)$ shown on the left.

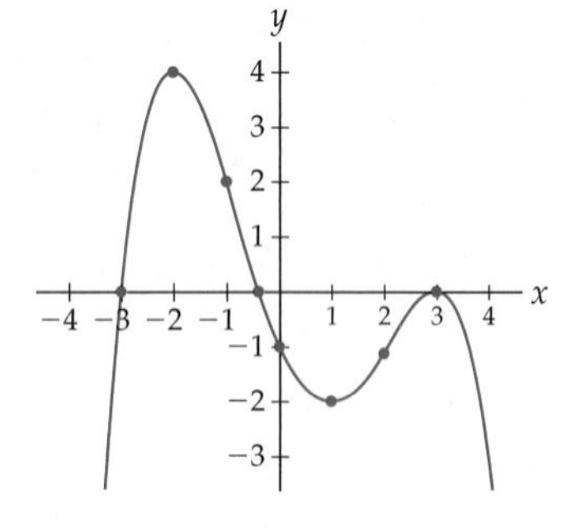

- roots at $x = -3$, $x \approx -0.4$, and $x = 3$
- y-intercept at $y = -1$
- local maxima at $x = -2$ and $x = 3$
- global maximum at $x = -2$
- inflection points at $x = -1$ and $x = 2$
- positive on $(-3, -0.4)$
- increasing on $(-\infty, -2)$ and $(1, 3)$
- concave up on $(-1, 2)$

In fact, technically the function f graphed at the left is increasing on the larger intervals $(-\infty, -2]$ and $[1, 3]$. This is because we do have $f(b) > f(a)$ for all values $b > a$ in these *closed* intervals. Most of the time we will be concerned only with the open intervals on which a function is increasing or decreasing, but there will be a few times in later chapters when we need to consider closed or half-closed intervals. For now, we will use open intervals unless we require otherwise.

The increasing and/or decreasing behavior of a function is related to its ***average rate of change*** on various intervals. The average rate of change of a function f on an interval $[a, b]$ measures how much the output $f(x)$ changes over that interval. Average rates of change will be extremely important in Chapter 2 when we study the derivative.

DEFINITION 0.6

Average Rate of Change

The ***average rate of change*** of a function f on an interval $[a, b]$ is the slope of the line from $(a, f(a))$ to $(b, f(b))$, which is given by the quotient

$$\frac{f(b) - f(a)}{b - a}.$$

For example, the function whose properties we just enumerated is increasing on the interval $(1, 3)$, moving up from $(1, f(1)) = (1, -2)$ to $(3, f(3)) = (3, 0)$. The average rate of change tells us how much the function increased per unit change in the input, on average:

$$\frac{f(3) - f(1)}{3 - 1} = \frac{0 - (-2)}{2} = 1$$

unit up for every unit across. We can also measure average rate of change over intervals where the function both increases and decreases; for example, with the same function, on the interval $[-3, 3]$ there is an average rate of change of

$$\frac{f(3) - f(-3)}{3 - (-3)} = \frac{0 - 0}{6} = 0$$

units up for every unit across; look at the graph to see why this makes sense.

Sometimes a graph gets closer and closer to a horizontal or vertical line, or ***asymptote***. In Chapter 1, we will define asymptotes precisely, using limits. For now, we will use the following definition: A line l is an ***asymptote*** of a function f if the difference between the graph of l and the graph of f gets as small as we want as either x or y increases in magnitude. For example, the following graph of a function f has vertical asymptotes at $x = -2$ and $x = 2$, and a horizontal asymptote at $y = 1$:

A function with three asymptotes

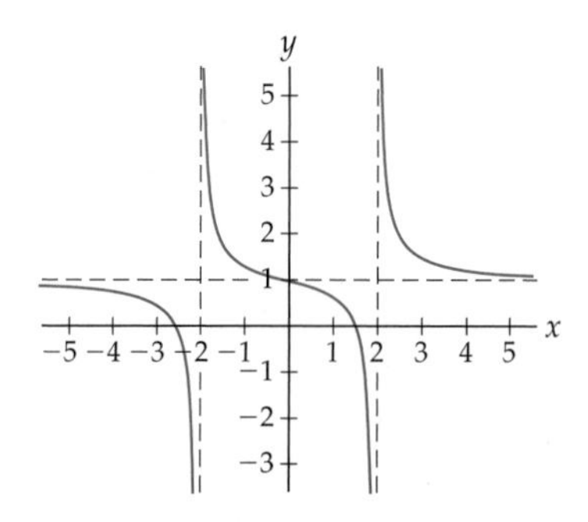

Notice that a graph *can* cross one of its horizontal asymptotes; the preceding graph above does so at the point $(0, 1)$. This is just one of the reasons that we are avoiding using the overly loose definition of asymptote that you may have heard in previous courses ("an asymptote is a line that the graph gets infinitely close to, but never reaches").

Examples and Explorations

EXAMPLE 1

Identifying functions and their domains and ranges

Determine whether or not each of the following relationships is a function. For each relationship that is a function, describe its natural domain and range and determine whether or not it is one-to-one. For each relationship that is not a function, describe the parts of the definition of a function that are violated.

(a) The rule $g\colon \mathbb{R} \to \mathbb{R}$ that assigns each real number x to the numbers whose square is x.

(b) The relationship defined by this table:

x	1	2	3	4	5	6
$P(x)$	5	2	9	−1	0	9

(c) Let P be the set of all living people in the world, and let W be the set of all women that have ever lived. Define $f\colon P \to W$ so that each person is assigned to his or her biological mother.

(d) $f(x) = 2 - \sqrt{x+5}$

(e) $h(x) = \dfrac{1}{x^2 - 4}$

SOLUTION

(a) This rule is not a function, for two reasons. First of all, negative real numbers do not have real square roots, so g is not defined on the given domain of $\mathbb{R}$. Second, each positive number x has two numbers whose square is x, namely, $\sqrt{x}$ and $-\sqrt{x}$, so this rule would not send each domain element to exactly one output.

(b) The relationship $P(x)$ defined by the table is a function, because the table assigns each value in the domain $\{1, 2, 3, 4, 5, 6\}$ to exactly one element of the range $\{-1, 0, 2, 5, 9\}$. This function is not one-to-one because $P(3)$ and $P(6)$ are both equal to 9.

(c) This relationship is a function because each person has one exactly one woman who is his or her biological mother. No person is without a biological mother, and no person has more than one biological mother. Here the domain is P and the range is the subset of W consisting of women that have had biological children. This function is not one-to-one, since there are examples of different people that have the same biological mother.

(d) This rule is a function because for each value x for which the formula makes sense, there is exactly one real number described by $2 - \sqrt{x+5}$. For x to be in the domain, we must have $x+5 \geq 0$ (since $x+5$ is under a square-root sign), and thus we must have $x \geq -5$. Therefore the domain of f is $[-5, \infty)$. The range of $y = f(x)$ is the set of y-values that can occur as outputs. Since $\sqrt{x+5}$ can take on any value greater than or equal to 0, the expression $2 - \sqrt{x+5}$ can take on any value less than or equal to 2. Therefore the range of f is $(-\infty, 2]$. This function is one-to-one because if $f(a) = f(b)$, then

$$\begin{aligned} 2 - \sqrt{a+5} = 2 - \sqrt{b+5} \quad &\Longrightarrow \quad \sqrt{a+5} = \sqrt{b+5} \\ &\Longrightarrow \quad a+5 = b+5 \\ &\Longrightarrow \quad a = b. \end{aligned}$$

(e) The rule $h(x)$ is a function because for each value x at which $\frac{1}{x^2-4}$ is defined, there is exactly one real number that h describes. The domain of $h(x)$ is the set of all x-values for which $x^2 - 4 \neq 0$ (since $x^2 - 4$ is in a denominator). Therefore the domain of h is everything except $x = \pm 2$. To find the range of $h(x)$ we must find the y-values that can be expressed in the form $y = h(x)$ for some x. Solving for x in terms of y we obtain $x = \sqrt{\frac{1}{y} + 4}$. This means we can find an x that maps via f to y as long as $y \neq 0$ and $\frac{1}{y} + 4 \geq 0$. It can be shown that the solution of the latter inequality is

$\left(-\infty, -\frac{1}{4}\right] \cup [0, \infty)$. Therefore the range of $h(x)$ is $\left(-\infty, -\frac{1}{4}\right] \cup (0, \infty)$. This function is not one-to-one because, for example, $h(1)$ and $h(-1)$ are both equal to $-\frac{1}{3}$. □

The following functions f and h have domains marked in blue on the x-axis and ranges marked in red on the y-axis:

f has domain $[-5, \infty)$ and range $(-\infty, 2]$

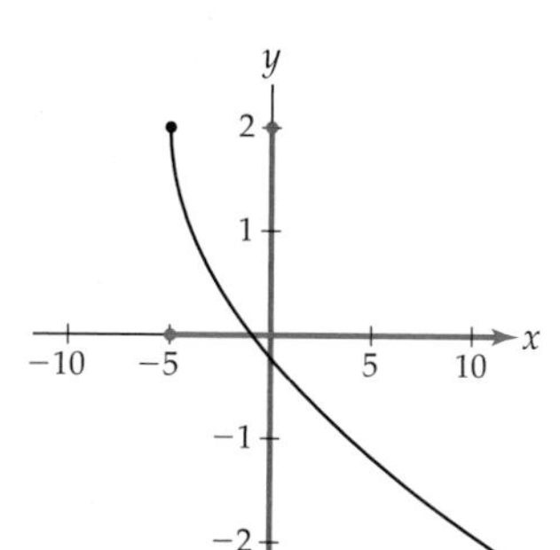

h has domain $x \neq \pm 2$ and range $\left(-\infty, -\frac{1}{4}\right] \cup (0, \infty)$

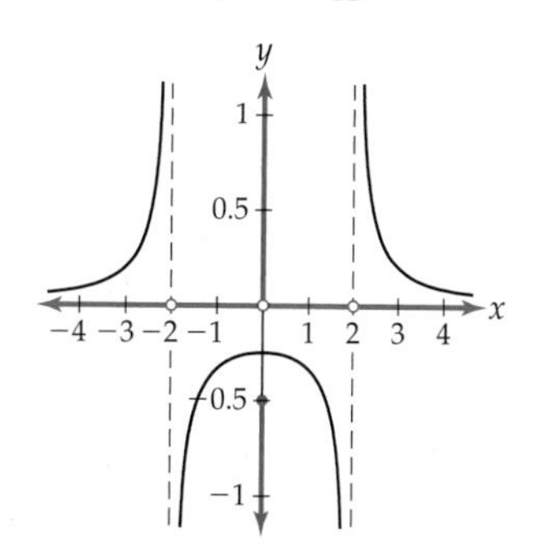

EXAMPLE 2

Evaluating function notation

Given that $f(x) = x\sqrt{3-x}$, evaluate $f(2)$, $f(a)$, $f(x+1)$, and $f(f(x))$.

SOLUTION

To evaluate $f(x) = x\sqrt{3-x}$ at a given input, simply replace x in the formula with whatever the input is:

$$f(2) = 2\sqrt{3-2} = 2\sqrt{1} = 2;$$

$$f(a) = a\sqrt{3-a};$$

$$f(x+1) = (x+1)\sqrt{3-(x+1)};$$

$$f(f(x)) = f(x)\sqrt{3-f(x)} = (x\sqrt{3-x})\sqrt{3-x\sqrt{3-x}}.$$

□

EXAMPLE 3

Finding a "good" graphing window

Use a graphing utility to find a graphing window that accurately represents the key features of the graph of the function $f(x) = x^3 - 6x^2 - x + 6$.

SOLUTION

The three graphs that follow show $y = f(x)$ in various graphing windows. Each of these windows is "bad" in the sense that the true behavior of the graph of f is not represented.

$f(x) = x^3 - 6x^2 - x + 6$
$x \in [-3, 3]$, $y \in [-10, 10]$

$f(x) = x^3 - 6x^2 - x + 6$
$x \in [-100, 100]$, $y \in [-50, 50]$

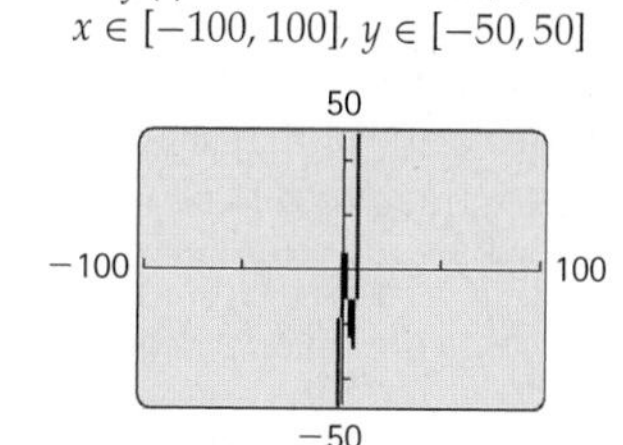

$f(x) = x^3 - 6x^2 - x + 6$
$x \in [-20, 20]$, $y \in [-1000, 1000]$

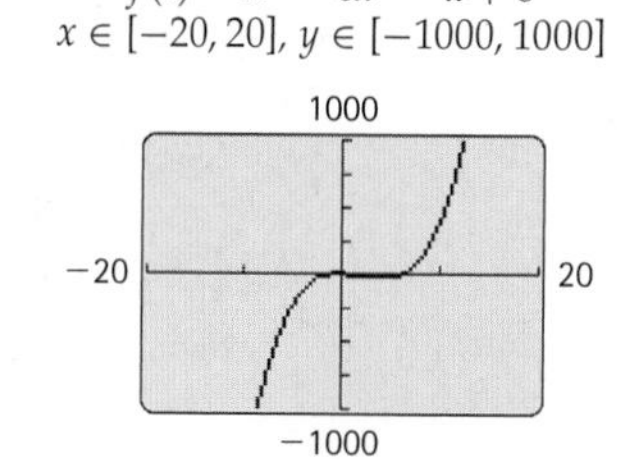

A "good" window (if one exists) is a window in which the local behavior of the graph of f is clear and the global behavior is accurately represented (the "ends" of the graph keep going in the direction indicated). The following figure shows the graph of f in a "good" window:

$f(x) = x^3 - 6x^2 - x + 6$
$x \in [-3, 7]$, $y \in [-40, 40]$

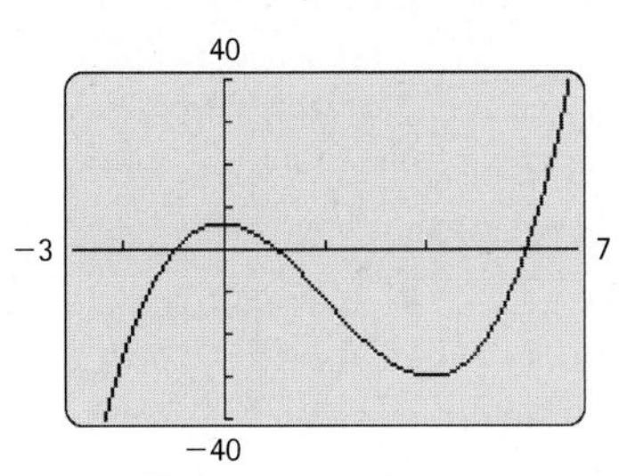

For now we will use trial and error to find an effective graphing window. We will be able to be more systematic after we have learned more about derivatives and function behavior. □

EXAMPLE 4 **Function behavior at points and on intervals**

(a) Sketch the graph of a function that has the following characteristics:

- roots at $x = -2$, $x = 1$, and $x = 3$
- local minimum at $x = -1$
- horizontal asymptote at $y = -2$
- local maximum at $x = 2$

(b) Approximate the locations of the inflection points on your graph.

(c) Given the graph that you sketched, find the intervals on which f is

- positive
- increasing
- decreasing
- concave up

SOLUTION

(a) By plotting the points $(-2, 0)$, $(1, 0)$, and $(3, 0)$, drawing a dashed asymptote at $y = -2$, and plotting some low point for the function at $x = -1$ and some high point at $x = 2$ (the information in the problem does not tell us exactly how high or how low), one might make the following sketch:

One possible graph of f

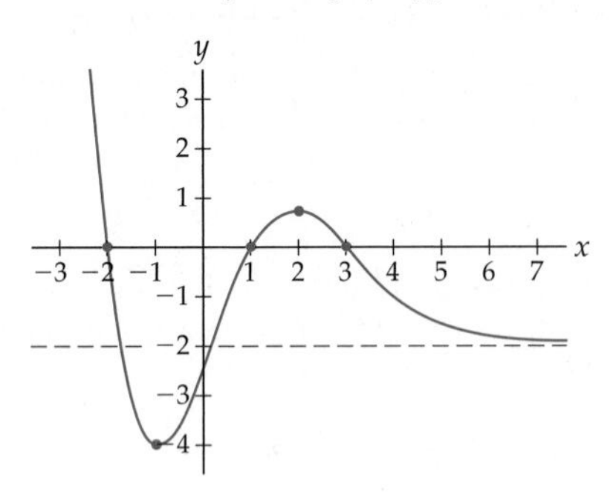

(b) The inflection points on this graph occur where the concavity of the graph changes from a ∪ shape to a ∩ shape, or vice versa. These are the locations where the graph "flexes." In our graph, these points occur at approximately $x = \frac{1}{2}$ and $x = 4$. Note that these points are also the locations of the steepest parts of the graph.

(c) Given our graph, the function f is positive on $(-\infty, -2)$ and $(1, 3)$, increasing on $(-1, 2)$, decreasing on $(-\infty, -1)$ and $(2, \infty)$, and concave up on $\left(-\infty, \frac{1}{2}\right)$ and $(4, \infty)$. □

EXAMPLE 5 **Finding the domain of a function by using equalities and inequalities**

Find the domain of the function $f(x) = \sqrt{\dfrac{1-2x}{x+1}}$.

SOLUTION

To find the domain of f, we ask which values of x can be validly plugged into the equation that defines $f(x)$. In order for the value of f to be defined for a real number x, that value of x must make the quotient underneath the square-root sign nonnegative and the denominator nonzero. Thus the domain of $f(x)$ is the set of real numbers that simultaneously satisfy the following:

$$\frac{1-2x}{x+1} \geq 0 \quad \text{and} \quad x+1 \neq 0.$$

The only x-values at which the quotient in the inequality can change sign are $x = \frac{1}{2}$ and $x = -1$, since those are the values that make either the numerator or denominator equal to zero. To determine the intervals on which the quotient is positive or negative we need only check its sign between the possible change points $x = \frac{1}{2}$ and $x = -1$. For example, evaluating the expression at $x = -2$, $x = 0$, and $x = 1$ gives

$$\frac{1-2(-2)}{-2+1} = \frac{\text{pos}}{\text{neg}} = \text{negative},$$
$$\frac{1-2(0)}{0+1} = \frac{\text{pos}}{\text{pos}} = \text{positive},$$
$$\frac{1-2(1)}{1+1} = \frac{\text{neg}}{\text{pos}} = \text{negative}.$$

We can record this information on a number line with a ***sign chart*** as follows:

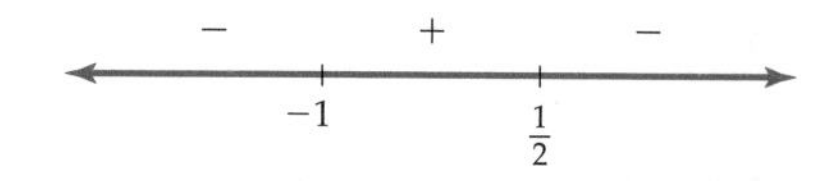

Since the quotient in question is negative on $(-\infty, -1)$ and $\left(\frac{1}{2}, \infty\right)$, the function $f(x)$ is not defined on those intervals. Note that $x = \frac{1}{2}$ is in the domain of f because there is no problem taking the square root of zero. However, since we cannot divide by zero, the function is not defined at $x = -1$. Therefore the domain of f is the half-open interval $\left(-1, \frac{1}{2}\right]$. □

EXAMPLE 6 **A review of factoring techniques**

Find the solution sets of each of the following equations:

(a) $2x^3 - 5x^2 - 3x = 0$ **(b)** $3x^2 = 7x - 1$ **(c)** $2x^5 - 32x = 0$

SOLUTION

(a) The number of real number solutions of a polynomial equation is at most the value of the ***degree***, or highest power, of the polynomial. Therefore we can expect the equation $2x^3 - 5x^2 - 3x = 0$ to have at most three real number solutions. The right-hand side of this equation can be easily factored as:

$$2x^3 - 5x^2 - 3x = x(2x^2 - 5x - 3) = x(2x+1)(x-3).$$

The expression $2x^3 - 5x^2 - 3x$ is zero if and only if one of x, $2x+1$, or $x-3$ is zero. In other words, the solutions of the equation $2x^3 - 5x^2 - 3x = 0$ are $x = 0$, $x = -\frac{1}{2}$, and $x = 3$.

(b) We first need to write the equation $3x^2 = 7x - 1$ in the general form of a quadratic equation: $3x^2 - 7x + 1 = 0$. This equation cannot be easily factored with the reverse "FOIL" (first-outside-inside-last) method, so we'll apply the ***quadratic formula***, which says that the roots of a quadratic equation of the form $ax^2 + bx + c = 0$ are

$$x = \frac{-b \pm \sqrt{b^2 - 4ac}}{2a}.$$

In this example we have $a = 3$, $b = -7$, and $c = 1$, so the solutions of $3x^2 - 7x + 1 = 0$ are

$$x = \frac{-(-7) \pm \sqrt{(-7)^2 - 4(3)(1)}}{2(3)} = \frac{7 \pm \sqrt{49 - 12}}{6} = \frac{7 \pm \sqrt{37}}{6}.$$

Therefore, the solutions of $3x^2 - 7x + 1 = 0$ are $x = \frac{1}{6}(7 + \sqrt{37})$ and $x = \frac{1}{6}(7 - \sqrt{37})$. Clearly we could not have easily figured that out by doing the "FOIL" method backwards!

(c) This time the factoring will involve two applications of the well-known factoring formula $a^2 - b^2 = (a+b)(a-b)$ for the difference of two squares:

$$\begin{aligned} 2x^5 - 32x &= 0 \\ 2x(x^4 - 16) &= 0 \\ 2x(x^2 - 4)(x^2 + 4) &= 0 && \leftarrow \text{formula for } a^2 - b^2 \text{ with } a = x^2 \text{ and } b = 4 \\ 2x(x-2)(x+2)(x^2+4) &= 0 && \leftarrow \text{formula for } a^2 - b^2 \text{ with } a = x \text{ and } b = 2 \end{aligned}$$

Thus $2x^5 - 32x = 0$ whenever $2x = 0$, $x - 2 = 0$, $x + 2 = 0$, or $x^2 + 4 = 0$. Note that $x^2 + 4 = 0$ has no real solutions, because there is no real number that satisfies $x^2 = -4$. Therefore the real-number solution set of the original equation $2x^5 - 32x = 0$ is $\{-2, 0, 2\}$. □

CHECKING THE ANSWER

To check the answers in Example 6, simply substitute each proposed solution into the original equation. Each solution should satisfy the equation. For example, to check that $x = -2$, $x = 0$, and $x = 2$ are solutions in part (c) of the example we note that

$$\begin{aligned} 2(-2)^5 - 32(-2) &= 0, && \leftarrow \text{evaluate equation at } x = -2 \\ 2(0)^5 - 32(0) &= 0, && \leftarrow \text{evaluate equation at } x = 0 \\ 2(2)^5 - 32(2) &= 0. && \leftarrow \text{evaluate equation at } x = 2 \end{aligned}$$

Of course, this will not tell you whether you have *missed* any solutions, but it will tell you whether the solutions you found are correct.

EXAMPLE 7 Finding the average rate of change of a function on an interval

Calculate the average rate of change of the function $f(x) = x^2 - x + 1$ (a) on the interval $[0, 3]$ and (b) on the interval $[-1, 1]$. Then (c) illustrate these average rates of change graphically.

SOLUTION

(a) Using the formula from Definition 0.6 we find that the average rate of change of f on $[0, 3]$ is

$$\frac{f(3) - f(0)}{3 - 0} = \frac{(3^2 - 3 + 1) - (0^2 - 0 + 1)}{3 - 0} = \frac{7 - 1}{3 - 0} = \frac{6}{3} = 2.$$

(b) Using the same formula again, we find that the average rate of change of f on $[-1, 1]$ is

$$\frac{f(1) - f(-1)}{1 - (-1)} = \frac{(1^2 - 1 + 1) - ((-1)^2 - (-1) + 1)}{1 - (-1)} = \frac{1 - 3}{1 - (-1)} = \frac{-2}{2} = -1.$$

Notice that the average rate of change of $f(x) = x^2 - x + 1$ is different, depending on what interval we consider.

(c) Graphically, the two average rates we found can be represented as slopes of line segments, as follows:

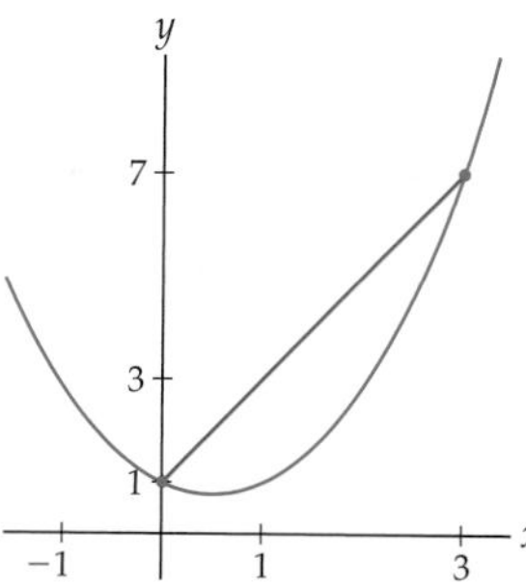

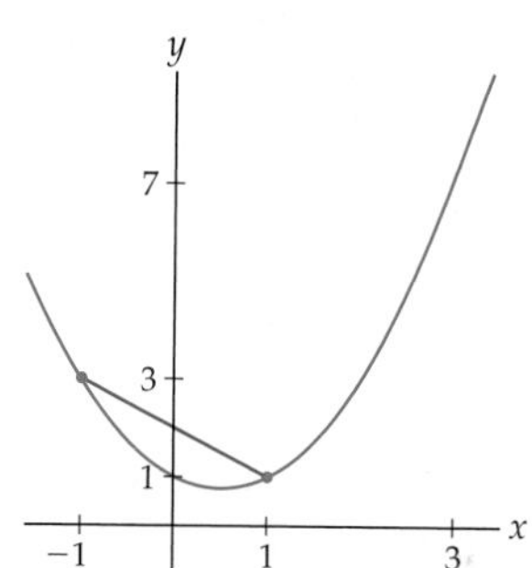

□

EXAMPLE 8 **A function that is defined in pieces**

A ***piecewise-defined function*** is a function that is defined in pieces, with different formulas on different parts of its domain. Let f be the function defined piecewise by

$$f(x) = \begin{cases} x^2, & \text{if } x \le -1 \\ 2x, & \text{if } x > -1. \end{cases}$$

Find $f(-5)$, $f(-1)$, and $f(3)$, and then sketch a graph of $y = f(x)$.

SOLUTION

Since $-5 \le -1$, we have $f(-5) = (-5)^2 = 25$. Since $-1 \le -1$, we have $f(-1) = (-1)^2 = 1$. In contrast, since $3 > -1$, we have $f(3) = 2(3) = 6$. To graph f, we begin by graphing the functions $y = x^2$ and $y = 2x$ that are used in the definition of f, as shown in the first two figures that follow:

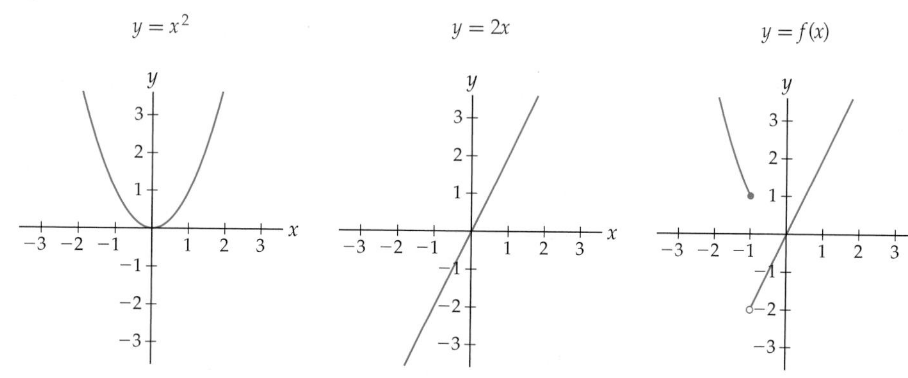

To graph f, we must restrict the graph of $y = x^2$ to the interval $(-\infty, -1]$ and restrict $y = 2x$ to $(-1, \infty)$. Whether these intervals are open or closed is important. To find $f(-1)$ we use the *first* equation $y = x^2$: $f(-1) = (-1)^2 = 1$. Note that when we sketched the graph of f in the third figure, we used open and closed dots to represent the function values corresponding to the ends of open and closed intervals in the domain, respectively. □

EXAMPLE 9 **Functions involving pairs or triples of real numbers***

In this example we examine functions with ***multivariable*** inputs and/or outputs. Let $\mathbb{R}^2$ be the set of ordered pairs of real numbers, and $\mathbb{R}^3$ the set of ordered triples of real numbers. We will not need to work with such functions until much later in this book, but we present them here to illustrate the variety of functions we will be encountering in this course.

(a) Explain why the rule $p: \mathbb{R} \to \mathbb{R}^3$ defined by $p(t) = (3t, t^2 - 1, t)$ is a function. Then find $p(2)$.

(b) Explain why the rule $q: \mathbb{R}^2 \to \mathbb{R}^2$ defined by $q(x, y) = (x - y, -3x)$ is a function. Then find $q(3, 2)$.

SOLUTION

(a) The rule $p(t)$ is a function because every real number t gets sent to exactly one triple of numbers. For example, $p(2) = (3(2), 2^2 - 1, 2) = (6, 3, 2)$.

(b) Similarly, the rule $q(x, y)$ is a function because each pair of numbers (x, y) in the domain $\mathbb{R}^2$ gets sent to exactly one pair of numbers in the range $\mathbb{R}^2$. For example, $q(3, 2) = (3 - 2, -3(3)) = (1, -9)$. □

? TEST YOUR UNDERSTANDING

- Why might the notation $f(x)$ be wrongly confused with the notation for a product? How is it different from the notation for a product? What is the difference between the notation f and the notation $f(x)$?
- If a rule f assigns both 4 and 8 to the same output value, can that rule be a function? Why or why not?
- The vertical line test states that each vertical line has to intersect the graph in *at most* one point. Is it okay for a vertical line to pass through the graph at *no* points, that is, for the line not to intersect the graph?
- How do we evaluate a function that is defined in pieces? That is, given a function $f(x)$ defined piecewise, how do we go about finding, say, $f(2)$?
- Define each of the following with mathematical notation: function, one-to-one, global maximum, asymptote, root.

EXERCISES 0.1

Thinking Back

Interval notation: Describe each of the following subsets of the real numbers in interval notation.

- $x \neq \pm 3$
- $x > -2$ and $x \neq 5$
- $x < 0$ or $x \geq 10$
- $x \not< 3$ and $x \neq 4$

Solving equations and inequalities: Find the solution sets of the equations and inequalities that follow. Write your answers in interval notation (or, if the solution is a discrete set of points, a list of those points).

- $\frac{x}{x-2} = 0$
- $x^3 - 5x^2 + 6x < 0$
- $\frac{x}{x-2} > 0$
- $x^3 - 5x^2 + 6x = 0$

Concepts

0. *Problem Zero:* Read the section and make your own summary of the material.

1. *True/False:* Determine whether each of the statements that follow is true or false. If a statement is true, explain why. If a statement is false, provide a counterexample.

(a) *True or False:* Functions are the same as equations.

(b) *True or False:* The domain of every function is a subset of $\mathbb{R}$.

(c) *True or False:* The function that for each x has output $f(x) = 1$ is a one-to-one function.

(d) *True or False:* Every global maximum of a function is also a local maximum.

(e) *True or False:* Every local minimum of a function is also a global minimum.

(f) *True or False:* The graph of a function can never cross one of its asymptotes.

(g) *True or False:* Average rates of change can be thought of as slopes.

(h) *True or False:* A function can have different average rates of change on different intervals.

2. *Examples:* Construct examples of the thing(s) described in the following. Try to find examples that are different than any in the reading.

(a) A function that is defined with a formula.

(b) A function that is not defined with a formula.

(c) A formula that does not define a function.

3. State the mathematical definition of a function, and describe its meaning in your own words. Support your answer with an example of something that is a function and an example of something that is not.

4. Suppose P is the set of people alive today and C is the set of possible eye colors. Let $f\colon P \to C$ be the rule that assigns to each person his or her eye color. Is f a function? Why or why not?

5. Use set notation to define the domain of a function. Then use the same notation to express the domain of the function $f(x) = \sqrt{x}$.

6. Use set notation to define the range of a function. Then use the same notation to express the range of the function $f(x) = x^2$.

7. Determine whether the points (a) $(3, 2)$, (b) $(1, 1)$, and (c) $(-5, 2)$ lie on the graph of $f(x) = \sqrt{x+1}$, without referring to a picture of the graph of f.

8. Describe the graph of the function $f(x) = 3x + 2$ as a set of ordered pairs.

9. Consider the function $f(x) = x^2 + 1$.

(a) Explain why $y = 5$ is in the range of f.

(b) Explain why $y = 0$ is *not* in the range of f.

(c) Argue that the range of $f(x) = x^2 + 1$ is $[1, \infty)$.

10. Determine whether or not each diagram that follows represents a function. If it does, find its domain and range, and determine whether it is one-to-one. If it does not, explain what goes wrong.

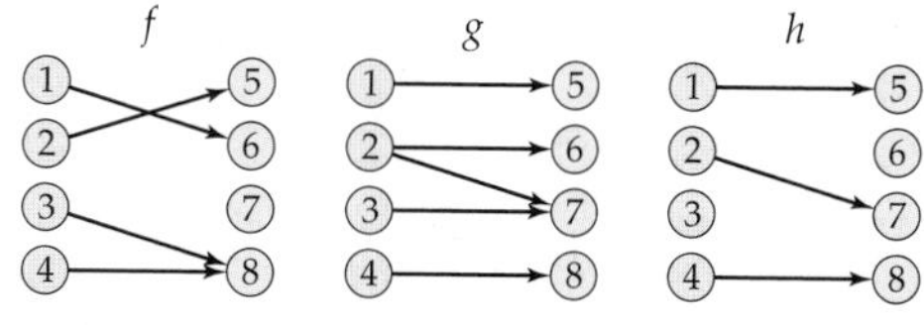

11. Construct a rule $f\colon \{2, 4, 6, 8, 10\} \to \{1, 2, 3, 4\}$ that is a function. Express this function three ways: as a list, as a table, and as a diagram. Is your function one-to-one? What is its range?

12. Construct a rule $f\colon \{2, 4, 6, 8, 10\} \to \{1, 2, 3, 4\}$ that is not a function. Justify your answer.

13. If the graph of a rule $y = f(x)$ passes through $(-2, 1)$ and $(2, 1)$, could that rule be a function? Why or why not?

14. A ***constant function*** is a function $f\colon A \to B$ with the property that there is some $b \in B$ for which $f(x) = b$ for all $x \in A$. (The output of the function is *constantly* the same.) Describe

(a) a constant function $f\colon \mathbb{R} \to \mathbb{R}$

(b) a constant function $g\colon \mathbb{R} \to [-5, -2]$

(c) a constant function $h\colon \mathbb{R} \to \mathbb{R}^2$

15. The ***identity function*** for a set A is the function $f\colon A \to A$ defined by $f(x) = x$ (so called because the output is *identical* to the input). For which of the following domains and ranges is there a well-defined identity function? Why or why not?

(a) $f\colon \mathbb{R} \to \mathbb{R}$

(b) $g\colon \mathbb{R}^2 \to \mathbb{R}$

(c) $h\colon \mathbb{R}^3 \to \mathbb{R}^3$

16. Let P be the set of all people living in the United States. Give examples of each of the following functions and state their ranges:

(a) the identity function $f\colon P \to P$

(b) two different constant functions $g\colon P \to P$

(c) a non-constant, non-identity function $g\colon P \to P$

(d) a constant function $h\colon P \to \mathbb{R}$

(e) a non-constant function $h\colon P \to \mathbb{R}$

17. Explain in your own words why the vertical line test determines whether a graph is a function.

18. Explain in your own words why the horizontal line test determines whether a function is one-to-one.

19. Show that $f(x) = x^2 + 1$ is not one-to-one, using values of f (*not* the horizontal line test).

20. Define what it means for a function f with domain $\mathbb{R}$ to have (a) a global minimum at $x = c$ and (b) a local minimum at $x = c$.

21. Define what it means for a function f with domain $\mathbb{R}$ to be (a) negative on an interval I and (b) decreasing on an interval I.

22. Make a labeled graph that illustrates why it makes sense that a function is increasing on an interval I if, for all $b > a$ in I, we have $f(b) > f(a)$. Include labels for a, b, $f(a)$, and $f(b)$, and for the interval I.

23. How is the formula for average rate of change related to the formula for computing slope?

24. Illustrate on a graph of $f(x) = 1 - x^2$ that the average rate of change of f on $[-1, 3]$ is -2.

25. For each local maximum $x = c$ in the following graph, approximate the largest possible $\delta > 0$ so that $f(c) \geq f(x)$ for all $x \in (c - \delta, c + \delta)$. Similarly, for the one local minimum $x = b$, find the largest δ so that $f(b) \leq f(x)$ for all $x \in (b - \delta, b + \delta)$.

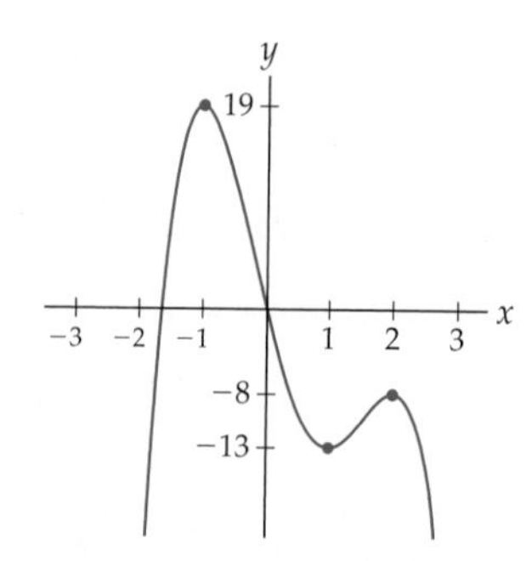

26. Use the definition of a local maximum to explicitly argue why the function graphed in Exercise 25 does *not* have a local maximum at $x = 0$.

Skills

Find the domain and range of each function in Exercises 27–32. Use a graphing utility or plot points to sketch a graph of the function, and illustrate the domain and range on the graph.

27. $f(x) = \sqrt{x-1}$

28. $f(x) = \sqrt{x} - 1$

29. $f(x) = \dfrac{1}{x+2}$

30. $f(x) = \dfrac{1}{\sqrt{5-x}}$

31. $f(x) = \dfrac{1}{x^2+1}$

32. $f(x) = \dfrac{1}{x^2-1}$

Find the domain of each function in Exercises 33–42.

33. $f(x)=\sqrt{x(x-2)}$

34. $f(x)=\dfrac{3x+1}{2x-1}$

35. $f(x)=\sqrt{(x-1)(x+3)}$

36. $f(x)=\dfrac{1}{\sqrt{x^2-4}}$

37. $f(x)=\dfrac{1}{\sqrt{(x-1)(x+3)}}$

38. $f(x)=\sqrt{\dfrac{1}{x^2}-4}$

39. $f(x)=\dfrac{\sqrt{x^2-1}}{x^2-9}$

40. $f(x)=\dfrac{x^{3/4}}{3x-5}$

41. $f(x)=\dfrac{1}{\sqrt{x-1}} - \dfrac{\sqrt{x}}{x-2}$

42. $f(x)=\dfrac{\sqrt{x^2-1}}{\sqrt{x^2-9}}$

Evaluate each function in Exercises 43–47 at the values indicated. Simplify your answers if possible. *(Note: We will not need to work with such functions until much later in the book, but we present them here to illustrate the variety of functions we will be encountering in this course.)*

43. If $f(x) = x^2 + 1$, find

(a) $f(-4)$ (b) $f(a^3)$ (c) $f(f(x))$

44. If $k(x) = \dfrac{x^2}{x+1}$, find

(a) $k(5)$ (b) $k(x+h)$ (c) $k(k(x))$

45. If $l(a,b,c) = \sqrt{a^2+b^2+c^2}$, find:

(a) $l(5,3,2)$ (b) $l(3,0,4)$ (c) $l(x,y,z)$

46. If $g(v) = (v-1, v, v^2)$, find

(a) $g(0)$ (b) $g(1)$ (c) $g(x+1)$

47. If $F(u,v,w) = (3u+v, u-w, v+2w)$, find

(a) $F(2,3,5)$ (b) $F(5,2,3)$ (c) $F(a,b,0)$

For each piecewise-defined function in Exercises 48–50, (a) calculate $f(-1)$, $f(0)$, $f(1)$, and $f(2)$, and (b) sketch a graph of f.

48. $f(x) = \begin{cases} x^2+3, & \text{if } x < 0 \\ 3-x, & \text{if } x \geq 0 \end{cases}$

49. $f(x) = \begin{cases} 3x+1, & \text{if } x \leq 0 \\ 4, & \text{if } 0 < x \leq 1 \\ x^3, & \text{if } x > 1 \end{cases}$

50. $f(x) = \begin{cases} 4x-1, & \text{if } x < 0 \\ 2, & \text{if } x = 0 \\ -3x+5, & \text{if } x > 0 \end{cases}$

Use a graphing utility to sketch a graph of each function in Exercises 51–56. Use trial and error to find a graphing window so that your graph represents the local and global behavior of the function. Include the x and y ranges of your window in your answer.

51. $f(x) = x^2 - 0.1$

52. $f(x) = (x^2-5)^7$

53. $f(x) = x^5 - 3x^4 - 7x$

54. $f(x) = x^3 - 11x^2 + 10x$

55. $f(x) = x^2 - 17x - 18$

56. $f(x) = \dfrac{2x^2-2}{x^2-3x-5}$

Describe the key properties of each graph in Exercises 57–62, including the following:

- domain and range;
- locations of roots, intercepts, local and global maxima and minima, and inflection points;
- intervals on which the function is positive or negative, increasing or decreasing, and concave up or down;
- any horizontal or vertical asymptotes.

57.

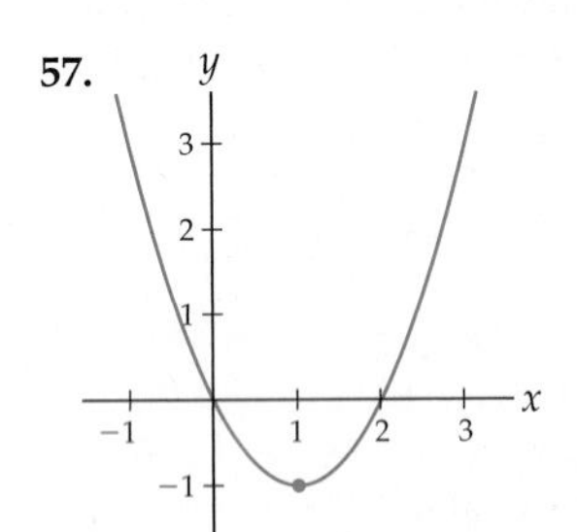

58.

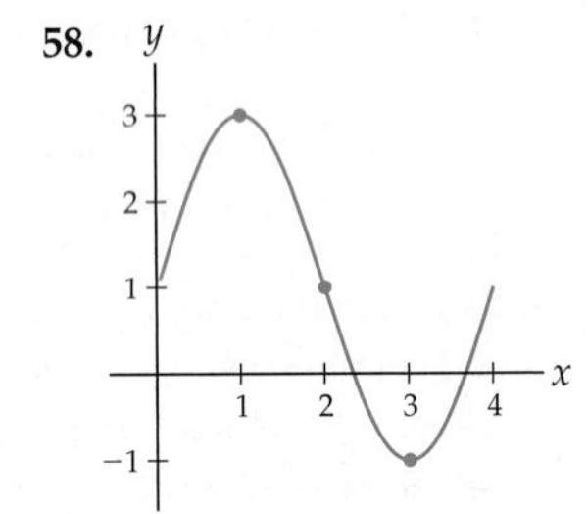

59.

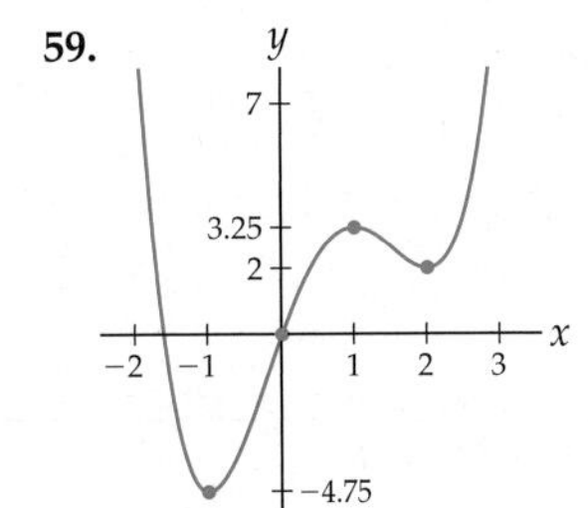

60.

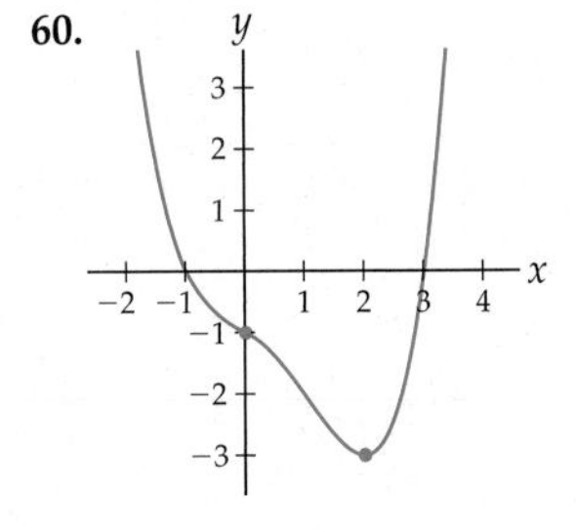

61.

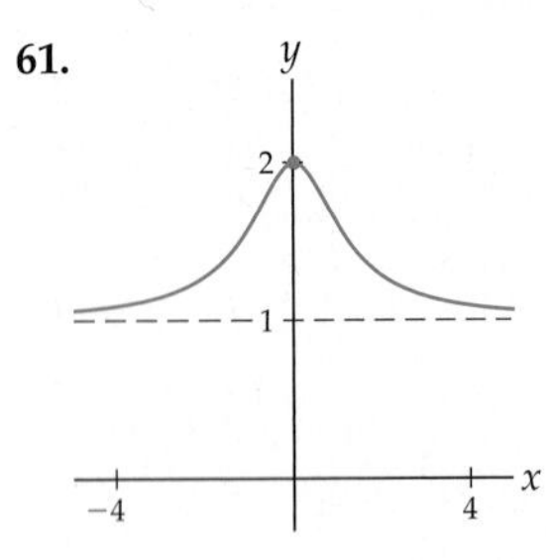

62.

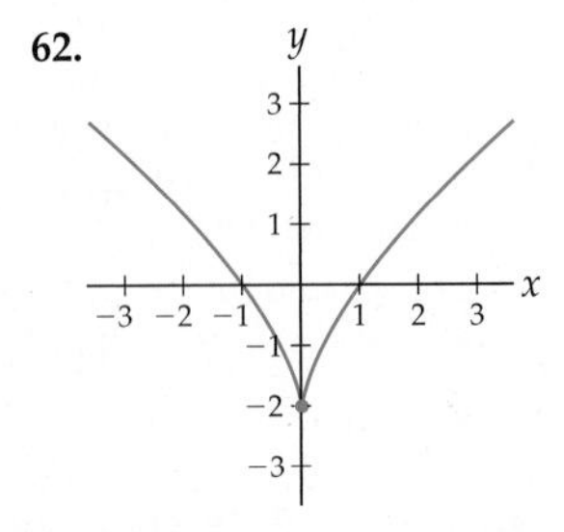

Sketch the graph of functions f that satisfy the lists of conditions given in Exercises 63–72, if possible.

63. Domain $\mathbb{R}$, concave up everywhere, and decreasing everywhere.

64. Domain $\mathbb{R}$, concave down everywhere, and decreasing everywhere.

65. Domain $\mathbb{R}$, concave up everywhere, increasing everywhere, and negative everywhere.

66. Domain $\mathbb{R}$, concave down everywhere, increasing everywhere, and negative everywhere.

67. Always increasing, with two horizontal asymptotes, one at $y = -2$ and one at $y = 2$.

68. Domain $(0, \infty)$, always negative, and always increasing.

69. Four roots but no y-intercept.

70. Concave down on $(-\infty, 2)$, concave up on $(2, \infty)$, and always increasing.

71. Concave down on $(-\infty, 0)$ and concave up on $(0, \infty)$ but *without* an inflection point at $x = 0$.

72. Average rate of change of 3 on $[0, 2]$, average rate of change of -1 on $[0, 1]$, and average rate of change of 0 on $[-2, 2]$.

In Exercises 73–78, find the average rate of change of the function f on the interval $[a, b]$.

73. $f(x) = -0.5 + 4.2x, \quad [a, b] = [1, 3.5]$

74. $f(x) = 3, \quad [a, b] = [-100, 100]$

75. $f(x) = \sqrt{x+1}, \quad [a, b] = [1, 9]$

76. $f(x) = \dfrac{1-x}{1+x^3}, \quad [a, b] = [0, 0.5]$

77. $f(x) = \dfrac{1}{x}, \quad [a, b] = [0.9, 1.1]$

78. $f(x) = (x-2)^2 + \dfrac{3}{x}, \quad [a, b] = [-2, 2]$

Applications

In Exercises 79–82, sketch and label the graph of a function that describes the given situation.

79. An island warthog population initially grows quickly, but as space and food become sparse on the island, the population growth slows down. Eventually the population of the island levels off at 512 warthogs.

80. Susie is late for calculus class and leaves her dorm in a panic. She hurries towards the math building, but about halfway there, she realizes she has left her notebook in her room. She sprints back to her dorm and gets her notebook. Coming out of the dorm, she sprains her ankle, so the best she can do is limp as fast as she can to her classroom.

81. Suppose that after you drink a cup of coffee the amount of caffeine in your body rises sharply and then decreases by half every hour. You have one cup of coffee in the morning and then no more.

82. On a dare, you go skydiving. Gravity causes you to fall faster and faster as you plummet towards the ground. When you open your parachute, your speed is drastically reduced. After opening your parachute you approach the ground at a constant speed.

(a) Graph your distance from the ground as a function of time.

(b) Graph your velocity as a function of time.

83. For each situation described, identify any independent and dependent variables, and express their relationship as an equation in multivariable function notation (see Example 9):

(a) H is the length of the hypotenuse of the right triangle with legs of length a and b.

(b) V is the volume of a rectangular prism ("box") with dimensions x, y, and z.

84. If your rain-catching bucket starts empty and collects 3 inches of rain during a 6-hour rainstorm, what is the average rate of change of the level of rainwater in the bucket over the 6 hours that it rained? Did the rain necessarily collect in the bucket at a constant rate?

85. A disgruntled pet store owner abandoned an unknown number of groundhogs on a small island in 1996. Since then it has been determined that the average rate of change of the groundhog population was 4 groundhogs per year and that the groundhog population was a linear function of time. When the abandoned groundhogs were discovered in 2001, there were 376 groundhogs on the island. How many groundhogs did the disgruntled pet store owner originally leave on the island?

86. The number N of operating drive-in movie theatres in the state of Virginia in various years y is given in the table below.

y	1958	1967	1972	1977	1982	1999
N	143	90	102	87	56	9

(a) Find the average rate of change in the number of drive-in movie theatres in Virginia over each time interval between table entries.

(b) Describe the units and real-world significance of these average rates of change.

(c) Over which time period was the average rate of change the most drastic? On average, assuming that no new theatres were built, how many drive-ins closed per year during that period?

87. In your first job after graduating from college you make \$36,000 a year before taxes. After four years you get a raise of \$2,500. Two years after that you change jobs and go to work for a company that pays you \$49,000 a year.

(a) Construct a piecewise-defined function that describes your pretax income in the year that is t years after you graduate from college.

(b) Write down a function that describes the total amount of money you will have earned t years after graduating from college.

(c) How many years after graduating from college will you have earned a total of one million pre-tax dollars?

88. The following table shows the year 2000 Federal Tax Rate Schedule for single filers:

Taxable income: The federal tax owed is:

Over	Not over	Amount	Plus %	Of amt. over
\$0	\$26,250	\$0	15%	\$0
\$26,250	\$63,550	\$3,937	28%	\$26,250
\$63,550	\$132,600	\$14,381	31%	\$63,550
\$132,600	\$288,350	\$35,787	36%	\$132,600
\$288,350	——	\$91,857	39.6%	\$288,350

(a) How much tax would you owe if you made \$18,000 of taxable income? What if you made \$180,000?

(b) What percentage of your taxable income did you owe in taxes if your taxable income was \$18,000? What if your taxable income was \$180,000?

(c) Construct a piecewise-defined function describing the dollar amount of tax T owed by a single person with m dollars of taxable yearly income. Each piece of your function will be linear. Do the pieces "match up"? Does this make financial sense?

Proofs

89. Use Definition 0.2 to prove that the range of the function $f(x) = 3x - 1$ is $\mathbb{R}$.

90. Use Definition 0.1 to prove that a graph represents a function if and only if it passes the vertical line test.

91. Use Definition 0.5 to prove that a function is one-to-one if and only if its graph passes the horizontal line test.

92. Use the contrapositive form of Definition 0.5 to prove that the function $f(x) = 3x + 1$ is one-to-one.

93. Use the definition of decreasing to prove that the function $f(x) = 1 - 3x$ is decreasing on $(-\infty, \infty)$.

94. Use the definition of increasing to prove that the function $f(x) = \frac{1}{3-x}$ is increasing on $(-\infty, 3)$.

95. Prove that the average rate of change of the linear function $f(x) = -2x + 4$ on *any* interval I is always equal to -2.

96. Show that the average rate of change of every linear function $f(x) = mx + b$ is constant, that is, the same over any choice of interval. *(Hint: Use $[c, d]$ to denote the interval, since the letter b is already used in the equation for $f(x)$.)*

Thinking Forward

Evaluations for slopes and derivatives: Evaluate each function at the values indicated. Simplify your answers if possible.

▶ If $f(x) = 4 - x^2$, find

(a) $\frac{f(1+0.1) - f(1)}{0.1}$ (b) $\frac{f(1+0.001) - f(1)}{0.001}$

▶ If $f(x) = \sqrt{x}$, find

(a) $\frac{f(1+h) - f(1)}{h}$ (b) $\frac{f(x) - f(1)}{x - 1}$

▶ If $q(x, h) = \frac{(x+h)^2 - x^2}{h}$, find

(a) $q(3, 0.5)$ (b) $q(3, h)$ (c) $q(x, 0.5)$

Evaluations for series: Evaluate each function at the values indicated. Simplify your answers if possible.

▶ If $S(x, n) = x - \frac{x^2}{2} + \frac{x^3}{3} - \frac{x^4}{4} + \cdots + (-1)^{n+1}\frac{x^n}{n}$, find

(a) $S\left(\frac{1}{2}, 5\right)$ (b) $S\left(\frac{1}{2}, n\right)$ (c) $S(x, 5)$

▶ If $c(x, n) = 1 - \frac{x^2}{2!} + \frac{x^4}{4!} - \frac{x^6}{6!} + \cdots + (-1)^n \frac{x^{2n}}{(2n)!}$, find

(a) $c(\pi, 3)$ (b) $c(\pi, n)$ (c) $c(x, 3)$

Tangent lines: The **tangent line** to the graph of a function f at $x = c$ is the line that passes through the point $(c, f(c))$ and has slope determined by the "direction" that the graph is moving. If you imagine a graph of $y = f(x)$ as a hilly curve that a small car is driving on, then the tangent line is the line determined by the car's headlights at time $x = c$. For example, the graph that follows shows the tangent line for $f(x) = x^2$ at $x = 2$.

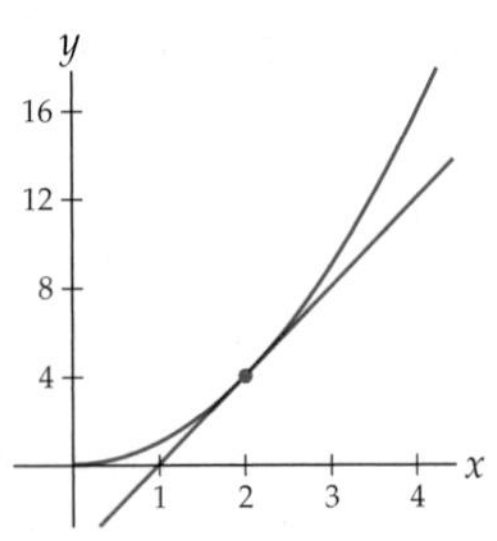

▶ Find the equation of the tangent line for $f(x) = 3x + 1$ at $x = 2$. *(Hint: Think about the graph.)* Why is this not surprising?

▶ Find the equation of the tangent line for $f(x) = 4 - x^2$ at $x = 0$. Again, think about the shape of the graph before you attempt to answer this question.

▶ Use a graph to visually estimate the slope of the tangent line for $f(x) = x^2$ at $x = 2$. Use this slope estimate to write down an approximate formula for the tangent line to $f(x) = x^2$ at $x = 2$.

0.2 OPERATIONS, TRANSFORMATIONS, AND INVERSES

- ▶ Constant multiples, sums, products, quotients, and compositions of functions
- ▶ Translations, stretches, compressions, and reflections of graphs
- ▶ Inverse functions and their properties

Combinations of Functions

So far we have been thinking of functions as *operators*, that is, a function f takes an input x and operates on it to produce an output $f(x)$. We now want to think of functions in a different way, as *objects* that can be added to, subtracted from, and multiplied or divided by one another. Of course, sums, differences, products, and quotients of functions f and g will be defined in terms of sums, differences, products, and quotients of their outputs $f(x)$ and $g(x)$.

For example, given two functions f and g, we can add them together and get a new function $f+g$. How this new function operates on inputs will depend on how the original two functions operated. In other words, to define the function $f+g$, we must say what $f+g$ does to each input x. If x is in the domain of f and in the domain of g, the obvious choice is to define $(f+g)(x)$ to be the sum of $f(x)$ and $g(x)$. For example, if $f(x) = x^2$ and $g(x) = 3x+1$, then for all values of x we define $(f+g)(x)$ to be $f(x)+g(x) = x^2 + 3x + 1$. In particular, this means that $(f+g)(2)$ is equal to the sum of $f(2) = 2^2 = 4$ and $g(2) = 3(2)+1 = 7$, so that $(f+g)(2) = 4+7 = 11$. The other arithmetic operations work similarly on functions:

DEFINITION 0.7

Arithmetic Combinations of Functions

Suppose f and g are functions and k is a real number.

(a) The ***constant multiple*** of f by k is the function kf defined by $(kf)(x) = kf(x)$ for all x in the domain of f.

(b) The ***sum*** of f and g is the function $f+g$ defined by $(f+g)(x) = f(x)+g(x)$ for all x in the domains of both f and g.

(c) The ***product*** of f and g is the function $f \cdot g$ defined by $(f \cdot g)(x) = f(x)g(x)$ for all x in the domains of both f and g.

(d) The ***quotient*** of f and g is the function $\frac{f}{g}$ defined by $\left(\frac{f}{g}\right)(x) = \frac{f(x)}{g(x)}$ for all x in the domains of both f and g with $g(x) \neq 0$.

There is an additional operation on functions that we do not have for numbers, called ***composition***. We ***compose*** two functions f and g by taking the output from one function as the input for the other:

DEFINITION 0.8

The Composition of Two Functions

The ***composition*** of two functions f and g is the function $f \circ g$ defined by

$$(f \circ g)(x) = f(g(x))$$

for all x in the domain of g such that $g(x)$ is in the domain of f.

For example, if $f(4) = 6$ and $g(10) = 4$, then $(f \circ g)(10) = f(g(10)) = f(4) = 6$. You should think of compositions as nestings of functions. The notation $(f \circ g)$ is pronounced "f composed with g" or sometimes "f circle g." The notation $f(g(x))$ is pronounced "f of g of x."

If $g\colon X \to Y$ and $f\colon Y \to Z$, then their composition is a function $(f \circ g)\colon X \to Z$ that takes an input x first to $g(x)$ and then to $f(g(x))$. For example, if $f(x) = x^2$ and $g(x) = 3x + 1$ then

$$(f \circ g)(x) = f(g(x)) = f(3x + 1) = (3x + 1)^2.$$

Notice that although the function f appears first (i.e., on the left) in the notation, it is the function g that gets applied to the input x first. Composition is not a ***commutative*** operation, which means that $f \circ g$ is not necessarily the same function as $g \circ f$. With the same example of $f(x) = x^2$ and $g(x) = 3x + 1$, if we compose in the other order, we get

$$(g \circ f)(x) = g(f(x)) = g(x^2) = 3(x^2) + 1.$$

You may have noticed that the notation for composition looks a bit like multiplication notation, but there is a key difference. When we want to denote multiplication we will use a small closed dot or no dot at all. To denote composition of functions we will always use an open circle.

Transformations and Symmetry

Another way we can obtain new functions from old is through ***transformations***. Given a function $f(x)$ and constants C and k, we could consider such modifications as $f(x) + C$, $f(x + C)$, $kf(x)$, and $f(kx)$. Each of these transformations changes $f(x)$ graphically and algebraically.

For example, transforming $f(x)$ to $f(x) \pm C$ clearly adds C units to every output of $f(x)$. This means that the graph of $y = f(x)$ shifts up or down vertically C units everywhere, to become the graph of $y = f(x) + C$ or $y = f(x) - C$, as illustrated by the red and green graphs, respectively, shown in the figure next at the left. If we instead add a constant to the independent variable and transform $f(x)$ to $f(x \pm C)$, the graph shifts left or right horizontally by C units, as illustrated in the green and red graphs shown in the figure at the right. (Note that the shift to the left for $f(x + 2)$ and to the right for $f(x - 2)$ might be the opposite of what we might initially expect.) These additive transformations are called ***translations***.

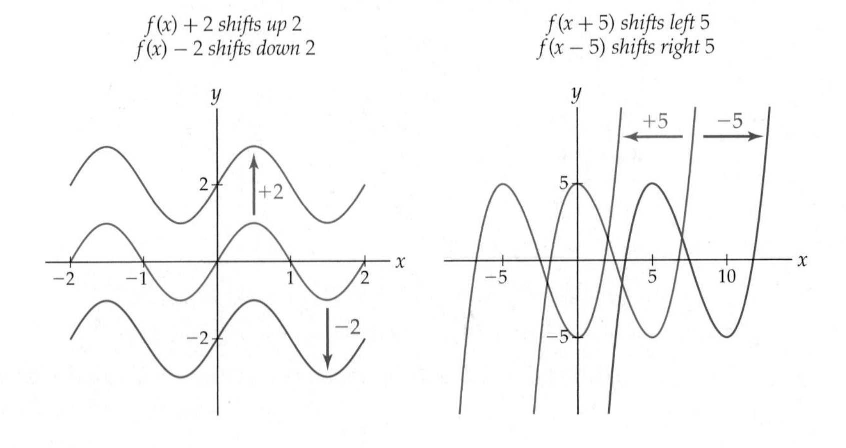

If we instead transform $f(x)$ by multiplication to $kf(x)$, then the graph of $y = f(x)$ expands or contracts vertically by a factor of k to become the graph of $y = kf(x)$, as shown in the red and green graphs next at the left. In contrast, if we do the same transformation to the independent variable and transform $f(x)$ to $f(kx)$, this contracts or expands the graph of $y = f(x)$ by a factor of k in the horizontal direction, as illustrated in the red and green graphs next at the right.

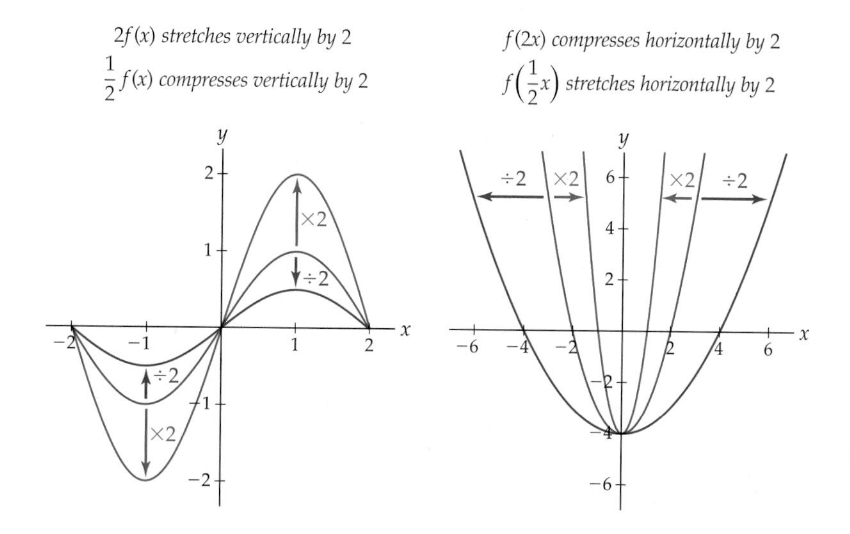

What happens if we multiply x or $f(x)$ by a negative number? We can answer that question by just looking at what happens when we multiply by -1. Changing $f(x)$ to $-f(x)$ transforms all positive outputs into negative outputs, and vice versa. The graph of $y = f(x)$ is then reflected across the x-axis to become the graph of $y = -f(x)$, as shown in the red graph in the figure that follows. If we instead multiply the independent variable by -1, then we obtain a reflection across the y-axis, as shown in the green graph.

$-f(x)$ reflects across the x-axis
$f(-x)$ reflects across the y-axis

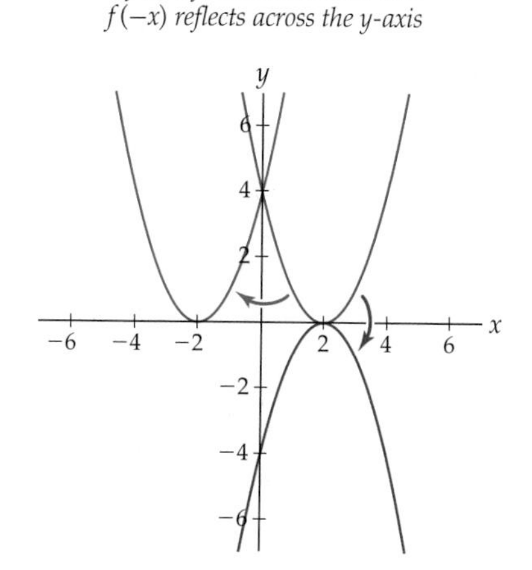

Now if we want to transform $f(x)$ to $f(-2x)$, for example, we can transform $f(x)$ first to $f(2x)$ and then by reflection to $f(-2x)$.

The table that follows summarizes the graphical and algebraic effects of the transformations just discussed. You will prove these general results in Exercises 86–88.

Transformation	Graphical Result	Algebraic Result
$f(x)+C$	shifts up C units if $C > 0$ shifts down C units if $C < 0$	$(x, y) \to (x, y+C)$
$f(x+C)$	shifts left C units if $C > 0$ shifts right C units if $C < 0$	$(x, y) \to (x-C, y)$
$kf(x)$	vertical stretch by k if $k > 1$ vertical compression by k if $0 < k < 1$	$(x, y) \to (x, ky)$
$f(kx)$	horizontal compression by k if $k > 1$ horizontal stretch by k if $0 < k < 1$	$(x, y) \to \left(\frac{1}{k}x, y\right)$
$-f(x)$	graph reflects across the x-axis	$(x, y) \to (x, -y)$
$f(-x)$	graph reflects across the y-axis	$(x, y) \to (-x, y)$

Some graphs do not change under certain transformations. For example, the graph of $f(x) = x^2$ shown next at the left remains the same if we reflect it across the y-axis. We say that this function has ***y*-axis symmetry**. As another example, the graph of $g(x) = x^3$ shown at the right remains the same if we reflect it first across the y-axis and then across the x-axis.

$f(x) = x^2$ preserved under y-axis reflection

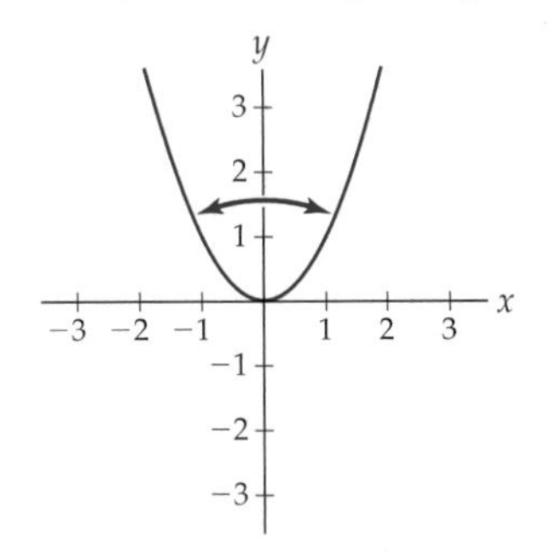

$g(x) = x^3$ preserved under 180° rotation

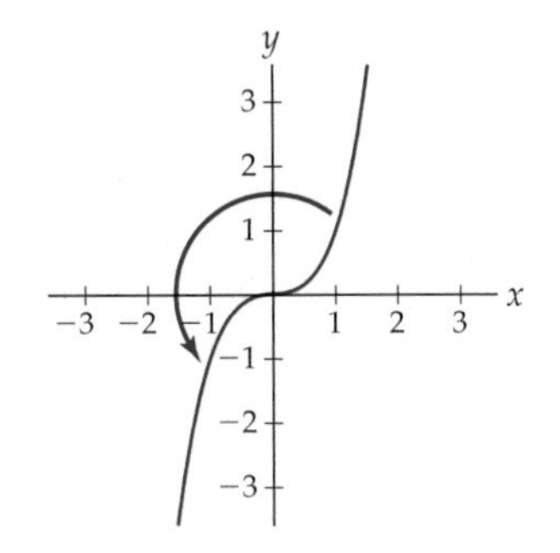

It turns out that the double-reflection we just described for $g(x) = x^3$ is equivalent to rotation around the origin by 180°. You can try this equivalence out for yourself by physically double-reflecting and rotating the book while looking at the preceding graph of $g(x) = x^3$. You can also see the equivalence by using a piece of paper with a smiley-face drawn on the front: Flipping the paper vertically and then horizontally is equivalent to rotating the paper by 180 degrees. A function that is preserved under the transformation of 180° rotation is said to have 180° ***rotational symmetry***.

These types of symmetries are also called ***even symmetry*** and ***odd symmetry***, since power functions with even powers all have y-axis symmetry and power functions with odd powers all have rotational symmetry. Because graphical reflections correspond to multiplication by -1, we can describe functions with even and odd symmetry algebraically as follows:

DEFINITION 0.9

Even and Odd Functions

A function f is an ***even function*** if $f(-x) = f(x)$ for all x in the domain of f.

A function f is an ***odd function*** if $f(-x) = -f(x)$ for all x in the domain of f.

For example, the function $f(x) = x^2$ is even because for all x we have

$$f(-x) = (-x)^2 = x^2 = f(x).$$

In contrast, the function $g(x) = x^3$ is odd because for all x we have

$$g(-x) = (-x)^3 = -(x^3) = -g(x).$$

Note that some functions are neither even nor odd; for example $h(x) = x^2 + x$ is one such function, because $h(-x) = (-x)^2 + (-x) = x^2 - x$, which is equal neither to $h(x)$ nor to $-h(x)$. Consequently, the function $h(x) = x^2 + x$ has neither y-axis symmetry nor rotational symmetry.

Inverse Functions

The ***inverse*** of a function f is a function that undoes the action of f. For example, the function that adds 1 to each real number can be undone by subtracting 1 from each real number. If two functions undo each other, then composing them results in the identity function. This property suggests the following definition:

DEFINITION 0.10 **The Inverse of a Function**

If f and g are functions such that

$$g(f(x)) = x, \quad \text{for all } x \text{ in the domain of } f$$
$$f(g(x)) = x, \quad \text{for all } x \text{ in the domain of } g$$

then g is the ***inverse*** of f and we denote g by f^{-1}.

Note that the two conditions in this definition guarantee that if a function g is the inverse of a function f, then f is the inverse of the function g.

For example, the functions $f(x) = x^3$ and $g(x) = \sqrt[3]{x}$ are inverses of each other. Intuitively, taking the cube of a number is undone by taking the cube root, and vice versa. It is easy to verify that these two functions satisfy the condition in Definition 0.10:

$$g(f(x)) = g(x^3) = \sqrt[3]{x^3} = x,$$
$$f(g(x)) = f(\sqrt[3]{x}) = (\sqrt[3]{x})^3 = x.$$

It is important to note that although we use the notation x^{-1} to denote the reciprocal $\frac{1}{x}$, the notation f^{-1} does *not* stand for the reciprocal $\frac{1}{f}$ of f. The notation f^{-1} used in Definition 0.10 is pronounced "f inverse." We are now using the same notation for two very different things, but it should be clear from the context which one we mean.

Not all functions have inverses. For example, consider the squaring function $f(x) = x^2$ on the domain $\mathbb{R}$. Since $f(2) = 4$ and $f(-2) = 4$, an inverse of f would have to send the input 4 to both 2 and -2; this is clearly not a function. As the following theorem asserts, a function has an inverse only if no two inputs are ever sent to the same output:

THEOREM 0.11 **Invertible Functions are One-to-One**

A function f has an inverse if and only if f is one-to-one.

Since a function is one-to-one if and only if it passes the horizontal line test, this theorem implies that a function has an inverse if and only if it passes the horizontal line test. The proof follows directly from the properties of inverse functions:

Proof. We first prove that if a function is invertible, then it must be one-to-one. If f is an invertible function, then it has an inverse f^{-1}. Suppose two domain values a and b are sent by f to the same output $f(a) = f(b)$. Applying f^{-1} to both sides, we have

$$f(a) = f(b) \implies f^{-1}(f(a)) = f^{-1}(f(b)) \implies a = b.$$

For the converse, suppose f is one-to-one. Then each element b in the range of f is the output of *exactly one* element a from the domain. We can define $f^{-1}(b)$ to be this element a. Since f is one-to-one, this new relationship f^{-1} will be a function and we will have $f^{-1}(b) = a$ if and only if $f(a) = b$. Therefore if f is one-to-one, then f has an inverse. ■

The properties of inverses given in the next theorem follow directly from our definition of an inverse function, that is, from the fact that f^{-1} undoes the function f. Functions that have inverses are said to be ***invertible***.

THEOREM 0.12

Properties of Inverses

If f is an invertible function with inverse f^{-1}, then the following statements hold.

(a) $\text{Domain}(f^{-1}) = \text{Range}(f)$ and $\text{Range}(f^{-1}) = \text{Domain}(f)$.

(b) $f^{-1}(b) = a$ if and only if $f(a) = b$.

(c) The graph of $y = f^{-1}(x)$ is the graph of $y = f(x)$ reflected across the line $y = x$.

Proof. The proofs of parts (a) and (c) are left to Exercises 91 and 92. To prove part (b), suppose $f^{-1}(b) = a$. Applying f to both sides, we have $f(f^{-1}(b)) = f(a)$. Since f and f^{-1} are inverses, their composition is the identity function. Therefore we have $b = f(a)$, as desired. With an entirely similar argument we can show that if $f(a) = b$, then $f^{-1}(b) = a$; see Exercise 93. ■

For example, consider the one-to-one function $f(x) = \sqrt{x+1}$. To find a function that undoes $f(x)$ we solve $y = \sqrt{x+1}$ for y, obtaining $x = y^2 - 1$. Changing notation so that x is again the independent variable, we see that the inverse of $f(x)$ is $f^{-1}(x) = x^2 - 1$. The following three figures illustrate the properties from Theorem 0.12 for $f(x) = \sqrt{x+1}$.

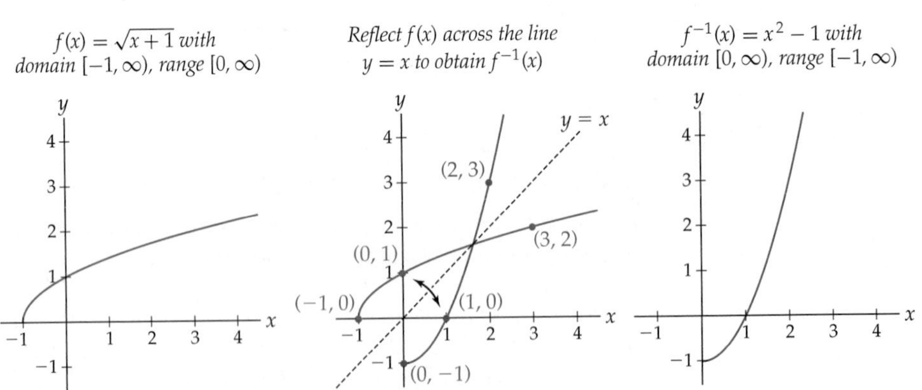

Sometimes a function is not invertible on its largest domain, but *is* invertible on some smaller, "restricted" domain. For example, the function $g(x) = x^2 - 1$ is not invertible on its usual domain $\mathbb{R}$, but it is invertible on the restricted domain $[0, \infty)$. In general, to find a restricted domain on which a function is one-to-one, we choose a smaller domain on which the graph passes the horizontal line test; see Example 4.

Examples and Explorations

EXAMPLE 1

Combinations of functions and their domains

Describe the domains of each of the combinations of $f(x) = \frac{1}{x}$ and $g(x) = \sqrt{x+1}$ that follow. Then find an expression for the combination function.

(a) $3f$ **(b)** $f+g$ **(c)** $\frac{f}{g}$ **(d)** $f \circ g$ **(e)** $g \circ f$

SOLUTION

(a) The domain of $3f$ is the same as the domain of f, which is $x \neq 0$. We have $(3f)(x) = 3f(x) = 3\left(\frac{1}{x}\right) = \frac{3}{x}$.

(b) The domain of f is $x \neq 0$, and the domain of g is $x \geq -1$. The domain of their sum $f+g$ is the intersection of these domains, or $[-1, 0) \cup (0, \infty)$. For values of x in this domain we have $(f+g)(x) = f(x) + g(x) = \frac{1}{x} + \sqrt{x+1}$.

(c) We have $g(x) = 0$ only when $x = -1$, so the quotient $\frac{f}{g}$ is defined on the intersection of the domains of f and g with the point $x = -1$ removed, or $(-1, 0) \cup (0, \infty)$. On this domain we have

$$\left(\frac{f}{g}\right)(x) = \frac{f(x)}{g(x)} = \frac{1/x}{\sqrt{x+1}} = \frac{1}{x\sqrt{x+1}}.$$

(d) For a value x to be in the domain of $f \circ g$, it must first be in the domain $[-1, \infty)$ of g. Then the value of $g(x) = \sqrt{x+1}$ must be in the domain of f, so we must have $\sqrt{x+1} \neq 0$, or in other words, $x \neq -1$. Therefore the domain of $f \circ g$ is $(-1, \infty)$. For values of x in this domain we have $(f \circ g)(x) = f(g(x)) = f(\sqrt{x+1}) = \frac{1}{\sqrt{x+1}}$. Notice that this equation is consistent with our calculation of the domain.

(e) For a value x to be in the domain of $g \circ f$, it must first be in the domain $(-\infty, 0) \cup (0, \infty)$ of f. Then the value of $f(x) = \frac{1}{x}$ must be in the domain of g, so we must have $\frac{1}{x} \in [-1, \infty)$. Since $\frac{1}{x} \geq -1$ when $x \leq -1$ or $x > 0$, the domain of $g \circ f$ is $(-\infty, -1] \cup (0, \infty)$. For values of x in this domain we have $(g \circ f)(x) = g(f(x)) = g\left(\frac{1}{x}\right) = \sqrt{\frac{1}{x} + 1}$. Again notice that the domain that we found does make sense with this equation. □

EXAMPLE 2

Vertical and horizontal translations, stretches, and reflections

The figure that follows shows a piece of the graph of $f(x) = 3 + 2x - x^2$ with five marked points. Find equations and graphs for each of the given transformations. On each graph, mark the new coordinates of the five marked points.

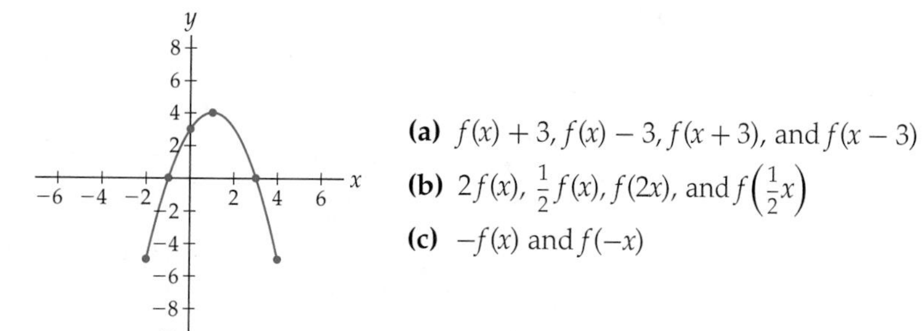

(a) $f(x) + 3$, $f(x) - 3$, $f(x+3)$, and $f(x-3)$

(b) $2f(x)$, $\frac{1}{2}f(x)$, $f(2x)$, and $f\left(\frac{1}{2}x\right)$

(c) $-f(x)$ and $f(-x)$

SOLUTION

(a) The equations for the four transformations are

$$f(x) + 3 = (3 + 2x - x^2) + 3 = 6 + 2x - x^2,$$
$$f(x) - 3 = (3 + 2x - x^2) - 3 = 2x - x^2,$$
$$f(x + 3) = 3 + 2(x + 3) - (x + 3)^2 = -4x - x^2, \text{ and}$$
$$f(x - 3) = 3 + 2(x - 3) - (x - 3)^2 = -12 + 8x - x^2.$$

The graphs of the transformations are shifts up, down, left, and right, respectively, of the original graph by 3 units, as shown in the red graphs in the following figures:

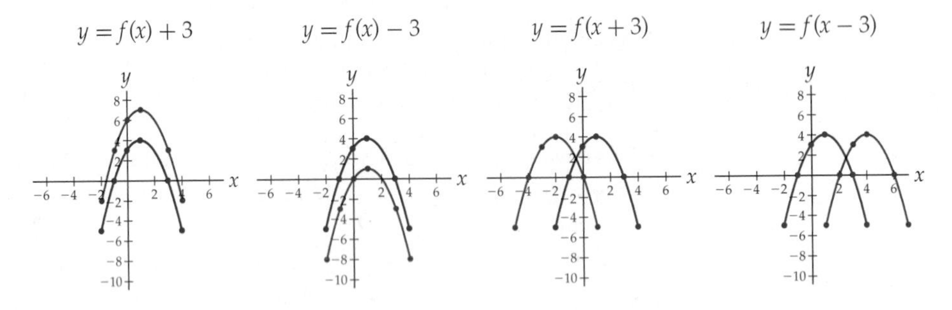

In each case, every point on the original graph is shifted in some direction by 3 units. For example, the point $(1, 4)$ on the original graph becomes $(1, 7)$ on the graph of $f(x) + 3$, $(1, 1)$ on the graph of $f(x) - 3$, $(-2, 4)$ on the graph of $f(x + 3)$, and $(4, 4)$ on the graph of $f(x - 3)$. The other four marked points move in a similar fashion.

(b) Algebraically, the four transformations are given by the equations

$$2f(x) = 2(3 + 2x - x^2) = 6 + 4x - 2x^2,$$
$$\frac{1}{2}f(x) = \frac{1}{2}(3 + 2x - x^2) = \frac{3}{2} + x - \frac{1}{2}x^2,$$
$$f(2x) = 3 + 2(2x) - (2x)^2 = 3 + 4x - 4x^2, \text{ and}$$
$$f\left(\frac{1}{2}x\right) = 3 + 2\left(\frac{1}{2}x\right) - \left(\frac{1}{2}x\right)^2 = 3 + x - \frac{1}{4}x^2.$$

The first two transformations cause the graph of f to stretch or compress vertically, and the last two cause the graph to compress or stretch horizontally, as shown in the red graphs in the next four figures.

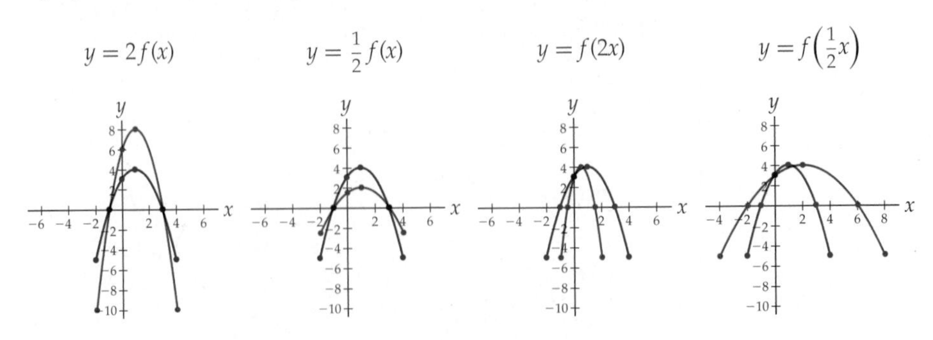

In the first two transformations it is the y-coordinate that changes; for example, the point $(1, 4)$ on the original graph moves to $(1, 8)$ for the first transformation and to $(1, 2)$ for the second. For the last two transformations, it is the x-coordinate that changes; for example, $(1, 4)$ moves to $\left(\frac{1}{2}, 4\right)$ for the third transformation and to $(2, 4)$ for the fourth. The other marked points move in a similar fashion, with either the x- or the y-coordinate being multiplied or divided by 2.

(c) The equations for the two transformations are

$$-f(x) = -(3 + 2x - x^2) = -3 - 2x + x^2 \text{ and}$$
$$f(-x) = 3 + 2(-x) - (-x)^2 = 3 - 2x - x^2.$$

The first transformation gives a vertical reflection across the x-axis, with each marked point (x, y) moving to the point $(x, -y)$, as shown next at the left. The second transformation causes a horizontal reflection across the y-axis, with each marked point (x, y) moving to the point $(-x, y)$, as shown at the right.

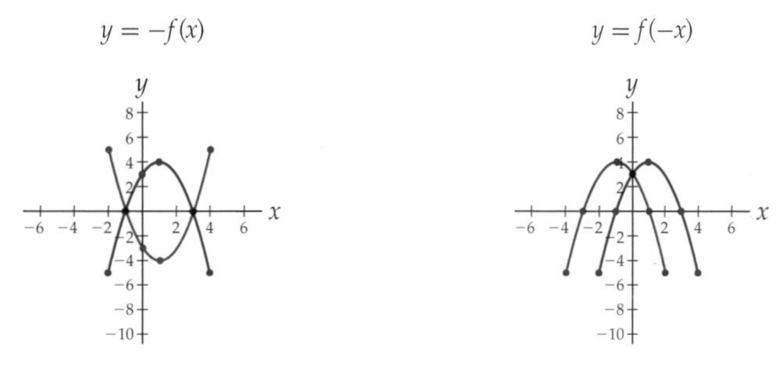

□

EXAMPLE 3

Testing if functions are even or odd

Determine whether each of the following functions is even, odd, or neither:

(a) $f(x) = \dfrac{1}{x}$ **(b)** $g(x) = x^4 - x^2$ **(c)** $h(x) = \dfrac{2 + x}{1 + x^2}$

SOLUTION

(a) To determine whether f is even or odd (or neither) we must calculate $f(-x)$ and determine if it is equal to $f(x)$, $-f(x)$, or neither. We have

$$f(-x) = \frac{1}{-x} = -\frac{1}{x} = -f(x),$$

so $f(x) = \frac{1}{x}$ is an odd function.

(b) The function $g(x)$ is even because

$$g(-x) = (-x)^4 - (-x)^2 = x^4 - x^2 = g(x).$$

(c) Since we have

$$h(-x) = \frac{2 + (-x)}{1 + (-x)^2} = \frac{2 - x}{1 + x^2},$$

which is equal neither to $h(x)$ nor to $-h(x)$, this function is neither even nor odd. □

The graphs of the functions f, g, and h are shown next. Note that f has rotational symmetry about the origin, g has y-axis reflectional symmetry, and h does not have either type of symmetry.

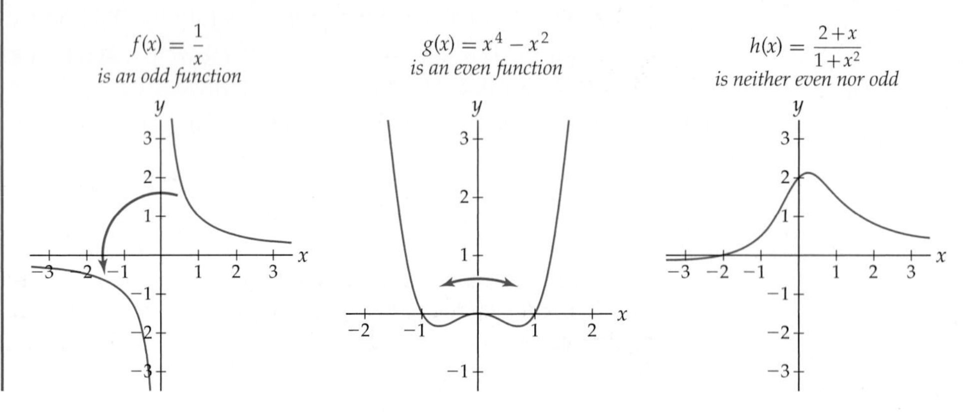

EXAMPLE 4

Graphically finding a restricted domain on which a function is invertible

Explain why the function f graphed here is not invertible on its domain. Then find three restricted domains on which the function *does* have an inverse.

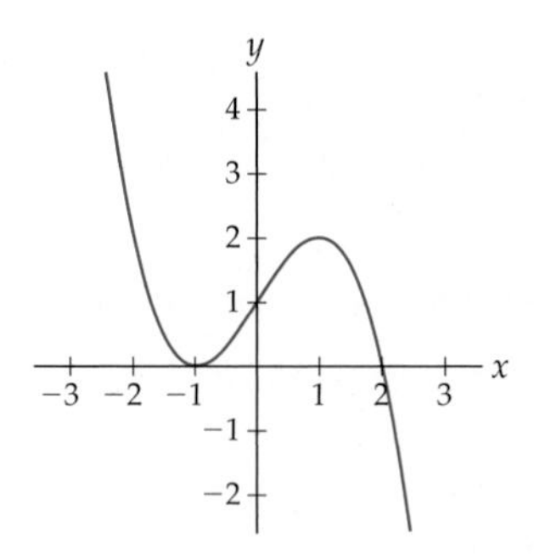

SOLUTION

This function is not invertible on the domain $\mathbb{R}$, because it does not pass the horizontal line test and therefore fails to be a one-to-one function. However, small enough pieces of the graph of f do pass the horizontal line test and therefore have an inverse on that restricted domain. The three graphs that follow show the graph of f on the restricted-domain domains $[-1, 1]$, $(-\infty, -1]$, and $[1, \infty)$, respectively. All three of these restricted graphs pass the horizontal line test and thus are invertible.

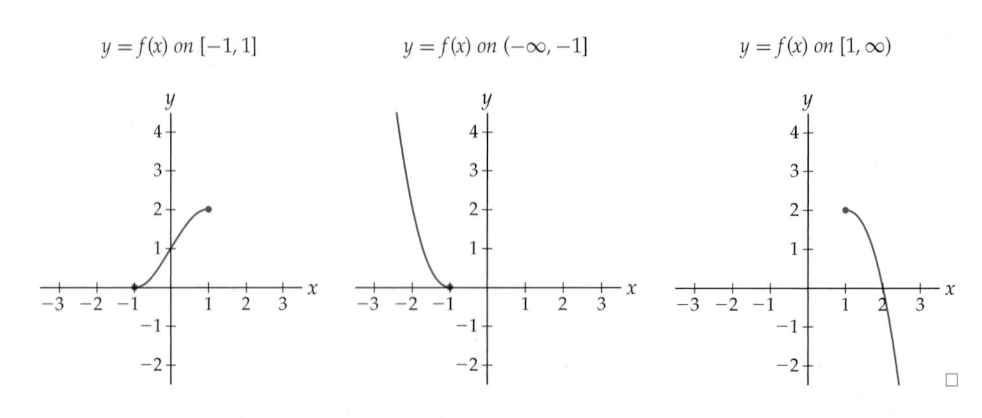

□

EXAMPLE 5 **The graphical and algebraic relationships of inverse functions**

Consider the function $f(x) = 2x - 2$.

(a) Explain why f must have an inverse. Then sketch graphs of $y = f(x)$ and $y = f^{-1}(x)$ and explain the relationship between these two graphs.

(b) Find a formula for $f^{-1}(x)$.

(c) Algebraically verify that $f(f^{-1}(x)) = x$ and $f^{-1}(f(x)) = x$.

SOLUTION

(a) $f(x) = 2x - 2$ has an inverse because it is a non-constant linear function and thus passes the horizontal line test and is one-to-one. By reflecting the graph of $y = f(x)$ across the line $y = x$, we obtain the graph of this inverse; see the following figures:

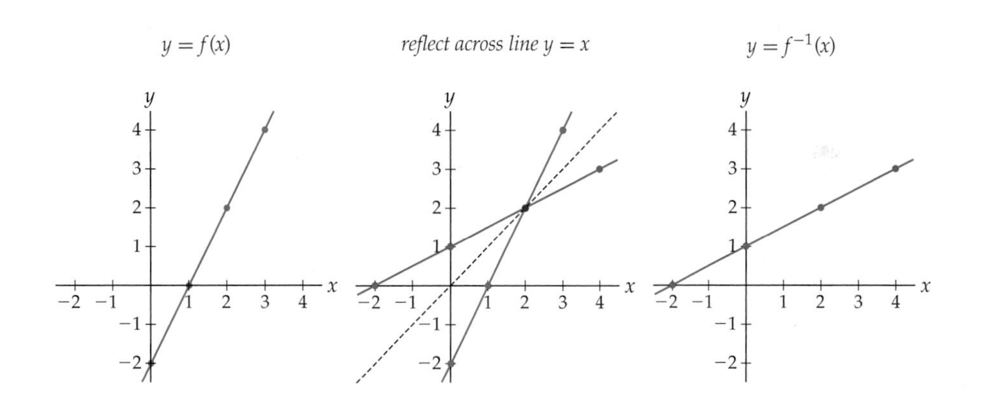

Whenever (x, y) is a point on the graph of f, the point (y, x) is on the graph of f^{-1}. For example, in the preceding graphs we see that $(2, 2)$ and $(3, 4)$ are on the graph of f while $(2, 2)$ and $(4, 3)$ are on the graph of f^{-1}.

(b) We could find a formula for f^{-1} by using the two-point form of a line and the two points shown in the rightmost figure. However, we will find it by solving the equation $y = 2x - 2$ for x:

$$y = 2x - 2 \implies y + 2 = 2x \implies x = \frac{1}{2}(y + 2) = \frac{1}{2}y + 1.$$

If $y = f(x)$, then $x = f^{-1}(y)$; thus we have shown that $f^{-1}(y) = \frac{1}{2}y + 1$. Since we would rather represent the independent variable of f^{-1} by the traditional letter x, replace the y's in the equation with x's to get $f^{-1}(x) = \frac{1}{2}x + 1$. Notice that this equation does appear reasonable, given the slope and y-intercept in the rightmost graph.

(c) For $f(x) = 2x - 2$ and $f^{-1}(x) = \frac{1}{2}x + 1$, we have

$$f(f^{-1}(x)) = f\left(\frac{1}{2}x + 1\right) = 2\left(\frac{1}{2}x + 1\right) - 2 = x + 2 - 2 = x,$$

$$f^{-1}(f(x)) = f^{-1}(2x - 2) = \frac{1}{2}(2x - 2) + 1 = x - 1 + 1 = x.$$

□

? TEST YOUR UNDERSTANDING

- When we write $(f+g)(x) = f(x) + g(x)$, what kind of objects is the first "+" symbol adding together? What kind of objects is the second "+" adding together?
- Why does it make sense that adding or multiplying a constant to the dependent variable of $y = f(x)$ would cause a vertical change in the graph?
- Why does it make sense that the graph of $y = f(x-5)$ would be obtained by shifting the graph of $y = f(x)$ to the right, rather than to the left?
- In Definition 0.10 we talk about *the* inverse of f. If a function has an inverse it has *only one*. Can you explain why?
- What is the difference between the inverse and the reciprocal of the function $f(x) = 2x - 2$ from Example 5? Why might someone wrongly confuse these two functions?

EXERCISES 0.2

Thinking Back

Evaulating functions: Find each of the following evaluations of the function $f(x) = \frac{x}{x-1}$:

- $f(1.5)$
- $f(x^2)$
- $f(x^2+1)$
- $f\left(\frac{1}{x}\right)$
- $f(f(x))$
- $f(f(x^2))$

Solving equations: Each equation that follows expresses s in terms of r. Solve for r (i.e., write r in terms of s).

- $s = \frac{r-2}{3}$
- $s = \frac{3}{r-2}$
- $s = \frac{r+1}{r}$
- $s = \frac{r+2}{r+3}$

Concepts

0. *Problem Zero:* Read the section and make your own summary of the material.

1. *True/False:* Determine whether each of the statements that follow is true or false. If a statement is true, explain why. If a statement is false, provide a counterexample.

(a) *True or False:* The domain of a function of the form $f+g$ is equal to the domain of f or the domain of g, whichever is smaller.

(b) *True or False:* The domain of a function of the form $f \cdot g$ is equal to the intersection of the domains of f and g.

(c) *True or False:* If the graph of $y = f(x)$ contains the point (a, b), then the graph of $y = f(x) + C$ must contain the point $(a, b+C)$.

(d) *True or False:* If the graph of $y = f(x)$ contains the point (a, b), then the graph of $y = f(x+C)$ must contain the point $(a+C, b)$.

(e) *True or False:* The inverse of the one-to-one function $f(x) = x^5$ is $f^{-1}(x) = x^{-5}$.

(f) *True or False:* If f is an invertible function, then $f^{-1} = \frac{1}{f}$.

(g) *True or False:* Every even function is a function that involves only even exponents.

(h) *True or False:* If f is an even function and $(0, b)$ is a point on the graph of $y = f(x)$, then $(0, -b)$ must also be on the graph of $y = f(x)$.

2. *Examples:* Construct examples of the thing(s) described in the following. Try to find examples that are different than any in the reading.

(a) A pair of functions f and g for which $f \circ g$ happens to be equal to $g \circ f$.

(b) A function f that is its own inverse.

(c) A function f that is both even and odd.

3. Explain what the definition of $(f-g)(x)$ ought to be. Show that this definition is just a combination of the definitions of $(f+g)(x)$ and $kf(x)$.

4. Suppose $(2, 5)$ is a point on the graph of $y = f(x)$. Fill in the blanks with the transformed coordinates of this point under each of the following transformations:

(a) ________ is on the graph of $f(x) - 4$.

(b) ________ is on the graph of $f(x-4)$.

(c) ________ is on the graph of $-7f(x)$.

(d) ________ is on the graph of $f\left(\frac{1}{3}x\right)$.

(e) ________ is on the graph of $3f(x+1)$.

(f) ________ is on the graph of $f(3x+1)$.

5. Fill in the blanks as appropriate. There may be more than one possible answer.

(a) If the point ________ is on the graph of $y = f(x)$, then $(4, 2)$ is on the graph of $y = f(x-3)$.

(b) If $(3, -2)$ is on the graph of $y = f(x)$, then $(6, -2)$ is on the graph of the function ________ .

(c) If $(1, 4)$ is on the graph of an even function $y = f(x)$, then ________ is also on the graph of $y = f(x)$.

(d) If $(-2, 5)$ is on the graph of an odd function $y = f(x)$, then ________ is also on the graph of $y = f(x)$.

6. Suppose f has domain $[1, \infty)$ and g has domain $[-4, 4]$. Suppose also that $f(x)$ is nonzero *except* at $x = 2$ and $x = 5$ and that $g(x)$ is nonzero *except* at $x = -1$ and $x = 1$. Find the domains of the following functions (if possible):

(a) $3f + 4g$ (b) $\dfrac{1}{fg}$ (c) $\dfrac{1}{f+g}$

7. Suppose f is a function with domain $[2, \infty)$ and range $[-3, 3]$, and let g be a function with domain $[-10, \infty)$ and range $[0, \infty)$.

(a) What is the domain of the composition $g \circ f$? Justify your answer.

(b) It is not possible to determine the domain of $f \circ g$ in this example; explain why not. What extra information would you have to know to be able to determine the domain of $f \circ g$?

(c) Is there enough information here to determine the domain of the composition $f \circ f$? What about the function $g \circ g$?

8. Use compositions to answer each of the following:

(a) If $g(x) = x^2$ and $f(g(x)) = \dfrac{1}{x^2+1}$, what is $f(x)$?

(b) If $h(x) = x^2 - 1$ and $h(l(x)) = \dfrac{1}{x^4} - 1$, what is $l(x)$?

(c) If $u(x) = \dfrac{1}{1-x}$ and $y(u(x)) = \dfrac{1}{1-x}$, what is $y(x)$?

9. Given that $y = f(x)$ has the graph on the left, use transformations to find a formula in terms of $f(x)$ for the function graphed on the right.

$y = f(x)$

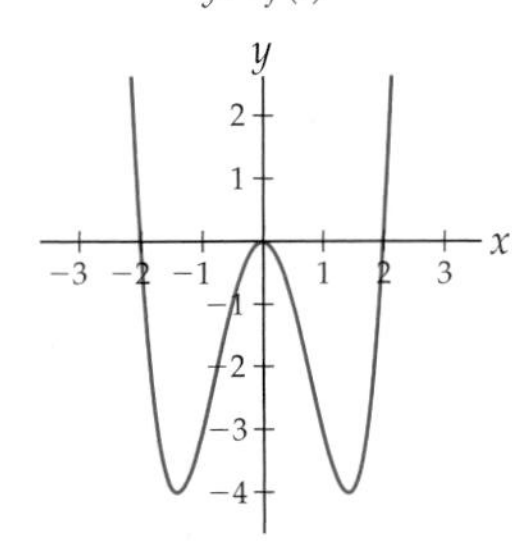

A transformation of $y = f(x)$

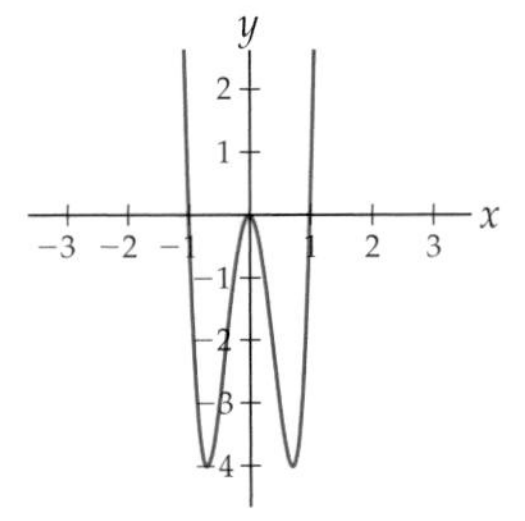

10. Given that $y = f(x)$ has the graph on the left, use transformations to find a formula in terms of $f(x)$ for the function graphed on the right.

$y = f(x)$

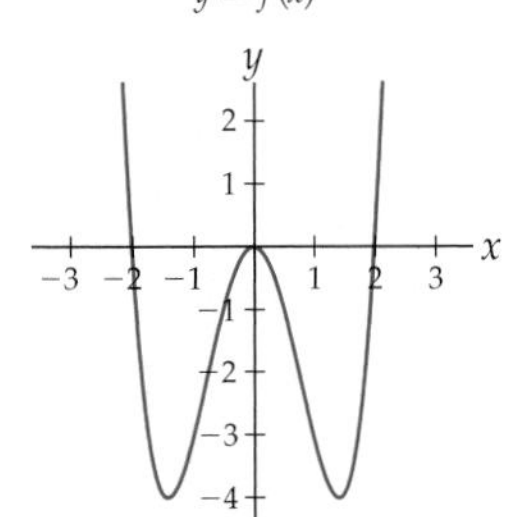

A transformation of $y = f(x)$

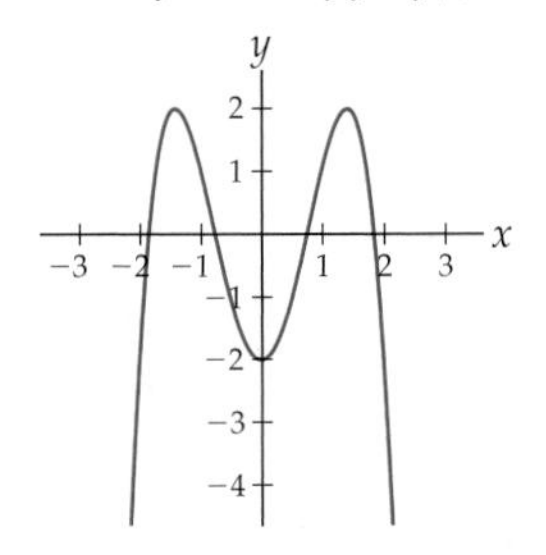

11. If $f(0) = 2$, can f be an odd function? What if $f(0)$ is undefined? Explain your answers.

12. Determine graphically whether each of the following four functions is even, odd, or neither.

(a)

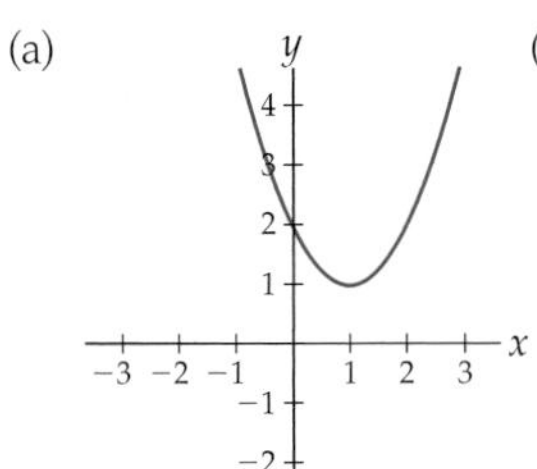

(b)

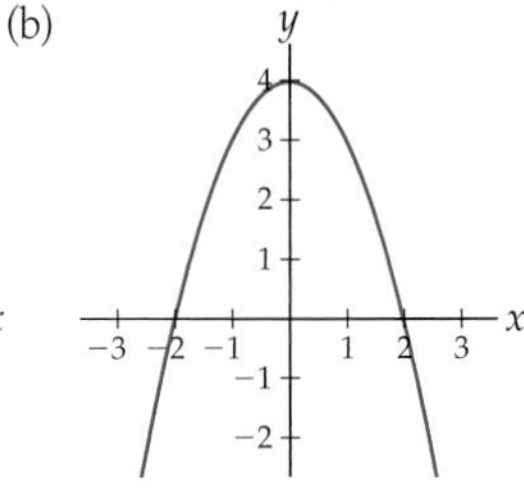

(c)

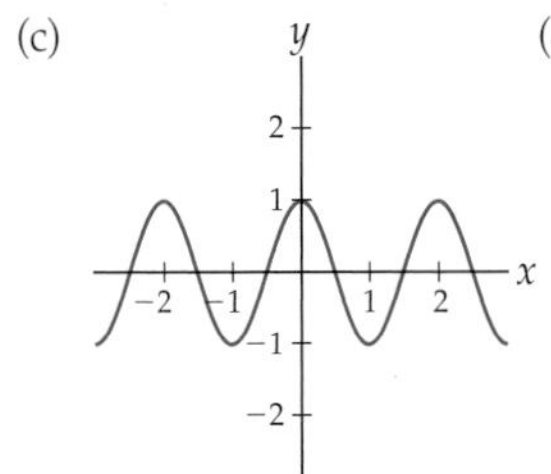

(d)

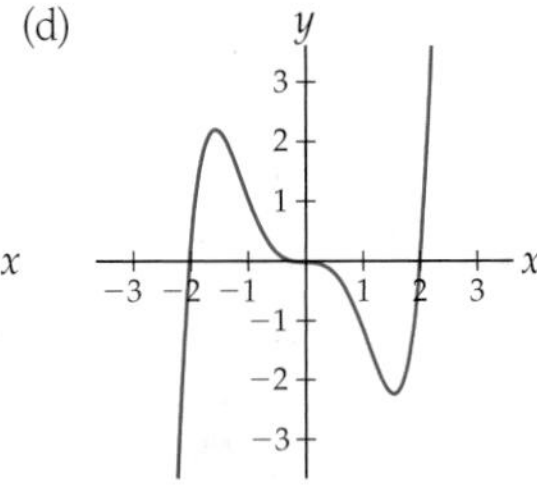

13. Complete the entries in the following table two ways: (a) to make an even function and (b) to make an odd function:

x	-3	-2	-1	0	1	2	3
$f(x)$	4				-2	1	

14. Suppose f is a function with domain $\mathbb{R}$ whose right-hand side is as shown here. Sketch the left-hand side of the graph so that (a) f is an even function, (b) f is an odd function, and (c) f is neither even nor odd.

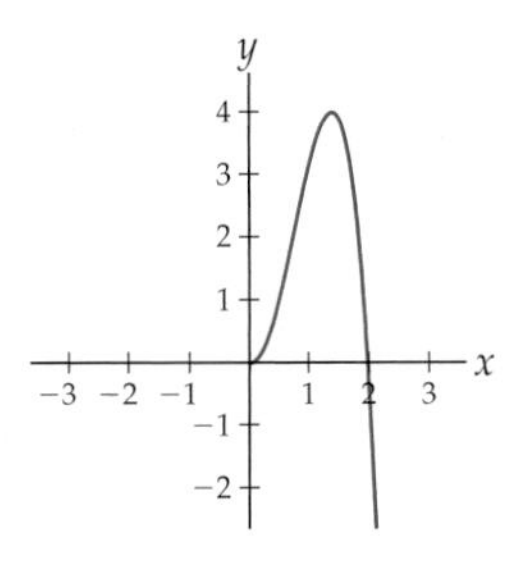

15. Suppose f is an invertible function with inverse f^{-1}. What is $(f^{-1})^{-1}$? Explain your answer.

16. Given that f is an invertible function, fill in the blanks.

(a) If $f(-1) = 0$, then $f^{-1}(0) =$________ .

(b) If $(2, 3)$ is on the graph of f, then ________ is on the graph of f^{-1}.

(c) If ________ is on the graph of f, then $(-2, 4)$ is on the graph of f^{-1}.

17. If an invertible function f has domain $[-1, 1)$ and range $(-\infty, 3]$, what are the domain and range of f^{-1}? Sketch the graph of a function f with the domain and range given, and on the same set of axes sketch the graph of its inverse f^{-1}.

18. Consider the function $f(x) = x^5 + 4x^3 + 2x + 1$. Alina says that f is invertible, because its graph appears to be one-to-one. Linda says that f is not invertible, because she cannot figure out how to solve $y = x^5 + 4x^3 + 2x + 1$ for x in terms of y. Who is correct, and why?

19. Use the values given in the table to fill in the missing values. There is only one correct way to fill in the table.

x	0	1	2
$f(x)$	1		
$g(x)$	0		-2
$h(x)$		1	
$(-2f)(x)$		4	
$(f + 2g - h)(x)$	4		
$(fg)(x)$			2
$\left(\frac{g}{h}\right)(x)$		3	-1

20. Use the values given in the table to fill in the missing values. There is only one correct way to fill in the table.

x	1	2	3	4
$f(x)$	4	3	1	2
$g(x)$	3		1	2
$h(x)$		2	4	
$(f \circ g)(x)$		4		
$(h \circ f)(x)$	3			2
$(g \circ g)(x)$				
$(g \circ f \circ h)(x)$	2			

Skills

Given that $f(x) = x^2 + 1$, $g(x) = \frac{1}{x-2}$, and $h(x) = \sqrt{x}$, find the domain of each function in Exercises 21–29. Then find an equation for the function and calculate its value at $x = 1$.

21. $(f + g)(x)$ **22.** $(3fh)(x)$ **23.** $\left(\frac{g}{h}\right)(x)$

24. $(g \circ f)(x)$ **25.** $(g \circ g)(x)$ **26.** $(g \circ h \circ f)(x)$

27. $g(x - 5)$ **28.** $h(3x) + 1$ **29.** $h(3x + 1)$

The table that follows defines three functions f, g, and h. Create additional rows for the table for each function in Exercises 30–38. (For some transformations, you may have to use different x-values than the ones in the table.)

x	0	1	2	3	4	5	6
$f(x)$	0	1	3	2	3	0	2
$g(x)$	1	0	1	1	0	1	0
$h(x)$	3	2	0	3	2	3	1

30. $(f - g)(x)$ **31.** $2f(x) + 3$ **32.** $(gh)(x)$

33. $(h \circ g)(x)$ **34.** $(g \circ h)(x)$ **35.** $(f \circ f \circ f)(x)$

36. $-h(-x)$ **37.** $g(x - 1)$ **38.** $f(2x)$

Use the graphs of f and g given here to sketch the graphs of the functions in Exercises 39–50. Label at least four points on each graph. Don't find or use equations for the given graphs.

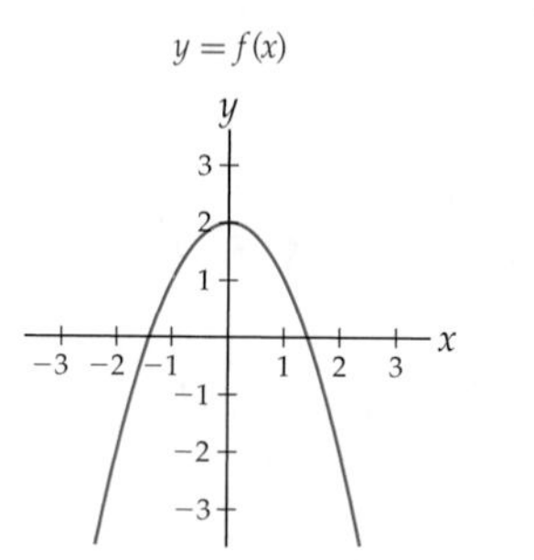

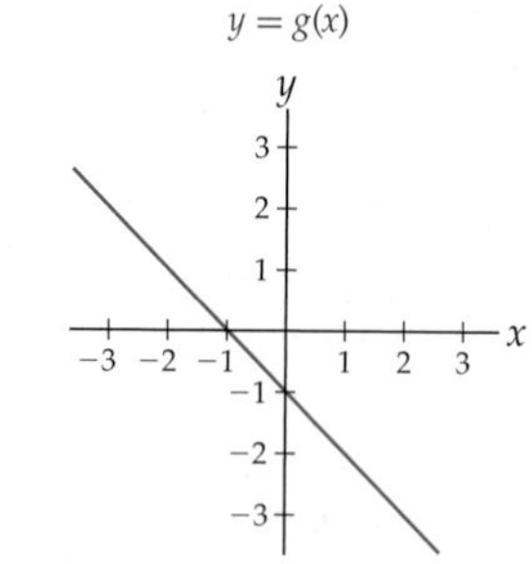

39. $(f + g)(x)$ **40.** $2 - 3f(x)$ **41.** $g\left(\frac{1}{2}x\right)$

42. $g(x - 2) + 1$ **43.** $(-.5f)(x)$ **44.** $(fg)(x)$

45. $(f \circ g)(x)$ **46.** $(g \circ f)(x)$ **47.** $(f \circ f)(x)$

48. $g^{-1}(x)$ **49.** $\left(\frac{1}{f}\right)(x)$ **50.** $\left(\frac{1}{g}\right)(x)$

Suppose f and g are the piecewise-defined functions defined here. For each combination of functions in Exercises 51–56, (a) find its values at $x=-1$, $x=0$, $x=1$, $x=2$, and $x=3$, (b) sketch its graph, and (c) write the combination as a piecewise-defined function.

$$f(x)=\begin{cases} 2x+1, & \text{if } x\le 0 \\ x^2, & \text{if } x>0 \end{cases} \qquad g(x)=\begin{cases} -x, & \text{if } x<2 \\ 5, & \text{if } x\ge 2 \end{cases}$$

51. $(f+g)(x)$ **52.** $3f(x)$ **53.** $(g\circ f)(x)$

54. $f(x)-1$ **55.** $f(x-1)$ **56.** $g(3x)$

Find two nontrivial ways to write each of the functions in Exercises 57–60 as a composition $f=g\circ h$. "Nontrivial" means that you should *not* choose $g(x)=x$ or $h(x)=x$.

57. $f(x)=3x^2+5$ **58.** $f(x)=(3x+5)^2$

59. $f(x)=\dfrac{6}{x+1}$ **60.** $f(x)=\dfrac{x^2}{\sqrt{1+x^2}}$

Determine algebraically whether the functions in Exercises 61–66 are even, odd, or neither. Afterwards, verify your answers by inspecting the symmetries of the graphs.

61. $f(x)=x^4+1$ **62.** $f(x)=1-2x$

63. $f(x)=x^3+x^2$ **64.** $f(x)=-\dfrac{2}{5x^3}$

65. $f(x)=\dfrac{x^3}{x^2+1}$ **66.** $f(x)=\dfrac{1}{x+1}$

Use Definition 0.10 to show that each pair of functions in Exercises 67–70 are inverses of each other.

67. $f(x)=2-3x$ and $g(x)=-\dfrac{1}{3}x+\dfrac{2}{3}$

68. $f(x)=x^2$ restricted to $[0,\infty)$ and $g(x)=\sqrt{x}$

69. $f(x)=\dfrac{x}{1-x}$ and $g(x)=\dfrac{x}{1+x}$

70. $f(x)=\dfrac{1}{2x}$ and $g(x)=\dfrac{1}{2x}$

For each invertible function in Exercises 71–74, sketch the graph of f^{-1} and label three points on its graph. If a function is not invertible, then find a restricted domain on which it is invertible and sketch a graph of the restricted inverse. Don't find or use equations for the given graphs.

71.

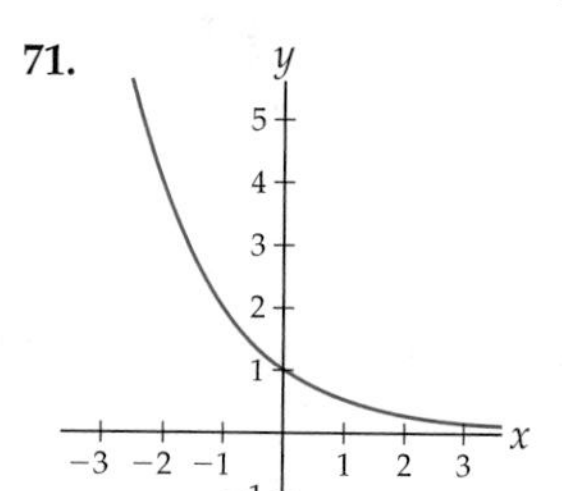

72.

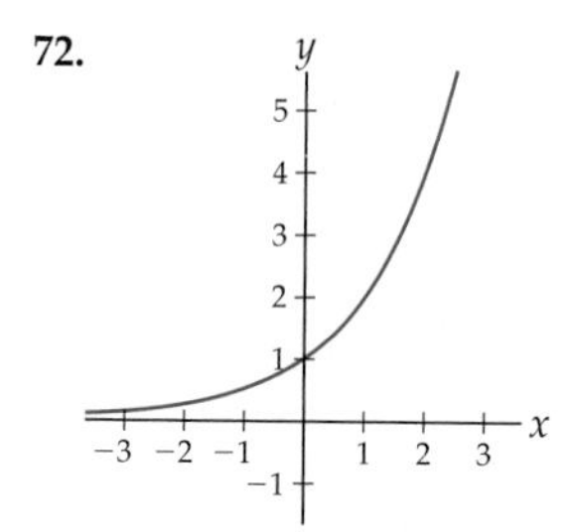

73.

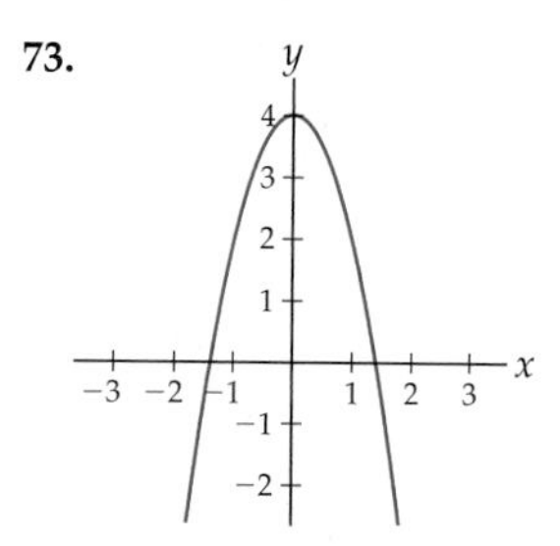

74.

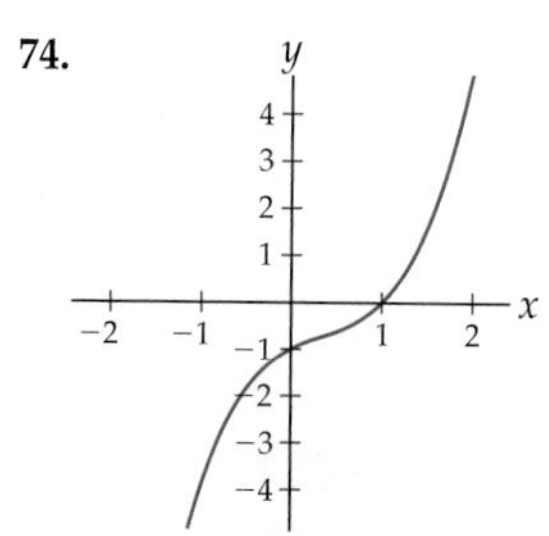

Find the inverse of each function in Exercises 75–80. Then check your answers by graphing f and f^{-1}.

75. $f(x)=\dfrac{1-5x}{2}$ **76.** $f(x)=1.2-3.5x$

77. $f(x)=1+\dfrac{1}{x}$ **78.** $f(x)=\dfrac{2}{x+1}$

79. $f(x)=\dfrac{x-1}{x+1}$ **80.** $f(x)=\dfrac{x-2}{x}$

Applications

81. CarpetKing charges $4.25 per square foot for its Deluxe ThriftySoft carpet, plus a flat fee of $200.00 for delivery and installation.

(a) If a square room measures x feet on a side, and S is the number of square feet of floor in the room, write down S as a function of x.

(b) Write down a function that describes the cost C of carpeting a room enclosing S square feet.

(c) Write down a function that describes the cost C of carpeting a square room that measures x feet on a side. Explain how this function is a composition.

82. The first table that follows shows the number of deer in Happyland Forest Park during 1990–1995. The number of deer seems to affect the number of ticks in the park; the second table shows the number of ticks that can be expected for various numbers of deer in the park.

Year	1990	1991	1992	1993	1994	1995
Deer	183	180	177	179	184	181

Deer	177	178	179	180	181	182	183	184
Ticks	850	855	860	865	870	875	880	885

(a) Use the tables to estimate the number of ticks that were in the park in 1995.

(b) Make a table that predicts the number of ticks in the park during 1990–1995.

(c) Explain how your table represents a composition of the deer and tick tables.

83. Juri's custom printing t-shirt shop charges $12.00 per t-shirt plus a $20.00 setup fee.

(a) Write a function that describes the cost C of printing n t-shirts at Juri's store.

(b) Show that $C(n)$ is an invertible function.

(c) Find the inverse of $C(n)$. What does this inverse function $n(C)$ represent in terms of C and n?

(d) Use the function you found in part (c) to calculate the number of shirts you can produce with $150.

84. The math teachers at Pinnacle High School have discovered a relationship between the number of hours their students watch television each day and their students' scores on math tests. The following table shows this relationship:

Hours of T.V.	0	1	2	3	4	5	6
Grade on test	92	86	80	74	68	62	56

(a) Find a linear function that describes the grade g on a student's math test as a function of the number of hours t of television that the student watches every day.

(b) Show that your function $g(t)$ is invertible.

(c) Find the inverse of $g(t)$. What does this function $t(g)$ represent in terms of t and g?

(d) Use $t(g)$ to predict the number of hours that Calvin watched television each day given that he scored a 40 on his math test. How much television can he be allowed to watch each day if his mother wants him to get an 85 on his next test?

Proofs

85. Prove algebraically that if $f(x) = x^k$, where k is a positive integer, then the graphs of $y = f(2x)$ and $y = 2^k f(x)$ are the same.

86. Prove that if (x, y) is a point on the graph of $y = f(x)$ and C is a real number, then

(a) $(x, y + C)$ is a point on the graph of $y = f(x) + C$.

(b) $(x - C, y)$ is a point on the graph of $y = f(x + C)$.

87. Prove that if (x, y) is a point on the graph of $y = f(x)$ and $k \neq 0$, then

(a) (x, ky) is a point on the graph of $y = kf(x)$.

(b) $\left(\frac{1}{k}x, y\right)$ is a point on the graph of $y = f(kx)$.

88. Prove that if (x, y) is a point on the graph of $y = f(x)$, then

(a) $(x, -y)$ is a point on the graph of $y = -f(x)$.

(b) $(-x, y)$ is a point on the graph of $y = f(-x)$.

89. Prove that every odd function that is defined at $x = 0$ must pass through the origin.

90. Prove that the algebraic definitions of even and odd functions imply even and odd graphical symmetry, by showing that:

(a) If (x, y) is a point on the graph of an even function $y = f(x)$, then $(-x, y)$ is also on the graph.

(b) If (x, y) is a point on the graph of an odd function $y = f(x)$, then $(-x, -y)$ is also on the graph.

91. Prove Theorem 0.12 (a) by using Definition 0.10 to argue that the domain of f^{-1} is the range of f and that the range of f^{-1} is the domain of f.

92. Prove Theorem 0.12 (c) by showing that if f is a function with an inverse f^{-1} and (x, y) is on the graph of $y = f(x)$, then (y, x) is on the graph of $y = f^{-1}(x)$. Why does this conclusion imply that the graph of $f^{-1}(x)$ is the reflection of the graph of $f(x)$ across the line $y = x$?

93. Suppose f is invertible with inverse f^{-1}. Prove that if $f(a) = b$, then $f^{-1}(b) = a$.

94. Prove that an invertible function f can have only *one* inverse. *(Hint: Suppose g and h are both inverses of a function f, and suppose also that there is some real number x_0 for which $g(x_0) \neq h(x_0)$. Show that these suppositions together produce a contradiction.)*

Thinking Forward

Transformations of trigonometric graphs: Consider the function $f(x) = \sin x$ shown in the following figure:

$f(x) = \sin x$

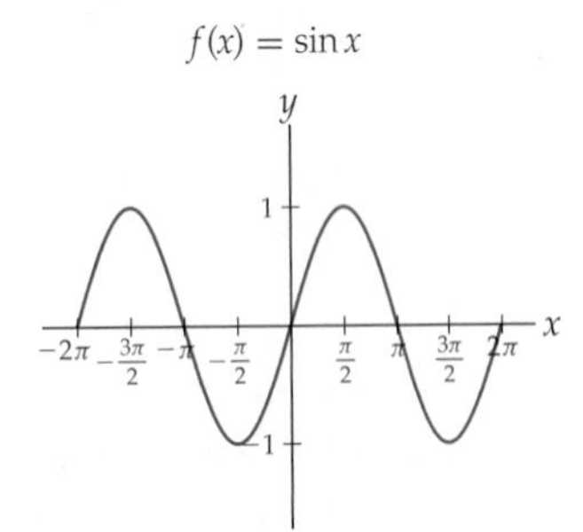

▶ Use transformations to sketch the graph of the function $f\left(x + \frac{\pi}{2}\right)$. (The resulting graph will be the graph of $\cos x$, so this problem illustrates the identity $\sin\left(x + \frac{\pi}{2}\right) = \cos x$.)

▶ Use transformations to sketch the graph of the function $f(x + 2\pi)$. What do you notice about this new graph? What does this mean about the function $f(x) = \sin x$?

▶ Sketch the graph of the reciprocal $\frac{1}{f(x)}$ of $f(x)$. Be especially careful when considering what happens at the places where $f(x)$ has zeros. (The resulting graph will be the graph of $\csc x$, so this problem illustrates that $\frac{1}{\sin x} = \csc x$.)

▶ The function $f(x) = \sin x$ is not one-to-one and therefore is not invertible. However, $f(x) = \sin x$ is invertible on the restricted domain $\left[-\frac{\pi}{2}, \frac{\pi}{2}\right]$. Sketch this restricted function, and then sketch its inverse $f^{-1}(x)$. (The graph of $f^{-1}(x)$ is the graph of the function known as $\sin^{-1} x$.)

0.3 ALGEBRAIC FUNCTIONS

- Power functions, polynomial functions, and rational functions
- Functions that involve absolute values
- Domains, properties, and graphs of algebraic functions

Power Functions

A function is ***algebraic*** if it can be expressed in terms of constants and a variable x by using only arithmetic operations ($+$, $-$, $\times$, and $\div$) and rational constant powers of the variable. For example,

$$f(x) = \frac{3x^2 - 5}{1 - x} \quad \text{and} \quad g(x) = (\sqrt{x} + \sqrt[3]{x})^{7/2}$$

are algebraic functions, but $h(x) = 2^x$ and $k(x) = \sin x$ are not.All algebraic functions are combinations and/or compositions of ***power functions***:

DEFINITION 0.13 **Power Functions**

A ***power function*** is a function that can be written in the form

$$f(x) = Ax^k$$

for some nonzero real number A and some rational number k.

In this definition, the constant k is called the ***power*** or ***exponent***, and the constant A is called the ***coefficient***. Note that the exponent k must be a rational number and the variable must be in the base; this means that, for example, $f(x) = x^{\pi}$ and $f(x) = 10^x$ are not considered power functions.

Although power functions all have the simple form $f(x) = Ax^k$, they vary greatly for different values of exponent k. The following graphs illustrate eight common power functions:

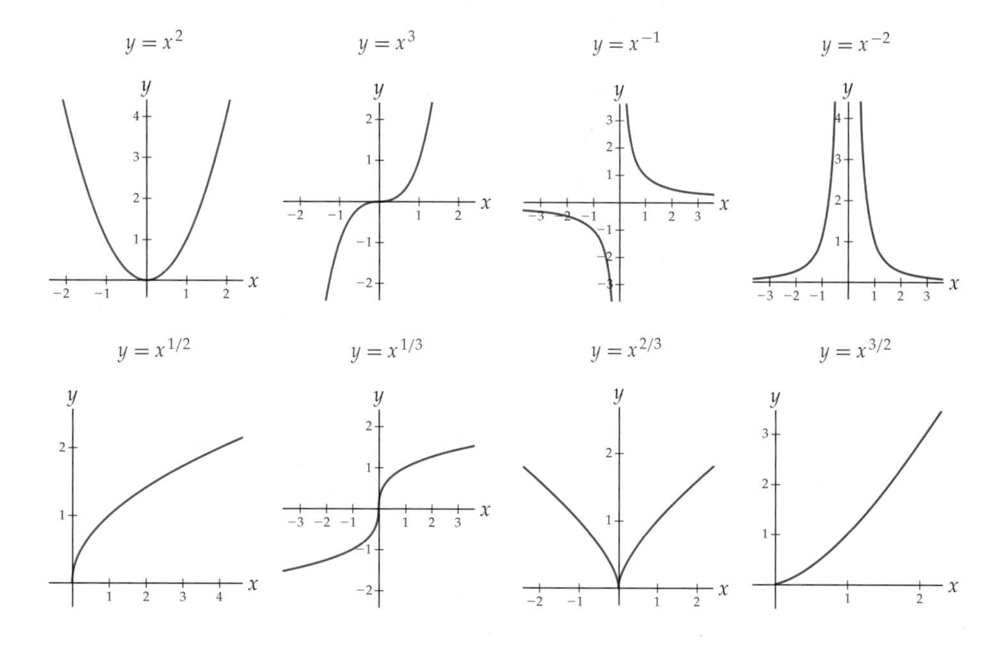

Although all eight of these graphs represent power functions, they are very different. They have different domains and ranges, some have asymptotes and some do not, some have sharp corners and some do not, and so on. However, all power functions with integer powers or positive rational powers have graphs that look similar to one of these eight graphs. For example, if k is a positive odd integer, then the graph of $f(x) = x^k$ looks a lot like the graph of x^3. If $\frac{p}{q}$ is a positive rational number with p even and q odd, and if $\frac{p}{q} < 1$, then the graph of $g(x) = x^{p/q}$ looks a lot like the graph of $x^{2/3}$.

Polynomial Functions

A ***polynomial function*** is a finite sum of power functions that have nonnegative integer powers. The following definition provides a general notation:

DEFINITION 0.14

Polynomial Functions

A ***polynomial function*** of degree n is a function that can be written in the form

$$f(x) = a_n x^n + a_{n-1} x^{n-1} + a_{n-2} x^{n-2} + \cdots + a_2 x^2 + a_1 x + a_0$$

for some integer $n \geq 0$ and some real numbers $a_0, a_1, \ldots a_n$ with $a_n \neq 0$.

As a matter of convention, we also say that the constant zero function $f(x) = 0$ is a polynomial function (and that its degree is undefined).

The numbers a_i (for $i = 0, 1, 2, \ldots, n$) are called the ***coefficients*** of the polynomial. Note that the coefficient belonging to the term containing the power x^i is conveniently named a_i; for example, the coefficient of the x^2 term is called a_2. The coefficient a_n belonging to the highest power of x is called the ***leading coefficient***, and the term $a_n x^n$ containing the highest power of x with $a_n \neq 0$ is called the ***leading term***. The coefficient a_0 is called the ***constant term***.

We have special names for some polynomials according to their degrees. For example, polynomials of degrees 0, 1, 2, 3, 4, and 5 are called ***constant***, ***linear***, ***quadratic***, ***cubic***, ***quartic***, and ***quintic*** polynomials, respectively. (Note: We will require non-zero leading coefficients for quadratics and higher degrees, but not for linear functions; in other words, we will consider constant functions to be linear.) Higher degrees can sometimes result in more roots and more turning points in the graph of a polynomial; for example, examine the following cubic, quartic, and quintic polynomials:

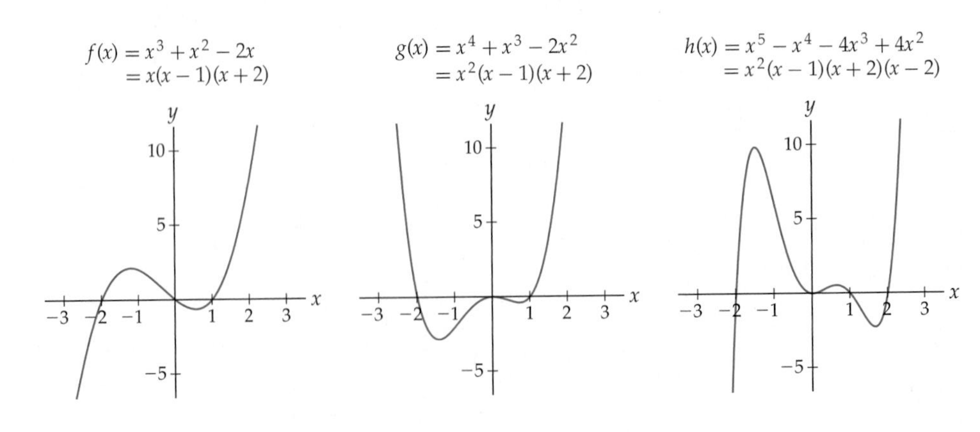

A quadratic expression is ***irreducible*** if it cannot be factored with real-number coefficients; that is, $ax^2 + bx + c$ is irreducible if it cannot be written in the form $a(x - r_1)(x - r_2)$ for some real numbers r_1 and r_2. A quadratic expression $ax^2 + bx + c$ is irreducible if and only if its ***discriminant*** $b^2 - 4ac$ is negative; think about the quadratic formula to see why. For example, the quadratic expressions $x^2 + 5$ and $x^2 + x + 7$ are irreducible.

Note that every linear factor $(x - r)$ of a polynomial corresponds to a root $x = r$ of that polynomial, since if a polynomial $f(x)$ can be factored as $f(x) = (x - r)g(x)$ for some other polynomial $g(x)$, then $f(r) = (r - r)g(r) = 0$ and thus $x = r$ is a root of $f(x)$. It happens that every polynomial function can be factored into linear factors (which correspond to real-number roots) and/or irreducible quadratic factors (which do not correspond to real-number roots). However, just because a polynomial *has* a factorization doesn't mean that we have an easy way to actually factor that polynomial!

The next theorem describes four key graphical properties of polynomial functions. The first part of Theorem 0.15 is related to the ***Fundamental Theorem of Algebra***, and the proof of this deep theorem is beyond the scope of this course. We will have the tools to prove the last three parts of Theorem 0.15 once we study limits and derivatives.

THEOREM 0.15 **Graphical Properties of Polynomial Functions**

If f is a polynomial function of degree n, then the graph of f

(a) has at most n real roots;

(b) has at most $n - 1$ local extrema;

(c) is "smooth" and "unbroken" on $\mathbb{R}$ and has no asymptotes;

(d) behaves like the graph of its leading term at the "ends" of the graph.

The last part tells us that a polynomial function $f(x) = a_n x^n + a_{n-1} x^{n-1} + \cdots + a_1 x + a_0$ with $a_n \neq 0$ will behave like the power function $a_n x^n$ at its "ends." This means that the "ends" of the graph of a polynomial always looks like one of the four graphs that follow, depending on whether the degree n is even or odd and whether the leading coefficient a_n is positive or negative. The dashed part of each graph indicates that this part of the theorem does not tell us the behavior in the middle of the graph, only at the ends.

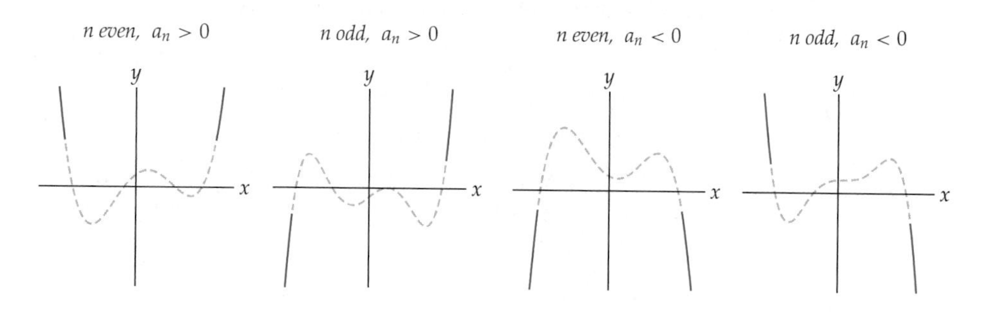

Rational Functions

A rational number is a number that can be written as a quotient of the simplest possible numbers you can imagine, namely, integers. Similarly, a rational *function* is a quotient of very simple functions, namely, polynomials.

DEFINITION 0.16

Rational Functions are Quotients of Polynomials

A ***rational function*** is a function that can be written as the quotient of two polynomial functions

$$f(x) = \frac{p(x)}{q(x)} = \frac{a_n x^n + a_{n-1}x^{n-1} + \cdots + a_1 x + a_0}{b_m x^m + b_{m-1}x^{m-1} + \cdots + b_1 x + b_0},$$

for any x such that $q(x) \neq 0$.

Graphs of rational functions are highly dependent on small changes in their numerators and denominators; for example, examine the three following rational functions:

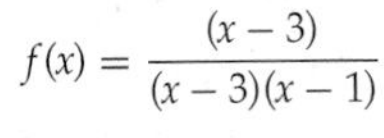

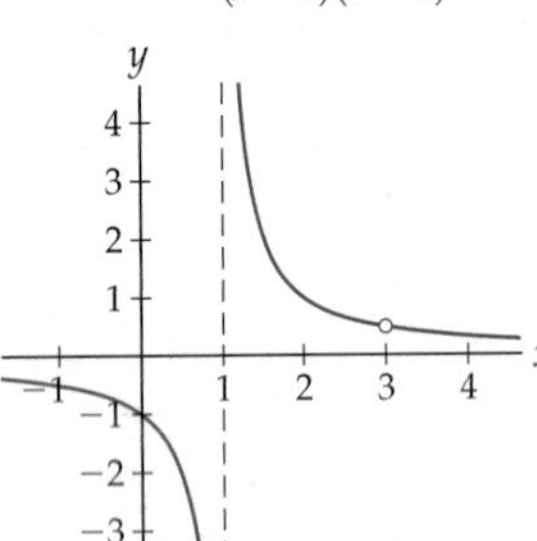

$$g(x) = \frac{(x-3)^2}{(x-3)(x-1)}$$

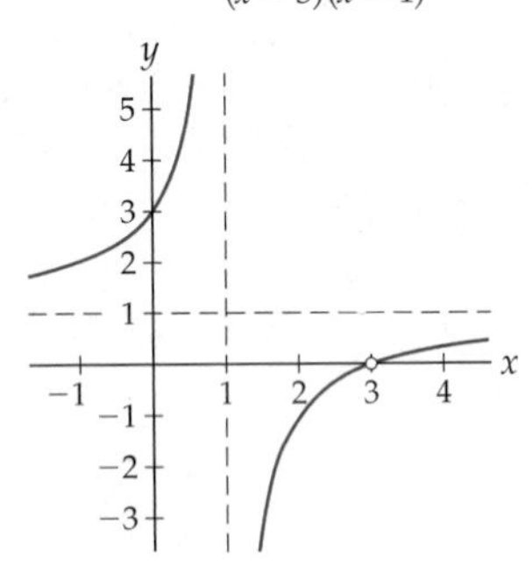

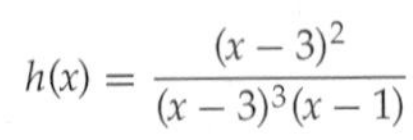

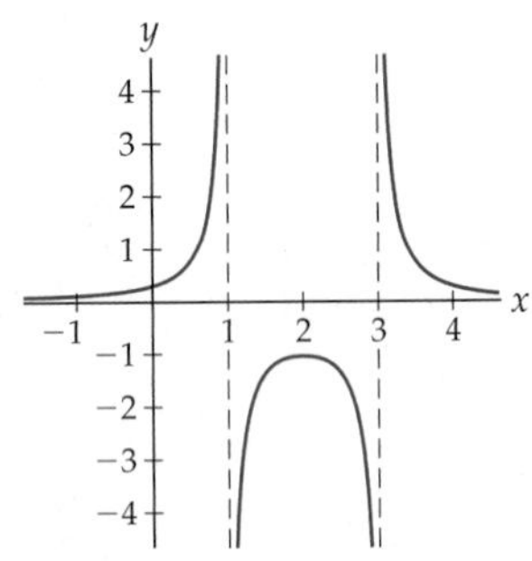

We say that the graph of a function f has a ***hole*** at $x = c$ if the graph of f is a simple unbroken curve near $x = c$, but at $x = c$ there is a point missing from the graph. We will have a more exact definition of holes in Chapter 1. For example, the first graph shown has a hole at $\left(3, \frac{1}{2}\right)$, and the second graph has a hole at $(3, 0)$.

As the preceding examples suggest, the domain, roots, and holes in the graph of a rational function are determined by the roots of its numerator and denominator. These relationships are made precise in Theorems 0.17 and 0.18.

THEOREM 0.17

Graphical Properties of Rational Functions

If $f(x) = \frac{p(x)}{q(x)}$ is a rational function, then the following are true.

(a) f is not defined at the roots of $q(x)$.

(b) f has roots at the points that are roots of $p(x)$ but not roots of $q(x)$.

(c) f has holes at the points that are roots of both $p(x)$ and $q(x)$, provided that the roots have higher or equal multiplicity in $p(x)$ than in $q(x)$.

The ***multiplicity*** of a root $x = c$, as mentioned in the theorem, is the number of times that $(x - c)$ is a factor of the numerator or of the denominator. The first two parts of this theorem follow directly from properties of polynomials, roots, and quotients; see Exercises 92 and 93. The proof of the third part of the theorem will have to wait until we cover limits in Chapter 1.

The next theorem describes the asymptotes of rational functions in terms of their numerator and denominator polynomials. Its proof is necessarily postponed until we have studied limits in Chapter 1.

THEOREM 0.18 **Vertical and Horizontal Asymptotes of Rational Functions**

Suppose $f(x) = \frac{p(x)}{q(x)}$ is a rational function in which $p(x)$ has degree n and $q(x)$ has degree m.

(a) f has vertical asymptotes at the roots of $q(x)$, provided that the roots have a higher multiplicity in $q(x)$ than in $p(x)$.

(b) If $n < m$, then f has a horizontal asymptote at $y = 0$.

(c) If $n = m$, then f has a horizontal asymptote at $y = \frac{a_n}{b_n}$, the ratio of the leading coefficients of $p(x)$ and $q(x)$.

(d) If $n > m$, then the graph of f does not have any horizontal asymptotes.

For example, looking back at our three graphs of rational functions, we see that the first and third functions are "bottom heavy" (since the degree of the denominator is greater than that of the numerator) and have horizontal asymptotes at $y = 0$. The second is "balanced" (since the degrees of the numerator and denominator are the same) and has a horizontal asymptote at $y = 1$.

Absolute Value Functions

Recall that the absolute value $|x|$ of a real number x is equal to x if x is positive (or zero) and is equal to $-x$ if x is negative. For example, $|2| = 2$ while $|-2| = -(-2) = 2$. Thus we can write the function $f(x) = |x|$ as a piecewise-defined function by splitting the definition of $|x|$ into two cases: $x \geq 0$ and $x < 0$.

DEFINITION 0.19 **The Absolute Value Function**

The ***absolute value function*** is the piecewise-defined function $|x| = \begin{cases} x, & \text{if } x \geq 0 \\ -x, & \text{if } x < 0. \end{cases}$

The graph of $y = |x|$ is a combination of the graph of $y = x$ on $[0, \infty)$ and the graph of $y = -x$ on $(-\infty, 0)$:

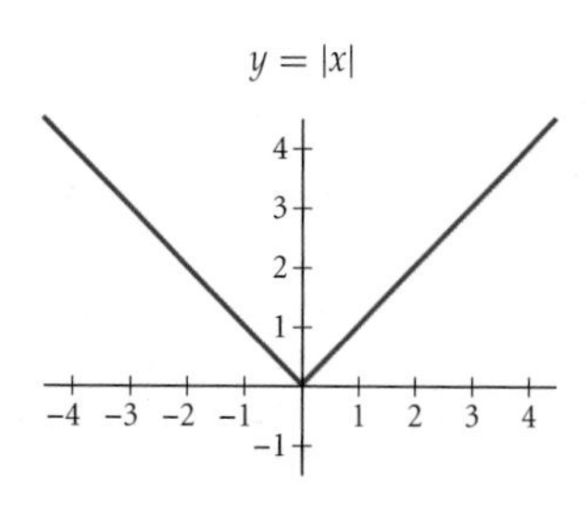

Of course, in general we might wish to take the absolute value of a more complicated expression. In the more general case we do exactly the same thing: The absolute value will leave positive quantities untouched, but flip the sign of negative quantities.

DEFINITION 0.20 **The Absolute Value of a Function**

The ***absolute value of a function*** $g(x)$ is $|g(x)| = \begin{cases} g(x), & \text{for all } x \text{ with } g(x) \geq 0 \\ -g(x), & \text{for all } x \text{ with } g(x) < 0. \end{cases}$

For example, consider the function $g(x) = 2x - 1$ and its absolute value $f(x) = |g(x)| = |2x - 1|$. When $2x - 1$ is positive or zero (i.e., when $x \geq 1/2$) the absolute value remains $2x-1$. But when $2x-1$ is negative (i.e., when $x < 1/2$) the absolute value of $g(x)$ is $-(2x-1)$. The graph of $y = |2x - 1|$ is a combination of the graphs of $y = 2x - 1$ and $y = -(2x - 1)$, switching between graphs at $x = 1/2$, as shown in the following figures:

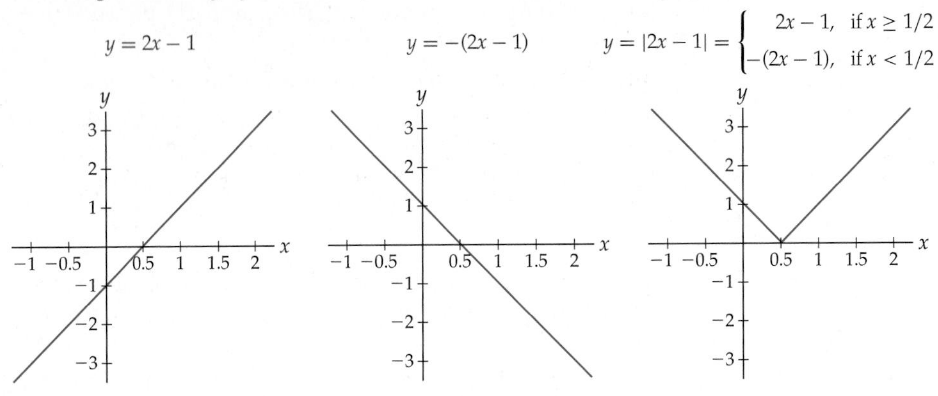

Examples and Explorations

EXAMPLE 1

Finding domains of power functions

Find the domains of the following power functions:

(a) $f(x) = 3x^{-2}$ **(b)** $g(x) = \frac{1}{2}x^{3/4}$ **(c)** $h(x) = 2x^{4/5}$ **(d)** $k(x) = 8x^{-1/2}$

SOLUTION

(a) If we rewrite the power function so that any roots or denominators are visible, the calculation of the domain is obvious: $f(x) = 3x^{-2} = \frac{3}{x^2}$ is defined everywhere but at $x = 0$, so its domain is $(-\infty, 0) \cup (0, \infty)$.

(b) The function $g(x) = \frac{1}{2}x^{3/4} = \frac{(\sqrt[4]{x})^3}{2}$ is defined everywhere except $x < 0$, so its domain is $[0, \infty)$.

(c) The function $h(x) = 2x^{-4/5} = \frac{2}{(\sqrt[5]{x})^4}$ is defined for all $x \neq 0$, so its domain is $(-\infty, 0) \cup (0, \infty)$.

(d) The function $k(x) = 8x^{-1/2} = \frac{8}{\sqrt{x}}$ fails to be defined for $x = 0$ and for $x < 0$, so its domain is $(0, \infty)$. □

EXAMPLE 2

Modeling a graph with a polynomial

Explain why the graph shown here could be modeled with a polynomial function f. Then say what you can about the degree and leading coefficient of f, and find a possible equation for $f(x)$.

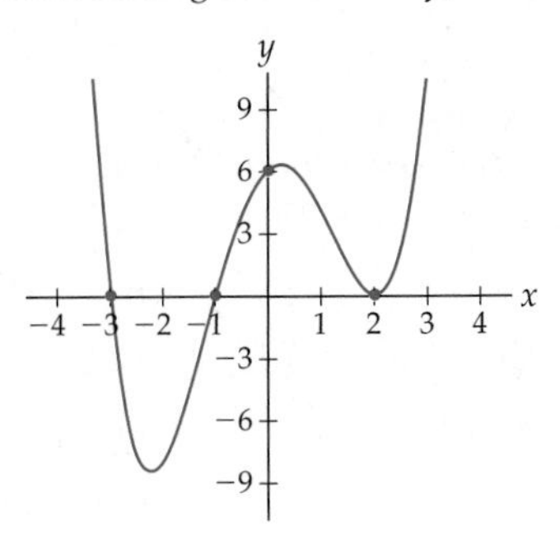

SOLUTION

This graph is defined on all of $\mathbb{R}$, is smooth and unbroken everywhere, and has no asymptotes, so it could be part of the graph of a polynomial function f. We will now find a polynomial whose graph is like the one pictured. Since both ends of the graph point upwards, we know that the degree of f must be even and the leading coefficient of f must be positive. The fact that the graph has three roots means that degree of f is at least 3. The fact that the graph has three local extrema means that the degree of f is in fact at least 4.

The roots of f are $x = -3$, $x = -1$, and $x = 2$. Note that the function f behaves differently at the root $x = 2$ than at the other roots. Near $x = 2$ the graph appears to have a quadratic type of shape. This means that $x = 2$ is a ***repeated root*** of f and therefore that $(x-2)$ is a factor of f more than once. Given all of this information, we see that one possible form for $f(x)$ is

$$f(x) = A(x+3)(x+1)(x-2)^2,$$

where A is some positive constant. The y-intercept $(0, 6)$ marked on the graph can help us determine A. Since $f(0) = 6$, we have

$$6 = A(0+3)(0+1)(0-2)^2 \quad \Longrightarrow \quad 6 = A(3)(1)(4) \quad \Longrightarrow \quad 6 = 12A \quad \Longrightarrow \quad A = \frac{1}{2}.$$

Therefore a function that could have the given graph is the quartic polynomial function $f(x) = \frac{1}{2}(x+3)(x-1)(x-2)^2$. □

EXAMPLE 3

Making a rough graph of a rational function

Without a calculator, sketch a rough graph of the function $f(x) = \dfrac{2(x-1)^2(x+2)}{(x-1)(x+1)(x-2)}$.

SOLUTION

By Theorems 0.17 and 0.18, we can see immediately from the factors in the numerator and denominator that the graph of f will have

- a hole at $x = 1$;
- a root at $x = -2$;
- vertical asymptotes at $x = -1$ and $x = 2$;
- a horizontal asymptote at $y = \frac{a_n}{b_n} = \frac{1}{2} = 2$.

A quick sign analysis tells us where the graph of f is above or below the x-axis. We must check the sign of f on each subinterval between the roots and non-domain points:

+ − DNE + − DNE +

−2 −1 1 2 f

Note that on this number line we include tick-marks only at the locations where $f(x)$ is zero or does not exist. The unlabeled tick-marks are the locations where $f(x)$ is zero, and the ones labeled "DNE" are the locations where $f(x)$ is not defined.

Plotting a few key points will help us make a more accurate graph. The y-intercept of the graph is $f(0) = 2$. The height of the hole at $x = 1$ will be the value of f at $x = 1$ after cancelling common factors:

$$\frac{2(1-1)(1+2)}{(1+1)(1-2)} = \frac{2(0)(3)}{(2)(-1)} = 0.$$

Finally, just to get a value to the right side of the graph, we also calculate $f(3) = 5$. The figure that follows at the left shows some of the information we have collected. The figure at the right uses this information and the sign analysis to fill in a rough sketch of the graph.

Some information about the graph

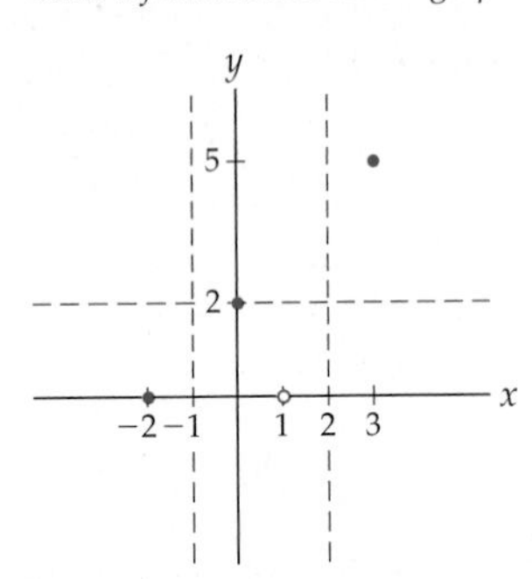

A rough sketch of f

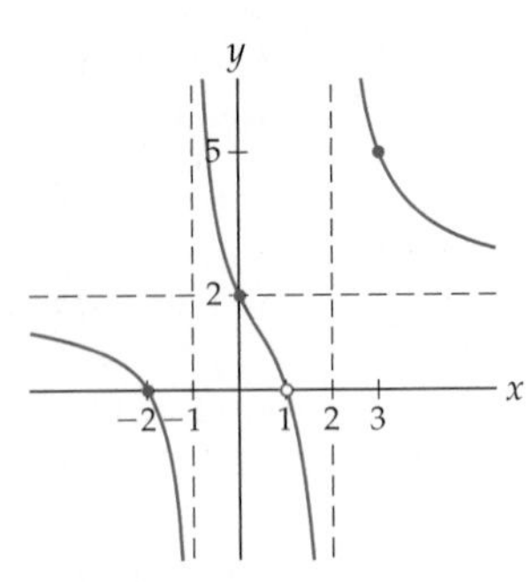

□

EXAMPLE 4

Writing an absolute value expression as a piecewise-defined function

Write the function $f(x) = |x^2 - 1|$ as a piecewise-defined function, and use this piecewise-defined function to calculate $f(-2)$, $f(0)$, and $f(1)$. Then sketch a graph of f.

SOLUTION

When $x^2 - 1$ is positive or zero, its absolute value will remain $x^2 - 1$. When $x^2 - 1$ is negative, its absolute value will be $-(x^2 - 1)$. Therefore we have

$$f(x) = |x^2 - 1| = \begin{cases} x^2 - 1, & \text{for all } x \text{ with } x^2 - 1 \geq 0 \\ -(x^2 - 1), & \text{for all } x \text{ with } x^2 - 1 < 0. \end{cases}$$

Although this is one way to write $|x^2 - 1|$ as a piecewise-defined function, it is difficult to work with. For example, to evaluate $f(-2)$ we would need to know whether $x^2 - 1 \geq 0$ or $x^2 - 1 \leq 0$ when $x = -2$. To simplify this piecewise-defined function, we need to rewrite the conditions as intervals of x-values. Since $x^2 - 1 \geq 0$ when $x \geq 1$ or $x \leq -1$, and $x^2 - 1 < 0$ when $-1 < x < 1$, we have

$$f(x) = |x^2 - 1| = \begin{cases} x^2 - 1, & \text{if } x \leq -1 \\ -(x^2 - 1), & \text{if } -1 < x < 1 \\ x^2 - 1, & \text{if } x \geq 1. \end{cases}$$

Now we can easily use our formula to calculate f when $x = -2$, $x = 0$, and $x = 1$:

$$f(-2) = (-2)^2 - 1 = 3 \qquad \text{(use first case, since } -2 \leq -1\text{)}$$
$$f(0) = -((0)^2 - 1) = 1 \qquad \text{(use middle case, since } -1 < 0 < 1\text{)}$$
$$f(1) = (1)^2 - 1 = 0 \qquad \text{(use last case, since } 1 \geq 1\text{).}$$

Of course, we could have just substituted our x-values into the expression $|x^2 - 1|$, but we are practicing writing absolute value functions as piecewise-defined functions. This is a skill that will come in handy when we later try to differentiate or integrate functions that involve absolute values.

The graph of $y = |x^2 - 1|$ is the same as the graph of $y = x^2 - 1$ on the intervals $(-\infty, -1]$ and $[1, \infty)$. On the interval $(-1, 1)$, the quantity $|x^2 - 1|$ has the sign opposite that of $x^2 - 1$, so its graph is the reflection of the negative parts of $x^2 - 1$ across the x-axis, as the following figure shows:

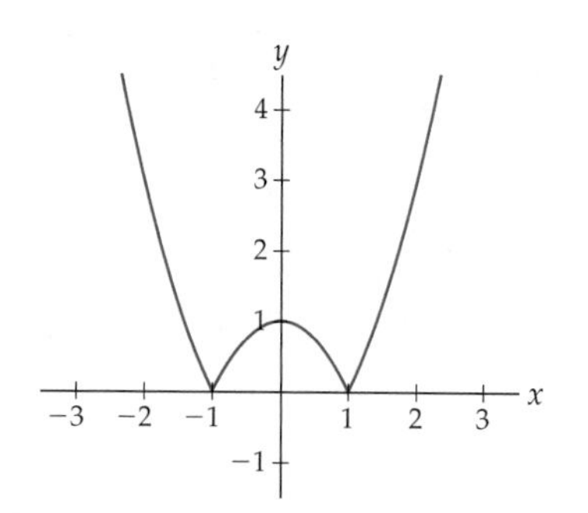

□

? TEST YOUR UNDERSTANDING

- Why is not every power function also a polynomial function?
- Why is every polynomial function also a rational function?
- Given that the degree of a polynomial function is the integer n that represents the highest power of x with a nonzero coefficient in the polynomial, why do you think we say that the degree of the zero polynomial (the function f defined by $f(x) = 0$ for *all* values of x) is undefined?
- Calculators are notoriously bad at graphing rational functions. Sometimes they connect a graph over its vertical asymptotes, and most times the holes of a rational function are not immediately clear from a calculator graph. What do you think causes calculators to make these errors?
- In Example 4, why does the function $f(x) = |x^2 - 1|$ have break points at $x = -1$ and $x = 1$?

EXERCISES 0.3

Thinking Back

Basic algebra review: The following exercises will help you review your skill with exponents, fractions, and factoring.

- Calculate the value of $\left(\frac{8}{27}\right)^{-2/3}$ by hand.
- Write $\frac{x^{-3} - x^{-2}}{x^{-1} - 1}$ in the form Ax^k for some real numbers A and k.
- Factor $16x^6 - 81x^2$ as much as possible.
- Solve the equation $\frac{x^3 + x^2 - 2x}{x^2 - 4x + 3} = 0$.
- Solve the equation $\frac{1}{3x^2 - 2x - 1} = 2x - 1$.

Factoring after root-guessing: Factor as much as possible each of the polynomial and rational functions that follow. In these exercises it is necessary to guess a root and use synthetic division to get started with factoring.

- $f(x) = 2x^4 + 6x^2 - 8$
- $f(x) = x^3 + 4x^2 - 11x + 6$
- $f(x) = \frac{x^3 + 6x^2 + 3x - 10}{x^5 + 3x^4 - x - 3}$

Concepts

0. *Problem Zero:* Read the section and make your own summary of the material.

1. *True/False:* Determine whether each of the statements that follow is true or false. If a statement is true, explain why. If a statement is false, provide a counterexample.

(a) *True or False:* The sum of any two algebraic functions is itself an algebraic function.

(b) *True or False:* Every power function is a polynomial function.

(c) *True or False:* Every polynomial function is a rational function.

(d) *True or False:* For any real numbers a and b, the polynomial function $f(x) = x^4 - ax^3 + bx + 7$ has exactly four real roots.

(e) *True or False:* If f is a polynomial function with k real roots, then the degree of f must be at most k.

(f) *True or False:* If f is a polynomial function with k turning points, then the degree of f must be at least $k+1$.

(g) *True or False:* A rational function is a quotient of two polynomials, where the denominator polynomial is not the zero polynomial.

(h) *True or False:* Every rational function has a horizontal asymptote.

2. *Examples:* Construct examples of the thing(s) described in the following. Try to find examples that are different than any in the reading.

(a) Three functions that are algebraic.

(b) Three functions that are algebraic and involve quotients, but are not rational.

(c) Three functions that fail to be algebraic.

3. Sketch graphs of $y = -2x^3$ and $y = 3x^{-2}$ by hand, without using a calculator. Label three points on each graph.

4. Sketch graphs of $y = -3x^{1/2}$ and $y = x^{2/3}$ by hand, without using a calculator. Label three points on each graph.

5. Sketch graphs of $y = x^2$, $y = x^4$, and $y = x^6$ by hand, without a calculator, on the same set of axes. Label three points on each graph.

6. Sketch graphs of $y = x^3$, $y = x^5$, and $y = x^7$ by hand, without a calculator, on the same set of axes. Label three points on each graph.

7. If $f(x) = Ax^k$ is a power function, is its reciprocal $\frac{1}{f(x)}$ also a power function? If so, write the reciprocal in the form Cx^r for some real numbers C and r.

8. Suppose $f(x) = Ax^k$ is a power function, where k is an integer.

(a) What must be true about k for f to be one-to-one, and why?

(b) If f is one-to-one, is its inverse f^{-1} also a power function? If so, write the inverse in the form $f^{-1}(x) = Cx^r$ for some real numbers C and r.

9. For the polynomial $f(x) = x(3x+1)(x-2)^2$, determine the leading coefficient, leading term, degree, constant term, and coefficients a_1 and a_3.

10. For the polynomial $f(x) = x^5(1-2x)(1-x^3)$, determine the leading coefficient, leading term, degree, constant term, and coefficients a_1 and a_5.

11. Use the quadratic formula to explain why a quadratic polynomial function $f(x) = ax^2 + bx + c$ is irreducible if, and only if, the discriminant $b^2 - 4ac$ is negative. Then use the discriminant to show that $f(x) = 3x^2 + 2x + 6$ is irreducible.

12. Give an example of each of the following types of polynomials:

(a) A quintic polynomial with only one real root.

(b) A polynomial, all of whose roots are rational, but non-integer, numbers.

(c) A polynomial with integer coefficients that has four real roots, only two of which are integers.

13. Explain why the graph of $f(x) = \frac{(x-1)(x+2)}{(x+2)^2}$ does not have a hole in it at $x = -2$.

14. Use a calculator to graph the function $f(x) = \frac{x^2+x-2}{x^2-4}$. This graph has a hole, where the function is not defined; determine the location of the hole, and trace along the graph on your calculator until you find it there.

15. Construct an equation of a rational function whose graph has no roots, no holes, vertical asymptotes at $x = \pm 2$, and a horizontal asymptote at $y = 3$.

16. Construct an equation of a rational function whose graph has a hole at the coordinates $(-2, 0)$, vertical asymptotes at $x = 1$ and $x = -3$, and a horizontal asymptote at $y = -5$.

17. Suppose f is a rational function with roots at $x = 1$ and $x = 3$, a hole at $x = -1$, a vertical asymptote at $x = 2$, and a horizontal asymptote at $y = -1$.

(a) Sketch three possible graphs of f. Make the graphs as different as you can while still having the given characteristics.

(b) Write down the equations of three functions f that have the given properties. (Your equations do not have to match your graphs from part (a).)

18. Suppose f is a rational function with root at $x = -2$, holes at $x = 0$ and $x = 3$, a vertical asymptote at $x = 1$, and no horizontal asymptote.

(a) Sketch three possible graphs of f. Make the graphs as different as you can while still having the given characteristics.

(b) Write down the equations of three functions f that have the given properties. (Your equations do not have to match your graphs from part (a).)

For each graph $y = f(x)$ shown, sketch the graph of $y = |f(x)|$.

19.

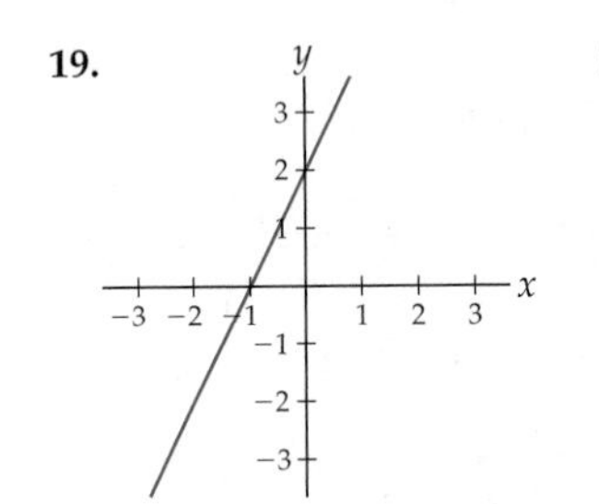

20.

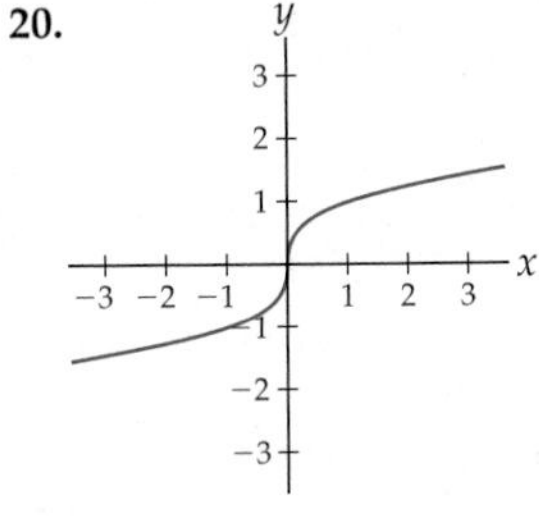

21.

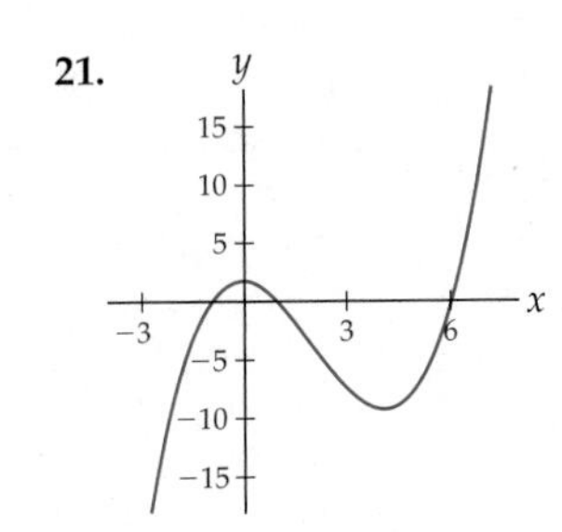

22.

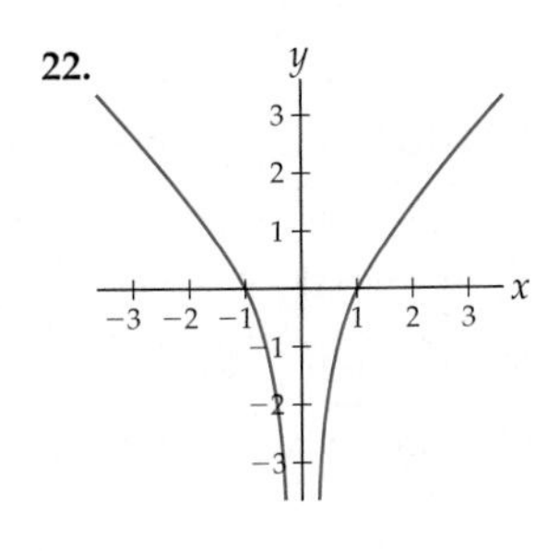

23. An alternative definition for the absolute value function is $|x| = \sqrt{x^2}$. Explain in your own words why this definition is equivalent to that of the absolute value function given in Definition 0.19. This exercise shows that the absolute value function is algebraic, not just piecewise algebraic.

24. Suppose that f is a function defined in pieces, using some functions g and h, as follows:

$$f(x) = \begin{cases} g(x), & \text{if } x \le 2 \\ h(x), & \text{if } x > 2. \end{cases}$$

Will the piece of the graph on $(-\infty, 2]$ always "match up" with the piece of the graph on $(2, \infty)$? Why or why not? Use graphs to illustrate your answer.

Skills

Find the domain and the zeroes of each function in Exercises 25–34. Check your answers afterwards with graphs.

25. $f(x) = -\frac{1}{4}x^{2/3}$

26. $f(x) = \frac{-4}{\sqrt[6]{x^{7/3}}}$

27. $f(x) = 2x^{-7/4}$

28. $f(x) = \frac{x^3 + x^2 - 2x}{2x^3 - x^2 - 6x}$

29. $f(x) = (x^2 - 1)^{-1/4}$

30. $f(x) = |x^2 - 9|^{-3/4}$

31. $f(x) = \frac{\sqrt{3 - x}}{\sqrt[3]{x^2 - 3x - 4}}$

32. $f(x) = \sqrt[4]{x^{-2} + 1}$

33. $f(x) = \sqrt{\frac{x^2 - 1}{x^3 - 7x + 6}}$

34. $f(x) = \frac{3x^{-3}}{x^{-1/4} - x^{3/4}}$

Find equations for each of the functions described in Exercises 35–42.

35. The linear function whose graph has slope -1 and passes through the point $(3, -2)$.

36. The linear function whose graph passes through the points $(-2, 1)$ and $(3, -4)$.

37. The linear function whose graph is parallel to $y = 2x + 1$ and passes through the point $(-1, 4)$.

38. A power function whose graph passes through $(0, 0)$ and $(1, 3)$.

39. A power function whose graph passes through $(0, 0)$ and $(2, 8)$.

40. A polynomial function whose graph passes through $(-2, 0)$, $(1, 0)$, and $(3, 0)$.

41. A polynomial function whose graph passes through $(0, 0)$, $(2, 0)$, and $(4, 0)$.

42. A polynomial function whose graph passes through $(0, 0)$, $(2, 0)$, $(4, 0)$, and $(1, 2)$.

Sketch rough graphs of the functions in Exercises 43–56 without using a calculator or graphing utility. Be as accurate as you can, and identify any roots, holes, or asymptotes.

43. $f(x) = \sqrt[3]{x + 1}$

44. $f(x) = (x + 3)^{2/3} - 2$

45. $f(x) = \frac{1}{16 - x^4}$

46. $f(x) = x^4 - 6x^2 + 9$

47. $f(x) = 2x^4 - x^3 - x^2$

48. $f(x) = x^4 - 2x^2 + 1$

49. $f(x) = x^3 - 2x^2 - 4x + 8$

50. $f(x) = x(x + 2)(x - 3)^2$

51. $f(x) = \frac{2x^3 + 4x^2 - 6x}{x^2 - 4}$

52. $f(x) = \frac{(x^2 - 4)^2}{2x^2 - 3x - 2}$

53. $f(x) = \frac{(x + 1)(x - 3)(x + 2)}{(x + 2)(x - 3)}$

54. $f(x) = \frac{(x - 1)(x + 2)}{(x + 1)(x - 1)}$

55. $f(x) = \frac{2x^3 + 3x^2 - 2x - 3}{x^2 - 2x - 3}$

56. $f(x) = \left| \frac{x^2(x - 3)(x + 2)}{x - 1} \right|$

Given that the graph of $f(x) = \sin x$ is as shown here, sketch graphs of each of the following absolute value transformations $g(x)$:

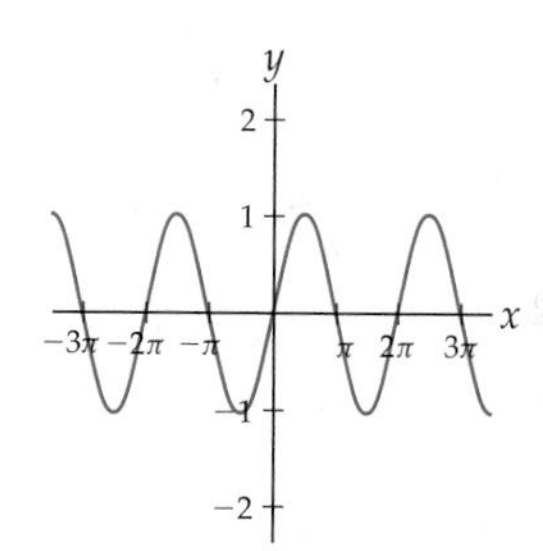

57. $g(x) = |\sin x|$

58. $g(x) = \sin |x|$

59. $g(x) = -2 + |\sin x|$

60. $g(x) = |-2 + \sin x|$

For each graph in Exercises 61–72, find a function whose graph looks like the one shown. When you are finished, use a graphing utility to check that your function f has the properties and features of the given graph.

61.

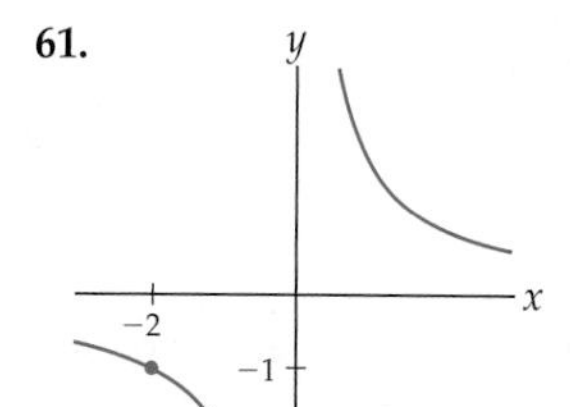

62.

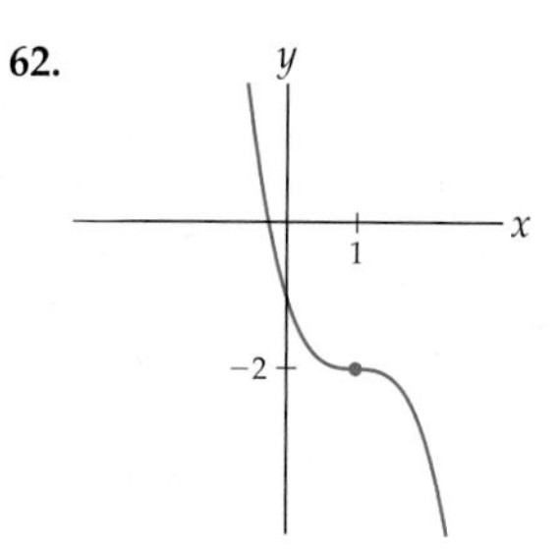

63.

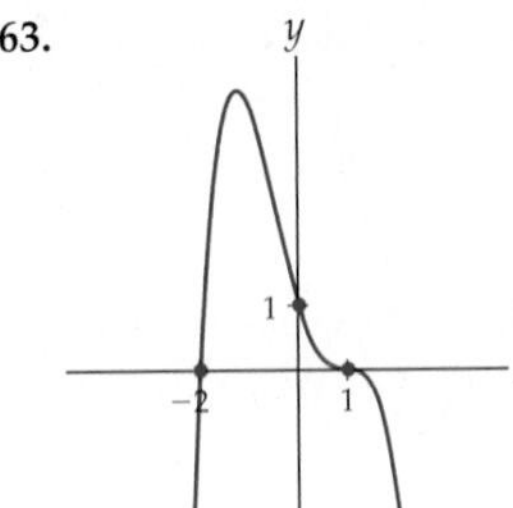

64.

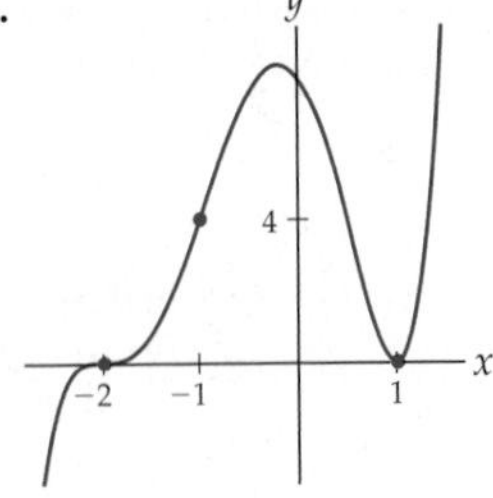

65.

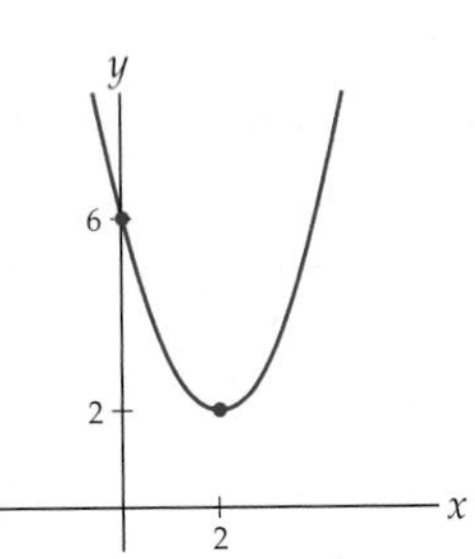

66.

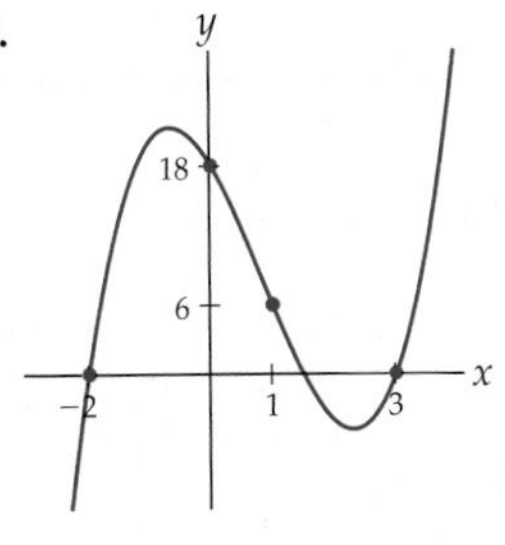

67.

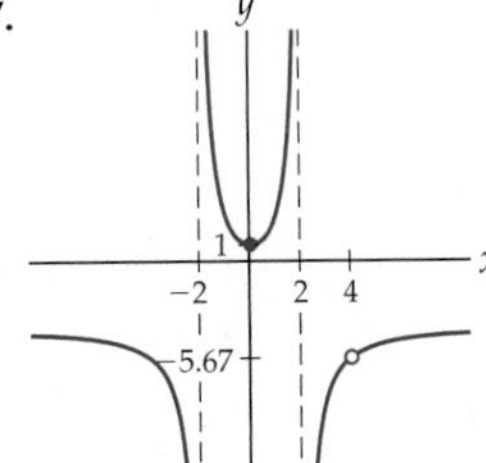

68.

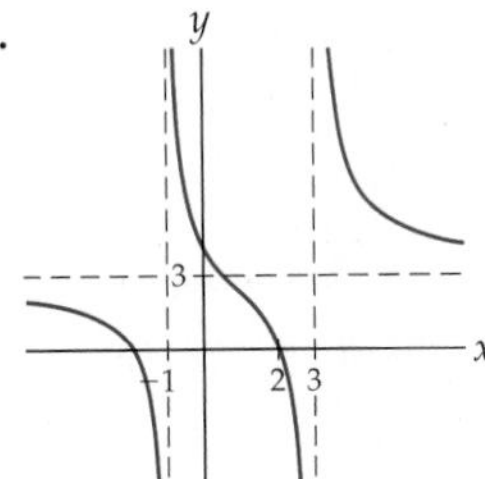

69.

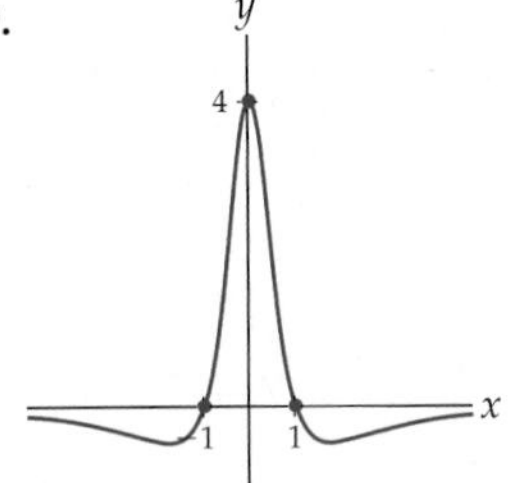

70.

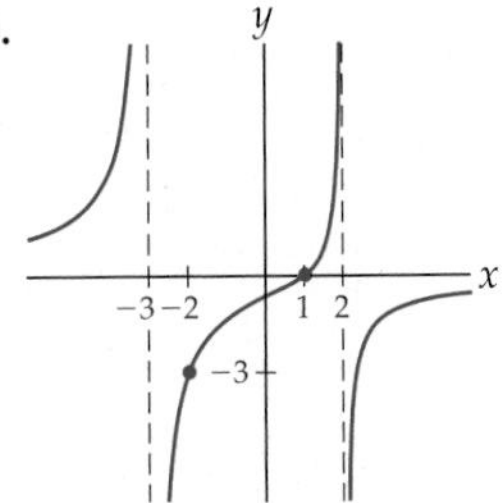

71.

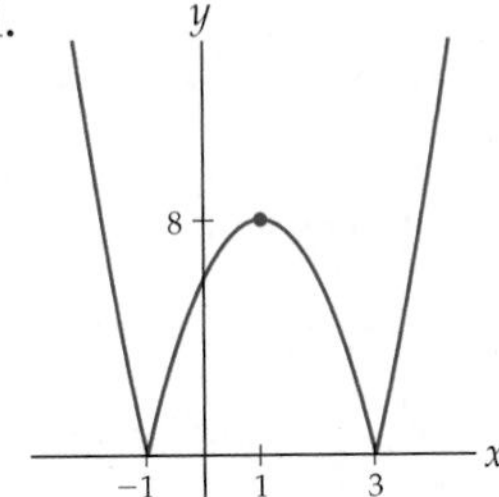

72.

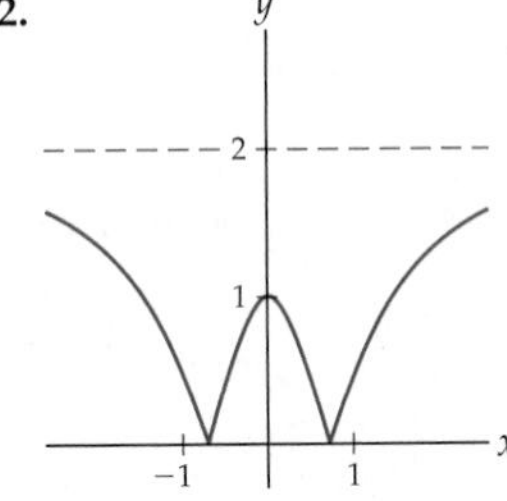

Write each function in Exercises 73–80 as a piecewise-defined function where each piece is defined on an interval of x-values. Then sketch a labeled graph of each piecewise-defined function.

73. $f(x) = |5 - 3x|$

74. $f(x) = |1.5x + 2.3|$

75. $f(x) = |x^2 + 1|$

76. $f(x) = |x^2 - 4|$

77. $f(x) = |9 - x^2|$

78. $f(x) = |3 - 4x + x^2|$

79. $f(x) = |x^2 - 3x - 4|$

80. $f(x) = |x^3 + x^2 - 2x|$

Applications

The following table for Exercises 81 and 82 describes the number of cars that were on a particular 1-mile stretch of Route 97 at t hours after 6:00 A.M. Monday morning:

t (hours after 6:00 A.M.)	0	1	2	3
N (cars on 1 mile of road)	0	28	12	18

81. Make a plot of the data given in the preceding table. Suppose you wanted to find a polynomial function $N(t)$ that passes through each of the four data points. What is the minimum degree that this polynomial function could be, and why?

82. Given the same table of car densities,

(a) Find a polynomial function $N(t)$ that goes through all four data points in the table. *(Hint: Use the data to solve for the coefficients.)*

(b) Overlay a plot of the function $N(t)$ you found with the plot of the data. Does $N(t)$ look the way you expected it to?

(c) Use your function $N(t)$ to predict the number of cars that were on that 1-mile stretch at 7:30 a.m. How many cars does your function predict will be there at 3 p.m.? On what time interval does your model function make practical sense?

For Exercises 83 and 84, suppose that Emmy is investigating a release of toxins from a tank farm on the Hanford nuclear reservation into groundwater. The groundwater eventually forms a spring that runs into the Columbia River. The following table describes the amount of toxins in the spring:

Years after 2006	0	1	2
Concentration of toxin (ppm)	0.53	0.65	0.74

83. Emmy believes that the leak is not getting larger, so that the concentration of toxin in the spring will approach some steady constant value. She wants to make estimates of the date on which the leak started and of the eventual steady concentration of toxin. Is a polynomial the best choice of a function to fit to the data for this purpose? Why or why not?

84. Emmy wants to fit a rational function of the form $T(t) = \dfrac{at+b}{t+d}$ to the data, where t is the number of years after 2006.

(a) What values should Emmy use for the coefficients a, b and d?

(b) Use your rational function model to estimate the date that the leak started.

(c) What asymptotic value will the concentration approach as time t increases without bound?

(d) What are the potential problems in using this model?

Proofs

85. Prove that the composition of two power functions is a power function.
86. Prove that the product of two power functions is a power function.
87. Prove that the composition of two linear functions is also a linear function.
88. Prove that the sum of two polynomial functions is a polynomial.
89. Prove that the product of two cubic polynomials is a polynomial of degree six.
90. Prove that the sum or product of two rational functions is a rational function.
91. Prove that (a) every constant function is linear and (b) every linear function is a polynomial.
92. Prove that the domain of a rational function $f(x) = \frac{p(x)}{q(x)}$ is the set $\{x \mid q(x) \neq 0\}$.
93. Prove that the graph of a rational function $f(x) = \frac{p(x)}{q(x)}$ has a root at $x = c$ if and only if $p(c) = 0$ but $q(c) \neq 0$.

Thinking Forward

Algebra for derivatives: Simplify and rewrite each expression until you can cancel the h in the denominator:

- ▶ $\dfrac{(x+h)^3 - x^3}{h}$
- ▶ $\dfrac{(x+h)^{1/2} - x^{1/2}}{h}$
- ▶ $\dfrac{(x+h)^{-2} - (x)^{-2}}{h}$
- ▶ $\dfrac{(x+h)^{-1/2} - x^{-1/2}}{h}$

Alternative algebra for derivatives: Simplify and rewrite each expression until you can cancel a common factor from the numerator and denominator:

- ▶ $\dfrac{t^3 - x^3}{t - x}$
- ▶ $\dfrac{t^{1/2} - x^{1/2}}{t - x}$
- ▶ $\dfrac{t^{-2} - x^{-2}}{t - x}$
- ▶ $\dfrac{t^{-1/2} - x^{-1/2}}{t - x}$

0.4 EXPONENTIAL AND TRIGONOMETRIC FUNCTIONS

- ▶ Definitions and properties of exponential and logarithmic functions
- ▶ Definitions and properties of trigonometric and inverse trigonometric functions
- ▶ Graphs and equations involving transcendental functions

Exponential Functions

Functions that are not algebraic are called ***transcendental*** functions. In this book we will investigate four basic types of transcendental functions: exponential, logarithmic, trigonometric, and inverse trigonometric functions. Exponential functions are similar to power functions, but with the roles of constant and variable reversed in the base and exponent:

DEFINITION 0.21 **Exponential Functions**

An ***exponential function*** is a function that can be written in the form

$$f(x) = Ab^x$$

for some real numbers A and b such that $A \neq 0$, $b > 0$, and $b \neq 1$.

There is an important technical problem with this definition: We know what it means to raise a number to a rational power by using integer roots and powers, but we don't know what it means to raise a number to an irrational power. We need to be able to raise numbers to irrational powers to talk about exponential functions; for example, if $f(x) = 2^x$, then we need to be able to compute $f(\pi) = 2^\pi$. One way to think of b^x where x is irrational is as a limit:

$$b^x = \lim_{\substack{r \to x \\ r \text{ rational}}} b^r.$$

The "lim" notation will be explored more in Chapter 1. For now you can just imagine that if x is rational, then we can approximate b^x by looking at quantities b^r for various rational numbers r that get closer and closer to the irrational number x. For example, 2^π can be approximated by 2^r for rational numbers r that are close to π:

$$2^\pi \approx 2^{3.14} = 2^{314/100} = \sqrt[100]{2^{314}}.$$

As we consider rational numbers r that are closer and closer to π, the expression 2^r will get closer and closer to 2^π; see Exercise 4. In Chapter 7 we will give a more rigorous definition of exponential functions as the inverses of certain accumulation integrals.

We will assume that you are familiar with the basic algebraic rules of exponents, for example that $b^{x+y} = b^x b^y$, that $b^0 = 1$ for any nonzero b, and that $(b^x)^y = b^{xy}$. Proving those rules requires the more rigorous definition of exponential functions that we will see in Chapter 5, so for the moment we will take these algebraic rules as given. From those basic rules it follows that an exponential function $f(x) = b^x$ is one-to-one, and that b^x is never zero for any value of x. (See Exercises 85 and 86.)

Interestingly, the most natural base b to use for an exponential function isn't a simple integer, like $b = 2$ or $b = 3$. Instead, for reasons that will become clear when we study derivatives, the most natural base is the irrational number known as e, and the function e^x is therefore called the ***natural exponential function***. An approximation of the number e to 65 digits is:

$$2.71828182845904523536028747135266249775724709369995957496696762 77\ldots.$$

Of course, since e is an irrational number, we cannot define it just by writing an approximation of e in decimal notation; we will define e properly once we cover limits in Chapter 1.

In Exercise 88 you will prove that every exponential function can be written so that its base is the natural number e, as the next theorem states:

THEOREM 0.22

Natural Exponential Functions

Every exponential function can be written in the form

$$f(x) = Ae^{kx}$$

for some real number A and some nonzero real number k.

Every exponential function has a graph similar to either the ***exponential growth*** graph that follows at the left or the ***exponential decay*** graph at the right, depending on the values of k and b. Of course, if the coefficient A is negative, then the graph of $f(x) = Ae^{kx}$ or $f(x) = Ab^x$ will be a reflection of one of these two graphs over the x-axis.

$f(x) = e^{kx}$ *with* $k > 0$,
$f(x) = b^x$ *with* $b > 1$

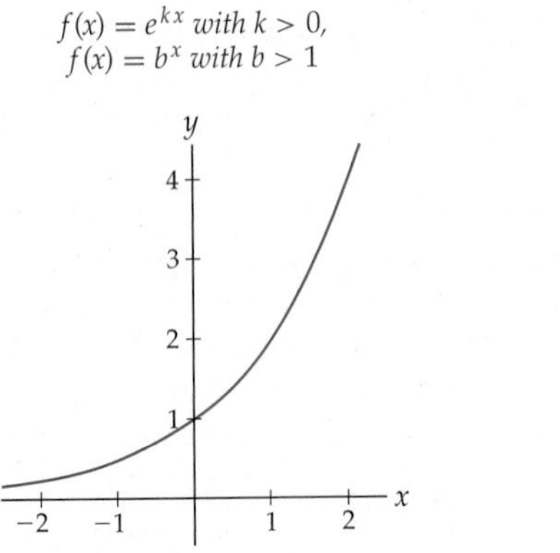

$f(x) = e^{kx}$ *with* $k < 0$,
$f(x) = b^x$ *with* $0 < b < 1$

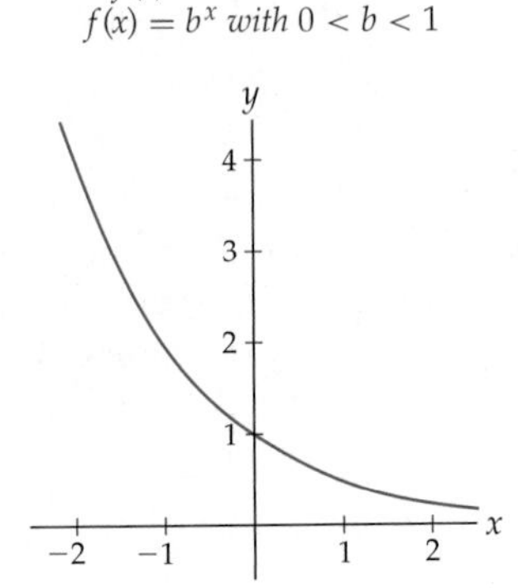

Logarithmic Functions

Since every exponential function b^x is one-to-one, every exponential function has an inverse. These inverses are what we call the ***logarithmic functions***:

DEFINITION 0.23

Logarithmic Functions as Inverses of Exponential Functions

The inverse of the exponential function b^x is the ***logarithmic function***

$$\log_b x.$$

As a special case, the inverse of the natural exponential function e^x is the ***natural logarithmic function***

$$\log_e x = \ln x.$$

We require that the base b satisfy $b > 0$ and $b \neq 1$, because these are exactly the conditions we must have for $y = b^x$ to be an exponential function. In Section 7.7 we will define logarithms another way, in terms of integrals and accumulation functions.

You should already be familiar with the algebraic rules of logarithms, but we restate them here in case you need a refresher; see Exercises 90–94 for proofs.

THEOREM 0.24

Algebraic Rules for Logarithmic Functions

For all values of x, y, b, and a for which these expressions are defined, we have

(a) $\log_b x = y$ if and only if $b^y = x$

(b) $\log_b(b^x) = x$

(c) $b^{\log_b x} = x$

(d) $\log_b(x^a) = a\log_b x$

(e) $\log_b(xy) = \log_b x + \log_b y$

(f) $\log_b\left(\frac{1}{x}\right) = -\log_b x$

(g) $\log_b\left(\frac{x}{y}\right) = \log_b x - \log_b y$

(h) $\log_b x = \frac{\log_a x}{\log_a b}$

The first three properties follow from properties of inverse functions, and tell us that $\log_b x$ is the exponent to which you have to raise b in order to get x. For example, $\log_2 8$ is the power to which you have to raise 2 to get 8; since $2^3 = 8$, we have $\log_2 8 = 3$. All of these rules also apply to the natural exponential function, because $\ln x$ is just $\log_b x$ with base $b = e$.

Properties (d) and (e) follow from the algebraic rules of exponents, and properties (f) and (g) are their immediate consequences. The final property in Theorem 0.24 is called the ***base conversion formula***, because it allows us to translate from one logarithmic base to another. The base conversion formula is especially helpful for converting to base e or base 10 so that we can calculate logarithms on a calculator. For example, $\log_7 2$ is equal to $\frac{\ln 2}{\ln 7}$, which we can approximate using the built-in "ln" key on a calculator.

The graphs of logarithmic functions can be obtained easily from the graphs of exponential functions by reflection over the line $y = x$, resulting in the following graphs:

$y = b^x$ *and* $y = \log_b x$ *with* $b > 1$

$y = b^x$ *and* $y = \log_b x$ *with* $0 < b < 1$

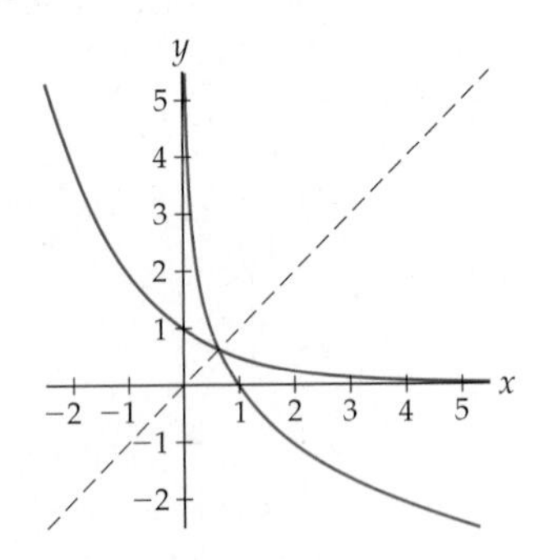

Trigonometric Functions

There are six ***trigonometric functions*** defined as ratios of side lengths of right triangles, or, more generally, as ratios of coordinate lengths on the unit circle. We now provide a quick review of the definitions of these functions and their graphical and algebraic properties.

We can place any angle in a ***standard position*** in the xy-plane by placing its vertex at the origin and its ***initial edge*** along the positive side of the x-axis, as shown next at the left. The angle then opens up in either a counterclockwise or clockwise direction until it reaches its ***terminal edge***. A ***positive angle*** is measured counterclockwise from its initial edge, while a ***negative angle*** is measured clockwise from its initial edge.

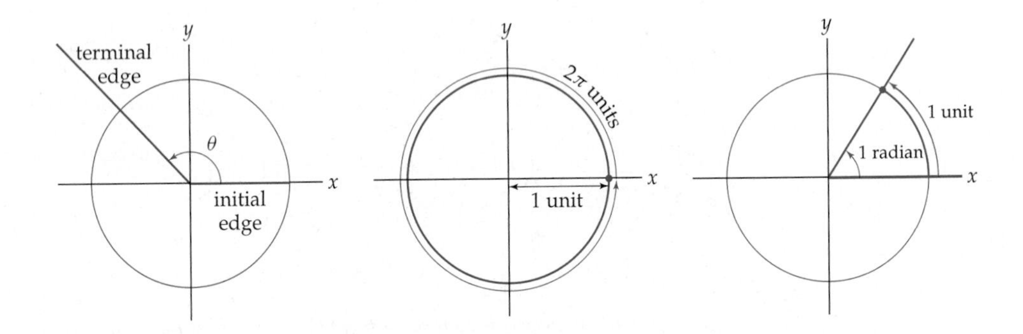

Now consider the ***unit circle*** shown in the center diagram. Since the unit circle has radius $r = 1$ unit, its circumference is $C = 2\pi r = 2\pi(1) = 2\pi$ units. That's a circumference of approximately 6.283185 units, which certainly is not as nice as a number like 360 that we can easily divide into integer-sized pieces. However, we can still measure everything in terms of this circumference by defining a new unit of angle measure called a ***radian*** that represents the size of an angle in standard position whose terminal edge intersects the unit circle after an arc length of 1 unit, as shown in the diagram at the right.

Since the distance all the way around the circle is 2π units, the distance halfway around is π units and the distance one-quarter of the way around the circle is $\frac{\pi}{2}$ units. This means that an angle of 90° measures $\frac{\pi}{2}$ radians, an angle of 180° measures π radians, and an angle of 360° measures, of course, 2π radians. The following three diagrams illustrate some common positive angles in radian measure around the unit circle:

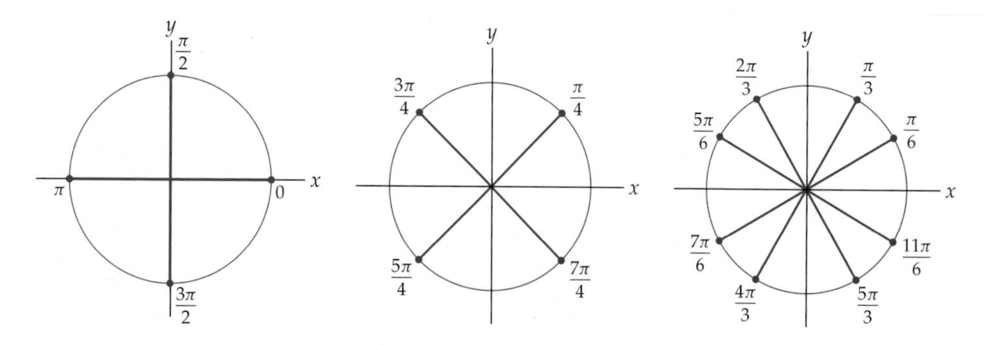

Of course we can also consider negative angles; for example, the angle that opens up in the clockwise direction for one quarter of the distance around the bottom half of the circle has radian measure $-\frac{\pi}{4}$. Its terminal edge intersects the unit circle in the same location as the angle $\frac{7\pi}{4}$ shown in the middle figure. We can also consider angles that go more than once around the circle; for example, the angle $\frac{5\pi}{2} = 2\pi + \frac{\pi}{2}$ intersects the unit circle at the same point as the angle $\frac{\pi}{2}$ in the diagram at the left.

Given any angle θ in standard position, the terminal edge of θ intersects the unit circle at some point (x, y) in the xy-plane. We will define the height y of that point to be the ***sine*** of θ, while the ***cosine*** of θ will be defined as the x-coordinate of that point.

DEFINITION 0.25

Trigonometric Functions for Any Angle

Given any angle θ measured in radians and in standard position, let (x, y) be the point where the terminal edge of θ intersects the unit circle. The six ***trigonometric functions*** of an angle θ are the six possible ratios of the coordinates x and y for θ:

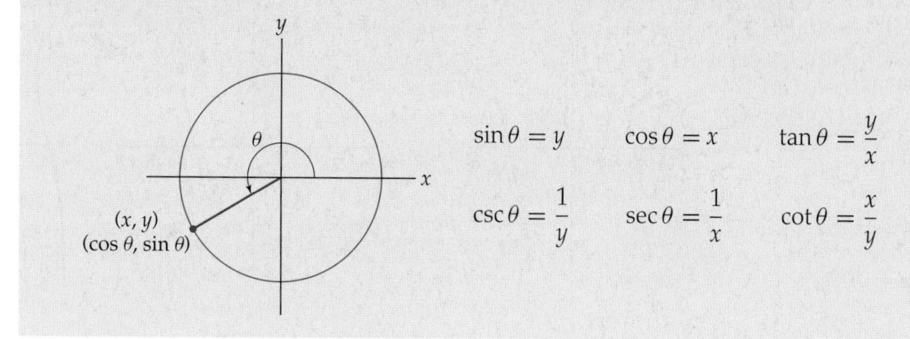

$$\sin\theta = y \qquad \cos\theta = x \qquad \tan\theta = \frac{y}{x}$$

$$\csc\theta = \frac{1}{y} \qquad \sec\theta = \frac{1}{x} \qquad \cot\theta = \frac{x}{y}$$

Notice that the sine and cosine functions determine the remaining four trigonometric functions, since $\tan\theta$ is the ratio $\frac{\sin\theta}{\cos\theta}$, and the last three trigonometric functions are the reciprocals of the first three.

You should already be familiar with the basic trigonometric identities, but they are repeated next for your review; see Exercises 95–100 for proofs. The first Pythagorean identity, the even–odd identities, and the shift identities follow easily from the definitions of the trigonometric functions. The sum identities follow from a geometric argument that we will not get into here. The remaining identities can all be proved from the previous identities. In the following identities we are using the notation $\sin^2 x$ as shorthand for $(\sin x)^2$.

THEOREM 0.26 **Basic Trigonometric Identities**

Pythagorean Identities	Even–Odd Identities	Shift Identities
$\sin^2\theta + \cos^2\theta = 1$	$\sin(-\theta) = -\sin\theta$	$\cos\left(\theta - \frac{\pi}{2}\right) = \sin\theta$
$\tan^2\theta + 1 = \sec^2\theta$	$\cos(-\theta) = \cos\theta$	$\sin\left(\theta + \frac{\pi}{2}\right) = \cos\theta$
$1 + \cot^2\theta = \csc^2\theta$	$\tan(-\theta) = -\tan\theta$	$\sin(\theta + 2\pi) = \sin\theta$
		$\cos(\theta + 2\pi) = \cos\theta$

Sum Identities	Difference Identities
$\sin(\alpha + \beta) = \sin\alpha\cos\beta + \sin\beta\cos\alpha$	$\sin(\alpha - \beta) = \sin\alpha\cos\beta - \sin\beta\cos\alpha$
$\cos(\alpha + \beta) = \cos\alpha\cos\beta - \sin\alpha\sin\beta$	$\cos(\alpha - \beta) = \cos\alpha\cos\beta + \sin\alpha\sin\beta$

Double-Angle Identities	Alternative Forms	Alternative Forms
$\sin 2\theta = 2\sin\theta\cos\theta$	$\cos 2\theta = 1 - 2\sin^2\theta$	$\sin^2\theta = \dfrac{1 - \cos 2\theta}{2}$
$\cos 2\theta = \cos^2\theta - \sin^2\theta$	$\cos 2\theta = 2\cos^2\theta - 1$	$\cos^2\theta = \dfrac{1 + \cos 2\theta}{2}$

The graphs of the six trigonometric functions are shown next. Each of the graphs in the second row is the reciprocal of the graph immediately above it. Remember that you can use the graph of a function f to sketch the graph of its reciprocal $\frac{1}{f}$. In particular, the zeros of f will be vertical asymptotes of $\frac{1}{f}$, large heights on the graph of f will become small heights on the graph of $\frac{1}{f}$, and vice versa.

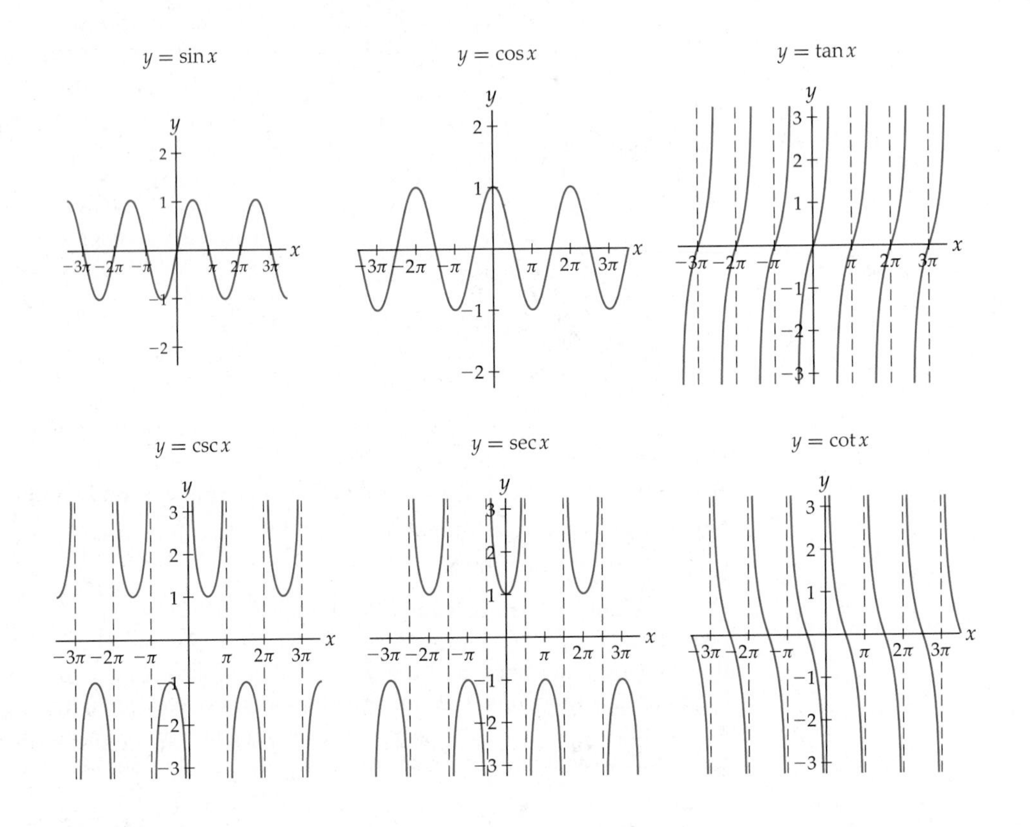

Inverse Trigonometric Functions

None of the six trigonometric functions are one-to-one, but after restricting domains we can construct the so-called inverse trigonometric functions. In this section we will focus on the inverses of only three of the six inverse trigonometric functions: Those for sine, tangent, and secant. (The inverses of these functions will be particularly useful to us in Chapter 5 when we study integration techniques, and the inverses of the remaining three trigonometric functions would add no more to that discussion.) There are many different restricted domains that we could use to obtain partial inverses to these three functions. We need to pick one restricted domain for each function and stick with it. In this text we will use the restricted domains shown below.

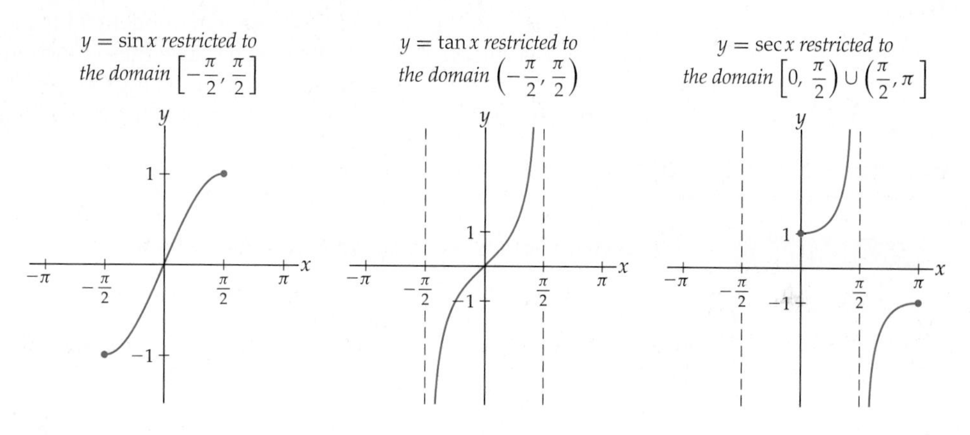

Each of these restricted functions is one-to-one and thus invertible. The inverses of these restricted functions, respectively, are the inverse sine, inverse tangent, and inverse secant functions.

DEFINITION 0.27

The Inverse Trigonometric Functions

(a) The ***inverse sine function*** $\sin^{-1} x$ is the inverse of the restriction of $\sin x$ to the interval $\left[-\frac{\pi}{2}, \frac{\pi}{2}\right]$.

(b) The ***inverse tangent function*** $\tan^{-1} x$ is the inverse of the restriction of $\tan x$ to the interval $\left(-\frac{\pi}{2}, \frac{\pi}{2}\right)$.

(c) The ***inverse secant function*** $\sec^{-1} x$ is the inverse of the restriction of $\sec x$ to the interval $\left[0, \frac{\pi}{2}\right) \cup \left(\frac{\pi}{2}, \pi\right]$.

Notice that since the inputs to the trigonometric functions are angles, it is the *outputs* of the inverse trigonometric functions that are angles. We will interchangeably use the alternative notations $\arcsin x$, $\arctan x$, and $\operatorname{arcsec} x$ for these inverse trigonometric functions.

CAUTION

Although we use the notation $\sin^2 x$ to represent $(\sin x)^2$ and the notation x^{-1} to represent $\frac{1}{x}$, the notation $\sin^{-1} x$ does *not* represent $\frac{1}{\sin x}$. Inverse functions in general have nothing to do with reciprocals, despite what one might imagine from the notation.

All of the properties of $\sin^{-1} x$, $\tan^{-1} x$, and $\sec^{-1} x$ come from the fact that they are the inverses of the restricted functions $\sin x$, $\tan x$, and $\sec x$. For example, we can graph the inverse trigonometric functions simply by reflecting the graphs of the restricted trigonometric functions over the line $y = x$, as follows:

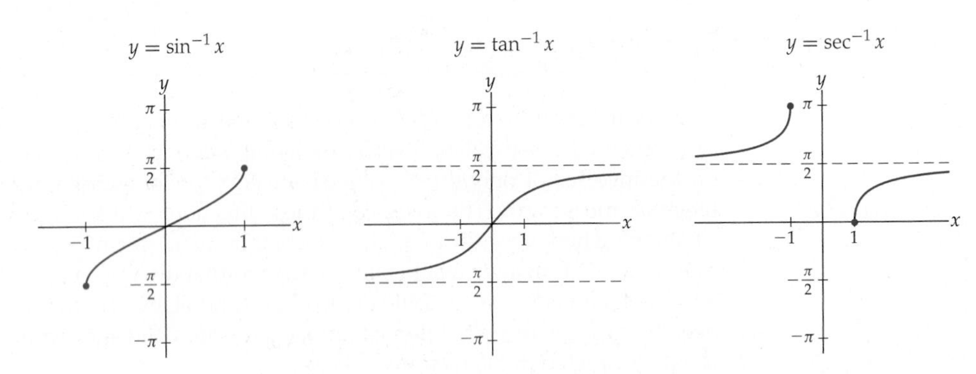

Although $\sin^{-1} x$ and (restricted) $\sin x$ are transcendental functions, their composition $\sin^{-1}(\sin x) = x$ is algebraic. This is obvious because these functions are inverses of each other. However, something more general and surprising is true: The composition of any inverse trigonometric function with any trigonometric function is algebraic; see Example 4.

Examples and Explorations

EXAMPLE 1

Finding values of transcendental functions by hand

Calculate each of the following by hand, without a calculator:

(a) $\log_6 3 + \log_6 12$ **(b)** $\cos \frac{5\pi}{6}$ **(c)** $\sin^{-1} \frac{1}{2}$

SOLUTION

(a) $\log_6 3$ is the exponent to which we would have to raise 6 to get 3; think $6^? = 3$. It is not immediately apparent what this exponent is. Similarly, it is not clear how to calculate $\log_6 12$ without a calculator. However, using the additive property of logarithms we can write

$$\log_6 3 + \log_6 12 = \log_6(3 \cdot 12) = \log_6 36 = 2.$$

The final equality holds because $6^2 = 36$.

(b) The diagram that follows at the left shows where the angle $\frac{5\pi}{6}$ lies on the unit circle. If we draw a line from the point (x, y) where the angle meets the unit circle to the x-axis, we obtain a triangle whose ***reference angle*** is 30°. Using the known side lengths of a 30–60–90 triangle with hypotenuse of length 1, we can label the side lengths of our reference triangle, as shown in the middle figure. This in turn means that we know the coordinates $(x, y) = \left(-\frac{\sqrt{3}}{2}, \frac{1}{2}\right)$ of the point at which the terminal edge of θ intersects the unit circle. Therefore $\cos \frac{5\pi}{6} = -\frac{\sqrt{3}}{2}$.

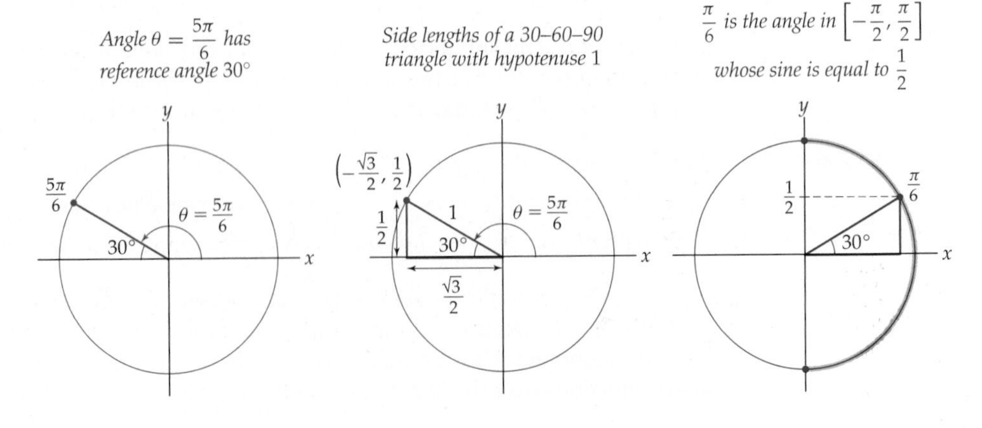

(c) If $\theta = \sin^{-1}\frac{1}{2}$, then we must have $\sin\theta = \frac{1}{2}$. There are infinitely many angles whose sine is $\frac{1}{2}$, but only one of those angles is in the restricted domain $\left[-\frac{\pi}{2}, \frac{\pi}{2}\right]$. Thus $\theta = \sin^{-1}\left(\frac{1}{2}\right)$ is the unique angle in $\left[-\frac{\pi}{2}, \frac{\pi}{2}\right]$ whose sine is $\frac{1}{2}$, as shown in the figure at the right. Notice that the triangle must be a 30–60–90 triangle (since its height is $\frac{1}{2}$), and therefore the angle θ we are looking for must be 30° (i.e., $\frac{\pi}{6}$ radians). Therefore $\sin^{-1}\frac{1}{2} = \frac{\pi}{6}$. □

EXAMPLE 2

Solving equations that involve transcendental functions

Solve each of the following equations:

(a) $3.25(1.72)^x = 1000$ **(b)** $\sin\theta = \cos\theta$ **(c)** $\sec^{-1}x = \frac{\pi}{6}$

SOLUTION

(a) To solve for x we will isolate the expression $(1.72)^x$ and then apply the natural logarithm so that we can get x out of the exponent:

$$3.25(1.72)^x = 1000 \implies \ln((1.72)^x) = \ln\left(\frac{1000}{3.25}\right) \implies x\ln(1.72) = \ln\left(\frac{1000}{3.25}\right).$$

It is now a simple matter to solve for

$$x = \frac{\ln(1000/3.25)}{\ln(1.72)} \approx 10.564.$$

(b) If $\sin\theta = \cos\theta$, then θ is an angle whose terminal edge intersects the unit circle at a point (x, y) with $x = y$. The only such points on the unit circle are $(\sqrt{2}/2, \sqrt{2}/2)$ and $(-\sqrt{2}/2, -\sqrt{2}/2)$, as shown in the left-hand diagram that follows. The angles that end at these points are all of the form $\theta = \frac{\pi}{4} + \pi k$ for some integer k. Thus the solution set for the equation is $\left\{\ldots, -\frac{3\pi}{4}, \frac{\pi}{4}, \frac{5\pi}{4}, \frac{9\pi}{4}, \ldots\right\}$.

Diagram to solve $\sin\theta = \cos\theta$

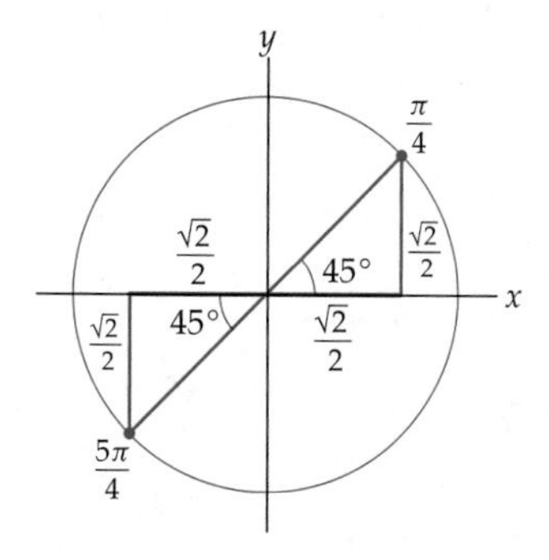

Diagram to solve $\sec^{-1}x = \frac{\pi}{6}$

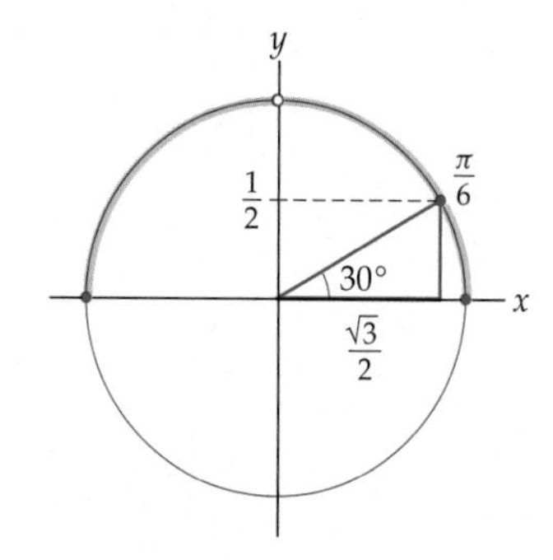

(c) If $\sec^{-1}x = \frac{\pi}{6}$, then

$$x = \sec\frac{\pi}{6} = \frac{1}{\cos\frac{\pi}{6}} = \frac{1}{\sqrt{3}/2} = \frac{2}{\sqrt{3}}.$$

The angle $\frac{\pi}{6}$ and the reference triangle we used for this calculation are shown in the right-hand diagram. □

EXAMPLE 3

Domains and graphs of transcendental functions

Find the domain of each of the functions that follow. Then use transformations to sketch careful graphs of each function by hand, without a graphing utility.

(a) $f(x) = 5 - 3e^{1.7x}$ **(b)** $g(x) = \dfrac{1}{\ln(x-2)}$ **(c)** $h(x) = 3\sec 2x$

SOLUTION

(a) The domain of $f(x) = 5-3e^{1.7x}$ is $\mathbb{R}$, and the graph of f is a transformation of the exponential growth function $e^{1.7x}$ shown in the left-hand figure that follows. $y = -3e^{1.7x}$ can be obtained by reflecting the leftmost graph over the x-axis and then stretching vertically by a factor of 3, as shown in the middle figure. The graph of $f(x) = 5 - 3e^{1.7x}$ can now be obtained by shifting the middle graph up five units, as shown at the right.

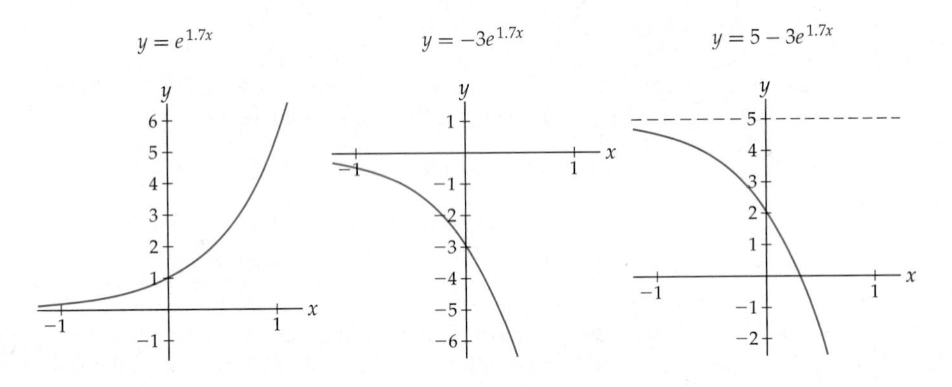

(b) For the function $g(x) = \frac{1}{\ln(x-2)}$ to be defined at a value x, we must have $x - 2 > 0$, and thus $x > 2$. We must also have $\ln(x-2) \neq 0$, which means that $x - 2 \neq 1$, and thus $x \neq 3$. Therefore the domain of $g(x)$ is $(2, 3) \cup (3, \infty)$. To sketch the graph of $g(x) = \frac{1}{\ln(x-2)}$ we start with the graph of $y = \ln x$ in the left-hand figure that follows, translate to the right two units as shown in the middle figure, and then sketch the reciprocal as shown at the right.

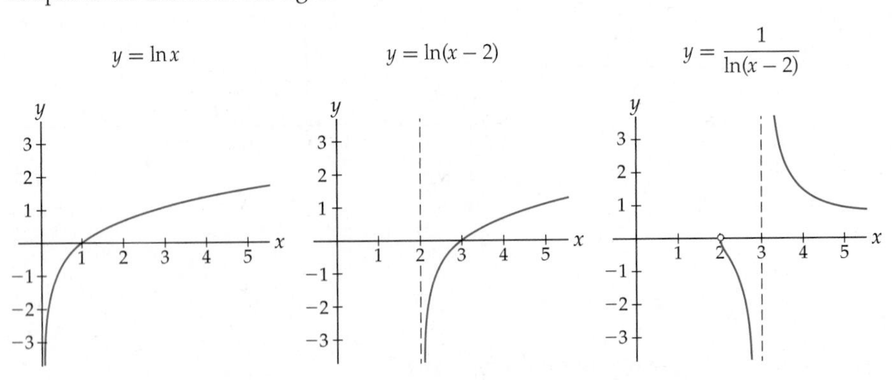

(c) The function $h(x) = 3\sec 2x = \frac{3}{\cos 2x}$ is defined when $\cos 2x \neq 0$. The latter condition occurs when $2x$ is not an odd multiple of $\frac{\pi}{2}$ and thus when x is not an odd multiple of $\frac{\pi}{4}$. Therefore the domain of $h(x)$ is $x \neq \frac{\pi}{4}(2k+1)$ for positive integers k. To sketch the graph of $h(x)$, we start with the graph of $y = \sec x$ as follows at the left, stretch vertically by a factor of 3 as shown in the middle figure, and then compress horizontally by a factor of 2 as shown at the right. (See Section 0.2 for a review of transformations of graphs.)

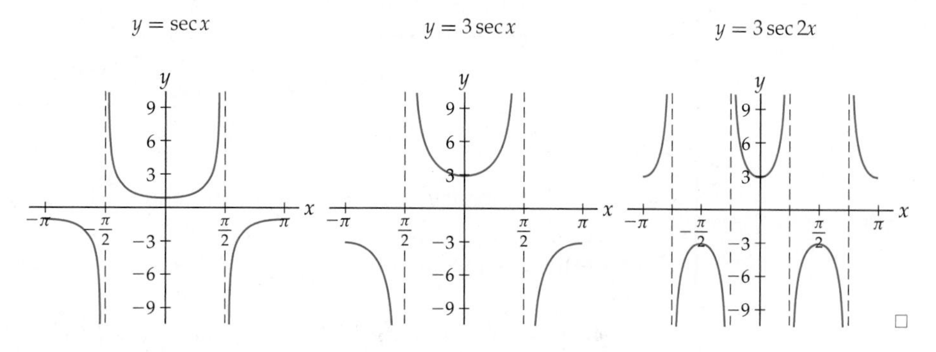

□

EXAMPLE 4

Simplifying compositions of inverse trigonometric and trigonometric functions

Write $\cos(\sin^{-1} x)$ as an algebraic function, that is, a function that involves only arithmetic operations and rational powers.

SOLUTION

If we define $\theta = \sin^{-1} x$, then $\sin\theta = x$ and θ must be in the interval $\left[-\frac{\pi}{2}, \frac{\pi}{2}\right]$. Let's first consider the case where θ is in the first quadrant $\left[0, \frac{\pi}{2}\right]$; the reference triangle for such a θ is shown next at the left. If we wish θ to have a sine of x, then the length of the vertical leg of the triangle must be x. The hypotenuse of the triangle is length 1, since we are on the *unit* circle. We could also have considered that the sine of θ is "opposite over hypotenuse"; thus one triangle involving our angle θ could have an opposite side of length x and a hypotenuse of length 1. Using the Pythagorean theorem, we find that the length of the remaining leg of the triangle is $\sqrt{1-x^2}$, as shown at the right:

Reference triangle for an angle θ in $\left[0, \frac{\pi}{2}\right]$

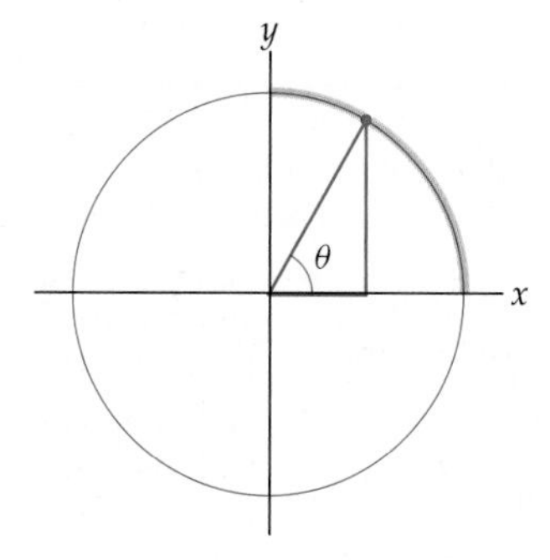

Use Pythagorean theorem to determine length of remaining leg

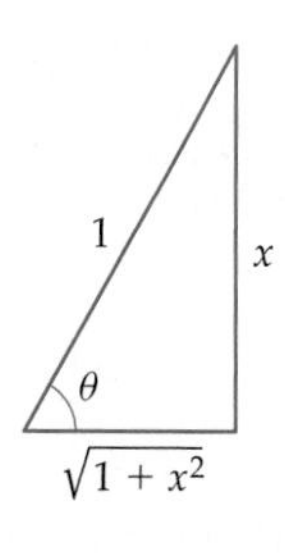

Now $\cos\theta$ is the horizontal coordinate of the point on the unit circle corresponding to θ, or, in terms of "adjacent over hypotenuse," we have

$$\cos\theta = \frac{\sqrt{1-x^2}}{1} = \sqrt{1-x^2}.$$

The case where θ is in the fourth quadrant, that is, where $\theta \in \left[-\frac{\pi}{2}, 0\right]$, is similar and also shows that $\cos\theta = \sqrt{1-x^2}$. Therefore we have shown that $\cos(\sin^{-1} x)$ is equal to the algebraic function $\sqrt{1-x^2}$. □

CHECKING THE ANSWER

To verify the strange fact that $\cos(\sin^{-1} x) = \sqrt{1-x^2}$, try evaluating both sides at some simple x-values. While looking at just a few x-values will not prove that the two expressions are equal for all x, it will at least give us some evidence that the equality is reasonable. For example, at $x = 0$ we have

$$\cos(\sin^{-1} 0) = \cos 0 = 1 \quad \text{and} \quad \sqrt{1-0^2} = \sqrt{1} = 1,$$

and at $x = 1$ we have

$$\cos(\sin^{-1} 1) = \cos\left(\frac{\pi}{2}\right) = 0 \quad \text{and} \quad \sqrt{1-1^2} = \sqrt{0} = 0.$$

As a less trivial example, consider $x = \frac{1}{2}$. At this value we have

$$\cos\left(\sin^{-1}\left(\frac{1}{2}\right)\right) = \cos\left(\frac{\pi}{6}\right) = \frac{\sqrt{3}}{2} \quad \text{and} \quad \sqrt{1-\left(\frac{1}{2}\right)^2} = \sqrt{1-\frac{1}{4}} = \sqrt{\frac{3}{4}} = \frac{\sqrt{3}}{2}.$$

TEST YOUR UNDERSTANDING

- Why do we require that $A \neq 0$ and $b > 0$, $b \neq 1$ in the definition of exponential functions? What would the graphs look like when $A = 0$, when $b < 0$, $b = 0$, or $b = 1$?
- In the reading we calculated $\log_7 2$ by finding $\frac{\ln 2}{\ln 7}$ with a calculator. Would we get the same answer if we computed $\frac{\log_{10} 2}{\log_{10} 7}$?
- How do you convert from radians to degrees, or vice versa?
- How is the graph of the reciprocal of a function related to the graph of that function? How can that information be useful for remembering the graphs of $y = \csc x$, $y = \sec x$, and $y = \cot x$?
- How are the unit-circle definitions of the trigonometric functions related to the right-triangle definitions of trigonometric functions?

EXERCISES 0.4

Thinking Back

Algebra with exponents: Write each of the following expressions in the form Ab^x for some real numbers A and b:

(a) 3^{2x+1} (b) $5^x 2^{3-x}$ (c) $(2^{3x-5})^4$

(d) $\dfrac{1}{2(3^{x-4})}$ (e) $\dfrac{4(3^x)^2}{2^x}$ (f) $\dfrac{(1/8)^x}{3(2^{3x+1})}$

Inverse functions: Suppose f and g are inverses of each other.

- What can you say about $f(g(x))$ and $g(f(x))$?
- If f has a horizontal asymptote at $y = 0$, what can you say about g?
- If f has a y-intercept at $y = 1$, what can you say about g?

Famous triangles, degrees, and radians: The following exercises will help you review and recall basic trigonometry.

- Suppose a right triangle has angles 30°, 60°, and 90° and a hypotenuse of length 1. What are the lengths of the remaining legs of the triangle?
- Suppose a right triangle has angles 45°, 45°, and 90° and a hypotenuse of length 1. What are the lengths of the remaining legs of the triangle?
- What is a radian? Is it larger or smaller than a degree? Compare an angle of 1 degree with an angle of 1 radian, with both angles in standard position.
- Show each of the following angles in standard position on the unit circle, in radians:

(a) $\dfrac{3\pi}{4}$ (b) $-\dfrac{4\pi}{3}$ (c) $\dfrac{17\pi}{6}$ (d) 21π

Concepts

0. *Problem Zero:* Read the section and make your own summary of the material.

1. *True/False:* Determine whether each of the statements that follow is true or false. If a statement is true, explain why. If a statement is false, provide a counterexample.

(a) *True or False:* The function $f(x) = 3e^{0.5x} - 2$ is an exponential function.

(b) *True or False:* Every exponential function $f(x) = Ae^{kx}$ has a horizontal asymptote at $y=0$.

(c) *True or False:* For all $x > 0$, $\ln(x^3) = 3\ln x$.

(d) *True or False:* For all $x > 0$, $\dfrac{\log_2 x}{\log_2 3} = \dfrac{\log_6 x}{\log_6 3}$.

(e) *True or False:* If (x, y) is the point on the unit circle corresponding to the angle $-\dfrac{7\pi}{3}$, then x is positive and y is negative.

(f) *True or False:* The sine of an angle θ is always equal to the sine of the reference angle for θ.

(g) *True or False:* For any x, $1 - \cos^2(5x^3) = \sin^2(5x^3)$.

(h) *True or False:* $\sec^{-1} x = \dfrac{1}{\cos^{-1} x}$.

2. *Examples:* Construct examples of the thing(s) described in the following. Try to find examples that are different than any in the reading.

(a) Two exponential functions and their inverses.

(b) Two x-values at which $\tan x$ is not defined.

(c) Two x-values at which $\sec^{-1} x$ is not defined.

3. What is the definition of an exponential function, and how is such a function different from a power function? Is the function $f(x) = x^x$ a power function, an exponential function, or neither, and why?

4. In this exercise we will examine two ways to think of a^b when b is an irrational number, and in particular, we will consider what the quantity 2^π represents.

(a) One way to define 2^π is to think of it as a limit. If we take a sequence $a_1, a_2, a_3, \ldots$ of rational numbers that approaches π, then the sequence $2^{a_1}, 2^{a_2}, 2^{a_3}, \ldots$ should approach 2^π. Said in terms of limits, this means that

$$2^\pi = \lim_{a \to \pi} 2^a,$$

where each a is assumed to be a rational number. Can you think of a sequence of rational numbers that gets closer and closer to π? (*Hint: Think about the decimal expansion of* π.)

(b) Another way to consider 2^π is to write it as an infinite product:

$$2^\pi = 2^3\, 2^{1/10}\, 2^{4/100}\, 2^{1/1000}\, 2^{5/10000}\, 2^{9/100000} \cdots.$$

What will the next term in the product be? How could 2^π equal the product of infinitely many numbers? Wouldn't that make 2^π infinitely large? Calculate some of the later terms in the product (for example, $2^{5/10000}$ or $2^{9/100000}$), and use these calculations to argue that even though 2^π can be written as a product of infinitely many numbers, it is not necessarily infinitely large.

5. Approximate $2^{\sqrt{3}}$ by calculating 2^r for rational values r that get closer and closer to $\sqrt{3}$. (*Hint: You can use the decimal expansion of* $\sqrt{3}$ *to get a sequence of rational numbers that approaches* $\sqrt{3}$.)

6. Why can't we define the number e just by writing it down in decimal notation to lots of decimal places?

7. Write the exponential function $f(x) = 3e^{-2x}$ in the form Ab^x for some real numbers A and b. Then write the exponential function $g(x) = -2(3^x)$ in the form Ae^{kx} for some real numbers A and k.

8. Fill in each blank with an interval of real numbers.

(a) An exponential function $f(x) = Ab^x$ represents exponential growth if $b \in$ ____ and exponential decay if $b \in$ ____ .

(b) An exponential function $f(x) = Ae^{kx}$ represents exponential growth if $k \in$ ____ and exponential decay if $k \in$ ____ .

(c) Suppose that $e^{kx} = b^x$ for some real numbers k and b. Then $k \in (0, \infty)$ if and only if $b \in$ ____ .

(d) Suppose that $e^{kx} = b^x$ for some real numbers k and b. Then $k \in (-\infty, 0)$ if and only if $b \in$ ____ .

9. In the definition of the logarithmic function $\log_b x$, what are the allowable values for the base b, and why?

10. Fill in the blanks in each of the following statements.

(a) For all $x \in$ ____ , $\log_2 x = y$ if and only if $x =$ ____ .

(b) For all $x \in$ ____ , $3^{\log_3 x} =$ ____ .

(c) For all $x \in$ ____ , $\log_4(4^x) =$ ____ .

(d) $\log_2 3$ is the exponent to which you have to raise ____ to get ____ .

11. The graphs of $y = \log_2 x$ and $y = \log_4 x$ are shown here. Determine which graph is which, without using a calculator. (*Hint: Think about the graphs* $y = 2^x$ *and* $y = 4^x$, *and then reflect those graphs over the line* $y = x$.)

$y = \log_2 x$ *and* $y = \log_4 x$

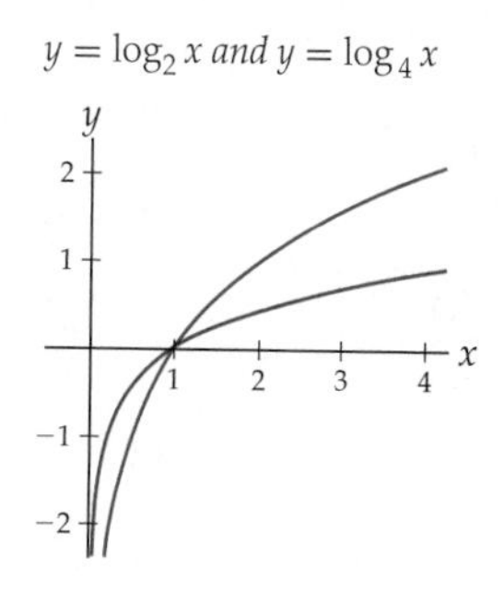

12. State the algebraic properties of the natural logarithm function that correspond to the eight properties of logarithmic functions in Theorem 0.24.

13. Use algebraic properties of logarithms, the graph of $y = \ln x$, and your knowledge of transformations to sketch graphs of $f(x) = \ln(x^2)$ and $g(x) = \ln\left(\frac{1}{x}\right)$.

14. Solve the inequality $\ln\left(\frac{x+1}{x-1}\right) \geq 0$.

15. Give a mathematical definition of $\sin\theta$ for any angle θ. Your definition should include the words "unit circle," "standard position," "terminal," and "coordinate."

16. Give a mathematical definition of $\tan\theta$ for any angle θ. Your definition should include the words "unit circle," "standard position," "terminal," and "coordinate."

17. Use the definition of the sine function to explain why $\sin\left(\frac{\pi}{4}\right)$ is equal to both $\sin\left(\frac{9\pi}{4}\right)$ and $\sin\left(-\frac{7\pi}{4}\right)$.

18. Fill in each blank with an interval of real numbers.

(a) The function $f(x) = \cos x$ has domain ____ and range ____.

(b) The function $f(x) = \csc x$ has domain ____ and range ____.

(c) The restricted tangent function has domain ____ and range ____.

(d) The function $f(x) = \sec^{-1} x$ has domain ____ and range ____.

19. Suppose θ is an angle in standard position whose terminal edge intersects the unit circle at the point (x, y). If $y = -\frac{1}{3}$, what are the possible values of $\cos\theta$? If you know that the terminal edge of θ is in the third quadrant, what can you say about $\cos\theta$? What if the terminal edge of θ is in the fourth quadrant? Could the terminal edge of θ be in the first or second quadrant?

20. Show that $-\sqrt{3}$ is in the range of the tangent function by finding an angle θ for which $\tan\theta = -\sqrt{3}$.

21. Describe restricted domains for $\sin x$, $\tan x$, and $\sec x$ on which each function is invertible. Then describe the corresponding domains and ranges for $\arcsin x$, $\arctan x$, and $\operatorname{arcsec} x$.

22. Fill in the blanks:

(a) $\sin^{-1} x$ is the angle in the interval ____ whose ____ is x.

(b) $y = \arcsin x$ if and only if $\sin y =$ ____, for all $x \in$ ____ and $y \in$ ____.

(c) If $\tan^{-1} x = \theta$ and $\tan\theta$ is positive, then θ is in the ____ quadrant.

(d) If $\arctan x = \theta$ and $\sin\theta = \frac{1}{3}$, then $\cos\theta =$ ____.

23. Which of the following expressions are defined? Why or why not?

(a) $\sin^{-1}\left(-\frac{1}{25}\right)$ (b) $\sin^{-1}\frac{3}{2}$

(c) $\tan^{-1} 100$ (d) $\sec^{-1}\frac{\pi}{4}$

24. Sketch a graph of the restricted cosine function on the domain $[0, \pi]$, and argue that this restricted function is one-to-one. Then sketch a graph of $\cos^{-1} x$, and list the domains and ranges of the inverse $\cos^{-1} x$ of this restricted function.

25. Without calculating the exact or approximate values of the following quantities, use the unit circle to determine whether each of those quantities is positive or negative:

(a) $\sin^{-1}\left(-\frac{1}{5}\right)$ (b) $\sin^{-1}\left(-\frac{2}{3}\right)$

(c) $\tan^{-1} 2$ (d) $\sec^{-1}(-5)$

26. Find all angles whose secant is 2, and then find $\sec^{-1}(2)$.

Skills

Find the domains of the functions in Exercises 27–32.

27. $f(x) = \dfrac{\ln(x+1)}{\ln(x-2)}$

28. $f(x) = \dfrac{1}{e^x - e^{2x}}$

29. $f(x) = \dfrac{1}{\sqrt{\ln(x-1)}}$

30. $f(x) = \dfrac{1}{1 - \tan\theta}$

31. $f(x) = \sqrt{\sec\theta}$

32. $f(x) = 2\sin^{-1}(x-3)$

Find the exact values of each of the quantities in Exercises 33–44. Do not use a calculator.

33. $\ln\left(\frac{1}{e^2}\right)$

34. $\log_{1/2} 4$

35. $4\log_2 6 - 2\log_2 9$

36. $\dfrac{\log_7 9}{\log_7 1/3} + \log_3 1$

37. $\tan\left(-\frac{\pi}{4}\right)$

38. $\cos\left(\frac{48\pi}{3}\right)$

39. $\csc\left(-\frac{5\pi}{4}\right)$

40. $\sin(201\pi)$

41. $\cos^{-1}(-1)$

42. $\sin^{-1}(-1)$

43. $\operatorname{arcsec}\left(-\frac{2}{\sqrt{2}}\right)$

44. $\arctan\left(-\frac{1}{\sqrt{3}}\right)$

Solve the equations in Exercises 45–50 by hand. When you are finished, check your answers either by testing your solutions or by graphing an appropriate function.

45. $2^x = 3^{x-1}$

46. $2 = 10\left(1 + \frac{0.19}{12}\right)^{12x}$

47. $\log_2\left(\frac{x-1}{x+1}\right) = 4$

48. $\sin x = \frac{1}{2}$

49. $\cos 2x = 1$

50. $\sec^{-1} x = \pi$

Suppose that $\cos(\theta) = \frac{1}{6}$, $\sin(\theta) > 0$, $\sin(\phi) = \frac{3}{5}$, and $\cos(\phi) < 0$. Use trigonometric identities to identify the quantities in Exercises 51–56.

51. $\sin(\theta)$

52. $\sin(-\phi)$

53. $\cos(2\theta)$

54. $\sin\left(\theta + \frac{\pi}{2}\right)$

55. the sign of $\cos(\theta + \phi)$

56. the sign of $\tan(\theta + \pi)$

Write each of the expressions in Exercises 57–60 as an algebraic expression that does not involve trigonometric or inverse trigonometric functions.

57. $\sin(\cos^{-1} x)$

58. $\tan(\tan^{-1} 2x)$

59. $\sec^2(\tan^{-1} x)$

60. $\sin^2(\tan^{-1} x)$

61. $\sin\left(\sec^{-1} \frac{3}{x}\right)$

62. $\csc(2\tan^{-1} x)$

63. $\cos(2\sin^{-1} 5x)$

64. $\tan^2\left(2\sec^{-1} \frac{x}{3}\right)$

Sketch graphs of the functions in Exercises 65–72 by hand, without using a calculator or graphing utility. Indicate any roots, intercepts, and asymptotes on your graphs.

65. $f(x) = -\left(\frac{1}{2}\right)^x + 10$

66. $f(x) = -0.25(3^{x-2})$

67. $f(x) = 20 - 5e^{-2x}$

68. $f(x) = \log_{1/2} x$

69. $f(x) = -\log_2(x - 3)$

70. $f(x) = \sin(2x) + 4$

71. $f(x) = 2\cos\left(x - \frac{\pi}{4}\right)$

72. $f(x) = \tan^{-1}(x - 2) + \pi$

For each graph in Exercises 73–78, find a function whose graph looks like the one shown. When you are finished, use a graphing utility to check that your function f has the properties and features of the given graph.

73.

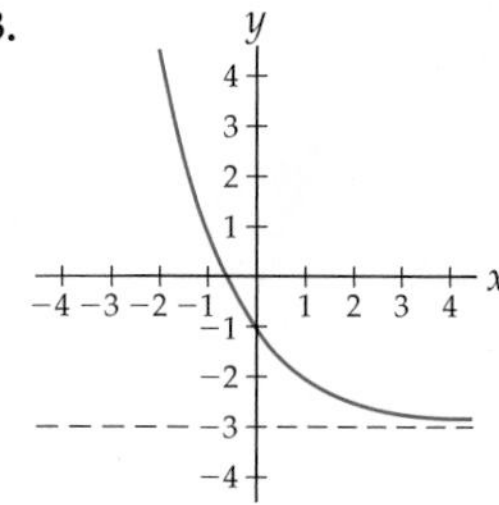

74.

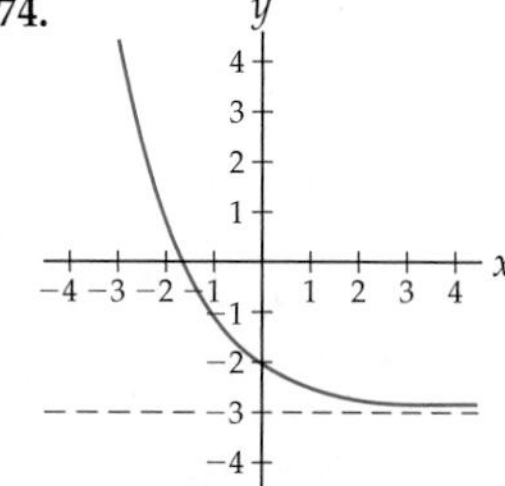

75.

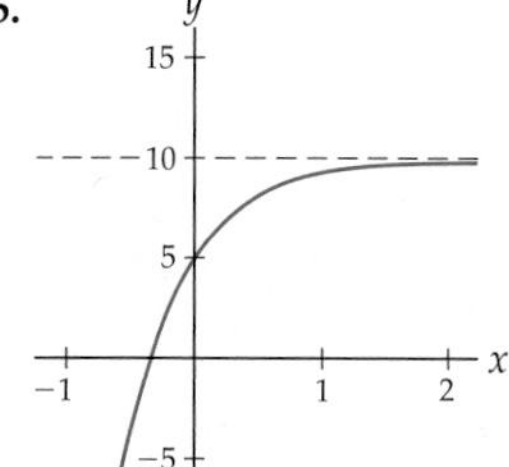

76.

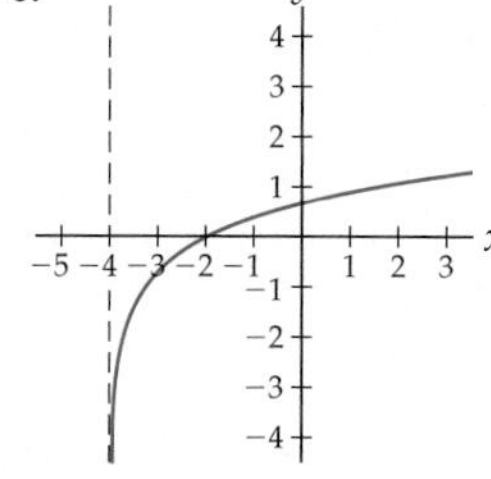

77.

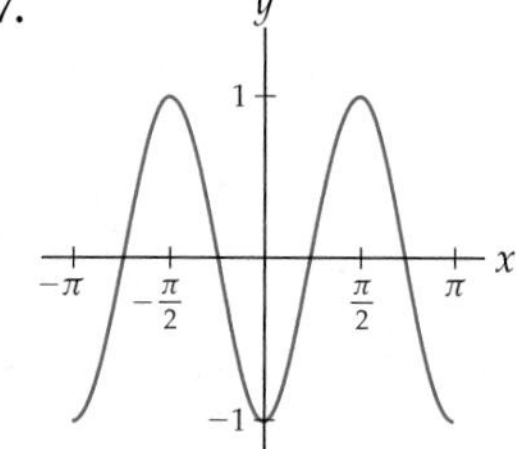

78.

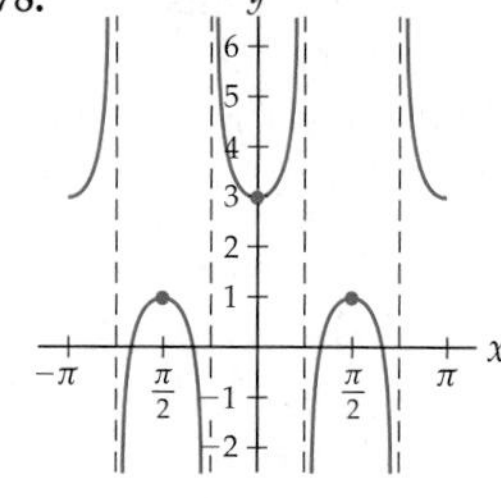

Applications

79. Ten years ago, Jenny deposited \$10,000 into an investment account. Her investment account now holds \$22,609.80. Her accountant tells her that her investment account balance $I(t)$ is an exponential function.

(a) Find an exponential function of the form $I(t) = Ae^{kt}$ to model Jenny's investment account balance.

(b) Find an exponential function of the form $I(t) = Ab^t$ to model Jenny's investment account balance.

80. Suppose there were 500 rats on a certain island in 1973 and 1,697 rats on the same island 10 years later. Assume that the number $R(t)$ of rats on the island t years after 1973 is an exponential function.

(a) Find an equation for the exponential function $R(t)$ that describes the number of rats on the island. Let $t = 0$ represent the year 1973.

(b) According to your function $R(t)$, how many rats will be on the island in 2020?

(c) How long did it take for the population of rats to double from its 1973 amount? How long did it take for it to double again? And again?

81. Suppose a rock sample initially contains 250 grams of the radioactive substance unobtainium, and that the amount of unobtainium after t years is given by an exponential function of the form $S(t) = Ae^{kt}$. The half-life of unobtainium is 29 years, which means that it takes 29 years for the amount of the substance to decrease by half.

(a) Find an equation for the exponential function $S(t)$.

(b) What percentage of unobtainium decays each year?

(c) How long will it be before the rock sample contains only 6 grams of unobtainium?

82. Again considering the rock sample described in Exercise 81, answer the following questions:

(a) At one point the rock sample contained 900 grams of unobtainium; how long ago?

(b) What percentage of the unobtainium will be left in 300 years?

(c) How long will it be before 95% of the unobtainium has decayed?

83. Alina is flying a kite and has managed to get her kite so high in the air that she has let out 400 feet of kite string. If the angle made by the ground and the line of kite string is 32 degrees, how high is the kite?

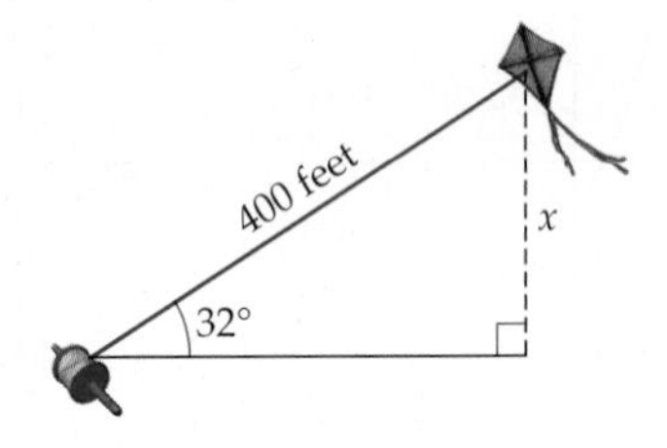

84. Suppose two stars are each 60 light-years away from Earth. The angle between the line of sight to the first star and the line of sight to the second star is 2 degrees. In other words, if you look at the first star, then turn your head to look at the second star, your head will move through an angle of 2 degrees. How far apart are the stars?

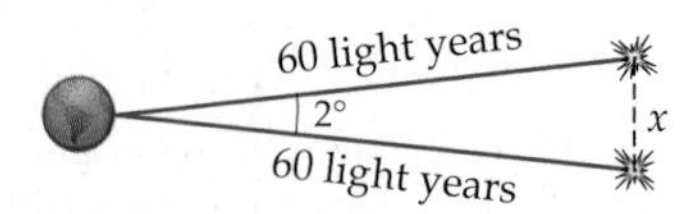

Proofs

85. Prove by contradiction that every exponential function $f(x) = Ab^x$ has the property that $f(x)$ is never zero. *(Hint: Use what you know about the algebraic properties of exponential functions, and the fact that if $f(x) = Ab^x$ is an exponential function, then neither A nor b is zero.)*

86. Use the definition of a one-to-one function to prove that every exponential function $f(x) = Ab^x$ is one-to-one. *(Hint: Use the fact that $b^x = 1$ only when $x = 0$.)*

87. Use the base conversion formula for logarithms to prove that the function $f(x) = \log_2 x$ is equal to the function $g(x) = \log_3 x$ only when $x = 1$.

88. Use logarithms to prove that every exponential function of the form $f(x) = Ab^x$ can be written in the form $f(x) = Ae^{kx}$, and vice versa.

89. Use the definition of a logarithmic function $y = \log_b x$ to prove that for any $b > 0$ with $b \neq 1$, the quantity $\log_b 1$ is equal to zero.

In Exercises 90–94, assume that x, y, a, and b are values which make sense in the expressions involved.

90. Use the fact that logarithmic functions are the inverses of exponential functions to prove that:

(a) $\log_b x = y$ if and only if $b^y = x$

(b) $\log_b(b^x) = x$

(c) $b^{\log_b x} = x$

91. Prove that $\log_b(x^a) = a \log_b x$. *(Hint: Start with $\log_b(x^a)$ and replace x with $b^{\log_b x}$.)*

92. Prove that $\log_b(xy) = \log_b x + \log_b y$. *(Hint: Show that this statement is equivalent to the statement $xy = b^{\log_b x + \log_b y}$, and prove the new statement instead.)*

93. Use the results of the two exericses above to prove that:

(a) $\log_b\left(\frac{1}{x}\right) = -\log_b x$

(b) $\log_b\left(\frac{x}{y}\right) = \log_b x - \log_b y$

94. Prove the base conversion formula $\log_b x = \dfrac{\log_a x}{\log_a b}$. *(Hint: Set $y = \log_b x$ and then show that $b^y = x$.)*

95. Use the unit-circle definitions of sine and cosine to prove the identity $\sin^2\theta + \cos^2\theta = 1$.

96. Use the first Pythagorean identity $\sin^2\theta + \cos^2\theta = 1$ to prove the second and third Pythagorean identities listed in Theorem 0.26. *(Hint: To prove the second identity, divide both sides of the first identity by $\cos^2 x$. A similar strategy will prove the third identity.)*

97. Use the unit-circle definitions of the trigonometric functions to prove the even-odd identities and the shift identities listed in Theorem 0.26.

98. Use the sum identities and the even-odd identities to prove the difference identities listed in Theorem 0.26.

99. Use the sum identities to prove the double-angle identities listed in Theorem 0.26. *(Hint: Note that 2θ is equal to $\theta + \theta$.)*

100. The four identities listed as alternative forms in Theorem 0.26 are alternative ways of writing the double-angle identity $\cos 2\theta = \cos^2\theta - \sin^2\theta$. Use this double-angle identity, algebra, and the Pythagorean identities to prove these four alternative forms.

Thinking Forward

- *A special exponential limit:* Use a calculator to approximate $\frac{e^h - 1}{h}$ for the following values of h: (a) $h = 0.1$; (b) $h = 0.01$; (c) $h = 0.001$. As h gets closer to zero, what number does your approximations seem to approach?
- *Logarithms with absolute values:* Sketch a graph of the function $f(x) = \ln|x|$. What is the domain of this function? Is the function even, odd, or neither, and why?
- *Rewriting trigonometric expressions:* Use the double-angle identity $\sin^2 x = \frac{1-\cos 2x}{2}$ to rewrite the expression $\sin^4 x \cos^2 x$ in terms of a sum of expressions of the form $A \cos kx$. (Note: You'll have to multiply out some expressions, and use the double-angle identity more than once.)

0.5 LOGIC AND MATHEMATICAL THINKING*

- Logical statements that involve quantifiers or implications
- Using a counterexample to show that a statement is false
- Mathematical proof techniques, including direct proof and proof by contradiction

From Definitions to Theorems

Throughout this chapter we have learned a lot of the mathematical language that we will use throughout the book. It is now time to start thinking about how to build on our basic definitions and develop the theory of calculus. Developing such a theory is like building a skyscraper. With our library of mathematical definitions and language we have set the foundation. Throughout the rest of this book we will build on that foundation by using logical, mathematical arguments to develop new theorems. Each new theorem will be the springboard to a new set of definitions and theorems and will form a new level of our skyscraper. In this section we present an overview of the logic and proof techniques needed for our construction of calculus.

In what follows we will keep the definitions simple so that we can focus on the logic and proofs. Most of the mathematical statements we will discuss will concern real numbers and whether they are positive, negative, or zero, or will build upon the following five definitions:

- An ***integer*** is a whole number that is positive, negative, or zero. The set of all integers is
$$\{\ldots, -5, -4, -3, -2, -1, 0, 1, 2, 3, 4, 5, \ldots\}.$$
- An integer n is ***divisible*** by an integer m if we can write $n = km$ for some integer k.
- An integer n is ***even*** if we can write $n = 2k$ for some integer k.
- An integer n is ***odd*** if we can write $n = 2k + 1$ for some integer k.
- A ***rational number*** is a real number that can be written in the form $\frac{p}{q}$ for some integers p and q, where $q \neq 0$.

Quantifiers

We will often be interested in stating that a property is true all of the time, some of the time, at least once, or none of the time. Logical ***quantifiers*** are the key to making such statements precisely.

DEFINITION 0.28 **The Quantifiers "For All" and "There Exists"**

Suppose P is a property that depends on a value x.

(a) ***For all x, property P*** means that property P holds for all possible values of x.

(b) ***There exists x such that property P*** means that property P is true for at least one value of x.

For example, consider the following four quantified statements:

For all real numbers x, $2x + 2 = 2(x + 1)$.

For all integers x, $\frac{1}{x}$ *is a rational number.*

There exists an integer x such that x is even.

There exists a real number x such that $\frac{1}{x} = 0$.

The first statement is true because its claim that $2x + 2 = 2(x + 1)$ is indeed true for *every* real number x. The second statement is false because its claim that $\frac{1}{x}$ is rational fails for $x = 0$. The third statement is true because many even integers exist; for example, $x = 2$. The fourth statement is false because there is not even one real number x for which $\frac{1}{x} = 0$.

For statements with more than one quantifier, the order in which the quantifiers appear can make a big difference. For example, consider the following two statements, which differ only in the order that their quantifiers are listed:

"For all $x > 0$*, there exists* $y > 0$ *such that* $y > x$*."*

"There exists $y > 0$ *such that for all* $x > 0$*,* $y > x$*."*

The first statement claims that given any positive number x, we can find some number y that is greater than x. This is clearly true, since we can always choose y as large as we like to ensure that it is larger than the given number x. The second statment claims that there is some positive number y with the property that y is greater than *every* number x. This is clearly false, since no real number is greater than all other real numbers.

Implications

Statements of the form "If A, then B" are called ***implications.*** Most of the theorems in this book have the form of an implication. The statement A is called the ***hypothesis,*** and the statement B is called the ***conclusion.*** We will use implications in this book so often that we have a shorthand notation for them, namely, "$A \Rightarrow B$" (pronounced "A implies B").

DEFINITION 0.29 Implications

An *implication* is a statement of the form ***if A, then B*** (also written $A \Rightarrow B$). Such an implication is true if, whenever statement A is true, statement B must also be true.

For example, the statement "If $x > 2$, then $x > 0$" is an implication. In the arrow notation we would write this as "$x > 2 \Rightarrow x > 0$," and as a quantified statement we could equivalently write "For all $x > 2$, $x > 0$." The hypothesis of the statement is "$x > 2$," and the conclusion is "$x > 0$." Thus, if we know that x is greater than 2, then we can conclude that x must also be greater than 0.

Suppose we have statements A and B, and the statement that A implies B. Does this mean that B also implies A? Not necessarily; for example, it is true that $x > 2$ implies $x > 0$, but the reverse implication is not true: $x > 0$ does not imply $x > 2$. When we switch the roles of the hypothesis and the conclusion of an implication, we have a new implication called the ***converse*** of the original.

DEFINITION 0.30 The Converse of an Implication

The ***converse*** of the implication $A \Rightarrow B$ is the implication $B \Rightarrow A$.

If an implication $A \Rightarrow B$ and its converse $B \Rightarrow A$ happen to both be true, then we have a two-way implication. We write this as $A \Longleftrightarrow B$, which is pronounced ***A if and only if B***. For example, the statements "x is even" and "x is divisible by 2" are equivalent; each implies the other. Thus we can say that "x is even if and only if x is divisible by 2."

The converse of an implication is obtained when the hypothesis and conclusion switch places. This results in a very different statement from the original. However, if the hypothesis and conclusion switch places *and* are negated, then surprisingly, we end up with a statement that is equivalent to the original.

DEFINITION 0.31

The Contrapositive of an Implication

The ***contrapositive*** of the implication $A \Rightarrow B$ is the statement (Not B) $\Rightarrow$ (Not A).

The contrapositive of an implication is *always* logically equivalent to the original implication. For example, the contrapositive of the statement "If x is an integer, then x is a rational number" is the statement "If x is not a rational number, then x is not an integer." These two statements are logically equivalent (and happen to be true).

Counterexamples

A statement of the form "for all x, property P" is false if there is even one instance x for which P is false. Such an instance is called a ***counterexample***. For example, the statement "all cats are grey" means that *every* cat is grey. If there is one cat that is not grey, then *some* cats are *not* grey, and this shows that the statement "all cats are grey" is false.

DEFINITION 0.32

Counterexamples

A ***counterexample*** is an example of a value that makes a statement false.

THEOREM 0.33

Counterexamples to "For All" Statements

Suppose P is a property that depends on a value x. Then the statement "For all x, we have P" is false if and only if there is a counterexample for which P is false.

For example, consider the statement "For all integers x, $\frac{1}{x}$ is a rational number." This statement is false, because we can find a counterexample: when $x = 0$, $\frac{1}{x}$ is not a rational number. In contrast, the statement "For all real numbers x, $2x+2 = 2(x+1)$" is true, since there are no counterexamples x that do not have the property $2x+2 = 2(x+1)$. Finding a counterexample is a fast and easy way to prove that a "for all" statement is false. To show that a "for all" statement is true, however, requires substantially more work, as we will see shortly in this section.

What about counterexamples to implications? An implication "$A \Rightarrow B$" implies nothing about the truth or falsehood of A. Moreover, if A is false, then the implication "$A \Rightarrow B$" does not imply anything about statement B. For example, consider the statement "If I am elected, then I will lower your taxes." A politician who says this, gets elected, but then does not lower taxes is a liar. But that same politician is *not* a liar if he fails to get elected, whether he raises taxes or not. A statement of the form "If A, then B" is false only if there is an instance when its hypothesis A is true and yet its conclusion B is false.

THEOREM 0.34 **Counterexamples to Implications**

Suppose A and B are statements that depend on a value x. Then the statement "For all x, if A, then B" is false if and only if we can find a counterexample for which A is true but B is false.

For example, the quantified implication "For all x, if x is positive, then x is even" is false, because we can exhibit a counterexample, say, $x = 3$, that is positive but not even. Notice that there are *some* values of x that are positive and even, but not *all* values of x that are positive are also even.

Simple Mathematical Proofs

A mathematical ***proof*** is a logical argument. Every theorem in this book can be proved with the use of previous theorems or definitions. Calculus, like mathematics in general, is about building up a logical system of definitions, facts, and theorems that can be used to investigate functions and describe real-world phenomena. In mathematics, the "building-up" is just as important as any eventual application, and it is very important that each new theorem rest on a foundation of previous theorems and definitions. It is not enough simply to rely on our intuition of what *ought* to be true; we must make sure that every statement we assert is true and that every theorem that we state is mathematically and logically sound.

If you're wondering "when you'll ever use this stuff," perhaps the best answer is that you probably won't, at least not directly. You probably won't need to directly use your studies of Shakespeare or American history, either. If you pursue a career in science, then you might use calculus to model or analyze real-world situations. If you become a literary critic, then you probably won't need to solve equations, find derivatives, or solve integrals. However, learning calculus and the theory behind it will help teach you how to *think*. No matter what you choose to do, the ability to think logically and solve problems will be an invaluable asset. That is yet another reason that it is so important not only to learn the calculational mechanics of calculus or how to apply calculus to real-world problems, but also to understand the theory—and thus the proofs—of calculus.

A proof of a statement or theorem of the form "if A, then B" is a logical argument that starts by assuming the hypothesis A and then argues that the conclusion B must follow. We assume that A is true, and then make a clear, concise, logical argument that B must also be true. We indicate that a proof is over by making a box "■" or by writing "QED," which represents the Latin phrase *quod erat demonstrandum*, meaning "which was to be demonstrated."

As a simple example, we present a proof that every integer that is divisible by 10 must be an even number. Notice that the proof is not much more than a tour through the definitions of divisibility and even numbers:

> ***Proof.*** Suppose n is an integer that is divisible by 10. By the definition of divisibility, this means that we can write $n = 10k$ for some integer k. Rewriting this equation we have $n = 2(5k)$. Since $5k$ is also an integer, n satisfies the definition of an even integer. Therefore n is even. ■

Notice that the proof has a beginning (state the hypothesis), a middle (make a logical argument), and an end (make the conclusion). This proof is an example of a ***direct proof***, which means that the conclusion follows from the hypothesis via a fairly straightforward logical argument. In the examples and exercises we will also explore more involved methods of proof, such as ***proof by contradiction***, in which we show that something is true by proving that it cannot be false.

Examples and Explorations

EXAMPLE 1

Determining the truth or falsehood of quantified statements

Determine whether each of the quantified statements that follow are true or false. If a statement is true, explain why. If a statement is false, provide a counterexample.

(a) For all real numbers x, $x \leq 12$.

(b) For all real numbers x, $x^2 \geq 0$.

(c) There exists a real number x such that $x^2 = -1$.

(d) For all integers x, there is some integer y such that $y = x + 1$.

(e) There exists some integer x such that for all integers y, $y = x + 1$.

(f) For all $e > 0$, there exists $d > 0$ such that for all $x > 0$, if $x < d$, then $x^2 < e$.

SOLUTION

(a) This statement is false, because not all real numbers are less than or equal to 12. The real number $x = 20$ is a counterexample.

(b) This statement is true, because the square of any real number is nonnegative. Therefore there are no counterexamples to this statement.

(c) This statement is false, since no real number has a square that is negative. Although the complex number $i = \sqrt{-1}$ satisfies $x^2 = -1$, it is not a real number.

(d) Given any integer x, we can always find some other integer y that is 1 greater than x. For example, given $x = 3$, we can choose $y = 3 + 1 = 4$; given $x = 4$, we can choose $y = 4 + 1 = 5$, and so on. The given statement is true.

(e) There is no integer x for which *every* other integer y is one greater than x. For example, for $x = 3$, there is an integer y that is 1 greater (namely, $3 + 1 = 4$), but not *all* integers y are one greater than 3. The given statement is false.

(f) This one takes some parsing, but it will be worth it since the given statement is similar to many of those which we will be studying in Chapter 1. Let's try an example. If $e = 9$, can we find some d such that $x < d$ guarantees that $x^2 < 9$ for all positive values of x? Yes; if $d = 3$ and $x > 0$, then $x < 3$ guarantees that $x^2 < 9$. In fact, for *any* value of e, the value $d = \sqrt{e}$ (as well as many other values of d) will make the implication "if $x < \sqrt{e}$, then $x^2 < e$" true, since $0 < x < \sqrt{e}$ guarantees that $x^2 < e$. Therefore the given statement is true. □

EXAMPLE 2

Finding counterexamples to false implications

Each of the implications that follow is false. Provide counterexamples.

(a) If x is even, then $x \geq 0$.

(b) The converse of the statement in part (a).

(c) The contrapositive of the statement in part (a).

SOLUTION

(a) This statement is false because not all even integers are nonnegative. One counterexample is $x = -2$, since -2 is even but -2 is not greater than or equal to 0.

(b) The converse of the given statement is "If $x \geq 0$, then x is even." This statement is false because not all nonnegative integers are even. One counterexample is $x = 3$, since $3 \geq 0$ but 3 is not even. In this example the original statement in part (a) and the converse here in part (b) both happened to be false. In general, however, a statement and its converse may or may not have the same truth value, since they are logically different statements.

(c) The contrapositive of the statement in part (a) is "If $x < 0$, then x is odd." Notice that the negation of the statement $x \geq 0$ is *not* the statement $x \leq 0$; why? The contrapositive statement is false, because not all negative numbers are odd. One counterexample is $x = -2$, since $-2 < 0$ but -2 is not odd. Notice that we can use the same counterexample for the contrapositive as for the statement in part (a), since the contrapositive is logically equivalent to the original statement. □

EXAMPLE 3

A simple calculational proof

Prove that for all real numbers a and b, $a^3 - b^3 = (a - b)(a^2 + ab + b^2)$.

SOLUTION

Sometimes a proof is nothing more than a calculation, written out with justifications for the steps:

Proof. For any real numbers a and b,

$$\begin{aligned} (a-b)(a^2+ab+b^2) &= a^3 + a^2b + ab^2 - a^2b - ab^2 - b^3 && \leftarrow \text{multiply out} \\ &= a^3 - b^3. && \leftarrow \text{simplify} \end{aligned}$$

Therefore, $a^3 - b^3 = (a - b)(a^2 + ab + b^2)$. ■

EXAMPLE 4

A direct proof

Prove that the sum of any two rational numbers is a rational number.

SOLUTION

Before writing a proof, it is helpful to write down the hypothesis that you are given and the conclusion that you are trying to show. While doing this, give names to the variables involved.

Given: a and b are any rational numbers.

Show: $a + b$ is a rational number.

The first line of your proof will be the "given," the last will be the "show," and the middle will be an argument that uses definitions and logical inferences to connect the two. Since rational numbers are involved, we should remind ourselves of their definition: Recall that a number is ***rational*** if it can be written as the quotient of two integers. This gives us an updated and more descriptive version of our "given" and "show":

Given: $a = \frac{p}{q}$ and $b = \frac{r}{s}$ for some integers p, q, r, and s with $q \neq 0$ and $s \neq 0$.

Show: $a + b$ can be written in the form $\frac{u}{v}$ for some integers u and v with $v \neq 0$.

We now have a very clear road map that indicates how we should prove the implication; we just have to get from the "given" to the "show" by calculating $a + b$:

Proof. Suppose a and b are rational numbers. Then we can write a and b as quotients of integers, say, $a = \frac{p}{q}$ and $b = \frac{r}{s}$, for some integers p, q, r, and s, where q and s are nonzero. With this notation, the sum of a and b is

$$a + b = \frac{p}{q} + \frac{r}{s} = \frac{ps + qr}{qs}.$$

Since p, q, r, and s are integers, so are $ps + qr$ and qs. Moreover, since q and s are nonzero, so is qs. Therefore we have written $a + b$ as a valid quotient of two integers, and thus $a + b$ is a rational number. ■

EXAMPLE 5

A proof by contradiction

Prove that the sum of a rational number and an irrational number is irrational.

SOLUTION

Let's begin by writing out the "given" and "show:"

Given: r is a rational number and x is an irrational number.

Show: $r + x$ is irrational.

A number is irrational if it cannot be written as the quotient of two integers. It can be difficult to show that we *can't* write a number as a quotient of two integers, so instead of using a direct proof we will use the method of ***proof by contradiction***. This means that we will show that $r + x$ is irrational by showing that it cannot possibly be rational. More precisely, we will suppose that $r + x$ is rational and then show that this supposition causes a logical contradiction.

Proof. Suppose r is a rational number and x is an irrational number. Seeking a contradiction, suppose that the sum $r + x$ is rational. In the previous example we proved that the sum of two rational numbers is rational. In addition, if a number r is rational, then so is $-r$. This means that the sum of the rational numbers $r + x$ and $-r$ must also be rational, so $(r + x) + (-r) = x$ must be rational. But this conclusion contradicts our assumption that x is irrational. If $r + x$ is rational, we are led to a contradiction; therefore $r + x$ must be an irrational number. ■

? TEST YOUR UNDERSTANDING

- What does it mean to say that a statement A "implies" a statement B? How is it different than asserting that A and B are true?
- What can you say about an implication if its hypothesis is always false? What about if its conclusion is always true?
- Why is $x = 3$ *not* a counterexample to the implication "If x is even, then $x \geq 0$"?
- What does it mean to say that an implication and its contrapositive are logically equivalent statements?
- Define each of the following: counterexample, implication, converse, contrapositive.

EXERCISES 0.5

Thinking Back

Basic definitions: Recall the definitions of each of the following terms or quantities:

- nonnegative
- integer
- rational number
- irrational number
- $|x|$
- $\text{dist}(a, b)$

- *Types of numbers:* Are all integers rational? Are all rational numbers integers? Is 0 a rational number? Why or why not?
- *Inequality opposites:* If $x > 9$ is false for a particular value of x, does this necessarily mean that $x < 9$? Why or why not? What is the logical opposite of the statement "$x > 9$"?

Concepts

0. *Problem Zero:* Read the section and make your own summary of the material.

1. *True/False:* Determine whether each of the statements that follow is true or false. If a statement is true, explain why. If a statement is false, provide a counterexample.

(a) *True or False:* You can show that "For all x, P" is true by exhibiting just one value of x that makes P true.

(b) *True or False:* You can show that "For all x, P" is false by exhibiting just one value of x that makes P false.

(c) *True or False:* You can show that "There exists x such that we have P" is true by exhibiting just one value of x that makes P true.

(d) *True or False:* You can show that "There exists x such that we have P" is false by exhibiting just one value of x that makes P false.

(e) *True or False:* The converse of an implication is also an implication.

(f) *True or False:* When A is true and B is false, the implication $A \Rightarrow B$ is false.

(g) *True or False:* When A is false and B is true, the implication $A \Rightarrow B$ is false.

(h) *True or False:* When A is false and B is false, the implication $A \Rightarrow B$ is false.

2. *Examples:* Construct examples of the thing(s) described in the following. Try to find examples that are different than any in the reading.

(a) A true statement involving two quantifiers, and a false statement involving two quantifiers.

(b) A statement of the form "For all x, $A \Rightarrow B$" that has just one counterexample, and another that has many counterexamples.

(c) An implication whose converse is false, and an implication whose converse is true.

3. Suppose the implication "$C \Rightarrow D$" is true. If C is true, what can you say about D? If C is false, what can you say about D?

4. Suppose the implication "$R \Rightarrow S$" is false. What does this mean about statements R and S?

5. Consider the statement "Every positive real number is greater than -2." Write this statement using the quantifier "for all." Then write a statement that is logically equivalent but uses "if ..., then ..." instead of quantifiers.

6. Consider the statement "The square of any real number is nonnegative." Write this statement using the quantifier "for all." Then write a statement that is logically equivalent but uses "if ..., then ..." instead of quantifiers.

7. Consider the statement "Every square is a rectangle." Is this statement true? Write down the converse and the contrapositive of the statement, and determine whether they are true or false.

8. What is the converse of the statement $C \Rightarrow D$? Is the converse logically equivalent to the original statement? Why or why not?

9. What is the contrapositive of the statement $P \Rightarrow Q$? Is the contrapositive logically equivalent to the original statement? Why or why not?

10. Prove, by exhibiting examples, that the sum of two irrational numbers can be either rational or irrational. Why is it okay to prove "by example" here, whereas it is not okay to prove "by example" in general?

For Exercises 11–16, suppose you know the following (no more and no less) about a function $f(x)$ at values of x:

$$\text{If } 0 < |x-3| < 0.1, \text{ then } |f(x)+5| < 0.2.$$

Note that this implication means that *if* we know that x is a solution of the double inequality $0 < |x-3| < 0.1$, *then* we can conclude that $f(x)$ is a solution of the inequality $|f(x)+5| < 0.2$. Use the meaning of the given implication to determine whether or not each of the following related implications is guaranteed to be true. *(Hint: You may have to solve inequalities or sketch number lines and think about distances to determine the meanings of the hypotheses and conclusions of the implications.)*

11. If $0 < |x-3| < 0.1$, then $|f(x)+5| < 0.1$.

12. If $0 < |x-3| < 0.1$, then $|f(x)+5| < 0.3$.

13. If $0 < |x-3| < 0.05$, then $|f(x)+5| < 0.2$.

14. If $0 < |x-3| < 0.05$, then $|f(x)+5| < 0.1$.

15. If $0 < |x-3| < 0.05$, then $|f(x)+5| < 0.4$.

16. If $0 < |x-3| < 0.2$, then $|f(x)+5| < 0.4$.

Skills

Determine whether each statement in Exercises 17–46 is true or false. Justify your answers with reasoning, examples, or counterexamples, as appropriate.

17. There is some real number between 2 and 3.

18. Every real number is a rational number.

19. No real number is both rational and irrational.

20. Every real number is either rational or irrational.

21. For all real numbers x, there is some real number y with $y = x^2$.

22. For all real numbers x, there is some real number y with $x = y^2$.

23. If x is an integer greater than 1, then $x \geq 2$.

24. All real numbers are either greater than zero or less than zero.

25. For all real numbers x, either x is even or x is odd.

26. For all real numbers x, if $x < -2$, then $x^2 > 4$.

27. There exists a real number x such that $x \leq 1$ and $x \geq 2$.

28. There exist real numbers x and y such that $x + y = 4$.

29. For all real numbers x, if x is negative, then x is irrational.

30. For all real numbers x, if x is an integer, then x is rational.

31. There exists a real number x such that $x \leq 1$ or $x \geq 2$.

32. There exists a real number x such that $x > 0$ and $x^2 > 10$.

33. For all real numbers a and b, if $a < b$, then $3a+1 < 3b+1$.

34. No rational number is both less than $\frac{1}{2}$ and greater than $\frac{1}{3}$.

35. For all real numbers x, $x^2 \geq 0$ and $|x| \geq 0$.

36. For all real numbers x, either $x \geq 2$ or $x \leq 1$.

37. There exist real numbers $x < 0$ and $y < 0$ such that $xy < 0$.

38. There exist real numbers $x > 0$ and $y > 0$ such that $xy = 0$.

39. For all real numbers x and y, if $x < y$, then $2x - 1 < 2y - 1$.

40. For all real numbers x and y, if $x < y$, then $x^2 < y^2$.

41. There exists a real number x such that for all real numbers y, $|y| > x$.

42. For all real numbers x and y, $xy = 0$ if and only if $x = 0$ or $y = 0$.

43. For all real numbers x, there exists some y such that $x < y$.

44. For all real numbers x, there exists some y such that $x = y^2$.

45. For all real numbers x and y, $\frac{x}{y} = 0$ if and only if $x = 0$.

46. There exists a real number x such that for all real numbers y, $y > x$.

Suppose A and B represent logical statements. In Exercises 47–54, write (a) the converse and (b) the contrapositive of the given statement. Simplify each of your statements if possible.

47. (Not A) $\Rightarrow B$

48. $A \Rightarrow$ (Not B)

49. (Not B) $\Rightarrow$ (Not A)

50. (Not A) $\Rightarrow$ (Not B)

51. (A and B) $\Rightarrow C$

52. (A or B) $\Rightarrow C$

53. $A \Rightarrow$ (B and C)

54. $A \Rightarrow$ (B or C)

In Exercises 55–66, write (a) the converse and (b) the contrapositive of each statement. Simplify your statements as much as possible. (c) Provide counterexamples if the original, the converse, and/or the contrapositive statements are false.

55. If x is a real number, then x is rational.

56. If $x \geq 2$, then $x \geq 3$.

57. If $x > 2$, then $x \geq 3$.

58. If $x \geq 2$, then $x \geq 1$.

59. If x is negative, then $\sqrt{x}$ is not a real number.

60. If x is rational, then x is not irrational.

61. If $x \leq 0$, then $|x| = -x$.

62. If $x < -2$, then $|x| = -x$.

63. If x is not zero, then $x^2 > x$.

64. If x is positive and rational, then $x - 1$ is positive and rational.

65. If x is odd, then there is some integer n such that $x = 2n + 1$.

66. If x is even, then there is some integer n such that $x = 2n$.

Applications

Use logic to solve the puzzles in Exercises 67 and 68. Then write proofs to argue that your solutions are correct.

67. Linda, Alina, Phil, and Stuart are wearing different-colored hats: either red, yellow, green, or blue. From the following statements, determine which hat each person is wearing:

- Neither boy wears a red hat.
- The oldest person wears a green hat.
- Linda is older than Alina.
- Alina never wears yellow or red.
- Stuart is the youngest and hates blue.

68. Xena, Yolanda, and Zeke each have different favorite fruits: either apples, bananas, or cantaloupes. Use the statements that follow to determine which person prefers which fruit. Then write a proof that argues that your solution is correct.

- Xena likes bananas better than apples.
- Zeke is allergic to cantaloupes.
- Bananas are the favorite fruit of one of the girls.
- Yolanda likes bananas better than cantaloupes.

Consider an Island X where there are exactly two types of people: those who always tell the truth ("truth-tellers") and those who always lie ("liars"). Given the statements in Exercises 69–71, determine which people are truth-tellers and which are liars. Then write a proof which argues that your solution is correct. *(These types of puzzles, based on those made popular by Raymond Smullyan, are a great avenue to developing your logical thinking skills and your proof-writing abilities.)*

69. You meet Liz, Rein, and Zubin, who say the following:

Liz: "We all tell the truth."
Rein: "Exactly two of us tell the truth."
Zubin: "Liz and Rein always lie."

70. You meet Anita, Bill, and Chris, who say the following:

Anita: "At least one of Bill or Chris tells the truth."
Bill: "Anita is a liar."
Chris: "Bill is a liar."

71. You meet Hyun, Jaan, and Kate, and only Kate and Hyun have something to say:

Kate: "Hyun and Jaan are liars."
Hyun: "Kate always tells the truth."

Proofs

72. Prove that $x^2 - y^2 = (x - y)(x + y)$ for all real numbers x and y.

73. Prove that if x is an irrational number and r is a rational number, then the difference $x - r$ must be an irrational number. You may assume that the sum of two rational numbers is a rational number.

74. Prove that for any real numbers a and b, $|b - a| = |a - b|$.

75. Prove that the sum of an even integer and an odd integer is odd.

76. Prove that the sum of an odd integer and an odd integer is even.

77. Prove that the product of an even integer and an odd integer is even.

78. The Pythagorean theorem states that if a right triangle has legs of length a and b, and a hypotenuse of length c, then $a^2 + b^2 = c^2$. Use the Pythagorean theorem and the definition of the distance between two real numbers to prove that the distance between any two points $P = (x_1, y_1)$ and $Q = (x_2, y_2)$ in the plane is given by the "distance formula" $\sqrt{(x_2 - x_1)^2 + (y_2 - y_1)^2}$. (*Hint: Draw an example of two points P and Q in the plane, label their coordinates, and use an appropriate right triangle.*)

79. Prove that the midpoint $\left(\frac{x_1 + x_2}{2}, \frac{y_1 + y_2}{2}\right)$ between the points $P = (x_1, y_1)$ and $Q = (x_2, y_2)$ is equidistant from P and Q.

80. Prove that the numbers

$$x = \frac{-b + \sqrt{b^2 - 4ac}}{2a} \quad \text{and} \quad x = \frac{-b - \sqrt{b^2 - 4ac}}{2a}$$

found by the quadratic formula are solutions of the quadratic equation $ax^2 + bx + c = 0$.

81. Use the fact that $\frac{(a/b)}{(c/d)} = \frac{ad}{bc}$ to prove that $\frac{(a/b)}{c} = \frac{a}{bc}$. You may assume that all denominators are nonzero.

82. Use the fact that $\left(\frac{a}{b}\right)\left(\frac{c}{d}\right) = \frac{ac}{bd}$ to prove that $c\left(\frac{a}{b}\right) = \frac{ac}{b}$. You may assume that any denominators are nonzero.

83. Follow the steps outlined here to prove the ***triangle inequality***: $|a + b| \leq |a| + |b|$ for any real numbers a and b:

(a) Argue that for any real number x, $|x| = \sqrt{x^2}$.

(b) Show that $(a + b)^2 \leq (|a| + |b|)^2$. (*Hint: Start on the left-hand side, multiply out the expression, and use the fact that $a \leq |a|$ and $b \leq |b|$.*)

(c) Take the square root of both sides of the inequality from part (b) (this is valid since both sides are positive), and use part (a) to show that $|a + b| \leq |a| + |b|$.

Use the triangle inequality to prove the following two inequalities, for any real numbers a and b:

84. $|a - b| \leq |a| + |b|$

85. $|a - b| \geq |a| - |b|$

For Exercises 86 and 87, use the definition of absolute value and systems of inequalities to prove that for any real numbers x and c, and any positive real number δ, the given statement is true:

86. $|x - c| < \delta \iff x \in (c - \delta, c + \delta)$

87. $|x - c| > \delta \iff x \in (-\infty, c - \delta) \cup (c + \delta, \infty)$

88. The "Monty Hall Problem" is a good example of a problem about which people's initial intuition is often incorrect. On the game show *Let's Make a Deal*, the host, Monty Hall, presents you with a choice of three doors. Behind one door is a lot of money. Behind the other two doors are worthless gag prizes. You pick a door and point at it. Monty Hall knows which door conceals the prize, and he opens one of the two doors you didn't pick to show you a gag prize. Then he gives you the option of keeping your original choice or switching your choice to the remaining door.

(a) Is it better to switch or to stick with your original choice? Or are both choices equally likely to lead to the money? Think about this problem for awhile and convince yourself of an answer before you go on to the next part.

(b) Come up with a proof or argument that would convince another person of the correct answer.

Thinking Forward

Quantified statements about distance: Show that each of the statements that follow is true by exhibiting a value of δ that satisfies each statement. These types of statements will be the backbone of our study of limits in Chapter 1. (Note: The symbols ϵ and δ are Greek letters that represent real numbers that are usually positive and quite small. In the third statement you will need to write δ in terms of ϵ.)

- There exists a $\delta > 0$ such that for all x, if x is within distance δ of 2, then $3x + 1$ is within distance 1 of 7.
- There exists a $\delta > 0$ such that for all x, if $|x - 2| < \delta$, then $|3x - 6| < 0.3$.
- For all $\epsilon > 0$, there exists a $\delta > 0$ such that if x is within distance δ of 3, then $2x$ is within distance ϵ of 6.

Negating quantified statements about distance: Write down the negation of each statement that follows. We will see these types of statements again when we formally define the concept of a limit in Chapter 1.

- For all $M > 0$, there exists an $N > 0$ such that for all x, if $x > N$, then $x^2 > M$.
- For all $\epsilon > 0$, there exists a $\delta > 0$ such that for all x, if $0 < |x - 2| < \delta$, then $|x^2 - 4| < \epsilon$.
- For all $\epsilon > 0$, there exists a $\delta > 0$ such that for all x, if $0 < |x - 4| < \delta$, then $|\sqrt{x} - 2| < \epsilon$.

CHAPTER REVIEW, SELF-TEST, AND CAPSTONES

Before you progress to the next chapter, be sure you are familiar with the definitions, concepts, and basic skills outlined here. The capstone exercises at the end bring together ideas from this chapter and look forward to future chapters.

Definitions

Give precise mathematical definitions or descriptions of each of the concepts that follow. Then illustrate the definition with a graph or algebraic example, if possible.

- a *function* from a set A to a set B
- the *independent variable* and the *dependent variable* of a function
- the *domain* and *range* of a function, in set notation
- the *graph* of a function, in set notation and as a picture
- a *one-to-one* function
- a *root* of a function and a *y-intercept* of a function
- a *local maximum, local minimum, global maximum,* or *global minimum* of a function
- an *inflection point* of a function (in rough terms)
- what it means for a function f to be *positive, negative, increasing,* or *decreasing* on an interval I
- what it means (roughly) for a function f to be *concave up* or *concave down* on an interval I
- the *average rate of change* of a function f on an interval $[a, b]$
- a *piecewise-defined function*
- the form of a *power function* and the form of a *polynomial function*
- the *leading coefficient,* the *leading term,* and the *constant term* of a polynomial
- the forms of *constant, linear, quadratic, cubic, quartic,* and *quintic* polynomials
- what it means for a quadratic polynomial to be *irreducible,* and how this is related to the *discriminant*
- the form of a *rational function*
- what it means for a function to be *algebraic*
- the piecewise-defined form of $f(x) = |x|$, and more generally of a function $f(x) = |g(x)|$
- what it means for a function to be *transcendental*
- the form of an *exponential function,* and when this type of function represents *exponential growth* or *exponential decay*
- the *natural exponential function* and the *natural logarithmic function*
- the algebraic definitions of *even functions* and *odd functions,* and the graphical meaning of *y-axis symmetry* and *rotational symmetry*
- what it means for functions f and g to be *inverses* of each other
- the *sine* and *cosine* of an radian angle θ, in terms of coordinates on the unit circle.
- the *tangent, cotangent, secant,* and *cosecant* of a radian angle θ, in terms of the sine and cosine of θ
- the *inverse sine, inverse tangent,* and *inverse secant* functions and their domains and ranges
- the meaning of the quantified statement *"For all x, P"*
- the meaning of the quantified statement *"There exists x such that P"*
- the meaning of the implication *"If A, then B"*
- the *hypothesis* and the *conclusion* of an implication
- the *converse* and the *contrapositive* of an implication $A \Rightarrow B$
- a *counterexample* to a statement

Theorems

Fill in the blanks to complete each of the following statements of theorems:

- $y = f(x) + C$ is shifted C units ____ from $y = f(x)$ if $C > 0$ and C units ____ from $y = f(x)$ if $C < 0$.
- $y = f(x + C)$ is shifted C units ____ from $y = f(x)$ if $C > 0$ and C units ____ from $y = f(x)$ if $C < 0$.
- $y = kf(x)$ is vertically stretched from $y = f(x)$ by a factor of k if ____ and vertically compressed by a factor of k if ____.
- $y = f(kx)$ is horizontally stretched from $y = f(x)$ by a factor of k if ____ and horizontally compressed by a factor of k if ____.
- If f is an invertible function with inverse f^{-1}, then the domain of f^{-1} is ____ and the range of f^{-1} is ____.
- If f is an invertible function with inverse f^{-1}, then $f^{-1}(b) = a$ if and only if ____, and the graph of $y = f^{-1}(x)$ can be obtained from the graph of $y = f(x)$ by ____.
- A function has an inverse if and only if the function is ____.
- If f is a polynomial function of degree n, then the graph of f has at most ____ real roots and at most ____ local extrema.
- If f is a polynomial function, then the graph of f behaves like the graph of ____ at its "ends."

- If $f(x) = \frac{p(x)}{q(x)}$ is a rational function, then f is not defined at the roots of ____, and f has roots at the points that are roots of ____ but not roots of ____ .
- If $f(x) = \frac{p(x)}{q(x)}$ is a rational function, then f has holes at the points that are roots of ____, provided that ____ .
- If $f(x) = \frac{p(x)}{q(x)}$ is a rational function, then f has vertical asymptotes at the points that are roots of ____, provided that ____ .
- Suppose $f(x) = \frac{p(x)}{q(x)}$ is a rational function with $\deg(p(x)) = n$ and $\deg(q(x)) = m$. If $n < m$, then f has a horizontal asymptote at ____ ; if $n = m$, then f has a horizontal asymptote at ____ ; and if $n > m$, then f ____ .
- A statement of the form "For all x, P" is false if and only if there is a counterexample in which ____ .
- A statement of the form "For all x, if A, then B" is false if and only if there is a counterexample in which ____ but ____ .

Notation and Algebraic Rules

Notation: Describe the meanings of each of the following mathematical expressions.

- $f: A \to B$
- $y = f(x)$
- $f(x) = |x|$
- $(kf)(x)$
- $(f + g)(x)$
- $(f \cdot g)(x)$
- $\left(\frac{f}{g}\right)(x)$
- $(f \circ g)(x)$
- $(f \circ g \circ h)(x)$
- $A \Rightarrow B$
- $A \Longleftrightarrow B$
- ■

Logarithms: Fill in the blanks to complete each of the algebraic rules that follow. You may assume that x, y, b, and a are real numbers whose values make the expressions well-defined.

- $\log_b x = y \Longleftrightarrow$ ____
- $\log_b(b^x) =$ ____
- $b^{\log_b x} =$ ____
- $\log_b(x^a) =$ ____
- $\log_b(xy) =$ ____
- $\log_b\left(\frac{1}{x}\right) =$ ____
- $\log_b\left(\frac{x}{y}\right) =$ ____
- $\frac{\log_a x}{\log_b x} =$ ____

Trigonometric identities: Fill in the blanks to complete each of the following trigonometric identities, where θ, α, and β are angles measured in radians.

- $\sin^2\theta + \cos^2\theta =$ ____
- $\tan^2\theta + 1 =$ ____
- $1 + \cot^2\theta =$ ____
- $\sin(-\theta) =$ ____
- $\cos(-\theta) =$ ____
- $\tan(-\theta) =$ ____
- $\cos\left(\theta - \frac{\pi}{2}\right) =$ ____
- $\sin\left(\theta + \frac{\pi}{2}\right) =$ ____
- $\sin(\theta + 2\pi) =$ ____
- $\cos(\theta + 2\pi) =$ ____
- $\sin(\alpha + \beta) =$ ____
- $\cos(\alpha + \beta) =$ ____
- $\sin(\alpha - \beta) =$ ____
- $\cos(\alpha - \beta) =$ ____
- $\sin 2\theta =$ ____
- $\cos 2\theta =$ ____

Skill Certification: Algebra and Functions

Simplifying expressions: Simplify each expression as much as possible.

1. $\frac{x^3 - 2^3}{x - 2}$

2. $\frac{\frac{1}{2+h} - \frac{1}{2}}{h}$

3. $\frac{x - 2}{x^3 - x^2 - 4x + 4}$

4. $\frac{x^4 + 27x}{x^2 + 5x + 6}$

5. $|-2(x^2 + 1)|$

6. $\frac{|4 - 2x|}{x - 2}$

7. $f(x) = \frac{x^{1/4} + x^{1/3}}{x^2}$

8. $f(x) = \frac{x^{-2/5}\sqrt{4x}}{\sqrt[3]{x}}$

9. $f(x) = e^{2\ln x}$

10. $f(x) = \log_2(8(4^x))$

11. $f(x) = \tan\left(\frac{\pi}{3}\right) + \tan\left(\frac{\pi}{4}\right)$

12. $f(x) = \left(\sin^{-1}\left(-\frac{1}{2}\right)\right)^2$

Solving inequalities: Solve each of the inequalities, and express each solution set in interval notation.

13. $2x^2 - 7x + 3 > 0$

14. $\frac{x^2 - 9}{x - 1} \le 0$

15. $\frac{3}{x - 2} < 1$

16. $|3x - 4| < \frac{1}{2}$

17. $|5x - 2| > 1$

18. $|x^2 - 4| < 2$

Finding zeros and undefined values: Determine the x-values for which each function is zero, and the x-values for which each function does not exist.

19. $f(x) = \frac{2x^2 - 5x + 3}{x}$

20. $f(x) = 3x^4 - 6x^3 + x^2$

21. $f(x) = |x - 2| - 5$

22. $f(x) = \frac{1}{x} - \frac{1}{1 - x}$

23. $f(x) = \dfrac{2(x-1)^2 - 4}{x-1}$

24. $f(x) = \dfrac{2x^3 - 1}{2x^2 - 5x - 3}$

25. $f(x) = e^x(1 - 2e^x)$

26. $f(x) = \dfrac{x \ln x}{e^x - 1}$

27. $f(x) = \dfrac{\sin(\pi x)}{x}$

28. $f(x) = \dfrac{1}{\sec x + 1}$

29. $f(x) = \dfrac{1}{\sin^{-1} x}$

30. $f(x) = \arctan(x^2 + 1)$

Finding domains: Find the domain of each function, and express the domain in interval notation.

31. $f(x) = \dfrac{\sqrt{x+2}}{x}$

32. $f(x) = \sqrt{x^2 - 2x - 3}$

33. $f(x) = \dfrac{x+2}{x^2 - x - 6}$

34. $f(x) = \dfrac{1}{\sqrt{x^2 - 3x}}$

35. $f(x) = \sqrt{x} + \sqrt{2-x}$

36. $f(x) = \sqrt{\dfrac{x-3}{1-x}}$

37. $f(x) = \dfrac{1}{e^{-1/2x}}$

38. $f(x) = \ln\left(\dfrac{x+1}{x-1}\right)$

39. $f(x) = \dfrac{1}{\frac{1}{2} - \sin x}$

40. $f(x) = \dfrac{\tan^{-1} x}{\pi - \sec^{-1} x}$

Graphs of basic functions: Sketch the graph of each function by hand, using your knowledge of simple graphs and transformations. Label any important points or features.

41. $f(x) = 3 - 2x$

42. $f(x) = 4(x-1) + 2$

43. $f(x) = \sqrt{x-3}$

44. $f(x) = (x-3)^2 + 1$

45. $f(x) = 2x^3 - 1$

46. $f(x) = \sqrt{1-x}$

47. $f(x) = \dfrac{1}{x} - 2$

48. $f(x) = \dfrac{1}{x-2}$

49. $f(x) = |4x - 3|$

50. $f(x) = |x^2 - 9|$

51. $f(x) = -2x^{1/3}$

52. $f(x) = 2x^{-1/3}$

53. $f(x) = -x(x+1)^2$

54. $f(x) = 3x^3 + x^2 - 3x - 1$

55. $f(x) = \dfrac{(x-1)(x+2)}{(x-1)^2}$

56. $f(x) = \dfrac{x^3 + x^2 + x + 1}{2x^2 + x - 1}$

57. $f(x) = 2^x$

58. $f(x) = \left(\dfrac{1}{2}\right)^x$

59. $f(x) = e^{3x}$

60. $f(x) = e^{-3x}$

61. $f(x) = 1 - 5e^x$

62. $f(x) = -3e^{x-1}$

63. $f(x) = \ln x$

64. $f(x) = \log_2 x$

65. $f(x) = \log_{1/2} x$

66. $f(x) = 1 - 3\ln x$

67. $f(x) = \sin x$

68. $f(x) = \tan x$

69. $f(x) = \sec x$

70. $f(x) = 2\sin\left(x - \dfrac{\pi}{4}\right)$

71. $f(x) = \sin^{-1} x$

72. $f(x) = \tan^{-1} x$

Capstone Problems

A. *Transformations of cubic functions:* Prove algebraically that a vertical or horizontal shift or stretch of a cubic function is also a cubic function. That is, prove that if $f(x)$ is a cubic function, then so are $f(x) + C$, $f(x + C)$, $kf(x)$, and $f(kx)$.

B. *Peeking forward to derivatives:* Suppose $f(x) = \dfrac{1}{x}$, and consider the two-variable function

$$q(x, h) = \frac{f(x+h) - f(x)}{h}.$$

(a) Simplify $q(3, h)$ as much as possible, and argue that it approaches $-\frac{1}{9}$ as h gets closer to 0.

(b) Simplify $q(x, h)$ as much as possible, and argue that it approaches $-\frac{1}{x^2}$ as h gets closer to 0.

C. *Approximating the area of a region in the plane:* Sketch the region that lies between the graph of $f(x) = 20 - 2^x$, the x-axis, and the lines $y = 2$ and $y = 4$. Use geometric figures to approximate upper and lower bounds for the area of this region.

D. *Optimizing the area of a region given its perimeter:* Elizabeth wants to build a rectangular pen for her dogs with 100 feet of spare fencing.

(a) Write down an equation in terms of length l and width w for the perimeter P of the pen.

(b) Use the perimeter equation and the constraint on fencing material to construct a one-variable equation for the area A of the enclosure.

(c) Sketch a graph of the one-variable function A, and use the graph to argue that its maximum occurs when $l = w = 25$ and the enclosure is square.

CHAPTER 1

Limits

1.1 An Intuitive Introduction to Limits

Examples of Limits
Limits of Functions
Infinite Limits, Limits at Infinity, and Asymptotes
Examples and Explorations

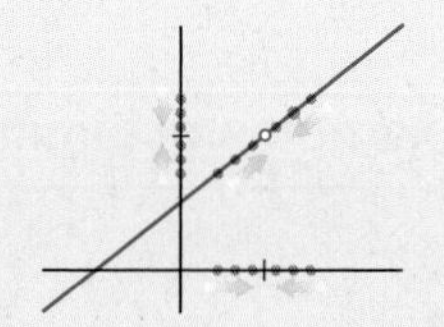

1.2 Formal Definition of Limit

Formalizing the Intuitive Definition of Limit
Uniqueness of Limits
One-Sided Limits
Infinite Limits and Limits at Infinity
Examples and Explorations

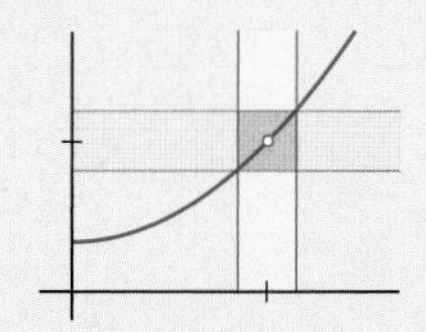

1.3 Delta-Epsilon Proofs*

Describing Limits with Absolute Value Inequalities
Finding a Delta for Every Epsilon
Writing Delta-Epsilon Proofs
Examples and Explorations

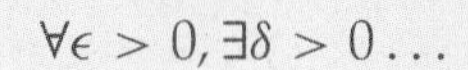

1.4 Continuity and Its Consequences

Defining Continuity with Limits
Types of Discontinuities
Continuity of Very Basic Functions
Extreme and Intermediate Values of Continuous Functions
Examples and Explorations

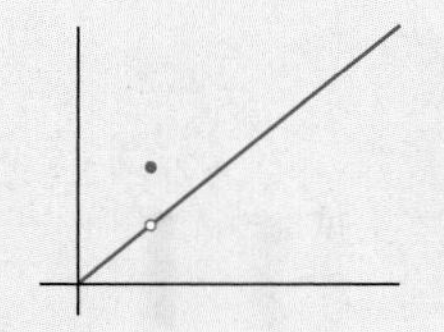

1.5 Limit Rules and Calculating Basic Limits

Limits of Combinations of Functions
Limits of Algebraic Functions
Finding Limits by Cancelling or Squeezing
Defining the Number *e*
Continuity of Exponential and Trigonometric Functions
Delta-Epsilon Proofs of the Limit Rules
Examples and Explorations

1.6 Infinite Limits and Indeterminate Forms

Infinite Limits
Limits at Infinity
Indeterminate and Non-Indeterminate Forms
Special Trigonometric Limits
Examples and Explorations

$$\frac{\infty}{\infty}$$

Chapter Review, Self-Test, and Capstones

1.1 AN INTUITIVE INTRODUCTION TO LIMITS

- Examples of the types of limits that are important in calculus
- Intuitive descriptions of two-sided and one-sided limits
- Infinite limits, limits at infinity, and horizontal and vertical asymptotes

Examples of Limits

Limits are the backbone of calculus. Limits are the key to investigating the local behavior of functions, giving meaning to the slope of a curve, finding areas under curves and volumes inside surfaces, and determining the long-term behavior of infinite sequences and sums. In the next section we will formally and mathematically define limits; in this section we focus on intuitive examples. Let's start with three examples that illustrate how limits can arise.

Limits of sequences: As a starting point for thinking about the concept of a limit, consider the following ***sequence*** of numbers:

$$\frac{1}{2}, \frac{1}{4}, \frac{1}{8}, \frac{1}{16}, \frac{1}{32}, \frac{1}{64}, \ldots, \frac{1}{2^k}, \ldots$$

If the pattern of this sequence continues, then the numbers will continue to get smaller and smaller, approaching zero. We say that 0 is the ***limit*** of the sequence $\left\{\frac{1}{2^k}\right\}$ as k approaches infinity. We can never actually let k be equal to infinity, because infinity is not a real number. However, we can let k get as large as we like. For each large value of k, the value of $\frac{1}{2^k}$ is very small, but not actually zero. When we "take the limit," we make an important theoretical transition: Instead of evaluating $\frac{1}{2^k}$ at a particular value of k, we think about the behavior of the sequence as k gets larger and larger. We think about what the sequence *approaches*, even if it never actually gets there for any real number k.

Limits of sequences of sums: Now consider the sequence defined by adding up more and more terms from the previous sequence:

$$\frac{1}{2}, \quad \frac{1}{2}+\frac{1}{4}, \quad \frac{1}{2}+\frac{1}{4}+\frac{1}{8}, \quad \frac{1}{2}+\frac{1}{4}+\frac{1}{8}+\frac{1}{16}, \quad \frac{1}{2}+\frac{1}{4}+\frac{1}{8}+\frac{1}{16}+\frac{1}{32}, \quad \ldots$$

After computing the sums, this sequence is equal to

$$\frac{1}{2}, \frac{3}{4}, \frac{7}{8}, \frac{15}{16}, \frac{31}{32}, \frac{63}{64}, \ldots$$

The terms get closer and closer to 1 as we go further and further out in this ***sequence of sums***. You may have noticed a pattern in the sequence: It turns out that for any given k, the value of $\frac{1}{2}+\frac{1}{4}+\cdots+\frac{1}{2^k}$ will be equal to the quantity $\frac{2^k-1}{2^k}$. The larger the value of k, the closer this quantity gets to 1. Moreover, we can get the sum to be as close as we like to 1 by choosing a sufficiently large value of k. Again, we can't plug in infinity for k, since we can't in real life add up infinitely many numbers, even if those numbers are getting infinitesimally small, as they are here. However, mathematically we can use the concept of a limit to notice that this sequence of sums *approaches* the quantity 1. We will study limits of sequences of sums in depth in Chapter 7; for now we present this type of limit only as an interesting example to consider.

Limits of average rates of change: Now let's switch gears and think of something a little more practical. Suppose you drop a bowling ball from the top of a 100-foot parking deck and want to know how fast it is falling when it hits the ground. Suppose that the height of the ball t seconds after it is dropped is given by $s(t) = 100 - 16t^2$ feet and thus that the ball hits the ground after 2.5 seconds. By the old formula "distance equals rate times time," the average rate that the ball falls over any time period Δt will be $r = \frac{\Delta s}{\Delta t}$, where Δs is the elapsed distance over the period Δt. A good approximation of the final speed of the bowling ball is the average rate in the last half-second, which is

$$r = \frac{s(2.5) - s(2)}{2.5 - 2} = \frac{0 - 36}{0.5} = -72 \text{ feet per second}$$

(negative since the ball is falling downwards). To get a better approximation for the ball's final ***instantaneous velocity***, we could calculate the rate over the last quarter-second. The following table records a sequence of better and better approximations for the final velocity of the bowling ball:

Time interval	[2, 2.5]	[2.25, 2.5]	[2.4, 2.5]	[2.49, 2.5]	[2.4999, 2.5]
Average rate	$-72\,\frac{\text{ft}}{\text{sec}}$	$-76\,\frac{\text{ft}}{\text{sec}}$	$-78.4\,\frac{\text{ft}}{\text{sec}}$	$-79.84\,\frac{\text{ft}}{\text{sec}}$	$-79.9984\,\frac{\text{ft}}{\text{sec}}$

We can't compute the actual final velocity of the ball this way, because we can't use $r = \frac{\Delta s}{\Delta t}$ when Δt is zero. But the average rates seem to *approach* -80 feet per second as Δt gets smaller and smaller. This sounds like another limit, and it is. In fact it is a very famous and useful type of limit called a ***derivative*** that we will introduce in Chapter 2. In general, limits help us discuss what happens when we let things get infinitesimally small, infinitely large, or arbitrarily close to some number.

Limits of Functions

Intuitively, a limit is what the output of a function approaches as we let the input of that function approach some value. In the previous examples, we saw that:

- As k approaches infinity, the quantity $\frac{1}{2^k}$ approaches 0.
- As k approaches infinity, the sum $\frac{1}{2} + \frac{1}{4} + \cdots + \frac{1}{2^k}$ approaches 1.
- As t approaches 2.5, the average rate $\frac{s(2.5) - s(t)}{2.5 - t}$ approaches -80 feet per second.

If the values of a function $f(x)$ approach some number L as x gets closer and closer to some value $x = c$, we will write

$$\lim_{x \to c} f(x) = L.$$

We can also consider limits of functions as $x \to \infty$, that is, as x grows without bound. For example, in this notation we have

$$\lim_{k \to \infty} \frac{1}{2^k} = 0 \quad \text{and} \quad \lim_{t \to 2.5} \frac{s(2.5) - s(t)}{2.5 - t} = -80.$$

When considering a limit, it only matters what happens as x gets closer and closer to c, not what happens when it actually gets there. This means that $\lim_{x \to c} f(x)$ may or may not in general be the same as the value $f(c)$ of the function at $x = c$. For example, the functions $f(x) = x + 1$ and $g(x) = \frac{x^2 - 1}{x - 1}$ shown in the following figures are not equal *at* the point $x = 1$, but they do *approach* the same value as $x \to 1$:

$\lim_{x \to 1} f(x) = 2$ *and* $f(1) = 2$

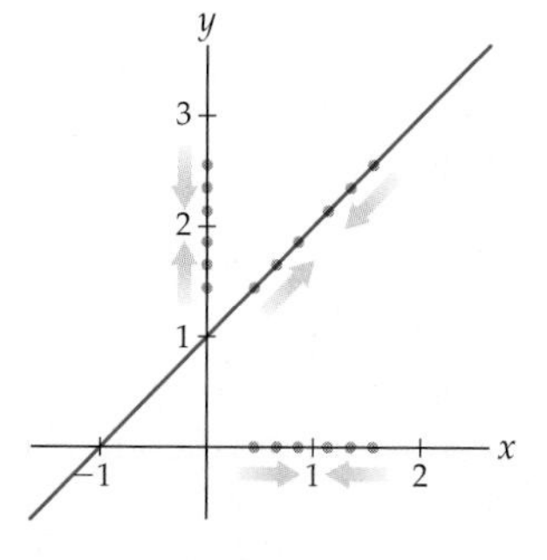

$\lim_{x \to 1} g(x) = 2$ *but* $g(1)$ *is undefined*

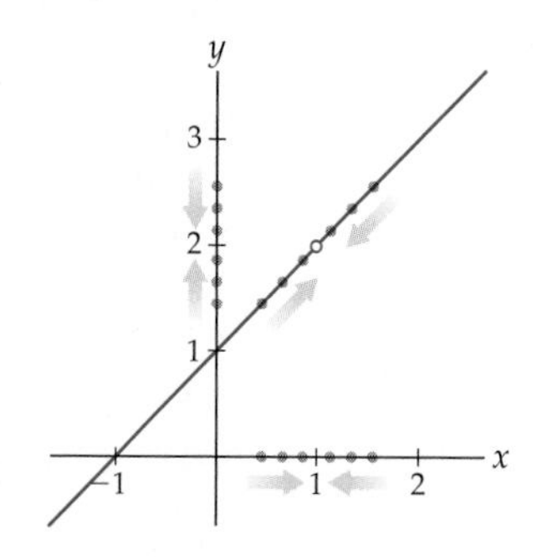

The dots on the x-axis represent a sequence of values of x that approach $x = 1$. When we evaluate the functions f and g at these values, in both cases we get a sequence of values of y that gets closer and closer to $y = 2$. Although $f(1) \neq g(1)$, we do have $\lim_{x \to 1} f(x) = \lim_{x \to 1} g(x) = 2$.

We can also consider limits as x grows without bound, and/or as $f(x)$ grows without bound. The following definition summarizes the notation we will use in each case:

DEFINITION 1.1

Intuitive Description of and Notation for Limits

Suppose f is a function and L and c are real numbers.

(a) ***Limit:*** If the values of a function $f(x)$ approach L as x approaches c, then we say that L is the ***limit*** of $f(x)$ as x approaches c and we write

$$\lim_{x \to c} f(x) = L.$$

(b) ***Limit at Infinity:*** If the values of a function $f(x)$ approach L as x grows without bound, then we say that L is the ***limit*** of $f(x)$ as x approaches ∞ and we write

$$\lim_{x \to \infty} f(x) = L.$$

(c) ***Infinite Limit:*** If the values of a function $f(x)$ grow without bound as x approaches c, then we say that ∞ is the ***limit*** of $f(x)$ as x approaches c and we write

$$\lim_{x \to c} f(x) = \infty.$$

(d) ***Infinite Limit at Infinity:*** If the values of a function $f(x)$ grow without bound as x grows without bound, then we say that ∞ is the ***limit*** of $f(x)$ as x approaches ∞ and we write

$$\lim_{x \to \infty} f(x) = \infty.$$

When a limit approaches a real number, we say that the limit ***exists***. When a limit approaches ∞ or $-\infty$ we say that the limit ***does not exist*** (because ∞ and $-\infty$ are not real numbers), but we will always be as specific as possible and describe the sign of infinity in such cases. Later we will see more pathological limits that "do not exist" in a way other than being infinite.

When we say that x approaches a real number c, we implicitly mean to consider values of x that are close to c from either the right or the left. In other words, when trying to find $\lim_{x \to c} f(x)$, we consider both values of x that are slightly less than c as well as values of x that are slightly greater than c. Sometimes it is convenient to consider these two cases separately:

DEFINITION 1.2

Intuitive Description of One-Sided Limits

If the values of a function $f(x)$ approach a value L as x approaches c from the left, we say that L is the ***left-hand limit*** of $f(x)$ as x approaches c and we write

$$\lim_{x \to c^-} f(x) = L.$$

If the values of a function $f(x)$ approach a value R as x approaches c from the right, we say that R is the ***right-hand limit*** of $f(x)$ as x approaches c and we write

$$\lim_{x \to c^+} f(x) = R.$$

Note that the notation $x \to c^-$ does not mean anything about whether c is a positive or negative number, only that x approaches c from the left. The two-sided limit of $f(x)$ as $x \to c$

exists if and only if the left and right limits as x approaches c exist and are equal. This means that both the left and right limits approach the same real number.

For example, the function graphed here has a different limit from the left than from the right as x approaches 1:

$$\lim_{x\to 1^-} f(x) = 2 \textit{ but } \lim_{x\to 1^+} f(x) = 3$$

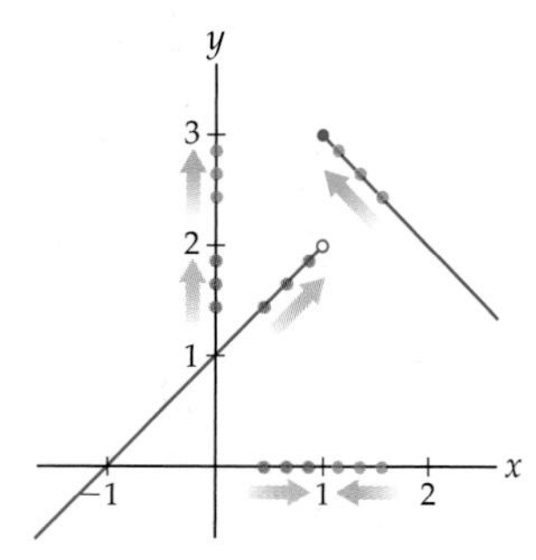

The purple sequence of values of x that approach $x = 1$ from the left determines a sequence of values of $f(x)$ that approach $y = 2$, while the red sequence determines values that approach $y = 3$. The value of the function at $x = 1$ happens to be $f(1) = 3$, but that is not relevant to either limit calculation. Since the limits from the left and right are not the same, there is no one real number that the function approaches as $x \to c$ and we say the two-sided limit ***does not exist***.

Infinite Limits, Limits at Infinity, and Asymptotes

Armed with the concept of limits, we can now give proper definitions for horizontal and vertical asymptotes. If a function f increases or decreases without bound as x approaches a real number c from either the right or the left, then f has a vertical asymptote at $x = c$:

DEFINITION 1.3 Vertical Asymptotes

A function f has a ***vertical asymptote*** at $x = c$ if one or more of the following are true:

$$\lim_{x\to c^+} f(x) = \infty, \quad \lim_{x\to c^-} f(x) = \infty, \quad \lim_{x\to c^+} f(x) = -\infty, \quad \text{or} \quad \lim_{x\to c^-} f(x) = -\infty.$$

If $f(x)$ approaches ∞ from both the left and the right as $x \to c$, then we say that $\lim_{x\to c} f(x) = \infty$, as happens in the leftmost graph that follows. If $f(x)$ approaches $-\infty$ from both the left and the right,then we say that $\lim_{x\to c} f(x) = -\infty$. If $f(x)$ approaches different signs of infinity from the left and the right, then the two-sided limit $\lim_{x\to c} f(x)$ does not exist, as happens in the middle graph.

$\lim_{x\to 1} f(x) = \infty$ $\qquad$ $\lim_{x\to 1^-} f(x) = -\infty, \lim_{x\to 1^+} f(x) = \infty$ $\qquad$ $\lim_{x\to \infty} f(x) = 10$

If the values of a function $f(x)$ approach a real-number value as x increases or decreases without bound, then f has a horizontal asymptote. For example, the rightmost graph of the three just presented shows a function with a horizontal asymptote and its corresponding limit. In general, we have the following definition:

DEFINITION 1.4

Horizontal Asymptotes

A nonconstant function f has a ***horizontal asymptote*** at $y = L$ if one or both of the following are true:

$$\lim_{x\to\infty} f(x) = L, \quad \text{or} \quad \lim_{x\to-\infty} f(x) = L.$$

Note that by convention, if $f(x)$ is actually *equal* to L as $x \to \infty$ or as $x \to -\infty$, then we do not consider f to have a horizontal asymptote at $y = L$. For example, the constant function $f(x) = 2$ has $\lim_{x\to\infty} f(x) = 2$ and yet does *not* have a horizontal asymptote at $y = 2$, since $f(x)$ is constantly equal to 2 as $x \to \infty$.

Examples and Explorations

EXAMPLE 1

Determining limits with tables of values

Use tables of values to find (a) $\lim_{x\to1}(x+1)$ and (b) $\lim_{x\to\infty} \frac{x}{x-1}$.

SOLUTION

(a) To see what happens to $x + 1$ as $x \to 1$, we choose a sequence of values approaching $x = 1$ from the left and a sequence approaching $x = 1$ from the right, and record the corresponding values of $x + 1$:

x	.9	.99	.999	1	1.001	1.01	1.1
$x+1$	1.9	1.99	1.999	*	2.001	2.01	2.1

From both the left and the right, the values of of $x + 1$ approach 2. Assuming that this pattern continues for values of x that are even closer to 1, we have $\lim_{x\to1}(x+1) = 2$.

(b) To see what happens to $\frac{x}{x-1}$ as $x \to \infty$, we choose a sequence of values of x that gets larger and larger, and record the corresponding (rounded) values of $\frac{x}{x-1}$:

x	25	50	100	1000	10,000
$\frac{x}{x-1}$	1.04167	1.02041	1.0101	1.001	1.0001

As x grows larger, the quantity $\frac{x}{x-1}$ approaches 1, so, assuming that the pattern in the table continues for even larger values of x, we have $\lim_{x\to\infty} \frac{x}{x-1} = 1$. □

EXAMPLE 2

Graphically identifying limits

Determine the limits at any holes, corners, or asymptotes on the graphs of the functions (a) f, (b) g, and (c) h:

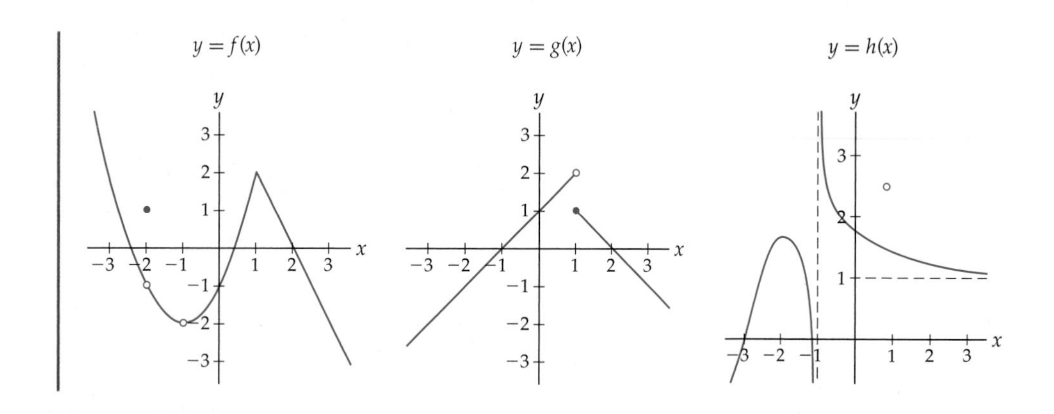

SOLUTION

(a) Observe that the graph in question has holes at $x = -2$ and $x = -1$ and a corner at $x = 1$, so we will examine limits at those three points. As x approaches -2, the height of the graph of $y = f(x)$ approaches -1. The value of $f(x)$ at $x = -2$ is $f(-2) = 1$, but this is not relevant to what $f(x)$ *approaches* as $x \to -2$ and thus does not affect the limit. Therefore $\lim_{x \to -2} f(x) = -1$.

As x approaches -1, the height of the graph approaches -2. The value $f(-1)$ is not defined, but that is not relevant to the limit as $x \to -1$. Thus $\lim_{x \to -1} f(x) = -2$.

As x approaches 1, the height of the graph approaches 2. The value of $f(x)$ at $x = 1$ happens to also equal 2, although this is irrelevant to the value of the limit as $x \to 1$. We have $\lim_{x \to 1} f(x) = 2$.

(b) The function $g(x)$ approaches different values as we approach $x = 1$ from the left and the right. As x approaches 1 from the left, the height of the graph approaches $y = 2$. As x approaches 1 from the right, the height of the graph approaches $y = 1$. The value of $g(x)$ at $x = 1$ happens to be $g(1) = 1$, but that is not relevant to either limit. Therefore we have $\lim_{x \to 1^-} g(x) = 2$ and $\lim_{x \to 1^+} g(x) = 1$, but the two-sided limit $\lim_{x \to 1} g(x)$ does not exist.

(c) The function $h(x)$ has a vertical asymptote at $x = -1$, with the height of the function decreasing without bound as we approach from the left and increasing without bound as we approach from the right. Therefore we have $\lim_{x \to -1^-} h(x) = -\infty$ and $\lim_{x \to -1^+} h(x) = \infty$, but the limit $\lim_{x \to -1} h(x)$ does not exist.

Let's also investigate the limits at the ends of the graph of $h(x)$. On the left side, as $x \to -\infty$, the graph decreases without bound; therefore $\lim_{x \to -\infty} h(x) = -\infty$. On the right side, as $x \to \infty$, the height of the function approaches $y = 1$. Therefore $h(x)$ has a horizontal asymptote on the right, and $\lim_{x \to \infty} h(x) = 1$. □

EXAMPLE 3 **A function with infinitely many oscillations as x approaches 0**

Use a table of values and various graphing windows on a calculator or other graphing utility to investigate the limit of the function $f(x) = \sin\left(\frac{1}{x}\right)$ as x approaches 0.

SOLUTION

Be sure that your calculator or other graphing utility is set to radian mode, rather than degree mode, for this example. To see what happens to the quantity $\sin\left(\frac{1}{x}\right)$ as x approaches 0, we choose points progressively closer to $x = 0$ from both the left and the right and record the corresponding (rounded) values of $\sin\left(\frac{1}{x}\right)$ in a table:

x	-0.001	-0.0001	-0.00000001	0	0.00000001	0.0001	0.001
$\sin\left(\frac{1}{x}\right)$	-0.827	0.306	-0.932	??	0.932	-0.306	0.827

From the table we see that as x approaches 0, the values of $f(x)$ seem to jump around! It is not clear whether or not the function $f(x) = \sin\left(\frac{1}{x}\right)$ will eventually approach any particular value as $x \to 0$. It is impossible to make an educated guess for $\lim_{x\to 0} \sin\left(\frac{1}{x}\right)$ from this table.

The reason that this is happening is that as $x \to 0$ the function $f(x) = \sin\left(\frac{1}{x}\right)$ oscillates faster and faster between -1 and 1, and never settles down. The graph on the left that follows shows this function on $[-3, 3]$, and the graph on the right shows the same function after reducing by a factor of 10 on the x-scale (but keeping the y-scale the same). No matter how much we "zoom in" towards $x = 0$, this function will keep oscillating. Therefore $\lim_{x\to 0} \sin\left(\frac{1}{x}\right)$ does not exist.

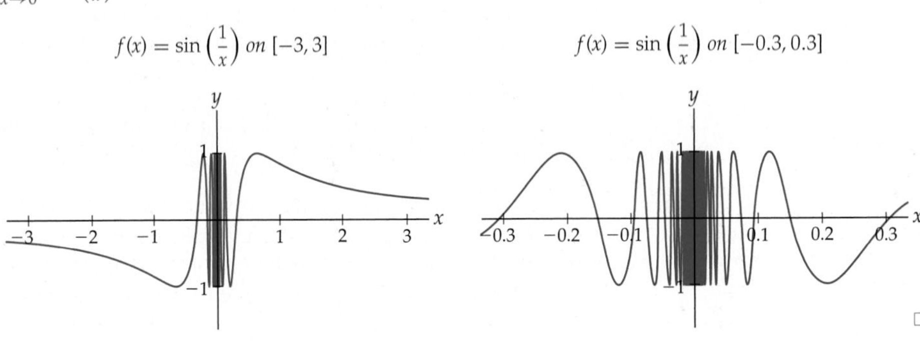

□

EXAMPLE 4

Areas under curves as another application of limits

Consider the area between the graph of $f(x) = x^2$ and the x-axis from $x = 0$ to $x = 1$, as shown in the leftmost figure.

(a) Find the value of the four-rectangle approximation shown in the middle figure.

(b) Find the value of the eight-rectangle approximation shown in the rightmost figure.

(c) Describe what would happen if we were to do similar approximations with more and more rectangles.

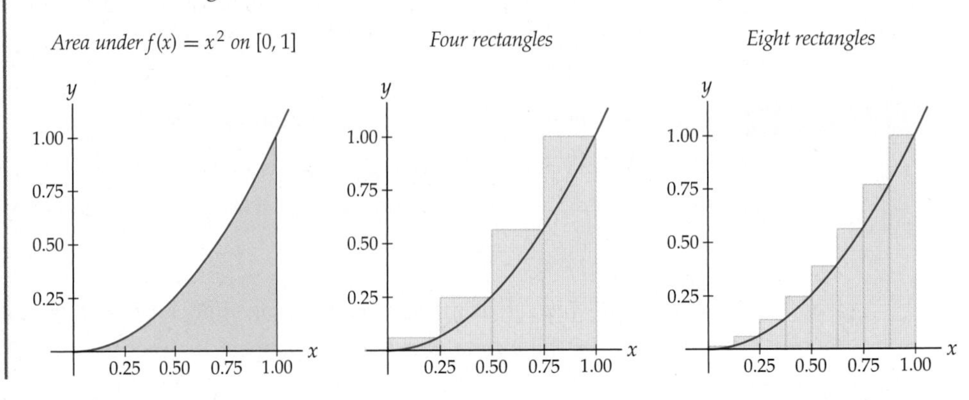

SOLUTION

(a) The four rectangles in the given middle graph each have width $\frac{1}{4}$, and their heights are given by $f\left(\frac{1}{4}\right), f\left(\frac{2}{4}\right), f\left(\frac{3}{4}\right)$, and $f\left(\frac{4}{4}\right)$, from left to right. Therefore, since $f(x) = x^2$, the sum of the areas of the four rectangles is

$$\left(\frac{1}{4}\right)^2\left(\frac{1}{4}\right) + \left(\frac{2}{4}\right)^2\left(\frac{1}{4}\right) + \left(\frac{3}{4}\right)^2\left(\frac{1}{4}\right) + \left(\frac{4}{4}\right)^2\left(\frac{1}{4}\right) = \frac{15}{32} \approx 0.4688.$$

(b) Similarly, the eight rectangles in the given rightmost graph each have width $\frac{1}{8}$, and their heights are given by $f\left(\frac{1}{8}\right), f\left(\frac{2}{8}\right), f\left(\frac{3}{8}\right), \ldots, f\left(\frac{8}{8}\right)$, from left to right. Therefore the sum of their eight areas is

$$\left(\frac{1}{8}\right)^2\left(\frac{1}{8}\right) + \left(\frac{2}{8}\right)^2\left(\frac{1}{8}\right) + \left(\frac{3}{8}\right)^2\left(\frac{1}{8}\right) + \cdots + \left(\frac{8}{8}\right)^2\left(\frac{1}{8}\right) = \frac{51}{128} \approx 0.3984.$$

(c) So what happens when we do similar approximations with more rectangles? Consider the following three figures, where we consider more and more rectangles:

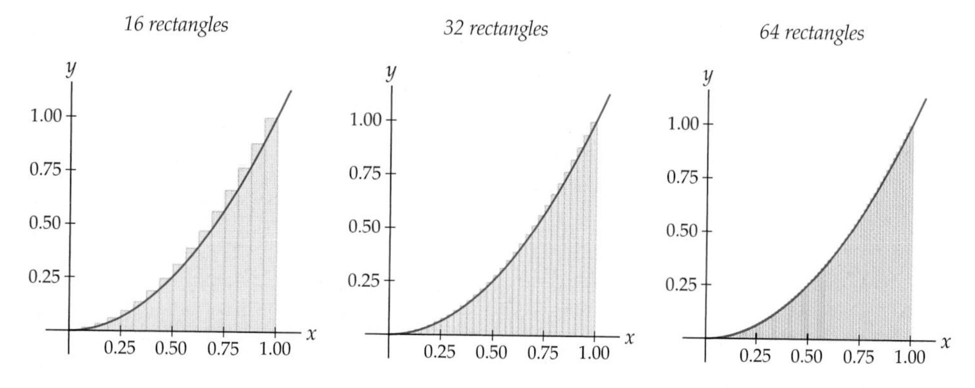

Clearly as we let the number N of rectangles get larger and larger, the sum of the areas of the rectangles gets closer and closer to a particular real number, namely, the actual area under the graph of $f(x) = x^2$ on $[0, 1]$. In limit notation we have

$$\lim_{N\to\infty} (N\text{-rectangle area approximation}) = (\text{actual area under graph}).$$

It turns out that the area approximations corresponding to the 16-, 32-, and 64-rectangle figures are 0.3652, 0.3491, and 0.3412, respectively. In Chapter 7 we will develop theory that will allow us to show that these area approximations have a limit of $\frac{1}{3}$ as $N \to \infty$. □

? TEST YOUR UNDERSTANDING

- What is the difference between taking a limit of some quantity as, say, x approaches 2 and actually computing the value of the quantity at $x = 2$?
- What can you say about the two-sided limit of a function as $x \to c$ if the left and right limits as $x \to c$ both exist but are not equal to each other?
- Can a function have more than one horizontal asymptote? More than two? Use Definition 1.4 to support your answer.
- Explain why the table in Example 1 cannot guarantee that $\lim_{x\to 1}(x+1)$ is actually 2 rather than some other number that is close to 2, such as 2.000035.
- Why does the table in Example 3 suggest that $\lim_{x\to 0} \sin\left(\frac{1}{x}\right)$ does not exist?

EXERCISES 1.1

Thinking Back

Finding the pattern in a sequence: For each sequence shown, find the next two terms. Then write a general form for the kth term of the sequence.

- $2, 6, 10, 14, 18, 22, \ldots$
- $1, 8, 27, 64, 125, 216, \ldots$
- $3, \frac{3}{4}, \frac{3}{9}, \frac{3}{16}, \frac{3}{25}, \frac{3}{36}, \ldots$
- $1, \frac{1}{3}, \frac{1}{9}, \frac{1}{27}, \frac{1}{81}, \frac{1}{243}, \ldots$
- $\frac{3}{5}, \frac{4}{7}, \frac{5}{9}, \frac{6}{11}, \frac{7}{13}, \frac{8}{15}, \ldots$
- $\frac{3}{2}, \frac{5}{5}, \frac{7}{10}, \frac{9}{17}, \frac{11}{26}, \frac{13}{37}, \ldots$

Distance, rate, and time: A watermelon dropped from the top of a 50-foot building has height given by $s(t) = 50 - 16t^2$ feet after t seconds. Calculate each of the following:

- The average rate of change of the watermelon over its entire fall, over the first half of its fall, and over the second half of its fall.
- The average rate of change over the last second, the last half-second, and the last quarter-second of its fall.

Concepts

0. *Problem Zero:* Read the section and make your own summary of the material.

1. *True/False:* Determine whether each of the statements that follow is true or false. If a statement is true, explain why. If a statement is false, provide a counterexample.

(a) *True or False:* A limit exists if there is some real number that it is equal to.

(b) *True or False:* The limit of $f(x)$ as $x \to c$ is the value $f(c)$.

(c) *True or False:* The limit of $f(x)$ as $x \to c$ might exist even if the value $f(c)$ does not.

(d) *True or False:* The two-sided limit of $f(x)$ as $x \to c$ exists if and only if the left and right limits of $f(x)$ exist as $x \to c$.

(e) *True or False:* If the graph of f has a vertical asymptote at $x = 5$, then $\lim_{x\to 5} f(x) = \infty$.

(f) *True or False:* If $\lim_{x\to 5} f(x) = \infty$, then the graph of f has a vertical asymptote at $x = 5$.

(g) *True or False:* If $\lim_{x\to 2} f(x) = \infty$, then the graph of f has a horizontal asymptote at $x = 2$.

(h) *True or False:* If $\lim_{x\to -\infty} f(x) = 2$, then the graph of f has a horizontal asymptote at $y = 2$.

2. *Examples:* Construct examples of the thing(s) described in the following. Try to find examples that are different than any in the reading.

(a) The graph of a function f for which $f(2)$ does not exist but $\lim_{x\to 2} f(x)$ does exist.

(b) The graph of a function f for which $f(2)$ exists and $\lim_{x\to 2} f(x)$ exists, but the two are not equal.

(c) The graph of a function f for which neither $f(2)$ nor $\lim_{x\to 2} f(x)$ exist.

3. If $\lim_{x\to 1^-} f(x) = 5$ and $\lim_{x\to 1^+} f(x) = 5$, what can you say about $\lim_{x\to 1} f(x)$? What can you say about $f(1)$?

4. If $\lim_{x\to 0^+} f(x) = -2$, $\lim_{x\to 0^-} f(x) = 3$, and $f(0) = -2$, what can you say about $\lim_{x\to 0} f(x)$?

5. If $\lim_{x\to 2^+} f(x) = 8$ but $\lim_{x\to 2} f(x)$ does not exist, what can you say about $\lim_{x\to 2^-} f(x)$?

6. If $\lim_{x\to -1^+} f(x) = -\infty$ and $\lim_{x\to -1^-} f(x) = -\infty$, what can you say about $\lim_{x\to -1} f(x)$?

7. If $\lim_{x\to -\infty} f(x) = \infty$, $\lim_{x\to \infty} f(x) = 3$, and $\lim_{x\to 1^+} f(x) = \infty$, what can you say about any horizontal and vertical asymptotes of f?

8. Consider the sequence $\frac{1}{2}, \frac{2}{3}, \frac{3}{4}, \frac{4}{5}, \ldots, \frac{k}{k+1}, \ldots$.

(a) What happens to the terms of this sequence as k gets larger and larger? Express your answer in limit notation.

(b) Use a calculator to find a sufficiently large value of k so that every term past the kth term of this sequence will be within 0.01 unit of 1.

9. Consider the sequence $\frac{1}{3}, \frac{1}{9}, \frac{1}{27}, \frac{1}{81}, \ldots, \frac{1}{3^k}, \ldots$.

(a) What happens to the terms of this sequence as k gets larger and larger? Express your answer in limit notation.

(b) Find a sufficiently large value of k so that every term past the kth term of this sequence will be less than 0.0001.

10. Consider the sequence of sums $\frac{1}{3}, \frac{1}{3} + \frac{1}{9}, \frac{1}{3} + \frac{1}{9} + \frac{1}{27}, \frac{1}{3} + \frac{1}{9} + \frac{1}{27} + \frac{1}{81}, \ldots$.

(a) What happens to the terms of this sequence of sums as k gets larger and larger?

(b) Find a sufficiently large value of k which will guarantee that every term past the kth term of this sequence of sums is in the interval $(0.49999, 0.5)$.

11. Consider the sequence of sums 1, 1 + 2, 1 + 2 + 3, 1 + 2 + 3 + 4, 1 + 2 + 3 + 4 + 5, ...

(a) What happens to the terms of this sequence of sums as k gets larger and larger?

(b) Find a sufficiently large value of k that will guarantee that every term past the kth term of this sequence of sums is greater than 1000.

12. An orange falling from 20 feet has a height of $s(t) = 20 - 16t^2$ feet when it has fallen for t seconds.

(a) Graph the position function $s(t)$ and find the time that the orange will hit the ground.

(b) Make a table to record the average rates that the orange is falling during the last second, half-second, quarter-second, and eighth-of-a-second of its fall.

(c) From the data in your table, make a guess for the instantaneous final velocity of the orange at the moment it hits the ground.

13. If you are on the moon, then an orange falling from 20 feet has a height of $s(t) = 20 - 2.65t^2$ feet when it has fallen for t seconds.

(a) Graph the position function $s(t)$ and find the time that the orange will hit the surface of the moon.

(b) Make a table to record the average rates that the orange is falling during the last second, half-second, quarter-second, and eighth-of-a-second of its fall on the moon.

(c) From the data in your table, make a guess for the instantaneous final velocity of the orange at the moment it hits the surface of the moon.

14. Consider the area between the graph of $f(x) = \sqrt{x}$ and the x-axis on $[0, 4]$.

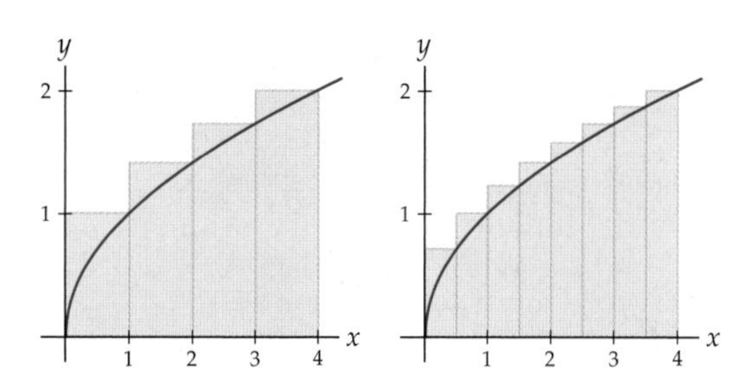

(a) Use the four rectangles shown on the left to approximate the given area, and then use the eight rectangles shown on the right to obtain another approximation of that area. Be sure to use the fact that the graph shown is that of the function $f(x) = \sqrt{x}$ in your calculations.

(b) Describe what would happen if we did similar approximations with more and more rectangles, and make a guess for the resulting limit.

15. Consider the area between the graph of $f(x) = 4 - x^2$ and the x-axis on $[0, 2]$.

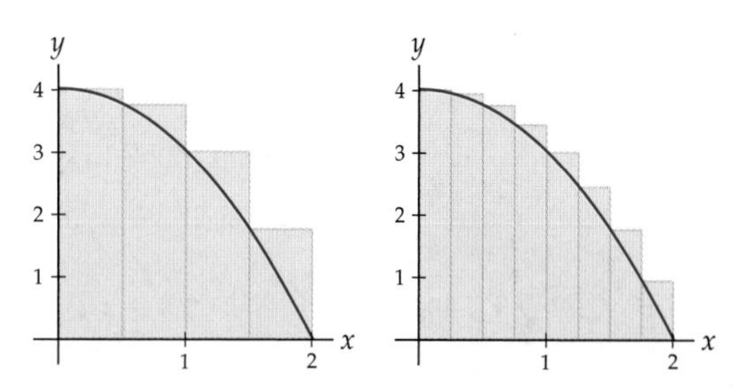

(a) Use the four rectangles shown on the left to approximate the given area, and then use the eight rectangles shown on the right to obtain another approximation of that area. Be sure to use the fact that the graph shown is that of the function $f(x) = 4 - x^2$ in your calculations.

(b) Describe what would happen if we did similar approximations with more and more rectangles, and make a guess for the resulting limit.

16. Sketch a function that has the following table of values, but whose limit as $x \to \infty$ is equal to $-\infty$:

x	100	200	500	1,000	10,000
$f(x)$	50	55	56.2	56.89	56.99

17. Sketch a function that has the following table of values, but whose limit as $x \to 2$ does not exist:

x	1.9	1.99	1.999	2	2.001	2.01	2.1
$f(x)$	3.12	3.09	3.01	-	2.99	2.92	2.87

18. Use a calculator or other graphing utility to graph the function $f(x) = \frac{x-2}{x^2-x-2}$.

(a) Show that $f(x)$ is not defined at $x = 2$. How is this reflected in your calculator graph?

(b) Use the graph to argue that even though $f(2)$ is undefined, we have $\lim_{x\to 2} f(x) = \frac{1}{3}$.

19. Use a calculator or other graphing utility to graph the function $g(x) = \frac{x^2-2x+1}{x-1}$.

(a) Show that $g(x)$ is not defined at $x = 1$. How is this reflected in your calculator graph?

(b) Use the graph to argue that even though $g(1)$ is undefined, we have $\lim_{x\to 1} g(x) = 0$.

20. Use a calculator or other graphing utility to investigate the graph $f(x) = x \sin\left(\frac{1}{x}\right)$ near $x = 0$. Be sure to have your calculator set to radian mode. Use the graphs to make an educated guess for $\lim_{x\to 0} f(x)$.

Skills

Sketch the graphs of functions that have the given limits and values in Exercises 21–30. (There are multiple correct answers.)

21. $\lim_{x\to -\infty} f(x) = 3$ and $\lim_{x\to \infty} f(x) = -\infty$

22. $\lim_{x\to 2} f(x) = -4$ and $\lim_{x\to -\infty} f(x) = -\infty$

23. $\lim_{x\to 0^+} f(x) = \infty$ and $\lim_{x\to 0^-} f(x) = \infty$

24. $\lim_{x\to 5^-} f(x) = 3$ and $\lim_{x\to 5^+} f(x) = 1$

25. $\lim_{x\to -\infty} f(x) = 2$ and $\lim_{x\to \infty} f(x) = 2$

26. $\lim_{x\to 2^-} f(x) = \infty$, $\lim_{x\to 2^+} f(x) = -\infty$, and $f(2) = 1$

27. $\lim_{x\to 3^-} f(x) = 2$, $\lim_{x\to 3^+} f(x) = 2$, but $f(3)$ does not exist

28. $\lim_{x\to -\infty} f(x) = -2$ and $\lim_{x\to 3} f(x) = \infty$, $f(0) = -5$

29. $\lim_{x\to 2^-} f(x) = 2$, $\lim_{x\to 2^+} f(x) = -1$, and $f(2) = 2$

30. $\lim_{x\to 2^-} f(x) = 3$, $\lim_{x\to 2^+} f(x) = 3$, and $f(2) = 0$

For the function f graphed as follows, approximate each of the limits and values in Exercises 31–34:

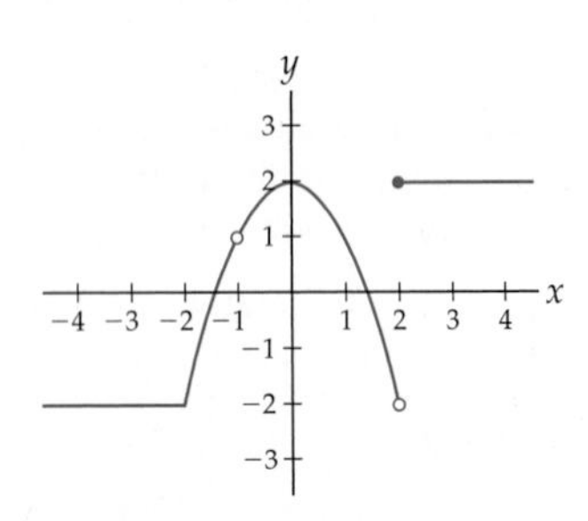

31. $\lim_{x\to -2^-} f(x)$, $\lim_{x\to -2^+} f(x)$, $\lim_{x\to -2} f(x)$, and $f(-2)$.

32. $\lim_{x\to -1^-} f(x)$, $\lim_{x\to -1^+} f(x)$, $\lim_{x\to -1} f(x)$, and $f(-1)$.

33. $\lim_{x\to 2^-} f(x)$, $\lim_{x\to 2^+} f(x)$, $\lim_{x\to 2} f(x)$, and $f(2)$.

34. $\lim_{x\to 0} f(x)$, $\lim_{x\to 1} f(x)$, $\lim_{x\to -\infty} f(x)$, and $\lim_{x\to \infty} f(x)$.

For the function $g(x)$ graphed as follows, approximate each of the limits and values in Exercises 35–38:

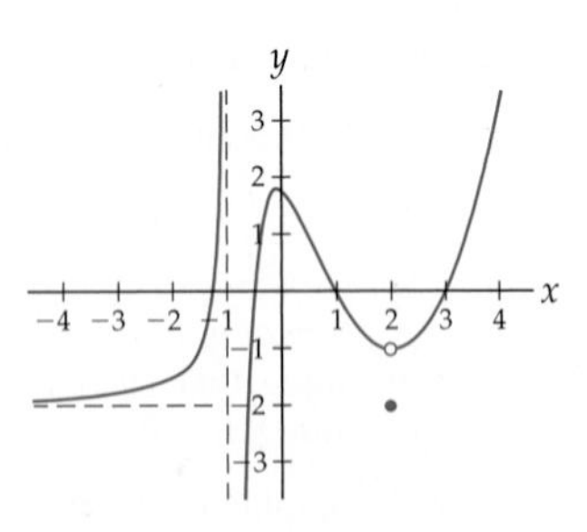

35. $\lim_{x\to -1^-} g(x)$, $\lim_{x\to -1^+} g(x)$, $\lim_{x\to -1} g(x)$, and $g(-1)$.

36. $\lim_{x\to 1^-} g(x)$, $\lim_{x\to 1^+} g(x)$, $\lim_{x\to 1} g(x)$, and $g(1)$.

37. $\lim_{x\to 2^-} g(x)$, $\lim_{x\to 2^+} g(x)$, $\lim_{x\to 2} g(x)$, and $g(2)$.

38. $\lim_{x\to 0} g(x)$, $\lim_{x\to 3} g(x)$, $\lim_{x\to -\infty} g(x)$, and $\lim_{x\to \infty} g(x)$.

Use tables of values to make educated guesses for each of the limits in Exercises 39–52.

39. $\lim_{x\to 2^-} (x^2 + x + 1)$

40. $\lim_{x\to 3^+} (1 - 3x + x^2)$

41. $\lim_{x\to 2} \frac{1}{x^2 - 4}$

42. $\lim_{x\to 1} \frac{1}{1 - x}$

43. $\lim_{x\to 3} \frac{x - 3}{(x^2 - 2)(x - 3)}$

44. $\lim_{x\to 5} \frac{x - 5}{x^2 - 25}$

45. $\lim_{x\to 2} \frac{3}{4 - 2^x}$

46. $\lim_{x\to \infty} (3e^{-2x} + 1)$

47. $\lim_{x\to -\infty} \frac{3x + 1}{1 - x}$

48. $\lim_{x\to \infty} \frac{1 + 2x}{x - 1}$

49. $\lim_{x\to \infty} \frac{x + 1}{x^2 - 1}$

50. $\lim_{x\to \infty} \left(1 + \frac{1}{2x + 1} + \frac{1}{x^2}\right)$

51. $\lim_{x\to \infty} \sin x$

52. $\lim_{x\to \infty} \sin\left(\frac{1}{x}\right)$

Sketch graphs by hand and use them to make approximations for each of the limits in Exercises 53–66. If a two-sided limit does not exist, describe the one-sided limits.

53. $\lim_{x\to 0} \frac{1}{x}$

54. $\lim_{x\to -1} (x^3 - 2)$

55. $\lim_{x\to 1} \frac{x^2 - 1}{x - 1}$

56. $\lim_{x\to -2} \frac{x^2 + x - 2}{x + 2}$

57. $\lim_{x\to 1} \frac{x - 1}{x^2 - 1}$

58. $\lim_{x\to \infty} \frac{x - 4}{x^2 - 4}$

59. $\lim_{x\to \infty} (1 - e^{-x})$

60. $\lim_{x\to -\infty} (3e^{4x} + 1)$

61. $\lim_{x\to \pi/2} \tan x$

62. $\lim_{x\to \pi} \csc x$

63. $\lim_{x\to 2} f(x)$, for $f(x) = \begin{cases} x^2, & \text{if } x < 2 \\ 1 - 3x, & \text{if } x \geq 2 \end{cases}$

64. $\lim_{x\to 0} f(x)$, for $f(x) = \begin{cases} 2x + 1, & \text{if } x \leq 0 \\ 2x - 1, & \text{if } x > 0 \end{cases}$

65. $\lim_{x\to 1} f(x)$, for $f(x) = \begin{cases} x^2 + 1, & \text{if } x < 1 \\ 3, & \text{if } x = 1 \\ 3 - x, & \text{if } x > 1 \end{cases}$

66. $\lim_{x\to -1} f(x)$, for $f(x) = \begin{cases} x + 1, & \text{if } x < -1 \\ 2, & \text{if } x = -1 \\ -x^2, & \text{if } x > -1 \end{cases}$

Use calculator graphs to make approximations for each of the limits in Exercises 67–74.

67. $\lim_{x\to 4} (3 - 4x - 5x^2)$

68. $\lim_{x\to \infty} (-0.2x^5 + 100x)$

69. $\lim_{x\to 1} \frac{3 - x}{x - 1}$

70. $\lim_{x\to 2} \frac{x + 1}{x - 2}$

71. $\lim_{x\to \infty} \frac{x^{100}}{2^x}$

72. $\lim_{x\to \infty} \frac{\ln x}{x}$

73. $\lim_{x\to 0} \frac{\sin x}{x}$

74. $\lim_{x\to 0} \frac{1 - \cos x}{x}$

Applications

75. There are four squirrels currently living in Linda's attic. If she does nothing to evict these squirrels, the number of squirrels in her attic after t days will be given by the formula $S(t) = \frac{12+5.5t}{3+0.25t}$.

(a) Verify that there are four squirrels in Linda's attic at time $t = 0$.

(b) Determine the number of squirrels in Linda's attic after 30 days, 60 days, and one year.

(c) Approximate $\lim_{t\to\infty} S(t)$ with a table of values. What does this limit mean in real-world terms?

(d) Graph $S(t)$ with a graphing utility, and use the graph to verify your answer to part (c).

76. The following graph describes the temperature $T(t)$ of a yam in an oven, where temperature T is measured in degrees Fahrenheit and time t is measured in minutes:

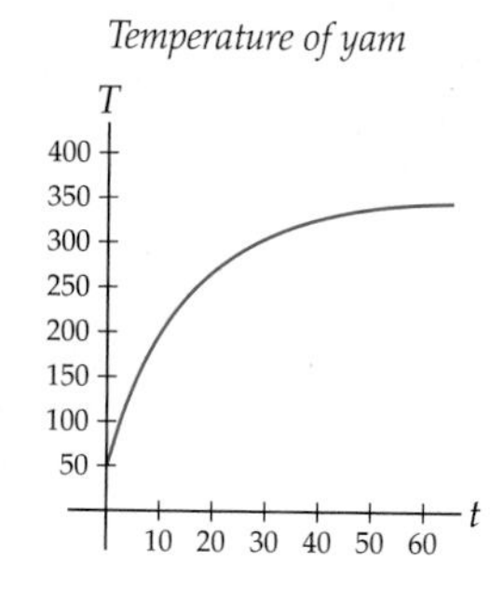

(a) Use the graph to approximate the temperature of the yam when it is first put in the oven.

(b) Use the graph to approximate $\lim_{t\to\infty} T(t)$.

(c) What is the temperature of the oven, and why?

77. In 1960, H. von Foerster suggested that the human population could be measured by the function

$$P(t) = \frac{179 \times 10^9}{(2027 - t)^{0.99}}.$$

Here P is the size of the human population. The time t is measured in years, where $t = 1$ corresponds to the year 1 A.D., time $t = 1973$ corresponds to the year 1973 A.D., and so on.

(a) Use a graphing utility to graph this function. You will have to be very careful when choosing a graphing window!

(b) Use the graph you found in part (a) to approximate $\lim_{t\to 2027^-} P(t)$.

(c) This population model is sometimes called the ***doomsday model.*** Why do you think this is? What year is doomsday, and why?

(d) In part (b), we considered only the *left* limit of $P(t)$ as $x \to 2027$. Why? What is the real-world meaning of the part of the graph that is to the right of $t = 2027$?

Proofs

78. Prove that for all $k > 100$, the quantity $\frac{1}{k^2}$ is in the interval $(0, 0.0001)$. What does this have to do with the limit of the sequence $\left\{\frac{1}{k^2}\right\}$ as $k \to \infty$?

79. For any positive integer k, the following equation holds: $1+2+3+\cdots+k = \frac{k(k+1)}{2}$. Use this fact to prove that for all $k > 100$, the value of the sum of the first k integers is greater than 5000. What does this have to do with the limit of a sequence of sums as $k \to \infty$?

80. Prove that for all x within 0.01 of the value $x = 1$, the quantity $(x-1)^2$ is within the interval $(0, 0.0001)$. What does this have to do with $\lim_{x\to 1}(x-1)^2$?

81. Prove that for all x within 0.01 of the value $x = 1$, the quantity $\frac{1}{(x-1)^2}$ is greater than $10{,}000$. What does this have to do with $\lim_{x\to 1} \frac{1}{(x-1)^2}$?

Thinking Forward

Convergence and divergence of sequences: If a sequence $a_1, a_2, a_3, \ldots, a_k, \ldots$ approaches a real-number limit as $k \to \infty$, then we say that the sequence $\{a_k\}$ ***converges.*** If the terms of the sequence do not get arbitrarily close to some real number, then we say that the sequence ***diverges.*** Write out enough terms of each sequence to make an educated guess as to whether it converges or diverges.

▶ $\left\{\left(\frac{1}{4}\right)^k\right\}$ ▶ $\left\{\left(\frac{5}{4}\right)^k\right\}$ ▶ $\left\{\frac{k}{k+2}\right\}$ ▶ $\left\{\frac{k+1}{k}\right\}$

Convergence and divergence of series: A **series** can be thought of as an infinite sum $a_1 + a_2 + a_3 + a_4 + \cdots + a_k + \cdots$. A series ***converges*** if this sum gets closer and closer to some real number limit as we add up more and more terms. Otherwise, the series is said to ***diverge.***

▶ As you will see in Chapter 8, the series $1 + \frac{1}{4} + \frac{1}{9} + \frac{1}{16} + \cdots + \frac{1}{k^2} + \cdots$ converges. Calculate partial sums including more and more terms until you are convinced that the sum eventually approaches a real-number limit and does not grow without bound.

▶ Although you might think that the series $1 + \frac{1}{2} + \frac{1}{3} + \frac{1}{4} + \cdots + \frac{1}{k} + \cdots$ converges because its terms get smaller and smaller, you will see in Chapter 8 that it does not. Calculate partial sums including more and more terms until you are convinced that this sum diverges and in fact grows without bound, never approaching a real-number limit. *(A calculator will come in handy here!)*

1.2 FORMAL DEFINITION OF LIMIT

- Moving from an intuitive concept of limit to a formal mathematical definition
- Uniqueness and existence of limits
- Limits from the left and right, limits at infinity, and infinite limits

Formalizing the Intuitive Definition of Limit

In the previous section we gave an intuitive description of limits. Now that we understand the basic concept, we are ready to give a precise, rigorous mathematical definition. Let's start with our intuitive description: For real numbers c and L and a function f, we have $\lim_{x \to c} f(x) = L$ if the values of $f(x)$ get closer and closer to L as x gets closer and closer to c. For example, $\lim_{x \to 2} x^2 = 4$ because the values of $f(x) = x^2$ approach 4 as x approaches 2. From the left, $f(1.9) = (1.9)^2 = 3.61$, $f(1.99) = (1.99)^2 = 3.9601$, $f(1.999) = (1.999)^2 \approx 3.996$, and so on, getting closer and closer to 4. A similar thing happens as x approaches 2 from the right.

Note that to be able to discuss $\lim_{x \to c} f(x)$, we must know how to calculate $f(x)$ near, but not necessarily at, the point $x = c$. Throughout this section we will assume that $f(x)$ is defined on a ***punctured interval*** $(c - \delta, c) \cup (c, c + \delta)$, where $\delta > 0$ represents a small distance to the left and right of $x = c$, as shown on the number line that follows. Notice that in our discussion of limits we will never be concerned with what happens *at* the point $x = c$, only *near* the point $x = c$.

Punctured δ-interval around c

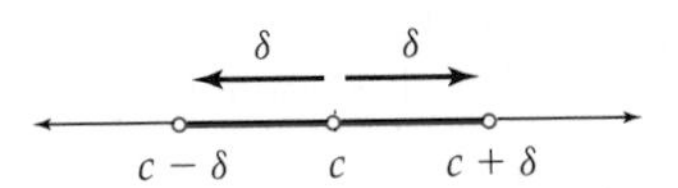

To make the definition of limit precise, we have to be very clear about what we mean when we say that $f(x)$ "approaches" L. We want to capture the idea that we can make the values of $f(x)$ not just close to L, but *as close as we like* to L if only we choose values of x that are sufficiently close to c. For example, we can guarantee that $f(x) = x^2$ is within 0.05 unit of 4 if we choose values of x that are within 0.01 unit of 2. Note that $f(2.01) = (2.01)^2 = 4.0401$ and $f(1.99) = (1.99)^2 = 3.9601$ are both within 0.05 unit of 4, and values of x that are closer to 2 will result in values of $f(x)$ that are even closer to 4. If we want values of $f(x)$ that are even closer to $L = 4$, then we can just choose values of x that are even closer to $c = 2$.

In general, suppose we want to guarantee that the values of $f(x)$ are within some very small distance ϵ above or below limit value L, as shown at the left in the graphs that follow. To do this we must choose values of x that are sufficiently close to c, say, some distance $\delta > 0$ left or right of c, as shown in the middle graph. The Greek letters ***delta*** (δ) and ***epsilon*** (ϵ) are the traditional letters used for these small distances. The figure at the right illustrates that a choice of x-value inside the blue punctured δ-interval $(c - \delta, c) \cup (c, c + \delta)$ determines an $f(x)$-value within the beige ϵ-interval $(L - \epsilon, L + \epsilon)$. In these figures we have omitted the point at $x = c$ to emphasize that we are not concerned with the actual value of $f(x)$ at the point $x = c$, only with the behavior of $f(x)$ at points near $x = c$.

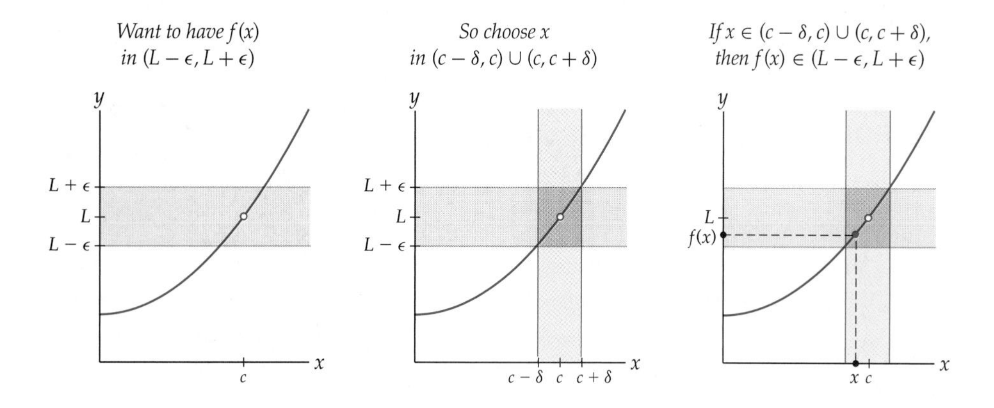

If the values of $f(x)$ get arbitrarily close to L as x approaches c, then we can choose smaller and smaller beige ϵ-intervals $(L - \epsilon, L + \epsilon)$ and in each case always find some blue punctured δ-interval $(c-\delta, c) \cup (c, c+\delta)$ that determines values of $f(x)$ which are within ϵ of the limit L. The following three figures illustrate this idea:

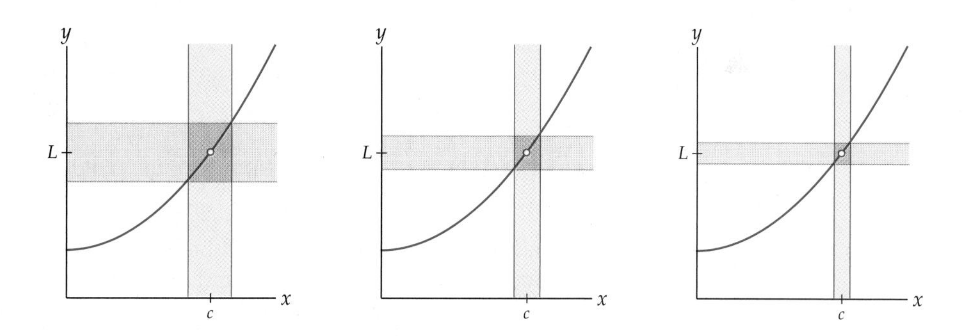

We want the limit statement $\lim_{x \to c} f(x) = L$ to mean that no matter how small a distance we choose for ϵ, we can find some δ so that values of x that are within δ of $x = c$ will yield values of $f(x)$ that are within ϵ of $y = L$. Writing this in terms of intervals gives us the following definition:

DEFINITION 1.5 Formal Definition of Limit

The ***limit*** $\lim_{x \to c} f(x) = L$ means that for all $\epsilon > 0$, there exists $\delta > 0$ such that

$$\text{if } x \in (c-\delta, c) \cup (c, c+\delta), \text{ then } f(x) \in (L-\epsilon, L+\epsilon).$$

Stop and think about that for a minute until it makes sense. Understanding this definition is the key to understanding limits, and limits are the foundation of everything in calculus. So take a few minutes, have some tea, and get everything straight in your head before you continue reading.

Uniqueness of Limits

A limit $\lim_{x \to c} f(x)$ ***exists*** if it is equal to some real number L. If a limit exists, then it can be equal to one and only one number. That sounds obvious, since the values of a function $f(x)$ cannot approach two different values L and M as x approaches c. However, as we are about to see, to prove uniqueness of limits we must carefully apply the formal definition of limits.

THEOREM 1.6 **Uniqueness of Limits**

If $\lim_{x\to c} f(x) = L$ and $\lim_{x\to c} f(x) = M$, then $L = M$.

Proof. Suppose to the contrary that somehow $\lim_{x\to c} f(x) = L$ and $\lim_{x\to c} f(x) = M$, with $L \neq M$. Let's suppose that $L > M$, since, if not, then we can just reverse the roles of L and M. If $L > M$, then we must have $L = M + k$ for some positive real number k. Now consider $\epsilon = \frac{k}{2}$; note that with this choice of ϵ, the intervals $(L - \epsilon, L + \epsilon)$ and $(M - \epsilon, M + \epsilon)$ do not overlap.

Since $\lim_{x\to c} f(x) = L$, we can find $\delta_1 > 0$ such that for all $x \in (c - \delta_1, c) \cup (c, c + \delta_1)$, we have $f(x) \in (L - \epsilon, L + \epsilon)$. Similarly, since $\lim_{x\to c} f(x) = M$, we can find $\delta_2 > 0$ such that for all $x \in (c - \delta_2, c) \cup (c, c + \delta_2)$, we have $f(x) \in (M - \epsilon, M + \epsilon)$. Now if we let δ be the smaller of δ_1 and δ_2, we can say that for any $x \in (c - \delta, c) \cup (c, c + \delta)$, we can guarantee that both $f(x) \in (L - \epsilon, L + \epsilon)$ and $f(x) \in (M - \epsilon, M + \epsilon)$. But this cannot be, since the intervals $(L - \epsilon, L + \epsilon)$ and $(M - \epsilon, M + \epsilon)$ do not overlap. Therefore we could not have initially had $f(x)$ approaching two different limits L and M; we must have $L = M$. ■

One-Sided Limits

We can consider each limit $\lim_{x\to c} f(x) = L$ from two different directions: from the left and from the right. We say that we have a left limit $\lim_{x\to c^-} f(x) = L$ if, given an ϵ-interval $(L - \epsilon, L + \epsilon)$, we can always find a sufficiently small half-neighborhood $(c - \delta, c)$ to the left of $x = c$ so that values of x that are in that left hand δ-interval yield values of $f(x)$ that are in the ϵ-interval, as shown in the left-hand graph that follows. We define right limits similarly, as shown in the right-hand graph:

Can get $f(x)$ in $(L - \epsilon, L + \epsilon)$ by choosing x in $(c - \delta, c)$

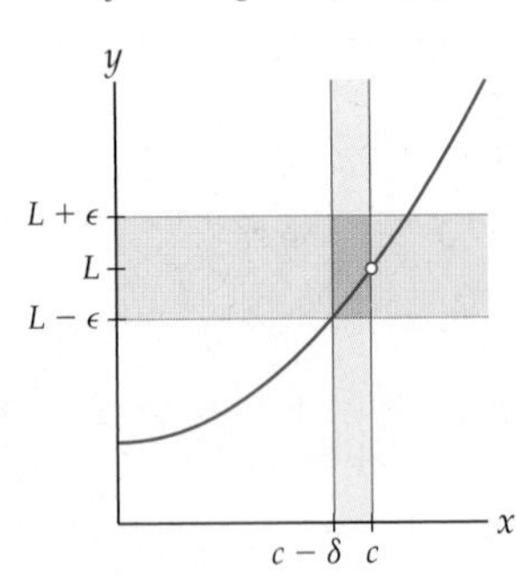

Can get $f(x)$ in $(L - \epsilon, L + \epsilon)$ by choosing x in $(c, c + \delta)$

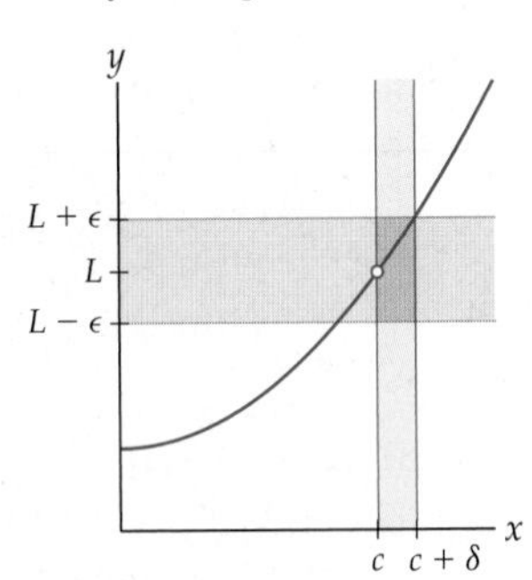

DEFINITION 1.7 **One-Sided Limits**

The ***left limit*** $\lim_{x\to c^-} f(x) = L$ means that for all $\epsilon > 0$, there exists $\delta > 0$ such that

$$\text{if } x \in (c - \delta, c), \text{ then } f(x) \in (L - \epsilon, L + \epsilon).$$

The ***right limit*** $\lim_{x\to c^+} f(x) = L$ means that for all $\epsilon > 0$, there exists $\delta > 0$ such that

$$\text{if } x \in (c, c + \delta), \text{ then } f(x) \in (L - \epsilon, L + \epsilon).$$

A two-sided limit $\lim_{x \to c} f(x)$ is equal to some real number L if and only if the corresponding left and right limits exist and are also equal to that same number L.

THEOREM 1.8 **For a Limit to Exist, the Left and Right Limits Must Exist and Be Equal**

$\lim_{x \to c} f(x) = L$ if and only if $\lim_{x \to c^-} f(x) = L$ and $\lim_{x \to c^+} f(x) = L$.

The proof of this theorem is a straightforward application of the definitions of two-sided and one-sided limits.

Proof. Suppose $\lim_{x \to c} f(x) = L$. Then for all $\epsilon > 0$, there is some $\delta > 0$ such that if $x \in (c - \delta, c) \cup (c, c + \delta)$, then $f(x) \in (L - \epsilon, L + \epsilon)$. In particular this means that if $x \in (c - \delta, c)$ or if $x \in (c, c + \delta)$, then we will have $f(x) \in (L - \epsilon, L + \epsilon)$. Therefore $\lim_{x \to c^-} f(x) = L$ and $\lim_{x \to c^+} f(x) = L$.

For the converse, suppose $\lim_{x \to c^-} f(x) = L$ and $\lim_{x \to c^+} f(x) = L$. Then for all $\epsilon > 0$, there exist numbers $\delta_1 > 0$ and $\delta_2 > 0$ such that for either $x \in (c-\delta_1, c)$ or $x \in (c, c+\delta_2)$, we have $f(x) \in (L-\epsilon, L+\epsilon)$. If we let δ be the smaller of δ_1 and δ_2, then we can say that for $x \in (c - \delta, c) \cup (c, c + \delta)$, we can guarantee that $f(x) \in (L - \epsilon, L + \epsilon)$. Therefore $\lim_{x \to c} f(x) = L$. ■

Infinite Limits and Limits at Infinity

So far we have formalized the definition of limit only in the case where both x and $f(x)$ are approaching real numbers. Now we consider what happens if one or both of x and $f(x)$ approach $\pm\infty$. For example, we want $\lim_{x \to c} f(x) = \infty$ to capture the idea that as x approaches c, the values of $f(x)$ grow without bound. In other words, $\lim_{x \to c} f(x) = \infty$ should guarantee that values of $f(x)$ will lie above any given large number M as long as we choose values of x that are sufficiently close to c; see the figure that follows at the left.

$\lim_{x \to c} f(x) = \infty$

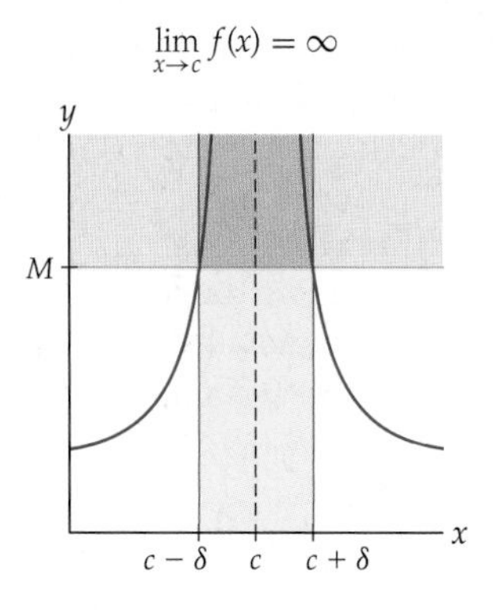

$\lim_{x \to \infty} f(x) = L$

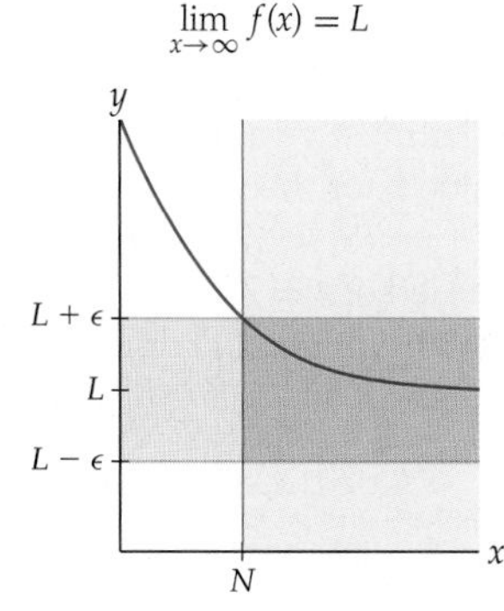

$\lim_{x \to \infty} f(x) = \infty$

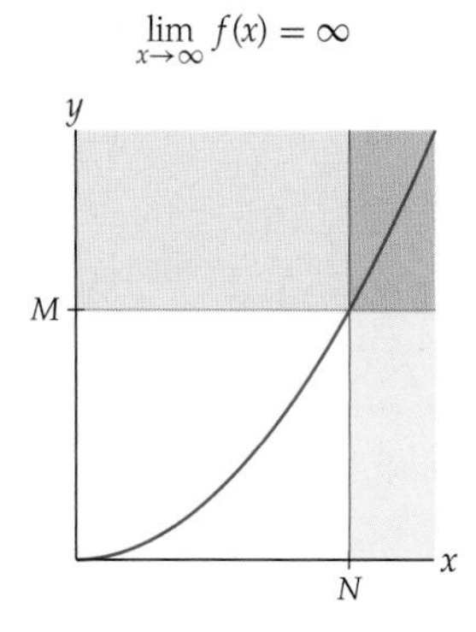

Similarly, we want the limit statement $\lim_{x \to \infty} f(x) = L$ to indicate that given any ϵ-interval around L, we can choose values of x sufficiently large so that $f(x)$ is in the ϵ-interval; see the middle graph. And $\lim_{x \to \infty} f(x) = \infty$ should mean that given any large number M, we can get values of $f(x)$ that are greater than M if we choose sufficiently large values of x, say, larger than some big number N, as in the graph at the right. The definition that follows expresses these three limits in terms of intervals; compare the three parts with the preceding three figures. You will define other related types of limits in Exercises 37–42.

DEFINITION 1.9 Limits Involving Infinity

The ***infinite limit*** $\lim_{x \to c} f(x) = \infty$ means that for all $M > 0$, there exists $\delta > 0$ such that

$$\text{if } x \in (c - \delta, c) \cup (c, c + \delta), \text{ then } f(x) \in (M, \infty).$$

The ***limit at infinity*** $\lim_{x \to \infty} f(x) = L$ means that for all $\epsilon > 0$, there exists $N > 0$ such that

$$\text{if } x \in (N, \infty), \text{ then } f(x) \in (L - \epsilon, L + \epsilon).$$

The ***infinite limit at infinity*** $\lim_{x \to \infty} f(x) = \infty$ means that for all $M > 0$, there exists $N > 0$ such that

$$\text{if } x \in (N, \infty), \text{ then } f(x) \in (M, \infty).$$

Examples and Explorations

EXAMPLE 1 **Approximating δ given ϵ for a limit**

Use a graph to illustrate and approximate

(a) the largest δ which guarantees that if $x \in (2 - \delta, 2) \cup (2, 2 + \delta)$, then $x^2 \in (3, 5)$.

(b) the largest δ which guarantees that if $x \in (2 - \delta, 2) \cup (2, 2 + \delta)$, then $x^2 \in (3.5, 4.5)$.

What limit statement do these problems have to do with, and why?

SOLUTION

These problems concern the limit statement $\lim_{x \to 2} x^2 = 4$, which by definition means that for all $\epsilon > 0$, there exists $\delta > 0$ such that if $x \in (2 - \delta, 2) \cup (2, 2 + \delta)$, then $x^2 \in (4 - \epsilon, 4 + \epsilon)$. Therefore parts (a) and (b) of this example ask us to find corresponding values of δ for $\epsilon = 1$ and $\epsilon = 0.5$, respectively.

(a) To find the largest δ corresponding to $\epsilon = 1$, we begin by drawing $f(x) = x^2$ and the beige ϵ-interval of width 1 around $y = 4$, as shown in the graph that follows at the left. We then draw the vertical blue band shown in the figure, to represent the range of values of x which determine values of $f(x)$ that are within the horizontal beige band. The leftmost x-value a in the blue band is a solution of $f(a) = a^2 = 3$, and the rightmost x-value b for the blue band is a solution of $f(b) = b^2 = 5$. Therefore $a = \sqrt{3} \approx 1.732$ and $b = \sqrt{5} \approx 2.236$. Now, what is δ in this case? We can move $2 - 1.732 = 0.268$ unit to the left of $x = 2$ and $2.236 - 2 = 0.236$ unit to the right of $x = 2$. We need the largest δ that will work in *both* directions, which is the smaller of the two distances we just found: $\delta = 0.236$. Then if $x \in (2 - 0.236, 2) \cup (2, 2 + 0.236)$, we can guarantee that $x^2 \in (3, 5)$.

If $x \in (1.732, 2.236)$, then $x^2 \in (3, 5)$

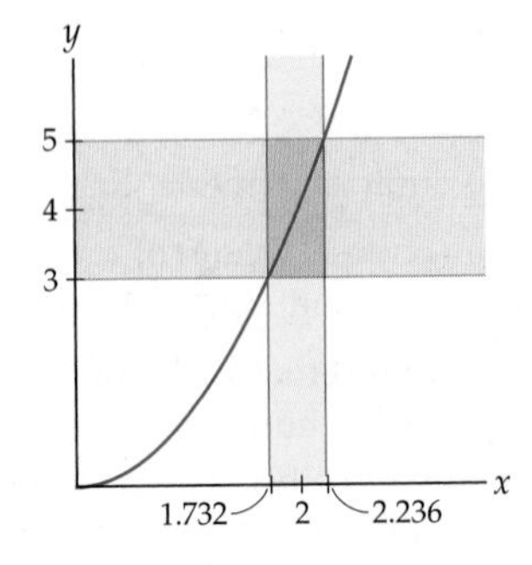

If $x \in (1.87, 2.12)$, then $x^2 \in (3.5, 4.5)$

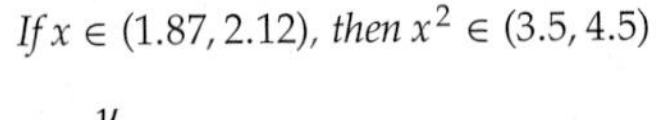

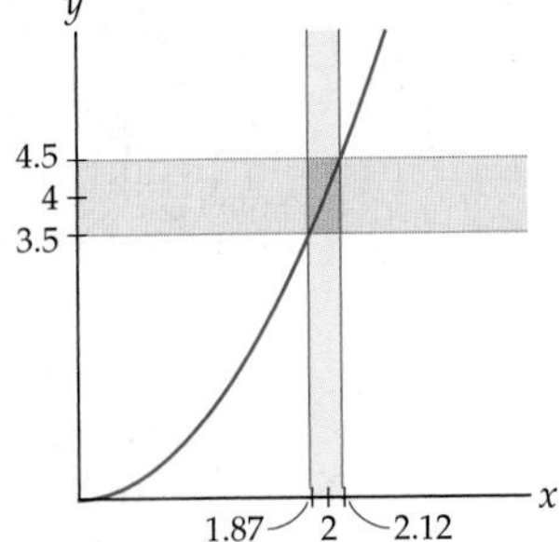

(b) We now repeat the problem with a smaller value of ϵ. For $\epsilon = 0.5$ we draw the smaller horizontal beige ϵ-bar shown in the preceding graph at the right, and the corresponding vertical blue bar. Solving $f(a) = a^2 = 3.5$ and $f(b) = b^2 = 4.5$, we get $a \approx 1.87$ and $b \approx 2.12$ as the leftmost and rightmost values of x contained in the vertical blue bar. Therefore values of x that are at most $2 - 1.87 = 0.13$ unit to the left of $x = 2$ or at most $2.12 - 2 = 0.12$ unit to the right of $x = 1$ will determine values of $f(x)$ that are within 0.5 unit of $y = 4$. The smaller of these two distances is the largest δ that will work in both directions, namely, $\delta = 0.12$. If $x \in (2 - 0.12, 2) \cup (2, 2 + 0.12)$, then we can guarantee that $x^2 \in (3.5, 4.5)$. □

EXAMPLE 2

Approximating *N* given ϵ for a limit at infinity

Use a graph to illustrate and approximate

(a) the smallest N which guarantees that if $x \in (N, \infty)$, then $\frac{x+1}{x} \in (0.75, 1.25)$.

(b) the smallest N which guarantees that if $x \in (N, \infty)$, then $\frac{x+1}{x} \in (0.9, 1.1)$.

What limit statement do these problems have to do with, and why?

SOLUTION

These problems are about the limit statement $\lim_{x\to\infty} \frac{x+1}{x} = 1$, which by definition means that for all $\epsilon > 0$, there is some $N > 0$ such that if $x \in (N, \infty)$, then $\frac{x+1}{x} \in (1 - \epsilon, 1 + \epsilon)$. Therefore parts (a) and (b) of this example ask us to find the corresponding values of N for $\epsilon = 0.25$ and $\epsilon = 0.1$, respectively.

(a) The figure that follows at the left shows $f(x) = \frac{x+1}{x}$ and a beige bar representing all the heights within 0.25 unit of $y = 1$. The blue area shows all of the values of x for which the corresponding values of $f(x)$ lie within the beige bar. According to this graph, to find the leftmost point $x = a$ of the blue area we must solve $f(a) = 1.25$:

$$f(a) = 1.25 \implies \frac{a+1}{a} = 1.25 \implies a + 1 = 1.25a \implies 1 = 0.25a \implies a = 4.$$

Therefore if $x \in (4, \infty)$, then we can guarantee that $\frac{x+1}{x} \in (0.75, 1.25)$.

If $x \in (4, \infty)$, *then* $\frac{x+1}{x} \in (0.75, 1.25)$

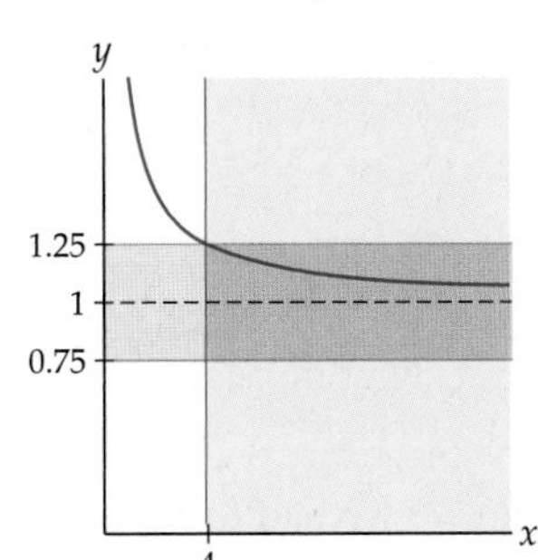

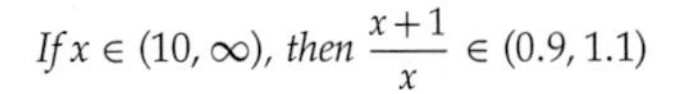
If $x \in (10, \infty)$, *then* $\frac{x+1}{x} \in (0.9, 1.1)$

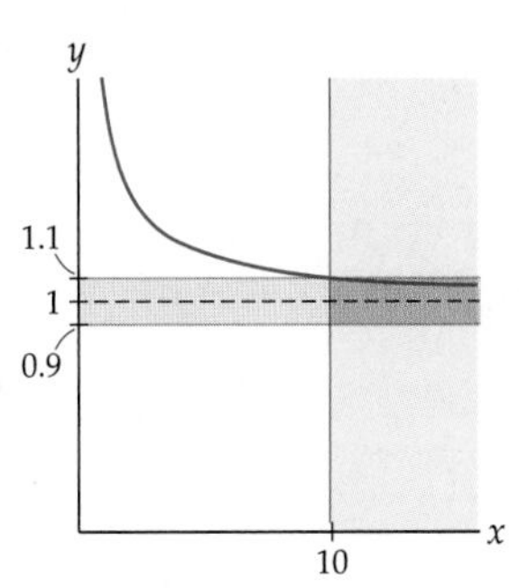

(b) We now do the same thing but for $\epsilon = 0.1$. For this smaller value of ϵ we must draw a smaller beige bar around $y = 1$, which in turn requires a different blue area of values of x for which the corresponding values of $f(x)$ lie within the ϵ-bar, as shown in the preceding graph at the right. To find the leftmost point $x = a$ of the blue area, we solve $f(a) = 1.1$:

$$f(a) = 1.1 \implies \frac{a+1}{a} = 1.1 \implies a + 1 = 1.1a \implies 1 = 0.1a \implies a = 10.$$

Thus for $x \in (10, \infty)$ we can guarantee that $\frac{x+1}{x} \in (0.9, 1.1)$. □

EXAMPLE 3

A real–world example of finding δ given ϵ

Fuel efficiency depends on driving speed. A typical car runs at 100% fuel efficiency when driven at 55 miles per hour. Suppose that the fuel efficiency percentage at speed s (in MPH) is given by $E(s) = -0.033058(s^2 - 110s)$. If you want your car to run with at least 95% fuel efficiency, how close to 55 miles per hour do you have to drive?

SOLUTION

In the language of limits, we are asking: For the limit $\lim_{s \to 55} E(s) = 100$, if $\epsilon = 5$, what is δ? The corresponding ϵ-bar and δ-bar are shown below in the following graph of $E(s)$:

Staying within 5% fuel efficiency

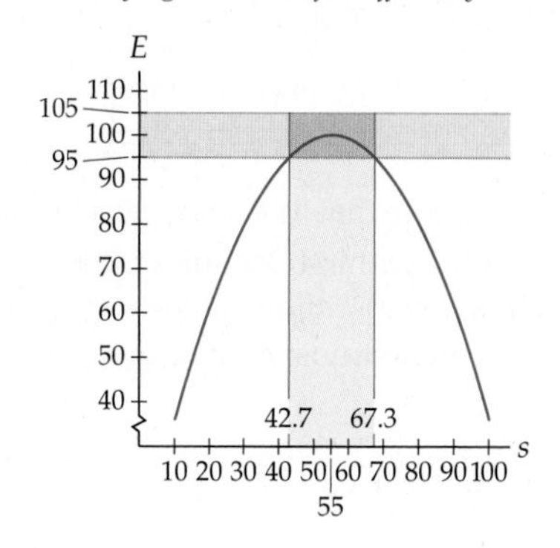

The leftmost and rightmost values $x = a$ and $x = b$ of the blue δ-interval shown can be found by using the quadratic formula to find the two solutions of the equation $E(s) = 95$ or by using a calculator graph to approximate values. In either case, we find $a \approx 42.7$ and $b \approx 67.3$. Therefore, according to the graph you can drive anywhere between 42.7 and 67.3 miles per hour and get at least 95% fuel efficiency. □

? TEST YOUR UNDERSTANDING

- When discussing limits as $x \to c$, why do we consider *punctured* intervals around $x = c$?
- In the definition of limit, why do we need the statement to be true for *all* values $\epsilon > 0$?
- In the proof of Theorem 1.6 we had $L = M + k$ and $\epsilon = \frac{k}{2}$. Why does this mean that $(L - \epsilon, L + \epsilon)$ and $(M - \epsilon, M + \epsilon)$ do not overlap?
- How are left limits and right limits related to two-sided limits?
- How is a limit at infinity different from an infinite limit?

EXERCISES 1.2

Thinking Back

Logical quantifiers: Determine whether each of the following statements about real numbers is true or false, and why.

- For all a, there exists some b such that $b = a^2$.
- For all a, there exists some b such that $a = b^2$.
- For all a, there exists some b such that $b = a + 1$.
- For all integers a, there exists some integer b such that if $x \geq a$, then $x > b$.
- For all integers a, there exists some integer b such that if $x > a$, then $x = b$.

Solving function equations: Solve each of the following equations for x, and illustrate these solutions on a graph of $y = f(x)$.

- If $f(x) = x^3$, solve $f(x) = 7.5$ and $f(x) = 8.5$.
- If $f(x) = \sqrt{x-1}$, solve $f(x) = 1.8$ and $f(x) = 2.2$.
- If $f(x) = -0.033058(x^2 - 110x)$, solve $f(x) = 90$.
- If $f(x) = 2 - x^2$ and $x > 0$, solve $f(x) = -7.01$ and $f(x) = -6.99$.
- If $f(x) = \dfrac{x^2 - 2x - 3}{x - 1}$ and $x < 0$, solve $f(x) = -2$ and $f(x) = 0$.

Concepts

0. *Problem Zero:* Read the section and make your own summary of the material.

1. *True/False:* Determine whether each of the statements that follow is true or false. If a statement is true, explain why. If a statement is false, provide a counterexample.

(a) *True or False:* For $\lim_{x \to c} f(x)$ to be defined, the function f must be defined at $x = c$.

(b) *True or False:* We can calculate a limit of the form $\lim_{x \to c} f(x)$ simply by finding $f(c)$.

(c) *True or False:* If $\lim_{x \to c} f(x) = 10$, then $f(c) = 10$.

(d) *True or False:* If $f(c) = 10$, then $\lim_{x \to c} f(x) = 10$.

(e) *True or False:* A function can approach more than one limit as x approaches c.

(f) *True or False:* If $\lim_{x \to 4} f(x) = 10$, then we can make $f(x)$ as close to 4 as we like by choosing values of x sufficiently close to 10.

(g) *True or False:* If $\lim_{x \to 6} f(x) = \infty$, then we can make $f(x)$ as large as we like by choosing values of x sufficiently close to 6.

(h) *True or False:* If $\lim_{x \to \infty} f(x) = 100$, then we can find values of $f(x)$ between 99.9 and 100.1 by choosing values of x that are sufficiently large.

2. *Examples:* Construct examples of the thing(s) described in the following. Try to find examples that are different than any in the reading.

(a) A function f and a value c such that $\lim_{x \to c} f(x)$ happens to be equal to $f(c)$.

(b) A function f and a value c such that $\lim_{x \to c} f(x)$ is not equal to $f(c)$.

(c) A function f and a value c such that $\lim_{x \to c} f(x)$ exists but $f(c)$ does not exist.

3. What are punctured intervals, and why do we need to use them when discussing limits?

4. Describe the punctured interval around $x = 2$ that has a radius of 3 and the punctured interval around $x = 4$ that has a radius of 0.25.

5. Find punctured intervals on which the function $f(x) = \dfrac{1}{x^2 - x}$ is defined, centered around

(a) $x = 1.5$ (b) $x = 0.25$ (c) $x = 1$

6. Find punctured intervals on which the function $f(x) = \dfrac{1}{x \ln(x+2)}$ is defined, centered around

(a) $x = 0$ (b) $x = -1$ (c) $x = -1.5$

Use interval notation to fill in the blanks that follow. Your answers will involve δ, ϵ, N, and/or M.

7. If $\lim_{x \to 2} f(x) = 5$, then for all $\epsilon > 0$, there is some $\delta > 0$ such that if $x \in$ ____, then $f(x) \in$ ____.

8. If $\lim_{x \to 3^-} f(x) = 1$, then for all $\epsilon > 0$, there is some $\delta > 0$ such that if $x \in$ ____, then $f(x) \in$ ____.

9. If $\lim_{x \to \infty} f(x) = 2$, then for all $\epsilon > 0$, there is some $N > 0$ such that if $x \in$ ____, then $f(x) \in$ ____.

10. If $\lim_{x \to \infty} f(x) = -\infty$, then for all $M > 0$, there is some $N > 0$ such that if $x \in$ ____, then $f(x) \in$ ____.

11. If $\lim_{x \to 1^+} f(x) = \infty$, then for all $M > 0$, there is some $\delta > 0$ such that if $x \in$ ____, then $f(x) \in$ ____.

12. Sketch a labeled graph that illustrates what is going on in the proof of Theorem 1.6 in the reading. Your graph should include two different ϵ-bars and a graphical reason that they cannot overlap.

13. Sketch a labeled graph that illustrates what is going on in the proof of Theorem 1.8 in the reading. Your graph should include two different δ-bars and a graphical reason why they combine to make a punctured delta-interval.

14. Suppose f is a function with $f(2) = 5$ where for all $\epsilon > 0$, there is some $\delta > 0$ such that if $x \in (2 - \delta, 2) \cup (2, 2 + \delta)$, then $f(x) \in (3 - \epsilon, 3 + \epsilon)$. Sketch a possible graph of f.

15. If $x \in (1.5, 2.5)$, what is the largest interval $I = (4 - \epsilon, 4 + \epsilon)$ for which we can guarantee that $x^2 \in I$?

16. It is *false* that $\lim_{x \to 1}(x + 1.01) = 2$. Express this fact in a mathematical sentence involving δ and ϵ, to show how the formal definition of limit fails in this case.

17. It is *false* that $\lim_{x\to\infty} \frac{1000}{x} = \infty$. Express this fact in a mathematical sentence involving M and N, to show how the formal definition of limit fails in this case.

18. Show that the limit as $x \to 2$ of $f(x) = \sqrt{x - 1.1}$ is *not* equal to 1, by finding an $\epsilon > 0$ for which there is no corresponding $\delta > 0$ satisfying the formal definition of limit.

Skills

Write each limit in Exercises 19–42 as a formal statement involving δ, ϵ, N, and/or M, and sketch a graph that illustrates the roles of these constants. In the last six exercises you may take f to be any appropriate graph.

19. $\lim_{x\to -3} \sqrt{x+7} = 2$

20. $\lim_{x\to 2} \frac{x^2-4}{x+2} = 0$

21. $\lim_{x\to -1} (x^3 - 2) = -3$

22. $\lim_{x\to 1^-} \sqrt{1-x} = 0$

23. $\lim_{x\to 2} \frac{x^2-4}{x-2} = 4$

24. $\lim_{x\to 1} (x^2 - 3) = -2$

25. $\lim_{x\to 0^+} \sqrt{x} = 0$

26. $\lim_{x\to 3^-} (4 - x^2) = -5$

27. $\lim_{x\to -2^+} \frac{1}{x+2} = \infty$

28. $\lim_{x\to 0^-} \frac{1}{x} = -\infty$

29. $\lim_{x\to 2^+} \frac{1}{2^x - 4} = \infty$

30. $\lim_{x\to 0^+} \frac{1}{1-e^x} = -\infty$

31. $\lim_{x\to \infty} \frac{x}{1-2x} = -.5$

32. $\lim_{x\to -\infty} \frac{1}{x^2+1} = 0$

33. $\lim_{x\to \infty} (x^3 + x + 1) = \infty$

34. $\lim_{x\to -\infty} (1 - 3x) = \infty$

35. $\lim_{h\to 0} \frac{(2+h)^2 - 4}{h} = 4$

36. $\lim_{h\to 0} \frac{\frac{1}{2+h} - \frac{1}{2}}{h} = -\frac{1}{4}$

37. $\lim_{x\to c^+} f(x) = -\infty$

38. $\lim_{x\to c^-} f(x) = \infty$

39. $\lim_{x\to -\infty} f(x) = L$

40. $\lim_{x\to -\infty} f(x) = -\infty$

41. $\lim_{x\to -\infty} f(x) = \infty$

42. $\lim_{x\to \infty} f(x) = -\infty$

For each limit $\lim_{x\to c} f(x) = L$ in Exercises 43–54, use graphs and algebra to approximate the largest value of δ such that if $x \in (c-\delta, c) \cup (c, c+\delta)$, then $f(x) \in (L-\epsilon, L+\epsilon)$.

43. $\lim_{x\to 2} x^3 = 8, \quad \epsilon = 0.5$

44. $\lim_{x\to 2} x^3 = 8, \quad \epsilon = 0.25$

45. $\lim_{x\to 5} \sqrt{x-1} = 2, \quad \epsilon = 1$

46. $\lim_{x\to 5} \sqrt{x-1} = 2, \quad \epsilon = 0.2$

47. $\lim_{x\to \pi} \sin x = 0, \quad \epsilon = \frac{\sqrt{2}}{2}$

48. $\lim_{x\to \pi} \sin x = 0, \quad \epsilon = \frac{1}{2}$

49. $\lim_{x\to 0} \frac{1-\cos x}{x} = 0, \quad \epsilon = \frac{1}{2}$

50. $\lim_{x\to 0} \frac{1-\cos x}{x} = 0, \quad \epsilon = \frac{1}{4}$

51. $\lim_{x\to 3} (2 - x^2) = -7, \quad \epsilon = 0.01$

52. $\lim_{x\to 3} (2 - x^2) = -7, \quad \epsilon = 0.001$

53. $\lim_{x\to -1} \frac{x^2-2x-3}{x+1} = -4, \quad \epsilon = 1$

54. $\lim_{x\to -1} \frac{x^2-2x-3}{x+1} = -4, \quad \epsilon = 0.1$

For each limit in Exercises 55–64, use graphs and algebra to approximate the largest δ or smallest-magnitude N that corresponds to the given value of ϵ or M, according to the appropriate formal limit definition.

55. $\lim_{x\to 1^+} \frac{1}{x^2-1} = \infty, \quad M = 1000$, find largest $\delta > 0$

56. $\lim_{x\to 1^+} \frac{1}{x^2-1} = \infty, \quad M = 10{,}000$, find largest $\delta > 0$

57. $\lim_{x\to \infty} \frac{3x}{x+1} = 3, \quad \epsilon = 0.5$, find smallest $N > 0$

58. $\lim_{x\to \infty} \frac{3x}{x+1} = 3, \quad \epsilon = 0.1$, find smallest $N > 0$

59. $\lim_{x\to \infty} \ln x = \infty, \quad M = 100$, find smallest $N > 0$

60. $\lim_{x\to \infty} \ln x = \infty, \quad M = 100{,}000$, find smallest $N > 0$

61. $\lim_{x\to -\infty} 3^x = 0, \quad \epsilon = \frac{1}{2}$, find smallest-magnitude $N < 0$

62. $\lim_{x\to -\infty} 3^x = 0, \quad \epsilon = \frac{1}{4}$, find smallest-magnitude $N < 0$

63. $\lim_{x\to \infty} (4 - x^2) = -\infty, \quad M = -100$, find smallest $N > 0$

64. $\lim_{x\to \infty} (4 - x^2) = -\infty, \quad M = -10{,}000$, find smallest $N > 0$

Applications

65. Every month, Jack hides \$50 under a broken floorboard to save up for a new boat. After t months of saving, he will have $F(t) = 50t$ dollars.

(a) The boat Jack wants costs at least \$7,465. How many months does Jack have to save money before he will have enough to pay for the boat? Illustrate this information on a graph of $F(t)$.

(b) Suppose a different boat costs M dollars. Will there be a value $t = N$ for which $F(N) > M$? What does this mean in real-world terms? Illustrate the roles of M and N on a graph of $F(t)$.

Money saved for Jack's boat

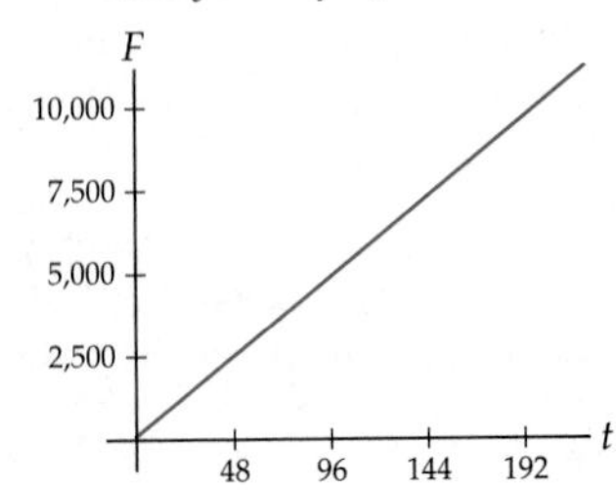

66. Len's company produces different-sized cylindrical cans that are each 6 inches tall. The cost to produce a can with radius r is $C(r) = 10\pi r^2 + 24\pi r$ cents.

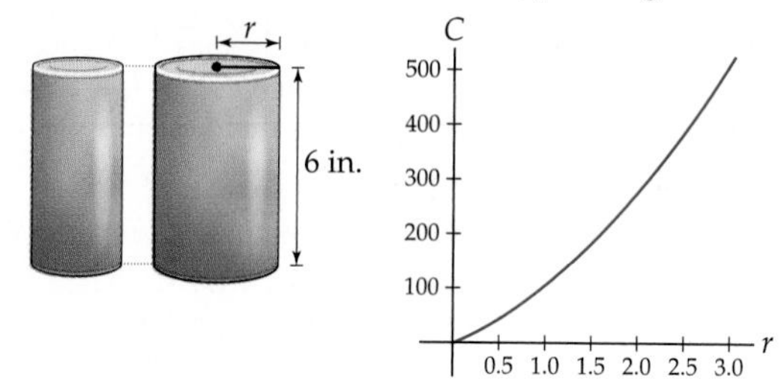

(a) Len's boss wants him to construct the cans so that the cost of each can is within 25 cents of \$4.00. Given these cost requirements, what is the acceptable range of values for r?

(b) Len's boss now says that he wants the cans to cost within 10 cents of \$4.00. Under these new cost requirements, what is the acceptable range of values for r?

(c) Interpret this problem in terms of δ and ϵ ranges. Specifically, what is c? What is L? What is ϵ for part (a) and part (b)? What are the corresponding values of δ? Illustrate these values of c, L, ϵ, and δ on a graph of $C(r)$.

67. You work for a company that sells velvet Elvis paintings. The function $N(p) = 9.2p^2 - 725p + 16,333$ predicts the number N of velvet Elvis paintings that your company will sell if they are priced at p dollars each, and is shown in the following graph. The Presley estate does not allow you to charge more than \$50 per painting.

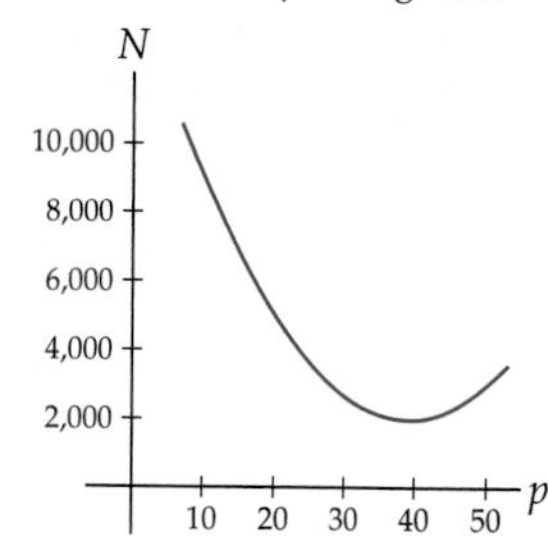

(a) Use a graphing utility to estimate the price your company should charge per painting if it wishes to sell 6000 velvet Elvis paintings.

(b) Find the range of prices that would enable your company to sell between 5000 and 7000 velvet Elvis paintings.

(c) Interpret this problem in terms of δ and ϵ ranges. Specifically, what is c? What is L? What is ϵ? What is the corresponding value of δ? Illustrate these values of c, L, ϵ, and δ on a graph of $N(p)$.

Proofs

Prove the four limit statements in Exercises 68–71. In the next section we will present a systematic method for such proofs.

68. Prove that $\lim_{x\to 1} 3x = 3$, with these steps:

(a) What is the δ–ϵ statement that must be shown to prove that $\lim_{x\to 1} 3x = 3$?

(b) Argue that $x \in (1-\delta, 1) \cup (1, 1+\delta)$ if and only if $-\delta < x - 1 < \delta$, with $x \neq 1$. Then use algebra to show that this means that $0 < |x-1| < \delta$.

(c) Argue that $3x \in (3-\epsilon, 3+\epsilon)$ if and only if $-\epsilon < 3(x-1) < \epsilon$. Then use algebra to show that this means that $3|x-1| < \epsilon$.

(d) Given any particular $\epsilon > 0$, what value of δ would guarantee that if $0 < |x-1| < \delta$, then $3|x-1| < \epsilon$? Your answer will depend on ϵ.

(e) Put the previous four parts together to prove the limit statement.

69. Prove that $\lim_{x\to 2}(7-x) = 5$, with these steps:

(a) What is the δ–ϵ statement that must be shown to prove that $\lim_{x\to 2} 7 - x = 5$?

(b) Argue that $x \in (2-\delta, 2) \cup (2, 2+\delta)$ if and only if $-\delta < x - 2 < \delta$. Then use algebra to show that this means that $0 < |x-2| < \delta$.

(c) Argue that $7 - x \in (5-\epsilon, 5+\epsilon)$ if and only if $-\epsilon < 2 - x < \epsilon$. Then use algebra to show that this means that $|x-2| < \epsilon$.

(d) Given any particular $\epsilon > 0$, what value of δ would guarantee that if $0 < |x-2| < \delta$, then $|x-2| < \epsilon$? Your answer will depend on ϵ.

(e) Put the previous four parts together to prove the limit statement.

70. Prove that $\lim_{x\to 0^+} \frac{1}{x} = \infty$, with these steps:

(a) What is the M–δ statement that must be shown to prove that $\lim_{x\to 0^+} \frac{1}{x} = \infty$?

(b) Argue that $x \in (0, 0+\delta)$ if and only if $0 < x < \delta$.

(c) Argue that $\frac{1}{x} \in (M, \infty)$ if and only if $x < \frac{1}{M}$. You may assume that $M > 0$.

(d) Given any particular $M > 0$, what value of δ would guarantee that if $0 < x < \delta$, then $x < \frac{1}{M}$? Your answer will depend on M.

(e) Put the previous four parts together to prove the limit statement.

71. Prove that $\lim_{x\to\infty} \frac{1}{x} = 0$, with these steps:

(a) What is the ϵ–N statement that must be shown to prove that $\lim_{x\to\infty} \frac{1}{x} = 0$?

(b) Argue that $x \in (N, \infty)$ if and only if $x > N$.

(c) Argue that $\frac{1}{x} \in (0-\epsilon, 0+\epsilon)$ if and only if $-\epsilon < \frac{1}{x} < \epsilon$. Then argue that for this limit, it suffices to consider $0 < \frac{1}{x} < \epsilon$.

(d) Given any particular ϵ, what value of $N > 0$ would guarantee that if $x > N$, then $0 < \frac{1}{x} < \epsilon$? Your answer will depend on ϵ.

(e) Put the previous four parts together to prove the limit statement.

Thinking Forward

Continuity: As you have already seen, sometimes $\lim_{x\to c} f(x)$ is equal to $f(c)$, and sometimes it is not. As we will see in Section 1.4, when the limit of a function f as $x \to c$ does happen to be equal to the value of $f(x)$ at $x = c$, we say that the function f is ***continuous*** at $x = c$.

- You may have heard the following loose, only partially accurate definition of continuity in a previous class: A function is continuous if you "can draw it without picking up your pencil." Why does it make sense that this would be related to the definition just presented of continuity in terms of limits?
- State the limit-definition of continuity with a formal δ–ϵ statement.
- The function $f(x) = x^2$ is continuous at every point. Use this fact and the formal definition of continuity to calculate $\lim_{x\to 2} x^2$, $\lim_{x\to 5} x^2$, and $\lim_{x\to -4} x^2$.

Limits of Sequences: We say that an infinite sequence of real numbers $a_1, a_2, a_3, \dots, a_k, \dots$ ***converges to a limit*** L, and we write $\lim_{k\to\infty} s_k = L$, if for all $\epsilon > 0$, there exists some $N > 0$ such that if $k > N$, then $|a_k - L| < \epsilon$.

- Use algebra to solve the inequality $|a_k - L| < \epsilon$ for a_k. Your answer should be in the form $a_k \in$ ____, where the blank is filled in with interval notation.
- Relate the definition of the convergence of a sequence to the definition of a limit at infinity.

1.3 DELTA-EPSILON PROOFS*

- Developing an equivalent algebraic definition of limits from our geometric definition
- Finding delta in terms of epsilon so that we can prove a limit statement
- The formal logic of writing delta–epsilon proofs

Describing Limits with Absolute Value Inequalities

In Definition 1.5, we formally defined the limit statement $\lim_{x\to c} f(x) = L$ to mean that for all $\epsilon > 0$, there exists $\delta > 0$ such that whenever $x \in (c - \delta, c) \cup (c, c + \delta)$, we can guarantee that $f(x) \in (L - \epsilon, L + \epsilon)$. This definition of limit has a very geometric flavor, since it is stated in the language of ϵ-intervals and punctured δ-intervals. That kind of language is useful when looking at specific values of ϵ or δ, but not as useful when trying to prove that *every* value of ϵ has a corresponding value of δ. For the purposes of proving limit statements, we give the following algebraic definition of limit and prove that it is equivalent to our previous geometric definition:

DEFINITION 1.10 Algebraic Definition of Limit

The ***limit*** $\lim_{x\to c} f(x) = L$ means that for all $\epsilon > 0$, there exists $\delta > 0$ such that

$$\text{if } 0 < |x - c| < \delta, \text{ then } |f(x) - L| < \epsilon.$$

THEOREM 1.11 **Equivalence of Geometric and Algebraic Definitions of Limit**

The two definitions of limit in Definition 1.5 and 1.10 are equivalent. Specifically,

(a) $x \in (c - \delta, c) \cup (c, c + \delta)$ if and only if $0 < |x - c| < \delta$;

(b) $f(x) \in (L - \epsilon, L + \epsilon)$ if and only if $|f(x) - L| < \epsilon$.

Proof. We will begin by proving part (b), since it is slightly easier. The statement $f(x) \in (L - \epsilon, L + \epsilon)$ means that $L - \epsilon < f(x) < L + \epsilon$. Subtracting L from all three parts of this double inequality, we get $-\epsilon < f(x) - L < \epsilon$. This is precisely the solution of the inequality $|f(x) - L| < \epsilon$.

For part (a) we have a similar situation, except that we must deal with a punctured interval. The statement $x \in (c - \delta, c) \cup (c, c + \delta)$ means that $c - \delta < x < c + \delta$ and $x \neq c$. Subtracting c from all three parts of the double inequality, we get $-\delta < x - c < \delta$, which is the solution set for the inequality $|x - c| < \delta$. The fact that $x \neq c$ means that $x - c \neq 0$. This is equivalent to saying that $|x - c| \neq 0$, and since the absolute value of a number is always positive or zero, it is also equivalent to saying that $|x - c| > 0$. Therefore $x \in (c - \delta, c) \cup (c, c + \delta)$ if and only if $0 < |x - c| < \delta$. ■

Finding a Delta for Every Epsilon

With our new algebraic definition of limit, we have the final tool we need to be able to effectively *prove* limit statements. This is the first step towards being able to *calculate* limits, something that, perhaps surprisingly, we do not yet know how to do.

Consider for example the limit statement $\lim_{x \to 2}(3x - 1) = 5$. We can examine this limit with a table of values, noticing that as the values of x approach $x = 2$ from the left and the right, the corresponding values of $3x - 1$ approach 5:

x	1.9	1.99	1.999	2	2.001	2.01	2.1
$3x - 1$	4.7	4.97	4.997	-	5.003	5.03	5.3

We can also investigate the limit statement $\lim_{x \to 2}(3x - 1) = 5$ with the following graph of $f(x) = 3x - 1$ at the left, noticing that a sequence of values of x approaching $x = 2$ from either the left or the right determines a sequence of values of $f(x)$ that approach $y = 5$:

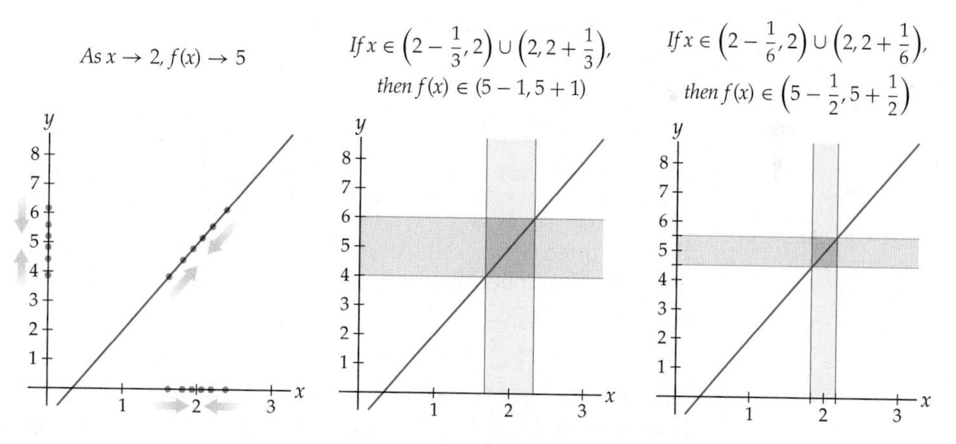

We can even show that the formal geometric definition of limit holds for a particular ϵ, by finding some δ for which $x \in (2 - \delta, 2) \cup (2, 2 + \delta)$ guarantees that $3x - 1 \in (5 - \epsilon, 5 + \epsilon)$. For example, it turns out that if $\epsilon = 1$, then $\delta = \frac{1}{3}$ will work, as illustrated in the middle figure shown, and if $\epsilon = \frac{1}{2}$, then $\delta = \frac{1}{6}$ will work, as in the rightmost figure.

Although the preceding investigations are encouraging evidence, none of them *prove* that $\lim_{x \to 2}(3x - 1)$ is equal to 5. In the table, perhaps we did not consider values of x close enough to 2 to see the true behavior of $f(x) = 3x - 1$, or perhaps the limit is actually 5.0001 and not 5. The three graphs shown each have the same possible problems. To prove that $\lim_{x \to 2}(3x - 1) = 5$ we would need to know for certain that we could get values of $f(x)$ *arbitrarily close* to $y = 5$ by choosing values of x that are sufficiently close to $x = 2$. For *every* epsilon we would have to be able to find a delta.

Specifically, in this example, we need to show that for any choice of $\epsilon > 0$, we can find some $\delta > 0$ such that whenever a real number x satisfies $0 < |x - 2| < \delta$, we can also guarantee that x satisfies $|(3x-1)-5| < \epsilon$. Notice that $|(3x-1)-5| = |3x-6| = 3|x-2|$, so what we really need is to conclude that $3|x-2| < \epsilon$, i.e., that $|x-2| < \frac{\epsilon}{3}$. But that is easy if we are allowed to choose a different δ for each ϵ: Given any $\epsilon > 0$, simply choose $\delta = \frac{\epsilon}{3}$. Then whenever $0 < |x-2| < \delta$, we also have $|x-2| < \frac{\epsilon}{3}$, as desired. Notice that in both the second and third of the graphs shown previously, we did in fact choose δ to be one-third of ϵ. We have just *proved* the limit statement $\lim\limits_{x \to 2}(3x-1) = 5$ by showing that *every* value of ϵ has a corresponding value of $\delta = \frac{\epsilon}{3}$ for which the formal algebraic definition of limit holds.

Writing Delta-Epsilon Proofs

We have just proved that $\lim\limits_{x \to 2}(3x-1) = 5$, but our proof meandered about in a paragraph of discussion. We now present a systematic way to write up delta-epsilon proofs for limit statements. Remember that in our example what we must show is the following doubly quantified logical implication:

"For all $\epsilon > 0$, there exists a $\delta > 0$ such that if $0 < |x-2| < \delta$, then $|(3x-1)-5| < \epsilon$."

Logically, to prove such a statement we must first let ϵ be an arbitrary positive number and then choose a value for δ in terms of ϵ. We must then show that for all values of x with $0 < |x-2| < \delta$, we can also say that $|(3x-1)-5| < \epsilon$.

In our example, we already know that for any $\epsilon > 0$ we should choose $\delta = \frac{\epsilon}{3}$. Using this fact, we could arrange our proof very concisely as follows:

Proof. Given $\epsilon > 0$, choose $\delta = \frac{\epsilon}{3}$. For all x with $0 < |x-2| < \delta$, we have

$$|(3x-1)-5| = |3x-6| = |3(x-2)| = 3|x-2| < 3\delta \overset{\star}{=} 3\left(\frac{\epsilon}{3}\right) = \epsilon.$$

Therefore whenever $0 < |x-2| < \delta$, we also have $|(3x-1)-5| < \epsilon$. ■

The first three equals signs in the proof are simple algebra. The less-than step followed from our assumption that $|x-2| < \delta$. The starred equality followed from the fact that $\delta = \frac{\epsilon}{3}$.

In general we will not know at the outset what to choose for δ, as we did just now. In those cases we can either do a side-calculation to find δ in advance or just leave a blank space for δ and continue with the proof. When we get to the starred equality, after using the assumption $0 < |x-c| < \delta$, we will be able to see what to choose for δ and can fill in the blank as if we knew it all along; see Example 2. Throughout the examples we will try our hand at proving all kinds of limit statements, including one-sided limits, infinite limits, and limits at infinity.

Examples and Explorations

EXAMPLE 1 **Finding delta as a function of epsilon**

Find formulas for δ in terms of ϵ for each of the following limit statements:

(a) $\lim\limits_{x \to 4}(2x+1) = 9$ **(b)** $\lim\limits_{x \to 2}(x^2 - 4x + 5) = 1$

Then use those formulas to find punctured δ-intervals for $\epsilon = 1$ and $\epsilon = 0.5$.

SOLUTION

(a) The limit statement $\lim_{x\to 4}(2x+1) = 9$ means that given any $\epsilon > 0$, there is some $\delta > 0$ so that if $0 < |x-4| < \delta$, then we can conclude that $|(2x+1)-9| < \epsilon$. We have

$$|(2x+1)-9| = |2x-8| = |2(x-4)| = 2|x-4|,$$

so $|(2x+1)-9| < \epsilon$ when $2|x-4| < \epsilon$, i.e., when $|x-4| < \frac{\epsilon}{2}$. Therefore we should choose $\delta = \frac{\epsilon}{2}$. In particular, when $\epsilon = 1$, we have $\delta = \frac{1}{2} = 0.5$ and punctured δ-interval $(3.5, 4) \cup (4, 4.5)$. When $\epsilon = 0.5$, we choose $\delta = \frac{0.5}{2} = 0.25$, which gives us the punctured δ-interval $(3.75, 4) \cup (4, 4.25)$.

(b) The limit statement $\lim_{x\to 2}(x^2-4x+5) = 1$ means that for any $\epsilon > 0$, we can find $\delta > 0$ so that whenever $0 < |x-2| < \delta$, we can conclude that $|(x^2-4x+5)-1| < \epsilon$. Notice that

$$|(x^2-4x+5)-1| = |x^2-4x+4| = |(x-2)^2| = |x-2|^2,$$

which is clearly very closely related to our δ-inequality $0 < |x-2| < \delta$. In fact, the inequality $|x-2|^2 < \epsilon$ is equivalent to the inequality $|x-2| < \sqrt{\epsilon}$. Therefore we should choose $\delta = \sqrt{\epsilon}$. In particular, when $\epsilon = 1$, we should choose $\delta = \sqrt{1} = 1$ and thus punctured δ-interval $(1,2)\cup(2,3)$. When $\epsilon = 0.5$, we should choose $\delta = \sqrt{0.5} \approx 0.707$ and thus punctured δ-interval $(1.293, 2.707)$. □

✔ CHECKING THE ANSWER

To check that our formulas for δ in terms of ϵ are reasonable, we can graph the functions and the punctured δ-intervals that we found in each case. For the limit $\lim_{x\to 4}(2x+1) = 9$, $\delta = 0.5$ looks right for $\epsilon = 1$ and $\delta = 0.25$ looks right for $\epsilon = 0.5$:

If $x \in (4-0.5, 4) \cup (4, 4+0.5)$, then $f(x) \in (9-1, 9+1)$

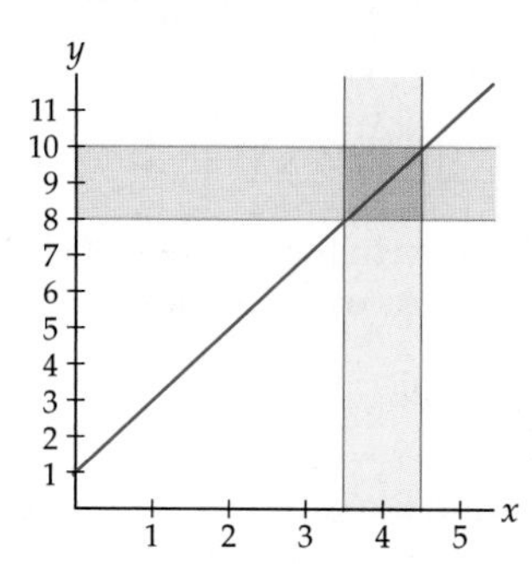

If $x \in (4-0.25, 4) \cup (4, 4+0.25)$, then $f(x) \in (9-0.5, 9+0.5)$

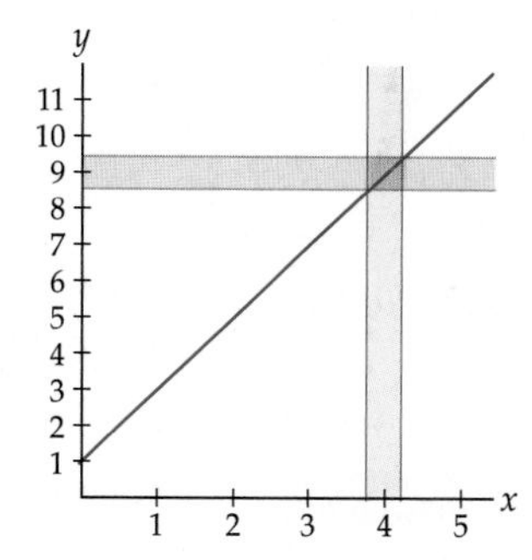

For $\lim_{x\to 2}(x^2-4x+5) = 1$, $\delta = 1$ looks right for $\epsilon = 1$ and $\delta \approx 0.707$ looks right for $\epsilon = 0.5$:

If $x \in (2-1, 2) \cup (2, 2+1)$, then $f(x) \in (1-1, 1+1)$

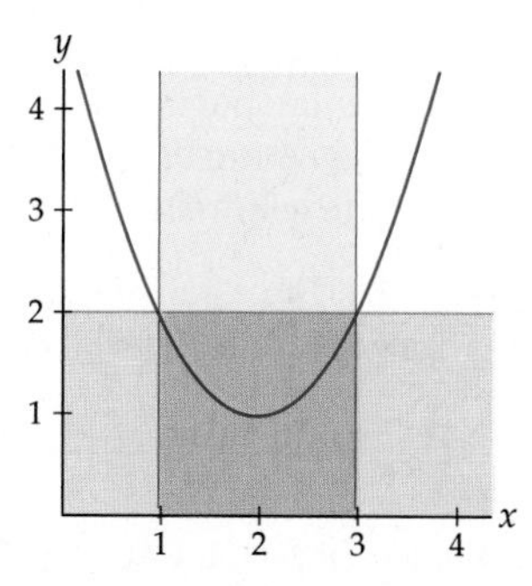

If $x \in (2-0.707, 2) \cup (2, 2+0.707)$, then $f(x) \in (1-0.5, 1+0.5)$

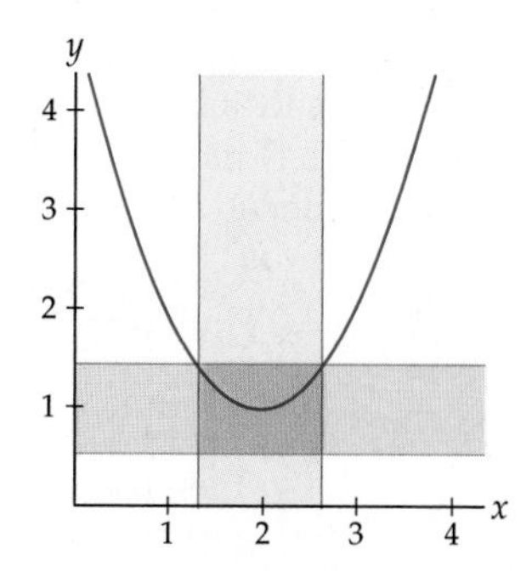

EXAMPLE 2 **Writing basic delta–epsilon proofs**

Write delta–epsilon proofs for each of the following limit statements:

(a) $\lim_{x \to 4} (2x + 1) = 9$ **(b)** $\lim_{x \to 2} (x^2 - 4x + 5) = 1$

SOLUTION

(a) To show that $\lim_{x \to 4}(2x + 1) = 9$ we must start with an arbitrary $\epsilon > 0$, choose some $\delta > 0$, and then show that for all x with $0 < |x - 4| < \delta$, we also have $|(2x+1) - 9| < \epsilon$. From the previous example we know that we should choose $\delta = \frac{\epsilon}{2}$.

Proof. Given $\epsilon > 0$, choose $\delta = \frac{\epsilon}{2}$. For all x with $0 < |x - 4| < \delta$, we have

$$|(2x + 1) - 9| = |2x - 8| = |2(x - 4)| = 2|x - 4| < 2\delta \stackrel{\star}{=} 2\left(\frac{\epsilon}{2}\right) = \epsilon.$$

Therefore whenever $0 < |x - 4| < \delta$, we also have $|(2x + 1) - 9| < \epsilon$. ■

If we had not already known that we should choose $\delta = \frac{\epsilon}{2}$, we could have determined that choice when we reached the starred equality. At that point we had shown that $|(2x + 1) - 9|$ was less than 2δ. What we were trying to show was that $|(2x + 1) - 9|$ was less than ϵ. If 2δ were equal to ϵ, then we would be done; solving $2\delta = \epsilon$ for δ, we arrive at the choice $\delta = \frac{\epsilon}{2}$.

(b) To show that $\lim_{x \to 2}(x^2 - 4x + 5) = 1$ we must start with an arbitrary $\epsilon > 0$, choose some $\delta > 0$, and then show that for all x with $0 < |x - 2| < \delta$, we also have $|(x^2 - 4x + 5) - 1| < \epsilon$. From the previous example we know that we should choose $\delta = \sqrt{\epsilon}$.

Proof. Given $\epsilon > 0$, choose $\delta = \sqrt{\epsilon}$. For all x with $0 < |x - 2| < \delta$, we have

$$|(x^2 - 4x + 5) - 1| = |x^2 - 4x + 4| = |(x - 2)^2| = |x - 2|^2 < \delta^2 \stackrel{\star}{=} (\sqrt{\epsilon})^2 = \epsilon.$$

Thus we can conclude that whenever $0 < |x - 2| < \delta$, we also have $|(x^2 - 4x + 5) - 1| < \epsilon$. ■

Again, if we had not already known that we should choose $\delta = \sqrt{\epsilon}$, then at the starred equality we would have δ^2 where we wish to have ϵ. Solving $\delta^2 = \epsilon$ for δ, we get our choice of $\delta = \sqrt{\epsilon}$. □

EXAMPLE 3 **Proofs for one-sided and infinite limits**

Give formal proofs for each of the following limit statements:

(a) $\lim_{x \to 1^+} \sqrt{x - 1} = 0$ **(b)** $\lim_{x \to \infty} \frac{1}{x} = 0$ **(c)** $\lim_{x \to 3^-} \frac{1}{3 - x} = \infty$

SOLUTION

(a) The limit statement $\lim_{x \to 1^+} \sqrt{x - 1} = 0$ means that for all $\epsilon > 0$, there exists $\delta > 0$ such that if $x \in (1, 1 + \delta)$, then $|\sqrt{x - 1} - 0| < \epsilon$. Other than a small bit of extra work to translate the meaning of $x \in (1, 1 + \delta)$, this is a standard delta-epsilon proof.

Proof. Given $\epsilon > 0$, choose $\delta = \epsilon^2$. (The reason for that choice of δ will become clear at the starred equality that follows.) Suppose $x \in (1, 1 + \delta)$. Then $1 < x < 1 + \delta$, which means that $0 < x - 1 < \delta$. We then have

$$|\sqrt{x - 1} - 0| = |\sqrt{x - 1}| = \sqrt{x - 1} < \sqrt{\delta} = \sqrt{\epsilon^2} = \epsilon.$$

Thus we can conclude that whenever $x \in (1, 1 + \delta)$, we also have $|\sqrt{x - 1} - 0| < \epsilon$. ■

(b) Geometrically, the limit statement $\lim_{x\to\infty} \frac{1}{x} = 0$ means that for all $\epsilon > 0$, there exists $N > 0$ such that if $x \in (N, \infty)$, then $\frac{1}{x} \in (0-\epsilon, 0+\epsilon)$. Algebraically, the implication after the quantifiers can be rewritten as follows: If $x > N$, then $\left|\frac{1}{x} - 0\right| < \epsilon$. This is what we shall prove.

> ***Proof.*** Given $\epsilon > 0$, choose $N = \frac{1}{\epsilon}$. (The reason for that choice of N will become clear at the starred equality that follows.) If $x > N$, then $\frac{1}{x} < \frac{1}{N}$ and x is positive, and therefore
>
> $$\left|\frac{1}{x} - 0\right| = \left|\frac{1}{x}\right| = \frac{1}{x} < \frac{1}{N} = \frac{1}{\frac{1}{\epsilon}} = \epsilon.$$
>
> Thus we can conclude that whenever $x > N$, we also have $\left|\frac{1}{x} - 0\right| < \epsilon$. ■

(c) Geometrically, the limit $\lim_{x\to 3^-} \frac{1}{3-x} = \infty$ means that for all $M > 0$, there exists $\delta > 0$ such that if $x \in (3-\delta, 3)$, then $\frac{1}{3-x} \in (M, \infty)$. Note that $x \in (3-\delta, 3)$ means that $3 - \delta < x < 3$ and therefore that $-\delta < x - 3 < 0$. Multiplying by -1 and flipping inequalities, this becomes $\delta > -(x-3) > 0$, or equivalently, $0 < 3 - x < \delta$. Hence the implication in our limit statement can be expressed as follows: If $0 < 3 - x < \delta$, then $\frac{1}{3-x} > M$. This is what we will prove:

> ***Proof.*** Given $M > 0$, choose $\delta = \frac{1}{M}$. (As usual, the reason for this choice will become clear at the starred equality that follows.) For all x with $0 < 3 - x < \delta$, we have $\frac{1}{3-x} > \frac{1}{\delta}$, and therefore
>
> $$\frac{1}{3-x} > \frac{1}{\delta} = \frac{1}{\frac{1}{M}} = M.$$
>
> Thus we can conclude that whenever $0 < 3 - x < \delta$, we also have $\frac{1}{3-x} > M$. ■

EXAMPLE 4

A delta–epsilon proof where it is necessary to bound delta from above

Write a delta-epsilon proof for the limit statement $\lim_{x\to 2} 5x^4 = 80$.

SOLUTION

To prove the limit statement $\lim_{x\to 2} 5x^4 = 80$, we must show that for all $\epsilon > 0$, there exists a choice of $\delta > 0$ such that whenever $0 < |x - 2| < \delta$, we also have $|5x^4 - 80| < \epsilon$. In our previous delta-epsilon proofs, the algebra has always magically worked out nicely. In this example there will be a point at which we get stuck. What will get us out of the jam will be assuming that δ is no larger than 1. This assumption will allow us to put a bound on an expression that would otherwise be in our way. For that reason, our choice of δ in this proof will be the minimum of 1 and an expression that depends on epsilon.

> ***Proof.*** Given $\epsilon > 0$, choose $\delta = \min\left(1, \frac{\epsilon}{325}\right)$, i.e., the smaller of 1 and $\frac{\epsilon}{325}$. (The reason for this elaborate choice of δ will be made clear at the two starred inequalities that follow.) If x is a real number with $0 < |x - 2| < \delta$, then we have
>
> $$\begin{aligned} |5x^4 - 80| &= |5(x^4 - 16)| = 5|(x^2 - 4)(x^2 + 4)| \\ &= 5|(x-2)(x+2)(x^2+4)| \\ &= 5|x-2| \cdot |(x+2)(x^2+4)| \\ &< 5\delta \cdot |(x+2)(x^2+4)|. \end{aligned}$$

Because our choice of δ ensures that δ is at most 1, we can say that $0 < |x-2| \leq 1$ and therefore that x is between 1 and 3. This means that we can bound the troublesome quantity $|(x+2)(x^2+4)|$ from above; it is largest when $x = 3$, so we know that $|(x+2)(x^2+4)|$ is at most $|(3+2)(3^2+4)| = 65$. Combining this result with the work we just did, we have (using our choice of $\delta = \min\left(1, \frac{\epsilon}{325}\right)$ at the two starred inequalities)

$$|5x^4 - 80| \overset{\star}{<} 5\delta \cdot 65 = 325\,\delta \overset{\star}{\leq} 325\left(\frac{\epsilon}{325}\right) = \epsilon.$$

Thus we can conclude that whenever $0 < |x-2| < \delta$, we also have $|5x^4 - 80| < \epsilon$. ■

? TEST YOUR UNDERSTANDING

- Why do we say that Definition 1.5 from the previous section is geometric in nature while we say that Definition 1.10 is algebraic? In what situations might one definition be more useful than another?
- How can we express the formal definitions for one-sided limits, infinite limits, and limits at infinity "algebraically," in the spirit of Definition 1.10?
- Suppose that we have a limit of the form $\lim_{x \to c} f(x) = L$ and that we can find a value of δ for some given value of ϵ. Why does this not prove definitively that $\lim_{x \to c} f(x) = L$?
- In a delta-epsilon proof, how do we come up with a choice for δ in terms of ϵ?
- Why is it sometimes necessary to require that $\delta \leq 1$ in a delta-epsilon proof?

EXERCISES 1.3

Thinking Back

Inequalities: Find the solution sets of each of the following inequalities.

- $0 < |x-2| < 0.5$
- $0 < |x+5| < 0.1$
- $|x^2 - 4| < 0.5$
- $|(3x+1) - 10| < 1$
- $\dfrac{1}{x^2} < 0.01$
- $\dfrac{1}{x^2} > 1000$

Logical implications: Suppose A and B are statements and that the implication "If A, then B" holds.

- If A is true, then what, if anything, can you conclude about B, and why?
- If A is false, then what, if anything, can you conclude about B, and why?
- If B is true, then what, if anything, can you conclude about A, and why?

Concepts

0. *Problem Zero:* Read the section and make your own summary of the material.

1. *True/False:* Determine whether each of the statements that follow is true or false. If a statement is true, explain why. If a statement is false, provide a counterexample.

(a) *True or False:* If $x \neq c$, then $|x - c|$ is strictly greater than zero.

(b) *True or False:* If $|x-c|$ is strictly greater than zero, then $x \neq c$.

(c) *True or False:* x is a solution of $0 < |x-c| < \delta$ if and only if $c - \delta < x < c + \delta$.

(d) *True or False:* If $0 < |x-c| < \delta$, then $x \in (c-\delta, c) \cup (c, c+\delta)$.

(e) *True or False:* If $|f(x)-L| < \epsilon$, then $L-\epsilon < f(x) < L+\epsilon$.

(f) *True or False:* If $f(x) \in (L-\epsilon, L+\epsilon)$, then $0 < f(x) < |L+\epsilon|$.

(g) *True or False:* The fact that $0 < |x-3| < 0.25$ guarantees that $|(2x-1)-5| < 0.5$ proves that $\lim_{x \to 3}(2x-1) = 5$.

(h) *True or False:* $\lim_{x \to 3}(2x-1) = 5$ means that for all $\delta > 0$ there is some $\epsilon > 0$ such that if $0 < |x-c| < \delta$, then $|(2x-1)-5| < \epsilon$.

2. *Examples:* Construct examples of the thing(s) described in the following. Try to find examples that are different than any in the reading.

(a) A function f with values given in the following table but whose limit as $x \to 2$ is *not* equal to 5:

x	1.9	1.99	1.999	2	2.001	2.01	2.1
$f(x)$	4.7	4.97	4.997	-	5.003	5.03	5.3

(b) An inequality involving absolute values whose solution set is $(2.75, 3) \cup (3, 3.25)$.

(c) An inequality involving absolute values whose solution set is $(-0.01, 0) \cup (0, 0.01)$.

3. Write down the formal delta-epsilon statement you would have to prove in order to prove the limit statement $\lim_{x \to -2} \frac{3}{x+1} = -3$.

4. Suppose you show that $|(1-2x) - (-5)| < 0.05$ for all x with $0 < |x-3| < 0.025$. Explain why this does *not* prove that $\lim_{x \to 3}(1-2x) = -5$.

5. Write down a mathematical equation that expresses the sentence "x is not equal to 5, and the distance between x and 5 is less than 0.01." Then write an equation that means "the distance between $f(x)$ and -2 is less than 0.5."

6. Why do we have $0 < |x-c| < \delta$ instead of just $|x-c| < \delta$ in Definition 1.10?

Write each of the following inequalities in interval notation:

7. $0 < |x-2| < 0.1$

8. $0 < |x+3| < 0.05$

9. $|(x^2-1)+3| < 0.5$

10. $|(3x+1)-2| < 0.1$

11. $|f(x) - L| < \epsilon$

12. $0 < |x-c| < \delta$

Determine whether each implication that follows is true or false. Use graphs to justify any implications that are true, and counterexamples for any implications that are false.

13. If $0 < |x-2| < 1$, then $|x^2-4| < 0.5$.

14. If $0 < |x-2| < 0.2$, then $|x^2-4| < 1$.

15. If $0 < |x-0| < 0.75$, then $|x^2-0| < 0.5$.

16. If $0 < |x+2| < 0.1$ then $|x^2-4| < 0.4$.

17. If $0 < |x+2| < 0.075$, then $|x^2-4| < 0.4$.

18. In Example 2 we proved that $\lim_{x \to 2}(x^2-4x+5) = 1$. Use the proof to find values of δ corresponding to (a) $\epsilon = 1$, (b) $\epsilon = 0.1$, and (c) $\epsilon = 0.01$. Illustrate that your choices of δ work by examining a graph of $f(x) = x^2 - 4x + 5$ and sketching appropriate ϵ and δ intervals.

19. In Example 4 we proved that $\lim_{x \to 2} 5x^4 = 80$. Use the proof to find values of δ corresponding to (a) $\epsilon = 5$, (b) $\epsilon = 0.01$, and (c) $\epsilon = 350$. Illustrate that your choices of δ work by examining a graph of $f(x) = 5x^4$ and sketching appropriate ϵ and δ intervals.

20. Use algebra to solve the inequality $0 < |x-c| < \delta$ and show that its solution set is $x \in (c-\delta, c) \cup (c, c+\delta)$.

21. Use algebra to solve the inequality $|f(x) - L| < \epsilon$ and show that its solution set is $f(x) \in (L-\epsilon, L+\epsilon)$.

22. Suppose $f(x) = mx + b$ is a linear function with $m \neq 0$, and let c be any real number.

(a) Show that for all $\epsilon > 0$, if $0 < |x-c| < \frac{\epsilon}{|m|}$, then $|f(x) - f(c)| < \epsilon$.

(b) What does the implication in part (a) have to do with limits?

(c) Illustrate the implication in part (a) with a labeled graph. Explain in terms of slopes why it makes sense that given $\epsilon > 0$, the corresponding $\delta > 0$ is $\delta = \frac{\epsilon}{|m|}$.

Skills

Use algebra to find the largest possible value of δ or smallest possible value of N that makes each implication in Exercises 23–28 true. Then verify and support your answers with labeled graphs.

23. If $0 < |x-2| < \delta$, then $|(3x-1) - 5| < 0.25$.

24. If $0 < |x-3| < \delta$, then $\left|\frac{1}{x} - \frac{1}{3}\right| < 0.2$.

25. If $x \in (1, 1+\delta)$, then $\left|\sqrt{x-1} - 0\right| < 0.5$.

26. If $x \in (3-\delta, 3)$, then $\frac{1}{3-x} > 1000$.

27. If $x > N$, then $\left|\frac{1}{x^2} - 0\right| < 0.001$.

28. If $x > N$, then $1 - 2x < -500$.

For each limit statement $\lim_{x \to c} f(x) = L$ in Exercises 29–40, use algebra to find $\delta > 0$ in terms of $\epsilon > 0$ so that if $0 < |x-c| < \delta$, then $|f(x) - L| < \epsilon$.

29. $\lim_{x \to 3}(x+5) = 8$

30. $\lim_{x \to -2}(4-2x) = 8$

31. $\lim_{x \to 0}(3-4x) = 3$

32. $\lim_{x \to 1}(3x+8) = 11$

33. $\lim_{x \to 0}(5x^2-1) = -1$

34. $\lim_{x \to 3}(x^2-6x+5) = -4$

35. $\lim_{x \to 2}(x^2-4x+6) = 2$

36. $\lim_{x \to 0}(x^3+1) = 1$

37. $\lim_{x \to 2} \frac{1}{x} = \frac{1}{2}$; you may assume $\delta \leq 1$

38. $\lim_{x \to 3} \frac{1}{x} = \frac{1}{3}$; you may assume $\delta \leq 1$

39. $\lim_{x \to 3}(x^2-2x-3) = 0$; you may assume $\delta \leq 1$

40. $\lim_{x \to 1} 2x^4 = 2$; you may assume $\delta \leq 1$

For each limit statement in Exercises 41–44, use algebra to find δ or N in terms of ϵ or M, according to the appropriate formal limit definition.

41. $\lim_{x \to -2^+}(1 + \sqrt{x+2}) = 1$, find δ in terms of ϵ

42. $\lim_{x \to \infty} \frac{x-1}{x} = 1$, find N in terms of ϵ

43. $\lim_{x \to 1^-} \frac{1}{1-x} = \infty$, find δ in terms of M

44. $\lim_{x \to \infty}(x^2+2) = \infty$, find N in terms of M

Applications

For Exercises 45 and 46, suppose you work for a company that manufactures gourmet soup cans. The material for the curved sides of the cans costs 0.25 cent (a quarter of a cent) per square

Cost of materials for producing a soup can

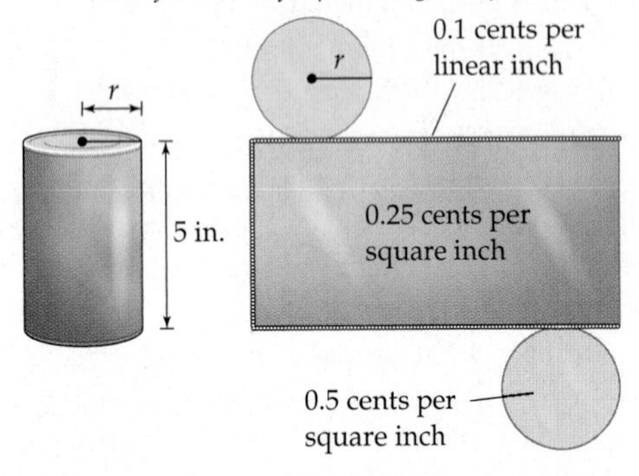

inch, the material for the top and bottom costs 0.5 cent per square inch, and the reinforcing weld around the seams costs 0.1 cent per linear inch. The seams run around the edges of the top and bottom and also in a straight line from the top to the bottom of the curved side.

45. Find a formula for the cost $C(r)$ of producing a gourmet soup can with radius r and height 5 inches, and answer the following questions:
(a) What is the radius of a can that is 5 inches tall and costs 30 cents to produce?
(b) Your manager wants you to produce 5-inch-tall cans that cost between 20 and 40 cents. Write this requirement as an absolute value inequality.
(c) What range of radii would satisfy your manager? Write an absolute value inequality whose solution set lies inside this range of radii.

46. Find a formula for the cost $C(h)$ of producing a gourmet soup can with height h and radius 2 inches, and answer the following questions:
(a) What is the height of a can that has radius 2 inches and costs 45 cents to produce?
(b) Your manager wants you to produce 2-inch-radius cans that cost between 40 and 50 cents. Write this requirement as an absolute value inequality.
(c) What range of heights would satisfy your manager? Write an absolute value inequality whose solution set lies within this range of heights.

Proofs

Write delta-epsilon proofs for each of the limit statements $\lim_{x\to c} f(x) = L$ in Exercises 47–60.

47. $\lim_{x\to 1}(2x+4) = 6$

48. $\lim_{x\to 2}(3-4x) = -5$

49. $\lim_{x\to -6}(x+2) = -4$

50. $\lim_{x\to -3}(1-x) = 4$

51. $\lim_{x\to 4}(6x-1) = 23$

52. $\lim_{x\to 8}(3x-11) = 13$

53. $\lim_{x\to 0}(3x^2+1) = 1$

54. $\lim_{x\to 3}(x^2-6x+11) = 2$

55. $\lim_{x\to 1}(2x^2-4x+3) = 1$

56. $\lim_{x\to 2}(3x^2-12x+15) = 3$

57. $\lim_{x\to 1}\dfrac{x^2-1}{x-1} = 2$

58. $\lim_{x\to 2}\dfrac{x^2-3x+2}{x-2} = 1$

59. $\lim_{x\to 5^+}\sqrt{x-5} = 0$

60. $\lim_{x\to 2^+} 3\sqrt{2x-4} = 0$

For each of the limit statements in Exercises 61–66, write a δ–M, N–ϵ, or N–M proof, according to the type of limit statement.

61. $\lim_{x\to -2^+}\dfrac{1}{x+2} = \infty$

62. $\lim_{x\to -2^-}\dfrac{1}{x+2} = -\infty$

63. $\lim_{x\to \infty}\dfrac{2x-1}{x} = 2$

64. $\lim_{x\to -\infty}\dfrac{2x-1}{x} = 2$

65. $\lim_{x\to \infty}(3x-5) = \infty$

66. $\lim_{x\to -\infty}(3x-5) = -\infty$

Prove each of the limit statements in Exercises 67–72. You will have to bound δ.

67. $\lim_{x\to 3}(x^2-2x-3) = 0$

68. $\lim_{x\to -1}(x^2-2x-3) = 0$

69. $\lim_{x\to 5}(x^2-6x+7) = 2$

70. $\lim_{x\to 1}(x^2-6x+7) = 2$

71. $\lim_{x\to 2}\dfrac{4}{x^2} = 1$

72. $\lim_{x\to 3}\dfrac{18}{x^2} = 2$

Thinking Forward

Calculating limits: We still do not have a way to calculate limits easily. In the following problems you will develop rules for calculating limits of some very simple functions.

▶ Explain why it makes intuitive sense that $\lim_{x\to c} x = c$ for any real number c. Then use a delta–epsilon argument to prove it.

▶ Explain why it makes intuitive sense that $\lim_{x\to c} x^2 = c^2$ for any real number c. Then use a delta–epsilon argument to prove it. *(Hint: You will need to assume that $\delta \le 1$.)*

▶ Use the preceding two problems and the result of Exercise 22 to calculate the following limits:

- $\lim_{x\to -1} x$
- $\lim_{x\to 4} x$
- $\lim_{x\to \pi} x$
- $\lim_{x\to 0} x^2$
- $\lim_{x\to 5} x^2$
- $\lim_{x\to -2} x^2$
- $\lim_{x\to 0}(2x-3)$
- $\lim_{x\to 1}(1-x)$
- $\lim_{x\to 3}(3x+1)$

▶ When calculating each of these limits $\lim_{x\to c} f(x)$, you simply used the value of $f(c)$. Will that method always work for any limit? Why or why not?

1.4 CONTINUITY AND ITS CONSEQUENCES

- Continuity of functions at points and on intervals, and basic types of discontinuities
- Simple functions that are continuous on their domains
- The Extreme Value Theorem and the Intermediate Value Theorem

Defining Continuity with Limits

Intuitively, a function is ***continuous*** if its graph has no breaks, jumps, or holes. Loosely speaking, you can sketch the graph of a continuous function "without picking up your pencil." We can make the notion of continuity more precise by using limits. For example, consider the following four graphs:

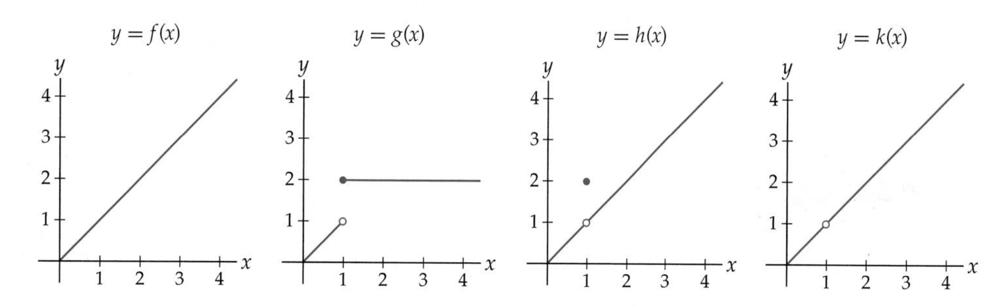

While the first graph has no breaks or holes, the remaining three graphs all have some sort of bad behavior at $x = 1$. It turns out that limits as $x \to 1$ detect exactly this bad behavior: In each case, the limit as $x \to 1$ is not the same as the value at $x = 1$. For example, $\lim_{x\to 1} g(x)$ does not exist, but $g(1) = 2$. For $h(x)$, $\lim_{x\to 1} h(x)$ is equal to 1, while the value $h(1)$ is equal to 2. Finally, for $k(x)$, the limit is $\lim_{x\to 1} k(x) = 1$ but the value $k(1)$ does not exist. On the other hand, for $f(x)$, we have both $\lim_{x\to 1} f(x) = 1$ and $f(1) = 1$.

The preceding examples suggest the following definition: A function is continuous at a point $x = c$ if its limit as $x \to c$ is equal to a real number that is the same as the value of the function at $x = c$.

DEFINITION 1.12

Continuity of a Function at a Point

A function f is ***continuous at*** $x = c$ if $\lim_{x\to c} f(x) = f(c)$.

By considering one-sided limits, we can get a more detailed picture of continuity. For example, with the previous function $g(x)$, the left limit as $x \to 1$ is not equal to the value $g(1)$ but the right limit is. We say that $g(x)$ is ***right continuous*** at $x = 1$ but not ***left continuous***.

DEFINITION 1.13

Left and Right Continuity at a Point

A function f is ***left continuous at*** $x = c$ if $\lim_{x\to c^-} f(x) = f(c)$ and is ***right continuous at*** $x = c$ if $\lim_{x\to c^+} f(x) = f(c)$.

Sometimes it is convenient to talk about continuity of a function on an interval. We say that a function is continuous on an open interval if it is continuous at each point in the interval. For non-open intervals we also require one-sided continuity as we approach any closed endpoints.

DEFINITION 1.14 Continuity of a Function on an Interval

A function f is ***continuous on an interval*** I if it is continuous at every point in the interior of I, right continuous at any closed left endpoint, and left continuous at any closed right endpoint.

The graphs that follow provide examples of continuity on the four possible types of bounded intervals. For example, a function f is continuous on $I = (1, 3]$ if it is continuous at every point in the interior $(1, 3)$ and left continuous at the right endpoint $x = 3$, as shown in the third figure. In terms of limits this means that $\lim_{x \to c} f(x) = f(c)$ for all $c \in (1, 3)$ and $\lim_{x \to 3^-} f(x) = f(3)$.

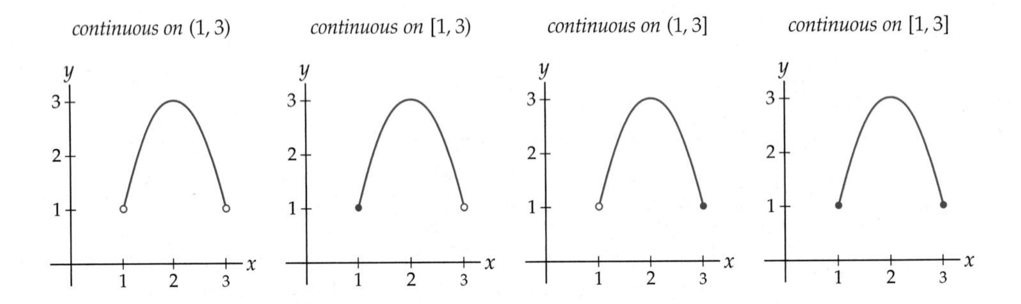

Types of Discontinuities

When a function is not continuous at a point $x = c$, we say that it is ***discontinuous*** at $x = c$. In terms of limits this means that the limit $\lim_{x \to c} f(x)$ is not equal to the value $f(c)$. The three most basic types of discontinuities that a function can have are illustrated as follows:

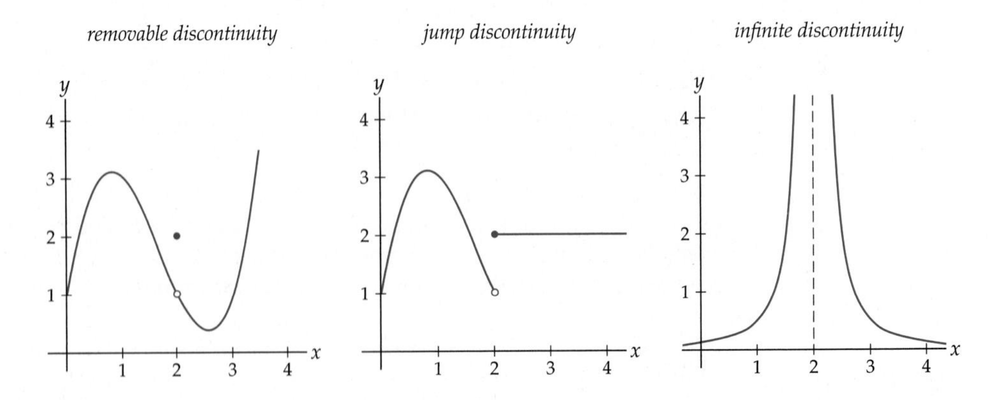

Intuitively, we say that a discontinuity is ***removable*** if we could remove it just by changing one function value. At a ***jump discontinuity***, the function jumps from one value to another, and at an ***infinite discontinuity*** the function has a vertical asymptote.

These types of discontinuities can be described precisely in terms of limits as follows:

DEFINITION 1.15

Removable, Jump, and Infinite Discontinuities

Suppose f is discontinuous at $x = c$. We say that $x = c$ is a

(a) ***removable discontinuity*** if $\lim_{x \to c} f(x)$ exists but is not equal to $f(c)$;

(b) ***jump discontinuity*** if $\lim_{x \to c^-} f(x)$ and $\lim_{x \to c^+} f(x)$ both exist but are not equal;

(c) ***infinite discontinuity*** if one or both of $\lim_{x \to c^-} f(x)$ and $\lim_{x \to c^+} f(x)$ is infinite.

For example, in the first figure shown, the limit $\lim_{x \to 2} f(x) = 1$ exists but is not equal to $f(2) = 2$, and therefore $f(x)$ has a removable discontinuity at $x = 2$. The function $g(x)$ in the second figure has left and right limits $\lim_{x \to 2^-} g(x) = 1$ and $\lim_{x \to 2^+} g(x) = 2$, respectively; the limits from both sides exist, but they are not equal to each other, and therefore $h(x)$ has a jump discontinuity at $x = 2$. Finally, the function $h(x)$ in the third graph has an infinite limit from both the left and the right at $x = 2$, and therefore has an infinite discontinuity at that point.

Continuity of Very Basic Functions

We say that a function is ***continuous on its domain*** if it is continuous on every interval on which it is defined. The following theorem proves that, unsurprisingly, our simplest examples of functions are continuous on their domains:

THEOREM 1.16

Continuity of Simple Functions

(a) Constant, identity, and linear functions are continuous everywhere. In terms of limits, for every k, c, m, and b in $\mathbb{R}$ we have

$$\lim_{x \to c} k = k, \quad \lim_{x \to c} x = c, \quad \text{and} \quad \lim_{x \to c}(mx + b) = mc + b.$$

(b) Power functions are continuous on their domains. In terms of limits, if A is real and k is rational, then for all values $x = c$ at which x^k is defined we have

$$\lim_{x \to c} Ax^k = Ac^k.$$

This is a powerful theorem, because it tells us that we can calculate limits of certain simple functions at domain points just by evaluating the functions at those points. For example, $f(x) = \frac{1}{x} = x^{-1}$ is a power function, so by the preceding theorem, it is continuous on its domain $(-\infty, 0) \cup (0, \infty)$. This means that at any point $c \neq 0$ we can calculate $\lim_{x \to c} \frac{1}{x}$ by simply calculating the value $f(c) = \frac{1}{c}$. However, the theorem does not tell us how to calculate $\lim_{x \to 0} \frac{1}{x}$; we will discuss such limits in a later section.

Technical point: In part (b) of the theorem the limit may sometimes be only one-sided. For example, $f(x) = x^{1/2}$ is defined at $x = 0$ and to the right of $x = 0$, but not for $x < 0$. Therefore the corresponding limit statement is one-sided: $\lim_{x \to 0^+} x^{1/2} = 0$.

Although it seems graphically obvious that the simple types of functions described in Theorem 1.16 are continuous everywhere they are defined, to actually prove continuity we need to appeal to the definition of limit. We will prove part (a) here and discuss the proof of part (b) after we learn about limit rules in the next section. You will prove that $f(x) = x^k$ is continuous on its domain for $k = 2, 3, -1, -2, \frac{1}{2}$, and $-\frac{1}{2}$ in Exercises 89–92.

Proof. *(This proof requires material covered in optional Section 1.3.)*
The limit $\lim_{x \to c} k = k$ makes intuitive sense because as x approaches c, the number k should simply remain k; there is no x involved. To prove this limit statement we must show that for all $\epsilon > 0$, there is some $\delta > 0$ such that if $x \in (c - \delta, c) \cup (c, c + \delta)$, then $k \in (k - \epsilon, k + \epsilon)$. But k is *always* in the interval $(k - \epsilon, k + \epsilon)$, so the implication is trivially true for all values of ϵ and δ. This is illustrated in the leftmost figure that follows.

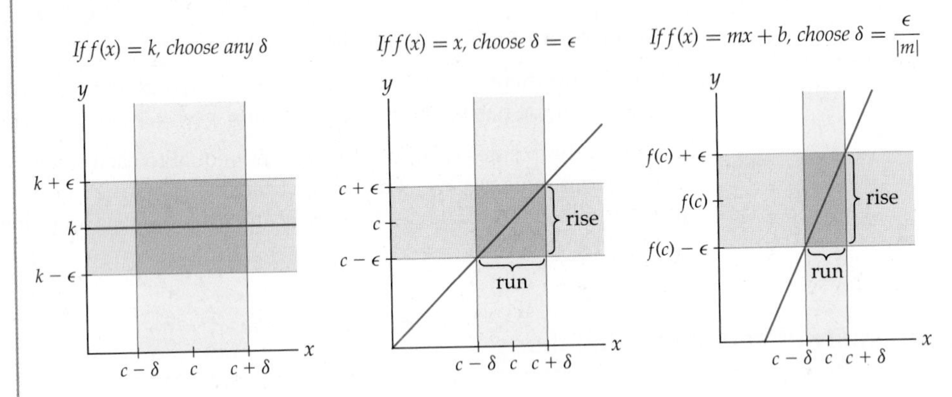

To prove that $\lim_{x \to c} x = c$ we must show that for all $\epsilon > 0$, there exists $\delta > 0$ such that if $x \in (c - \delta, c) \cup (c, c + \delta)$, then $x \in (L - \epsilon, L + \epsilon)$. This clearly holds if we choose δ to be equal to ϵ, as shown in the middle figure.

To prove that $\lim_{x \to c}(mx + b) = (mc + b)$ we will use the definition of limit in terms of absolute value inequalities from Definition 1.10. In the case when $m = 0$, the linear function $f(x) = mx + b$ is the constant function $f(x) = b$; we have already proved that case. If $m \neq 0$, then given $\epsilon > 0$, choose $\delta = \frac{\epsilon}{|m|}$. Then for all x satisfying $0 < |x - c| < \delta$ we also have

$$|(mx + b) - (mc + b)| = |mx - mc| = |m||x - c| < |m|\delta = |m|\left(\frac{\epsilon}{|m|}\right) = \epsilon.$$

It makes intuitive sense that δ should be $\frac{\epsilon}{m}$, since the ratio $\frac{\epsilon}{\delta}$ is equal to the ratio $\frac{\text{rise}}{\text{run}}$, or the slope of $f(x) = mx + b$, as shown in the rightmost figure. ■

Extreme and Intermediate Values of Continuous Functions

In this section we examine two important consequences of continuity. First, a continuous function on a closed interval must be bounded and attain its upper and lower bounds. Second, if f is continuous between two values $x = a$ and $x = b$, then the corresponding values of $f(x)$ go through every possible intermediate value between the y-values $f(a)$ and $f(b)$. Both of these consequences are intuitively obvious if we think of continuous functions as having "unbroken" graphs.

For example, consider the function f in the first figure that follows. This function is continuous on $[a, b]$. In the second graph we see that f attains its maximum on that interval at $x = M$ and its minimum at $x = m$. In the third graph we see that the function attains every intermediate value K between $f(a)$ and $f(b)$.

f is continuous on $[a, b]$

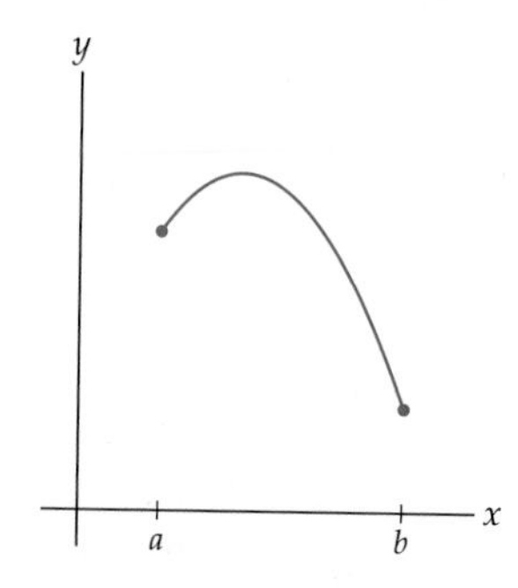

Maximum at M, minimum at m

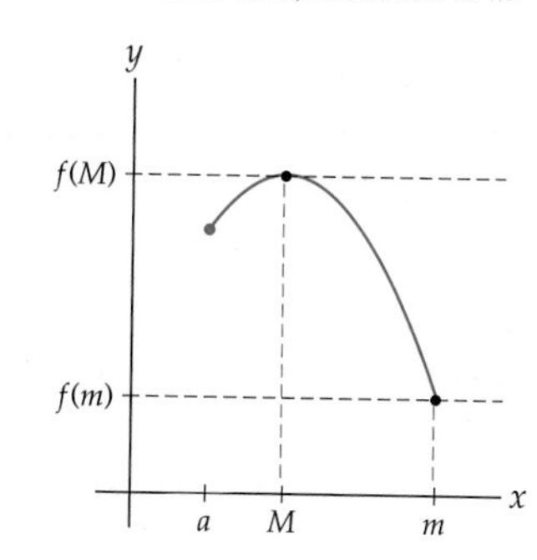

Height of $y = K$ at $x = c$

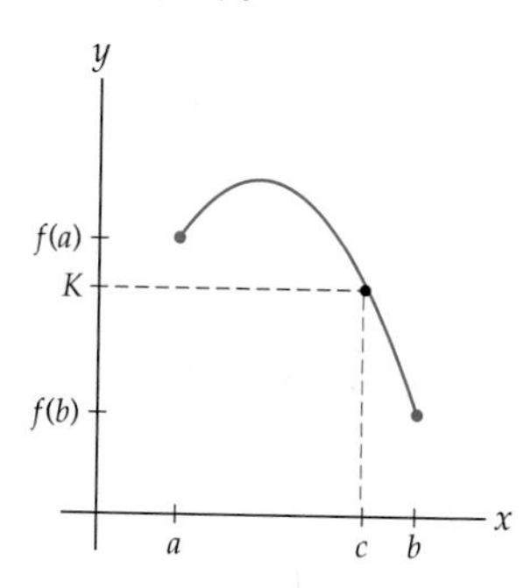

In general, attaining extreme values and passing through all intermediate values are properties that hold for every function that is continuous on a closed interval $[a, b]$:

THEOREM 1.17

The Extreme Value Theorem

If f is continuous on a closed interval $[a, b]$, then there exist values M and m in the interval $[a, b]$ such that $f(M)$ is the maximum value of $f(x)$ on $[a, b]$ and $f(m)$ is the minimum value of $f(x)$ on $[a, b]$.

THEOREM 1.18

The Intermediate Value Theorem

If f is continuous on a closed interval $[a, b]$, then for any K strictly between $f(a)$ and $f(b)$, there exists at least one $c \in (a, b)$ such that $f(c) = K$.

These two important consequences of continuity may seem obvious, but in fact they rely on a subtle mathematical property of the real numbers called the ***Least Upper Bound Axiom***. Properly explaining the proofs of these theorems is outside of the scope of this book.

In the Extreme and Intermediate Value Theorems, the hypothesis that f be continuous on a closed interval $[a, b]$ is essential. If f either fails to be continuous on the interior of the interval or fails to be continuous at a closed endpoint, then the conclusions of these theorems do not necessarily hold. For example, each of the following three functions fails to be continuous on $[a, b]$ and also fails to satisfy at least one of the conclusions of the two theorems.

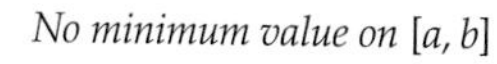

No minimum value on $[a, b]$

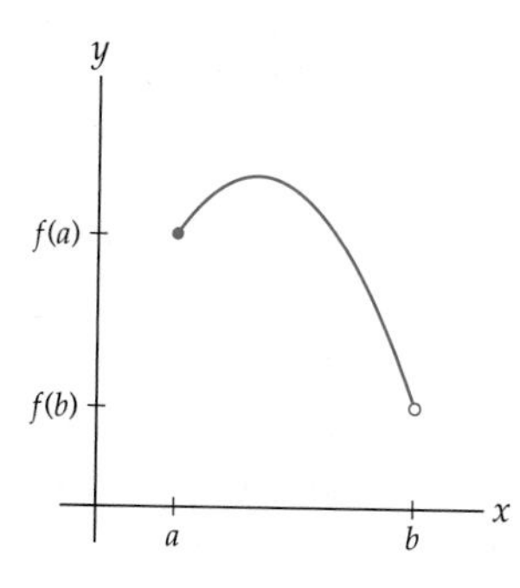

No c with $f(c) = K$

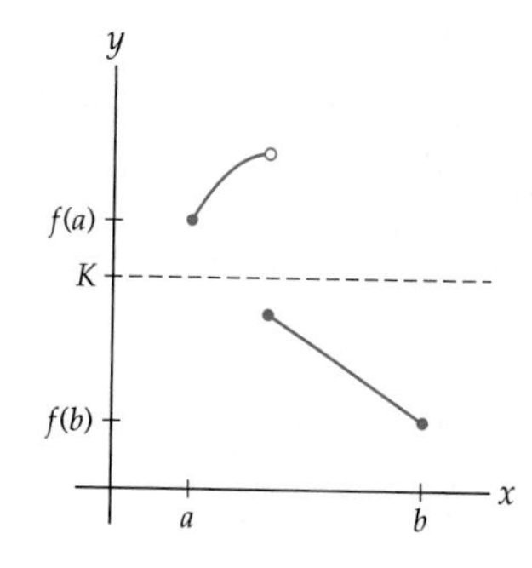

No maximum value on $[a, b]$

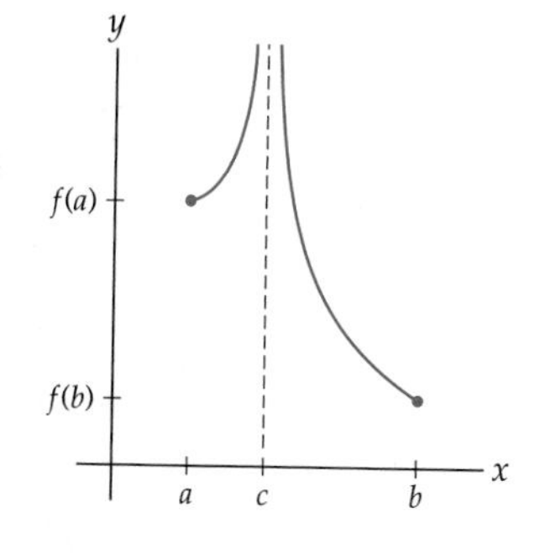

An extremely useful special case of the Intermediate Value Theorem is the case when we consider the intermediate value $K = 0$. In this case, the Intermediate Value Theorem says that if f is continuous on $[a, b]$ and $f(a)$ and $f(b)$ have opposite signs, then there exists at least one $c \in (a, b)$ where $f(c) = 0$. We use an equivalent variant of this special case every time we solve an inequality by checking signs between roots and discontinuities:

THEOREM 1.19 **A Function Can Change Sign Only at Roots and Discontinuities**

A function f can change sign (from positive to negative or vice versa) at a point $x = c$ only if $f(x)$ is zero, undefined, or discontinuous at $x = c$.

The graph that follows at the left shows a function f that is continuous on $[a, b]$, and changes sign only at its roots c_1, c_2, and c_3. The graph at the right is discontinuous somewhere in $[a, b]$ and therefore can change sign as we move from left to right without ever crossing the y-axis.

$f(x)$ can change sign at roots

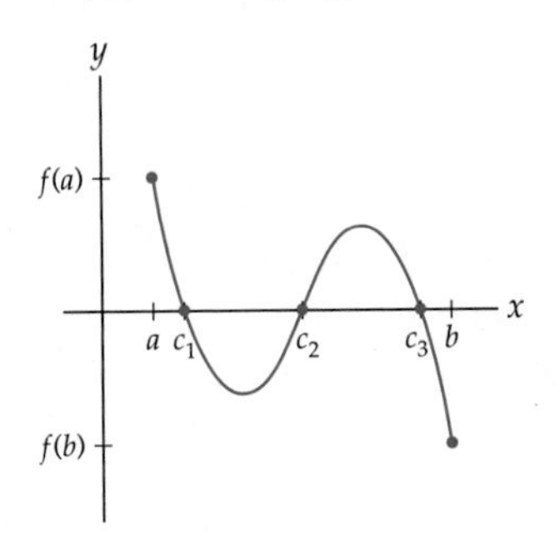

$f(x)$ can change sign at a discontinuity

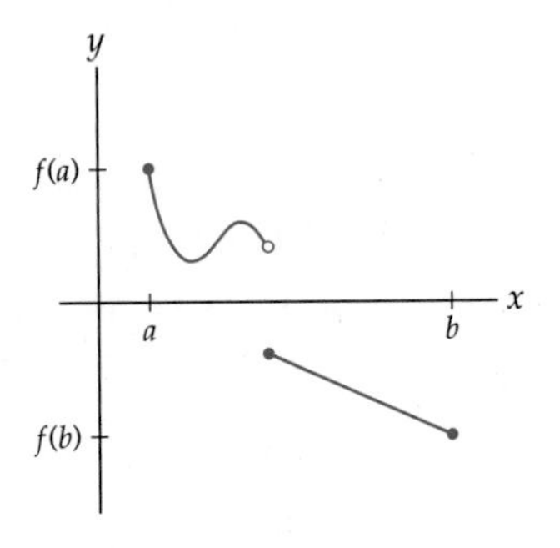

Examples and Explorations

EXAMPLE 1 **Limits and continuity of piecewise-defined functions**

Describe the continuity or discontinuity of each piecewise-defined function that follows at $x = 1$ by using graphs to determine the left, right, and two-sided limits at $x = 1$. Then describe the intervals on which each function is continuous.

(a) $f(x) = \begin{cases} x + 1, & \text{if } x < 1 \\ 3 - x^2, & \text{if } x \geq 1 \end{cases}$ **(b)** $g(x) = \begin{cases} 4 - x^2, & \text{if } x \leq 1 \\ x - 1, & \text{if } x > 1 \end{cases}$

SOLUTION

(a) The graph of f looks like $y = x + 1$ to the left of $x = 1$ and like $y = 3 - x^2$ to the right of $x = 1$, as shown next at the left. From the graph we see that $\lim_{x \to 1^-} f(x) = 2$ and $\lim_{x \to 1^+} f(x) = 2$, and thus $\lim_{x \to 1} f(x) = 2$. We also have $f(1) = 3 - 1^2 = 2$. Since the limit of $f(x)$ as $x \to 1$ is equal to the value of $f(x)$ at $x = 1$, we can conclude that f is continuous at $x = 1$. In fact, according to the graph, the function f is continuous on all of $(-\infty, \infty)$.

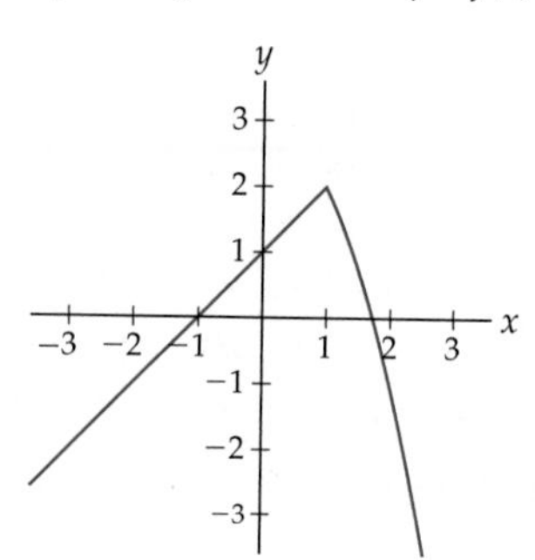

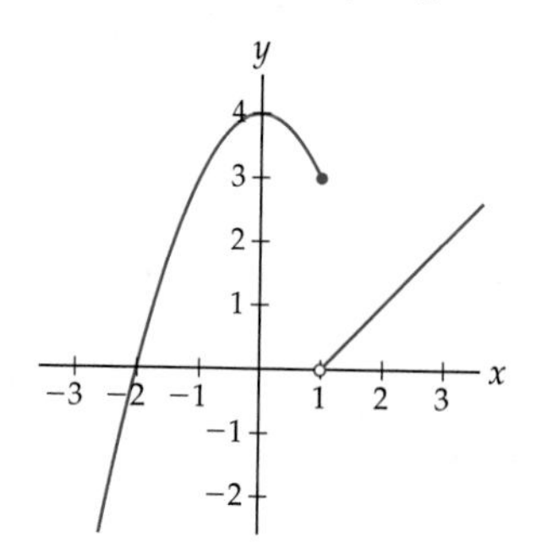

(b) The graph of $g(x)$ looks like $4 - x^2$ to the left of $x = 1$ and like $x - 1$ to the right of $x = 1$, with value $g(1) = 4 - 1^2 = 3$, as shown previously at the right. From the graph we see that $\lim_{x\to 1^-} g(x) = 3$ while $\lim_{x\to 1^+} g(x) = 0$. The left and right limits both exist, but they are not equal; thus $g(x)$ has a jump discontinuity at $x = 1$. Since $\lim_{x\to 1^-} g(x) = 3 = g(1)$ but $\lim_{x\to 1^+} g(x) = 0 \neq g(1)$, we can also say that $g(x)$ is left continuous, but not right continuous, at $x = 1$. According to the graph, the function $g(x)$ is continuous on $(-\infty, 1]$ and on $(1, \infty)$. □

EXAMPLE 2 **Continuity of a function that is defined separately for rationals and irrationals**

Determine graphically whether or not the rather exotic functions that follow are continuous at $x = 0$. Although these types of functions are not going to be a major focus in this course, this example helps get at the root of what continuity really means. You might be surprised by the solution!

(a) $f(x) = \begin{cases} 1, & \text{if } x \text{ is rational} \\ -1, & \text{if } x \text{ is irrational} \end{cases}$ **(b)** $g(x) = \begin{cases} 1, & \text{if } x \text{ is rational} \\ x+1, & \text{if } x \text{ is irrational} \end{cases}$

SOLUTION

In the graphs of f and g that follow, the lighter dotted line represents the values of the function at rational-number inputs and the darker dotted line represents the values at irrational-number inputs. Note that the graphs of both f and g pass the vertical line test, since every input x is either rational or irrational and never both.

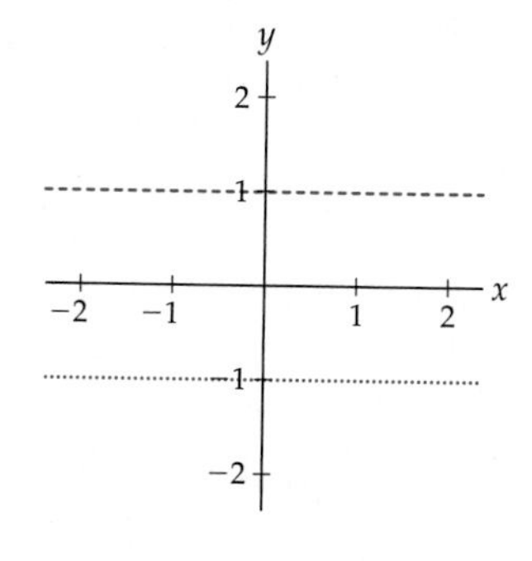

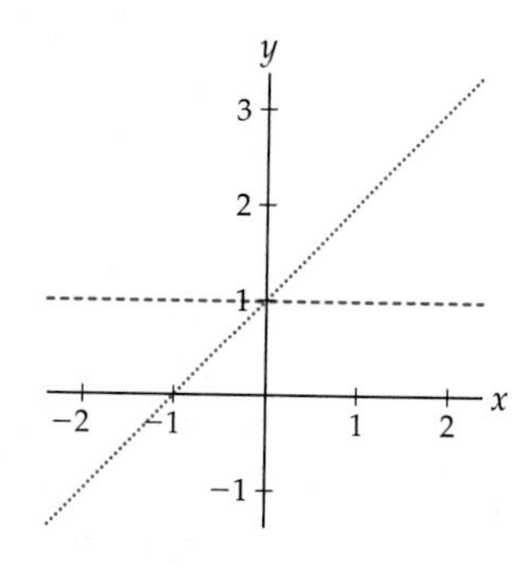

(a) We must consider the limit of f as $x \to 0$ separately for rational and irrational values of x. From the graph at the left, we see that for rational values of x we have $\lim_{x\to 0} f(x) = 1$

while for irrational values of x we have $\lim_{x \to 0} f(x) = -1$. Note that every punctured interval $(-\delta, 0) \cup (0, \delta)$ around $x = 0$ contains both rational and irrational numbers. Since the limit of $f(x)$ as $x \to 0$ is different depending on whether we choose rational or irrational values of x, the overall limit does not exist. Therefore the function f is not continuous at $x = 0$.

(b) On the other hand, looking at the graph of g at the right, we see that for rational values of x we have $\lim_{x \to 0} g(x) = 1$ and for irrational values of x we also have $\lim_{x \to 0} g(x) = 1$. Therefore $\lim_{x \to 0} g(x)$ exists and is equal to 1. Since $g(0)$ is also equal to 1, the function g is in fact continuous at $x = 0$, as strange as that may seem. □

EXAMPLE 3

Calculating limits of very basic functions

Use continuity to calculate each of the limits that follow, if possible. If we do not yet have enough information to calculate a limit, explain why not.

(a) $\lim_{x \to 3} 2$ **(b)** $\lim_{x \to 3} \frac{1}{x}$ **(c)** $\lim_{x \to 0} \frac{1}{x}$ **(d)** $\lim_{x \to 4} \sqrt{x - 1}$

SOLUTION

By Theorem 1.16 we know that the constant function $f(x) = 2$ and the power function $g(x) = \frac{1}{x}$ are continuous on their domains. Therefore for parts (a) and (b) we can calculate the limits just by evaluating at $x = 3$: $\lim_{x \to 3} 2 = 2$ and $\lim_{x \to 3} \frac{1}{x} = \frac{1}{3}$.

We do not yet know how to calculate the remaining two limits algebraically. In part (c), the point $x = 0$ is not in the domain of $g(x) = \frac{1}{x}$, so we cannot apply Theorem 1.16. In part (d), the function $h(x) = \sqrt{x - 1}$ is not a constant, identity, linear, or power function, and thus at this point we cannot conclude anything about its continuity or its limits. □

EXAMPLE 4

A real-world illustration of the Extreme and Intermediate Value Theorems

Consider the function $w(t)$ that describes a particular person's weight at t years of age between the ages of 18 and 45. Why does it make sense that this function is continuous on $[18, 45]$? What do the Extreme Value Theorem and the Intermediate Value Theorem say about $w(t)$?

SOLUTION

The weight function $w(t)$ should be continuous on $[18, 45]$ because a person's weight changes continuously over time and cannot jump from one value to another. (We are assuming typical circumstances, so that a person does not get a serious haircut, lose a limb, or somehow otherwise get their weight to change drastically in an instant.)

The Extreme Value Theorem tells us that there is some time $M \in [18, 45]$ at which the person's weight was greatest and some time $m \in [18, 45]$ at which that person weighed the least. In other words, at some time between 18 and 45 years of age, the person must have had a maximum weight and a minimum weight.

The Intermediate Value Theorem tells us that for every weight K between $w(18)$ and $w(45)$, there is some time $c \in (18, 45)$ for which $w(c) = K$. For example, if the person weighed $w(18) = 130$ pounds at age 18 and $w(45) = 163$ pounds at age 45, then there must be some age between 18 and 45 at which the person weighed, say, exactly 144 pounds. □

EXAMPLE 5

Applying the Intermediate Value Theorem to a Continuous Function

The function $f(x) = x^3 - 3x + 1$ is continuous everywhere. (We will see this later in Section 1.5.) Use the Intermediate Value Theorem to conclude that there is some point c for which $f(c) = 2$. Then use a graph of f to approximate at least one such value of c.

SOLUTION

To show that there is some c with $f(c) = 2$ we need to find values a and b such that $K = 2$ is between $f(a)$ and $f(b)$, and apply the Intermediate Value Theorem. By trial and error we can find such values a and b, by testing different values of $f(x)$ until we find one that is less than and one that is greater than 2. For example,

$$f(0) = 0^3 - 3(0) + 1 = 1 < 2,$$

$$f(2) = 2^3 - 3(2) + 1 = 3 > 2.$$

Since f is continuous on $[0, 2]$ and $f(0) < 2 < f(2)$, by the Intermediate Value Theorem there is some value $c \in (0, 2)$ for which $f(c) = 2$. Note that the Intermediate Value Theorem doesn't tell us where c is, only that such a c exists somewhere in the interval $(0, 2)$.

We can approximate some values of c for which $f(c) = 2$ by approximating the values of x for which the graph of $f(x) = x^3 - 3x + 1$ intersects the line $y = 2$:

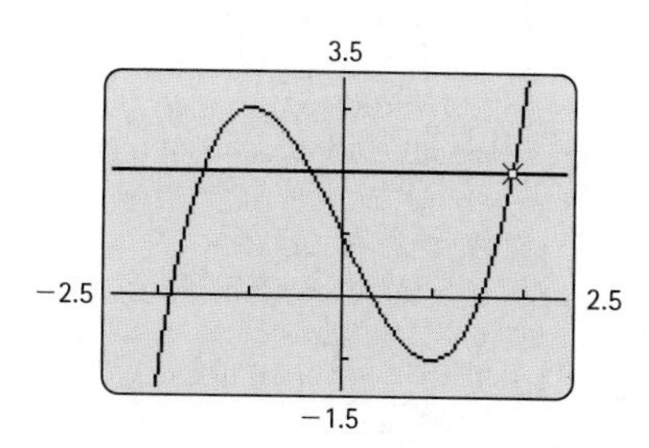

From this graph we can conclude that $f(c) = 2$ at $c \approx -1.5$, $c \approx -0.4$, and $c \approx 1.9$. To get better approximations we could trace along the graph on a calculator or other graphing utility. □

EXAMPLE 6

Determining sign information between zeroes and discontinuities

Determine the intervals on which the functions that follow are positive or negative. You may assume that the function in part (a) is continuous everywhere and the function in part (b) is continuous on each piece.

(a) $f(x) = 3x^3 + 3x^2 - 6x$

(b) $g(x) = \begin{cases} x + 1, & \text{if } x < -2 \\ (x+1)^2, & \text{if } -2 \leq x \leq 1 \\ 2 - x, & \text{if } x > 1 \end{cases}$

SOLUTION

(a) The roots of $f(x) = 3x^3 + 3x^2 - 6x = 3x(x^2 + x - 2) = 3x(x-1)(x+2)$ are $x = 0$, $x = 1$, and $x = -2$. By Theorem 1.19, in each of the intervals between these points the function f is either always positive or always negative. We need to test the sign of

$f(x)$ at one point in each interval. For example, $f(-3) = -36 < 0$, $f(-1) = 6 > 0$, $f(0.5) = -1.875 < 0$, and $f(2) = 24 > 0$. Reading from the resulting sign chart shown at the left, we can see that $f(x) = 3x^3 + 3x^2 - 6x$ is negative on $(-\infty, -2) \cup (0, 1)$ and positive on $(-2, 0) \cup (1, \infty)$:

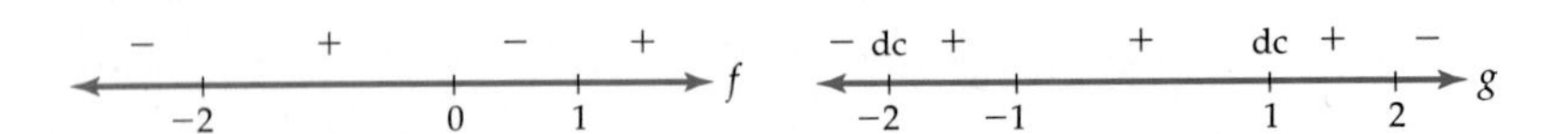

(b) The piecewise-defined function g can be discontinuous only at its break points $x = -2$ and $x = 1$. Furthermore, its first component $x+1$ is never zero on $(-\infty, -2)$, its second component $(x + 1)^2$ is zero only at $x = -1$, and its third component $2 - x$ is zero when $x = 2$. By Theorem 1.19 the function g can change sign only at the roots and discontinuities at $-2, -1, 1$, and 2. All that now remains is to check the sign of $g(x)$ one time between each of these points; the results are recorded on the preceding sign chart at the right. We marked the discontinuous points with "dc" to distinguish them from the zeros. Reading from this sign chart and keeping careful track of the sign of $g(x)$ at the break points, we see that $g(x)$ is negative on $(-\infty, -2) \cup (2, \infty)$ and positive on $[-2, -1) \cup (-1, 2]$. □

CHECKING THE ANSWER

We can graph f and g with a calculator or other graphing utility to verify that the sign charts we found are reasonable. Notice that the intervals where $f(x)$ or $g(x)$ are positive are the intervals on which their graphs are above the x-axis.

$f(x)$ is above the x-axis on $(-2, 0) \cup (1, \infty)$

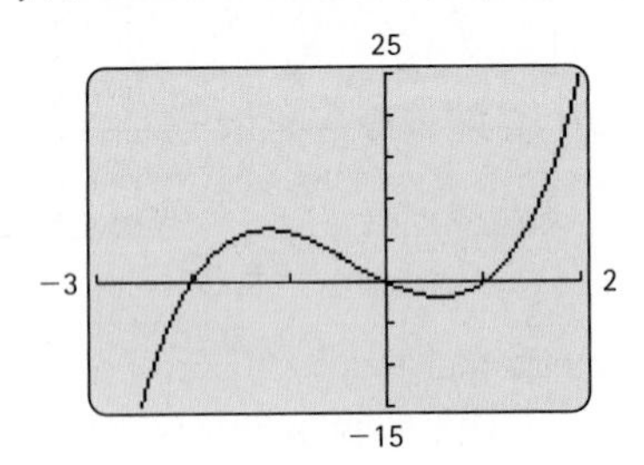

$g(x)$ is above the x-axis on $[-2, -1) \cup (-1, 2)$

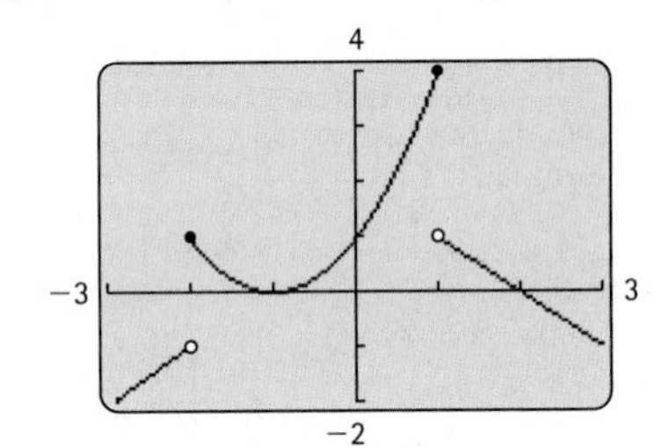

? TEST YOUR UNDERSTANDING

- Use limits to give definitions of each of the following: continuity at a point, continuity on an interval, left and right continuity.
- Use limits to give definitions of each of the following: removable discontinuity, jump discontinuity, infinite discontinuity.
- In the reading is a graph of a function with a removable discontinuity. We could make this graph continuous if we could change just one function value. What value?
- Why would it be difficult to prove at this point that all power functions, no matter what kind of power, are continuous on their domains?
- The conclusion of the Intermediate Value Theorem tells us of the existence of a point $c \in (a, b)$ with $f(c) = K$. Why do we need only the open interval (a, b), and not the closed interval $[a, b]$, in this conclusion?

EXERCISES 1.4

Thinking Back

Finding roots of piecewise-defined functions: For each function f that follows, find all values $x = c$ for which $f(c) = 0$. Check your answers by sketching a graph of f.

- $f(x) = \begin{cases} 4 - x^2, & \text{if } x < 0 \\ x + 1, & \text{if } x \geq 0 \end{cases}$
- $f(x) = \begin{cases} x + 1, & \text{if } x < 0 \\ 4 - x^2, & \text{if } x \geq 0 \end{cases}$
- $f(x) = \begin{cases} 2x - 1, & \text{if } x \leq 1 \\ 2x^2 + x - 3, & \text{if } x > 1 \end{cases}$

Logical existence statements: Determine whether each of the statements that follow are true or false. Justify your answers.

- If x is an integer, then there exists some positive integer y such that $|y| = x$.
- If x is a positive integer, then there exists some negative integer y such that $|y| = x$.
- If $x \in [-2, 2]$, then there exists some $y \in (0, 4)$ such that $y = x^2$.
- If $x \in [0, 100]$, then there exists some $y \in [-10, 10]$ such that $x = y^2$.

Concepts

0. *Problem Zero:* Read the section and make your own summary of the material.

1. *True/False:* Determine whether each of the statements that follow is true or false. If a statement is true, explain why. If a statement is false, provide a counterexample.

(a) *True or False:* If f is both left and right continuous at $x = c$, then f is continuous at $x = c$.

(b) *True or False:* If f is continuous on the open interval $(0, 5)$, then f is continuous at every point in $(0, 5)$.

(c) *True or False:* If f is continuous on the closed interval $[0, 5]$, then f is continuous at every point in $[0, 5]$.

(d) *True or False:* If f is continuous on the interval $(2, 4)$, then f must have a maximum value and a minimum value on $(2, 4)$.

(e) *True or False:* If $f(3) = -5$ and $f(9) = -2$, then there must be a value c at which $f(c) = -3$.

(f) *True or False:* If f is continuous everywhere, and if $f(-2) = 3$ and $f(1) = 2$, then $f(x)$ must have a root somewhere in $(-2, 1)$.

(g) *True or False:* If f is continuous everywhere, and if $f(0) = -2$ and $f(4) = 3$, then $f(x)$ must have a root somewhere in $(0, 4)$.

(h) *True or False:* If $f(0) = f(6) = 0$ and $f(2) > 0$, then $f(x)$ is positive on the entire interval $(0, 6)$.

2. *Examples:* Construct examples of the thing(s) described in the following. Try to find examples that are different than any in the reading.

(a) The graph of a function with $f(4) = 2$ that has a removable discontinuity at $x = 4$.

(b) The graph of a function that is continuous on its domain but not continuous at $x = 0$.

(c) The graph of a function that is continuous on $(0, 2]$ and $(2, 3)$ but not on $(0, 3)$.

3. If f is a continuous function, what can you say about $\lim_{x \to 1} f(x)$?

4. Explain what it means for a function f to be continuous at a point $x = c$, with a sentence that includes the words "approaches" and "value."

5. In our proof that constant functions are continuous, we used the fact that given any $\epsilon > 0$, a choice of *any* $\delta > 0$ will work in the formal definition of limit. Use a graph to explain why this makes intuitive sense. *(This exercise depends on Section 1.3.)*

6. In our proof that linear functions are continuous, we used the fact that given any $\epsilon > 0$, the choice of $\delta = \frac{\epsilon}{|m|}$ will work in the formal definition of limit. Use a graph to explain why this makes intuitive sense. *(This exercise depends on Section 1.3.)*

7. Given the following function f, define $f(1)$ so that f is continuous at $x = 1$, if possible:

$$f(x) = \frac{x^2 - 2x + 1}{x^2 - 6x + 5}.$$

8. Given the following function f, define $f(1)$ so that f is continuous at $x = 1$, if possible:

$$f(x) = \begin{cases} 3x - 1, & \text{if } x < 1 \\ x^2 + 1, & \text{if } x > 1. \end{cases}$$

Each function in Exercises 9–12 is discontinuous at some value $x = c$. Describe the type of discontinuity and any one-sided continuity at $x = c$, and sketch a possible graph of f.

9. $\lim_{x \to -1^-} f(x) = 2, \ \lim_{x \to -1^+} f(x) = 2, \ f(-1) = 1.$

10. $\lim_{x \to 2^-} f(x) = 2, \ \lim_{x \to 2^+} f(x) = 1, \ f(2) = 1.$

11. $\lim_{x \to 0^-} f(x) = -1, \ \lim_{x \to 0^+} f(x) = 1, \ f(0) = 0.$

12. $\lim_{x \to 2^-} f(x) = -\infty, \ \lim_{x \to 2^+} f(x) = \infty, \ f(2) = 3.$

13. State what it means for a function f to be continuous at a point $x = c$, in terms of the delta–epsilon definition of limit. *(This exercise depends on Section 1.3.)*

14. State what it means for a function f to be left continuous at a point $x = c$, in terms of the delta–epsilon definition of limit. *(This exercise depends on Section 1.3.)*

15. State what it means for a function f to be right continuous at a point $x = c$, in terms of the delta–epsilon definition of limit. *(This exercise depends on Section 1.3.)*

16. Sketch a labeled graph of a function that satisfies the hypothesis of the Extreme Value Theorem, and illustrate on your graph that the conclusion of the Extreme Value Theorem follows.

17. Sketch a labeled graph of a function that satisfies the hypothesis of the Intermediate Value Theorem, and illustrate on your graph that the conclusion of the Intermediate Value Theorem follows.

18. Sketch a labeled graph of a function that fails to satisfy the hypothesis of the Intermediate Value Theorem, and illustrate on your graph that the conclusion of the Intermediate Value Theorem does not necessarily hold.

19. Sketch a labeled graph of a function that fails to satisfy the hypothesis of the Extreme Value Theorem, and illustrate on your graph that the conclusion of the Extreme Value Theorem does not necessarily hold.

20. Explain why the Intermediate Value Theorem allows us to say that a function can change sign only at discontinuities and zeroes.

For each of the following sign charts, sketch the graph of a function f that has the indicated signs, zeros, and discontinuities:

21. Signs: $-$ $+$ $+$ $-$; marked points: 0, 1, 3; f

22. Signs: $+$ $-$ dc $-$ $+$ dc $-$; marked points: -4, -2, 0, 2; f

Skills

For each function f graphed in Exercises 23–26, describe the intervals on which f is continuous. For each discontinuity of f, describe the type of discontinuity and any one-sided continuity. Justify your answers about discontinuities with limit statements.

23. 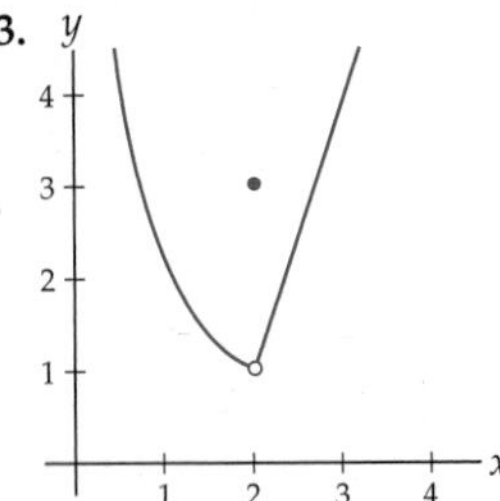

24. 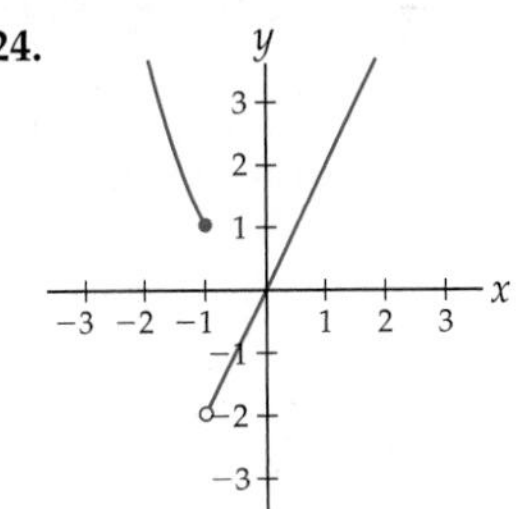

25. 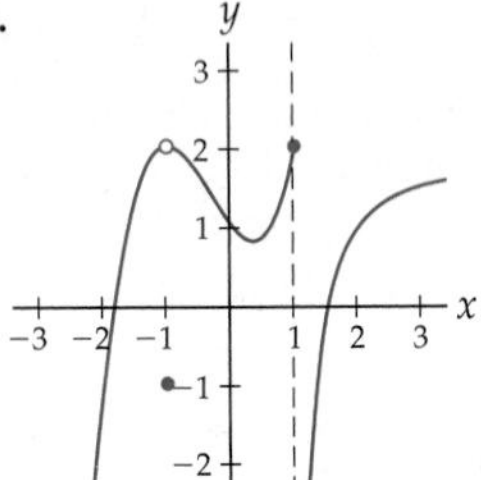

26. 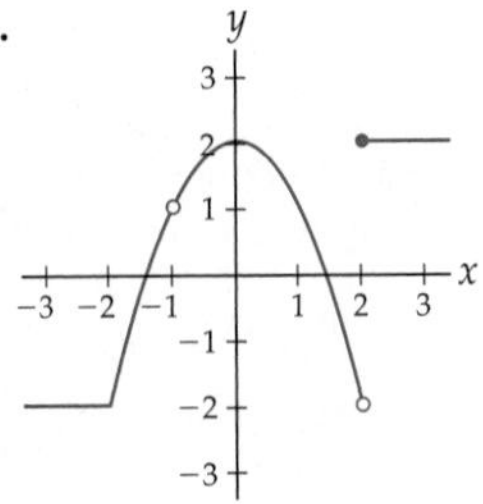

Sketch the graph of a function f described in Exercises 27–32, if possible. If it is not possible, explain why not.

27. f is left continuous at $x = 1$ and right continuous at $x = 1$, but is not continuous at $x = 1$, and $f(1) = -2$.

28. f is left continuous at $x = 2$ but not continuous at $x = 2$, and $f(2) = 3$.

29. f has a jump discontinuity at $x = -1$ and is left continuous at $x = -1$, and $f(-1) = 2$.

30. f has an infinite discontinuity at 0 but is right continuous at 0, and $f(0) = 1$.

31. f has a removable discontinuity at $x = -2$ and is right continuous at $x = -2$, and $f(-2) = 0$.

32. f is continuous on $[0, 2)$ but not on $[0, 2]$.

For each limit in Exercises 33–38, either use continuity to calculate the limit or explain why Theorem 1.16 does not apply.

33. $\lim_{x \to -1} 6$

34. $\lim_{x \to -1} x$

35. $\lim_{x \to -5} (3x - 2)$

36. $\lim_{x \to 3} x^4$

37. $\lim_{x \to 0} x^{-3}$

38. $\lim_{x \to -5} \sqrt{x}$

In Exercises 39–44, use Theorem 1.16 and left and right limits to determine whether each function f is continuous at its break point(s). For each discontinuity of f, describe the type of discontinuity and any one-sided discontinuity.

39. $f(x) = \begin{cases} x - 3, & \text{if } x < 3 \\ -(x - 3), & \text{if } x \geq 3 \end{cases}$

40. $f(x) = \begin{cases} x - 3, & \text{if } x < 0 \\ -(x - 3), & \text{if } x \geq 0 \end{cases}$

41. $f(x) = \begin{cases} x^2, & \text{if } x < 2 \\ 4, & \text{if } x = 2 \\ 2x + 1, & \text{if } x > 2 \end{cases}$

42. $f(x) = \begin{cases} \sqrt{-x}, & \text{if } x < 0 \\ 2, & \text{if } x = 0 \\ \sqrt{x}, & \text{if } x > 0 \end{cases}$

43. $f(x) = \begin{cases} x + 1, & \text{if } x < 1 \\ 3x - 1, & \text{if } 1 \leq x < 2 \\ x + 2, & \text{if } x \geq 2 \end{cases}$

44. $f(x) = \begin{cases} x^3, & \text{if } x \leq 0 \\ 1 - x, & \text{if } 0 < x < 3 \\ x - 5, & \text{if } x \geq 3 \end{cases}$

Use graphs to determine if each function f in Exercises 45–48 is continuous at the given point $x = c$.

45. $f(x) = \begin{cases} 2 - x, & \text{if } x \text{ rational} \\ x^2, & \text{if } x \text{ irrational,} \end{cases} \quad c = 2$

46. $f(x) = \begin{cases} x^2 - 3, & \text{if } x \text{ rational} \\ 3x + 1, & \text{if } x \text{ irrational,} \end{cases} \quad c = 0$

47. $f(x) = \begin{cases} 2 - x, & \text{if } x \text{ rational} \\ x^2, & \text{if } x \text{ irrational,} \end{cases} \quad c = 1$

48. $f(x) = \begin{cases} x^2 - 3, & \text{if } x \text{ rational} \\ 3x + 1, & \text{if } x \text{ irrational,} \end{cases} \quad c = 4$

Use the Extreme Value Theorem to show that each function f in Exercises 49–54 has both a maximum and a minimum value on $[a, b]$. Then use a graphing utility to approximate values M and m in $[a, b]$ at which f has a maximum and a minimum, respectively. You may assume that these functions are continuous everywhere.

49. $f(x) = x^4 - 3x^2 - 2, \; [a, b] = [-2, 2]$

50. $f(x) = x^4 - 3x^2 - 2, \; [a, b] = [0, 2]$

51. $f(x) = x^4 - 3x^2 - 2, \; [a, b] = [-1, 1]$

52. $f(x) = 3 - 2x^2 + x^3, \; [a, b] = [-1, 2]$

53. $f(x) = 3 - 2x^2 + x^3, \; [a, b] = [0, 2]$

54. $f(x) = 3 - 2x^2 + x^3, \; [a, b] = [-1, 1]$

Use the Intermediate Value Theorem to show that for each function f, interval $[a, b]$, and value K in Exercises 55–60, there is some $c \in (a, b)$ for which $f(c) = K$. Then use a graphing utility to approximate all such values c. You may assume that these functions are continuous everywhere.

55. $f(x) = 5 - x^4, \; [a, b] = [0, 2], \; K = 0$

56. $f(x) = 5 - x^4, \; [a, b] = [-2, -1], \; K = 0$

57. $f(x) = x^3 - 3x^2 - 2, \; [a, b] = [-2, 4], \; K = -4$

58. $f(x) = x^3 - 3x^2 - 2, \; [a, b] = [0, 2], \; K = -4$

59. $f(x) = x^3 - 3x^2 - 2, \; [a, b] = [2, 4], \; K = -4$

60. $f(x) = 2 + x + x^3, \; [a, b] = [-1, 2], \; K = 3$

Use the Intermediate Value Theorem to show that for each function f and value K in Exercises 61–66, there must be some $c \in \mathbb{R}$ for which $f(c) = K$. You will have to select an appropriate interval $[a, b]$ to work with. Then find or approximate one such value of c. You may assume that these functions are continuous everywhere.

61. $f(x) = x^3 + 2, \; K = -15$

62. $f(x) = -2x^2 + 4, \; K = 0$

63. $f(x) = \sin x, \; K = \frac{1}{2}$

64. $f(x) = \sin x, \; K = \frac{\sqrt{3}}{2}$

65. $f(x) = |3x + 1|, \; K = 1$

66. $f(x) = |2 - 3x|, \; K = 2$

Find the intervals on which each function in Exercises 67–74 is positive or negative. Make clear how your work uses the Intermediate Value Theorem and continuity. You may assume that polynomials and their quotients are continuous on the intervals on which they are defined.

67. $f(x) = 2 + 5x + 2x^2$

68. $f(x) = x^3 - 2x^2 - 3x$

69. $f(x) = \dfrac{x^2 - 4}{x^2 - 1}$

70. $f(x) = \dfrac{(x + 4)(x - 1)}{2x + 3}$

71. $f(x) = \begin{cases} x - 4, & \text{if } x \le 1 \\ x^2 - 4, & \text{if } x > 1 \end{cases}$

72. $f(x) = \begin{cases} 3x + 1, & \text{if } x < 0 \\ x, & \text{if } x \ge 0 \end{cases}$

73. $f(x) = \begin{cases} x^3, & \text{if } x \le 2 \\ 4x - x^3, & \text{if } x > 2 \end{cases}$

74. $f(x) = \begin{cases} x^2 - 9, & \text{if } x \le -2 \\ x^2 + x - 2, & \text{if } x > -2 \end{cases}$

Applications

Explain in practical terms what the Extreme Value Theorem says about each continuous function defined in Exercises 75–77. Then explain in practical terms what the Intermediate Value Theorem says in each situation.

75. Alina hasn't cut her hair for six years. Six years ago her hair was just 2 inches long. Now her hair is 42 inches long. Let $H(t)$ be the function that describes the length, in inches, of Alina's hair t years after she stopped cutting it.

76. Linda collects rain in a bucket outside her back door. Since the first day of April she has been keeping track of how the amount of water in the bucket changes as it fills with rain and evaporates. On April 1 the bucket was empty, and today it contains 4 inches of water. Let $w(t)$ be the height, in inches, of rainwater in the bucket t days after the first day of April.

77. The number of gallons of gas in Phil's new station wagon t days after he bought it is given by the function $g(t)$. When he purchased the station wagon one year ago, the tank had 19 gallons of gas in it. Today he ran out of gas.

78. Lars was 20 inches tall when he was born, and six foot one when he died at age 83. Use the Intermediate Value Theorem to show that there must have been some point in Lars's life at which his height in inches was equal to his age in years. *(Hint: Think about when the difference between his height and age is zero.)*

79. As a vacuum cleaner salesman, Alex earns a salary of \$8,500 a year, whether he sells any vacuum cleaners or not. In addition, for every 30 vacuum cleaners he sells, he earns a \$1,500 commission.

(a) Construct a piecewise-defined function $M(v)$ that describes the amount of money M that Alex will make in a year if he sells v vacuum cleaners over the course of the year. Assume he sells between 0 and 90 vacuum cleaners in a year.

(b) Check that your function makes sense by using it to calculate $M(0)$, $M(30)$, $M(59)$, $M(61)$, and $M(90)$. Then sketch a graph of $M(v)$ on the interval $0 \le v \le 90$.

(c) The piecewise-defined function $M(v)$ is *not* continuous. List all the values at which $M(v)$ fails to be continuous, and use the definition of continuity to support your answers.

80. One immediate application of the Intermediate Value Theorem is the method of finding roots called the ***Bisection Method***. In this problem you will develop this method and then use it to approximate the square root of 2.

(a) Suppose f is continuous on $\mathbb{R}$ and that a and b are some real numbers for which $f(a)$ is negative and $f(b)$ is positive. Explain why the Intermediate Value Theorem guarantees that there must be some point in (a, b) where $f(x)$ has a root.

(b) Consider the function $f(x) = x^2 - 2$. Show that $f(0)$ is negative and $f(2)$ is positive. What conclusion can we draw from the Intermediate Value Theorem?

(c) We can bisect the interval $(0, 2)$ by finding the midpoint of the interval, which in this case is $x = 1$. Is $f(1)$ positive or negative? Does the Intermediate Value Theorem say anything about $f(x) = x^2 - 2$ on the interval $(0, 1)$? What about on the interval $(1, 2)$?

(d) Your answer to part (b) tells you that $f(x) = x^2 - 2$ must have a root somewhere in the interval $(0, 2)$ of length 2. Your answer to part (c) tells you that $f(x) = x^2 - 2$ must have a root in a shorter interval of length 1. Now repeat! Bisect the interval of length 1 to find an interval of length $\frac{1}{2}$ on which $f(x)$ must have a root.

(e) Describe why this Bisection Method will in general give better and better approximations for finding a root of a given function. In this particular example, with $f(x) = x^2 - 2$, why does the Bisection Method give us an approximation for $\sqrt{2}$?

Proofs

81. Write a delta–epsilon proof that shows that the function $f(x) = 3x - 5$ is continuous at $x = 2$. *(This exercise depends on Section 1.3.)*

82. Write a delta–epsilon proof that shows that the function $f(x) = 2x + 1$ is continuous at $x = 5$. *(This exercise depends on Section 1.3.)*

83. Write a delta–epsilon proof that shows that the function $f(x) = |x|$ is continuous. You may find the following inequality useful: For any real numbers a and b, $||a| - |b|| \le |a - b|$. *(This exercise depends on Section 1.3.)*

84. Use what you know about one-sided limits to prove that a function f is continuous at a point $x = c$ if and only if it is both left and right continuous at $x = c$.

For each function f in Exercises 85–86, use the delta–epsilon definition of continuity to argue that f is or is not continuous at the indicated point $x = c$.

85. $f(x) = \begin{cases} 2 - x, & \text{if } x \text{ rational} \\ x^2, & \text{if } x \text{ irrational}, \end{cases} \quad c = 2$

86. $f(x) = \begin{cases} 2 - x, & \text{if } x \text{ rational} \\ x^2, & \text{if } x \text{ irrational}, \end{cases} \quad c = 1$

87. Use the Intermediate Value Theorem to prove that every cubic function $f(x) = Ax^3 + Bx^2 + Cx + D$ has at least one real root. You will have to first argue that you can find real numbers a and b so that $f(a)$ is negative and $f(b)$ is positive.

For each power function f in Exercises 89–93, write a delta–epsilon proof which proves that f is continuous on its domain. In each case you will need to assume that δ is less than or equal to 1. *(These exercises depend on Section 1.3.)*

88. $f(x) = x^{-1}$

89. $f(x) = x^2$

90. $f(x) = x^3$

91. $f(x) = x^{-2}$

92. $f(x) = x^{-1/2}$

93. $f(x) = x^{1/2}$

Thinking Forward

Interesting trigonometric limits: For each of the functions that follow, use a calculator or other graphing utility to examine the graph of f near $x = 0$. Does it appear that f is continuous at $x = 0$? Make sure your calculator is set to radian mode.

- $f(x) = \begin{cases} \frac{1}{x}\sin(x), & \text{if } x \ne 0 \\ 1, & \text{if } x = 0 \end{cases}$

- $f(x) = \begin{cases} \sin\left(\frac{1}{x}\right), & \text{if } x \ne 0 \\ 0, & \text{if } x = 0 \end{cases}$

- $f(x) = \begin{cases} x\sin\left(\frac{1}{x}\right), & \text{if } x \ne 0 \\ 0, & \text{if } x = 0 \end{cases}$

- $f(x) = \begin{cases} x^2\sin\left(\frac{1}{x}\right), & \text{if } x \ne 0 \\ 0, & \text{if } x = 0 \end{cases}$

1.5 LIMIT RULES AND CALCULATING BASIC LIMITS

- Rules for calculating limits of arithmetic combinations and compositions of functions
- Continuity of algebraic and transcendental functions at domain points
- The Cancellation Theorem and the Squeeze Theorem for calculating limits

Limits of Combinations of Functions

Although we now understand in depth what limit statements *mean*, at this point we do not have many tools for calculating limits. We can calculate limits of continuous functions at domain points by evaluation, and we know that very simple functions, such as constant and linear functions, are continuous. What about more complicated functions? For example, we already know from the continuity of power functions that

$$\lim_{x\to 2} x^2 = 2^2 = 4 \quad \text{and} \quad \lim_{x\to 2} x^3 = 2^3 = 8.$$

Can we use these results to say something about $\lim_{x\to 2}(x^2+x^3)$ of the sum of these functions?

The key theorem that follows will help us answer this question; it says that limits behave well with respect to all of the arithmetic operations and even with respect to composition.

THEOREM 1.20 **Rules for Calculating Limits of Combinations**

If $\lim_{x\to c} f(x)$ and $\lim_{x\to c} g(x)$ exist, then the following rules hold for their combinations:

Constant Multiple Rule: $\lim_{x\to c} kf(x) = k\lim_{x\to c} f(x)$, for any real number k.

Sum Rule: $\lim_{x\to c}(f(x)+g(x)) = \lim_{x\to c} f(x) + \lim_{x\to c} g(x)$

Difference Rule: $\lim_{x\to c}(f(x)-g(x)) = \lim_{x\to c} f(x) - \lim_{x\to c} g(x)$

Product Rule: $\lim_{x\to c}(f(x)g(x)) = (\lim_{x\to c} f(x))(\lim_{x\to c} g(x))$

Quotient Rule: $\lim_{x\to c}\frac{f(x)}{g(x)} = \frac{\lim_{x\to c} f(x)}{\lim_{x\to c} g(x)}$, if $\lim_{x\to c} g(x) \neq 0$

Composition Rule: $\lim_{x\to c} f(g(x)) = f(\lim_{x\to c} g(x))$, if f is continuous at $\lim_{x\to c} g(x)$

This theorem is a powerful tool for calculating limits, since it tells us how to find limits of compound functions in terms of the limits of their components. For example, we can calculate the limit of the sum $x^2 + x^3$ as $x \to 2$ by taking the sum of the limits of x^2 and x^3 as $x \to 2$:

$$\lim_{x\to 2}(x^2+x^3) = \lim_{x\to 2} x^2 + \lim_{x\to 2} x^3 = 4 + 8 = 12.$$

We will postpone the proofs of the limit rules in Theorem 1.20 until the end of this section so that we can first explore their consequences and practical uses. For example, an immediate consequence of Theorem 1.20 is that constant multiples, sums, differences, products, quotients, and compositions of continuous functions are continuous:

THEOREM 1.21 **Combinations of Continuous Functions Are Continuous**

If f and g are continuous at $x = c$ and k is any constant, then the functions kf, $f+g$, $f-g$, and fg are also continuous at $x = c$.

Moreover, if $g(c) \neq 0$, then $\frac{f}{g}$ is continuous at $x = c$, and if f is also continuous at $g(c)$, then $f \circ g$ is continuous at $x = c$.

For example, since $f(x) = x^2$ and $g(x) = x^3$ are continuous at $x = 2$, Theorem 1.21 tells us that $(f + g)(x) = x^2 + x^3$ must be also be continuous at $x = 2$. This makes sense given Theorem 1.20 because

$$\lim_{x\to 2}(f+g)(x) = \lim_{x\to 2}(x^2 + x^3) = \lim_{x\to 2} x^2 + \lim_{x\to 2} x^3 = 2^2 + 2^3 = 4 + 8 = 12 = (f+g)(2).$$

Limits of Algebraic Functions

With the limit rules we can now prove that most of the functions we will use in this book are continuous on their domains. We will start with the algebraic functions. Recall that a function is ***algebraic*** if it can be expressed with the use of only arithmetic operations (+, −, ×, and ÷) and rational constant powers. Power functions, polynomial functions, and rational functions are all examples of algebraic functions.

THEOREM 1.22 **Continuity of Algebraic Functions**

All algebraic functions are continuous on their domains. In particular, if $x = c$ is in the domain of an algebraic function f, then we can calculate $\lim_{x\to c} f(x)$ by evaluating $f(c)$.

With this theorem we can do lots of basic limit calculations. For example, $\lim_{x\to 4} x^{1/2} = \sqrt{4} = 2$, $\lim_{x\to 1}(3x^4 - 2x) = 3(1)^4 - 2(1) = 1$, and $\lim_{x\to 2} \frac{1+x}{3-x} = \frac{1+2}{3-2} = 3$. For certain special cases the limits are only one-sided; for example, $\lim_{x\to 2^+} \sqrt{x-2} = 0$. Note that the theorem does not tell us how to calculate limits at non-domain points; for example, we still do not know how to calculate $\lim_{x\to 2} \frac{1}{x-2}$.

Proof. Algebraic functions are by definition built out of rational powers and arithmetic combinations of real numbers and the variable x. We already know how to handle limits of constant multiples, sums, products, quotients, and compositions by using the limit rules. We also know from Theorem 1.16 that $\lim_{x\to c} k = k$ and $\lim_{x\to c} x = c$. Therefore to show that every algebraic function is continuous on its domain, it suffices to show that every function of the form $f(x) = x^k$ is continuous on its domain.

We must show that for any rational number k, if $x = c$ is in the domain of x^k, then $\lim_{x\to c} x^k = c^k$. There are a few cases to consider. If k is a positive integer, then we just repeatedly apply the product rule for limits so that we can use the known limit $\lim_{x\to c} x = c$:

$$\lim_{x\to c} x^k = \underbrace{(\lim_{x\to c} x)(\lim_{x\to c} x)\cdots(\lim_{x\to c} x)}_{k \text{ times}} = \underbrace{(c)(c)\cdots(c)}_{k \text{ times}} = c^k.$$

For negative integer powers, we apply the quotient rule for limits and the result for positive integer powers. In this case we must require $c \neq 0$ so that c will be in the domain of x^{-k}, and we obtain the following limit:

$$\lim_{x\to c} x^{-k} = \lim_{x\to c} \frac{1}{x^k} = \frac{\lim_{x\to c} 1}{\lim_{x\to c} x^k} = \frac{1}{c^k} = c^{-k}.$$

Although we will not prove so here, it can be shown that $\lim_{x\to c} x^{1/q} = c^{1/q}$ when c is in the domain of $x^{1/q}$. Given these facts, the composition rule for limits allows us to prove that $x^{p/q}$ is continuous at domain points for any rational power $\frac{p}{q}$:

$$\lim_{x\to c} x^{p/q} = \lim_{x\to c} \sqrt[q]{x^p} = \sqrt[q]{\lim_{x\to c} x^p} = \sqrt[q]{c^p} = c^{p/q}.$$

■

Finding Limits by Cancelling or Squeezing

The continuity of algebraic functions and the limit rules can help us calculate a great many limits, but only at domain points. One thing that can help us at non-domain points is the cancellation of common factors. For example, consider the limit $\lim_{x\to 1} \frac{x^2-1}{x-1}$. At $x = 1$ we have $x^2 - 1 = 0$ and $x - 1 = 0$, and therefore the limit is of the form $\frac{0}{0}$. Limits of this form are said to be ***indeterminate***, which means that they may or may not exist, depending on the situation. We will examine indeterminate forms in depth in Section 1.6. In the example we are considering, we can determine the limit by simple cancellation:

$$\lim_{x\to 1} \frac{x^2-1}{x-1} = \lim_{x\to 1} \frac{(x-1)(x+1)}{x-1} = \lim_{x\to 1}(x+1) = 2.$$

The cancellation of the common factor $x - 1$ is valid because $\frac{(x-1)(x+1)}{x-1}$ and $x + 1$ differ only when $x = 1$, and when we take the limit, we are not concerned with what happens *at* the point $x = 1$. In general, by definition, limits as $x \to c$ never have anything to do with what happens at $x = c$, which proves the following theorem:

THEOREM 1.23 **The Cancellation Theorem for Limits**

If $\lim_{x\to c} g(x)$ exists, and f is a function that is equal to g for all x sufficiently close to c except possibly at c itself, then $\lim_{x\to c} f(x) = \lim_{x\to c} g(x)$.

Another useful tool for calculating new types of limits is the Squeeze Theorem. This theorem says that if the output of a function $f(x)$ is always bounded between a lower function $l(x)$ and an upper function $u(x)$, and if the lower and upper functions approach the same value L as $x \to c$, then $f(x)$ gets squeezed between the lower and upper functions and also approaches L as $x \to c$.

THEOREM 1.24 **The Squeeze Theorem for Limits**

If $l(x) \le f(x) \le u(x)$ for all x sufficiently close to c, but not necessarily at $x = c$, and if $\lim_{x\to c} l(x)$ and $\lim_{x\to c} u(x)$ are both equal to L, then $\lim_{x\to c} f(x) = L$.

Similar results hold for limits at infinity and one-sided limits.

For example, the figure that follows shows a function f that is sandwiched between two functions u and l as $x \to 0$. Since $u(x)$ and $l(x)$ have the same limit at $x \to 0$ and $f(x)$ is squeezed between them, we know that $f(x)$ must share that same limit as $x \to 0$.

$l(x) \le f(x) \le u(x)$

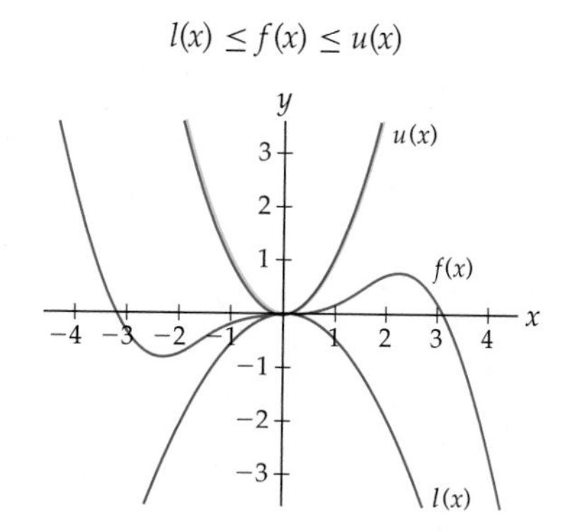

Proof. Given $\epsilon > 0$, we can choose $\delta_1 > 0$ to get $u(x)$ within ϵ of L and also choose δ_2 to get $l(x)$ within ϵ of L. If $\delta = \min(\delta_1, \delta_2)$, then whenever $x \in (c - \delta, c) \cup (c, c + \delta)$, we also have

$$L - \epsilon < l(x) \le f(x) \le u(x) < L + \epsilon.$$

Similar arguments prove that the Squeeze Theorem holds for limits as $x \to \infty$ and $x \to c^+$ or $x \to c^-$. ■

For example, we can use the Squeeze Theorem to calculate the limits of $\sin\theta$ and $\cos\theta$ at $x = 0$. Consider the following diagrams of portions of the unit circle, with angles measured in radians and $0 < \theta < \frac{\pi}{4}$:

$0 < \sin\theta < \theta$

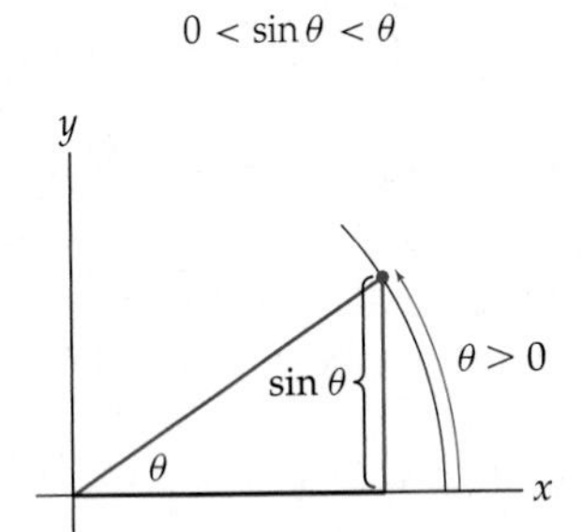

$0 < 1 - \cos\theta < \theta$

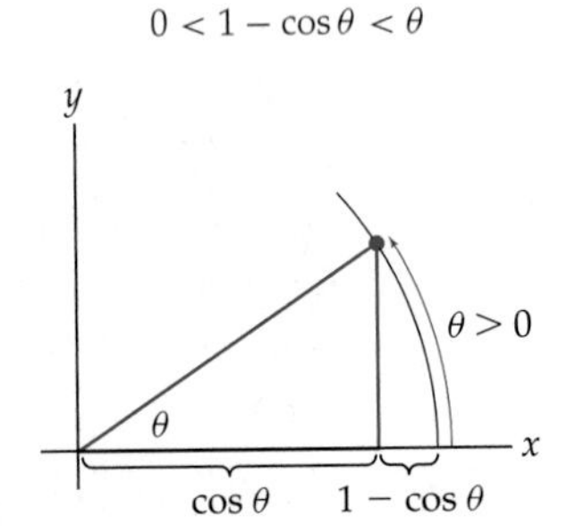

According to the leftmost figure, we clearly have $0 < \sin\theta < \theta$. Since $\lim\limits_{\theta\to 0^+} 0 = 0$ and $\lim\limits_{\theta\to 0^+} \theta = 0$, by the Squeeze Theorem we must also have $\lim\limits_{\theta\to 0^+} \sin\theta = 0$. In the second figure we can see that $0 < 1 - \cos\theta < \theta$. Thus again by the Squeeze Theorem we have $\lim\limits_{\theta\to 0^+} (1 - \cos\theta) = 0$. Rewriting this limit with some simple limit rules, we see that $\lim\limits_{x\to 0^+} \cos\theta = 1$. We can illustrate similar limits as $\theta \to 0^-$ by drawing pictures of small negative values of θ in the fourth quadrant (see Exercises 22 and 23).

Defining the Number e

Before we can discuss continuity and limits of exponential functions, we must have a proper definition for the irrational number e that we have so far been approximating as $e \approx 2.71828$. It turns out that this definition itself involves a limit.

DEFINITION 1.25 **The Number e**

We define e to be the number that $(1 + h)^{1/h}$ approaches as h approaches 0:

$$e = \lim_{h\to 0} (1 + h)^{1/h}.$$

It is important to note that this is a weak definition, because we have not proven that the limit in this definition exists. That is, we have not shown that $(1 + h)^{1/h}$ converges to a real number as $h \to 0$.

For small values of h it is easy to see that the quantity $(1 + h)^{1/h}$ is close to the approximation $e \approx 2.71828$ we have been using so far. For example, when $h = 0.0001$, we have

$$(1 + 0.0001)^{1/0.0001} \approx 2.71815.$$

Proving that the limit in Definition 1.25 does converge to a real number is beyond the scope of this chapter. In Example 5 and Exercise 24 we will use tables of values to show that this limit is reasonable. We will use accumulation integrals to give another definition of the number e in Section 4.7.

As we will see in the next chapter, our definition of the number e is partially motivated by derivatives. Specifically, the reason that e is the natural base for exponential functions has to do with the slope of the graph of $y = e^x$ at $x = 0$. In the next chapter we will see that the slope of the graph of an exponential function $y = b^x$ at $x = 0$ is given by the limit $\lim_{h \to 0} \frac{b^h - 1}{h}$. We can use our definition of e to show that when $b = e$, this slope is equal to 1:

THEOREM 1.26

Another Characterization of the Number e

The number e satisfies the following limit statement:

$$\lim_{h \to 0} \frac{e^h - 1}{h} = 1.$$

Proof. For a proper proof of this limit, we need a technique that we will not cover until Section 3.6. Thus we give here only a convincing argument that uses approximations. Given that $e = \lim_{h \to 0}(1 + h)^{1/h}$ as in Definition 1.25, for sufficiently small values of h we have

$$e \approx (1 + h)^{1/h} \iff e^h \approx 1 + h \iff e^h - 1 \approx h \iff \frac{e^h - 1}{h} \approx 1$$

Since the preceding approximations get better as $h \to 0$, it is reasonable that $\lim_{h \to 0} \frac{e^h - 1}{h} = 1$. ■

Continuity of Exponential and Trigonometric Functions

We can use the definition of e to prove that exponential and logarithmic functions are continuous everywhere they are defined. This allows us to calculate limits such as $\lim_{x \to 2} 3^x = 3^2 = 9$ and $\lim_{x \to 1} \ln x = \ln 1 = 0$ by simple evaluation.

THEOREM 1.27

Continuity of Exponential and Logarithmic Functions

All exponential and logarithmic functions are continuous on their domains.

Proof. The proof hinges entirely on algebra, limit rules, and the definition of e in Definition 1.25. We will prove continuity for (a) the natural exponential function and (b) the natural logarithmic function here, and use limit rules to extend these results to general exponential and logarithmic functions in Exercises 94–96. The proofs are a little technical, but without them we would not be able to compute even the simplest limits of exponential and logarithmic functions!

(a) To prove that e^x is continuous on its domain $\mathbb{R}$ we must show that for all $c \in \mathbb{R}$ we have $\lim_{x \to c} e^x = e^c$. If we want to use the definition of e, then we need to have a limit as $h \to 0$, so we define $h = x - c$. Then $x = c + h$, and as $x \to c$, we have $h \to 0$. This makes our limit equal to

$$\lim_{x \to c} e^x = \lim_{h \to 0} e^{c+h} = \lim_{h \to 0} e^c e^h = e^c \lim_{h \to 0} e^h.$$

The last step follows from the constant multiple rule for limits, since e^c is a constant. At this point we would be done if we could show that $\lim_{h \to 0} e^h = e^0 = 1$. In other words, the proof that e^x is continuous for all x essentially boils down to showing that it is continuous at one point, namely, 0. To finish the proof we employ a series of algebraic manipulations followed by some limit rules and the definition of e:

$$\lim_{h\to 0} e^h = \lim_{h\to 0}(e^h - 1 + 1) = \lim_{h\to 0}\left(\frac{e^h - 1}{h}\cdot h + 1\right) \qquad \leftarrow \text{algebra}$$

$$= \left(\lim_{h\to 0}\frac{e^h - 1}{h}\right)\left(\lim_{h\to 0} h\right) + \left(\lim_{h\to 0} 1\right) \qquad \leftarrow \text{limit rules}$$

$$= (1)(0) + (1) = 1. \qquad \leftarrow \text{Theorem 1.26}$$

One technical point: In the second line of the above calculation we applied the product rule for limits, which is only valid when the limits involved are known to exist. Therefore this proof hinges on knowing that $\lim_{h\to 0}\frac{e^h-1}{h}$ exists, which is a nontrivial fact that we will not prove here.

(b) To show that $\ln x$ is continuous on its domain, we must show that for all $c \in (0, \infty)$ we have $\lim_{x\to c} \ln x = \ln c$. We will use what we just proved about the continuity of e^x. By the composition rule for limits and the fact that e^x is continuous everywhere, for $c > 0$ we have

$$\lim_{x\to c} e^{\ln x} = e^{\lim_{x\to c} \ln x}.$$

Since $\ln x$ is the inverse of e^x, we know that $e^{\ln x} = x$. Therefore for $c > 0$ we also have

$$\lim_{x\to c} e^{\ln x} = \lim_{x\to c} x = c = e^{\ln c}.$$

Since e^x is a one-to-one-function, putting these two calculations together we see that $\lim_{x\to c} \ln x$ must be equal to $\ln c$. (Note that once again, we are making an important assumption here, that $\lim_{x\to c} \ln x$ is equal to some real number. If that number does not exist then we cannot apply the composition rule for limits here.) ■

Using similar techniques, we can prove that trigonometric and inverse trigonometric functions are continuous on their domains. This again allows us to calculate limits of trigonometric and inverse trigonometric functions at domain points by simple evaluation. For example, $\cos x$ is defined at $x = \pi$, and this theorem says that the limit $\lim_{x\to\pi} \cos x$ is equal to the value $\cos(\pi) = -1$. The theorem does not tell us how to calculate $\lim_{x\to\pi} \sec x$, however, because π is not in the domain of $\sec x$.

THEOREM 1.28

Continuity of Trigonometric Functions

All trigonometric and inverse trigonometric functions are continuous on their domains.

Proof. We will prove Theorem 1.28 for the functions (a) $\sin x$ and (b) $\sin^{-1} x$. Proofs for other basic transcendental functions will be covered in Exercises 94–100.

(a) To show that $\sin x$ is continuous on its domain, we must show that $\lim_{x\to c} \sin x = \sin c$ for all $c \in \mathbb{R}$. Following the same technique as for e^x, we change variables with $h = x - c$ and relate to some limits that we already know. Recall that in the discussion after the Squeeze Theorem we showed that $\lim_{x\to 0} \sin x = 0$ and $\lim_{x\to 0} \cos x = 1$. We thus have

$$\lim_{x\to c} \sin x = \lim_{h\to 0} \sin(c + h) \qquad \leftarrow \text{change variables}$$

$$= \lim_{h\to 0}(\sin c \cos h + \sin h \cos c) \qquad \leftarrow \text{sum identity for sine}$$

$$= \sin c \left(\lim_{h\to 0} \cos h\right) + \cos c \left(\lim_{h\to 0} \sin h\right) \qquad \leftarrow \text{sum and constant multiple rules}$$

$$= (\sin c)(1) + (\cos c)(0) = \sin c. \qquad \leftarrow \text{known limits}$$

(b) Finally, to show that $\sin^{-1} x$ is continuous on its domain $[-1, 1]$ we use the fact that it is the inverse of $\sin x$ restricted to $\left[-\frac{\pi}{2}, \frac{\pi}{2}\right]$, with the same method we used for $\ln x$. Before we do this, a technical point: this proof will show that if $\lim_{x\to c} \sin^{-1}$ is equal to some real number, then that real number must be $\sin^{-1} c$. We will assume that $\lim_{x\to c} \sin^{-1}$ exists, a fact that is necessary for the application of limit rules. By the composition rule for limits and the fact that $\sin x$ is continuous everywhere, for $c \in [-1, 1]$ we have

$$\lim_{x\to c} \sin(\sin^{-1} x) = \sin(\lim_{x\to c} \sin^{-1} x).$$

On the other hand, by properties of inverses, for $c \in [-1, 1]$ we also have

$$\lim_{x\to c} \sin(\sin^{-1} x) = \lim_{x\to c} x = c = \sin(\sin^{-1} c).$$

Because $\sin x$ is a one-to-one function on $\left[-\frac{\pi}{2}, \frac{\pi}{2}\right]$, we can put the preceding two calculations together to conclude that $\lim_{x\to c} \sin^{-1} x = \sin^{-1} c$. ■

Delta-Epsilon Proofs of the Limit Rules

The limit rules seem almost obvious; for example, if $f(x)$ approaches L and $g(x)$ approaches M as $x \to c$, it is reasonable to expect that $f(x) + g(x)$ approaches $L + M$ as $x \to c$. To *prove* the limit rules in Theorem 1.20, however, we must appeal to the delta–epsilon definition of limit.

Proof. *(This proof requires material covered in optional Section 1.3.)*
We will prove the (a) sum, (b) product, and (c) composition rules and leave the proofs of the remaining rules to Exercises 91, 92, and 93.

(a) To prove the sum rule for limits, we must show that we can get $f(x) + g(x)$ as close as we like to $L + M$ by choosing δ so that $f(x)$ and $g(x)$ are each half of that distance from L and M, as illustrated in the graph that follows at the left. Given $\epsilon > 0$, choose δ_1 to get $f(x)$ within $\frac{\epsilon}{2}$ of L and choose δ_2 to get $g(x)$ within $\frac{\epsilon}{2}$ of M. Then for $\delta = \min(\delta_1, \delta_2)$ and $x \in (c - \delta, c) \cup (c, c + \delta)$, we have

$$L - \frac{\epsilon}{2} < f(x) < L + \frac{\epsilon}{2} \quad \text{and} \quad M - \frac{\epsilon}{2} < g(x) < M + \frac{\epsilon}{2}.$$

Adding these two double inequalities together, we get our desired conclusion:

$$(L + M) - \epsilon < f(x) + g(x) < (L + M) + \epsilon.$$

Given ϵ for $f+g$, choose δ to get $\frac{\epsilon}{2}$ for f and g

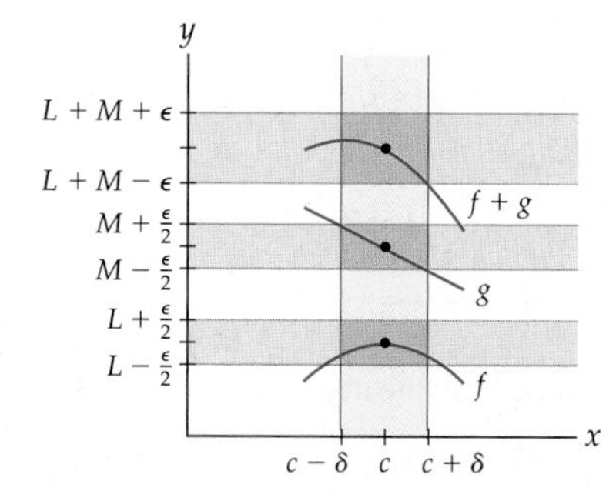

Given ϵ for fg, choose δ to get ϵ' for f and g

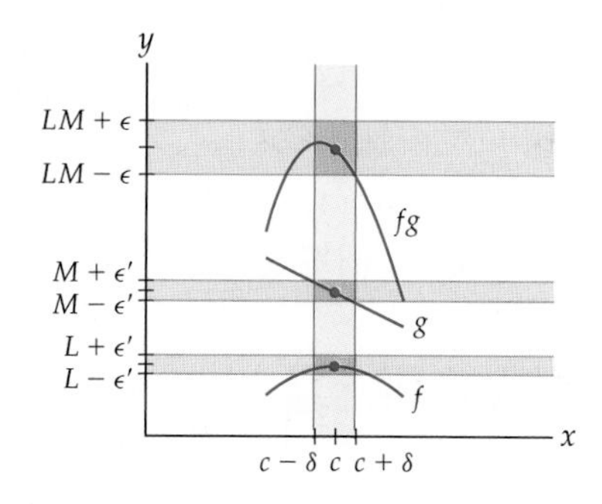

(b) The proof for the product rule for limits is similar: We will get $f(x)g(x)$ as close as we like to LM by choosing δ so that $f(x)$ and $g(x)$ are sufficiently close enough to L and M, respectively, as illustrated in the preceding graph at the right. We will assume here that both L and M are positive; the other cases are similar. Given $\epsilon > 0$, choose $\epsilon' > 0$ sufficiently small so that $\epsilon' < L$, $\epsilon' < M$, and $(L + M)\epsilon' + (\epsilon')^2 < \epsilon$. The reason for this odd choice of ϵ' will become clear in a moment. Now choose δ_1 and δ_2 to get $f(x)$ and $g(x)$ within ϵ' of L and M, respectively. Then for $\delta = \min(\delta_1, \delta_2)$ and $x \in (c - \delta, c) \cup (c, c + \delta)$, we have

$$L - \epsilon' < f(x) < L + \epsilon' \quad \text{and} \quad M - \epsilon' < g(x) < M + \epsilon'.$$

Since L and M are assumed to be positive, we can assume that δ has been chosen small enough so that for $x \in (c - \delta, c) \cup (c, c + \delta)$ the values of $f(x)$ and $g(x)$ are also positive. Therefore, multiplying our two double inequalities together gives

$$(L - \epsilon')(M - \epsilon') < f(x)g(x) < (L + \epsilon')(M + \epsilon')$$

By our choice of ϵ' it is easy to show that $(L + \epsilon')(M + \epsilon') < LM + \epsilon$ and that $LM - \epsilon < (L - \epsilon')(M - \epsilon')$. Putting these two inequalities together with our double inequality for $f(x)g(x)$, we can say that for $x \in (c - \delta, c) \cup (c, c + \delta)$ we have

$$LM - \epsilon < f(x)g(x) < LM + \epsilon.$$

(c) To prove the composition rule for limits, let $\lim_{x \to c} g(x) = L$. We will show that we can get $f(g(x))$ as close as we like to $f(L)$ by choosing δ so that $g(x)$ is sufficiently close to L. Since f is continuous at L, we know that $\lim_{u \to L} f(u) = L$. Thus given $\epsilon > 0$, we can choose $\delta' > 0$ so that whenever $u \in (L - \delta', L) \cup (L, L + \delta')$, we also have $f(u) \in (f(L) - \epsilon, f(L) + \epsilon)$. In fact, if $u = L$, then $f(u) = f(L)$, so we can say a little bit more:

$$\text{if } u \in (L - \delta', L + \delta'), \text{ then } f(u) \in (f(L) - \epsilon, f(L) + \epsilon).$$

Now, $\lim_{x \to c} g(x) = L$ allows us to choose $\delta > 0$ so that

$$\text{if } x \in (c - \delta, c) \cup (c, c + \delta), \text{ then } g(x) \in (L - \delta', L + \delta').$$

Given ϵ for $f(u)$, choose δ' for u

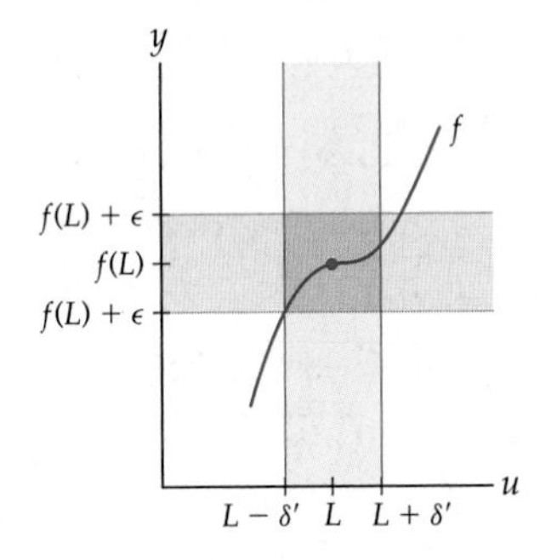

Then given δ' for u, choose δ for x

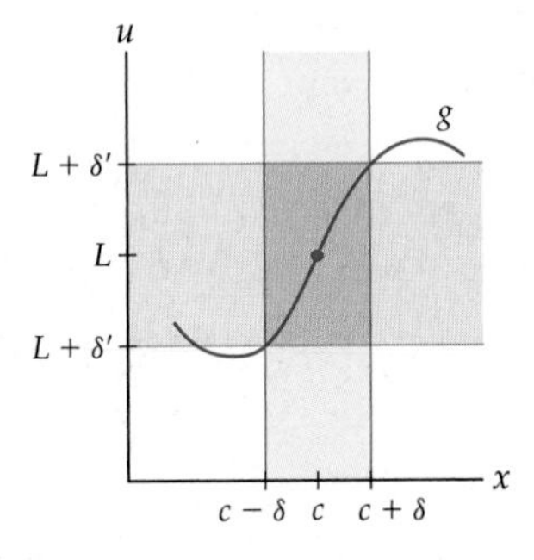

The two figures above illustrate our choices for δ' and δ. Now given our choice of δ we can let $g(x) = u$ and string together the last two displayed implications above to conclude that

$$\text{if } x \in (c - \delta, c) \cup (c, c + \delta), \text{ then } f(g(x)) \in (f(L) - \epsilon, f(L) + \epsilon). \quad \blacksquare$$

Examples and Explorations

EXAMPLE 1

Calculating limits by using continuity and limit rules

Calculate the limits that follow, using only the continuity of linear and power functions and the limit rules in Theorem 1.20. Cite each limit rule that you apply.

(a) $\lim_{x \to 3} (3x^2 - 2x + 1)$

(b) $\lim_{x \to 1} (3x - 1)^{12}$

SOLUTION

(a) We apply the rules for limits of combinations of functions until we have reduced the problem to limits of functions that we know to be continuous:

$$\begin{aligned}\lim_{x\to3}(3x^2-2x+1) &= \lim_{x\to3}3x^2-\lim_{x\to3}2x+\lim_{x\to3}1 && \leftarrow \text{sum and difference rules}\\ &= 3\lim_{x\to3}x^2-2\lim_{x\to3}x+\lim_{x\to3}1 && \leftarrow \text{constant multiple rule}\\ &= 3(3^2)-2(3)+1=22. && \leftarrow \text{limits of continuous functions}\end{aligned}$$

(b) We could find this limit by multiplying out $(3x-1)^{12}$ and then applying the sum and constant multiple rules as we did in part (a), but it is much faster to instead apply the composition rule for limits. We know that the power function x^{12} is continuous everywhere and, in particular, at $\lim_{x\to1}(3x-1)=2$. Therefore we have

$$\begin{aligned}\lim_{x\to1}(3x-1)^{12} &= \left(\lim_{x\to1}(3x-1)\right)^{12} && \leftarrow \text{composition rule for limits}\\ &= (3(1)-1)^{12}=2^{12}=4096. && \leftarrow \text{linear functions are continuous}\end{aligned}$$

□

EXAMPLE 2 **Limits and continuity of piecewise-defined functions**

Describe the continuity or discontinuity of each of the following piecewise-defined functions at $x=1$ by algebraically calculating the left, right, and two-sided limits at $x=1$:

(a) $f(x)=\begin{cases}x+1, & \text{if } x<1\\ 3-x^2, & \text{if } x\geq1\end{cases}$ **(b)** $g(x)=\begin{cases}4-x^2, & \text{if } x\leq1\\ x-1, & \text{if } x>1\end{cases}$

SOLUTION

(a) These two functions are the same ones we investigated graphically in Example 1 of Section 1.4. Here, for each function we must compute the limit as $x\to1$ and compare it with the value at $x=1$. Since the piecewise-defined function f has a break point at $x=1$ we must compute its left and right limits separately. Because the component functions $x+1$ and $3-x^2$ are continuous, we can calculate the left and right limits by evaluation:

$$\lim_{x\to1^-}f(x)=\lim_{x\to1^-}(x+1)=1+1=2 \quad\text{and}$$
$$\lim_{x\to1^+}f(x)=\lim_{x\to1^+}(3-x^2)=3-(1)^2=2.$$

Since the left and right limits both exist and are equal to 2, we have $\lim_{x\to1}f(x)=2$, which is equal to the value $f(1)=2$. Therefore, f is continuous at $x=2$.

(b) Once again we must calculate the left and right limits of $g(x)$ separately. By continuity of the component functions we have

$$\lim_{x\to1^-}g(x)=\lim_{x\to1^-}(4-x^2)=4-(1)^2=3 \quad\text{and}$$
$$\lim_{x\to1^+}g(x)=\lim_{x\to1^+}(x-1)=1-1=0.$$

Since the left and right limits both exist but are not equal as $x\to1$, the function $g(x)$ has a jump discontinuity at $x=1$. Because $\lim_{x\to1^-}g(x)=3=g(1)$ but $\lim_{x\to1^+}g(x)=0\neq g(1)$, we can also say that $g(x)$ is left continuous, but not right continuous, at $x=1$. □

EXAMPLE 3

Calculating limits

Calculate each of the following limits:

(a) $\lim\limits_{x\to 0}\sqrt{\sin\left(x+\frac{\pi}{2}\right)}$ **(b)** $\lim\limits_{x\to 2}\frac{2-x}{4-x^2}$ **(c)** $\lim\limits_{x\to 0}\frac{e^{2x}+e^x-2}{e^x-1}$

SOLUTION

(a) $f(x) = \sqrt{\sin\left(x+\frac{\pi}{2}\right)}$ is a combination of functions that are continuous on their domains, and thus is continuous on its domain. Since $x = 0$ is in the domain of f we can solve this limit by simple evaluation:

$$\lim_{x\to 0}\sqrt{\sin\left(x+\frac{\pi}{2}\right)} = \sqrt{\sin\left(0+\frac{\pi}{2}\right)} = \sqrt{1} = 1.$$

(b) The function $f(x) = \frac{2-x}{4-x^2}$ is algebraic and thus continuous on its domain, but unfortunately $x = 2$ is not in that domain. As $x \to 2$, both the numerator and the denominator approach 0, and therefore this limit is of the form $\frac{0}{0}$, which is indeterminate. This means that we don't know at this point whether or not the limit exists or not, or if it does, what it might be equal to. We can solve this indeterminacy by doing some preliminary algebra; after cancellation we get a limit that is no longer indeterminate and, in fact, that we can find by evaluation at $x = 2$:

$$\lim_{x\to 2}\frac{2-x}{4-x^2} = \lim_{x\to 2}\frac{2-x}{(2-x)(2+x)} = \lim_{x\to 2}\frac{1}{2+x} = \frac{1}{4}.$$

(c) As $x \to 0$ both the numerator and the denominator approach zero, so this limit is of the form $\frac{0}{0}$, which is indeterminate. After factoring and cancelling we can resolve this problem:

$$\lim_{x\to 0}\frac{e^{2x}+e^x-2}{e^x-1} = \lim_{x\to 0}\frac{(e^x-1)(e^x+2)}{e^x-1} = \lim_{x\to 0}(e^x+2) = e^0+2 = 3.$$ □

EXAMPLE 4

Finding a Limit with the Squeeze Theorem

Use the Squeeze Theorem to find $\lim\limits_{x\to 0} x^2 \sin\frac{1}{x}$.

SOLUTION

The graph of $f(x) = x^2 \sin\frac{1}{x}$ follows at the left.

$f(x) = x^2 \sin\frac{1}{x}$

y
0.10
0.05
−0.4
−0.2
0.2
0.4
x
−0.05
−0.10

Squeezed between $y = x^2$ and $y = -x^2$

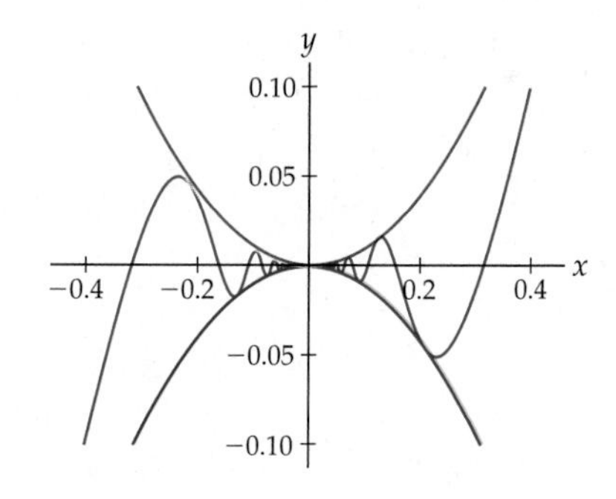

From this graph it seems reasonable that $\lim_{x \to 0} x^2 \sin \frac{1}{x} = 0$. We can use the Squeeze Theorem to find this limit algebraically. The sine function has outputs that are always between -1 and 1. Therefore $-1 < \sin \frac{1}{x} < 1$, which means that $-x^2 < x^2 \sin \frac{1}{x} < x^2$, as shown in the graph at the right. We know that $\lim_{x \to 0} -x^2 = 0$ and $\lim_{x \to 0} x^2 = 0$. Therefore by the Squeeze Theorem we must also have $\lim_{x \to 0} x^2 \sin \frac{1}{x} = 0$. □

EXAMPLE 5

Using tables of values to approximate limits related to e

Use tables of values to approximate each of the limits that follow. In the second limit you will have to use a calculator approximation of e to perform the calculations.

(a) $\lim_{h \to 0} \frac{3^h - 1}{h}$ **(b)** $\lim_{h \to 0} \frac{e^h - 1}{h}$ **(c)** $\lim_{h \to 0} (1 + h)^{1/h}$

SOLUTION

(a) This limit is similar to the one in Theorem 1.26, but with the base e replaced by 3. Limits of the form $\lim_{h \to 0} \frac{b^h - 1}{h}$ converge to different numbers, depending on the base b, but such a limit converges to 1 only for the number $b = e$. Thus we would expect that $\lim_{h \to 0} \frac{3^h - 1}{h}$ should converge to some number other than 1. The following table of values of $\frac{3^h - 1}{h}$ suggests that this is indeed the case:

h	-0.1	-0.01	-0.001	0	0.001	0.01	0.1
$\frac{3^h - 1}{h}$	1.040415	1.092600	1.098009	?	1.099216	1.104669	1.161232

If the pattern in the table continues, then we would expect this limit to converge to some number between 1.098009 and 1.099216. The numbers in that range are close to 1, but none of them are equal to 1. This is not unexpected, because the base in our limit, which is 3, is close to the base $e \approx 2.71828$ that would cause the limit we seek to approach exactly 1.

(b) From Theorem 1.26 we know that $\lim_{h \to 0} \frac{e^h - 1}{h}$ should approach 1. Let's see if that is the case. For small values of h approaching 0 we have

h	-0.1	-0.01	-0.001	0	0.001	0.01	0.1
$\frac{e^h - 1}{h}$	0.951626	0.995017	0.999500	?	1.000500	1.005017	1.051709

As expected, this limit does seem to be approaching 1 as h approaches 0.

(c) We can approximate the limit in Definition 1.25 (and thus the value of e) by using a table of approximate values of $(1 + h)^{1/h}$ for small values of h:

h	-0.1	-0.01	-0.001	0	0.001	0.01	0.1
$(1 + h)^{1/h}$	2.867972	2.731999	2.719642	?	2.716924	2.704814	2.593742

Although $(1 + h)^{1/h}$ does not seem to approach a very *nice* number as $h \to 0$, it appears that the limit does exist and is somewhere between 2.719642 and 2.716923. If we evaluate $(1+h)^{1/h}$ at an extremely small value of h, say, $h = 0.000001$, then we can get a relatively accurate approximation for the limit in Definition 1.25 and thus for the number e:

$$e \approx (1 + 0.000001)^{1/0.000001} \approx 2.718280. \qquad \square$$

? TEST YOUR UNDERSTANDING

- The sum rule for limits says that the limit of a sum is the sum of the limits. In English, what do the other limit rules say?
- In the proof of the sum rule for limits, in order to get $f(x)+g(x)$ within ϵ of $L+M$, how close do we have to get $f(x)$ and $g(x)$ to L and M, respectively?
- What is the rule for the limit of a constant? What is the rule for the limit of a constant multiple of a function? How are these two rules different?
- Why does it make sense that cancellation would be a valid operation when dealing with a limit as $x \to c$, even if what is being cancelled approaches zero as $x \to c$?
- In the Squeeze Theorem for limits, why do we require that the upper and lower functions $u(x)$ and $l(x)$ have the same limit as $x \to c$?

EXERCISES 1.5

Thinking Back

Values of transcendental functions: Without a calculator, find each of the function values that follow. For some values the answer may be undefined.

- If $f(x) = \csc x$, find $f(\pi)$ and $f\left(\frac{\pi}{2}\right)$.
- If $f(x) = \tan^2 x$, find $f(\pi)$ and $f\left(\frac{\pi}{2}\right)$.
- If $f(x) = \sin^{-1} x$, find $f(-1)$ and $f\left(\frac{1}{2}\right)$.
- If $f(x) = \tan^{-1}\sqrt{x}$, find $f(1)$ and $f(3)$.

The δ–ϵ definition of limit: Write each limit statement that follows in terms of the δ–ϵ definition of limit. Then approximate the largest value of δ corresponding to $\epsilon = 0.5$, and illustrate this choice of δ on a graph of f.

- $\lim_{x \to 2}(3x - 2) = 4$
- $\lim_{x \to 1}(x^3 - 1) = 0$
- $\lim_{x \to 1} \frac{x^2 - 1}{x + 3} = 0$
- $\lim_{x \to 0^+} \sqrt{x + 4} = 2$

Concepts

0. *Problem Zero:* Read the section and make your own summary of the material.

1. *True/False:* Determine whether each of the statements that follow is true or false. If a statement is true, explain why. If a statement is false, provide a counterexample.

(a) *True or False:* The limit of a difference of functions as $x \to c$ is equal to the difference of the limits of those functions as $x \to c$, provided that all limits involved exist.

(b) *True or False:* If $f(x)$ is within 0.25 unit of 7 and $g(x)$ is within 0.25 unit of 2, then $f(x) + g(x)$ is within 0.5 unit of 9.

(c) *True or False:* If $f(x)$ is within 0.25 unit of 7 and $g(x)$ is within 0.25 unit of 2, then $f(x)g(x)$ is within 0.5 unit of 9.

(d) *True or False:* Every algebraic function f is continuous at every real number $x = c$.

(e) *True or False:* Every power function $f(x) = Ax^k$ is continuous at the point $x = 2$.

(f) *True or False:* The function $f(x) = \sec x$ is continuous at $x = \frac{\pi}{2}$.

(g) *True or False:* The value of $\frac{(x-c)f(x)}{(x-c)g(x)}$ at $x = c$ is equal to the limit of $\frac{f(x)}{g(x)}$ at $x = c$.

(h) *True or False:* The limit of $\frac{(x-c)f(x)}{(x-c)g(x)}$ as $x \to c$ is equal to the limit of $\frac{f(x)}{g(x)}$ as $x \to c$.

2. *Examples:* Construct examples of the thing(s) described in the following. Try to find examples that are different than any in the reading.
   (a) Two limits that are initially in an indeterminate form but can be solved with the Cancellation Theorem.
   (b) Two limits that can be solved with the Squeeze Theorem.
   (c) Two limits that we do not yet know how to calculate.
3. State the constant multiple rule, sum rule, product rule, quotient rule, and composition rule for limits.
4. Explain in your own words the types of functions whose limits we can calculate with the limit rules in this section.
5. Explain why we can't calculate every limit $\lim_{x \to c} f(x)$ just by evaluating $f(x)$ at $x = c$. Support your argument with the graph of a function f for which $\lim_{x \to c} f(x) \neq f(c)$.
6. Find functions f and g and a real number c such that $\lim_{x \to c} f(x) + \lim_{x \to c} g(x) \neq \lim_{x \to c}(f(x) + g(x))$. Does this example contradict the sum rule for limits? Why or why not?
7. Find functions f and g and a real number c such that $(\lim_{x \to c} f(x))(\lim_{x \to c} g(x)) \neq \lim_{x \to c}(f(x)g(x))$. Does this example contradict the product rule for limits? Why or why not?
8. Write the constant multiple rule for limits in terms of delta–epsilon statements.
9. Write the difference rule for limits in terms of delta–epsilon statements.
10. Write the product rule for limits in terms of delta–epsilon statements.
11. Explain how the algebraic function
$$f(x) = (\sqrt{x} + 1)^3$$
is a combination of identity, constant, and power functions. Why does this mean that we can calculate limits of this function at domain points by evaluation?
12. Explain how the algebraic function
$$f(x) = \frac{(x^2 + 1)(4 - 3x)}{3x^2}$$
is a combination of identity, constant, and power functions. Why does this mean that we can calculate limits of this function at domain points by evaluation?

Suppose f and g are functions such that $\lim_{x \to 3} f(x) = 5$, $\lim_{x \to 4} f(x) = 2$, and $\lim_{x \to 3} g(x) = 4$. Given this information, calculate the limits that follow, if possible. If it is not possible with the given information, explain why.

13. $\lim_{x \to 3}(2f(x) - 3g(x))$
14. $\lim_{x \to 3} -2f(x)$
15. $\lim_{x \to 7} f(x)$
16. $\lim_{x \to 4} f(x)g(x)$
17. $\lim_{x \to 3} \frac{f(x) - 3}{g(x)}$
18. $\lim_{x \to 3} f(g(x))$
19. Graph the functions $f(x) = x + 1$ and $g(x) = \frac{x^2 - 1}{x - 1}$, and show that they are equal everywhere except at one point. Then show that $f(x)$ and $g(x)$ have different values, but the same limit, at this point.
20. Graph the functions $f(x) = 2 - x$ and $g(x) = \frac{4 - x^2}{x + 2}$, and show that they are equal everywhere except at one point. Then show that $f(x)$ and $g(x)$ have different values, but the same limit, at this point.
21. In the Squeeze Theorem for limits, we require that $l(x) \le f(x) \le u(x)$ for all x sufficiently close to c, but we do not require this inequality to hold *at* the point $x = c$. Why not?
22. Use a geometric argument and the Squeeze Theorem for limits to argue that
$$\lim_{\theta \to 0^-} \sin\theta = 0$$
for sufficiently small negative angles θ.
23. Use a geometric argument and the Squeeze Theorem for limits to argue that
$$\lim_{\theta \to 0^-} \cos\theta = 1$$
for sufficiently small negative angles θ.
24. In this exercise you will use a calculator to investigate the number e.
   (a) Make a table of values that describes the behavior of the quantity $(1 + h)^{1/h}$ as $h \to 0$.
   (b) Make a table of values that describes the behavior of the quantity $\frac{e^h - 1}{h}$ as $h \to 0$.
   (c) What do your tables of values have to do with Definition 1.25 and Theorem 1.26?

Skills

Calculate the limits in Exercises 25–28, using only the continuity of linear and power functions and the limit rules. Cite each limit rule that you apply.

25. $\lim_{x \to 1} 15(3 - 2x)$
26. $\lim_{x \to -1} \frac{x - 1}{(x + 4)(x + 2)}$
27. $\lim_{x \to 3}(3x + x^2(2x + 1))$
28. $\lim_{x \to 0} \frac{3}{2x^2 - 4x + 1}$

Calculate each of the limits in Exercises 29–70.

29. $\lim_{x \to 0}(x^2 - 1)$
30. $\lim_{x \to 2}(x - 1)(x + 1)(x + 5)$
31. $\lim_{x \to -1}\left(x^2 - \frac{3x}{x + 2}\right)$
32. $\lim_{x \to 1.7}(3.1x^2 - 4x + 0.8)$
33. $\lim_{x \to 2} \frac{4 + 2x}{x + 2}$
34. $\lim_{x \to -2} \frac{4 + 2x}{x + 2}$
35. $\lim_{x \to 1} \frac{x^2 - 1}{x - 1}$
36. $\lim_{x \to 0} \frac{x^2 - 1}{x - 1}$
37. $\lim_{x \to -3} \frac{x + 3}{3x^2 + 8x - 3}$
38. $\lim_{x \to -3} \frac{3x^2 + 8x - 3}{x + 3}$
39. $\lim_{x \to 1/2} \frac{4x - 2}{6x^2 + x - 2}$
40. $\lim_{x \to -2/3} \frac{6x^2 + x - 2}{3x + 2}$

41. $\lim_{x\to 1^+} \frac{x-1}{\sqrt{x-1}}$

42. $\lim_{x\to 1^+} \frac{1-\sqrt{x}}{1-x}$

43. $\lim_{x\to 1} \frac{x+x^2-2x^3}{x-x^2}$

44. $\lim_{x\to 0} \frac{x+x^2-2x^3}{x-x^2}$

45. $\lim_{h\to 0} \frac{(1+h)^2-1}{h}$

46. $\lim_{h\to 0} \frac{(-1+h)^2-1}{h}$

47. $\lim_{x\to 0} \frac{2^x-3^x}{4^x}$

48. $\lim_{x\to 2^+} \sqrt{2^x-4}$

49. $\lim_{x\to 4}(3e^{1.7x}+1)$

50. $\lim_{x\to 0^+} \ln(1+\sqrt{x})$

51. $\lim_{x\to 0} \frac{e^x-1}{e^{2x}+2e^x-3}$

52. $\lim_{x\to 1} \frac{e^x-1}{e^{2x}+2e^x-3}$

53. $\lim_{x\to 0} \frac{2x}{e^x-1}$

54. $\lim_{x\to 0} \frac{e^x-1}{x}$

55. $\lim_{x\to \pi} \frac{1}{\csc(x-\pi)}$

56. $\lim_{x\to 1} \frac{1-\cos(x-1)}{x}$

57. $\lim_{x\to 0} \frac{\sin x}{\tan x}$

58. $\lim_{x\to \pi/2} \frac{\cot x}{\cos x}$

59. $\lim_{x\to 1} x\sin^{-1}\frac{x}{2}$

60. $\lim_{x\to 3} \frac{\sqrt{x}}{\tan^{-1}\sqrt{x}}$

61. $\lim_{x\to 1^-} \frac{1}{\sin^{-1}x}$

62. $\lim_{x\to 0} \frac{1}{\sec^{-1}x}$

63. $\lim_{h\to 0} \frac{(3+h)^2-3^2}{h}$

64. $\lim_{h\to 0} \frac{(2+h)^2-2^2}{h}$

65. $\lim_{h\to 0} \frac{(1+h)^3-1^3}{h}$

66. $\lim_{h\to 0} \frac{(-1+h)^3-(-1)^3}{h}$

67. $\lim_{h\to 0} \frac{\frac{1}{1+h}-1}{h}$

68. $\lim_{h\to 0} \frac{\frac{1}{2+h}-\frac{1}{2}}{h}$

69. $\lim_{h\to 0} \frac{\frac{4}{(2+h)^2}-1}{h}$

70. $\lim_{h\to 0} \frac{\frac{4}{(1+h)^2}-4}{h}$

Describe the intervals on which each function f in Exercises 71–78 is continuous. At each point where f fails to be continuous, use limits to determine the type of discontinuity and any left- or right-continuity.

71. $f(x) = \begin{cases} x^2+1, & \text{if } x \le 0 \\ 1-x, & \text{if } x > 0 \end{cases}$

72. $f(x) = \begin{cases} 3x+2, & \text{if } x < -1 \\ 5+4x^3, & \text{if } x \ge -1 \end{cases}$

73. $f(x) = \begin{cases} x^2-3x-1, & \text{if } x \ne -2 \\ 3, & \text{if } x = -2 \end{cases}$

74. $f(x) = \begin{cases} \frac{3x^2}{4+x}, & \text{if } x < 2 \\ x^2-2, & \text{if } x \ge 2 \end{cases}$

75. $f(x) = \begin{cases} \frac{x^2-1}{x-1}, & \text{if } x < 1 \\ 0, & \text{if } x = 1 \\ 3x-1, & \text{if } x > 1 \end{cases}$

76. $f(x) = \begin{cases} x+1, & \text{if } x < 3 \\ 2, & \text{if } x = 3 \\ x^2-9, & \text{if } x > 3 \end{cases}$

77. $f(x) = \begin{cases} \sin x, & \text{if } x < \pi \\ \cos x, & \text{if } x \ge \pi \end{cases}$

78. $f(x) = \begin{cases} 2^x-1, & \text{if } x \le 0 \\ \frac{4^x-1}{2^x-1}, & \text{if } x > 0 \end{cases}$

Use the Squeeze Theorem to find each of the limits in Exercises 79–86. Explain exactly how the Squeeze Theorem applies in each case.

79. $\lim_{x\to 0} x\sin\frac{1}{x}$

80. $\lim_{x\to 0} x\sin\frac{1}{x^2}$

81. $\lim_{x\to 0}(e^x-1)\sin\frac{1}{x}$

82. $\lim_{x\to 0} \sin x \sin\frac{1}{x}$

83. $\lim_{x\to 1}(x-1)\cos\frac{1}{x-1}$

84. $\lim_{x\to 1}(x-1)^2\cos\frac{1}{x-1}$

85. $\lim_{x\to 0} x\tan^{-1}\frac{1}{x}$

86. $\lim_{x\to 0} x^2\tan^{-1}\frac{1}{x}$

Applications

In Exercise 88 in Section 0.1, you constructed a piecewise-defined function from the 2000 Federal Tax Rate Schedule that you will use in the next two problems. Specifically, you found that a person who makes m dollars a year will pay $T(m)$ dollars in tax, given by the function

$$T(m) = \begin{cases} 0.15m, & \text{if } 0 \le m \le 26{,}250 \\ 3{,}937 + 0.28(m - 26{,}250), & \text{if } 26{,}250 < m \le 63{,}550 \\ 14{,}381 + 0.31(m - 63{,}550), & \text{if } 63{,}550 < m \le 132{,}600 \\ 35{,}787 + 0.36(m - 132{,}600), & \text{if } 132{,}600 < m \le 288{,}350 \\ 91{,}857 + 0.396(m - 288{,}350), & \text{if } m > 288{,}350. \end{cases}$$

87. Suppose you make $63,550 a year and pay taxes according to the given formula.

(a) Calculate the value of $T(63{,}550)$ and the limit of $T(m)$ as m approaches 63,550 from the left and from the right.

(b) Use part (a) to argue that the function $T(m)$ is continuous at $m = 63{,}550$. What does this mean in real-world terms?

88. Suppose you make $288,350 a year and pay taxes according to the given formula.

(a) Calculate the value of $T(288{,}350)$ and the limit of $T(m)$ as m approaches 288,350 from the left and from the right.

(b) Use part (a) to argue that the function $T(m)$ is continuous at $m = 288{,}350$. What does this mean in real-world terms?

Proofs

89. Use limit rules and the continuity of power functions to prove that every polynomial function is continuous everywhere.

90. Use limit rules and the continuity of polynomial functions to prove that every rational function is continuous on its domain.

91. Prove the constant multiple rule for limits: If $\lim_{x \to c} f(x) = L$ and $k \in \mathbb{R}$, then $\lim_{x \to c} kf(x) = kL$.

92. Prove the difference rule for limits by applying the sum and constant multiple rules for limits.

93. Suppose that we know the reciprocal rule for limits: If $\lim_{x \to c} g(x) = M$ exists and is nonzero, then $\lim_{x \to c} \frac{1}{g(x)} = \frac{1}{M}$. This limit rule is tedious to prove and we do not include it here. Use the reciprocal rule and the product rule for limits to prove the quotient rule for limits.

94. Use algebra, limit rules, and the continuity of e^x to prove that every exponential function of the form $f(x) = Ae^{kx}$ is continuous everywhere.

95. Use algebra, limit rules, and the continuity of e^x to prove that every exponential function of the form $f(x) = Ab^x$ is continuous everywhere.

96. Use algebra, limit rules, and the continuity of $\ln x$ on $(0, \infty)$ to prove that every logarithmic function of the form $f(x) = \log_b x$ is continuous on $(0, \infty)$.

97. In the reading we used the Squeeze Theorem to prove that $\lim_{h \to 0} \sin h = 0$ and $\lim_{h \to 0} \cos h = 1$. Use these facts, the sum identity for cosine, and limit rules to prove that $f(x) = \cos x$ is continuous everywhere.

98. Use the quotient rule for limits and the continuity of $\sin x$ and $\cos x$ to prove that $f(x) = \tan x$ is continuous on its domain.

99. Use the quotient rule for limits and the continuity of $\cos x$ to prove that $f(x) = \sec x$ is continuous on its domain.

100. Use the composition rule for limits and the fact that $\tan x$ is continuous on its domain to prove that $\tan^{-1} x$ is continuous everywhere.

Thinking Forward

Limits for Derivatives: In Chapter 2 we will define the ***derivative*** of a function f at a point $x = c$ to be the slope of the line that points in the direction of the graph of f at $x = c$. Algebraically, the derivative of f at c is given by the following limit:

$$\lim_{h \to 0} \frac{f(c+h) - f(c)}{h}.$$

Such limits are always of the indeterminate form $\frac{0}{0}$, so we must do algebra before we can resolve the limit.

- Calculate the derivative of $f(x) = x^2$ at $c = 0$.
- Calculate the derivative of $f(x) = x^2$ at $c = 2$.
- Calculate the derivative of $f(x) = x^2$ at $c = 4$.
- Sketch a graph of $f(x) = x^2$, and sketch the lines that point in the direction of the curve at $(0, f(0))$, $(2, f(2))$, and $(4, f(4))$. Relate the slopes of these lines to the answers to the last three exercises.
- Use the definition of the derivative to calculate the derivative of $f(x) = \sqrt{x}$ at $c = 4$. At some point you will need to multiply numerator and denominator by the ***conjugate*** of $\sqrt{4+h} - 2$, which is $\sqrt{4+h} + 2$.
- Use the definition of the derivative to calculate the derivative of $f(x) = x^{-1/2}$ at $c = 4$. As in the previous calculation, you will need to multiply numerator and denominator by a conjugate at some point.
- Calculate the derivative of $f(x) = e^x$ at $c = 0$. At some point you should need the characterization of e given in Theorem 1.26.

1.6 INFINITE LIMITS AND INDETERMINATE FORMS

- Calculating limits at infinity and infinite limits
- Recognizing non-indeterminate forms and dealing with indeterminate forms
- Special trigonometric limits

Infinite Limits

The utility of continuity is that it enables us to calculate limits by evaluation. However, we can't solve limits by evaluation if continuity fails. For example, consider the limit $\lim_{x\to 1} \frac{1}{x-1}$. The function $\frac{1}{x-1}$ is neither defined nor continuous at $x = 1$, so we cannot find its limit by evaluation. If we try to evaluate $\frac{1}{x-1}$ at $x = 1$, we would not get a real number, because the denominator would be 0. In terms of limits, we say that

$$\lim_{x\to 1} \frac{1}{x-1} \text{ is of the form } \frac{1}{0}, \text{ and thus } \lim_{x\to 1} \frac{1}{x-1} \text{ is not a real number.}$$

Can we be more specific than just pointing out that limits of type $\frac{1}{0}$ "do not exist" (i.e., are not real numbers)? Consider the behavior of $\frac{1}{x-1}$ as x approaches 1 from the right, as shown in the table below. As $x \to 1^+$ (see the first row of the table) the values of $x - 1$ approach 0^+ (see the second row). The reciprocals of these values (see the third row) then approach ∞:

x	1.1	1.01	1.001	1.0001	1.00001	$\to 1^+$
$x-1$	0.1	0.01	0.001	0.0001	0.00001	$\to 0^+$
$\frac{1}{x-1}$	10	100	1000	10,000	100,000	$\to \infty$

In symbols, this means that $\frac{1}{x-1} \to \infty$ as $x \to 1^+$. In terms of limits we can express this behavior from the right by saying that

$$\lim_{x\to 1^+} \frac{1}{x-1} \text{ is of the form } \frac{1}{0^+}, \text{ and thus } \lim_{x\to 1^+} \frac{1}{x-1} = \infty.$$

We will prove a more general version of this statement in Theorem 1.29.

From the left, we have a similar situation: As $x \to 1^-$, we have $x - 1 \to 0^-$, so the quantity $\frac{1}{x-1}$ will remain negative as $x \to 1^-$, and as the magnitude of the denominator $x - 1$ gets smaller and smaller, the magnitude of $\frac{1}{x-1}$ gets larger and larger. In terms of limits we can express this behavior from the left by saying that

$$\lim_{x\to 1^-} \frac{1}{x-1} \text{ is of the form } \frac{1}{0^-}, \text{ and thus } \lim_{x\to 1^-} \frac{1}{x-1} = -\infty.$$

We have now shown that $\lim_{x\to 1} \frac{1}{x-1}$ does not exist, and more specifically, that the limit is ∞ from the right and $-\infty$ from the left.

If an expression approaches ∞ from both the right and the left, then we say that the two-sided limit is ∞. For example, since $\frac{1}{(x-1)^2}$ approaches ∞ from both the right and the left as $x \to 1$, we would write $\lim_{x\to 1} \frac{1}{(x-1)^2}$ is of the form $\frac{1}{0^+}$, and thus $\lim_{x\to 1} \frac{1}{(x-1)^2} = \infty$.

The theoretical basis for the discussion above is summarized in the following theorem:

THEOREM 1.29 **Limits Whose Denominators Approach Zero from the Right or the Left**

(a) If $\lim_{x \to c} \frac{f(x)}{g(x)}$ is of the form $\frac{1}{0^+}$, then $\lim_{x \to c} \frac{f(x)}{g(x)} = \infty$.

(b) If $\lim_{x \to c} \frac{f(x)}{g(x)}$ is of the form $\frac{1}{0^-}$, then $\lim_{x \to c} \frac{f(x)}{g(x)} = -\infty$.

Theorem 1.29 also applies to one-sided limits and to limits as $x \to \infty$ or as $x \to -\infty$. We will prove only the case for limits from the right:

Proof. We will prove the case where $\lim_{x \to c^+} \frac{f(x)}{g(x)}$ is of the form $\frac{1}{0^+}$. The other cases are similar; you will handle another in Exercise 87. Since $f(x) \to 1$ as $x \to 0$, it follows that for any $\epsilon_1 > 0$, we can find $\delta_1 > 0$ such that

$$\text{if } c < x < c + \delta_1, \text{ then } 1 - \epsilon_1 < f(x) < 1 + \epsilon_1.$$

Similarly, since $g(x) \to 0^+$ as $x \to 0$, it follows that for any $\epsilon_2 > 0$, we can find $\delta_2 > 0$ such that

$$\text{if } c < x < c + \delta_2, \text{ then } 0 < g(x) < 0 + \epsilon_2.$$

Now to prove that $\lim_{x \to c^+} \frac{f(x)}{g(x)} = \infty$, take any $M > 0$. Choose δ to be the minimum of the δ_1 corresponding to $\epsilon_1 = \frac{1}{2}$ and the δ_2 corresponding to $\epsilon_2 = \frac{1}{2M}$. With this choice of δ, we have

$$\text{if } c < x < c + \delta, \text{ then } \frac{f(x)}{g(x)} > \frac{1 - \epsilon_1}{\epsilon_2} = \frac{\frac{1}{2}}{\frac{1}{2M}} = M.$$

■

Limits at Infinity

We have just seen that limits of the form $\frac{1}{0^+}$ and $\frac{1}{0^-}$ are always infinite. A sort of reverse of this is also true, and is the subject of our next theorem: Limits of the form $\frac{1}{\infty}$ and $\frac{1}{-\infty}$ are always zero. This makes intuitive sense because the reciprocal of a number of large magnitude is a number of small magnitude. Note that even though the expression "$\frac{1}{\infty}$" does not represent a real number, a limit of that form *will* be equal to a real number, namely, 0.

THEOREM 1.30 **Limits Whose Denominators Become Infinite Approach Zero**

(a) If $\lim_{x \to \infty} \frac{f(x)}{g(x)}$ is of the form $\frac{1}{\infty}$, then $\lim_{x \to \infty} \frac{f(x)}{g(x)} = 0$.

(b) If $\lim_{x \to \infty} \frac{f(x)}{g(x)}$ is of the form $\frac{1}{-\infty}$, then $\lim_{x \to \infty} \frac{f(x)}{g(x)} = 0$.

For example, as $x \to \infty$ we have $x + 1 \to \infty$ and thus $\frac{1}{x+1} \to 0$; therefore $\lim_{x \to \infty} \frac{1}{x+1} = 0$. Similarly, as $x \to -\infty$ we have $x + 1 \to -\infty$ and thus $\frac{1}{x+1} \to 0$; therefore $\lim_{x \to -\infty} \frac{1}{x+1} = 0$. Theorem 1.30 also applies for limits as $x \to -\infty$ and as $x \to c$, although we will not prove that here.

Proof. We prove the first part and leave the second part to Exercise 88. Since $f(x) \to 1$, it follows that for any $\epsilon_1 > 0$, we can find $N_1 > 0$ such that

$$\text{if } x > N_1, \text{ then } 1 - \epsilon < f(x) < 1 + \epsilon.$$

Similarly, since $g(x) \to \infty$, it follows that for any $M > 0$, we can find $N_2 > 0$ such that

$$\text{if } x > N_2, \text{ then } g(x) > M.$$

Now to prove that $\lim_{x \to \infty} \frac{f(x)}{g(x)} = 0$, take any $\epsilon > 0$. Choose N to be the maximum of the N_1 corresponding to $\epsilon_1 = 1$ and the N_2 corresponding to $M = \frac{2}{\epsilon}$. With this choice of N, we have

$$\text{if } x > N, \text{ then } 0 < \frac{f(x)}{g(x)} < \frac{1 + \epsilon_1}{M} = \frac{2}{\left(\frac{2}{\epsilon}\right)} = \epsilon.$$

■

The next theorem lists the limits at infinity of some simple functions. You will prove a selection of these limits in Exercises 90 and 91.

THEOREM 1.31

Limits of Some Basic Functions at Infinity

(a) If $k > 0$, then $\lim_{x \to \infty} x^k = \infty$ and $\lim_{x \to \infty} x^{-k} = 0$.

(b) If $k > 0$, then $\lim_{x \to \infty} e^{kx} = \infty$ and $\lim_{x \to \infty} e^{-kx} = 0$.

(c) $\lim_{x \to \infty} \ln x = \infty$.

(d) The functions $\sin x$, $\cos x$, $\tan x$, $\sec x$, $\csc x$, and $\cot x$ all have periodic behavior as $x \to \infty$, and thus their limits as $x \to \infty$ do not exist.

(e) $\lim_{x \to \infty} \tan^{-1} x = \frac{\pi}{2}$ and $\lim_{x \to -\infty} \tan^{-1} x = -\frac{\pi}{2}$.

Rather than memorizing these limits, it is better to remember the behavior on the right side of the graphs of these basic functions, as in the following examples:

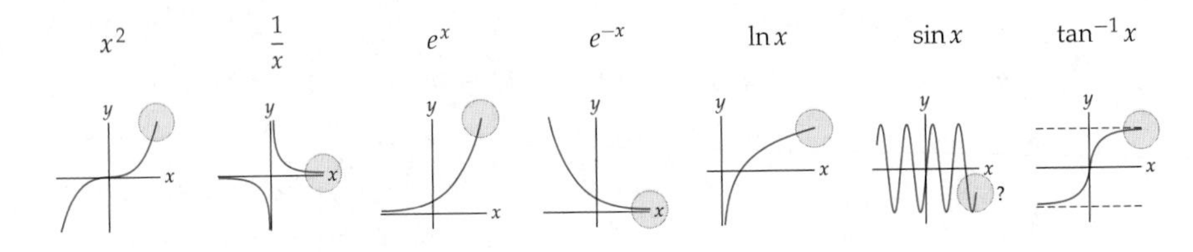

We can say similar things about limits as $x \to -\infty$. In fact, in most cases limits as $x \to -\infty$ can be rewritten with algebra as limits as $x \to \infty$. It is also helpful to remember the behavior on the left side of common graphs to determine limits as $x \to -\infty$. Although we will not prove so in general here, the limit rules from Theorem 1.20 also apply when $x \to \infty$ and when $x \to -\infty$. You will prove this for the sum rule for limits in Exercise 89. For example, given that $\lim_{x \to \infty} x^{-2} = 0$ and that $\lim_{x \to \infty} x^{-3} = 0$, we can conclude that $\lim_{x \to \infty} (x^{-2} + x^{-3}) = 0 + 0 = 0$.

As we will see in Example 4 and Exercise 85, polynomials behave like their leading terms as $x \to \infty$ and as $x \to -\infty$. For example, the function $f(x) = 2x^3 - 5x - 1$ will be dominated by its leading term $2x^3$ as x takes on larger and larger magnitudes. Therefore f approaches ∞ as $x \to \infty$, and approaches $-\infty$ as $x \to -\infty$. Since rational functions are by definition quotients of polynomial functions, we can use what we know about the global behavior of polynomials to determine the global behavior of rational functions:

THEOREM 1.32 **Horizontal Asymptote Theorem for Rational Functions**

If $f(x) = \frac{p(x)}{q(x)}$ is a rational function in which the polynomials $p(x)$ and $q(x)$ have leading terms $a_n x^n$ and $b_m x^m$, respectively, then

(a) if $n < m$, then the graph of $y = f(x)$ has a horizontal asymptote at $y = 0$.

(b) if $n = m$, then the graph of $y = f(x)$ has a horizontal asymptote at $y = \frac{a_n}{b_m}$.

(c) if $n > m$, then the graph of $y = f(x)$ does not have a horizontal asymptote.

Indeterminate and Non-Indeterminate Forms

If a limit is ***indeterminate***, then we cannot initially say whether or not it exists—or if it exists, what real number it is equal to. Many indeterminate limits can be resolved with algebra such as cancellation or factoring. For example, the four limits that follow are all initially of the indeterminate form $\frac{0}{0}$. After some simple algebra, we see that three limits exist and one limit becomes infinite. The three that exist are each equal to different real numbers.

$$\lim_{x\to 1} \frac{(x-1)^2}{x-1} = \lim_{x\to 1} \frac{x-1}{1} = \frac{0}{1} = 0 \qquad \lim_{x\to 1} \frac{x-1}{(x-1)^3} = \lim_{x\to 1} \frac{1}{(x-1)^2} = \infty$$

$$\lim_{x\to 1} \frac{x-1}{x-1} = \lim_{x\to 1} 1 = 1 \qquad \lim_{x\to 1} \frac{x-1}{3(x-1)} = \lim_{x\to 1} \frac{1}{3} = \frac{1}{3}$$

When a limit is indeterminate, it is essentially because there is a "fight" going on between two parts of the limit. For example, limits of the form $\frac{0}{0}$ approach different things depending on whether the numerator or the denominator approaches 0 faster. If the numerator does, then it "wins" the fight and the limit is equal to zero. If the denominator approaches 0 faster, then the limit will become infinite. If the numerator and denominator are balanced appropriately, then they cancel each other out and the limit will approach a nonzero real number. Over time you will develop an intuition for what types of expressions are likely to win such fights, and the examples at the end of this section illustrate algebraic techniques for resolving such indeterminacies.

The following theorem identifies seven common indeterminate forms for limits:

THEOREM 1.33 **Indeterminate Forms for Limits**

Each of the following is an ***indeterminate form***, meaning that a limit in one of these forms may or may not exist, depending on the situation:

$$\frac{0}{0} \qquad \frac{\infty}{\infty} \qquad 0\cdot\infty \qquad \infty-\infty \qquad 0^0 \qquad 1^\infty \qquad \infty^0$$

To prove this theorem, for each indeterminate form we need only exhibit an example of a limit of that form that exists and an example of a limit of that form that does not, as we did before for limits of the indeterminate form $\frac{0}{0}$. Identifying such examples for the remaining six forms is left to you in Exercises 15–21. Limits that have the first four types of indeterminate forms listed in Theorem 1.33 can often be solved after simple factoring or cancelling. Limits of the remaining three types pose more of a challenge; see Section 3.6.

CAUTION Of course, the indeterminate expression "$\frac{0}{0}$" is not a real number and we cannot actually divide the number 0 by the number 0. Theorem 1.33 tells us that limits of this form are indeterminate, and thus cannot be determined until we somehow rewrite or re-examine the limit, perhaps by factoring, cancelling, or some other method. Note that this is very different than saying that a limit "does not exist."

We have already seen that a limit of the form $\frac{1}{\infty}$ is equal to zero and a limit of the form $\frac{1}{0^+}$ is infinite. The following limit forms also always either approach 0 or become infinite and thus are not indeterminate:

THEOREM 1.34 **Non-Indeterminate Forms for Limits**

(a) A limit in any of these forms must be equal to 0:

$$\frac{1}{\infty} \qquad \frac{0}{\infty} \qquad \frac{0}{1} \qquad 0^{\infty} \qquad 0^1$$

(b) A limit in any of these forms must be ∞:

$$\frac{1}{0^+} \qquad \frac{\infty}{0^+} \qquad \frac{\infty}{1} \qquad \infty + \infty \qquad \infty \cdot \infty \qquad \infty^{\infty} \qquad \infty^1$$

We are *not* suggesting that you should memorize Theorem 1.34. Each of the limit forms in the theorem can be easily determined by investigation. We will not give a formal proof of this theorem, but rather, we can argue that in each case there is no "fight"; it is clear what limits of each form must approach. For example, in a limit of the form $\frac{0}{\infty}$, the numerator approaches zero, making the quotient smaller and smaller; at the same time the denominator grows without bound, which also makes the quotient smaller and smaller. Thus for the form $\frac{0}{\infty}$, the behavior of both the numerator and the denominator causes the limit to approach 0. You will see some of the other non-indeterminate forms in Exercises 11–14.

Special Trigonometric Limits

Certain indeterminate limits can be reduced with algebra to two specific trigonometric limits. These limits expand the library of limits that we can compute and will be vital tools for determining the derivatives of sine and cosine in the next chapter.

THEOREM 1.35 **Two Useful Trigonometric Limits**

(a) $\displaystyle\lim_{\theta \to 0} \frac{\sin\theta}{\theta} = 1$ **(b)** $\displaystyle\lim_{\theta \to 0} \frac{1 - \cos\theta}{\theta} = 0$

Notice that the limits in this theorem give us a way to determine a number of related limits that are initially of the form $\frac{0}{0}$. For example, we can use the first limit to show that

$$\lim_{x \to 0} \frac{\sin 2x}{x} = \lim_{x \to 0} \frac{\sin 2x}{2x}(2) = 1(2) = 2,$$

since as $x \to 0$, we also have $2x \to 0$. As another example, since $x - \pi \to 0$ when $x \to \pi$, we have

$$\lim_{x \to \pi} \frac{\sin(x - \pi)}{x - \pi} = 1.$$

Proof. We can intuitively see that the two special trigonometric limits make sense based on the following figures of portions of the unit circle, with angles measured in radians:

$\sin\theta \approx \theta$ $\qquad\qquad$ $1 - \cos\theta$ *much smaller than* θ

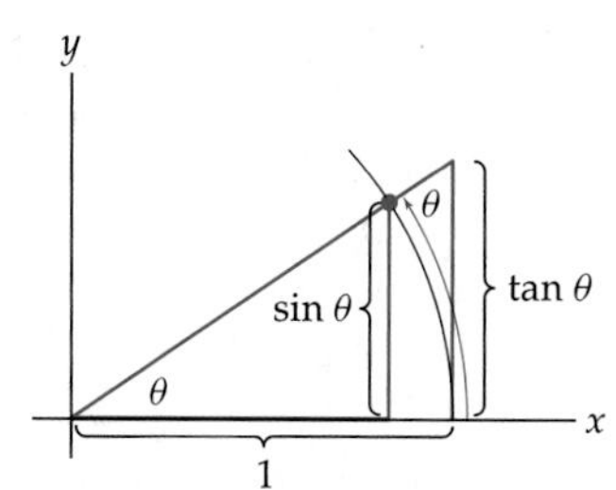

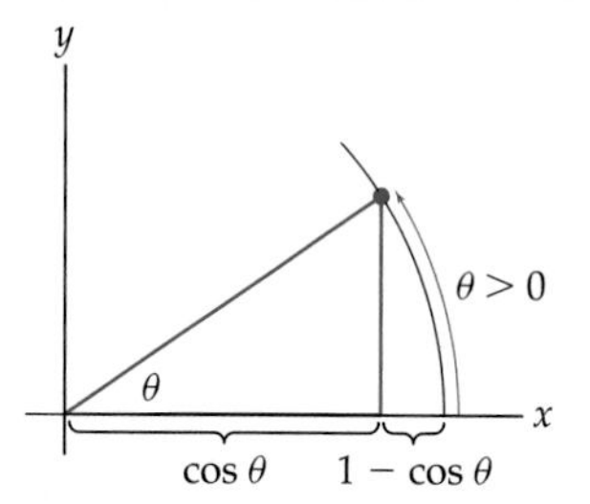

For small positive values of θ, the picture on the left suggests that $\sin\theta \approx \theta$ and therefore that $\frac{\sin\theta}{\theta} \approx 1$. Similarly, for small positive θ, the picture on the right suggests that $1 - \cos\theta$ approaches 0 much faster than θ and therefore that $\frac{1-\cos\theta}{\theta} \approx 0$.

More formally, from the earlier figure on the left, we can see that $\sin\theta \leq \theta \leq \tan\theta$. Dividing all expressions in this chain of inequalitites by $\sin\theta$ and using the fact that $\tan\theta = \frac{\sin\theta}{\cos\theta}$, we have

$$1 \leq \frac{\theta}{\sin\theta} \leq \frac{1}{\cos\theta}.$$

Since all of the expressions in this chain of inequalities are positive, we can take reciprocals to obtain:

$$\cos\theta \leq \frac{\sin\theta}{\theta} \leq 1.$$

Now by applying the Squeeze Theorem to these two inequalities we have $\lim_{\theta\to 0^+} \frac{\sin\theta}{\theta} = 1$. The argument for $\theta \to 0^-$ is similar, with a picture in the fourth quadrant.

The second trigonometric limit can be proved from the first by using the double-angle formula $\cos 2\theta = 1 - 2\sin^2\theta$; see Exercise 94. ■

The two figures that follow show the functions $y = \frac{\sin x}{x}$ and $y = \frac{1-\cos x}{x}$. Notice that neither function is defined at $x = 0$, but both approach a specific real-number value as $x \to 0$.

$\lim_{x\to 0} \frac{\sin x}{x} = 1$

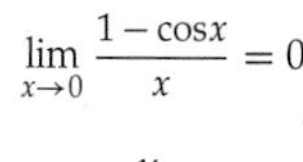

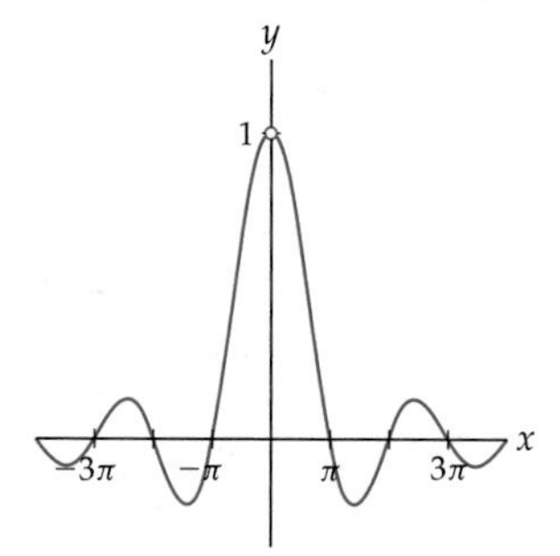

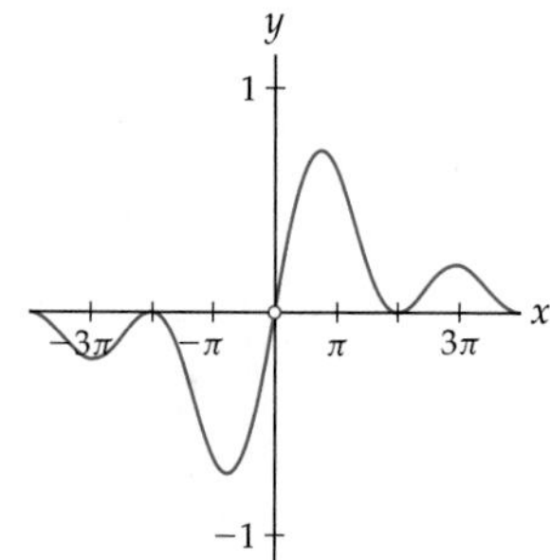

Examples and Explorations

EXAMPLE 1 **Calculating limits to determine horizontal asymptotes**

Use limits to find any horizontal asymptotes of the following functions:

(a) $f(x) = \frac{x}{x-1}$ $\qquad$ **(b)** $g(x) = \frac{x^2-1}{x^3-4x}$ $\qquad$ **(c)** $h(x) = \frac{\sin x}{x}$

SOLUTION

(a) To find horizontal asymptotes we must examine limits as $x \to \pm\infty$. As $x \to \infty$, we have $x \to \infty$ and $x - 1 \to \infty$, and therefore

$$\lim_{x\to\infty} \frac{x}{x-1} \text{ is of the indeterminate form } \frac{\infty}{\infty}.$$

Limits of the indeterminate form $\frac{\infty}{\infty}$ can often be resolved by dividing the numerator and denominator by the highest power of x that appears, as follows:

$$\lim_{x\to\infty} \frac{x}{x-1} = \lim_{x\to\infty} \frac{x}{x-1}\left(\frac{1/2x}{1/2x}\right) = \lim_{x\to\infty} \frac{1}{1-(1/x)} = \frac{1}{1-0} = 1.$$

In a similar fashion we can show that $\lim_{x\to-\infty} \frac{x}{x-1} = 1$. Therefore $f(x) = \frac{x}{x-1}$ has a two-sided horizontal asymptote at $y = 1$.

(b) As $x \to \infty$ we have $x^2 - 1 \to \infty$ in the numerator and $x^3 \to \infty$ and $4x \to \infty$ in the denominator. Since $\infty - \infty$ is an indeterminate form, we cannot even be certain what the denominator $x^3 - 4x$ approaches at this point. To resolve this limit we will once again divide the numerator and denominator by the highest power of x:

$$\lim_{x\to\infty} \frac{x^2-1}{x^3-4x} = \lim_{x\to\infty} \frac{x^2-1}{x^3-4x}\left(\frac{1/x^3}{1/x^3}\right) = \lim_{x\to\infty} \frac{1/x - 1/x^3}{1-(4/x^2)} = \frac{0-0}{1-0} = \frac{0}{1} = 0.$$

Similarly, $\lim_{x\to-\infty} \frac{x^2-1}{x^3-4x}$ is equal to 0. Therefore $g(x) = \frac{x^2-1}{x^3-4x}$ has a two-sided horizontal asymptote at $y = 0$.

(c) Looking at $h(x) = \frac{\sin x}{x}$ as $x \to \infty$, we see that the numerator $\sin x$ oscillates between -1 and 1 while the denominator x gets infinitely large. A bounded quantity divided by a quantity that increases without bound must approach zero; in other words, as $x \to \infty$ we have

$$\frac{\sin x}{x} \to \frac{\text{bounded}}{\infty} \to 0.$$

The same is true as $x \to -\infty$, and thus $\lim_{x\to\infty} \frac{\sin x}{x}$ and $\lim_{x\to-\infty} \frac{\sin x}{x}$ are both equal to 0. Therefore the function $h(x) = \frac{\sin x}{x}$ has a two-sided horizontal asymptote at $y = 0$. □

✔ CHECKING THE ANSWER

We can use calculator graphs to verify the horizontal asymptotes that we just found. These graphs also provide verification for the vertical asymptotes in the next example.

f has horizontal asymptote at $y = 1$

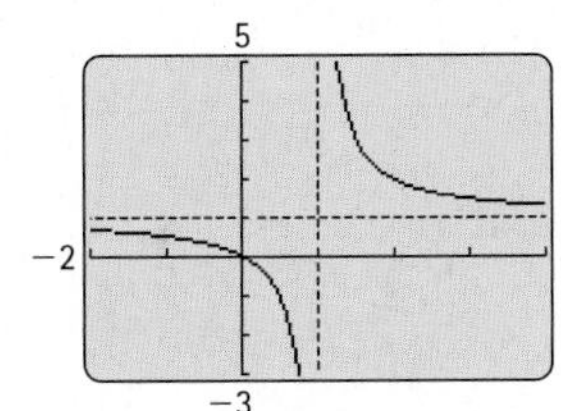

g has horizontal asymptote at $y = 0$

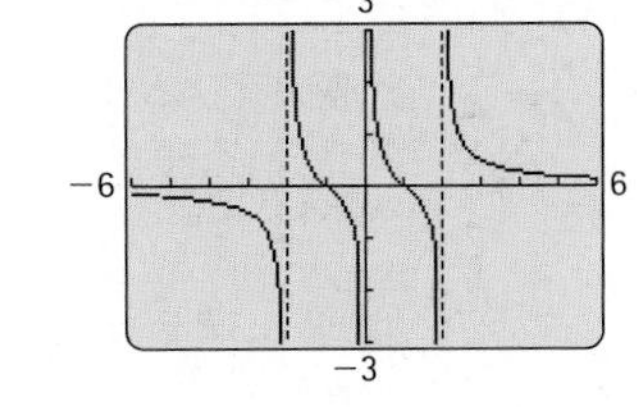

h has horizontal asymptote at $y = 0$

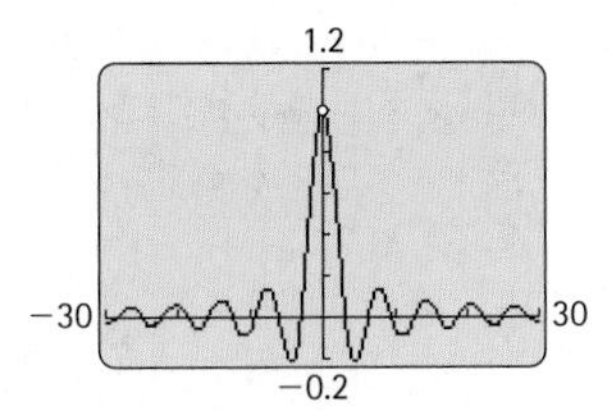

EXAMPLE 2 **Calculating limits to determine vertical asymptotes**

Use limits to describe any vertical asymptotes of the following functions:

(a) $f(x) = \frac{x}{x-1}$ **(b)** $g(x) = \frac{x^2-1}{x^3-4x}$ **(c)** $h(x) = \frac{\sin x}{x}$

SOLUTION

(a) From the formula for $f(x)$, we see that $x = 1$ is the only serious candidate for a vertical asymptote. As $x \to 1$ we have $x - 1 \to 0$ and therefore

$$\lim_{x \to 1} \frac{x}{x-1} \text{ is of the form } \frac{1}{0}.$$

This tells us that $f(x) = \frac{x}{x-1}$ has a vertical asymptote at $x = 1$. If we want to describe the behavior of f near this asymptote more precisely, we can calculate the right and left limits separately. As $x \to 1^-$ we have $x - 1 \to 0^-$, and as $x \to 1^+$ we have $x - 1 \to 0^+$; therefore

$$\lim_{x \to 1^-} \frac{x}{x-1} = -\infty \text{ and } \lim_{x \to 1^+} \frac{x}{x-1} = \infty.$$

This means that the vertical asymptote at $x = 1$ is downward-pointing on the left and upward-pointing on the right; see the leftmost graph from the previous "Checking the Answer" figures.

(b) The function $g(x)$ factors as

$$\frac{x^2 - 1}{x^3 - 4x} = \frac{(x-1)(x+1)}{x(x-2)(x+2)}.$$

The values of x that cause the denominator of this quotient to approach zero are $x = 0$, $x = 2$, and $x = -2$. None of these values cause the numerator to approach zero, so in each case we will get a limit that is either ∞ or $-\infty$ from the left and/or the right. In any case, we know that $g(x)$ has vertical asymptotes at $x = 0$, $x = -2$, and $x = 2$. If we want to know the precise behavior of $g(x)$ at one of these vertical asymptotes, we can look from the left and the right. For example, as $x \to 2^-$ we have $x(x-2)(x+2) \to 2(0^-)(4) \to 0^-$, and as $x \to 2^+$ we have $x(x-2)(x+2) \to 2(0^+)(4) \to 0^+$. Therefore the left and right limits at $x = 2$ are

$$\lim_{x \to 2^-} \frac{x^2 - 1}{x^3 - 4x} = \lim_{x \to 2^-} \frac{(x-1)(x+1)}{x(x-2)(x+2)} = -\infty$$

and

$$\lim_{x \to 2^+} \frac{x^2 - 1}{x^3 - 4x} = \lim_{x \to 2^+} \frac{(x-1)(x+1)}{x(x-2)(x+2)} = \infty.$$

Notice that in these calculations we kept track only of the left/right "$\pm$" directions when we encountered 0. This is because whether a multiplicative factor in the limit is approaching, say, 1^+ or 1^-, will not affect the overall sign. On the other hand, the difference between a factor of 0^+ and 0^- does affect the sign, which in turn will determine whether the limit approaches ∞ or $-\infty$. Notice also that we used the factored form of $g(x)$ to determine the sign of infinity in each case, since it would have been difficult to determine the sign of the denominator as $x \to 2^+$ and as $x \to 2^-$ in the unfactored expression for $g(x)$.

(c) The denominator of $h(x) = \frac{\sin x}{x}$ is zero only when $x = 0$, so $x = 0$ is the only candidate for a vertical asymptote. However, at $x = 0$ we have one of our special limits from Theorem 1.35:

$$\lim_{x \to 0} \frac{\sin x}{x} = 1.$$

Therefore $h(x)$ does not have a vertical asymptote at $x = 0$. Instead, the graph has a hole, since $\lim_{x \to 0} h(x) = 1$ exists but $h(0)$ does not exist; see the third graph in the Checking the Answer discussion before this example. □

EXAMPLE 3 **Indeterminate forms**

Determine whether each of the limits that follows is initially in indeterminate form or non-indetermine form. Then calculate each limit.

(a) $\lim_{x\to\infty} \frac{3^x - 2^x}{1+4^x}$ **(b)** $\lim_{x\to\infty} (x^{2/3} - x^{3/4})$ **(c)** $\lim_{x\to\infty} \left(1 + \frac{3}{x}\right)^{2x}$

SOLUTION

(a) By Theorem 1.31, as $x \to \infty$ the expressions 3^x, 2^x, and 4^x all approach ∞. Therefore the limit in question is an indeterminate form. An analog of the method of dividing by the highest power works in this case, but this time we divide numerator and denominator by the exponential function with the largest base:

$$\lim_{x\to\infty} \frac{3^x - 2^x}{1+4^x} = \lim_{x\to\infty} \frac{3^x - 2^x}{1+4^x}\left(\frac{1/4^x}{1/4^x}\right) = \lim_{x\to\infty} \frac{(3/4)^x - (1/2)^x}{(1/4)^x + 1} = \frac{0-0}{0+1} = \frac{0}{1} = 0.$$

(b) Since $\frac{2}{3}$ and $\frac{3}{4}$ are positive powers, as $x \to \infty$ we have $x^{2/3} \to \infty$ and $x^{3/4} \to \infty$. Therefore the limit in question is of the indeterminate form $\infty - \infty$. Limits of this form can often be resolved by factoring. We have

$$x^{2/3} - x^{3/4} = x^{2/3}(1 - x^{1/12}),$$

and as $x \to \infty$ we have $x^{2/3} \to \infty$ and $(1 - x^{1/12}) \to -\infty$. Therefore the limit in question is now of the form $(\infty)(-\infty)$ which means that

$$\lim_{x\to\infty} (x^{2/3} - x^{3/4}) = \lim_{x\to\infty} x^{2/3}(1 - x^{1/12}) = -\infty.$$

(c) As $x \to \infty$, the base $1 + \frac{3}{x}$ approaches $1 + 0 = 1$ and the exponent $2x$ approaches ∞. Therefore this limit is of the form 1^∞, which is indeterminate. Fortunately, with a substitution we can rewrite the limit in such a way that allows us to apply the definition of e from Definition 1.25. Let $h = \frac{3}{x}$. Then as $x \to \infty$, we have $h \to 0^+$. Using this relationship and the fact that $x = \frac{3}{h}$, we have

$$\lim_{x\to\infty} \left(1 + \frac{3}{x}\right)^{2x} = \lim_{h\to 0^+} (1+h)^{2(3/h)} = (\lim_{h\to 0^+} (1+h)^{1/h})^6 = e^6.$$

□

CHECKING THE ANSWER

We can use intuition to verify that these answers seem reasonable. Remember that each time a limit has an indeterminate form, two parts of the limit are fighting against each other. In part (a) of the preceding example, $\frac{3^x - 2^x}{1+4^x}$ approaches an indeterminate form as $x \to \infty$. Since 4^x is the exponential function with the largest base in the expression, it is reasonable to expect that it will dominate the expression as $x \to \infty$, dragging the whole limit down to zero.

In part (b) we saw that $\lim_{x\to\infty} x^{2/3} - x^{3/4}$ was of the indeterminate form $\infty - \infty$. Since $\frac{3}{4} > \frac{2}{3}$, $x^{3/4}$ should approach ∞ faster than $x^{2/3}$ does. Thus it makes sense to expect that $x^{3/4}$ should win the battle and $x^{2/3} - x^{3/4}$ should eventually approach $-\infty$.

It is difficult to use intuition to verify the limit in part (c), except that we might expect in this case that as $x \to \infty$ the 1 and the ∞ are balanced in the indeterminate form 1^∞, due to the fact that each involve a single power of x. This might lead us to suspect that the answer is neither 1, nor ∞, but rather some number in between.

EXAMPLE 4 **The global behavior of a polynomial is determined by its leading term**

Use limits to show that the polynomial $f(x) = x^4 - x^3 - 11x^2 + 9x + 18$ behaves like its leading term x^4 as $x \to \infty$ and $x \to -\infty$. Then use graphs to compare the graph of f with the graph of $y = x^4$ in different graphing windows.

SOLUTION

It is not immediately obvious how to calculate this limit, because, as $x \to \infty$, the terms x^4 and $9x$ approach ∞ while the terms $-x^3$ and $-11x^2$ approach $-\infty$. Therefore $\lim_{x\to\infty} x^4 - x^3 - 11x^2 + 9x + 18$ is indeterminate.

However, with some simple algebra we can change this sum and difference of infinities into a product that is easier to work with. Specifically, we can factor out the largest power of x:

$$\lim_{x\to\infty}(x^4 - x^3 - 11x^2 + 9x + 18) = \lim_{x\to\infty} x^4\left(1 - \frac{1}{x} - \frac{11}{x^2} + \frac{9}{x^3} + \frac{18}{x^4}\right).$$

Since as $x \to \infty$ we have $x^4 \to \infty$ and the remainder of the expression approaching $1 - 0 - 0 + 0 + 0 = 1$, we can say that the limit is equal to ∞.

Similarly, the limit as $x \to -\infty$ is also ∞. Notice that the only term which ended up being relevant in the limit calculation was the leading term. The figures that follow show the function $f(x) = x^4 - x^3 - 11x^2 + 9x + 18$ in blue and its leading term $y = x^4$ in red, in three different viewing windows. The more we enlarge the graphing window, the more the graph of the function $y = f(x)$ looks like the graph of $y = x^4$.

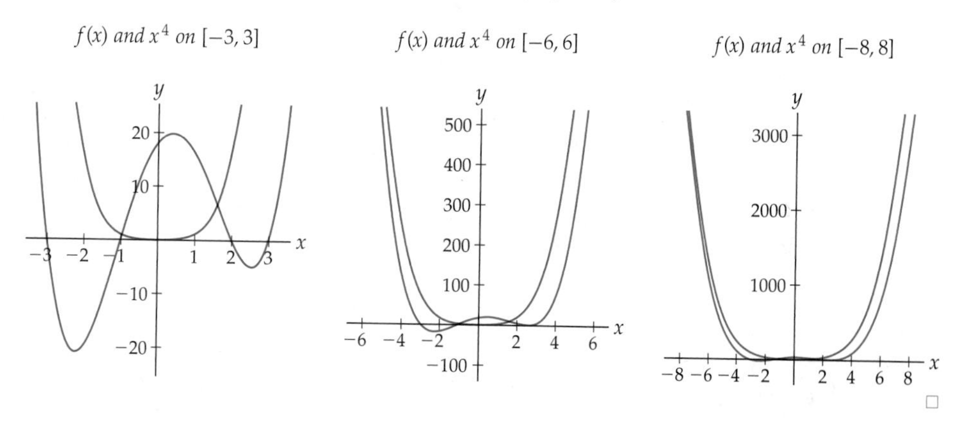

□

? TEST YOUR UNDERSTANDING

- In terms of large and small numbers, why does it make intuitive sense that limits of the form $\frac{1}{0^+}$ must always equal ∞?
- If a limit is of the form $\frac{1}{0}$ as $x \to c$, why should we examine the corresponding left and right limits separately?
- Why does it make sense that limits of the form $\frac{1}{\infty}$ and of the form $\frac{1}{-\infty}$ are always equal to zero?
- Why does it make intuitive sense that limits of the form $\frac{0}{0}$ are indeterminate? What "fight" is happening between the numerator and the denominator? What will happen if the numerator "wins"? The denominator? If there is a tie?
- Can a function have two different horizontal asymptotes? Can you think of a specific example?

EXERCISES 1.6

Thinking Back

Behavior of transcendental functions: Determine whether each function approaches 0, approaches a nonzero real number, or becomes infinite as x approaches each indicated value.

- $f(x) = \csc x$, with $x \to 0$ and $x \to \pi$.
- $f(x) = \tan^2 x$, with $x \to 0$ and $x \to \pi$.
- $f(x) = \sin^{-1} x$, with $x \to 0$ and $x \to 1$.
- $f(x) = \tan^{-1} \sqrt{x}$, with $x \to 0$ and $x \to 3$.

The definition of infinite limits and limits at infinity: Write each limit statement that follows in terms of the formal definition of limit. Then approximate the largest value of δ or N corresponding to $\epsilon = 0.5$ or $M = 100$, as appropriate, and illustrate this choice of δ or N on a graph of f.

- $\lim\limits_{x \to \infty} \dfrac{2x}{x-1} = 2$
- $\lim\limits_{x \to 2^+} \dfrac{1}{x^2 - 4} = \infty$
- $\lim\limits_{x \to \infty} \sqrt{x-1} = \infty$
- $\lim\limits_{x \to -\infty} \dfrac{1}{x} = 0$

Concepts

0. *Problem Zero:* Read the section and make your own summary of the material.

1. *True/False:* Determine whether each of the statements that follow is true or false. If a statement is true, explain why. If a statement is false, provide a counterexample.

(a) *True or False:* If $f(x) \to 0^+$, then $\dfrac{1}{f(x)} \to \infty$.

(b) *True or False:* If $f(x) \to \infty^+$, then $\dfrac{1}{f(x)} \to 0^+$.

(c) *True or False:* If a limit initially has an indeterminate form, then it can never be solved.

(d) *True or False:* A limit "does not exist" if there is no real number that it approaches.

(e) *True or False:* As limit forms, $\infty^2 \to \infty$.

(f) *True or False:* As limit forms, $2^\infty \to \infty$.

(g) *True or False:* As limit forms, $\infty - \infty \to 0$.

(h) *True or False:* The limit of a function f as $x \to c$ is always equal to the value $f(c)$, provided that $f(c)$ exists.

2. *Examples:* Construct examples of the thing(s) described in the following. Try to find examples that are different than any in the reading.

(a) A limit of the form $\frac{1}{0}$ that approaches ∞ as $x \to c^+$ and $-\infty$ as $x \to c^-$.

(b) Two limits that can be solved with the special trigonometric limits from Theorem 1.35.

(c) Formulas for three functions that are discontinuous at $x = 3$: one removable, one jump, and one infinite discontinuity.

In Exercises 3–6, $\lim\limits_{x \to c} f(x) = L$ and $\lim\limits_{x \to c} g(x) = M$ for some real numbers L and M. What, if anything, can you say about $\lim\limits_{x \to c} \dfrac{f(x)}{g(x)}$ in each case?

3. $L \neq 0$ and $M \neq 0$

4. $L = 0$ and $M \neq 0$

5. $L \neq 0$ and $M = 0$

6. $L = 0$ and $M = 0$

7. Determine which of the given forms are indeterminate. For each form that is not indeterminate, describe the behavior of a limit of that form.

$$\infty + \infty \qquad \infty - \infty \qquad \infty + 1 \qquad 0 + \infty$$

8. Determine which of the given forms are indeterminate. For each form that is not indeterminate, describe the behavior of a limit of that form.

$$\infty \cdot \infty \qquad 0 \cdot \infty \qquad 5 \cdot \infty \qquad 5 \cdot 0 \qquad 0 \cdot 0$$

9. Determine which of the given forms are indeterminate. For each form that is not indeterminate, describe the behavior of a limit of that form.

$$\frac{0}{0} \qquad \frac{0}{\infty} \qquad \frac{\infty}{0} \qquad \frac{1}{0} \qquad \frac{0}{1} \qquad \frac{\infty}{1} \qquad \frac{1}{\infty} \qquad \frac{\infty}{\infty}$$

10. Determine which of the given forms are indeterminate. For each form that is not indeterminate, describe the behavior of a limit of that form.

$$0^1 \qquad 0^0 \qquad 0^\infty \qquad 1^\infty \qquad \infty^1 \qquad \infty^0 \qquad \infty^\infty$$

11. Describe in terms of large and small numbers why it makes intuitive sense that limits of the form (a) $\frac{1}{\infty}$, (b) $\frac{0}{\infty}$, and (c) $\frac{0}{1}$ must equal 0.

12. Describe in terms of large and small numbers why it makes intuitive sense that limits of the form (a) 0^∞ and (b) 0^1 must equal 0.

13. Describe in terms of large and small numbers why it makes intuitive sense that limits of the form (a) $\frac{1}{0^+}$, (b) $\frac{\infty}{0^+}$, and (c) $\frac{\infty}{1}$ must be infinite.

14. Describe in terms of large and small numbers why it makes intuitive sense that limits of the form (a) $\infty + \infty$, (b) $\infty \cdot \infty$, (c) ∞^∞, and (d) ∞^1 must be infinite.

To prove that the limit forms in Theorem 1.33 are indeterminate, we need only list explicit examples of limits that do and do not exist for each form. Do so for each of the limit forms from Exercises 15–21. For the last three forms you may want to experiment with a graphing utility to find your examples.

15. $\frac{0}{0}$ that approaches (a) 0, (b) 2, (c) ∞.

16. $\frac{\infty}{\infty}$ that approaches (a) 1, (b) 6, (c) ∞.

17. $0 \cdot \infty$ that approaches (a) 0 (b) 1, (c) ∞.

18. $\infty - \infty$ that approaches (a) 0, (b) 5, (c) ∞.

19. 0^0 that approaches (a) 1, (b) 0, (c) ∞.

20. 1^∞ that approaches (a) 1, (b) e, (c) ∞.

21. ∞^0 that approaches (a) 1, (b) 2, (c) ∞.

22. Find the equation of a rational function that could have the graph shown. Take into account roots, holes, and vertical and horizontal asymptotes when constructing your function.

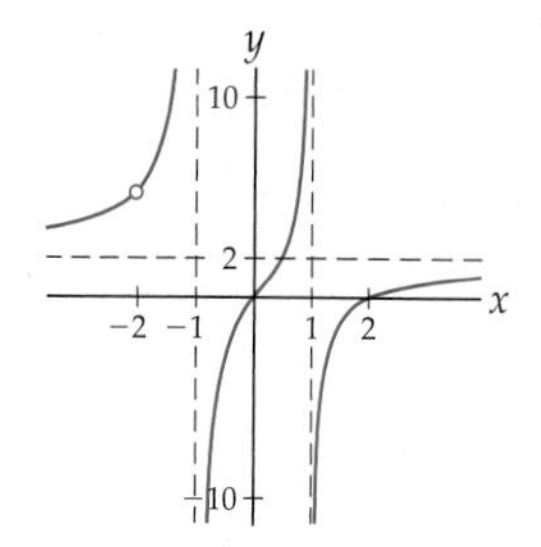

Skills

Find the roots, discontinuities, and horizontal and vertical asymptotes of the functions in Exercises 23–34. Support your answers by explicitly computing any relevant limits.

23. $f(x) = \frac{x^2 - 2x - 3}{x - 3}$

24. $f(x) = \frac{2x^2 - 1}{x^2 - 2x + 1}$

25. $f(x) = \frac{(x+1)(x-2)}{(x-2)(x+2)}$

26. $f(x) = \frac{(x+1)(x-2)^2}{(x-2)(x+2)}$

27. $f(x) = \frac{(x+1)(x-2)}{(x-2)^2(x+2)}$

28. $f(x) = \frac{(x+1)(x-2)}{(x-2)(x+2)^2}$

29. $f(x) = \frac{2}{4 + e^{-2x}}$

30. $f(x) = \frac{1}{2 + 3^x}$

31. $f(x) = \frac{2^x - 4^x}{3^x}$

32. $f(x) = \frac{4^x - 6(2^x) + 5}{1 - 2^x}$

33. $f(x) = \tan^{-1}(3x) + 1$

34. $f(x) = \frac{1}{\tan^{-1} x}$

Calculate each limit in Exercises 35–80.

35. $\lim_{x \to 0} -4x^{-3}$

36. $\lim_{x \to 0} 2x^{-3/4}$

37. $\lim_{x \to \infty} 2x^{-4/3}$

38. $\lim_{x \to -\infty} -5x^{3/5}$

39. $\lim_{x \to \infty} (\sqrt{x} - x)$

40. $\lim_{x \to -\infty} (x^4 - x^5)$

41. $\lim_{x \to \infty} (-3x^5 + 4x + 11)$

42. $\lim_{x \to -\infty} (5 - 2x + 3x^3)$

43. $\lim_{x \to -4} \frac{x^2 + 8x + 16}{(x + 4)^2(x + 1)}$

44. $\lim_{x \to 2} \frac{x + 1}{(x - 2)^2}$

45. $\lim_{x \to 0} \frac{x^2 + 1}{x(x - 1)}$

46. $\lim_{x \to -4} \frac{x + 4}{x^2 + 8x + 16}$

47. $\lim_{x \to 0} \frac{x}{x^2 - x}$

48. $\lim_{x \to 1} \frac{x - 1}{x^2 - 2x + 1}$

49. $\lim_{x \to \infty} \frac{(3x + 1)^2(x - 1)}{1 - x^3}$

50. $\lim_{x \to -\infty} \frac{1 - 2x^2}{(3 - x)(3 + 4x)}$

51. $\lim_{x \to 0^+} (x^{-1/3} - x^{-1/2})$

52. $\lim_{x \to \infty} (x^{-1/3} - x^{-1/2})$

53. $\lim_{x \to \infty} \frac{x^{-3}}{x^2 - x^{-1}}$

54. $\lim_{x \to 0^+} \frac{x^{-3}}{x^2 - x^{-1}}$

55. $\lim_{x \to 0^+} \frac{x^{7/2} - x^{8/3}}{x^2}$

56. $\lim_{x \to \infty} \frac{x^{7/2} - x^{8/3}}{x^2}$

57. $\lim_{x \to \infty} \frac{4^x - 3^x}{5^x}$

58. $\lim_{x \to \infty} \frac{2^x - 4^{-x}}{3^x}$

59. $\lim_{x \to -\infty} \frac{3^x - 5^x}{4^x}$

60. $\lim_{x \to \infty} \frac{4(3^x)}{2 + 3^x}$

61. $\lim_{x \to \infty} \frac{2e^{1.5x}}{3e^{2x} + e^{1.5x}}$

62. $\lim_{x \to \infty} \frac{1 - 5e^{2x}}{3e^x + 4e^{2x}}$

63. $\lim_{x \to 3^+} \ln(x^2 - 9)$

64. $\lim_{x \to 0^+} \ln\left(\frac{1}{x}\right)$

65. $\lim_{x \to \infty} (\ln x^2 - \ln(2x + 1))$

66. $\lim_{x \to \infty} (\ln 3x - \ln 2x)$

67. $\lim_{x \to 0} \frac{1 - \cos 2x}{7x}$

68. $\lim_{x \to 0} \frac{\sin 3x}{5x}$

69. $\lim_{x \to 0} \frac{x}{1 - \cos x}$

70. $\lim_{x \to 0} \frac{3 \sin x + x}{x}$

71. $\lim_{x \to 0} \frac{\sin^2 3x}{x^3 - x}$

72. $\lim_{x \to 0} \frac{\sin(3x^2)}{x^3 - x}$

73. $\lim_{x \to 0^+} \frac{x^2 \csc 3x}{1 - \cos 2x}$

74. $\lim_{x \to 0} \frac{x^2 \cot x}{\sin x}$

75. $\lim_{x \to 0} \frac{\sec x \tan x}{x}$

76. $\lim_{x \to 0} 3x^2 \cot^2 x$

77. $\lim_{x \to 0} (1 + x)^{2/x}$

78. $\lim_{x \to 0} (1 + 2x)^{3/x}$

79. $\lim_{x \to \infty} \left(1 + \frac{1}{x}\right)^{3x}$

80. $\lim_{x \to \infty} \left(1 - \frac{5}{x}\right)^{x}$

Applications

81. In 1960, H. von Foerster suggested that the human population could be measured by the function

$$P(t) = \frac{179 \times 10^9}{(2027 - t)^{0.99}}.$$

The time t is measured in years, where $t = 1$ corresponds to the year 1 A.D., $t = 1973$ corresponds to the year 1973 A.D., and so on. (We saw this "doomsday model" for population in Problem 77 of Section 1.1, on page 89.) Use limit techniques to calculate $\lim_{t \to 2027^-} P(t)$. What does this limit mean in real-world terms?

82. Suppose instead we consider the population model

$$Q(t) = \frac{44 \times 10^{10}}{1 + (2027 - t)^{4/3}},$$

with t measured in years as in the previous problem.

(a) Use limit techniques to calculate $\lim_{t \to \infty} Q(t)$. What does this limit mean in real-world terms? What happens in this model in the year 2027?

(b) Use calculator graphs to compare the population models in this exercise with those in the previous exercise. Describe the long-term population growth scenarios that are suggested by these models.

83. Consider a mass hanging from the ceiling at the end of a spring. If you pull down on the mass and let go, it will oscillate up and down according to the equation

$$s(t) = A \sin\left(\sqrt{\frac{k}{m}}\, t\right) + B \cos\left(\sqrt{\frac{k}{m}}\, t\right),$$

where $s(t)$ is the distance of the mass from its equilibrium position, m is the mass of the bob on the end of the spring, and k is a "spring coefficient" that measures how tight or stiff the spring is. The constants A and B depend on initial conditions—specifically, how far you pull down the mass (s_0) and the velocity at which you release the mass (v_0). This equation does *not* take into effect any friction due to air resistance.

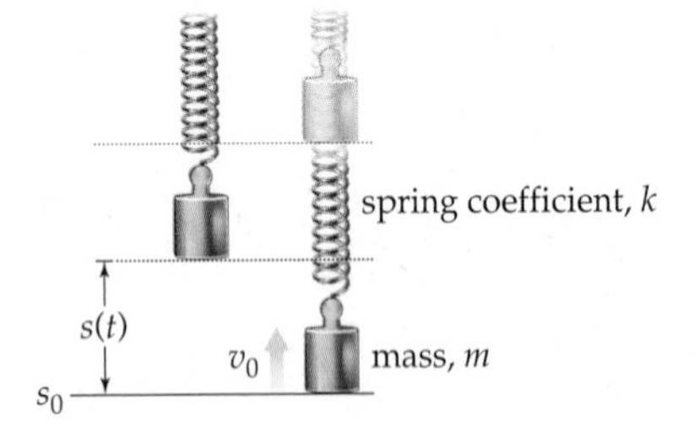

(a) Determine whether or not the limit of $s(t)$ as $t \to \infty$ exists. What does this say about the long-term behavior of the mass on the end of the spring?

(b) Explain how this limit relates to the fact that the equation for $s(t)$ does *not* take friction due to air resistance into account.

(c) Suppose the bob at the end of the spring has a mass of 2 grams and that the coefficient for the spring is $k = 9$. Suppose also that the spring is released in such a way that $A = \sqrt{2}$ and $B = 2$. Use a graphing utility to graph the function $s(t)$ that describes the distance of the mass from its equilibrium position. Use your graph to support your answer to part (a).

84. In the previous exercise we gave an equation describing spring motion without air resistance. If we take into account friction due to air resistance, the mass will oscillate up and down according to the equation

$$s(t) = e^{(-f/2m)t}\left(A \sin\left(\frac{\sqrt{4km - f^2}}{2m}\, t\right) + B \cos\left(\frac{\sqrt{4km - f^2}}{2m}\, t\right)\right),$$

where m, k, A, and B are the constants described in Problem 83 and f is a positive "friction coefficient" that measures the amount of friction due to air resistance.

(a) Find the limit of $s(t)$ as $t \to \infty$. What does this say about the long-term behavior of the mass on the end of the spring?

(b) Explain how this limit relates to the fact that the new equation for $s(t)$ *does* take friction due to air resistance into account.

(c) Suppose the bob at the end of the spring has a mass of 2 grams, the coefficient for the spring is $k = 9$, and the friction coefficient is $f = 6$. Suppose also that the spring is released in such a way that $A = 4$ and $B = 2$. Use a graphing utility to graph the function $s(t)$ that describes the distance of the mass from its equilibrium position. Use your graph to support your answer to part (a).

Proofs

85. Use limits to prove that the limits of a polynomial $f(x) = a_n x^n + a_{n-1}x^{n-1} + a_1 x + a_0$ are the same as the limits of its leading term $a_n x^n$ as $x \to \infty$ and as $x \to -\infty$. *(Hint: Show that $\lim_{x\to\infty} f(x)$ is equal to $\lim_{x\to\infty} a_n x^n$ by factoring out $a_n x^n$ from $f(x)$.)*

86. Use limit techniques to prove that a rational function $f(x) = \frac{p(x)}{q(x)}$ will have

(a) a horizontal asymptote at $y = 0$, if the degree of $p(x)$ is less than the degree of $q(x)$;

(b) a horizontal asymptote at $y = \frac{a_n}{b_m}$, where a_n and b_m are the leading terms of $p(x)$ and $q(x)$, respectively, if $p(x)$ and $q(x)$ have the same degree;

(c) no horizontal asymptote, if the degree of $p(x)$ is greater than the degree of $q(x)$.

87. Prove the second part of Theorem 1.29: If $\lim_{x\to c} \frac{f(x)}{g(x)}$ is of the form $\frac{1}{0^-}$, then $\lim_{x\to c} \frac{f(x)}{g(x)} = -\infty$.

88. Prove the second part of Theorem 1.30: If $\lim_{x\to\infty} \frac{f(x)}{g(x)}$ is of the form $\frac{1}{-\infty}$, then $\lim_{x\to\infty} \frac{f(x)}{g(x)} = 0$.

89. Prove that the sum rule for limits also applies for limits as $x \to \infty$: If $\lim_{x\to\infty} f(x) = L$ and $\lim_{x\to\infty} g(x) = M$, then $\lim_{x\to\infty} (f(x) + g(x)) = L + M$.

90. Prove the first part of Theorem 1.31(a): If $k > 0$, then $\lim_{x\to\infty} x^k = \infty$. *(Hint: Given $M > 0$, choose $N = M^{1/k}$. Then show that for $x > N$ it must follow that $x^k > M$.)*

91. Prove the second part of Theorem 1.31(a): If $k > 0$, then $\lim_{x\to\infty} x^{-k} = 0$.

92. Prove the $k = 1$ case of the first part of Theorem 1.31(b): that $\lim_{x\to\infty} e^x = \infty$. *(Hint: Given $M > 0$, choose $N = \ln M$. Then if $x > N = \ln M$, we must have $x = \ln M + c$ for some positive number c. Use this to show that $e^x > M$.)*

93. Prove the $k = 1$ case of the second part of Theorem 1.31(b): that $\lim_{x\to\infty} e^{-x} = 0$.

94. Prove that $\lim_{\theta\to 0} \frac{1-\cos\theta}{\theta} = 0$ by using the double-angle identity $\cos 2\theta = 1 - 2\sin^2\theta$ and the other special trigonometric limit $\lim_{\theta\to 0} \frac{\sin\theta}{\theta} = 1$.

Thinking Forward

A limit representing an instantaneous rate of change: After t seconds, a bowling ball dropped from 350 feet has height $h(t) = 350 - 16t^2$, measured in feet.

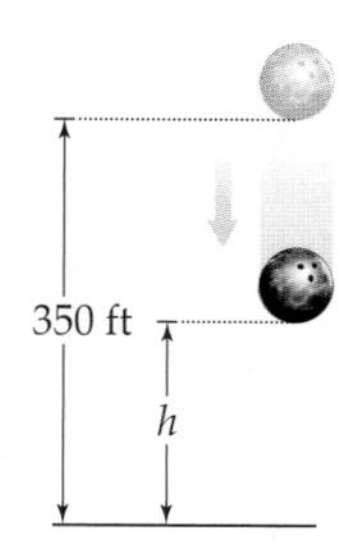

- Calculate the average rate of change of the height of the bowling ball from $t = 3$ to $t = 3 + h$ seconds in the cases where h is equal to 0.5, 0.25, 0.1, and 0.01.
- Write down a formula for the average rate of change of the height of the bowling ball from time $t = 3$ to time $t = 3 + h$, assuming that $h > 0$. The only letter in your formula should be h.
- Take the limit as $h \to 0^+$ of the formula you found for average rate of change in the previous problem. What does this limit represent in real–world terms?

Taylor Series: In this section we learned that e can be thought of as the following limit:

$$\lim_{h\to 0}(1 + h)^{1/h} = e.$$

In the following exercise you will investigate the convergence of this limit and also get a preview of ***Taylor series***, which we will see in Chapter 8.

- Use the substitution $n = \frac{1}{h}$ to show that the preceding limit statement is equivalent to the limit statement

$$\lim_{n\to\infty}\left(1 + \frac{1}{n}\right)^n = e.$$

- The ***Binomial Theorem*** says that an expression of the form $(a + b)^n$ can be expanded to

$$\binom{n}{0}a^n b^0 + \binom{n}{1}a^{n-1}b^1 + \binom{n}{2}a^{n-1}b^2 + \cdots + \binom{n}{n}x^0 y^n,$$

where, for any $0 \le k \le n$, the symbol $\binom{n}{k}$ is equal to $\frac{n!}{k!(n-k)!}$. Here $n!$ is n ***factorial***, the product of the integers from 1 to n. By convention we set $0! = 1$. Apply this expansion to the expression $\left(1 + \frac{1}{n}\right)^n$.

- Show that as $n \to \infty$ we would expect the preceding expansion to approach

$$1 + \frac{1}{1!} + \frac{1}{2!} + \frac{1}{3!} + \frac{1}{4!} + \frac{1}{5!} + \cdots.$$

(Hint: Think about limits of rational functions and ratios of leading coefficients.)

- Use a calculator to find the sum of the first six terms of the sum from the previous problem, and compare this sum with your calculator's best approximation of the number e.

CHAPTER REVIEW, SELF-TEST, AND CAPSTONES

Before you progress to the next chapter, be sure you are familiar with the definitions, concepts, and basic skills outlined here. The capstone exercises at the end bring together ideas from this chapter and look forward to future chapters.

Definitions

Give precise mathematical definitions or descriptions of each of the concepts that follow. Then illustrate the definition with a graph or algebraic example, if possible.

- the intuitive meaning of the limit statements $\lim_{x \to c} f(x) = L$, $\lim_{x \to c^-} f(x) = L$, and $\lim_{x \to c^+} f(x) = L$
- the intuitive meaning of the limit statements $\lim_{x \to c} f(x) = \infty$ and $\lim_{x \to \infty} f(x) = L$
- the formal δ–ϵ definition of the limit statements $\lim_{x \to c} f(x) = L$, $\lim_{x \to c^-} f(x) = L$, and $\lim_{x \to c^+} f(x) = L$
- the formal δ–M, N–ϵ, and N–M definitions of the limit statements $\lim_{x \to c} f(x) = \infty$, $\lim_{x \to \infty} f(x) = L$, and $\lim_{x \to \infty} f(x) = \infty$, respectively
- what we mean when we say that a limit *exists*, or that a limit *does not exist*
- what it means, in terms of limits, for a function f to have a *vertical asymptote* at $x = c$ or a *horizontal asymptote* at $y = L$
- what it means, in terms of limits, for a function f to be *continuous at a point* $x = c$, *left continuous* at $x = c$, and *right continuous* at $x = c$
- what it means for a function f to be *continuous on a closed interval* $[a, b]$, or *continuous on an open interval* (a, b), or *continuous on a half-closed interval* $[a, b)$
- what it means, in terms of limits, for a function to have a *removable discontinuity*, a *jump discontinuity*, or an *infinite discontinuity* at $x = c$
- the definition of the number e in terms of a limit

Theorems

Fill in the blanks to complete each of the following theorem statements:

- If $\lim_{x \to c} f(x) = L$ and $\lim_{x \to c} f(x) = M$, then ____ .
- $\lim_{x \to c} f(x) = L$ if and only if $\lim_{x \to c^-} f(x) =$ ____ and $\lim_{x \to c^+} f(x) =$ ____ .
- For $\delta > 0$, $x \in (c - \delta, c) \cup (c, c + \delta)$ if and only if $0 <$ ____ $< \delta$.
- For $\epsilon > 0$, $f(x) \in (L - \epsilon, L + \epsilon)$ if and only if ____ $< \epsilon$.
- *The Extreme Value Theorem:* If f is ____ on a closed interval $[a, b]$, then there exist values M and m in the interval $[a, b]$ such that $f(M)$ is ____ and $f(m)$ is ____ .
- *The Intermediate Value Theorem:* If f is ____ on a closed interval $[a, b]$, then for any K strictly between ____ and ____ , there exists at least one $c \in (a, b)$ such that ____ .
- A function f can change sign from positive to negative, or vice versa, at $x = c$ only if $f(x)$ is ____ , ____ , or ____ at $x = c$.
- Constant, identity, and linear functions are continuous everywhere, which means in terms of limits that ____ , ____ , and ____ .
- Power functions are continuous everywhere, which means in terms of limits that ____ .
- All algebraic functions are ____ on their domains, which means in terms of limits that if $x = c$ is in the domain of an algebraic function f, then ____ .
- All basic transcendental functions are ____ on their domains, which means in terms of limits that if $x = c$ is in the domain of a basic exponential, logarithmic, trigonometric, or inverse trigonometric function f, then ____ .
- If $\lim_{x \to c} g(x)$ exists and $f(x)$ is a function that is ____ to $g(x)$ for all x sufficiently close to ____ , but not necessarily at ____ , then ____ .
- *The Squeeze Theorem for Limits:* If $l(x) \le f(x) \le u(x)$ for all x sufficiently close to ____ , but not necessarily at ____ , and if $\lim_{x \to c} l(x)$ and $\lim_{x \to c} u(x)$ are both equal to L, then ____ .
- Suppose $\frac{p(x)}{q(x)}$ is a rational function with $\deg(p(x)) = n$ and $\deg(q(x)) = m$. If $n < m$, then $\lim_{x \to \infty} \frac{p(x)}{q(x)} =$ ____ ; if $n = m$, then $\lim_{x \to \infty} \frac{p(x)}{q(x)} =$ ____ ; and if $n > m$, then $\lim_{x \to \infty} \frac{p(x)}{q(x)} =$ ____ .

Limit Rules and Indeterminate Forms

Limits of basic functions: Fill in the blanks to complete the limit rules that follow. You may assume that k is positive.

- $\lim_{x\to c} k =$ ____
- $\lim_{x\to c} x =$ ____
- $\lim_{x\to c}(mx+b) =$ ____
- $\lim_{x\to c} Ax^n =$ ____
- $\lim_{x\to\infty} x^k =$ ____
- $\lim_{x\to\infty} x^{-k} =$ ____
- $\lim_{x\to c} b^x =$ ____
- $\lim_{x\to c} \sin x =$ ____
- $\lim_{x\to\infty} e^{kx} =$ ____
- $\lim_{x\to\infty} e^{-kx} =$ ____
- $\lim_{x\to\infty} 2^x =$ ____
- $\lim_{x\to\infty} (0.75)^x =$ ____
- $\lim_{x\to 0^+} \ln x =$ ____
- $\lim_{x\to\infty} \ln x =$ ____
- $\lim_{x\to\infty} \tan^{-1} x =$ ____
- $\lim_{x\to -\infty} \tan^{-1} x =$ ____
- $\lim_{x\to 0} \dfrac{e^x - 1}{x} =$ ____
- $\lim_{x\to 0}(1+x)^{1/x} =$ ____
- $\lim_{x\to 0} \dfrac{\sin x}{x} =$ ____
- $\lim_{x\to 0} \dfrac{1-\cos x}{x} =$ ____

Limits of combinations: Fill in the blanks to complete the limit rules that follow. You may assume that k and c are any real numbers and that both $\lim_{x\to c} f(x)$ and $\lim_{x\to c} g(x)$ exist.

- $\lim_{x\to c} kf(x) =$ ____
- $\lim_{x\to c}(f(x)+g(x)) =$ ____
- $\lim_{x\to c}(f(x)-g(x)) =$ ____
- $\lim_{x\to c}(f(x)g(x)) =$ ____
- $\lim_{x\to c} \dfrac{f(x)}{g(x)} =$ ____, provided that ____
- $\lim_{x\to c} f(g(x)) =$ ____, provided that ____

Indeterminate forms: Identify which of the limit forms listed here are indeterminate. For each form that is not indeterminate, describe the behavior of a limit of that form.

- $\dfrac{1}{0^+}$
- $\dfrac{1}{0^-}$
- $\dfrac{1}{\infty}$
- $\dfrac{1}{-\infty}$
- $\dfrac{0}{1}$
- $\dfrac{0}{0}$
- $\dfrac{\infty}{1}$
- $\dfrac{0}{\infty}$
- $\dfrac{\infty}{\infty}$
- $\dfrac{\infty}{0^+}$
- $0\cdot\infty$
- $\infty\cdot\infty$
- $\infty(-\infty)$
- $\infty+\infty$
- $\infty-\infty$
- 0^0
- 0^1
- 0^∞
- 1^∞
- ∞^0
- ∞^∞
- ∞^1
- $0^{-\infty}$
- $\infty^{-\infty}$

Skill Certification: Basic Limits

Calculating limits: Find each limit by hand.

1. $\lim_{x\to 0} 3x^{-4}$
2. $\lim_{x\to\infty} -2x^{-1/2}$
3. $\lim_{x\to 2} \dfrac{1}{2-x}$
4. $\lim_{x\to 1} \dfrac{1}{x^2-1}$
5. $\lim_{x\to 1} \dfrac{2x^3-x^2-2x+1}{x^2-2x+1}$
6. $\lim_{x\to -\infty} \dfrac{x^3+2x-1}{1-x^4}$
7. $\lim_{x\to 0} \dfrac{3^x-4^x}{3^x}$
8. $\lim_{x\to 0} \dfrac{e^x-1}{3e^{2x}-2e^x-1}$
9. $\lim_{x\to 0^+} \dfrac{\ln x}{x}$
10. $\lim_{x\to\infty} \ln\left(\dfrac{x-1}{1-3x^2}\right)$
11. $\lim_{x\to\infty} \dfrac{1-e^x}{e^{2x}}$
12. $\lim_{x\to \pi/2} \dfrac{\sin x}{x}$
13. $\lim_{x\to\infty} (\sqrt{x}-x)$
14. $\lim_{x\to 0^+} \dfrac{\sqrt{x}-x^3}{x^2}$
15. $\lim_{x\to 3} \dfrac{\frac{1}{x-3}-\frac{1}{x}}{x-3}$
16. $\lim_{x\to 4} \dfrac{2-\sqrt{x}}{4-x}$
17. $\lim_{x\to\infty} (-2x^3+x^2-10)$
18. $\lim_{x\to 0}(x^{-3}-2x^{-1})$
19. $\lim_{x\to\infty} \dfrac{(2x-1)(x^2+1)}{x^2-4}$
20. $\lim_{x\to\infty} \dfrac{(x-1)(3x+1)^3}{(x-2)^4}$
21. $\lim_{x\to\infty} \dfrac{\sqrt{x}}{1-\sqrt{x}}$
22. $\lim_{x\to -\infty} e^x \tan^{-1} x$
23. $\lim_{x\to 0} \dfrac{3x}{\sin 2x}$
24. $\lim_{x\to 0} \dfrac{\sin^2 3x}{x}$
25. $\lim_{x\to 0} \dfrac{1-\cos x}{\sin x}$
26. $\lim_{x\to\infty} \sin(\tan^{-1} x)$
27. $\lim_{x\to\infty} \left(1+\dfrac{1}{x}\right)^x$
28. $\lim_{x\to 0} \dfrac{x^2}{e^x-1}$
29. $\lim_{x\to\infty} \sin x$
30. $\lim_{x\to\infty} \sin \dfrac{1}{x}$
31. $\lim_{x\to\infty} \dfrac{1}{x} \sin x$
32. $\lim_{x\to 0} x \sin \dfrac{1}{x}$

Capstone Problems

A. *Continuity of piecewise-defined functions:* For each given function f, find a real number a that makes f continuous at $x = 0$, if possible.

(a) $f(x) = \begin{cases} 3x+1, & \text{if } x < 0 \\ 2x+a, & \text{if } x \geq 0 \end{cases}$

(b) $f(x) = \begin{cases} \dfrac{a}{x+2}, & \text{if } x < 0 \\ 3, & \text{if } x = 0 \\ ax+1, & \text{if } x > 0 \end{cases}$

B. *Limits that define derivatives:* In the next chapter we will be interested in ***derivatives***, which we will define as limits of the form

$$\lim_{h \to 0} \frac{f(c+h) - f(c)}{h}.$$

(a) Calculate this limit for $f(x) = x^3$ and $c = 0$.
(b) Calculate this limit for $f(x) = x^3$ and $c = 2$.
(c) Calculate this limit for $f(x) = x^3$ and general $c = x$. This time your answer will be a function of x instead of a number.

C. *The limit of a model at infinity:* Leila is interested in the effect of a stabilized wolf population on the eventual population of beavers in Idaho. The following table gives estimated beaver populations $B(t)$ for $t = 0, 1, 2, 3, 4,$ and 5 years after 2005:

t	0	1	2	3	4	5
$B(t)$	48,112	42,256	47,088	43,684	46,320	44,704

(a) Leila makes a plot of these values of $B(t)$ and notes that the population of beavers is cyclical with diminishing amplitude. She finds that the quadratic function

$$M(t) = -51x^2 + 918x + 41,389$$

is a good model for the relative maximum data points at $t = 0$, 2, and 4, and that

$$m(t) = 33.25x^2 - 583.5x + 48,122$$

is a good model for the relative minimum data points at $t = 1$, 3, and 5. Verify that these functions do in fact pass through the relevant data points, and graph the data for $B(t)$ along with the two functions.

(b) Do the two quadratics $M(t)$ and $m(t)$ ever meet? If so, where? What conclusion could Leila make concerning the eventual steady population $\lim_{t \to \infty} B(t)$ of beavers in Idaho?

D. *The limit of a rational function model at infinity:* Upon further reflection, Leila decides that the quadratics used in the previous problem are unreasonable, since the quadratic model for the relative maximum values could be interpreted as indicating that the eventual number of beavers would be unbounded. She decides to change her model for the relative maximum beaver populations to

$$M(t) = \frac{40944\,t^2 + 454512\,t - 1732032}{t^2 + 9t - 36}.$$

(a) Verify that this function does pass through the data points at $t = 0$, 2, and 4. Is this function continuous everywhere? *(Hint: Consider* $\lim_{t \to 3} M(t)$*.)*

(b) Compute $\lim_{t \to \infty} M(t)$. What is the significance of this number?

CHAPTER 2

Derivatives

2.1 An Intuitive Introduction to Derivatives

Slope Functions
Position and Velocity
Approximating the Slope of a Tangent Line
Approximating an Instantaneous Rate of Change
Examples and Explorations

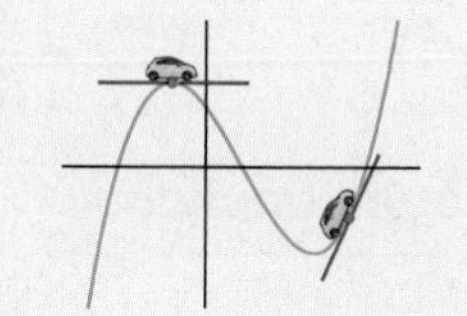

2.2 Formal Definition of the Derivative

The Derivative at a Point
The Derivative as a Function
Differentiability
Tangent Lines and Local Linearity
Leibniz Notation and Differentials
Examples and Explorations

$$\lim_{h \to 0} \frac{f(x+h) - f(x)}{h}$$

2.3 Rules for Calculating Basic Derivatives

Derivatives of Linear Functions
The Power Rule
The Constant Multiple and Sum Rules
The Product and Quotient Rules
Examples and Explorations

$$\frac{d}{dx}(x^k) = kx^{k-1}$$

2.4 The Chain Rule and Implicit Differentiation

Differentiating Compositions of Functions
Implicit Differentiation
Examples and Explorations

$$\frac{df}{dx} = \frac{df}{du}\frac{du}{dx}$$

2.5 Derivatives of Exponential and Logarithmic Functions

Derivatives of Exponential Functions
Exponential Functions Grow Proportionally to Themselves
Derivatives of Logarithmic Functions
Derivatives of Inverse Functions*
Examples and Explorations

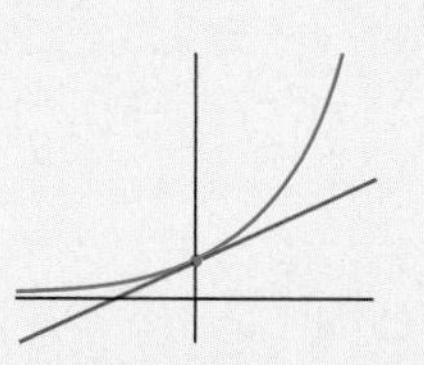

2.6 Derivatives of Trigonometric and Hyperbolic Functions

Derivatives of Trigonometric Functions
Derivatives of Inverse Trigonometric Functions
Hyperbolic Functions and Their Derivatives*
Inverse Hyperbolic Functions and Their Derivatives*
Examples and Explorations

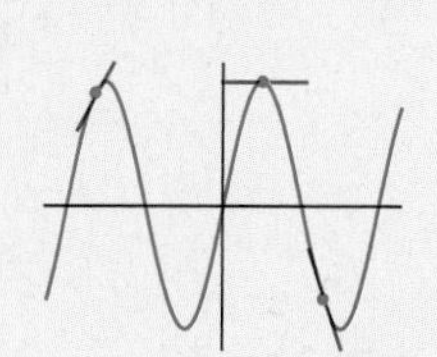

Chapter Review, Self-Test, and Capstones

2.1 AN INTUITIVE INTRODUCTION TO DERIVATIVES

- Associated slope functions, tangent lines, and secant lines
- Velocity and other instantaneous rates of change
- Approximating slopes of tangent lines and instantaneous rates of change

Slope Functions

We begin our study of the derivative with an intuitive introduction in terms of slopes and rates of change. We also start thinking about how one might calculate, or at least approximate, derivatives. In Section 2.2 we will give a formal mathematical definition of the derivative in terms of limits.

Intuitively speaking, if the graph of a function f is ***smooth*** on an interval (a, b)—meaning that it does not have any corners, cusps, jumps, or holes—then at every point $(x, f(x))$ on the the graph of f on the interval (a, b) we can consider the direction, or ***slope***, of the function at that point. For example, in the figure that follows at the left, the ***tangent line*** drawn at $x = -1$ points in the same direction as the function f at the point $(-1, f(-1))$. More precisely, if you imagine yourself in a tiny car driving along the graph of f with your headlights on, then that line represents the direction that your headlights are pointing when you reach the point $(-1, f(-1))$ from the right or the left. Similarly, the line drawn at $x = 4$ represents the direction of the graph of f at the point $(4, f(4))$.

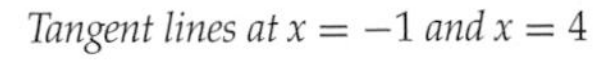
Tangent lines at $x = -1$ and $x = 4$

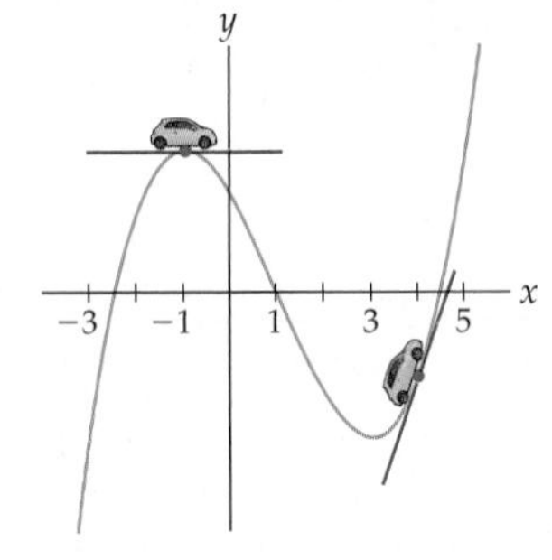

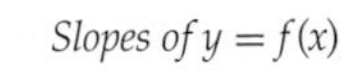
Slopes of $y = f(x)$

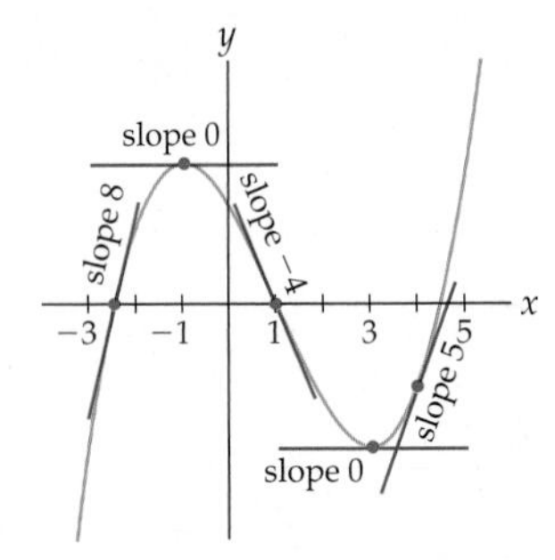

Heights of $y = f'(x)$

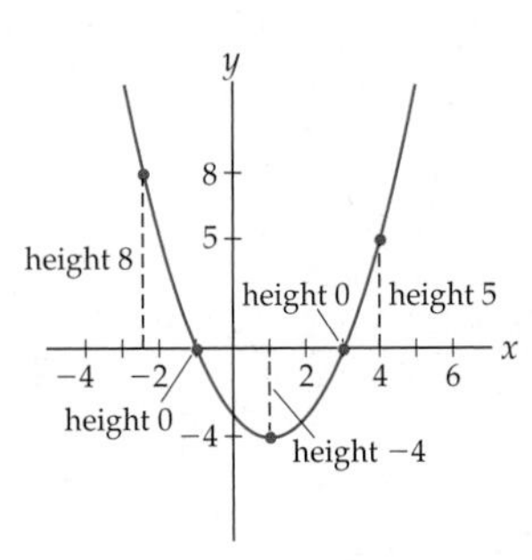

We can use the graph of a smooth function f to define a new function whose output at each point x is the slope of the tangent line at x. For example, the slopes shown on the graph of f in the middle figure are used to define heights on the graph of the ***associated slope function*** shown at the right. This associated slope function is what we will define in Section 2.2 as the ***derivative*** of $f(x)$ and denote as $f'(x)$ (pronounced "f prime of x"). Notice the following relationships between a function f and its derivative f':

- For each x, the slope of $f(x)$ is the height of $f'(x)$.
- Where f has a horizontal tangent line, the derivative f' has a root.
- Where the graph of f is increasing, the derivative f' is above the x-axis.
- Where the graph of f is decreasing, the derivative f' is below the x-axis.
- Where f has steep slope, the derivative f' has large magnitude.
- Where f has shallow slope, the derivative f' has small magnitude.

Position and Velocity

Suppose an object moves in a straight path so that after t seconds it is a distance of $s(t)$ units from its starting point. We will call the function $s(t)$ describing the motion of the object a ***position function***. The moving object has a speed and a direction at any time t, and the combination of these two measurements defines a ***velocity function*** for the object. Specifically, if we consider one direction on the straight path as the "positive" direction and the other as the "negative" direction, then the velocity of the object at time t is the speed of the object times either $+1$ or -1, depending on the direction in which the object is moving.

The velocity $v(t)$ of such a moving object is a measurement of how the position function of the object is changing over time. Intuitively, because the way position changes at a particular moment in time is measured by the slope of its graph, velocity is the associated slope function for position. In other words, velocity is the derivative of position, or in symbols, $v(t) = s'(t)$. Similarly, acceleration $a(t)$ measures how velocity changes, and thus $a(t) = v'(t)$. We will examine these relationships more precisely in Section 2.2.

For example, suppose you throw a grapefruit straight up into the air, releasing it at a height of 4 feet and an upwards velocity of 32 feet per second, as illustrated in the figure that follows on the left. On the right is a plot of the height of the grapefruit over time. The points A, B, C, D, and E show the height $s(t)$, in feet, of the grapefruit at $t = 0$, $t = 0.6$, $t = 1$, $t = 1.75$, and $t = 2.118$ seconds. At A, the grapefruit is moving upwards quickly. Because of the downwards pull of gravity, the grapefruit is moving upwards more slowly at B. At C the grapefruit is at the top of its flight and about to fall to the ground. Gravity then causes the grapefruit to fall faster and faster through D and then finally E when it hits the ground.

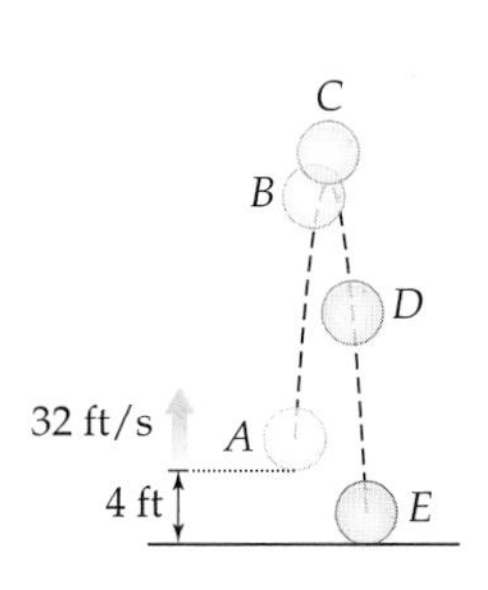

Position increases, then decreases

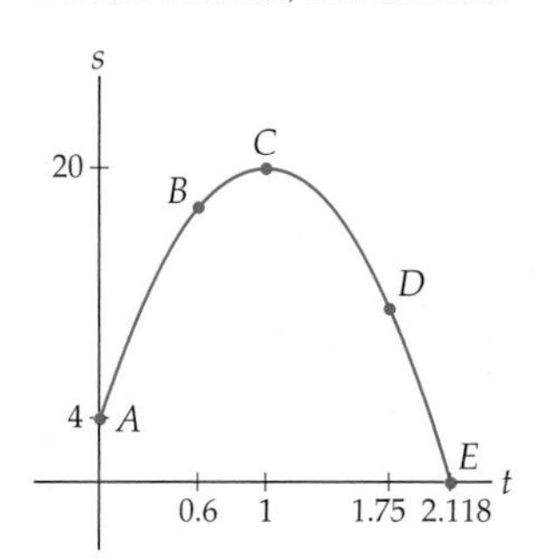

The next leftmost figure shows the velocity $v(t)$ of the grapefruit. Notice that at times A and B, when the grapefruit is moving upwards, its velocity is positive; at C, when the grapefruit turns around at the top of its flight, its velocity is zero; and at D and E, when the grapefruit is falling to the ground, its velocity is negative. The rightmost figure shows the constant acceleration of the grapefruit due to gravity.

Velocity is positive, then negative

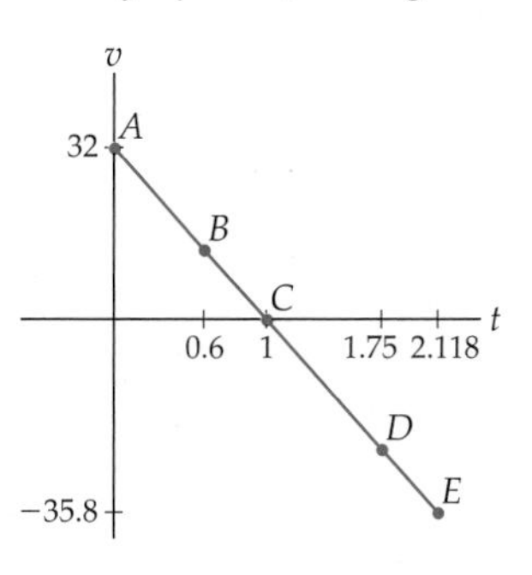

Acceleration is constant

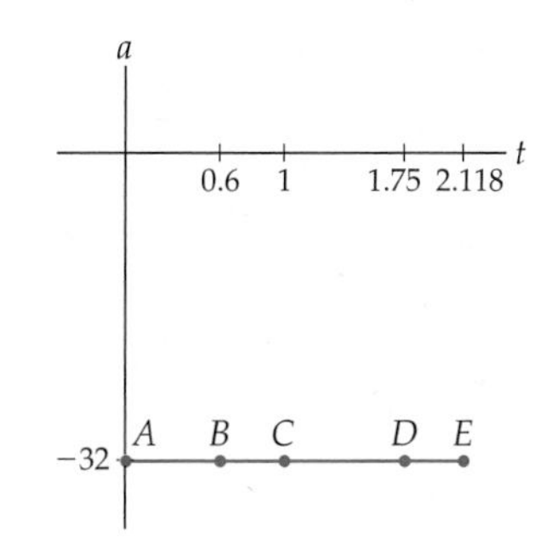

In general, our earlier list of relationships between a function $f(x)$ and its derivative $f'(x)$ translates into a list of relationships between position $s(t)$ and velocity $v(t) = s'(t)$ as follows:

- For each t, the way position $s(t)$ is changing is measured by velocity $v(t)$.
- When position $s(t)$ is not changing, velocity $v(t)$ is zero.
- When position $s(t)$ is increasing, velocity $v(t)$ is positive.
- When position $s(t)$ is decreasing, velocity $v(t)$ is negative.
- When position $s(t)$ is changing rapidly, velocity $v(t)$ has large magnitude.
- When position $s(t)$ is changing slowly, velocity $v(t)$ has small magnitude.

Approximating the Slope of a Tangent Line

Usually, calculating the slope of a line is a simple matter: Simply take two points (x_0, y_0) and (x_1, y_1) on the line and calculate the "rise over run," which is equal to the average rate of change $\frac{\Delta y}{\Delta x} = \frac{y_1 - y_0}{x_1 - x_0}$. With tangent lines the situation is more complicated, because we know only *one* point on a tangent line, namely, the point $(c, f(c))$ where it touches the function. The slope of the tangent line measures the "direction" of the function, but how do we calculate that from only one point? The key will be to use nearby points on the function to approximate nearby slopes.

The ***secant line*** from a to b for a function f is the line that passes through the points $(a, f(a))$ and $(b, f(b))$. If f is a smooth function and $x = z$ is a point that is close to $x = c$, then the slope of the secant line from $x = c$ to $x = z$ will be close to the slope of the tangent line to f at $x = c$, as shown in the middle graph that follows:

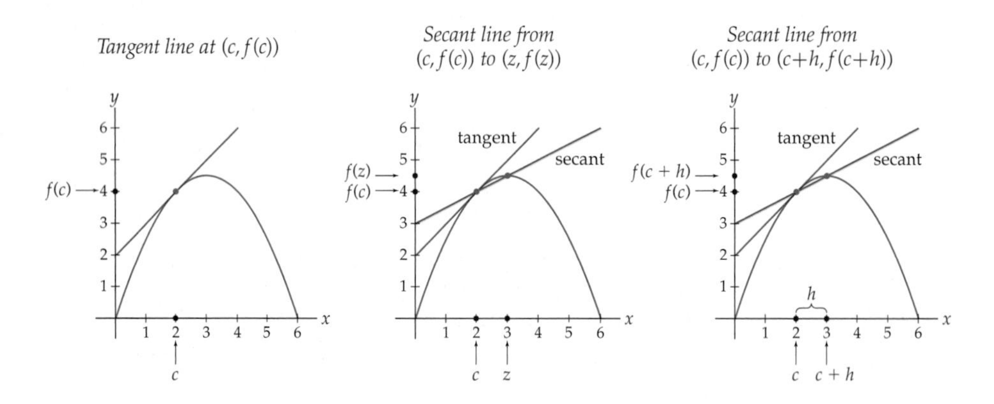

If we choose points z that are closer and closer to the point $x = c$, we will get secant lines that get closer and closer to the tangent line we are interested in. Equivalently, we could think of the second point z as "c plus a little bit," where the little bit is called h. In other words, $z = c + h$, as in the rightmost graph shown.

Since we know two points on a secant line, we can easily calculate its slope. The slope $f'(c)$ of the tangent line to f at $x = c$ can be approximated by the slope of a nearby secant line from $x = c$ to $x = z$, or equivalently, from $x = c$ to $x = c + h$:

$$f'(c) \approx \frac{f(z) - f(c)}{z - c}, \quad \text{or equivalently,} \quad f'(c) \approx \frac{f(c+h) - f(c)}{h}.$$

The preceding expressions are often called ***difference quotients***, and when we find the derivative of a function f, we say that we are ***differentiating*** the function. If the graph of f is smooth, then as z gets closer to c, or as h gets closer to 0, these approximations get closer and closer to $f'(c)$. In Section 2.2 we will in fact take the limit as $z \to c$, or, equivalently, as $h \to 0$, to define the derivative exactly.

Approximating an Instantaneous Rate of Change

You may have noticed that the approximations we have been using for $f'(c)$ are closely related to the formula for average rate of change. This is no coincidence, since average rates of change are in fact the same as slopes of secant lines. As we choose points $z = c+h$ closer and closer to $x=c$, these average rates of change approach the ***instantaneous rate of change*** of the function at $x = c$. For a general function f, this instantaneous rate of change at $x = c$ is the derivative, that is, the slope $f'(c)$ of the tangent line to f at $x = c$. In the case of a position function $s(t)$, the instantaneous rate of change is the velocity $v(c) = s'(c)$.

We can approximate instantaneous rates of change in a position or velocity context in much the same way as we approximated slopes of tangent lines in the previous discussion. Suppose an object is moving along a straight path. The ***distance formula*** says that for such a moving object, the distance travelled, average rate, and time elapsed are related by the formula $d = rt$ ("distance equals rate times time"). We can also write this formula as $r = \frac{d}{t}$, or more accurately, as $r = \frac{\Delta d}{\Delta t}$, since we want to consider the change in distance over a corresponding change in time. If an object starts at position s_0 at time t_0 and ends at position s_1 at time t_1, then we have

$$\begin{matrix}\text{average velocity} \\ \text{from } t_0 \text{ to } t_1\end{matrix} = \begin{matrix}\text{average rate of change of} \\ \text{position from } t_0 \text{ to } t_1\end{matrix} = \frac{\Delta d}{\Delta t} = \frac{s_1 - s_0}{t_1 - t_0}.$$

Now suppose $s(t)$ describes the position of the object at time t and we are interested in finding the velocity $v(c)$ at some time $t = c$. This instantaneous velocity can be approximated by the average velocity over a small time interval $[c, z]$, or equivalently, $[c, c+h]$:

$$v(c) \approx \frac{s(z) - s(c)}{z - c}, \quad \text{or equivalently,} \quad v(c) \approx \frac{s(c+h) - s(c)}{h}.$$

Notice that this is just a special case of what we did earlier for a general function $f(x)$ and its derivative $f'(x)$ at a point $x = c$.

We can use derivatives to examine instantaneous rates of change in many contexts. In general, the derivative of a function $y(x)$ represents the instantaneous rate of change of the variable y as the variable x varies. The units for the derivative $y'(x)$ are the units for the variable y divided by the units for the variable x. For example, if time t is measured in hours and position $s(t)$ is measured in miles, then the velocity $v(t) = s'(t)$ is measured in miles per hour. As another example, if $Q(t)$ is the amount of money in a savings account after t years, measured in dollars, then $Q'(t)$ is the rate at which the savings balance changes over time, with units measured in dollars per year.

Examples and Explorations

EXAMPLE 1

Sketching the graph of an associated slope function

Given the following graph of the smooth function f, sketch the graph of its associated slope function f':

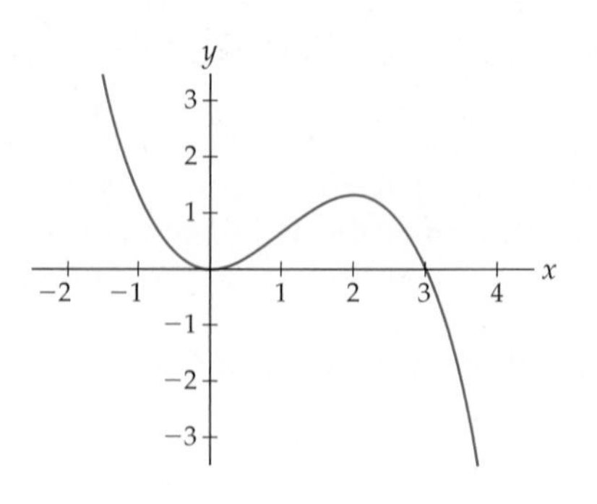

SOLUTION

A good place to begin is by marking all the locations on the graph of f where the tangent line is horizontal and thus has slope 0. In this case that happens at $x = 0$ and at $x = 2$, as shown next at the left. Thus the associated slope function f' has zeroes at $x = 0$ and $x = 2$, as shown next at the right.

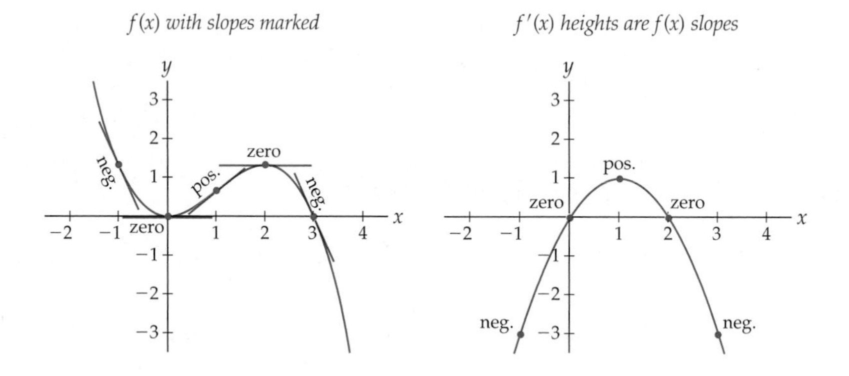

Looking again at the graph of f, we see that its tangent lines have positive slope between $x = 0$ and $x = 2$ (see, for example, the positive slope marked at $x = 1$). This means that, in the graph of f', the heights will be positive between $x = 0$ and $x = 2$. Similarly, the negative slopes on the graph of f to the left of $x = 0$ and to the right of $x = 2$ correspond to negative heights on the graph of f'. □

EXAMPLE 2

Graphing velocity from the graph of position

Suppose the graph that follows describes your distance from home one morning as you drive back and forth from your sister's house. Describe a possible scenario for your travels that morning. Then sketch the corresponding graph of your velocity.

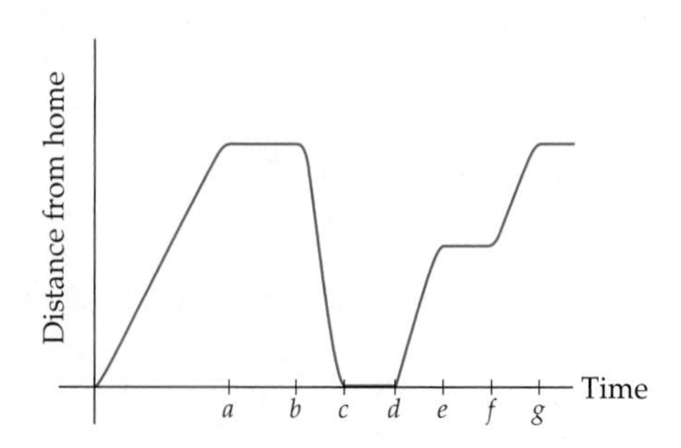

SOLUTION

One possible scenario is this: You drive to your sister's house for a visit. After talking to her for a few minutes, you realize you forgot something at home and race back to get it. In the middle of returning to your sister's house, you have to stop at a red light for a couple of minutes. Following the times marked on the time axis, we see that from 0 to a you drive to your sister's house, you talk until b, race home from b to c, leave your house at d and get stopped at the light at e, and move on at f until you get back to your sister's house at g.

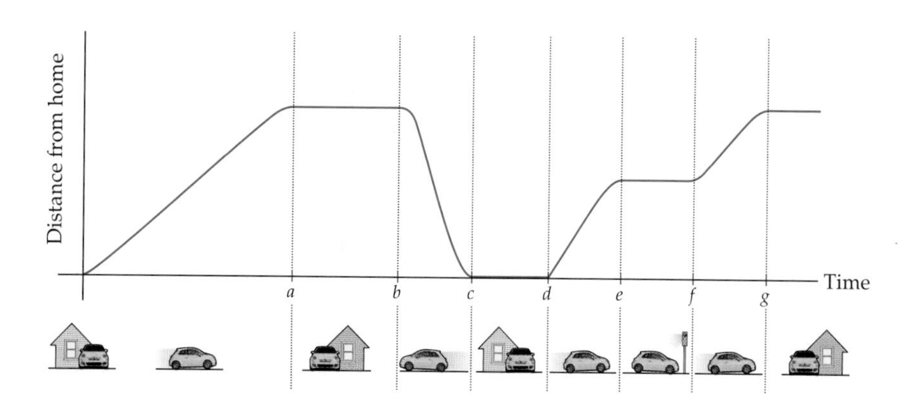

The graph of your velocity that morning is the graph of the associated slope function for the given position graph. The slope of the position graph is positive from 0 to a, zero from a to b, negative and steep from b to c, zero from c to d, positive and steep from d to e, zero from e to f, positive and steep from f to g, and finally zero again after g. The previous sentence also describes the height of the corresponding velocity graph, where steep slope values correspond to large magnitudes of velocity:

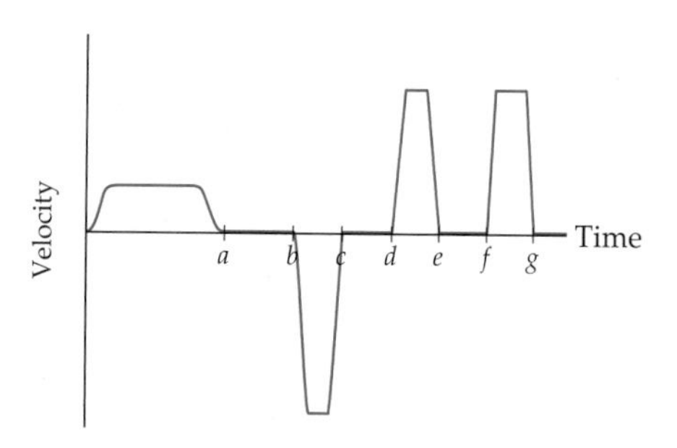

□

EXAMPLE 3

Estimating the slope of a tangent line with a sequence of secant lines

Estimate the slope of the line tangent to the graph of $f(x) = -\frac{1}{2}x^2 + 3x$ at the point $(2, f(2))$ by calculating a sequence of slopes of secant lines.

SOLUTION

The tangent line passes through the point $(2, f(2)) = \left(2, -\frac{1}{2}(2)^2 + 3(2)\right) = (2, 4)$ and is shown in red in each of the graphs that follow. Since we only know one point on this line, we cannot compute its slope directly. We will approximate its slope by considering a sequence the slopes of secant lines on smaller and smaller intervals, namely, $[2, 3]$, $[2, 2.5]$, $[2, 2.25]$, and $[2, 2.1]$, as shown in the following four graphs below:

Secant line on [2, 3] *Secant line on* [2, 2.5] *Secant line on* [2, 2.25] *Secant line on* [2, 2.1]

In our difference quotient notation, these intervals correspond to a sequence of points $z = 3$, $z = 2.5$, $z = 2.25$, and $z = 2.1$ that approach $c = 2$. Equivalently, we can think of this sequence as a series of h-values $h = 1$, $h = 0.5$, $h = 0.25$, and $h = 0.1$ approaching zero.

The slope of the secant line from $x = 2$ to $x = 3$ in the leftmost graph is given by the difference quotient:

$$\frac{f(3) - f(2)}{3 - 2} = \frac{\left(-\frac{1}{2}(3)^2 + 3(3)\right) - \left(-\frac{1}{2}(2)^2 + 3(2)\right)}{3 - 2} = \frac{4.5 - 4}{3 - 2} = 0.5.$$

Similarly, the slopes of the remaining three secant lines are given by the difference quotients:

$$\frac{f(2.5) - f(2)}{2.5 - 2} = 0.75, \quad \frac{f(2.25) - f(2)}{2.25 - 2} = 0.875, \quad \text{and} \quad \frac{f(2.1) - f(2)}{2.1 - 2} = 0.95.$$

Each of these slopes is an approximation to the slope of the red tangent line. As the graphs shown suggest, we would expect this sequence of slopes to be getting closer and closer to the actual slope of the red tangent line; notice for example that, in the last figure shown, the green secant line is almost indistinguishable from the red tangent line.

In a similar fashion we can calculate the slopes of secant lines from the left of $x = 2$. For example, the slope of the secant line from $x = 1$ to $x = 2$ is given by the difference quotient:

$$\frac{f(1) - f(2)}{1 - 2} = \frac{\left(-\frac{1}{2}(1)^2 + 3(1)\right) - \left(-\frac{1}{2}(2)^2 + 3(2)\right)}{1 - 2} = \frac{2.5 - 4}{1 - 2} = 1.5.$$

Over the smaller intervals [1.5, 2], [1.75, 2], and [1.9, 2] we have secant lines with slopes given by

$$\frac{f(1.5) - f(2)}{1.5 - 2} = 1.25, \quad \frac{f(1.75) - f(2)}{1.7 - 2} = 1.125, \quad \text{and} \quad \frac{f(1.9) - f(2)}{1.9 - 2} = 1.05.$$

Putting all this information together, we obtain the following table:

Interval	[1, 2]	[1.5, 2]	[1.75, 2]	[1.9, 2]	*	[2, 2.1]	[2, 2.25]	[2, 2.5]	[2, 3]
Slope	1.5	1.25	1.125	1.05	*	0.95	0.875	0.75	0.5

From this table, we might guess that the slope of the tangent line is 1. This guess is only an estimate; the slope of the tangent line might instead be something like 0.97 or 1.02, but we don't have enough information to say otherwise at this point. □

EXAMPLE 4 **Estimating instantaneous velocity with a sequence of average velocities**

It can be shown that a watermelon dropped from a height of 100 feet will be $s(t) = -16t^2 + 100$ feet off the ground t seconds after it is dropped. Approximate the instantaneous velocity of the watermelon at time $t = 1$ by calculating a sequence of average velocities. Then interpret these average velocities graphically as slopes of secant lines.

SOLUTION

To estimate the instantaneous velocity at $t = 1$ we will look at a sequence of small time intervals near $t = 1$ and consider the corresponding average velocities. The time intervals we choose to consider are $[1, 2]$, $[1, 1.5]$, $[1, 1.25]$, and $[1, 1.1]$. These intervals correspond to $z = 2, 1.5, 1.25$, and 1.1, or equivalently, to $h = 1, 0.5, 0.25$, and 0.1.

Let's look first at the interval $[1, 2]$. At $t = 1$ the watermelon is $s(1) = 84$ feet from the ground, and at $t = 2$ the watermelon is $s(2) = 36$ feet from the ground, as illustrated here:

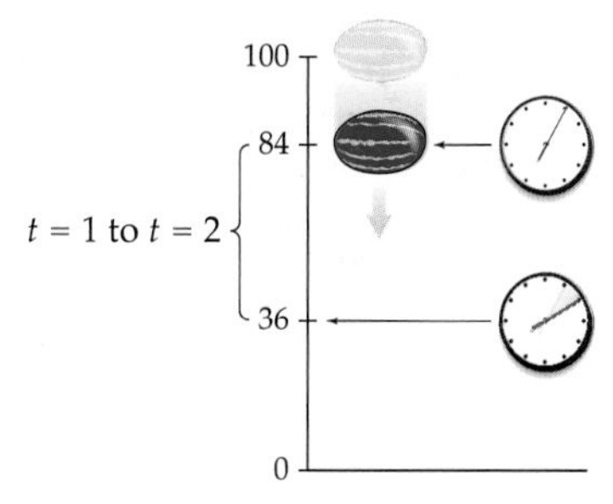

The average velocity over the first interval is therefore given by the difference quotient:

$$\frac{s(2) - s(1)}{2 - 1} = \frac{(-16(2)^2 + 100) - (-16(1)^2 + 100)}{1} = 36 - 84$$
$$= -48 \text{ feet per second.}$$

Similarly, the average velocities over the remaining three time intervals, in feet per second, are

$$\frac{s(1.5) - s(1)}{1.5 - 1} = -40, \quad \frac{s(1.25) - s(1)}{1.25 - 1} = -36, \quad \text{and} \quad \frac{s(1.1) - s(1)}{1.1 - 1} = -33.6.$$

Each of these average velocities is an approximation to the instantaneous velocity of the watermelon at time $t = 1$. Since the approximations should be improving as z gets closer to 1 (or, equivalently, as h gets closer to 0), we might guess that the instantaneous velocity of the watermelon at time $t = 1$ is some value greater than, but close to, -33.6 feet per second. For example, we might estimate that the instantaneous velocity at time $t = 1$ is -33 feet per second.

Each average velocity just calculated is an average rate of change of position, and thus can be thought of as the slope of a secant line, as in the four graphs shown next. As we consider smaller and smaller time intervals, we see that the slopes corresponding to these average velocities approach the slope of the red tangent line to $s(t)$ at $t = 1$, which in turn represents the instantaneous velocity at $t = 1$.

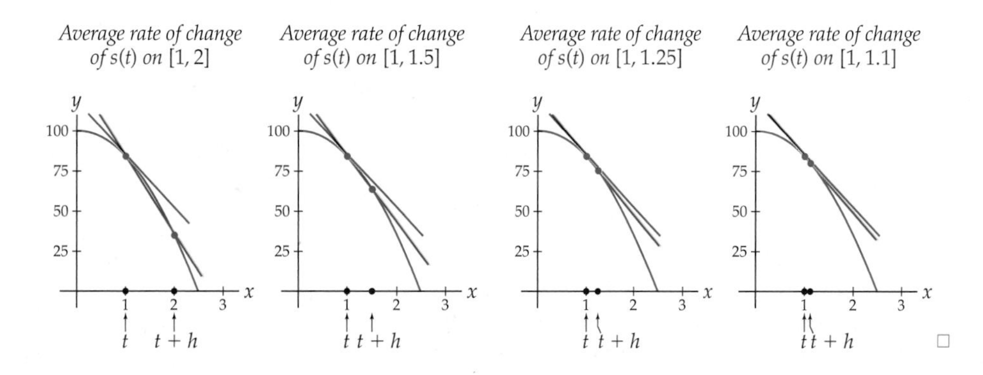

□

? TEST YOUR UNDERSTANDING

- Why does it make intuitive sense that when the graph of a smooth function f has a horizontal tangent line, the graph of its associated slope function will have a root?
- Why does it make intuitive sense that when the graph of a smooth function f is increasing, the graph of its associated slope function will be above the x-axis?
- In our initial discussion of position and velocity in this section, we showed graphs for the position, velocity, and acceleration of a grapefruit thrown into the air. Why does it make sense that the graph of acceleration is constant, given that the graph of velocity is a straight line?
- How can we use a sequence of slopes of secant lines to estimate the slope of a tangent line? Why is considering a sequence of secant lines better than considering just one secant line?
- In real–world examples, how are the units of a derivative of a function related to the units of the independent and dependent variables of that function?

EXERCISES 2.1

Thinking Back

- *Slope and linear functions:* If f is a linear function with slope -3 such that $f(2) = 1$, find the following, *without* first finding an equation for $f(x)$.
  - $f(4)$
  - $f(7)$
  - $f(-2)$
- *Approximating limits:* Use sequences of approximations to estimate the values of
  - $\lim\limits_{x \to 2} \dfrac{4 - x^2}{2 - x}$
  - $\lim\limits_{z \to 3} \dfrac{z^3 - 27}{z - 3}$
- *Identifying increasing and decreasing behavior:* Use a graphing utility to determine the intervals on which $f(x) = -4x^5 + 25x^4 - 40x^3$ is increasing or decreasing.
- *Interpreting distance graphically:* When flying home for the holidays, Eva often flies between Denver International Airport (DIA) and Chicago O'Hare (ORD). Suppose Eva's plane takes off from DIA and 50 miles from ORD the plane has to circle the airport because of snow. The plane circles ORD four times and then lands.
  (a) Draw a graph depicting the distance from DIA to Eva's plane as a function of time.
  (b) Draw a graph depicting the distance from ORD to Eva's plane as a function of time.

Concepts

0. *Problem Zero:* Read the section and make your own summary of the material.

1. *True/False:* Determine whether each of the statements that follow is true or false. If a statement is true, explain why. If a statement is false, provide a counterexample.

(a) *True or False:* The slope of the tangent line to a function f at the point $x = 4$ is given by $f'(4)$.

(b) *True or False:* The instantaneous rate of change of a function f at the point $x = -3$ is given by $f'(-3)$.

(c) *True or False:* The instantaneous rate of change of a function f at a point $x = a$ can be represented as the slope of a secant line.

(d) *True or False:* Where a function f is positive, its associated slope function f' is increasing.

(e) *True or False:* Where a function f is decreasing, its associated slope function f' is negative.

(f) *True or False:* When a function f has a steep slope at a point on its graph, its instantaneous rate of change at that point will have a large magnitude.

(g) *True or False:* When the graph of a function f is decreasing with a steep slope, the graph of the associated slope function f' is negative with a large magnitude.

(h) *True or False:* Suppose an object is moving in a straight path with position function $s(t)$. If $s(t)$ is positive and decreasing, then the velocity $v(t)$ is negative.

2. *Examples:* Construct examples of the thing(s) described in the following. Try to find examples that are different than any in the reading.

(a) The graph of a function whose associated slope function f' is positive on $(-\infty, 2)$ and negative on $(2, \infty)$.

(b) The graph of a function with the following three properties: The average rate of change of f on $[0, 2]$ is 3, the average rate of change of f on $[0, 1]$ is -1, and the average rate of change of f on $[-2, 2]$ is 0.

(c) The graph of a function f with the following three properties: The instantaneous rate of change of f at $x = 2$ is zero, the average rate of change of f on $[1, 2]$ is 2, and the average rate of change of f on $[2, 4]$ is 1.

3. Explain why it is not a simple task to calculate the slope of the tangent line to a function f at a point $x = c$. Shouldn't calculating the slope of a line be really easy? What goes wrong here?

4. Let l be the line connecting two points $(a, f(a))$ and $(b, f(b))$ on the graph of a function f. What does this line l have to do with the average rate of change of f on the interval $[a, b]$, and why?

5. Given that $s(t)$ measures the distance an object has travelled over time, explain what the expression $\frac{s(b)-s(a)}{b-a}$ has to do with the distance formula $d = rt$.

6. How is velocity different from speed? What does it mean if velocity is negative?

7. What is the relationship between the derivative of a function f at a point $x = c$, the slope of the tangent line to the graph of f at $x = c$, and the instantaneous rate of change of f at $x = c$?

8. On a graph of $f(x) = x^2$,

(a) draw the tangent line to the graph of f at the point $(2, f(2))$;

(b) draw the secant line from $(2, f(2))$ to $(2.75, f(2.75))$;

(c) draw the secant line from $(1.75, f(1.75))$ to $(2, f(2))$.

(d) Which secant line is a better approximation to the tangent line, and why?

9. In Example 3 we estimated the slope of the tangent line to $f(x) = -\frac{1}{2}x^2 + 3x$ at $x = 2$. Get a better estimate by calculating the slopes of secant lines with values of z even closer to $x = 2$—for example, $z = 2.01$, $z = 2.001$, and $z = 2.0001$.

10. In Example 3 we estimated the slope of the tangent line to $f(x) = -\frac{1}{2}x^2 + 3x$ at $x = 2$ by finding slopes of secant lines from $x = 2$ to various points $x = z$ with $z > 2$. Draw a sequence of graphs that illustrates how to do this for $z < 2$, and then make specific calculations for $z = 1$, $z = 1.5$, $z = 1.75$, and $z = 1.9$. What are the corresponding values of h in this example?

11. For the graph of f appearing next at the left, label each of the following quantities to illustrate that $f'(c) \approx \frac{f(c+h)-f(c)}{h}$:

(a) the locations c, $c + h$, $f(c)$, and $f(c + h)$

(b) the distances h and $f(c + h) - f(c)$

(c) the slopes $\frac{f(c+h)-f(c)}{h}$ and $f'(c)$

12. For the graph of $g(x)$ appearing next at the right, label each of the following quantities to illustrate that $g'(c) \approx \frac{g(c+h)-g(c)}{h}$:

(a) the locations c, $c + h$, $g(c)$, and $g(c + h)$

(b) the distances h and $g(c + h) - g(c)$

(c) the slopes $\frac{g(c+h)-g(c)}{h}$ and $g'(c)$

f(x), Exercises 11 and 13

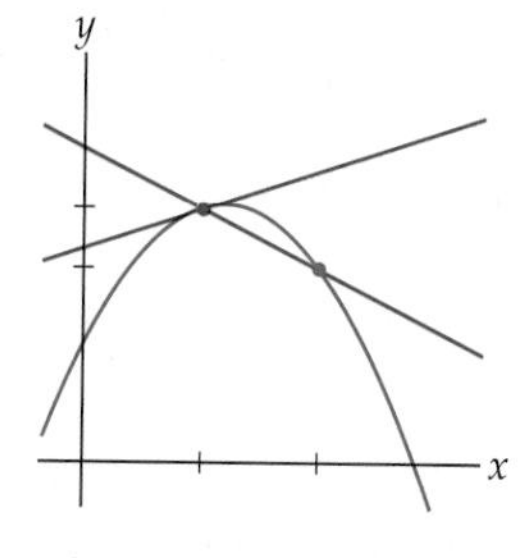

g(x), Exercises 12 and 14

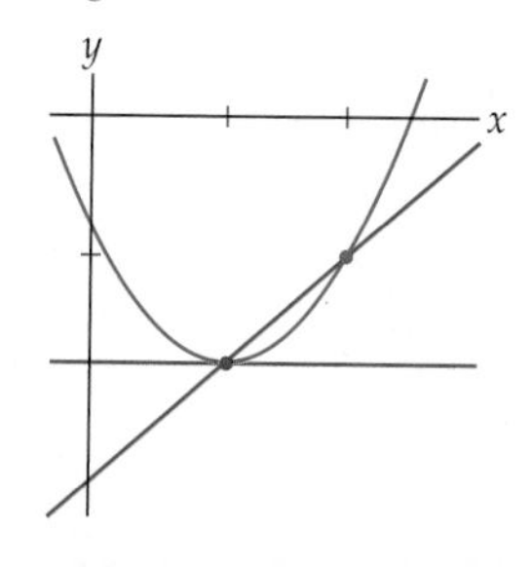

13. Consider again the graph of f at the left. Label each of the following quantities to illustrate that $f'(c) \approx \frac{f(z)-f(c)}{z-c}$:

(a) the locations c, z, $f(c)$, and $f(z)$

(b) the distances $z - c$ and $f(z) - f(c)$

(c) the slopes $\frac{f(z)-f(c)}{z-c}$ and $f'(c)$

14. Consider again the graph of $g(x)$ at the right. Label each of the following quantities to illustrate that $g'(c) \approx \frac{g(z)-g(c)}{z-c}$:

(a) the locations c, z, $g(c)$, and $g(z)$

(b) the distances $z - c$ and $g(z) - g(c)$

(c) the slopes $\frac{g(z)-g(c)}{z-c}$ and $g'(c)$

15. For the graph of f shown next at the left, list the following quantities in order from least to greatest:

(a) the average rate of change of f on $[-1, 1]$

(b) the instantaneous rate of change of f at $x = 1$

(c) $f'(-1)$

(d) $\frac{f(2)-f(-1)}{2-(-1)}$

16. For the graph of $g(x)$ shown next at the right, list the following quantities in order from least to greatest:

(a) the average rate of change of g on $[0, 1]$

(b) the instantaneous rate of change of g at $x = 1$

(c) $\frac{g(-1+0.1)-g(-1)}{0.1}$

(d) $\frac{g(1)-g(-1)}{1-(-1)}$

f(x), Exercises 15 and 17

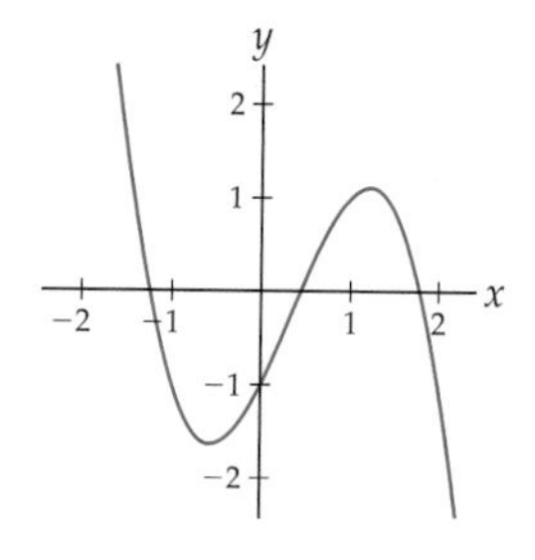

g(x), Exercises 16 and 18

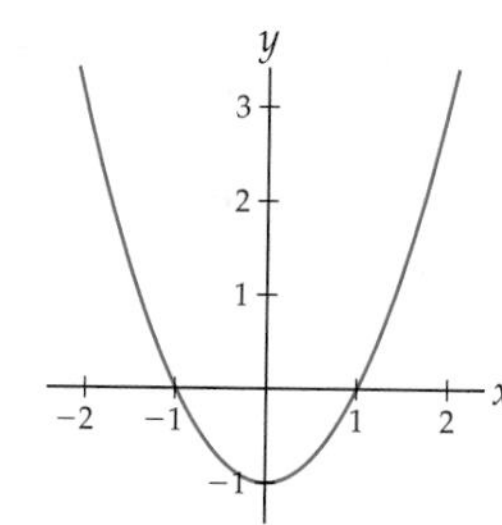

17. Consider again the function f graphed at the left. At which values of x does f have the greatest instantaneous rate of change? The least? At which values of x is the instantaneous rate of change of f equal to zero?

18. Consider again the function $g(x)$ graphed at the right. For which values of x does $g(x)$ have a positive instantaneous rate of change? Negative? Zero?

19. Make a copy of the graph of f used in Exercises 11 and 13, and sketch additional secant lines to illustrate that as $h \to 0$ (or equivalently, as $z \to c$) the slopes of the secant line get closer and closer to the slope of the tangent line to f at $x = c$.

20. The derivative of a smooth function f at a point $x = c$ can also be approximated with a ***symmetric difference quotient***:

$$f'(c) \approx \frac{f(c+h) - f(c-h)}{2h}.$$

(a) Use a graph to illustrate what the symmetric difference measures. Why would it be reasonable to use the two-sided symmetric difference to approximate $f'(c)$? *(Hint: Your answer should involve a certain kind of secant line and a discussion of what happens as h gets close to 0.)*

(b) Use a sequence of symmetric difference approximations to estimate the derivative of $f(x) = x^2$ at $x = 3$. Illustrate your answer with a sequence of graphs.

Skills

In Exercises 21–24, sketch the graph of a function f that has the listed characteristics.

21. $f(1) = 2,\ f'(1) = 0,\ f'(3) = 2$

22. $f'(-3) = 0,\ f'(-1) = 0,\ f'(2) = 0$

23. $f(-1) = 2,\ f'(-1) = 3,\ f(1) = -2,\ f'(1) = 3$

24. $f'(-2) = 2,\ f'(0) = 1,\ f(1) = -5$

Sketch a graph of the associated slope function f' for each function f in Exercises 25–30.

25.

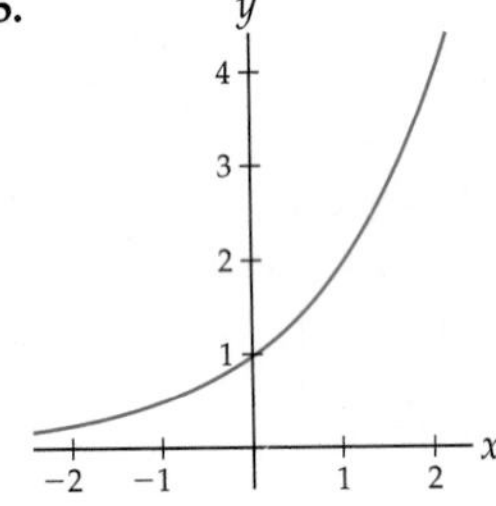

26.

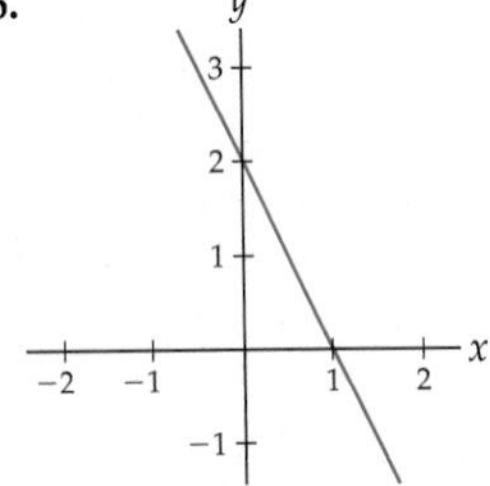

27.

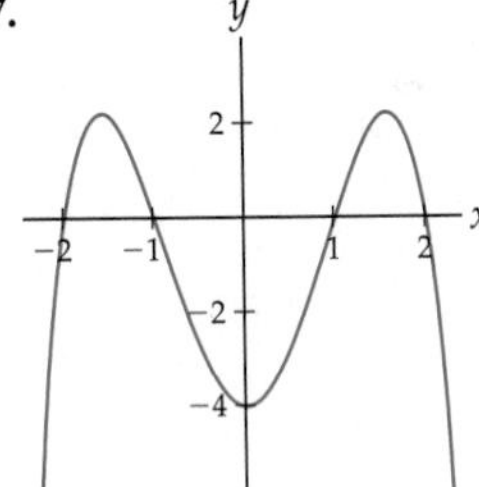

28.

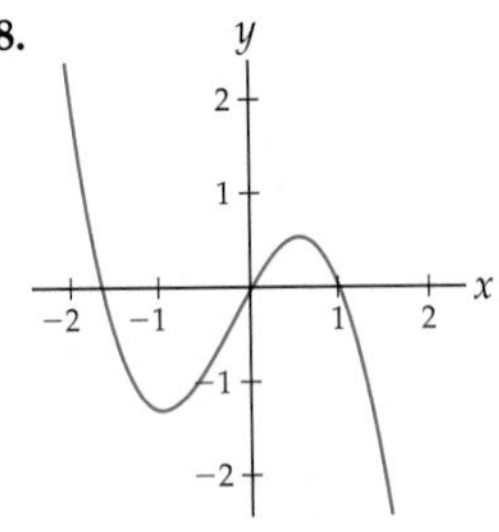

29.

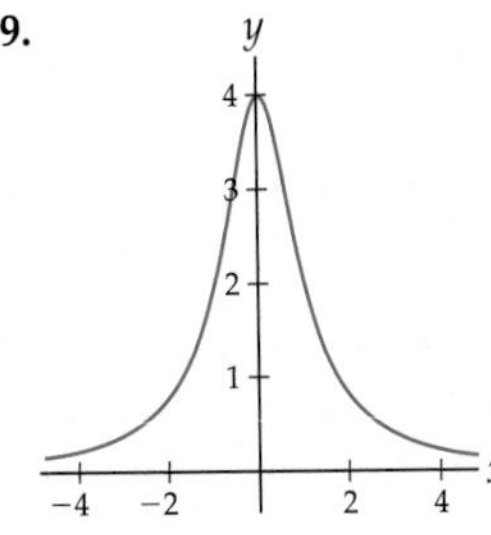

30.

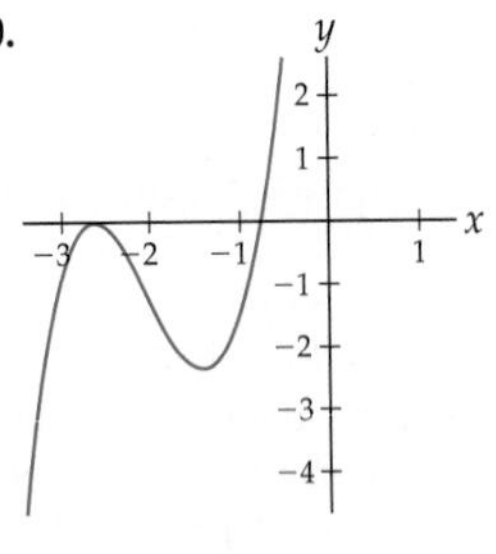

Now go the other way! Each graph in Exercises 31–34 can be thought of as the associated slope function f' for some unknown function f. In each case sketch a possible graph of f.

31.

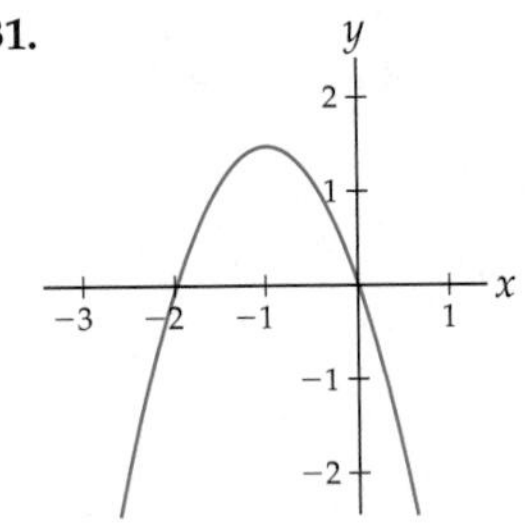

32.

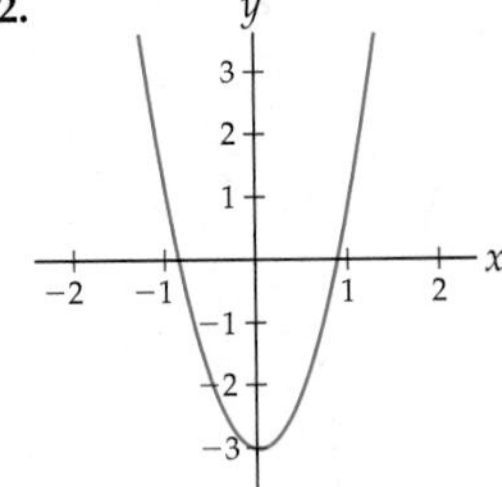

33.

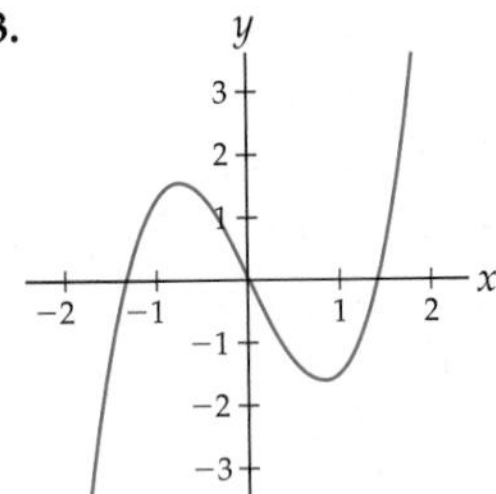

34.

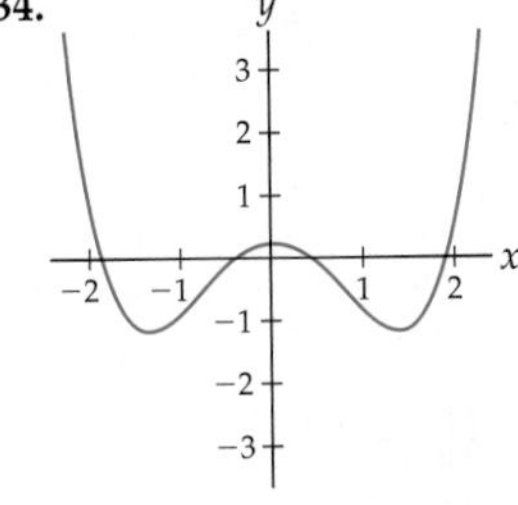

For each function f and value $x = c$ in Exercises 35–44, use a sequence of approximations to estimate $f'(c)$. Illustrate your work with an appropriate sequence of graphs of secant lines.

35. $f(x) = 4 - x^2,\ c = 1$

36. $f(x) = 4 - x^2,\ c = 0$

37. $f(x) = x + x^3,\ c = 0$

38. $f(x) = x + x^3,\ c = 1$

39. $f(x) = \ln(x^2+1), c=0$

40. $f(x) = e^x,\ c = 0$

41. $f(x) = \sin x,\ c = \dfrac{\pi}{2}$

42. $f(x) = \arctan x,\ c = 0$

43. $f(x) = |x - 1|,\ c = 3$

44. $f(x) = |x^2 - 4|,\ c = 1$

Applications

A bowling ball dropped from a height of 400 feet will be $s(t) = 400 - 16t^2$ feet from the ground after t seconds. Use a sequence of average velocities to estimate the instantaneous velocities described in Exercises 45–48.

45. When the bowling ball is first dropped, with $h = 0.5$, $h = 0.25$, and $h = 0.1$

46. After $t = 1$ seconds, with $h = 0.5$, $h = 0.25$, $h = -0.5$, and $h = -0.2$

47. After $t = 2$ seconds, with $h = 0.1$, $h = 0.01$, $h = -0.1$, and $h = -0.01$

48. When the bowling ball hits the ground, with $h = -0.5$, $h = -0.2$, and $h = -0.1$

49. Think about what you did today and how far north you were from your house or dorm throughout the day. Sketch a graph that represents your distance north from your house or dorm over the course of the day, and explain how the graph reflects what you did today. Then sketch a graph of your velocity.

50. Stuart left his house at noon and walked north on Pine Street for 20 minutes. At that point he realized he was late for an appointment at the dentist, whose office was located *south* of Stuart's house on Pine Street; fearing he would be late, Stuart sprinted south on Pine Street, past his house, and on to the dentist's office. When he got there, he found the office closed for lunch; he was 10 minutes early for his 12:40 appointment. Stuart waited at the office for 10 minutes and then found out that his appointment was actually for the next day, so he walked back to his house. Sketch a graph that describes Stuart's position over time. Then sketch a graph that describes Stuart's velocity over time.

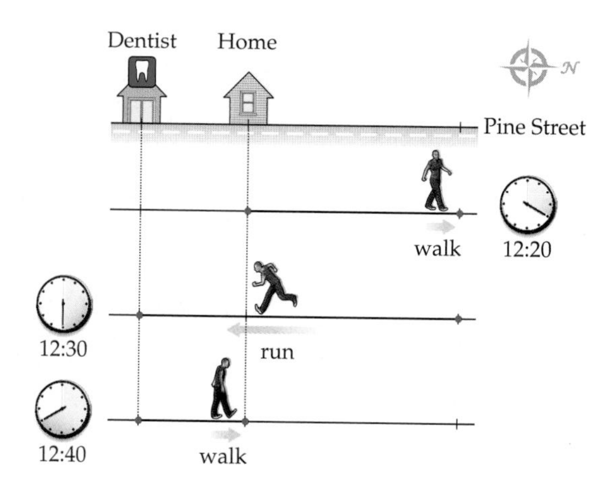

51. Every morning Linda takes a thirty-minute jog in Central Park. Suppose her distance s in feet from the oak tree on the north side of the park t minutes after she begins her jog is given by the function $s(t)$ shown that follows at the left, and suppose she jogs on a straight path leading into the park from the oak tree.

(a) What was the average rate of change of Linda's distance from the oak tree over the entire thirty-minute jog? What does this mean in real-world terms?

(b) On which ten-minute interval was the average rate of change of Linda's distance from the oak tree the greatest: the first 10 minutes, the second 10 minutes, or the last 10 minutes?

(c) Use the graph of $s(t)$ to estimate Linda's average velocity during the 5-minute interval from $t = 5$ to $t = 10$. What does the sign of this average velocity tell you in real-world terms?

(d) Approximate the times at which Linda's (instantaneous) velocity was equal to zero. What is the physical significance of these times?

(e) Approximate the time intervals during Linda's jog that her (instantaneous) velocity was negative. What does a negative velocity mean in terms of this physical example?

Distance from the oak tree

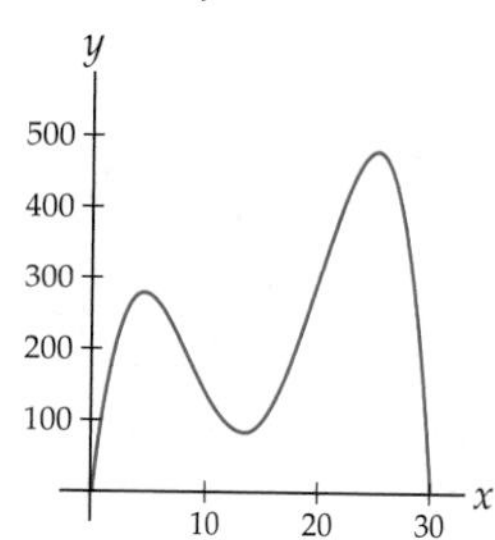

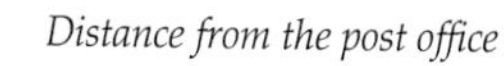
Distance from the post office

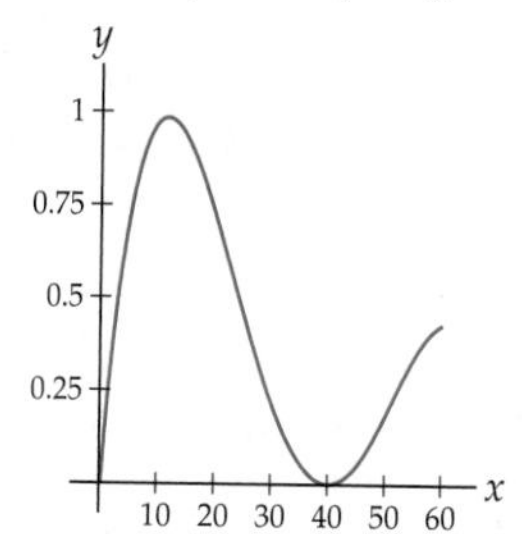

52. Last night Phil went jogging along Main Street. His distance from the post office t minutes after 6:00 P.M. is shown in the preceding graph at the right.

(a) Give a narrative (that matches the graph) of what Phil did on his jog.

(b) Sketch a graph that represents Phil's instantaneous velocity t minutes after 6:00 P.M. Make sure you label the tick marks on the vertical axis as accurately as you can.

(c) When was Phil jogging the fastest? The slowest? When was he the farthest away from the post office? The closest to the post office?

53. Suppose $h(t)$ represents the average height, in feet, of a person who is t years old.

(a) In real-world terms, what does $h(12)$ represent and what are its units? What does $h'(12)$ represent, and what are its units?

(b) Is $h(12)$ positive or negative, and why? Is $h'(12)$ positive or negative, and why?

(c) At approximately what value of t would $h(t)$ have a maximum, and why? At approximately what value of t would $h'(t)$ have a maximum, and why?

54. A tomato plant given x ounces of fertilizer will successfully bear $T(x)$ pounds of tomatoes in a growing season.

(a) In real-world terms, what does $T(5)$ represent and what are its units? What does $T'(5)$ represent and what are its units?

(b) A study has shown that this fertilizer encourages tomato production when less than 20 ounces are used, but inhibits production when more than 20 ounces are used. When is $T(x)$ positive and when is $T(x)$ negative? When is $T'(x)$ positive and when is $T'(x)$ negative?

55. If Katie walked at 3 miles per hour for 20 minutes and then sprinted at 10 miles an hour for 8 minutes, how fast would Dave have to walk or run to go the same distance as Katie did in the same time while moving at a constant speed? Sketch a graph of Katie's position over time and a graph of Dave's position over time on the same set of axes.

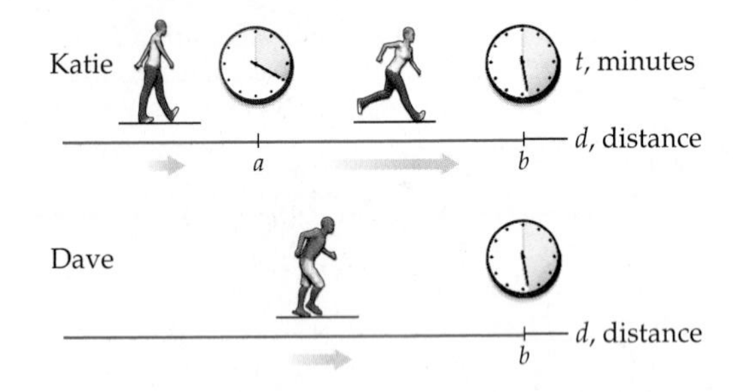

56. Velocity $v(t)$ is the derivative of position $s(t)$. It is also true that acceleration $a(t)$ (the rate of change of velocity) is the derivative of velocity. If a race car's position in miles t hours after the start of a race is given by the function $s(t)$,what are the units of $s(1.2)$? What are the units and real-world interpretation of $v(1.2)$? What are the units and real-world interpretation of $a(1.2)$?

57. The total yearly expenditures by public colleges and universities from 1990 to 2000 can be modeled by the function $E(t) = 123(1.025)^t$, where expenditures are measured in billions of dollars and time is measured in years since 1990.

(a) Estimate the total yearly expenditures by these colleges and universities in 1995.

(b) Compute the average rate of change in yearly expenditures between 1990 and 2000.

(c) Compute the average rate of change in yearly expenditures between 1995 and 1996.

(d) Estimate the rate at which yearly expenditures of public colleges and universities were increasing in 1995.

Proofs

58. Show that if f is a function and $z = x + h$, then

$$\frac{f(z) - f(x)}{z - x} = \frac{f(x+h) - f(x)}{h}.$$

59. Suppose f is a linear function with positive slope. Show that the average rate of change of f on any interval $[a, b]$ is positive, and then use this fact to show that f is always increasing.

Thinking Forward

Taking the limit: We have seen that if f is a smooth function, then $f'(c) \approx \frac{f(c+h) - f(c)}{h}$. This approximation should get better as h gets closer to zero. In fact, in the next section we will define the derivative in terms of such a limit.

$$f'(c) = \lim_{h \to 0} \frac{f(c+h) - f(c)}{h}.$$

- ▶ Use the limit just defined to calculate the exact slope of the tangent line to $f(x) = x^2$ at $x = 4$.
- ▶ Instead of choosing small values of h, we could have chosen values of z close to c. What limit involving z instead of h is equivalent to the one involving h?
- ▶ Use the limit you just found to calculate the exact slope of the tangent line to $f(x) = x^2$ at $x = 4$. Obviously you should get the same final answer as you did earlier.

2.2 FORMAL DEFINITION OF THE DERIVATIVE

- Using limits to define derivatives, tangent lines, and instantaneous rates of change
- Differentiability at a point, from one side, and on intervals
- Leibniz notation and multiple derivatives

The Derivative at a Point

In the previous section we examined derivatives intuitively, by discussing tangent lines and instantaneous rates of change. We will now use limits to make these ideas precise. We have seen that the slope $f'(c)$ of the tangent line to f at $x = c$ can be approximated by the slope of a nearby secant line from $x = c$ to $x = c + h$, or equivalently, from $x = c$ to $x = z$:

$$f'(c) \approx \frac{f(c+h) - f(c)}{h}, \quad \text{or equivalently,} \quad f'(c) \approx \frac{f(z) - f(c)}{z - c}.$$

If the limit of these quantities approaches a real number as $h \to 0$, or as $z \to c$, then we will define that real number to be the ***derivative*** of f at the point $x = c$.

DEFINITION 2.1

The Derivative of a Function at a Point

The ***derivative at*** $x = c$ of a function f is the number

$$f'(c) = \lim_{h \to 0} \frac{f(c+h) - f(c)}{h}, \quad \text{or equivalently,} \quad f'(c) = \lim_{z \to c} \frac{f(z) - f(c)}{z - c},$$

provided that this limit exists.

The derivative of a function f at a point $x = c$ measures the ***instantaneous rate of change*** of the function at that point. Notice that this instantaneous rate of change is a limit of average rates of change.

For example, consider the function $f(x) = x^2$. We can calculate the derivative of this function at the point $x = 3$ with a limit as $h \to 0$ or with a limit as $z \to 3$. Using the $h \to 0$ definition of the derivative, we have

$$f'(3) = \lim_{h \to 0} \frac{(3+h)^2 - 3^2}{h} = \lim_{h \to 0} \frac{6h + h^2}{h} = \lim_{h \to 0} \frac{h(6+h)}{h} = \lim_{h \to 0} (6 + h) = 6.$$

Using the $z \to c$ definition of derivative we obtain the same answer:

$$f'(3) = \lim_{z \to c} \frac{z^2 - 3^2}{z - 3} = \lim_{z \to c} \frac{(z+3)(z-3)}{z - 3} = \lim_{z \to c} (z + 3) = 6.$$

In both calculations above we have shown the instantaneous rate of change of $f(x) = x^2$ at $x = 3$ is equal to 6. At the instant that we have $x = 3$, the function $f(x) = x^2$ is changing at a rate of 6 vertical units for each horizontal unit. This is equivalent to saying that the slope of the tangent line is equal to 6, as shown here:

Tangent line to $f(x) = x^2$ at $x = 3$ has slope $f'(3) = 6$

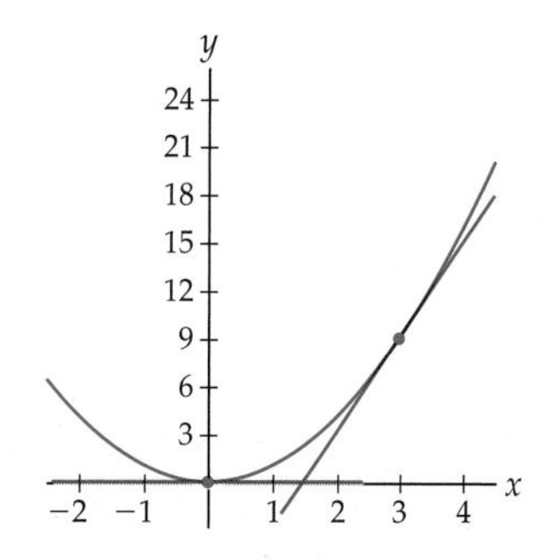

The Derivative as a Function

By putting all of the point-derivatives of a function together, we can define a function f' whose output at any value of x is defined to be the derivative, or instantaneous rate of change, of f at that point.

DEFINITION 2.2 **The Derivative of a Function**

The ***derivative*** of a function f is the function f' defined by

$$f'(x) = \lim_{h\to 0} \frac{f(x+h) - f(x)}{h}, \quad \text{or equivalently,} \quad f'(x) = \lim_{z\to x} \frac{f(z) - f(x)}{z - x}.$$

The domain of f' is the set of values x for which the defining limit of f' exists.

The function f' is the ***associated slope function*** that we investigated in the previous section, since at each point x its value is the slope of the graph of f. In addition, the function f' represents the instantaneous rate of change at every point x. In particular, if $s(t)$ is a position function, then its instantaneous rate of change is the velocity function $v(t) = s'(t)$. Similarly, the instantaneous rate of change of velocity $v(t)$ is the the acceleration function $a(t) = v'(t)$.

For example, we can calculate the derivative of $f(x) = x^2$ for *all* values of x with either a limit as $h \to 0$ or a limit as $z \to x$; again we choose the first method:

$$f'(x) = \lim_{h\to 0} \frac{(x+h)^2 - x^2}{h} = \lim_{h\to 0} \frac{2xh + h^2}{h} = \lim_{h\to 0} \frac{h(2x+h)}{h} = \lim_{h\to 0}(2x+h) = 2x.$$

Finding $f'(x)$ for general x is like calculating $f'(c)$ for all possible values $x = c$ at the same time. Once we have a formula for $f'(x)$, we can easily calculate any particular value $f'(c)$. For example, evaluating $f'(x) = 2x$ at $x = 3$ does give us $f'(3) = 2(3) = 6$, as we calculated before. The following figures show slopes on the graph of $f(x) = x^2$ together with heights on the graph of $f'(x) = 2x$, for $x = -2$, $x = 0$, and $x = 3$.

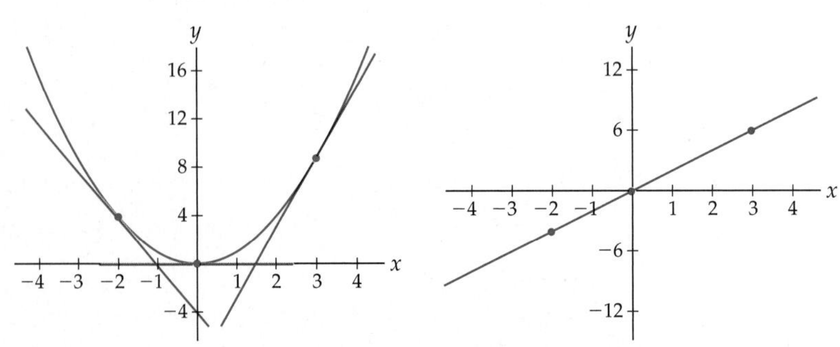

In this book we will most often use the $h \to 0$ version of the derivative, but will use the $z \to x$ version when it suits our purposes or makes a calculation easier. You will verify that these two versions of the derivative are equivalent in Exercise 5.

Differentiability

Given a function f and a value $x = c$, there may or may not be a well-defined tangent line to the graph of f at $(c, f(c))$. When there is a well-defined tangent line with a finite slope, we say that f is ***differentiable*** at $x = c$:

DEFINITION 2.3 **Differentiability at a Point**

A function f is ***differentiable at*** $x = c$ if $\displaystyle\lim_{h\to 0} \frac{f(c+h) - f(c)}{h}$ exists.

We used the $h \to 0$ definition of the derivative in Definition 2.3, but everything would work equally well with the equivalent $z \to c$ definition of the derivative. If a point-derivative $f'(c)$ is infinite, then we say that f has a ***vertical tangent line*** at $x = c$. In this case a line exists that is tangent to the graph of the function, but since that line is vertical, its slope is undefined and the function fails to be differentiable at that point.

Like continuity, differentiability can be considered from the left or from the right. A function is ***left differentiable at*** $x = c$ if its left derivative exists, and ***right differentiable at*** $x = c$ if its right derivative exists, where the left and right derivatives are defined with left and right limits as follows:

DEFINITION 2.4

One-sided Differentiability at a Point

The ***left derivative*** and ***right derivative*** of a function f at a point $x = c$ are, respectively, equal to the following, if they exist:

$$f'_-(c) = \lim_{h \to 0^-} \frac{f(c+h) - f(c)}{h}, \qquad f'_+(c) = \lim_{h \to 0^+} \frac{f(c+h) - f(c)}{h}.$$

We could also use the $z \to c$ definition of the derivative to define left and right derivatives, by considering limits of difference quotients as $z \to c^-$ and as $z \to c^+$.

For example, consider the function $f(x) = |x|$. This function has a sharp corner at $x = 0$, and therefore we would not expect it to have a well-defined tangent line at that point. Indeed, when we try to calculate $f'(0)$, we encounter the following limit:

$$f'(0) = \lim_{h \to 0} \frac{|0 + h| - |0|}{h} = \lim_{h \to 0} \frac{|h|}{h}.$$

This limit is initially in the indeterminate form $\frac{0}{0}$, but we cannot cancel anything in it until we get rid of the absolute value. Recall that if $h \geq 0$, then $|h| = h$, but if $h < 0$, then $|h| = -h$. Looking from the left and the right, we have the following limits:

$$f'_-(0) = \lim_{h \to 0^-} \frac{|h|}{h} = \lim_{h \to 0^-} \frac{-h}{h} = \lim_{h \to 0^-} -1 = -1,$$

$$f'_+(0) = \lim_{h \to 0^+} \frac{|h|}{h} = \lim_{h \to 0^+} \frac{h}{h} = \lim_{h \to 0^+} 1 = 1.$$

Since $f'_-(0)$ and $f'_+(0)$ exist but are not equal, $f'(0)$ does not exist. The first two graphs shown next illustrate the left and right derivatives at 0, for small negative h and small positive h, respectively. The third graph shows the graph of the derivative function f'; note that this function has a jump discontinuity and is not defined at $x = 0$.

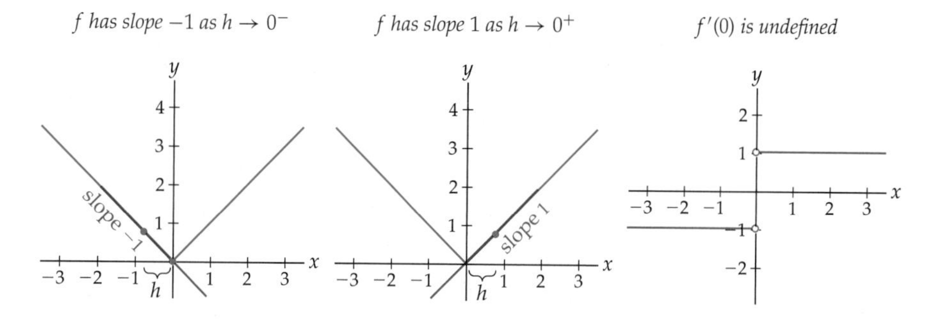

As with continuity, we say that a function is ***differentiable on an interval*** I if it is differentiable at every point in the interior of I, right differentiable at any closed left endpoint, and left differentiable at any closed right endpoint. For example, the first graph shown next is differentiable on $[2, 3]$, since it is right differentiable at $x = 2$, even though it is not differentiable at $x = 2$. The second and third graphs show two more ways that a function could fail to be differentiable.

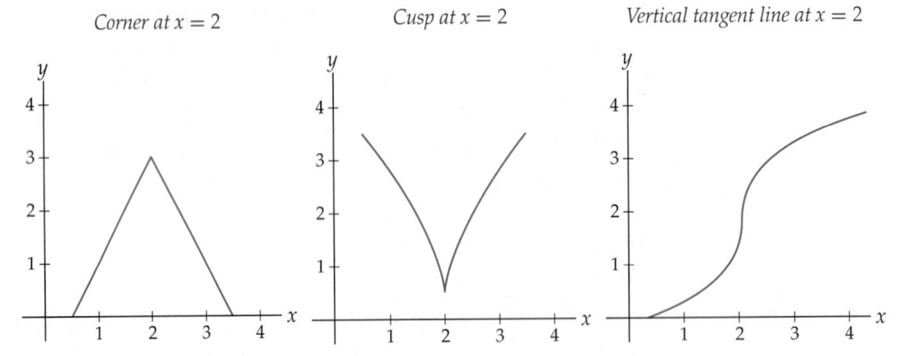

In the first and second graphs, the left and right derivatives exist at $x = 2$ but the two-sided derivative at $x = 2$ does not. In the third graph, the left, right, and two-sided derivatives at $x = 2$ are all infinite, because the tangent line is vertical.

Another way that a function f can fail to be differentiable at a point $x = c$ is if f fails to be continuous at $x = c$. Intuitively, if a function is not continuous, then it has absolutely no chance of being differentiable; think for a minute about secant lines from the left and the right at a jump or removable discontinuity, for example. (See also Example 6.) What this means is that although not every continuous function is differentiable, every differentiable function is continuous:

THEOREM 2.5 **Differentiability Implies Continuity**

If f is differentiable at $x = c$, then f is continuous at $x = c$.

Proof. If f is differentiable at $x = c$, then $\lim_{x \to c} \frac{f(x) - f(c)}{x - c} = f'(c)$ exists; that is, the limit, and thus $f'(c)$, is equal to some real number. We will use this fact to show that f is continuous at $x = c$, by showing that $\lim_{x \to c} f(x) = f(c)$. By the sum rule for limits, it is equivalent to show that $\lim_{x \to c} (f(x) - f(c)) = 0$, which we can do by using the expression for the derivative:

$$\lim_{x \to c} (f(x) - f(c)) = \lim_{x \to c} \left(\frac{f(x) - f(c)}{x - c} (x - c) \right) \qquad \leftarrow \frac{x - c}{x - c} = 1 \text{ if } x \neq c$$

$$= \left(\lim_{x \to c} \frac{f(x) - f(c)}{x - c} \right) (\lim_{x \to c} (x - c)) \qquad \leftarrow \text{product rule for limits}$$

$$= f'(c) \lim_{x \to c} (x - c) \qquad \leftarrow \text{definition of the derivative}$$

$$= f'(c)(0) = 0. \qquad \leftarrow \text{limit rules}$$

■

Tangent Lines and Local Linearity

Although we have an intuitive sense of the tangent line to a graph at a point, up until now we did not have a formal mathematical definition for this tangent line. Our geometrical intuition helped us arrive at an algebraic definition for the derivative, but it is the algebraic definition that now allows us to define the tangent line precisely. Specifically, we define the ***tangent line*** through a point on the graph of a function to be the line whose slope is given by the derivative of the function at that point.

THEOREM 2.6 **Equation of the Tangent Line to a Function at a Point**

The ***tangent line*** to the graph of a function f at a point $x = c$ is defined to be the line passing through $(c, f(c))$ with slope $f'(c)$, provided that the derivative $f'(c)$ exists. This line has equation

$$y = f(c) + f'(c)(x - c).$$

Proof. We need only calculate the form of the line that passes through the point $(c, f(c))$ and has slope $f'(c)$. Using the point-slope form $y - y_0 = m(x - x_0)$, we see that this line has equation

$$y - f(c) = f'(c)(x - c),$$

and solving for y, we obtain the desired equation $y = f(c) + f'(c)(x - c)$. ■

For example, we saw earlier that the derivative of $f(x) = x^2$ at $x = 3$ is equal to $f'(3) = 6$. This means that the tangent line to the graph of $f(x) = x^2$ at $x = 3$ has slope $f'(3) = 6$ and passes through $(3, f(3)) = (3, 9)$. This line has equation $y = 9 + 6(x - 3)$, which, in slope-intercept form, is the equation $y = 6x - 9$.

The tangent line to a function f at a point $x = c$ is the unique line that "agrees with" both the height of the function and the slope of the function at $x = c$. This means that near $x = c$ the graph of a function is very close to the graph of its tangent line. Therefore we can use the tangent line as a rough approximation to the function f itself near $x = c$.

DEFINITION 2.7 **Local Linearity**

If f has a well-defined derivative $f'(c)$ at the point $x = c$, then, for values of x near c, the function $f(x)$ can be approximated by the tangent line to f at $x = c$ with the ***linearization of f around*** $x = c$ given by

$$f(x) \approx f(c) + f'(c)(x - c).$$

Note that this definition does not assert how good an approximation one can make by using the tangent line. What constitutes "near" and what constitutes "good" will be determined by the context of the problem at hand.

For example, since the tangent line to $f(x) = x^2$ at $x = 3$ is the line $y = 6x - 9$, the line $y = 6x - 9$ can be used as a rough approximation to the graph of the function $f(x) = x^2$, at least for values of x close to 3. This approximation will be better the closer we are to $x = 3$. As the following figure shows, the two graphs have nearly the same height at the point $x = 3.5$ and are still relatively close even at $x = 4$.

$y = 6x - 9$ is close to $f(x) = x^2$ near $x = 3$

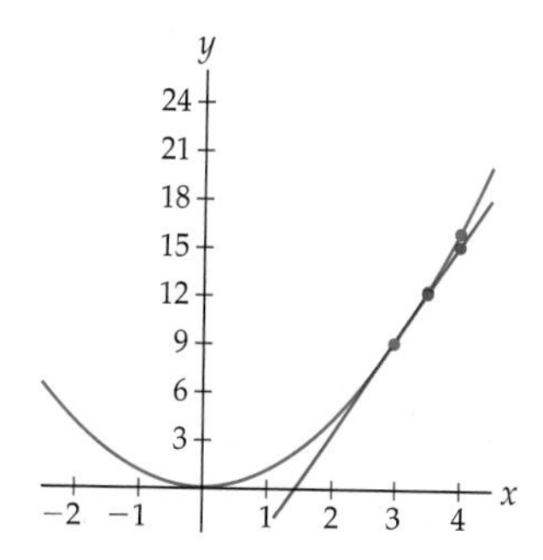

The concept of local linearity will be particularly helpful for approximating roots with Newton's method in Example 8 and Exercises 81–86, as well as for approximating solutions of differential equations with Euler's method in Section 7.5.

Leibniz Notation and Differentials

Derivatives are used in so many different fields of study that they are represented with a wide variety of notations. In this book we will focus on two types: first, the "prime" notation $f'(x)$ that we have already established, and second, the ***Leibniz notation*** $\frac{df}{dx}$. If $y = f(x)$, then we can also write $y'(x)$ or $\frac{dy}{dx}$ to represent the derivative.

Leibniz notation is named for one of the founders of calculus and is intentionally structured to remind us of the connection between derivatives and average rates of change. (The "prime" notation we have been using so far is due to Lagrange.) Intuitively, the expression dx represents an infinitesimally small change in x, just as Δx represents a small finite change in x. In Leibniz notation the definition of the derivative as a limit of average rates of change is strikingly clear:

$$\frac{dy}{dx} = \lim_{\Delta x \to 0} \frac{\Delta y}{\Delta x}.$$

It is sometimes convenient to think of the expressions dy and dx as ***differentials*** in a functional relationship, with dy as a function of dx, in the following sense: We know that $\frac{\Delta y}{\Delta x}$ represents the slope of a line with vertical change of Δy for a given horizontal change Δx, as in the figure shown next at the left. In the same way we might try to think of $\frac{dy}{dx}$ as the slope of a line with a vertical change of dy for a given horizontal change dx, as in the figure at the right.

Δy depends on Δx and gives the height of the function

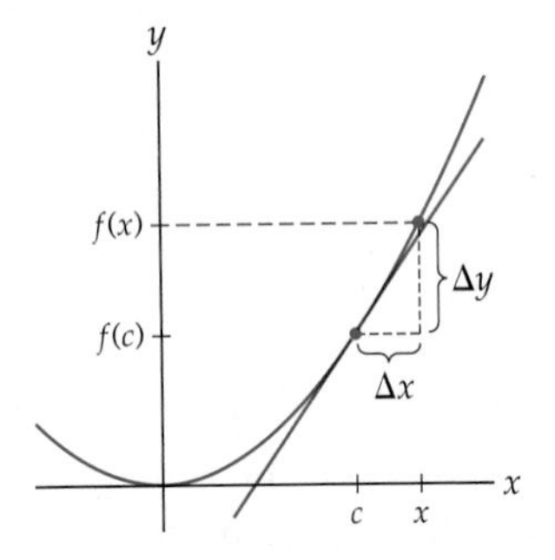

As differentials, dy depends on dx and give the height of the tangent line

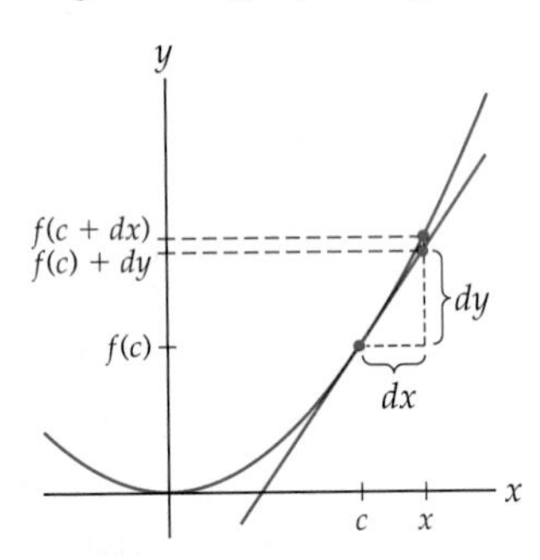

With this interpretation of dy as a differential we can recast local linearity (Definition 2.7) as saying that when x is sufficiently close to a point c, we can approximate $f(x)$ by adding dy to $f(c)$. If $x = c + dx$ as in the rightmost figure, this approximation can be expressed as

$$f(c + dx) \approx f(c) + dy.$$

CAUTION It is important to note that we will *not* be thinking of dx and dy as numbers, but rather thinking of $\frac{dy}{dx}$ as a formal symbol that represents the derivative $f'(x)$. This formal symbol suggests meanings such as those given to the differentials in the preceding figure, but $\frac{dy}{dx}$ is not an actual quotient of numbers; it is a limit of slopes of secant lines as defined in Definition 2.2.

We can also write Leibniz notation in a slightly different way, called ***operator notation***, as follows:

$$\frac{dy}{dx} = \frac{d}{dx}(y(x)).$$

Here we are thinking of $\frac{d}{dx}$ as a sort of metafunction that operates on functions instead of numbers: It takes functions as inputs and returns the derivatives of those functions as outputs. Using operator notation, we can express the statement "if $y = x^2$, then $\frac{dy}{dx} = 2x$" in the more compact form "$\frac{d}{dx}(x^2) = 2x$."

Although Leibniz notation is sometimes more convenient or informative than the usual "prime" notation, it has one drawback: It is cumbersome to write down the point-derivative of a function in Leibniz notation. For example, as we showed earlier, the derivative of $f(x) = x^2$ is $f'(x) = 2x$. This derivative is easy to express in Leibniz notation as $\frac{d}{dx}(x^2) = 2x$. Now suppose we wish to consider the same derivative at the point $x = 3$. In "prime notation" it is easily expressed as $f'(3) = 2(3) = 6$. In Leibniz notation it is more difficult to work the $x = 3$ evaluation into the notation. We could write any of the following:

$$\left.\frac{df}{dx}\right|_3 = 6, \quad \left.\frac{d}{dx}(x^2)\right|_{x=3} = 6, \quad \text{or} \quad \left.\frac{d}{dx}(x^2)\right|_3 = 6.$$

If f is a function, then its derivative f' is a function as well. This means that we can also consider *its* derivative, which we call f'', the ***second derivative*** of f. In Leibniz notation we write the second derivative as $\frac{d^2f}{dx^2}$. This notation is supposed to suggest the fact that we are differentiating twice, that is, finding $\frac{d}{dx}\left(\frac{d}{dx}(f(x)\right)$. Similarly, we could find the third, fourth, or fifth derivatives of a function f, and so on. For example, the third derivative of $f(x)$ can be written as $f'''(x)$. For larger values of n we will replace the primes with a parenthetical notation: for example, $f''''''(x) = f^{(6)}(x)$. In general, the ***nth derivative*** of a function f is denoted by:

$$f^{(n)}(x) = \frac{d^nf}{dx^n} = \underbrace{\frac{d}{dx}\left(\frac{d}{dx}\left(\frac{d}{dx}\cdots\left(\frac{d}{dx}\right.\right.\right.}_{n \text{ times}}(f(x))\Big)\cdots\Big)\Big).$$

Examples and Explorations

EXAMPLE 1

Calculating the derivative at a point

Consider the function $f(x) = x^3$.

(a) Use the $h \to 0$ definition of the derivative to find $f'(2)$.

(b) Use the $z \to x$ definition of the derivative to find $f'(2)$.

(c) Find the equation of the tangent line to $f(x) = x^3$ at $x = 2$, and graph $f(x) = x^3$ and this line on the same set of axes.

SOLUTION

(a) Using the $h \to 0$ definition of the derivative, we have

$$\begin{aligned} f'(2) &= \lim_{h\to 0}\frac{f(2+h)-f(2)}{h} = \lim_{h\to 0}\frac{(2+h)^3-2^3}{h} && \leftarrow \text{derivative with } x=2 \\ &= \lim_{h\to 0}\frac{(8+12h+6h^2+h^3)-8}{h} && \leftarrow \text{multiply out } (2+h)^3 \\ &= \lim_{h\to 0}\frac{12h+6h^2+h^3}{h} = \lim_{h\to 0}\frac{h(12+6h+h^2)}{h} && \leftarrow \text{algebra} \\ &= \lim_{h\to 0}(12+6h+h^2) = 12+6(0)+(0)^2 = 12. && \leftarrow \text{cancellation, limit rules} \end{aligned}$$

(b) With the $z \to 2$ definition of the derivative the algebra is different, but the final answer will be the same. Along the way we will need to use the factoring formula $a^3 - b^3 = (a-b)(a^2+ab+b^2)$, as follows:

$$\begin{aligned} f'(x) &= \lim_{z\to 2}\frac{f(z)-f(2)}{z-2} = \lim_{z\to 2}\frac{z^3-2^3}{z-2} && \leftarrow \text{derivative with } x=2 \\ &= \lim_{z\to 2}\frac{(z-2)(z^2+2z+4)}{z-2} && \leftarrow \text{factoring formula} \\ &= \lim_{z\to 2}(z^2+2z+4) = 2^2+2(2)+4 = 12. && \leftarrow \text{cancellation, limit rule} \end{aligned}$$

(c) We have just shown two ways that if $f(x) = x^3$, then $f'(2) = 12$. Thus the tangent line to $f(x)$ at $x = 2$ has slope 12 and passes through the point $(2, f(2)) = (2, 8)$. The equation of this line is

$$y - 8 = 12(x - 2) \implies y = 12x - 24 + 8 \implies y = 12x - 16.$$

When we graph the line along with the original function, we see that indeed the line $y = 12x - 16$ is the tangent line to $f(x) = x^3$ at $x = 2$:

$y = 12x - 16$ is the tangent line to $f(x) = x^2$ at $x = 2$

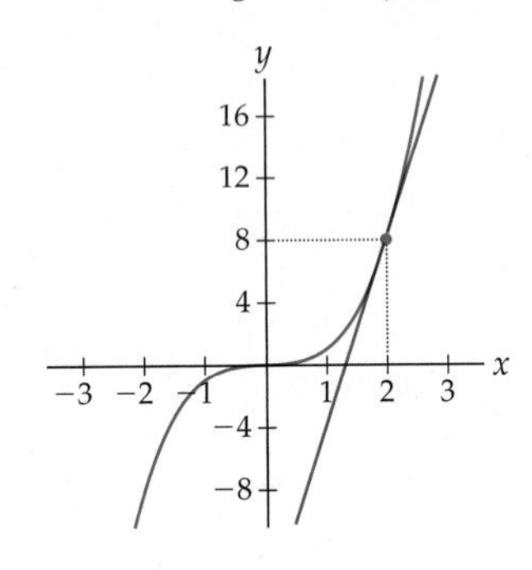

□

EXAMPLE 2

Calculating the derivative of a function

Consider the function $f(x) = x^3$.

(a) Use the $h \to 0$ definition of the derivative to find $f'(x)$.

(b) Use the $z \to x$ definition of the derivative to find $f'(x)$.

(c) Graph the function $f(x) = x^3$ and the derivative $f'(x)$ you found in parts (a) and (b), and argue that one graph is the slope function of the other.

SOLUTION

(a) The calculations in this example will be similar to those in the previous example, except that we will not specify a specific value of x here. Using the $h \to 0$ definition of the derivative, we have

$$f'(x) = \lim_{h\to 0} \frac{f(x+h) - f(x)}{h} = \lim_{h\to 0} \frac{(x+h)^3 - x^3}{h} \qquad \leftarrow \text{derivative}$$

$$= \lim_{h\to 0} \frac{(x^3 + 3x^2h + 3xh^2 + h^3) - x^3}{h} \qquad \leftarrow \text{multiply out } (x+h)^3$$

$$= \lim_{h\to 0} \frac{3x^2h + 3xh^2 + h^3}{h} = \lim_{h\to 0} \frac{h(3x^2 + 3xh + h^2)}{h} \qquad \leftarrow \text{algebra}$$

$$= \lim_{h\to 0} (3x^2 + 3xh + h^2) = 3x^2 + 3x(0) + (0)^2 = 3x^2. \qquad \leftarrow \text{cancellation, limit rules}$$

Notice that before we did any algebra, the limit was of the indeterminate form $\frac{0}{0}$. For derivative calculations with the $h \to 0$ definition, the goal is often to expand and simplify until a common factor of h can be cancelled, as we did here.

(b) With the $z \to x$ definition of the derivative the algebra is different, but the final answer is the same:

$$f'(x) = \lim_{z\to x} \frac{f(z) - f(x)}{z - x} = \lim_{z\to x} \frac{z^3 - x^3}{z - x} \qquad \leftarrow \text{derivative}$$

$$= \lim_{z\to x} \frac{(z - x)(z^2 + zx + x^2)}{z - x} \qquad \leftarrow \text{factoring formula}$$

$$= \lim_{z\to x} (z^2 + zx + x^2) = x^2 + x^2 + x^2 = 3x^2. \qquad \leftarrow \text{cancellation, limit rules}$$

Again, notice that the limit was initially of the indeterminate form $\frac{0}{0}$. For derivative calculations with the $z \to x$ definition, the goal is often to factor and simplify until a common factor such as $z - x$ can be cancelled, as we just did.

(c) We have just shown two ways that if $f(x) = x^3$, then $f'(x) = 3x^2$. When we graph these two functions, we can see that at each value of x, the slope of the graph of $f(x) = x^3$ is equal to the height of the graph of $f'(x) = 3x^2$. In particular notice that the slopes of $f(x) = x^3$ are always positive and the graph of $f'(x) = 3x^2$ is always positive.

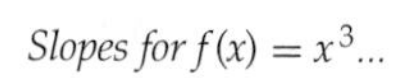

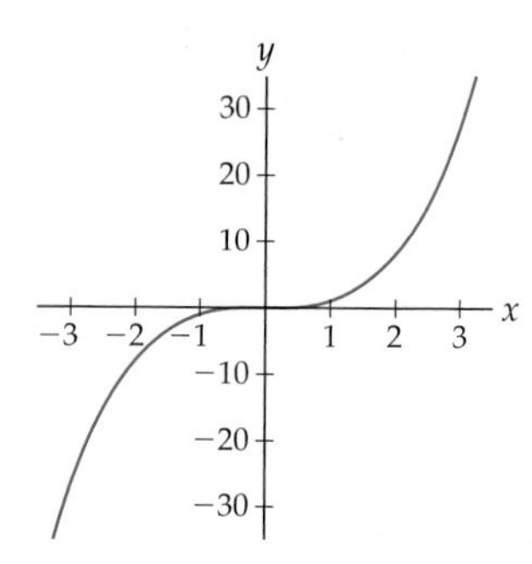

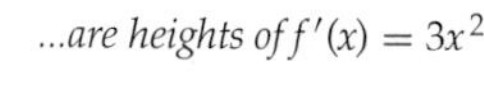

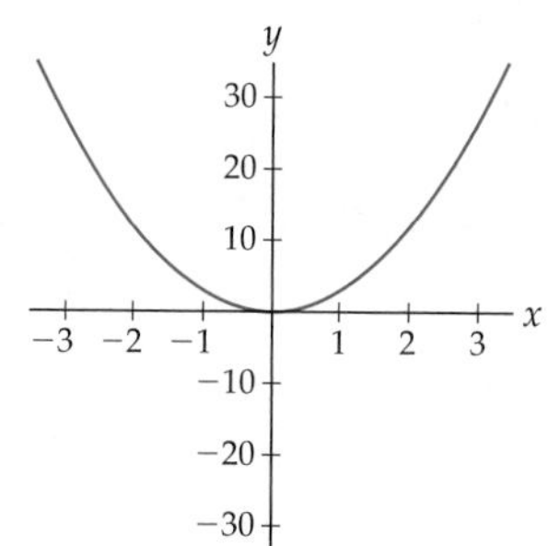

□

EXAMPLE 3

Finding the equation of a tangent line

Suppose $f(x) = \frac{2}{x}$. Find the equation of the tangent line to f at $x = 1$.

SOLUTION

By definition, the tangent line to f at $x = 1$ has slope $f'(1)$ and passes through the point $(1, f(1)) = \left(1, \frac{2}{1}\right) = (1, 2)$. Before we can find the equation of this line, we must calculate the value of $f'(1)$. A lot of algebra will be needed before we can cancel a common factor of h and solve the limit:

$$f'(1) = \lim_{h\to 0} \frac{f(1+h) - f(1)}{h} = \lim_{h\to 0} \frac{\frac{2}{1+h} - \frac{2}{1}}{h} \qquad \leftarrow \text{derivative with } x = 1$$

$$= \lim_{h\to 0} \frac{\left(\frac{2 - 2(1+h)}{1+h}\right)}{h} = \lim_{h\to 0} \frac{2 - 2 - 2h}{h(1+h)} \qquad \leftarrow \text{algebra}$$

$$= \lim_{h\to 0} \frac{-2h}{h(1+h)} = \lim_{h\to 0} \frac{-2}{1+h} = \frac{-2}{1+0} = -2. \qquad \leftarrow \text{cancellation, limit rules}$$

We now use the point-slope form of a line to find the equation of the line that has slope $f'(1) = -2$ and that passes through the point $(1, f(1)) = (1, 2)$:

$$y - 2 = -2(x-1) \implies y = -2(x-1) + 2 \implies y = -2x + 4.$$

□

CHECKING THE ANSWER We can verify the reasonableness of the answer we just found by graphing the function $f(x) = \frac{2}{x}$ and the line $y = -2x + 4$ on the same set of axes and checking that the line appears to be tangent to the graph of f at the point $(1, f(1))$:

$f(x) = \frac{2}{x}$ *and* $y = -2x + 4$

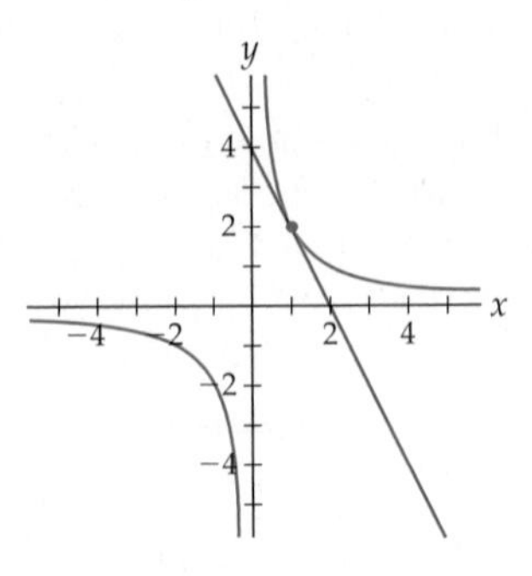

EXAMPLE 4

Finding derivatives of functions that involve roots

(a) Use the $z \to x$ definition of the derivative to find $\frac{d}{dx}(\sqrt{x})$.

(b) Use the $h \to 0$ definition of the derivative to find $\frac{d}{dx}(\sqrt{x})$.

SOLUTION

(a) In the calculation that follows we will use the factoring formula $a^2 - b^2 = (a - b)(a + b)$ in the case where $a = \sqrt{z}$ and $b = \sqrt{x}$. In other words we will apply the formula $z - x = (\sqrt{z} - \sqrt{x})(\sqrt{z} + \sqrt{x})$. Remember that our goal in this calculation of the derivative is to factor and simplify until we can cancel something in the numerator with the same thing in the denominator and take the limit:

$$\frac{d}{dx}(\sqrt{x}) = \lim_{z \to x} \frac{\sqrt{z} - \sqrt{x}}{z - x} \qquad \leftarrow \text{derivative}$$

$$= \lim_{z \to x} \frac{\sqrt{z} - \sqrt{x}}{(\sqrt{z} - \sqrt{x})(\sqrt{z} + \sqrt{x})} \qquad \leftarrow \text{factoring formula}$$

$$= \lim_{z \to x} \frac{1}{\sqrt{z} + \sqrt{x}} \qquad \leftarrow \text{cancellation}$$

$$= \frac{1}{\sqrt{x} + \sqrt{x}} = \frac{1}{2\sqrt{x}}. \qquad \leftarrow \text{take limit, algebra}$$

(b) We encounter different algebra when we use the $h \to 0$ definition of the derivative. The ***conjugate*** of an expression of the form $\sqrt{a} - \sqrt{b}$ is $\sqrt{a} + \sqrt{b}$, and vice versa. Notice that the product of such an expression and its conjugate does not involve any square roots, since $(\sqrt{a} - \sqrt{b})(\sqrt{a} + \sqrt{b}) = a - b$. In the calculation that follows we will simplify our limit by multiplying numerator and denominator by a conjugate, to clear square roots. Remember that our goal in this calculation is to cancel a common factor of h so that we can take the limit:

$$\frac{d}{dx}(\sqrt{x}) = \lim_{h \to 0} \frac{\sqrt{x+h} - \sqrt{x}}{h} \qquad \leftarrow \text{derivative}$$

$$= \lim_{h \to 0} \frac{\sqrt{x+h} - \sqrt{x}}{h} \left(\frac{\sqrt{x+h} + \sqrt{x}}{\sqrt{x+h} + \sqrt{x}} \right) \qquad \leftarrow \text{multiply by conjugate}$$

$$= \lim_{h \to 0} \frac{(x+h) - x}{h(\sqrt{x+h} + \sqrt{x})} = \lim_{h \to 0} \frac{h}{h(\sqrt{x+h} + \sqrt{x})} \qquad \leftarrow \text{algebra}$$

$$= \lim_{h \to 0} \frac{1}{\sqrt{x+h} + \sqrt{x}} = \frac{1}{\sqrt{x+0} + \sqrt{x}} = \frac{1}{2\sqrt{x}}. \qquad \leftarrow \text{cancel, take limit}$$

□

EXAMPLE 5 **The derivative of a piecewise-defined function**

Consider the piecewise-defined function $f(x) = \begin{cases} 3x+2, & \text{if } x < 1 \\ 6-x, & \text{if } x \geq 1. \end{cases}$

(a) Calculate $f'_+(1)$ and $f'_-(1)$. What can you say about $f'(1)$?

(b) Write down a formula for $f'(x)$ as a piecewise-defined function.

SOLUTION

(a) We start by finding the right derivative of f at $x = 1$. In this case we examine $h \to 0^+$, which means that $h > 0$, and thus $1 + h > 1$. Therefore we will use the second part of the piecewise-defined function f to evaluate $f(1 + h)$ in this case:

$$f'_+(1) = \lim_{h\to 0^+} \frac{f(1+h) - f(1)}{h} = \lim_{h\to 0^+} \frac{(6-(1+h)) - (6-1)}{h}$$
$$= \lim_{h\to 0^+} \frac{6-1-h-6+1}{h} = \lim_{h\to 0^+} \frac{-h}{h} = \lim_{h\to 0^+} (-1) = -1.$$

In contrast, when we calculate the left derivative of f at $x = 1$, we will have $h \to 0^-$, and thus $h < 0$. This means that $1+h < 1$, so we will use the first part of the piecewise-defined function f to evaluate $f(1 + h)$. Of course we still have $1 \geq 1$, so we still use the *second* part of f to evaluate $f(1)$:

$$f'_-(1) = \lim_{h\to 0^-} \frac{f(1+h) - f(1)}{h} = \lim_{h\to 0^-} \frac{(3(1+h)+2) - (6-1)}{h}$$
$$= \lim_{h\to 0^-} \frac{3+3h+2-6+1}{h} = \lim_{h\to 0^-} \frac{3h}{h} = \lim_{h\to 0^-} 3 = 3.$$

Since the left and right derivatives of f at $x = 1$ are not equal to each other, the derivative $f'(1)$ of f at $x = 1$ is undefined. Note that f is left differentiable and right differentiable at $x = 1$, but *not* differentiable at $x = 1$.

(b) We have just calculated that $f'(1)$ does not exist. It now remains to determine $f'(x)$ for values of x that are less than or greater than 1. For $x < 1$ the value of $f(x)$ is equal to $3x + 2$. Using the definition of the derivative, we see that for $x < 1$ we have

$$f'(x) = \lim_{h\to 0} \frac{(3(x+h)+2) - (3x+2)}{h} = \lim_{h\to 0} \frac{3x+3h+2-3x-2}{h}$$
$$= \lim_{h\to 0} \frac{3h}{h} = \lim_{h\to 0} 3 = 3.$$

For $x > 1$ the value of $f(x)$ is equal to $6 - x$. A similar calculation shows that for $x > 1$ we have

$$f'(x) = \lim_{h\to 0} \frac{(6-(x+h)) - (6-x)}{h} = \lim_{h\to 0} \frac{6-x-h-6+x}{h}$$
$$= \lim_{h\to 0} \frac{-h}{h} = \lim_{h\to 0} -1 = -1.$$

Therefore the derivative of the piecewise-defined function f is

$$f'(x) = \begin{cases} 3, & \text{if } x < 1 \\ \text{undefined}, & \text{if } x = 1 \\ -1, & \text{if } x > 1, \end{cases}$$

where "DNE" stands for "does not exist," representing the fact that the function f' is not defined at $x = 1$ (i.e., that there is no real number assigned to $f'(1)$). □

The figure that follows at the left shows the graphs of the function f and its derivative f' from the previous example. We can see from the graph of f that the right derivative of the function at $x = 1$ is negative (think about the secant lines when $h > 0$) while the left derivative at $x = 1$ is positive and steeper (think about the secant lines when $h < 0$). In fact, since the pieces of this graph are linear, we can clearly see from the graph that the slopes of the secant lines from the right and the left are -1 and 3, respectively.

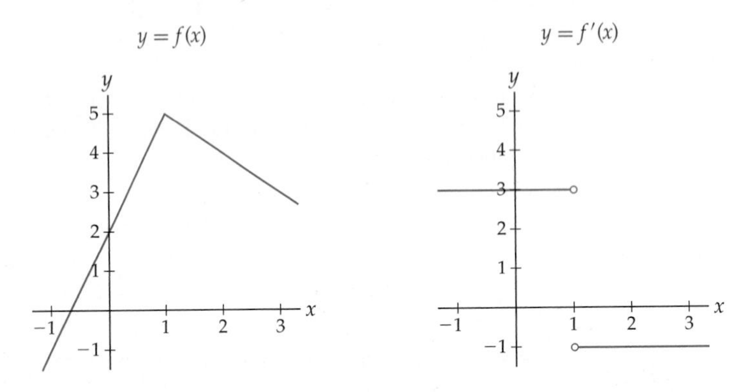

Because the left and right derivatives at $x = 1$ do not agree, there is no well-defined tangent line at $x = 1$; the (two-sided) derivative of f at $x = 1$ does not exist. Graphically, there is no *unique* tangent line that passes through the point $(1, f(1))$. In some sense, the graph of f has *two* directions at $x = 1$, one with slope -1 and one with slope 3. In the graph of f' at the right we see that the slopes of f are recorded correctly, with a slope of 3 for $x < 1$, a slope of -1 for $x > 1$, and an undefined slope for $x = 1$.

EXAMPLE 6

A function that fails to be continuous also fails to be differentiable

Consider the piecewise-defined function f given by the equation

$$f(x) = \begin{cases} x^2, & \text{if } x < 3 \\ 12, & \text{if } x \geq 3. \end{cases}$$

Argue graphically that f is not differentiable at $x = 3$ by examining secant lines on the graph of f from both the left and the right of $x = 3$.

SOLUTION

The graph of f looks like this:

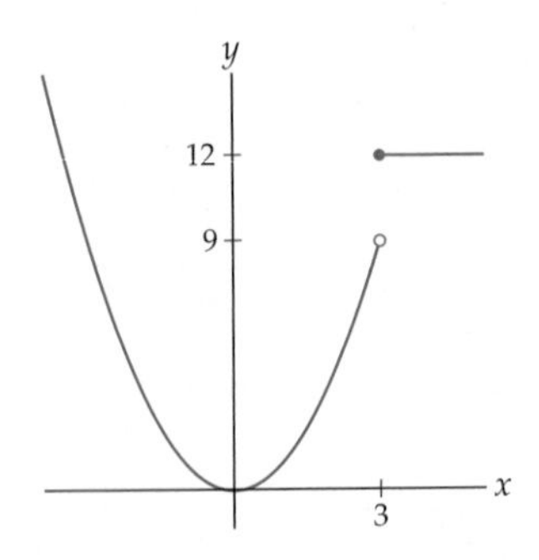

On the one hand, for $h > 0$, the secant lines over intervals of the form $[3, 3 + h]$ all have slope zero, since, on the interval $[3, \infty)$, the graph of f is a horizontal line. Therefore the right derivative of f at $x = 3$ must be $f'_+(3) = 0$. On the other hand, for $h < 0$, the secant lines over intervals of the form $[3+h, 3]$ behave as shown in the following three figures:

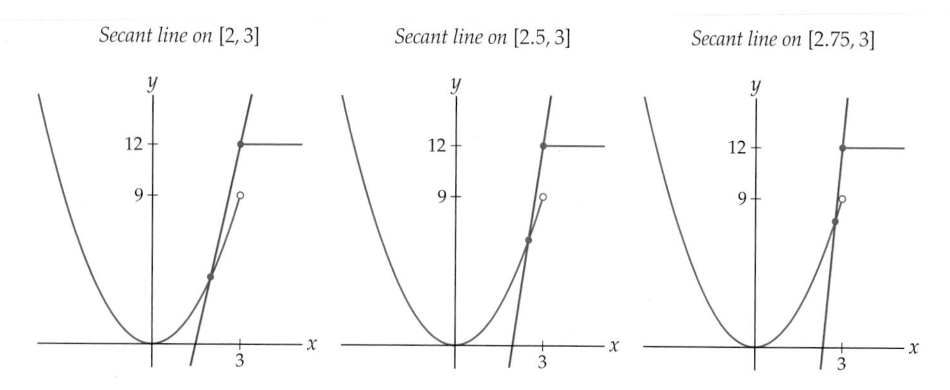

As $h \to 0^-$, we get a sequence of secant lines that are becoming more and more vertical. The slopes of these secant lines are large and positive, and approach ∞ as $h \to 0^-$. Therefore the left derivative $f'_-(3)$ does not exist. This means that f is not differentiable at $x = 3$, since its left derivative does not exist. □

EXAMPLE 7

Approximating roots with local linearity and Newton's method

Use local linearity to approximate a root of $f(x) = -x^3 - x + 1$.

SOLUTION

First of all, notice that by the Intermediate Value Theorem and the fact that $f(0) = 1$ is positive and $f(1) = -1$ is negative, the continuous function $f(x) = -x^3 - x + 1$ must have at least one root between $x = 0$ and $x = 1$. However, since we cannot factor $f(x)$, we are unable to solve for the root directly. Our approximation strategy will be to guess a location for a root, find the tangent line to the function at that point, and then look at the root of that tangent line. If the tangent line is a good local approximation for the function, then the root of the tangent line will be close to the root of the function if our initial guess is close enough. We then repeat the process, using the root of the tangent line as our next guess for the root of the function, to get better and better approximations of the root we are looking for. This strategy is known as ***Newton's method*** for approximating roots.

We'll start with $x_1 = 0$ as our first guess for a root of $f(x)$. Clearly this is not actually a root of the function, because $f(0) = 1$ is not equal to zero. Hopefully the tangent line at $x_1 = 0$ will help us make a better guess for a root. Before we compute the tangent line, we must find the derivative of $f(x) = -x^3 - x + 1$. We will leave the bulk of the computational details to the reader (see Exercise 42):

$$\begin{aligned} f'(x) &= \lim_{h \to 0} \frac{f(x+h) - f(x)}{h} \\ &= \lim_{h \to 0} \frac{(-(x+h)^3 - (x+h) + 1) - (-x^3 - x + 1)}{h} \\ &= \cdots = -3x^2 - 1. \end{aligned}$$

Therefore the derivative of f at $x_1 = 0$ is $f'(0) = -3(0)^2 - 1 = -1$, and thus the tangent line to $f(x)$ at $(0, f(0)) = (0, 1)$ has equation

$$y = f(0) + f'(0)(x - 0) = 1 + (-1)(x - 0) = 1 - x.$$

Although it is not easy to calculate a root of our original cubic $f(x) = -x^3 - x + 1$, it is easy to calculate that 1 is a root of the linear function $y = 1 - x$. Although $x_2 = 1$ is still not a root of $f(x)$ (since $f(1) = -1$), the hope is that repeating this process will get us closer and closer to a root. The following figure shows our initial guess of $(x_1, f(x_1)) = (0, 1)$, the root at $x = 1$ of the tangent line to $f(x)$ at $x_1 = 0$, and the new guess of $(x_2, f(x_2)) = (1, -1)$.

$f(x) = -x^3 - x + 1$ *with* $x_1 = 0$ *and* $x_2 = 1$

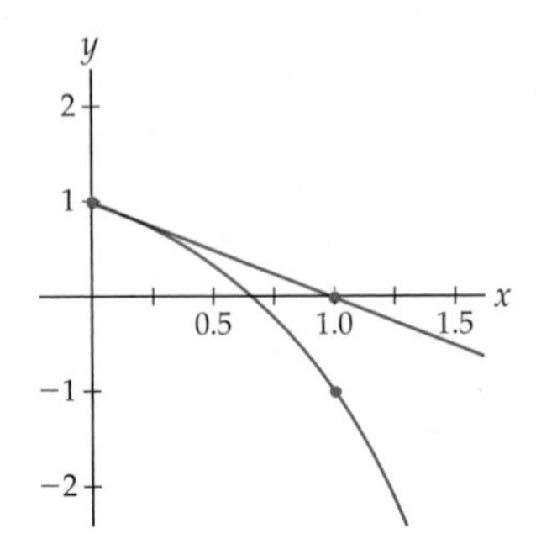

Repeating this process, we can use a root of the tangent line to $f(x)$ at x_2 to obtain a revised guess x_3 for a root. The tangent line passes through $(1, -1)$ and has slope $f'(1) = -3(1) - 1 = -4$, so its equation is

$$y = f(1) + f'(1)(x - 1) = -1 + (-4)(x - 1) = -4x + 3.$$

The root of this line, and our new guess for a root of $f(x)$, is $x_3 = \frac{3}{4}$; see the graph that follows. Note that already our third approximation, x_3, appears quite close to the actual root of $f(x)$:

$f(x) = -x^3 - x + 1$ *with* $x_2 = 1$ *and* $x_3 = \frac{3}{4}$

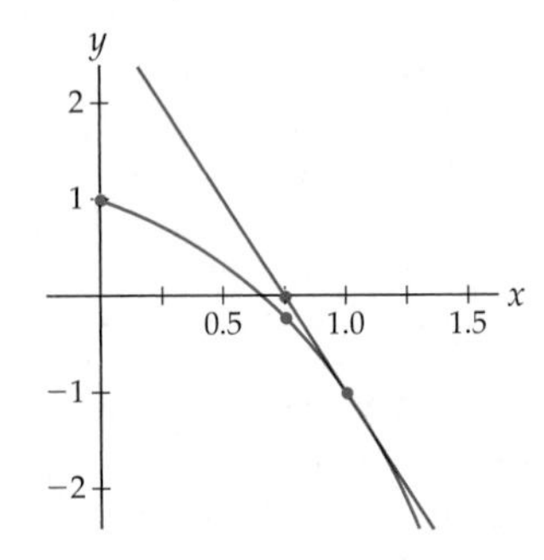

Repeating this process one more time, we obtain $x_4 \approx 0.686047$, which is in fact extremely close to the actual root of $x \approx 0.682328$. In Exercise 22 you will compare the accuracy of this method with that of the Bisection Method of finding roots that we used in Exercise 80 of Section 1.4. □

? TEST YOUR UNDERSTANDING

- What is the definition of the derivative of a function f at a point $x = c$? Give two answers, one with a limit that involves h and one with a limit that involves z.
- When calculating a derivative with the $h \to 0$ definition of the derivative, why is the limit always initially of the form $\frac{0}{0}$? What about when we use the $z \to x$ definition of the derivative?
- Is a function that is continuous at $x = c$ necessarily differentiable at $x = c$? Is a function that is differentiable at $x = c$ necessarily continuous at $x = c$?

- What is the connection between the Leibniz notation $\frac{dy}{dx}$ for the derivative and the difference quotient $\frac{\Delta y}{\Delta x}$?
- How can we express the fifth derivative of a function f at the point $x = 2$ in Leibniz notation? In "prime" notation?

EXERCISES 2.2

Thinking Back

Simplifying quotients: Simplify and rewrite the following expressions until you can cancel a common factor in the numerator and the denominator.

- $\dfrac{(x+h)^4 - x^4}{h}$
- $\dfrac{(x+h)^{-2} - x^{-2}}{h}$
- $\dfrac{z^4 - x^4}{z - x}$
- $\dfrac{z^{-2} - x^{-2}}{z - x}$

Limit calculations: Find each of the following limits.

- $\lim\limits_{h\to 0} \dfrac{(1-h) - 1}{h}$
- $\lim\limits_{h\to 0} \dfrac{(3(-1+h)^2 + 1) - 4}{h}$
- $\lim\limits_{h\to 0} \dfrac{\frac{1}{2+h} - 0.5}{h}$
- $\lim\limits_{z\to 2} \dfrac{z^2 - 4}{z - 2}$
- $\lim\limits_{z\to 4} \dfrac{(1-3z) + 11}{z - 4}$
- $\lim\limits_{z\to 1} \dfrac{\frac{1}{z} - 1}{z - 1}$

Concepts

0. *Problem Zero:* Read the section and make your own summary of the material.

1. *True/False:* Determine whether each of the statements that follow is true or false. If a statement is true, explain why. If a statement is false, provide a counterexample.

(a) *True or False:* $f'(x) = \dfrac{f(x+h) - f(x)}{h}$.

(b) *True or False:* $f'(x) = \lim\limits_{x\to 0} \dfrac{f(x+h) - f(x)}{h}$.

(c) *True or False:* $f'(x) = \lim\limits_{z\to 0} \dfrac{f(z) - f(x)}{z - x}$.

(d) *True or False:* If $f(x) = x^3$, then $f(x+h) = x^3 + h$.

(e) *True or False:* If $f(x) = x^3$, then $f'(x) = \lim\limits_{h\to 0} \dfrac{f(x^3+h) - f(x)}{h}$.

(f) *True or False:* A function f is differentiable at $x = c$ if and only if $f'_-(c)$ and $f'_+(c)$ both exist.

(g) *True or False:* If f is continuous at $x = c$, then f is differentiable at $x = c$.

(h) *True or False:* If f is not continuous at $x = c$, then f is not differentiable at $x = c$

2. *Examples:* Construct examples of the thing(s) described in the following. Try to find examples that are different than any in the reading.

(a) The graph of a function that is continuous, but not differentiable, at $x = 2$.

(b) The graph of a function that is left and right differentiable, but not differentiable, at $x = 3$.

(c) The graph of a function that is differentiable on the interval $[-1, 1]$ but not differentiable at the point $x = 1$.

3. Use limits to give mathematical definitions for each of the following derivatives, first with the $h \to 0$ definition of the derivative, and then with the $z \to x$ definition:

(a) the derivative of a function f at the point $x = 5$

(b) the derivative of a function f

(c) the right derivative of a function f at the point $x = -2$

4. Use limits to give mathematical definitions for:

(a) the slope of the line tangent to the graph of a function f at the point $x = 4$

(b) the line tangent to the graph of a function f at the point $x = 4$

(c) the instantaneous rate of change of a function f at the point $x = 1$

(d) the acceleration at time $t = 1.65$ of an object that moves with position function $s(t)$

5. Explain why the limits $\lim\limits_{h\to 0} \dfrac{f(x+h) - f(x)}{h}$ and $\lim\limits_{z\to x} \dfrac{f(z) - f(x)}{z - x}$ are the same for any function f. *(Hint: Consider the substitution $z = x + h$.)*

6. Explain why the limits $\lim\limits_{h\to 0} \dfrac{f(x+h) - f(x)}{h}$ and $\lim\limits_{z\to x} \dfrac{f(z) - f(x)}{z - x}$ are each initially in the form $\frac{0}{0}$. Why would cancelling a common factor of h or $z - x$ be likely to resolve this indeterminate form?

7. The function $f(x) = 4x^3 - 5x + 1$ is both continuous and differentiable at $x = 2$. Write these facts as limit statements.

8. The function $f(x) = 4 - x^2$ is both continuous and differentiable at $x = 1$. Write these facts as limit statements.

9. If $\lim\limits_{x\to c} \dfrac{f(x) - f(c)}{x - c}$ exists, what can you say about the differentiability of f at $x = c$? What can you say about the continuity of f at $x = c$?

10. Suppose $f(0) = 1$, $\lim\limits_{x\to 0^+} f(x) = 1$, $\lim\limits_{x\to 0^-} f(x) = 1$, $\lim\limits_{x\to 0^+} \dfrac{f(x) - f(0)}{x} = 3$, and $\lim\limits_{x\to 0^-} \dfrac{f(x) - f(0)}{x} = -2$.

(a) Is f continuous and/or differentiable at $x = 0$? What about from the left or right?

(b) Sketch a possible graph of f.

11. Suppose $f(1) = 3$, $\lim\limits_{x\to 1^-} f(x) = 3$, $\lim\limits_{x\to 1^+} f(x) = 3$, $\lim\limits_{h\to 0^-} \dfrac{f(1+h) - f(1)}{h} = 2$, and $\lim\limits_{h\to 0^+} \dfrac{f(1+h) - f(1)}{h} = 0$.

(a) Is f continuous and/or differentiable at $x = 0$? What about from the left or right?
(b) Sketch a possible graph of f.

12. Consider the function f graphed here:

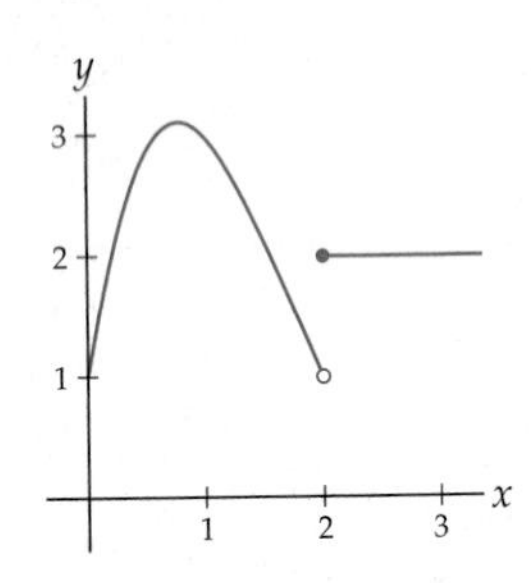

(a) Sketch secant lines from $(2, f(2))$ to $(2 + h, f(2 + h))$ on the graph of f, for the following values of h: $h = 0.5$, $h = 0.25$, $h = 0.1$, $h = -0.5$, $h = -0.25$, and $h = -0.1$.
(b) Use the secant lines you sketched in part (a) to graphically evaluate $\lim_{h \to 0^+} \frac{f(2+h) - f(2)}{h}$ and $\lim_{h \to 0^-} \frac{f(2+h) - f(2)}{h}$.
(c) Use your answer from part (b) to show that f is not differentiable at $x = 2$.

13. Sketch secant lines on a graph of $f(x) = |x|$, and use them to argue that the absolute value function is not differentiable at $x = 0$.

14. The two-sided symmetric difference approximation for the slope of a tangent line (see Exercise 20 in Section 2.1) can sometimes be misleading. Use a sequence of symmetric difference approximations to estimate the derivative of $f(x) = |x|$ at $x = 0$. What does your sequence of approximations suggest about $f'(0)$? Does this seem right?

15. Express the sentence "the derivative of $f(x) = 3x^2 - 1$ at $x = -4$ is -24" in each of the following notations:
(a) prime (b) Leibniz (c) operator

16. Express the sentence "the fourth derivative of $f(x) = 3x^2 - 1$ is equal to 0" in each of the following notations:
(a) prime (b) Leibniz (c) operator

Suppose $f(x) = x^3 - 2x + 1$, and let $y = f(x)$. It can be shown that $f'(x) = 3x^2 - 2$. Use this information to determine the expressions in Exercises 17 and 18. *(Note: No differentiation will be necessary, since the derivatives are given. These problems are just a test of your ability to interpret derivative notation.)*

17. $\frac{dy}{dx}$, $\frac{df}{dx}$, $\frac{d}{dx}(f(x))$, and $\frac{dy}{dx}(f(x))$.

18. $\left.\frac{dy}{dx}\right|_{x=2}$, $\left.\frac{df}{dx}\right|_{x=2}$, $\left.\frac{d}{dx}(f(x))\right|_{x=2}$, and $\frac{dy}{dx}(f(2))$

Express each of the following limit statements as delta–epsilon statements:

19. $f'(3) = \lim_{h \to 0} \frac{(3+h)^2 - 9}{h}$

20. $f'(3) = \lim_{x \to 3} \frac{x^2 - 9}{x - 3}$

21. Suppose that you know that the derivative of $f(x) = \sqrt{x}$ is equal to $f'(x) = \frac{1}{2\sqrt{x}}$. Use this fact, local linearity, and the fact that $\sqrt{4} = 2$ to approximate the value of $\sqrt{4.1}$. How close is your approximation with the approximation of $\sqrt{4.1}$ that you can find with a calculator? *(Hint: Consider the tangent line to $f(x)$ at $x = 4$, and use it to approximate the function nearby.)*

22. Suppose you wish to find a root of $f(x) = x^3 + x^2 + x + 1$ in the interval $[-3, 2]$. Compare the accuracy of Newton's method as applied in Example 8 with the accuracy of the Bisection Method used in Exercise 80 of Section 1.4 for the same function. Which method gets closest to the root in three iterations?

Skills

Use (a) the $h \to 0$ definition of the derivative and then (b) the $z \to c$ definition of the derivative to find $f'(c)$ for each function f and value $x = c$ in Exercises 23–38.

23. $f(x) = x^2$, $x = -3$

24. $f(x) = x^3$, $x = 1$

25. $f(x) = \frac{1}{x}$, $x = -1$

26. $f(x) = \frac{1}{x^2}$, $x = 2$

27. $f(x) = 1 - x^3$, $x = -1$

28. $f(x) = x^4 + 1$, $x = 2$

29. $f(x) = x^{1/2}$, $x = 9$

30. $f(x) = x^{-1/2}$, $x = 9$

31. $f(x) = \frac{x - 1}{x + 3}$, $x = 2$

32. $f(x) = \frac{x^2 - 3x}{x + 1}$, $x = 0$

33. $f(x) = e^x$, $x = 0$

34. $f(x) = 2e^x$

35. $f(x) = \sin x$, $x = 0$

36. $f(x) = \cos x$, $x = 0$

37. $f(x) = \tan x$, $x = 0$

38. $f(x) = \sec x$, $x = 0$

Use the definition of the derivative to find f' for each function f in Exercises 39–54.

39. $f(x) = -2x^2$

40. $f(x) = 4 + x^2$

41. $f(x) = x^3 + 2$

42. $f(x) = -x^3 - x + 1$

43. $f(x) = \frac{2}{x + 1}$

44. $f(x) = \frac{2}{1 - x}$

45. $f(x) = \frac{1}{x^2}$

46. $f(x) = \frac{1}{x^3}$

47. $f(x) = 3\sqrt{x}$

48. $f(x) = \frac{3}{\sqrt{x}}$

49. $f(x) = \sqrt{2x + 1}$

50. $f(x) = \frac{1}{\sqrt{2x + 1}}$

51. $f(x) = \frac{x - 1}{x + 3}$

52. $f(x) = \frac{1}{x^2 - 1}$

53. $f(x) = \frac{x^3}{x + 1}$

54. $f(x) = \frac{x^2 - 1}{x^2 - x - 2}$

Use the definition of the derivative to find the derivatives described in Exercises 55–58.

55. Find $\frac{d}{dx}(2x^3)$, $\frac{d^2}{dx^2}(2x^3)$, and $\frac{d^3}{dx^3}(2x^3)$.

56. Find $\left.\frac{d}{dx}(2x^3)\right|_3$, $\left.\frac{d^2}{dx^2}(2x^3)\right|_1$, and $\left.\frac{d^3}{dx^3}(2x^3)\right|_{-2}$.

57. Given $f^{(3)}(x) = 3x^2 + 1$, find $\dfrac{d^4f}{dx^4}$ and $\left.\dfrac{d^4f}{dx^4}\right|_2$.

58. Given $\dfrac{d^2f}{dx^2} = 2 - x^2$, find $f^{(3)}(4)$ and $\dfrac{d^4}{dx^4}(f(x))$.

Use the definition of the derivative to find the equations of the lines described in Exercises 59–64.

59. The tangent line to $f(x) = x^2$ at $x = -3$.

60. The tangent line to $f(x) = x^2$ at $x = 0$.

61. The line tangent to the graph of $y = 1 - x - x^2$ at the point $(1, -1)$.

62. The line tangent to the graph of $y = 4x + 3$ at the point $(-2, -5)$.

63. The line that passes through the point $(3, 2)$ and is parallel to the tangent line to $f(x) = \dfrac{1}{x}$ at $x = -1$.

64. The line that is perpendicular to the tangent line to $f(x) = x^4 + 1$ at $x = 2$ and also passes through the point $(-1, 8)$.

For each function f graphed in Exercises 65–68, determine the values of x at which f fails to be continuous and/or differentiable. At such points, determine any left or right continuity or differentiability. Sketch secant lines supporting your answers.

65.

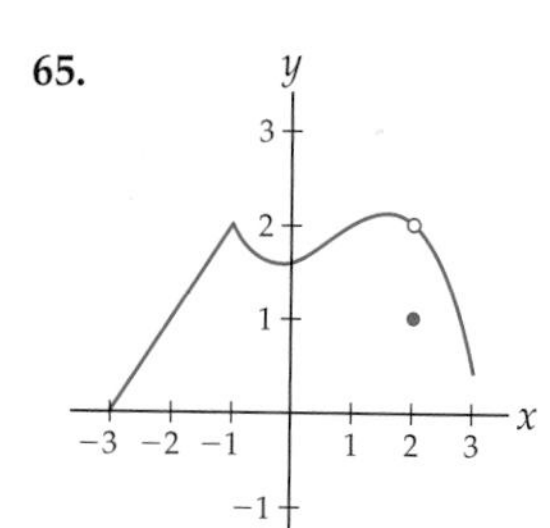

66.

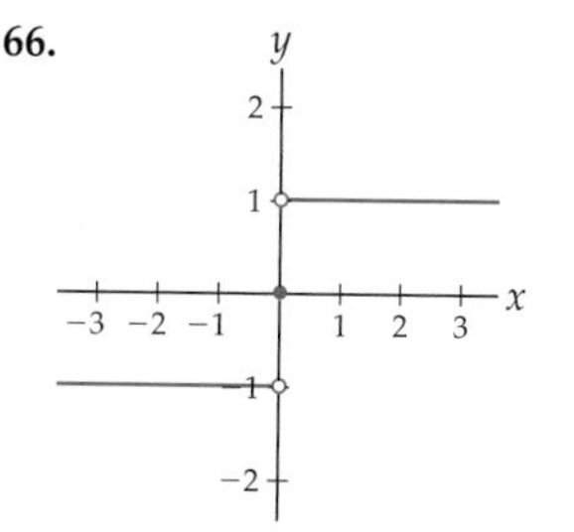

67.

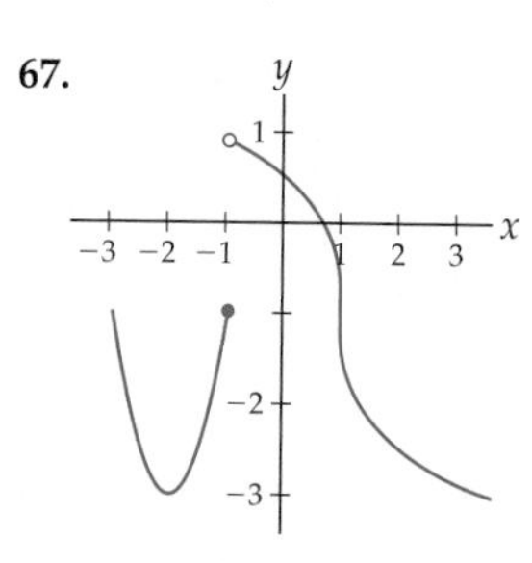

68.

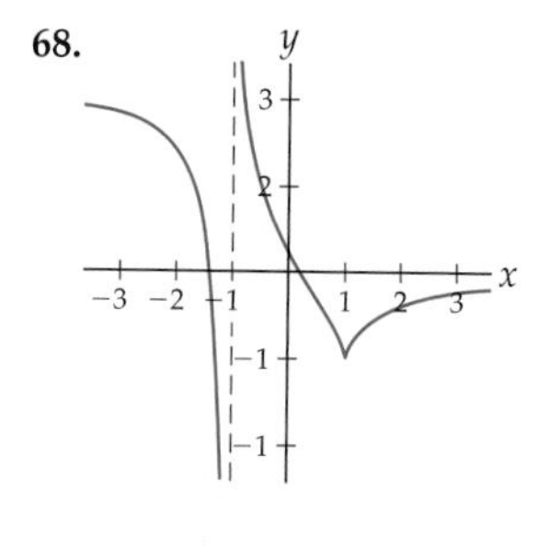

In Exercises 69–80, determine whether or not f is continuous and/or differentiable at the given value of x. If not, determine any left or right continuity or differentiability. For the last four functions, use graphs instead of the definition of the derivative.

69. $f(x) = \dfrac{1}{x}$, $x = 0$

70. $f(x) = x^{2/3}$, $x = 0$

71. $f(x) = |x^2 - 4|$, $x = 2$

72. $f(x) = |x^2 - 4|$, $x = -2$

73. $f(x) = \begin{cases} x + 4, & \text{if } x < 2 \\ 3x, & \text{if } x \geq 2, \end{cases}$ $x = 2$

74. $f(x) = \begin{cases} x^2 - 3, & \text{if } x < 3 \\ x + 2, & \text{if } x \geq 3, \end{cases}$ $x = 3$

75. $f(x) = \begin{cases} x^2, & \text{if } x \leq 1 \\ 2x - 1, & \text{if } x > 1, \end{cases}$ $x = 1$

76. $f(x) = \begin{cases} x^2, & \text{if } x \leq 1 \\ 2x + 4, & \text{if } x > 1, \end{cases}$ $x = 1$

77. $f(x) = \begin{cases} x \sin\left(\dfrac{1}{x}\right), & \text{if } x \neq 0 \\ 0, & \text{if } x = 0, \end{cases}$ $x = 0$

78. $f(x) = \begin{cases} x^2 \sin\left(\dfrac{1}{x}\right), & \text{if } x \neq 0 \\ 0, & \text{if } x = 0, \end{cases}$ $x = 0$

79. $f(x) = \begin{cases} 1, & \text{if } x \text{ rational} \\ x + 1, & \text{if } x \text{ irrational}, \end{cases}$ $x = 1$

80. $f(x) = \begin{cases} x^2, & \text{if } x \text{ rational} \\ 2x - 1, & \text{if } x \text{ irrational}, \end{cases}$ $x = 1$

For each function $f(x)$ and interval $[a, b]$ in Exercises 81–86, use the Intermediate Value Theorem to argue that the function must have at least one real root on $[a, b]$. Then apply Newton's method to approximate that root.

81. $f(x) = x^2 - 5$, $[a, b] = [1, 3]$

82. $f(x) = x^2 - 2$, $[a, b] = [1, 2]$

83. $f(x) = x^3 - 3x + 1$, $[a, b] = [0, 1]$

84. $f(x) = x^3 - 3x + 1$, $[a, b] = [1, 2]$

85. $f(x) = x^3 + 1$, $[a, b] = [-2, 1]$

86. $f(x) = x^4 - 2$, $[a, b] = [1, 2]$

Applications

87. In Example 4 of Section 2.1, we saw that a watermelon dropped from a height of 100 feet will be $s(t) = -16t^2 + 100$ feet above the ground t seconds after it is dropped. In that example, we approximated the velocity of such a watermelon at time $t = 1$ by calculating a sequence of average rates of change. Now we can calculate this velocity exactly, using the definition of the derivative. Do so, and compare the exact answer to the approximation we found earlier.

88. On a long road trip you are driving along a straight portion of Route 188. Suppose that t hours after entering Nevada your distance from the Donut Hole is $s(t) = -10t^2 - 40t + 120$ miles.

(a) How long will it take you to reach the Donut Hole after entering Nevada?

(b) Find your velocity $v(t)$ as you drive toward the Donut Hole.

(c) Are you accelerating or decelerating as you approach the Donut Hole? At what rate?

89. Suppose your position $s(t)$ as you drive north along a straight highway is as shown in the graph that follows at the left, with t measured in hours and s measured in miles.

(a) Sketch a graph of your velocity $v(t)$, and use the graph to describe your velocity over the course of the two-hour drive.

(b) Sketch a graph of your acceleration $a(t)$, and use the graph to describe your acceleration over the course of the two-hour drive.

Position on highway

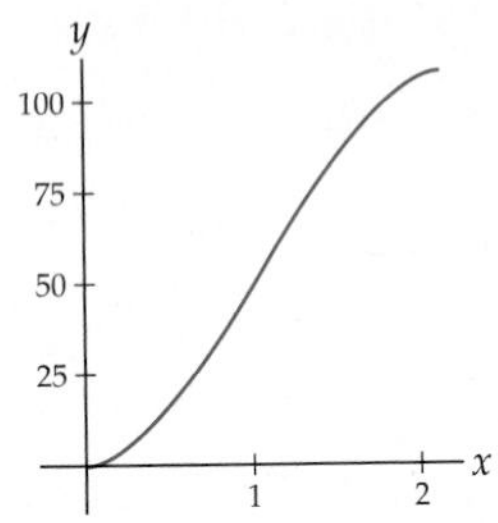

Carol's velocity

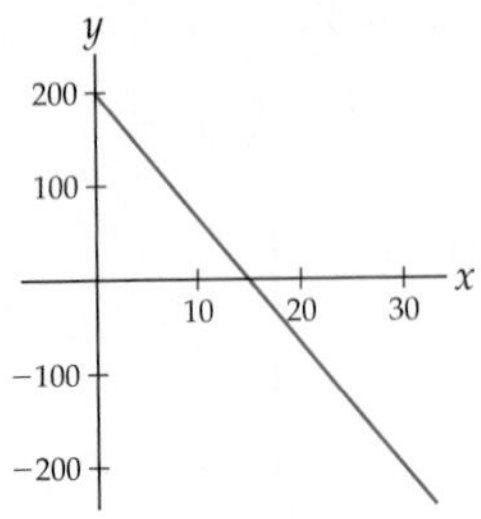

90. While Carol is walking along a straight north-south forest path, her velocity, in feet per minute, after t minutes is given by the preceding graph at the right. Suppose the "positive" direction is north.

(a) Describe the sort of walk Carol must have taken to have this velocity graph. Be sure your description explains the physical significance of the fact that her velocity is zero at $t = 15$ minutes and the fact that her velocity is negative for the second half of her walk.

(b) Find Carol's average acceleration over the 30-minute walk. Was her acceleration constant over the duration of her walk? Why or why not?

(c) What was Carol's average velocity over her entire walk? Why does your answer make sense?

91. To save up for a car, you take a job working 10 hours a week at the school library. For the first six weeks the library pays you \$8.00 an hour. After that you earn \$11.50 an hour. You put all of the money you earn each week into a savings account. On the day you start work your savings account already holds \$200.00. Let $S(t)$ be the function that describes the amount in your savings account t weeks after your library job begins.

(a) Find the values of $S(3)$, $S(6)$, $S(8)$, $S'(3)$, $S'(6)$, and $S'(8)$, if possible, and describe their meanings in practical terms. If it is not possible to find one or more of these values, explain why.

(b) Write an equation for the function $S(t)$. *(Hint: $S(t)$ will be a piecewise-defined function.)* Be sure that your equation correctly produces the values you calculated in part (a).

(c) Sketch a labeled graph of $S(t)$. By looking at this graph, determine whether $S(t)$ is continuous and whether $S(t)$ is differentiable. Explain the practical significance of your answers.

(d) Show algebraically that $S(t)$ is a continuous function, but not a differentiable function.

Proofs

92. Use the definition of the derivative to prove that every quadratic function $f(x) = ax^2 + bx + c$ has the property that its graph has a horizontal tangent line at the point $x = -\dfrac{b}{2a}$.

93. Use the definition of the derivative to prove that our concept of slope for linear functions matches the slope that is defined by the derivative. In other words, show that if $f(x) = mx + b$ is any linear function, then $f'(x) = m$.

94. Use Problem 93 to prove that a linear function is its own tangent line at every point. In other words, show that if $f(x) = mx + b$ is any linear function, then the tangent line to f at any point $x = c$ is given by $y = mx + b$.

95. Use the mathematical definition of a tangent line and the point-slope form of a line to show that if f is differentiable at $x = c$, then the tangent line to f at $x = c$ is given by the equation $y = f'(c)(x - c) + f(c)$.

96. Prove that if a function f is differentiable at $x = c$, then f is continuous at $x = c$.

(a) We are given that f is differentiable at $x = c$. Use the alternative definition of the derivative to write down what that statement means.

(b) We want to show that f is continuous at $x = c$. Use the definition of continuity to show that this statement is equivalent to the statement $\lim_{x \to c} (f(x) - f(c)) = 0$.

(c) Now use part (a) to show that $\lim_{x \to c} (f(x) - f(c)) = 0$. *(Hint: Multiply $(f(x) - f(c))$ by $\frac{x-c}{x-c}$ and use the product rule for limits.)*

97. Use the definition of two-sided and one-sided derivatives, together with properties of limits, to prove that $f'(c)$ exists if and only if $f'_-(c)$ and $f'_+(c)$ exist and are equal.

98. Show that if a function $y = f(x)$ is differentiable at x_0 and

$$\Delta y = f(x_0 + \Delta x) - f(x_0),$$

then

$$\Delta y = f'(x_0)\Delta x + \epsilon \Delta x,$$

where ϵ is a function satisfying $\lim_{\Delta x \to 0} \epsilon = 0$.

Thinking Forward

Derivatives of power functions: After differentiating a few power functions, you may have noticed a pattern emerging. In the following exercises we will investigate a possible formula for differentiating power functions.

- ▶ Use the $z \to x$ definition of the derivative to show that $\frac{d}{dx}(x^4) = 4x^3$.
- ▶ Use the $z \to x$ definition of the derivative to show that $\frac{d}{dx}(x^8) = 8x^7$. (*Hint: The following factoring formula will come in handy: If n is a positive integer, then $z^n - x^n = (z-x)(z^{n-1} + z^{n-2}x + z^{n-3}x^2 + \cdots + z^2x^{n-3} + zx^{n-2} + x^{n-1})$.*)
- ▶ Use the preceding two derivative formulas to make a conjecture about a formula for $\frac{d}{dx}(x^n)$, where n is a positive integer.

Derivatives of combinations of functions: We have already seen that the limit of a sum is the sum of the limits and that the limit of a product is the product of the limits. Do derivatives also interact well with sums and products?

- ▶ Use the definition of the derivative (or exercises done previously in this section) to find (a) $\frac{d}{dx}(3x)$, (b) $\frac{d}{dx}(x^2)$, and (c) $\frac{d}{dx}(3x + x^2)$. Use your answers to make a conjecture as to whether or not $\frac{d}{dx}(f(x) + g(x)) = \frac{df}{dx} + \frac{dg}{dx}$.
- ▶ Use the definition of the derivative (or exercises done previously in this section) to find (a) $\frac{d}{dx}(x - 3)$, (b) $\frac{d}{dx}(2x + 1)$, and (c) $\frac{d}{dx}((x - 3)(2x + 1))$. Use your answers to make a conjecture as to whether or not $\frac{d}{dx}(f(x)g(x)) = \left(\frac{df}{dx}\right)\left(\frac{dg}{dx}\right)$.

2.3 RULES FOR CALCULATING BASIC DERIVATIVES

- ▶ Formulas for differentiating constant, identity, linear, and power functions
- ▶ Rules for differentiating constant multiples, sums, products, and quotients
- ▶ Using differentiation rules to quickly calculate derivatives and antiderivatives

Derivatives of Linear Functions

You may have noticed by now that using the definition of the derivative to calculate derivatives can be rather tedious. Since derivatives will be used often throughout this course, we need to develop a faster method of calculating them. Let's start with linear functions. If f is a linear function, then it has slope m everywhere, and therefore its derivative is constantly m. Since constant and identity functions are linear functions with slopes 0 and 1, their derivatives are constantly 0 and 1, respectively. These are our first differentiation rules.

THEOREM 2.8 **Derivatives of Constant, Identity, and Linear Functions**

For any real numbers k, m, and b,

(a) $\frac{d}{dx}(k) = 0$ **(b)** $\frac{d}{dx}(x) = 1$ **(c)** $\frac{d}{dx}(mx + b) = m$

With this differentiation rule we can find derivatives of linear functions very quickly, without having to consider the definition of the derivative. For example, $\frac{d}{dx}(3) = 0$, $\frac{d}{dx}(3x + 1) = 3$, and $\frac{d}{dx}(2x - 99) = 2$.

Proof. We need only prove the third formula, since first two are special cases of that formula. Our proof is just a general calculation with the definition of the derivative: If m and b are any constants and $f(x) = mx + b$, then

$$f'(x) = \lim_{h\to 0} \frac{f(x+h) - f(x)}{h} \qquad \leftarrow \text{definition of the derivative}$$

$$= \lim_{h\to 0} \frac{(m(x+h)+b) - (mx+b)}{h} \qquad \leftarrow \text{use formula for } f(x)$$

$$= \lim_{h\to 0} \frac{mx + mh + b - mx - b}{h} \qquad \leftarrow \text{algebra}$$

$$= \lim_{h\to 0} \frac{mh}{h} = \lim_{h\to 0} m = m. \qquad \leftarrow \text{limit of a constant}$$

■

The Power Rule

You may have already noticed a particular pattern for the derivatives of power functions. In various examples and exercises in the previous section, we have seen that

$$\frac{d}{dx}(x^3) = 3x^2, \quad \frac{d}{dx}(x^4) = 4x^3, \quad \frac{d}{dx}(\sqrt{x}) = \frac{1}{2\sqrt{x}}, \quad \text{and} \quad \frac{d}{dx}\left(\frac{1}{x^3}\right) = \frac{-3}{x^4}.$$

The pattern becomes clear if we write these derivative formulas in exponent notation:

$$\frac{d}{dx}(x^3) = 3x^2, \quad \frac{d}{dx}(x^4) = 4x^3, \quad \frac{d}{dx}(x^{1/2}) = \frac{1}{2}x^{-1/2}, \quad \text{and} \quad \frac{d}{dx}(x^{-3}) = -3x^{-4}.$$

From these examples it appears that, to take the derivative of x^k, we bring down the exponent k to the front of the expression and then decrease the exponent by one, to get kx^{k-1}. This is in fact the case in general:

THEOREM 2.9 **The Power Rule**

For any nonzero rational number k, $\frac{d}{dx}(x^k) = kx^{k-1}$.

Although we require that k be a rational number in this formula, the power rule is actually true for any nonzero real number k. If $k = 0$, then x^0 is the constant function 1, which we already know how to differentiate.

Considering the algebra involved in applying the definition of the derivative, it is a relief to have such a simple formula for finding derivatives of power functions! With this formula we can quickly say, for example, that $\frac{d}{dx}(x^{15}) = 15x^{14}$ or that $\frac{d}{dx}(x^{-1000}) = -1000x^{-1001}$, or even that $\frac{d}{dx}(x^{17/12}) = \frac{17}{12}x^{5/12}$. We can also find higher derivatives very easily; for example, if $f(x) = x^4$, then $f'(x) = 4x^3$, $f''(x) = 12x^2$, $f'''(x) = 24x$, $f^{(4)}(x) = 24$, and $f^{(5)}(x) = 0$. These functions are graphed next; each one is the associated slope function for the one before.

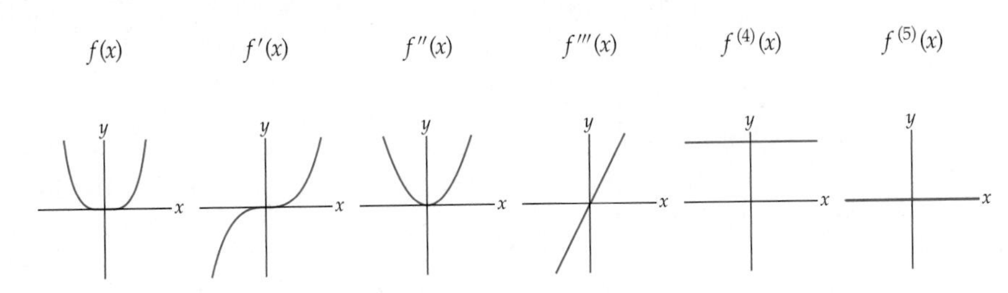

Proof. We will prove the power rule by cases. The case where k is a positive integer is proved in what follows. The case where k is a negative integer is left for Exercise 86. The case where k is a general rational number $\frac{p}{q}$ must be put off until after we know about implicit differentiation in Section 2.4.

Given any positive integer k, we need to apply the definition of the derivative to find $\frac{d}{dx}(x^k)$. It turns out that the $z \to x$ definition of the derivative is easier to use in this case. Along the way we will require the following factoring formula: If k is a positive integer, then

$$z^k - x^k = (z-x)(z^{k-1} + z^{k-2}x + z^{k-3}x^2 + \cdots + z^2x^{k-3} + zx^{k-2} + x^{k-1}).$$

(You can verify this formula by simply multiplying out the right-hand side of the equation; everything except the first and last terms will cancel, and you will obtain the left side of the equation.) Now using the definition of the derivative, we have

$$\begin{aligned}
\frac{d}{dx}(x^k) &= \lim_{z\to x} \frac{z^k - x^k}{z - x} && \leftarrow \text{definition of the derivative} \\
&= \lim_{z\to x} \frac{(z-x)(z^{k-1} + z^{k-2}x + z^{k-3}x^2 + \ldots + zx^{k-2} + x^{k-1})}{z-x} && \leftarrow \text{factoring formula} \\
&= \lim_{z\to x}(z^{k-1} + z^{k-2}x + z^{k-3}x^2 + \ldots + zx^{k-2} + x^{k-1}) && \leftarrow \text{cancellation} \\
&= x^{k-1} + x^{k-2}x + x^{k-3}x^2 + \ldots + xx^{k-2} + x^{k-1}. && \leftarrow \text{evaluate limit} \\
&= x^{k-1} + x^{k-1} + x^{k-1} + \ldots + x^{k-1} + x^{k-1} && \leftarrow \text{there are } k \text{ of these} \\
&= kx^{k-1}. &&
\end{aligned}$$

■

The Constant Multiple and Sum Rules

In Chapter 1 we saw that limits behave well with respect to arithmetic combinations of functions. For example, limits commute with sums: The limit of a sum is the sum of the limits, provided that the limits involved exist. Is the same thing true for derivatives? For constant multiples and sums, the answer is yes:

THEOREM 2.10

Derivatives of Constant Multiples and Sums of Functions

If f and g are functions and k is a constant, then for all x where the functions involved are differentiable, we have the following differentiation formulas:

Constant Multiple Rule: $(kf)'(x) = kf'(x)$

Sum Rule: $(f+g)'(x) = f'(x) + g'(x)$

Difference Rule: $(f-g)'(x) = f'(x) - g'(x)$

The notation $(kf)'(x)$ indicates that we are differentiating the function kf with respect to x. In Leibniz notation we would write this differentiation as $\frac{d}{dx}(kf(x))$. The difference rule is of course just a combination of the sum and constant multiple rules, since $f(x) - g(x) = f(x) + (-g(x))$.

These rules mean that we can factor out constants and split up sums when calculating derivatives. For example, derivatives of sums or constant multiples of the functions $f(x) = x^2$ and $g(x) = x^4$ can be expressed as sums or constant multiples of their derivatives

$f'(x) = 2x$ and $g'(x) = 4x^3$, as illustrated here:

$$\begin{aligned}
\frac{d}{dx}(5x^2 + 2x^4) &= \frac{d}{dx}(5x^2) + \frac{d}{dx}(2x^4) && \leftarrow \text{sum rule} \\
&= 5\frac{d}{dx}(x^2) + 2\frac{d}{dx}(x^4) && \leftarrow \text{constant multiple rule} \\
&= 5(2x) + 2(4x^3) && \leftarrow \text{power rule} \\
&= 10x + 8x^3.
\end{aligned}$$

Proof. We will use the definition of the derivative to prove (a) the constant multiple and (b) sum rules in what follows. The difference rule can also be proved with the definition of the derivative, or by applying the sum and constant multiple rules; see Exercise 87.

(a) The proof of the constant multiple rule uses the definition of the derivative and the constant multiple rule for limits. Given a function $f(x)$ and a constant k, we wish to show that the derivative of the function $kf(x)$ is the same as k times the derivative of $f(x)$. We'll start from the left and work to the right:

$$\begin{aligned}
(kf)'(x) &= \lim_{h\to 0} \frac{kf(x+h) - kf(x)}{h} && \leftarrow \text{definition of the derivative for } kf(x) \\
&= \lim_{h\to 0} \frac{k(f(x+h) - f(x))}{h} && \leftarrow \text{factor out } k \text{ from numerator} \\
&= \lim_{h\to 0} \left(k\left(\frac{f(x+h) - f(x)}{h}\right)\right) && \leftarrow \text{factor out } k \text{ from quotient} \\
&= k\left(\lim_{h\to 0} \frac{f(x+h) - f(x)}{h}\right) && \leftarrow \text{constant multiple rule for limits} \\
&= kf'(x). && \leftarrow \text{definition of the derivative for } f(x)
\end{aligned}$$

(b) Similarly, the proof of the sum rule uses the definition of the derivative and the sum rule for limits. We wish to show that the derivative of the function $f+g$ is the sum of the derivative of f and the derivative of g. We will work from the left to the right; our goal is to use the definition of the derivative to write the left-hand statement as a limit and then use algebra and the sum rule for limits to split this limit into two limits, one of which will be the derivative of f, and one the derivative of g:

$$\begin{aligned}
(f+g)'(x) &= \lim_{h\to 0} \frac{(f(x+h) + g(x+h)) - (f(x) + g(x))}{h} && \leftarrow \text{definition of derivative} \\
&= \lim_{h\to 0} \frac{f(x+h) + g(x+h) - f(x) - g(x)}{h} && \leftarrow \text{simplify} \\
&= \lim_{h\to 0} \frac{(f(x+h) - f(x)) + (g(x+h) - g(x))}{h} && \leftarrow \text{reordering terms} \\
&= \lim_{h\to 0} \left(\frac{f(x+h) - f(x)}{h} + \frac{g(x+h) - g(x)}{h}\right) && \leftarrow \text{algebra} \\
&= \left(\lim_{h\to 0} \frac{f(x+h) - f(x)}{h}\right) + \left(\lim_{h\to 0} \frac{g(x+h) - g(x)}{h}\right) && \leftarrow \text{sum rule for limits} \\
&= f'(x) + g'(x). && \leftarrow \text{definition of derivative} \ \blacksquare
\end{aligned}$$

The Product and Quotient Rules

We now know that derivatives interact nicely with constant multiples and sums; for example, the derivative of a sum is the sum of the derivatives. Do derivatives also commute with products and quotients? Sadly, the answer is no. For example, consider the functions $f(x)=x^2+1$ and $g(x)=x^3$, whose derivatives are $f'(x)=2x$ and $g'(x)=3x^2$. Now consider the product $f(x)g(x) = (x^2+1)(x^3) = x^5+x^3$; by the power and sum rules, the derivative of that product is $5x^4+3x^2$. However, this is clearly not equal to the product $f'(x)g'(x) = (2x)(3x^2) = 6x^3$.

In general, the derivative of a product is not necessarily the product of its component derivatives. Similarly, the derivative of a quotient is not in general the quotient of its component derivatives. However, we can write the derivative of a product fg or quotient $\frac{f}{g}$ in terms of f, g, f', and g', with the following, somewhat surprising, formulas:

THEOREM 2.11

Derivatives of Products and Quotients of Functions

If f and g are functions, then for all x such that both f and g are differentiable, we have the following differentiation formulas:

$$\textbf{\textit{Product Rule:}}\quad (fg)'(x) = f'(x)\,g(x) + f(x)\,g'(x)$$

$$\textbf{\textit{Quotient Rule:}}\quad \left(\frac{f}{g}\right)'(x) = \frac{f'(x)\,g(x) - f(x)g'(x)}{(g(x))^2}$$

Some people remember the product rule by remembering the pattern of differentiating one piece and not the other in both possible ways and then taking the sum. Some people remember the quotient rule with a phrase like "lo d-hi minus hi d-lo over lo lo," although it might be easier just to remember the quotient rule than to try and remember that! In any case, you will have to memorize these differentiation formulas, because they will be needed often throughout this book.

Armed with the product and quotient rules, we can now differentiate a lot more functions. For example, with our earlier example of $f(x) = x^2+1$ and $g(x) = x^3$, we can differentiate $f(x)g(x)$ without multiplying it out first:

$$\begin{aligned}
\frac{d}{dx}(f(x)g(x)) &= f'(x)\,g(x) + f(x)\,g'(x) && \leftarrow \text{product rule} \\
&= (2x)(x^3) + (x^2+1)(3x^2) && \leftarrow f'(x) = 2x,\ g(x) = 3x^2 \\
&= 2x^4 + 3x^4 + 3x^2 && \leftarrow \text{algebra} \\
&= 5x^4 + 3x^2.
\end{aligned}$$

This is exactly what we found earlier by multiplying out $f(x)g(x)$ first and then taking the derivative.

Proof. The proof of the product rule is yet another general calculation using the definition of the derivative, with a small twist that we will explain shortly. A similar method of proof works for the quotient rule and is left to Exercise 89. We will have an easier way to prove the quotient rule once we learn the chain rule in Section 2.4.

Suppose f and g are differentiable functions. We wish to show that $(fg)' = f'g + fg'$. We will apply the definition of the derivative to the left side of that equation and then use a long string of algebra and limit rules to rewrite it in terms of f', g, f, and g'. This will be made possible in the second step of the following calculation by adding and subtracting the same expression, $f(x)g(x+h)$, from the numerator.

$$
\begin{aligned}
(fg)'(x) &= \lim_{h\to 0} \frac{f(x+h)g(x+h) - f(x)g(x)}{h} && \leftarrow \text{derivative} \\
&= \lim_{h\to 0} \frac{f(x+h)g(x+h) - f(x)g(x+h) + f(x)g(x+h) - f(x)g(x)}{h} && \leftarrow \text{algebra} \\
&= \lim_{h\to 0} \frac{(f(x+h) - f(x))g(x+h) + f(x)(g(x+h) - g(x))}{h} && \leftarrow \text{factoring} \\
&= \lim_{h\to 0} \left(\frac{f(x+h) - f(x)}{h} \cdot g(x+h) + f(x) \cdot \frac{g(x+h) - g(x)}{h} \right) && \leftarrow \text{algebra} \\
&= \left(\lim_{h\to 0} \frac{f(x+h) - f(x)}{h} \right) \lim_{h\to 0} g(x+h) + \lim_{h\to 0} f(x) \left(\lim_{h\to 0} \frac{g(x+h) - g(x)}{h} \right) && \leftarrow \text{limit rules} \\
&= \left(\lim_{h\to 0} \frac{f(x+h) - f(x)}{h} \right) g(x) + f(x) \left(\lim_{h\to 0} \frac{g(x+h) - g(x)}{h} \right) && \leftarrow \text{continuity} \\
&= f'(x)\, g(x) + f(x)\, g'(x). && \leftarrow \text{derivative}
\end{aligned}
$$

The step labeled "continuity" follows from the fact that $g(x)$ is differentiable and therefore continuous, because that is what allows us to say that $\lim_{h\to 0} g(x+h) = g(x)$. Notice that throughout the entire calculation, our goal was to extract the expressions that represent the derivatives of f and g. The algebra steps were not meant to "simplify"; they were meant to get us closer to the final form of $f'g + fg'$. ■

Examples and Explorations

EXAMPLE 1

Determining whether the differentiation rules apply

Find the derivatives of each of the functions that follow, using the differentiation rules from this section, if possible. If it is not possible, explain why not.

(a) $f(x) = 2(5^x)$ **(b)** $g(x) = 3^4$ **(c)** $h(x) = (2x+3)^2$ **(d)** $k(x) = \sqrt{3x^2+1}$

SOLUTION

(a) We cannot differentiate $f(x) = 2(5^x)$ with any of the rules from this section. In particular, the power rule applies only to power functions of the form x^k, where the variable x is in the base and a constant k is in the exponent. The function 5^x is not a power function, but an exponential function, which we will see how to differentiate in Section 2.5.

(b) Since $g(x) = 3^4 = 81$ is a constant function, $\frac{d}{dx}(3^4) = 0$.

(c) $h(x) = (2x+3)^2$ is a composition of functions, and we do not yet have a rule for differentiating compositions. Therefore we must use algebra to expand the function first. We could write the function as $(2x+3)(2x+3)$ and use the product rule, or we could just expand $(2x+3)^2$ entirely and then need only the sum and constant multiple rules. We present the latter approach here:

$$
\begin{aligned}
\frac{d}{dx}((2x+3)^2) &= \frac{d}{dx}(4x^2 + 12x + 9) && \leftarrow \text{algebra} \\
&= \frac{d}{dx}(4x^2) + \frac{d}{dx}(12x) + \frac{d}{dx}(9) && \leftarrow \text{sum rule} \\
&= 4\frac{d}{dx}(x^2) + 12\frac{d}{dx}(x) + \frac{d}{dx}(9) && \leftarrow \text{constant multiple rule} \\
&= 4(2x^1) + 12(1) + 0 \;=\; 8x + 12. && \leftarrow \text{power rule}
\end{aligned}
$$

(d) The function $k(x) = \sqrt{3x^2 + 1}$ is a composition that cannot be simplified, and we do not yet have a rule for differentiating compositions. (Note in particular that this function is *not* equal to $\sqrt{3}x + 1$, because square roots do not distribute over sums; there is no useful simplification that can be done to this function.) We will see how to differentiate such functions in Section 2.4. □

EXAMPLE 2

Finding derivatives with and without simplifying first

Differentiate each of the following functions with and without the quotient rule:

(a) $f(x) = \dfrac{4}{3x^2}$ **(b)** $g(x) = \dfrac{x^2}{x^{-3}\sqrt{x}}$ **(c)** $h(x) = \dfrac{x^7 + \sqrt{x}}{x^3}$

SOLUTION

(a) Applying the quotient rule gives us

$$\frac{d}{dx}\left(\frac{4}{3x^2}\right) = \frac{\frac{d}{dx}(4) \cdot (3x^2) - (4) \cdot \frac{d}{dx}(3x^2)}{(3x^2)^2} \qquad \leftarrow \text{quotient rule}$$

$$= \frac{(0)(3x^2) - (4)(6x)}{9x^4} \qquad \leftarrow \text{constant and power rules}$$

$$= \frac{-24x}{9x^4} = -\frac{8}{3}x^{-3}. \qquad \leftarrow \text{algebra}$$

Alternatively, we could simplify f first and then apply the constant multiple and power rules. We of course will get the same final answer:

$$\frac{d}{dx}\left(\frac{4}{3x^2}\right) = \frac{d}{dx}\left(\frac{4}{3}x^{-2}\right) \qquad \leftarrow \text{algebra}$$

$$= \frac{4}{3}\frac{d}{dx}(x^{-2}) \qquad \leftarrow \text{constant multiple rule}$$

$$= \frac{4}{3}(-2x^{-3}) = -\frac{8}{3}x^{-3}. \qquad \leftarrow \text{power rule, algebra}$$

(b) If we apply differentiation rules immediately without simplifying first, we need to use both the quotient and product rules:

$$\frac{d}{dx}\left(\frac{x^2}{x^{-3}\sqrt{x}}\right) = \frac{\frac{d}{dx}(x^2) \cdot (x^{-3}\sqrt{x}) - (x^2) \cdot \frac{d}{dx}(x^{-3}\sqrt{x})}{(x^{-3}\sqrt{x})^2} \qquad \leftarrow \text{quotient rule}$$

$$= \frac{(2x)(x^{-3}\sqrt{x}) - (x^2)\left(-3x^{-4}\sqrt{x} + x^{-3}\left(\frac{1}{2}x^{-1/2}\right)\right)}{(x^{-3}\sqrt{x})^2}. \qquad \leftarrow \text{more rules}$$

However, if we do some preliminary algebra, then the differentiation steps and final simplification will both be much, much simpler:

$$\frac{d}{dx}\left(\frac{x^2}{x^{-3}\sqrt{x}}\right) = \frac{d}{dx}(x^2x^3x^{-1/2}) \qquad \leftarrow \text{algebra}$$

$$= \frac{d}{dx}(x^{9/2}) \qquad \leftarrow \text{algebra}$$

$$= \frac{9}{2}x^{7/2}. \qquad \leftarrow \text{power rule}$$

In Exercise 17 you will show that the preceding two answers are equal. Students often want to differentiate as soon as possible, since the differentiation rules can be easier to implement than algebraic simplification. However, in many cases, in the long run, simplifying before differentiating ends up saving a lot of even messier algebra later on. Look at the end result of the two calculations we just did, and ask yourself which one you would rather work with!

(c) Applying the quotient rule first, we have

$$\frac{d}{dx}\left(\frac{x^7+\sqrt{x}}{x^3}\right) = \frac{\frac{d}{dx}(x^7+\sqrt{x})\cdot(x^3) - (x^7+\sqrt{x})\cdot\frac{d}{dx}(x^3)}{(x^3)^2} \qquad \leftarrow \text{quotient rule}$$

$$= \frac{\left(7x^6 + \frac{1}{2}x^{-1/2}\right)(x^3) - (x^7+\sqrt{x})(3x^2)}{x^6}. \qquad \leftarrow \text{sum, power rules}$$

Alternatively, we could do some algebra before differentiating and then apply the sum and power rules. This gives us a simpler but equivalent answer:

$$\frac{d}{dx}\left(\frac{x^7+\sqrt{x}}{x^3}\right) = \frac{d}{dx}\left(\frac{x^7}{x^3} + \frac{x^{1/2}}{x^3}\right) \qquad \leftarrow \text{algebra}$$

$$= \frac{d}{dx}(x^4 + x^{-5/2}) \qquad \leftarrow \text{algebra}$$

$$= 4x^3 - \frac{5}{2}x^{-7/2}. \qquad \leftarrow \text{sum, power rules}$$

Doing algebra before differentiating almost always means that there is much less algebra needed to simplify your answer after differentiating. In Exercise 18 you will show that the two answers we just found are in fact equal. □

✔ CHECKING THE ANSWER

You can always check the reasonableness of the answer to a differentiation problem by graphing both f and f' and verifying that f' appears to be the associated slope function for f. Do this for the following graphs, for part (c) of the example:

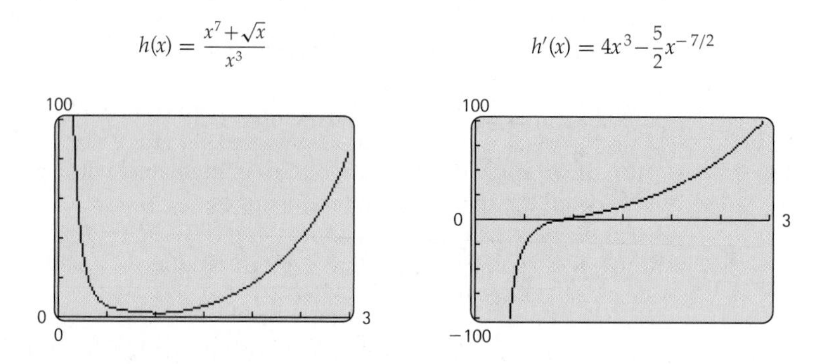

EXAMPLE 3

Differentiating a piecewise-defined function

Find the derivative of the function $f(x) = \begin{cases} x^2, & \text{if } x \le 1 \\ \frac{1}{x}, & \text{if } x > 1. \end{cases}$

SOLUTION

For $x < 1$ we use the first case of the function to find the derivative:

$$f'(x) = \frac{d}{dx}(x^2) = 2x.$$

Similarly, for $x > 1$ we use the second case:

$$f'(x) = \frac{d}{dx}\left(\frac{1}{x}\right) = \frac{d}{dx}(x^{-1}) = -x^{-2} = \frac{-1}{x^2}.$$

It now remains only to determine what the derivative is at the breakpoint $x = 1$, if it exists. The function f will be differentiable at $x = 1$ if it is continuous at $x = 1$ and the derivatives of each of its pieces are equal at $x = 1$. Since $(1)^2 = \frac{1}{1}$, f is continuous at $x = 1$. However, since $2(1) \neq \frac{-1}{(1)^2}$, the derivatives of the left and right pieces are not equal at $x = 1$. Therefore f is not differentiable at $x = 1$, that is, $f'(1)$ does not exist. Thus the derivative of f is

$$f'(x) = \begin{cases} 2x, & \text{if } x < 1 \\ \text{DNE}, & \text{if } x = 1 \\ \frac{-1}{x^2}, & \text{if } x > 1. \end{cases}$$

□

CHECKING THE ANSWER

Consider the graphs of f and f' that follow. Note that f is indeed continuous, but not differentiable, at $x = 1$. In the graph of f' we have a hole at $x = 1$. We can see in *both* graphs the fact that $f'_{-}(1) = 2$ but $f'_{+}(1) = -1$. We can also see that, except at $x = 1$, the slopes of $f(x)$ do seem to be the same as the heights of $f'(x)$.

f is not differentiable at $x = 1$

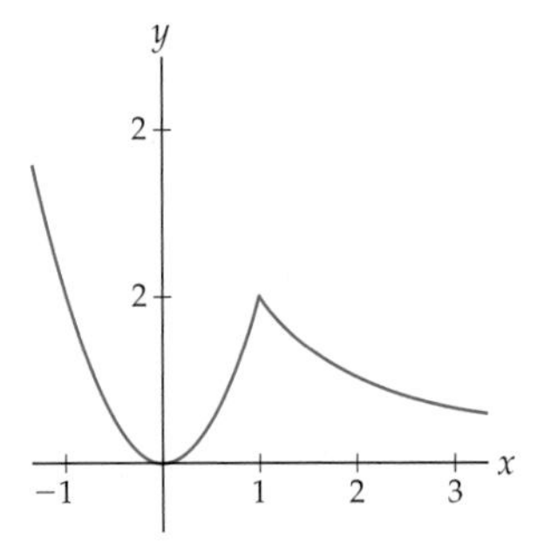

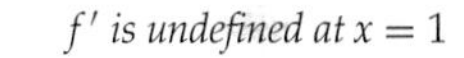
f' is undefined at $x = 1$

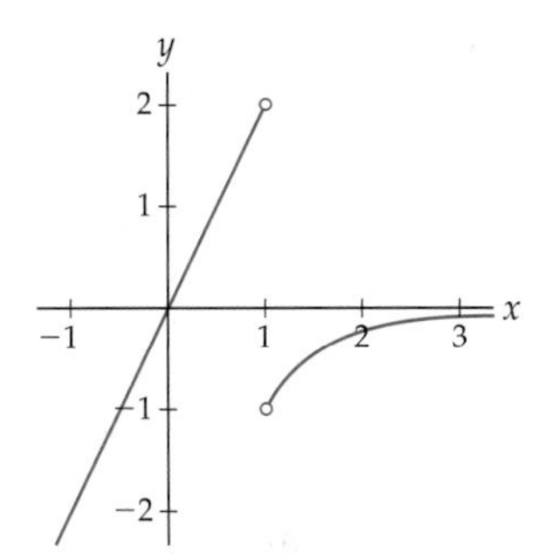

EXAMPLE 4

Finding antiderivatives

An ***antiderivative*** of a function $g(x)$ is a function whose derivative is $g(x)$. In this problem you may assume that any two functions with the same derivative must differ by a constant. (We will prove this fact in Section 3.2.)

(a) Find one antiderivative of $g(x) = 8x^3$.

(b) Describe the set of antiderivatives $g(x) = 8x^3$.

(c) Find the one antiderivative of $g(x) = 8x^3$ that passes through $(1, 4)$.

SOLUTION

(a) We can use a targeted guess-and-check method to find an antiderivative of $g(x) = 8x^3$. Since differentiating a power function decreases its power by one, we might start with the function $f(x) = x^4$, whose derivative is $f'(x) = 4x^3$; this is almost what we want, but off by a factor of two. A good second guess would be $f(x) = 2x^4$, and indeed its derivative is $f'(x) = 8x^3$, as desired.

(b) We now have one function whose derivative is $f'(x) = 8x^3$. Since any two functions with that derivative must differ by a constant, the functions whose derivative is $f'(x) = 8x^3$ are all the functions of the form $f(x) = 2x^4 + C$ for some real number

C, as shown in the next figure at the left. The graph corresponding to our previous choice of $f(x) = 2x^4$, with $C = 0$, is shown in black.

One antiderivative has $C = 0$

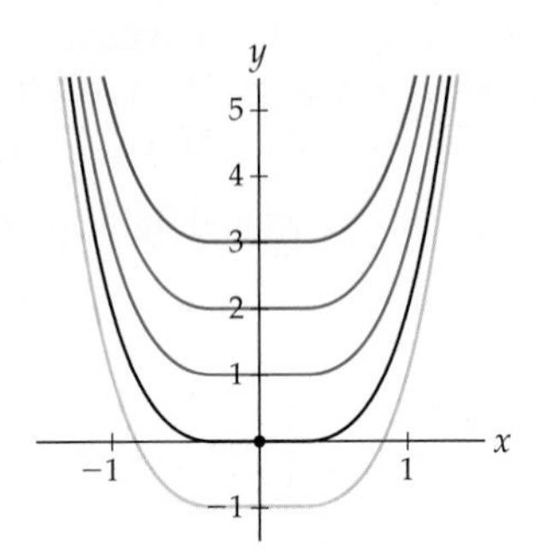

One antiderivative has $f(1) = 4$

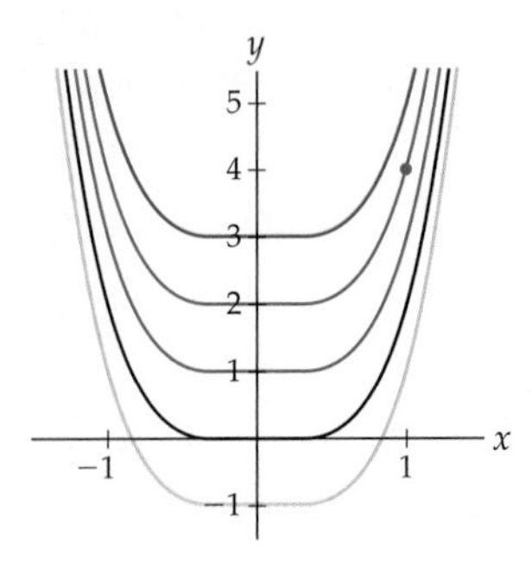

(c) Only one of the antiderivatives of $f'(x) = 8x^3$ passes through the point $(1, 4)$, as shown in red at the right. To find this antiderivative we set $f(1) = 4$ and solve for C:

$$f(1) = 4 \implies 2(1)^4 + C = 4 \implies C = 2.$$

Therefore the only antiderivative of $f'(x) = 8x^3$ that passes through $(1, 4)$ is the function $f(x) = 2x^4 + 2$. □

? TEST YOUR UNDERSTANDING

- Why are the rules for differentiating constants and the identity function special cases of the rule for differentiating a linear function?
- What is the difference between the rule for differentiating a constant function and the constant multiple rule?
- Can you find examples of functions f and g such that the derivative of their quotient is not the same as the quotient of their derivatives?
- Why do we add and subtract $f(x)g(x + h)$ in the calculation in the proof of the product rule? What does that enable us to do in the calculation?
- What sorts of functions can we differentiate with the rules in this section? Are there any functions that we can't differentiate with these rules?

EXERCISES 2.3

Thinking Back

Factoring formulas: Pull out a linear factor from each of the following expressions.

- $z^2 - 100$
- $z^3 - 27$
- $z^6 - 64$

Definition-of-derivative calculations: Use the definition of the derivative to find f' for each function.

- $f(x) = x^4$
- $f(x) = x^{-2}$
- $f(x) = x^2(x + 1)$
- $f(x) = \dfrac{x^2}{x + 1}$

Associated slope functions: For each of the following two function graphs, sketch a careful, labeled graph of its associated slope function.

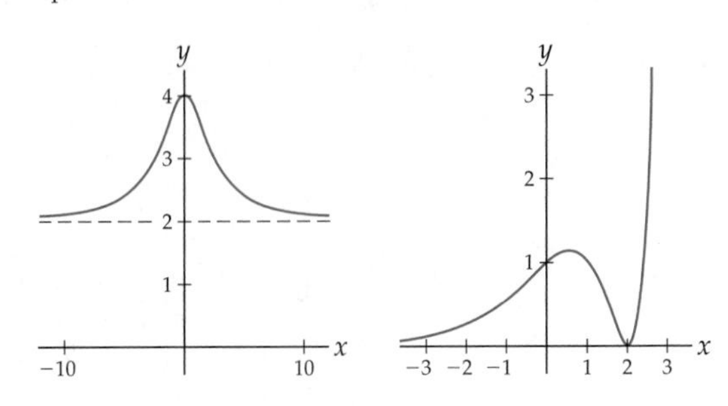

Concepts

0. *Problem Zero:* Read the section and make your own summary of the material.

1. *True/False:* Determine whether each of the statements that follow is true or false. If a statement is true, explain why. If a statement is false, provide a counterexample.

(a) *True or False:* $\frac{d}{dx}(5) = 0$.

(b) *True or False:* $\frac{d}{dr}(ks + r) = k$.

(c) *True or False:* $\frac{d}{ds}(ks + r) = k$.

(d) *True or False:* $\frac{d}{dx}(3x + 1)^k = k(3x + 1)^{k-1}$.

(e) *True or False:* $\frac{d}{dx}\left(\frac{1}{x^3}\right) = \frac{1}{3x^2}$.

(f) *True or False:* If f and g are differentiable functions, then $(f(x)g(x))' = g'(x)f(x) + f'(x)g(x)$.

(g) *True or False:* If g and h are differentiable functions, then $\left(\frac{g(x)}{h(x)}\right)' = \frac{h(x)g'(x) - g(x)h'(x)}{(h(x))^2}$.

(h) *True or False:* Proving the sum rule for differentiation involves the definition of the derivative, a lot of algebraic manipulation, and the sum rule for limits.

2. *Examples:* Construct examples of the thing(s) described in the following. Try to find examples that are different than any in the reading.

(a) Functions f and g, which illustrate that, in general, derivatives and products do not commute.

(b) Functions f and g, which illustrate that, in general, derivatives and quotients do not commute.

(c) Three functions whose derivatives we cannot calculate by using the differentiation rules we have developed so far.

3. Express the constant multiple, sum, and difference rules in Leibniz/operator notation.

4. Express the product and quotient rules in Leibniz/operator notation.

5. Why does it make graphical sense that the derivative of a constant is zero? That the derivative of the identity function is constantly equal to 1? That the derivative of a linear function $f(x) = mx + b$ is equal to m?

6. Why does the proof of the power rule (Theorem 2.9) in this text work only when k is an integer? Also, at which point in the proof do we use the fact that k is positive?

7. Explain why the power rule does *not* say that the derivative of 3^x is $x3^{x-1}$. Specifically, why doesn't 3^x fit the pattern required for the power rule to apply?

8. Explain why the power rule cannot be used to differentiate the function $(2 - x)^{1/3}$.

9. Use the product rule to find and state the rule for differentiating a product of *three* functions f, g, and h. In other words, fill in the blank: $(f(x)g(x)h(x))' = $ ________ . Then use your rule to differentiate the function $y = (2x - 1)(x^2 + x + 1)(1 - 3x^4)$. Check your answer by differentiating the function $y(x)$ another way.

10. Two operations commute if they can be done in either order. Does multiplying a function by a constant commute with differentiation? Does adding two functions commute with differentiation? What about products and quotients and the operation of differentiation?

Given that f, g, and h are functions with values $f(2) = 1$, $g(2) = -4$, and $h(2) = 3$ and point-derivatives $f'(2) = 3$, $g'(2) = 0$, and $h'(2) = -1$, calculate

11. $(3f + 4h)'(2)$

12. $(2f + 3g - h)'(2)$

13. $(fh)'(2)$

14. $\left(\frac{f}{g}\right)'(2)$

15. In the text of this section we displayed graphs of $f(x) = x^4$ and its first five derivatives. Use the slope-height behavior of the graphs to verify that each is the associated slope function of the one before.

16. Sketch graphs of $f(x) = \frac{1}{x}$ and its first five derivatives. Then use the slope-height behavior of the graphs to verify that each is the associated slope function of the one before.

17. In Example 2(b) we calculated a derivative two different ways. Use algebra to simplify the first answer and show that it is equal to the second.

18. In Example 2(c) we calculated a derivative two different ways. Use algebra to simplify the first answer and show that it is equal to the second.

19. Suppose f is a piecewise-defined function, equal to $g(x)$ if $x < 2$ and $h(x)$ if $x \geq 2$, where g and h are continuous and differentiable everywhere. If $g(2) = h(2)$, is the function f necessarily differentiable at $x = 2$? Why or why not?

20. Suppose f is a piecewise-defined function, equal to $g(x)$ if $x < 2$ and $h(x)$ if $x \geq 2$, where g and h are continuous and differentiable everywhere. If $g'(2) = h'(2)$, is the function f necessarily differentiable at $x = 2$? Why or why not?

21. If possible, find constants a and b so that the function f that follows is continuous and differentiable everywhere. If it is not possible, explain why not.

$$f(x) = \begin{cases} 3x + a, & \text{if } x < 1 \\ x^{b/2}, & \text{if } x \geq 1. \end{cases}$$

22. If possible, find constants a and b so that the function f that follows is continuous and differentiable everywhere. If it is not possible, explain why not.

$$f(x) = \begin{cases} ax - b, & \text{if } x < 2 \\ bx^2 + 1, & \text{if } x \geq 2. \end{cases}$$

Skills

Suppose $g(x)$, $h(x)$, and $j(x)$ are differentiable functions with the values of the function and its derivative given in the following table:

x	$g(x)$	$h(x)$	$j(x)$	$g'(x)$	$h'(x)$	$j'(x)$
−1	3	0	1	−1	−2	−2
0	2	3	0	−2	3	−2
1	0	−1	−2	−2	−2	−1
2	−2	−2	−3	−1	0	2
3	−3	0	1	0	1	2

Use the table to calculate the values of the derivatives listed in Exercises 23–28.

23. If $f(x) = 3j(x) - 2h(x)$, find $f'(3)$.

24. If $f(x) = g(x)h(x)$, find $f'(0)$.

25. If $f(x) = g(x)(h(x) + j(x))$, find $f'(2)$.

26. If $f(x) = h(x)g(x)j(x)$, find $f'(1)$.

27. If $f(x) = \dfrac{3h(x)}{g(x) + j(x)}$, find $f'(0)$.

28. If $f(x) = \dfrac{g(x)h(x) + j(x)}{h(x)}$, find $f'(-1)$.

Differentiate each of the functions in Exercises 29–34 in two different ways: first with the product and/or quotient rules and then without these rules. Then use algebra to show that your answers are the same.

29. $f(x) = x^2(x + 1)$

30. $f(x) = \dfrac{3x + 1}{x^4}$

31. $f(x) = x^{7/2}(2 - 5x^3)$

32. $f(x) = \sqrt{x}\,(x^{-1} + 1)$

33. $f(x) = \dfrac{x^2 - x^3}{\sqrt{x}}$

34. $f(x) = \dfrac{\sqrt{x}}{x^{-2}x^3}$

Use the differentiation rules developed in this section to find the derivatives of the functions in Exercises 35–64. Note that it may be necessary to do some preliminary algebra before differentiating.

35. $f(x) = 4 - 3x^7$

36. $f(x) = 2x - 3 + 4x^2$

37. $f(x) = 2(1 + 3x^2)$

38. $f(x) = 5x^3 - 2x^2 + 7$

39. $f(x) = 2(3x+1) - 4x^5$

40. $f(x) = x^2 + x(2 - 3x^2)$

41. $f(x) = (x + 2)(x - 1)$

42. $f(x) = (x^2 - 3)^2$

43. $f(x) = (3x + 2)^3$

44. $f(x) = (3 - x)^2 + 5$

45. $f(x) = \dfrac{1 - 6x^3}{3}$

46. $f(x) = \dfrac{x}{x + 1}$

47. $f(x) = \dfrac{x^2 - 1}{x + 1}$

48. $f(x) = \dfrac{x^4 - 7x^3}{2x}$

49. $f(x) = \pi^2$

50. $f(x) = (\sqrt{x} - \sqrt[3]{x})^2$

51. $f(x) = \dfrac{x^2}{\sqrt[5]{x^2}}$

52. $f(x) = \dfrac{x - 1}{(x + 1)(x + 2)}$

53. $f(x) = \dfrac{\sqrt[5]{x^7} - 2x^4}{x^3}$

54. $f(x) = (3x\sqrt{x})^{-2}$

55. $f(x) = \dfrac{x^7 - 3x^5 + 4}{1 - 3x^4}$

56. $f(x) = \dfrac{x^3 + x - 1}{x^2 - 7}$

57. $f(x) = \sqrt{\dfrac{1}{x^3}} + \left(\dfrac{1}{\sqrt{x}}\right)^3$

58. $f(x) = \dfrac{x^2 - 3x}{x^2 - 2x + 1}$

59. $f(x) = \dfrac{2x - 3}{5x + 4}$

60. $f(x) = x^3\sqrt{x}\,(x^{2/3})$

61. $f(x) = \dfrac{1}{(x - 2)(x - 3)}$

62. $f(x) = \dfrac{1}{(x + 1)^3}$

63. $f(x) = \dfrac{x^2}{x^3 + 5x^2 - 3x}$

64. $f(x) = \dfrac{(x - 2)^2}{(x^2 + 1)(x - 3)}$

Find the derivatives of each of the absolute value and piecewise-defined functions in Exercises 65–72.

65. $f(x) = |x|$

66. $f(x) = |3x + 1|$

67. $f(x) = |1 - 2x|$

68. $f(x) = |x^2 - 1|$

69. $f(x) = \begin{cases} x^3, & \text{if } x < 1 \\ x, & \text{if } x \geq 1 \end{cases}$

70. $f(x) = \begin{cases} 1, & \text{if } x \leq -1 \\ x^{2/3}, & \text{if } x > -1 \end{cases}$

71. $f(x) = \begin{cases} -x^2, & \text{if } x \leq 0 \\ x^2, & \text{if } x > 0 \end{cases}$

72. $f(x) = \begin{cases} 3x + 1, & \text{if } x \leq 1 \\ x^3, & \text{if } x > 1 \end{cases}$

In Exercises 73–78, find a function that has the given derivative and value. In each case you can find the answer with an educated guess-and-check process. It may be helpful to do some preliminary algebra.

73. $f'(x) = 3x^5 - 2x^2 + 4$, $f(0) = 1$

74. $f'(x) = 7x^2 + 8x^{11} - 18$, $f(0) = -2$

75. $f'(x) = 1 - 4x^6$, $f(1) = 3$

76. $f'(x) = x(4 - 2x)$, $f(0) = 0$

77. $f'(x) = (x^4 - 8)(1 - 3x^5)$, $f(0) = 2$

78. $f'(x) = (3x + 1)^3$, $f(2) = 1$

Applications

79. A spaceship is moving along a straight path from Venus into the heart of the Sun. The velocity of the spaceship t hours after leaving Venus is $v(t) = 0.012t^2 + 400$ thousands of miles per hour. (To simplify matters we will pretend that Venus is not moving with respect to the sun; you may assume that everything is fixed in place in this exercise.)

(a) Say what you can about the initial values s_0, v_0, and a_0, and then use derivatives and antiderivatives to find equations for the position and acceleration of the spaceship.

(b) Is the spaceship always moving towards the sun? How can you tell?

(c) Is the spaceship travelling at a constant acceleration? Is it speeding up or slowing down, or neither? How can you tell?

(d) The distance between Venus and the sun is about 67 million miles. How long will it take the spaceship to reach the sun? How fast will the spaceship be going when it gets there?

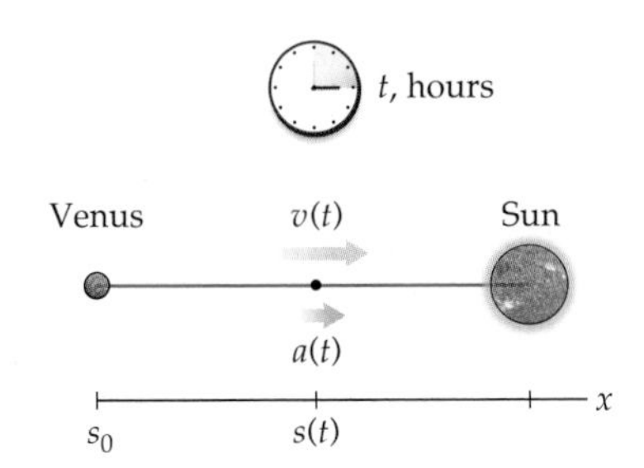

80. A bowling ball is thrown down from a 20th-story window. After 3 seconds, the bowling ball is 26 feet from the ground and falling at a rate of -106 feet per second (downwards). You may assume that gravity causes a constant downward acceleration of -32 feet per second.

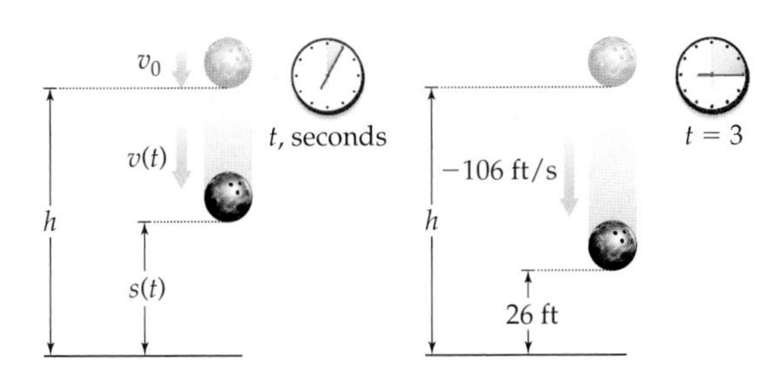

(a) If the height $s(t)$ of the bowling ball t seconds after being thrown is given by a quadratic polynomial function, use $s(3)$, $s'(3)$, and $s''(3)$ to find an equation for $s(t)$.

(b) How high is the 20th-story window from which the bowling ball was thrown?

(c) How fast was the bowling ball initially thrown?

81. On Earth, a falling object has a downward acceleration of -32 feet per second per second due to gravity. Suppose an object falls from an initial height of s_0 feet, with an initial velocity of v_0 feet per second. Use antiderivatives to show that the equations for the position and velocity of the falling object after t seconds are, respectively, $s(t) = -16t^2 + v_0 t + s_0$ and $v(t) = -32t + v_0$.

82. In a science fiction novel, gravity on the planet XV-37 acts very differently than it does in our universe. An object dropped from a 1000-foot building on planet XV-37 will have a downward gravitational acceleration of $a(t) = -6t$ feet per second per second after falling for t seconds. Use antiderivatives to find equations for the position and velocity of such a falling object. What might be the consequences of living on this strange planet?

83. In another science fiction novel, gravity on planet Xillian again acts very strangely. The height $s(t)$ of a falling object on Xillian is always a cubic polynomial function. Suppose a kiwi fruit is dropped (with initial velocity of zero) from the top of a Xillian radio tower. After 5 seconds, the kiwi fruit is 100 feet from the ground and falling at a rate of -200 feet per second. The acceleration of the kiwi fruit at that moment is -46 feet per second per second.

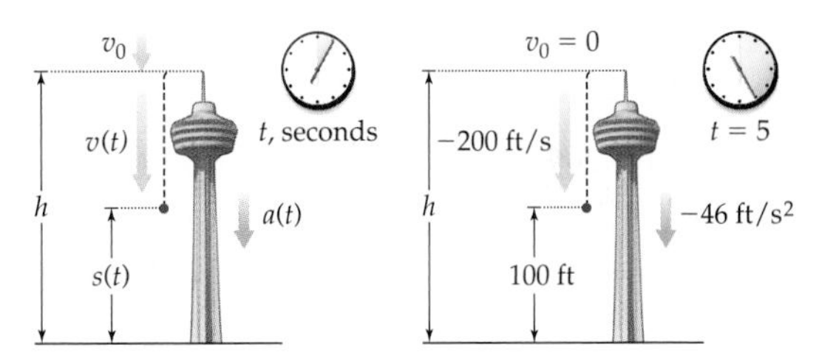

(a) Use the values of $s'(0)$ (note the kiwi initially has velocity zero), $s(5)$, $s'(5)$, and $s''(5)$ given in the preceding description to find a formula for the height $s(t)$ of the kiwi fruit t seconds after being dropped from the Xillian radio tower. Specifically, use these four data points to solve for the coefficients of a cubic polynomial $s(t) = at^3 + bt^2 + ct + d$.

(b) Verify that the function $s(t)$ you just found produces the correct values for $s'(0)$, $s(5)$, $s'(5)$, and $s''(5)$ in this exercise.

(c) How high is the Xillian radio tower from which the kiwi fruit was dropped?

(d) On Earth, acceleration due to gravity is given by a constant -32 feet per second per second, that is, on Earth we always have $a(t) = -32$ for falling objects. What is the function $a(t)$ for acceleration due to Xillian "gravity"? Is this acceleration constant? What are the physical implications of gravity on Xillian?

Proofs

84. Use the definition of the derivative to directly prove the differentiation rules for constant and identity functions.

85. Use the $h \to 0$ definition of the derivative to prove that the power rule holds for positive integer powers.

86. Prove, in two ways, that the power rule holds for negative integer powers:

(a) By using the $z \to x$ definition of the derivative.

(b) By using the $h \to 0$ definition of the derivative.

87. Prove the difference rule in two ways:

(a) Using the definition of the derivative.

(b) Using the sum and constant multiple rules.

88. Use the definition of the derivative to prove the following special case of the product rule:

$$\frac{d}{dx}(x^2 f(x)) = 2xf(x) + x^2 f'(x).$$

89. Use the definition of the derivative to prove the quotient rule. *(Hint: At some point you will have to add and subtract the quantity $f(x)g(x)$ from the numerator.)*

90. The following ***reciprocal rule*** tells us how to differentiate the reciprocal of a function:

$$\frac{d}{dx}\left(\frac{1}{f(x)}\right) = -\frac{f'(x)}{(f(x))^2}.$$

Prove the reciprocal rule in two ways:

(a) By using the definition of the derivative.

(b) By using the quotient rule.

91. Consider the following formula for antidifferentiating power functions: If $f'(x) = x^k$ and $k \neq -1$, then $f(x) = \frac{1}{k+1}x^{k+1} + C$ for some constant C.

(a) Prove this antidifferentiation formula. You may assume that any two functions with the same derivative differ by a constant, as we will prove in Section 3.2.

(b) What part of your argument from part (a) breaks down when $f(x) = x^{-1}$?

92. Consider the piecewise-defined function

$$f(x) = \begin{cases} g(x), & \text{if } x \leq c \\ h(x), & \text{if } x > c. \end{cases}$$

Prove that if $g(x)$ and $h(x)$ are continuous and differentiable at $x = c$, and if $g(c) = h(c)$ and $g'(c) = h'(c)$, then f is differentiable at $x = c$.

Thinking Forward

For each function f that follows, find all of the x-values in the domain of f for which $f'(x) = 0$ and all values for which $f'(x)$ does not exist. In later sections we will call these values the ***critical points*** of f.

- $f(x) = x^3 - 2x$
- $f(x) = \sqrt{x} - x$
- $f(x) = \dfrac{1}{1+\sqrt{x}}$
- $f(x) = \dfrac{x^2(x-1)}{(x-2)^2}$

Taylor polynomials: In the exercises that follow, you will investigate the relationship between the coefficients of a polynomial and its higher order derivatives. In later chapters this same sort of idea will be used to locally approximate differentiable functions with polynomials.

- Prove that if f is any quadratic polynomial function $f(x) = ax^2 + bx + c$, then the coefficients of f are completely determined by the values of $f(x)$ and its derivatives at $x = 0$, as follows:

$$a = \frac{f''(0)}{2}, \quad b = f'(0), \quad \text{and} \quad c = f(0).$$

  In particular, this means that if f is any quadratic polynomial function, then $f(x) = \frac{f''(0)}{2}x^2 + f'(0)x + f(0)$.

- Prove that if f is any cubic polynomial function $f(x) = ax^3 + bx^2 + cx + d$, then the coefficients of f are completely determined by the values of $f(x)$ and its derivatives at $x = 0$, as follows:

$$a = \frac{f'''(0)}{6}, \quad b = \frac{f''(0)}{2}, \quad c = f'(0), \quad \text{and} \quad d = f(0).$$

  This means that every cubic polynomial function can be written $f(x) = \frac{f'''(0)}{6}x^3 + \frac{f''(0)}{2}x^2 + f'(0)x + f(0)$.

- Suppose f is any cubic polynomial function $f(x) = ax^3 + bx^2 + cx + d$. Prove that the coefficients a, b, c, and d of f can be expressed in terms of the values of $f(x)$ and its derivatives at the point $x = 2$. *(Hint: In other words, show that you can write the coefficients a, b, c, and d in terms of $f(2)$, $f'(2)$, $f''(2)$, and $f'''(2)$.)*

- Suppose f is a polynomial of degree n, and let k be some integer with $0 \leq k \leq n$. Prove that if $f(x)$ is of the form

$$f(x) = a_n x^n + a_{n-1}x^{n-1} + \cdots + a_k x^k + \cdots + a_1 x + a_0,$$

  then $a_k = \frac{f^{(k)}(0)}{k!}$, where $f^{(k)}(x)$ is the kth derivative of $f(x)$ and $k! = k(k-1)\cdots(2)(1)$. *(Hint: Find $f^{(k)}(x)$, and use it to show that $f^{(k)}(0) = k!a_k$.)*

2.4 THE CHAIN RULE AND IMPLICIT DIFFERENTIATION

- The chain rule for differentiating compositions of functions
- Implicit functions and implicit differentiation
- Using implicit differentiation to prove derivative formulas

Differentiating Compositions of Functions

With the rules we have developed so far, we can differentiate any arithmetic combination of functions whose components have known derivatives. There is, however, one other way that functions can be combined: by composition. We still don't know how to differentiate, say, the composition $y = (x^2 + 1)^{1/2}$. Is the derivative of a composition $f(g(x))$ somehow related to the derivatives of f and g?

Let's consider a simple example involving rates of change. Suppose you own a small factory that makes 30 widgets an hour and you make a profit of \$10.00 for each widget made. This means that you can make a profit of \$300.00 an hour by making widgets at your factory. We arrive at this answer by using the product:

$$300\,\frac{\text{dollars}}{\text{hour}} = \left(10\,\frac{\text{dollars}}{\text{widget}}\right)\left(30\,\frac{\text{widgets}}{\text{hour}}\right).$$

What does this have to do with the derivative of a composition? Let $w(t)$ be the number of widgets you have t hours after starting production, and let $p(w)$ be the profit made from producing w widgets. Then the composition $p(w(t))$ gives the profit made after t hours. We are interested in the rate of change of profit per hour, which is the derivative $\frac{dp}{dt}$. The fact that you make a profit of $\frac{dp}{dw} = 10$ dollars per widget and the fact that your factory can make $\frac{dw}{dt} = 30$ widgets per hour are also statements about derivatives. Repeating our previous calculation, we can relate the derivative of the composition $p(w(t))$ to the derivatives of $p(w)$ and $w(t)$:

$$\frac{dp}{dt} = \frac{dp}{dw}\frac{dw}{dt}.$$

This example suggests that the derivative of a composition is the product of the derivatives of the component functions. That is in fact the case in general:

THEOREM 2.12 **The Chain Rule**

Suppose $f(u(x))$ is a composition of functions. Then for all values of x at which u is differentiable at x and f is differentiable at $u(x)$, the derivative of f with respect to x is equal to the product of the derivative of f with respect to u and the derivative of u with respect to x.

In Leibniz notation, we write this as

$$\frac{df}{dx} = \frac{df}{du}\frac{du}{dx}.$$

In "prime" notation, we write it as

$$(f \circ u)'(x) = f'(u(x))\,u'(x).$$

To use the chain rule, we must first recognize a function as a composition $f(u(x))$ and identify the "outside" and "inside" functions f and u. For example, the function $y = (x^2 + 1)^{1/2}$ can be thought of as a composition $y = f(u(x))$ with inside function $u(x) = x^2 + 1$ and outside function $f(u) = u^{1/2}$. The chain rule says that we should differentiate the outside function f with respect to u and then multiply the result by the derivative of the inside function:

$$(f(u(x)))' = f'(u(x))\,u'(x) = \frac{1}{2}(u(x))^{-1/2}u'(x) = \frac{1}{2}(x^2 + 1)^{-1/2}(2x).$$

For more examples of how to apply the chain rule, see Examples 1 and 2. Most of the examples of the chain rule in this section will be fairly simple. We will do more complicated examples in later sections after we have seen how to differentiate exponential, logarithmic, and trigonometric functions.

The "prime" notation in Theorem 2.12 makes clear why we want u to be differentiable at x and f to be differentiable at $u(x)$. When it suits our purposes, we can also write $(f(u(x)))'$ for $(f \circ u)'(x)$. The Leibniz notation suggests why we call it the "chain" rule: We are taking the derivative of a chain of functions by multiplying a chain of derivatives. If we had a longer chain of functions, then we would have a longer chain of derivatives; for example, the derivative of the composition $f(u(v(x))$ with respect to x is given by the chain of derivatives

$$\frac{df}{dx} = \frac{df}{du}\frac{du}{dv}\frac{dv}{dx}.$$

The chain rule seems sensible if you consider its analog with difference quotients, since we can cancel Δu's to justify the equation $\frac{\Delta f}{\Delta x} = \frac{\Delta f}{\Delta u}\frac{\Delta u}{\Delta x}$. However, this cancellation does not automatically apply to the Leibniz notation, since the differentials df, dx, and du do not represent numerical quantities. They cannot be cancelled just because the notation makes it look tempting to do so. The proof of the chain rule requires more work than simply "cancelling," but except for a certain technical point, is not that difficult.

Proof. Our proof will start with the definition of the derivative for $(f \circ g)$. After some algebra and a limit rule, a change of variables will give us the result we desire. In the middle of the calculation we will make a simplifying substitution:

$$\begin{aligned}
(f \circ g)'(x) &= \lim_{h\to 0} \frac{f(g(x+h)) - f(g(x))}{h} && \leftarrow \text{derivative}\\
&= \lim_{h\to 0} \left(\frac{f(g(x+h)) - f(g(x))}{g(x+h) - g(x)}\,\frac{g(x+h) - g(x)}{h}\right) && \leftarrow \text{algebra}\\
&= \left(\lim_{h\to 0} \frac{f(g(x+h)) - f(g(x))}{g(x+h) - g(x)}\right)\left(\lim_{h\to 0} \frac{g(x+h) - g(x)}{h}\right) && \leftarrow \text{product rule for limits}\\
&= \left(\lim_{k\to 0} \frac{f(g(x) + k) - f(g(x))}{k}\right)(g'(x)) && \leftarrow \text{see below}\\
&= f'(g(x))\,g'(x) && \leftarrow \text{derivative}
\end{aligned}$$

There is one technical point to consider in this proof. In the fourth step we applied the substitution $k = g(x+h) - g(x)$. Since $g(x)$ is differentiable, it is also continuous; therefore as $h \to 0$, we also have $k = g(x+h) - g(x) \to 0$. However, the preceding calculation assumes that $k = g(x+h) - g(x)$ is nonzero for small enough h, which in some situations may not be the case. In the case when this happens because $g(x+h) = g(x)$ as $h \to 0$, it is easy to show that $(f \circ g)'(x)$ and $f'(g(x))g'(x)$ are both zero and therefore equal. There are other cases when this can happen, but handling the details for those functions is beyond the scope of this course, so we will assume here that $g(x)$ is nice enough that we can avoid those cases. ∎

Implicit Differentiation

Consider the equation $x^2+y^2=1$ that describes the circle of radius 1 centered at the origin, as shown next at the left. Clearly, this graph does not represent a function, since it does not pass the vertical line test. However, *locally*, that is, in small pieces, the graph does define y as a function of x. For example, the top half of the graph shown in the middle figure does represent a function, as does the graph of the bottom half shown at the right.

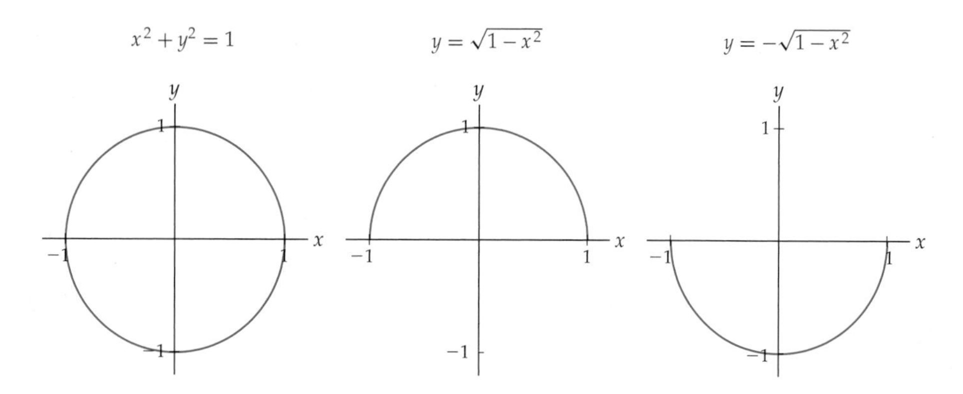

Although we cannot solve the equation $x^2+y^2=1$ for y and obtain a single well-defined function, we can still think of the x-values as inputs and the y-values as outputs; the only difference is that there may be more than one y-value for each x-value. In cases such as these, we say that y is an ***implicit function*** of x.

Thinking locally, we can use our usual differentiation techniques to show that the unit circle has horizontal tangent lines at $(0,1)$ and at $(0,-1)$. Looking at the top of the circle, we have $y=\sqrt{1-x^2}$, with derivative

$$y'=\frac{1}{2}(1-x^2)^{-1/2}(-2x)=\frac{-x}{\sqrt{1-x^2}},$$

which is clearly zero only when $x=0$. Therefore the unit circle has a horizontal tangent line at the point $(0,\sqrt{1-0^2})=(0,1)$. In a similar fashion we could use the equation $y=-\sqrt{1-x^2}$ for the bottom half of the circle to find the other horizontal tangent line.

We will not always be able to divide an implicit function into pieces given by actual functions whose equations we know. That is, we will not always be able to "solve for y." However, given an equation that defines an implicit function, we can still find information about slopes and derivatives simply by differentiating both sides of the equation with respect to x. This technique is known as ***implicit differentiation***. The key to implementing the technique will be applying the chain rule appropriately.

For example, we can differentiate both sides of the equation $x^2+y^2=1$ from the previous example. Along the way we will have to remember that we are thinking of $y=y(x)$ as a function of x and apply the chain rule:

$$x^2+y^2=1 \qquad \leftarrow \text{given equation}$$

$$\frac{d}{dx}(x^2+(y(x))^2)=\frac{d}{dx}(1) \qquad \leftarrow \text{differentiate both sides}$$

$$2x+2y\frac{dy}{dx}=0 \qquad \leftarrow \text{power and chain rules}$$

$$\frac{dy}{dx}=\frac{-2x}{2y}=-\frac{x}{y}. \qquad \leftarrow \text{solve for } \frac{dy}{dx}$$

The use of the chain rule in the differentiation step is crucial to these types of calculations. Notice that the derivative of x^2 with respect to x is the familiar $2x$, but the derivative of y^2 with respect to x uses the chain rule to arrive at $2y\,\frac{dy}{dx}$. Here y is thought of as an implicit function of x, so the expression y^2 is really the composition of the implicit function $y(x)$ with the squaring function. That is why the chain rule must come into play when we differentiate y^2 with respect to x.

The fact that $\frac{dy}{dx} = -\frac{x}{y}$ means that for any point (a, b) that lies on the graph of the circle $x^2 + y^2 = 1$, the slope of the tangent line to the circle at (a, b) is given by $-\frac{a}{b}$. For example, the point $\left(\frac{1}{2}, \frac{\sqrt{3}}{2}\right)$ is on the graph of the circle because $\left(\frac{1}{2}\right)^2 + \left(\frac{\sqrt{3}}{2}\right)^2 = 1$ and the slope of the line tangent to the circle at that point is $\frac{-1/2}{\sqrt{3}/2} = -\frac{1}{\sqrt{3}}$. Similarly, at the point $\left(-\frac{1}{2}, -\frac{\sqrt{3}}{2}\right)$ the slope of the tangent line is also $-\frac{1}{\sqrt{3}}$, as shown next at the left. At the point $(0, 1)$ on the circle, the tangent line has slope $-\frac{0}{1} = 0$ and is thus horizontal, and at the point $(1, 0)$ the tangent line has undefined slope $-\frac{1}{0}$ and thus is vertical, as shown at the right.

Slope at (a, b) given by $-\frac{a}{b}$

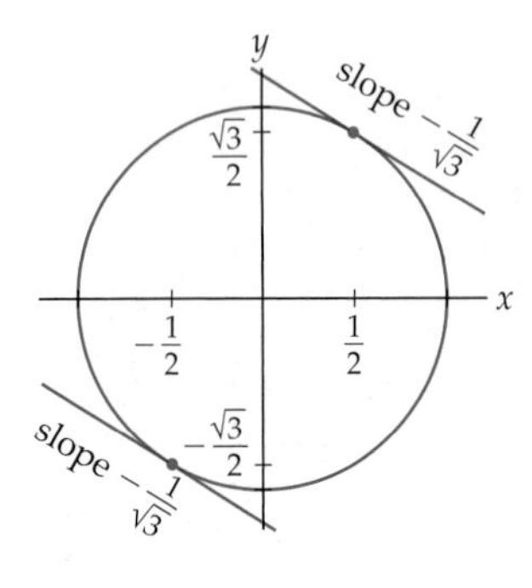

Horizontal and vertical tangent lines

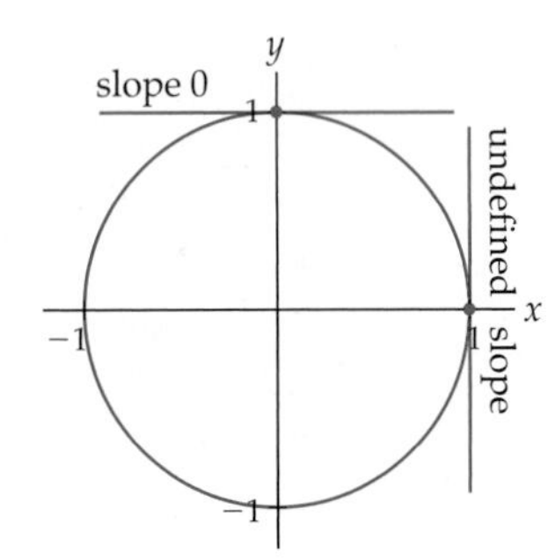

Examples and Explorations

EXAMPLE 1

Applying the chain rule one step at a time

Use the chain rule to differentiate $h(x) = \left(\frac{x}{1 - 3x^2}\right)^4$, writing out all steps.

SOLUTION

We can think of $h(x)$ as a composition $f(u(x))$ with inside function $u(x) = \frac{x}{1 - 3x^2}$ and outside function $f(u) = u^4$. Applying the chain rule, we have

$$\begin{aligned}
h'(x) &= f'(u(x))u'(x) && \leftarrow \text{chain rule} \\
&= 4(u(x))^3 u'(x) && \leftarrow \text{derivative of outside function} \\
&= 4\left(\frac{x}{1 - 3x^2}\right)^3 u'(x) && \leftarrow \text{evaluate at inside function} \\
&= 4\left(\frac{x}{1 - 3x^2}\right)^3 \left(\frac{1(1 - 3x^2) - x(-6x)}{(1 - 3x^2)^2}\right). && \leftarrow \text{derivative of inside function}
\end{aligned}$$

When you become comfortable using the chain rule, you won't write out each step separately as we just did; instead you will simply think something like "the derivative of the outside with the inside plugged in, times the derivative of the inside," and write out the derivative calculation in one step. Note that once again we have not simplified the derivative that we found. In this text we will not simplify such answers unless we have a specific use for the derivative in question. □

EXAMPLE 2

Determining which differentiation rule to apply first

For each function f, find the derivative f'. What is the *first* differentiation rule that you must use in each case?

(a) $f(x) = \left(\sqrt[3]{x} - \frac{1}{x}\right)^{-2}$ **(b)** $g(x) = \dfrac{\sqrt{x^2-1}}{1-\frac{1}{x}}$ **(c)** $h(x)=(x^2-1)^5(3x-1)^2$

SOLUTION

(a) The function f is at its outermost layer a composition of the form $(u(x))^{-2}$, where $u(x) = \sqrt[3]{x} - \frac{1}{x}$. Therefore, the first rule to use is the chain rule. The power rule is the *second* differentiation rule we must use, since we need to use the chain rule to even begin differentiating the function. Before applying the chain rule, we use algebra to rewrite the function f in a more convenient exponent form:

$$\frac{d}{dx}\left(\left(\sqrt[3]{x} - \frac{1}{x}\right)^{-2}\right) = \frac{d}{dx}((x^{1/3} - x^{-1})^{-2}) \qquad \leftarrow \text{algebra}$$

$$= -2(x^{1/3} - x^{-1})^{-3} \cdot \frac{d}{dx}(x^{1/3} - x^{-1}) \qquad \leftarrow \text{chain, power rules}$$

$$= -2(x^{1/3} - x^{-1})^{-3}\left(\frac{1}{3}x^{-2/3} - (-1)x^{-2}\right). \qquad \leftarrow \text{difference, power rules}$$

(b) The function $g(x)$ is at its outermost layer a quotient, so we will begin by using the quotient rule. Again we write roots and fractions as exponents first, so that the differentiation steps will be easier:

$$\frac{d}{dx}\left(\frac{\sqrt{x^2-1}}{1-(1/x)}\right) = \frac{d}{dx}\left(\frac{(x^2-1)^{1/2}}{1-x^{-1}}\right) \qquad \leftarrow \text{algebra}$$

$$= \frac{\frac{d}{dx}((x^2-1)^{1/2})(1-x^{-1}) - (x^2-1)^{1/2}\frac{d}{dx}(1-x^{-1})}{(1-x^{-1})^2} \qquad \leftarrow \text{quotient rule}$$

$$= \frac{\frac{1}{2}(x^2-1)^{-1/2}(2x)(1-x^{-1}) - (x^2-1)^{1/2}(0-(-1)x^{-2})}{(1-x^{-1})^2}. \qquad \leftarrow \text{more rules}$$

(c) The function $h(x)$ is at its outermost layer a product, so we begin by applying the product rule:

$$\frac{d}{dx}((x^2-1)^5(3x-1)^2) = \frac{d}{dx}((x^2-1)^5)(3x-1)^2 + (x^2-1)^5\frac{d}{dx}((3x-1)^2) \qquad \leftarrow \text{product rule}$$

$$= 5(x^2-1)^4(2x)(3x-1)^2 + (x^2-1)^5(2)(3x-1)(3). \leftarrow \text{more rules}$$

□

EXAMPLE 3

Implicit differentiation and the chain rule

Each of the equations that follow defines $y = y(x)$ as an implicit function of x. Use implicit differentiation to find $y' = y'(x)$.

(a) $3x^4 + 2y^4 + x + y = 0$ **(b)** $x^2y + xy^2 = 1$

SOLUTION

(a) Before we begin, note that $y = y(x)$ is an implicit function of x, so when we differentiate any expression involving y with respect to x, we will need the chain rule. We will write y and y' to avoid any confusion with multiplication here, but each time we do so we will think of the (implicit) functions $y(x)$ and $y'(x)$. Differentiating both sides of the given equation and then solving for y' gives

$$\frac{d}{dx}(3x^4 + 2y^4 + x + y) = \frac{d}{dx}(0) \qquad \leftarrow \text{differentiate both sides}$$

$$3(4x^3) + 2(4y^3)y' + 1 + y' = 0 \qquad \leftarrow \text{note } x \text{ and } y \text{ are dealt with differently}$$

$$8y^3y' + y' = -12x^3 - 1 \qquad \leftarrow \text{isolate } y' \text{ terms to one side}$$

$$y'(8y^3 + 1) = -12x^3 - 1 \qquad \leftarrow \text{factor out } y' \text{ on the left}$$

$$y' = \frac{-12x^3 - 1}{8y^3 + 1}. \qquad \leftarrow \text{divide to finish solving for } y'$$

(b) This time we will need the product rule, since x^2y is the product of the function x^2 and the implicit function y, and xy^2 is the product of the function x and the implicit function y^2. For the latter product we will also require the chain rule. The overall process is much the same as in the previous calculation:

$$\frac{d}{dx}(x^2y + xy^2) = \frac{d}{dx}(1) \qquad \leftarrow \text{differentiate both sides}$$

$$(2xy + x^2y') + (1y^2 + x(2yy')) = 0 \qquad \leftarrow \text{product and chain rules}$$

$$x^2y' + 2xyy' = -2xy - y^2 \qquad \leftarrow \text{isolate } y' \text{ terms to one side}$$

$$y'(x^2 + 2xy) = -2xy - y^2 \qquad \leftarrow \text{factor out } y' \text{ on the left}$$

$$y' = \frac{-2xy - y^2}{x^2 + 2xy}. \qquad \leftarrow \text{divide to finish solving for } y'$$

□

EXAMPLE 4

Finding values and tangent lines on the graph of an implicit function

Consider the equation $y^3 + xy + 2 = 0$ that defines y as an implicit function of x.

(a) Show that $y^3 - 5y + 2$ factors as $(y - 2)(y^2 + 2y - 1)$.

(b) If $x = -5$, find all the possible values for y.

(c) Find the slope of the line tangent to $y^3 + xy + 2 = 0$ at the point $(-5, 2)$.

SOLUTION

(a) We could factor $y^3 - 5y + 2$ by using synthetic division, which you may have seen in a previous course, but since the factorization is already given in the problem, all we need to do is multiply out and check:

$$(y - 2)(y^2 + 2y - 1) = y^3 + 2y^2 - y - 2y^2 - 4y + 2 = y^3 - 5y + 2.$$

(b) Given the factorization from part (a), we substitute $x = -5$ into the equation $y^3 + xy + 2 = 0$ and solve for all corresponding values of y:

$$y^3 - 5y + 2 = 0 \qquad \leftarrow \text{the equation with } x = -5$$

$$(y - 2)(y^2 + 2y - 1) = 0 \qquad \leftarrow \text{factorization in part (a)}$$

$$y = 2,\ y = -1 + \sqrt{2},\ \text{or } y = -1 - \sqrt{2}. \qquad \leftarrow \text{quadratic formula}$$

Thus $(-5, 2)$, $(-5, -1 + \sqrt{2})$, and $(-5, -1 - \sqrt{2})$ are points on the graph of $y^3 + xy + 2 = 0$.

(c) To find the slope of the tangent line we must first calculate $\frac{dy}{dx}$. Using implicit differentiation and the chain rule, we have

$$y^3 + xy + 2 = 0 \qquad \leftarrow \text{the given equation}$$

$$\frac{d}{dx}((y(x))^3 + x \cdot y(x) + 2) = \frac{d}{dx}(0) \qquad \leftarrow \text{differentiate both sides}$$

$$3y^2\frac{dy}{dx} + (1)(y) + (x)\left(\frac{dy}{dx}\right) + 0 = 0 \qquad \leftarrow \text{chain and product rules}$$

$$3y^2\frac{dy}{dx} + x\frac{dy}{dx} = -y \qquad \leftarrow \text{start solving for } \frac{dy}{dx}$$

$$\frac{dy}{dx}(3y^2 + x) = -y \qquad \leftarrow \text{algebra}$$

$$\frac{dy}{dx} = \frac{-y}{3y^2 + x}. \qquad \leftarrow \text{algebra}$$

This means that if a point (a, b) is on the graph of $y^3 + xy + 2 = 0$, then the line tangent to the graph of $y^3 + xy + 2 = 0$ at (a, b) has slope

$$\left.\frac{dy}{dx}\right|_{\substack{x=a\\y=b}} = \frac{-b}{3b^2 + a}.$$

Therefore the slope of the tangent line at the point $(a, b) = (-5, 2)$ is

$$\left.\frac{dy}{dx}\right|_{\substack{x=-5\\y=2}} = \frac{-2}{3(2)^2 + (-5)} = \frac{-2}{12 - 5} = \frac{-2}{7}.$$

□

✔ CHECKING THE ANSWER

We can use a graphing utility to sketch the graph of the implicit function from the previous example, as shown next at the left. Notice that for $x = -5$ there are three corresponding y-values, as we showed in part (b). At each of these points $(-5, y)$ there is a tangent line to the graph. From the graph at the right it does appear (taking axes scales into account) that the slope of the tangent line at $(-5, 2)$ could be approximately $-\frac{2}{7}$.

$y^3 + xy + 2 = 0$ and the points at $x = -5$

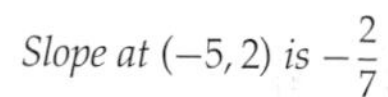
Slope at $(-5, 2)$ is $-\frac{2}{7}$

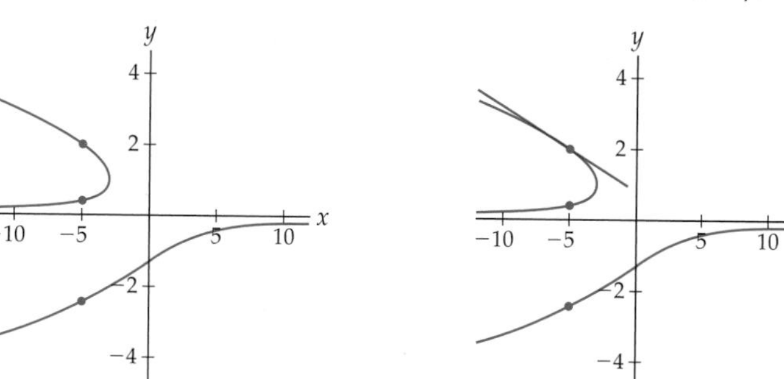

EXAMPLE 5

Using implicit differentiation to prove the general power rule

Although we have been using the power rule for general rational powers, we have in fact proved it only for integer powers.

(a) Use implicit differentiation and the power rule for integer powers to prove that $\frac{d}{dx}(x^{5/3}) = \frac{5}{3}x^{2/3}$.

(b) Use implicit differentiation and the power rule for integer powers to prove the power rule for rational powers.

SOLUTION

(a) If $y = x^{5/3}$, then $y^3 = x^5$. With this equation we can use implicit differentiation and the power rule for integer powers to find y':

$$y^3 = x^5 \qquad \leftarrow \text{since } y = x^{5/3}$$

$$\frac{d}{dx}((y(x))^3) = \frac{d}{dx}(x^5) \qquad \leftarrow \text{differentiate both sides}$$

$$3y^2y' = 5x^4 \qquad \leftarrow \text{chain rule and integer power rule}$$

$$y' = \frac{5x^4}{3y^2} = \frac{5x^4}{3(x^{5/3})^2} \qquad \leftarrow \text{solve for } y', \text{ substitute } y = x^{5/3}$$

$$y' = \frac{5}{3}x^{4-(10/3)} = \frac{5}{3}x^{2/3}. \qquad \leftarrow \text{algebra}$$

(b) Suppose $y = x^{p/q}$, where $\frac{p}{q}$ is a rational number. Generalizing the method used in part (a), we use implicit differentiation on the equation $y^q = x^p$ to solve for y':

$$y^q = x^p \qquad \leftarrow \text{since } y = x^{p/q}$$

$$\frac{d}{dx}((y(x))^q) = \frac{d}{dx}(x^p) \qquad \leftarrow \text{differentiate both sides}$$

$$qy^{q-1}y' = px^{p-1} \qquad \leftarrow \text{chain rule, integer power rule}$$

$$y' = \frac{px^{p-1}}{qy^{q-1}} = \frac{px^{p-1}}{q(x^{p/q})^{q-1}} \qquad \leftarrow \text{solve for } y', \text{ substitute } y = x^{p/q}$$

$$y' = \frac{px^{p-1}}{qx^{p(q-1)/q}} = \frac{p}{q}x^{((p-1)-(p(q-1)/q))} \qquad \leftarrow \text{algebra}$$

$$y' = \frac{p}{q}x^{p/q-1}. \qquad \leftarrow \text{even more algebra}$$

□

? TEST YOUR UNDERSTANDING

- What is the difference between $f'(g(x))$ and $(f(g(x)))'$?
- The chain rule formula $f'(u(x))u'(x)$ is not exactly the same as the product of the derivatives of f and u. What is the difference?
- What is the difference between saying that y is a function of x and saying that y is an implicit function of x?
- Suppose y is an implicit function of x. Would it be correct or incorrect to say that $\frac{d}{dx}(y^3) = 3y^2$, and why?
- If f is an invertible function with inverse f^{-1}, then what is the relationship between the derivatives of f and f^{-1}?

EXERCISES 2.4

Thinking Back

Differentiation review: Without using the chain rule, find the derivatives of each of the functions f that follow. Some algebra may be required before differentiating.

- $f(x) = (3x+1)^4$
- $f(x) = \left(\frac{1}{x} + \frac{1}{x^2}\right)^2$
- $f(x) = (x+1)^2\sqrt{x}$
- $f(x) = \frac{(x+1)^2}{\sqrt{x}}$

Decomposing functions: For each function k that follows, find functions f, g, and h so that $k = f \circ g \circ h$. There may be more than one possible answer.

- $k(x) = (x^2+1)^3$
- $k(x) = \sqrt{1+(x-2)^2}$
- $k(x) = \sqrt{\frac{1}{3x+1}}$
- $k(x) = \frac{1}{\sqrt{3x+1}}$

Concepts

0. *Problem Zero:* Read the section and make your own summary of the material.

1. *True/False:* Determine whether each of the statements that follow is true or false. If a statement is true, explain why. If a statement is false, provide a counterexample.

(a) *True or False:* The chain rule is used to differentiate compositions of functions.

(b) *True or False:* If f and g are differentiable functions, then the derivative of $f \circ g$ is equal to the derivative of $g \circ f$.

(c) *True or False:* If f and g are differentiable functions, then $\frac{d}{dx}(f(g(x))) = f'(x)g'(x)$.

(d) *True or False:* If u and v are differentiable functions, then $\frac{d}{dx}(u(v(x))) = u'(v'(x))$.

(e) *True or False:* If h and k are differentiable functions, then $\frac{d}{dx}(k(h(x))) = k'(h(x))h'(x)$.

(f) *True or False:* If y is an implicit function of x, then there can be more than one y-value corresponding to a given x-value.

(g) *True or False:* The graph of an implicit function can have vertical tangent lines.

(h) *True or False:* If y is an implicit function of x and $\frac{dy}{dx}\big|_{x=2} = 0$, then the graph of the implicit function has a horizontal tangent line at $(2, 0)$.

2. *Examples:* Construct examples of the thing(s) described in the following. Try to find examples that are different than any in the reading.

(a) Three functions that we could not have differentiated before learning the chain rule, even after algebraic simplification.

(b) An equation that defines y as an implicit function of x, but not as a function of x.

(c) The graph of an implicit function with three horizontal tangent lines and two vertical tangent lines.

3. State the chain rule for differentiating a composition $g(h(x))$ of two functions expressed (a) in "prime" notation and (b) in Leibniz notation.

4. In the text we noted that if $f(u(v(x)))$ was a composition of three functions, then its derivative is $\frac{df}{dx} = \frac{df}{du}\frac{du}{dv}\frac{dv}{dx}$. Write this rule in "prime" notation.

5. Write down a rule for differentiating a composition $f(u(v(w(x))))$ of four functions (a) in "prime" notation and (b) in Leibniz notation.

6. Suppose $u(x) = \sqrt{3x^2+1}$ and $f(u) = \frac{u^2+3u^5}{1-u}$. Use the chain rule to find $\frac{d}{dx}(f(u(x)))$ without first finding the formula for $f(u(x))$.

7. Differentiate $f(x) = (3x + \sqrt{x})^2$ in three ways. When you have completed all three parts, show that your three answers are the same:

(a) with the chain rule

(b) with the product rule but not the chain rule

(c) without the chain or product rules

8. Differentiate $f(x) = \left(\frac{x^4-2}{\sqrt{x}}\right)^3$ in three ways. When you have completed all three parts, show that your three answers are the same:

(a) with the chain rule

(b) with the quotient rule but not the chain rule

(c) without the chain or quotient rules

Suppose g, h, and j are differentiable functions with the values for the function and derivative given in the following table:

x	$g(x)$	$h(x)$	$j(x)$	$g'(x)$	$h'(x)$	$j'(x)$
−3	0	3	1	1	0	2
−2	1	2	3	2	−3	0
−1	3	0	1	−1	−2	−2
0	2	3	0	−2	3	−2
1	0	−1	−2	−2	−2	−1
2	−2	−2	−3	−1	0	2
3	−3	0	1	0	1	2

Use the table to calculate the values of the derivatives listed in Exercises 9–16.

9. If $f(x) = g(h(x))$, find $f'(3)$.

10. If $f(x) = h(g(x))$, find $f'(3)$.

11. If $f(x) = (g(x))^3$, find $f'(-2)$.

12. If $f(x) = g(x^3 - 6)$, find $f'(2)$.

13. If $f(x) = h(g(j(x)))$, find $f'(1)$.

14. If $f(x) = j(2x)$, find $f'(-1)$.

15. If $f(x) = h(g(x)j(x))$, find $f'(0)$.

16. If $f(x) = h(h(h(x)))$, find $f'(1)$

17. If y is a function of x, then how is the chain rule involved in differentiating y^3 with respect to x, and why?

18. Show that, for any integers p and q (with $q \neq 0$),

$$(p-1) - \frac{p(q-1)}{q} = \frac{p}{q} - 1.$$

What does this equation have to do with the current section?

19. Match the two graphs shown here to the equations $(x+1)(y^2+y-1)=1$ and $xy^2+y=1$. Explain your choices.

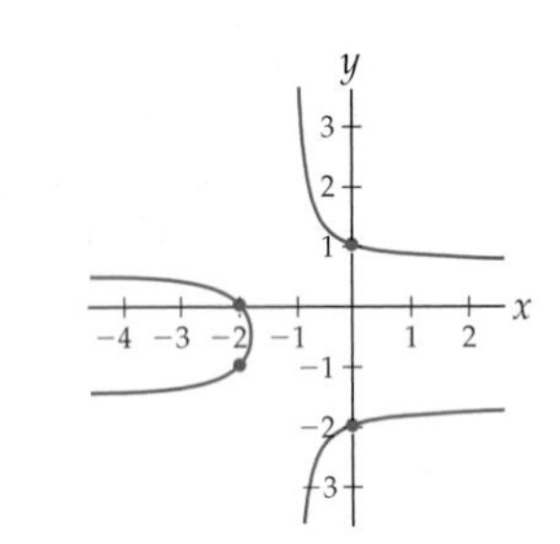

20. Match the two graphs shown here to the equations $xy^2+x=1$ and $1+x+xy^2=0$. Explain your choices.

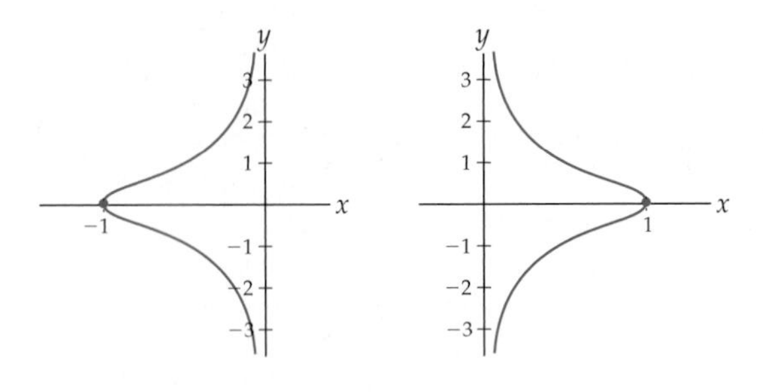

Skills

Find the derivatives of the functions in Exercises 21–46. Keep in mind that it may be convenient to do some preliminary algebra before differentiating.

21. $f(x)=\dfrac{1}{x^3+1}$

22. $f(x)=(\sqrt{x}+4)^5$

23. $f(x)=x(3x^2+1)^9$

24. $f(x)=\sqrt{x^2+1}$

25. $f(x)=\dfrac{1}{\sqrt{x^2+1}}$

26. $f(x)=\dfrac{3x+1}{\sqrt{x^2+1}}$

27. $f(x)=(x\sqrt{x+1})^{-2}$

28. $f(x)=\dfrac{(1+\sqrt{x})^2}{3x^2-4x+1}$

29. $f(x)=\dfrac{\frac{1}{x}-3x^2}{x^5-\frac{1}{\sqrt{x}}}$

30. $f(x)=\dfrac{(x+1)(3x-4)}{\sqrt{x^3-27}}$

31. $f(x)=(x^{1/3}-2x)^{-1}$

32. $f(x)=\sqrt{2-\sqrt{3x+1}}$

33. $f(x)=x^{-1/2}(x^2-1)^3$

34. $f(x)=(x^{-1/2}(x^2-1))^3$

35. $f(x)=\sqrt{3x-4(2x+1)^6}$

36. $f(x)=\dfrac{x^2-\sqrt[3]{x}+5x^9}{\sqrt{x^{-1}}}$

37. $f(x)=(5(3x^4-1)^3+3x-1)^{100}$

38. $f(x)=(1-4x)^2(3x^2+1)^9$

39. $f(x)=3((x^2+1)^8-7x)^{-2/3}$

40. $f(x)=\dfrac{(3x+1)(x^4-3)^4}{(x+5)^{-2}(1+x^2)^5}$

41. $f(x)=(5x^4-3x^2)^7(2x^3+1)$

42. $f(x)=x\sqrt{3x^2+1}\sqrt[3]{2x+5}$

43. $f(x)=((2x+1)^{-5}-1)^{-9}$

44. $f(x)=x(x^2)(\sqrt{x})(x^{2/3})$

45. $f(x)=(x^4-\sqrt{3-4x})^8+5x$

46. $f(x)=\sqrt{1+\sqrt{1+\sqrt{1-2x}}}$

Calculate each of the derivatives or derivative values in Exercises 47–52.

47. $\dfrac{d^2}{dx^2}((x\sqrt{x+1})^{-2})$

48. $\dfrac{d^2}{dx^2}\left(\dfrac{\frac{1}{x}-3x^2}{x^5-\frac{1}{\sqrt{x}}}\right)$

49. $\dfrac{d^2}{dx^2}\left(\dfrac{3}{x^{-3/2}\sqrt{x}}\right)\Bigg|_{x=2}$

50. $\dfrac{d^2}{dx^2}((5x^4-3x^2)^7(2x^3+1))$

51. $\dfrac{d^3}{dx^3}(x(3x^2+1)^9)|_{x=0}$

52. $\dfrac{d}{dx}\left(\dfrac{x^2+1}{(x^2+4)(3x-2)}\right)\Bigg|_{x=-1}$

Suppose that r is an independent variable, s is a function of r, and q is a constant. Calculate the derivatives in Exercises 53–58. Your answers may involve r, s, q, or their derivatives.

53. $\dfrac{d}{dr}(s^3)$

54. $\dfrac{d}{dr}(r^3)$

55. $\dfrac{d}{dr}(q^3)$

56. $\dfrac{d}{dr}(sr^2)$

57. $\dfrac{d}{dr}(rs^2)$

58. $\dfrac{d}{dr}(qs^2)$

For each of the equations in Exercises 59–62, y is defined as an implicit function of x. Solve for y, and use what you find to sketch a graph of the equation.

59. $\dfrac{1}{4}x^2+y^2=9$

60. $(x-1)^2+(y+2)^2=4$

61. $x^2-3y^2=16$

62. $4y^2-x^2+25=0$

In Exercises 63–68, find a function that has the given derivative and value. In each case you can find the answer with an educated guess-and-check process while thinking about the chain rule.

63. $f'(x)=5(x^2+1)^4(2x)$, $f(0)=1$

64. $f'(x)=5(x^2+1)^4(2x)$, $f(1)=25$

65. $f'(x)=x(x^2+1)^4$, $f(0)=0$

66. $f'(x)=-6x(x^2+1)^4$, $f(0)=1$

67. $f'(x) = \sqrt{3x+1}$, $f(0) = 1$

68. $f'(x) = \dfrac{x^2}{\sqrt{x^3+1}}$, $f(1) = 2$

Each of the equations in Exercises 69–80 defines y as an implicit function of x. Use implicit differentiation (*without* solving for y first) to find $\frac{dy}{dx}$.

69. $4x^2 - y^2 = 9$

70. $x^2 + y^2 = 4$

71. $xy^2 + 3x^2 = 4$

72. $y^6 - 3x + 4 = 0$

73. $(3x+1)(y^2 - y + 6) = 0$

74. $x^2y - y^2x = x^2 + 3$

75. $\sqrt{3y-1} = 5xy$

76. $(3y^2 + 5xy - 2)^4 = 2$

77. $\dfrac{y^2+1}{3y-1} = x$

78. $3y = 5x^2 + \sqrt[3]{y-2}$

79. $\dfrac{1}{y} - \dfrac{1}{x} = \dfrac{x^2}{y+1}$

80. $\dfrac{x+1}{y^2-3} = \dfrac{1}{xy}$

In Exercises 81–84, use implicit differentiation to algebraically find each quantity or location related to the given implicit function.

81. Consider the circle of radius 1 centered at the origin, that is, the solutions of the equation $x^2 + y^2 = 1$.
   (a) Find all points on the graph with an x-coordinate of $x = \frac{1}{2}$, and then find the slope of the tangent line at each of these points.
   (b) Find all points on the graph with a y-coordinate of $y = \frac{\sqrt{2}}{2}$, and then find the slope of the tangent line at each of these points.
   (c) Find all points on the graph where the tangent line is vertical.
   (d) Find all points on the graph where the tangent line has a slope of -1.

82. Consider the graph of the solutions of the equation $4y^2 - x^2 + 2x = 2$.
   (a) Find all points on the graph with an x-coordinate of $x = 3$, and then find the slope of the tangent line at each of these points.
   (b) Find all points on the graph with a y-coordinate of $y = 3$, and then find the slope of the tangent line at each of these points.
   (c) Find all points where the graph has a horizontal tangent line.
   (d) Find all points where the graph has a vertical tangent line.

83. Consider the graph of the solutions of the equation $y^3 + xy + 2 = 0$.
   (a) Find all points on the graph with an x-coordinate of $x = 1$, and then find the slope of the tangent line at each of these points.
   (b) Find all points on the graph with a y-coordinate of $y = 1$, and then find the slope of the tangent line at each of these points.
   (c) Find all points where the graph has a horizontal tangent line.
   (d) Find all points where the graph has a vertical tangent line.

84. Consider the graph of the solutions of the equation $y^3 - 3y - x = 1$.
   (a) Find all points on the graph with an x-coordinate of $x = -1$, and then find the slope of the tangent line at each of these points.
   (b) Find all points on the graph with a y-coordinate of $y = 2$, and then find the slope of the tangent line at each of these points.
   (c) Find all points where the graph has a horizontal tangent line.
   (d) Find all points where the graph has a vertical tangent line.

Applications

85. Linda can sell 12 magazine subscriptions per week and makes \$4.00 for each magazine subscription she sells. Obviously this means that Linda will make (12)(\$4.00) = \$48.00 per week from magazine subscriptions. Explain this result mathematically, using mathematical notation and the chain rule.

86. If you drop a pebble into a large lake, you will cause a circle of ripples to expand outward. The area $A = A(t)$ and radius $r = r(t)$ are clearly functions of t (they change over time) and are related by the formula $A = \pi r^2$.

Area $A = A(t)$

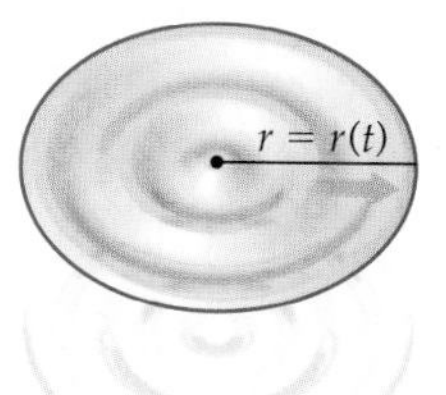

t, seconds

   (a) If r is measured in inches and t is measured in seconds, what are the units of $\frac{dA}{dt}$? What are the units of $\frac{dA}{dr}$?
   (b) Find $\frac{dA}{dr}$ and explain the practical meaning of $\left.\frac{dA}{dr}\right|_{r=2}$.
   (c) Find $\frac{dA}{dt}$ and explain the practical meaning of $\left.\frac{dA}{dt}\right|_{r=2}$.

87. The area of a circle can be written in terms of its radius as $A = \pi r^2$, where both A and r are functions of time. Suppose a circular area from a spotlight on a stage floor is slowly expanding.
   (a) Find $\frac{dA}{dr}$ and explain its meaning in practical terms.
   (b) Does the rate $\frac{dA}{dr}$ depend on how fast the radius of the circle is increasing? Does it depend on the radius of the circle? Why or why not?

(c) Find $\frac{dA}{dt}$ and explain its meaning in practical terms.

(d) Does the rate $\frac{dA}{dt}$ depend on how fast the radius of the circle is increasing? Does it depend on the radius of the circle?

(e) If the radius of the circle of light is increasing at a constant rate of 2 inches per second, how fast is the area of the circle of light increasing at the moment that the spotlight has a radius of 24 inches?

Proofs

88. Use the chain rule twice to prove that $\frac{d}{dx}(f(u(v(x)))) = f'(u(v(x)))u'(v(x))v'(x)$.

89. In Exercise 89 of the Section 2.3 you used the definition of derivative to prove the quotient rule. Prove it now another way: by writing a quotient $\frac{f}{g}$ as a product and applying the product, power, and chain rules. Point out where you use each rule.

90. Use implicit differentiation and the fact that $\frac{d}{dx}(x^4) = 4x^3$ to prove that $\frac{d}{dx}(x^{-4}) = -4x^{-5}$.

91. Use implicit differentiation and the fact that $\frac{d}{dx}(x^3) = 3x^2$ and $\frac{d}{dx}(x^5) = 5x^4$ to prove that $\frac{d}{dx}(x^{3/5}) = \frac{3}{5}x^{-2/5}$.

92. Use implicit differentiation and the power rule for integer powers (*not* the general power rule) to prove that $\frac{d}{dx}(x^{2/3}) = \frac{2}{3}x^{-1/3}$.

93. Use implicit differentiation, the product rule, and the power rule for positive integer powers to prove the power rule for negative integer powers.

94. Use implicit differentiation and the power rule for integer powers to prove the power rule for rational powers.

Thinking Forward

Finding critical points: For each of the following functions f, find all of the x-values for which $f'(x) = 0$ and all of the x-values for which $f'(x)$ does not exist.

- $f(x) = x^3\sqrt{3x+1}$
- $f(x) = (1-x^4)^7$
- $f(x) = (x^2+3)(x-2)^{3/2}$
- $f(x) = 3x\left(x+\frac{1}{x}\right)$
- $f(x) = \frac{\sqrt{x}}{x\sqrt{x-1}}$
- $f(x) = (x\sqrt{x+1})^{-2}$

Finding antiderivatives by undoing the chain rule: For each function f that follows, find a function F with the property that $F'(x) = f(x)$. You may have to guess and check to find such a function.

- $f(x) = x\sqrt{1+x^2}$
- $f(x) = x^2\sqrt{1+x^3}$
- $f(x) = \frac{1}{(2-5x)^3}$
- $f(x) = \frac{1}{\sqrt{1+3x}}$

2.5 DERIVATIVES OF EXPONENTIAL AND LOGARITHMIC FUNCTIONS

- Formulas for differentiating exponential and logarithmic functions
- Rates of growth of exponential functions
- The method of logarithmic differentiation

Derivatives of Exponential Functions

The power rule tells us that the derivative of x^k is kx^{k-1}. This rule works only for power functions, where the base is the variable x and the power is a constant k; it does not tell us how to differentiate an exponential function like 2^x or e^{3x}, where the variable is in the exponent. To determine such derivatives we must return to the definition of the derivative:

$$\frac{d}{dx}(b^x) = \lim_{h\to 0}\frac{b^{x+h}-b^x}{h} = \lim_{h\to 0}\frac{b^x b^h - b^x}{h} = \lim_{h\to 0}\frac{b^x(b^h-1)}{h} = b^x\left(\lim_{h\to 0}\frac{b^h-1}{h}\right).$$

Although we have simplified as much as possible, this is a limit that we do not yet know how to calculate.

In one special case we do already know how to evaluate this limit. In Theorem 1.26 we saw that e is the unique number such that $\lim_{h \to 0} \frac{e^h - 1}{h} = 1$. Therefore when $b = e$, the preceding calculation looks like this:

$$\frac{d}{dx}(e^x) = \lim_{h \to 0} \frac{e^{x+h} - e^x}{h} = \cdots = e^x \left(\lim_{h \to 0} \frac{e^h - 1}{h} \right) = e^x(1) = e^x.$$

We have just shown that the function $f(x) = e^x$ is its own derivative! This is in fact the exact reason that we defined the number e the way that we did. As illustrated here, $y = e^x$ is its own associated slope function:

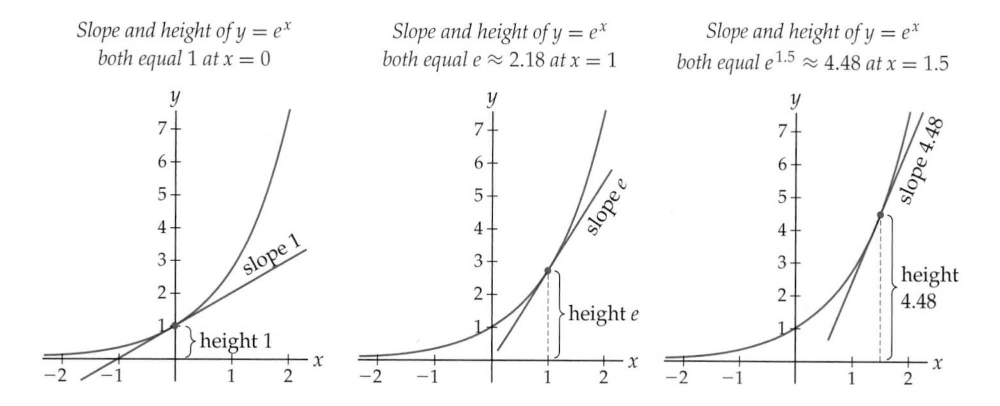

Other exponential functions of the form e^{kx} or b^x have graphs similar to the graph of e^x, but only e^x is scaled in exactly the right way to be its own derivative. With the chain rule, we can use the derivative of e^x to find the derivatives of general exponential functions:

THEOREM 2.13

Derivatives of Exponential Functions

For any constant k, any constant $b > 0$ with $b \neq 1$, and all $x \in \mathbb{R}$,

(a) $\frac{d}{dx}(e^x) = e^x$ **(b)** $\frac{d}{dx}(b^x) = (\ln b)\, b^x$ **(c)** $\frac{d}{dx}(e^{kx}) = ke^{kx}$

Proof. The previous discussion proves that $\frac{d}{dx}(e^x) = e^x$. To prove the second rule we rewrite b^x as $(e^{\ln b})^x$ and then apply the chain rule:

$$\begin{aligned} \frac{d}{dx}(b^x) &= \frac{d}{dx}((e^{\ln b})^x) && \leftarrow b = e^{\ln b} \\ &= \frac{d}{dx}(e^{(\ln b)x}) && \leftarrow \text{algebra of exponents} \\ &= e^{(\ln b)x} \frac{d}{dx}((\ln b)x) && \leftarrow \text{chain rule and derivative of } e^x \\ &= (e^{\ln b})^x(\ln b) && \leftarrow \text{algebra and derivative of linear function} \\ &= (\ln b)\, b^x. && \leftarrow \text{algebra} \end{aligned}$$

The proof for e^{kx} is similar, with a simpler application of the chain rule, and is left for Exercise 69. ■

By combining our new differentiation rules for exponential functions with the chain rule, we obtain the following rules for exponential compositions:

$$\frac{d}{dx}(e^{u(x)}) = e^{u(x)}u'(x),$$

$$\frac{d}{dx}(b^{u(x)}) = (\ln b)b^{u(x)}u'(x).$$

For example, we have

$$\frac{d}{dx}\left(e^{x^2+1}\right) = e^{x^2+1} \cdot \frac{d}{dx}(x^2+1) = e^{x^2+1}(2x),$$

$$\frac{d}{dx}\left(2^{x^2+1}\right) = (\ln 2)2^{x^2+1} \cdot \frac{d}{dx}(x^2+1) = (\ln 2)2^{x^2+1}(2x).$$

Exponential Functions Grow Proportionally to Themselves

Notice that all exponential functions have the property that their derivatives are constant multiples of the original function; for example, $\frac{d}{dx}(e^{2x}) = 2e^{2x}$ is just 2 times the original function e^{2x}. This means that the rate of change of an exponential function f is proportional to the function f itself. In fact, the converse is also true, although we will not have the tools to prove it until Section 7.5. Together, both proportionalities give us the following two-sided statement, which will prove useful in many application problems:

THEOREM 2.14 **Rates of Change and Exponential Functions**

$f'(x) = kf(x)$ for some constant k if and only if f is an exponential function of the form $f(x) = Ae^{kx}$.

Recall that one quantity y is ***proportional*** to another quantity x if y is a constant multiple of x. Therefore the preceding theorem gives us a nice characterization of exponential functions: *All* exponential functions f, and *only* exponential functions, have the property that f' is proportional to f.

If a word problem states that the rate of change of a function is constant, we immediately know that the function is linear. Similarly, by Theorem 2.14, if a word problem states that the rate of change of a function is proportional to the function itself, we immediately know that that function is exponential. Then it is just a question of finding values A and k to determine the function $f(x) = Ae^{kx}$ that models the situation. Exponential functions often arise when modeling populations, for example, since a larger population can produce greater numbers of offspring than a smaller population and can therefore grow at a faster rate.

Derivatives of Logarithmic Functions

Up to this point, the derivatives that we have seen are similar to their original functions. For example, derivatives of power functions are power functions, derivatives of polynomials are polynomials, and derivatives of exponential functions are exponential. Now something surprising happens: Derivatives of logarithmic functions are not logarithmic. Even more surprisingly, logarithmic functions are transcendental but their derivatives are algebraic!

THEOREM 2.15 **Derivatives of Logarithmic Functions**

For any constant $b > 0$ with $b \neq 1$, and all appropriate values of x,

(a) $\dfrac{d}{dx}(\log_b x) = \dfrac{1}{(\ln b)x}$ **(b)** $\dfrac{d}{dx}(\ln x) = \dfrac{1}{x}$ **(c)** $\dfrac{d}{dx}(\ln |x|) = \dfrac{1}{x}$

The second rule is a special case of the first, with $b = e$. Because $\ln x$ has domain $(0, \infty)$, when we say that $\frac{d}{dx}(\ln x) = \frac{1}{x}$, we are also restricting $\frac{1}{x}$ to the domain $(0, \infty)$. In the third rule we generalize the second rule so that we are considering the full domain $(-\infty, 0) \cup (0, \infty)$ of $\frac{1}{x}$.

Proof. We will prove the second rule and leave the proofs of the first and third rules to Exercises 72 and 73, respectively. Since $y = \ln x$ and $y = e^x$ are inverses, $e^{\ln x} = x$ for all x in the domain $(0, \infty)$ of $y = \ln x$. Differentiating both sides of the equation gives

$$e^{\ln x} = x \qquad \leftarrow \text{property of inverses}$$

$$\frac{d}{dx}(e^{\ln x}) = \frac{d}{dx}(x) \qquad \leftarrow \text{differentiate both sides}$$

$$e^{\ln x}\frac{d}{dx}(\ln x) = 1 \qquad \leftarrow \text{exponential and chain rules}$$

$$\frac{d}{dx}(\ln x) = \frac{1}{e^{\ln x}} \qquad \leftarrow \text{solve for } \frac{d}{dx}(\ln x)$$

$$\frac{d}{dx}(\ln x) = \frac{1}{x}. \qquad \leftarrow \text{since } e^{\ln x} = x$$

■

The two graphs that follow illustrate that the associated slope function for $\ln |x|$ is the function $\frac{1}{x}$. For negative values of x, as we move from left to right, the slopes of $\ln |x|$ are negative with larger and larger magnitude while the heights of $\frac{1}{x}$ behave the same way. For positive values of x, as we move from left to right, the slopes of $\ln x$ are positive but getting smaller while the heights of $\frac{1}{x}$ do the same.

Slopes of $f(x) = \ln |x|$

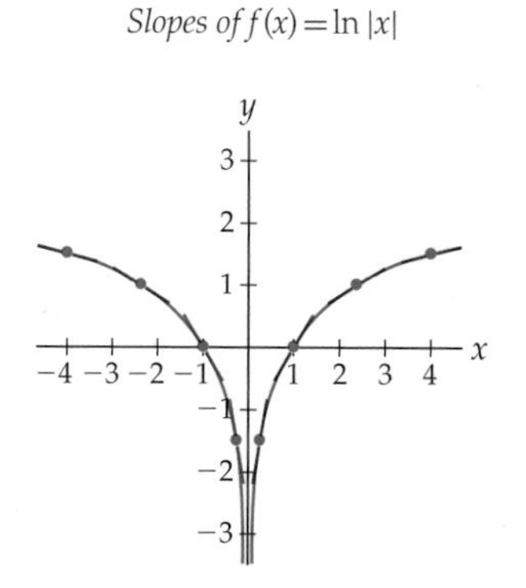

Heights of $f'(x) = \frac{1}{x}$

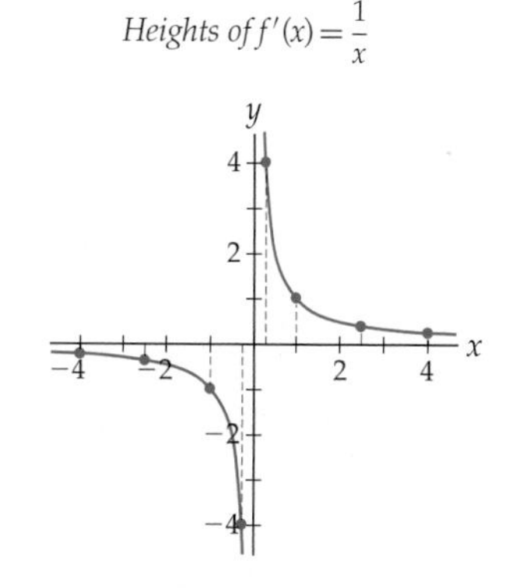

The rule for differentiating $\ln |x|$ will come in handy for calculating certain derivatives. Functions that involve many products or quotients or that have variables in both a base and an exponent can be difficult to differentiate. One strategy for differentiating such functions is to apply $\ln|x|$ to both sides of the equation $y = f(x)$ and then differentiate both sides. This process is called ***logarithmic differentiation*** and is illustrated in Examples 4 and 5.

By combining our new differentiation rules for logarithmic functions with the chain rule, we obtain the following rules for logarithmic compositions:

$$\frac{d}{dx}(\ln u(x)) = \frac{1}{u(x)} \cdot u'(x) = \frac{u'(x)}{u(x)},$$

$$\frac{d}{dx}(\log_b u(x)) = \frac{1}{(\ln b)u(x)} \cdot u'(x) = \frac{u'(x)}{(\ln b)u(x)}.$$

For example, we have

$$\frac{d}{dx}(\ln(x^2+1)) = \frac{1}{x^2+1} \cdot 2x = \frac{2x}{x^2+1},$$

$$\frac{d}{dx}(\log_b(x^2+1)) = \frac{1}{(\ln b)(x^2+1)} \cdot 2x = \frac{2x}{(\ln b)(x^2+1)}.$$

Derivatives of Inverse Functions*

We can generalize the technique used in the proof of Theorem 2.15 to obtain a formula for the derivative of the inverse of any function whose derivative we already know. If f is an invertible function, then the graph of its inverse $y = f^{-1}(x)$ can be obtained by reflecting the graph of $y = f(x)$ over the line $y = x$. This reflection yields the reciprocals of all slopes on the graph, as shown in the following figure:

The slope of $f^{-1}(x)$ at (b, a) is the reciprocal of the slope of $f(x)$ at (a, b)

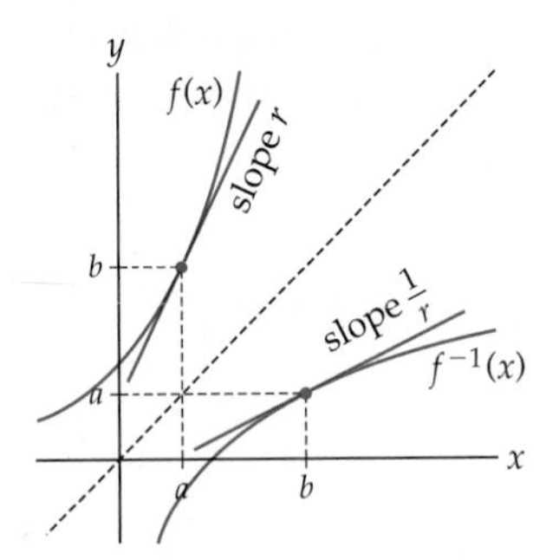

From this example we might expect that the derivatives of f and f^{-1} would have some sort of reciprocal relationship. In particular, we would expect that if $f(a) = b$ and thus $a = f^{-1}(b)$, then $(f^{-1})'(b) = \frac{1}{f'(a)}$. Using implicit differentiation and the chain rule, we can show that this is indeed the case in general:

THEOREM 2.16 **Derivatives of Inverse Functions**

If f and f^{-1} are inverse functions and are both differentiable, then, for all appropriate values of x,

$$(f^{-1})'(x) = \frac{1}{f'(f^{-1}(x))}.$$

For example, if $f(x) = x^3$, then its inverse is the function $f^{-1}(x) = x^{1/3}$, since $y = x^3$ if and only if $y^{1/3} = x$. By Theorem 2.16, the derivative of $f^{-1}(x) = x^{1/3}$ must be

$$(f^{-1})'(x) = \frac{1}{f'(f^{-1}(x))} = \frac{1}{3(f^{-1}(x))^2} = \frac{1}{3(x^{1/3})^2} = \frac{1}{3x^{2/3}} = \frac{1}{3}x^{-2/3},$$

just as we would expect from the power rule.

Proof. Since f and f^{-1} are inverses, we know that their composition in either order is the identity function. Starting from this fact and applying implicit differentiation, we have

$$f(f^{-1}(x)) = x \qquad \leftarrow \text{definition of inverses}$$

$$\frac{d}{dx}(f(f^{-1}(x))) = \frac{d}{dx}(x) \qquad \leftarrow \text{differentiate both sides}$$

$$f'(f^{-1}(x))\,\frac{d}{dx}(f^{-1}(x)) = 1 \qquad \leftarrow \text{chain rule}$$

$$\frac{d}{dx}(f^{-1}(x)) = \frac{1}{f'(f^{-1}(x))}. \qquad \leftarrow \text{solve for } \frac{d}{dx}(f^{-1}(x))$$

This result holds whenever x is in the domain of f^{-1} and f is differentiable at $f^{-1}(x)$. ■

Examples and Explorations

EXAMPLE 1

Differentiating combinations of exponential and logarithmic functions

Find the derivatives of each of the following functions:

(a) $f(x) = e^{x^2}\ln x$ **(b)** $f(x) = \dfrac{7e^{3x} - x^2 2^x}{\log_5 x}$ **(c)** $f(x) = \ln\left(\left(\dfrac{x^2-1}{1-2x}\right)^2\right)$

SOLUTION

(a) Using the product and chain rules, we have

$$\frac{d}{dx}\left(e^{x^2}\ln x\right) = \frac{d}{dx}\left(e^{x^2}\right)\cdot\ln x + e^{x^2}\cdot\frac{d}{dx}(\ln x) = 2xe^{x^2}\ln x + e^{x^2}\left(\frac{1}{x}\right).$$

(b) The function $f(x)$ is a quotient, so we begin by applying the quotient rule:

$$f'(x) = \frac{\frac{d}{dx}(7e^{3x} - x^2 2^x)\cdot\log_5 x - (7e^{3x} - x^2 2^x)\cdot\frac{d}{dx}(\log_5 x)}{(\log_5 x)^2} \qquad \leftarrow \text{quotient rule}$$

$$= \frac{\left(21e^{3x} - \frac{d}{dx}(x^2 2^x)\right)(\log_5 x) - (7e^{3x} - x^2 2^x)\left(\frac{1}{(\ln 5)x}\right)}{(\log_5 x)^2} \qquad \leftarrow \text{other rules}$$

$$= \frac{(21e^{3x} - (2x2^x + x^2(\ln 2)2^x))(\log_5 x) - (7e^{3x} - x^2 2^x)\left(\frac{1}{(\ln 5)x}\right)}{(\log_5 x)^2} \qquad \leftarrow \text{product rule}$$

In the preceding calculation, we saved the product rule calculation for the last step. For lengthy derivative problems it is often helpful to postpone some calculations to later steps.

(c) There are a number of ways we could find $f'(x)$. One way is to jump right in and start differentiating; although it is possible to do this all in one step, we'll do the calculation in a few separate steps to be as clear as possible:

$$\frac{d}{dx}\left(\ln\left(\left(\frac{x^2-1}{1-2x}\right)^2\right)\right) = \frac{1}{\left(\frac{x^2-1}{1-2x}\right)^2}\cdot\frac{d}{dx}\left(\left(\frac{x^2-1}{1-2x}\right)^2\right) \quad \leftarrow \text{chain rule}$$

$$= \frac{1}{\left(\frac{x^2-1}{1-2x}\right)^2}(2)\left(\frac{x^2-1}{1-2x}\right)^1\cdot\frac{d}{dx}\left(\frac{x^2-1}{1-2x}\right) \quad \leftarrow \text{chain rule}$$

$$= \frac{1}{\left(\frac{x^2-1}{1-2x}\right)^2}(2)\left(\frac{x^2-1}{1-2x}\right)^1\left(\frac{(2x)(1-2x)-(x^2-1)(-2)}{(1-2x)^2}\right). \quad \leftarrow \text{quotient rule}$$

The first step used the chain rule and the logarithmic rule, the second step used the chain rule and the power rule, and the last step used the quotient rule.

If we do a little bit of algebra first, the differentiation step becomes much easier:

$$\frac{d}{dx}\left(\ln\left(\left(\frac{x^2-1}{1-2x}\right)^2\right)\right) = \frac{d}{dx}\left(2\ln\left(\frac{x^2-1}{1-2x}\right)\right) \quad \leftarrow \text{algebra}$$

$$= \frac{d}{dx}(2(\ln(x^2-1)-\ln(1-2x))) \quad \leftarrow \text{algebra}$$

$$= 2\left(\frac{1}{x^2-1}(2x)-\frac{1}{1-2x}(-2)\right). \quad \leftarrow \text{differentiate}$$

In the preceding calculation, the first two steps were algebra; only the final step involved differentiation. Although the two answers obtained look very different, they are in fact the same. (As an exercise, use algebra to show this.) □

EXAMPLE 2

Differentiating a piecewise-defined function

Find the derivative of the piecewise-defined function $f(x) = \begin{cases} e^x, & \text{if } x < 0 \\ e^{-x}, & \text{if } x \geq 0. \end{cases}$

SOLUTION

From the differentiation rules we have developed, we know that $\frac{d}{dx}(e^x) = e^x$ and $\frac{d}{dx}(e^{-x}) = -e^{-x}$. These are the expressions for the derivative of f when $x < 0$ and when $x > 0$, respectively. It remains only to determine what happens at the breakpoint $x = 0$.

If $g(x) = e^x$ and $h(x) = e^{-x}$, then we must first check that $g(0) = h(0)$; since $e^0 = 1$ and $e^{-0} = 1$, this is true. It follows that the function f is continuous at $x = 0$. Second, we must check that $g'(0) = h'(0)$. Here $g'(0) = e^0 = 1$ but $h'(0) = -e^{-0} = -1$, so although f is continuous at $x = 0$, it is not differentiable. The derivative of f is therefore

$$f'(x) = \begin{cases} e^x, & \text{if } x < 0 \\ \text{undefined}, & \text{if } x = 0 \\ -e^{-x}, & \text{if } x > 0. \end{cases}$$

□

EXAMPLE 3

A real-world problem that can be modeled with an exponential function

Suppose a population of wombats on a small island is growing at a rate proportional to the number of wombats on the island. If there were 12 wombats on the island in 1990 and 37 wombats on the island in 1998, how many wombats were on the island in the year 2010?

SOLUTION

We are given that the rate of change $W'(t)$ of the population of wombats is proportional to the population $W(t)$ of wombats at time t; in other words, $W'(t) = kW(t)$ for some real number k. By Theorem 2.14, this means that the population $W(t)$ of wombats on the island must be an exponential function $W(t) = Ae^{kt}$. We now only need to find constants A and k that match the information given in the problem.

If we let $t = 0$ represent 1990 (so that $t = 8$ will represent 1998), then from the information given in the problem, we have $W(0) = 12$ and $W(8) = 37$. We will use these two data points to find A and k. Using the first data point, we have

$$W(0) = 12 \implies Ae^{k(0)} = 12 \implies Ae^0 = 12 \implies A(1) = 12 \implies A = 12,$$

so $W(t) = 12e^{kt}$ for some k. Using the second data point, we can now solve for the value of k:

$$W(8) = 37 \implies 12e^{k(8)} = 37 \implies e^{8k} = \frac{37}{12} \implies 8k = \ln\left(\frac{37}{12}\right) \implies k = \frac{\ln\left(\frac{37}{12}\right)}{8} \approx 0.14.$$

Thus $W(t) = 12e^{0.14t}$. Using this function, we can now easily calculate the number of wombats that were on the island in the year 2010. Since 2010 is 20 years after 1990, we need to find

$$W(20) = 12e^{0.14(20)} \approx 197.34.$$

In the year 2010, there were approximately 197 wombats on the island, assuming of course that we cannot have "parts" of wombats. □

EXAMPLE 4

Using logarithmic differentiation when products and quotients are involved

Use logarithmic differentiation to calculate the derivative of the function

$$f(x) = \frac{\sqrt{x}\,(x^2-1)^5}{(x+2)(x-4)^3}.$$

SOLUTION

Obviously it would take a great deal of work to differentiate this function as it is currently written. We would have to apply the quotient rule once and the product rule twice, among other things. We could do some algebra and multiply out some of the factors in the numerator and the denominator of $f(x)$ to make our job easier, but that is in itself a pretty nasty calculation. However, it happens to be not nearly as difficult to differentiate the related function $\ln|f(x)|$. Taking the logarithm of both sides and then differentiating both sides, we have

$$
\begin{aligned}
y &= \frac{\sqrt{x}\,(x^2-1)^5}{(x+2)(x-4)^3} && \leftarrow \text{set } y = f(x)\\
\ln|y| &= \ln\left|\frac{\sqrt{x}(x^2-1)^5}{(x+2)(x-4)^3}\right| && \leftarrow \text{apply } \ln|x|\\
\ln|y| &= \ln|\sqrt{x}| + \ln|(x^2-1)^5| - \ln|x+2| - \ln|(x-4)^3| && \leftarrow \text{algebra}\\
\ln|y| &= \frac{1}{2}\ln|x| + 5\ln|x^2-1| - \ln|x+2| - 3\ln|x-4| && \leftarrow \text{algebra}\\
\frac{d}{dx}(\ln|y|) &= \frac{d}{dx}\left(\frac{1}{2}\ln|x| + 5\ln|x^2-1| - \ln|x+2| - 3\ln|x-4|\right) && \leftarrow \text{differentiate}\\
\frac{1}{y}y' &= \frac{1}{2x} + \frac{5(2x)}{x^2-1} - \frac{1}{x+2} - \frac{3}{x-4} && \leftarrow \text{derivative rules}\\
y' &= y\left(\frac{1}{2x} + \frac{10x}{x^2-1} - \frac{1}{x+2} - \frac{3}{x-4}\right) && \leftarrow \text{solve for } y'\\
y' &= \frac{\sqrt{x}(x^2-1)^5}{(x+2)(x-4)^3}\left(\frac{1}{2x} + \frac{10x}{x^2-1} - \frac{1}{x+2} - \frac{3}{x-4}\right). && \leftarrow \text{definition of } y
\end{aligned}
$$

Although the calculation as a whole is algebraically complicated, the differentiation step was simple. The most difficult part was to remember to use the chain rule when we differentiated $\frac{d}{dx}(\ln y) = \frac{1}{y}y'$. Note also that we used the fact that $|ab| = |a||b|$ and the fact that $|a^b| = |a|^b$ in our calculations. □

EXAMPLE 5

Using logarithms to differentiate a function with variables in both base and exponent

Use logarithmic differentiation to find $\frac{d}{dx}((x^2 - 3)^x)$.

SOLUTION

In this example using logarithmic differentiation is not a choice, but a necessity: Neither the product rule nor the exponential rule applies to this function. When taking the derivative of a function that involves the variable in both the base and the exponent, we must use logarithmic differentiation. After setting $y = (x^2 - 3)^x$ and then applying the natural logarithm to both sides, we will be able to use algebra to remove the variable x from the exponent and then differentiate both sides:

$$
\begin{aligned}
y &= (x^2 - 3)^x && \leftarrow \text{set } y = f(x) \\
\ln y &= \ln((x^2 - 3)^x) && \leftarrow \text{apply } \ln(x) \text{ to both sides} \\
\ln y &= x \ln(x^2 - 3) && \leftarrow \text{algebra} \\
\frac{d}{dx}(\ln y) &= \frac{d}{dx}(x \ln(x^2 - 3)) && \leftarrow \text{differentiate both sides} \\
\frac{1}{y}y' &= (1)\ln(x^2 - 3) + (x)\frac{2x}{x^2 - 3} && \leftarrow \text{chain, product rules} \\
y' &= y\left(\ln(x^2 - 3) + \frac{2x^2}{x^2 - 3}\right) && \leftarrow \text{solve for } y' \\
y' &= (x^2 - 3)^x\left(\ln(x^2 - 3) + \frac{2x^2}{x^2 - 3}\right). && \leftarrow \text{since } y = (x^2 - 3)^x
\end{aligned}
$$

Note that we did not need any absolute values in this calculation, because $(x^2 - 3)^x$ is always positive where it is defined. □

? TEST YOUR UNDERSTANDING

- What differentiation fact is the consequence of the limit statement $\lim_{h \to 0} \frac{e^h - 1}{h} = 1$ that characterizes the number e?
- In the proof that $\frac{d}{dx}(\ln x) = \frac{1}{x}$, we used the fact that $e^{\ln x} = x$. It is also true that $\ln(e^x) = x$; could we have started with this equality instead? Why or why not?
- What can you say about a quantity that grows at a rate proportional to the amount of the quantity that is present?
- Although there are absolute values in the fifth line of the calculation in Example 4, there are no absolute values in the sixth line; what happened to the absolute values and why?
- Why was logarithmic differentiation necessary in Example 5? In particular, why did neither the power nor the exponential rule apply?

EXERCISES 2.5

Thinking Back

Solving exponential and logarithmic equations: Use rules of exponents and logarithms to solve each of the following equations.

- ▶ $3(1.2)^x = 500$
- ▶ $\ln(x^2 + x - 5) = 0$
- ▶ $\dfrac{\ln(x+1)}{\ln(x-2)} = 0$
- ▶ $\dfrac{2^x + 1}{3^x - 5} = 0$

Compositions: For each function k, find functions f, g, and h such that $k = f \circ g \circ h$. There may be more than one possible answer.

- ▶ $k(x) = \ln(\sqrt{x^2 + 5})$
- ▶ $k(x) = e^{1/\sqrt{x+1}}$
- ▶ $k(x) = (\ln 3x)^2$
- ▶ $k(x) = 3\ln(5^x + 2)$

Concepts

0. *Problem Zero:* Read the section and make your own summary of the material.

1. *True/False:* Determine whether each of the statements that follow is true or false. If a statement is true, explain why. If a statement is false, provide a counterexample.

(a) *True or False:* $\dfrac{d}{dx}(e^{\pi}) = 0$.

(b) *True or False:* $\dfrac{d}{dz}(e^z) = e^z$.

(c) *True or False:* $\dfrac{d}{dx}\left(\dfrac{1}{x}\right) = \ln x$.

(d) *True or False:* $\dfrac{d}{dx}(\ln|x|) = \dfrac{1}{|x|}$.

(e) *True or False:* If f is an exponential function, then f' is a constant multiple of f.

(f) *True or False:* If f' is a constant multiple of f, then f is an exponential function.

(g) *True or False:* Logarithmic differentiation is required in order to differentiate complicated products and quotients.

(h) *True or False:* Logarithmic differentiation is required in order to differentiate expressions that have a variable in both the base and the exponent.

2. *Examples:* Construct examples of the thing(s) described in the following. Try to find examples that are different than any in the reading.

(a) Three functions f whose derivatives are just constant multiples of f.

(b) Three functions that are transcendental, but whose derivatives are algebraic.

(c) A function whose derivative would be difficult or impossible to find without the method of logarithmic differentiation.

3. Does the exponential rule apply to the function $f(x) = x^x$? What about the power rule? Explain your answers.

4. The natural exponential function is its own derivative. Explain what this means graphically. (Use words like "height" and "slope.")

5. Explain how the formula for differentiating the natural exponential function is a special case of the formula for differentiating exponential functions of the form e^{kx}. Then explain why it is a special case of the formula for differentiating functions of the form b^x.

6. The function $f(x) = e^x$ is its own derivative. Are there other functions with this property? If not, explain why not. If so, give three examples.

7. Explain how the formula for differentiating the natural logarithm function is a special case of the formula for differentiating logarithmic functions of the form $\log_b x$.

8. When we say that $\dfrac{d}{dx}(\ln x) = \dfrac{1}{x}$, we really mean to consider the function $\dfrac{1}{x}$ on the restricted domain $(0, \infty)$. Why?

9. The graphs of the exponential functions $y = 2^x$, $y = 4^x$, and $y = 2(2^x)$ are shown in the figure at the left. Use your knowledge of transformations to determine which graph is which without using a graphing calculator.

$y = 2^x,\ y = 4^x,\ y = 2(2^x)$

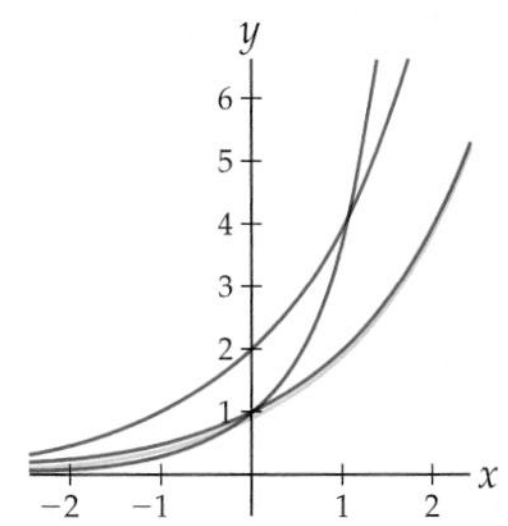

$y = e^x,\ y = e^{3x},\ y = e^{-2x}$

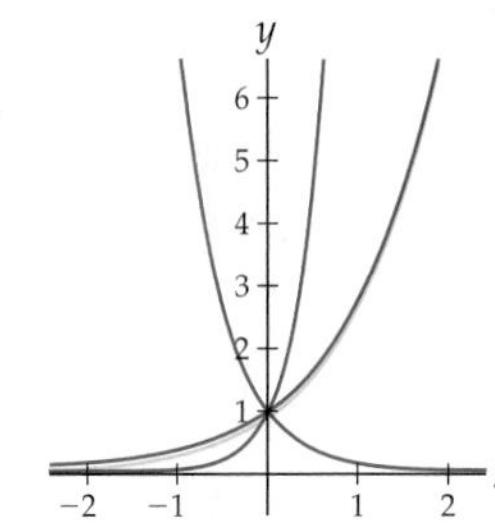

10. The graphs of the exponential functions $y = e^x$, $y = e^{3x}$, and $y = e^{-2x}$ are shown in the preceding graph at the right. Use your knowledge of transformations to determine which graph is which without using a graphing calculator.

11. The functions $f(x) = 2^{(x^2)}$ and $g(x) = (2^x)^2$ look similar, but are very different functions. (This is why we try to avoid using the ambiguous notation 2^{x^2}.) Calculate $f(3)$ and $g(3)$, and show that they are not the same (and thus that $f(x)$ and $g(x)$ are not the same function). Then find all the values for which $f(x) = g(x)$.

12. Every exponential function of the form $f(x) = b^x$ (with $b > 0$ and $b \neq 1$) is one-to-one. Explain why this fact implies that $b^x = b^y$ if and only if $x = y$.

13. Explain how we know that logarithmic functions are one-to-one. Why does this mean that $A = B$ if and only if $\log_b A = \log_b B$ (assuming that A and B are positive)?

14. What is the definition of the number e? What does this definition tell you about $\lim_{h\to 0} \frac{e^h - 1}{h}$? Why is this limit relevant to calculating the derivative of the function $f(x) = e^x$?

15. Describe the process called logarithmic differentiation. What types of differentiation problems is logarithmic differentiation useful for?

16. Why do we need to consider absolute values when we apply logarithmic differentiation to $f(x) = xe^x \sin x$? In contrast, why do we not need to consider absolute values when we apply logarithmic differentiation to $f(x) = x^x$?

Skills

Find the derivatives of each of the functions in Exercises 17–44. In some cases it may be convenient to do some preliminary algebra.

17. $f(x) = \dfrac{1}{2 - e^{5x}}$

18. $f(x) = \log_2(3x^2 - 5)$

19. $f(x) = 3x^2 e^{-4x}$

20. $f(x) = e^{3x} \ln(x^2 + 1)$

21. $f(x) = \dfrac{1 - x}{e^x}$

22. $f(x) = \ln\left(\dfrac{x^3}{x^2 + x + 1}\right)$

23. $f(x) = e^x(x^2 + 3x - 1)$

24. $f(x) = \dfrac{e^x \ln x}{x^2 - 1}$

25. $f(x) = e^{3 \ln x}$

26. $f(x) = 3^x + \log_3 x$

27. $f(x) = e^{(e^x)}$

28. $f(x) = e^{(x^e)}$

29. $f(x) = (e^x)^e$

30. $f(x) = (x^e)^e$

31. $f(x) = \dfrac{\ln(x^5)}{\ln(x^4)}$

32. $f(x) = 11 + e^{\pi} - \ln 2$

33. $f(x) = x^{-3} e^{2x}$

34. $f(x) = \ln(x^2 2^x)$

35. $f(x) = x^2 \log_2(x2^x)$

36. $f(x) = \sqrt{\ln(x^2 + 1)}$

37. $f(x) = \ln(x^2 + e^{\sqrt{x}})$

38. $f(x) = \dfrac{\sqrt{e^x}}{\ln \sqrt{x}}$

39. $f(x) = \sqrt{\log_2(3^x - 5)}$

40. $f(x) = \ln(x^2+1)(e^x)^{-1/3}$

41. $f(x) = x^2 \ln(\ln x)$

42. $f(x) = \ln(x^x)$

43. $f(x) = (2^x)^x$

44. $f(x) = 2^{1-3^x}$

Describe the derivatives of each of the piecewise-defined functions in Exercises 45–48.

45. $f(x) = \begin{cases} 2^x, & \text{if } x \le -2 \\ \dfrac{1}{x^2}, & \text{if } x > -2 \end{cases}$

46. $f(x) = \begin{cases} \ln(-x), & \text{if } x < 0 \\ \ln x, & \text{if } x \ge 0 \end{cases}$

47. $f(x) = \begin{cases} x^2, & \text{if } x < 1 \\ 1 + \ln x, & \text{if } x \ge 1 \end{cases}$

48. $f(x) = \begin{cases} \dfrac{1}{2} - \dfrac{1}{4}x, & \text{if } x \le 0 \\ \dfrac{1}{1 + e^x}, & \text{if } x > 0 \end{cases}$

Use logarithmic differentiation to find the derivatives of each of the functions in Exercises 49–58.

49. $f(x) = \sqrt{x \ln |2^x + 1|}$

50. $f(x) = 12x\sqrt[3]{1 - x}\sqrt{x + 1}$

51. $f(x) = \dfrac{2^x \sqrt{x^3 - 1}}{\sqrt{x}(2x - 1)}$

52. $f(x) = \dfrac{e^{2x}(x^3 - 2)^4}{x(3e^{5x} + 1)}$

53. $f(x) = x^{\ln x}$

54. $f(x) = (2x + 1)^{3x}$

55. $f(x) = \left(\dfrac{x}{x - 1}\right)^x$

56. $f(x) = (\ln x)^x$

57. $f(x) = (\ln x)^{\ln x}$

58. $f(x) = \left(\dfrac{1}{x + 1}\right)^x$

In Exercises 59–63, find a function f that has the given derivative f'. In each case you can find the answer with an educated guess-and-check process.

59. $f'(x) = \dfrac{4e^{4x}(3x^5 - 1) - e^{4x}(15x^4)}{(3x^5 - 1)^2}$

60. $f'(x) = x^2 e^{x^3}$

61. $f'(x) = \dfrac{x}{x^2 + 3}$

62. $f'(x) = e^x(1 + e^x)$

63. $f'(x) = \dfrac{e^x}{1 + e^x}$

Applications

64. An abandoned building contained 45 rats on the first day of the year and 53 rats 30 days later. Let $r(t)$ be the function that describes the number of rats in the building t days after the first of the year.

(a) Find a formula for $r(t)$ given that the rate of change of the rat population is constant, and use this formula to predict the number of rats in the building on the 100th day of the year ($t = 99$).

(b) Find a formula for $r(t)$ given that the rate of change of the rat population is proportional to the number of rats in the building, and use this formula to predict the number of rats in the building on the 100th day of the year ($t = 99$).

65. Alina started an investment account with an initial deposit of one thousand dollars. Following the initial deposit, the amount of money increased at a rate proportional to her investment account balance. After three years her balance was $1,260.

(a) Write down a function $A(t)$ that describes the amount of money in Alina's investment account t years after her initial deposit.

(b) How much money will Alina have in her investment account after 30 years?

(c) How long will it be before Alina's initial investment quadruples?

66. The temperature T, in degrees Fahrenheit, of a yam after sitting in a hot oven for t minutes is given by the function

$$T(t) = 350 - 280e^{-0.2t}.$$

(a) What is the initial temperature of the yam, before it is put in the oven?

(b) Given that over time the temperature of the yam will approach the temperature inside the oven, use a limit to determine the temperature of the oven.

(c) How long will it take for the yam to be within 5 degrees Fahrenheit of the temperature of the oven?

(d) The first derivative of $T(t)$ measures the rate of change of the temperature of the yam. The second derivative of $T(t)$ measures the rate of change of the rate of change of the temperature of the yam. Use $T'(t)$ and $T''(t)$ to argue that the temperature of the yam increases at a decreasing rate. This statement is related to the odd saying "Cold water boils faster." How?

67. A political candidate starts an advertising campaign in Hamtown, Virginia. The number of people in Hamtown that have heard of him t days after the start of his campaign is given by

$$P(t) = \frac{45{,}000}{1 + 35e^{-0.12t}}.$$

(a) How many people knew about the candidate before the start of his advertising campaign?

(b) Given that over time the advertising will eventually reach everyone in the town, use a limit to determine the population of Hamtown.

(c) How many days will it take for all but one person in Hamtown to have heard of the candidate?

(d) Find $P'(t)$ and use it to argue that in this model, the number of people who have heard of the candidate is always increasing. Does this make sense in the context of this problem?

Proofs

68. Use the definition of the derivative and the definition of the number e to prove that $f(x) = e^x$ is its own derivative.

69. Use the chain rule to prove that $\frac{d}{dx}(e^{kx}) = ke^{kx}$.

70. Use the fact that $\frac{d}{dx}(b^x) = (\ln b)\, b^x$ to prove that $\frac{d}{dx}(e^{kx}) = ke^{kx}$.

71. Prove that if f is an exponential function, then $f'(x)$ is proportional to $f(x)$.

72. Use implicit differentiation and the fact that $\log_b x$ is the inverse of b^x to prove that $\frac{d}{dx}(\log_b x) = \frac{1}{(\ln b)x}$.

73. Use the definition of $|x|$, the chain rule, and the fact that $\frac{d}{dx}(\ln x) = \frac{1}{x}$ for $x > 0$ to prove that $\frac{d}{dx}(\ln |x|) = \frac{1}{x}$ for all $x \neq 0$.

74. Use a direct application of the fact that $\frac{d}{dx}(f^{-1}(x)) = \frac{1}{f'(f^{-1}(x))}$ to prove that $\frac{d}{dx}(\ln x) = \frac{1}{x}$.

Thinking Forward

L'Hôpital's rule: At the end of Chapter 3 we will see that, under certain conditions, the limit of a quotient of functions is equal to the limit of the quotient of the derivatives of those functions. Specifically, if $f(x)$ and $g(x)$ both approach zero as $x \to c$, then

$$\lim_{x\to c} \frac{f(x)}{g(x)} = \lim_{x\to c} \frac{f'(x)}{g'(x)}.$$

Show that each of the following limits is of the form $\frac{0}{0}$ and then use L'Hôpital's rule to calculate the limit:

- $\lim_{x\to 0} \frac{x^3}{1 - 2^x}$
- $\lim_{x\to 1} \frac{3^x - 3}{1 - x^2}$
- $\lim_{x\to 1} \frac{\ln x}{x - 1}$
- $\lim_{x\to 3} \frac{(x-3)^2}{1 - e^{x-3}}$

Differential equations: A function $y(x)$ is exponential if and only if its derivative is proportional to itself. This means that exponential functions are solutions of ***differential equations*** of the form $\frac{dy}{dx} = ky$. A solution of a differential equation is a function $y(x)$ that makes the equation true.

- Show that $y(x) = 4e^{3x}$ is a solution of the differential equation $\frac{dy}{dx} = 3y$.
- Show that $y(x) = 1.7e^{-2.1x}$ is a solution of the differential equation $\frac{dy}{dx} = -2.1y$.
- Show that $y(x) = 3(2^x)$ is a solution of the differential equation $\frac{dy}{dx} = (\ln 2)y$.
- Describe all of the solutions of the differential equation $\frac{dy}{dx} = 3y$.
- Describe all of the solutions of the differential equation $\frac{dy}{dx} = 3y$ that satisfy $y(0) = 2$.

2.6 DERIVATIVES OF TRIGONOMETRIC AND HYPERBOLIC* FUNCTIONS

- Derivatives of the six trigonometric functions
- Derivatives of inverse trigonometric functions
- Hyperbolic functions, inverse hyperbolic functions, and their derivatives

Derivatives of Trigonometric Functions

Because trigonometric functions have periodic oscillating behavior, and their slopes also have periodic oscillating behavior, it would make sense if the derivatives of trigonometric functions were trigonometric. For example, the two graphs that follow show the function $f(x) = \sin x$ and its derivative $f'(x) = \cos x$. As we will prove in Theorem 2.17, it turns out that, at each value of x, the slope of the graph of $f(x) = \sin x$ is given by the height of the graph of $f'(x) = \cos x$. Before we tackle this fact algebraically, take a minute to verify that it is the case with these graphs for the values $x = -5.2$, $x = \frac{\pi}{2}$, and $x = 4$, as shown in the following figures:

Slopes of $f(x) = \sin x$ at three points

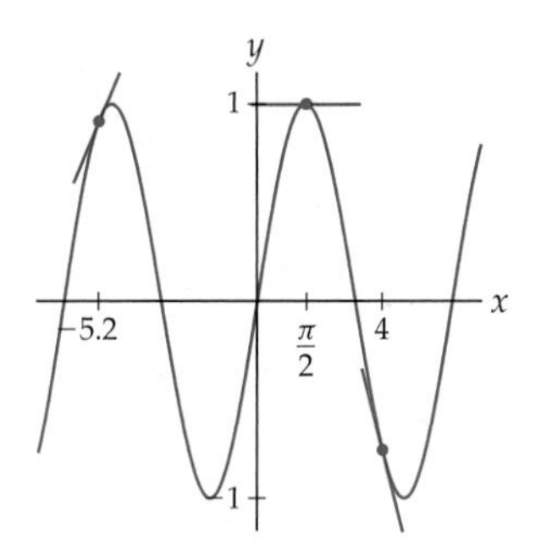

Heights of $f'(x) = \cos x$ at three points

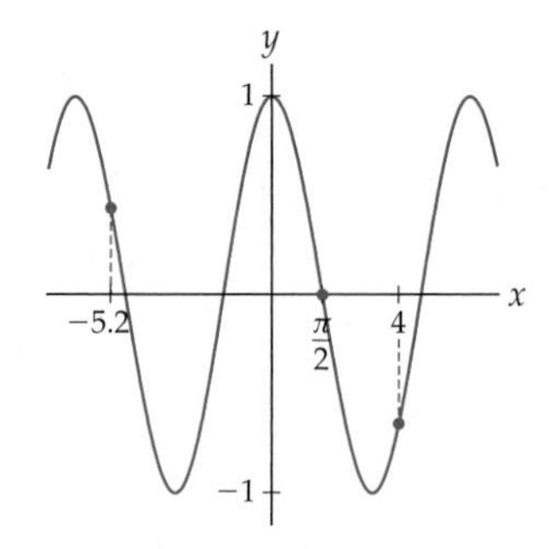

The six trigonometric functions have the following derivatives:

THEOREM 2.17 Derivatives of the Trigonometric Functions

For all values of x at which the following functions below are defined,

(a) $\frac{d}{dx}(\sin x) = \cos x$ **(c)** $\frac{d}{dx}(\tan x) = \sec^2 x$ **(e)** $\frac{d}{dx}(\cot x) = -\csc^2 x$

(b) $\frac{d}{dx}(\cos x) = -\sin x$ **(d)** $\frac{d}{dx}(\sec x) = \sec x \tan x$ **(f)** $\frac{d}{dx}(\csc x) = -\csc x \cot x$

It is important to note that these formulas for derivatives are true only if angles are measured in radians; see Exercise 5.

Proof. We will prove the formulas for $\sin x$ and $\tan x$ from parts (a) and (c) and leave the proofs of the remaining four formulas to Exercises 83–86.

(a) The proof of the first formula is nothing more than an annotated calculation using the definition of the derivative. To simplify the limit we obtain, we will rewrite $\sin(x + h)$ with a trigonometric identity. Our goal after that will be to rewrite the limit so that we can apply the two trigonometric limits from Theorem 1.35 in Section 1.6. The calculation is as follows:

$$\begin{aligned}
\frac{d}{dx}(\sin x) &= \lim_{h\to 0}\frac{\sin(x+h)-\sin x}{h} && \leftarrow \text{definition of the derivative}\\
&= \lim_{h\to 0}\frac{(\sin x\cos h+\sin h\cos x)-\sin x}{h} && \leftarrow \text{sum identity for sine}\\
&= \lim_{h\to 0}\frac{\sin x(\cos h-1)+\sin h\cos x}{h} && \leftarrow \text{algebra}\\
&= \lim_{h\to 0}\left(\sin x\,\frac{\cos h-1}{h}+\cos x\,\frac{\sin h}{h}\right) && \leftarrow \text{algebra}\\
&= \sin x\left(\lim_{h\to 0}\frac{\cos h-1}{h}\right)+\cos x\left(\lim_{h\to 0}\frac{\sin h}{h}\right) && \leftarrow \text{limit rules}\\
&= (\sin x)(0)+(\cos x)(1)=\cos x. && \leftarrow \text{trigonometric limits}
\end{aligned}$$

(c) We do not have to resort to the definition of the derivative in order to prove the formula for differentiating $\tan x$. Instead we can use the quotient rule, the fact that $\tan x = \frac{\sin x}{\cos x}$, and the formulas for differentiating $\sin x$ and $\cos x$:

$$\begin{aligned}
\frac{d}{dx}(\tan x) &= \frac{d}{dx}\left(\frac{\sin x}{\cos x}\right)\\
&= \frac{\frac{d}{dx}(\sin x)\cdot(\cos x)-(\sin x)\cdot\frac{d}{dx}(\cos x)}{(\cos x)^2} && \leftarrow \text{quotient rule}\\
&= \frac{(\cos x)(\cos x)-(\sin x)(-\sin x)}{\cos^2 x} && \leftarrow \text{derivatives of } \sin x \text{ and } \cos x\\
&= \frac{\cos^2 x+\sin^2 x}{\cos^2 x}=\frac{1}{\cos^2 x}=\sec^2 x. && \leftarrow \text{algebra and identities}
\end{aligned}$$

■

Derivatives of Inverse Trigonometric Functions

We can use the formulas for the derivatives of the trigonometric functions to prove the following formulas for the derivatives of the inverse trigonometric functions:

THEOREM 2.18 **Derivatives of Inverse Trigonometric Functions**

For all values of x at which the following functions are defined,

(a) $\frac{d}{dx}(\sin^{-1} x)=\frac{1}{\sqrt{1-x^2}}$ **(b)** $\frac{d}{dx}(\tan^{-1} x)=\frac{1}{1+x^2}$ **(c)** $\frac{d}{dx}(\sec^{-1} x)=\frac{1}{|x|\sqrt{x^2-1}}$

It is extremely important and surprising to note that although inverse trigonometric functions are transcendental, their derivatives are algebraic! This property makes these derivative formulas particularly useful for finding certain antiderivatives, and in Chapter 6 they will be part of our arsenal of integration techniques. Of course, all of these rules can be used in combination with the sum, product, quotient, and chain rules. For example,

$$\frac{d}{dx}(\sin^{-1}(3x^2-1)) = \frac{1}{\sqrt{1-(3x^2-1)^2}}(6x).$$

Proof. We will prove the rule for $\sin^{-1} x$ and leave the remaining two rules to Exercises 87 and 88. We could apply Theorem 2.16 here, but it is just as easy to do the implicit differentiation by hand. Since $\sin(\sin^{-1} x) = x$ for all x in the domain of $\sin^{-1} x$, we have

$$\sin(\sin^{-1} x) = x \qquad \leftarrow \sin^{-1} x \text{ is the inverse of } \sin x$$

$$\frac{d}{dx}(\sin(\sin^{-1} x)) = \frac{d}{dx}(x) \qquad \leftarrow \text{differentiate both sides}$$

$$\cos(\sin^{-1} x) \cdot \frac{d}{dx}(\sin^{-1} x) = 1 \qquad \leftarrow \text{chain rule}$$

$$\frac{d}{dx}(\sin^{-1} x) = \frac{1}{\cos(\sin^{-1} x)} \qquad \leftarrow \text{algebra}$$

$$\frac{d}{dx}(\sin^{-1} x) = \frac{1}{\sqrt{1 - \sin^2(\sin^{-1} x)}} \qquad \leftarrow \text{since } \sin^2 x + \cos^2 x = 1$$

$$\frac{d}{dx}(\sin^{-1} x) = \frac{1}{\sqrt{1 - x^2}}. \qquad \leftarrow \sin x \text{ is the inverse of } \sin^{-1} x$$

We could also have used triangles and the unit circle to show that the composition $\cos(\sin^{-1} x)$ is equal to the algebraic expression $\sqrt{1 - x^2}$, as we did in Example 4 of Section 0.4. ■

An interesting fact about the derivatives of the inverse sine and inverse secant functions is that their domains are slightly smaller than the domains of the original functions. The graphs of the inverse trigonometric functions are as follows (note their domains):

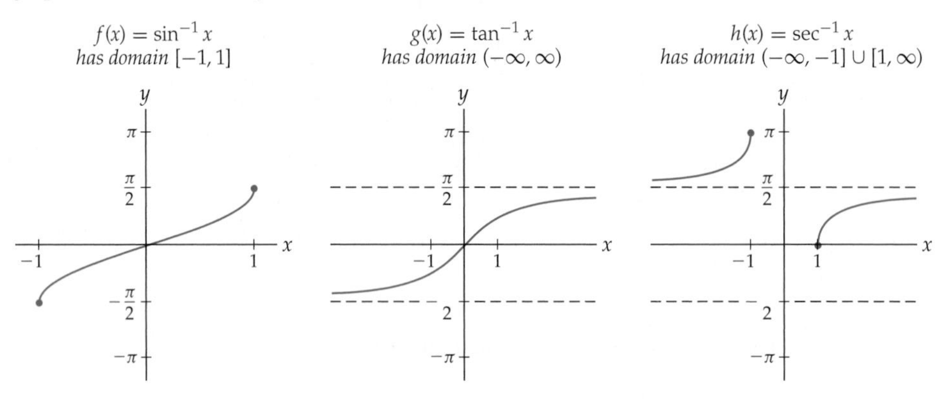

If you look closely at the first and third graphs, you should notice that at the ends of the domains the tangent lines will be vertical. Since a vertical line has undefined slope, the derivative does not exist at these points. This means that the derivatives of $\sin^{-1} x$ and $\sec^{-1} x$ are not defined at $x = 1$ or $x = -1$; see the first and third graphs shown next:

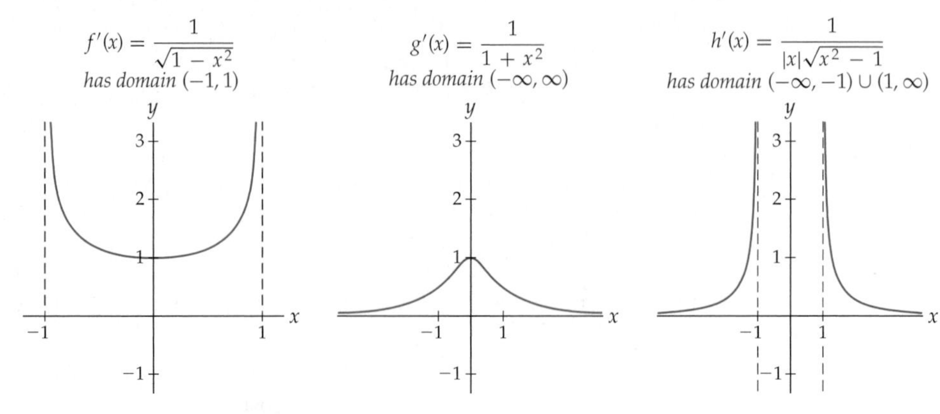

Hyperbolic Functions and Their Derivatives*

The trigonometric functions sine and cosine are ***circular functions*** in the sense that they are defined to be the coordinates of a ***parameterization*** of the unit circle. This means that the circle defined by $x^2+y^2 = 1$ is the path traced out by the coordinates $(x, y) = (\cos t, \sin t)$ as t varies; see the following figure at the left:

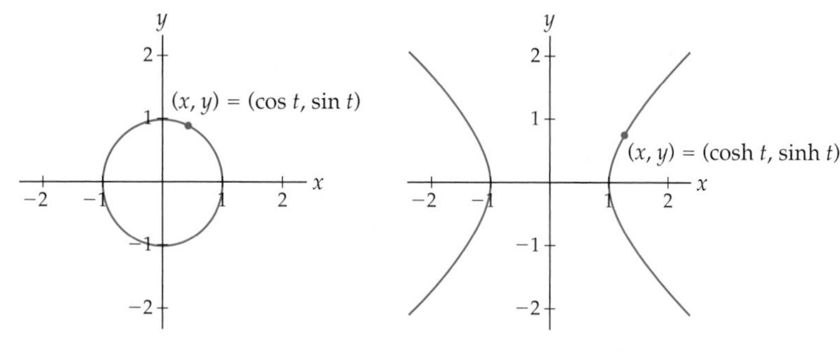

Now let's consider the path traced out by the hyperbola $x^2 - y^2 = 1$ as shown at the right. One parameterization of the right half of this hyperbola is traced out by the ***hyperbolic functions*** $(\cosh t, \sinh t)$ that we will spend the rest of this section investigating.

The hyperbolic functions are nothing more than simple combinations of the exponential functions e^x and e^{-x}:

DEFINITION 2.19 **Hypberbolic Sine and Hyperbolic Cosine**

For any real number x, the ***hyperbolic sine function*** and the ***hyperbolic cosine function*** are, respectively, defined as the following combinations of exponential functions:

$$\sinh x = \frac{e^x - e^{-x}}{2} \qquad \cosh x = \frac{e^x + e^{-x}}{2}$$

The hyperbolic sine function is pronounced "sinch" and the hyperbolic cosine function is pronounced "cosh." The "h" is for "hyperbolic." As we will soon see, the properties and interrelationships among the hyperbolic functions are similar to the properties and interrelationships among the trigonometric functions. These properties will be particularly useful in Chapter 6 when we attempt to solve certain forms of integrals.

It is a simple matter to use Definition 2.19 to verify that, for any value of t, the point $(x, y) = (\cosh t, \sinh t)$ lies on the hyperbola $x^2 - y^2 = 1$; see Exercise 89. We will usually think of this fact rewritten so that the independent variable is x, as follows:

$$\cosh^2 x - \sinh^2 x = 1.$$

Here we are using the familiar convention that, for example, $\sinh^2 x$ is shorthand for $(\sinh x)^2$. Note the similarity between the hyperbolic identity $\cosh^2 t - \sinh^2 t = 1$ and the Pythagorean identity for sine and cosine. Hyperbolic functions also satisfy many other algebraic identities that are reminiscent of those that hold for trigonometric functions, as you will see in Exercises 90–92.

Just as we can define four additional trigonometric functions from sine and cosine, we can define four additional hyperbolic functions from hyperbolic sine and hyperbolic cosine. We will be interested primarily in the ***hyperbolic tangent function***:

$$\tanh x = \frac{\sinh x}{\cosh x} = \frac{e^x - e^{-x}}{e^x + e^{-x}}.$$

We can also define csch x, sech x, and coth x as the reciprocals of sinh x, cosh x, and tanh x, respectively.

The graphs of sinh x, cosh x, and tanh x are shown next. In Exercises 13–16 you will investigate various properties of these graphs.

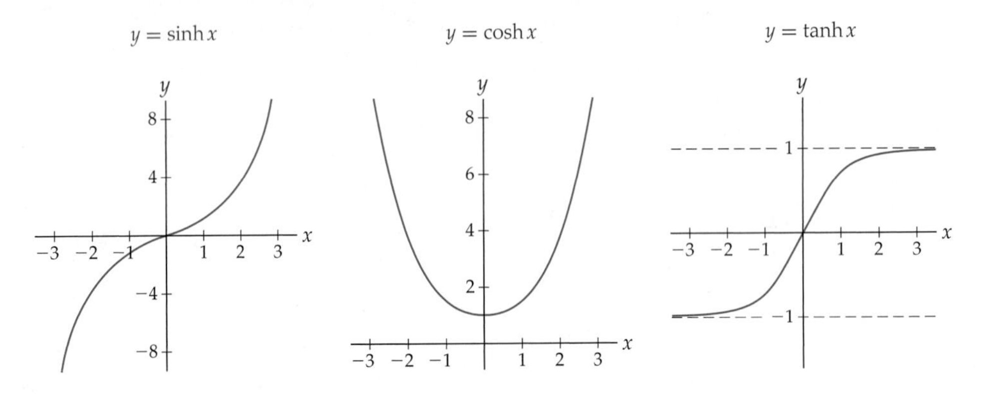

In Chapter 4 we will see that the graph of $y = \cosh x$ is an example of a ***catenary*** curve (see also Exercise 82), which is the shape formed by a hanging chain or cable.

As with any functions that we study, we are interested in finding formulas for the derivatives of sinh x, cosh x, and tanh x. The similarity between hyperbolic functions and trigonometric functions continues here. These derivatives follow a very familiar pattern, differing from the pattern for trigonometric functions only by a sign change.

THEOREM 2.20 **Derivatives of Hyperbolic Functions**

For all real numbers x,

(a) $\frac{d}{dx}(\sinh x) = \cosh x$ **(b)** $\frac{d}{dx}(\cosh x) = \sinh x$ **(c)** $\frac{d}{dx}(\tanh x) = \operatorname{sech}^2 x$

If you prefer to stay away from the hyperbolic secant function sech x, you can write the derivative in part (c) as $\frac{1}{\cosh^2 x}$.

Proof. The proofs of these differentiation formulas follow immediately from the definitions of the hyperbolic functions as simple combinations of exponential functions. For example,

$$\frac{d}{dx}(\sinh x) = \frac{d}{dx}\left(\frac{1}{2}(e^x - e^{-x})\right) = \frac{1}{2}(e^x + e^{-x}) = \cosh x.$$

The proofs of parts (b) and (c) are left to Exercises 93 and 94. ∎

Although hyperbolic functions may seem somewhat exotic, they work with the other differentiation rules just as any other functions do. For example, with the product and chain rules we can calculate

$$\frac{d}{dx}(5x \sinh^3 x^2) = 5\sinh^3 x^2 + 5x(3\sinh^2 x^2)(\cosh x^2)(2x).$$

The derivatives of the hyperbolic cotangent, secant, and cosecant functions are also similar to those of their trigonometric cousins, but at the moment we will be focusing only on hyperbolic sine, cosine, and tangent.

Inverse Hyperbolic Functions and Their Derivatives*

For a function to have an inverse, it must be one-to-one. Looking back at the graphs of $\sinh x$, $\cosh x$, and $\tanh x$, we see that only $\cosh x$ fails to be one-to-one. Just as we did when we defined the trigonometric inverses, we will restrict the domain of $\cosh x$ to a smaller domain on which it is one-to-one. We will choose the restricted domain of $\cosh x$ to be $x \geq 0$. The notation we will use for the inverses of these three functions is what you would expect: $\sinh^{-1} x$, $\cosh^{-1} x$ and $\tanh^{-1} x$.

Since the hyperbolic functions are defined as combinations of exponential functions, it would seem reasonable to expect that their inverses could be expressed in terms of logarithmic functions. This is in fact the case, as you will see in Exercises 97–99. However, our main concern here is to find formulas for the derivatives of the inverse hyperbolic functions, which we can do directly from identities and properties of inverses.

THEOREM 2.21 **Derivatives of Inverse Hyperbolic Functions**

For all x for which the following are defined,

(a) $\frac{d}{dx}(\sinh^{-1} x) = \frac{1}{\sqrt{x^2+1}}$ **(b)** $\frac{d}{dx}(\cosh^{-1} x) = \frac{1}{\sqrt{x^2-1}}$ **(c)** $\frac{d}{dx}(\tanh^{-1} x) = \frac{1}{1-x^2}$

Similar formulas can be developed for the inverse hyperbolic cotangent, secant, and cosecant functions. Notice the strong similarities between these derivatives and the derivatives of the inverse trigonometric functions.

Proof. We will prove the rule for the derivative of $\sinh^{-1} x$ and leave the remaining two rules to Exercises 95 and 96. Starting from the fact that $\sinh(\sinh^{-1} x) = x$ for all x, we can apply implicit differentiation:

$$\sinh(\sinh^{-1} x) = x \qquad \leftarrow \sinh^{-1} x \text{ is the inverse of } \sinh x$$

$$\frac{d}{dx}(\sinh(\sinh^{-1} x)) = \frac{d}{dx}(x) \qquad \leftarrow \text{differentiate both sides}$$

$$\cosh(\sinh^{-1} x) \cdot \frac{d}{dx}(\sinh^{-1} x) = 1 \qquad \leftarrow \text{chain rule, derivative of } \sinh x$$

$$\frac{d}{dx}(\sinh^{-1} x) = \frac{1}{\cosh(\sinh^{-1} x)}. \qquad \leftarrow \text{algebra}$$

$$\frac{d}{dx}(\sinh^{-1} x) = \frac{1}{\sqrt{1+\sinh^2(\sinh^{-1} x)}} \qquad \leftarrow \text{since } \cosh^2 x - \sinh^2 x = 1$$

$$\frac{d}{dx}(\sinh^{-1} x) = \frac{1}{\sqrt{1+x^2}}. \qquad \leftarrow \sinh x \text{ is the inverse of } \sinh^{-1} x$$

Compare this proof with our proof earlier in this section for the derivative of $\sin^{-1} x$; the two are similar. ■

Examples and Explorations

EXAMPLE 1 **Differentiating combinations of trigonometric functions**

Find the derivatives of each of the following functions:

(a) $f(x) = \frac{\tan x}{x^3 - 2}$ **(b)** $f(x) = x \sin^{-1}(3x+1)$ **(c)** $f(x) = \sec^2 e^x$

SOLUTION

(a) The function $f(x) = \frac{\tan x}{x^3-2}$ is a quotient of two functions. By the quotient rule and the rule for differentiating tangent, we have

$$\frac{d}{dx}\left(\frac{\tan x}{x^3-2}\right) = \frac{\frac{d}{dx}(\tan x)\cdot(x^3-2) - (\tan x)\cdot\frac{d}{dx}(x^3-2)}{(x^3-2)^2}$$
$$= \frac{(\sec^2 x)(x^3-2) - (\tan x)(3x^2)}{(x^3-2)^2}.$$

(b) The function $f(x) = x\sin^{-1}(3x+1)$ is a product of two functions, and thus we begin with the product rule. We will also need the chain rule to differentiate the composition $\sin^{-1}(3x+1)$:

$$f'(x) = (1)\cdot\sin^{-1}(3x+1) + x\cdot\frac{1}{\sqrt{1-(3x+1)^2}}(3)$$
$$= \sin^{-1}(3x+1) + \frac{3x}{\sqrt{1-(3x+1)^2}}.$$

(c) The function $f(x) = \sec^2 e^x$ is a composition of three functions, and thus we need to apply the chain rule twice:

$$\frac{d}{dx}(\sec^2 e^x) = \frac{d}{dx}((\sec(e^x))^2) \quad \leftarrow \text{rewrite so compositions are clear}$$
$$= 2(\sec e^x)^1\cdot\frac{d}{dx}(\sec e^x) \quad \leftarrow \text{first application of chain rule}$$
$$= 2(\sec e^x)(\sec e^x)(\tan e^x)\cdot\frac{d}{dx}(e^x) \quad \leftarrow \text{second application of chain rule}$$
$$= 2(\sec e^x)(\sec e^x)(\tan e^x)e^x \quad \leftarrow \text{derivative of } e^x$$

Perhaps the most difficult part of the preceding calculation is that the derivative of $\sec x$ has *two* instances of the independent variable: $\frac{d}{dx}(\sec x) = \sec \underline{x} \tan \underline{x}$. This means that we needed to put the "inside" function e^x into *both* of the slots for variables. □

EXAMPLE 2 **Differentiating combinations of hyperbolic functions***

Find the derivatives of each of the following functions:

(a) $f(x) = \ln(\tanh^2(x^3+2x+1))$ **(b)** $f(x) = \sqrt{\cosh^{-1}(e^{3x})}$

SOLUTION

(a) This is a nested chain-rule problem, since $f(x)$ is a composition of multiple functions. We will work from the outside to the inside, one step at a time:

$$f'(x) = \frac{1}{\tanh^2(x^3+2x+1)}\frac{d}{dx}(\tanh^2(x^3+2x+1))$$
$$= \frac{1}{\tanh^2(x^3+2x+1)}(2\tanh(x^3+2x+1))\frac{d}{dx}(\tanh(x^3+2x+1))$$
$$= \frac{1}{\tanh^2(x^3+2x+1)}(2\tanh(x^3+2x+1))(\operatorname{sech}^2(x^3+2x+1))(3x^2+2).$$

(b) Once again we have a nested chain-rule situation. Notice in particular how e^{3x} works with the derivative of the inverse hyperbolic cosine function:

$$f'(x) = \frac{1}{2}(\cosh^{-1}(e^{3x}))^{-1/2}\frac{d}{dx}(\cosh^{-1}(e^{3x})) = \frac{1}{2}(\cosh^{-1}(e^{3x}))^{-1/2}\left(\frac{1}{\sqrt{(e^{3x})^2-1}}\right)(3e^{3x}).$$
□

EXAMPLE 3 **Finding antiderivatives that involve inverse trigonometric functions**

Find a function f whose derivative is $f'(x) = \dfrac{1}{1+4x^2}$.

SOLUTION

Since the derivative of $\tan^{-1} x$ is $\frac{1}{1+x^2}$, we might suspect that the function we are looking for is related to the inverse tangent function. We will use an intelligent guess-and-check method to find f. Clearly $f(x) = \tan^{-1} x$ isn't exactly right, since its derivative is missing the "4." A good guess might be $f(x) = \tan^{-1}(4x)$; let's try that:

$$\frac{d}{dx}(\tan^{-1}(4x)) = \frac{1}{1+(4x)^2}(4) = \frac{4}{1+16x^2}.$$

Obviously that wasn't quite right either; but by examining the results we can make a new guess. We might try $\tan^{-1}(2x)$, since the "$2x$" will be squared in the derivative and become the "$4x^2$" we are looking for in the denominator:

$$\frac{d}{dx}(\tan^{-1}(2x)) = \frac{1}{1+(2x)^2}(2) = \frac{2}{1+4x^2}.$$

Now we are getting somewhere; this result differs by a multiplicative constant from the derivative $f'(x)$ we are looking for, and that is easy to fix. We need only divide our guess by that constant. Try the function $f(x) = \frac{1}{2}\tan^{-1}(2x)$:

$$\frac{d}{dx}\left(\frac{1}{2}\tan^{-1}(2x)\right) = \left(\frac{1}{2}\right)\frac{1}{1+(2x)^2}(2) = \frac{1}{1+4x^2}.$$

We now know that $f(x) = \frac{1}{2}\tan^{-1}(2x)$ is a function whose derivative is $f'(x) = \frac{1}{1+4x^2}$. Of course, we could also add any constant to $f(x)$ and not change its derivative; for example, $f(x) = \frac{1}{2}\tan^{-1}(2x) + 5$ would work as well. In fact, any function of the form $f(x) = \frac{1}{2}\tan^{-1}(2x) + C$ will have $f'(x) = \frac{1}{1+4x^2}$. □

EXAMPLE 4 **Finding antiderivatives that involve hyperbolic functions***

Find a function f whose derivative is $f'(x) = \dfrac{e^x}{\sqrt{e^{2x}-1}}$.

SOLUTION

Until we learn more specific antidifferentiation techniques in Chapter 6, a problem like this is best done by an intelligent guess-and-check procedure. Given that we have the inverse hyperbolic functions in mind, the best match of the three is the derivative of $\cosh^{-1} x$. Since the expression for $f'(x)$ also involves an e^x, let's revise that guess right away to $\cosh^{-1} e^x$. Now we check by differentiating with the chain rule:

$$\frac{d}{dx}(\cosh^{-1} e^x) = \frac{1}{\sqrt{(e^x)^2-1}} \cdot e^x = \frac{e^x}{\sqrt{e^{2x}-1}}.$$

We guessed it on the first try! We have just shown that $f(x) = \cosh^{-1} e^x$ has the desired derivative. □

? TEST YOUR UNDERSTANDING

- What trigonometric limits were used to find the derivative of $\sin x$?
- How can we obtain the derivative of $\sec x$ from the derivative of $\cos x$?
- What is the graphical reason that the domains of the derivatives of $\sin^{-1} x$ and $\sec^{-1} x$ are slightly smaller than the domains of the functions themselves?

- How are hyperbolic functions similar to trigonometric functions? How are they different?
- How can we obtain the derivative of $\sinh^{-1} x$ from the derivative of $\sinh x$?

EXERCISES 2.6

Thinking Back

Trigonometric and inverse trigonometric values: Without using a calculator, find the exact values of each of the following quantities.

- $\sin\left(-\frac{\pi}{3}\right)$
- $\tan\left(-\frac{\pi}{4}\right)$
- $\sec\left(\frac{5\pi}{6}\right)$
- $\sin^{-1} 1$
- $\tan^{-1}(\sqrt{3})$
- $\sec^{-1}(-2)$

Compositions: For each function k that follows, find functions f, g, and h such that $k = f \circ g \circ h$. There may be more than one possible answer.

- $k(x) = \dfrac{1}{\sin(x^3)}$
- $k(x) = \sin^{-1}(\cos^2 x)$
- $k(x) = \tan^2(3x + 1)$
- $k(x) = \sec(x^3)\tan(x^3)$

Writing trigonometric compositions algebraically: Prove each of the following equalities, which rewrite compositions of trigonometric and inverse trigonometric functions as algebraic functions.

- $\cos(\sin^{-1} x) = \sqrt{1 - x^2}$
- $\sin(\cos^{-1} x) = \sqrt{1 - x^2}$
- $\sec^2(\tan^{-1} x) = 1 + x^2$
- $\tan(\sec^{-1} x) = |x|\sqrt{1 - \frac{1}{x^2}}$

Concepts

0. *Problem Zero:* Read the section and make your own summary of the material.

1. *True/False:* Determine whether each of the statements that follow is true or false. If a statement is true, explain why. If a statement is false, provide a counterexample.

(a) *True or False:* To find the derivative of $\sin x$ we have to use the definition of the derivative.

(b) *True or False:* To find the derivative of $\tan x$ we have to use the definition of the derivative.

(c) *True or False:* The derivative of $\frac{x^4}{\sin x}$ is $\frac{4x^3}{\cos x}$.

(d) *True or False:* If a function is algebraic, then so is its derivative.

(e) *True or False:* If a function is transcendental, then so is its derivative.

(f) *True or False:* If f is a trigonometric function, then f' is also a trigonometric function.

(g) *True or False:* If f is an inverse trigonometric function, then f' is also an inverse trigonometric function.

(h) *True or False:* If f is a hyperbolic function, then f' is also a hyperbolic function.

2. *Examples:* Construct examples of the thing(s) described in the following. Try to find examples that are different than any in the reading.

(a) A function that is its own fourth derivative.

(b) A function whose domain is larger than the domain of its derivative.

(c) Three non-logarithmic functions that are transcendental, but whose derivatives are algebraic.

3. What limit facts and trigonometric identities are used in the proof that $\frac{d}{dx}(\sin x) = \cos x$?

4. Sketch graphs of $\sin x$ and $\cos x$ on $[-2\pi, 2\pi]$.

(a) Use the graph of $\sin x$ to determine where $\sin x$ is increasing and and where it is decreasing.

(b) Use the graph of $\cos x$ to determine where $\cos x$ is positive and where it is negative.

(c) Explain why your answers to parts (a) and (b) suggest that $\cos x$ is the derivative of $\sin x$.

5. The differentiation formula $\frac{d}{dx}(\sin x) = \cos x$ is valid only if x is measured in radians. In this exercise you will explore why this derivative relationship does not hold if x is measured in degrees.

(a) Set your calculator to degree mode, and sketch a graph of $\sin x$ that shows at least two periods. If the derivative of $\sin x$ is $\cos x$, then the slope of your graph at $x = 0$ should be equal to $\cos 0 = 1$. Use your graph to explain why this is not the case when we use degrees. *(Hint: Think about your graphing window scale.)*

(b) Now set your calculator back to radians mode!

6. Suppose you wish to differentiate $g(x) = \sin^2(x) + \cos^2(x)$. What is the fastest way to do this, and why?

7. The derivatives of the function $f(x) = \cos(3x^2)$ that follow are incorrect. What misconception occurs in each case?

(a) *Incorrect:* $f'(x) = (-\sin x)(3x^2) + (\cos x)(6x)$.

(b) *Incorrect:* $f'(x) = -\sin(6x)$.

8. The derivatives of the function $f(x) = \cos(3x^2)$ that follow are incorrect. What misconception occurs in each case?

(a) *Incorrect:* $f'(x) = -\sin(3x^2)$.

(b) *Incorrect:* $f'(x) = -\sin(3x^2)(6x)(6)$.

9. In the proof that $\frac{d}{dx}(\sin^{-1} x) = \frac{1}{\sqrt{1-x^2}}$, we used the fact that $\sin(\sin^{-1} x) = x$. It is also true that $\sin^{-1}(\sin x) = x$; could we have started with that inequality instead? Why or why not?

10. Both of the following equations are true: $\tan(\tan^{-1} x) = x$ and $\tan^{-1}(\tan x) = x$. We can find the derivative of $\tan^{-1} x$ by differentiating both sides of one of these equations and solving for $\frac{d}{dx}(\tan^{-1} x)$. Which one of the equations should we use, and why?

11. How can the derivative of $\sin^{-1} x$ be equal to *both* $\frac{1}{\sqrt{1-x^2}}$ and $\frac{1}{\cos(\sin^{-1} x)}$? Which expression is easier to use, and why?

12. The function $\sin^{-1} x$ is defined on $[-1, 1]$, but its derivative $\frac{1}{\sqrt{1-x^2}}$ is defined only on $(-1, 1)$. Explain why the tangent lines to the graph of $y = \sin^{-1} x$ do not exist at $x = \pm 1$. *(Hint: Think about the corresponding tangent lines on the graph of the restricted sine function.)*

13. The figure that follows at the left shows the graphs of $y = \sinh x$, $y = \cosh x$, and $y = \frac{1}{2}e^x$. For each of the statements that follows, explain graphically why the statement is true. Then justify the statement algebraically, using the definitions of the hyperbolic functions.

(a) $\sinh x \le \frac{1}{2}e^x \le \cosh x$ for all x

(b) $\lim_{x\to\infty} \frac{\sinh x}{(1/2)e^x} = 1$ and $\lim_{x\to\infty} \frac{\cosh x}{(1/2)e^x} = 1$

Graph for Exercise 13 *Graph for Exercise 14*

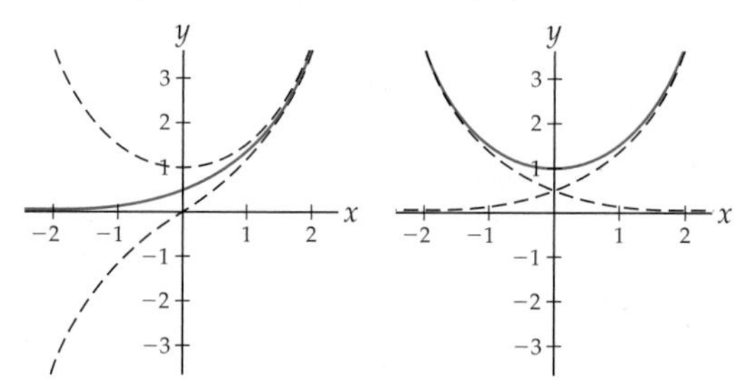

14. The preceding figure at the right shows the graphs of $y = \cosh x$, $y = \frac{1}{2}e^x$, and $y = \frac{1}{2}e^{-x}$. For each of the statements that follows, explain graphically why the statement is true. Then justify the statement algebraically, using the definitions of the hyperbolic functions.

(a) $\cosh x = \frac{1}{2}e^x + \frac{1}{2}e^{-x}$

(b) $\lim_{x\to -\infty} \frac{\cosh x}{(1/2)e^{-x}} = 1$

15. The figure that follows at the left shows the graphs of $y = \sinh x$, $y = \frac{1}{2}e^x$, and $y = -\frac{1}{2}e^{-x}$. For each of the statements that follows, explain graphically why the statement is true. Then justify the statement algebraically, using the definitions of the hyperbolic functions.

(a) $\sinh x = \frac{1}{2}e^x - \frac{1}{2}e^{-x}$

(b) $\lim_{x\to -\infty} \frac{\sinh x}{-(1/2)e^{-x}} = 1$

Graph for Exercise 15 *Graph for Exercise 16*

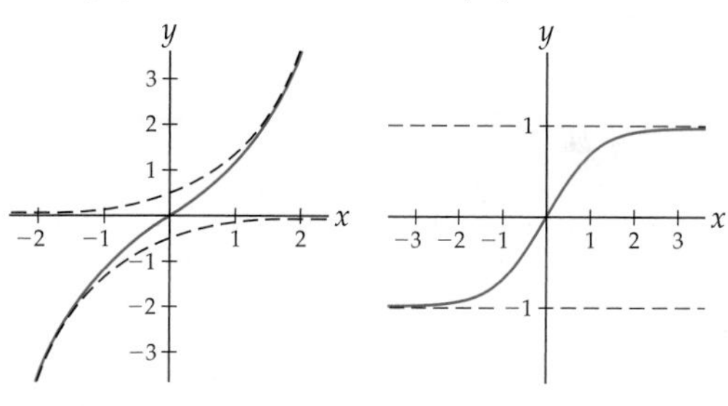

16. The preceding figure at the right shows the graphs of $y = \tanh x$, $y = 1$, and $y = -1$. For each of the statements that follows, explain graphically why the statement is true. Then justify the statement algebraically, using the definitions of the hyperbolic functions.

(a) $-1 \le \tanh x \le 1$

(b) $\lim_{x\to\infty} \tanh x = 1$ and $\lim_{x\to -\infty} \tanh x = -1$

Skills

Find the derivatives of each of the functions in Exercises 17–50. In some cases it may be convenient to do some preliminary algebra.

17. $f(x) = \frac{x^2+1}{\cos x}$

18. $f(x) = 2\cos(x^3)$

19. $f(x) = \cot x - \csc x$

20. $f(x) = \tan^2(3x+1)$

21. $f(x) = 4\sin^2 x + 4\cos^2 x$

22. $f(x) = \sec^2 x^{-1}$

23. $f(x) = 3\sec x \tan x$

24. $f(x) = 3^x \sec x + 17$

25. $f(x) = \sin(\cos(\sec(x)))$

26. $f(x) = \csc^2(e^x)$

27. $f(x) = e^{\csc^2 x}$

28. $f(x) = e^x \csc^2 x$

29. $f(x) = \frac{-2^x}{5x \sin x}$

30. $f(x) = \frac{\log_3(3^x)}{\sin^2 x + \cos^2 x}$

31. $f(x) = x\sqrt{\sin x \cos x}$

32. $f(x) = \frac{\sin x \csc x}{\cot x \cos x}$

33. $f(x) = \frac{3x^2 \ln x}{\tan x}$

34. $f(x) = \frac{\ln(3x^2)}{\tan x}$

35. $f(x) = \sin(\ln x)$

36. $f(x) = \ln(x \sin x)$

37. $f(x) = \sin^{-1}(3x^2)$

38. $f(x) = 3(\sin^{-1} x)^2$

39. $f(x) = x^2 \arctan x^2$

40. $f(x) = \tan^{-1}(\ln x)$

41. $f(x) = \sec^{-1} x^2$

42. $f(x) = \sin(\sin^{-1} x)$

43. $f(x) = \sin^{-1}(\sec^2 x)$

44. $f(x) = \sin^2(\sec^{-1} x)$

45. $f(x) = \frac{\sin^{-1} x}{\tan^{-1} x}$

46. $f(x) = \frac{\sin^{-1} x}{\sec^{-1} x}$

47. $f(x) = \ln(\text{arcsec}(\sin^2 x))$

48. $f(x) = x^{-2} e^{4x} \sin^{-1} x$

49. $f(x) = \sec(1 + \tan^{-1} x)$

50. $f(x) = \frac{\sin(\arcsin x)}{\arctan x}$

Find the derivatives of each of the functions in Exercises 51–62. In some cases it may be convenient to do some preliminary algebra. *(These exercises involve hyperbolic functions and their inverses.)*

51. $f(x) = x \sinh x^3$

52. $f(x) = x \sinh^3 x$

53. $f(x) = \cosh(\ln(x^2 + 1))$

54. $f(x) = 3 \tanh^2 e^x$

55. $f(x) = \sqrt{\cosh^2 x + 1}$

56. $f(x) = \dfrac{\tanh \sqrt{x}}{\sqrt{\sinh x}}$

57. $f(x) = \sinh^{-1}(x^3)$

58. $f(x) = \tanh^{-1}(\tan x^2)$

59. $f(x) = \dfrac{\sinh^{-1} x}{\cosh^{-1} x}$

60. $f(x) = x\sqrt{\tanh^{-1} x}$

61. $f(x) = \sin(e^{\sinh^{-1} x})$

62. $f(x) = \cosh^{-1}(\cosh^{-1} x)$

Use logarithmic differentiation to find the derivatives of each of the functions in Exercises 63–65.

63. $(\sin x)^x$

64. $(\sec x)^x$

65. $(\sin x)^{\cos x}$

In Exercises 66–71, find a function f that has the given derivative f'. In each case you can find the answer with an educated guess-and-check process.

66. $f'(x) = \dfrac{2x}{\sqrt{1 - 4x^2}}$

67. $f'(x) = \dfrac{2}{\sqrt{1 - 4x^2}}$

68. $f'(x) = \dfrac{1}{1 + 9x^2}$

69. $f'(x) = \dfrac{3x}{1 + 9x^2}$

70. $f'(x) = \dfrac{1}{9 + x^2}$

71. $f'(x) = \dfrac{3}{\sqrt{4 - 9x^2}}$

In Exercises 72–77, find a function f that has the given derivative f'. In each case you can find the answer with an educated guess-and-check process. *(Some of these exercises involve hyperbolic functions.)*

72. $f'(x) = \dfrac{2x}{\sqrt{1 + 4x^2}}$

73. $f'(x) = \dfrac{2}{\sqrt{1 + 4x^2}}$

74. $f'(x) = \dfrac{1}{1 - 9x^2}$

75. $f'(x) = \dfrac{3x}{1 - 9x^2}$

76. $f'(x) = \dfrac{1}{9 - x^2}$

77. $f'(x) = \dfrac{3}{\sqrt{4 + 9x^2}}$

Applications

78. In Exercise 83 from Section 1.6 we saw that the oscillating position of a mass hanging from the end of a spring, neglecting air resistance, is given by the following equation, where A, B, k, and m are constants:

$$s(t) = A \sin\left(\sqrt{\frac{k}{m}}\, t\right) + B \cos\left(\sqrt{\frac{k}{m}}\, t\right).$$

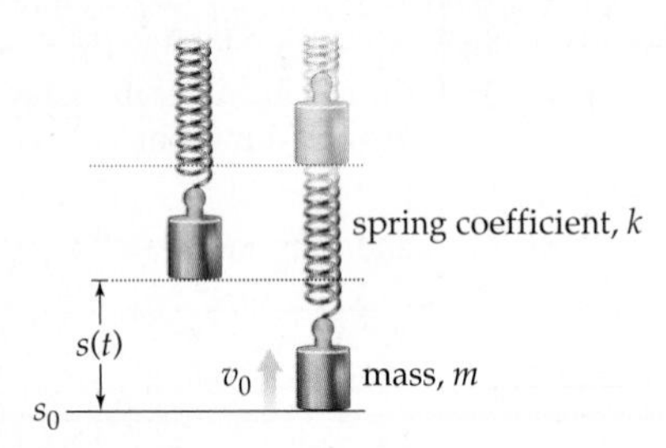

(a) Show that the function $s(t)$ has the property that $s''(t) + \frac{k}{m}s(t) = 0$. This is the differential equation for the spring motion, an equation involving derivatives that describes the motion of the bob on the end of the spring.

(b) Suppose the spring is released from an initial position of s_0 and with an initial velocity of v_0. Show that $A = v_0\sqrt{\frac{m}{k}}$ and $B = s_0$.

79. In Exercise 84 from Section 1.6 we learned that the oscillating position of a mass hanging from the end of a spring, taking air resistance into account, is given by the following equation, where A, B, k, f, and m are constants:

$$s(t) = e^{-f/2m\,t}\left(A \sin\left(\frac{\sqrt{4km - f^2}}{2m}\, t\right) + B \cos\left(\frac{\sqrt{4km - f^2}}{2m}\, t\right)\right).$$

(a) Show that the function $s(t)$ has the property that $s''(t) + \frac{f}{m}s'(t) + \frac{k}{m}s(t) = 0$ for some constant f. This is the differential equation for spring motion, taking air resistance into account. *(Hint: Find the first and second derivatives of $s(t)$ first, and then show that $s(t)$, $s'(t)$, and $s''(t)$ have the given relationship.)*

(b) Suppose the spring is released from an initial position of s_0 with an initial velocity of v_0. Show that $A = \frac{2mv_0 + fs_0}{\sqrt{4km - f^2}}$ and $B = s_0$.

80. Suppose your friend Max drops a penny from the top floor of the Empire State Building, 1250 feet from the ground. After t seconds, the penny is a distance of $s(t) = -16t^2 + 1250$ from the ground. You are standing about a block away, 250 feet from the base of the building.

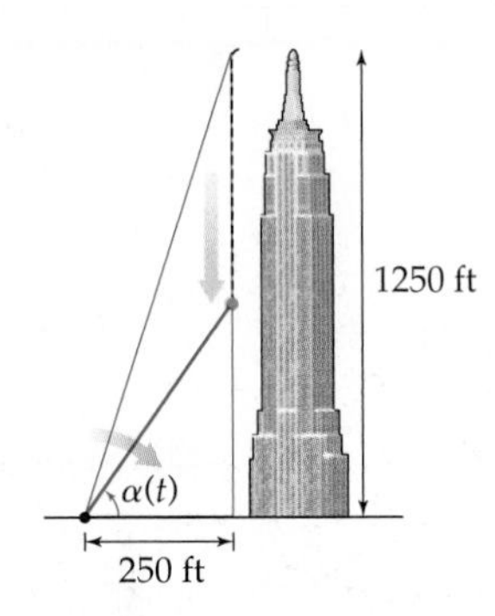

(a) Find a formula for the angle of elevation $\alpha(t)$ from the ground at your feet to the height of the penny t seconds after Max drops it. Multiply by an appropriate constant so that $\alpha(t)$ is measured in degrees.

(b) Find a formula for the rate at which the angle of elevation $\alpha(t)$ is changing at time t, and use the formula to determine the rate of change of the angle of elevation at the time the penny hits the ground.

81. The Gateway Arch in St. Louis is designed as an inverted catenary curve. The arch is a complex three-dimensional structure, but some sources model it simply, using the hyperbolic function

$$A(x) = 693.8 - 68.8\cosh\left(\frac{1}{99.7}(x - 299.22)\right),$$

where x denotes the distance in feet from one base of the arch as you approach the other. *(This exercise involves hyperbolic functions.)*

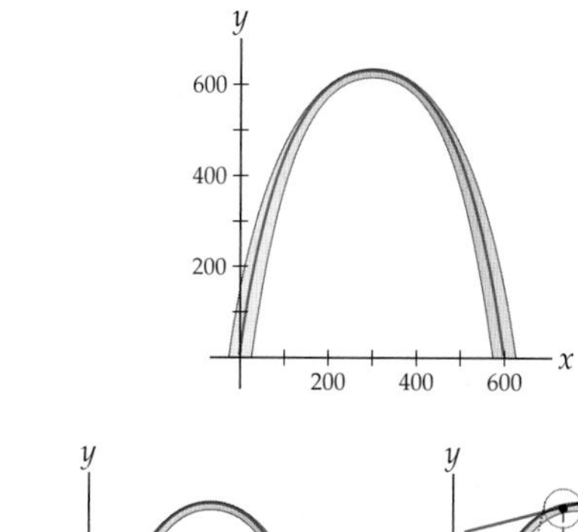

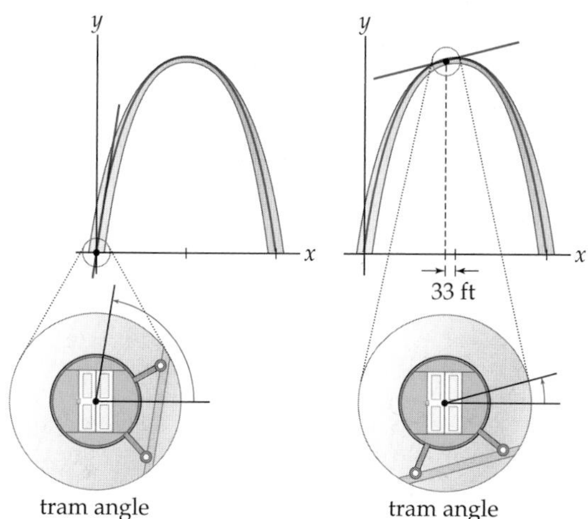

(a) How tall is the arch, according to this model?

(b) There is a tram that takes visitors to an observation deck in the top of the arch. The cabin of the tram rotates rather like cars on a Ferris wheel to keep visitors upright, but the outer part of the tram changes angle with the curve of the arch. What angle does the tram make with the ground at the bottom of the arch?

(c) Visitors leave the tram at the observation deck 33 feet from the center of the arch. What angle does the tram make with the ground there?

82. Ian has climbed a pinnacle that is detached from the main peak by roping down into the notch dividing them and then climbing the pinnacle. He pulled an extra rope behind him so that he could get back to the main peak by using a *Tyrolean traverse*, meaning that he would use the rope to go directly back to the peak instead of descending and then climbing on rock again. When he anchors the rope, it hangs in a catenary curve, with equation

$$r(x) = 125\cosh(0.008x - 0.6528).$$

The point $x = 0$ is where the rope attaches to the main peak, while $x = 136$ is where it attaches to the pinnacle. Heights are measured in feet above the notch. *(This exercise involves hyperbolic functions.)*

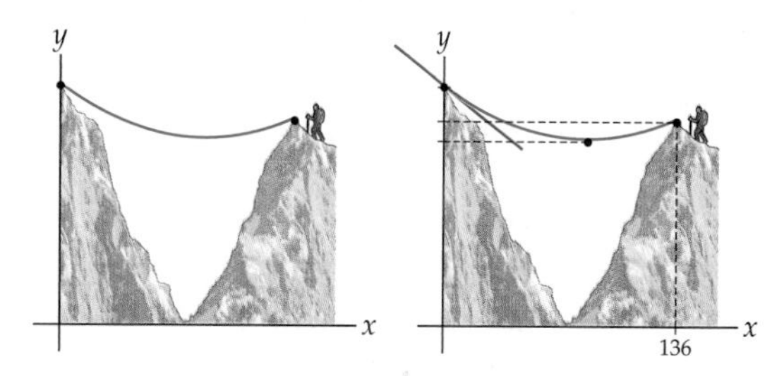

(a) How much higher is the main peak than the detached pinnacle?

(b) Where is the low point of the rope as it hangs loosely? How high is the rope above the notch at that point?

(c) What angle does the rope make with the horizontal where it attaches to the main peak?

Proofs

83. Use the definition of the derivative, a trigonometric identity, and known trigonometric limits to prove that $\frac{d}{dx}(\cos x) = -\sin x$.

84. Use the quotient rule and the derivative of the cosine function to prove that $\frac{d}{dx}(\sec x) = \sec x \tan x$.

85. Use the quotient rule and the derivative of the sine function to prove that $\frac{d}{dx}(\csc x) = -\csc x \cot x$.

86. Use the quotient rule and the derivatives of the sine and cosine functions to prove that $\frac{d}{dx}(\cot x) = -\csc^2 x$.

87. Use implicit differentiation and the fact that $\tan(\tan^{-1} x) = x$ for all x in the domain of $\tan^{-1} x$ to prove that $\frac{d}{dx}(\tan^{-1} x) = \frac{1}{1+x^2}$.

88. Use implicit differentiation and the fact that $\sec(\sec^{-1} x) = x$ for all x in the domain of $\sec^{-1} x$ to prove that $\frac{d}{dx}(\sec^{-1} x) = \frac{1}{|x|\sqrt{x^2-1}}$. You will have to consider the cases $x > 1$ and $x < -1$ separately.

89. Prove that for any value of t, the point $(x, y) = (\cosh t, \sinh t)$ lies on the hyperbola $x^2 - y^2 = 1$. Bonus question: In fact, these points will always lie on the right-hand side of the hyperbola; why? *(This exercise involves hyperbolic functions.)*

Use the definitions of the hyperbolic functions to prove that each of the identities in Exercises 90–92 hold for all values of x and y. Note how similar these identities are to those which hold for trigonometric functions. *(These exercises involve hyperbolic functions.)*

90. (a) $\sinh(-x) = -\sinh x$, and (b) $\cosh(-x) = \cosh x$

91. $\sinh(x + y) = \sinh x \cosh y + \cosh x \sinh y$

92. $\cosh(x + y) = \cosh x \cosh y + \sinh x \sinh y$

Prove each of the differentiation formulas in Exercises 93–96. *(These exercises involve hyperbolic functions.)*

93. $\frac{d}{dx}(\cosh x) = \sinh x$

94. $\frac{d}{dx}(\tanh x) = \operatorname{sech}^2 x$

95. $\frac{d}{dx}(\cosh^{-1} x) = \frac{1}{\sqrt{x^2 - 1}}$

96. $\frac{d}{dx}(\tanh^{-1} x)\frac{1}{1 - x^2}$

Prove that the inverse hyperbolic functions can be written in terms of logarithms as shown in Exercises 97–99. *(Hint for the first problem: Solve $\sinh y = x$ for y by using algebra to get an expression that is quadratic in e^y (i.e., of the form $ae^{2y} + be^y + c$) and then applying the quadratic formula.)*
(These exercises involve hyperbolic functions.)

97. $\sinh^{-1} x = \ln(x + \sqrt{x^2 + 1})$, for any x.

98. $\cosh^{-1} x = \ln(x + \sqrt{x^2 - 1})$, for $x \geq 1$.

99. $\tanh^{-1} x = \frac{1}{2} \ln\left(\frac{1 + x}{1 - x}\right)$, for $-1 < x < 1$.

Thinking Forward

Local extrema and inflection points: In the exercises that follow, you will investigate how derivatives can help us find the locations of the maxima and minima of a function.

- Suppose f has a maximum or minimum value at $x = c$. If f is differentiable at $x = c$, what must be true of $f'(c)$, and why?
- If f is a differentiable function, then the values $x = c$ at which the sign of the derivative $f'(x)$ changes are the locations of the local extrema of f. Use this information to find the local extrema of the function $f(x) = \sin x$. Illustrate your answer on a graph of $y = \sin x$.
- If f is a differentiable function, then the values $x = c$ at which the sign of the second derivative $f''(x)$ changes are the locations of the inflection points of f. Use this information to find the inflection points of the function $f(x) = \sin x$. Illustrate your answer on a graph of $y = \sin x$.

CHAPTER REVIEW, SELF-TEST, AND CAPSTONES

Before you progress to the next chapter, be sure you are familiar with the definitions, concepts, and basic skills outlined here. The capstone exercises at the end bring together ideas from this chapter and look forward to future chapters.

Definitions

Give precise mathematical definitions or descriptions of each of the concepts that follow. Then illustrate the definition with a graph or algebraic example, if possible.

- the graphical interpretations of a *tangent line* and a *secant line* to a graph
- the real–world interpretations of *position*, *velocity*, and *acceleration*
- the real–world interpretations of *average rate of change* and *instantaneous rate of change*
- the formal definition of the *derivative* of a function f at a point $x = c$ (both $z \to x$ form and $h \to 0$ form)
- the formal definition of the *derivative* of a function f, as a function (both $z \to x$ form and $h \to 0$ form)
- the formal definitions of the *tangent line* and the *instantaneous rate of change* of the graph of a function f at a point $x = c$
- what it means for a function f to be *differentiable*, *left differentiable*, and *right differentiable* at a point $x = c$.
- what it means for a function f to be *differentiable* on an open or closed interval I
- what it means to say that y is an *implicit function* of x, and the meaning of *implicit differentiation*
- the definitions of the *hyperbolic functions* $\sinh x$, $\cosh x$, and $\tanh x$ in terms of exponential functions

Theorems

Fill in the blanks to complete each of the following theorem statements:

- If a function f is differentiable at $x = c$, then f is ____ at $x = c$.
- If f and f^{-1} are inverse functions, then for all appropriate values of x we can write the derivative of $f^{-1}(x)$ in terms of the derivative of $f'(x)$ as follows: ____ .
- $f'(x) = kf(x)$ for some constant k if and only if f is a/an ____ function.
- For all real numbers x, $\cosh^2 x - \sinh^2 x =$ ____ .

Notation and Differentiation Rules

Leibniz notation: Describe the meanings of each of the mathematical expressions that follow. Translate expressions written in Leibniz notation to "prime" notation, and vice versa.

- $f'(x)$
- $f'(2)$
- $f''(x)$
- $\dfrac{df}{dx}$
- $\dfrac{dy}{dx}$
- $\dfrac{d}{dx}(y(x))$
- $\left.\dfrac{dg}{dt}\right|_{t=3}$
- $\dfrac{d}{dx}(x^2)$
- $\left.\dfrac{d}{dx}\right|_{-1}(x^2)$
- $\dfrac{d^2y}{dx^2}$
- $\dfrac{d}{dx}\left(\dfrac{d}{dx}(x^2)\right)$
- $f^{(5)}(x)$

Derivatives of basic functions: Fill in the blanks to differentiate each of the given basic functions. You may assume that k, m, and b are appropriate constants.

- $\dfrac{d}{dx}(k) =$ ____
- $\dfrac{d}{dx}(x) =$ ____
- $\dfrac{d}{dx}(mx + b) =$ ____
- $\dfrac{d}{dx}(x^k) =$ ____
- $\dfrac{d}{dx}(\sqrt{x}) =$ ____
- $\dfrac{d}{dx}\left(\dfrac{1}{x}\right) =$ ____
- $\dfrac{d}{dx}(e^x) =$ ____
- $\dfrac{d}{dx}(b^x) =$ ____
- $\dfrac{d}{dx}(e^{kx}) =$ ____
- $\dfrac{d}{dx}(\log_b x) =$ ____
- $\dfrac{d}{dx}(\ln x) =$ ____
- $\dfrac{d}{dx}(\ln |x|) =$ ____
- $\dfrac{d}{dx}(\sin x) =$ ____
- $\dfrac{d}{dx}(\cos x) =$ ____
- $\dfrac{d}{dx}(\tan x) =$ ____
- $\dfrac{d}{dx}(\sec x) =$ ____
- $\dfrac{d}{dx}(\cot x) =$ ____
- $\dfrac{d}{dx}(\csc x) =$ ____
- $\dfrac{d}{dx}(\sin^{-1} x) =$ ____
- $\dfrac{d}{dx}(\tan^{-1} x) =$ ____
- $\dfrac{d}{dx}(\sec^{-1} x) =$ ____
- $\dfrac{d}{dx}(\sinh x) =$ ____
- $\dfrac{d}{dx}(\cosh x) =$ ____
- $\dfrac{d}{dx}(\tanh x) =$ ____
- $\dfrac{d}{dx}(\sinh^{-1} x) =$ ____
- $\dfrac{d}{dx}(\cosh^{-1} x) =$ ____
- $\dfrac{d}{dx}(\tanh^{-1} x) =$ ____

Derivatives of combinations: Fill in the blanks to complete each of the given differentiation rules. You may assume that f and g are differentiable everywhere.

- $(kf)'(x) =$ ____
- $(f + g)'(x) =$ ____
- $(f - g)'(x) =$ ____
- $(fg)'(x) =$ ____
- $\left(\dfrac{f}{g}\right)'(x) =$ ____
- $(f \circ g)'(x) =$ ____

Skill Certification: Basic Derivatives

Basic definition-of-derivative calculations: Find the derivatives of the functions that follow, using (a) the $h \to 0$ definition of the derivative and (b) the $z \to x$ definition of the derivative.

1. $f(x) = x^2$
2. $f(x) = x^3$
3. $f(x) = \dfrac{1}{x}$
4. $f(x) = \dfrac{1}{x^2}$
5. $f(x) = \sqrt{x}$
6. $f(x) = \dfrac{1}{\sqrt{x}}$

Calculating basic derivatives: Find the derivatives of the functions that follow, using the differentiation rules developed in this chapter. *(The last two exercises involve hyperbolic functions.)*

7. $f(x) = \dfrac{1}{x^4 - 5x^3 + 2}$
8. $f(x) = \dfrac{x - 2}{(x - 1)(x + 3)}$
9. $f(x) = (x\sqrt{2x - 1})^{-3}$
10. $f(x) = x^{-1}(4 - x)^2$

11. $f(x) = \dfrac{1 - \frac{1}{x}}{\sqrt{x}}$

12. $f(x) = \dfrac{3x\sqrt{2x+1}}{1-x}$

13. $f(x) = |x|$

14. $f(x) = |3x+1|$

15. $f(x) = e^x \sin x$

16. $f(x) = \dfrac{e^x}{\sin x}$

17. $f(x) = \sin(e^x)$

18. $f(x) = e^{\sin x}$

19. $f(x) = \ln(\tan^2 x)$

20. $f(x) = x^3 \sec x$

21. $f(x) = \sqrt{\sin^{-1}(x^2)}$

22. $f(x) = \dfrac{\sin^2 x + \cos^2 x}{\csc x}$

23. $f(x) = x2^{3x+1}$

24. $f(x) = x^x$

25. $f(x) = \sqrt{\tanh^3(x^5)}$

26. $f(x) = \dfrac{\sinh^{-1} x}{\tanh^{-1} x}$

Calculating antiderivatives: For each exercise that follows, find a function f that has the given derivative f' and value $f(c)$. In each case you can find the answer with an educated guess-and-check process. The last exercise involves an inverse hyperbolic function.

27. $f'(x) = -32,\ f(0) = 4$

28. $f'(x) = -32x + 4,\ f(0) = 100$

29. $f'(x) = x(3x+1),\ f(2) = 4$

30. $f'(x) = \dfrac{3x^2}{\sqrt{x^3+1}},\ f(2) = 6$

31. $f'(x) = 8e^{4x} + 1,\ f(0) = 3$

32. $f'(x) = 2x \sec x^2 \tan x^2,\ f(0) = 2$

33. $f'(x) = \dfrac{1}{1+4x},\ f(0) = 1$

34. $f'(x) = \dfrac{1}{1+4x^2},\ f(0) = 1$

35. $f'(x) = \dfrac{x}{1+4x^2},\ f(0) = 1$

36. $f'(x) = \dfrac{1}{1-4x^2},\ f(0) = 0$

Differentiating with respect to different variables: Find each derivative described.

37. If $3v^2 + xv - 1 = 0$, find $\dfrac{dv}{dx}$.

38. If $3v^2 + xv - 1 = 0$, find $\dfrac{dx}{dv}$.

39. If $A = \pi r^2$, find $\dfrac{dA}{dr}$.

40. If $A = \pi r^2$, and A and r are both functions of time t, find $\dfrac{dA}{dt}$.

41. If $y = 3x^2 t - t^k$, where x and k are constant, find $\dfrac{dy}{dt}$.

42. If $y = 3x^2 t - t^k$, where t and k are constant, find $\dfrac{dy}{dx}$.

Capstone Problems

A. *The sum rule for differentiation:* Use the definition of the derivative to prove that the derivative of a sum of functions $f(x) + g(x)$ is equal to the sum of their derivatives $f'(x) + g'(x)$.

B. *The power rule for differentiation:* Use the definition of the derivative and factoring formulas to prove that for any positive integer k, the derivative of x^k is kx^{k-1}.

C. *Rates of change from data:* The following table lists the consumption of gasoline in billions of gallons in the United States from 1994 to 2000:

Year	1994	1995	1996	1997	1998	1999	2000
Gas	109	111	113	117	118	121	122

(a) Compute the average rate of change in gasoline consumption in the United States for each year from 1994 to 2000.

(b) During which year was gasoline consumption increasing most rapidly? Least rapidly? Estimate the instantaneous rates of change in gasoline consumption during those years.

D. *Derivatives and graphical behavior:* In the next chapter we will see that we can get a lot of information about the graph of a function f by looking at the signs of $f(x)$ and its first and second derivatives. Let's do this for the function $f(x) = x^3 - 3x^2 - 9x + 27$.

(a) Find the roots of f, and then determine the intervals on which f is positive or negative.

(b) Find the roots of f', and then determine the intervals on which f' is positive or negative.

(c) Find the roots of f'', and then determine the intervals on which f'' is positive or negative.

(d) The graph of f will be above the x-axis when $f(x)$ is positive and below the x-axis when $f(x)$ is negative. Moreover, the graph of f will be increasing when f' is positive and decreasing when f' is negative. Finally, the graph of f will be concave up when f'' is positive and concave down when f'' is negative. Given this information and your answers from parts (a)–(c), sketch a careful, labeled graph of f.

Applications of the Derivative

3.1 The Mean Value Theorem

The Derivative at a Local Extremum
Rolle's Theorem
The Mean Value Theorem
Examples and Explorations

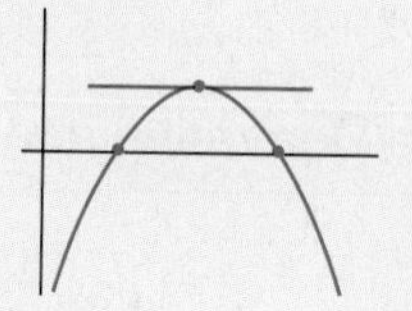

3.2 The First Derivative and Curve Sketching

Derivatives and Increasing/Decreasing Functions
Functions with the Same Derivative
The First-Derivative Test
Examples and Explorations

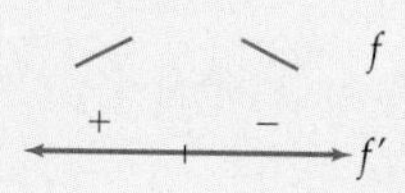

3.3 The Second Derivative and Curve Sketching

Derivatives and Concavity
Inflection Points
The Second-Derivative Test
Curve-Sketching Strategies
Examples and Explorations

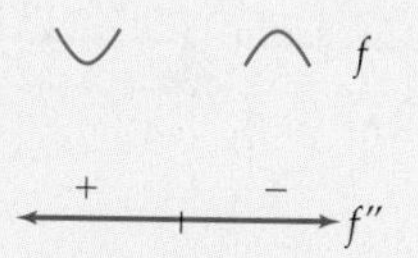

3.4 Optimization

Finding Global Extrema
Translating Word Problems into Mathematical Problems
Examples and Explorations

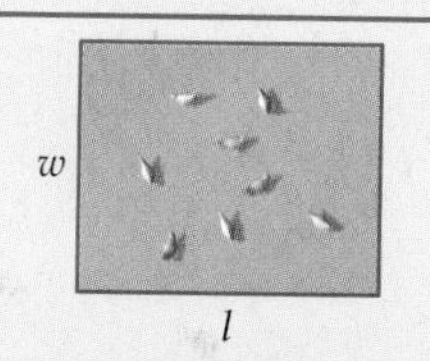

3.5 Related Rates

Related Quantities Have Related Rates
Volumes and Surface Areas of Geometric Objects
Similar Triangles
Examples and Explorations

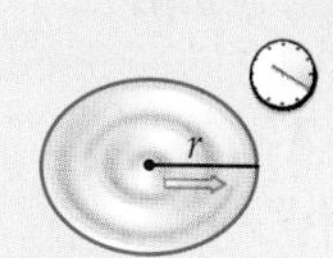

3.6 L'Hôpital's Rule

Geometrical Motivation for L'Hôpital's Rule
L'Hôpital's Rule for the Indeterminate Forms $\frac{0}{0}$ and $\frac{\infty}{\infty}$
Using Logarithms for the Indeterminate Forms 0^0, 1^∞, and ∞^0
Examples and Explorations

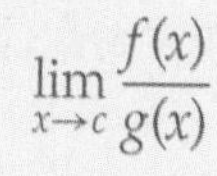

Chapter Review, Self-Test, and Capstones

3.1 THE MEAN VALUE THEOREM

- Local extrema, critical points, and their relationships
- Rolle's Theorem and the Mean Value Theorem
- Using critical points to find local extrema

The Derivative at a Local Extremum

Suppose a function f has a local maximum at some point $x = c$. This means that the value $f(c)$ is greater than or equal to all other nearby $f(x)$ values. The following definition makes this notion precise:

DEFINITION 3.1

Local Extrema of a Function

(a) f has a ***local maximum*** at $x = c$ if there exists some $\delta > 0$ such that $f(c) \geq f(x)$ for all $x \in (c - \delta, c + \delta)$.

(b) f has a ***local minimum*** at $x = c$ if there exists some $\delta > 0$ such that $f(c) \leq f(x)$ for all $x \in (c - \delta, c + \delta)$.

Intuitively, at a local extremum, the tangent line of a function must be either horizontal or undefined; for example, consider the following three graphs:

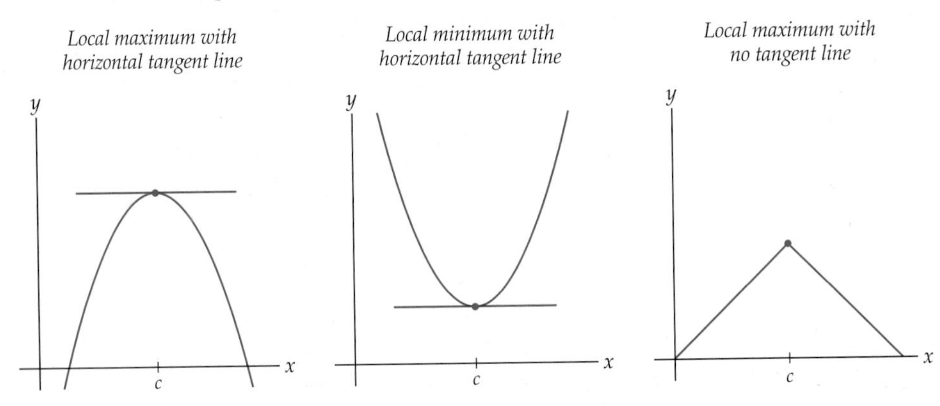

When a function has a horizontal or an undefined tangent line at a point, its derivative at that point is either zero or undefined. We call such points ***critical points*** of the function:

DEFINITION 3.2

Critical Points of a Function

A point $x = c$ in the domain of f is called a ***critical point*** of f if $f'(c) = 0$ or $f'(c)$ does not exist.

Notice that only points in the domain of f are considered critical points. For example, consider the function $f(x) = \frac{1}{x}$, whose derivative is $f'(x) = -\frac{1}{x^2}$. Clearly $f'(0)$ does not exist; however, since $x = 0$ is not in the domain of f, it is not called a critical point.

The preceding graphs suggest that every local extremum is also a critical point. This seemingly obvious relationship between critical points and extrema turns out to be the foundation on which we will build two more theorems that are key in our development of calculus:

THEOREM 3.3 **Local Extrema are Critical Points**

If $x = c$ is the location of a local extremum of f, then $x = c$ is a critical point of f.

The converse of this theorem is not true. That is, not every critical point is a local extremum of f; see Example 1. Although the implication in Theorem 3.3 is intuitively obvious just by thinking about graphs and the behavior of the derivative at local maxima and minima, actually proving it requires a somewhat subtle argument. The key is to look at the left and right derivatives at a local extremum. Theorem 3.3 is also known as *Fermat's Theorem for Local Extrema*, when formulated equivalently as saying that if $x = c$ is a local extremum and f is differentiable at $x = c$, then $f'(c)$ must be zero.

Proof. We will prove the case for local maxima and leave the similar proof for local minima to Exercise 64. Suppose $x = c$ is the location of a local maximum of f. If $f'(c)$ does not exist, then $x = c$ is a critical point and we are done. It now suffices to show that if $f'(c)$ exists, then it must be equal to 0. We will do so by examining the right and left derivatives at $x = c$.

Since $x = c$ is the location of a local maximum of f, there is some $\delta > 0$ such that for all $x \in (c - \delta, c + \delta)$, $f(c) \geq f(x)$, and thus $f(x) - f(c) \leq 0$. In the case where $x \in (c, c + \delta)$, it follows that $x > c$, which means that $x - c$ is positive. Thus in this case $\frac{f(x) - f(c)}{x - c} \leq 0$, and therefore

$$f'_+(c) = \lim_{x \to c^+} \frac{f(x) - f(c)}{x - c} \leq 0.$$

By a similar argument for $x \in (c - \delta, c)$, we have $x - c < 0$ and $f(x) - f(c) \leq 0$, and therefore

$$f'_-(c) = \lim_{x \to c^-} \frac{f(x) - f(c)}{x - c} \geq 0.$$

Since we are assuming that $f'(c)$ exists, we know that both $f'_+(c)$ and $f'_-(c)$ must exist and be equal to $f'(c)$. We have just shown both that $f'(c) \leq 0$ and that $f'(c) \geq 0$. Therefore, we must have $f'(c) = 0$, as desired. ■

Rolle's Theorem

Suppose a differentiable function f has two roots $x = a$ and $x = b$. What can you say about the graph of f between a and b? The three graphs that follow next provide a clue; if the graph of a function is smooth and unbroken, then somewhere between each root of f the function must turn around, and at that turning point it must have a local extremum with a horizontal tangent line:

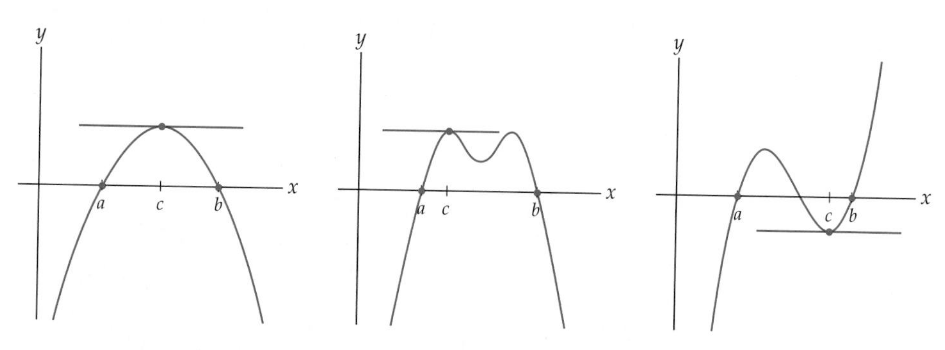

The preceding discussion is a summary of both the statement and the proof of the following key theorem:

THEOREM 3.4 **Rolle's Theorem**

If f is continuous on $[a, b]$ and differentiable on (a, b), and if $f(a) = f(b) = 0$, then there exists at least one value $c \in (a, b)$ for which $f'(c) = 0$.

Actually, Rolle's Theorem also holds in the more general case where $f(a)$ and $f(b)$ are equal to each other (not necessarily both zero). For example, Rolle's Theorem is also true if $f(a) = f(b) = 5$, or if $f(a) = f(b) = -3$, and so on, because vertically shifting a function by adding a constant term does not change its derivative. However, the classic way to state Rolle's Theorem is with $f(a)$ and $f(b)$ both equal to zero.

Proof. Rolle's Theorem is an immediate consequence of the Extreme Value Theorem from Section 1.4 and the fact that every extremum is a critical point. Suppose f is continuous on the closed interval $[a, b]$ and differentiable on the open interval (a, b), with $f(a) = f(b) = 0$. By the Extreme Value Theorem, we know that f attains both a maximum and a minimum value on $[a, b]$. If one of these extreme values occurs at a point $x = c$ in the interior (a, b) of the interval, then $x = c$ is a local extremum of f. By the previous theorem, this means that $x = c$ is a critical point of f. Since f is assumed to be differentiable at $x = c$, it follows that $f'(c) = 0$ and we are done.

It remains to consider the special case where all of the maximum and minimum values of f on $[a, b]$ occur at the endpoints of the interval (i.e., at $x = a$ or at $x = b$). In this case, since $f(a) = f(b) = 0$, the maximum and minimum values of $f(x)$ must both equal zero. For all x in $[a, b]$ we would have $0 \leq f(x) \leq 0$, which means that f would have to be the constant function $f(x) = 0$ on $[a, b]$. Since the derivative of a constant function is always zero, in this special case we have $f'(x) = 0$ for *all* values of c in (a, b), and we are done. ■

Just as the Intermediate Value Theorem and the Extreme Value Theorem illustrate basic properties of continuous functions, Rolle's Theorem illustrates a basic property of functions that are both continuous and differentiable. Like those two theorems before, Rolle's Theorem is a theorem about *existence*, not calculation; it tells you that there must exist some value $c \in (a, b)$ where the derivative of f is zero, but it does not tell you what that value is. It is important to note that the continuity and the differentiability hypotheses of Rolle's Theorem are essential: If a function f fails to be continuous on $[a, b]$ or fails to be differentiable on (a, b), then the conclusion of Rolle's Theorem does not necessarily follow; see Example 2.

The Mean Value Theorem

The Mean Value Theorem is a generalization of Rolle's Theorem to the case where $f(a)$ and $f(b)$ are not necessarily equal. Suppose f is a continuous, differentiable function. What can we say about the derivative of f between two points $x = a$ and $x = b$? The three graphs that follow suggest an answer: Somewhere between a and b the slope of the tangent line must be the same as the slope of the line from $(a, f(a))$ to $(b, f(b))$. If you turn your head so that the green line is horizontal in each figure, you can see that these figures are similar to rotated versions of the earlier figures used in illustrating Rolle's Theorem.

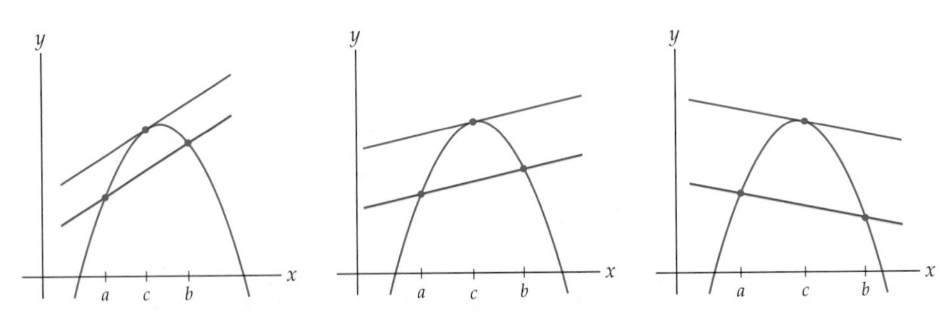

Algebraically, this means that there must be some $c \in (a, b)$ whose derivative value $f'(c)$ is equal to the average rate of change of f on $[a, b]$:

THEOREM 3.5

The Mean Value Theorem

If f is continuous on $[a, b]$ and differentiable on (a, b), then there exists at least one value $c \in (a, b)$ such that

$$f'(c) = \frac{f(b) - f(a)}{b - a}.$$

The "mean" in the "Mean Value Theorem" refers to an ***average***. Basically, this theorem says that if a function is continuous on a closed interval and differentiable on its interior, then there is always at least one place in the interval where the instantaneous rate of change of the function is equal to its average rate of change over the whole interval. As a real–world example, suppose you drove at an average speed of 50 miles per hour on a short road trip. The Mean Value Theorem guarantees that at some point along your journey you must have been travelling at exactly 50 miles per hour.

The Mean Value Theorem is intuitively clear if you believe that you can just "turn your head to the side" and see Rolle's Theorem. In fact, the proof of the Mean Value Theorem is based on an algebraic version of this intuition:

Proof. Suppose f is a function that is continuous on $[a, b]$ and differentiable on (a, b), and let $l(x)$ be the secant line from $(a, f(a))$ to $(b, f(b))$. The idea of the proof is to "turn our heads" algebraically. To rotate so that the secant line $l(x)$ plays the role of the x-axis, we will consider the function $g(x) = f(x) - l(x)$. The graph of this new function $g(x)$ will have roots at $x = a$ and $x = b$, and we will be able to apply Rolle's Theorem.

Since the secant line $l(x)$ has slope $\frac{f(b)-f(a)}{b-a}$ and passes through the point $(a, f(a))$, its equation is

$$l(x) = \frac{f(b) - f(a)}{b - a}(x - a) + f(a).$$

This means that the function $g(x) = f(x) - l(x)$ is equal to

$$g(x) = f(x) - \frac{f(b) - f(a)}{b - a}(x - a) - f(a).$$

If we want to apply Rolle's Theorem to $g(x)$, then we must first verify that $g(x)$ satisfies all the hypotheses of Rolle's Theorem. First, $g(x)$ is continuous on $[a, b]$ because it is a combination of continuous functions. Second, $g(x)$ is differentiable on (a, b) because f is differentiable on (a, b). Finally, $g(a) = 0$ and $g(b) = 0$:

$$g(a) = f(a) - \frac{f(b) - f(a)}{b - a}(a - a) - f(a) = f(a) - 0 - f(a) = 0,$$

$$g(b) = f(b) - \frac{f(b) - f(a)}{b - a}(b - a) - f(a) = f(b) - (f(b) - f(a)) - f(a) = 0.$$

Since Rolle's Theorem applies to the function $g(x)$, we can conclude that there exists some $c \in (a, b)$ for which $g'(c) = 0$. How does this conclusion relate to our original problem? To answer that, we must first calculate $g'(x)$:

$$g'(x) = \frac{d}{dx}\left(f(x) - \frac{f(b) - f(a)}{b - a}(x - a) - f(a)\right) \quad \leftarrow \text{definition of } g(x)$$

$$= f'(x) - \frac{f(b) - f(a)}{b - a}(1) - 0 \quad \leftarrow \frac{f(b) - f(a)}{b - a} \text{ and } f(a) \text{ are constants}$$

$$= f'(x) - \frac{f(b) - f(a)}{b - a}. \quad \leftarrow \text{simplify}$$

Rolle's Theorem now guarantees that there exists a $c \in (a, b)$ for which $g'(c) = 0$. By our previous calculation, for this value of c we have $g'(c) = f'(c) - \frac{f(b)-f(a)}{b-a} = 0$, and therefore $f'(c) = \frac{f(b)-f(a)}{b-a}$, as desired. ■

Examples and Explorations

EXAMPLE 1

Not every critical point is a local extremum

Show that $x = 1$ is a critical point of $f(x) = x^3 - 3x^2 + 3x$. Then use a graph to show that $x = 1$ is not a local extremum of f.

SOLUTION

If $f(x) = x^3 - 3x^2 + 3x$, then $f'(x) = 3x^2 - 6x + 3$ and thus $f'(1) = 3(1)^2 - 6(1) + 3 = 0$. Therefore $x = 1$ is a critical point of f. However, looking at the following graph of f, we can see that f has neither a local minimum nor a local maximum at $x = 1$:

$x = 1$ is a critical point but not an extremum

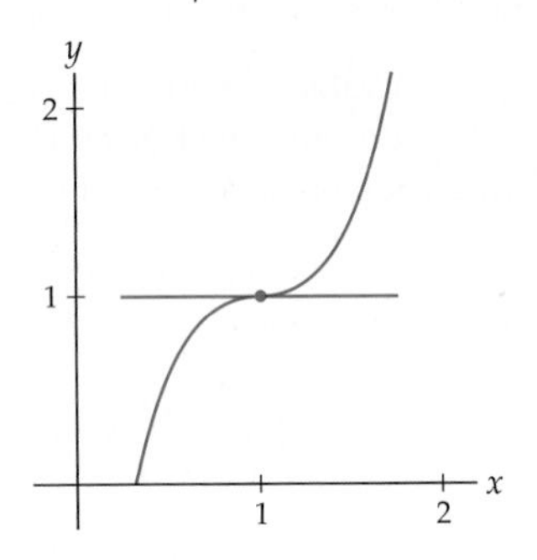

□

EXAMPLE 2

The hypotheses of Rolle's Theorem are important

Sketch graphs of three functions that fail to satisfy the hypotheses of Rolle's Theorem, for which the conclusion of Rolle's Theorem does not follow.

SOLUTION

The function f in the first graph that follows at the right fails to be differentiable on $(1, 3)$, and therefore can "turn around" at $x = 2$ without having a horizontal tangent line. For this function, there is no value of $c \in (a, b)$ with $f'(c) = 0$.

f is not differentiable on $(1, 3)$

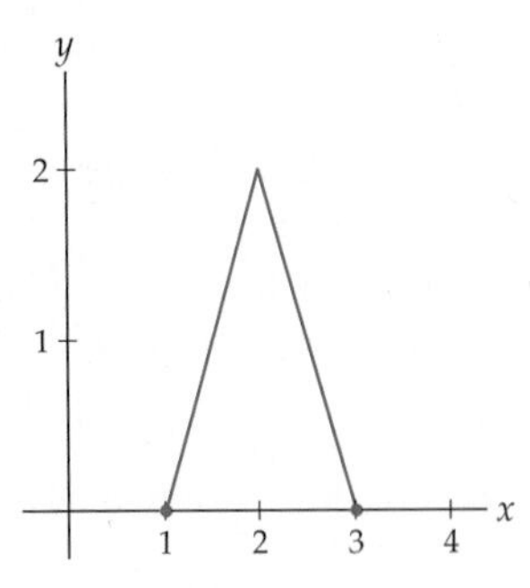

g is not continuous on $[1, 3]$

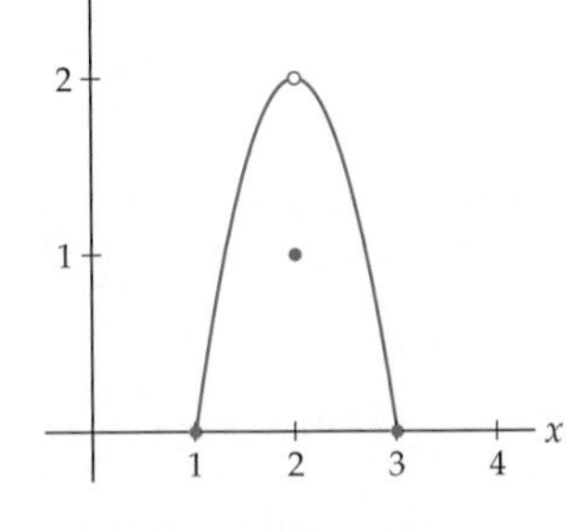

h is not continuous on $[1, 3]$

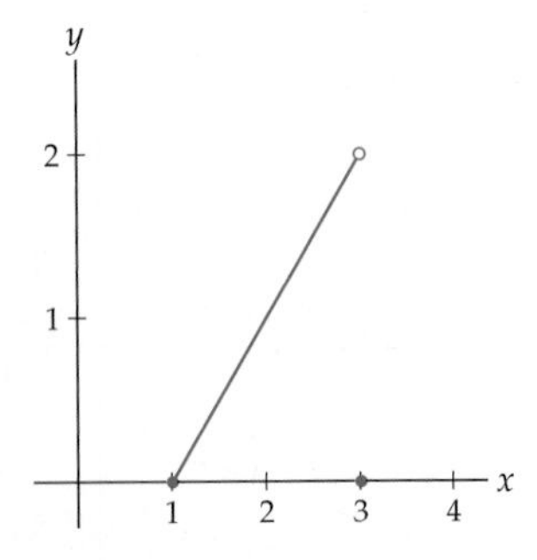

In the second graph, the function $g(x)$ fails to be continuous at the very place where we would have expected its derivative to be zero. Since this function is not continuous at $x = 2$, it is also not differentiable at $x = 2$, and there is no value $c \in (1, 3)$ with $g'(c) = 0$.

The third graph illustrates a function $h(x)$ that fails to be continuous at the right endpoint $x = 3$ of the interval. There is no value $c \in (1, 3)$ with $h'(c) = 0$; the function never has to "turn around," since it just jumps down to the root at $x = 3$. □

EXAMPLE 3

Applying Rolle's Theorem

Use Rolle's Theorem to show that there must exist some value of c in $(-2, 5)$ at which the function $f(x) = x^2 - 3x - 10$ has a horizontal tangent line. Then use f' to find such a value c algebraically, and verify your answer with a graph.

SOLUTION

First notice that $f(x) = x^2 - 3x - 10 = (x + 2)(x - 5)$ has roots at $x = -2$ and at $x = 5$. Since f is a polynomial, it is continuous and differentiable. In particular, it is continuous on $[-2, 5]$ and differentiable on $(-2, 5)$. Therefore Rolle's Theorem applies to the function f, and we can conclude that there must exist some value of $c \in (-2, 5)$ for which $f'(c) = 0$. At this value of c the graph of f will have a horizontal tangent line.

Rolle's Theorem tells us that there exists some $c \in (-2, 5)$ where $f'(c) = 0$, but it doesn't tell us exactly where. We can find such a c by solving the equation $f'(x) = 0$. Since $f(x) = x^2 - 3x - 10$, we have $f'(x) = 2x - 3$, which is equal to zero when $x = \frac{3}{2}$. Therefore f has a horizontal tangent line at $c = \frac{3}{2}$, which is in the interval $(-2, 5)$. The following graph illustrates that $f(x) = x^2 - 3x - 10$ does appear to have a horizontal tangent line at $x = \frac{3}{2}$.

Horizontal tangent at $x = \frac{3}{2}$

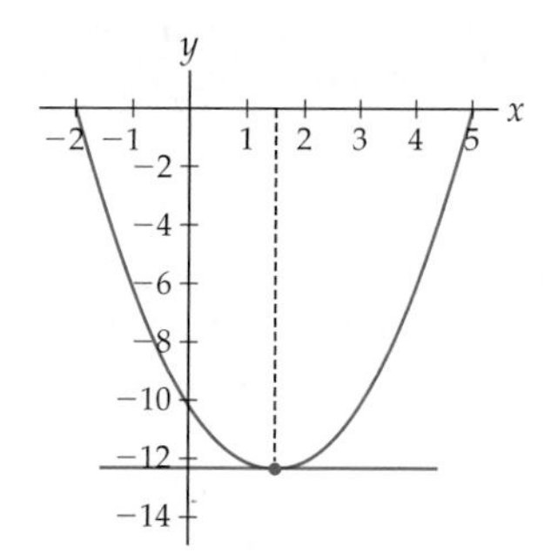

□

EXAMPLE 4

Using critical points and Rolle's Theorem to find local extrema

The function $f(x) = x(x - 1)(x - 3)$ is a cubic polynomial with one local maximum and one local minimum. Use Rolle's Theorem to identify intervals on which these extrema exist. Then use derivatives to find the exact locations of these extrema.

SOLUTION

The roots of $f(x) = x(x - 1)(x - 3)$ are $x = 0$, $x = 1$, and $x = 3$. Since f is a polynomial, it is continuous and differentiable everywhere. Therefore Rolle's Theorem applies on the intervals $[0, 1]$ and $[1, 3]$, and it tells us that at least one critical point must exist inside each of these intervals.

The critical points of f are the possible locations of the local extrema that we seek. To find the critical points we must solve the equation $f'(x) = 0$. It is simpler to do some algebra before differentiating:

$$f'(x) = \frac{d}{dx}(x(x - 1)(x - 3)) = \frac{d}{dx}(x^3 - 4x^2 + 3x) = 3x^2 - 8x + 3.$$

By the quadratic formula, we have $f'(x) = 0$ at the points

$$x = \frac{-(-8) \pm \sqrt{8^2 - 4(3)(3)}}{2(3)} = \frac{8 \pm \sqrt{28}}{6} = \frac{4 \pm \sqrt{7}}{3}.$$

These x-values are approximately $x \approx 0.451$ and $x \approx 2.215$. If we look at the graph of f, then we can see that the smaller of these two x-values is the location of the local maximum and the larger is the location of the local minimum; see the figure that follows. □

CHECKING THE ANSWER

The graph of $f(x) = x(x-1)(x-3)$ is shown next. Notice that the local extrema do seem to occur at the values we just found.

Extrema at $x \approx 0.451$ and $x \approx 2.215$

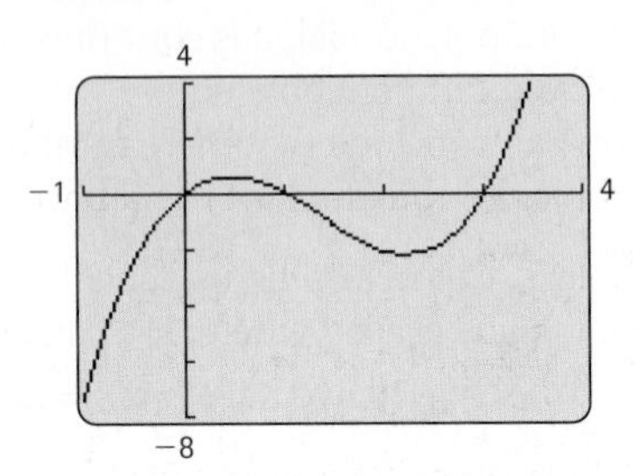

EXAMPLE 5

Applying the Mean Value Theorem

Use the Mean Value Theorem to show that there is some value $c \in (0, 2)$ at which the tangent line to the function $f(x) = x^2 - 2$ has slope 2. Then use f' to find such a value c algebraically, and verify your answer with a graph.

SOLUTION

The function $f(x) = x^2 - 2$ is a polynomial and thus is continuous and differentiable; in particular it is continuous on $[0, 2]$ and differentiable on $(0, 2)$. Therefore, the Mean Value Theorem applies to this function on the interval $[0, 2]$. The slope of the line from $(0, f(0))$ to $(2, f(2))$ is

$$\frac{f(2) - f(0)}{2 - 0} = \frac{(2^2 - 2) - (0^2 - 2)}{2} = \frac{2 - (-2)}{2} = \frac{4}{2} = 2.$$

By the Mean Value Theorem, there must exist at least one point $c \in (0, 2)$ with $f'(c) = 2$.

To find such a value of c algebraically, observe that the derivative of $f(x) = x^2 - 2$ is $f'(x) = 2x + 0 = 2x$. We want to find $c \in (0, 2)$ with $f'(c) = 2$, so we solve:

$$f'(c) = 2 \implies 2c = 2 \implies c = 1.$$

The point $c = 1$ is indeed in the interval $(0, 2)$, and $f'(1) = 2$, so we are done. The following figure illustrates that $f(x) = x^2 - 2$ does appear to have the same slope at $x = 1$ as the secant line from $(0, f(0))$ to $(2, f(2))$.

$f'(1)$ equals average rate of change on $[0, 2]$

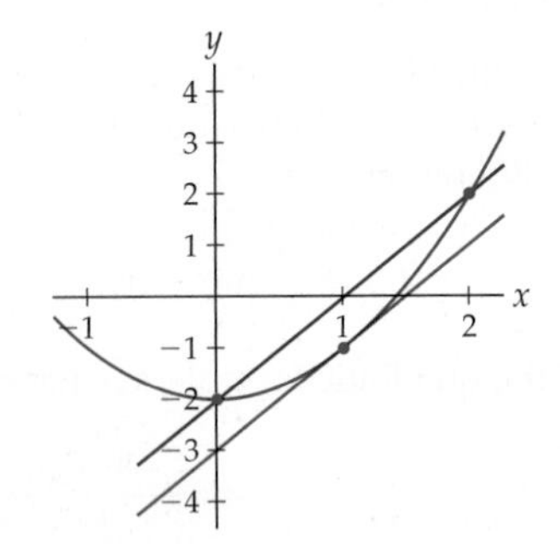

□

? TEST YOUR UNDERSTANDING

- Is every critical point a local extremum? Is every local extremum a critical point?
- What is the role of δ in the definitions of local extrema given in Definition 3.1?
- Why do you think $f(a) = f(b)$ would be a sufficient hypothesis in the statement of Rolle's Theorem? Think about the situation graphically.
- Can Rolle's Theorem tell you the exact location of a root of f'?
- How is the Mean Value Theorem related to Rolle's Theorem?

EXERCISES 3.1

Thinking Back

Review of definitions and theorems: State each theorem or definition that follows in precise mathematical language. Then give an illustrative graph or example, as appropriate.

- f has a local maximum at $x = c$
- f has a local minimum at $x = c$
- f is continuous on $[a, b]$
- f is differentiable on (a, b)
- The secant line from $(a, f(a))$ to $(b, f(b))$
- The right derivative $f'_+(c)$ at a point $x = c$
- The left derivative $f'_-(c)$ at a point $x = c$
- The Extreme Value Theorem
- The Intermediate Value Theorem

Concepts

0. *Problem Zero:* Read the section and make your own summary of the material.

1. *True/False:* Determine whether each of the statements that follow is true or false. If a statement is true, explain why. If a statement is false, provide a counterexample.

(a) *True or False:* Rolle's Theorem is a special case of the Mean Value Theorem.

(b) *True or False:* The Mean Value Theorem is so named because it concerns the average (or "mean") rate of change of a function on an interval.

(c) *True or False:* If f is differentiable on $\mathbb{R}$ and has an extremum at $x = -2$, then $f'(-2) = 0$.

(d) *True or False:* If f has a critical point at $x = 1$, then f has a local minimum or maximum at $x = 1$.

(e) *True or False:* If f is any function with $f(2) = 0$ and $f(8) = 0$, then there is some c in the interval $(2, 8)$ such that $f'(c) = 0$.

(f) *True or False:* If f is continuous and differentiable on $[-2, 2]$ with $f(-2) = 4$ and $f(2) = 0$, then there is some $c \in (-2, 2)$ with $f'(c) = -1$.

(g) *True or False:* If f is continuous and differentiable on $[0, 10]$ with $f'(5) = 0$, then f has a local maximum or minimum at $x = 5$.

(h) *True or False:* If f is continuous and differentiable on $[0, 10]$ with $f'(5) = 0$, then there are some values a and b in $(0, 10)$ for which $f(a) = 0$ and $f(b) = 0$.

2. *Examples:* Construct examples of the thing(s) described in the following. Try to find examples that are different than any in the reading.

(a) A function with a local minimum at $x = 3$ that is continuous but not differentiable at $x = 3$.

(b) A function with a local maximum at $x = -2$ that is not differentiable at $x = -2$ because of a removable discontinuity.

(c) A function with a local minimum at $x = 1$ that is not differentiable at $x = 1$ because of a jump discontinuity.

3. If f has a local maximum at $x = 1$, then what can you say about $f'(1)$? What if you also know that f is differentiable at $x = 1$?

4. If f has a local minimum at $x = 0$ and a local maximum at $x = 2$, what can you say about $f'(0)$ and $f'(2)$? Is there anything else you can say about f'?

5. Suppose that f is defined on $(-\infty, \infty)$ and differentiable everywhere except at $x = -2$ and $x = 4$, and that $f'(x) = 0$ only at $x = 0$ and $x = 5$. List all the critical points of f and sketch a possible graph of f.

6. Suppose that f is defined for $x \neq 0$ and differentiable everywhere except at $x = 0$ and $x = 1$, and that $f'(x) = 0$ only at $x = \pm 2$. List all the critical points of f and sketch a possible graph of f.

7. If a continuous, differentiable function f has zeroes at $x = -4$, $x = 1$, and $x = 2$, what can you say about f' on $[-4, 2]$?

8. If a continuous, differentiable function f is equal to 2 at $x = 3$ and at $x = 5$, what can you say about f' on $[3, 5]$?

9. If a continuous, differentiable function f has values $f(-2) = 3$ and $f(4) = 1$, what can you say about f' on $[-2, 4]$?

10. Restate Theorem 3.3 so that its conclusion has to do with tangent lines.

11. Restate Rolle's Theorem so that its conclusion has to do with tangent lines.

12. Restate the Mean Value Theorem so that its conclusion has to do with tangent lines.

In Exercises 13–22, sketch the graph of a function that satisfies the given description. Label or annotate your graph so that it is clear that it satisfies each part of the description.

13. A function that satisfies the hypothesis, and therefore the conclusion, of Rolle's Theorem on $[2, 6]$.

14. A function that satisfies the hypothesis, and therefore the conclusion, of the Mean Value Theorem.

15. A function f that satisfies the hypotheses of Rolle's Theorem on $[-2, 2]$ and for which there are exactly three values $c \in (-2, 2)$ that satisfy the conclusion of the theorem.

16. A function f that satisfies the hypothesis of the Mean Value Theorem on $[0, 4]$ and for which there are exactly three values $c \in (0, 4)$ that satisfy the conclusion of the theorem.

17. A function f that is defined on $[-2, 2]$ with $f(-2) = f(2) = 0$ such that f is continuous everywhere, differentiable everywhere except at $x = -1$, and fails the conclusion of Rolle's Theorem.

18. A function f defined on $[1, 5]$ with $f(1) = f(5) = 0$ such that f is continuous everywhere except for $x = 2$, differentiable everywhere except at $x = 2$, and fails the conclusion of Rolle's Theorem.

19. A function f defined on $[-3, -1]$ with $f(-3) = f(-1) = 0$ such that f is continuous everywhere except at $x = -1$ and differentiable everywhere except at $x = -1$, and fails the conclusion of Rolle's Theorem.

20. A function f defined on $[0, 4]$ such that f is continuous everywhere, differentiable everywhere except at $x = 2$, and fails the conclusion of the Mean Value Theorem with $a = 0$ and $b = 4$.

21. A function f defined on $[-3, 3]$ such that f is continuous everywhere except at $x = 1$, differentiable everywhere except at $x = 1$, and fails the conclusion of the Mean Value Theorem with $a = -3$ and $b = 3$.

22. A function f defined on $[-2, 0]$ such that f is continuous everywhere except at $x = -2$, differentiable everywhere except at $x = -2$, and fails the conclusion of the Mean Value Theorem with $a = -2$ and $b = 0$.

Skills

For each graph of f in Exercises 23–26, approximate all the values $x \in (0, 4)$ for which the derivative of f is zero or does not exist. Indicate whether f has a local maximum, minimum, or neither at each of these critical points.

23.

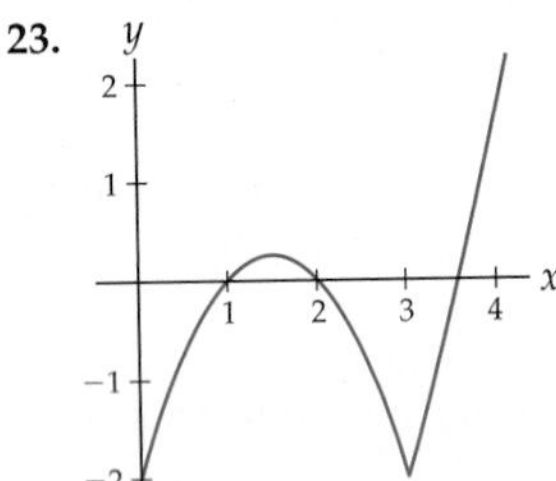

24.

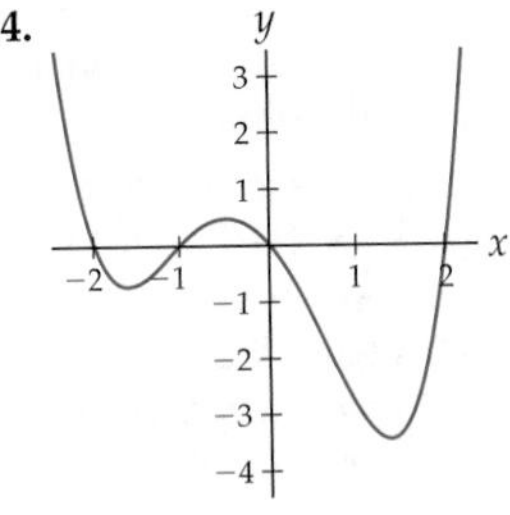

25.

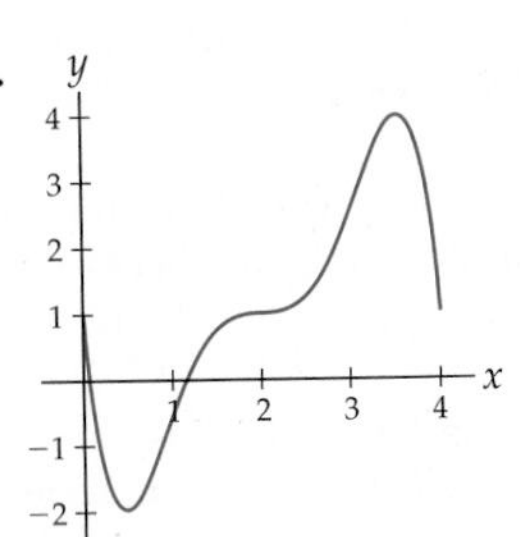

26.

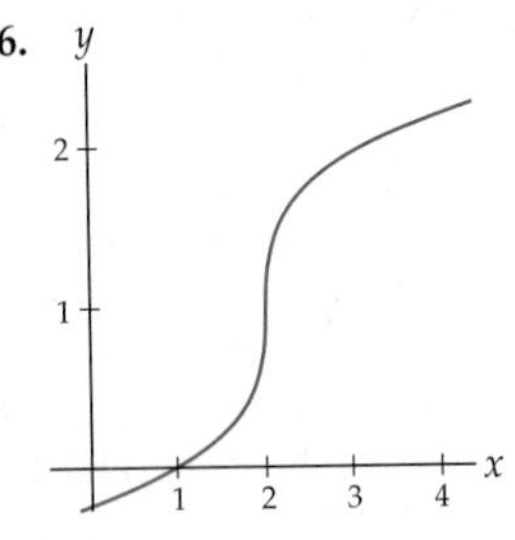

Find the critical points of each function f in Exercises 27–36. Then use a graphing utility to determine whether f has a local minimum, a local maximum, or neither at each of these critical points.

27. $f(x) = (x - 1.7)(x + 3)$

28. $f(x) = x^3 + x^2 + 1$

29. $f(x) = 3x^4 + 8x^3 - 18x^2$

30. $f(x) = (2x - 1)^5$

31. $f(x) = 3x - 2e^x$

32. $f(x) = 3^x - 2^x$

33. $f(x) = \dfrac{\ln 2x}{x}$

34. $f(x) = 2^{1 - \ln x}$

35. $f(x) = \cos x$

36. $f(x) = \sec x$

For each graph of f in Exercises 37–40, explain why f satisfies the hypotheses of Rolle's Theorem on the given interval $[a, b]$. Then approximate any values $c \in (a, b)$ that satisfy the conclusion of Rolle's Theorem.

37. 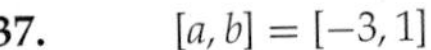 $[a, b] = [-3, 1]$

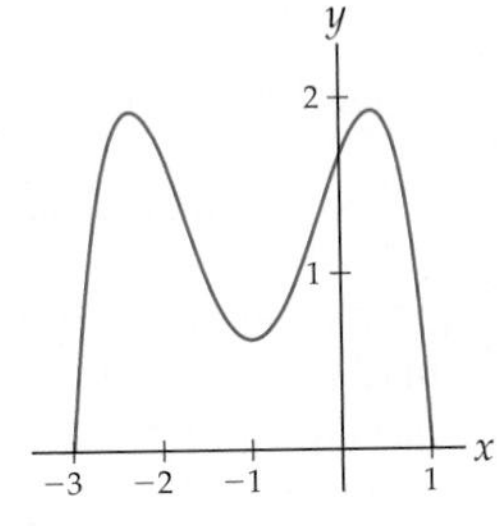

 38. $[a, b] = [-3, 3]$

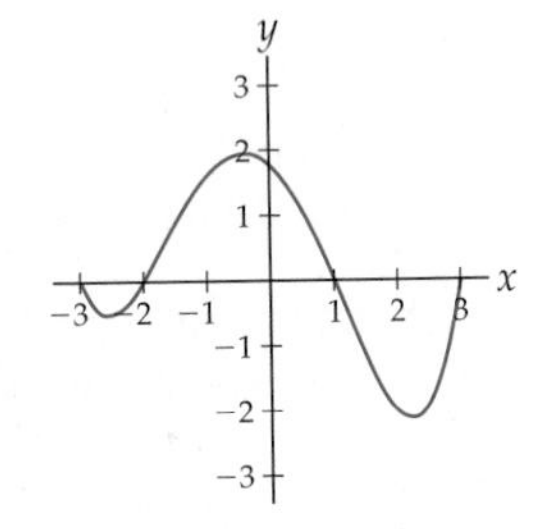

39. 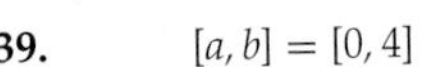 $[a, b] = [0, 4]$

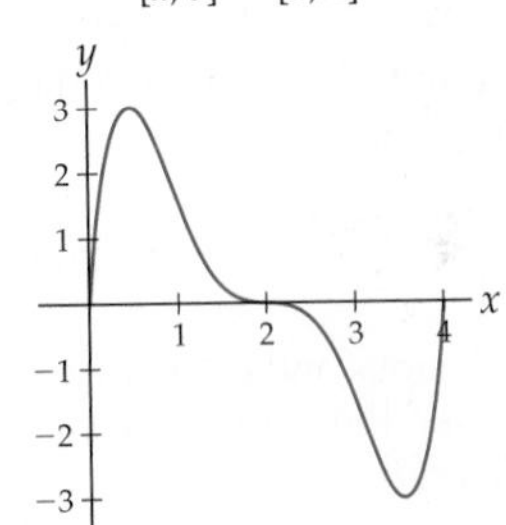

40. $[a, b] = [-1, 1]$

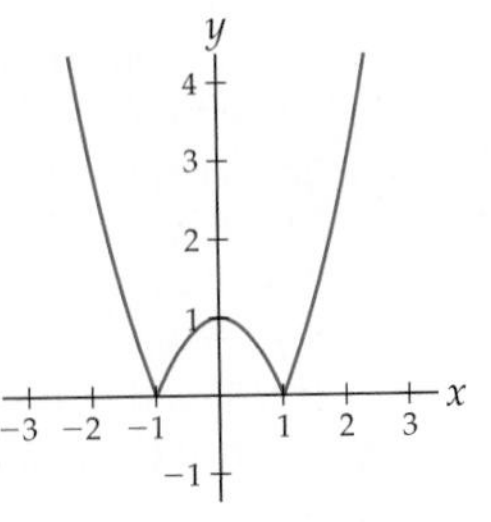

Determine whether or not each function f in Exercises 41–48 satisfies the hypotheses of Rolle's Theorem on the given interval $[a, b]$. For those that do, use derivatives and algebra to find the exact values of all $c \in (a, b)$ that satisfy the conclusion of Rolle's Theorem.

41. $f(x) = x^3 - 4x^2 + 3x$, $[a, b] = [0, 3]$

42. $f(x) = x^3 - 4x^2 + 3x$, $[a, b] = [1, 3]$

43. $f(x) = x^4 - 3.24x^2 - 3.04$, $[a, b] = [-2, 2]$

44. $f(x) = \dfrac{x^2 - 4x}{x^2 - 4x + 3}$, $[a, b] = [0, 4]$

45. $f(x) = \cos x$, $[a, b] = \left[-\dfrac{\pi}{2}, \dfrac{3\pi}{2}\right]$

46. $f(x) = \sin 2x$, $[a, b] = [0, 2\pi]$

47. $f(x) = e^x(x^2 - 2x)$, $[a, b] = [0, 2]$

48. $f(x) = \ln|x^2 - 1|$, $[a, b] = [-\sqrt{2}, \sqrt{2}]$

For each graph of f in Exercises 49–52, explain why f satisfies the hypotheses of the Mean Value Theorem on the given interval $[a, b]$ and approximate any values $c \in (a, b)$ that satisfy the conclusion of the Mean Value Theorem.

49. $[a, b] = [0, 2]$

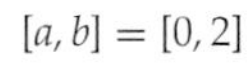

50. $[a, b] = [-1, 3]$

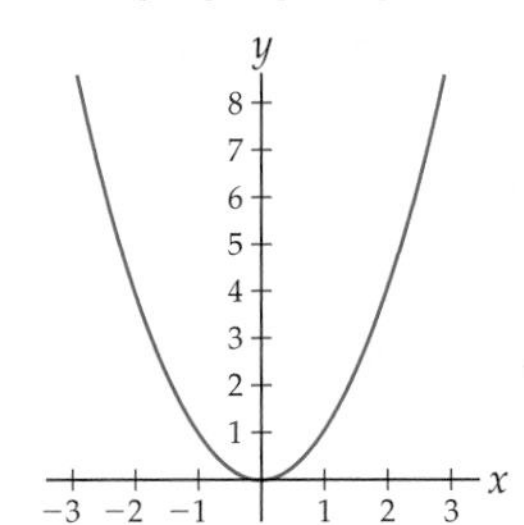

51. $[a, b] = [-3, 0]$

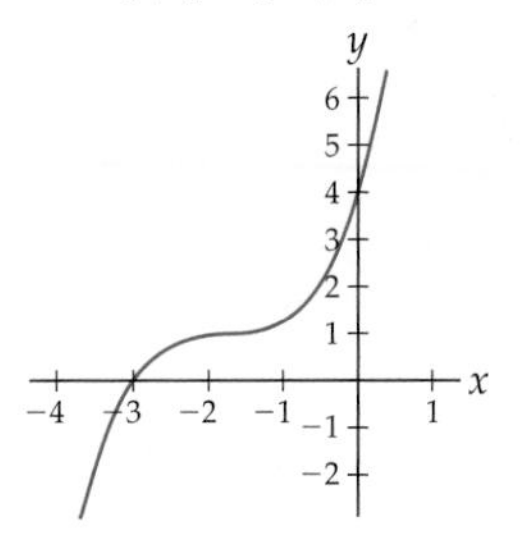

52. $[a, b] = [0, 4]$

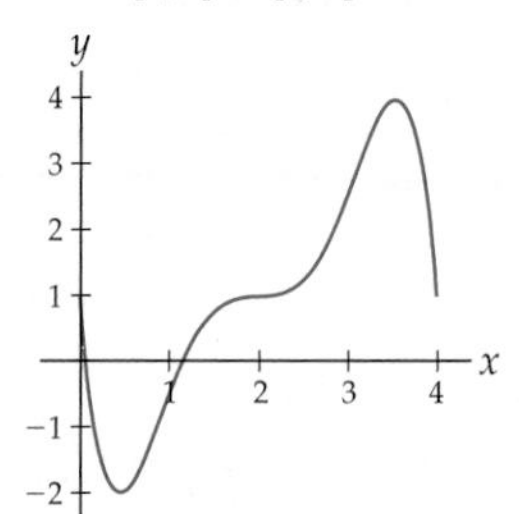

Determine whether or not each function f in Exercises 53–60 satisfies the hypotheses of the Mean Value Theorem on the given interval $[a, b]$. For those that do, use derivatives and algebra to find the exact values of all $c \in (a, b)$ that satisfy the conclusion of the Mean Value Theorem.

53. $f(x) = x^2 + \dfrac{1}{x}$, $[a, b] = [-3, 2]$

54. $f(x) = x^3 - 2x + 1$, $[a, b] = [0, 6]$

55. $f(x) = -x^3 + 3x^2 - 7$, $[a, b] = [-2, 3]$

56. $f(x) = (x^2 - 1)(x^2 - 4)$, $[a, b] = [-3, 3]$

57. $f(x) = \ln(x^2 + 1)$, $[a, b] = [0, 1]$

58. $f(x) = 2^x$, $[a, b] = [0, 3]$

59. $f(x) = \sin x$, $[a, b] = \left[0, \dfrac{\pi}{2}\right]$

60. $f(x) = \tan x$, $[a, b] = [-\pi, \pi]$

Applications

61. The cost of manufacturing a container for frozen orange juice is $C(h) = h^2 - 7.4h + 13.7$ cents, where h is the height of the container in inches. Your boss claims that the containers will be cheapest to make if they are 4 inches tall. Use Theorem 3.3 to quickly show that he is wrong.

62. Last night at 6 P.M., Linda got up from her blue easy chair. She did not return to her easy chair until she sat down again at 8 P.M. Let $s(t)$ be the distance between Linda and her easy chair t minutes after 6 P.M. last night.
    (a) Sketch a possible graph of $s(t)$, and describe what Linda did between 6 P.M. and 8 P.M. according to your graph. (Questions to think about: Will Linda necessarily move in a continuous and differentiable way? What are good ranges for t and s?)
    (b) Use Rolle's Theorem to show that at some point between 6 P.M. and 8 P.M., Linda's velocity $v(t)$ with respect to the easy chair was zero. Find such a place on the graph of $s(t)$.

63. It took Alina half an hour to drive to the grocery store that is 20 miles from her house.
    (a) Use the Mean Value Theorem to show that, at some point during her trip, Alina must have been travelling exactly 40 miles per hour.
    (b) Why does what you have shown in part (a) make sense in real-world terms?

Proofs

64. Prove the part of Theorem 3.3 that was not proved in the reading: If a function f has a local minimum at $x = c$, then either $f'(c)$ does not exist or $f'(c) = 0$.

65. Prove Rolle's Theorem: If f is continuous on $[a, b]$ and differentiable on (a, b), and if $f(a) = f(b) = 0$, then there is some value $c \in (a, b)$ with $f'(c) = 0$.

66. Prove the Mean Value Theorem: If f is continuous on $[a, b]$ and differentiable on (a, b), then there is some value $c \in (a, b)$ with $f'(c) = \dfrac{f(b) - f(a)}{b - a}$.

67. Use Rolle's Theorem to prove that if f is continuous and differentiable everywhere and has three roots, then its derivative f' has at least two roots.

68. Follow the method of proof that we used for Rolle's Theorem to prove the following slightly more general theorem: If f is continuous on $[a, b]$ and differentiable on (a, b), and if $f(a) = f(b)$, then there is some value $c \in (a, b)$ with $f'(c) = 0$.

69. Use Rolle's Theorem to prove the slightly more general theorem from Exercise 68: If f is continuous on $[a, b]$ and differentiable on (a, b), and if $f(a) = f(b)$, then there is some value $c \in (a, b)$ with $f'(c) = 0$. *(Hint: Apply Rolle's Theorem to the function $g(x) = f(x) - f(a)$.)*

Thinking Forward

Sign analyses for derivatives: For each function f that follows, find the derivative f'. Then determine the intervals on which the derivative f' is positive and the intervals on which the derivative f' is negative. Record your answers on a sign chart for f', with tick-marks only at the x-values where f' is zero or undefined.

- $f(x) = \dfrac{x}{x^2+1}$
- $f(x) = x^2 3^x$
- $f(x) = \dfrac{\sin x}{e^x}$
- $f(x) = \ln(\ln x)$

Sign analyses for second derivatives: Repeat the instructions of the previous block of problems, except find sign intervals for the second derivative f'' instead of the first derivative.

- $f(x) = \dfrac{x}{x^2+1}$
- $f(x) = x^2 3^x$
- $f(x) = \dfrac{\sin x}{e^x}$
- $f(x) = \ln(\ln x)$

3.2 THE FIRST DERIVATIVE AND CURVE SKETCHING

- The relationship between the derivative and increasing/decreasing functions
- Proving that all antiderivatives of a function differ by a constant
- Using the first-derivative test to determine whether critical points are maxima, minima, or neither

Derivatives and Increasing/Decreasing Functions

In Section 0.4 we defined a function f to be ***increasing*** on an interval if, for all a and b in the interval with $b > a$, $f(b) > f(a)$. In other words, the height of f at points farther to the right are higher. Similarly, f is ***decreasing*** on an interval if, for all $b > a$ in the interval, $f(b) < f(a)$. These definitions can be difficult to work with if we wish to find the intervals on which a given function is increasing or decreasing. Luckily, the derivative will provide us with an easier method.

We have seen that the first derivative f' in some sense measures the direction of the graph of a function f at each point, since f' is the associated slope function for f. In particular, if f' is positive at a point $x = c$, then the graph of f must be moving in an upwards direction, that is, increasing, as it passes $x = c$. Similarly, if $f'(c)$ is negative, then the graph of f must be decreasing at $x = c$. For example, we can divide the real-number line into intervals according to where the function $f(x) = x^3 - 3x^2 - 9x + 11$ is increasing or decreasing, as shown in the figure that follows at the left. This same division into subintervals describes where the derivative $f'(x) = 3x^2 - 6x - 9$ is positive and negative, as shown at the right.

Intervals where f increases/decreases

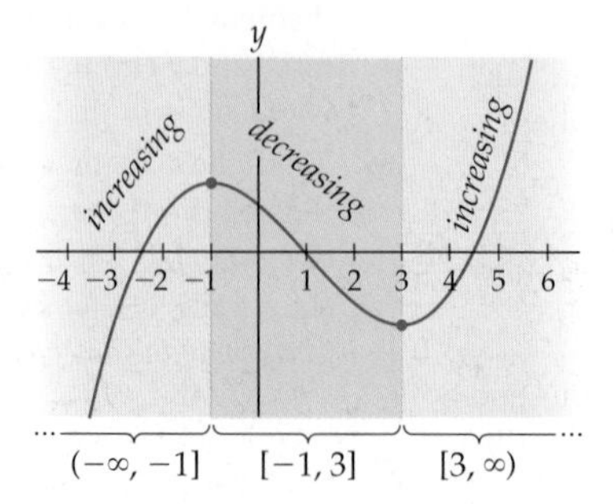

Intervals where f' is positive/negative

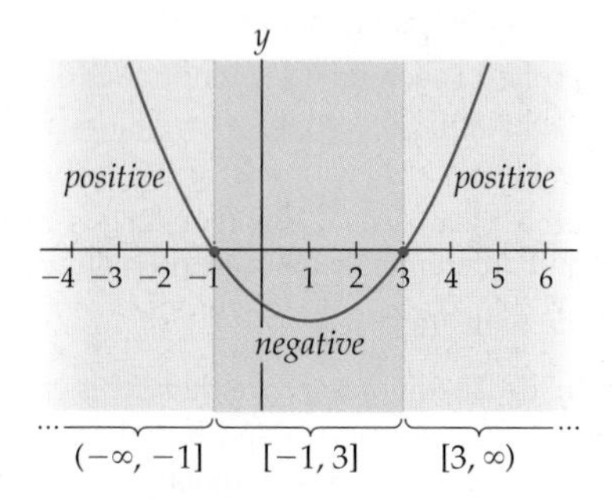

This relationship between intervals on which f is increasing or decreasing exactly when f' is positive or negative, respectively, holds in general. The only wrinkle is at the extremum of a function f; for example, in the graph at the left we say that f is increasing on $(-\infty, -1]$, which includes the extremum at $x = -1$, because it is true that for all $b > a$ less than *or equal to* $x = -1$ we do have $f(b) > f(a)$. However, the derivative at $x = -1$ is not positive, but zero. Therefore f' is positive only on the interior $(-\infty, -1)$ of the interval, but the function f is increasing on the entire interval $(-\infty, -1]$.

THEOREM 3.6 **The Derivative Measures Where a Function is Increasing or Decreasing**

Let f be a function that is differentiable on an interval I.

(a) If f' is positive in the interior of I, then f is increasing on I.

(b) If f' is negative in the interior of I, then f is decreasing on I.

(c) If f' is zero in the interior of I, then f is constant on I.

Theorem 3.6 is intuitively obvious if we consider the slopes of tangent lines on increasing and decreasing graphs. To prove this theorem formally we require the Mean Value Theorem.

Proof. We'll prove part (a) here and leave the similar proofs of parts (b) and (c) to Exercises 89 and 90, respectively. The key to the proof will be the Mean Value Theorem.

Let f be a function that is differentiable on an interval I and whose derivative f' is positive on the interior of that interval. Suppose also that $a, b \in I$ with $b > a$. By the definition of increasing, we must show that $f(b) > f(a)$. Since f is differentiable, and thus also continuous, on the interval I, and since $[a, b]$ is contained in the interval I, f satisfies the hypotheses of the Mean Value Theorem on $[a, b]$. Therefore we can conclude that there exists some $c \in (a, b)$ such that

$$f'(c) = \frac{f(b) - f(a)}{b - a}.$$

To show that $f(b) > f(a)$ it suffices to show that $f(b) - f(a) > 0$; with this in mind we can rearrange the preceding equation as

$$f(b) - f(a) = f'(c)\,(b - a).$$

Since $c \in (a, b)$, it follows that c is in the interior of I, and thus by hypothesis $f'(c) > 0$. Furthermore, since $b > a$, we have $(b - a) > 0$. Therefore $f(b) - f(a)$ is the product of two positive numbers and must itself be positive, which is what we wanted to show. ■

Up to this point we could only graphically approximate the intervals on which a function is increasing or decreasing. With Theorem 3.6 we can now find these intervals algebraically, by examining the sign of f'. We have thus reduced the difficult problem of finding the intervals on which a function is increasing or decreasing to the much simpler problem of finding the intervals on which an associated function—the derivative—is positive or negative.

Functions with the Same Derivative

If two functions differ by a constant, then obviously they will have the same derivative, because the derivative of a constant is zero. For example, $f(x) = x^3$ and $g(x) = x^3 + 10$ differ by a constant because their difference $g(x) - f(x)$ is equal to the constant 10, and their derivatives $f'(x) = 3x^2$ and $g'(x) = 3x^2 + 0 = 3x^2$ are equal.

Although it is less obvious, the converse is also true: Any two functions that have the same derivative must differ by a constant. Algebraically, this means that if you find one antiderivative of a function, then all other antiderivatives of that function differ from the one that you found by a constant. For example, one antiderivative of $3x^2$ is x^3, and thus all antiderivatives of $3x^2$ are of the form $x^3 + C$ for some constant C. Graphically, this means that if two functions have the same derivative, then one is a vertical shift of the other. For example, the graph of $3x^2$ is shown next at the left and five of its antiderivatives $x^3 + C$ are shown at the right, for $C = 0$, $C = \pm 10$, and $C = \pm 20$. The red graph of $y = 3x^2$ yields information about every one of the blue graphs $y = x^3 + C$, regardless of the vertical shift C.

Graph of $y = 3x^2$

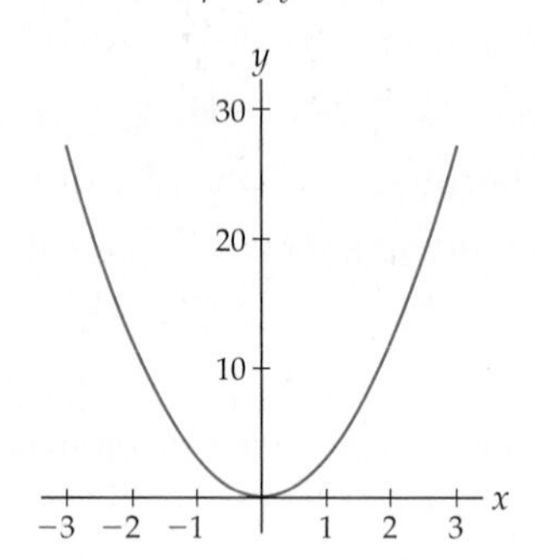

All antiderivatives of $3x^2$ are of the form $x^3 + C$

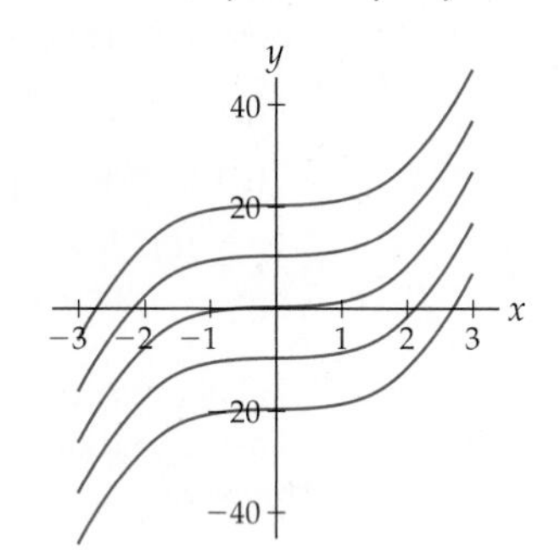

THEOREM 3.7 Functions with the Same Derivative Differ by a Constant

If $f'(x) = g'(x)$ for all $x \in [a, b]$, then, for some constant C, $f(x) = g(x) + C$ for all $x \in [a, b]$.

Proof. Suppose $f'(x) = g'(x)$ for all $x \in [a, b]$. Then $f'(x) - g'(x) = 0$, which by the difference rule means that $\frac{d}{dx}(f(x) - g(x)) = 0$ on $[a, b]$. By the third part of Theorem 3.6 this means that the function $f(x) - g(x)$ is constant for all $x \in [a, b]$, say, $f(x) - g(x) = C$ for some real number C. Therefore $f(x) = g(x) + C$ for all $x \in [a, b]$. ■

The First-Derivative Test

In the previous section we saw that the set of critical points of a function, that is, the values of x for which f' is zero or does not exist—is a complete list of all the possible local extrema of f. We now develop a method for using the first derivative to determine which critical points are local maxima, which are local minima, and which are neither.

Suppose $f'(c) = 0$ and that f is differentiable near c. Then if f is not constant, there are four different ways that f can behave near $x = c$:

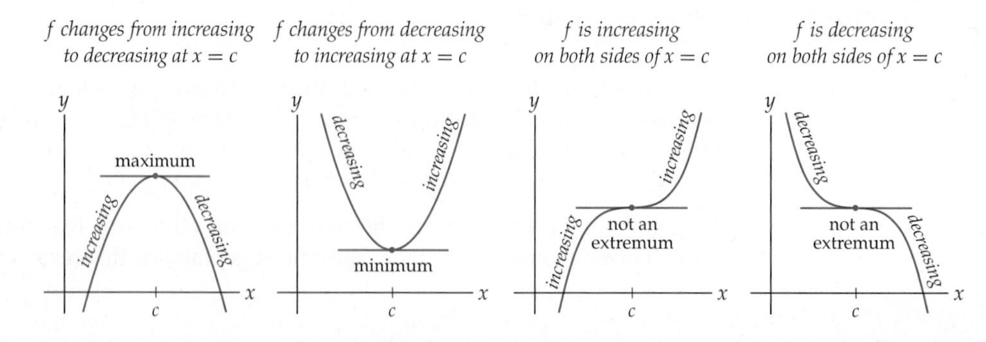

On the one hand, if f changes from increasing to decreasing at $x = c$, then f has a local maximum at $x = c$. If f changes from decreasing to increasing at the point $x = c$, then f has a local minimum at $x = c$. On the other hand, if f does not change direction at $x = c$, then f does not have a local extremum at $x = c$. Since a function f is increasing or decreasing according to whether its derivative is positive or negative, we can record the preceding four situations on sign charts for f', with the corresponding information for f recorded as the following sign analyses:

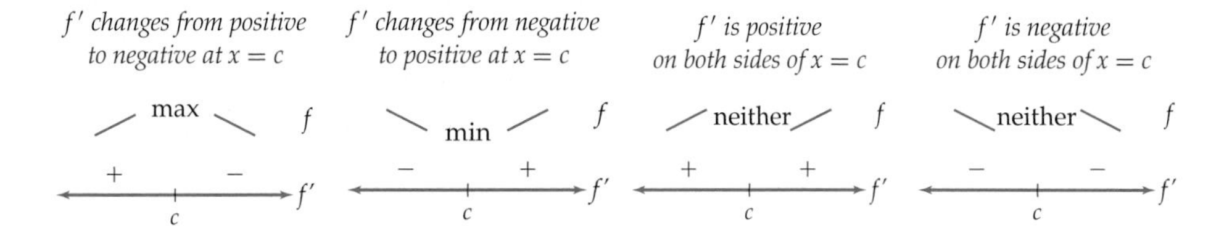

CAUTION Please recall that, throughout this book, we will make tick-marks on sign charts *only* at the locations where the function is zero or fails to exist. We will mark with "DNE" any locations where the function does not exist, but we will not explicitly mark the zeroes. You should assume that any unlabeled tick-marks on a sign chart are zeroes of the function in question. Do not make additional tick-marks on your sign charts if you follow this convention.

Using a sign chart for the first derivative f' to test whether critical points of f are local maxima, minima, or neither is quite sensibly known as the ***first-derivative test***. The statement of the test is wordy and its proof is somewhat technical, but its meaning is equivalent to information obtained from the previous sign chart analyses.

THEOREM 3.8 **The First-Derivative Test**

Suppose $x = c$ is the location of a critical point of a function f, and let (a, b) be an open interval around c that is contained in the domain of f and does not contain any other critical points of f. If f is continuous on (a, b) and differentiable at every point of (a, b) except possibly at $x = c$, then the following statements hold.

(a) If $f'(x)$ is positive for $x \in (a, c)$ and negative for $x \in (c, b)$, then f has a local maximum at $x = c$.

(b) If $f'(x)$ is negative for $x \in (a, c)$ and positive for $x \in (c, b)$, then f has a local minimum at $x = c$.

(c) If $f'(x)$ is positive for both $x \in (a, c)$ and $x \in (c, b)$, then f does not have a local extremum at $x = c$.

(d) If $f'(x)$ is negative for both $x \in (a, c)$ and $x \in (c, b)$, then f does not have a local extremum at $x = c$.

Proof. We will prove parts (a) and (c) here and leave the similar proofs of parts (b) and (d) to Exercises 91 and 92, respectively. The proof will be an application of Theorem 3.6. Suppose $x = c$ is a critical point of f, and let (a, b) be an interval around $x = c$ satisfying the hypotheses of the theorem.

To prove part (a), suppose $f'(x) > 0$ for $x \in (a, c)$ and $f'(x) < 0$ for $x \in (c, b)$, that is, suppose that f is increasing on $(a, c]$ and decreasing on $[c, b)$. We will show that $f(c) \geq f(x)$ for all $x \in (a, b)$, which will tell us that f has a local maximum at $x = c$. Given any $x \in (a, b)$, there are three cases to consider. First, if $x = c$, then clearly $f(c) = f(x)$. Second, if $a < x < c$, then since f is increasing on $(a, c]$, we have $f(x) < f(c)$. Third, if $c < x < b$, then since f is decreasing on $[c, b)$, we have $f(c) > f(x)$. In all three cases we have $f(c) \geq f(x)$, and therefore f has a local maximum at $x = c$.

To prove part (c), suppose that $f'(x) > 0$ for all $x \in (a, c) \cup (c, b)$. Then by Theorem 3.6, f must be increasing on all of (a, b). Now, the point $x = c$ cannot be the location of a local maximum of f, because, for all $x > c$ in (a, b), $f(x) > f(c)$, since f is increasing on (a, b). But neither can $x = c$ be a local minimum of f, because, for all $x < c$ in (a, b), $f(x) < f(c)$. Therefore f has neither a local minimum nor a local maximum at $x = c$. ■

Now, to find the local extrema of a function f, we need only find the critical points of f and then test each one with the first-derivative test. In other words, we find the derivative f', determine where f' is zero or does not exist, and then make a sign chart for f' around these critical points to determine whether f has a local maximum, a local minimum, or neither at each critical point. This method will find all local extrema for functions f that are defined on open intervals. For functions defined on closed or half-closed intervals, we will also have to consider endpoint extrema, which we will discuss in Section 3.4.

Examples and Explorations

EXAMPLE 1 **Using the derivative to determine where a function is increasing or decreasing**

Use Theorem 3.6 to determine the intervals on which each of the following functions are increasing or decreasing:

(a) $f(x) = x^3$ **(b)** $g(x) = x^2 - 2x + 1$ **(c)** $h(x) = \dfrac{x^2 + 10x + 1}{x - 2}$

SOLUTION

(a) If $f(x) = x^3$, then $f'(x) = 3x^2$, which is positive as long as $x \neq 0$. Therefore f' is positive on $(-\infty, 0)$ and $(0, \infty)$. By Theorem 3.6 we can say that f is increasing on the entire half-closed intervals $(-\infty, 0]$ and $[0, \infty)$. Thus f is increasing on all of $\mathbb{R}$. Notice that, as shown in the graph that follows, the tangent line to $f(x) = x^3$ is horizontal at $x = 0$. However, the function $f(x) = x^3$ is still increasing everywhere: For all real numbers $b > a$ we have $b^3 > a^3$, even if one of a or b is zero.

$f(x) = x^3$ *is increasing everywhere*

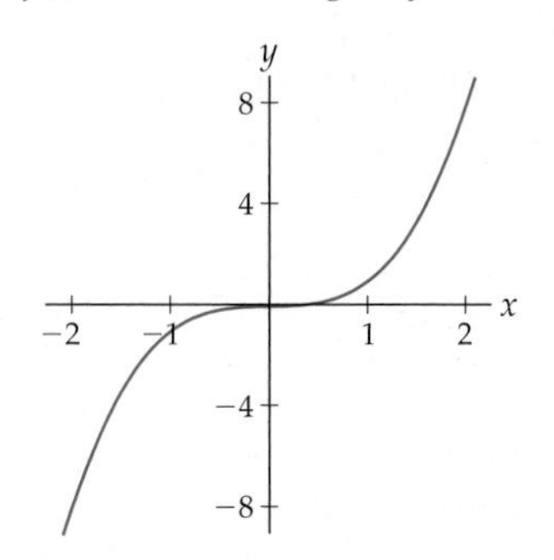

(b) If $g(x) = x^2 - 2x + 1$, then $g'(x) = 2x - 2$. Recall that functions can change signs only at roots, non-domain points, and discontinuities. The derivative $g'(x) = 2x - 2$ is always defined and continuous, and $g'(x) = 0$ when $2x - 2 = 0$ (i.e., when $x = 1$). Therefore we need only check the sign of $g'(x)$ to the left and right of $x = 1$. For example, $g'(0) = 2(0) - 2 = -2$ is negative, and $g'(2) = 2(2) - 2 = 2$ is positive, as the following sign chart shows:

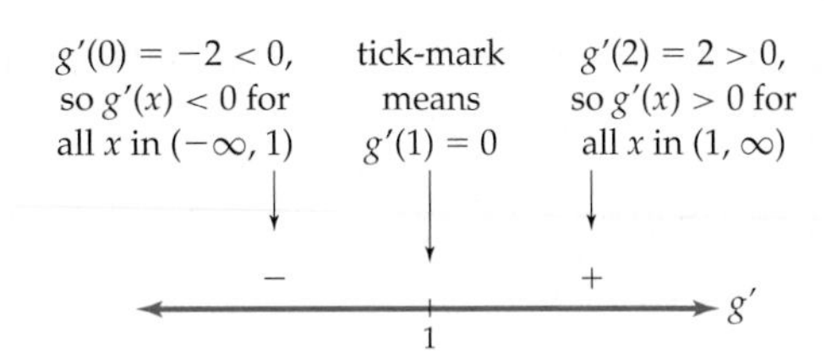

By Theorem 3.6, the information on the sign chart for $g'(x)$ shows that $g(x)$ is decreasing on $(-\infty, 1]$ and increasing on $[1, \infty)$. However, from this point forward we will record the closed-endpoint information only when we need it and instead record only the open-interval information. Therefore here we would say simply that $g(x)$ is decreasing on $(-\infty, 1)$ and increasing on $(1, \infty)$.

(c) By the quotient rule, the derivative of $h(x)$ is

$$h'(x) = \frac{(2x+10)(x-2) - (x^2+10x+1)(1)}{(x-2)^2}$$
$$= \frac{x^2-4x-21}{(x-2)^2} = \frac{(x+3)(x-7)}{(x-2)^2}.$$

Notice that we simplified the preceding equation so that it would be immediately clear that $h'(x) = 0$ when $x = -3$ and $x = 7$ and that $h'(x)$ does not exist when $x = 2$, a point that is not in the domain of $h(x)$ in the first place. To determine the sign chart for $h'(x)$ we need only check the sign of $h'(x)$ one time between each of these critical points and non-domain points. For example, using the factored form of $h'(x)$ we have:

$$h'(-5) : \frac{(-)(-)}{(-)^2} \text{ (positive)} \qquad h'(0) : \frac{(+)(-)}{(-)^2} \text{ (negative)}$$
$$h'(5) : \frac{(+)(-)}{(+)^2} \text{ (negative)} \qquad h'(10) : \frac{(+)(+)}{(+)^2} \text{ (postiive)}$$

We can record this information on a sign chart for $h'(x)$ as follows, with annotations above the chart for the corresponding increasing/decreasing behavior of $h(x)$:

h
+ − DNE − + h'
−3 2 7

Reading off the sign chart, we see that $h(x)$ is increasing on $(-\infty, -3)$ and $(7, \infty)$ and decreasing on $(-3, 2)$ and on $(2, 7)$. □

To check our answers to parts (b) and (c) of the example, we can simply graph $g(x)$ and $h(x)$. Notice that the graph of $g(x)$ shown next at the left does appear to be decreasing on $(-\infty, 1)$ and increasing on $(1, \infty)$. The graph of $h(x)$ at the right does appear to be increasing on $(-\infty, -3)$ and $(7, \infty)$ and decreasing elsewhere.

$g(x) = x^2 - 2x + 1$ $\qquad$ $h(x) = \dfrac{x^2+10x+1}{x-2}$

4, −1, 3, 0 $\qquad$ 40, −10, 15, −40

EXAMPLE 2 **Interpreting the sign chart for the derivative of a function**

Let f be a function whose derivative f' has the signs indicated on the sign chart shown. Recall our convention that tick-marks on this sign chart indicate the locations where f' is zero (if unlabeled) or does not exist (if marked with "DNE"). Use the sign chart to sketch possible graphs for f and f'.

SOLUTION

This sign chart says that f' is positive on the intervals $(-\infty, -1)$ and $(2, \infty)$, and negative on the interval $(-1, 2)$. It also says that $f'(-1) = 0$ and that $f'(2)$ does not exist. Given this information, one possible sketch of f' is shown next at the left.

We can determine a lot about the graph of a function f from the graph of its derivative f'. The function f must be increasing on $(-\infty, -1)$ and $(2, \infty)$, and decreasing on $(-1, 2)$, with a horizontal tangent line at $x = -1$, and a non-differentiable point at $x = 2$. One possible sketch of a function f that has these characteristics is shown next at the right. Notice that the information about f' does not tell us how high or low to sketch the graph of f. In fact, any vertical shift of f would do just as well, since every antiderivative of f' is of the form $f(x) + C$.

Possible graph of f'

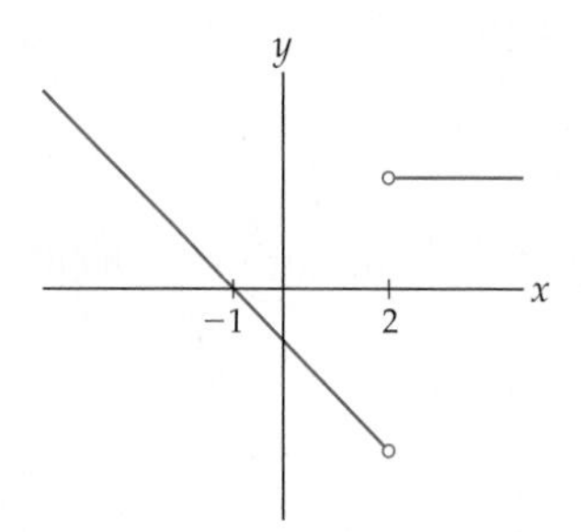

Possible graph of f

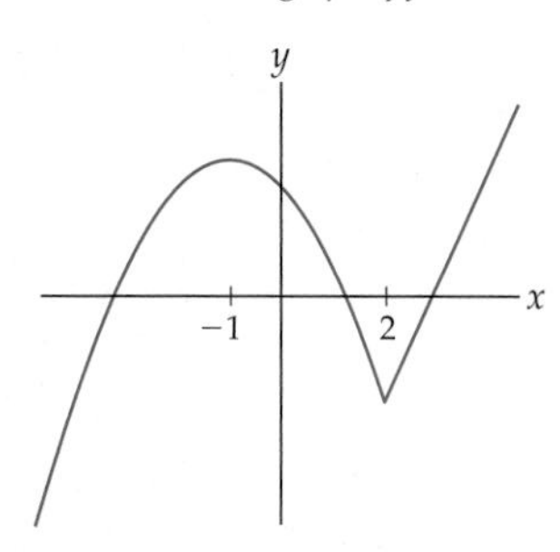

□

EXAMPLE 3 **Detailed curve-sketching analyses using the first derivative**

Sketch the graph of each of the given functions. Along the way, determine all local extrema and important features of the graph.

(a) $f(x) = x^3 - 3x + 2$ **(b)** $f(x) = \dfrac{x^2}{x^2 - 2x + 1}$

SOLUTION

(a) For $f(x) = x^3 - 3x + 2$ we have $f'(x) = 3x^2 - 3$. This derivative is always defined and continuous, so the critical points of f are just the places where $f'(x) = 0$:

$$3x^2 - 3 = 0 \implies 3x^2 = 3 \implies x^2 = 1 \implies x = \pm 1.$$

These critical points divide the real-number line into three intervals, namely, $(-\infty, -1)$, $(-1, 1)$, and $(1, \infty)$. We'll test the sign of f' on each interval by testing the sign of $f'(x)$ at one point in that interval, say, at $x = -2$, $x = 0$, and $x = 2$:

$$f'(-2) = 3(-2)^2 - 3 = 12 - 3 = 9 > 0,$$
$$f'(0) = 3(0)^2 - 3 = 0 - 3 = -3 < 0,$$
$$f'(2) = 3(2)^2 - 3 = 12 - 3 = 9 > 0.$$

Recording this information on a sign chart for f', with consequences for f illustrated above the chart, we have

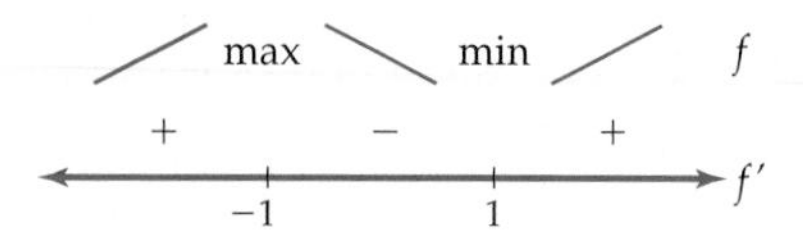

The maximum and minimum of f listed on the chart were identified by the first-derivative test: Since f' changes from positive to negative at $x = -1$, f changes from increasing to decreasing there; thus the point $x = -1$ must be a local maximum of f. Since f' changes from negative to positive at $x = 1$, f changes from decreasing to increasing there; thus f must have a local minimum at the point $x = 1$.

With all the first-derivative information we have collected, we can sketch a reasonable graph of f by plotting just a few points. As a general rule it is a good idea to plot the points $(c, f(c))$ for each critical point $x = c$ and then use the information about the derivative to connect the dots accordingly. In this example, the critical points are $x = -1$ and $x = 1$, so we calculate:

$$f(-1) = (-1)^3 - 3(-1) + 2 = -1 + 3 + 2 = 4,$$
$$f(1) = (1)^3 - 3(1) + 2 = 1 - 3 + 2 = 0.$$

Using this information, we can make a sketch of the graph of $f(x) = x^3 - 3x + 2$, as shown next. Compare the features of this graph with the information in the sign chart for f':

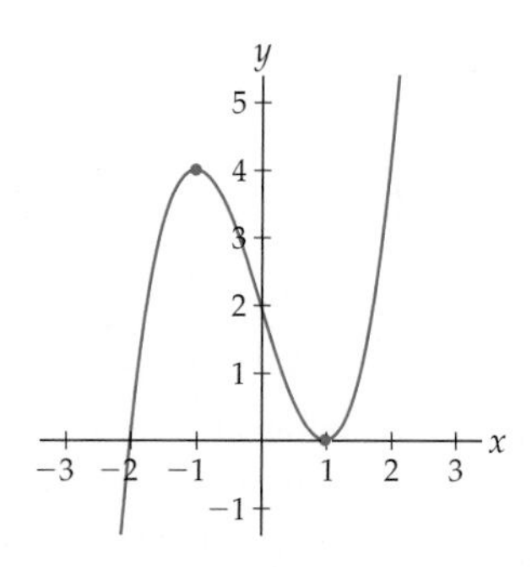

(b) Before we do anything else, we should note that the domain of

$$f(x) = \frac{x^2}{x^2 - 2x + 1} = \frac{x^2}{(x-1)^2}$$

is all points except $x = 1$. The function f is zero only at $x = 0$, and at all other points in its domain it is positive, since both its numerator and denominator are perfect squares. So far we know that the entire graph of f will be above the x-axis, except at the root $(0, 0)$ and at the non-domain point $x = 1$, where something interesting may occur.

To determine local extrema and increasing/decreasing behavior, we must find the derivative of $f(x) = \frac{x^2}{x^2 - 2x + 1}$. Because we will be interested in where f' is zero or does not exist, we also simplify as much as possible:

$$f'(x) = \frac{(2x)(x^2 - 2x + 1) - (x^2)(2x - 2)}{(x^2 - 2x + 1)^2} = \frac{-2x(x-1)}{(x-1)^4} = \frac{-2x}{(x-1)^3}.$$

This derivative is zero when its numerator is zero but its denominator is not, that is, when $x = 0$. The derivative is undefined when $x = 1$, which makes sense because the original function f is also undefined at $x = 1$. The critical point $x = 0$ and the non-domain point $x = 1$ are the only places at which the sign of f' can change.

Now we make a sign chart representing the sign of the derivative f' between each of these critical points. For example, we can find that $f'(-1) < 0$, $f'(0.5) > 0$, and $f'(2) < 0$:

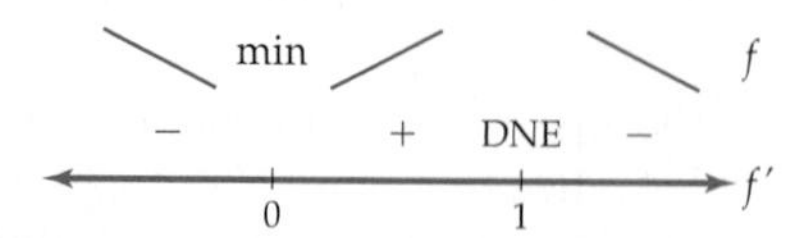

We can identify the local minimum at $x = 0$ indicated on the chart with the first-derivative test, since $f'(x)$ changes sign from negative to positive at $x = 0$. However, the same type of argument cannot be applied to the point $x = 1$: Even though $f'(x)$ changes sign at $x = 1$, the function f does not have a local extremum at $x = 1$, because it is not defined at that point.

Information about limits can be important to figuring out the key features of a graph, so we will examine limits of $f(x) = \frac{x^2}{x^2 - 2x + 1}$ at any "interesting" places. In this case the interesting limits to consider are at the point $x = 1$, where the function does not exist, and at the ends as $x \to \pm\infty$. As $x \to 1$ we have $x^2 \to 1$ and $x^2 - 2x + 1 = (x - 1)^2 \to 0^+$, and therefore the limit in question is of the form $\frac{1}{0^+}$. Thus

$$\lim_{x \to 1} \frac{x^2}{x^2 - 2x + 1} = \infty.$$

We can calculate the limit of f as $x \to \infty$ with the familiar strategy of dividing numerator and denominator by the highest power of x:

$$\lim_{x \to \infty} \frac{x^2}{x^2 - 2x + 1} \left(\frac{1/x^2}{1/x^2} \right) = \lim_{x \to \infty} \frac{1}{1 - \frac{2}{x} + \frac{1}{x^2}} = \frac{1}{1 - 0 + 0} = 1.$$

A similar argument shows that $\lim\limits_{x \to -\infty} \frac{x^2}{x^2 - 2x + 1}$ is also equal to 1. (We could also appeal to Theorem 1.32 and notice that the numerator and denominator of the function in question are polynomials of the same degree, and thus that the limit of the quotient as $x \to \pm\infty$ is the ratio $\frac{1}{1} = 1$ of the leading coefficients of those polynomials.) Since $\lim\limits_{x \to 1} f(x) = \infty$, we can see that f has a vertical asymptote at the non-domain point $x = 1$, with the function approaching ∞ on both sides of this asymptote. Since $\lim\limits_{x \to \infty} f(x) = 1$ and $\lim\limits_{x \to -\infty} f(x) = 1$, f has a horizontal asymptote on both the left and the right at $y = 1$.

Putting all of this information together, we get the following graph:

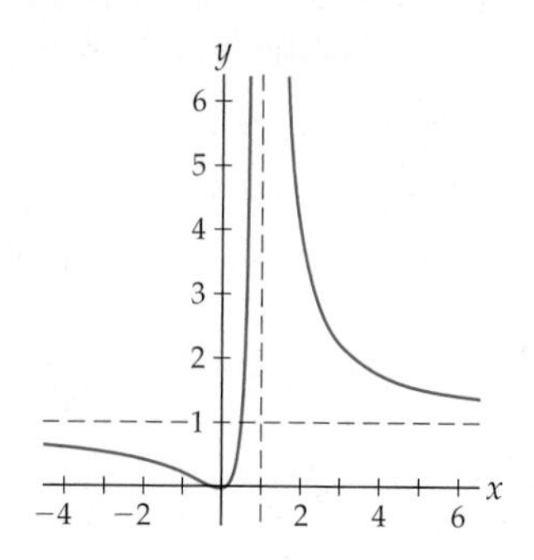

□

? TEST YOUR UNDERSTANDING

- If f' is positive on $(0, 3)$, why does it make sense that we can say that f is increasing on all of $[0, 3]$?
- Can you give an example of a function f for which f' is zero on all of $(-1, 1)$?
- If $x^3 - 4x^2 - 2$ is one antiderivative of some function f, what can you say about the other antiderivatives of f?
- If f is defined, continuous, and differentiable at $x = 3$, and if f' changes sign at $x = 3$, then what can you say about f at $x = 3$?
- Can you sketch an example of a function f with a critical point at $x = 2$ but no local extremum at $x = 2$?

EXERCISES 3.2

Thinking Back

Differentiation: Find the derivative of each function f, and then simplify as much as possible.

- $f(x) = (2x - 1)^3(3x + 1)^2$
- $f(x) = \dfrac{(2x-1)^3}{(3x+1)^2}$
- $f(x) = 3x^2e^{-4x}$
- $f(x) = \sin(\ln x)$

Solving equations: For each of the following functions $g(x)$, find the solutions of $g(x) = 0$ and also find the values of x for which $g(x)$ does not exist.

- $g(x) = \dfrac{\frac{1}{2}x^{-1/2}(1+5x) - \sqrt{x}(5)}{(1+5x)^2}$
- $g(x) = 2x\sqrt{x-1} + x^2\left(\frac{1}{2}(x-1)^{-1/2}\right)$
- $g(x) = \dfrac{e^x(1-e^x) - e^x(-e^x)}{(1-e^x)^2}$

Sign analyses: For each of the following functions $g(x)$, use algebra and a sign chart to find the intervals on which $g(x)$ is positive and the intervals on which $g(x)$ is negative.

- $g(x) = 6x^2 - 18x$
- $g(x) = 2 - x - 2x^2 + x^3$
- $g(x) = \dfrac{(x+2)(x-1)^4}{e^x}$
- $g(x) = \dfrac{3x^2 - 5x - 2}{\sin^2 x}$

Concepts

0. *Problem Zero:* Read the section and make your own summary of the material.

1. *True/False:* Determine whether each of the statements that follow is true or false. If a statement is true, explain why. If a statement is false, provide a counterexample.

(a) *True or False:* If $f'(x) < 0$ for all $x \in (0, 3)$, then f is decreasing on $[0, 3]$.

(b) *True or False:* If f is increasing on $(-2, 2)$, then $f'(x) \geq 0$ for all $x \in (-2, 2)$.

(c) *True or False:* If $f'(x) = 2x$, then $f(x) = x^2 + C$ for some constant C.

(d) *True or False:* If f' is continuous on $(1, 8)$ and $f'(3)$ is negative, then f' is negative on all of $(1, 8)$.

(e) *True or False:* If f' changes sign at $x = 3$, then $f'(3) = 0$.

(f) *True or False:* If $f'(-2) = 0$, then f has either a local maximum or a local minimum at $x = -2$.

(g) *True or False:* If $x = 1$ is the only critical point of f and $f'(0)$ is positive, then $f'(2)$ must be negative.

(h) *True or False:* If $f'(1)$ is negative and $f'(3)$ is positive, then f has a local minimum at $x = 2$.

2. *Examples:* Construct examples of the thing(s) described in the following. Try to find examples that are different than any in the reading.

(a) A function that is decreasing on $(-\infty, 0)$, increasing on $(0, \infty)$, and undefined at $x = 0$.

(b) A function that is decreasing on $(-\infty, 0]$ and increasing on $[0, \infty)$.

(c) A function that is always positive and always decreasing, on all of $\mathbb{R}$.

3. State the definition of what it means for a function f to be increasing on an interval I and what it means for a function f to be decreasing on an interval I.

4. Describe what a critical point is, intuitively and in mathematical language. Then describe what a local extremum is. How are these two concepts related?

5. Can a point $x = c$ be both a local extremum and a critical point of a differentiable function f? Both an inflection point and a critical point? Both an inflection point and a local extremum? Sketch examples, or explain why such a point cannot exist.

6. Suppose f is a function that may be non-differentiable at some points. Can a point $x = c$ be both a local extremum and a critical point of such a function f? Both an inflection point and a critical point? Both an inflection point and a local extremum? Sketch examples, or explain why such a point cannot exist.

7. Suppose f is defined and continuous everywhere. Why is testing the sign of the derivative f' at just one point sufficient to determine the sign of f' on the whole interval between critical points of f?

8. Describe what the first-derivative test is for and how to use it. Sketch graphs and sign charts to illustrate your description.

9. Sketch the graph of a function f with the following properties:
   - f is continuous and defined on $\mathbb{R}$;
   - $f(0) = 5$;
   - $f(-2) = -3$ and $f'(-2) = 0$;
   - $f'(1)$ does not exist;
   - f' is positive only on $(-2, 1)$.

10. Sketch the graph of a function f with the following properties:
    - f is continuous and defined on $\mathbb{R}$;
    - f has critical points at $x = -3, 0$, and 5;
    - f has inflection points at $x = -3, -1$, and 2.

11. Use the definitions of *increasing* and *decreasing* to argue that $f(x) = x^4$ is decreasing on $(-\infty, 0]$ and increasing on $[0, \infty)$. Then use derivatives to argue the same thing.

12. Use the definition of *increasing* to argue that $f(x) = x^5$ is increasing on all of $\mathbb{R}$. Then use derivatives to argue the same thing.

13. Suppose f is a function that is continuous and differentiable everywhere and that the derivative of f is

$$f'(x) = \sqrt{1 + x^2} - 4.$$

What are the critical points of f?

14. Suppose f is a function that is continuous and differentiable everywhere and that the derivative of f is

$$f'(x) = \frac{(x-1)(x-2)}{x-3}.$$

What are the critical points of f?

15. If $g(x)$ and $h(x)$ are both antiderivatives of some function $f(x)$, then what can you say about the function $g(x) - h(x)$?

16. If $g(x)$ is an antiderivative of $f(x)$, then what is the relationship between the functions $g(x) + 10$ and $f(x)$?

17. One of the graphs shown is a function f and the other is its derivative f'. Which one is which, and why?

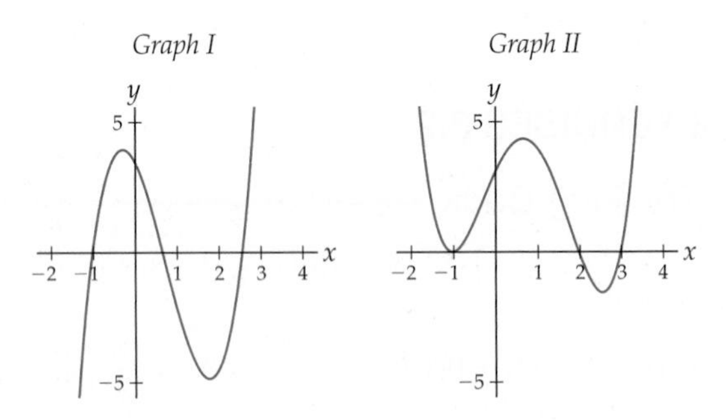

18. One of the graphs shown is a function f and the other is its derivative f'. Which one is which, and why?

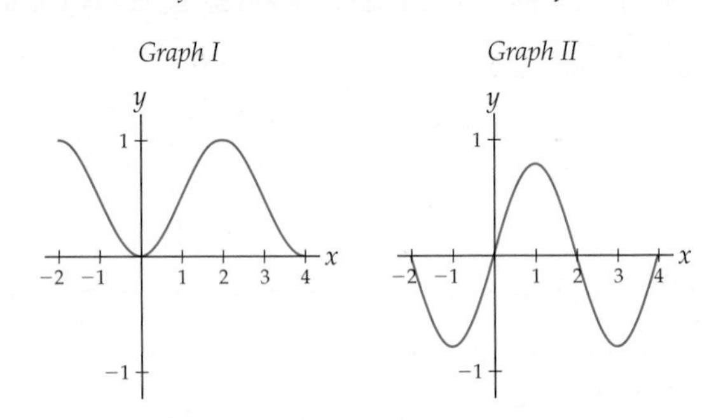

Skills

For each function f graphed in Exercises 19–22, sketch a possible graph of its derivative f'.

19.

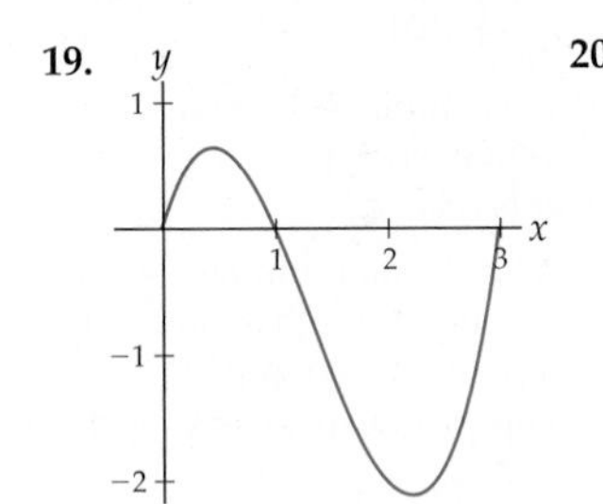

20.

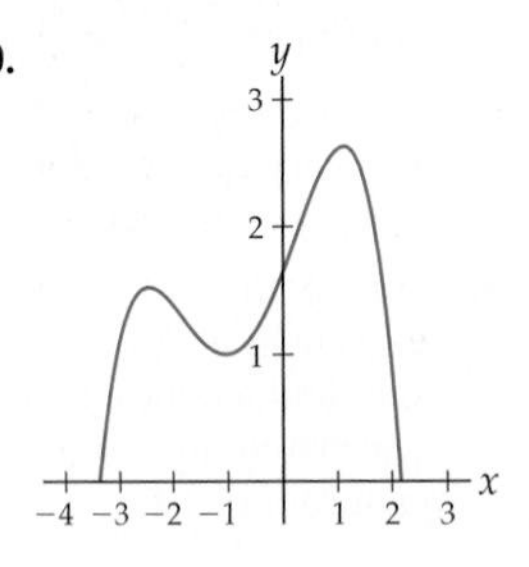

21. 22.

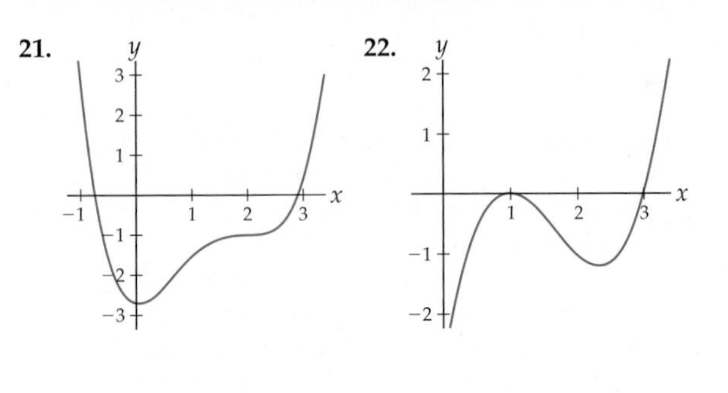

Each graph in Exercises 23–26 represents the derivative f' of some function f. Use the given graph of f' to sketch a possible graph of f.

23.

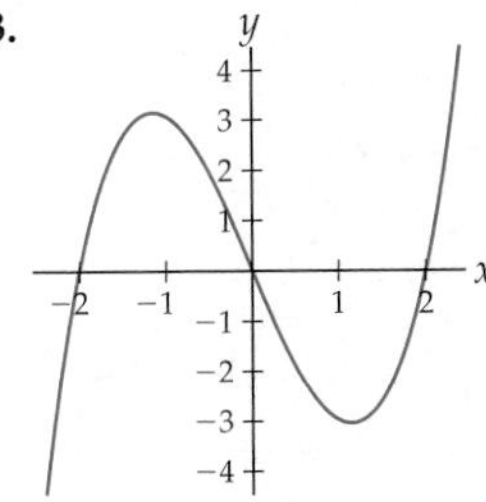

24.

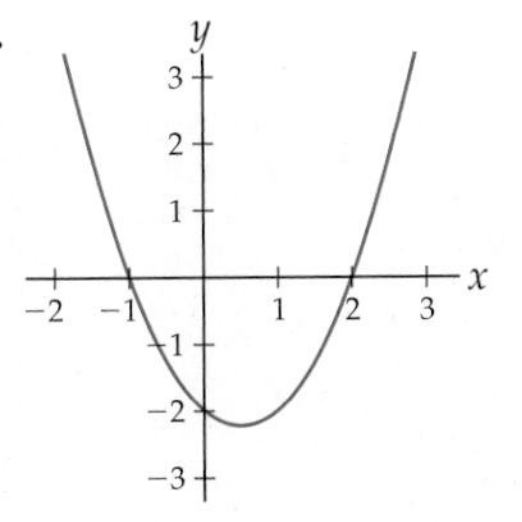

25.

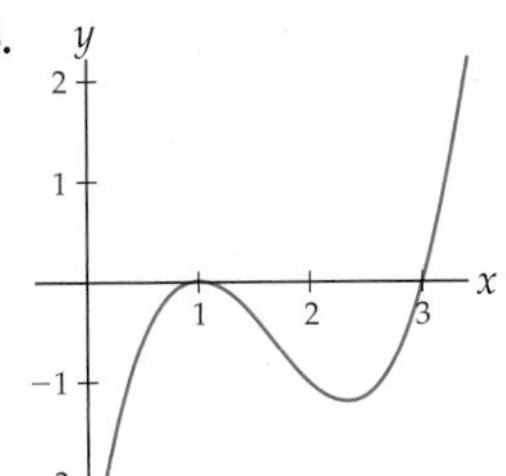

26.

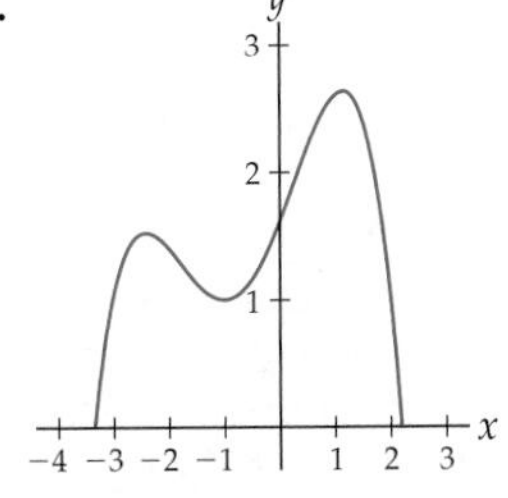

Use a sign chart for f' to determine the intervals on which each function f in Exercises 27–38 is increasing or decreasing. Then verify your algebraic answers with graphs from a calculator or graphing utility.

27. $f(x) = 2x^3 - 9x^2 + 1$

28. $f(x) = x^3 + 4x^2 + 4x - 5$

29. $f(x) = \dfrac{x}{x^2 + 4}$

30. $f(x) = \dfrac{3x + 1}{x^2 - 1}$

31. $f(x) = e^x(x - 2)$

32. $f(x) = \dfrac{e^x}{1 + e^x}$

33. $f(x) = \ln(x^2 + 1)$

34. $f(x) = \ln(\ln x)$

35. $f(x) = \sin\left(\dfrac{\pi}{2}x\right)$

36. $f(x) = \cos^2(x)$

37. $f(x) = \sin x \cos x$

38. $f(x) = \dfrac{1}{\sin x}$

Use the first-derivative test to determine the local extrema of each function f in Exercises 39–50. Then verify your algebraic answers with graphs from a calculator or graphing utility.

39. $f(x) = (x - 2)^2(1 + x)$

40. $f(x) = x^2(x - 1)(x + 1)$

41. $f(x) = \dfrac{1 + x + x^2}{x^2 + x - 2}$

42. $f(x) = \dfrac{(x - 1)^2}{x + 2}$

43. $f(x) = \dfrac{1}{3 - 2e^x}$

44. $f(x) = e^x(x^2 - x - 1)$

45. $f(x) = \cos(\pi(x + 1))$

46. $f(x) = \cos(\pi x)$

47. $f(x) = \arctan x$

48. $f(x) = \sin^{-1} x^2$

49. $f(x) = \sin(\cos^{-1} x)$

50. $f(x) = \cos(\sin^{-1} x)$

For each sign chart for f' in Exercises 51–56, sketch possible graphs of both f' and f. On each sign chart, unlabeled tick-marks are locations where $f'(x)$ is zero and x-values where $f'(x)$ does not exist are indicated by tick-marks labeled "DNE."

51. f': + | 0 | − | 5 | +

52. f': + | −2 | −

53. f': − | −3 | − | 3 | +

54. f': − | −2 | + | 0 | + | 2 | −

55. f': + | 1 | − | 3 (DNE) | + | 8 | +

56. f': − | −1 (DNE) | + | 4 (DNE) | +

Sketch careful, labeled graphs of each function f in Exercises 57–82 by hand, without consulting a calculator or graphing utility. As part of your work, make sign charts for the signs, roots, and undefined points of f and f', and examine any relevant limits so that you can describe all key points and behaviors of f.

57. $f(x) = (x - 2)(3x + 1)$

58. $f(x) = x^2 - x + 100$

59. $f(x) = x^3 - x^2 - x + 1$

60. $f(x) = x^3 - 9x + 1$

61. $f(x) = x^3 - 6x^2 + 12x$

62. $f(x) = x(x^2 - 4)$

63. $f(x) = (2x + 11)(x^2 + 10)$

64. $f(x) = 3x^5 - 10x^4$

65. $f(x) = (1 - x^4)^7$

66. $f(x) = \sqrt{x} - \dfrac{1}{\sqrt{x}}$

67. $f(x) = \dfrac{x + 1}{x - 1}$

68. $f(x) = \dfrac{1}{x} + \dfrac{1}{x^2}$

69. $f(x) = \sqrt{x^2 + 1}$

70. $f(x) = \dfrac{1}{\sqrt{x^2 - 1}}$

71. $f(x) = \dfrac{(x - 1)^2}{x^2 + x - 6}$

72. $f(x) = \dfrac{(x - 1)^2}{x^2 - 1}$

73. $f(x) = \dfrac{x^2(x - 1)}{(x - 2)^2}$

74. $f(x) = \dfrac{x^3}{x^2 - 3x + 2}$

75. $f(x) = x \ln x$

76. $f(x) = \dfrac{2^x}{1 - 2^x}$

77. $f(x) = x^2 3^x$

78. $f(x) = x^3 e^x$

79. $f(x) = \dfrac{\ln x}{x}$

80. $f(x) = \ln((x-1)(x-2))$

81. $f(x) = e^{x^3 - 3x^2 + 2x}$

82. $f(x) = e^{\ln e^x}$

Applications

83. Dr. Alina is interested in the behavior of rats trapped in a long tunnel. Her rat Bubbles is released from the left-hand side of the tunnel and runs back and forth in the tunnel for 4 minutes. Bubbles' velocity $v(t)$, in feet per minute, is given by the following graph.

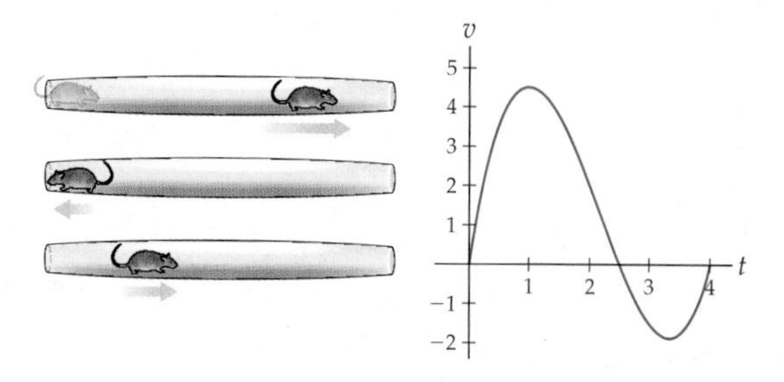

(a) On which time intervals is Bubbles moving towards the right-hand side of the tunnel?

(b) At which point in time is Bubbles farthest away from the left-hand side of the tunnel, and why? Do you think that Bubbles ever comes back to the left-hand side of the tunnel?

(c) On which time intervals does Bubbles have a positive acceleration?

(d) Find an interval on which Bubbles has a negative velocity but a positive acceleration. Describe what Bubbles is doing during this period.

84. Calvin uses a slingshot to launch an orange straight up in the air to see what will happen. The distance in feet between the orange and the ground after t seconds is given by the equation $s(t) = -16t^2 + 90t + 5$. Use this equation to answer the following questions:

(a) What is the initial height of the orange? What is the initial velocity of the orange? What is the initial acceleration of the orange?

(b) What is the maximum height of the orange?

(c) When will the orange hit the ground?

For Exercises 85 and 86, suppose that Annie is planning a kayak trip around Orcas Island in August. The tides create strong currents in several places on the coast of that island.

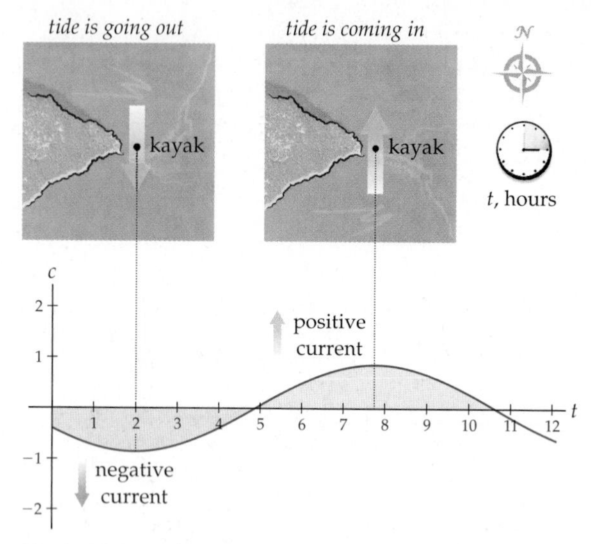

85. Annie has information from the gauge at Point Lawrence to help her decide the best time to round that point. The tidal current velocity (in nautical miles per hour) around Point Lawrence this August can be modeled very simply as

$$c(t) = 0.86 \cos(0.51t + 2.04).$$

A positive sign on the current indicates that it is going roughly north (the tide is coming in), while a negative sign indicates southward motion of the current. The time t is given in hours after midnight of the morning of August 1.

(a) When is $c(t)$ equal to zero? What is the significance of these times for the tides?

(b) When do the high tides occur? When are the low tides?

(c) Suppose Annie does not want to have to fight the current. Approximately when would be good time intervals for Annie to lead her party southward around the point?

86. After a bad experience on one trip, Annie models the tidal current velocity around Point Lawrence in a more accurate way as

$$C(t) = 0.90 \cos(0.51t + 2.02) + 0.49 \cos\left(\frac{0.51t}{2} + 2.13\right).$$

(a) Plot this function $C(t)$ together with the function $c(t)$ from the previous exercise on the same axes.

(b) The maximum currents given by $c(t)$ occurred when $t = -4 \pm 12.32n$ hours, for n an integer. Demonstrate that $C(t)$ does not have maxima at these points.

Proofs

87. Prove that every nonconstant linear function is either always increasing or always decreasing.

88. Prove that every quadratic function has exactly one local extremum.

89. Prove part (b) of Theorem 3.6: Suppose f is differentiable on an interval I; if f' is negative on the interior of I, then f is decreasing on I.

90. Prove part (c) of Theorem 3.6: Suppose f is differentiable on an interval I; if f' is zero on the interior of I, then f is constant on I. *(Hint: use the Mean Value Theorem to show that any two numbers a and b in I must be equal.)*

91. Prove part (b) of Theorem 3.8: With hypotheses as stated in the theorem, if $x = c$ is a critical point of f, where $f'(x) < 0$ to the left of c and $f'(x) > 0$ to the right of c, then f has a local minimum at $x = c$.

92. Prove part (d) of Theorem 3.8: With hypotheses as stated in the theorem, if $x = c$ is a critical point of f, where $f'(x) < 0$ to the left and to the right of c, then $x = c$ is not a local extremum of f.

Thinking Forward

▶ *Second-derivative graphs:* The three graphs shown are graphs of a function f and its first and second derivatives f' and f'', in no particular order. Identify which graph is which.

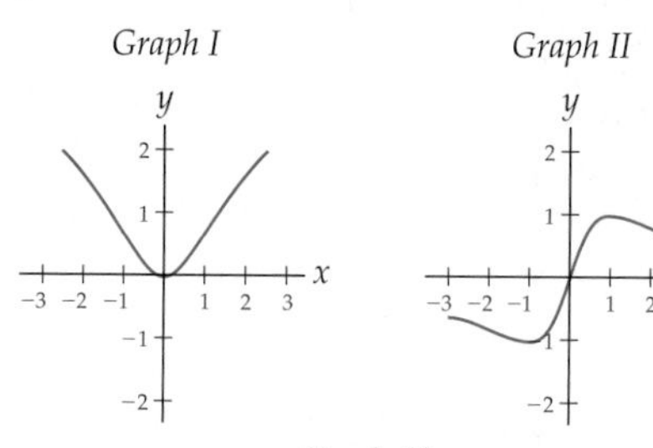

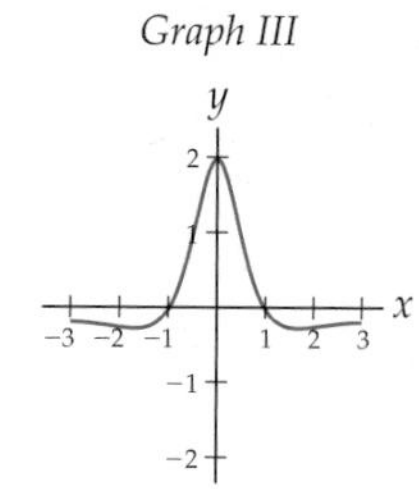

▶ *More second-derivative graphs:* The three graphs shown are graphs of a function f and its first and second derivatives f' and f'', in no particular order. Identify which graph is which.

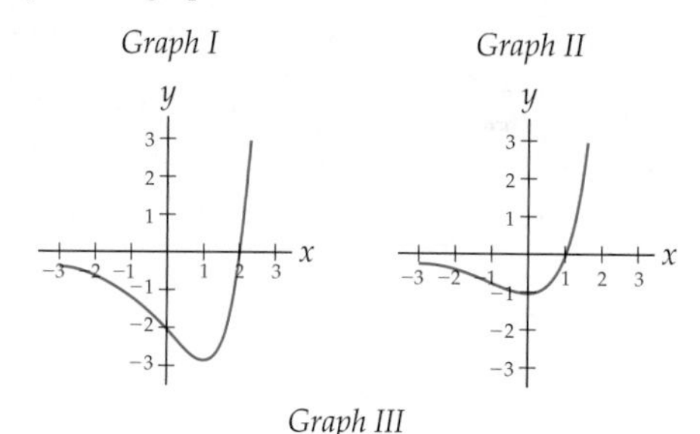

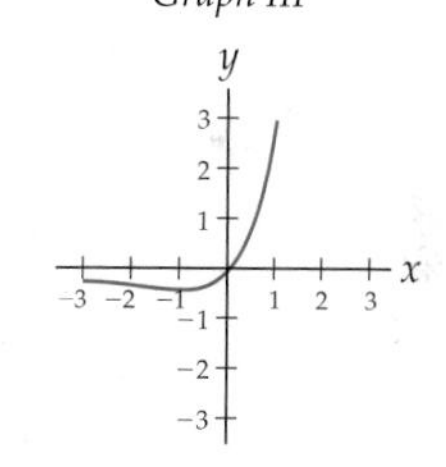

3.3 THE SECOND DERIVATIVE AND CURVE SKETCHING

- Using first and second derivatives to define and detect concavity
- The behavior of the first and second derivatives at inflection points
- Using the second-derivative test to determine whether critical points are maxima, minima, or neither

Derivatives and Concavity

In Section 0.4 we gave an informal definition of concavity: The graph of a function is ***concave up*** if it "curves upward" and ***concave down*** if it "curves downward." This is equivalent to saying that the graph of a concave-up function lies below its secant lines and above its tangent lines, and the graph of a concave-down function lies above its secant lines and below its tangent lines. Now that we know about derivatives, we are finally able to give a more precise definition of concavity.

DEFINITION 3.9 **Formally Defining Concavity**

Suppose f and f' are both differentiable on an interval I.

(a) f is ***concave up*** on I if f' is increasing on I.

(b) f is ***concave down*** on I if f' is decreasing on I.

How does this formal definition of concavity correspond with our intuitive notion of concavity? Consider the functions graphed next. On each graph four slopes are illustrated and estimated. Notice that when f is concave up, its slopes increase from left to right, and when f is concave down, its slopes decrease from left to right.

Slopes increase when f is concave up

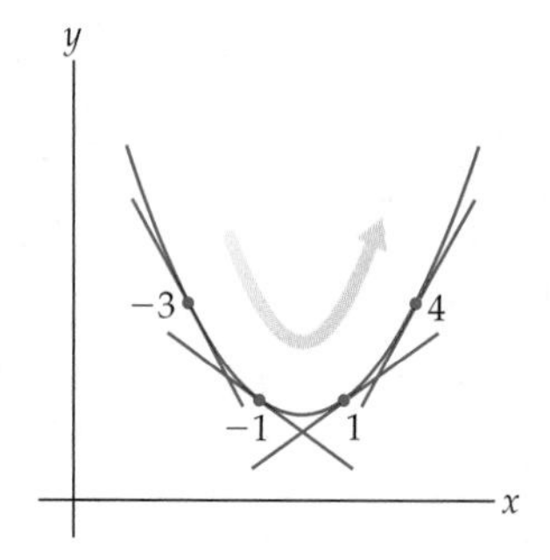

Slopes decrease when f is concave down

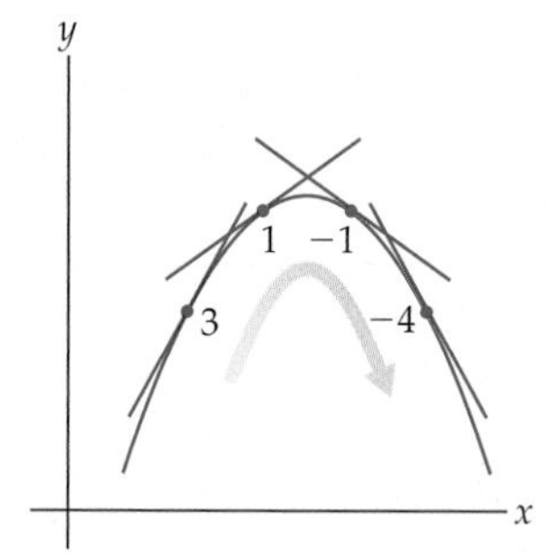

As we have already seen, a function increases where its derivative is positive. Taking this up one level, we see that the derivative function f' is increasing where *its* derivative function f'' is positive. Therefore we can check whether a function f is concave up or concave down by looking at the sign of its second derivative:

THEOREM 3.10 **The Second Derivative Determines Concavity**

Suppose both f and f' are differentiable on an interval I.

(a) If f'' is positive on I, then f is concave up on I.

(b) If f'' is negative on I, then f is concave down on I.

Proof. We will prove part (a) and leave part (b) to Exercise 91. Suppose that f and f' are differentiable on I and that $f''(x) > 0$ for all x in I. Then since the derivative of f' is f'', it follows from Theorem 3.6 that f' is increasing on I. By the definition of concavity this means that f is concave up on the interval I. ■

For example, we can divide the real-number line into intervals according to where the function $f(x) = x^3$ is concave up or concave down. This same division into subintervals describes where the derivative $f'(x) = 3x^2$ is increasing or decreasing and where the second derivative $f''(x) = 6x$ is positive or negative:

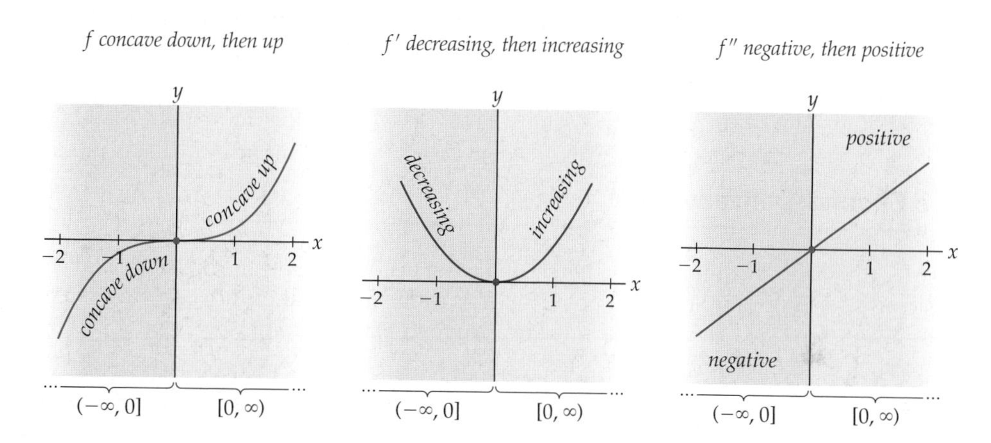

Inflection Points

Recall from Section 0.1 that the ***inflection points*** of a function f are the points in the domain of f at which its concavity changes. Since the sign of f'' measures the concavity of f, we can find inflection points by looking for the places where f'' changes sign. For example, if $f(x) = x^3$, then $f''(x) = 6x$, which is zero only when $x = 0$. The sign of f'' changes from negative to positive at $x = 0$, and therefore the concavity of f changes from down to up at that inflection point.

If $x = c$ is an inflection point of f and $f''(c) = 0$, then the graph of f could look one of the following four ways near $x = c$, depending on how f changes concavity and whether f is increasing or decreasing:

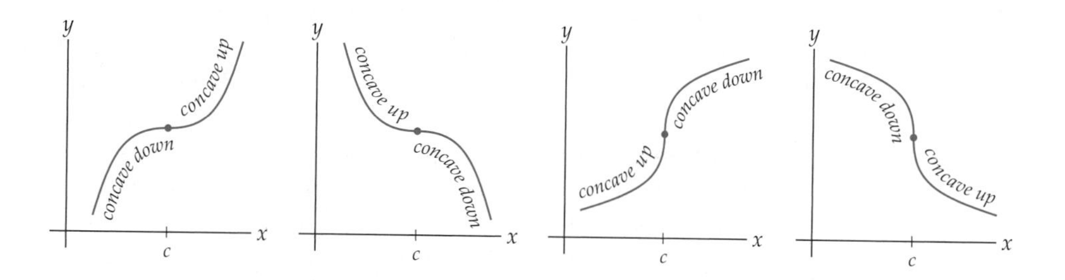

We know that at an inflection point of a function f, the function changes concavity and the second derivative f'' changes sign. What happens to the first derivative f'? The answer lies in the fact that f'' is the derivative of f'. Thus we know that If f'' is positive, then f' is increasing, and If f'' is negative, then f' is decreasing. The four possible scenarios corresponding to the preceding graphs are shown here:

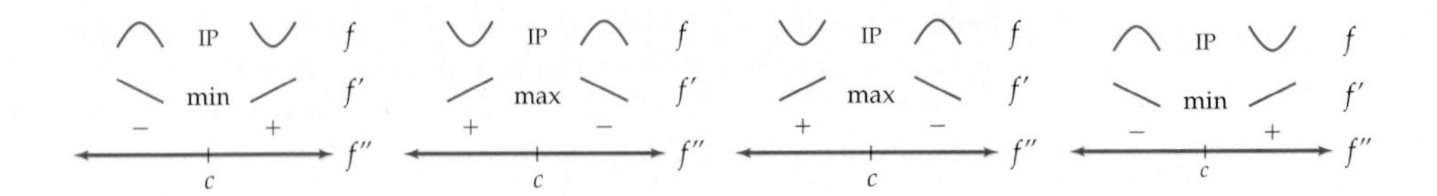

Notice that in each case the sign of f'' changes at $x = c$, causing f to have an inflection point at $x = c$ and in addition causing f' to have a local maximum or minimum at $x = c$. If you sketch tangent lines on the four graphs shown, you should be able to see that the slopes are at their maximum or minimum values at the inflection points. For example, in the first graph, the slopes start out large and positive, decrease to a minimum of zero at $x = c$, and then increase to larger and larger positive slopes as we move from left to right.

The Second-Derivative Test

In the previous section we saw how to apply the first-derivative test to determine whether the critical points of a function were local maxima, local minima, or neither. We can also use the second derivative to test critical points, by examining the concavity of the function at each critical point.

Suppose f is a differentiable function and $x = c$ is a critical point of f with $f'(c) = 0$. Then there are four possible ways that f can behave near $x = c$, as shown in the figures that follow. In each case we can examine the second derivative at the point $x = c$:

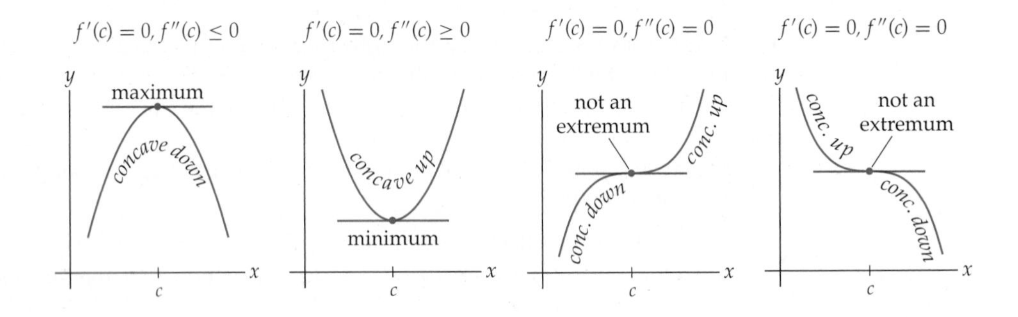

In the first graph, the second derivative is negative at $x = c$, so f curves downwards and has a local maximum at that point. In the second graph, the second derivative is positive at $x = c$, so f curves upwards and has a local minimum at that point.

In general, knowing that $f''(c) = 0$ does not tell us whether f has a maximum, minimum, or neither at $x = c$. The reason is that $f''(c) = 0$ is a possibility in all four of the cases just graphed. It is obvious that the last two graphs must have $f''(c) = 0$ because in those cases the function changes concavity at $x = c$. Perhaps less obvious is that we could have $f''(c) = 0$ in the first two graphs, where no inflection points occur; for example see the function $f(x) = x^4$ that we will examine later in part (a) of Example 1.

What we have just illustrated with the preceding four graphs is the ***second-derivative test***. When we applied the first-derivative test to a critical point, we were interested in the sign of $f'(x)$ to the left and the right of the critical point. When we use the second derivative to test a critical point, we will look at the sign of the second derivative *at* the critical point, as follows:

THEOREM 3.11

The Second-Derivative Test

Suppose $x = c$ is the location of a critical point of a function f with $f'(c) = 0$, and suppose both f and f' are differentiable and f'' is continuous on an interval around $x = c$.

(a) If $f''(c)$ is positive, then f has a local minimum at $x = c$.

(b) If $f''(c)$ is negative, then f has a local maximum at $x = c$.

(c) If $f''(c) = 0$, then this test says nothing about whether or not f has an extremum at $x = c$.

Proof. Suppose both f and f' are differentiable in a neighborhood of a critical point $x = c$ with $f'(c) = 0$. To prove part (a), suppose $f''(c) > 0$. Since f'' is assumed to be continuous near c, f'' must be positive in a small neighborhood $(c - \delta, c + \delta)$ of c. Because f'' is the derivative of f', it follows from Theorem 3.6 that f' is increasing on $(c - \delta, c + \delta)$. Since f' is increasing near c, and is zero at c, we must have $f'(x) < 0$ to the immediate left of c and $f'(x) > 0$ to the immediate right of c. Therefore, by the first-derivative test, f has a local minimum at $x = c$. The proof of part (b) is similar and is left to Exercise 92.

To prove part (c), it suffices to exhibit three functions with $f''(c) = 0$ at some point c where one function has a local maximum at $x = c$, one function has a local minimum at $x = c$, and one function does not have a local extremum at $x = c$. You will do this in Exercise 10. ■

Curve-Sketching Strategies

When we graph a function by hand, a good place to start is to use algebra, derivatives, and sign charts to determine the intervals where the function is positive or negative, increasing or decreasing, and concave up or down, as well as the coordinates of any roots, extrema, and inflection points of f. Armed with this information and occasionally a few strategic function values and limits, we can often sketch a fairly accurate graph.

Of course, actually sketching a graph based on the information just described can take a little bit of practice. One useful thing to notice is that a continuous, differentiable function can change sign only at it roots, change direction only at its local extrema, and change concavity only at its inflection points. This means that on the intervals between such points the graph of the function is relatively homogeneous. In fact, if f is a sufficiently well-behaved function, then we can connect each adjacent pair of dots determined by the coordinates of its local extrema and inflection points with one of the following four types of arcs:

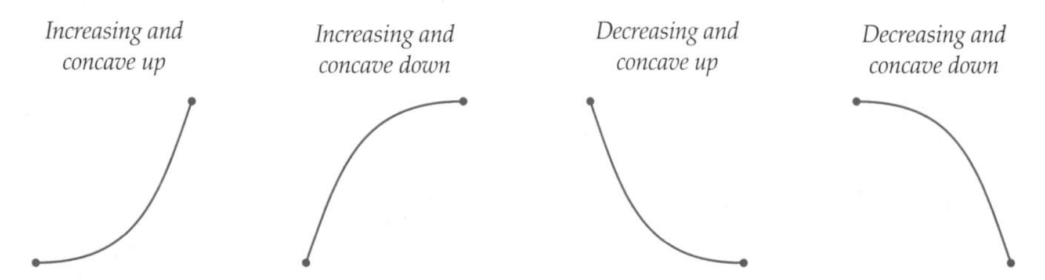

Here is a summary of what you might consider when sketching the graph of a function f by hand:

- Find the domain of f.
- Determine the roots of f and the intervals on which f is positive or negative. Record this information on a sign chart for f.

- Find f' and determine the points where f' is zero or does not exist. Each of these points that is in the domain of f is a critical point of f and therefore a possible local extremum of f.
- Determine the intervals on which f' is positive or negative. These are the same as the intervals on which f is increasing or decreasing, respectively. Record this information on a sign chart for f'.
- Determine whether f has a local minimum, local maximum, or neither at each critical point.
- Find f'' and determine the points where f'' is zero or does not exist. Each of these points that is in the domain of f is a critical point of f' and therefore a possible inflection point of f.
- Determine the intervals on which f'' is positive or negative. These are the same as the intervals on which f is concave up or concave down, respectively. Record this information on a sign chart for f''.
- Determine whether or not each critical point of f' is an inflection point of f.
- At each local extremum or inflection point $x = c$, determine the value of $f(c)$. Plot these key points $(c, f(c))$.
- At each non-continuous or non-differentiable point $x = c$ in the domain of f, determine the value of $f(c)$. Use limits of f and f' to determine the type of discontinuity or non-differentiable point. Plot these key points $(c, f(c))$.
- Calculate limits of f at any non-domain points and as $x \to \pm\infty$. Determine any horizontal or vertical asymptotes and the long-term behavior of the graph.
- Between key points and non-domain points, use the sign charts for f' and f'' to determine which of the four types of arc shapes the graph will have.
- Graph the function by connecting the key points with arcs.

For some functions you will not be able to obtain all of the information you want, but in every case you should explore as much as you can until you can confidently determine the behavior of the graph of f.

Why would we want to spend time using derivatives and algebra to sketch the graph of a function by hand when we could make essentially the same sketch with a graphing calculator in just a few keystrokes? One reason is that graphing by hand provides more specific information, such as the exact locations and values of the key points on the graph. Moreover, if we do the work by hand, then we can determine a graphing window that captures all of the important features of the graph and possibly also discern features of the graph that are not apparent with a graphing calculator, such as holes and asymptotes. Derivatives, limits, and algebra not only enable us to sketch graphs by hand, but also help us correctly interpret graphs made by calculators.

Examples and Explorations

EXAMPLE 1 **Using the second derivative to determine concavity and inflection points**

Use second derivatives to determine the intervals on which each of the functions that follow are concave up or concave down. Then determine any inflection points.

(a) $f(x) = x^4$ **(b)** $g(x) = x^3 - 3x + 2$

SOLUTION

(a) We start by calculating the first and second derivatives of f:

$$f(x) = x^4 \implies f'(x) = 4x^3 \implies f''(x) = 12x^2.$$

Therefore $f''(x) = 0$ only when $x = 0$, and f'' is positive to both the left and right of $x = 0$, as shown in the following sign chart:

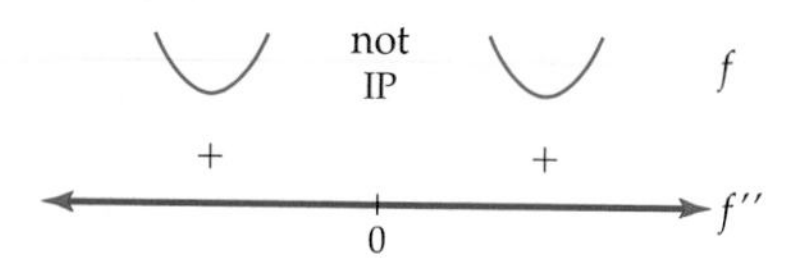

Since the sign of $f''(x) = 12x^2$ does not change at $x = 0$, the function $f(x) = x^4$ does not have an inflection point at $x = 0$. Therefore $f(x) = x^4$ has no inflection points and is concave up on both $(-\infty, 0)$ and $(0, \infty)$.

(b) Again we start by calculating derivatives:

$$g(x) = x^3 - 3x + 2 \quad \Longrightarrow \quad g'(x) = 3x^2 - 3 \quad \Longrightarrow \quad g''(x) = 6x.$$

Therefore $g''(x) = 0$ only when $x = 0$. Testing the sign of $g''(x)$ to the left and right of $x = 0$, we have $g''(-1) = -6 < 0$ and $g''(1) = 6 > 0$. This information about $g''(x)$ is summarized in the following sign chart:

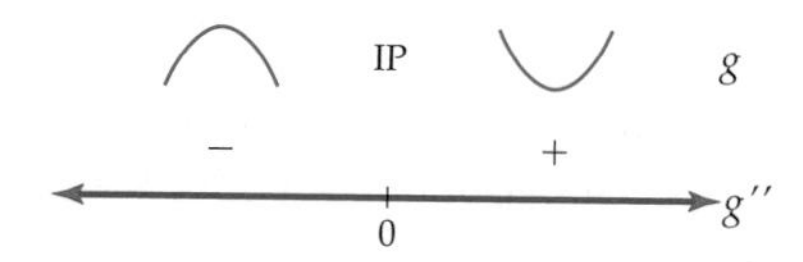

Thus g is concave down on $(-\infty, 0)$, is concave up on $(0, \infty)$, and has an inflection point at $x = 0$. □

✔ CHECKING THE ANSWER

To verify the preceding calculations, we sketch graphs of f and g. In the graph that follows at the left we see that the curve $f(x) = x^4$ is flat enough to have zero curvature at $x = 0$, but has positive curvature to both the left and the right of the origin. In the graph at the right we see that $g(x) = x^3 - 3x + 2$ does have an inflection point at $x = 0$, where its concavity changes from concave-down to concave-up. Note that $g(x)$ is the same function we examined in Example 3(a) of the previous section.

$y = x^4$ has no inflection points

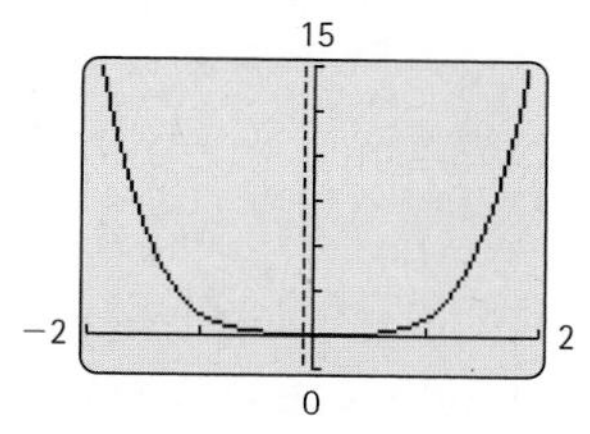

$g(x)$ has an inflection point at $x = 0$

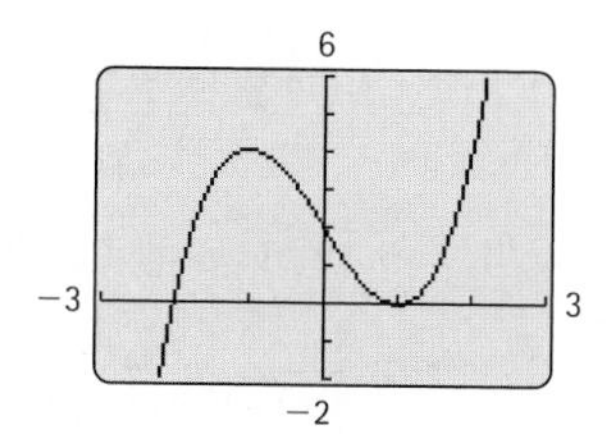

EXAMPLE 2

Comparing the first- and second-derivative tests

Determine the local extrema of the function $f(x) = x^3 - x^2 + 2$, using

(a) the first-derivative test **(b)** the second-derivative test

SOLUTION

(a) The derivative of $f(x) = x^3 - x^2 + 2$ is $f'(x) = 3x^2 - 2x = x(3x - 2)$.

This derivative f' is zero at the points $x = 0$ and $x = \frac{2}{3}$, and always exists. Thus the only critical points of f are $x = 0$ and $x = \frac{2}{3}$. To apply the first-derivative test we must find the sign of the derivative between each of these points. For example, we could calculate:

$$f'(-1) = 5 > 0, \quad f'\left(\tfrac{1}{2}\right) = -\tfrac{1}{4} < 0, \quad \text{and} \quad f'(1) = 1 > 0.$$

The following sign chart summarizes this information about f':

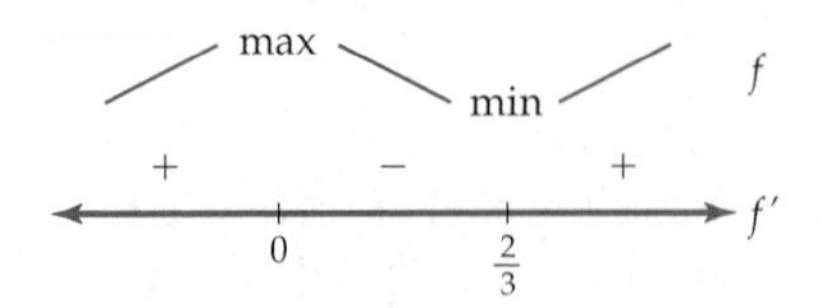

By the first-derivative test, f has a local maximum at $x = 0$ and a local minimum at $x = \frac{2}{3}$.

(b) The calculation for the second derivative starts out the same way, by computing the derivative $f'(x) = 3x^2 - 2x$ and then finding the critical points $x = 0$ and $x = \frac{2}{3}$. The difference is that we will test the sign of the second derivative $f''(x) = 6x - 2$ at each of these critical points:

$$f''(0) = -2 < 0 \quad \text{and} \quad f''\left(\tfrac{2}{3}\right) = 2 > 0.$$

By the second-derivative test, since f is concave down at the critical point $x = 0$, f has a local maximum at $x = 0$. Similarly, since f is concave up at the critical point $x = \frac{2}{3}$, we know that f has a local minimum at $x = \frac{2}{3}$. This is of course the same conclusion we reached when we applied the first-derivative test. □

EXAMPLE 3 **A detailed curve-sketching analysis**

Use derivatives, algebra, and sign charts to sketch the graph of $f(x) = x^5 - 15x^3$. Identify the coordinates of each root, local extremum, and inflection point.

SOLUTION

We'll start by finding and then simplifying the functions f, f', and f''. We will factor each function as much as possible so that we can easily identify its roots:

$$f(x) = x^5 - 15x^3 = x^3(x^2 - 15) = x^3(x - \sqrt{15})(x + \sqrt{15});$$
$$f'(x) = 5x^4 - 45x^2 = 5x^2(x^2 - 9) = 5x^2(x - 3)(x + 3);$$
$$f''(x) = 20x^3 - 90x = 10x(2x^2 - 9) = 10x(\sqrt{2}x - 3)(\sqrt{2}x + 3).$$

From these factorizations we can see that f has roots at $x = 0$ and $x = \pm\sqrt{15} \approx \pm 3.87$, that the only possible local extrema of f are $x = 0$ and $x = \pm 3$, and that the only possible inflection points of f are $x = 0$ and $x = \pm\frac{3}{\sqrt{2}} \approx \pm 2.12$. By checking the signs of $f(x)$, $f'(x)$, and $f''(x)$ at appropriate values, we obtain the following sign charts:

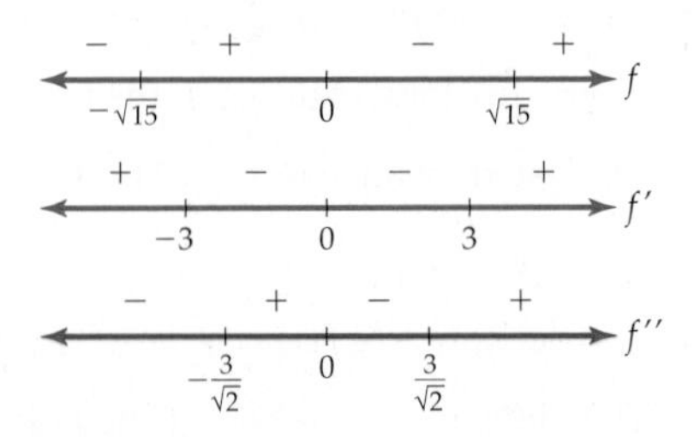

From this information we can see that f has a maximum at $x = -3$, a minimum at $x = 3$, and inflection points at $x = 0$ and $x = \pm\frac{3}{\sqrt{2}}$. We can also identify the intervals on which

f is positive or negative, increasing or decreasing, and concave up or concave down, as summarized in the following three sign charts:

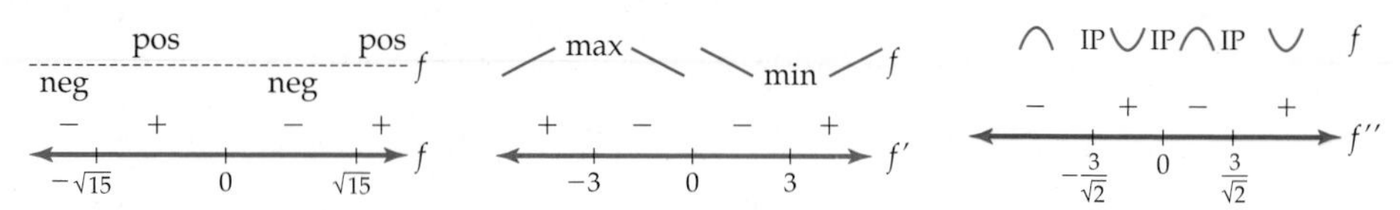

By evaluating $f(x) = x^5 - 15x^3$ at its key points we can obtain the coordinates of its roots, extrema, and inflection points:

- roots at $(-\sqrt{15}, 0)$, $(0, 0)$, and $(\sqrt{15}, 0)$;
- local maximum at $(-3, 162)$, local minimum at $(3, -162)$;
- inflection points at $\left(-\frac{3}{\sqrt{2}}, 100.232\right)$, $(0, 0)$, and $\left(\frac{3}{\sqrt{2}}, -100.232\right)$.

Now we need only plot these points and connect the dots with appropriate arcs according to the sign information in the sign charts. For example, we can see from the sign charts for f, f', and f'' that between $x = -3$ and $x = -\frac{3}{\sqrt{2}}$ the graph of f should be positive, decreasing, and concave down.

We already have all of the information we need about the derivatives, but for some people it helps to collect all this information in one place. The arc shapes on each subinterval between key points are recorded on the combined number-line chart that follows. Note that on this number line the tick-marks represent the locations of all interesting points on the graph of f, meaning that they are a composite of the tick-marks from the three sign charts for f, f', and f''.

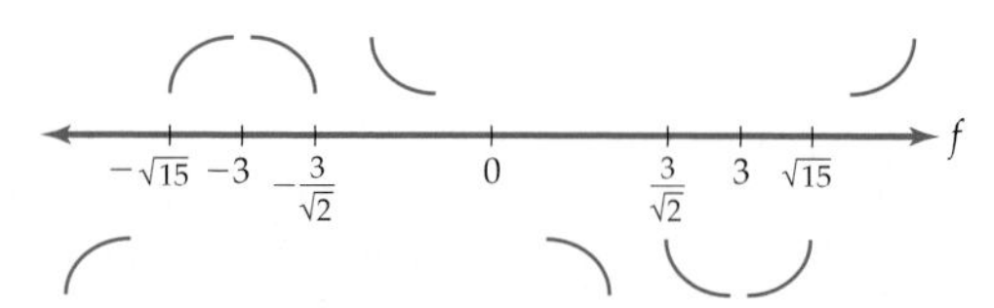

Although at this point the shape of the graph is pretty clear, for completeness we should compute limits at any interesting points. The function $f(x) = x^5 - 15x^3$ has no non-domain points, discontinuities, or non-differentiable points, so the only limits to check are those as $x \to \pm\infty$. We can use what we know about the behavior of fifth-degree polynomials, or we can just compute these limits directly:

$$\lim_{x\to\infty} (x^5 - 15x^3) = \lim_{x\to\infty} (x^3)(x^2 - 15) = \infty,$$

$$\lim_{x\to-\infty} (x^5 - 15x^3) = \lim_{x\to-\infty} (x^3)(x^2 - 15) = -\infty.$$

This information tells us that the graph has no horizontal asymptotes and indicates what happens at the "ends" of the graph of f.

Putting all of the information together into a labeled graph, we have

$f(x) = x^5 - 15x^3$

□

We can verify the graph we constructed in the preceding example by using a calculator or graphing utility to graph $f(x) = x^5 - 15x^3$ in a similar graphing window. Notice that we did indeed capture the major features of this graph. In addition, because we did the work by hand, we know the exact values of every key point on the graph of f.

Calculator graph of $f(x) = x^5 - 15x^3$

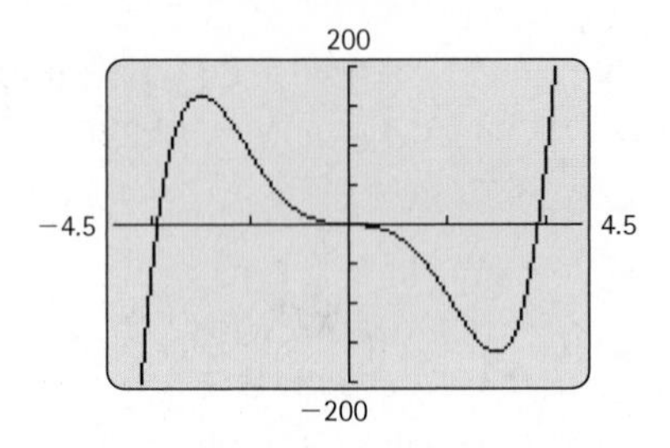

EXAMPLE 4

A curve-sketching analysis with asymptotes

Sketch an accurate, labeled graph of the function $f(x) = \frac{6}{4-2^x}$. Include complete sign analyses of f, f', and f'', and calculate any relevant limits.

SOLUTION

Let's begin by finding and simplifying the first and second derivatives of $f(x)$. The first derivative of $f(x) = 6(4 - 2^x)^{-1}$ is

$$f'(x) = 6(-1)(4-2^x)^{-2}(-(\ln 2)2^x) = \frac{6(\ln 2)2^x}{(4-2^x)^2}.$$

Differentiating that result and then simplifying as much as possible so that we can easily identify roots, we find that the second derivative of f is

$$f''(x) = \frac{6(\ln 2)(\ln 2)2^x(4-2^x)^2 - (6(\ln 2)2^x)(2)(4-2^x)^1(-(\ln 2)2^x)}{(4-2^x)^4}$$

$$= \frac{6(\ln 2)^2 2^x(4-2^x)((4-2^x)+2\cdot 2^x)}{(4-2^x)^4} = \frac{6(\ln 2)^2 2^x(4+2^x)}{(4-2^x)^3}.$$

To determine the intervals on which f, f', and f'' are positive and negative we must first locate the values of x for which these functions are zero or do not exist. The function $f(x) = \frac{6}{4-2^x}$ is never zero, but is undefined where its denominator is zero, at $x = 2$. Looking at the formula for $f'(x)$, we can easily see that it is never zero (since 2^x is never zero), but is undefined if $x = 2$ (since the denominator $4 - 2^x$ is zero for $x = 2$). Thus $x = 2$ is the only critical point of f and the only point we will mark on the number line for f'. Similarly, from the formula for $f''(x)$, it is clear that $f''(x)$ is never zero (since neither 2^x nor $4 + 2^x$ can ever be zero), but is undefined at $x = 2$. Checking signs on either side of $x = 2$ for each function, we obtain the following set of number lines:

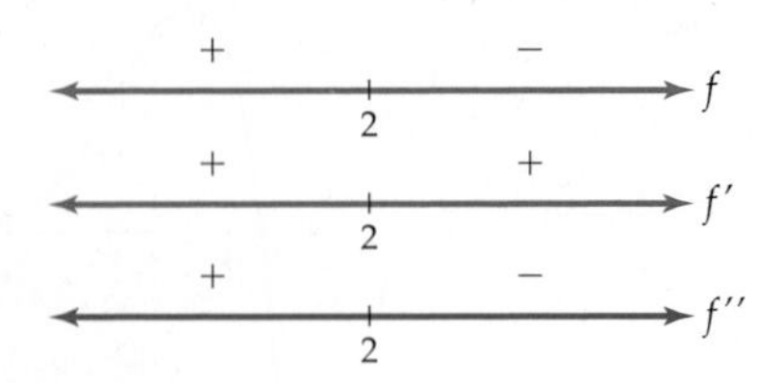

We now know that $f(x)$ is positive, increasing, and concave up on $(-\infty, 2)$ and negative, increasing, and concave down on $(2, \infty)$. The graph has no roots, no extrema, and no inflection points.

It remains to calculate any interesting limits. Since the domain of f is $(-\infty, 2) \cup (2, \infty)$, we must investigate the limits of $f(x) = \frac{6}{4-2^x}$ as $x \to \pm\infty$ and as $x \to 2$ from the left and the right. As $x \to \infty$ the denominator $4 - 2^x$ of $f(x)$ approaches $-\infty$, and thus

$$\lim_{x\to\infty} \frac{6}{4-2^x} = 0.$$

As $x \to -\infty$, the denominator $4 - 2^x$ approaches 4, and thus

$$\lim_{x\to-\infty} \frac{6}{4-2^x} = \frac{6}{4} = \frac{3}{2}.$$

This information tells us that the graph of f has *two* horizontal asymptotes: at $y = 0$ on the right and at $y = \frac{3}{2}$ on the left.

As $x \to 2^-$ we have $4 - 2^x \to 0^+$ and thus

$$\lim_{x\to 2^-} \frac{6}{4-2^x} = \infty,$$

and as $x \to 2^+$ we have $4 - 2^x \to 0^-$ and thus

$$\lim_{x\to 2^+} \frac{6}{4-2^x} = -\infty.$$

Thus the graph of f has a vertical asymptote at $x = 2$, where the graph approaches ∞ to the left of 2 and approaches $-\infty$ to the right of 2.

Putting all of this information together, we can now sketch the graph:

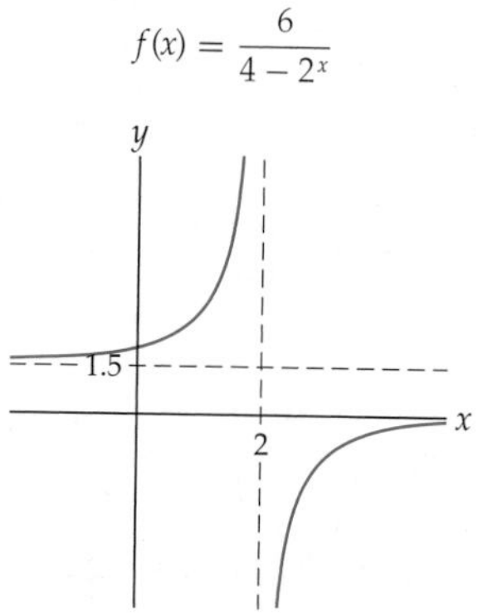

$$f(x) = \frac{6}{4-2^x}$$

□

? TEST YOUR UNDERSTANDING

- Why could we not give a precise mathematical definition of concavity before this section?
- The domain points $x = c$ where $f''(c) = 0$ or where $f''(c)$ does not exist are the critical points of the function f'. Why?
- Why is it not clear to say a sentence such as "Because it is positive, it is concave up"? How could this information be conveyed more precisely?
- Why does it make sense that f' is increasing when f'' is positive?
- Suppose $x = c$ is a critical point with $f'(c) = 0$. Why does it make graphical sense that f has a local minimum at $x = c$ when f is concave up in a neighborhood around $x = c$?

EXERCISES 3.3

Thinking Back

Finding the second derivative: For each of the following functions f, calculate and simplify the second derivative f''.

- ▶ $f(x) = \dfrac{x-1}{3x^2 - 4x + 2}$
- ▶ $f(x) = \dfrac{x}{\sqrt{x^2+1}}$
- ▶ $f(x) = \dfrac{1-x}{e^x}$
- ▶ $f(x) = e^{3x}\ln(x^2+1)$

Solving for zeroes and non-domain points: For each of the following expressions, find all values of x for which $g(x)$ is zero or does not exist.

- ▶ $g(x) = \dfrac{3x^2 - x - 2}{x^4 + 2x^2 - 3}$
- ▶ $g(x) = \dfrac{1}{x-2} - \dfrac{3x+1}{x+2}$
- ▶ $g(x) = \dfrac{\sin x}{\cos x}$
- ▶ $g(x) = \dfrac{e^{3x}(x-1)}{\ln x}$

Concepts

0. *Problem Zero:* Read the section and make your own summary of the material.

1. *True/False:* Determine whether each of the statements that follow is true or false. If a statement is true, explain why. If a statement is false, provide a counterexample.

(a) *True or False:* If $f''(2) = 0$, then $x = 2$ is an inflection point of f.

(b) *True or False:* If f'' is concave up on an interval I, then it is positive on I.

(c) *True or False:* If f is concave up on an interval I, then f'' is positive on I.

(d) *True or False:* If $f''(2)$ does not exist and $x = 2$ is in the domain of f, then $x = 2$ is a critical point of the function f'.

(e) *True or False:* If f has an inflection point at $x = 3$ and f is differentiable at $x = 3$, then the derivative f' has a local minimum or maximum at $x = 3$.

(f) *True or False:* If $f'(1) = 0$ and $f''(1) = -2$, then f has a local minimum at $x = 1$.

(g) *True or False:* The second-derivative test involves checking the sign of the second derivative on each side of every critical point.

(h) *True or False:* The second-derivative test always produces exactly the same information as the first-derivative test.

2. *Examples:* Construct examples of the thing(s) described in the following. Try to find examples that are different than any in the reading.

(a) The graph of a function f for which f' is positive everywhere, $f''(x) > 0$ for $x < -2$, and $f''(x) < 0$ for $x > -2$.

(b) The graph of a function f for which $f(3) = 0$, $f'(3) = 0$, and $f''(3) = 0$.

(c) The graph of a function f for which $f(x)$ is zero at $x = -1$, $x = 2$, and $x = 4$; $f'(x)$ is zero at $x = -1$, $x = 1$, and $x = 3$; and $f''(x)$ is zero at $x = 0$ and $x = 2$.

3. Sketch the graph of a function f that is concave up everywhere. Then draw five tangent lines on the graph, and explain how you can see that the derivative of f is increasing.

4. Sketch the graph of a function f that is concave down everywhere. Then draw five tangent lines on the graph, and explain how you can see that the derivative of f is decreasing.

5. State the converse of Theorem 3.10(a). Is the converse true? If so, explain why; if not, provide a counterexample.

6. State the contrapositive of Theorem 3.10(a). Is the contrapositive true? If so, explain why; if not, provide a counterexample.

7. Sketch the graph of a function f that has an inflection point at $x = c$ in such a way that the derivative f' has a local maximum at $x = c$. Add tangent lines to your sketch to illustrate that f' does have a local maximum at $x = c$.

8. Sketch the graph of a function f that has an inflection point at $x = c$ in such a way that the derivative f' has a local minimum at $x = c$. Add tangent lines to your sketch to illustrate that f' does have a local minimum at $x = c$.

9. Show that for $f(x) = x^6$ we have $f''(0) = 0$ but the point $x = 0$ is not an inflection point of f.

10. In this problem we will verify part (c) of Theorem 3.11.

(a) For $f(x) = x^3$, show that $f'(0) = 0$ and $f''(0) = 0$ while f does not have a local extremum at $x = 0$.

(b) For $g(x) = x^4$, show that $g'(0) = 0$ and $g''(0) = 0$ while g has a local minimum at $x = 0$.

(c) For $h(x) = -x^4$, show that $h'(0) = 0$ and $h''(0) = 0$ while h has a local maximum at $x = 0$.

(d) Explain how the parts (a)–(c) show that the second-derivative test does not tell us anything when the second derivative at a critical point is zero.

11. We could use part (c) from Theorem 3.6 to add a third part to Theorem 3.10 that would tell us what it means when f'' is zero in the interior of an interval I. Fill in the blank accordingly: If f'' is zero on I, then f is ________ on I.

12. Describe what the second-derivative test is for and how to use it. Sketch graphs and sign charts to illustrate your description.

13. If a function f has four critical points, how many calculations after finding derivatives are required in order to apply the first-derivative test? The second-derivative test?

14. Describe in words, and then illustrate in pictures, the four types of arcs that are the building blocks of most continuous, differentiable graphs.

For Exercises 15–20,sketch the graph of a function f that has the indicated characteristics. If a graph is not possible, explain why.

15. f positive, f' negative, and f'' positive on $[0, 3]$.

16. f negative, f' negative, and f'' negative on $[0, 3]$.

17. f negative, f' positive, and f'' positive on $[0, 3]$.

18. f positive, f' negative, and f'' positive on $\mathbb{R}$.

19. f negative, f' negative, and f'' negative on $\mathbb{R}$.

20. f negative, f' positive, and f'' positive on $\mathbb{R}$.

Skills

In Exercises 21–28, graphs of f, f', or f'' are given. Whichever is shown, sketch graphs of the remaining two functions. Label the locations of any roots, extrema, and inflection points on each graph.

21. *Graph of f*

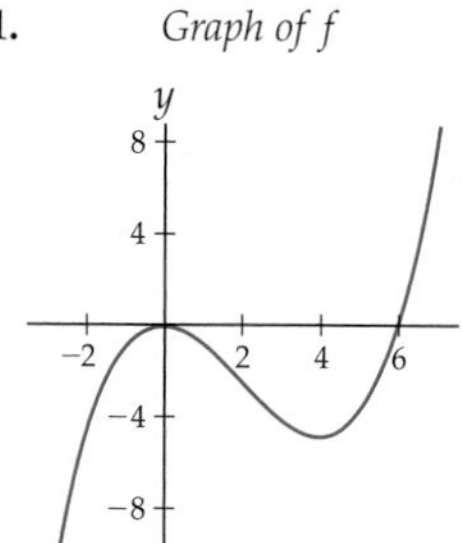

22. *Graph of f*

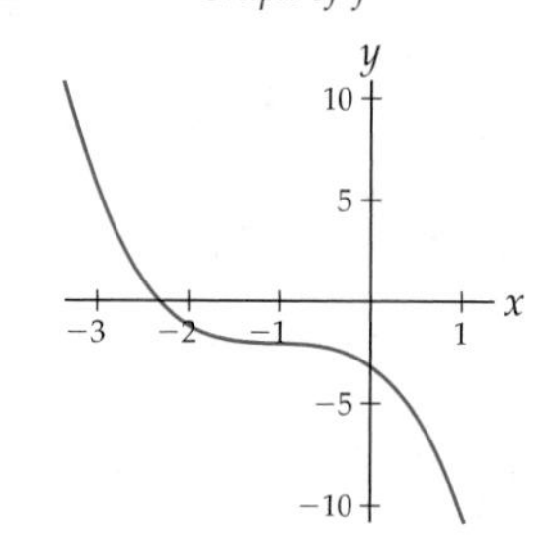

23. *Graph of f*

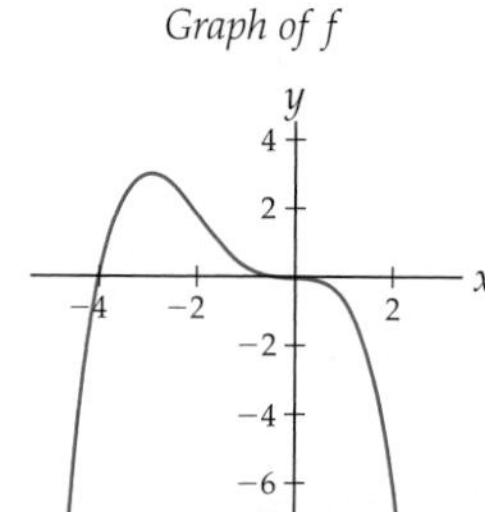

24. *Graph of f′*

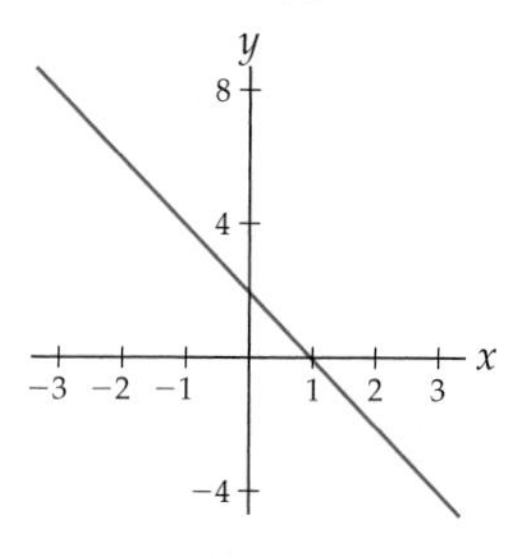

25. *Graph of f′*

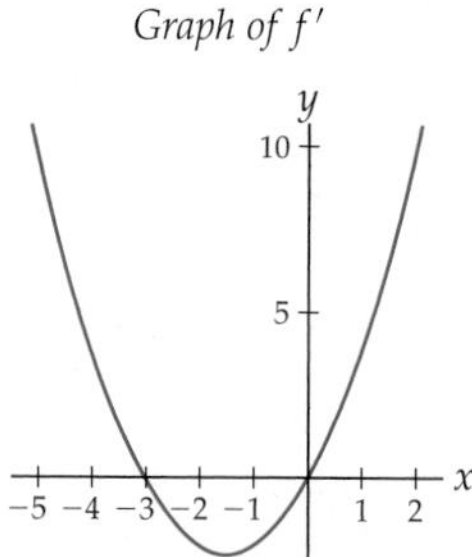

26. *Graph of f′*

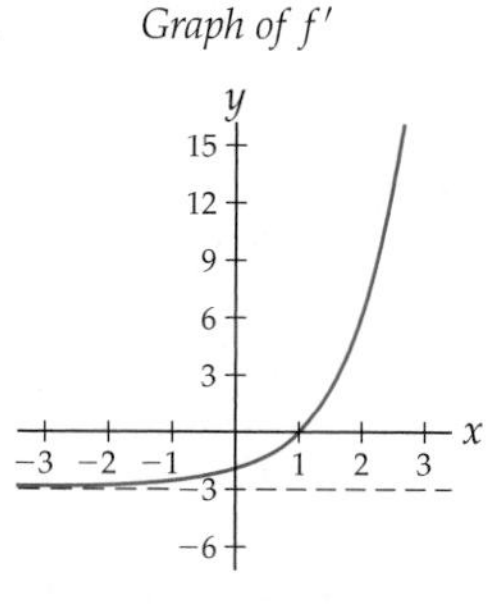

27. *Graph of f″*

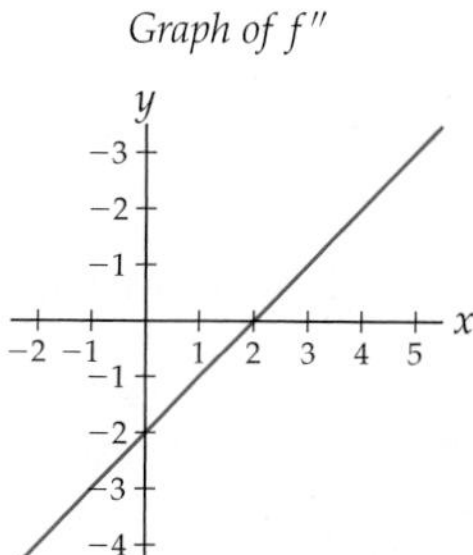

28. *Graph of f″*

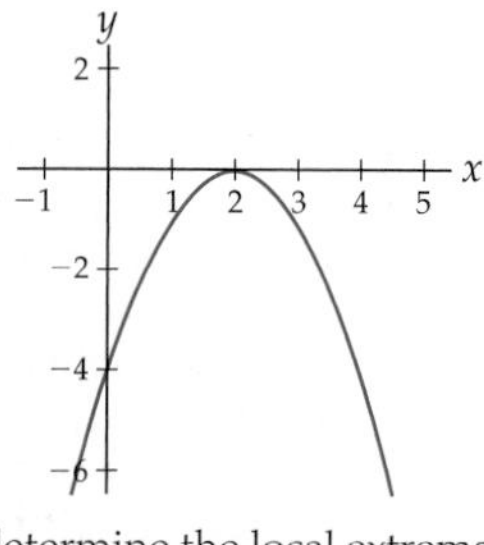

Use the second-derivative test to determine the local extrema of each function f in Exercises 29–40. If the second-derivative test fails, you may use the first-derivative test. Then verify your algebraic answers with graphs from a calculator or graphing utility. *(Note: These are the same functions that you examined with the first-derivative test in Exercises 39–50 of Section 3.2.)*

29. $f(x) = (x-2)^2(1+x)$

30. $f(x) = x^2(x-1)(x+1)$

31. $f(x) = \dfrac{1+x+x^2}{x^2+x-2}$

32. $f(x) = \dfrac{(x-1)^2}{x+2}$

33. $f(x) = \dfrac{1}{3-2e^x}$

34. $f(x) = e^x(x^2-x-1)$

35. $f(x) = \cos(\pi(x+1))$

36. $f(x) = \cos(\pi x)$

37. $f(x) = \arctan x$

38. $f(x) = \sin^{-1} x^2$

39. $f(x) = \sin(\cos^{-1} x)$

40. $f(x) = \cos(\sin^{-1} x)$

Use a sign chart for f'' to determine the intervals on which each function f in Exercises 41–52 is concave up or concave down, and identify the locations of any inflection points. Then verify your algebraic answers with graphs from a calculator or graphing utility.

41. $f(x) = (x-2)^4$

42. $f(x) = (x-3)^3(x-1)$

43. $f(x) = x^4 - 2x^3 - 5$

44. $f(x) = \dfrac{1}{1+x^2}$

45. $f(x) = \dfrac{1}{x^2+x+1}$

46. $f(x) = \dfrac{\sqrt{x}}{x-2}$

47. $f(x) = e^{3x}(1-e^x)$

48. $f(x) = \dfrac{x}{\ln x}$

49. $f(x) = e^{1+2x-x^2}$

50. $f(x) = \dfrac{2^x}{1-2^x}$

51. $f(x) = \sin\left(x - \dfrac{\pi}{4}\right)$

52. $f(x) = 3\cos\left(\dfrac{\pi}{2}x\right) + 5$

For each set of sign charts in Exercises 53–62, sketch a possible graph of f.

53.

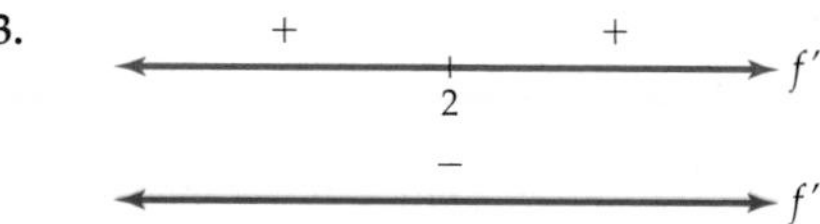

54.

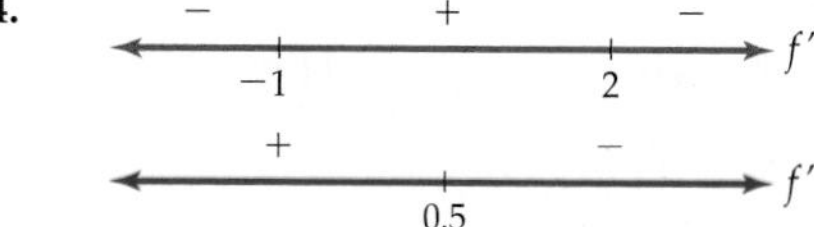

55.

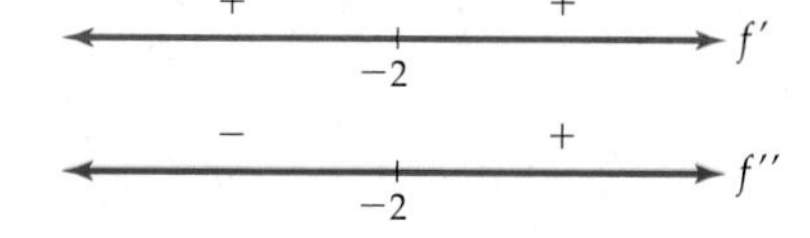

56.

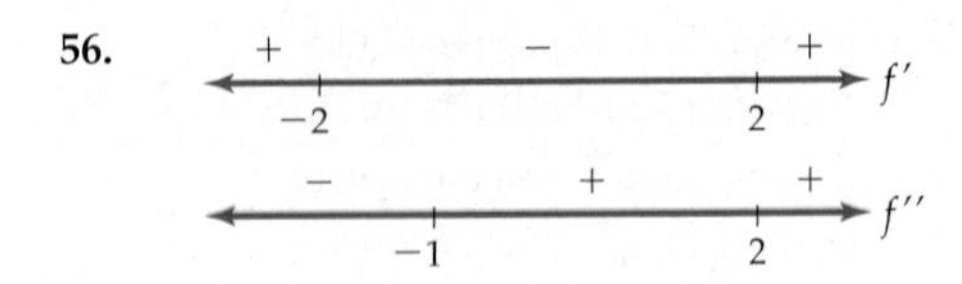

57.

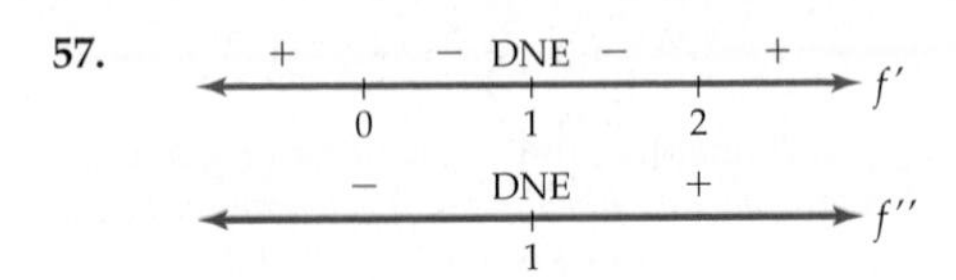

58.

− + DNE + −
−1 1 2
f'
+ DNE −
1
f''

59.

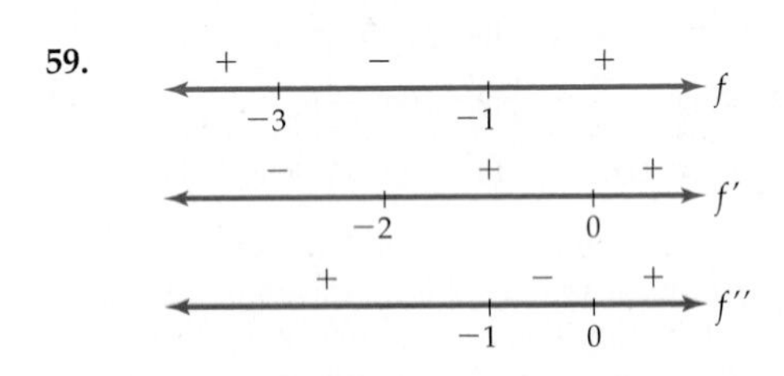

60.

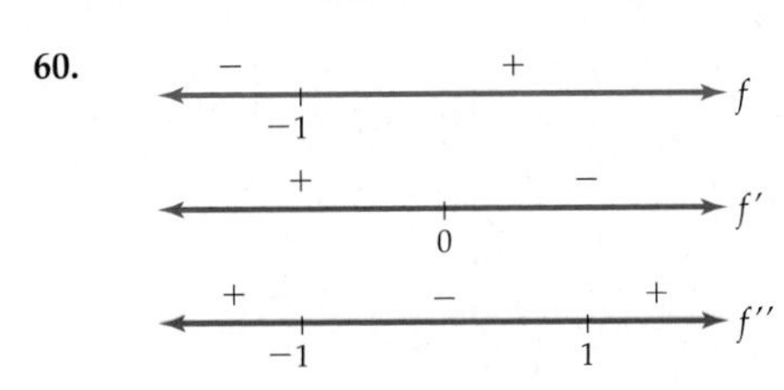

61.

− + − +
−3 1 4
f
+ − − +
−2 1 3
f'
− + − +
−1 1 2
f''

62.

− + + DNE − +
−2 0 1 2
f
+ DNE − + DNE +
−1 0 1
f'
+ DNE + DNE −
−1 1
f''

Sketch careful, labeled graphs of each function f in Exercises 63–82 by hand, without consulting a calculator or graphing utility. As part of your work, make sign charts for the signs, roots, and undefined points of f, f', and f'', and examine any relevant limits so that you can describe all key points and behaviors of f.

63. $f(x) = x^2 + 3x$

64. $f(x) = (x-2)(x+2)$

65. $f(x) = x^3 + 3x^2$

66. $f(x) = \dfrac{1}{x^2+3}$

67. $f(x) = x^3 - 2x^2 + x$

68. $f(x) = (1-x)^4 - 2$

69. $f(x) = x^3(x+2)$

70. $f(x) = \dfrac{x}{x^2+1}$

71. $f(x) = \sqrt{x}\,(4-x)$

72. $f(x) = \dfrac{1}{(x-1)^2(x-2)}$

73. $f(x) = \dfrac{x^2-1}{x^2-5x+4}$

74. $f(x) = \dfrac{x^2-x}{x^2-3x+2}$

75. $f(x) = \dfrac{1-x}{e^x}$

76. $f(x) = \ln(x^2+1)$

77. $f(x) = \dfrac{e^x}{x}$

78. $f(x) = e^{3x} - e^{2x}$

79. $f(x) = x^{2/3} - x^{1/3}$

80. $f(x) = \cos\left(3\left(x - \frac{\pi}{2}\right)\right)$

81. $f(x) = (\ln x)^2 + 1$

82. $f(x) = \sin(\tan^{-1} x)$

In Exercises 83–86, use the given derivative f' to find any local extrema and inflection points of f and sketch a possible graph without first finding an formula for f.

83. $f'(x) = x^3 - 3x^2 + 3x$

84. $f'(x) = x^4 - 1$

85. $f'(x) = \dfrac{1}{x}$

86. $f'(x) = e^x(x+4)$

Applications

87. Jason's distance in miles north from the corner of Main Street and High Street t minutes after noon on Tuesday is given by the following function $s(t)$:

Distance north of Main and High

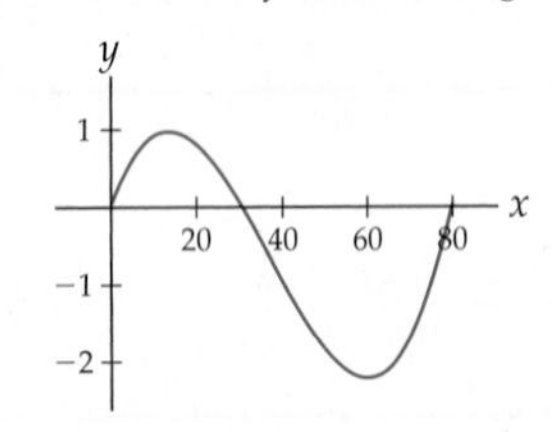

(a) Find an interval on which Jason's velocity is positive and decreasing. Describe what Jason is doing over this time interval.

(b) Find a time interval on which Jason is moving north and his velocity is increasing. Describe what Jason is doing over this time interval.

(c) Find a time interval on which Jason's acceleration and velocity are both negative. Describe what Jason is doing over this time interval.

(d) At which time is Jason's velocity at a minimum? What is he doing at that moment?

88. Suppose Juri drives for two hours and that his distance from home in miles is given by the function $s(t)$ shown in the following figure.

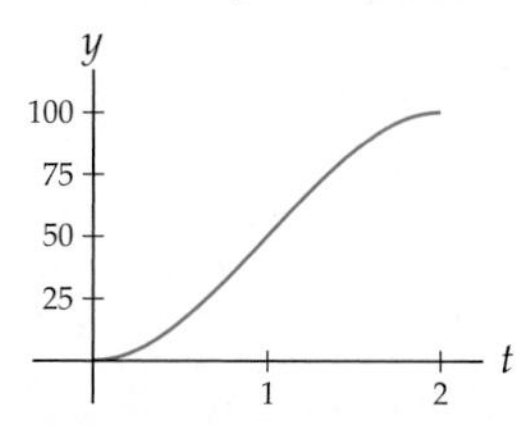

(a) Find a time interval on which Juri's acceleration is positive. Is his velocity positive or negative on this interval? Describe what Juri is doing over this time interval.

(b) The graph $y = s(t)$ of Juri's position has an inflection point at $t = 1$ hour. What does this inflection point say about Juri's velocity at $t = 1$? About his acceleration at $t = 1$?

89. For Exercises 89 and 90, suppose that Ian is a climber who is planning a path over a glacier in Canada's Icefield Range. Glaciers are a little like gelatin: They tend to form cracks (crevasses) when their surfaces are concave down. Cracks close up and travel is easy when they are concave up, as shown in the figure.

Using a map, Ian approximates the elevation of the glacier on a line that runs up through its center as

$$h(x) = 1.2 + .0095x + 0.037x^2 - 0.0072x^3 + 0.00046x^4,$$

where both x and $h(x)$ are measured in miles. Find the areas of the glacier where it is concave down and hence where Ian will need to move away from the center to avoid the crevasses.

90. Ian is a bit worried about taking a fall into a crevasse while carrying a heavy pack and towing a heavy sled. He does some tests on an old rope, dropping from a tree in his backyard as shown below.

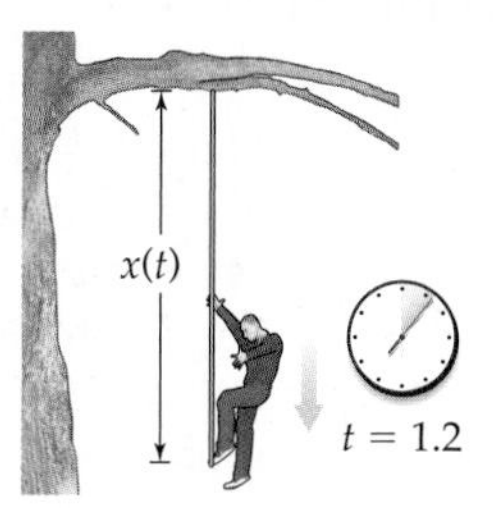

He cannot measure the force on the rope, but he takes a video, from which he can find his position (in feet below the tree limb) at time $t > 1.2$ seconds as

$$x(t) = 20 + 7e^{-0.25t}\sin(4.7t - 5.8).$$

(a) Weight is a force, given by mass times acceleration. Ian weighs 160 pounds, and the acceleration due to gravity that causes his weight is 32 feet per square second. What is Ian's mass? (The units are called "slugs.")

(b) Recalling that acceleration is the second derivative of position $x(t)$, what is the force on Ian at any time $t > 1.2$?

(c) Use a graphing calculator or other graphing utility to make a graph of Ian's acceleration over time. When is the *upward* force on Ian the greatest? Note that since we are measuring distance below the tree limb, in this situation an upward force is negative. What is that force?

Proofs

91. Prove part (b) of Theorem 3.10: If both f and f' are differentiable on an interval I, and f'' is negative on I, then f is concave down on I.

92. Prove Theorem 3.11 (b): If $x = c$ is a critical point of f, both f and f' are differentiable near $x = c$, and if $f''(c)$ is negative, then f has a local maximum at $x = c$.

93. Prove that every quadratic function is either always concave up or always concave down.

94. Prove that every cubic function (i.e., every function of the form $f(x) = ax^3 + bx^2 + cx + d$ for some constants a, b, c, and d) has exactly one inflection point. (*Note: It is not enough just to show that the second derivative of any cubic function has exactly one zero; you must also show that the sign of the second derivative changes.*)

95. Prove that if f'' is zero on an interval, then f is linear on that interval.

Thinking Forward

▶ *Global extrema on an interval:* The first-derivative test can be used to show that the function $f(x) = x^2 + 3x$ has a local minimum at $x = -\frac{3}{2}$. Is this a global minimum of the function? Is there a global maximum? What are the global extrema (if any) if we consider the function restricted to the interval $[-3, 3]$?

▶ *Global extrema and derivatives:* The first-derivative test can be used to show that the function $f(x) = x^3 - x^2 + x$ has a local maximum at $x = \frac{1}{3}$ and a local minimum at $x = 1$. Are either of these local extrema also global extrema? Can the first or second derivative tell you whether or not a local extremum is a global extremum?

3.4 OPTIMIZATION

- Comparing local extrema and endpoint behavior to find global extrema
- Strategies for translating real-world problems into mathematical problems
- Optimizing real-world quantities by solving global extrema problems

Finding Global Extrema

Many real-world problems involve optimization, that is, finding the global maximum and minimum values of a function on an interval. For example, a company might be interested in the maximum amount of profit it can make or in the minimum cost of producing an object. In this section we will learn how to use limits, derivatives, and values to find the global extrema of a function on an interval, so that we can solve such real-world problems.

Recall from Section 0.4 that $x = c$ is the location of a ***global maximum*** of a function f on an interval I if $f(c) \geq f(x)$ for all $x \in I$. Sometimes local extrema are also global extrema, and sometimes they are not. For example, the following four functions each have a local maximum at the critical point $x = 2$:

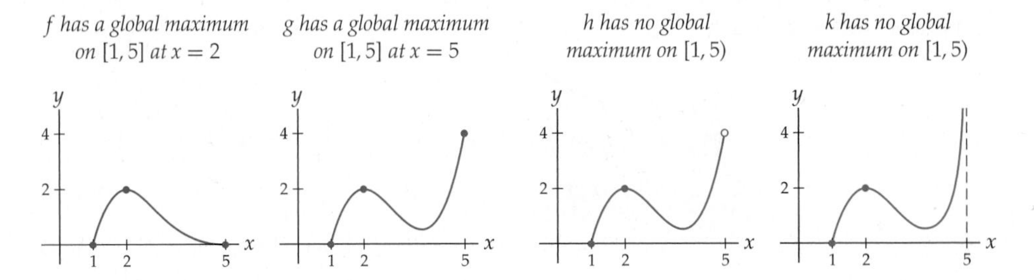

Only in the first graph is the local maximum also a global maximum. In the graph of g the global maximum on $[1, 5]$ is instead at the right endpoint $x = 5$. In the last two graphs there is no global maximum on the interval, because there is no point on either graph that is higher than all other points on its graph.

Although derivatives can help us locate local extrema, they do not always provide enough information to detect global extrema. The previous figures suggest that to find a global extremum of a continuous function we must compare the following:

- the values $f(c)$ for each interior local extremum $c \in I$;
- the values $f(c)$ at any closed endpoints $x = c$ of I;
- the limits $\lim_{x \to c} f(x)$ at any open endpoints or non-domain points $x = c$ of I.

Whichever of these three values is largest determines the location, if any, of the global maximum of f on I. Whichever is smallest determines the location, if any, of the global minimum. For example, in the graph of f that we just looked at, the value at the local extremum is $f(2) = 2$ and at the left and right endpoints we have $f(1) = 0$ and $f(5) = 0$. Since the value is highest at $x = 2$, that point is the global maximum of f on $[1, 5]$. A different thing happens with the graph of h; we again have $h(2) = 2$ at the local extremum and $h(1) = 0$ at the left endpoint, but at the open right endpoint we have $\lim_{x \to 5^-} h(x) = 4$. Since the limit at the open right endpoint is larger than the value at the left endpoint and the values at any local extrema, the function h has no global maximum on $[1, 5)$.

Translating Word Problems into Mathematical Problems

Finding the global extrema of a function f on an interval I is a mathematical problem that we can tackle with function values, limits, and derivatives. Real–world problems are less straightforward, since they are expressed in full sentences in which the variables, constants, functions, and relationships are described in words instead of in mathematical notation. To solve a real–world optimization problem we must first translate it into a mathematical global extrema problem. Once we solve the mathematical problem, we can translate the solution back into the real–world context. This translating procedure is illustrated in the following diagram:

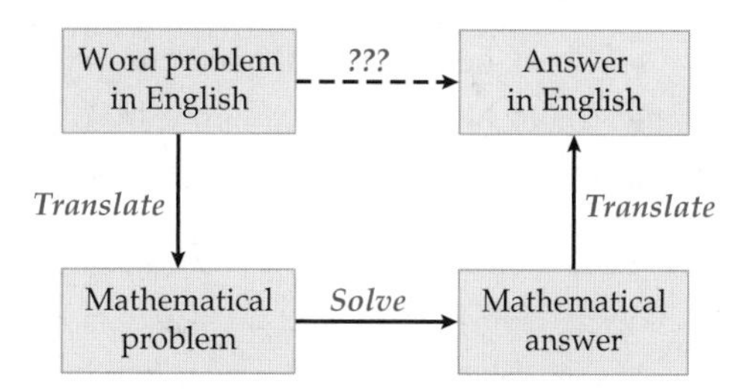

Many students have difficulty with word problems because of the translation step, not because of the mathematical solving step. As a simple example, suppose we wanted to solve the following real–world problem:

> *Calvin throws a baseball straight into the air at 50 feet per second, releasing the ball when his hand is 5 feet above the ground. How high did Calvin throw the baseball?*

To figure out how to answer this problem with calculus, we need to get some equations and variables to work with. A good place to start is to list what you know and set variable names as you go along:

- Let $s(t)$ be the height/position of the baseball at time t, in feet.
- We know that the position equation will be of the form $s(t) = -16t^2 + v_0t + s_0$.
- $v_0 = 50$ feet per second is the initial velocity of the baseball.
- $s_0 = 5$ feet is the initial position of the baseball.
- We seek the greatest height of the baseball during the time that it is in the air.
- Solving $s(t) = 0$, we see that the ball hits the ground at $t \approx 3.22$ seconds.

At some point while developing this list of facts it is good to collect the information into a figure, such as this one:

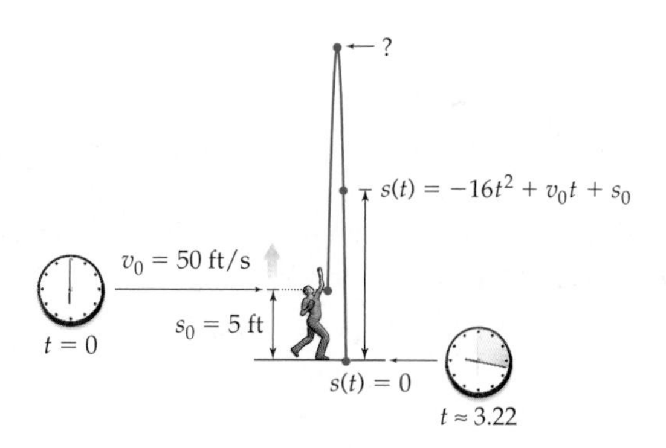

Putting this all together, we see that the mathematical problem that we wish to solve is

Find the global maximum value of $s(t) = -16t^2 + 50t + 5$ *on the interval* $[0, 3.22]$.

Compare this mathematical statement with our earlier real–world statement. After the hard work of translating, we now have a fairly straightforward calculus problem to solve. Here is an outline of the steps we could use to solve this optimization problem:

- Take the derivative: $s'(t) = -32t + 50$.
- Find critical points: $s'(t) = 0$ when $t = \frac{50}{32} = 1.5625$.
- Test critical points: $s''(1.5625) = -32 < 0$, so $s(t)$ has a local maximum at $x = 1.5625$.
- Check height at local maximum: $s(1.5625) \approx 44.06$.
- Check heights at endpoint values: $s(0) = 0$ and $s(3.22) \approx 0$.
- Conclusion: The maximum value of $s(t)$ on $[0, 3.22]$ is approximately 44.06.

Translating back into the real–world context, we can say that Calvin threw the baseball to a height of just over 44 feet.

In general, when translating a word problem we must identify any variables or functions and express them in mathematical notation with letters and symbols. With those same letters and symbols we can construct labeled diagrams, formulas, and relationships in mathematical notation. We must keep translating until we form a well-posed mathematical problem that we know how to solve with the tools and techniques of calculus. In the context of optimization problems, this means that we must identify a specific function and a specific interval on which we wish to find a global maximum or minimum value. At this point we only know how to find global extrema of one-variable functions, so we will sometimes have to use constraint equations to reduce multivariable functions to single-variable functions; see Example 2.

Examples and Explorations

EXAMPLE 1 **Finding global extrema of a function on an interval**

Find the global maximum and minimum values, if any, of the function $f(x) = 2x^3 - 15x^2 + 24x + 20$ on the interval $(0, 6)$.

SOLUTION

Our goal will be to use the derivative to identify local extrema on the interior of the interval and then compare those extrema with values or limits at the ends of the interval. In this case the interval is open, so we will compare with limits at $x \to 0$ and $x \to 6$. Step by step, we have

- Take the derivative: $f'(x) = 6x^2 - 30x + 24 = 6(x^2 - 5x + 4) = 6(x - 1)(x - 4)$.
- Find critical points: $f'(x) = 0$ at $x = 1$ and $x = 4$.
- Test critical points: $f''(x) = 12x - 30$, so $f''(1) = -18 < 0$ and $f''(4) = 18 > 0$; thus f has a local maximum at $x = 1$ and a local minimum at $x = 4$.
- Check height at local extrema: $f(1) = 31$ and $f(4) = 4$.
- Check limits at open endpoints:

$$\lim_{x\to 0^+} f(x) = \lim_{x\to 0^+} (2x^3 - 15x^2 + 24x + 20) = 20,$$

$$\lim_{x\to 6^-} f(x) = \lim_{x\to 6^-} (2x^3 - 15x^2 + 24x + 20) = 56.$$

The local maximum value of $f(1) = 31$ is exceeded by the limit of f at the right endpoint, so the function f has no global maximum on $(0, 6)$. The local minimum value of $f(4) = 4$ is less than the limit at either end of the interval and therefore is also a global minimum of f on $(0, 6)$. The graph of f on $(0, 6)$ looks like this:

Global minimum at $x = 4$,
no global maximum on $(0, 6)$

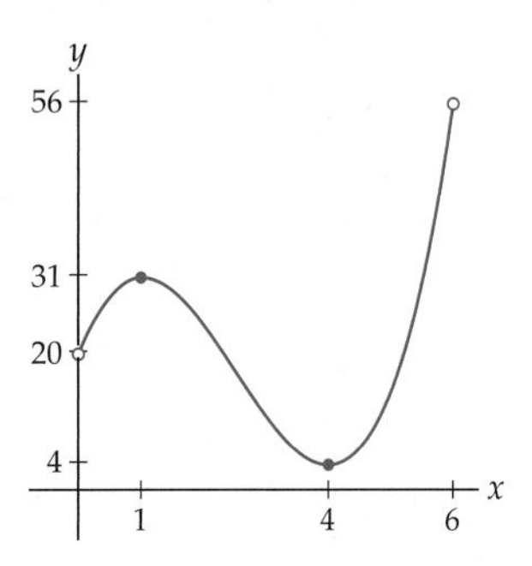

□

EXAMPLE 2

Optimizing a function on an interval given a constraint

Farmer Joe wants to build a rectangular chicken pen for his chickens. He wants to build the pen so that it has the largest area possible, and he has only 100 feet of chicken-wire fencing. What dimensions should he use for the pen?

SOLUTION

Let l and w represent the length and width of the rectangular pen, in feet. We know that the perimeter P of the pen must be 100 feet, since that is how much chicken-wire fencing Farmer Joe has; therefore $P = 2w + 2l = 100$. We are interested in maximizing the area $A = lw$ of the pen. The following diagram summarizes this information:

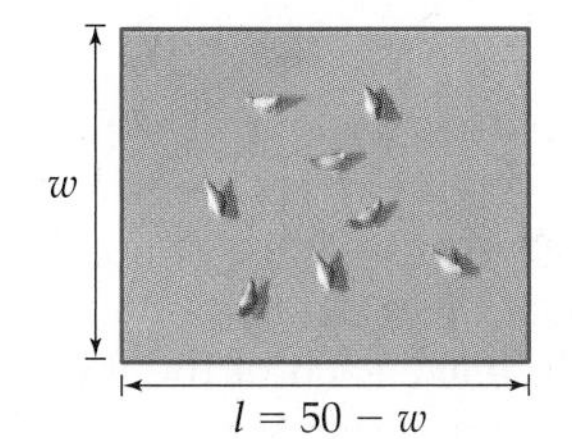

Constraint: $P = 2w + 2l = 100$

Maximize: $A = lw$

At this point the function A that we wish to maximize is written in terms of two variables, but our calculus techniques work only for functions of one variable. Luckily, we can use the constraint $P = 2w + 2l = 100$ to solve for one variable in terms of the other; this will reduce A to a one-variable function that we can optimize. Solving for l in terms of w, we have

$$P = 2w + 2l = 100 \quad \Longrightarrow \quad 2l = 100 - 2w \quad \Longrightarrow \quad l = 50 - w.$$

Using this equation, we can write the area function entirely in terms of w:

$$A = lw = (50 - w)w = 50w - w^2.$$

We now have a one-variable function $A(w) = 50w - w^2$ to maximize, but on what interval? We need to know which values of w we are willing to consider in this problem. Clearly we must have $w \geq 0$, since the width of the chicken pen cannot be negative. What is the upper bound on w? The key is to realize that the width w is the largest when the length l is the smallest. The smallest value that l could have is $l = 0$, and when $l = 0$, we have $0 = 50 - w$, or $w = 50$. Therefore we must have $w \leq 50$. We have now determined that the underlying mathematical problem for this word problem is

Find the global maximum of $A(w) = 50w - w^2$ *on the interval* $[0, 50]$.

Now we just have a simple mathematical problem to solve, and we follow the usual procedure for finding a global extremum on an interval:

- Take the derivative: $A'(w) = 50 - 2w$.
- Find critical points: $A'(w) = 0$ at $w = 25$.
- Test critical points: $A''(w) = -2$, so $A''(25) = -2$, and by the second-derivative test, $A(w)$ has a local maximum at $w = 25$.
- Check height at local extremum: $A(25) = 625$.
- Check heights at closed endpoints: $A(0) = 0$ and $A(50) = 0$.

From this work we see that $w = 25$ is indeed the global maximum of $A(w)$ on the interval $[0, 50]$. Thus the width of the pen should be $w = 25$ feet long. Since $l = 50 - w$, the length of the pen should also be $l = 25$ feet long, and the pen is square. The final answer to the word problem is therefore that Farmer Joe should build a square chicken pen where each side is 25 feet long. □

EXAMPLE 3

Modeling and minimizing a cost function

You work for a company that makes jewelry boxes. Your boss tells you that each jewelry box must have a square base and an open top and that you can spend \$3.75 on the materials for each box. The people in production tell you that the material for the sides of the box costs 2 cents per square inch while the reinforced material for the base of the box costs 5 cents per square inch. What is the largest volume jewelry box that you can make and still stay within budget?

SOLUTION

From a quick reading of the problem we can see that we wish to maximize the volume of a jewelry box given certain monetary constraints. Let's begin with a simple picture, assign variable names, and list what we know in mathematical language. Suppose x is the length of the sides of the square base and y is the height of the jewelry box, both measured in inches. The material to make each of the four sides of the box will cost $0.02(xy)$ dollars, and the material to make the base of the box will cost $0.05(x^2)$ dollars. Therefore we must have $0.02(4xy) + 0.05(x^2) = 3.75$, since we have \$3.75 to spend on materials. We wish to maximize the volume $V = x^2y$ of the box:

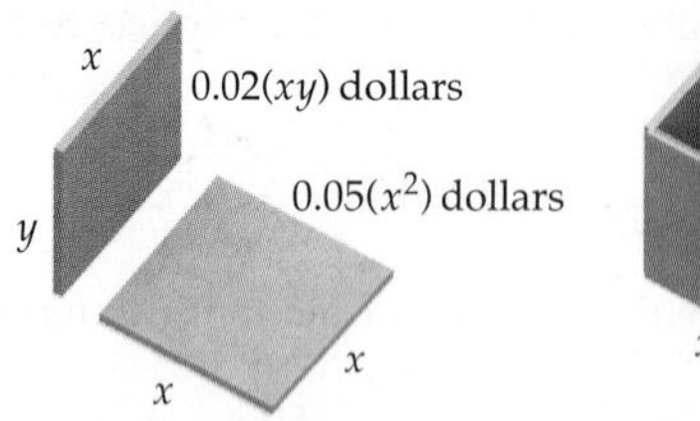

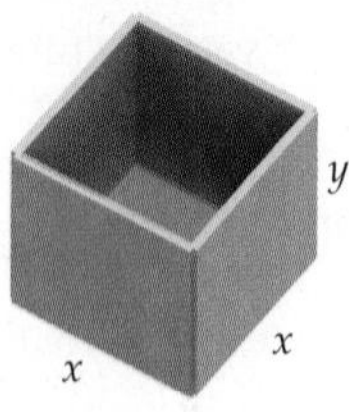

Constraint: $0.02(4xy) + 0.05(x^2) = 3.75$

Maximize: $V = x^2y$

Again, the function to be optimized is written in terms of two variables at this point, but we can use the constraint to solve for one variable in terms of the other. It is easiest to solve for y in terms of x:

$$0.02(4xy) + 0.05(x^2) = 3.75 \implies 0.08xy = 3.75 - 0.05x^2 \implies y = \frac{3.75 - 0.05x^2}{0.08x}.$$

We can use this equation for y to write the volume function V in terms of one variable. Doing this and then simplifying as much as possible, we have

$$V = x^2y = x^2\left(\frac{3.75 - 0.05x^2}{0.08x}\right) = \frac{1}{0.08}x(3.75 - 0.05x^2) = 46.875x - 0.625x^3.$$

Now we must determine the appropriate domain for $V(x) = 46.875x - 0.625x^3$ in the context of this problem. Clearly the smallest that the length x can be is zero. To find the largest that the base side length x can be, consider that the smallest possible value of the height of the box is $y = 0$. When $y = 0$, we have

$$0 = \frac{3.75 - 0.05x^2}{0.08x} \implies 0 = 3.75 - 0.05x^2 \implies x = \sqrt{\frac{3.75}{0.05}} = \sqrt{75}.$$

Note that we do not consider the negative square root of 75, since we know that the length x must be nonnegative. We have now completely translated the original word problem into the following mathematical problem:

Find the global maximum value of $V(x) = 46.875x - 0.625x^3$ *on the interval* $[0, \sqrt{75}\,]$.

To solve this problem we will find all the local interior extrema of $V(x)$ in the interval $[0, \sqrt{75}\,]$ and compare their values with the values of V at the endpoints of the interval. The steps are the same as those from previous examples:

- Take the derivative: $V'(x) = 46.875 - 3(0.625)x^2$.
- Find critical points: $V'(x) = 0$ when $x = \pm 5$, but only $x = 5$ is in the interval $[0, \sqrt{75}\,]$.
- Test critical points: $V''(5) = -2(3)(0.625)(5)$ is negative, so by the second-derivative test, $V(x)$ has a local maximium at $x = 5$.
- Check height at local extremum: $V(5) = 156.25$.
- Check heights at closed endpoints: $V(0) = 0$ and $V(\sqrt{75}\,) = 0$.

We now see that $x = 5$ is not only the location of a local maximum, but in fact the location of the *global* maximum of $V(x)$ on $[0, \sqrt{75}\,]$. Therefore the largest jewelry box that we can make with the given cost restrictions has volume 156.25 cubic inches. □

✔ CHECKING THE ANSWER

When $x = 5$, we must have $y = \frac{3.75 - 0.05(5)^2}{0.08(5)} = 6.25$. To algebraically check some of the work we have done, we can verify that, with these values for x and y, the cost of producing the jewelry box is indeed $0.02(4xy) + 0.05(x^2) = \$3.75$. We can check our work regarding the optimization of V and the choice of endpoints for the interval by graphing $V(x) = 46.875x - 0.625x^3$, as shown next. From this graph, it does seem reasonable that at $x = 5$ we have a global maximum of 156.25 and that the point $x = \sqrt{75} \approx 8.66$ should be the right end of the interval for the problem.

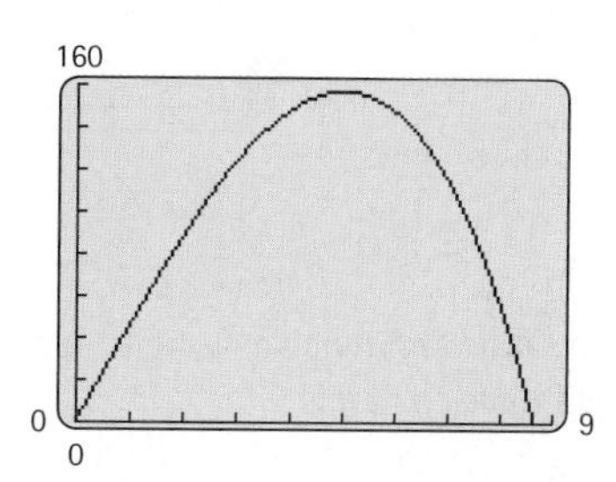

EXAMPLE 4

The fastest way to put out a flaming tent

While you are on a camping trip, your tent accidentally catches fire. Luckily, you happen to be standing right at the edge of a stream with a bucket in your hand. The stream runs east–west, and the tent is 40 feet north of the stream and 100 feet farther east than you are, as shown here:

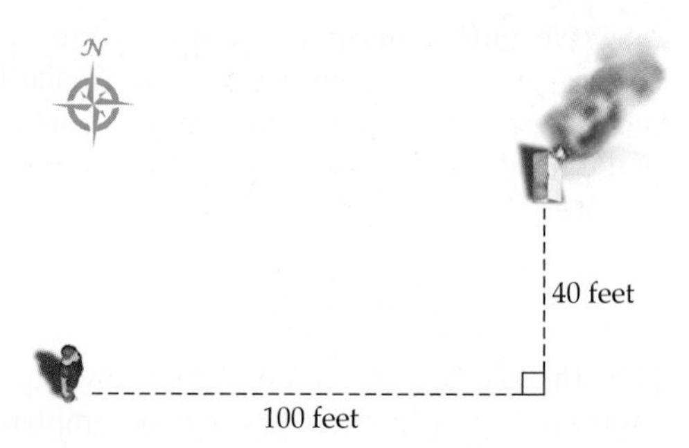

You can run only half as fast while carrying the full bucket as you can empty handed, and thus any distance travelled with the full bucket is effectively twice as long. What is the fastest way for you to get water to the tent?

SOLUTION

On the one hand, notice that if you were to get water immediately and run diagonally to the tent, you would run the entire distance with a full bucket of water at half your normal speed. On the other hand, if you were to run along the side of the stream until you were directly south of the tent, then get water and run north to the tent, you would have a lot of total distance to run. Clearly, it would be more efficient for you to run along the side of the stream for a while, fill the bucket, and then run diagonally to the tent. The question is, how far should you run along the side of the stream before you stop to fill up the bucket?

We need some mathematical notation to get our heads around this problem. Suppose x is the distance you will run along the stream before filling your bucket. The distance you will run with the full bucket is then the hypotenuse of a right triangle with legs of lengths 40 feet and $100 - x$ feet. By the Pythagorean theorem, you will have to run $\sqrt{40^2 + (100 - x)^2}$ feet with a full bucket of water, as shown here:

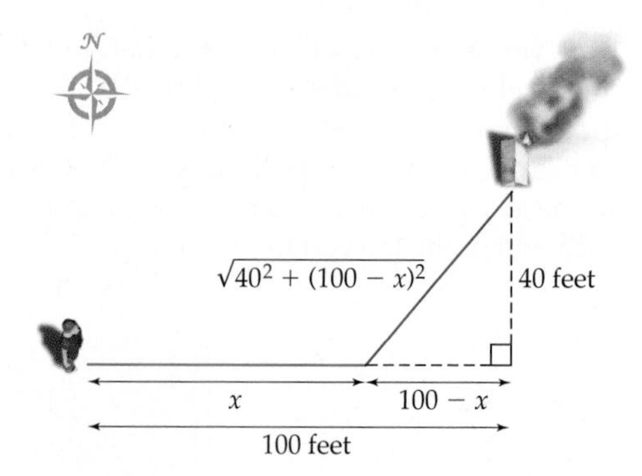

Since it takes you twice as long to run with a full bucket of water, the total *effective* distance you will have to run is

$$D(x) = x + 2\sqrt{40^2 + (100 - x)^2}.$$

This is the function we want to minimize. What is the interval of x-values that we are interested in? From the diagram we can see that x must be between 0 and 100 feet, since clearly you wouldn't want to run away from the tent or past the tent. The endpoint cases $x = 0$ and $x = 100$ correspond to the two special cases we discussed earlier: The case when you get water immediately and the case when you run the full 100 feet before getting water. We have now translated the word problem into the following mathematical optimization problem:

Find the global minimum value of $D(x) = x + 2\sqrt{40^2 + (100 - x)^2}$ *on the interval* $[0, 100]$.

Once again this is a straightforward global extremum problem, and we follow the same steps as previously; but this time the mathematics is a little more involved:

- Take the derivative: By applying the chain rule twice, we see that the derivative of $D(x)$ is

$$D'(x) = 1 + 2\left(\frac{1}{2}\right)(40^2 + (100 - x)^2)^{-1/2}(2(100 - x)(-1)).$$

- Find critical points: We need to simplify $D'(x)$ before we attempt to find its zeroes or the values where it does not exist. With a bit of algebra we can write $D'(x)$ in the form of a quotient:

$$D'(x) = 1 + \frac{-2(100 - x)}{\sqrt{40^2 + (100 - x)^2}} = \frac{\sqrt{40^2 + (100 - x)^2} - 2(100 - x)}{\sqrt{40^2 + (100 - x)^2}}.$$

From this quotient form it is easy to pick out any places where $D'(x)$ is zero or undefined. The denominator of $D'(x)$ is never zero, and thus $D'(x)$ always exists. To find the values of x for which $D'(x) = 0$, we set the numerator equal to 0 and solve:

$$\begin{aligned}
\sqrt{40^2 + (100 - x)^2} - 2(100 - x) &= 0 \\
\sqrt{40^2 + (100 - x)^2} &= 2(100 - x) \\
40^2 + (100 - x)^2 &= 4(100 - x)^2 && \leftarrow \text{square both sides} \\
1600 + 10,000 - 200x + x^2 &= 40,000 - 800x + 4x^2 \\
0 &= 3x^2 - 600x + 28400 \\
x &= 100 \pm \frac{40}{3}\sqrt{3}. && \leftarrow \text{quadratic formula}
\end{aligned}$$

Note that $x = 100 + \frac{40}{3}\sqrt{3} \approx 123.1$ is not in the interval $[0, 100]$, but $x = 100 - \frac{40}{3}\sqrt{3} \approx 76.9$ is. In addition, recall that squaring both sides of an equation sometimes leads to extraneous, or false, solutions; interestingly, $x = 100 + \frac{40}{3}\sqrt{3}$ happens to be one of those false solutions. Therefore the only critical point of $D(x)$ is $x = 100 - \frac{40}{3}\sqrt{3} \approx 76.9$.

- Test critical points: The first-derivative test can be applied to show that $D(x)$ has a local minimum at $x = 100 - \frac{40}{3}\sqrt{3} \in [0, 100]$.
- Check height at local extremum: $D(100 - \frac{40}{3}\sqrt{3}) \approx 169.282$.
- Check heights at closed endpoints: $D(0) \approx 215.407$, $D(100) = 180$.

The work that we just did shows that the global minimum of $D(x)$ on $[0, 100]$ is at $x = 100 - \frac{40}{3}\sqrt{3}$. This means that the quickest way to bring water to the burning tent is to run for $100 - \frac{40}{3}\sqrt{3} \approx 76.9$ feet along the side of the stream, then fill the bucket with water and run diagonally to the tent. □

? TEST YOUR UNDERSTANDING

- Why can't the derivative necessarily detect the global minimum of a function f on an interval $[a, b]$ if that minimum happens to occur at an endpoint of the interval?
- Suppose a function f satisfies $\lim_{x \to 3^-} f(x) = -\infty$. Why does this mean that f has no global minimum on $[0, 3]$. What can you say about any global minima of f on $[0, 6]$?
- Suppose a function f satisfies $\lim_{x \to 1^+} f(x) = \infty$. What can you say about any global maxima of f on $[-2, 4]$?
- What types of real–world problems translate into mathematical problems in which we must find the global maximum or minimum of a function on an interval?
- What are the general steps for solving an optimization word problem?

EXERCISES 3.4

Thinking Back

Local and global extrema: Use mathematical notation, including inequalities as used in the definition of local and global extrema, to express each of the following statements.

- On the interval $[-3, 5]$, f has a local maximum at $x = 2$.
- On the interval $[0, 2]$, f has a global maximum at $x = -1$.
- On the interval $[-4, 4]$, f has a global minimum at $x = 0$.
- On the interval $[0, 5]$, f has no global minimum.

Critical points: Find the critical points of each of the following functions.

- $f(x) = \dfrac{3x - 2}{x - 1}$
- $f(x) = \dfrac{1}{1 + \sqrt{x}}$
- $f(x) = \sin\left(\dfrac{\pi}{2}x\right)$
- $f(x) = e^x(x - 2)$

Concepts

0. *Problem Zero:* Read the section and make your own summary of the material.

1. *True/False:* Determine whether each of the statements that follow is true or false. If a statement is true, explain why. If a statement is false, provide a counterexample.

(a) *True or False:* Every local maximum is a global maximum.

(b) *True or False:* Every global minimum is a local minimum.

(c) *True or False:* If f has a global maximum at $x = 2$ on the interval $(-\infty, \infty)$, then the global maximum of f on the interval $[0, 4]$ must also be at $x = 2$.

(d) *True or False:* If f has a global maximum at $x = 2$ on the interval $[0, 4]$, then the global maximum of f on the interval $(-\infty, \infty)$ must also be at $x = 2$.

(e) *True or False:* If f is continuous on an interval I, then f has both a global maximum and a global minimum on I.

(f) *True or False:* Suppose f has two local minima on the interval $[0, 10]$, one at $x = 2$ with a value of 4 and one at $x = 7$ with a value of 1. Then the global minimum of f on $[0, 10]$ must be at $x = 7$.

(g) *True or False:* If f has no local maxima on $(-\infty, \infty)$, then it will have no global maximum on the interval $[0, 5]$.

(h) *True or False:* If $f'(3) = 0$, then f has either a local minimum or a local maximum at $x = 3$.

2. *Examples:* Construct examples of the thing(s) described in the following. Try to find examples that are different than any in the reading.

(a) The graph of a function with a local minimum at $x = 2$ but no global minimum on $[0, 4]$.

(b) The graph of a function with no local or global extrema on $(-3, 3)$.

(c) The graph of a function whose global maximum on $[2, 6]$ does not occur at a critical point.

3. When you try to find the local extrema of a function f on an interval I, one of the first steps is to find the critical points of f. Explain why these critical points of f won't help you locate any "endpoint" extrema.

4. Explain why you can't find the global maximum of a function f on an interval I just by finding all the local extrema of f and then checking to see which one has the highest value $f(c)$.

5. Suppose f is a function that is defined and continuous on an open interval I. Will the endpoints of I always be local extrema of f? Will f necessarily have a global maximum or minimum in the interval I? Justify your answers.

6. Suppose f is a function that is defined and continuous on a closed interval I. Will the endpoints of I always be local extrema of f? Will f necessarily have a global maximum or minimum in the interval I? Justify your answers.

7. Suppose f is a function that is discontinuous somewhere on an interval I. Explain why comparing the values of any local extrema of f on I and the values or limits of f at the endpoints of I is not in general sufficient to determine the global extrema of f on I.

8. Given the following graph of f, graphically estimate the global extrema of f on each of the six intervals listed:

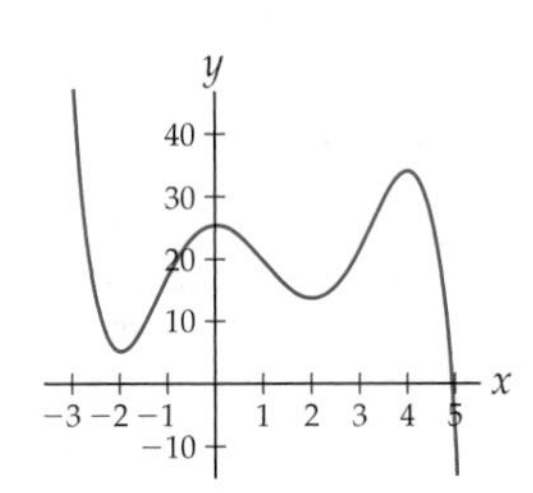

(a) $[-2, 4]$ (b) $(-2, 4)$ (c) $(-1, 1)$
(d) $(0, 4]$ (e) $[0, 4)$ (f) $(-\infty, \infty)$

9. Given the following graph of f, graphically estimate the global extrema of f on each of the six intervals listed:

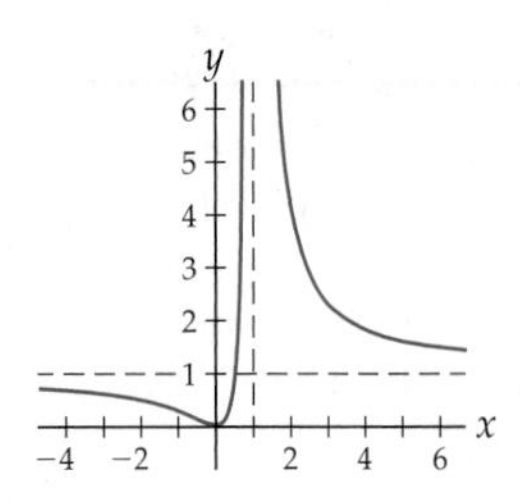

(a) $[0, 4]$ (b) $[2, 5]$ (c) $(-2, 1)$
(d) $[0, \infty)$ (e) $(-\infty, 0]$ (f) $(-\infty, \infty)$

10. Given the following graph of f, graphically estimate the global extrema of f on each of the six intervals listed:

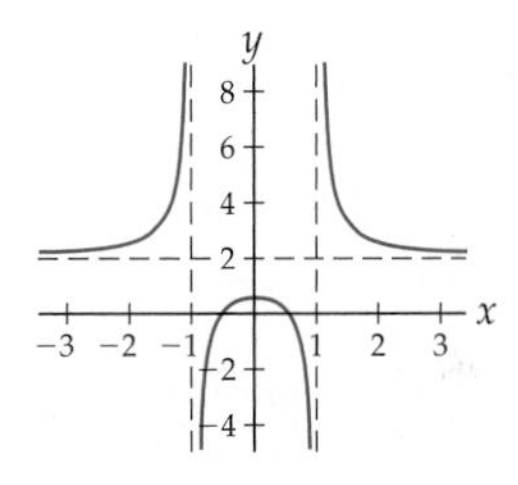

(a) $(-1, 1)$ (b) $[2, \infty)$ (c) $[-2, -1)$
(d) $[0, 2]$ (e) $(1, \infty)$ (f) $(-\infty, \infty)$

Skills

Find the locations and values of any global extrema of each function f in Exercises 11–20 on each of the four given intervals. Do all work by hand by considering local extrema and endpoint behavior. Afterwards, check your answers with graphs.

11. $f(x) = 2x^3 - 3x^2 - 12x$, on the intervals

(a) $[-3, 3]$ (b) $[0, 3]$ (c) $(-1, 2]$ (d) $(-2, 1)$

12. $f(x) = 3x^4 + 4x^3 - 36x^2$, on the intervals

(a) $[-5, 5]$ (b) $[-2, 2]$ (c) $(-3, 1]$ (d) $(-1, 3)$

13. $f(x) = -12x + 6x^2 + 4x^3 - 3x^4$, on the intervals

(a) $[-1, 1]$ (b) $(-1, 1)$ (c) $(-3, 0]$ (d) $[0, 3]$

14. $f(x) = \dfrac{3x - 2}{x - 1}$, on the intervals

(a) $[0, 2]$ (b) $[-2, 0]$ (c) $[-1, 1]$ (d) $(-1, 1)$

15. $f(x) = \dfrac{1}{1 + \sqrt{x}}$, on the intervals

(a) $[0, 3]$ (b) $(0, 3)$ (c) $[1, 2]$ (d) $[0, \infty)$

16. $f(x) = \dfrac{4}{\sqrt{x^2 + 1}} + 3$, on the intervals

(a) $(-5, 2]$ (b) $[-5, 2)$ (c) $[1, 10]$ (d) $[0, 20]$

17. $f(x) = x^{3/2}(3x - 5)$, on the intervals

(a) $[0, 4]$ (b) $[0, 4)$ (c) $[0, 1)$ (d) $(0, 1)$

18. $f(x) = \sin\left(\frac{\pi}{2}x\right)$, on the intervals

(a) $[-2, 2]$ (b) $(-2, 2)$ (c) $[-1, 1)$ (d) $[0, \infty)$

19. $f(x) = (x^2 - 4x + 3)^{-1/2}$, on the intervals

(a) $[0, 4]$ (b) $[0, 10]$ (c) $[0, 3.5]$ (d) $(3, \infty)$

20. $f(x) = e^x(x - 2)$, on the intervals

(a) $[-2, 2]$ (b) $(0, 3)$ (c) $[0, \infty)$ (d) $(-\infty, 0]$

Find dimensions for each shape in Exercises 21–24 so that the total area enclosed is as large as possible, given that the total edge length is 120 inches. The rounded shapes are half-circles, and the triangles are equilateral.

21. 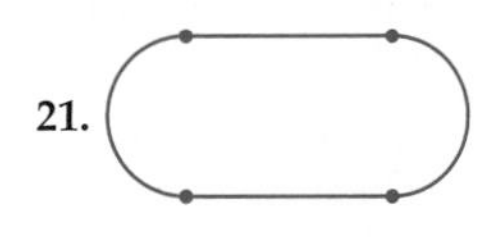

22. 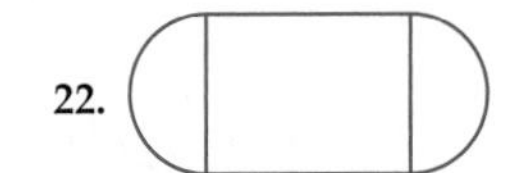

23. 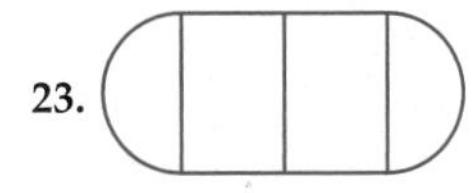

24. 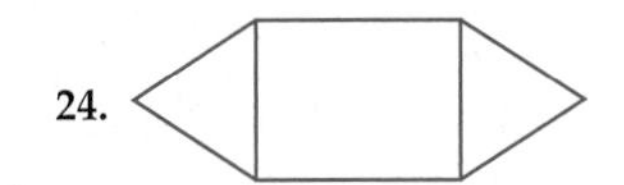

Use optimization techniques to answer the questions in Exercises 25–30.

25. Find two real numbers x and y whose sum is 36 and whose product is as large as possible.

26. Find two real numbers x and y whose sum is 36 and whose product is as small as possible.

27. Find real numbers a and b whose sum is 100 and for which the sum of the squares of a and b is as small as possible.

28. Find the area of the largest rectangle that fits inside a circle of radius 4.

29. Find the area of the largest rectangle that fits inside a circle of radius 10.

30. Find the volume of the largest cylinder that fits inside a sphere of radius 10.

In Exercises 31–34, find the point on the graph of the function f that is closest to the point (a, b) by minimizing the square of the distance from the graph to the point.

31. $f(x) = 3x + 1$ and the point $(-2, 1)$

32. $f(x) = x^2$ and the point $(0, 3)$

33. $f(x) = x^2 - 2x + 1$ and the point $(1, 2)$

34. $f(x) = \sqrt{x^2 + 1}$ and the point $(2, 0)$

Applications

A farmer wants to build four fenced enclosures on his farmland for his free-range ostriches. To keep costs down, he is always interested in enclosing as much area as possible with a given amount of fence. For the fencing projects in Exercises 35–38, determine how to set up each ostrich pen so that the maximum possible area is enclosed, and find this maximum area.

35. A rectangular ostrich pen built with 350 feet of fencing material.

36. A rectangular ostrich pen built along the side of a river (so that only three sides of fence are needed), with 540 feet of fencing material.

37. A rectangular ostrich pen built with 1000 feet of fencing material, divided into three equal sections by two interior fences that run parallel to the exterior side fences, as shown next at the left.

Ostrich pen with three sections

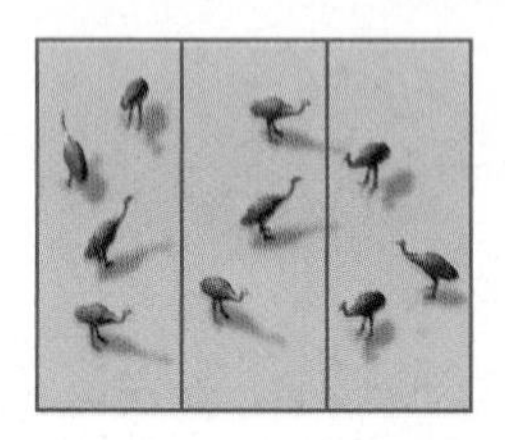

Ostrich pen with six sections

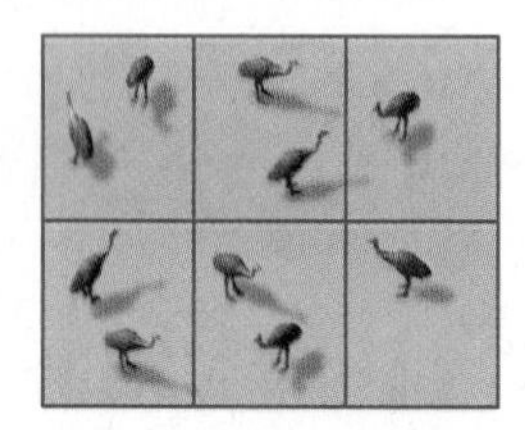

38. A rectangular ostrich pen that is divided into six equal sections by two interior fences that run parallel to the east and west fences, and another interior fence running parallel to the north and south fences, as shown previously at the right. The farmer has allotted 2400 feet of fencing material for this important project.

You are in charge of constructing a zoo habitat for prairie dogs, with the requirement that the habitat must enclose 2500 square feet of area and use as little border fencing as possible. For each of the habitat designs described in Exercises 39–42, find the amount of border fencing that the project will require.

39. A rectangular habitat with a 20-foot-wide nesting hutch along one side (so that fencing is not needed along those 20 feet), as shown next at the left.

Rectangular habitat with hutch

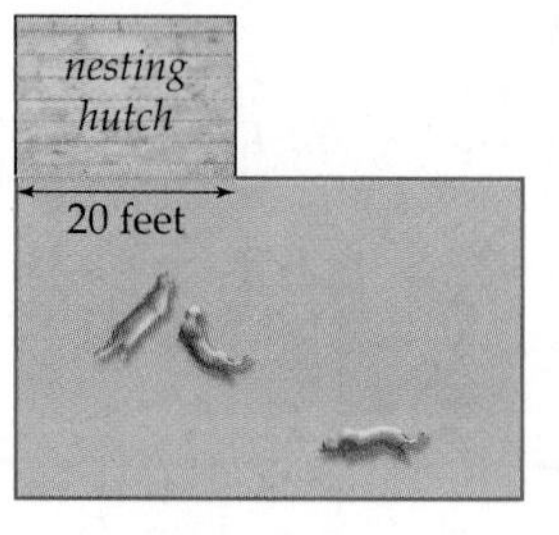

Arena-style habitat

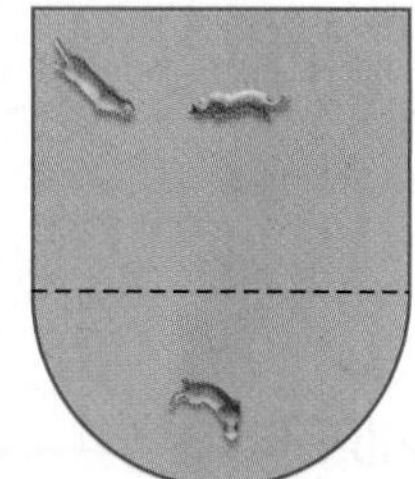

40. An arena-style habitat whose front area is a semicircle and whose back area is rectangular, as shown previously at the right.

41. A trapezoid-shaped habitat whose angled side is an enclosed walkway for zoo patrons (so that no fencing is needed along the walkway), where the walkway makes an angle of 60° with the right fence, as shown next at the left.

Habitat along walkway

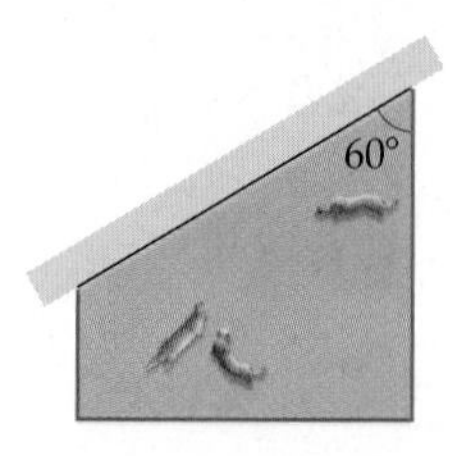

Habitat with mural

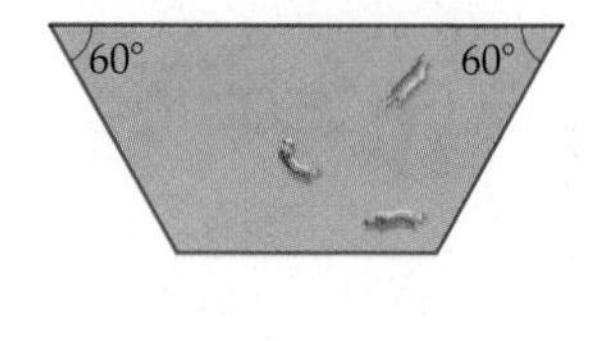

42. An arena-style trapezoid-shaped habitat whose long back side is a wall with a landscape mural (so no fencing is needed along the back wall), where the back wall makes an angle of 60° with the slanted side fences, as shown previously at the right.

Alina wants to make keepsake boxes for her two best friends. She doesn't have a lot of money, so she wants to make each box described in Exercises 43–44 so that it holds as much as possible with a limited amount of material.

43. For Jen, Alina wants to make a box with a square base whose sides and base are made of wood and whose top is made of metal. The wood she wants to use costs 5 cents per square inch, while the material for the metal top costs 12 cents per square inch. What is the largest possible box that Alina can make for Jen if she only has \$20.00 to spend on materials?

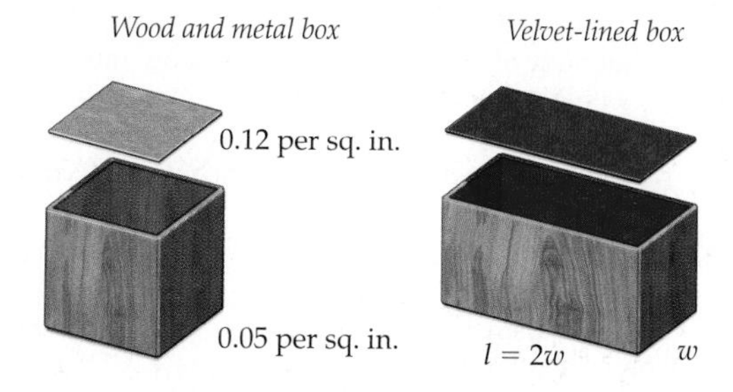

44. For Eliza, Alina wants to make a rectangular box whose base is twice as long as it is wide. This box will be lined on the entire inside with velvet and in addition the outside of the top of the box is to be lined in velvet. If Alina has 240 square inches of velvet, how can she make Eliza's box so that it holds as many keepsakes as possible?

The U.S. Postal Service ships a package under large-package rates if the sum of the length and the girth of the package is greater than 84 inches and less than or equal to 108 inches. The *length* of a package is considered to be the length of its longest side, and the *girth* of the package is the distance around the package perpendicular to its length. In each of Exercises 45–47, Linda wants to ship packages under the USPS large-package rates.

45. Linda needs to mail a rectangular package with square ends (in other words, with equal width and height). What is the largest volume that her package can have? What is the largest surface area that her package can have?

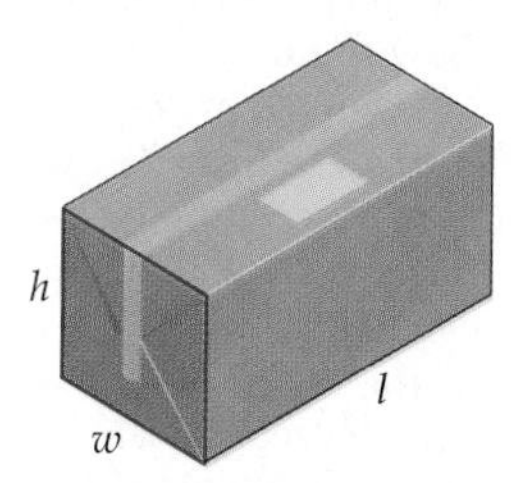

46. Linda's second package must be rectangular and 40 inches in length. What is the largest volume that her package can have? What is the largest surface area that her package can have?

47. Linda also needs to mail some architectural plans, which must be shipped in a cylindrical container. What is the largest volume that her package can have? What is the largest surface area that her package can have? *(Hints: The volume of a right circular cylinder with radius r and height h is $V = \pi r^2 h$; the total surface area of such a cylinder is $SA = 2\pi rh + 2\pi r^2$.)*

For Exercises 48–50, consider a toy car that moves back and forth on a long, straight track for 4 minutes in such a way that the function $s(t) = 96t - 84t^2 + 28t^3 - 3t^4$ describes how far the car is to the right of the starting point, in centimeters, after t minutes.

48. When is the toy car farthest away from the starting point? Does it ever return to the starting point? For how long does this model make sense?

49. Within the domain of your model, when is the toy car moving fastest to the right? When is the toy car moving fastest to the left?

50. When is the toy car accelerating the fastest to the right? When is the toy car accelerating fastest to the left?

Suppose you have a 10-inch length of wire that you wish to cut and form into shapes. In each of Exercises 51–53 you will determine how to cut the wire to minimize or maximize the area of the resulting shapes.

51. Suppose you wish to make one cut in the wire and use the two pieces to form a square and a circle. Determine how to cut the wire so that the combined area enclosed by the square and the circle is (a) as small as possible and (b) as large as possible.

52. Suppose you wish to make one cut in the wire and use the two pieces to form a square and an equilateral triangle. Determine how to cut the wire so that the combined area is (a) as small as possible and (b) as large as possible.

53. Suppose you wish to make one cut in the wire and use the two pieces to form a circle and an equilateral triangle. Determine how to cut the wire so that the combined area of these two shapes is (a) as small as possible and (b) as large as possible.

In each situation described in Exercises 54–62, set up and solve a global extrema problem that solves the given real–world optimization problem.

54. Alina needs to make a flyer for her band's concert. The flyer must contain 20 square inches of printed material and for design purposes should have side margins of 1 inch and top and bottom margins of 2 inches. What size paper should Alina use in order to use the least amount of paper per flyer as possible?

55. Your company produces cylindrical metal oil drums that must each hold 40 cubic feet of oil. How should the oil drums be constructed so that they use as little metal as possible? Can they be constructed to use as *much* metal as possible?

56. An airplane leaves Chicago at noon and travels south at 500 miles per hour. Another airplane is travelling east towards Chicago at 650 miles per hour and arrives at 2:00 P.M.. When were these two airplanes closest to each other, and how far apart were they at that time?

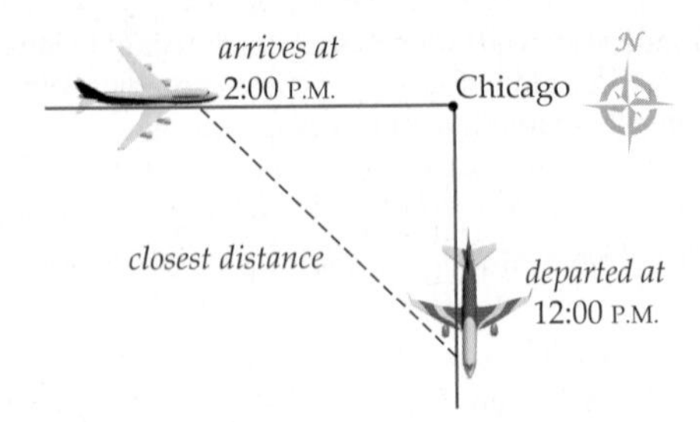

57. The cost of the material for the top and bottom of a cylindrical can is 5 cents per square inch. The material for the rest of the can costs only 2 cents per square inch. If the can must hold 400 cubic inches of liquid, what is the cheapest way to make the can? What is the most expensive way?

58. Consider the can-making situation in the previous exercise, but suppose that the cans are made with open tops. If each can must hold 400 cubic inches of liquid, what is the cheapest way to make the cans? What is the most expensive way?

59. A steam pipe must be buried underground to reach from one corner of a rectangular parking lot to the diagonally opposite corner. The dimensions of the parking lot are 500 feet by 800 feet. It costs 5 dollars per foot to lay steam pipe under the pavement but only 3 dollars per foot to lay the pipe along one of the long edges of the parking lot. Because of nearby sidewalks, the pipe cannot be laid along the 500-foot sides of the parking lot. How should the steam pipe be buried so as to cost as little as possible?

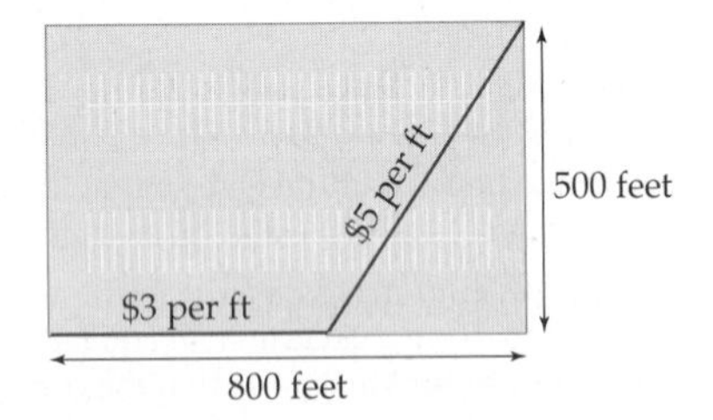

60. Suppose you want to make an open-topped box out of a 4×6 index card by cutting a square out of each corner and then folding up the edges, as shown in the figure. How large a square should you cut out of each corner in order to maximize the volume of the resulting box?

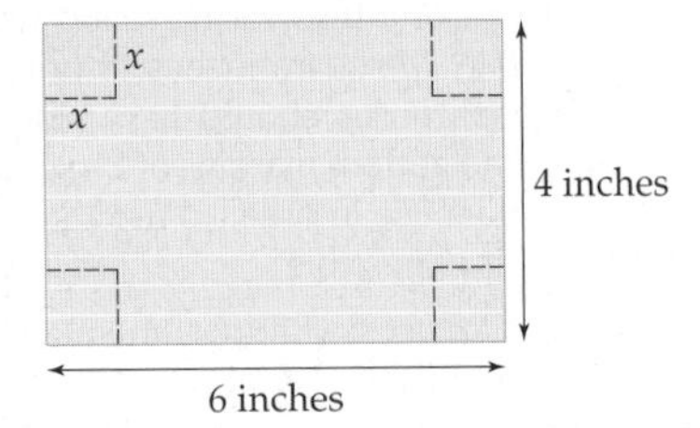

61. Your family makes and sells velvet Elvis paintings. After many years of research you have found a function that predicts how many paintings you will sell in a year, based on the price that you charge per painting. You always charge between \$5.00 and \$55.00 per painting. Specifically, if you charge c dollars per painting, then you can sell $N(c) = 0.6c^2 - 54c + 1230$ paintings in a year.

(a) What price should you charge to sell the *greatest* number of velvet Elvis paintings, and how many could you sell at that price? For what price would you sell the *least* number of paintings, and how many would you sell?

(b) Write down a function that predicts the revenue $R(c)$, in dollars, that you will earn in a year if you charge c dollars per painting. *(Hint: Try some examples first; for example, what would your yearly revenue be if you charged* \$10.00 *per painting? What about* \$50.00*? Then write down a function that works for all values of c.)*

(c) What price should you charge to earn the *most* money, and how much money would you earn? What price per painting would cause you to make the *least* amount of money in a year, and how much money would you make in that case?

(d) Explain why you do not make the most money at the same price per painting for which you sell the most paintings.

62. While you are on a camping trip, your tent accidentally catches fire. At the time, you and the tent are both 50 feet from a stream and you are 200 feet away from the tent, as shown in the diagram. You have a bucket with you, and need to run to the stream, fill the bucket, and run to the tent as fast as possible. You can run only half as fast while carrying the full bucket as you can empty handed, and thus any distance travelled with the full bucket is effectively twice as long. Complete parts (a)–(f) to determine how you can put out the fire as quickly as possible.

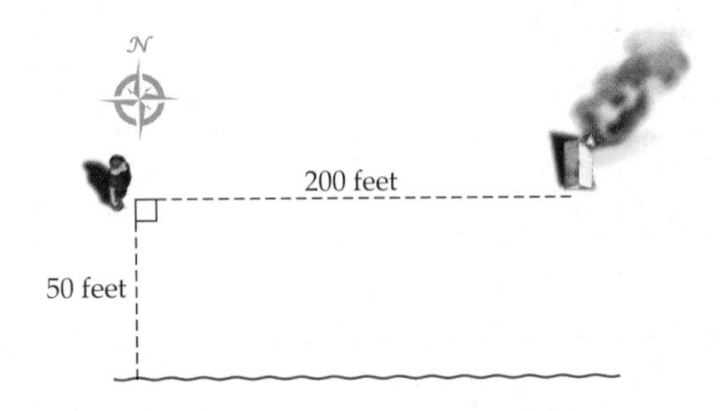

(a) Let x represent the distance from the point on the stream directly "below" you to the point on the stream that you run to. Sketch the path that you would follow to run from your starting position, to the point x along the stream, to the tent.

(b) Let $D(x)$ represent the effective distance (counting twice any distance travelled while carrying a full bucket) you have to run in order to collect water and get to the tent. Write a formula for $D(x)$.

(c) Determine the interval I of x-values on which $D(x)$ should be minimized. Explain in practical terms what happens at the endpoints of this interval, and calculate the value of $D(x)$ at these endpoints.

(d) Find $D'(x)$, and simplify as much as possible. Are there any points (in the interval I) at which $D'(x)$ is undefined?

(e) It is difficult to find the zeroes of $D'(x)$ by hand. Use a graphing utility to approximate any zeroes of $D'(x)$ in the interval I, and test these zeroes by evaluating D' at each one.

(f) Use the preceding information to determine the minimum value of $D(x)$, and then use this value to answer the original word problem.

Proofs

63. Prove that the rectangle with the largest possible area given a fixed perimeter P is always a square.

64. Prove that the most efficient way to build a rectangular fenced area along a river—so that only three sides of fencing are needed—is to make the side parallel to the river twice as long as the other sides. You may assume that you have a fixed amount of fencing material.

Thinking Forward

Consider the graph of the function f shown next. Define $A(x)$ to be the area of the region between the graph of f and the x-axis from 0 to x. We will count areas of regions above the x-axis positively and areas of regions below the x-axis negatively.

$A(x)$ is area under this graph on $[0, x]$

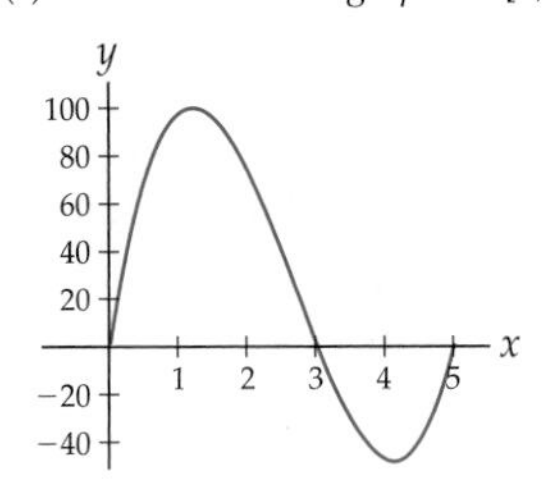

- Use the graph to approximate the values of $A(0)$, $A(1)$, $A(2)$, $A(3)$, $A(4)$, and $A(5)$.
- From the graph of f, estimate all local maxima and minima of $A(x)$.
- From the graph of f, estimate all global maxima and minima of $A(x)$, if any.
- It turns out that the function f whose graph is shown is given by the formula $f(x) = 12x^3 - 96x^2 + 180x$ and that the area function A is given by the formula $A(x) = 3x^4 - 32x^3 + 90x^2$. What surprising relationship do f and A have?
- Show that your answers for the local and global extrema of $A(x)$ are reasonable by using optimization techniques on the area function $A(x) = 3x^4 - 32x^3 + 90x^2$.

3.5 RELATED RATES

- Using implicit differentiation to obtain relationships between rates
- Formulas for volume, surface area, and relationships between side lengths of triangles
- Techniques for solving related-rates problems

Related Quantities Have Related Rates

If two quantities that change over time are related to each other, then their rates of change over time will also be related to each other. For example, consider an expanding circle. Clearly the radius $r = r(t)$ of the circle and the area $A = A(t)$ of the circle are related: If you know one of these quantities at some time t, then you also know the other, via the formula $A = \pi r^2$. As the circle expands over time, the rate $\frac{dr}{dt}$ at which its radius increases is related to the rate $\frac{dA}{dt}$ at which its area increases. We can find an equation that relates these rates by applying implicit differentiation to the formula that relates the quantities r and A:

$$A = \pi r^2 \qquad \leftarrow \text{relationship between } A \text{ and } r$$

$$\frac{d}{dt}(A(t)) = \frac{d}{dt}(\pi(r(t))^2) \qquad \leftarrow \text{differentiate both sides with respect to } t$$

$$\frac{dA}{dt} = \pi\left(2r\frac{dr}{dt}\right) \qquad \leftarrow \text{chain rule}$$

$$\frac{dA}{dt} = 2\pi r\frac{dr}{dt}. \qquad \leftarrow \text{relationship between } \frac{dA}{dt} \text{ and } \frac{dr}{dt}$$

Notice that in this calculation we use A interchangeably with $A(t)$ and r interchangeably with $r(t)$, employing the explicit function notation only when we need to be reminded that A and r are functions of t.

According to the formula we just found, the rate $\frac{dA}{dt}$ depends on both the size of the radius r at time t and the rate $\frac{dr}{dt}$ at which the radius is increasing. For example, suppose that our expanding circle has an initial radius of $r(0) = 2$ inches at time $t = 0$, and that the radius then increases at a constant rate of $\frac{dr}{dt} = 3$ inches per second. Then the formula we found tells us that at the instant the circle has a radius of 4 inches, the area of the circle is increasing at a rate of

$$\left.\frac{dA}{dt}\right|_{r=4} = 2\pi(4)(3) = 24\pi \text{ square inches per second.}$$

In contrast, when the radius of the circle is 20 inches, the area of the circle is increasing at the faster rate of

$$\left.\frac{dA}{dt}\right|_{r=20} = 2\pi(20)(3) = 120\pi \text{ square inches per second.}$$

Notice that although the rate of change of the radius of the circle is constant, the rate at which the area is changing is not: As the circle gets larger and larger, changes to the radius create larger and larger changes to the area. We could also determine the rate of change of the area at a particular time, say, after $t = 3$ seconds. Since $r(t) = 2 + 3t$ in our example, at time $t = 3$ seconds, we have a radius of $r(3) = 2 + 3(3) = 11$ inches. The rate of change of the area at this moment is

$$\left.\frac{dA}{dt}\right|_{t=3} = \left.\frac{dA}{dt}\right|_{r=11} = 2\pi(11)(3) = 66\pi \text{ square inches per second.}$$

Most related-rates problems involve two rates, one of which is known and one of which you are asked to find. These rates will be related in some way that is determined by the way the corresponding quantities are related. Translating a related-rates problem should result in an equation that relates the two quantities whose rates you are interested in. You can then implicitly differentiate the equation relating the quantities to get an equation relating the rates, as we did at the outset of this section.

Volumes and Surface Areas of Geometric Objects

Many related-rates problems involve geometric quantities such as volume, area, and surface area. Many of the formulas for these quantities should already be familiar to you. For example, here are some formulas for two-dimensional objects: A circle with radius r has area $A = \pi r^2$ and circumference $C = 2\pi r$, a rectangle with length l and width w has area $A = lw$ and perimeter $P = 2l + 2w$, and any triangle with base b and height h has area $A = \frac{1}{2}bh$.

We now gather some three-dimensional geometric formulas for reference, although we will not have the tools to prove them until much later in the book.

THEOREM 3.12

Volume and Surface Area Formulas

The formulas that follow describe the volume V and surface area S of a rectangular box, sphere, right circular cylinder, and right circular cone. The lateral (side) surface area L is also given for the cylinder and cone.

(a) The volume and surface area of a ***rectangular box*** of length x, width y, and height z are

$$V = xyz$$
$$S = 2xy + 2yz + 2xz$$

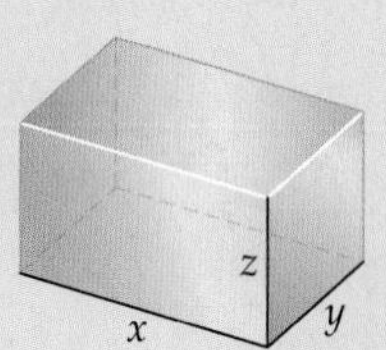

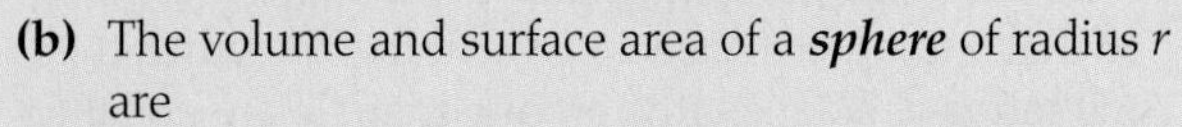

(b) The volume and surface area of a ***sphere*** of radius r are

$$V = \frac{4}{3}\pi r^3$$
$$S = 4\pi r^2$$

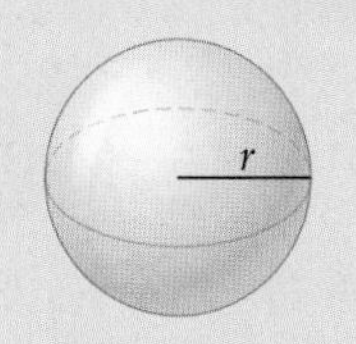

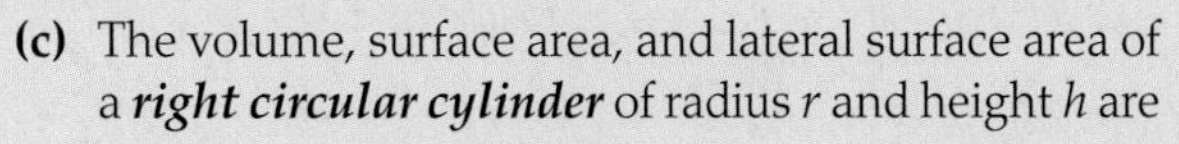

(c) The volume, surface area, and lateral surface area of a ***right circular cylinder*** of radius r and height h are

$$V = \pi r^2 h$$
$$S = 2\pi rh + 2\pi r^2$$
$$L = 2\pi rh$$

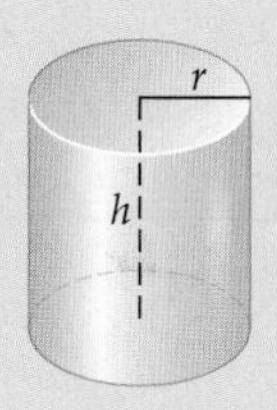

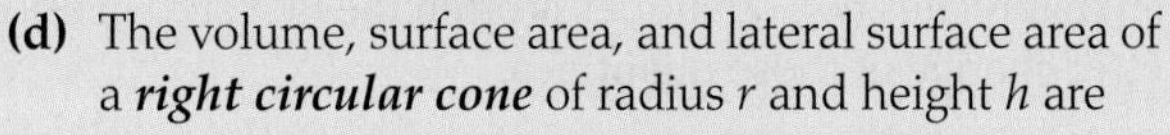

(d) The volume, surface area, and lateral surface area of a ***right circular cone*** of radius r and height h are

$$V = \frac{1}{3}\pi r^2 h$$
$$S = \pi r\sqrt{r^2 + h^2} + \pi r^2$$
$$L = \pi r\sqrt{r^2 + h^2}$$

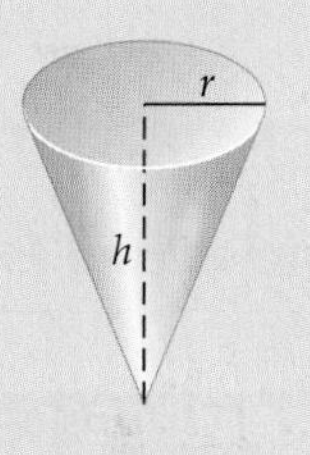

As always, whenever possible you should *think* rather than memorize. For example, let's walk through why the volume and surface area formulas for the cylinder make sense. The volume of a cylinder is the product of the area πr^2 of its top and its height h. The surface area of a cylinder is the sum of two things: the area of its curvy side and the areas of its top and bottom circles. Notice that if we unrolled the curvy side, it would be a rectangle whose width is the circumference $2\pi r$ of the top and bottom circles and whose height is h. Therefore the lateral, or side, surface area of the cylinder is the product $2\pi rh$. Of course, the area of the top and bottom circles each have area πr^2, so the top and bottom together have area $2\pi r^2$. Thus the total surface area of the cylinder is $2\pi rh + 2\pi r^2$.

Similar Triangles

It is also common for related-rates word problems to involve the following two well-known theorems concerning right triangles, which we present here for reference, without proof:

THEOREM 3.13 **Two Theorems About Right Triangles**

The following two theorems describe well-known relationships between the side lengths of right triangles:

(a) The ***Pythagorean theorem*** states that if a right triangle has legs of lengths a and b and hypotenuse of length c, then:

$$a^2 + b^2 = c^2$$

(b) The ***law of similar triangles*** states that if two right triangles have the same three angle measures, so that one is just a scaled-up version of the other, then the ratios of side lengths on one triangle are equal to the ratios of corresponding side lengths on the other. Specifically, with the side lengths shown in the diagram at the right, we have

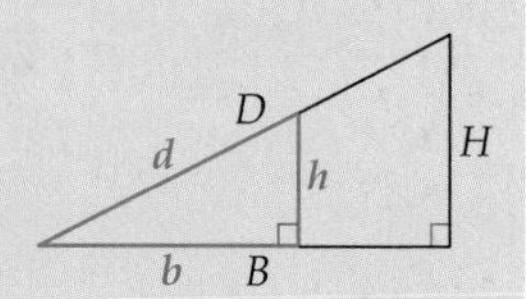

$$\frac{h}{b} = \frac{H}{B}, \qquad \frac{d}{b} = \frac{D}{B}, \qquad \frac{d}{h} = \frac{D}{H}$$

The reason these theorems about triangles arise in related-rates problems is that both theorems give us ways to relate quantities that might change together over time. Finding an equation that relates two quantities is often the first step in finding an equation that relates the rates of change of those quantities.

Examples and Explorations

EXAMPLE 1 **Relating quantities and rates**

In each part that follows, write down an equation that relates the two given quantities. Then use implicit differentiation to obtain a relationship between the rates at which the following quantities change over time:

(a) the circumference C and the area A of a circle;

(b) the surface area S and the radius r of a cylinder with a fixed height of 4 units;

(c) the lengths a and b of the legs of a right triangle with hypotenuse of fixed length 7 units.

SOLUTION

(a) We know that a circle of radius r has circumference $C = 2\pi r$ and area $A = \pi r^2$. We need to find an equation relating C and A. Since $C = 2\pi r$, we have $r = \frac{C}{2\pi}$; substituting the right-hand side into $A = \pi r^2$ gives

$$A = \pi \left(\frac{C}{2\pi}\right)^2 = \frac{1}{4\pi} C^2.$$

Now suppose that the circle is expanding or contracting, so that its area $A = A(t)$ and circumference $C = C(t)$ are changing over time. By implicitly differentiating the preceding equation with respect to t we have

$$\frac{dA}{dt} = \frac{1}{4\pi} 2C \frac{dC}{dt} = \frac{1}{2\pi} C \frac{dC}{dt}.$$

(b) The formula for the surface area of a cylinder with radius r and height 4 is

$$S = 2\pi r(4) + 2\pi r^2 = 8\pi r + 2\pi r^2.$$

By differentiating both sides we can express the relationship between the radius $r = r(t)$ and surface area $S = S(t)$ if the cylinder changes size over time:

$$\frac{d(S)}{dt} = 8\pi \frac{dr}{dt} + 2\pi\, 2r \frac{dr}{dt} = (8\pi + 4\pi r)\frac{dr}{dt}.$$

(c) By the Pythagorean theorem, if a right triangle has legs of length a and b and hypotenuse of length 7, then

$$a^2 + b^2 = 7^2.$$

If the triangle is changing shape or size over time in such a way that the hypotenuse remains 7 units in length, then by implicit differentiation the rates of change of the leg lengths $a = a(t)$ and $b = b(t)$ are related as follows:

$$2a \frac{da}{dt} + 2b \frac{db}{dt} = 0.$$ □

EXAMPLE 2

Relating the changing radius and area of an expanding circle

Suppose a rock dropped into a pond causes a circular wavefront of ripples whose radius increases at 3 inches per second. How fast is the area of the circle of ripples expanding at the instant that the circle has a radius of 12 inches?

SOLUTION

Like many related-rates problems, this situation involves (1) two quantities that are related and (2) known information about the rate of change of one of these quantities; we are then asked to find information about the rate of change of the other quantity. In this case the related quantities are the radius $r = r(t)$ and area $A = A(t)$ of the circle, which are related by the formula $A = \pi r^2$. Both of these quantities change over time as the circle expands. We are given that the radius changes constantly at a rate of $\frac{dr}{dt} = 3$ inches per second and asked to find the rate of change $\frac{dA}{dt}$ of the area at a particular moment. This information is summarized as follows:

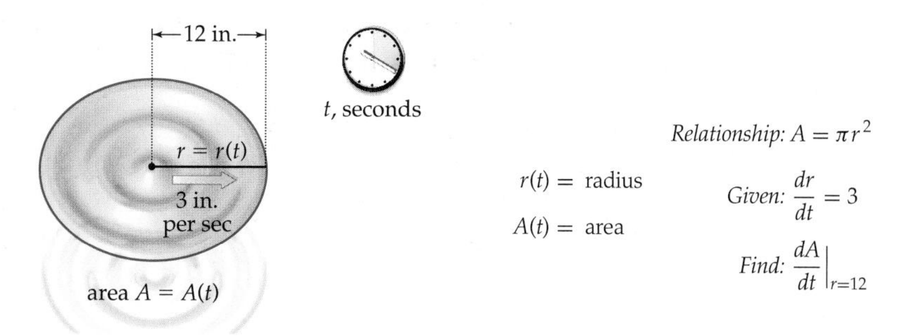

At this point we are done translating the original real–world problem into a mathematical related-rates problem that we know how to solve. To get a formula relating the rates of change of $r(t)$ and $A(t)$ we differentiate both sides of the area formula, yielding $\frac{dA}{dt} = 2\pi r \frac{dr}{dt}$. By evaluating this equation for $\frac{dA}{dt}$ at $r = 12$ and using the given information that $\frac{dr}{dt} = 3$, we see that at the moment that the circle has a 12-inch radius, it is expanding at a rate of

$$\left.\frac{dA}{dt}\right|_{r=12} = 2\pi(12)\frac{dr}{dt} = 2\pi(12)(3) = 72\pi \text{ square inches per second.} \qquad \square$$

EXAMPLE 3 **Relating the changing volume and radius of an inflating balloon**

Suppose a pink spherical party balloon is being inflated at a constant rate of 44 cubic inches per second.

(a) How fast is the radius of the balloon increasing at the instant that the balloon has a radius of 4 inches?

(b) How fast is the radius of the balloon increasing at the instant that the balloon contains 100 cubic inches of air?

SOLUTION

This is a related-rates problem because it involves two rates, namely, the rate at which the balloon is being inflated and the rate of change of the radius of the balloon. We know something about the first rate and wish to say something about the second. Suppose $r = r(t)$ is the radius of the balloon in inches after t seconds and $V = V(t)$ is the volume of the balloon in cubic inches after t seconds. The quantities r and V are related by the volume equation $V = \frac{4}{3}\pi r^3$. We are given that the rate of change of the volume is constantly $\frac{dV}{dt} = 44$, and we want to find the rate of change $\frac{dr}{dt}$ of the radius when $r = 4$ and when $V = 100$. The following diagram summarizes this translation of the problem into mathematical notation:

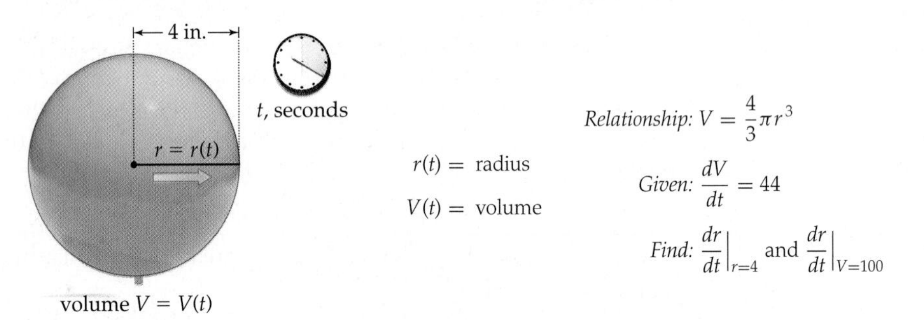

To get an equation relating the rates $\frac{dV}{dt}$ and $\frac{dr}{dt}$ we differentiate both sides of the equation that relates V and r with respect to t:

$$\frac{dV}{dt} = \frac{4}{3}\pi\, 3r^2\frac{dr}{dt} = 4\pi r^2\frac{dr}{dt}.$$

We know from the given information that $\frac{dV}{dt}$ is always equal to 44. Therefore we have the following formula concerning the rate of change of the radius of the balloon:

$$44 = 4\pi r^2\frac{dr}{dt}.$$

(a) To find the rate of change of the radius at the instant that the ballon has a 4-inch radius, we evaluate the preceding equation at $r = 4$ and then solve for $\left.\frac{dr}{dt}\right|_{r=4}$:

$$44 = 4\pi(4^2)\left.\frac{dr}{dt}\right|_{r=4} \implies \left.\frac{dr}{dt}\right|_{r=4} = \frac{44}{4\pi(16)} \approx 0.2188 \text{ inch per second.}$$

(b) For the second part we must find the rate at which the radius is changing at the moment that the volume of the balloon is 100 cubic inches. Our formula for the rate of change of the radius of the balloon depends on r, not V, so we must first use the equation $V = \frac{4}{3}\pi r^3$ to determine the value of r when $V = 100$:

$$100 = \frac{4}{3}\pi r^3 \quad \Longrightarrow \quad r^3 = \frac{100}{(4/3)\pi} \quad \Longrightarrow \quad r = \left(\frac{100}{(4/3)\pi}\right)^{1/3} \approx 2.879.$$

Using our previously developed formula, we find that when the volume of the balloon is $V = 100$ cubic inches and thus the radius is approximately $r \approx 2.879$ inches, the rate of change of the radius of the balloon is

$$44 = 4\pi(2.879)^2 \left.\frac{dr}{dt}\right|_{V=100} \quad \Longrightarrow \quad \left.\frac{dr}{dt}\right|_{V=100} = \frac{44}{4\pi(2.879)^2} \approx 0.422 \text{ inch per second.} \quad \square$$

EXAMPLE 4 **The shadow of a person walking away from a streetlight**

Matt is 6 feet tall and is walking away from a 10-foot streetlight at a rate of 3 feet per second. As he walks away from the streetlight, his shadow gets longer. How fast is the length of Matt's shadow increasing when he is 8 feet from the streetlight?

SOLUTION

We are given the rate at which Matt walks away from the streetlight, and we wish to find the rate of change of the length of Matt's shadow. To find a relationship between these two rates we will find a relationship between their underlying variables: the distance s between Matt and the streetlight and the length l of Matt's shadow. By the law of similar triangles, s and l are related by the equation $\frac{10}{s+l} = \frac{6}{l}$, as shown in the following diagram:

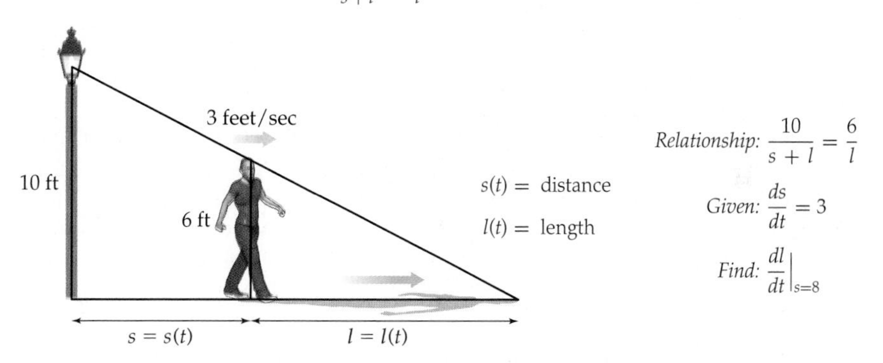

To find the relationship between $\frac{ds}{dt}$ and $\frac{dl}{dt}$ we must implicitly differentiate the equation relating s and l. We will simplify the equation first to make our lives easier:

$$\frac{10}{s+l} = \frac{6}{l} \quad \Longrightarrow \quad 10l = 6(s+l) \quad \Longrightarrow \quad 4l = 6s.$$

Differentiating both sides and then using the fact that $\frac{ds}{dt} = 3$ gives us

$$4\frac{dl}{dt} = 6\frac{ds}{dt} \quad \Longrightarrow \quad 4\frac{dl}{dt} = 6(3) \quad \Longrightarrow \quad \frac{dl}{dt} = 4.5.$$

Interestingly, we have just discovered that Matt's shadow is increasing at a *constant* rate of 4.5 feet per second. In particular, when Matt is 8 feet from the streetlight, the length of his shadow is increasing at a rate of $\left.\frac{dl}{dt}\right|_{s=8} = 4.5$ feet per second. $\square$

? TEST YOUR UNDERSTANDING

- Suppose $r = r(t)$ is a function of t. Why is $\frac{d}{dt}(r^3)$ not equal to $3r^2$?
- What are the formulas for the volumes of spheres, cylinders, and cones? What about the formulas for surface area?
- What do we mean when we say that two triangles are similar? What does the law of similar triangles say?
- In Example 2, why did we not label the diagram of the circle with the number 12? Why did we use the variable r instead?
- In Example 3 the radius of the balloon increases at a faster rate when the balloon is smaller. Why does this make sense?

EXERCISES 3.5

Thinking Back

Using the chain rule: Given that $r = r(t)$, $s = s(t)$, and $u = u(t)$ are functions of t and that c and k are constants, find each of the following derivatives.

- $\frac{d}{dt}(\pi u^2)$
- $\frac{d}{dt}(3r + 2s)$
- $\frac{d}{dt}(cu + rs)$
- $\frac{d}{dt}(k + cu^3)$
- $\frac{d}{dt}(cr^2 u)$
- $\frac{d}{dt}\left(\frac{c+s}{k+u}\right)$

Evaluation in Leibniz notation: Given that $r = r(t)$, $s = s(t)$, and $u = u(t)$ are functions of t, answer each of the following.

- If $\frac{ds}{dt} = 3s^2 - 4$, find $\left.\frac{ds}{dt}\right|_{s=2}$.
- If $r^2\frac{dr}{dt} - 2r = 0$, find $\left.\frac{dr}{dt}\right|_{r=3}$.
- If $4 = 2u\frac{du}{dt}$ and $u = 2 + 3t$, find $\left.\frac{du}{dt}\right|_{t=4}$.

Concepts

0. *Problem Zero:* Read the section and make your own summary of the material.

1. *True/False:* Determine whether each of the statements that follow is true or false. If a statement is true, explain why. If a statement is false, provide a counterexample.

(a) *True or False:* If a square grows larger, so that its side length increases at a constant rate, then its area will also increase at a constant rate.

(b) *True or False:* If a square grows larger, so that its side length increases at a constant rate, then its perimeter will also increase at a constant rate.

(c) *True or False:* If a circle grows larger, so that its radius increases at a constant rate, then its circumference will also increase at a constant rate.

(d) *True or False:* If a sphere grows larger, so that its radius increases at a constant rate, then its volume will also increase at a constant rate.

(e) *True or False:* The volume of a right circular cone is one-third of the volume of the right circular cylinder with the same radius and height.

(f) *True or False:* If $V(r)$ is the volume of a sphere as a function of its radius, and $S(r)$ is the surface area of a sphere as a function of its radius, then $V'(r) = S(r)$.

(g) *True or False:* If you unroll the side of a right circular cylinder with radius r and height h, you get a flat rectangle with height h and width $2\pi r$.

(h) *True or False:* Given a right triangle with side lengths a and c and hypotenuse of length b, we must have $a^2 + b^2 = c^2$.

2. *Examples:* Construct examples of the thing(s) described in the following. Try to find examples that are different than any in the reading.

(a) Two triangles that are similar to the right triangle with legs of length 1 and hypotenuse $\sqrt{2}$.

(b) Two triangles with a hypotenuse of length 5.

(c) Two cylinders with a volume of 100 cubic units.

3. Give formulas for the volume and the surface area of a cylinder with radius y and height s.

4. Give formulas for the volume and the surface area of a cone with radius u and height w.

5. Give formulas for the volume and the surface area of a cylinder whose radius r is half of its height h.

6. Give formulas for the volume and the surface area of a cone whose height h is three times its radius r.

7. State the Pythagorean theorem and give an example of a triangle that illustrates the theorem.

8. State the law of similar triangles and give an example of a pair of triangles that illustrate this law.

9. If the volume and radius of a sphere are functions of time, what is the relationship between the rate of change of the volume of the sphere and the rate of change of the radius of the sphere?

10. If the volume and radius of a cone are functions of time, what is the relationship between the rate of change of the volume of the cone and the rate of change of the radius of the cone?

11. If the volume and equator of a sphere are functions of time, what is the relationship between the rate of change of the volume of the sphere and the rate of change of the equator of the sphere?

12. Suppose the side lengths x, y, and z of a rectangular box are each functions of time.
    (a) How is the rate of change of the volume of the box related to the rates of change of x, y, and z?
    (b) How is the rate of change of the surface area of the box related to the rates of change of x, y, and z?

13. Suppose the radius r, volume V, and surface area S of a sphere are functions of time t.
    (a) How are $\frac{dV}{dt}$ and $\frac{dr}{dt}$ related?
    (b) How are $\frac{dS}{dt}$ and $\frac{dr}{dt}$ related?

14. Suppose the radius r, volume V, and surface area S of a sphere are functions of time t. How are $\frac{dV}{dt}$ and $\frac{dS}{dt}$ related?

15. Suppose the radius r, height h, and volume V of a cylinder are functions of time t. How is $\frac{dV}{dt}$ related to $\frac{dr}{dt}$ if the height of the cylinder is constant?

16. Suppose the radius r, height h, and volume V of a cylinder are functions of time t. How is $\frac{dV}{dt}$ related to $\frac{dh}{dt}$ if the radius of the cylinder is constant?

17. Suppose the radius r, height h, and volume V of a cylinder are functions of time t, and further suppose that the height of the cylinder is always twice its radius. Write $\frac{dV}{dt}$ in terms of h and $\frac{dh}{dt}$.

18. Suppose the radius r, height h, and volume V of a cylinder are functions of time t, and further suppose that the volume of the cylinder is always constant. Write $\frac{dr}{dt}$ in terms of r, h, and $\frac{dh}{dt}$.

Skills

In Exercises 19–26, write down an equation that relates the two quantities described. Then use implicit differentiation to obtain a relationship between the rates at which the quantities change over time.

19. The area A and perimeter P of a square.

20. The area A and perimeter P of an equilateral triangle.

21. The surface area S and height h of a cylinder with a fixed radius of 2 units.

22. The volume V and radius r of a cylinder with a fixed height of 10 units.

23. The surface area S and radius r of a cone with a fixed height of 5 units.

24. The volume V and height h of a cone with a fixed radius of 3 units.

25. The area A and hypotenuse c of an isosceles right triangle.

26. The area A and hypotenuse c of a triangle that is similar to a right triangle with legs of lengths 3 and 4 units and hypotenuse of length 5 units.

Given that $u = u(t)$, $v = v(t)$, and $w = w(t)$ are functions of t and that k is a constant, calculate the derivative $\frac{df}{dt}$ of each function $f(t)$ in Exercises 27–36. Your answers may involve u, v, w, $\frac{du}{dt}$, $\frac{dv}{dt}$, $\frac{dw}{dt}$, k, and/or t.

27. $f(t) = u^2 + kv$

28. $f(t) = u + v + w$

29. $f(t) = tv + kv$

30. $f(t) = kuvw$

31. $f(t) = 2v\sqrt{u + w}$

32. $f(t) = 3u^2v + vt$

33. $f(t) = w(u + t)^2$

34. $f(t) = \frac{w}{uv}$

35. $f(t) = \frac{ut + w}{k}$

36. $f(t) = \frac{k}{u^2w}$

Applications

37. A rock dropped into a pond causes a circular wave of ripples whose radius increases at 4 inches per second. How fast is the area of the circle of ripples expanding at the instant that the radius of the circle is 12 inches? 24 inches? 100 inches? Explain why it makes sense that the rate of change of the area increases as the radius increases.

38. A rock dropped into a pond causes a circular wave of ripples whose radius increases at 6 inches per second. How fast is the area of the circle of ripples expanding at the instant that the area of the circle is 100 square inches? 200 square inches? 1000 square inches? Explain why it makes sense that the rate of change of the area increases as the area increases.

In Exercises 39–42, suppose the sides of a cube are expanding at a rate of 2 inches per minute.

39. How fast is the volume of the cube changing at the moment that the cube has a side length of 8 inches?

40. How fast is the volume of the cube changing at the moment that the cube has a side length of 20 inches?

41. How fast is the volume of the cube changing at the moment that the cube's volume is 55 cubic inches?

42. How fast is the volume of the cube changing at the moment that the area of the cube's base is 10 square inches?

In Exercises 43–45, consider a large helium balloon that is being inflated at the rate of 120 cubic inches per second.

43. How fast is the radius of the balloon increasing at the instant that the balloon has a radius of 12 inches?

44. How fast is the radius of the balloon increasing at the instant that the balloon contains 300 cubic inches of air?

45. How fast is the surface area of the balloon increasing at the instant that the radius of the balloon is 15 inches?

In Exercises 46–48, suppose that Stuart is 6 feet tall and is walking towards a 20-foot streetlight at a rate of 4 feet per second. As he walks towards the streetlight, his shadow gets shorter.

46. How fast is the length of Stuart's shadow changing? Does it depend on how far Stuart is from the streetlight?

47. How fast is the tip of Stuart's shadow moving? Does it depend on how far Stuart is from the streetlight?

48. How fast is the area of the triangle made up of Stuart's legs and his shadow changing? Is it increasing or decreasing as Stuart walks towards the streetlight?

In Exercises 49–51, Alina props a 12-foot ladder against the side of her house so that she can sneak into her upstairs bedroom window. Unfortunately, the ground is muddy because of a recent rainstorm, and the base of the ladder slides away from the house at a rate of half a foot per second.

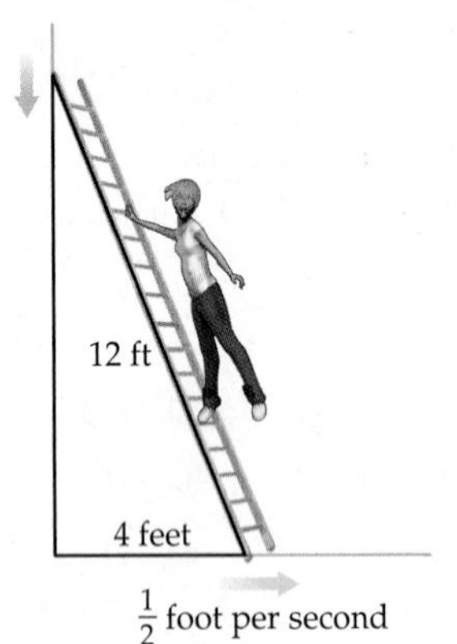

49. How fast is the top of the ladder moving down the side of the house when the base of the ladder is 4 feet from the house?

50. How fast is the top of the ladder moving down the side of the house when the base of the ladder is 10 feet from the house?

51. How fast is the area of the triangle formed by the ladder, the house, and the ground changing when the top of the ladder is 6 feet from the ground?

In Exercises 52–55, Linda is bored and decides to pour an entire container of salt into a pile on the kitchen floor. She pours 3 cubic inches of salt per second into a conical pile whose height is always two-thirds of its radius.

52. How fast is the radius of the conical salt pile changing when the radius of the pile is 2 inches?

53. How fast is the radius of the conical salt pile changing when the height of the pile is 4 inches?

54. How fast is the height of the conical salt pile changing when the radius of the pile is 2 inches?

55. How fast is the height of the conical salt pile changing when the height of the pile is 4 inches?

56. Linda is still bored and is now pouring sugar onto the floor. The poured-out sugar forms a conical pile whose height is three-quarters of its radius and whose height is growing at a rate of 1.5 inches per second. How fast is Linda pouring the sugar at the instant that the pile of sugar is 3 inches high?

57. Riley is holding an ice cream cone on a hot summer day. As usual, the cone has a small hole at the bottom, and ice cream is melting and dripping through the hole at a rate of half a cubic inch per minute. The cone has a radius of 2 inches and a height of 5 inches. How fast is the height of the ice cream changing when the height of the ice cream in the cone is 3 inches?

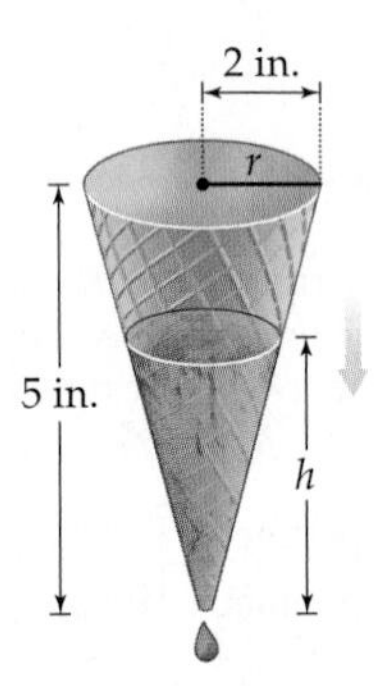

58. Suppose the width w of a rectangle is decreasing at a rate of 3 inches per second while the height h of the rectangle is increasing at a rate of 3 inches per second. The rectangle initially has a width of 100 inches and a height of 75 inches.

(a) Find the rate of change of the area of the rectangle in terms of its width and height.

(b) On what intervals do the variables w and h make sense in this problem? On what time interval does the problem make sense?

(c) When will the area of the rectangle be increasing, and when will it be decreasing? Answer these questions both in terms of the width and height of the rectangle and in terms of time.

59. Suppose the length a of one leg of a right triangle is increasing at a rate of 4 inches per second while the length b of its other leg is decreasing at a rate of 2 inches per second. The triangle initially has legs of width $a = 1$ inch and $b = 10$ inches.

(a) Find the rate of change of the area of the triangle over time, in terms of its width and height.

(b) On what intervals do the variables a and b make sense in this problem? On what time interval does the problem make sense?

(c) When will the area of the triangle be increasing, and when will it be decreasing? Answer these questions both in terms of the width and height of the triangle and in terms of time.

60. Annie is paddling her kayak through the San Juan Islands and is a quarter of a mile away from where she wants to cross a channel. She sees a ferry in the channel approaching fast from her left, about 2 miles away. The ferry travels at about 20 mph, while Annie can do about 3 mph if she jams.

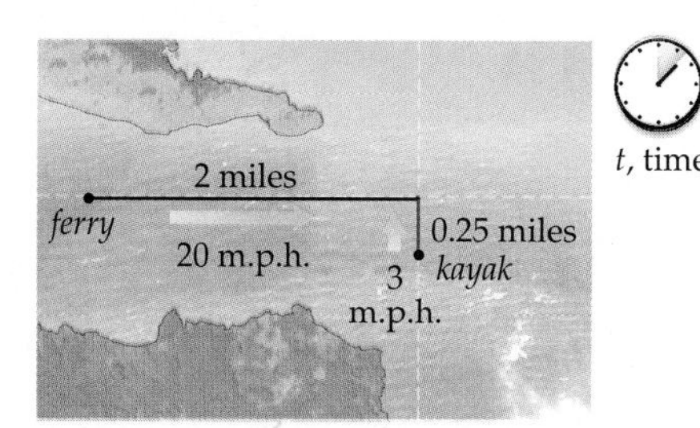

(a) To set up a model for this problem, suppose Annie is travelling on the x-axis and is approaching the origin from the right. Suppose also that the ferry is travelling on the y-axis and is approaching the origin from above. Then $x = x(t)$ represents Annie's position at time t and $y = y(t)$ represents the ferry's position. Given what you know about this problem, what are $x'(t)$ and $y'(t)$?

(b) Construct an equation in terms of $x = x(t)$ and $y = y(t)$ that describes the distance between Annie and the ferry at time t.

(c) Use implicit differentiation to determine how fast the distance between Annie and the ferry is decreasing when she first sees the ferry.

(d) If Annie decides to jam across the channel, will the ferry hit her?

61. The sun goes down at a rate of about 11 degrees per hour in Colorado in the middle of summer. Ian finds himself contemplating this fact one evening while sitting at Chasm Lake, below Long's Peak in Colorado, watching the sun descend behind the peak. The point on the ridge where the sun is descending is at 13,200 feet. The lake is at 11,710 feet. Ian is sitting 3,100 horizontal feet from the ridge.

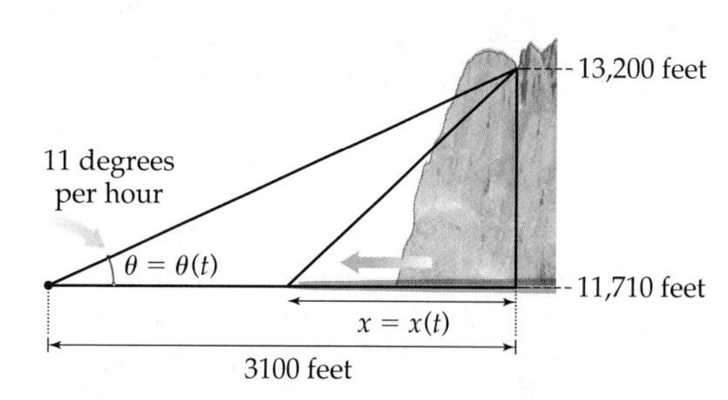

(a) Suppose $x = x(t)$ is the distance of the edge of the shadow from the ridge at time t. This distance is related to the angle $\theta = \theta(t)$ shown in the figure. Find a formula for the speed $x'(t)$ of the shadow.

(b) Use the model from part (a) to determine how fast the shadow of the ridge is moving when it reaches Ian.

Proofs

62. Prove that the lateral surface area of a right circular cone is equal to πrl, where r is the radius of the cone and l is the length of the diagonal of the cone, that is, the distance from the vertex of the cone to a point on its circumference.

63. Prove that the rate of change of the volume of a sphere with respect to its radius r is equal to the surface area of the sphere. Why does it make geometric sense that the surface area would be related to this rate of change?

64. Prove that the rate of change of the volume of a cylinder with fixed height with respect to its radius r is equal to the lateral surface area of the cylinder. Why does it make geometric sense that the lateral surface area would be related to this rate of change?

65. Suppose a right triangle has legs of lengths a and b and a hypotenuse of length h and that this triangle is changing size, so that the length of its hypotenuse does not change. Prove that the ratio of the rates of change $\frac{da}{dt}$ and $\frac{db}{dt}$ is equal to $-\frac{b}{a}$.

Thinking Forward

Parametric curves: Imagine the curve traced in the xy-plane by the coordinates $(x, y) = (3z + 1, z^2 - 4)$ as z varies, where the parameter z is a function of time t.

- ▶ Plot the points (x, y) in the plane that correspond to $z = -3, -2, -1, 0, 1, 2,$ and 3.
- ▶ If the parameter z moves at 3 units per second and $z = 0$ when $t = 0$, plot the points (x, y) in the plane that correspond to $t = 0, 1, 2, 3,$ and 4.
- ▶ If the parameter z moves at 5 units per second, find the instantaneous rate of change of the x- and y-coordinates as the curve passes through the point $(7, 0)$.
- ▶ If the x-coordinate moves at 5 units per second, find the instantaneous rate of change of the y-coordinate as the curve passes through the point $(7, 0)$.

3.6 L'HÔPITAL'S RULE

- L'Hôpital's rule for calculating limits of the indeterminate forms $\frac{0}{0}$ and $\frac{\infty}{\infty}$
- Rewriting limits of the indeterminate form $0 \cdot \infty$ so that L'Hôpital's rule applies
- Using logarithms to calculate limits of the indeterminate forms 0^0, 1^∞, and ∞^0

Geometrical Motivation for L'Hôpital's Rule

As we have already seen, limits of the form $\frac{0}{0}$ are indeterminate. At first glance it is not clear whether such a limit exists or what it might be equal to. In some cases we can resolve the indeterminate form $\frac{0}{0}$ with some algebra, such as in this example:

$$\lim_{x\to 0} \frac{x^2}{x^3 - x} = \lim_{x\to 0} \frac{x}{x^2 - 1} = \frac{0}{0^2 - 1} = 0.$$

Other limits of the indeterminate form $\frac{0}{0}$ are not so easy to simplify. In particular, limits of quotients that involve a mixture of different types of functions are usually more resistant to algebra. For example, consider the limit

$$\lim_{x\to 0} \frac{x^2}{2^x - 1}.$$

As $x \to 0$ we have $x^2 \to 0$ and $2^x - 1 \to 1 - 1 = 0$, and thus this limit is of the indeterminate form $\frac{0}{0}$. This indeterminate limit cannot be simplified with algebra. So what can we do?

Let's approach the problem graphically. The graphs of $f(x) = x^2$ and $g(x) = 2^x - 1$ are shown next at the left. Since we are interested in a limit as $x \to 0$, we should focus on what happens as we look at smaller and smaller graphing windows around $x = 0$, as shown in the second and third graphs.

$f(x) = x^2$ and $g(x) = 2^x - 1$

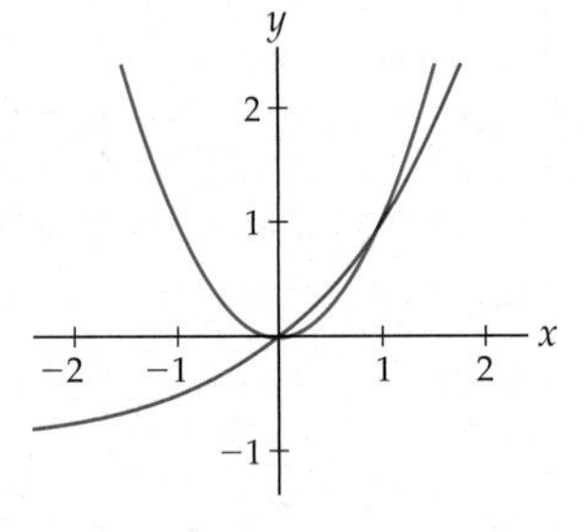

Same graph but in smaller window

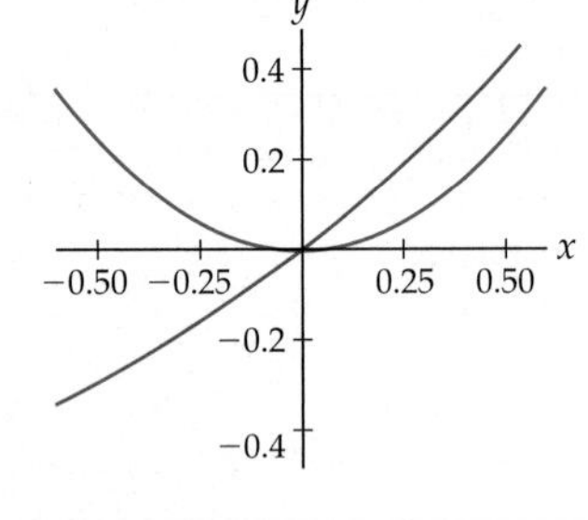

Graphs are almost linear here

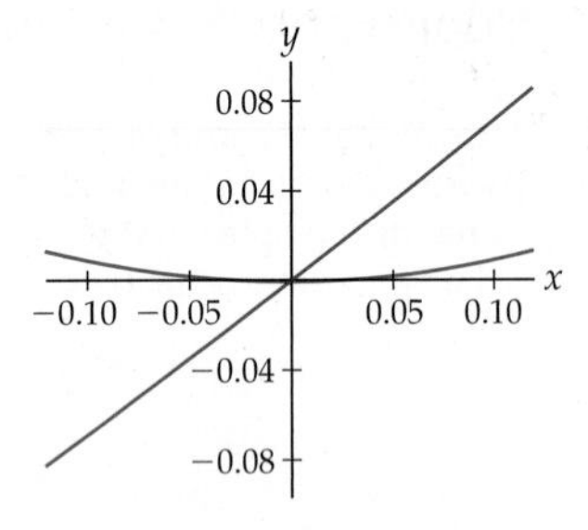

Near $x = 0$, the graph of $f(x) = x^2$ looks a lot like its horizontal tangent line $y = 0$, and the graph of $g(x) = 2^x - 1$ looks a lot like its tangent line $y = x$. Thus we would anticipate the behavior of the quotient $\frac{f(x)}{g(x)} = \frac{x^2}{2^x - 1}$ as $x \to 0$ to be similar to that of the quotient $\frac{0}{x}$ of the corresponding tangent lines at $x = 0$. From this information it would be reasonable to guess that $\lim_{x\to 0} \frac{x^2}{2^x - 1} = 0$. As we are about to see, that is in fact the case. Indeed, we will soon see that, in general, limits of the indeterminate form $\frac{0}{0}$ are related to the limit of the quotient of the slopes, or derivatives, of those numerator and denominator functions.

We can also use tangent lines to examine limits of the indeterminate form $\frac{\infty}{\infty}$. For example, consider the limit

$$\lim_{x\to\infty} \frac{x^2}{2^x - 1}.$$

This is the same quotient $\frac{x^2}{2^x-1}$ of functions as before, but with a limit as $x \to \infty$ instead of as $x \to 0$. As $x \to \infty$ we have $x^2 \to \infty$ and $2^x - 1 \to \infty$, and therefore this limit is of the indeterminate form $\frac{\infty}{\infty}$.

Again, we cannot simplify this expression with algebra. Let's examine what happens as we look at larger graphing windows to see the behavior of $f(x) = x^2$ and $g(x) = 2^x - 1$ as $x \to \infty$:

$f(x) = x^2$ and $g(x) = 2^x - 1$

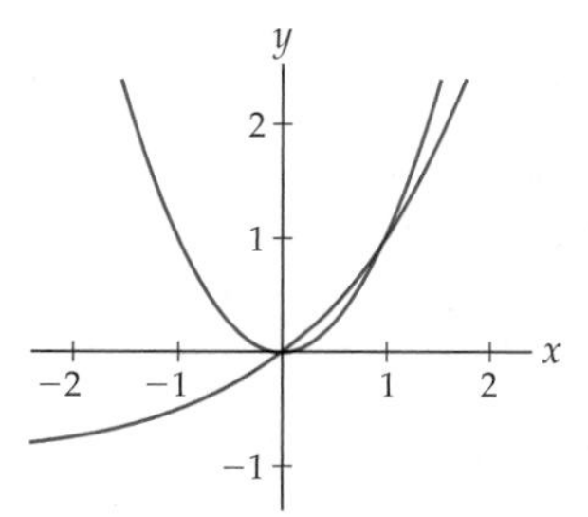

Same graph but in larger window

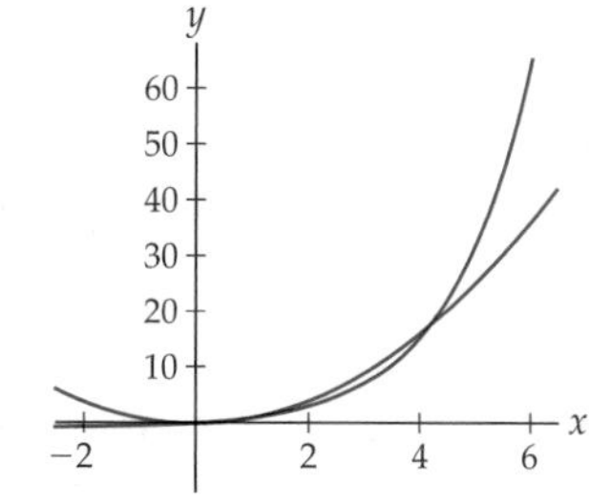

Graphs are much different out here

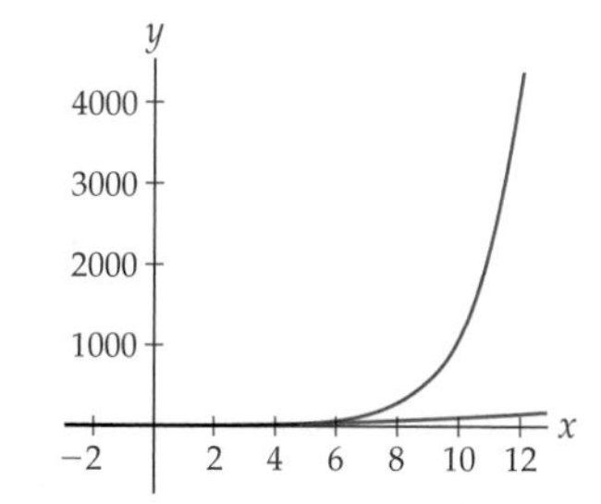

From the rightmost graph we can see that the heights on the graph of $g(x) = 2^x - 1$ are in some sense approaching ∞ in a fundamentally faster way than the heights on the graph of $f(x) = x^2$. Specifically, the slopes of $g(x) = 2^x - 1$ are significantly steeper than the slopes of $f(x) = x^2$ for large values of x. Accordingly, we would anticipate that as $x \to \infty$ the values of $2^x - 1$ would win the race to ∞ and that the limit $\lim_{x\to\infty} \frac{x^2}{2^x-1}$ would therefore be equal to zero. Once again a ratio of slopes has helped us to guess the value of the limit of a quotient. This geometric intuition is the basis for the powerful limit technique known as ***L'Hôpital's rule***.

L'Hôpital's Rule for the Indeterminate Forms $\frac{0}{0}$ and $\frac{\infty}{\infty}$

The ideas we have developed regarding limits of quotients suggest the theorem that follows. This key theorem will allow us to solve some limits of the indeterminate form $\frac{0}{0}$ or $\frac{\infty}{\infty}$ by relating them to the quotients of the corresponding derivatives:

THEOREM 3.14 **L'Hôpital's Rule**

Suppose f and g are differentiable functions on some punctured interval around $x = c$ on which $g(x)$ is nonzero. If $\lim_{x\to c} \frac{f(x)}{g(x)}$ is of the indeterminate form $\frac{0}{0}$ or $\frac{\infty}{\infty}$, then

$$\lim_{x\to c} \frac{f(x)}{g(x)} = \lim_{x\to c} \frac{f'(x)}{g'(x)},$$

as long as the second limit exists or is infinite.

The conclusion holds also if $x \to \infty$ (or $x \to -\infty$), as long as f and g are differentiable on some interval (N, ∞) (or $(-\infty, N)$) on which $g(x)$ is nonzero.

There are a few technical points to notice about the hypotheses of this theorem. First, notice that the functions f and g must be differentiable with $g(x)$ nonzero for x-values near, but not necessarily at, $x = c$. Second, notice that the conclusion does not necessarily hold if the limit on the right does not exist for some reason other than being infinite. In practice, most of the functions we look at will satisfy the two conditions we just mentioned. The vital thing to check when applying L'Hôpital's rule is that the limit is of the indeterminate form $\frac{0}{0}$ or $\frac{\infty}{\infty}$. If the limit is not in one of those two indeterminate forms, then L'Hôpital's rule cannot be applied.

The proof of L'Hôpital's rule requires a more general version of the Mean Value Theorem called the ***Cauchy Mean Value Theorem***. This theorem—and the resulting proof of L'Hôpital's rule—are not that difficult, but proper proofs would take us too far afield for our purposes, so we leave that for future mathematics courses. To simplify matters we will prove L'Hôpital's rule only in the special case where f and g are very well behaved and the limit is of the indeterminate form $\frac{0}{0}$:

Proof. We prove the theorem in the special case where f and g are continuous and differentiable on an interval around and including $x = c$, and $f(c) = g(c) = 0$ but $g'(c) \neq 0$. We wish to show that

$$\lim_{x \to c} \frac{f(x)}{g(x)} = \frac{f'(c)}{g'(c)}.$$

We will work backwards from the right-hand side of the equality to the left-hand side. Applying the definition of the derivative to the numerator and the denominator, we have

$$\frac{f'(c)}{g'(c)} = \frac{\lim\limits_{x \to c} \frac{f(x) - f(c)}{x - c}}{\lim\limits_{x \to c} \frac{g(x) - g(c)}{x - c}} = \lim_{x \to c} \frac{f(x) - f(c)}{g(x) - g(c)} = \lim_{x \to c} \frac{f(x) - 0}{g(x) - 0} = \lim_{x \to c} \frac{f(x)}{g(x)}.$$

■

L'Hôpital's rule can be an extremely powerful tool for resolving indeterminate limits of the indeterminate form $\frac{0}{0}$ or $\frac{\infty}{\infty}$. In Example 2 we will use the rule to find the limits that we examined graphically at the start of this section. L'Hôpital's rule can also be useful for resolving other indeterminate forms, provided that we can rewrite them so that they are of the form $\frac{0}{0}$ or $\frac{\infty}{\infty}$; see Example 3.

Using Logarithms for the Indeterminate Forms 0^0, 1^∞, and ∞^0

Recall from Section 1.6 that limits of the form 0^0, 1^∞, and ∞^0 are indeterminate. For example, all three of the following limits are of the indeterminate form 1^∞, but each one of them approaches something different.

$$\lim_{x \to \infty} \left(\frac{x}{x-1}\right)^{x^2} = \infty; \qquad \lim_{x \to 1^+} x^{1/(x^2-1)} = \sqrt{e}; \qquad \lim_{x \to \infty} \left(1 + \frac{1}{x}\right)^x = e.$$

In each of these limits there is a race between how fast the base approaches 1 and how fast the exponent approaches ∞, and in some sense the winner of that race determines the limit. But how can we determine who wins this race?

One difficulty with limits of the indeterminate forms 0^0, 1^∞, and ∞^0 is that such limits involve a variable in *both* the base *and* the exponent. Fortunately, logarithms have the power to change exponentiation into multiplication, in the sense that $\ln(a^b) = b \ln a$. The key to using logarithms to calculate limits is the following theorem:

THEOREM 3.15 **Relating the Limit of a Function to the Limit of Its Logarithm**

(a) If $\lim_{x \to c} \ln(f(x)) = L$, then $\lim_{x \to c} f(x) = e^L$.

(b) If $\lim_{x \to c} \ln(f(x)) = \infty$, then $\lim_{x \to c} f(x) = \infty$.

(c) If $\lim_{x \to c} \ln(f(x)) = -\infty$, then $\lim_{x \to c} f(x) = 0$.

To use this theorem to calculate a limit of the form $\lim_{x \to c} u(x)^{v(x)}$, we consider instead the limit

$$\lim_{x \to c} \ln(u(x)^{v(x)}) = \lim_{x \to c} v(x) \ln(u(x)).$$

Notice that the logarithm allows us to consider the limit of a product rather than a limit involving an exponent. Once we find this limit L, the theorem tells us that the answer to our original limit $\lim_{x \to c} u(x)^{v(x)}$ must be e^L. If instead of L we get $\pm\infty$, our original limit must be equal to $e^{\pm\infty}$.

Proof. We will prove only the first part of the theorem. The proof will follow directly from the composition rule for limits of continuous functions from Section 1.5. Suppose f is a function that is positive as x approaches c, so that $\ln(f(x))$ is defined near $x = c$. For functions f with an exponent and base both involving the variable x, domain restrictions will ensure that this will always be the case. Since the function $f(x) = \ln x$ is continuous on $(0, \infty)$, by the rule for limits of compositions of continuous functions, we have

$$L = \lim_{x \to c} \ln(f(x)) = \ln(\lim_{x \to c} f(x)).$$

Since $L = \ln(A)$ if and only if $A = e^L$, this equation implies that, as desired,

$$\lim_{x \to c} f(x) = e^L. \quad \blacksquare$$

Examples and Explorations

EXAMPLE 1 **Checking to see if L'Hôpital's rule applies**

Determine whether or not L'Hôpital's rule applies to each of the following limits as they are written here (without any preliminary algebra or simplification):

(a) $\lim_{x \to 1} \frac{(x-1)^2}{x-1}$ **(b)** $\lim_{x \to 0} \frac{x^2}{1+2^x}$ **(c)** $\lim_{x \to \infty} (2^x - x^3)$ **(d)** $\lim_{x \to \infty} \frac{2^x}{1-3^x}$

SOLUTION

(a) L'Hôpital's rule does apply, since as $x \to 1$, both the numerator $(x-1)^2$ and the denominator $x - 1$ approach zero, and therefore the limit $\lim_{x \to 1} \frac{(x-1)^2}{x-1}$ is of the indeterminate form $\frac{0}{0}$. The more technical hypotheses of L'Hôpital's rule are also satisfied by the numerator and denominator, as they are in most common examples. It is also possible to use algebra to solve this limit.

(b) L'Hôpital's rule does not apply, since $\frac{x^2}{1+2^x}$ approaches $\frac{0^2}{1+2^0} = \frac{0}{2} = 0$ as $x \to 0$.

(c) L'Hôpital's rule does not apply, since $2^x - x^3$ is not a quotient.

(d) L'Hôpital's rule does apply, since as $x \to \infty$ we have $2^x \to \infty$ and $1 - 3^x \to -\infty$ and therefore this limit is of the indeterminate form $\frac{\infty}{-\infty}$. The negative sign in the denominator could be factored out, but L'Hôpital's rule will work even if we do not extract the negative sign. □

EXAMPLE 2

Applying L'Hôpital's rule

Use L'Hôpital's rule to calculate **(a)** $\lim_{x \to 0} \frac{x^2}{2^x - 1}$ and **(b)** $\lim_{x \to \infty} \frac{x^2}{2^x - 1}$.

SOLUTION

(a) Since the limit $\lim_{x \to 0} \frac{x^2}{2^x - 1}$ is of the indeterminate form $\frac{0}{0}$, L'Hôpital's rule applies and says that we can calculate it by considering instead the limit of the quotient of the derivatives of the numerator and denominator:

$$\lim_{x \to 0} \frac{x^2}{2^x - 1} \stackrel{\text{L'H}}{=} \lim_{x \to 0} \frac{\frac{d}{dx}(x^2)}{\frac{d}{dx}(2^x - 1)} = \lim_{x \to 0} \frac{2x}{(\ln 2)2^x} = \frac{2(0)}{(\ln 2)2^0} = \frac{0}{\ln 2} = 0.$$

Notice that we wrote "L'H" above the equals sign where we applied L'Hôpital's rule, to indicate our reasoning in that step. Notice also that we did not apply the quotient rule to differentiate the quotient $\frac{f(x)}{g(x)}$, because that is not the way that L'Hopital's rule works. Instead, following L'Hôpital's rule, we differentiated the numerator and denominator individually.

(b) The limit $\lim_{x \to \infty} \frac{x^2}{2^x - 1}$ is of the indeterminate form $\frac{\infty}{\infty}$, so L'Hôpital's rule applies. Again we replace the numerator and denominator of our quotient with their derivatives:

$$\lim_{x \to \infty} \frac{x^2}{2^x - 1} \stackrel{\text{L'H}}{=} \lim_{x \to \infty} \frac{\frac{d}{dx}(x^2)}{\frac{d}{dx}(2^x - 1)} = \lim_{x \to \infty} \frac{2x}{(\ln 2)2^x}.$$

Unfortunately, our application of L'Hôpital's rule was not sufficient to resolve the limit here, because if we let $x \to \infty$ we have $2x \to \infty$ and $(\ln 2)2^x \to \infty$, so our limit is still in the indeterminate form $\frac{\infty}{\infty}$. However, this means that we can apply L'Hôpital's rule again!

$$\lim_{x \to \infty} \frac{x^2}{2^x - 1} \stackrel{\text{L'H}}{=} \lim_{x \to \infty} \frac{2x}{(\ln 2)2^x} \stackrel{\text{L'H}}{=} \lim_{x \to \infty} \frac{2}{(\ln 2)(\ln 2)2^x}$$

As $x \to \infty$ the denominator of this limit approaches ∞, while the numerator is equal to 2. Therefore we have

$$\lim_{x \to \infty} \frac{2}{(\ln 2)(\ln 2)2^x} = 0.$$

Notice that both of the preceding answers agree with what we guessed from a graphical analysis at the start of this section. □

EXAMPLE 3

Rewriting limits in the form $\frac{0}{0}$ or $\frac{\infty}{\infty}$ so that L'Hôpital's rule applies

Use L'Hôpital's rule to calculate each of the following limits:

(a) $\lim_{x \to \infty} x^2 e^{-3x}$ **(b)** $\lim_{x \to 0^+} x \ln x$ **(c)** $\lim_{x \to 0} \left(\frac{1}{x} - \frac{1}{\sin x}\right)$

SOLUTION

(a) As $x \to \infty$ we have $x^2 \to \infty$ and $e^{-3x} \to 0$, so the limit $\lim_{x\to\infty} x^2 e^{-3x}$ is in the indeterminate form $\infty \cdot 0$. Therefore, this limit is not yet in a form to which we can apply L'Hôpital's rule. Luckily, limits of the indeterminate form $\infty \cdot 0$ can always be rewritten as a quotient of the form $\frac{\infty}{\infty}$ or as a quotient of the form $\frac{0}{0}$, simply by inverting one of the factors and placing it in the denominator. We can then choose whichever of these two indeterminate forms we prefer and apply L'Hôpital's rule. One way we could rewrite the limit is

$$\lim_{x\to\infty} x^2 e^{-3x} = \lim_{x\to\infty} \frac{x^2}{1/e^{-3x}} = \lim_{x\to\infty} \frac{x^2}{e^{3x}},$$

which is of the indeterminate form $\frac{\infty}{\infty}$. Another way we could write the limit is

$$\lim_{x\to\infty} x^2 e^{-3x} = \lim_{x\to\infty} \frac{e^{-3x}}{1/x^2},$$

which is of the indeterminate form $\frac{0}{0}$.

The first way of rewriting seems like it would be easier to deal with, so we apply L'Hôpital's rule to that version:

$$\begin{aligned}
\lim_{x\to\infty} x^2 e^{-3x} &= \lim_{x\to\infty} \frac{x^2}{e^{3x}} && \leftarrow \text{rewrite to form } \tfrac{\infty}{\infty} \\
&\overset{\text{L'H}}{=} \lim_{x\to\infty} \frac{2x}{3e^{3x}} && \leftarrow \text{apply L'Hôpital's rule; still of form } \tfrac{\infty}{\infty} \\
&\overset{\text{L'H}}{=} \lim_{x\to\infty} \frac{2}{9e^{3x}} && \leftarrow \text{apply L'Hôpital's rule} \\
&= 0. && \leftarrow \text{since } 9e^{3x} \to \infty \text{ as } x \to \infty
\end{aligned}$$

(b) Note that we consider only the limit from the right, since $\ln x$ is not defined for negative numbers. As $x \to 0^+$ we have $x \to 0$ and $\ln x \to -\infty$, and therefore the limit $\lim_{x\to 0^+} x \ln x$ is in the indeterminate form $0(-\infty)$. In order to apply L'Hôpital's rule we must rewrite the limit in the form $\frac{0}{0}$ or $\frac{\infty}{\infty}$. In this case it is easier to leave $\ln x$ in the numerator; as $x \to 0^+$ we have

$$\begin{aligned}
\lim_{x\to 0^+} x \ln x &= \lim_{x\to 0^+} \frac{\ln x}{1/x} && \leftarrow \text{limit is now in the form } \tfrac{-\infty}{\infty} \\
&\overset{\text{L'H}}{=} \lim_{x\to 0^+} \frac{1/x}{-1/x^2} && \leftarrow \text{apply L'Hôpital's rule} \\
&= \lim_{x\to 0^+} \frac{-x^2}{x} && \leftarrow \text{use algebra to simplify} \\
&= \lim_{x\to 0^+} (-x) = 0. && \leftarrow \text{simplify more and evaluate limit}
\end{aligned}$$

Note that we could have applied L'Hôpital's rule a second time in this problem, since immediately after the first application of the rule the limit was again in the indeterminate form $\frac{\infty}{\infty}$. However, simplifying instead resulted in a very simple limit that we could easily evaluate.

(c) As $x \to 0$, both $\frac{1}{x}$ and $\frac{1}{\sin x}$ become infinite, so the limit $\lim_{x\to 0} \left(\frac{1}{x} - \frac{1}{\sin x}\right)$ is potentially of the indeterminate form $\infty - \infty$.

To be honest, we are playing pretty fast and loose here: Specifically, we are not bothering to examine whether the terms $\frac{1}{x}$ and $\frac{1}{\sin x}$ approach ∞ or $-\infty$. Without knowing that, we don't know for sure whether this is a limit of the indeterminate form $\infty - \infty$ or the non-indeterminate form $\infty + \infty$. We could look from the left and right and do this more precisely if we cared to, but it is easier to instead do some algebra so that we can apply L'Hôpital's rule:

$$\lim_{x\to 0}\left(\frac{1}{x} - \frac{1}{\sin x}\right) = \lim_{x\to 0}\frac{\sin x - x}{x\sin x} \qquad \leftarrow \text{combine fractions; form is now } \frac{0}{0}$$

$$\overset{\text{L'H}}{=} \lim_{x\to 0}\frac{\cos x - 1}{\sin x + x\cos x} \qquad \leftarrow \text{apply L'Hôpital's rule; still form } \frac{0}{0}$$

$$\overset{\text{L'H}}{=} \lim_{x\to 0}\frac{-\sin x}{\cos x + \cos x - x\sin x} \qquad \leftarrow \text{apply L'Hôpital's rule again}$$

$$= \frac{-\sin 0}{\cos 0 + \cos 0 - 0\sin 0} \qquad \leftarrow \text{evaluate limit}$$

$$= \frac{0}{2} = 0. \qquad \leftarrow \text{use trigonometric values}$$

□

All three limits we just calculated happened to be equal to zero, the first as $x \to \infty$ and the last two as $x \to 0$. We can check these limits with calculator graphs:

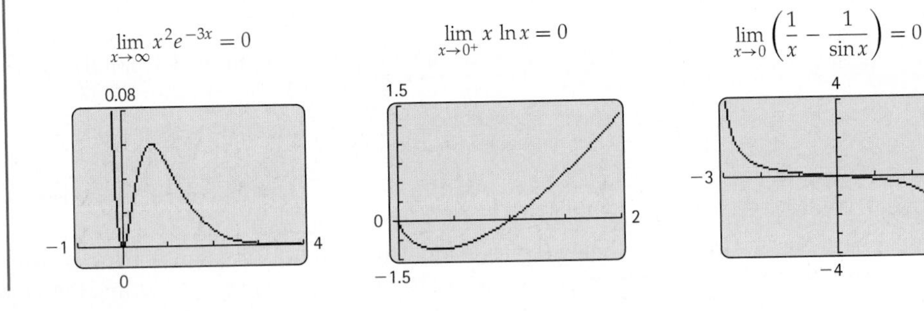

EXAMPLE 4

Using logarithms to calculate a limit

Use logarithms and L'Hôpital's rule to calculate each of the following limits:

(a) $\lim_{x\to\infty} x^{1/x}$ **(b)** $\lim_{x\to 0^+} (\sin x)^x$

SOLUTION

(a) Since $x \to \infty$ and $\frac{1}{x} \to 0$ as x approaches ∞, this limit is of the indeterminate form ∞^0. Let's calculate the related, but *different*, limit $\lim_{x\to\infty} \ln(x^{1/x})$ and see what we get:

$$\lim_{x\to\infty} \ln(x^{1/x}) = \lim_{x\to\infty}\frac{1}{x}\ln x \qquad \leftarrow \text{algebra; now the limit is of the form } 0\cdot\infty$$

$$= \lim_{x\to\infty}\frac{\ln x}{x} \qquad \leftarrow \text{algebra; now the limit is of the form } \frac{\infty}{\infty}$$

$$\overset{\text{L'H}}{=} \lim_{x\to\infty}\frac{1/x}{1} \qquad \leftarrow \text{apply L'Hôpital's rule}$$

$$= \lim_{x\to\infty}\frac{1}{x} = 0. \qquad \leftarrow \text{simplify and evaluate the limit}$$

But wait! This is not the answer to our *original* limit. Since this limit is equal to 0, we now know by Theorem 3.15 that our original limit must be equal to $e^0 = 1$.

(b) Once again we have a limit that involves a variable in both the base and the exponent, and we will need to use logarithms to resolve this problem. As $x \to 0^+$ we have $\sin x \to 0$ and $x \to 0$, so $\lim_{x\to 0^+} (\sin x)^x$ is in the indeterminate form 0^0. Let's look at the related limit obtained by first taking the logarithm:

$$\begin{aligned}
\lim_{x\to 0^+} \ln((\sin x)^x) &= \lim_{x\to 0^+} x\ln(\sin x) && \leftarrow \text{algebra; limit is now of form } 0(-\infty) \\
&= \lim_{x\to 0^+} \frac{\ln(\sin x)}{1/x} && \leftarrow \text{algebra; limit is now of form } \frac{-\infty}{\infty} \\
&\stackrel{\text{L'H}}{=} \lim_{x\to 0^+} \frac{(\cos x)/(\sin x)}{-1/x^2} && \leftarrow \text{apply L'Hôpital's rule} \\
&= \lim_{x\to 0^+} \frac{-x^2\cos x}{\sin x} && \leftarrow \text{algebra; limit is now of form } \frac{0}{0} \\
&\stackrel{\text{L'H}}{=} \lim_{x\to 0^+} \frac{-2x\cos x + x^2\sin x}{\cos x} && \leftarrow \text{apply L'Hôpital's rule again} \\
&= \frac{0+0}{1} = 0. && \leftarrow \text{evaluate the limit}
\end{aligned}$$

Since this limit is equal to 0, our original limit $\lim_{x\to 0^+} (\sin x)^x$ must be equal to $e^0 = 1$. □

? TEST YOUR UNDERSTANDING

- What sorts of limits does L'Hôpital's rule help us calculate?
- How do you explain in words what L'Hôpital's rule says we can do to solve limits?
- When calculating a limit by using L'Hôpital's rule multiple times, how do you know when to stop applying L'Hôpital's rule and evaluate the limit?
- How can L'Hôpital's rule sometimes be used to solve limits of the indeterminate form $0 \cdot \infty$?
- How can we use logarithms to solve limits of the indeterminate forms 0^0, 1^∞, and ∞^0?

EXERCISES 3.6

Thinking Back

Indeterminate forms: Determine which of the given forms are indeterminate. For each form that is not indeterminate, describe the behavior of a limit of that form.

- $\infty \cdot \infty$
- $\frac{\infty}{\infty}$
- 0^0
- ∞^∞
- $\frac{0}{0}$
- 1^∞
- $\infty - \infty$
- $\frac{0}{\infty}$
- 0^∞
- $\infty \cdot 0$
- $\frac{\infty}{0}$
- ∞^0

Simple limit calculations: Determine each of the limits that follow. You should be able to solve all of these very quickly by thinking about the graphs of the functions.

- $\lim_{x\to\infty} 2^x$
- $\lim_{x\to\infty} x^{-5}$
- $\lim_{x\to-\infty} e^{3x}$
- $\lim_{x\to 0} \frac{1}{x^2}$
- $\lim_{x\to 0^+} \ln x$
- $\lim_{x\to\infty} \left(\frac{1}{2}\right)^x$
- $\lim_{x\to\pi/2} \tan x$
- $\lim_{x\to\infty} \sin x$
- $\lim_{x\to\infty} \log_{1/2} x$

Concepts

0. *Problem Zero:* Read the section and make your own summary of the material.

1. *True/False:* Determine whether each of the statements that follow is true or false. If a statement is true, explain why. If a statement is false, provide a counterexample.

   (a) *True or False:* If a limit has an indeterminate form, then that limit does not have a real number as its solution.

   (b) *True or False:* L'Hôpital's rule can be used to find the limit of any quotient $\frac{f(x)}{g(x)}$ as $x \to c$.

   (c) *True or False:* When using L'Hôpital's rule, you need to apply the quotient rule in the differentiation step.

   (d) *True or False:* L'Hôpital's rule applies only to limits as $x \to 0$ or as $x \to \infty$.

   (e) *True or False:* L'Hôpital's rule applies only to limits of the indeterminate form $\frac{0}{0}$ or $\frac{\infty}{\infty}$.

   (f) *True or False:* If $\lim_{x\to 2} \ln(f(x)) = 4$, then $\lim_{x\to 2} f(x) = \ln 4$.

   (g) *True or False:* If $\lim_{x\to 2} \ln(f(x)) = \infty$, then $\lim_{x\to 2} f(x) = \infty$.

   (h) *True or False:* If $\lim_{x\to 2} \ln(f(x)) = -\infty$, then $\lim_{x\to 2} f(x) = -\infty$.

2. *Examples:* Construct examples of the thing(s) described in the following. Try to find examples that are different than any in the reading.

   (a) Three limits of the indeterminate form $\frac{0}{0}$, one that approaches ∞, one that equals 0, and one that equals 3.

   (b) Three limits of the indeterminate form $\frac{\infty}{\infty}$, one that approaches ∞, one that equals 0, and one that equals 3.

   (c) Three limits of the indeterminate form $0 \cdot \infty$, one that approaches ∞, one that equals 0, and one that equals 3.

3. Explain graphically why it makes sense that a limit $\lim_{x\to c} \frac{f(x)}{g(x)}$ of the indeterminate form $\frac{0}{0}$ would be related to $\lim_{x\to c} \frac{f'(x)}{g'(x)}$.

4. Explain graphically why it might make sense that a limit $\lim_{x\to\infty} \frac{f(x)}{g(x)}$ of the indeterminate form $\frac{\infty}{\infty}$ would be related to $\lim_{x\to\infty} \frac{f'(x)}{g'(x)}$.

5. Suppose $f(x) = x^2 - 1$ and $g(x) = \ln x$. Find the equations of the tangent lines to these functions at $x = 1$. Then argue graphically that it would be reasonable to think that the limit of the quotient $\frac{f(x)}{g(x)}$ as $x \to 1$ might be equal to the limit of the quotient of these tangent lines as $x \to 1$.

6. Suppose $f(x) = 2^x - 4$ and $g(x) = x - 2$. Find the equations of the tangent lines to these functions at $x = 2$. Then argue graphically that it would be reasonable to think that the limit of the quotient $\frac{f(x)}{g(x)}$ as $x \to 2$ might be equal to the limit of the quotient of these tangent lines as $x \to 2$.

Each of the limits in Exercises 7–12 is of the indeterminate form $0 \cdot \infty$ or $\infty \cdot 0$. Rewrite each limit so that it is (a) in the form $\frac{0}{0}$ and then (b) in the form $\frac{\infty}{\infty}$. Then (c) determine which of these indeterminate forms would be easier to work with when applying L'Hôpital's rule.

7. $\lim_{x\to\infty} 2^{-x}x$

8. $\lim_{x\to 0} (2^x - 1)x^{-2}$

9. $\lim_{x\to\infty} x^{-2}\ln x$

10. $\lim_{x\to 0} x^3 \ln x$

11. $\lim_{x\to 0} x \csc x$

12. $\lim_{x\to 1} \sqrt{x-1}\,\ln(x-1)$

13. Find the error in the following *incorrect* calculation, and then calculate the limit correctly:

$$\lim_{x\to 0} \frac{x^2 - x}{2^x - 1} \overset{\text{L'H}}{=} \lim_{x\to 0} \frac{2x - 1}{(\ln 2)2^x} \overset{\text{L'H}}{=} \lim_{x\to 0} \frac{2}{(\ln 2)^2 2^x} = \frac{2}{(\ln 2)^2 2^0} = \frac{2}{(\ln 2)^2}.$$

14. Find the error in the following *incorrect* calculation. Then calculate the limit correctly.

$$\lim_{x\to\infty} \frac{e^{-x}}{x^2} \overset{\text{L'H}}{=} \lim_{x\to\infty} \frac{-e^{-x}}{2x} \overset{\text{L'H}}{=} \lim_{x\to\infty} \frac{e^{-x}}{2} = \frac{0}{2} = 0.$$

Skills

Calculate each of the limits in Exercises 15–20 (a) using L'Hôpital's rule and (b) without using L'Hôpital's rule.

15. $\lim_{x\to 1} \frac{x^2 + x - 2}{x - 1}$

16. $\lim_{x\to 2} \frac{x^2 - 4x + 4}{x - 2}$

17. $\lim_{x\to\infty} \frac{x - 1}{2 - 3x^2}$

18. $\lim_{x\to\infty} \frac{3^x}{1 - 4^x}$

19. $\lim_{x\to\infty} \frac{e^{3x}}{1 - e^{2x}}$

20. $\lim_{x\to 0} \frac{2^x - 1}{4^x - 1}$

Calculate each of the limits in Exercises 21–48. Some of these limits are made easier by L'Hôpital's rule, and some are not.

21. $\lim_{x\to 3} \frac{2^x - 8}{3 - x}$

22. $\lim_{x\to 1} \frac{x - 1}{e^{x-1} - 1}$

23. $\lim_{x\to 2} \frac{e^{2x-4} - 1}{x^2 - 4}$

24. $\lim_{x\to\infty} \frac{e^x - 7}{8x^2 + 12x + 5}$

25. $\lim_{x\to\infty} \frac{2^{-x}}{x^2 + 1}$

26. $\lim_{x\to\infty} \frac{x + 3^x}{3 - 2^x}$

27. $\lim_{x\to 0} \frac{x^{-2}}{x^{-3}+1}$

28. $\lim_{x\to\infty} x^2e^{-x}$

29. $\lim_{x\to\infty} \left(e^x - \frac{e^x}{x+1}\right)$

30. $\lim_{x\to 0^+} \left(x^{-1} - \frac{1}{2^x-1}\right)$

31. $\lim_{x\to\infty} x\left(\frac{1}{2}\right)^x$

32. $\lim_{x\to\infty} \frac{xe^{-3x}}{x^2+3x+1}$

33. $\lim_{x\to\infty} \frac{xe^x}{e^{2x}+1}$

34. $\lim_{x\to 0^+} \left(\frac{2^x-1}{x^2} - \frac{1}{1-e^x}\right)$

35. $\lim_{x\to\infty} \frac{e^{-3x}}{x^2+3x+1}$

36. $\lim_{x\to\infty} \frac{\ln x}{\ln(2x+1)}$

37. $\lim_{x\to 0^+} \frac{\log_2 x}{\log_2 3x}$

38. $\lim_{x\to 2^+} \ln\left(\frac{2^x-4}{x-2}\right)$

39. $\lim_{x\to 2^+} \frac{\ln(x-2)}{\ln(x^2-4)}$

40. $\lim_{x\to 0^+} \frac{\ln x}{\ln x - x + 1}$

41. $\lim_{x\to 0} \frac{\cos x - 1}{\sin x}$

42. $\lim_{x\to 0} \frac{\sin x}{x+\sin x}$

43. $\lim_{x\to 0} \frac{1-\cos x}{\tan x}$

44. $\lim_{x\to \pi/2} \frac{\sin(\cos x)}{\cos x}$

45. $\lim_{x\to 0} \frac{x\cos x}{1-e^x}$

46. $\lim_{x\to 1} \frac{\sin(\ln x)}{x-1}$

47. $\lim_{x\to 0} \frac{\tan^{-1} x}{\sin x}$

48. $\lim_{x\to 0} \frac{\tan^{-1} x}{\sin^{-1} x}$

Calculate each of the limits in Exercises 49–64. Some of these limits are made easier by considering the logarithm of the limit first, and some are not.

49. $\lim_{x\to 0^+} x^{\ln x}$

50. $\lim_{x\to\infty} x^{\ln x}$

51. $\lim_{x\to 2^+} (x-2)^{x^2-4}$

52. $\lim_{x\to 0^+} (x^2+1)^x$

53. $\lim_{x\to 1^+} x^{1/(x-1)}$

54. $\lim_{x\to 0^+} x^{2x}$

55. $\lim_{x\to\infty} x^{1/x}$

56. $\lim_{x\to\infty} \left(\frac{1}{x}\right)^x$

57. $\lim_{x\to\infty} \left(\frac{1}{x+1}\right)^x$

58. $\lim_{x\to\infty} \left(\frac{x}{x-1}\right)^x$

59. $\lim_{x\to 1^+} (x-1)^{\ln x}$

60. $\lim_{x\to 1^+} (\ln x)^{x-1}$

61. $\lim_{x\to 0^+} x^{\sin x}$

62. $\lim_{x\to 0^+} (\sin 3x)^{2x}$

63. $\lim_{x\to 0^+} (\cos x)^{1/x}$

64. $\lim_{x\to 0} (1-\cos x)^x$

A function f ***dominates*** another function g as $x \to \infty$ if $f(x)$ and $g(x)$ both grow without bound as $x \to \infty$ and if

$$\lim_{x\to\infty} \frac{f(x)}{g(x)} = \infty.$$

Intuitively, f dominates g as $x \to \infty$ if $f(x)$ is very much larger than $g(x)$ for very large values of x. In Exercises 65–74, use limits to determine whether $u(x)$ dominates $v(x)$, or $v(x)$ dominates $u(x)$, or neither.

65. $u(x) = x + 100,\ v(x) = x$

66. $u(x) = 5x^2 + 1,\ v(x) = x^3$

67. $u(x) = 100x^2,\ v(x) = 2x^{100}$

68. $u(x) = x^2,\ v(x) = 2^x$

69. $u(x) = 2^x,\ v(x) = e^x$

70. $u(x) = x^{10} + 3,\ v(x) = 10^x + 3$

71. $u(x) = \log_2 x,\ v(x) = \log_{30} x$

72. $u(x) = \ln(x^2+1),\ v(x) = x^2 + 1$

73. $u(x) = 0.001e^{0.001x},\ v(x) = 100x^{100}$

74. $u(x) = 0.001x^2 - 100x,\ v(x) = 100\log_3 x$

As you will prove in Exercises 93 and 94, exponential growth functions e^{kx} always dominate power functions x^r, and power functions x^r with positive powers always dominate logarithmic functions $\log_b x$. Use these facts to quickly determine each of the limits in Exercises 75-80.

75. $\lim_{x\to\infty} \frac{x^{101}+500}{e^x}$

76. $\lim_{x\to\infty} \frac{e^x}{x^{101}+500}$

77. $\lim_{x\to\infty} \frac{\sqrt{x}}{300\ln x}$

78. $\lim_{x\to\infty} \frac{\ln x^{100}}{x^8}$

79. $\lim_{x\to\infty} 2^x x^{-100}$

80. $\lim_{x\to\infty} e^{-x}\ln x$

Now that we know L'Hôpital's rule, we can apply it to solve more sophisticated global optimization problems. Consider domains, limits, derivatives, and values to determine the global extrema of each function f in Exercises 81–86 on the given intervals I and J.

81. $f(x) = x\ln x,\ I = (0, 1],\ J = (0, \infty)$

82. $f(x) = x^2\ln(0.2x),\ I = (0, 4],\ J = (0, \infty)$

83. $f(x) = x^3e^{-x},\ I = [0, \infty),\ J = (-\infty, \infty)$

84. $f(x) = \frac{\ln x}{\ln 2x},\ I = [0, 1], J = [1, \infty)$

85. $f(x) = \frac{\sin x}{1-\cos x},\ I = (0, \pi],\ J = (0, 2\pi)$

86. $f(x) = \frac{e^x}{1+x^2},\ I = [0, \infty), J = (-\infty, \infty)$

Applications

In Exercises 87–89, suppose that Leila is a population biologist with the Idaho Fish and Game Service. Wolves were introduced formally into Idaho in 1994, but there were some wolves in the state before that. Leila has been assigned the task of estimating the rate at which the number of wolves in Idaho increased naturally, before the animals were introduced. The only information she has is a population model, which indicates that the wolf population currently satisfies the formula

$$w(t) = 835(1 - e^{-.006t}),$$

where t is the number of years since 1994.

87. Leila reasons that the average rate of change of wolves from 1994 to a given time t is given by $\frac{w(t)}{t}$. Why does her reasoning make sense?

88. To approximate the rate of increase of wolves per year at the beginning of 1994, Leila decides to take the limit of $\frac{w(t)}{t}$ as $t \to 0$. Why does this approach make sense, and what is the value of that limit? Is there another way she could find the same number?

89. Leila must also determine hunting policies to sustain a population W_0 of wolves that satisfy federal guidelines, while maximizing the sustained elk population E_0 for which the state can sell hunting tags. Her predator-prey models approximate the number of elk over time as

$$E(t) = \frac{E_0 + 72e^{-0.006t}\sin(\pi t/4) + 8te^{-0.006t}\sin(\pi t/4)}{1 + 0.2W_0}$$

(a) Use the Squeeze Theorem for Limits to show that the population goes toward $\frac{E_0}{1+0.2W_0}$ as $t \to \infty$.

(b) Explain why L'Hôpital's Rule is *not* a good method for calculating this limit.

Proofs

90. In your own words, prove the special case of L'Hôpital's rule that is proved in the reading. Explain each step in detail.

Exercises 91–94 concern dominance of functions as defined earlier in Exercises 65–74.

91. Use L'Hôpital's rule to prove that every exponential growth function dominates the power function $g(x) = x^2$.

92. Use L'Hôpital's rule to prove that every power function with a positive power dominates the logarithmic function $g(x) = \ln x$.

93. Use L'Hôpital's rule to prove that exponential growth functions always dominate power functions.

94. Use L'Hôpital's rule to prove that power functions with positive powers always dominate logarithmic functions.

95. Use logarithms to prove that for any real number r, $\lim_{x\to\infty}\left(1 + \frac{r}{x}\right)^x = e^r$.

Thinking Forward

Convergence and divergence of sequences: If a sequence $a_1, a_2, a_3, \ldots, a_k, \ldots$ approaches a real-number limit as $k \to \infty$, then the sequence $\{a_k\}$ ***converges***. If the terms of the sequence do not get arbitrarily close to some real number, then the sequence ***diverges***. Determine the general form $\{a_k\}$ for each of the following sequences, and then use L'Hôpital's rule to determine whether that sequence converges or diverges.

- ▶ $\frac{1}{2}, \frac{4}{4}, \frac{9}{8}, \frac{16}{16}, \frac{25}{32}, \frac{36}{64}, \ldots$
- ▶ $\frac{\ln 1}{\ln 2}, \frac{\ln 2}{\ln 3}, \frac{\ln 3}{\ln 4}, \frac{\ln 4}{\ln 5}, \frac{\ln 5}{\ln 6}, \frac{\ln 6}{\ln 7}, \ldots$
- ▶ $\frac{1}{10}, \frac{3}{100}, \frac{7}{1000}, \frac{15}{10{,}000}, \frac{31}{100{,}000}, \frac{63}{1{,}000{,}000}, \ldots$
- ▶ $\frac{1}{301}, \frac{8}{304}, \frac{27}{309}, \frac{64}{316}, \frac{125}{325}, \frac{216}{336}, \ldots$

CHAPTER REVIEW, SELF-TEST, AND CAPSTONES

Before you progress to the next chapter, be sure you are familiar with the definitions, concepts, and basic skills outlined here. The capstone exercises at the end bring together ideas from this chapter and look forward to future chapters.

Definitions

Give precise mathematical definitions or descriptions of each of the concepts that follow. Then illustrate the definition with a graph or algebraic example, if possible.

- f has a *local maximum* or a *local minimum* at $x = c$
- f has a *critical point* at $x = c$
- f is *increasing* or *decreasing* on an interval I
- f is *concave up* or *concave down* on on interval I
- f has a *global maximum* or a *global minimum* at $x = c$
- f does not have any *global maximum* or does not have any *global minimum*

Theorems

Fill in the blanks to complete each of the following theorem statements:

- If $x = c$ is a local extremum of f, then $f'(c)$ is either ____ or ____ .
- *Rolle's Theorem:* If f is ____ on $[a, b]$ and ____ on (a, b), and if ____ , then there exists at least one value $c \in (a, b)$ for which $f'(c) =$ ____ .
- *The Mean Value Theorem:* If f is ____ on $[a, b]$ and ____ on (a, b), then there exists at least one value $c \in (a, b)$ for which $f'(c) =$ ____ .
- If f is differentiable on an interval I and f' is positive in the interior of I, then f is ____ on I.
- If f is differentiable on an interval I and f' is ____ in the interior of I, then f is decreasing on I.
- If f is differentiable on an interval I and f' is zero in the interior of I, then f is ____ on I.
- If $f'(x) = g'(x)$ for all $x \in [a, b]$, then for some constant C, $f(x) =$ ____ for all $x \in [a, b]$.
- *The first-derivative test:* Suppose $x = c$ is a ____ of a differentiable function f. If ____ , then f has a local maximum at $x = c$. If ____ , then f has a local minimum at $x = c$. If ____ , then f has neither a local maximum nor a local minimum at $x = c$.
- Suppose f and f' are differentiable on an interval I. If ____ is positive on I, then f is concave up on I. If ____ is ____ on I, then f is concave down on I.
- *The second-derivative test:* Suppose $x = c$ is a ____ of a twice-differentiable function f. If ____ , then f has a local maximum at $x = c$. If ____ , then f has a local minimum at $x = c$. If ____ , then this test is inconclusive.
- *L'Hôpital's Rule:* If f and g are ____ and $g(x)$ is nonzero near $x = c$, and if $\lim_{x\to c} \frac{f(x)}{g(x)}$ is of the indeterminate form ____ or ____ , then $\lim_{x\to c} \frac{f(x)}{g(x)} =$ ____ .
- If $\lim_{x\to c} \ln(f(x)) = L$, then $\lim_{x\to c} f(x) =$ ____ .
- If $\lim_{x\to c} \ln(f(x)) =$ ____ , then $\lim_{x\to c} f(x) = \infty$.
- If $\lim_{x\to c} \ln(f(x)) =$ ____ , then $\lim_{x\to c} f(x) = 0$.

Geometric Formulas and Theorems

Volume and Surface Area Formulas: Write a formula for (a) the volume and (b) the surface area of each solid described below.

- A rectangular box of width w, length l, and height h
- A box of height h with a square base of area A
- A sphere of radius r
- A sphere of circumference C
- A right circular cylinder of radius R and height y
- A right circular cylinder whose radius r is half of its height
- A cone of height y and circular base of area A

Right-Triangle Theorems: Write an equation that describes the relationships between the variables given for each theorem that follows. Draw a picture to illustrate the theorem and the roles of the variables in your equation.

- *The Pythagorean Theorem:* If a right triangle has legs of lengths x and y and a hypotenuse of length h, then ____ .
- *The Law of Similar Triangles:* Suppose two right triangles have the same angle measures as each other (i.e., they are similar), where the first has legs of lengths x_1 and y_1 and a hypotenuse of length h_1 and the second has corresponding legs of lengths x_2 and y_2 and a hypotenuse of length h_2. Then we have the following three equations involving ratios: ____ , ____ , and ____ .

Skill Certification: Curve Sketching and L'Hôpital's Rule

Intervals of behavior: For each of the following functions f, determine the intervals on which f is positive, negative, increasing, decreasing, concave up, and concave down.

1. $f(x) = x^3 + 3x^2 - 9x - 27$
2. $f(x) = x^{4/3}$
3. $f(x) = \frac{x-1}{x+3}$
4. $f(x) = \frac{1}{(x-1)(x+2)}$
5. $f(x) = e^{1+x^2}$
6. $f(x) = \sin\left(\frac{x}{4}\right)$
7. $f(x) = 2^x(2^x - 1)$
8. $f(x) = \sec^2 x$

Important points: Find all roots, local maxima and minima, and inflection points of each function f. In addition, determine whether any local extrema are also global extrema on the domain of f.

9. $f(x) = 3x^4 - 8x^3$
10. $f(x) = x^3 - 15x^2 - 33x$
11. $f(x) = \frac{x}{x^2+1}$
12. $f(x) = \frac{1}{1+\sqrt{x}}$
13. $f(x) = x \ln x$
14. $f(x) = x^{4/3} - x^{1/3}$
15. $f(x) = \frac{e^x}{1-e^x}$
16. $f(x) = \tan^{-1} x^2$

Curve sketching: For each function f that follows, construct sign charts for f, f', and f'', if possible. Examine function values or limits at any interesting values and at $\pm\infty$. Then interpret this information to sketch a labeled graph of f.

17. $f(x) = x^3 - 2x^2 - 4x + 8$
18. $f(x) = \frac{1}{3x+1}$
19. $f(x) = x\sqrt{x+1}$
20. $f(x) = \sqrt{x^2+2x+10}$
21. $f(x) = xe^x$
22. $f(x) = \ln(x^2+1)$
23. $f(x) = \cos x$
24. $f(x) = \tan^{-1} x$

L'Hôpital's Rule limit calculations: Calculate each of the limits that follow. Some of these limits are easier to calculate by using L'Hôpital's rule, and some are not.

25. $\lim_{x\to\infty} \frac{x^{-2}}{x^{-3}+1}$
26. $\lim_{x\to 0} \frac{3^x - 1}{2^x - 1}$
27. $\lim_{x\to 0} \frac{\sin x}{\cos x - 1}$
28. $\lim_{x\to\infty} x^3 e^{-x}$
29. $\lim_{x\to\infty} \frac{\ln x}{\ln(x+1)}$
30. $\lim_{x\to 0} \frac{\tan^{-1} x}{1-\cos x}$
31. $\lim_{x\to 1^+} x^{1/(x-1)}$
32. $\lim_{x\to 0^+} x^{1-\cos x}$

Capstone Problems

A. *Critical points, extrema, and inflection points:* Find examples of differentiable functions which illustrate that not every critical point is an extremum and that not every zero of the second derivative is an inflection point. More specifically, find the following:
   (a) A function f with $f'(2) = 0$ and an extremum at $x = 2$, and a function g with $g'(2) = 0$ but no extremum at $x = 2$.
   (b) A function k with $k''(2) = 0$ and an inflection point at $x = 2$, and a function h with $h''(2) = 0$ but no inflection point at $x = 2$.

B. *Optimizing perimeter, given area:* Suppose that you want to cut a rectangular shape with a particular area A from a sheet of material, and that you want the perimeter of the shape to be as small as possible. Use techniques of optimization to argue that the smallest possible perimeter will be achieved if the rectangular shape that you cut out is a square.

C. *The Mean Value Theorem:* Recap the development of the Mean Value Theorem as follows:
   (a) Prove that if f is a differentiable function, then every extremum $x = c$ of f is also a critical point of f. *(Hint: Show that $f'(c) = 0$ by proving that $f'_+(c) \le 0$ and $f'_-(c) \ge 0$.)*
   (b) Use part (a) and the Extreme Value Theorem to prove Rolle's Theorem. *(Hint: Consider the case where f has an extremum on the interior of the interval first.)*
   (c) Explain how the Mean Value Theorem is essentially a rotated version of Rolle's Theorem and how the proof in the reading makes use of that fact.

D. *Area accumulation functions:* Suppose f is the function pictured here, and $A(x)$ is the associated function whose value at any $x \ge 0$ is equal to the area between the graph of f and the x-axis from 0 to x. The quantity $A(2.5)$ is shaded in the figure. We will count area below the x-axis negatively, so that in this example $A(5)$ is less than $A(4)$.

Area function $A(x)$ is defined by the shaded region as x varies

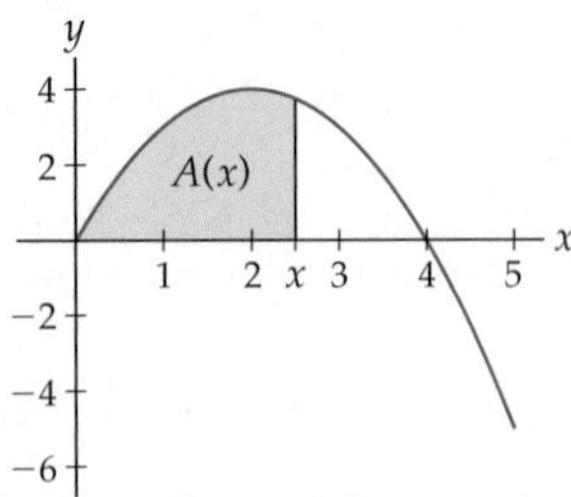

   (a) On what interval of x-values is $A(x)$ an increasing function? On what interval is $A(x)$ decreasing?
   (b) On what interval of x-values is the function f positive? On what interval is f negative?
   (c) Here is a surprising fact: One of these functions is the derivative of the other! Use your answers to parts (a) and (b) to determine whether $A' = f$ or $f' = A$.

CHAPTER 4

Definite Integrals

4.1 Addition and Accumulation

Accumulation Functions
Sigma Notation
The Algebra of Sums in Sigma Notation
Formulas for Common Sums
Examples and Explorations

$$\sum_{k=1}^{n} a_k$$

4.2 Riemann Sums

Subdivide, Approximate, and Add Up
Approximating Area with Rectangles
Riemann Sums
Signed Area
Types of Riemann Sums
Examples and Explorations

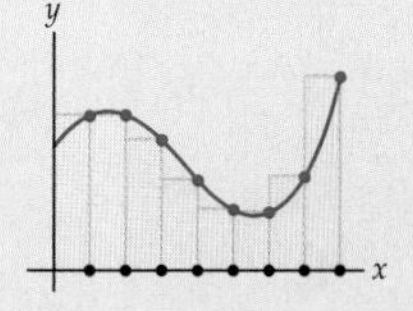

4.3 Definite Integrals

Defining the Area Under a Curve
Properties of Definite Integrals
Formulas for Three Simple Definite Integrals
Examples and Explorations

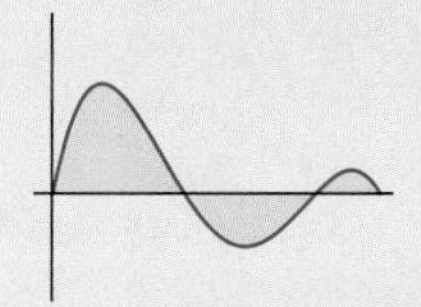

4.4 Indefinite Integrals

Antiderivatives and Indefinite Integrals
Antidifferentiation Formulas
Antidifferentiating Combinations of Functions
Examples and Explorations

$$\int f(x)\,dx$$

4.5 The Fundamental Theorem of Calculus

The Fundamental Theorem
Using the Fundamental Theorem of Calculus
The Net Change Theorem
The Proof of the Fundamental Theorem
Examples and Explorations

$$\int_a^b f(x)\,dx = F(b) - F(a)$$

4.6 Areas and Average Values

The Absolute Area Between a Graph and the x-axis
Areas Between Curves
The Average Value of a Function on an Interval
The Mean Value Theorem for Integrals
Examples and Explorations

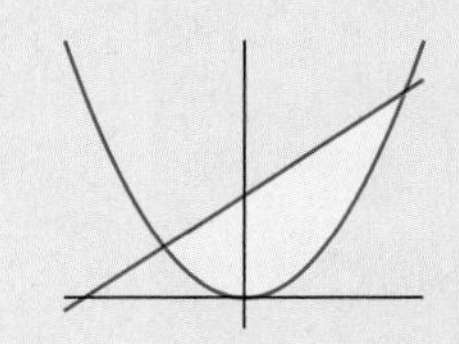

4.7 Functions Defined by Integrals
Area Accumulation Functions
The Second Fundamental Theorem of Calculus
Defining the Natural Logarithm Function with an Integral
The Proof of the Second Fundamental Theorem
Examples and Explorations

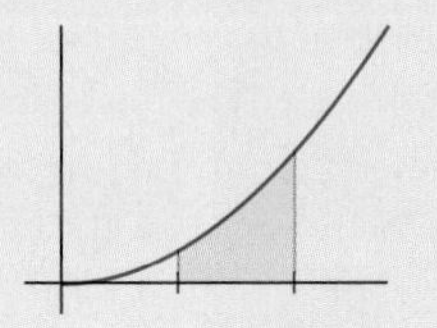

Chapter Review, Self-Test, and Capstones

4.1 ADDITION AND ACCUMULATION

- An introduction to accumulation functions and the area under a curve
- Sigma notation and properties of sums
- A preview of limits of sums, the backbone of the definition of the definite integral

Accumulation Functions

With this chapter we begin the study of integrals. While derivatives describe the rates at which functions change, integrals can describe how functions accumulate. As we will see throughout this chapter, the concepts of accumulation, area, and differentiation are fundamentally intertwined.

For a simple example of how these concepts are intertwined, imagine that you are driving down a straight road that has stoplights. Suppose that you start from a full stop at one stoplight, increase your velocity for 20 seconds to reach 60 miles per hour (88 feet per second), then decrease your velocity for 20 seconds until you come to a full stop at the next stoplight, as illustrated next at the left. Your speedometer works, so you can tell how fast you are going at any time. However, your odometer is broken, so you have no idea how far you have travelled. Using data from your speedometer, you can show that your velocity on this trip is given by $v(t) = -0.22t^2 + 8.8t$ feet per second, from $t = 0$ to $t = 40$ seconds, as illustrated in the graph shown at the right.

0 mph 60 mph 0 mph

$v = 88$ ft/s

$t = 0$ $t = 20$ $t = 40$

$v(t) = -0.22t^2 + 8.8t$

v 88 t 10 20 30 40

So, how far was it between the two stoplights? If you had travelled at a constant velocity, this would be an easy application of the "distance equals average rate times time" formula. For example, driving at exactly 30 miles per hour (44 feet per second) for 40 seconds would accumulate a distance of $d = (44)(40) = 1760$ feet.

Unfortunately in our example, velocity varies. However, we can approximate the distance travelled by assuming a constant velocity over small chunks of time. For example, we could use $v(5)$, $v(15)$, $v(25)$, and $v(35)$ as constant velocities over the intervals $[0, 10]$, $[10, 20]$, $[20, 30]$, and $[30, 40]$, respectively. Using the $d = rt$ formula over each of the four time intervals would give us an approximate distance travelled of

$$\begin{aligned} d &\approx d_1 + d_2 + d_3 + d_4 = r_1 t_1 + r_2 t_2 + r_3 t_3 + r_4 t_4 \\ &= v(5)(10) + v(15)(10) + v(25)(10) + v(35)(10) \\ &= (38.5)(10) + (82.5)(10) + (82.5)(10) + (38.5)(10) = 2420 \text{ feet.} \end{aligned}$$

Notice that we have just estimated the distance between the two stoplights by means of only the readings on your speedometer at $t = 5$, $t = 15$, $t = 25$, and $t = 35$. Despite having used only this small amount of information, our estimate is fairly close to the actual distance, which in this example happens to be just over 2346 feet.

We can think of the preceding four 10-second distance approximations as areas of four rectangles, as shown next at the left. For example, over the first time interval we have a rectangle of width $t_1 = 10$, height $r_1 = v(5) = 38.5$, and area $d_1 = (t_1)(r_1) = (38.5)(10) = 385$.

Distance was approximated with a sum of areas of rectangles

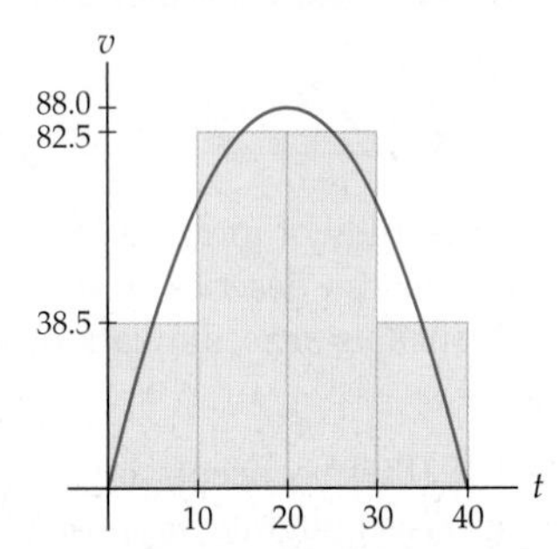

Thinner rectangles give a better approximation

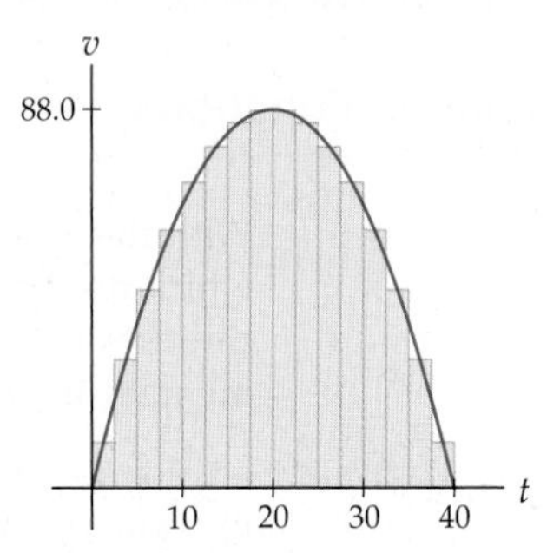

Notice that the sum of the areas of these rectangles is a pretty good approximation for the distance between the stoplights and also a pretty good approximation for the area under the velocity curve in the figure at the left. Using smaller time intervals would give us thinner rectangles (as shown at the right), whose combined areas would be a better approximation for the distance travelled, as well as a better approximation for the area under $v(t)$. These figures suggest that the exact distance travelled might equal the exact area under the velocity curve. Furthermore, since we already know that velocity is the derivative of position, the figures also suggest connections between accumulation, area, and derivatives. Over the rest of this chapter we will make these notions and connections more precise.

Sigma Notation

From the stoplight example we just discussed, it seems that we will have to do a lot of adding in order to investigate area and accumulation functions. Our better approximation with smaller, but more, rectangles involved 16 rectangles with areas to calculate and add up—and for an even better approximation we might consider 100 or more rectangles. We now develop a compact notation to represent the sum of a sequence of numbers—in particular, a sequence of numbers that has a recognizable pattern. This notation is called ***sigma notation,*** since it uses the letter "sigma" (written Σ), which is the Greek counterpart of the letter "S" (for "sum").

DEFINITION 4.1 Sigma Notation

If a_k is a function of k, and m and n are nonnegative integers with $m \leq n$, then

$$\sum_{k=m}^{n} a_k = a_m + a_{m+1} + a_{m+2} + \cdots + a_{n-1} + a_n.$$

As an initial example, consider our earlier sum of four distances. This sum can be written in sigma notation as follows:

$$d \approx \sum_{k=1}^{4} d_k = \sum_{k=1}^{4} r_k t_k = 2420 \text{ feet.}$$

The expression in Definition 4.1 is pronounced "the sum from $k = m$ to n of a_k." The "$k = m$" below the sigma shows where we should begin the sum, and it also specifies that the ***index***, or stepping variable, for the sum is k. With sigma notation, the step from one value of k to the next is always equal to 1. If we start with $k = 1$, then the next value of k will be $k = 2$, the next will be $k = 3$, and so on. We end at the value $k = n$, where n is the number appearing above the sigma in the notation. The terms a_k represent the pattern of the items to be summed. When we say that a_k is a function of k, we mean that, for every integer index k, the expression for a_k determines a unique real number. Since the index k can have only integer values, it is traditional to use a_k rather than the usual function notation $a(k)$.

When printed in a line of text, sigma notation is more compact and looks like this: $\sum_{k=m}^{n} a_k$. Sigma notation always follows the general pattern

$$\sum_{k=\text{starting value}}^{\text{ending value}} (\text{function of } k).$$

To find the value of a sum in sigma notation $\sum_{k=m}^{n} a_k$ you would evaluate the function a_k at $k = m$, $k = m + 1$, $k = m + 2$, and so on until $k = n$, and then add up all of these values. To put a given sum into sigma notation you must identify a pattern for the function a_k, as well as starting and ending values for k. For example, we could represent the sum

$$1 + \frac{1}{2} + \frac{1}{3} + \frac{1}{4} + \frac{1}{5} + \frac{1}{6} + \frac{1}{7} + \frac{1}{8} + \frac{1}{9} + \frac{1}{10}$$

with the sigma notation $\sum_{k=1}^{10} \frac{1}{k}$, since each of the numbers in our sum is of the form $\frac{1}{k}$ for some integer k, with $k = 1$ for the first term and $k = 10$ for the last term.

The Algebra of Sums in Sigma Notation

The next theorem expresses two common properties of sums in sigma notation. For the moment we will restrict our attention to sums whose index k starts at $k=1$. However, Theorem 4.2 is true for sums that begin at any integer $k=m$.

THEOREM 4.2 **Constant-Multiple and Sum Rules for Sums**

If a_k and b_k are functions defined for nonnegative integers k, and c is any real number, then

(a) $\displaystyle\sum_{k=1}^{n} ca_k = c\sum_{k=1}^{n} a_k.$ **(b)** $\displaystyle\sum_{k=1}^{n} (a_k + b_k) = \sum_{k=1}^{n} a_k + \sum_{k=1}^{n} b_k.$

Part (a) of this theorem is a general version of the distributive property $c(x + y) = cx + cy$. Part (b) is a general version of the associative and commutative properties of addition (i.e., the fact that we can reorder and regroup the numbers in a sum). A simple example is the fact that $(a_1 + b_1) + (a_2 + b_2) = (a_1 + a_2) + (b_1 + b_2)$. The proof of Theorem 4.2 consists mostly of translating the sigma notation.

Proof. To prove part (a), we translate the sigma notation to its expanded form and then apply the distributive property:

$$\begin{aligned} \sum_{k=1}^{n} ca_k &= ca_1 + ca_2 + ca_3 + \cdots + ca_{n-1} + ca_n && \leftarrow \text{write out the sum} \\ &= c(a_1 + a_2 + a_3 + \cdots + a_{n-1} + a_n) && \leftarrow \text{factor out } c \\ &= c\sum_{k=1}^{n} a_k. && \leftarrow \text{write in sigma notation} \end{aligned}$$

The proof of part (b) is similar; we will write the sigma notation in expanded form, reorder and regroup the terms in the sum, and then write the reordered sum in sigma notation:

$$\begin{aligned}\sum_{k=1}^{n}(a_k + b_k) &= (a_1 + b_1) + (a_2 + b_2) + (a_3 + b_3) + \cdots (a_{n-1} + b_{n-1}) + (a_n + b_n)\\ &= (a_1 + a_2 + a_3 + \cdots + a_{n-1} + a_n) + (b_1 + b_2 + b_3 + \cdots + b_{n-1} + b_n)\\ &= \sum_{k=1}^{n} a_k + \sum_{k=1}^{n} b_k.\end{aligned}$$

■

We can strip off terms from the beginning or end of a sum, or split a sum into two pieces, by applying the following theorem:

THEOREM 4.3

Splitting a Sum

Given any function a_k defined for nonnegative integers k, and given any integers m, n, and p such that $0 \le m < p < n$,

$$\sum_{k=m}^{n} a_k = \sum_{k=m}^{p-1} a_k + \sum_{k=p}^{n} a_k.$$

Proof. The proof of this theorem is fairly simple; we need only write out the terms of the sum. If $m < p < n$, we have

$$\begin{aligned}\sum_{k=m}^{n} a_k &= a_m + a_{m+1} + \cdots + a_{p-1} + a_p + a_{p+1} + \cdots + a_{n-1} + a_n && \leftarrow \text{expand}\\ &= (a_m + a_{m+1} + \cdots + a_{p-1}) + (a_p + a_{p+1} + \cdots + a_{n-1} + a_n) && \leftarrow \text{regroup}\\ &= \sum_{k=m}^{p-1} a_k + \sum_{k=p}^{n} a_k. && \leftarrow \text{sigma notation}\end{aligned}$$

■

Formulas for Common Sums

It can be tedious to calculate a long sum by hand; for example, consider the sum $\sum_{k=1}^{1000} k^2$. To calculate this sum directly, we would have to write out all 1000 of its terms and then add them together. Luckily, the next theorem provides formulas for the values of a few simple sums. Using these formulas, we can quickly calculate any sum whose general term is a constant, linear, quadratic, or cubic polynomial in k.

THEOREM 4.4

Sum Formulas

If n is a positive integer, then

(a) $\displaystyle\sum_{k=1}^{n} 1 = n$

(b) $\displaystyle\sum_{k=1}^{n} k = \frac{n(n+1)}{2}$

(c) $\displaystyle\sum_{k=1}^{n} k^2 = \frac{n(n+1)(2n+1)}{6}$

(d) $\displaystyle\sum_{k=1}^{n} k^3 = \frac{n^2(n+1)^2}{4}$

Proof. To prove the first formula we only have to write out the sum

$$\sum_{k=1}^{n} 1 = \underbrace{1+1+1+\cdots+1}_{n\text{ times}} = n.$$

You will prove the second formula by using two cases in Exercises 61 and 62. The third and fourth formulas are simple to prove by a method called ***induction***, but that method of proof is best left for a future mathematics course, so we do not include proofs of the third and fourth formulas at this time. ■

Examples and Explorations

EXAMPLE 1

Converting from sigma notation to an expanded sum

Write the sum represented by the sigma notation $\sum_{k=2}^{8} 2^k$ in expanded form, and find the value of the sum.

SOLUTION

The given sigma notation is a compact way of writing "the sum of all numbers of the form 2^k, where k is an integer greater than or equal to 2 and less than or equal to 8." This means we must find the values of 2^k for $k = 2$, $k = 3$, and so on until $k = 8$, and then add all of these values:

$$\sum_{k=2}^{8} 2^k = 2^2 + 2^3 + 2^4 + 2^5 + 2^6 + 2^7 + 2^8 = 508.$$

□

EXAMPLE 2

Converting from an expanded sum to sigma notation

Write the following sum in sigma notation:

$$\frac{3}{4} + \frac{4}{5} + \frac{5}{6} + \frac{6}{7} + \frac{7}{8} + \frac{8}{9} + \frac{9}{10}.$$

SOLUTION

First, we look for a pattern in the numbers of the sum, so we can determine the function a_k. One possible pattern is that each number in the sum is of the form $\frac{k}{k+1}$ for some non-negative integer k. Therefore, we can set $a_k = \frac{k}{k+1}$. The first number, $\frac{3}{4}$, corresponds to the value $k = 3$. The last number, $\frac{9}{10}$, corresponds to the value $k = 9$ (when $k = 9$, we have $\frac{k}{k+1} = \frac{9}{10}$). The sum can be written in sigma notation as

$$\frac{3}{4} + \frac{4}{5} + \frac{5}{6} + \frac{6}{7} + \frac{7}{8} + \frac{8}{9} + \frac{9}{10} = \sum_{k=3}^{9} \frac{k}{k+1}.$$

□

EXAMPLE 3

Algebraically manipulating sums in sigma notation

Given that $\sum_{k=1}^{4} a_k = 7$ and $\sum_{k=1}^{4} b_k = 10$, find $\sum_{k=1}^{4} (a_k + 3b_k)$.

SOLUTION

Notice that we do not know what the functions a_k and b_k are in this problem, and thus we cannot directly compute $\sum_{k=1}^{4}(a_k + 3b_k)$. However, using Theorem 4.2, we have

$$\begin{aligned}\sum_{k=1}^{4}(a_k + 3b_k) &= \sum_{k=1}^{4} a_k + \sum_{k=1}^{4} 3b_k && \leftarrow \text{Part (a) of Theorem 4.2}\\ &= \sum_{k=1}^{4} a_k + 3\sum_{k=1}^{4} b_k && \leftarrow \text{Part (b) of Theorem 4.2}\\ &= 7 + 3(10) = 37. && \leftarrow \text{using the values of the sums given}\end{aligned}$$

□

EXAMPLE 4

Sums can be combined only if they start and end at the same value

Use the fact that $\sum_{k=2}^{20} \frac{1}{k} \approx 2.5977$ and $\sum_{k=0}^{19} \sqrt{k} \approx 57.1938$ to estimate $\sum_{k=2}^{19} \left(\frac{2}{k} + \sqrt{k}\right)$.

SOLUTION

We could of course just directly calculate the desired sum by adding up terms, but instead we will combine the two given sums to obtain the information we need. The key is to rewrite the two sums we were given so that they each begin and end and the same value; we will rewrite so that they begin at $k = 2$ and end at $k = 19$.

We are given that $\sum_{k=2}^{20} \frac{1}{k} = \frac{1}{2} + \frac{1}{3} + \cdots + \frac{1}{19} + \frac{1}{20}$ is approximately 2.5977. To get the sum of the $k = 2$ through $k = 19$ terms, all we have to do is subtract the extra term $\frac{1}{20}$:

$$\sum_{k=2}^{19} \frac{1}{k} = \left(\sum_{k=2}^{20} \frac{1}{k}\right) - \frac{1}{20} \approx 2.5977 - \frac{1}{20} = 2.5477.$$

Similarly, we can find the sum from $k = 2$ to $k = 19$ of $\sqrt{k}$ by subtracting the two extra terms (at $k = 0$ and $k = 1$) at the beginning of the sum $\sum_{k=0}^{19} \sqrt{k}$:

$$\sum_{k=2}^{19} \sqrt{k} = \left(\sum_{k=0}^{19} \sqrt{k}\right) - \sqrt{0} - \sqrt{1} \approx 57.1938 - 0 - 1 = 56.1938.$$

We can now use Theorem 4.2 to combine these sums and find the value of the desired sum:

$$\begin{aligned}\sum_{k=2}^{19}\left(\frac{2}{k} + \sqrt{k}\right) &= \sum_{k=2}^{19} \frac{2}{k} + \sum_{k=2}^{19} \sqrt{k} && \leftarrow \text{sum of sums}\\ &= 2\sum_{k=2}^{19} \frac{1}{k} + \sum_{k=2}^{19} \sqrt{k} && \leftarrow \text{constant times a sum}\\ &\approx 2(2.5477) + 56.1938 = 61.2892. && \leftarrow \text{add previously computed values}\end{aligned}$$

□

EXAMPLE 5

Using sum formulas to calculate the value of a sum

Find the value of the sum $\sum_{k=1}^{300} (k^3 - 4k^2 + 2)$.

SOLUTION

By writing the sum as three simple sums and using the formulas in Theorem 4.4, we have

$$\begin{aligned}\sum_{k=1}^{300}(k^3-4k^2+2) &= \left(\sum_{k=1}^{300}k^3\right)-4\left(\sum_{k=1}^{300}k^2\right)+2\left(\sum_{k=1}^{300}1\right) && \leftarrow \text{properties of sums}\\ &= \frac{300^2(301)^2}{4}-4\frac{300(301)(601)}{6}+2(300) && \leftarrow \text{Theorem 4.4}\\ &= 2,002,342,900.\end{aligned}$$

Of course that was a whole lot easier than expanding the sum and then adding all 300 terms! □

EXAMPLE 6

Using a sum formula to calculate the limit of a sum as $n \to \infty$

Find the values of the sum $\sum_{k=1}^{n} \frac{k^2}{n^3}$ for $n = 3$, $n = 4$, $n = 100$, and $n = 1000$. Then investigate what happens as n approaches infinity.

SOLUTION

To find the sums for $n = 3$ and $n = 4$, it is easy to expand and calculate the sums directly:

$$\sum_{k=1}^{3}\frac{k^2}{n^3}=\frac{1^2}{3^3}+\frac{2^2}{3^3}+\frac{3^2}{3^3}=\frac{1}{27}+\frac{4}{27}+\frac{9}{27}=\frac{14}{27};$$

$$\sum_{k=1}^{4}\frac{k^2}{n^3}=\frac{1^2}{4^3}+\frac{2^2}{4^3}+\frac{3^2}{4^3}+\frac{4^2}{4^3}=\frac{1}{64}+\frac{4}{64}+\frac{9}{64}+\frac{16}{64}=\frac{30}{64}=\frac{15}{32}.$$

Notice that the denominator of each term in the first sum is constantly equal to n^3 and does not change as k changes. Notice also that the sum from $n = 1$ to $n = 4$ *cannot* be obtained from the sum from $n = 1$ to $n = 3$ just by adding a fourth term; *all* the terms in the second sum are different from the terms in the first sum. In fact, the second sum is actually *smaller* than the first sum, since $\frac{14}{27} \approx 0.5185$ and $\frac{15}{32} = 0.46875$. Although we are adding more terms in the second sum, each of those terms is smaller than the terms in the first sum (since their denominators are larger).

To find the sum for $n = 100$, we will use properties of sums and one of the sum formulas from Theorem 4.4:

$$\begin{aligned}\sum_{k=1}^{100}\frac{k^2}{n^3} &= \sum_{k=1}^{100}\frac{k^2}{100^3}=\frac{1}{100^3}\sum_{k=1}^{100}k^2 && \leftarrow \text{pull out the constant } \tfrac{1}{100^3}\\ &= \frac{1}{100^3}\frac{100(101)(201)}{6} && \leftarrow \text{sum formula from Theorem 4.4}\\ &= \frac{2,030,100}{6,000,000}=0.33835.\end{aligned}$$

Similarly, the sum from $k = 1$ to $k = 1000$ is

$$\begin{aligned}\sum_{k=1}^{1000}\frac{k^2}{n^3} &= \sum_{k=1}^{1000}\frac{k^2}{1000^3}=\frac{1}{1000^3}\sum_{k=1}^{1000}k^2\\ &= \frac{1}{1000^3}\left(\frac{1000(1001)(2001)}{6}\right)=\frac{2,003,001,000}{6,000,000,000}\approx 0.33383.\end{aligned}$$

Notice that even though we just summed up a thousand terms, the sum is a very small number. In fact, the more terms we sum up, the smaller that number seems to get! It looks like if we let n continue to grow, the sum from $k = 1$ to $k = n$ will approach a number that is approximately 0.33 or so. We can show that this is indeed the case by taking a limit of the formula we used to compute the two previous sums:

$$\lim_{n\to\infty} \sum_{k=1}^{n} \frac{k^2}{n^3} = \lim_{n\to\infty} \frac{1}{n^3} \sum_{k=1}^{n} k^2 \qquad \leftarrow n \text{ is constant with respect to } k$$

$$= \lim_{n\to\infty} \frac{1}{n^3}\left(\frac{n(n+1)(2n+1)}{6}\right) \qquad \leftarrow \text{sum formula from Theorem 4.4}$$

$$= \lim_{n\to\infty} \frac{(n+1)(2n+1)}{6n^2} \qquad \leftarrow \text{algebra}$$

$$= \lim_{n\to\infty} \frac{2n^2+3n+1}{6n^2} = \frac{2}{6} = \frac{1}{3}. \qquad \leftarrow \text{take the limit}$$

□

? TEST YOUR UNDERSTANDING

- Considering the stoplight example at the beginning of this section, how can the approximate distances travelled be interpreted as rectangles? How is exact distance travelled related to the area under a velocity curve?
- Can you find an example in which $\sum_{k=1}^{n}(a_k b_k)$ is *not* equal to the product of $\sum_{k=1}^{n} a_k$ and $\sum_{k=1}^{n} b_k$?
- Do two sums in sigma notation have to start and end at the same index value in order to be combined into one sum? Why or why not?
- How can we express the sum in Example 2 by using sigma notation that starts at $k = 4$ and ends at $k = 10$?
- How could a sum of infinitely many things ever be a small finite number, as happened in Example 6?

EXERCISES 4.1

Thinking Back

- *Approximations and limits:* Describe in your own words how the slope of a tangent line can be approximated by the slope of a nearby secant line. Then describe how the derivative of a function at a point is defined as a limit of slopes of secant lines. What is the approximation/limit situation described in this section?
- *Properties of addition:* State the associative law for addition, the commutative law for addition, and the distributive law for multiplication over addition of real numbers. (You may have to think back to a previous algebra course.)
- *Sum and constant-multiple rules:* State the sum and constant-multiple rules for (a) derivatives and (b) limits.

Concepts

0. *Problem Zero:* Read the section and make your own summary of the material.

1. *True/False:* Determine whether each of the statements that follow is true or false. If a statement is true, explain why. If a statement is false, provide a counterexample.

(a) *True or False:* The sum formulas in Theorem 4.4 can be applied only to sums whose starting index value is $k = 1$.

(b) *True or False:* $\sum_{k=0}^{n} \frac{1}{k+1} + \sum_{k=1}^{n} k^2$ is equal to $\sum_{k=0}^{n} \frac{k^3+k^2+1}{k+1}$.

(c) *True or False:* $\sum_{k=1}^{n} \frac{1}{k+1} + \sum_{k=0}^{n} k^2$ is equal to $\sum_{k=1}^{n} \frac{k^3+k^2+1}{k+1}$.

(d) *True or False:* $\left(\sum_{k=1}^{n} \frac{1}{k+1}\right)\left(\sum_{k=1}^{n} k^2\right)$ is equal to $\sum_{k=1}^{n} \frac{k^2}{k+1}$.

(e) *True or False:* $\sum_{k=0}^{m} \sqrt{k} + \sum_{k=m}^{n} \sqrt{k}$ is equal to $\sum_{k=0}^{n} \sqrt{k}$.

(f) *True or False:* $\sum_{k=0}^{n} a_k = -a_0 - a_n + \sum_{k=1}^{n-1} a_k$.

(g) *True or False:* $\left(\sum_{k=1}^{10} a_k\right)^2 = \sum_{k=1}^{10} a_k^2$.

(h) *True or False:* $\sum_{k=1}^{n} (e^x)^2 = \dfrac{e^x(e^x+1)(2e^x+1)}{6}$.

2. *Examples:* Construct examples of the thing(s) described in the following. Try to find examples that are different than any in the reading.
   (a) A sum that would not be suitable for expressing in sigma notation.
   (b) Two different sigma notation expressions of the same sum.
   (c) A sum from $k = 1$ to $k = n$ that is smaller for $n = 10$ than it is for $n = 5$.

3. Explain why it would be difficult to write the sum $\frac{1}{3} + \frac{1}{4} + \frac{1}{5} + \frac{1}{8} + \frac{1}{11} + \frac{1}{12} + \frac{1}{13}$ in sigma notation.

4. Use a sentence to describe what the notation $\sum_{k=2}^{100} \sqrt{k}$ means. (*Hint: Start with "The sum of...."*)

5. Use a sentence to describe what the notation $\sum_{k=3}^{87} k^2$ means. (*Hint: Start with "The sum of...."*)

6. Consider the general sigma notation $\sum_{k=m}^{n} a_k$. What do we mean when we say that a_k is a function of k?

7. Consider the sum $\sum_{i=p}^{q} b_i$.
   (a) Write out this sum in expanded form (i.e., without sigma notation).
   (b) What is the index of the sum? What is the starting value? What is the ending value? Which part of the notation describes the form of each of the terms in the sum?
   (c) Do p and q have to be integers? Can they be negative? What about b_i? What else can you say about p and q?

8. Consider the sum $\sum_{k=2}^{5} \frac{k}{1-k}$. Identify the terms a_2, a_3, a_4, and a_5.

9. Consider the sum $\sum_{k=m}^{n} a_k = 9 + 16 + 25 + 36 + 49$. What is a_k? What is m? What is n?

10. Show that $\sum_{k=3}^{9} \frac{k}{k+1}$ is equal to $\sum_{k=4}^{10} \frac{k-1}{k}$ by writing out the terms in each sum.

11. Show that $\sum_{k=0}^{8} \frac{1}{k^2+1}$ is equal to $2\sum_{k=0}^{8} \frac{1}{2k^2+2}$ by writing out the terms in each sum.

12. Write the sum $\frac{4}{7} + \frac{5}{8} + \frac{6}{9} + \frac{7}{10} + \frac{8}{11}$ in sigma notation in three ways: with a starting value of (a) $k = 4$, (b) $k = 7$, and (c) $k = 5$.

13. Write the sum $2 + \frac{2}{4} + \frac{2}{9} + \frac{2}{16} + \frac{2}{25}$ in sigma notation in three ways: with a starting value of (a) $k = 1$, (b) $k = 2$, and (c) $k = 0$.

14. Split the sum $\sum_{k=4}^{11} \sqrt{k}$ into three sums, each in sigma notation, where the first sum has two terms and the last two sums each have three terms.

15. Verify that $\sum_{k=1}^{n} k$ is equal to $\frac{n(n+1)}{2}$ for the cases (a) $n = 2$, (b) $n = 8$, and (c) $n = 9$.

16. Verify that $\sum_{k=1}^{n} k^2$ is equal to $\frac{n(n+1)(2n+1)}{6}$ for the cases (a) $n = 1$, (b) $n = 5$, and (c) $n = 10$.

17. State algebraic formulas that express the following sums, where n is a positive integer:

    (a) $\displaystyle\sum_{k=1}^{n} 1$ (b) $\displaystyle\sum_{k=1}^{n} k$ (c) $\displaystyle\sum_{k=1}^{n} k^2$ (d) $\displaystyle\sum_{k=1}^{n} k^3$

18. Explain why terms in the sum in Example 6 with $n = 4$ are completely different from the terms in the sum when $n = 3$. How can the sum from $k = 1$ to $k = 4$ be *smaller* than the sum from $k = 1$ to $k = 3$? What will happen as n gets larger in this example?

19. Considering the discussion at the end of the stoplight example in the reading, would you expect that the area under the graph of a function f is related to the derivative f'? Or would you expect that the area under the graph of a derivative function f' is related to the function f?

20. Consider again the stoplight example from the reading. In making an approximation for distance travelled, why do we assume that velocity is constant on small subintervals? What are some different ways that we could choose which velocity to use on each subinterval? Illustrate a couple of these ways with graphs that involve rectangles.

Skills

Write each of the sums in Exercises 21–28 in sigma notation. Identify m, n, and a_k in each problem.

21. $3 + 3 + 3 + 3 + 3 + 3 + 3 + 3$

22. $\frac{4}{3} + \frac{5}{4} + \frac{6}{5} + \frac{7}{6} + \frac{8}{7} + \frac{9}{8} + \frac{10}{9} + \frac{11}{10}$

23. $3 + \frac{4}{8} + \frac{5}{27} + \frac{6}{64} + \frac{7}{125}$

24. $\frac{1}{4} + \frac{1}{9} + \frac{1}{16} + \frac{1}{25} + \frac{1}{36} + \frac{1}{49} + \frac{1}{64}$

25. $5 + 10 + 17 + 26 + 37 + 50 + 65 + 82 + 101$

26. $9 + 12 + 15 + 18 + 21 + 24 + 27$

27. $\frac{1}{n} + \frac{2}{n} + \frac{3}{n} + \cdots + \frac{n}{n}$

28. $-2^n - 1^n + 0^n + 1^n + \cdots + n^n$

Write out each sum in Exercises 29–34 in expanded form, and then calculate the value of the sum.

29. $\displaystyle\sum_{k=4}^{9} k^2$

30. $\displaystyle\sum_{k=0}^{6} \frac{2}{k+1}$

31. $\displaystyle\sum_{k=0}^{5} \left(\frac{1}{2}k\right)^2 \left(\frac{1}{2}\right)$

32. $\displaystyle\sum_{k=3}^{10} \ln k$

33. $\displaystyle\sum_{k=0}^{9} \sqrt{3 + \frac{1}{10}k} \left(\frac{1}{10}\right)$

34. $\displaystyle\sum_{k=1}^{4} ((2+k)^2 + 1)$

Find a formula for each of the sums in Exercises 35–40, and then use these formulas to calculate each sum for $n = 100$, $n = 500$, and $n = 1000$.

35. $\sum_{k=1}^{n}(3-k)$

36. $\sum_{k=1}^{n}(k^3 - 10k^2 + 2)$

37. $\sum_{k=3}^{n}(k+1)^2$

38. $\sum_{k=1}^{n}\frac{k^3-1}{4}$

39. $\sum_{k=1}^{n}\frac{k^3-1}{n^4}$

40. $\sum_{k=1}^{n}\frac{k^2+k+1}{n^3}$

Write each expression in Exercises 41–43 in one sigma notation (with some extra terms added to or subtracted from the sum, as necessary).

41. $2\sum_{k=0}^{100} a_k - \sum_{k=3}^{101} a_k$

42. $\sum_{k=1}^{40}\frac{1}{k} - \sum_{k=0}^{39}\frac{1}{k+1}$

43. $3\sum_{k=2}^{25} k^2 + 2\sum_{k=2}^{24} k - \sum_{k=0}^{25} 1$

In Exercises 44–46, find the sum or quantity without completely expanding or calculating any sums.

44. Given $\sum_{k=3}^{10} a_k = 12$ and $\sum_{k=2}^{10} a_k = 23$, find a_2.

45. Given $\sum_{k=1}^{4} a_k = 7$, $\sum_{k=0}^{4} b_k = 10$, and $a_0 = 2$, find the value of $\sum_{k=0}^{4}(2a_k + 3b_k)$.

46. Given $\sum_{k=0}^{25} k = 325$ and $\sum_{k=3}^{28}(k-3)^2 = 14,910$, find the value of $\sum_{k=3}^{25}(k^2 - 5k + 9)$.

Determine which of the limit of sums in Exercises 47–52 are infinite and which are finite. For each limit of sums that is finite, compute its value.

47. $\lim_{n\to\infty}\sum_{k=1}^{n}\frac{k^2+k+1}{n^3}$

48. $\lim_{n\to\infty}\sum_{k=1}^{n}(k^2+k+1)$

49. $\lim_{n\to\infty}\sum_{k=1}^{n}\frac{(k+1)^2}{n^3-1}$

50. $\lim_{n\to\infty}\sum_{k=1}^{n}\frac{k^2+k+1}{n^2}$

51. $\lim_{n\to\infty}\sum_{k=1}^{n}\left(1+\frac{k}{n}\right)^2\cdot\frac{1}{n}$

52. $\lim_{n\to\infty}\sum_{k=1}^{n}\frac{k^3}{n^4+n+1}$

Applications

53. Considering the stoplight example in the reading with velocity $v(t) = -0.22t^2 + 8.8t$ as shown next at the left, approximate the distance travelled by dividing the time interval $[0, 40]$ into eight pieces and assuming constant velocity on each piece. Interpret this distance in terms of rectangles on the graph of $v(t)$.

Velocity of car
$v(t) = -0.22t^2 + 8.8t$

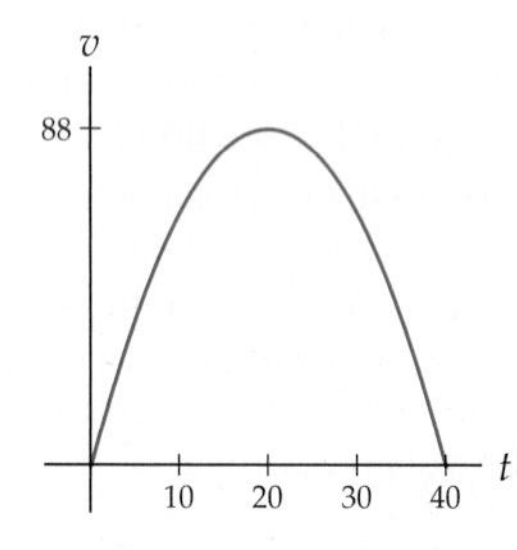

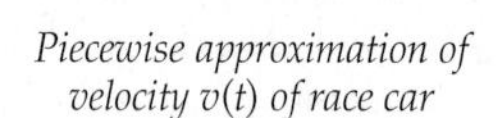
Piecewise approximation of velocity $v(t)$ of race car

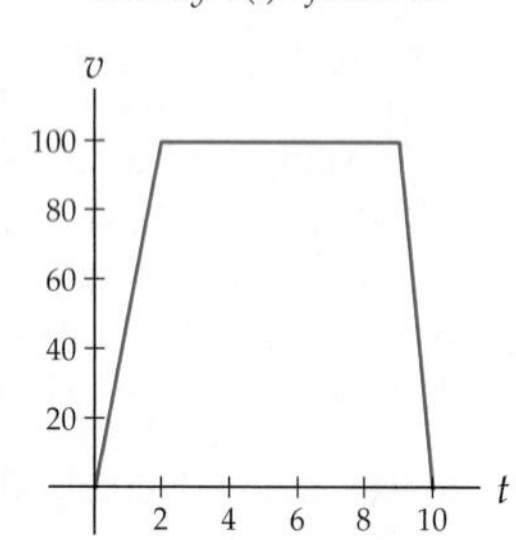

54. Suppose you drive on a racetrack for 10 minutes with velocity as shown in the graph at the right.

(a) Describe in words the behavior of your race car over the 10 minutes as shown in the graph.

(b) Find a piecewise-defined formula for your velocity $v(t)$, in miles per hour, t hours after you start from rest. (Note that 1 minute is $\frac{1}{60}$ of an hour.)

(c) Approximate the distance you travelled over the 10 minutes by using 10 subintervals of 1 minute over which you assume a constant velocity. Illustrate this approximation by showing rectangles on the graph of $v(t)$.

(d) Given that distance travelled is the area under the velocity graph, use triangles and squares to calculate the exact distance travelled.

55. The table that follows describes the activity in a college tuition savings account over four years. Notice that 2008 was a particularly bad year for investing! Let $I(t)$ be the amount by which your account increased or decreased in year t, and let $B(t)$ be the balance of your account at the end of year t.

Activity in tuition savings account

Year	2005	2006	2007	2008
Deposited	\$ 600	\$ 1200	\$ 1200	\$ 1200
Earnings	\$ 10	\$ 183	\$ 317	\$ −1650
Increase	\$ 610	\$ 1383	\$ 1517	\$ −450
Balance	\$ 610	\$ 1994	\$ 3512	\$ 3061

(a) Describe in your own words how $B(t)$ is the accumulation function for $I(t)$.

(b) Plot a step-function graph of $I(t)$, and describe how $B(t)$ relates to the area under that graph.

(c) What, if anything, can you say about $B(t)$ when $I(t)$ is positive? Negative? If you had to guess that one of these functions was related to the derivative of the other, which one would it be?

56. Suppose 100 mg of a drug is administered to a patient each morning in pill form and it is known that after 24 hours the body processes 80% of the drug from such a pill, leaving 20% of the drug in the body. The amount of the drug in the body right after the first pill is taken is $A(1) = 100$ mg. 24 hours later, after the second pill has been taken, the amount in the body is $A(2) = 100(0.2) + 100 = 120$ mg. 48 hours later, the amount in the body right after taking the third pill is $A(3) = 100(0.2)(0.2) + 100(0.2) + 100 = 124$ mg.

Repeated doses of a drug

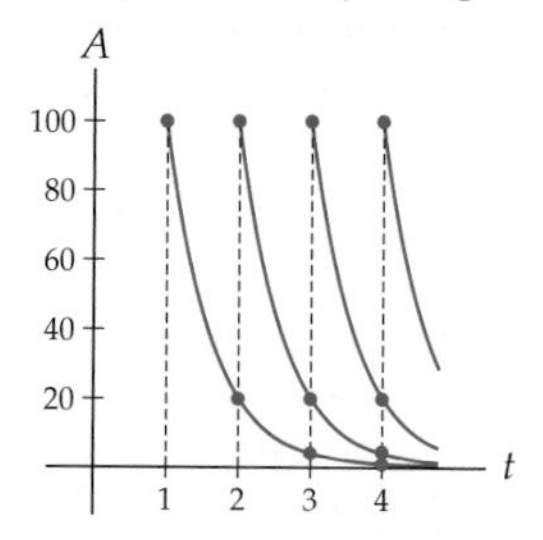

(a) Explain the calculations for $A(1)$, $A(2)$, and $A(3)$ described in the exercise. Which term in $A(3)$ corresponds to the drug left from the first pill?
(b) Interpret the given graph in the context of this problem. What do the marked points represent?
(c) Express $A(n)$ in sigma notation.
(d) Calculate the amount of drug in the body after the 4th through 10th pills. Do you notice anything special about $A(n)$ as n gets larger?

57. Suppose the government enacts a \$10 billion tax cut and that the people who save money from this tax cut will spend 70% of it and save the rest. This generates $\$10(0.7) = \$7$ billion of extra income for other people. Assume these people also spend 70% of their extra income, and that these transactions continue. Let $A(n)$ be the accumulated amount of spending, in billions of dollars, that has occurred after n such transactions. For example, $A(1)$ is the amount of spending that has occurred after the first group of people has spent its money, so $A(1) = \$10(0.7) = \$7$ billion. $A(2)$ is the amount of spending that has occurred after the first and second groups of people have spent their money, so $A(2) = \$10(0.7) + \$7(0.7) = \$11.9$ billion, as shown in the following graph:

Accumulation of tax cut spending

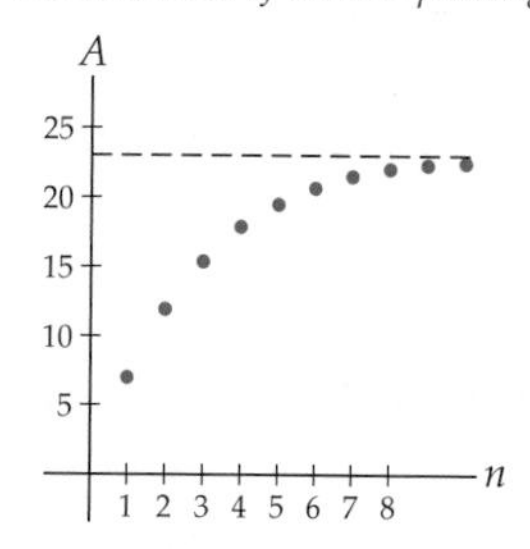

(a) Express $A(n)$ in sigma notation.
(b) Calculate $A(3)$, $A(4)$, and $A(5)$.
(c) Estimate the total spending created by this tax cut by calculating the accumulated spending for at least 10 of these transactions. Interpret your answer in terms of the given graph.

Proofs

58. Give a simple proof that $\sum_{k=5}^{n}(a_k + b_k) = \sum_{k=5}^{n} a_k + \sum_{k=5}^{n} b_k$.

59. Give a simple proof that $\sum_{k=0}^{n} 3a_k = 3\sum_{k=0}^{n} a_k$.

60. Give a simple proof that if n is a positive integer and c is any real number, then $\sum_{k=1}^{n} c = cn$.

61. Prove part (b) of Theorem 4.4 in the case when n is even: If n is a positive even integer, then $\sum_{k=1}^{n} k = \frac{n(n+1)}{2}$. (*Hint: Try some examples first, such as $n = 6$ and $n = 10$, and think about how to group the terms to get the sum quickly.*)

62. Prove part (b) of Theorem 4.4 in the case when n is odd: If n is a positive odd integer, then $\sum_{k=1}^{n} k = \frac{n(n+1)}{2}$. (*Hint: Use a method similar to the one for the previous exercise, but take note of what happens with the extra middle term of the sum.*)

Thinking Forward

Functions defined by area accumulation: Let f be the function that is shown here at the left, and define a new function A so that for every $c \geq 1$, $A(c)$ is the area of the region between the graph of f and the x-axis over the interval $[1, c]$. For example, $A(2)$ is the area of the shaded region in the graph at the right.

- Use the graph of f to estimate the values of $A(1)$, $A(2)$, and $A(3)$. (*Hint: Consider the grid lines in the graph shown at the right.*)
- Describe the intervals on which the function f is positive, negative, increasing, and decreasing. Then describe the intervals on which the function A is positive, negative, increasing, and decreasing.
- From the figures, we can see that f is increasing and positive on $[1, \infty)$ and A is also increasing and positive on $[1, \infty)$. What would you be able to say about the area accumulation function A if f were instead decreasing and positive? Or increasing and negative? Draw some pictures in your investigation.

Graph of $y = f(x)$

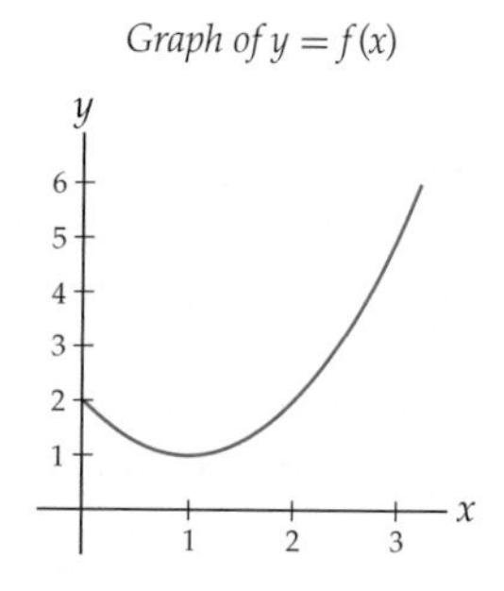

The shaded area is $A(2)$

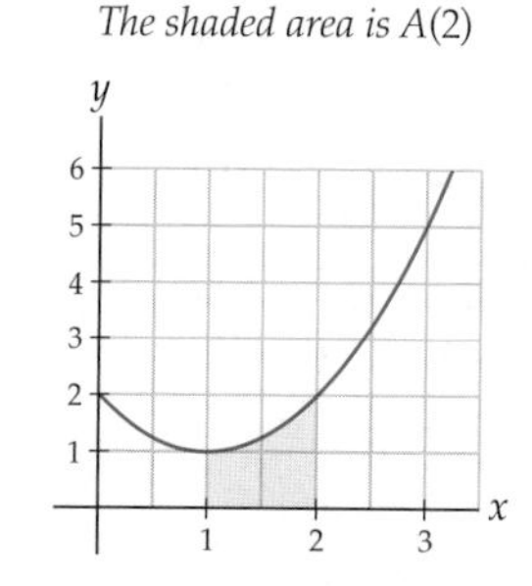

Approximating the area under a curve with rectangles: Suppose you want to find the area between the graph of a positive function f and the x-axis from $x = a$ to $x = b$. We can approximate such an area by using a sum of areas of small rectangles whose heights depend on the height $f(x)$ at various points. For the problems that follow you should choose rectangles so that each rectangle has the same width and the top left corner of each rectangle intersects the graph of f.

- Approximate the area between the graph of $f(x) = x^2$ and the x-axis from $x = 0$ to $x = 4$, by using four rectangles. Include a picture of the rectangles that you are using.
- Approximate the same area as earlier, but this time with eight rectangles. Is this an over-approximation or an under-approximation of the exact area under the graph?

Sequences of partial sums: In Exercise 57 we saw a function $A(n)$ that was defined as a sum of n terms, $A(n) = \sum_{k=1}^{n} 10(0.7)^k$. What happens as n approaches infinity? The sum $A(n)$ is called a *partial sum* because it represents part of the sum that accumulates if you let n approach infinity.

- Consider the sequence $A(1)$, $A(2)$, $A(3)$, $\ldots$, $A(n)$. Write out this sequence up to $n = 10$. What do you notice?
- As n approaches infinity, this sequence of partial sums could either *converge*, meaning that the terms eventually approach some finite limit, or it could *diverge* to infinity, meaning that the terms eventually grow without bound. Which do you think is the case here, and why?

4.2 RIEMANN SUMS

- Geometric approximation by the process of subdividing, approximating, and adding up
- Using rectangles to approximate the area under a curve
- Definition and types of Riemann sums in formal mathematical notation

Subdivide, Approximate, and Add Up

As you well know, the formula for finding the area of a circle of radius r is $A = \pi r^2$. In particular, a circle of radius 2 units has area $A = \pi 2^2 = 4\pi$. But wait a moment; where does this area formula come from? Why is it true? Suppose for a moment that we don't know the area formula. How could we find, or at least approximate, the area of a circle of radius 2? The three diagrams that follow suggest an answer.

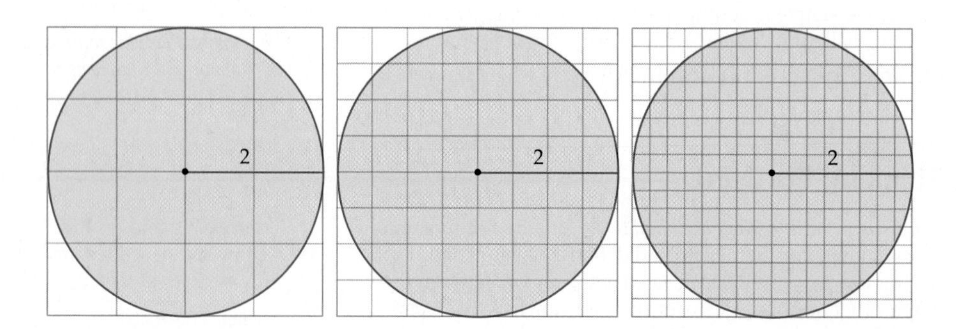

In the figure at the left, a circle of radius 2 units is shown with a grid where each square has side length 1 unit and thus area 1 square unit. We need only count up the approximate number of squares to approximate the area of the circle. The circle encloses four full squares, and 12 partial squares. We will approximate by counting each partial square as *half* of a square. This is just one of many approximation methods we could use. This method produces the approximation

$$A \approx 4(1) + 12\left(\frac{1}{2}\right) = 10 \text{ square units.}$$

As we know, the actual area of the circle is $4\pi \approx 12.5664$, so the approximation we found is not very accurate. If instead we use the grid in the second figure shown previously, we can obtain a better approximation. In this case, the circle covers 32 full squares (each of area $\left(\frac{1}{2}\right)^2 = \frac{1}{4}$) and 28 partial squares (which we will count as half-squares, or $\frac{1}{8}$ square unit each). We now have the approximation

$$A \approx 32\left(\frac{1}{4}\right) + 28\left(\frac{1}{8}\right) = 11.5 \text{ square units.}$$

This approximation is closer to the area we expected, since the squares we used were smaller. An even better approximation can be obtained by using the rightmost figure shown earlier, in which there are 164 full squares of area $\left(\frac{1}{4}\right)^2 = \frac{1}{16}$ and 60 partial squares that we'll count as having an area of $\frac{1}{32}$ square units each. This gives us the even better approximation

$$A \approx 164\left(\frac{1}{16}\right) + 60\left(\frac{1}{32}\right) = 12.125 \text{ square units.}$$

We just used a grid to subdivide a circle into a number of smaller pieces, approximated the area of each of those pieces (by counting each piece as either a full square or a half-square), and then added up each of the small approximated areas. This process of "subdividing, approximating, and adding up" is the cornerstone of the ***definite integral***, which we will introduce in Section 4.3. After we have learned the theory of integrals, we will be able to *prove* that the area formula $A = \pi r^2$ for a circle of radius r is correct. (See Section 5.5.)

Before moving on, notice that we could think of each one of our approximations of the area of a circle as the output of a sort of approximation function whose inputs are the possible grid sizes. Smaller grid sizes should produce more accurate approximations. As the grid sizes get smaller, if those approximations somehow stabilize at a real number, then we will say that the area of the object is the number to which the approximations stabilize. This is yet another place where our study of limits will pay off! We will make the ideas in this paragraph more precise in Section 4.3.

Approximating Area with Rectangles

In the rest of this chapter we will be concerned primarily with finding or approximating the area under a curve, that is, the area enclosed between the graph of some function and the x-axis on some interval $[a, b]$. To keep things simple we will begin by restricting our attention to positive functions. For example, we might be interested in calculating the area between the graph of the function f and the x-axis from $x = 0$ to $x = 2$ as shown here:

Area between f and the x-axis on $[0, 2]$

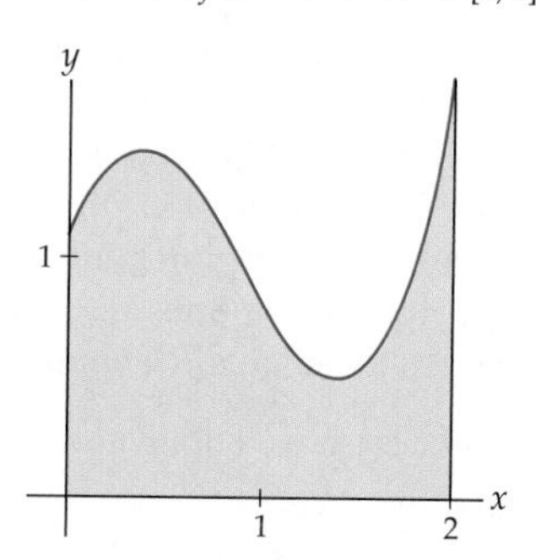

We could use a grid of squares to approximate this area, but it is more efficient to use a set of rectangles whose heights depend on the height of the function f. For example, we could subdivide the interval $[0, 2]$ into four smaller intervals of width $\frac{1}{2}$ as shown next at the left. Then we could use the heights of f at the leftmost points of the subintervals to define four rectangles, as shown in the middle and rightmost figures.

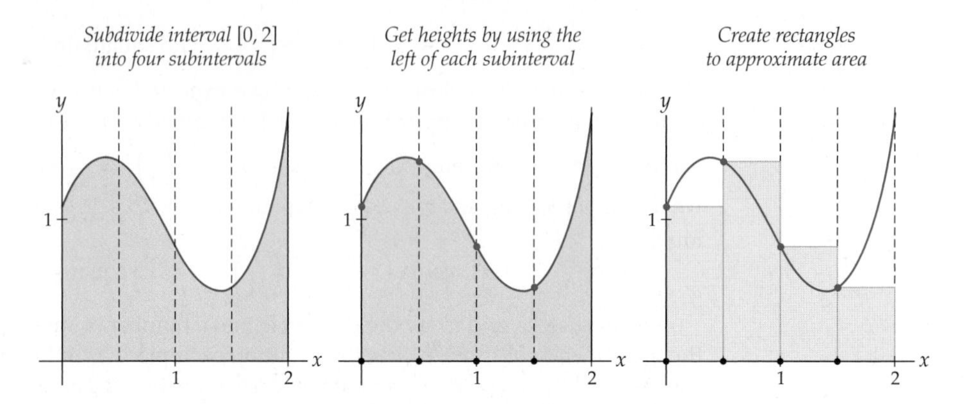

In the third figure, the first rectangle has height $f(0)$ and width $\frac{1}{2}$. Therefore, the area of the first rectangle is $f(0)\left(\frac{1}{2}\right)$. The remaining three rectangles have heights $f(0.5)$, $f(1)$, and $f(1.5)$, respectively, all with width $\frac{1}{2}$. Therefore the sum of the areas of these four rectangles is

$$f(0)\left(\frac{1}{2}\right) + f(0.5)\left(\frac{1}{2}\right) + f(1)\left(\frac{1}{2}\right) + f(1.5)\left(\frac{1}{2}\right).$$

The function f we have been working with happens to have the unwieldy equation

$$f(x) = \frac{1}{80}(125x^3 - 325x^2 + 175x + 89).$$

Evaluating $f(x)$ at $x = 0$, $x = 0.5$, $x = 1$, and $x = 1.5$, we see that the sum of the areas of the four rectangles is approximately

$$(1.11)\left(\frac{1}{2}\right) + (1.39)\left(\frac{1}{2}\right) + (0.80)\left(\frac{1}{2}\right) + (0.53)\left(\frac{1}{2}\right) \approx 1.915.$$

The combined area of the rectangles represents a rough approximation of the area under the graph of f on $[0, 2]$. Notice that this answer is reasonable, since the region whose area we are approximating looks to be about half of the size of the square defined by $0 \le x \le 2$ and $0 \le y \le 2$, a square with area 4.

Riemann Sums

We will now develop mathematical notation to formalize the method of using rectangles to approximate the area under a curve. This formalization will eventually enable us to apply limits so that we can find the *actual* area under a curve. Suppose f is a function that is nonnegative on an interval $[a, b]$. Consider the area under the graph of f on this interval, that is, the area of the region bounded above by the graph of f, below by the x-axis, and to the left and right by the lines $x = a$ and $x = b$. We will use the *subdivide, approximate, and add up* process to define a ***Riemann sum*** that will help us to approximate this area.

Subdivide: First, we subdivide the interval $[a, b]$ into n subintervals of equal width Δx. This means that $\Delta x = \frac{b-a}{n}$. We will give names to the subdivision points between the subintervals as shown next at the left. Notice that $x_0 = a$, $x_1 = a + \Delta x$, $x_2 = a + 2\Delta x$, and so on, until we end at $x_n = b$. In general, the kth subdivision point in the subdivision x_k is equal to $a + k\Delta x$.

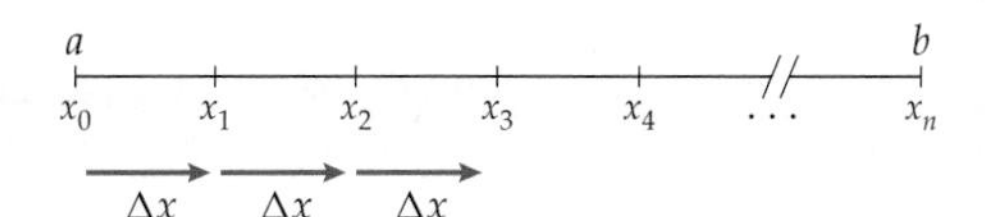

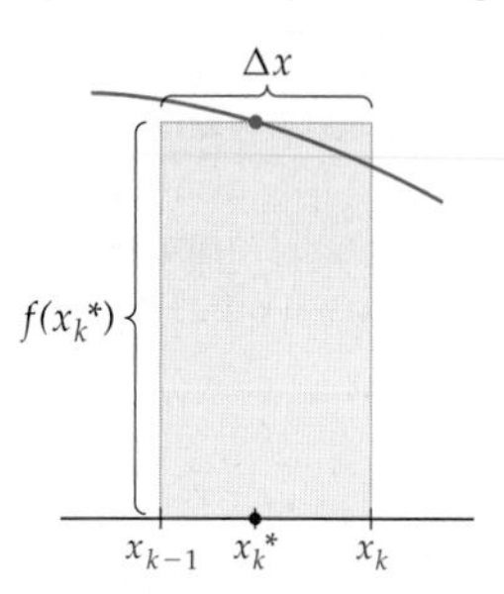

Approximate: We now need to define, for each subinterval $[x_{k-1}, x_k]$, a rectangle that approximates the area under the graph of f on that subinterval. We need to choose a height for the rectangle so that the height is somehow related to the height of the graph of f. In our previous example we chose the leftmost point of each subinterval, but in general we can choose *any* point x_k^* in $[x_{k-1}, x_k]$. The star in the x_k^* is supposed to suggest that we are choosing any point we like in that subinterval. Now we can define a rectangle over the kth subinterval $[x_{k-1}, x_k]$ such that the height of the rectangle is $f(x_k^*)$, as shown previously at the right.

Add up: If we choose n large enough—or equivalently, make Δx small enough—then the combined areas of the rectangles we define with the points x_k^* will be a decent approximation for the area under the curve. It now remains only to add up the areas of these rectangles. The area of the kth rectangle is its height times its width, or $f(x_k^*)\Delta x$, so the total area under the curve is approximated by the sum

$$f(x_1^*)\Delta x + f(x_2^*)\Delta x + f(x_3^*)\Delta x + \cdots + f(x_k^*)\Delta x + \cdots + f(x_n^*)\Delta x.$$

This sum can be expressed very compactly in sigma notation:

$$\sum_{k=1}^{n} f(x_k^*)\Delta x.$$

Note that to find this sum, we followed a "subdivide, approximate, and add" process. Sums of this form are called ***Riemann sums***, named for the prolific mathematician Bernhard Riemann, who developed the notion. The following definition summarizes the notation that we have developed:

DEFINITION 4.5

Riemann Sums

A ***Riemann sum*** for a function f on an interval $[a, b]$ is a sum of the form

$$\sum_{k=1}^{n} f(x_k^*)\,\Delta x,$$

where $\Delta x = \frac{b-a}{n}$, $x_k = a + k\Delta x$, and x_k^* is some point in the interval $[x_{k-1}, x_k]$.

As we have seen, a Riemann sum for a function f on an interval $[a, b]$ approximates the area under the graph of f between $x = a$ and $x = b$. If f is continuous on $[a, b]$, then as the number of rectangles n approaches infinity, the approximation will get better and better and approach the actual area under the graph of f on $[a, b]$. We will make this idea mathematically precise in Section 4.3.

It is possible to consider Riemann sums for which the interval $[a, b]$ is partitioned into n subintervals by points $a = x_0, x_1, x_2, \ldots, x_{n-1}, x_n = b$ that are not equal distances apart. In such subdivisions we would have a different width $(\Delta x)_k$ for each subinterval $[x_{k-1}, x_k]$. This more general theory of Riemann sums defined with arbitrary partitions is important in later mathematics courses, but is not needed for our study of calculus here.

Signed Area

When the graph of a function lies below the x-axis, we can still use rectangles to approximate area. For example, consider the following Riemann sum with $n = 4$ rectangles that approximates the area of the region between the graph of $f(x) = x^2 - 1$ and the x-axis on $[0, 1]$:

A Riemann sum with $n = 4$ for $f(x) = x^2 - 1$ on $[0, 1]$

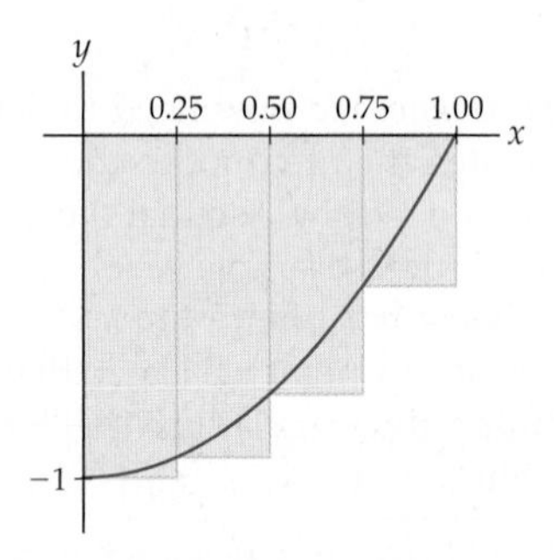

Note that the "height" of each of the rectangles in this Riemann sum is *negative*. For example, the first rectangle has a height of $f(0) = 0^2 - 1 = -1$ and is counted as having an "area" of -0.25. The left sum in this case is the sum of the areas of these four "negative" rectangles, $-0.25 - 0.234375 - 0.1875 - 0.109375 = -0.78125$.

Since, by construction, Riemann sums automatically count the area between a negative function and the x-axis as a negative number and the area between a positive function and the x-axis as a positive number, we will say that Riemann sums measure ***signed area***, also known as ***net area***. For example, if f is the following function graphed on $[0, 5]$, then the right sum, with 25 rectangles, will have some rectangles counting area positively and some rectangles counting area negatively, depending on whether f is above or below the x-axis:

Area is counted positively on $[0, 2]$ and $[4, 5]$, and negatively on $[2, 4]$

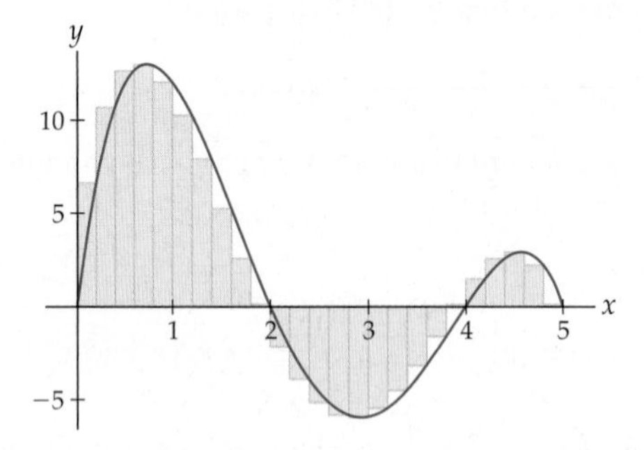

Types of Riemann Sums

Depending on how we choose the point x_k^* in each subinterval $[x_{k-1}, x_k]$, we get different types of Riemann sums. For example, we could choose each x_k^* to be the leftmost point of the kth subinterval, as shown in the first graph that follows; this is called a ***left sum***. Alternatively, we could construct a ***right sum*** by choosing x_k^* to be the rightmost point of the kth subinterval, or a ***midpoint sum*** by choosing x_k^* to be the midpoint, as shown in the following middle and rightmost graphs, respectively:

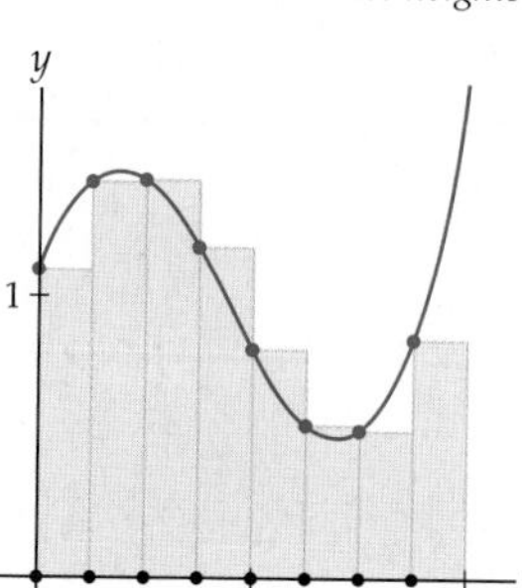

Leftmost point of each subinterval determines heights

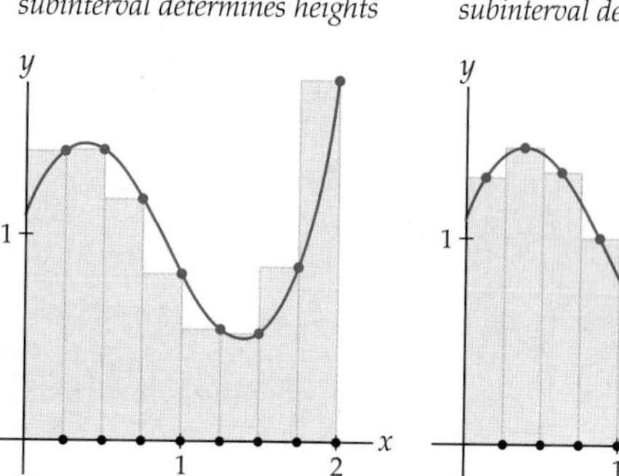

Rightmost point of each subinterval determines heights

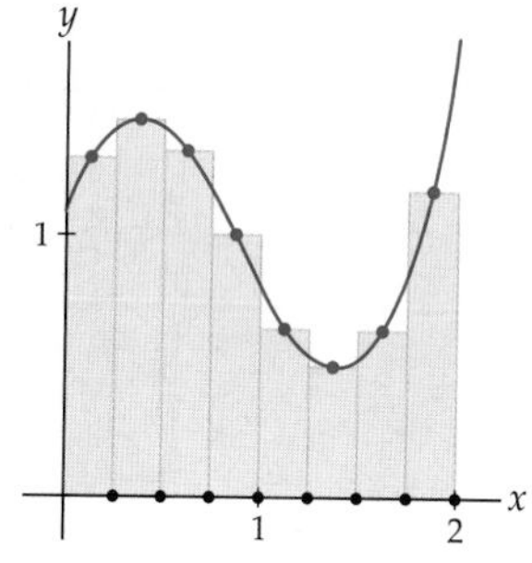

Midpoint of each subinterval determines heights

Algebraically, these three ways of choosing x_k^* from each subinterval $[x_{k-1}, x_k]$ yield the following three types of Riemann sums:

DEFINITION 4.6 **Left, Right, and Midpoint Sums**

Suppose f is a function defined on the interval $[a, b]$. Given a positive integer n, let $\Delta x = \frac{b-a}{n}$ and $x_k = a + k\Delta x$. Then

(a) The n-rectangle ***left sum*** for f on $[a, b]$ is $\sum_{k=1}^{n} f(x_{k-1})\,\Delta x$.

(b) The n-rectangle ***right sum*** for f on $[a, b]$ is $\sum_{k=1}^{n} f(x_k)\,\Delta x$.

(c) The n-rectangle ***midpoint sum*** for f on $[a, b]$ is $\sum_{k=1}^{n} f\left(\frac{x_{k-1}+x_k}{2}\right)\Delta x$.

Note that given any interval $[a, b]$ and number n of rectangles, we can write Δx and x_k in terms of a, b, and n. In practice, we will always need to use the explicit expressions $\Delta x = \frac{b-a}{n}$ and $x_k = a + k\Delta x$ (as well as using the definition of the function f) when evaluating a Riemann sum. For example, the right sum expressed earlier is equal to

$$\sum_{k=1}^{n} f\left(a + k\left(\frac{b-a}{n}\right)\right)\left(\frac{b-a}{n}\right).$$

We can also set up Riemann sums that we can guarantee will be over-approximations or under-approximations of the actual area under the graph of a continuous function. The first figure that follows shows the ***upper sum,*** where each x_k^* is chosen so that $f(x_k^*)$ is the maximum value of f on $[x_{k-1}, x_k]$. The upper sum is always greater than or equal to the actual signed area. Similarly, in the ***lower sum*** shown in the second figure we choose x_k^* so that $f(x_k^*)$ is the minimum value of f on $[x_{k-1}, x_k]$. The lower sum is always less than or equal to the actual signed area.

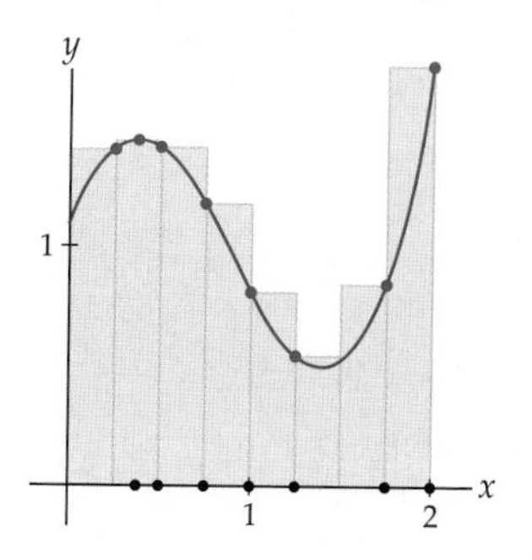

Maximum value of f on each subinterval determines heights

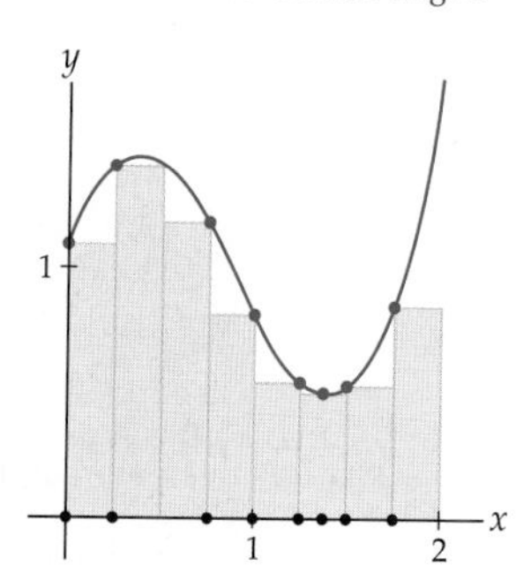

Minimum value of f on each subinterval determines heights

Algebraically, we can express these two types of Riemann sums as follows:

DEFINITION 4.7 **Upper and Lower Sums**

Suppose f is a function that is continuous on the interval $[a, b]$. Given a positive integer n, let $\Delta x = \frac{b-a}{n}$ and $x_k = a + k\Delta x$. Then

(a) The n-rectangle ***upper sum*** for f on $[a, b]$ is $\sum_{k=1}^{n} f(M_k)\,\Delta x$, where each M_k is chosen so that $f(M_k)$ is the maximum value of f on $[x_{k-1}, x_k]$.

(b) The n-rectangle ***lower sum*** for f on $[a, b]$ is $\sum_{k=1}^{n} f(m_k)\,\Delta x$, where each m_k is chosen so that $f(m_k)$ is the minimum value of f on $[x_{k-1}, x_k]$.

The reason we require that f be continuous on $[a, b]$ is that this is necessary in order for the Extreme Value Theorem to guarantee that f attains maximum and minimum values on each subinterval.

The upper sum and the lower sum can be more complicated to calculate than the left, right, or midpoint sums, but they can be useful if we wish to not only approximate the area under a curve, but get a bound on how much error is involved in our approximation. For example, if an upper sum approximation for the area under a curve was calculated to be 4.4 and a lower sum approximation for the same area was calculated to be 4.1, then we would know that the actual area was in the interval $[4.1, 4.4]$. In particular, this would mean that both the upper sum and the lower sum were within 0.3 square unit of the actual area under the curve.

We also don't necessarily have to use rectangles to approximate the area under a curve. For example, we could use trapezoids instead, as shown here:

Trapezoids on each subinterval approximate area under a curve

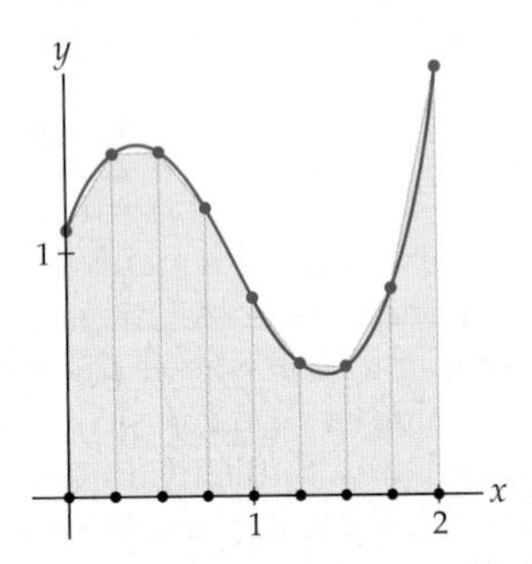

The ***trapezoid sum*** uses a trapezoid with width Δx, left height $f(x_{k-1})$, and right height $f(x_k)$ to approximate the slice of area in the subinterval $[x_{k-1}, x_k]$. Recall that the area of a trapezoid with base b and heights h_1 and h_2 is $\left(\frac{h_1+h_2}{2}\right) b$. This formula for the area suggests the following algebraic definition:

DEFINITION 4.8 **Trapezoid Sums**

Suppose f is a function defined on the interval $[a, b]$. Given a positive integer n, let $\Delta x = \frac{b-a}{n}$ and $x_k = a + k\Delta x$. Then

The n-rectangle ***trapezoid sum*** for f on $[a, b]$ is $\displaystyle\sum_{k=1}^{n} \frac{f(x_{k-1})+f(x_k)}{2}\,\Delta x$.

TECHNICAL POINT To be honest, in the form of Definition 4.8, the trapezoid sum is not technically a Riemann sum for f; recall that the terms in a Riemann sum must be of the form $f(x_k^*)\Delta x$ for some $x_k^* \in [x_{k-1}, x_k]$. However, if f is a continuous function, then in each subinterval we can use the Intermediate Value Theorem to find some x_k^* in for which $f(x_k^*)$ is equal to the average of $f(x_{k-1})$ and $f(x_k)$. This means that when f is continuous, the trapezoid sum really is a Riemann sum "in disguise."

Examples and Explorations

EXAMPLE 1

Approximating area with a right sum

Approximate the area between the graph of $f(x) = x^2 - 2x + 2$ and the x-axis from $x = 1$ to $x = 3$, using a right sum with four rectangles.

SOLUTION

We begin by subdividing the interval $[1, 3]$ into four subintervals: $[1, 1.5]$, $[1.5, 2]$, $[2, 2.5]$, and $[2.5, 3]$, each of width $\frac{1}{2}$. The rightmost points of these four subintervals are, respectively, 1.5, 2, 2.5, and 3. The values of f at these points will be the heights of our four rectangles, as shown here:

Right sum with four rectangles

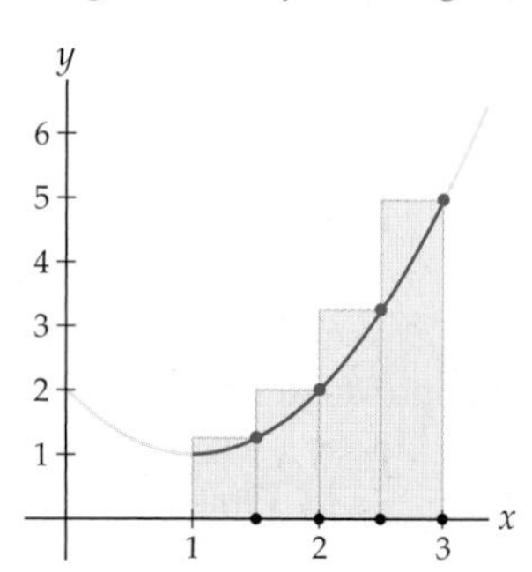

It is now a simple matter to add up the areas of the four rectangles:

$$\begin{aligned}\text{Area} &\approx f(1.5)\left(\tfrac{1}{2}\right) + f(2)\left(\tfrac{1}{2}\right) + f(2.5)\left(\tfrac{1}{2}\right) + f(3)\left(\tfrac{1}{2}\right) \\ &= (1.25)\left(\tfrac{1}{2}\right) + (2)\left(\tfrac{1}{2}\right) + (3.25)\left(\tfrac{1}{2}\right) + (5)\left(\tfrac{1}{2}\right) \quad \leftarrow \text{plug into } f(x) = x^2 - 2x + 2 \\ &= 5.75.\end{aligned}$$

□

EXAMPLE 2

Understanding the notation used in Riemann sums

Looking back at Example 1, identify each of the following:

$$a, b; \quad n; \quad \Delta x; \quad x_0, x_1, x_2, x_3, x_4; \quad x_1^*, x_2^*, x_3^*, x_4^*.$$

SOLUTION

We have $a = 1$, $b = 3$, $n = 4$, and $\Delta x = \frac{1}{2}$. The subdivision points are $x_0 = a = 1$, $x_1 = 1.5$, $x_2 = 2$, $x_3 = 2.5$, and $x_4 = b = 3$. The chosen points in each subinterval $[x_{k-1}, x_k]$ are taken to be the rightmost points: $x_1^* = 1.5$, $x_2^* = 2$, $x_3^* = 2.5$, and $x_4^* = 3$. Notice that since this is a right sum, each x_k^* is just equal to x_k. □

EXAMPLE 3

Expressing a right sum in sigma notation

Consider again the area between the graph of $f(x) = x^2 - 2x + 2$ and the x-axis from $x = 1$ to $x = 3$. Express the n-rectangle right-sum approximation in sigma notation. Simplify until k and n are the only variables that appear in your expression. Then use your expression to calculate the right-sum approximations with $n = 4$ and $n = 8$ rectangles.

SOLUTION

We first subdivide $[1, 3]$ into n subintervals, each of width $\Delta x = \frac{b-a}{n} = \frac{3-1}{n} = \frac{2}{n}$. The subdivision points x_k are of the form $x_k = a + k\Delta x = 1 + \frac{2k}{n}$. Since we are constructing a right sum, in each subinterval $[x_{k-1}, x_k]$ we will choose x_k^* to be the rightmost point x_k. The kth rectangle in our approximation will have height $f(x_k^*)$ and width Δx. Using the notation we have developed, we find that the sum of the areas of the n rectangles is

$$\sum_{k=1}^{n} f(x_k^*)\Delta x = \sum_{k=1}^{n} f\left(1 + \frac{2k}{n}\right)\left(\frac{2}{n}\right)$$

$$= \sum_{k=1}^{n}\left(\left(1 + \frac{2k}{n}\right)^2 - 2\left(1 + \frac{2k}{n}\right) + 2\right)\left(\frac{2}{n}\right) = \sum_{k=1}^{n}\left(1 + \frac{4k^2}{n^2}\right)\left(\frac{2}{n}\right).$$

Substituting $n = 4$ into this expression and then writing out the sum gives us the same four-rectangle right-sum approximation that we found in Example 1:

$$\sum_{k=1}^{4}\left(1 + \frac{4k^2}{4^2}\right)\left(\frac{2}{4}\right) = \left(1 + \frac{4(1)^2}{16}\right)\left(\frac{1}{2}\right) + \left(1 + \frac{4(2)^2}{16}\right)\left(\frac{1}{2}\right)$$
$$+ \left(1 + \frac{4(3)^2}{16}\right)\left(\frac{1}{2}\right) + \left(1 + \frac{4(4)^2}{16}\right)\left(\frac{1}{2}\right)$$
$$= (1.25)\left(\frac{1}{2}\right) + (2)\left(\frac{1}{2}\right) + (3.25)\left(\frac{1}{2}\right) + (5)\left(\frac{1}{2}\right)$$
$$= 5.75.$$

Similarly, we can substitute $n = 8$ and then expand the sum to calculate the eight-rectangle right-sum approximation:

$$\sum_{k=1}^{8}\left(1 + \frac{4k^2}{(8)^2}\right)\left(\frac{2}{8}\right) = \left(1 + \frac{4(1)^2}{64}\right)\left(\frac{1}{4}\right) + \left(1 + \frac{4(2)^2}{64}\right)\left(\frac{1}{4}\right) + \left(1 + \frac{4(3)^2}{64}\right)\left(\frac{1}{4}\right)$$
$$+ \left(1 + \frac{4(4)^2}{64}\right)\left(\frac{1}{4}\right) + \left(1 + \frac{4(5)^2}{64}\right)\left(\frac{1}{4}\right) + \left(1 + \frac{4(6)^2}{64}\right)\left(\frac{1}{4}\right)$$
$$+ \left(1 + \frac{4(7)^2}{64}\right)\left(\frac{1}{4}\right) + \left(1 + \frac{4(8)^2}{64}\right)\left(\frac{1}{4}\right)$$
$$= (1.0625)\left(\frac{1}{4}\right) + (1.25)\left(\frac{1}{4}\right) + (1.5625)\left(\frac{1}{4}\right) + (2)\left(\frac{1}{4}\right)$$
$$+ (2.5625)\left(\frac{1}{4}\right) + (3.25)\left(\frac{1}{4}\right) + (4.0625)\left(\frac{1}{4}\right) + (5)\left(\frac{1}{4}\right)$$
$$= 5.1875.$$

Notice that sigma notation does not in general make it easier to calculate the areas of the rectangles in our approximations. However, the general n-rectangle sigma notation for a Riemann sum will be very important when we discuss the theory of definite integrals in the next section. It would also be very useful if you wanted to write a computer program to approximate the area under a curve, especially with a large number of rectangles. □

EXAMPLE 4

Using the left, midpoint, and trapezoid sums to approximate area

Approximate the area between the graph of $f(x) = x^2 - 2x + 2$ and the x-axis on $[1, 3]$, using $n = 4$ subintervals with (a) the left sum; (b) the midpoint sum; and (c) the trapezoid sum.

SOLUTION

(a) As in Example 1, we have $\Delta x = \frac{1}{2}$ and $x_k = 1 + k\Delta x = 1 + \frac{k}{2}$. For a left sum we choose x_k^* to be the leftmost point in the interval $[x_{k-1}, x_k]$, and therefore we choose $x_k^* = x_{k-1} = 1 + \frac{k-1}{2}$. The left sum with four rectangles is therefore equal to

$$\sum_{k=1}^{4} f\left(1 + \frac{k-1}{2}\right)\left(\frac{1}{2}\right) = f(1)\left(\frac{1}{2}\right) + f(1.5)\left(\frac{1}{2}\right) + f(2)\left(\frac{1}{2}\right) + f(2.5)\left(\frac{1}{2}\right) = 3.75.$$

(b) For a midpoint sum we choose x_k^* to be the midpoint in the interval $[x_{k-1}, x_k]$, and therefore we choose $x_k^* = \frac{x_{k-1}+x_k}{2} = \frac{(1+\frac{k-1}{2})+(1+\frac{k}{2})}{2} = \frac{3+2k}{4}$. The midpoint sum with four rectangles is therefore equal to

$$\sum_{k=1}^{4} f\left(\frac{3+2k}{4}\right)\left(\frac{1}{2}\right) = f(1.25)\left(\frac{1}{2}\right) + f(1.75)\left(\frac{1}{2}\right) + f(2.25)\left(\frac{1}{2}\right) + f(2.75)\left(\frac{1}{2}\right) = 4.625.$$

(c) For the trapezoid sum we do something completely different and use the heights at both $x_{k-1} = 1 + \frac{k-1}{2}$ and $x_k = 1 + \frac{k}{2}$ to calculate the area of the kth trapezoid. The trapezoid sum with four trapezoids is equal to

$$\sum_{k=1}^{4} \frac{f\left(1 + \frac{k-1}{2}\right) + f\left(1 + \frac{k}{2}\right)}{2}\left(\frac{1}{2}\right) = \frac{f(1) + f(1.5)}{2}\left(\frac{1}{2}\right) + \frac{f(1.5) + f(2)}{2}\left(\frac{1}{2}\right)$$
$$+ \frac{f(2) + f(2.5)}{2}\left(\frac{1}{2}\right) + \frac{f(2.5) + f(3)}{2}\left(\frac{1}{2}\right) = 4.75.$$

The three figures that follow show the three area approximations we just calculated. Which one do you think is the most accurate, and why? (See Exercise 5.)

Left sum

Trapezoid sum

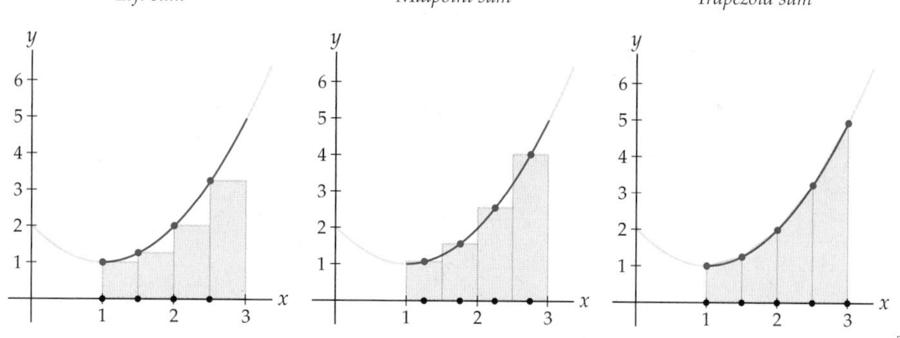

□

? TEST YOUR UNDERSTANDING

- Recalling the discussion of the circle at the start of this section, can you think of a different way of approximating the area of the circle?
- The Riemann sums presented in this section are not the only sums we could use to approximate the area under a curve. For example, we could use a "one-third" sum, which would define x_k^* to be the point one-third of the way across the interval $[x_{k-1}, x_k]$. How would you write x_k^* in terms of x_{k-1} and x_k for the one-third sum? What other methods for choosing x_k^* can you think of?
- Both the midpoint sum and the trapezoid sum use an average somewhere in their formula. What is the average that is being taken in each case?
- Can you label x_2, x_2^*, and $f(x_2^*)$ in each of the Riemann sum figures in this section?
- In general, the left sum will always underestimate the area under an increasing function and overestimate the area under a decreasing function; why? What about the right sum?

EXERCISES 4.2

Thinking Back

Simple area formulas: Give formulas for the areas of each of the following geometric figures.

- A circle of radius r
- A semicircle of radius r
- A right triangle with legs of lengths a and b
- A triangle with base b and altitude h
- A rectangle with sides of lengths w and l
- A trapezoid with width w and heights h_1 and h_2

Concepts

0. *Problem Zero:* Read the section and make your own summary of the material.

1. *True/False:* Determine whether each of the statements that follow is true or false. If a statement is true, explain why. If a statement is false, provide a counterexample.

(a) *True or False:* The n-rectangle lower sum for f on $[a, b]$ is always equal to the corresponding right sum.

(b) *True or False:* If f is positive and increasing on $[a, b]$, then any left sum for f on $[a, b]$ will be an under-approximation.

(c) *True or False:* If $f(a) > f(b)$, then any right-sum approximation for f on $[a, b]$ will be an under-approximation.

(d) *True or False:* A midpoint sum is always a better approximation than a left sum.

(e) *True or False:* If f is positive and concave up on all of $[a, b]$, then any left-sum approximation for f on $[a, b]$ will be an under-approximation.

(f) *True or False:* If f is positive and concave up on all of $[a, b]$, then any trapezoid sum approximation for f on $[a, b]$ will be an over-approximation.

(g) *True or False:* An upper sum approximation for f on $[a, b]$ can never be an under-approximation.

(h) *True or False:* For every function f on $[a, b]$, the left sum is always a better approximation with 10 rectangles than with 5 rectangles.

2. *Examples:* Construct examples of the thing(s) described in the following. Try to find examples that are different than any in the reading.

(a) A graph of a function f on an interval $[a, b]$ for which the left sum with four rectangles finds the *exact* area under the graph of f from $x = a$ to $x = b$.

(b) A graph of a function f on an interval $[a, b]$ for which *all* trapezoid sums (regardless of the size of n) will find the *exact* area under the graph of f from $x = a$ to $x = b$.

(c) A graph of a function f on an interval $[a, b]$ for which the upper sum with four rectangles is a much better approximation than the lower sum with four rectangles.

3. In the reading we used objects whose area we knew (squares) to approximate the area of a more complicated object (a circle). The same kind of technique can be used to approximate the area of the blob pictured here.

(a) Given that each of the squares in the grid has a side length of $\frac{1}{2}$ square unit, approximate the area of the blob.

(b) How could you get a better approximation?

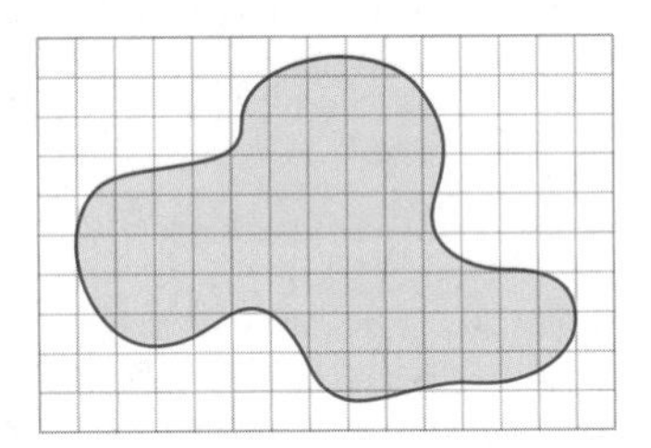

4. Consider the area between the graph of a function f and the x-axis from $x = 0$ to $x = 2$, as shown in each of the two figures that follow.

(a) Use the grid on the left and whatever method you like to approximate this area. Then use the grid on the right and the same method to make another approximation. Which approximation is likely more accurate, and why?

(b) Use the grid at the left to get an upper bound on the area of the region. In other words, make an approximation that you *know* is greater than the actual area. Repeat for the grid at the right.

(c) Use the grid at the left to get a lower bound on the area of the region. In other words, make an approximation that you *know* is less than the actual area. Repeat for the grid at the right.

(d) Use your answers to parts (a)–(c) to come up with your best possible guess for the actual area under the curve.

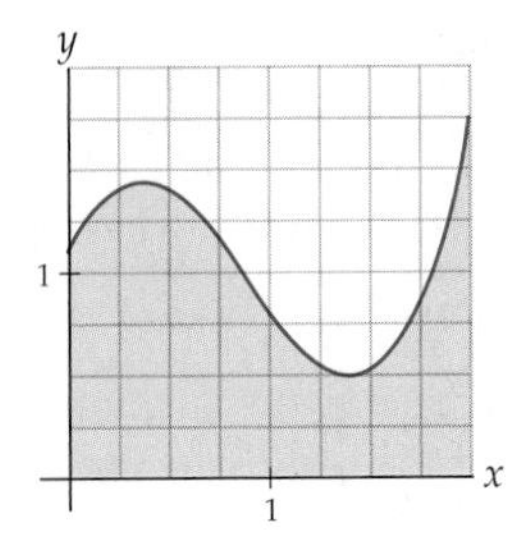

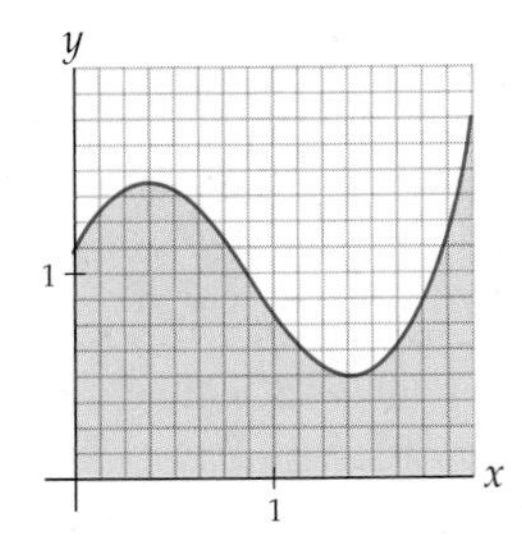

5. In Examples 1 and 4 we found four different approximations for the area between the graph of $f(x) = x^2 - 2x + 2$ and the x-axis on $[1, 3]$.
   (a) Based on the pictures of these approximations given in the reading, which are over-approximations? Which are under-approximations? Which approximation looks like it might be closest to the actual area under the curve?
   (b) The actual area (which we won't know how to calculate until Section 4.5) under the graph of $f(x) = x^2 - 2x + 2$ on $[1, 3]$ is equal to $\frac{14}{3} \approx 4.6667$. Given this information, which of the approximations in Examples 1 and 4 was the most accurate? Was that what you expected?
6. Do you think that one type of Riemann sum (right, left, midpoint, upper, lower, trapezoid) is usually more accurate than the others? Why or why not?
7. Suppose that the left-sum approximation with eight rectangles for the area between the graph of a function f and the x-axis from $x = a$ to $x = b$ is equal to 8.2 and that the corresponding right sum approximation is equal to 7.5.
   (a) What is the corresponding trapezoid sum approximation for this area?
   (b) Is the corresponding midpoint sum for this area necessarily between 7.5 and 8.2? If so, explain why. If not, sketch an example of a function f on an interval $[a, b]$ whose midpoint sum is *not* between the left sum and the right sum.
   (c) What can you say about the corresponding upper sum for this area? The corresponding lower sum?
   (d) Is it necessarily true that f is decreasing on the entire interval $[a, b]$? If so, explain why. If not, sketch an counterexample in which the left sum is greater than the right sum but f is *not* decreasing on all of $[a, b]$.
   (e) Could the function f be increasing on the entire interval $[a, b]$? If not, explain why not. If so, sketch a possible example in which the left sum is greater than the right sum and f is increasing on all of $[a, b]$.
8. In the reading we mentioned that the trapezoid sum is the average of the left sum and the right sum. Use the solutions of Examples 1 and 4 to show that for $f(x) = x^2 - 2x + 2$, $[a, b] = [1, 3]$, and $n = 4$, the trapezoid sum is indeed the average of the left sum and the right sum.
9. Explain why the upper sum approximation for the area between the graph of a function f and the x-axis on $[a, b]$ must always be larger than or equal to any other type of Riemann sum approximation with the same number n of rectangles.
10. Consider the area between the graph of a positive function f and the x-axis on an interval $[a, b]$. Explain why the upper sum approximation for this area with $n = 8$ boxes must be smaller than or equal to the upper sum approximation with $n = 4$ boxes. It may help to sketch some examples.
11. Suppose you wanted to calculate the upper sum approximation for the area between the graph of $f(x) = (x - 1)^2$ and the x-axis from $x = 0$ to $x = 2$. List all of the values M_k used for (a) $n = 2$ rectangles, (b) $n = 3$ rectangles, and (c) $n = 4$ rectangles. Sketch graphs of your rectangles to illustrate your answers.
12. Repeat Exercise 11, using the lower sum approximation and the values m_k.
13. Suppose $v(t)$ is the velocity of a particle moving on a straight path, where v is measured in meters per second and t is measured in seconds. The particle starts moving at time t_0 and moves for Δt seconds.
   (a) What are the units of $v(t_0)\Delta t$?
   (b) Geometrically, what does $v(t_0)\Delta t$ represent?
   (c) What do these questions have to do with this section?
14. Explain in no more than three sentences how we can approximate the derivative of a function f at a point c if we know the graph of f. Then, in no more than three additional sentences, discuss how the method for approximating area is similar to, and how is it different from, approximating the derivative. (Both descriptions should involve *multiple* approximations, each better than the last.)
15. Explain why the sum $\sum_{k=1}^{20} f(-3 + \frac{k}{2})(0.25)$ can't be a right sum for f on $[a, b] = [-3, 2]$.
16. Explain why the sum $\sum_{k=1}^{100} f(2 + 0.1(k - 1))(0.1)$ can't be a left sum for f on $[a, b] = [2, 5]$.

Each of the sums in Exercises 17–20 approximates the area between the graph of some function f and the x-axis from $x = a$ to $x = b$. Do some "reverse engineering" to determine the type of approximation (left sum, midpoint sum, etc.) and identify $f(x)$, a, b, n, Δx, and x_k. Then sketch the approximation described.

17. $\displaystyle\sum_{k=1}^{4} \left(1 + \frac{k}{2}\right)^2 \left(\frac{1}{2}\right)$
18. $\displaystyle\sum_{k=0}^{2} \ln\left(2 + \frac{k}{3}\right) \left(\frac{1}{3}\right)$
19. $\displaystyle\sum_{k=1}^{100} \frac{\sin(0.05(k-1)) + \sin(0.05k)}{2}(0.05)$
20. $\displaystyle\sum_{k=1}^{9} \sqrt{\frac{(1 + \frac{k-1}{3}) + (1 + \frac{k}{3})}{2}} \left(\frac{1}{3}\right)$

Skills

Your calculator should be able to approximate the area between a graph and the x-axis. Determine how to do this on your particular calculator, and then, in Exercises 21–26, use the method to approximate the signed area between the graph of each function f and the x-axis on the given interval $[a, b]$.

21. $f(x) = \sqrt{x - 1}$, $[a, b] = [2, 3]$
22. $f(x) = x^2$, $[a, b] = [0, 3]$
23. $f(x) = e^x$, $[a, b] = [1, 4]$
24. $f(x) = \sin x$, $[a, b] = [0, \pi]$

25. $f(x) = (x-2)^2 + 1$, $[a, b] = [1, 3]$

26. $f(x) = 1 - 2^x$, $[a, b] = [-3, 1]$

For each function f and interval $[a, b]$ in Exercises 27–33, use the given approximation method to approximate the signed area between the graph of f and the x-axis on $[a, b]$. Determine whether each of your approximations is likely to be an over-approximation or an under-approximation of the actual area.

27. $f(x) = x^2$, $[a, b] = [0, 3]$, left sum with
(a) $n = 3$ (b) $n = 6$

28. $f(x) = \sin x$, $[a, b] = [0, \pi]$, $n = 3$, with
(a) trapezoid sum (b) upper sum

29. $f(x) = \sqrt{x-1}$, $[a, b] = [2, 3]$, $n = 4$, with
(a) left sum (b) right sum

30. $f(x) = 1 - 2^x$, $[a, b] = [-3, 1]$, $n = 8$, with
(a) left sum (b) right sum

31. $f(x) = e^x$, $[a, b] = [1, 4]$, $n = 6$, with
(a) midpoint sum (b) trapezoid sum

32. $f(x) = 9 - x^2$, $[a, b] = [0, 5]$, $n = 5$, with
(a) midpoint sum (b) lower sum

33. $f(x) = (x-2)^2 + 1$, $[a, b] = [1, 3]$, lower sum with
(a) $n = 2$ (b) $n = 3$ (c) $n = 4$

For each function f and interval $[a, b]$ in Exercises 34–38, it is possible to find the *exact* signed area between the graph of f and the x-axis on $[a, b]$ geometrically by using the areas of circles, triangles, and rectangles. Find this exact area, and then calculate the left, right, midpoint, upper, lower, and trapezoid sums with $n = 4$. Which approximation rule is most accurate?

34. $f(x) = 5$, $[a, b] = [-2, 2]$

35. $f(x) = 3x + 1$, $[a, b] = [3, 5]$

36. $f(x) = 4 - x$, $[a, b] = [0, 6]$

37. $f(x) = \sqrt{1 - x^2}$, $[a, b] = [-1, 1]$

38. $f(x) = 3 + \sqrt{4 - x^2}$, $[a, b] = [-2, 2]$

In Exercises 39–44, write out the sigma notation for the Riemann sum described in such a way that the only letter which appears in the general term of the sum is k. Don't calculate the value of the sum; just write it down in sigma notation.

39. $f(x) = \sqrt{x-1}$, $[a, b] = [2, 3]$, right sum, $n = 4$.

40. $f(x) = x^2$, $[a, b] = [0, 3]$, left sum, $n = 3$.

41. $f(x) = e^x$, $[a, b] = [1, 4]$, midpoint sum, $n = 6$.

42. $f(x) = \ln x$, $[a, b] = [2, 5]$, left sum, $n = 100$.

43. $f(x) = \sin x$, $[a, b] = [0, \pi]$, trapezoid sum, $n = 4$.

44. $f(x) = \sqrt{1 - x^2}$, $[a, b] = [-1, 1]$, midpoint sum, $n = 20$.

Applications

45. Suppose that, as in Section 4.1, you drive in a car for 40 seconds with velocity $v(t) = -0.22t^2 + 8.8t$ feet per second, as shown in the graph that follows. If your total distance travelled is equal to the area under the velocity curve on $[0, 40]$, then find lower and upper bounds for your distance travelled by using
(a) the lower sum with $n = 4$ rectangles;
(b) the upper sum with $n = 4$ rectangles.

Velocity of car
$v(t) = -0.22t^2 + 8.8t$

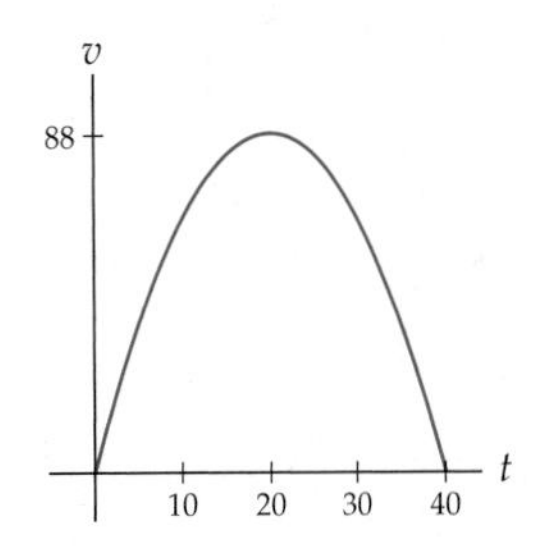

46. Repeat Exercise 45, using
(a) the midpoint sum with $n = 4$ rectangles;
(b) the trapezoid sum with $n = 4$ trapezoids.
If the exact distance travelled is just over 2,346 feet, then which of these approximations is the most accurate?

47. Dad's casserole surprise is hot out of the oven, and its temperature after t minutes is given by the function $T(t)$, measured in degrees. The casserole cools by changing at a rate of $T'(t) = -15e^{-0.5t}$ degrees per minute.

Rate of change of casserole temperature
$T'(t) = -15e^{-0.5t}$

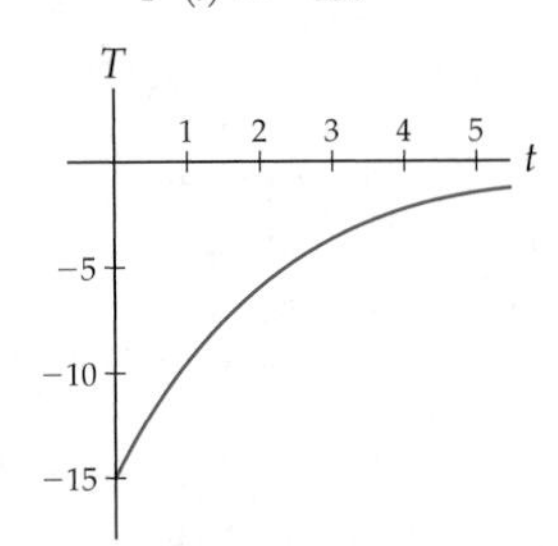

(a) Discuss what the graph of $T'(t)$ says about the behavior of the casserole's temperature after the casserole is taken out of the oven.
(b) Just as we can approximate distance travelled by approximating the area under the corresponding velocity curve, we can approximate the change in temperature T of the casserole by approximating the area under the graph of T'. Why?
(c) Estimate the change in temperature of the casserole over the first 5 minutes it is out of the oven, using any Riemann sum you like, with $n = 10$.

48. The National Oceanic and Atmospheric Administration tabulates flow data from many American rivers. From this data they compute and plot median annual flows. The flows are given by functions whose closed form is not known, but for which we can read off values for any day of the year we like. The following table describes the flow $f(t)$ in cubic feet per second for Idaho's Lochsa River, t days after January 1:

t	0	60	120	180	240	300	360
$f(t)$	700	1000	6300	4000	500	650	700

(a) Keeping in mind that the data are periodic with period 365, compute the left Riemann sum for the function that these data have sampled.

(b) What is the relation between the numbers you computed and the total amount of water that flows down the Lochsa annually? Estimate the total amount of water that flows down the Lochsa each year.

(c) Most of the flow down the river takes place from April to July. We can get a better idea of the total flow if we add a few data points. Recompute the left Riemann sum, adding the data points $(90, 2100)$, $(150, 11000)$, and $(210, 1000)$.

(d) What would you need to do to get an even better estimate of the total flow?

49. To approximate the flow $f(t)$ of the Lochsa River in its flood stage, we can use a function of the form

$$g(t) = c_1 + c_2\left(\sin\left(\frac{(t-90)\pi}{105}\right) - \frac{2}{\pi}\right),$$

where the coefficients c_1 and c_2 are found by evaluating the following two integrals:

$$c_1 = \frac{1}{105}\int_{90}^{195} f(t)\,dt,$$

$$c_2 = \frac{1}{9.95}\int_{90}^{195} f(t)\left(\sin\left(\frac{(t-90)\pi}{105}\right) - \frac{2}{\pi}\right)dt.$$

(a) Use the data points $(t, f(t)) = (90, 2100)$, $(120, 6300)$, $(150, 11000)$, and $(180, 4000)$, and left Riemann sums to approximate the values of the integrals for c_1 and c_2.

(b) Now that you have found c_1 and c_2, plot the resulting function $g(t)$ against the data points from Exercise 48.

Proofs

50. Use Definition 4.7 to prove that for any function f and interval $[a, b]$, the upper sum with n rectangles is greater than or equal to the lower sum with n rectangles.

51. Use Definition 4.6 and the definition of *increasing* to prove that if a function f is positive and increasing on $[a, b]$, then the left sum with n rectangles is less than or equal to the right sum with n rectangles.

52. Use Definitions 4.6 and 4.8 to prove that for any function f and interval $[a, b]$, the trapezoid sum with n trapezoids is always the average of the left sum and the right sum with n rectangles.

53. Use Definition 4.8 to prove that if a function f is positive and concave up on $[a, b]$, then the trapezoid sum with n trapezoids is an always an over-approximation for the actual area.

Thinking Forward

Approximating the length of a curve: Suppose you want to calculate the driving distance between New York City and Dallas, Texas.

- Print out a highway map of the United States, and highlight a route, snaking along with the paths of the major highways.
- How could we use the method of "subdivide, approximate, and add up" to approximate this driving distance?
- Make an actual approximation of the driving distance from New York City to Dallas, Texas, using the route and method you just described.
- How accurate do you think your approximation is? Is it an over-approximation or an under-approximation?

Limits of Riemann sums: In the reading we saw that the area between the graph of $f(x) = x^2 - 2x + 2$ and the x-axis on $[1, 3]$ could be approximated with the right sum $\sum_{k=1}^{n}\left(1 + \frac{4k^2}{n^2}\right)\left(\frac{2}{n}\right)$. Let $A(n)$ be equal to this n-rectangle right-sum approximation. The following table describes various values of $A(n)$:

n	10	100	1000	10,000
$A(n)$	5.08	4.7068	4.6707	4.66707

- Describe the meaning of the entries in this table, and verify that the entry for $A(10)$ is correct.
- Use the table to make a graph of $A(n)$, and discuss what happens to this graph as n approaches infinity.
- What does your graph tell you about the right-sum approximations of the area under the graph of f as n approaches infinity?

Approximations and error: In Section 4.5 we will see that definite integrals can be computed by taking differences of antiderivatives; in particular, the *Fundamental Theorem of Calculus* will reveal that if f is continuous on $[a, b]$, then $\int_a^b f(x)\,dx = F(b) - F(a)$, where F is any antiderivative of f. Armed with this fact, we can check the exact error of Riemann sum approximations for integrals of functions that we can antidifferentiate.

- Use the given antiderivative fact to find the exact value of $\int_1^4 \frac{1}{x}\,dx$. (*Hint: What is an antiderivative of* $\frac{1}{x}$*?*

In other words, what is a function F whose derivative is $f(x) = \frac{1}{x}$*?)*

- What is the actual error that results from a right-sum approximation with $n = 4$ for $\int_1^4 \frac{1}{x}\,dx$?
- By trial and error with smaller values of n, find the smallest value of n for which a right sum will approximate $\int_1^4 \frac{1}{x}\,dx$ to within 0.25.
- Repeat the preceding steps, but with the midpoint sum in place of the right sum.

4.3 DEFINITE INTEGRALS

- The exact area under the graph of a function f on $[a,b]$
- Defining the definite integral as a limit of Riemann sums
- Properties of the definite integral

Defining the Area Under a Curve

We have seen how to approximate areas with Riemann sums, but how do we find the *exact* area between the graph of a function f and the x-axis on an interval $[a, b]$? Consider the area of the region bounded by the curve $f(x) = x^2$, the x-axis, and the line $x = 1$ as shown here:

Area under $f(x) = x^2$ *on* $[0, 1]$

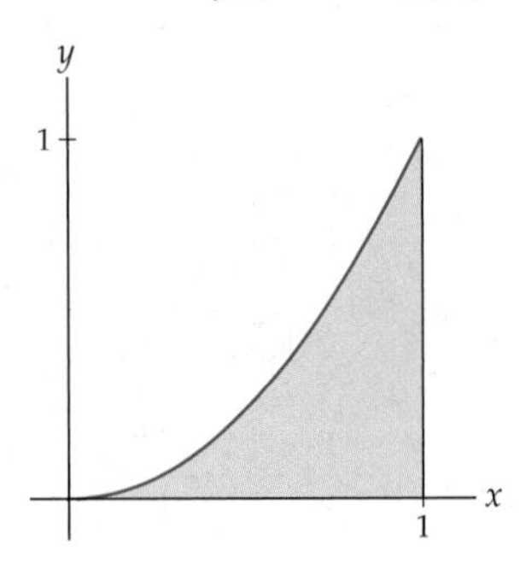

We don't yet have a rigorous mathematical method for finding this area exactly. Using Riemann sums, we can get approximations for the area. For example, the three figures that follow illustrate upper-sum approximations using $n = 10$, $n = 50$, and $n = 100$. Notice that in this case, since $f(x) = x^2$ is increasing on $[0, 1]$, the upper sum happens to be the same as the right sum.

Upper sum with $n = 10$ *Upper sum with* $n = 50$ *Upper sum with* $n = 100$

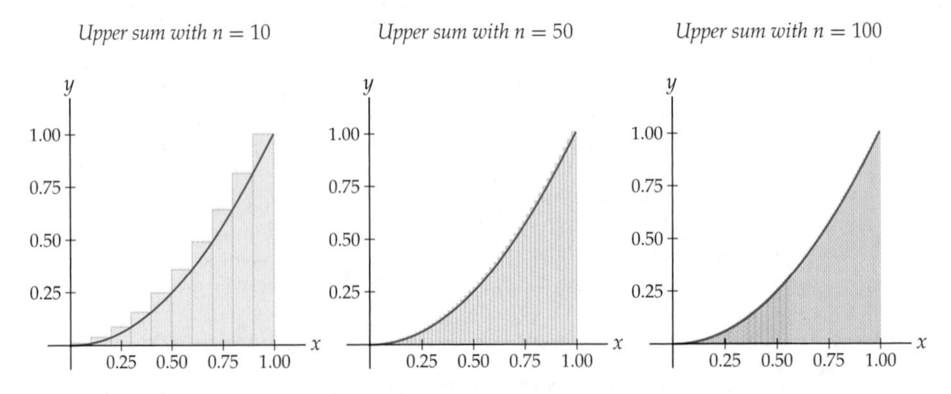

As the number n of rectangles gets larger, the upper sum approximation gets closer to the area under $f(x) = x^2$ on $[0, 1]$. The same thing happens if we consider any other Riemann sum, say, the lower sum, as pictured here for $n = 10$, $n = 50$, and $n = 100$:

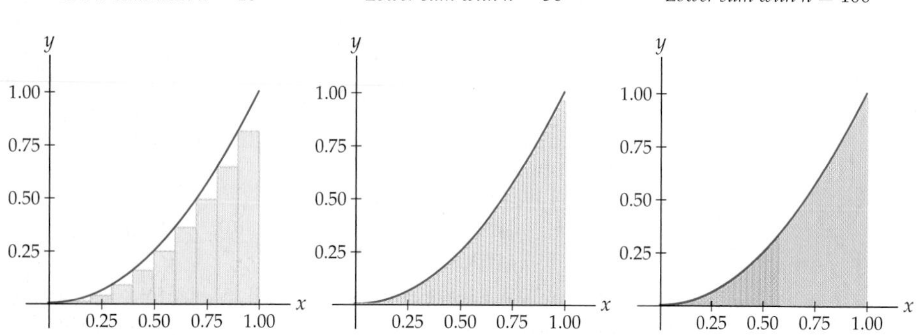

If a sequence of Riemann sum approximations converges to a real number when we take the limit as n approaches infinity, then we call this number the "area" under the graph. Notice that up until this point we did not actually have a truly rigorous definition for even the *concept* of such an area! The following definition describes this limit and the notation we will use to refer to it.

DEFINITION 4.9

The Definite Integral of a Function on an Interval

Let f be a function defined on an interval $[a, b]$. The ***definite integral*** of f from $x = a$ to $x = b$ is defined to be the number

$$\int_a^b f(x)\,dx = \lim_{n\to\infty} \sum_{k=1}^{n} f(x_k^*)\,\Delta x,$$

if this limit exists, where $\Delta x = \frac{b-a}{n}$, $x_k = a + k\,\Delta x$, and x_k^* is any choice of point in $[x_{k-1}, x_k]$.

The "$\int$" symbol should remind you of the letter "S" for "Sum." A Riemann sum describes a discrete sum of areas of rectangles, while in a loose sense an integral represents a continuous sum of the areas of infinitely many rectangles, each of which is infinitely thin. The a and b below and above the integral symbol are called the ***limits of integration***, where here "limits" is used in the sense of "ends," not in the sense of limits of functions. It may help you remember this definition if you think of it in the following way: As $n \to \infty$, the finite sum $(\sum)$ of n things becomes an integral $(\int)$ that accumulates everything from $x = a$ to $x = b$. The discrete list of values $f(x_k^*)$ becomes the continuous function $f(x)$, and the small change Δx becomes an "infinitesimal" change dx.

CAUTION

The "dx" portion of the definite integral is not optional; it must always be included with the notation of the definite integral. Just as

$$\frac{d}{dx}(\_\_\_\_\_)$$

is how we denote "the derivative of _____," the notation

$$\int_a^b \_\_\_\_\_\; dx$$

is how we will denote "the definite integral of _____ on $[a, b]$," which represents "the signed area between the graph of _____ and the x-axis on $[a, b]$." Note that in both cases there is the presence of a "dx" that represents the result of a limiting process on a change Δx in the independent variable with respect to which the differentiation or integration/accumulation/area process is taking place.

Consider again the example with $f(x) = x^2$ and $[a, b] = [0, 1]$. For any given n, we have

$$\Delta x = \frac{b-a}{n} = \frac{1}{n} \quad \text{and} \quad x_k = a + k\Delta x = \frac{k}{n}.$$

We will use the right sum and choose each x_k^* to be the rightmost point $x_k = \frac{k}{n}$ of the subinterval $[x_{k-1}, x_k]$. When we take the limit as the number n of subdivisions approaches infinity, we obtain

$$\int_0^1 x^2\,dx = \lim_{n\to\infty} \sum_{k=1}^{n} \left(\frac{k}{n}\right)^2 \left(\frac{1}{n}\right).$$

If, for a given function f and interval $[a, b]$, the limit defining the integral exists for any choice of points x_k^*, then we say that f is ***integrable*** on $[a, b]$. It turns out that every continuous function is integrable on $[a, b]$. (The proof of this theorem is beyond the scope of the text and will not be presented here.) However, *some* discontinuous functions are still integrable; for example, functions with removable or jump discontinuities are still integrable, but some functions with vertical asymptotes may not be. You will investigate various examples of discontinuous functions in Exercises 19 and 20. In Section 5.6 you will see the somewhat surprising spectacle of certain functions that have vertical asymptotes and yet are integrable.

Properties of Definite Integrals

Recall from the previous section that Riemann sums measure *signed* areas: Areas of regions above the x-axis are counted positively, and areas of regions below the x-axis are counted negatively. Because definite integrals are defined to be limits of Riemann sums, this means that definite integrals also automatically count signed areas:

THEOREM 4.10 **Definite Integrals Measure Signed Areas**

Definite integrals count areas above the x-axis positively and areas below the x-axis negatively. Algebraically, this means that

(a) If $f(x) \geq 0$ on all of an interval $[a, b]$, then $\int_a^b f(x)\,dx \geq 0$.

(b) If $f(x) \leq 0$ on all of an interval $[c, d]$, then $\int_c^d f(x)\,dx \leq 0$.

For example, if f is the function graphed next, then the definite integral $\int_0^5 f(x)\,dx$ will count the areas of the regions marked A and C positively and count the area of the region marked B negatively:

$\int_0^5 f(x)\,dx$ *measures the signed area* $A - B + C$

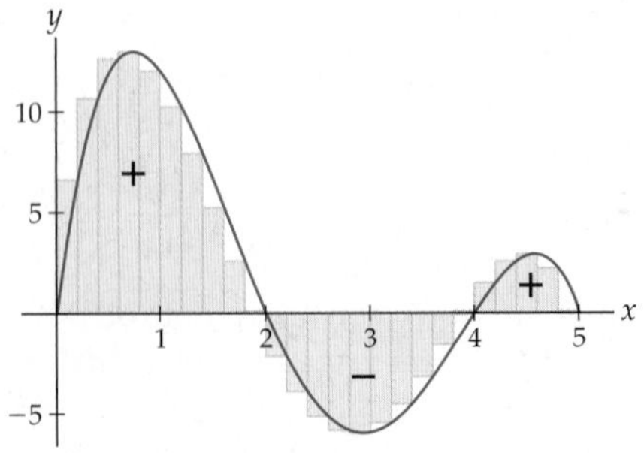

The fact that definite integrals are defined in terms of sums and limits allows us to state many nice properties of definite integrals. In particular, because definite integrals are just limits of sums, they behave well with sums and constant multiples, as follows:

THEOREM 4.11

Sum and Constant-Multiple Rules for Definite Integrals

For any functions f and g that are integrable on $[a, b]$ and any real number k,

(a) $\displaystyle\int_a^b (f(x) + g(x))\,dx = \int_a^b f(x)\,dx + \int_a^b g(x)\,dx.$

(b) $\displaystyle\int_a^b kf(x)\,dx = k\int_a^b f(x)\,dx.$

Before we present a formal proof of this theorem, let's think about the first part of it graphically. The first graph that follows shows a left sum for the area between the graph of $f(x) = x^2$ and the x-axis on $[0, 2]$. The second graph shows a left sum (with the same n) for the area between $g(x) = x$ and the x-axis on the same interval. In the third graph we see that the sum of these two left sums is itself a left sum for the graph of $f(x)+g(x) = x^2+x$ on $[0, 2]$. As n approaches infinity, the area of the beige rectangles will approach the area under f, the area of the blue rectangles will approach the area under g, and the sum of the areas of the beige and blue rectangles will approach the area under $f + g$.

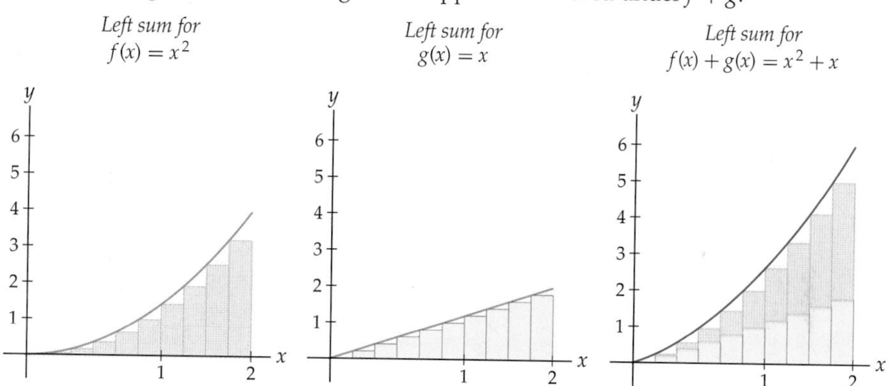

The key to the algebraic proof of Theorem 4.11 is that sums and constant multiples commute with limits and with Riemann sums. Since the definite integral is a limit of Riemann sums, it must be that sums and constant multiples commute with definite integrals.

Proof. Given a positive integer n, define $\Delta x = \dfrac{b-a}{n}$ and $x_k = a + k\Delta x$, and let x_k^* be any point in the subinterval $[x_{k-1}, x_k]$. Then by the definition of the definite integral, we have

$$\begin{aligned}
\int_a^b (f(x) + g(x))\,dx &= \lim_{n\to\infty} \sum_{k=1}^{n} (f(x_k^*) + g(x_k^*))\,\Delta x && \leftarrow \text{definite integral definition} \\
&= \lim_{n\to\infty} \sum_{k=1}^{n} (f(x_k^*)\,\Delta x + g(x_k^*)\,\Delta x) && \leftarrow \text{algebra} \\
&= \lim_{n\to\infty} \left(\sum_{k=1}^{n} f(x_k^*)\,\Delta x + \sum_{k=1}^{n} g(x_k^*)\,\Delta x \right) && \leftarrow \text{split into two sums} \\
&= \lim_{n\to\infty} \sum_{k=1}^{n} f(x_k^*)\,\Delta x + \lim_{n\to\infty} \sum_{k=1}^{n} g(x_k^*)\,\Delta x && \leftarrow \text{sum rule for limits} \\
&= \int_a^b f(x)\,dx + \int_a^b g(x)\,dx. && \leftarrow \text{definite integral definition}
\end{aligned}$$

The proof of part (b) is similar: Write the definite integral of $y = kf(x)$ as a limit of Riemann sums, use properties of sums and limits to pull out the constant k, and then use the definition of the definite integral again to get k times the definite integral of f. You will write out the details in Exercise 55. ■

The following theorem describes what happens if we take a definite integral over a zero-length interval, or from right to left, or in two pieces:

THEOREM 4.12

Properties Concerning Limits of Integration

If $a < b$ are real numbers and f is integrable on $[a, b]$, then

(a) $\int_a^a f(x)\,dx = 0$ **(b)** $\int_b^a f(x)\,dx = -\int_a^b f(x)\,dx$

Moreover, if c is any real number in $[a, b]$, then

(c) $\int_a^b f(x)\,dx = \int_a^c f(x)\,dx + \int_c^b f(x)\,dx$

Again, the properties stated in this theorem follow directly from the definition of the definite integral as a limit of Riemann sums. The first property makes sense graphically because the area under the graph of a function f from $x = a$ to $x = a$ would have a width of zero. You will prove this property algebraically in Exercise 56. The second property enables us to consider definite integrals like $\int_4^1 x^2\,dx$, in which the starting x-value is greater than the ending x-value. If we integrate "backwards," say, from $x = 4$ to $x = 1$, then we count the area given by the definite integral negatively. This property can also be proved from the definition of the definite integral, where in this situation we would have a *negative* change $\Delta x = \frac{a-b}{n}$. You will prove the property in Exercise 57.

The third part of Theorem 4.12 can also be proved by using the definition of the definite integral, but instead we present a "convincing argument by picture" in the three figures shown next. In this picture we have $a = 0$, $b = 3$, and $c = 1$. The area from $x = 0$ to $x = 3$ is clearly the sum of the area from $x = 0$ to $x = 1$ and the area from $x = 1$ to $x = 3$.

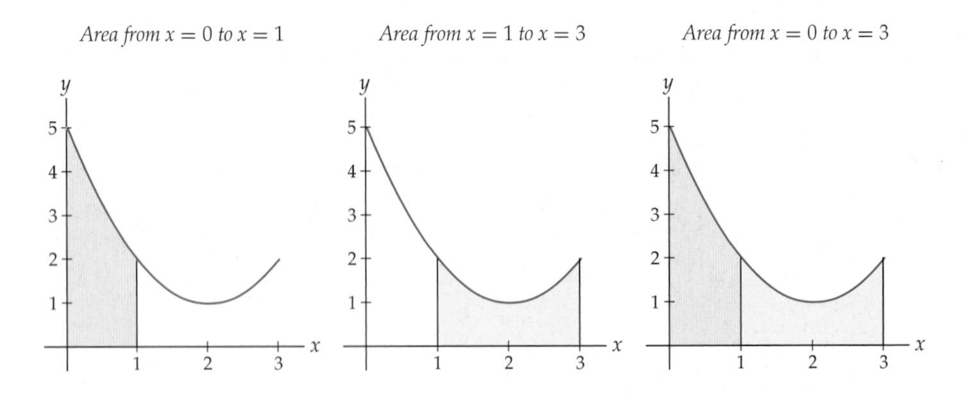

Formulas for Three Simple Definite Integrals

In Section 4.5 we will learn a method for quickly calculating many definite integrals. In the meantime, it will be useful to have a small set of integration formulas that we can use to solve simple problems. The next theorem describes definite integrals of three common functions: constant functions, the identity function, and the squaring function. Using these formulas, together with the sum and constant-multiple rules for definite integrals, we will be able to quickly evaluate definite integrals for any quadradic functions.

THEOREM 4.13 **Definite Integral Formulas**

For any real numbers a, b, and c,

(a) $\int_a^b c\,dx = c(b-a)$ **(b)** $\int_a^b x\,dx = \frac{1}{2}(b^2 - a^2)$ **(c)** $\int_a^b x^2\,dx = \frac{1}{3}(b^3 - a^3)$

These formulas are just the tip of the iceberg; in Section 4.5 we will show that we can solve many definite integrals by using antiderivatives, and in Chapter 5 we will develop methods for quickly evaluating many more types of definite integrals.

TECHNICAL POINT The definite integral formulas in Theorem 4.13 are true even when the definite integral in question turns out to be a negative number. For example,

$$\int_{-2}^{1} x\,dx = \frac{1}{2}(1^2 - (-2)^2) = \frac{1}{2}(1-4) = -\frac{3}{2}.$$

In Section 4.6 we will discuss what it means when a definite integral on an interval $[a, b]$ is negative.

Proof. We will use the definition of the definite integral to prove part (a) of the theorem. The proofs of parts (b) and (c) are similar and are left to Exercises 60 and 61. The first two parts of the theorem can also be argued geometrically; see Exercises 58 and 59. Suppose $f(x) = c$ is a constant function on $[a, b]$. Then by the definition of the definite integral, we have

$$\int_a^b f(x)\,dx = \lim_{n\to\infty} \sum_{k=1}^{n} f(x_k^*)\,\Delta x \qquad \leftarrow \text{definition of definite integral}$$

$$= \lim_{n\to\infty} \sum_{k=1}^{n} c\left(\frac{b-a}{n}\right) \qquad \leftarrow f(x) = c \text{ and } \Delta x = \frac{b-a}{n}$$

$$= \lim_{n\to\infty} c\left(\frac{b-a}{n}\right)\sum_{k=1}^{n} 1 \qquad \leftarrow \text{pull constants out of sum}$$

$$= \lim_{n\to\infty} c\left(\frac{b-a}{n}\right)(n) \qquad \leftarrow \text{sum formula}$$

$$= \lim_{n\to\infty} c(b-a) = c(b-a) \qquad \leftarrow \text{limit of a constant}$$

■

Examples and Explorations

EXAMPLE 1 **Using a sum formula to approximate an area with a large number of rectangles**

Find the n-rectangle right-sum approximation for $\int_2^4 (x^2+1)\,dx$, and then use it to calculate approximations with $n = 100$ rectangles and with $n = 1000$ rectangles.

SOLUTION

We'll find a formula for the n-rectangle approximation and then evaluate at $n = 100$ and $n = 1000$ at the end. Since $a = 2$ and $b = 4$, we have $\Delta x = \frac{b-a}{n} = \frac{2}{n}$, and therefore $x_k = a + k\Delta x = 2 + \frac{2k}{n}$. For the right sum with $x_k^* = x_k$, we have

$$\begin{aligned}
\sum_{k=1}^{n} f(x_k^*)\Delta x &= \sum_{k=1}^{n} \left(\left(2+\tfrac{2k}{n}\right)^2 + 1\right)\left(\tfrac{2}{n}\right) \\
&= \sum_{k=1}^{n} \left(\frac{8k^2}{n^3} + \frac{16k}{n^2} + \frac{10}{n}\right) && \leftarrow \text{simple algebra} \\
&= \frac{8}{n^3}\sum_{k=1}^{n} k^2 + \frac{16}{n^2}\sum_{k=1}^{n} k + \frac{10}{n}\sum_{k=1}^{n} 1 && \leftarrow \text{separate the sums} \\
&= \frac{8}{n^3}\left(\frac{n(n+1)(2n+1)}{6}\right) + \frac{16}{n^2}\left(\frac{n(n+1)}{2}\right) + \frac{10}{n}(n) && \leftarrow \text{sum formulas} \\
&= \frac{8n(n+1)(2n+1)}{6n^3} + \frac{16n(n+1)}{2n^2} + 10. && \leftarrow \text{simplify}
\end{aligned}$$

Notice that our expression for the n-rectangle right sum no longer has any sigma notation in it. This means that it is now a simple matter to evaluate the expression at $n = 100$ and at $n = 1000$. At $n = 100$ we have

$$\frac{800(101)(201)}{6,000,000} + \frac{1600(101)}{20,000} + 10 = 20.7868,$$

and at $n = 1000$ we have

$$\frac{8000(1001)(2001)}{6,000,000,000} + \frac{16000(1001)}{2,000,000} + 10 = 20.6787.$$

□

✔ CHECKING THE ANSWER

After all that work it is good to do a reality check. The figure that follows shows a graphing calculator plot of the area that we just approximated. The area we are considering should be just a little less than the area of the trapezoid with base from $(2, 0)$ to $(4, 0)$ and top side connecting $(2, 5)$ and $(4, 17)$. This trapezoid has area $\frac{5+17}{2}(2) = 21$ square units, just slightly more than our earlier approximations.

Area under $f(x) = x^2 + 1$ on $[2, 4]$

EXAMPLE 2 **Calculating a definite integral exactly**

Take the limit as $n \to \infty$ of an n-rectangle right sum to calculate $\int_2^4 (x^2+1)\,dx$ exactly.

SOLUTION

We already did the hard work in the previous example, and all that remains is to take the limit as $n \to \infty$. The exact area between the graph of $f(x) = x^2+1$ and the x-axis on $[2,4]$ is

$$\int_2^4 (x^2+1)\,dx = \lim_{n\to\infty} \sum_{k=1}^{n} f(x_k^*)\Delta x \qquad \leftarrow \text{definite integral definition}$$

$$= \lim_{n\to\infty}\left(\frac{8n(n+1)(2n+1)}{6n^3} + \frac{16n(n+1)}{2n^2} + 10\right) \qquad \leftarrow \text{Example 1}$$

$$= \frac{16}{6} + \frac{16}{2} + 10 = \frac{62}{3}. \qquad \leftarrow \text{ratios of leading coefficients}$$

In the step where we evaluated the limit, we used the fact that the limit of a "balanced" rational function with the same degree in the numerator and denominator is equal to the ratio of leading coefficients.

Notice that we would not have been able to calculate this limit if we did not have a *formula* for the Riemann sum from Example 1 that was expressed in terms of n, but with no "Σ". Note also that our previous approximation with $n = 1000$ was quite accurate; it was only approximately 0.012 square unit larger than the actual area, $\frac{62}{3} \approx 20.6667$. □

EXAMPLE 3 **Using definite integral formulas**

Use properties of definite integrals and the definite integral formulas in Theorem 4.13 to verify the calculation of $\int_2^4 (x^2+1)\,dx$ from the previous example.

SOLUTION

Using the sum rule for definite integrals from Theorem 4.11, we can indeed calculate the same answer that we did in Example 2:

$$\int_2^4 (x^2+1)\,dx = \int_2^4 x^2\,dx + \int_2^4 1\,dx \qquad \leftarrow \text{sum rule for definite integrals}$$

$$= \frac{1}{3}(4^3 - 2^3) + 1(4-2) = \frac{62}{3}. \qquad \leftarrow \text{definite integral formulas}$$ □

EXAMPLE 4 **Using the algebraic properties of definite integrals**

Given that $\int_1^3 f(x)\,dx = 4$ and $\int_5^3 2f(x)\,dx = -3$, find $\int_1^5 f(x)\,dx$. Identify the theorems or properties that allow each of your steps.

SOLUTION

We pick apart the definite integral we are seeking until we can express it in terms of the definite integrals that we are given:

$$\int_1^5 f(x)\,dx = \int_1^3 f(x)\,dx + \int_3^5 f(x)\,dx \qquad \leftarrow \text{Theorem 4.12(c)}$$

$$= \int_1^3 f(x)\,dx - \int_5^3 f(x)\,dx \qquad \leftarrow \text{Theorem 4.12(b)}$$

$$= \int_1^3 f(x)\,dx - \frac{1}{2}\int_5^3 2f(x)\,dx \qquad \leftarrow \text{Theorem 4.11(b)}$$

$$= 4 - \frac{1}{2}(-3) = \frac{11}{2}. \qquad \leftarrow \text{using what was given}$$ □

EXAMPLE 5

Interpreting definite integrals as signed areas

Use a graph and properties of definite integrals to argue that $\int_1^5 (2-x)\,dx$ is negative. Then find a number a so that $\int_a^5 (2-x)\,dx$ is exactly zero.

SOLUTION

The region between the graph of $f(x) = 2 - x$ and the x-axis on $[1, 5]$ consists of one small triangular region above the x-axis and one larger triangular region below the x-axis:

The region between $f(x) = 2 - x$ and the x-axis on $[1, 5]$

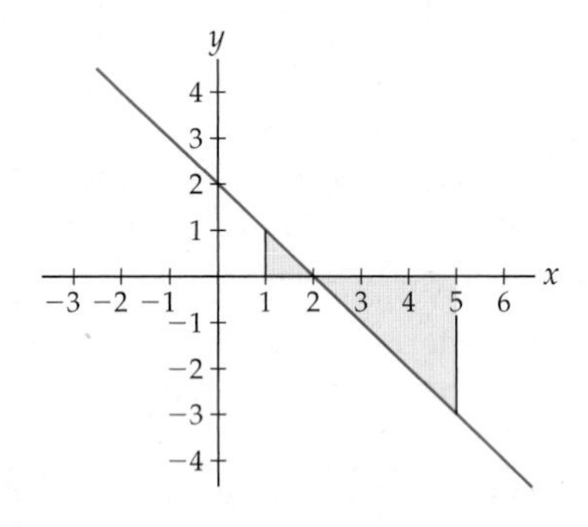

The definite integral will count the region above the x-axis positively and the region below the x-axis negatively; that is,

$$\int_1^2 (2-x)\,dx > 0 \quad \text{and} \quad \int_2^5 (2-x)\,dx < 0.$$

Since the region above the x-axis is smaller than the region below the x-axis, we must have

$$\int_1^5 (2-x)\,dx = \int_1^2 (2-x)\,dx + \int_2^5 (2-x)\,dx < 0.$$

We will have $\int_a^5 (2-x)\,dx$ equal to zero when $a = -1$, because this is the value for which the triangle above the x-axis on the interval $[a, 1]$ is the same size as the triangle below the x-axis on the interval $[2, 5]$. We could compute the areas of the triangles in question exactly from the formula for the area of a triangle, but it is sufficient in this case to see the graph:

The region between $f(x) = 2 - x$ and the x-axis on $[-1, 5]$

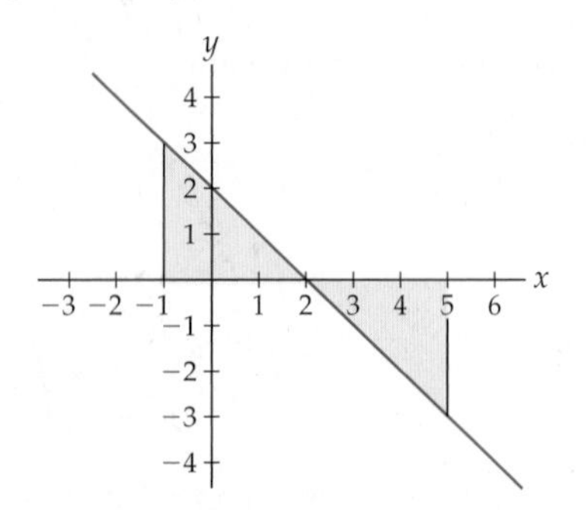

□

? TEST YOUR UNDERSTANDING

- How is a definite integral different than a Riemann sum?
- What do definite integrals have to do with areas under curves?
- What types of functions are integrable?
- Why does it make sense that if $a < c < b$, then $\int_a^b f(x)\,dx = \int_a^c f(x)\,dx + \int_c^b f(x)\,dx$?
- Why is it not surprising that definite integrals behave well with respect to sums and constant multiples?

EXERCISES 4.3

Thinking Back

- *Commuting with sums:* What does it mean to say that derivatives commute with sums? That limits commute with sums? That sums written in sigma notation commute with sums? Express your answers in words and algebraically.
- *Commuting with constant multiples:* What does it mean to say that derivatives commute with constant multiples? That limits commute with constant multiples? That sums written in sigma notation commute with constant multiples? Express your answers in words and algebraically.

Concepts

0. *Problem Zero:* Read the section and make your own summary of the material.

1. *True/False:* Determine whether each of the statements that follow is true or false. If a statement is true, explain why. If a statement is false, provide a counterexample.

(a) *True or False:* The left-sum and right-sum approximations are the same if the number n of rectangles is very large.

(b) *True or False:* $\int_{-2}^{5}(x+2)^3\,dx$ is a real number.

(c) *True or False:* $\int(5x^2 - 3x + 2)\,dx$ is exactly 26.167.

(d) *True or False:* $\int_{-3}^{-2} f(x)\,dx = -\int_2^3 f(x)\,dx$.

(e) *True or False:* If $\int_0^2 f(x)\,dx = 3$ and $\int_0^2 g(x)\,dx = 2$, then $\int_0^2 f(g(x)) = 6$.

(f) *True or False:* If $\int_0^2 f(x)\,dx = 3$ and $\int_0^2 g(x)\,dx = 2$, then $\int_0^2 f(x)g(x)\,dx = 6$.

(g) *True or False:* If $\int_0^1 f(x)\,dx = 3$ and $\int_{-1}^0 f(x)\,dx = 4$, then $\int_{-1}^1 f(x)\,dx = 7$.

(h) *True or False:* If $\int_0^2 f(x)\,dx = 3$ and $\int_2^4 g(x)\,dx = 4$, then $\int_0^4 (f(x) + g(x))\,dx = 7$.

2. *Examples:* Construct examples of the thing(s) described in the following. Try to find examples that are different than any in the reading.

(a) A function that is not integrable on $[2, 10]$.

(b) A function f for which we currently know how to calculate $\int_{-2}^2 f(x)\,dx$ exactly.

(c) A function f for which we currently do not know how to calculate $\int_{-2}^2 f(x)\,dx$ exactly.

3. Fill in the blanks: The signed area between the graph of a continuous function f and the x-axis on $[a, b]$ is represented by the notation ________ and is called the ________ .

4. Explain why it makes sense that *every* Riemann sum for a continuous function f on an interval $[a, b]$ approaches the same number as the number n of rectangles approaches infinity. Illustrate your argument with graphs.

5. Fill in the blanks: The definite integral of an integrable function f from $x = a$ to $x = b$ is defined to be

$$\int_a^b f(x)\,dx = \lim_{\square} \sum_{\square}^{\square} \boxed{},$$

where $\Delta x =$ ____, $x_k =$ ____, and x_k^* is ____ .

6. Explain geometrically what the definition of the definite integral as a limit of Riemann sums represents. Include a labeled picture of a Riemann sum (for a particular n) that illustrates the roles of n, Δx, x_k, x_k^*, and $f(x_k^*)$. What happens in the picture as $n \to \infty$?

7. If $f(x)$ is defined at $x = a$, then $\int_a^a f(x)\,dx = 0$. Explain why this makes sense in terms of area.

8. Draw pictures illustrating the fact that if $a \le c \le b$, then

$$\int_a^c f(x)\,dx + \int_c^b f(x)\,dx = \int_a^b f(x)\,dx.$$

Use graphs to determine whether each of the following definite integrals is equal to a positive number, a negative number, or zero:

9. $\displaystyle\int_{-3}^{3} (x^2 - 4)\,dx$

10. $\displaystyle\int_{-3}^{3} |x^2 - 4|\,dx$

11. $\displaystyle\int_0^{2\pi} \cos x\,dx$

12. $\displaystyle\int_{-3\pi/4}^{3\pi/4} \cos x\,dx$

13. Consider the function f graphed here. Shade in the regions between f and the x-axis on (a) $[-2, 6]$ and (b) $[-4, 2]$. Are the signed areas on these intervals positive or negative, and why?

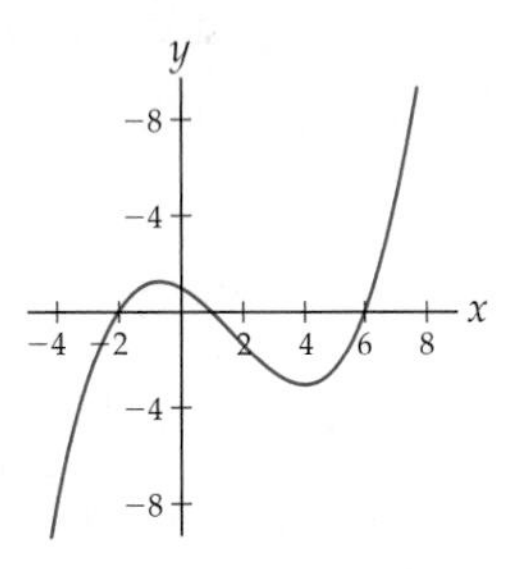

14. In a right sum with $n = 4$ rectangles for the area between the graph of $f(x) = x^2 - 1$ and the x-axis on $[0, 2]$, how many rectangles have negative heights? What if $n = 7$? In each case, is the estimate for the signed area positive or negative? Is the exact value of the signed area positive or negative?

15. Although the definite integral of a sum of functions is equal to the sum of the definite integrals of those functions, the definite integral of a product of functions is *not* the product of two definite integrals.

(a) Use mathematical notation to write the preceding sentence in this form:

$$\_\_\_\_\_ = \_\_\_\_\_, \quad \text{but} \quad \_\_\_\_\_ \neq \_\_\_\_\_.$$

(b) Choose two simple functions f and g so that you can calculate the definite integrals of f, g, and $f + g$ on $[0, 1]$, and show that the sum of the first two definite integrals is equal to the third.

(c) Find two simple functions f and g such that $\int_0^1 f(x)g(x)\,dx$ is not equal to the product of $\int_0^1 f(x)\,dx$ and $\int_0^1 g(x)\,dx$. (*Hint: Choose f and g so that you can calculate the definite integrals involved.*)

16. Suppose f is an integrable function $[a, b]$ and k is a real number. Use pictures of Riemann sums to illustrate that the right sum for the function $kf(x)$ on $[a, b]$ is k times the value of the right sum (with the same n) for f on $[a, b]$. What happens as $n \to \infty$? What does this exercise say about the definite integrals $\int_a^b f(x)\,dx$ and $\int_a^b kf(x)\,dx$?

17. Consider the definite integral $\int_0^3 x^2\,dx$.

(a) Write down an n-rectangle right sum for $\int_0^3 x^2\,dx$, and use algebra and a sum formula to write this sum as a formula in terms of n.

(b) Write down an n-rectangle left sum for $\int_0^3 x^2\,dx$, and use algebra and a sum formula to write this sum as a formula in terms of n.

(c) Use your answers to (a) and (b) to show that the right sum and the left sum for $\int_0^3 x^2\,dx$ are different for $n = 100$ and $n = 1000$.

(d) Use your answers to (a) and (b) to show that the right sum and the left sum for $\int_0^3 x^2\,dx$ approach the same quantity as $n \to \infty$. What does this quantity represent geometrically?

18. The definite integral of a function f on an interval $[a, b]$ is defined as a limit of Riemann sums. How can it be that the sum of the areas of infinitely many rectangles that are each "infinitely thin" is a finite number? On the one hand, shouldn't it be infinite, since we are adding up infinitely many rectangles? On the other hand, shouldn't it always be zero, since the width of each of the rectangles is approaching zero as $n \to \infty$?

19. Some discontinuous functions are not integrable. For example, consider the function $f(x) = \dfrac{1}{x^2}$ on $[-2, 2]$.

(a) Sketch a graph of f on $[-2, 2]$. What kind of discontinuity does f have, and where?

(b) Why might it be reasonable to think that f is not integrable on $[-2, 2]$ because of this discontinuity?

(c) What do you think happens to a Riemann sum approximation for the area between f and the x-axis on $[-2, 2]$ as $n \to \infty$?

(d) Although your intuition probably told you that the area between the graph of $f(x) = \dfrac{1}{x^2}$ and the x-axis on $[-2, 2]$ was likely to be infinite, this is not always the case for functions with vertical asymptotes. Surprisingly, as we will see in Section 5.6, the function $g(x) = \dfrac{1}{x^{2/3}}$ has an asymptote at $x = 0$ and yet its area on $[-2, 2]$ is actually finite! Compare the graphs of $f(x) = \dfrac{1}{x^2}$ and $g(x) = \dfrac{1}{x^{2/3}}$, and speculate as to why this might be the case.

20. Some discontinuous functions are still "nice" enough to be integrable. For example, consider the function

$$f(x) = \begin{cases} x, & \text{if } x \le 2 \\ x + 3, & \text{if } x > 2. \end{cases}$$

(a) Sketch a graph of f on $[0, 5]$. What kind of discontinuity does f have, and where?

(b) Why is it reasonable that f is integrable on $[0, 5]$ despite this discontinuity?

(c) What happens to a Riemann sum approximation for the area between a function f and the x-axis on $[0, 5]$ as $n \to \infty$?

Skills

Use geometry (i.e., areas of triangles, rectangles, and circles) to find the exact values of each of the definite integrals in Exercises 21–28.

21. $\displaystyle\int_0^2 (2 - x)\,dx$

22. $\displaystyle\int_2^1 0(4x - 3)\,dx$

23. $\displaystyle\int_{-3}^8 24\,dx$

24. $\displaystyle\int_{-2}^4 |3x + 1|\,dx$

25. $\displaystyle\int_{-1}^1 \sqrt{1 - x^2}\,dx$

26. $\displaystyle\int_{-r}^r \sqrt{r^2 - x^2}\,dx$

27. $\int_0^3 (1 - |x - 1|)\, dx$

28. $\int_{-2}^2 (3 + \sqrt{4 - x^2})\, dx$

If $\int_{-2}^3 f(x)\, dx = 4$, $\int_{-2}^6 f(x)\, dx = 9$, $\int_{-2}^3 g(x)\, dx = 2$, and $\int_3^6 g(x)\, dx = 3$, then find the values of each definite integral in Exercises 29–40. If there is not enough information, explain why.

29. $\int_3^6 f(x)\, dx$

30. $\int_{-2}^6 g(x)\, dx$

31. $\int_{-2}^6 (f(x) + g(x))\, dx$

32. $\int_3^6 (2f(x) - g(x))\, dx$

33. $\int_{-2}^3 f(x)g(x)\, dx$

34. $\int_{-2}^6 (g(x) + x)\, dx$

35. $\int_3^6 (g(x))^2\, dx$

36. $\int_{-2}^3 x f(x)\, dx$

37. $\int_6^3 (f(x) + g(x))\, dx$

38. $\int_{-2}^6 (4f(x) - 2)\, dx$

39. $\int_3^{-2} (2x^2 - 3g(x))\, dx$

40. $\int_{-2}^{-2} x(f(x) + 3)^2\, dx$

For each definite integral in Exercises 41–46, (a) find the general n-rectangle right sum and simplify your answer with sum formulas. Then (b) use your answer to approximate the definite integral with $n = 100$ and $n = 1000$. Finally, (c) take the limit as $n \to \infty$ to find the exact value.

41. $\int_2^5 (5 - x)\, dx$

42. $\int_{-3}^3 (2x + 1)\, dx$

43. $\int_0^1 2x^2\, dx$

44. $\int_{-3}^2 x^2\, dx$

45. $\int_2^3 (x + 1)^2\, dx$

46. $\int_{-1}^2 (1 - x^2)\, dx$

Calculate the exact value of each definite integral in Exercises 47–52 by using properties of definite integrals and the formulas in Theorem 4.13.

47. $\int_2^4 (x^2 + 1)\, dx$

48. $\int_0^6 (3x + 2)\, dx$

49. $\int_5^2 (9 + 10x - x^2)\, dx$

50. $\int_1^3 (x + 1)^2\, dx$

51. $\int_0^4 ((2x - 3)^2 + 5)\, dx$

52. $\int_6^1 (3(1 - 2x)^2 + 4x)\, dx$

Applications

53. Suppose that once again you drive in a car for 40 seconds with velocity $v(t) = -0.22t^2 + 8.8t$ feet per second, as shown in the graph that follows. Suppose also that your total distance travelled is equal to the area under the velocity curve on $[0, 40]$.

Velocity of car
$v(t) = -0.22t^2 + 8.8t$

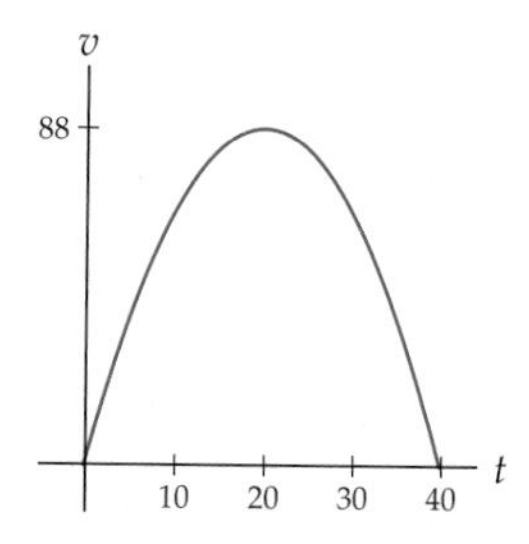

(a) What definite integral would you have to compute in order to find your exact distance travelled over the 40 seconds of your trip?
(b) Find the exact value of that definite integral by taking a limit of Riemann sums.

54. The function for the ***standard normal distribution*** is

$$f(x) = \frac{1}{\sqrt{2\pi}} e^{-x^2/2}$$

Its graph is that of the **bell curve**. Probability and statistics books often have tables like the one following, which lists some approximate areas under the bell curve:

Areas under the bell curve

b	$\frac{1}{\sqrt{2\pi}} \int_{-b}^{b} e^{-x^2/2}\, dx$
0.5	0.3829
1	0.6827
1.5	0.8664
2	0.9545
2.5	0.9876

Use the information given in the table, properties of definite integrals, and symmetry to find

(a) $\frac{1}{\sqrt{2\pi}} \int_{-0.5}^{1.5} e^{-x^2/2}\, dx$ (b) $\frac{1}{\sqrt{2\pi}} \int_{1.5}^{2} e^{-x^2/2}\, dx$

Proofs

55. Use the definition of the definite integral as a limit of Riemann sums to prove Theorem 4.11(b): For any function f that is integrable on $[a, b]$ and any real number c, $\int_a^b cf(x)\, dx = c\int_a^b f(x)\, dx$.

56. Use the definition of the definite integral as a limit of Riemann sums to prove Theorem 4.12(a): For any function f and real number a, $\int_a^a f(x)\, dx = 0$.

57. Use the definition of the definite integral as a limit of Riemann sums to prove Theorem 4.12(b): For any function f that is integrable on $[a, b]$, $\int_b^a f(x)\, dx = -\int_a^b f(x)\, dx$.

58. Give a geometric argument to prove Theorem 4.13(a): For any real numbers a, b, and c, $\int_a^b c\, dx = c(b - a)$. (*Hint: Use a rectangle.*)

59. Give a geometric argument to prove Theorem 4.13(b): For any real numbers $0 < a < b$, $\int_a^b x\,dx = \frac{1}{2}(b^2 - a^2)$. (*Hint: Use a trapezoid.*)

60. Prove Theorem 4.13(b): For any real numbers a and b, we have $\int_a^b x\,dx = \frac{1}{2}(b^2 - a^2)$. Use the proof of Theorem 4.13(a) as a guide.

61. Prove Theorem 4.13(c): For any real numbers a and b, $\int_a^b x^2\,dx = \frac{1}{3}(b^3 - a^3)$. Use the proof of Theorem 4.13(a) as a guide.

62. Prove that $\int_1^3 (3x+4)\,dx = 20$ in three different ways:

(a) algebraically, by calculating a limit of Riemann sums;

(b) geometrically, by recognizing the region in question as a trapezoid and calculating its area;

(c) with formulas, by using properties and formulas of definite integrals.

Thinking Forward

Functions defined by area accumulation: We know that for fixed real numbers a and b and an integrable function f, the definite integral $\int_a^b f(x)\,dx$ is a real number. For different real values of b, we get (potentially) different values for the integral $\int_a^b f(x)\,dx$.

- Make a table of the values of the integral $\int_0^b 2x\,dx$ corresponding to the values $-3, -2, -1, 0, 1, 2$, and 3 for b. Conjecture a formula for the relationship between the values of b and the corresponding value of the integral.
- What is the word that describes the kind of relationship that exists between the values of b and the corresponding value of the integral?
- Now make a table of the values of the integral $\int_1^b 2x\,dx$ corresponding to the values $-3, -2, -1, 0, 1, 2$, and 3 for b. Conjecture a formula for the relationship between the values of b and the corresponding value of the integral.
- What is the relationship between the formula that describes $\int_0^b 2x\,dx$ and the formula that describes $\int_1^b 2x\,dx$?

4.4 INDEFINITE INTEGRALS

- The definition of the indefinite integral as a notation for antidifferentiation
- Formulas for integrating some basic functions
- Guessing and checking to solve integrals of combinations of functions

Antiderivatives and Indefinite Integrals

As we have seen throughout the previous chapters, an ***antiderivative*** of a function f is a function F whose derivative is f. For example, an antiderivative of $f(x) = 2x$ is the function $F(x) = x^2$. Another antiderivative of $f(x) = 2x$ is the function $G(x) = x^2 + 3$. In fact, any function of the form $x^2 + C$ is an antiderivative of $f(x) = 2x$, and these are the only possible antiderivatives of $f(x) = 2x$. In general, the antiderivatives of a given function are all related to each other, so we call the set of all antiderivatives of f the ***family of antiderivatives*** of f. As we showed in Theorem 3.7 of Section 3.2, any two antiderivatives of a function must differ by a constant. For convenience we restate that theorem now:

THEOREM 4.14 **Functions with the Same Derivative Differ by a Constant**

If F and G are differentiable functions, then $F'(x) = G'(x)$ if and only if $G(x) = F(x) + C$ for some constant C.

In the previous section we defined the ***definite integral*** of a function f on an interval $[a, b]$ as a limit of Riemann sums used in calculating the signed area between the graph of a function f and the x-axis on an interval $[a, b]$. We will now define a completely different object called the ***indefinite integral*** of a function f. This new object will be a family of functions, not a number, but we will see in Section 4.5 that these two types of integrals are related, which is why we'll call them by similar names.

The indefinite integral of a function is a notation for expressing the collection of all possible antiderivatives of that function:

DEFINITION 4.15

The Indefinite Integral of a Function

The ***indefinite integral*** of a continuous function f is defined to be the family of antiderivatives

$$\int f(x)\,dx = F(x) + C,$$

where F is an antiderivative of f, that is, a function for which $F' = f$.

For example, since an antiderivative of $f(x) = 2x$ is $F(x) = x^2$, it follows that all antiderivatives of $f(x) = 2x$ are of the form $x^2 + C$, and therefore that $\int 2x\,dx = x^2 + C$. Note that it would be equally accurate to use a different antiderivative, such as $G(x) = x^2 + 3$, and say that $\int 2x\,dx = (x^2 + 3) + C$.

CAUTION

Although the notation and terminology used for indefinite integrals in Definition 4.15 is similar to what we used for definite integrals in Section 4.3, it is important to note that at this point we have no proof that the two types of integrals are related. When we see the Fundamental Theorem of Calculus in Section 4.5, we will make the surprising discovery that the area under a curve is in fact related to families of antiderivatives, and this relationship will justify why we use such similar notation for two different kinds of objects.

The "dx" in the notation of Definition 4.15 represents the fact that we are antidifferentiating with respect to the variable x. The constant C represents an arbitrary constant. The function f inside the integral notation is called the ***integrand,*** and when we find $\int f(x)\,dx$, we say that we are ***integrating*** the function f. The indefinite integral of a function will often be called simply the ***integral*** of that function. The continuity hypothesis is important (see Exercises 18–20), and we will assume throughout this section that we are working with intervals where our functions are continuous.

Antidifferentiation Formulas

All of the rules that we have developed for differentiating functions can be used to develop *anti*differentiation rules, which in turn will give us formulas for some common indefinite integrals. For example, the rule for differentiating power functions says that for any constant k, $\frac{d}{dx}(x^k) = kx^{k-1}$. The rule for antidifferentiating a power function should "undo" this process and is given in the following theorem:

THEOREM 4.16

Integrals of Power Functions

(a) If $k \neq -1$, then $\displaystyle\int x^k\,dx = \frac{1}{k+1}x^{k+1} + C.$

(b) $\displaystyle\int \frac{1}{x}\,dx = \ln|x| + C.$ *(See Exercise 18 for a technical point.)*

Proof. To prove the first part of the theorem, it suffices to show that if $k \neq -1$, then $\frac{1}{k+1}x^{k+1}$ is an antiderivative of x^k. (Note that if $k = -1$, then $\frac{1}{k+1}$ is not even defined.) In other words, we need only show that the derivative of $\frac{1}{k+1}x^{k+1}$ is x^k. This is a simple application of the power and constant-multiple rules of differentiation:

$$\frac{d}{dx}\left(\frac{1}{k+1}x^{k+1}\right) = \frac{1}{k+1}(k+1)x^k = x^k.$$

The second integration formula in the theorem covers the case when $k = -1$, since $x^{-1} = \frac{1}{x}$. This formula follows immediately from the fact that $\frac{d}{dx}(\ln|x|) = \frac{1}{x}$. ■

The next three theorems describe formulas for antidifferentiating–and thus integrating–other common types of functions. Each of these formulas can be proved by differentiating; you will do so in Exercises 73–75.

THEOREM 4.17 **Integrals of Exponential Functions**

(a) If $k \neq 0$, then $\displaystyle\int e^{kx}\,dx = \frac{1}{k}e^{kx} + C.$

(b) If $b > 0$ and $b \neq 1$, then $\displaystyle\int b^x\,dx = \frac{1}{\ln b}b^x + C.$

For example, $\int 2^x\,dx = \frac{1}{\ln 2}2^x$ because $\frac{d}{dx}\left(\frac{1}{\ln 2}2^x\right) = \frac{1}{\ln 2}(\ln 2)2^x = 2^x$. Notice that both of the rules in Theorem 4.17 imply that the integral of e^x is itself, that is, that $\int e^x\,dx = e^x + C$.

THEOREM 4.18 **Integrals of Certain Trigonometric Expressions**

(a) $\displaystyle\int \sin x\,dx = -\cos x + C$ **(b)** $\displaystyle\int \cos x\,dx = \sin x + C$

(c) $\displaystyle\int \sec^2 x\,dx = \tan x + C$ **(d)** $\displaystyle\int \csc^2 x\,dx = -\cot x + C$

(e) $\displaystyle\int \sec x \tan x\,dx = \sec x + C$ **(f)** $\displaystyle\int \csc x \cot x\,dx = -\csc x + C$

THEOREM 4.19 **Integrals Whose Solutions Are Inverse Trigonometric Functions**

(a) $\displaystyle\int \frac{1}{\sqrt{1-x^2}}\,dx = \sin^{-1} x + C.$

(b) $\displaystyle\int \frac{1}{1+x^2}\,dx = \tan^{-1} x + C.$

(c) $\displaystyle\int \frac{1}{|x|\sqrt{x^2-1}}\,dx = \sec^{-1} x + C.$

Finally, if you are covering hyperbolic and inverse hyperbolic functions in your course (see the last two subsections of Section 2.6), then you should also be familiar with the next two sets of antidifferentiation formulas, which follow directly from the differentiation formulas in Theorems 2.20 and 2.21.

THEOREM 4.20 **Integrals of Hyperbolic Functions***

(a) $\displaystyle\int \sinh x\,dx = \cosh x + C$

(b) $\displaystyle\int \cosh x\,dx = \sinh x + C$

(c) $\displaystyle\int \operatorname{sech}^2 x\,dx = \tanh x + C$

THEOREM 4.21 **Integrals Whose Solutions Are Inverse Hyperbolic Functions***

(a) $\displaystyle\int \frac{1}{\sqrt{x^2+1}}\,dx = \sinh^{-1} x + C$

(b) $\displaystyle\int \frac{1}{\sqrt{x^2-1}}\,dx = \cosh^{-1} x + C$

(c) $\displaystyle\int \frac{1}{1-x^2}\,dx = \tanh^{-1} x + C$

Antidifferentiating Combinations of Functions

We have rules for differentiating constant multiples, sums, products, quotients, and compositions of functions. Only the constant-multiple rule and the sum rule translate directly into antidifferentiation rules:

THEOREM 4.22 **Constant Multiple and Sum Rules for Indefinite Integrals**

(a) $\displaystyle\int kf(x)\,dx = k\int f(x)\,dx.$

(b) $\displaystyle\int (f(x)+g(x))\,dx = \int f(x)\,dx + \int g(x)\,dx.$

Proof. Suppose F is any antiderivative of f, that is, $F'(x) = f(x)$. Then, by the constant-multiple rule, $kF'(x) = kf(x)$ for any constant k. Furthermore, for any constant D, $F(x) + D$ is also an antiderivative of f, so $(F(x) + D)' = f(x)$. Therefore

$$\int kf(x)\,dx = kF'(x) + C = k\left(F'(x) + \frac{C}{k}\right) = k(F'(x) + D) = k\int f(x)\,dx.$$

Note that in the calculation we just did, $D = \frac{C}{k}$ is just a constant.

Similarly, if $F'(x) = f(x)$ and $G'(x) = g(x)$, then, by the sum rule, $(F(x) + G(x))' = f(x) + g(x)$, and therefore

$$\int f(x)\,dx + \int g(x)\,dx = (F'(x) + C_1) + (G'(x) + C_2) = (F'(x) + G'(x)) + C = \int (f(x) + g(x))\,dx,$$

where $C = C_1 + C_2$. ■

There are no general product, quotient, and chain rules for antidifferentiation. However, if we think of these differentiation rules "backwards," then we can say something about certain types of integrands:

THEOREM 4.23 **Reversing the Product, Quotient, and Chain Rules**

(a) $\int (f'(x)\,g(x) + f(x)\,g'(x))\,dx = f(x)\,g(x) + C$

(b) $\int \frac{f'(x)\,g(x) - f(x)\,g'(x)}{(g(x))^2}\,dx = \frac{f(x)}{g(x)} + C$

(c) $\int f'(g(x))\,g'(x)\,dx = f(g(x)) + C$

Of course, the trouble is recognizing when an integrand is in one of these special forms. If we are lucky enough to recognize an integrand as the result of a product, quotient, or chain rule calculation, then we can make an educated guess at the integral and check our answer by differentiating. Repeated guess-and-check is at this point our best strategy for integrating combinations of functions. The first and third parts of Theorem 4.23 will form the basis for the methods of integration by parts and integration by substitution that we will see in Chapter 5.

In general, antidifferentiation is much more difficult than differentiation. We can differentiate *every* function that we currently know how to write down; however, at this point we cannot integrate even very simple functions like $\ln x$ or $\sec x$. You should think of integration as a puzzle, not a procedure. Unlike differentiation, where it is always clear which rules to apply, and in which order, it is *not* always immediately clear how to find a given integral. We will learn some more methods for calculating integrals in Chapter 5, but even then we will not be able to calculate all integrals. In fact, as we will see in Section 4.7, some functions have no elementary antiderivative, which means that their antiderivatives cannot even be written down in terms of the functions we now know.

Examples and Explorations

EXAMPLE 1 **Identifying antiderivatives**

Which of the following are antiderivatives of $f(x) = x^4$, and why?

(a) $4x^3$ **(b)** $\frac{1}{5}x^5$ **(c)** $\frac{1}{5}x^5 - 2$ **(d)** $\frac{1}{5}(x^5 - 2)$

SOLUTION

An antiderivative of $f(x) = x^4$ is a function whose derivative is x^4. Choice (a) is clearly the derivative, not an antiderivative, of $f(x) = x^4$. The remaining three choices are *all* antiderivatives of $f(x) = x^4$, since each of those functions has derivative x^4. □

EXAMPLE 2 **Using algebra to identify and then integrate power functions**

Find $\int \frac{1}{x^2}\,dx$ and $\int \sqrt{x^3}\,dx$.

SOLUTION

Solving these integrals is an easy application of Theorem 4.16, once we write the integrands in the form x^k:

$$\int \frac{1}{x^2}\,dx = \int x^{-2}\,dx = \frac{1}{-2+1}x^{-2+1} = \frac{1}{-1}x^{-1} + C = -\frac{1}{x} + C;$$

$$\int \sqrt{x^3}\,dx = \int x^{3/2}\,dx = \frac{1}{\frac{3}{2}+1}x^{3/2+1} = \frac{1}{\frac{5}{2}}x^{5/2} + C = \frac{2}{5}\sqrt{x^5} + C.$$

□

EXAMPLE 3

Using the sum and constant-multiple rules for indefinite integrals

Find $\int (5x^3 - \sin x)\,dx$, and show explicitly how the sum and constant-multiple rules for indefinite integrals are applied.

SOLUTION

The function $f(x) = 5x^3 - \sin x$ is a sum of constant multiples of functions whose integrals we know, namely, the functions x^3 and $\sin x$. By Theorem 4.22, we have

$$\begin{aligned}
\int (5x^3 - \sin x)\,dx &= \int (5x^3 + (-\sin x))\,dx && \leftarrow \text{write the difference as a sum} \\
&= \int 5x^3\,dx + \int (-\sin x)\,dx && \leftarrow \text{sum rule} \\
&= 5\int x^3\,dx - \int \sin x\,dx && \leftarrow \text{constant-multiple rule} \\
&= 5\left(\tfrac{1}{4}x^4\right) - (-\cos x) + C && \leftarrow \text{antidifferentiation formulas} \\
&= \frac{5}{4}x^4 + \cos x + C.
\end{aligned}$$

Notice that we added only *one* constant C in this calculation. Technically there are *two* such constants, since the family of antiderivatives for x^3 is $\frac{1}{4}x^4 + C_1$, and the family of antiderivatives for $\sin x$ is $-\cos x + C_2$. The C we are using is really the sum $C_1 + C_2$. There is no need to write the two constants C_1 and C_2 separately, since the sum of two arbitrary constants is also an arbitrary constant. □

EXAMPLE 4

Recognizing integrands in product, quotient, and chain rule forms

Solve the following indefinite integrals:

(a) $\displaystyle\int (e^x + xe^x)\,dx$ **(b)** $\displaystyle\int 5(x^2+1)^4(2x)\,dx$ **(c)** $\displaystyle\int \frac{2x\tan x - x^2\sec^2 x}{\tan^2 x}\,dx$

SOLUTION

(a) This integrand looks like it could be the result of a product rule calculation where x and e^x are the factors in the product. We check by differentiating, and indeed $\frac{d}{dx}(xe^x) = 1e^x + xe^x$. Therefore $\int (e^x + xe^x)\,dx = xe^x + C$.

(b) The integrand $5(x^2+1)^4(2x)$ looks like it could be the result of a chain rule calculation, with outside function x^5 and inside function $x^2 + 1$. We can check whether this is the case by differentiating to find $\frac{d}{dx}((x^2+1)^5) = 5(x^2+1)^4(2x)$, which is what we had hoped for. Therefore $\int 5(x^2+1)^4(2x)\,dx = (x^2+1)^5 + C$.

(c) Finally, this integrand could be the result of differentiating a quotient $\frac{f(x)}{g(x)}$ where the denominator is $g(x) = \tan x$. Looking at the numerator of the integrand, we might guess that $f(x) = x^2$. We now check: $\frac{d}{dx}\left(\frac{x^2}{\tan x}\right) = \frac{(2x)\tan x - x^2\sec^2 x}{(\tan x)^2}$, which is what we wanted for the integrand. Therefore $\int \frac{2x\tan x - x^2\sec^2 x}{\tan^2 x}\,dx = \frac{x^2}{\tan x} + C$. □

EXAMPLE 5 **Integrating by educated guess-and-check**

Solve the following indefinite integrals:

(a) $\int \sin 3x \, dx$ **(b)** $\int xe^{x^2+1} \, dx$ **(c)** $\int \frac{1}{1+4x^2} \, dx$

SOLUTION

(a) At first glance the function $\sin(3x)$ does not appear to be of the form $f'(g(x))g'(x)$. Specifically, if we choose $f'(x) = \sin x$, so that the inside function is $g(x) = 3x$, then the derivative $g'(x) = 3$ does not appear in the integrand. However, since an antiderivative of $\sin x$ is $-\cos x$, we might try $F(x) = -\cos 3x$ as a first guess for an antiderivative of $\sin 3x$. Testing this guess, we find that $\frac{d}{dx}(-\cos 3x) = -(-\sin 3x)(3) = 3 \sin 3x$. This is *almost* the integrand we are looking for, but with an extra constant multiple of 3. Accordingly, we update our guess to $F(x) = -\frac{1}{3}\cos 3x$, so that the $\frac{1}{3}$ at the front will cancel out the factor of 3 that appears when we apply the chain rule. We now have

$$\frac{d}{dx}\left(-\frac{1}{3}\cos 3x\right) = -\frac{1}{3}(-\sin 3x)(3) = \sin 3x,$$

and therefore $\int \sin 3x \, dx = -\frac{1}{3}\cos 3x + C$.

(b) The first thing to recognize about this problem is that the function $x^2 + 1$ is an inside function in the integrand while its derivative $2x$ is *almost* part of the integrand: We have an x instead of a $2x$, so we're missing only a constant multiple. Since the function e^x is its own antiderivative, a good first guess for an antiderivative of xe^{x^2+1} might be $F(x) = e^{x^2+1}$. Checking this guess, we find that $\frac{d}{dx}(e^{x^2+1}) = e^{x^2+1}(2x)$, which is *almost* what we want. To get rid of the extra 2, we update our guess to $F(x) = \frac{1}{2}e^{x^2+1}$; with this guess we have

$$\frac{d}{dx}\left(\frac{1}{2}e^{x^2+1}\right) = \frac{1}{2}e^{x^2+1}(2x) = xe^{x^2+1}.$$

It then follows that $\int xe^{x^2+1} \, dx = \frac{1}{2}e^{x^2+1} + C$. It is important to note that the method used in this example, whereby we update our guess by dividing by what is "missing," works only if what is missing is a *constant*.

(c) In this example we are searching for a function whose derivative is $\frac{1}{1+4x^2}$. Our integrand is similar in structure to the second integrand in Theorem 4.19, so we will start the guessing process with $\tan^{-1}(x)$. We guess, check, and re-guess until we get the derivative we seek:

$$\frac{d}{dx}(\tan^{-1} x) = \frac{1}{1+x^2};$$

$$\frac{d}{dx}(\tan^{-1}(4x)) = \frac{1}{1+(4x)^2}(4) = \frac{4}{1+16x^2};$$

$$\frac{d}{dx}(\tan^{-1}(2x)) = \frac{1}{1+(2x)^2}(2) = \frac{2}{1+4x^2};$$

$$\frac{d}{dx}\left(\frac{1}{2}\tan^{-1}(2x)\right) = \left(\frac{1}{2}\right)\frac{1}{1+(2x)^2}(2) = \frac{1}{1+4x^2}.$$

Therefore $\int \frac{1}{1+4x^2} \, dx = \frac{1}{2}\tan^{-1}(2x) + C$. □

? TEST YOUR UNDERSTANDING

- How is an indefinite integral related to antiderivatives? In what way is a definite integral different from an antiderivative?
- Which of the six trigonometric functions do we know how to integrate at this point, and why?
- Is the integral of a difference of two functions always equal to the difference of the integrals of those two functions? Why or why not?
- Is the integral of a product of two functions equal to the product of the integrals of those two functions? Why or why not?
- How can we solve integrals of combinations of functions by guessing and checking? Why is this a valid method?

EXERCISES 4.4

Thinking Back

- *Antiderivatives:* If $g'(x) = h(x)$, then is g an antiderivative of h or is h an antiderivative of g?
- *Definite integrals:* State the definition of the definite integral of a function f on an interval $[a, b]$.
- *The Mean Value Theorem:* State the Mean Value Theorem.

Derivatives and antiderivatives: Find the derivative and an antiderivative of each of the following functions:

- $f(x) = \pi^2$
- $f(x) = (\sqrt[11]{x})^{-2}$
- $f(x) = \dfrac{1}{5x}$
- $f(x) = \sec^2(3x + 1)$

Concepts

0. *Problem Zero:* Read the section and make your own summary of the material.

1. *True/False:* Determine whether each of the statements that follow is true or false. If a statement is true, explain why. If a statement is false, provide a counterexample.

(a) *True or False:* If $f(x) - g(x) = 2$ for all x, then f and g have the same derivative.

(b) *True or False:* If $f(x) - g(x) = 2$ for all x, then f and g have the same antiderivative.

(c) *True or False:* If $f(x) = 2x$, then $F(x) = x^2$.

(d) *True or False:* $\int (x^3 + 1)\, dx = \left(\frac{1}{4}x^4 + x + 2\right) + C$.

(e) *True or False:* $\int \sin(x^2)\, dx = -\cos(x^2) + C$.

(f) *True or False:* $\int e^x \cos x\, dx = e^x \sin x + C$.

(g) *True or False:* $\int \frac{1}{x^2+1}\, dx = \ln|x^2 + 1| + C$.

(h) *True or False:* $\int \frac{1}{x^2+1}\, dx = \frac{1}{2x} \ln|x^2 + 1| + C$.

2. *Examples:* Construct examples of the thing(s) described in the following. Try to find examples that are different than any in the reading.

(a) An integral that can be solved by recognizing the integrand as the result of a product rule calculation.

(b) An integral that can be solved by recognizing the integrand as the result of a chain rule calculation.

(c) Six functions that we do not currently know how to integrate (and why we cannot integrate each of them given what we currently know).

3. What is the difference between an antiderivative of a function and the indefinite integral of a function?

4. Explain why we call the collection of antiderivatives of a function f a family. How are the antiderivatives of a function related?

5. Compare the definitions of the definite and indefinite integrals. List at least three things that are different about these mathematical objects.

6. Fill in each of the blanks:

(a) $\int$ _____ $dx = x^6 + C$.

(b) x^6 is an antiderivative of _____ .

(c) The derivative of x^6 is _____ .

7. Fill in each of the blanks:

(a) $\int x^6\, dx =$ _____ $+ C$.

(b) _____ is an antiderivative of x^6.

(c) The derivative of _____ is x^6.

8. Explain why the formula for the integral of x^k does not apply when $k = -1$. What *is* the integral of x^{-1}?

9. Explain why at this point we don't have an integration formula for the function $f(x) = \sec x$ whereas we *do* have an integration formula for $f(x) = \sin x$.

10. Why don't we bother to state an integration formula that has to do with $\cos^{-1} x$? *(Hint: Think about the derivatives of $\cos^{-1} x$ and $\sin^{-1} x$.)* What would the integration formula be? Why is it "redundant," given the integration formula that has to do with $\sin^{-1} x$?

11. Write out all the integration formulas and rules that we know at this point.

12. Show that $F(x) = \sin x - x\cos x + 2$ is an antiderivative of $f(x) = x\sin x$.

13. Show that $F(x) = \frac{1}{2}e^{(x^2)}(x^2 - 1)$ is an antiderivative of $f(x) = x^3 e^{(x^2)}$.

14. Verify that $\int \cot x\, dx = \ln(\sin x) + C$. (Do not try to solve the integral from scratch.)

15. Verify that $\int \ln x\, dx = x(\ln x - 1) + C$. (Do not try to solve the integral from scratch.)

16. Show by exhibiting a counterexample that, in general, $\int \frac{f(x)}{g(x)}\, dx \neq \frac{\int f(x)\,dx}{\int g(x)\,dx}$. In other words, find two functions f and g such that the integral of their quotient is not equal to the quotient of their integrals.

17. Show by exhibiting a counterexample that, in general, $\int f(x)g(x)\, dx \neq (\int f(x)\, dx)(\int g(x)\, dx)$. In other words, find two functions f and g so that the integral of their product is not equal to the product of their integrals.

In the definition of the definite integral we required integrands to be continuous. If an integrand fails to be continuous everywhere, then a kind of branching can occur in its antiderivatives. In Exercises 18–19 we investigate some discontinuous examples where we choose nonstandard antiderivatives. Outside of this set of examples we will restrict our attention to the standard antiderivatives in Theorems 4.16–4.19.

18. Consider the function

$$F(x) = \begin{cases} \ln|x|, & \text{if } x < 0 \\ \ln|x| + 4, & \text{if } x > 0. \end{cases}$$

Show that the derivative of this function is the function $f(x) = \frac{1}{x}$. Compare the graphs of $F(x)$ and $\ln|x|$, and discuss how this exercise relates to the second part of Theorem 4.16.

19. Consider the function

$$F(x) = \begin{cases} -\cot x, & \text{if } x < 0 \\ -\cot x + 100, & \text{if } x > 0. \end{cases}$$

Show that the derivative of this function is the function $f(x) = \csc^2 x$. Compare the graphs of $F(x)$ and $-\cot x$, and discuss how this exercise relates to the fourth part of Theorem 4.18.

20. Consider the function

$$F(x) = \begin{cases} \sec^{-1} x, -\pi & \text{if } x < -1 \\ \sec^{-1} x + \pi, & \text{if } x > 1. \end{cases}$$

Show that the derivative of this function is the function $f(x) = \frac{1}{|x|\sqrt{x^2-1}}$. Compare the graphs of $F(x)$ and $\sec^{-1} x$, and discuss how this exercise relates to the second part of Theorem 4.19.

Skills

Use integration formulas to solve each integral in Exercises 21–62. You may have to use algebra, educated guess-and-check, and/or recognize an integrand as the result of a product, quotient, or chain rule calculation. Check each of your answers by differentiating. *(Hint for Exercise 54:* $\tan x = \frac{\sin x}{\cos x}$*).*

(Exercises 59–62 involve hyperbolic functions and their inverses; see Section 2.6.)

21. $\int (x^2 - 3x^5 - 7)\, dx$

22. $\int (x^3 + 4)^2\, dx$

23. $\int (x^2 - 1)(3x + 5)\, dx$

24. $\int \left(\frac{2}{3x} + 1\right) dx$

25. $\int (x^2 + 2^x + 2^2)\, dx$

26. $\int \frac{3}{x^2+1}\, dx$

27. $\int \frac{x+1}{\sqrt{x}}\, dx$

28. $\int \left(\frac{3}{x^2} - 4\right) dx$

29. $\int (3.2(1.43)^x - 50)\, dx$

30. $\int 5\sin(2x+1)\, dx$

31. $\int 4(e^{x-3})^2\, dx$

32. $\int 4e^{2x-3}\, dx$

33. $\int \sec(3x)\tan(3x)\, dx$

34. $\int 3\csc(\pi x)\cot(\pi x)\, dx$

35. $\int 3\tan^2 x\, dx$

36. $\int (\sin^2 x + \cos^2 x)\, dx$

37. $\int \pi\, dx$

38. $\int \frac{x^{-1}}{1 - x^{-1}}\, dx$

39. $\int \frac{7}{\sqrt{1-(2x)^2}}\, dx$

40. $\int \frac{-3}{1+16x^2}\, dx$

41. $\int \frac{1}{|3x|\sqrt{9x^2-1}}\, dx$

42. $\int \frac{1}{4+x^2}\, dx$

43. $\int \frac{2x}{1+x^2}\, dx$

44. $\int 2\sin x\cos x\, dx$

45. $\int (x^2 e^x + 2xe^x)\, dx$

46. $\int 3x^2\sin(x^3+1)\, dx$

47. $\int \frac{6x}{3x^2+1}\, dx$

48. $\int x^3 e^{3x^4-2}\, dx$

49. $\int \frac{e^{3x} - 2e^{4x}}{e^{2x}}\, dx$

50. $\int \sec^2 x + \csc^2 x\, dx$

51. $\int x^2(x^3+1)^5\, dx$

52. $\int (\tan x + x\sec^2 x)\, dx$

53. $\int \frac{2x\ln x - x}{(\ln x)^2}\, dx$

54. $\int \tan x\, dx$

55. $\int \left(e^x\sqrt{x} + \frac{e^x}{2\sqrt{x}}\right) dx$

56. $\int \frac{1}{2\sqrt{x}} e^{\sqrt{x}}\, dx$

57. $\displaystyle\int \left(\frac{1}{x}\cos x - \sin x \ln x\right) dx$

58. $\displaystyle\int \frac{2x\sin x - (x^2+1)\cos x}{\sin^2 x}\, dx$

59. $\displaystyle\int 3\cosh 2x\, dx$

60. $\displaystyle\int \frac{1}{2\sqrt{x^2-1}}\, dx$

61. $\displaystyle\int (\tanh x + x\,\mathrm{sech}^2 x)\, dx$

62. $\displaystyle\int \frac{1}{(1+x)(1-x)}\, dx$

Solve each of the integrals in Exercises 63–68, where a, b, and c are real numbers with $a \neq 0$, $b \neq 0$, $c > 0$, and $c \neq 1$.

63. $\displaystyle\int ae^{bx+c}\, dx$

64. $\displaystyle\int \frac{a}{bx^c}\, dx$

65. $\displaystyle\int \frac{\sec(ax)\tan(ax)}{b}\, dx$

66. $\displaystyle\int (b(c^x) + a)\, dx$

67. $\displaystyle\int \frac{a}{1+(bx)^2}\, dx$

68. $\displaystyle\int (a\sin(bx) - c)\, dx$

Applications

69. Suppose you throw a ball straight up into the air at a velocity of $v_0 = 42$ feet per second, initially releasing the ball at a height of 4 feet. Acceleration due to gravity is constantly $a(t) = -32$ feet per second squared (negative since it pulls downward).

Position, velocity, and acceleration of a ball thrown upwards

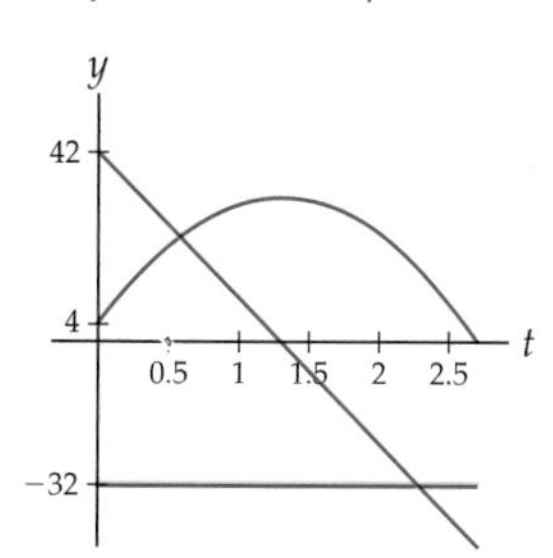

(a) Calculate $\int a(t)\, dt$. Then, given that acceleration is the derivative of velocity and that the initial velocity was $v_0 = 42$, find a formula for the velocity $v(t)$ of the ball after t seconds.

(b) Calculate $\int v(t)\, dt$. Then, given that velocity is the derivative of position and that the initial position was $s_0 = 4$ feet, find a formula for the velocity $s(t)$ of the ball after t seconds.

(c) Discuss the relationships among the graphs of $s(t)$, $v(t)$, and $a(t)$ shown in this exercise.

70. Dad's casserole surprise is for dinner again, and from previous experiments we know that after the casserole is taken out of the oven, the temperature of the casserole will cool by changing at a rate of $T'(t) = -15e^{-0.5t}$ degrees Fahrenheit per minute.

Rate of change of casserole temperature $T'(t) = -15e^{-0.5t}$

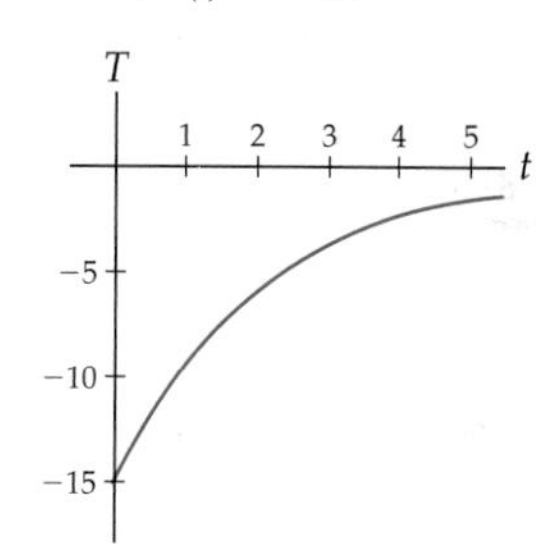

(a) Calculate $\int T'(t)\, dt$. Then, given that the casserole was $T_0 = 350$ degrees when it was removed from the oven, find a formula for the temperature $T(t)$ of the casserole after t minutes.

(b) Calculate $T(5) - T(0)$. Compare your answer with the answer to the last part of Exercise 47 from Section 4.2. Write a few sentences about the connection that this comparison suggests between Riemann sums and antidifferentiation.

Proofs

71. Use differentiation to prove the two formulas from Theorem 4.16 for integrating power functions.

72. Prove that $\int e^x\, dx = e^x + C$ in two ways: (a) by using the first part of Theorem 4.17 and (b) by using the second part of Theorem 4.17.

73. Use differentiation to prove the two formulas from Theorem 4.17 for integrating exponential functions.

74. Use differentiation to prove the six formulas from Theorem 4.18 for integrating certain trigonometric functions.

75. Use differentiation to prove the three formulas from Theorem 4.19 for integrals whose solutions are inverse trigonometric functions.

76. Prove the first part of Theorem 4.22 in your own words: $\int kf(x), dx = k\int f(x)\, dx$.

77. Prove the second part of Theorem 4.22 in your own words: $\int (f(x) + g(x))\, dx = \int f(x)\, dx + \int g(x)\, dx$.

78. Prove the "forward" direction of Theorem 4.14: If F and G are differentiable and $F'(x) = G'(x)$, then $G(x) = F(x) + C$ for some constant C. *(Hint: Show that if $F'(x) = G'(x)$, then the function $F(x) - G(x)$ is constant. Recall that if the derivative of a function is zero, then that function must be constant.)*

79. Prove the "backwards" direction of Theorem 4.14: If F and G are differentiable and $G(x) = F(x) + C$ for some constant C, then $F'(x) = G'(x)$. *(Hint: Differentiate both sides of the equality $G(x) = F(x) + C$.)*

Thinking Forward

Initial-Value Problems: Find a function f that has the given derivative f' and value $f(c)$, if possible.

- $f'(x) = x^8 - x^2 + 1,\ f(0) = -1$
- $f'(x) = \dfrac{2}{\sqrt[3]{x}},\ f(1) = 2$
- $f'(x) = 3e^{2x} + 1,\ f(0) = 2$
- $f'(x) = 3\left(\dfrac{4}{5}\right)^x,\ f(1) = 0$
- $f'(x) = \dfrac{5}{3x - 2},\ f(1) = 8$
- $f'(x) = 3\sin(-2x) - 5\cos(3x),\ f(0) = 0$

Preview of Differential Equations: Given each of the following equations involving a function f, find a possible formula for $f(x)$.

- $f'(x) = f(x)$
- $f'(x) = 2f(x)$
- $f'(t) = \sin(bt)$
- $f''(t) = \cos(t)$

4.5 THE FUNDAMENTAL THEOREM OF CALCULUS

- The statement, meaning, and proof of the Fundamental Theorem of Calculus
- Other equivalent formulations of the Fundamental Theorem
- Using the Fundamental Theorem of Calculus to calculate definite integrals exactly

The Fundamental Theorem

So far we have two very different mathematical objects that we call integrals, namely, the definite integral $\int_a^b f(x)\,dx$ and the indefinite integral $\int f(x)\,dx$. The reason these objects have such similar names and notation is that they are related by two very important theorems collectively called the ***Fundamental Theorem of Calculus***.

We have already seen that our intuition about average velocity suggests that the area under a velocity curve is related to the distance travelled. In other words, we suspect that

$$\int_a^b v(t)\,dt = s(b) - s(a).$$

We have seen hints of this on multiple occasions in previous sections: The signed area under the graph of a function such as $v(t)$ on an interval is related to a difference of values of its antiderivative $s(t)$. This incredible relationship between signed area and antiderivatives is so important, in fact so *fundamental*, to the study of calculus that we call it the Fundamental Theorem of the subject:

THEOREM 4.24 **The Fundamental Theorem of Calculus**

If f is continuous on $[a, b]$ and F is any antiderivative of f, then

$$\int_a^b f(x)\,dx = F(b) - F(a).$$

CAUTION If f is not continuous on $[a, b]$, then the Fundamental Theorem may not apply to f on that interval, because we can't even be sure that its signed area on the interval $[a, b]$ is well defined. For example, consider the function $f(x) = \frac{1}{(x-1)^2}$ on the interval $[0, 3]$. The discontinuity of f at $x = 1$ is a problem that we won't learn how to handle until Section 5.6.

The Fundamental Theorem of Calculus expresses a deep fact about the relationship between antiderivatives and signed area and moreover provides a very powerful tool for calculating definite integrals. We will prove the Fundamental Theorem of Calculus at the end of this section, after we have discussed the various ways that we can use this powerful tool. In Section 4.7 we will discuss a related theorem called the Second Fundamental Theorem of Calculus.

Using the Fundamental Theorem of Calculus

When we apply the Fundamental Theorem in practice, we need to identify an antiderivative of the integrand, evaluate that antiderivative at two points, and then subtract. For example, since an antiderivative of $f(x) = x^2$ is $F(x) = \frac{1}{3}x^3$, the Fundamental Theorem tells us that

$$\int_1^4 x^2\, dx = F(4) - F(1) = \frac{1}{3}(4)^3 - \frac{1}{3}(1)^3 = 21.$$

The following notation gives us a more efficient way of working with the Fundamental Theorem in practice:

DEFINITION 4.25 **Evaluation Notation**

For any function F on an interval $[a, b]$, the difference $F(b) - F(a)$ will be called ***the evaluation of*** $F(x)$ ***on*** $[a, b]$ and will be denoted by

$$\left[F(x)\right]_a^b = F(b) - F(a).$$

This notation is also sometimes written as $F(x)|_a^b$. We call this the "evaluation" of $F(x)$ on $[a, b]$ because $[F(x)]_a^b$ is the difference of $F(x)$ evaluated at $x = b$ and $F(x)$ evaluated at $x = a$. With this new notation, there is no longer a need to give a letter name to the antiderivative if we want to write the antidifferentiation step separately from the evaluation step:

$$\int_1^4 x^2\, dx = \left[\frac{1}{3}x^3\right]_1^4 = \frac{1}{3}(4)^3 - \frac{1}{3}(1)^3 = 21.$$

If F and G are two antiderivatives of a function $f(x)$ on $[a, b]$, then by Theorem 4.14, for some constant C we have $G(x) = F(x) + C$ for all values of x. This means that the evaluations of F and G on $[a, b]$ will be equal:

$$\left[G(x)\right]_a^b = \left[F(x) + C\right]_a^b = (F(b) + C) - (F(a) + C) = F(b) - F(a) = \left[F(x)\right]_a^b.$$

For an integrable function f, the indefinite integral $\int f(x)\,dx$ is the family $F(x) + C$ of antiderivatives of $f(x)$. Using evaluation notation, we can reword the Fundamental Theorem of Calculus to see clearly how definite and indefinite integrals are related:

THEOREM 4.26

The Fundamental Theorem of Calculus in Evaluation Notation

If f is continuous on $[a, b]$, then

$$\int_a^b f(x)\,dx = \left[\int f(x)\,dx\right]_a^b$$

In other words, the definite integral of a function f on an interval $[a, b]$ is equal to the indefinite integral of f evaluated on that interval. The proof of this theorem follows directly from the Fundamental Theorem, evaluation notation, and the definition of the indefinite integral, and is left to Exercise 80.

The Net Change Theorem

We can rephrase the Fundamental Theorem of Calculus in terms of a function and its derivative, rather than in terms of a function and its antiderivative. In this alternative form of the Fundamental Theorem (which you will prove in Exercise 78), the function f plays the role of the antiderivative while its derivative f' plays the role of the integrand.

THEOREM 4.27

The Net Change Theorem

If f is differentiable on $[a, b]$, then

$$\int_a^b f'(x)\,dx = f(b) - f(a).$$

In other words, the ***net change*** $f(b) - f(a)$ of a function f on $[a, b]$ is equal to the signed area between the graph of its derivative f' and the x-axis on that interval.

In this form, the Fundamental Theorem of Calculus says that in some sense definite integration "undoes" differentiation, in that if you integrate the derivative of a function from $x = a$ to $x = b$, you get something that has to do with the original function (namely, $f(b) - f(a)$). This form of the Fundamental Theorem can also be interpreted as saying that if you accumulate all the instantaneous rates of change of $f(x)$ by integrating its derivative $f'(x)$ on the interval $[a, b]$, you get the total change $f(b) - f(a)$ of the values of $f(x)$ on that interval.

The Net Change Theorem is useful in a variety of practical applications. For example, at the start of this section we saw that if $s(t)$ and $v(t)$ are the position and velocity functions of an object travelling in a straight line, then

$$\int_a^b v(t)\,dt = s(b) - s(a).$$

Since $v(t) = s'(t)$, we can rephrase this as

$$\int_a^b s'(t)\,dt = s(b) - s(a).$$

This formula says that we can find the net change in distance $s(t)$ on $[a, b]$ by finding the net area between the graph of $v(t) = s'(t)$ and the x-axis on $[a, b]$. In any situation where

$Q(t)$ is a quantity that varies over time and whose rate of change is given by the derivative $r(t) = Q'(t)$, we can say that the net change in Q on an interval is the area under r on that interval:

$$\int_a^b r(t)\,dt = \int_a^b Q'(t)\,dt = Q(b) - Q(a)$$

The Proof of the Fundamental Theorem

To prove the Fundamental Theorem of Calculus, we must show that the Riemann sums that define the definite integral of a function f on an interval can be written in terms of an antiderivative F of f. The key to doing this will be applying the Mean Value Theorem to the antiderivative F. This will allow us to relate the average rate of change of F to the instantaneous rate of change–or derivative–of F. Before reading this proof, you may want to review the definition of, and notation for, the definite integral of a function (Section 4.3), as well as the Mean Value Theorem (Section 3.1).

Proof. Suppose f is a continuous function on $[a, b]$, and let F be any antiderivative of f, so that $F' = f$. Since f is continuous, it must be integrable; in other words, its definite integral from $x = a$ to $x = b$ is well defined by a limit of Riemann sums, namely,

$$\int_a^b f(x)\,dx = \lim_{n\to\infty} \sum_{k=1}^{n} f(x_k^*)\,\Delta x = \lim_{n\to\infty} \sum_{k=1}^{n} F'(x_k^*)\,\Delta x,$$

where, for each n, we define $\Delta x = \frac{b-a}{n}$ and $x_k = a + k\Delta x$, and where each x_k^* is a point in the subinterval $[x_{k-1}, x_k]$. Note that we could write $f(x_k^*) = F'(x_k^*)$, because $f = F'$ by hypothesis.

We can choose each x_k^* to be *any* point we like in the subinterval $[x_{k-1}, x_k]$. The key to this proof is to use the Mean Value Theorem to choose the points x_k^* in a very special way. Since, by hypothesis, F is differentiable on $[x_{k-1}, x_k]$, it is also continuous on that subinterval. Therefore the Mean Value Theorem applies to F on $[a, b] = [x_{k-1}, x_k]$ and says that there exists some point $c_k \in [x_{k-1}, x_k]$ such that

$$F'(c_k) = \frac{F(x_k) - F(x_{k-1})}{x_k - x_{k-1}}.$$

Because we can choose x_k^* to be any point we like in $[x_{k-1}, x_k]$, we will choose it to be the point c_k that we defined for that subinterval. Since $x_k - x_{k-1} = \Delta x$, when we substitute this expression for $F'(x_k^*)$ into our Riemann sum expression for $\int_a^b f(x)\,dx$, we have

$$\int_a^b f(x)\,dx = \lim_{n\to\infty} \sum_{k=1}^{n} \frac{F(x_k) - F(x_{k-1})}{\Delta x}\,\Delta x = \lim_{n\to\infty} \sum_{k=1}^{n} (F(x_k) - F(x_{k-1})).$$

For every n, this sum is what is known as a ***telescoping sum***; this means that if we write out the terms of this sum, then, after rearranging terms, we find that most will cancel each other out and, luckily, the sum will collapse like a folding telescope:

$$\begin{aligned}\sum_{k=1}^{n} (F(x_k) - F(x_{k-1})) &= (F(x_1) - F(x_0)) + (F(x_2) - F(x_1)) + (F(x_3) - F(x_2)) \\ &\quad + \cdots + (F(x_{n-1}) - F(x_{n-2})) + (F(x_n) - F(x_{n-1})) \\ &= -F(x_0) + (F(x_1) - F(x_1)) + (F(x_2) - F(x_2)) \\ &\quad + \cdots + (F(x_{n-1}) - F(x_{n-1})) + F(x_n) \\ &= -F(x_0) + 0 + 0 + \cdots + 0 + F(x_n) = F(b) - F(a).\end{aligned}$$

Putting this computation together with our Riemann sum calculation of the definite integral and taking what is now a very simple limit, we have the equality that we seek:

$$\int_a^b f(x)\,dx = \lim_{n\to\infty} \sum_{k=1}^{n} (F(x_k) - F(x_{k-1})) = \lim_{n\to\infty} (F(b) - F(a)) = F(b) - F(a).$$

■

Examples and Explorations

EXAMPLE 1

Distance travelled is the definite integral of velocity

Suppose a car travelled 100 miles along a straight path in two hours. If $s(t)$ describes the car's position in feet after t hours and $v(t) = s'(t)$ describes its velocity in miles per hour after t hours, find the signed area between the graph of $v(t)$ and the t-axis from $t = 0$ to $t = 2$.

SOLUTION

If $s(0) = 0$ miles, then $s(2) = 100$ miles. By the Fundamental Theorem of Calculus and the fact that $s(t)$ is an antiderivative of $v(t)$, the signed area under the graph of $v(t)$ on $[0, 2]$ is

$$\int_0^2 v(t)\,dt = s(2) - s(0) = 100 - 0 = 100.$$

Note that we didn't need to know anything about the velocity function to do this calculation; all we had to know was the change in the position of the car over the time interval $[0, 2]$. Neither did we have to calculate any limits or Riemann sums to find the definite integral $\int_0^2 v(t)\,dt$. □

EXAMPLE 2

Calculating definite integrals with the Fundamental Theorem of Calculus

Use the Fundamental Theorem of Calculus to calculate the exact values of the following definite integrals:

(a) $\displaystyle\int_2^4 (x^2 + 1)\,dx$ **(b)** $\displaystyle\int_{-\pi}^{\pi} \sin x\,dx$ **(c)** $\displaystyle\int_0^2 \frac{1}{x-1}\,dx$

SOLUTION

(a) Calculating this definite integral exactly with the Fundamental Theorem of Calculus is as easy as antidifferentiating $x^2 + 1$ and then evaluating from 2 to 4:

$$\int_2^4 (x^2 + 1)\,dx = \left[\tfrac{1}{3}x^3 + x\right]_2^4 = \left(\tfrac{1}{3}(4)^3 + 4\right) - \left(\tfrac{1}{3}(2)^3 + 2\right) = \frac{62}{3}.$$

Notice that the Fundamental Theorem just allowed us to calculate a definite integral quickly without having to make a lengthy computation of limits of Riemann sums. Compare this solution with the work we did to investigate the same definite integral in Examples 1 and 2 from Section 4.3.

(b) Since an antiderivative of $\sin x$ is $-\cos x$ (check this by differentiating), $\cos(\pi) = -1$, and $\cos(-\pi) = -1$, we have

$$\int_{-\pi}^{\pi} \sin x\,dx = \left[-\cos x\right]_{-\pi}^{\pi} = -\cos(\pi) - (-\cos(-\pi)) = -(-1) - (-(-1)) = 0.$$

(c) At first glance we might be tempted to apply the Fundamental Theorem and say that $\int_0^2 \frac{1}{x-1}\,dx$ is equal to $\left[\ln|x-1|\right]_0^2 = \ln|1| - \ln|-1| = 0$. But this is not the correct answer, because $\frac{1}{x-1}$ is not continuous at $x = 1$, which is part of the interval $[0, 2]$. The Fundamental Theorem of Calculus can be used *only* if a function is continuous on the entire interval $[a, b]$. At this point we do not have the tools to discuss the area under the graph of a function with a vertical asymptote in the interval (although we will in Section 5.6). □

CHECKING THE ANSWER

We can use calculator graphs to check that our definite integral calculations are reasonable. The graphs shown next illustrate the three areas we discussed in the previous example. In the example, we found that the first area was $\frac{62}{3}$, which is just over 20 units; this seems reasonable given the leftmost graph. The middle graph clearly shows an area of 0, since the regions below and above the x-axis are the same size; this answer matches with the calculation we made. The rightmost graph shows that the function $\frac{1}{x-1}$ has an asymptote at $x = 1$ over which we cannot integrate.

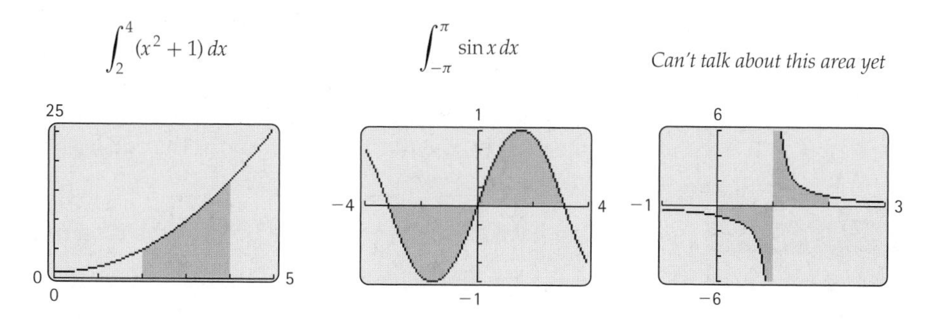

EXAMPLE 3

Using the Fundamental Theorem of Calculus to find a signed area

Find the signed area between the graph of $f(x) = 3e^{2x-5} - 3$ and the x-axis from $x = 1$ to $x = 3$. Use a graph to verify that the sign of your answer is reasonable.

SOLUTION

By definition, the signed area is the definite integral of f on $[1, 3]$. Using integration formulas and a little guess and check, we find that $F(x) = \frac{3}{2}e^{2x-5} - 3x$ is an antiderivative of $f(x) = 3e^{2x-5} - 3$. By the Fundamental Theorem of Calculus, the signed area is then

$$\begin{aligned}\int_1^3 (3e^{2x-5} - 3)\,dx &= \left[\frac{3}{2}e^{2x-5} - 3x\right]_1^3 \\ &= \left(\frac{3}{2}e^{2(3)-5} - 3(3)\right) - \left(\frac{3}{2}e^{2(1)-5} - 3(1)\right) \\ &= \frac{3}{2}e - \frac{3}{2}e^{-3} - 6 \approx -1.99726.\end{aligned}$$

To determine whether it is reasonable that our answer is negative, we must first know where the function $f(x) = 3e^{2x-5} - 3$ is positive and where it is negative. The graph of f crosses the x-axis when $f(x) = 0$. Solving $3e^{2x-5} - 3 = 0$, we get $x = \frac{5}{2} = 2.5$. As you can see in the following graph of f, $f(x)$ is negative on $[1, 2.5]$ and positive on $[2.5, 3]$:

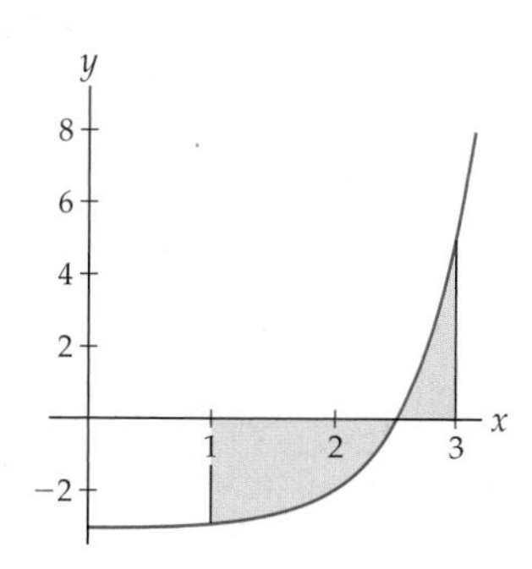

Since the part of the region in question that is below the x-axis is larger than the part of the region that is above the x-axis, it does make sense that the signed area of the region is negative. □

There are many ways that we can check calculations that use the Fundamental Theorem of Calculus. First, we can check the antidifferentiation step by differentiating. For instance, for the previous example we can verify that

$$\frac{d}{dx}\left(\frac{3}{2}e^{2x-5} - 3x\right) = \frac{3}{2}(2)e^{2x-5} - 3 = 3e^{2x-5} - 3.$$

Second, we can check the numerical answer of a definite integral calculation by approximating the value of the definite integral with a graphing calculator or other graphing utility. We can also quickly check that the size and sign of the numerical answer make sense in terms of the graph of the function, as we did at the end of the previous example.

EXAMPLE 4 **Computing a value of f given a value of f and the area under f'**

Suppose f is a function whose *derivative* f' is shown next. Given that $f(1) = 3$, use the Fundamental Theorem of Calculus and the graph of f' to approximate $f(4)$.

Graph of f' and the area $\int_1^4 f'(x)\,dx$

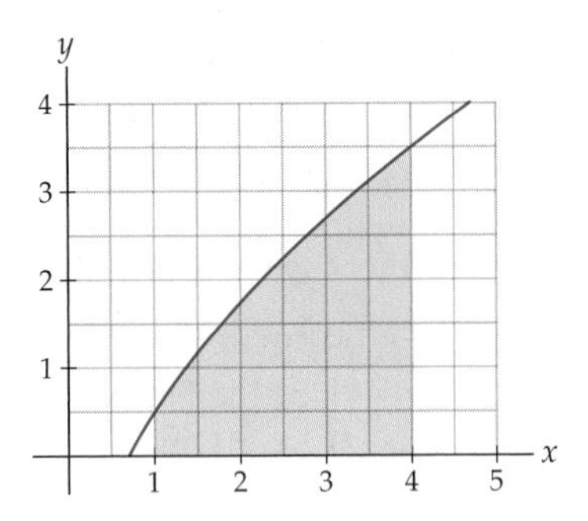

SOLUTION

By the Fundamental Theorem of Calculus, we know that the value $f(1)$, the value $f(4)$, and the area under the graph of f' on $[1, 4]$ are related. Specifically, we know that

$$\int_1^4 f'(x)\,dx = f(4) - f(1).$$

Solving for $f(4)$ gives us

$$f(4) = f(1) + \int_1^4 f'(x)\,dx.$$

We are given that $f(1) = 3$, so all that remains is to approximate the value of the definite integral. We can do this by using the grid in the given figure. Each box in the grid has a length and height of half a unit, and therefore each box has an area of $\left(\frac{1}{2}\right)\left(\frac{1}{2}\right) = \frac{1}{4}$ square units. By counting boxes (and estimating "partial" boxes), we see that the area under the graph of f' from $x = 1$ to $x = 4$ covers approximately 26.5 boxes. Therefore the area in question is approximately $26.5\left(\frac{1}{4}\right) \approx 6.6$. This means that the value of $f(4)$ must be approximately

$$f(4) = f(1) + \int_1^4 f'(x)\,dx \approx 3 + 6.6 = 9.6.$$

Notice that we were able to find an approximate value of $f(4)$ even though we were never given a graph or equation for the function f. □

EXAMPLE 5

Finding net change with a definite integral

A termite colony is slowly dying off in such a way that the number of termites $P(t)$ after t weeks is changing at a rate of $P'(t) = -576.68(0.794)^t$. Determine the number of termites that die in the first three weeks, and interpret your answer as an area under a curve.

SOLUTION

By the Net Change Theorem, we can find the net change in the number of termites over the first three weeks by calculating a definite integral:

$$\int_0^3 -576.68(0.794)^t = \left[\frac{-576.68}{\ln 0.794}(0.794)^t\right]_0^3$$

$$= \frac{-576.68}{\ln 0.794}(0.794)^3 - \frac{-576.68}{\ln 0.794}(0.794)^0$$

$$\approx -1248.58.$$

Therefore about 1249 termites in the colony die in the first three weeks. To visualize this answer as the area under a curve, consider the graph of the rate $P'(t)$ shown next at the left. Notice that this rate is negative, since the termites are dying, and that the rate has a large magnitude at first and then tapers towards zero. The net change in termites is shown at the right, as the signed area between the graph of $P'(t)$ and the t-axis on the interval $[0, 3]$.

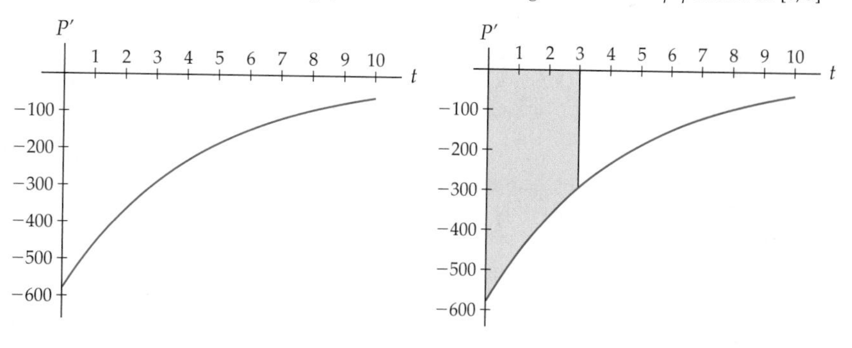

□

? TEST YOUR UNDERSTANDING

- How does the Fundamental Theorem of Calculus connect definite and indefinite integrals?
- What important theorem is the key to proving the Fundamental Theorem of Calculus?
- What is the identifying characteristic of a telescoping sum?
- Does the Fundamental Theorem of Calculus necessarily apply to a discontinuous function? Why or why not?
- How can you check your answer after applying the Fundamental Theorem of Calculus to calculate a definite integral?

EXERCISES 4.5

Thinking Back

- *Definite integrals:* State the definition of the definite integral of an integrable function f on $[a, b]$.
- *Indefinite integrals:* State the definition of the indefinite integral of an integrable function f.
- *Approximating with a Riemann sum:* Use a right sum with 10 rectangles to approximate the area under the graph of $f(x) = x^2$ on $[0, 5]$.
- *Limits of Riemann sums:* Use a limit of Riemann sums to calculate the exact area under the graph of $f(x) = x^2$ on $[0, 5]$.

Concepts

0. *Problem Zero:* Read the section and make your own summary of the material.

1. *True/False:* Determine whether each of the statements that follow is true or false. If a statement is true, explain why. If a statement is false, provide a counterexample.

(a) *True or False:* Definite integrals and indefinite integrals are the same.

(b) *True or False:* If f is integrable on $[a, b]$ and F is any other function, then $\int_a^b f(x)\,dx = F(b) - F(a)$.

(c) *True or False:* The Fundamental Theorem of Calculus applies to $f(x) = \sin x$ on $[0, \pi]$.

(d) *True or False:* The Fundamental Theorem of Calculus applies to $f(x) = \tan x$ on $[0, \pi]$.

(e) *True or False:* If f is an even function, a is a real number, and f is integrable on $[-a, a]$, then $\int_{-a}^{a} f(x)\,dx = 0$.

(f) *True or False:* If f is an odd function, a is a real number, and f is integrable on $[-a, a]$, then $\int_{-a}^{a} f(x)\,dx = 0$.

(g) *True or False:* If f is an even function, a is a real number, and f is integrable on $[-a, a]$, then $\int_{-a}^{0} f(x)\,dx = \int_0^a f(x)\,dx$.

(h) *True or False:* If f is an odd function, a is a real number, and f is integrable on $[-a, a]$, then $\int_{-a}^{0} f(x)\,dx = \int_0^a f(x)\,dx$.

2. *Examples:* Construct examples of the thing(s) described in the following. Try to find examples that are different than any in the reading.

(a) A function f for which the Fundamental Theorem of Calculus does not apply on $[0, 10]$.

(b) Three functions f that we do not yet know how to antidifferentiate.

(c) A non-constant function f for which $\int_0^4 f(x)\,dx = 0$.

3. Why is the Fundamental Theorem of Calculus such an incredible, *fundamental* theorem?

4. State the Fundamental Theorem of Calculus (a) in its original form, (b) in its alternative form, and (c) by using an indefinite integral and evaluation notation.

5. What important theorem is the key to proving the Fundamental Theorem of Calculus?

6. Why must the integrand be an integrable function in the Fundamental Theorem of Calculus?

7. Intuitively, the average value of a velocity function should be the same as the average rate of change of the corresponding position function. Explain why this intuition suggests that the signed area under the velocity graph on an interval is equal to the difference in the position function on that interval, and tell what this has to do with the Fundamental Theorem of Calculus.

8. Why is it not necessary to write down an antiderivative family when using the Fundamental Theorem of Calculus to calculate definite integrals? In other words, why don't we have to use "+C"?

Suppose f is a function whose *derivative* f' is given by the graph shown next. In Exercises 9–12, use the given value of f, an area approximation, and the Fundamental Theorem to approximate the requested value.

Graph of f', the derivative of f

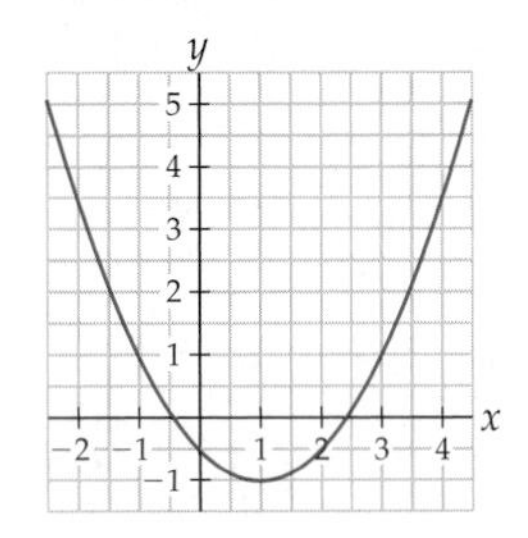

9. Given that $f(3) = 2$, approximate $f(4)$.

10. Given that $f(0) = -1$, approximate $f(2)$.

11. Given that $f(2) = 3$, approximate $f(-2)$.

12. Given that $f(-1) = 2$, approximate $f(1)$.

Calculate each definite integral in Exercises 13–14, using (a) the definition of the definite integral as a limit of Riemann sums, (b) the definite integral formulas from Theorem 4.13, and (c) the Fundamental Theorem of Calculus. Then show that your three answers are the same.

13. $\displaystyle\int_1^2 3x^2\,dx$

14. $\displaystyle\int_0^4 (3x + 2)^2\,dx$

15. In the proof of the Fundamental Theorem of Calculus we encounter a telescoping sum. Find the values of the following sums, which are also telescoping.

(a) $\displaystyle\sum_{k=1}^{100} \left(\frac{1}{k} - \frac{1}{k+1}\right)$

(b) $\displaystyle\sum_{k=1}^{100} \frac{k^2 - (k-1)^2}{2}$

16. Determine whether or not each statement that follows is equivalent to the Fundamental Theorem of Calculus. Assume that all functions here are integrable.

(a) If $f'(x) = F(x)$, then $\int_a^b F(x)\,dx = \left[f(x)\right]_a^b$.

(b) $\int_a^b G(x)\,dx = [G'(x)]_a^b$.

(c) If $h(x)$ is the derivative of $g(x)$, then $\int_a^b h(x)\,dx = g(b) - g(a)$.

(d) $\left[\int w'(x)\,dx\right]_a^b = \int_a^b w(x)\,dx$.

(e) $\left[\int s'(x)\,dx\right]_a^b = [s''(x)]_a^b$.

17. Determine whether or not each statement that follows is equivalent to the Fundamental Theorem of Calculus. Assume that all functions here are integrable.

(a) $\int_a^b f(x)\,dx = g(b) - g(a)$, where $g'(x) = f(x)$.

(b) If $f'(x) = F(x)$, then $\int_a^b f(x)\,dx = [F(x)]_a^b$.

(c) If $g(x)$ is any antiderivative of $h(x)$, then $\int_a^b h(x)\,dx = g(a) - g(b)$.

(d) $\int_a^b h''(x)\,dx = [h'(x)]_a^b$.

(e) $\left[\int p(x)\,dx\right]_a^b = \int_a^b p(x)\,dx$.

18. In the proof of the Fundamental Theorem of Calculus, the Mean Value Theorem is used to choose values x_k^* in each subinterval $[x_{k-1}, x_k]$. Use the Mean Value Theorem in the same way to find the corresponding values x_k^* for a Riemann sum approximation of $\int_0^2 x^2\,dx$ with four rectangles.

Skills

Use the Fundamental Theorem of Calculus to find the exact values of each of the definite integrals in Exercises 19–64. Use a graph to check your answer. *(Hint: The integrands that involve absolute values will have to be considered piecewise.)*

19. $\int_{-1}^{1} (x^4 + 3x + 1)\,dx$

20. $\int_{-3}^{3} (x - 1)(x + 3)\,dx$

21. $\int_{1}^{4} 3(2^x)\,dx$

22. $\int_{0}^{3} (x^2 - 4)^2\,dx$

23. $\int_{2}^{5} \frac{1}{\sqrt{x^5}}\,dx$

24. $\int_{-\pi}^{\pi} \sin(3x)\,dx$

25. $\int_{1}^{3} \frac{\sqrt{x} - 1}{\sqrt{x}}\,dx$

26. $\int_{1}^{3} 2x\sqrt{x^2 + 1}\,dx$

27. $\int_{0}^{\pi/4} \sec x \tan x\,dx$

28. $\int_{0}^{\pi/4} \sec^2 x\,dx$

29. $\int_{0}^{\sqrt{2}/4} \frac{1}{\sqrt{1 - 4x^2}}\,dx$

30. $\int_{0}^{1/2} \frac{1}{1 + 4x^2}\,dx$

31. $\int_{-\pi}^{\pi} (1 + \sin x)\,dx$

32. $\int_{2}^{4} 3e^{2x-4}\,dx$

33. $\int_{0}^{1} \frac{1}{2e^x}\,dx$

34. $\int_{2}^{4} \frac{1}{4 - 3x}\,dx$

35. $\int_{0}^{1} \frac{1}{1 + x^2}\,dx$

36. $\int_{0}^{1} \frac{x}{1 + x^2}\,dx$

37. $\int_{-1}^{1} \frac{2^x}{4^x}\,dx$

38. $\int_{0}^{\pi} (\sin^2 x + \cos^2 x)\,dx$

39. $\int_{-3}^{2} \frac{1}{x + 5}\,dx$

40. $\int_{-3}^{2} \frac{1}{(x + 5)^2}\,dx$

41. $\int_{-3}^{2} \frac{2x}{x^2 + 5}\,dx$

42. $\int_{-3}^{2} \frac{x^2 + 5}{2x}\,dx$

43. $\int_{1}^{4} \frac{2x + 3}{x^2 + 3x + 4}\,dx$

44. $\int_{\pi/4}^{\pi/2} \csc x \cot x\,dx$

45. $\int_{0}^{3/2} \frac{1}{\sqrt{9 - x^2}}\,dx$

46. $\int_{-3}^{3} 2\cos(\pi x)\,dx$

47. $\int_{1}^{e-3} \frac{3\pi}{2x + 6}\,dx$

48. $\int_{1}^{e} 2(\ln x)\left(\frac{1}{x}\right)\,dx$

49. $\int_{0}^{1} (x^{2/3} - x^{1/3})\,dx$

50. $\int_{0}^{4} \frac{1}{(2x + 1)^{5/2}}\,dx$

51. $\int_{0}^{\pi/2} \sin x(1 + \cos x)\,dx$

52. $\int_{2}^{3} \frac{\ln x}{x}\,dx$

53. $\int_{-1}^{1} \frac{e^x - xe^x}{e^{2x}}\,dx$

54. $\int_{1}^{2} \frac{x^2 e^x - 2xe^x}{x^4}\,dx$

55. $\int_{\pi/4}^{\pi/2} x\csc^2(x^2)\,dx$

56. $\int_{-\pi/4}^{\pi/4} x^2 \sec^2(x^3)\,dx$

57. $\int_{2}^{4} \frac{3x}{1 - x^2}\,dx$

58. $\int_{0}^{1} \frac{2^x - 3^x}{4^x}\,dx$

59. $\int_{-2}^{5} |x - 2|\,dx$

60. $\int_{-1}^{3} |4 - x^2|\,dx$

61. $\int_{-2}^{4} |2x^2 - 5x - 3|\,dx$

62. $\int_{0}^{6} |(x - 1)(x - 4)|\,dx$

63. $\int_{0}^{5\pi/4} |\sin x|\,dx$

64. $\int_{-1}^{1} |e^{2x} - 1|\,dx$

Applications

65. Suppose a large oil tank develops a hole that causes oil to leak from the tank at a rate of $r(t) = 0.05t$ gallons per hour. Construct and then solve a definite integral to determine the amount of oil that has leaked from the tank after one day.

66. Suppose another oil tank with a hole in it takes three days to leak 11 gallons of oil. Assuming that the rate of leakage at time t is given by $r(t) = kt$ for some constant k, set up and solve an equation involving a definite integral to find k.

67. A figurine that is part of the display in a mechanical music box moves back and forth along a straight track as shown next at the left. The mechanics of the music box are programmed to move the figurine back and forth with velocity

$$v(t) = 0.00015t(t-5)(t-10)(t-15)(t-20)$$

inches per second over the 20 seconds that the music program plays; see the graph next at the right.

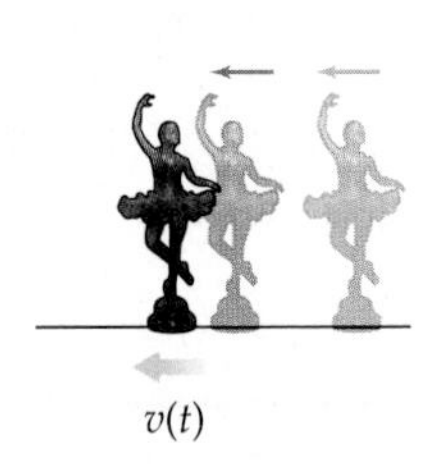

Velocity $v(t)$ of the figurine

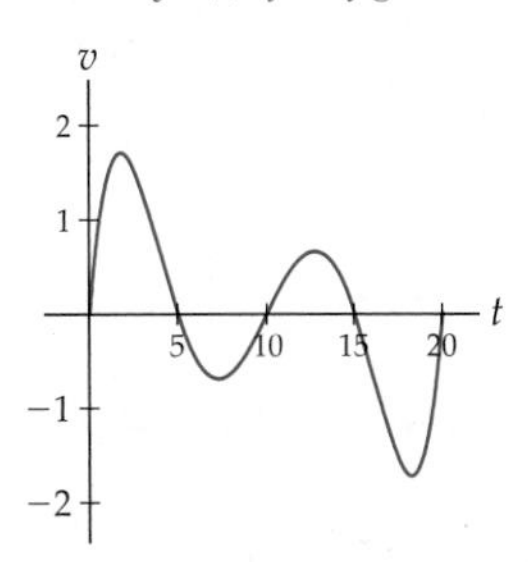

(a) Use the graph just shown to describe the direction of motion of the figurine during various parts of the music program.

(b) Determine the net distance travelled by the figurine in the first 5 seconds, the first 10 seconds, and the full 20-second music program.

(c) Determine the total amount of distance travelled by the figurine in either direction along the track.

68. Water is evaporating from an open vase. The rate of evaporation changes according to the amount of exposed surface area at the top of the container in such a way that the rate of change of the volume of water in the vase is given by

$$V'(t) = -0.05\pi(10-0.2t)^2$$

cubic centimeters per hour. The vase initially contains 200 cubic centimeters of water. Use the Net Change Theorem to determine the amount of time it will take for all of the water in the vase to evaporate.

69. Many email filters can be trained how to recognize spam by having a user identify spam messages from a lineup. Suppose you have looked at 12 messages with the word "opportunity" in the subject line, and identified 8 of them to the filter as spam. Now you get a new message with that word in the subject line, and your filter must compute the probability p that the message is spam. One way to do that is to identify p as the expected value of the Beta distribution

$$f(x) = 6435x^8(1-x)^4.$$

The expected value of this distribution is

$$\int_0^1 x f(x)\,dx.$$

Evaluate the probability that the next message you receive with the word "opportunity" in the subject line will be spam.

70. During World War II, a handful of German submarines were captured by the Allies to get access to German codes. On one of these submarines, the submarine's depth had been plotted carefully as $D(t)$, in meters, where t measures the number of hours after noon on that day. Unfortunately, during the capture, the information about $D(t)$ was lost. However, the following graph of the rate of change $D'(t)$ of the submarine's depth was recovered:

Rate of change $D'(t)$ of the depth of a submarine

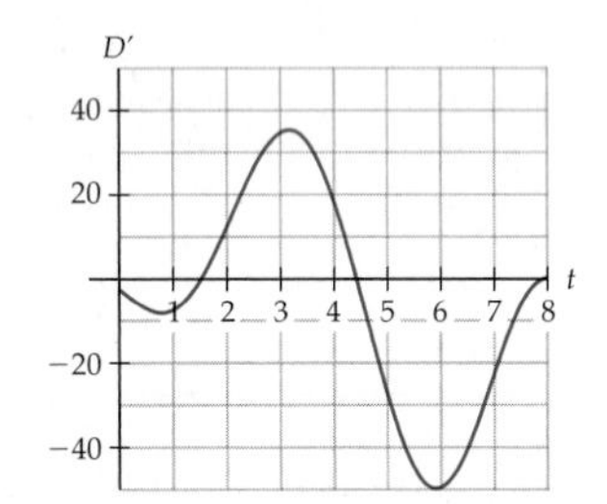

(a) At 4:00 P.M., was the submarine rising or falling? Was it speeding up or slowing down?

(b) Was the submarine closer to the surface at 2:00 P.M. or at 5:00 P.M.?

(c) At the end of the first two hours of the voyage, had the submarine risen or fallen from its position at the beginning? By how much?

(d) Approximately how much did the submarine rise or fall in the two minutes after 6:00 P.M.?

(e) At what time was the submarine diving at the fastest rate?

(f) When was the submarine at its highest point?

(g) Suppose the submarine was on the surface when it was at its highest point. What was the greatest depth of the submarine during its voyage that day?

71. A specialty bookshop is open from 8:00 A.M. to 6:00 P.M. Customers arrive at the shop at a rate $r(t)$ arrivals per hour. A graph of $r(t)$ for the duration of the business day is shown here, where t is the number of hours after the shop opens at 8:00 A.M. As a group, the salespeople in the bookshop can serve customers at a rate of 30 customers per hour. If there are no salespeople available, customers have to wait in line until they can be served, and the customers are willing to wait as long as is necessary in this line. Use the graph of $r(t)$ to estimate answers to the questions that follow.

Rate $r(t)$ of arrivals at a bookshop

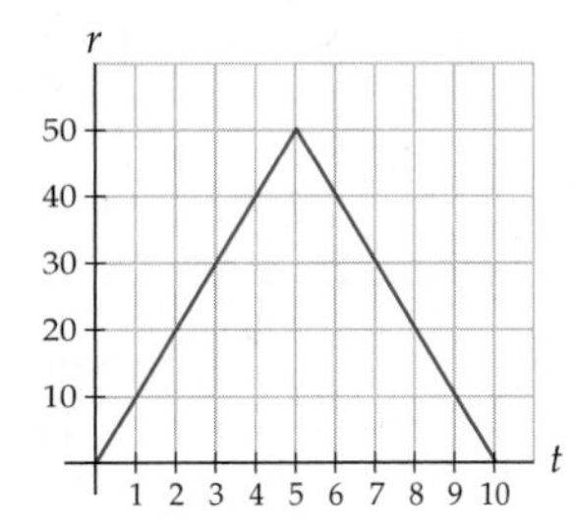

(a) At noon, approximately how many people had entered the bookshop so far? Is there a line yet, and if so, how long is it?

(b) At what rate is the line at the bookshop growing (or shrinking) at 1:00 P.M.?

(c) At approximately what time of day is the line the longest, and how many people are in the line at that time?

(d) At 6:00 P.M. when the bookshop closes, has every customer been served or is there still a line at closing time?

Proofs

Use the Fundamental Theorem of Calculus to give alternative proofs of the integration facts shown in Exercises 72–76. You may assume that all functions here are integrable.

72. $\displaystyle\int_a^b c\,dx = c(b-a)$

73. $\displaystyle\int_a^b x\,dx = \frac{1}{2}(b^2-a^2)$

74. $\displaystyle\int_a^b x^2\,dx = \frac{1}{3}(b^3-a^3)$

75. $\displaystyle\int_a^b kf(x)\,dx = k\int_a^b f(x)\,dx$

76. $\displaystyle\int_a^b (f(x)+g(x))\,dx = \int_a^b f(x)\,dx + \int_a^b g(x)\,dx$

77. Prove the Fundamental Theorem of Calculus in your own words. Use the proof in this section as a guide.

78. Prove that the Net Change Theorem (Theorem 4.27) is equivalent to the Fundamental Theorem of Calculus.

79. Prove that if two functions F and G differ by a constant, then $[F(x)]_a^b = [G(x)]_a^b$.

80. Use evaluation notation and the Fundamental Theorem of Calculus to prove Theorem 4.26: $\int_a^b f(x)\,dx = \left[\int f(x)\,dx\right]_a^b$.

Thinking Forward

Functions Defined by Integrals: Suppose $A(x)$ is the function that for each $x > 0$ is equal to the area under the graph of $f(t) = t^2$ from 0 to x.

- Write an equation that defines $A(x)$ in terms of a definite integral. *(Hint: Let the variable inside the integral be t, so as not to confuse it with the variable x.)*
- Calculate $A(0)$, $A(1)$, $A(2)$, $A(3)$, and $A(4)$.
- Sketch a graph of $f(x) = x^2$ and a graph of $A(x)$ on the same set of axes.
- Use the Fundamental Theorem of Calculus to find an equation for $A(x)$ that does not involve an integral.
- Use the equation for $A(x)$ that you just found to find $A'(x)$. What do you notice about $A'(x)$ and $f(x)$?
- Use the graphs of $A(x)$ and $f(x)$ that you drew earlier to illustrate the relationship between $A'(x)$ and $f(x)$ that you just discovered. Use the words *increasing, decreasing, positive,* and *negative* in your discussion.

4.6 AREAS AND AVERAGE VALUES

- Definite integrals of functions whose values are sometimes negative
- Using definite integrals to find the area between two curves
- Average value as a definite integral, and the Mean Value Theorem for Integrals

The Absolute Area Between a Graph and the x-axis

We have seen that Riemann sums, and thus definite integrals, automatically count area above the x-axis positively and area below the x-axis negatively. But what if we want to count the true area of a region, counting both area above and below the x-axis positively? The definite integral will not do this automatically, but we can use absolute values to express such an ***absolute area*** with a definite integral.

THEOREM 4.28

Signed Area and Absolute Area

For any integrable function f on an interval $[a, b]$,

(a) The signed area between the graph of f and the x-axis from $x = a$ to $x = b$ is given by the definite integral $\int_a^b f(x)\, dx$.

(b) The absolute area between the graph of f and the x-axis from $x = a$ to $x = b$, counting all areas positively, is given by the definite integral $\int_a^b |f(x)|\, dx$.

The first part of this theorem follows directly from the definition of signed area and the definite integral. The second part of the theorem follows from the fact that the graph of the absolute value of a function can be obtained by reflecting all the negative portions of the graph over the x-axis.

For example, consider the region between the graph of $f(x) = x - 1$ and the x-axis on $[0, 3]$. This region consists of a small triangle below the x-axis with area $1/2$ and a larger triangle above the x-axis with area 2; see the figure next at the left. The signed area of this region is $(-1/2) + 2 = 3/2$, and the absolute area of the region is $(1/2) + 2 = 5/2$. The figure next at the right shows the graph of $y = |x - 1|$, in which the triangle in the fourth quadrant has been reflected into the first quadrant.

$$\int_0^3 (x-1)\, dx = (-1/2) + 2 = 3/2$$

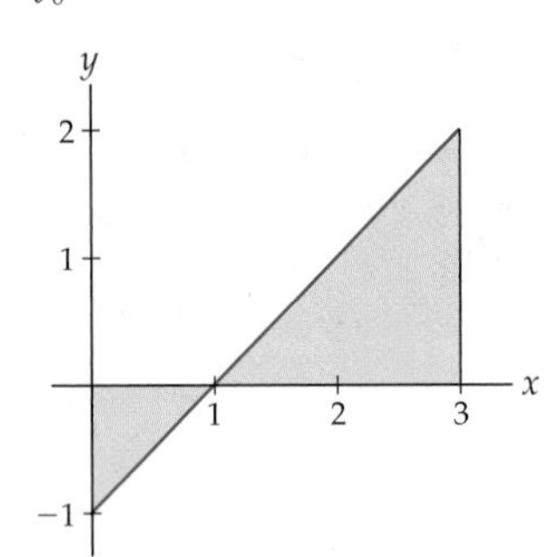

$$\int_0^3 |x-1|\, dx = (1/2) + 2 = 5/2$$

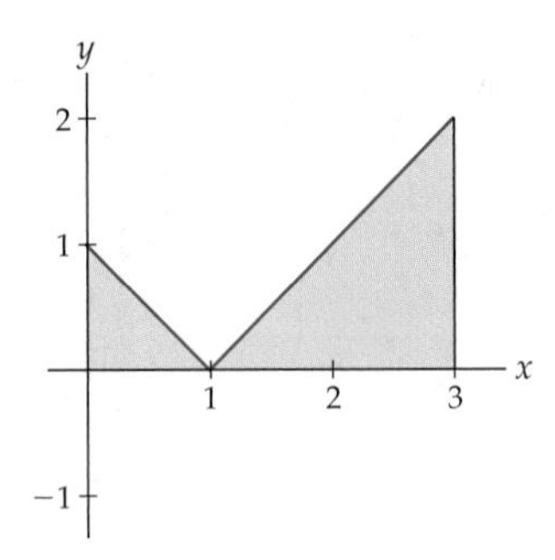

Although the second part of Theorem 4.28 gives us an expression for absolute area in terms of a definite integral, we cannot immediately calculate this type of definite integral with the Fundamental Theorem of Calculus. The reason is that we do not in general know how to antidifferentiate functions that involve absolute values. However, since we can write absolute values in terms of piecewise-defined functions, we write any definite integral of the form $\int_a^b |f(x)|\, dx$ as a sum of definite integrals that do not involve absolute values. Specifically, to calculate the integral of $|f(x)|$, we will split up the interval $[a, b]$ into smaller intervals on which f is either always above or always below the x-axis and then add negative signs as needed to turn negative values into positive values. For example, the definite integral described earlier can be split as

$$\int_0^3 |x-1|\, dx = -\int_0^1 (x-1)\, dx + \int_1^3 (x-1)\, dx = -(-1/2) + 2 = 5/2.$$

Areas Between Curves

We can think of the area between the graph of a function f and the x-axis on an interval $[a, b]$ as the area between two graphs, namely, the graph of f and the graph of $y = 0$, on the interval $[a, b]$. What about the area between the graph of f and the graph of some other function g on $[a, b]$? For example, consider the region between the curves $f(x) = x + 2$ and $g(x) = x^2$ on the interval $[-1, 2]$, as shown here:

Region between $f(x) = x + 2$ and $g(x) = x^2$ on $[-1, 2]$

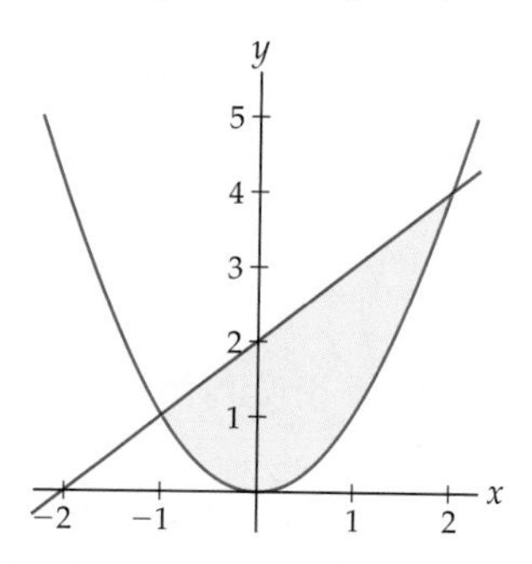

One way to find the area of this region is to calculate the area under the upper graph ($f(x) = x+2$) on $[-1, 2]$ and subtract the area under the lower graph ($f(x) = x^2$) on $[-1, 2]$. These two areas are shown in the first and second figures here:

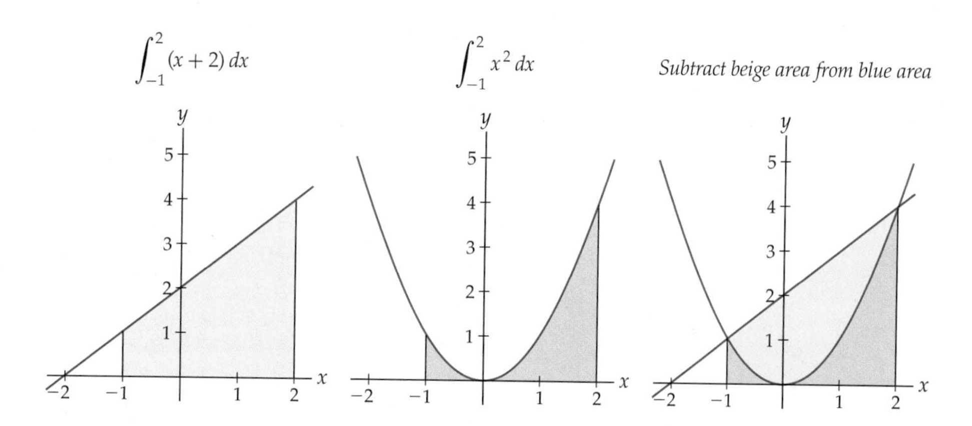

From these pictures and the properties of definite integrals, we can express the area between f and g as a definite integral that we could easily calculate:

$$\text{Area between } f(x) = x + 2 \text{ and } g(x) = x^2 \text{ on } [-1, 2] = \int_{-1}^{2} (x + 2)\, dx - \int_{-1}^{2} x^2\, dx = \int_{-1}^{2} ((x + 2) - x^2)\, dx.$$

The following definition generalizes this method of finding the area between two curves:

DEFINITION 4.29 The Area Between Two Graphs

If f and g are integrable functions, then the ***area between the graphs of f and g from $x = a$ to $x = b$*** is

$$\int_a^b |f(x) - g(x)|\, dx.$$

On any interval $[c, d]$ where $f(x) \geq g(x)$, we will have $|f(x) - g(x)| = f(x) - g(x)$. The area between the graphs of f and g on such an interval will be simply $\int_c^d (f(x) - g(x))\, dx$.

Similarly, on any interval $[r, s]$ where $g(x) \geq f(x)$, we will have $|f(x) - g(x)| = g(x) - f(x)$, and the area between the graphs of f and g on this interval will be $\int_r^s (g(x) - f(x))\, dx$. In general, sometimes $f(x)$ will be greater than $g(x)$, and sometimes it will be less than $g(x)$. In such cases we will have to split up the interval $[a, b]$ into smaller intervals on which $f(x)$ is always greater than $g(x)$–or vice versa–so we can express the area without using any absolute values, as will be shown in Example 2.

CAUTION The phrase "area between the graphs of f and g" always refers to an absolute area. The definite integral in Definition 4.29 does *not* consider the area between f and g to be positive or negative depending on whether it is, respectively, above or below the x-axis.

The Average Value of a Function on an Interval

So far we have used definite integrals to find areas; we will now investigate how definite integrals can be used to find the ***average value*** of a function on an interval. We will motivate our development of the definition of average value with a discrete approximation. Suppose we wish to approximate the average height of a plant during the first four days of its growth, given that the function $f(x) = 0.375x^2$ describes the height of the plant, in centimeters, x days after it breaks through the soil. We can approximate the average height of this plant by choosing a discrete number of x-values, for example $x = 1$, $x = 2$, $x = 3$, and $x = 4$, and averaging the heights $f(x)$ of the plant at these times:

$$\frac{f(1) + f(2) + f(3) + f(4)}{4} = \frac{0.375(1)^2 + 0.375(2)^2 + 0.375(3)^2 + 0.375(4)^2}{4}$$
$$= 2.8125 \text{ cm.}$$

This answer seems reasonable given the graph of the plant height function $f(x)$ shown next; the height $y = 2.8125$ does seem to be approximately the average height of the graph over $[0, 4]$.

Height of the plant after x days is $f(x) = 0.375x^2$

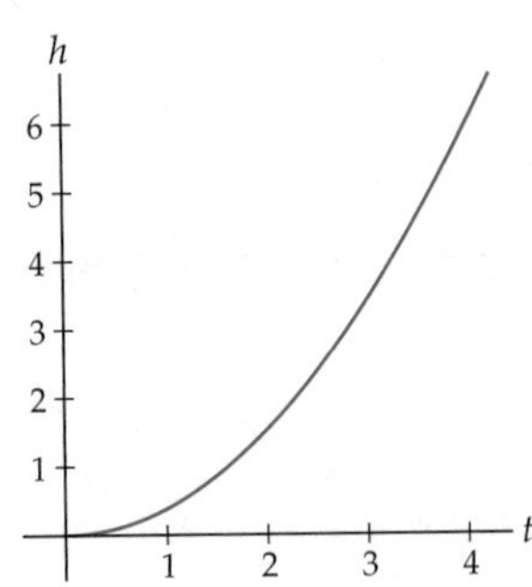

Of course, during the first four days the plant has more than just four heights. The height of the plant changes *continuously* as x changes, and we took into account only a *discrete* number of plant heights. To make a better approximation of the average height of the plant, we could average over a larger number of heights; for example, we could consider the height of the plant every quarter day (at $x = 0.25$, $x = 0.5$, $x = 0.75$, and so on until $x = 4$) and average those 16 heights:

$$\frac{f(0.25) + f(0.5) + f(0.75) + \cdots + f(3.75) + f(4)}{16} \approx 2.1914 \text{ cm.}$$

We can make as good an approximation as we like for the average value by choosing a large enough sample size n. To make n height measurements at equally spaced times, we would measure every $\frac{b-a}{n} = \frac{4}{n}$ days, at times $x_k = 0 + k\left(\frac{4}{n}\right)$ for $k = 1$ to $k = n$. The average value of the height function over those n measurements is

$$\frac{f(x_1) + f(x_2) + f(x_3) + \cdots + f(x_n)}{n} = \frac{\sum_{k=1}^{n} f(x_k)}{n} = \frac{1}{b-a}\sum_{k=1}^{n} f(x_k)\Delta x.$$

In the last step of the calculation, we used the fact that $\Delta x = \frac{b-a}{n}$. This equation should look awfully familiar to you. As we let n approach infinity we obtain the *exact* average value on $[a, b]$. But as $n \to \infty$ we are taking a limit of Riemann sums, which is a definite integral!

This discussion suggests the following general definition of average value:

DEFINITION 4.30

The Average Value of a Function on an Interval

The ***average value*** of an integrable function f on an interval $[a, b]$ is defined to be $\frac{1}{b-a}$ times the definite integral of f on $[a, b]$:

$$\begin{array}{c}\text{Average value of } f \\ \text{from } x = a \text{ to } x = b\end{array} = \frac{1}{b-a}\int_a^b f(x)\,dx.$$

Definition 4.30 gives us another way to utilize the power of the Fundamental Theorem of Calculus. Because we can express the exact average value of a function f on an interval in terms of a definite integral, we can find that average value as long as we can antidifferentiate the function f; see part (a) of Example 3.

The Mean Value Theorem for Integrals

The Mean Value Theorem (Theorem 3.5) guaranteed that for any function $f(x)$ that is continuous on a closed interval $[a, b]$ and differentiable on (a, b), there is always some point c in $[a, b]$ where the slope of f is equal to the average rate of change of f on $[a, b]$. We can now say a similar thing about finding a point c where the height of a function f is equal to its average value on an interval $[a, b]$.

Intuitively, the average value of a function f on $[a, b]$ should be between the highest and lowest values of f on that interval. By the Intermediate Value Theorem from Section 1.4, the function f has to achieve that average value at some point $c \in (a, b)$. This property suggests the following theorem, which you will prove in Exercise 82:

THEOREM 4.31

The Mean Value Theorem for Integrals

If f is continuous on a closed interval $[a, b]$, then there exists some $c \in (a, b)$ such that

$$f(c) = \frac{1}{b-a}\int_a^b f(x)\,dx.$$

Note that the Mean Value Theorem for Integrals tells us only that a certain value $x = c$ must exist, not what that value *is*. If we want to actually find the values $x = c$ for which the height of a function is equal to its absolute value, we have to use the Fundamental Theorem of Calculus and solve the equation in Theorem 4.31 for $x = c$.

Our construction in Definition 4.30 means that the average value of a function on an interval is related to the signed area under the same function on the same interval. Specifically, if $f(c) = K$ is the average value of f on an interval $[a, b]$, then

$$K = \frac{1}{b-a}\int_a^b f(x)\,dx \quad \Rightarrow \quad K(b-a) = \int_a^b f(x)\,dx.$$

This means that the signed area under the graph of f from $x = a$ to $x = b$ is equal to $K(b-a)$, which is the area of a rectangle of height K and width $b - a$. One way to think of this area is to imagine the region between the graph of f and the x-axis as the cross section of a wave of water, as if we are looking, from the side, at water sloshing in a glass tank. When the water settles, it will have a height of K, which means that its height will be the average value of the function f on the interval.

For example, consider the following graphs:

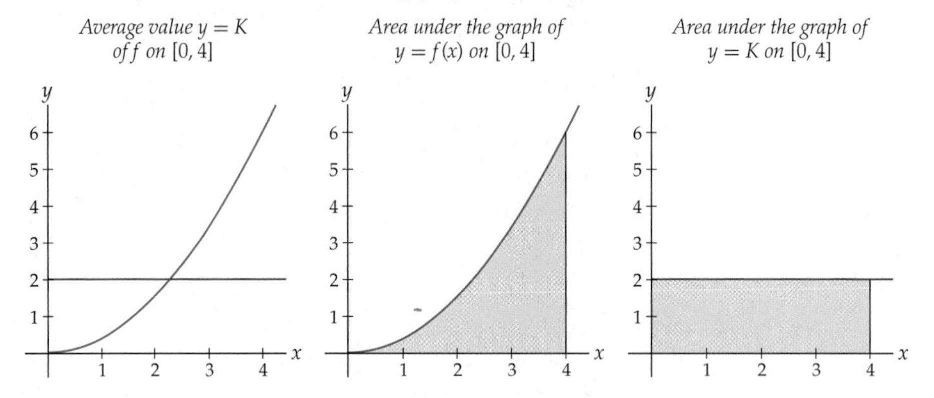

Notice that the area of the shaded region in the middle figure is the same as the area of the shaded region in the last figure. In addition, the value $x = c$ guaranteed by the Mean Value Theorem for Integrals is approximately $x = 2.3$ in this example, since, according to the figure at the left this is the value at which the height of the function f is equal to the average value shown at height $y = 2$.

Examples and Explorations

EXAMPLE 1 **Using definite integrals to find signed and absolute areas**

Consider the region between the graph of a function f and the x-axis on $[-2, 4]$ as shown here. Express (a) the signed area and (b) the absolute area of this region in terms of definite integrals of the function f that do not involve absolute values.

Region between f and x-axis on $[-2, 4]$

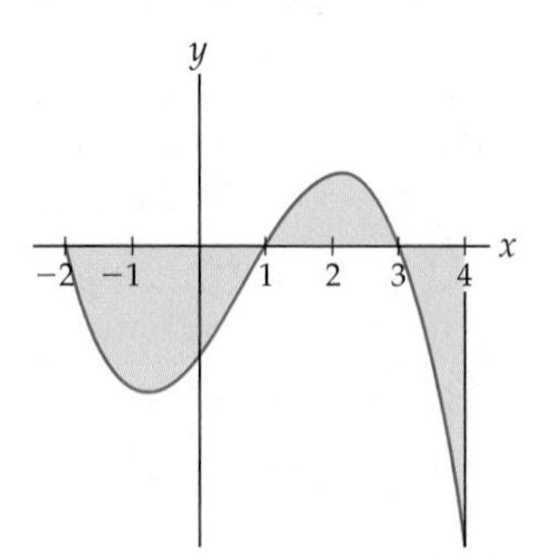

SOLUTION

(a) The signed area of this region is simply the definite integral of f from $x = -2$ to $x = 4$; in other words,

$$\text{signed area} = \int_{-2}^{4} f(x)\, dx.$$

(b) The absolute area of the region (counting all areas positively) is $\int_{-2}^{4} |f(x)|\,dx$. Unfortunately, we do not know how to calculate a definite integral of a function with an absolute value. However, as usual we can get rid of an absolute value if we are willing to work piecewise. We will express the absolute area without using absolute values by using three definite integrals. From $x = -2$ to $x = 1$ the function f is negative, and thus on the interval $[-2, 1]$ we have $|f(x)| = -f(x)$. From $x = 1$ to $x = 3$ the function is positive, so $|f(x)| = f(x)$ on this interval. Finally, from $x = 3$ to $x = 4$ the function is negative, so $|f(x)| = -f(x)$ on $[3, 4]$. Therefore the absolute area of the region is

$$\int_{-2}^{4} |f(x)|\,dx = -\int_{-2}^{1} f(x)\,dx + \int_{1}^{3} f(x)\,dx - \int_{3}^{4} f(x)\,dx.$$

Notice that in this expression, $\int_{-2}^{1} f(x)\,dx$ is a *negative* number, and thus $-\int_{-2}^{1} f(x)\,dx$ will be a *positive* number. We have essentially just flipped the signs of the definite integrals that measure area negatively. □

CAUTION In general, absolute values do *not* commute with definite integrals. In other words, $\int_a^b |f(x)|\,dx$ is not in general the same as $\left|\int_a^b f(x)\,dx\right|$. This means that to find the absolute area it is not enough just to take the absolute value of the signed area.

EXAMPLE 2 **Using definite integrals to find the area between two graphs**

Find the area of the region between the graphs of $f(x) = x^2 - x - 2$ and $g(x) = 4 - x^2$ on the interval $[-2, 3]$.

SOLUTION

According to Definition 4.29, we need to calculate

$$\int_{-2}^{3} |(x^2 - x - 2) - (4 - x^2)|\,dx.$$

However, we do not know how to calculate a definite integral that involves an absolute value. Once again if we are willing to work piecewise, we can get around this problem. Our first task is to find the points where f and g intersect. The region whose area we are interested in calculating is as follows (with f shown in blue and g in red):

Region between $f(x) = x^2 - x - 2$ and $g(x) = 4 - x^2$ on $[-2, 3]$

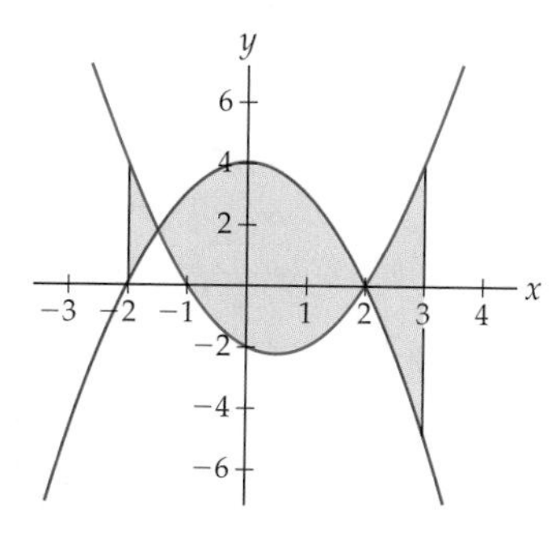

The solutions of the equation $x^2 - x - 2 = 4 - x^2$ are $x = -\frac{3}{2}$ and $x = 2$, so these are the two points of intersection shown in the graph. Notice that f is above g on $\left[-2, -\frac{3}{2}\right]$, g is above f on $\left[-\frac{3}{2}, 2\right]$, and f is again above g on $[2, 3]$. On each of these smaller intervals we can use a definite integral to calculate the area between the graphs without requiring an absolute value. Each such definite integral is easy to calculate with the Fundamental Theorem of Calculus. Using the fact that $f(x) - g(x) = 2x^2 - x - 6$, we see that the total area of the region between the graphs of f and g on $[-2, 3]$ is

$$\int_{-2}^{3} |f(x) - g(x)|\, dx = \int_{-2}^{-3/2} (f(x) - g(x))\, dx + \int_{-3/2}^{2} (g(x) - f(x))\, dx + \int_{2}^{3} (f(x) - g(x))\, dx$$

$$= \int_{-2}^{-3/2} (2x^2 - x - 6)\, dx + \int_{-3/2}^{2} -(2x^2 - x - 6)\, dx + \int_{2}^{3} (2x^2 - x - 6)\, dx$$

$$= \left[\tfrac{2}{3}x^3 - \tfrac{1}{2}x^2 - 6x\right]_{-2}^{-3/2} + \left[-\left(\tfrac{2}{3}x^3 - \tfrac{1}{2}x^2 - 6x\right)\right]_{-3/2}^{2} + \left[\tfrac{2}{3}x^3 - \tfrac{1}{2}x^2 - 6x\right]_{2}^{3}$$

$$= \frac{23}{24} + \frac{343}{24} + \frac{25}{6} = \frac{466}{24}.$$

□

EXAMPLE 3

The difference between average value and average rate of change

Suppose the function $f(x) = 0.375x^2$ describes the height, in centimeters, of a growing plant after x days.

(a) Find the average height of the plant during the first four days of its growth.

(b) Find the average rate of growth of the plant over those four days.

SOLUTION

(a) The average height of the plant during the first four days is the average value of the function $f(x) = 0.375x^2$ from $x = 0$ to $x = 4$, which by Definition 4.30 is

$$\frac{1}{4-0}\int_0^4 0.375x^2\, dx = \frac{1}{4}\left[\frac{0.375}{3}x^3\right]_0^4 = \frac{1}{4}\left(\frac{0.375}{3}(4)^3 - \frac{0.375}{3}(0)^3\right) = 2 \text{ cm}.$$

Notice that we reduced the problem of finding the average height of f to the problem of solving a definite integral. This is something that is simple to do with the Fundamental Theorem of Calculus, as long as we can find an antiderivative of f. Compare this result with our earlier approximations for the average height of the plant.

(b) In contrast, the average rate of growth of the plant over the first four days is the average rate of change of $f(x) = 0.375x^2$ on $[0, 4]$, which is equal to

$$\frac{f(4) - f(0)}{4 - 0} = \frac{0.375(4)^2 - 0.375(0)^2}{4} = 1.5\frac{\text{cm}}{\text{day}}.$$

Note that this answer makes sense, since the plant grew six centimeters in four days, and thus its average rate of growth was $\frac{6}{4} = 1.5$ centimeters per day. □

EXAMPLE 4

Interpreting the Mean Value Theorem for Integrals

Answer the following questions about the function $f(x) = x^2$ on the interval $[1, 3]$:

(a) Why does the Mean Value Theorem for Integrals apply to this function on this interval? What can we conclude from it? Interpret this conclusion in terms of an area and in terms of an average value.

(b) Find a value c that satisfies the conclusion of the Mean Value Theorem for Integrals in this case.

SOLUTION

(a) The theorem applies because $f(x) = x^2$ is continuous on $[1, 3]$. The conclusion is that there is some point $c \in (1, 3)$ for which $f(c) = \frac{1}{3-1}\int_1^3 x^2\,dx$. This means that the average value of $f(x) = x^2$ on $[1, 3]$ is equal to the height c^2 of the function at the point c. Multiplying by $3 - 1$ we have $(3 - 1)f(c) = \int_1^3 x^2\,dx$, which says that the area under the graph of $f(x) = x^2$ on $[1, 3]$ is equal to the area of a rectangle with height c^2 and width $3 - 1$.

(b) Of course the Mean Value Theorem for Integrals does not tell us how to find a point $c \in (1, 3)$ at which f achieves its average value, only that such a point must exist. However, we can easily find the average value by applying the Fundamental Theorem of Calculus to the expression of average value as a definite integral:

$$\frac{1}{3-1}\int_1^3 x^2\,dx = \frac{1}{2}\left[\frac{1}{3}x^3\right]_1^3 = \frac{1}{2}\left(\frac{1}{3}(3)^3 - \frac{1}{3}(1)^3\right) = \frac{13}{3}.$$

Now it is easy to find a number $c \in (1, 3)$ so that $f(c) = c^2$ is equal to this average value:

$$c^2 = \frac{13}{3} \quad \Longrightarrow \quad c = \pm\sqrt{\frac{13}{3}}.$$

Note that $-\sqrt{\frac{13}{3}}$ is *not* in the interval $(1, 3)$, so the only value $c \in (1, 3)$ for which $f(c)$ is the average value $\frac{13}{3}$ is $c = \sqrt{\frac{13}{3}}$. □

CHECKING THE ANSWER

In the previous example, we found that the average value of $f(x) = x^2$ on $[1, 3]$ should occur at $x = \sqrt{\frac{13}{3}} \approx 2.08$, and should be $f(x) = \frac{13}{3} \approx 4.33$. This means that the area under the graph of f on $[1, 3]$ should be equal to the area of the rectangle on $[1, 3]$ with height approximately 4.33, a result that looks reasonable considering the following two graphs:

Area under $f(x) = x^2$ on $[1, 3]$

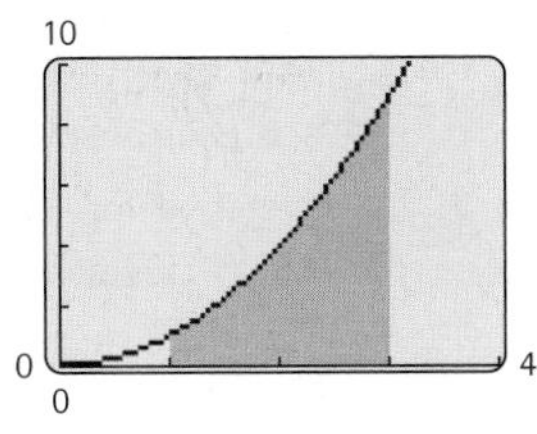

Area under height $f(2.08) = 4.33$ on $[1, 3]$

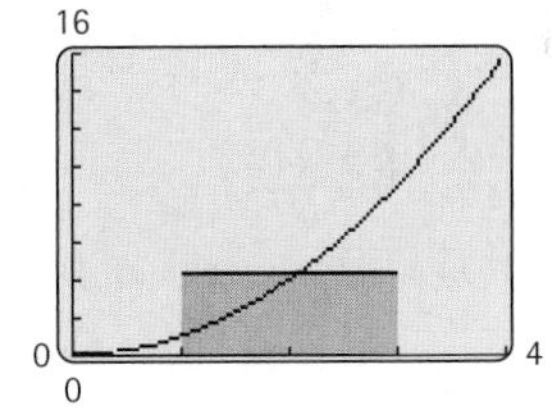

? TEST YOUR UNDERSTANDING

- What is the difference between signed area and absolute area? Which one does the definite integral automatically calculate?
- What can you say about f if the signed area between f and the x-axis is zero on an interval $[a, b]$? What can you say if the *absolute* area is zero on that interval?
- Why do we have to know where two graphs f and g intersect in order to calculate the area between f and g on an interval?
- How can it be that the plant described in the reading grows from 0 to 6 centimeters in the first 4 days and yet has an average height of only 2 centimeters, rather than 3? *(Hint: Use the shape of the graph.)*
- What is the difference between the average value of a function on an interval and the average rate of change of a function on an interval?

EXERCISES 4.6

Thinking Back

- *Finding sign intervals:* Consider the function $f(x) = x^3 - 4x^2 - 12x$. Find the intervals on which f is positive and the intervals on which f is negative.
- *Finding intersection points:* Find the x-coordinates of the points where the graphs of $f(x) = x^3 - 9x^2$ and $g(x) = x - 9$ intersect.
- *The Intermediate Value Theorem:* State the Intermediate Value Theorem, and illustrate the theorem with a graph.
- *The Mean Value Theorem:* State the Mean Value Theorem, and illustrate the theorem with a graph.

Absolute values as piecewise-defined functions: Write out each of the following functions as a piecewise-defined function that does not involve absolute values.

- $f(x) = |x|$
- $f(x) = |2 - 3x|$
- $f(x) = |x^2 - 3x - 4|$
- $f(x) = ||x - 1| - 1|$

Concepts

0. *Problem Zero:* Read the section and make your own summary of the material.

1. *True/False:* Determine whether each of the statements that follow is true or false. If a statement is true, explain why. If a statement is false, provide a counterexample.
   (a) *True or False:* The absolute area between the graph of f and the x-axis on $[a, b]$ is equal to $|\int_a^b f(x)\,dx|$.
   (b) *True or False:* The area of the region between $f(x) = x - 4$ and $g(x) = -x^2$ on the interval $[-3, 3]$ is negative.
   (c) *True or False:* The signed area between the graph of f on $[a, b]$ is always less than or equal to the absolute area on the same interval.
   (d) *True or False:* The area between any two graphs f and g on an interval $[a, b]$ is given by $\int_a^b (f(x) - g(x))\,dx$.
   (e) *True or False:* The average value of the function $f(x) = x^2 - 3$ on $[2, 6]$ is $\frac{f(6)+f(2)}{2} = \frac{33+1}{2} = 17$.
   (f) *True or False:* The average value of the function $f(x) = x^2 - 3$ on $[2, 6]$ is $\frac{f(6)-f(2)}{4} = \frac{33-1}{4} = 8$.
   (g) *True or False:* The average value of f on $[1, 5]$ is equal to the average of the average value of f on $[1, 2]$ and the average value of f on $[2, 5]$.
   (h) *True or False:* The average value of f on $[1, 5]$ is equal to the average of the average value of f on $[1, 3]$ and the average value of f on $[3, 5]$.

2. *Examples:* Construct examples of the thing(s) described in the following. Try to find examples that are different than any in the reading.
   (a) A function f for which the signed area between f and the x-axis on $[0, 4]$ is zero, and a different function g for which the absolute area between g and the x-axis on $[0, 4]$ is zero.
   (b) A function f whose signed area on $[0, 5]$ is less than its signed area on $[0, 3]$.
   (c) A function f whose average value on $[-1, 6]$ is negative while its average rate of change on the same interval is positive.

3. Without using absolute values, how many definite integrals would we need in order to calculate the absolute area between $f(x) = \sin x$ and the x-axis on $[-\frac{\pi}{2}, 2\pi]$? Will the absolute area be positive or negative, and why? Will the signed area will be positive or negative, and why?

4. Without using absolute values, how many definite integrals would we need in order to calculate the area between the graphs of $f(x) = \sin x$ and $g(x) = \frac{1}{2}$ on $[-\frac{\pi}{2}, 2\pi]$?

5. Suppose f is positive on $(-\infty, -1]$ and $[2, \infty)$ and negative on the interval $[-1, 2]$. Write (a) the signed area and (b) the absolute area between the graph of f and the x-axis on $[-3, 4]$ in terms of definite integrals that do not involve absolute values.

6. Suppose $f(x) \geq g(x)$ on $[1, 3]$ and $f(x) \leq g(x)$ on $(-\infty, 1]$ and $[3, \infty)$. Write the area of the region between the graphs of f and g on $[-2, 5]$ in terms of definite integrals without using absolute values.

7. If f is negative on $[-3, 2]$, is the definite integral $\int_{-3}^{2} f(x)\,dx$ positive or negative? What about the definite integral $-\int_{-3}^{2} f(x)\,dx$?

8. Shade in the regions between the two functions shown here on the intervals (a) $[-2, 3]$; (b) $[-1, 2]$; and (c) $[1, 3]$. Which of these regions has the largest area? The smallest?

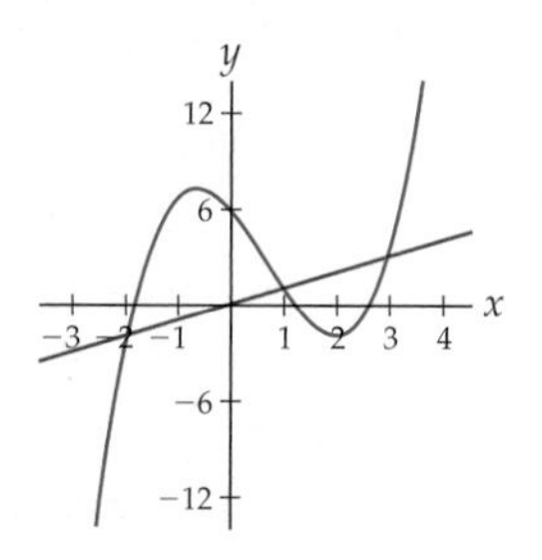

9. Describe an example that illustrates that $\int_a^b |f(x)|\,dx$ is not equal to $|\int_a^b f(x)\,dx|$.

10. Use the definition of absolute value to explain why the absolute area $\int_a^b |f(x)|\, dx$ counts area positively regardless of whether it is above or below the x-axis. Include graphs that support your explanation.

11. Consider the region between f and the x-axis on $[-2, 4]$ as in the graph next at the left. (a) Draw the rectangles of the left-sum approximation for the area of this region, with $n = 8$. Then express (b) the signed area and (c) the absolute area of the region with definite integrals that do not involve absolute values.

Graph for Exercise 11

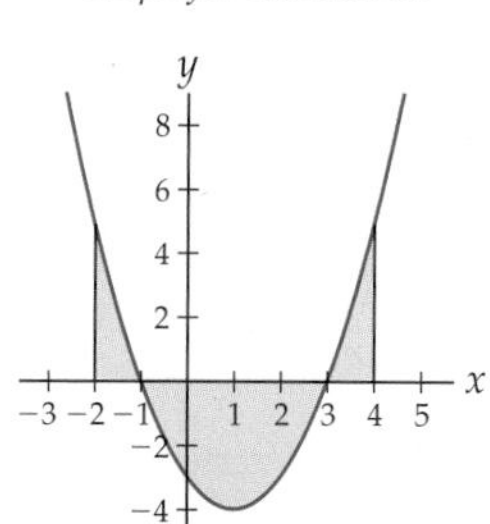

Graph for Exercise 12

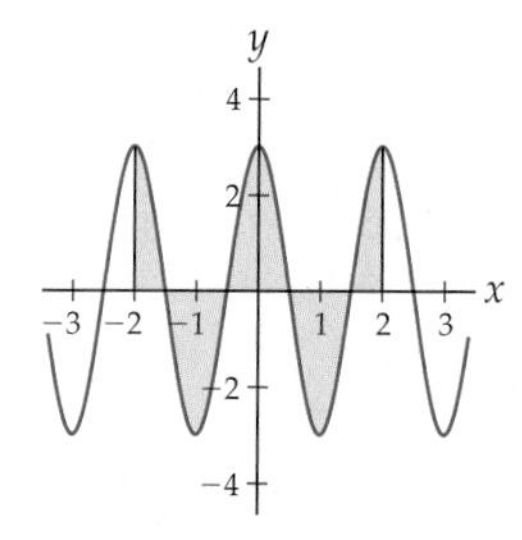

12. Repeat Exercise 11 for the function f shown above at the right, on the interval $[-2, 2]$.

13. Consider the region between f and g on $[0, 4]$ as in the graph next at the left. (a) Draw the rectangles of the left-sum approximation for the area of this region, with $n = 8$. Then (b) express the area of the region with definite integrals that do not involve absolute values.

Graph for Exercise 13

Graph for Exercise 14

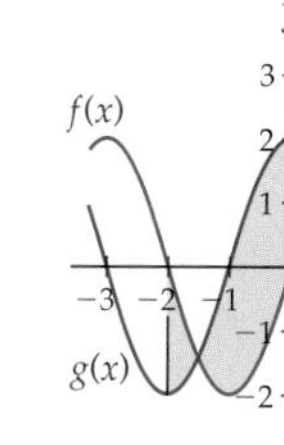

14. Repeat Exercise 13 for the function f shown above at the right, on the interval $[-2, 2]$.

15. Consider the function f shown in the graph next at the right. Use the graph to make a rough estimate of the average value of f on $[-4, 4]$, and illustrate this average value as a height on the graph.

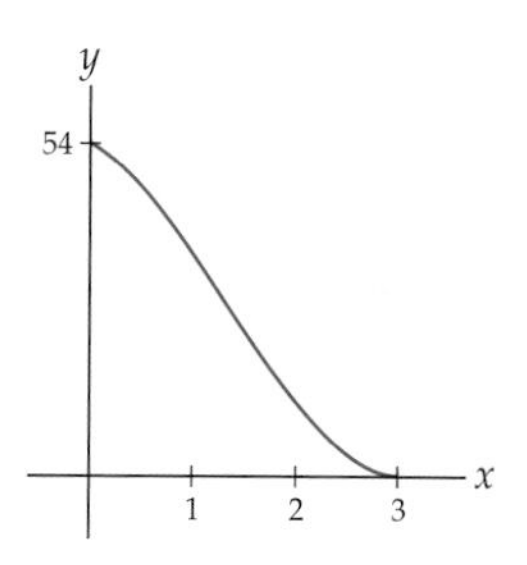

16. Without calculating any sums or definite integrals, determine the values of the described quantities. *(Hint: Sketch graphs first.)*
    (a) The signed area between the graph of $f(x) = \cos x$ and the x-axis on $[-\pi, \pi]$.
    (b) The average value of $f(x) = \cos x$ on $[0, 2\pi]$.
    (c) The area of the region between the graphs of $f(x) = \sqrt{4 - x^2}$ and $g(x) = -\sqrt{4 - x^2}$ on $[-2, 2]$.

17. Suppose f is a function whose average value on $[-2, 5]$ is 10 and whose average rate of change on the same interval is -3. Sketch a possible graph for f. Illustrate the average value and the average rate of change on your graph of f.

18. Suppose f is a function whose average value on $[-3, 1]$ is -2 and whose average rate of change on the same interval is 4. Sketch a possible graph for f. Illustrate the average value and the average rate of change on your graph of f.

19. State the Mean Value Theorem for Integrals, and explain what this theorem means. Include a picture with your explanation. What does the Mean Value Theorem for Integrals have to do with average values?

20. Explain what the Mean Value Theorem for Integrals has to do with the Intermediate Value Theorem.

21. Explain why the Mean Value Theorem for Integrals applies to the function $f(x) = x(x - 6)$ on the interval $[1, 5]$. Next state the conclusion of the Mean Value Theorem for Integrals in this particular case, and sketch a graph illustrating your conclusion. Then find all the values $c \in [1, 5]$ for which $f(c)$ is equal to the average value of f on $[1, 5]$, and indicate these values on your graph.

22. Explain why the Mean Value Theorem for Integrals applies to the function $f(x) = x^2$ on the interval $[-2, 5]$. Next state the conclusion of the Mean Value Theorem for Integrals in this particular case, and sketch a graph illustrating your conclusion. Then find all the values $c \in [-2, 5]$ for which $f(c)$ is equal to the average value of f on $[-2, 5]$, and indicate these values on your graph.

Skills

For each function f and interval $[a, b]$ in Exercises 23–25, use at least eight rectangles to approximate (a) the signed area and (b) the absolute area between the graph of f and the x-axis from $x = a$ to $x = b$. Your work should include a graph of f together with the rectangles that you used.

23. $f(x) = x^2 - 3x - 4$, $[-5, 5]$

24. $f(x) = \cos x$, $[-2\pi, 2\pi]$

25. $f(x) = 1 - e^x$, $[-1, 3]$

For each function f and interval $[a, b]$ in Exercises 26–37, use definite integrals and the Fundamental Theorem of Calculus to find the exact values of (a) the signed area and (b) the absolute area of the region between the graph of f and the x-axis from $x = a$ to $x = b$.

26. $f(x) = x^2$, $[-2, 2]$

27. $f(x) = 3 - x$, $[0, 5]$

28. $f(x) = x^2 - 1$, $[-1, 3]$

29. $f(x) = 2x^2 - 7x + 3$, $[0, 4]$

30. $f(x) = 1 - x^2$, $[0, 3]$

31. $f(x) = 3x^2 + 5x - 2$, $[-3, 2]$

32. $f(x) = \cos x$, $[a, b] = [-\pi, \pi]$

33. $f(x) = (2x - 1)^2 - 4$, $[a, b] = [-2, 4]$

34. $f(x) = 2 - e^{-x}$, $[a, b] = [-1, 0]$

35. $f(x) = \dfrac{1}{2x^3}$, $[a, b] = [1, 3]$

36. $f(x) = \dfrac{3 - 2x}{x^2}$, $[a, b] = [1, 4]$

37. $f(x) = \dfrac{1}{1 + x^2}$, $[a, b] = [-1, 1]$

For each pair of functions f and g and each interval $[a, b]$ in Exercises 38–40, use at least eight rectangles to approximate the area of the region between the graphs of f and g on the interval $[a, b]$. Your work should include graphs of f and g together with the rectangles that you used.

38. $f(x) = \sin x$, $g(x) = \cos x$, $[0, \pi]$

39. $f(x) = x$, $g(x) = 2$, $[-4, 4]$

40. $f(x) = e^x$, $g(x) = \ln x$, $[0.5, 2.5]$

For each pair of functions f and g and interval $[a, b]$ in Exercises 41–52, use definite integrals and the Fundamental Theorem of Calculus to find the exact area of the region between the graphs of f and g from $x = a$ to $x = b$.

41. $f(x) = 1 + x$, $g(x) = 2 - x$, $[0, 3]$

42. $f(x) = 1 + x$, $g(x) = x + 2$, $[0, 3]$

43. $f(x) = x^2$, $g(x) = x + 2$, $[-2, 2]$

44. $f(x) = x^2$, $g(x) = x - 2$, $[-2, 2]$

45. $f(x) = x^2$, $g(x) = x + 2$, $[-3, 3]$

46. $f(x) = x^2 - x - 1$, $g(x) = 5 - x^2$, $[-2, 3]$

47. $f(x) = x - 1$, $g(x) = x^2 - 2x - 1$, $[a, b] = [-1, 3]$

48. $f(x) = \sin x$, $g(x) = \cos x$, $[a, b] = \left[-\frac{\pi}{2}, \frac{\pi}{2}\right]$

49. $f(x) = x^3$, $g(x) = (x - 2)^2$, $[a, b] = [-1, 2]$

50. $f(x) = 2^x$, $g(x) = x^2$, $[a, b] = [0, 3]$

51. $f(x) = \dfrac{2}{1 + x^2}$, $g(x) = 1$, $[a, b] = [0, \sqrt{3}]$

52. $f(x) = \dfrac{3}{x - 2}$, $g(x) = 6 - x$, $[a, b] = [3, 7]$

For each function f and interval $[a, b]$ in Exercises 53–55, approximate the average value of f from $x = a$ to $x = b$, using a sample size of at least eight.

53. $f(x) = 2^x$, $[-1, 1]$

54. $f(x) = \ln x$, $[1, 5]$

55. $f(x) = \sin x$, $[0, 2\pi]$

For each function f and interval $[a, b]$ in Exercises 56–67, use definite integrals and the Fundamental Theorem of Calculus to find the exact average value of f from $x = a$ to $x = b$. Then use a graph of f to verify that your answer is reasonable.

56. $f(x) = x - 1$, $[-1, 3]$

57. $f(x) = 3x + 1$, $[0, 4]$

58. $f(x) = 4 - x^2$, $[-2, 2]$

59. $f(x) = 4$, $[-37.2, 103.75]$

60. $f(x) = (x + 2)^2 - 5$, $[-5, 0]$

61. $f(x) = x^2 - 2x - 1$, $[0, 3]$

62. $f(x) = 4x^{3/2}$, $[a, b] = [0, 2]$

63. $f(x) = (e^x)^2$, $[a, b] = [-1, 1]$

64. $f(x) = \dfrac{1}{3x + 1}$, $[a, b] = [2, 5]$

65. $f(x) = 2 - \sqrt[3]{x}$, $[a, b] = [1, 8]$

66. $f(x) = x^2 \sin(x^3 + 1)$, $[a, b] = [-1, 2]$

67. $f(x) = \sin x + x \cos x$, $[a, b] = [-\pi, \pi]$

For each function f and interval $[a, b]$ given in Exercises 68–73, find a real number $c \in (a, b)$ such that $f(c)$ is the average value of f on $[a, b]$. Then use a graph of f to verify that your answer is reasonable.

68. $f(x) = 3x + 1$, $[a, b] = [2, 6]$

69. $f(x) = 2x^2 + 1$, $[a, b] = [-1, 2]$

70. $f(x) = 9 - x^2$, $[a, b] = [0, 3]$

71. $f(x) = 1 + x + 2x^2$, $[a, b] = [-2, 2]$

72. $f(x) = 100(1 - x)$, $[a, b] = [0, 10]$

73. $f(x) = (x + 1)^2$, $[a, b] = [-1, 2]$

Applications

74. You slam on your brakes and come to a full stop exactly at a stop sign. Your distance from the sign after t seconds is $s(t) = 3t^3 - 12t^2 - 9t + 54$ feet, as shown in the following graph:

Distance from a stop sign
$s(t) = 3t^3 - 12t^2 - 9t + 54$

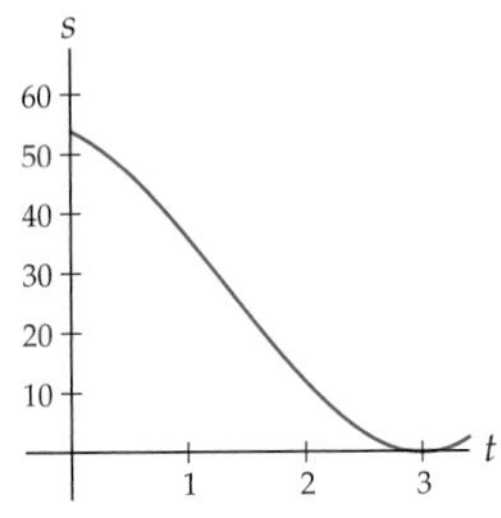

(a) What does the Mean Value Theorem from Section 3.1 say about your distance and/or velocity from the stop sign during the time that you are applying your brakes?

(b) What does the Mean Value Theorem for Integrals say about your distance from the stop sign during the time that you are applying the brakes?

(c) What does the Mean Value Theorem for Integrals say about your velocity during the time that you are applying the brakes?

75. Suppose the height, in centimeters, of a growing plant t days after it breaks through the soil is given by $f(t) = 0.36t^2$, as shown next at the left.

Height of a growing plant
$h(t) = .36t^2$

Height of a growing tree
$h(t) = .25t^2 + 1$

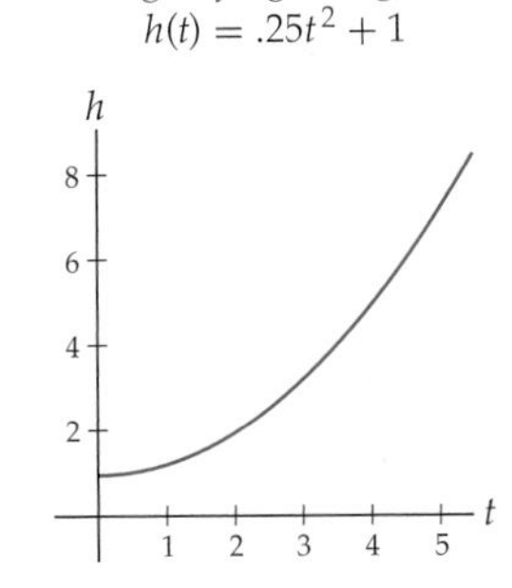

(a) What does the Mean Value Theorem from Section 3.1 say about the height of the plant during the first four days of its growth?

(b) What does the Mean Value Theorem for Integrals say about the height of the plant during the first four days of its growth?

76. Suppose the height of a growing tree t years after it is planted is $h(t) = 0.25t^2 + 1$ feet, as shown earlier at the right.

(a) Approximate the average height of the tree during the first five years of its growth, using a sample size of $n = 5$ times and then $n = 10$ times.

(b) Find the exact average height of the tree during the first five years of its growth.

(c) What was the average rate of growth of the tree over the first five years?

77. Wally's Burger Shack wants to put up a giant sign by Interstate 81. According to local sign ordinances, any sign visible from the interstate must have a frontal square footage of 529 feet or less. The entire sign will be a gigantic W cut out from billboard material, as shown in the graph that follows, where the top edges of the W are at a height of 55 feet and the boundaries of the W are given by the functions

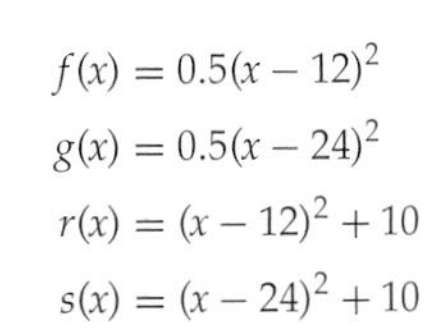

$f(x) = 0.5(x - 12)^2$

$g(x) = 0.5(x - 24)^2$

$r(x) = (x - 12)^2 + 10$

$s(x) = (x - 24)^2 + 10$

Shape of Wally's sign

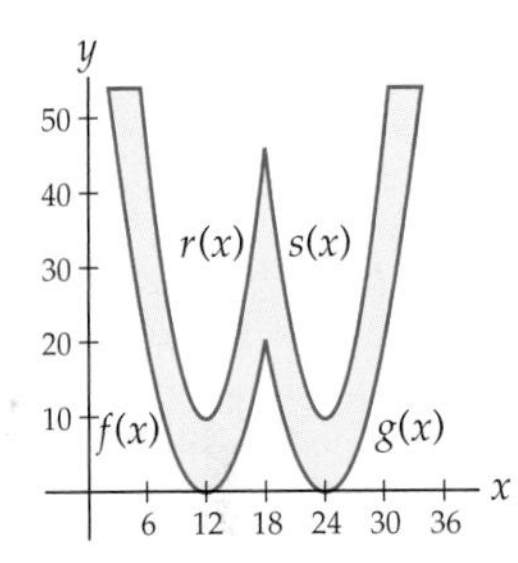

(a) Write the total area of the front of the W sign in terms of definite integrals. You will need to find the solutions of $f(x) = 55$, $r(x) = 55$, $s(x) = 55$, and $g(x) = 55$ as part of your work. Be careful about how you divide up the region.

(b) Use your answer to part (a) to calculate the exact frontal square footage of the W sign. Will Wally's sign meet the local square footage requirements?

Proofs

78. Prove that for the region between the graph of a function f and the x-axis on an interval $[a, b]$, the absolute area is always greater than or equal to the signed area.

79. Prove that for all real numbers a and b with $a < b$, we have $|\int_a^b f(x)\,dx| \le \int_a^b |f(x)|\,dx$.

80. In this exercise you will use two different methods to prove that, for any real numbers a, b, c, and k,

$$\int_{-k}^{k} (ax^2 + c)\,dx = 2\int_0^k (ax^2 + c)\,dx.$$

(a) Prove this equality by using a geometric argument that involves signed area.

(b) Now prove the equality a different way, by using an algebraic argument and the Fundamental Theorem of Calculus.

81. Suppose f is an integrable function on $[a, b]$ and $x_k = a + k\left(\frac{b-a}{n}\right)$.

(a) Use the definition of the definite integral as a limit of Riemann sums to show that

$$\lim_{n\to\infty} \frac{\sum_{k=1}^{n} f(x_k)}{n} = \frac{1}{b-a}\int_a^b f(x)\,dx.$$

(b) Why is it algebraically sensible that the left-hand side of the equation is a calculation of average value?

(c) Why is it graphically sensible that the right-hand side of the equation is a calculation of average value?

82. Prove the Mean Value Theorem for Integrals by following these steps:

(a) Use the Extreme Value Theorem to argue that f has a maximum value M and a minimum value m on the interval $[a, b]$.

(b) Use an upper sum and a lower sum with one rectangle to argue that $m(b-a) \le \int_a^b f(x)\,dx \le M(b-a)$ and thus that the average value of f on $[a, b]$ is between m and M.

(c) Use the Intermediate Value Theorem to argue that there is some point $c \in (a, b)$ for which $f(c)$ is equal to the average value of f on $[a, b]$.

Thinking Forward

The Second Fundamental Theorem of Calculus: Just as you are driving past the big oak tree on Main Street, you notice a new stop sign 54 feet ahead of you. You slam on your brakes and end up coming to a full stop exactly at the stop sign.

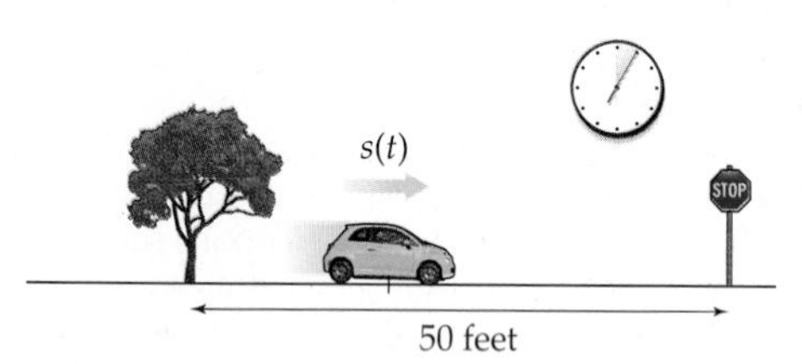

Suppose that your distance from the stop sign (in feet) t seconds after stepping on the brakes is given by the function $s(t) = 3t^3 - 12t^2 - 9t + 54$. By working through the following five problems you will see another argument that the distance travelled is related to the signed area under the velocity curve; this set of problems previews the Second Fundamental Theorem of Calculus, which we will see in Section 4.7.

- How long did it take you to come to a full stop?
- What was your average distance from the stop sign from the time that you first saw it to the time that you came to a stop?
- Use an average rate of change to find your average velocity during the time that you were trying to stop the car. (Assume that the average velocity is the average rate of change of position.)
- Find your average velocity from the oak tree to the stop sign another way, as follows: Differentiate the formula for position to get a formula $v(t)$ for your velocity, in feet per second, t seconds after hitting the brakes. Then use a definite integral to find the average velocity during the time that you were trying to stop the car. (Your final answer should of course be the same as the average velocity you found in the last part!)
- You have just calculated your average velocity in two ways, once using the formula $\frac{s(b)-s(a)}{b-a}$ for the average rate of change of position and once using the definition $\frac{1}{b-a}\int_a^b v(t)\,dt$ of the average value of $v(t)$ on $[a, b]$, where $a = 0$ and b is the amount of time it took you to stop the car. Use the fact that these two quantities are equal to discuss the relationship between the area under the graph of your velocity $v(t)$ on $[a, b]$ and the total distance that you travelled while trying to stop the car.

4.7 FUNCTIONS DEFINED BY INTEGRALS

- Functions that measure accumulated area
- The Second Fundamental Theorem of Calculus and constructing antiderivatives
- Defining the natural logarithm function with an integral

Area Accumulation Functions

We have seen many types of functions in this course, most of them being arithmetic combinations and/or compositions of powers of x, exponential functions, logarithmic functions, trigonometric functions, and inverse trigonometric functions (and hyperbolic and inverse hyperbolic functions, if you covered them in Section 2.6). Of course there are many more functions than the ones we have named and notated throughout this course; any assignment of real numbers that associates exactly one real number to each number in the domain is a function. For example, we could define a function f so that for each real number x, the value of $f(x)$ is given by the greatest integer less than x. This function is called the ***floor function***, as it rounds down to the next-lowest integer. As another example, we could define a function g so that for each real number x, the value of $g(x)$ is given by the the first perfect square greater than x. We can't express these functions in terms of combinations of simple functions, but they are functions nonetheless.

In this section we examine a new way of defining functions, where for each input x the output is given by an accumulation of area. For example, consider the function $f(t)$ on $[a, b]$ shown next. We can associate with each x in $[a, b]$ a unique area, namely, the area under the curve on $[a, x]$. Since each input x corresponds to exactly one area, this relationship defines a function whose value at any real number x is given by $A(x) = \int_a^x f(t)\,dt$.

Each input x gives a unique area A(x) as output

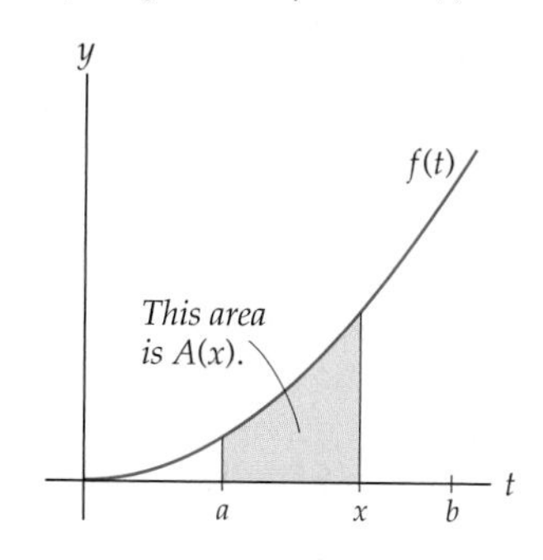

DEFINITION 4.32 **Area Accumulation Functions**

Suppose f is a continuous function on $[a, b]$. Then the ***area accumulation function*** for f on $[a, b]$ is the function that, for $x \in [a, b]$, is equal to the signed area between the graph of f and the x-axis on $[a, x]$:

$$A(x) = \int_a^x f(t)\,dt.$$

Notice that we have named the variable in the integrand t to distinguish it from the variable x that represents the input of the area accumulation function. The variable t is called a "dummy variable," because it will not appear in the output of our function $A(x)$. We are using the function f only as a means to define our new function $A(x)$.

For example, for $f(t) = \sqrt{t}$ on $[0, 4]$, the area accumulation function is

$$A(x) = \int_0^x \sqrt{t}\,dt = \left[\frac{2}{3}t^{3/2}\right]_0^x = \frac{2}{3}x^{3/2} - \frac{2}{3}0^{3/2} = \frac{2}{3}x^{3/2}.$$

Using this equation for $A(x)$, we can easily calculate values such as $A(1)$, $A(2)$, and $A(3)$, illustrated, respectively, in the three graphs that follow. As x moves from left to right, $A(x)$ accumulates more and more area under the graph of $f(t) = \sqrt{t}$.

$$A(1) = \int_0^1 \sqrt{t}\,dt \approx 0.67 \qquad A(2) = \int_0^2 \sqrt{t}\,dt \approx 1.89 \qquad A(3) = \int_0^3 \sqrt{t}\,dt \approx 3.46$$

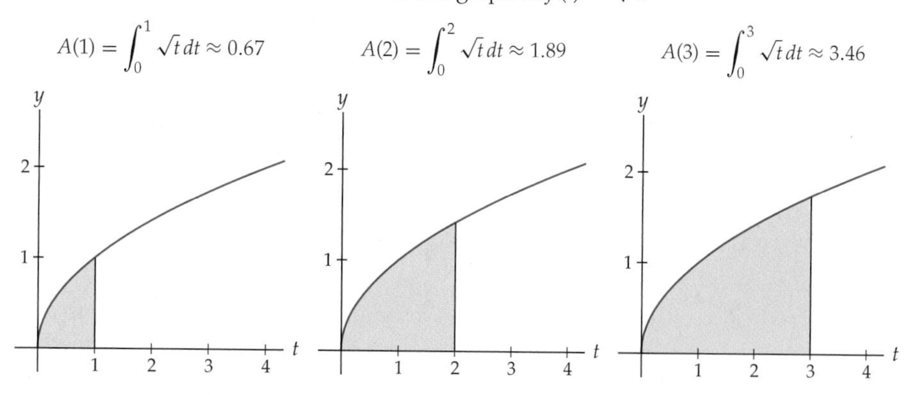

In this example, the area accumulation function turned out to be equal to one of the elementary functions we have studied in this course (namely, $\frac{2}{3}x^{3/2}$). This is not always the case. For example, the area accumulation function $A(x) = \int_0^x e^{-t^2}\,dt$ does not have such an elementary representation; note that in fact we do not even know how to antidifferentiate e^{-t^2}, so even if there were such an elementary representation of $A(x)$, we would be unable to identify it.

The Second Fundamental Theorem of Calculus

In the example we just discussed, you may have noticed something interesting about how A and f are related. Since $A(x) = \frac{2}{3}x^{3/2}$, we have $A'(x) = \left(\frac{2}{3}\right)\left(\frac{3}{2}\right)x^{1/2} = x^{1/2}$, which is equal to $f(x)$. In other words, the derivative of the area accumulation function is the function whose area we were considering! Said another way, the area accumulation function A is an antiderivative of the function f. This is true in general: Every area accumulation function for a continuous function f is an antiderivative of f. The relationship is so important that it is called the Second Fundamental Theorem of Calculus.

THEOREM 4.33 **The Second Fundamental Theorem of Calculus**

If f is continuous on $[a, b]$, and we define $F(x) = \int_a^x f(t)\,dt$ for all $x \in [a, b]$, then

(a) F is continuous on $[a, b]$ and differentiable on (a, b);

(b) F is an antiderivative of f, that is, $F'(x) = f(x)$.

We will postpone the proof of the Second Fundamental Theorem until the end of this section, so that we can focus on some of its consequences first.

The Second Fundamental Theorem guarantees that we can *always* construct an antiderivative for any continuous function f simply by defining an area accumulation function for f. For example, consider the function $f(x) = \sin(e^x)$. It is not immediately apparent how we would antidifferentiate this function. However, we can construct an antiderivative of f by using the Second Fundamental Theorem: If we define $F(x) = \int_0^x \sin(e^t)\,dt$, then we can see that this function F is an antiderivative of f, since $F'(x) = f(x)$ by part (b) of the theorem. We can construct other antiderivatives of f by choosing different values of a; for example, $G(x) = \int_3^x \sin(e^t)\,dt$ is also an antiderivative of $f(x) = \sin(e^x)$.

The second part of the Second Fundamental Theorem immediately gives us a formula for differentiating certain functions defined by integrals:

THEOREM 4.34 **Differentiating Area Accumulation Functions**

If f is continuous on $[a, b]$, then for all $x \in [a, b]$,

$$\frac{d}{dx}\int_a^x f(t)\,dt = f(x).$$

This theorem illustrates yet another way that integration and differentiation can be thought of as inverse operations: Given a function f and a real number a, if we integrate f from a to x and then differentiate the resulting function with respect to x, we arrive back at the function

f we started with. Said another way, the derivative of an area accumulation function for any function f is equal to the original function f.

The part of the Second Fundamental Theorem expressed in Theorem 4.34 follows immediately from the (first) Fundamental Theorem from Section 4.5, since if F is a function with $F' = f$, then

$$\frac{d}{dx}\int_a^x f(t)\,dt = \frac{d}{dx}\left[F(t)\right]_a^x = \frac{d}{dx}(F(x) - F(a)) = F'(x) = f(x).$$

The nontrivial part of the Second Fundamental Theorem is that such an antiderivative function F exists in the first place and that moreover it is continuous and differentiable.

We must be careful when using Theorem 4.34. In particular, in order for us to apply the theorem, the upper limit of the integral *must* be x and the lower limit of the integral must be a constant. If the upper limit is instead a function of x, then we will need to apply the chain rule. The following theorem illustrates how to do this:

THEOREM 4.35 **Differentiating a Composition Involving an Area Accumulation Function**

If f is continuous on $[a, b]$ and $u(x)$ is differentiable on $[a, b]$, then for all $x \in [a, b]$,

$$\frac{d}{dx}\int_a^{u(x)} f(t)\,dt = f(u(x))u'(x).$$

The right-hand side of the equation in Theorem 4.35 looks like the chain rule, with the important exception that it begins with $f(u(x))$ rather than $f'(u(x))$. In fact, it *is* the chain rule, and $f(x)$ is the derivative of the area accumulation function $F(x) = \int_a^x f(t)\,dt$.

Proof. The proof involves recognizing $\int_a^{u(x)} f(t)\,dt$ as a composition and then applying the chain rule. If $F(x) = \int_a^x f(t)\,dt$, then $\int_a^{u(x)} f(t)\,dt$ is the composition $F(u(x))$. By the Second Fundamental Theorem we know that $F'(x) = f(x)$. Thus by the chain rule, we have

$$\frac{d}{dx}\int_a^{u(x)} f(t)\,dt = \frac{d}{dx}(F(u(x))) = F'(u(x))u'(x) = f(u(x))u'(x).$$

■

Defining the Natural Logarithm Function with an Integral

At the beginning of this section we saw that the function $A(x) = \int_0^x \sqrt{t}\,dt$ is equal to the elementary function $\frac{2}{3}x^{3/2}$. Many area accumulation functions can be written as elementary functions. One in particular can be used as the *definition* of a function we already know quite well:

DEFINITION 4.36 **The Natural Logarithm Function**

The ***natural logarithm function*** is the function that for $x > 0$ is defined by

$$\ln x = \int_1^x \frac{1}{t}\,dt.$$

For example, we define $\ln 2$ to be the area between the graph of $\frac{1}{t}$ and the x-axis from $t = 1$ to $t = 2$, or the number $\int_1^2 \frac{1}{t}\,dt$. Since $\frac{1}{t}$ is a continuous function, its definite integral on $[1, 2]$ is a well-defined real number that we can choose to call "ln 2."

Of course we have already discussed logarithmic functions as the inverses of exponential functions in Section 0.5. However, our previous development of exponential functions was not entirely rigorous. In particular, our definition of irrational powers relied on a loose argument that we could consider them to be limits of rational powers. We can get around this problem by instead defining $\ln x$ with an integral, as just shown, and then defining exponential functions as the inverses of logarithmic functions. All of the properties of $\ln x$ that you have come to know and love follow directly from this new definition of $\ln x$ as an area accumulation function:

THEOREM 4.37 **Properties of the Natural Logarithm Function**

(a) $\ln x$ is continuous on $(0, \infty)$

(b) $\ln x$ is differentiable on $(0, \infty)$

(c) $\frac{d}{dx}(\ln x) = \frac{1}{x}$

(d) $\ln 1 = 0$

(e) $\ln x < 0$ on $(0, 1)$, $\ln x > 0$ on $(1, \infty)$

(f) $\ln x$ is increasing on $(0, \infty)$

(g) $\ln x$ is concave down on $(0, \infty)$

(h) $\ln x$ is one-to-one on $(0, \infty)$

Proof. Properties (a)–(c) follow from the Second Fundamental Theorem of Calculus. Properties (d) and (e) are easily proved by using properties of definite integrals. Properties (f) and (g) can be proved by considering the first and second derivatives of $\ln x$. We leave the proofs of these properties to Exercises 71–74 and prove only property (h) here.

To show that $F(x) = \ln x$ is one-to-one, we must show that for all $a, b \in (0, \infty)$, if $F(a) = F(b)$, then $a = b$. Now if $F(a) = F(b)$, then $\ln a = \ln b$, and by Definition 4.36 we have $\int_1^a \frac{1}{t}\,dt = \int_1^b \frac{1}{t}\,dt$. Subtracting all terms to one side and applying parts (b) and (c) of Theorem 4.12 gives

$$0 = \int_1^a \frac{1}{t}\,dt - \int_1^b \frac{1}{t}\,dt = -\int_a^1 \frac{1}{t}\,dt - \int_1^b \frac{1}{t}\,dt = -\int_a^b \frac{1}{t}\,dt.$$

Since $f(t) = \frac{1}{t}$ is positive for all $t > 0$, the integral $\int_a^b \frac{1}{t}\,dt$ can be zero only if a is equal to b, which is what we wanted to show. ■

Because $\ln x$ is one-to-one on its domain, it is invertible. As you might expect, the name that we will give to its inverse is e^x. This means that $e^x = y$ if and only if $\ln y = x$. Since $\ln x$ has domain $(0, \infty)$ and range $\mathbb{R}$ (as you will argue in Exercise 21), its inverse e^x has domain $\mathbb{R}$ and range $(0, \infty)$. We will define general logarithmic functions $\log_b x$ and general exponential functions b^x or e^{kx} in a similar fashion, in Exercises 24–26. Every single property of logarithmic and exponential functions can be derived from this new definition of logarithms. You will explore some of these properties in Exercises 75–78.

The Proof of the Second Fundamental Theorem

The proof of the Second Fundamental Theorem of calculus is lengthy, but involves little more than the definition of the derivative and the Squeeze Theorem for limits.

Proof. Suppose f is continuous on $[a, b]$, and define $F(x) = \int_a^x f(t)\,dt$ for $x \in [a, b]$. We'll leave the proof that F is continuous on $[a, b]$ to Exercise 66 and show that for all $x \in (a, b)$, F is differentiable with derivative $F'(x) = f(x)$. We calculate the right derivative of F (the calculation for the left derivative is similar):

$$F'(x) = \lim_{h\to 0^+} \frac{F(x+h) - F(x)}{h} = \lim_{h\to 0^+} \frac{\int_a^{x+h} f(t)\,dt - \int_a^x f(t)\,dt}{h} = \lim_{h\to 0^+} \frac{\int_x^{x+h} f(t)\,dt}{h}.$$

Note that since we are considering only the right derivative here, h is positive. The last step in the calculation follows from the fact that $a < x < x + h$, and thus

$$\int_a^{x+h} f(t)\,dt = \int_a^x f(t)\,dt + \int_x^{x+h} f(t)\,dt.$$

Let's examine the quantity $\int_x^{x+h} f(t)\,dt$ for a small value h. This definite integral represents the signed area between the graph of $f(t)$ and the x-axis from $t = x$ to $t = x + h$, for example, as shown in the first figure that follows. For each $h > 0$, define m_h to be a point in $[x, x + h]$ for which $f(t)$ has a minimum value and define M_h to be a point in $[x, x + h]$ for which $f(t)$ has a maximum value, as shown in the second and third figures. Note that the values m_h and M_h depend on h and may change as $h \to 0^+$.

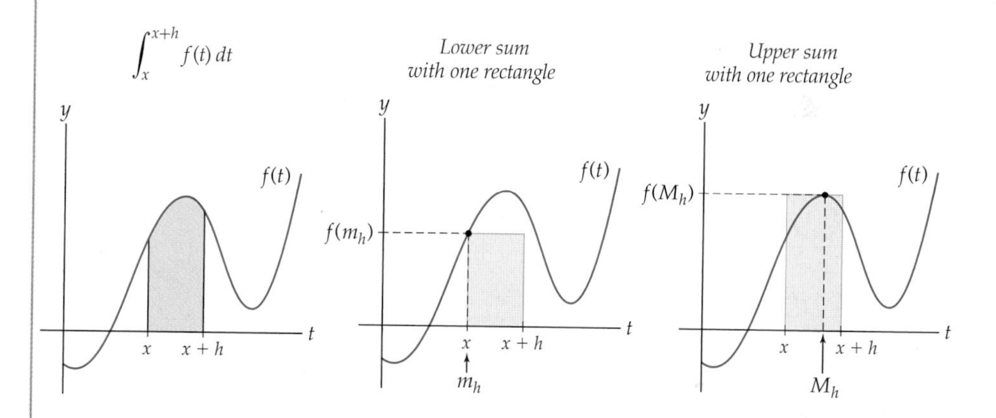

The area of the rectangle in the second of these figures is $f(m_h)h$, and the area of the rectangle in the third figure is $f(M_h)h$. Because $m_h \le f(x) \le M_h$ for all $x \in [x, x + h]$, we have

$$f(m_h)h \le \int_x^{x+h} f(t)\,dt \le f(M_h)h.$$

Dividing both sides by h (which in this case is positive), this inequality becomes

$$f(m_h) \le \frac{\int_x^{x+h} f(x)\,dt}{h} \le f(M_h).$$

Note that the quantity in the middle is exactly the quantity whose limit we need to find to calculate $F'(x)$. Since our results hold for all $h > 0$, we have

$$\lim_{h\to 0^+} f(m_h) \le \lim_{h\to 0^+} \frac{\int_x^{x+h} f(t)\,dt}{h} \le \lim_{h\to 0^+} f(M_h).$$

Because f is continuous, both $f(m_h)$ and $f(M_h)$ approach $f(x)$ as $h \to 0^+$ and the interval $[x, x + h]$ shrinks. Therefore both sides of the double inequality approach $f(x)$ as $h \to 0^+$, so by the Squeeze Theorem,

$$\lim_{h\to 0^+} \frac{\int_x^{x+h} f(t)\,dt}{h} = f(x).$$

Since we started this proof by showing that $F'(x)$ is equal to this limit, we have now shown that $F'(x) = f(x)$, as desired. ■

Examples and Explorations

EXAMPLE 1 **Graphically interpreting an area accumulation function**

Suppose f is the function shown here, and define the area accumulation function $A(x) = \int_1^x f(t)\,dt$. Use the graph of f to answer the following questions:

(a) Which is larger, $A(2)$ or $A(3)$? Which is larger, $A(3)$ or $A(6)$?

(b) List the intervals on which $A(x)$ is increasing and decreasing.

(c) Sketch a rough graph of the function $A(x)$ on $[1, 8]$.

(d) Use the graphs of f and A to verify that one of these functions is the derivative of the other.

The function f on $[1, 8]$

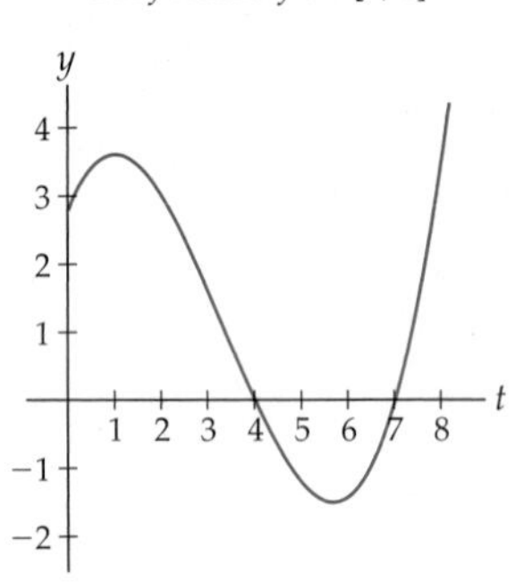

SOLUTION

(a) $A(2)$ is the signed area between the graph of f and the x-axis on $[1, 2]$, while $A(3)$ is the signed area from $t = 1$ to $t = 3$; see the first two graphs that follow. Clearly $A(3) > A(2)$.

$A(2) = \int_1^2 f(t)\,dt$

$A(3) = \int_1^3 f(t)\,dt$

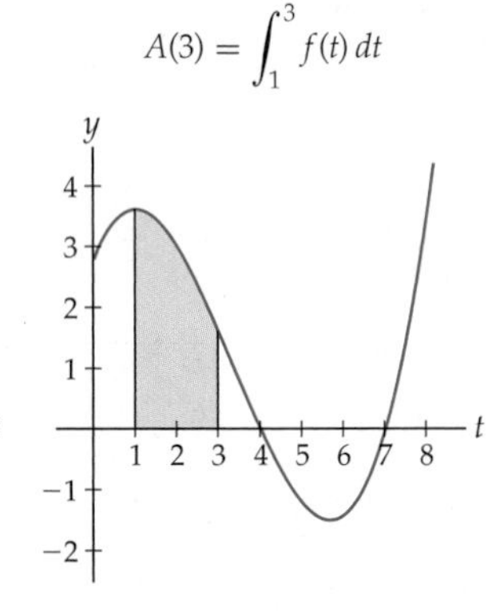

$A(6) = \int_1^6 f(t)\,dt$

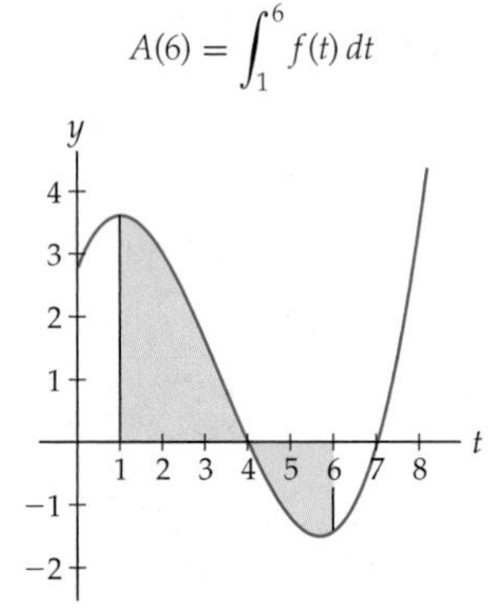

Similarly, $A(6)$ is the signed area between the graph of f and the x-axis on $[1, 6]$, as shown at the right, and differs from $A(3)$ by the area accumulated between $t = 3$ and $t = 6$. The area on $[3, 4]$ is positive, but the area on $[4, 6]$ is negative and larger. Therefore between $t = 3$ and $t = 6$ the accumulation function $A(x)$ actually *decreases*, and thus $A(6) < A(3)$.

(b) $A(x)$ is the accumulation of signed area as x moves from the left to the right. When f is positive, the area function A will increase; when f is negative, the area function A will decrease. From the graph of f we see that f is positive on $(1, 4) \cup (7, 8)$ and negative on $(4, 7)$. Therefore $A(x)$ is increasing on $(1, 4) \cup (7, 8)$ and decreasing on $(4, 7)$.

(c) To sketch the graph of $A(x)$, first notice that $A(1) = \int_1^1 f(t)\,dt = 0$. The height $A(x)$ is the signed area between the graph of f and the x-axis on $[1, x]$, as shown next at the left. Using the information about where A is increasing and where it is decreasing, we can sketch the rough graph of A shown at the right.

The function $f(t)$ on $[1, 8]$

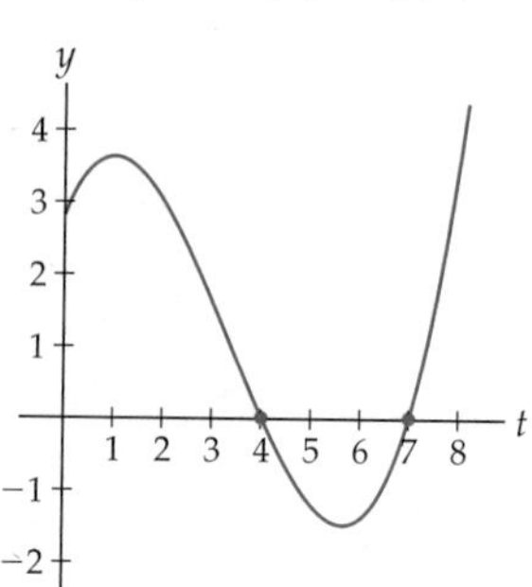

The area accumulation function $A(x)$ on $[1, 8]$

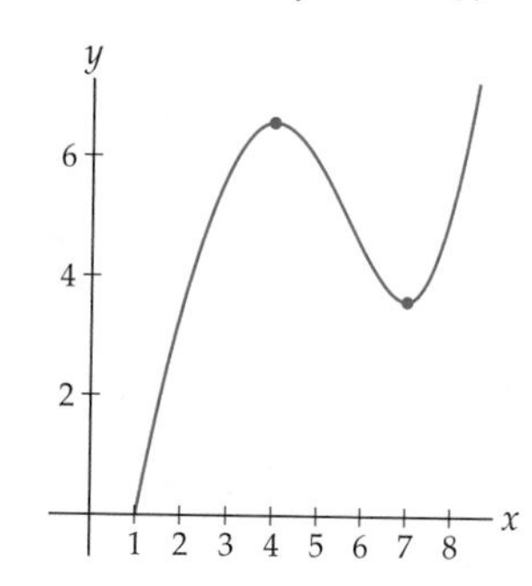

(d) By the Second Fundamental Theorem of Calculus, A is an antiderivative of f, that is, $A' = f$. In the graphs just shown, we can see that for each x, the slope of A at x is equal to the height of f at x. In particular, A has horizontal tangent lines at approximately $x = 4$ and $x = 7$, and its derivative f has roots at those same points. When A is increasing, its derivative f is above the x-axis, and when A is decreasing, f is below the x-axis. □

EXAMPLE 2 **Differentiating area accumulation functions**

Find the derivatives of each of the following functions:

(a) $F(x) = \int_1^x \frac{\cos(\ln t)}{e^{\sin(t^2)}}\,dt$ **(b)** $F(x) = \int_x^2 \ln t\,dt$

SOLUTION

(a) In this case we can apply Theorem 4.34 directly:

$$F'(x) = \frac{d}{dx}\int_1^x \frac{\cos(\ln t)}{e^{\sin(t^2)}}\,dt = \frac{\cos(\ln x)}{e^{\sin(x^2)}}.$$

(b) Here we cannot apply Theorem 4.34 immediately, because the lower limit of integration is the variable x, rather than a constant. However, we can easily fix this problem before differentiating:

$$F'(x) = \frac{d}{dx}\int_x^2 \ln t\,dt = \frac{d}{dx}\left(-\int_2^x \ln t\,dt\right) = -\frac{d}{dx}\int_2^x \ln t\,dt = -\ln x.$$ □

EXAMPLE 3 **Differentiating compositions that involve area accumulation functions**

Find the derivatives of each of the following functions:

(a) $F(x) = \int_0^{x^2} \sec t\,dt$ **(b)** $F(x) = \int_0^{3x} \frac{x}{t^2+1}\,dt$

SOLUTION

(a) We cannot apply Theorem 4.34 directly to this function, because the upper limit of integration is x^2, not x. Instead we need to use the chain rule since the function is a composition, with outside function $A(x) = \int_0^x \sec t\,dt$ and inside function $u = x^2$. In other words, $F(x) = A(x^2)$. By the chain rule and the fact that $A'(x) = \sec x$,

Theorem 4.35 tells us that

$$F'(x) = \frac{d}{dx}\int_0^{x^2} \sec t \, dt = \frac{d}{dx}(A(x^2)) = A'(x^2)(2x) = \sec(x^2)(2x).$$

It may have occurred to you that we could have solved the integral $\int_0^x \sec t \, dt$ first and then found the derivative afterwards. However, in this example we cannot do that, because we do not know how to integrate the function $\sec t$. Our only option is to use Theorem 4.35.

(b) In this example the integrand $\frac{x}{t^2+1}$ involves *both* the variable x and the dummy variable t. Since the integral varies with t, x can be factored out of the integral:

$$F(x) = \int_0^{3x} \frac{x}{t^2+1} \, dt = x \int_0^{3x} \frac{1}{t^2+1} \, dt.$$

It is now clear that F is really the *product* of two functions, namely, x and $\int_0^{3x} \frac{1}{t^2+1} \, dt$. We must use the product rule to differentiate F. We will also need the chain rule, since $\int_0^{3x} \frac{1}{t^2+1} \, dt$ is the composition of $A(x) = \int_0^x \frac{1}{t^2+1}$ with $3x$. We have

$$F'(x) = \frac{d}{dx}\left(x \int_0^{3x} \frac{1}{t^2+1} \, dt\right) \qquad \leftarrow \text{factor out } x$$

$$= 1\left(\int_0^{3x} \frac{1}{t^2+1} \, dt\right) + x\left(\frac{d}{dx}\int_0^{3x} \frac{1}{t^2+1} \, dt\right) \qquad \leftarrow \text{product rule}$$

$$= \int_0^{3x} \frac{1}{t^2+1} \, dt + x\left(\frac{1}{(3x)^2+1}\right)(3). \qquad \leftarrow \text{chain rule}$$

□

EXAMPLE 4

Getting comfortable with the integral definition of ln x

Use pictures and properties of definite integrals to illustrate that $\ln(0.5)$ is negative, $\ln 3$ is positive, and $\ln 4$ is greater than $\ln 3$.

SOLUTION

By Definition 4.36, $\ln(0.5) = \int_1^{0.5} \frac{1}{t} \, dt = -\int_{0.5}^1 \frac{1}{t} \, dt$, and since the left-hand figure that follows shows that the signed area under $\frac{1}{t}$ on $[0.5, 1]$ is positive, it follows that $\ln(0.5)$ is negative. Similarly, the middle figure shows that $\ln 3 = \int_1^3 \frac{1}{t} \, dt$ is positive, and the right-hand figure shows that $\ln 4$ represents a larger area than $\ln 3$ does.

$\ln(0.5) = -\int_{0.5}^1 \frac{1}{t} \, dt$ $\qquad$ $\ln 3 = \int_1^3 \frac{1}{t} \, dt$ $\qquad$ $\ln 4 = \int_1^4 \frac{1}{t} \, dt$

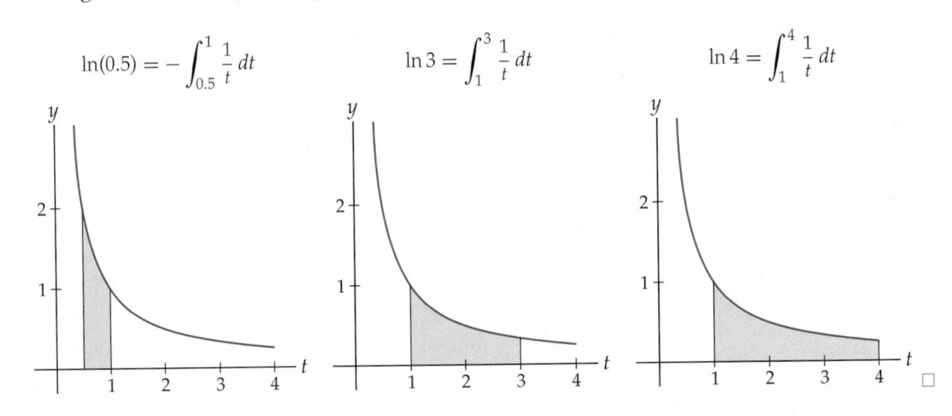

□

EXAMPLE 5

Approximating logarithms with Riemann sums

Use Riemann sums with four rectangles to find upper and lower bounds for $\ln 3$.

SOLUTION

Since $\frac{1}{t}$ decreases as t increases, a left sum for $\ln 3 = \int_1^3 \frac{1}{t}\,dt$ will be an over-approximation and a right sum an under-approximation, as shown here:

Left sum with four rectangles

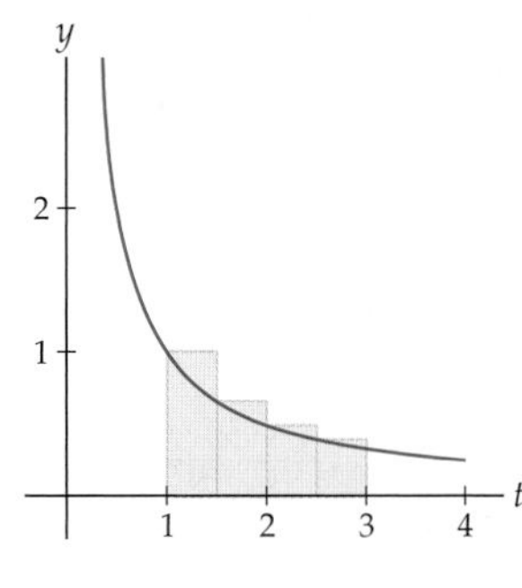

This area is ln 3

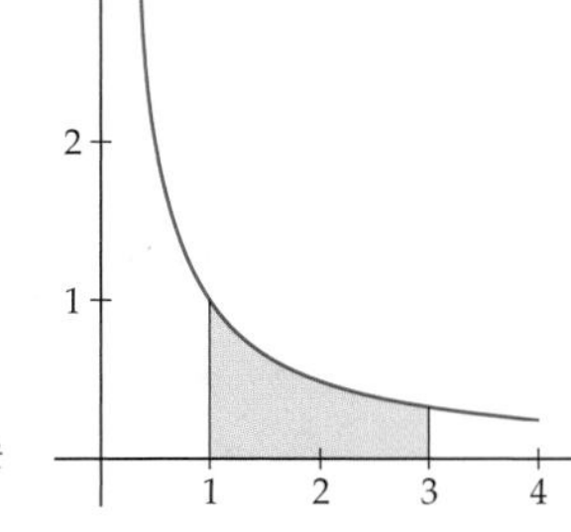

Right sum with four rectangles

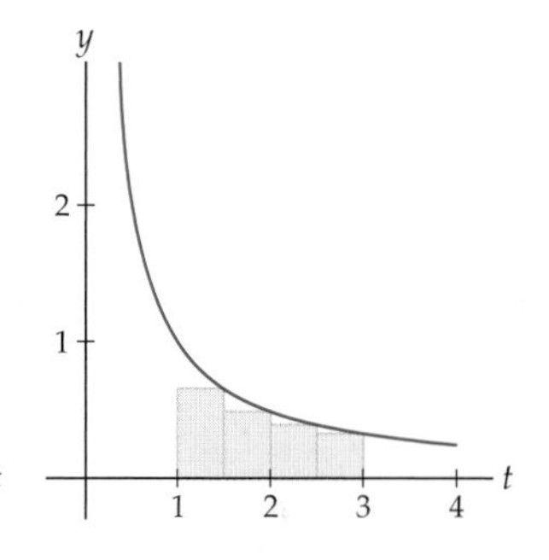

The left sum just shown uses the values of $\frac{1}{t}$ at $t = 1, 1.5, 2,$ and 2.5 to define the heights of the rectangles and will be greater than the actual area:

$$\ln 3 = \int_1^3 \frac{1}{t}\,dt \approx \frac{1}{1}(0.5) + \frac{1}{1.5}(0.5) + \frac{1}{2}(0.5) + \frac{1}{2.5}(0.5) \approx 1.28333.$$

In contrast, the right sum just shown is less than the actual area:

$$\ln 3 = \int_1^3 \frac{1}{t}\,dt \approx \frac{1}{1.5}(0.5) + \frac{1}{2}(0.5) + \frac{1}{2.5}(0.5) + \frac{1}{3}(0.5) = 0.95.$$

Therefore $0.95 < \ln 3 < 1.28333$. We could, of course, get better upper and lower bounds for $\ln 3$ by using more rectangles. □

? TEST YOUR UNDERSTANDING

- What is the difference between the variable x and the variable t in Definition 4.32?
- If $A(x) = \int_1^x t\,dt$, does it make sense to calculate $A(0) = \int_1^0 f(t)\,dt$?
- If $F(x) = \int_\pi^x \cos t\,dt$ and $u(x) = x^2$, what is the composition $F(u(x))$?
- In Example 1, why is $A(x)$ always positive, even though f is sometimes negative? Why do you think the graph of A is concave down until about $x = 5.8$?
- What is the difference between the development of $\ln x$ and e^x in Section 0.5 and the development presented in this section?

EXERCISES 4.7

Thinking Back

The relationship between a function and its derivative: Fill in the blanks if a relationship exists. If there is nothing of substance to say, write "not applicable."

- If f has a zero, then f' ________ .
- If f' has a zero, then f ________ .
- If f is increasing, then f' is ________ .
- If f is positive, then f' is ________ .
- If f' is negative, then f is ________ .
- If f'' is negative, then f is ________ .
- If f'' is negative, then f' is ________ .

Functions and their properties: Give formal mathematical definitions for each of the following functions.

- A *function* f from a set A to a set B
- A *positive* function on $[a, b]$
- An *increasing* function on $[a, b]$

Properties of logarithms and exponents: Determine whether each of the following statements is true or false.

- $\ln(a + b) = \ln(ab)$
- $\ln(a^x) = x \ln a$
- $\ln\left(\frac{a}{b}\right) = \ln a - \ln b$
- $e^{a/b} = e^a - e^b$
- $e^{\ln x} = x$
- $\ln(e^x) = 1$

Concepts

0. *Problem Zero:* Read the section and make your own summary of the material.

1. *True/False:* Determine whether each of the statements that follow is true or false. If a statement is true, explain why. If a statement is false, provide a counterexample.

(a) *True or False:* $\frac{d}{dv} \int_2^u f(v)\, du = f(u)$

(b) *True or False:* $\frac{d}{dx} \int_a^{x^2} \sin t\, dt = \sin x^2$

(c) *True or False:* If f is continuous on $[a, b]$, then there is exactly one $c \in [a, b]$ with $f(c) = \frac{1}{b-a} \int_a^b f(x)\, dx$.

(d) *True or False:* The function $A(x) = \int_0^x (3 - t)\, dt$ is positive and increasing on $[0, 3]$.

(e) *True or False:* The function $A(x) = \int_0^x (3 - t)\, dt$ is concave up.

(f) *True or False:* Every continuous function has an antiderivative.

(g) *True or False:* Every continuous function has more than one antiderivative.

(h) *True or False:* $\ln 5 = \int_0^5 \frac{1}{t}\, dt$.

2. *Examples:* Construct examples of the thing(s) described in the following. Try to find examples that are different than any in the reading.

(a) A function f whose area accumulation function is negative on $[0, 5]$.

(b) A function f whose area accumulation function is decreasing on $[0, 5]$.

(c) Three antiderivatives of $e^{\sin^2 x}$.

3. Use a definite integral to express the function whose output at any real number x is the signed area between the graph of $f(t) = t^2$ and the t-axis on $[2, x]$.

4. What, if anything, is wrong with the expression for a function F given by $F(x) = \int_2^x x^2\, dx$?

5. Consider the function $A(x) = \int_a^x f(t)\, dt$. What is the independent variable of this function? What does the dependent variable represent? Explain why we say that t is a "dummy" variable.

6. Suppose $A(x) = \int_0^x f(t)\, dt$, where f is positive and decreasing for $x > 0$. Is $A(x)$ increasing or decreasing, and why?

7. Suppose $A(x) = \int_0^x f(t)\, dt$, where f is positive and decreasing for $x > 0$.

(a) Explain graphically why the graph of $A(x)$ is concave down.

(b) Find $A''(x)$ and argue that if $f(x)$ is decreasing, then $A(x)$ must be concave down.

8. State the Second Fundamental Theorem of Calculus. Why is this theorem important?

9. The functions $A(x) = \int_0^x t^2\, dt$ and $B(x) = \int_3^x t^2\, dt$ differ by a constant. Explain why this is so in three ways, as follows:

(a) By comparing the derivatives of $A(x)$ and $B(x)$.

(b) By showing algebraically that $A(x) - B(x)$ is a constant.

(c) By interpreting $A(x) - B(x)$ graphically.

10. The functions $A(x) = \int_{-\pi}^x \cos x\, dx$ and $B(x) = \int_\pi^x \cos x\, dx$ differ by a constant. Explain why this is so in three ways, using Exercise 9 as a guide.

11. Write the function $F(x) = \int_3^{x^2} \sin t\, dt$ as a composition of two functions. What is the inside function? What is the outside function?

12. Let f be the function shown here, and define $A(x) = \int_0^x f(t)\, dt$. List the following quantities in order from smallest to largest:

(a) $A(1)$ (b) $A(3)$ (c) $A(6)$ (d) $A(7)$

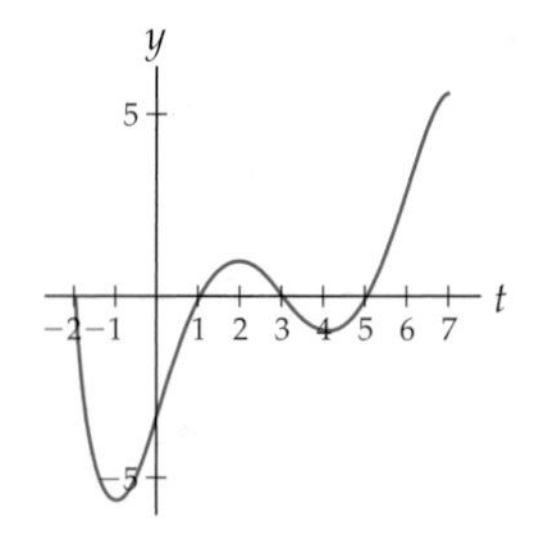

13. Let f be the function graphed in Exercise 12, and define $A(x) = \int_0^x f(t)\,dt$. List the following quantities in order from smallest to largest:

(a) $A(0)$ (b) $A(-1)$ (c) $A(-2)$ (d) $A(5)$

14. Let f be the function shown next at the left, and define $A(x) = \int_0^x f(t)\,dt$. On which interval(s) is A positive? Negative? Increasing? Decreasing? Sketch a rough graph of A.

Graph for Exercise 14

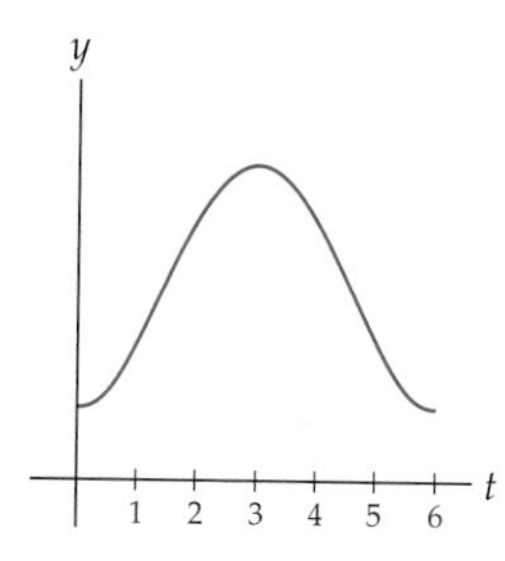

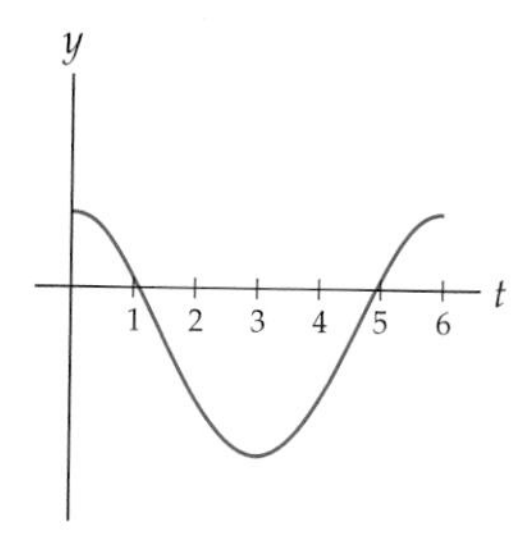

15. Let f be the function shown earlier at the right, and define $A(x) = \int_0^x f(t)\,dt$. On which interval(s) is A positive? Negative? Increasing? Decreasing? Sketch a rough graph of A.

16. Explain how we get the inequality

$$f(m_h)\,h \le \int_x^{x+h} f(t)\,dt \le f(M_h)\,h$$

in the proof of the Second Fundamental Theorem of Calculus. Make sure you define m_h and M_h clearly.

17. Explain precisely how the formula $\frac{d}{dx}\int_a^{u(x)} f(t)\,dt = f(u(x))u'(x)$ in Theorem 4.35 is an application of the chain rule.

18. Are indefinite integrals the "inverse" of differentiation? In other words, does one undo the other? Simplify each of the following to answer this question:

(a) $\int f'(x)\,dx$ (b) $\frac{d}{dx}\int f(x)\,dx$

19. Are definite integrals the "inverse" of differentiation? In other words, does one undo the other? Simplify each of the following to answer this question:

(a) $\int_a^b f'(x)\,dx$ (b) $\frac{d}{dx}\int_a^b f(x)\,dx$

20. Are area accumulation functions the "inverse" of differentiation? In other words, does one undo the other? Simplify each of the following to answer this question:

(a) $\int_a^x f'(t)\,dt$ (b) $\frac{d}{dx}\int_a^x f(t)\,dt$

21. Use the new definition of $\ln x$ from Definition 4.36 to argue that

(a) $\ln x$ has domain $(0, \infty)$ and range $\mathbb{R}$.
(b) e^x has domain $\mathbb{R}$ and range $(0, \infty)$.

22. Express the signed area between the graph of $y = \frac{1}{x}$ and the x-axis from $x = 0.25$ to $x = 1$ in terms of logarithms.

23. Express the signed area between the graph of $y = \frac{1}{x}$ and the x-axis from $x = e$ to $x = 10$ in terms of logarithms.

Now that we have defined $\ln x$ with an integral, we can define a general logarithm with base b as $\log_b x = \frac{1}{\ln b}\ln x$. In Exercises 24–26 you will investigate this definition of $\log_b x$.

24. For which b does the definition of $\log_b x$ just given make sense, and why? Is this the same allowable range of values for b that we saw in our old definition of $\log_b x$ from Chapter 0?

25. Use the definition of $\log_b x$ just given and the new definition of $\ln x$ from Definition 4.36 to express $\log_b x$ as an integral.

26. We can now define general exponential functions b^x as the inverses of the general logarithmic functions $\log_b x$. What can you say about $\log_b(b^y)$? Use this information to simplify $\frac{1}{\ln b}\int_1^{b^y} \frac{1}{t}\,dt$ so that it is written without an integral.

Skills

For each area accumulation function A in Exercises 27–30, (a) illustrate $A(2)$ graphically, (b) calculate $A(2)$ and $A(5)$, and (c) find an explicit elementary formula for $A(x)$.

27. $A(x) = \int_1^x (t^2 + 1)\,dt$

28. $A(x) = \int_{-\pi}^x \sin t\,dt$

29. $A(x) = \int_3^x \frac{1}{t}\,dt$

30. $A(x) = \int_{-x}^x t^3\,dt$

Use the Second Fundamental Theorem of Calculus to write down three antiderivatives of each function f in Exercises 31–34.

31. $f(x) = \sin^2(3^x)$

32. $f(x) = \frac{1}{\sqrt{1 + e^{4x}}}$

33. $f(x) = e^{(x^2)}$

34. $f(x) = \frac{\ln(\sin x)}{x^2 - 1}$

Use the Second Fundamental Theorem of Calculus, if needed, to calculate each the derivatives expressed in Exercises 35–48. (Recall that the operator $\frac{d^2}{dx^2}$ indicates that you should find the second derivative.)

35. $\frac{d}{dx}\int_x^4 e^{t^2+1}\,dt$

36. $\frac{d}{dx}\int_2^x \frac{\sin t}{t}\,dt$

37. $\frac{d}{dx}\int_0^{x^2} \cos t\,dt$

38. $\frac{d}{dx}\int_0^x e^{-t^2}\,dt$

39. $\frac{d}{dx}\int_1^{\sqrt{x}} x \ln t\,dt$

40. $\frac{d}{dx}\left(\left(\int_{\sin x}^{\pi} \frac{t}{\cos t}\,dt\right)^3\right)$

41. $\frac{d}{dx}\int_1^8 \ln t\,dt$

42. $\frac{d}{dx}\int_1^x \ln t\,dt$

43. $\dfrac{d^2}{dx^2}\displaystyle\int_{x^2}^{1} \ln|t|\,dt$

44. $\dfrac{d^2}{dx^2}\displaystyle\int_{1}^{e^x} f(t)g(t)\,dt$

45. $\dfrac{d^2}{dx^2}\displaystyle\int_{2}^{3x} (t^2+1)\,dt$

46. $\dfrac{d}{dx}\left(x^2\displaystyle\int_{0}^{\sin x} \sqrt{t}\,dt\right)$

47. $\dfrac{d}{dx}\displaystyle\int_{x}^{x+2} \sin(t^2)\,dt$

48. $\dfrac{d}{dx}\displaystyle\int_{x}^{\sqrt{x}} \frac{e^x}{x}\,dt$

In Exercises 49–54, find a function f that has the given derivative f' and value $f(c)$. Find an antiderivative of f' by hand, if possible; if it is not possible to antidifferentiate by hand, use the Second Fundamental Theorem of Calculus to write down an antiderivative.

49. $f'(x) = \dfrac{1}{2x-1}, \quad f(1) = 3$

50. $f'(x) = \dfrac{1}{x^2+1}, \quad f(-1) = 0$

51. $f'(x) = \dfrac{1}{x^3+1}, \quad f(2) = 0$

52. $f'(x) = 2\sin(\pi x), \quad f(2) = 4$

53. $f'(x) = e^{-x^2}, \quad f(1) = 0$

54. $f'(x) = \sin(x^2), \quad f(0) = 0$

For each logarithmic value in Exercises 55–58, use Riemann sums with at least four rectangles to find an over-approximation and an under-approximation for the value. *(Note: In Exercise 58, you will need to use the fact that* $\log_2 x = \frac{1}{\ln 2}\ln x$.*)*

55. $\ln 10$

56. $\ln 0.4$

57. $\ln e$

58. $\log_2 4$

Applications

59. Suppose a very strange particle moves back and forth along a straight path in such a way that its velocity after t seconds is given by $v(t) = \sin(0.1t^2)$, measured in feet per second. Consider positions left of the starting position to be in the negative direction and positions right of the starting position in the positive direction.

Velocity of very strange particle
$v(t) = \sin(0.1t^2)$

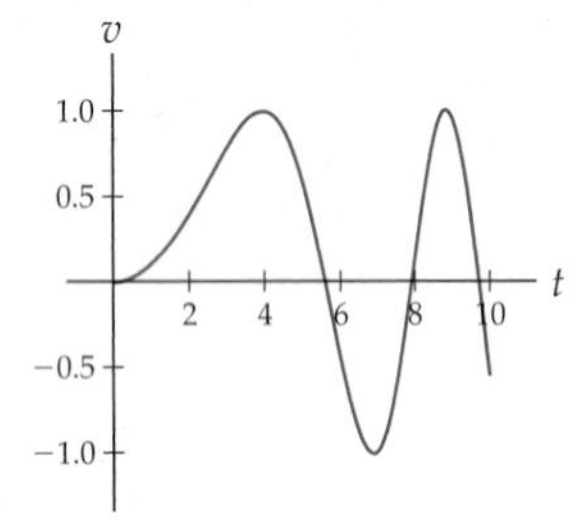

(a) Write down an expression for the position $s(t)$ of the very strange particle, measured in feet left or right of the starting position. Your expression for $s(t)$ will involve an integral.

(b) Use your answer to part (a) and a Riemann sum with 10 rectangles to approximate the position of the particle after 10 seconds of motion.

(c) Verify that the units in your calculation to part (b) make sense. Why is it clear that the Riemann sum will have units measured in feet?

(d) What happens to the velocity of the very strange particle after a long time? What does this mean about how the particle is moving after, say, about 100 seconds?

60. Jimmy is doing some arithmetic problems. As the evening wears on, he gets less and less effective, so that $t \geq 1$ minutes after the start of his study session he can solve $r(t) = 1 + \frac{1}{t}$ problems per minute. Assume that it takes 1 minute for Jimmy to set up his work area; thus he is not doing arithmetic problems in the first minute.

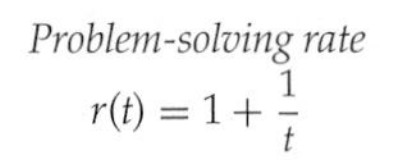

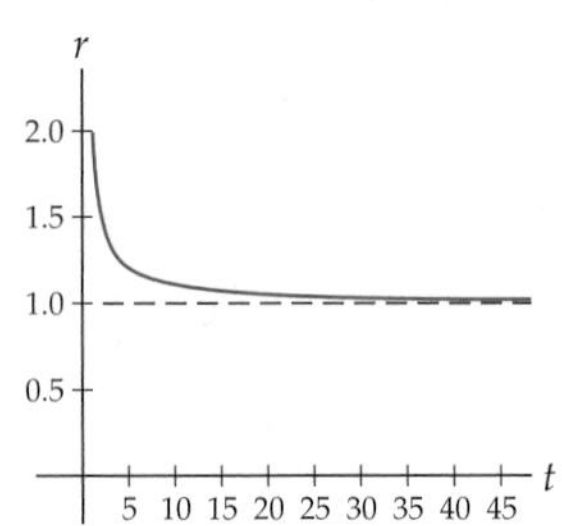

(a) How many arithmetic problems per minute can Jimmy do when he first begins his studies at time $t = 1$? What about at $t = 4$ and $t = 20$?

(b) Make a rough estimate of the number of arithmetic problems Jimmy will have completed after 4 minutes.

(c) Use an integral to express the number of arithmetic problems completed after t minutes, and interpret this definite integral as a logarithm. Calculate the number of problems Jimmy can complete in 10 minutes and in 20 minutes.

(d) Approximately how long will it take for Jimmy to finish his arithmetic homework if he must complete 40 problems?

61. The median rate of flow of the Lochsa River (in Idaho) at the Lowell gauge can be modeled surprisingly accurately as

$$f(t) = 800 + 3.1p(t),$$

where $p(t)$ is a function given by $(t-90)(195-t)$ for $t \in (90, 195)$ and is zero otherwise. The time t is measured in days after the beginning of the year, and the flow is measured in cubic feet of water per second.

(a) Given a starting day t_0, use the Fundamental Theorem of Calculus to find the function $F_{t_0}(t)$ that gives the total amount of water that has flowed past the Lowell gauge since that day. Note that there are 86400 seconds per day.

(b) In a non-leap year, March 31 is day 90 of the year and July 14 is day 195. Use the function you found in part (a) to compute how many cubic feet of water flowed past the gauge between those days.

(c) Use your function to compute how much water flows down the Lochsa River in a full year. What fraction of that amount flows by during the span of time you examined in part (b)?

62. Ian is climbing every day, using a camp at the base of a snowfield. His only supply of water is a trickle that comes out of the the snowfield. The trickle dries at night, because the temperature drops and the snow stops melting. Since he has run out of books, to entertain himself he uses measurements of the water in his cooking pot to model the flow as

$$w(t) = 15\left(1 + \cos\left(\frac{2\pi(t-16)}{24}\right)\right),$$

where t is the time in hours after midnight and $w(t)$ is the rate at which snow is melting into his pot, in gallons per hour.

(a) About how long does it take Ian to fill his 2-quart pot at 4 in the afternoon?

(b) What is the total amount of water that flows out of the snowfield in a single day?

(c) Write an expression for the total amount of water that flows from the snowfield between any starting time t_0 and an ending time t.

(d) If Ian stashes his pot under the trickle at 5 in the morning, how long must he wait until he comes back to a full pot? *(Hint: You will need to do this numerically.)*

Proofs

63. Prove that if f has an antiderivative (say, G), then the function $A(x) = \int_0^x f(t)\,dt$ must also be an antiderivative of f. *(Hint: Use the Fundamental Theorem of Calculus.)*

64. Suppose f is continuous on all of $\mathbb{R}$. Prove that for all real numbers a and b, the functions $A(x) = \int_a^x f(t)\,dt$ and $B(x) = \int_b^x f(t)\,dt$ differ by a constant. Interpret this constant graphically.

65. The proof of the latter part of the Second Fundamental Theorem of Calculus in the reading covered only the case $h \to 0^+$. Rewrite this proof in your own words, and then write a proof of what happens as $h \to 0^-$.

66. Prove that if f is continuous on $[a, b]$ and we define $F(x) = \int_a^x f(t)\,dt$, then F is continuous on the closed interval $[a, b]$. *(Hint: Argue that F is continuous on (a, b), left-continuous at a, and right-continuous at b.)*

67. Prove Theorem 4.34: If f is continuous on $[a, b]$, then for all $x \in [a, b]$, $\frac{d}{dx}\int_a^x f(t)\,dt = f(x)$. The proof follows directly from the Second Fundamental Theorem of Calculus.

68. Prove Theorem 4.35 in your own words: If f is continuous on $[a, b]$ and $u(x)$ is a differentiable function, then for all $x \in [a, b]$, $\int_a^{u(x)} f(t)\,dt = f(u(x))u'(x)$. Be especially clear about how you use the chain rule.

69. Follow the given steps to give an alternative proof of the Mean Value Theorem for Integrals (Theorem 4.31): If f is continuous on a closed interval $[a, b]$, then there exists some $c \in (a, b)$ such that:

$$f(c) = \frac{1}{b-a}\int_a^b f(x)\,dx.$$

(a) Define $F(x) = \int_a^x f(t)\,dt$. What three things does the Second Fundamental Theorem of Calculus say about F?

(b) Why does the Mean Value Theorem apply to F on $[a, b]$, and what conclusion can we obtain from the Mean Value Theorem?

(c) Show that, for the value c that is guaranteed by the Mean Value Theorem, $f(c) = \frac{1}{b-a}\int_a^b f(t)\,dt$.

70. Prove in your own words the last part of Theorem 4.37: If we define $\ln x = \int_1^x \frac{1}{t}\,dt$ for $x > 0$, then $\ln x$ is one-to-one on $(0, \infty)$.

Prove each statement in Exercises 71–78, using the new definition of $\ln x$ as an integral and e^x as the inverse of $\ln x$.

71. Prove that $\ln x$ is continuous and differentiable on its entire domain $(0, \infty)$.

72. Prove that $\frac{d}{dx}(\ln x) = \frac{1}{x}$.

73. Prove that $\ln x$ is zero if $x = 1$, negative if $0 < x < 1$, and positive if $x > 1$.

74. Prove that $\ln x$ is increasing and concave down on its entire domain $(0, \infty)$.

75. Prove that $\ln(ab) = \ln a + \ln b$ for any $a, b > 0$ by following these steps:

(a) Use Theorem 4.35 to show that $\frac{d}{dx}(\ln ax) = \frac{1}{x}$ for any real number $a > 0$.

(b) Use your answer to part (a) to argue that $\ln ax = \ln x + C$. *(Hint: Think about Theorem 4.14.)*

(c) Solve for the constant C in the equation from the previous part, by evaluating the equation at $x = 1$. Use your answer to show that $\ln ax = \ln x + \ln a$. Why does this argument complete the proof?

76. Prove that $\ln x^a = a \ln x$ for any $x > 0$ and any a by following these steps:

(a) Use Theorem 4.35 to show that $\frac{d}{dx}(\ln x^a) = \frac{a}{x}$.

(b) Compare the derivatives of $\ln x^a$ and $a \ln x$ to argue that $\ln x^a = a \ln x + C$.

(c) Use part (b) with $x = 1$ and $a = 1$ to show that $C = 0$, and then complete the proof.

77. Use the previous two exercises to prove that for any a, $b > 0$, $\ln\left(\frac{a}{b}\right) = \ln a - \ln b$.

78. Prove that $e^{a+b} = e^a e^b$ for any real numbers a and b by following the steps in parts (a)–(c). You may use any properties of logarithms proved in the reading or the exercises.

(a) Show that if $y = e^{a+b}$, then $\ln y = a + b$.

(b) Show that if $z = e^a e^b$, then $\ln z = a + b$.

(c) Use parts (a) and (b) to prove that $y = z$, and therefore that $e^{a+b} = e^a e^b$.

Thinking Forward

Differential Equations: A ***differential equation*** is simply an equation that involves derivatives. The solution of a differential equation is the family of functions that make the differential equation true.

- Consider the differential equation $f'(x) = g(x)$. What has to be true about $g(x)$ for the Second Fundamental Theorem of Calculus to guarantee that a function f exists that satisfies $f'(x) = g(x)$?
- Solve the differential equation $f'(x) = \sin x$. Remember, the solution will be a *family* of functions.
- Solve the differential equation $f'(x) = \sin(e^x)$. The solution will be a family of nonelementary functions.
- What family of functions satisfies the differential equation $f'(x) = f(x)$? Notice that a function f satisfying this differential equation has the property that it is its own derivative.

Initial-Value Problems: An ***initial-value problem*** is a differential equation together with an initial condition that specifies one value of a function.

- If $f_1(x)$ and $f_2(x)$ are both solutions of the differential equation $f'(x) = g(x)$, where $g(x)$ is continuous, then what can you say about the relationship between f_1 and f_2? What could you say if, in addition, f_1 and f_2 agreed on any given value?
- Find the solution f to the initial-value problem $f'(x) = \sin x$, $f(\pi) = 0$.
- Find the solution f to the initial-value problem $f'(x) = \sin(e^x)$, $f(\pi) = 0$.
- Find the solution f to the initial-value problem $f'(x) = f(x)$, $f(0) = 3$.

CHAPTER REVIEW, SELF-TEST, AND CAPSTONES

Before you progress to the next chapter, be sure you are familiar with the definitions, concepts, and basic skills outlined here. The capstone exercises at the end bring together ideas from this chapter and look forward to future chapters.

Definitions

Give precise mathematical definitions or descriptions of each of the concepts that follow. Then illustrate the definition with a graph or algebraic example, if possible.

- a *Riemann sum* for a function f on an interval $[a, b]$, including the definitions of Δx, x_k, and x_k^*
- the geometric interpretations of the n-rectangle *left sum*, *right sum*, and *midpoint sum* for a function f on an interval $[a, b]$
- the geometric interpretations of the n-rectangle *upper sum* and *lower sum* for a function f on an interval $[a, b]$
- the geometric interpretations of the n-rectangle *trapezoid sum* for a function f on an interval $[a, b]$
- the *definite integral* of a function f on an interval $[a, b]$ as a limit of Riemann sums, including the definitions of Δx, x_k, and x_k^*
- the *indefinite integral* of a continuous function f
- what we mean when we refer to an *integrand*
- the *signed area* between the graph of a function f and the x-axis on $[a, b]$
- the *absolute area* between the graph of a function f and the x-axis on $[a, b]$
- the *area between two curves* f and g on $[a, b]$
- the *average value* of a function f on $[a, b]$
- the *area accumulation function* for a function f on an interval $[a, b]$
- the formal definition of the *natural logarithm function* in terms of an accumulation integral

Theorems

Fill in the blanks to complete each of the following theorem statements:

- If F and G are ____ functions, then $F'(x) = G'(x)$ if and only if $G(x) =$ ____ .
- *The Fundamental Theorem of Calculus*: If f is ____ on $[a, b]$, and if F is any ____ of f, then
$$\int_a^b f(x)\,dx = \underline{\quad\quad}.$$
- *The Net Change Theorem*: If f is ____ on $[a, b]$ and its derivative f' is ____ on $[a, b]$, then
$$\int_a^b f'(x)\,dx = \underline{\quad\quad}.$$
- *The Fundamental Theorem of Calculus in Evaluation Notation*: If f is ____ on $[a, b]$, then
$$\int_a^b f(x)\,dx = [\underline{\quad\quad}]_a^b.$$
- *The Mean Value Theorem for Integrals*: If f is ____ on $[a, b]$, then there is some $c \in$ ____ such that $f(c) =$ ____ .
- *The Second Fundamental Theorem of Calculus*: Suppose f is ____ on $[a, b]$ and, for all $x \in [a, b]$, we define
$$F(x) = \int_a^x f(t)\,dt.$$
Then F is ____ on $[a, b]$ and ____ on (a, b), and F is an ____ of f, or in other words, ____ = ____ .
- If f is ____ on $[a, b]$, then for all $x \in [a, b]$,
$$\frac{d}{dx}\int_a^x f(t)\,dt = \underline{\quad\quad}.$$
- If f is ____ on $[a, b]$ and $u(x)$ is ____ on $[a, b]$, then for all $x \in [a, b]$,
$$\frac{d}{dx}\int_a^{u(x)} f(t)\,dt = \underline{\quad\quad}.$$
- The function
$$\ln x = \int_1^x \frac{1}{t}\,dt$$
is continuous and differentiable on the interval ____ . Moreover, the following properties follow from this definition (fill in the blanks or select the correct options):
  - $\frac{d}{dx}(\ln x) =$ ____ and $\ln 1 =$ ____ .
  - $\ln x < 0$ on the interval ____ , and $\ln x > 0$ on the interval ____ .
  - $\ln x$ is (increasing)/(decreasing) on its entire domain and (concave up)/(concave down) on its entire domain
  - $\ln x$ is (one-to-one)/(not one-to-one) on its domain

Notation, Formulas, and Algebraic Rules

Notation: Describe the meanings of each of the following mathematical expressions or how they are commonly used in this chapter:

- $\sum_{k=m}^{n} a_k$
- $\sum_{i=1}^{m} b_i$
- Δx
- x_k
- $[x_{k-1}, x_k]$
- x_k^*
- $f(x_k^*)$
- $f(x_k^*)\Delta x$
- $\int_a^b f(x)\,dx$
- $\int f(x)\,dx$
- $\int_a^b f(x)\,dx$
- $[F(x)]_a^b$

The algebra of sums: Fill in the blanks to complete the sum rules that follow. You may assume that a_k and b_k are functions defined for nonnegative integers k and that c is any real number.

- $\sum_{k=1}^{n} c a_k =$ ____
- $\sum_{k=1}^{n} (a_k + b_k) =$ ____
- $\sum_{k=m}^{p-1} a_k + \sum_{k=p}^{n} a_k =$ ____

Sum formulas: Fill in the blanks to complete each sum formula:

- $\sum_{k=1}^{n} 1 =$ ____
- $\sum_{k=1}^{n} k =$ ____
- $\sum_{k=1}^{n} k^2 =$ ____
- $\sum_{k=1}^{n} k^3 =$ ____

Formulas for Riemann sums: Express each of the types of Riemann sums that follow in general sigma notation and also as an expanded sum. You may assume that f is a function defined on $[a, b]$, n is a positive integer, $\Delta x = \frac{b-a}{n}$, and $x_k = a + k\Delta x$.

- left sum
- right sum
- midpoint sum
- trapezoid sum
- upper sum
- lower sum

The algebra of definite integrals: Fill in the blanks to complete the definite integral rules that follow. You may assume that f and g are integrable functions on $[a, b]$, that $c \in [a, b]$, and that k is any real number.

- $\int_a^b kf(x)\,dx =$ ____
- $\int_a^b (f(x)+g(x))\,dx =$ ____

- $\int_a^a f(x)\,dx =$ ____
- $\int_b^a f(x)\,dx =$ ____
- $\int_a^c f(x)\,dx + \int_c^b f(x)\,dx =$ ____

Integral Formulas: Fill in the blanks to complete each of the following integration formulas.
(The last six formulas involve hyperbolic functions and their inverses.)

- For $k \neq -1$, $\int x^k\,dx =$ ____
- $\int \frac{1}{x}\,dx =$ ____
- For $k \neq 0$, $\int e^{kx}\,dx =$ ____
- For $b > 0$ and $b \neq 1$, $\int b^x\,dx =$ ____
- $\int \sin x\,dx =$ ____
- $\int \cos x\,dx =$ ____
- $\int \sec^2 x\,dx =$ ____
- $\int \csc^2 x\,dx =$ ____
- $\int \sec x \tan x\,dx =$ ____
- $\int \csc x \cot x\,dx =$ ____
- $\int \frac{1}{1-x^2}\,dx =$ ____
- $\int \frac{1}{1+x^2}\,dx =$ ____
- $\int \frac{1}{|x|\sqrt{x^2-1}}\,dx =$ ____
- $\int \sinh x\,dx =$ ____
- $\int \cosh x\,dx =$ ____
- $\int \operatorname{sech}^2 x\,dx =$ ____
- $\int \frac{1}{\sqrt{x^2+1}}\,dx =$ ____
- $\int \frac{1}{\sqrt{x^2-1}}\,dx =$ ____
- $\int \frac{1}{1-x^2}\,dx =$ ____

Indefinite integrals of combinations: Fill in the blanks to complete the integration rules that follow. You may assume that f and g are continuous functions and that k is any real number.

- $\int kf(x)\,dx =$ ____
- $\int (f(x)+g(x))\,dx =$ ____
- $\int (f'(x)g(x)+f(x)g'(x))\,dx =$ ____
- $\int \frac{f'(x)g(x)-f(x)g'(x)}{(g(x))^2}\,dx =$ ____
- $\int f'(g(x))g'(x)\,dx =$ ____

Skill Certification: Sums, Integrals, and the Fundamental Theorem

Calculating sums: Determine the value of each of the sums that follow. Some can be computed directly, some require the use of sum formulas, and for some you will have to also compute a limit.

1. $\sum_{k=1}^{10} \frac{1}{k^2}$
2. $\sum_{k=1}^{4} \left(1+\frac{k}{2}\right)^2 \left(\frac{1}{2}\right)$
3. $\sum_{k=1}^{200} (k^2+1)$
4. $\sum_{k=5}^{50} (k+1)^3$
5. $\lim_{n\to\infty} \sum_{k=1}^{n} \frac{k^2+1}{n^3}$
6. $\lim_{n\to\infty} \sum_{k=1}^{n} \frac{(k+1)^3}{n^4+n+1}$

Riemann sums: Calculate each of the following Riemann sum approximations for the definite integral of f on $[a,b]$, using the given value of n.

7. The left sum for $f(x)=\sqrt{x}$ on $[0,4]$, $n=4$.
8. The right sum for $f(x)=\sqrt{x}$ on $[0,4]$, $n=8$.
9. The midpoint sum for $f(x)=9-x^2$ on $[0,3]$, $n=3$.
10. The trapezoid sum for $f(x)=9-x^2$ on $[0,3]$, $n=6$.
11. The upper sum for $f(x)=4x-x^2$ on $[0,3]$, $n=3$.
12. The lower sum for $f(x)=4x-x^2$ on $[0,3]$, $n=6$.

Calculating definite integrals with limits of Riemann sums: Calculate the exact value of each the following definite integrals by setting up a general Riemann sum and then taking the limit as $n \to \infty$.

13. $\int_3^5 (2x+4)\,dx$
14. $\int_0^4 x^2\,dx$
15. $\int_{-2}^3 (4-x^2)\,dx$
16. $\int_1^4 (3x+1)^2\,dx$

Indefinite integrals: Use integration formulas, algebra, and educated guess-and-check strategies to find the following integrals.

17. $\int (3x^4 - x^3 + 2)\,dx$
18. $\int \frac{1-x+x^7}{x^2}\,dx$
19. $\int e^x(1-e^{3x})\,dx$
20. $\int 2\cos 3x\,dx$
21. $\int \frac{3}{1+4x}\,dx$
22. $\int \frac{3}{1+4x^2}\,dx$
23. $\int (2x3^x + x^2(\ln 3)3^x)\,dx$
24. $\int x^3 \sin x^4\,dx$
25. $\int 3\operatorname{sech}^2 x\,dx$
26. $\int \frac{1}{1-2x^2}\,dx$

Definite integrals: Use the Fundamental Theorem of Calculus to find the exact values of each of the definite integrals that follow. Sketch the areas described by these definite integrals to determine whether your answers are reasonable.

27. $\displaystyle\int_{-2}^{2} (x^2 - 1)\, dx$

28. $\displaystyle\int_{1}^{3} \frac{1}{x}\, dx$

29. $\displaystyle\int_{-\pi}^{\pi} \sin 2x\, dx$

30. $\displaystyle\int_{0}^{2} e^{-3x}\, dx$

31. $\displaystyle\int_{0}^{4} (x-1)(x-2)\, dx$

32. $\displaystyle\int_{-\pi/4}^{\pi/4} \sec^2 x\, dx$

33. $\displaystyle\int_{-1}^{1} \frac{x}{1+x^2}\, dx$

34. $\displaystyle\int_{-1}^{1} \frac{-1}{1+x^2}\, dx$

Calculating areas and average values: Express each of the following areas or average values with a definite integral, and then use definite integral formulas to compute the exact value of the area or average value.

35. The signed area between the graph of $f(x) = 3x^2 - 7x + 2$ and the x-axis on $[0, 4]$.

36. The absolute area between the graph of $f(x) = 3x^2 - 7x + 2$ and the x-axis on $[0, 4]$.

37. The average value of the function $f(x) = \sin 2x$ on $[0, \pi]$.

38. The area between the graphs of $f(x) = 4 - x^2$ and $1 - 2x$ on $[-4, 4]$.

Combining derivatives and integrals: Simplify each of the following as much as possible.

39. $\displaystyle\frac{d}{dx}\int x^3\, dx$

40. $\displaystyle\frac{d}{dx}\int_{0}^{x} t^3\, dt$

41. $\displaystyle\int_{-2}^{2} \frac{d}{dx}(\ln(x^2+1))\, dx$

42. $\displaystyle\int_{-2}^{x} \frac{d}{dx}(\ln(x^2+1))\, dx$

43. $\displaystyle\frac{d}{dx}\int_{0}^{3} e^{-t^2}\, dt$

44. $\displaystyle\frac{d}{dx}\int_{0}^{x} e^{-t^2}\, dt$

45. $\displaystyle\frac{d}{dx}\int_{0}^{\ln x} \sin^3 t\, dt$

46. $\displaystyle\frac{d}{dx}\int_{x}^{3} \sin^3 t\, dt$

Capstone Problems

A. *The area under a velocity curve:* Return to the very start of this chapter, and review the discussion of driving down a straight road with stoplights. Describe in your own words the relationship velocity, distance, and accumulation functions illustrated in that discussion. Then use what you know from the material you learned in this chapter to calculate the exact distance travelled in that situation.

B. *Is integration the opposite of differentiation?* In what sense do derivatives "undo" integrals? In what sense do integrals "undo" derivatives? In what sense do they not? In your answer, be sure to consider indefinite integrals, definite integrals, and accumulation functions defined by integrals.

C. *The Fundamental Theorem of Calculus:* Why is the Fundamental Theorem of Calculus so fundamental? What does it allow us to calculate, and what concepts does it relate? Give an overview outline of the proof of this important theorem.

D. *Defining logarithms with integrals:* In this chapter we defined the natural logarithm function as the accumulation integral

$$\ln x = \int_{0}^{x} \frac{1}{t}\, dt.$$

(a) Use the graph of $y = \frac{1}{x}$ and this definition to describe the graphical features of $y = \ln x$.

(b) Given this definition of $\ln x$, how would we define the natural exponential function e^x? Why is this a better definition for e^x than the one we introduced in Definition 1.25?

CHAPTER 5

Techniques of Integration

5.1 Integration by Substitution

The Art of Integration
Undoing the Chain Rule
Choosing a Useful Substitution
Finding Definite Integrals by Using Substitution
Examples and Explorations

$$\int f'(u)du$$

5.2 Integration by Parts

Undoing the Product Rule
Strategies for Applying Integration by Parts
Finding Definite Integrals by Using Integration by Parts
Examples and Explorations

$$\int u\,dv = uv - \int v\,du$$

5.3 Partial Fractions and Other Algebraic Techniques

Proper and Improper Rational Functions
Partial Fractions
Algebraic Techniques
Examples and Explorations

$$\frac{A}{x-2} + \frac{B}{x-3}$$

5.4 Trigonometric Integrals

Using Pythagorean Identities to Set Up a Substitution
Using Double-Angle Formulas to Reduce Powers
Integrating Secants and Cosecants
Examples and Explorations

$$\int \sin^2 x \cos^3 x\,dx$$

5.5 Trigonometric Substitution

Backwards Trigonometric Substitutions
Domains and Simplifications with Trigonometric Substitutions
Rewriting Trigonometric Compositions
Examples and Explorations

$$x = a \sin u$$

5.6 Improper Integrals

Integrating over an Unbounded Interval
Integrating Unbounded Functions
Improper Integrals of Power Functions
Determining Convergence or Divergence with Comparisons
Examples and Explorations

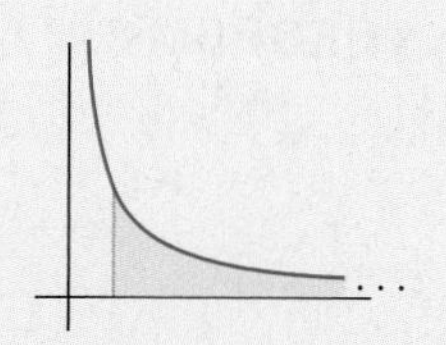

5.7 Numerical Integration*

Approximations and Error
Error in Left and Right Sums
Error in Trapezoid and Midpoint Sums
Simpson's Rule
Examples and Explorations

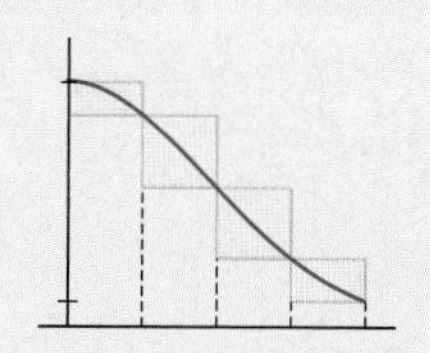

Chapter Review, Self-Test, and Capstones

5.1 INTEGRATION BY SUBSTITUTION

- Reversing the chain rule to obtain a formula for integration by substitution
- Strategies for choosing a useful substitution
- Two ways of finding definite integrals with integration by substitution

The Art of Integration

In this chapter we will explore a series of techniques for solving integrals. At this point we know how to calculate simple integrals involving basic antidifferentiation, such as

$$\int x^3\,dx = \frac{1}{4}x^4 + C, \quad \int \cos x\,dx = \sin x + C, \quad \text{and} \quad \int \frac{1}{1+x^2}\,dx = \tan^{-1} x + C.$$

We have also seen how to use algebra and intelligent guess-and-check methods to calculate certain types of integrals. However, at this point we do not know how to calculate the vast majority of integrals.

For example, consider the integrals

$$\int \frac{e^x}{1+e^x}\,dx, \quad \int x^2 \cos x\,dx, \quad \text{and} \quad \int \sin^3 x \cos^3 x\,dx.$$

Although differentiating any of these integrands would be quite simple, it is not clear how to *anti*differentiate them. Throughout the chapter we will discuss techniques for calculating integrals such as the three we just listed, by rewriting the integrals in terms of simpler ones that we can solve by straightforward antidifferentiation. Even after learning these techniques, it will still take creativity and problem-solving skills to solve many integrals. Integration is not a rote calculational activity like differentiation; it is more of an *art*.

Undoing the Chain Rule

We will begin with a simple technique for rewriting integrals that is based on the chain rule. Since the derivative of a composition $f(u(x))$ is $f'(u(x))u'(x)$, we have the following theorem:

THEOREM 5.1 **A Formula for Integration by Substitution**

If f and u are functions such that $f'(u(x))u'(x)$ is integrable, then

$$\int f'(u(x))\,u'(x)\,dx = f(u(x)) + C.$$

For example, with $f(x) = \sin x$ and $u(x) = x^2$, we have $\frac{d}{dx}(\sin x^2) = (\cos x^2)(2x)$, and therefore $\int (\cos x^2)(2x)\,dx = \sin x^2 + C$. The trouble is, the pattern $f'(u(x))u'(x)$ is not always easy to recognize, so it is not always clear when or how to apply Theorem 5.1 to an integral. In this section we will formalize the procedure of undoing the chain rule with a method known as ***integration by substitution***, also known as ***u*-substitution**.

The key to integration by substitution is a ***change of variables***, a technique by which we change an integral in terms of a variable x into an integral in terms of a different variable u. To do this, we must pay close attention to the ***differential*** dx that appears in integrals. Consider, for example, a function f whose independent variable is called u. Since the derivative of $f(u)$ with respect to u is $f'(u)$, we have

$$\int f'(u)\,du = f(u) + C.$$

Notice that this is exactly the familiar equation $\int f'(x)\,dx = f(x) + C$, but with the variable x renamed as u. Now if $u = u(x)$ is a function of x, comparing the equation we just displayed with the one in Theorem 5.1 tells us that we must have $du = u'(x)\,dx$. This is exactly what we would expect from the notation $\frac{du}{dx} = u'(x)$ if we imagine multiplying both sides by dx. In other words, the differentials dx and du act precisely the way the Leibniz notation $\frac{du}{dx}$ would suggest, even though Leibniz notation does not represent an actual numerical quotient. This property motivates the following definition:

DEFINITION 5.2

The Differential of a Function

If $u(x)$ is a differentiable function of x, then the ***differential*** du is equal to

$$du = u'(x)\,dx.$$

In particular, note that we *cannot* just replace dx with du when we change variables from x to $u = u(x)$. When changing variables in an integral, we will also have to change the differentials, and this will involve the derivative $u'(x)$. As a simple example, consider again the integral $\int (\cos x^2)(2x)\,dx = \int 2x \cos x^2\,dx$, and let $u(x) = x^2$. We must differentiate u to find du:

$$\boxed{u = x^2} \quad \Longrightarrow \quad \frac{du}{dx} = 2x \quad \Longrightarrow \quad \boxed{du = 2x\,dx}.$$

Using the boxed equations to substitute u for x^2 and du for $2x\,dx$, integrating with respect to u, and then writing again in terms of our original variable x, we have

$$\int 2x \cos x^2\,dx = \int \cos u\,du = \sin u + C = \sin x^2 + C.$$

Notice that in this example, the substitution $u = x^2$ changed our original integral into one that was easier to integrate. Integration by substitution is not a method for antidifferentiating; it is a method for changing a difficult integral into a simpler one.

Choosing a Useful Substitution

The first step in trying to solve an integral is to choose a method of attack. Very few integrals can be solved immediately from memory with antidifferentiation formulas. Other integrals can be simplified with an appropriate u-substitution, recognized as backwards quotient or product rule problems, or simplified with some clever algebra. If we choose to attack an integral with the method of integration by substitution, our next step will be to try to find a useful choice for the substitution variable u.

So what makes a good choice for u? The first thing to remember is that we are trying to match the integrand to the pattern $f'(u(x))u'(x)$. Therefore it is a good idea to try choices for $u(x)$ that are inside a composition of functions and whose derivatives $u'(x)$ appear elsewhere in the integrand as a multiplicative factor. For these reasons, we would expect $u = \sin x$ to be a good choice for each of the following integrals:

$$\int \cos x\, e^{\sin x}\,dx, \quad \int \sin^5 x\, \cos x\,dx, \quad \text{and} \quad \int \frac{\cos x}{\sin x}\,dx.$$

With this choice of u we have

$$\boxed{u = \sin x} \quad \Longrightarrow \quad \frac{du}{dx} = \cos x \quad \Longrightarrow \quad \boxed{du = \cos x\,dx},$$

and these integrals would be transformed into, respectively:

$$\int e^u \, du, \quad \int u^5 \, du, \quad \text{and} \quad \int \frac{1}{u} \, du,$$

each of which can be solved immediately with antidifferentiation formulas. In contrast, $u = \sin x$ would *not* be a good choice for the integral $\int \frac{\sin x}{\cos x} \, dx$, because in this case the derivative $u' = \cos x$ would not be a multiplicative part of the integrand. With $u = \sin x$ this integral has the form $\int \frac{u(x)}{u'(x)} \, dx$, which does not match the pattern in Theorem 5.1.

Another good rule of thumb is to choose $u(x)$ to be as much of the integrand as possible while still having things work out with the change in differential from dx to du. This helps change the integral into one that is as simple as possible. For example, with the integral

$$\int x^2 \sin(x^3 + 1) \, dx,$$

a choice of $u = x^3 + 1$ is better than a choice of $u = x^3$. Once we choose u, we must change the entire rest of the integral so that it is in terms of the new variable. After substituting $u = x^3 + 1$, we will still have $x^2 \, dx$ to replace, which we can do by considering the differentials as follows:

$$\boxed{u = x^3 + 1} \quad \Longrightarrow \quad \frac{du}{dx} = 3x^2 \quad \Longrightarrow \quad du = 3x^2 \, dx \quad \Longrightarrow \quad \boxed{\frac{1}{3} du = x^2 \, dx}.$$

Now by substituting with the boxed expressions, our integral is changed into something that is easy to integrate:

$$\int x^2 \sin(x^3 + 1) \, dx = \int \sin u \cdot \frac{1}{3} \, du = \frac{1}{3} \int \sin u \, du = \frac{1}{3}(-\cos u) + C = -\frac{1}{3} \cos(x^3 + 1) + C.$$

It is simple to check this calculation by differentiating:

$$\frac{d}{dx}\left(-\frac{1}{3}\cos(x^3 + 1)\right) = -\frac{1}{3}(-\sin(x^3 + 1))(3x^2) = x^2 \sin(x^3 + 1).$$

Note also that in this example, the original integrand was not *exactly* in the form $f'(u(x))u'(x)$, since we had $u'(x) = 3x^2$ but the constant multiple 3 did not appear in our original integral. However, the method of u-substitution helped us determine the constant $\frac{1}{3}$ that was needed to make everything work out.

Occasionally some preliminary algebra can help us find a useful u-substitution. This is the case for the following two basic integral formulas:

THEOREM 5.3 **Integrals of the Tangent and Cotangent Functions**

(a) $\int \tan x \, dx = \ln |\sec x| + C$ **(b)** $\int \cot x \, dx = -\ln |\csc x| + C$

Proof. The integral $\int \tan x \, dx$ looks too simple for applying integration by substitution, until we rewrite $\tan x$ as $\frac{\sin x}{\cos x}$. We can then choose

$$\boxed{u = \cos x} \quad \Longrightarrow \quad \frac{du}{dx} = -\sin x \quad \Longrightarrow \quad \boxed{-du = \sin x \, dx}.$$

This substitution changes the integral into

$$\int \tan x \, dx = \int \frac{\sin x}{\cos x} \, dx = -\int \frac{1}{u} \, du = -\ln |u| + C = -\ln |\cos x| + C.$$

Although we have found the integral of $\tan x$, it is not in the form we see in Theorem 5.3. Using properties of logarithms and trigonometric functions, we can rewrite our answer as desired:

$$-\ln|\cos x| + C = \ln\left|\frac{1}{\cos x}\right| + C = \ln|\sec x| + C.$$

The proof of part (b) is similar and is left to Exercise 92. ■

Finding Definite Integrals by Using Substitution

There are two methods for calcuating a definite integral $\int_a^b f(x)\,dx$ with the use of integration by substitution. Perhaps the most straightforward way is to use the Fundamental Theorem of Calculus as expressed in Theorem 4.26, which, when considered with the method of u-substitution, says the following:

THEOREM 5.4

Evaluating a Definite Integral After a Substitution

If $f(x) = g(u(x))$ is continuous on $[a, b]$, $u(x)$ is differentiable on (a, b), and $G(u)$ is an antiderivative of $g(u)$, then

$$\int_a^b f(x)\,dx = \int_{x=a}^{x=b} g(u)\,du = \left[G(u)\right]_{x=a}^{x=b} = \left[G(u(x))\right]_a^b = G(u(b)) - G(u(a)).$$

Theorem 5.4 tells us that we can simply use u-substitution (or any integration method, for that matter) to find $\int f(x)\,dx$ in terms of x and then evaluate this antiderivative from $x = a$ to $x = b$. For example, repeating our first example with a definite integral, we have

$$\int_0^{\pi} 2x\cos x^2\,dx = \int_{x=0}^{x=\pi} \cos u\,du = \left[\sin u\right]_{x=0}^{x=\pi} = \left[\sin x^2\right]_0^{\pi} = \sin(\pi^2) - \sin 0.$$

Notice that we are careful to write the limits of integration as $x = 0$ and $x = \pi$ when we change variables to u, so that we remember to substitute back $u = x^2$ before evaluating from 0 to π.

Alternatively, we could change the limits of integration so that they themselves are in terms of the new variable u and then not have to substitute back $u = u(x)$ at the end of the calculation:

THEOREM 5.5

Another Way to Evaluate a Definite Integral After a Substitution

If $f(x) = g(u(x))$ is continuous on $[a, b]$, $u(x)$ is differentiable on (a, b), and $G(u)$ is an antiderivative of $g(u)$, then

$$\int_a^b f(x)\,dx = \int_{u(a)}^{u(b)} g(u)\,du = \left[G(u)\right]_{u(a)}^{u(b)} = G(u(b)) - G(u(a)).$$

For example, if we repeat our earlier calculation with this second method, we have $u = x^2$ and thus $u(0) = 0^2 = 0$ and $u(\pi) = \pi^2$. Calculating the same definite integral gives

$$\int_0^{\pi} 2x\cos x^2\,dx = \int_0^{\pi^2} \cos u\,du = \left[\sin u\right]_0^{\pi^2} = \sin(\pi^2) - \sin 0,$$

which of course is the same as the answer we got earlier. In general, you may choose either of the two methods according to your preference. In some cases the first method will be easier, and in some cases the second. Both will always give the same answer.

Examples and Explorations

EXAMPLE 1 **Using integration by substitution when a constant multiple is missing**

Use integration by substitution to find $\int x^6 e^{5x^7-2}\,dx$.

SOLUTION

Since $5x^7 - 2$ is the inside function of a composition in the integrand and the x^6 part of its derivative $35x^6$ is a multiplicative part of the integrand, we'll try setting $u = 5x^7 - 2$ and see what happens. With this choice of u we have

$$\boxed{u = 5x^7 - 2} \implies \frac{du}{dx} = 35x^6 \implies du = 35x^6\,dx \implies \boxed{\frac{1}{35}\,du = x^6\,dx}.$$

Using the boxed equations, we can now change variables and differentials to rewrite the integral in terms of u and du; moreover, this new integral is easy to solve:

$$\int x^6 e^{5x^7-2}\,dx = \int e^{5x^7-2}(x^6\,dx) = \int e^u\left(\frac{1}{35}\,du\right) = \frac{1}{35}\int e^u\,du = \frac{1}{35}e^u + C = \frac{1}{35}e^{5x^7-2} + C.$$

Although it was not necessary, we rewrote the integral in the first step of the calculation to make sure that the substitution of $x^6\,dx$ by $\frac{1}{35}\,du$ was clear. The antidifferentiation step in the calculation is the step $\int e^u\,du = e^u + C$, which is true because $\frac{d}{du}(e^u) = e^u$. Notice that u-substitution helped us rewrite the original integral into something that was very easy to antidifferentiate. □

CHECKING THE ANSWER

We can differentiate with the chain rule to check the answer we just found. Since we can see that

$$\frac{d}{dx}\left(\frac{1}{35}e^{5x^7-2}\right) = \frac{1}{35}e^{5x^7-2}(35x^6) = x^6 e^{5x^7-2},$$

we know that we have integrated correctly.

EXAMPLE 2 **Deciding when to use integration by substitution**

For each integral, determine whether integration by substitution is appropriate, and if so, determine a useful substitution $u(x)$.

(a) $\int x^4\sqrt{x^5+1}\,dx$ **(c)** $\int \cos x \sin^2 x\,dx$ **(e)** $\int \frac{e^x}{1+e^x}\,dx$

(b) $\int x^3\sqrt{x^5+1}\,dx$ **(d)** $\int \cos x \tan x\,dx$ **(f)** $\int \frac{1+e^x}{e^x}\,dx$

SOLUTION

(a) Integration by substitution will work for this integral. If we choose $u = x^5 + 1$, then $\frac{du}{dx} = 5x^4$, so $x^4\,dx = \frac{1}{5}du$. We can rewrite the integral as $\frac{1}{5}\int \sqrt{u}\,du$, which we know how to integrate.

(b) Integration by substitution won't work here. The most reasonable choice for u would be $u = x^5 + 1$, which makes $\frac{du}{dx} = 5x^4$. We cannot write the integral entirely in terms of u and du with this choice of u. For this integral, no choice of u will work.

(c) If we choose $u = \sin x$, then $\frac{du}{dx} = \cos x$, so $du = \cos x\, dx$. This substitution changes the integral into $\int u^2\, du$, which is easy to integrate. The choice $u = \cos x$ would *not* work in this example; why not?

(d) Neither $u = \cos x$ nor $u = \tan x$ work as substitutions here, because their derivatives are not present in the integrands. However, $\cos x \tan x = \cos x\left(\frac{\sin x}{\cos x}\right)$ is equal to $\sin x$, which we know how to integrate.

(e) At first glance, it does not appear that integration by substitution will work for this problem. However, remember that the derivative of e^x is itself—and in fact the derivative of $1 + e^x$ is also e^x. Thus we can let u be the denominator $u = 1 + e^x$ and get $\frac{du}{dx} = e^x$, that is, $du = e^x\, dx$. This substitution changes the integral into $\int \frac{1}{u}\, du$, which is easy to integrate.

(f) The method we used in part (e) won't work here: We cannot use $u = 1 + e^x$ and $du = e^x\, dx$ to change this integral into a new integral involving only u and du. The problem with this substitution is that the derivative e^x is in the denominator; it is not a *multiplicative* factor of the integrand. This integral can be solved by using algebra to write the integrand $\frac{1+e^x}{e^x}$ as $\frac{1}{e^x} + \frac{e^x}{e^x} = e^{-x} + 1$. Try it! □

EXAMPLE 3

Choosing a substitution

Use integration by substitution to find (a) $\displaystyle\int \frac{\ln x}{x}\, dx$ and (b) $\displaystyle\int \frac{\sin\sqrt{x}}{\sqrt{x}}\, dx$.

SOLUTION

(a) Clearly $u = x$ will not be a good substitution, since it would not simplify the integral. Perhaps $u = \ln x$ will work. With this choice of u, we have

$$\boxed{u = \ln x} \quad \Longrightarrow \quad \frac{du}{dx} = \frac{1}{x} \quad \Longrightarrow \quad \boxed{du = \frac{1}{x}\, dx}.$$

At first you might think that there is no $\frac{1}{x}$ in the original integrand; we only have $\ln x$ and x. But notice that the quotient $\frac{\ln x}{x}$ is actually equal to the product $(\ln x)\left(\frac{1}{x}\right)$; we *do* have $\frac{1}{x}$ in the integrand. Changing variables gives us

$$\int \frac{\ln x}{x}\, dx = \int (\ln x)\left(\frac{1}{x}\, dx\right) = \int u\, du = \frac{1}{2}u^2 + C = \frac{1}{2}(\ln x)^2 + C.$$

As always, it is a good idea to check by differentiating. Since $\frac{d}{dx}\left(\frac{1}{2}(\ln x)^2\right) = \frac{1}{2}(2\ln x)\left(\frac{1}{x}\right) = \frac{\ln x}{x}$, we know we did the integration correctly.

(b) A sensible first choice for u would be $u = \sqrt{x}$, since this is the inside function of a composition in the integrand. With this choice of u we have

$$\boxed{u = \sqrt{x}} \quad \Longrightarrow \quad \frac{du}{dx} = \frac{1}{2}x^{-1/2} = \frac{1}{2\sqrt{x}} \quad \Longrightarrow \quad \boxed{2\, du = \frac{1}{\sqrt{x}}\, dx}.$$

We solved for the expression $\frac{1}{\sqrt{x}}\, dx$ because that is the part of the integrand we need to rewrite in terms of du. With this change of variables we have

$$\begin{aligned} \int \frac{\sin\sqrt{x}}{\sqrt{x}}\, dx &= \int (\sin\sqrt{x})\left(\frac{1}{\sqrt{x}}\, dx\right) = \int (\sin u)(2\, du) \\ &= 2\int \sin u\, du = 2(-\cos u) + C = -2\cos\sqrt{x} + C. \end{aligned}$$

Again, we can easily check this answer: $\frac{d}{dx}(-2\cos\sqrt{x}) = -2(-\sin\sqrt{x})\left(\frac{1}{2}x^{-1/2}\right) = \frac{\sin\sqrt{x}}{\sqrt{x}}$. □

EXAMPLE 4

Integration by substitution followed by back-substitution

Use integration by substitution and algebra to find $\int x\sqrt{x-1}\,dx$.

SOLUTION

This is an example in which integration by substitution works even though the integrand is not at all in the form $f'(u(x))u'(x)$. A clever change of variables will allow us to rewrite the integral so that it can be algebraically simplified. We'll start by choosing $u = x - 1$, which is sensible because this choice of u is the inside function of a composition in the integrand:

$$\boxed{u = x - 1} \quad \Rightarrow \quad \frac{du}{dx} = 1 \quad \Rightarrow \quad \boxed{du = dx}.$$

We can now replace $\sqrt{x-1}$ with $\sqrt{u}$ and dx with du. However, if we do this, there will be an x left over in our integrand that we haven't written in terms of u or du. Luckily, we can use back-substitution to solve for x; since $u = x - 1$, we have

$$\boxed{x = u + 1}.$$

By using the three boxed equations, we have

$$\begin{aligned}
\int x\sqrt{x-1}\,dx = \int (u+1)\sqrt{u}\,du = \int (u^{3/2} + u^{1/2})\,du & \quad \leftarrow \text{substitute and simplify} \\
= \frac{2}{5}u^{5/2} + \frac{2}{3}u^{3/2} + C & \quad \leftarrow \text{antidifferentiate} \\
= \frac{2}{5}(x-1)^{5/2} + \frac{2}{3}(x-1)^{3/2} + C. &
\end{aligned}$$

The original integrand could not be algebraically expanded, but after our clever change of variables we had an integral that we could multiply out, since the sum was then outside, rather than inside, the square root. □

CHECKING THE ANSWER

As usual we can check our integration work by differentiating, and in particular, we know that the chain rule will be involved. In this case a bit of algebra is needed after differentiating, to show that $\frac{d}{dx}\left(\frac{2}{5}(x-1)^{5/2} + \frac{2}{3}(x-1)^{3/2}\right)$ is equal to $x\sqrt{x-1}$. Try it!

EXAMPLE 5

Using integration by substitution to solve a definite integral

Use integration by substitution to find $\int_1^4 \frac{x}{3x^2+1}\,dx$.

SOLUTION

A sensible substitution to try first is

$$\boxed{u = 3x^2 + 1} \quad \Longrightarrow \quad \frac{du}{dx} = 6x \quad \Longrightarrow \quad \boxed{\frac{1}{6}\,du = x\,dx}.$$

A definite integral consists of an integrand *and* limits of integration. When making our substitution we have to be mindful of both:

$$\int_1^4 \frac{x}{3x^2+1}\,dx = \int_1^4 \frac{1}{3x^2+1}(x\,dx) = \int_{x=1}^{x=4} \frac{1}{u}\left(\frac{1}{6}\,du\right) = \frac{1}{6}\int_{x=1}^{x=4} \frac{1}{u}\,du.$$

Note that this is not the same as writing 1 and 4 as the limits of integration on the new integral: Since the new integral is in terms of u and du, that would implicitly mean $u = 1$ and $u = 4$, which is not what we want. Our new definite integral is easy to evaluate:

$$\frac{1}{6}\int_{x=1}^{x=4} \frac{1}{u}\,du = \frac{1}{6}\big[\ln|u|\big]_{x=1}^{x=4} = \frac{1}{6}\big[\ln|3x^2+1|\big]_1^4 = \frac{1}{6}(\ln 49 - \ln 4).$$ □

EXAMPLE 6

Changing the limits of integration when using substitution

Repeat the calculation of $\int_1^4 \frac{x}{3x^2+1}\,dx$, this time replacing the limits of integration during the substitution.

SOLUTION

The choice of u will still be the same as in Example 5. Again we have

$$\boxed{u = 3x^2+1} \implies \frac{du}{dx} = 6x \implies \boxed{\frac{1}{6}\,du = x\,dx}.$$

We will now write the limits of integration ($x = 1$ and $x = 4$) in terms of the new variable u. When $x = 1$ we have $u = 3(1)^2 + 1 = 4$, and when $x = 4$ we have $u = 3(4)^2 + 1 = 49$; in other words,

$$\boxed{u(1) = 4} \quad \text{and} \quad \boxed{u(4) = 49}.$$

Using the information in all four boxed equations, we can change variables completely:

$$\int_1^4 \frac{x}{3x^2+1}\,dx = \int_1^4 \frac{1}{3x^2+1}(x\,dx) = \int_4^{49} \frac{1}{u}\left(\frac{1}{6}\,du\right) = \frac{1}{6}\int_4^{49} \frac{1}{u}\,du.$$

This time we do not have to write the antiderivative in terms of x before the evaluation step, because the limits of integration are already written in terms of u:

$$\frac{1}{6}\int_4^{49} \frac{1}{u}\,du = \frac{1}{6}\big[\ln|u|\big]_4^{49} = \frac{1}{6}(\ln 49 - \ln 4).$$ □

✔ CHECKING THE ANSWER

As usual, we can check our antidifferentiation steps by differentiating. In addition, we can check the numerical answer of a definite integral by remembering that it represents a signed area under a curve. In both examples, our numerical answer was $\frac{1}{6}(\ln 49 - \ln 4) \approx 0.417588$, which we can see is reasonable by using a calculator to approximate the signed area between the graph of $f(x) = \frac{x}{3x^2+1}$ and the x-axis on $[1, 4]$.

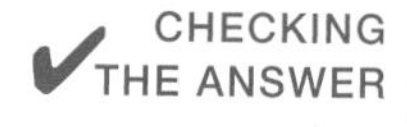

- What would go wrong if we tried to use the substitution $u = x^2$ to solve the integral $\int x^2 \cos x^2\,dx$?
- When we change variables in an integral with a substitution $u = u(x)$, how is the differential dx related to the new differential du?

- Why is $u = \sin x$ not a good choice of substitution for solving the integral $\int \frac{\sin x}{\cos x}\,dx$?
- Show that $\frac{1}{6}\int_1^4 \frac{1}{u}\,du = \frac{\ln 4}{6} \approx 0.23104$. Why is this answer different from the value of the definite integral $\frac{1}{6}\int_{x=1}^{x=4} \frac{1}{u}\,du$ that we computed in Example 5?
- Compare the calculations in Examples 5 and 6. How are they the same? How are they different? What work did you have to do in Example 5 that you did not have to do in Example 6? Where was the equivalent work done in Example 6?

EXERCISES 5.1

Thinking Back

Algebra review: Simplify each algebraic expression as much as possible, until the expression is something that would be easy to antidifferentiate.

- $\sqrt{x}(x^{-2} - 5x^{2/3})$
- $\ln(5x^3) + 3\ln(2^x)$
- $\dfrac{x+1}{2\sqrt[3]{x}}$
- $\dfrac{e^{3x} - 5e^{-2x}}{3e^{2x}}$
- $\sec x \tan x \cos^2 x$
- $\sqrt{1 - \sin^2 x}$
- $\dfrac{2\sqrt{x} - 5}{3\sqrt{x}}$
- $2^x(1 - 3^x)$

Differentiation review: Differentiate each of the functions that follow. Simplify your answers as much as possible.

- $f(x) = -\frac{1}{2}\cot x^2$
- $f(x) = \frac{1}{3}\ln|3x + 1|$
- $f(x) = \frac{1}{2\pi}\sin^2 \pi x$
- $f(x) = \frac{2}{3}(\ln x)^{3/2}$
- $f(x) = (\sin x^2)^{1/2}$
- $f(x) = \frac{1}{2}(\sin^{-1} x)^2$
- $f(x) = \cos(\sin^2 x^3)$
- $f(x) = \ln(\ln(\ln x))$

Antidifferentiation review: Antidifferentiate each of the following functions.

- $f(x) = \sqrt{x} + \sqrt[3]{x}$
- $f(x) = \sec^2 5x$
- $f(x) = \dfrac{2}{3x}$
- $f(x) = \dfrac{\sin \pi x}{2}$
- $f(x) = \dfrac{3}{x^2 + 1}$
- $f(x) = \dfrac{-2}{\sqrt{1 - x^2}}$

Expressing geometric quantities with integrals: Express each of the given geometric quantities in terms of definite integrals. You do not have to solve the integrals.

- The signed area of the region between the graph of $f(x) = \sin x$ and the x-axis on $\left[0, \frac{3\pi}{2}\right]$.
- The absolute area of the region between the graph of $f(x) = \sin x$ and the x-axis on $\left[0, \frac{3\pi}{2}\right]$.
- The signed area of the region between the graphs of x^2 and 2^x on $[-5, 5]$.
- The average value of the function $f(x) = \frac{1}{x}$ from $x = 1$ to $x = 10$.

Concepts

0. *Problem Zero:* Read the section and make your own summary of the material.

1. *True/False:* Determine whether each of the statements that follow is true or false. If a statement is true, explain why. If a statement is false, provide a counterexample.

(a) *True or False:* $\int g'(h(x))\,h'(x)\,dx = g(h(x)) + C$.

(b) *True or False:* If $v = u^2 + 1$, then $\int \sqrt{u^2+1}\,du = \int \sqrt{v}\,dv$.

(c) *True or False:* If $u = x^3$, then $\int x\sin(x^3)\,dx = \frac{1}{3x}\int \sin u\,du$.

(d) *True or False:* $\int_0^3 u^2\,du = \int_{x=0}^{x=3} (u(x))^2\,du$.

(e) *True or False:* $\int_0^1 x^2\,dx = \int_0^1 u^2\,du$.

(f) *True or False:* $\int_2^4 xe^{x^2-1}\,dx = \frac{1}{2}\int_2^4 e^u\,du$.

(g) *True or False:* $\int_2^3 f(u(x))u'(x)dx = \int_{u(2)}^{u(3)} f(u)du$.

(h) *True or False:* $\int_0^6 f(u(x))u'(x)dx = \left[\int f(u)du\right]_0^6$.

2. *Examples:* Construct examples of the thing(s) described in the following. Try to find examples that are different than any in the reading.

(a) Five integrals that can be solved with the method of integration by substitution.

(b) Five integrals that cannot be solved with the method of integration by substitution.

(c) Three relatively simple integrals that we cannot solve with any of the methods we now know.

3. Explain why $\int \frac{2x}{x^2+1}\,dx$ and $\int \frac{1}{x\ln x}\,dx$ are essentially the same integral after a change of variables.

4. List some things which would suggest that a certain substitution $u(x)$ could be a useful choice. What do you look for when choosing $u(x)$?

For each integral in Exercises 5–8, write down three integrals that will have that form after a substitution of variables.

5. $\displaystyle\int \sin u\,du$

6. $\displaystyle\int u^2\,du$

7. $\displaystyle\int \frac{1}{\sqrt{u}}\,du$

8. $\displaystyle\int e^u\,du$

For each function $u(x)$ in Exercises 9–12, write the differential du in terms of the differential dx.

9. $u(x) = x^2 + x + 1$

10. $u(x) = x^2 + 1$

11. $u(x) = \sin x$

12. $u(x) = \dfrac{1}{x}$

13. Suppose $u(x) = x^2$. Calculate and compare the values of the following definite integrals:

$$\int_{-1}^{5} u^2\,du,\quad \int_{x=-1}^{x=5} u^2\,du,\quad \text{and} \quad \int_{u(-1)}^{u(5)} u^2\,du.$$

14. Find three integrals in Exercises 21–70 in which the denominator of the integrand is a good choice for a substitution $u(x)$.

15. Find three integrals in Exercises 21–70 that we can antidifferentiate immediately after algebraic simplification.

16. Consider the integral $\int \sin x \cos x\,dx$.
   (a) Solve this integral by using u-substitution with $u = \sin x$ and $du = \cos x\,dx$.
   (b) Solve the integral another way, using u-substitution with $u = \cos x$ and $du = -\sin x\,dx$.
   (c) How must your two answers be related? Use algebra to prove this relationship.

17. Consider the integral $\int x(x^2-1)^2\,dx$.
   (a) Solve this integral by using u-substitution.
   (b) Solve the integral another way, using algebra to multiply out the integrand first.
   (c) How must your two answers be related? Use algebra to prove this relationship.

18. Consider the integral $\int \frac{x^{-2}-4}{x^3}$.
   (a) Solve this integral by using u-substitution.
   (b) Solve the integral another way, using algebra to simplify the integrand first.
   (c) How must your two answers be related? Use algebra to prove this relationship.

19. Consider the integral $\int_{-2}^{2} e^{5-3x}\,dx$.
   (a) Solve this integral by using u-substitution while keeping the limits of integration in terms of x.
   (b) Solve the integral again with u-substitution, this time changing the limits of integration to be in terms of u.

20. Consider the integral $\int_2^4 \frac{x}{x^2-1}\,dx$.
   (a) Solve this integral using u-substitution while keeping the limits of integration in terms of x.
   (b) Solve this integral again with u-substitution, this time changing the limits of integration to be in terms of u.

Skills

Solve each of the integrals in Exercises 21–70. Some integrals require substitution, and some do not. *(Exercise 69 involves a hyperbolic function.)*

21. $\displaystyle\int (3x+1)^2\,dx$

22. $\displaystyle\int x(x^2-1)^2\,dx$

23. $\displaystyle\int \frac{8x}{x^2+1}\,dx$

24. $\displaystyle\int \left(\frac{2}{x-1} - \frac{3}{2x+1}\right)dx$

25. $\displaystyle\int \frac{\sqrt{x}+3}{2\sqrt{x}}\,dx$

26. $\displaystyle\int \frac{x^{2/3}+1}{\sqrt[3]{x}}\,dx$

27. $\displaystyle\int x\csc^2 x^2\,dx$

28. $\displaystyle\int 3\cos(\pi x)\,dx$

29. $\displaystyle\int \frac{1}{3x+1}\,dx$

30. $\displaystyle\int x^3\cos x^4\,dx$

31. $\displaystyle\int \sin \pi x \cos \pi x\,dx$

32. $\displaystyle\int \frac{x}{\sqrt{3x^2+1}}\,dx$

33. $\displaystyle\int \sec 2x \tan 2x\,dx$

34. $\displaystyle\int \frac{\sqrt{x}+1}{x}\,dx$

35. $\displaystyle\int \cot^5 x \csc^2 x\,dx$

36. $\displaystyle\int \sin x\, e^{\cos x}\,dx$

37. $\displaystyle\int x^4(x^3+1)^2\,dx$

38. $\displaystyle\int 2xe^{3x^2}\,dx$

39. $\displaystyle\int x^{1/4}\sin x^{5/4}\,dx$

40. $\displaystyle\int \frac{\cos(\ln x)}{x}\,dx$

41. $\displaystyle\int x(2^{x^2+1})dx$

42. $\displaystyle\int x\ln(e^{x^2+1})\,dx$

43. $\displaystyle\int \frac{\sqrt{\ln x}}{x}\,dx$

44. $\displaystyle\int \frac{e^{\sqrt{x}}}{\sqrt{x}}\,dx$

45. $\displaystyle\int \frac{\ln\sqrt{x}}{x}\,dx$

46. $\displaystyle\int \frac{(\cos x+1)^{3/2}}{\csc x}\,dx$

47. $\displaystyle\int \frac{3x+1}{x^2+1}\,dx$

48. $\displaystyle\int \frac{1-x}{1+3x^2}\,dx$

49. $\displaystyle\int \frac{e^x}{2-e^x}\,dx$

50. $\displaystyle\int (x^2+1)\sqrt{x}\,dx$

51. $\displaystyle\int \frac{2-e^x}{e^x}\,dx$

52. $\displaystyle\int x^2\sqrt{x+1}\,dx$

53. $\displaystyle\int \sin x\cos^7 x\,dx$

54. $\displaystyle\int \tan^3 x\sec^2 x\,dx$

55. $\displaystyle\int \frac{x\cos x^2}{\sqrt{\sin x^2}}\,dx$

56. $\displaystyle\int \frac{x^2}{\sqrt{x+1}}\,dx$

57. $\displaystyle\int \frac{\sin(1/x)}{x^2}\,dx$

58. $\displaystyle\int \frac{1}{x\ln x}\,dx$

59. $\displaystyle\int x\sqrt{x^2+1}\,dx$

60. $\displaystyle\int \frac{\sec^2 x}{\tan x+1}\,dx$

61. $\displaystyle\int x\sqrt{x+1}\,dx$

62. $\displaystyle\int x(x+1)^{1/3}\,dx$

63. $\displaystyle\int \frac{x}{\sqrt{3x+1}}\,dx$

64. $\displaystyle\int \frac{1}{\sqrt{x}\sec\sqrt{x}}\,dx$

65. $\displaystyle\int \frac{e^x}{\csc e^x}\,dx$

66. $\displaystyle\int \cot x\ln(\sin x)\,dx$

67. $\displaystyle\int x^2(x-1)^{100}\,dx$

68. $\displaystyle\int \frac{\sin^{-1} x}{\sqrt{1-x^2}}\,dx$

69. $\displaystyle\int x\sinh x^2\,dx$

70. $\displaystyle\int \frac{e^x}{\sqrt{e^{2x}-1}}\,dx$

Solve each of the definite integrals in Exercises 71–78. Some integrals require substitution, and some do not.

71. $\displaystyle\int_0^3 x(x^2+1)^{1/3}\,dx$

72. $\displaystyle\int_{-3}^{-1} \frac{1}{\sqrt{1-3x}}\,dx$

73. $\displaystyle\int_{-\pi}^{\pi} \sin x\cos x\,dx$

74. $\displaystyle\int_0^{\sqrt{4\pi}} x\sin x^2\,dx$

75. $\displaystyle\int_2^3 \frac{1}{x\sqrt{\ln x}}\,dx$

76. $\displaystyle\int_2^4 \frac{e^{2x}}{1+e^{2x}}\,dx$

77. $\displaystyle\int_0^1 \frac{x}{1+x^4}\,dx$

78. $\displaystyle\int_0^1 x\sqrt{1-x}\,dx$

79. Consider the function $f(x) = \frac{x}{x^2+1}$.

(a) Find the signed area between the graph of $f(x)$ and the x-axis on $[-1, 3]$ shown next at the left.

(b) Find the area between the graph of $f(x)$ and the graph of $g(x) = \frac{1}{4}x$ on $[-1, 3]$ shown next at the right.

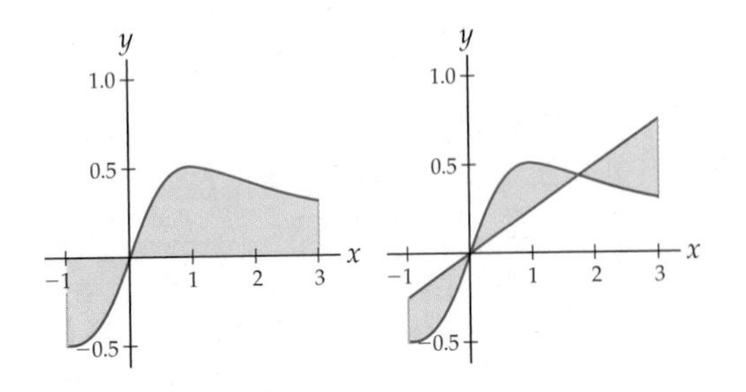

80. Consider the function $f(x) = 4xe^{-x^2}$.

(a) Find the signed area of the region between the graph of $f(x)$ and the x-axis on $[-1, 2]$ shown here:

(b) Find the absolute area of the same region.

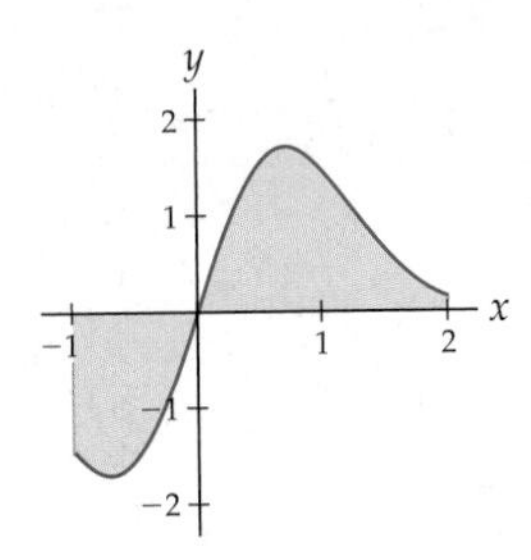

81. Consider the function $f(x) = \frac{\ln x}{x}$ shown here:

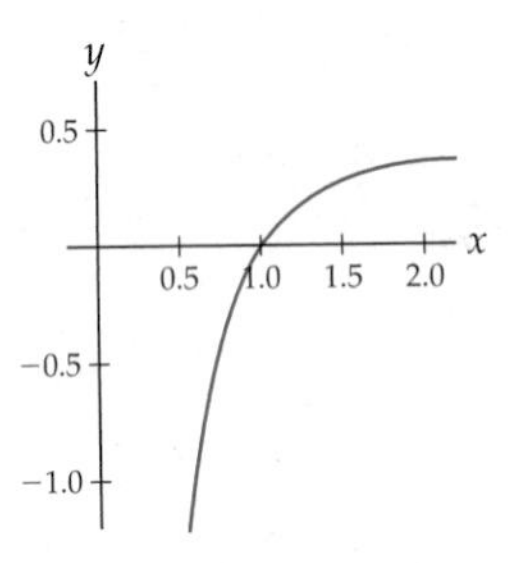

(a) Find the average value of $f(x)$ on $\left[\frac{1}{2}, 2\right]$.

(b) Find a value $c \in \left[\frac{1}{2}, 2\right]$ at which $f(x)$ achieves its average value.

82. Consider the function $f(x) = \sin x\cos x$.

(a) Find the area between the graphs of $f(x)$ and $g(x) = \sin x$ on $[0, \pi]$ shown next at the left.

(b) Find the area between the graphs of $f(x)$ and $h(x) = \sin 2x$ on $[0, \pi]$ shown next at the right.

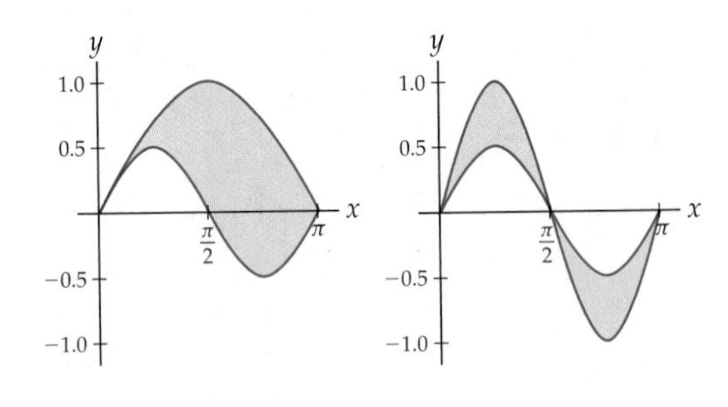

Applications

83. Your local weatherman predicts that there will be a rainstorm this afternoon for three hours and that the rate of rainfall will be $r(t) = 0.6t^2\sqrt{3-t}$ inches per hour from $t = 0$ to $t = 3$, as shown here. Find the total amount of rainfall predicted for the storm.

Rate of rainfall
$r(t) = 0.6t^2\sqrt{3-t}$

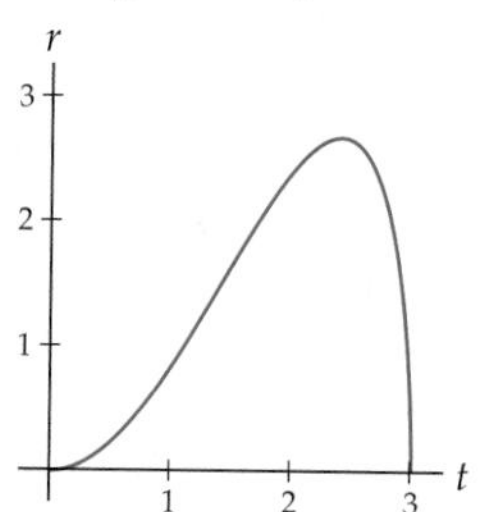

84. Rajini bends a metal rod into a curve that lines up with the graph of $y = 4x^{3/2}$ from $x = 0$ to $x = 2$, as shown here. Given that the length of a curve $y = f(x)$ from $x = a$ to $x = b$ can be calculated with the formula $\int_a^b \sqrt{1 + (f'(x))^2}\,dx$, find the length of the thin metal rod.

Shape of metal rod
$y = 4x^{3/2}$ *on* $[0, 2]$

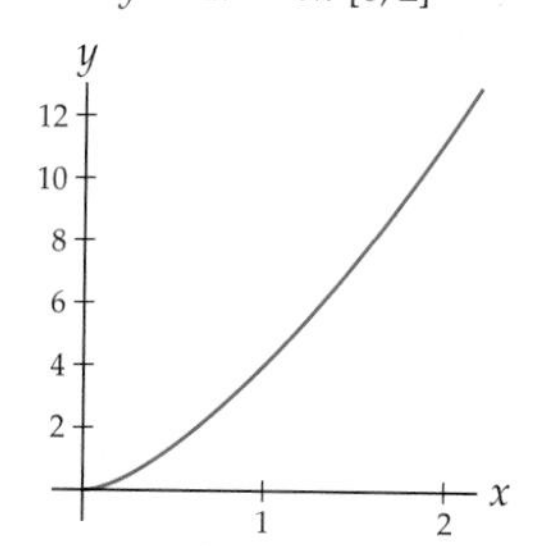

85. One of Dr. Geek's favorite beakers is exactly like the shape obtained by revolving the graph of

$$y = 2\left(\frac{\ln x}{x}\right)^{1/2}$$

from $x = 1$ to $x = 10$ around the x-axis, as shown in the figure and measured in inches. Given that the volume of the shape obtained by revolving f around the x-axis on $[a, b]$ can be calculated with the formula $\pi \int_a^b (f(x))^2\,dx$, about how much liquid can the beaker hold?

Shape of beaker

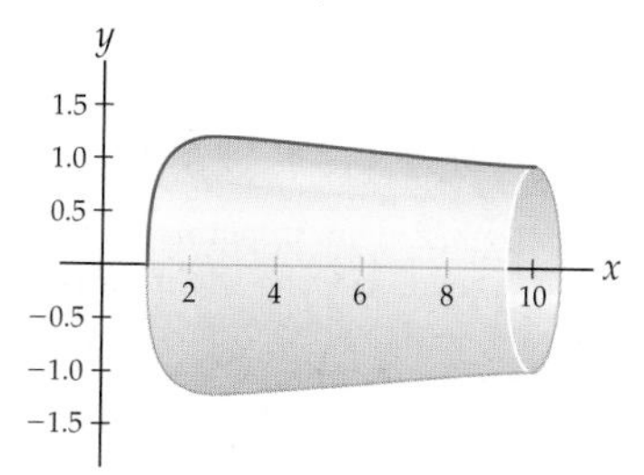

86. A mass hanging at the end of a spring oscillates up and down from its equilibrium position with velocity

$$v(t) = 3\cos\left(\frac{3t}{\sqrt{2}}\right) - 3\sqrt{2}\sin\left(\frac{3t}{\sqrt{2}}\right)$$

centimeters per second, as shown in the figure. The mass is at its equilibrium at $t = 0$. Use definite integrals to determine whether the mass will be above or below its equilibrium position at times $t = 4$ and $t = 5$.

Velocity $v(t)$ of spring

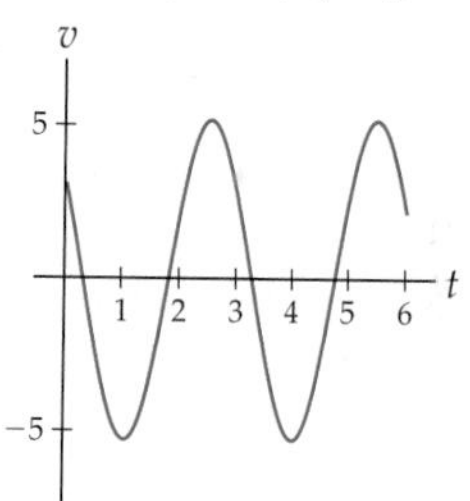

Proofs

87. Prove, in the following two ways, that for any integer k, the signed area under the graph of the function $f(x) = \sin(2(x - (\pi/4)))$ on the interval $[0, k\pi]$ is always zero:

(a) by calculating a definite integral;

(b) by considering the period and graph of the function $f(x) = \sin(2(x - (\pi/4)))$.

88. Prove, in the following two ways, that the signed area under the graph of the function $f(x) = \sin x \cos^2 x$ on an interval $[-a, a]$ centered about the origin is always zero:

(a) by calculating a definite integral;

(b) by considering the symmetry of the graph of the function $f(x) = \sin x \cos^2 x$.

89. Use the chain rule to prove the formula for integration by substitution:

$$\int f'(u(x))u'(x)\,dx = f(u(x)) + C.$$

90. Use the chain rule and the Fundamental Theorem of Calculus to prove the integration-by-substitution formula for definite integrals:

$$\int_a^b f'(u(x))u'(x)\,dx = f(u(b)) - f(u(a)).$$

91. Prove the integration formula

$$\int \tan x\,dx = \ln|\sec x| + C$$

(a) by using algebra and integration by substitution to find $\int \tan x\,dx$;
(b) by differentiating $\ln|\sec x|$.

92. Prove the integration formula

$$\int \cot x\,dx = -\ln|\csc x| + C$$

(a) by using algebra and integration by substitution to find $\int \cot x\,dx$;
(b) by differentiating $-\ln|\csc x|$.

Thinking Forward

Trigonometric integrals: The integrals that follow can be solved by using algebra to write the integrands in the form $f'(u(x))u'(x)$ so that u-substitution will apply.

- Solve $\int \sin^5 x\,dx$ by using the Pythagorean identity $\sin^2 x + \cos^2 x = 1$ to rewrite the integrand as $(1 - \cos^2 x)^2 \sin x$ and then applying substitution with $u = \cos x$.
- Solve $\int \sin^3 x \cos^4 x\,dx$ by using the Pythagorean identity $\sin^2 x + \cos^2 x = 1$ to rewrite the integrand as $(1 - \cos^2 x)\cos^4 x \sin x$ and then applying substitution with $u = \cos x$.
- Solve $\int \sec^4 x \tan^3 x\,dx$ by using the Pythagorean identity $\tan^2 x + 1 = \sec^2 x$ to rewrite the integrand as $(\tan^2 x + 1)\tan^3 x \sec^2 x$ and then applying substitution with $u = \tan x$.

5.2 INTEGRATION BY PARTS

- Reversing the product rule to arrive at the formula for integration by parts
- Strategies for deciding when and how to apply integration by parts
- Using integration by parts to calculate definite integrals

Undoing the Product Rule

As we saw in Theorem 4.23, since the derivative of a product $u(x)v(x)$ of differentiable functions is $u'(x)v(x) + u(x)v'(x)$, we have the following theorem:

THEOREM 5.6 **Reversing the Product Rule**

If u and v are functions such that $u'(x)v(x) + u(x)v'(x)$ is integrable, then

$$\int (u'(x)v(x) + u(x)v'(x))\,dx = u(x)v(x) + C.$$

For example, with $u(x) = x$ and $v(x) = \sin x$, we have $\frac{d}{dx}(x \sin x) = \sin x + x\cos x$, and thus $\int (\sin x + x \cos x)\,dx = x \sin x + C$.

The pattern $u'(x)v(x) + u(x)v'(x)$ does not often appear in the integrals we will be concerned with; however, products will routinely appear in them. With a little bit of rearranging we can rewrite the formula in Theorem 5.6 to obtain a formula for a technique called ***integration by parts***, which will help us solve integrals that involve products. By splitting up the integral in the theorem and then solving for one of the smaller integrals, we obtain

$$\int u(x)v'(x)\,dx = u(x)v(x) - \int v(x)u'(x)\,dx.$$

You may have noticed that we just dropped the "$+C$" that had appeared in the formula from Theorem 5.6. This is allowable because there are indefinite integrals on both sides of the new equation, and thus both sides already represent families of antiderivatives.

Moreover, since we can write the differentials for u and v as $du = u'(x)\,dx$ and $dv = v'(x)\,dx$, we can write the new formula as follows:

THEOREM 5.7 **Formula for Integration by Parts**

If $u = u(x)$ and $v = v(x)$ are differentiable functions, then

$$\int u\,dv = uv - \int v\,du.$$

The parts referred to in Theorem 5.7 are u and dv. When applying integration by parts we will choose u and dv, use them to find du and v, and then apply the integration by parts formula. For example, to solve $\int x\cos x\,dx$ we could choose $u = x$ and $dv = \cos x\,dx$. We can find $du = u'(x)\,dx$ by differentiating u, and we can find v by antidifferentiating $dv = v'(x)\,dx$. Since differentiation intuitively seems like a forward operation, whereas antidifferentiation goes backwards, a sensible way to organize this work is as follows:

$$\begin{array}{lcl} u = x & \longrightarrow & du = 1\,dx \\ v = \sin x & \longleftarrow & dv = \cos x\,dx \end{array}$$

Now the formula for integration by parts tells us that $\int x\cos x\,dx = x\sin x - \int \sin x\,dx$. This is fortunate because it gives us a way to relate an integral that we cannot directly antidifferentiate to an easier integral that is simple to antidifferentiate. Notice that like integration by substitution, integration by parts does not solve integrals; rather, it is a method for rewriting a difficult integral into what is presumably a simpler integral.

Strategies for Applying Integration by Parts

Deciding whether or not an integral can be improved by applying the integration-by-parts formula for some choices of u and dv can sometimes be difficult. In general it is best to try algebraic simplification and u-substitution before attempting integration by parts. However, integrands that are certain simple products, such as $x\cos x$, x^2e^x, or $e^x\sin x$, are often good candidates for integration by parts.

Note that integrals that look similar can sometimes require different methods of solution. For example, consider the integrals

$$\int xe^x\,dx, \quad \int x^2e^x\,dx, \quad \int xe^{(x^2)}\,dx, \quad \text{and} \quad \int e^{(x^2)}\,dx.$$

The first integral can be integrated by parts, with $u = x$ and $v = e^x$, and the second can be integrated by applying parts twice. However, the third integral is better suited for u-substitution with $u = x^2$, and the fourth integrand does not even have an elementary antiderivative.

Once we do decide to use integration by parts, how do we decide what to choose for u and dv? A good rule of thumb is that u should be something that gets simpler when differentiated, and/or dv should be something that gets simpler (or at least not more complicated) when integrated. In other words, we want to choose u and dv so that the associated du and v are simpler than what we started with. This makes sense if you remember that the original integral in terms of u and dv will be replaced by an integral in terms of v and du that we hope will be simpler.

Let's return again to the example $\int x\cos x\,dx = \int u\,dv$. There are four possible ways to split the integrand into the parts u and dv. The possible choices for u are $u = x$, $u = \cos x$, $u = x\cos x$, and $u = 1$. The following table lists each of these choices, the corresponding dv, the resulting du and v, and the expressions for $uv - \int v\,du$:

Choice of u and dv and resulting du and v	New integral $uv - \int v\,du$
$u = x \longrightarrow du = 1\,dx$ $v = \sin x \longleftarrow dv = \cos x\,dx$	$x\sin x - \int \sin x\,dx$
$u = \cos x \longrightarrow du = -\sin x\,dx$ $v = \frac{1}{2}x^2 \longleftarrow dv = x\,dx$	$\frac{1}{2}x^2\cos x + \frac{1}{2}\int x^2\sin x\,dx$
$u = x\cos x \longrightarrow du = (\cos x - x\sin x)\,dx$ $v = x \longleftarrow dv = dx$	$x^2\cos x - \int (x\cos x - x^2\sin x)\,dx$
$u = 1 \longrightarrow du = 0\,dx$ $v = ??? \longleftarrow dv = x\cos x\,dx$	*not possible to apply formula*

At the beginning of this section we solved the integral $\int x\cos x\,dx$ by choosing $u = x$ and $dv = \cos x\,dx$, as shown in the first row of the table, and that worked out well. If we had chosen instead $u = \cos x$ and $dv = x\,dx$, as shown in the second row, we would not have been so lucky: In that case we would have rewritten our integral $\int x\cos x\,dx$ in terms of the more complicated integral $\int x^2\sin x\,dx$. The third possible case is to choose u to be everything but the dx. In this example that does not work out so well. The fourth case is to choose dv to be the entire integrand. This choice is *never* going to work, because we would have to antidifferentiate dv to find v, and antidifferentiating the integrand is what we couldn't do in the first place.

CAUTION However you choose u and dv, you must do it so that the integrand is the *product* of u and dv. For example, with $\int \frac{x}{e^x}\,dx$, we could *not* choose $u = x$ and $dv = e^x\,dx$, since, with this second choice, the integrand is not the product of u and dv. We could, however, choose $u = x$ and $dv = \frac{1}{e^x}\,dx$.

There are instances in which choosing u to be the entire integrand and $dv = dx$ is a useful choice. In fact, the method of choosing u to be the entire integrand yields a formula for integrating the natural logarithm function:

THEOREM 5.8 **Integral of the Natural Logarithm Function**

$$\int \ln x\,dx = x\ln x - x + C.$$

Proof. At first glance the integral of $\ln x$ might not look like it could be a good candidate for integration by parts, since $\ln x$ does not look like a product. If, however, we choose $u = \ln x$ to be the entire integrand and $dv = dx$, then we have

$$u = \ln x \longrightarrow du = \frac{1}{x}\,dx$$
$$v = x \longleftarrow dv = dx.$$

Now the formula for integration by parts gives us

$$\int \ln x\,dx = (\ln x)(x) - \int x\left(\frac{1}{x}\right)dx = x\ln x - \int dx = x\ln x - x + C. \quad \blacksquare$$

As you will show in Exercises 90 and 91, a similar method yields formulas for integrating two inverse trigonometric functions:

THEOREM 5.9 **Integrals of Inverse Sine and Inverse Tangent**

(a) $\int \sin^{-1} x\, dx = x\sin^{-1} x + \sqrt{1-x^2} + C$

(b) $\int \tan^{-1} x\, dx = x\tan^{-1} x - \frac{1}{2}\ln(x^2+1) + C$

Finding Definite Integrals by Using Integration by Parts

When using integration by parts to solve a definite integral $\int_a^b f(x)\,dx$, we must evaluate the *entire* right-hand side of the integration-by-parts formula from $x = a$ to $x = b$. The integration-by-parts formula for definite integrals follows directly from the formula for indefinite integrals, the Fundamental Theorem of Calculus, and the definition of evaluation notation:

THEOREM 5.10 **Integration by Parts for Definite Integrals**

If $u = u(x)$ and $v = v(x)$ are differentiable functions on $[a, b]$, then

$$\int_a^b u\,dv = \left[\int u\,dv\right]_a^b = [uv]_a^b - \int_a^b v\,du.$$

For example, we have already seen that $\int x\cos x\,dx = x\sin x - \int \sin x\,dx$. This means that, for instance, on $[0, \pi]$ we have

$$\int_0^\pi x\cos x\,dx = \left[x\sin x\right]_0^\pi - \int_0^\pi \sin x\,dx = \left[x\sin x\right]_0^\pi - \left[-\cos x\right]_0^\pi.$$

We now have a choice: We can evaluate the expressions $x\sin x$ and $-\cos x$ on $[0, \pi]$ separately, or we can combine them first. Here we choose the latter approach for finishing the calculation:

$$\left[x\sin x + \cos x\right]_0^\pi = (\pi\sin\pi + \cos\pi) - (0\sin 0 + \cos 0) = -1 - 1 = -2.$$

Examples and Explorations

EXAMPLE 1 **Choosing *u* and *dv* when applying integration by parts**

Use integration by parts to find

(a) $\int xe^{2x}\,dx$ **(b)** $\int x^2 \ln x\,dx$

SOLUTION

(a) Since x gets simpler when differentiated and e^{2x} at least does not get worse when antidifferentiated, a sensible choice for parts is $u = x$ and $dv = e^{2x}\,dx$. We can use u

and dv to solve for du and v:

$$u = x \quad \longrightarrow \quad du = dx$$

$$v = \frac{1}{2}e^{2x} \quad \longleftarrow \quad dv = e^{2x}\,dx$$

We can now apply the formula for integration by parts to write our integral in terms of an integral that we hope will be easier to antidifferentiate:

$$\int x e^{2x}\,dx = x\left(\tfrac{1}{2}e^{2x}\right) - \int \tfrac{1}{2}e^{2x}\,dx \qquad \leftarrow \text{integration-by-parts formula}$$

$$= \frac{1}{2}x e^{2x} - \frac{1}{2}\int e^{2x}\,dx \qquad \leftarrow \text{simplify}$$

$$= \frac{1}{2}x e^{2x} - \frac{1}{4}e^{2x} + C. \qquad \leftarrow \text{antidifferentiate}$$

(b) This time there are six different ways to choose u and dv. For example we could choose $u = x^2$, which does indeed get simpler when differentiated. But with that choice of u we would have $dv = \ln x\,dx$, whose antiderivative is the relatively messy function $x \ln x - x$. A better choice is

$$u = \ln x \quad \longrightarrow \quad du = \frac{1}{x}\,dx$$

$$v = \frac{1}{3}x^3 \quad \longleftarrow \quad dv = x^2\,dx.$$

Notice that $u = \ln x$ gets *much* simpler when differentiated: It goes from being a transcendental function to being an algebraic function! Even though $v = \frac{1}{3}x^3$ is a little more complicated than $dv = x^2\,dx$, it is worth it to get rid of the logarithm. There is also some nice cancelling in the integral $\int v\,du$:

$$\int x^2 \ln x\,dx = (\ln x)\left(\tfrac{1}{3}x^3\right) - \int \left(\tfrac{1}{3}x^3\right)\left(\tfrac{1}{x}\right)\,dx \qquad \leftarrow \text{integration-by-parts formula}$$

$$= \frac{1}{3}x^3 \ln x - \frac{1}{3}\int x^2\,dx \qquad \leftarrow \text{simplify}$$

$$= \frac{1}{3}x^3 \ln x - \frac{1}{3}\left(\tfrac{1}{3}x^3\right) + C. \qquad \leftarrow \text{antidifferentiate}$$

□

EXAMPLE 2

Deciding when to use integration by parts

For each integral, determine whether integration by parts is appropriate, and if so, determine useful choices for u and dv.

(a) $\int x \sin x^2\,dx$ **(c)** $\int x \ln x\,dx$ **(e)** $\int \frac{x+1}{e^x}\,dx$

(b) $\int x^2 \sin x\,dx$ **(d)** $\int \frac{\ln x}{x}\,dx$ **(f)** $\int \frac{e^x}{e^x+1}\,dx$

SOLUTION

(a) Although the integrand here is a product, integration by parts is not the best strategy for this integral. This is a classic u-substitution problem, since $u = x^2$ is the inside of a composition and its derivative $2x$ (without the constant 2) is a multiplicative factor in the integrand. This substitution gives us the integral $\frac{1}{2}\int \sin u\,du$, which is easy to solve.

(b) This time u-substitution cannot help us, because the x^2 is outside of the sine function while its derivative is inside. A good choice for integration by parts here is $u = x^2$ and $dv = \sin x\,dx$, since the x^2 will become simpler upon differentiating. In this case we will have to apply integration by parts twice, as we will do in part (a) of Example 3.

(c) This integrand is well suited for integration by parts because it is the product of two unrelated functions. Although we often like to choose $u = x$ in such integrals (e.g., with $\int x \sin x\,dx$ or $\int xe^x\,dx$), in this case that would force us to choose $dv = \ln x\,dx$, which becomes much more complicated after antidifferentiating. If we instead choose $u = \ln x$ and $dv = x\,dx$, then we can successfully apply integration by parts.

(d) Although we could certainly think of the integrand as a product $(\ln x)\left(\frac{1}{x}\right)$, and there is a choice of parts that will work here (see Exercise 24), this integral is best solved with u-substitution. With $u = \ln x$, we have $du = \frac{1}{x}\,dx$ and therefore can change variables to get the very simple integral $\int u\,du$.

(e) To apply integration by parts it is best to rewrite the integrand as the product $e^{-x}(x + 1)$. The most sensible choice of parts is $u = x + 1$ (since this expression gets simpler after differentiating) and $dv = e^{-x}\,dx$ (since this expression stays about the same after antidifferentiating).

(f) This integral is best solved with u-substitution, with $u = e^x + 1$. Then we have $du = e^x\,dx$ and can transform to the simpler integral $\int \frac{1}{u}\,du$. □

EXAMPLE 3 **Applying integration by parts multiple times**

Use integration by parts to find

(a) $\displaystyle\int x^2 e^x\,dx$ **(b)** $\displaystyle\int e^x \sin x\,dx$

SOLUTION

(a) A sensible choice of parts here is $u = x^2$ and $dv = e^x\,dx$, since differentiation will improve x^2 and integration will not make e^x any worse. This choice of parts results in a simpler integral, but one that itself requires an application of integration by parts to solve:

$$\int x^2 e^x\,dx = x^2 e^x - 2\int xe^x\,dx \qquad \leftarrow \text{parts with } u = x^2 \text{ and } dv = e^x\,dx$$

$$= x^2 e^x - 2\left(xe^x - \int e^x\,dx\right) \qquad \leftarrow \text{parts with } u = x \text{ and } dv = e^x\,dx$$

$$= x^2 e^x - 2(xe^x - e^x) + C. \qquad \leftarrow \text{antidifferentiation}$$

The first application of integration by parts takes us from an integral involving x^2e^x to one involving the simpler xe^x. The second application of integration by parts further reduces the power of x until we are left with an integral involving simply e^x. If the original integral had been $\int x^5 e^x\,dx$, then we could use the same method, integrating by parts five times to eventually end up with $\int e^x\,dx$.

(b) In some cases, integrating by parts twice brings us back to the original integral. Although sometimes this is no help at all, in this case it will enable us to construct an equation involving the original integral, and it will be a simple matter to solve that equation. Applying parts to the resulting integral with $u = e^x$ and $dv = \sin x\,dx$, and then applying parts again with $u = e^x$ and $dv = \cos x\,dx$, we get

$$\int e^x \sin x\,dx = -e^x \cos x + \int e^x \cos x\,dx = -e^x \cos x + \left(e^x \sin x - \int e^x \sin x\,dx\right).$$

It might seem as if we have gone in a circle, but a sign change has made this into a very useful equation. For a moment let I represent the integral $\int e^x \sin x\,dx$ that we are trying to find. Then the calculation we just did implies that $I = -e^x \cos x + e^x \sin x - I$. It is now a simple matter to solve this equation for $I = \frac{1}{2}(-e^x \cos x + e^x \sin x) + C$. The "$+C$" must be there because I represents a family of antiderivatives. We have solved our integral, without even having to do a final antidifferentiation step! □

✔ CHECKING THE ANSWER

As usual, we can check our answers by differentiating. In both cases of the previous example we applied parts twice, so to differentiate we need to apply the product rule twice. We have

$$\frac{d}{dx}(x^2e^x - 2(xe^x - e^x)) = 2xe^x + x^2e^x - 2(e^x + xe^x - e^x) = x^2e^x \text{ and}$$

$$\frac{d}{dx}\left(\frac{1}{2}(-e^x \cos x + e^x \sin x)\right) = \frac{1}{2}(-e^x \cos x + e^x \sin x + e^x \sin x + e^x \cos x) = e^x \sin x.$$

EXAMPLE 4

Solving a definite integral with integration by parts

Calculate $\displaystyle\int_0^{\pi/4} x \sec^2 x\,dx$.

SOLUTION

Notice that $\sec^2 x$ is easy to antidifferentiate, since it is the derivative of $\tan x$. Therefore we choose

$$\begin{array}{ccc} u = x & \longrightarrow & du = dx \\ v = \tan x & \longleftarrow & dv = \sec^2 x\,dx \end{array}$$

Integrating by parts, we have

$$\int_0^{\pi/4} x \sec^2 x\,dx = \left[x \tan x\right]_0^{\pi/4} - \int_0^{\pi/4} \tan x\,dx.$$

We can solve the new integral $\int_0^{\pi/4} \tan x\,dx = \int_0^{\pi/4} \frac{\sin x}{\cos x}\,dx$ with integration by substitution. To avoid confusion, since we've already used the letter u, let's use the letter w here:

$$\boxed{w = \cos x} \implies \frac{dw}{dx} = -\sin x \implies \boxed{-dw = \sin x\,dx}.$$

Continuing the calculation, we have

$$\begin{aligned}
\int_0^{\pi/4} x \sec^2 x\,dx &= \left[x \tan x\right]_0^{\pi/4} + \int_{x=0}^{x=\pi/4} \frac{1}{w}\,dw \\
&= \left[x \tan x\right]_0^{\pi/4} + \left[\ln|w|\right]_{x=0}^{x=\pi/4} && \leftarrow \text{antidifferentiate} \\
&= \left[x \tan x\right]_0^{\pi/4} + \left[\ln|\cos x|\right]_0^{\pi/4} && \leftarrow \text{since } w = \cos x \\
&= \left(\frac{\pi}{4} \tan \frac{\pi}{4} - 0 \tan 0\right) + \left(\ln\left|\cos \frac{\pi}{4}\right| - \ln|\cos 0|\right) && \leftarrow \text{evaluate} \\
&= \left(\frac{\pi}{4} - 0\right) + \left(\ln \frac{\sqrt{2}}{2} - 0\right) \\
&= \frac{\pi}{4} + \ln\left(\frac{\sqrt{2}}{2}\right).
\end{aligned}$$

□

CHECKING THE ANSWER

You can check the integration steps in the previous example by differentiating:

$$\frac{d}{dx}(x\tan x + \ln|\cos x|) = \tan x + x\sec^2 x + \frac{1}{\cos x}(-\sin x) = x\sec^2 x.$$

You can check the numerical answer by using a graphing calculator to verify that the signed area under the graph of $y = x\sec^2 x$ from $x = 0$ to $x = \frac{\pi}{4} \approx 0.785$ is approximately $\frac{\pi}{4} + \ln\left(\frac{\sqrt{2}}{2}\right) \approx 0.439$:

$$\int_0^{\pi/4} x\sec^2 x\,dx \approx 0.439$$

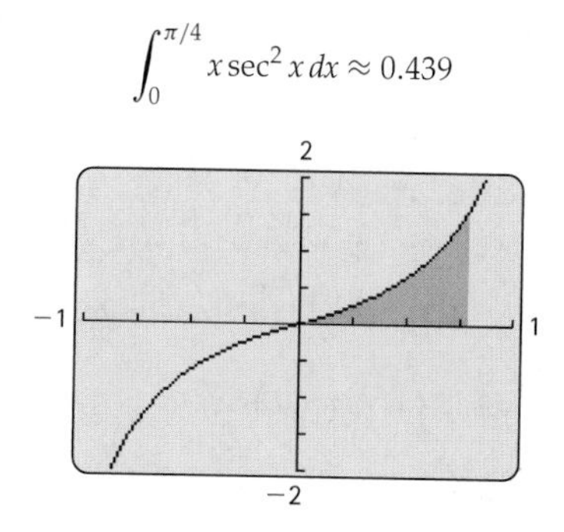

? TEST YOUR UNDERSTANDING

- What types of functions get simpler when differentiated? What types of functions get simpler when antidifferentiated? What types of functions have derivatives and/or antiderivatives of about the same complexity?
- What types of integrals are usually amenable to integration by parts? How does integration by parts help you calculate an integral?
- After choosing $u = e^x$ and $dv = \sin x\,dx$ in Example 3(b), we could have chosen $u = \cos x$ and $dv = e^x\,dx$ in the second application of integration by parts. Would this work?
- In Example 3(b) we could have instead chosen $u = \sin x$ and $dv = e^x\,dx$ in the first integration by parts and then a similar u and dv in the second integration by parts. Would this work?
- Now that we know how to integrate $\ln x$, are there any other elementary functions that we do not yet know how to antidifferentiate?

EXERCISES 5.2

Thinking Back

Differentiation review: Differentiate each of the functions that follow. Simplify your answers as much as possible.

- $f(x) = xe^x - e^x$
- $f(x) = \frac{1}{2}x^2\ln x - \frac{1}{4}x^2$
- $f(x) = -e^{-x}(x^2+1) - 2xe^{-x} - 2e^{-x}$
- $f(x) = \frac{1}{3}x^3 - 2(xe^x - e^x) + \frac{1}{2}e^{2x}$
- $f(x) = -x\cot x + \ln|\sin x|$
- $f(x) = x^3\sin x + 3x^2\cos x - 6x\sin x - 6\cos x$

Review of integration by substitution: Use u-substitution to find each of the following integrals.

- $\int e^{3x+1}\,dx$
- $\int xe^{x^2+1}\,dx$
- $\int \frac{\ln x}{x}\,dx$
- $\int \frac{1}{x\ln x}\,dx$
- $\int \tan x\sec^2 x\,dx$
- $\int \frac{1}{1+x^2}\,dx$
- $\int \sin x\sqrt{\cos x}\,dx$
- $\int e^x\sin e^x\,dx$

Concepts

0. *Problem Zero:* Read the section and make your own summary of the material.

1. *True/False:* Determine whether each of the statements that follow is true or false. If a statement is true, explain why. If a statement is false, provide a counterexample.

(a) *True or False:* If $u = x^2 + 1$, then $du = 2x$.

(b) *True or False:* If $dv = x^2\,dx$, then $v = \frac{1}{3}x^3$.

(c) *True or False:* We can apply integration by parts with $u = \ln x$ and $dv = x\,dx$ to the integral $\int \frac{\ln x}{x}\,dx$.

(d) *True or False:* We can apply integration by parts with $u = x$ and $dv = \ln x\,dx$ to the integral $\int \frac{\ln x}{x}\,dx$.

(e) *True or False:* Integration by parts has to do with reversing the product rule.

(f) *True or False:* Integration by parts is a good method for any integral that involves a product.

(g) *True or False:* In applying integration by parts, it is sometimes a good idea to choose u to be the entire integrand and let $dv = dx$.

(h) *True or False:* $\int_0^3 xe^x\,dx = xe^x - \int_0^3 e^x\,dx$.

2. *Examples:* Construct examples of the thing(s) described in the following. Try to find examples that are different than any in the reading.

(a) Five integrals for which u-substitution is a better strategy than integration by parts. List a good choice for u in each case.

(b) Five integrals for which integration by parts is a better strategy than u-substitution. List good choices for u and dv in each case.

(c) Three integrals that we cannot integrate with only the techniques we have learned so far.

3. State the integration-by-parts formula for indefinite integrals, (a) using the notation du and dv and (b) without using the notation du and dv.

4. State the integration-by-parts formula for definite integrals, (a) using the notation du and dv and (b) without using the notation du and dv.

5. Write down an integral that can be solved by using integration by parts with $u = x$ and another integral that can be solved by using integration by parts with $dv = x\,dx$.

6. Write down an integral that can be solved by using integration by parts with $u = \sin x$ and another integral that can be solved by using integration by parts with $dv = \sin x\,dx$.

7. Write down an integral that can be solved with integration by parts by choosing u to be the entire integrand and $dv = dx$.

8. Suppose $v(x)$ is a function of x. Explain why the integral of dv is equal to v (up to a constant).

9. Explain why choosing $u = 1$ (and thus choosing dv to be the entire integrand, including dx) is *never* a good choice for integration by parts.

10. Find three integrals in Exercises 27–70 for which either algebra or u-substitution is a better strategy than integration by parts.

11. Find three integrals in Exercises 27–70 for which a good strategy is to use integration by parts with $u = x$ (and dv the remaining part).

12. Find three integrals in Exercises 27–70 for which a good strategy is to apply integration by parts twice.

13. If $u(x) = \sin 3x$ and $v(x) = x$, what are du and dv? Write down $\int u\,dv$ and $\int v\,du$ in this situation. Which of these integrals would be easier to find? What does this exercise have to do with integration by parts?

14. Provide a justification for each equality in the statement of the integration-by-parts formula for definite integrals from Theorem 5.10.

15. Explain why $[g(x)]_a^b - [h(x)]_a^b = [g(x) - h(x)]_a^b$. What does this equation have to do with calculations of definite integrals with integration by parts?

For each pair of functions $u(x)$ and $v(x)$ in Exercises 16–18, fill in the blanks to complete each of the following:

(a) $\frac{d}{dx}(u(x)v(x)) = \_\_\_\_$

(b) $\int \_\_\_\_\ dx = u(x)v(x) + C$

(c) $\int u\,dv = \_\_\_\_$

16. $u(x) = x,\ v(x) = \cos 2x$

17. $u(x) = \ln x,\ v(x) = x$

18. $u(x) = x^3,\ v(x) = e^{3x}$

Each expression that follows is the result of a calculation that uses integration by parts. That is, each is an expression of the form $uv - \int v\,du$ for some functions u and v. Identify u, v, du, and dv, and determine the original integral $\int u\,dv$.

19. $\frac{x\,2^x}{\ln 2} - \frac{1}{\ln 2}\int 2^x\,dx$

20. $-x^2\cos x + 2\int x\cos x\,dx$

21. $-\frac{\ln x}{x^2} + \int \frac{1}{x^3}\,dx$

22. $\frac{1}{3}xe^{3x} - \frac{1}{3}\int e^{3x}\,dx$

23. $x\tan^{-1}x - \int \frac{x}{x^2+1}\,dx$

24. Consider the integral $\int \frac{\ln x}{x}\,dx$.

(a) Solve this integral by using integration by parts with $u = \ln x$ and $dv = \frac{1}{x}\,dx$.

(b) Now solve the integral another way, by using u-substitution with $u = \ln x$.

(c) How must your answers to parts (a) and (b) be related? Use algebra to prove that this is so.

25. Consider the integral $\int x^2 \ln(x^3)\, dx$.

(a) Solve this integral by using integration by parts with $u = \ln(x^3)$ and $dv = x^2\, dx$.

(b) Now solve the integral another way, by using u-substitution with $u = x^3$.

(c) How must your answers to parts (a) and (b) be related? Use algebra to prove this relationship.

26. Consider the integral $\int x(x+1)^{100}\, dx$.

(a) Solve this integral by using integration by parts with $u = x$ and $dv = (x+1)^{100}\, dx$.

(b) Now solve the integral another way, by using u-substitution with $u = x + 1$ and back-substitution.

(c) How must your answers to parts (a) and (b) be related? Use graphs to prove this relationship.

Skills

Solve each of the integrals in Exercises 27–70. Some integrals require integration by parts, and some do not. *(The last two exercises involve hyperbolic functions.)*

27. $\int xe^x\, dx$

28. $\int x \sin x\, dx$

29. $\int x \ln x\, dx$

30. $\int x \cos x\, dx$

31. $\int x \sin x^2\, dx$

32. $\int x^2 \cos x\, dx$

33. $\int x^2 e^{3x}\, dx$

34. $\int (3 - x)e^{2x}\, dx$

35. $\int \frac{x}{e^x}\, dx$

36. $\int \frac{x^2}{e^x}\, dx$

37. $\int 3xe^{x^2}\, dx$

38. $\int \ln(3x)\, dx$

39. $\int \ln(x^3)\, dx$

40. $\int x \ln(x^2)\, dx$

41. $\int \frac{x^2+1}{e^x}\, dx$

42. $\int \frac{x}{e^{x^2+1}}\, dx$

43. $\int \frac{3x+1}{x}\, dx$

44. $\int \frac{3x+1}{\sec x}\, dx$

45. $\int (x - e^x)^2\, dx$

46. $\int \ln \sqrt{x}\, dx$

47. $\int \sqrt{x} \ln x\, dx$

48. $\int x \ln \sqrt{x}\, dx$

49. $\int x^3 e^{x^2}\, dx$

50. $\int x^3 e^{-x}\, dx$

51. $\int x \csc^2 x\, dx$

52. $\int x^5 \cos x^3\, dx$

53. $\int x^3 \cos x\, dx$

54. $\int \frac{x^2}{\sqrt{x-1}}\, dx$

55. $\int x^3 \sin x^2\, dx$

56. $\int \arcsin 2x\, dx$

57. $\int \tan^{-1} 3x\, dx$

58. $\int \sin \pi x \cos \pi x\, dx$

59. $\int e^{2x} \sin x\, dx$

60. $\int e^x \ln(e^{2x})\, dx$

61. $\int \frac{2^x}{3^x}\, dx$

62. $\int \frac{\sin x}{3^x}\, dx$

63. $\int e^{-x} \cos x\, dx$

64. $\int x^3 e^{x^2}\, dx$

65. $\int \frac{x}{\cos^2 x}\, dx$

66. $\int \sec^4 x\, dx$

67. $\int x \tan^{-1} x\, dx$

68. $\int x^2 \tan^{-1} x\, dx$

69. $\int \frac{x^2}{\sqrt{x^2+1}}\, dx$

70. $\int x^3 \cosh 2x\, dx$

Solve each of the definite integrals in Exercises 71–78. Some integrals require integration by parts, and some do not.

71. $\int_1^2 \ln x\, dx$

72. $\int_0^{\pi} x \sin x\, dx$

73. $\int_{-1}^1 \frac{x}{e^x}\, dx$

74. $\int_{-\pi/4}^{-\pi/4} x \sec x \tan x\, dx$

75. $\int_0^{\pi} e^x \sin x\, dx$

76. $\int_{-1}^1 2^x \cos x\, dx$

77. $\int_{\pi/4}^{\pi/2} x \csc^2 x\, dx$

78. $\int_{-1/2}^{1/2} \sin^{-1} x\, dx$

79. Consider the function $f(x) = xe^{-x}$.

(a) Find the signed area of the region between the graph of f and the x-axis on $[-1, 1]$ shown in the next figure.

(b) Find the absolute area of the same region.

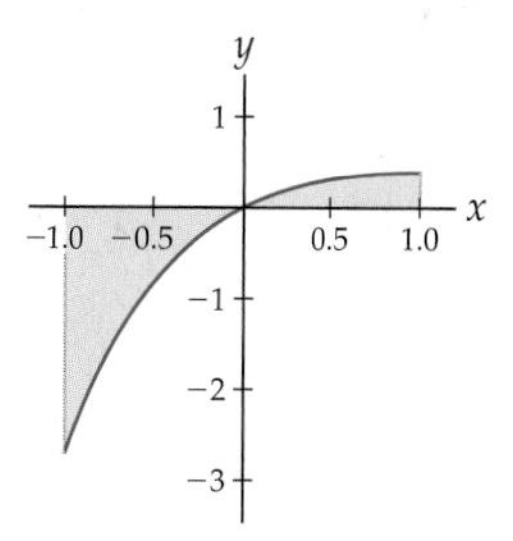

80. Consider the function $f(x) = 2 + \tan^{-1} x$ shown in the next figure.

(a) Find the average value of $f(x)$ on $[-4, 4]$.

(b) Find a value $c \in [-4, 4]$ at which $f(x)$ achieves its average value.

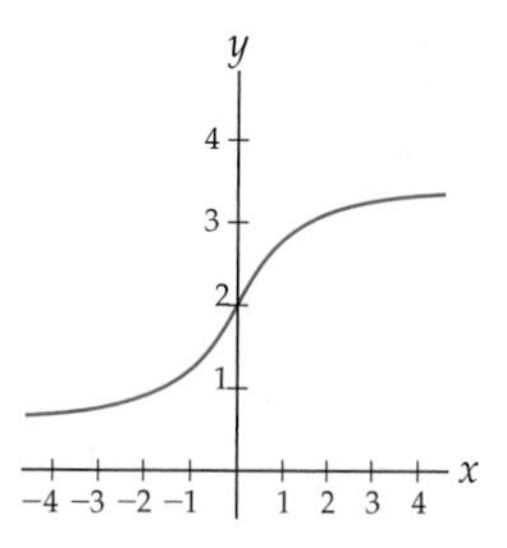

81. Consider the function $f(x) = x\cos 2x$.

(a) Find the signed area between the graph of $f(x)$ and the x-axis on $[0, 3\pi/4]$ shown in the figure next at the left.

(b) Find the area between the graph of $f(x)$ and the graph of $g(x) = -\cos 2x$ on $[0, 3\pi/4]$ shown in the figure next at the right.

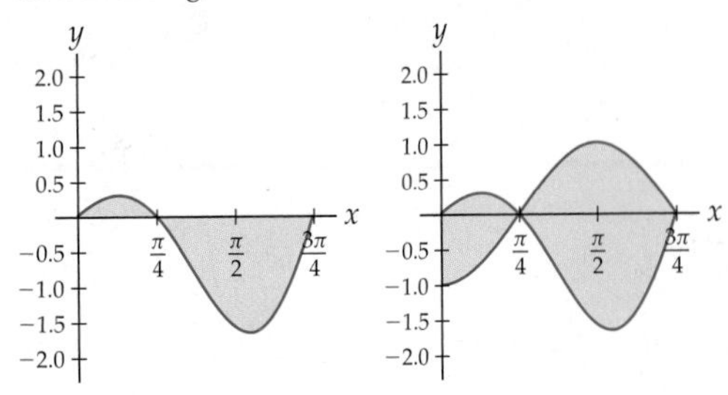

82. Consider the function $f(x) = \ln x$.

(a) Find the signed area between the graph of $f(x)$ and the x-axis on $[1/2, 4]$ shown in the figure next at the left.

(b) Find the area between the graph of $f(x)$ and the graph of $g(x) = 1$ on $[1/2, 4]$ shown in the figure next at the right.

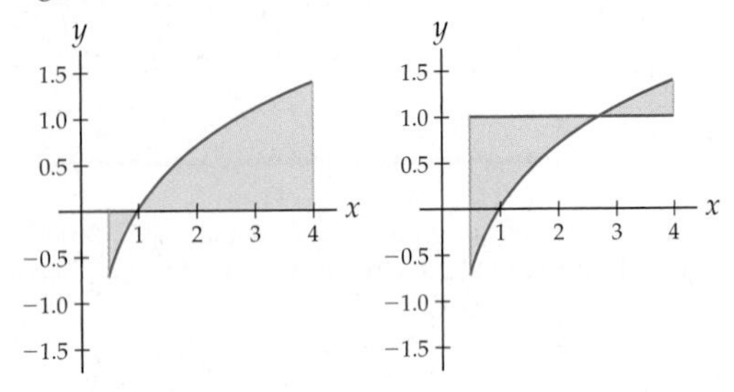

Applications

83. Your local weatherman predicts that there will be a rainstorm this afternoon for three hours and that the rate of rainfall will be given by $r(t) = 0.6t^2\sqrt{3-t}$ for those three hours from $t = 0$ to $t = 3$, as shown in the figure. Use integration by parts (twice) to solve the definite integral that represents the total amount of rainfall predicted for the storm. Compare this exercise with Exercise 83 in Section 5.1, where you used u-substitution to solve the same problem.

Rate of rainfall
$r(t) = 0.6t^2\sqrt{3-t}$

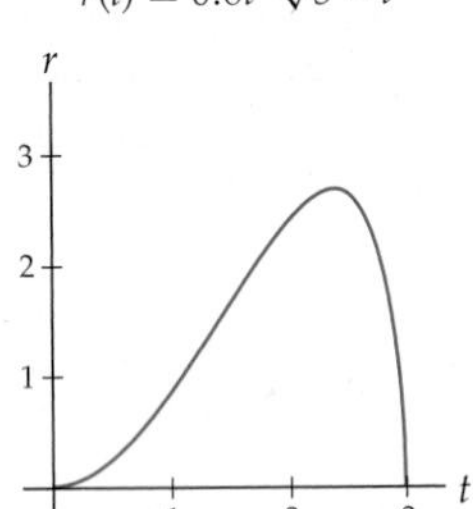

84. Suppose the rate at which a population of 100 weasels catches weasel-mumps is $r(t) = 8 - 2\ln(t+1)$ weasels per day, where $t > 0$ measures the number of days after the initial outbreak. Find a formula that represents the number $W(t)$ of weasels that have contracted weasel-mumps t days after the initial outbreak. Approximately how long will it be until all but one of the weasels are infected?

Rate of weasel-mump infection
$r(t) = 8 - 2\ln(t+1)$

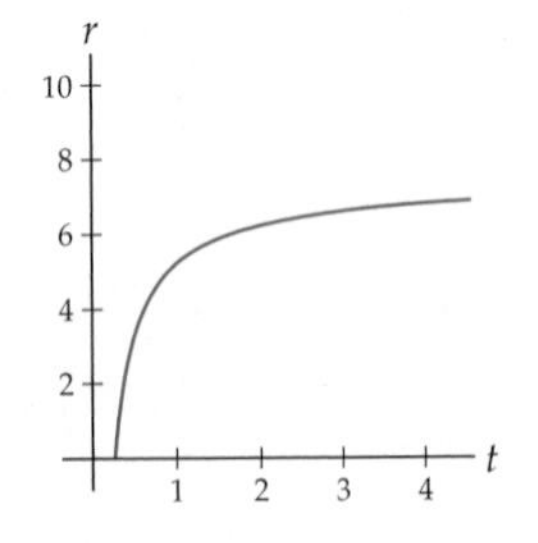

85. At Dr. Geek's house, the wine goblets are shaped exactly like the shape obtained by revolving the graph of $y = \ln x$ from $x = 0.5$ to $x = 5$ around the x-axis, as shown in the figure and measured in inches. Given that the volume of the shape obtained from revolving f around the x-axis on $[a, b]$ can be calculated with the formula $\pi\int_a^b (f(x))^2\,dx$, about how much wine can each glass hold?

Shape of wine goblet

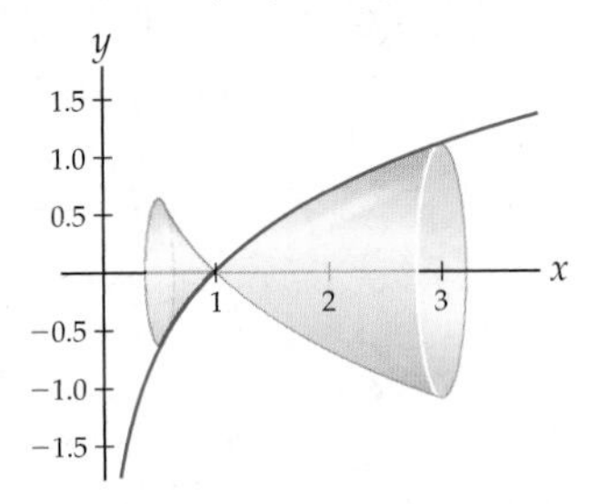

86. Dr. Geek also owns some champagne flutes, which are shaped exactly like the shape obtained by revolving the graph of $y = \dfrac{\ln x}{x}$ from $x = 0.6$ to $x = 5$ around the x-axis, as shown in the figure and measured in inches. Given that the volume of the shape obtained from revolving f around the x-axis on $[a, b]$ can be calculated with the formula $\pi\int_a^b (f(x))^2\,dx$, about how much champagne can each glass hold?

Shape of champagne flute

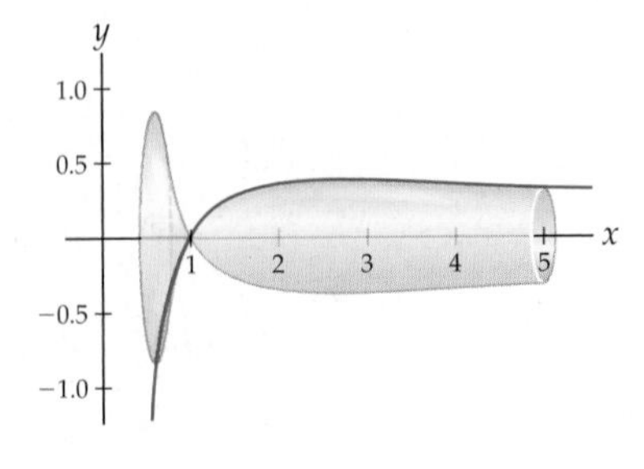

87. Suppose a very odd particle moves back and forth along a straight path in such a way that its velocity after t seconds is given by $v(t) = 5t\sin(\pi t)$, measured in feet per second, as shown in the figure. Consider positions left of the starting position to be in the negative direction and positions right of the starting position to be in the positive direction.

Velocity of odd particle
$v(t) = 5t\sin(\pi t)$

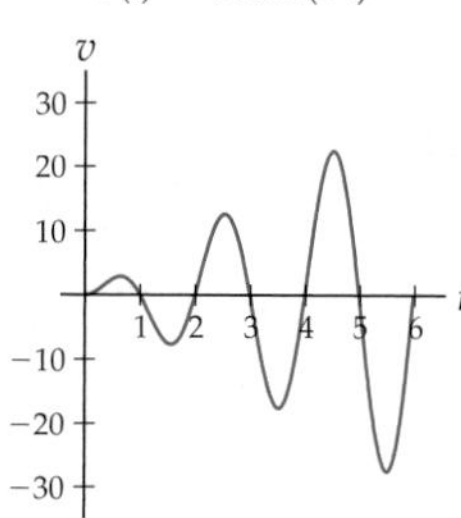

(a) Use integration by parts to find an equation for the position $s(t)$ of the very odd particle, measured in feet left or right of the starting position.

(b) Sketch a graph of $s(t)$, and describe the behavior of the particle over the first 3 seconds.

(c) Where is the particle at $t = 3$ seconds, and what it is doing at that time?

(d) What happens to the position and velocity of the very odd particle after a long time, say, 100 seconds?

Proofs

88. Use the product rule to derive the formula for integration by parts in Theorem 5.7.

89. Prove the integration formula

$$\int \ln x\,dx = x\ln x - x + C$$

(a) by applying integration by parts to $\int \ln x\,dx$;

(b) by differentiating $x\ln x - x$.

90. Prove the integration formula

$$\int \sin^{-1} x\,dx = x\sin^{-1} x + \sqrt{1-x^2} + C$$

(a) by applying integration by parts to $\int \sin^{-1} x\,dx$;

(b) by differentiating $\sqrt{1-x^2} + x\sin^{-1} x$.

91. Prove the integration formula

$$\int \tan^{-1} x\,dx = x\tan^{-1} x - \frac{1}{2}\ln(x^2+1) + C$$

(a) by applying integration by parts to $\int \tan^{-1} x\,dx$;

(b) by differentiating $x\tan^{-1} x - \frac{1}{2}\ln(x^2+1)$.

92. Show that choosing a different antiderivative $v + C$ of dv will yield an equivalent formula for integration by parts, as follows:

(a) Explain why what we need to show is that $uv - \int v\,du = u(v+C) - \int (v+C)\,du$.

(b) Rewrite the equation from part (a), substituting $u = u(x)$, $v = v(x)$, and $du = u'(x)\,dx$.

(c) Prove the equality you wrote down in part (b).

Thinking Forward

Volumes of surfaces of revolution: Consider the region between the graph of $f(x) = \ln x$ and the x-axis on $[1, 3]$ shown next at the left. If we revolve this region around the x-axis, then we get the three-dimensional solid shown at the right. In the problems that follow you will investigate two different methods of finding the volume of this solid.

$f(x) = \ln x$ *on* $[1, 3]$

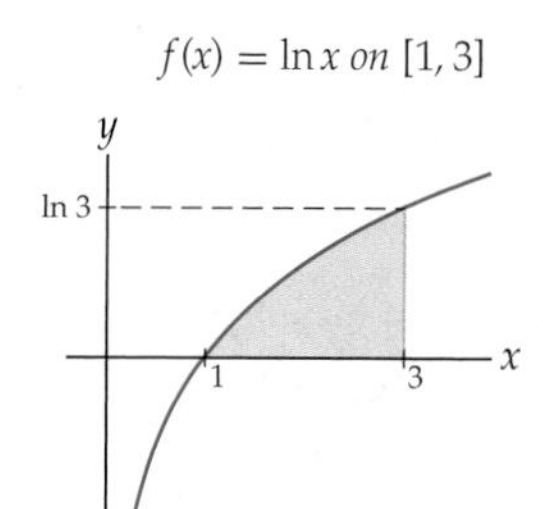

Surface of revolution

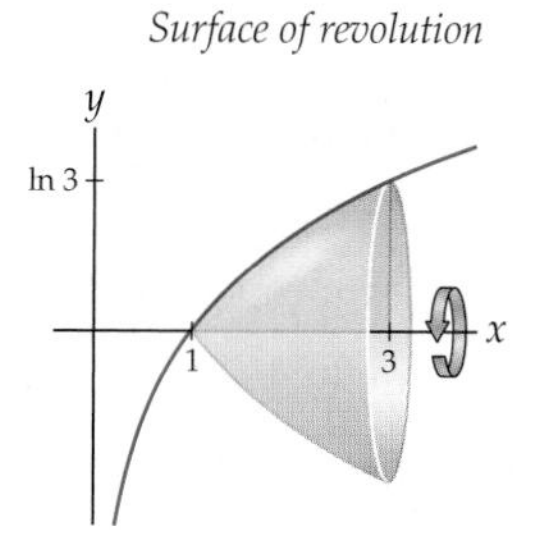

▶ One way to think of this volume is as an accumulation of disks as x varies from 1 to 3, as shown next at the left. The disk at a given $x \in [1, 3]$ has radius $r = f(x)$ and thus area $\pi(f(x))^2$. As we will see in Section 6.1, the definite integral

$$\pi \int_1^3 (f(x))^2\,dx$$

calculates the volume of the solid. Use integration by parts to calculate this volume.

A disk at $x \in [1, 3]$ *A shell at* $y \in [0, \ln 3]$

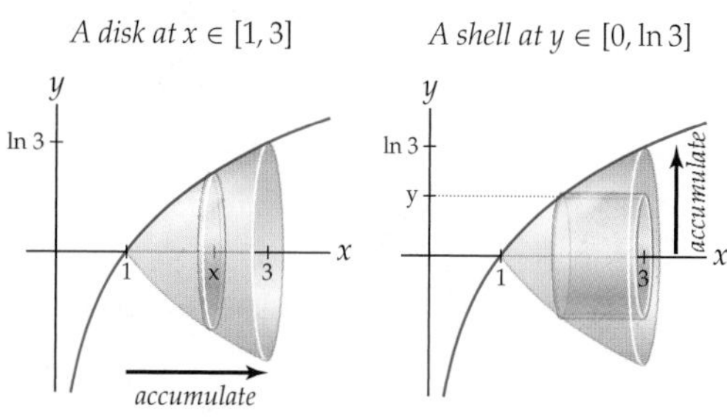

- Another way of thinking of the volume in question is as an accumulation of nested open cylinders, or "shells," as y varies from 0 to ln 3, as shown previously at the right. The shell at a given $y \in [0, \ln 3]$ has radius $r = y$ and length $h = 3 - f^{-1}(y)$, and thus surface area $2\pi y(3 - f^{-1}(y))$, where $f^{-1}(y) = e^y$ here is the inverse function of $f(x) = \ln x$. As we will see in Section 6.2, the definite integral

$$2\pi \int_0^{\ln 3} y\,(3 - f^{-1}(y))\, dy$$

calculates the volume of the same solid shown in the previous problem. Use integration by parts to calculate this volume.

- Since we just calculated the same volume two different ways, obviously we should have obtained the same answer both times. Verify that this is the case for the previous two volume calculations.

Centroid of a shape: Suppose you cut a shape out of a piece of cardboard and attempt to balance the shape on the end of a pencil. The point in the interior at which it balances perfectly is called the ***centroid***. If f is positive and continuous on $[a, b]$, then the centroid of the region between the graph of $f(x)$ and the x-axis on $[a, b]$ is the point $(\bar{x}, \bar{y})$, where

$$\bar{x} = \frac{\int_a^b x f(x)\, dx}{\int_a^b f(x)\, dx} \quad \text{and} \quad \bar{y} = \frac{(1/2)\int_a^b (f(x))^2\, dx}{\int_a^b f(x)\, dx}.$$

- Find the coordinates of the centroid of the region between the graph of $f(x) = x^2$ and the x-axis on $[0, 3]$.
- Find the coordinates of the centroid of the region between $f(x) = \cos x$ and the x-axis on $[-\pi/2, \pi/2]$.

5.3 PARTIAL FRACTIONS AND OTHER ALGEBRAIC TECHNIQUES

- Writing improper rational functions by using polynomial long division
- The method and form of partial fractions
- Rewriting quadratics by the method of completing the square

Proper and Improper Rational Functions

The methods of integration by substitution and integration by parts are useful, but they can't help us solve every integration problem. Sometimes what we need is some good old-fashioned algebra. We have already used algebra extensively as part of our integration toolbox; for example the integral $\int \frac{x^3+1}{x}\, dx$ is easy to solve if we first apply algebra to rewrite it as $\int \left(x^2 + \frac{1}{x}\right) dx$. In this section we will examine some fancier algebraic tools for rewriting and simplifying integrals that involve rational functions.

Let's start by considering the integral

$$\int \frac{3x^3 + x^2 - 14x - 5}{x^2 - 5}\, dx.$$

The integrand is an example of an ***improper*** rational function, meaning that the degree of its numerator is greater than or equal to the degree of its denominator. It turns out that for many integration problems it is easier to work with ***proper*** rational functions, that is, rational functions for which the degree of the numerator is strictly less than the degree of the denominator.

Luckily, in much the same way that an improper fraction like $\frac{3}{2}$ can be written as the sum of an integer and a proper fraction, namely, $1 + \frac{1}{2}$, we can rewrite any improper rational function as the sum of a polynomial and a proper rational function. As we will see in Example 1, we have

$$\int \frac{3x^3 + x^2 - 14x - 5}{x^2 - 5}\, dx = \int \left((3x+1) + \frac{x}{x^2 - 5}\right) dx.$$

This puts the integral into a form that we know how to deal with, since $3x+1$ is easy to integrate and $\frac{x}{x^2-5}$ can be integrated with the simple substitution of $u = x^2-5$, $du = 2x\,dx$.

It is easy to combine fractions to show that $(3x+1) + \frac{x}{x^2-5}$ is equal to $\frac{3x^3+x^2-14x-5}{x^2-5}$. But of course, what we need to be able to do is go the other way and turn an improper rational function into the sum of a polynomial and a proper rational function. The technique for doing this is called ***polynomial long division***, a more general version of the *synthetic division* algorithm that you have seen in previous courses. The polynomial long-division algorithm provides a method to divide any polynomial $p(x)$ by another polynomial $q(x)$. This algorithm always terminates with a remainder $r(x)$ whose degree is strictly less than that of $q(x)$, enabling us to rewrite any improper rational function as the sum of a polynomial and a proper rational function:

THEOREM 5.11 **Rewriting Improper Rational Functions**

Suppose $\frac{p(x)}{q(x)}$ is an improper rational function where $p(x)$ and $q(x)$ are polynomial functions with $\deg(p(x)) = n$ and $\deg(q(x)) = m < n$. Then

$$\frac{p(x)}{q(x)} = s(x) + \frac{r(x)}{q(x)},$$

for some polynomial $s(x)$ of degree $n-m$ and some proper rational function $\frac{r(x)}{q(x)}$ with $\deg(r(x)) < m$.

Partial Fractions

Let's think for a moment about the types of proper rational functions that are easy for us to integrate. Here are some examples of integrals that we can solve:

$$\int \frac{1}{x-3}\,dx, \quad \int \frac{1}{(x-3)^4}\,dx, \quad \int \frac{1}{x^2+1}\,dx, \quad \int \frac{x}{x^2+1}\,dx, \quad \int \frac{x}{(x^2+1)^3}\,dx.$$

The first is an integral involving a logarithm, the second can be done with a simple linear substitution and a power rule, the third concerns the inverse tangent, and the fourth and fifth can be solved after the substitution $u = x^2+1$ and $du = 2x\,dx$. There are many other examples, but in these five we see a pattern: Each integrand shown has a denominator that is some power of a linear or quadratic function.

What about more complicated rational functions? For example, we do not yet have a method that helps us solve the integral

$$\int \frac{8}{(x-1)^2(x^2+3)}\,dx.$$

What would be helpful at this point is if we could use algebra to rewrite the integrand $\frac{8}{(x-1)^2(x^2+3)}$ in terms of proper rational functions that we do know how to integrate, that is, in terms of proper rational functions whose denominators are powers of linear or quadratic functions. Obvious candidates for these denominators in this example are $x-1$, $(x-1)^2$, and x^2+3, since those are factors of the denominator of our integrand.

As you will see in Example 3, we can rewrite our integral as follows:

$$\int \frac{8}{(x-1)^2(x^2+3)}\,dx = \int \left(\frac{-1}{x-1} + \frac{2}{(x-1)^2} + \frac{x-1}{x^2+3} \right) dx.$$

Again note that it is easy to verify that this decomposition works, simply by combining the three terms on the right side to show that they are equal to our original proper rational function on the left. The question is how to find this rewritten form in the first place! We will examine two different ways to do that in Examples 3 and 4. Note that we know how to integrate each of the three parts of the decomposition; the first uses a logarithm, the second is a power rule calculation, and the third can be split into an integral we can integrate by substitution and an integral involving the inverse tangent.

It turns out that this type of rewriting is always possible if we use what are called ***partial fractions***:

THEOREM 5.12

Rewriting Proper Rational Functions with Partial Fractions

Every proper rational function $\frac{p(x)}{q(x)}$ can be written as a sum of ***partial fractions***, as follows:

(a) If $q(x)$ has a linear factor $x - c$ with multiplicity m, then for some constants $A_1, A_2, \ldots, A_m$, the sum will include terms of the form

$$\frac{A_1}{x-c} + \frac{A_2}{(x-c)^2} + \cdots + \frac{A_m}{(x-c)^m}.$$

(b) If $q(x)$ has an irreducible quadratic factor $x^2 + bx + c$ with multiplicity m, then for some constants $B_1, B_2, \ldots, B_m$ and $C_1, C_2, \ldots, C_m$, the sum will include terms of the form

$$\frac{B_1 x + C_1}{x^2+bx+c} + \frac{B_2 x + C_2}{(x^2+bx+c)^2} + \cdots + \frac{B_m x + C_m}{(x^2+bx+c)^m}.$$

Proving that such a decomposition always exists requires some detailed algebra that is not within the scope of this course, so we will omit the proof. Such general theory is not our concern here; we are interested only in finding these partial-fraction decompositions in particular examples. Given a particular proper rational function $\frac{r(x)}{q(x)}$, we need only factor $q(x)$ and then set up systems of equations to solve for the constants A_i or B_i and C_i. Theorem 5.12 guarantees that in each case we will be able to succeed in finding such a decomposition.

Algebraic Techniques

To rewrite an improper rational function as a sum of a polynomial and a proper rational function as in Theorem 5.11 we must know how to divide one polynomial by another. Let's start by reviewing what we do when we apply the long-division algorithm to divide one integer by another. For example, suppose we use long division to divide 131 by 4:

$$\begin{array}{r} 32 \\ 4\,\overline{)131} \\ \underline{12} \\ 11 \\ \underline{8} \\ 3 \end{array}$$

We have shown that 4 goes into 131 thirty-two times, with a remainder of 3; in other words we have shown that $131 = 4(32) + 3$. We can also rewrite this equality by dividing both sides by 4, to obtain $\frac{131}{4} = 32 + \frac{3}{4}$.

The algorithm for long division of polynomials is exactly the same, but with powers of x determining place value instead of powers of 10:

THEOREM 5.13

Polynomial Long Division

To use polynomial long division to divide a polynomial $p(x)$ by a polynomial $q(x)$ of equal or lower degree, we fill in the following diagram as described next:

$$\begin{array}{r} m(x) \\ q(x)\;\overline{)\,p(x)} \\ \vdots \\ R(x) \end{array}$$

- Find an expression ax^k whose product with the leading term of $q(x)$ is equal to the leading term of $p(x)$. Record this expression above the line.
- Multiply $q(x)$ by the expression ax^k you found in the first step, and *subtract* the product from $p(x)$. For the purposes of the next step, call the resulting polynomial the "remainder polynomial."
- Repeat, with $p(x)$ replaced by the remainder polynomial, until the degree of the final remainder polynomial $R(x)$ is less than the degree of $q(x)$.

At the end of the algorithm, we will have found $m(x)$ and $R(x)$ so that

$$p(x) = q(x)\,m(x) + R(x).$$

In other words,

$$\frac{p(x)}{q(x)} = m(x) + \frac{R(x)}{q(x)}.$$

A formal proof that the polynomial long-division algorithm always works is more appropriate for a course in abstract algebra, and we will not include the proof here. (See Example 1 for an example of the implementation of polynomial long division.)

Another technique that can be useful when attempting to integrate rational functions is the method known as ***completing the square***. The goal of this method is to write a quadratic of the form $f(x) = x^2 + bx + c$ in the form $(x-k)^2 + C$ for some numbers k and C; note that this rewritten form has a perfect square in it, from where the method gets its name. The completed-square form is sometimes more suitable for integration problems.

THEOREM 5.14

Completing the Square

Every quadratic function $x^2 + bx + c$ can be rewritten in the form

$$x^2 + bx + c = (x-k)^2 + C,$$

where $k = -\frac{b}{2}$ and $C = c - \frac{b^2}{4}$.

In practice, we do not actually memorize the forms of k and C in this theorem. Instead, when confronted with a quadratic x^2+bx+c, we simply add and subtract the quantity $(b/2)^2$ and then rewrite in the form of a perfect square. For example, the quadratic $x^2 + 6x + 10$ has $b = 6$ and thus $(b/2)^2 = 9$; to complete the square we calculate

$$x^2 + 6x + 10 = (x^2 + 6x + 10) + 9 - 9 = (x^2 + 6x + 9) + (10 - 9) = (x+3)^2 + 1.$$

The proof of Theorem 5.14 is precisely like this calculation; we leave the details to Exercise 63. This particular example is used in an integration problem in Example 5.

You have most likely noticed by now that many integrals need some kind of rewriting before we can antidifferentiate or apply substitution or integration by parts. Deciding what kind of algebra to apply is a skill that can be developed only by doing many practice problems. One thing to keep in mind as you work through such problems is that when we use algebra it is always our goal to rewrite the integrand in a form that is easier to integrate than the original. This means that we don't just apply algebra because we *can;* we do it to satisfy a certain goal. Make sure you have such a goal in mind before you apply any algebra.

Examples and Explorations

EXAMPLE 1

Using polynomial long division to rewrite an improper rational function

Solve the integral

$$\int \frac{3x^3+x^2-14x-5}{x^2-5}\,dx$$

by writing the improper rational function in the integrand as the sum of a polynomial and a proper rational function.

SOLUTION

Following the algorithm for polynomial long division to divide $p(x)=3x^3+x^2-14x-5$ by $q(x)=x^2-5$, we have

$$\begin{array}{r|l} & \quad\quad\quad\quad\quad 3x+1 \\ x^2+0x-5 & 3x^3+\ x^2-14x-5 \\ & \underline{-3x^3+\ 0x^2+15x} \\ & \quad\quad\ \ x^2+\ \ x-5 \\ & \quad\quad\ \underline{-x^2+\ 0x+5} \\ & \quad\quad\quad\quad\quad\ \ x \end{array}$$

We know to stop the algorithm at this point because the degree of the remainder polynomial x is less than the degree of the divisor polynomial $q(x)=x^2+5$. Using the result of this calculation, we can rewrite our integral and then perform the integration steps:

$$\int \frac{3x^3+x^2-14x-5}{x^2-5}\,dx = \int \left((3x+1)+\frac{x}{x^2-5}\right)dx \quad \leftarrow \text{polynomial long division}$$

$$= \frac{3}{2}x^2+x+\frac{1}{2}\ln|x^2-5|+C. \quad \leftarrow \text{substitution } u=x^2-5 \quad \square$$

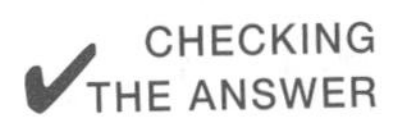
CHECKING THE ANSWER

As usual, we can check our work by differentiating. To verify that we arrive back at the original integrand, we will also have to do some algebra:

$$\frac{d}{dx}\left(\frac{3}{2}x^2+x+\frac{1}{2}\ln|x^2-5|\right) = \left(\frac{3}{2}\right)2x+1+\left(\frac{1}{2}\right)\frac{1}{x^2-5}(2x) \quad \leftarrow \text{differentiate}$$

$$= \frac{(3x+1)(x^2-5)}{x^2-5}+\frac{x}{x^2-5} \quad \leftarrow \text{algebra}$$

$$= \frac{3x^3-15x+x^2-5+x}{x^2-5} \quad \leftarrow \text{algebra}$$

$$= \frac{3x^3+x^2-14x-5}{x^2-5}. \quad \leftarrow \text{algebra}$$

EXAMPLE 2

Determining the form of a partial-fraction decomposition

Write each of the proper rational functions that follow as a sum of partial fractions with coefficients A_i, B_i, and/or C_i. Do not solve for these coefficients; just write out the form of the partial-fraction decomposition.

(a) $\dfrac{2x}{(x-1)(3x^2-2x-1)}$ **(b)** $\dfrac{1}{x^3-6x^2+11x-6}$ **(c)** $\dfrac{x^2-1}{x^7+2x^5+x^3}$

SOLUTION

(a) To find the partial-fraction decomposition we must first split the denominator $(x-1)(3x^2-2x-1)$ into linear and irreducible quadratic factors. Although the denominator is currently expressed as a product of a linear and a quadratic function, the quadratic $3x^2-2x-1$ is not irreducible and in fact can be factored as $(x-1)(3x+1)$. Therefore we have

$$\frac{2x}{(x-1)(3x^2-2x-1)} = \frac{2x}{(x-1)(x-1)(3x+1)} = \frac{2x}{(x-1)^2(3x+1)}.$$

By Theorem 5.12, the right-hand side has partial-fraction decomposition of the form

$$\frac{2x}{(x-1)^2(3x+1)} = \frac{A_1}{x-1} + \frac{A_2}{(x-1)^2} + \frac{A_3}{3x+1}.$$

(b) In this case the factorization of the denominator is not particularly easy, but by guessing and verifying that $x=1$ is a root (since $1^3-6(1)^2+11(1)-6=0$) and then using either synthetic division or polynomial long division, we see that

$$\frac{1}{x^3-6x^2+11x-6} = \frac{1}{(x-1)(x^2-5x+6)} = \frac{1}{(x-1)(x-2)(x-3)}.$$

From the right-hand fraction we obtain the partial-fraction decomposition

$$\frac{1}{(x-1)(x-2)(x-3)} = \frac{A_1}{x-1} + \frac{A_2}{x-2} + \frac{A_3}{x-3}.$$

(c) Again we must begin by factoring the denominator:

$$\frac{x^2-1}{x^7+2x^5+x^3} = \frac{x^2-1}{x^3(x^4+2x^2+1)} = \frac{x^2-1}{x^3(x^2+1)^2}.$$

The fraction to the right of the equals sign can be written as a partial-fraction decomposition of the form

$$\frac{x^2-1}{x^3(x^2+1)^2} = \frac{A_1}{x} + \frac{A_2}{x^2} + \frac{A_3}{x^3} + \frac{B_1x+C_1}{x^2+1} + \frac{B_2x+C_2}{(x^2+1)^2}.$$

□

EXAMPLE 3

Solving for the coefficients in a partial-fractions decomposition

Solve for the coefficients A_1, A_2, B, and C that satisfy the partial-fraction decomposition

$$\frac{8}{(x-1)^2(x^2+3)} = \frac{A_1}{x-1} + \frac{A_2}{(x-1)^2} + \frac{Bx+C}{x^2+3}.$$

SOLUTION

By using a common denominator to combine fractions on the right-hand side of the equation, we have

$$\frac{8}{(x-1)^2(x^2+3)} = \frac{A_1(x-1)(x^2+3) + A_2(x^2+3) + (Bx+C)(x-1)^2}{(x-1)^2(x^2+3)}.$$

After some tedious algebra to expand and then collect terms in the numerator according to the degrees of x, we can show that the fraction on the right is equal to

$$\frac{(A_1+B)x^3+(-A_1+A_2-2B+C)x^2+(3A_1+B-2C)x+(-3A_1+3A_2+C)}{(x-1)^2(x^2+3)}.$$

Since this fraction must be equal to our original proper rational function $\frac{8}{(x-1)^2(x^2-3)}$, its numerator must be equal to 8. This means that the terms in the numerator have to be equal to the terms in $0x^3+0x^2+0x+8$. Equating the corresponding coefficients of these terms gives us a system of equations that we can solve for A_1, A_2, B, and C:

$$\begin{cases} A_1+B=0 \\ -A_1+A_2-2B+C=0 \\ 3A_1+B-2C=0 \\ -3A_1+3A_2+C=8 \end{cases} \implies \begin{cases} A_1=-B \\ A_2=2B \\ C=-B \\ 8B=8 \end{cases} \implies \begin{cases} A_1=-1 \\ A_2=2 \\ C=-1 \\ B=1. \end{cases}$$

Putting these coefficients back into the partial-fractions decomposition, we have

$$\frac{8}{(x-1)^2(x^2+3)} = \frac{-1}{x-1}+\frac{2}{(x-1)^2}+\frac{x-1}{x^2+3.}$$ □

EXAMPLE 4 **Another way to solve for the coefficients in a partial-fractions decomposition**

Solve for the coefficients A_1, A_2, and A_3 that satisfy the partial-fraction decomposition

$$\frac{1}{(x-1)(x-2)(x-3)} = \frac{A_1}{x-1}+\frac{A_2}{x-2}+\frac{A_3}{x-3}.$$

SOLUTION

Again using a common denominator to combine fractions, we have

$$\frac{1}{(x-1)(x-2)(x-3)} = \frac{A_1(x-2)(x-3)+A_2(x-1)(x-3)+A_3(x-1)(x-2)}{(x-1)(x-2)(x-3)}.$$

At this point we could solve for the coefficients A_1, A_2, and A_3 by the same method we used in the previous example: expanding the numerator and collecting terms according to powers of x. However, for partial fractions with many linear factors there is a faster way, which we outline here.

In order for this partial-fractions decomposition to be valid, we must have $A_1(x-2)(x-3)+A_2(x-1)(x-3)+A_3(x-1)(x-2)=1$ for all values of x. By examining this equality for clever values of x, namely, those which make certain factors equal to zero, we can quickly solve for the coefficients A_i. For example, when $x=1$, all of the $x-1$ terms are zero and thus we must have

$$A_1(-1)(-2)+A_2(0)(-2)+A_3(0)(-1)=1 \implies 2A_1=1 \implies A_1=\frac{1}{2}.$$

Now, for $x=2$, we have

$$A_1(0)(-1)+A_2(-1)(1)+A_3(1)(0)=1 \implies -A_2=1 \implies A_2=-1.$$

And finally, for $x=3$, we have

$$A_1(1)(0)+A_2(2)(0)+A_3(2)(1)=1 \implies 2A_3=1 \implies A_3=\frac{1}{2}.$$

Therefore the partial-fraction decomposition is

$$\frac{1}{(x-1)(x-2)(x-3)} = \frac{1/2}{x-1}+\frac{-1}{x-2}+\frac{1/2}{x-3}.$$

Notice that each of the terms in this decomposition is easy to integrate. In fact we can very quickly compute the integral to be $\frac{1}{2}\ln|x-1|-\ln|x-2|+\frac{1}{2}\ln|x-3|+C$. □

✔ CHECKING THE ANSWER

It is easy to check the answers to the previous two examples by using common denominators to combine the terms in the partial-fraction decompositions. For example,

$$\frac{1/2}{x-1} + \frac{-1}{x-2} + \frac{1/2}{x-3} = \frac{(1/2)(x-2)(x-3) - (x-1)(x-3) + (1/2)(x-1)(x-2)}{(x-1)(x-2)(x-3)}$$
$$= \frac{(1/2)(x^2-5x+6) - (x^2-4x+3) + (1/2)(x^2-3x+2)}{(x-1)(x-2)(x-3)}$$
$$= \frac{1}{(x-1)(x-2)(x-3)}.$$

EXAMPLE 5

Combining algebraic techniques to integrate a rational function

Solve the integral $\int \frac{x^3+4x^2-21}{x^2+6x+10}\,dx$.

SOLUTION

Since neither substitution nor integration by parts seems to be a useful approach to solving this integral, and there is no obvious algebraic simplification trick, we will begin by rewriting the improper rational integrand with the use of polynomial long division. After using the polynomial long-division algorithm to divide x^3+4x^2-21 by $x^2+6x+10$, we obtain

$$\int \frac{x^3+4x^2-21}{x^2+6x+10}\,dx = \int \left((x-2) + \frac{2x-1}{x^2+6x+10}\right) dx.$$

Obviously the $x-2$ portion of the integrand is easy to deal with, so let's focus on the proper rational function $\frac{2x-1}{x^2+6x+10}$. Substitution does not seem to be immediately useful here, so we might try partial fractions. However, $x^2+6x+10$ is an irreducible quadratic, since its discriminant $b^2-4ac = 36-4(1)(10) = -4$ is negative; thus, because the denominator does not factor any further, we cannot decompose this rational function into partial fractions.

Let's go back to substitution for a moment. The obvious choice to try is $u = x^2+6x+1$, which gives $du = (2x+6)\,dx$. This du is not exactly what we need for our numerator of $2x-1$, but we can rewrite the numerator by adding and subtracting 6 so that the integral becomes

$$\int (x-2)\,dx + \int \frac{2x+6}{x^2+6x+10}\,dx + \int \frac{-7}{x^2+6x+10}\,dx.$$

Now the first two terms are easy to integrate, so we need only to deal with the third. We can use the method of completing the square to rewrite this third integral in a form whose antiderivative is related to the inverse tangent function. By the example following Theorem 5.14, we have $x^2+6x+10 = (x+3)^2+1$. Therefore the integral expression we are working with is equal to

$$\int (x-2)\,dx + \int \frac{2x+6}{x^2+6x+10}\,dx + \int \frac{-7}{(x+3)^2+1}\,dx.$$

Finally, with the substitution $u = x^2+6x+10$ and $du = (2x+6)\,dx$ in the second integral, and the substitution $w = x+3$ and $dw = dx$ in the third, we can integrate to obtain

$$\frac{1}{2}x^2 - 2x + \ln|x^2+6x+10| - 7\tan^{-1}(x+3) + C.$$

□

CHECKING THE ANSWER

We can check by differentiating the function $\frac{1}{2}x^2 - 2x + \ln|(x+3)^2+1| - 7\tan^{-1}(x+3)$ and algebraically simplifying to get

$$x - 2 + \frac{2(x+3)}{(x+3)^2+1} - \frac{7}{(x+3)^2+1} = \frac{(x-2)((x+3)^2+1) + 2(x+3) - 7}{(x+3)^2+1}$$
$$= \frac{x^3+4x^2-21}{x^2+6x+10}.$$

TEST YOUR UNDERSTANDING

- Is $\frac{x^2+1}{x^2-1}$ a proper rational function or an improper rational function? Why?
- What is the form for the partial-fraction decomposition of the rational function $\frac{x^2-4}{(x-1)^6}$? What about $\frac{x^2-4}{(x^2+x+1)^6}$?
- What are two ways that we could solve for the coefficients in a partial-fractions decomposition?
- Why is the method of partial fractions a useful technique for solving some integrals? What do we hope to gain by rewriting an integrand into the form of a partial-fraction decomposition?
- What does it mean to complete the square of a quadratic expression?

EXERCISES 5.3

Thinking Back

Factoring polynomials: Factor each of the given polynomials into a product of linear and irreducible quadratic factors. For the third and fourth polynomials you will have to guess a root and use synthetic division or other factoring methods.

- $x^3 - 2x^2 - x + 2$
- $x^3 - 1$
- $x^3 - 4x^2 + x + 6$
- $x^3 - x^2 - 5x - 3$

Integration: Calculate the following integrals.

- $\int \frac{1}{(x-1)^2}\,dx$
- $\int \frac{x-1}{x^2-1}\,dx$
- $\int \frac{1}{x^2+1}\,dx$
- $\int \frac{x}{x^2+1}\,dx$
- $\int \frac{x^2+1}{x^3}\,dx$
- $\int \frac{x+1}{x^2+1}\,dx$

Concepts

0. *Problem Zero:* Read the section and make your own summary of the material.

1. *True/False:* Determine whether each of the statements that follow is true or false. If a statement is true, explain why. If a statement is false, provide a counterexample.

(a) *True or False:* $f(x) = \frac{x+1}{x-1}$ is a proper rational function.

(b) *True or False:* Every improper rational function can be expressed as the sum of a polynomial and a proper rational function.

(c) *True or False:* After polynomial long division of $p(x)$ by $q(x)$, the remainder $r(x)$ has a degree strictly less than the degree of $q(x)$.

(d) *True or False:* Polynomial long division can be used to divide two polynomials of the same degree.

(e) *True or False:* If a rational function is improper, then polynomial long division must be applied before using the method of partial fractions.

(f) *True or False:* The partial-fraction decomposition of $\frac{x^2+1}{x^2(x-3)}$ is of the form $\frac{A}{x^2} + \frac{B}{x-3}$.

(g) *True or False:* The partial-fraction decomposition of $\frac{x^2+1}{x^2(x-3)}$ is of the form $\frac{Bx+C}{x^2} + \frac{A}{x-3}$.

(h) *True or False:* Every quadratic function can be written in the form $A(x-k)^2 + C$.

2. *Examples:* Construct examples of the thing(s) described in the following. Try to find examples that are different than any in the reading.

(a) A rational function that is its own partial-fraction decomposition.

(b) A rational function whose partial-fraction decomposition has five terms.

(c) A rational function $\frac{p(x)}{q(x)}$ that is of the form $(3x+1) + \frac{1}{x^2-1}$ after polynomial long division.

3. What is a rational function? What does it mean for a rational function to be proper? Improper?

4. Explain how to use long division to write the improper fraction $\frac{172}{5}$ as the sum of an integer and a proper fraction.

5. Suppose you use polynomial long division to divide $p(x)$ by $q(x)$, and after doing your calculations you end up with the polynomial $x^2 - x + 3$ as the quotient above the top line, and the polynomial $3x - 1$ at the bottom as the remainder. Then $p(x) =$ ________ and $\frac{p(x)}{q(x)} =$ ________ .

6. Describe two ways in which the long-division algorithm for polynomials is similar to the long-division algorithm for integers and then two ways in which the two algorithms are different.

7. Show that the equation $p(x) = q(x)m(x) + R(x)$ is equivalent to the equation $\frac{p(x)}{q(x)} = m(x) + \frac{R(x)}{q(x)}$ for all x for which $q(x)$ is nonzero.

8. Verify that $(x^2 - 1) + \frac{x^2}{x^4+3x^2+4}$ is a valid result for the polynomial long-division calculation performed to divide the polynomial $x^6 + 2x^4 + 2x^2 - 4$ by the polynomial $x^4 + 3x^2 + 4$.

In Exercises 9–11, suppose that you apply polynomial long division to divide a polynomial $p(x)$ by a polynomial $q(x)$ with the goal of obtaining an expression of the form

$$\frac{p(x)}{q(x)} = m(x) + \frac{R(x)}{q(x)}.$$

9. If the degree of $p(x)$ is two greater than the degree of $q(x)$, what can you say about the degrees of $m(x)$ and $R(x)$?

10. If the degree of $p(x)$ is equal to the degree of $q(x)$, what can you say about the degrees of $m(x)$ and $R(x)$?

11. If the degree of $p(x)$ is less than the degree of $q(x)$, what happens? Does polynomial long division work? Why or why not?

In Exercises 12–14, suppose that you want to obtain a partial-fraction decomposition of a rational function $\frac{p(x)}{q(x)}$ according to Theorem 5.12.

12. What happens if the degree of $p(x)$ is greater than or equal to the degree of $q(x)$? Try to perform a partial-fraction decomposition of the improper rational function $\frac{x^3}{(x-1)(x-2)}$, and see what goes wrong.

13. If $q(x)$ is an irreducible quadratic, what can you say about its partial-fraction decomposition? What if $q(x)$ is a reducible quadratic? Consider the functions $q(x) = x^2 + 1$ and $q(x) = (x - 1)(x - 2)$ to find your answer.

14. Why do we need to put linear numerators of the form $B_i x + C_i$ in every term of a partial-fraction decomposition which involves irreducible quadratics? Think about the example $\frac{p(x)}{q(x)} = \frac{1}{(x^2+1)^2}$, and figure out what goes wrong if we attempt to make a decomposition of the form $\frac{C_1}{x^2+1} + \frac{C_2}{(x^2+1)^2}$.

15. Use the method of completing the square and what we know about transformations of graphs to show that the vertex of the parabola $f(x) = x^2 - 3x + 5$ is the point $(3/2, 11/4)$.

16. Theorem 5.14 tells us that if a quadratic $ax^2 + bx + c$ is *monic*, meaning that its leading coefficient is $a = 1$, then we can express that quadratic in the form $(x-k)^2 + C$ with $k = -\frac{b}{2}$ and $C = c - \frac{b^2}{4}$. What happens in the general case for non-monic quadratics? If $ax^2 + bx + c = (Ax - k)^2 + C$, then how can we express A, k and C in terms of a, b, and c?

Skills

Calculate each of the integrals in Exercises 17–46. For some integrals you may need to use polynomial long division, partial fractions, factoring or expanding, or the method of completing the square.

17. $\int \frac{3}{(x-2)(x+1)}\,dx$

18. $\int \frac{6(x+1)}{x(x-3)}\,dx$

19. $\int \frac{x+1}{(x-1)^2}\,dx$

20. $\int \frac{x+1}{(x-1)^3}\,dx$

21. $\int \frac{(x+3)(x-2)}{x^2}\,dx$

22. $\int \frac{x^2}{(x+3)(x-2)}\,dx$

23. $\int \frac{5x^3}{(x+3)(x-2)}\,dx$

24. $\int \frac{25x^2}{(x+3)(x-2)^2}\,dx$

25. $\int \frac{x}{(x^2+3)(x-2)}\,dx$

26. $\int \frac{7(x^2+1)}{(x^2+3)(x-2)}\,dx$

27. $\int \frac{3+3x}{x^3-1}\,dx$

28. $\int \frac{3x^3}{x^3-1}\,dx$

29. $\int \frac{2(x-1)}{x^2-4x+5}\,dx$

30. $\int \frac{2(x-1)}{x^2-2x+5}\,dx$

31. $\int \frac{3x^2-13x+13}{(x-2)^3}\,dx$

32. $\int \frac{x+7}{x^2-x-2}\,dx$

33. $\int \frac{-5x-5}{3x^2-8x-3}\,dx$

34. $\int \frac{x^2+4x+1}{x^3+x^2}\,dx$

35. $\int \frac{x^2+2x+3}{x^4+4x^2+3}\,dx$

36. $\int \frac{x^2+x+1}{(x^2+1)^2}\,dx$

37. $\int \frac{1-2x}{x^2(x^2+1)}\,dx$

38. $\int \frac{16x^2(x^2+1)}{1-2x}\,dx$

39. $\int \frac{2x^2+4x}{x^3+x^2+x+1}\,dx$

40. $\int \frac{x^3}{x^3+x^2+x+1}\,dx$

41. $\int \frac{2x(x^2-1)}{x^2-4}\,dx$

42. $\int \frac{(x-1)(7x-11)}{(x-2)(x-3)(x+1)}\,dx$

43. $\displaystyle\int \frac{2x^3 - 9x^2 + 7x + 6}{2x + 1}\,dx$

44. $\displaystyle\int \frac{(x-1)(x^2 - 2x - 2)}{x - 3}\,dx$

45. $\displaystyle\int \frac{2x^4 - 8x^3 + 9x - 4}{x^2 - 4x + 1}\,dx$

46. $\displaystyle\int \frac{2x^4 - 3x^3 + x^2 - 4x - 5}{x^3 + 2x + 1}\,dx$

Calculate each of the definite integrals in Exercises 47–52. Some integrals require partial fractions or polynomial long division, and some do not.

47. $\displaystyle\int_3^4 \frac{4}{x^2 - 4}\,dx$

48. $\displaystyle\int_{-1}^1 \frac{4(x^2 - 1)}{x^2 - 4}\,dx$

49. $\displaystyle\int_1^3 \frac{1}{x(x+1)}\,dx$

50. $\displaystyle\int_1^3 \frac{1}{x^2(x+1)}\,dx$

51. $\displaystyle\int_{-1}^1 \frac{x^2 - 2x - 4}{x - 3}\,dx$

52. $\displaystyle\int_1^2 \frac{1}{x(x+2)(x-3)}\,dx$

Calculate each of the integrals in Exercises 53–56. Each integral requires substitution or integration by parts as well as the algebraic methods described in this section.

53. $\displaystyle\int \frac{\sin x}{\cos x + \cos^2 x}\,dx$

54. $\displaystyle\int \frac{e^x}{e^{3x} - 2e^{2x}}\,dx$

55. $\displaystyle\int \frac{\ln x}{x(1 + \ln x)^2}\,dx$

56. $\displaystyle\frac{\ln x + 1}{x((\ln x)^2 - 4)}$

57. Consider the function $f(x) = \frac{1}{x^2 - 4}$.

(a) Find the signed area between the graph of $f(x)$ and the x-axis on $[-1, 1]$ shown next at the left.

(b) Find the area between the graph of $f(x)$ and the graph of $g(x) = \frac{1}{4}(x - 1)$ on $[-1, 1]$ shown next at the right.

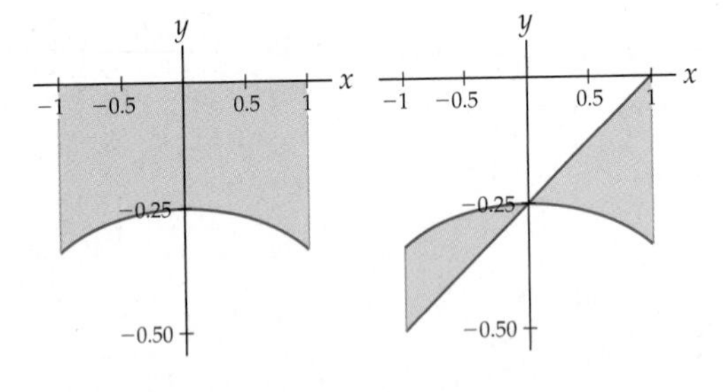

58. Consider the function $f(x) = \frac{2}{x^3 - 1} + 1$.

(a) Find the signed area of the region between the graph of $f(x)$ and the x-axis on $\left[-2, \frac{1}{2}\right]$ shown in the figure.

(b) Find the absolute area of the same region.

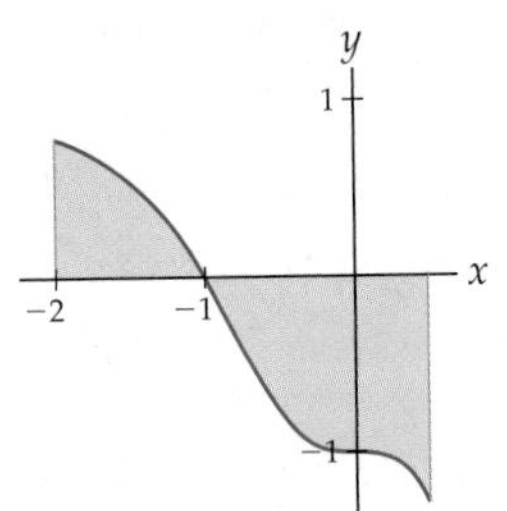

59. Consider the function $f(x) = \frac{2x^3 + x}{(x^2 + 1)^2}$.

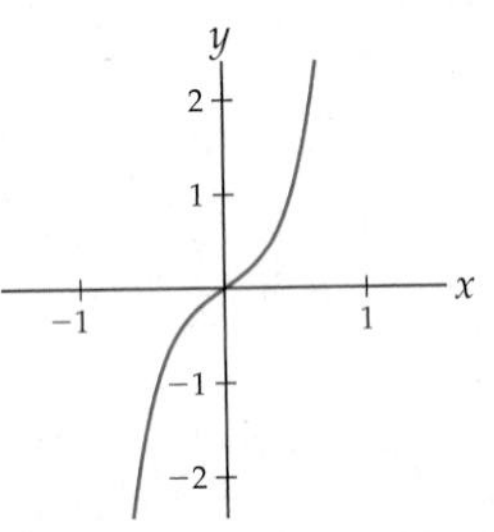

(a) Find the average value of $f(x)$ on $[0, 2]$.

(b) Graphically approximate a value of x for which $f(x)$ is equal to its average value on $[0, 2]$, and use this value to verify that your answer from part (a) is reasonable.

60. Consider the function $f(x) = \frac{16}{(x+1)(x-3)^2}$.

(a) Find the area between the graphs of $f(x)$ and $g(x) = 4 - 2x$ on $[0, 2]$ shown next at the left.

(b) Find the area between the graphs of $f(x)$ and $h(x) = \frac{4}{x+1}$ on $[0, 2]$ shown next at the right.

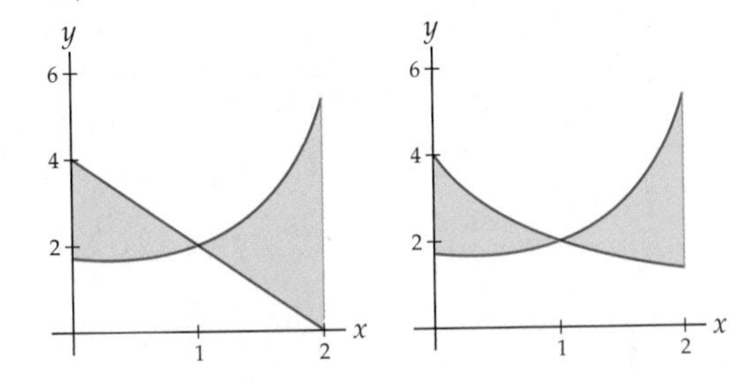

Applications

61. In her work as a population biologist for Idaho Fish and Game, Leila is assigned the task of modeling a population $p(t)$ of prairie dogs that ranchers in the eastern part of the state are complaining about. She uses the so-called ***logistic growth model***, which quickly results in an expression of the form

$$r(t) = \int_{p_0}^{p(t)} \frac{1}{q\left(1 - \frac{q}{K}\right)}\,dq,$$

where K represents the carrying capacity of the land where the prairie dogs live, r is the rate of reproduction, and p_0 is the initial population of the prairie dogs. Use the Fundamental Theorem of Calculus to calculate $r(t)$ and then use your answer to solve for $p(t)$ in terms of $r(t)$, assuming that p_0 and $p(t)$ both lie between 0 and K.

62. Leila's model for the prairie dogs in the previous exercise was so successful that she was invited to assist a University of Idaho professor in modeling sage grouse populations. In contrast to prairie dogs, sage grouse show evidence of an inverse density dependence: If the number g of grouse falls too low, then the population is no longer sustainable and dies off. This relationship can be represented in a model that satisfies

$$-rt = \int_{g(0)}^{g(t)} \frac{1}{y(1-y/K)(1-y/m)}\,dy,$$

where $r > 0$, $0 < m < K$, and $g(t)$ represents the population of grouse at time t.

(a) Evaluate the integral in the case that t satisfies $0 < g(t) < g(0) < m$.

(b) It is difficult to solve for $g(t)$, but given that t satisfies $0 < g(t) < g(0) < m < K$, differentiate both sides of the equation with respect to t and use the Fundamental Theorem of Calculus in solving for $g'(t)$, to show that $\lim_{t\to\infty} g(t) = 0$. You may assume that all of the limits involved exist. This means that if the grouse are hunted to a population lower than m, then the grouse eventually all die off.

Proofs

63. Prove that every quadratic function of the form $f(x) = x^2 + bx + c$ can be written in the form $f(x) = (x-k)^2 + C$, with $k = -\frac{b}{2}$ and $C = c - \frac{b^2}{4}$.

64. Use the method of completing the square and what we know about transformations of graphs to prove that for all real numbers b and c, the vertex of the parabola with equation $f(x) = x^2 + bx + c$ is located at the point $\left(-\frac{b}{2}, c - \frac{b^2}{4}\right)$.

Thinking Forward

Separable differential equations: Suppose a population $P = P(t)$ grows in such a way that its rate of growth $\frac{dP}{dt}$ obeys the equation

$$\frac{dP}{dt} = P(100 - P).$$

This is called a ***differential equation*** because it is an equation that involves a derivative. In the series of steps that follow, you will find a function $P(t)$ that behaves according to this differential equation.

- As we will see in Section 6.5, a differential equation of this form is ***separable*** to the following equivalent equation involving integrals:

$$\int \frac{1}{P(100-P)}\,dP = \int 1\,dt.$$

- Use partial fractions to show that the integral on the left side of this equation is equal to

$$\frac{1}{100}(\ln|P| - \ln|100 - P|) + C_1.$$

- Use properties of logarithms to simplify the answer you found in the previous step.
- Set the answer from the previous step equal to $\int 1\,dt = t + C_2$, and solve for $P = P(t)$. Along the way you can combine unknown constants into new constants; for example, if you encounter $C_2 - C_1$, then you could just rename that constant C and proceed from there. At the end of your calculations you should have

$$P(t) = \frac{100Ae^{100t}}{1 + Ae^{100t}}$$

for some constant A.

- What happens to the function $P(t)$ as $t \to \infty$? What does that mean about this particular population?

5.4 TRIGONOMETRIC INTEGRALS

- Using Pythagorean identities to set up trigonometric integrals for simple u-substitutions
- Using trigonometric identities to simplify integrals involving powers of trigonometric functions
- The integrals of secant and cosecant

Using Pythagorean Identities to Set Up a Substitution

Integration by substitution is a powerful tool, but as we have seen, it can be difficult to recognize an integrand as being in the form $f'(u(x))u'(x)$. If we are lucky, we can use algebra to rewrite an integrand to set it up for a particular substitution. In the case of trigonometric integrands, this algebra often involves the Pythagorean identities, which we repeat here for convenience:

THEOREM 5.15 **Pythagorean Identities**

For all values of x at which the following equations are defined,

(a) $\sin^2 x + \cos^2 x = 1$ **(b)** $\tan^2 x + 1 = \sec^2 x$ **(c)** $1 + \cot^2 x = \csc^2 x$

It will help you remember these formulas if you notice that the second identity follows from the first by dividing both sides by $\cos^2 x$ and that the third identity follows from the first by dividing both sides by $\sin^2 x$.

With the Pythagorean identities we can freely write even powers of $\sin x$, $\tan x$, and $\cot x$ in terms of even powers of $\cos x$, $\sec x$, and $\csc x$, and vice versa. For example, using the first Pythagorean identity, we can write $\cos^4 x$ in terms of sine as

$$\cos^4 x = (\cos^2 x)^2 = (1 - \sin^2 x)^2.$$

With this equation in mind, consider the integral $\int \cos^5 x\,dx$. As it is written, we do not know how to solve the integral. However, we can convert four of the cosines to sines and have one cosine left over:

$$\int \cos^5 x\,dx = \int \cos^4 x\,\cos x\,dx = \int (\cos^2 x)^2\,\cos x\,dx = \int (1 - \sin^2 x)^2\,\cos x\,dx.$$

In this new form we have an integrand that can be easily solved with u-substitution with $u = \sin x$ and $du = \cos x\,dx$:

$$\int (1 - \sin^2 x)^2\,\cos x\,dx = \int (1 - u^2)^2\,du = \int (1 - 2u^2 + u^4)\,du$$
$$= u - \frac{2}{3}u^3 + \frac{1}{5}u^5 + C = \sin x - \frac{2}{3}\sin^3 x + \frac{1}{5}\sin^5 x + C.$$

In general, we would like to rewrite trigonometric integrands so that they are expressed in terms of a single trigonometric function multiplied by the derivative of that trigonometric function. Whether or not we can do this for a particular integral depends on the trigonometric functions and powers involved. For example, we can use the method for some integrals of the form $\int \sin^m x\,\cos^n x\,dx$ when either m or n is odd, since in that case we can convert all but one sine or cosine into the other function and apply a substitution. For instance, we could set up $\int \sin^8 x\,\cos^3 x\,dx$ for substitution with $u = \sin x$ by writing

$$\int \sin^8 x\,\cos^3 x\,dx = \int \underbrace{\sin^8 x(1 - \sin^2 x)}\,\cos x\,dx.$$

We could now perform the substitution $u = \sin x$, which would change the bracketed portion of the integral to simply $u^8(1 - u^2)$, which is easy to multiply out and integrate.

Similarly, we can use this method for some integrals of the form $\int \sec^m x \tan^n x\,dx$. When m is even, all but two of the secants can be converted into tangents. When n is odd, we can save a $\sec x \tan x$ and rewrite the rest of the integrand in terms of the secant. Both of these conditions hold for the integral $\int \sec^6 x \tan^5 x\,dx$, so we can rewrite this integral in two ways:

$$\int \sec^6 x \tan^5 x\,dx = \int \underbrace{(\tan^2 x + 1)^2 \tan^5 x}\, \sec^2 x\,dx,$$

$$\int \sec^6 x \tan^5 x\,dx = \int \underbrace{\sec^5 x(\sec^2 x - 1)^2}\, \sec x \tan x\,dx.$$

In both cases the integral is set up for a simple substitution, with $u = \tan x$ in the first case and $u = \sec x$ in the second case. The bracketed portions of the integrands become functions of u that are easy to antidifferentiate.

The table that follows summarizes some common situations in which we can use Pythagorean identities to set up a simple substitution in a trigonometric integral. In this table it is assumed that m and n are positive integers.

Form	Rewritten Form	Choose
$\int \sin^n x\,dx$, n odd	$\int$ (expression in $\cos x$) $\sin x\,dx$	$u = \cos x$, $du = -\sin x\,dx$
$\int \sin^m x \cos^n x\,dx$, n odd	$\int$ (expression in $\sin x$) $\cos x\,dx$	$u = \sin x$, $du = \cos x\,dx$
$\int \sin^m x \cos^n x\,dx$, m odd	$\int$ (expression in $\cos x$) $\sin x\,dx$	$u = \cos x$, $du = -\sin x\,dx$
$\int \sec^m x \tan^n x\,dx$, m even	$\int$ (expression in $\tan x$) $\sec^2 x\,dx$	$u = \tan x$, $du = \sec^2 x\,dx$
$\int \sec^m x \tan^n x\,dx$, n odd	$\int$ (expression in $\sec x$) $\sec x \tan x\,dx$	$u = \sec x$, $du = \sec x \tan x\,dx$

Similar rows could be made for integrals that involve powers of cosine or products of powers of cosecant and cotangent.

You do not need to memorize the preceding table! Instead, notice that in each case we rewrite the original integral in such a way as to save part of the integrand for the du that corresponds to our desired choice of u-substitution. Whether or not we can do this depends whether the exponents involved in the original integral are even or odd. Integrals that are not in one of the forms in the table will require different solution techniques that are the focus of the remainder of this section.

The Pythagorean identities can also be applied in less obvious cases. For example, we can effectively reduce the integral of a power of a tangent by converting two copies of the tangent at a time. As an illustration of this procedure, consider the integral $\int \tan^5 x\,dx$. We can write

$$\tan^5 x = \tan^3 x \tan^2 x = \tan^3 x (\sec^2 x - 1) = \tan^3 x \sec^2 x - \tan^3 x.$$

Notice that we have written $\tan^5 x$ as the sum of a function we can integrate with the techniques discussed earlier and a function that is a lower power of the tangent. Repeating this kind of reduction will eventually solve this integral, as you will see in Example 4.

Using Double-Angle Formulas to Reduce Powers

Not all integrals involving trigonometric functions can be solved by using Pythagorean identities and u-substitution. For example, that technique would not work for the integrals $\int \sin^6 x\,dx$, $\int \sin^2 x \cos^2 x\,dx$, and $\int \tan^5 x\,dx$, because of the parity of the exponents involved, that is, whether certain exponents are even or odd. The first two of these integrals can be simplified with the use of the double-angle identities, which we repeat here:

THEOREM 5.16 **Double-Angle Identities**

For all real numbers x,

(a) $\sin^2 x = \frac{1}{2}(1 - \cos 2x)$ **(b)** $\cos^2 x = \frac{1}{2}(1 + \cos 2x)$

For example, we can write the integral $\int \sin^2 x\, dx$ as

$$\int \sin^2 x\, dx = \frac{1}{2}\int (1 - \cos 2x)\, dx = \frac{1}{2}\left(x - \frac{1}{2}\sin 2x\right) + C.$$

For higher even powers of sine or cosine we apply double-angle identities repeatedly, reducing the power by half each time. For example, since $\sin^4 x = (\sin^2 x)^2$, we have

$$\sin^4 x = \left(\frac{1}{2}(1 - \cos 2x)\right)^2 = \frac{1}{4}(1 - 2\cos 2x + \cos^2 2x) = \frac{1}{4}\left(1 - 2\cos 2x + \frac{1}{2}(1 + \cos 4x)\right).$$

Although it seems like we made the expression $\sin^4 x$ much more complicated, in fact the messy expression on the right-hand side of the equation is easy to integrate, piece by piece.

Integrating Secants and Cosecants

Believe it or not, at this point we still do not have formulas for antidifferentiating two of the six basic trigonometric functions. To finish our investigation of trigonometric integrals, we give formulas for these integrals:

THEOREM 5.17 **Integrals of Secant and Cosecant**

(a) $\displaystyle\int \sec x\, dx = \ln|\sec x + \tan x| + C$ **(b)** $\displaystyle\int \csc x\, dx = -\ln|\csc x + \cot x| + C$

Proof. We prove part (a) and leave the proof of part (b) to Exercise 83. Neither integration by substitution nor integration by parts can immediately improve the integral $\int \sec x\, dx$. However, there is a particular technique that happens to apply to this integral. If we multiply the integrand by a certain very special form of 1, we will have an integral that can be evaluated by substitution:

$$\int \sec x\, dx = \int \sec x\left(\frac{\sec x + \tan x}{\sec x + \tan x}\right) dx = \int \frac{\sec^2 x + \sec x \tan x}{\sec x + \tan x}\, dx.$$

It was not at all obvious why we multiplied by $\frac{\sec x + \tan x}{\sec x + \tan x}$ just now. However, now that we have done so, the following choice of substitution works out surprisingly well:

$$\boxed{u = \sec x + \tan x} \implies \frac{du}{dx} = \sec x \tan x + \sec^2 x \implies \boxed{du = (\sec x \tan x + \sec^2 x)\, dx}$$

With this change of variables, the integral becomes

$$\int \sec x\, dx = \int \frac{1}{u}\, du = \ln|u| + C = \ln|\sec x + \tan x| + C.$$

■

What about integrating powers of $\sec x$ and $\csc x$? Of course, it is easy to integrate $\sec^2 x$, since it is the derivative of $\tan x$. This fact can be exploited when integrating higher powers of $\sec x$, since we will be able to choose $dv = \sec^2 x\, dx$ and integrate it to get $v = \tan x$. For example, consider the integral $\int \sec^4 x\, dx$, with parts chosen as follows:

$$u = \sec^2 x \quad \longrightarrow \quad du = 2\sec x\,(\sec x \tan x)\, dx$$

$$v = \tan x \quad \longleftarrow \quad dv = \sec^2 x\, dx.$$

Applying the formula for integration by parts and simplifying a bit, we have

$$\int \sec^4 x\,dx = \sec^2 x \tan x - 2\int \tan^2 x \sec^2 x\,dx.$$

The new integral on the right can be solved by substitution, with $u = \tan x$ and $du = \sec^2 x\,dx$. This method of choosing parts with $dv = \sec^2 x\,dx$ is effective for simplifying any integrand that is a power of a secant, since it reduces the power each time the method is applied.

Examples and Explorations

EXAMPLE 1

Setting up substitutions with Pythagorean identities

Use algebra and u-substitution to find each of the following integrals:

(a) $\int \sin^3 x \cos^4 x\,dx$ **(b)** $\int \csc^4 x \cot^3 x\,dx$ **(c)** $\int \sec^3 x \tan^3 x\,dx$

SOLUTION

(a) Using the Pythagorean identity $\sin^2 x = 1 - \cos^2 x$, we can write the integrand so that it involves only cosines (the bracketed portion) except for a single copy of $\sin x$:

$$\int \sin^3 x \cos^4 x\,dx = \int \sin^2 x \sin x \cos^4 x\,dx = \int \underbrace{(1-\cos^2 x)\cos^4 x}\,\sin x\,dx$$

Now with the substitution $u = \cos x$ we have $-du = \sin x\,dx$, and our integral becomes

$$-\int (1-u^2)\,u^4\,du = -\int (u^4 - u^6)\,du = -\frac{1}{5}\cos^5 x + \frac{1}{7}\cos^7 x + C.$$

Notice that this method worked only because the original power of $\sin x$ was odd. If both powers had been even, we would have needed to use another method, as we will in Example 2.

(b) We will set up the integral so that we can apply the substitution $u = \cot x$. This means that we must write the integral in terms of cotangents (the bracketed portion), with one copy of $\csc^2 x$ saved for the differential du:

$$\int \csc^4 x \cot^3 x\,dx = \int \csc^2 x \csc^2 x \cot^3 x\,dx = \int \underbrace{(1+\cot^2 x)\cot^3 x}\,\csc^2 x\,dx$$

Now with $u = \cot x$ and thus $-du = \csc^2 x\,dx$, substitution gives us the new integral

$$-\int (u^2+1)\,u^3\,du = -\int (u^5 + u^3)\,du = -\frac{1}{6}\cot^6 x - \frac{1}{4}\cot^4 x + C.$$

This method worked only because the original power of $\csc x$ was even, which allowed us to apply a Pythagorean identity to convert all but two of the cosecants into cotangents.

(c) This time the power of secant is odd, so the method used in part (b) will not work. Instead, we will convert everything to secants (the bracketed portion), with one copy of $\sec x \tan x$ saved for the differential:

$$\int \sec^3 x \tan^3 x\,dx = \int \sec^2 x \sec x \tan^2 x \tan x\,dx = \int \underbrace{\sec^2 x(\sec^2 x - 1)}\,\sec x \tan x\,dx.$$

Now with the substitution $u = \sec x$ and thus $du = \sec x \tan x\,dx$, our integral becomes

$$\int u^2(u^2-1)\,du = \int (u^4 - u^2)\,du = \frac{1}{5}\sec^5 x - \frac{1}{3}\sec^3 x + C.$$

The method we just applied works only if the power of secant (respectively, cosecant) is odd. In part (b) we saw a method that works only if the power of tangent (respectively, cotangent) is even. If neither of these conditions holds, then we can apply a reduction method; see Example 4. □

EXAMPLE 2

Reducing powers with double-angle identities

Use double-angle identities to find each of the following integrals:

(a) $\displaystyle\int \cos^6 x\,dx$ **(b)** $\displaystyle\int \sin^2 x\cos^2 x\,dx$

SOLUTION

(a) Notice that we cannot use the method of Pythagorean identities as in the previous example, because the power of cosine is even. Instead, we apply the double-angle identity for cosine to rewrite $\cos^6 x$ as

$$\cos^6 x = (\cos^2 x)^3 = \left(\frac{1}{2}(1+\cos 2x)\right)^3 = \frac{1}{8} + \frac{3}{8}\cos 2x + \frac{3}{8}\cos^2 2x + \frac{1}{8}\cos^3 2x.$$

We have now traded the problem of integrating $\cos^6 x$ for the problem of integrating four simpler functions: $\frac{1}{8}$ and $\frac{3}{8}\cos 2x$, which are easy to antidifferentiate; $\frac{3}{8}\cos^2 2x$, which requires another application of the double-angle formula; and $\frac{1}{8}\cos^3 2x$, which can be solved with a Pythagorean identity and the substitution $u = \sin 2x$, $du = 2\cos 2x\,dx$. We have

$$\begin{aligned}\int \cos^6 x\,dx &= \int \frac{1}{8}\,dx + \int \frac{3}{8}\cos 2x\,dx + \int \frac{3}{8}\cos^2 2x\,dx + \int \frac{1}{8}\cos^3 2x\,dx \\ &= \frac{1}{8}x + \frac{3}{8}\left(\frac{1}{2}\sin 2x\right) + \frac{3}{8}\int \frac{1}{2}(1+\cos 4x)\,dx + \frac{1}{8}\int (1-\sin^2 2x)\cos 2x\,dx \\ &= \frac{1}{8}x + \frac{3}{8}\left(\frac{1}{2}\sin 2x\right) + \frac{3}{16}\left(x + \frac{1}{4}\sin 4x\right) + \frac{1}{8}\left(\frac{1}{2}\right)\int (1-u^2)\,du \\ &= \frac{1}{8}x + \frac{3}{8}\left(\frac{1}{2}\sin 2x\right) + \frac{3}{16}\left(x + \frac{1}{4}\sin 4x\right) + \frac{1}{16}\left(u - \frac{1}{3}u^3\right) + C \\ &= \frac{1}{8}x + \frac{3}{8}\left(\frac{1}{2}\sin 2x\right) + \frac{3}{16}\left(x + \frac{1}{4}\sin 4x\right) + \frac{1}{16}\left(\sin 2x - \frac{1}{3}\sin^3 2x\right) + C.\end{aligned}$$

(b) One way to simplify this integral is to apply double-angle identities to write both $\sin^2 x$ and $\cos^2 x$ in terms of $\cos 2x$ and then apply a double-angle identity *again* to reduce powers even further:

$$\begin{aligned}\int \sin^2 x\cos^2 x\,dx &= \int \left(\frac{1}{2}(1-\cos 2x)\right)\left(\frac{1}{2}(1+\cos 2x)\right)dx && \leftarrow \text{double-angle identities} \\ &= \frac{1}{4}\int (1-\cos^2 2x)\,dx && \leftarrow \text{simplify} \\ &= \frac{1}{4}\int \left(1 - \frac{1}{2}(1+\cos 4x)\right)dx && \leftarrow \text{double-angle identity} \\ &= \frac{1}{4}\int \left(\frac{1}{2} - \frac{1}{2}\cos 4x\right)dx && \leftarrow \text{simplify} \\ &= \frac{1}{4}\left(\frac{1}{2}x - \frac{1}{8}\sin 4x\right) + C. && \leftarrow \text{antidifferentiate}\end{aligned}$$

□

EXAMPLE 3 **Using integration by parts to integrate a power of the secant**

Find $\int \sec^3 x\,dx$.

SOLUTION

In this case, Pythagorean identities would not be helpful. However, since we know how to integrate $\sec^2 x$, we can apply integration by parts with $dv = \sec^2 x\,dx$:

$$u = \sec x \quad \longrightarrow \quad du = \sec x \tan x\,dx$$
$$v = \tan x \quad \longleftarrow \quad dv = \sec^2 x\,dx.$$

Applying the integration-by-parts formula and then a Pythagorean identity, we have

$$\begin{aligned}
\int \sec^3 x\,dx &= \sec x \tan x - \int \sec x \tan^2 x\,dx. && \leftarrow \text{integration by parts} \\
&= \sec x \tan x - \int \sec x(\sec^2 x - 1)\,dx && \leftarrow \text{Pythagorean identity} \\
&= \sec x \tan x - \int \sec^3 x\,dx + \int \sec x\,dx && \leftarrow \text{expand} \\
&= \sec x \tan x - \int \sec^3 x\,dx + \ln|\sec x + \tan x| && \leftarrow \text{Theorem 5.17}
\end{aligned}$$

We are now in a fortunate situation, since all we have to do is solve for $\int \sec^3 x\,dx$ in the equation $\int \sec^3 x\,dx = \sec x \tan x - \int \sec^3 x\,dx + \ln|\sec x + \tan x|$. Doing so gives us

$$\int \sec^3 x\,dx = \frac{1}{2}(\sec x \tan x + \ln|\sec x + \tan x|) + C.$$

□

EXAMPLE 4 **Using Pythagorean identities to reduce to simpler cases and smaller powers**

Find each of the following integrals:

(a) $\int \tan^5 x\,dx$ **(b)** $\int \sec x \tan^2 x\,dx$

SOLUTION

(a) The key here is to repeatedly apply a Pythagorean identity to decrease the powers of the tangent until we have a sum of expressions that we can integrate. We can write

$$\begin{aligned}
\tan^5 x &= \tan^3 x\,(\sec^2 x - 1) && \leftarrow \text{Pythagorean identity} \\
&= \tan^3 x\,\sec^2 x - \tan^3 x && \leftarrow \text{expand} \\
&= \tan^3 x\,\sec^2 x - \tan x\,(\sec^2 x - 1) && \leftarrow \text{Pythagorean identity} \\
&= \tan^3 x\,\sec^2 x - \tan x\,\sec^2 x + \tan x && \leftarrow \text{expand}
\end{aligned}$$

The first two expressions can be integrated with the substitution $u = \tan x$ and $du = \sec^2 x\,dx$, and the third is an expression we already know how to antidifferentiate, as follows:

$$\begin{aligned}
\int \tan^5 x\,dx &= \int u^3\,du - \int u^2\,du + \int \tan x\,dx \\
&= \frac{1}{4}u^4 - \frac{1}{3}u^3 + \ln|\cos x| + C = \frac{1}{4}\tan^4 x - \frac{1}{3}\tan^3 x + \ln|\cos x| + C.
\end{aligned}$$

(b) Again we will apply Pythagorean identities to reduce the powers of the tangents to expressions that we know how to integrate. We have

$$\sec x \tan^2 x = \sec x(\sec^2 x - 1) = \sec^3 x - \sec x.$$

We saw how to integrate $\sec^3 x$ in Example 3 and how to integrate $\sec x$ in Theorem 5.17. Putting these results together, we have

$$\int \sec x \tan^2 x \, dx = \frac{1}{2} \sec x \tan x - \frac{1}{2} \ln|\sec x + \tan x| + C.$$ □

EXAMPLE 5

Hyperbolic integrals*

Adapt the techniques of this section to solve the integral $\int \sinh^2 x \cosh^3 x \, dx$.

SOLUTION

Since the hyperbolic functions $\sinh x$, $\cosh x$, and $\tanh x$ behave so much like their trigonometric counterparts, the techniques of this section can also be used to solve integrals involving products and powers of hyperbolic functions. Recall from Section 2.6 that $(\sinh t, \cosh t)$ is a point on the hyperbola $x^2 - y^2 = 1$ and thus that we have the identity

$$\cosh^2 t - \sinh^2 t = 1.$$

We can use this identity to set up a convenient substitution:

$$\int \sinh^2 x \cosh^3 x \, dx = \int \underbrace{\sinh^2 x(1 + \sinh^2 x)} \cosh x \, dx.$$

Now with the substitution $u = \sinh x$ and $du = \cosh x$, our integral becomes

$$\int u^2(1 + u^2) \, du = \int (u^2 + u^4) \, du = \frac{1}{3}u^3 + \frac{1}{5}u^5 + C = \frac{1}{3} \sinh^3 x + \frac{1}{5} \sinh^5 x + C.$$ □

? TEST YOUR UNDERSTANDING

- Why can't we use Pythagorean identities and u-substitution to solve the integral $\int \sin^4 x \cos^4 x \, dx$? What method could we use instead?
- In Example 1(b) we used a Pythagorean identity followed by substitution with $u = \cot x$ to solve the integral $\int \csc^4 x \cot^3 x \, dx$. Can you find another way to solve this integral?
- When we integrated $\int \sec^3 x \, dx$ in Example 3, we used integration by parts and identities to return to the original integral. Would the same thing happen if the power of the secant were even? Why or why not?
- The technique used in Example 4 also works if the power of the tangent is even. Use this technique to find $\int \tan^6 x \, dx$. What is different about the last steps of the computation when k is even?
- In solving trigonometric integrals by the techniques in this section, it is often difficult to check answers by differentiating. Why?

EXERCISES 5.4

Thinking Back

Pythagorean identities: Prove each of the following Pythagorean identities in the manner specified.

- ▶ Prove that $\sin^2 x + \cos^2 x = 1$ for acute angles x, using the right-triangle definitions of sine and cosine. Why is this called a *Pythagorean* identity?
- ▶ Prove that $\sin^2 x + \cos^2 x = 1$ for any angle x, using the unit-circle definitions of sine and cosine.
- ▶ Use the fact that $\sin^2 x + \cos^2 x = 1$ to prove that $\tan^2 x + 1 = \sec^2 x$.
- ▶ Use the fact that $\sin^2 x + \cos^2 x = 1$ to prove that $1 + \cot^2 x = \csc^2 x$.

Double-angle identities: Prove each of the following double-angle identities by applying the sum identity for the cosine followed by a Pythagorean identity.

- ▶ $\sin^2 x = \frac{1}{2}(1 - \cos 2x)$
- ▶ $\cos^2 x = \frac{1}{2}(1 + \cos 2x)$

Concepts

0. *Problem Zero:* Read the section and make your own summary of the material.

1. *True/False:* Determine whether each of the statements that follow is true or false. If a statement is true, explain why. If a statement is false, provide a counterexample.

(a) *True or False:* $\int \cos^8 x\,dx = \frac{1}{9}\cos^9 x + C$.

(b) *True or False:* $\int \cos^8 x\,dx = \frac{1}{9}\sin^9 x + C$.

(c) *True or False:* $\int \cos^8 x\,dx = -\frac{1}{9}\sin^9 x + C$.

(d) *True or False:* If k is positive and odd, then $\int \sin^k x\,dx$ can be rewritten so that integration by substitution is possible with $u = \sin x$.

(e) *True or False:* The best method for integrating $\int \sin^3 x \cos^2 x\,dx$ is to apply double-angle identities.

(f) *True or False:* The best method for integrating $\int \sec^4 x \cos^5 x\,dx$ is to apply Pythagorean identities.

(g) *True or False:* The best method for integrating $\int \csc^2 x \cot^3 x\,dx$ is to apply substitution with $u = \cot x$.

(h) *True or False:* We now know how to integrate all six of the basic trigonometric functions.

2. *Examples:* Construct examples of the thing(s) described in the following. Try to find examples that are different than any in the reading.

(a) Three integrals that can be rewritten by using the Pythagorean identity $\sin^2 x + \cos^2 x = 1$ so that integration by substitution applies.

(b) Three integrals that can be rewritten by using the Pythagorean identity $\tan^2 x + 1 = \sec^2 x$ so that integration by substitution applies.

(c) Three integrals that can be improved by using one or both of the double-angle identities.

3. In Example 1(a) we showed that $\int \sin^3 x \cos^4 x\,dx = -\frac{1}{5}\cos^5 x + \frac{1}{7}\cos^7 x + C$. Check that this answer is correct by differentiating and applying trigonometric identities.

4. Compile lists of (a) the derivatives and (b) the integrals of the six basic trigonometric functions: $\sin x$, $\cos x$, $\tan x$, $\sec x$, $\csc x$, and $\cot x$.

5. Given the algebraic trick for integrating $\sec x$ used in the proof of Theorem 5.17, what do you think is the algebraic trick used for integrating $\csc x$?

6. Find three integrals in Exercises 21–66 that can be solved by applying Pythagorean identities to set up simple u-substitutions.

7. Find three integrals in Exercises 21–66 that can be solved by the application of double-angle formulas.

Describe strategies for solving the types of integrals given in Exercises 8–18.

8. $\displaystyle\int \cos^k x\,dx,\ k$ odd

9. $\displaystyle\int \cos^k x\,dx,\ k$ even

10. $\displaystyle\int \cot^k x\,dx,\ k$ odd

11. $\displaystyle\int \cot^k x\,dx,\ k$ even

12. $\displaystyle\int \csc^k x\,dx,\ k = 2$

13. $\displaystyle\int \csc^k x\,dx,\ k > 2$

14. $\displaystyle\int \sin^m x \cos^n x\,dx$, one of m and n odd

15. $\displaystyle\int \sin^m x \cos^n x\,dx$, both m and n even

16. $\displaystyle\int \sec^m x \tan^n x\,dx$, m even

17. $\displaystyle\int \sec^m x \tan^n x\,dx$, n odd

18. $\displaystyle\int \sec^m x \tan^n x\,dx$, m odd and n even

19. Consider the integral $\int \sin^2 x \cos^2 x\,dx$. We solved this integral in Example 2(b) by applying double-angle identities at the very beginning.

(a) Solve this integral by applying the identity $\sin^2 x = 1 - \cos^2 x$.

(b) Solve this integral another way, by applying the identity $\sin x \cos x = \frac{1}{2}\sin 2x$.

20. Consider the integral $\int \sec^2 x \tan^3 x\,dx$.

(a) Solve this integral by using integration by substitution with $u = \tan x$.

(b) Solve this integral another way, by using integration by substitution with $u = \sec x$.

Skills

Solve each of the integrals in Exercises 21–66. Some of the integrals require the methods presented in this section, and some do not. *(The last four exercises involve hyperbolic functions.)*

21. $\int \sin^2 3x\, dx$

22. $\int \cos^2 x\, dx$

23. $\int \cos^5 x\, dx$

24. $\int \sin^3 x\, dx$

25. $\int \tan 2x\, dx$

26. $\int \csc \pi x\, dx$

27. $\int \sec 4x\, dx$

28. $\int \tan^6 x\, dx$

29. $\int \cot^5 x\, dx$

30. $\int \csc^2 x\, dx$

31. $\int \sec^3 2x\, dx$

32. $\int \sec^6 x\, dx$

33. $\int \cos^4 x\, dx$

34. $\int \sin^6 x\, dx$

35. $\int \csc^4 x\, dx$

36. $\int \sin x \cos^4 x\, dx$

37. $\int \cos^3 x \sin^4 x\, dx$

38. $\int \sin^5 x \cos^2 x\, dx$

39. $\int \sec x \tan^3 x\, dx$

40. $\int \sin^2 x \cos^3 x\, dx$

41. $\int \sin^2 3x \cos^2 3x\, dx$

42. $\int \sec^2 x \tan^5 x\, dx$

43. $\int \sec^8 x \tan x\, dx$

44. $\int \tan^2 4x \sec 4x\, dx$

45. $\int \sec^3 x \tan^5 x\, dx$

46. $\int \csc^4 x \cot^2 x\, dx$

47. $\int \sin^{13} x \cos^5 x\, dx$

48. $\int \sec^3 x \tan^2 x\, dx$

49. $\int \csc x \cot^2 x\, dx$

50. $\int (\cot x \csc x)^3\, dx$

51. $\int \frac{\tan x}{\cos^3 x}\, dx$

52. $\int \frac{\cos^3 x}{\csc^8 x}\, dx$

53. $\int \sin x \sec x\, dx$

54. $\int \cos^6 x \sin^3 x\, dx$

55. $\int \cos^3 x \sec^2 x\, dx$

56. $\int \sin x \tan x\, dx$

57. $\int \tan^2 x \csc x\, dx$

58. $\int \tan x \cos^5 x\, dx$

59. $\int \frac{\cos x \ln(\cos x)}{\csc x}\, dx$

60. $\int \frac{3}{\sin^2(\pi x)}\, dx$

61. $\int (\sin x \sqrt{\cos x})^3\, dx$

62. $\int \frac{\sin^3 x}{\cos x}\, dx$

63. $\int \sinh x \cosh^2 x\, dx$

64. $\int \sinh^5 x \cosh^2 x\, dx$

65. $\int \tanh x \cosh^7 x\, dx$

66. $\int \sinh^2 x \cosh^2 x\, dx$

Solve each of the definite integrals in Exercises 67–76.

67. $\int_0^{\pi} \sin^5 x\, dx$

68. $\int_0^{3\pi} \sin^5 x\, dx$

69. $\int_{-\pi/2}^{\pi/2} \cos^3 x\, dx$

70. $\int_{-\pi}^{\pi} \sin^2 x\, dx$

71. $\int_0^{\pi} \sin^3 x \cos^2 x\, dx$

72. $\int_{-\pi}^{\pi} \sin^2 x \cos^2 x\, dx$

73. $\int_0^{\pi/4} \sec^4 x \tan^2 x\, dx$

74. $\int_{\pi/4}^{\pi/2} \csc x \cot^3 x\, dx$

75. $\int_0^{\pi/4} \frac{\sin^2 x}{\cos^4 x}\, dx$

76. $\int_0^{\pi/4} \frac{\sin x}{\cos^2 x}\, dx$

77. Consider the function $f(x) = 4\sin^3 x \cos^2 x$.

(a) Find the signed area between the graph of $f(x)$ and the x-axis on $[-\pi, \pi]$, shown next at the left.

(b) Find the absolute area between the graph of $f(x)$ and the x-axis on $[-\pi, \pi]$.

(c) Find the average value of $f(x)$ on $\left[0, \frac{\pi}{2}\right]$, shown next at the right.

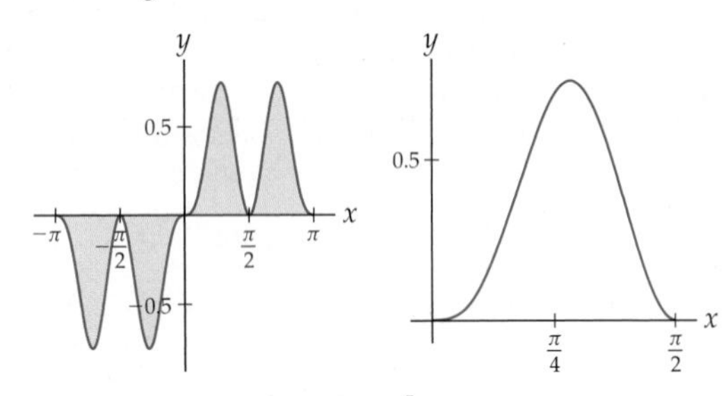

78. Consider the function $f(x) = \sin^5 x$.

(a) Find the area between the graphs of $f(x)$ and $g(x) = \sin x$ on $[0, \pi]$, shown next at the left.

(b) Find the area between the graphs of $f(x)$ and $h(x) = \sin^5 2x$ on $[0, \pi]$, shown next at the right. Note that the intersection point of the two graphs is at $x = \frac{\pi}{3}$.

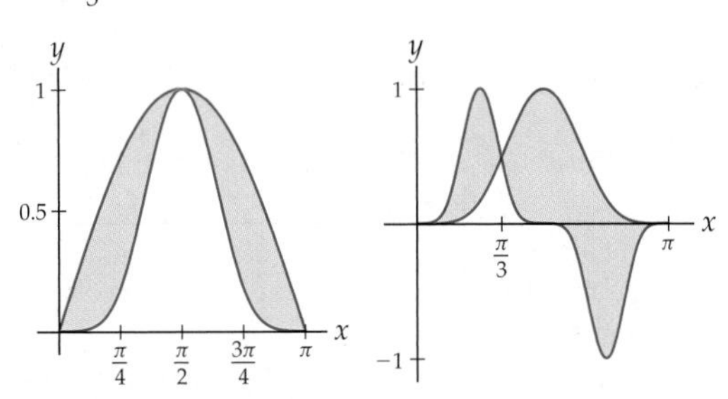

79. Consider the function $f(x) = \sin^2(\pi x)$.

(a) Find the area between the graphs of $f(x)$ and $f(x) + \frac{1}{2}$ on $[0, 3]$, shown next at the left.

(b) Find the area between the graphs of $f(x)$ and $-f(x) + \frac{1}{2}$ on $[0, 3]$, shown next at the right.

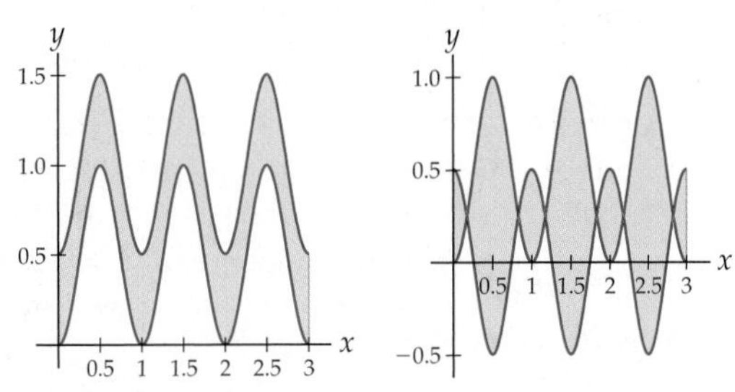

Applications

80. An office building has a thermostat that keeps the temperature between 20 and 26 degrees Celsius for 24 hours a day. As the heating and cooling units alternate, the temperature starting from midnight is given by

$$T(t) = 20\sin^3\left(\frac{\pi}{3}x\right)\cos^3\left(\frac{\pi}{3}x\right) + 22.5,$$

as shown in the figure. Find the average temperature in the office building between the hours of 9am and 5pm.

Temperature of office building

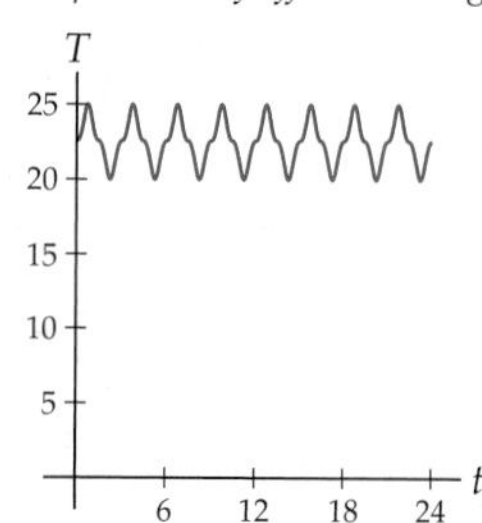

81. A wire filament is shaped to match the graph of the function of $y = \ln(2\cos x)$ from $x = -\frac{\pi}{3}$ to $x = \frac{\pi}{3}$, measured in inches, as shown in the figure. Given that the length of a curve $y = f(x)$ from $x = a$ to $x = b$ can be calculated with the formula

$$\int_a^b \sqrt{1 + (f'(x))^2}\,dx,$$

find the length of the filament.

Shape of wire filament

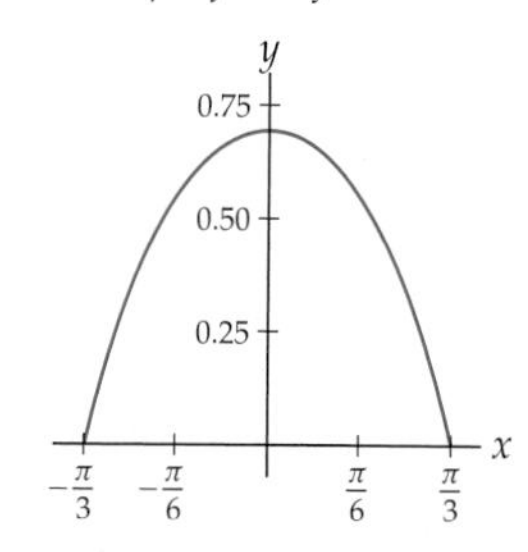

Proofs

82. Prove the integration formula

$$\int \sec x\,dx = \ln|\sec x + \tan x| + C$$

(a) by multiplying the integrand by a form of 1 and then applying a substitution;

(b) by differentiating $\ln|\sec x + \tan x|$.

83. Prove the integration formula

$$\int \csc x\,dx = -\ln|\csc x + \cot x| + C$$

(a) by multiplying the integrand by a form of 1 and then applying a substitution;

(b) by differentiating $-\ln|\csc x + \cot x|$.

84. In this problem we will see how to solve integrals of the form $\int \sin mx \cos nx\,dx$ by using the trigonometric identity $\sin\alpha\cos\beta = \frac{1}{2}(\sin(\alpha-\beta) + \sin(\alpha+\beta))$.

(a) Explain why the techniques covered earlier in the reading do not apply to $\int \sin 2x \cos 3x\,dx$.

(b) Use the given trigonometric identity to solve $\int \sin 2x \cos 3x\,dx$.

(c) Use the sum and difference identities for the sine function to prove the given trigonometric identity.

85. It can be tedious to integrate powers of the sine, especially for powers that are large and even. The following ***reduction formula*** can help: For $k > 2$,

$$\int \sin^k x\,dx = -\frac{1}{k}\sin^{k-1} x\cos x + \frac{k-1}{k}\int \sin^{k-2} x\,dx.$$

(a) Use the given reduction formula to find $\int \sin^4 x\,dx$ and $\int \sin^8 x\,dx$. You may have to apply the formula more than once.

(b) Why is this formula called a *reduction* formula?

(c) Use integration by parts to prove the reduction formula. (*Hint: Choose* $u = \sin^{n-1} x$.)

86. It can be tedious to integrate powers of the secant, especially for powers that are large. The following reduction formula can help: For $k > 2$,

$$\int \sec^k x\,dx = \frac{1}{k-1}\sec^{k-2} x\tan x + \frac{k-2}{k-1}\int \sec^{k-2} x\,dx.$$

(a) Use the given reduction formula to find $\int \sec^3 x\,dx$ and $\int \sec^7 x\,dx$. You may have to apply the formula more than once.

(b) Use integration by parts to prove the reduction formula. (*Hint: Choose* $dv = \sec^2 x\,dx$.)

Thinking Forward

Trigonometric substitution: Some integrals that don't start out involving trigonometric functions can be cleverly converted into ones that do, and then we can bring the techniques of this section to bear. In the next few problems this is how we will solve the integral $\int \frac{1}{x\sqrt{1-x^2}}\,dx$.

- The non-intuitive substitution $u = \sin^{-1} x$ is equivalent to the backwards substitution $x = \sin u$. How are dx and du related?
- Rewrite the given integral by substituting $x = \sin u$ and $dx = \cos u$.
- Use algebra to show that your new integral is equivalent to $\int \csc u\,dx$.
- Solve the integral and write your answer in terms of x. Your answer will involve both trigonometric and inverse trigonometric functions.

5.5 TRIGONOMETRIC SUBSTITUTION

- Using inverse trigonometric functions as substitutions to solve integrals
- Domain considerations when applying trigonometric substitution
- Writing a trigonometric and inverse trigonometric composition as an algebraic function

Backwards Trigonometric Substitutions

As we have seen, integration is not always a simple, straightforward process. Sometimes, similar integrals require different solution strategies. For example, consider the four integrals

$$\int \frac{1}{\sqrt{1-x^2}}\,dx, \quad \int \frac{x}{\sqrt{1-x^2}}\,dx, \quad \int \frac{x^3}{\sqrt{1-x^2}}\,dx, \quad \text{and} \quad \int \frac{1}{x^2\sqrt{1-x^2}}\,dx.$$

The first three can be solved with methods we have used so far: the first by antidifferentiating, the second with the substitution $u = 1 - x^2$, and the third with the same substitution along with a back-substitution. The fourth integral, however, is more difficult. Fortunately, a nonintuitive substitution will save the day:

$$\boxed{u = \sin^{-1} x} \quad \Longrightarrow \quad \frac{du}{dx} = \frac{1}{\sqrt{1-x^2}} \quad \Longrightarrow \quad \boxed{du = \frac{1}{\sqrt{1-x^2}}\,dx}.$$

The differential du looks promising, since $\frac{1}{\sqrt{1-x^2}}$ does appear in our integrand. The choice of $u = \sin^{-1} x$ does not appear in the integrand, but we will be able to use back-substitution to fix that problem, with $x = \sin u$. With this back-substitution we can rewrite our integral as one that involves trigonometric functions but is *much* simpler:

$$\int \frac{1}{x^2\sqrt{1-x^2}}\,dx = \int \frac{1}{x^2}\left(\frac{1}{\sqrt{1-x^2}}\,dx\right) = \int \frac{1}{\sin^2 u}\,du = \int \csc^2 u\,du.$$

We can achieve the same result with a slightly different method of back-substitution called ***trigonometric substitution***. Instead of starting with the substitution $u = \sin^{-1} x$, we will start immediately with the substitution $x = \sin u$. Starting over with this choice, we have

$$\boxed{x = \sin u} \quad \Longrightarrow \quad \frac{dx}{du} = \cos u \quad \Longrightarrow \quad \boxed{dx = \cos u\,du}.$$

Compare this substitution with the one we did previously; the two substitutions are equivalent, but the one starting with $x = \sin u$ is simpler. Notice also that we are thinking of x as a function of u. With this new version of our substitution and a Pythagorean identity, we obtain the same simplification as before:

$$\int \frac{1}{x^2\sqrt{1-x^2}}\,dx = \int \frac{\cos u}{\sin^2 u\sqrt{1-\sin^2 u}}\,du = \int \frac{1}{\sin^2 u}\,du = \int \csc^2 u\,du.$$

Although the second calculation involves more algebra, at least it does not involve inverse trigonometric functions. The rewritten integral is easy to antidifferentiate, and since we have $x = \sin u$ and thus $u = \sin^{-1} x$, the integral is equal to $-\cot u + C = -\cot(\sin^{-1} x) + C$.

Domains and Simplifications with Trigonometric Substitutions

One key thing that made the substitution $x = \sin u$ work in our earlier calculation was that the Pythagorean identity $\sin^2 x + \cos^2 x = 1$ allowed us to simplify the expression

$\sqrt{1-x^2} = \sqrt{1-\sin^2 u}$. The other two Pythagorean identities allow similar simplifications. In each case we must be very careful concerning the domains of any variables that we use. Every substitution we make must be compatible with the domain of the integrand.

The three types of trigonometric substitutions and their allowable domains are described in the following theorem:

THEOREM 5.18

Three Types of Trigonometric Substitutions

Suppose a is a positive real number. The following substitutions can be useful in solving certain integrals:

(a) If $x = a\sin u$, then $\sqrt{a^2 - x^2} = a\cos u$.
This substitution is valid for $x \in [-a, a]$ and $u \in \left[-\frac{\pi}{2}, \frac{\pi}{2}\right]$.

(b) If $x = a\tan u$, then $x^2 + a^2 = a^2\sec^2 u$.
This substitution is valid for $x \in (-\infty, \infty)$ and $u \in \left(-\frac{\pi}{2}, \frac{\pi}{2}\right)$.

(c) If $x = a\sec u$, then $\sqrt{x^2 - a^2} = \begin{cases} a\tan u, & \text{if } x > a \\ -a\tan u, & \text{if } x < -a. \end{cases}$
This substitution is valid for $x \in (-\infty, -a] \cup [a, \infty)$ and $u \in \left[0, \frac{\pi}{2}\right) \cup \left(\frac{\pi}{2}, \pi\right]$.

The domain considerations in Theorem 5.18 are essential. For example, if $x = a\sin u$, then x must be between $-a$ and a. Therefore the substitution $x = a\sin u$ makes sense only for integrands whose domain is contained in $[-a, a]$. Notice that the domain of $\sqrt{a^2 - x^2}$ is precisely $[-a, a]$, so integrals involving $\sqrt{a^2 - x^2}$ are well suited for trigonometric substitution with $x = a\sin u$. Furthermore, if $x = a\sin u$ and we want to be able to solve for $u = \sin^{-1}(x/a)$, then we must have u in the range of $\sin^{-1} x$. This is why we require that $u \in [-\pi/2, \pi/2]$ in part (a). The domain restrictions for u in the theorem come from the restricted domains of the sine, secant, and tangent, respectively.

Proof. To prove part (a), suppose $x \in [-a, a]$, $u \in [-\pi/2, \pi/2]$, and $x = a\sin u$. Using the Pythagorean identity $a^2\sin^2 u + a^2\cos^2 u = a^2$, we have

$$\sqrt{a^2 - x^2} = \sqrt{a^2 - a^2\sin^2 u} = \sqrt{a^2\cos^2 u} = |a\cos u| = a\cos u.$$

The last couple of steps in this calculation need some technical discussion. An absolute value appears in the second-to-last step because for any number A, we have $\sqrt{A^2} = |A|$. Luckily, in this case we can drop the absolute value sign because we are assuming $a > 0$ and because $\cos u$ is always positive for $u \in [-\pi/2, \pi/2]$.

The proof of part (b) is similar to the proof of part (a) and is left to Exercise 94. In this case there is no domain restriction on x, because $x = a\tan u$ can be any real number. This means that trigonometric substitution with $x = a\tan u$ can be used even for integrands that do not involve a square root.

When using secant for a trigonometric substitution as in part (c), we run into a technical point. Suppose $x \in (-\infty, a] \cup [a, \infty)$, $u \in [0, \pi/2) \cup (\pi/2, \pi]$, and $x = a\sec u$. Using the Pythagorean identity $a^2\tan^2 u + a^2 = a^2\sec^2 u$, we can write

$$\sqrt{x^2 - a^2} = \sqrt{a^2\sec^2 u - a^2} = \sqrt{a^2\tan^2 u} = |a\tan u| = \begin{cases} -a\tan u, & \text{if } x < -a \\ a\tan u, & \text{if } x > a. \end{cases}$$

Once again, absolute values appear because we are taking the square root of a square. However, this time the absolute values are significant, since $\tan u$ is positive for u in the first quadrant $[0, \pi/2)$ and negative for u in the second quadrant $(\pi/2, \pi]$. The former case happens when $x \in [a, \infty)$, and the latter case when $x \in (-\infty, -a]$. The piecewise-defined expression for $\sqrt{x^2 - a^2}$ means that we will have to consider the cases $x > a$ and $x < -a$ separately when using substitutions of the form $x = a \sec u$. ■

The point of all the work we just did boils down essentially to this: When making trigonometric substitutions with the sine or tangent, we need not worry about quadrants. But when making a trigonometric substitution with secant, there will be cases to consider.

Rewriting Trigonometric Compositions

Previously, we used trigonometric substitution to show that

$$\int \frac{1}{x^2\sqrt{1-x^2}}\,dx = -\cot(\sin^{-1} x) + C.$$

Although we have solved this integral, the form of its answer can be simplified. In fact, as we will see in a moment, we can always rewrite compositions of trigonometric functions and inverse trigonometric functions as algebraic functions.

To solve the integral listed above, we used the substitution $x = \sin u$. The domain of our integrand is $x \in (-1, 1)$, and considering this and the restricted domain of sine, we have $u \in (-\pi/2, \pi/2)$. Since $x = \sin u$, u is an angle in the first or fourth quadrant whose reference triangle has altitude x, as shown in the figure that follows. By the Pythagorean theorem, the length of the remaining side of the reference triangle is $\sqrt{1-x^2}$. This expression should look familiar: It's part of our original integrand!

Reference triangles for $x = \sin u$ in the first and fourth quadrants

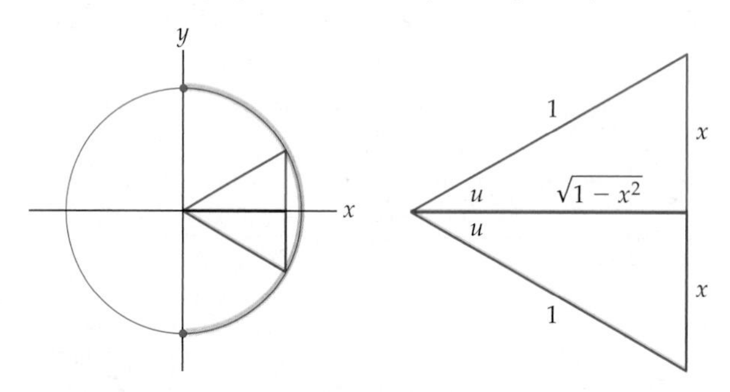

The quantity x is positive for the first-quadrant triangle and negative for the fourth-quadrant triangle, but in either case the cotangent of the angle $u = \sin^{-1} x$ is given by the ratio of the adjacent side to the opposite side.

$$\cot(\sin^{-1} x) = \cot u = \frac{\text{adj}}{\text{opp}} = \frac{\sqrt{1-x^2}}{x}.$$

Combining this with our original work solving the integral, we have

$$\int \frac{1}{x^2\sqrt{1-x^2}}\,dx = -\cot(\sin^{-1} x) + C = \frac{-\sqrt{1-x^2}}{x} + C.$$

Luckily, it turns out that when using the substitutions $x = a \sin u$ and $x = a \tan u$, choice of quadrants will never be an issue and we will not have to consider the unit circle.

From this point forward we will only need to look at one reference triangle in those cases. Unfortunately, we will have to take quadrants into account when we use the substitution $x = a \sec u$; we will see this in Example 4, where the choice of quadrant will make a sign difference.

Examples and Explorations

EXAMPLE 1

Deciding when to use trigonometric substitution

Which of these integrals would be good candidates for trigonometric substitution?

(a) $\displaystyle\int \frac{x}{1+x^2}\,dx$ **(b)** $\displaystyle\int \frac{x^2}{1+x^2}\,dx$ **(c)** $\displaystyle\int \frac{1+x^2}{x^2}\,dx$

SOLUTION

(a) Trigonometric substitution is not necessary here because we can apply regular u-substitution with $u = 1 + x^2$. This technique simplifies the integral to $\frac{1}{2}\int \frac{1}{u}\,du$, which is easy to solve.

(b) Neither u-substitution nor integration by parts seems useful here. This integral is a good candidate for trigonometric substitution, with $x = \tan u$. This choice simplifies the integral to $\int \tan^2 u\,du$, which we can solve by the techniques of Section 5.4. Alternatively, we could use polynomial long division or other algebraic methods to write $\frac{x^2}{1+x^2}$ as $1 - \frac{1}{1+x^2}$, which is easy to antidifferentiate.

(c) Trigonometric substitution is not necessary here. Applying algebra first, we can write $\frac{1+x^2}{x^2}$ as $\frac{1}{x^2} + \frac{x^2}{x^2} = x^{-2} + 1$, which is easy to antidifferentiate. □

EXAMPLE 2

An integral involving $\sqrt{a^2 - x^2}$ and the substitution $x = a \sin u$

Use trigonometric substitution to solve the integral $\displaystyle\int (4 - x^2)^{-3/2}\,dx$.

SOLUTION

First, notice that the integrand can be expressed as $\frac{1}{(\sqrt{4-x^2})^3}$ and therefore does involve the expression $\sqrt{a^2 - x^2}$, where $a = 2$. Using the substitution $x = 2\sin u$ and the Pythagorean identity $\sin^2 u + \cos^2 u = 1$, we have

$$(4 - x^2)^{-3/2} = (4 - 4\sin^2 u)^{-3/2} = \left(\sqrt{4\cos^2 u}\right)^{-3} = (2|\cos u|)^{-3} = \frac{1}{8\cos^3 u}.$$

Theorem 5.18 explains why the absolute values can be ignored in the preceding calculation. Now with the substitution $x = 2\sin u$, we have $\sqrt{4 - x^2} = 2\cos u$ and the differential $dx = 2\cos u\,du$, and thus we can write

$$\int (4 - x^2)^{-3/2}\,dx = \int \frac{1}{8\cos^3 u}(2\cos u)\,du$$
$$= \frac{1}{4}\int \sec^2 u\,du = \frac{1}{4}\tan u + C = \frac{1}{4}\tan\left(\sin^{-1}\left(\frac{x}{2}\right)\right) + C.$$

The last step in the calculation is a back-substitution obtained by solving $x = 2\sin u$ for $u = \sin^{-1}(x/2)$.

It now remains only to rewrite the answer as an algebraic function. Because we already know that quadrants are not relevant when using trigonometric substitution with the sine, we need to consider only one reference triangle. Since $x = 2\sin u$, we have $\sin u = \frac{x}{2}$. Thinking about "opposite over hypotenuse," we label the angle u and label two sides of the triangle with x and 2. The remaining leg of the triangle is determined by the Pythagorean theorem to be $\sqrt{4 - x^2}$, as follows:

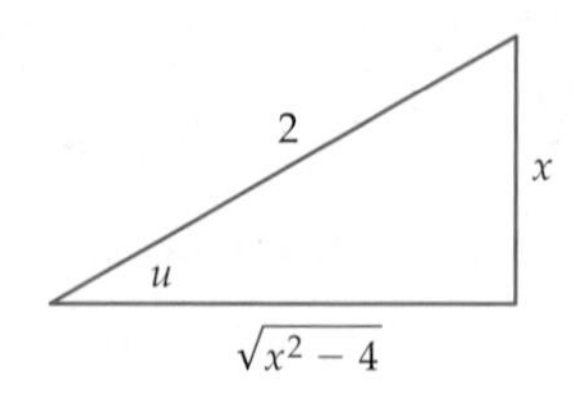

Now thinking about "opposite over adjacent," we can easily calculate that

$$\tan\left(\sin^{-1}\left(\frac{x}{2}\right)\right) = \tan u = \frac{x}{\sqrt{4 - x^2}}.$$

Therefore the integral in question is equal to

$$\int (4 - x^2)^{-3/2}\, dx = \frac{1}{4}\tan\left(\sin^{-1}\left(\frac{x}{2}\right)\right) + C = \frac{x}{4\sqrt{4 - x^2}} + C.$$

□

Since we took the time to rewrite the trigonometric composition as an algebraic function, we can check the answer to the previous example by using the quotient rule and some algebra:

$$\frac{d}{dx}\left(\frac{x}{4\sqrt{4 - x^2}}\right) = \frac{(1)(4\sqrt{4 - x^2}) - (x)\left(4\left(\frac{1}{2}\right)(4 - x^2)^{-1/2}(-2x)\right)}{16(4 - x^2)}$$

$$= \frac{4\sqrt{4 - x^2} + \frac{4x^2}{\sqrt{4 - x^2}}}{16(4 - x^2)}\left(\frac{\sqrt{4 - x^2}}{\sqrt{4 - x^2}}\right)$$

$$= \frac{4(4 - x^2) + 4x^2}{16(4 - x^2)^{3/2}} = \frac{16}{16(4 - x^2)^{3/2}} = (4 - x^2)^{-3/2}.$$

EXAMPLE 3

An integral involving $x^2 + a^2$ and the substitution $x = a\tan u$

Use trigonometric substitution to solve the integral $\int (x^2 + 5)^{-3/2}\, dx$.

SOLUTION

This example is similar to the previous one, except that this time the substitution $x = \sqrt{5}\tan u$ is suggested. Using a Pythagorean identity and some basic algebra, we have

$$(x^2 + 5)^{-3/2} = (5\tan^2 u + 5)^{-3/2} = \sqrt{5\sec^2 u}^{\,-3} = (\sqrt{5}|\sec u|)^{-3} = \frac{1}{(\sqrt{5})^3\sec^3 u}.$$

With the substitution $x = \sqrt{5}\tan u$ we also have $dx = \sqrt{5}\sec^2 u\, du$ and $u = \tan^{-1}(x/\sqrt{5})$, and thus can write

$$\int (x^2 + 5)^{-3/2}\, dx = \int \frac{1}{(\sqrt{5})^3\sec^3 u}(\sqrt{5}\sec^2 u)\, du$$

$$= \frac{1}{5}\int \cos u\, du = \frac{1}{5}\sin u + C = \frac{1}{5}\sin\left(\tan^{-1}\left(\frac{x}{\sqrt{5}}\right)\right) + C.$$

Once again, we construct a reference triangle to rewrite our answer as an algebraic function. We have $\tan u = \frac{x}{\sqrt{5}}$, so relative to the angle u we label the opposite side x and the adjacent side $\sqrt{5}$. By the Pythagorean theorem, the corresponding hypotenuse has length $\sqrt{x^2+5}$:

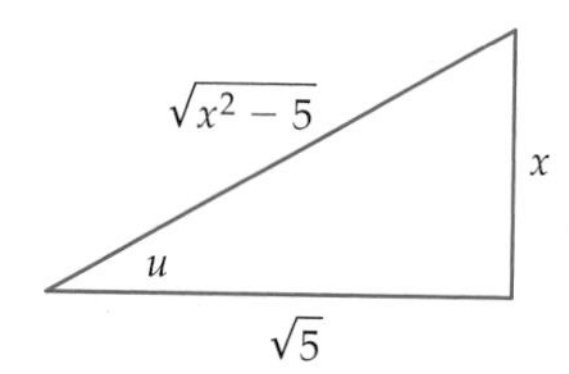

Using this triangle we have

$$\sin\left(\tan^{-1}\left(\frac{x}{\sqrt{5}}\right)\right) = \sin u = \frac{x}{\sqrt{x^2+5}}.$$

Therefore our integral is equal to

$$\int (x^2+5)^{-3/2}\,dx = \frac{1}{5}\sin\left(\tan^{-1}\left(\frac{x}{\sqrt{5}}\right)\right) + C = \frac{x}{5\sqrt{x^2+5}} + C.$$

The astute reader will notice that this example is quite similar to the previous one. To get a better understanding, compare and contrast the two examples. □

EXAMPLE 4

An integral involving $\sqrt{x^2-a^2}$ and the substitution $x = a\sec u$

Use trigonometric substitution to solve the integral $\displaystyle\int \frac{1}{\sqrt{x^2-4}}\,dx$.

SOLUTION

With the substitution $x = 2\sec u$ we have

$$\frac{1}{\sqrt{x^2-4}} = \frac{1}{\sqrt{4\sec^2 u - 4}} = \frac{1}{\sqrt{4\tan^2 u}} = \frac{1}{2|\tan u|}.$$

Unfortunately, at this point a problem arises that we didn't encounter in the previous two examples. The difficulty is that in this example the integrand $\frac{1}{\sqrt{x^2-4}}$ is defined for $x > 2$ and for $x < -2$, but not in between. Since we are using the substitution $x = 2\sec u$, we need u to be in the restricted domain of the secant, which is the first two quadrants. Let's examine the two cases.

If $x > 2$, then $x = 2\sec u$ means that $\sec u > 1$, which happens when u is in the first quadrant. The tangent function is positive in that quadrant, so if $x > 2$, then $\tan u > 0$ and thus $|\tan u| = \tan u$. However, if $x < -2$, then $\sec u < 1$ and thus u is in the second quadrant, where the tangent is negative. In that case we have $|\tan u| = -\tan u$. Unlike the solution in the previous two examples, the choice of quadrant will be significant. The problem here is that our original integrand was defined on a disconnected domain, and we have to deal with the two pieces of the domain separately.

If $x > 2$, then $|\tan u| = \tan u$, and thus with $x = 2\sec u$, $dx = 2\sec u\tan u\,du$, and a known antidifferentiation formula, we have

$$\int \frac{1}{\sqrt{x^2-4}}\,dx = \int \frac{1}{2\tan u}(2\sec u\,\tan u)\,du = \int \sec u\,du$$

$$= \ln|\sec u + \tan u| + C = \ln\left|\sec\left(\sec^{-1}\left(\frac{x}{2}\right)\right) + \tan\left(\sec^{-1}\left(\frac{x}{2}\right)\right)\right| + C.$$

Still considering the case where $x > 2$, we see that our reference triangle is in the first quadrant. Since $\sec u = \frac{x}{2}$, we have a hypotenuse of length x and an adjacent leg of length 2, relative to the angle u. By the Pythagorean theorem, the remaining side has length $\sqrt{x^2 - 4}$, as follows:

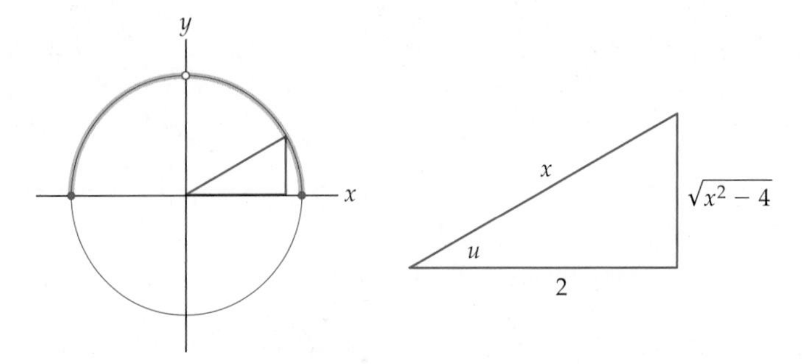

We clearly have $\sec(\sec^{-1}(x/2)) = x/2$. Using the triangle shown we see that

$$\tan\left(\sec^{-1}\left(\frac{x}{2}\right)\right) = \tan u = \frac{\sqrt{x^2 - 4}}{2}.$$

Therefore in the case where $x > 2$, we have

$$\int \frac{1}{\sqrt{x^2 - 4}}\, dx = \ln\left|\sec\left(\sec^{-1}\left(\frac{x}{2}\right)\right) + \tan\left(\sec^{-1}\left(\frac{x}{2}\right)\right)\right| + C = \ln\left|\frac{x}{2} + \frac{\sqrt{x^2 - 4}}{2}\right| + C.$$

In the case where $x < -2$, there are two places where our answer could differ from the answer for $x > 2$. First, if $x < -2$ and $x = 2\sec u$, then $\sec u$ is negative, which means that u is an angle terminating in the second quadrant. The tangent function is negative in that quadrant, so in this case $|\tan u| = -\tan u$. This means that before integrating, we have to insert a negative sign. By the end of the integration step we still have a negative sign, so for $x < -2$, we have

$$\int \frac{1}{\sqrt{x^2 - 4}}\, dx = -\ln\left|\sec\left(\sec^{-1}\left(\frac{x}{2}\right)\right) + \tan\left(\sec^{-1}\left(\frac{x}{2}\right)\right)\right| + C.$$

Second, when we use a reference triangle to write the answer as an algebraic function, we will have to consider quadrants. The diagrams that follow show the angle u in the unit circle and its corresponding reference triangle v.

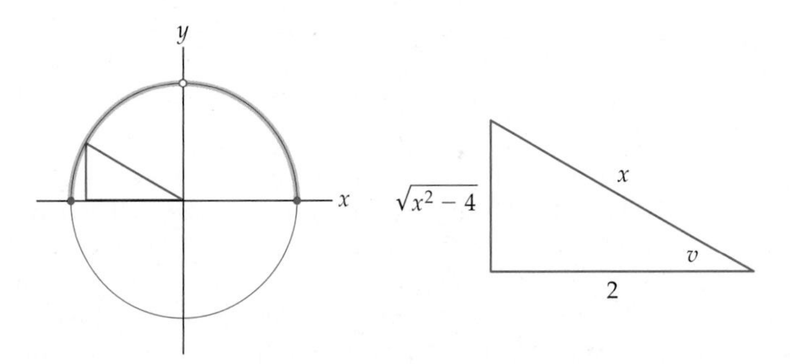

Since the tangent is negative in the second quadrant,

$$\tan\left(\sec^{-1}\left(\frac{x}{2}\right)\right) = \tan u = -\frac{\sqrt{x^2 - 4}}{2}.$$

Therefore for $x < -2$, the solution of our integral is

$$\int \frac{1}{\sqrt{x^2 - 4}}\, dx = -\ln\left|\sec\left(\sec^{-1}\left(\frac{x}{2}\right)\right) + \tan\left(\sec^{-1}\left(\frac{x}{2}\right)\right)\right| + C = -\ln\left|\frac{x}{2} - \frac{\sqrt{x^2 - 4}}{2}\right| + C.$$

□

In the previous example we got different answers depending on whether we had $x > 2$ or $x < -2$. We can see graphically why this has to be the case, by plotting the integrand $f(x)$ and one of its antiderivatives $F(x)$ (we chose $C = 0$ in each case), as shown in the following graphs:

$$f(x) = \frac{1}{\sqrt{x^2 - 4}}$$

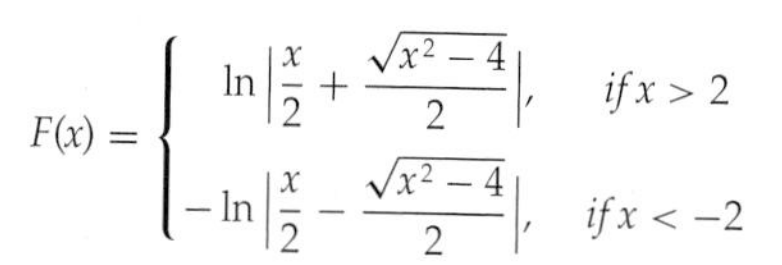

$$F(x) = \begin{cases} \ln\left|\frac{x}{2} + \frac{\sqrt{x^2-4}}{2}\right|, & \text{if } x > 2 \\ -\ln\left|\frac{x}{2} - \frac{\sqrt{x^2-4}}{2}\right|, & \text{if } x < -2 \end{cases}$$

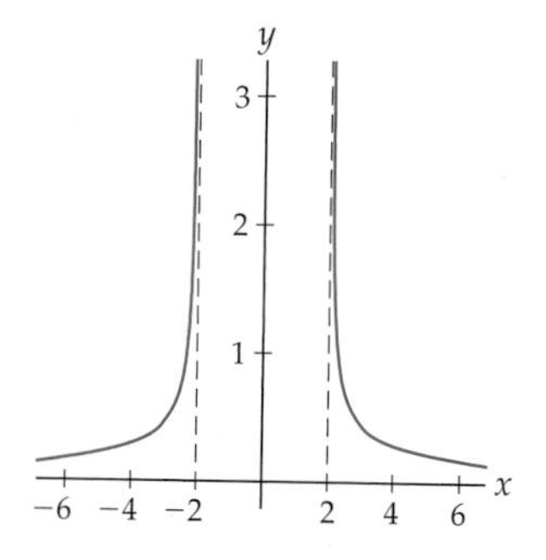

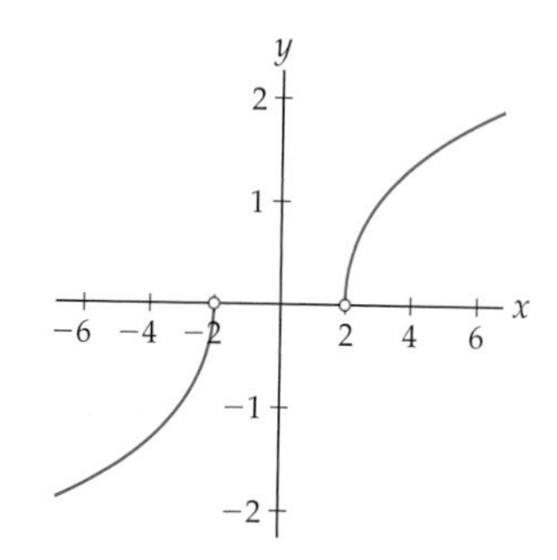

You should be able to convince yourself from looking at these graphs that it is at least reasonable that the derivative of $F(x)$ is equal to $f(x)$. Notice also that it makes sense that we had to look at two separate cases to integrate $f(x)$, since that function is defined on two different intervals (and not between).

EXAMPLE 5

Choosing an appropriate trigonometric substitution

Determine appropriate trigonometric substitutions for each of the integrals that follow. Apply the substitution and simplify until the integral is one that you know how to solve.

(a) $\int \sqrt{x^2 - 7}\, dx$ **(b)** $\int \sqrt{4x^2 + 9}\, dx$

SOLUTION

(a) This integrand has the form $\sqrt{x^2 - a^2}$, with $a = \sqrt{7}$. According to Theorem 5.18, this suggests that we apply trigonometric substitution with $x = \sqrt{7} \sec u$. With that substitution we have $dx = \sqrt{7} \sec u \tan u\, du$, so we can rewrite the integral as

$$\int \sqrt{x^2 - 7}\, dx = \int \sqrt{7 \sec^2 u - 7}\, \sqrt{7} \sec u \tan u\, du.$$

Notice that the domain of the integrand $\sqrt{x^2 - 7}$ is $(-\infty, -\sqrt{7}] \cup [\sqrt{7}, \infty)$. According to Theorem 5.18, we must handle the cases $x < -\sqrt{7}$ and $x > \sqrt{7}$ separately. Keeping this in mind, we can apply a Pythagorean identity to write

$$\sqrt{7 \sec^2 u - 7} = \sqrt{7 \tan^2 u} = \sqrt{7}\, |\tan u| = \begin{cases} -\sqrt{7} \tan u, & \text{if } x < -\sqrt{7} \\ \sqrt{7} \tan u, & \text{if } x > \sqrt{7}. \end{cases}$$

This means that for $x < -\sqrt{7}$ our integral becomes $7 \int \sec u \tan^2 u\, du$, and for $x > \sqrt{7}$ our integral becomes $-7 \int \sec u \tan^2 u\, du$. Both of these integrals can be solved with the methods of Section 5.4.

(b) This integrand involves an expression that is the sum of two squares: $(2x)^2 + 3^2$. This suggests that we should apply trigonometric substitution with $2x = 3\tan u$, or equivalently, $x = \frac{3}{2}\tan u$. With this choice we have $dx = \frac{3}{2}\sec^2 u\,du$, so our integral can be written as

$$\int \sqrt{4x^2 + 9}\,dx = \int \sqrt{4\left(\frac{3}{2}\tan u\right)^2 + 9}\,\left(\frac{3}{2}\sec^2 u\right)\,du.$$

After a little bit of algebra the latter integral simplifies to $\frac{9}{2}\int \sec^3 u\,du$, which we know how to solve by the methods of the previous section. □

EXAMPLE 6

Completing the square before trigonometric substitution

Determine an appropriate trigonometric substitution for the following integral:

$$\int \frac{1}{\sqrt{x^2 - 6x + 13}}\,dx$$

Then apply the substitution and simplify until the integral is one that you know how to solve.

SOLUTION

To use trigonometric substitution, we need to have an expression of the form $\sqrt{a^2 - x^2}$, $x^2 + a^2$, or $\sqrt{x^2 - a^2}$ in the integrand. To obtain a sum or difference of perfect squares we will apply the method of ***completing the square*** to the quadratic $x^2 - 6x + 13$:

$$x^2 - 6x + 13 = (x^2 - 6x + 9) - 9 + 13 = (x - 3)^2 + 4.$$

We added and subtracted 9 from the quadratic so that it would involve the perfect square $x^2 - 6x + 9 = (x - 3)^2$. In general, recall from Section 5.3 that to complete the square of a quadratic $x^2 + bx + c$, we add and subtract $(b/2)^2$. This enables us to write the quadratic in terms of the perfect square $(x - (b/2))^2$ and some other constant. Now that we have completed the square, we have

$$\int \frac{1}{\sqrt{x^2 - 6x + 13}}\,dx = \int \frac{1}{\sqrt{(x-3)^2 + 4}}\,dx.$$

The part of the integrand inside the radical is *almost* of the form $x^2 + a^2$. Instead, it is of the form $(x - 3)^2 + a^2$ (with $a = 2$). Therefore we will choose the substitution $x - 3 = 2\tan u$. This means that $dx = 2\sec^2 u\,du$, and our integral becomes

$$\int \frac{1}{\sqrt{x^2 - 6x + 13}}\,dx = \int \frac{1}{\sqrt{(x-3)^2 + 4}} = \int \frac{1}{\sqrt{4\tan^2 u + 4}}(2\sec^2 u)\,du,$$

which simplifies easily to $\int \sec u\,du$, an integral that we know how to solve. □

? TEST YOUR UNDERSTANDING

- In the first subsection we did one example two different ways. In the first way we had $du = \frac{1}{\sqrt{1-x^2}}\,dx$, and in the second we had $dx = \cos u\,du$. Why are these the same if $x = \sin u$?
- What are the three trigonometric identities that come in handy for simplifying after performing a trigonometric substitution?
- Why would trigonometric substitution not be a good strategy for solving the integral $\int \frac{1}{x^2(1-x^2)}\,dx$?

- Why do we have to consider cases for different domains and quadrants when doing a trigonometric substitution with secant, but not when we use trigonometric substitution with sine or tangent?
- We can use the unit circle and reference triangles to write certain trigonometric compositions as algebraic functions. What will go wrong if we use that technique to simplify $\sin(2\cos^{-1} x)$, and how can we fix it?

EXERCISES 5.5

Thinking Back

Domains and ranges of inverse trigonometric functions: For each function that follows, (a) list the domain and range, (b) sketch a labeled graph, and (c) discuss the domains and ranges in the context of the unit circle.

- $f(x) = \sin^{-1} x$
- $f(x) = \tan^{-1} x$
- $f(x) = \sec^{-1} x$
- $f(x) = \sin^{-1}\left(\frac{x}{3}\right)$

Working with inverse trigonometric functions: Solve each of the following problems by hand, using reference triangles in the unit circle.

- If $x = \tan u$ and $u \in \left(-\frac{\pi}{2}, \frac{\pi}{2}\right)$, find $\sin u$.
- If $x = 3\sin u$ and $u \in \left[-\frac{\pi}{2}, \frac{\pi}{2}\right]$, find $\cot u$.
- If $x = \sec u$ and $u \in \left[0, \frac{\pi}{2}\right)$, find $\tan u$.
- If $x = \sec u$ and $u \in \left(\frac{\pi}{2}, \pi\right]$, find $\tan u$.

Review of trigonometric integrals: Use the methods of the previous section to find each of the following integrals.

- $\int \sec x\, dx$
- $\int \sec^2 x\, dx$
- $\int \sec^3 x\, dx$
- $\int \sec x \tan^2 x\, dx$

Review of other integration methods: Use any method *except* trigonometric substitution to find each of the following integrals.

- $\int \frac{1}{\sqrt{1-x^2}}\, dx$
- $\int \frac{x}{\sqrt{x^2-9}}\, dx$
- $\int \frac{9+x^2}{x}\, dx$
- $\int \frac{1+x^4}{1+x^2}\, dx$
- $\int \frac{1}{1+4x^2}\, dx$
- $\int \frac{x^3}{\sqrt{x^2+1}}\, dx$

Concepts

0. *Problem Zero:* Read the section and make your own summary of the material.

1. *True/False:* Determine whether each of the statements that follow is true or false. If a statement is true, explain why. If a statement is false, provide a counterexample.

(a) *True or False:* The substitution $x = 2\sec u$ is a suitable choice for solving $\int \frac{1}{x^2-4}\, dx$.

(b) *True or False:* The substitution $x = 2\sec u$ is a suitable choice for solving $\int \frac{1}{\sqrt{x^2-4}}\, dx$.

(c) *True or False:* The substitution $x = 2\tan u$ is a suitable choice for solving $\int \frac{1}{x^2+4}\, dx$

(d) *True or False:* The substitution $x = 2\sin u$ is a suitable choice for solving $\int (x^2+4)^{-5/2}\, dx$

(e) *True or False:* Trigonometric substitution is a useful strategy for solving any integral that involves an expression of the form $\sqrt{x^2-a^2}$.

(f) *True or False:* Trigonometric substitution doesn't solve an integral; rather, it helps you rewrite integrals as ones that are easier to solve by other methods.

(g) *True or False:* When using trigonometric substitution with $x = a\sin u$, we must consider the cases $x > a$ and $x < -a$ separately.

(h) *True or False:* When using trigonometric substitution with $x = a\sec u$, we must consider the cases $x > a$ and $x < -a$ separately.

2. *Examples:* Construct examples of the thing(s) described in the following. Try to find examples that are different than any in the reading.

(a) An integral to which we could reasonably apply trigonometric substitution with $x = \tan u$.

(b) An integral to which we could reasonably apply trigonometric substitution with $x = 4\sec u$.

(c) An integral to which we could reasonably apply trigonometric substitution with $x - 2 = 3\sin u$.

3. Show that if $x = \tan u$, then $dx = \sec^2 u\, du$, in the following two ways: (a) by using implicit differentiation, thinking of u as a function of x and (b) by thinking of x as a function of u.

4. Show by differentiating (and then using algebra) that $-\cot(\sin^{-1} x)$ and $-\frac{\sqrt{1-x^2}}{x}$ are both antiderivatives of $\frac{1}{x^2\sqrt{1-x^2}}$. How can these two very different-looking functions be antiderivatives of the same function?

5. Consider the integral $\int \frac{1}{x^2\sqrt{1-x^2}}\,dx$ from the reading at the beginning of the section.

(a) Use the inverse trigonometric substitution $u = \sin^{-1} x$ to solve this integral.

(b) Use the trigonometric substitution $x = \sin u$ to solve the integral.

(c) Compare and contrast the two methods used in parts (a) and (b).

6. Give an example of an integral for which trigonometric substitution is possible but an easier method is available. Then give an example of an integral that we still don't know how to solve given the techniques we know at this point.

7. Why don't we ever have cause to use the trigonometric substitution $x = \cos u$?

8. Explain how to know when to use the trigonometric substitutions $x = a \sin u$, $x = a \tan u$, and $x = a \sec u$. Describe the trigonometric identity and the triangle that will be needed in each case. What are the possible values for x and u in each case?

9. Why don't we need to have a square root involved in order to apply trigonometric substitution with the tangent? In other words, why can we use the substitution $x = a \tan u$ when we see x^2+a^2, even though we can't use the substitution $x = a \sin u$ unless the integrand involves the square root of $a^2 - x^2$? *(Hint: Think about domains.)*

10. Explain why it makes sense to try the trigonometric substitution $x = \sec u$ if an integrand involves the expression $\sqrt{x^2 - 1}$.

11. Which of the integrals that follow would be good candidates for trigonometric substitution? If trigonometric substitution is a good strategy, name the substitution. If another method is a better strategy, explain that method.

(a) $\displaystyle\int \frac{4+x^2}{x}\,dx$ (b) $\displaystyle\int \frac{x}{4+x^2}\,dx$

(c) $\displaystyle\int \frac{x^2}{4+x^2}\,dx$ (d) $\displaystyle\int \frac{16-x^4}{4+x^2}\,dx$

12. Explain why using trigonometric substitution with $x = a \tan u$ often involves a triangle with side lengths a and x and hypotenuse of length $\sqrt{x^2 + a^2}$.

13. Explain why, if $x = a \sin u$, then $\sqrt{a^2 - x^2} = a \cos u$. Your explanation should include a discussion of domains and absolute values.

14. Explain why, if $x = a \tan u$, then $\sqrt{x^2 + a^2} = a \sec u$. Your explanation should include a discussion of domains and absolute values.

15. Explain why, if $x = a \sec u$, then $\sqrt{a^2 \sec^2 u - a^2}$ is $-a \tan u$ if $x < -a$ and is $a \tan u$ if $x > a$. Your explanation should include a discussion of domains and absolute values.

16. Why is it okay to use a triangle *without* thinking about the unit circle when simplifying expressions that result from a trigonometric substitution with $x = a \sin u$ or $x = a \tan u$? Why do we need to think about the unit circle after trigonometric substitution with $x = a \sec u$?

17. Why doesn't the definite integral $\int_2^3 \sqrt{1-x^2}\,dx$ make sense? *(Hint: Think about domains.)*

18. Find three integrals in Exercises 39–74 that can be solved by using a trigonometric substitution of the form $x = a \tan u$.

19. Find three integrals in Exercises 39–74 that can be solved *without* using trigonometric substitution.

In Exercises 20–27, use reference triangles and the unit circle to write the given trigonometric compositions as algebraic functions.

20. $\cot(\sec^{-1} x)$ **21.** $\cos(\sin^{-1} x)$

22. $\tan\left(\sin^{-1}\left(\frac{x}{3}\right)\right)$ **23.** $\sin\left(\tan^{-1}\left(\frac{x}{2}\right)\right)$

24. If $x + 3 = 4 \tan u$, then write $\sec u$ as an algebraic function of x.

25. If $x - 5 = 3 \sin u$, then write $\tan^2 u$ as an algebraic function of x.

26. Write $\tan(2 \sec^{-1} x)$ as an algebraic function.

27. Write $\sin(2 \cos^{-1} x)$ as an algebraic function.

Complete the square for each quadratic in Exercises 28–33. Then describe the trigonometric substitution that would be appropriate if you were solving an integral that involved that quadratic.

28. $x^2 - 4x - 8$ **29.** $x^2 + 6x - 2$

30. $2x^2 - 4x + 1$ **31.** $x^2 - 5x + 1$

32. $x - 3x^2$ **33.** $2(x+2)^2$

34. Solve $\int \frac{x^2}{1+x^2}\,dx$ the following two ways:

(a) with the substitution $u = \tan^{-1} x$;

(b) with the trigonometric substitution $x = \tan u$.

35. Solve $\int \frac{1}{x^2+9}\,dx$ the following two ways:

(a) with the trigonometric substitution $x = 3 \tan u$;

(b) with algebra and the derivative of the arctangent.

36. Solve $\int \frac{x+2}{(x^2+4x)^{3/2}}\,dx$ the following two ways:

(a) with the substitution $u = x^2 + 4x$;

(b) by completing the square and then applying the trigonometric substitution $x + 2 = 2 \sec u$.

37. Solve $\int \frac{x}{4+x^2}\,dx$ the following two ways:

(a) with the substitution $u = 4 + x^2$;

(b) with the trigonometric substitution $x = 2 \tan u$.

38. Solve the integral $\int x^3\sqrt{x^2-1}\,dx$ three ways:

(a) with the substitution $u = x^2 - 1$, followed by back-substitution;

(b) with integration by parts, choosing $u = x^2$ and $dv = x\sqrt{x^2-1}\,dx$;

(c) with the trigonometric subsitution $x = \sec u$.

Skills

Solve each of the integrals in Exercises 39–74. Some integrals require trigonometric substitution, and some do not. Write your answers as algebraic functions whenever possible.

39. $\int \frac{\sqrt{4-x^2}}{x^2}\,dx$

40. $\int x^3\sqrt{x^2+1}\,dx$

41. $\int \frac{x}{\sqrt{x^2-1}}\,dx$

42. $\int x\sqrt{x^2+1}\,dx$

43. $\int \frac{1}{\sqrt{3-x^2}}\,dx$

44. $\int \frac{1}{(x^2+4)^{3/2}}\,dx$

45. $\int \frac{\sqrt{9-x^2}}{x}\,dx$

46. $\int \frac{x^2+9}{x^{3/2}}\,dx$

47. $\int \sqrt{3-x}\sqrt{x-1}\,dx$

48. $\int \sqrt{4-x}\sqrt{x-2}\,dx$

49. $\int \frac{1}{x^2\sqrt{x^2-9}}\,dx$

50. $\int \frac{\sqrt{x^2-4}}{2x}\,dx$

51. $\int (1-x^2)^{-3/2}\,dx$

52. $\int \frac{1}{x^2\sqrt{x^2+3}}\,dx$

53. $\int \frac{1}{x^2-4x+13}\,dx$

54. $\int \frac{x}{\sqrt{1-4x^2}}\,dx$

55. $\int \frac{3+x}{\sqrt{9-4x^2}}\,dx$

56. $\int \frac{1}{\sqrt{x^2+9}}\,dx$

57. $\int \sqrt{x^2-8x+25}\,dx$

58. $\int \sqrt{x^2+6x+18}\,dx$

59. $\int (3-x^2)^{3/2}\,dx$

60. $\int \frac{x^3}{3x^2+5}\,dx$

61. $\int \sqrt{2-x^2}\,dx$

62. $\int \frac{1}{(x^2+1)^{5/2}}\,dx$

63. $\int \frac{\sqrt{x^2-1}}{x}\,dx$

64. $\int \frac{1}{x^2\sqrt{4-x^2}}\,dx$

65. $\int \frac{1}{\sqrt{x^2-2}}\,dx$

66. $\int \sqrt{x^2-2}\,dx$

67. $\int x^3\sqrt{x^2+1}\,dx$

68. $\int x^5\sqrt{x^2-1}\,dx$

69. $\int \ln(x^2+1)\,dx$

70. $\int \ln(4+x^2)\,dx$

71. $\int e^{2x}\sqrt{e^{2x}+1}\,dx$

72. $\int e^{2x}(1-e^{4x})^{3/2}\,dx$

73. $\int \frac{1}{e^x\sqrt{e^{2x}+3}}\,dx$

74. $\int x\sin^{-1}x\,dx$

Solve each of the integrals in Exercises 75–78 by using polynomial long division to rewrite the integrand. This is one way that you can sometimes avoid using trigonometric substitution; moreover, sometimes it works when trigonometric substitution does not apply.

75. $\int \frac{x^3}{x^2+4}\,dx$

76. $\int \frac{x^2-1}{x^2+1}\,dx$

77. $\int \frac{x^4-3}{2+3x^2}\,dx$

78. $\int \frac{x^3-3x^2+2x-3}{x^2+1}\,dx$

Solve each of the definite integrals in Exercises 79–86. Some integrals require trigonometric substitution, and some do not.

79. $\int_0^4 x\sqrt{x^2+4}\,dx$

80. $\int_3^5 \sqrt{x^2-9}\,dx$

81. $\int_4^5 \frac{1}{x\sqrt{x^2+9}}\,dx$

82. $\int_1^2 \frac{1}{x\sqrt{9-x^2}}\,dx$

83. $\int_{1/4}^{1/2} \frac{1}{x^2\sqrt{1-x^2}}\,dx$

84. $\int_{-1}^{1} x^3\sqrt{9x^2-1}\,dx$

85. $\int_1^2 \frac{3}{(9-2x^2)^{3/2}}\,dx$

86. $\int_0^4 x^3\sqrt{x^2+4}\,dx$

We can extend the technique of trigonometric substitution to the hyperbolic functions. Use Theorem 2.20 and the identity $\cosh^2 x - \sinh^2 x = 1$ to solve each integral in Exercises 87–90 with an appropriate hyperbolic substitution $x = a\sinh u$ or $x = a\cosh u$. *(These exercises involve hyperbolic functions.)*

87. $\int \frac{1}{\sqrt{x^2+4}}\,dx$

88. $\int \frac{1}{\sqrt{2x^2-3}}\,dx$

89. $\int \frac{x^2}{(x^2+1)^{3/2}}\,dx$

90. $\int \frac{x^3}{(x^2-4)^{3/2}}\,dx$

Applications

91. A pharmaceutical company is designing a new drug whose shape in tablet form is obtained by rotating the graph of $f(x) = 2(42-x^2)^{1/4}$ on the interval $[-6, 6]$ around the x-axis, where units are measured in millimeters.

Shape of a tablet

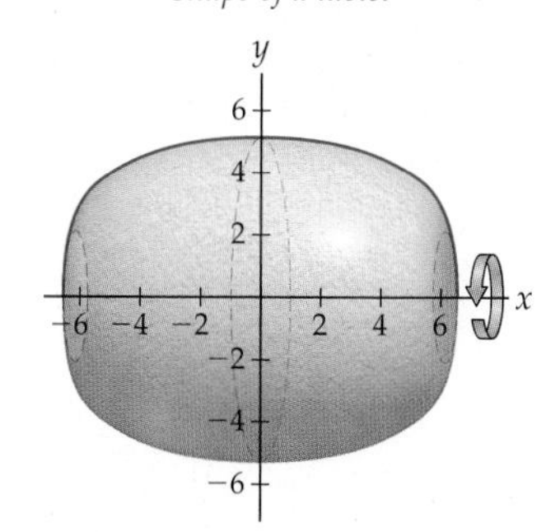

As we will see in Section 6.1, the volume of such a solid is given by $\pi\int_a^b (f(x))^2\,dx$, where a and b are the places where the graph of $f(x)$ meets the x-axis.

(a) Write down a specific definite integral that represents the amount of material, in cubic millimeters, required to make the tablet.

(b) Use trigonometric substitution to solve the definite integral and determine the amount of material needed.

92. The main cable on a certain suspension bridge follows a parabolic curve with equation $f(x) = (0.025x)^2$, measured in feet, as shown in the following figure:

Main cable of a suspension bridge

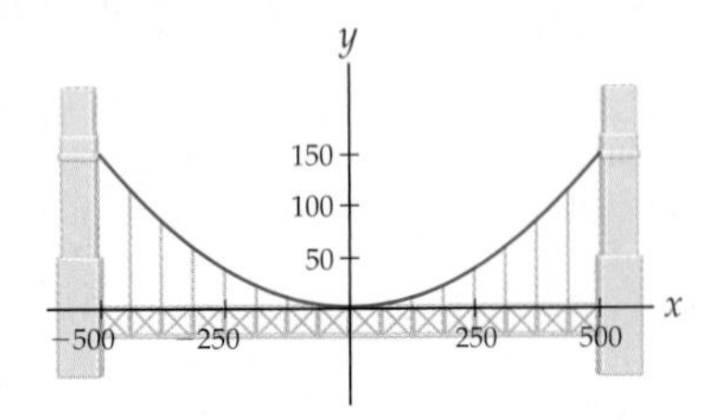

As we will see in Section 6.3, the length of a curve $f(x)$ from $x = a$ to $x = b$ can be calculated from the formula $\int_a^b \sqrt{1 + (f'(x))^2}\, dx$.

(a) Write down a specific definite integral that represents the length of the main cable of the suspension bridge.

(b) Use trigonometric substitution to solve the definite integral and determine the length of the cable.

Proofs

93. Prove that the area of a circle of radius r is πr^2, as follows:

(a) Write down a definite integral that represents the area of the circle of radius r centered at the origin. *(Hint: The equation of such a circle is $x^2 + y^2 = r^2$.)*

(b) Use trigonometric substitution to solve the definite integral you found in part (a). *(Hint: Change the limits of integration to match your substitution.)*

94. Prove part (b) of Theorem 5.18: For $x \in (-\infty, \infty)$ and $u \in (-\pi/2, \pi/2)$, the substitution $x = a \tan u$ gives $x^2 + a^2 = a \sec u$. Your proof should include a discussion of domains and a consideration of absolute values.

Thinking Forward

Arc length of a curve: Suppose we want to find the length of the curve traced out by the graph of a function $f(x)$ on an interval $[a, b]$. The obvious way to approximate this length is to use a series of straight line segments such as the one shown in the figure.

Approximating length with a line segment

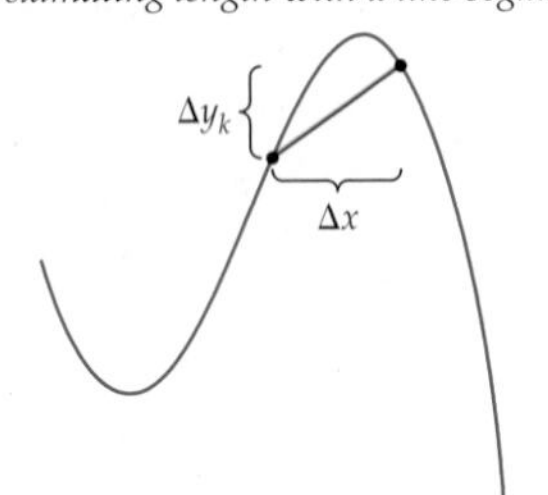

By the distance formula, the length of the kth segment is $\sqrt{(\Delta x)^2 + (\Delta y_k)^2}$, where Δy_k will be different for each segment we use.

- Use algebra to show that the distance of the kth segment can be rewritten as

$$\sqrt{1 + \left(\frac{\Delta y_k}{\Delta x}\right)^2}\, \Delta x.$$

- Approximate the length of the graph of $f(x) = \sqrt{9 - x^2}$ on $[-3, 3]$, using four line segments and the distance formula. Then make a better approximation with eight line segments.

- As we will see in Section 6.3, the definite integral

$$L = \int_a^b \sqrt{1 + (f'(x))^2}\, dx$$

calculates the length of the curve of $f(x)$ from $x = a$ to $x = b$. Explain intuitively what this might have to do with the distance formula calculation.

- Use the formula just given for arc length to write down a specific definite integral that represents the length of $f(x) = \sqrt{9 - x^2}$ on $[-3, 3]$. Then use trigonometric substitution to solve that definite integral.

- Verify your previous answer by using the formula for the circumference of a circle.

Surface area of a solid of revolution: If $f(x)$ is a differentiable function with a continuous derivative, then the surface area of the solid of revolution obtained by rotating the region under the graph of $f(x)$ on the interval $[a, b]$ around the x-axis is given by the definite integral

$$S = 2\pi \int_a^b f(x)\sqrt{1 + (f'(x))^2}\, dx.$$

- Calculate the surface area of the solid of revolution obtained by revolving the region between $f(x) = 2x^3$ and the x-axis on $[0, 3]$ around the x-axis.

- Use the formula just given for surface area to write down and solve a specific definite integral to show that the surface area of a sphere of radius r is given by the formula $S = 4\pi r^2$.

5.6 IMPROPER INTEGRALS

- Defining improper integrals over unbounded intervals and for unbounded functions
- Determining the convergence or divergence of improper integrals of power functions
- The comparison test for improper integrals

Integrating over an Unbounded Interval

So far we have examined definite integrals $\int_a^b f(x)\,dx$ only for continuous functions $f(x)$ over finite intervals $[a, b]$, since those are the conditions under which the Fundamental Theorem of Calculus applies. If $f(x)$ fails to be continuous on the interval or if the interval itself is unbounded, then the Fundamental Theorem of Calculus does not apply and we say that the integral is ***improper***. In this section we develop definitions and strategies for examining such improper integrals.

For example, consider the functions $\frac{1}{\sqrt{x}}$, $\frac{1}{x}$, and $\frac{1}{x^2}$ on the interval $[1, \infty)$, as shown in the three figures that follow. Perhaps unsurprisingly, for the first two graphs there is an infinite amount of area between the graph of the function and the x-axis on this unbounded interval. However, you might be shocked to learn that for the third graph this area turns out to be finite!

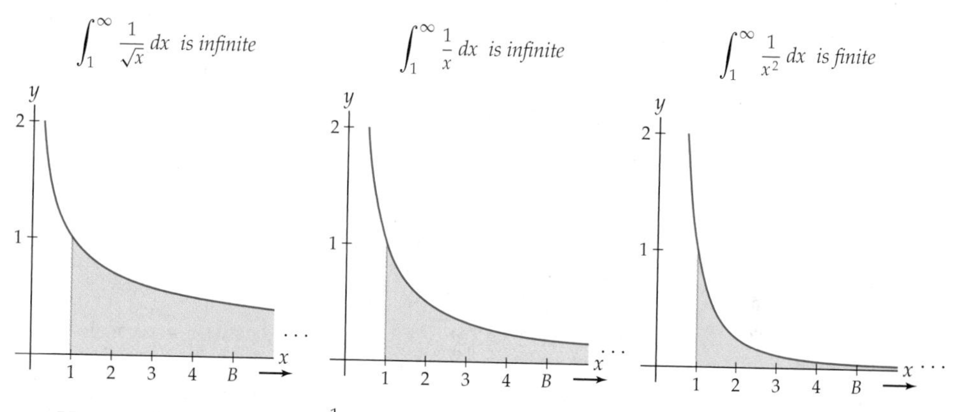

How could the graph of $\frac{1}{x^2}$ be positive on all of $[1, \infty)$ and yet collect only a finite amount of area on this interval? Let's look at the area under each graph for intervals of the form $[1, B)$, with B getting larger and larger:

	$B = 5$	$B = 10$	$B = 50$	$B = 100$	$B = 500$	$B = 1000$	$B \to \infty$
$\int_1^B \frac{1}{\sqrt{x}}\,dx$	6.787	20.415	235.036	666.000	7,452.893	21,081.184	$\to \infty$?
$\int_1^B \frac{1}{x}\,dx$	1.609	2.302	3.912	4.605	6.215	6.908	$\to \infty$??
$\int_1^B \frac{1}{x^2}\,dx$	0.800	0.900	0.980	0.990	0.998	0.999	$\to 1$???

As B increases, the area under the graph of $\frac{1}{\sqrt{x}}$ on $[1, B)$ seems to grow without bound. The same seems true of $\frac{1}{x}$, but the growth is much slower. For $\frac{1}{x^2}$, however, the area under the graph on $[1, B)$ seems to stay below 1 no matter how large a value we take for B. The graph of $\frac{1}{x^2}$ decreases in some fundamentally faster way than the other two graphs, so much so that it collects only a finite amount of area.

We can investigate these areas algebraically over finite intervals $[1, B)$ by using the Fundamental Theorem of Calculus. For $B > 1$ we have

$$\int_1^B \frac{1}{\sqrt{x}}\,dx = \left[2\sqrt{x}\right]_1^B = 2\sqrt{B} - 2\sqrt{1} = 2\sqrt{B} - 2,$$

$$\int_1^B \frac{1}{x}\,dx = \left[\ln|x|\right]_1^B = \ln B - \ln 1 = \ln B,$$

$$\int_1^B \frac{1}{x^2}\,dx = \left[-\frac{1}{x}\right]_1^B = -\frac{1}{B} + \frac{1}{1} = 1 - \frac{1}{B}.$$

We can now see that the values of the first two definite integrals will grow without bound as $B \to \infty$, since $\lim_{B\to\infty}(2\sqrt{B} - 2) \to 2(\infty) - 2 = \infty$ and $\lim_{B\to\infty} \ln B = \infty$. In contrast, since $\lim_{B\to\infty}(1 - (1/B)) = 1 - \frac{1}{\infty} = 1$, the value of the third definite integral approaches 1.

In general, we define integrals over unbounded intervals as limits of integrals over ever-larger finite intervals:

DEFINITION 5.19

Improper Integrals over Unbounded Intervals

Suppose a and b are any real numbers and $f(x)$ is a function.

(a) If f is continuous on $[a, \infty)$, then $\int_a^\infty f(x)\,dx = \lim_{B\to\infty} \int_a^B f(x)\,dx.$

(b) If f is continuous on $(-\infty, b]$, then $\int_{-\infty}^b f(x)\,dx = \lim_{A\to-\infty} \int_A^b f(x)\,dx.$

(c) If f is continuous on $(-\infty, \infty)$, then define the integral over $(-\infty, \infty)$ as a sum of the first two cases, splitting at any real number c:

$$\int_{-\infty}^\infty f(x)\,dx = \lim_{A\to-\infty} \int_A^c f(x)\,dx + \lim_{B\to\infty} \int_c^B f(x)\,dx$$

In all three cases, if the limits involved exist, then we say that the improper integral ***converges*** to the value determined by those limits, and if the limits are infinite or fail to exist, then we say that the improper integral ***diverges***.

Remember that a limit exists if it is equal to a finite number, so an improper integral converges if it is equal to some finite number and diverges if it is infinite or for some other reason does not have a well-defined limit. In Exercise 81 you will use properties of definite integrals to prove that we can in fact split the integral in the third part of the theorem at any point c that we like. In the terminology of this theorem, the improper integrals $\int_1^\infty \frac{1}{\sqrt{x}}\,dx$ and $\int_1^\infty \frac{1}{x}\,dx$ diverge while the improper integral $\int_1^\infty \frac{1}{x^2}\,dx$ converges to 1.

Integrating Unbounded Functions

We now know how to extend our theory of definite integrals to unbounded intervals. What about over finite intervals, but where the function itself is unbounded? The Fundamental Theorem of Calculus does not apply to integrands that have discontinuities such as vertical asymptotes, but we can again use limits to determine whether integrals of unbounded functions converge or diverge.

For example, consider the same three functions we were working with earlier, but over a different graphing window to illustrate their behavior on $[0, 1]$, as shown in the next three figures. Each graph has a vertical asymptote at $x = 0$. In the second and third figures the area between the graph and the x-axis on $[0, 1]$ will turn out to be infinite, and in the first figure the graph of $\frac{1}{\sqrt{x}}$ will turn out to have a finite area over $[0, 1]$.

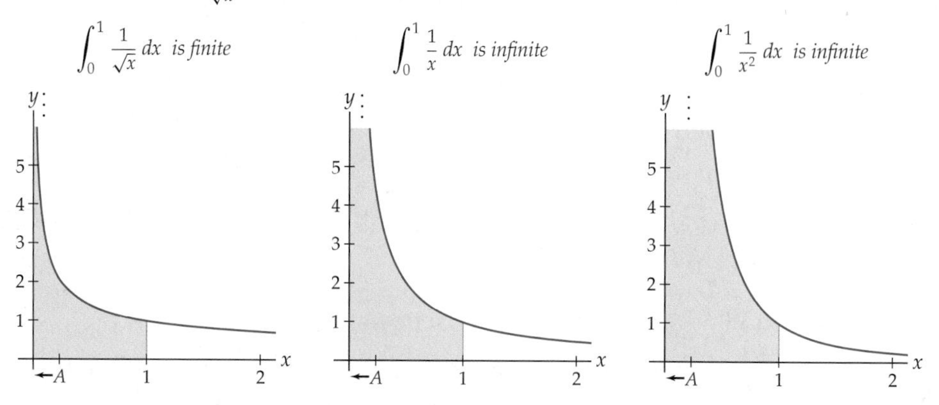

As suggested by the labels under the axes in each figure, what we should think about is the area under these curves on intervals of the form $(A, 1)$ as A gets closer and closer to 0 from the right. For $0 < A < 1$ the Fundamental Theorem of Calculus tells us that

$$\int_A^1 \frac{1}{\sqrt{x}}\,dx = \left[2\sqrt{x}\right]_A^1 = 2\sqrt{1} - 2\sqrt{A} = 2 - 2\sqrt{A},$$

$$\int_A^1 \frac{1}{x}\,dx = \left[\ln|x|\right]_A^1 = \ln 1 - \ln A = -\ln A,$$

$$\int_A^1 \frac{1}{x^2}\,dx = \left[-\frac{1}{x}\right]_A^1 = -\frac{1}{1} + v1A = \frac{1}{A} - 1.$$

As $A \to 0^+$, we have $\sqrt{A} \to 0$, $\ln A \to -\infty$, and $\frac{1}{A} \to \infty$, so the first integral will approach 2 and the second and third will both approach ∞. Therefore, of these three functions, only $\frac{1}{\sqrt{x}}$ has a finite area over $[0, 1]$.

The preceding example suggests how we should define improper integrals $\int_a^b f(x)\,dx$ for which $f(x)$ has a discontinuity at an endpoint of $[a, b]$. In cases where $f(x)$ is discontinuous somewhere in the interior of $[a, b]$, we have to split the integral at the discontinuity:

DEFINITION 5.20

Improper Integrals over Discontinuities

Suppose a and b are any real numbers and $f(x)$ is a function.

(a) If f is continuous on $(a, b]$ but not at $x = a$, then $\displaystyle\int_a^b f(x)\,dx = \lim_{A\to a^+} \int_A^b f(x)\,dx.$

(b) If f is continuous on $[a, b)$ but not at $x = b$, then $\displaystyle\int_a^b f(x)\,dx = \lim_{B\to b^-} \int_a^B f(x)\,dx.$

(c) If f is continuous on $[a, c) \cup (c, b]$ but not at $x = c$, then we can define the integral over $[a, b]$ as the sum of the first two cases:

$$\int_a^b f(x)\,dx = \lim_{B\to c^-} \int_a^B f(x)\,dx + \lim_{A\to c^+} \int_A^b f(x)\,dx$$

In all three cases, if the limits involved exist, then we say that the improper integral ***converges*** to the value determined by those limits, and if the limits are infinite or fail to exist, then we say that the improper integral ***diverges***.

Improper Integrals of Power Functions

The function $\frac{1}{x}$ is a kind of boundary between power functions whose integrals converge and power functions whose integrals diverge. On both $[0, 1]$ and $[1, \infty)$ the improper integral of $\frac{1}{x}$ diverges. A power function of the form $\frac{1}{x^p}$ has an improper integrals that converges or diverges on an interval according to whether that power function is less than or greater than $\frac{1}{x}$ on that interval. We use the letter p in our notation here because the integrals we are studying now are closely related to certain infinite sums that we will later call ***p*-series** in Chapter 7.

More precisely, in Exercises 74 and 75 you will use limits of proper definite integrals to show that, for improper integrals over $[1, \infty)$, we have the following theorem:

THEOREM 5.21 **Improper Integrals of Power Functions on $[1, \infty)$**

(a) If $0 < p \le 1$, then $\frac{1}{x^p} \ge \frac{1}{x}$ for $x \in [1, \infty)$ and $\int_1^\infty \frac{1}{x^p}\,dx$ diverges.

(b) If $p > 1$, then $\frac{1}{x^p} < \frac{1}{x}$ for $x \in [1, \infty)$ and $\int_1^\infty \frac{1}{x^p}\,dx$ converges to $\frac{1}{p-1}$.

Notice that we do not bother to consider power functions like x^2 or x^3 in this theorem, since those functions are increasing on $[1, \infty)$ and therefore have improper integrals that obviously diverge over that interval. The statements in the theorem about whether $\frac{1}{x^p}$ is greater or less than $\frac{1}{x}$ are there to help us remember whether we have convergence or divergence in each case. Theorem 5.21 will be a key tool in our investigation both of improper integrals in this section and of power series in Chapter 7.

In addition, in Exercises 76 and 77 you will prove that improper integrals of power functions over $[0, 1]$ also either converge or diverge according to how closely they compare to $\frac{1}{x}$, as follows:

THEOREM 5.22 **Improper Integrals of Power Functions on $[0, 1]$**

(a) If $0 < p < 1$, then $\frac{1}{x^p} < \frac{1}{x}$ for $x \in [0, 1]$ and $\int_0^1 \frac{1}{x^p}\,dx$ converges to $\frac{1}{1-p}$.

(b) If $p \ge 1$, then $\frac{1}{x^p} \ge \frac{1}{x}$ for $x \in [0, 1]$ and $\int_0^1 \frac{1}{x^p}\,dx$ diverges.

Again we do not consider power functions such as x^2 or x^3 in the theorem because their integrals over $[0, 1]$ are not improper. (See Example 2 for an illustration of the four cases in the two previous theorems.)

Determining Convergence or Divergence with Comparisons

We now have a complete understanding of the convergence or divergence of improper integrals of power functions. These special examples of improper integrals will be the yardsticks against which we compare many other types of improper integrals. This comparison will enable us to determine the convergence or divergence of many improper integrals without actually doing any integration or antidifferentiation.

Loosely speaking, the key idea that will allow us to compare two improper integrals is this: Improper integrals that are smaller than convergent ones will also converge, and improper integrals that are larger than divergent ones will also diverge. This idea makes intuitive sense because a (positive) quantity that is smaller than a finite number will also be finite and a quantity that is larger than an infinite quantity will also be infinite. More precisely, we have the following theorem:

THEOREM 5.23

Comparison Test for Improper Integrals

Suppose f and g are functions that are continuous on an interval I.

(a) If the improper integral of f on I converges and $0 \le g(x) \le f(x)$ for all $x \in I$ where the functions are defined, then the improper integral of g on I also converges.

(b) If the improper integral of f on I diverges and $0 \le f(x) \le g(x)$ for all $x \in I$ where the functions are defined, then the improper integral of g on I also diverges.

Theorem 5.23 applies to all the types of improper integrals we have studied. For example, for improper integrals on an unbounded interval $[a, \infty)$, the theorem says that

$$\begin{array}{l} \int_a^\infty f(x)\,dx \text{ converges and} \\ 0 \le g(x) \le f(x) \text{ for all } x \in [a,\infty) \end{array} \quad \Longrightarrow \quad \int_a^\infty g(x)\,dx \text{ also converges;}$$

$$\begin{array}{l} \int_a^\infty f(x)\,dx \text{ diverges and} \\ g(x) \ge f(x) \text{ for all } x \in [a,\infty) \end{array} \quad \Longrightarrow \quad \int_a^\infty g(x)\,dx \text{ also diverges.}$$

It is important to note that Theorem 5.23 says nothing about improper integrals that are larger than convergent ones or about improper integrals that are smaller than divergent ones.

For example, consider the improper integrals $\int_1^\infty \frac{1}{x^2}\,dx$ and $\int_1^\infty \frac{\sin^2 x}{x^2}\,dx$. We already know that the first of these converges. Considering the fact that $-1 \le \sin x \le 1$, we have $0 \le \sin^2 x \le 1$ and therefore $0 \le \frac{\sin^2 x}{x^2} \le \frac{1}{x^2}$. These facts taken together tell us that $\int_1^\infty \frac{\sin^2 x}{x^2}\,dx$ must also converge, even without our having to integrate the function $\frac{\sin^2 x}{x^2}$.

Proof. We will prove the theorem in the case for convergence of improper integrals over unbounded intervals of the form $[a, \infty)$ and leave similar proofs of other cases to Exercises 78–80. Suppose that $f(x)$ is a function for which the improper integral $\int_a^\infty f(x)\,dx$ converges. If $g(x)$ is a function satisfying $0 \le g(x) \le f(x)$ for all $x \in [a, \infty)$, then for all $B > a$, we have the following inequality of definite integrals:

$$0 \le \int_a^B g(x)\,dx \le \int_a^B f(x)\,dx.$$

Now taking the limit as $B \to \infty$ on each side, we get

$$0 \le \lim_{B\to\infty} \int_a^B g(x)\,dx \le \lim_{B\to\infty} \int_a^B f(x)\,dx.$$

By the definition of improper integrals on unbounded intervals, this double inequality is equivalent to

$$0 \le \int_a^\infty g(x)\,dx \le \int_a^\infty f(x)\,dx.$$

We are given that $\int_a^\infty f(x)\,dx$ converges, say, to some real number L. Then the inequality we found shows that $\int_a^\infty g(x)\,dx$ must converge to a real number between 0 and L. ■

Examples and Explorations

EXAMPLE 1

Writing improper integrals as limits of definite integrals

Express each improper integral that follows as a sum of limits of proper definite integrals. Do not calculate any integrals or limits; just write them down.

(a) $\displaystyle\int_0^1 (3x-1)^{-2/3}\,dx$ **(b)** $\displaystyle\int_0^1 \frac{1}{x\ln x}\,dx$ **(c)** $\displaystyle\int_1^\infty \frac{1}{(x-2)^2}\,dx$

SOLUTION

The graphs of the three functions we are attempting to integrate are shown here:

$(3x-1)^{-2/3}$ *on* $[0,1]$ $\qquad \dfrac{1}{x\ln x}$ *on* $[0,1]$ $\qquad \dfrac{1}{(x-2)^2}$ *on* $[1,\infty)$

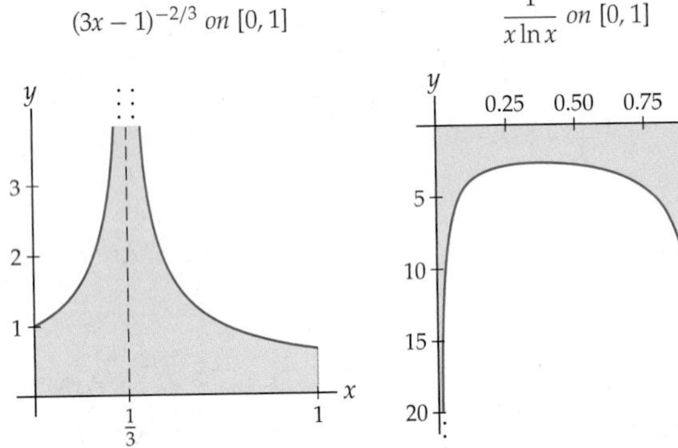
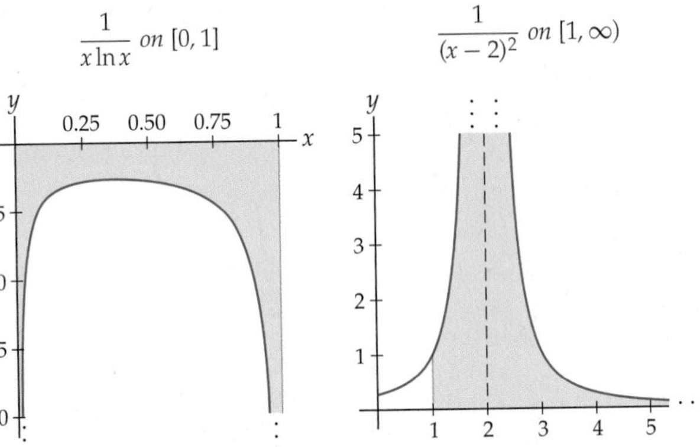

We have to split each improper integral into a sum of integrals so that in each integral only one endpoint requires a limit. In parts (a) and (b) this will mean splitting into two integrals, and in part (c) we will require three integrals.

(a) The function $y = (3x-1)^{-2/3}$ is continuous everywhere except at $x = \frac{1}{3}$, where it has a vertical asymptote. Since $x = \frac{1}{3}$ is in the interior of the interval $[0,1]$, we must split the integral into two improper integrals:

$$\int_0^1 (3x-1)^{-2/3}\,dx = \int_0^{1/3} (3x-1)^{-2/3}\,dx + \int_{1/3}^1 (3x-1)^{-2/3}\,dx.$$

Each of these improper integrals can be written as a limit of definite integrals, which gives us

$$\int_0^1 (3x-1)^{-2/3}\,dx = \lim_{B\to 1/3^-}\int_0^B (3x-1)^{-2/3}\,dx + \lim_{A\to 1/3^+}\int_A^1 (3x-1)^{-2/3}\,dx.$$

If either of these limits of integrals diverges, then the entire improper integral diverges. If both limits of integrals converge, then the original improper integral is equal to the sum of those limits. (See Exercise 29.)

(b) The function $y = \frac{1}{x\ln x}$ has vertical asymptotes at both endpoints of the interval $[0,1]$, so we have to split this improper integral somewhere between $x = 0$ and $x = 1$. The point $x = \frac{1}{2}$ is as good a place as any to make the split:

$$\int_0^1 \frac{1}{x\ln x}\,dx = \int_0^{1/2} \frac{1}{x\ln x}\,dx + \int_{1/2}^1 \frac{1}{x\ln x}\,dx.$$

Now taking limits at the endpoints where there are vertical asymptotes, we have

$$\int_0^1 \frac{1}{x \ln x}\,dx = \lim_{A\to 0^+} \int_A^{1/2} \frac{1}{x \ln x}\,dx + \lim_{B\to 1^-} \int_{1/2}^{B} \frac{1}{x \ln x}\,dx.$$

You will determine whether this improper integral converges or diverges in Exercise 47.

(c) The graph of $y = \frac{1}{(x-2)^2}$ has a vertical asymptote at $x = 2$, and we need to express the improper integral over $[1, \infty)$. This means that we will need to split the integral at $x = 2$, as well as at some later point, say, $x = 3$, in order to separately consider the limit as $x \to \infty$. This approach forces a split into three integrals:

$$\int_1^\infty \frac{1}{(x-2)^2}\,dx = \int_1^2 \frac{1}{(x-2)^2}\,dx + \int_2^3 \frac{1}{(x-2)^2}\,dx + \int_3^\infty \frac{1}{(x-2)^2}\,dx.$$

Taking limits at the points where we have vertical asymptotes and as $x \to \infty$, we have

$$\int_1^\infty \frac{1}{(x-2)^2}\,dx = \lim_{B\to 2^-} \int_1^B \frac{1}{(x-2)^2}\,dx + \lim_{A\to 2^+} \int_A^3 \frac{1}{(x-2)^2}\,dx + \lim_{C\to\infty} \int_3^C \frac{1}{(x-2)^2}\,dx.$$

We used the letter C in the last limit only to distinguish it from the variable B that we used in the first limit. For calculations involving these improper integrals, see Exercises 30 and 31. □

EXAMPLE 2

Calculating improper integrals of power functions

For each power function $f(x)$ that follows, use limits of definite integrals to determine whether $\int_0^1 f(x)\,dx$ and $\int_1^\infty f(x)\,dx$ converge or diverge. Find the values of those improper integrals which converge.

(a) $f(x) = \frac{1}{\sqrt[3]{x}}$ **(b)** $f(x) = \frac{1}{x}$ **(c)** $f(x) = \frac{1}{x^3}$

SOLUTION

(a) Since $f(x) = \frac{1}{\sqrt[3]{x}}$ is continuous on $(0, 1]$ but has a vertical asymptote at $x = 0$, Definition 5.20 and the Fundamental Theorem of Calculus tell us that the improper integral on $[0, 1]$ converges:

$$\int_0^1 \frac{1}{\sqrt[3]{x}}\,dx = \lim_{A\to 0^+} \int_A^1 \frac{1}{\sqrt[3]{x}}\,dx = \lim_{A\to 0^+} \left[\frac{3}{2}x^{2/3}\right]_A^1 = \lim_{A\to 0^+} \left(\frac{3}{2} - \frac{3}{2}A^{2/3}\right) = \frac{3}{2}.$$

$f(x) = \frac{1}{\sqrt[3]{x}}\,dx$ is also continuous on $[1, \infty)$, and therefore Definition 5.19 and the Fundamental Theorem of Calculus tell us that the improper integral on $[1, \infty)$ diverges:

$$\int_1^\infty \frac{1}{\sqrt[3]{x}}\,dx = \lim_{B\to\infty} \int_1^B \frac{1}{\sqrt[3]{x}}\,dx = \lim_{B\to\infty} \left[\frac{3}{2}x^{2/3}\right]_1^B = \lim_{B\to\infty} \left(\frac{3}{2}B^{2/3} - \frac{3}{2}\right) = \infty.$$

(b) Earlier in the reading we used limits of definite integrals to determine that $\int_0^1 \frac{1}{x}\,dx$ and $\int_1^\infty \frac{1}{x}\,dx$ both diverge.

(c) Again we apply the definitions of improper integrals to get

$$\lim_{A\to 0^+}\int_A^1 \frac{1}{x^3}\,dx = \lim_{A\to 0^+}\left[-\tfrac{1}{2}x^{-2}\right]_A^1 = \lim_{A\to 0^+}\left(-\tfrac{1}{2}+\tfrac{1}{2}A^{-2}\right) = -\frac{1}{2}+\frac{1}{2(0^+)} = \infty,$$

$$\lim_{B\to\infty}\int_1^B \frac{1}{x^3}\,dx = \lim_{B\to\infty}\left[-\tfrac{1}{2}x^{-2}\right]_1^B = \lim_{B\to\infty}\left(-\tfrac{1}{2}B^{-2}+\tfrac{1}{2}\right) = -\frac{1}{2(\infty)}+\frac{1}{2} = \frac{1}{2}. \quad \square$$

CHECKING THE ANSWER

First, notice that the preceding two answers do match with the statements in Theorems 5.21 and 5.22. For $p = \frac{1}{3}$ only the integral on $[0, 1]$ converges, and it converges to $\frac{1}{1-p} = \frac{1}{1-(1/3)} = \frac{3}{2}$; for $p = 1$ neither improper integral converges; and for $p = 3$ only the integral on $[1, \infty)$ converges, to $\frac{1}{p-1} = \frac{1}{3-1} = \frac{1}{2}$.

In addition, we can examine the graphs of $\frac{1}{\sqrt[3]{x}}$, $\frac{1}{x}$, and $\frac{1}{x^3}$. Of course we cannot tell whether the areas of regions are infinite or finite just by visual inspection, but when we look at all three graphs together, the answers do seem plausible:

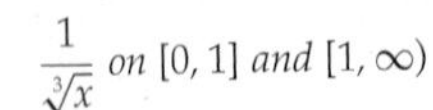

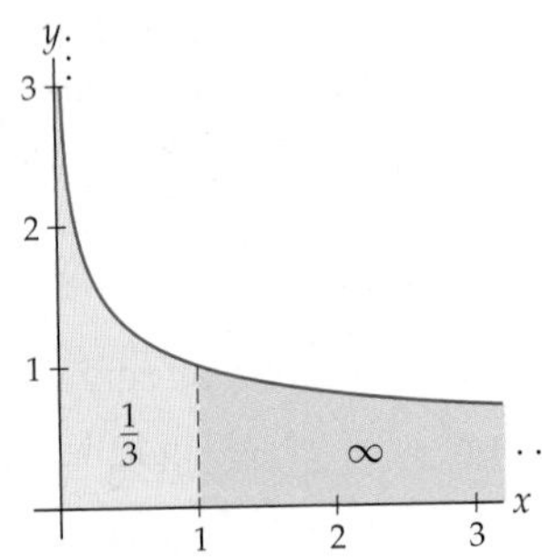

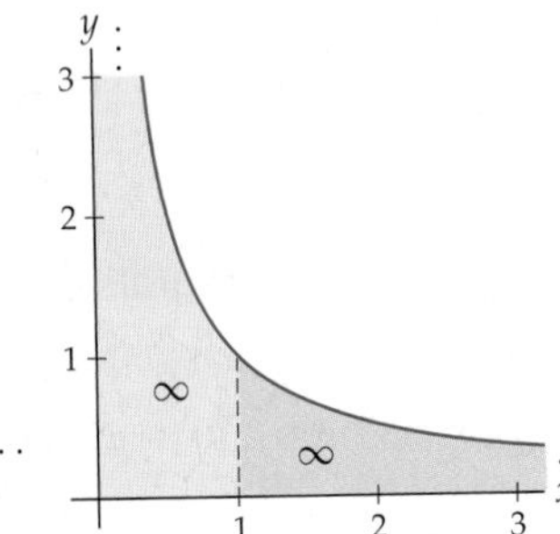

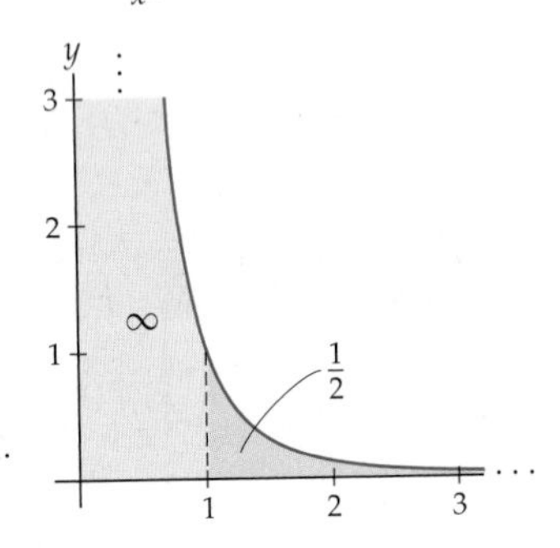

EXAMPLE 3

Calculating improper integrals

Use limits of definite integrals to calculate each of the following improper integrals:

(a) $\displaystyle\int_0^{\infty} e^{-x}\,dx$ **(b)** $\displaystyle\int_0^{5} \frac{1}{(x-3)^{4/5}}\,dx$ **(c)** $\displaystyle\int_{-\infty}^{\infty} \frac{1}{1+x^2}\,dx$

SOLUTION

(a) The function $y = e^{-x}$ has no vertical asymptotes and is in fact continuous on $[0, \infty)$. Therefore we have

$$\begin{aligned}\int_0^{\infty} e^{-x}\,dx &= \lim_{B\to\infty}\int_0^B e^{-x}\,dx && \leftarrow \text{definition of improper integral}\\ &= \lim_{B\to\infty}\left[-e^{-x}\right]_0^B && \leftarrow \text{antidifferentiate}\\ &= \lim_{B\to\infty}(-e^{-B}+1) && \leftarrow \text{evaluate}\\ &= -\frac{1}{\infty}+1 = 1. && \leftarrow \text{calculate limit}\end{aligned}$$

(b) The function $y = \frac{1}{(x-3)^{4/5}}$ has a vertical asymptote at $x = 3$ but is continuous on $[0, 3) \cup (3, 5]$. We can solve this improper integral by splitting at $x = 3$:

$$\int_0^5 \frac{1}{(x-3)^{4/5}}\,dx = \int_0^3 \frac{1}{(x-3)^{4/5}}\,dx + \int_3^5 \frac{1}{(x-3)^{4/5}}\,dx \qquad \leftarrow \text{split at } x = 3$$

$$= \lim_{B\to 3^-} \int_0^B \frac{1}{(x-3)^{4/5}}\,dx + \lim_{A\to 3^+} \int_A^5 \frac{1}{(x-3)^{4/5}}\,dx \qquad \leftarrow \text{use limits}$$

$$= \lim_{B\to 3^-} \left[5(x-3)^{1/5}\right]_0^B + \lim_{A\to 3^+} \left[5(x-3)^{1/5}\right]_A^5 \qquad \leftarrow \text{antidifferentiate}$$

$$= \lim_{B\to 3^-} (5(B-3)^{1/5} - 5(-3)^{1/5}) + \lim_{A\to 3^+} (5(2)^{1/5} - 5(A-3)^{1/5}) \qquad \leftarrow \text{evaluate}$$

$$= 0 + 5(3^{1/5}) + 5(2^{1/5}) - 0 = 5(3^{1/5} + 2^{1/5}). \qquad \leftarrow \text{calculate limit}$$

(c) The function $y = \frac{1}{1+x^2}$ is continuous on all of $(-\infty, \infty)$, so we need only split this improper integral into two pieces so that we can handle the limits to ∞ and $-\infty$ separately. We might as well make the split at $x = 0$:

$$\int_{-\infty}^{\infty} \frac{1}{1+x^2}\,dx = \int_{-\infty}^{0} \frac{1}{1+x^2}\,dx + \int_0^{\infty} \frac{1}{1+x^2}\,dx \qquad \leftarrow \text{split at } x = 0$$

$$= \lim_{A\to -\infty} \int_A^0 \frac{1}{1+x^2}\,dx + \lim_{B\to\infty} \int_0^B \frac{1}{1+x^2}\,dx \qquad \leftarrow \text{use limits}$$

$$= \lim_{A\to -\infty} \left[\tan^{-1} x\right]_A^0 + \lim_{B\to\infty} \left[\tan^{-1} x\right]_0^B \qquad \leftarrow \text{antidifferentiate}$$

$$= \lim_{A\to -\infty} (0 - \tan^{-1} A) + \lim_{B\to\infty} (\tan^{-1} B - 0) \qquad \leftarrow \text{evaluate}$$

$$= \frac{\pi}{2} + \frac{\pi}{2} = \pi. \qquad \leftarrow \text{calculate limit}$$

□

EXAMPLE 4

Using the comparison test to determine convergence or divergence

Determine whether each of the first two improper integrals listed here converges or diverges by applying the comparison test with simpler improper integrals. For the third improper integral, explain why the comparison test with $\frac{1}{x^2}$ does not give information; then determine convergence or divergence by using limits of definite integrals.

(a) $\displaystyle\int_1^{\infty} \frac{1}{xe^x}\,dx$ **(b)** $\displaystyle\int_0^1 \frac{1+x}{x^3}\,dx$ **(c)** $\displaystyle\int_2^{\infty} \frac{1}{x^2 - x}\,dx$

SOLUTION

(a) To use the comparison test we have two choices: If we think that this improper integral converges, then we should try to find a function larger than $\frac{1}{xe^x}$ on $[1, \infty)$, which we know converges on that interval. If instead we think that the improper integral diverges, then we should try to find a function smaller than $\frac{1}{xe^x}$ on $[1, \infty)$, which we know diverges.

A good first try might be to notice that $xe^x \geq x$ for all $x \in [1, \infty)$, from which we can deduce that $0 < \frac{1}{xe^x} \leq \frac{1}{x}$ on $[1, \infty)$. But we know from Theorem 5.21 that $\int_1^{\infty} \frac{1}{x}\,dx$ diverges, and knowing that our improper integral is less than something infinite does not tell us whether it is infinite or finite.

As another try, notice that since $e^x \geq x$ for all $x \in [1, \infty)$, we must have $xe^x \geq x^2$ and thus $0 < \frac{1}{xe^x} \leq \frac{1}{x^2}$. Now we are getting somewhere: Since $\int_1^\infty \frac{1}{x^2}\,dx$ converges by Theorem 5.21, we know from Theorem 5.23 that the smaller improper integral $\int_1^\infty \frac{1}{xe^x}\,dx$ must also converge. In symbols, we have

$$\int_1^\infty \frac{1}{xe^x}\,dx \leq \int_1^\infty \frac{1}{x^2}\,dx,$$

which is finite, and thus $\int_1^\infty \frac{1}{xe^x}\,dx$ must also be finite. Note that the comparison test has not told us the value of our improper integral, only that it must converge to some finite value.

(b) For all $x \in (0, 1]$ we have

$$1 + x \geq x \quad \Longrightarrow \quad \frac{1+x}{x^3} \geq \frac{x}{x^3} \quad \Longrightarrow \quad \frac{1+x}{x^3} \geq \frac{1}{x^2}.$$

(Alternatively, we could obtain this inequality by noticing that $\frac{1+x}{x^3} = \frac{1}{x^3} + \frac{1}{x^2} \geq \frac{1}{x^2}$ on $(0, 1]$.) We also know by Theorem 5.22 that the improper integral of $\frac{1}{x^2}$ diverges on $[0, 1]$. Combining these facts, we have

$$\int_0^1 \frac{1+x}{x^3}\,dx \geq \int_0^1 \frac{1}{x^2}\,dx = \infty,$$

and thus, by the comparison test, $\int_0^1 \frac{1+x}{x^3}\,dx$ is also infinite and thus diverges.

(c) First notice that since $0 \leq x^2 - x \leq x^2$ for all $x \in [2, \infty)$, it follows that $\frac{1}{x^2-x} \geq \frac{1}{x^2}$ on that interval. But a comparison with this function whose improper integral on $[2, \infty)$ converges can tell us only that our improper integral is greater than something that is finite, which does not tell us whether our improper integral is finite or infinite.

Resorting to calculation instead, we can use a limit of definite integrals that we can solve by the method of partial fractions:

$$\begin{aligned}
\int_2^\infty \frac{1}{x^2 - x}\,dx &= \lim_{B\to\infty} \int_2^B \frac{1}{x^2 - x}\,dx && \leftarrow \text{definition of improper integral} \\
&= \lim_{B\to\infty} \int_2^B \frac{1}{x(x-1)}\,dx && \leftarrow \text{algebra} \\
&= \lim_{B\to\infty} \int_2^B \left(-\frac{1}{x} + \frac{1}{x-1}\right) dx && \leftarrow \text{partial fractions} \\
&= \lim_{B\to\infty} \left[-\ln|x| + \ln|x-1|\right]_2^B && \leftarrow \text{antidifferentiate} \\
&= \lim_{B\to\infty} (-\ln B + \ln(B-1) + \ln 2 - \ln 1) && \leftarrow \text{evaluate} \\
&= \lim_{B\to\infty} \left(\ln\left(\frac{B-1}{B}\right) + \ln 2\right) && \leftarrow \text{properties of logarithms} \\
&= \ln 1 + \ln 2 = \ln 2. && \leftarrow \text{calculate limit}
\end{aligned}$$

From this calculation we see that the improper integral converges. □

? TEST YOUR UNDERSTANDING

- What types of power functions have improper integrals that converge on $[1, \infty)$? That diverge? What about on $[0, 1]$?
- Why do we have to split the improper integral $\int_{-\infty}^\infty \frac{1}{1+x^2}\,dx$ at some point $x = c$ in order to calculate it in terms of limits of definite integrals?
- Why do we have to split the improper integral $\int_0^2 \frac{1}{(x-1)^2}\,dx$ at $x = 1$ in order to calculate it in terms of limits of definite integrals?

- Why does the Fundamental Theorem of Calculus not apply directly to improper integrals? What hypotheses of the Fundamental Theorem of Calculus fail to be true for improper integrals?
- Why is it not helpful to know that an improper integral has a larger value than a convergent improper integral or a smaller value than a divergent improper integral?

EXERCISES 5.6

Thinking Back

Limits: Solve each of the following limits.

- $\lim_{x\to\infty} x^{-2/3}$
- $\lim_{x\to\infty} x^{-4/3}$
- $\lim_{x\to 0^+} x^{-2/3}$
- $\lim_{x\to 0^+} x^{-4/3}$
- $\lim_{x\to 1^-} \frac{1}{x\ln x}$
- $\lim_{x\to 0^+} \frac{1}{x\ln x}$
- $\lim_{x\to\infty} e^{-x^2}$
- $\lim_{x\to\infty} xe^{-x^2}$
- $\lim_{x\to\pi/2^-} \sec x$

Integration: Solve each of the following indefinite integrals.

- $\int (3x-1)^{-2/3}\,dx$
- $\int \frac{1}{x\ln x}\,dx$
- $\int xe^{-x^2}\,dx$
- $\int \frac{\ln x}{x}\,dx$
- $\int \frac{1}{x(x-1)}\,dx$
- $\int \sec x\,dx$

The Fundamental Theorem of Calculus: Solve each of the following definite integrals.

- $\int_1^{100} \frac{1}{\sqrt[3]{x^4}}\,dx$
- $\int_1^{100} \frac{1}{\sqrt[4]{x^3}}\,dx$
- $\int_{0.01}^{1} \frac{1}{\sqrt[3]{x^4}}\,dx$
- $\int_{0.01}^{1} \frac{1}{\sqrt[4]{x^3}}\,dx$

Concepts

0. *Problem Zero:* Read the section and make your own summary of the material.

1. *True/False:* Determine whether each of the statements that follow is true or false. If a statement is true, explain why. If a statement is false, provide a counterexample.

(a) *True or False:* The Fundamental Theorem of Calculus applies to $\int_0^2 \frac{1}{1-x}\,dx$.

(b) *True or False:* The Fundamental Theorem of Calculus applies to $\int_2^\infty \frac{1}{1-x}\,dx$.

(c) *True or False:* If $f(x)$ is positive and decreasing, then the area between the graph of $f(x)$ and the x-axis on $[1,\infty)$ must be finite.

(d) *True or False:* If $f(x)$ is positive and decreasing, then the area between the graph of $f(x)$ and the x-axis on $[1,\infty)$ must be infinite.

(e) *True or False:* If $p > 1$, then $\int_1^\infty \frac{1}{x^p}\,dx = \frac{1}{p-1}$.

(f) *True or False:* If $p < 1$, then $\int_0^1 \frac{1}{x^p}\,dx = \frac{1}{1-p}$.

(g) *True or False:* If $\int_1^\infty f(x)\,dx$ converges and $0 \le f(x) \le g(x)$ for all $x \in [1,\infty)$, then $\int_1^\infty g(x)\,dx$ must diverge.

(h) *True or False:* If $\int_1^\infty f(x)\,dx$ diverges and $0 \le f(x) \le g(x)$ for all $x \in [1,\infty)$, then $\int_1^\infty g(x)\,dx$ must diverge.

2. *Examples:* Construct examples of the thing(s) described in the following. Try to find examples that are different than any in the reading.

(a) Three power functions whose improper integrals on $[1,\infty)$ converge.

(b) Three power functions whose improper integrals on $[0,1]$ converge.

(c) A function with the property that both its improper integral on $[1,\infty)$ and its improper integral on $[0,1]$ diverge.

3. What sorts of situations should you look for in order to determine whether an integral is improper?

4. What do we mean when we say that an improper integral converges? What do we mean when we say that an improper integral diverges?

5. Why is it very easy to conclude that $\int_0^1 \frac{1}{e^x}\,dx$ converges without making any integration calculations?

6. Why is it very easy to conclude that $\int_1^\infty x^2\,dx$ diverges without making any integration calculations?

7. The Fundamental Theorem of Calculus does not apply to the integral $\int_0^5 (x-3)^{-4/3}\,dx$; why not? What would the Fundamental Theorem of Calculus calculate for this integral if we applied it anyway? Would it give the correct answer?

8. Why does it make sense that $\int_1^\infty \frac{1}{x^p}\,dx$ diverges when $0 < p \le 1$? Consider how $\frac{1}{x^p}$ compares with $\frac{1}{x}$ in this case.

9. Why does it make sense that $\int_0^1 \frac{1}{x^p}\,dx$ diverges when $p \ge 1$? Consider how $\frac{1}{x^p}$ compares with $\frac{1}{x}$ in this case.

10. Draw pictures to illustrate why the comparison test for improper integrals makes intuitive sense for both convergence comparisons and for divergence comparisons.

11. What, if anything, does the divergence of $\int_1^\infty \frac{1}{x}\,dx$ and the comparison test tell you about the convergence or divergence of $\int_1^\infty \frac{1}{x+1}\,dx$, and why?

12. What, if anything, does the convergence of $\int_1^\infty \frac{1}{x^2}\,dx$ and the comparison test tell you about the convergence or divergence of $\int_1^\infty \frac{1}{x^2+1}\,dx$, and why?

13. Why can't the comparison test make conclusions about $\int_1^\infty \frac{1}{x+1}\,dx$ based on the divergence of $\int_1^\infty \frac{1}{x}\,dx$? What graphical way could you argue for divergence instead? *(Hint: Think about transformations of graphs.)*

14. Why can't the comparison test make conclusions about $\int_2^\infty \frac{1}{(x-1)^2}\,dx$ based on the convergence of $\int_2^\infty \frac{1}{x^2}\,dx$? What graphical way could you argue for the convergence instead?

Express each improper integral in Exercises 15–20 as a sum of limits of proper definite integrals. Do not calculate any integrals or limits; just write them down.

15. $\displaystyle\int_1^\infty \frac{1}{x-1}\,dx$

16. $\displaystyle\int_1^\infty \frac{1}{x-2}\,dx$

17. $\displaystyle\int_1^5 \frac{1}{2x^2-10x+12}\,dx$

18. $\displaystyle\int_0^4 \frac{1}{3x^2-4x+1}\,dx$

19. $\displaystyle\int_{-\infty}^\infty \frac{e^{-x^2}}{x^2}\,dx$

20. $\displaystyle\int_2^\infty \frac{1}{x\ln(x-2)}\,dx$

Skills

Use limits of definite integrals to calculate each of the improper integrals in Exercises 21–56.

21. $\displaystyle\int_1^\infty \frac{1}{\sqrt[3]{x^4}}\,dx$

22. $\displaystyle\int_1^\infty \frac{1}{\sqrt[4]{x^3}}\,dx$

23. $\displaystyle\int_0^1 \frac{1}{\sqrt[3]{x^4}}\,dx$

24. $\displaystyle\int_0^1 \frac{1}{\sqrt[4]{x^3}}\,dx$

25. $\displaystyle\int_0^1 x^{-0.99}\,dx$

26. $\displaystyle\int_0^1 x^{-1.01}\,dx$

27. $\displaystyle\int_1^\infty x^{-0.99}\,dx$

28. $\displaystyle\int_1^\infty x^{-1.01}\,dx$

29. $\displaystyle\int_0^1 (3x-1)^{-2/3}\,dx$

30. $\displaystyle\int_3^\infty \frac{1}{(x-2)^2}\,dx$

31. $\displaystyle\int_1^\infty \frac{1}{(x-2)^2}\,dx$

32. $\displaystyle\int_1^\infty \frac{1+x}{x^2}\,dx$

33. $\displaystyle\int_0^4 \frac{2x-4}{x^2-4x+3}\,dx$

34. $\displaystyle\int_4^\infty \frac{2x-4}{x^2-4x+3}\,dx$

35. $\displaystyle\int_0^\infty xe^{-x^2}\,dx$

36. $\displaystyle\int_0^\infty xe^{-x}\,dx$

37. $\displaystyle\int_0^\infty x^2e^{-x}\,dx$

38. $\displaystyle\int_0^1 \frac{1}{\sqrt{x}\,e^{\sqrt{x}}}\,dx$

39. $\displaystyle\int_0^\infty \frac{1}{\sqrt{x}\,e^{\sqrt{x}}}\,dx$

40. $\displaystyle\int_0^\infty \frac{x}{(x^2+1)^2}\,dx$

41. $\displaystyle\int_{-\infty}^\infty \frac{x}{(x^2+1)^2}\,dx$

42. $\displaystyle\int_0^\infty \frac{x}{(x^2+1)}\,dx$

43. $\displaystyle\int_0^{\pi/2} \sec x\,dx$

44. $\displaystyle\int_0^{\pi/2} \tan x\,dx$

45. $\displaystyle\int_0^1 \frac{\ln x}{x}\,dx$

46. $\displaystyle\int_1^\infty \frac{\ln x}{x}\,dx$

47. $\displaystyle\int_0^1 \frac{1}{x\ln x}\,dx$

48. $\displaystyle\int_1^\infty \frac{1}{x\ln x}\,dx$

49. $\displaystyle\int_e^\infty \frac{1}{x(\ln x)^2}\,dx$

50. $\displaystyle\int_1^e \frac{1}{x(\ln x)^2}\,dx$

51. $\displaystyle\int_{1/\pi}^\infty \frac{\sin(1/x)}{x^2}\,dx$

52. $\displaystyle\int_0^1 \frac{1}{x(x-1)}\,dx$

53. $\displaystyle\int_1^2 \frac{1}{x(x-1)}\,dx$

54. $\displaystyle\int_2^\infty \frac{1}{x(x-1)}\,dx$

55. $\displaystyle\int_0^2 \frac{x-2}{x(x-1)}\,dx$

56. $\displaystyle\int_2^\infty \frac{x-2}{x(x-1)}\,dx$

Determine the convergence or divergence of each improper integral in Exercises 57–64 by comparing it to simpler improper integrals whose convergence or divergence is known or can be found directly.

57. $\displaystyle\int_1^\infty \frac{x}{1+\sin^2 x}\,dx$

58. $\displaystyle\int_1^\infty \frac{1+x}{x^2}\,dx$

59. $\displaystyle\int_1^\infty \frac{1}{x^2+1}\,dx$

60. $\displaystyle\int_1^\infty \frac{x+2}{x^2}\,dx$

61. $\displaystyle\int_1^\infty \frac{1+\ln x}{x}\,dx$

62. $\displaystyle\int_1^\infty \frac{x+\ln x}{x^3}\,dx$

63. $\displaystyle\int_1^\infty \frac{1+\ln x}{x^3}\,dx$

64. $\displaystyle\int_1^\infty \frac{1+\ln x}{x^2}\,dx$

The Limit Comparison Test: If two functions $f(x)$ and $g(x)$ behave in a similar way on $[a,\infty)$, then their improper integrals on $[a,\infty)$ should converge or diverge together. More specifically, the ***limit comparison test*** says that if

$$\lim_{x\to\infty} \frac{f(x)}{g(x)} = L$$

for some positive real number L, then the improper integrals $\int_a^\infty f(x)\,dx$ and $\int_a^\infty g(x)\,dx$ either both converge or both diverge. We will learn more about limit comparison tests in Chapter 7. Use this test to determine the convergence or divergence of the integrals in Exercises 65–70.

65. $\displaystyle\int_2^\infty \frac{1}{(x-1)^2}\,dx$

66. $\displaystyle\int_1^\infty \frac{1}{x+0.001}\,dx$

67. $\displaystyle\int_1^\infty \frac{1}{3x^2-1}\,dx$

68. $\displaystyle\int_1^\infty \frac{x^2-x+3}{x^4+1}\,dx$

69. $\displaystyle\int_1^\infty \frac{x}{\sqrt[4]{x^3+100}}\,dx$

70. $\displaystyle\int_1^\infty \frac{1+e^x}{xe^x}\,dx$

Applications

For Exercises 71 and 72, suppose that in the course of Emmy's job as a civil engineer at Hanford, she needs to find out how much radioactive cesium was present in the mixture in a certain tank when it was first filled exactly 12 years ago. Suppose $Q(0)$ is the number of pounds of radioactive cesium that was initially in the tank 12 years ago, at time $T = 0$. The amount of cesium in the tank at any time $T \geq 0$ is the amount that remains to decay, which for some constant k is equal to

$$Q(t) = \int_T^\infty ke^{-0.023t}\,dt.$$

71. If Emmy measures 0.95 lb of cesium in the tank now, how much cesium was in the tank when it was first filled 12 years ago?

72. In another location in the tank farm, there is a tank that was filled with a different, known mixture that contained 1.5 lb of radioactive cesium when the tank was filled. However, there is no record of when this tank was filled. Emmy measures 0.85 lb of cesium in the tank now. How long ago was the tank filled?

73. Frank is evaluating electric motors to drive automated mixing for some waste tanks that he must maintain. One pump is advertised to have a probability of failure that follows the exponential distribution

$$f(t) = 0.31e^{-0.31t},$$

where the time $t > 0$ is measured in years. Frank knows that the expected time of failure for something following this distribution is

$$\int_0^\infty tf(t)dt.$$

How long can he expect one of these pumps to last?

Proofs

Prove each statement in Exercises 74–77, using limits of definite integrals for general values of p.

74. If $0 < p \leq 1$, then $\int_1^\infty \frac{1}{x^p}\,dx$ diverges.

75. If $p > 1$, then $\int_1^\infty \frac{1}{x^p}\,dx$ converges to $\frac{1}{p-1}$.

76. If $0 < p < 1$, then $\int_0^1 \frac{1}{x^p}\,dx$ converges to $\frac{1}{1-p}$.

77. If $p \geq 1$, then $\int_0^1 \frac{1}{x^p}\,dx$ diverges.

Prove each statement in Exercises 78–80, using the definition of improper integrals as limits of proper definite integrals.

78. If $\int_a^\infty f(x)\,dx$ diverges and $0 \leq f(x) \leq g(x)$ for all $x \in [a, \infty)$, then $\int_a^\infty g(x)\,dx$ also diverges.

79. Suppose $f(x)$ and $g(x)$ are both continuous on $(a, b]$ but not at $x = a$. If $\int_a^b f(x)\,dx$ converges and $0 \leq g(x) \leq f(x)$ for all $x \in (a, b]$, then $\int_a^b g(x)\,dx$ also converges.

80. Suppose $f(x)$ and $g(x)$ are both continuous on $(a, b]$ but not at $x = a$. If $\int_a^b f(x)\,dx$ diverges and $0 \leq f(x) \leq g(x)$ for all $x \in (a, b]$, then $\int_a^b g(x)\,dx$ also diverges.

81. Suppose $f(x)$ is continuous on $\mathbb{R}$ and that for some real number c, both $\int_{-\infty}^c f(x)\,dx$ and $\int_c^\infty f(x)\,dx$ exist. Use properties of definite integrals to prove that for all real numbers d,

$$\int_{-\infty}^c f(x)\,dx + \int_c^\infty f(x)\,dx$$

is equal to

$$\int_{-\infty}^d f(x)\,dx + \int_d^\infty f(x)\,dx.$$

Thinking Forward

Gabriel's Horn: If we rotate the graph of $f(x) = \frac{1}{x}$ on $[1, \infty)$ around the x-axis, we get what is known as ***Gabriel's Horn***, as shown in the figure.

Graph of $f(x) = \frac{1}{x}$ rotated around x-axis

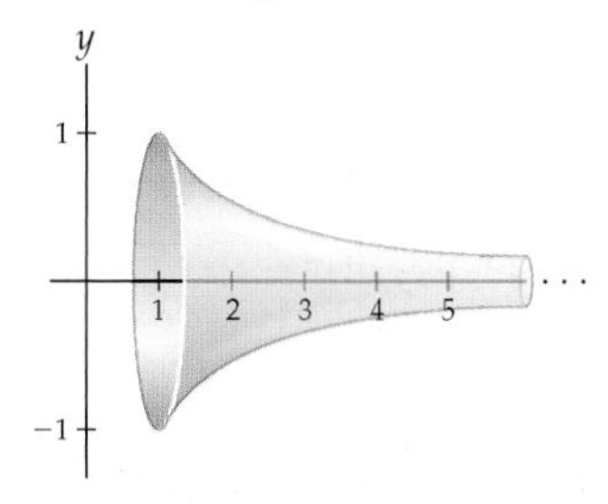

We already know that the area $\int_1^\infty \frac{1}{x}\,dx$ under the graph of $\frac{1}{x}$ on $[1, \infty)$ is infinite. Does this mean that the volume of Gabriel's Horn will also be infinite? The answer, surprisingly, is no. Let's find out why.

It turns out that the volume of the solid of revolution obtained by rotating the graph of a function $y = f(x)$ around the x-axis on an interval $[a, b]$ is given by the formula

$$\text{Volume} = \pi \int_a^b (f(x))^2\,dx.$$

It also turns out that the surface area of such a solid of revolution is given by the formula

$$\text{Surface Area} = 2\pi \int_a^b f(x)\sqrt{1 + (f'(x))^2}\,dx.$$

- Set up an improper integral that describes the volume of Gabriel's Horn. Then solve the improper integral to show that the volume of Gabriel's Horn is not infinite, but rather is equal to π.
- Get ready to be surprised again: Although the volume of Gabriel's Horn is finite, the surface area of Gabriel's Horn is infinite! That means that it would take an infinite amount of paint to coat the outside of Gabriel's Horn, even though it takes only a finite amount of paint to fill up the inside. Set up an improper integral that describes the surface area of Gabriel's Horn. Then use the limit comparison test (see Exercises 65–70) with $\frac{1}{x}$ to show that the surface area of Gabriel's Horn is infinite.

5.7 NUMERICAL INTEGRATION*

- Error bounds for certain left, right, trapezoid, and midpoint sums
- Simpson's Rule for approximating areas under curves
- Using error bounds to calculate definite integrals to within specified degrees of accuracy

Approximations and Error

Even with a whole chapter of integration techniques under our belt, there are many integrals that we do not know how to solve. For example, the definite integrals

$$\int_0^{1.5} e^{-x^2}\,dx, \quad \int_1^2 \frac{x}{e^x - 1}\,dx, \quad \text{and} \quad \int_0^{\pi/2} \cos(\cos x)\,dx$$

have relatively simple integrands, although it turns out that their antiderivatives cannot be expressed in ***closed form***, that is, as combinations of the types of functions we have studied in this course. However, each of the integrands in these integrals is continuous and thus is integrable, which means that each of the given definite integrals is equal to some real number that we can approximate by using Riemann sums.

Approximating is easy; the tough part is knowing when your approximation is "good." For example, suppose we want to get a rough estimate of the number of seconds in a day. The actual answer is of course that there are (60 seconds)(60 minutes)(24 hours) = 86,400 seconds in each day, but if you wanted to calculate quickly in your head you might instead consider the simpler product (60)(60)(20) = 72,000. This approximation is good for knowing roughly what order of magnitude we are talking about, but not good enough for most practical applications. In this case we can see that our simple approximation is off by an error of 4,400 seconds. But how do we know how good or bad an approximation is without knowing the exact value we are approximating?

The ***error*** of an approximation is the difference between the approximated value and the actual value of what is being calculated. When we are interested only in the magnitude of such error and not in whether our approximation is more or less than the actual value, we consider the absolute value of the error:

$$|\text{error}| = |\text{true value} - \text{approximation}|.$$

When we use a Riemann sum, we are approximating an exact signed area under a curve. The resulting error will typically be large if we use only a few rectangles, but becomes smaller when we use a large number of rectangles. In general, the error $E(n)$ incurred by using a given n-rectangle Riemann sum is the difference between the actual area and the approximated area:

$$|E(n)| = \left| \int_a^b f(x)\,dx - \sum_{k=1}^{n} f(x_k^*)\,\Delta x \right|.$$

But how can we possibly find this error without knowing the exact value of $\int_a^b f(x)\,dx$ in the first place? It turns out that in specific situations we can get bounds on the error even without knowing the value of the definite integral. This means that in those situations we will be able to approximate definite integrals to whatever degree of accuracy we choose.

Error in Left and Right Sums

Since we will be calculating and comparing many types of Riemann sums in this section, it will be useful to have some descriptive notation for such sums. Suppose we wish to examine the signed area between the graph of a function f and the x-axis on $[a, b]$. We will denote the left-sum and right-sum n-rectangle approximations for this area, respectively, as

$$\text{LEFT}(n) = \sum_{k=1}^{n} f(x_{k-1})\Delta x = (f(x_0) + f(x_1) + f(x_2) + \cdots + f(x_{n-1}))\,\Delta x,$$

$$\text{RIGHT}(n) = \sum_{k=1}^{n} f(x_k)\Delta x = (f(x_1) + f(x_2) + f(x_3) + \cdots + f(x_n))\,\Delta x.$$

Sometimes these sums are over-approximations of the actual area, and sometimes they are under-approximations, depending on the behavior of the function f. If f happens to be ***monotonic*** on $[a, b]$, in other words either always increasing on $[a, b]$ or always decreasing on $[a, b]$, we can say more. For example, consider the monotonically decreasing function f shown following at the left. For this function, the four-rectangle left sum is an over-approximation and the four-rectangle right sum is an under-approximation.

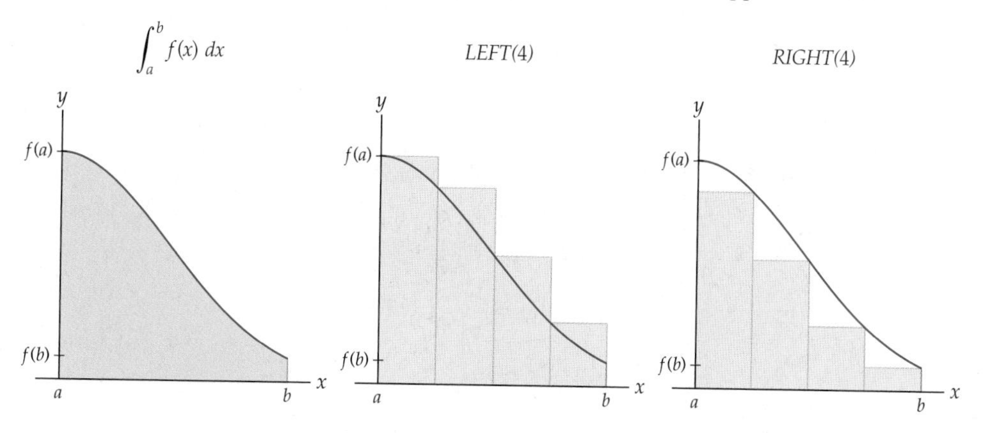

In fact, for any number n of rectangles, the area under a monotonically increasing or decreasing graph will be between the left-sum and right-sum approximations, as described in the following theorem:

THEOREM 5.24 **Left and Right Sums of Monotonic Functions**

Suppose f is integrable on $[a, b]$ and let n be a positive integer.

(a) If f is increasing on $[a, b]$, then $\text{LEFT}(n) \leq \int_a^b f(x)\,dx \leq \text{RIGHT}(n)$.

(b) If f is decreasing on $[a, b]$, then $\text{RIGHT}(n) \leq \int_a^b f(x)\,dx \leq \text{LEFT}(n)$.

Proof. We will prove part (b) and leave the entirely similar proof of part (a) to Exercise 58. Suppose we partition $[a, b]$ into n rectangles over subintervals of the form $[x_{k-1}, x_k]$. If f is monotonically decreasing, then for each k we have $f(x_{k-1}) \geq f(x) \geq f(x_k)$ for all $x \in [x_{k-1}, x_k]$, and thus the area under f on $[x_{k-1}, x_k]$ will be less than the area of the kth left-sum rectangle and greater than the area of the kth right-sum rectangle. ■

The fact that the actual area under a monotonic function is always sandwiched between the left sum and the right sum can actually tell us something about the amount of error in those Riemann sum approximations. The shaded rectangles in the figures that follow represent the combined error from the left sum and the right sum. These errors join up to form a rectangle that can be stacked into a larger rectangle whose area $|f(b) - f(a)|\Delta x$ is the difference $|\text{LEFT}(n) - \text{RIGHT}(n)|$, regardless of the value of n:

Difference between LEFT(4) and RIGHT(4)

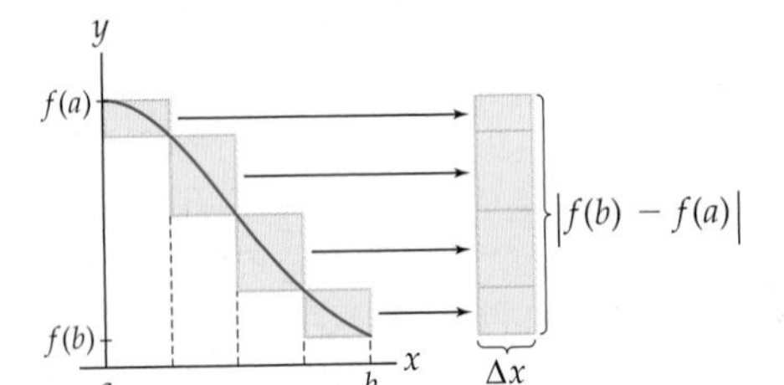

Difference between LEFT(8) and RIGHT(8)

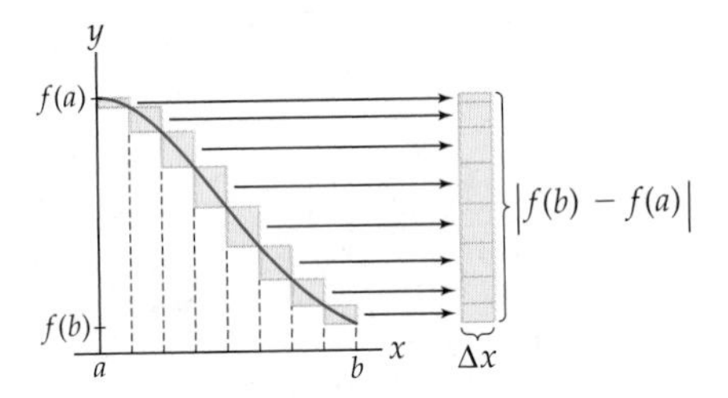

Since the stack of rectangles represents the combined error from the left and right sums, it follows that the error from either of these approximations must itself be less than $|f(b) - f(a)|\Delta x$. This argument holds for any monotonically decreasing function, and we can say similar things when f is monotonically increasing. Therefore we have proved the following theorem:

THEOREM 5.25 **Error Bounds for Left and Right Sums**

Suppose f is integrable and monotonic on $[a, b]$ and n is a positive integer. Then we can bound the error $E_{\text{LEFT}(n)}$ of the left sum as follows:

$$|E_{\text{LEFT}(n)}| \leq |\text{LEFT}(n) - \text{RIGHT}(n)| = |f(b) - f(a)|\,\Delta x.$$

The error $|E_{\text{RIGHT}(n)}|$ of the corresponding right sum has the same bound.

As we will see in Example 1, we can use these error bounds to ensure that our left- or right-sum approximations of area are to within any desired degree of accuracy.

Error in Trapezoid and Midpoint Sums

What about bounding error for other types of Riemann sums? Recall from Section 4.2 the trapezoid and midpoint sums for n subdivisions, which can be expressed, respectively, as

$$\text{TRAP}(n) = \sum_{k=1}^{n} \frac{f(x_{k-1}) + f(x_k)}{2} \Delta x = (f(x_0) + 2f(x_1) + 2f(x_2) + \cdots + f(x_n))\left(\frac{\Delta x}{2}\right),$$

$$\text{MID}(n) = \sum_{k=1}^{n} f\left(\frac{x_{k-1} + x_k}{2}\right) \Delta x, = \left(f\left(\frac{x_0 + x_1}{2}\right) + f\left(\frac{x_1 + x_2}{2}\right) + \cdots + f\left(\frac{x_{n-1} + x_n}{2}\right)\right) \Delta x.$$

It turns out that we can guarantee that these sums are either over-approximations or under-approximations depending on whether the function f is concave up or concave down. For example, consider the concave-up function f shown following at the left. It is immediately obvious that the trapezoid sum approximation must be greater than the actual area under the curve, as shown in the middle figure.

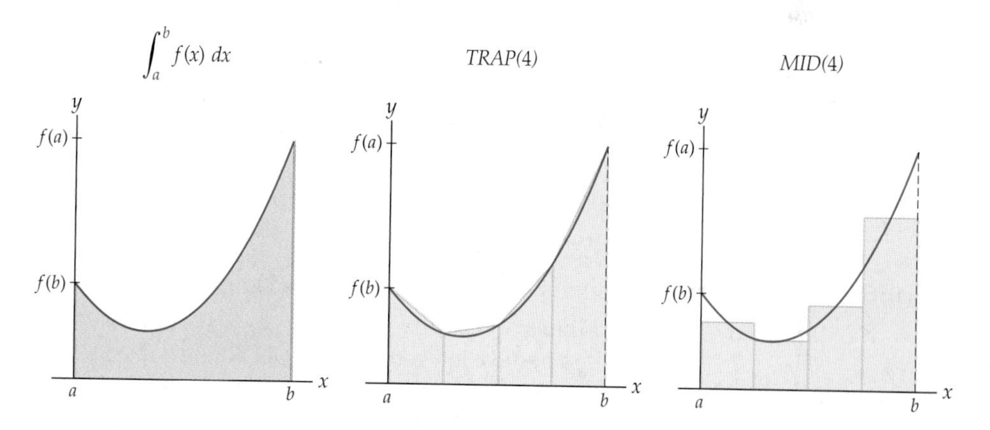

What is less obvious is that the midpoint sum of a concave-up function, as shown in the third figure, will always be an under-approximation of the actual area. For concave-down functions the situation is reversed, with the actual area sandwiched between the smaller trapezoid sum and the larger midpoint sum, as stated in the following theorem:

THEOREM 5.26 **Trapezoid and Midpoint Sums of Functions with Consistent Concavity**

Suppose f is integrable on $[a, b]$ and let n be a positive integer.

(a) If f is concave up on $[a, b]$, then $\text{MID}(n) \le \int_a^b f(x)\,dx \le \text{TRAP}(n)$.

(b) If f is concave down on $[a, b]$, then $\text{TRAP}(n) \le \int_a^b f(x)\,dx \le \text{MID}(n)$.

Proof. We will use geometric arguments to prove part (a) and leave the similar proof of part (b) to Exercise 59. That the trapezoid sum is an over-approximation for concave-up functions is clear from our earlier middle diagram. Specifically, upwards concavity guarantees that the secant line from $(x_{k-1}, f(x_{k-1}))$ to $(x_k, f(x_k))$ will always be completely above the graph of f on $[x_{k-1}, x_k]$.

That the midpoint sum is an under-approximation for concave-up functions becomes clear if we reinterpret each midpoint rectangle as a trapezoid. Given a subinterval $[x_{k-1}, x_k]$, consider the ***midpoint trapezoid*** on the interval to be the trapezoid whose top is formed by the line tangent to the graph of f at the midpoint of the subinterval. From the figure it is clear that the midpoint trapezoid will have the same area as the midpoint rectangle whose height is given by the height of $f(x)$ at the midpoint of the subinterval. Now look at the figure on the right; since f is concave up, the midpoint tangent line on each subinterval will be below the graph. Therefore the area of each midpoint trapezoid will be less than the area under the curve on that subinterval.

Midpoint trapezoid

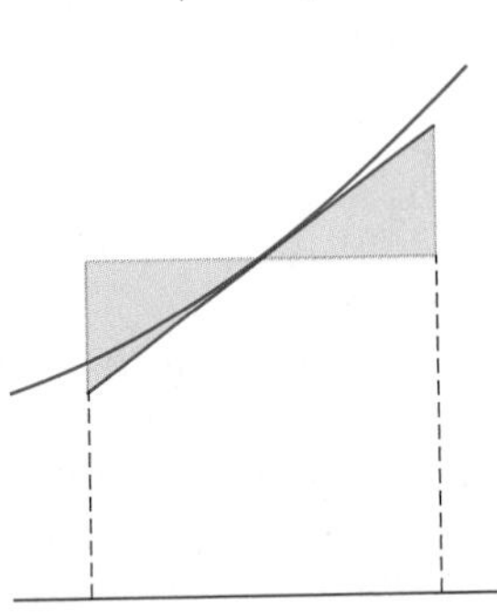

Reinterpretation of MID(4) with trapezoids

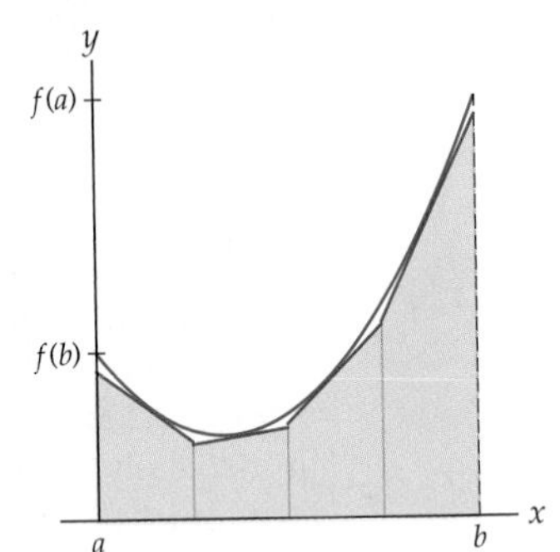

■

Because the area under the graph of a consistently concave-up or concave-down function will always be between the trapezoid sum and the midpoint sum, the error for either of those sums is less than the difference of those sums. This is entirely analogous to what we saw in Theorem 5.25. We can actually get an even better bound if we are able to restrict the behavior of the second derivative:

THEOREM 5.27

Error Bounds for Trapezoid and Midpoint Sums

Suppose f is integrable and either always concave up or always concave down on $[a, b]$, and let n be a positive integer. Then we can bound both $E_{\text{TRAP}(n)}$ and $E_{\text{MID}(n)}$ as follows:

(a) $|E_{\text{TRAP}(n)}| \le |\text{TRAP}(n) - \text{MID}(n)|$ **(b)** $|E_{\text{MID}(n)}| \le |\text{TRAP}(n) - \text{MID}(n)|$

If we suppose further that f'' is bounded on $[a, b]$, that is, that there is some positive real number M such that $|f''(x)| \le M$ for all $x \in [a, b]$, then we can say that:

(c) $|E_{\text{TRAP}(n)}| \le \dfrac{M(b-a)^3}{12n^2}$ **(d)** $|E_{\text{MID}(n)}| \le \dfrac{M(b-a)^3}{24n^2}$

The proof of the second part of this theorem can be found in many numerical analysis texts. You will explore why these bounds are reasonable in Exercise 18. In Example 2 you will use the same bounds to calculate trapezoid and midpoint sums to specified degrees of accuracy.

Simpson's Rule

So far all of our approximations of a definite integral $\int_a^b f(x)\,dx$ have utilized piecewise-linear approximations of f. In other words, after subdividing $[a, b]$, we approximated each piece of f with a line. For example, the left sum just represents the area under the graph of a piecewise-linear step function that agrees with f on the left-hand side of each subinterval. The trapezoid sum uses a continuous piecewise-linear function whose pieces are secant lines of the graph of f, connecting the heights of the function at the subdivision points.

One way to get a more accurate approximation of a definite integral is to use a non-linear approximation instead. For example, we could approximate each piece of f with a parabola instead of a line. The first figure that follows shows some function f on $[a, b]$, and the second figure shows how we could approximate the area under the graph of f with three parabola-topped rectangles. We choose the parabola $p(x)$ on each subinterval to be the unique parabola that passes through the leftmost height, the midpoint height, and the rightmost height on that subinterval; see the third figure.

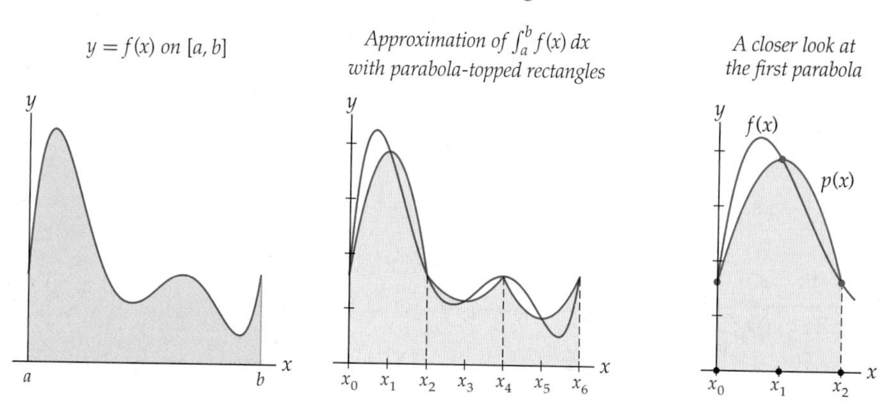

The example we just discussed was specifically chosen so that the parabolas would be easily distinguishable from the original function. In other words, the example does not show a very accurate approximation! However, in general, approximation by parabolas can be extremely accurate; imagine using 20 parabolas along the graph of f, and consider how closely such an approximation would match the function.

So how do we calculate a particular parabola-based area approximation? One way would be to actually find the equations of each little quadratic parabola piece and then use definite integral formulas for quadratics to calculate the area under each such piece. Doing this in practice, however, would be quite time-consuming. The next theorem can help make the process faster; it provides a specific formula for calculating the area under a quadratic on an interval that uses only the values of the quadratic at the endpoints and the midpoint of the interval. This nice theorem will allow us to bypass the step of actually finding the equation of each parabola.

THEOREM 5.28 **The Definite Integral of a Quadratic in Terms of Endpoints and Midpoint**

If $p(x)$ is a quadratic function on $[A, B]$, then

$$\int_A^B p(x)\,dx = \frac{B-A}{6}\left(p(A) + 4p\left(\frac{A+B}{2}\right) + p(B)\right).$$

The proof of this convenient formula is just an application of known definite integral formulas and a bunch of algebra and is left to Exercise 62.

It turns out to be convenient to construct our parabolas on *pairs* of intervals. For example, in the middle figure just shown we divided $[a, b]$ into $n = 6$ subintervals and then constructed three parabolas to approximate the curve $y = f(x)$. In Exercise 63 you will show that this process leads to the following summation formula:

THEOREM 5.29

Simpson's Rule

Suppose f is integrable on $[a, b]$ and n is a positive even integer. Let $\Delta x = \frac{b-a}{n}$ and $x_k = a + k\Delta x$. Then we can approximate $\int_a^b f(x)\,dx$ with $\frac{n}{2}$ parabola-topped rectangles by using the following sum, which is known as ***Simpson's Rule***:

$$\text{SIMP}(n) = (f(x_0) + 4f(x_1) + 2f(x_2) + 4f(x_3) + \cdots + 2f(x_{n-2}) + 4f(x_{n-1}) + f(x_n))\left(\frac{\Delta x}{3}\right).$$

Intuitively, it makes sense that, for most functions, an approximation using pieces of parabolas could be more accurate than an approximation with pieces of lines. The following theorem backs up this intuition with a bound on the error from Simpson's Rule that is clearly better than the error bounds we found for left, right, trapezoid, and midpoint sums:

THEOREM 5.30

Error Bound for Simpson's Rule

Suppose f is an integrable function on $[a, b]$ whose fourth derivative is bounded. Then there is some positive real number M such that $|f^{(4)}(x)| \le M$ for all $x \in [a, b]$. In this situation we can bound the error $E_{\text{SIMP}(n)}$ from Simpson's Rule as follows:

$$|E_{\text{SIMP}(n)}| \le \frac{M(b-a)^5}{180n^4}.$$

The proof of Theorem 5.30 can be found in many numerical analysis texts, and we will not concern ourselves with it here.

Notice that all of the error bounds we have discussed have a requirement that either f or some derivative of f be bounded. For the left and right sums we required f to be monotonic, which implies that f is bounded. This restriction in turn puts some kind of limit on how much the height function f can change on $[a, b]$. For the trapezoid and midpoint sums we required f'' to be bounded, which in a sense puts a limit on the curviness of f on $[a, b]$. Notice that curvier functions will tend to have less accurate trapezoid sum approximations. For Simpson's Rule we required a bound on the fourth derivative, which also puts some limits on how wildly behaved f can be on $[a, b]$ and therefore limits how much error can arise by using parabolas to approximate the function on that interval.

Examples and Explorations

EXAMPLE 1

Guaranteeing a level of accuracy from a right sum

Suppose we wish to approximate the definite integral $\int_0^{1.5} e^{-x^2}\,dx$ with an n-rectangle right sum so that we can be sure that our estimate is within 0.005 of the actual answer. How large will we have to make n?

SOLUTION

As we mentioned at the start of this section, the integral $\int_0^{1.5} e^{-x^2}\,dx$ has no closed-form solution. This means that we cannot calculate its actual value with the Fundamental Theorem of Calculus, because we would be unable to perform the antidifferentiation step. Therefore in this example we have no choice but to approximate the value of the definite integral, and in addition we do not know the value of the actual answer we are attempting to approximate.

The graph of f on $[a, b]$ that appears first in this section happens to be the graph of $f(x) = e^{-x^2}$ on the interval $[a, b] = [0, 1.5]$ that we are concerned with in this example. The function $f(x) = e^{-x^2}$ is monotonically decreasing on $[0, 1.5]$ because its derivative $f'(x) = -2xe^{-x^2}$ is always negative on that interval. Therefore Theorem 5.25 applies. We have $a = 0$, $b = 1.5$, and $\Delta x = \frac{1.5-0}{n} = \frac{1.5}{n}$, so the magnitude of the error from using an n-rectangle right sum has the following bound:

$$|E_{\text{RHS}(n)}| \le |f(b) - f(a)|\Delta x = \left|e^{-(1.5)^2} - e^{-0^2}\right|\left(\frac{1.5}{n}\right) \approx \frac{1.3419}{n}.$$

We want to find n such that $|E_{\text{RHS}(n)}|$ is less than 0.005. To do this we just have to find n such that the error bound is less than 0.005, so we need:

$$\frac{1.3419}{n} < 0.005 \implies \frac{1.3419}{0.005} < n \implies n > 268.38.$$

This means that $n = 269$ is the first positive integer for which the n-rectangle right sum will be within 0.005 of the actual area under the curve.

In case you're interested, the actual value of the definite integral is approximately 0.856188, and the right-sum approximation with 269 rectangles is approximately 0.853693; notice that our approximation is indeed within 0.005 of the actual answer. □

EXAMPLE 2

Finding bounds for the error from trapezoid and midpoint sums

Consider the signed area between the graph of $f(x) = x^2 - 2x + 2$ and the x-axis on the interval $[0, 3]$.

(a) Find trapezoid and midpoint sum approximations for this area with $n = 4$.

(b) Describe error bounds for the approximations from part (a) in two different ways (with Theorem 5.26 and with Theorem 5.27).

(c) Use the Fundamental Theorem of Calculus to find the actual area, and then verify that the error bounds computed in part (b) are accurate.

SOLUTION

(a) The function $f(x) = x^2 - 2x + 2$ on $[a, b] = [0, 3]$ just happens to be the function f on $[a, b]$ that we used in our discussion of the trapezoid and midpoint sums before Theorem 5.26. We begin by dividing the interval $[0, 3]$ into $n = 4$ subintervals of width $\Delta x = \frac{3}{4}$, with subdivision points $x_0 = 0$, $x_1 = \frac{3}{4}$, $x_2 = \frac{3}{2}$, $x_3 = \frac{9}{4}$, and $x_4 = 3$. Using the equation $f(x) = x^2 - 2x + 2$, we find that the $n = 4$ trapezoid sum approximation is

$$\text{TRAP}(4) = \left(f(0) + 2f\left(\frac{3}{4}\right) + 2f\left(\frac{3}{2}\right) + 2f\left(\frac{9}{4}\right) + f(3)\right)\left(\frac{3/4}{2}\right) = \frac{201}{32} \approx 6.28125.$$

To find MID(4), we must first find the midpoints of the four subintervals; these midpoints are $x_1^* = \frac{3}{8}$, $x_2^* = \frac{9}{8}$, $x_3^* = \frac{15}{8}$, and $x_4^* = \frac{21}{8}$. Therefore the $n = 4$ midpoint sum approximation is

$$\text{MID}(4) = \left(f\left(\frac{3}{8}\right) + f\left(\frac{9}{8}\right) + f\left(\frac{15}{8}\right) + f\left(\frac{21}{8}\right)\left(\frac{3}{4}\right)\right) = \frac{375}{64} \approx 5.85938.$$

(b) Theorem 5.26 applies because $f'(x) = 2x - 2$ is positive on all of $[0, 3]$ and thus f is concave up on that entire interval. Therefore the actual area under f on $[0, 3]$ is greater than $\text{MID}(4) = \frac{375}{64} \approx 5.85938$ and smaller than $\text{TRAP}(4) = \frac{201}{32} \approx 6.28125$. In particular, by the first part of Theorem 5.27 this means that

$$|E_{\text{TRAP}(4)}| \text{ and } |E_{\text{MID}(4)}| \le \left|\frac{201}{32} - \frac{375}{64}\right| = \frac{27}{64} \approx 0.421875.$$

Accordingly, we can expect both the $n = 4$ trapezoid sum and the $n = 4$ midpoint sum to be within about 0.421875 of the actual area under f on $[0, 3]$.

The second part of Theorem 5.27 has the potential to give us a much better error bound in the sense that we should be confident in our $n = 4$ approximations to an even greater degree of accuracy. Let's see if that is the case in this example. We first have to get a bound M on the second derivative that is as good as possible. The second derivative of $f(x) = x^2 - 2x + 2$ is $f''(x) = 2$, and thus clearly $|f''(x)| \leq 2$ for all values $x \in [0, 3]$, so we can take $M = 2$. Note that we are lucky to have an *extremely* good bound for f'' in this example, since we know that it is constantly 2. We can now say that

$$|E_{\text{TRAP}(4)}| \leq \frac{M(b-a)^3}{12n^2} = \frac{2(3-0)^3}{12(4^2)} = \frac{9}{32} \approx 0.28125$$

and

$$|E_{\text{MID}(4)}| \leq \frac{M(b-a)^3}{24n^2} = \frac{2(3-0)^3}{24(4^2)} = \frac{9}{64} \approx 0.140625.$$

From the $n = 4$ trapezoid sum approximation, it follows that the actual value of $\int_0^3 f(x)\,dx$ is within $\frac{9}{32}$ of $\frac{201}{32}$ (i.e., between $\frac{192}{32} = 6$ and $\frac{210}{32} \approx 6.5625$). With the $n = 4$ midpoint sum we can guarantee even better accuracy, namely, that the actual value of $\int_0^3 f(x)\,dx$ is within $\frac{9}{64}$ of $\frac{375}{64}$ (i.e., between $\frac{366}{64} \approx 5.71875$ and $\frac{384}{64} = 6$).

(c) In general, we use approximations with error bounds to investigate definite integrals that we are unable to solve exactly. However, in this example the definite integral $\int_0^3 (x^2 - 2x + 2)\,dx$ is easy to solve exactly, so we can actually verify that the error bounds we just found are accurate. Using the Fundamental Theorem of Calculus, we obtain

$$\int_0^3 (x^2 - 2x + 2)\,dx = \left[\frac{1}{3}x^3 - x^2 + 2x\right]_0^3 = 6.$$

In part (b) we used Theorem 5.26 to argue that the value of this definite integral was within 0.421875 of both the trapezoid sum approximation of 6.28125 and the midpoint sum approximation of 5.85938. This meant that we could guarantee that the value of the definite integral was both in the interval $[5.859375, 6.703125]$ and in the interval $[5.437505, 6.281255]$, which is true, because we just showed that the exact value was 6. We then used Theorem 5.27 to obtain better error bounds, which allowed us to guarantee that the value of the definite integral was in both the interval $[6, 6.5625]$ and the interval $[5.71875, 6]$. In fact, notice that the only number in both of those intervals happens to be the exact answer of 6. □

EXAMPLE 3

Arriving at the Simpson's Rule formula when $n = 6$

Show that the Simpson's Rule formula in Theorem 5.29 results from repeated application of Theorem 5.28 in the case where $n = 6$ with $\frac{n}{2} = 3$ parabolas.

SOLUTION

Suppose f is an integrable function on an interval $[a, b]$. In this case we have three quadratic functions $p_1(x)$, $p_2(x)$, and $p_3(x)$ that approximate $f(x)$ on the double subintervals $[x_0, x_2]$, $[x_2, x_4]$, and $[x_4, x_6]$, and agreeing with $f(x)$ at the endpoints and midpoints of these subintervals. Note that the midpoints of the parabola pieces are the odd subdivision points x_1, x_3, and x_5. We can approximate the area under f on $[a, b]$ by adding up the areas under these three parabola pieces $p_1(x)$, $p_2(x)$, and $p_3(x)$:

$$\int_a^b f(x)\,dx \approx \int_{x_0}^{x_2} p_1(x)\,dx + \int_{x_2}^{x_4} p_2(x)\,dx + \int_{x_4}^{x_6} p_3(x)\,dx.$$

By applying Theorem 5.28 to the first double subinterval $[x_0, x_2]$ we see that

$$\int_{x_0}^{x_2} p_1(x)\,dx = \frac{x_2 - x_0}{6}(p_1(x_0) + 4p_1(x_1) + p_1(x_2)).$$

Similarly, on the second and third double subintervals $[x_2, x_4]$ and $[x_4, x_6]$ we have

$$\int_{x_2}^{x_4} p_1(x)\,dx = \frac{x_4 - x_2}{6}(p_1(x_2) + 4p_1(x_3) + p_1(x_4)),$$

$$\int_{x_4}^{x_6} p_1(x)\,dx = \frac{x_6 - x_4}{6}(p_1(x_4) + 4p_1(x_5) + p_1(x_6)).$$

By construction, the values of $p_1(x)$ at x_0, x_1, and x_2 are the same as the corresponding values of $f(x)$. Similarly, the values of $p_2(x)$ and $p_3(x)$ agree with the values of $f(x)$ at the corresponding endpoints and midpoints. Moreover, the width of each double subinterval is $2\Delta x$, and thus $\frac{x_2-x_0}{6}$, $\frac{x_4-x_2}{6}$, and $\frac{x_6-x_4}{6}$ are each equal to $\frac{\Delta x}{3}$. Therefore $\int_a^b f(x)\,dx$ is approximated with Simpson's Rule by the sum

$$(f(x_0) + 4f(x_1) + f(x_2))\frac{\Delta x}{3} + (f(x_2) + 4f(x_3) + f(x_4))\frac{\Delta x}{3} + (f(x_4) + 4f(x_5) + f(x_6))\frac{\Delta x}{3}.$$

A little algebra shows that Simpson's Rule indeed says what is predicted by Theorem 5.29:

$$\int_a^b f(x)\,dx \approx (f(x_0) + 4f(x_1) + 2f(x_2) + 4f(x_3) + 2f(x_4) + 4f(x_5) + f(x_6))\frac{\Delta x}{3}.$$

Note that if we knew the function f and the interval $[a, b]$, then we could easily calculate the preceding sum. □

? TEST YOUR UNDERSTANDING

- What does it mean for a function to be monotonic on an interval?
- Why do we require that f be monotonic on all of $[a, b]$ in Theorem 5.24?
- Why do we require that f be either always concave up or always concave down on $[a, b]$ in Theorem 5.26?
- As we saw in the reading, a trapezoid sum with n trapezoids can be written in the form $(f(x_0) + 2f(x_1) + 2f(x_2) + \cdots + f(x_n))\left(\frac{\Delta x}{2}\right)$. Why are $f(x_0)$ and $f(x_n)$ not multiplied by 2 in this expression?
- Simpson's Rule approximates the area under a curve by approximating a function with pieces of parabolas. How do we define the parabolas that are used on each subinterval?

EXERCISES 5.7

Thinking Back

Monotonicity and concavity: For each function f and interval $[a, b]$, use derivatives to determine whether or not f is monotonic on $[a, b]$ and whether or not f has consistent concavity on $[a, b]$.

- $f(x) = x^3 - 2x^2$, $[a, b] = [1, 3]$
- $f(x) = x^3 - 2x^2$, $[a, b] = [0, 1]$
- $f(x) = \sin\left(\frac{\pi}{2}x\right)$, $[a, b] = [-1, 1]$

Bounding the second derivative: Consider the function $f(x) = x^2(x - 3)$.

- To the nearest tenths place, find the smallest number M so that $|f''(x)| \le M$ for all $x \in [0, 2]$.
- To the nearest tenths place, find the smallest number M so that $|f''(x)| \le M$ for all $x \in [0, 4]$.

Determining quadratics: For each of the following problems, find the equation of the unique quadratic function $p(x)$ with the given properties.

- The graph of $p(x)$ is a parabola that agrees with $f(x) = x^3 + 1$ at $f(-1), f(0)$, and $f(1)$.
- The graph of $p(x)$ is a parabola that agrees with the function $f(x) = \frac{1}{x}$ at $x = 1$, $x = 2$, and $x = 3$.
- The graph of $p(x)$ is a parabola that agrees with the function $f(x) = \sin x$ at $x = 0$, $x = \frac{\pi}{2}$, and $x = \pi$.

Concepts

0. *Problem Zero:* Read the section and make your own summary of the material.

1. *True/False:* Determine whether each of the statements that follow is true or false. If a statement is true, explain why. If a statement is false, provide a counterexample.
   (a) *True or False:* If we knew the value of the error from a given Riemann sum calculation, then we would know the actual value of the signed area under the curve.
   (b) *True or False:* We can sometimes get a bound on the error from a given Riemann sum calculation even without knowing the actual value of the signed area under the curve.
   (c) *True or False:* A left sum with $n = 10$ rectangles will always have a larger error than the corresponding left sum with $n = 100$ rectangles.
   (d) *True or False:* A right sum with $n = 10$ rectangles for a monotonically decreasing function will always have a larger error than the corresponding right sum with $n = 100$ rectangles.
   (e) *True or False:* If f is monotonically increasing on $[a, b]$, then the error incurred from using the left sum with n rectangles will always be less than the difference of the right and left sums with n rectangles.
   (f) *True or False:* If f is concave down on $[a, b]$, then every trapezoid sum on that interval will be an under-approximation.
   (g) *True or False:* If f is concave down on $[a, b]$, then every midpoint sum on that interval will be an over-approximation.
   (h) *True or False:* Simpson's Rule with $n = 10$ subdivisions (and thus 5 parabolas) will always be more accurate than the trapezoid sum with $n = 10$ trapezoids.

2. *Examples:* Construct examples of the thing(s) described in the following. Try to find examples that are different than any in the reading.
   (a) A function f on an interval $[a, b]$ for which the left sum and the trapezoid sum are over-approximations for every n.
   (b) A function f on an interval $[a, b]$ for which the right sum and the trapezoid sum are over-approximations for every n.
   (c) A function f on an interval $[a, b]$ for which Simpson's Rule has no error at all.

3. Suppose f is monotonicically increasing on $[a, b]$. Use pictures to explain why the error incurred from using the left sum with n rectangles to approximate $\int_a^b f(x)\,dx$ will always be less than or equal to the difference of the right sum with n rectangles and the left sum with n rectangles.

4. Suppose f is concave down on $[a, b]$. Use pictures to explain why the error incurred from using the trapezoid sum with n trapezoids to approximate $\int_a^b f(x)\,dx$ will always be less than or equal to the difference of the trapezoid sum with n trapezoids and the midpoint sum with n rectangles.

5. Sketch an example that shows that the inequality $\text{LEFT}(n) \le \int_a^b f(x)\,dx \le \text{RIGHT}(n)$ is not necessarily true if f is not monotonically increasing on $[a, b]$.

6. Sketch an example that shows that the inequality $\text{MID}(n) \le \int_a^b f(x)\,dx \le \text{TRAP}(n)$ is not necessarily true if f is not concave up on all of $[a, b]$.

7. Sketch an example that shows that the left-sum error bound $|E_{\text{LEFT}(n)}| \le |f(b) - f(a)|\,\Delta x$ does not necessarily hold for functions f that fail to be monotonic on $[a, b]$.

8. If f is monotonically increasing on $[a, b]$, which of the given approximations is *guaranteed* to be an over-approximation for $\int_a^b f(x)\,dx$? (Select *all* that apply.)
   (a) left sum (b) right sum
   (c) trapezoid sum (d) midpoint sum

9. If f is concave down on all of on $[a, b]$, which of the given approximations is *guaranteed* to be an over-approximation for $\int_a^b f(x)\,dx$? (Select *all* that apply.)
   (a) left sum (b) right sum
   (c) trapezoid sum (d) midpoint sum

10. The error bounds for right and left sums in Theorem 5.25 apply only to monotonic functions. Suppose f is a positive integrable function that is increasing on $[a, c]$ and decreasing on $[c, b]$, with $a < c < b$. How could you use a right or left sum to make an estimate of $\int_a^b f(x)\,dx$ and still get a bound on the error? Draw a picture to help illustrate your answer.

11. The error bounds for trapezoid and midpoint sums in Theorem 5.27 apply only to functions with consistent concavity. Suppose f is a positive integrable function that is concave down on $[a, c]$ and concave up on $[c, b]$, with $a < c < b$. How could you use a trapezoid or midpoint sum to make an estimate of $\int_a^b f(x)\,dx$ and still get a bound of the error? Draw a picture to help illustrate your answer.

12. Explain what we mean when we say that the right, left, trapezoid, and midpoint sum approximations for $\int_a^b f(x)\,dx$ involve piecewise-linear approximations of the function f.

13. What is a "midpoint trapezoid," and what does it have to do with the midpoint sum? How do midpoint trapezoids help us determine whether the midpoint sum will be an over-approximation or under-approximation?

14. Theorem 5.27 allows us to estimate $\text{MID}(n)$ with twice the accuracy (i.e., half the error) of $\text{TRAP}(n)$. Is this what you would have expected from looking at graphical examples? Why do you think the midpoint sum is so accurate? *(Hint: Think about "midpoint trapezoids" and how they are defined.)*

15. Suppose f is a function with $f(x) > 0$, $f'(x) \le 0$, and $f''(x) \le 0$.
   (a) Put the right-sum, left-sum, trapezoid-sum, and midpoint-sum approximations of $\int_a^b f(x)\,dx$ in order from smallest to largest. (The order will not depend

on the number n of rectangles.) Explain your answer with a picture.

(b) The true value of $\int_a^b f(x)\,dx$ must lie between two of the approximations that you listed in part (a). Which two, and why?

16. The following table describes 12 different approximations of the value of $\int_1^2 \frac{1}{x}\,dx$:

Method	$n=4$	$n=8$	$n=16$
Right sum	0.634524	0.662872	0.677766
Left sum	0.759524	0.725372	0.709016
Trapezoid sum	0.697024	0.694122	0.693391
Midpoint sum	0.693147	0.693147	0.693147

(a) The actual value of $\int_1^2 \frac{1}{x}\,dx$ is $\ln 2 \approx 0.69314718$. Create a table showing the errors corresponding to each of the 12 approximations. What do you notice as n increases in each row?

(b) What do you notice about the sign of the errors of the right and left sums? What about the signs of the errors of the trapezoid and midpoint sums?

(c) How do the errors of the right and left sums compare with the errors of the trapezoid and midpoint sums for each n? How do the errors from the trapezoid sums compare with the errors from the midpoint sums?

17. The table that follows describes four different approximations of the value of $\int_2^5 f(x)\,dx$ for some differentiable function f. Use the weighted average formula given in Exercise 61 to find an estimate of $\int_2^5 f(x)\,dx$ that you would expect to be more accurate than any of the estimates in the table, and explain your answer.

Method	$n=1000$
Right sum	1.09794591
Left sum	1.09927925
Trapezoid sum	1.09861258
Midpoint sum	1.09861214

18. Suppose f is a positive, increasing, concave-up function on an interval $[a, b]$, and let c be the midpoint of the interval. Let A be the point $(a, 0)$, B be the point $(b, 0)$, C be the point $(c, f(c))$, D be the point $(b, f(b))$, E be the point $(a, f(a))$, R be the point $(b, f(c))$, and S be the point $(a, f(c))$.

(a) Draw a picture illustrating the situation described. What does the area of trapezoid $ABDE$ represent?

(b) Draw a line through point C that is parallel to the line connecting D and E. Let P be the point on the line with x-coordinate b. Let Q be the point on the line with x-coordinate a. What does the area of the rectangle $ABRS$ represent? What can you say about the areas of the rectangle $ABRS$ and the trapezoid $ABPQ$?

(c) Let A_1 be the area between the line PQ and the graph of f. Let A_2 be the area between the line DE and the graph of f. How do these areas compare in size? What do the areas represent? How do your answers to these questions help motivate the statement of Theorem 5.27?

19. Verify that the formula in Theorem 5.28 is true in the case where $p(x) = 3x^2 - x + 4$ and $[A, B] = [1, 5]$.

20. Sketch the graph of some function f on $[2, 5]$ for which SIMP(6) (i.e., with $\frac{n}{2} = 3$ parabola-topped rectangles) is an under-approximation for $\int_a^b f(x)\,dx$.

21. Show that the summation expression for Simpson's Rule in Theorem 5.29 follows from repeated applications of the formula for the definite integral of a quadratic in Theorem 5.28, in the case where $n = 4$ (i.e., where there are $\frac{n}{2} = 2$ parabolas).

22. In this exercise we investigate two cases where an approximation with Simpson's Rule happens to find the exact value of a definite integral.

(a) Explain why Simpson's Rule finds the *exact* value of $\int_a^b f(x)\,dx$ if f is a quadratic function. *(Hint: Think about Theorem 5.28.)*

(b) Explain why Simpson's Rule finds the *exact* value of $\int_a^b f(x)\,dx$ if f is a cubic function. *(Hint: Think about M in the formula for the error bound in Theorem 5.30.)*

Skills

Calculate each definite integral approximation in Exercises 23–40, and then find an error bound for your approximation. If it is possible to calculate the definite integral exactly, then do so and verify that the error bounds you found are accurate.

23. $\displaystyle\int_0^4 (x^2 + x)\,dx$, right sum, $n = 4$

24. $\displaystyle\int_0^4 (x^2 + x)\,dx$, right sum, $n = 8$

25. $\displaystyle\int_1^3 e^{-x}\,dx$, left sum with $n = 6$

26. $\displaystyle\int_1^3 e^{-x}\,dx$, left sum, $n = 10$

27. $\displaystyle\int_1^3 e^{-x}\,dx$, trapezoid sum, $n = 6$

28. $\displaystyle\int_1^3 e^{-x}\,dx$, midpoint sum, $n = 6$

29. $\displaystyle\int_1^7 \ln x\,dx$, midpoint sum, $n = 12$

30. $\displaystyle\int_1^7 \ln x\,dx$, trapezoid sum, $n = 12$

31. $\int_0^2 \sqrt{1+x^3}\,dx$, trapezoid sum with $n=8$

32. $\int_0^2 \sqrt{1+x^3}\,dx$, right sum, $n=8$

33. $\int_0^3 \cos(x^2)\,dx$, left sum, $n=9$

34. $\int_0^3 \cos(x^2)\,dx$, midpoint sum, $n=9$

35. $\int_{-2}^4 x(x+1)(x-2)\,dx$, Simpson's Rule, $n=12$

36. $\int_{-2}^4 x(x+1)(x-2)\,dx$, Simpson's Rule, $n=24$

37. $\int_{-1}^2 (x^4-4x^3+4x^2)\,dx$, Simpson's Rule, $n=4$

38. $\int_{-1}^2 (x^4-4x^3+4x^2)\,dx$, Simpson's Rule, $n=8$

39. $\int_{-\pi}^{\pi} \cos x\,dx$, Simpson's Rule, $n=8$

40. $\int_{-\pi}^{\pi} \cos x\,dx$, Simpson's Rule, $n=12$

Approximate each definite integral in Exercises 41–52 with the indicated method and to the given degree of accuracy. Then calculate the definite integral exactly, and verify that the error bounds you found are accurate.

41. $\int_{-1}^1 xe^x\,dx$, left sum, within 0.5

42. $\int_{-1}^1 xe^x\,dx$, right sum, within 0.5

43. $\int_2^4 \frac{1}{x}\,dx$, right sum, within 0.05

44. $\int_2^4 \frac{1}{x}\,dx$, left sum, within 0.05

45. $\int_0^3 x^2\,dx$, midpoint sum, within 0.1

46. $\int_0^3 x^2\,dx$, trapezoid sum, within 0.1

47. $\int_1^4 x^{2/3}$, trapezoid sum, within 0.005

48. $\int_1^4 x^{2/3}$, midpoint sum, within 0.005

49. $\int_0^2 10e^{-x}\,dx$, Simpson's Rule, within 0.01

50. $\int_0^2 10e^{-x}\,dx$, Simpson's Rule, within 0.001

51. $\int_0^{\pi} \sin x\,dx$, Simpson's Rule, within 0.0005

52. $\int_0^{\pi} \sin x\,dx$, Simpson's Rule, within 0.00005

Applications

53. Dad's casserole surprise is hot out of the oven. It cools at a rate of $T'(t) = 15e^{-0.5t}$ degrees per minute. The change in temperature $T(t)$ over an interval is equal to the area under the graph of $T'(t)$ over that interval.

(a) Estimate the change in temperature of the casserole during its first 4 minutes out of the oven, using LEFT(4), RIGHT(4), MID(4), TRAP(4), and SIMP(4).

(b) Give a bound on the errors involved in each approximation.

54. The table that follows shows the velocity in meters per second of a parachutist at various times. The distance travelled by the parachutist over an interval of time is equal to the area under the velocity curve on the same interval.

t	1	2	3	4
$v(t)$	9	17	23	28

(a) The parachutist's acceleration is decreasing because of air resistance. What does this fact imply about $v(t)$?

(b) Assuming that the velocity curve is increasing and concave down as the data suggests, find the best possible upper and lower bounds on the distance the parachutist fell from time $t=0$ to $t=4$ seconds.

55. The rate r of increase of the gross domestic product (GDP), in percent per year, for the United States from 1993 to 2001 is recorded in the table shown here. For example, in 1993 the GDP rose 2.7% from January 1 to December 31. You may assume that there is no growth from December 31 of one year to January 1 of the next year.

Year	1993	1994	1995	1996	1997	1998	1999	2000	2001
r	2.7	4.0	2.7	3.6	4.4	4.3	4.1	3.8	0.3

(a) Use Simpson's Rule to estimate the total percentage change in GDP from January 1, 1993, to December 31, 2001.

(b) If the GDP was $6,880 billion on January 1, 1993, what was the GDP on December 31, 2001? Use the best approximating tool available to you.

(c) Estimate a bound for the fourth derivative of r over the period from 1993 to 2001, and use that information to determine an error bound for your estimate of the total change in r.

56. Annie is building a kayak using wood and fabric. She needs to know the length of a sheet of polyurethane fabric to buy, which depends on the arc length of the longest surface of her kayak. The bottom of her kayak follows the curve

$$h(x) = (6 \times 10^{-4})\,x^4,$$

where h is the height of the bottom of the kayak above the ground when it is resting upright, and x is the distance from the center of the kayak. The kayak is 13 feet long; that is, it extends 6.5 feet from the center in either direction.

(a) Approximate the length of the curve along the bottom of Annie's kayak by using nine equally spaced points along the bottom of the kayak, and straight line segments between each adjacent pair of points.

(b) Make another approximation of the length of the curve along the bottom of Annie's kayak by Simpson's rule on the same nine equally spaced points, given that the length of this curve is given by the definite integral

$$\int_{-6.5,6.5} \sqrt{1 + (h'(x))^2}\,dx.$$

57. The flow (in cubic feet per second) down the Lochsa River over the course of a year is given in the table that follows, where t measures days after January 1 and $r(t)$ measures the rate of flow at time t. Use Simpson's Rule to approximate the total amount of water that flows down the river in a year. What did you do to account for the fact that a year has 365 days in it?

t	0	60	120	180	240	300	360
$r(t)$	700	1000	6300	4000	500	650	700

Proofs

58. Prove part (a) of Theorem 5.24: If f is integrable and monotonically increasing on $[a, b]$, then, for any positive integer n, $\text{LEFT}(n) \le \int_a^b f(x)\,dx \le \text{RIGHT}(n)$.

59. Use geometric arguments to support part (b) of Theorem 5.26: If f is integrable and concave down on $[a, b]$ and n is any positive integer, then $\text{TRAP}(n) \le \int_a^b f(x)\,dx \le \text{MID}(n)$.

60. Prove that the trapezoid sum is equivalent to the average of the left and right sums:

$$\text{TRAP}(n) = \frac{\text{LEFT}(n) + \text{RIGHT}(n)}{2}.$$

61. Prove that Simpson's Rule is equivalent to the following weighted average of the trapezoid and midpoint sums:

$$\text{SIMP}(2n) = \frac{1}{3}\text{TRAP}(n) + \frac{2}{3}\text{MID}(n).$$

62. Use steps (a) and (b) that follow to prove Theorem 5.28: if $p(x)$ is a quadratic function on $[A, B]$, then

$$\int_A^B p(x)\,dx = \frac{B-A}{6}\left(p(A) + 4p\left(\frac{A+B}{2}\right) + p(B)\right).$$

(a) First, calculate the integral of an arbitrary quadratic $p(x) = a_0 + a_1x + a_2x^2$ over an interval $[A, B]$.

(b) Now use the fact that the parabola $p(x)$ passes through the points $(A, p(A))$, $\left(\frac{A+B}{2}, p\left(\frac{A+B}{2}\right)\right)$, and $(B, p(B))$ to show that what you found in part (a) is equivalent to the desired formula.

63. Prove that by repeated application of the formula for the definite integral of a quadratic in Theorem 5.28, we can arrive at the Simpson's Rule approximation given in Theorem 5.29. (*Hint: Apply Theorem 5.28 to the subintervals* $[x_k, x_{k+2}]$ *where* k *is even.*)

Thinking Forward

Approximating with power series: For values of x in the interval $(-1, 1)$, the rational function $f(x) = \frac{1}{1-x}$ turns out to be equal to

$$\lim_{n\to\infty} \sum_{k=1}^{n} x^k.$$

- Calculate the preceding sum when $x = \frac{1}{2}$ for $n = 5$, $n = 10$, and $n = 20$, and compare these numbers with the quantity $f\left(\frac{1}{2}\right)$. What is the error in each case?
- Repeat the previous problem for $x = 2$, and argue that the sum does *not* provide a good approximation of $f(2)$.

CHAPTER REVIEW, SELF-TEST, AND CAPSTONES

Before you progress to the next chapter, be sure you are familiar with the definitions, concepts, and basic skills outlined here. The capstone exercises at the end bring together ideas from this chapter and look forward to future chapters.

Definitions

Give precise mathematical definitions or descriptions of each of the concepts that follow. Then illustrate the definition with a graph or algebraic example, if possible.

- the *differential* of a differentiable function $u(x)$
- what it means for a rational function to be *improper*
- what it means for a definite integral to be *improper*
- what it means for an improper integral to *converge* or to *diverge*
- the *error* $|E(n)|$ of an approximation of a definite integral by a Riemann sum
- the geometric interpretations of the $\frac{n}{2}$-parabola *Simpson's Rule* approximation for the definite integral of a function f on an interval $[a, b]$

Theorems

Fill in the blanks to complete each of the following theorem statements:

- If f and u are functions such that $f'(u(x))u'(x)$ is ____ , then we can use integration by substitution to solve

$$\int f'(u(x))u'(x)\,dx = \_\_\_\_ .$$

- If $f(x) = g(u(x))$ is ____ on $[a, b]$, $u(x)$ is ____ on (a, b), and $G(u)$ is an ____ of $g(u)$, then integration by substitution for definite integrals gives the following string of equalities (fill in the upper and lower limits that are marked by asterisks; there are two possible ways to do this):

$$\int_a^b f(x)\,dx = \int_*^* g(u)\,du = \left[G(u)\right]_*^*$$
$$= \left[G(u(x))\right]_*^* = G(u(b)) - G(u(a)).$$

- If u and v are functions such that $u'(x)v(x) + u(x)v'(x)$ is ____ , then we can use the product rule in reverse to obtain

$$\int (u'(x)v(x) + u(x)v'(x))\,dx = \_\_\_\_ .$$

- If $u = u(x)$ and $v = v(x)$ are differentiable functions, then the following integration-by-parts formula holds:

$$\int u\,dv = \_\_\_\_ - \_\_\_\_ .$$

- If $u = u(x)$ and $v = v(x)$ are differentiable functions on $[a, b]$, then the following integration-by-parts formula holds for definite integrals:

$$\int_a^b u\,dv = [\_\_\_\_]_a^b = \_\_\_\_ - \_\_\_\_ .$$

- Suppose $\frac{p(x)}{q(x)}$ is an ____ rational function with $\deg(p(x)) = n$ and $\deg(q(x)) = m < n$. Then we can write

$$\frac{p(x)}{q(x)} = \_\_\_\_ + \_\_\_\_ ,$$

for some polynomial $s(x)$ of degree $n-m$ and some proper rational function $\frac{r(x)}{q(x)}$ with $\deg(r(x)) <$ ____ .

- In a partial-fractions decomposition of a proper rational function $\frac{p(x)}{q(x)}$, if $q(x)$ has a linear factor $x - c$ with multiplicity m, then for some constants $A_1, A_2, \ldots, A_m$, the sum will include terms of the forms ____ .
- In a partial-fractions decomposition of a proper rational function $\frac{p(x)}{q(x)}$, if $q(x)$ has an irreducible quadratic factor $x^2 + bx + c$ with multiplicity m, then for some constants $B_1, B_2, \ldots, B_m$ and $C_1, C_2, \ldots, C_m$, the sum will include terms of the forms ____ .
- By applying polynomial long division to divide a polynomial $p(x)$ by a polynomial $q(x)$ of equal or lower degree, we can obtain polynomials $m(x)$ and $R(x)$ such that the following equation is true:

$$p(x) = \_\_\_\_ + \_\_\_\_ .$$

Equivalently, we have the following equation:

$$\frac{p(x)}{q(x)} = \_\_\_\_ + \_\_\_\_ .$$

- Every quadratic function $x^2 + bx + c$ can be rewritten in the form $x^2 + bx + c = (x - k)^2 + C$, where $k =$ ____ and $C =$ ____ .
- If a is a positive real number, and if $x \in$ ____ and $u \in$ ____ , then the substitution $x = a \sin u$ gives us $\sqrt{a^2 - x^2} =$ ____ .
- If a is a positive real number, and if $x \in$ ____ and $u \in$ ____ , then the substitution $x = a \tan u$ gives us $x^2 + a^2 =$ ____ .
- If a is a positive real number, and if $x \in$ ____ and $u \in$ ____ , then the substitution $x = a \sec u$ gives us $\sqrt{x^2 - a^2} =$ ____ if $x < -a$ and $\sqrt{x^2 - a^2} =$ ____ if $x > a$.
- If the improper integral of a function f on an interval I converges, and if ____ for all $x \in I$ where the functions f and g are defined, then the improper integral of $g(x)$ on I also converges.

- If the improper integral of f on I diverges, and $0 \le f(x) \le g(x)$ for all $x \in I$ where the functions are defined, then the improper integral of g on I must ____ .

- If f is integrable and increasing on $[a, b]$ and n is a positive integer, then

$$\text{LEFT}(n) \le \_\_\_\_ \le \text{RIGHT}(n).$$

- If f is integrable and monotonic on $[a, b]$ and n is a positive integer, then

$$|E_{\text{LEFT}(n)}| \le |\_\_\_\_ - \_\_\_\_| = |\_\_\_\_|\Delta x.$$

- If f is integrable and either always concave up or always concave down on $[a, b]$ and n is a positive integer, then

$$|E_{\text{TRAP}(n)}| \le \_\_\_\_ \quad \text{and} \quad |E_{\text{MID}(n)}| \le \_\_\_\_ .$$

- If f is integrable and either always concave up or always concave down on $[a, b]$, n is a positive integer, and f'' is bounded on $[a, b]$ so that there is some positive real number M such that ____ $\le M$ for all $x \in [a, b]$, then

$$|E_{\text{TRAP}(n)}| \le \_\_\_\_ \quad \text{and} \quad |E_{\text{MID}(n)}| \le \_\_\_\_ .$$

- If $p(x)$ is a quadratic function on $[A, B]$, then we can express $\int_A^B p(x)\,dx$ in terms of $p(A)$, $p\left(\frac{A+B}{2}\right)$, and $p(B)$ as follows: ____ .

- If f is integrable and $f^{(4)}$ is bounded on $[a, b]$ so that there is some positive real number M such that ____ $\le M$ for all $x \in [a, b]$, then

$$|E_{\text{SIMP}(n)}| \le \_\_\_\_ .$$

Rules of Algebra and Integration

Pythagorean identities: Use Pythagorean identities to rewrite each of the following trigonometric expressions.

- $\sin^2 x =$ ____
- $\cos^4 x =$ ____
- $\tan^2 x =$ ____
- $\sec^6 x =$ ____
- $\cot^2 x =$ ____
- $\csc^8 x =$ ____

Double-angle identities: Use double-angle identities to rewrite each of the following trigonometric expressions until no exponents are involved.

- $\sin^2 x =$ ____
- $\cos^2 x =$ ____
- $\sin^4 x =$ ____
- $\cos^6 x =$ ____

Integration formulas: Fill in the blanks to complete each of the following integration formulas.

- $\int \tan x\,dx =$ ____
- $\int \cot x\,dx =$ ____
- $\int \ln x\,dx =$ ____
- $\int \sin^{-1} x\,dx =$ ____
- $\int \tan^{-1} x\,dx =$ ____
- $\int \sec x\,dx =$ ____
- $\int \csc x\,dx =$ ____

Defining improper integrals: Fill in the blanks, using limits and proper definite integrals to express each of the following types of improper integral.

- If f is continuous on $[a, \infty)$, then

$$\int_a^\infty f(x)\,dx = \_\_\_\_ .$$

- If f is continuous on $(-\infty, b]$, then

$$\int_{-\infty}^b f(x)\,dx = \_\_\_\_ .$$

- If f is continuous on $(-\infty, \infty)$, then for any real number c,

$$\int_{-\infty}^\infty f(x)\,dx = \_\_\_\_ + \_\_\_\_ .$$

- If f is continuous on $(a, b]$ but not at $x = a$, then

$$\int_a^b f(x)\,dx = \_\_\_\_ .$$

- If f is continuous on $[a, b)$ but not at $x = b$, then

$$\int_a^b f(x)\,dx = \_\_\_\_ .$$

- If f is continuous on $[a, c) \cup (c, b]$ but not at $x = c$, then

$$\int_a^b f(x)\,dx = \_\_\_\_ + \_\_\_\_ .$$

Convergence and divergence of basic improper integrals: Determine whether each of the given improper integrals converges or diverges. For those that converge, give the exact solution of the integral.

- $\int_0^1 \frac{1}{x^p}\,dx$, for $0 < p < 1$
- $\int_1^\infty \frac{1}{x^p}\,dx$, for $0 < p < 1$
- $\int_0^1 \frac{1}{x^p}\,dx$, for $p = 1$
- $\int_1^\infty \frac{1}{x^p}\,dx$, for $p = 1$
- $\int_0^1 \frac{1}{x^p}\,dx$, for $p > 1$
- $\int_1^\infty \frac{1}{x^p}\,dx$, for $p > 1$

Skill Certification: Integration Techniques

Integration techniques: Use whatever method you like to solve each of the given definite and indefinite integrals. These integrals are neither in order of difficulty nor in order of technique. Many of the integrals can be solved in more than one way.

1. $\int \cos^7 x\,dx$

2. $\int (e^{3\ln x})^2\,dx$

3. $\int x \sec x \tan x\,dx$

4. $\int \frac{x^2 \sin x + 1}{x}\,dx$

5. $\int \frac{\ln 2x}{x^2}\,dx$

6. $\int \sin^{-1} x\,dx$

7. $\int \cot^6 x\,dx$

8. $\int x^3 \sin x\,dx$

9. $\int 2\pi x(8 - x^{3/2})\,dx$

10. $\int \csc^4 x \cot^4 x\,dx$

11. $\int_0^1 \ln(1+3x)\,dx$

12. $\int_0^{\pi/2} \sin^2 x \cos^5 x\,dx$

13. $\int \frac{\sqrt{x}}{1 - x\sqrt{x}}\,dx$

14. $\int e^x \cos 2x\,dx$

15. $\int \sin(\ln x)\,dx$

16. $\int \sin^4 3x\,dx$

17. $\int \ln\left(\frac{x^3}{2x+1}\right)dx$

18. $\int \frac{\sin^3 x}{\sqrt{\cos x}}\,dx$

19. $\int (\ln x)^3\,dx$

20. $\int \csc^3 x\,dx$

21. $\int \tan^3 x \sec^3 x\,dx$

22. $\int (x \ln x)^{-1}\,dx$

23. $\int_1^2 \frac{x^2 + 3x + 7}{\sqrt{x}}\,dx$

24. $\int \tan x \ln(\cos x)\,dx$

25. $\int \csc 3x\,dx$

26. $\int (\sin x + e^x)^2\,dx$

27. $\int_1^3 \frac{e^{1/x}}{x^2}\,dx$

28. $\int x^3 \cos x^2\,dx$

29. $\int x(x+5)^{7/2}\,dx$

30. $\int e^x \tan e^x\,dx$

31. $\int \frac{x+1}{x^4 + 3x^2}\,dx$

32. $\int \sec^5 x \tan^3 x\,dx$

33. $\int \frac{x^3}{e^{x^2}}\,dx$

34. $\int x\sqrt{1 - x\sqrt{x}} - 5\,dx$

35. $\int 2^x e^x\,dx$

36. $\int \frac{x^4}{(x-1)^2(x+2)}\,dx$

37. $\int \frac{7x^2 + 6x + 5}{x^3 + x^2 + x - 3}\,dx$

38. $\int \frac{1}{e^x\sqrt{4 + e^{2x}}}\,dx$

39. $\int \frac{1}{(9 - 4x^2)^{3/2}}\,dx$

40. $\int \frac{2x^3 + x^2 + 2}{x^3 + 1}\,dx$

41. $\int \frac{\ln(\tan^{-1} x)}{x^2 + 1}\,dx$

42. $\int \frac{\sqrt{x}}{1 + x}\,dx$

43. $\int x \operatorname{sech}^2 x^2\,dx$

44. $\int \sinh^2 x \cosh^3 x\,dx$

45. $\int \frac{\sinh^{-1} x}{\sqrt{x^2 + 1}}\,dx$

46. $\int \frac{x^2}{(x^2 - 4)^{3/2}}\,dx$

Areas and average values: Set up and solve definite integrals to calculate the exact values of each of the given geometric quantities. Verify that your answers are reasonable with graphs.

47. The signed area between the graph of $f(x) = \frac{x^2-1}{x^2+1}$ and the x-axis on the interval $[-4, 4]$.

48. The absolute area between the graph of $f(x) = \sin^2 x \cos x$ and the x-axis on the interval $[0, \pi]$.

49. The area between the graphs of $f(x) = \sin^2 x$ and $g(x) = \frac{1}{4}$ on the interval $[0, \pi]$.

50. The area between the graphs of $f(x) = \ln x$ and $g(x) = (\ln x)^2$ on the interval $\left[\frac{1}{2}, 5\right]$.

51. The average value of the function $f(x) = xe^{-x}$ on the interval $[-1, 3]$.

52. The average value of the function $f(x) = \frac{2x-1}{(x+2)(x^2+1)}$ on the interval $[0, 10]$.

Improper integrals: Set up and solve limits of definite integrals to calculate each of the following improper integrals.

53. $\int_0^8 x^{-1/3}\,dx$

54. $\int_1^\infty x^{-1/4}\,dx$

55. $\int_0^\infty x \ln x\,dx$

56. $\int_0^1 \frac{1}{x(\ln x)^2}\,dx$

57. $\int_0^{\pi/2} \sec x \tan^3 x\,dx$

58. $\int_1^\infty \frac{1}{x^2\sqrt{x^2 - 1}}\,dx$

Numerical integration: Approximate each of the following definite integrals with the indicated method and to the given degree of accuracy. *(These problems assume that you have covered Section 5.7)*

59. $\int_{-1}^1 e^{-x^2}\,dx$, right sum, within 0.5

60. $\int_0^\pi \sin x^2\,dx$, midpoint sum, within 0.1

61. $\int_0^3 \sqrt{1 + x^3}\,dx$, Simpson's Rule, within 0.01

Capstone Problems

A. *Integration strategies:* Make a flowchart or diagram that describes a strategy for choosing integration techniques. Given a particular integral, what techniques might you try first? What features of the integrand would lead you to choose one technique over another?

B. *Impossible integrals:* Even with all of the techniques of this chapter, there are still plenty of integrals that we cannot solve.

(a) List three functions $f(x)$ that we cannot integrate with the techniques that we have learned so far. Describe why each of our integration techniques fails or does not apply to each of your three integrals.

(b) For each of your functions $f(x)$, use the Second Fundamental Theorem of Calculus (Theorem 4.33) to write down an antiderivative of $f(x)$.

(c) For each of your functions $f(x)$, choose an interval $[a, b]$ over which $f(x)$ is defined and continuous and consider the definite integral $\int_a^b f(x)\,dx$. Use Riemann sums to approximate the values of each of these definite integrals to within an accuracy of 0.05. (See Section 5.7.)

C. *Comparing polynomial long division and synthetic division:* There are two methods that we can use to divide a polynomial by a linear function. In this exercise you will compare those two methods.

(a) Find the quotient that results by dividing the polynomial $x^3 - 4x^2 - 7x + 10$ by the linear function $x - 5$, using polynomial long division.

(b) Repeat the same calculation, this time using synthetic division.

(c) Use a comparison of the coefficients that appear in your two calculations to investigate how synthetic division is just a shorthand for polynomial long division.

D. *Improper integrals and infinite sums:* Consider the improper integral

$$\int_1^\infty \frac{x}{e^x}\,dx$$

(a) Show, by calculating a limit of definite integrals, that this improper integral converges.

(b) Sketch a right sum with $\Delta x = 1$ for the given improper integral.

(c) Use the sketch you made in part (b) to argue that

$$\sum_{k=2}^{\infty} \frac{k}{e^k}$$

converges, that is, that the sum of infinitely many numbers $\frac{2}{e^2} + \frac{3}{e^3} + \frac{4}{e^4} + \cdots$ is a finite real number.

CHAPTER 6

Applications of Integration

6.1 Volumes by Slicing

Approximating Volume by Slicing
Volume as a Definite Integral of Cross-Sectional Area
Volumes by Disks and Washers
Distances Defined by a Point on a Graph
Examples and Explorations

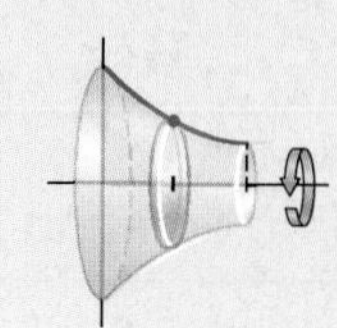

6.2 Volumes by Shells

Approximating Volume by Shells
Finding a Riemann Sum for Volumes by Shells
A Definite Integral for Volume by Shells
Examples and Explorations

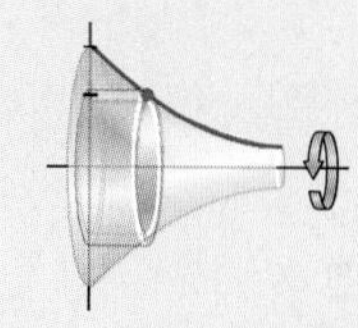

6.3 Arc Length and Surface Area

Approximating Arc Length
A Definite Integral for Arc Length
Approximating Surface Area
A Definite Integral for Surface Area
Examples and Explorations

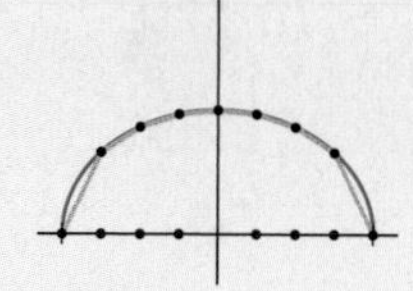

6.4 Real-World Applications of Integration

Mass and Density
Work and Force
Hydrostatic Force
Representing Varying Quantities with Definite Integrals
Centroids and Centers of Mass
Examples and Explorations

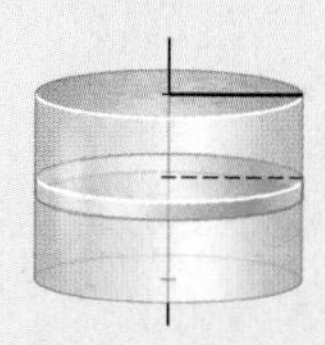

6.5 Differential Equations*

Differential Equations and Initial-Value Problems
Separable Differential Equations
Slope Fields
Euler's Method
Applications of Initial-Value Problems
Examples and Explorations

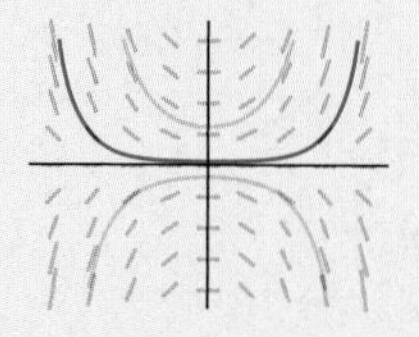

Chapter Review, Self-Test, and Capstones

6.1 VOLUMES BY SLICING

- Approximating volumes of solids of revolution with disks and washers
- Finding exact volumes by accumulating cross-sectional area
- Using definite integrals to describe volumes of solids of revolution

Approximating Volume by Slicing

Some geometric quantities are easy to calculate, such as the area of a rectangle, length of a line segment, or volume of a cylinder. Harder to calculate are things like the area of a region with curved boundary, length of a curve, or volume of a curvy solid. In Chapter 4 we calculated the area under a curve by using a "subdivide, approximate, and add" strategy and then taking a limit to obtain a definite integral. We now apply the same strategy to volumes. Let's start by trying to find the volume of a sphere.

Of course, you probably already know that the formula for the volume V of a sphere with radius r is $V = \frac{4}{3}\pi r^3$. But why? Let's approximate the volume of a sphere of radius 2 to find out. We can construct this sphere by considering the region bounded by the graph of $\sqrt{4-x^2}$ and the x-axis on $[-2, 2]$ and rotating this region around the x-axis, as shown in the first figure that follows. In the second figure we show an approximation of this region with rectangles. If we rotate these rectangles around the x-axis along with the region as shown in the third figure, then they become cylinders that approximate the sphere.

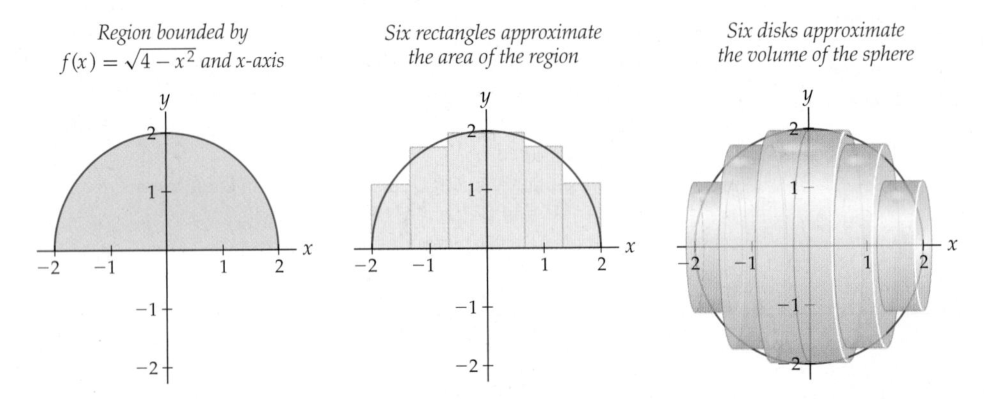

We can approximate the volume of the sphere by adding up the volumes of these cylinders, which we will call ***disks***. We have $n = 6$ disks, each of width $\Delta x = \frac{2}{3}$. These values define subdivision points $x_k = -2 + k\left(\frac{2}{3}\right)$, for $k = 0, 1, 2, 3, 4,$ and 5. The heights of the rectangles in the middle figure are defined by the values of $f(x)$ at the midpoints x_k^* of each subinterval $[x_{k-1}, x_k]$. Each disk has radius $f(x_k^*)$ and thus volume $\pi f(x_k^*)^2\left(\frac{2}{3}\right)$. The volume of the sphere is therefore approximately

$$\sum_{k=1}^{6} \pi f(x_k^*)^2\left(\frac{2}{3}\right) = \sum_{k=1}^{6} \pi\left(\sqrt{4-(x_k^*)^2}\right)^2\left(\frac{2}{3}\right) = \pi\sum_{k=1}^{6}(4-(x_k^*)^2)\left(\frac{2}{3}\right).$$

To make our approximation better, we could start with more rectangles and get more cylinders, each of which is a better approximation to its part of the sphere. For n disks and

$\Delta x = \frac{2-(-2)}{n} = \frac{4}{n}$, we have

$$\sum_{k=1}^{n} \pi f(x_k^*)^2 \left(\frac{4}{n}\right) = \sum_{k=1}^{n} \pi \left(\sqrt{4-(x_k^*)^2}\right)^2 \left(\frac{4}{n}\right) = \pi \sum_{k=1}^{n} (4-(x_k^*)^2) \left(\frac{4}{n}\right).$$

To find the exact volume we would try to take the limit as n approaches infinity. This should sound very familiar, because it is a limit of Riemann sums. That is fantastic news, because it means that we will be able to write down definite integrals that represent volumes, and definite integrals are much easier to calculate than limits of sums.

Volume as a Definite Integral of Cross-Sectional Area

As a warm-up, recall the definition of the area under a curve. For any continuous, nonnegatively-valued function $f(x)$ on $[a, b]$, we defined the area under the graph of f on $[a, b]$ to be

$$\text{Area} = \lim_{n\to\infty} \sum_{k=1}^{n} f(x_k^*)\Delta x = \int_a^b f(x)\,dx,$$

where $\Delta x = \frac{b-a}{n}$, $x_k = a + k\Delta x$, and x_k^* is a point in $[x_{k-1}, x_k]$. Here $f(x^*)$ represents a height, Δx is a width, and their product $f(x_k^*)\Delta x$ is the area of a rectangle that approximates a slice of the area under the graph of f, as shown here in the leftmost figure:

Accumulate heights $f(x_k^*)$ to obtain area

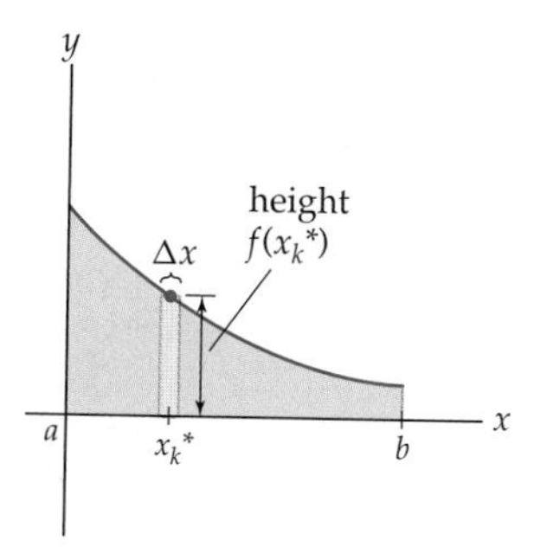

Accumulate cross-sectional areas $A(x_k^*)$ to obtain volume

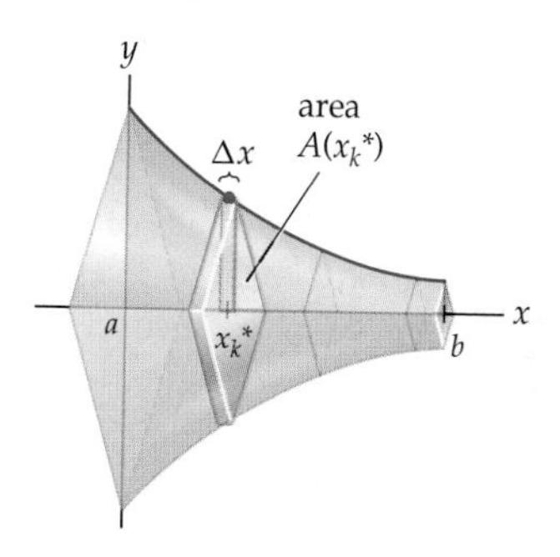

So how should we define volume? This is a similar situation. Suppose we have a three-dimensional solid with cross sections that are defined in some homogeneous way, such as that shown in the rightmost figure. We subdivide the solid by slicing it, and then we approximate the volume of each slice by multiplying a cross-sectional area $A(x_k^*)$ at some point x_k^* by the width Δx. If the cross sections of the solid vary continuously, then we can get a formula for $A(x_k^*)$. In the preceding rightmost graph, the cross-sectional areas happen to be squares whose diagonals depend on $f(x_k^*)$.

DEFINITION 6.1 **Volume as a Definite Integral of Cross-Sectional Areas**

If S is a solid whose cross sections between $x = a$ and $x = b$ have areas given by a continuous function $A(x)$ on $[a, b]$, then the ***volume*** of S is

$$\text{Volume} = \lim_{n\to\infty} \sum_{k=1}^{n} A(x_k^*)\Delta x = \int_a^b A(x)\,dx,$$

where $\Delta x = \frac{b-a}{n}$, $x_k = a + k\Delta x$, and $x_k^* \in [x_{k-1}, x_k]$.

Of course if we sliced instead in the y-direction, then we could make a similar definition, but with area as a function of y instead of x and integrating from some point $y = A$ to some point $y = B$ instead of from $x = a$ to $x = b$.

Volumes by Disks and Washers

A ***solid of revolution*** is a three-dimensional object obtained by rotating a planar region around an axis, or line. For example, consider the planar region $\mathbb{R}$ bounded by $f(x) = x^2$, the x-axis, and the line $x = 2$, as shown in the first figure that follows. By rotating this region around different lines we can obtain many different solids of revolution. For example, the second figure shows the solid of revolution obtained by rotating $\mathbb{R}$ around the x-axis, and the third figure shows the solid of revolution obained by rotating $\mathbb{R}$ around the y-axis.

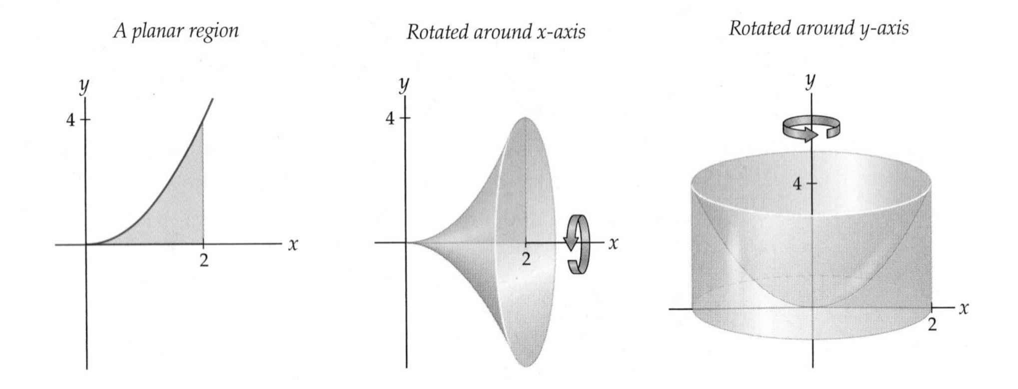

Solids of revolution usually have cross sections that are either disks or ***washers***, which are large disks with a smaller disk removed from the center. For example, the solid shown in the second figure has disk cross sections if we slice along the x-axis, and the solid in the third figure has washer cross sections if we slice along the y-axis. If the cross sections are disks, then we can define volume as the accumulation of the areas of those disks:

DEFINITION 6.2 **Volume as a Definite Integral Using Disks**

Suppose S is a solid of revolution obtained by rotating a region on $[a, b]$ around a horizontal axis. If the cross sections of S are disks with radii given by a continuous function $r(x)$ on $[a, b]$, then the ***volume*** of S is

$$\text{Volume of solid with disk cross sections} = \lim_{n\to\infty} \sum_{k=1}^{n} \pi (r_k^*)^2 \Delta x = \pi \int_a^b (r(x))^2\, dx,$$

where $\Delta x = \frac{b-a}{n}$, $x_k = a + k\Delta x$, $x_k^* \in [x_{k-1}, x_k]$, and $r_k^* = r(x_k^*)$.

In this definition, the cross section at each x_k^* is a disk with radius r_k^* as shown next at the left, so its cross-sectional area is $A(x_k^*) = \pi (r_k^*)^2$. If the cross section of a solid of revolution at each x_k^* is a washer with large radius R_k^* and small radius r_k^*, as shown next at the right, then its cross-sectional area is $A(x_k^*) = \pi ((R_k^*)^2 - (r_k^*)^2)$ and its volume can be defined as the accumulation of those areas.

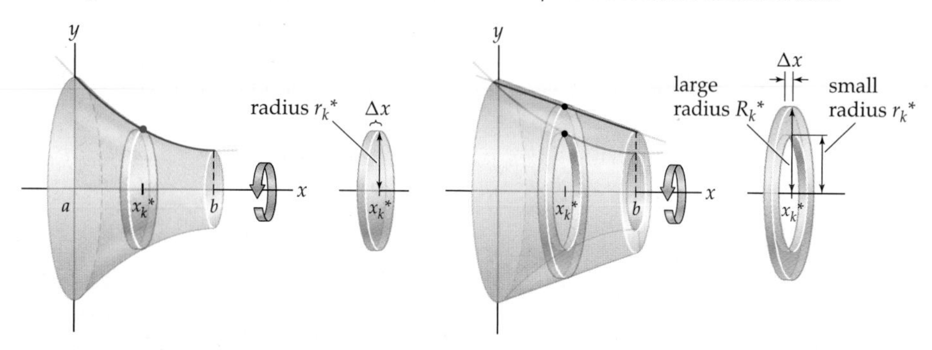

DEFINITION 6.3 **Volume as a Definite Integral Using Washers**

Suppose S is a solid of revolution obtained by rotating a region on $[a, b]$ around a horizontal axis. If the cross sections of S are washers with large radii given by a continuous funtion $R(x)$ and small radii given by a continuous function $r(x)$ on $[a, b]$, then the ***volume*** of S is

$$\text{Volume of solid with washer cross sections} = \lim_{n\to\infty} \sum_{k=1}^{n} \pi\,((R_k^*)^2 - (r_k^*)^2)\Delta x = \pi \int_a^b ((R(x))^2 - (r(x))^2)\,dx,$$

where $\Delta x = \frac{b-a}{n}$, $x_k = a + k\Delta x$, $x_k^* \in [x_{k-1}, x_k]$, $R_k^* = R(x_k^*)$, and $r_k^* = r(x_k^*)$.

We have similar definitions for solids of revolution obtained by rotation around a vertical axis on an interval $[A, B]$ of y-values. The only differences in that case are that the radius functions will be functions of y rather than x and that we will integrate from A to B with respect to y instead of from a to b with respect to x, as shown in the following two figures:

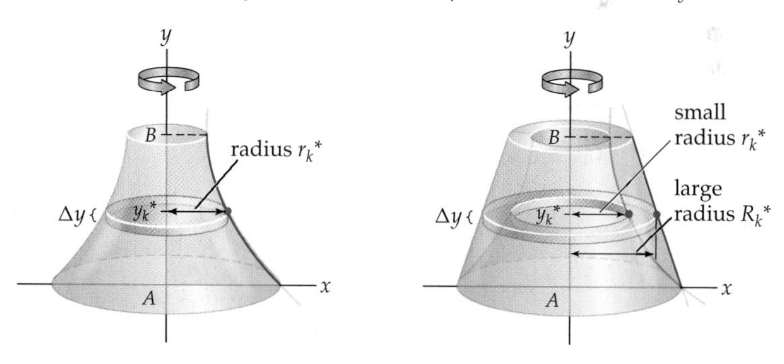

The analogs of Definitions 6.2 and 6.3 for integrating along the y-axis are then

$$\text{Volume of solid with disk cross sections} = \lim_{n\to\infty} \sum_{k=1}^{n} \pi\,(r_k^*)^2 \Delta y = \pi \int_A^B (r(x))^2\,dy,$$

$$\text{Volume of solid with washer cross sections} = \lim_{n\to\infty} \sum_{k=1}^{n} \pi\,((R_k^*)^2 - (r_k^*)^2)\Delta y = \pi \int_A^B ((R(x))^2 - (r(x))^2)\,dy.$$

Distances Defined by a Point on a Graph

In setting up definite integrals to represent volumes by disks or washers, the first step is to determine the radius functions. When we slice across the x direction, radius functions often depend on the function f that was rotated to obtain the solid or on some boundary x-value. When we slice across the y direction, radius functions often depend on f^{-1} or a boundary y-value. As a handy reference, the following two figures show the vertical and horizontal distances defined by a point on a graph:

Vertical and horizontal distances in terms of $y = f(x)$

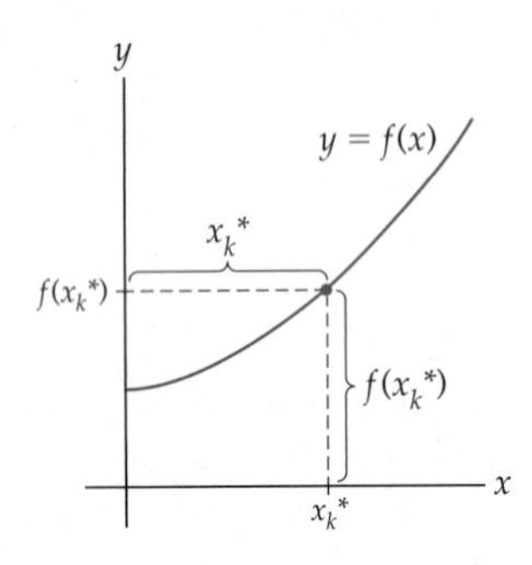

Vertical and horizontal distances in terms of $x = f^{-1}(y)$

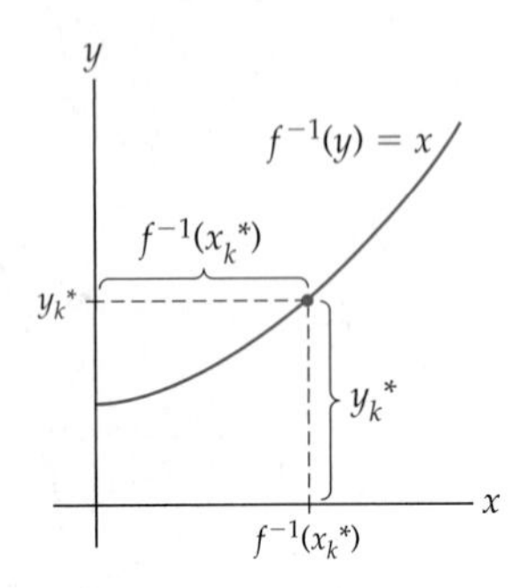

Notice that to find the distance between the y-axis and a point on the graph in the second figure, we need to use $f^{-1}(y_k^*)$. This means that f has to be an invertible function (at least on a restricted domain) and that we have to be able to actually figure out what its inverse is.

Examples and Explorations

EXAMPLE 1

Using disks to approximate volume

Approximate the volume of a sphere of radius 2 with six disks.

SOLUTION

Let's use the six disks that were defined in the reading at the start of this section. Recall that a sphere of radius 2 can be obtained by rotating the region under the graph of $f(x) = \sqrt{4 - x^2}$ around the x-axis. For reference, the diagrams of these disks and the rectangles that define them are repeated here:

Six rectangles approximate the area under $f(x) = \sqrt{4 - x^2}$

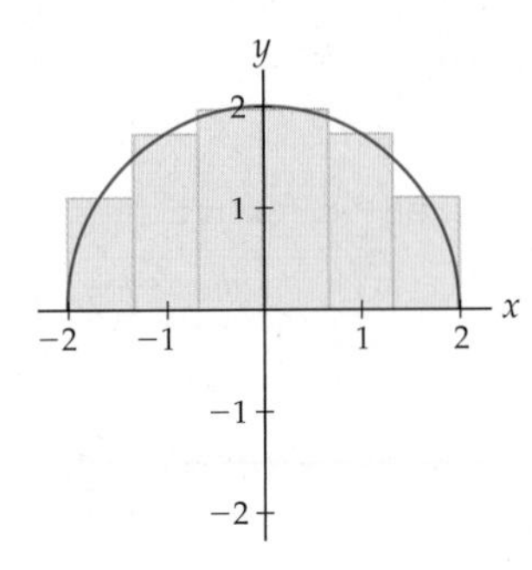

Six disks approximate the volume of the sphere

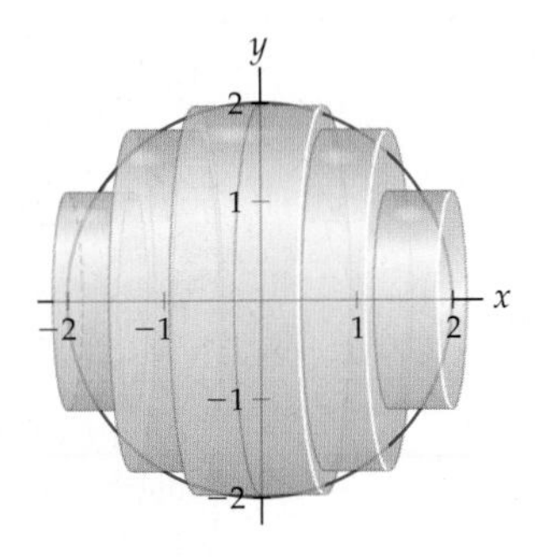

We have $n = 6$ disks, and thus $\Delta x = \frac{2}{3}$. The six subintervals are

$$\left[-2, -\frac{4}{3}\right], \left[-\frac{4}{3}, -\frac{2}{3}\right], \left[-\frac{2}{3}, 0\right], \left[0, \frac{2}{3}\right], \left[\frac{2}{3}, \frac{4}{3}\right], \text{ and } \left[\frac{4}{3}, 2\right].$$

The radii of the six disks shown in the figure earlier at the right are the heights of the six rectangles in the graph shown earlier at the left, which are obtained by evaluating the function $f(x) = \sqrt{4 - x^2}$ at the midpoints of the six subintervals. Thus the radii of the six disks are

$$f\left(-\frac{5}{3}\right),\ f(-1),\ f\left(-\frac{1}{3}\right),\ f\left(\frac{1}{3}\right),\ f(1),\ \text{and}\ f\left(\frac{5}{3}\right).$$

The volume of a cylinder with radius r and height h is $\pi r^2 h$, and our disks are cylinders with radii as given earlier and heights (for us they appear as widths) each $\Delta x = \frac{2}{3}$. The sum of their volumes is

$$\pi\left(f\left(-\frac{5}{3}\right)\right)^2\left(\frac{2}{3}\right) + \pi(f(-1))^2\left(\frac{2}{3}\right) + \pi\left(f\left(-\frac{1}{3}\right)\right)^2\left(\frac{2}{3}\right)$$
$$+\pi\left(f\left(\frac{1}{3}\right)\right)^2\left(\frac{2}{3}\right) + \pi(f(1))^2\left(\frac{2}{3}\right) + \pi\left(f\left(\frac{5}{3}\right)\right)^2\left(\frac{2}{3}\right).$$

Evaluating with $f(x) = \sqrt{4 - x^2}$ and simplifying, we find that the combined volume of the six disks is

$$\frac{22}{27}\pi + 2\pi + \frac{70}{27}\pi + \frac{70}{27}\pi + 2\pi + \frac{22}{27}\pi \approx 33.9758 \text{ cubic units.}$$

This approximation is close to what our old formula for the volume of a sphere predicts: $\frac{4}{3}\pi r^3 = \frac{4}{3}\pi(2)^3 \approx 33.5103$ cubic units. □

EXAMPLE 2

Using disks to construct a definite integral for volume

Find the exact volume of a sphere of radius 2 by using a definite integral to calculate the volume of the solid of revolution obtained by rotating the area under the graph of $f(x) = \sqrt{4 - x^2}$ on $[-2, 2]$ around the x-axis.

SOLUTION

We can rotate a representative rectangle at x_k^* to obtain a representative disk at x_k^*. The radius of this disk is given by the height $f(x_k^*)$ of the function, as shown in the following figure:

A representative disk

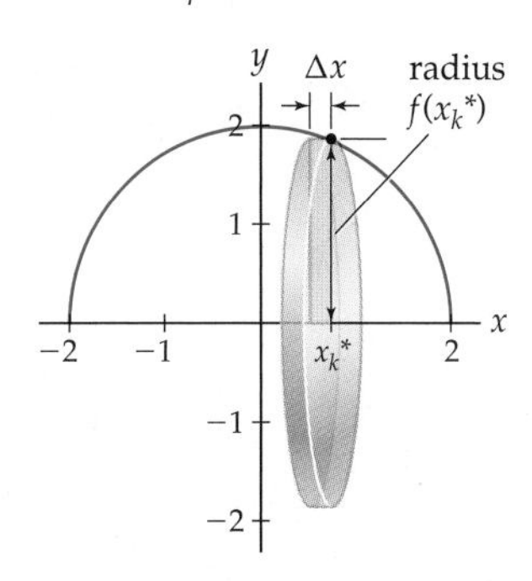

Therefore the volume of this representative disk is

$$\pi(f(x_k^*))^2\Delta x = \pi\left(\sqrt{4 - (x_k^*)^2}\right)^2\Delta x = \pi(4 - (x_k^*)^2)\Delta x.$$

We need to accumulate the volumes of these disks from $x = -2$ to $x = 2$, so, converting to definite integral form, we find that the volume of the sphere is

$$\pi\int_{-2}^{2}(4 - x^2)\,dx = \pi\left[4x - \frac{1}{3}x^3\right]_{-2}^{2} = \pi\left(\left(4(2) - \frac{1}{3}(2)^3\right) - \left(4(-2) - \frac{1}{3}(-2)^3\right)\right) = \frac{32\pi}{3}.$$

Note that of course our answer here exactly matches the answer we get from the formula for the volume of a sphere: $V = \frac{4}{3}\pi(2)^3 = \frac{32\pi}{3}$. □

EXAMPLE 3

Using washers to construct a definite integral for volume

Find the volume of the solid of revolution obtained by rotating the region between the graphs of $f(x) = x + 1$ and $g(x) = x^2 - 4x + 5$ on $[1, 4]$ around the x-axis.

SOLUTION

The leftmost figure that follows shows the region in question, and the middle figure shows the solid of revolution whose volume we wish to find. Once again we can take a representative rectangle for the region at x_k^* and rotate it around the x-axis to obtain a representative slice of the solid at x_k^*, which in this case is washer shaped, as shown in the rightmost figure.

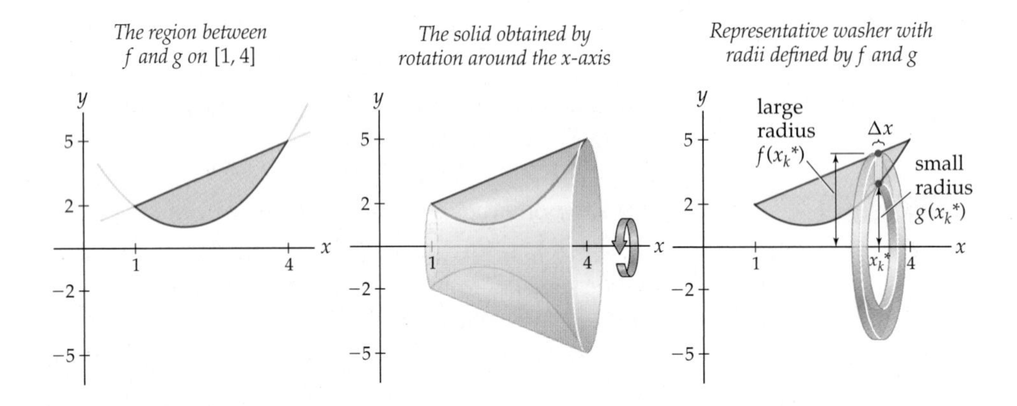

The representative washer has large radius given by $f(x_k^*)$ and small radius given by $g(x_k^*)$. Since the volume of a washer with large radius R, small radius r, and thickness Δx is $\pi R^2 \Delta x - \pi r^2 \Delta x = \pi(R^2 - r^2)\Delta x$, the volume of the representative washer at x_k^* is given by $\pi((f(x_k^*))^2 - (g(x_k^*))^2)\Delta x$.

We must accumulate these washers from $x = 1$ to $x = 4$, so, using the volume formula we just constructed and the equations for $f(x)$ and $g(x)$, we find that the volume of the solid is

$$\pi\int_1^4 ((x+1)^2 - (x^2 - 4x + 5)^2)\,dx = \pi\int_1^4 (-x^4 + 8x^3 - 25x^2 + 42x - 24)\,dx$$

$$= \pi\left[-\frac{1}{5}x^5 + \frac{8}{4}x^4 - \frac{25}{3}x^3 + \frac{42}{2}x^2 - 24x\right]_1^4$$

$$= \frac{117}{5}\pi \approx 73.513 \text{ cubic units.}$$ □

EXAMPLE 4

Finding volume by integrating along the y-axis

Find the volume of the cone obtained by rotating the region between $f(x) = \frac{3}{2}x$ and $g(x) = 3$ on $[0, 2]$ around the y-axis.

SOLUTION

The first figure that follows shows the region bounded by $f(x) = \frac{3}{2}x$ and $g(x) = 3$ from $x = 0$ to $x = 2$, and the middle figure shows this region rotated around the y-axis. Since we rotated around the y-axis rather than the x-axis, we will choose a representative rectangle in the horizontal rather than the vertical direction. Then this rectangle will become a representative disk when we rotate it around the y-axis, as shown in the rightmost figure.

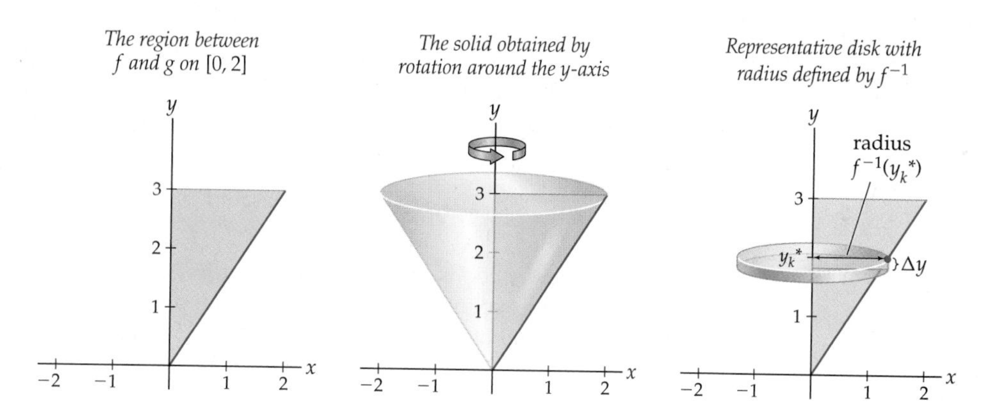

The representative disk at height y_k^* has thickness Δy and radius $f^{-1}(y_k^*)$. Since $f(x) = \frac{3}{2}x$, we have $y = \frac{3}{2}x$ and thus $x = \frac{2}{3}y$. Therefore $f^{-1}(y) = \frac{2}{3}y$, and the volume of the representative disk at y_k^* is

$$\pi(f^{-1}(y_k^*))^2 \Delta y = \pi\left(\frac{2}{3}y_k^*\right)^2 \Delta y = \frac{4}{9}\pi(y_k^*)^2 \Delta y.$$

To find the volume of the cone we must accumulate disks like the one just described, *vertically* from $y = 0$ to $y = 3$. This gives us a definite integral in terms of y and dy instead of x and dx, but all of our previous solving techniques still work, no matter what letter we happen to be using. The volume of the cone is therefore given by

$$\frac{4}{9}\pi \int_0^3 y^2\, dy = \frac{4}{9}\pi\left[\frac{1}{3}y^3\right]_0^3 = \frac{4\pi}{9}\left(\frac{1}{3}(3)^3 - \frac{1}{3}(0)^3\right) = 4\pi.$$

Notice that 4π is the volume predicted by the formula for the volume of a cone of radius $r = 2$ and height $h = 3$: $V = \frac{1}{3}\pi r^2 h = \frac{1}{3}\pi(2)^2(3) = 4\pi$. In Exercise 63 you will use a general version of the preceding calculation to prove the cone volume formula. □

EXAMPLE 5 **Setting up more complicated volume integrals**

Use definite integrals to express the volumes of the following solids of revolution:

(a) The solid obtained from the region bounded by the graph of $f(x) = 3\ln x$, the line $y = 2$, and the x- and y-axes by rotating around the x-axis.

(b) The solid obtained from the region bounded by the graph of $f(x) = x^2 + 2$ and the line $y = 2$ on the interval $[0, 1]$ by rotating around the line $y = 1$.

SOLUTION

(a) The region and the solid it defines when rotated around the x-axis are shown in the first figure. The graph of $f(x) = 3\ln x$ intersects the line $y = 0$ when $x = 1$ and intersects the line $y = 2$ when $x = e^{2/3}$. The vertical cross sections of this solid are sometimes disks and sometimes washers, as shown in the second and third figures.

The solid obtained by rotating the region around the x-axis

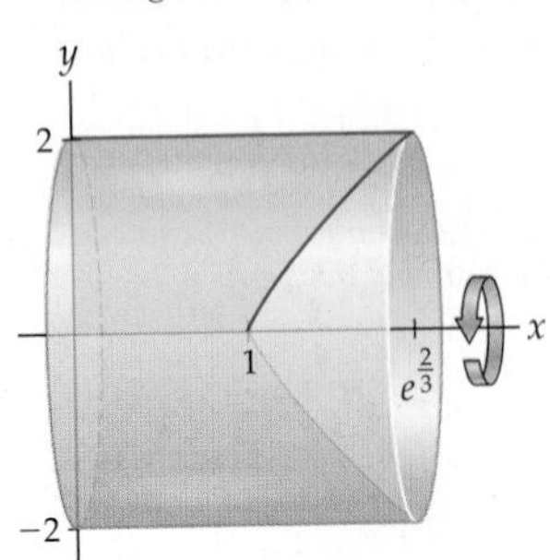

Disk cross sections on $[0, 1]$

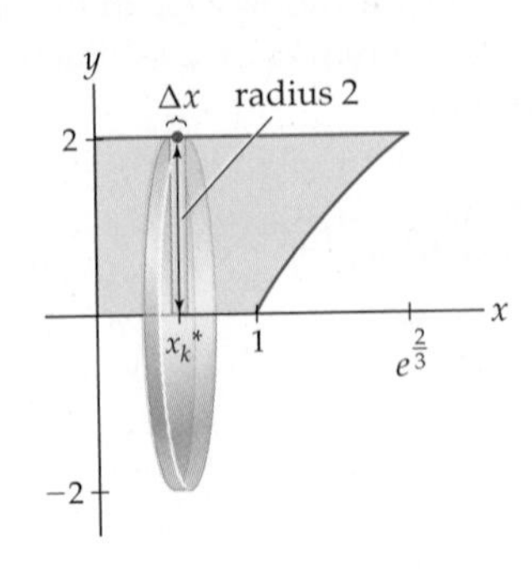

Washer cross sections on $[1, e^{2/3}]$

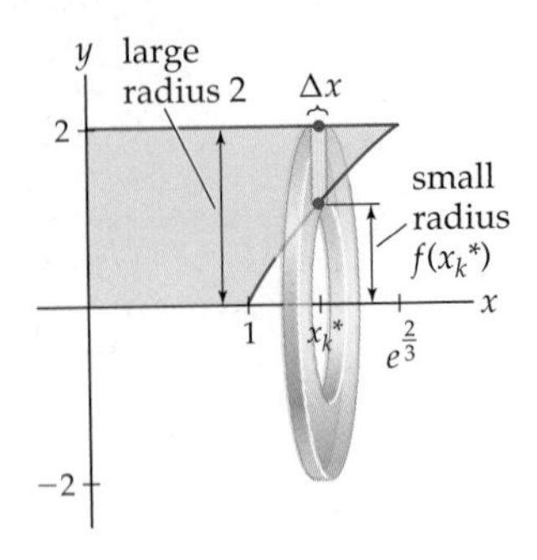

Since the cross sections change character at $x = 1$, we have to use two definite integrals to compute this volume. We must accumulate disks of radius 2 from $x = 0$ to $x = 1$ and then accumulate washers with large radius 2 and small radius $f(x_k^*) = 3\ln x_k^*$ from $x = 1$ to $x = e^{2/3}$. The volume of the solid is therefore represented by

$$\pi\int_0^1 2^2\,dx + \pi\int_1^{e^{2/3}} (2^2 - (3\ln x)^2)\,dx = \pi\int_0^1 4\,dx + \pi\int_1^{e^{2/3}} (4 - 9(\ln x)^2)\,dx.$$

(b) The region and the line that we plan to rotate about are shown in the first figure that follows. The second figure shows the solid of revolution that results. To get a good picture of a representative washer, draw the mirror image of the region over the line $y = 1$. Then draw a representative rectangle in the region, and imagine rotating this rectangle down to its mirror image. The resulting washer is shown in the third figure.

The region and the axis of rotation

The solid obtained by rotation around $y = 1$

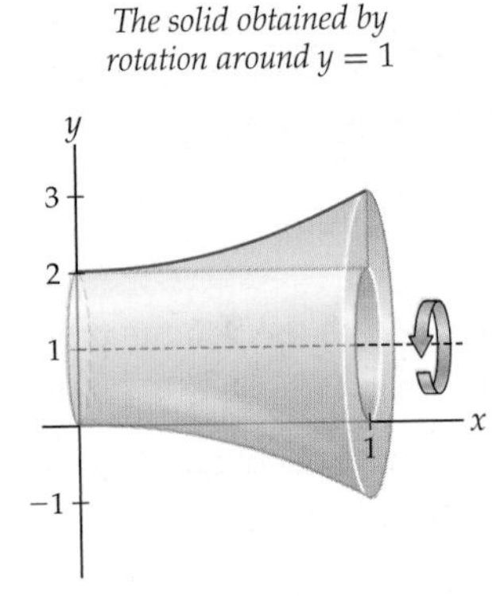

Representative washer with radii defined by f and $y = 1$

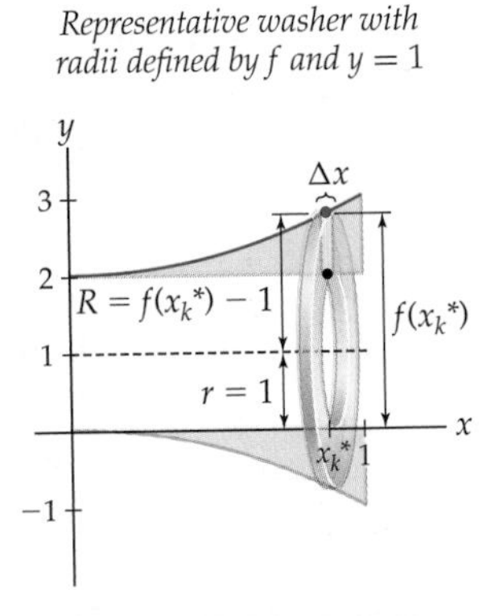

Because we rotated around a line other than one of the coordinate axes, determining the large and small radii of the representative washer takes some concentration. In the figure at the right we have labeled the associated distances very carefully. Notice that the large radius of the washer is $R = f(x_k^*) - 1$ and the small radius is constantly $r = 1$. Accumulating these washers from $x = 0$ to $x = 1$, we see that the volume of the solid is represented by the definite integral

$$\pi\int_0^1 ((f(x) - 1)^2 - 1^2)\,dx = \pi\int_0^1 ((x^2 + 1)^2 - 1)\,dx = \pi\int_0^1 (x^4 + 2x^2)\,dx. \qquad \square$$

? TEST YOUR UNDERSTANDING

- For a solid S defined by revolving a region $\mathbb{R}$ around an axis or a line, how is an approximation of R by rectangles related to an approximation of S by disks and/or washers?
- If S is a solid whose cross-sectional area at $x \in [a, b]$ is $A(x)$, what has to be true about $A(x)$ for volume approximations to converge to the definite integral $\int_a^b A(x)\,dx$?

- What type of situation would require us to integrate along the y-axis instead of the x-axis to use disks or washers, and why?
- What is the vertical distance between the x-axis and a point on the graph of $y = f(x)$? What is the horizontal distance between the y-axis and a point on the graph?
- In Example 3, would we get the same answer if we subtracted the volume of the solid obtained by rotating $g(x) = x^2 - 4x + 5$ around the x-axis from the volume of the solid obtained by rotating $f(x) = x + 1$ around the x-axis? Why or why not?

EXERCISES 6.1

Thinking Back

Definite integrals: Calculate each of the following definite integrals, using integration techniques and the Fundamental Theorem of Calculus.

- $\int_{-3}^{3} (9 - x^2)\, dx$
- $\int_{0}^{1} (x^4 + 2x^2)\, dx$
- $\int_{0}^{1} 4\, dx$
- $\int_{1}^{e^{2/3}} (4 - 9(\ln x)^2)\, dx$
- $\int_{0}^{2} (4y^2 - y^4)\, dy$
- $\int_{0}^{1} (2 - \sqrt{y})^2\, dy$

Definite integrals for geometric quantities: Let $f(x) = \sqrt{x}$ and $g(x) = 2 - x$. Express each of the given quantities in terms of definite integrals. Do not solve the integrals; just set them up.

- The area under the graph of $f(x)$ on $[0, 4]$.
- The absolute area under the graph of $g(x)$ on $[0, 4]$.
- The area between $f(x)$ and $g(x)$ on $[0, 4]$.
- The average value of $f(x)$ on $[0, 4]$.

Concepts

0. *Problem Zero:* Read the section and make your own summary of the material.

1. *True/False:* Determine whether each of the statements that follow is true or false. If a statement is true, explain why. If a statement is false, provide a counterexample.

(a) *True or False:* Every sum is a Riemann sum and can be turned into a definite integral.

(b) *True or False:* Every sum involving only continuous functions is a Riemann sum and can be turned into a definite integral.

(c) *True or False:* The volume of a disk can be obtained by multiplying its thickness by the circumference of a circle of the same radius.

(d) *True or False:* The volume of a disk can be obtained by multiplying its thickness by the area of a circle of the same radius.

(e) *True or False:* The volume of a cylinder can be obtained by multiplying the height of the cylinder by the area of a circle of the same radius.

(f) *True or False:* The volume of a washer can be expressed as the difference of the volume of two disks.

(g) *True or False:* The volume of a right cone is exactly one third of the volume of a cylinder with the same radius and height.

(h) *True or False:* The volume of a sphere of radius r is $V = \frac{4}{3}\pi r^3$.

2. *Examples:* Construct examples of the thing(s) described in the following. Try to find examples that are different than any in the reading.

(a) A region that, when revolved around the x-axis, has both disk and washer cross sections.

(b) A region that, when revolved around the y-axis, has both disk and washer cross sections.

(c) A solid of revolution for which it is not possible to use the disk or washer method.

3. Consider the rectangle bounded by $y = 3$ and $y = 0$ on the x-interval $[2, 2.25]$.

(a) What is the volume of the disk obtained by rotating this rectangle around the x-axis?

(b) What is the volume of the washer obtained by rotating this rectangle around the line $y = 5$?

4. Consider the rectangle bounded by $x = 1$ and $x = 4$ on the y-interval $[3, 3.5]$.

(a) What is the volume of the disk obtained by rotating this rectangle around the line $x = 4$?

(b) What is the volume of the washer obtained by rotating this rectangle around the y-axis?

5. Consider the region between $f(x) = 5 - x^2$ and the x-axis between $x = 0$ and $x = 4$. Draw a Riemann sum approximation of the area of this region, using a midpoint sum with four rectangles, and explain how it is related to a four-disk approximation of the solid obtained by rotating the region around the x-axis.

6. For a four-disk approximation of the volume of the solid obtained from the region between $f(x) = 5 - x^2$ and the x-axis between $x = 0$ and $x = 4$ by rotating around the x-axis, illustrate and calculate

(a) Δx and each x_k;

(b) some x_k^* in each subinterval $[x_{k-1}, x_k]$;

(c) each $f(x_k^*)$;

(d) the volume of the second disk.

7. For a four-washer approximation of the volume of the solid obtained from the region between $f(x) = x^2$ and the y-axis between $y = 0$ and $y = 4$ by rotating around the x-axis, illustrate and calculate
   (a) Δx and each x_k;
   (b) some x_k^* in each subinterval $[x_{k-1}, x_k]$;
   (c) each $f(x_k^*)$;
   (d) the volume of the second washer.

8. For a four-disk approximation of the volume of the solid obtained from the region between $f(x) = x^2$ and the y-axis between $y = 0$ and $y = 4$ by rotating around the y-axis, illustrate and calculate
   (a) Δy and each y_k;
   (b) some y_k^* in each subinterval $[y_{k-1}, y_k]$;
   (c) each $f^{-1}(y_k^*)$;
   (d) the volume of the second disk.

Write each of the limits in Exercises 9–11 in terms of definite integrals, and identify a solid of revolution whose volume is represented by that definite integral.

9. $\displaystyle\lim_{n\to\infty} \sum_{k=1}^{n} \pi(1 + x_k^*)^2 \left(\frac{1}{n}\right)$, with $x_k^* = x_k = 2 + k\left(\frac{1}{n}\right)$

10. $\displaystyle\lim_{n\to\infty} \sum_{k=1}^{n} \pi(1 + y_k^*)^2 \left(\frac{3}{n}\right)$, with $y_k^* = y_k = 1 + k\left(\frac{3}{n}\right)$

11. $\displaystyle\lim_{n\to\infty} \sum_{k=1}^{n} \pi(4 - (x_k^*)^2) \left(\frac{2}{n}\right)$, with $x_k^* = x_k = k\left(\frac{2}{n}\right)$

12. Suppose that for some $b > 0$, the region between $y = \sqrt{x}$ and $y = 0$ on $[0, b]$, rotated around the x-axis, has volume $V = 8\pi$. Without solving any integrals, find the volume of solid obtained by rotating the region bounded by $y = \sqrt{x}$, $y = \sqrt{b}$, and $x = 0$ around the x-axis.

For each pair of definite integrals in Exercises 13–18, decide which, if either, is larger, without computing any integrals.

13. $\displaystyle\pi \int_0^{\pi/2} \cos^2 x\,dx$ and $\displaystyle\pi \int_{\pi/2}^{\pi} \cos^2 x\,dx$

14. $\displaystyle\pi \int_0^1 e^{2x}\,dx$ and $\displaystyle\pi \int_0^1 (e^{2x} - 1)\,dx$

15. $\displaystyle\pi \int_0^2 x^4\,dx$ and $\displaystyle\pi \int_0^2 (16 - x^4)\,dx$

16. $\displaystyle\pi \int_0^3 x^2\,dx$ and $\displaystyle\pi \int_0^3 (9 - x^2)\,dx$

17. $\displaystyle\pi \int_0^2 x^4\,dx$ and $\displaystyle\pi \int_0^4 y\,dy$

18. $\displaystyle\pi \int_0^4 y\,dy$ and $\displaystyle\pi \int_4^8 (8 - y)\,dy$

Each of the definite integrals in Exercises 19–24 represents the volume of a solid of revolution obtained by rotating a region around either the x- or y-axis. Find this region.

19. $\displaystyle\pi \int_1^3 (16 - (x + 1)^2)\,dx$

20. $\displaystyle\pi \int_0^{\pi} \sin^2 x\,dx$

21. $\displaystyle\pi \int_1^3 (x^2 - 2x + 1)\,dx$

22. $\displaystyle\pi \int_0^2 y\,dy$

23. $\displaystyle\pi \int_1^5 \left(\frac{y - 1}{2}\right)^2 dy$

24. $\displaystyle\pi \int_0^4 (2^2 - (\sqrt{y})^2)\,dy$

Write the volume of the two solids of revolution that follow in terms of definite integrals that represent accumulations of disks and/or washers. Do not compute the integrals.

25.

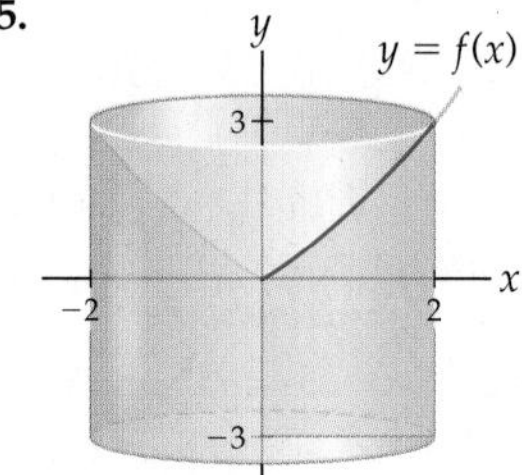

26.

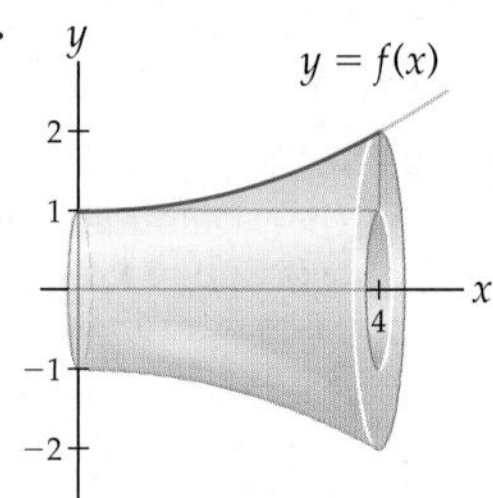

Skills

Consider the region between $f(x) = \sqrt{x}$ and the x-axis on $[0, 4]$. For each line of rotation given in Exercises 27–30, use four disks or washers based on the given rectangles to approximate the volume of the resulting solid.

27. *Around the x-axis*

28. *Around the y-axis*

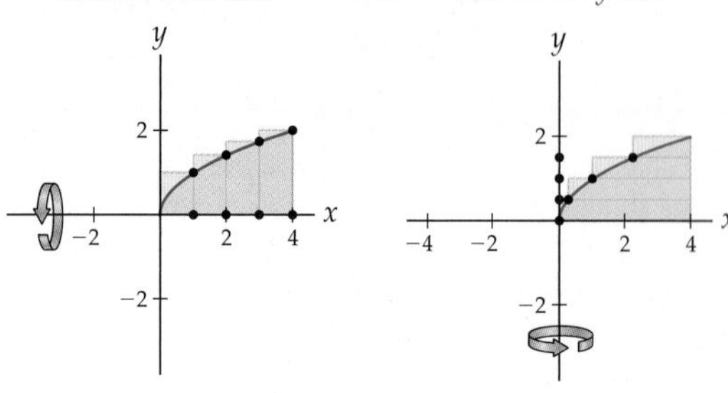

29. *Around the line $y = -1$*

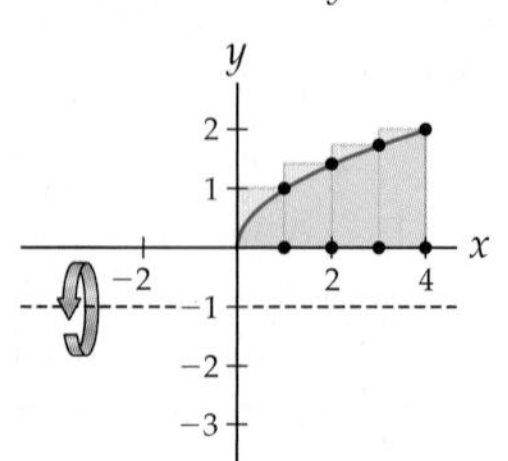

30. *Around the line $x = 5$*

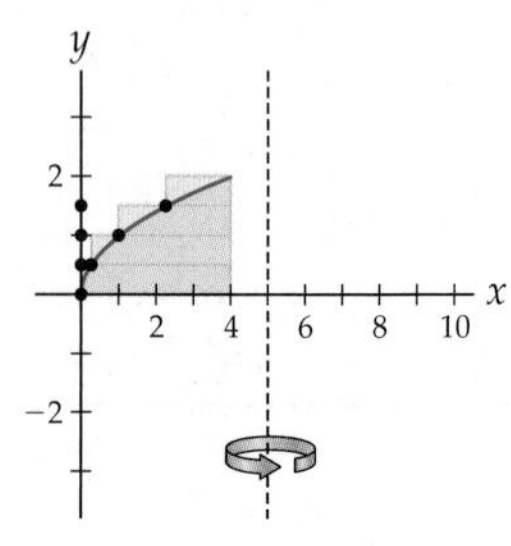

Consider the region between the graph of $f(x) = \frac{3}{x}$ and the x-axis on $[1, 3]$. For each line of rotation given in Exercises 31–34, use definite integrals to find the volume of the resulting solid.

31. *Around the x-axis*

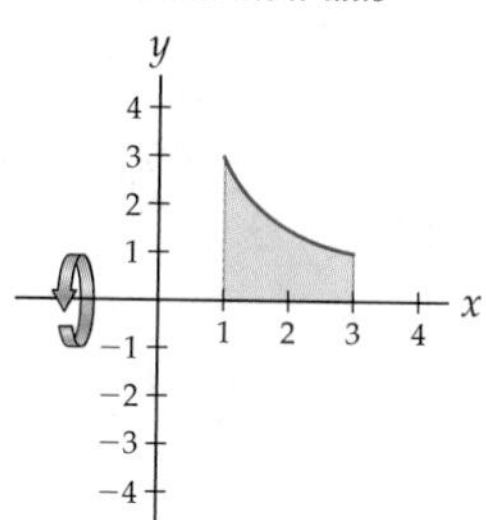

32. *Around the line $y = -1$*

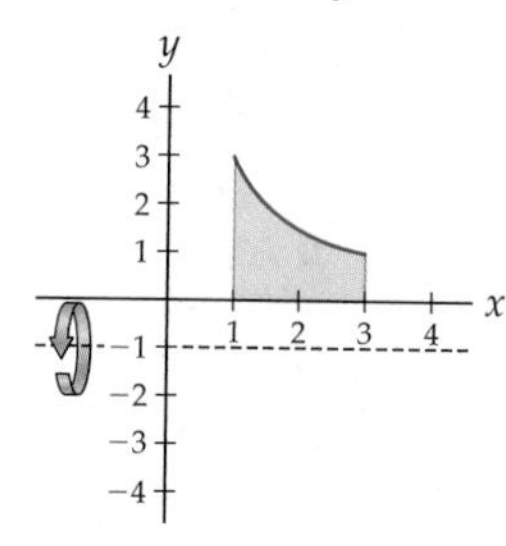

33. *Around the y-axis*

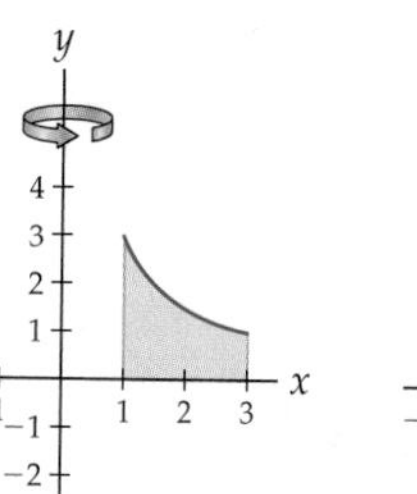

34. *Around the line $x = 1$*

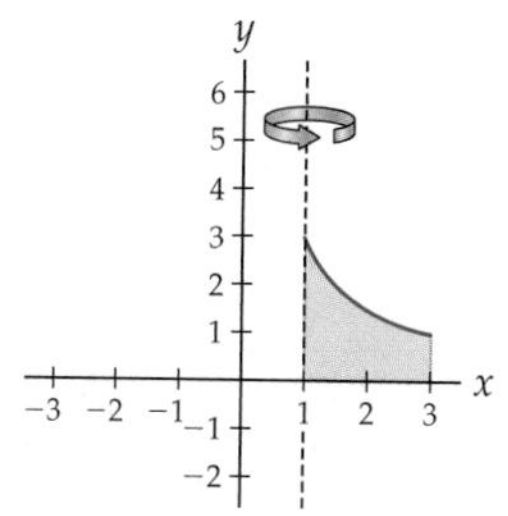

Consider the region between the graph of $f(x) = x - 2$ and the x-axis on $[2, 5]$. For each line of rotation given in Exercises 35–40, use definite integrals to find the volume of the resulting solid.

35. *Around the y-axis*

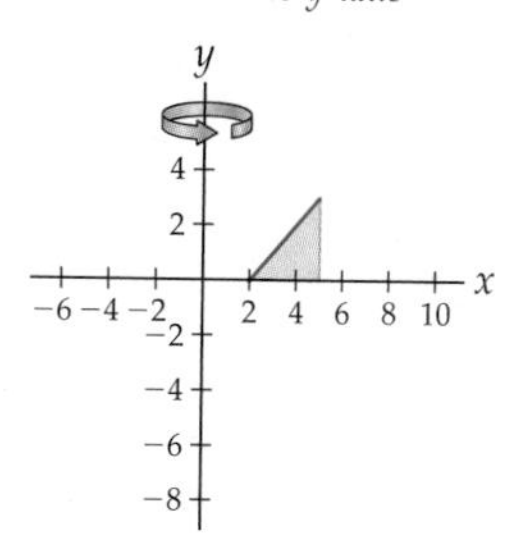

36. *Around the x-axis*

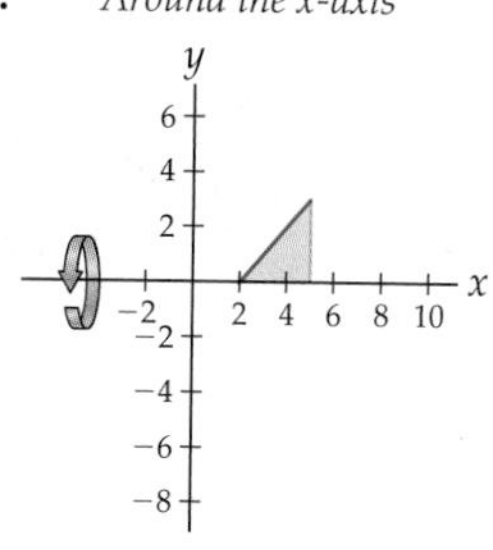

37. *Around the line $y = 3$*

38. *Around the line $y = -2$*

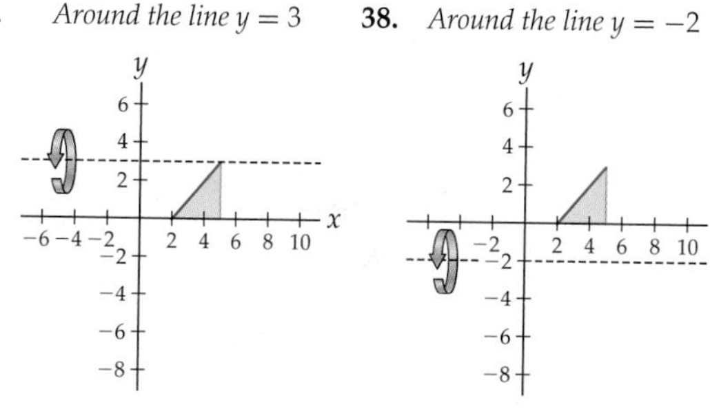

39. *Around the line $x = 2$*

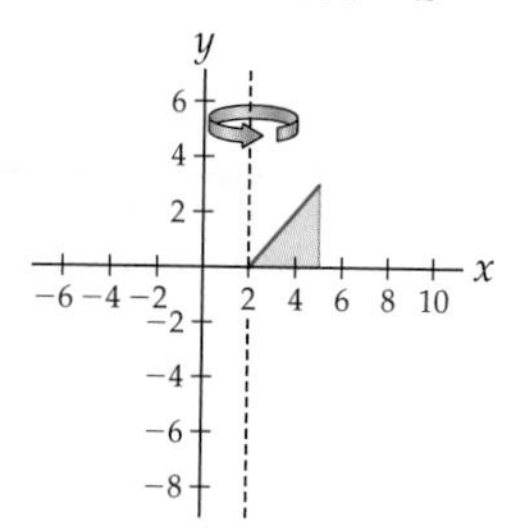

40. *Around the line $x = 6$*

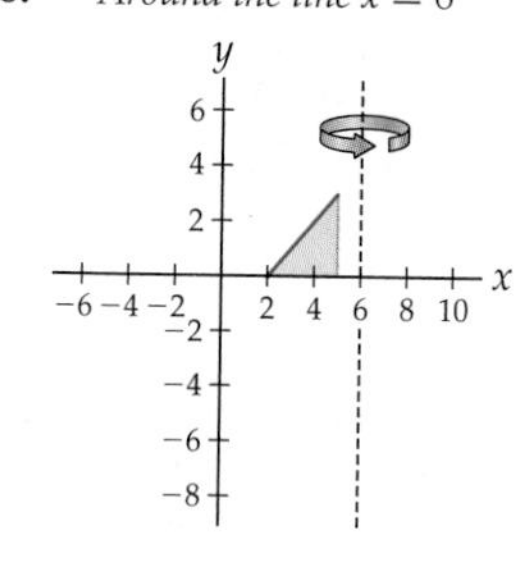

Consider the region between the graph of $f(x) = 4 - x^2$ and the line $y = 5$ on $[0, 2]$. For each line of rotation given in Exercises 41–44, use definite integrals to find the volume of the resulting solid.

41. *Around the y-axis*

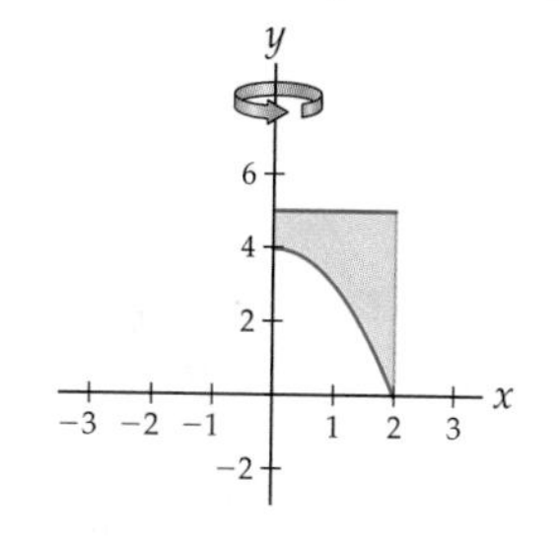

42. *Around the line $x = 2$*

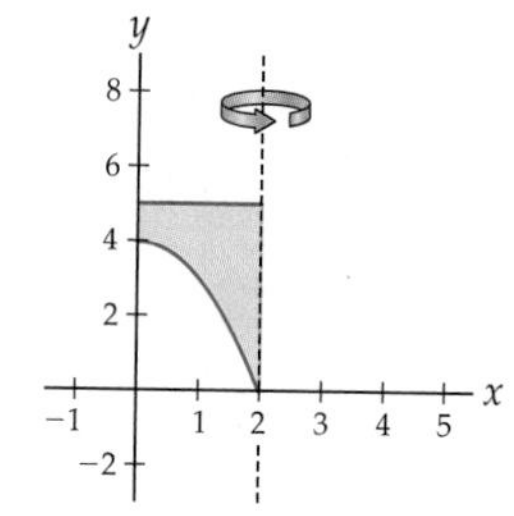

43. *Around the line $x = 3$*

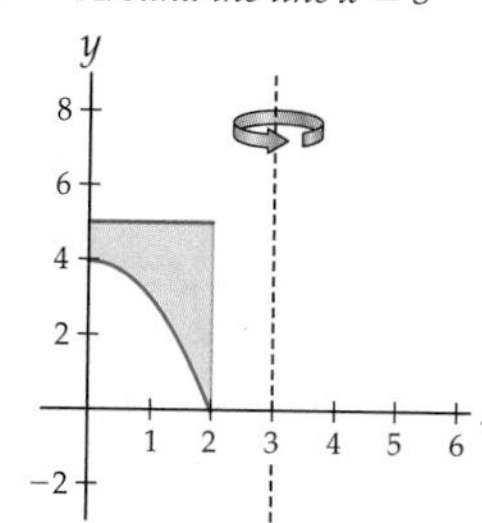

44. *Around the line $y = 5$*

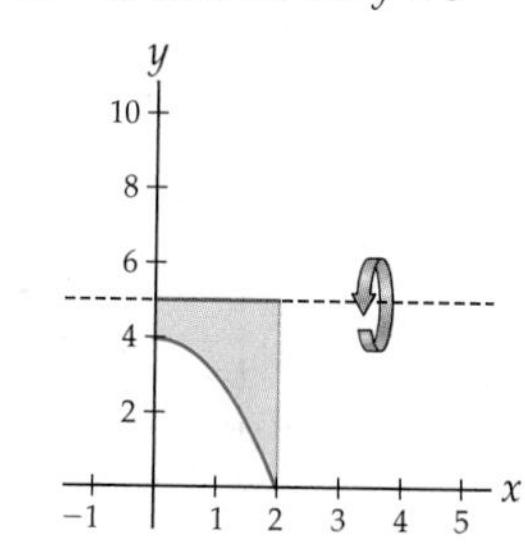

Consider the region between the graphs of $f(x) = 5 - x$ and $g(x) = 2x$ on $[1, 4]$. For each line of rotation given in Exercises 45 and 46, use definite integrals to find the volume of the resulting solid.

45. *Around the x-axis*

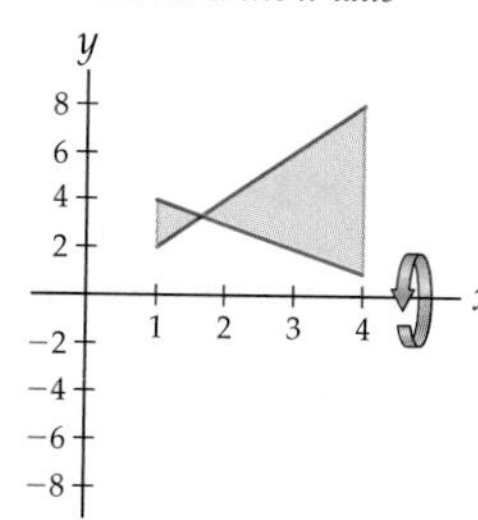

46. *Around the line $x = 1$*

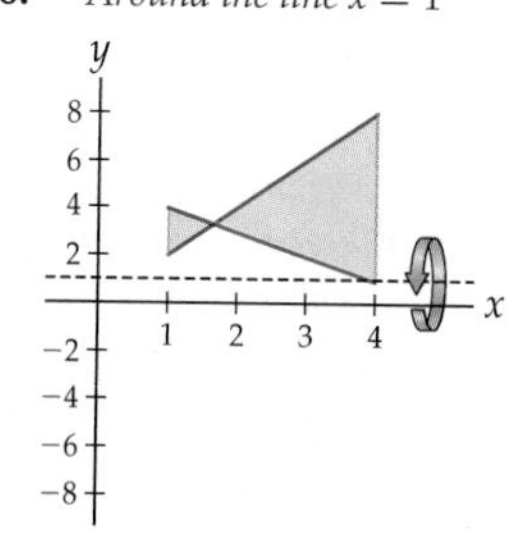

Consider the region between the graphs of $f(x) = x^2$ and $g(x) = 2x$ on $[0, 2]$. For each line of rotation given in Exercises 47–50, use definite integrals to find the volume of the resulting solid.

47. *Around the y-axis*

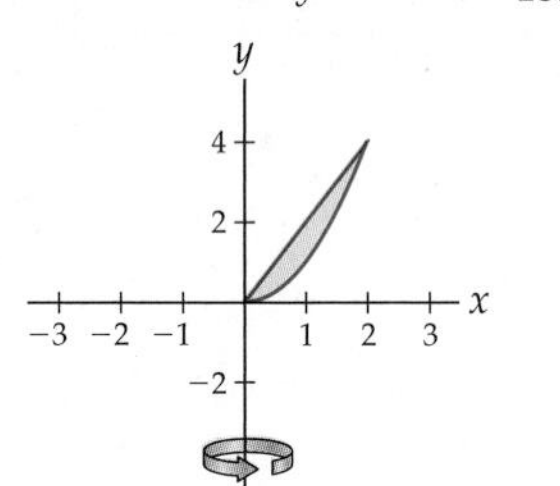

48. *Around the x-axis*

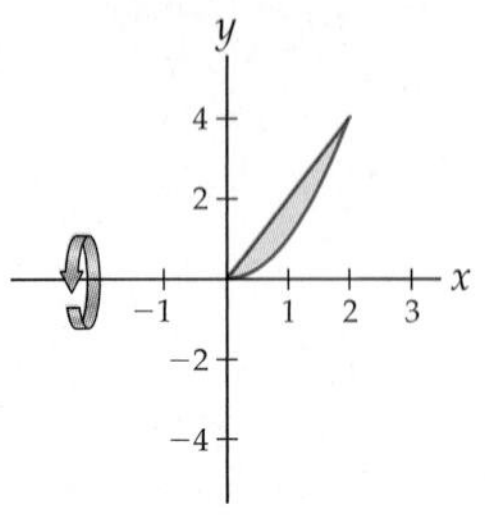

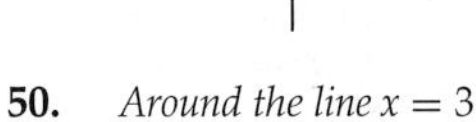

49. *Around the line $y = -1$*

50. *Around the line $x = 3$*

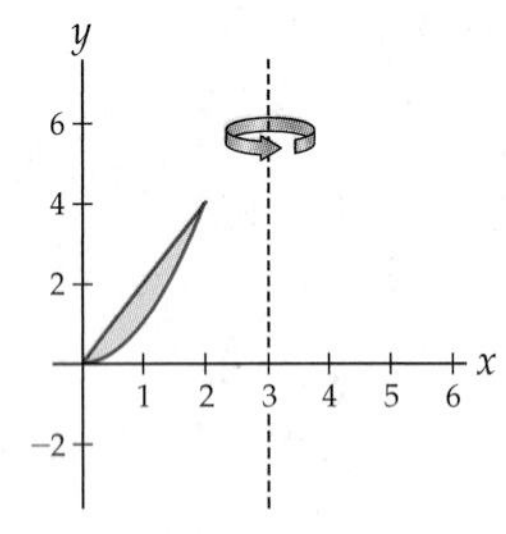

Consider the region between the graphs of $f(x) = (x - 2)^2$ and $g(x) = x$ on $[1, 4]$. For each line of rotation given in Exercises 51–54, use definite integrals to find the volume of the resulting solid.

51. *Around the x-axis*

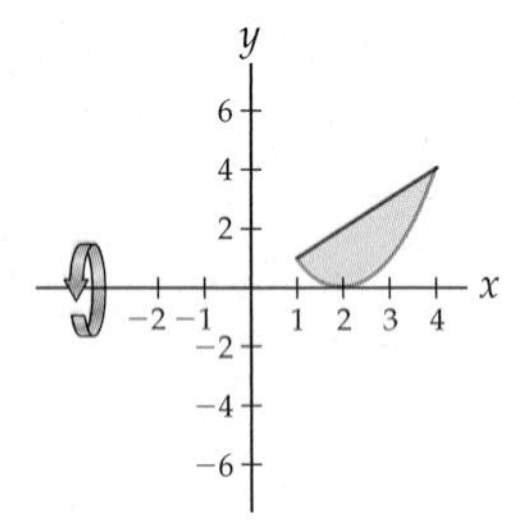

52. *Around the y-axis*

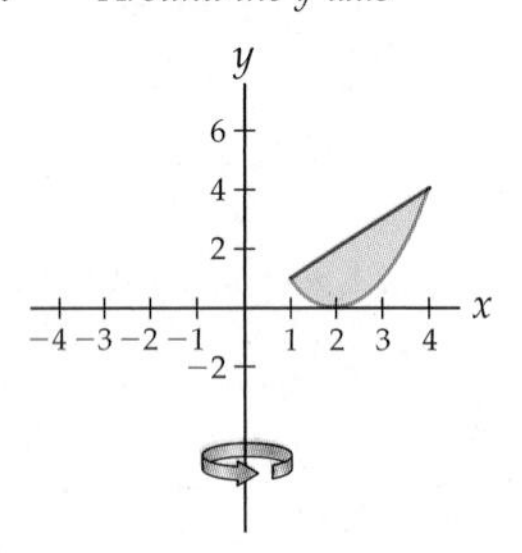

53. *Around the line $y = -1$*

54. *Around the line $x = 1$*

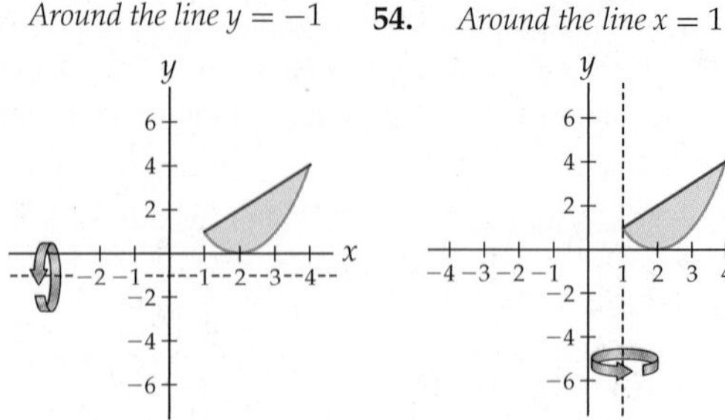

Consider the region between the graph of $f(x) = 1 - \cos x$ and the x-axis on $[0, \pi]$. For each line of rotation given in Exercises 55–58, write down definite integrals that represent the volume of the resulting solid and then use a calculator or computer to approximate the integrals.

55. *Around the x-axis*

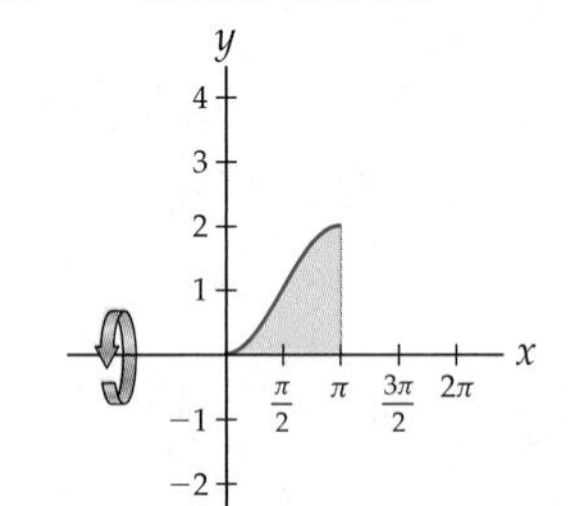

56. *Around the y-axis*

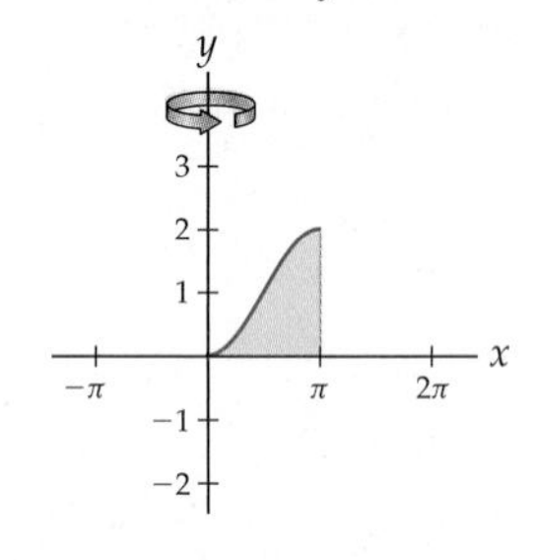

57. *Around the line $x = \pi$*

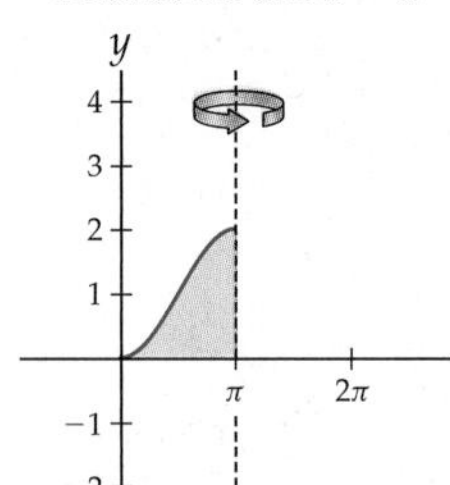

58. *Around the line $y = 2$*

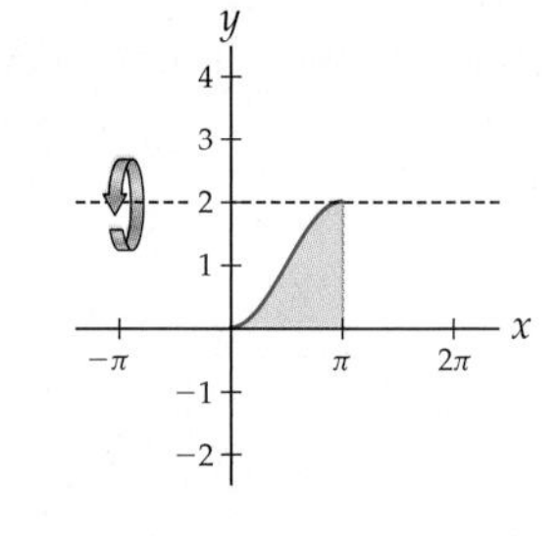

Applications

59. Rebecca plans to visit the Great Pyramid of Giza in Cairo, Egypt. The Great Pyramid has a square base with each side approximately 756 feet in length.

(a) An interesting fact about the Great Pyramid of Giza is that the perimeter of its base is equal to the circumference of a circle whose radius is equal to the height of the pyramid. Use this fact to find the height of the Great Pyramid, rounded to the nearest integer.

(b) The volume of a square-based pyramid with base area A and height h is given by the formula $V = \frac{1}{3}Ah$. Use this formula to find the volume of the Great Pyramid.

(c) Find the volume of the Great Pyramid another, much more difficult way: Set up a definite integral that represents the volume, and solve it. Use the given diagram as a starting point, and justify your answer.

Great Pyramid of Giza

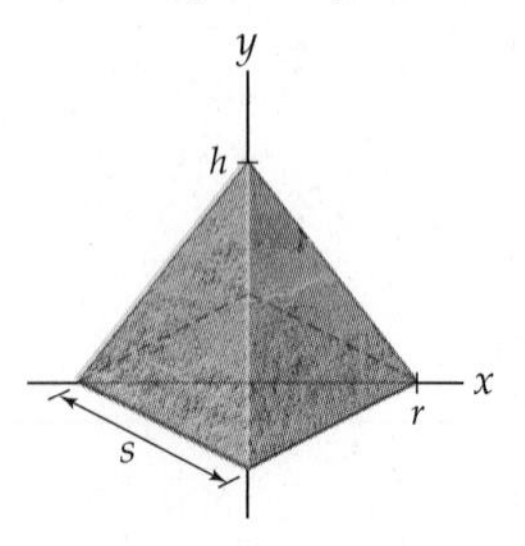

60. Two of Dr. Geek's friends got married this year, one in June and one in July. His traditional gift for such occasions is a fancy crystal bowl, whose volume corresponds to how much he likes the future spouse of the recipient.

Crystal hexagonal bowl

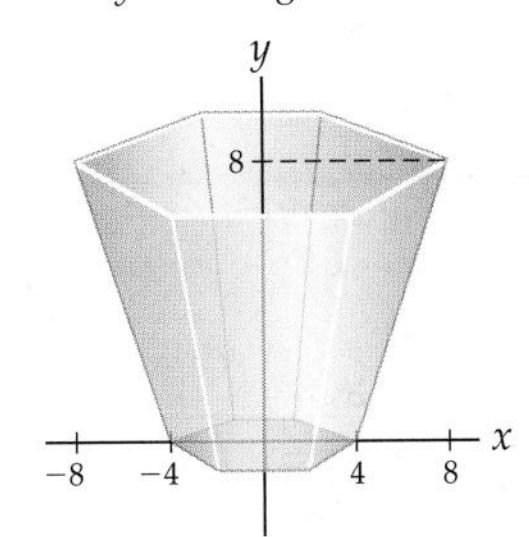

Crystal octagonal bowl

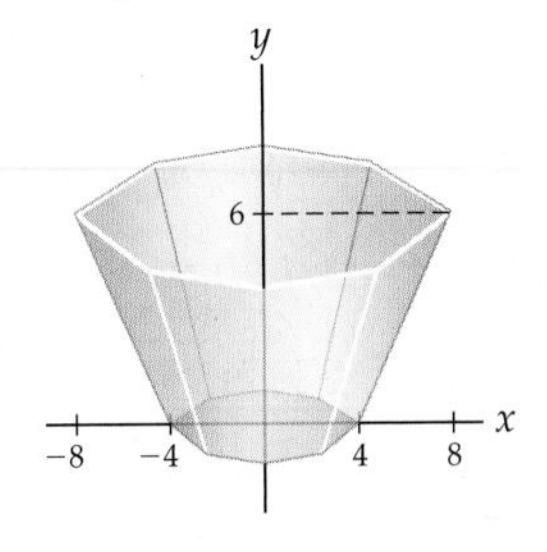

(a) For the June wedding, Dr. Geek gave a fancy hexagonal crystal bowl with base hexagon of radius 4 inches, top hexagon of radius 8 inches, and height of 8 inches. The area of a hexagon with radius r is $A = \frac{3\sqrt{3}}{2}r^2$. What is the volume of this bowl, in cubic inches?

(b) For the July wedding, Dr. Geek gave a fancy octagonal crystal bowl with base radius 4, top radius 8, and height 6 inches. The area of an octagon with radius r is $A = 2\sqrt{2}r^2$. What is the volume of this bowl, in cubic inches?

Proofs

61. Use a definite integral to prove that the volume formula $V = \frac{1}{3}\pi r^2 h$ holds for a cone of radius 3 and height 5.

62. Use a definite integral to prove that the volume formula $V = \frac{4}{3}\pi r^3$ holds for a sphere of radius 3.

63. Use a definite integral to prove that a cone of radius r and height h has volume given by the formula $V = \frac{1}{3}\pi r^2 h$.

64. Use a definite integral to prove that a sphere of radius r has volume given by the formula $V = \frac{4}{3}\pi r^3$.

Thinking Forward

Slicing to obtain a definite integral for work: The "subdivide, approximate, and add" strategy works for more than just geometric quantities like area, length, and volume. In this problem you will investigate the use of a slicing process to write work as a definite integral.

- The ***work*** involved in lifting an object is the product of the weight of the object and the distance it must be lifted. Suppose a flat, wide cylinder that weighs 50 pounds must be lifted from ground level to a height of 10 feet. How much work does this require, in foot-pounds?
- Now suppose we want to pump all of the water out of the top of a 4-foot-high cylindrical tank with radius 10 feet. It takes more work to pump out the water from the bottom of the tank than it does to pump it out from the top of the tank. With the given figure as a guide, use horizontal slices to set up and solve a definite integral that represents the work required to pump all of the water out of the top of the tank.

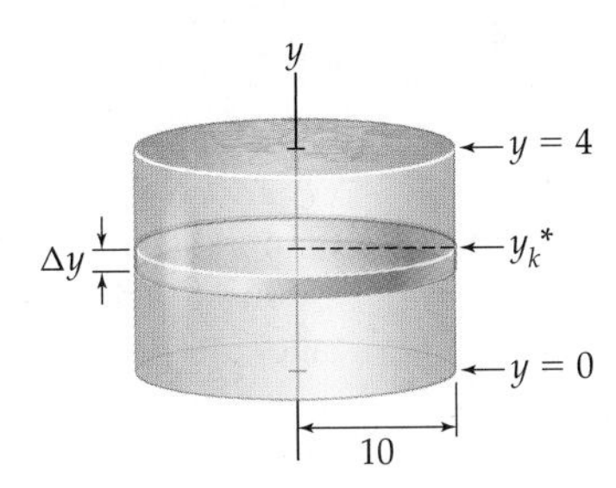

6.2 VOLUMES BY SHELLS

- Approximating volumes of solids of revolution with shells
- Setting up Riemann sums for shells by using an alternative volume formula
- Using definite integrals to describe volumes of solids of revolution with shells

Approximating Volume by Shells

We now develop an alternative method for calculating the volume of a solid of revolution, using "shells" rather than slices. This new method will help us find some volumes that are inconvenient to find by using disks or washers.

For example, consider the region bounded by the graph of $f(x) = -x^2 - x + 8$ and the x-axis from $x = 0$ to $x = 2$, rotated about the y-axis. To use disks to approximate the volume of this solid we could slice the region into horizontal rectangles, for example as shown in the figure that follows at the left. When we revolve these rectangles about the y-axis, they produce four disks stacked on top of one another, as shown in the figure at the right.

Four horizontal rectangles

A stack of four disks

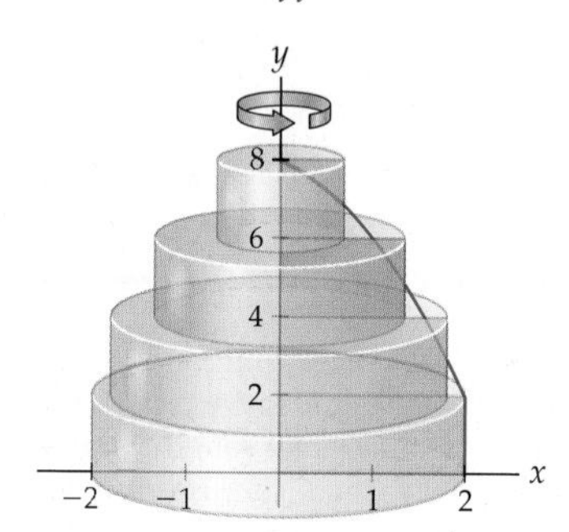

Unfortunately, it is difficult to find the radii of these disks, since it is not easy to solve the equation $y = -x^2 - x + 8$ for x. This makes it difficult to use the slicing/disk method here.

What happens if we instead try to use vertical rectangles, as shown next at the left? When we revolve these rectangles around the y-axis, each rectangle makes a ***shell*** rather than a disk or washer, as shown at the right. The green shell is a washer that is wrapped around the other shells. The blue shell is an even taller washer that wraps around an even taller purple washer, which in turn wraps around a very tall pink cylinder. Instead of slicing the solid horizontally or vertically, we have sliced it in layers from the inside out, like the layers of an onion.

Four vertical rectangles

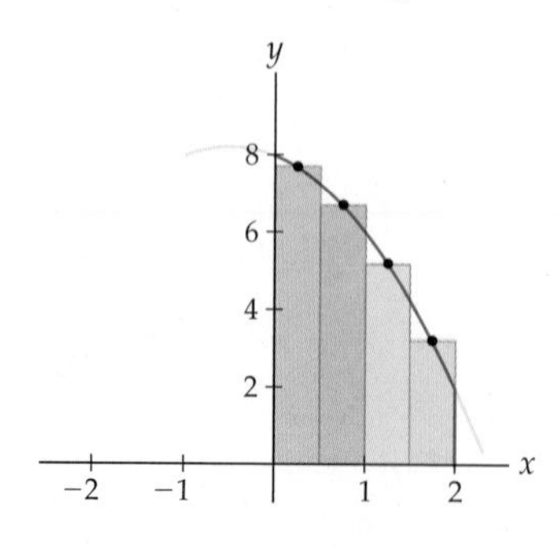

Four nested shells

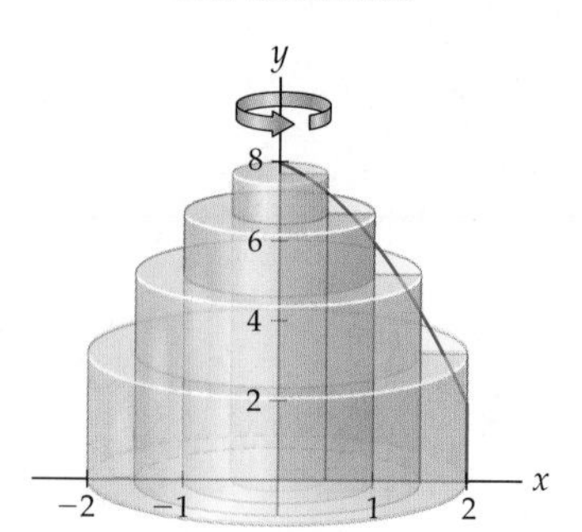

To get an approximation for the volume of the solid we just add up the volumes of these four shells. In this example, shells are easier to work with than disks because the heights of the shells are determined by $f(x)$ rather than the hard-to-compute $f^{-1}(x)$.

Finding a Riemann Sum for Volumes by Shells

Before we can construct a definite integral to represent the exact volume of a solid of revolution with the shell method, we need to know how to calculate the volume of a representative shell such as the one shown in the next figure. This shell has height $f(x_k^*)$ and thickness Δx.

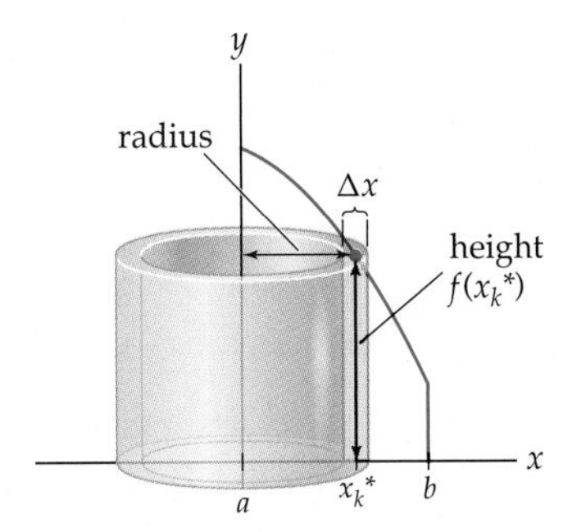

So what is the radius of this shell? Really, there are two radii, an inner radius r_1 and an outer radius r_2, where $r_2 - r_1$ is the thickness Δx of the shell. The shell can be thought of as a cylinder of radius r_2 and height $h = f(x_k^*)$ with a smaller cylinder of radius r_2 and height $h = f(x_k^*)$ removed. Therefore its volume is

$$V = \pi r_2^2 h - \pi r_1^2 h = \pi(r_2^2 - r_1^2) f(x_k^*).$$

This formula is not in the right form for constructing a definite integral, because it does not explicitly involve Δx, which we need in our expression if we want to construct a Riemann sum. We can get around this problem by calculating the volume of a shell in terms of its average radius.

THEOREM 6.4 **The Volume of a Shell**

A shell with inner radius r_1, outer radius r_2, and height h is said to have ***average radius*** $r = \frac{r_1 + r_2}{2}$ and ***thickness*** $\Delta x = r_2 - r_1$. Such a shell has volume $V = 2\pi r h \,\Delta x$.

The volume formula in this theorem makes intuitive sense, because a shell is a thickened cylinder and the formula is the product of the lateral surface area $2\pi r h$ and the thickness Δx.

Proof. Given a shell with inner radius r_1, outer radius r_2, and height h, define $r = \frac{r_1 + r_2}{2}$ and $\Delta x = r_2 - r_1$. With this notation we have

$$2\pi r h \,\Delta x = 2\pi \left(\frac{r_1 + r_2}{2}\right) h(r_2 - r_1) = \pi(r_1 + r_2)(r_1 - r_2)h = \pi(r_2^2 - r_1^2)h,$$

which is the volume of the shell. ∎

Now suppose we are given a partition $x_0, x_1, \ldots, x_n$ of $[a, b]$. If the kth shell has width Δx, average radius $r_k = \frac{x_k + x_{k-1}}{2}$, and height $f(x_k^*)$, then the sum of the volumes of the n shells is given by

$$\sum_{k=1}^{n} 2\pi\, r_k f(x_k^*)\, \Delta x.$$

This is almost a Riemann sum, since the summand is the product of a function $2\pi r_k f(x_k^*)$ of the partition points and a factor of Δx. As $n \to \infty$ and thus $\Delta x \to 0$, the radius r_k that is the midpoint of each subinterval will get closer and closer to x_k^*, and the limit of this sum will converge to a definite integral.

A Definite Integral for Volume by Shells

In general, the heights of a representative shell may be given by something more complicated than just the height $f(x_k^*)$ of some function, depending on the region that is rotated. But in any case, if the shell height function is continuous, then we can use shells to express the volume of the solid of revolution as a definite integral.

DEFINITION 6.5 **Volume as a Definite Integral Using Shells**

Suppose S is a solid of revolution obtained by rotating a region on $[a, b]$ around a vertical axis. If S can be represented with nested shells whose average radii and heights are given by continuous functions $r(x)$ and $h(x)$ on $[a, b]$, then the ***volume*** of S is

$$\begin{matrix}\text{Volume of solid} \\ \text{with nested shells}\end{matrix} = \lim_{n\to\infty} \sum_{k=1}^{n} 2\pi\, r_k h_k^*\, \Delta x = 2\pi \int_a^b r(x)\, h(x)\, dx,$$

where $\Delta x = \frac{b-a}{n}$, $x_k = a + k\Delta x$, $x_k^* \in [x_{k-1}, x_k]$, $r_k = \frac{x_k + x_{k-1}}{2}$, and $h_k^* = h(x_k^*)$.

Of course, there is a similar definition for the volume of a solid of revolution obtained by rotating around a horizontal axis and integrating along an interval $[A, B]$ of y-values. In this case the radius and height functions for the shells will be functions of y rather than x. The following two figures illustrate examples of shells to be integrated in the x and y directions:

Shells in the x direction

Shells in the y direction

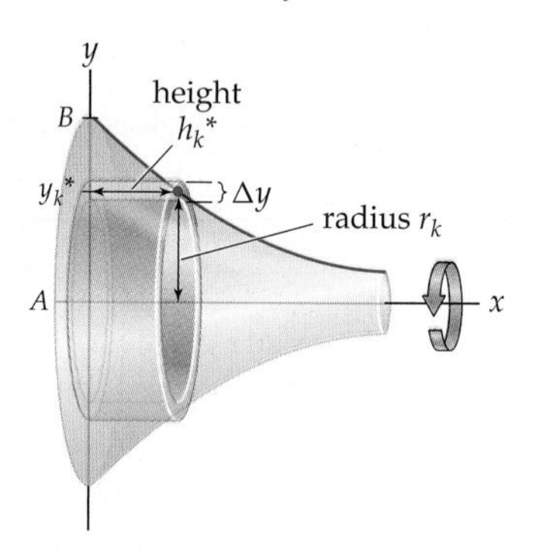

When integrating by shells in the x direction, we will have the radius function in terms of x and the height function in terms of some function of x. When integrating by shells in the y direction, we will have the radius function in terms of y and the height function in terms of some function of y, usually given by boundary values or some inverse function. The formula that is analogous to Definition 6.5 for integrating along the y-axis is

$$\begin{matrix}\text{Volume of solid} \\ \text{with nested shells}\end{matrix} = \lim_{n\to\infty} \sum_{k=1}^{n} 2\pi\, r_k h_k^*\, \Delta y = 2\pi \int_A^B r(x)\, h(x)\, dy.$$

Examples and Explorations

EXAMPLE 1

Approximating a volume with shells

Consider the solid obtained by rotating the region between $f(x) = -x^2 - x + 8$ and the x-axis on $[0, 2]$ around the y-axis. Approximate the volume of this region with four shells.

SOLUTION

Let's use the four shells that were defined at the start of the reading, whose heights were defined at the midpoints of each subinterval. For reference, we repeat the diagrams here:

Four rectangles approximate the area under $f(x) = -x^2 - x + 8$

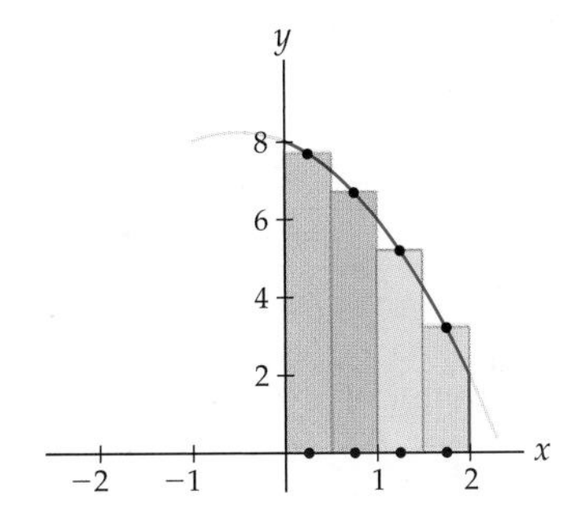

Four shells approximate the volume of the solid

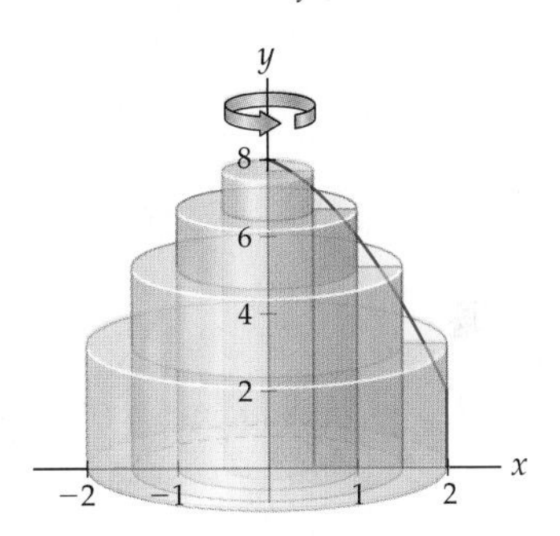

The centermost shell is the tall pink disk defined on $[0, 0.5]$. This disk has radius $r = 0.5$ and height defined by the value of the function at the midpoint $x = 0.25$ of the interval: $h = f(0.25) = -(0.25)^2 - 0.25 + 8 = 7.8125$. Therefore the first, innermost shell has volume

$$\pi r^2 h = \pi(0.5)^2 f(0.25) \approx 6.04 \text{ cubic units.}$$

The next shell is the purple shell defined on $[0.5, 1]$ with outer radius $r_2 = 1$, inner radius $r_1 = 0.5$, and height defined by the function at the midpoint $x = 0.75$ of the interval: $h = f(0.75) = -(0.75)^2 - (0.75) + 8 = 6.6875$. Therefore the volume of the second, purple shell is

$$\pi(r_2^2 - r_1^2)h = \pi(1^2 - 0.5^2) f(0.75) \approx 15.76 \text{ cubic units.}$$

We could have instead used the shell volume formula described just before Theorem 6.4 to compute the volume of the solid. The purple shell has average radius $r = 0.75$, thickness $\Delta x = \frac{1}{2}$, and height $h = f(0.75) = 6.6875$. Using the new formula, we see that once again the volume of the purple shell is

$$2\pi r h\, \Delta x = 2\pi(0.75) f(0.75)\left(\frac{1}{2}\right) \approx 15.76 \text{ cubic units.}$$

Using similar methods, we can calculate that the volumes of the blue and green shells are approximately 20.37 and 17.52 cubic units, respectively. The volume V of the solid of revolution is approximately the sum of the volumes of the four shells:

$$V \approx 6.04 + 15.76 + 20.37 + 17.52 = 59.69 \text{ cubic units.}$$

□

To verify that our answer is reasonable, we can compare the solid (shown next at the left) with a very rough approximation whose volume is easy to calculate (shown at the right).

Solid of revolution

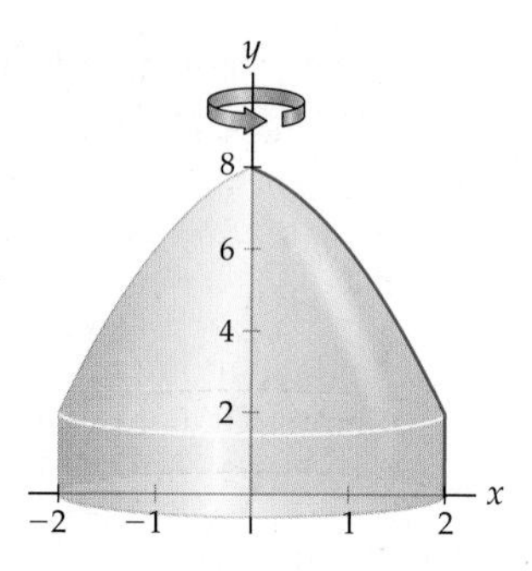

Very rough approximation

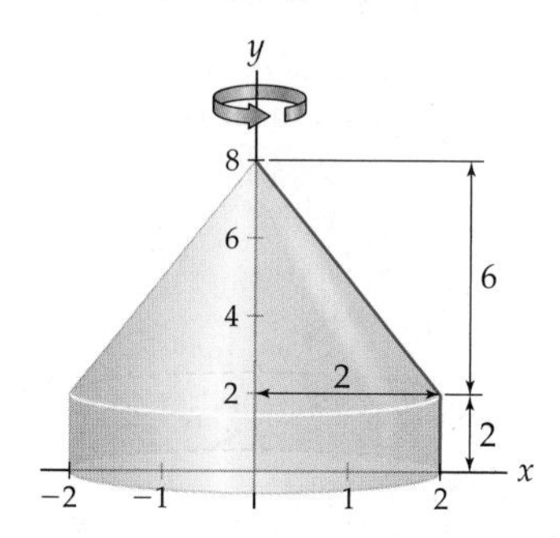

The solid in the second figure is just a disk of radius 2 and height 2 with a cone of radius 2 and height 6 on top, so its volume is

$$\pi 2^2(2) + \frac{1}{3}\pi 2^2(6) = 16\pi \approx 50.2655.$$

This number does make our earlier approximation of 59.69 seem reasonably accurate, since the actual solid is somewhat larger than our cone-on-disk approximation.

EXAMPLE 2

Using shells to construct a definite integral for volume

Find the exact volume of the solid obtained by rotating the region between the graph of $f(x) = -x^2 - x + 8$ and the x-axis on $[0, 2]$ around the y-axis.

SOLUTION

The first figure that follows shows the region in question, and the second figure shows the resulting solid of revolution. If we take a vertical representative rectangle in the region and rotate it around the y-axis, then we obtain a representative shell for the solid at x_k^*, as shown in the third figure.

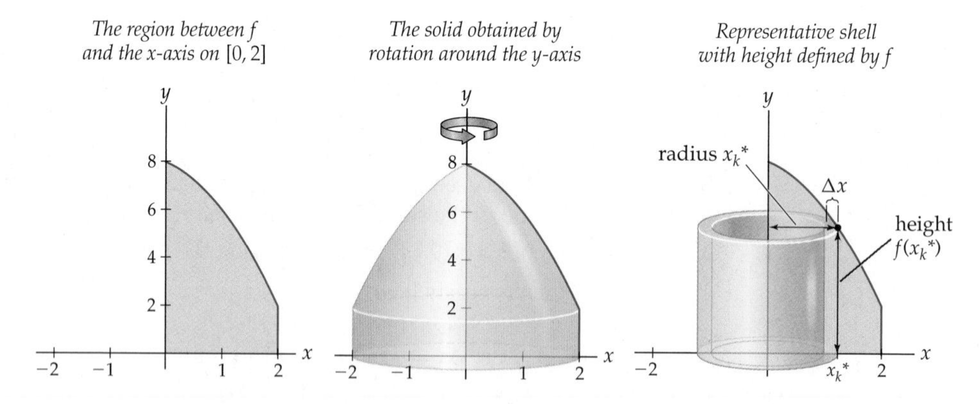

If we imagine x_k^* to be the midpoint of the subinterval $[x_{k-1}, x_k]$, then this representative shell has average radius x_k^*, height $f(x_k^*)$ and thickness Δx. Therefore the volume of the representative shell at x_k^* is $2\pi x_k^* f(x_k^*)\Delta x$.

Accumulating these shells from the inside out, from $x = 0$ at the center to $x = 2$ on the outside, and applying the function $f(x) = -x^2 - x + 8$, we can write the volume of the solid as the definite integral

$$2\pi \int_0^2 x(-x^2 - x + 8)\,dx = 2\pi \int_0^2 (-x^3 - x^2 + 8x)\,dx$$
$$= 2\pi \left[-\frac{1}{4}x^4 - \frac{1}{3}x^3 + \frac{8}{2}x^2\right]_0^2$$
$$= 2\pi \left(\left(-\frac{1}{4}(2)^4 - \frac{1}{3}(2)^3 + 4(2)^2\right) - \left(-\frac{1}{4}(0)^4 - \frac{1}{3}(0)^3 + 4(0)^2\right)\right)$$
$$= \frac{56}{3}\pi.$$

Notice that this volume of $\frac{56}{3}\pi \approx 58.6431$ cubic units is not far off from the four-shell approximation we did in the previous example. □

EXAMPLE 3

Finding volume by integrating along the *y*-axis with shells

Find the volume of the solid obtained by rotating the region bounded by the graph of $f(x) = 3\ln x$, the line $y = 2$, and the x- and y-axes around the x-axis.

SOLUTION

The three figures that follow illustrate, respectively, the region in question, the resulting solid, and a representative shell obtained by rotating a horizontal rectangle around the x-axis. Note that this is the same solid whose volume we computed by using disks and washers in the first part of Example 5 of Section 6.1.

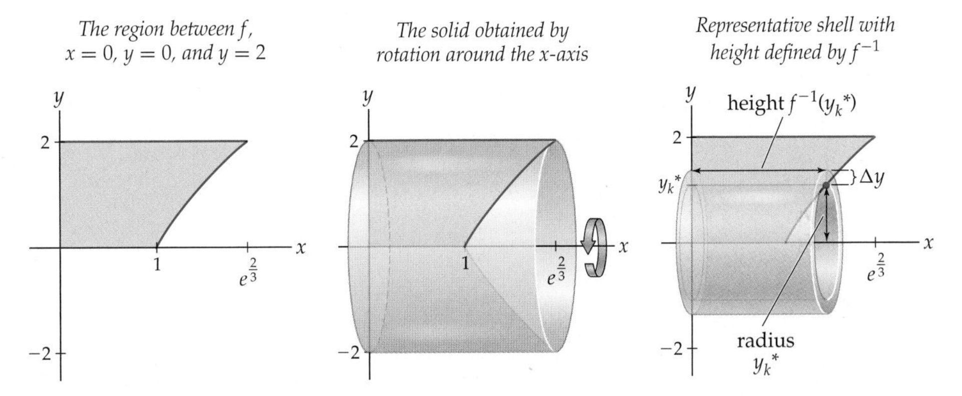

The representative shell at y_k^* has average radius y_k^*, (horizontal) height $f^{-1}(y_k^*)$, and thickness Δy. Therefore the volume of this representative shell is $2\pi y_k^* f^{-1}(y_k^*)\,\Delta y$. Since $f(x) = y = 3\ln x$, we have $f^{-1}(y) = e^{y/3}$. Accumulating the shells from the inside at $y = 0$ to the outside at $y = 2$, we get the volume integral

$$2\pi \int_0^2 y\,e^{y/3}\,dy.$$

This definite integral can be solved by using integration by parts with $u = y$ and $dv = e^{y/3}\,dy$:

$$2\pi \int_0^2 y\,e^{y/3}\,dy = 2\pi \left(\left[3ye^{y/3}\right]_0^2 - 3\int_0^2 e^{y/3}\,dy\right)$$
$$= 2\pi \left(\left[3ye^{y/3}\right]_0^2 - \left[9e^{y/3}\right]_0^2\right)$$
$$= 2\pi((6e^{2/3} - 0) - (9e^{2/3} - 9e^0))$$
$$= 2\pi(-3e^{2/3} + 9).$$

The volume of $2\pi(-3e^{2/3} + 9) \approx 19.83$ cubic units does seem appropriate given the dimensions of the solid in question and given that the volume of the cylinder that encloses our solid is $\pi(2)^2 e^{2/3} \approx 24.48$ cubic units. □

EXAMPLE 4

Setting up a variety of volume integrals

Consider the region bounded by the graph of $f(x) = e^x$ and the lines $y = 0$, $y = 3$, $x = 0$, and $x = 2$. Use definite integrals to express the volumes of the following solids of revolution:

(a) The solid obtained by rotating the region around the x-axis. Do this problem in two ways: first by integrating in the x direction and then by integrating in the y direction.

(b) The solid obtained by rotating the region around the y-axis. Again, do this problem in two ways: first by integrating in the x direction and then by integrating in the y direction.

SOLUTION

The region in question is shown in the first figure that follows. This region changes character at $x = \ln 3$, since that is the solution of the equation $e^x = 3$. The second figure shows the solid resulting from rotating this region around the x-axis, and the third figure shows the result of rotating the region around the y-axis.

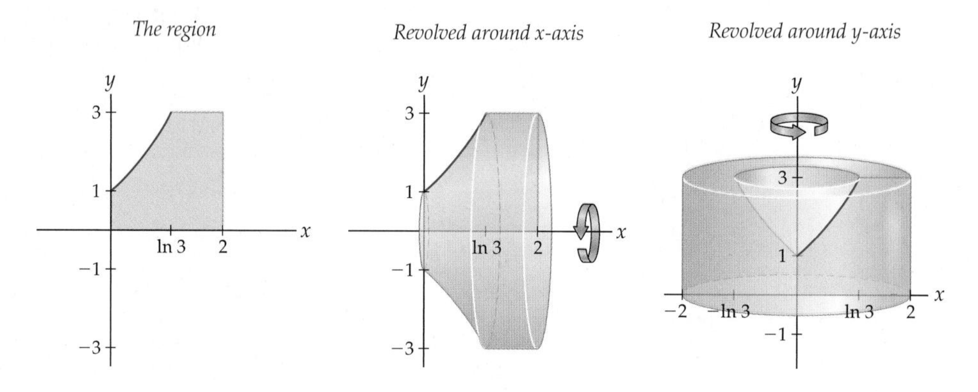

(a) Integrating in the x direction, we can use disks to write the volume of the solid in the center graph as the sum of two definite integrals. From $x = 0$ to $x = \ln 3$ the radius of the disk at x_k^* will be given by $e^{x_k^*}$, and from $x = \ln 3$ to $x = 2$ the radius of the disk at x_k^* is 3. The volume can be written as

$$\pi \int_0^{\ln 3} (e^x)^2\, dx + \pi \int_{\ln 3}^2 (3)^2\, dx.$$

Integrating in the y direction, we can use shells. From $y = 0$ to $y = 1$, each shell will have a (horizontal) height of $2 - 0 = 2$. From $y = 1$ to $y = 3$, the (horizontal) height of the shell at y_k^* will be the difference $2 - \ln y_k^*$. Therefore the volume of the solid is

$$2\pi \int_0^1 y\,(2)\, dy + 2\pi \int_1^3 y\,(2 - \ln y)\, dy.$$

Of course, the two definite integral expressions we just constructed should find the same volume. Calculate them and see!

(b) We now turn to the solid shown earlier at the right. Integrating in the x direction produces shells in this case. From $x = 0$ to $x = \ln 3$ the shell at x_k^* will have a height of $e^{x_k^*}$, and from $x = \ln 3$ to $x = 2$ the shell will have a height of 3. Therefore the volume of the solid can be written as the following sum of definite integrals:

$$V_2 = 2\pi \int_0^{\ln 3} xe^x \, dx + 2\pi \int_{\ln 3}^{2} x(3) \, dx.$$

To integrate in the y direction, we have to use disks and washers. From $y = 0$ to $y = 1$ the disk at y_k^* will have a radius of 2, and from $y = 1$ to $y = 3$ a rectangle at y_k^* will produce a washer with inner radius given by $\ln(y_k^*)$ (since $y = e^x$, we have $x = \ln y$) and an outer radius of 2. Therefore, we can write the volume of the solid as

$$V_2 = \pi \int_0^1 (2)^2 \, dy + \pi \int_1^3 (2^2 - (\ln y)^2) \, dy.$$

Again, the two definite integral expressions we found should give the same volume. Calculating a volume in two ways, when possible, can be a good method for checking your work. □

? TEST YOUR UNDERSTANDING

- ▶ Why was the shell volume formula $2\pi rh \, \Delta x$ more useful than the shell volume formula $\pi(r_2^2 - r_1^2)h$ for constructing a Riemann sum?
- ▶ With shells, why do we integrate from the center of a solid to the outside, instead of from one side to the other?
- ▶ What is the analog of Definition 6.5 for using the shell method in the y direction?
- ▶ Why, geometrically, is the answer found in Example 3 reasonable?
- ▶ What do the disks, washers, and shells used in the various parts of Example 4 look like?

EXERCISES 6.2

Thinking Back

Definite integrals: Calculate each of the following definite integrals, using integration techniques and the Fundamental Theorem of Calculus.

- ▶ $\int_0^1 (3 - x)(x^2 + 1) \, dx$
- ▶ $\int_0^4 x(2 - \sqrt{x}) \, dx$
- ▶ $\int_0^6 x((x - 3)^2 + 2) \, dx$
- ▶ $\int_1^2 xe^{x/3} \, dx$
- ▶ $\int_0^1 (e^x)^2 \, dx$
- ▶ $\int_2^4 x \ln x \, dx$
- ▶ $\int_1^5 ye^y \, dy$
- ▶ $\int_1^e (\ln y)^2 \, dy$
- ▶ $\int_0^{\pi} y \sin y \, dy$
- ▶ $\int_2^4 y\sqrt{6 - y} \, dy$

Finding volumes geometrically: Use volume formulas to find the exact volumes of the solids pictured here.

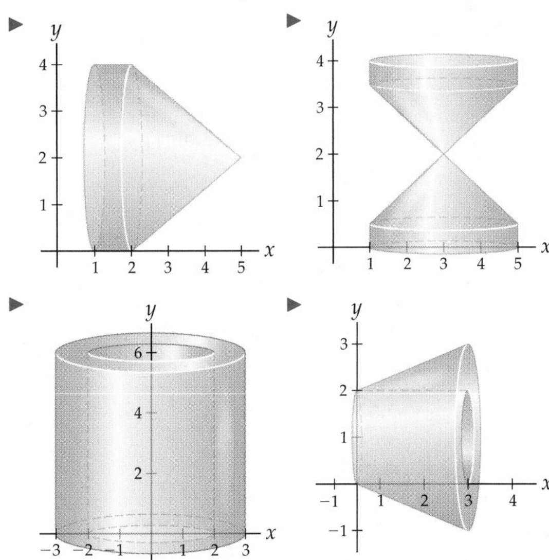

Concepts

0. *Problem Zero:* Read the section and make your own summary of the material.

1. *True/False:* Determine whether each of the statements that follow is true or false. If a statement is true, explain why. If a statement is false, provide a counterexample.

(a) *True or False:* The thickness of a shell with inner radius r and outer radius R is $R - r$.

(b) *True or False:* The average radius of a shell with inner radius r and outer radius R is $\frac{r+R}{2}$.

(c) *True or False:* A shell with height h and average radius r has volume $\pi r^2 h \Delta x$.

(d) *True or False:* A shell with height h, average radius r, and thickness Δx has volume $\pi r^2 h \Delta x$.

(e) *True or False:* A shell with height h, outer radius r, and thickness Δx has volume $2\pi r h \Delta x$.

(f) *True or False:* Rotating a vertical rectangle around the y-axis results in a shell.

(g) *True or False:* Rotating a horizontal rectangle around the y-axis results in a shell.

(h) *True or False:* Rotating a horizontal rectangle around the x-axis results in a shell.

2. *Examples:* Construct examples of the thing(s) described in the following. Try to find examples that are different than any in the reading.

(a) A solid of revolution around the x-axis whose volume can be computed with disks/washers but not shells.

(b) A solid of revolution around the y-axis whose volume can be computed with shells but not disks/washers.

(c) A solid of revolution whose volume can be computed both with shells and with disks/washers.

3. Consider the rectangle bounded by $y = 1$ and $y = 4$ on the x-interval $[1, 1.5]$. Calculate the volume of the geometric object obtained by rotating this rectangle around

(a) the x-axis
(b) the y-axis
(c) the line $x = 1$
(d) the line $y = 1$
(e) the line $x = 2$
(f) the line $y = -1$

4. Consider the rectangle bounded by $x = 1$ and $x = 3$ on the y-interval $[2, 2.5]$. Calculate the volume of the geometric object obtained by rotating this rectangle around

(a) the x-axis
(b) the y-axis
(c) the line $x = 1$
(d) the line $y = 2$
(e) the line $x = -1$
(f) the line $y = 3$

5. In what way are shells the same as washers? How are they different?

6. Consider the region between $f(x) = x^2$ and the x-axis between $x = 0$ and $x = 2$. Draw a Riemann sum approximation of the area of this region, using a midpoint sum with four rectangles, and explain how this Riemann sum approximation is related to a four-shell approximation of the solid obtained by rotating the region around the y-axis.

7. For a four-shell approximation of the volume of the solid obtained from the region between $f(x) = x^2$ and the x-axis between $x = 0$ and $x = 2$ by rotating the region around the y-axis, illustrate and calculate

(a) Δx and each x_k;
(b) the midpoint x_k^* of each subinterval $[x_{k-1}, x_k]$;
(c) each $f(x_k^*)$;
(d) the volume of the second shell.

8. For a four-shell approximation of the volume of the solid obtained from the region between $f(x) = x^2$ and the x-axis between $x = 0$ and $x = 2$ by rotating the region around the x-axis, illustrate and calculate

(a) Δy and each y_k;
(b) the midpoint y_k^* of each subinterval $[y_{k-1}, y_k]$;
(c) each $f^{-1}(y_k^*)$;
(d) the volume of the second shell.

Write each of the limits in Exercises 9 and 10 as a definite integral, and identify the corresponding solid of revolution, given that x_k^* and y_k^* represent midpoints of subintervals.

9. $\lim_{n\to\infty} \sum_{k=1}^{n} 2\pi (y_k^*)^3 \left(\frac{2}{n}\right)$, with $y_k = \frac{2k}{n}$

10. $\lim_{n\to\infty} \sum_{k=1}^{n} 2\pi (x_k^*)^{3/2} \left(\frac{4}{n}\right)$, with $x_k = \frac{4k}{n}$

Each of the definite integrals in Exercises 11–16 represents the volume of a solid of revolution obtained by rotating a region around either the x- or y-axis, computed with the shell method. Find this region.

11. $2\pi \int_0^{\pi/4} x \cos x \, dx$

12. $2\pi \int_0^1 x(x^2)\, dx$

13. $2\pi \int_4^5 y\sqrt{y-4}\, dy$

14. $2\pi \int_0^1 (x^2 - x^3)\, dx$

15. $2\pi \int_1^e y \ln y \, dy$

16. $2\pi \int_0^1 (3y - 3y^2)\, dy$

Write the volumes of the solids of revolution shown in Exercises 17–20 in terms of definite integrals that represent accumulations of shells. Do not solve the integrals.

17.

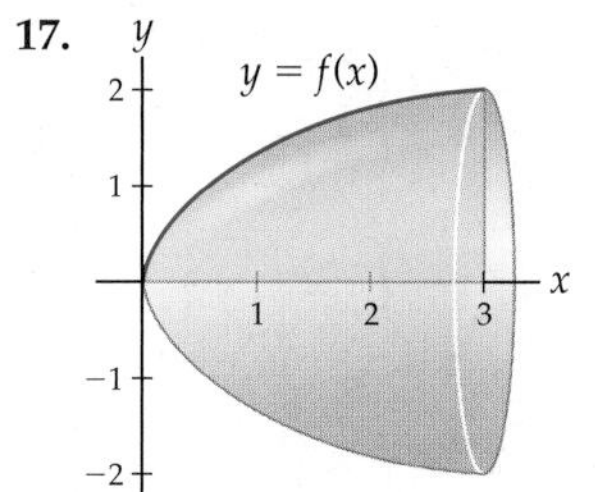

18.

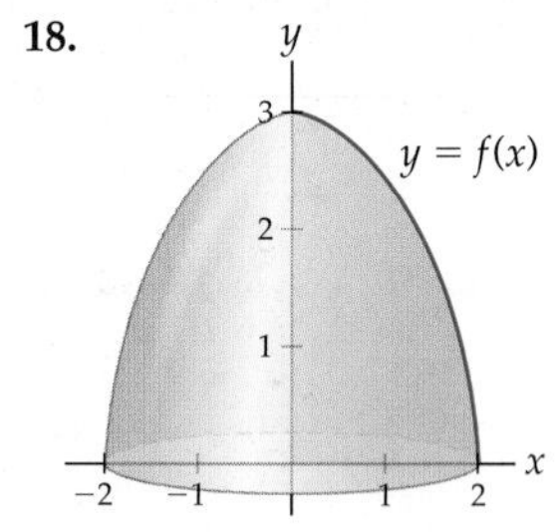

19.

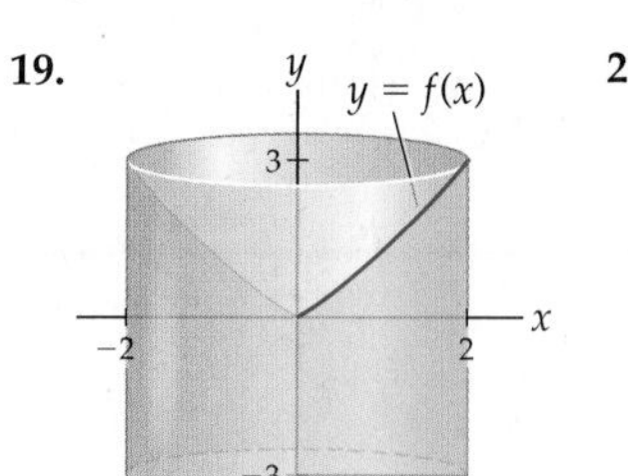

20.

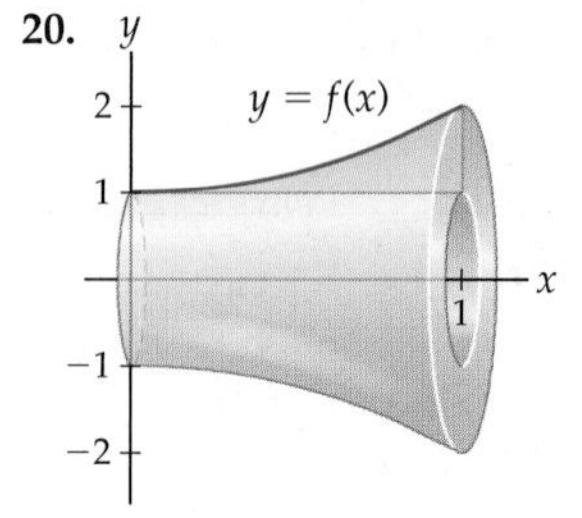

For each solid described in Exercises 21–24, set up volume integrals using both the shell and disk/washer methods. Which method produces an easier integral in each case, and why? Do not solve the integrals.

21. The region between the graph of $f(x) = |x|$ and the x-axis on $[-2, 2]$, revolved around the x-axis.

22. The region between the graph of $f(x) = \frac{1}{x}$ and the lines $y = 0$, $y = 1$, $x = 0$, and $x = 2$, revolved around the y-axis.

23. The region between the graph of $f(x) = x^2 + 1$ and the x-axis on $[0, 1]$, revolved around the line $x = 3$.

24. The region between the graph of $f(x) = x^2 + 1$ and the x-axis on $[0, 1]$, revolved around the line $y = -2$.

Skills

Consider the region between $f(x) = \sqrt{x}$ and the x-axis on $[0, 4]$. For each line of rotation given in Exercises 25–28, use four shells based on the given rectangles to approximate the volume of the resulting solid.

25. *Around the x-axis* **26.** *Around the y-axis*

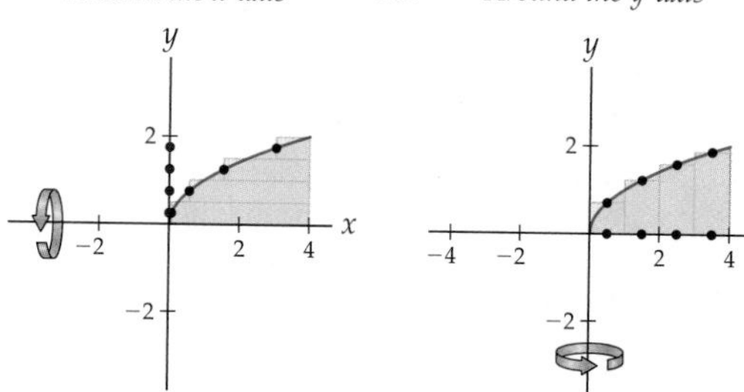

27. *Around the line $y = -1$* **28.** *Around the line $x = 5$*

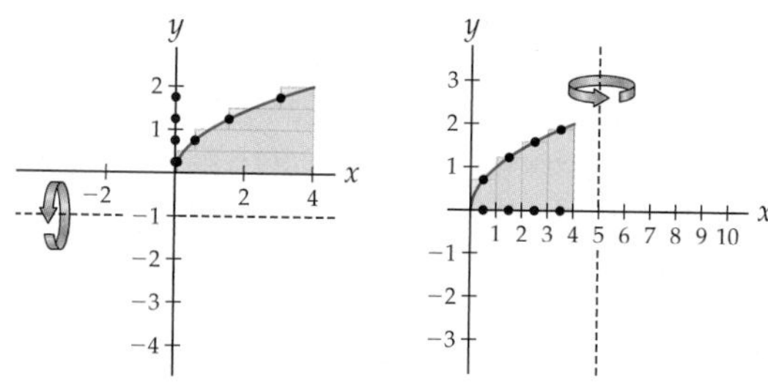

Consider the region between $f(x) = 4 - x^2$ and the x-axis on $[0, 2]$. For each line of rotation given in Exericses 29–32, use the shell method to construct definite integrals to find the volume of the resulting solid.

29. *Around the y-axis*

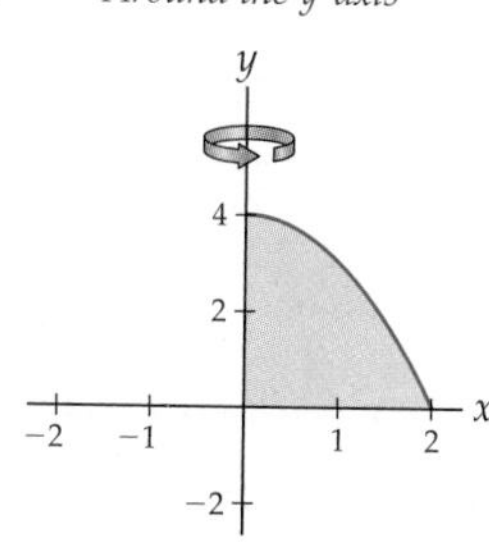

30. *Around the x-axis*

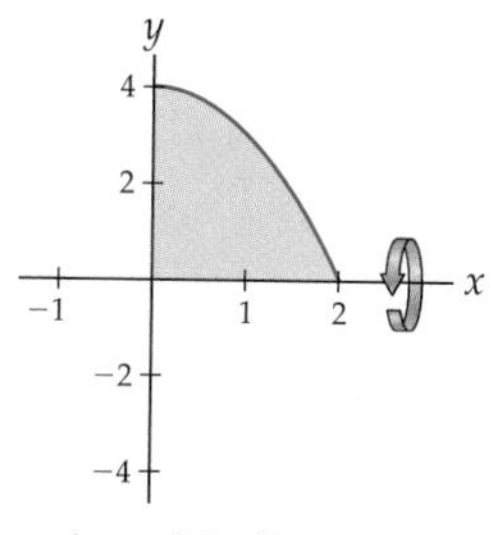

31. *Around the line $x = 2$*

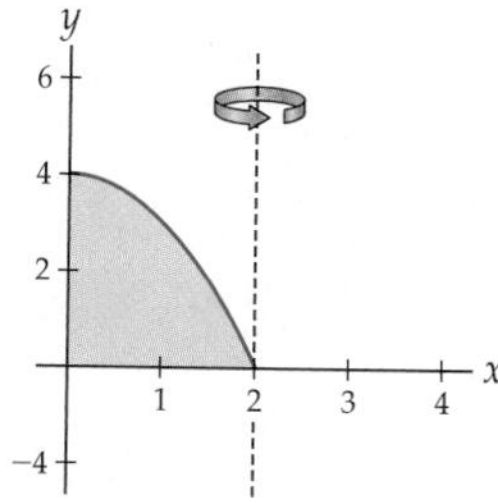

32. *Around the line $y = 4$*

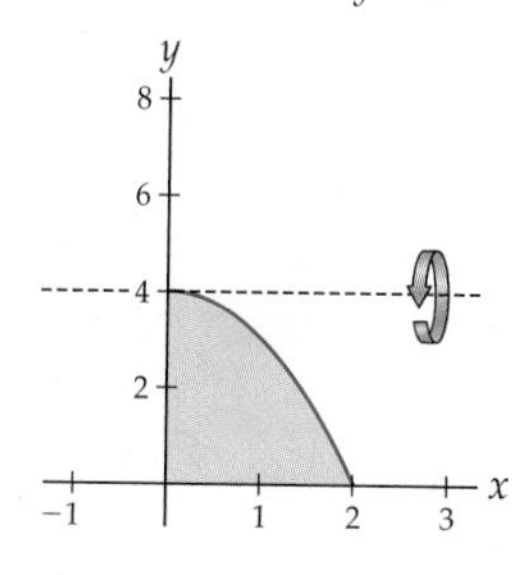

Consider the region between $f(x) = x - 2$ and the x-axis on $[2, 5]$. For each line of rotation given in Exercises 33–38, use the shell method to construct definite integrals to find the volume of the resulting solid.

33. *Around the y-axis*

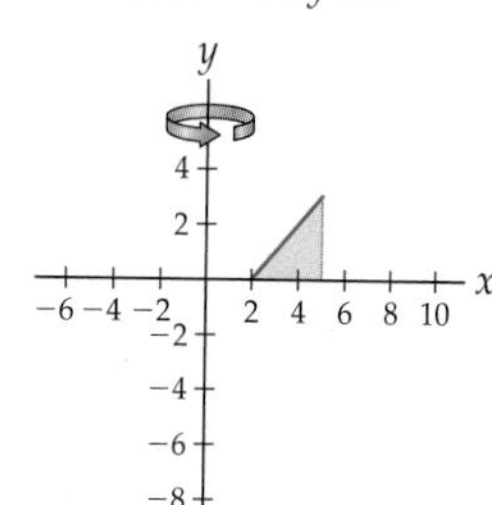

34. *Around the x-axis*

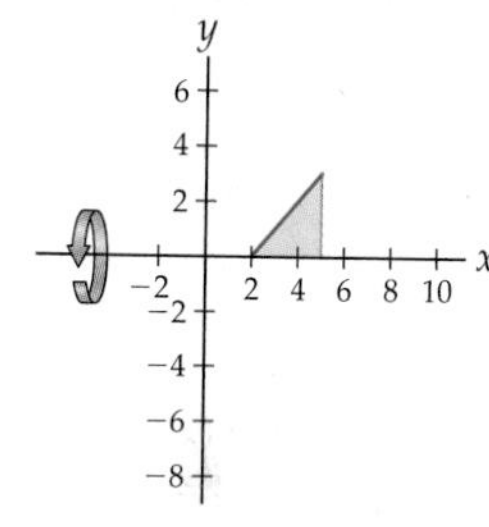

35. *Around the line $y = 3$*

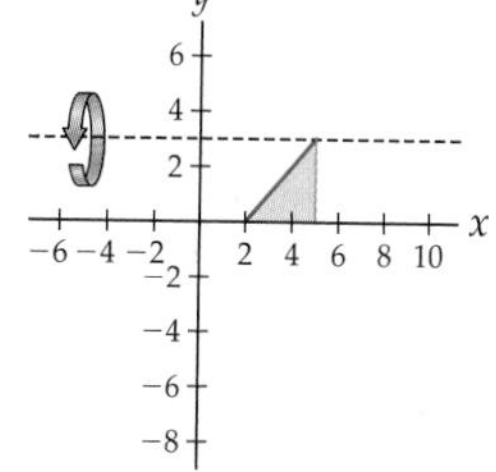

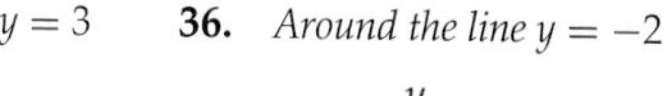

36. *Around the line $y = -2$*

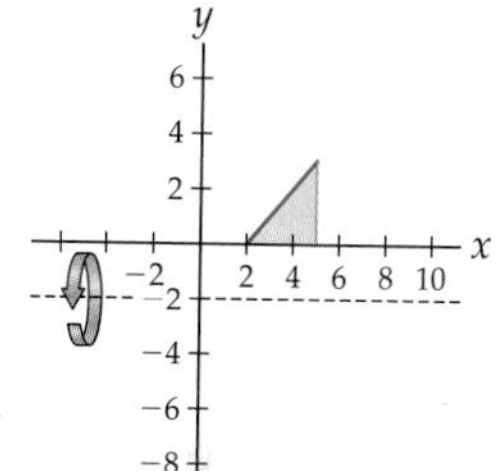

37. *Around the line $x = 2$*

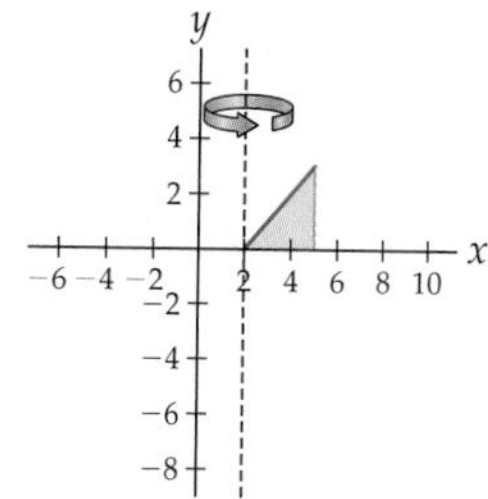

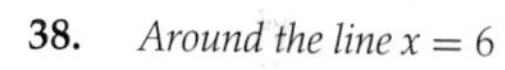

38. *Around the line $x = 6$*

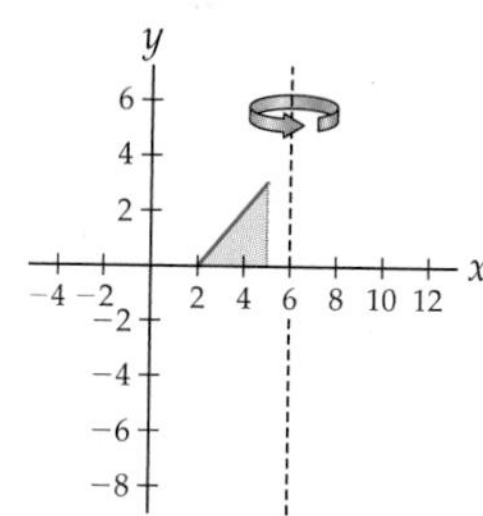

Consider the region between $f(x) = 4 - x^2$ and the line $y = 5$ on $[0, 2]$. For each line of rotation given in Exercises 39–42, use the shell method to construct definite integrals to find the volume of the resulting solid.

39. *Around the y-axis*

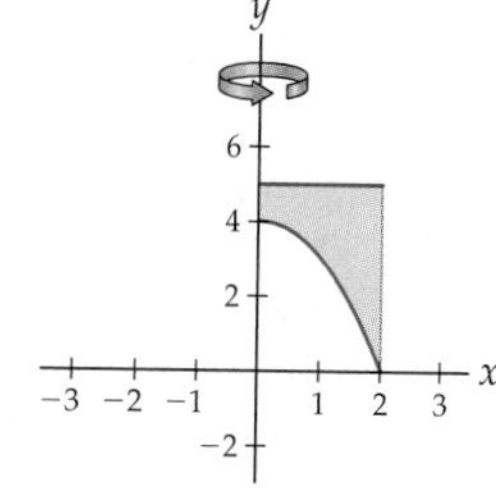

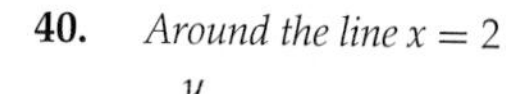

40. *Around the line $x = 2$*

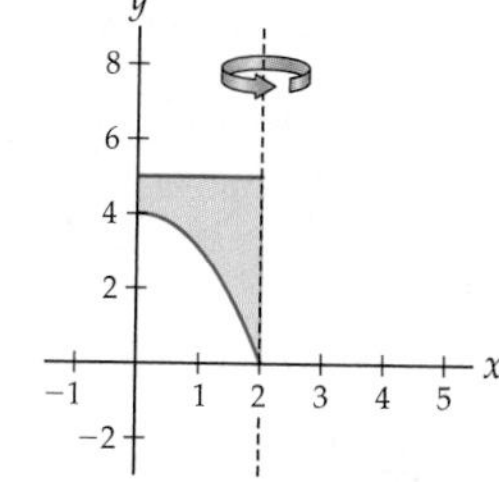

41. *Around the line $x = 3$*

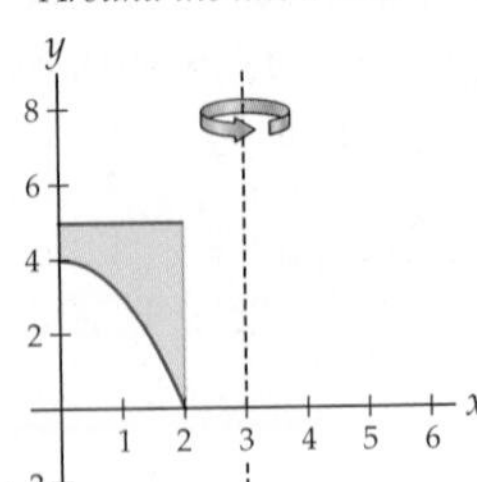

42. *Around the line $y = 5$*

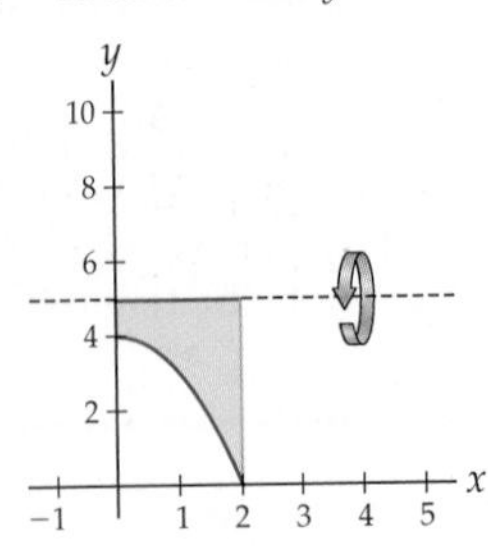

Consider the region between $f(x) = x+1$ and the line $y = -2$ on $[1, 5]$. For each line of rotation given in Exercises 43 and 44, use the shell method to construct definite integrals to find the volume of the resulting solid.

43. *Around the line $x = 5$*

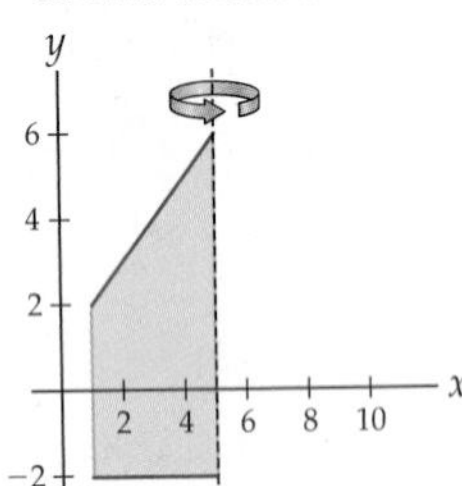

44. *Around the y-axis*

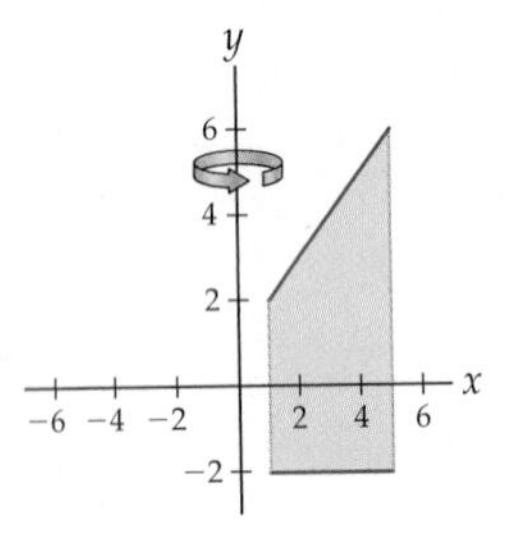

Consider the region between $f(x) = (x-2)^2$ and $g(x) = x$ on $[1, 4]$. For each line of rotation given in Exercises 45–48, use the shell method to construct definite integrals to find the volume of the resulting solid.

45. *Around the x-axis*

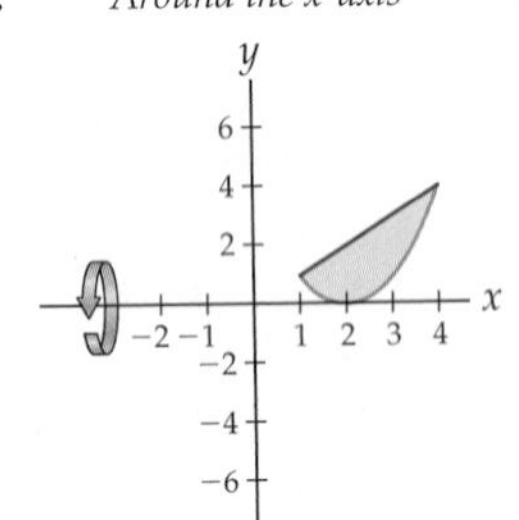

46. *Around the y-axis*

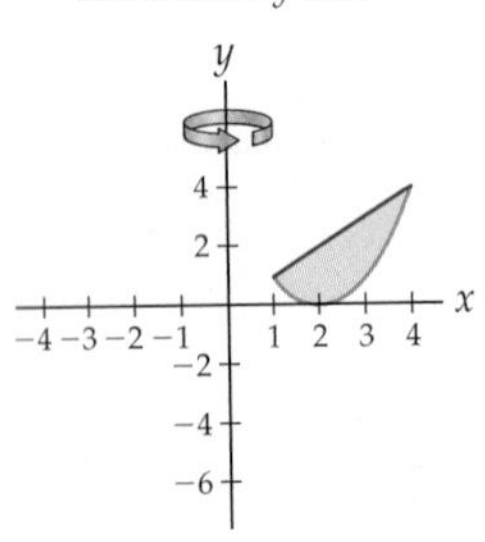

47. *Around the line $y = -1$*

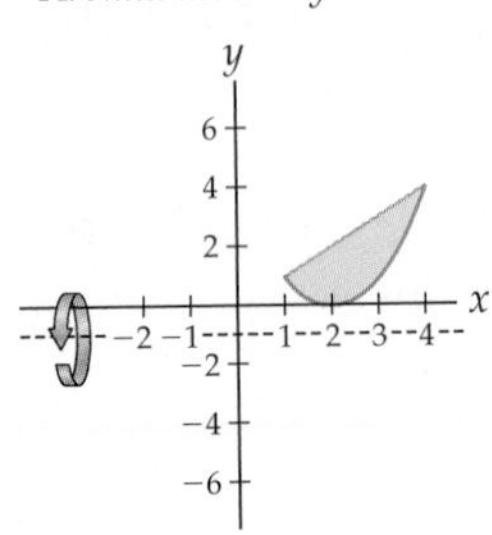

48. *Around the line $x = 1$*

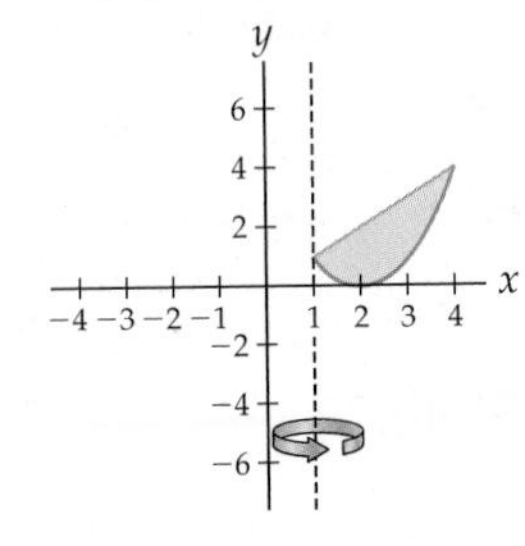

Use definite integrals to find the volume of each solid of revolution described in Exercises 49–61. (It is your choice whether to use disks/washers or shells in these exercises.)

49. The region between the graph of $f(x) = 4 - x^2$ and the line $y = 4$ on $[0, 2]$, revolved around the y-axis.

50. The region between the graph of $f(x) = e^{2x}$ and the x-axis on $[0, 4]$, revolved around the x-axis.

51. The region between the graph of $f(x) = (x-3)^2 + 2$ and the x-axis on $[0, 6]$, revolved around the y-axis.

52. The region between the graph of $f(x) = x^2$ and the x-axis on $[0, 3]$, revolved around the line $y = 10$.

53. The region between the graphs of $f(x) = \sqrt{x}$ and $g(x) = x^2$ on $[0, 1]$, revolved around the x-axis.

54. The region between the graph of $f(x) = e^x$ and the line $y = e$ on $[0, 1]$, revolved around the x-axis.

55. The region between the graph of $f(x) = 9 - x^2$ and the x-axis on $[0, 3]$, revolved around the x-axis.

56. The region between the graph of $f(x) = (x-2)^2$ and the x-axis on $[0, 4]$, revolved around the x-axis.

57. The region between the graph of $f(x) = x^2 - 4x + 4$ and the x-axis on $[0, 2]$, revolved around the y-axis.

58. The region bounded by the graphs of $f(x) = 2 \ln x$, $y = 0$, $y = 3$, and $x = 0$, revolved around the x-axis.

59. The region bounded by the graphs of $f(x) = \frac{2}{x}$, $y = 0$, $y = 3$, $x = 0$, and $x = 2$, revolved around the x-axis.

60. The region between the graph of $f(x) = x^2 + 2$ and the x-axis on $[-1, 3]$, revolved around the line $x = -1$.

61. The region bounded by the graphs of $f(x) = |x|$ and $y = 2$ on $[-2, 2]$, revolved around the line $x = 3$.

Applications

62. Mark carves napkin rings out of wooden spheres. His napkin rings have height $h = 8$ and radius $R = 5$, measured in centimeters, as shown next at the left. Alina also carves napkin rings out of spheres, but she likes bigger napkins, so she carves them with height $h = 8$ and radius $R = 6$, as shown at the right. Use the shell method to show that Mark's and Alina's napkin rings use exactly the same amount of wood. (This is an example of the famous *Napkin Ring Theorem*; see Exercise 70).

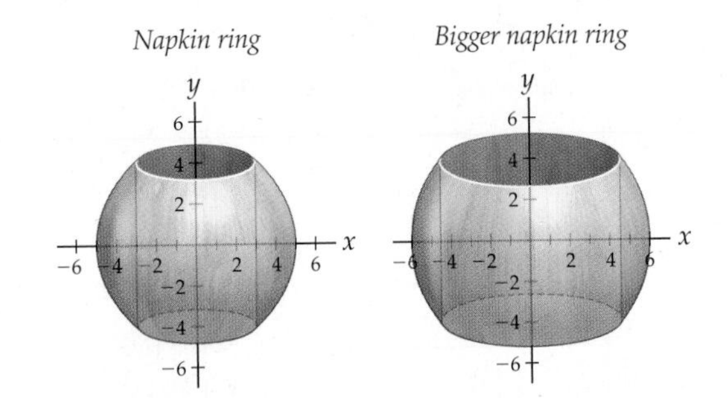

63. Alan makes beads that have the shape of flattened spheres with holes drilled through for stringing. The beads can be modeled by revolving the curve $f(x) = 3 - 0.15x^2$ on $[-4, 4]$ around the x-axis, as shown in the next figure, in which units are in centimeters. Use the shell method to determine the amount of material required to make 100 beads of this shape.

Shape of a bead

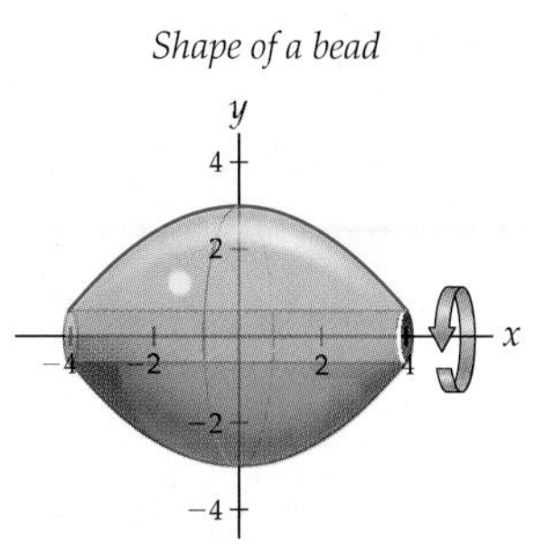

Proofs

64. The disk and shell methods for calculating volume of solids of revolution always give the same answer (when they are both computable). Prove that this is so for any solid of revolution defined by revolving the region between a linear function $y = mx + c$ and the x-axis on $[a, b]$, where m, c, a, and b are positive.

65. Prove that a shell with average radius r, height h, and thickness Δx has volume $V = 2\pi rh\, \Delta x$.

66. Use a definite integral with the shell method to prove that the volume formula $V = \frac{1}{3}\pi r^2 h$ holds for a cone of radius 3 and height 5.

67. Use a definite integral with the shell method to prove that the volume formula $V = \frac{4}{3}\pi r^3$ holds for a sphere of radius 3.

68. Use a definite integral with the shell method to prove that a cone of radius r and height h has volume given by the formula $V = \frac{1}{3}\pi r^2 h$.

69. Use a definite integral with the shell method to prove that a sphere of radius r has volume given by the formula $V = \frac{4}{3}\pi r^3$.

70. Prove the *Napkin Ring Theorem:* If a napkin ring is made by removing a cylinder of height h from a sphere, then the volume of the resulting shape does not depend on the radius of the sphere! Look at the figures in Exercise 62 to see why this result is surprising.

Thinking Forward

More "subdivide, approximate, and add": We have already seen how to use limits and a "subdivide, approximate, and add" strategy to calculate area by approximating with rectangles and to calculate volume by using disks, washers, or shells. What else can we calculate in this way?

- ▶ Consider the curve traced out by the graph of a function $y = f(x)$ on an interval $[a, b]$. Come up with a proposal for approximating the length of such a curve with a "subdivide, approximate, and add" strategy. What would play the role of rectangles, disks, washers, or shells in this case?
- ▶ Consider the surface of revolution obtained by revolving the graph of a function $y = f(x)$ around the x-axis on an interval $[a, b]$. Come up with a proposal for approximating the surface area of such an object with a "subdivide, approximate, and add" strategy. What do you think would play the role of rectangles, disks, washers, or shells in this case?
- ▶ Suppose you wish to find the mass of a metal rod that is made up of a material whose density increases linearly from the left end to the right end. Mass is equal to density times volume, but since density is varying in this example, we need to somehow use a "subdivide, approximate, and add" strategy. Come up with a proposal for how to use this strategy to calculate the mass of the rod. Again, consider what the best choice would be for the role of the pieces in this example.

6.3 ARC LENGTH AND SURFACE AREA

- Approximating arc length with a sum of line-segment lengths
- Approximating surface area with a sum of frustum surface areas
- Using definite integrals to calculate exact arc lengths and surface areas

Approximating Arc Length

We have now used the "subdivide, approximate, and add" strategy to define definite integrals for areas under curves and for volumes of solids of revolution. We will now use the same strategy to find lengths of curves and surface areas of solids of revolution. To illustrate the process in the context of lengths of curves, let's try to find the length of part of a circle.

You probably already know that the circumference C of a circle of radius r is given by $C = 2\pi r$. But where does that magical formula come from? If we didn't know this formula, how could we approximate the circumference of a circle, say, of radius 2? Since such a circle is given by the equation $x^2 + y^2 = 4$, the function $f(x) = \sqrt{4 - x^2}$ on $[-2, 2]$ traces out its top half, as shown here:

$f(x) = \sqrt{4 - x^2}$ on $[-2, 2]$

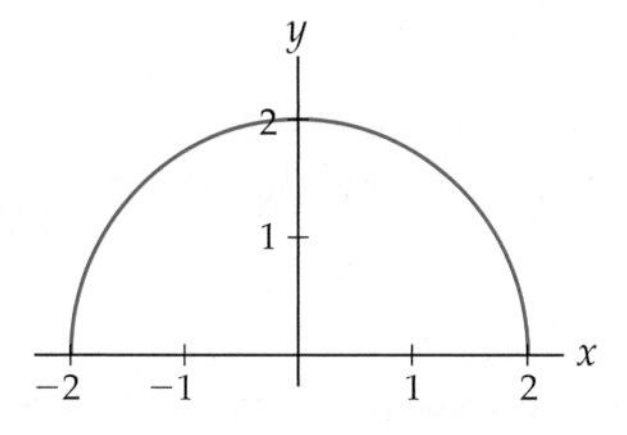

The most obvious way to approximate the length of this curve is with line segments, much like the way that you would approximate the length of a curvy highway on a map by using straight segments. For example, we could divide the interval $[-2, 2]$ into four subintervals and approximate the length of the curve on each subinterval with a line segment, as shown in the first figure that follows. To get a better approximation of the length of the curve, we could use more line segments, as shown, for example, in the second figure.

Four-segment approximation

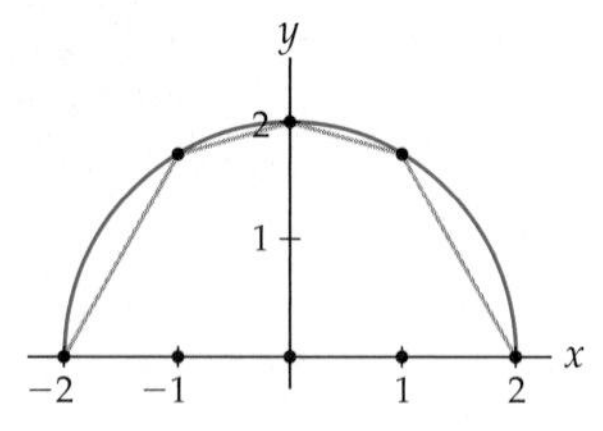

Eight-segment approximation

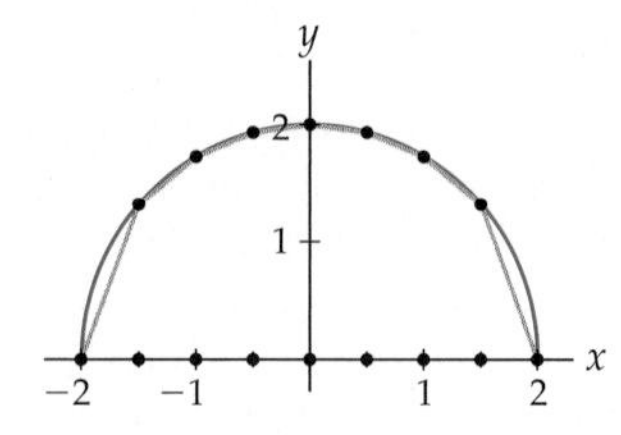

In order to use our "subdivide, approximate, and add" strategy to find the *exact* length of a curve, we will need to take a limit, letting the size of the subintervals get smaller and smaller. Before we can apply such a limit, we need to set up some general notation, much like the notation we used when we defined the area under a curve.

Suppose we wish to approximate the length of a curve $f(x)$ on an interval $[a, b]$. If we divide $[a, b]$ into n subintervals of equal width, then each subinterval will have width

$\Delta x = \frac{b-a}{n}$. If we define $x_k = a + k\Delta x$ for $k = 0, 1, \ldots, n$, then the kth subinterval is the interval $[x_{k-1}, x_k]$. Each subinterval will have width Δx, and the line segment on the kth subinterval will change vertically by $\Delta y_k = f(x_k) - f(x_{k-1})$ units on that subinterval, as illustrated in the next figure. Notice that Δx does not depend on the choice of k, but Δy_k does, since the graph of f increases and decreases at different rates as we move from $x = a$ to $x = b$.

Notation for the kth line segment

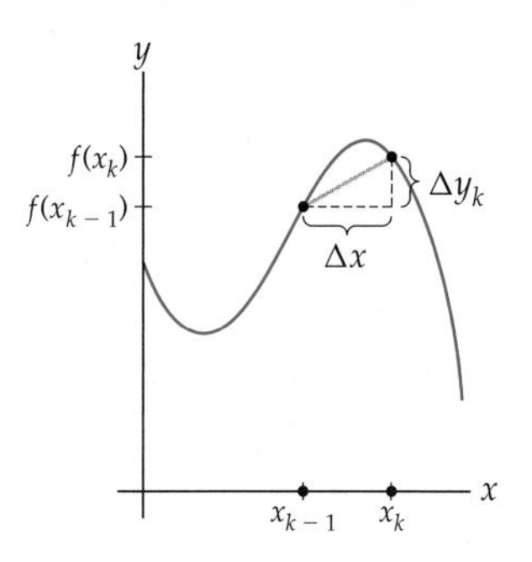

We can use the distance formula to calculate the length l_k of each line segment:

$$l_k = \text{length of } k\text{th line segment} = \sqrt{(\Delta x)^2 + (\Delta y_k)^2}.$$

With just a little bit of algebra we can factor out a Δx, as follows:

$$l_k = \sqrt{(\Delta x)^2 + (\Delta y_k)^2} = \sqrt{(\Delta x)^2\left(1 + \frac{(\Delta y_k)^2}{(\Delta x)^2}\right)} = \sqrt{1 + \left(\frac{\Delta y_k}{\Delta x}\right)^2}\,\Delta x.$$

As we will see in Theorem 6.7, this second form for the distance will be much more useful. Now given a function $f(x)$, an interval $[a, b]$, and a number of subdivisions n, we have defined n subintervals, each with a line segment that approximates a small piece of the curve. Adding up the lengths of these line segments gives us an approximation for the length of the curve:

$$\text{length of } f(x) \text{ on } [a, b] \approx \sum_{k=1}^{n} l_k = \sum_{k=1}^{n} \sqrt{1 + \left(\frac{\Delta y_k}{\Delta x}\right)^2}\,\Delta x.$$

You will calculate such an approximation by hand in Example 1. For any individual approximation all this notation is not particularly necessary, but as we will see in a moment, such notation will enable us to obtain a Riemann sum, and thus a definite integral, for arc length.

A Definite Integral for Arc Length

So what is the *exact* length of the curve? For sufficiently well behaved functions the approximation we just developed gets better as we let the number n of line segments get larger and larger, since this makes the width Δx of the subintervals get smaller and smaller. Therefore to get the exact length we need to take the limit of our n-subinterval approximation as n approaches infinity. For nice-enough functions this limit will converge to a real number, and we define that number to be the ***arc length*** of $f(x)$ on $[a, b]$.

DEFINITION 6.6

The Arc Length of a Function on an Interval

Suppose $f(x)$ is a differentiable function with a continuous derivative. Then the ***arc length*** of $f(x)$ from $x = a$ to $x = b$ is

$$\begin{array}{c}\text{Arc length of } f(x) \\ \text{from } x = a \text{ to } x = b\end{array} = \lim_{n\to\infty} \sum_{k=1}^{n} \sqrt{1 + \left(\frac{\Delta y_k}{\Delta x}\right)^2}\, \Delta x,$$

where $\Delta x = \frac{b-a}{n}$, $x_k = a + k\Delta x$, and $\Delta y_k = f(x_k) - f(x_{k-1})$ for all $k = 0, 1, 2, \ldots, n$.

The limit of sums in this definition should look familiar to you. It is a kind of Riemann sum, and therefore it can be written as a definite integral.

THEOREM 6.7

Arc Length as a Definite Integral

Suppose $f(x)$ is a differentiable function with a continuous derivative on $[a, b]$. Then the arc length of $f(x)$ from $x = a$ to $x = b$ can be represented by the definite integral

$$\begin{array}{c}\text{Arc length of } f(x) \\ \text{from } x = a \text{ to } x = b\end{array} = \int_a^b \sqrt{1 + (f'(x))^2}\, dx.$$

If you think about it for a moment, you will see that Theorem 6.7 says something very interesting: The length of the curve traced out by the graph of a function $f(x)$ on an interval $[a, b]$ is the same as the *area* under the graph of $\sqrt{1 + (f'(x))^2}$ on that same interval. There is a nice consequence to this: We already know how to calculate definite integrals, and therefore Theorem 6.7 makes it simple for us to calculate certain exact arc lengths.

Considering the definition of the definite integral, we can see how the limit of sums in Definition 6.6 becomes the definite integral in Theorem 6.7, at least if we believe that $\frac{\Delta y_k}{\Delta x}$ becomes $f'(x)$ as $n \to \infty$. The key to showing the latter, and thus the key to the proof of Theorem 6.7, is the Mean Value Theorem.

Proof. If $f(x)$ is a differentiable function, then the Mean Value Theorem applies to the function $f(x)$ on each subinterval $[x_{k-1}, x_k]$ and guarantees that there exists some point $x_k^* \in (x_{k-1}, x_k)$ at which

$$f'(x_k^*) = \frac{f(x_k) - f(x_{k-1})}{x_k - x_{k-1}} = \frac{\Delta y_k}{\Delta x},$$

where $\Delta y_k = f(x_k) - f(x_{k-1})$. Therefore, there is a point x_k^* in each subinterval such that the definition of arc length can be written as

$$\lim_{n\to\infty} \sum_{k=1}^{n} \sqrt{1 + \left(\frac{\Delta y_k}{\Delta x}\right)^2}\, \Delta x = \lim_{n\to\infty} \sum_{k=1}^{n} \sqrt{1 + (f'(x_k^*))^2}\, \Delta x.$$

Since the derivative $f'(x)$ is assumed to be continuous, so is the function $\sqrt{1 + (f'(x))^2}$. In addition, $\Delta x = \frac{b-a}{n}$ and $x_k = a + k\,\Delta x$. Therefore this limit of sums represents the definite integral of $\sqrt{1 + (f'(x))^2}$ on the interval $[a, b]$. In other words,

$$\lim_{n\to\infty} \sum_{k=1}^{n} \sqrt{1 + (f'(x_k^*))^2}\, \Delta x = \int_a^b \sqrt{1 + (f'(x))^2}\, dx. \quad \blacksquare$$

Approximating Surface Area

The tools we have developed for calculating arc length can be adapted to define the area of a ***surface of revolution***. Given the graph of a function $f(x)$ on an interval $[a, b]$, we can revolve just the graph–not the area under the graph–around some axis to obtain a surface. As we did for arc length, we can use a piecewise-linear approximation of the graph of $y = f(x)$ and a "subdivide, approximate, and add" strategy to find the area of such a surface.

For example, consider the graph shown next at the left. Revolving this graph around the x-axis produces the surface in the middle figure. One way to approximate the area of this surface is by approximating the graph with two straight line segments and then revolving these line segments around the x-axis to obtain truncated cones, as illustrated in the rightmost figure.

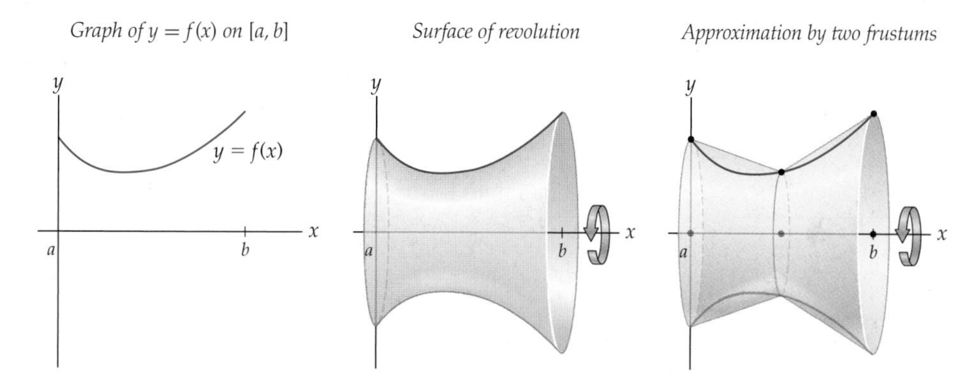

The truncated cones in this example are called ***frustums***. In Exercise 81 you will use the fact that a frustum is obtained by removing the top of a cone to prove that the surface area of a frustum is given by the following formula:

THEOREM 6.8

The Surface Area of a Frustum

A frustum with radii p and q and ***slant length*** s, as shown in the figure, has surface area

$$S = 2\pi rs,$$

where $r = \frac{p+q}{2}$ is the ***average radius*** of the frustum.

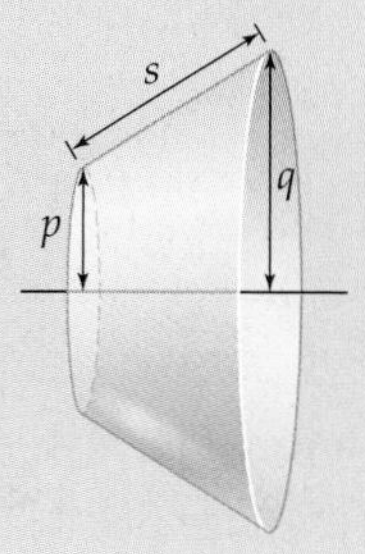

To remember the formula $S = 2\pi rs$, it may help to notice that it is a lot like the formula $S = 2\pi rl$ for the surface area of a cylinder, except that in the case of a frustum the length is measured at a slant and the radius is an average radius.

Now suppose we wish to use frustums to approximate the surface area of the surface obtained by revolving the graph of a function $y = f(x)$ around the x-axis on the interval

$[a, b]$. Just as we did with arc length, we subdivide $[a, b]$ into n subintervals $[x_{k-1}, x_k]$, each of width $\Delta x = \frac{b-a}{n}$, with corresponding vertical changes $\Delta y_k = f(x_k) - f(x_{k-1})$ on each subinterval as shown here:

Notation for the kth frustum

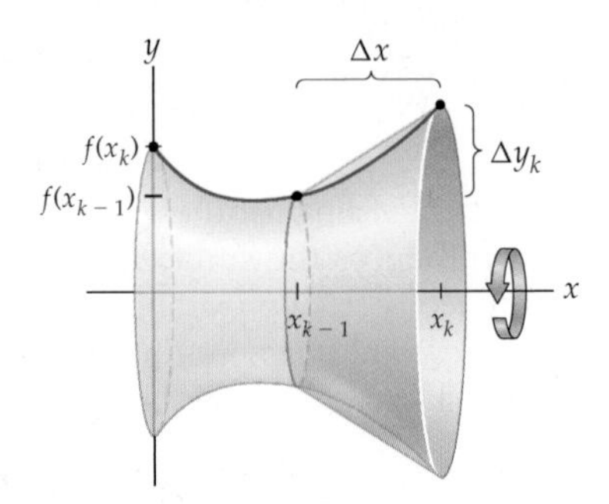

We must find the surface area of each frustum and then add up all those surface areas. Following the same algebra as we used for arc length, we can calculate the slant length of each frustum:

$$s_k = \text{slant length of } k\text{th frustum} = \sqrt{(\Delta x)^2 + (\Delta y_k)^2} = \sqrt{1 + \left(\frac{\Delta y_k}{\Delta x}\right)^2}\,\Delta x.$$

The average radius of each frustum depends on the function $y = f(x)$:

$$r_k = \text{average radius of } k\text{th frustum} = \frac{f(x_{k-1}) + f(x_k)}{2}.$$

Using the formula for the surface area of a frustum and adding up these surface areas, we obtain an approximation for the surface area of the surface of revolution:

$$\text{surface area of } S \approx \sum_{k=1}^{n} 2\pi r_k s_k = \sum_{k=1}^{n} 2\pi \left(\frac{f(x_{k-1}) + f(x_k)}{2}\right)\sqrt{1 + \left(\frac{\Delta y_k}{\Delta x}\right)^2}\,\Delta x.$$

You will calculate such an approximation by hand in Example 5, and just as happened for arc length, we can use the notation we have developed to obtain a definite integral for surface area.

A Definite Integral for Surface Area

We have already done the "legwork" needed to define surface area as a limit of sums of frustums:

DEFINITION 6.9 The Area of a Surface of Revolution

Suppose $f(x)$ is a positive, differentiable function with a continuous derivative. The ***surface area*** of the solid of revolution obtained by revolving $f(x)$ around the x-axis from $x = a$ to $x = b$ is

$$\begin{array}{c}\text{Surface area under } f(x) \\ \text{from } x = a \text{ to } x = b\end{array} = \lim_{n\to\infty} \sum_{k=1}^{n} 2\pi \left(\frac{f(x_{k-1}) + f(x_k)}{2}\right)\sqrt{1 + \left(\frac{\Delta y_k}{\Delta x}\right)^2}\,\Delta x,$$

where $\Delta x = \frac{b-a}{n}$, $x_k = a + k\Delta x$, and $\Delta y_k = f(x_k) - f(x_{k-1})$ for all $k = 0, 1, 2, \ldots, n$.

Intuitively, as $n \to \infty$, the average radius $\frac{f(x_{k-1}) + f(x_k)}{2}$ will be given by the height $f(x)$ of the function and the quotient $\frac{\Delta y_k}{\Delta x}$ will be given by the derivative $f'(x)$. This gives us the following definite integral formula for surface area:

THEOREM 6.10

Surface Area as a Definite Integral

Suppose $f(x)$ is a positive, differentiable function with a continuous derivative. The ***surface area*** of the solid of revolution obtained by revolving $f(x)$ around the x-axis from $x = a$ to $x = b$ is

$$\begin{array}{c}\text{Surface area under } f(x)\\ \text{from } x = a \text{ to } x = b\end{array} = 2\pi \int_a^b f(x)\sqrt{1 + (f'(x))^2}\, dx.$$

Proof. To obtain a definite integral we will attempt to write the expression in Definition 6.9 as a Riemann sum. That means that we need the terms of the sum to be in the form $g(x_k^*)\Delta x$ for some function g and some choice of point x_k^* in each subinterval. This will be a two-step process.

As we saw in the proof of Theorem 6.7, the Mean Value Theorem guarantees that on each subinterval $[x_{k-1}, x_k]$ we can find some point x_k^* at which $f'(x_k^*) = \frac{\Delta y_k}{\Delta x}$.

Now consider the expression $\frac{f(x_{k-1}) + f(x_k)}{2}$ for the average radius. Since f is continuous and this average lies between $f(x_{k-1})$ and $f(x_k)$, the Intermediate Value Theorem guarantees that there is some point x_k^{**} at which $f(x_k^{**})$ is equal to $\frac{f(x_{k-1}) + f(x_k)}{2}$.

Although in general x_k^* and x_k^{**} will not be the same number in $[x_{k-1}, x_k]$, as $n \to \infty$ and thus $\Delta x \to 0$ these two numbers will get closer and closer to each other. Thus we *almost* have a Riemann sum. There are some technical details here that we are omitting, but given that $x_k^* \approx x_k^{**}$ and using the fact that $f(x)$ and $f'(x)$ are assumed to be continuous, we obtain

$$\lim_{n\to\infty} \sum_{k=1}^{n} 2\pi f(x_k^{**})\sqrt{1 + (f'(x_k^*))^2}\, \Delta x = 2\pi \int_a^b f(x)\sqrt{1 + (f'(x))^2}\, dx.$$

■

We now have a way to use a definite integral to calculate the exact surface area of a surface of revolution. However, this method is practical only if the resulting definite integral is one that we know how to solve. In the examples and exercises we will choose functions very carefully so that the definite integrals are solvable with the techniques we have covered in this course.

Examples and Explorations

EXAMPLE 1

Using line segments to approximate the length of a curve

Approximate the length of the curve $f(x) = \sqrt{4 - x^2}$ from $x = -2$ to $x = 2$ with four line segments. Then find a better approximation that uses eight line segments.

SOLUTION

We begin by dividing the interval $[-2, 2]$ into four subintervals of equal size: $[-2, -1]$, $[-1, 0]$, $[0, 1]$, and $[1, 2]$. The four line segments in question are shown next at the left.

Four-segment approximation

Eight-segment approximation

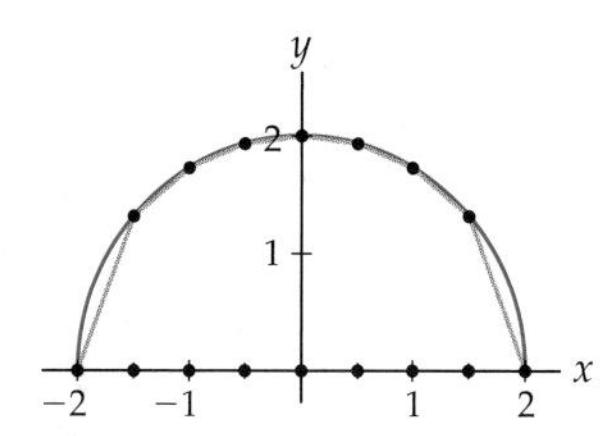

It is easy to calculate the lengths of these line segments with the distance formula; for example, the first line segment starts at the point $(-2, f(-2)) = (-2, 0)$ and ends at the point $(-1, f(-1)) = (-1, \sqrt{3})$ and therefore has length

$$\sqrt{(-1-(-2))^2 + (\sqrt{3}-0)^2} = \sqrt{1+3} = 2.$$

Similarly, by using the coordinates of $(-1, f(-1))$, $(0, f(0))$, $(1, f(1))$, and $(2, f(2))$ we can calculate that the lengths of the remaining three line segments are 1.035, 1.035, and 2. The sum of these four lengths approximates the length of the curve:

$$2 + 1.035 + 1.035 + 2 = 6.070.$$

Clearly this approximation should be an under-approximation of the actual length, since the lengths of the lines are shorter than the lengths of the curves they approximate. Because the curve in question is half of the circumference of a circle of radius 2, the length of the curve should be $\frac{1}{2}(2\pi(2)) = 2\pi \approx 6.28319$. Our four-segment approximation is an under-approximation by about 0.21263 unit.

To get a better approximation with the eight line segments shown earlier at the right, we need to make eight distance formula calculations. The first line segment extends from $(-2, 0)$ to $(-1.5, \sqrt{1.75})$ and thus has length

$$\sqrt{(-1.5-(-2))^2 + (\sqrt{1.75}-0)^2} = \sqrt{0.25+1.75} = \sqrt{2} \approx 1.41421.$$

The remaining seven line segments can be calculated similarly, and the sum of the eight lengths gives us an eight-segment approximation for the arc length of the curve:

$$1.414 + 0.646 + 0.540 + 0.504 + 0.504 + 0.540 + 0.646 + 1.414 \approx 6.208.$$

Note that the eight-segment approximation is indeed closer to the length we would expect from the circumference formula and in fact under-approximates the exact value of $2\pi \approx 6.28319$ by only about 0.07519 unit. □

EXAMPLE 2

Approximating arc length in general notation

Approximate the length of one period of the function $f(x) = \sin x$ by using the formula and notation in Definition 6.6, with $n = 4$. For each k, describe x_k, y_k, and Δy_k.

SOLUTION

We will approximate the arc length of $f(x) = \sin x$ from $x = 0$ to $x = 2\pi$. Since we are using $n = 4$ line segments, as shown in the following figure, we have $\Delta x = \frac{\pi}{2}$:

Arc length of $f(x) = \sin x$ on $[0, 2\pi]$ approximated by four line segments

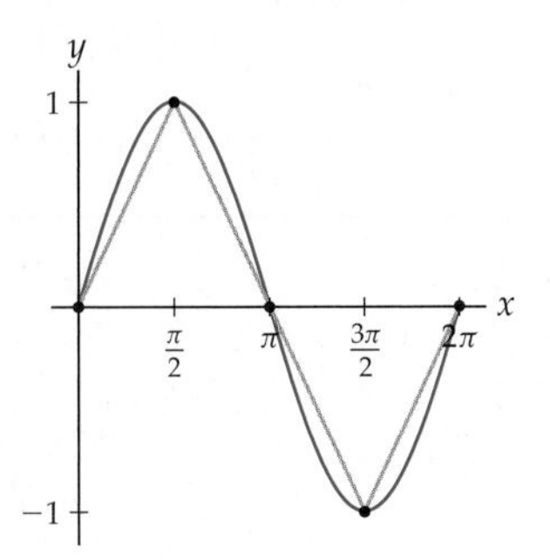

We could use the distance formula to find the length of each of these line segments without bothering with notation at all, as in the previous example. However, the purpose of this example is to practice using the notation from Definition 6.6. We have

$$x_0 = 0, \quad x_1 = \frac{\pi}{2}, \quad x_2 = \pi, \quad x_3 = \frac{3\pi}{2}, \quad \text{and} \quad x_4 = 2\pi.$$

Evaluating the function $f(x) = \sin x$ at these points gives

$$f(x_0) = f(0) = \sin 0 = 0,$$
$$f(x_1) = f\left(\frac{\pi}{2}\right) = \sin\frac{\pi}{2} = 1,$$
$$f(x_2) = f(\pi) = \sin\pi = 0,$$
$$f(x_3) = f\left(\frac{3\pi}{2}\right) = \sin\frac{3\pi}{2} = -1, \text{ and}$$
$$f(x_4) = f(2\pi) = \sin 2\pi = 0.$$

Although $\Delta x = \frac{\pi}{2}$ is the same on each subinterval, the changes in the y-values vary:

$$\Delta y_1 = f(x_1) - f(x_0) = 1 - 0 = 1,$$
$$\Delta y_2 = f(x_2) - f(x_1) = 0 - 1 = -1,$$
$$\Delta y_3 = f(x_3) - f(x_2) = -1 - 0 = -1, \text{ and}$$
$$\Delta y_4 = f(x_4) - f(x_3) = 0 - (-1) = 1.$$

In the notation of Definition 6.6, the lengths of the four line segments are thus

$$\sqrt{1 + \left(\frac{\Delta y_1}{\Delta x}\right)^2}\,\Delta x = \sqrt{1 + \left(\frac{1}{\pi/2}\right)^2}\left(\frac{\pi}{2}\right) \approx 1.8621,$$

$$\sqrt{1 + \left(\frac{\Delta y_2}{\Delta x}\right)^2}\,\Delta x = \sqrt{1 + \left(\frac{-1}{\pi/2}\right)^2}\left(\frac{\pi}{2}\right) \approx 1.8621,$$

$$\sqrt{1 + \left(\frac{\Delta y_3}{\Delta x}\right)^2}\,\Delta x = \sqrt{1 + \left(\frac{-1}{\pi/2}\right)^2}\left(\frac{\pi}{2}\right) \approx 1.8621, \text{ and}$$

$$\sqrt{1 + \left(\frac{\Delta y_4}{\Delta x}\right)^2}\,\Delta x = \sqrt{1 + \left(\frac{1}{\pi/2}\right)^2}\left(\frac{\pi}{2}\right) \approx 1.8621.$$

In this example, all four of the line segments happen to be the same length; of course that is not usually the case! The reason it is here is that $\sin x$ just happens to have the same shape of arc on each of the four pieces we are examining. Adding up all of these lengths, we get the following approximation for the length of one period of the sine function:

$$\sum_{k=1}^{4} \sqrt{1 + \left(\frac{\Delta y_k}{\Delta x}\right)^2}\,\Delta x \approx 1.8621 + 1.8621 + 1.8621 + 1.8621 \approx 7.4484.$$

□

✔ CHECKING THE ANSWER

The answer of 7.4484 we just found does seem roughly reasonable for the length of one period of the sine function, considering its graph. To get a better check on our approximation, we could compute the definite integral

$$\int_0^{2\pi} \sqrt{1 + (f'(x))^2}\,dx = \int_0^{2\pi} \sqrt{1 + \left(\frac{d}{dx}(\sin x)\right)^2}\,dx = \int_0^{2\pi} \sqrt{1 + \cos^2 x}\,dx.$$

Unfortunately, we do not know how to calculate this integral. However, it represents an area, and we can use a graphing calculator to approximate area. The area in question is shown in the following figure:

The area under the graph of
$\sqrt{1+(f'(x))^2} = \sqrt{1+\cos^2 x}$ *on* $[0, 2\pi]$

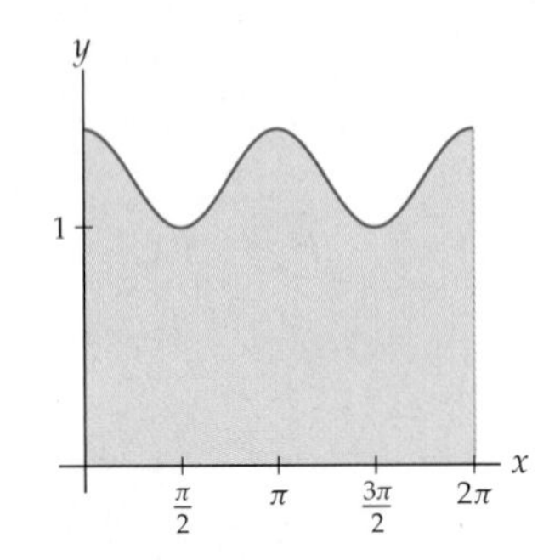

This area is *not* the area under the graph of $f(x) = \sin x$. Rather, it is the area under the graph of $\sqrt{1+(f'(x))^2} = \sqrt{1+\cos^2 x}$ from $x = 0$ to $x = 2\pi$. Using a graphing calculator, we can approximate this area as 7.6404; therefore the arc length of $f(x) = \sin x$ from $x = 0$ to $x = 2\pi$ is approximately 7.6404. This verifies that 7.4484 was indeed a reasonable approximation of the arc length of one period of $f(x) = \sin x$.

EXAMPLE 3

Using a definite integral to calculate exact arc length

Construct and solve a definite integral to find the exact length of the curve traced out by $f(x) = x^{3/2}$ on the interval $[0, 3]$.

SOLUTION

By the arc length formula in Theorem 6.7, the arc length of $f(x) = x^{3/2}$ on $[0, 3]$ is given by

$$\int_0^3 \sqrt{1+\left(\frac{d}{dx}(x^{3/2})\right)^2}\,dx = \int_0^3 \sqrt{1+\left(\frac{3}{2}x^{1/2}\right)^2}\,dx = \int_0^3 \sqrt{1+\frac{9}{4}x}\,dx.$$

Applying the simple substitution $u = 1 + \frac{9}{4}x$, $du = \frac{9}{4}\,dx$, we can write this definite integral as

$$\frac{4}{9}\int_{x=0}^{x=3} \sqrt{u}\,du = \frac{4}{9}\left[\frac{2}{3}u^{3/2}\right]_{x=0}^{x=3} = \frac{4}{9}\left[\frac{2}{3}\left(1+\frac{9}{4}x\right)^{3/2}\right]_{x=0}^{x=3}$$
$$= \frac{8}{27}\left(\left(1+\frac{9}{4}(3)\right)^{3/2} - \left(1+\frac{9}{4}(0)\right)^{3/2}\right)$$
$$= \frac{8}{27}\left(\left(\frac{31}{4}\right)^{3/2} - 1\right) \approx 6.096.$$

□

The answer we just found appears reasonable if we consider a rough estimate of the length of the curve traced out by the graph of $f(x) = x^{3/2}$ from $x = 0$ to $x = 3$, as shown in the next figure. By the distance formula, the diagonal line from $(0, 0)$ to $(3, 5)$ has length $\sqrt{3^2 + 5^2} \approx 5.83$, just a little less than our approximation of 6.096 for the length of the curve.

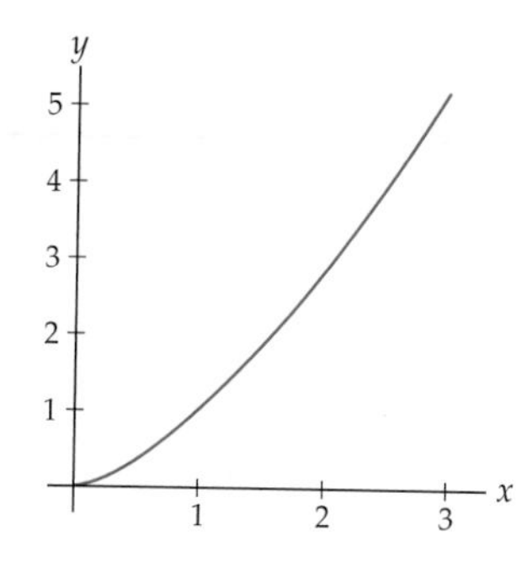

EXAMPLE 4 **Recovering the formula for the circumference of a circle**

Construct and solve a definite integral to find the circumference of a circle of radius 2.

SOLUTION

As we have already discussed in this section, to calculate the circumference of a circle of radius 2 we can instead calculate the arc length of $f(x) = \sqrt{4-x^2}$ on the interval $[-2, 2]$ and then multiply that answer by 2. Using the definite integral formula for arc length, we have

$$\int_{-2}^{2} \sqrt{1+(f'(x))^2}\,dx = \int_{-2}^{2} \sqrt{1+\left(\frac{d}{dx}(\sqrt{4-x^2})\right)^2}\,dx \quad \leftarrow \text{definition of } f(x)$$

$$= \int_{-2}^{2} \sqrt{1+\left(\frac{1}{2}(4-x^2)^{-1/2}(-2x)\right)^2}\,dx \quad \leftarrow \text{differentiate } f(x)$$

$$= \int_{-2}^{2} \sqrt{1+\frac{x^2}{4-x^2}}\,dx = \int_{-2}^{2} \sqrt{\frac{(4-x^2)+x^2}{4-x^2}}\,dx \quad \leftarrow \text{algebra}$$

$$= \int_{-2}^{2} \sqrt{\frac{4}{4-x^2}}\,dx = 2\int_{-2}^{2} \frac{1}{\sqrt{4-x^2}}\,dx. \quad \leftarrow \text{algebra}$$

Fortunately, $\frac{1}{\sqrt{4-x^2}}$ is a function whose integral we can find by using either algebra and the antiderivative of the arcsine (Sections 2.6 and 4.4) or trigonometric substitution (Section 5.5). With those techniques we find that

$$\int \frac{1}{\sqrt{4-x^2}}\,dx = \sin^{-1}\frac{x}{2} + C.$$

Less fortunately, since $\frac{1}{\sqrt{4-x^2}}$ is not defined at $x = \pm 2$, the definite integral in question is actually an improper integral. To solve it we will have to split the integral and consider limits at the endpoints $x = \pm 2$ (see Section 5.6):

$$2\int_{-2}^{2} \frac{1}{\sqrt{4-x^2}}\,dx = 2\left(\lim_{A\to -2^+} \int_{A}^{0} \frac{1}{\sqrt{4-x^2}}\,dx + \lim_{B\to 2^-} \int_{0}^{B} \frac{1}{\sqrt{4-x^2}}\,dx\right)$$

$$= 2\left(\lim_{A\to -2^+} \left[\sin^{-1}\frac{x}{2}\right]_A^0 + \lim_{B\to 2^-} \left[\sin^{-1}\frac{x}{2}\right]_0^B\right)$$

$$= 2\left(\lim_{A\to -2^+} \left(\sin^{-1} 0 - \sin^{-1}\frac{A}{2}\right) + \lim_{B\to 2^-} \left(\sin^{-1}\frac{B}{2} - \sin^{-1} 0\right)\right).$$

Since $\sin^{-1} 0 = 0$, $\sin^{-1} 1 = \frac{\pi}{2}$, and $\sin^{-1} -1 = -\frac{\pi}{2}$, we have

$$2\int_{-2}^{2} \frac{1}{\sqrt{4-x^2}}\,dx = 2\left(\left(0 - \left(-\frac{\pi}{2}\right)\right) + \left(\frac{\pi}{2} - 0\right)\right) = 2\pi.$$

Therefore, the length of the graph of $f(x) = \sqrt{4 - x^2}$ from $x = -2$ to $x = 2$ is 2π units. This arc length is *half* of the arc length of the circumference of a circle with radius 2, so the circumference of a circle of radius 2 is $2(2\pi) = 4\pi$ units. This is exactly the answer we expected, since, by the formula for the circumference of a circle, we should have $C = 2\pi r = 2\pi(2) = 4\pi$. □

TECHNICAL POINT Although it may seem as if the preceding example could be used to re-derive the formula for the circumference of a circle, note that solving the definite integral involved the use of inverse trigonometric functions, which are defined from trigonometric functions, which are in turn defined with the use of the unit circle and the fact that the unit circle has circumference 2π.

EXAMPLE 5

Approximating the surface area of a sphere with frustums

Use four frustums to approximate the surface area of the sphere of radius 2 centered at the origin. This sphere can be realized by rotating the curve $f(x) = \sqrt{4 - x^2}$ from $x = -2$ to $x = 2$ around the x-axis.

SOLUTION

Subdividing the interval $[-2, 2]$ into four subintervals and then rotating a corresponding piecewise-linear approximation of $y = \sqrt{4 - x^2}$ around the x-axis yields four frustums (two are actually cones):

Approximating a sphere with four frustums

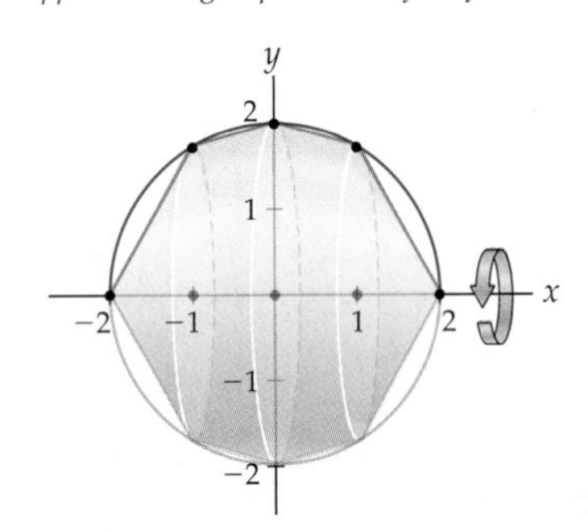

In our subdivision notation we have $x_0 = -2$, $x_1 = -1$, $x_2 = 0$, $x_3 = 1$, and $x_4 = 2$, with $\Delta x = 1$. We can treat the cones at the end as cones or as frustums; either surface area formula will work. For simplicity we will treat them as frustums.

Since $f(-2) = \sqrt{4 - (-2)^2} = 0$ and $f(-1) = \sqrt{4 - (-1)^2} = \sqrt{3}$, the first frustum (a cone) has average radius and slant length given by

$$r_1 = \frac{f(-2) + f(-1)}{2} = \frac{\sqrt{3}}{2},$$

$$s_1 = \sqrt{(-2 - (-1))^2 + (f(-2) - f(-1))^2} = \sqrt{1 + 3} = 2.$$

Therefore the surface area of the first frustum is $2\pi r_1 s_1 = 2\pi\left(\frac{\sqrt{3}}{2}\right)(2)$.

In a similar fashion we can find the volumes of the remaining three frustums to obtain the following approximation for the surface area S of a sphere:

$$\begin{aligned} S &= 2\pi r_1 s_1 + 2\pi r_2 s_2 + 2\pi r_3 s_3 + 2\pi r_4 s_4 \\ &= 2\pi\left(\frac{\sqrt{3}}{2}\right)(2) + 2\pi\left(\frac{\sqrt{3}+2}{2}\right)\left(\sqrt{8-4\sqrt{3}}\right) \\ &\quad + 2\pi\left(\frac{\sqrt{3}+2}{2}\right)\left(\sqrt{8-4\sqrt{3}}\right) + 2\pi\left(\frac{\sqrt{3}}{2}\right)(2) \\ &\approx 46.04. \end{aligned}$$

□

CHECKING THE ANSWER

In this case we happen to know the exact value of the surface area of a sphere of radius 2, since the surface area of a sphere of radius r is given by the formula $S = 4\pi r^2$. Thus the exact answer to the preceding problem is $4\pi(2)^2 = 16\pi \approx 50.265$.

EXAMPLE 6

Using a definite integral to calculate exact surface area

Construct and solve a definite integral to find the exact surface area of the surface of revolution obtained by revolving the graph of $f(x) = x^3$ around the x-axis on $[0, 1]$.

SOLUTION

Since the derivative of $f(x) = x^3$ is $f'(x) = 3x^2$, the surface area formula in Theorem 6.10 gives us the following definite integral to solve:

$$2\pi \int_0^1 x^3\sqrt{1+9x^4}\,dx.$$

This integral is simple to solve by using substitution, with $u = 1+9x^4$ and $du = 36x^3$. With this substitution the integral becomes

$$\begin{aligned} 2\pi\left(\frac{1}{36}\right)\int_{x=0}^{x=1} \sqrt{u}\,du &= \frac{\pi}{18}\left[\frac{2}{3}u^{3/2}\right]_{x=0}^{x=1} \\ &= \frac{\pi}{18}\left(\frac{2}{3}(1+9(1)^4)^{3/2} - \frac{2}{3}(1+9(0)^4)^{3/2}\right) \\ &= \frac{\pi}{18}\left(\frac{2}{3}\sqrt{1000} - \frac{2}{3}\right). \end{aligned}$$

This number is approximately equal to 3.56, which is a reasonable size for the surface area obtained by revolving $f(x) = x^3$ around the x-axis on $[0, 1]$.

□

? TEST YOUR UNDERSTANDING

- In the four-segment approximation of the arc length of $f(x) = \sqrt{4-x^2}$ on $[-2, 2]$, what is it about the graph of the function that causes some line segments to be longer than others, even though we are dividing the interval $[-2, 2]$ into equal pieces?
- Why would you expect an arc length approximation with six segments to be a better approximation than one with three segments?
- What is the difference between Definition 6.6 and Theorem 6.7? What needs to be proved to show Theorem 6.7?
- What parts of the construction of the definite integral for surface area are generalized from the construction of the definite integral for arc length?
- How are the Mean Value Theorem and the Intermediate Value Theorem involved in the proof of Theorem 6.10?

EXERCISES 6.3

Thinking Back

Approximating area with rectangles: Approximate the area represented by each of the following definite integrals by using a right sum with $n = 4$ rectangles.

▶ $\int_0^2 (\sqrt{x} + 1)\, dx$

▶ $\int_3^4 \frac{1}{x-2}\, dx$

Riemann sums: For each given definite integral, write the integral as a Riemann sum of the form $\lim_{n\to\infty} \sum_{k=1}^{n} f(x_k^*)\, \Delta x$, using a right sum (i.e., using the rightmost point of each subinterval as x_k^*).

▶ $\int_1^e \ln x\, dx$

▶ $\int_{-\pi}^{\pi} \sin x\, dx$

Review of integration techniques: Solve each of the following definite integrals.

▶ $\int_0^4 \sqrt{3 + 16x}\, dx$

▶ $\int_{-2}^{2} \sqrt{26}\, dx$

▶ $\int_0^3 \frac{3}{\sqrt{9 - x^2}}\, dx$

▶ $\int_{\pi/4}^{\pi/2} \csc x\, dx$

Approximating definite integrals: Use a calculator or computer algebra system to approximate the values of the following definite integrals.

▶ $\int_{-1}^{3} \sqrt{1 + e^{2x}}\, dx$

▶ $\int_0^{2\pi} \sqrt{1 + \sin^2 x}\, dx$

Concepts

0. *Problem Zero:* Read the section and make your own summary of the material.

1. *True/False:* Determine whether each of the statements that follow is true or false. If a statement is true, explain why. If a statement is false, provide a counterexample.

(a) *True or False:* If we subdivide an interval $[a, b]$ into n subintervals, then each subinterval has width $\Delta x = \frac{b-a}{n}$.

(b) *True or False:* If we subdivide an interval $[a, b]$ into n subintervals at points $x_0 = a, x_1, \dots, x_{n-1}, x_n = b$, then the kth subdivision point x_k is given by $a + k$.

(c) *True or False:* The distance formula says that if $\Delta x = x_k - x_{k-1}$ and $\Delta y_k = f(x_k) - f(x_{k-1})$, then the distance from $(x_{k-1}, f(x_{k-1}))$ to $(x_k, f(x_k))$ is given by $\sqrt{(\Delta x)^2 + (\Delta y_k)^2}$.

(d) *True or False:* For any real numbers a and b, $\sqrt{a^2 + b^2} = a^2\sqrt{1 + \left(\frac{b}{a}\right)^2}$.

(e) *True or False:* We cannot use the definite integral formula for arc length for a function $f(x)$ on $[a, b]$ if $f(x)$ has a discontinuity on an interval $[a, b]$.

(f) *True or False:* An approximation of arc length by line segments can sometimes be an over-approximation.

(g) *True or False:* An approximation of surface areas by frustums can sometimes be an over-approximation.

(h) *True or False:* The area of a frustum can be expressed in terms of average radius and slant length.

2. *Examples:* Construct examples of the thing(s) described in the following. Try to find examples that are different than any in the reading.

(a) A graph of a function $f(x)$ on $[a, b]$ whose arc length is poorly approximated by four line segments.

(b) An equation of a function $f(x)$ on $[a, b]$ that gives rise to an arc length integral that we do not know how to solve.

(c) An equation of a surface of revolution defined by revolving a function $f(x)$ on $[a, b]$ around the x-axis that gives rise to a surface area integral that we do not know how to solve.

3. What is the definition of the arc length of a continuous function $f(x)$ on an interval $[a, b]$? Explain this definition in words in terms of a limit of sums, and explain the meaning of the notation used in the definition.

4. Suppose $f(x)$ is a differentiable function with a continuous derivative. Describe the arc length of $f(x)$ on the interval $[a, b]$ in terms of a definite integral. Why is this definite integral equal to the definition of arc length for $f(x)$ on $[a, b]$?

5. How is the Mean Value Theorem involved in proving that the arc length of a function on an interval can be represented by a definite integral?

6. Fill in the blank: The length of the graph of $y = x^2$ on $[2, 4]$ is equal to the area under the graph of the function ____________ on the same interval.

7. Fill in the blank: The area under the graph of $y = \sqrt{1 + e^{2x}}$ on $[0, 2]$ is equal to the length of the graph of the function ____________ on the same interval.

8. In this exercise you will investigate whether approximating arc length with line segments results in an under-approximation or an over-approximation.

(a) Sketch graphs of the functions $f(x) = 4 - x^2$, $g(x) = e^x$, and $h(x) = 3x + 1$ on $[-2, 2]$.

(b) Suppose you wanted to approximate the arc length of each of these functions on $[-2, 2]$ by using eight line segments. Draw these line segments on each graph.

(c) Determine graphically whether each approximation is greater than, less than, or equal to the actual arc length for each function listed in part (a).

(d) Explain why approximating the length of a curve with line segments will *always* yield either an under-approximation or an exact answer.

9. Suppose you want to approximate the arc length of $f(x) = \tan x$ on the interval $[0, \pi]$. Would you get a good approximation if you subdivided the interval $[0, \pi]$ into five subintervals and added up the lengths of the line segments over each subinterval? Why or why not? *(Hint: Draw the line segments.)*

10. Use definite integrals to show that the graphs of $f(x) = \frac{1}{x}$ and $g(x) = -\frac{1}{x}$ have the same arc length on $[1, 3]$. (*Hint: You do not have to solve the integrals!*) Why does this equality make sense in terms of graphical transformations?

11. Use definite integrals to show that $f(x) = x^2 - 3$ and $g(x) = 5 - x^2$ have the same arc length on $[0, 2]$. (*Hint: You do not have to solve the integrals!*) Why does this equality make sense in terms of transformations?

12. Do you think that the function $f(x) = 2x^2$ has twice the arc length of $g(x) = x^2$ on the interval $[0, 3]$? Why or why not?

13. Write down a definite integral that describes the circumference of a circle of radius 5. Don't try to solve the integral; just write it down.

14. Write down a definite integral that describes the arc length around an ellipse with horizontal radius of 3 units and vertical radius of 2 units. The ellipse in this exercise extends 6 units across and 4 units vertically and has equation $\frac{x^2}{9} + \frac{y^2}{4} = 1$. Don't try to solve the integral; just write it down.

15. In this exercise you will approximate the arc length of $f(x) = \cos x$ on $[0, \pi]$ in two ways and compare your answers.

(a) By using four line segments and the distance formula.

(b) By using a right sum with four rectangles to approximate the area under the graph of the function $y = \sqrt{1 + \sin^2 x}$ on $[0, 2]$.

(c) Why do the approximations you found in parts (a) and (b) both approximate the arc length of $f(x) = \cos x$ on $[0, \pi]$? Which, if either, do you think might be a better approximation?

16. In this exercise you will approximate the arc length of $f(x) = x^2$ on $[0, 2]$ in two ways and compare your answers.

(a) By using four line segments and the distance formula.

(b) By using a right sum with four rectangles to approximate the area under the graph of the function $y = \sqrt{1 + 4x^2}$ on $[0, 2]$.

(c) Why do the approximations you found in parts (a) and (b) both approximate the arc length of $f(x) = x^2$ on $[0, 2]$? Which, if either, do you think might be a better approximation?

17. Suppose you wish to use n frustums to approximate the area of the surface obtained by revolving the graph of a function $y = f(x)$ around the x-axis on $[a, b]$. Use a labeled graph to explain why the slant length s_k of the kth frustum is given by

$$s_k = \sqrt{(\Delta x)^2 + (\Delta y_k)^2}.$$

18. Suppose you wish to use n frustums to approximate the area of the surface obtained by revolving the graph of a function $y = f(x)$ around the x-axis on $[a, b]$. Use a labeled graph to explain why the average radius r_k of the kth frustum is given by

$$r_k = \frac{f(x_{k-1}) + f(x)}{2}.$$

Skills

In Exercises 19–24, approximate the arc length of $f(x)$ on $[a, b]$, using n line segments and the distance formula. Include a sketch of $f(x)$ and the line segments.

19. $f(x) = x^3, \quad [a, b] = [-2, 2], \quad n = 4$

20. $f(x) = \cos x, \quad [a, b] = [0, \pi], \quad n = 3$

21. $f(x) = \ln x, \quad [a, b] = [1, 3], \quad n = 4$

22. $f(x) = e^x, \quad [a, b] = [-1, 3], \quad n = 4$

23. $f(x) = \sqrt{9 - x^2}, \quad [a, b] = [-3, 3], \quad n = 3$

24. $f(x) = \sqrt{9 - x^2}, \quad [a, b] = [-3, 3], \quad n = 6$

In Exercises 25–30, approximate the arc length of $f(x)$ on $[a, b]$, using the approximation $\sum_{k=1}^{n} \sqrt{1 + \left(\frac{\Delta y_k}{\Delta x}\right)^2}\,\Delta x$ with the given value of n. In each problem list the values of Δy_k for $k = 1, 2, \ldots, n$.

25. $f(x) = x^3, \quad [a, b] = [-2, 2], \quad n = 4$

26. $f(x) = \cos x, \quad [a, b] = [0, \pi], \quad n = 3$

27. $f(x) = \ln x, \quad [a, b] = [1, 3], \quad n = 4$

28. $f(x) = e^x, \quad [a, b] = [-1, 3], \quad n = 4$

29. $f(x) = \sqrt{9 - x^2}, \quad [a, b] = [-3, 3], \quad n = 3$

30. $f(x) = \sqrt{9 - x^2}, \quad [a, b] = [-3, 3], \quad n = 6$

Find the *exact* value of the arc length of each function $f(x)$ in Exercises 31–46 on $[a, b]$ by writing the arc length as a definite integral and then solving that integral.

31. $f(x) = 3x + 1, \quad [a, b] = [-1, 4]$

32. $f(x) = 4 - x, \quad [a, b] = [2, 5]$

33. $f(x) = 2x^{3/2} + 1, \quad [a, b] = [1, 3]$

34. $f(x) = x^{3/2}, \quad [a, b] = [0, 2]$

35. $f(x) = (2x + 3)^{3/2}, \quad [a, b] = [-1, 1]$

36. $f(x) = 2(1 - x)^{3/2} + 3, \quad [a, b] = [-2, 0]$

37. $f(x) = \sqrt{9 - x^2}, \quad [a, b] = [-3, 3]$

38. $f(x) = \sqrt{1 - x^2}, \quad [a, b] = [-1, 1]$

39. $f(x) = \frac{1}{3}x^{3/2} - x^{1/2}, \quad [a, b] = [0, 1]$

40. $f(x) = (1 - x^{2/3})^{3/2}, \quad [a, b] = [0, 1]$

41. $f(x) = (4 - x^{2/3})^{3/2}, \quad [a, b] = [0, 2]$

42. $f(x) = x^2, \quad [a, b] = [-1, 1]$

43. $f(x) = x^2 - \frac{1}{8}\ln x, \quad [a, b] = [1, 2]$

44. $f(x) = \ln(\cos x), \quad [a, b] = \left[0, \frac{\pi}{4}\right]$

45. $f(x) = \ln(\sin x), \quad [a, b] = \left[\frac{\pi}{4}, \frac{3\pi}{4}\right]$

46. $f(x) = \frac{x^4 + 3}{6x}, \quad [a, b] = [1, 3]$

Each definite integral in Exercises 47–52 represents the arc length of a function $f(x)$ on an interval $[a, b]$. Determine the function and interval.

47. $\int_{-2}^{5} \sqrt{1 + 9e^{6x}}\,dx$

48. $\int_{0}^{\pi} \sqrt{1 + \sin^2 x}\,dx$

49. $\int_{2}^{3} \sqrt{\frac{x^4 + 1}{x^4}}\,dx$

50. $\int_{0}^{2} \sqrt{1 + 36x}\,dx$

51. $\int_0^{\pi/4} \sec x \, dx$

52. $\int_0^1 \sqrt{1+9x^4} \, dx$

You may have noticed that even very simple functions give rise to arc length integrals that we have no idea how to compute. In Exercises 53–56, use a graphing calculator to approximate a definite integral that represents the arc length of the given function $f(x)$ on the interval $[a, b]$.

53. $f(x) = x^3, \quad [a, b] = [-1, 1]$

54. $f(x) = x^2 + 1, \quad [a, b] = [0, 4]$

55. $f(x) = 3x^2 - 1, \quad [a, b] = [1, 2]$

56. $f(x) = \ln x, \quad [a, b] = [1, 3]$

In Exercises 57–62, use n frustums to approximate the area of the surface of revolution obtained by revolving the curve $y = f(x)$ around the x-axis on the interval $[a, b]$.

57. $f(x) = x^2, \quad [a, b] = [0, 4], \quad n = 2$

58. $f(x) = x^2, \quad [a, b] = [0, 4], \quad n = 4$

59. $f(x) = x^3, \quad [a, b] = [1, 3], \quad n = 4$

60. $f(x) = \ln x, \quad [a, b] = [1, 4] \quad n = 3$

61. $f(x) = \sin x, \quad [a, b] = [0, \pi], \quad n = 2$

62. $f(x) = \sin x, \quad [a, b] = [0, \pi], \quad n = 4$

In Exercises 63–72, set up and solve a definite integral to find the exact area of each surface of revolution obtained by revolving the curve $y = f(x)$ around the x-axis on the interval $[a, b]$.

63. $f(x) = x, \quad [a, b] = [0, 2]$

64. $f(x) = \frac{1}{x}, \quad [a, b] = [1, 10]$

65. $f(x) = \sqrt{x}, [a, b] = [0, 4]$

66. $f(x) = \sqrt{25 - x^2}, \quad [a, b] = [-5, 5]$

67. $f(x) = \sqrt{3x + 1}, \quad [a, b] = [0, 3]$

68. $f(x) = e^{-x}, \quad [a, b] = [-1, 1]$

69. $f(x) = e^{4x}, \quad [a, b] = [-1, 1]$

70. $f(x) = x^2, \quad [a, b] = [-1, 1]$

71. $f(x) = \sin x, \quad [a, b] = [0, \pi]$

72. $f(x) = \cos x, \quad [a, b] = \left[-\frac{\pi}{2}, \frac{\pi}{2}\right]$

Applications

73. Suppose a drinking fountain expels water so that it falls in the shape of the parabola $f(x) = 3 - \frac{1}{2}x^2$, measured in inches, as shown in the given figure. Set up and solve a definite integral that measures the distance that water travels starting from when it leaves the fountain (at $x = 0$) and ending when it hits the surface of the fountain (when $f(x) = 0$).

Path of water from a drinking fountain

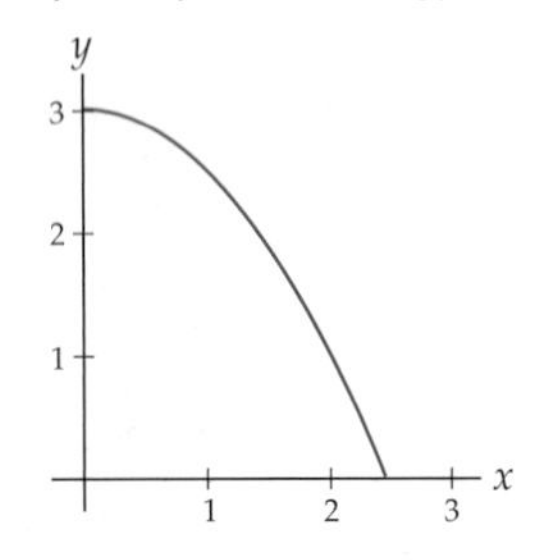

74. Dr. Geek built a skate ramp in his backyard modeled after the function $f(x) = 11 \sin\left(\frac{1}{42}x^2\right)$, measured in feet, as shown in the next figure. Set up a definite integral that measures the distance required to skate from one side of the ramp to the other. The starting and ending x-values should both satisfy $f(x) = 11$ feet. Use a graphing calculator to approximate the value of this definite integral.

Curve of skate ramp

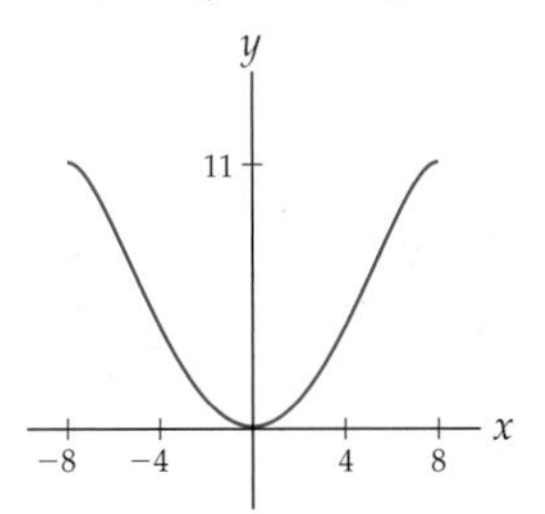

Proofs

75. Suppose $f(x)$ is a continuous function on an interval $[a, b]$, and let n be a positive integer. With the notation Δx, x_k, and Δy_k given in Definition 6.6, prove that the sum $\sum_{k=1}^{n} \sqrt{(x_k - x_{k-1})^2 + (f(x_k) - f(x_{k-1}))^2}$ is equal to the sum $\sum_{k=1}^{n} \sqrt{1 + \left(\frac{\Delta y_k}{\Delta x}\right)^2} \, \Delta x$.

76. Use Definition 6.6, the Mean Value Theorem, and the definition of the definite integral to prove Theorem 6.7: The arc length of a sufficiently well behaved function $f(x)$ on an interval $[a, b]$ can be represented by the definite integral $\int_a^b \sqrt{1 + (f'(x))^2} \, dx$.

77. Prove, in two ways, that the arc length of a linear function $f(x) = mx + c$ on an interval $[a, b]$ is equal to $(b - a)\sqrt{1 + m^2}$:
(a) by using the distance formula;
(b) by using Theorem 6.7.

78. Use Theorem 6.7 to prove that a circle of radius 5 has circumference 10π.

79. Use Theorem 6.7 to prove that a circle of radius r has circumference $2\pi r$.

80. Prove that if f is continuous on $[a, b]$ and C is any real number, then $f(x)$ and $f(x) + C$ have the same arc length on $[a, b]$. Then explain why this makes sense graphically.

81. In this exercise you will prove that the surface area of a frustum with radii p and q and slant length s is equal to $SA = 2\pi rs$, where $r = \frac{p+q}{2}$ is the average radius of the frustum. The argument will hinge on the fact that a frustum is a cone with its top removed, as shown here:

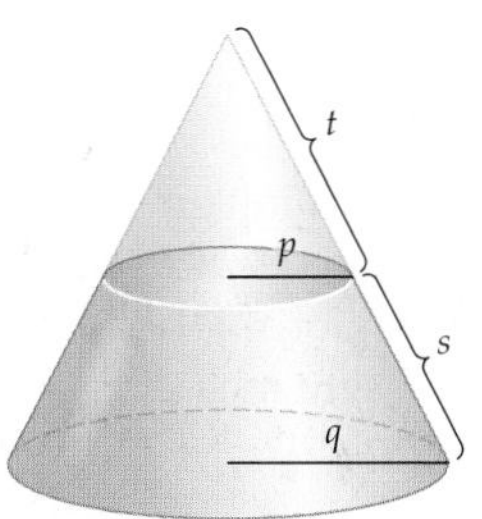

(a) Use similar triangles to prove that, in the notation of the given diagram, $pt + ps = qt$.

(b) We know from Theorem 3.12 that the surface area of a cone with radius A and height B is given by $\pi A\sqrt{A^2 + B^2}$. Use the Pythagorean theorem to prove that this expression implies that the surface area of a cone with radius A and slant height C is given by πAC.

(c) Express the surface area of the frustum in the figure as the difference of the area of the larger cone and the area of the smaller cone, and use the previous two parts of this problem to prove that the surface area of the frustum is $SA = \pi(p + q)s$.

(d) Finally, use the relationship $r = \frac{p+q}{2}$ to prove that the surface area of the frustum is $S = 2\pi rs$.

82. Use Theorem 6.10 to prove that a sphere of radius 5 has surface area 100π.

83. Use Theorem 6.10 to prove that a sphere of radius r has surface area $4\pi r^2$.

84. Prove the surprising fact that the arc length of the catenary curve traced out by the hyperbolic function $f(x) = \cosh x$ on any interval $[a, b]$ is equal to the area under the same graph on $[a, b]$.

Thinking Forward

Arc length of parametric curves: A curve in the plane is defined by *parametric equations* if we can express the points on the curve in the form $(x(t), y(t))$ for some pair of continuous functions $x(t)$ and $y(t)$. You should think of t as a *parameter* and $x(t)$ and $y(t)$ as tracing out the x- and y-coordinates of the curve as t changes. A classic example of a parametric curve is $(\cos t, \sin t)$, which traces out the unit circle as t increases from 0 to 2π.

- Calculate the coordinates $(\cos t, \sin t)$ for the values $t = 0, \frac{\pi}{4}, \frac{5\pi}{6}, \frac{3\pi}{2}$, and 2π, and verify that these coordinates define the points pictured in the following diagram:

$(\cos t, \sin t)$ traces out a circle

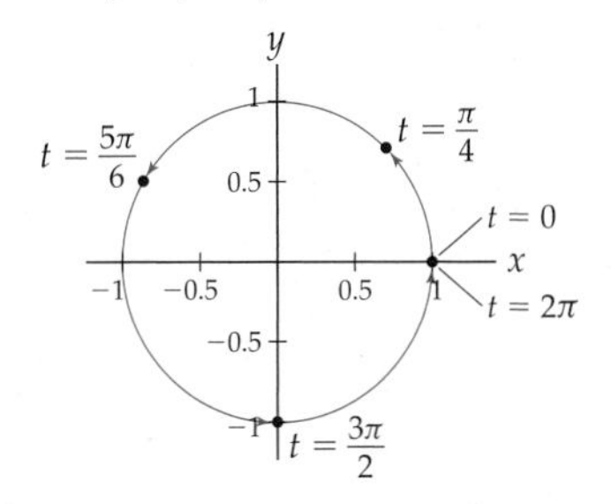

- To approximate the arc length of a parametric curve on some interval from $t = a$ to $t = b$, we subdivide $[a, b]$ into n pieces to define $t_1 = a, t_2, t_3, \ldots, t_n$, and consider the line segments connecting $(x(t_{k-1}), y(t_{k-1}))$ and $(x(t_k), y(t_k))$ for each $k = 1, 2, \ldots, n$. Define Δx_k and Δy_k for the kth line segment in terms of the coordinates shown in the following figure:

Notation for the kth line segment

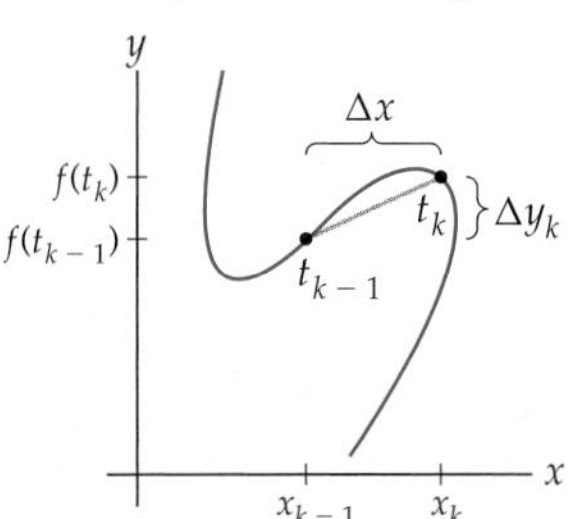

- Mimic the argument in the reading for this section to argue that a reasonable definition for the arc length of a parametric curve $(x(t), y(t))$ from $t = a$ to $t = b$ is

$$\lim_{n\to\infty} \sum_{k=1}^{n} \sqrt{(\Delta x_k)^2 + (\Delta y_k)^2},$$

where Δx_k and Δy_k are defined as in the previous problem.

- Use algebra and the concept of Riemann sums to argue that the arc length of a parametric curve $(x(t), y(t))$ from $t = a$ to $t = b$ should be represented by the definite integral

$$\int_a^b \sqrt{(x'(t))^2 + (y'(t))^2}\, dt.$$

- Use this new description of arc length to show that the circumference of the unit circle is 2π, by thinking of the unit circle as the parametric curve $(\cos t, \sin t)$ from $t = 0$ to $t = 2\pi$ and applying the definite integral formula from the previous problem.

6.4 REAL-WORLD APPLICATIONS OF INTEGRATION

- Definitions of mass, weight, density, work, force, and hydrostatic force
- Using definite integrals to represent accumulations of varying quantities
- Examples of real-world physics problems

Mass and Density

We now examine applications from physics that can be solved by our "subdivide, approximate, and add" strategy, and thus with definite integrals. The first concept that will concern us is ***mass***. Mass is different from, but related to, ***weight***. Mass is an inherent quantity that does not depend on which gravitational force is acting on the object. Mass measures the amount of matter in an object, while weight measures the force of gravity on the object.

A small object that is very dense (like a small rock) can have the same mass as a larger object that is less dense (for example, a very large serving of cotton candy). ***Density*** is mass per unit volume, and is traditionally labeled by the letter ρ, the lowercase Greek letter "rho." The mass of an object is then the product of its density and its volume:

DEFINITION 6.11 Mass Given a Constant Density

The ***mass*** of an object with density ρ and volume V is $m = \rho V$.

For example, gold has a density of approximately 19.3 grams per cubic centimeter, so the mass of a gold coin of radius 1.5 centimeters and thickness 0.2 centimeter is

$$m = \rho V = \left(19.3\,\frac{\text{grams}}{\text{cm}^3}\right)(\pi(1.5)^2(0.2)\text{cm}^3) \approx 27.28 \text{ grams}.$$

A silver coin of the same dimensions has a smaller mass, since silver has a lower density of approximately 10.5 grams per cubic centimeter:

$$m = \rho V = \left(10.5\,\frac{\text{grams}}{\text{cm}^3}\right)(\pi(1.5)^2(0.2)\text{cm}^3) \approx 14.84 \text{ grams}.$$

This simple mass formula works only for objects that are homogeneous, with same density everywhere. When density varies, finding mass is more complicated. The key will be to slice the object in such a way that for each piece, we can assume that the density is approximately constant, as we will soon see.

Work and Force

The second concept from physics that we will be interested in is ***work***. Clearly it takes more work to lift a heavy object (say, a car) than it takes to lift a lighter object (say, a book). It also takes more work to lift an object a long distance (such as lifting a car 30 feet) than a short distance (such as lifting a car 1 inch off the ground). The work required to lift an object is thus the product of its weight and the distance that it must be moved.

DEFINITION 6.12

Work Given a Constant Distance

The ***work*** required to lift an object with a weight of F through a distance d is $W = Fd$.

We use the letter F for weight because the weight of an object is the force exerted on it by gravity. This force is the product $F = \omega V$ of the object's weight per unit volume, or its ***weight-density***, and its volume.

For example, water has a weight-density of $\omega = 62.4$ pounds per cubic foot. How much work does it take to lift a full bucket of water with radius 1 foot and height 2 feet to a height of 3 feet off the ground? The water in the bucket weighs

$$F = \omega V = 62.4(\pi(1)^2(2)) = 124.8\pi \text{ pounds}.$$

Assuming that the weight of the bucket is negligible, the work required to lift such a full bucket of water a distance of 3 feet off the ground is

$$W = Fd = (124.8\pi \text{ pounds})(3 \text{ feet}) \approx 1176 \text{ foot-pounds}.$$

Note that work involves units called *foot-pounds*; this is because work is the product of force (measured in pounds) and distance (measured in feet). In physics, one might see force measured in *newtons* and distance measured in meters, in which case work would be measured in *newton-meters* which are also called *joules*.

The simple work formula just presented applies only when force and displacement are constant. If either the weight of an object or the distance that it is to be moved is variable, then finding work will require integrals, as we will soon see.

Hydrostatic Force

Our final physics concept is ***hydrostatic force***, which is the force exerted on a surface by a volume of water or some other liquid. The force exerted on the bottom, horizontal surface of any tank or container is equal to the weight of the water above it. Force due to gravity is equal to weight, which is equal to weight-density times volume, which, for water in a container, is equal to weight-density, times area, times depth and thus:

DEFINITION 6.13

Hydrostatic Force Given a Constant Depth

The ***hydrostatic force*** exerted by a liquid of weight-density ω and depth d on a horizontal wall of area A is given by $F = \omega A d$.

For example, in a rectangular swimming pool measuring 10 feet by 30 feet that has a constant depth of 5 feet, the hydrostatic force on the bottom surface of the pool is

$$F = \omega A d = \left(62.4 \frac{\text{lbs}}{\text{ft}^3}\right)(300 \text{ ft}^2)(5 \text{ ft}) = 93{,}600 \text{ pounds}.$$

In fact, the same principle applies to the force exerted against *any* wall (horizontal, vertical, or otherwise) by a mass of water. The reason is that the pressure, or force per square foot of area, exerted by the water is the same in all directions. However, a non-horizontal wall will be at different depths at various points on the wall, and the force from the water will be different at these various points. The water in a pool will exert a great deal of force near the bottom of a side wall, but less force near the top of that side wall. We will have to use a slicing technique–and thus, eventually, definite integrals–to find the hydrostatic force exerted on nonhorizontal walls.

Representing Varying Quantities with Definite Integrals

Mass, work, and hydrostatic force have simple formulas when all the associated quantities are constant. But what happens when things vary? How do we calculate the mass of an object of varying density, or the work required to pump water of varying depths out of a tank, or the hydrostatic force on the side wall of a pool?

The key, of course, is to consider slices on which any varying quantities can be approximated with *constant* values. Then the simple formulas we developed earlier will apply on each slice, and we can add up all the slice-approximations to find an approximation for the total mass, work, or hydrostatic force. Subdivide, approximate, and add... and of course if we set things up right, then we will be able to express our approximation as a Riemann sum, take a limit, and obtain a definite integral for the exact answer!

Let's walk through this process with a particular example. Consider the task of pumping all the water out of the top of a cylindrical hot tub that is 4 feet deep and has a radius of 3.5 feet. The water at the bottom of the tub has to be moved farther than the water at the top of the tub, so the distance involved is variable. If we consider a thin horizontal slice of the hot tub, we can assume without too much error that all the water in that slice has to be moved the same upward distance; see the following figure:

Horizontal slice of hot tub

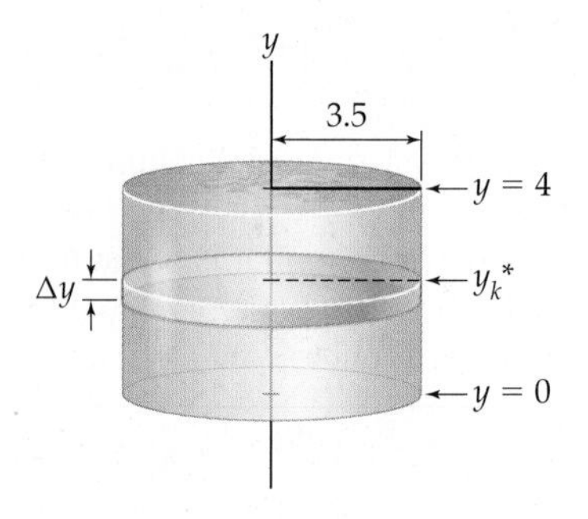

This figure introduces some needed notation. We are imagining that the bottom of the hot tub is at $y = 0$ and and the top is at $y = 4$. We divide the interval from $y = 0$ to $y = 4$ into n subintervals $[y_{k-1}, y_k]$ of thickness Δy. A representative slice of the hot tub at some height $y_k^* \in [y_{k-1}, y_k]$ is shown in the figure.

The slice at y_k^* needs to be moved upwards $d_k = 4 - y_k^*$ units in order to be pumped out of the tub. This slice is a disk with volume $V_k = \pi r^2 h = \pi(3.5)^2 \Delta y$ cubic feet. Water weighs $\omega = 62.4$ pounds per cubic foot. Since for constant quantities we have $W = Fd = \omega V d$, the work W_k required to pump the kth slice of water out of the hot tub is

$$W_k = F_k d_k = \omega V_k d_k = 62.4(\pi(3.5)^2 \Delta y)(4 - y_k^*) \text{ foot-pounds.}$$

By repeating this approximation for each horizontal slice and then adding all these approximations, we can get an approximation for the total work W required to pump all of the water out of the hot tub:

$$W \approx \sum_{k=1}^{n} (62.4)(\pi)(3.5)^2(4 - y_k^*)\Delta y.$$

You should recognize this as a Riemann sum. As n gets larger, we make thinner and thinner slices and better approximations. In the limit, as $n \to \infty$, this approximation approaches the actual work required. Accumulating slices of work from $y = 0$ to $y = 4$, the actual amount of work required is given by the definite integral

$$W = \int_0^4 (62.4)(\pi)(3.5)^2(4 - y)\, dy.$$

With basic integration techniques it is easy to solve this integral to find that the total amount of work required to pump all of the water out of the top of the hot tub is $6115.2\pi \approx 19{,}211.5$ foot-pounds.

Let's review what just happened from a more general perspective. The sum we found earlier was a Riemann sum that converged to a definite integral because it was of the form $\sum_{k=1}^{n} f(y_k^*)\Delta y$ for some continuous function $f(y)$. In this case the function was

$$f(y) = 62.4\pi(3.5)^2(4 - y) = \text{(weight-density)(cross-sectional area)(distance)} = \omega A d,$$

where $d = d(y)$ is a function of y. Compare this with the formula $W = Fd = \omega V d$ for work. The difference is that $V = A\Delta y$; we extracted this Δy so that we could form a Riemann sum. In the definite integral, Δy becomes dy.

In other examples we will follow the same basic strategy: Subdivide, or slice, so that quantities are constant; use physics formulas to approximate quantities on each slice; extract a Δx or Δy to write the sum as a Riemann sum; and then take the limit to get a definite integral that we can solve.

Centroids and Centers of Mass

As a geometric application of the techniques of this section, we will investigate the ***centroid***, otherwise known as the ***center of mass***, of a thin plate of material. Let's consider the simple case where the plate is defined by the region between a positive function $f(x)$ and the x-axis, as shown next at the left. The centroid of a thin plate of material in the shape of this region is the point $(\bar{x}, \bar{y})$ where the plate would balance horizontally, as shown at the right.

A region in the plane and its centroid

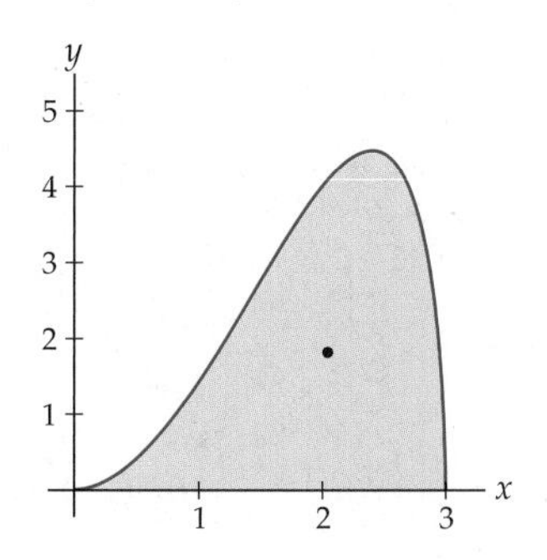

Balancing horizontally on the centroid

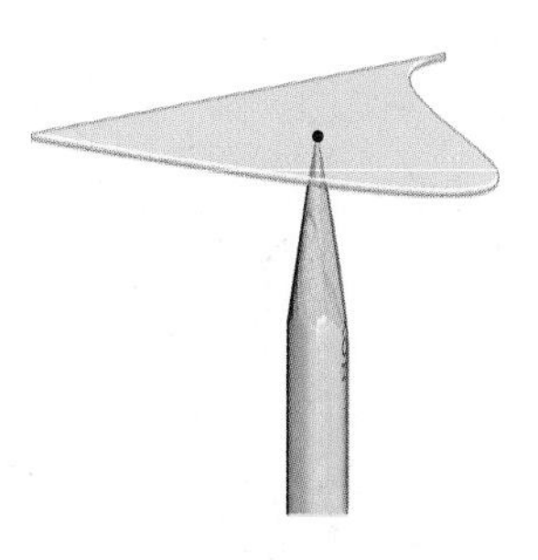

As one might expect, the centroid of a rectangular region is easy to find; it is found at the center of the rectangle. It is also easy to find the centroid of a collection of rectangles, because the centroid of a composite object is a weighted average of the centroids of its pieces. Specifically, suppose a region with area A is divided into a collection of regions with areas $A_1, A_2, \dots A_n$ and centroids $(\bar{x}_1, \bar{y}_1), (\bar{x}_2, \bar{y}_2), \dots, (\bar{x}_n, \bar{y}_n)$. Then the centroid $(\bar{x}, \bar{y})$ of the combined region can be obtained by taking the average of the centroids of the rectangles, weighted by their areas, with coordinates as follows:

$$\bar{x} = \frac{\bar{x}_1 A_1 + \bar{x}_2 A_2 + \cdots + \bar{x}_n A_n}{A},$$

$$\bar{y} = \frac{\bar{y}_1 A_1 + \bar{y}_2 A_2 + \cdots + \bar{y}_n A_n}{A}.$$

An example of this situation with three rectangular regions is shown next at the left, with the centroid of each rectangle marked with an "x" and the centroid of the overall shape marked with a dot. Note that in our weighted average, the third "x" will count the most, since it is the centroid of the largest rectangle. This particular set of rectangles was not arbitrary; the figure at the right shows the same collection of rectangles set against the region we illustrated at the start of this subsection. It is reasonable to imagine that the centroid of the collection of rectangles is an approximation for the centroid of this region. (See Example 6.)

The centroid of a collection of rectangles

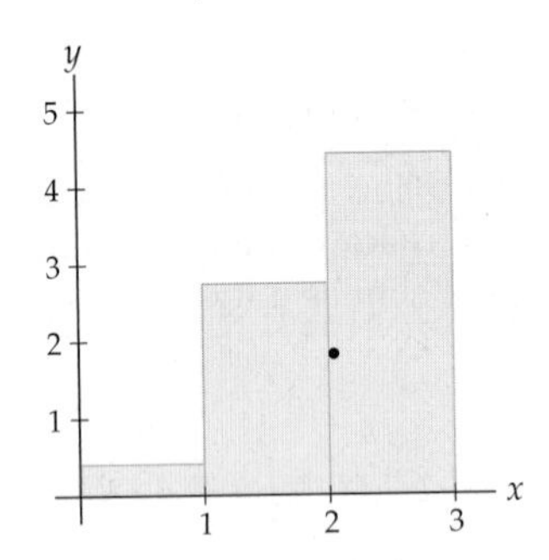

Approximating the centroid of a region

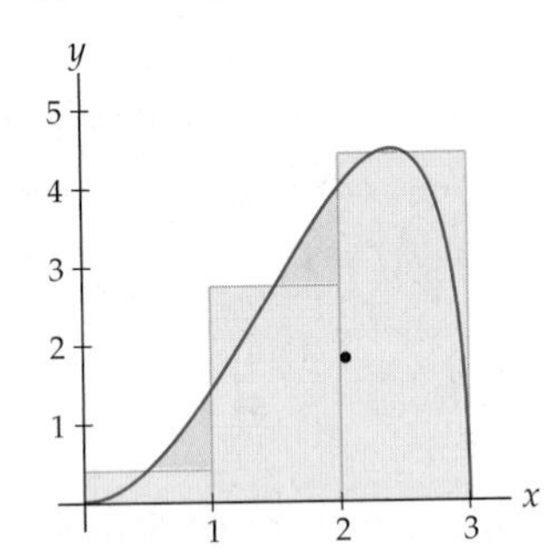

We are now on very familiar ground. Suppose we want to find the centroid of a region between the graph of some function $f(x)$ and the x-axis on an interval $[a, b]$. Following our usual "subdivide, approximate, and add" strategy, we begin by subdividing the interval $[a, b]$ into n subintervals of the form $[x_{k-1}, x_k]$, where $x_k = a + k\,\Delta x$ and $\Delta x = \frac{b-a}{n}$. For each k, we then choose x_k^* to be the midpoint of the subinterval $[x_{k-1}, x_k]$ and consider the rectangle on that subinterval whose height is $f(x_k^*)$ as shown here:

The centroid of a representative rectangle

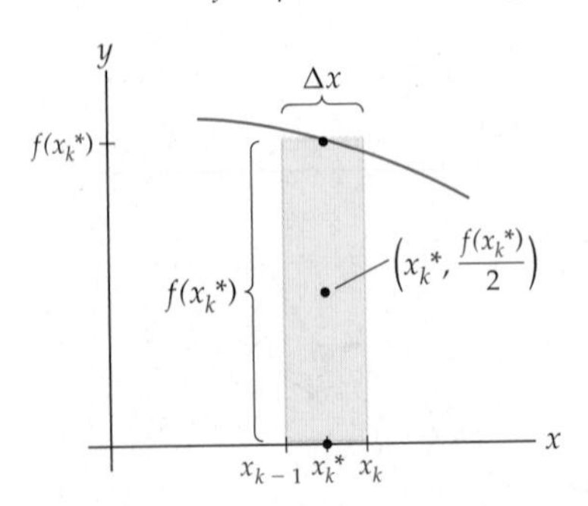

The centroid of this rectangle is at its center, which is the point

$$(\bar{x}_k, \bar{y}_k) = \left(x_k^*, \frac{1}{2}f(x_k^*)\right).$$

The area of this kth rectangle is

$$A_k = f(x_k^*)\Delta x.$$

Using our weighted average formula from earlier, we find that the centroid of the collection of all n rectangles has coordinates

$$\bar{x} = \frac{\sum_{k=1}^n \bar{x}_k A_k}{\sum_{k=1}^n A_k} = \frac{\sum_{k=1}^n x_k^* f(x_k^*)\Delta x}{\sum_{k=1}^n f(x_k^*)\Delta x},$$

$$\bar{y} = \frac{\sum_{k=1}^n \bar{y}_k A_k}{\sum_{k=1}^n A_k} = \frac{\sum_{k=1}^n \frac{1}{2} f(x_k^*) f(x_k^*)\Delta x}{\sum_{k=1}^n f(x_k^*)\Delta x}.$$

Finally, we notice that the numerators and denominators of $\bar{x}$ and $\bar{y}$ are Riemann sums! Taking the limit as $n \to \infty$ allows us to describe the exact centroid of the region in terms of definite integrals:

THEOREM 6.14

The Centroid of a Region Under a Curve

Let f be an positive, integrable function on $[a, b]$. Then the centroid $(\bar{x}, \bar{y})$ of the region between the graph of $f(x)$ and the x-axis on the interval $[a, b]$ is the point

$$(\bar{x}, \bar{y}) = \left(\frac{\int_a^b x f(x)\, dx}{\int_a^b f(x)\, dx}, \frac{\frac{1}{2}\int_a^b f(x)^2\, dx}{\int_a^b f(x)\, dx}\right).$$

By construction, this formula works only for finding centroids of regions between the graph of some function $f(x)$ and the x-axis. A similar construction allows us to describe the centroid between two graphs $f(x)$ and $g(x)$ on an interval $[a, b]$ (see Example 7 for this construction):

THEOREM 6.15

The Centroid of a Region Between Two Curves

Let f and g be integrable functions on $[a, b]$. The centroid $(\bar{x}, \bar{y})$ of the region between the graphs of $f(x)$ and $g(x)$ on the interval $[a, b]$ is the point

$$(\bar{x}, \bar{y}) = \left(\frac{\int_a^b x|f(x) - g(x)|\, dx}{\int_a^b |f(x) - g(x)|\, dx}, \frac{\frac{1}{2}\int_a^b |f(x)^2 - g(x)^2|\, dx}{\int_a^b |f(x) - g(x)|\, dx}\right).$$

The absolute values in these expressions play the same role as those we saw in the calculations of the area between two curves as those we saw in Section 4.6. Specifically, they allow us to write the centroid as one formula regardless of whether $f(x) \geq g(x)$ or $f(x) \leq g(x)$. Just as in Section 4.6, when calculating centroids you must divide the interval $[a, b]$ into subintervals according to these inequalities.

Examples and Explorations

EXAMPLE 1

Finding work when distance and volume vary

Find the work involved in pumping all of the water out of the top of a cone-shaped tank that is 4 feet tall and has a radius of 3.5 feet at the top.

SOLUTION

The first thing we have to do is establish some notation. Suppose that the bottom of the tank is at height $y = 0$ and the top of the tank is at height $y = 4$. A representative horizontal slice at some point y_k^* is shown next at the left.

Horizontal slice of conical tank

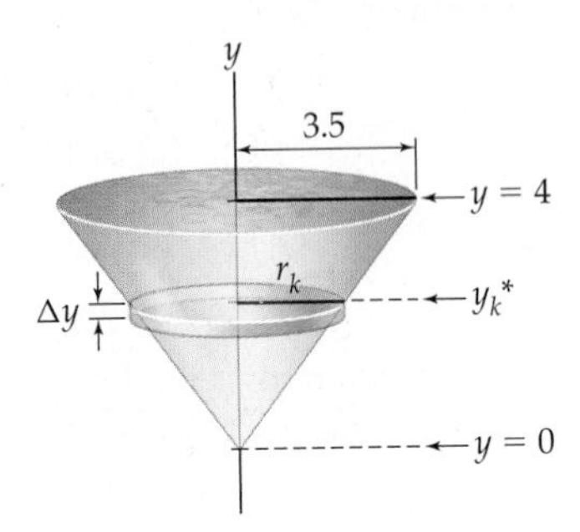

Use the law of similar triangles

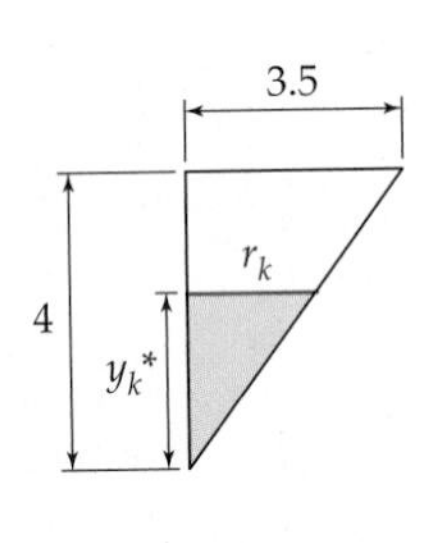

We can use the law of similar triangles (see the diagram at the right) to find the radius r_k of this slice:

$$\frac{r_k}{y_k^*} = \frac{3.5}{4} \implies r_k = \frac{3.5}{4} y_k^*.$$

The volume of the slice is $V_k = \pi r^2 \Delta y$, the weight-density of water is $\omega = 62.4$ pounds per cubic foot, and the vertical distance that the water in the kth slice must move is approximately $d_k = 4 - y_k^*$. Since for constant quantities we have $W = Fd = \omega V d$, the work involved in pumping out the representative slice of water is

$$W_k = F_k d_k = \omega V_k d_k = (62.4)(\pi)\left(\frac{3.5}{4} y_k^*\right)^2 \Delta y \cdot (4 - y_k^*) \text{ foot-pounds.}$$

Thus the work W required to pump all of the water out of the top of the conical tank is approximately

$$W \approx \sum_{k=1}^{n} (62.4)(\pi)\left(\frac{3.5}{4}\right)^2 (y_k^*)^2 (4 - y_k^*) \Delta y.$$

This is a Riemann sum, so as $n \to \infty$ we obtain a definite integral. Accumulating slices from $y = 0$ to $y = 4$, we find that the amount of work required to pump all of the water out of the top of the tank is

$$\begin{aligned} W &= \int_0^4 (62.4)(\pi)\left(\frac{3.5}{4}\right)^2 y^2 (4 - y)\, dy \\ &= 47.775\pi \int_0^4 (4y^2 - y^3)\, dy = 47.775\pi \left[\frac{4}{3} y^3 - \frac{1}{4} y^4\right]_0^4 \\ &= 3201.91\pi \text{ foot-pounds.} \end{aligned}$$

□

EXAMPLE 2 **Finding mass when density varies**

Suppose a cylindrical rod with a radius of 2 centimeters and a length of 24 centimeters is made of a combination of silver and gold in such a way that the density of the rod x centimeters from the left end is given by the function $\rho(x) = 10.5 + 0.01527x^2$ grams per cubic centimeter. Find the total mass of the rod.

SOLUTION

The following diagram shows a thin representative slice of the rod at some point x_k^* centimeters from the left end:

Representative slice of rod with varying mass

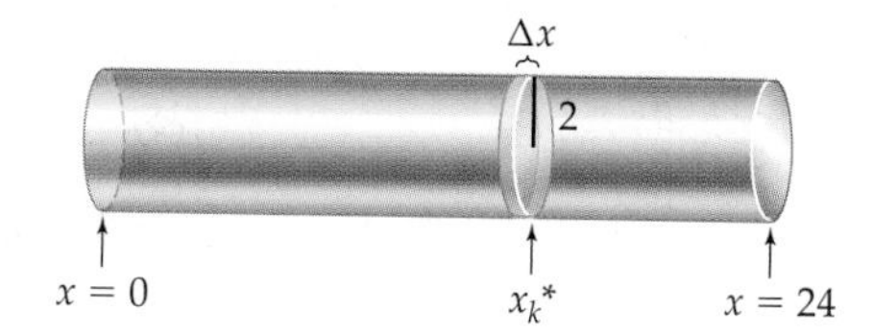

Each representative slice is a disk of radius 2 and thickness Δx, and therefore volume $\pi(2)^2\Delta x$. We can assume without too much error that the density throughout the entire thin slice is $\rho_k = \rho(x_k^*) = 10.5 + 0.01527(x_k^*)^2$ grams per cubic centimeter. Since for constant values we have $m = \rho V$, the mass of the slice is

$$m_k = \rho_k V_k = (10.5 + 0.01527(x_k^*)^2)(\pi(2)^2\Delta x).$$

The approximate mass of the entire rod is thus given by the sum

$$m \approx \sum_{k=1}^{n} 4\pi(10.5 + 0.01527(x_k^*)^2)\,\Delta x.$$

As $n \to \infty$, this Riemann sum approaches the actual mass of the rod. Accumulating mass from $x = 0$ to $x = 24$, we see that the total mass of the rod is

$$m = \int_0^{24} 4\pi(10.5 + 0.01527x^2)\,dx$$

$$= 4\pi\left[10.5x + \frac{0.01527}{3}x^3\right]_0^{24} = 1289.46\pi \approx 4050.95 \text{ grams.}$$

□

EXAMPLE 3

Finding mass when density and volume vary

Alina has made a gelatin mold that is 4.5 inches tall, 6 inches across, and in the shape of the top of a downwards-pointing parabola that has been revolved around the y-axis. She put strawberry pieces into the gelatin mold, but they have tended to settle down to the bottom. This causes the density of the gelatin mold to vary linearly with height in such a way that the density at the very top of the mold is 0.25 ounce per cubic inch, while the density at the very bottom of the mold is 1.3 ounces per cubic inch. Find the mass of the entire gelatin mold.

SOLUTION

We first need to find functions that describe the shape and density of the gelatin mold. This is not difficult to do, using the information given in the problem. It turns out that the shape of the gelatin mold is the solid of revolution obtained by rotating the graph of $y = 4.5 - 0.5x^2$ on $[0, 3]$ around the x-axis and that the density of the gelatin mold at height y is approximately given by the function $\rho(y) = 1.3 - 0.233y$. (See Exercises 10 and 11.)

In this example, both the density and the volume vary with the slice of the object. The next figure shows a representative slice at some height y_k^*. We can assume without introducing too much error that the entire slice has a constant density of $\rho_k = \rho(y_k^*) = 1.3 - 0.233y_k^*$ ounces per cubic inch.

Strawberries settle to the bottom

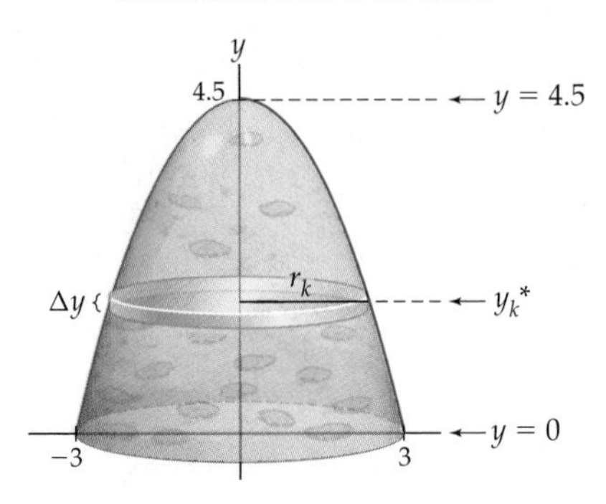

According to this graph, if r_k is the radius corresponding to the height y_k^*, then $y_k^* = 4.5 - 0.5r_k^2$, and thus $r_k = \sqrt{9 - 2y_k^*}$. The volume of the representative slice is therefore

$$V_k = \pi r_k^2 \,\Delta y = \pi \left(\sqrt{9 - 2y_k^*}\right)^2 \Delta y = \pi(9 - 2y_k^*)\Delta y.$$

Since for constant values we have $m = \rho V$, the mass of our representative slice is

$$m_k = \rho_k V_k = (1.3 - 0.233y_k^*)(\pi)(9 - 2y_k^*)\Delta y.$$

The total mass m of the gelatin mold can be approximated by the sum of the approximate masses of each of the slices:

$$m \approx \sum_{k=1}^{n} \pi(1.3 - 0.233y_k^*)(9 - 2y_k^*)\Delta y.$$

This is a Riemann sum, and as $n \to \infty$ the sum converges to a definite integral. Accumulating slices from $y = 0$ to $y = 4.5$, we get the total mass of the gelatin mold:

$$\int_0^{4.5} \pi(1.3 - 0.233y)(9 - 2y)\,dy = \pi \int_0^{4.5} (11.7 - 4.697y + 0.466y^2)\,dy$$

$$= \pi\left[11.7y - \frac{4.697}{2}y^2 + \frac{0.466}{3}y^3\right]_0^{4.5} \approx 60.4682 \text{ ounces.} \quad \square$$

EXAMPLE 4

Finding hydrostatic force when depth varies

Consider a swimming pool that measures 10 feet by 30 feet, with a constant depth of 5 feet. Find the hydrostatic force exerted on one of the shorter side walls of the pool.

SOLUTION

The force exerted by the water on the side walls varies with depth. However, if we consider a very thin horizontal slice of the wall, then we can assume without too much error that the entire slice is at the same depth. The following diagram shows a representative slice of the side wall at some height y_k^* from the bottom of the pool:

Representative slice of side wall

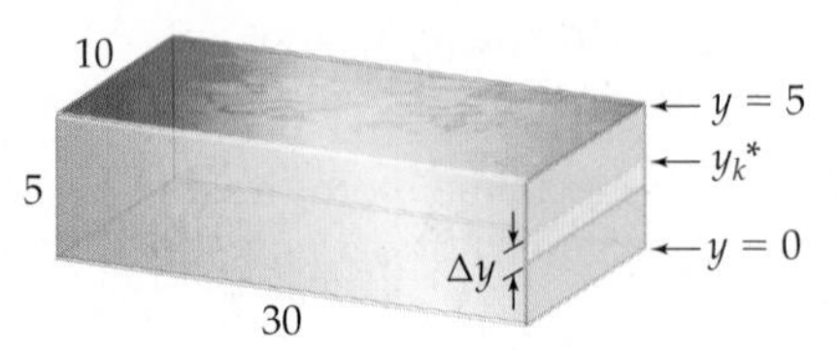

The area of the representative wall slice is $A_k = 10\Delta y$ square feet. Water has a weight-density of $\omega = 62.4$ pounds per cubic foot. When values are constant, hydrostatic force is given by $F = \omega A d$. Therefore, assuming that the entire thin slice of wall is at a depth of $d_k = 5 - y_k^*$ feet, we find that the hydrostatic force on that slice of wall is given by

$$F_k = \omega A_k d_k = 62.4(10\Delta y)(5 - y_k^*).$$

The hydrostatic force on the entire side wall is thus approximately

$$F \approx \sum_{k=1}^{n} 62.4(10)(5 - y_k^*)\,\Delta y.$$

As n approaches ∞ and we accumulate slices from $y = 0$ to $y = 5$, this Riemann sum approaches a definite integral. The hydrostatic force on a shorter side wall of the pool is

$$\int_0^5 62.4(10)(5 - y)\,dy = (62.4)(10)\left[5y - \frac{1}{2}y^2\right]_0^5 = 7800 \text{ pounds.}$$ □

EXAMPLE 5

Finding hydrostatic force when depth and area vary

Consider a rectangular swimming pool that is 30 feet long and 10 feet wide, with a depth of 4 feet at the shallow end and 8 feet at the deep end. Suppose the shallow end of the pool makes up the first 10 feet of the pool and is followed by 10-foot linear ramp along which the depth of the pool increases from 4 feet to 8 feet and then a final length of 10 feet for the deep end. Find the hydrostatic force exerted on one of the long sides of the pool.

SOLUTION

The next diagram shows a side view of the pool, as well as two types of horizontal slices. If y_k^* is between $y = 0$ and $y = 4$, then the length of the slice varies with the depth of the pool. If y_k^* is between $y = 4$ and $y = 8$, then the length of the slice is always 30 feet.

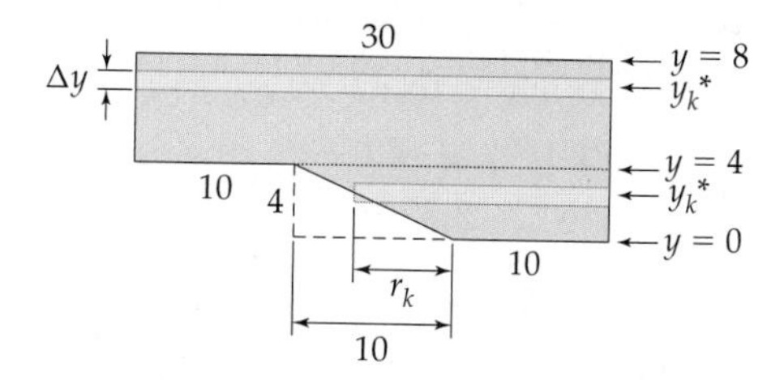

For y_k^* between $y = 0$ and $y = 4$, the kth slice has width Δy and length $10 + r_k$, where r_k is the distance that the slice extends into the ramp part of the pool. By the law of similar triangles and the triangles whose legs are shown as dotted lines in the figure, we have

$$\frac{r_k}{y_k^*} = \frac{10}{4} \quad \Longrightarrow \quad r_k = \frac{5}{2}y_k^*.$$

The area of the kth slice, between $y = 0$ and $y = 4$, is $A_k = \left(10 + \frac{5}{2}y_k^*\right)\Delta y$. Again, the weight-density of water is $\omega = 62.4$ pounds per cubic foot. When values are constant, hydrostatic force is given by $F = \omega A d$. Since we are assuming that the entire slice is at depth $d_k = 8 - y_k^*$, the hydrostatic force on each slice between $y = 0$ and $y = 4$ is

$$F_k = \omega A_k d_k = 62.4\left(\left(10 + \frac{5}{2}y_k^*\right)\Delta y\right)(8 - y_k^*).$$

The total hydrostatic force on the part of the side wall from $y = 0$ to $y = 4$ (the lower half of the wall) is thus approximately

$$F_{\text{lower}} \approx \sum_{k=1}^{n} 62.4\left(10 + \frac{5}{2}y_k^*\right)(8 - y_k^*)\Delta y.$$

As $n \to \infty$ and we accumulate slices from $y = 0$ to $y = 4$, this sum converges to a definite integral that represents the hydrostatic force F_{lower} on the bottom half of the wall:

$$F_{\text{lower}} = \int_0^4 62.4\left(10 + \frac{5}{2}y\right)(8 - y)\,dy = 21,632 \text{ pounds}.$$

By an argument similar to the one used in the previous example, we can show that the hydrostatic force from $y = 4$ to $y = 8$ (the upper half of the side wall) is given by the definite integral

$$F_{\text{upper}} = \int_4^8 62.4(30)(8 - y)\,dy = 59,904 \text{ pounds}.$$

(We leave it to you to solve these integrals.) The total hydrostatic force exerted on a long side wall is the sum of these two forces:

$$F = F_{\text{lower}} + F_{\text{upper}} = 21,632 \text{ pounds} + 59,904 \text{ pounds} = 81,536 \text{ pounds}.$$

Notice that in this problem we never needed to use the fact that the pool was 10 feet wide! Surprisingly, the answer is the same whether the pool is 10 feet wide or 100 feet wide; the hydrostatic force exerted on the side wall does not depend on how far the pool extends from that wall. □

EXAMPLE 6

Approximating the centroid of a region

Use three rectangles to approximate the centroid of the region between the graph of $f(x) = x^2\sqrt{3 - x}$ and the x-axis on $[0, 3]$.

SOLUTION

Subdividing $[0, 3]$ into three intervals and then defining rectangles whose heights match the function at the midpoints of those subintervals gives the following picture:

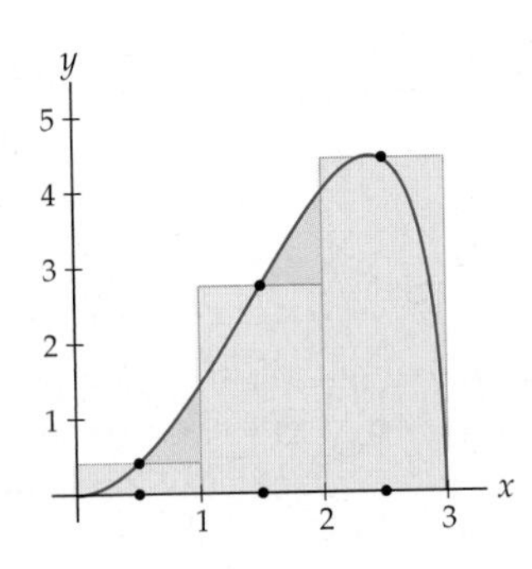

The centroids of the rectangles are at their centers, and, given the graph and equation for $f(x)$, these centroids are

$$(\bar{x}_1, \bar{y}_1) = \left(0.5, \frac{1}{2}f(0.5)\right) = \left(0.5, \frac{1}{2}((0.5)^2\sqrt{3 - 0.5})\right) \approx (0.5, 0.1976),$$
$$(\bar{x}_2, \bar{y}_2) = \left(1.5, \frac{1}{2}f(1.5)\right) = \left(1.5, \frac{1}{2}((1.5)^2\sqrt{3 - 1.5})\right) \approx (1.5, 1.3778),$$
$$(\bar{x}_3, \bar{y}_3) = \left(2.5, \frac{1}{2}f(2.5)\right) = \left(2.5, \frac{1}{2}((2.5)^2\sqrt{3 - 2.5})\right) \approx (2.5, 2.2097).$$

Note that these coordinates do seem reasonable compared with the points marked "x" in the preceding figure.

The three rectangles each have $\Delta x = 1$ and height defined by $f(x)$ at their midpoints $x = 0.5$, 1.5, and 2.5, and therefore have the following areas:

$$A_1 = f(0.5)\,\Delta x = ((0.5)^2\sqrt{3-0.5}\,)(1) \approx 0.3953,$$
$$A_2 = f(1.5)\,\Delta x = ((1.5)^2\sqrt{3-1.5}\,)(1) \approx 2.7557,$$
$$A_3 = f(2.5)\,\Delta x = ((2.5)^2\sqrt{3-2.5}\,)(1) \approx 4.4194.$$

Note again that these values seem reasonable compared with the three rectangles we graphed earlier. The area of the three rectangles together is naturally $A = A_1 + A_2 + A_3 \approx 7.5704$.

Finally, to find the coordinates of the centroid of the collection of three rectangles we must take the average of the three centroids we found, weighted by the areas of the corresponding rectangles. This gives us the following approximation for the centroid $(\bar{x}, \bar{y})$ of the region:

$$\bar{x} \approx \frac{\bar{x}_1 A_1 + \bar{x}_2 A_2 + \bar{x}_3 A_3}{A} \approx \frac{0.5(0.3953) + 1.5(2.7557) + 2.5(4.4194)}{7.5704} \approx 2.0316,$$
$$\bar{y} \approx \frac{\bar{y}_1 A_1 + \bar{y}_2 A_2 + \bar{y}_3 A_3}{A} \approx \frac{0.1976(0.3953) + 1.3778(2.7557) + 2.2097(4.4194)}{7.5704} \approx 1.8018.$$

Therefore the centroid of the region between $f(x) = x^2\sqrt{3-x}$ and the x-axis on $[0, 3]$ is approximately $(\bar{x}, \bar{y}) \approx (2.0316, 1.8018)$. You will find a finer approximation with $n = 6$ rectangles in Exercise 32 and find this centroid exactly in Exercise 36. □

EXAMPLE 7

Expressing a centroid in terms of Riemann sums and definite integrals

Consider the region between the graphs of $f(x) = e^{1-x}$ and $g(x) = x^2$ on the interval $[0, 1]$. Express the centroid of this region in terms of (a) Riemann sums and (b) definite integrals.

SOLUTION

(a) The region whose centroid we wish to calculate is shown next at the left. We can already visually judge that this centroid should be close to, say, $(0.3, 1)$. The figure at the right shows a representative rectangle at some point x_k^* in $[0, 1]$.

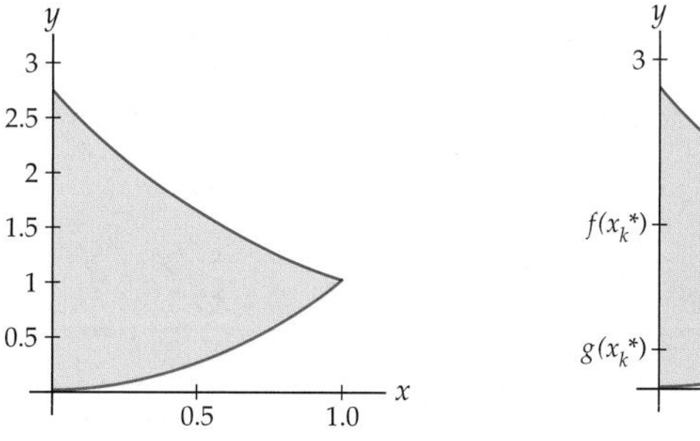

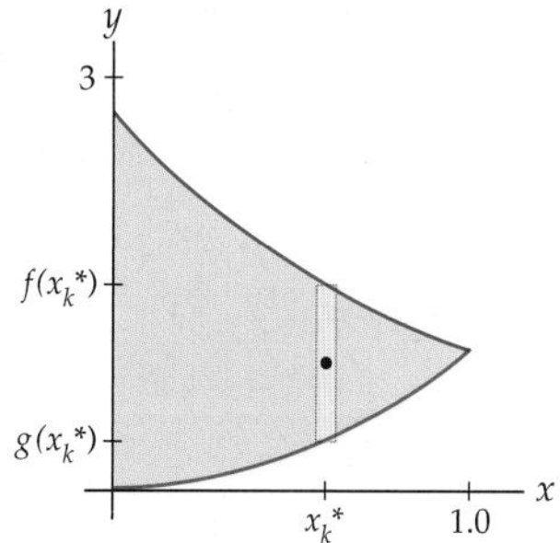

Note that the top of this rectangle is at height $f(x_k^*) = e^{1-x_k^*}$ and the bottom is at height $g(x_k^*) = (x_k^*)^2$. Thus the height of the marked point in the center of the rectangle is the average $\frac{f(x_k^*)+g(x_k^*)}{2}$, and therefore the centroid of the rectangle is

$$(\bar{x}_k, \bar{y}_k) = \left(x_k^*, \frac{f(x_k^*)+g(x_k^*)}{2}\right).$$

The height of the rectangle is the difference $f(x_k^*) - g(x_k^*)$, and therefore the area of the rectangle is

$$A_k = (f(x_k^*) - g(x_k^*))\Delta x.$$

We can now approximate the coordinates of the centroid of the region in terms of Riemann sums by taking a weighted average of the centroids of the rectangles:

$$\bar{x} \approx \frac{\sum_{k=1}^{n} x_k^*\left(f(x_k^*) - g(x_k^*)\right)\Delta x}{\sum_{k=1}^{n}\left(f(x_k^*) - g(x_k^*)\right)\Delta x},$$

$$\bar{y} \approx \frac{\sum_{k=1}^{n} \frac{f(x_k^*)+g(x_k^*)}{2}\left(f(x_k^*) - g(x_k^*)\right)\Delta x}{\sum_{k=1}^{n}\left(f(x_k^*) - g(x_k^*)\right)\Delta x} = \frac{\sum_{k=1}^{n} \frac{1}{2}\left(f(x_k^*)^2 - g(x_k^*)^2\right)\Delta x}{\sum_{k=1}^{n}\left(f(x_k^*) - g(x_k^*)\right)\Delta x}.$$

(b) Taking the limit as $n \to \infty$ and using the equations $f(x) = e^{1-x}$ and $g(x) = x^2$, we obtain an expression for the coordinates of the centroid of the region in terms of definite integrals:

$$\bar{x} = \frac{\int_0^1 x(f(x) - g(x))\,dx}{\int_0^1 (f(x) - g(x))\,dx} = \frac{\int_0^1 x(e^{1-x} - x^2)\,dx}{\int_0^1 (e^{1-x} - x^2)\,dx},$$

$$\bar{y} = \frac{\frac{1}{2}\int_0^1 (f(x)^2 - g(x)^2)\,dx}{\int_0^1 (f(x) - g(x))\,dx} = \frac{\frac{1}{2}\int_0^1 ((e^{1-x})^2 - (x^2)^2)\,dx}{\int_0^1 (e^{1-x} - x^2)\,dx}.$$

Calculating these definite integrals (some require preliminary algebra or integration by parts) gives us a centroid of $(\bar{x}, \bar{y}) \approx (0.338, 1.081)$, which does seem compatible with our initial visual approximation. □

? TEST YOUR UNDERSTANDING

- Would the work required to pump the water out of the top of a hot tub be the same as the work required to pump the water out of the bottom? Why or why not?
- Would the work required to pump the water out of a buried conical tank be the same as the work required to lift the entire conical tank up to ground level? Why or why not?
- Why did we have to consider two types of horizontal slices in Example 5?
- Where does the expression $|f(x)^2 - g(x)^2|$ come from in the formula for the centroid of a region between two curves in Theorem 6.15?
- Each example in this section followed the same general procedure. Can you explain this procedure in your own words?

EXERCISES 6.4

Thinking Back

Definite integrals: Calculate each of the following definite integrals, using integration techniques and the Fundamental Theorem of Calculus.

- $\int_0^{10} 4\pi(11.2 + 0.072x^2)\,dx$
- $\int_0^4 (62.4\pi)(3.5)^2(4 - y)\,dy$
- $\int_0^4 62.4\left(10 + \frac{5}{2}y\right)(8 - y)\,dy$
- $\pi\int_0^{100} (0.75x + 250)(200 - x)\,dx$
- $\pi\int_0^6 (-2y + 1.12)\left(100 - \frac{25}{2}y\right)dy$

- *Setting up and solving a word problem involving arc length:* Calvin wants to hang some Christmas lights along the edge of the front side of his garage roof. The edge of the front side of the roof of his garage is a curve in the shape of a downwards-pointing parabola extending 3 feet above the ceiling of the garage and 12 feet across. How long a string of Christmas lights does Calvin need?

- *Setting up and solving a word problem involving volume:* A specialty glass flask has a bulb-shaped bottom half and a cylindrical top half. The bulb consists of a sphere with a radius of 10 centimeters whose top and bottom are truncated by 2 centimeters. The cylindrical tube has a radius just large enough to connect to the bulb and is 7 centimeters tall. Find the volume of the flask.

Concepts

0. *Problem Zero:* Read the section and make your own summary of the material.

1. *True/False:* Determine whether each of the statements that follow is true or false. If a statement is true, explain why. If a statement is false, provide a counterexample.

(a) *True or False:* Mass and weight are the same.

(b) *True or False:* A large object can have less mass than a smaller object.

(c) *True or False:* An object can have one mass on Earth and another mass on the moon.

(d) *True or False:* An object can have one weight on Earth and another weight on the moon.

(e) *True or False:* Water has a weight-density of 62.4 foot-pounds.

(f) *True or False:* It always takes more work to lift a heavy object than a light object for the same distance.

(g) *True or False:* Hydrostatic force can be measured only on horizontal walls.

(h) *True or False:* Hydrostatic force is stronger at greater depths than it is at lesser depths.

2. *Examples:* Construct examples of the thing(s) described in the following. Try to find examples that are different than any in the reading.

(a) An object whose density is constant throughout, and another whose density varies.

(b) A situation where work can be calculated without an integral, and another situation where a definite integral is needed to calculate work.

(c) An example of something that is measured in foot-pounds.

3. What is the difference between mass and weight?

4. What is the difference between density and weight-density?

5. Why is a definite integral needed to calculate the hydrostatic force exerted on a vertical wall of a swimming pool, but no integral is needed to calculate the hydrostatic force on the bottom of the pool?

6. Describe the "subdivide, approximate, and add" strategy that is used in each of the examples in this section. Why is it necessary? How does it work? What do Riemann sums and definite integrals have to do with it?

7. Explain how the values of Δx and x_k (or Δy and y_k) in a Riemann sum determine the limits of integration on the definite integral obtained by taking the limit of the Riemann sum as $n \to \infty$.

8. What do we mean when we say that a sum is a Riemann sum? What has to be true about a Riemann sum for us to be able to guarantee that it converges to a definite integral when we take the limit as $n \to \infty$?

9. Why is the task of pumping all of the water out of the top of a hot tub different from the task of lifting the entire hot tub 4 feet in the air (even assuming that the hot tub itself doesn't weigh anything)? Which task requires more work?

10. Use the information in the description of the problem in Example 3 to show that the height of the gelatin mold is given by the function $y = 4.5 - 0.5x^2$ and that the density of the gelatin mold at height y is given by the linear function $\rho(y) = 1.3 - 0.233y$.

11. If the approximating sum in Example 3 runs from $k = 1$ to $k = n$, why does the definite integral have 0 and 4.5 as its limits of integration?

12. Make a list of all of the physics formulas introduced in this section. Give an example of an application of each formula, together with appropriate units.

13. Strangely, the hydrostatic force exerted on the side wall of a pool does not depend on how much water is behind that wall, but only on the depth of the water. Explain why this makes sense.

14. Mass is the product of density and volume. When we integrate to find accumulated mass, we integrate the product of density and cross-sectional area. Express the previous two sentences in mathematical notation. Then explain why volume is used in one, but area in the other.

15. Work is the product of weight-density, volume, and distance. When we integrate to find accumulated work, we integrate the product of weight-density, cross-sectional area, and distance. Express the previous two sentences in mathematical notation. Then explain why volume is used in one, but area in the other.

16. Hydrostatic force is the product of weight-density, area, and depth. When we integrate to find accumulated hydrostatic force, we integrate the product of weight-density, length, and depth. Express the previous two sentences in mathematical notation. Then explain why area is used in one, but length in the other.

17. Consider a region in the plane divided into disjoint rectangles with areas $A_1, A_2, \dots, A_n$ and individual centroids $(\bar{x}_1, \bar{y}_1), (\bar{x}_2, \bar{y}_2), \dots, (\bar{x}_n, \bar{y}_n)$. The centroid of this region is

$$(\bar{x}, \bar{y}) = \left(\frac{\sum_{k=1}^{n} x_k A_k}{\sum_{k=1}^{n} A_k}, \frac{\sum_{k=1}^{n} y_k A_k}{\sum_{k=1}^{n} A_k} \right).$$

Explain what it means to say that this is a "weighted" average of the coordinates of the centroids of the rectangles.

18. In Theorem 6.14 there are four definite integrals in the expression for the centroid of a region under a curve. Explain where each of these definite integrals comes from in the construction of this centroid by Riemann sums. Then give a similar explanation for the four definite integrals in the expression for the centroid of a region between two curves given in Theorem 6.15. In this second theorem, why are there absolute values?

Skills

Use physics formulas to calculate each quantity in Exercises 19–24.

19. The mass of an object with density 12 grams per cubic centimeter in the shape of a cylinder with radius 4 centimeters and height 10 centimeters.

20. The mass of an object with density 600 grams per cubic centimeter in the shape of a cone with radius 0.5 meters and height 1 meter.

21. The work required to lift an object that weighs 20 pounds to a height of 100 feet.

22. The work required to lift an object 2 feet into the air, given that the object has a weight-density of 50 pounds per cubic foot and that the object is a cylinder with radius 2 inches and height 6 inches.

23. The hydrostatic force exerted by water on the bottom of a full, 2-foot deep rectangular tank with a 3×4 foot base.

24. The hydrostatic force exerted by a liquid of weight-density 70 pounds per cubic foot on the bottom of a cylindrical tank with height 8 feet and radius 2 feet.

Use four slices to construct an approximation of each of the quantities described in Exercises 25–30. In each case include a labeled diagram and an explicit list of the values of Δx and each x_k and x_k^* (or Δy and each y_k and y_k^*).

25. The mass of a cylindrical rod with radius of 2 centimeters and length of 24 centimeters whose density at a point x centimeters from the left end is $\rho(x) = 10.5 + 0.01527x^2$ grams per cubic centimeter.

26. The mass of the solid of revolution obtained by rotating the graph of $y = 4.5 - 0.5x^2$ on $[0, 3]$ around the y-axis and whose density at height y is $\rho(y) = 1.3 - 0.233y$ ounces per cubic inch.

27. The work required to pump all of the water out of a cone with top radius 3.5 feet and height 4 feet.

28. The work required to pump all of the water out of a cylinder with radius 3.5 feet and depth 4 feet.

29. The hydrostatic force exerted by water on one of the sides of a small cubical tank that is 4 feet to a side.

30. The hydrostatic force exerted by water on one of the shorter side walls of a rectangular container that is 3 feet long, 4 feet wide, and 2 feet deep.

In Exercises 31–34, use a weighted average over n rectangles to approximate the centroid of the region described. *(Hint: It may help to draw a picture.)*

31. The region between $f(x) = \sqrt{x}$ and the x-axis on $[a, b] = [1, 9]$, with $n = 2$.

32. The region between $f(x) = x^2\sqrt{3-x}$ and the x-axis on $[a, b] = [0, 3]$, with $n = 6$.

33. The region between $f(x) = x^3$ and the line $y = 8$ on $[a, b] = [0, 2]$, with $n = 4$.

34. The region between $f(x) = x^2$ and $g(x) = 64 - x^2$ on $[a, b] = [0, 8]$, with $n = 4$.

In Exercises 35–40, use definite integrals to calculate the centroid of the region described. Use graphs to verify that your answers are reasonable.

35. The region between $f(x) = \sqrt{x}$ and the x-axis on $[a, b] = [1, 9]$. (Compare with Exercise 31.)

36. The region between $f(x) = x^2\sqrt{3-x}$ and the x-axis on $[a, b] = [0, 3]$. (Compare with Exercise 32.)

37. The region between $f(x) = x^3$ and the line $y = 8$ on $[a, b] = [0, 2]$. (Compare with Exercise 33.)

38. The region between $f(x) = \ln x$ and $g(x) = 2 - \ln x$ on $[a, b] = [1, e^2]$.

39. The region between $f(x) = \sin x$ and the line $y = \frac{1}{2}$ on $[a, b] = [0, \pi]$.

40. The region between $f(x) = x^2$ and $g(x) = 64 - x^2$ on $[a, b] = [0, 8]$. (Compare with Exercise 34.)

Applications

Find each mass described in Exercises 41–48.

41. The mass of a block of copper that has a square base with sides measuring 20 centimeters and a height of 13 centimeters. Copper has a density of 8.93 grams per cubic centimeter.

42. The mass of a gold walking stick in the shape of a tall, thin cylinder that is 4 feet high and 3 inches wide. Gold has a density of 19.3 grams per cubic centimeter. *(Caution: Watch your units! You will need to use the conversion* 2.54 cm = 1 inch.*)*

43. The mass of a 12-inch rod with a square cross section of side length 1.5 inches, with density x inches from the left end given by $\rho(x) = 4.2 + 0.4x - 0.03x^2$ grams per cubic inch.

44. The mass of a cylindrical rod with a radius of 3 centimeters and a length of 20 centimeters, made of a combination of copper and aluminum in such a way that the density of the rod x centimeters from the left end is $\rho(x) = 8.93 - 0.015x$ grams per cubic centimeter.

45. The mass of a 10-centimeter rod with square cross section of side length 1 centimeter, made of a combination of metals in such a way that the density of the rod x centimeters from the left end is $\rho(x) = 15 - 0.5x$ grams per cubic centimeter.

46. The mass of a rectangular block of metal that is 30 centimeters wide, 21 centimeters long, and 14 centimeters tall, given that its density y centimeters from its base is $\rho(y) = 6.8 + 0.14y$ grams per cubic centimeter.

47. Last week Jack made an upside-down pudding cake that was 8 inches tall and 10 inches across, in the shape of the top of a downwards-pointing parabola that has been revolved around the y-axis. The raisins he put in the cake have tended to drift towards the bottom of the cake over the past week (nobody seems to want to eat this cake!) in such a way that the density of the cake at the very top of the mold is 0.15 ounce per cubic inch, while the density at the very bottom of the cake is 1.12 ounces per cubic inch.

(a) Describe Jack's upside-down pudding cake as a solid of revolution, by identifying the region to be revolved and the axis of revolution.

(b) Describe the density of Jack's upside-down pudding cake as a linear function of the height y of a point in the cake.

(c) Find the weight of Jack's upside-down pudding cake.

48. A solid cone of metal with a radius of 10 centimeters and a height of 15 centimeters is made of a combination of metals in such a way that, if the cone is pictured with its point up, then the density of the metal in the cone increases by 0.35 gram per cubic centimeter for each centimeter of height.

(a) Describe the metal cone as a solid of revolution, by identifying the region to be revolved and the axis of revolution.

(b) Given that the density at the bottom of the cone is 7.25 grams per cubic centimeter, find a function $\rho(y)$ that describes the density of the cone y centimeters above its base.

(c) Using the density function you found in part (b), find the total mass of the cone.

Find the work described in each of Exercises 49–60.

49. The work required to lift a full cylindrical soda fountain cup 1 foot into the air, given that the cup is full and has a radius of 3 inches and a height of 6 inches. You may assume that soda has the same weight-density as water. *(Caution: Watch your units!)*

50. The work required to lift a bathtub full of water 30 feet into the air, given that the bathtub itself weighs 150 pounds and it holds a rectangular volume of water measuring 5 feet by 2 feet by 2 feet.

51. The work required to pump all of the water out of the top of a giant rectangular hot tub that is 5 feet wide, 8 feet long, and 3 feet deep.

52. The work required to pump all of the water out of the top of a spherical tank with a radius of 27 feet.

53. The work required to pump all of the water out of the top of an upright conical tank that is 10 feet high and has a radius of 4 feet at the top.

54. The work required to pump all of the water out of the top of an upright conical tank that is 12 feet high and has a radius of 6 feet at the top.

55. The work required to pump all the water out of the top of a tank and up to ground level, given that the tank is a large upright cylinder with a radius of 4 feet and a height of 13 feet, buried so that its top is 3 feet below the surface.

56. The work required to pump all the liquid out of the top of a tank and up to 4 feet above ground level, given that the tank is a giant upright cylinder with a radius of 8 feet and a height of 20 feet, buried so that its top is 2 feet below the surface. The tank is only two-thirds full and contains a strange liquid that weighs 71.8 pounds per cubic foot.

57. The work required to pump all of the liquid out of the top of a cylindrical tank on a tanker truck (so that the cylinder is on its side) with a length of 100 feet and a radius of 8 feet. The tank is full of a liquid that weighs 42.3 pounds per cubic foot. Be careful; in this problem the slices are not cylinders!

58. Buffy has tied a rope to a 50-pound sword so that she can hoist it up to her upstairs bedroom window, 23 feet off the ground. How much work will it take for Buffy to lift the sword up to her window, assuming that the weight of the rope is negligible?

59. Willow has tied a rope to a 20-pound mallet so that she can hoist it up to her upstairs bedroom window, 18 feet off the ground. How much work will it take for Willow to lift the mallet up to her window, assuming that the rope weighs 0.25 pound per foot?

60. Xander has tied a rope to a 100-pound block of ice so that he can hoist it up to his upstairs bedroom window, 22 feet off the ground. How much work will it take for Xander to lift the block of ice up to his window, assuming that the weight of the rope is negligible and the block of ice melts off 2 pounds of water for each foot it is lifted?

Find the hydrostatic force described in each of Exercises 61–69.

61. The hydrostatic force on the bottom of a full cylindrical water glass with a radius of 2 inches and a height of 7 inches.

62. The hydrostatic force on the bottom of a full cylindrical beer stein with a radius of 6 centimeters and a height of 20 centimeters. You may assume that the weight-density of beer is the same as for water.

63. The hydrostatic force exerted on one of the long sides of a giant rectangular hot tub that is 5 feet wide, 8 feet long, and 3 feet deep.

64. The hydrostatic force exerted on one of the long sides of a rectangular swimming pool that is 8 feet long, 6 feet wide, and 4.5 feet deep.

65. The hydrostatic force exerted by the water in a birdbath on its cylindrical side, given that the birdbath is in the shape of a cylinder that is 1 foot deep with a radius of 0.4 foot.

66. The hydrostatic force exerted on a dam in the shape of an isosceles triangle whose top is 200 feet wide and whose total height is 100 feet, given that the dam is holding back a body of water that reaches to the top of the dam.

67. The hydrostatic force exerted on a dam in the shape of a trapezoid whose top is 400 feet long and whose base is 250 feet long, given that the dam is holding back a body of water that reaches all 200 feet from the bottom to the top of the dam.

68. The hydrostatic force exerted on one of the long sides of a swimming pool that is 24 feet long and 10 feet wide, with a shallow end that has a depth of 3 feet (for the first 4 feet of the pool) and a deep end that has a depth of 15 feet (for the last 8 feet of the pool). The middle portion of the bottom of the pool is a straight ramp leading down from the shallow depth to the depth at the deep end.

69. The hydrostatic force exerted on one of the long sides of a pool that is 20 feet long and 8 feet wide, with a shallow end that has a depth of 4 feet (for the first 6 feet of the pool) and a deep end that has a depth of 12 feet (for the last 6 feet of the pool), and with a middle portion that is a straight ramp.

Use centroids to solve the real-world problems in Exercises 70–71.

70. Ian is making some snow flukes, which are metal sheets that can be used to prevent a fall on steep snow. The flukes are in the shape of a polygon with vertices $(0, 8)$, $(6, 8)$, $(6, 2)$, $(3, 0.5)$, and $(0, 8)$ with all distances given in inches. Ian needs to clip a cable to the centroid of the fluke. Where should he attach the cable?

71. Leila needs to find a location for a radio repeater that will monitor the signals from a pack of radio-collared wolves. In order to get the strongest signal, she wants to locate the repeater at the centroid of the range of the wolf pack. The valley in which the wolves live encompasses roughly the union of the rectangle $0 \le x \le 8, 0 \le y \le 4$ and the part of the ellipse

$$\frac{(x-6)^2}{4} + \frac{(y-2)^2}{100} = 1$$

that lies above the x-axis, where all distances are in miles. Where should Leila put the repeater?

Proofs

72. Use the physics formulas given in this text to prove that the work required to lift a cube of liquid with weight-density ω and side length s through a distance of d feet is $W = \omega s^2 d$.

73. Suppose an object has horizontal cross sections for $x \in [a, b]$, so that the functions for density $\rho(x)$ and area $A(x)$ are continuous. Use physics formulas and the definition of the definite integral to prove that the accumulated mass of such an object is

$$m = \int_a^b \rho(x)\, A(x)\, dx.$$

74. Suppose an object whose particles are to be displaced has vertical cross sections for $y \in [a, b]$, so that the functions for weight-density $\omega(y)$, area $A(y)$, and displacement $d(y)$ are continuous. Use physics formulas and the definition of the definite integral to prove that the accumulated work required is

$$W = \int_a^b \omega(y)\, A(y)\, d(y)\, dy.$$

75. Suppose the wall of a container has vertical cross sections for $y \in [a, b]$, so that the functions for weight-density $\rho(y)$, length $l(y)$, and depth $d(x)$ are continuous. Use physics formulas and the definition of the definite integral to prove that the accumulated hydrostatic force on such a wall is

$$F = \int_a^b \omega(y)\, l(y)\, d(y)\, dy.$$

76. Use a Riemann sum argument to prove that the centroid $(\bar{x}, \bar{y})$ of the region between two curves $f(x)$ and $g(x)$ on an interval $[a, b]$ has coordinates

$$\bar{x} = \frac{\int_a^b x|f(x) - g(x)|\, dx}{\int_a^b |f(x) - g(x)|\, dx},$$

$$\bar{y} = \frac{\frac{1}{2}\int_a^b |f(x)^2 - g(x)^2|\, dx}{\int_a^b |f(x) - g(x)|\, dx}.$$

As part of your argument you will need to explain the reason for the four definite integrals in these expressions, as well as the absolute value signs that appear in the expression. Your argument should also contain the words "rectangle," "weighted average," and "limit."

Thinking Forward

- *The volume of a torus:* A **torus** is a donut-shaped object that is obtained by rotating a circle around an axis. For example, we could revolve the circle with equation $y^2 + (x-3)^2 = 1$ (i.e., the circle with radius 1 and center at $(0, 3)$) around the y-axis to obtain a torus. Use a "subdivide, approximate, and add" strategy to find the volume of this torus. (*Hint: The functions $\sqrt{1-y^2}+3$ and $-\sqrt{1-y^2}+3$ will be involved; why? You can use disks/washers or shells to find this volume.*)

- *Pappus's Centroid Theorem:* The volume of a solid obtained by revolving a region in the plane around an axis that does not intersect that region is equal to $V = 2\pi r A$, where r is the distance from the centroid of the region to the axis and A is the area of the region. Use this theorem to find the volume of the torus described in the previous problem without having to set up any definite integrals. Then discuss why the theorem makes intuitive sense in terms of an alternative "subdivide, approximate, and add" strategy for finding volume in this case.

6.5 DIFFERENTIAL EQUATIONS*

- Using antidifferentiation and separation of variables to solve differential equations
- Using slope fields and Euler's method to approximate solutions of differential equations
- Common forms of initial-value problems and their applications

Differential Equations and Initial-Value Problems

Recall that an ***equation*** is nothing more than two mathematical expressions connected by an equals sign. For example, $x^2-4=0$ is a single-variable equation, and its set of ***solutions*** is the set the values of x that make the equation true, namely, $x=2$ and $x=-2$. Similarly, $y=x^2-4$ is a two-variable equation, and its set of solutions is the set of ordered pairs (x,y) that make the equation true, namely, the ordered pairs of the form $(x,y)=(x,x^2-4)$ that, when graphed together, make up the parabola that is the graph of $y=x^2-4$.

If an equation involves the derivative $\frac{dy}{dx}$ of some function $y(x)$, then we say that the equation is a ***differential equation***. For example, two examples of differential equations are

$$\frac{dy}{dx}=x^2 \quad \text{and} \quad \frac{dy}{dx}=y.$$

Once again, the solutions of these equations are the objects that make the equations true. For example, the solutions of $\frac{dy}{dx}=x^2$ are the functions $y(x)$ whose derivatives equal x^2, namely, the functions of the form $y(x)=\frac{1}{3}x^3+C$. The solutions of $\frac{dy}{dx}=y$ are the functions $y(x)$ that are equal to their own derivatives, namely, the functions $y(x)=Ae^x$. In general, a ***solution*** of a differential equation is a function that makes the equation true.

A ***first-order*** differential equation is an equation that involves the first derivative $\frac{dy}{dx}$, a function y, and independent variable x. Higher order differential equations involve higher order derivatives. Many of the first-order differential equations we will study are of the form $\frac{dy}{dx}=g(x,y)$, where $g(x,y)$ is some function of the variables x and y. The simplest case occurs when g involves only the variable x. Such a differential equation can be solved by antidifferentiation:

THEOREM 6.16 **Differential Equations Whose Solutions Are Antiderivatives**

The solution of a differential equation of the form $\frac{dy}{dx}=g(x)$ is the family $y(x)=G(x)+C$ of antiderivatives of $g(x)$.

Proof. If $G(x)$ is an antiderivative of $g(x)$, then $G'(x)=g(x)$, which means that $y(x)=G(x)$ is a solution of the differential equation $\frac{dy}{dx}=g(x)$. Moreover, we know that two functions of x have the same derivative precisely when they differ by a constant. This means that *every* solution of the differential equation $\frac{dy}{dx}=f(x)$ must be of the form $y(x)=G(x)+C$. ■

A differential equation of the form $\frac{dy}{dx}=g(x,y)$ with initial condition $y(x_0)=y_0$ is called an ***initial-value problem***. Provided that various continuity, differentiability, and boundedness conditions are satisfied, an initial-value problem will have one and only one solution $y(x)$. The details can be found in any standard differential equations textbook. For our purposes it is sufficient to say that when we consider an initial-value problem in this book, it will be one for which these conditions are met and it will have one and only one solution.

For example, the differential equation $\frac{dy}{dx}=x^2$ has many solutions, all of the form $y(x)=\frac{1}{3}x^3+C$. Only one of these solutions also satisfies the initial condition $y(0)=2$, namely, the function $y(x) = \frac{1}{3}x^3 + 2$. Therefore $y(x) = \frac{1}{3}x^3 + 2$ is the unique solution of the initial-value problem $\frac{dy}{dx} = x^2$ and $y(0) = 2$. This solution is shown next at the left, highlighted amongst other solutions of the differential equation $\frac{dy}{dx} = x^2$. Similarly, the function $y(x) = 2e^x$ is the unique solution of the initial-value problem $\frac{dy}{dx} = y$ and $y(0) = 2$, as illustrated at the right.

$y(x) = \frac{1}{3}x^3 + 2$ is the only solution of $\frac{dy}{dx} = x^2$ that satisfies $y(0) = 2$

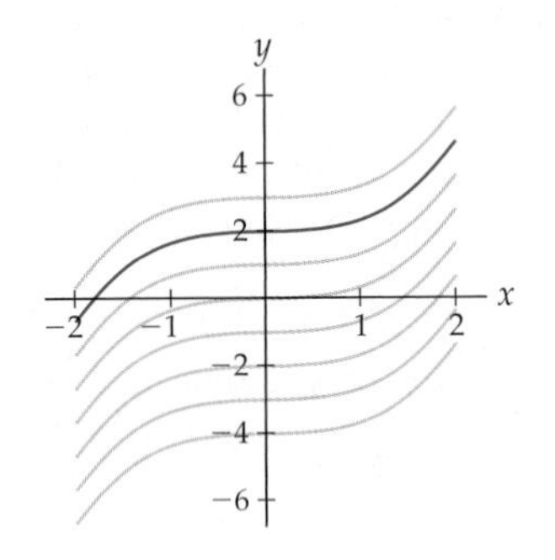

$y(x) = 2e^x$ is the only solution of $\frac{dy}{dx} = y$ that satisfies $y(0) = 2$

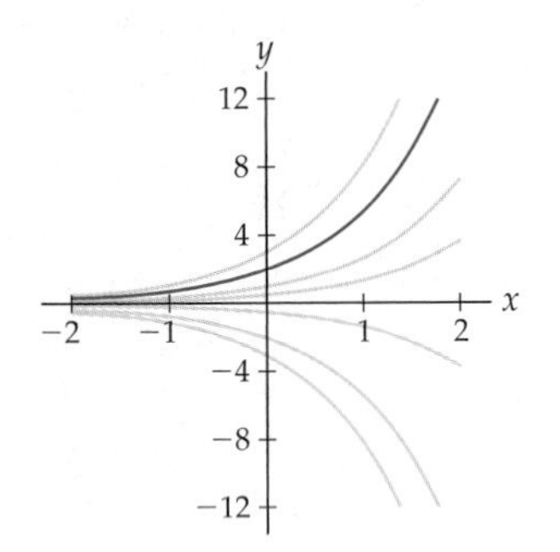

Separable Differential Equations

We have seen that differential equations of the form $\frac{dy}{dx} = g(x)$ can be solved if we can antidifferentiate the function $g(x)$. How do we solve differential equations that involve the independent variable y? The answer is that sometimes we can and sometimes we cannot. One particular technique is useful for differential equations for which the y-related information can be separated from the x-related information. Note that in this section we will be assuming that the functions $y(x)$ we work with are continuous on their domains.

Let's consider the simple example $\frac{dy}{dx} = y$. Our strategy will be to move all the y-related parts of the differential equation to the side with the derivative $\frac{dy}{dx}$ and then integrate both sides:

$$\frac{dy}{dx} = y \quad \Longrightarrow \quad \frac{1}{y}\frac{dy}{dx} = 1 \quad \Longrightarrow \quad \int \frac{1}{y}\frac{dy}{dx}\,dx = \int 1\,dx \quad \Longrightarrow \quad \int \frac{1}{y}\,dy = \int 1\,dx.$$

The last step followed from integration by substitution, with $y(x)$ playing the role of what we would normally call $u(x)$ when substituting. Notice that Leibniz notation makes it easy for us to remember this substitution, since it has the effect of "cancelling the dx" in the notation.

After considering the work we just did, we now have a way to move towards finding the solutions $y(x)$ to the differential equation: We simply perform the integration and solve for $y(x)$. This gives us

$$\ln|y| + C_1 = x + C_2 \quad \Longrightarrow \quad \ln|y| = x + C \quad \Longrightarrow \quad |y| = e^{x+C} = e^C e^x \quad \Longrightarrow \quad y = Ae^x.$$

Along the way we just defined C_1 and C_2 as arbitrary constants, with $C = C_2 - C_1$ as their difference and $A = \pm e^C$. All that matters by the end of this calculation is that A is some constant. If we knew a particular initial value for $y(x)$, then we could solve uniquely for this constant A.

For a more complicated differential equation like $\frac{dy}{dx} = y + xy$, we can use a similar technique, rearranging the equation so that all the y-related expressions are on the left side with a multiplier of $\frac{dy}{dx}$ and all the x-related expressions are on the right, as follows:

$$\frac{dy}{dx} = y + xy \quad \Longrightarrow \quad \frac{dy}{dx} = y(1+x) \quad \Longrightarrow \quad \frac{1}{y}\frac{dy}{dx} = 1 + x.$$

We could then integrate both sides and solve for y to find the family of solutions of the differential equation; see Example 1.

This technique is called ***separation of variables***, because the initial step involves separating the y-related and x-related parts of the differential equation. We say that a differential equation $\frac{dy}{dx} = g(x, y)$ is ***separable*** if we can write it in the form $\frac{dy}{dx} = p(x)q(y)$ for some functions $p(x)$ and $q(x)$. With such a separable differentiable equation we have

$$\frac{1}{q(y)}\frac{dy}{dx} = p(x).$$

Integrating both sides with respect to x and applying u-substitution with $u = y$ and $du = dy = y'(x)\,dx = \frac{dy}{dx}\,dx$, we have the following theorem:

THEOREM 6.17 **Solving a Differential Equation by Separation of Variables**

Given a separable differential equation $\frac{dy}{dx} = p(x)q(y)$, it follows that

$$\int \frac{1}{q(y)}\,dy = \int p(x)\,dx.$$

Solving both of these integrals and then solving for $y(x)$, if possible, yields a solution of the differential equation.

Of course, in some cases one or both integration steps will be impossible, or perhaps after integrating we might be unable to solve for the function $y(x)$. But if everything works out, then this method gives us a nice way to solve more complicated differential equations with our antidifferentiation skills.

Slope Fields

Every differential equation $\frac{dy}{dx} = g(x, y)$ is at its heart a statement about the slopes of some function $y(x)$ at various points (x, y) in the plane. For example, a solution of the differential equation $\frac{dy}{dx} = x^2$ is a function $y(x)$ whose slope at each of its graphed points (a, b) is equal to a^2. If we draw small line segments at each point (a, b) in the plane, each with slope a^2, we get the ***slope field*** shown next at the left. The solutions of $\frac{dy}{dx} = x^2$ are all of the form $y(x) = \frac{1}{3}x^3 + C$, and each of these functions flows along with the small line segments in the slope field. Similarly, a slope field for the differential equation $\frac{dy}{dx} = y$ consists of line segments whose slope at (a, b) is equal to b, as shown at the right. The curves that flow through this slope field are the solutions $y(x) = Ae^{kx}$.

Slope at (a, b) is equal to a^2

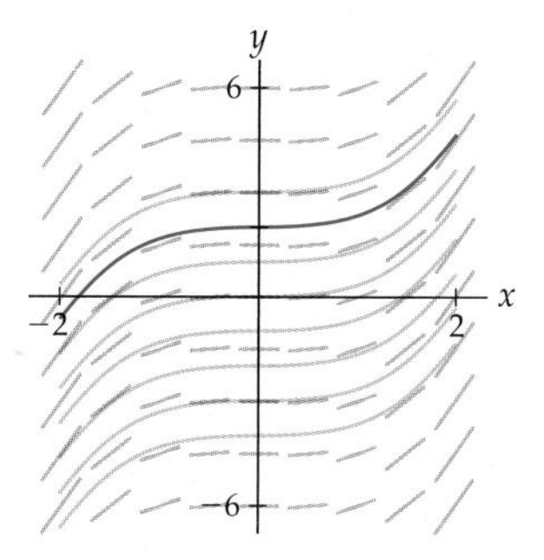

Slope at (a, b) is equal to b

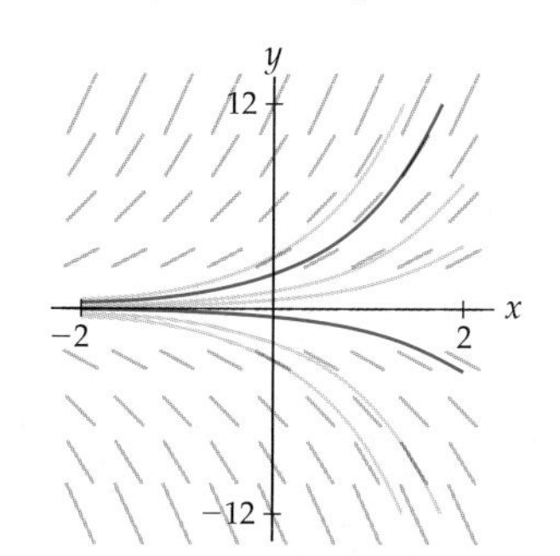

In the first slope field, the slopes are the same across each column because the differential equation $\frac{dy}{dx} = x^2$ does not involve the variable y. Similarly, in the second slope field the slopes are the same in each row because $\frac{dy}{dx} = y$ does not involve the variable x. In Example 2 you will construct a slope field that depends on both x and y. In general, we can sketch a slope field for any differential equation:

DEFINITION 6.18 The Slope Field of a Differential Equation

Given a differential equation $\frac{dy}{dx} = g(x, y)$ and a lattice of points in the plane, the associated ***slope field*** is a collection of line segments drawn at each point (x, y) in the lattice, with the property that the slope of the segment drawn at (x, y) is equal to $g(x, y)$.

Euler's Method

The method of tracing through a slope field to approximate a solution of a differential equation can be put into algebraic form with the technique known as ***Euler's method***. Consider for example the initial-value problem consisting of the differential equation $\frac{dy}{dx} = y$ and the initial value $y(0) = 2$. We will use slope field information to construct a sequence of points going forward from this initial condition that approximate values of the solution of the given differential equation.

By construction, the solution $y(x)$ we would like to approximate must pass through the initial point $(x_0, y_0) = (0, 2)$. We also know from the differential equation $\frac{dy}{dx} = y$ that the slope of the function $y(x)$ at this point is given by $\frac{dy}{dx}\big|_{(0,2)} = 2$. Near the point $(0, 2)$ we would expect the function $y(x)$ to behave a lot like its tangent line, so we will follow this slope for a step of $\Delta x = 0.5$ to the right, as illustrated here:

Getting from (x_0, y_0) to (x_1, y_1)

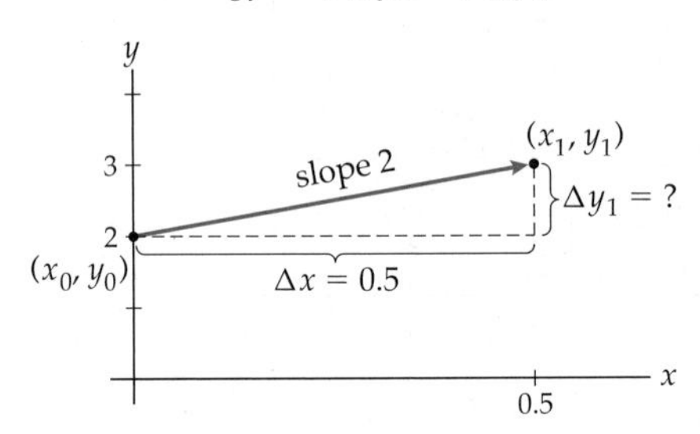

So what are the coordinates of our new points (x_1, y_1)? Clearly we have $x_1 = x_0 + \Delta x = 0 + 0.5 = 0.5$. To find the new y-coordinate y_1 we must find the change Δy_1 between y_0 and y_1. Using the slope information, we have

$$2 = \frac{\Delta y_1}{0.5} \quad \Longrightarrow \quad \Delta y_1 = 1.$$

Since $y_0 = 2$, it follows that $y_1 = y_0 + \Delta y_1 = 2 + 1 = 3$.

We have just followed the slope $\frac{dy}{dx}\big|_{(x_0,y_0)}$ from our initial point (x_0, y_0) to a new point (x_1, y_1). This new point will in general not be on the actual graph of the solution $y(x)$ to the differential equation, but it will be a reasonable approximation to a point on the graph of the solution. We can repeat the process, following the slope $\frac{dy}{dx}\big|_{(x_1,y_1)}$ to get from the point (x_1, y_1) to another new point (x_2, y_2), and so on. This produces a sequence of points that we can connect to form an approximation of the solution function $y(x)$. In the next figure at the left is the result of continuing the process with $\Delta x = 0.5$ for four steps. At the right is the result of following the same process with $\Delta x = 0.25$ for eight steps; see Example 3 for the details.

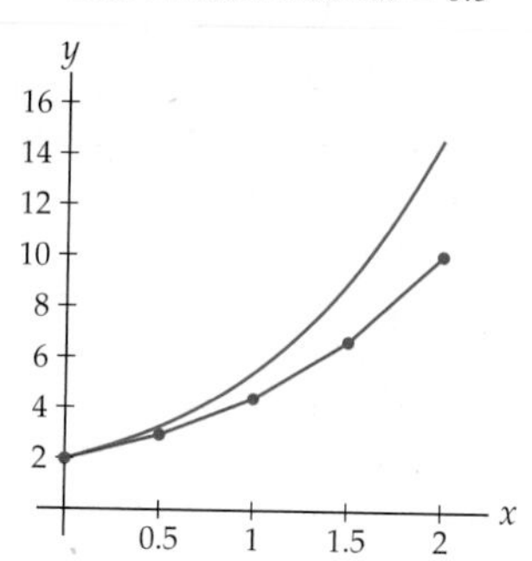

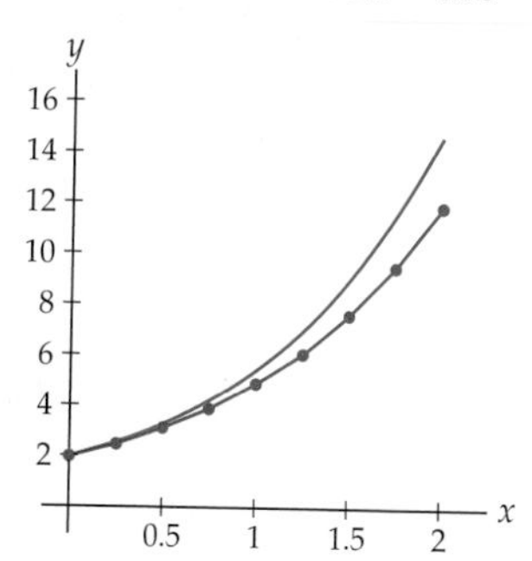

In each of these figures the Euler's method approximation is shown, compared with the actual solution $y(x) = 2e^x$ of the initial-value problem $\frac{dy}{dx} = y$, $y(0) = 2$. Notice that the approximation is better when we use a smaller value of Δx and therefore update our slope information more frequently. Notice also that Euler's method approximations start out close to the solution graph but then tend to drift away as errors compound in each step of the process. In general, Euler's method provides us with a useful way of approximating solutions of initial-value problems that are too complicated to solve algebraically:

DEFINITION 6.19 **Euler's Method of Approximating the Solution of an Initial-Value Problem**

Given an initial-value problem with differential equation $\frac{dy}{dx} = g(x, y)$ and initial condition $y(x_0) = y_0$, and given a value $\Delta x > 0$, define a sequence of points (x_k, y_k) by the iterative formula

$$(x_{k+1}, y_{k+1}) = (x_k + \Delta x, y_k + \Delta y_k), \quad \text{where } \Delta y_k = g(x_k, y_k)\,\Delta x.$$

This sequence of points forms an ***Euler's method*** approximation to the solution $y(x)$ of the initial-value problem.

Applications of Initial-Value Problems

Many real-world situations can be modeled by initial-value problems. When we have information about the rate of change and initial value of a quantity, we can often set up a model of an initial-value problem whose solution will be the function that describes that quantity. We will describe three such models next, each of which results in a differential equation that can be solved exactly by the technique of separation of variables.

One of the simplest situations is when a quantity $Q(t)$ grows at a rate proportional to itself. This happens often with simple growth and decay models, as you will see in Exercises 65–67.

THEOREM 6.20 **Exponential Growth and Decay Models**

If a quantity $Q(t)$ changes over time at a rate proportional to its value, then that quantity is modeled by a differential equation of the form

$$\frac{dQ}{dt} = kQ$$

for some proportionality constant k, where $k > 0$ represents growth and $k < 0$ represents decay. The solution of this differential equation that satisfies the initial value $Q(0) = Q_0$ is

$$Q(t) = Q_0 e^{kt}.$$

The proof of this theorem requires that we solve the differential equation $\frac{dQ}{dt} = kQ$ with initial condition $Q(0) = Q_0$. The work is similar to what we did in our motivating example for the technique of separation of variables, so the details of the proof are left to Exercise 75.

Of course, real-life situations are often more complicated. For example, it makes sense to imagine that a population of animals on an island might grow at a rate proportional to the number of animals on the island; when there are more animals on the island, there are also more animals around to make offspring. However, at some point there may be too many animals on the island to be supported by the available food and shelter. In this case we say that the island has a ***carrying capacity*** of some maximum number L of animals. The following ***logistic model*** describes such a situation, where for small population values the growth rate is approximately exponential with some ***natural growth rate*** k, but for population values near the carrying capacity L the growth rate levels off to zero:

THEOREM 6.21

Logistic Growth Model

If a population $P(t)$ has natural growth rate k and is restricted by a carrying capacity L, then it can be modeled by the differential equation

$$\frac{dP}{dt} = kP\left(1 - \frac{P}{L}\right).$$

The solution of this differential equation that satisfies the initial value $P(0) = P_0$ is

$$P(t) = \frac{LP_0}{P_0 + (L - P_0)e^{-kt}}.$$

To prove this theorem we must solve the initial-value problem that represents logistic growth and show that we obtain the given solution. This can be done by separation of variables followed by partial fractions, either in general (see Exercise 76) or on a case-by-case basis (see Exercises 68 and 69).

We can also use differential equations to model the changing temperature of an object. If a hot or cold object is in an environment with a constant ambient temperature, then the object will cool down or heat up until its temperature matches that of its environment. In an ideal situation this happens in such a way that the rate of change of the temperature of the object is proportional to the difference of the ambient temperature and the current temperature of the object. In this situation we have the following differential equation, attributed to Newton:

THEOREM 6.22

Newton's Law of Cooling and Heating

If an object with temperature $T(t)$ is placed in a location with a constant ambient temperature A, then the temperature of the object can be modeled by a differential equation of the form

$$\frac{dT}{dt} = k(A - T)$$

for some proportionality constant k. The solution of this differential equation that satisfies the initial value $T(0) = T_0$ is

$$T(t) = A - (A - T_0)e^{-kt}.$$

It is not difficult to solve this initial-value problem by the technique of separation of variables; you will work out the details of the proof in general in Exercise 78 and in specific cases in Exercises 70 and 71. Note that in practice we might encounter a situation where we know the temperature of the object not at time $t = 0$, but at some other point in time. In those situations the equation for $T(t)$ in Theorem 6.22 will involve different constants, which are not difficult to solve for after performing separation of variables to solve the differential equation; see Example 4.

Examples and Explorations

EXAMPLE 1

Using separation of variables to solve initial-value problems

Solve each of the following initial-value problems:

(a) $\frac{dy}{dx} = \sin x,\; y(0) = 2$ **(b)** $\frac{dy}{dx} = y^2,\; y(0) = 3$ **(c)** $\frac{dy}{dx} = y + xy,\; y(0) = 2$

SOLUTION

(a) The differential equation $\frac{dy}{dx} = \sin x$ does not involve the dependent variable at all, so technically it is already separated. By antidifferentiating, the solutions of this differential equation are of the form

$$y(x) = -\cos x + C.$$

Using the initial condition $y(0) = 2$, we get

$$2 = -\cos 0 + C \quad\Longrightarrow\quad 2 = -1 + C \quad\Longrightarrow\quad C = 3.$$

Therefore the solution of the initial-value problem is the function $y(x) = -\cos x + 3$.

(b) We can write this differential equation in separable form as $\frac{dy}{dx} = (1)(y^2)$; in the notation of Theorem 6.17 we have $p(x) = 1$ and $q(x) = y^2$. Dividing both sides by the y^2 expression and integrating both sides, we obtain

$$\frac{dy}{dx} = y^2 \quad\Longrightarrow\quad \frac{1}{y^2}\frac{dy}{dx} = 1 \quad\Longrightarrow\quad \int y^{-2}\,dy = \int 1\,dx \quad\Longrightarrow\quad -y^{-1} = x + C.$$

Notice that we did not add a constant after integrating the left side of the equation; this is because any constant on that side can be absorbed into the constant on the right side. All that remains now is to solve for y:

$$-\frac{1}{y} = x + C \quad\Longrightarrow\quad y = -\frac{1}{x + C}.$$

We can use the initial condition $y(0) = 3$ to solve for C:

$$y(0) = 3 \quad\Longrightarrow\quad 3 = -\frac{1}{0 + C} \quad\Longrightarrow\quad C = -\frac{1}{3}.$$

Therefore the solution of the initial-value problem is $y(x) = -\frac{1}{x - (1/3)}$.

(c) The differential equation $\frac{dy}{dx} = y + xy$ involves both x and y. As we saw earlier in the reading, we can separate the variables and then integrate:

$$\frac{dy}{dx} = y + xy \quad\Longrightarrow\quad \frac{dy}{dx} = y(1 + x) \quad\Longrightarrow\quad \int \frac{1}{y}\,dy = \int 1 + x\,dx$$

$$\Longrightarrow\quad \ln|y| = x + \frac{1}{2}x^2 + C.$$

Now solving for y, we have

$$|y| = e^{x + (1/2)x^2 + C} \quad\Longrightarrow\quad y = Ae^{x + (1/2)x^2}.$$

Notice that C is an arbitrary constant from the integration step, and we have defined $A = \pm e^C$ to simplify the expression and take care of the absolute value. We are given the initial value $y(0) = 2$, which yields:

$$y(0) = 2 \quad \Longrightarrow \quad 2 = Ae^0 \quad \Longrightarrow \quad A = 2.$$

Therefore the solution we seek is $y(x) = 2e^{x+(1/2)x^2}$. □

CHECKING THE ANSWER

To check our answers to the previous example we need only differentiate and plug in values to compare against the original initial-value problems. Let's do this for part (c). If $y(x) = 2e^{x+(1/2)x^2}$, then we have $\frac{dy}{dx} = 2e^{x+(1/2)x^2}(1+x) = y(1+x) = y + xy$, as we started with in that problem. Moreover, $y(0) = 2e^{0+(1/2)(0)} = 2(1) = 2$, as desired.

EXAMPLE 2

Sketching a slope field and tracing solution curves

Sketch a slope field for the differential equation $\frac{dy}{dx} = 2xy$ on a lattice of points contained within $-2 \le x \le 2$ and $-10 \le y \le 10$. Then use the slope field to sketch graphs of three approximate solutions of the differential equation.

SOLUTION

Let's start by looking at the slopes at a few chosen points in the lattice. We have $\frac{dy}{dx} = g(x, y)$ with $g(x, y) = 2xy$. At each point (a, b) in the lattice we want to sketch a line segment with slope $g(a, b) = 2ab$. For example, at the point $(0, 0)$ we draw a segment with slope $2(0)(0)$, and at the point $(1, 2)$ we draw a segment with slope $2(1)(2) = 4$. Clearly, for positive a and b, the slope $2ab$ will be positive. Likewise, if a and b are both negative, then the slope $2ab$ will be positive, and if a and b have opposite signs, then the slope $2ab$ will be negative. In all cases, the magnitude of the slope will grow larger as a and/or b increase, and grow smaller as a and/or b decrease. After plotting a few of these line segments to get an initial idea, we can sketch the slope field shown here at the left:

Slope field for $\frac{dy}{dx} = 2xy$

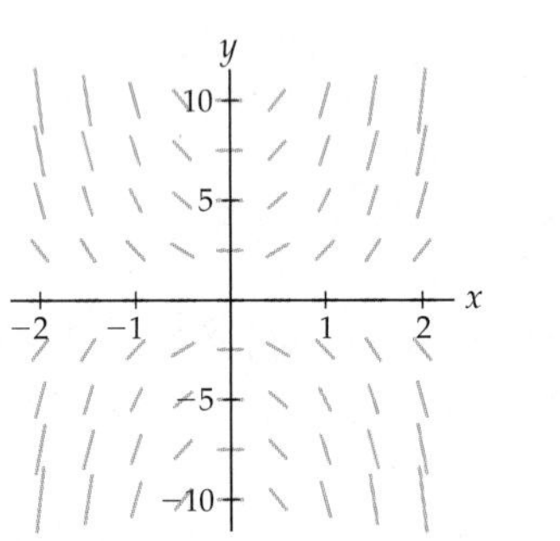

Three solution curves in the slope field

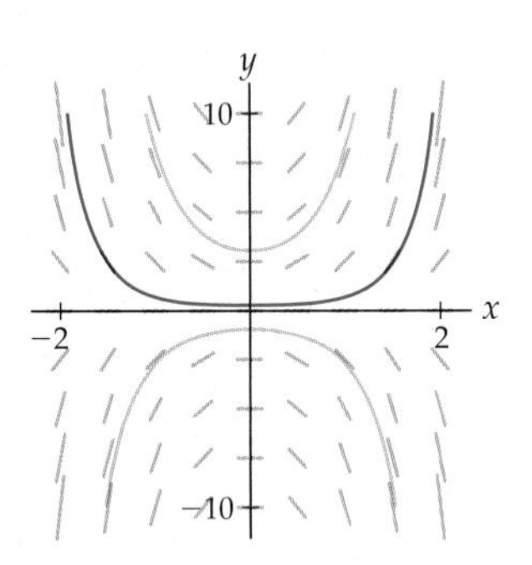

Three curves through the slope field are shown in the figure at the right. Each of these is an approximate solution of the differential equation $\frac{dy}{dx} = 2xy$. □

CHECKING THE ANSWER

Sometimes we sketch a slope field because we cannot solve a differential equation by hand. That is not the case in Example 2, and so we can check whether our slope field is reasonable by actually solving the differential equation $\frac{dy}{dx} = 2xy$. A straightforward separation-of-variables calculation (see Exercise 21) shows that the solutions are each of the form $y(x) = Ae^{x^2}$ for some constant A. This means that the graphs of, for example, $y(x) = 3e^{x^2}$, $y(x) = \frac{1}{4}e^{x^2}$, and $y(x) = -e^{x^2}$ should flow through the slope field. And in fact, these are precisely the three functions that we traced through the slope field earlier at the right.

EXAMPLE 3

Applying Euler's method

Consider the differential equation $\frac{dy}{dx} = y$ with initial condition $y(0) = 2$. Use Euler's method with $\Delta x = 0.25$ to approximate the solution $y(x)$ to this initial-value problem on the interval $[0, 2]$.

SOLUTION

We want to approximate the solution of $\frac{dy}{dx} = y$ that passes through the point $(x_0, y_0) = (0, 2)$. Euler's method with $\Delta x = 0.25$ allows us to construct a sequence of eight more coordinate points $(x_1, y_1), \ldots, (x_8, y_8)$ over the interval $[0, 2]$. Each (x_{k+1}, y_{k+1}) is obtained from the previous point (x_k, y_k) by stepping over Δx and up or down $g(x_k, y_k)\Delta x$ according to the value of $\frac{dy}{dx}\big|_{(x_k, y_k)} = g(x_k, y_k) = y_k$:

$$(x_0,\ y_0) = (0, 2);$$

$$(x_1,\ y_1) = (x_0 + \Delta x,\ y_0 + g(x_0, y_0)\Delta x) = (0 + 0.25,\ 2 + 2(0.25)) = (0.25,\ 2.5);$$

$$(x_2,\ y_2) = (x_1 + \Delta x,\ y_1 + g(x_1, y_1)\Delta x) = (0.25 + 0.25,\ 2.5 + 2.5(0.25)) = (0.5,\ 3.125);$$

$$(x_3,\ y_3) = (x_2 + \Delta x,\ y_2 + g(x_2, y_2)\Delta x) = (0.5 + 0.25,\ 3.125 + 3.125(0.25)) = (0.75,\ 3.91);$$

$$(x_4,\ y_4) = (x_3 + \Delta x,\ y_3 + g(x_3, y_3)\Delta x) = (0.75 + 0.25,\ 3.91 + 3.91(0.25)) = (1,\ 4.89);$$

$$(x_5,\ y_5) = (x_4 + \Delta x,\ y_4 + g(x_4, y_4)\Delta x) = (1 + 0.25,\ 4.89 + 4.89(0.25)) = (1.25,\ 6.11);$$

$$(x_6,\ y_6) = (x_5 + \Delta x,\ y_5 + g(x_5, y_5)\Delta x) = (1.25 + 0.25,\ 6.11 + 6.11(0.25)) = (1.5,\ 7.64);$$

$$(x_7,\ y_7) = (x_6 + \Delta x,\ y_6 + g(x_6, y_6)\Delta x) = (1.5 + 0.25,\ 7.64 + 7.64(0.25)) = (1.75,\ 9.55);$$

$$(x_8,\ y_8) = (x_7 + \Delta x,\ y_7 + g(x_7, y_7)\Delta x) = (1.75 + 0.25,\ 9.55 + 9.55(0.25)) = (2, 11.94).$$

When plotted and connected by line segments, this sequence of points produces a piecewise-linear approximation to the solution of the initial-value problem $\frac{dy}{dx} = y$ with initial condition $y(0) = 2$. This is the same piecewise-linear approximation that we sketched in the rightmost figure preceding Definition 6.19. □

EXAMPLE 4

Using Newton's Law of Cooling

Suppose your teacher sets down a fresh cup of coffee onto a table in a classroom that is kept at a constant temperature of 68° Fahrenheit. After giving a fascinating 45-minute lecture on differential equations she remembers her cup of coffee, which has now cooled to a lukewarm 75° Fahrenheit. Assuming that the proportionality constant for Newton's Law of Cooling in this case is $k = 0.065$, how hot was the coffee before the lecture started?

SOLUTION

Newton's Law of Cooling tells us that the coffee should get cooler until it matches the temperature of the 68° room. We are told that for this particular cup of coffee and room environment, the rate of this cooling will be proportional to the difference in temperatures of the coffee and the room, with proportionality constant $k = 0.065$. In other words, if $T(t)$ is the temperature of the coffee after t minutes, then we have the differential equation

$$\frac{dT}{dt} = 0.065(68 - T).$$

Notice that T starts out hotter than the 68° room, so the rate of change $\frac{dT}{dt} = 0.065(68 - T)$ is negative, as we would expect.

Let us start by solving this differential equation. After applying separation of variables, we have

$$\int \frac{1}{68-T}\,dT = \int 0.065\,dt.$$

Integrating both sides gives us the equation

$$-\ln|68-T| = 0.065t + C,$$

and after solving this equation for T, we have

$$T = 68 - Be^{-0.065t},$$

where $B = e^{-C}$ is some constant.

We do not know the initial temperature $T(0)$ of the coffee at the start of the teacher's lecture, but we do know that the temperature of the coffee after 45 minutes is $T(45) = 75°$. Substituting this number into our equation for $T(t)$ results in

$$75 = 68 - Be^{-0.065(45)} \quad \Rightarrow \quad B = \frac{-7}{e^{-0.065(45)}} \approx -130.44.$$

Therefore the temperature of the coffee is given by the equation $T(t) = 68 + 130.44e^{-0.065t}$.

Now that we have an equation for the temperature $T(t)$ of the coffee, we can easily calculate $T(0)$ to obtain the initial temperature of the teacher's coffee. This initial temperature was approximately $T(0) = 68 + 130.44e^{-0.065(0)} = 198.44$ degrees Fahrenheit. □

CHECKING THE ANSWER

As a reality check, let's graph the function $T(t) = 68 + 130.44e^{-0.065t}$ and verify that it is a good model for the temperature of the coffee as it cools:

Temperature T(t) of coffee

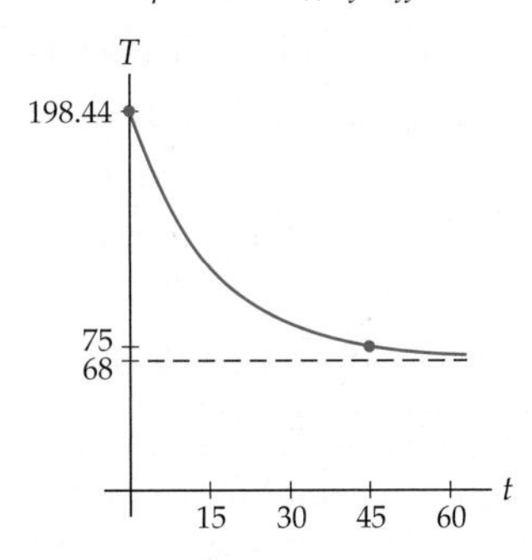

Notice that the coffee starts out at nearly 200°, and then its temperature declines, first quickly and then more slowly, approaching the ambient temperature of the room. According to this model the coffee's temperature after $t = 45$ minutes is 75°, as we wanted. Notice also that as $t \to \infty$ the temperature of the coffee becomes asymptotically close to the temperature of the 68° classroom, as we would expect. This model function $T(t)$ does indeed seem to match with what we would expect from the teacher's cooling cup of coffee.

TEST YOUR UNDERSTANDING

- What does it mean for a function $y(x)$ to be a solution of a differential equation? What about a solution of an initial-value problem?
- What does it mean for a differential equation to be separable? Can you give an example of a differential equation that is not separable?
- What is a slope field associated with a differential equation, and how are the slopes of the line segments in a slope field related to the differential equation?

- When using Euler's method, how do we use slope information to determine how to get each point (x_{k+1}, y_{k+1}) from the previous point (x_k, y_k)?
- Why does the differential equation for Newton's Law of cooling make real-world sense?

EXERCISES 6.5

Thinking Back

Integration: Solve each of the following indefinite integrals.

- $\displaystyle\int \frac{1}{y^2}\,dy$
- $\displaystyle\int \frac{1}{1-y}\,dy$
- $\displaystyle\int \frac{1}{y^2+1}\,dy$
- $\displaystyle\int \frac{y}{y^2+1}\,dy$
- $\displaystyle\int \frac{1}{y(1-y)}\,dy$
- $\displaystyle\int \frac{1}{y(1-3y)}\,dy$

Local linearity: In each problem that follows we are given values for $f(c)$ and $f'(c)$ for some c. Use this information to approximate $f(c+h)$ for the given value of h.

- $f(0) = 2,\ f'(0) = 1,\ h = 1$
- $f(1) = -1,\ f'(1) = -3,\ h = 0.5$
- $f(-2) = -3,\ f'(-2) = 2,\ h = 0.25$
- $f(1.5) = 6.2,\ f'(1.5) = 2.2,\ h = 0.1$

Concepts

0. *Problem Zero:* Read the section and make your own summary of the material.

1. *True/False:* Determine whether each of the statements that follow is true or false. If a statement is true, explain why. If a statement is false, provide a counterexample.

(a) *True or False:* A solution of an initial-value problem is a function that satisfies both the differential equation and the initial value.

(b) *True or False:* If $y_1(x)$ and $y_2(x)$ are both solutions of some differential equation $\frac{dy}{dx} = g(x, y)$, then $y_1(x) - y_2(x) = C$ for some constant C.

(c) *True or False:* Solving a differential equation $\frac{dy}{dx} = g(x)$ by antidifferentiating is just a special case of solving by separation of variables.

(d) *True or False:* The differential equation $\frac{dy}{dx} = x + y$ is separable.

(e) *True or False:* In a slope field for a differential equation $\frac{dy}{dx} = g(x, y)$, the slope of the line segment at (a, b) is given by $\left.\frac{dy}{dx}\right|_{(a,b)}$.

(f) *True or False:* In a slope field for a differential equation $\frac{dy}{dx} = g(y)$, the slope at $(2, b)$ will be the same as the slope at $(3, b)$.

(g) *True or False:* Euler's method is a way to construct a piecewise-linear approximation of the solution of an initial-value problem.

(h) *True or False:* For populations well below carrying capacity, the logistic growth model is similar to an exponential model.

2. *Examples:* Construct examples of the thing(s) described in the following. Try to find examples that are different than any in the reading.

(a) Two differential equations that are separable and two that are not separable.

(b) A slope field that describes solution functions of the form $f(x) = x^2 + C$.

(c) Three real-world situations that could be modeled with initial-value problems.

3. What is the difference between a solution of a differential equation and a solution of an initial-value problem?

4. Suppose you solve an initial-value problem of the form $\frac{dy}{dx} = g(x, y)$ with $y(0) = y_0$ and obtain an explict solution function $y(x)$. How could you check that your function $y(x)$ is indeed a solution of the initial-value problem?

5. Verify that $y(x) = \sqrt{x + C}$ is a solution of the differential equation $\frac{dy}{dx} = \frac{1}{2y}$ for every constant C.

6. Verify that $y(x) = \sqrt{t + 9}$ is a solution of the initial-value problem $\frac{dy}{dx} = \frac{1}{2y}$ and $y(0) = 3$.

7. A falling object accelerates downwards due to gravity at a rate of 9.8 meters per second squared. Suppose an object is dropped and falls to the ground from a height of 100 meters.

(a) Set up and solve a first-order initial-value problem whose solution is the velocity $v(t)$ of the object at time t.

(b) Use your answer to part (a) to set up and solve a first-order initial-value problem whose solution is the position $s(t)$ of the object at time t.

8. Explain, using the chain rule and/or u-substitution, why

$$\int \frac{1}{q(y)} \frac{dy}{dx}\,dx = \int \frac{1}{q(y)}\,dy.$$

9. Why do we use the terminology "separable" to describe a differential equation that can be written in the form $\frac{dy}{dx} = p(x)q(y)$? Once we have a differential equation in this form, how does the technique of separation of variables work?

10. In the process of solving $\frac{dy}{dx} = y$ by separation of variables, we obtain the equation $|y| = e^{x+C}$. After solving for y, this equation becomes $y = Ae^x$. How is A related to C? What happened to the absolute value?

11. In the process of solving the differential equation $\frac{dy}{dx} = 1 - y$ by separation of variables, we obtain the equation $-\ln|1-y| = x + C$. After solving for y, this equation becomes $y = 1 - Ae^{-x}$. Given that $y > 1$, how is A related to C?

12. Explain in your own words how the slopes of the line segments in a slope field for a differential equation are related to the differential equation.

13. How does a slope field help us to understand the solutions of a differential equation? How can a slope field help us sketch an approximate solution of an initial-value problem?

14. Given an initial-value problem, we can apply Euler's method to generate a sequence of points (x_0, y_0), (x_1, y_1), (x_2, y_2), and so on. How are these coordinate points related to the solution of the initial-value problem?

15. Explain how the equality $\left.\frac{dy}{dx}\right|_{(x_k, y_k)} = \frac{\Delta y_k}{\Delta x}$ is relevant to Euler's method.

16. Suppose your bank account grows at 3 percent interest yearly, so that your bank balance after t years is $B(t) = B_0(1.03)^t$.

(a) Show that your bank balance grows at a rate proportional to the amount of the balance.

(b) What is the proportionality constant for the growth rate, and what is the corresponding differential equation for the exponential growth model of $B(t)$?

17. Suppose a population $P(t)$ of animals on a small island grows according to a logistic model of the form $\frac{dP}{dt} = kP\left(1 - \frac{P}{500}\right)$ for some constant k.

(a) What is the carrying capacity of the island under this model?

(b) Given that the population $P(t)$ is growing and that $0 < P(t) < 500$, is the constant k positive or negative, and why?

(c) Explain why $\frac{dP}{dt} \approx kP$ for small values of P.

(d) Explain why $\frac{dP}{dt} \approx 0$ for values of P that are close to the carrying capacity.

18. Suppose an object is heating up according to a model for Newton's Law of Cooling with temperature satisfying $\frac{dT}{dt} = k(350 - T)$ for some constant k.

(a) What is the ambient temperature of the environment under this model?

(b) Given that the temperature $T(t)$ is increasing and that $0 < T < 350$, is the constant k positive or negative, and why?

(c) Use the differential equation to argue that the object's temperature changes faster when it is much cooler than the ambient temperature than when it is close to the ambient temperature.

(d) Part (c) is the basis for the oft-misunderstood saying "Cold water boils faster." Why?

Skills

Use antidifferentiation and/or separation of variables to solve each of the differential equations in Exercises 19–28. Your answers will involve unsolved constants.

19. $\frac{dy}{dx} = (x^3 + 4)^2$

20. $\frac{dy}{dx} = \frac{x+1}{\sqrt{x}}$

21. $\frac{dy}{dx} = 2xy$

22. $\frac{dy}{dx} = -3xy$

23. $\frac{dy}{dx} = x^2 y$

24. $\frac{dy}{dx} = xy^2$

25. $\frac{dy}{dx} = \frac{3x+1}{xy}$

26. $\frac{dy}{dx} = y \sin x$

27. $\frac{dy}{dx} = xe^{-y}$

28. $\frac{dy}{dx} = x^2 e^{-y}$

Use antidifferentiation and/or separation of variables to solve each of the initial-value problems in Exercises 29–52.

29. $\frac{dy}{dx} = 3y,\ y(0) = 4$

30. $\frac{dy}{dx} = 3x,\ y(0) = 4$

31. $\frac{dy}{dx} = -6x^2,\ y(1) = 3$

32. $\frac{dy}{dx} = -5y,\ y(0) = 2$

33. $\frac{dy}{dx} = 1 - y,\ y(0) = 4$

34. $\frac{dy}{dx} = \frac{1}{y},\ y(0) = 3$

35. $\frac{dy}{dx} = \frac{3}{y},\ y(0) = 2$

36. $\frac{dy}{dx} = \frac{6}{x},\ y(1) = 7$

37. $\frac{dy}{dx} = xy,\ y(0) = -1$

38. $\frac{dy}{dx} = \sqrt{4y},\ y(1) = 1$

39. $\frac{dy}{dx} = e^{x+y},\ y(0) = 2$

40. $\frac{dy}{dx} = \frac{2x}{y^2},\ y(0) = 4$

41. $\frac{dy}{dx} = \frac{x}{1+x^2},\ y(0) = 4$

42. $\frac{dy}{dx} = 10 - 7y,\ y(0) = \frac{1}{11}$

43. $\frac{dy}{dx} = 9.8 - \frac{0.3}{150}y,\ y(0) = 1000$

44. $\frac{dy}{dx} = 0.4y - 100,\ y(0) = 0$

45. $\frac{dy}{dx} = -2(1 - 3y),\ y(0) = 2$

46. $\frac{dy}{dx} = -2y(1 - 3y),\ y(0) = 2$

47. $\frac{dy}{dx} = 3y(1 - y),\ y(0) = 1$

48. $\frac{dy}{dx} = xy + x + y + 1,\ y(0) = c$

49. $\frac{dy}{dx} = y^2 \cos x,\ y(\pi) = 1$

50. $\frac{dy}{dx} = y^3 \cos x,\ y(\pi) = 1$

51. $\dfrac{dy}{dx} = \dfrac{x^2-2}{y^2}$, $y(0) = 8$

52. $\dfrac{dy}{dx} = \dfrac{y+1}{x^2+1}$, $y(0) = 1$

For each initial-value problem $\frac{dy}{dx} = g(x, y)$ with $y(x_0) = y_0$ in Exercises 53–58, use Euler's method with the given Δx to approximate four additional points on the graph of the solution $y(x)$. Use these points to sketch a piecewise-linear approximation of the solution.

53. $\dfrac{dy}{dx} = -2y$, $y(0) = 3$; $\Delta x = 0.25$

54. $\dfrac{dy}{dx} = \dfrac{1}{x}$, $y(1) = 1$; $\Delta x = 0.25$

55. $\dfrac{dy}{dx} = x + y$, $y(0) = 0$; $\Delta x = 0.5$

56. $\dfrac{dy}{dx} = \dfrac{x}{1-y}$, $y(0) = 2$; $\Delta x = 0.1$

57. $\dfrac{dy}{dx} = \dfrac{x}{y}$, $y(0) = 2$; $\Delta x = 0.5$.

58. $\dfrac{dy}{dx} = x^2 - y$, $y(1) = 0$; $\Delta x = 0.25$

Sketch slope fields for each of the differential equations in Exercises 59–64, and within each slope field sketch four different approximate solutions of the differential equation.

59. $\dfrac{dy}{dx} = -y$

60. $\dfrac{dy}{dx} = 1 - x$

61. $\dfrac{dy}{dx} = x + y$

62. $\dfrac{dy}{dx} = x - y$

63. $\dfrac{dy}{dx} = y - 2x$

64. $\dfrac{dy}{dx} = e^x - y$

Applications

The situations in Exercises 65–67 concern exponential growth models of the form $\frac{dQ}{dt} = kQ$. When the constant of proportionality k (also called the ***continuous growth rate***, when expressed as a percentage) happens to be negative rather than positive, this differential equation models exponential decay.

65. Suppose that the current population of Freedonia is 1.08 million people and that the continuous growth rate of the population is 1.39%.

(a) Set up a differential equation describing $\frac{dP}{dt}$, and solve it to get a formula for the population $P(t)$ of Freedonia t years from now.

(b) According to your model, how long ago was it that the population of Freedonia was just half a million people?

(c) How long will it take for the population of Freedonia to double from its current size?

66. Suppose $X(t)$ is the number of milligrams of the drug Xenaphoril that is present in the body t hours after it is ingested. As the drug is absorbed, the quantity of the drug decreases at a rate proportional to the amount of the drug in the body.

(a) Set up a differential equation describing $\frac{dX}{dt}$, and solve it to get a formula for $X(t)$. Your answer will involve two constants.

(b) The ***half-life*** of a drug is the number of hours that it takes for the quantity of the drug to decrease by half. In an exponential decay model, the half-life will be the same no matter when we start measuring the amount of the drug. If Xenaphoril has a half-life of 4 hours, what is the constant of proportionality for this model?

(c) Given the constant of proportionality you found in part (b), how much of a 20-mg dose of Xenaphoril remains after 10 hours?

67. The amount of the radioactive isotope carbon-14 present in small quantities can be measured with a Geiger counter. Carbon-14 is replenished in live organisms, and after an organism dies the carbon-14 in it decays at a rate proportional to the amount of carbon-14 present in the body. Suppose $C(t)$ is the amount of carbon-14 in a dead organism t years after it dies.

(a) Set up a differential equation describing $\frac{dC}{dt}$, and solve it to get a formula for $C(t)$. Your answer will involve two constants.

(b) The half-life of carbon-14 is 5730 years. (See part (b) of the previous problem for the definition of half-life.) Use this half-life to find the value of the proportionality constant for the model you found in part (a).

(c) Suppose you find a bone fossil that has 10% of its carbon-14 left. How old would you estimate the fossil to be?

The situations in Exercises 68 and 69 concern logistic growth models of the form $\frac{dP}{dt} = kP\left(1 - \frac{P}{L}\right)$ for some constant of proportionality k (also called the ***natural growth rate***) and some carrying capacity L.

68. Suppose that the country of Freedonia has a carrying capacity of 5 million people, with natural growth rate and initial population as given in Exercise 65.

(a) Set up a differential equation describing $\frac{dP}{dt}$, and solve it to get a formula for the population $P(t)$ of Freedonia t years from now.

(b) How long will it be before the population of Freedonia is half of the carrying capacity?

(c) How fast is Freedonia's population changing when the population is at half of carrying capacity? What about when the population has reached 90% of carrying capacity?

69. Suppose 100 rabbits are shipwrecked on a deserted island and their population $P(t)$ after t years is determined by a logistic growth model, where the natural growth rate of the rabbits is $k = 0.1$ and the carrying capacity of the island is 1000 rabbits.

(a) Set up a differential equation describing $\frac{dP}{dt}$, and solve it to get a formula for the population $P(t)$ of rabbits on the island in t years.

(b) Sketch a graph of the population $P(t)$ of rabbits on the island over the next 100 years.

(c) It turns out that a population governed by a logistic model will be growing fastest when the population is equal to exactly half of the carrying capacity. In how many years will the population of rabbits be growing the fastest?

The situations in Exercises 70 and 71 concern Newton's Law of Cooling and Heating and models of the form $\frac{dT}{dt} = k(A - T)$ for some proportionality constant k and constant ambient temperature A.

70. A cold drink is heating up from an initial temperature $T(0) = 2°C$ to room temperature of 22°C according to Newton's Law of Heating with constant of proportionality 0.05°C.

(a) Set up a differential equation describing $\frac{dT}{dt}$, and solve it to get a formula for the temperature of the drink after t minutes.

(b) Use the differential equation and/or its solution to determine the units of the constant of proportionality.

(c) How long will it take for the drink to warm up to within 1 degree of room temperature?

71. A crime scene investigator finds a body at 8 P.M., in a room that is kept at a constant temperature of 70°F. The temperature of the body is 88.8°F at that time. Thirty minutes later the temperature of the body is 87.5°F.

(a) Set up a differential equation describing $\frac{dT}{dt}$, and solve it to get a formula for the temperature of the body t minutes after the time of death. Your answer will involve a proportionality constant k.

(b) Use the information in the problem to determine k.

(c) Assuming that the body had a normal temperature of 98.6°F at the time of death, estimate the time of death of the victim.

In Exercises 72 and 73 you will set up and solve differential equations to model different population growth scenarios.

72. A colony of bacteria is growing in a large petri dish in such a way that the shape of the colony is a disk whose area $A(t)$ after t days grows at rate proportional to the diameter d of the disk. Suppose the colony has an area of 2 cm^2 on the first day and 5 cm^2 on the third day.

(a) Set up a differential equation that describes this situation, and solve it to get an equation for the area $A(t)$ of the colony after t days. Your answer will involve a proportionality constant k.

(b) Use the information in the problem to determine k.

(c) Given that the petri dish has a diameter of 6 inches and that the colony started in the exact center of the dish, how long will it take for the colony to fill the entire petri dish?

73. Another model for population growth is what is called *supergrowth*. It assumes that the rate of change in a population is proportional to a higher power of P than P^1. For example, suppose the rate of change of the world's human population $P(t)$ is proportional to $P^{1.1}$.

(a) Set up a differential equation that describes $\frac{dP}{dt}$, and solve it to get an equation for the world population $P(t)$. Your answer will involve two unsolved constants.

(b) Given that the world population was estimated to be 6.451 billion in 2005 and 6.775 billion in 2010, what is the value of the proportionality constant for this supergrowth model?

(c) According to your model, when will the world population be 7 billion?

Proofs

74. Show that if $y_1(x)$ and $y_2(x)$ are both solutions of the differential equation $\frac{dy}{dx} = ky$, then the sum $y_1(x) + y_2(x)$ is also a solution of the differential equation.

75. Prove Theorem 6.20 by solving the initial-value problem $\frac{dQ}{dt} = kQ$ with $Q(0) = Q_0$, where k is a constant.

76. Prove Theorem 6.21 by solving the initial-value problem $\frac{dP}{dt} = rP\left(1 - \frac{P}{K}\right)$ with $P(0) = P_0$, where r and K are constants. *(Hint: After separation of variables you will need to use partial fractions to solve the integral that concerns P. Before solving for P, use properties of logarithms to simplify the expression.)*

77. Use the solution of the logistic model $\frac{dP}{dt} = kP\left(1 - \frac{P}{L}\right)$ to prove that as $t \to \infty$, the population $P(t)$ approaches the carrying capacity L. Assume that the constant k is positive.

78. Prove Theorem 6.22 by solving the initial-value problem $\frac{dT}{dt} = k(A - T)$ with $T(0) = T_0$, where k and A are constants.

79. Use the solution of the differential equation $\frac{dT}{dt} = k(A-T)$ for the Newton's Law of Cooling and Heating model to prove that as $t \to \infty$, the temperature $T(t)$ of an object approaches the ambient temperature A of its environment. The proof requires that we assume that k is positive. Why does this make sense regardless of whether the model represents heating or cooling?

Thinking Forward

Implicit solution curves: Sometimes the solution of a differential equation will be an implicit solution, meaning that y will be given as an implicit function of x, not an explicit function of x. For each differential equation $\frac{dy}{dx} = g(x, y)$ that follows, find an equation that gives y as an implicit function of x. Then draw the continuous curve that satisfies this differential equation and passes through the point $(2, 0)$.

▶ $\frac{dy}{dx} = \frac{-x}{y}$

▶ $\frac{dy}{dx} = \frac{x}{y}$

CHAPTER REVIEW, SELF-TEST, AND CAPSTONES

Before you progress to the next chapter, be sure you are familiar with the definitions, concepts, and basic skills outlined here. The capstone exercises at the end bring together ideas from this chapter and look forward to future chapters.

Definitions

Give precise mathematical definitions or descriptions of each of the concepts that follow. Then illustrate the definition with a graph or algebraic example, if possible.

- the *volume of a solid with cross-sectional area function* given by $A(x)$ on $[a, b]$, both as a limit of Riemann sums and as a definite integral
- the *volume of a solid with disk cross sections* whose radii are given by $r(x)$ on $[a, b]$, both as a limit of Riemann sums and as a definite integral
- the *volume of a solid with washer cross sections* whose outer and inner radii are given, respectively, by $R(x)$ and $r(x)$ on $[a, b]$, as a limit of Riemann sums
- the *volume of a solid with nested shell sections* whose heights and average radii are given, respectively, by $h(x)$ and $r(x)$ on $[a, b]$, as a limit of Riemann sums
- the *arc length* of the graph of a continuous, differentiable function $y = f(x)$ whose derivative is also continuous, from $x = a$ to $x = b$, as a limit of Riemann sums
- if f is a continuous, differentiable function with a continuous derivative, the *area of the surface of revolution* obtained by revolving $y = f(x)$ around the x-axis on $[a, b]$, as a limit of Riemann sums
- the definition of a *differential equation* and what it means for a function to be a *solution of a differential equation*
- the definition of an *initial-value problem* and what it means for a function to be a *solution of an initial-value problem*
- what it means for a differential equation to be *separable*
- the definition of a *slope field* for a differential equation, including the specific property that holds at each lattice point (x, y)
- the method of approximation known as *Euler's method*, including the iterative formula that describes a new coordinate point (x_{k+1}, y_{k+1}) in terms of the previous point (x_k, y_k)

Theorems

Fill in the blanks to complete each of the following theorem statements.

- The volume of a shell with height h, average radius r, and thickness Δx is $V =$ ____ .
- The surface area of a frustum with average radius r and slant length s is $S =$ ____ .
- If a quantity $Q(t)$ changes over time at a rate proportional to its value, then that quantity is modeled by a differential equation of the form $\frac{dQ}{dt} =$ ____ , with solution $Q(t) =$ ____ .
- If a population $P(t)$ changes over time with natural growth rate k and carrying capacity L, then it can be modeled by the differential equation $\frac{dP}{dt} =$ ____ , with solution $P(t) =$ ____ .
- If an object with temperature $T(t)$ is in a location with constant ambient temperature A, then the temperature of the object can be modeled by a differential equation of the form $\frac{dT}{dt} =$ ____ , with solution $T(t) =$ ____ .

Formulas and Geometric Quantities

Definite integral formulas for geometric quantities: Write down definite integrals to express each of the given geometric quantities. You may assume that $f(x)$ is continuous and differentiable, with a continuous derivative.

- The volume of the solid obtained by revolving $f(x)$ on $[a, b]$ around the x-axis, by the disk method.
- The volume of the solid obtained by revolving $f(x)$ on $[a, b]$ around the y-axis, by the disk method.
- The volume of the solid obtained by revolving $f(x)$ on $[a, b]$ around the x-axis, by the shell method.
- The volume of the solid obtained by revolving $f(x)$ on $[a, b]$ around the y-axis, by the shell method.
- The arc length of the curve formed by $y = f(x)$ on $[a, b]$.
- The area of the surface obtained by revolving $f(x)$ on $[a, b]$ around the x-axis.
- The centroid $(\bar{x}, \bar{y})$ of the region between the graph of an integrable function f and the x-axis on an interval $[a, b]$.
- The centroid $(\bar{x}, \bar{y})$ of the region between the graphs of two integrable functions f and g on an interval $[a, b]$.

Finding distances related to graphs: In setting up definite integrals for volume problems, it is often necessary to describe various distances in terms of f, f^{-1}, x_k^*, and/or y_k^*. Describe these distances for each figure that follows.

▶ For the following figure, write the distances A and B in terms of f or f^{-1} and the point x_k^*.

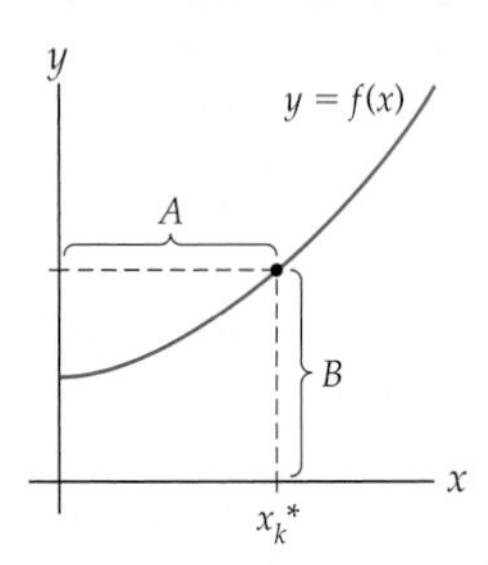

▶ For the following figure, write the distances A and B in terms of f or f^{-1} and the point y_k^*.

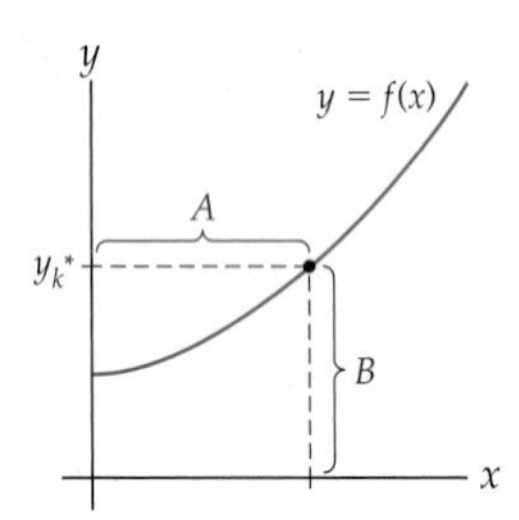

▶ For the following figure, write the distances A, B, and C in terms of f or f^{-1}, the point x_k^*, and K.

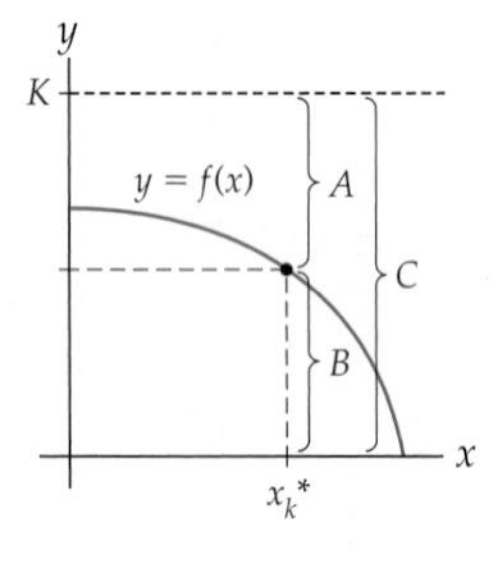

▶ For the following figure, write the distances A, B, and C in terms of f or f^{-1}, the point y_k^*, and K.

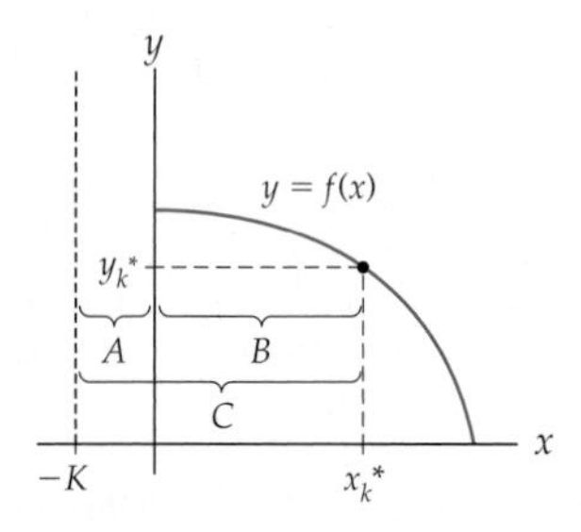

Concepts from physics: Fill in the blanks to complete the descriptions of each of the following physical quantities and units.

▶ The *mass* of an object with density ρ and volume V is given by $m =$ ____ .

▶ If density is measured in grams per cubic centimeter and volume is measured in cubic centimeters, then mass is measured in units of ____ .

▶ The *work* required to lift an object with a weight F through a distance d is given by $W =$ ____ .

▶ If weight is measured in pounds and distance is measured in feet, then work is measured in units of ____ .

▶ The *hydrostatic force* exerted by a liquid of weight-density ω and depth d on a horizontal wall of area A is given by $F =$ ____ .

▶ If weight-density is measured in pounds per cubic foot and distance is measured in feet, then hydrostatic force is measured in units of ____ .

Skill Certification: Definite Integrals for Geometry and Applications

Sketching disks, washers, and shells: Sketch the three disks, washers, or shells that result from revolving the rectangles shown in the given figures around the given lines.

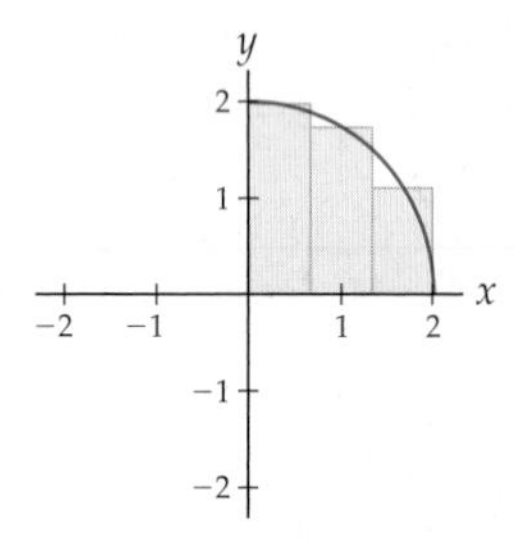

1. the x-axis **2.** the y-axis

3. the line $x = -1$ **4.** the line $y = -1$

Sketching a representative disk, washer, or shell: Sketch a representative disk, washer, or shell for the solid obtained by revolving the regions shown in the given figures around the given lines.

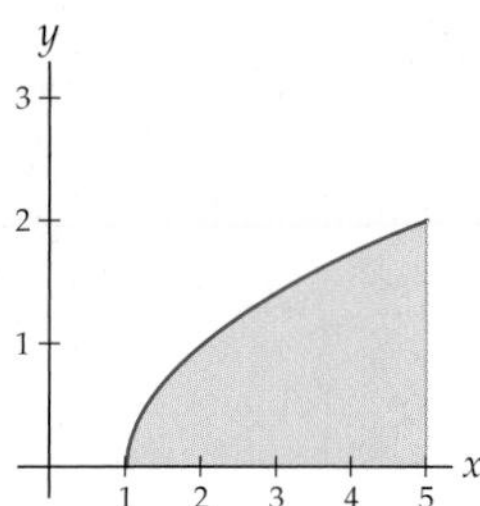

5. the x-axis **6.** the y-axis

7. the line $x = 7$ **8.** the line $y = 3$

Finding geometric quantities with definite integrals: Set up and solve definite integrals to find each volume, surface area, or arc length that follows. Solve each volume problem both with disks/washers and with shells, if possible.

9. The volume of the solid obtained by revolving the region between the graph of $f(x) = x^2$ and the y-axis for $0 \le x \le 2$ around (a) the x-axis and (b) the y-axis.

10. The volume of the solid obtained by revolving the region between the graph of $f(x) = 9 - x^2$ and the x-axis on $[-3, 3]$ around (a) the x-axis and (b) the line $y = -3$.

11. The volume of the solid obtained by revolving the region between the graphs of $f(x) = \sqrt{x}$ and $g(x) = x^3$ on $[0, 1]$ around (a) the y-axis and (b) the line $x = 2$.

12. The arc length of the curve traced out by the graph of $f(x) = \ln(\csc x)$ on the interval $\left[\frac{\pi}{4}, \frac{\pi}{2}\right]$.

13. The area of the surface obtained by revolving the curve $f(x) = \sin(\pi x)$ around the x-axis on $[-1, 1]$.

14. The centroid of the region between the graph of $f(x) = x^2$ and the x-axis on $[0, 2]$.

15. The centroid of the region between the graphs of $f(x) = \sqrt{x}$ and $g(x) = 4 - x$ on $[0, 4]$.

Real-world applications of definite integrals: Set up and solve definite integrals to answer each of the following questions.

16. Find the mass of a 30-centimeter rod with square cross sections of side length 2 centimeters, given that the density of the rod x centimeters from the left end is $\rho(x) = 10.5 + 0.01527x^2$ grams per cubic centimeter.

17. Find the work required to pump all of the water out of the top of a cylindrical hot tub that is 8 feet in diameter and 3 feet deep.

18. Find the hydrostatic force exerted on one of the long sides of a rectangular swimming pool that is 20 feet long, 12 feet wide, and 6 feet deep.

Differential equations: Solve each of the following initial-value problems. *(These problems assume that you covered Section 6.5.)*

19. $\dfrac{dy}{dx} = 5y, \quad y(0) = 1$

20. $\dfrac{dy}{dx} = 3 - 4y, \quad y(0) = 2$

21. $\dfrac{dy}{dx} = 2y(1 - 5y), \quad y(0) = 1$

22. $\dfrac{dy}{dx} = y^2 \sin x, \quad y(\pi) = 1$

Capstone Problems

A. *Proving geometric properties of spheres:* Set up and solve definite integrals to prove each of the following geometric formulas concerning the sphere with radius r.

(a) Volume: $V = \frac{4}{3}\pi r^3$ (with disks)

(b) Volume: $V = \frac{4}{3}\pi r^3$ (with shells)

(c) Equator: $C = 2\pi r$

(d) Surface area: $S = 4\pi r^2$

B. *Approximating volumes with frustums:* In Section 6.3 we used frustums to approximate surface areas. We could also use frustums, instead of disks, to approximate volumes.

(a) Consider the solid of revolution obtained by revolving the graph of $f(x) = x^2$ around the x-axis on $[0, 2]$. Use four frustums to approximate the volume of this solid.

(b) Suppose we used a limit of volumes of frustums to develop a definite integral formula for volume. Would we obtain a different definite integral formula? Why or why not?

C. *Center of mass when density varies:* In Section 6.4 we saw both how to find the mass of a rod whose density varies and how to find the center of mass of a planar region with constant density. Combine these two ideas in a "subdivide, approximate, and add" strategy to construct a formula for determining the center of mass of a planar region whose density varies. You may assume that the density at any point in the region is given by some function $\rho(x)$ that depends only on the x-coordinate.

D. *A differential equation for escape velocity:* To get a satellite up into space, it needs to escape the gravitational pull of the earth. The velocity it must attain to break free from the earth is called its ***escape velocity***. Let $s = s(t)$ be the position of the satellite above the surface of the earth. As an object moves away from the earth, the acceleration due to gravity is reduced. In fact, the acceleration due to gravity on an object s kilometers from the surface of the earth is

$$a = a(t) = \frac{gR^2}{(R+s)^2},$$

where R is the radius of the earth and g is the acceleration due to gravity on the surface of the earth. The sign on g is negative, pointing in the direction toward the earth. Knowing that $a = \frac{dv}{dt}$, we can rewrite our equation for acceleration due to gravity as a differential equation:

$$\frac{dv}{dt} = \frac{gR^2}{(R+s)^2}.$$

(a) Use the chain rule to show that

$$v\frac{dv}{ds} = \frac{gR^2}{(R+s)^2}.$$

(b) Use separation of variables to solve the differential equation.

(c) Find the escape velocity of the satellite, that is, the smallest value of $v(0)$ for which the velocity $v(t)$ will never be zero.

CHAPTER 7

Sequences and Series

7.1 Sequences

Sequences of Numbers
Recursively Defined Sequences
Geometric and Arithmetic Sequences
Monotonic Sequences
Bounded Sequences
Examples and Explorations

7.2 Limits of Sequences

Convergence or Divergence of a Sequence
Theorems About Convergent Sequences
Convergence and Divergence of Basic Sequences
Bounded Monotonic Sequences
Examples and Explorations

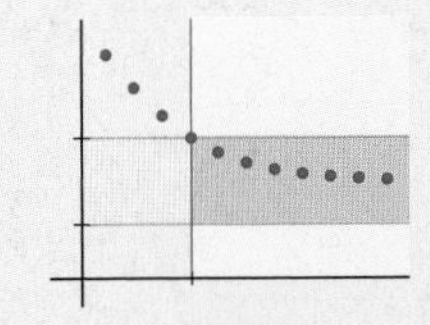

7.3 Series

Adding Up Sequences to Get Series
Convergence and Divergence of Series
The Algebra of Series
Geometric Series
Examples and Explorations

7.4 Introduction to Convergence Tests

An Overview of Convergence Tests for Series
The Divergence Test
The Integral Test
Convergence and Divergence of p-Series and the Harmonic Series
Approximating a Convergent Series
Examples and Explorations

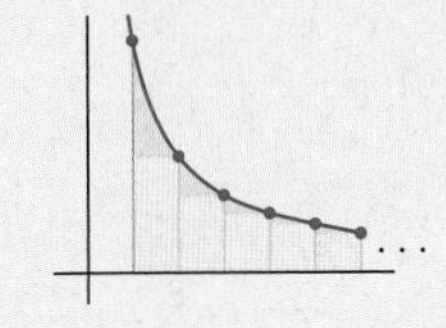

7.5 Comparison Tests

The Comparison Test
The Limit Comparison Test
Examples and Explorations

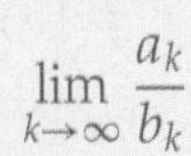

7.6 The Ratio and Root Tests

The Ratio Test
The Root Test
Examples and Explorations

$$\lim_{k\to\infty} \frac{a_{k+1}}{a_k}$$

7.7 Alternating Series

Alternating Series
Absolute and Conditional Convergence
The Curious Behavior of a Conditionally Convergent Series
The Ratio Test for Absolute Convergence
A Summary of Convergence Tests
Examples and Explorations

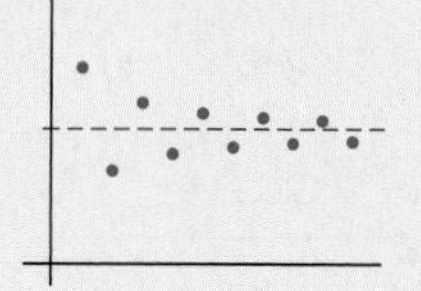

Chapter Review, Self-Test, and Capstones

7.1 SEQUENCES

- Sequences as functions whose domains are subsets of the nonnegative integers
- Common sequences defined with formulas and with recursion
- Monotonic sequences and bounded sequences

Sequences of Numbers

In this chapter we will discuss two distinct, yet related, infinite processes: sequences and series. Informally, a sequence is an infinite list of numbers and a series is a sum of an infinite list of numbers. To examine series in more generality we must first make our idea of a sequence more precise.

Here are a few examples of sequences:

$$3, 4, 5, 6, 7, \ldots$$
$$1, 1, 2, 3, 5, 8, \ldots$$
$$2, 3, 5, 7, 11, 13, \ldots$$
$$1, \frac{1}{2}, \frac{1}{3}, \frac{1}{4}, \frac{1}{5}, \frac{1}{6}, \ldots$$
$$1, -\frac{1}{3}, \frac{1}{7}, -\frac{1}{13}, \frac{1}{21}, -\frac{1}{31}, \ldots$$
$$1, -1, 1, -1, 1, -1, \ldots$$
$$\pi, \pi, \pi, \pi, \pi, \pi, \ldots$$

It is not hard to see the pattern in, say, the fourth sequence shown; the kth term in the list is of the form $\frac{1}{k}$. We can think of this sequence as a function that, for each positive integer k, returns a real number $\frac{1}{k}$. This is how we will define sequences in general:

DEFINITION 7.1 Sequences of Real Numbers

A ***sequence of real numbers*** $\{a_k\}$ is a function from the positive integers to the real numbers. For each positive integer ***index*** k, the output a_k is called the ***kth term*** of the sequence.

To start a sequence at some integer r other than $k = 1$, we can use the more general notation $\{a_k\}_{k=r}^{\infty}$. For example, the sequence $3, 4, 5, 6, 7, \ldots$ can be expressed in many different ways depending on where we choose to start the index: $\{k+2\} = \{k+2\}_{k=1}^{\infty}$, or $\{k\}_{k=3}^{\infty}$, or even $\{k-5\}_{k=8}^{\infty}$.

If we let $\mathbb{Z}^+$ represent the positive integers, then a sequence $\{a_k\}$ defined on $\mathbb{Z}^+$ can be expressed in function notation as

$$a : \mathbb{Z}^+ \to \mathbb{R}.$$

Although in regular function notation we would represent the output for a given $k \in \mathbb{Z}^+$ as $a(k)$, with sequences the convention is instead to use a subscripted letter such as a_k. This subscript notation, and the fact that we are using k rather than x as our input variable, stresses that the inputs make up a discrete list of positive integers, not arbitrary real numbers.

Sometimes it is easy to find a straightforward formula for the general term of a sequence, and sometimes it is not. For example, the sixth sequence in our previous list is

$$\{1, -1, 1, -1, 1, -1, \ldots\} = \{(-1)^{k+1}\},$$

since $(-1)^{k+1}$ is equal to 1 if k is odd and equal to -1 if k is even. An equation that allows us to compute the value of each term of a sequence from its index is called a ***closed formula***. The fifth sequence in our list also has a closed formula, although it is one that is more difficult to identify:

$$\left\{1, -\frac{1}{3}, \frac{1}{7}, -\frac{1}{13}, \frac{1}{21}, -\frac{1}{31}, \ldots\right\} = \left\{\frac{(-1)^{k+1}}{k^2 - k + 1}\right\}.$$

In some cases we cannot find a simple algebraic formula; for example, the third sequence in our list was $\{2, 3, 5, 7, 11, 13, \ldots\}$, which is the sequence whose kth term a_k is equal to the kth prime number.

Recursively Defined Sequences

Some sequences are best described in terms of previous terms instead of with a closed formula. For example, the second sequence in our previous list is the ***Fibonacci sequence***

$$\{f_k\} = \{1, 1, 2, 3, 5, 8, 13, 21, 34, \ldots\}.$$

This famous sequence is obtained by starting with $f_1 = 1$ and $f_2 = 1$ and then defining each subsequent term to be the sum of the two previous terms in the sequence. Thus $f_3 = f_1 + f_2 = 1 + 1 = 2$ and $f_7 = f_5 + f_6 = 5 + 8 = 13$. We can completely describe all of the terms in the Fibonacci sequence as follows:

$$f_1 = 1,\ f_2 = 1; \quad \text{and for } k \geq 3, f_k = f_{k-2} + f_{k-1}.$$

In general, sequences that are defined by specifying one or more initial terms and then defining remaining terms as functions of previous terms are called ***recursive*** sequences.

DEFINITION 7.2 **Recursive Sequences**

A sequence $\{a_k\}$ is said to be ***recursively defined*** if there is some index K so that for $k > K$, the value of a_k is determined by $a_1, a_2, \ldots, a_{k-1}$. The equation defining a_k in terms of $a_1, a_2, \ldots, a_{k-1}$ is called a ***recursion formula***.

As another example of a recursively defined sequence, recall that the ***factorial*** of a positive integer k, denoted by $k!$, is the product of all the positive integers from 1 to k:

$$k! = 1(2)(3)\cdots(k-1)(k).$$

By convention we also say that the factorial of 0 is equal to $0! = 1$; see Exercise 6. We can recursively define the ***factorial sequence***

$$\{k!\}_{k=0}^{\infty} = \{1, 1, 2, 6, 24, 120, 720, \ldots\}$$

as follows:

$$0! = 1; \quad \text{and for } k \geq 1, k! = k \cdot (k-1)!$$

A closed formula and a recursion formula give us different types of information about a sequence. For example, consider this sequence of nonnegative powers of two:

$$\{2^k\} = \{1, 2, 4, 8, 16, \ldots\}_{k=0}^{\infty}.$$

The closed formula $a_k = 2^k$ tells us exactly how to compute the kth term of the sequence, no matter how large a value of k we choose. We can easily see that the 100th term of this sequence is 2^{99}. The same sequence can also be recursively defined as follows:

$$a_0 = 1; \quad \text{and for } k \geq 1, a_k = 2a_{k-1}.$$

Although it is obvious from looking at the sequence $\{1, 2, 4, 8, 16, \ldots\}$ that every term is twice the previous term, the recursion formula makes it explicit. One computational advantage of a recursively defined sequence is that the recursion formula relates new terms to terms we've already computed. One disadvantage is that in order to compute a new term we must have already computed all previous terms. For example, to use the recursive definition to compute the 100th term , we need to have previously computed a_{99}, and to have computed a_{99}, we need to have also computed a_{98}, and so on.

Geometric and Arithmetic Sequences

The sequence $\{2^k\}$ is an example of a type of sequence that is common enough to deserve its own definition.

DEFINITION 7.3

Geometric Sequences

A ***geometric sequence*** is a sequence in which each term differs from the previous term by a constant multiplicative factor r. We will commonly desire to start such sequences at $k = 0$:

$$\left\{cr^k\right\}_{k=0}^{\infty} = \left\{c, cr, cr^2, cr^3, \ldots\right\}.$$

For example $\{3 \cdot 2^k\}_{k=0}^{\infty} = \{3, 3 \cdot 2, 3 \cdot 2^2, 3 \cdot 2^3, \ldots\}$ is a geometric sequence starting at 3 and with multiplicative factor 2. Note that each term of the sequence is two times the previous term. In general, we can define a geometric sequence $\{g_k\}$ recursively as follows:

$$g_0 = c; \quad \text{and for } k \geq 1, g_k = r g_{k-1}.$$

If the terms are nonzero, then the ratio of successive terms in a geometric sequence is constant, since $\frac{g_k}{g_{k-1}} = r$.

The additive "cousin" of the geometric sequence is the arithmetic sequence, where instead of multiplying by a constant to get the next term, we add a constant:

DEFINITION 7.4

Arithmetic Sequences

An ***arithmetic sequence*** is a sequence in which each term differs from the previous term by a constant additive increment d:

$$\left\{c + dk\right\}_{k=0}^{\infty} = \{c, c + d, c + 2d, c + 3d, \ldots\}.$$

For example, $\{a_k\} = \{5, 7, 9, 11, 13, \ldots\}$ is an arithmetic sequence with starting value $a_0 = 5$ and additive difference $d = 2$. This example immediately suggests how to write an additive sequence recursively:

$$a_0 = c; \quad \text{and for } k \geq 1, a_k = a_{k-1} + d.$$

The difference of successive terms in an arithmetic series is constant, since $a_k - a_{k-1} = d$.

Monotonic Sequences

In Section 7.2 we will be interested in the long-term behavior of sequences—specifically, whether sequences converge or diverge as $k \to \infty$. One important idea that will help us answer such questions is monotonicity. A ***monotonic*** sequence is a sequence whose terms either always increase as the index grows or always decrease. For example, the sequence $\{2, 4, 6, 8, \dots\}$ is monotonically increasing, while the sequence $\{1, -1, 1, -1, \dots\}$ is not monotonic. More precisely, we have the following definition:

DEFINITION 7.5

Monotonic Sequences

A ***monotonic*** sequence $\{a_k\}$ is a sequence that is either always increasing or always decreasing, where

(a) a sequence $\{a_k\}$ is ***increasing*** if $a_{k+1} \ge a_k$ for all $k \ge 1$;

(b) a sequence $\{a_k\}$ is ***decreasing*** if $a_{k+1} \le a_k$ for all $k \ge 1$.

Note that the definitions of increasing and decreasing are just the discrete versions of our definitions from Section 0.1 where we said that a function $f(x)$ is increasing if $f(b) \ge f(a)$ when $a < b$, and is decreasing if $f(b) \le f(a)$ when $a < b$. A sequence can also satisfy the stronger condition of being ***strictly increasing*** or ***strictly decreasing***, which means that the inequalities ($\le$ and $\ge$) in the preceding definition become strict inequalities ($<$ or $>$).

When we study the long-term behavior of a sequence, we do not actually care if the sequence is always monotonic; we only need to know if it is ***eventually monotonic***. For example, we say that a sequence $\{a_k\}$ is ***eventually increasing*** if it is increasing after some index K, or more precisely, if there is some positive integer K for which $a_k \ge a_{k-1}$ for all $k \ge K$; see Exercises 11–14.

The graphs that follow show three examples of sequence behavior. The first is the sequence $\{a_k\} = \left\{\frac{1}{k}\right\}$, the second is $\{b_k\} = \left\{\frac{k^2}{2^k}\right\}$, and the third is $\{c_k\} = \left\{\frac{(-1)^k + k}{2k}\right\}$.

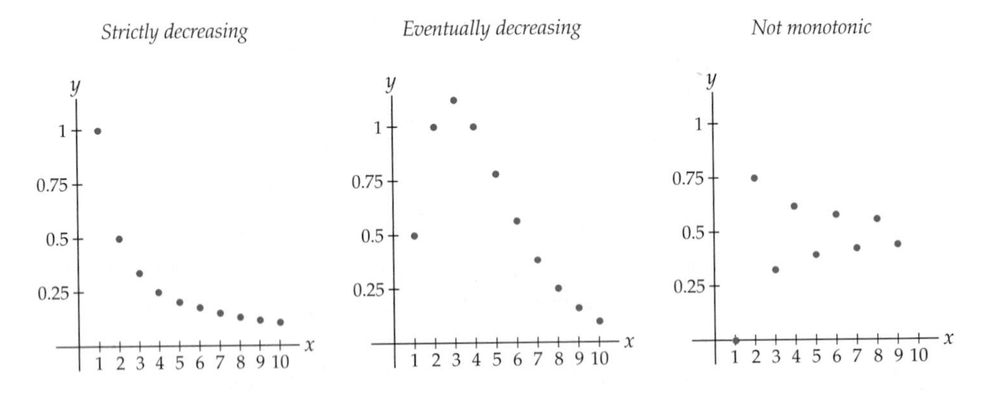

For some sequences, monotonicity is easy to determine; for example, we can see just by inspection that the sequence $\{k^2\} = \{1, 4, 9, 16, 25, \dots\}$ clearly has the property that $a_{k+1} > a_k$ for $k \ge 1$ and thus is strictly increasing. We can also prove it algebraically: Since $2k + 1 > 0$ whenever $k \ge 1$, we have

$$a_{k+1} = (k+1)^2 = k^2 + 2k + 1 > k^2 = a_k.$$

For more complicated sequences it is useful to have different tests for monotonicity. The following theorem provides three such tests:

THEOREM 7.6 **Three Tests for Monotonicity**

A sequence $\{a_k\}$ is increasing if it passes any of the following tests:

(a) The ***difference test***: $a_{k+1} - a_k \geq 0$ for all $k \geq 1$.

(b) The ***ratio test***: all terms are positive, and $\frac{a_{k+1}}{a_k} \geq 1$ for all $k \geq 1$.

(c) The ***derivative test***: $a'(x) \geq 0$ for all $x > 1$, given that $a(x)$ is a function that is differentiable on $[1, \infty)$ and whose value at any positive integer k is $a(k) = a_k$.

The obvious analogs of these tests work to determine other types of monotonicity (decreasing, strictly increasing, eventually increasing, etc.).

The difference test follows directly from the definition of monotonicity; the ratio test follows from the fact that a fraction $\frac{a}{b}$ is greater than or equal to 1 exactly when the numerator is greater than or equal to the denominator; and the derivative test is based on our usual test for increasing behavior in differentiable functions. You will formally prove these three tests in Exercises 94, 95, and 96.

As an illustration of the three monotonicity tests, we will use each of them to prove that the sequence $\{a_k\} = \{k^2\}$ is strictly increasing. This sequence passes the difference test for strictly increasing behavior because, for $k \geq 1$, we have $2k + 1 > 0$, and thus

$$a_{k+1} - a_k = (k+1)^2 - k^2 = k^2 + 2k + 1 - k^2 = 2k + 1 > 0.$$

The sequence $\{a_k\} = \{k^2\}$ has all positive terms, so we can also use the ratio test to show that it is strictly increasing. When $k \geq 1$, we have $\frac{2k+1}{k^2} > 0$, and thus

$$\frac{a_{k+1}}{a_k} = \frac{(k+1)^2}{k^2} = \frac{k^2 + 2k + 1}{k^2} = \frac{k^2}{k^2} + \frac{2k+1}{k^2} = 1 + \frac{2k+1}{k^2} > 1.$$

Finally, if we consider the function $a(x) = x^2$, then we clearly have $a'(x) = 2x$, which is positive for all $x > 1$. Together with the derivative test, this fact proves that the sequence $\{k^2\}$ is strictly increasing.

Not all three monotonicity tests are equally useful for all sequences. For example, the derivative test for monotonicity can be applied only if we can find a differentiable function $a(x)$ that agrees with the sequence $\{a_k\}$ on the positive integers. By contrast, sequences that involve exponential factors or factorials are particularly suited to the ratio test for monotonicity, because of the ease at which cancellation occurs in those cases; see Example 5.

Bounded Sequences

Another concept that will be useful in our later study of the long-term behavior of sequences is ***boundedness***. Loosely speaking, we are interested in whether the terms of a sequence always lie above or below a certain value. For example, the sequence

$$\left\{\frac{1}{k}\right\} = \left\{1, \frac{1}{2}, \frac{1}{3}, \frac{1}{4}, \dots\right\}$$

has terms that are clearly always less than or equal to 1 and always strictly greater than zero, so we say that $\left\{\frac{1}{k}\right\}$ is bounded above by 1 and bounded below by zero. In general we have the following definition:

DEFINITION 7.7

Bounded Sequences

(a) A sequence $\{a_k\}$ is ***bounded above*** if there exists a number M such that $a_k \leq M$ for all k.

(b) A sequence $\{a_k\}$ is ***bounded below*** if there exists a number m such that $a_k \geq m$ for all k.

If a sequence is both bounded above and bounded below, then we say that it is a ***bounded sequence***. If a sequence fails to be bounded above, fails to be bounded below, or both, then we say that it is an ***unbounded sequence***.

For example, the sequence $\left\{\frac{1}{k}\right\}$ is bounded above by 1 and bounded below by zero. The geometric sequence $\left\{2^k\right\}_{k=0}^{\infty}$ is bounded below by zero or any negative number, but is not bounded above. The sequence $\left\{(-1)^k k\right\}_{k=0}^{\infty}$ is bounded neither above nor below.

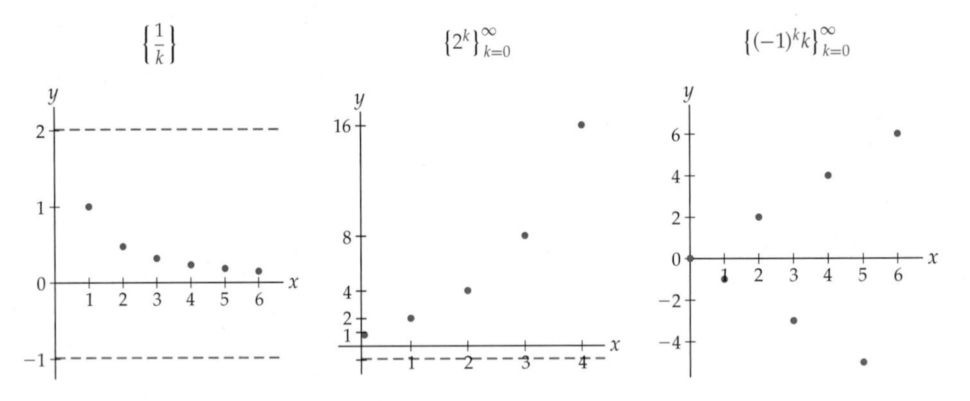

An upper bound M that is less than or equal to all other upper bounds for a sequence is called the ***least upper bound*** for the sequence. Likewise, a lower bound m that is greater than or equal to all other lower bounds for a sequence is called the ***greatest lower bound*** for the sequence. These are in some sense the "best" bounds for the sequence. As we just mentioned, the sequence $\left\{\frac{1}{k}\right\}$ is bounded above by 2 and below by -1, but its least upper bound is $M = 1$ and its greatest lower bound is $m = 0$.

The sequence $\left\{\frac{1}{k}\right\}$ has a least upper bound and a greatest lower bound

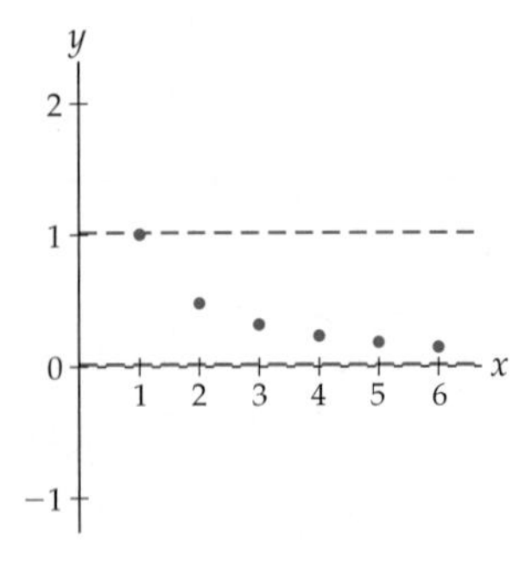

A deep assumption that we make about the structure of the real numbers $\mathbb{R}$ is that every subset of $\mathbb{R}$, which is bounded above, has a least upper bound. This is known as the ***Least Upper Bound Axiom*** and, in an upper-level math course, you will find that it is the key to proving the Extreme Value Theorem and the Intermediate Value Theorem, both of which we saw in Section 1.4. Since each sequence is an infinite set of real numbers, the Least Upper Bound Axiom says that every bounded sequence has both a least upper bound and (similarly) a greatest lower bound.

Examples and Explorations

EXAMPLE 1 **Using a closed formula to find the terms of a sequence**

Give the first four terms of the following sequences:

(a) $\left\{\dfrac{\sin k}{k}\right\}$ **(b)** $\left\{1+\dfrac{(-1)^k}{k}\right\}_{k=2}^{\infty}$ **(c)** $\{k^2+k\}_{k=3}^{\infty}$

SOLUTION

We evaluate each of these sequences for the four smallest integers in their domains and obtain

(a) $\sin 1, \dfrac{\sin 2}{2}, \dfrac{\sin 3}{3}, \dfrac{\sin 4}{4}$ **(b)** $\dfrac{3}{2}, \dfrac{2}{3}, \dfrac{5}{4}, \dfrac{4}{5}$ **(c)** $12, 20, 30, 42$ □

EXAMPLE 2 **Changing the indexing of a sequence**

Express the sequence

$$\left\{1+\frac{(-1)^k}{k}\right\}_{k=2}^{\infty}$$

so that the indexing begins with 1.

SOLUTION

It can be helpful to introduce a different letter for the index. We will use n. That is, we wish to express the given sequence in the form $\{a_n\}_{n=1}^{\infty}$. We need to find a linear relationship between k and n only for the first term of the sequence. When we find that, every other term will follow the pattern as well, since both k and n are incremented by 1. Thus, to change $\left\{1+\frac{(-1)^k}{k}\right\}_{k=2}^{\infty}$, we need $n=1$ when $k=2$. We obtain the relationship $k=n+1$. We use this equation and replace each k in $1+\frac{(-1)^k}{k}$ with $n+1$ to obtain

$$\left\{1+\frac{(-1)^{n+1}}{n+1}\right\}_{n=1}^{\infty}.$$

Note the index is a "dummy" variable: Any letter will suffice. In fact, if you wish to reuse k, you may, as long as you now replace each n with a k in the new expression. □

CHECKING THE ANSWER

We may easily check to ensure that our new expression for the sequence $\left\{1+\frac{(-1)^{n+1}}{n+1}\right\}_{n=1}^{\infty}$ gives the same first four terms we had in Example 1. Substituting 1, 2, 3, and 4 for n, we obtain the values

$$1+\frac{(-1)^{1+1}}{1+1}=\frac{3}{2},\quad 1+\frac{(-1)^{2+1}}{2+1}=\frac{2}{3},\quad 1+\frac{(-1)^{3+1}}{3+1}=\frac{5}{4},\quad 1+\frac{(-1)^{4+1}}{4+1}=\frac{4}{5}.$$

We see that our values agree.

EXAMPLE 3 **Finding a formula for a sequence**

Find a closed formula for each of the following sequences:

(a) $0, \frac{1}{2}, \frac{2}{3}, \frac{3}{4}, \frac{4}{5}, \frac{5}{6}, \ldots$ **(b)** $1, -\frac{1}{2}, \frac{1}{4}, -\frac{1}{8}, \frac{1}{16}, -\frac{1}{32}, \ldots$

(c) $0, \frac{1}{3}, 0, \frac{1}{5}, 0, \frac{1}{7}, 0, \frac{1}{9}, \ldots$

SOLUTION

There are many correct expressions for each of these sequences, but any answer requires discerning a pattern in the listed terms, assuming there is one.

(a) In the first list, we have a list of quotients, with each numerator 1 less than the matching denominator. (Note that we are thinking of 0 as $\frac{0}{1}$.) The denominators start with 1 and increase by 1 for each term in the sequence. Thus

$$\left\{\frac{k-1}{k}\right\} = 0, \frac{1}{2}, \frac{2}{3}, \frac{3}{4}, \frac{4}{5}, \frac{5}{6}, \ldots$$

We may also start the indexing with a different nonnegative integer. For example, $\left\{\frac{k-3}{k-2}\right\}_{k=3}^{\infty}$ represents the same sequence.

(b) Here we see that each term is a power of $\left(-\frac{1}{2}\right)$. We have

$$\left\{\left(-\frac{1}{2}\right)^k\right\}_{k=0}^{\infty} = 1, -\frac{1}{2}, \frac{1}{4}, -\frac{1}{8}, \frac{1}{16}, -\frac{1}{32}, \ldots$$

(c) For the third sequence, we will use a conditional definition for the terms of the sequence $\{a_k\}$:

$$a_k = \begin{cases} 0, & \text{if } k \text{ is odd} \\ \frac{1}{k+1}, & \text{if } k \text{ is even.} \end{cases}$$

□

EXAMPLE 4 **Building a recursion formula**

The sequence $1, 1, 1, 3, 5, 9, 17, 31, \ldots$ has the property that the first three terms are 1 and after that every term is the sum of the previous three terms. Express this sequence by using a recursion formula.

SOLUTION

The description of the given sequence is quite similar to the description of the Fibonacci sequence, but here we are adding three terms instead of two to obtain each new term. We will denote this sequence by $\{t_k\}$. Thus we have

$$t_1 = t_2 = t_3 = 1, \quad \text{and} \quad t_k = t_{k-3} + t_{k-2} + t_{k-1} \quad \text{for} \quad k \geq 4.$$

If we prefer, we may also define the sequence by

$$t_1 = t_2 = t_3 = 1, \quad \text{and} \quad t_{k+3} = t_k + t_{k+1} + t_{k+2} \quad \text{for} \quad k \geq 1.$$

The two recursion formulas tell us that, starting with the fourth term, each term in the sequence is the sum of the previous three. □

EXAMPLE 5

Checking for boundedness and monotonicity

Determine whether the following sequences are bounded and/or monotonic:

(a) $\left\{\dfrac{\sin k}{k}\right\}$ **(b)** $\left\{\dfrac{k^3}{e^k}\right\}$ **(c)** $\{k^2 - k\}$ **(d)** $\left\{\dfrac{10^k}{k!}\right\}$

SOLUTION

(a) The sequence $\left\{\frac{\sin k}{k}\right\}$ is bounded because $-1 \le \frac{\sin k}{k} \le 1$ for every $k \in \mathbb{Z}^+$. The sequence is not monotonic, because the signs of $\sin k$ vary as k increases. The changing signs of the terms mean that the sequence cannot be monotonic.

(b) To analyze the sequence $\left\{\frac{k^3}{e^k}\right\}$ we use the function $f(x) = \frac{x^3}{e^x}$ and the derivative test. You should verify that $f'(x) = x^2e^{-x}(3 - x)$. We see that the derivative is positive for $x < 3$ and negative for $x > 3$. Thus, this sequence is eventually decreasing. The derivative also tells us that the largest term in the sequence will occur when $k = 3$, so $\frac{27}{e^3}$ is an upper bound for the sequence. Since all terms in the sequence are positive, zero is a lower bound. The sequence is bounded and eventually decreasing.

(c) For variety we will use the difference test on our third sequence, although the derivative test would probably be more efficient. We let $a_k = k^2 - k$; thus

$$a_{k+1} - a_k = [(k+1)^2 - (k+1)] - [k^2 - k] = 2k.$$

Since $2k$ is positive when k is positive, this sequence is strictly increasing. The sequence is bounded below by its first term, 0, but because it is an increasing sequence, the more important issue concerns the existence of an upper bound. If we think of $a_k = k^2 - k$ as $a_k = k(k-1)$, we see that a_k may be expressed as a product of two factors, each of which increases without bound. Therefore, the given sequence increases without bound.

(d) The ratio test is particularly useful for gaining information about sequences that contain exponential factors or factorials. All of the terms are positive in our final sequence $\left\{\frac{10^k}{k!}\right\}$. We have $a_k = \frac{10^k}{k!}$ and $a_{k+1} = \frac{10^{k+1}}{(k+1)}$. Thus,

$$\frac{a_{k+1}}{a_k} = \frac{10^{k+1}/(k+1)!}{10^k/k!} = \frac{10^{k+1}k!}{10^k(k+1)!} = \frac{10}{k+1}.$$

This quotient is greater than 1 for $1 < k < 9$, but is less than 1 when $k > 10$. The sequence is eventually strictly monotonically decreasing. □

EXAMPLE 6

Analyzing a sequence with "factorial-like" factors

Discuss the monotonicity and boundedness of the sequence

$$\frac{2}{1}, \frac{2^2}{1\cdot 3}, \frac{2^3}{1\cdot 3\cdot 5}, \frac{2^4}{1\cdot 3\cdot 5\cdot 7}, \ldots$$

SOLUTION

Before testing a sequence for monotonicity, it is often helpful to find a nice, succinct form for the general term of the sequence a_k. The numerators in this sequence are powers of 2, but there is no commonly used notation for the denominators, which, although factorial-like, are not factorials. Here we will let

$$a_k = \frac{2^k}{1\cdot 3\cdot 5\cdots(2k-1)}.$$

That is, the denominator of the general term a_k is a product of all odd integers from 1 through $2k-1$. Because the terms are all positive, and because they involve exponential factors in the numerators and factorial-like factors in the denominators, the ratio test will be our choice to check for monotonicity. We have

$$a_k = \frac{2^k}{1 \cdot 3 \cdot 5 \cdots (2k-1)} \quad \text{and} \quad a_{k+1} = \frac{2^{k+1}}{1 \cdot 3 \cdot 5 \cdots (2k-1) \cdot (2k+1)}.$$

That is, the denominator of a_{k+1} contains one more odd factor, $2k+1$, than the denominator of a_k. Now, using the ratio test, we obtain

$$\frac{a_{k+1}}{a_k} = \frac{2^{k+1}/(1 \cdot 3 \cdot 5 \cdots (2k-1) \cdot (2k+1))}{2^k/(1 \cdot 3 \cdot 5 \cdots (2k-1))} = \frac{2^{k+1} \cdot 1 \cdot 3 \cdot 5 \cdots (2k-1)}{2^k \cdot 1 \cdot 3 \cdot 5 \cdots (2k-1) \cdot (2k+1)}.$$

You should verify that

$$\frac{a_{k+1}}{a_k} = \frac{2}{2k+1}.$$

This ratio is less than 1 for every positive integer k; thus the sequence is strictly decreasing.

Since all of the terms of the sequence are positive, it is bounded below by zero, and because the sequence is strictly decreasing, the first term, 2, is an upper bound for the sequence. Therefore, the sequence is bounded.

We are done analyzing this sequence, but we would like to note that its terms may be rewritten with factorials. If we multiply the numerator and denominator of $a_k = \frac{2^k}{1 \cdot 3 \cdot 5 \cdots (2k-1)}$ by the product $2 \cdot 4 \cdot 6 \cdots (2k)$, the new denominator will be $(2k)!$. It may seem that this just moves the problem of having factorial-like terms to the numerator, but since each of the factors of $2 \cdot 4 \cdot 6 \cdots (2k)$ is even, we may rewrite this product as

$$2 \cdot 4 \cdot 6 \cdots (2k) = 2^k k!$$

Thus, if we wish, we may express a_k as the quotient $\frac{2^{2k}k}{2k}$. As we've already seen, this manipulation is not necessary to analyze the sequence. In addition, not all factorial-like products can be handled in a similar fashion. However, this trick may be used when a quotient contains a product of all odd integers from 1 through $2k-1$, as we have here. □

EXAMPLE 7

Modeling drug levels with a sequence

Many prescribed drugs must reach a "maintenance level" in the bloodstream to be effective. Say a person takes D milligrams of wonder drug *Excellenté* per day and that whatever level of *Excellenté* is in the bloodstream, $p\%$ is eliminated in one 24-hour period.

(a) Find a recurrence relation that models the maintenance level of *Excellenté* in a patient's bloodstream.

(b) Assume that 40% of *Excellenté* is eliminated from a patient's body in every 24-hour period and that 100 milligrams are taken daily. Use the result from part (a) to find the level of *Excellenté* in the bloodstream during the first week. Round each answer to the nearest milligram.

SOLUTION

(a) Since the daily dosage is D, we let $L_1 = D$. That is, if we assume that the drug dissolves instantly in the patient's bloodstream, L_1 is the amount present immediately after the drug is taken. Twenty-four hours later $p\%$ has been eliminated from the bloodstream, but another dose is taken; therefore the amount of *Excellenté* in the blood immediately after the second dose will be

$$L_2 = \left(1 - \frac{p}{100}\right) L_1 + D.$$

The situation is similar after every 24-hour period; therefore we obtain the recursively defined sequence

$$L_1 = D \text{ and } L_{k+1} = \left(1 - \frac{p}{100}\right) L_k + D \text{ for } k \geq 1.$$

(b) Using the given quantities and our model from (a), we have, in milligrams,

$$L_1 = 100, \text{ and } L_{k+1} = 0.6L_k + 100 \text{ for } k \geq 1.$$

Thus,

$$L_2 = 0.6(100) + 100 = 160,\ L_3 = 0.6(160) + 100 = 196,$$

$$L_4 = 0.6(196) + 100 = 217.6 \approx 218,\ L_5 = 0.6(218) + 100 \approx 231,$$

$$L_6 = 0.6(231) + 100 \approx 239,\ L_7 = 0.6(239) + 100 \approx 243 \text{ milligrams.}$$ □

EXAMPLE 8

Finding the roots of a function by using Newton's method

Recall that in Example 7 of Section 2.2 we discussed Newton's method. Newton's method provides a recursive formula for approximating roots of equations of the form $f(x) = 0$. To use Newton's method, guess an approximate value, x_0, for the root of $f(x) = 0$ and then use

$$x_{k+1} = x_k - \frac{f(x_k)}{f'(x_k)} \text{ for } k \geq 0$$

to form other approximations that are often more accurate. When this method works, it usually works quickly. We usually terminate the procedure when the absolute value of the difference of two consecutive approximations, $|x_{k+1} - x_k|$, is smaller than a predetermined level of accuracy.

Use Newton's method to approximate a root for the functions

(a) $f(x) = x^2 - 5$ **(b)** $g(x) = e^x - x^3$

both with $x_0 = 1$. Terminate your sequence when $|x_{k+1} - x_k| < 0.01$.

SOLUTION

(a) Here, we have

$$x_{k+1} = x_k - \frac{f(x_k)}{f'(x_k)} = x_k - \frac{x_k^2 - 5}{2x_k} = \frac{x_k^2 + 5}{2x_k}.$$

Thus,

$$x_1 = \frac{1^2 + 5}{2 \cdot 1} = 3,\ x_2 = \frac{3^2 + 5}{2 \cdot 3} = \frac{7}{3},\ x_3 = \frac{(7/3)^2 + 5}{2 \cdot (7/3)} = \frac{47}{21},\ x_4 = \frac{(47/21)^2 + 5}{2 \cdot (47/21)} = \frac{2207}{987}.$$

Note that

$$|x_4 - x_3| = \left|\frac{2207}{987} - \frac{47}{21}\right| = \frac{2}{987} < 0.01,$$

so this is where we terminate our sequence. If we had wanted more accuracy in our approximation, we could have calculated a few more terms in the sequence.

Note also that the roots of the equation $x^2 - 5 = 0$ are $\pm\sqrt{5} \approx \pm 2.2360679$ and that $\frac{2207}{987} \approx 2.360688$, so Newton's method has produced a very good approximation for $\sqrt{5}$ in relatively few steps. More generally, Newton's method provides an effective algorithm for approximating $\sqrt{a}$ when applied to the function $f(x) = x^2 - a$ with $a > 0$.

(b) For our second function $g(x) = e^x - x^3$, we have

$$x_{k+1} = x_k - \frac{g(x_k)}{g'(x_k)} = x_k - \frac{e^{x_k} - x_k^3}{e^{x_k} - 3x_k^2}.$$

You should verify that

$$x_1 \approx 7.0993, \quad x_2 \approx 6.2942, \quad x_3 \approx 5.6031, \quad x_4 \approx 5.0646,$$
$$x_5 \approx 4.7154, \quad x_6 \approx 4.5640, \quad x_7 \approx 4.5372, \quad \text{and} \quad x_8 \approx 4.5364.$$

Since $|x_8 - x_7| < 0.01$, this is where we terminate our approximations. □

CHECKING THE ANSWER

Following is a plot of the graph of g:

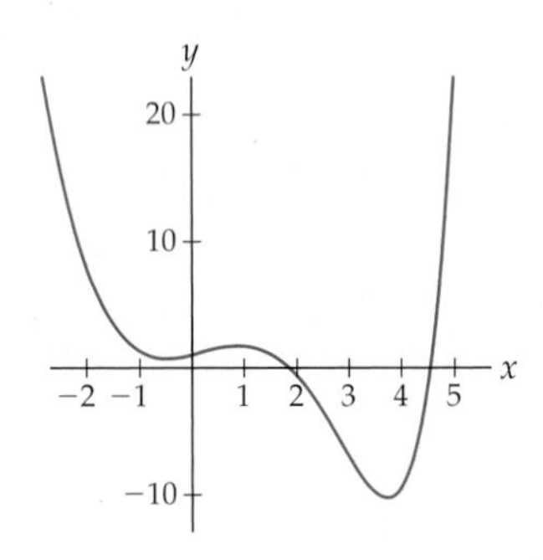

We have approximated one of the two roots of g, but interestingly, not the root closest to our seed value $x_0 = 1$. This is one of the pitfalls of Newton's method. In other cases, it may fail entirely; see Exercises 84 and 85. However, it is often an excellent tool for approximating roots of functions.

EXAMPLE 9 **Showing that a set of rational numbers does not have a least upper bound in $\mathbb{Q}$**

Show that the set $S = \{x \in \mathbb{Q} : 0 \le x^2 \le 2\}$ has a least upper bound in $\mathbb{R}$, but not in $\mathbb{Q}$.

SOLUTION

We will start by showing that $\sqrt{2}$ is the least upper bound of set S in $\mathbb{R}$. Set S consists of all the rational numbers that are in the interval $(-\sqrt{2}, \sqrt{2})$. So, $\sqrt{2}$ is certainly an upper bound for S in $\mathbb{R}$. If u is a positive real number such that $0 < u < \sqrt{2}$, then $0 < u^2 < 2$. So u is not an upper bound for S. Therefore, $\sqrt{2}$ is the least upper bound for the set S in $\mathbb{R}$.

We now show that S does not have a least upper bound in $\mathbb{Q}$. If it did, the least upper bound would be some rational number q such that $\sqrt{2} < q$. However, between every two distinct real numbers there is a rational number. So there is a rational number u such that $\sqrt{2} < u < q$. Therefore, $2 < u^2 < q^2$. This implies that u is another upper bound for the set S in $\mathbb{Q}$ and, therefore, contradicts the fact that q was the least upper bound for S in $\mathbb{Q}$. Our point here is that since every rational number is also a real number, every set of rationals will have a least upper bound, which may be irrational, as it is in this example. □

? TEST YOUR UNDERSTANDING

- What is a sequence? What is the difference between a sequence and any other function?
- What is recursion? How can recursion be used to define a sequence?
- What does it mean for a sequence to be monotonic? What is the distinction between an increasing sequence and a strictly increasing sequence?
- What does it mean for a sequence to be bounded? How can you recognize boundedness in the graph of a sequence? What is meant by a least upper bound? What is meant by a greatest lower bound?
- What are the tests for monotonicity? How do these tests work? Why do they work?

EXERCISES 7.1

Thinking Back

Functions: Provide definitions for each of the following:

- *function*
- the *domain* of a function
- the *codomain* of a function
- the function f is *increasing* on interval $[a, b]$
- the function f is *strictly increasing* on interval $[a, b]$
- the function f is *decreasing* on interval $[a, b]$
- the function f is *strictly decreasing* on interval $[a, b]$
- the function f is *constant* on interval $[a, b]$
- the function f is *bounded* on interval $[a, b]$

Concepts

0. *Problem Zero:* Read the section and make your own summary of the material.

1. *True/False:* Determine whether each of the statements that follow is true or false. If a statement is true, explain why. If a statement is false, provide a counterexample.

(a) *True or False:* Every sequence is a function.

(b) *True or False:* The third term of the sequence $\{k+1\}_{k=1}^{\infty}$ is 4.

(c) *True or False:* The third term of the sequence $\{k^2\}_{k=2}^{\infty}$ is 9.

(d) *True or False:* Every sequence of real numbers is either increasing or decreasing.

(e) *True or False:* Every sequence of numbers has a smallest term.

(f) *True or False:* Every recursively defined sequence has an infinite number of distinct outputs.

(g) *True or False:* Every sequence has an upper bound, a lower bound, or both an upper bound and a lower bound.

(h) *True or False:* Every monotonic sequence has an upper bound, a lower bound, or both an upper bound and a lower bound.

2. *Examples:* Construct examples of the thing(s) described in the following. Try to find examples that are different than any in the reading.

(a) A strictly increasing sequence with an upper bound.

(b) A decreasing sequence without a lower bound.

(c) A sequence of real numbers that is neither increasing nor decreasing.

3. What is a sequence?

4. What is a *term* of a sequence?

5. What is meant by the *index* of a term of a sequence?

6. Give a recursive definition for $K!$ for integers $k \geq 0$. Be sure you define $0!$ as part of your answer.

7. Give the first five terms of the following recursively defined sequence:

$$a_1 = 1, \text{ and } a_k = a_{k-1} + 2 \text{ for } k \geq 2.$$

Also, give a closed formula for the sequence.

8. Give the first five terms of the following recursively defined sequence:

$$a_1 = 2, \text{ and } a_k = a_{k-1} + 2 \text{ for } k \geq 2.$$

Also, give a closed formula for the sequence.

9. Give a recursive definition for the sequence $1, 2, 3, 4, \ldots$ of positive integers. (*Hint: Let $a_1 = 1$.*)

10. The Lucas numbers are defined recursively as follows:

$$L_1 = 1, L_2 = 3, \text{ and } L_k = L_{k-2} + L_{k-1} \text{ for } k \geq 3.$$

What are L_3, L_4, L_5, and L_6?

11. Define what it means for a sequence $\{a_k\}$ to be eventually strictly increasing.

12. Define what it means for a sequence $\{a_k\}$ to be eventually decreasing.

13. Define what it means for a sequence $\{a_k\}$ to be eventually strictly decreasing.

14. Define what it means for a sequence $\{a_k\}$ to be eventually monotonic.

15. What does it mean for a sequence $\{a_k\}$ to be bounded above? Bounded below? Bounded?

16. Explain why a sequence that is bounded above has infinitely many upper bounds.

17. Give an example of a sequence with neither an upper bound nor a lower bound.

18. Explain why every monotonic sequence has an upper bound, a lower bound, or both an upper bound and a lower bound.

19. Explain why we require the terms of the sequence $\{a_k\}$ to be positive when we use the ratio test from Theorem 7.6.

20. State a variation of the ratio test from Theorem 7.6 that would allow you to use ratios to test a sequence $\{a_k\}$ for monotonicity when each $a_k < 0$.

21. The Fibonacci numbers may be computed with the formula

$$f_k = \frac{(1+\sqrt{5})^k - (1-\sqrt{5})^k}{2^k\sqrt{5}}.$$

Use this formula to compute f_1, f_2, f_3, f_4, and f_5. (Imagine computing f_{100}.)

22. What is the least upper bound property for nonempty subsets of real numbers? Does the least upper bound property hold for subsets of the rational numbers? Does it hold for subsets of the integers?

23. Make a statement expressing a property analogous to the least upper bound property for nonempty subsets of real numbers that are bounded below.

24. Let $\{a_k\}$ be the sequence $a_1 = 3$, $a_2 = 3.1$, $a_3 = 3.14$, $a_4 = 3.141, \ldots$. That is, each term a_k contains the first k decimal digits of π.

(a) Explain why a_k is a rational number for each positive integer k.

(b) Explain why the sequence $\{a_k\}$ is increasing.

(c) Provide an upper bound for the sequence $\{a_k\}$.

(d) What is the least upper bound of the sequence $\{a_k\}$?

(e) Use this sequence to explain why the Least Upper Bound Axiom does not apply to the set of rational numbers.

Skills

In Exercises 25–30 find a plausible formula for the general term of the given sequence.

25. $\{0, 1, 0, 1, 0, 1, \ldots\}$

26. $\{1, 7, 13, 19, 25, \ldots\}$

27. $\left\{\frac{1}{3}, \frac{2}{9}, \frac{1}{9}, \frac{4}{81}, \frac{5}{243}, \ldots\right\}$

28. $\left\{5, -\frac{5}{2}, \frac{5}{4}, -\frac{5}{8}, \frac{5}{16}, \ldots\right\}$

29. $\left\{\frac{2}{3}, \frac{3}{5}, \frac{4}{7}, \frac{5}{9}, \ldots\right\}$

30. $\left\{1, \frac{1}{2}, \frac{1}{6}, \frac{1}{24}, \frac{1}{120}, \ldots\right\}$

In Exercises 31–36 provide the first five terms of the given sequence. Unless specified, assume that the first term has index 1.

31. $\left\{\frac{1-(-1)^k}{k}\right\}$

32. $\left\{\frac{n}{2n+1}\right\}$

33. $a_k = \frac{\cos(kx)}{x^k + k^2}$

34. $\left\{\frac{(-1)^{k-1}x^{2k}}{2k!}\right\}$

35. $\left\{\frac{\sqrt{k}}{k}\right\}$

36. $\left\{\frac{1}{2} + \frac{1}{4} + \frac{1}{8} + \cdots + \frac{1}{2^k}\right\}$

Find the least upper bound of the sequences in Exercises 37–42.

37. $\left\{2 - \frac{1}{k^2}\right\}$

38. $\left\{\frac{1}{2}, \frac{2}{3}, \frac{3}{4}, \frac{4}{5}, \ldots\right\}$

39. $\{-k\}$

40. $\{0, 1, 0, 1, 0, 1, \ldots\}$

41. $\left\{\frac{1}{2} + \frac{1}{4} + \frac{1}{8} + \cdots + \frac{1}{2^k}\right\}$

42. $\{2, 2.7, 2.71, 2.718, 2.7182, \ldots\}$

In Exercises 43–46 give the first five terms for a geometric sequence $\{cr^k\}_{k=0}^{\infty}$ with the specified values of c and r.

43. $c = 3,\ r = \frac{1}{2}$

44. $c = -2,\ r = -\frac{1}{3}$

45. $c = -2,\ r = -3$

46. $c = -1,\ r = -\frac{1}{2}$

Write each of the arithmetic sequences in Exercises 47–50 in the form $\{c + dk\}_{k=0}^{\infty}$.

47. $-3, 7, 17, 27, \ldots$

48. $-6, -7.1, -8.2, -9.3, \ldots$

49. $1, -1, \ldots$

50. $5, \pi, \ldots$

In Exercises 51–54 use the difference test in Theorem 7.6 to analyze the monotonicity of the given sequence.

51. $\{k^2 - 5k\}$

52. $\{\sqrt{k} - \sqrt{k+1}\}$

53. $\left\{\frac{k}{k+2}\right\}$

54. $\left\{\frac{1}{k!}\right\}$

In Exercises 55–58 use the ratio test in Theorem 7.6 to analyze the monotonicity of the given sequence.

55. $\left\{\frac{k^2}{k!}\right\}$

56. $\left\{\sqrt{1 - \frac{1}{k}}\right\}$

57. $\left\{\frac{3^k}{2 \cdot 4 \cdot 6 \cdots (2k)}\right\}$

58. $\left\{\frac{(k!)^2}{(2k)!}\right\}$

In Exercises 59–62 use the derivative test in Theorem 7.6 to analyze the monotonicity of the given sequence.

59. $\left\{\frac{\sqrt{k+1}}{k}\right\}$

60. $\{k^2 - k\}$

61. $\left\{\frac{\sin k}{k}\right\}$

62. $\left\{\frac{k!}{(k+1)!}\right\}$

Determine whether the sequences in Exercises 63–74 are monotonic or not. Also determine whether the given sequence is bounded or unbounded.

63. $\left\{2-\frac{k-1}{10}\right\}$

64. $\left\{\frac{1}{2k}\right\}$

65. $a_k = \frac{(-1)^k}{k}$

66. $a_k = \frac{k^2}{k!}$

67. $\{1, -1, 1, -1, 1, \ldots\}$

68. $\left\{1, \frac{3}{4}, \frac{8}{9}, \frac{15}{16}, \frac{24}{25}, \ldots\right\}$

69. $\left\{\frac{\cos k}{k}\right\}$

70. $\left\{\frac{\cos(2\pi k)}{k}\right\}$

71. $\left\{\frac{e^k}{k!}\right\}$

72. $\left\{\frac{(2k)!}{(k!)^2}\right\}$

73. $\{(-1)^k k\}$

74. $\left\{5\left(1-\left(\frac{1}{10}\right)^k\right)\right\}$

In Exercises 75–78 use Newton's method (see Example 8) to approximate a root for the given function with the specified value of x_0. Terminate your sequence when $|x_{n+1}-x_n| < 0.001$.

75. $f(x) = x^3 - 2,\ x_0 = 1$

76. $f(x) = e^x + \sin x,\ x_0 = 0$

77. $f(x) = e^x + \sin x,\ x_0 = -2$

78. $f(x) = \sqrt{x+1} - \frac{1}{x},\ x_0 = 1$

79. Use Newton's method to derive the recursion formula

$$x_{k+1} = \frac{1}{2}\left(x_k + \frac{a}{x_k}\right)$$

for approximating $\sqrt{a}$. (*Hint: Let* $f(x) = x^2 - a$.)

Use the result of Exercise 79 to approximate the square roots in Exercises 80–83. In each case, start with $x_0 = 1$ and stop when $|x_{k+1} - x_k| < 0.001$.

80. $\sqrt{2}$

81. $\sqrt{3}$

82. $\sqrt{4}$

83. $\sqrt{101}$

84. Explain why Newton's method will fail if you choose a value of x_0 for which $f'(x_0) = 0$.

85. Newton's method will also fail when the difference of successive approximations, $|x_{k+1} - x_k|$, does not decrease as k increases.

(a) Show that this happens for the function $f(x) = \sqrt[3]{x-2}$ when you choose $x_0 = 1$.

(b) What is the root of $f(x) = \sqrt[3]{x-2}$?

Applications

Use the result of Example 7 to approximate the levels of the drug *Excellenté* during the first week, assuming the dosages and decay rates in Exercises 86–88.

86. $L_1 = 200,\ p = 50$

87. $L_1 = 100,\ p = 25$

88. $L_1 = 300,\ p = 60$

89. Suppose you invest \$100.00 in a bank that pays you 5% interest compounded annually. The balance in the account after k years is given by $a_k = 100(1+0.05)^k$. To the nearest cent, determine the first five terms of the sequence, starting at $k = 0$. What does $k = 0$ mean in practical terms? Determine whether the sequence is bounded. Determine whether the sequence is increasing, decreasing, or not monotonic.

Proofs

90. Prove that the ratio of successive terms of a nonzero geometric sequence is constant.

91. Prove that a sequence $\{a_k\}$ that is both increasing and decreasing is constant.

92. Prove that every sequence of the form $\{a_k\}_{k=n}^{\infty}$ can be rewritten as a sequence of the form $\{\alpha_k\}_{k=1}^{\infty}$.

93. Prove that if $\{a_k\}_{k=1}^{\infty}$ is a sequence of positive real numbers, then the sequence $\{S_n\}_{n=1}^{\infty}$, where the sequence $S_n = a_1 + a_2 + \cdots + a_n$, is an increasing sequence.

94. Let $\{a_k\}$ be a sequence. Prove Theorem 7.6 (a) along with the following variations:

(a) Show that when $a_{k+1} - a_k \geq 0$ for every $k \geq 1$, the sequence is increasing.

(b) Show that when $a_{k+1} - a_k > 0$ for every $k \geq 1$, the sequence is strictly increasing.

(c) Show that when $a_{k+1} - a_k \leq 0$ for every $k \geq 1$, the sequence is decreasing.

(d) Show that when $a_{k+1} - a_k < 0$ for every $k \geq 1$, the sequence is strictly decreasing.

95. Let $\{a_k\}$ be a sequence of positive terms. Prove Theorem 7.6 (b) along with the following variations:

(a) Show that when $\frac{a_{k+1}}{a_k} \geq 1$ for every $k \geq 1$, the sequence is increasing.

(b) Show that when $\frac{a_{k+1}}{a_k} > 1$ for every $k \geq 1$, the sequence is strictly increasing.

(c) Show that when $\frac{a_{k+1}}{a_k} \leq 1$ for every $k \geq 1$, the sequence is decreasing.

(d) Show that when $\frac{a_{k+1}}{a_k} < 1$ for every $k \geq 1$, the sequence is strictly decreasing.

96. Let $a(x)$ be a differentiable function on the interval $[1, \infty)$, and let $a_k = a(k)$ for every positive integer k. Prove Theorem 7.6 (c) along with the following variations:

(a) Show that when $a'(x) \geq 0$ for $x > 1$, the sequence $\{a_k\}$ is increasing.

(b) Show that when $a'(x) > 0$ for $x > 1$, the sequence $\{a_k\}$ is strictly increasing.

(c) Show that when $a'(x) \leq 0$ for $x > 1$, the sequence $\{a_k\}$ is decreasing.

(d) Show that when $a'(x) < 0$, for $x > 1$, the sequence $\{a_k\}$ is strictly decreasing.

Thinking Forward

▶ *A sequence of sums:* Consider the sequence $\left\{\frac{1}{k!}\right\}_{k=0}^{\infty}$. The associated sequence $\{S_n\}_{n=0}^{\infty}$, where

$$S_n = 1 + 1 + \frac{1}{2!} + \cdots + \frac{1}{n!},$$

is a sequence of sums. In Chapter 8 we will see that this sequence converges to the number e. Evaluate S_n for $n = 1,\ 2,\ 3,\ 10$. How close is S_{10} to e?

▶ *A sequence of polynomials:* Consider the sequence of monomials $\left\{\frac{x^k}{k!}\right\}_{k=0}^{\infty}$. The associated sequence $\{p_n\}_{n=0}^{\infty}$, where $p_n = 1 + x + \cdots + \frac{x^n}{n!}$ is a sequence of polynomials. This latter sequence of polynomials has a very interesting connection with the graph of the exponential function $f(x) = e^x$. Graph various p_n versus $f(x)$ in a window containing the point $(0, 1)$. What do you notice? Take the derivatives of p_n and $f(x)$, evaluate them at $x = 0$, and take notice of what happens.

7.2 LIMITS OF SEQUENCES

- What it means for a sequence to converge or diverge
- Algebraic properties of convergent sequences
- The relationships among monotonicity, boundedness, and convergence for sequences

Convergence or Divergence of a Sequence

Since a sequence is essentially an infinite list of numbers indexed by $1, 2, 3, \ldots, k, \ldots$, one of the most fundamental questions we can ask about a sequence is what happens as $k \to \infty$. We have already set up machinery to answer such questions for functions $f(x)$ as $x \to \infty$. Recall from Definition 1.9 that we say $\lim\limits_{x\to\infty} f(x) = L$ when, for any $\epsilon > 0$, there exists some $N > 0$ such that whenever $x > N$, the value of $f(x)$ lies in the horizontal bar defined by $(L - \epsilon, L + \epsilon)$; see the following figure at the left:

$\lim\limits_{x\to\infty} a(x) = L$

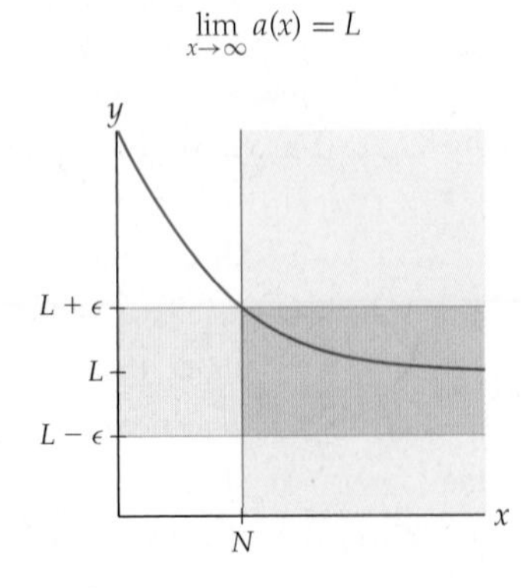

$\lim\limits_{k\to\infty} a_k = L$

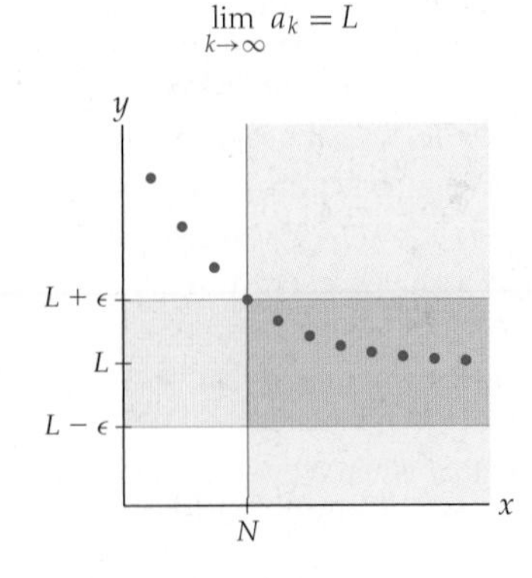

If we restrict our attention to integer inputs, as in the right-hand figure, then the same machinery can be used to define what we mean by $\lim_{k\to\infty} a_k = L$:

DEFINITION 7.8 **The Limit of a Convergent Sequence**

Suppose $\{a_k\}$ is a sequence of real numbers. We say that $\lim_{k\to\infty} a_k = L$ for some real number L, or equivalently that $a_k \to L$, if the following statement is true:

For any $\epsilon > 0$, there exists some $N > 0$ such that if $k > N$, then $a_k \in (L - \epsilon, L + \epsilon)$.

If $a_k \to L$ for some real number L, then we say that the sequence $\{a_k\}$ ***converges*** to L. If no such L exists, then we say that the sequence ***diverges***.

Informally, for any value of ϵ, we can find some N so that all terms a_k after a_N are "close enough" to L. We call the set of sequence terms after some index N a ***tail*** of the sequence. Thus, if a sequence is convergent, we can always find some N so that the entire tail of the sequence after index N is as close to L as we like.

Although not every sequence $\{a_k\}$ is the integer part of a continuous function $a(x)$ in a simple way, when that is the case, the limit of $a(x)$ as $x \to \infty$ determines the limit of a_k as $k \to \infty$.

THEOREM 7.9 **Limits of Sequences Defined by Continuous Functions**

Let $a(x)$ be a function that is continuous on $[1, \infty)$ and $\{a_k\}$ be the sequence defined by $a_k = a(k)$ for every $k \in \mathbb{Z}^+$. If $\lim_{x\to\infty} a(x) = L$, then $\lim_{k\to\infty} a_k = L$.

For example, since we already know that $\lim_{x\to\infty} \frac{1}{x} = 0$, we can say that the limit of the sequence $\left\{\frac{1}{k}\right\}$ is also zero. The proof of this theorem follows directly from the definitions of limits at infinity for functions and sequences and is left to Exercise 75.

When a sequence diverges, it can be for a number of reasons. For example, the sequence $\{(-1)^k\}$ diverges because it oscillates forever between -1 and 1, never approaching any one real number L. Other sequences diverge because their terms increase or decrease without bound; in this case we say that $a_k \to \infty$ or $a_k \to -\infty$. The next two figures illustrate the situation when $a_k \to \infty$. Once again we can define this notion by using the machinery of limits that we developed in Chapter 1 for functions.

$\lim_{x\to\infty} a(x) = \infty$

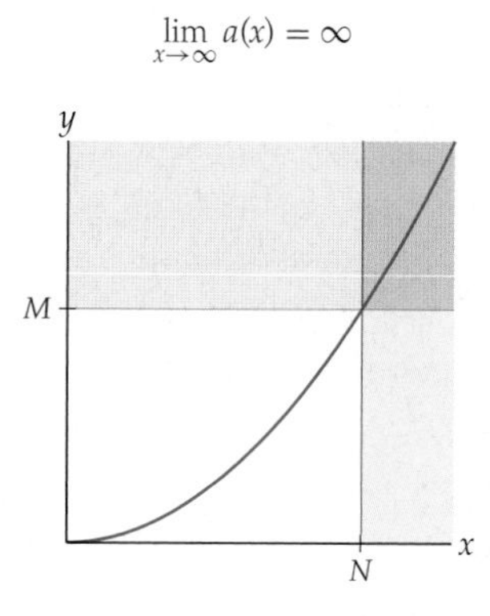

$\lim_{k\to\infty} a_k = \infty$

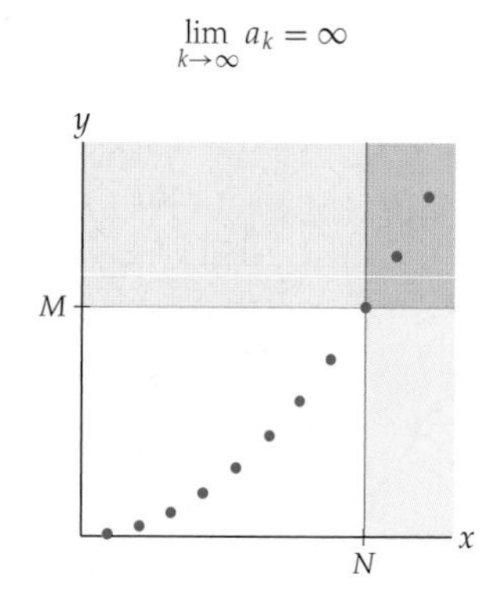

DEFINITION 7.10 **Sequences that Diverge to ∞ or $-\infty$**

Suppose $\{a_k\}$ is a sequence of real numbers.

(a) We say that $\lim_{k\to\infty} a_k = \infty$, or equivalently, that $a_k \to \infty$, if the following statement is true:

For any $M > 0$, there exists some $N > 0$ such that if $k > N$, then $a_k > M$.

(b) We say that $\lim_{k\to\infty} a_k = -\infty$, or equivalently, that $a_k \to -\infty$, if the following statement is true:

For any $M > 0$, there exists some $N > 0$ such that if $k > N$, then $a_k < -M$.

For example, the sequence of natural numbers $\{k\} = \{1, 2, 3, \ldots\}$ diverges to ∞ and the sequence of odd negative integers $\{-2k+1\} = \{-1, -3, -5, \ldots\}$ diverges to $-\infty$.

Theorems About Convergent Sequences

Sequences are functions, and therefore the properties of limits that we established in Section 1.5 for functions $f(x)$ as $x \to \infty$ also apply to sequences $\{a_k\}$ as $k \to \infty$. The following theorems recast the constant-multiple, sum, product, quotient, composition, and uniqueness rules for limits in terms of the notation of convergent sequences:

THEOREM 7.11 **Basic Limit Rules for Convergent Sequences**

If $\{a_k\}$ and $\{b_k\}$ are convergent sequences with $a_k \to L$ and $b_k \to M$ as $k \to \infty$, and if c is any constant, then

(a) $ca_k \to cL$

(b) $(a_k + b_k) \to L + M$

(c) $a_k b_k \to LM$

(d) If $M \neq 0$, then $\dfrac{a_k}{b_k} \to \dfrac{L}{M}$

The proofs of these limit rules for sequences are analogous to the proofs we did in Section 1.5; see Exercises 67–70.

THEOREM 7.12 **More Limit Rules for Convergent Sequences**

Suppose $\{a_k\}$ is a sequence.

(a) *Limits of Functions of Sequences:* If $a_k \to L$ and f is a function that is continuous at L, then $f(a_k) \to f(L)$.

(b) *Uniqueness of Limits for Sequences:* If $a_k \to L$ and $a_k \to M$, then $L = M$.

(c) *Squeeze Theorem for Sequences:* If $\{m_k\}$ and $\{M_k\}$ are sequences that both converge to L, and if $m_k \leq a_k \leq M_k$ for all k, then $a_k \to L$.

(d) *The Limit of the Absolute Value of a Sequence:* If $|a_k| \to 0$, then $a_k \to 0$.

Again, you may find the proofs of analogous properties in Section 1.5.

Part (a) of this theorem can be particularly useful. For example, since the sequence $\left\{\frac{1}{n}\right\}$ converges to 0, we can conclude that the sequence $\{e^{1/n}\}$ converges to $e^0 = 1$. The proofs of the four parts of this theorem are essentially the same as the proofs of the analogous theorems from Chapter 1; see Exercises 71–74.

One of the conditions specified in the Squeeze Theorem (Theorem 7.12 (c)) may be relaxed slightly. As long as the inequality $m_n \le a_n \le M_n$ holds eventually, the conclusion of the theorem will remain valid. For example, it may be shown that for all $k \ge 5$, $0 < \frac{3}{k^2-8} < \frac{1}{k}$. Therefore, since the constant sequence $\{0\}$ and the sequence $\left\{\frac{1}{k}\right\}$ both converge to 0, the sequence $\left\{\frac{3}{k^2-8}\right\}$ must also converge to zero.

Informally, a subsequence of a sequence $\{a_k\}$ is an (infinite) subset of $\{a_k\}$, which maintains the order of the terms of the sequence. For example, if we start with the sequence given by $\left\{\frac{(-1)^{k+1}}{k}\right\}$, then

$$\left\{\frac{(-1)^{k+1}}{k}\right\} = 1, -\frac{1}{2}, \frac{1}{3}, -\frac{1}{4}, \frac{1}{5}, -\frac{1}{6}, \ldots.$$

If we select every other term, starting with 1, we obtain the subsequence

$$1, \frac{1}{3}, \frac{1}{5}, \frac{1}{7}, \frac{1}{9}, \ldots.$$

More formally, we have Definition 7.13. Recall that $\mathbb{Z}^+$ denotes the positive integers.

DEFINITION 7.13 **Subsequence**

Let $\{a_k\}$ be a sequence and $f : \mathbb{Z}^+ \to \mathbb{Z}^+$ be a strictly increasing function. Then the sequence $\{a_{f(k)}\}$ is called a ***subsequence*** of $\{a_k\}$.

In the example immediately preceding the definition, the increasing function that selects the desired terms is $f(k) = 2k - 1$.

THEOREM 7.14 **Subsequences of Convergent Sequences Converge**

Let $\{a_k\}$ be a sequence that converges to L. Then every subsequence of $\{a_k\}$ also converges to L.

The converse of Theorem 7.14 is not true. For example, consider the sequence $\{a_k\}$, where

$$a_k = \begin{cases} k, & \text{if } k \text{ is odd} \\ \frac{1}{k}, & \text{if } k \text{ is even} \end{cases}$$
$$= 1, \frac{1}{2}, 3, \frac{1}{4}, 5, \frac{1}{6}, \ldots.$$

The subsequence $\{a_{2k}\} = \left\{\frac{1}{2k}\right\} \to 0$, but the sequence $\{a_k\}$ diverges. In Exercise 88 we ask you to prove Theorem 7.14.

Convergence and Divergence of Basic Sequences

Since we will encounter geometric sequences so frequently in this course, let us summarize their convergence and divergence properties:

THEOREM 7.15 **Convergence and Divergence of Geometric Sequences**

Let $\{r^k\}_{k=0}^{\infty}$ be a geometric sequence.

(a) If $|r| > 1$, then $\{r^k\}$ diverges.

(b) If $r = -1$, then $\{r^k\}$ diverges.

(c) If $r = 1$, then $\{r^k\}$ converges to 1.

(d) If $|r| < 1$, then $\{r^k\}$ converges to 0.

The results of this theorem are unsurprising if we consider some examples for various values of r:

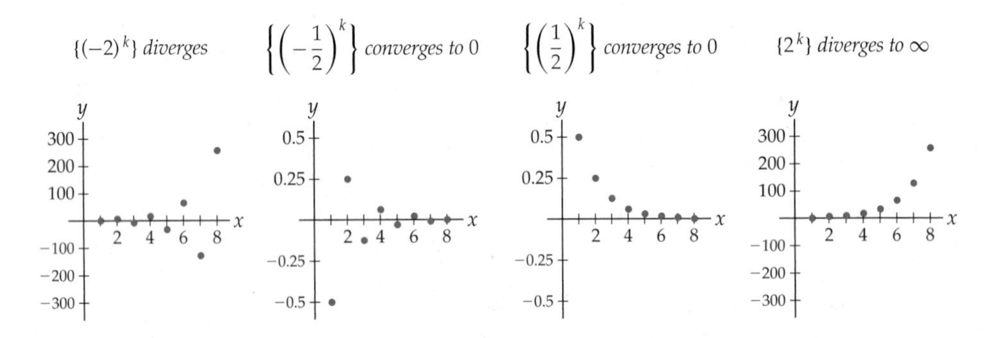

Proof. Parts of this theorem are equivalent to corresponding statements about limits that we have already proved. For example, if we already know that $r > 1$, then $\lim_{x \to \infty} r^x = \infty$. Similarly, if we already know that $0 \le r < 1$, then $\lim_{x \to \infty} r^x = 0$. Also when $r = 1$, $\lim_{x \to \infty} 1^x = 1$. By Theorem 7.9, this proves Theorem 7.15 for all positive values of r.

In Example 4 we provide a formal proof that the geometric sequence $\{(-1)^k\}_{k=0}^{\infty}$ diverges. In Exercise 63 we ask you to prove that $\{r^k\}_{k=0}^{\infty}$ diverges when $r < -1$. ■

In order to develop an intuition for the long-term behavior of sequences, it is important to understand a few simple examples. Many such examples rely on the concept of ***dominance*** that we defined in the exercises of Section 3.6. In the language of sequences, an increasing sequence $\{a_k\}$ dominates another increasing sequence $\{b_k\}$ if $\frac{a_k}{b_k} \to \infty$, or equivalently, if $\frac{b_k}{a_k} \to 0$. Intuitively, dominance essentially means that $\{a_k\}$ grows at a fundamentally faster rate than $\{b_k\}$, and we can represent this faster growth with the notation $a_k >> b_k$. The dominance relationships between various basic sequences are as follows:

THEOREM 7.16 **Dominance Relationships for Simple Sequences**

For any real numbers $a > 0$ and $b > 1$, the following string of dominances holds:

$$\ln k << k^a << b^k << k!, \text{ for } k \text{ sufficiently large.}$$

We already established in Section 3.6 that logarithmic functions are dominated by power functions, which in turn are dominated by exponential functions. The fact that factorial expressions dominate even exponential functions becomes clear when we use the ratio test from Theorem 7.6 to analyze the sequence of quotients $\left\{\frac{k!}{b^k}\right\}$. We ask you to provide the details of this analysis in Exercise 64.

The next theorem summarizes the convergence results that follow immediately from the dominance relationships presented in Theorem 7.16.

THEOREM 7.17 **Convergence of Sequences Based on Dominance Ratios**

For any real numbers $a > 0$ and $b > 1$, the following sequences converge to zero:

$$\left\{\frac{\ln k}{k^a}\right\}, \quad \left\{\frac{\ln k}{b^k}\right\}, \quad \left\{\frac{\ln k}{k!}\right\}, \quad \left\{\frac{k^a}{b^k}\right\}, \quad \left\{\frac{k^a}{k!}\right\}, \quad \left\{\frac{b^k}{k!}\right\}.$$

It is also useful to gain an intuition into the convergence and divergence of basic sequences that involve exponents. Sometimes this intuition is obvious; for example it is clear that $\{p^k\}$ and $\{k^p\}$ must diverge to ∞ for all $p > 1$. It is also clear that $\{k^{-p}\} = \left\{\frac{1}{k^p}\right\}$ must converge to zero. A few of the less obvious basic sequences involving exponents are summarized in the following theorem:

THEOREM 7.18 **Some Convergent Sequences Involving Exponents**

For any real number $p > 0$, the following sequences converge:

(a) $p^{1/k} \to 1$ **(b)** $k^{1/k} \to 1$ **(c)** $\left(\frac{1}{k}\right)^p \to 0$

We will prove the first part of Theorem 7.18 and leave the proofs of parts (b) and (c) to Exercises 65 and 66, but before we do, we will examine the first few terms of one of these sequences. The first five terms of the sequence $\{k^{1/k}\}$ are

$$1, 2^{1/2} \approx 1.414, \; 3^{1/3} \approx 1.442, \; 4^{1/4} \approx 1.414, \; \text{and } 5^{1/5} \approx 1.380.$$

When $k = 1000$, we have $1000^{1/1000} \approx 1.007$. Of course, this reasoning does not constitute a proof that $k^{1/k} \to 1$, but it does indicate that the result is plausible.

Proof. **(a)** We could prove this statement by using L'Hôpital's rule and the logarithm methods of Section 3.6 to show that $\lim_{x\to\infty} p^{1/x} = 1$, but we will follow a different path and use the first part of Theorem 7.12. We already know that the sequence $\left\{\frac{1}{k}\right\}$ converges to zero. Since the exponential function p^x is continuous for $p > 0$, part (a) of Theorem 7.12 tells us that the sequence $\{p^{1/k}\}$ converges to $p^0 = 1$. ■

Bounded Monotonic Sequences

In the previous section, we defined what it means for a sequence to be bounded and what it means for a sequence to be monotonic. These two concepts are important individually, but when we combine them, we get a very powerful convergence theorem:

THEOREM 7.19 **Bounded Monotonic Sequences Always Converge**

If $\{a_k\}$ is both bounded and (eventually) monotonic, then $\{a_k\}$ converges.

This theorem should make immediate intuitive sense. For example, if a monotonically increasing sequence never gets above a certain value M, then the values of its tail terms keep getting larger and yet never get above M. This means that the terms of the sequence must at some point level off at or before that upper bound M.

The proof of this theorem is fairly technical and requires the ***Least Upper Bound Axiom***, which we discussed in the previous section. This says that every nonempty subset of the real numbers that is bounded above has a least upper bound.

Proof. We will prove the theorem in the case where $\{a_k\}$ is bounded and increasing. The case where $\{a_k\}$ is bounded and decreasing is similar. The sequence is a set of real numbers. Since this set is assumed to be bounded, it must have an upper bound. In fact, by the Least Upper Bound Axiom there must be a real number that is its least upper bound; call that number M.

We now appeal to the definition of convergence in Definition 7.8. Given any $\epsilon > 0$, there must be some index N for which $a_N > M - \epsilon$. This statement is true because if there were not such an N, then every term in the tail of the sequence would be less than or equal to $M - \epsilon$, and this would contradict the fact that M is the *least* upper bound of $\{a_k\}$. Since $\{a_k\}$ is by assumption an increasing sequence, it must be that for all $k > N$,

$$a_k \geq a_N > M - \epsilon.$$

Since M is an upper bound of $\{a_k\}$, it follows that $M \geq a_k$ for all k. Combining this result with the double inequality we have

$$M - \epsilon < a_k \leq M < M + \epsilon,$$

which is precisely what we need to show that $a_k \in (M-\epsilon, M+\epsilon)$. By the definition of convergence for sequences we have shown that $\{a_k\}$ converges. In fact we have shown more than that: We have shown that an increasing bounded sequence converges to its least upper bound. ■

It is important to note that the converse of Theorem 7.19 is not true. That is, not every convergent sequence is necessarily bounded and (eventually) monotonic. For example, the sequence $\{(-1/2)^n\}$ converges to zero, but is not (even eventually) monotonic, since it alternates between positive- and negative-valued terms. However, we do have the following partial converse:

THEOREM 7.20 **Convergent Sequences are Bounded**

If a sequence converges, then it is bounded.

The proof of this theorem is left to Exercise 85.

Examples and Explorations

EXAMPLE 1 **Finding the limit of a sequence**

Use the theorems from this section to find the limit of the sequence $\left\{\dfrac{k-2}{k^2+k-7}\right\}$.

SOLUTION

With a quotient like the one given, we will often use Theorem 7.11(d). Notice that we need to know that the limits of the numerator and denominator exist prior to using the theorem. As the quotient is initially stated, neither the sequence in the numerator nor the sequence in the denominator has a limit. Therefore, our first step is algebraic, to ensure that the new numerator and new denominator are both convergent. To evaluate this limit, we use various parts of Theorem 7.11 provisionally and then verify at the end that all the hypotheses of the theorems have been satisfied:

$$\lim_{k\to\infty} \frac{k-2}{k^2+k-7} = \lim_{k\to\infty} \frac{(1/k)-(2/k^2)}{1+(1/k)-(7/k^2)} \qquad \leftarrow \text{multiply by } \frac{1/k^2}{1/k^2}$$

$$= \frac{\lim_{k\to\infty}((1/k)-(2/k^2))}{\lim_{k\to\infty}(1+(1/k)-(7/k^2))} \qquad \leftarrow \text{Theorem 7.11 (d)}$$

$$= \frac{\lim_{k\to\infty}(1/k) - \lim_{k\to\infty}(2/k^2)}{\lim_{k\to\infty} 1 + \lim_{k\to\infty}(1/k) - \lim_{k\to\infty}(7/k^2)} \qquad \leftarrow \text{Theorem 7.11 (a and b)}$$

$$= \frac{\lim_{k\to\infty}(1/k) - 2\lim_{k\to\infty}(1/k^2)}{\lim_{k\to\infty} 1 + \lim_{k\to\infty}(1/k) - 7\lim_{k\to\infty}(1/k^2)} \qquad \leftarrow \text{Theorem 7.11 (a)}$$

$$= \frac{0-2\cdot 0}{1+0-7\cdot 0} = 0 \qquad \leftarrow \text{Theorem 7.18 (c)}$$

□

EXAMPLE 2

Using the convergence theorems for sequences

Use Theorems 7.15–7.18 to analyze the following sequences:

(a) $\left\{\left(\frac{999}{1000}\right)^k\right\}$ **(b)** $\left\{\frac{100^k}{k!}\right\}$ **(c)** $\left\{\frac{k!}{10000^k}\right\}$ **(d)** $\left\{\left(\frac{1}{100}\right)^{1/k}\right\}$

SOLUTION

(a) The sequence $\left\{\left(\frac{999}{1000}\right)^k\right\}$ is geometric. Since $\left|\frac{999}{1000}\right| < 1$, the sequence converges to zero.

(b) The dominance of factorial growth over polynomial growth is one of the relationships expressed in Theorem 7.16. Therefore, $\frac{100^k}{k!} \to 0$.

(c) Here, again, since factorial growth dominates polynomial growth, $\frac{k!}{10000^k} \to \infty$.

(d) By Theorem 7.18 (a) $\left(\frac{1}{100}\right)^{1/k} \to 1$.

□

EXAMPLE 3

Using boundedness and monotonicity

Use the boundedness and monotonicity of the following sequences to analyze their convergence properties:

(a) $\left\{\frac{\sin k}{k}\right\}$ **(b)** $\left\{\frac{k^3}{e^k}\right\}$ **(c)** $\{k^2-k\}$

SOLUTION

In Example 5 of Section 7.1 we discussed the boundedness and monotonicity of these three sequences. We now extend the work we started there to determine whether those sequences converge or diverge.

(a) In Section 7.1 we saw that the sequence $\left\{\frac{\sin k}{k}\right\}$ is bounded but not monotonic. Such a sequence may or may not converge. Since $-1 \le \sin k \le 1$ and $k > 1$, we also have

$$-\frac{1}{k} \le \frac{\sin}{k} \le \frac{1}{k}.$$

But $\lim_{k\to\infty}\left(\pm\frac{1}{k}\right) = 0$; thus, by the Squeeze Theorem, the sequence $\left\{\frac{\sin k}{k}\right\}$ converges to zero.

(b) In Section 7.1 we saw that the sequence $\left\{\frac{k^3}{e^k}\right\}$ is both bounded and eventually monotonic. Therefore, by Theorem 7.19, this sequence converges. Theorem 7.19 does not tell us the limit of the sequence. In Exercise 19 we will ask you to show that the sequence converges to zero.

(c) The sequence $\{k^2 - k\}$ is monotonic, but not bounded. Theorem 7.20 implies that every unbounded sequence diverges. □

EXAMPLE 4

Showing that a geometric sequence diverges

Show that the geometric sequence $\{(-1)^k\}_{k=0}^{\infty}$ diverges.

SOLUTION

The sequence takes on only the values 1 and -1. We claim that there is no limit for this sequence. Suppose to the contrary that there was a limit L. Then for $\epsilon = 1$, there would be a positive integer N such that for all integers $n > N$, we would have $|a_n - L| < 1$. There are some $n > N$ that are even and some that are odd. For those that are even, $|a_n - L| = |1 - L|$. For those that are odd, $|a_n - L| = |(-1) - L|$. Now, using algebra, the triangle inequality, and the definition of absolute value, we see that

$$\begin{aligned} 2 = 1 - (-1) &= |1 - (-1)| = |1 - L + L - (-1)| \\ &\le |1 - L| + |L - (-1)| = |1 - L| + |(-1) - L| < 1 + 1 = 2. \end{aligned}$$

Since $2 < 2$ is not true, our assumption about the existence of the limit L is false. Therefore, the given sequence has no limit. □

EXAMPLE 5

Using the hypotheses of a convergence theorem

Show that $\lim_{k\to\infty} a_k + \lim_{k\to\infty} b_k$ does not necessarily equal $\lim_{k\to\infty}(a_k + b_k)$ for divergent sequences $\{a_k\}$ and $\{b_k\}$.

SOLUTION

Consider the sequences defined by $a_k = (-1)^k$ and $b_k = (-1)a_k = (-1)^{k+1}$. As shown in Example 4, the sequence $\{(-1)^k\}$ diverges, so $\lim_{k\to\infty} a_k$ does not exist. Similarly, $\lim_{k\to\infty} b_k$ does not exist. Therefore, the sum of these limits, $\lim_{k\to\infty} a_k + \lim_{k\to\infty} b_k$, does not exist. By contrast, $\lim_{k\to\infty}(a_k + b_k) = 0$, because for each $k \in \mathbb{Z}^+$, $a_k + b_k = 0$. □

? TEST YOUR UNDERSTANDING

- What is the definition of the limit of a sequence? How is this definition similar to and different from the limit of a function $f(x)$ as $x \to \infty$?
- What is the tail of a sequence? Why is it important?
- Is it true that for any sequences $\{a_k\}$ and $\{b_k\}$, $\lim_{k\to\infty}(a_k \pm b_k) = \lim_{k\to\infty} a_k \pm \lim_{k\to\infty} b_k$? If so, explain. If not, what is needed to make the statement true?

- What is the Squeeze Theorem? How can the Squeeze Theorem be used to help understand the convergence of a sequence?
- What is the Completeness Axiom for the real numbers? Does a nonempty subset of the rational numbers that is bounded above always have a rational least upper bound? Does a nonempty subset of the integers that is bounded above always have an integer as a least upper bound?

EXERCISES 7.2

Thinking Back

Analyzing the behavior of a continuous function: Consider the function $f(x) = \frac{x^3}{e^x}$ on the interval $[0, \infty)$.

- Where is f increasing and where is f decreasing? Is the function bounded above and/or below? Does f have a limit as $x \to \infty$?

The asymptotic behavior of a continuous function: Let $y = f(x)$ be an increasing function defined on the interval $[0, \infty)$.

- Provide a condition that f must satisfy in order for f to have a horizontal asymptote as $x \to \infty$. Use the words "upper bound" in your answer.
- Explain why the lack of an upper bound means that the graph of the function f does not have a horizontal asymptote.
- Explain the connection between the answers you just gave and the convergence or divergence of an increasing sequence.
- Provide an example of a continuous function f defined on the interval $[0, \infty)$ such that f is neither increasing nor decreasing on any interval of the form $[a, \infty)$, where $a > 0$, yet f has a horizontal asymptote at $y = 0$.

Concepts

0. *Problem Zero:* Read the section and make your own summary of the material.

1. *True/False:* Determine whether each of the statements that follow is true or false. If a statement is true, explain why. If a statement is false, provide a counterexample.

(a) *True or False:* If the tail of a sequence converges, the sequence converges.

(b) *True or False:* If $\{a_k\}$ and $\{b_k\}$ are two divergent sequences, then the sequence $\{a_k + b_k\}$ diverges.

(c) *True or False:* If $\{a_k\}$ is a convergent sequence of rational numbers, it must converge to a rational number.

(d) *True or False:* Every convergent sequence is bounded.

(e) *True or False:* Every bounded sequence is convergent.

(f) *True or False:* $\lim_{k\to\infty} \frac{k^{1000000}}{(1.000001)^k} = 0$

(g) *True or False:* Every increasing sequence of negative numbers converges.

(h) *True or False:* Every increasing sequence of positive numbers converges.

2. *Examples:* Construct examples of the thing(s) described in the following. Try to find examples that are different than any in the reading.

(a) A sequence $\{a_k\}$ of irrational numbers that converges to a rational number.

(b) A bounded sequence without a limit.

(c) A convergent sequence that is not eventually monotonic.

3. Explain why the convergence of a sequence depends only on the convergence of the tail of the sequence.

In Exercises 4–11, give examples of sequences satisfying the given conditions or explain why such an example cannot exist.

4. Two convergent sequences $\{a_k\}$ and $\{b_k\}$ such that the sequence $\{a_k + b_k\}$ converges.

5. Two convergent sequences $\{a_k\}$ and $\{b_k\}$ such that the sequence $\{a_k - b_k\}$ diverges.

6. Two divergent sequences $\{a_k\}$ and $\{b_k\}$ such that the sequence $\{a_k \cdot b_k\}$ converges.

7. Two divergent sequences $\{a_k\}$ and $\{b_k\}$ such that the sequence $\{a_k \cdot b_k\}$ diverges.

8. Two divergent sequences $\{a_k\}$ and $\{b_k\}$ such that the sequence $\left\{\frac{a_k}{b_k}\right\}$ converges.

9. Two divergent sequences $\{a_k\}$ and $\{b_k\}$ such that the sequence $\left\{\frac{a_k}{b_k}\right\}$ diverges.

10. Two convergent sequences $\{a_k\}$ and $\{b_k\}$ such that the sequence $\left\{\frac{a_k}{b_k}\right\}$ converges.

11. Two convergent sequences $\{a_k\}$ and $\{b_k\}$ such that the sequence $\left\{\frac{a_k}{b_k}\right\}$ diverges.

12. Explain what is important about monotonic and bounded sequences.

In Exercises 13–16, give examples of sequences satisfying the given conditions or explain why such an example cannot exist.

13. A convergent sequence that is not eventually monotonic.

14. An increasing sequence that is not strictly increasing.

15. A decreasing sequence that is bounded below but is not bounded above.

16. A bounded and convergent sequence that is not eventually monotonic.

17. Find a divergent sequence $\{a_k\}$ such that the sequence given by $\left\{\frac{k!}{a_k}\right\}$ converges.

18. What do we mean when we say *factorial growth dominates exponential growth*? Provide an example illustrating this fact.

19. Complete Example 3 by showing that the limit of the sequence $\left\{\frac{k^3}{e^k}\right\}$ is 0.

20. State the converse of Theorem 7.19. Explain why Theorem 7.20 is a partial converse.

21. Discuss the boundedness and monotonicity of the geometric sequence $\{cr^k\}_{k=0}^{\infty}$ with $c > 0$ and $r > 0$. Determine the values of r for which the sequence converges and the values of r for which the sequence diverges. If the sequence converges, find the limit of the sequence.

22. Discuss the boundedness and monotonicity of the geometric sequence $\{cr^k\}_{k=0}^{\infty}$ with $c > 0$ and $r < 0$. In addition, determine whether the sequence converges or diverges. If it converges, find the limit of the sequence.

Skills

For each of the sequences in Exercises 23–52 determine whether the sequence is monotonic or eventually monotonic and whether the sequence is bounded above and/or below. If the sequence converges, give the limit.

23. $\{11\}$

24. $\left\{1 + \frac{1}{k}\right\}$

25. $\left\{\sin\left(\frac{k\pi}{2}\right)\right\}$

26. $\{\cos(k\pi)\}$

27. $\left\{\frac{(-1)^k}{k+6}\right\}$

28. $\{k^{1/10}\}$

29. $\{(3^{-k} - 2)\}$

30. $\left\{\left(\frac{1}{5}\right)^k\right\}$

31. $\left\{\frac{k - 2^{-k}}{2^k}\right\}$

32. $\left\{\frac{2^k - 2^{-k}}{2^k}\right\}$

33. $\left\{\frac{k}{2+k}\right\}$

34. $\left\{\frac{k^2 - 2}{k^2 + 2k + 2}\right\}$

35. $\left\{\left(1 + \frac{1}{k}\right)^2\right\}$

36. $\left\{\left(1 + \frac{1}{k}\right)^k\right\}$

37. $\left\{\frac{2^k}{k!}\right\}$

38. $\left\{\frac{k^2 - 2}{k^2 + 2k + 2}\right\}$

39. $\{3^{1/k}\}$

40. $\{k!\}$

41. $\{(42 + (-1)^k)\}$

42. $\left\{\frac{k}{k^2 + 2k + 2}\right\}$

43. $\left\{\left(\frac{2k^2}{k^2 + 2k + 2}\right)^3\right\}$

44. $\left\{\frac{k^2 - 1}{10000k + 2}\right\}$

45. $\left\{\left(42 - \frac{k}{k^2 - 2k + 50}\right)\right\}$

46. $\left\{\tan\left(\frac{k\pi}{2}\right)\right\}$

47. $\left\{\frac{1 \cdot 3 \cdot 5 \cdots (2k-1)}{10^k}\right\}$

48. $\left\{\frac{1 \cdot 3 \cdot 5 \cdots (2k-1)}{1 \cdot 4 \cdot 7 \cdots (3k-2)}\right\}$

49. $\left\{\frac{\sin k}{k}\right\}$

50. $\left\{\frac{k+1}{k+7}\right\}$

51. $\left\{\frac{(k!)^2}{(2k)!}\right\}$

52. $\left\{\frac{(k!)^3}{(3k)!}\right\}$

For the sequences in Exercises 53 and 54, determine whether the sequence converges or diverges. If the sequence converges, give the limit.

53. $\{(k!)^{1/k}\}$

54. $\{k^{1/k!}\}$

Evaluate the limits in Exercises 55–60. Use the theorems in this section to justify each step of your work.

55. $\lim_{k\to\infty} \frac{\sin k}{2^k}$ (*Hint: Use the Squeeze Theorem.*)

56. $\lim_{k\to\infty} \left(1 + \frac{1}{k}\right)\left(2 - \frac{1}{k}\right)$

57. $\lim_{k\to\infty} \frac{k-7}{3k+5}$

58. $\lim_{k\to\infty} \frac{k^3+1}{k^2}$

59. $\lim_{k\to\infty} (\sqrt{k^2 + k} - k)$

60. $\lim_{k\to\infty} (\sqrt{k^3 + k + 1} - \sqrt{k^3 - k - 1})$ (*Hint: Write as a quotient and rationalize the numerator.*)

Applications

61. Suppose you invest \$100.00 in a bank that pays you a nominal annual interest rate of 6%. The bank offers you the option of compounding your money n times over the course of a year. Your balance in the account after one year is given by

$$a_k = 100.00\left(1 + \frac{0.06}{k}\right)^k.$$

Find the balance in the account, rounded to the nearest cent, if you compound the interest k times during the year, where k is 1, 2, 6, 12, 52, and 365. You should see that the balance at the end of the year is increasing. Do you believe the sequence is bounded? If so, by what value? Be as specific as possible.

62. We may use a recursively defined sequence to approximate the current amount of a radioactive element. For example, radioactive radium changes into lead over time. The rate of decay is proportional to the amount of radium present. Experimental data suggests that a gram of radium decays into lead at a rate of $\frac{1}{2337}$ gram per year. Let a_k be the amount of radium at the end of year k. Since the decay rate is constant, if we use a linear model to approximate the amount that remains after one year has passes, we have

$$a_1 = a_0 - \frac{1}{2337} a_0 = \frac{2336}{2337} a_0.$$

More generally, we obtain the recursion formula

$$a_{k+1} = \frac{2336}{2337} a_k.$$

Use this formula to estimate how much radium remains after 100 years if we start off with $a_0 = 10$ grams of radium.

Proofs

63. Prove that the geometric sequence $\{r^k\}_{k=0}^{\infty}$ diverges when $r < -1$.

64. Prove that the sequence of factorials $\{k!\}$ dominates every sequence of exponential functions $\{b^k\}$, where $b > 0$, by applying the ratio test from Theorem 7.6 to the sequence of quotients $\left\{\frac{k!}{b^k}\right\}$.

Complete the proof of Theorem 7.18 by evaluating the limits of the sequences in Exercises 65 and 66.

65. Explain why $\lim_{k\to\infty} k^{1/k}$ is an indeterminate form. Use L'Hôpital's Rule or another valid method to prove that $k^{1/k} \to 1$.

66. Prove that $\left(\frac{1}{k}\right)^p \to 0$, when $p > 0$.

In Exercises 67–70, let $\{a_k\}$ and $\{b_k\}$ be convergent sequences with $a_k \to L$ and $b_k \to M$ as $k \to \infty$ and let c be a constant. Prove the indicated basic limit rules from Theorem 7.11. You may wish to model your proofs on the proofs of the analogous statements from Section 1.5.

67. Prove that $ca_k \to cL$.

68. Prove that $(a_k + b_k) \to L + M$.

69. Prove that $a_k b_k \to LM$.

70. Prove that if $M \neq 0$, then $\frac{a_k}{b_k} \to \frac{L}{M}$.

In Exercises 71–74 let $\{a_k\}$ be a sequence. Prove the indicated limit rules from Theorem 7.12. You may wish to model your proofs on the proofs of the analogous statements from Section 1.5.

71. Prove that if $a_k \to L$ and $a_k \to M$, then $L = M$.

72. Prove that if $\{a_k\} \to L$ and f is a function that is continuous at L, then $f(a_k) \to f(L)$.

73. Prove that if $\{m_k\}$ and $\{M_k\}$ are sequences that both converge to L, and if $m_k \le a_k \le M_k$ for all k, then $a_k \to L$.

74. Prove that if $|a_k| \to 0$, then $a_k \to 0$.

75. Prove Theorem 7.9. That is, let $a : [1, \infty) \to \mathbb{R}$ be a continuous function and let $a_k = a(k)$ for every $k \in \mathbb{Z}^+$. Show that if $\lim_{x\to\infty} a(x) = L$, then $\{a_k\} \to L$.

76. Prove that the converse of Theorem 7.9 is not true by finding a continuous function $a : [1, \infty) \to \mathbb{R}$ such that $\lim_{x\to\infty} a(x)$ does not exist but $\{a(k)\}$ converges.

77. Prove that if $\{a_k\}$ is a sequence of nonzero terms with the property that $\lim_{k\to\infty} a_k = \infty$, then $\frac{1}{a_k} \to 0$.

Prove the statements about the convergence or divergence of sequences in Exercises 78–83, referring to theorems in the section as necessary. For each of these statements, assume that r is a real number and p is a positive real number.

78. $k^p \to \infty$

79. $\left(\frac{1}{k}\right)^p \to 0$

80. If $r = 1$, then $r^k \to 1$.

81. If $|r| > 1$, then the sequence $\{r^k\}$ diverges and $|r|^k \to \infty$.

82. If $|r| < 1$, then the sequence $r^k \to 0$.

83. $\frac{k^r}{(1+p)^k} \to 0$.

84. Prove that if $\lim_{k\to\infty} a_k = L$, then $\lim_{k\to\infty} a_{k+1} = L$.

85. Prove that every convergent sequence is bounded.

86. Consider the sequence $\{a_k\}$ defined recursively by $a_1 = 1$ and for $k > 1$, $a_k = \sqrt{2a_{k-1}}$. Prove that $a_k \to 2$ by first proving that the limit must exist. (*Hint: Use induction to show that the terms of the sequence may be expressed with the closed formula for* $a_k = 2^{1-(1/2)^{k-1}}$.)

87. Let $\{a_k\}$ be a sequence such that $a_{2k} \to L$ and $a_{2k+1} \to L$. Prove that $\{a_k\} \to L$. (What you will have proven is that if the even terms and odd terms of a sequence both converge to L, then the sequence converges to L.)

88. Prove Theorem 7.14. That is, show that if $\{a_k\}$ is a sequence that converges to L, then every subsequence of $\{a_k\}$ also converges to L.

Thinking Forward

▶ *A sequence of sums:* Consider the sequence $\{a_k\}_{k=0}^{\infty}$ where $a_k = 1 + \frac{1}{2} + \frac{1}{4} + \cdots + \frac{1}{2^k}$. Note that after the zeroth term, every term is given by a sum. As k increases, the sum increases, so we see that this is an increasing sequence. Find a closed formula for a_k to show that the sequence is bounded and therefore converges.

▶ *Another sequence of sums:* Consider the sequence $\{a_k\}_{k=1}^{\infty}$ where $a_k = 1 + \frac{1}{2} + \frac{1}{3} + \cdots + \frac{1}{k}$. Draw the graph of this sequence together with the graph of $f(x) = \frac{1}{x}$. Argue that $a_k \ge \int_1^k \frac{1}{x}\,dx$. What is $\lim_{k\to\infty} \int_1^k \frac{1}{x}\,dx$? What does the value of this improper integral imply about the convergence of $\{a_k\}$?

7.3 SERIES

- Series as limits of partial sums of sequences
- Convergence of series in terms of convergence of sequences of partial sums
- Geometric series and when they converge or diverge

Adding Up Sequences to Get Series

If we take an infinite sequence and add up all of its terms, we get what is known as a ***series***:

DEFINITION 7.21

Series of Real Numbers

A ***series*** $\sum_{k=1}^{\infty} a_k$ is the sum of the terms of a sequence $\{a_k\}$ of real numbers:

$$\sum_{k=1}^{\infty} a_k = a_1 + a_2 + a_3 + \cdots + a_k + \cdots.$$

For each positive integer index k, a_k is called the ***kth term of the series.***

Of course if we wish to, we can also choose to start a series at an index other than $k = 1$. For example, $\sum_{k=3}^{\infty} \frac{1}{k^2} = \frac{1}{9} + \frac{1}{16} + \frac{1}{25} + \cdots$ is a series.

Given a sequence, we can sum up the terms to get a series. We can also consider partial sums of a series to obtain another sequence:

DEFINITION 7.22

The Sequence of Partial Sums Associated with a Series

Given a series $\sum_{k=1}^{\infty} a_k$, for each positive integer n we can consider the ***partial sum***, S_n, of the first through nth terms. That is,

$$S_n = a_1 + a_2 + a_3 + \cdots + a_n.$$

Considered together these form a ***sequence of partial sums***:

$$\{S_n\} = \left\{\sum_{k=1}^{n} a_k\right\} = \{a_1, (a_1 + a_2), (a_1 + a_2 + a_3), \ldots, (a_1 + a_2 + a_3 + \cdots + a_n), \ldots\}.$$

In other words, every series is a limit of partial sums:

$$\sum_{k=1}^{\infty} a_k = \lim_{n\to\infty} \sum_{k=1}^{n} a_k = \lim_{n\to\infty} S_n.$$

We can determine the long-term behavior of a series by examining the long-term behavior of its partial sums. For example, the series

$$\sum_{k=0}^{\infty} 2^k = 1 + 2 + 2^2 + 2^3 + \cdots + 2^k + \cdots$$

is the sum of a sequence of infinitely many numbers, each larger than the one before it. The partial sums for this series grow as we add up more and more terms, and the sum of all the terms will be infinite. In other words, the sequence of partial sums clearly diverges to ∞:

$$\{S_n\} = \{1, 1+2, 1+2+2^2, 1+2+2^2+2^3, \ldots\} = \{1, 3, 7, 15, 31, \ldots\}.$$

In contrast, we will soon see that the series

$$\sum_{k=0}^{\infty} \left(\frac{1}{2}\right)^k = 1 + \frac{1}{2} + \left(\frac{1}{2}\right)^2 + \left(\frac{1}{2}\right)^3 + \cdots + \left(\frac{1}{2}\right)^k + \cdots$$

adds up to the real number 2. Just as with Riemann sums, in this example we are adding up infinitely many terms but the terms themselves are becoming infinitely small. Later in the section we will see that the sequence of partial sums for this sequence converges to 2; no matter how many terms in the series we add up, the sum never gets larger than 2:

$$\{S_n\} = \left\{1, 1+\frac{1}{2}, 1+\frac{1}{2}+\left(\frac{1}{2}\right)^2, 1+\frac{1}{2}+\left(\frac{1}{2}\right)^2+\left(\frac{1}{2}\right)^3, \ldots\right\}$$

$$= \left\{1, \frac{3}{2}, \frac{7}{4}, \frac{15}{8}, \frac{31}{16}, \ldots\right\}, \text{ which converges to 2.}$$

Convergence and Divergence of Series

In general, a series will converge to a real number, or not, depending on whether the limit of its partial sums converges to a real number:

DEFINITION 7.23

Convergence or Divergence of a Series

Suppose $\sum_{k=1}^{\infty} a_k$ is a series whose sequence of partial sums is $S_n = \sum_{k=1}^{n} a_k$. We say that the series ***converges to a limit*** L if $\{S_n\}$ converges to L; that is,

$$\sum_{k=1}^{\infty} a_k = L \text{ when } \lim_{n\to\infty} \left(\sum_{k=1}^{n} a_k\right) = \lim_{n\to\infty} S_n = L.$$

In this case we say that the ***sum*** of the series is L. We say that a series ***diverges*** if its sequence of partial sums diverges.

Although sequences of partial sums may seem to be an additional complication, basing the convergence of a series on such sequences is the most natural thing we can do. The sequence of partial sums simply considers adding one additional term at a time. If the limit of the sequence of partial sums converges as we add more and more terms, then we say that the series converges, too.

Unfortunately, not every sequence whose terms become infinitely small will have a corresponding series that converges. For example, we will soon see that the series

$$\sum_{k=1}^{\infty} \frac{1}{k} = 1 + \frac{1}{2} + \frac{1}{3} + \cdots + \frac{1}{k} + \cdots$$

does not add up to a finite real number, even though the terms of the sequence $\left\{\frac{1}{k}\right\}$ tend to zero. The fact that the series $\sum_{k=1}^{\infty} \frac{1}{k}$ diverges is not immediately clear from examining the first few terms of its sequence of its partial sums:

$$\{S_n\} = \left\{1, 1+\frac{1}{2}, 1+\frac{1}{2}+\frac{1}{3}, 1+\frac{1}{2}+\frac{1}{3}+\frac{1}{4}, \ldots\right\} = \left\{1, \frac{3}{2}, \frac{11}{6}, \frac{25}{12}, \frac{137}{60}, \ldots\right\}.$$

Although it is difficult to tell from just the first few terms, we will see in Section 7.4 that this sequence of partial sums eventually grows without bound, even though with each term we are adding smaller and smaller numbers to the partial sums.

Determining whether a given series converges or diverges is not always an easy task. In fact, most of the rest of this chapter is devoted to developing techniques for determining whether series converge or diverge.

The Algebra of Series

Convergent series obey most of the algebraic rules we might expect, although in Section 7.7 we will see that there are some surprises, too. The following theorem guarantees that series behave well with respect to sums and constant multiples:

THEOREM 7.24 **Sums and Constant Multiples of Convergent Series**

If $\sum_{k=1}^{\infty} a_k$ and $\sum_{k=1}^{\infty} b_k$ are convergent series and c is any real number, then

(a) $\displaystyle\sum_{k=1}^{\infty} c\,a_k = c\sum_{k=1}^{\infty} a_k$ **(b)** $\displaystyle\sum_{k=1}^{\infty}(a_k + b_k) = \sum_{k=1}^{\infty} a_k + \sum_{k=1}^{\infty} b_k$

This theorem is unsurprising, since we already know that both limits and sums commute with sums and constant multiples. We will prove part (b) and leave the proof of part (a) to Exercise 84.

Proof.

(b) The result follows from looking at limits of partial sums:

$$\begin{aligned}
\sum_{k=1}^{\infty}(a_k + b_k) &= \lim_{n\to\infty}\sum_{k=1}^{n}(a_k + b_k) && \leftarrow \text{definition of series}\\
&= \lim_{n\to\infty}\left(\sum_{k=1}^{n} a_k + \sum_{k=1}^{n} b_k\right) && \leftarrow \text{sums commute with sums}\\
&= \lim_{n\to\infty}\sum_{k=1}^{n} a_k + \lim_{n\to\infty}\sum_{k=1}^{n} b_k && \leftarrow \text{sums commute with limits}\\
&= \sum_{k=1}^{\infty} a_k + \sum_{k=1}^{\infty} b_k. && \leftarrow \text{definition of series}
\end{aligned}$$

■

Just as the tail of a sequence is what determines whether or not the sequence converges, the tail of a series is what determines the convergence or divergence of the series. This means that adding or removing any finite number of terms from the start of a series does not change its convergence properties. Of course if a series is convergent and we add or remove terms, then this does change the sum of the series. The proof involves splitting a sum from 1 to ∞ into two sums, one from 1 to $M-1$ and one from M to ∞; the details are left to Exercise 86.

THEOREM 7.25 **The Tail of a Series Determines the Convergence of the Series**

For any integer M, the series $\sum_{k=1}^{\infty} a_k$ and $\sum_{k=M}^{\infty} a_k$ either both converge or both diverge. In addition, if $\sum_{k=M}^{\infty} a_k$ converges to L, then $\sum_{k=1}^{\infty} a_k$ converges to $a_1 + a_2 + a_3 + \cdots + a_{M-1} + L$.

Geometric Series

In Section 7.1 we defined a ***geometric sequence*** as a sequence of the form $\{cr^k\}_{k=0}^{\infty}$ for some constants c and r. If we add up the terms of a geometric sequence, we get a ***geometric series*** $\sum_{k=0}^{\infty} cr^k$. The next theorem completely characterizes the behavior of all geometric series. Not only can we determine exactly which geometric series converge or diverge just by looking at r, but we can also determine the exact sum of any convergent geometric series.

Since we can always factor a constant multiple out of a series, it suffices to characterize the behavior of geometric series of the form $\sum_{k=0}^{\infty} r^k$:

THEOREM 7.26

Convergence and Divergence of Geometric Series

Suppose r is a nonzero real number. Then

(a) For $|r| < 1$, the geometric series $\sum_{k=0}^{\infty} r^k$ converges to the sum $\frac{1}{1-r}$.

(b) For $|r| \geq 1$, the geometric series $\sum_{k=0}^{\infty} r^k$ diverges.

For example, on the one hand, we can tell immediately from this theorem that the series $\sum_{k=0}^{\infty} 5(1.1)^k$ must diverge, since $|1.1| \geq 1$. On the other hand, we can immediately deduce not only that the series $\sum_{k=0}^{\infty} 5(0.9)^k$ converges (since $|0.9| < 1$), but also what sum it converges to:

$$\sum_{k=0}^{\infty} 5(0.9)^k = 5\sum_{k=0}^{\infty}(0.9)^k = 5\left(\frac{1}{1-0.9}\right) = 50.$$

To prove this useful theorem, we will perform a clever algebraic manipulation to determine a formula for the partial sums of a geometric series and then examine the limit of these partial sums for various values of r.

Proof. We first note that if $r = 1$, then we have the series $\sum_{k=0}^{\infty} 1 = 1+1+1+\cdots$, which diverges to ∞. Thus for the rest of the proof we may assume that $r \neq 1$.

Let $\sum_{k=0}^{\infty} r^k$ be a geometric series with $r \neq 1$. For any positive integer n, the nth partial sum of this series is

$$S_n = 1 + r + r^2 + r^3 + \cdots + r^n.$$

If we multiply both sides of the preceding equation by r, we obtain

$$rS_n = r + r^2 + r^3 + r^4 + \cdots + r^{n+1}.$$

Now an interesting thing happens. If we subtract the corresponding sides of the foregoing two equations, then all terms on the right-hand side cancel except for two terms, and we get

$$S_n - rS_n = 1 - r^{n+1}.$$

Factoring out S_n from the left side and then dividing both sides by $r - 1$ (note that we specifically assumed that $r \neq 1$ so that we could be sure that we were not dividing by zero in this step), we get a simple formula for the nth partial sum of the geometric series:

$$S_n = \frac{1-r^{n+1}}{1-r}.$$

To determine whether or not the series $\sum_{k=0}^{\infty} r^k$ converges or diverges, we need to examine $\lim_{n\to\infty} S_n$ in the cases $|r| < 1$, $|r| > 1$, and $r = -1$. Note that the only part of S_n that depends upon n is the r^{n+1} term in the numerator.

If $|r| < 1$, then $r^{n+1} \to 0$ as $n \to \infty$ and therefore we have a convergent geometric series whose sum is

$$\sum_{k=1}^{\infty} r^k = \lim_{n\to\infty} \frac{1 - r^{n+1}}{1 - r} = \frac{1}{1 - r}.$$

If $r > 1$, then $r^{n+1} \to \infty$ as $n \to \infty$ and therefore the geometric series diverges:

$$\sum_{k=1}^{\infty} r^k = \lim_{n\to\infty} \frac{1 - r^{n+1}}{1 - r} \to \infty.$$

If $r < -1$, then the magnitude of r^{n+1} increases without bound as $n \to \infty$ and the geometric series oscillates and diverges.

Finally, if $r = -1$, then the term $r^{n+1} = (-1)^{n+1}$ alternates between 1 and -1 for integer values of n as $n \to \infty$ and again the limit of partial sums diverges. ■

We will spend much of the next three sections studying various methods that help us determine whether or not a series converges; but most of the time we will not be able to compute the actual sum of a convergent series. Geometric series are a rare example of series for which we can determine not only convergence, but also an exact sum. There is only one other type of series whose sums we can calculate exactly, and that is ***telescoping series***; see Example 2.

Examples and Explorations

EXAMPLE 1

Analyzing series for convergence

Determine whether the following series converge or diverge:

(a) $\displaystyle\sum_{k=5}^{\infty} (0.8)^k$ **(b)** $\displaystyle\sum_{k=1}^{\infty} 1$ **(c)** $\displaystyle\sum_{k=0}^{\infty} (-1)^k$

If a series converges, find its sum.

SOLUTION

(a) Any series in which the quotient of every two consecutive terms is the same constant is a geometric series. That is the case here, since $\frac{(0.8)^{k+1}}{(0.8)^k} = 0.8$. When the absolute value of this quotient is less than 1, as we have here, the series will converge. The geometric series

$$\sum_{k=0}^{\infty} cr^k = c + cr + cr^2 + \cdots + cr^k + \cdots$$

converges to $\frac{c}{1-r}$ if $|r| < 1$. In our series, since the first term is $(0.8)^5$ and $r = 0.8$, the series converges to $\frac{(0.8)^5}{1-0.8} = \frac{(0.8)^5}{0.2} = 1.6384$.

(b) We next consider the series in which every term is the constant 1. That is,

$$\sum_{k=1}^{\infty} 1 = 1 + 1 + 1 + \cdots + 1 + \cdots$$

Here the sequence of partial sums is

$$\{S_n\} = 1,\ (1+1),\ (1+1+1), \ldots$$
$$= 1,\ 2,\ 3, \ldots$$

We see that the sequence of partial sums is $\mathbb{Z}^+$, which diverges. Thus, the series diverges as well.

(c) Not every divergent series diverges to infinity. For our third series, we have

$$\sum_{k=0}^{\infty}(-1)^k = 1 - 1 + 1 - 1 + \cdots + (-1)^k + \cdots$$

Here, the sequence of partial sums is

$$\begin{aligned}\{S_n\}_{n=0}^{\infty} &= 1,\ (1-1),\ (1-1+1),\ (1-1+1-1),\ \ldots\\ &= 1,\ 0,\ 1,\ 0, \ldots\end{aligned}$$

This sequence of partial sums diverges. Thus, the series diverges as well. □

EXAMPLE 2

A telescoping series

Use a sequence of partial sums to determine whether the series $\sum_{k=1}^{\infty} \frac{1}{k(k+1)}$ converges or diverges. If the series converges, find its value.

SOLUTION

You should verify that the kth term of this series has the partial-fraction decomposition

$$\frac{1}{k(k+1)} = \frac{1}{k} - \frac{1}{k+1}.$$

Therefore we may rewrite the series in the form

$$\sum_{k=1}^{\infty} \frac{1}{k(k+1)} = \sum_{k=1}^{\infty}\left(\frac{1}{k} - \frac{1}{k+1}\right).$$

Now using this decomposition, we construct the nth term in the sequence of partial sums:

$$S_n = \left(\frac{1}{1} - \frac{1}{2}\right) + \left(\frac{1}{2} - \frac{1}{3}\right) + \left(\frac{1}{3} - \frac{1}{4}\right) + \cdots + \left(\frac{1}{n} - \frac{1}{n+1}\right).$$

Note that between each two consecutive pairs, the second term of a pair cancels with the first term of the subsequent pair. The cancellation between these values gives rise to the term "telescoping" series in that each term in the sequence of partial sums collapses like a folding telescope. Because of the telescoping within the summation for S_n, we have

$$S_n = 1 - \frac{1}{n+1}.$$

We now take the limit of S_n as $n \to \infty$ and see that $\lim_{n\to\infty}\left(1 - \frac{1}{n+1}\right) = 1$. Therefore the series sums to 1.

Before we proceed to other examples, we note that telescoping series have the characteristic that each term of the series can be written as a sum of two or more summands and that once they are decomposed in this way, there will be cancellation between terms in the sequence of partial sums. That cancellation in a telescoping series is not always as simple as the cancellation in this example. You will see other telescoping series in the exercises. □

EXAMPLE 3

Finding terms of a sequence of partial sums

Find the first five terms in the sequence of partial sums for the series $\sum_{k=0}^{\infty} \frac{1}{k!}$.

SOLUTION

The series is $\sum_{k=0}^{\infty} \frac{1}{k!} = \frac{1}{0!} + \frac{1}{1!} + \frac{1}{2!} + \frac{1}{3!} + \frac{1}{4!} + \cdots$.

The first five terms in the sequence of partial sums are

$$\begin{aligned}
S_0 &= \frac{1}{0!} &&= 1 \\
S_1 &= \frac{1}{0!}+\frac{1}{1!} &&= 1+1 &&= 2 \\
S_2 &= \frac{1}{0!}+\frac{1}{1!}+\frac{1}{2!} &&= 1+1+\frac{1}{2} &&= \frac{5}{2} \\
S_3 &= \frac{1}{0!}+\frac{1}{1!}+\frac{1}{2!}+\frac{1}{3!} &&= 1+1+\frac{1}{2}+\frac{1}{6} &&= \frac{8}{3} \\
S_4 &= \frac{1}{0!}+\frac{1}{1!}+\frac{1}{2!}+\frac{1}{3!}+\frac{1}{4!} &&= 1+1+\frac{1}{2}+\frac{1}{6}+\frac{1}{24} &&= \frac{65}{24}.
\end{aligned}$$

In Chapter 8 we will see that this series converges to e. □

EXAMPLE 4 **Working with geometric series**

Show that each of the sums that follow defines a geometric series, and then determine which series converge and which diverge. Find the sum of each convergent series.

(a) $\sum_{k=0}^{\infty} 7(0.9)^k$ **(b)** $\sum_{k=5}^{\infty} \frac{1}{100}(-1.00001)^k$ **(c)** $-\frac{43}{5}+\frac{43}{25}-\frac{43}{125}+\frac{43}{625}\cdots$

SOLUTION

(a) The first of these series is in the standard form for a geometric series with $c = 7$ and $r = 0.9$. Since $|r| = |0.9| < 1$, by Theorem 7.26 the series converges to $\frac{c}{1-r} = \frac{7}{1-0.9} = 70$.

(b) The second series is almost in the standard form for a geometric series, but the index starts with 5, rather than 0. However, recall that the convergence of a series depends not upon the first few terms, but only upon the tail of the series. In this geometric series, $r = -1.00001$. Of course, $|-1.00001| > 1$, so by Theorem 7.26, this series diverges.

(c) A geometric series may be defined recursively as a sum of terms in which every term after the first term, c, is the same constant multiple, r, of the previous term. In the third series in this example we see that the first term is $c = -\frac{43}{5}$. Every term after that is $-\frac{1}{5}$ times the previous term. Since $|r| = \left|-\frac{1}{5}\right| < 1$, the given series is a convergent geometric series. Its sum is

$$\frac{-43/5}{1-(-1/5)} = -\frac{43}{6}.$$

□

EXAMPLE 5 **Expressing repeating decimals as geometric series**

Express the following repeating decimal expansions as geometric series and as the quotient of two integers:

(a) 0.123123123... **(b)** 6.2353535...

SOLUTION

(a) Every decimal that is eventually repeating may be expressed as a geometric series and as the quotient of two integers, that is, as rational numbers. Note that

$$\begin{aligned}
0.123123123\ldots &= 0.123(1+0.001+(0.001)^2+(0.001)^3+\cdots) \\
&= \sum_{k=0}^{\infty} 0.123(0.001)^k = \frac{0.123}{1-0.001} = \frac{0.123}{0.999} = \frac{123}{999} = \frac{41}{333}.
\end{aligned}$$

(b) The repetition in the expansion of our second decimal, 6.2353535..., does not start until after the tenths place. To mimic the technique from (a), we need the repetition of the decimal to start at the decimal point. To accomplish this we may simultaneously multiply and divide by 10 to obtain $\frac{1}{10}(62.353535\ldots)$. Now, working with the decimal 0.353535..., we have

$$0.353535\ldots = 0.35(1 + 0.01 + (0.01)^2 + (0.01)^3 + \cdots)$$
$$= \sum_{k=0}^{\infty} 0.35(0.01)^k = \frac{0.35}{1-0.01} = \frac{0.35}{0.99} = \frac{35}{99}.$$

Thus,

$$6.2353535\ldots = \frac{1}{10}\left(62 + \frac{35}{99}\right) = \frac{6173}{990}.$$

□

CHECKING THE ANSWER

These results may easily be checked with a calculator to see if our rational numbers have the correct decimal expansions. For example, when we divide 6173 by 990, we obtain 6.2353535....

EXAMPLE 6

Changing the index of a series

Rewrite the series $\sum_{k=5}^{\infty} \frac{k}{k^2+3}$ with the index starting at zero.

SOLUTION

We introduce a different letter, j, as the new index. We want $j = 0$ when $k = 5$, giving the relationship $k = j + 5$. If we replace each k with $j + 5$ in the original expression for the series, we will have achieved our goal. We have

$$\sum_{j=0}^{\infty} \frac{j+5}{(j+5)^2+3} = \sum_{j=0}^{\infty} \frac{j+5}{j^2+10j+28}.$$

□

CHECKING THE ANSWER

We may check the answer by evaluating the original expression of the series for the first few values of k and doing the same for the first few values of j in our new expression. From the original expression, we have

$$\sum_{k=5}^{\infty} \frac{k}{k^2+3} = \frac{5}{28} + \frac{6}{39} + \frac{7}{52} + \cdots.$$

For the new expression, we have

$$\sum_{j=0}^{\infty} \frac{j+5}{j^2+10j+28} = \frac{5}{28} + \frac{6}{39} + \frac{7}{52} + \cdots.$$

The first three terms of the two expressions agree.

? TEST YOUR UNDERSTANDING

- What is a series? What is a sequence of partial sums?
- How do we define convergence for a series?
- What is meant by the "tail" of a series? Why does the tail of a series determine the convergence of the series?
- What is a geometric series? Which geometric series converge and which diverge?
- What is a telescoping series? Why are telescoping series given this name?

EXERCISES 7.3

Thinking Back

Working with sequences: Determine which of the sequences that follow converge and which diverge. Explain your reasoning for each sequence.

- $\{1 - r^{n+1}\}$ for $0 < r < 1$
- $\{1 - r^{n+1}\}$ for $r > 1$
- $\{1 - r^{n+1}\}$ for $-1 < r < 0$
- $\{1 - r^{n+1}\}$ for $r < -1$
- $\{n^r\}$ for $r > 0$
- $\{n^r\}$ for $r < 0$
- *Convergence of an Improper Integral:* If the function $f : [1, \infty) \to \mathbb{R}$ is continuous on its domain, what does it mean for the improper integral $\int_1^\infty f(x)\,dx$ to converge? What does it mean for the integral to diverge?
- *Convergence of a Sequence:* Let $\{a_n\}$ be a sequence. What does it mean for the sequence to converge? What does it mean for the sequence to diverge?

Concepts

0. *Problem Zero:* Read the section and make your own summary of the material.

1. *True/False:* Determine whether each of the statements that follow is true or false. If a statement is true, explain why. If a statement is false, provide a counterexample.

(a) *True or False:* If the sequence $\{a_n\}$ converges, then the series $\sum_{n=1}^\infty a_n$ converges.

(b) *True or False:* If a series $\sum_{k=1}^\infty a_k$ diverges and $\{S_n\}$ is its sequence of partial sums, then $\lim_{n\to\infty} S_n = \infty$.

(c) *True or False:* If two series $\sum_{k=1}^\infty a_k$ and $\sum_{k=1}^\infty b_k$ both diverge, then the series $\sum_{k=1}^\infty (a_k + b_k)$ diverges.

(d) *True or False:* If $\sum_{k=1}^\infty a_k$ and $\sum_{k=1}^\infty b_k$ are two convergent series, then for any real numbers c and d, $\sum_{k=1}^\infty (ca_k + db_k) = c\sum_{k=1}^\infty a_k + d\sum_{k=1}^\infty b_k$.

(e) *True or False:* If the series $\sum_{k=0}^\infty a_{n+k}$ converges to 5, then the series $\sum_{k=100}^\infty a_k$ converges to a value $L < 5$.

(f) *True or False:* If the series $\sum_{k=N}^\infty a_k$ converges, then $\sum_{k=N}^\infty a_k = \sum_{k=0}^\infty a_{n+k}$.

(g) *True or False:* If a geometric series $\sum_{k=0}^\infty cr^k$ converges, then $\lim_{k\to\infty} cr^k = 0$.

(h) *True or False:* The series $\sum_{k=1}^\infty a_k$ where $a_1 = 4$ and $a_{k+1} = \frac{a_k}{2}$ for $k > 1$, converges to 7.

2. *Examples:* Construct examples of the thing(s) described in the following. Try to find examples that are different than any in the reading.

(a) A convergent geometric series $\sum_{k=0}^\infty cr^k$ with $r < 0$.

(b) A divergent geometric series $\sum_{k=0}^\infty cr^k$ with $r < 0$.

(c) A divergent series that is *not* a geometric series.

3. What is meant by the sequence of partial sums for a series $\sum_{k=1}^\infty a_k$? Why does it make sense to define the convergence of a series in terms of the convergence of its sequence of partial sums?

4. What is meant by the tail of a series? Why does it make sense that the tail of a series determines the convergence of the series?

5. What is the relationship between $\sum_{k=1}^\infty a_k$ and $\sum_{k=10}^\infty a_k$? If you knew that $\sum_{k=10}^\infty a_k$ converged to 42, what other information would you need to know to find the sum of $\sum_{k=1}^\infty a_k$?

6. Explain why the sequence of partial sums of the series $\sum_{k=1}^\infty a_k$ in which $a_k > 0$ for every $k \in \mathbb{Z}^+$ is strictly increasing.

7. Explain how to change the index of the series $\sum_{k=1}^\infty a_k$ to start with an initial value other than 1.

8. What is a geometric progression? What determines the convergence of a geometric progression?

9. What is a geometric series? What determines the convergence of a geometric series?

10. Let $\sum_{k=1}^\infty cr^k$ be a series with c and $r \in \mathbb{R}$. Explain why the convergence of this series depends only upon the magnitude of r and not on c.

11. Explain why all the terms of a divergent geometric series are nonzero.

12. What is a telescoping series? Give an example of a convergent telescoping series and an example of a divergent telescoping series.

13. Find a series $\sum_{k=1}^\infty a_k$ with all nonzero terms that converges to 1.

14. Let $\alpha \in \mathbb{R}$. Explain why you can find a series $\sum_{k=1}^\infty a_k$ with all nonzero terms that converges to α. You may wish to use your answer to Exercise 13.

15. Find two divergent series $\sum_{k=1}^\infty a_k$ and $\sum_{k=1}^\infty b_k$ such that the series $\sum_{k=1}^\infty (a_k + b_k)$ converges. Carefully use the sequence of partial sums for this new series to show that your answer is correct.

16. Find two divergent series $\sum_{k=1}^\infty a_k$ and $\sum_{k=1}^\infty b_k$ such that the series $\sum_{k=1}^\infty (a_k - b_k)$ converges. Carefully use the sequence of partial sums for this new series to show that your answer is correct.

17. Find two convergent geometric series $\sum_{k=0}^\infty a_k = L$ and $\sum_{k=0}^\infty b_k = M$ such that the series $\sum_{k=0}^\infty (a_k \cdot b_k)$ converges. Does this series converge to LM?

18. Find two convergent geometric series $\sum_{k=0}^{\infty} a_k = L$ and $\sum_{k=0}^{\infty} b_k = M$ with all positive terms such that $\sum_{k=0}^{\infty} \frac{a_k}{b_k}$ converges but whose sum is not $\frac{L}{M}$.

19. Find two divergent geometric series $\sum_{k=0}^{\infty} a_k$ and $\sum_{k=0}^{\infty} b_k$ with all positive terms such that $\sum_{k=0}^{\infty} \frac{a_k}{b_k}$ converges.

20. Find two convergent geometric series $\sum_{k=0}^{\infty} a_k = L$ and $\sum_{k=0}^{\infty} b_k = M$ with all positive terms such that $\sum_{k=0}^{\infty} \frac{a_k}{b_k}$ diverges.

Skills

In Exercises 21–28 provide the first five terms of the series.

21. $\displaystyle\sum_{k=1}^{\infty} \frac{1}{k^2+2}$ **22.** $\displaystyle\sum_{n=1}^{\infty} \frac{n+1}{n!}$

23. $\displaystyle\sum_{j=0}^{\infty} (-1)^{j+1}$ **24.** $\displaystyle\sum_{n=3}^{\infty} (2n+1)$

25. $\displaystyle\sum_{i=0}^{\infty} \frac{i!}{(i+1)!}$ **26.** $\displaystyle\sum_{k=0}^{\infty} \frac{k!}{(2k)!}$

27. $\displaystyle\sum_{n=0}^{\infty} \frac{(-1)^n}{(2n)!}$ **28.** $\displaystyle\sum_{n=1}^{\infty} (-1)^n \frac{n^2}{n!}$

In Exercises 29–36 provide the first five terms of the sequence of partial sums for the given series. You may find it useful to refer to your answers to Exercises 21–28.

29. $\displaystyle\sum_{k=1}^{\infty} \frac{1}{k^2+2}$ **30.** $\displaystyle\sum_{n=1}^{\infty} \frac{n+1}{n!}$

31. $\displaystyle\sum_{j=0}^{\infty} (-1)^{j+1}$ **32.** $\displaystyle\sum_{n=3}^{\infty} (2n+1)$

33. $\displaystyle\sum_{i=0}^{\infty} \frac{i!}{(i+1)!}$ **34.** $\displaystyle\sum_{k=0}^{\infty} \frac{k!}{(2k)!}$

35. $\displaystyle\sum_{n=0}^{\infty} \frac{(-1)^n}{(2n)!}$ **36.** $\displaystyle\sum_{n=1}^{\infty} (-1)^n \frac{n^2}{n!}$

Evaluate the finite sums in Exercises 37–42.

37. $\displaystyle\sum_{k=0}^{100} \frac{1}{2^k}$ **38.** $\displaystyle\sum_{k=1}^{50} \frac{1}{3^k}$

39. $\displaystyle\sum_{k=0}^{1000} 2^k$ **40.** $\displaystyle\sum_{j=1}^{60} 3^j$

41. $\displaystyle\sum_{k=6}^{150} \left(\frac{2}{3}\right)^k$ **42.** $\displaystyle\sum_{j=11}^{200} \left(\frac{3}{2}\right)^j$

Given that $a_0 = -3$, $a_1 = 5, a_2 = -4, a_3 = 2$, and $\sum_{k=2}^{\infty} a_k = 7$, find the values of the specified quantities in Exercises 43–46.

43. $\displaystyle\sum_{k=0}^{\infty} a_k$ **44.** $\displaystyle\sum_{k=1}^{\infty} a_k$

45. $\displaystyle\sum_{k=3}^{\infty} a_k$ **46.** $\displaystyle\sum_{k=4}^{\infty} a_k$

For each of the geometric series in Exercises 47–60, determine whether the series converges or diverges. Give the sum of each convergent series.

47. $\displaystyle\sum_{k=0}^{\infty} \frac{3}{2^k}$ **48.** $\displaystyle\sum_{n=1}^{\infty} \frac{(-1)^{n+1}}{3^n}$

49. $\displaystyle\sum_{j=2}^{\infty} \frac{2^j}{3}$ **50.** $\displaystyle\sum_{n=4}^{\infty} \frac{4}{11^n}$

51. $\displaystyle\sum_{k=0}^{\infty} \pi \left(\frac{e}{3}\right)^k$ **52.** $\displaystyle\sum_{k=0}^{\infty} e \left(\frac{\pi}{3}\right)^k$

53. $\displaystyle\sum_{k=2}^{\infty} \left(-\frac{3}{5}\right)^k$ **54.** $\displaystyle\sum_{k=0}^{\infty} \left(-\frac{9}{8}\right)^k$

55. $\displaystyle\sum_{k=0}^{\infty} \frac{2^{k+2}}{5^{k-1}}$ **56.** $\displaystyle\sum_{k=0}^{\infty} \frac{4^{k+1}}{3^{2k}}$

57. $\displaystyle\sum_{k=0}^{\infty} \frac{(-3)^{k+1}}{4^{k-2}}$ **58.** $\displaystyle\sum_{k=0}^{\infty} \frac{5^{k+1}}{(-6)^k}$

59. $\displaystyle\frac{80}{3} - \frac{20}{3} + \frac{5}{3} - \frac{5}{12} + \cdots$

60. $\displaystyle\frac{99}{40} - \frac{99}{20} + \frac{99}{10} - \frac{99}{5} + \cdots$

Show that each series in Exercises 61–66 is a telescoping series. For each series, provide the general term S_n in its sequence of partial sums and find the sum of the series if it converges.

61. $\displaystyle\sum_{k=1}^{\infty} \left(\frac{1}{k} - \frac{1}{k+2}\right)$ **62.** $\displaystyle\sum_{k=1}^{\infty} \frac{3}{k^2+3k+2}$

63. $\displaystyle\sum_{k=0}^{\infty} \left(\frac{1}{3^k} - \frac{1}{3^{k+1}}\right)$ **64.** $\displaystyle\sum_{k=1}^{\infty} \frac{1}{k(k+1)(k+2)}$

65. $\displaystyle\sum_{k=1}^{\infty} \ln \frac{k}{k+1}$ **66.** $\displaystyle\sum_{k=2}^{\infty} \ln \frac{1}{k^2-1}$

For each series in Exercises 67–70, find all values of x for which the series converges.

67. $\displaystyle\sum_{k=0}^{\infty} x^k$ **68.** $\displaystyle\sum_{k=0}^{\infty} \left(\frac{3}{x}\right)^k$

69. $\displaystyle\sum_{k=0}^{\infty} (\sin x)^k$ **70.** $\displaystyle\sum_{k=0}^{\infty} \left(\frac{\cos x}{2}\right)^k$

Express each of the repeating decimals in Exercises 71–78 as a geometric series and as the quotient of two integers reduced to lowest terms.

71. 0.237237237...

72. 3.454545...

73. 0.305130513051...

74. 2.2131313...

75. 1.272727...

76. 0.6345345...

77. 0.199999...

78. 0.99999...

Applications

79. Whenever a certain ball is dropped, it always rebounds to a height 60% of its original position. What is the total distance the ball travels before coming to rest when it is dropped from a height of 1 meter?

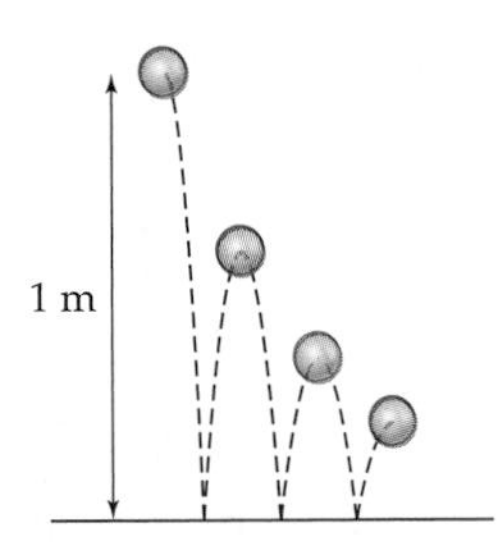

80. Whenever a certain ball is dropped, it always rebounds to a height $p\%$ $(0 < p < 100)$ of its original position. What is the total distance the ball travels before coming to rest when it is dropped from a height of h meters?

In Exercises 81–83, find the areas of ***fractals*** by using properties of geometric series.

81. The figure shown is drawn recursively and then shaded. The largest square has side length 1 unit. A smaller square is inscribed in the square so that each vertex is at the midpoint of the larger square. This process is repeated recursively, resulting in shading as depicted in the figure. What is the area of the shaded portion of the picture?

82. The figure shown is drawn recursively and then shaded. The largest square has side length 1 unit. A square whose side length is $r\%$ as long as the larger square is inscribed with one vertex on each edge of the larger square. This process is repeated recursively, resulting in shading as depicted in the figure. What is the area of the shaded portion of the picture?

83. The figure shown is called a Sierpinski triangle. It may be constructed recursively as follows: We start with a large black equilateral triangle. Every time we see a black equilateral triangle, we inscribe a white equilateral triangle with each vertex at the midpoint of a side of the black triangle. If the side of the largest triangle is 1 unit and this process is repeated recursively, what is the area of the white-shaded region?

Proofs

84. Prove Theorem 7.24 (a). That is, show that if c is a real number and $\sum_{k=1}^{\infty} a_k$ is a convergent series, then $\sum_{k=1}^{\infty} ca_k = c\sum_{k=1}^{\infty} a_k$.

85. Prove that if $\sum_{k=1}^{\infty} a_k$ converges to L and $\sum_{k=1}^{\infty} b_k$ converges to M, then the series $\sum_{k=1}^{\infty} (a_k + b_k) = L + M$.

86. Prove Theorem 7.25. That is, show that the series $\sum_{k=1}^{\infty} a_k$ and $\sum_{k=M}^{\infty} a_k$ either both converge or both diverge. In addition, show that if $\sum_{k=M}^{\infty} a_k$ converges to L, then $\sum_{k=1}^{\infty} a_k$ converges to $a_1 + a_2 + a_3 + \cdots + a_{M-1} + L$.

87. Let $\sum_{k=0}^{\infty} cr^k$ and $\sum_{k=0}^{\infty} bv^k$ be two convergent geometric series. Prove that $\sum_{k=0}^{\infty}(cr^k \cdot bv^k)$ converges. If neither c nor b is 0, could the sum of this series be $\left(\sum_{k=0}^{\infty} cr^k\right)\left(\sum_{k=0}^{\infty} bv^k\right)$?

88. Let $\sum_{k=1}^{\infty} a_k$ be a convergent series and $\sum_{k=1}^{\infty} b_k$ be a divergent series. Prove that the series $\sum_{k=1}^{\infty} (a_k + b_k)$ diverges.

89. Let $\sum_{k=0}^{\infty} cr^k$ and $\sum_{k=0}^{\infty} bv^k$ be two convergent geometric series. If b and v are both nonzero, prove that

$$\sum_{k=0}^{\infty} \frac{cr^k}{bv^k}$$

is a geometric series. What condition(s) must be met for this series to converge?

Thinking Forward

Improper Integrals: Determine whether the following improper integrals converge or diverge.

▶ $\int_1^\infty \frac{1}{x}\,dx$

▶ $\int_1^\infty \frac{1}{x^2}\,dx$

▶ *An Improper Integral and Infinite Series:* Sketch the function $f(x) = \frac{1}{x}$ for $x \geq 1$ together with the graph of the terms of the series $\sum_{k=1}^{\infty} \frac{1}{k}$. Argue that for every term S_n of the sequence of partial sums for this series, $S_n > \int_1^{n+1} \frac{1}{x}\,dx$. What does this result tell you about the convergence of the series?

7.4 INTRODUCTION TO CONVERGENCE TESTS

▶ Convergence tests for series

▶ The divergence test and the integral test

▶ Approximating series by using partial sums

An Overview of Convergence Tests for Series

The rest of this chapter will be dedicated to determining which series converge and which diverge. In some sense the convergence of a sum of a sequence is a question of how fast the terms of that sequence converge to zero. If the terms of a sequence converge to zero, then the sum of that sequence might converge or it might not. For example we will soon see that the sum $1+\frac{1}{2^2}+\frac{1}{3^2}+\frac{1}{4^2}+\cdots$ converges, but the sum $1+\frac{1}{2}+\frac{1}{3}+\frac{1}{4}+\cdots$ is infinite, even though both of the sequences $\left\{\frac{1}{k^2}\right\}$ and $\left\{\frac{1}{k}\right\}$ converge to zero. The first sequence tends to zero fast enough that its sum is a real number, but the second sequence somehow does not tend to zero fast enough. Determining exactly what "fast enough" means is a nontrivial question.

At this point we can determine the convergence or divergence of certain series by looking at limits of partial sums or by recognizing a series as a geometric series. For most series, we will need a much bigger toolbox. For the rest of this chapter we will discuss various tests to help us decide which series converge and which diverge. These tests are as follows:

▶ The Divergence Test (Section 7.4)

▶ The Integral Test (Section 7.4)

▶ The Comparison Test (Section 7.5)

▶ The Limit Comparison Test (Section 7.5)

▶ The Ratio Test (Section 7.6)

▶ The Root Test (Section 7.6)

▶ The Alternating Series Test (Section 7.7)

▶ The Ratio Test for Absolute Convergence (Section 7.7)

At the end of Section 7.7 you will find a table that compares and contrasts the eight tests and comments on when they are most useful.

The Divergence Test

If the terms of a sequence converge to zero, then the sum of those terms might be finite or might be infinite. But if the terms of a sequence do not converge to zero, then we can be certain that the sum of those terms will be infinite. For example, the sequences $\{2, 2^2, 2^3, \ldots\}$ and $\{1, 1, 1, \ldots\}$ clearly have the infinite sums $2 + 2^2 + 2^3 + \cdots$ and $1 + 1 + 1 + \cdots$, respectively.

THEOREM 7.27

The Divergence Test

If the sequence $\{a_k\}$ does not converge to zero, then the series $\sum_{k=1}^{\infty} a_k$ diverges.

The divergence test gives us a quick way to determine that certain series diverge, but it can never tell us if a series converges. If we notice that the terms of a series do not converge to zero, then the divergence test immediately tells us that the series diverges. For example, the series $\sum_{k=1}^{\infty} \frac{k}{k+1}$ must diverge because its sequence of terms $\left\{\frac{k}{k+1}\right\}$ converges to 1, not 0. Note that if the terms of a series do converge to zero, then the divergence test tells us nothing at all.

Proof. It is easier to prove the contrapositive of the divergence test, so we will prove that if a series $\sum_{k=1}^{\infty} a_k$ converges, then its sequence of terms $\{a_k\}$ must converge to zero.

Suppose $\sum_{k=1}^{\infty} a_k$ converges to a sum L. Then the sequence of partial sums

$$\{S_n\} = \{a_1 + a_2 + \cdots + a_n\}$$

converges to L. Notice that the difference between the nth and $(n-1)$th partial sums is equal to the nth term of the series:

$$S_n - S_{n-1} = (a_1 + a_2 + a_3 + \cdots + a_n) - (a_1 + a_2 + a_3 + \cdots + a_{n-1}) = a_n.$$

Note that $\{S_{n-1}\}$ is just the sequence $\{S_n\}$ with a different index. Therefore since $\{S_n\}$ converges to L, we also have $\{S_{n-1}\}$ converges to L. Now we can calculate the limit of the original terms a_n of our series:

$$\lim_{n\to\infty} a_n = \lim_{n\to\infty} (S_n - S_{n-1}) = L - L = 0.$$

This means that the sequence of terms $\{a_n\}$ converges to zero. ■

The Integral Test

Suppose $\{a_k\}$ is a sequence of positive numbers that converges to zero, and suppose that we are interested in its sum. Recall that we can often think of $\{a_k\}$ as a sequence of heights on the graph of some continuous function $a(x)$, as shown in the leftmost figure that follows. We can also think of each term a_k as the area of a rectangle with height a_k and width $\Delta x = 1$ in two different ways, as shown in the middle and rightmost figures. The sum of the areas of these rectangles is intimately related to the area under the graph of $a(x)$ on $[1, \infty)$, an area that we can often calculate with an improper integral.

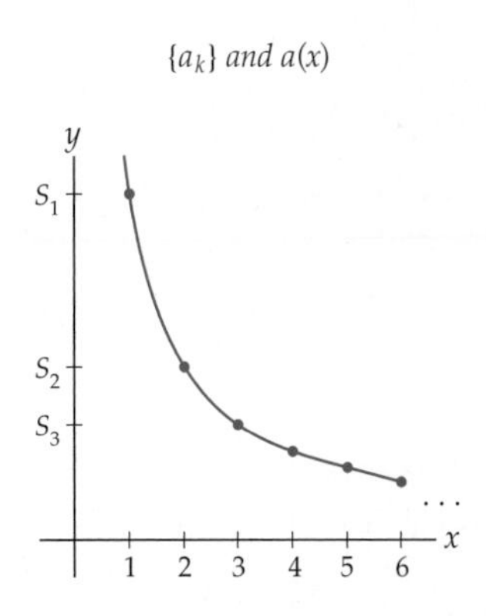

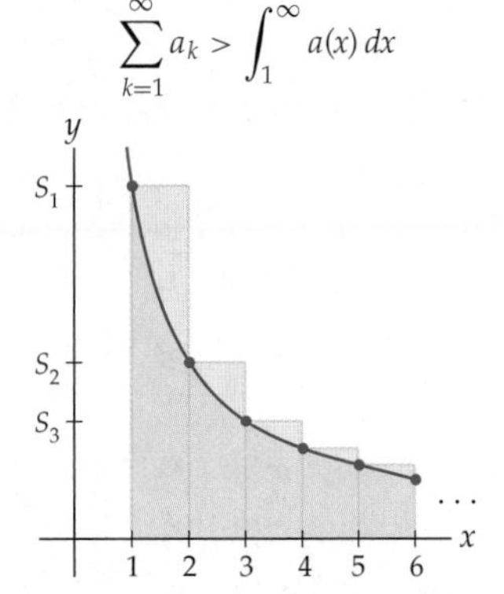

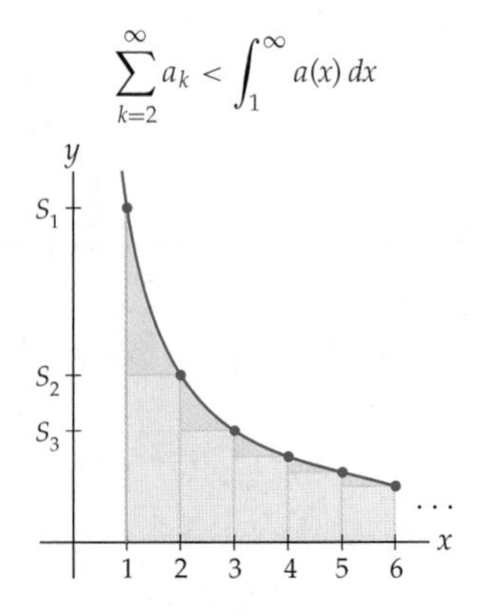

From the middle figure we see that if the integral of $a(x)$ on $[1, \infty)$ diverges, then so will the sum of $\{a_k\}$ from $k = 1$ to ∞, since the sum is larger than the area under the curve. From the rightmost figure we see that if the integral of $a(x)$ on $[1, \infty)$ converges, then so will the sum of $\{a_k\}$ from $k = 2$ to ∞, since that sum is smaller than the area under the curve. But then when $k = 1$, the term a_1 must be finite, so the sum of $\{a_k\}$ from $k = 1$ to ∞ will also converge.

In general, a series and its associated improper integral always converge or diverge together:

THEOREM 7.28 **The Integral Test**

If $a(x)$ is a function that is continuous, positive, and decreasing on $[1, \infty)$, and if $\{a_k\}$ is the sequence defined by $a_k = a(k)$ for every $k \in \mathbb{Z}^+$, then

$$\sum_{k=1}^{\infty} a_k \quad \text{and} \quad \int_1^{\infty} a(x)\, dx$$

either both converge or both diverge.

The integral test provides a way to test the convergence or divergence of a series on the basis of the convergence or divergence of a related improper integral. This can be a useful test when we recognize the terms a_k as a function $a(x)$ that we know how to integrate. It is important to note that a convergent series and its associated convergent improper integral do not in general converge to the same number L. The integral test tells us only whether or not a series converges; it does not tell us the actual sum of the series.

Proof. To prove this theorem it suffices to show that a series $\sum_{k=1}^{\infty} a_k$ converges if and only if the improper integral $\int_1^{\infty} a(x)\, dx$ converges, for some function $a(x)$ that agrees with a_k on the positive integers. We will show that convergence of the series implies convergence of the integral, and leave the converse to Exercise 54.

Suppose the series $\sum_{k=1}^{\infty} a_k$ converges to some value L. Since $a_k > 0$ for every k, the sequence of partial sums $S_n = a_1 + a_2 + \cdots + a_n$ is an increasing sequence that converges to L. Note that S_n is a left Riemann sum for $\int_1^{n+1} a(x)\, dx$. Since $a(x)$ is assumed to be continuous, positive, and decreasing, this left Riemann sum is an upper sum; see the previous middle figure. Therefore, for every positive integer n,

$$0 \le \int_1^{n+1} a(x)\, dx \le S_n \le L.$$

Now, by Theorem 7.19, $\left\{\int_1^n a(x)\, dx\right\}$ is an increasing sequence that is bounded above by L and therefore converges to some value less than or equal to L. Thus, the improper integral $\int_1^{\infty} a(x)\, dx = \lim_{n \to \infty} \int_1^n a(x)\, dx$ converges, which is what we set out to show. ■

Convergence and Divergence of *p*-Series and the Harmonic Series

We have already seen that we can completely classify the convergence and divergence of geometric series of the form $\sum_{k=1}^{\infty} r^k$. These series have exponential terms, in the sense that the variable index k is in the exponent, and a constant r is in the base, of each term. We can also consider series with power function terms, in the sense that the variable index k is in the base, and a constant p is in the exponent, of each term. Such series are called ***p*-series** and are of the form

$$\sum_{k=1}^{\infty} k^{-p} = \sum_{k=1}^{\infty} \frac{1}{k^p}.$$

The case when $p = 1$ is called the ***harmonic series***:

$$\sum_{k=1}^{\infty} \frac{1}{k}.$$

The integral test allows us to completely classify the convergence and divergence of p-series:

THEOREM 7.29 **Convergence and Divergence of p-series**

Suppose p is any real number. Then

(a) For $p > 1$, the p-series $\sum_{k=1}^{\infty} \frac{1}{k^p}$ converges.

(b) For $p = 1$, the harmonic series $\sum_{k=1}^{\infty} \frac{1}{k}$ diverges.

(c) For $p < 1$, the p-series $\sum_{k=1}^{\infty} \frac{1}{k^p}$ diverges.

Proof. In Theorem 5.21 of Section 5.6, we showed that

$$\int_1^{\infty} \frac{1}{x^p}\, dx$$

converges when $p > 1$ and diverges for $p \leq 1$. Therefore, by the integral test, a p-series also converges for $p > 1$ and diverges for $p \leq 1$. ■

Again, we point out that a convergent series does not in general have the same value as its associated improper integral. For example, when $p = 2$,

$$\int_1^{\infty} \frac{1}{x^2}\, dx = \lim_{b\to\infty} \left[-\frac{1}{x}\right]_1^b = \lim_{b\to\infty} \left(-\frac{1}{b} + 1\right) = 1.$$

The fact that this integral is finite means that the associated p-series $\sum_{k=1}^{\infty} \frac{1}{k^2}$ must also converge. However, this p-series converges, not to 1, but rather to the much more surprising value

$$\sum_{k=1}^{\infty} \frac{1}{k^2} = \frac{\pi^2}{6}.$$

Approximating a Convergent Series

Using the proof of the integral test, we may deduce that when the following series and related improper integrals both converge, their values are related by the inequalities

$$\int_2^{\infty} a(x)\, dx \leq \sum_{k=2}^{\infty} a(k) \leq \int_1^{\infty} a(x)\, dx \leq \sum_{k=1}^{\infty} a(k).$$

Thus, if we can evaluate these improper integrals, we can compute an upper and lower bound for the value of the series. (Similarly, but less likely, if we know the sum of the series in the preceding inequality, we may compute bounds for the value of the improper integral.)

This string of inequalities may be generalized to provide a method for bounding the error when we approximate the sum of a convergent series with one of its partial sums. We need a definition first, however:

DEFINITION 7.30

The Remainder of a Convergent Series

If $\sum_{k=1}^{\infty} a_k$ is a convergent series with sum L, then we may approximate L with the nth partial sum $S_n = \sum_{k=1}^{n} a_k$. The nth ***remainder*** is defined by $R_n = L - S_n = \sum_{k=n+1}^{\infty} a_k$.

The remainder is the error that arises from using S_n to approximate the sum of the series. The next theorem gives us a tool for understanding the magnitude of this error when we have a series that, by the integral test, we know converges. If we choose a "large enough" value of n to force the remainder to be small, the nth partial sum will be a good approximation.

THEOREM 7.31

Approximating the Remainder for a Series That Converges by the Integral Test

If a function a is continuous, positive, and decreasing, and if the improper integral $\int_1^{\infty} a(x)\,dx$ converges, then the nth remainder, R_n, for the series $\sum_{k=1}^{\infty} a(k)$ is bounded by

$$0 \le R_n \le \int_n^{\infty} a(x)\,dx.$$

Furthermore, if we define $B_n = \int_n^{\infty} a(x)\,dx$, then

$$S_n \le \sum_{k=1}^{\infty} a(k) \le S_n + B_n.$$

As we mentioned before, the proof of the first string of inequalities in Theorem 7.31 is a generalization of the proof of the inequalities we proved in the integral test. The proof is left for Exercise 55. The second string of inequalities in Theorem 7.31 follows from the fact that the sequence of partial sums for a series satisfying the conditions of the integral test monotonically increases as it converges to the sum of the series. We use Theorem 7.31 in Example 2.

Examples and Explorations

EXAMPLE 1

Choosing a convergence test

Determine whether the series $\sum_{k=1}^{\infty} \frac{k}{e^{k^2}}$ converges or diverges. Prove that the answer is correct by applying an appropriate convergence test.

SOLUTION

Frequently, more than one convergence test may be used to show that a given series converges or diverges. Right now, when you are presented with a new series, $\sum_{k=1}^{\infty} a_k$, you can quickly answer the following two questions:

- Does $a_k \not\to 0$ as $k \to \infty$? If so, then, by the divergence test, the series diverges.
- Does $a_k = a(k)$, where a is a continuous, positive, and decreasing function on $[1, \infty)$? If so, is the improper integral $\int_1^{\infty} a(x)\,dx$ "relatively easy" to evaluate? If it is, you may use the integral test.

After we introduce other convergence tests in the sections that follow, there will be new questions to add to this list. Every time you are presented with a new series, you can quickly run through these questions to help you decide which test might be applicable. After you analyze the behavior of a variety of series, it should become more apparent which test(s) may be applied to a given series.

For the series given in this example, we see that $\frac{k}{e^{k^2}} \to 0$ as $k \to \infty$. Therefore, the divergence test fails here.

Next, note that the function $a(x) = xe^{-x^2}$ is continuous, positive, and decreasing on $[1, \infty)$ and that $a(k) = \frac{k}{e^{k^2}}$. Thus, if we can evaluate the improper integral $\int_1^\infty xe^{-x^2}\,dx$, we will understand the behavior of the series. We now consider

$$\begin{aligned}
\int_1^\infty xe^{-x^2}\,dx &= \lim_{b\to\infty} \int_1^b xe^{-x^2}\,dx && \leftarrow \text{the definition of the improper integral} \\
&= \lim_{b\to\infty} -\frac{1}{2}e^{-x^2}\Big|_1^b && \leftarrow \text{the Fundamental Theorem of Calculus} \\
&= \lim_{b\to\infty} \left(\frac{1}{2e} - \frac{1}{2e^{b^2}}\right) && \leftarrow \text{evaluation of the antiderivative at the endpoints} \\
&= \frac{1}{2e} && \leftarrow \text{evaluation of the limit}
\end{aligned}$$

Since the improper integral converges, by the integral test the series converges as well. □

EXAMPLE 2

Approximating a convergent series and finding a bound for the remainder

Use the third term, S_3, in the sequence of partial sums to approximate the sum of the series $\sum_{k=1}^{\infty} \frac{k}{e^{k^2}}$, and use Theorem 7.31 to find a bound on the remainder.

SOLUTION

To use Theorem 7.31 we must show that the series converges and that it satisfies the hypotheses of the integral test. We did this in Example 1. We now compute S_3:

$$S_3 = \frac{1}{e} + \frac{2}{e^4} + \frac{3}{e^9} \approx 0.404881.$$

Theorem 7.31 tells us that the third remainder R_3 is bounded by

$$0 \le R_3 \le \int_3^\infty xe^{-x^2}\,dx = \frac{1}{2e^9} = B_3 \approx 0.000062.$$

Thus, if $\sum_{k=1}^{\infty} \frac{k}{e^{k^2}} = L$, then,

$$S_3 \le L \le S_3 + B_3.$$

That is, $L \in (0.404881, 0.404943)$. This approximation is quite good already.

We mentioned before that when we approximate a quantity, we usually begin with an understanding of how large an error we can tolerate. When we approximate a convergent series, this means that we should ensure that the remainder is smaller than the desired error. For the series in this example, if we wish the remainder to be less than, say, 10^{-10}, we need to find the smallest value of n such that $B_n \le 10^{-10}$. That is, we need to find n so that

$$\int_n^\infty xe^{-x^2}\,dx = \frac{1}{2e^{n^2}} \le 10^{-10}.$$

We leave it to you to show that $n = 5$ is the smallest positive integer satisfying the inequality. Thus, $S_5 = \frac{1}{e} + \frac{2}{e^4} + \frac{3}{e^9} + \frac{4}{e^{16}} + \frac{5}{e^{25}} \approx 0.4048813986$ with the desired accuracy. □

EXAMPLE 3 **Using the divergence test**

Determine whether the series $\sum_{k=1}^{\infty}\left(\frac{k}{k+1}\right)^k$ converges or diverges.

SOLUTION

We hope that the limit:

$$\lim_{x\to\infty}\left(1+\frac{1}{x}\right)^x$$

is familiar to you. Its value is e. If you do not recall this fact, review Chapter 1. Since we have $1+\frac{1}{x}=\frac{x+1}{x}$, we also have

$$\lim_{k\to\infty}\left(\frac{k}{k+1}\right)^k=\frac{1}{e}\neq 0.$$

By the divergence test, the series diverges. □

EXAMPLE 4 **Analyzing a category of series for convergence**

Find all values of p for which the series $\sum_{k=1}^{\infty}\frac{k}{(1+k^2)^p}$ converge.

SOLUTION

When $p<\frac{1}{2}$, $\frac{k}{(1+k^2)^p}\to\infty$ as $k\to\infty$. Therefore, by the divergence test, the series diverge. When $p=\frac{1}{2}$, $\frac{k}{(1+k^2)^p}\to 1$ as $k\to\infty$. Again, by the divergence test, the series diverges. For $p>\frac{1}{2}$, the function $\frac{x}{(1+x^2)^p}$ is continuous, positive, and decreasing for $x\geq 1$ and, therefore, satisfies the hypotheses of the integral test. Thus, when we know the positive values of p for which the improper integrals

$$\int_1^{\infty}\frac{x}{(1+x^2)^p}\,dx$$

converge, we will know the values of p for which the series converge. Using the substitution $u=1+x^2$, we obtain the integral

$$\frac{1}{2}\int_2^{\infty}\frac{du}{u^p}.$$

We have previously seen that this improper integral converges when $p>1$ and diverges when $0<p\leq 1$. Therefore, our series $\sum_{k=1}^{\infty}\frac{k}{(1+k^2)^p}$ converge when $p>1$ and diverge otherwise. □

? TEST YOUR UNDERSTANDING

- What is the divergence test? What does the divergence test tell you about a series? What can the divergence test not tell you about a series?
- What is the integral test? How is the integral test used? What conditions must be met for the integral test to be effective?
- How do we show that a series converges or diverges?
- What is the remainder of a convergent series? How is the remainder related to the sequence of partial sums?
- How is the integral test extended to find a bound on the error for an approximation to a series?

EXERCISES 7.4

Thinking Back

The contrapositive: What is the contrapositive of the implication "If A, then B."?

Find the contrapositives of the following implications:

- If a quadrilateral is a square, then it is a rectangle.
- If a positive integer is prime, then it has exactly two positive divisors.
- If $r < 0$, then $|r| = -r$.
- If a divides b and b divides c, then a divides c.

Concepts

0. *Problem Zero:* Read the section and make your own summary of the material.
1. *True/False:* Determine whether each of the statements that follow is true or false. If a statement is true, explain why. If a statement is false, provide a counterexample.
   (a) *True or False:* If $a_k \to 0$, then $\sum_{k=1}^{\infty} a_k$ converges.
   (b) *True or False:* If $\sum_{k=1}^{\infty} a_k$ converges, then $a_k \to 0$.
   (c) *True or False:* The improper integral $\int_1^{\infty} f(x)\,dx$ converges if and only if the series $\sum_{k=1}^{\infty} f(k)$ converges.
   (d) *True or False:* The harmonic series converges.
   (e) *True or False:* If $p > 1$, the series $\sum_{k=1}^{\infty} k^{-p}$ converges.
   (f) *True or False:* If $f(x) \to 0$ as $x \to \infty$, then $\sum_{k=1}^{\infty} f(k)$ converges.
   (g) *True or False:* If $\sum_{k=1}^{\infty} f(k)$ converges, then $f(x) \to 0$ as $x \to \infty$.
   (h) *True or False:* If $\sum_{k=1}^{\infty} a_k = L$ and $\{S_n\}$ is the sequence of partial sums for the series, then the sequence of remainders $\{L - S_n\}$ converges to 0.
2. *Examples:* Construct examples of the thing(s) described in the following. Try to find examples that are different than any in the reading.
   (a) A divergent series $\sum_{k=1}^{\infty} a_k$ in which $a_k \to 0$.
   (b) A divergent p-series.
   (c) A convergent p-series.
3. What is meant by a p-series?
4. Which p-series converge and which diverge?
5. Given a series $\sum_{k=1}^{\infty} a_k$, in general the divergence test is inconclusive when $a_k \to 0$. For a geometric series, however, if the limit of the terms of the series is zero, the series converges. Explain why.
6. Let $f(x)$ be a function that is continuous, positive, and decreasing on the interval $[1, \infty)$ such that $\lim_{x\to\infty} f(x) = \alpha > 0$. What can the divergence test tell us about the series $\sum_{k=1}^{\infty} f(k)$?
7. Let $f(x)$ be a function that is continuous, positive, and decreasing on the interval $[1, \infty)$ such that $\lim_{x\to\infty} f(x) = \alpha > 0$. What can the integral test tell us about the series $\sum_{k=1}^{\infty} f(k)$?
8. Explain how you could adapt the integral test to analyze a series $\sum_{k=1}^{\infty} f(k)$ in which the function $f : [1, \infty) \to \mathbb{R}$ is continuous, negative, and increasing.
9. Provide a more general statement of the integral test in which the function f is continuous and *eventually* positive, and decreasing. Explain why your statement is valid.
10. What is meant by the remainder R_n of a series $\sum_{k=1}^{\infty} a_k$?
11. For a convergent series satisfying the conditions of the integral test, why is every remainder R_n positive? How can R_n be used along with the term S_n from the sequence of partial sums to understand the quality of the approximation S_n?
12. Explain why, if n is an integer greater than 1, the series $\sum_{k=1}^{\infty} \frac{1}{\sqrt[n]{k}}$ diverges.
13. Explain why a function $a(x)$ has to be continuous in order for us to use the integral test to analyze a series $\sum_{k=1}^{\infty} a_k$ for convergence.
14. Sketch the graph of a function $a(x)$ that is continuous, positive, and decreasing on the interval $[1, \infty)$. Use your graph to explain the geometrical relationships between the improper integral $\int_1^{\infty} a(x)\,dx$ and the series $\sum_{k=1}^{\infty} a(k)$ and $\sum_{k=2}^{\infty} a(k)$.
15. Find an example of a continuous function $f : [1, \infty) \to \mathbb{R}$ such that $\int_1^{\infty} f(x)\,dx$ diverges and $\sum_{k=1}^{\infty} f(k)$ converges.

Skills

In Exercises 16–23 use the divergence test to analyze the given series. Each answer should either be *the series diverges* or *the divergence test fails,* along with the reason for your answer.

16. $\displaystyle\sum_{k=1}^{\infty} \frac{1}{k}$

17. $\displaystyle\sum_{k=1}^{\infty} \frac{1}{k^2}$

18. $\displaystyle\sum_{k=1}^{\infty} \frac{k}{3k^2 + 100}$

19. $\displaystyle\sum_{k=1}^{\infty} \frac{k}{3k + 100}$

20. $\displaystyle\sum_{k=1}^{\infty} \frac{1}{k^k}$

21. $\displaystyle\sum_{k=1}^{\infty} \frac{k!}{k^k}$

22. $\sum_{k=1}^{\infty} \left(\frac{k}{1+k}\right)^{k^2}$

23. $\sum_{k=1}^{\infty} \frac{k^2+1}{k!}$

In Exercises 24–31, explain why the integral test may be used to analyze the given series and then use the test to determine whether the series converges or diverges.

24. $\sum_{k=1}^{\infty} \frac{k}{e^k}$

25. $\sum_{k=0}^{\infty} e^{-k}$

26. $\sum_{k=2}^{\infty} \frac{1}{k(\ln k)^2}$

27. $\sum_{k=2}^{\infty} \frac{1}{k\sqrt{\ln k}}$

28. $\sum_{k=1}^{\infty} \frac{1+k}{k^2}$

29. $\sum_{k=4}^{\infty} \frac{2k-4}{k^2-4k+3}$

30. $\sum_{k=1}^{\infty} (3k-1)^{-2/3}$

31. $\sum_{k=3}^{\infty} \frac{1}{(k-2)^2}$

Use either the divergence test or the integral test to determine whether the series in Exercises 32–43 converge or diverge. Explain why the series meets the hypotheses of the test you select.

32. $\sum_{k=1}^{\infty} \frac{1}{2}k^{-3/4}$

33. $\sum_{k=1}^{\infty} k^{-3/2}$

34. $\sum_{k=1}^{\infty} \frac{\tan^{-1} k}{1+k^2}$

35. $\sum_{k=1}^{\infty} \frac{k}{2+k^2}$

36. $\sum_{k=1}^{\infty} \frac{k}{k^2+3}$

37. $\sum_{k=1}^{\infty} \frac{\ln k}{k}$

38. $\sum_{k=1}^{\infty} \frac{1}{3k+5}$

39. $\sum_{k=1}^{\infty} k^3 e^{-k^4}$

40. $\sum_{k=1}^{\infty} \frac{\sin(1/k)}{k^2}$

41. $\sum_{k=2}^{\infty} \frac{1}{k\,(k-1)}$

42. $\sum_{k=1}^{\infty} \frac{2^k}{k^5}$

43. $\sum_{k=1}^{\infty} \frac{k}{3^k}$

For each series in Exercises 44–47, do each of the following:

(a) Use the integral test to show that the series converges.
(b) Use the 10th term in the sequence of partial sums to approximate the sum of the series.
(c) Use Theorem 7.31 to find a bound on the tenth remainder, R_{10}.
(d) Use your answers from parts (b) and (c) to find an interval containing the sum of the series.
(e) Find the smallest value of n so that $R_n \leq 10^{-6}$.

44. $\sum_{k=1}^{\infty} \frac{k}{e^k}$

45. $\sum_{k=0}^{\infty} e^{-k}$

46. $\sum_{k=2}^{\infty} \frac{1}{k(\ln k)^2}$

47. $\sum_{k=1}^{\infty} \frac{1}{k^2}$

In Exercises 48–51 find all values of p so that the series converges.

48. $\sum_{k=1}^{\infty} \frac{\ln k}{k^p}$

49. $\sum_{k=2}^{\infty} \frac{1}{k\,(\ln k)^p}$

50. $\sum_{k=1}^{\infty} \frac{1}{(c+k)^p}$, where $c > 0$

51. $\sum_{k=1}^{\infty} \frac{k}{(1+k^2)^p}$

Applications

52. Leila, in her capacity as a population biologist in Idaho, is trying to figure out how many salmon a local hatchery should release annually in order to revitalize the fishery. She knows that if p_k salmon spawn in Redfish Lake in a given year, then only $0.2\,p_k$ fish will return to the lake from the offspring of that run, because of all the dams on the rivers between the sea and the lake. Thus, if she adds the spawn from h fish, from a hatchery, then the number of fish that return from that run k will be $p_{k+1} = 0.2(p_k + h)$.

(a) Show that the sustained number of fish returning approaches $p_\infty = h\sum_{k=1}^{\infty} 0.2^k$ as $k \to \infty$.
(b) Evaluate p_∞.
(c) How should Leila choose h, the number of hatchery fish to raise in order to hold the number of fish returning in each run at some constant P?

53. Leila finds that there are more factors affecting the number of salmon that return to Redfish Lake than the dams: There are good years and bad years. These happen at random, but they are more or less cyclical, so she models the number of fish q_k returning each year as $q_{k+1} = (0.14(-1)^k + 0.36)(q_k + h)$, where h is the number of fish whose spawn she releases from the hatchery annually.

(a) Show that the sustained number of fish returning in even-numbered years approaches approximately

$$q_e = 3h\sum_{k=1}^{\infty} 0.11^k.$$

(Hint: Make a new recurrence by using two steps of the one given.)

(b) Show that the sustained number of fish returning in odd-numbered years approaches approximately

$$q_o = \frac{61}{11}h\sum_{k=1}^{\infty} 0.11^k.$$

(c) How should Leila choose h, the number of hatchery fish to breed in order to hold the minimum number of fish returning in each run near some constant P?

Proofs

54. Let $a : [1, \infty) \to \mathbb{R}$ be a continuous, positive, and decreasing function. Complete the proof of the integral test (Theorem 7.28) by showing that if the improper integral $\int_1^{\infty} a(x)\,dx$ converges, then the series $\sum_{k=1}^{\infty} a(k)$ does too.

55. Prove Theorem 7.31. That is, show that if a function a is continuous, positive, and decreasing, and if the improper integral $\int_1^{\infty} a(x)\,dx$ converges, then the nth remainder, R_n, for the series $\sum_{k=1}^{\infty} a(k)$ is bounded by

$$0 \le R_n = \sum_{k=n+1}^{\infty} a(k) \le \int_n^{\infty} a(x)\,dx.$$

56. Use the divergence test to prove that a p-series $\sum_{k=1}^{\infty} \frac{1}{k^p}$ diverges when $p < 0$.

57. Use the divergence test to prove that a geometric series $\sum_{k=1}^{\infty} cr^k$ diverges when $|r| \ge 1$ and $c \ne 0$.

Thinking Forward

- *A series of monomials:* Find all values of x for which the series $\sum_{k=1}^{\infty} (4x)^k$ converges.
- *A series of monomials:* Find all values of x for which the series $\sum_{k=1}^{\infty} \left(\frac{x}{3}\right)^k$ converges.

7.5 COMPARISON TESTS

- Analyzing series by comparing them with series whose behavior we understand
- Using inequalities to compare series
- Using limits to compare series

The Comparison Test

Our first comparison is analogous to the comparison test for improper integrals we discussed in Section 5.6. In fact, if we combine the comparison test for improper integrals and the integral test, Theorem 7.28, we immediately obtain the next theorem.

THEOREM 7.32 **The Comparison Test**

Let $\sum_{k=1}^{\infty} a_k$ and $\sum_{k=1}^{\infty} b_k$ be two series with nonnegative terms such that $0 \le a_k \le b_k$ for every positive integer k. If the series $\sum_{k=1}^{\infty} b_k$ converges, then the series $\sum_{k=1}^{\infty} a_k$ converges.

We should also mention the contrapositive of Theorem 7.32. That is, given the same hypotheses for the two series, if the series $\sum_{k=1}^{\infty} a_k$ diverges, then the series $\sum_{k=1}^{\infty} b_k$ diverges.

As you may recall from your work with the comparison test for improper integrals, using a comparison test requires finesse. When we are trying to discover whether a given series $\sum_{k=1}^{\infty} c_k$ converges or diverges, we must think through a few questions:

- We fully understand the behavior of geometric series and p-series. Is the series a geometric series or a p-series? If it is, we are done.
- If the series is not a geometric series or a p-series, is it "similar enough" to one so that we can use the comparison test? If it is, we must ensure that the correct inequality holds. That is, if $\sum_{k=1}^{\infty} d_k$ represents a geometric series or p-series with terms similar to the terms of $\sum_{k=1}^{\infty} c_k$, and if the series $\sum_{k=1}^{\infty} d_k$ converges, we would need the inequality $0 \le c_k \le d_k$ to hold for every k. If it does, then the

series $\sum_{k=1}^{\infty} c_k$ converges as well. If, however, $\sum_{k=1}^{\infty} d_k$ diverges, then we need the inequality $0 \le d_k \le c_k$ to hold for every k. In this case we would deduce that the series $\sum_{k=1}^{\infty} c_k$ diverges.

- If the series $\sum_{k=1}^{\infty} c_k$ is not similar to a geometric series or a p-series, is it similar to another series whose behavior we can more easily analyze? If so, we may carry out the preceding comparison to analyze our series $\sum_{k=1}^{\infty} c_k$.

We use the comparison test in one example here and another later in our explorations.

Consider the series $\sum_{k=1}^{\infty} \frac{k-1}{k^3+k+1}$. This series is neither geometric nor a p-series. However, it is similar to the convergent p-series $\sum_{k=1}^{\infty} \frac{1}{k^2}$. We chose $\sum_{k=1}^{\infty} \frac{1}{k^2}$ because the dominant term in the numerator of $\frac{k-1}{k^3+k+1}$ is k, the dominant term in the denominator is k^3, and the quotient $\frac{k}{k^3} = \frac{1}{k^2}$. Now, since the p-series converges, we need the inequality

$$0 \le \frac{k-1}{k^3+k+1} \le \frac{1}{k^2}$$

to hold for every positive integer k. You should verify that this inequality is valid for every positive integer. Thus, the convergence of $\sum_{k=1}^{\infty} \frac{1}{k^2}$ implies the convergence of $\sum_{k=1}^{\infty} \frac{k-1}{k^3+k+1}$.

If the inequality had not held for every k, using the comparison test would not have been quite as simple. For example, if we start with the similar-looking series $\sum_{k=1}^{\infty} \frac{k+1}{k^3-k+1}$, we would still want to compare that series with the p-series $\sum_{k=1}^{\infty} \frac{1}{k^2}$, so we would still suspect that the series $\sum_{k=1}^{\infty} \frac{k+1}{k^3+k+1}$ converges. In this case, however, the required inequality

$$0 \le \frac{k+1}{k^3-k+1} \le \frac{1}{k^2}$$

does not hold for every positive integer. (For example, $\frac{2+1}{2^3-2+1} \not\le \frac{1}{2^2}$.) We could try to find a different convergent series to compare successfully against $\sum_{k=1}^{\infty} \frac{k+1}{k^3+k+1}$, using the comparison test, but we may use the limit comparison test, presented next, more easily.

The Limit Comparison Test

As we have just shown, although the comparison test can work effectively, we do not always have the required inequality between the terms of the given series and the most obvious comparison series. The limit comparison test provides another method of comparison, in which we do not have to consider inequalities. However, in exchange for avoiding inequalities, we must evaluate a limit.

THEOREM 7.33 **The Limit Comparison Test**

Let $\sum_{k=1}^{\infty} a_k$ and $\sum_{k=1}^{\infty} b_k$ be two series with positive terms.

(a) If $\lim_{k\to\infty} \frac{a_k}{b_k} = L$, where L is any positive real number, then either both series converge or both series diverge.

(b) If $\lim_{k\to\infty} \frac{a_k}{b_k} = 0$ and $\sum_{k=1}^{\infty} b_k$ converges, then $\sum_{k=1}^{\infty} a_k$ converges.

(c) If $\lim_{k\to\infty} \frac{a_k}{b_k} = \infty$ and $\sum_{k=1}^{\infty} b_k$ diverges, then $\sum_{k=1}^{\infty} a_k$ diverges.

The main result of Theorem 7.33 is found in part (a). Parts (b) and (c) are occasionally useful side results. Before we prove part (a) formally, consider the significance of $\lim_{n\to\infty} \frac{a_k}{b_k} = L$. This limit tells us that when k is "large," the terms of series $\sum_{k=1}^{\infty} a_k$ are roughly L times the size of the terms of the other series. But recall that multiplying the terms of a series by a nonzero constant does not change the convergence or divergence of the series. Thus, if one series converges, the other does also.

Proof. We will prove part (a) of Theorem 7.33 and leave parts (b) and (c) for Exercises 56 and 57.

Let $\lim_{k\to\infty} \frac{a_k}{b_k} = L$, where L is a positive real number. Thus, for any $\epsilon > 0$, there is an integer N such that $\left|\frac{a_k}{b_k} - L\right| < \epsilon$ when $k \geq N$.

We now assume that the series $\sum_{k=1}^{\infty} b_k$ converges and let $\epsilon = L$. We have $\frac{a_k}{b_k} < 2L$, or equivalently, $a_k < 2Lb_k$ for $k \geq N$. Therefore,

$$\sum_{k=N}^{\infty} a_k < \sum_{k=N}^{\infty} 2Lb_k = 2L \sum_{k=N}^{\infty} b_k.$$

Recall that removing finitely many terms from a series does not change its convergence. Thus, from the preceding inequality above we see that the convergence of the series $\sum_{k=1}^{\infty} b_k$ implies that the series $\sum_{k=1}^{\infty} a_k$ converges as well.

Next we assume that the series $\sum_{k=1}^{\infty} b_k$ diverges and let $\epsilon = \frac{L}{2}$. We now have $\frac{a_k}{b_k} > \frac{L}{2}$, or equivalently, $a_k > \frac{L}{2} b_k$ for $k \geq N$. Here,

$$\sum_{k=N}^{\infty} a_k > \sum_{k=N}^{\infty} \frac{L}{2} b_k = \frac{L}{2} \sum_{k=N}^{\infty} b_k.$$

Removing finitely many terms from a series does not change its divergence. Thus, from this new inequality we see that the divergence of the series $\sum_{k=1}^{\infty} b_k$ implies the divergence of $\sum_{k=1}^{\infty} a_k$. ■

As we mentioned earlier, to analyze the convergence of the series $\sum_{k=1}^{\infty} \frac{k+1}{k^3+k+1}$ the most "obvious" series to use for a comparison is the convergent p-series $\sum_{k=1}^{\infty} \frac{1}{k^2}$. We do not have the necessary inequality between the terms of the two series to successfully use the comparison test, but the limit comparison test will succeed. Consider the limit

$$\lim_{k\to\infty} \frac{(k+1)/(k^3+k+1)}{1/k^2} = \lim_{k\to\infty} \frac{k^3+k^2}{k^3+k+1} = 1.$$

Since this limit is positive and finite, and because we already know the the p-series converges, we have proven that the series $\sum_{k=1}^{\infty} \frac{k+1}{k^3+k+1}$ converges as well.

To use either of the comparison tests on a series $\sum_{k=1}^{\infty} a_k$ successfully, follow these steps:

1. If the general term of your series is a quotient, consider the *dominant* term in the numerator n_k and the *dominant* term in the denominator d_k.
2. Consider the series $\sum_{k=1}^{\infty} \frac{n_k}{d_k}$. Simplify the quotient $\frac{n_k}{d_k}$ as much as possible.
3. If the new series is a geometric series, a p-series, or another series whose behavior you can easily analyze, you now suspect that the original series $\sum_{k=1}^{\infty} a_k$ has the same behavior as your new series $\sum_{k=1}^{\infty} \frac{n_k}{d_k}$.
4. Complete the details of your analysis, using either the comparison test or limit comparison test. Be sure to show that you have the correct inequality if you use the comparison test. That is,

- If the series $\sum_{k=1}^{\infty} \frac{n_k}{d_k}$ converges, you must have
$$0 \le a_k \le \frac{n_k}{d_k} \text{ for every } k \in \mathbb{Z}^+.$$
- If the series $\sum_{k=1}^{\infty} \frac{n_k}{d_k}$ diverges, you must have
$$0 \le \frac{n_k}{d_k} \le a_k \text{ for every } k \in \mathbb{Z}^+.$$
- If you use the limit comparison test, you must compute the limit
$$\lim_{k\to\infty} \frac{a_k}{n_k/d_k}.$$

If this limit is positive and finite, the series $\sum_{k=1}^{\infty} a_k$ converges if and only if $\sum_{k=1}^{\infty} \frac{n_k}{d_k}$ does.

Examples and Explorations

EXAMPLE 1

Using the comparison test

Determine whether the series $\sum_{k=1}^{\infty} \frac{k^{3/2}-1}{5k^3+3}$ converges or diverges.

SOLUTION

Recall the list of questions we started in Example 1 of Section 7.4:

- Does $a_k \not\to 0$ as $k \to \infty$? If so, then, by the divergence test, the series diverges.
- Does $a_k = a(k)$, where a is a continuous, positive, and decreasing function on $[1, \infty)$? If so, is the improper integral $\int_1^{\infty} a(x)\,dx$ "relatively easy" to evaluate? If it is, you may use the integral test.

We now add the following question:

- Is there another related series whose terms are similar to those of the given series and whose behavior you either know or can more easily determine? If so, you may use one of the comparison tests.

You may show that the divergence test would fail on the series given here and that the integral test would be impractical. We will use the comparison test, comparing the given series with $\sum_{k=1}^{\infty} \frac{1}{5k^{3/2}}$. The latter series converges, since it is a multiple of a convergent p-series. For every positive integer k, we have
$$0 \le \frac{k^{3/2}-1}{5k^3+3} \le \frac{k^{3/2}}{5k^3} = \frac{1}{5}\frac{1}{k^{3/2}}.$$
Now, since the series $\sum_{k=1}^{\infty} \frac{1}{5k^{3/2}}$ converges, by the comparison test we know that the series $\sum_{k=1}^{\infty} \frac{k^{3/2}-1}{5k^3+3}$ converges as well. The comparison test worked well here because the inequalities are exactly those required to show that the given series converges. We could also have used the limit comparison test. You will be asked to supply the details for this in Exercise 19. □

EXAMPLE 2

Using the limit comparison test

Determine whether the series $\sum_{k=1}^{\infty} (3k^2+k+1)^{-1/2}$ converges or diverges.

SOLUTION

The divergence test would fail on this series, and again, the integral test would be impractical. We note that each term of the series is positive and that
$$(3k^2+k+1)^{-1/2} = \frac{1}{\sqrt{3k^2+k+1}}.$$

If we consider the dominant term in the denominator, we are led to compare the series with the harmonic series $\sum_{k=1}^{\infty} \frac{1}{k}$. To use the limit comparison test we evaluate

$$\lim_{k\to\infty} \frac{1/\sqrt{3k^2+k+1}}{1/k} = \lim_{k\to\infty} \frac{k}{\sqrt{3k^2+k+1}} = \frac{1}{\sqrt{3}}.$$

Since the harmonic series diverges and this limit is positive and finite, the given series diverges as well. Note that the initial quotient in the preceding equation is the result of the division of $\frac{1}{\sqrt{3k^2+k+1}}$ by $\frac{1}{k}$. We could also have used the reciprocal instead. If we had, the limit of that quotient would have been $\sqrt{3}$. Again, since the harmonic series diverges and this limit is positive and finite, the given series diverges. □

? TEST YOUR UNDERSTANDING

- What is the comparison test? How is the comparison test used?
- How is the comparison test for series related to the comparison test for improper integrals that we introduced in Section 5.6?
- What is the limit comparison test? How is the limit comparison test used?
- Which series are good to use in comparisons? How do you find a series to use for a comparison?
- What are the advantages and disadvantages of the comparison test versus the limit comparison test?

EXERCISES 7.5

Thinking Back

The comparison test: In Section 5.6 we discussed the comparison test for improper integrals.

- What is the comparison test for improper integrals?
- How is the comparison test for improper integrals used to analyze the convergence or divergence of an improper integral?
- Which improper integrals are used in the comparison test?

The limit comparison test: In Section 5.6 we discussed a limit comparison test for improper integrals.

- What is the limit comparison test for improper integrals?
- How is the limit comparison test for improper integrals used to analyze the convergence or divergence of an improper integral?
- Which improper integrals are used in the limit comparison test?

Concepts

0. *Problem Zero:* Read the section and make your own summary of the material.

1. *True/False:* Determine whether each of the statements that follow is true or false. If a statement is true, explain why. If a statement is false, provide a counterexample.

(a) *True or False:* If $0 \le f(x) \le g(x)$ for every $x \ge 0$ and the improper integral $\int_0^\infty g(x)\,dx$ converges, then the improper integral $\int_0^\infty f(x)\,dx$ converges.

(b) *True or False:* If $0 \le f(x) \le g(x)$ for every $x > 0$ and $\lim_{x\to\infty} \frac{f(x)}{g(x)} = 3$, then the improper integrals $\int_0^\infty g(x)\,dx$ and $\int_0^\infty f(x)\,dx$ both converge.

(c) *True or False:* If $0 \le a_k < \frac{1}{k}$ for every positive integer k, then the series $\sum_{k=1}^{\infty} a_k$ converges.

(d) *True or False:* If $\frac{1}{k^2} < b_k$ for every positive integer k, then the series $\sum_{k=1}^{\infty} b_k$ diverges.

(e) *True or False:* If $a_k \le b_k$ for every positive integer k and the series $\sum_{k=1}^{\infty} b_k$ converges, then the series $\sum_{k=1}^{\infty} a_k$ converges.

(f) *True or False:* If $\sum_{k=1}^{\infty} a_k$ and $\sum_{k=1}^{\infty} b_k$ both diverge, then $\sum_{k=1}^{\infty} (a_k \cdot b_k)$ diverges.

(g) *True or False:* If a_k and b_k are both positive for every positive integer k and $\lim_{k\to\infty} \frac{a_k}{b_k} = \frac{1}{2}$, then $\sum_{k=1}^{\infty} a_k$ and $\sum_{k=1}^{\infty} b_k$ both converge.

(h) *True or False:* If $\sum_{k=1}^{\infty} a_k$ and $\sum_{k=1}^{\infty} b_k$ both converge, then $\lim_{k\to\infty} \frac{a_k}{b_k}$ is finite.

2. *Examples:* Construct examples of the thing(s) described in the following. Try to find examples that are different than any in the reading.
   (a) A p-series you could use as a comparison to show that the series $\sum_{k=1}^{\infty} \frac{\sqrt{k+2}}{k^2+k+1}$ converges.
   (b) A p-series you could use as a comparison to show that the series $\sum_{k=1}^{\infty} \sin\left(\frac{1}{k}\right)$ diverges.
   (c) A p-series other than $\sum_{k=1}^{\infty} \frac{1}{k^2}$ you could use with the comparison test to show that the series $\sum_{k=1}^{\infty} \frac{k-1}{k^3+k+1}$ converges.
3. Explain how you could adapt the comparison test to analyze a series $\sum_{k=1}^{\infty} a_k$ in which all of the terms are negative.
4. Use the comparison test to explain why the series $\sum_{k=1}^{\infty} \frac{1}{\sqrt[\alpha]{k}}$ diverges when α is an integer greater than 1.
5. Provide a more general statement of the comparison test in which the inequality $0 \le a_k \le b_k$ holds only for integers $k > K$, where K is an arbitrary positive integer. Explain why your statement is valid.
6. Explain how you could adapt the limit comparison test to analyze a series $\sum_{k=1}^{\infty} a_k$ in which all of the terms are negative.
7. Provide a more general statement of the limit comparison test in which $\sum_{k=1}^{\infty} a_k$ and $\sum_{k=1}^{\infty} b_k$ are two series whose terms are eventually positive. Explain why your statement is valid.
8. If you suspect that a series $\sum_{k=1}^{\infty} a_k$ diverges, explain why you would need to compare the series with a divergent series, using either the comparison test or the limit comparison test.
9. If you suspect that a series $\sum_{k=1}^{\infty} a_k$ converges, explain why you would want to compare the series with a convergent series, using either the comparison test or the limit comparison test.
10. Briefly outline the advantages and disadvantages of using the two comparison tests to analyze the behavior of a series $\sum_{k=1}^{\infty} a_k$.

Exercises 11–13 concern the series $\sum_{k=2}^{\infty} \frac{1}{k \ln k}$.

11. Use the integral test to show that the series $\sum_{k=2}^{\infty} \frac{1}{k \ln k}$ diverges.
12. Let $0 < p < 1$. Show that $0 \le \frac{1}{k \ln k} \le \frac{1}{k^p}$ for large values of k. Explain why we cannot use a p-series with $0 < p < 1$ in a comparison test to verify the divergence of the series $\sum_{k=2}^{\infty} \frac{1}{k \ln k}$.
13. Let $0 < p < 1$. Evaluate the limit $\lim_{k \to \infty} \frac{1/(k \ln k)}{1/k^p}$. Explain why we cannot use a p-series with $0 < p < 1$ in a limit comparison test to verify the divergence of the series $\sum_{k=2}^{\infty} \frac{1}{k \ln k}$.

Let $\sum_{k=1}^{\infty} a_k$ and $\sum_{k=1}^{\infty} b_k$ be two series with positive terms. Answer the questions about the limit comparison test in Exercises 14–18.

14. If $\lim_{k \to \infty} \frac{a_k}{b_k} = L$, where L is a positive finite number, what may we conclude about the two series?
15. Fill in the blanks and select the correct word: If $\lim_{k \to \infty} \frac{a_k}{b_k} = 0$ and $\sum_{k=1}^{\infty}$ ___ converges, then $\sum_{k=1}^{\infty}$ ___ (converges/diverges).
16. If $\lim_{k \to \infty} \frac{a_k}{b_k} = 0$ and $\sum_{k=1}^{\infty} a_k$ converges, explain why we *cannot* draw any conclusions about the behavior of $\sum_{k=1}^{\infty} b_k$.
17. Fill in the blanks and select the correct word: If $\lim_{k \to \infty} \frac{a_k}{b_k} = \infty$ and $\sum_{k=1}^{\infty}$ ___ diverges, then $\sum_{k=1}^{\infty}$ ___ (converges/diverges).
18. If $\lim_{k \to \infty} \frac{a_k}{b_k} = \infty$ and $\sum_{k=1}^{\infty} a_k$ diverges, explain why we cannot draw any conclusions about the behavior of $\sum_{k=1}^{\infty} b_k$.
19. In Example 1 we used the comparison test to show that the series $\sum_{k=1}^{\infty} \frac{k^{3/2}-k-1}{5k^3+3}$ converges. Use the limit comparison test to prove the same result.
20. In Example 1 of Section 7.4 we used the integral test to show that the series $\sum_{k=1}^{\infty} \frac{k}{e^{k^2}}$ converges. Use the limit comparison test with the series $\sum_{k=1}^{\infty} \frac{1}{e^k}$ to prove the same result.

Skills

In Exercises 21–30 use one of the comparison tests to determine whether the series converges or diverges. Explain how the given series satisfies the hypotheses of the test you use.

21. $\sum_{k=0}^{\infty} \frac{3k^2+1}{k^3+k^2+5}$
22. $\sum_{k=1}^{\infty} \frac{3}{(1+2k)^2}$
23. $\sum_{k=2}^{\infty} \frac{3k-5}{k\sqrt{k^3-4}}$
24. $\sum_{k=1}^{\infty} \frac{\sqrt{k}}{1+k^2}$
25. $\sum_{k=1}^{\infty} \frac{1+\ln k}{k}$
26. $\sum_{k=1}^{\infty} \frac{1}{k+0.01}$
27. $\sum_{k=1}^{\infty} \frac{1+\ln k}{k^2}$
28. $\sum_{k=1}^{\infty} \frac{1}{(k-\pi)^2}$
29. $\sum_{k=1}^{\infty} \frac{1+\ln k}{k^3}$
30. $\sum_{k=1}^{\infty} \left(\frac{\sin k}{k}\right)^2$

Use any convergence test from this section or the previous section to determine whether the series in Exercises 31–48 converge or diverge. Explain how the series meets the hypotheses of the test you select.

31. $\sum_{k=1}^{\infty} k^{-1/2}$
32. $\sum_{k=1}^{\infty} k^{-4/3}$

33. $\sum_{k=1}^{\infty} \frac{k^{-3/4}}{k+2}$

34. $\sum_{k=1}^{\infty} \cos\left(\frac{1}{k}\right)$

35. $\sum_{k=1}^{\infty} \sin\left(\frac{1}{k}\right)$

36. $\sum_{k=1}^{\infty} \frac{1}{1+k^2}$

37. $\sum_{k=1}^{\infty} \frac{\ln k}{k}$

38. $\sum_{k=1}^{\infty} \sin\left(\frac{1}{k^2}\right)$

39. $\sum_{k=1}^{\infty} \frac{\sqrt{k+1}-\sqrt{k}}{k}$

40. $\sum_{k=1}^{\infty} \frac{1}{\sqrt{k+1}+\sqrt{k}}$

41. $\sum_{k=1}^{\infty} k^2 e^{-k^3}$

42. $\sum_{k=1}^{\infty} \frac{1}{\sqrt[3]{k+1}+\sqrt[3]{k}}$

43. $\sum_{k=1}^{\infty} \frac{k^2}{e^k}$

44. $\sum_{k=1}^{\infty} \frac{2^k}{k^2}$

45. $\sum_{k=1}^{\infty} \frac{1}{\sqrt{k}e^{\sqrt{k}}}$

46. $\sum_{k=3}^{\infty} \frac{k+1}{(k-2)^3}$

47. $\sum_{k=1}^{\infty} \frac{1}{2k+7}$

48. $\sum_{k=1}^{\infty} \frac{\sqrt{k}}{k^2+3}$

In Exercises 49 and 50 find all values of p so that the series converges.

49. $\sum_{k=1}^{\infty} \frac{k^p}{e^k}$

50. $\sum_{k=1}^{\infty} \frac{(\ln k)^p}{k^2}$

Applications

51. There are many ways to harness wave energy. One of them relies on a hollow platform that is anchored to the ocean floor. Waves wash over the top of the platform into the basin in its center, where the only way for water to get out is to pass through turbines, generating power.

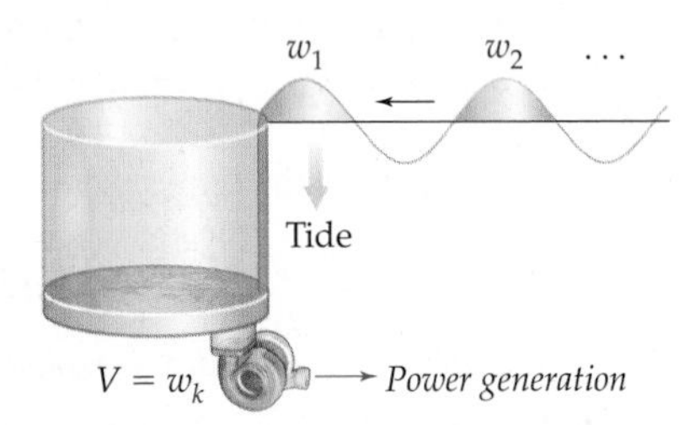

When the tide is going out, the amount of water each wave leaves in the central basin is given by a random w_k, where $0 \le w_k \le Wr^{\alpha(k)}$. Here, W is the amount of water deposited by the largest wave at high tide, $r < 1$, and $\alpha(k)$ is a polynomial in k with positive coefficients and no constant term. Although the number of waves for a given tide is actually finite, it is easiest to use the infinite series $V = \sum_{k=0}^{\infty} w_k$ to model the total water arriving in the basin. Does this series converge? Explain your answer.

52. Sandy is studying the relationship of light in a forest to the density of the canopy. She models this relationship simply by supposing that a fixed amount of radiance I_ω at a frequency ω passes through the vegetation canopy. Once light is below the vegetation, a fraction s is reflected from the surface while a different fraction, $c \ge s$, is reflected from the vegetation canopy. The total ambient light inside the forest is the sum of all these direct and reflected lights: the direct light, plus that reflected from the surface, plus that reflected from the surface and then reflected from the vegetation, and so on.

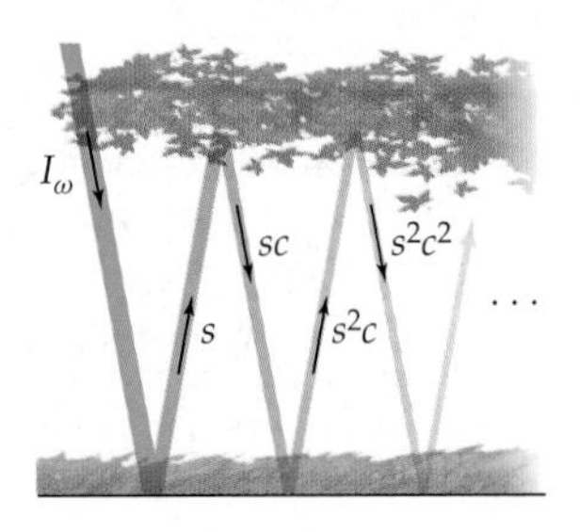

Sandy writes the relationship she has found as

$$I_{tot} = I_\omega(1 + s + sc + s^2c + s^2c^2 + \cdots).$$

Does this series converge? Why or why not?

Proofs

53. Prove that if $\sum_{k=1}^{\infty} a_k$ is a convergent series with $a_k \ge 0$ for every positive integer k, then the series $\sum_{k=1}^{\infty} a_k^2$ converges.

54. Prove that if $\sum_{k=1}^{\infty} a_k$ is a convergent series with $a_k \ge 0$ for every positive integer k, then the series $\sum_{k=1}^{\infty} \frac{a_k}{k}$ converges.

55. Prove that if $\sum_{k=1}^{\infty} a_k$ and $\sum_{k=1}^{\infty} b_k$ are two convergent series with $a_k \ge 0$ and $b_k \ge 0$ for every positive integer k, then the series $\sum_{k=1}^{\infty} (a_k \cdot b_k)$ converges.

In Exercises 56 and 57 we ask you to complete the proof of Theorem 7.33. For these exercises let $\sum_{k=1}^{\infty} a_k$ and $\sum_{k=1}^{\infty} b_k$ be two series with positive terms.

56. Show that if $\lim_{k\to\infty} \frac{a_k}{b_k} = \infty$ and $\sum_{k=1}^{\infty} b_k$ diverges, then $\sum_{k=1}^{\infty} a_k$ diverges.

57. Show that if $\lim_{k\to\infty} \frac{a_k}{b_k} = 0$ and $\sum_{k=1}^{\infty} b_k$ converges, then $\sum_{k=1}^{\infty} a_k$ converges.

Thinking Forward

- *A series of monomials:* Find all values of x for which the series $\sum_{k=1}^{\infty} \frac{x^{2k}}{k^2}$ converges.
- *A series of monomials:* Find all values of x for which the series $\sum_{k=0}^{\infty} \frac{|x|}{4^k}$ converges.

7.6 THE RATIO AND ROOT TESTS

- Series containing powers, factorials, and/or factorial-like terms
- Using the ratio test to determine the convergence of a series
- Using the root test to determine the convergence of a series

The Ratio Test

In this section we discuss two more convergence tests for series with positive terms: the ratio test and the root test. As we will see, the statements of these tests are quite similar. Both tests involve the evaluation of a limit. The magnitude of the appropriate limit can tell us whether the series converges or diverges. Unlike the limit we use in the limit comparison test, the limits we take for the tests in this section do not rely upon the terms of a secondary series. The ratio test, in particular, will be used extensively throughout the remainder of this chapter and in Chapter 8.

THEOREM 7.34

The Ratio Test

Let $\sum_{k=1}^{\infty} a_k$ be a series with positive terms, and assume $\rho = \lim_{k\to\infty} \frac{a_{k+1}}{a_k}$ exists.

(a) If $\rho < 1$, the series converges.

(b) If $\rho > 1$, the series diverges.

(c) If $\rho = 1$, the test is inconclusive. (Use a different test!)

The main concept used in the proof of the ratio test is that a series in which $\lim_{k\to\infty} \frac{a_{k+1}}{a_k} = \rho$ behaves like the geometric series $\sum \rho^k$.

Proof. Let $\sum_{k=1}^{\infty} a_k$ be a series with positive terms, and let $\rho = \lim_{k\to\infty} \frac{a_{k+1}}{a_k}$.

We first assume that $\rho < 1$. In this case, there is a positive real number r such that $\rho < r < 1$. Since $\rho = \lim_{k\to\infty} \frac{a_{k+1}}{a_k}$ and $r > \rho$, all values of the quotient $\frac{a_{k+1}}{a_k}$ will be less than r for sufficiently large values of k. That is, there is a positive integer N such that, for $k \geq N$, $\frac{a_{k+1}}{a_k} < r$. In particular,

$$a_{N+1} < a_N r.$$

The analogous inequalities hold for all subsequent successive pairs of terms of the series. For example,

$$a_{N+2} < a_{N+1} r$$

as well. Combining these two inequalities, we obtain $a_{N+2} < a_N r^2$. By induction, we may prove that

$$a_{N+n} < a_N r^n$$

for every positive integer n. (We leave this detail to Exercise 66.) We now consider the tail of the original series:

$$\sum_{k=N}^{\infty} a_k = a_N + a_{N+1} + a_{N+2} + \cdots + a_{N+n} + \cdots$$
$$< a_N + a_N r + a_N r^2 + \cdots + a_N r^n + \cdots \quad \leftarrow \text{since } a_{N+n} < a_N r^n \text{ for every positive integer } n$$
$$= a_N(1 + r + r^2 + \cdots + r^n + \cdots)$$

Note that since $a_N \in \mathbb{R}$ and $r < 1$, the preceding series is a convergent geometric series. Because this geometric series converges, by the comparison test and the fact that the tail of a series determines the convergence of the entire series, the series $\sum_{k=1}^{\infty} a_k$ converges.

We now show that if $\rho > 1$ the series diverges. Since $\rho = \lim\limits_{k\to\infty} \frac{a_{k+1}}{a_k}$ and $\rho > 1$, there is a positive integer N such that for every $k \geq N$, $a_{k+1} > a_k$. Thus, after index N, the terms of the series form an increasing sequence. Furthermore, since $a_N > 0$, it follows that $\lim\limits_{k\to\infty} a_k \neq 0$. The divergence test implies that when $\rho > 1$ the series $\sum_{k=1}^{\infty} a_k$ diverges.

Recall from Section 7.4 that a p-series $\sum_{k=1}^{\infty} \frac{1}{k^p}$ converges when $p > 1$ and diverges otherwise. In Example 1 we will show that for the (divergent) harmonic series $\sum_{k=1}^{\infty} \frac{1}{k}$ and the (convergent) p-series $\sum_{k=1}^{\infty} \frac{1}{k^2}$ the limit of the quotient of successive terms of the series is 1. This will complete the proof by showing that the ratio test is inconclusive when $\rho = 1$. ■

The ratio test works particularly well on series containing exponential factors, factorials, and/or factorial-like products. Since the terms of the series $\sum_{k=0}^{\infty} \frac{2^k}{k!}$ contain both exponentials and factorials, the series is a perfect candidate for the ratio test. Here $a_k = \frac{2^k}{k!}$ and $a_{k+1} = \frac{2^{k+1}}{(k+1)!}$, and we compute

$$\lim_{k\to\infty} \frac{a_{k+1}}{a_k} = \frac{2^{k+1}/(k+1)!}{2^k/k!} = \lim_{k\to\infty} \frac{2^{k+1}}{(k+1)!} \cdot \frac{k!}{2^k} = \lim_{k\to\infty} \frac{2}{k+1} = 0.$$

Since this limit is less than 1, the series converges.

The Root Test

THEOREM 7.35

The Root Test

Let $\sum_{k=1}^{\infty} a_k$ be a series with positive terms, and assume $\rho = \lim\limits_{k\to\infty} \sqrt[k]{a_k} = \lim\limits_{k\to\infty} (a_k)^{1/k}$ exists.

(a) If $\rho < 1$, the series converges.

(b) If $\rho > 1$, the series diverges.

(c) If $\rho = 1$, the test is inconclusive. (Use a different test!)

The main idea in the proof of the root test is the same as it was in the proof of the ratio test: A series in which $\lim\limits_{k\to\infty} \sqrt[k]{a_k} = \rho$ behaves like the geometric series $\sum \rho^k$. Therefore, the proof of Theorem 7.35 may be modeled on our proof of the ratio test. This is left for Exercise 68. The root test works particularly well on series containing exponential factors or other factors with kth powers. The series $\sum_{k=1}^{\infty} \left(\frac{2k}{3k-1}\right)^k$ is one such series. Here,

$$\lim_{k\to\infty} (a_k)^{1/k} = \lim_{k\to\infty} \left(\left(\frac{2k}{3k-1}\right)^k\right)^{1/k} = \lim_{k\to\infty} \frac{2k}{3k-1} = \frac{2}{3}.$$

Since this value is less than 1, the series converges.

Examples and Explorations

EXAMPLE 1 **Using the ratio test**

Use the ratio test to determine the convergence or divergence of the following series:

(a) $\displaystyle\sum_{k=0}^{\infty} \frac{10^k}{k!}$ (b) $\displaystyle\sum_{k=1}^{\infty} \frac{3^k}{k^5}$ (c) $\displaystyle\sum_{k=2}^{\infty} \frac{2^{3k-1}}{(2k-3)!}$

SOLUTION

(a) We have already mentioned that the ratio test is particularly effective on series containing exponential factors, factorials, or factorial-like products. Each of the given series in this example contains at least one of these. Also, note that the terms of these series are positive for every value of k. In the first series we let $a_k = \frac{10^k}{k!}$. Thus, $a_{k+1} = \frac{10^{k+1}}{(k+1)!}$ and

$$\rho = \lim_{k\to\infty} \frac{a_{k+1}}{a_k} = \lim_{k\to\infty} \frac{10^{k+1}/(k+1)!}{10^k/k!} = \lim_{k\to\infty} \frac{10^{k+1}}{(k+1)!}\frac{k!}{10^k} = \lim_{k\to\infty} \frac{10}{k+1} = 0.$$

Since $\rho < 1$, the series converges. Before we continue, be sure you understand the mechanics of the steps we just used. In particular, make sure you understand why $\frac{(k+1)!}{k!} = k+1$.

(b) Here we let $a_k = \frac{3^k}{k^5}$ and $a_{k+1} = \frac{3^{k+1}}{(k+1)^5}$. Then

$$\rho = \lim_{k\to\infty} \frac{a_{k+1}}{a_k} = \lim_{k\to\infty} \frac{3^{k+1}/(k+1)^5}{3^k/k^5} = \lim_{k\to\infty} \frac{3^{k+1}}{(k+1)^5}\frac{k^5}{3^k} = \lim_{k\to\infty} \frac{3k^5}{(k+1)^5} = 3.$$

Since $\rho > 1$, the series diverges. Before proceeding to the final series in this example we should note that we could also have used the divergence test to show that the given series diverges. Since exponential functions dominate polynomials, it follows that $\frac{3^k}{k^5} \to \infty$ as $k \to \infty$.

(c) For our final series we let $a_k = \frac{2^{3k-1}}{(2k-3)!}$ and

$$a_{k+1} = \frac{2^{3(k+1)-1}}{(2(k+1)-3)!} = \frac{2^{3k+2}}{(2k-1)!}.$$

Note that to form a_{k+1} we replaced each occurrence of k by $k+1$ in our formula for a_k and then simplified. Now,

$$\rho = \lim_{k\to\infty} \frac{a_{k+1}}{a_k} = \lim_{k\to\infty} \frac{2^{3k+2}/(2k-1)!}{2^{3k-1}/(2k-3)!} = \lim_{k\to\infty} \frac{2^{3k+2}}{(2k-1)!}\frac{(2k-3)!}{2^{3k-1}}.$$

We will need to simplify the rightmost quantity before we evaluate the limit. When we simplify the exponential factors in the numerator and denominator, we have

$$\frac{2^{3k+2}}{2^{3k-1}} = 2^{(3k+2)-(3k-1)} = 2^3 = 8.$$

From the definition of the factorial, we obtain

$$\frac{(2k-3)!}{(2k-1)!} = \frac{(2k-3)!}{(2k-1)(2k-2)(2k-3)!} = \frac{1}{(2k-1)(2k-2)}.$$

Therefore,

$$\rho = \lim_{k\to\infty} \frac{8}{(2k-1)(2k-2)} = 0.$$

By the ratio test, the series converges. □

EXAMPLE 2 **Showing that the ratio test is inconclusive when $\rho = 1$**

Show that the ratio test fails to determine the divergence of the harmonic series $\sum_{k=1}^{\infty} \frac{1}{k}$ or the convergence of the p-series $\sum_{k=1}^{\infty} \frac{1}{k^2}$.

SOLUTION

We consider the harmonic series first. For the harmonic series, $a_k = \frac{1}{k}$ and $a_{k+1} = \frac{1}{k+1}$. Thus,

$$\lim_{k\to\infty} \frac{a_{k+1}}{a_k} = \lim_{k\to\infty} \frac{1/(k+1)}{1/k} = \lim_{k\to\infty} \frac{k}{k+1} = 1.$$

We now consider $\sum_{k=1}^{\infty} \frac{1}{k^2}$. Here we have $a_k = \frac{1}{k^2}$ and $a_{k+1} = \frac{1}{(k+1)^2}$. Hence,

$$\lim_{k\to\infty} \frac{a_{k+1}}{a_k} = \lim_{k\to\infty} \frac{1/(k+1)^2}{1/k^2} = \lim_{k\to\infty} \frac{k^2}{(k+1)^2} = 1.$$

We already know that the harmonic series diverges and the p-series with $p = 2$ converges. This shows that the ratio test is inconclusive when $\rho = \lim_{k\to\infty} \frac{a_{k+1}}{a_k} = 1$.

Recall that we have already mentioned that the ratio test can be effective on series containing exponential factors, factorials, and/or factorial-like products. The two p-series in this example do not fall into this category. In particular, the ratio test will always be inconclusive for series whose terms are rational functions. □

EXAMPLE 3 **Analyzing a series containing products similar to factorials**

Determine the convergence or divergence of the following series:

(a) $\frac{2}{4} + \frac{2\cdot 4}{4\cdot 7} + \frac{2\cdot 4\cdot 6}{4\cdot 7\cdot 10} + \cdots + \frac{2\cdot 4\cdot 6\cdots(2k)}{4\cdot 7\cdot 10\cdots(3k+1)} + \cdots$

(b) $\frac{1}{2} + \frac{1\cdot 3}{2\cdot 4} + \frac{1\cdot 3\cdot 5}{2\cdot 4\cdot 6} + \cdots + \frac{1\cdot 3\cdot 5\cdots(2k-1)}{2\cdot 4\cdot 6\cdots(2k)} + \cdots$

SOLUTION

The terms of each of these series consist of quotients of factorial-like products. The ratio test will be the first convergence test we try on both series, but we will see that it succeeds only on the first.

(a) For the first series, we let $a_k = \frac{2\cdot 4\cdot 6\cdots(2k)}{4\cdot 7\cdot 10\cdots(3k+1)}$. Then the numerator of a_{k+1} contains all of the same factors as the numerator of a_k, along with the factor $2(k+1) = 2k+2$. Similarly, the denominator contains all of the same factors, along with $3(k+1)+1 = 3k+4$. Thus, $a_{k+1} = \frac{2\cdot 4\cdot 6\cdots(2k)(2k+2)}{4\cdot 7\cdot 10\cdots(3k+1)(3k+4)}$ and

$$\begin{aligned}
\rho &= \lim_{k\to\infty} \frac{a_{k+1}}{a_k} \\
&= \lim_{k\to\infty} \frac{2\cdot 4\cdot 6\cdots(2k)(2k+2)/(4\cdot 7\cdot 10\cdots(3k+1)(3k+4))}{2\cdot 4\cdot 6\cdots(2k)/(4\cdot 7\cdot 10\cdots(3k+1))} && \leftarrow \text{the definition of } \rho \\
&= \lim_{k\to\infty} \frac{2\cdot 4\cdot 6\cdots(2k)(2k+2)}{4\cdot 7\cdot 10\cdots(3k+1)(3k+4)} \frac{4\cdot 7\cdot 10\cdots(3k+1)}{2\cdot 4\cdot 6\cdots(2k)} && \leftarrow \text{invert and multiply} \\
&= \lim_{k\to\infty} \frac{2k+2}{3k+4} = \frac{2}{3}. && \leftarrow \text{simplify the expression and evaluate the limit}
\end{aligned}$$

Since $\rho < 1$, this series converges. Be sure that you understand the algebra in the preceding steps.

(b) The algebra required to use the ratio test on our second series is quite similar, so we will evaluate ρ with less explanation. However, check that you understand the following work. We have

$$\rho = \lim_{k\to\infty} \frac{1 \cdot 3 \cdot 5 \cdots (2k-1)(2k+1)/(2 \cdot 4 \cdot 6 \cdots (2k) \cdot (2k+2))}{1 \cdot 3 \cdot 5 \cdots (2k-1)/(2 \cdot 4 \cdot 6 \cdots (2k))} = \lim_{k\to\infty} \frac{2k+1}{2k+2} = 1.$$

As we predicted, the ratio test is inconclusive.

We regroup the factors of the kth term of the series:

$$\frac{1 \cdot 3 \cdot 5 \cdots (2k-1)}{2 \cdot 4 \cdot 6 \cdots (2k)} = \frac{1}{2k} \cdot \frac{3}{2} \cdot \frac{5}{4} \cdots \frac{2k-1}{2k-2}.$$

Then we note that the quotients $\frac{3}{2}, \frac{5}{4}, \ldots, \frac{2k-1}{2k-2}$ are all greater than 1. We now have

$$\frac{1 \cdot 3 \cdot 5 \cdots (2k-1)}{2 \cdot 4 \cdot 6 \cdots (2k)} \geq \frac{1}{2k}.$$

Next we use the comparison test to show that our series diverges. Since each term of the series $\sum_{k=1}^{\infty} \frac{1}{2k}$ is one-half of the corresponding term of the (divergent) harmonic series, this series diverges. Note also that, because $\frac{1 \cdot 3 \cdot 5 \cdots (2k-1)}{2 \cdot 4 \cdot 6 \cdots (2k)} \geq \frac{1}{2k}$, each term of our series is greater than the corresponding term of the divergent series, so by the comparison test our series diverges as well. □

EXAMPLE 4

Using the root test

Use the root test, if appropriate, to determine the convergence or divergence of the following series:

$$\sum_{k=0}^{\infty} \frac{10^k}{k!} \qquad \sum_{k=1}^{\infty} \frac{3^k}{k^5} \qquad \sum_{k=2}^{\infty} \frac{2^{3k-1}}{(2k-3)!}$$

SOLUTION

These are the same three series we analyzed in Example 1. We have mentioned that the root test is particularly effective on series containing exponential factors. In general, it should not be used on a series containing factorials or products like factorials. Therefore, of the three given series, only $\sum_{k=1}^{\infty} \frac{3^k}{k^5}$ is a candidate for the root test. We have

$$\rho = \lim_{k\to\infty} (a_k)^{1/k} = \lim_{k\to\infty} \left(\frac{3^k}{k^5}\right)^{1/k} = \lim_{k\to\infty} \frac{(3^k)^{1/k}}{(k^5)^{1/k}}.$$

The numerator in the limit simplifies to 3, and the denominator is equal to $(k^{1/k})^5$. Recall that $k^{1/k} \to 1$ as $k \to \infty$. (You will show this in the *Thinking Back* exercises at the end of this section.) Therefore, when we take the limit in the preceding equation, we obtain $\rho = 3$. Since $\rho > 1$, the series diverges.

Compare the analysis and algebra we did for this series in Example 1 with the work we did here. When the ratio and root tests may both be used successfully to analyze a series, sometimes the work required for one of these tests is significantly simpler than the work required for the other. For the series we just worked with, how do you think the tests compare? □

EXAMPLE 5

Showing that the root test is inconclusive when $\rho = 1$

Show that the root test fails to determine the divergence of the harmonic series $\sum_{k=1}^{\infty} \frac{1}{k}$ or the convergence of the p-series $\sum_{k=1}^{\infty} \frac{1}{k^2}$.

SOLUTION

We consider the harmonic series first. For the harmonic series, $a_k = \frac{1}{k}$. We have

$$\lim_{k\to\infty} (a_k)^{1/k} = \lim_{k\to\infty} \frac{1}{k^{1/k}} = 1.$$

Therefore, the root test fails.

For the series $\sum_{k=1}^{\infty} \frac{1}{k^2}$,

$$\lim_{k\to\infty} (a_k)^{1/k} = \lim_{k\to\infty} \frac{1}{(k^2)^{1/k}} = \lim_{k\to\infty} \frac{1}{k^{2/k}}.$$

Again, the limit of the denominator is indeterminate and of the type ∞^0. We leave it to you to show that $\lim_{k\to\infty} \frac{1}{k^{2/k}} = 1$.

As we mentioned in Example 1, we already know that the harmonic series diverges and the p-series with $p = 2$ converges. These two series show that the root test cannot discern convergence from divergence when $\rho = \lim_{k\to\infty} (a_k)^{1/k} = 1$.

We point out again that the root test is most effective on series with exponential factors or series with other types of factors involving kth powers. Since p-series do not contain factors of these types, we should not expect the root test to work here. Again, as with the ratio test, the root test will always be inconclusive for series whose terms are rational functions. □

? TEST YOUR UNDERSTANDING

- What is the statement of the ratio test?
- Given a series $\sum_{k=1}^{\infty} a_k$, to use the ratio test we let $\rho = \lim_{k\to\infty} \frac{a_{k+1}}{a_k}$. If $\rho < 1$, the series converges and if $\rho > 1$, the series diverges. Why does the value 1 mark the boundary between convergence and divergence?
- What characteristics should you look for in a series before you try the ratio test?
- What is the statement of the root test?
- What characteristics should you look for in a series before you try the root test?

EXERCISES 7.6

Thinking Back

- *An indeterminate form of the type ∞^0:* Explain why $\lim_{k\to\infty} k^{1/k}$ has an indeterminate form of the type ∞^0. Then show that this limit equals 1.
- *Basic functions:* What is a monomial? What is a power function? What is a polynomial? What is a rational function?

Concepts

0. *Problem Zero:* Read the section and make your own summary of the material.

1. *True/False:* Determine whether each of the statements that follow is true or false. If a statement is true, explain why. If a statement is false, provide a counterexample.

(a) *True or False:* If a_k is a nonzero polynomial function, the series $\sum_{k=1}^{\infty} a_k$ will diverge.

(b) *True or False:* The ratio test will be inconclusive for a series $\sum_{k=1}^{\infty} a_k$ where a_k is a nonzero polynomial function of k.

(c) *True or False:* The ratio test will be inconclusive for a series $\sum_{k=1}^{\infty} a_k$ where a_k is a rational function of k.

(d) *True or False:* If the function a_k contains factorials, the ratio test will be effective in determining the convergence or divergence of the series $\sum_{k=1}^{\infty} a_k$.

(e) *True or False:* The proof of the ratio test uses the fact that a geometric series $\sum_{k=1}^{\infty} r^k$ converges when $|r| < 1$.

(f) *True or False:* The root test can be used successfully to determine the convergence or divergence of every p-series.

(g) *True or False:* The root test is only used to analyze the convergence of a series of the form $\sum_{k=1}^{\infty}(a_k)^k$.

(h) *True or False:* $\dfrac{(n+2)!}{n!} = 2$.

2. *Examples:* Construct examples of the thing(s) described in the following. Try to find examples that are different than any in the reading.

(a) A series containing factorials on which the ratio test will be effective in determining convergence or divergence.

(b) A series containing factorials on which the ratio test will be ineffective in determining convergence or divergence.

(c) A series on which the root test will be effective in determining convergence or divergence.

3. Let $p(x)$ be a nonzero polynomial function. Evaluate $\lim\limits_{x\to\infty} \dfrac{p(x+1)}{p(x)}$.

4. Use Exercise 3 to explain why the ratio test will be inconclusive for every series $\sum_{k=1}^{\infty} a_k$ in which a_k is a polynomial.

5. Let $r(x)$ be a nonzero rational function. Evaluate $\lim\limits_{x\to\infty} \dfrac{r(x+1)}{r(x)}$.

6. Use Exercise 5 to explain why the ratio test will be inconclusive for every series $\sum_{k=1}^{\infty} a_k$ in which a_k is a rational function of k.

7. Explain how you could adapt the ratio test to analyze a series $\sum_{k=1}^{\infty} a_k$ in which the terms of the series are all negative.

8. Provide a more general statement of the ratio test in which the terms of the series are *eventually* all positive. Explain why your statement is valid.

9. Let $\sum_{k=1}^{\infty} a_k$ be a series in which all the terms are positive. If $\lim\limits_{k\to\infty} \dfrac{a_{k+1}}{a_k} > 1$, explain why both the ratio test and the divergence test could be used to show that the series diverges.

10. Explain why the ratio test does not work in determining the convergence or divergence of the series $\sum_{k=1}^{\infty} \dfrac{k!}{(k+2)!}$. What test would be more effective to analyze this series?

11. Let $m(x) = Ax^r$ be a power function. Evaluate $\lim\limits_{x\to\infty} m(x)^{1/x}$.

12. Use Exercise 11 to explain why the root test will be inconclusive for every series $\sum_{k=1}^{\infty} a_k$ in which a_k is a power function.

13. Explain how you could adapt the root test to analyze a series $\sum_{k=1}^{\infty} a_k$ in which the terms of the series are all negative.

14. Provide a more general statement of the root test in which the terms of the series are *eventually* all positive. Explain why your statement is valid.

15. Explain why it would be difficult to use the root test on the series $\sum_{k=1}^{\infty} \dfrac{1}{k!}$.

16. Explain why the series $\sum_{k=1}^{\infty} \dfrac{1}{k^3}$ converges. Which convergence tests could be used to prove this?

17. Explain why the series $\sum_{k=0}^{\infty} \dfrac{1}{3^k}$ converges. Which convergence tests could be used to prove this?

18. Explain why the series $\sum_{k=0}^{\infty} \dfrac{1}{k!}$ converges. Which convergence tests could be used to prove this?

19. Explain why the ratio test cannot be used on the series $1 + \frac{1}{5} + \frac{1}{10} + \frac{1}{50} + \frac{1}{100} + \frac{1}{500} + \cdots$. Then show that the series converges and find its sum.

20. Let α and β be two distinct positive numbers less than 1. Explain why the ratio test cannot be used on the series $\alpha + \beta + \alpha^2 + \beta^2 + \alpha^3 + \beta^3 + \cdots$. Then show that the series converges and find its sum.

Skills

Simplify the quotients in Exercises 21–28 without using a calculator.

21. $\dfrac{8!}{10!}$

22. $\dfrac{700!}{699!}$

23. $\dfrac{k!}{(k+2)!}$

24. $\dfrac{(2n-2)!}{(2n)!}$

25. $\dfrac{4!}{(2!)!}$

26. $\dfrac{6!}{(3!)!}$

27. $\dfrac{(n+1)!}{(n-2)!}$

28. $\dfrac{1\cdot 3\cdot 5\cdots(2k-1)}{1\cdot 3\cdot 5\cdots(2k+1)}$

In Exercises 29–34 use the ratio test to analyze whether the given series converges or diverges. If the ratio test is inconclusive, use a different test to analyze the series.

29. $\displaystyle\sum_{k=0}^{\infty} \frac{5^k}{k!}$

30. $\displaystyle\sum_{k=1}^{\infty} \frac{5^k}{k^5}$

31. $\displaystyle\sum_{k=0}^{\infty} \frac{1}{k^3+1}$

32. $\displaystyle\sum_{k=1}^{\infty} \frac{k!}{(2k)!}$

33. $\displaystyle\sum_{k=1}^{\infty} \frac{2\cdot 4\cdot 6\cdots(2k)}{1\cdot 3\cdot 5\cdots(2k-1)}$

34. $\displaystyle\sum_{k=1}^{\infty} \frac{1\cdot 4\cdot 7\cdots(3k-2)}{2\cdot 4\cdot 6\cdots(2k)}$

In Exercises 35–40 use the root test to analyze whether the given series converges or diverges. If the root test is inconclusive, use a different test to analyze the series.

35. $\sum_{k=1}^{\infty} \frac{5^k}{k^5}$

36. $\sum_{k=1}^{\infty} \frac{5^k}{k^2 4^{k+1}}$

37. $\sum_{k=0}^{\infty} \frac{1}{k^3+1}$

38. $\sum_{k=2}^{\infty} \left(\frac{1}{\ln k}\right)^k$

39. $\sum_{k=1}^{\infty} \left(\frac{1+4k}{5k-1}\right)^k$

40. $\sum_{k=1}^{\infty} \frac{k^k}{k!}$

Use any convergence test from Sections 7.4–7.6 to determine whether the series in Exercises 41–59 converge or diverge. Explain why each series that meets the hypotheses of the test you select does so.

41. $\sum_{k=1}^{\infty} \left(\frac{k+1}{3k+1}\right)^k$

42. $\sum_{k=1}^{\infty} \frac{3}{k(k+2)}$

43. $\sum_{k=1}^{\infty} \frac{4}{k+3}$

44. $\sum_{k=1}^{\infty} \frac{k^3}{k^2+7}$

45. $\sum_{k=0}^{\infty} \frac{k!}{(k+3)!}$

46. $\sum_{k=1}^{\infty} \frac{k^3}{2^k}$

47. $\sum_{k=1}^{\infty} \frac{\sqrt{k}}{k^2+3}$

48. $\sum_{k=1}^{\infty} \left(\frac{1}{k^2}\right)^{\sqrt{k}}$

49. $\sum_{k=1}^{\infty} \frac{3^k}{k^k}$

50. $\sum_{k=0}^{\infty} \frac{2^{k!}}{k!}$

51. $\sum_{k=1}^{\infty} \frac{(2k)!}{(k!)!}$

52. $\sum_{k=0}^{\infty} \frac{k!}{3^{k^2}}$

53. $\sum_{k=1}^{\infty} \frac{2 \cdot 5 \cdot 8 \cdots (3k-1)}{3 \cdot 5 \cdot 7 \cdots (2k+1)}$

54. $\sum_{k=0}^{\infty} \frac{2 \cdot 5 \cdot 8 \cdots (3k+2)}{1 \cdot 5 \cdot 9 \cdots (4k+1)}$

55. $\sum_{k=1}^{\infty} \left(\frac{k}{k+1}\right)^{k^2}$

56. $\sum_{k=1}^{\infty} \frac{e^{1/k}}{k^2}$

57. $\sum_{k=1}^{\infty} \frac{k!}{1 \cdot 3 \cdot 5 \cdot 7 \cdots (2k-1)}$

58. $\sum_{k=0}^{\infty} \frac{1 \cdot 4 \cdot 7 \cdots (3k+1)}{100^k}$

59. $\sum_{k=1}^{\infty} \frac{1 \cdot 4 \cdot 7 \cdots (3k-2)}{3 \cdot 6 \cdot 9 \cdots (3k)}$ (*Hint: See Example 3.*)

60. (a) Show that the series $\sum_{k=1}^{\infty} \frac{\ln k}{k^2}$ converges.

(b) Use the result from part (a) to show that $\sum_{k=1}^{\infty} \frac{\ln k}{k^n}$ converges for every integer $n > 2$.

61. Let n be a positive integer and let $r > 1$.

(a) Show that the series $\sum_{k=1}^{\infty} \frac{k^n}{r^k}$ converges.

(b) Explain why part (a) proves that $\lim_{k\to\infty} \frac{k^n}{r^k} = 0$.

(c) Explain why part (b) proves that exponential growth dominates polynomial growth.

62. Let $r > 1$.

(a) Show that the series $\sum_{k=1}^{\infty} \frac{r^k}{k!}$ converges.

(b) Explain why part (a) proves that $\lim_{k\to\infty} \frac{r^k}{k!} = 0$.

(c) Explain why part (b) proves that factorial growth dominates exponential growth.

63. (a) Show that the series $\sum_{k=1}^{\infty} \frac{k!}{k^k}$ converges.

(b) Explain why part (a) proves that $\lim_{k\to\infty} \frac{k!}{k^k} = 0$.

(c) Explain why part (b) proves that the function k^k dominates factorial growth.

Applications

In Exercises 64 and 65 we use the product notation $\prod_{j=1}^{n} r_j$. (The symbol Π is a capital π.) This notation is analogous to the summation notation $\sum_{k=1}^{n} a_k$, but means the product of the entries, rather than their sum. That is,

$$\prod_{j=1}^{n} r_j = r_1 \cdot r_2 \cdot r_3 \cdots r_n.$$

64. Leila needs to determine the number of trout to stock a lake with annually. She has evidence that between 65% and 95% of the trout persist in the lake from year to year. The actual percentage varies with the amount of fishing, as well as the weather, the number of insects available, and the fraction of the trout that are of a reproductive age. Denote the fraction of fish that persist from one year to the next by $r_k \in [0.65, 0.95]$. For simplicity, Leila assumes that $\lim_{k\to\infty} r_k = 0.80$. The number of fish she stocks annually is denoted s, with the steady number of fish in the lake given by

$$f_\infty = s\left(1 + \sum_{i=0}^{\infty} \left(\prod_{j=0}^{i} r_j\right)\right).$$

(a) Use the ratio test to show that this series converges.

(b) Can Leila use the given assumptions to find f_∞? Explain.

65. In Exercise 64, Leila actually needs to estimate the steady number of fish f_∞ that will be in the lake for a given annual stocking rate s. She supposes that $s = 420$ and that in the first 10 years of stocking the rates of persistence r_i are all equal to the mean: $r_i = 0.80$ for $i = 1, 2, \ldots, 10$. Note that she assumes that after the first 10 years the values of r_k can again vary between 0.65 and 0.95.

(a) Neglecting f_0, find $f_{10} = s\sum_{i=0}^{10} \prod_{j=1}^{i} r_j$ for the given values of s and r_i.

(b) Use Theorem 7.31 to find an upper bound for the error in this estimate for f_∞.

Proofs

66. Use the principle of mathematical induction to prove that if $a_{k+1} < a_k r$ for every $k \geq N$, then $a_{N+n} < a_N r^n$. Proving this implication completes our proof of the ratio test.

67. Prove that the ratio test will be inconclusive on every series of the form $\sum_{k=1}^{\infty} a_k$ where a_k is a rational function of k.

68. Prove the root test. You may model your proof on the proof of the ratio test.

Exercises 69–72 build to a proof that the root test is inconclusive on every series $\sum_{k=1}^{\infty} a_k$ where a_k is a rational function of k.

69. Let $a_k = ck^m$, where $c > 0$ and $m \in \mathbb{Z}$. Prove that $\lim_{k\to\infty} (ck^m)^{1/k} = 1$. Explain why this shows that the root test is inconclusive for every series of the form $\sum_{k=1}^{\infty} ck^m$, including every p-series.

70. Let c_1 be a real number, c_2 be a positive real number, and $m_1 \leq m_2$ be two nonnegative integers.

(a) Explain why, for large values of k, we have

$$0 < c_1 k^{m_1} + c_2 k^{m_2} \leq (c_1 + c_2)\, k^{m_2}.$$

(b) Use part (a), the result of Exercise 69, and the Squeeze Theorem to show that $\lim_{k\to\infty} (c_1 k^{m_1} + c_2 k^{m_2})^{1/k} = 1$. Explain why this shows that the root test is inconclusive for every series of the form $\sum_{k=1}^{\infty} (c_1 k^{m_1} + c_2 k^{m_2})$.

71. (a) Use the result of Exercise 70 and mathematical induction to show that if $c_m > 0$ and each m_j is a nonnegative integer with $m_1 \leq m_2 \leq \cdots \leq m_n$, then $\lim_{k\to\infty} (c_1 k^{m_1} + c_2 k^{m_2} + \cdots + c_n k^{m_n})^{1/k} = 1$.

(b) Explain why part (a) shows that the root test is inconclusive for every series of the form $\sum_{k=1}^{\infty} (c_1 k^{m_1} + c_2 k^{m_2} + \cdots + c_n k^{m_n})$.

72. Use the result of Exercise 71 to show that the root test is inconclusive for every series of the form $\sum_{k=1}^{\infty} a_k$ where a_k is a rational function of k.

Thinking Forward

▶ *A series of monomials:* Find all values of x for which the series $\sum_{k=1}^{\infty} \frac{x^{2k}}{k!}$ converges.

▶ *A series of monomials:* Find all values of x for which the series $\sum_{k=1}^{\infty} \left(\frac{x}{k}\right)^k$ converges.

7.7 ALTERNATING SERIES

- ▶ The alternating series test and approximating convergent alternating series
- ▶ The difference between absolute and conditional convergence
- ▶ A summary of convergence tests for series

Alternating Series

Unlike the divergence test, all of the convergence tests we discussed in Sections 7.4, 7.5 and 7.6 have required our series to have nonnegative terms. In this section we relax that requirement and discuss how to handle certain series in which the signs are arbitrary. We begin, however, with a discussion of series in which the signs alternate.

DEFINITION 7.36 Alternating Series

Let $\{a_k\}$ be a sequence of positive numbers. A series of the form

$$\sum_{k=1}^{\infty} (-1)^{k+1} a_k \qquad \text{or} \qquad \sum_{k=1}^{\infty} (-1)^k a_k$$

is called an ***alternating series***.

So, for an alternating series with each $a_k > 0$, we have one of two possible forms:

$$a_1 - a_2 + a_3 - a_4 + \cdots + (-1)^{k+1} a_k + \cdots$$

$$-a_1 + a_2 - a_3 + a_4 - \cdots + (-1)^k a_k + \cdots$$

We have already studied one type of alternating series. A geometric series $\sum_{k=0}^{\infty} cr^k$ is an alternating series exactly when $r < 0$. For example, we have the alternating geometric series

$$\sum_{k=0}^{\infty} 3\left(-\frac{1}{2}\right)^k = 3 - \frac{3}{2} + \frac{3}{4} - \frac{3}{8} + \cdots$$

An alternating series that we will see quite often is the ***alternating harmonic series***

$$\sum_{k=1}^{\infty} \frac{(-1)^{k+1}}{k} = 1 - \frac{1}{2} + \frac{1}{3} - \frac{1}{4} + \cdots.$$

THEOREM 7.37

The Alternating Series Test

Let $\{a_k\}$ be a sequence of positive numbers.

$$\text{If } a_{k+1} < a_k \text{ for every } k \geq 1 \quad \text{and} \quad \lim_{k\to\infty} a_k = 0,$$

then the alternating series

$$\sum_{k=1}^{\infty} (-1)^{k+1} a_k \qquad \text{and} \qquad \sum_{k=1}^{\infty} (-1)^k a_k$$

both converge.

Note that an arbitrary series that meets the two conditions (7.37) and (7.37) may or may not converge. For example, both the harmonic series and the p-series with $p = 2$ satisfy those two conditions, but the former series diverges and the latter series converges. The two conditions ensure convergence only when the series is alternating. For example, the alternating harmonic series converges because it meets the two conditions of Theorem 7.37 *and* it is an alternating series.

The figure shown next illustrates the main ideas of the proof of the alternating series test with the graph of the first 10 terms of the sequence of partial sums $\{S_n\}$ for the alternating harmonic series. In the graph we see that the subsequence $\{S_{2n-1}\}$ is strictly decreasing and is bounded below. Similarly, the subsequence $\{S_{2n}\}$ is strictly increasing and is bounded above. Since the terms of the series $a_k \to 0$, these two subsequences converge to the same value: the sum of the series, L. The graph of the sequence of partial sums for every alternating series of the form $\sum_{k=1}^{\infty} (-1)^{k+1} a_k$ satisfying the conditions of the alternating series test will be similar.

The first 10 *terms of the sequence of partial sums* $\{S_n\}$ *for the alternating harmonic series*

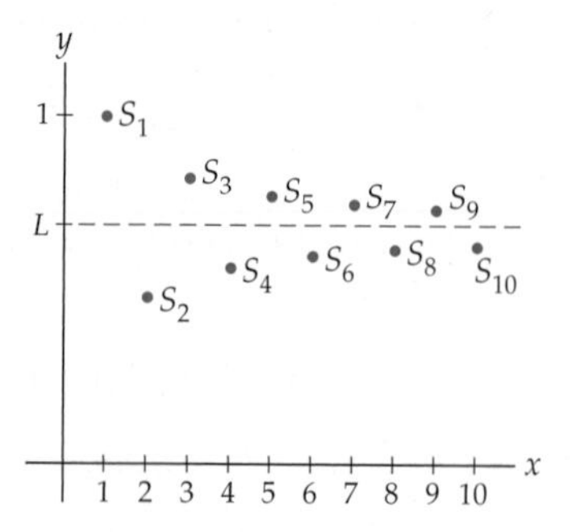

Proof. Let $\{a_k\}$ be a strictly decreasing sequence of positive numbers such that $a_k \to 0$ as $k \to \infty$. We will show that the alternating series $\sum_{k=1}^{\infty}(-1)^{k+1}a_k$ converges and leave the convergence of the alternating series $\sum_{k=1}^{\infty}(-1)^k a_k$ for Exercise 66. Consider the sequence of partial sums $\{S_n\}$ given by

$$\begin{aligned} S_1 &= a_1 \\ S_2 &= a_1 - a_2 \\ S_3 &= a_1 - a_2 + a_3 \\ S_4 &= a_1 - a_2 + a_3 - a_4 \\ S_5 &= a_1 - a_2 + a_3 - a_4 + a_5 \\ &\vdots \qquad\qquad \vdots \end{aligned}$$

We consider the subsequence $\{S_{2n-1}\}$ of $\{S_n\}$. That is,

$$\{S_{2n-1}\} = S_1,\ S_3,\ S_5,\ S_7, \ldots$$

We will show that $\{S_{2n-1}\}$ is bounded and strictly decreasing, and is, therefore, convergent. Consider the terms S_{2n-1} and S_{2n+1}. We have

$$\begin{aligned} S_{2n-1} &= a_1 - a_2 + a_3 - \cdots - a_{2n-2} + a_{2n-1} \\ S_{2n+1} &= a_1 - a_2 + a_3 - \cdots - a_{2n-2} + a_{2n-1} - a_{2n} + a_{2n+1} \end{aligned}$$

We will group the summands of these terms in two different ways to prove our assertions, but first we note that since the sequence $\{a_k\}$ is strictly decreasing, each difference

$$a_k - a_{k+1} > 0.$$

Now,

$$\begin{aligned} S_{2n-1} &= a_1 - (a_2 - a_3) - \cdots - (a_{2n-2} - a_{2n-1}) \\ S_{2n+1} &= a_1 - (a_2 - a_3) - \cdots - (a_{2n-2} - a_{2n-1}) - (a_{2n} - a_{2n+1}). \end{aligned}$$

Each of the paired differences in these sums is positive, and since

$$S_{2n+1} = a_1 - (a_2 - a_3) - \cdots - (a_{2n-2} - a_{2n-1}) - (a_{2n} - a_{2n+1}) = S_{2n-1} - (a_{2n} - a_{2n+1}),$$

we have $S_{2n-1} > S_{2n+1}$. Therefore, the sequence $\{S_{2n-1}\}$ is strictly decreasing.

Every decreasing sequence is bounded above by its first term. We now show that the sequence $\{S_{2n-1}\}$ is bounded below by zero. Consider the following grouping of the general term S_{2n-1}:

$$S_{2n-1} = (a_1 - a_2) + (a_3 - a_4) + \cdots + (a_{2n-3} - a_{2n-2}) + a_{2n-1}.$$

Again, each of the paired differences in the formula is positive, as is a_{2n-1}. Thus S_{2n-1}, the sum of positive numbers, is positive. Therefore, the sequence is bounded below by zero. We now know that the sequence $\{S_{2n-1}\}$ is bounded and strictly decreasing. Therefore, it converges to some value. Call this value L.

We will show that the subsequence $\{S_{2n}\}$ also converges to L. Consider

$$S_{2n} = a_1 - a_2 + a_3 - \cdots - a_{2n-2} + a_{2n-1} - a_{2n} = S_{2n-1} + a_{2n}.$$

The limit

$$\lim_{n\to\infty} S_{2n} = \lim_{n\to\infty}(S_{2n-1} + a_{2n}) = L + 0 = L,$$

since $S_{2n-1} \to L$ and $a_n \to 0$ as $n \to \infty$. Because the subsequences $\{S_{2n-1}\}$ and $\{S_{2n}\}$ both converge to L, the entire sequence of partial sums $\{S_n\}$ converges to L. Therefore the series $\sum_{k=1}^{\infty}(-1)^{k+1}a_k$ converges. ∎

In our proof of the alternating series test, we could have shown that the subsequence $\{S_{2n}\}$ is strictly increasing and bounded above. However, we did not need those two facts in the proof. You will prove these facts in Exercise 67.

Given any convergent series $\sum_{k=1}^{\infty} b_k$, the terms of the sequence of partial sums, $\{S_n\}$, may always be used to approximate the sum L of the series. Given an arbitrary convergent series, however, we cannot always be sure how good an approximation a term S_n in the sequence of partial sums is for L. The next theorem tells us that, in the sequence of partial sums for an alternating series satisfying the conditions of the alternating series test, the term S_n is within a_{n+1} of L.

THEOREM 7.38 **Approximating the Remainder for a Convergent Alternating Series**

Let L be the sum of an alternating series satisfying the hypotheses of the alternating series test. For any term S_n in the sequence of partial sums,

$$|L - S_n| < a_{n+1}.$$

Furthermore, the sign of the difference $L - S_n$ is the sign of the coefficient of the term a_{n+1}.

Proof. We consider an alternating series of the form $\sum_{k=1}^{\infty} (-1)^{k+1} a_k$ and leave the case of an alternating series of the form $\sum_{k=1}^{\infty} (-1)^k a_k$ for Exercise 68.

In our proof of the alternating series test we showed that the subsequence $\{S_{2n-1}\}$ of $\{S_n\}$ was strictly decreasing and converged to a limit L. Similarly, one may show that the subsequence $\{S_{2n}\}$ of $\{S_n\}$ is strictly increasing. In the proof, we did show that the subsequence $\{S_{2n}\}$ converges to L. Thus, for every n, either

$$S_n < L < S_{n+1} \qquad \text{or} \qquad S_{n+1} < L < S_n,$$

depending upon whether n is even or odd, respectively. In either case,

$$S_{n+1} - S_n = (-1)^{n+2} a_{n+1}.$$

If n is even, subtracting S_n throughout $S_n < L < S_{n+1}$, we obtain

$$0 < L - S_n < (-1)^{n+2} a_{n+1} = a_{n+1}.$$

If n is odd, when we subtract S_n throughout $S_{n+1} < L < S_n$, we obtain

$$(-1)^{n+2} a_{n+1} < L - S_n < 0.$$

Note that since n is odd, we have $(-1)^{n+2} a_{n+1} = -a_{n+1}$. If we multiply through by -1 in $(-1)^{n+2} a_{n+1} < L - S_n < 0$, we obtain

$$0 < S_n - L < a_{n+1}.$$

Considering the two inequalities $0 < L - S_n < a_{n+1}$ and $0 < S_n - L < a_{n+1}$ together, whether n is even or odd, we have the desired inequality

$$|L - S_n| < a_{n+1}.$$

Moreover, we note that the sign of the difference $L - S_n$ is the sign of the coefficient of the term a_{n+1}. ■

We have already mentioned that the alternating harmonic series $\sum_{k=1}^{\infty} \frac{(-1)^{k+1}}{k}$ satisfies the conditions of the alternating series test. If we add the first 10 terms of the series to obtain the 10th partial sum, S_{10}, we obtain

$$S_{10} = 1 - \frac{1}{2} + \frac{1}{3} - \frac{1}{4} + \cdots - \frac{1}{10} \approx 0.646.$$

This is an approximation for the sum L of the series. By Theorem 7.38, the error in using S_{10} is at most $\frac{1}{11} \approx 0.091$. That is, $|L - 0.646| < 0.091$. Furthermore, the approximation $S_{10} \approx 0.646$ is less than L, since the coefficient of a_{11} is positive. Therefore, we are ensured that the actual value of the sum L lies in the interval $[S_{10}, S_{10} + a_{11}]$, or, with decimal approximations,

$$0.646 < L < 0.737.$$

(In Chapter 8 we will prove that the sum of the alternating harmonic series is $\ln 2 \approx 0.693$. We see that $\ln 2$ is indeed in the interval we computed.) We also note that although the alternating harmonic series converges, it converges quite slowly, in that it takes many terms of the series to obtain a good approximation for the sum.

Absolute and Conditional Convergence

Consider the two alternating p-series

$$\sum_{k=1}^{\infty} \frac{(-1)^{k+1}}{k} \quad \text{and} \quad \sum_{k=1}^{\infty} \frac{(-1)^{k+1}}{k^2}.$$

We have already seen that the alternating series test may be used to prove that the alternating harmonic series converges. You should verify that the second alternating series shown also satisfies the conditions of the alternating series test and, therefore, converges. However, these two series are different in that, if you sum the absolute values of the terms, rather than the terms themselves, one of the resulting series diverges and one converges. That is,

$$\sum_{k=1}^{\infty} \left| \frac{(-1)^{k+1}}{k} \right| = \sum_{k=1}^{\infty} \frac{1}{k} \quad \text{and} \quad \sum_{k=1}^{\infty} \left| \frac{(-1)^{k+1}}{k^2} \right| = \sum_{k=1}^{\infty} \frac{1}{k^2},$$

and we have already determined that the harmonic series diverges and the p-series with $p = 2$ converges. This distinction is the basis for our definitions of absolute convergence and conditional convergence.

DEFINITION 7.39 Absolute Convergence

The series $\sum_{k=1}^{\infty} b_k$ is said to ***converge absolutely***, or be ***absolutely convergent***, if the series $\sum_{k=1}^{\infty} |b_k|$ converges.

We see that the series $\sum_{k=1}^{\infty} \frac{(-1)^{k+1}}{k^2}$ converges absolutely, since $\sum_{k=1}^{\infty} \left| \frac{(-1)^{k+1}}{k^2} \right| = \sum_{k=1}^{\infty} \frac{1}{k^2}$ converges.

DEFINITION 7.40 Conditional Convergence

A convergent series $\sum_{k=1}^{\infty} b_k$ is said to ***converge conditionally***, or be ***conditionally convergent***, if the series $\sum_{k=1}^{\infty} |b_k|$ diverges.

The alternating harmonic series is conditionally convergent because it converges, but the corresponding series of absolute values, $\sum_{k=1}^{\infty} \left| \frac{(-1)^{k+1}}{k} \right| = \sum_{k=1}^{\infty} \frac{1}{k}$, diverges.

The alternating series test was the first convergence test we studied that can directly handle series with some negative terms. However, as its name suggests, the alternating series test may be applied only to alternating series. The next theorem allows us to examine series in which the signs of the terms are less rigidly specified.

THEOREM 7.41

Absolute Convergence Implies Convergence

If the series $\sum_{k=1}^{\infty} b_k$ converges absolutely, then it converges.

The content of Theorem 7.41 may seem slightly odd to you, but recall that the definitions of *convergence* and *absolute convergence* are distinct, so there is something to prove here.

Proof. Since the series $\sum_{k=1}^{\infty} b_k$ converges absolutely, the series $\sum_{k=1}^{\infty} |b_k|$ converges. Consider the sum $|b_k| + b_k$, and note that for every $k \in \mathbb{Z}^+$,

$$0 \leq |b_k| + b_k \leq 2\,|b_k|\,.$$

Now, the series $\sum_{k=1}^{\infty} 2\,|b_k|$ converges, since $\sum_{k=1}^{\infty} 2\,|b_k| = 2\sum_{k=1}^{\infty} |b_k|$. Thus, by the comparison test, the series

$$\sum_{k=1}^{\infty} (|b_k| + b_k)$$

converges. Finally, using both parts of Theorem 7.24, the difference of two convergent series is convergent, and it follows that

$$\sum_{k=1}^{\infty} (|b_k| + b_k) - \sum_{k=1}^{\infty} |b_k| = \sum_{k=1}^{\infty} ((|b_k| + b_k) - |b_k|) = \sum_{k=1}^{\infty} b_k.$$

We have thus shown that the series $\sum_{k=1}^{\infty} b_k$ converges. ■

As we mentioned, Theorem 7.41 allows us to analyze the convergence of many (not necessarily alternating) series in which there are both positive and negative terms. For example, consider the series

$$1 - \frac{1}{4} + \frac{1}{9} - \frac{1}{16} - \frac{1}{25} + \frac{1}{36} + \frac{1}{49} - \frac{1}{64} - \frac{1}{81} + \cdots$$

The series of absolute values is

$$1 + \frac{1}{4} + \frac{1}{9} + \frac{1}{16} + \frac{1}{25} + \frac{1}{36} + \frac{1}{49} + \frac{1}{64} + \frac{1}{81} + \cdots = \sum_{k=1}^{\infty} \frac{1}{k^2}.$$

This is a convergent p-series. Therefore, the given series converges absolutely, and by Theorem 7.41 it also converges.

The Curious Behavior of a Conditionally Convergent Series

By this point in your study of mathematics you should be very familiar with the commutative and associative laws of addition. That is, for real numbers a, b, and c,

$$a + b = b + a \quad \text{and} \quad (a + b) + c = a + (b + c).$$

We also know that these properties may be combined to show that a finite sum of real numbers may be computed in any order we wish, without affecting the result. However, as we will see in a moment, the same may not be true when we rearrange the terms of an infinite sum. In particular, we will see that if we rearrange the terms of the alternating harmonic series, we can make it converge to a different value. Admittedly, we do have to rearrange infinitely many of the terms for the sum to change, but the sum does change.

Earlier in the section we showed that the alternating harmonic series

$$\sum_{k=1}^{\infty} \frac{(-1)^{k+1}}{k} = 1 - \frac{1}{2} + \frac{1}{3} - \frac{1}{4} + \frac{1}{5} - \frac{1}{6} + \cdots$$

converges to a positive number L in the interval $(0.646, 0.737)$ when the terms are in the order shown. For the argument we present next, we just need the fact that $L > 0$. Rearrange the terms of the alternating harmonic series as follows: Start with the first positive term, and then subtract the next two negative terms, add the next positive term and subtract the next two negative terms, etc. That is,

$$1 - \frac{1}{2} - \frac{1}{4} + \frac{1}{3} - \frac{1}{6} - \frac{1}{8} + \frac{1}{5} - \frac{1}{10} - \frac{1}{12} + \cdots$$

Take a moment to convince yourself that every term from the alternating harmonic series appears exactly once in our new sum, with the same sign it had in the original arrangement.

We now group the terms from the new arrangement as follows:

$$\begin{aligned}
&\left(1 - \frac{1}{2}\right) - \frac{1}{4} + \left(\frac{1}{3} - \frac{1}{6}\right) - \frac{1}{8} + \left(\frac{1}{5} - \frac{1}{10}\right) - \frac{1}{12} + \cdots \\
&= \frac{1}{2} - \frac{1}{4} + \frac{1}{6} - \frac{1}{8} + \frac{1}{10} - \frac{1}{12} + \cdots \\
&= \frac{1}{2}\left(1 - \frac{1}{2} + \frac{1}{3} - \frac{1}{4} + \frac{1}{5} - \frac{1}{6} + \cdots\right) \\
&= \frac{1}{2} \sum_{k=1}^{\infty} \frac{(-1)^{k+1}}{k} = \frac{1}{2} L
\end{aligned}$$

With this arrangement, the sum is $\frac{1}{2}L$, and since $L > 0$, the sum cannot equal L. This argument shows that the order of the summation is significant for the alternating harmonic series. Similarly, if we start with any conditionally convergent series, changing the order of (infinitely many) terms can result in a different sum. Perhaps more strangely, given a conditionally convergent series, we can find a rearrangement that will sum to any value we choose! For more on this, see Exercises 71–74.

The Ratio Test for Absolute Convergence

Now that we have studied absolute and conditional convergence, we should think about convergence tests that will tell us when a series converges absolutely or converges conditionally. Each of the convergence tests we have studied may be adapted to analyze a series for absolute convergence. When presented with a series $\sum_{k=1}^{\infty} b_k$ containing both positive and negative terms, we should always look at the series of absolute values, $\sum_{k=1}^{\infty} |b_k|$, first. If this series converges, then $\sum_{k=1}^{\infty} b_k$ converges absolutely. By Theorem 7.41, the series $\sum_{k=1}^{\infty} b_k$ also converges. Thus, we can often use a single convergence test to determine that a series converges and converges absolutely.

To determine that a series converges conditionally we *always* need to use two tests: one to show that the series $\sum_{k=1}^{\infty} b_k$ converges and a second to show that the series of absolute values, $\sum_{k=1}^{\infty} |b_k|$, diverges. A single test can never suffice to determine conditional convergence.

Although the convergence tests we have studied may be adapted to analyze a series for absolute convergence, the variant of the ratio test we studied in Section 7.6 warrants special mention. Recall that, to use the ratio test, we must have a series of the form $\sum_{k=1}^{\infty} a_k$ with each $a_k > 0$. To use the test, we examine $\lim_{k\to\infty} \frac{a_{k+1}}{a_k}$. In the ratio test for absolute convergence, we need a series with nonzero terms, $\sum_{k=1}^{\infty} b_k$, and instead examine $\lim_{k\to\infty} \left|\frac{b_{k+1}}{b_k}\right|$. We mention the ratio test for absolute convergence now because we will use it frequently in Chapter 8. The proof of the ratio test for absolute convergence is almost identical to the proof we saw in Section 7.6 and is left for Exercise 70.

THEOREM 7.42 **The Ratio Test for Absolute Convergence**

Let $\sum_{k=1}^{\infty} b_k$ be a series with nonzero terms, and assume $\rho = \lim_{k\to\infty} \left|\frac{b_{k+1}}{b_k}\right|$ exists.

(a) If $\rho < 1$, the series converges absolutely.

(b) If $\rho > 1$, the series diverges.

(c) If $\rho = 1$, the test is inconclusive. (Use a different test!)

The ratio test for absolute convergence works very well on many series containing exponential factors, factorials, and/or factorial-like products. The series $\sum_{k=0}^{\infty} \frac{(-3)^k}{(2k)!}$ is a good candidate. Here, $b_k = \frac{(-3)^k}{(2k)!}$ and $b_{k+1} = \frac{(-3)^{k+1}}{(2(k+1))!}$. Thus,

$$\lim_{k\to\infty} \left|\frac{b_{k+1}}{b_k}\right| = \lim_{k\to\infty} \left|\frac{(-3)^{k+1}/(2k+2)!}{(-3)^k/(2k)!}\right| = \lim_{k\to\infty} \left|\frac{(-3)^{k+1}}{(2k+2)!} \cdot \frac{(2k)!}{(-3)^k}\right|$$

$$= \lim_{k\to\infty} \frac{3}{(2k+2)(2k+1)} = 0.$$

Therefore, this series converges absolutely.

A Summary of Convergence Tests

The following table provides a compact summary of the statements of the convergence tests we've studied in Sections 7.4–7.7, along with a few notes about them:

Test	Theorem	Comments
Divergence Test *Section 7.4*	Given the series $\sum_{k=1}^{\infty} a_k$, if $\lim_{n\to\infty} a_k \neq 0$, the series diverges.	Can be used only to determine that a series diverges. Cannot be used to determine convergence.
Integral Test *Section 7.4*	If $f : [1, \infty) \to \mathbb{R}$ is a continuous, positive, and decreasing function, then the improper integral $\int_1^{\infty} f(x)\,dx$ and the series $\sum_{k=1}^{\infty} f(k)$ either both converge or both diverge.	The function f must be relatively easy to integrate. This test is not appropriate for series containing factorials or factors like factorials.
Comparison Test *Section 7.5*	Let $\sum_{k=1}^{\infty} a_k$ and $\sum_{k=1}^{\infty} b_k$ be two series with nonnegative terms such that $0 \le a_k \le b_k$ for every positive integer k. If the series $\sum_{k=1}^{\infty} b_k$ converges, then the series $\sum_{k=1}^{\infty} a_k$ converges.	Given a series to analyze, both comparison tests require you to find a related series different enough to be easy to analyze but similar enough to provide a meaningful comparison.
Limit Comparison Test *Section 7.5*	Let $\sum_{k=1}^{\infty} a_k$ and $\sum_{k=1}^{\infty} b_k$ be two series with positive terms. If $\lim_{k\to\infty} \frac{a_k}{b_k} = L$, where L is any positive real number, then either the series both converge or both diverge.	It is often easier to evaluate $\lim_{k\to\infty} \frac{a_k}{b_k}$, rather than ensure the required inequalities of the comparison test. There is an extension of the limit comparison test. See Section 7.4 for details.

Test	Theorem	Comments
Ratio Test *Section 7.6*	Let $\sum_{k=1}^{\infty} a_k$ be a series with nonnegative terms, and let $\rho = \lim_{k\to\infty} \frac{a_{k+1}}{a_k}$. If $\rho < 1$, the series converges. If $\rho > 1$, the series diverges. If $\rho = 1$, the test is inconclusive.	Works well on series containing exponential factors, factorials, and/or factors like factorials.
Root Test *Section 7.6*	Let $\sum_{k=1}^{\infty} a_k$ be a series with positive terms, and let $\rho = \lim_{k\to\infty} \sqrt[k]{a_k} = (a_k)^{1/k}$. If $\rho < 1$, the series converges. If $\rho > 1$, the series diverges. If $\rho = 1$, the test is inconclusive.	Works well on series containing exponential factors, or other factors with kth powers.
Alternating Series Test *Section 7.7*	Let $\{a_k\}$ be a strictly decreasing sequence of positive numbers such that $\lim_{k\to\infty} a_k = 0$. Then the alternating series $\sum_{k=1}^{\infty}(-1)^{k+1}a_k$ and $\sum_{k=1}^{\infty}(-1)^k a_k$ both converge.	There is an extension of this test that guarantees that for a series satisfying the conditions of the test, if S_n is any term in the sequence of partial sums, then $\lvert L - S_n\rvert < a_{n+1}$, where L is the sum of the series.
Absolute Convergence Implies Convergence *Section 7.7*	If the series $\sum_{k=1}^{\infty} b_k$ converges absolutely, then it converges.	Examine the series of absolute values, $\sum_{k=1}^{\infty} \lvert b_k\rvert$. If this series converges, the series $\sum_{k=1}^{\infty} b_k$ converges absolutely.
Ratio Test for Absolute Convergence *Section 7.7*	Let $\sum_{k=1}^{\infty} b_k$ be a series with nonzero terms, and let $\rho = \lim_{k\to\infty} \left\lvert\frac{b_{k+1}}{b_k}\right\rvert$. If $\rho < 1$, the series converges absolutely. If $\rho > 1$, the series diverges. If $\rho = 1$, the test is inconclusive.	This test can *never* be used by itself to determine that a series converges conditionally.

Examples and Explorations

EXAMPLE 1

Using the alternating series test

Use the alternating series test to show that the series $\sum_{k=0}^{\infty} \frac{(-1)^k}{k!}$ converges, and find an approximation to the sum within 10^{-6} of its value.

SOLUTION

The given series is an alternating series. Although the absolute values of the first two terms are both 1, after that the absolute values of the terms monotonically decrease to zero, so the series satisfies the conditions of the alternating series test. Therefore, the series converges. To calculate an approximation within 10^{-6} of the sum of the series, we find the first term whose absolute value is smaller than 10^{-6}. That is, we find the smallest value of k such that $\frac{1}{k!} \le 10^{-6}$. You may verify that $\frac{1}{9!} > 10^{-6}$ and $\frac{1}{10!} < 10^{-6}$. Thus, if we compute the ninth partial sum, S_9, we have

$$S_9 = \frac{1}{0!} - \frac{1}{1!} + \frac{1}{2!} - \frac{1}{3!} + \frac{1}{4!} - \frac{1}{5!} + \frac{1}{6!} - \frac{1}{7!} + \frac{1}{8!} - \frac{1}{9!} \approx 0.36787919.$$

In addition, Theorem 7.38 tells us that S_9 is less than the sum of the series, L, because $L - S_9$ has the same sign as the term "$+\frac{1}{10!}$". Furthermore, our approximation, S_9, is somewhat better than we specified, because $L - S_9 < a_{10} \approx 2.8 \times 10^{-7}$. We have $L \in (S_9, S_9 + a_{10})$. Therefore $L \in (0.36787919, 0.36787947)$. □

CHECKING THE ANSWER

In Chapter 8 we will see that the series $\sum_{k=0}^{\infty} \frac{(-1)^k}{k!}$ converges to $\frac{1}{e}$. The value $\frac{1}{e}$ is in the interval we obtained in Example 1.

EXAMPLE 2

Analyzing series for absolute convergence, conditional convergence, and divergence

Determine whether the series

(a) $\displaystyle\sum_{k=0}^{\infty} \frac{(-2)^k}{1+2^k}$ **(b)** $\displaystyle\sum_{k=2}^{\infty} \frac{(-1)^k}{k \ln k}$ **(c)** $\displaystyle\sum_{k=1}^{\infty} \frac{(-1)^k k^2}{k!}$

converge absolutely, converge conditionally, or diverge.

SOLUTION

(a) The divergence test often makes it (relatively) easy to see if a series diverges. The limit $\lim_{k\to\infty} \frac{(-2)^k}{1+2^k}$ does not exist, because when k is even, the quotient approaches 1, and when k is odd, it approaches -1. Therefore, the series $\sum_{k=0}^{\infty} \frac{(-2)^k}{1+2^k}$ diverges. Note that the limits of the terms of the other two series are both zero. Thus, the divergence test fails on those series and we must use other tests.

(b) If we can show that the series of absolute values converges, we will have proved both absolute convergence and convergence, by Theorem 7.41. Consider the series of absolute values

$$\sum_{k=2}^{\infty} \left| \frac{(-1)^k}{k \ln k} \right| = \sum_{k=2}^{\infty} \frac{1}{k \ln k}.$$

We see that this series satisfies the conditions of the integral test. We compute the improper integral

$$\int_2^{\infty} \frac{1}{x \ln x}\, dx = \lim_{b\to\infty} \ln|\ln x|\Big|_2^b = \lim_{b\to\infty} \left(\ln|\ln b| - \ln|\ln 2|\right) = \infty.$$

Since this improper integral diverges, the series of absolute values diverges. This tells us that the original series does not converge absolutely. However, if we can show that the original series converges, we will know that it converges conditionally. We see that the original series is an alternating series. We have already mentioned that the limit of the terms of the series is zero. By the alternating series test, since $\frac{1}{(k+1)\ln(k+1)} < \frac{1}{k \ln k}$ for every $k \geq 2$, the series $\sum_{k=2}^{\infty} \frac{(-1)^k}{k \ln k}$ converges, and, therefore, converges conditionally. Before we analyze the remaining series, we repeat:

Demonstrating conditional convergence for a series $\sum_{k=1}^{\infty} a_k$ always requires the use of (at least) two convergence tests; you must show that the series $\sum_{k=1}^{\infty} a_k$ converges and that the series $\sum_{k=1}^{\infty} |a_k|$ diverges.

(c) When presented with a series containing factorials or factors similar to factorials, one of the two ratio tests is usually the best choice. (This does not mean that these tests have to succeed; recall Example 3 from Section 7.6!) Since the series $\sum_{k=1}^{\infty} \frac{(-1)^k k^2}{k!}$ has both positive and negative terms, we use the ratio test for absolute convergence. We have $a_k = \frac{(-1)^k k^2}{k!}$. Thus, $a_{k+1} = \frac{(-1)^{k+1}(k+1)^2}{(k+1)!}$. We now evaluate

$$\lim_{k\to\infty} \left| \frac{(-1)^{k+1}(k+1)^2/(k+1)!}{(-1)^k k^2/k!} \right| = \lim_{k\to\infty} \frac{(k+1)^2}{(k+1)!} \cdot \frac{k!}{k^2} = \lim_{k\to\infty} \frac{k+1}{k^2} = 0.$$

Therefore, by the ratio test for absolute convergence, the series $\sum_{k=1}^{\infty} \frac{(-1)^k k^2}{k!}$ converges absolutely. □

EXAMPLE 3

Modifying the root test to check for absolute convergence

Modify and use the root test to analyze the series

$$\sum_{k=1}^{\infty} \left(-\frac{2}{k}\right)^k.$$

SOLUTION

Theorem 7.35, the root test, may be used on a series only if the terms are (eventually) all positive. However, just as we modified the ratio test, we may adapt the root test to check for absolute convergence. Given a series $\sum_{k=1}^{\infty} a_k$, instead of examining $\lim_{k\to\infty} (a_k)^{1/k}$, we let $\rho = \lim_{k\to\infty} |a_k|^{1/k}$. As with the ratio test for absolute convergence,

- if $\rho < 1$, the series converges absolutely;
- if $\rho > 1$, the series diverges;
- if $\rho = 1$, the test fails.

For our series we have

$$\rho = \lim_{k\to\infty} |a_k|^{1/k} = \lim_{k\to\infty} \left| \left(-\frac{2}{k}\right)^k \right|^{1/k} = \lim_{k\to\infty} \frac{2}{k} = 0.$$

So, the series converges absolutely. □

? TEST YOUR UNDERSTANDING

- What is an alternating series? What additional conditions must an alternating series meet in order to satisfy the conditions of the alternating series test?
- When a series meets the conditions of the alternating series test, how is the test extended to provide a bound on the remainder when a term from the sequence of partial sums is used to approximate the sum of the series?
- What is absolute convergence? What is the relationship between absolute convergence and convergence?
- What is conditional convergence? What is the relationship between conditional convergence and absolute convergence?
- How is it possible to use a single convergence test to show that a series converges absolutely? Why are two convergence tests required to show that a series converges conditionally?

EXERCISES 7.7

Thinking Back

- *A strictly increasing sequence of partial sums:* If $a_k > 0$ for $k \in \mathbb{Z}^+$, explain why the sequence of partial sums for the series $\sum_{k=1}^{\infty} a_k$ is strictly increasing.
- *A bounded sequence of partial sums:* If the series $\sum_{k=1}^{\infty} a_k$ converges, explain why the sequence of partial sums is bounded.

Concepts

0. *Problem Zero:* Read the section and make your own summary of the material.

1. *True/False:* Determine whether each of the statements that follow is true or false. If a statement is true, explain why. If a statement is false, provide a counterexample.
   (a) *True or False:* $\sum_{k=1}^{\infty}(-1)^{k+1}a_k$ is an alternating series.
   (b) *True or False:* If $a_k > 0$ for $k \in \mathbb{Z}^+$ and the alternating series $\sum_{k=1}^{\infty}(-1)^k a_k$ converges, then the series $\sum_{k=1}^{\infty}(-1)^{k+1}a_k$ converges.
   (c) *True or False:* If $a_k > 0$ for $k \in \mathbb{Z}^+$ and the series $\sum_{k=1}^{\infty} a_k$ converges, then the series also converges absolutely.
   (d) *True or False:* If a function f satisfies the hypotheses of the integral test, then the series $\sum_{k=1}^{\infty}(-1)^{k+1}f(k)$ converges.
   (e) *True or False:* If a series $\sum_{k=1}^{\infty} a_k$ converges conditionally, then the series $\sum_{k=1}^{\infty} |a_k|$ diverges.
   (f) *True or False:* If $\sum_{k=1}^{\infty} a_k$ is a series such that $\lim_{k\to\infty}\left|\frac{a_{k+1}}{a_k}\right| = 1$, then the series converges conditionally.
   (g) *True or False:* If $\sum_{k=1}^{\infty} a_k$ is a series such that $\lim_{k\to\infty}\sqrt[k]{|a_k|} < 1$, then the series converges absolutely.
   (h) *True or False:* If we rearrange infinitely many terms of the alternating harmonic series, we can change the value of its sum.

2. *Examples:* Construct examples of the thing(s) described in the following. Try to find examples that are different than any in the reading.
   (a) A series that converges absolutely.
   (b) A series that converges conditionally.
   (c) A series $\sum_{k=1}^{\infty} a_k$ such that $\lim_{k\to\infty}\left|\frac{a_{k+1}}{a_k}\right| = 1$, but the series is absolutely convergent.

3. What is an alternating series?

Exercises 4–8 concern the alternating series test. To use this test we need a series of the form

$$\sum_{k=1}^{\infty}(-1)^{k+1}a_k \quad \text{or} \quad \sum_{k=1}^{\infty}(-1)^k a_k,$$

where

(i) $\{a_k\}$ is a sequence of positive numbers;
(ii) the sequence $\{a_k\}$ is strictly decreasing;
(iii) $\lim_{k\to\infty} a_k = 0$.

4. Find an example of a divergent series of the form $\sum_{k=1}^{\infty}(-1)^{k+1}a_k$
   (a) that satisfies conditions (i) and (iii), but not condition (ii);
   (b) that satisfies conditions (i) and (ii), but not condition (iii).

5. A series can fail two of the three conditions, (i), (ii) and (iii), and still converge. Which of the three conditions must an alternating series pass in order to converge?

6. Explain why a series satisfying the hypotheses of the alternating series test has a sum with the same sign as the first term in the series.

7. Explain why the sum of a series satisfying the hypotheses of the alternating series test is between any two consecutive terms in its sequence of partial sums.

8. Outline the steps you would use to approximate the sum of a convergent series satisfying the hypotheses of the alternating series test to within ϵ of its value.

9. What condition(s) must a series $\sum_{k=1}^{\infty} a_k$ satisfy in order for the series to be absolutely convergent?

10. What condition(s) must a series $\sum_{k=1}^{\infty} a_k$ satisfy in order for the series to be conditionally convergent?

11. Explain why one convergence test can suffice to show that a series converges absolutely, even though it always requires two to show that a series converges conditionally.

12. Explain why you must use two convergence tests to show that a series $\sum_{k=1}^{\infty} a_k$ converges conditionally.

13. Explain why every convergent series consisting of positive terms is absolutely convergent.

14. Fill in each blank with an inequality involving p: The series $\sum_{k=1}^{\infty}\frac{(-1)^{k+1}}{k^p}$ converges absolutely if ________, converges conditionally if ________, and diverges if ________.

15. Fill in the blanks: Let f be a function with domain $[1, \infty)$. If the function $|f|$ is ________ and ________, and if________ converges, then the series ________ converges absolutely.

16. Fill in the blanks: Let $\sum_{k=1}^{\infty} a_k$ and $\sum_{k=1}^{\infty} b_k$ be two series such that $0 \le$ ________ $\le$ ________ for every ________. If the series ________ converges absolutely, then the series ________ converges absolutely.

17. Let $\{a_k\}$ be a sequence of positive numbers. Explain why, if the series $\sum_{k=1}^{\infty} a_{2k-1}$ and $\sum_{k=1}^{\infty} a_{2k}$ both converge absolutely, the series $\sum_{k=1}^{\infty} a_k$ converges absolutely.

18. Which convergence tests can be used only on series in which the terms eventually all have the same sign?

19. Give an example of divergent series $\sum_{k=1}^{\infty} a_k$ and $\sum_{k=1}^{\infty} b_k$ such that the series $\sum_{k=1}^{\infty} a_k b_k$ converges.

20. Give an example of convergent series $\sum_{k=1}^{\infty} a_k$ and $\sum_{k=1}^{\infty} b_k$ such that the series $\sum_{k=1}^{\infty} a_k b_k$ diverges.

Skills

Determine whether the series in Exercises 21–23 converge or diverge. Note that these series do *not* satisfy the criteria of the alternating series test.

21. $1 - \frac{1}{2} - \frac{1}{2} + \frac{1}{3} + \frac{1}{3} + \frac{1}{3} - \frac{1}{4} - \frac{1}{4} - \frac{1}{4} - \frac{1}{4} + \cdots$

22. $1 - \frac{1}{2} - \frac{1}{3} + \frac{1}{4} + \frac{1}{5} - \frac{1}{6} - \frac{1}{7} + \frac{1}{8} + \frac{1}{9} - \frac{1}{10} + \cdots$

23. $1 - \frac{1}{2^3} + \frac{1}{3^2} - \frac{1}{4^3} + \frac{1}{5^2} - \frac{1}{6^3} + \frac{1}{7^2} - \frac{1}{8^3} + \frac{1}{9^2} - \cdots$

Use the alternating series test to determine whether the series in Exercises 24–29 converge or diverge. If a series converges, determine whether it converges absolutely or conditionally.

24. $\sum_{k=1}^{\infty} \frac{(-k)^k}{k!}$

25. $\sum_{k=0}^{\infty} \frac{(-1)^{k+1}}{1+k^2}$

26. $\sum_{k=2}^{\infty} \frac{(-1)^{k+1}}{(\ln k)^2}$

27. $\sum_{k=0}^{\infty} \frac{(-2)^{k+1}}{1+3^k}$

28. $\sum_{k=0}^{\infty} \frac{(-1)^{k+1}k!}{(2k+1)!}$

29. $\sum_{k=0}^{\infty} \frac{(-1)^k k^2}{1+k^2}$

Use the ratio test for absolute convergence to determine whether the series in Exercises 30–35 converge absolutely or diverge.

30. $\sum_{k=1}^{\infty} \frac{(-k)^k}{k!}$

31. $\sum_{k=0}^{\infty} \frac{(-3)^{k+1}}{(2k)!}$

32. $\sum_{k=1}^{\infty} \frac{(-5)^k}{(k+1)! - k!}$

33. $\sum_{k=0}^{\infty} \frac{(-7)^k}{(2k+1)!}$

34. $\sum_{k=0}^{\infty} \frac{(-2)^k 1 \cdot 3 \cdot 5 \cdots (2k+1)}{1 \cdot 4 \cdot 7 \cdots (3k+1)}$

35. $\sum_{k=0}^{\infty} \frac{(-1)^k 1 \cdot 3 \cdot 5 \cdots (2k+1)}{1 \cdot 4 \cdot 7 \cdots (3k+1)}$

Use any convergence tests from Sections 7.4–7.7 to determine whether the series in Exercises 36–51 converge absolutely, converge conditionally, or diverge. Explain why the series meets the hypotheses of the test you select.

36. $\sum_{k=1}^{\infty} \frac{1}{(2k-1)(2k+1)}$

37. $\sum_{k=2}^{\infty} \frac{1}{k^2-1}$

38. $\sum_{k=2}^{\infty} \frac{(-1)^k}{k \ln k \ln(\ln k)}$

39. $\sum_{k=1}^{\infty} \frac{(-1)^k k}{k+1}$

40. $\sum_{k=2}^{\infty} \frac{(-1)^k \ln k}{k^3+1}$

41. $\sum_{k=2}^{\infty} \frac{(-1)^k \ln k}{k+1}$

42. $\sum_{k=1}^{\infty} (-1)^k \frac{\sqrt{k^3+3k^2-5}}{k^2+3}$

43. $\sum_{k=2}^{\infty} \frac{(-1)^k \ln k}{k^2-1}$

44. $\sum_{k=1}^{\infty} (-1)^k e^{-k}$

45. $\sum_{k=1}^{\infty} (-1)^k \frac{\sqrt[3]{k^2+3k+1}}{k+1}$

46. $\sum_{k=1}^{\infty} (-1)^k k e^{-k^2}$

47. $\sum_{k=1}^{\infty} (-1)^k e^{-k^2}$

48. $\sum_{k=1}^{\infty} \frac{\sin \sqrt{k}}{k^{3/2}}$

49. $\sum_{k=1}^{\infty} \frac{\cos k}{k^2}$

50. $\sum_{k=1}^{\infty} \cos^3\left(\frac{1}{k}\right)$

51. $\sum_{k=1}^{\infty} \sin^3\left(\frac{1}{k}\right)$

In Exercises 52–57 do each of the following:

(a) Show that the given alternating series converges.

(b) Compute S_{10} and use Theorem 7.38 to find an interval containing the sum L of the series.

(c) Find the smallest value of n such that Theorem 7.38 guarantees that S_n is within 10^{-6} of L.

52. $\sum_{k=1}^{\infty} \frac{(-1)^k}{k^3}$

53. $\sum_{k=1}^{\infty} \frac{(-1)^{k+1}}{k^2}$

54. $\sum_{k=0}^{\infty} \frac{(-1)^k}{(2k+1)!}$

55. $\sum_{k=0}^{\infty} \frac{(-1)^k}{(2k)!}$

56. $\sum_{k=0}^{\infty} \frac{(-1)^{k+1}k!}{(2k+1)!}$

57. $\sum_{k=0}^{\infty} \frac{(-1)^k(k+1)}{k!}$

In Exercises 58–61, find the values of p that make the series converge absolutely, the values that make the series converge conditionally, and the values that make the series diverge.

58. $\sum_{k=2}^{\infty} \frac{(-1)^{k+1}}{k^p \ln k}$

59. $\sum_{k=1}^{\infty} \frac{(-1)^{k+1} \ln k}{k^p}$

60. $\sum_{k=0}^{\infty} (-1)^k k^p e^{-k^2}$

61. $\sum_{k=0}^{\infty} (-1)^k k^p e^{-k}$

Applications

62. On the coast of Oregon, near Newport, there is a sandbar in a tidal area that blocks water going out to sea. Some water can escape from the side of the pool that develops behind the bar. As the tide is going out, waves wash over the bar into the pool behind it. The depth of water left by the kth wave is w_k, where $\lim_{k\to\infty} w_k = 0$. Between waves, a little water flows out the side outlet. The deeper the water in the pool, the deeper is the channel at the outlet, so that the amount of water flowing out of the pool between waves is $0.03d_k$, where d_k is the depth of the water in the pool after the kth wave. It follows that the depth of water in the pool is

$$d_0 - 0.03d_0 + w_1 - 0.03d_1 + w_2 - 0.03d_2 + \cdots.$$

Does this series converge? *(Hint: Notice that we have* $d_k = (1-0.03)d_{k-1} + w_k$*.)*

63. Alex is measuring nitrates from agricultural runoff from a new point source. The nitrates flow into a small pond, where Alex measures them. The contamination comes in pulses: The pond gains nitrates, then clean water flows for a while and the contaminated water flows away, and then another dump of nitrates occurs. Considering the flow of nitrates into the pond and the flow of the water and nitrate mixture out, Alex finds that the total amount of nitrate in the lake at his kth measurement satisfies $A_k = 8069 - \frac{A_{k-1}}{7}$. The water in the pond was initially clean, so $A_0 = 0$.

(a) Show that $\frac{A_k - A_{k-1}}{A_{k-1} - A_{k-2}} = -\frac{1}{7}$.

(b) Show that $A_N = \sum_{k=1}^{N}(A_k - A_{k-1})$.

(c) The lake started clean. Show that

$$\lim_{N\to\infty} A_N = \sum_{k=1}^{\infty}(A_k - A_{k-1})$$

converges.

Proofs

64. Prove that every convergent geometric series converges absolutely.

65. Show that if a series $\sum_{k=1}^{\infty} a_k$ converges absolutely, the series $\sum_{k=1}^{\infty} a_k^2$ also converges.

66. Let $\{a_k\}$ be a strictly decreasing sequence of positive numbers such that $a_k \to 0$ as $k \to \infty$. Show that the alternating series $\sum_{k=1}^{\infty}(-1)^k a_k$ converges. *(Hint: We have already shown that the hypotheses imply that the series $\sum_{k=1}^{\infty}(-1)^{k+1} a_k$ converges.)*

67. Let $\{a_k\}$ be a strictly decreasing sequence of positive numbers such that $a_k \to 0$ as $k \to \infty$. Show that the subsequence $\{S_{2n}\}$ of the sequence of partial sums $\{S_n\}$ for the alternating series $\sum_{k=1}^{\infty}(-1)^{k+1} a_k$ is strictly increasing and is bounded above by a_1.

68. Let $\sum_{k=1}^{\infty}(-1)^k a_k$ be an alternating series that satisfies the hypotheses of the alternating series test. Prove that if S_n is a term in the sequence of partial sums, then

$$|L - S_n| < a_{n+1}$$

and, furthermore, the sign of the difference, $L - S_n$, is the sign of the coefficient of the term a_{n+1}.

69. Prove that if the series $\sum_{k=1}^{\infty} a_k$ diverges, then the series $\sum_{k=1}^{\infty} |a_k|$ also diverges.

70. Prove the ratio test for absolute convergence. That is, let $\sum_{k=1}^{\infty} b_k$ be a series with nonzero terms, and let $\rho = \lim_{k\to\infty}\left|\frac{b_{k+1}}{b_k}\right|$.

(a) Show that if $\rho < 1$, the series converges absolutely.

(b) Show that if $\rho > 1$, the series diverges.

(c) Show that the test fails when $\rho = 1$, by finding a convergent series $\sum_{k=1}^{\infty} c_k$ such that $\left|\frac{c_{k+1}}{c_k}\right| \to 1$ as $k \to \infty$ and a divergent series $\sum_{k=1}^{\infty} d_k$ such that $\left|\frac{d_{k+1}}{d_k}\right| \to 1$ as $k \to \infty$.

Changing the order of the summands in a conditionally convergent series can change the value of the sum. We showed this earlier in the section for the alternating harmonic series $\sum_{k=1}^{\infty}(-1)^{k+1}/k$. In Exercises 71–74 we will continue working with the alternating harmonic series to prove a few other, related properties.

71. Prove that the series consisting of just the positive terms of the alternating harmonic series diverges. That is, show that $\sum_{k=1}^{\infty}\frac{1}{2k-1}$ diverges.

72. Prove that the series consisting of just the negative terms of the alternating harmonic series diverges. That is, show that $\sum_{k=1}^{\infty}\left(-\frac{1}{2k}\right)$ diverges.

73. Let α be any real number. Show that there is a rearrangement of the terms of the alternating harmonic series that converges to α. *(Hint: Argue that if you add up some finite number of the terms of $\sum_{k=1}^{\infty}\frac{1}{2k-1}$, the sum will be greater than α. Then argue that, by adding in some other finite number of the terms of $\sum_{k=1}^{\infty}\left(-\frac{1}{2k}\right)$, you can get the sum to be less than α. By alternately adding terms from these two divergent series as described in the preceding two steps, explain why the sequence of partial sums you are constructing will converge to α.)*

74. Show that there is a rearrangement of the terms of the alternating harmonic series that diverges to ∞. *(Hint: Argue that if you add up some finite number of the terms of $\sum_{k=1}^{\infty}\frac{1}{2k-1}$, the sum will be greater than 1. Then argue that, by adding in some other finite number of the terms of $\sum_{k=1}^{\infty}\left(-\frac{1}{2k}\right)$, you can get the sum to be less than 1. Next, explain why you can get the sum to be greater than 2 if you add in more terms of $\sum_{k=1}^{\infty}\frac{1}{2k-1}$. Then argue again that you can get the sum to be less than 2 by adding in more terms of $\sum_{k=1}^{\infty}\left(-\frac{1}{2k}\right)$. By alternately adding terms from these two divergent series in a similar fashion as just described, explain why the sum can then go above 3, then below 3, above 4, then below 4, etc. Explain why such a process results in a sequence of partial sums that diverges to ∞.)*

Thinking Forward

- *A series of monomials:* Use the ratio test for absolute convergence to find all values of x for which the series $\sum_{k=1}^{\infty}\frac{x^k}{k}$ converges.
- *A series of monomials:* Use the ratio test for absolute convergence to find all values of x for which the series $\sum_{k=1}^{\infty}\frac{x^k}{k!}$ converges.

CHAPTER REVIEW, SELF-TEST, AND CAPSTONES

Before you progress to the next chapter, be sure you are familiar with the definitions, concepts, and basic skills outlined here. The capstone exercises at the end bring together ideas from this chapter and look forward to future chapters.

Definitions

Give precise mathematical definitions or descriptions of each of the concepts that follow. Then illustrate the definition or description with a graph or an algebraic example.

- A *sequence of real numbers*, including the meaning of a *term* of the sequence
- A *geometric sequence*
- An *arithmetic sequence*
- A *monotonic sequence*, including the definitions of *increasing* sequences, *decreasing* sequences, *strictly increasing* sequences, *strictly decreasing* sequences, *eventually increasing* sequences, and *eventually decreasing* sequences
- A *bounded sequence*
- The *limit* of a sequence
- A *series*, including the meaning of a *term* of the series
- The *sequence of partial sums* for a given series
- The *sum* of a convergent series
- The *remainder* of a convergent series
- An *alternating series*
- *Absolute convergence* for a series
- *Conditional convergence* for a series

Theorems

Fill in the blanks to complete each of the following theorem statements:

Tests for Monotonicity: A sequence $\{a_k\}$ is increasing if it passes any of the following tests:

- The *Difference Test*: _____ ≥ 0 for all $k \geq 1$.
- The *Ratio Test*: All terms are positive and _____ ≥ 1 for all $k \geq 1$.
- The *Derivative Test*: _____ ≥ 0 for all $x \geq 1$, given that $a(x)$ is a function that is _____ on $[1, \infty)$ and whose value at any positive integer k is $a(k) =$_____.

Basic Limit Rules for Convergent Sequences: If $\{a_k\}$ and $\{b_k\}$ are convergent sequences with $a_k \to L$ and $b_k \to M$ as $k \to \infty$, and if c is any constant, then

- $ca_k \to$_____.
- $(a_k + b_k) \to$_____.
- $a_k b_k \to$_____.
- If $M \neq 0$, then $\frac{a_k}{b_k} \to$_____.
- If f is a function that is _____ at L, then $f(a_k) \to$_____.
- If $a_k \to L$ and $a_k \to M$, then _____.
- If $\{m_k\}$ and $\{M_k\}$ are convergent sequences that both converge to L, and if m_k _____ a_k _____ M_k for all k, then $a_k \to$_____.
- If $|a_k| \to$_____, then $a_k \to$_____.
- Let $\{a_k\}$ be a sequence that converges to L. Then every subsequence of $\{a_k\}$ _____.
- If $\{a_k\}$ is both bounded and (eventually) monotonic, then $\{a_k\}$ _____.

Sums and Constant Multiples of Convergent Series: If $\sum_{k=1}^{\infty} a_k$ and $\sum_{k=1}^{\infty} b_k$ are convergent series and c is any real number, then

- $\sum_{k=1}^{\infty} ca_k =$_____.
- $\sum_{k=1}^{\infty} (a_k + b_k) =$_____ + _____.

Sequences and Series

Geometric Sequences: Fill in the blanks.

- For r_____, the sequence $\{r^k\}$ diverges.
- For r_____, the sequence $\{r^k\}$ converges to _____.

Dominance Relationships for Sequences: Order the following sequences by dominance when $a > 0$ and $b > 1$:

- $\{k!\}$
- $\{\ln k\}$
- $\{b^k\}$
- $\{k^a\}$

Some Convergent Sequences Involving Exponents: For any real number $p > 0$, the following sequences converge. Fill in each blank with the appropriate value.

- $p^{1/k} \to$_____
- $k^{1/k} \to$_____
- $\left(\frac{1}{k}\right)^p \to$_____

Geometric Series and p-series: Suppose r is a nonzero real number and $p > 0$. Fill in the blanks.

- *Geometric Series:* For r_____, the geometric series $\sum_{k=0}^{\infty} r^k$ converges to the sum _____.

- *Geometric Series:* For r_____ the geometric series $\sum_{k=0}^{\infty} r^k$ diverges.
- *p-series:* For p_____, the p-series $\sum_{k=0}^{\infty} \frac{1}{k^p}$ converges and for p_____, the series diverges.

Convergence Tests for Series: Fill in the blanks.

- *The Divergence Test:* If the sequence $\{a_k\}$ does not converge to _____, then the series $\sum_{k=1}^{\infty} a_k$ _____.
- *The Integral Test:* If $a(x)$ is a function that is _____, _____, and _____ on $[1, \infty)$ and $\{a_k\}$ is the sequence defined by $a_k =$_____ for every $k \in \mathbb{Z}^+$, then $\sum_{k=1}^{\infty} a_k$ and $\int_1^{\infty} a(x)\,dx$ either both _____ or both _____.
- *The Comparison Test:* Let $\sum_{k=1}^{\infty} a_k$ and $\sum_{k=1}^{\infty} b_k$ be two series with _____ terms such that 0_____a_k_____b_k for every positive integer k. If the series $\sum_{k=1}^{\infty} b_k$ _____, then the series $\sum_{k=1}^{\infty} a_k$ _____.
- *The Limit Comparison Test:* Let $\sum_{k=1}^{\infty} a_k$ and $\sum_{k=1}^{\infty} b_k$ be two series with positive terms.

  If $\lim_{k\to\infty} \frac{a_k}{b_k} = L$, where L is _____, then either the series both converge or both diverge.

  If $\lim_{k\to\infty} \frac{a_k}{b_k} = 0$ and $\sum_{k=1}^{\infty} b_k$ _____, then $\sum_{k=1}^{\infty} a_k$ _____.

  (There are two correct ways to fill in the last two blanks.)
- *The Ratio Test:* Let $\sum_{k=1}^{\infty} a_k$ be a series with nonnegative terms, and let $\rho = \lim_{k\to\infty}$_____.

  If $\rho < 1$, the series _____.

  If $\rho > 1$, the series _____.

  If $\rho = 1$, _____.
- *The Root Test:* Let $\sum_{k=1}^{\infty} a_k$ be a series with nonnegative terms, and let $\rho = \lim_{k\to\infty}$_____.

  If $\rho < 1$, the series _____.

  If $\rho > 1$, the series _____.

  If $\rho = 1$, _____.
- *The Alternating Series Test:* Let $\{a_k\}$ be a sequence of positive numbers.

  If a_{k+1}_____a_k for every $k \geq 1$ and $\lim_{k\to\infty} a_k =$_____, then $\sum_{k=1}^{\infty} (-1)^{k+1} a_k$ _____.
- *The Ratio Test for Absolute Convergence:* Let $\sum_{k=1}^{\infty} a_k$ be a series with nonzero terms, and let $\rho = \lim_{k\to\infty}$_____.

  If $\rho < 1$, the series _____.

  If $\rho > 1$, the series _____.

  If $\rho = 1$, _____.

Skill Certification: Limits, Convergence, and Divergence

Limits of sequences: Determine whether the sequences that follow are bounded, monotonic and/or eventually monotonic. Determine whether each sequence converges or diverges. If the sequence converges, find its limit.

1. $\left\{\frac{e^k}{k}\right\}$

2. $\left\{e^{-k} \ln \frac{1}{k}\right\}$

3. $\left\{\frac{2k^2+1}{3k^2-1}\right\}$

4. $\left\{\frac{(2k)!}{5^k}\right\}$

5. $\left\{\frac{k^2}{k+3} - \frac{k^2}{k+4}\right\}$

6. $\left\{\sin\left(\frac{\pi}{2k}\right)\right\}$

7. $\left\{\frac{k!}{1 \cdot 3 \cdot 5 \cdots (2k-1)}\right\}$

8. $\left\{\frac{1 \cdot 3 \cdot 5 \cdots (2k-1)}{3 \cdot 6 \cdot 9 \cdots (3k)}\right\}$

Sequences of partial sums: For each of the series that follow, (a) provide the first five terms in the sequence of partial sums $\{S_k\}$, (b) provide a closed formula for S_k, and (c) find the sum of the series by evaluating $\lim_{k\to\infty} S_k$.

9. $\sum_{k=0}^{\infty} \left(\frac{1}{k+3} - \frac{1}{k+4}\right)$

10. $\sum_{k=0}^{\infty} \left(\frac{2^k}{(k+3)!} - \frac{2^{k+1}}{(k+4)!}\right)$

11. $\sum_{k=0}^{\infty} \left(\frac{(1/2)^k}{\ln(k+2)} - \frac{(1/2)^{k+1}}{\ln(k+3)}\right)$

12. $\sum_{k=2}^{\infty} \ln\left(\frac{k}{k+1}\right)$

Geometric series: For each of the series that follow, find the sum or explain why the series diverges.

13. $\sum_{k=4}^{\infty} \left(\frac{1}{3}\right)^{k+2}$

14. $\sum_{k=0}^{\infty} 3\left(\frac{-2}{5}\right)^k$

15. $\sum_{k=0}^{\infty} (1)^k$

16. $\sum_{k=0}^{\infty} \frac{5^{k+3}}{2^{3k+1}}$

Convergence or divergence of a series: For each of the series that follow, determine whether the series converges or diverges. Explain the criteria you are using and why your conclusion is valid.

17. $\sum_{k=1}^{\infty} \ln k$

18. $\sum_{k=1}^{\infty} \ln\left(\frac{1}{k^2}\right)$

19. $\sum_{k=1}^{\infty} \frac{k}{\ln k}$

20. $\sum_{k=1}^{\infty} \frac{\ln k}{k}$

21. $\sum_{k=1}^{\infty} \frac{k}{3^k}$

22. $\sum_{k=1}^{\infty} \left(\frac{1}{k^2}\right)^{\sqrt{k}}$

23. $\sum_{k=1}^{\infty} \frac{k}{1+k^2}$

24. $\sum_{k=0}^{\infty} \frac{5^k + k}{k! + 3}$

25. $\sum_{k=2}^{\infty} \frac{1}{\sqrt[k]{e}}$

26. $\sum_{k=0}^{\infty} \frac{k!10^k}{3^k}$

27. $\sum_{k=1}^{\infty} \frac{1}{k + \sqrt{k}}$

28. $\sum_{k=0}^{\infty} \frac{5^k}{(k+1)!}$

29. $\sum_{k=0}^{\infty} \frac{1}{2^k + \sin k}$

30. $\sum_{k=0}^{\infty} \frac{5^{2k+1}}{(2k+1)!}$

Conditional and absolute convergence: For each of the series that follow, determine whether the series converges absolutely, converges conditionally, or diverges. Explain the criteria you are using and why your conclusion is valid.

31. $\sum_{k=1}^{\infty} (-1)^k \ln k$

32. $\sum_{k=1}^{\infty} (-1)^k \frac{1}{\sqrt{k(1+k)}}$

33. $\sum_{k=1}^{\infty} (-1)^k e^{-k}$

34. $\sum_{k=1}^{\infty} (-1)^{k+1} \sin\left(\frac{\pi}{k}\right)$

35. $\sum_{k=0}^{\infty} \frac{(-2)^k}{1+2^k}$

36. $\sum_{k=1}^{\infty} (-1)^{k+1} \sin\left(\frac{\pi}{k}\right)$

37. $\sum_{k=0}^{\infty} \frac{(-1)^k}{3^k}$

38. $\sum_{k=1}^{\infty} (-1)^k \frac{k}{4^k}$

39. $\sum_{k=1}^{\infty} (-1)^k \frac{\sqrt{k}}{3k-1}$

40. $\sum_{k=1}^{\infty} (-1)^{k+1} \frac{k^{100}}{2^k}$

Capstone Problems

A. Find the limits of the sequences $\{(k!)^{1/k}\}$ and $\{k^{1/k!}\}$ if they converge.

B. Let $0 < a < b$. Does the sequence $\{(a^k + b^k)^{1/k}\}$ converge or diverge? If it converges, find its limit. If it diverges, explain why.

C. Determine whether the series $\sum_{k=0}^{\infty} ke^{-k^2}$ converges or diverges. Explain the criteria you are using and why your conclusion is valid. Confirm your conclusion by using as many different convergence tests as you can.

D. Consider the series $\sum_{k=1}^{\infty} \frac{\sin k}{k^p}$, where $p \geq 1$.

(a) Explain why the divergence test fails on this series and none of the other convergence tests discussed in Sections 7.4–7.6 apply to the series.

(b) Show that the series converges when $p > 1$.

(c) Show that the convergence tests from Section 7.7 fail to determine the convergence or divergence of this series when $p = 1$.

(d) Do you think the series $\sum_{k=1}^{\infty} \frac{\sin k}{k}$ converges or diverges?

CHAPTER 8

Power Series

8.1 Power Series

Power Series
The Interval of Convergence of a Power Series
Power Series in $x - x_0$
Examples and Explorations

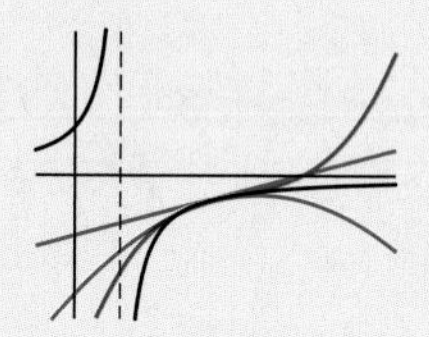

8.2 Maclaurin Series and Taylor Series

Maclaurin Polynomials and Taylor Polynomials
Taylor Series and Maclaurin Series
Examples and Explorations

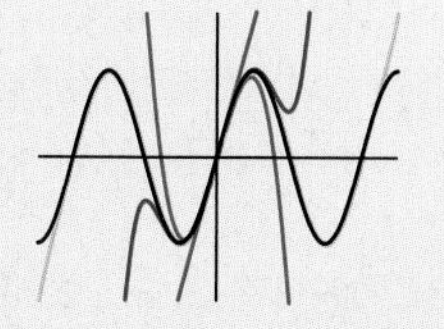

8.3 Convergence of Power Series

The Remainder
Basic Manipulations of Maclaurin Series
Examples and Explorations

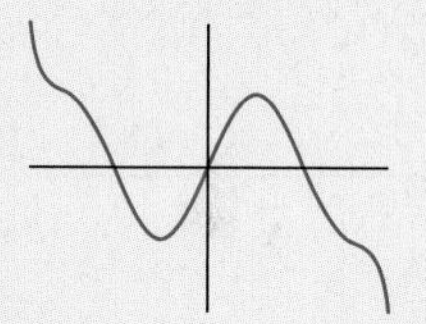

8.4 Differentiating and Integrating Power Series

Differentiating a Power Series
Integrating a Power Series
Examples and Explorations

$$\int \left(\sum_{k=0}^{\infty} a_k (x - x_0)^k \right) dx$$

Chapter Review, Self-Test, and Capstones

8.1 POWER SERIES

- Infinite sums of polynomials
- The power series in x and the power series in $x - x_0$
- The interval of convergence of a power series

Power Series

In this chapter we will extend our study of series to include series of monomials of the form $\sum_{k=0}^{\infty} a_k x^k$ and $\sum_{k=0}^{\infty} a_k (x - x_0)^k$. These are called power series. Before we start that discussion, we remind you that the tangent line to a differentiable function $f(x)$ at a point x_0 may be used to approximate f on an interval containing x_0. In this chapter we will discuss how higher degree polynomials can also be used to approximate functions. A *power series* is a generalization of these polynomials.

DEFINITION 8.1

Power Series in x

Let $\{a_k\}_{k=0}^{\infty}$ be a sequence of real numbers and let x be a variable. A ***power series in*** x is a series of the form $\sum_{k=0}^{\infty} a_k x^k$. That is,

$$\sum_{k=0}^{\infty} a_k x^k = a_0 + a_1 x + a_2 x^2 + a_3 x^3 + \cdots + a_k x^k + \cdots$$

If c is a real number at which the series of constants $\sum_{k=0}^{\infty} a_k c^k$ converges, the power series $\sum_{k=0}^{\infty} a_k x^k$ is said to ***converge*** at c.

For example, if $a_k = 1$ for every $k \geq 0$, we obtain the geometric series $\sum_{k=0}^{\infty} x^k$. We already know that this series converges to $\frac{1}{1-x}$ when $|x| < 1$ and diverges otherwise. That is,

$$\sum_{k=0}^{\infty} x^k = \frac{1}{1-x} \quad \text{when } |x| < 1.$$

Now consider this formula from the following perspective: On the interval $(-1, 1)$, the function $f(x) = \frac{1}{1-x}$ may be represented by the power series

$$\sum_{k=0}^{\infty} x^k = 1 + x + x^2 + x^3 + \cdots.$$

This means that as we add more terms to the series, the sequence of partial sums, given by the polynomials $1,\ 1+x,\ 1+x+x^2,\ 1+x+x^2+x^3, \cdots$ approximates the function $f(x) = \frac{1}{1-x}$ more closely on the interval $(-1, 1)$. We see this behavior in the following figure:

The polynomials $1+x$, $1+x+x^2$, and $1+x+x^2+x^3$ approximate the function $f(x) = \frac{1}{1-x}$

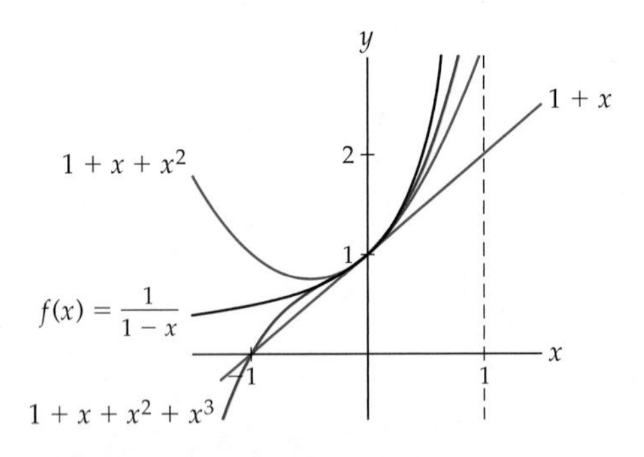

More generally:

- We say the power series in x, $\sum_{k=0}^{\infty} a_k x^k$, converges to a_0 when $x = 0$. Because of this behavior, every power series in x defines a function whose domain is a nonempty subset of the real numbers.
- A power series in x is a generalization of a polynomial, since if each $a_k = 0$ when $k > n$, we obtain the nth-degree polynomial $a_0 + a_1 x + a_2 x^2 + \cdots + a_n x^n$.

We already know that polynomials, from a computational standpoint, form a relatively simple class of functions; they are easy to evaluate, differentiate, and integrate. We will see that our most familiar functions can be expressed in terms of a power series, at least on some finite interval. In addition, if a function can be expressed as a convergent power series in x, we may use polynomials to approximate the function on an interval centered at $x = 0$, as we illustrated with the function $f(x) = \frac{1}{1-x}$.

The Interval of Convergence of a Power Series

We have already seen that every power series in x converges when $x = 0$. In a moment we will discuss a theorem that summarizes the possibilities for the types of intervals on which a power series in x can converge. Before we get to that theorem, however, we need the following one:

THEOREM 8.2 **The Convergence of a Power Series in x at a Nonzero Point Implies Convergence on an Interval**

Let $\sum_{k=0}^{\infty} a_k x^k$ be a power series in x.

(a) If the series converges at $c \neq 0$, then the series converges absolutely for all real numbers b such that $|b| < |c|$.

(b) If the series diverges at c, then the series diverges for all real numbers d such that $|c| < |d|$.

Proof. For Theorem 8.2 (a) we are given that the power series $\sum_{k=0}^{\infty} a_k x^k$ converges at $c \neq 0$. That is, the series of constants $\sum_{k=0}^{\infty} a_k c^k$ converges. Since this series converges, the Divergence Test implies that $\lim_{k\to\infty} a_k c^k = 0$. Furthermore, since every convergent sequence is bounded, there is a real number $M > 0$ such that $|a_k c^k| \leq M$ for every $k \geq 0$. We will now show that the series $\sum_{k=0}^{\infty} |a_k b^k|$ converges when $|b| < |c|$. This inequality implies that $\left|\frac{b}{c}\right| < 1$. For every $k \geq 0$, we have

$$0 \leq |a_k b^k| = \left|a_k c^k \left(\frac{b}{c}\right)^k\right| \leq M\left|\frac{b}{c}\right|^k.$$

Since $\left|\frac{b}{c}\right| < 1$, the geometric series $\sum_{k=0}^{\infty} M\left|\frac{b}{c}\right|^k$ converges. Thus, by the preceding inequality and the comparison test, the series $\sum_{k=0}^{\infty} |a_k b^k|$ converges. Hence, the series $\sum_{k=0}^{\infty} a_k b^k$ converges absolutely and we have proven (a).

The proof of part (b) is left for Exercise 61. ■

As a consequence of Theorem 8.2 we have the following theorem:

THEOREM 8.3 **The Convergence of a Power Series in x**

Let $\sum_{k=0}^{\infty} a_k x^k$ be a power series in x. Exactly one of the following occurs:

(a) The series converges only at $x = 0$.

(b) There exists a positive real number ρ such that the series converges absolutely for every $x \in (-\rho, \rho)$ and diverges if $x < -\rho$ or $x > \rho$.

(c) The series converges absolutely for every real number x.

Consider the following schematic illustrating the parts of Theorem 8.3:

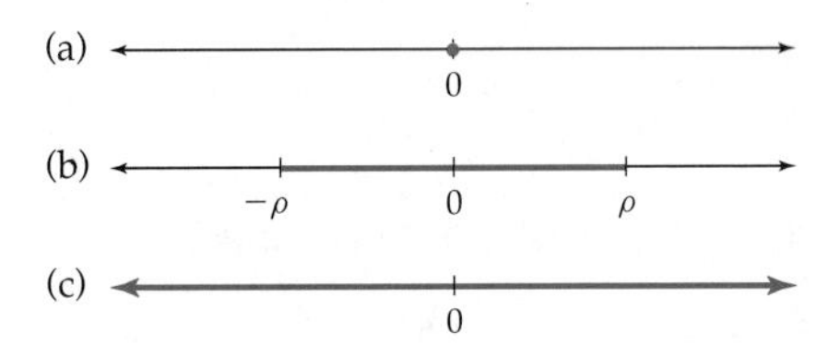

Each of the conditions (a), (b), and (c) of Theorem 8.3 defines both an ***interval of convergence*** and a ***radius of convergence*** for the power series:

- When a power series in x satisfies part (a) of Theorem 8.3, we say that the interval of convergence for the series is trivial and the radius of convergence is 0.
- Next, note that part (b) of the theorem does *not* say whether the power series converges at either ρ or $-\rho$. In fact, a power series in x satisfying part (b) may converge at both $\pm\rho$, at neither of these points, or at just one of these points. Thus, here the interval of convergence will have one of the forms

$$[-\rho, \rho], \ (-\rho, \rho), \ [-\rho, \rho) \text{ and } (-\rho, \rho].$$

  For any of these intervals, the radius of convergence is said to be ρ. When the series converges at ρ or $-\rho$, the convergence may be either conditional or absolute, depending upon the particular series. (See Exercise 62 for more on this.)
- Finally, in part (c) we say that the interval of convergence is $\mathbb{R}$ and that the radius of convergence is infinite.

We will discuss several examples later in the section.

Power Series in $x - x_0$

We wish to generalize the concept of a power series slightly. In the sections that follow we will be looking for power series representations for specified functions. At the beginning of the current section we saw that the function $f(x) = \frac{1}{1-x}$ may be represented by the following power series in x:

$$\frac{1}{1-x} = \sum_{k=0}^{\infty} x^k.$$

We also saw that the graphs of the terms of the sequence of partial sums

$$1, \ 1+x, \ 1+x+x^2, \ 1+x+x^2+x^3, \cdots$$

approximate the function with increasing accuracy on an interval centered at $x = 0$ as we include more summands. We may form similar power series at points other than $x = 0$. When we do, we obtain a power series in $x - x_0$. Just as we hoped to find power series in x that converge on an interval centered at $x = 0$, we wish to find power series in $x - x_0$ that converge on an interval centered at $x = x_0$.

DEFINITION 8.4 **Power Series in $x - x_0$**

Let $\{a_k\}_{k=0}^{\infty}$ be a sequence of real numbers, let x be a variable, and let x_0 be a real number. A ***power series in*** $x - x_0$ is a series of the form $\sum_{k=0}^{\infty} a_k (x - x_0)^k$. That is,

$$\sum_{k=0}^{\infty} a_k (x - x_0)^k = a_0 + a_1(x - x_0) + a_2(x - x_0)^2 + a_3(x - x_0)^3 + \cdots + a_k (x - x_0)^k + \cdots$$

If c is a real number at which the series of constants $\sum_{k=0}^{\infty} a_k (c - x_0)^k$ converges, the power series $\sum_{k=0}^{\infty} a_k (x - x_0)^k$ is said to ***converge*** at c.

For example, if $x_0 = 3$ and $a_k = \left(-\frac{1}{2}\right)^{k+1}$, we have the series

$$\sum_{k=0}^{\infty} \left(-\frac{1}{2}\right)^{k+1} (x - 3)^k = -\frac{1}{2} + \frac{1}{4}(x - 3) - \frac{1}{8}(x - 3)^2 + \frac{1}{16}(x - 3)^3 - \cdots .$$

In Section 8.3 we will see that this power series in $x - 3$ also converges to the function $f(x) = \frac{1}{1-x}$, but the series converges for every real number in the interval $(1, 5)$. In the following figure we graph the function $f(x) = \frac{1}{1-x}$ along with linear, quadratic, and cubic polynomials from the sequence of partial sums:

Polynomial approximations to the function $f(x) = \frac{1}{1-x}$ near $x = 3$

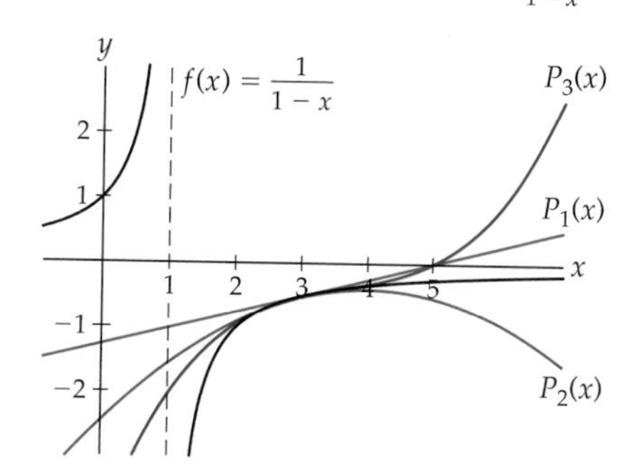

Note that if we start with a power series in $x - x_0$ and let $x_0 = 0$, we obtain a power series in x, so every power series in x is also a power series in $x - x_0$ where $x_0 = 0$. Thus, a power series in $x - x_0$ is a generalization of a power series in x. Theorem 8.3 outlined the possibilities for the radius of convergence and interval of convergence for a power series in x. The following theorem is the generalization of Theorem 8.3 for a power series in $x - x_0$:

THEOREM 8.5 **The Convergence of a Power Series in $x - x_0$**

Let $\sum_{k=0}^{\infty} a_k (x - x_0)^k$ be a power series in $x - x_0$. Exactly one of the following occurs:

(a) The series converges only at $x = x_0$.

(b) There exists a positive real number ρ such that the series converges absolutely for every $x \in (x_0 - \rho, x_0 + \rho)$ and diverges if $x < x_0 - \rho$ or $x > x_0 + \rho$.

(c) The series converges absolutely for every real number x.

The following schematic illustrates the three parts of Theorem 8.5 for an interval of convergence for a power series in $x - x_0$:

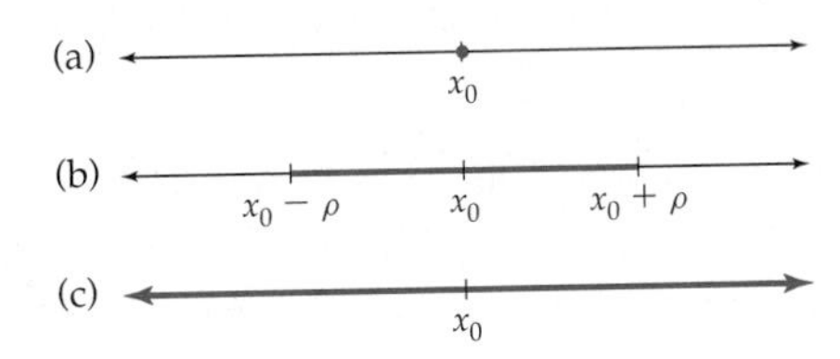

Each of the conditions (a), (b), and (c) of Theorem 8.5 defines both an interval of convergence and a radius of convergence for the power series:

- If the power series converges only at x_0, we say that the interval of convergence for the series is trivial and the radius of convergence is 0.
- If the series converges on a finite but nontrivial interval, we say that the radius of convergence is ρ. Part (b) of Theorem 8.5 does not tell us whether the series converges or diverges at either of the two points $x_0 + \rho$ and $x_0 - \rho$. Thus, in this case, the interval of convergence has one of the following forms:

$$[x_0 - \rho, x_0 + \rho], \ (x_0 - \rho, x_0 + \rho), \ [x_0 - \rho, x_0 + \rho) \text{ and } (x_0 - \rho, x_0 + \rho].$$

  When the series converges at $x_0 + \rho$ or $x_0 - \rho$, the convergence at either point may be either conditional or absolute, depending upon the particular series.
- If a power series in $x - x_0$ converges everywhere, we say that the interval of convergence is $\mathbb{R}$ and that the radius of convergence is infinite.

In the sections that follow, we will show how power series arise and how they may be used to help us understand the behavior of functions.

Examples and Explorations

EXAMPLE 1 **Finding the interval of convergence and radius of convergence for a power series in x**

Find the interval of convergence and radius of convergence for the following power series in x:

$$\sum_{k=0}^{\infty} \frac{1}{1 + 2^k} x^k = \frac{1}{2} + \frac{1}{3}x + \frac{1}{5}x^2 + \frac{1}{9}x^3 + \cdots$$

If the radius of convergence is finite, analyze the behavior of the series at the endpoints of the interval of convergence.

SOLUTION

Although occasionally it will be more convenient to use the root test, throughout this chapter our primary tool for finding the interval of convergence for a power series will be the ratio test for absolute convergence.

For this series we let $b_k = \frac{1}{1+2^k}x^k$. Therefore, $b_{k+1} = \frac{1}{1+2^{k+1}}\,x^{k+1}$, and thus

$$\lim_{k\to\infty}\left|\frac{b_{k+1}}{b_k}\right| = \lim_{k\to\infty}\left|\frac{x^{k+1}/(1+2^{k+1})}{x^k/(1+2^k)}\right| \qquad \leftarrow \text{the ratio test for absolute convergence}$$

$$= \lim_{k\to\infty}|x|\,\frac{1+2^k}{1+2^{k+1}} \qquad \leftarrow \text{algebraic simplification of the quotient}$$

$$= \lim_{k\to\infty}|x|\,\frac{(1/2^k)+1}{(1/2^k)+2} \qquad \leftarrow \text{division by } 2^k \text{ in the numerator and denominator}$$

$$= \frac{1}{2}\,|x|. \qquad \leftarrow \text{evaluation of the limit}$$

Before we proceed, notice that in the second line of the calculation we dropped the absolute value around the quotient $\frac{1+2^k}{1+2^{k+1}}$ because it would be redundant, since each term in the quotient is positive. However, the absolute value is necessary around the variable x, since that variable may take on both positive and negative values.

Now, from the ratio test for absolute convergence we know that the series will converge absolutely if the limit is less than 1. Therefore we set up the inequality

$$\frac{1}{2}\,|x| < 1,$$

or equivalently, $|x| < 2$. This inequality tells us that the radius of convergence for our series is 2 and that the series converges absolutely when $|x| < 2$ or equivalently when $-2 < x < 2$. Since the latter interval is finite, we will have to analyze the series at both endpoints of the interval, which we'll do in a moment. Before we do, we note that our results conform to part (b) of Theorem 8.3.

Now we consider the endpoints of the interval of convergence. We'll analyze the behavior of the power series when $x = 2$ and leave the analysis for $x = -2$ for Exercise 8. We evaluate our series when $x = 2$ and obtain

$$\sum_{k=0}^{\infty}\frac{1}{1+2^k}2^k = \sum_{k=0}^{\infty}\frac{2^k}{1+2^k}.$$

Consider the limit of the terms of this series as $k \to \infty$: $\lim_{k\to\infty}\frac{2^k}{1+2^k} = 1$. Thus, by the divergence test, the power series diverges when $x = 2$. In Exercise 8 you will show that the power series also diverges when $x = -2$. Therefore the interval of convergence for this power series is $(-2, 2)$.

Before we proceed to another example, we summarize a few of the main ideas we've used in this example. Our main tool for finding the radius of convergence and interval of convergence for any power series is the ratio test for absolute convergence. When you simplify $\frac{b_{k+1}}{b_k}$, remember to leave the variable x in an absolute value even if the other factors of the quotient are positive. When we are presented with a power series in x, if the radius of convergence ρ is positive and finite:

- The series converges absolutely for every value in the interval $(-\rho, \rho)$.
- You will need to analyze the behavior of the power series individually at both ρ and $-\rho$.

- The series may converge at one or both of the endpoints of the interval of convergence, but if it converges absolutely at either endpoint, then as you will see in Exercise 62, it will converge absolutely at the other endpoint.
- The ratio test for absolute convergence will *always fail* to uncover the convergence or divergence at the endpoints of a finite interval of convergence. We ask you to explain this phenomenon in Exercise 17. □

EXAMPLE 2 **More examples of finding the interval of convergence and radius of convergence for power series in x**

Find the interval of convergence and radius of convergence for the following power series in x:

(a) $\displaystyle\sum_{k=0}^{\infty} k!\,x^k$ **(b)** $\displaystyle\sum_{k=1}^{\infty} \left(\frac{2}{k}\right)^k x^{3k}.$

If the radius of convergence is finite, analyze the behavior of the series at the endpoints of the interval of convergence.

SOLUTION

(a) For the first series given, we again use the ratio test for absolute convergence. We let $b_k = k!\,x^k$. Thus, $b_{k+1} = (k+1)!\,x^{k+1}$, and we have

$$\lim_{k\to\infty}\left|\frac{b_{k+1}}{b_k}\right| = \lim_{k\to\infty}\left|\frac{(k+1)!\,x^{k+1}}{k!\,x^k}\right| = \lim_{k\to\infty}|x|(k+1).$$

Now, if $x = 0$, the preceding limit is zero, so the series converges absolutely at that point. However, if $x \neq 0$, the limit will be infinite. Therefore, by the ratio test for absolute convergence, we see that this series converges only when $x = 0$. The interval of convergence is trivial and the radius of convergence is 0.

(b) For variety we use the modified root test on the second series given. This is a reasonable choice for that series because the factors of the terms of the series involve kth powers. We let $b_k = \left(\frac{2}{k}\right)^k x^{3k}$ and evaluate

$$\lim_{k\to\infty}\sqrt[k]{|b_k|} = \lim_{k\to\infty}\sqrt[k]{|(2/k)^k x^{3k}|} = \lim_{k\to\infty}|x|^3\frac{2}{k}.$$

Note that we evaluate the preceding limit as $k \to \infty$. No matter what value the variable x takes on, this limit is zero. That is, $\lim_{k\to\infty}|x|^3\frac{2}{k} = 0$. Therefore, by the modified root test, the series converges absolutely for every value of x. Thus, the interval of convergence is $\mathbb{R}$ and the radius of convergence is infinite. □

EXAMPLE 3 **Finding the interval of convergence and radius of convergence for power series in $x - x_0$**

Find the interval of convergence and radius of convergence for the following power series:

(a) $\displaystyle\sum_{k=1}^{\infty}(-1)^{k+1}\frac{1}{k}(x+2)^k$ **(b)** $\displaystyle\sum_{k=0}^{\infty}\frac{1}{(2k+1)!}(x-\pi)^k$

If the radius of convergence is finite, analyze the behavior of the series at the endpoints of the interval of convergence.

SOLUTION

(a) For a power series in $x - x_0$, our primary tool for analyzing the interval of convergence is, again, the ratio test for absolute convergence. For the first series given, we let $b_k = (-1)^{k+1}\frac{1}{k}(x+2)^k$. Here, $b_{k+1} = (-1)^{k+1}\frac{1}{k+2}(x+2)^{k+1}$ and we evaluate

$$\lim_{k\to\infty}\left|\frac{b_{k+1}}{b_k}\right| = \lim_{k\to\infty}\left|\frac{(-1)^{k+2}(x+2)^{k+1}/(k+1)}{(-1)^{k+1}(x+2)^k/k}\right| = \lim_{k\to\infty}|x+2|\,\frac{k}{k+1}.$$

Since we are evaluating this limit as $k \to \infty$, its value is $|x+2|$. By the ratio test for absolute convergence we know that the series will converge absolutely when $|x+2| < 1$. We now know that the radius of convergence is 1, and the interval of convergence contains $(-3, -1)$. Since this a finite interval, we also analyze the behavior of the series at the endpoints of the interval. When $x = -1$,

$$\sum_{k=1}^{\infty}(-1)^{k+1}\frac{1}{k}(-1+2)^k = \sum_{k=1}^{\infty}(-1)^{k+1}\frac{1}{k}.$$

This is the alternating harmonic series that we already know converges conditionally. When $x = -3$,

$$\sum_{k=1}^{\infty}(-1)^{k+1}\frac{1}{k}(-3+2)^k = \sum_{k=1}^{\infty}(-1)^{k+1}\frac{1}{k}(-1)^k = -\sum_{k=1}^{\infty}\frac{1}{k}.$$

This is just a constant multiple of the harmonic series, which diverges. Thus, our interval of convergence is $(-3, -1]$.

(b) For our other given series, we let $b_k = \frac{1}{(2k+1)!}(x-\pi)^k$. Here, $b_{k+1} = \frac{1}{(2(k+)+1)!}(x-\pi)^{k+1}$ and we evaluate

$$\lim_{k\to\infty}\left|\frac{b_{k+1}}{b_k}\right| = \lim_{k\to\infty}\left|\frac{(x-\pi)^{k+1}/(2(k+1)+1)!}{(x-\pi)^k/(2k+1)!}\right| = \lim_{k\to\infty}|x-\pi|\,\frac{(2k+1)!}{(2k+3)!}$$

$$= \lim_{k\to\infty}|x-\pi|\,\frac{1}{(2k+2)(2k+3)}.$$

Make sure you understand the simplification of the quotients in this equation. Since we are evaluating the limit as $k \to \infty$, the limit is zero for any value of x. Thus, this power series converges for every real number, and the radius of convergence is infinite. □

EXAMPLE 4 **Analyzing a power series given in another form**

Show that if m and b are real numbers and $m \neq 0$, then a series of the form

$$\sum_{k=0}^{\infty} a_k(mx+b)^k$$

is a power series in $x + \frac{b}{m}$. Then find the interval of convergence and radius of convergence for the series

$$\sum_{k=0}^{\infty}(-1)^k\frac{k}{3k+1}(2x-3)^k.$$

If the radius of convergence is finite, analyze the behavior of the series at the endpoints of the interval of convergence.

SOLUTION

The series $\sum_{k=0}^{\infty} a_k(mx+b)^k$ is not in the form of a power series in $x - x_0$, but it can be put into that form. Since $(mx+b)^k = m^k\left(x + \frac{b}{m}\right)^k$, we have

$$\sum_{k=0}^{\infty} a_k(mx+b)^k = \sum_{k=0}^{\infty} a_k\, m^k\left(x+\frac{b}{m}\right)^k.$$

Thus, the series is a power series in $x + \frac{b}{m}$.

We do *not* need to put the series into this alternative form to find the radius of convergence or interval of convergence, however. We may use the ratio test for absolute convergence on the original form of the series. For the series $\sum_{k=0}^{\infty}(-1)^k \frac{k}{3k+1}(2x-3)^k$, we have $b_k = (-1)^k \frac{k}{3k+1}(2x-3)^k$ and $b_{k+1} = (-1)^{k+1}\frac{k+1}{3(k+1)+1}(2x-3)^{k+1}$. We evaluate

$$\lim_{k\to\infty}\left|\frac{b_{k+1}}{b_k}\right| = \lim_{k\to\infty}\left|\frac{(-1)^{k+1}(k+1)(2x-3)^{k+1}/(3k+4)}{(-1)^k k(2x-3)^k/(3k+1)}\right| = \lim_{k\to\infty}|2x-3|\,\frac{(k+1)(3k+1)}{k(3k+4)}.$$

Here the limit is $|2x-3|$. By the ratio test for absolute convergence, we know that the series will converge absolutely when $|2x-3|<1$ or, equivalently, when $\left|x-\frac{3}{2}\right| < \frac{1}{2}$. Thus, the radius of convergence is $\frac{1}{2}$, and the interval of convergence contains $(1,2)$. Since this is a finite interval, we also analyze the behavior of the series at the endpoints. When $x=1$,

$$\sum_{k=0}^{\infty}(-1)^k\frac{k}{3k+1}(2\cdot 1-3)^k = \sum_{k=0}^{\infty}\frac{k}{3k+1}.$$

Since $\lim_{k\to\infty}\frac{k}{3k+1} = \frac{1}{3}$, the power series diverges at $x=1$. In Exercise 9 we ask you to show that the power series also diverges when $x=2$. Therefore, the interval of convergence for the series is $(1,2)$. □

? TEST YOUR UNDERSTANDING

- What is a power series? What is the difference between a series of constants and a power series?
- What is the difference between a power series in x and a power series in $x - x_0$?
- What is the difference between a polynomial and a power series?
- What is meant by the interval of convergence of a power series? What is meant by the radius of convergence of a power series?
- How are the interval of convergence and radius of convergence of a series determined?

EXERCISES 8.1

Thinking Back

- *Translating the graph of a function:* What is the relationship between the graph of a function $f(x)$ and the graph of $f(x - x_0)$ when x_0 is a positive constant? When x_0 is a negative constant?
- *Odd and Even Functions:* What is the definition of an *odd* function? An *even* function?
- Fill in the blanks: The graph of every odd function is symmetric about ________. The graph of every even function is symmetric about ________.

Concepts

0. *Problem Zero:* Read the section and make your own summary of the material.
1. *True/False:* Determine whether each of the statements that follow is true or false. If a statement is true, explain why. If a statement is false, provide a counterexample.
   (a) *True or False:* $\sum_{k=1}^{\infty} \frac{1}{k} x^{-k}$ is a power series in x.
   (b) *True or False:* $\sum_{k=0}^{\infty} \frac{1}{k!}(x-2)^k$ is a power series in $x - 2$.
   (c) *True or False:* $\sum_{k=0}^{\infty} \frac{(-1)^k}{(2k+1)!} x^{2k+1}$ is a power series in x.
   (d) *True or False:* The series $\sum_{k=0}^{\infty} k!(x+3)^k$ converges when $x = 3$.
   (e) *True or False:* If a power series in x converges conditionally when $x = 3$, it converges absolutely at $x = -2$.
   (f) *True or False:* If a power series in x diverges when $x = 3$, it diverges at $x = -4$.
   (g) *True or False:* If a power series in $x - 4$ converges absolutely when $x = 10$, it converges absolutely at $x = -1$.
   (h) *True or False:* If a power series in $x + 5$ converges conditionally when $x = 3$, it converges conditionally at $x = -13$.
2. *Examples:* Construct examples of the thing(s) described in the following. Try to find examples that are different than any in the reading.
   (a) A power series in x with a finite radius of convergence.
   (b) A power series in $x - 4$ with an infinite radius of convergence.
   (c) A power series in $x + 3$ that converges only at $x = -3$.
3. What is a power series in x?
4. What is a power series in $x - x_0$?
5. Explain why $\sum_{k=0}^{\infty}(x-k)^k$ is *not* a power series.
6. What is meant by the interval of convergence for a power series in x? How is the interval of convergence determined? If a power series in x has a nontrivial interval of convergence, what types of intervals are possible?
7. What is meant by the interval of convergence for a power series in $x - x_0$? How is the interval of convergence determined? If a power series in $x - x_0$ has a nontrivial interval of convergence, what types of intervals are possible?
8. Show that $\sum_{k=0}^{\infty} \frac{1}{1+2^k} x^k$, the power series in x from Example 1, diverges when $x = -2$.
9. Complete Example 4 by showing that the power series $\sum_{k=0}^{\infty}(-1)^k \frac{k}{3k+1}(2x-3)^k$ diverges when $x = 2$.
10. Show that the power series $\sum_{k=0}^{\infty} \frac{(-1)^k}{2k+1} x^{2k+1}$ converges conditionally when $x = 1$ and when $x = -1$. What does this behavior tell you about the interval of convergence for the series?
11. Show that the power series $\sum_{k=1}^{\infty} \frac{(-1)^k}{k} x^k$ converges conditionally when $x = 1$ and diverges when $x = -1$. What does this behavior tell you about the interval of convergence for the series?
12. Show that the power series $\sum_{k=1}^{\infty} \frac{(-1)^k}{k^2} x^k$ converges absolutely when $x = 1$ and when $x = -1$. What does this behavior tell you about the interval of convergence for the series?
13. What is x_0 if the interval of convergence for the power series $\sum_{k=0}^{\infty} a_k (x - x_0)^k$ is $(2, 10]$?
14. What is x_0 if (p, q) is the interval of convergence for the power series $\sum_{k=0}^{\infty} a_k (x - x_0)^k$?
15. What is x_0 if the power series $\sum_{k=0}^{\infty} a_k (x - x_0)^k$ converges conditionally at both $x = -4$ and $x = 8$.
16. Is it possible for a power series to have $(0, \infty)$ as its interval of convergence? Explain your answer.
17. Let $\sum_{k=0}^{\infty} a_k x^k$ be a power series in x with a positive and finite radius of convergence ρ. Explain why the ratio test for absolute convergence will fail to determine the convergence of this power series when $x = \rho$ or when $x = -\rho$.
18. Let $\sum_{k=0}^{\infty} a_k x^k$ be a power series in x with radius of convergence ρ. What is the radius of convergence of the power series $\sum_{k=0}^{\infty} a_k (x - x_0)^k$? Make sure you justify your answer.
19. Let $a_k \neq 0$ for each value of k, and let $\sum_{k=0}^{\infty} a_k x^k$ be a power series in x with a positive and finite radius of convergence ρ. What is the radius of convergence of the power series $\sum_{k=0}^{\infty} \frac{1}{a_k} x^k$?
20. Let $\sum_{k=0}^{\infty} a_k x^k$ be a power series in x with interval of convergence $[-2, 2)$. What is the radius of convergence of the power series $\sum_{k=0}^{\infty} a_k (x-3)^k$? Justify your answer.

Skills

Find the interval of convergence for each power series in Exercises 21–48. If the interval of convergence is finite, be sure to analyze the convergence at the endpoints.

21. $\sum_{k=0}^{\infty} \frac{1}{k!}x^k$

22. $\sum_{k=0}^{\infty} \frac{(-2)^k}{k!}x^k$

23. $\sum_{k=0}^{\infty} \frac{(-1)^k}{(2k)!}x^{2k}$

24. $\sum_{k=0}^{\infty} \frac{(-1)^k}{(2k+1)!}x^{2k+1}$

25. $\sum_{k=1}^{\infty} \frac{(-1)^k}{k+1}x^k$

26. $\sum_{k=0}^{\infty} \frac{(-1)^k}{2k+1}x^{2k+1}$

27. $\sum_{k=1}^{\infty} \frac{1}{k}(x+2)^k$

28. $\sum_{k=1}^{\infty} \frac{1}{k^2}(x+3)^k$

29. $\sum_{k=1}^{\infty} \frac{2k+1}{k^3}(x-\pi)^k$

30. $\sum_{k=1}^{\infty} (-1)^k \frac{1}{k^k}(x-1)^k$

31. $\sum_{k=1}^{\infty} (-1)^k \frac{\sqrt{k}}{k^3}\left(x-\frac{3}{2}\right)^k$

32. $\sum_{k=1}^{\infty} \frac{2^k+1}{\sqrt{k}}x^k$

33. $\sum_{k=0}^{\infty} (-1)^k \frac{k^2}{3^k}\left(x-\frac{1}{2}\right)^k$

34. $\sum_{k=0}^{\infty} \frac{1}{3^k+5^k}x^k$

35. $\sum_{k=0}^{\infty} \frac{(k!)^2}{(2k)!}(x+2)^k$

36. $\sum_{k=0}^{\infty} \frac{1}{3^k+5^k}(x+3)^k$

37. $\sum_{k=0}^{\infty} k!x^k$

38. $\sum_{k=1}^{\infty} \frac{5}{k}x^k$

39. $\sum_{k=1}^{\infty} \frac{\ln k}{k^2}\left(x+\frac{5}{3}\right)^k$

40. $\sum_{k=1}^{\infty} \frac{\sqrt{k+1}}{k^2}\left(x+\frac{1}{2}\right)^k$

41. $\sum_{k=0}^{\infty} \frac{k+1}{3^k}(2x-5)^k$

42. $\sum_{k=0}^{\infty} \frac{k^2}{4^k+3}(3x+7)^k$

43. $\sum_{k=0}^{\infty} \frac{\sqrt{k}}{k!}(4x+7)^k$

44. $\sum_{k=0}^{\infty} \frac{(-1)^k}{(2k+1)!}(3x+7)^{2k+1}$

45. $\sum_{k=1}^{\infty} \frac{1}{2\cdot 4\cdot 6\cdots(2k)}x^k$

46. $\sum_{k=0}^{\infty} \frac{1}{1\cdot 3\cdot 5\cdots(2k+1)}x^k$

47. $\sum_{k=1}^{\infty} \frac{1}{k^k}(x-3)^k$

48. $\sum_{k=0}^{\infty} \frac{k^3}{1\cdot 3\cdot 5\cdots(2k+1)}(x+1)^k$

In Exercises 49–52 find the radius of convergence for the given series.

49. $\sum_{k=0}^{\infty} \frac{k!}{(k+m)!}x^k$, where $m \in \mathbb{Z}^+$

50. $\sum_{k=0}^{\infty} \frac{k!}{((k+m)!)^2}x^k$, where $m \in \mathbb{Z}^+$

51. $\sum_{k=0}^{\infty} \frac{k!}{(k+m)!(k+n)!}x^k$, where $m \in \mathbb{Z}^+$ and $n \in \mathbb{Z}^+$

52. $\sum_{k=0}^{\infty} \frac{(k!)^m}{(mk)!}x^k$, where $m \in \mathbb{Z}^+$

In Exercises 53–58 explain why the series is *not* a power series in $x - x_0$. Then use the ratio test for absolute convergence to find the values of x for which the given series converge.

53. $\sum_{k=0}^{\infty} \left(\frac{1}{x+1}\right)^k$

54. $\sum_{k=0}^{\infty} (-1)^k \frac{1}{k!}x^{-k}$

55. $\sum_{k=0}^{\infty} \frac{(\sin x)^k}{k!}$

56. $\sum_{k=1}^{\infty} \frac{(-1)^k}{k}\left(\frac{x}{x-1}\right)^k$

57. $\sum_{k=1}^{\infty} \frac{1}{k^3}\left(\frac{x+2}{x-3}\right)^k$

58. $\sum_{k=1}^{\infty} \frac{(-1)^k k!}{k^2}\left(\frac{3x}{x-2}\right)^k$

Applications

59. Leila, in her capacity as a wildlife biologist, is faced with concerns over the viability of the beaver population in Idaho in the presence of the wolves that have moved to the area. She has read papers that show that if the beaver population in an area stays above a certain solution $b(w)$ of a differential equation, where w represents the number of wolves in that area, then the population is viable. However, if the number of beavers falls below that function, then the population will collapse. Leila's solution of the differential equation for the function is

$$b(w) = 39w^{2/3}\sum_{n=0}^{\infty} \frac{w^n}{19200^n n!}.$$

(a) Where does this series converge?

(b) Plot an approximation to b on the interval $[0, 500]$. How do you choose how many terms to use?

60. Leila is managing elk hunting for one herd in a region of Idaho. She is responsible for picking a hunting rate h, where h represents the fraction of the herd remaining after hunting season is over. Each year, the herd size changes by a number Δ_k independently of the hunting rate. If P is the initial population of elk, then the number remaining after the first hunting season is Ph, the number remaining after the second is $(Ph + \Delta_1)h$, the number remaining after the third is $((Ph + \Delta_1)h + \Delta_2)h$, and so on. The number after n seasons is

$$Ph^{n+1} + \sum_{k=1}^{n} \Delta_k h^{n-k}.$$

Leila needs to prevent the size of the herd from going to zero or to infinity as $n \to \infty$. Given that $0 < h \le 1$, what conditions on Δ_k are required?

Proofs

61. Prove part (b) of Theorem 8.2. That is, let $\sum_{k=0}^{\infty} a_k x^k$ be a power series in x, and show that if the series diverges at c, then the series diverges for all real numbers d such that $|c| < |d|$. *(Hint: Use contradiction and part (a) of the theorem.)*

62. Let $\sum_{k=0}^{\infty} a_k x^k$ be a power series in x with a finite radius of convergence ρ. Prove that if the series converges absolutely at either $\pm\rho$, then the series converges absolutely at the other value as well.

63. Let $\sum_{k=0}^{\infty} a_k (x - x_0)^k$ be a power series in $x - x_0$ with a finite radius of convergence ρ. Prove that if the series converges absolutely at $x_0 \pm \rho$, then the series converges absolutely at the other value as well.

64. Prove that if $(-\rho, \rho]$ is the interval of convergence for the series $\sum_{k=0}^{\infty} a_k x^k$, then the series converges conditionally at ρ. *(Hint: Consider Exercise 62).*

65. Prove that if $(x_0 - \rho, x_0 + \rho]$ is the interval of convergence for the series $\sum_{k=0}^{\infty} a_k (x - x_0)^k$, then the series converges conditionally at $x_0 + \rho$.

66. Prove that if the power series $\sum_{k=0}^{\infty} a_k x^k$ has a positive and finite radius of convergence ρ, then the series $\sum_{k=0}^{\infty} a_k x^{2k}$ has radius of convergence $\sqrt{\rho}$.

67. Prove that if the power series $\sum_{k=0}^{\infty} a_k x^k$ and $\sum_{k=0}^{\infty} a_k x^{2k}$ have the same radius of convergence ρ, then ρ is 0, 1, or infinite. *(Hint: Use Exercise 66).*

68. Prove that if the power series $\sum_{k=0}^{\infty} a_k x^k$ has a positive and finite radius of convergence ρ, and if m is a positive integer greater than 1, then the series $\sum_{k=0}^{\infty} a_k x^{mk}$ has radius of convergence $\sqrt[m]{\rho}$.

69. Let b be a nonzero constant. Prove that the radius of convergence of the power series $\sum_{k=0}^{\infty} b^k x^k$ is $\dfrac{1}{|b|}$.

70. Let $\sum_{k=0}^{\infty} a_k x^k$ be a power series with a positive and finite radius of convergence ρ, and let $c \neq 0$. Prove that the series $\sum_{k=0}^{\infty} a_k (cx)^k$ has radius of convergence $\dfrac{\rho}{|c|}$.

71. A certain power series $\sum_{k=0}^{\infty} a_k x^k$ has the properties that a_0 and a_1 are nonzero and $a_{k+2} = a_k$ for every $k > 0$.
    (a) Prove that the radius of convergence ρ for the series is finite, and find the value of ρ.
    (b) Prove that the series diverges at the endpoints of the interval of convergence.
    (c) What is the sum of the series for $x \in (-\rho, \rho)$? Give your answer in terms of x, a_0, and a_1.

72. Recall that a rational function is the quotient of two polynomials. Prove that if a_k is a rational function of k, then the power series in x $\sum_{k=0}^{\infty} a_k x^k$ has radius of convergence $\rho = 1$.

Thinking Forward

▶ *Constructing a power series of the form* $\sum_{k=0}^{\infty} \dfrac{f^{(k)}(0)}{k!} x^k$*:* Using $f(x) = \sin x$, construct the power series $\sum_{k=0}^{\infty} \dfrac{f^{(k)}(0)}{k!} x^k$.

▶ *Graphing* $\sin x$ *and a polynomial approximation:* Graph the function $\sin x$, together with the polynomial formed from the first three nonzero terms you found in the previous problem, on the interval $[-\pi, \pi]$. What do you observe about the graphs of the two functions?

8.2 MACLAURIN SERIES AND TAYLOR SERIES

- Maclaurin polynomials and Taylor polynomials
- Taylor series
- Maclaurin series

Maclaurin Polynomials and Taylor Polynomials

We know that when a function f is differentiable at a point x_0, the tangent line to the graph of f at the point $(x_0, f(x_0))$ and with equation $y = f(x_0) + f'(x_0)(x - x_0)$ is the best linear approximation to the graph "close" to the point of tangency, as depicted in the figure that follows. Since this equation is that of a first-degree polynomial, we use the notation $P_1(x)$ to denote the tangent function. That is, $P_1(x) = f(x_0) + f'(x_0)(x - x_0)$.

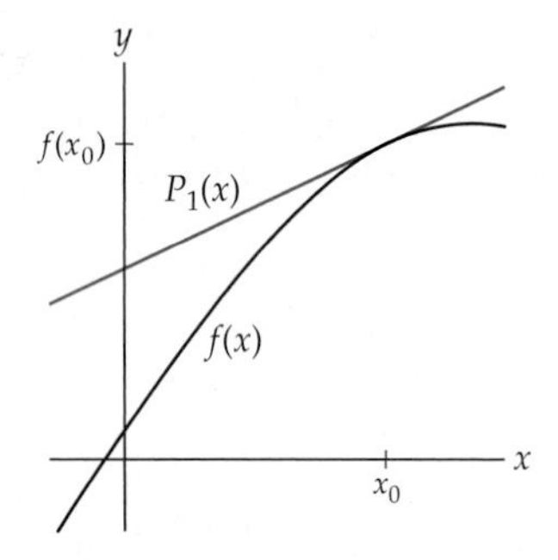

The tangent line has two properties that make it a good approximation for the function close to the point of tangency: It has the same value as the function at $x = x_0$, and it has the same derivative as the function at $x = x_0$. That is, $P_1(x_0) = f(x_0)$ and $P'_1(x_0) = f'(x_0)$.

For a function f with an nth-order derivative at the point x_0, we may similarly approximate the function with an nth-degree polynomial of the form

$$P_n(x) = a_0 + a_1(x - x_0) + a_2(x - x_0)^2 + \cdots + a_k(x - x_0)^k + \cdots + a_n(x - x_0)^n$$

if we insist that $P_n(x)$ and $f(x)$ share the following $n+1$ properties: P_n and f have the same value at x_0, P'_n and f' have the same value at x_0, P''_n and f'' have the same value at x_0, ..., and finally $P_n^{(n)}$ and $f^{(n)}$ have the same value at x_0. To construct such a polynomial, we must determine the constants $a_0, a_1, a_2, \ldots, a_n$ so that the polynomial $P_n(x)$ shares the necessary properties with the function $f(x)$.

Now consider the kth derivative of $P_n(x)$. Since the kth derivative of a polynomial of degree less than k is 0, the first nonzero term of $P_n^{(k)}(x)$ is $a_k k!$, and we have

$$\begin{aligned} P_n^{(k)}(x) &= a_k k! + a_{k+1}((k+1)k(k-1)\cdots 2)(x - x_0) + a_{k+2}((k+2)(k+1)k\cdots 3)(x - x_0)^2 \\ &\quad + \cdots + a_n(n(n-1)(n-2)\cdots(n-k+1))(x - x_0)^{n-k}. \end{aligned}$$

Now, if we evaluate this equation when $x = x_0$, we obtain $P_n^{(k)}(x_0) = a_k k!$. Since we wish to have $P_n^{(k)}(x_0) = f^{(k)}(x_0)$, we must have $a_k = \frac{f^{(k)}(x_0)}{k!}$. This holds for each integer $0 \le k \le n$, where we interpret $f^{(0)}$ to be the function f itself. Therefore, the nth-degree polynomial we hope to use to approximate f at x_0 is

$$\begin{aligned} P_n(x) &= f(x_0) + f'(x_0)(x - x_0) + \frac{f''(x_0)}{2!}(x - x_0)^2 + \cdots + \frac{f^{(n)}(x_0)}{n!}(x - x_0)^n \\ &= \sum_{k=0}^{n} \frac{f^{(k)}(x_0)}{k!}(x - x_0)^k. \end{aligned}$$

This important function is the basis for the following definition:

DEFINITION 8.6

Taylor Polynomials and Maclaurin Polynomials

Let f be a function with an nth-order derivative at the point x_0. We define the nth ***Taylor polynomial for*** f ***at*** x_0 to be the function

$$P_n(x) = \sum_{k=0}^{n} \frac{f^{(k)}(x_0)}{k!}(x - x_0)^k.$$

If $x_0 = 0$, we also call this the nth ***Maclaurin polynomial for*** f, which is of the form,

$$P_n(x) = \sum_{k=0}^{n} \frac{f^{(k)}(0)}{k!}x^k.$$

For example, the function $f(x) = \ln x$ has derivatives of all orders at every point in its domain, so we may find the third Taylor polynomial for $\ln x$ at $x = 1$. We have

$$f'(x) = \frac{1}{x}, \ f''(x) = -\frac{1}{x^2} \ \text{ and } \ f'''(x) = \frac{2}{x^3}.$$

To form the third Taylor polynomial P_3 at $x = 1$, we evaluate the function and these three derivatives at $x = 1$:

$$f(1) = 0, \ f'(1) = \frac{1}{1} = 1, \ f''(1) = -\frac{1}{1^2} = -1 \ \text{ and } \ f'''(1) = \frac{2}{1^3} = 2.$$

Therefore, $P_3(x) = (x-1) - \frac{1}{2}(x-1)^2 + \frac{1}{3}(x-1)^3$. Note that we have divided by the appropriate factorials to obtain the coefficients. We may also easily obtain the first and second Taylor polynomials for $\ln x$ at $x = 1$ by truncating P_3 appropriately. That is, $P_1(x) = x - 1$ is simply the tangent line to the function when $x = 1$, and $P_2(x) = (x-1) - \frac{1}{2}(x-1)^2$ is the second-order Taylor polynomial at $x = 1$. Following are plots of P_1, P_2, and P_3, along with a plot of the logarithm function:

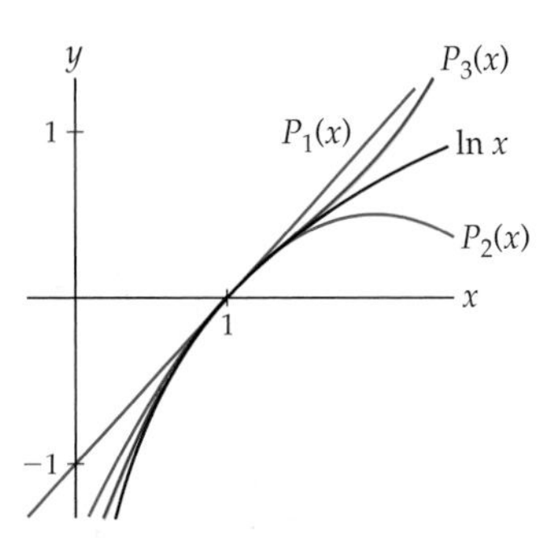

Note that each of these three polynomials approximates the natural logarithm function close to $x = 1$ and that the quality of the approximations improves as the degree of the Taylor polynomial increases.

Taylor Series and Maclaurin Series

Mathematicians love to go to extremes. Notice that our definitions of the Taylor polynomial at $x = x_0$ and the Maclaurin polynomial may be easily generalized to the following:

DEFINITION 8.7

Taylor Series and Maclaurin Series

Let f be a function with derivatives of all orders at the point x_0. We define the ***Taylor series for*** f ***at*** x_0 to be the following power series in $x - x_0$:

$$\sum_{k=0}^{\infty} \frac{f^{(k)}(x_0)}{k!}(x - x_0)^k.$$

If $x_0 = 0$, we also call this the ***Maclaurin series for*** f, which is of the form,

$$\sum_{k=0}^{\infty} \frac{f^{(k)}(0)}{k!} x^k.$$

This is a wonderful definition, but what relationship does a Taylor series at x_0 have to the original function f that is used to generate the series? Before answering that question, we will construct the Maclaurin series for the sine function:

n	$f^{(n)}(x)$	$f^{(n)}(0)$	$\frac{f^{(n)}(0)}{n!}$
0	$\sin x$	0	0
1	$\cos x$	1	1
2	$-\sin x$	0	0
3	$-\cos x$	-1	$-\frac{1}{3!}$
$\vdots$	$\vdots$	$\vdots$	$\vdots$
$2k$	$(-1)^k \sin x$	0	0
$2k+1$	$(-1)^k \cos x$	$(-1)^k$	$(-1)^k \frac{1}{(2k+1)!}$
$\vdots$	$\vdots$	$\vdots$	$\vdots$

The Maclaurin series for the sine function is

$$0 + x + 0x^2 - \frac{1}{3!}x^3 + 0x^4 + \frac{1}{5!}x^5 + 0x^6 - \cdots,$$

or $\sum_{k=0}^{\infty} \frac{(-1)^k}{(2k+1)!} x^{2k+1}$. In the following figure, we plot the sine function and the Maclaurin series for the sine, along with the 1st, 3rd, 5th and 17th Maclaurin polynomials for the sine function:

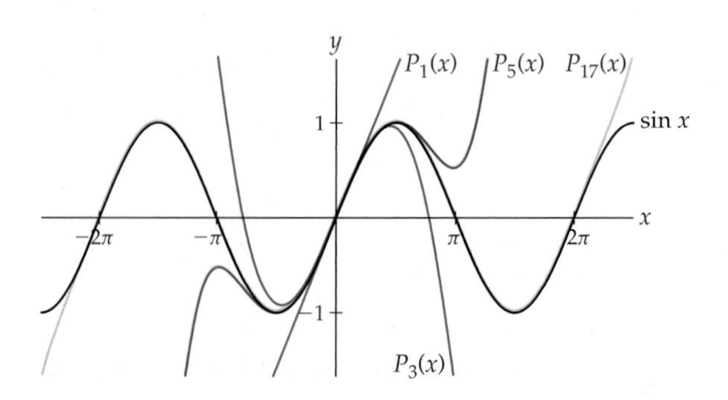

Note that the sine function and the 17th Maclaurin polynomial overlap almost perfectly on the interval $(-2\pi, 2\pi)$ and the sine function and its Maclaurin series *are identical* for every value of x, a fact that we will prove in Section 8.3.

We would like to repeat that a Maclaurin polynomial is just a Taylor polynomial when $x_0 = 0$. Taylor series and Maclaurin series bear the same relationship. Although the difference between the concepts is small, these are the traditional ways the names are used.

Examples and Explorations

EXAMPLE 1

Finding Maclaurin polynomials for e^x

Graph the exponential function e^x along with its first, second, and third Maclaurin polynomials.

SOLUTION

For any function f with a derivative of order 3 at $x = 0$, the first three Maclaurin polynomials are $P_1(x) = f(0) + f'(0)\,x$, $P_2(x) = f(0) + f'(0)\,x + \frac{f''(0)}{2!}x^2$, and

$$P_3(x) = f(0) + f'(0)\,x + \frac{f''(0)}{2!}x^2 + \frac{f'''(0)}{3!}x^3.$$

Since every derivative of $f(x) = e^x$ is e^x, we have $f^{(k)}(0) = 1$ for every nonnegative integer k. Therefore the first, second, and third Maclaurin polynomials for e^x are

$$P_1(x) = 1 + x,\ P_2(x) = 1 + x + \frac{1}{2}x^2 \text{ and } P_3(x) = 1 + x + \frac{1}{2}x^2 + \frac{1}{6}x^3.$$

Following are the graphs:

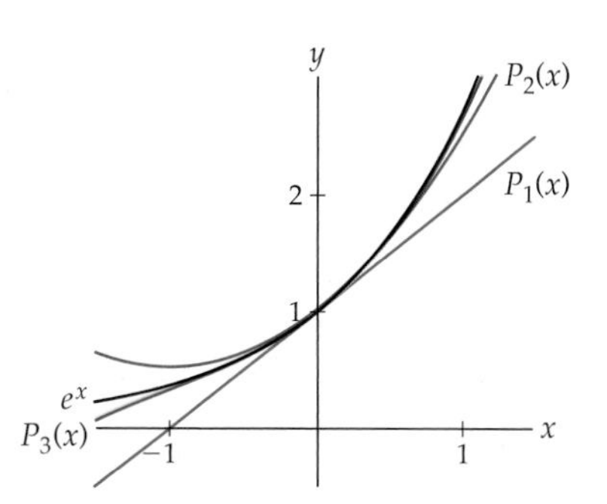

□

EXAMPLE 2

Finding the Maclaurin series for e^x

Find the Maclaurin series for the exponential function e^x, and find the interval of convergence for the series.

SOLUTION

In order to have a Maclaurin series, a function f must have a derivative of every order at $x = 0$. When this is the case, the Maclaurin series for f is given by $\sum_{k=0}^{\infty} \frac{f^{(k)}(0)}{k!}x^k$. We now build upon the work we started in Example 1. As we mentioned, when $f(x) = e^x$, $f^{(k)}(0) = 1$ for every nonnegative integer k. Thus, the Maclaurin series for e^x is $\sum_{k=0}^{\infty} \frac{1}{k!}x^k$.

To find the interval of convergence of the series, we use the ratio test for absolute convergence, with $b_k = \frac{1}{k!}x^k$ and $b_{k+1} = \frac{1}{(k+1)!}x^{k+1}$:

$$\lim_{k\to\infty}\left|\frac{b_{k+1}}{b_k}\right| = \lim_{k\to\infty}\left|\frac{x^{k+1}/(k+1)!}{x^k/k!}\right| \qquad \leftarrow \text{the ratio test for absolute convergence}$$

$$= \lim_{k\to\infty}|x|\frac{k!}{(k+1)!} \qquad \leftarrow \text{algebraic simplification of the quotient}$$

$$= \lim_{k\to\infty}|x|\frac{1}{k+1} \qquad \leftarrow \text{reduction of the factorials}$$

$$= 0. \qquad \leftarrow \text{evaluation of the limit}$$

Since the limit is zero for every value of x, the series has an infinite radius of convergence and the interval of convergence is $\mathbb{R}$. □

EXAMPLE 3

Finding Taylor series for $\ln x$ at $x = 2$

Find the Taylor series for $\ln x$ at $x = 2$ and the interval of convergence for the series. Also, graph the Taylor polynomials of degrees 1, 2, and 3 for $\ln x$ at $x = 2$.

SOLUTION

Since $f(x) = \ln x$ has derivatives of every order at every point in its domain, a Taylor series may be constructed for every $x > 0$. At $x = 2$ the Taylor series is given by $\sum_{k=0}^{\infty} \frac{f^{(k)}(2)}{k!}(x-2)^k$. To complete the series, we need to find the general form of the kth derivative of $\ln x$ at $x = 2$. Toward that end, we construct the following table:

k	$f^{(k)}(x)$	$f^{(k)}(2)$	$\frac{f^{(k)}(2)}{k!}$
0	$\ln x$	$\ln 2$	$\ln 2$
1	$\frac{1}{x}$	$\frac{1}{2}$	$\frac{1}{2}$
2	$-\frac{1}{x^2}$	$-\frac{1}{2^2}$	$-\frac{1}{2\cdot 2^2}$
3	$\frac{2}{x^3}$	$\frac{2}{2^3}$	$\frac{2}{3!2^3}$
4	$-\frac{3!}{x^4}$	$-\frac{3!}{2^4}$	$-\frac{3!}{4!2^4}$
⋮	⋮	⋮	⋮
k	$(-1)^{k-1}\frac{(k-1)!}{x^k}$	$(-1)^{k-1}\frac{(k-1)!}{2^k}$	$(-1)^{k-1}\frac{(k-1)!}{k!2^k} = (-1)^{k-1}\frac{1}{k2^k}$

(To be thorough, we should use the principle of mathematical induction to prove that

$$\frac{d^k}{dx^k}(\ln x) = (-1)^{k-1}\frac{(k-1)!}{x^k}.$$

We leave this step for Exercise 69.) Using the information in the table, we see that the Taylor series for $\ln x$ at $x = 2$ is

$$\sum_{k=0}^{\infty} \frac{f^{(k)}(2)}{k!}(x-2)^k = \ln 2 + \sum_{k=1}^{\infty} \frac{(-1)^{k-1}(k-1)!/2^k}{k!}(x-2)^k = \ln 2 + \sum_{k=1}^{\infty} \frac{(-1)^{k-1}}{k2^k}(x-2)^k.$$

Note that the "zeroth" derivative does not fall into the same pattern as every other derivative. Therefore, we split that term, ln 2, off from the summation of the rest of the terms.

To find the interval of convergence for this series, we let

$$b_k = \frac{(-1)^{k-1}}{k2^k}(x-2)^k \quad \text{and} \quad b_{k+1} = \frac{(-1)^k}{(k+1)2^{k+1}}(x-2)^{k+1}$$

and evaluate

$$\begin{aligned}
\lim_{k\to\infty}\left|\frac{b_{k+1}}{b_k}\right| &= \lim_{k\to\infty}\left|\frac{(-1)^k(x-2)^{k+1}/((k+1)2^{k+1})}{(-1)^{k-1}(x-2)^k/(k2^k)}\right| && \leftarrow \text{ratio test for absolute convergence}\\
&= \lim_{k\to\infty}|x-2|\frac{k2^k}{(k+1)2^{k+1}} && \leftarrow \text{algebraic simplification of the quotient}\\
&= \lim_{k\to\infty}|x-2|\frac{k}{2(k+1)} && \leftarrow \text{reduction}\\
&= \frac{1}{2}|x-2|. && \leftarrow \text{evaluation of the limit}
\end{aligned}$$

By the ratio test for absolute convergence, the series will converge absolutely when $\frac{1}{2}|x-2| < 1$ or, equivalently, when $|x-2| < 2$. The radius of convergence is 2, and the series converges absolutely for $0 < x < 4$. In Exercise 3 you will show that the series diverges when $x = 0$ and converges conditionally when $x = 4$. Thus, the interval of convergence for this series is $(0, 4]$.

Finally, in the following figure, we graph the natural logarithm function along with the Taylor polynomials $P_1(x) = \ln 2 + \frac{1}{2}(x-2)$, $P_2(x) = \ln 2 + \frac{1}{2}(x-2) - \frac{1}{8}(x-2)^2$, and $P_3(x) = \ln 2 + \frac{1}{2}(x-2) - \frac{1}{8}(x-2)^2 + \frac{1}{24}(x-2)^3$:

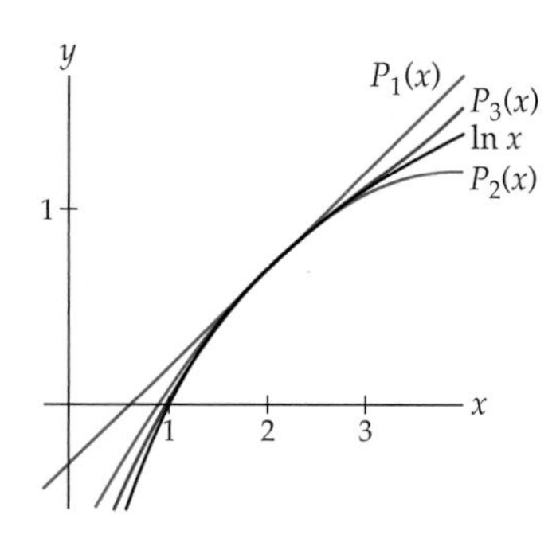

□

EXAMPLE 4 **Constructing binomial series**

For a nonzero constant p, the Maclaurin series for $(1+x)^p$ is called the ***binomial series***. Show that the binomial series is given by $\sum_{k=0}^{\infty}\binom{p}{k}x^k$, where

$$\binom{p}{k} = \begin{cases} \dfrac{p(p-1)(p-2)\cdots(p-k+1)}{k!}, & \text{if } k > 0 \\ 1, & \text{if } k = 0. \end{cases}$$

SOLUTION

Let $f(x) = (1+x)^p$. We begin by constructing the following table:

k	$f^{(k)}(x)$	$f^{(k)}(0)$
0	$(1+x)^p$	1
1	$p(1+x)^{p-1}$	p
2	$p(p-1)(1+x)^{p-2}$	$p(p-1)$
3	$p(p-1)(p-2)(1+x)^{p-3}$	$p(p-1)(p-2)$
$\vdots$	$\vdots$	$\vdots$
k	$p(p-1)(p-2)\cdots(p-k+1)(1+x)^{p-k}$	$p(p-1)(p-2)\cdots(p-k+1)$

Using the information in this table, we see that the Maclaurin series for $(1+x)^p$ is

$$1 + px + \frac{p(p-1)}{2!}x^2 + \frac{p(p-1)(p-2)}{3!}x^2 + \cdots + \frac{p(p-1)(p-2)\cdots(p-k+1)}{k!}x^k + \cdots.$$

We define the ***binomial coefficient*** $\binom{p}{k}$ by

$$\binom{p}{k} = \begin{cases} \dfrac{p(p-1)(p-2)\cdots(p-k+1)}{k!}, & \text{if } k > 0 \\ 1, & \text{if } k = 0. \end{cases}$$

The binomial coefficients $\binom{p}{k}$ just defined are generalizations of the binomial coefficients you may have previously studied, for the case where p is a nonnegative integer and $p \geq k$. For example, if $p = 3$, then

$$\binom{3}{0} = 1, \quad \binom{3}{1} = \frac{3}{1} = 3, \quad \binom{3}{2} = \frac{3 \cdot 2}{2!} = 3, \quad \binom{3}{3} = \frac{3 \cdot 2 \cdot 1}{3!} = 1.$$

Note that $\binom{3}{k} = 0$ for all integers $k > 3$, since the numerator of $\binom{3}{k}$ would contain a factor of 0. Thus, here we would have the binomial series $(1+x)^3 = 1 + 3x + 3x^2 + x^3$, as expected.

In Exercise 58 we ask you to find the interval of convergence for binomial series. □

EXAMPLE 5 **Constructing the binomial series for $\sqrt{1+x}$**

Use the result from Example 4 to find the binomial series for $f(x) = \sqrt{1+x}$.

SOLUTION

We have $f(x) = (1+x)^{1/2}$. Thus,

$$\begin{aligned}\sqrt{1+x} &= \sum_{k=0}^{\infty} \binom{1/2}{k} x^k \\ &= 1 + \frac{1}{2}x + \frac{\frac{1}{2}\left(-\frac{1}{2}\right)}{2!}x^2 + \frac{\frac{1}{2}\left(-\frac{1}{2}\right)\left(-\frac{3}{2}\right)}{3!}x^3 + \frac{\frac{1}{2}\left(-\frac{1}{2}\right)\left(-\frac{3}{2}\right)\left(-\frac{5}{2}\right)}{4!}x^4 + \cdots \\ &= 1 + \frac{1}{2}x - \frac{1}{8}x^2 + \frac{1}{16}x^3 - \frac{5}{128}x^4 + \cdots.\end{aligned}$$

□

? TEST YOUR UNDERSTANDING

- What is a Maclaurin polynomial and what is a Taylor polynomial? What is the difference between a Maclaurin polynomial and a Taylor polynomial?
- What is the relationship between the tangent line to a function f at $x = x_0$ and a Taylor polynomial for f at $x = x_0$?
- What is a Maclaurin series and what is a Taylor series? What is the difference between a Maclaurin series and a Taylor series?
- What is the relationship between a Maclaurin polynomial of degree n for a function f and the Maclaurin series for f?
- What is the relationship between a Taylor polynomial of degree n for a function f at $x = x_0$ and the Taylor series for f at $x = x_0$?

EXERCISES 8.2

Thinking Back

- *Polynomials:* What is a polynomial? What is the domain of every polynomial function?
- *Derivatives and linear approximations:* If a function f is differentiable at the point x_0, what is the equation of the line tangent to the graph of f at x_0? Why is this line a good approximation for f near x_0?

Concepts

0. *Problem Zero:* Read the section and make your own summary of the material.

1. *True/False:* Determine whether each of the statements that follow is true or false. If a statement is true, explain why. If a statement is false, provide a counterexample.

(a) *True or False:* If a function f is differentiable at 0, then f has a first-order Maclaurin polynomial.

(b) *True or False:* If a function f has derivatives of every order at $x = 5$, then f has a Taylor series at $x = 5$.

(c) *True or False:* The first-order Maclaurin polynomial for $f(x) = |x|$ is $P_1(x) = x$.

(d) *True or False:* If a function f has a Taylor polynomial of order n at $x = 3$, then f has a Taylor polynomial of order k at $x = 3$ for every $k < n$.

(e) *True or False:* If a function f has an nth-order Taylor polynomial $P_n(x)$ at $x = \pi$, then $f^{(n)}(\pi) = P_n(\pi)$.

(f) *True or False:* If a function f has a Maclaurin series, then f has derivatives of every order at every point in its domain.

(g) *True or False:* If a function f has a Taylor series at a point x_0, then f has derivatives of every order at x_0.

(h) *True or False:* If a function f has a Taylor series at a every point in its domain, then f has a Maclaurin series.

2. *Examples:* Construct examples of the thing(s) described in the following. Try to find examples that are different than any in the reading.

(a) A function with a Maclaurin polynomial of order 3.

(b) A function with domain $\mathbb{R}$ without a Maclaurin polynomial.

(c) A function with a Taylor series for every $x_0 \in \mathbb{R}$.

3. Show that the series

$$\ln 2 + \sum_{k=1}^{\infty} \frac{(-1)^{k-1}}{k2^k}(x-2)^k$$

from Example 3 diverges when $x = 0$ and converges conditionally when $x = 4$.

4. What is a Taylor polynomial for a function f at a point x_0?

5. Let f be a twice-differentiable function at a point x_0. Using the words *value, slope,* and *concavity*, explain why the second Taylor polynomial $P_2(x)$ might be a good approximation for f close to x_0.

6. Let f be a twice-differentiable function at a point x_0. Explain why the sum

$$f(x) + f'(x)(x - x_0) + \frac{f''(x)}{2}(x - x_0)^2$$

is *not* the second-order Taylor polynomial for f at x_0.

7. What is a difference between the Maclaurin polynomial of order n and the Taylor polynomial of order n for a function f?

8. What is a difference between a Maclaurin polynomial and the Maclaurin series for a function f?

9. What is a difference between a Taylor polynomial and the Taylor series for a function f at a point x_0?

10. What is the relationship between a Maclaurin series and a power series in x?

11. If a function f has a Maclaurin series, what are the possibilities for the interval of convergence for that series?

12. If a function f has a Taylor series at x_0, what are the possibilities for the interval of convergence for that series?

13. Let $f(x) = 3x^2 - 2x + 5$. Find the first-, second-, and third-order Maclaurin polynomials, $P_1(x)$, $P_2(x)$, and $P_3(x)$, for f. Explain why $f(x) = P_2(x) = P_3(x)$. Graph $f(x)$, $P_1(x)$, and $P_2(x)$.

14. Let $f(x) = 4x^3 - 5x^2 - 6x + 7$. Find the first- through fourth-order Maclaurin polynomials, $P_1(x)$, $P_2(x)$, $P_3(x)$, and $P_4(x)$, for f. Explain why $f(x) = P_3(x) = P_4(x)$. Graph $f(x)$, $P_1(x)$, $P_2(x)$, and $P_3(x)$.

15. Let $f(x) = 3x^2 - 2x + 5$. Find the first-, second-. and third-order Taylor polynomials, $P_1(x)$, $P_2(x)$, and $P_3(x)$, for f at 1. Explain why $f(x) = P_2(x) = P_3(x)$.

16. Let $f(x) = 4x^3 - 5x^2 - 6x + 7$. Find the first- through fourth-order Taylor polynomials, $P_1(x)$, $P_2(x)$, $P_3(x)$, and $P_4(x)$, for f at 1. Explain why $f(x) = P_3(x) = P_4(x)$.

17. Let $f(x) = ax^3 + bx^2 + cx + d$, where a, b, c, and d are constants. Find the first- through fourth-order Taylor polynomials, $P_1(x)$, $P_2(x)$, $P_3(x)$, and $P_4(x)$, for f at x_0. Explain why $f(x) = P_3(x) = P_4(x)$.

18. Let $f(x)$ be an nth-degree polynomial function.
    (a) Explain why f has a Taylor series for every value of x_0.
    (b) Explain why $P_m(x)$, the mth-order Taylor polynomial for f at x_0, equals $f(x)$ for every integer $m \geq n$.

19. Let $f(x) = x^{7/3}$.
    (a) Find $P_1(x)$ and $P_2(x)$, the Maclaurin polynomials of orders 1 and 2 for f. Graph $f(x)$ together with $P_1(x)$ and $P_2(x)$.
    (b) Explain why f does *not* have a Maclaurin polynomial of order 3.

20. Use the ideas in Exercise 19 to find a function that has a Maclaurin polynomial of order 3 but does not have a Maclaurin polynomial of order 4. Find a function f such that f has a Maclaurin polynomial of order n but does not have a Maclaurin polynomial of order $n + 1$.

Skills

In Exercises 21–30 find the fourth Maclaurin polynomial $P_4(x)$ for the specified function.

21. $\cos x$
22. e^x
23. $\sin x$
24. $\ln(1 + x)$
25. $\cos 2x$
26. $\sin 3x$
27. $\sqrt{1 - x}$
28. $\dfrac{1 + x}{1 - x}$
29. $x \sin x$
30. $x^2 e^x$

In Exercises 31–40 find the Maclaurin series for the specified function. Note: These are the same functions as in Exercises 21–30.

31. $\cos x$
32. e^x
33. $\sin x$
34. $\ln(1 + x)$
35. $\cos 2x$
36. $\sin 3x$
37. $\sqrt{1 - x}$
38. $\dfrac{1 + x}{1 - x}$
39. $x \sin x$
40. $x^2 e^x$

In Exercises 41–48 find the fourth Taylor polynomial $P_4(x)$ for the specified function and the given value of x_0.

41. $\cos x$, $\dfrac{\pi}{2}$
42. e^x, 1
43. $\sin x$, π
44. $\sqrt{x}$, 1
45. $\ln x$, 3
46. $\sqrt[3]{x}$, 1
47. $\cos 2x$, $\dfrac{\pi}{4}$
48. $\sin 3x$, $\dfrac{\pi}{6}$

In Exercises 49–56 find the Taylor series for the specified function and the given value of x_0. Note: These are the same functions and values as in Exercises 41–48.

49. $\cos x$, $\dfrac{\pi}{2}$
50. e^x, 1
51. $\sin x$, π
52. $\sqrt{x}$, 1
53. $\ln x$, 3
54. $\sqrt[3]{x}$, 1
55. $\cos 2x$, $\dfrac{\pi}{4}$
56. $\sin 3x$, $\dfrac{\pi}{6}$

Exercises 57–62 concern the binomial series discussed in Example 4. In Exercises 59–62 use the binomial series to find the Maclaurin series for the given function.

57. Show that if p is a positive integer, then the binomial series for $f(x) = (1 + x)^p$ is a polynomial.

58. Show that the radius of convergence for the binomial series is 1 when p is not a positive integer. What is the radius of convergence when p is a positive integer? *(Hint: Consider Exercise 57.)*

59. $\sqrt[3]{1 + x}$
60. $\dfrac{1}{(1 + x)^2}$
61. $\dfrac{1}{\sqrt{1 + x}}$
62. $(1 + x)^{2/3}$

63. Let $f(x) = \begin{cases} 0, & \text{if } x = 0 \\ e^{-1/x^2}, & \text{if } x \neq 0 \end{cases}$
    (a) Use the definition of the derivative at a point to show that f has derivatives of all orders at $x = 0$.
    (b) Show that $f^{(k)}(0) = 0$ for every nonnegative integer k.
    (c) What is the Maclaurin series for f?
    (d) Explain why the Maclaurin series does not converge to $f(x)$.

Applications

The second-order differential equation

$$x^2y'' + xy' + (x^2 - p^2) = 0,$$

where p is a nonnegative integer, arises in many applications in physics and engineering, including one model for the vibration of a beaten drum. The solution of the differential equation is called the ***Bessel function of order p***, denoted by $J_p(x)$. It may be shown that $J_p(x)$ is given by the following power series in x:

$$J_p(x) = \sum_{k=0}^{\infty} \frac{(-1)^k}{k!(k+p)!2^{2k+p}} x^{2k+p}.$$

Exercises 64–68 concern these Bessel functions.

64. What is the interval of convergence for $J_0(x)$?

65. Find and graph the first four terms in the sequence of partial sums of $J_0(x)$.

66. What is the interval of convergence for $J_1(x)$?

67. Graph the first four terms in the sequence of partial sums of $J_1(x)$.

68. What is the interval of convergence for $J_p(x)$ where p is a nonnegative integer?

Proofs

69. Use the principle of mathematical induction to prove that $\frac{d^k}{dx^k}(\ln x) = (-1)^{k-1}\frac{(k-1)!}{x^k}$, for $k \geq 1$.

70. Let f be a function with an nth-order derivative at a point x_0, and let $P_n(x) = \sum_{k=0}^{n} \frac{f^{(k)}(x_0)}{k!}(x - x_0)^k$. Prove that $f^{(k)}(x_0) = P_n^{(k)}(x_0)$ for every nonnegative integer $k \leq n$.

Thinking Forward

- *Graphing differences:* Let $P_1(x)$, $P_2(x)$, and $P_3(x)$ be the first three Maclaurin polynomials for the function e^x. Make separate graphs of the functions $e^x - P_1(x)$, $e^x - P_2(x)$, and $e^x - P_3(x)$ on the interval $[-2, 2]$. What do you observe? In particular, for each of these functions, for what values of x is the graph of the function within 0.1 of the x-axis?

- *Graphing more differences:* Let $P_1(x)$, $P_2(x)$, and $P_3(x)$ be the first three Maclaurin polynomials for the function $\ln(1+x)$. *(Hint: You may use your answer from Exercise 24 to construct these polynomials.)*

  Make separate graphs of the functions $\ln(1+x) - P_1(x)$, $\ln(1+x) - P_2(x)$, and $\ln(1+x) - P_3(x)$ on the interval $(-1, 3)$. What do you observe? In particular, for each of these functions, for what values of x is the graph of the function within 0.1 of the x-axis?

8.3 CONVERGENCE OF POWER SERIES

- The remainder of a power series
- Taylor's Theorem and Lagrange's form for the remainder
- Algebraic manipulations of power series

The Remainder

In Section 7.4 we defined the remainder of a convergent series of constants to be the difference of the sum L of the series and a term s_n from the sequence of partial sums. The remainder measures the error incurred in using s_n to approximate L. We now define the analogous concept for Taylor series and Maclaurin series.

DEFINITION 8.8 **The Remainder of a Function**

Let f be a function with an nth-order derivative at every point on an open interval containing the point x_0, and let $P_n(x) = \sum_{k=0}^{n} \frac{f^{(k)}(x_0)}{k!}(x - x_0)^k$, the nth Taylor polynomial for f at x_0. We define the nth ***remainder*** for f, denoted by $R_n(x)$, to be

$$R_n(x) = f(x) - P_n(x).$$

The remainder measures the error incurred in using the Taylor polynomial, $P_n(x)$, to approximate the function $f(x)$. As the notation implies, the remainder is a function of x. By the definition of the Taylor polynomial at $x = x_0$, the remainder will always be zero at x_0 and generally is greater the farther x is from the point x_0. For example, we have seen that the fifth Maclaurin polynomial for the sine function is $P_5(x) = x - \frac{x^3}{3!} + \frac{x^5}{5!}$. Therefore, the fifth remainder for the sine function at $x = 0$ is $R_5(x) = \sin x - \left(x - \frac{x^3}{3!} + \frac{x^5}{5!}\right)$, shown next:

$R_5(x)$ for $\sin x$ at $x = 0$

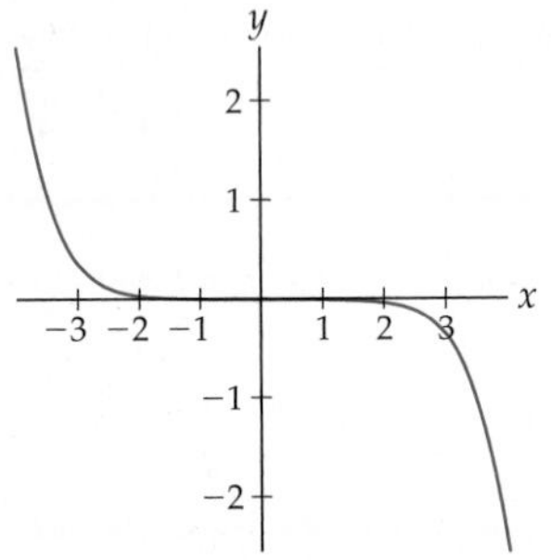

y
2
1
−3 −2 −1
1 2 3
x
−1
−2

Since the remainder is very close to zero on the interval $(-1.5, 1.5)$, $P_5(x)$ is an excellent approximation for the sine function on that interval.

We wish to understand the behavior of $R_n(x)$ as $n \to \infty$. If $\lim_{n\to\infty} R_n(x) = 0$ on some interval I, then the Taylor polynomials become better approximations for the function f as we include terms of higher degree, and, furthermore, the Taylor series is an alternative representation for f on the interval I. We will see that a Taylor series may converge to the function that generated it for every real number, converge to the function on a finite interval centered around the point $x = x_0$, or converge only at the point x_0. To see this, we need a tool to analyze the remainder.

THEOREM 8.9 **Taylor's Theorem**

Let f be a function that can be differentiated $n+1$ times in some open interval I containing the point x_0, and let $R_n(x)$ be the nth remainder for f at $x = x_0$. The nth remainder for f is given by

$$R_n(x) = \int_{x_0}^{x} \frac{(x-t)^n}{n!} f^{(n+1)}(t)\, dt.$$

Proof. By the Fundamental Theorem of Calculus, we have

$$\int_{x_0}^{x} f'(t)\, dt = f(x) - f(x_0).$$

The bulk of our proof of Theorem 8.9 comes from repeatedly applying integration by parts to integrals derived from this equation. Using integration by parts on the integral, with $u = f'(t)$ and $dv = dt$, we have $du = f''(t)\, dt$ and may let $v = t - x$. (Our variable here is t. Think of x as a constant of integration.) We obtain

$$\begin{aligned}\int_{x_0}^{x} f'(t)\, dt &= f'(t)(t-x)\Big|_{x_0}^{x} - \int_{x_0}^{x} (t-x) f''(t)\, dt \\ &= f'(x_0)(x-x_0) + \int_{x_0}^{x} (x-t) f''(t)\, dt.\end{aligned}$$

We apply integration by parts again, this time to the integral we just obtained with $u = f''(t)$ and $dv = (x - t)\,dt$. Then $du = f'''(t)\,dt$ and $v = -\frac{(x-t)^2}{2}$. (Remember that our variable is t.) Thus,

$$\int_{x_0}^{x} (x - x_0)f''(t)\,dt = \left[-f''(t)\frac{(x-t)^2}{2}\right]_{x_0}^{x} + \int_{x_0}^{x} \frac{(x-t)^2}{2} f'''(t)\,dt$$

$$= \frac{f''(x_0)}{2}(x - x_0)^2 + \int_{x_0}^{x} \frac{(x-t)^2}{2} f'''(t)\,dt.$$

We now combine the preceding equations to obtain

$$f(x) = f(x_0) + f'(x_0)(x - x_0) + \frac{f''(x_0)}{2}(x - x_0)^2 + \int_{x_0}^{x} \frac{(x-t)^2}{2} f'''(t)\,dt.$$

In Exercise 87 you will use the principle of mathematical induction to prove that

$$f(x) = f(x_0) + f'(x_0)(x - x_0) + \frac{f''(x_0)}{2}(x - x_0)^2 + \cdots$$

$$+ \frac{f^{(n)}(x_0)}{n!}(x - x_0)^n + \int_{x_0}^{x} \frac{(x-t)^n}{n!} f^{(n+1)}(t)\,dt$$

$$= P_n(x) + \int_{x_0}^{x} \frac{(x-t)^n}{n!} f^{(n+1)}(t)\,dt.$$

If we now compare this with our definition of $R_n(x)$, we have our result. ■

Unfortunately, it is usually difficult to analyze the remainder by using the integral in Theorem 8.9. Instead, the next theorem gives us a more useful tool for understanding the remainder.

THEOREM 8.10 **Lagrange's Form for the Remainder**

Let f be a function that can be differentiated $n + 1$ times in some open interval I containing the point x_0, and let $R_n(x)$ be the nth remainder for f at $x = x_0$. For each point $x \in I$, there is at least one c between x_0 and x such that

$$R_n(x) = \frac{f^{(n+1)}(c)}{(n+1)!}(x - x_0)^{n+1}.$$

Theorem 8.10 follows from Theorem 8.9 and Theorem 4.31 of Section 4.6. We ask you to prove Theorem 8.10 in Exercise 88.

We now use Theorem 8.10 to prove that the sine function and its Maclaurin series are identical for every real number; that is,

$$\sin x = \sum_{k=0}^{\infty} \frac{(-1)^k}{(2k+1)!} x^{2k+1}.$$

We let $f(x) = \sin x$. We know that for every $n \geq 0$, $f^{(n+1)}(x)$ is one of the four functions $\pm \sin x$ and $\pm \cos x$. For any of these functions, $\left|f^{(n+1)}(c)\right| \leq 1$ for every value of x, so, using Lagrange's form for the remainder, we have

$$\left|R_n(x)\right| = \left|\frac{f^{(n+1)}(c)}{(n+1)!} x^{n+1}\right| \leq \frac{|x|^{n+1}}{(n+1)!}.$$

We now show that $\lim_{n\to\infty} |R_n(x)| = 0$ for every value of x. Note that we are taking the limit as $n \to \infty$, so we may regard x as being fixed. Previously we proved that for every value of x, $\frac{|x|^n}{n!} \to 0$ as $n \to \infty$. Therefore by the Squeeze Theorem, $\lim_{n\to\infty} |R_n(x)| = 0$, which implies that $\lim_{n\to\infty} R_n(x) = 0$. Since $\lim_{n\to\infty} R_n(x) = 0$ for every value of x, the sine function is equal to its Maclaurin series.

There are several important Maclaurin series that we will be using regularly. All of these series are summarized in the following table:

Function	Series	Interval of Convergence
$\sin x$	$\sum_{k=0}^{\infty} \frac{(-1)^k}{(2k+1)!} x^{2k+1}$	$\mathbb{R}$
$\cos x$	$\sum_{k=0}^{\infty} \frac{(-1)^k}{(2k)!} x^{2k}$	$\mathbb{R}$
e^x	$\sum_{k=0}^{\infty} \frac{1}{k!} x^k$	$\mathbb{R}$
$\frac{1}{1-x}$	$\sum_{k=0}^{\infty} x^k$	$(-1, 1)$
$\ln(1+x)$	$\sum_{k=1}^{\infty} \frac{(-1)^{k+1}}{k} x^k$	$(-1, 1]$
$\tan^{-1} x$	$\sum_{k=0}^{\infty} \frac{(-1)^k}{2k+1} x^{2k+1}$	$[-1, 1]$
$\cosh x$	$\sum_{k=0}^{\infty} \frac{1}{(2k)!} x^{2k}$	$\mathbb{R}$
$\sinh x$	$\sum_{k=0}^{\infty} \frac{1}{(2k+1)!} x^{2k+1}$	$\mathbb{R}$

We have already proven that the sine function and its Maclaurin series are equal for every value of x and that the function $\frac{1}{1-x}$ and its Maclaurin series are equal on the interval $(-1, 1)$. In Exercise 89 you will prove that cosine and its Maclaurin series are equal for every x. We will prove that e^x and its Maclaurin series are equal in Example 1. We derive the Maclaurin series for the hyperbolic cosine in Example 2 and leave the derivation of the hyperbolic sine function to Exercise 91. The convergence of the remaining two series in this table is proven in Section 8.4.

Basic Manipulations of Maclaurin Series

In this subsection we will discuss three methods for obtaining a new Taylor series from known series relatively quickly: substitution, multiplication by a monomial, and the multiplication of two series. Given a function f that has derivatives of all orders at a point x_0, Definition 8.7 provides the pattern for constructing the Taylor series for the function. To form the series, we compute $f^{(k)}(x)$ for every k and then evaluate $f^{(k)}(x_0)$. Unfortunately, finding higher order derivatives even for a relatively simple function can become quite complicated. For example, think about constructing the Maclaurin series for $f(x) = \sin(x^2)$. Since the sine function and x^2 both have derivatives of all orders everywhere, the composition function has derivatives of all orders at $x = 0$. We need to find a pattern for the derivatives of f. When we start taking the derivatives, we find that

$$f'(x) = 2x\cos(x^2),\ f''(x) = 2\cos(x^2) - 4x^2\sin(x^2),\ f'''(x) = -12x\sin(x^2) - 8x^3\cos(x^2), \ldots$$

Certainly, a pattern is developing, but the derivatives become increasingly unwieldy with increasing order. Fortunately, there is a simpler way to find the Maclaurin series for this function. We have already shown that $\sin x = \sum_{k=0}^{\infty} \frac{(-1)^k}{(2k+1)!} x^{2k+1}$ and that this series converges for every real number. To construct the Maclaurin series for $\sin(x^2)$, all we have to do is substitute x^2 for each occurrence of x in the Maclaurin series for the sine. That is, since

$$\sin x = \sum_{k=0}^{\infty} \frac{(-1)^k}{(2k+1)!} x^{2k+1}$$

the Maclaurin series for $\sin(x^2)$ is

$$\sin(x^2) = \sum_{k=0}^{\infty} \frac{(-1)^k}{(2k+1)!} (x^2)^{2k+1} = \sum_{k=0}^{\infty} \frac{(-1)^k}{(2k+1)!} x^{4k+2}.$$

That was fast! Because the Maclaurin series for the sine converges for every real number, this new series will also converge for all real numbers.

More generally, given the Maclaurin series for a function f, $\sum_{k=0}^{\infty} \frac{f^{(k)}(0)}{k!} x^k$, and a monomial cx^m, where $c \in \mathbb{R}$ and $m \in \mathbb{Z}^+$ we have the Maclaurin series for $f(cx^m)$,

$$\sum_{k=0}^{\infty} \frac{f^{(k)}(0)}{k!} (cx^m)^k = \sum_{k=0}^{\infty} \frac{f^{(k)}(0)}{k!} c^k x^{mk}.$$

If the Maclaurin series $\sum_{k=0}^{\infty} \frac{f^{(k)}(0)}{k!} x^k$ converges for every real number, so will the new series $\sum_{k=0}^{\infty} \frac{f^{(k)}(0)}{k!} c^k x^{mk}$. If the Maclaurin series $\sum_{k=0}^{\infty} \frac{f^{(k)}(0)}{k!} x^k$ converges for every real number in the interval defined by $|x| < \rho$, the new series will converge on the interval defined by the inequality $|cx^m| < \rho$, that is $|x| < \sqrt[m]{\rho/|c|}$.

Next, when we have a Maclaurin series for $f(x)$ that converges on an interval I, we may also quickly find the Maclaurin series for $cx^m f(x)$, where $c \in \mathbb{R}$ and m is a nonnegative integer. We already know that for a convergent series of constants,

$$c\sum_{k=0}^{\infty} a_k = \sum_{k=0}^{\infty} ca_k.$$

We may similarly multiply a Maclaurin series by the monomial cx^m. That is, when $f(x) = \sum_{k=0}^{\infty} \frac{f^{(k)}(0)}{k!} x^k$ converges on some interval I, the Maclaurin series for $cx^m f(x)$ is

$$cx^m f(x) = cx^m \sum_{k=0}^{\infty} \frac{f^{(k)}(0)}{k!} x^k = \sum_{k=0}^{\infty} c\frac{f^{(k)}(0)}{k!} x^{k+m},$$

with interval of convergence I.

For example, we may immediately find the Maclaurin series for $\frac{x^3}{1-x}$, since this function is just $x^3 \cdot \frac{1}{1-x}$. So

$$\frac{x^3}{1-x} = x^3 \sum_{k=0}^{\infty} x^k = \sum_{k=0}^{\infty} x^{k+3} = \sum_{k=3}^{\infty} x^k.$$

Make sure you understand each of the preceding steps. Multiplication by a nonzero monomial does not change the interval of convergence, so this new Maclaurin series converges to $\frac{x^3}{1-x}$ on the interval $(-1, 1)$.

Finally, given two functions f and g with convergent Taylor series at x_0, $f(x) = \sum_{k=0}^{\infty} a_k(x-x_0)^k$ and $g(x) = \sum_{k=0}^{\infty} b_k(x-x_0)^k$, and with intervals of convergence I_1 and I_2, respectively, we may form the Taylor series at x_0 for the product function $f \cdot g$ by multiplying the two known series. That is,

$$f(x)\,g(x) = \sum_{k=0}^{\infty} c_k(x-x_0)^k,$$

where

$$c_k = \sum_{m=0}^{k} a_m b_{k-m} = a_0\, b_k + a_1 b_{k-1} + \cdots + a_{k-1} b_1 + a_k\, b_0.$$

The interval of convergence for the new series is $I_1 \cap I_2$. See Example 7.

Examples and Explorations

EXAMPLE 1 **Showing that the exponential function and its Maclaurin series are identical for every x**

Use Lagrange's form for the remainder to prove that

$$e^x = \sum_{k=0}^{\infty} \frac{1}{k!} x^k$$

for every real number.

SOLUTION

Lagrange's form for the remainder is $R_n(x) = \frac{f^{(n+1)}(c)}{(n+1)!}(x-x_0)^{n+1}$, where c is between x_0 and x. Here, since $f(x) = e^x$, we have $f^{(n+1)}(c) = e^c$ for every $n \geq 0$. Furthermore, since $x_0 = 0$, we also have $e^c < e^{|x|}$, so

$$\left|R_n(x)\right| = \left|\frac{e^c}{(n+1)!} x^{n+1}\right| \leq e^{|x|} \frac{|x|^{n+1}}{(n+1)!}.$$

We now take the limit

$$\lim_{n\to\infty} \left|R_n(x)\right| \leq \lim_{n\to\infty} e^{|x|} \frac{|x|^{n+1}}{(n+1)!} = 0.$$

The limit is zero because the quotient $\frac{|x|^{n+1}}{(n+1)!} \to 0$ as $n \to \infty$ and we are taking the limit with respect to n, not x. This limit implies that $\lim_{n\to\infty} R_n(x) = 0$ as well. Thus, the Maclaurin series for the exponential function converges to e^x for every value of x. □

EXAMPLE 2 **Deriving the Maclaurin series for the hyperbolic cosine**

Show that $\cosh x = \sum_{k=0}^{\infty} \frac{1}{(2k)!} x^{2k}$.

SOLUTION

We saw in Section 2.6 that $\cosh x = \frac{e^x+e^{-x}}{2}$. Since $e^x = \sum_{k=0}^{\infty} \frac{1}{k!}x^k = 1+x+\frac{x^2}{2!}+\frac{x^3}{3!}+\frac{x^4}{4!}+\cdots$, by substitution we have

$$e^{-x} = \sum_{k=0}^{\infty} \frac{1}{k!}(-x)^k = 1 - x + \frac{x^2}{2!} - \frac{x^3}{3!} + \frac{x^4}{4!} - \cdots.$$

When we add the series for e^x and e^{-x}, all of the terms with odd exponents sum to zero. That is,

$$e^x + e^{-x} = 2 + \frac{2x^2}{2!} + \frac{2x^4}{4!} + \frac{2x^6}{6!} + \cdots.$$

Dividing the resulting series by 2, we obtain the series

$$\cosh x = \sum_{k=0}^{\infty} \frac{1}{(2k)!}x^{2k}.$$

□

EXAMPLE 3

Evaluating Maclaurin series at points in their intervals of convergence

Find the sums of the following series:

(a) $\sum_{k=0}^{\infty} \frac{(-5)^k}{k!}$ **(b)** $\sum_{k=2}^{\infty} \frac{(-1)^k \pi^{2k+1}}{(2k+1)!}$

SOLUTION

In Chapter 7 we would only have asked whether these series converge or diverge. Since both series involve factorials, the most obvious convergence tests to try would be the ratio test or the ratio test for absolute convergence. However, we do not have to use any convergence test to show that the series converge.

(a) Note that the first series is the Maclaurin series for e^x when $x = -5$. Since the Maclaurin series for e^x converges for every value of x, we have $\sum_{k=0}^{\infty} \frac{(-5)^k}{k!} = e^{-5}$.

(b) Similarly, the Maclaurin series for $\sin x$ also converges for every value of x. The second series is almost the Maclaurin series for $\sin x$ when $x = \pi$. Since

$$0 = \sin \pi = \sum_{k=0}^{\infty} \frac{(-1)^k \pi^{2k+1}}{(2k+1)!} = \pi - \frac{\pi^3}{3!} + \sum_{k=2}^{\infty} \frac{(-1)^k \pi^{2k+1}}{(2k+1)!},$$

we have

$$\sum_{k=2}^{\infty} \frac{(-1)^k \pi^{2k+1}}{(2k+1)!} = \frac{\pi^3}{6} - \pi.$$

□

EXAMPLE 4

Using the Maclaurin series for $f(x)$ to find a Maclaurin series for $f(cx^m)$

Use the Maclaurin series for $f(x) = \frac{1}{1-x} = \sum_{k=0}^{\infty} x^k$ to find the Maclaurin series for the function $\frac{1}{4+x^2}$ and find the interval of convergence for the series.

SOLUTION

We begin with a small algebraic manipulation. To be able to use the Maclaurin series for $\frac{1}{1-x}$ to find a Maclaurin series for $\frac{1}{4+x^2}$ we first rewrite this function in the form

$$\frac{1}{4+x^2} = \frac{1}{4} \cdot \left(\frac{1}{1+(x/2)^2}\right) = \frac{1}{4} \cdot \left(\frac{1}{1-(-(x/2)^2)}\right) \quad \leftarrow \text{algebra}$$

$$= \frac{1}{4} f(-(x/2)^2) \quad \leftarrow \text{substitution}$$

Thus, if we substitute $-\left(\frac{x}{2}\right)^2$ for x in the series for $f(x) = \frac{1}{1-x}$ and multiply by $\frac{1}{4}$, we get our series for $\frac{1}{4+x^2}$. That is,

$$\frac{1}{4+x^2} = \frac{1}{4}\sum_{k=0}^{\infty}\left(-\left(\frac{x}{2}\right)^2\right)^k = \frac{1}{4}\sum_{k=0}^{\infty}\left(-\frac{1}{4}\right)^k x^{2k}.$$

To find the interval of convergence for this new series, we make the same substitution in the inequality that defines the interval of convergence for our original series. That is, we replace x by $-\left(\frac{x}{2}\right)^2$ in the inequality $|x| < 1$. We obtain

$$\left|-\left(\frac{x}{2}\right)^2\right| < 1 \text{ or, equivalently, } x^2 < 4.$$

The solution of this inequality is $|x| < 2$. Therefore, the series for $\frac{1}{4+x^2}$ converges to the function on the interval $(-2, 2)$. □

EXAMPLE 5

Manipulating a known Maclaurin series to find a new Maclaurin series

Find the Maclaurin series and its interval of convergence for $f(x) = \frac{x^2}{27-x^3}$.

SOLUTION

We know that the function $\frac{1}{1-x}$ has the Maclaurin series $\sum_{k=0}^{\infty} x^k$ with interval of convergence $(-1, 1)$. We rewrite the given function $f(x)$ as

$$f(x) = \frac{x^2}{27-x^3} = \frac{x^2}{27}\cdot\left(\frac{1}{1-(x/3)^3}\right).$$

You should check that the algebra is correct. We may find the Maclaurin series for $\frac{1}{1-(x/3)^3}$ by substituting $\left(\frac{x}{3}\right)^3$ into the series $\sum_{k=0}^{\infty} x^k$. We obtain

$$\frac{1}{1-(x/3)^3} = \sum_{k=0}^{\infty}\left(\left(\frac{x}{3}\right)^3\right)^k = \sum_{k=0}^{\infty}\left(\frac{x}{3}\right)^{3k}.$$

This series will converge when

$$\left|\left(\frac{x}{3}\right)^3\right| < 1 \text{ or, equivalently, } |x| < 3.$$

To find the Maclaurin series for $f(x)$, we multiply the series we just found by $\frac{x^2}{27}$. Thus,

$$\frac{x^2}{27-x^3} = \frac{x^2}{27}\sum_{k=0}^{\infty}\left(\frac{x}{3}\right)^{3k} = \sum_{k=0}^{\infty}\frac{x^{3k+2}}{3^{3k+3}}.$$

Since multiplication by a monomial does not affect the interval of convergence, this series converges on the interval $(-3, 3)$. □

EXAMPLE 6

Deriving a Taylor series from a Maclaurin series

Find the Taylor series for $f(x) = \sin(x - \pi)$ at π.

SOLUTION

First note that we could use Definition 8.7 to find the desired Taylor series. However, since we already know that $\sin t = \sum_{k=0}^{\infty} \frac{(-1)^k}{(2k+1)!} t^{2k+1}$, we may obtain the Taylor series by substituting $x - \pi$ for t in both versions of the function. We have

$$\sin(x - \pi) = \sum_{k=0}^{\infty} \frac{(-1)^k}{(2k+1)!}(x - \pi)^{2k+1}.$$

Furthermore, because the interval of convergence for the Maclaurin series is $\mathbb{R}$, the interval of convergence for this Taylor series is also $\mathbb{R}$. □

EXAMPLE 7 **Multiplying two Maclaurin series**

Find the fourth Maclaurin polynomial, $P_4(x)$, for $e^x \ln(1 + x)$.

SOLUTION

We know the Maclaurin series for both e^x and $\ln(1 + x)$; namely, these are:

$$e^x = \sum_{k=0}^{\infty} \frac{x^k}{k!} = 1 + x + \frac{x^2}{2} + \frac{x^3}{6} + \frac{x^4}{24} + \cdots \quad \text{and}$$

$$\ln(1 + x) = \sum_{k=1}^{\infty} \frac{(-1)^{k+1}}{k} x^k = x - \frac{x^2}{2} + \frac{x^3}{3} - \frac{x^4}{4} + \cdots$$

We build the series for $e^x \ln(1 + x)$ term by term, by multiplying the preceding two series together. The constant term of the new series will be the product of the constant terms of the series for e^x and $\ln(1 + x)$. These constants are 1 and 0, respectively, so the new series has 0 as its constant term. The coefficient of the x term is

$$1 \cdot 1 + 1 \cdot 0 = 1.$$

The coefficient of the x^2 term is

$$1 \cdot \left(-\frac{1}{2}\right) + 1 \cdot 1 + \frac{1}{2} \cdot 0 = \frac{1}{2}.$$

Similarly, the coefficient of the x^3 term is

$$1 \cdot \frac{1}{3} + 1 \cdot \left(-\frac{1}{2}\right) + \frac{1}{2} \cdot 1 + \frac{1}{6} \cdot 0 = \frac{1}{3}.$$

Finally, the coefficient of the x^4 term is

$$1 \cdot \left(-\frac{1}{4}\right) + 1 \cdot \frac{1}{3} + \frac{1}{2} \cdot \left(-\frac{1}{2}\right) + \frac{1}{6} \cdot 1 + \frac{1}{24} \cdot 0 = 0.$$

Since the coefficient of the quartic term is 0, it follows that $P_4(x) = P_3(x)$, and we have

$$P_4(x) = x + \frac{1}{2}x^2 + \frac{1}{3}x^3.$$

We were not asked to find the entire Maclaurin series for $e^x \ln(1+x)$, nor were we asked for the interval of convergence for the series. Nonetheless, the interval of convergence for the product function is the intersection of the intervals of convergence of the two series. Since the Maclaurin series for e^x converges everywhere, the interval of convergence for the Maclaurin series for $e^x \ln(1+x)$ would be the same as the interval of convergence for the Maclaurin series for $\ln(1+x)$, namely, $(-1, 1]$. □

EXAMPLE 8

Using a Maclaurin polynomial to approximate the graph of a function

Find a Maclaurin polynomial for $\sin x$ whose graph differs by less than 0.001 from the graph of $\sin x$ on the interval $[-\pi, \pi]$.

SOLUTION

We wish to find a value of n to ensure that $|R_n(x)| \leq 0.001$ on the interval $[-\pi, \pi]$. Using Theorem 8.10, we know that

$$R_n(x) = \frac{f^{(n+1)}(c)}{(n+1)!}(x - x_0)^{n+1} \text{ where } c \text{ is between } x_0 \text{ and } x.$$

When $f(x) = \sin x$, we know that $f^{(n+1)}(x)$ is one of the functions $\pm \sin x$ and $\pm \cos x$. For each of these functions, $|f^{(n+1)}(x)| \leq 1$. Letting $x_0 = 0$, we have

$$|R_n(x)| = \left| \frac{f^{(n+1)}(c)}{(n+1)!} x^{n+1} \right| \leq \frac{|x|^{n+1}}{(n+1)!}.$$

To ensure that $|R_n(x)| \leq 0.001$ on $[-\pi, \pi]$, we find the smallest value of n such that

$$\frac{|\pi|^{n+1}}{(n+1)!} \leq 0.001.$$

We leave it to you to check that when $n = 11$, the quotient $\frac{|\pi|^{11+1}}{(11+1)!} > 0.001$, but $\frac{|\pi|^{12+1}}{(12+1)!} \leq 0.001$. Therefore, $P_{12}(x) = x - \frac{x^3}{3!} + \frac{x^5}{5!} - \frac{x^7}{7!} + \frac{x^9}{9!} - \frac{x^{11}}{11!}$ approximates $\sin x$ to within 0.001 of its actual value on the interval $[-\pi, \pi]$. We graph this polynomial next and see that it does indeed look like an excellent approximation to the sine function on the interval $[-\pi, \pi]$:

$P_{12}(x)$ for $\sin x$

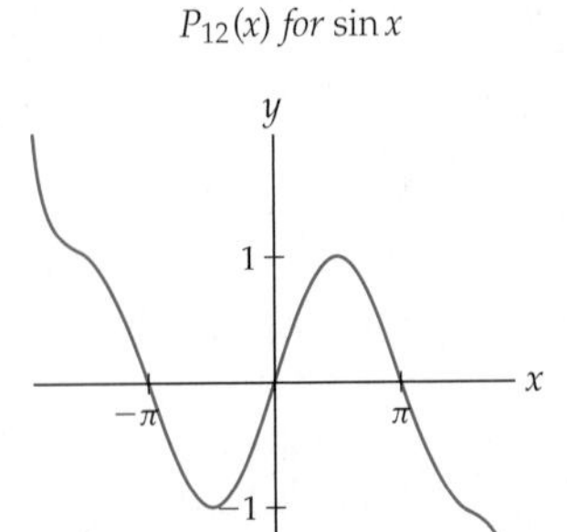

When $|x| > \pi$, $P_{12}(x)$ ceases to be a good approximation for $\sin x$. If we want to approximate the sine function over a larger interval or find a closer approximation over the same interval, we would need to use a Maclaurin polynomial of a higher degree. □

? TEST YOUR UNDERSTANDING

- What is meant by the remainder of a function? What does the remainder measure?
- What does Taylor's Theorem tell us about the remainder? Why is Lagrange's form for the remainder more useful for analyzing the remainder?
- What are the Maclaurin series for $\sin x$, $\cos x$, e^x, $\frac{1}{1-x}$, $\ln(1+x)$, and $\tan^{-1} x$?
- If you know the Maclaurin series for a function $f(x)$, how can you use that series to quickly find the Maclaurin series for $f(cx^m)$, where $c \neq 0$ and $m \in \mathbb{Z}^+$? How can the interval of convergence for the new series be found from the interval of convergence for the original series?
- If you know the Maclaurin series for a function $f(x)$, how can you use that series to quickly find the Maclaurin series for $cx^m f(x)$, where $c \neq 0$ and $m \in \mathbb{Z}^+$? What is the relationship between the interval of convergence for the new series and the interval of convergence for the original series?

EXERCISES 8.3

Thinking Back

- *Analyzing the quality of an approximation for a definite integral:* How is Theorem 5.27 used to analyze the quality of a midpoint approximation M_n for a definite integral $\int_a^b f(x)\,dx$?
- *Analyzing the quality of an approximation for the sum of an alternating series:* If $a_k > 0$ for every positive integer k and the series $\sum_{k=1}^{\infty}(-1)^{k+1}a_k$ satisfies the conditions of the alternating series test, how can a_{n+1} be used to analyze the quality of the partial sum $\sum_{k=1}^{n}(-1)^{k+1}a_k$ as an approximation for the sum of the series?

Concepts

0. *Problem Zero:* Read the section and make your own summary of the material.

1. *True/False:* Determine whether each of the statements that follow is true or false. If a statement is true, explain why. If a statement is false, provide a counterexample.

(a) *True or False:* Every Maclaurin series converges to the function used to generate the series.

(b) *True or False:* Every Taylor series converges to the function used to generate the series at at least one point.

(c) *True or False:* If f is a function with derivatives of all orders at the point x_0, the nth remainder of f is given by $R_n(x) = \sum_{k=n+1}^{\infty}\frac{f^{(k)}(x_0)}{k!}(x-x_0)^k$.

(d) *True or False:* If f is a function with derivatives of all orders at the point x_0, the nth remainder of f is given by $R_n(x) = \int_{x_0}^{x}\frac{(x-t)^n}{n!}f^{(n+1)}(t)\,dt$.

(e) *True or False:* If f is a function with derivatives of all orders at the point x_0, the nth remainder of f is given by $R_n(x) = \frac{f^{(n+1)}(c)}{(n+1)!}(x-x_0)^{n+1}$, where c is between x_0 and x.

(f) *True or False:* The nth remainder measures the error in using the Taylor polynomial $P_n(x)$ to approximate the function that generated the polynomial.

(g) *True or False:* If the Maclaurin series for $f(x)$ converges to $f(x)$ on the interval $(-\rho, \rho)$, the Maclaurin series for $f(x^3)$ converges to $f(x^3)$ on the interval $(-\rho^3, \rho^3)$.

(h) *True or False:* If the Maclaurin series for $f(x)$ converges to $f(x)$ on the interval $(-\rho, \rho)$, the Maclaurin series for $x^3 f(x)$ converges to $x^3 f(x)$ on the interval $(-\rho, \rho)$.

2. *Examples:* Construct examples of the thing(s) described in the following. Try to find examples that are different than any in the reading.

(a) The third remainder $R_3(x)$ for $\sin x$ when $P_3(x)$ is the third Taylor polynomial for $\sin x$ at $\frac{\pi}{4}$.

(b) The integral given by Taylor's Theorem equal to the third remainder $R_3(x)$ for $\sin x$ when $P_3(x)$ is the third Taylor polynomial for $\sin x$ at $\frac{\pi}{4}$.

(c) Lagrange's form for the third remainder $R_3(x)$ for $\sin x$ when $P_3(x)$ is the third Taylor polynomial for $\sin x$ at $\frac{\pi}{4}$.

3. If the series $\sum_{k=0}^{\infty} a_k x^k$ converges to the function $f(x)$ on the interval $(-2, 2)$, provide a formula for a_k in terms of the function f.

4. If the series $\sum_{k=0}^{\infty} b_k(x-3)^k$ converges to the function $g(x)$ on the interval $(-2, 8)$, provide a formula for b_k in terms of the function g.

5. If the series $\sum_{k=0}^{\infty} c_k(x - x_0)^k$ converges to the function $h(x)$ for every real number, provide a formula for c_k in terms of the function h.

6. Explain why $\lim_{n\to\infty} \frac{|x|^n}{n!} = 0$ for every value of x.

7. Let $f(x) = \frac{1}{1-x}$ and $g(x) = \frac{1}{ax+b}$. Show that if $b \neq 0$, then $g(x) = \frac{1}{b} f\left(-\frac{a}{b}x\right)$.

8. Given a function f and a Taylor polynomial for f at x_0, what is meant by the nth remainder $R_n(x)$? What does $R_n(x)$ measure?

9. If $f(x) = 4x^3 - 5x^2 + 6x + 1$ and $P_3(x)$ is the third Taylor polynomial for f at -1, what is the third remainder $R_3(x)$? What is $R_4(x)$? *(Hint: You can answer this question* without *finding any derivatives.)*

10. If $f(x)$ is an nth-degree polynomial and $P_n(x)$ is the nth Taylor polynomial for f at x_0, what is the nth remainder $R_n(x)$? What is $R_{n+1}(x)$?

11. What is Taylor's Theorem?

12. What is Lagrange's form for the remainder? Why is Lagrange's form usually more useful for analyzing the remainder than the definition of the remainder or the integral provided by Taylor theorem?

13. Why is it helpful to know the Maclaurin series for a few basic functions?

14. If m is a positive integer, how can we find the Maclaurin series for the function $f(cx^m)$ if we already know the Maclaurin series for the function $f(x)$? How do you find the interval of convergence for the new series?

15. If m is a positive integer, how can we find the Maclaurin series for the function $cx^m f(x)$ if we already know the Maclaurin series for the function $f(x)$? How do you find the interval of convergence for the new series?

16. How may we find the Maclaurin series for $f(x)g(x)$ if we already know the Maclaurin series for the functions $f(x)$ and $g(x)$? How do you find the interval of convergence for the new series?

Use an appropriate Maclaurin series to find the values of the series in Exercises 17–22.

17. $\sum_{k=0}^{\infty} \frac{(-\pi)^k}{k!}$

18. $\sum_{k=0}^{\infty} \frac{(-1)^k \pi^{2k}}{(2k)!}$

19. $\sum_{k=0}^{\infty} \frac{(-1)^k}{2k+1}\left(\frac{1}{\sqrt{3}}\right)^{2k+1}$

20. $\sum_{k=1}^{\infty} \frac{(-1)^k}{k}\left(\frac{1}{\pi}\right)^k$

21. $\sum_{k=0}^{\infty} \frac{(-1)^k \pi^{2k+1}}{(2k+1)!}$

22. $\sum_{k=1}^{\infty} \left(-\frac{1}{\pi}\right)^k$

Skills

In Exercises 21–30 in Section 8.2 you were asked to find the fourth Maclaurin polynomial $P_4(x)$ for the specified function. In Exercises 23–32 we ask you to give Lagrange's form for the corresponding remainder, $R_4(x)$.

23. $\cos x$

24. e^x

25. $\sin x$

26. $\ln(1+x)$

27. $\tan^{-1} x$

28. $\tan x$

29. $\sqrt{1-x}$

30. $\frac{1+x}{1-x}$

31. $x \sin x$

32. $x^2 e^x$

In Exercises 31–34 in Section 8.2 you were asked to find the Maclaurin series for the specified function. Now find the Lagrange's form for the remainder $R_n(x)$, and show that $\lim_{n\to\infty} R_n(x) = 0$ on the specified interval.

33. $\cos x$, $\mathbb{R}$

34. e^x, $\mathbb{R}$

35. $\sin x$, $\mathbb{R}$

36. $\ln(1+x)$, $(-1/2, 1/2)$

In Exercises 41–48 in Section 8.2, you were asked to find the fourth Taylor polynomial $P_4(x)$ for the specified function and the given value of x_0. In Exercises 37–44 give Lagrange's form for the remainder $R_4(x)$.

37. $\cos x$, $\frac{\pi}{2}$

38. e^x, 1

39. $\sin x$, π

40. $\sqrt{x}$, 1

41. $\ln x$, 3

42. $\sqrt[3]{x}$, 1

43. $\tan x$, $\frac{\pi}{4}$

44. $\tan^{-1} x$, 1

In Exercises 49–54 in Section 8.2 you were asked to find the Taylor series for the specified function at the given value of x_0. In Exercises 45–50 find the Lagrange's form for the remainder $R_n(x)$, and show that $\lim_{n\to\infty} R_n(x) = 0$ on the specified interval.

45. $\cos x$, $\frac{\pi}{2}$, $\mathbb{R}$

46. e^x, 1, $\mathbb{R}$

47. $\sin x$, π, $\mathbb{R}$

48. $\sqrt{x}$, 1, $(1/2, 3/2)$

49. $\ln x$, 3, $(2, 4)$

50. $\sqrt[3]{x}$, 1, $(1/2, 3/2)$

Find the Maclaurin series for the functions in Exercises 51–60 by substituting into a known Maclaurin series. Also, give the interval of convergence for the series.

51. e^{x^3}

52. e^{-3x^2}

53. $\sin(-5x^2)$

54. $x\cos(x^2)$

55. $\frac{1}{8+x^3}$

56. $\frac{x}{9-x^2}$

57. $\dfrac{e^x + e^{-x}}{2}$

58. $\dfrac{e^x - e^{-x}}{2}$

59. $\cos^2 x$ *(Hint: Use the identity* $\cos^2 x = \dfrac{1}{2}(1 + \cos 2x)$*).*

60. $\sin^2 x$ *(Hint: Use the identity* $\cos^2 x = \dfrac{1}{2}(1 - \cos 2x)$*).*

Use appropriate Maclaurin series to find the first four nonzero terms in the Maclaurin series for the product functions in Exercises 61–66. Also, give the interval of convergence for the series.

61. $e^x \sin x$

62. $e^x \cos x$

63. $e^x \ln(1 + x^3)$

64. $(\sin 2x)(\tan^{-1} x^3)$

65. $\dfrac{\sin x}{1 - x^2}$

66. $\dfrac{\tan^{-1} x}{1 - x^3}$

Use appropriate Maclaurin series to express the quantities in Exercises 67–76 as alternating series. Then use Theorem 7.38 to approximate the value of the specified quantities to within 0.001 of their actual value. How many terms in each series would be needed to approximate the given quantity to within 10^{-6} of its value? In Exercises 73–76 be sure to convert to radian measure first.

67. $e^{-0.3}$

68. e^{-2}

69. $\ln 1.5$

70. $\ln 1.3$

71. $\tan^{-1} 0.4$

72. $\tan^{-1}(-0.6)$

73. $\sin 2^\circ$

74. $\sin 1^\circ$

75. $\cos 5^\circ$

76. $\cos 10^\circ$

In Exercises 77–82,

(a) Use appropriate Maclaurin series to express the quantities in series form.

(b) Use Lagrange's form for the remainder to bound the error in using the 5th degree Maclaurin polynomial to approximate the given quantity.

(c) Find the smallest value of n so that the nth degree Maclaurin polynomial approximation to the given quantity is guaranteed to be accurate to within 10^{-6}.

77. $e^{0.3}$

78. $e^{0.9}$

79. $\ln 0.5$

80. $\ln 0.7$

81. $\sin 1$

82. $\cos 2$

83. Let

$$f(x) = \begin{cases} \dfrac{\sin x}{x}, & \text{if } x \neq 0 \\ 1, & \text{if } x = 0 \end{cases}$$

(a) Use the definition of the derivative to prove that f is differentiable at 0.

(b) Use the Maclaurin series for $\sin x$ to find a Maclaurin series for f.

84. Let

$$f(x) = \begin{cases} \dfrac{1 - \cos x}{x}, & \text{if } x \neq 0 \\ 0, & \text{if } x = 0 \end{cases}$$

(a) Use the definition of the derivative to prove that f is differentiable at 0.

(b) Use the Maclaurin series for $\cos x$ to find a Maclaurin series for f.

Applications

85. Emmy is a civil engineer who has to approximate the slope of the water table along a certain line in the Hanford Nuclear Reservation. She can dig only three test holes to evaluate the depth of groundwater at certain points. She finds that the water table lies at (0, 35), (300, 38), and (600, 42), with all distances given in feet and positive vertical distances representing distances below the surface.

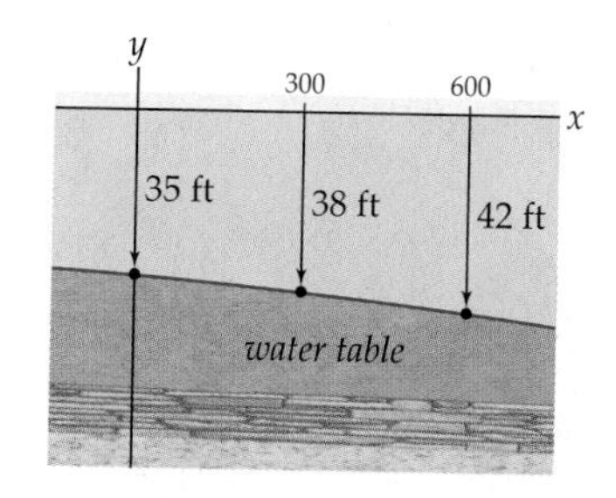

(a) Using the data from the first two wells, estimate the slope of the water table. Use this slope to write an equation of a line that describes the water table.

(b) Considering the equation of the line from part (a) as the first two terms of a Maclaurin series for the function $w(x)$ describing the water table, write an expression for the error in this linear approximation.

(c) Verify that the graph of the quadratic

$$w_2(x) = \frac{1}{180000}x^2 + \frac{1}{120}x + 35$$

passes through the three data points describing the water table.

(d) Use the quadratic from part (c) to estimate the error in the linear approximation to the water table.

86. In 2009, Mirec was a civil engineer in Fargo, North Dakota. He was charged with monitoring the Red River as it flooded catastrophically in March of that year. By measuring the flow of the river very precisely, he was able to estimate the flow $f(t)$ and many of its derivatives as of March 8, which he considered to be time $t = 0$. He found that $f(0) = 988.9$, $f'(0) = -853.8$, $f''(0) = 1026.9$, $f'''(0) = -528.1$, $f^{(4)}(0) = 141.3$, and $f^{(5)}(0) = -15.0$.

(a) Use these data to compute a Maclaurin polynomial of degree 5 for the flow.

(b) Mirec had to predict when the maximum flow would occur for $t > 0$. What day did he predict?

Proofs

87. Use the principle of mathematical induction to complete the proof of Theorem 8.9. That is, show that

$$f(x) = \sum_{k=0}^{n} \frac{f^{(k)}(x_0)}{k!}(x - x_0)^k + \int_{x_0}^{x} \frac{(x-t)^n}{n!} f^{(n+1)}(t)\, dt.$$

88. Use Theorem 8.9 and Theorem 4.31 of Section 4.6 to prove Theorem 8.10. That is, show that if f is a function that can be differentiated $n + 1$ times in some open interval I containing the point x_0, and if $R_n(x)$ is the nth remainder for f at $x = x_0$, then for each point $x \in I$ there is at least one c between x_0 and x such that $R_n(x) = \frac{f^{(n+1)}(c)}{(n+1)!}(x - x_0)^{n+1}$.

89. Use Lagrange's form for the remainder to prove that the Maclaurin series for the cosine,

$$\cos x = \sum_{k=0}^{\infty} \frac{(-1)^k}{(2k)!} x^{2k},$$

converges for every real number.

90. Let f be a function with the property that $\left|f^{(n)}(x)\right| < M$ for every nonnegative integer n and every real number x, where M is a positive number. Use Lagrange's form for the remainder to prove that the Taylor series for f at any point x_0 converges to f for every real number.

91. Use the Maclaurin series for e^x and e^{-x} to prove that

$$\sinh x = \sum_{k=0}^{\infty} \frac{1}{(2k+1)!} x^{2k+1}.$$

Thinking Forward

▶ *The derivative of a series:* Show that if you take the derivative of the Maclaurin series for e^x term by term, you get the same series back again. That is, show that

$$\frac{d}{dx}(e^x) = \sum_{k=0}^{\infty} \frac{x^k}{k!}.$$

▶ *The derivative of a series:* Show that if you take the derivative of the Maclaurin series for $\sin x$ term by term, you get the series for $\cos x$. That is, show that

$$\frac{d}{dx}(\sin x) = \sum_{k=0}^{\infty} \frac{(-1)^k}{(2k)!} x^{2k} = \cos x.$$

8.4 DIFFERENTIATING AND INTEGRATING POWER SERIES

- Differentiating a power series term by term
- Integrating a power series term by term
- Using a Taylor polynomial to approximate a function

Differentiating a Power Series

In Section 8.3 we discussed two methods for altering a Maclaurin series to obtain a new series: function composition and multiplication. We extend this idea by discussing the calculus of power series. The next two theorems tell us that we may differentiate and integrate power series term by term.

THEOREM 8.11 **Differentiating a Power Series**

If $\sum_{k=0}^{\infty} a_k(x-x_0)^k$ is a power series in $x - x_0$ that converges to a function $f(x)$ on an interval I, then the derivative of f is given by

$$f'(x) = \frac{d}{dx}\left(\sum_{k=0}^{\infty} a_k(x-x_0)^k\right) \stackrel{*}{=} \sum_{k=0}^{\infty} a_k \frac{d}{dx}((x-x_0)^k) = \sum_{k=1}^{\infty} k a_k (x-x_0)^{k-1}.$$

The interval of convergence for this new series is I, with the possible exception that if I is a finite interval, the convergence at the endpoints of I may be different than it was for the initial series.

The starred equality in Theorem 8.11 is the most significant. The first equality just expresses the fact that the function f and its power series are equal. The third equality says that $\frac{d}{dx}((x-x_0)^k) = k(x-x_0)^{k-1}$. In Chapter 2 we proved that the derivative of a finite sum is the sum of the derivatives, but here we have an infinite sum. The second equality says that this summation property of the derivative in finite situations also holds for a convergent power series. The proof of Theorem 8.11 is beyond the scope of this course.

Integrating a Power Series

THEOREM 8.12 **Integration of Power Series**

Let $\sum_{k=0}^{\infty} a_k(x-x_0)^k$ be a power series in $x - x_0$ that converges to a function $f(x)$ on an interval I.

(a) The indefinite integral of f is given by

$$\int f(x)\,dx = \int \left(\sum_{k=0}^{\infty} a_k(x-x_0)^k\right) dx$$

$$\stackrel{*}{=} \sum_{k=0}^{\infty} a_k \int (x-x_0)^k dx = \sum_{k=0}^{\infty} \frac{a_k}{k+1}(x-x_0)^{k+1} + C,$$

where C is a constant. The interval of convergence for this new series is I, with the possible exception that if I is a finite interval, the convergence at the endpoints of I may be different than it was for the initial series.

(b) If c and d are two points in I, then the definite integral of f on the interval $[c, d]$ is given by

$$\int_c^d f(x)\,dx = \int_c^d \left(\sum_{k=0}^{\infty} a_k(x-x_0)^k\right) dx \stackrel{*}{=} \sum_{k=0}^{\infty} a_k \int_c^d (x-x_0)^k dx$$

$$= \sum_{k=0}^{\infty} \frac{a_k}{k+1}\left[(x-x_0)^{k+1}\right]_c^d = \sum_{k=0}^{\infty} \frac{a_k}{k+1}((d-x_0)^{k+1} - (c-x_0)^{k+1}).$$

As in Theorem 8.11, the starred equalities in Theorem 8.12 are the most significant. The first equalities in parts (a) and (b) express the fact that the power series converges to the function f. All of the other equalities indicate basic facts about antiderivatives and definite integrals. In Chapter 4 we proved that the indefinite integral of a finite sum is the sum of the indefinite integrals, but in part (a) we have an infinite sum. Part (b) makes the analogous statement for definite integrals. The starred equalities say that these properties of antiderivatives and definite integrals are also valid for convergent power series. As with Theorem 8.11, the proof of Theorem 8.12 is beyond the scope of this course.

Examples and Explorations

EXAMPLE 1

Using the derivative to find the Maclaurin series for $\frac{1}{(1-x)^2}$

Find the Maclaurin series for $\frac{1}{(1-x)^2}$.

SOLUTION

We begin by noting that $\frac{d}{dx}\left(\frac{1}{1-x}\right) = \frac{1}{(1-x)^2}$. Since the Maclaurin series for $\frac{1}{1-x}$ is $\sum_{k=0}^{\infty} x^k$, we may use Theorem 8.11 to find the Maclaurin series for $\frac{1}{(1-x)^2}$:

$$\frac{1}{(1-x)^2} = \frac{d}{dx}\left(\frac{1}{1-x}\right) = \frac{d}{dx}\left(\sum_{k=0}^{\infty} x^k\right) = \sum_{k=0}^{\infty} \frac{d}{dx}(x^k) = \sum_{k=1}^{\infty} kx^{k-1} = \sum_{k=0}^{\infty}(k+1)x^k.$$

Note that we changed the indexing of the series in the final step. This series has the same radius of convergence as the original series, namely 1, but the new series may converge at ± 1 even though the original series diverges at both endpoints.

When $x = 1$, we have the series $\sum_{k=0}^{\infty}(k+1)$, which diverges by the divergence test. Similarly, when $x = -1$, we obtain the series $\sum_{k=0}^{\infty}(-1)^k(k+1)$, which also diverges by the divergence test. Thus the interval of convergence for the series is $(-1, 1)$. □

EXAMPLE 2

Using integration to find the Maclaurin series for $\ln(1+x)$

Find the Maclaurin series for $\ln(1+x)$.

SOLUTION

We could of course use Definition 8.7 to find the Maclaurin series for $\ln(1+x)$. However, we already know that

$$\frac{1}{1-x} = \sum_{k=0}^{\infty} x^k \quad \text{on the interval } (-1, 1).$$

If we replace each x in this equation with $-x$, we obtain the Maclaurin series for $\frac{1}{1+x}$: That is,

$$\frac{1}{1-(-x)} = \frac{1}{1+x} = \sum_{k=0}^{\infty}(-x)^k = \sum_{k=0}^{\infty}(-1)^k x^k.$$

This series has the same interval of convergence, $(-1, 1)$.

Now, since

$$\int_0^x \frac{1}{1+t}\,dt = \ln(1+t)\big|_0^x = \frac{1}{1+x},$$

if we integrate our series for $\frac{1}{1+t}$, we will have our Maclaurin series for $\ln(1+x)$. That is,

$$\ln(1+x) = \int_0^x \frac{1}{1+t}\,dt = \int_0^x \sum_{k=0}^{\infty}(-1)^k t^k\,dt = \sum_{k=0}^{\infty}(-1)^k\left(\int_0^x t^k\,dt\right)$$

$$= \sum_{k=0}^{\infty}\left[\frac{(-1)^k}{k+1}t^{k+1}\right]_0^x = \sum_{k=0}^{\infty}\frac{(-1)^k}{k+1}x^{k+1} x - \frac{1}{2}x^2 + \frac{1}{3}x^3 - \cdots.$$

The new series is guaranteed to have the same interval of convergence as the series for the function $\frac{1}{1+x}$, with the possible exception of the endpoints, which for this series are ± 1. By changing the indexing of the series, we may also express this as

$$\ln(1+x) = \sum_{k=1}^{\infty}\frac{(-1)^{k+1}}{k}x^k.$$

Finally, we need to see if the series converges at the endpoints of the interval. When $x = 1$, we have the alternating harmonic series $\sum_{k=1}^{\infty}\frac{(-1)^{k+1}}{k}$. In Section 7.7 we showed that the alternating harmonic series converges conditionally. At that time we mentioned that the alternating harmonic series converges to $\ln 2$, and we now see why. When $x = -1$, we obtain the series $\sum_{k=1}^{\infty}\frac{(-1)^{k+1}}{k}(-1)^k = -\sum_{k=1}^{\infty}\frac{1}{k}$. This is the (negative of the) harmonic series, which we know diverges. Thus the interval of convergence for the series is $(-1, 1]$. □

EXAMPLE 3

Using integration to find the Maclaurin series for $\tan^{-1} x$

Outline the steps for finding the Maclaurin series for $\tan^{-1} x$.

SOLUTION

The procedure for finding the Maclaurin series for $\tan^{-1} x$ is quite similar to the work we did in Example 2. Since $\frac{d}{dx}(\tan^{-1} x) = \frac{1}{1+x^2}$, we will perform the following steps:

- Substitute $-x^2$ for x in the series for $\frac{1}{1-x}$ to find the Maclaurin series for $\frac{1}{1+x^2}$.
- Use the same substitution in the inequality $|x| < 1$ to find the interval of convergence for the new series.
- Integrate the latter series to find the series for $\tan^{-1} x$. Be sure to include a single constant of integration. The resulting new series will have the same interval of convergence as the series for $\frac{1}{1+x^2}$, with the possible change at the endpoints of the interval.
- Evaluate the function $\tan^{-1} x$ and the series at $x = 0$ to determine the value of the constant of integration.
- Determine whether the series converges absolutely, converges conditionally, or diverges at each endpoint of the interval of convergence. □

EXAMPLE 4 **Using a Maclaurin series to approximate a definite integral**

Approximate the definite integral $\int_0^1 e^{-x^2}\,dx$ to within 0.001 of its value.

SOLUTION

The function e^{-x^2} does not have an elementary antiderivative. Therefore, we cannot use the Fundamental Theorem to evaluate the integral. We could use the techniques of Chapter 4 to approximate the integral, but Theorem 8.12 gives us another effective tool for approximating its value. Since

$$e^x = \sum_{k=0}^{\infty} \frac{x^k}{k!},$$

if we substitute $-x^2$ for x in this series, we immediately obtain the Maclaurin series for e^{-x^2}:

$$e^{-x^2} = \sum_{k=0}^{\infty} \frac{(-x^2)^k}{k!} = \sum_{k=0}^{\infty} (-1)^k \frac{x^{2k}}{k!}.$$

Now, using Theorem 8.12, we have

$$\int_0^1 e^{-x^2}\,dx = \int_0^1 \left(\sum_{k=0}^{\infty} (-1)^k \frac{x^{2k}}{k!}\right) dx = \sum_{k=0}^{\infty} \frac{(-1)^k}{k!}\left(\int_0^1 x^{2k}\,dx\right)$$

$$= \sum_{k=0}^{\infty} \left[\frac{(-1)^k}{k!}\frac{x^{2k+1}}{2k+1}\right]_0^1 = \sum_{k=0}^{\infty} \frac{(-1)^k}{k!}\frac{1}{2k+1}$$

$$= 1 - \frac{1}{3} + \frac{1}{10} - \frac{1}{42} + \frac{1}{216} - \frac{1}{1320} + \cdots.$$

This is a convergent alternating series that satisfies the conditions of the alternating series test. By Theorem 7.38 of Section 7.7, to approximate the value of the integral to within 0.001 of its value, we look for the first term of the series whose value is less than 0.001 and sum all the terms that precede it. Thus, the sum $1 - \frac{1}{3} + \frac{1}{10} - \frac{1}{42} + \frac{1}{216} \approx 0.7475$ is within $\frac{1}{1320} \approx 0.00076$ of the actual value of the integral. In fact, by Theorem 7.38,

$$0.7467 \approx \sum_{k=0}^{5} \frac{(-1)^k}{k!}\frac{1}{2k+1} < \int_0^1 e^{-x^2}\,dx < \sum_{k=0}^{4} \frac{(-1)^k}{k!}\frac{1}{2k+1} \approx 0.7475.$$

The terms of this series decrease in magnitude so quickly, that including a few extra terms would provide a much more accurate approximation for the integral. □

? TEST YOUR UNDERSTANDING

- If you know the Maclaurin series for a function f, how can you use that series to quickly find the Maclaurin series for f'? What is the relationship between the interval of convergence for the Maclaurin series for f and the interval of convergence for the Maclaurin series for f'?
- If you know the Taylor series for a function f at $x = x_0$, how can you use that series to quickly find the Taylor series for f' at $x = x_0$? What is the relationship between the interval of convergence for the Taylor series for f at $x = x_0$ and the interval of convergence for the Taylor series for f' at $x = x_0$?

- If you know the Maclaurin series for a function f, how can you use that series to quickly find the Maclaurin series for $\int f(x)\,dx$? What is the relationship between the interval of convergence for the Maclaurin series for f and the interval of convergence for the Maclaurin series for $\int f(x)\,dx$?
- If you know the Taylor series for a function f at $x = x_0$, how can you use that series to quickly find the Taylor series for $\int f(x)\,dx$ at $x = x_0$? What is the relationship between the interval of convergence for the Taylor series for f at $x = x_0$ and the interval of convergence for the Taylor series for $\int f(x)\,dx$ at $x = x_0$?
- If you know the Maclaurin series for the function $f(x)$, how can you combine techniques introduced in this section with the techniques from Section 8.3 to quickly find the Maclaurin series for the function $\int c_1 x^{m_1} f(c_2 x^{m_2})\,dx$, where c_1 and c_2 are real numbers and m_1 and m_2 are positive integers? How would you determine the interval of convergence for the new series?

EXERCISES 8.4

Thinking Back

- *The derivative of a sum:* Prove that $\frac{d}{dx}(f_1(x) + f_2(x)) = \frac{d}{dx}(f_1(x)) + \frac{d}{dx}(f_2(x))$. How would you prove that

$$\frac{d}{dx}\left(\sum_{k=1}^{n} f_k(x)\right) = \sum_{k=1}^{n} \frac{d}{dx}(f_k(x))?$$

- *The integral of a sum:* Prove that $\int (f_1(x) + f_2(x))dx = \int f_1(x)\,dx + \int f_2(x)\,dx$. How would you prove that

$$\int \left(\sum_{k=1}^{n} f_k(x)\right)dx = \sum_{k=1}^{n} \left(\int f_k(x)\,dx\right)?$$

Concepts

0. *Problem Zero:* Read the section and make your own summary of the material.

1. *True/False:* Determine whether each of the statements that follow is true or false. If a statement is true, explain why. If a statement is false, provide a counterexample.

(a) *True or False:* If a function f has a Taylor series at $x = 5$ that converges to f for every value of x, then the function f' has a Taylor series at $x = 5$ that converges to f' for every value of x.

(b) *True or False:* If the series $\sum_{k=0}^{\infty} a_k (x - 5)^k$ converges to the function $f(x)$ for every value of x, then the series $\sum_{k=0}^{\infty} \frac{a_k}{k+1}(x - 5)^{k+1}$ converges to a function $G(x)$ where $G'(x) = f(x)$ for every value of x.

(c) *True or False:* If the series $\sum_{k=0}^{\infty} a_k x^k$ converges absolutely on the interval $(-\rho, \rho)$, then for every $x \in (-\rho, \rho)$, $\frac{d}{dx}\left(\sum_{k=0}^{\infty} a_k x^k\right) = \sum_{k=0}^{\infty} (k + 1)a_{k+1} x^k$.

(d) *True or False:* If the interval of convergence for the series $\sum_{k=0}^{\infty} a_k x^k$ contains the points c and d, then

$$\int_c^d \left(\sum_{k=0}^{\infty} a_k x^k\right)dx = \sum_{k=0}^{\infty} \frac{a_k}{k+1}(d^{k+1} - c^{k+1}).$$

(e) *True or False:* If the series $\sum_{k=0}^{\infty} a_k x^k$ converges absolutely on the interval $(-\rho, \rho)$, then for every $x \in (-\rho, \rho)$, $\int \left(\sum_{k=0}^{\infty} a_k x^k\right)\,dx = \sum_{k=0}^{\infty} \frac{a_k}{k+1} x^{k+1}$.

(f) *True or False:* If the interval of convergence of the power series $\sum_{k=0}^{\infty} a_k x^k$ is $(-3, 3)$, then the interval of convergence of the power series $\sum_{k=1}^{\infty} k a_k x^{k-1}$ is $(-3, 3)$.

(g) *True or False:* If the interval of convergence of the power series $\sum_{k=0}^{\infty} a_k x^k$ is $(-3, 3)$, then the interval of convergence of the power series $\sum_{k=0}^{\infty} \frac{a_k}{k+1} x^{k+1}$ is $[-3, 3]$.

(h) *True or False:* If the interval of convergence of the power series $\sum_{k=0}^{\infty} a_k x^k$ is $(-3, 3)$, then the interval of convergence of the power series $\sum_{k=0}^{\infty} \frac{a_k}{k+1} x^{k+1}$ contains $(-3, 3)$.

2. *Examples:* Construct examples of the thing(s) described in the following. Try to find examples that are different than any in the reading.

(a) A function f with a Maclaurin series that has $[-2, 2]$ as its interval of convergence such that the Maclaurin series for f' converges only on the interval $(-2, 2)$.

(b) A function f such that both f and f' have Taylor series with $(0, 6]$ as their interval of convergence.

(c) A function f such that both f and f' have Taylor series in $(x - \pi)$ with $\mathbb{R}$ as their interval of convergence.

3. If $f(x) = \sum_{k=0}^{\infty} a_k (x - x_0)^k$, what is $f(x_0)$? What is $f'(x_0)$? What is $f''(x_0)$? What is $f^{(k)}(x_0)$?

4. Let $f(x) = \sum_{k=0}^{\infty} a_k(x - x_0)^k$ and let G be an antiderivative for f. Explain why we do not have enough information to determine $G(x_0)$. What is $G'(x_0)$? What is $G''(x_0)$? What is $G^{(k)}(x_0)$?

5. Let $f(x) = \sum_{k=0}^{\infty} a_k(x - x_0)^k$ and let G be the antiderivative for f with the property that $G(x_0) = 7$. Find the Taylor series in x_0 for G.

6. Let $f(x)$ be a function such that the power series in $x - x_0$, $\sum_{k=0}^{\infty} a_k (x - x_0)^k$ converges absolutely to f on the interval I. If G_1 and G_2 are two antiderivatives for f, explain why the power series in $x - x_0$ for G_1 and G_2 have the same interval of convergence.

7. In Example 1 we used Theorem 8.11 to find the Maclaurin series for $\frac{1}{(1-x)^2}$. Explain how Theorem 8.11 could be used to find the Maclaurin series for $\frac{1}{(1-x)^k}$, where k is a positive integer greater than 2.

8. If f is a function such that $f(0) = 1$ and $f'(x) = f(x)$ for every value of x, find the Maclaurin series for f.

9. If f is a function such that $f(0) = 2$ and $f'(x) = -3f(x)$ for every value of x, find the Maclaurin series for f.

10. If f is a function such that $f(0) = -3$ and $f'(x) = 2f(x)$ for every value of x, find the Maclaurin series for f.

Perform the following steps for the power series in $x - x_0$ in Exercises 11–16:

(a) Find the interval of covergence, I, for the series.

(b) Let f be the function to which the series converges on I. Find the power series in $x - x_0$ for f'.

(c) Find the power series in $x - x_0$ for $F(x) = \int_{x_0}^{x} f(t)\,dt$.

11. $\sum_{k=0}^{\infty} 2^k x^k$

12. $\sum_{k=0}^{\infty} \frac{5^k}{k!}(x-3)^k$

13. $\sum_{k=0}^{\infty} \frac{(-1)^k}{(2k)!} x^k$

14. $\sum_{k=0}^{\infty} \frac{1}{k+2} x^{3k+1}$

15. $\sum_{k=1}^{\infty} \frac{(-1)^k}{k}(x+5)^k$

16. $\sum_{k=0}^{\infty} (-1)^k \frac{k!}{(2k)!}(x-7)^{2k}$

17. (a) Explain why the definite integral $I = \int_0^{0.5} \frac{dx}{1+x^3}$ exists.

(b) Explain how to use Simpson's method to approximate I to within 0.001 of its actual value.

(c) Use substitution in the Maclaurin series for $\frac{1}{1-x}$ to find a Maclaurin series for $\frac{1}{1+x^3}$.

(d) Explain how to use Theorems 7.38 and 8.12 to approximate I to within 0.001 of its actual value.

18. (a) Explain why the definite integral $I = \int_0^2 \frac{dx}{1+x^3}$ exists.

(b) Explain how to use Simpson's method to approximate the value of I to within 0.001 of its actual value.

(c) Explain why you cannot use the method of Exercise 17 (b) and (c) to approximate I.

19. (a) Explain why the definite integral $I = \int_0^{0.5} \sin(x^2)\,dx$ exists.

(b) Explain how to use Simpson's method to approximate I to within 0.001 of its actual value.

(c) Use substitution in the Maclaurin series for $\sin x$ to find a Maclaurin series for $\sin(x^2)$.

(d) Explain how to use Theorems 7.38 and 8.12 to approximate I to within 0.001 of its actual value.

20. (a) Explain why the definite integral $I = \int_0^2 \sin(x^2)\,dx$ exists.

(b) Explain how to use Simpson's method to approximate the value of I to within 0.001 of its actual value.

(c) Use your answer from Exercise 19 (b) to explain how to use Theorems 7.38 and 8.12 to approximate I to within 0.001 of its actual value.

21. Show that when you take the derivative of the Maclaurin series for the sine function term by term you obtain the Maclaurin series for cosine.

22. Show that when you take the derivative of the Maclaurin series for the cosine function term by term you obtain the negative of the Maclaurin series for the sine.

23. Show that when you take the derivative of the Maclaurin series for the exponential function term by term you obtain the same series you started with. Why does that make sense?

24. (a) Verify that

$$\frac{d}{dx}(x\sin(x^3)) = \sin(x^3) + 3x^3\cos(x^3).$$

(b) Use multiplication and/or substitution in the Maclaurin series for the sine and the cosine to find the Maclaurin series for $x\sin(x^3)$, $\sin(x^3)$, and $3x^3\cos(x^3)$.

(c) Use Theorem 8.11 to find the Maclaurin series for $\frac{d}{dx}(x\sin(x^3))$, and show that this series is the sum of the Maclaurin series for $\sin(x^3)$ and $3x^3\cos(x^3)$ you obtained in part (b).

Skills

In Exercises 25–30, find Maclaurin series for the given pairs of functions, using these steps:

(a) Use substitution and/or multiplication and the Maclaurin series for $\frac{1}{1-x}$ to find the Maclaurin series for the given function. Also, provide the interval of convergence for the series you found.

(b) Use Theorem 8.11 and your answer from part (a) to find the Maclaurin series for the given function. Also, provide the interval of convergence for the series you found.

25. **(a)** $\dfrac{1}{1+x^2}$ **(b)** $\dfrac{x}{(1+x^2)^2}$

26. **(a)** $\dfrac{1}{1-x^3}$ **(b)** $\dfrac{x^2}{(1-x^3)^2}$

27. **(a)** $\dfrac{1}{1+8x^3}$ **(b)** $\dfrac{x^2}{(1+8x^3)^2}$

28. **(a)** $\dfrac{1}{3-x}$ **(b)** $\dfrac{1}{(3-x)^2}$

29. **(a)** $\dfrac{1}{4-x^2}$ **(b)** $\dfrac{x}{(4-x^2)^2}$

30. **(a)** $\dfrac{1}{(3-x)^2}$ **(b)** $\dfrac{1}{(3-x)^3}$

(Hint: See Exercise 28).

In Exercises 31–34, find Maclaurin series for the given pairs of functions, using these steps:

(a) Use substitution in the appropriate Maclaurin series to find the Maclaurin series for the given function.

(b) Use Theorem 8.11 and your answer from part (a) to find the Maclaurin series for the given function.

(c) Find the Maclaurin series for the function in (b), using multiplication and substitution with the appropriate Maclaurin series. Compare your answers from (b) and (c).

31. **(a)** $\sin(x^2)$ **(b)** $x\cos(x^2)$

32. **(a)** $\cos(4x^3)$ **(b)** $x^2\sin(4x^3)$

33. **(a)** e^{-x^2} **(b)** xe^{-x^2}

34. **(a)** $\tan^{-1}\left(\dfrac{x^2}{3}\right)$ **(b)** $\dfrac{x}{9+x^4}$

In Exercises 35–40, explore the Taylor series for the given pairs of functions, using these steps:

(a) Find the Taylor series for the given function at the specified value of x_0 and determine the interval of convergence for the series.

(b) Use Theorem 8.11 and your answer from part (a) to find the Taylor series for the given function for the same value of x_0. Also, find the interval of convergence for your series.

35. **(a)** $\dfrac{1}{1-x}$, $x_0 = 3$ **(b)** $\dfrac{1}{(1-x)^2}$

36. **(a)** $\dfrac{1}{1-x}$, $x_0 = -3$ **(b)** $\dfrac{1}{(1-x)^2}$

37. **(a)** $\dfrac{1}{2-x}$, $x_0 = 3$ **(b)** $\dfrac{1}{(2-x)^2}$

38. **(a)** $\dfrac{1}{2-x}$, $x_0 = -3$ **(b)** $\dfrac{1}{(2-x)^2}$

39. **(a)** $\dfrac{1}{1-3x}$, $x_0 = 3$ **(b)** $\dfrac{1}{(1-3x)^2}$

40. **(a)** $\dfrac{1}{1-3x}$, $x_0 = -3$ **(b)** $\dfrac{1}{(1-3x)^2}$

In Exercises 41–50, find Maclaurin series for the given pairs of functions, using these steps:

(a) Use substitution and/or multiplication and the appropriate Maclaurin series to find the Maclaurin series for the given function f.

(b) Use Theorem 8.12 and your answer from part (a) to find the Maclaurin series for the antiderivative $F = \int f$ that satisfies the specified initial condition.

41. **(a)** $f(x) = \sin(x^3)$ **(b)** $F(0) = 2$

42. **(a)** $f(x) = x\cos(x^3)$ **(b)** $F(0) = -1$

43. **(a)** $f(x) = e^{-x^2/3}$ **(b)** $F(0) = 0$

44. **(a)** $f(x) = \tan^{-1}(x^2)$ **(b)** $F(0) = \pi$

45. **(a)** $f(x) = \dfrac{1}{1+x^3}$ **(b)** $F(0) = -5$

46. **(a)** $f(x) = \ln(4+x^2)$ **(b)** $F(0) = -2$

47. **(a)** $f(x) = x^2\sin(5x^2)$ **(b)** $F(0) = 1$

48. **(a)** $f(x) = x^3\cos\left(\dfrac{x}{2}\right)$ **(b)** $F(0) = -4$

49. **(a)** $f(x) = x^2e^{-3x^2}$ **(b)** $F(0) = 1$

50. **(a)** $f(x) = x^4\tan^{-1}(3x^3)$ **(b)** $F(0) = 0$

Use Theorem 8.12 and the results from Exercises 41–50 to find series equal to the definite integrals in Exercises 51–60.

51. $\displaystyle\int_0^2 \sin(x^3)\,dx$

52. $\displaystyle\int_0^1 x\cos(x^3)\,dx$

53. $\displaystyle\int_{-1}^1 e^{-x^2/3}\,dx$

54. $\displaystyle\int_{0.1}^{0.2} \tan^{-1}(x^2)\,dx$

55. $\displaystyle\int_0^{0.3} \frac{dx}{1+x^3}$

56. $\displaystyle\int_{0.5}^1 \ln(4+x^2)\,dx$

57. $\displaystyle\int_1^2 x^2\sin(5x^2)\,dx$

58. $\displaystyle\int_{0.5}^2 x^3\cos\left(\frac{x}{2}\right)dx$

59. $\int_{0.5}^{1} x^2 e^{-3x^2}\,dx$

60. $\int_{0}^{0.3} x^4 \tan^{-1}(3x^3)\,dx$

Use the results from Exercises 51–60 and Theorem 7.38 to approximate the values of the definite integrals in Exercises 61–70 to within 0.001 of their values.

61. $\int_{0}^{2} \sin(x^3)\,dx$

62. $\int_{0}^{1} x\cos(x^3)\,dx$

63. $\int_{-1}^{1} e^{-x^2/3}\,dx$

64. $\int_{0.1}^{0.2} \tan^{-1}(x^2)\,dx$

65. $\int_{0}^{0.3} \frac{dx}{1+x^3}$

66. $\int_{0.5}^{1} \ln(4+x^2)\,dx$

67. $\int_{1}^{2} x^2\sin(5x^2)\,dx$

68. $\int_{0.5}^{2} x^3\cos\left(\frac{x}{2}\right)dx$

69. $\int_{0.5}^{1} x^2 e^{-3x^2}\,dx$

70. $\int_{0}^{0.3} x^4 \tan^{-1}(3x^3)\,dx$

Applications

71. Airy's equation $y'' = xy$ was developed to describe the patterns of light that pass through a circular aperture and that are caused by diffraction at the edges of the aperture. Show that the function

$$y_0(x) = 1 + \sum_{k=1}^{\infty} \frac{x^{3k}}{(2\cdot 3)(5\cdot 6)\cdots((3k-1)\cdot 3k)}$$

is a solution of Airy's equation.

72. Annie throws a rock into the water near her camp on a very calm morning and watches the ripples spread from the point where the rock splashed. She knows that those ripples can be described radially by Bessel functions. The Bessel function of order p is given by

$$J_p(x) = \sum_{k=0}^{\infty} \frac{(-1)^k x^{2k+p}}{k!(k+p)!2^{2k+p}}.$$

Show that

$$xJ_0(x) = \frac{d}{dx}(xJ_1(x)).$$

Proofs

73. Prove that if the series $\sum_{k=0}^{\infty} a_k x^k$ and $\sum_{k=0}^{\infty} b_k x^k$ both converge to the same sum for every value of x in some nontrivial interval, then $a_k = b_k$ for every nonnegative integer k.

74. Let f be an even function with Maclaurin series representation $\sum_{k=0}^{\infty} a_k x^k$. Prove that $a_{2k+1} = 0$ for every nonnegative integer k.

75. Let f be an odd function with Maclaurin series representation $\sum_{k=0}^{\infty} a_k x^k$. Prove that $a_{2k} = 0$ for every nonnegative integer k.

Thinking Forward

▸ *The complex number i:* If we define $i = \sqrt{-1}$, show that $i^2 = -1$, $i^3 = -i$, and $i^4 = 1$.

▸ *Euler's Formula:* Use the results of the previous exercise and the Maclaurin series for the sine, the cosine, and the exponential function to derive ***Euler's formula***

$$e^{ix} = \cos x + i\sin x.$$

CHAPTER REVIEW, SELF-TEST, AND CAPSTONES

Before you progress to the next chapter, be sure you are familiar with the definitions, concepts, and basic skills outlined here. The capstone exercises at the end bring together ideas from this chapter and look forward to future chapters.

Definitions

Give precise mathematical definitions or descriptions of each of the concepts that follow. Then illustrate the definition or description with a graph or an algebraic example.

- *power series in* x
- *power series in* $x - x_0$
- *interval of convergence*
- *radius of convergence*
- *Maclaurin polynomial*
- *Taylor polynomial*
- *Maclaurin series*
- *Taylor series*
- *the remainder of a function*

Theorems

Fill in the blanks to complete each of the following theorem statements:

- Let $\sum_{k=0}^{\infty} a_k x^k$ be a power series in x. Exactly one of the following is true:

  The series converges only at _____.

  There exists a positive real number ρ such that the series converges absolutely for every $x \in$_____ and diverges if x_____ or x_____.

  The series converges absolutely for every _____.

- Let $\sum_{k=0}^{\infty} a_k (x - x_0)^k$ be a power series in $x - x_0$. Exactly one of the following is true:

  The series converges only at _____.

  There exists a positive real number ρ such that the series converges absolutely for every $x \in$_____ and diverges if x_____ or x_____.

  The series converges absolutely for every _____.

- *Lagrange's Form for the Remainder*: Let f be a function that can be differentiated $n + 1$ times in some open interval I containing the point x_0, and let $R_n(x)$ be the nth remainder for f at $x = x_0$. For each point $x \in I$, there is at least one c between _____ and _____ such that $R_n(x) =$_____.

Derivatives and Integrals of Power Series

The Calculus of Power Series: Let $\sum_{k=0}^{\infty} a_k (x - x_0)^k$ be a power series in $x - x_0$ that converges to a function $f(x)$ on an interval I.

- The derivative of f is given by $f'(x) =$_____.

  The interval of convergence for this new series is _____, with the possible exception that _____.

- The indefinite integral of f is given by $\int f(x)\, dx =$ _____.

  The interval of convergence for this new series is _____, with the possible exception that _____.

- If c and d are two points in I, then the definite integral of f on the interval $[c, d]$ is given by $\int_c^d f(x)\, dx =$_____.

Skill Certification: Working with Power Series

Interval of convergence and radius of convergence: Find the interval of convergence and radius of convergence for each of the given power series. If the interval of convergence is finite, test the series for convergence at each of the endpoints of the interval.

1. $\displaystyle\sum_{k=2}^{\infty} \frac{1}{\ln k} x^k$

2. $\displaystyle\sum_{k=2}^{\infty} (-1)^k \frac{1}{k (\ln k)^2} x^k$

3. $\displaystyle\sum_{k=1}^{\infty} \ln k (x - 3)^k$

4. $\displaystyle\sum_{k=1}^{\infty} k^k (x + 1)^k$

5. $\sum_{k=0}^{\infty} \frac{5^k}{(2k)!}(x+2)^k$

6. $\sum_{k=1}^{\infty} \frac{k!}{k^k}(x-4)^k$

7. $\sum_{k=1}^{\infty} \frac{(-1)^k}{k2^k}(x-2)^k$

8. $\sum_{k=1}^{\infty}(x-3)^k$

Maclaurin and Taylor polynomials: Find third-order Maclaurin or Taylor polynomial for the given function about the indicated point.

9. $\tan^{-1} x, x_0 = 0$

10. $\tan x, x_0 = \frac{\pi}{3}$

11. $x \sin x, x_0 = 0$

12. $\sin x, x_0 = \frac{\pi}{2}$

13. $(x^2+1)^{3/2}, x_0 = 0$

14. $\sqrt{x}, x_0 = 1$

Maclaurin and Taylor series: Find the indicated Maclaurin or Taylor series for the given function about the indicated point, and find the radius of convergence for the series.

15. $\sin x, x_0 = 0$

16. $\sin x, x_0 = \frac{\pi}{2}$

17. $e^x, x_0 = 0$

18. $e^x, x_0 = 1$

19. $\frac{1}{1-x}, x_0 = 0$

20. $\frac{1}{1-x}, x_0 = 2$

21. $\frac{1}{x}, x_0 = 1$

22. $\frac{1}{x}, x_0 = -1$

23. $\sqrt{x}, x_0 = 1$

24. $\sqrt{x}, x_0 = 2$

Recognizing power series: Use the Maclaurin series for $\sin x$, $\cos x$, and e^x to find the values of the following series.

25. $\pi - \frac{\pi^3}{3!} + \frac{\pi^5}{5!} - \frac{\pi^7}{7!} + \cdots$

26. $\pi - \frac{\pi^2}{2!} + \frac{\pi^3}{3!} - \frac{\pi^4}{4!} + \frac{\pi^5}{5!} + \cdots$

27. $1 - \frac{\pi^2}{2^2 \cdot 2!} + \frac{\pi^4}{2^4 \cdot 4!} - \frac{\pi^6}{2^6 \cdot 6!} + \cdots$

28. $e - \frac{e^3}{3!} + \frac{e^5}{5!} - \frac{e^7}{7!} + \cdots$

Manipulating power series: Find power series representations for the indicated functions.

29. Use the Maclaurin series for e^x to find power series representations for $g(x) = x^2 e^x$, $g'(x)$, and $\int g(x)\,dx$.

30. Use the Maclaurin series for $\frac{1}{1-x}$ to find a power series representation for $\frac{x^2}{(1-x^3)^2}$.

31. Use the Maclaurin series for $\frac{1}{1-x}$ to find power series representations for $\frac{1}{1+x}$, $\ln(1+x)$, $\int \ln(1+x)\,dx$, and $\int_0^2 \ln(1+x)\,dx$.

32. Use the Maclaurin series for $\cos x$ to find series representations for $\cos(x^3)$, $\int \cos(x^3)\,dx$, and $\int_0^1 \cos(x^3)\,dx$.

Approximations using power series: Use the indicated power series to approximate each of the following quantities.

33. Find the Maclaurin series for e^{-x}, and use it to approximate $\frac{1}{\sqrt{e}}$ to within 0.001 of its value. How many terms would you need to approximate $\frac{1}{\sqrt{e}}$ to within 10^{-6} of its value?

34. Find the Maclaurin series for e^{-x^2}, and use it to approximate $\int_0^1 e^{-x^2}\,dx$ to within 0.001 of its value. How many terms would you need to approximate the integral to within 10^{-6} of its value?

35. Find the Maclaurin series for $\sin(x^2)$, and use it to approximate $\int_0^3 \sin(x^2)\,dx$ to within 0.001 of its value. How many terms would you need to approximate the integral to within 10^{-6} of its value?

36. Find the Maclaurin series for $\cos(x^4)$, and use it to approximate $\int_0^1 \cos(x^4)\,dx$ to within 0.001 of its value. How many terms would you need to approximate the integral to within 10^{-6} of its value?

Capstone Problems

A. Use the Maclaurin series for $\sin x$, $\cos x$, and e^x to prove that

(a) $\frac{d}{dx}(\sin x) = \cos x$

(b) $\frac{d}{dx}(\cos x) = -\sin x$

(c) $\frac{d}{dx}(e^x) = e^x$

B. Consider the definite integral $\int_0^1 \tan^{-1} x\,dx$.

(a) Evaluate the integral using integration by parts.

(b) Recall that

$$\tan^{-1} x = \sum_{k=0}^{\infty} \frac{(-1)^k}{2k+1} x^{2k+1},$$

with interval of convergence $[-1, 1]$. Use Theorem 8.12 to evaluate the integral by integrating the Maclaurin series.

(c) Assuming that you can rearrange the terms of the series, show that your answers to (a) and (b) are equivalent.

C. Consider the first-order linear initial-value problem $y' - y = 1$, $y(0) = 2$.

(a) Use the techniques of Section 6.5 to show that $y = 3e^x - 1$ is the solution to the initial-value problem.

(b) Follow these steps to find a series solution to the problem:

- Assume that y may be represented by a power series in x. That is, $y = \sum_{k=0}^{\infty} a_k x^k$. Explain why $a_0 = 2$.
- Use Theorem 8.11 to explain why the power series for y' is $y' = \sum_{k=1}^{\infty} k a_k x^{k-1}$.
- Reindex the series for y' to show that $y' = \sum_{k=0}^{\infty} (k+1) a_{k+1} x^k$.
- Use the differential equation $y' - y = 1$ to combine the series for y' and y, and thereby show that

$$\sum_{k=0}^{\infty} ((k+1)a_{k+1} - a_k)\, x^k = 1.$$

- Explain why the preceding series implies that $a_1 - a_0 = 1$ and therefore $a_1 = 3$.
- Explain why we have the recursion formula $a_{k+1} = \frac{a_k}{k+1}$ for every $k \geq 1$.
- Use the recursion formula to determine the values of a_k for $k \geq 2$.
- Use the values for $a_0, a_1, a_2, \ldots$ to find a series solution of the differential equation.
- Show that your series solution is equal to the function $3e^x - 1$.

D. Consider the initial-value problem $y' - 2xy = x$, $y(0) = 1$.

(a) Use the techniques of Section 6.5 to solve the initial-value problem.

(b) Use the series technique outlined in Problem D to solve the initial-value problem.

(c) Show that your two solutions are equivalent.

CHAPTER 9

Parametric Equations, Polar Coordinates, and Conic Sections

9.1 Parametric Equations

Parametric Equations
Graphing Parametric Equations
Direction of Motion and Tangent Lines
Arc Length
Examples and Explorations

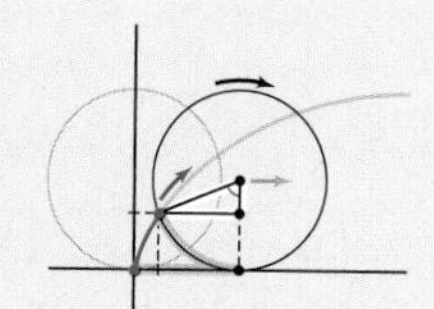

9.2 Polar Coordinates

Plotting Points in Polar Coordinates
Converting Between Polar and Rectangular Coordinates
The Graphs of Some Simple Polar Coordinate Equations
Examples and Explorations

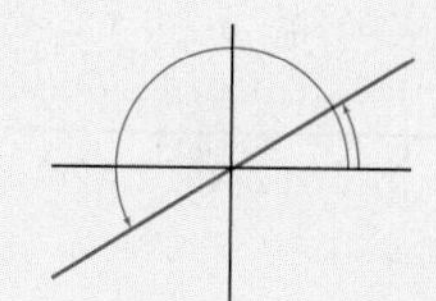

9.3 Graphing Polar Equations

Using the θr-plane to Get Information About a Polar Graph
Symmetry in Polar Graphs
Examples and Explorations

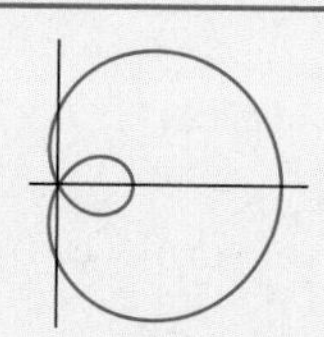

9.4 Computing Arc Length and Area with Polar Functions

Computing the Area Bounded by Polar Curves
Computing Arc Length in Polar Coordinates
Examples and Explorations

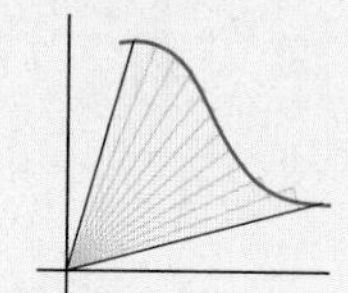

9.5 Conic Sections*

Circles and Ellipses
Parabolas
Hyperbolas
Conic Sections in Polar Coordinates
Examples and Explorations

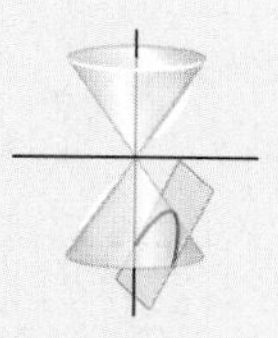

Chapter Review, Self-Test, and Capstones

9.1 PARAMETRIC EQUATIONS

- Defining and graphing parametric equations
- Finding slopes of tangent lines on parametric curves
- Finding the arc length of a parametric curve

Parametric Equations

In most of this book up to now we have studied functions of the form $y = f(x)$, where y is explicitly expressed as a function of x. There are advantages to using equations of this form. In particular, they are often relatively easy to analyze and graph with the tools of calculus. However, there are disadvantages as well. One fundamental limitation is that not every curve of interest has a simple representation in this form. As we saw in Section 2.4, this limitation can be partially addressed by using implicitly defined functions of the form $F(x, y) = c$. In this section we discuss a different method for expressing relationships in the plane: parametric equations. When we use parametric equations, each variable is expressed independently as a function of a new variable, called a parameter. The knobs on an Etch A Sketch® toy show how each variable of a graph can be controlled individually. In a more complex setting, machine tools are also guided parametrically.

DEFINITION 9.1 **Parametric Equations in Two Variables**

Parametric equations in two variables are a pair of functions

$$x = f(t) \quad \text{and} \quad y = g(t),$$

where the ***parameter*** t is defined on some interval I of real numbers.

A ***parametric curve*** is the set of points in the coordinate plane:

$$\{(x(t), y(t)) \mid t \in I\}.$$

For example, we will soon show that the pair of equations

$$x = t + 1, \quad y = t^2 - 4 \text{ for } t \geq -2$$

has the graph shown next on the left, and that the pair of equations

$$x = t - \sin t, \quad y = 1 - \cos t \text{ for } t \in [0, 4\pi]$$

has the graph shown on the right.

The graph of $x = t + 1$, $y = t^2 - 4$ *The graph of $x = t - \sin t$, $y = 1 - \cos t$*

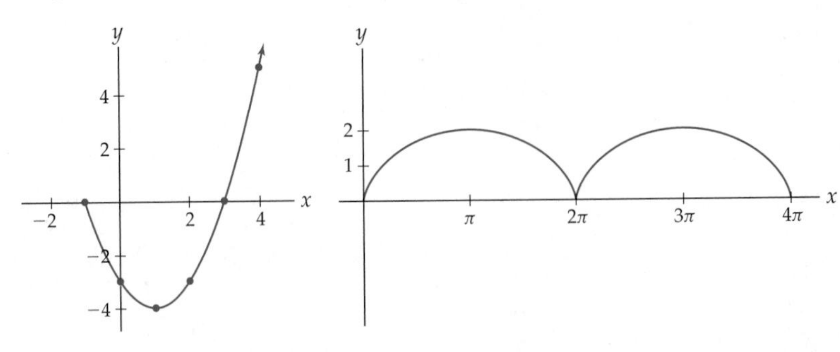

In these equations we used the letter t as the parameter, so we say that the curves are ***parametrized*** by t. The particular letter that we use is not important. However, we will often think of t as representing time. When this is the case, we say that the curve is ***parametrized by time***. On other occasions, the parameter may represent a rotation; in that case we might use the angles θ or ϕ as our parameter.

There are two important reasons for introducing parametric equations. First, some curves are most naturally expressed with parametric equations. Second, such equations can be generalized to express curves in three-dimensional space. (See Sections 10.5 and 11.1.) In many applications, we will have a curve or phenomenon we wish to understand and parametric equations will be the easiest way to describe the motion involved.

To gain a basic understanding of parametric equations, however, we continue by ignoring these important reasons, at least for now. Instead we will start with some parametric equations and look for the curves they describe.

Graphing Parametric Equations

The most basic way to plot any curve is to plot points and then "connect the dots." This is probably the first technique you learned when you started graphing functions. It is also the way your graphing calculator and most computer algebra systems plot curves. In fact, this is the very reason that graphs of curves with discontinuities often are not plotted well by a calculator. To start we will use the "connect the dots" approach to plot the parametric curve given by

$$x = t + 1 \quad \text{and} \quad y = t^2 - 4 \quad \text{for} \quad t \in [-2, \infty).$$

Using these equations, we may generate the following table that contains the coordinates of several points on the curve:

t	-2	-1	0	1	2	3
$x = t + 1$	-1	0	1	2	3	4
$y = t^2 - 4$	0	-3	-4	-3	0	5

We then plot the points (x, y) and connect the dots with a curve, as follows:

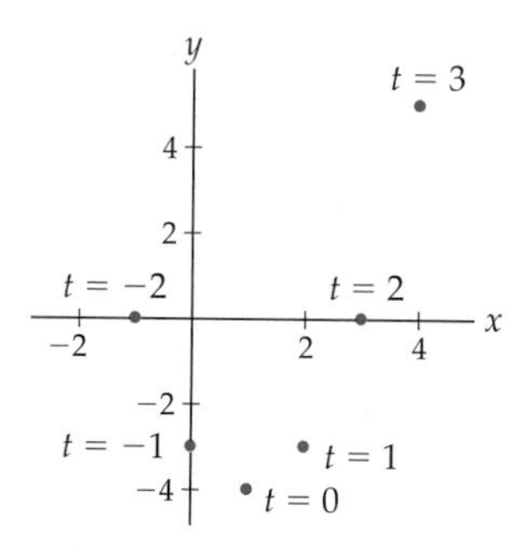

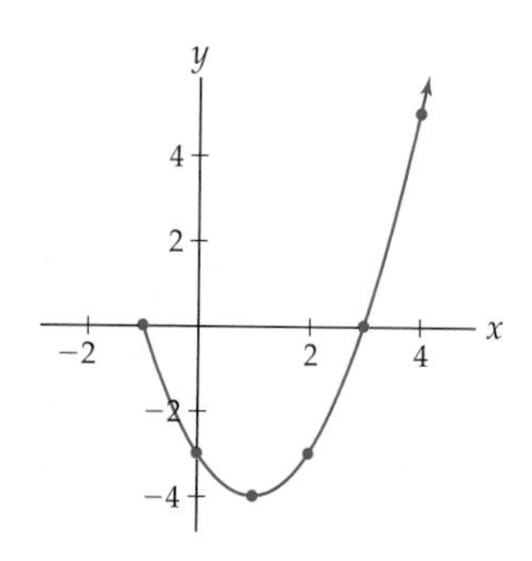

Observe that we have added an arrowhead to one end of the curve in the second figure. This is because, as the parameter t increases, it imposes a direction on the curve. For the values of t that we happened to consider, the curve begins at the smallest value, $t = -2$, and the arrow indicates the direction of motion as t increases. It is important to note that a given curve in the plane could be parametrized many different ways. Different parametric equations could traverse the same curve in the opposite direction. Other equations could traverse the same curve, but double back on portions of the curve over certain subintervals.

For example, in Exercises 14 and 15 we will ask you to show that the parametrization

$$x = 2t + 1 \quad \text{and} \quad y = 4t^2 - 4 \quad \text{for} \quad t \in [-1, \infty)$$

traces exactly the same curve as the one shown, and that the equations

$$x = \sin t + 1 \quad \text{and} \quad y = \sin^2 t - 4 \quad \text{for} \quad t \in (-\infty, \infty)$$

repeatedly trace a portion of the same parabola.

If the parametric equations are simple enough, we may also graph the curve by ***eliminating the parameter***. The process of eliminating the parameter usually takes one of two forms. If the function $x = f(t)$ is invertible and we can find a simple expression for the inverse, $t = f^{-1}(x)$, then we may obtain $y = g(f^{-1}(x))$. For example, this is possible with the preceding equations by solving $x = t + 1$ for t, yielding $t = x - 1$. We then replace t with $x - 1$ in the equation for y, giving the equation

$$y = (x - 1)^2 - 4.$$

This equation defines an upwards-opening parabola whose vertex is at the point $(1, -4)$, just as we saw earlier. However, note that when we eliminated the parameter we lost information: The equation $y = (x - 1)^2 - 4$ neither contains information about the direction of motion imposed by the parameter nor tells us that the parametric equations specify only a portion of the parabola.

We may use a variation of this procedure when $y = g(t)$ is invertible and we can find a simple expression for the inverse, $t = g^{-1}(y)$. In this case, we are expressing either x or y as a function of the other variable, and that allows us to use the techniques of earlier chapters to analyze the resulting function.

Another way to eliminate the parameter results in a relationship in which y is expressed implicitly as a function of x. For example, if $x = 2\sin t$ and $y = 2\cos t$, then when we square both equations, add the two left-hand sides, and add the two right-hand sides, we obtain $x^2 + y^2 = 4\sin^2 t + 4\cos^2 t$. Using one of the Pythagorean identities, we have $x^2 + y^2 = 4$. Thus, the graph of these parametric equations lies on the circle with radius 2 and centered at the origin.

Direction of Motion and Tangent Lines

In general, the parametric equations $x = f(t)$ and $y = g(t)$ do not have to be either continuous or differentiable. For the time being we will assume that the functions are differentiable, and, to allow us to use the chain rule, we will also assume that y is locally a differentiable function of x for a point at which x is a differentiable function of t.

We may use the derivatives of the parametric equations to obtain information about the direction of motion along the parametric curve with increasing values of t. When we compute $\frac{dx}{dt}$ and $\frac{dy}{dt}$ and examine the signs of these derivatives, we easily determine the direction in which the curve is being drawn. For example, for the parametric equation $x = t + 1$ and $y = t^2 - 4$ for $t \geq -2$, we have

$$\frac{dx}{dt} = 1 \text{ and } \frac{dy}{dt} = 2t.$$

Thus, for all values of t, since $\frac{dx}{dt} > 0$, the curve is being traced from left to right. Since $\frac{dy}{dt} < 0$ for $-2 \leq t < 0$ and $\frac{dy}{dt} > 0$ for $t > 0$, the curve is moving down at first, but then starts rising.

To understand the slope of a parametric curve we may use the chain rule,

$$\frac{dy}{dt} = \frac{dy}{dx}\frac{dx}{dt}.$$

Thus, if $\frac{dx}{dt} \neq 0$, then

$$\frac{dy}{dx} = \frac{dy/dt}{dx/dt} = \frac{y'(t)}{x'(t)}.$$

For example, the slope of the tangent line to the curve given by $x = t+1$ and $y = t^2 - 4$ when $t = 4$ is

$$\left.\frac{dy/dt}{dx/dt}\right|_{t=4} = \left.\frac{2t}{1}\right|_{t=4} = 8.$$

More generally, we use this idea to define the tangent line to a parametric curve.

DEFINITION 9.2

The Slope of a Parametric Curve and the Tangent Line to a Parametric Curve

Let $x = x(t),\ y = y(t)$ be parametric equations for t in some interval I, and let t_0 be a point in I at which $x'(t_0)$ and $y'(t_0)$ both exist.

(a) If $x'(t_0) \neq 0$, we define the ***slope*** of the parametric curve at the point $(x(t_0), y(t_0))$ to be

$$m = \frac{y'(t_0)}{x'(t_0)}.$$

If $x'(t_0) = 0$, we define the slope to be

$$m = \lim_{t \to t_0} \frac{y'(t)}{x'(t)},$$

provided that the limit exists.

(b) If the slope m is defined, the ***tangent line*** to the parametric curve at the point t_0 is given by the equation

$$y - y(t_0) = m(x - x(t_0)).$$

Note that certain tangents to parametric curves are vertical lines. As an extension to the preceding definition we may say that if $\lim_{t \to t_0} \frac{y'(t)}{x'(t)} = \pm\infty$, then the parametric curve has a ***vertical tangent line*** at $(x(t_0), y(t_0))$ whose equation is $x = x(t_0)$.

Even when the functions f and g are differentiable, the curve defined by the parametric equations $x = f(t),\ y = g(t)$ may have cusps or other anomalies. For example, the function in the figure that follows at the left is not differentiable at the cusps, although we will soon see that the parametric equations we use to define the curve are differentiable everywhere. In addition, a given parametric curve may have points with two or more tangent lines; see, for example, the following figure at the right:

Cusps are points of non-differentiability

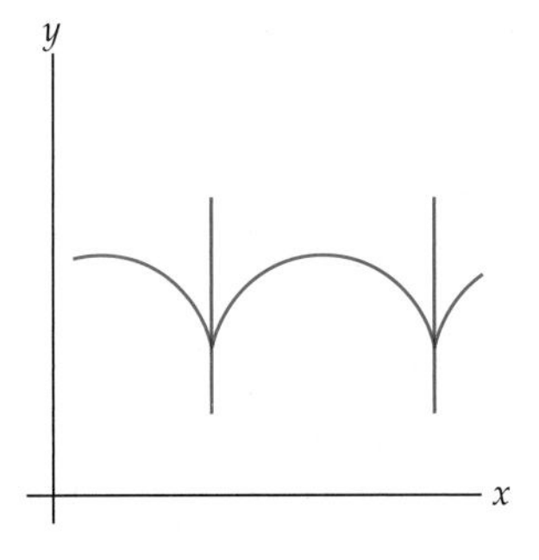

A point on a parametric curve with two tangents

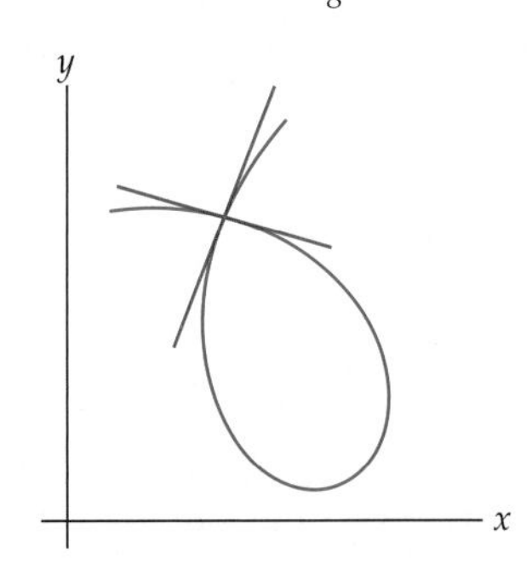

Arc Length

In Section 6.3 we saw that the arc length of a continuous function $f(x)$ on an interval $[x_1, x_2]$ is given by the formula

$$\int_{x_1}^{x_2} \sqrt{1 + (f'(x))^2}\, dx.$$

We will extend this definition to functions defined parametrically.

Suppose C is a parametric curve defined by $x = f(t)$ and $y = g(t)$ for $t \in [a, b]$. We can approximate the curve C with line segments:

Curve C can be approximated with line segments

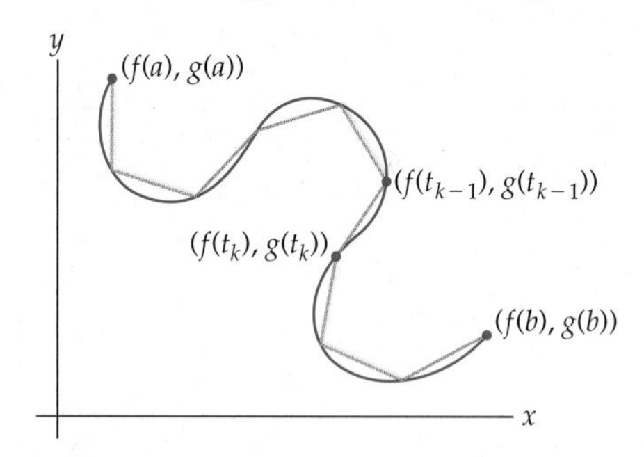

We need to use general notation to compute the sum of the lengths of the approximating line segments. Toward that end we subdivide the interval $[a, b]$ into n equal pieces and define $\Delta t = \frac{b-a}{n}$. Then, for $k = 0, 1, 2, \ldots, n$, we define $t_k = a + k\Delta t$. The length of C can be approximated by the sum of the lengths of the segments extending from the points $(f(t_{k-1}), g(t_{k-1}))$ to the points $(f(t_k), g(t_k))$. Notice that this is the same process we followed in Section 6.3, except that here we are subdividing the interval on which the parameter is defined. Using the distance formula and then adding up the lengths of each of the line segments, we arrive at the following approximation for the length of the curve C as the parameter t varies from a to b:

$$\sum_{k=1}^{n} \sqrt{(f(t_k) - f(t_{k-1}))^2 + (g(t_k) - g(t_{k-1}))^2}.$$

As in Section 6.3, the length of the curve C for $t \in [a, b]$ is defined as the limit of the approximate quantity as the number of pieces in our subdivision increases without bound.

DEFINITION 9.3 **The Arc Length of a Parametric Curve**

Let C be a curve in the plane with parametrization $x = f(t),\ y = g(t)$ for $t \in [a, b]$, where f and g are differentiable functions of t such that the parametrization is a one-to-one function from the interval $[a, b]$ to the curve C. Then the ***length of the curve*** C is

$$\lim_{n\to\infty} \sum_{k=1}^{n} \sqrt{(f(t_k) - f(t_{k-1}))^2 + (g(t_k) - g(t_{k-1}))^2},$$

where $\Delta t = \frac{b-a}{n}$ and $t_k = a + k\Delta t$.

The requirement in Definition 9.3 that the parametrization be a one-to-one function from the interval $[a, b]$ to the curve C ensures that C is traversed exactly once as t increases from a to b. This means that the parametrization does not trace any part of the curve more than once. (Actually, this requirement can be relaxed somewhat; we can allow isolated points to be duplicated, and the definition will still hold.)

You may have noticed that the expression in Definition 9.3 looks almost like a limit of Riemann sums. The following theorem tells us how to interpret this quantity as a definite integral:

THEOREM 9.4 **The Arc Length of a Parametric Curve**

Let C be a curve in the plane with parametrization $x = f(t)$, $y = g(t)$ for $t \in [a, b]$ such that the parametrization is a one-to-one function from the interval $[a, b]$ to the curve C. If $x = f(t)$ and $y = g(t)$ are differentiable functions of t such that $f'(t)$ and $g'(t)$ are continuous on $[a, b]$, then the length of the curve C is given by

$$\int_a^b \sqrt{(f'(t))^2 + (g'(t))^2}\, dt.$$

Proof. By Definition 9.3, the length of the curve C is

$$\lim_{n\to\infty} \sum_{k=1}^{n} \sqrt{(f(t_k) - f(t_{k-1}))^2 + (g(t_k) - g(t_{k-1}))^2}.$$

Some simple algebra then shows that this is equivalent to

$$\lim_{n\to\infty} \sum_{k=1}^{n} \sqrt{\left(\frac{f(t_k) - f(t_{k-1})}{\Delta t}\right)^2 + \left(\frac{g(t_k) - g(t_{k-1})}{\Delta t}\right)^2}\, \Delta t,$$

where $\Delta t = \dfrac{b-a}{n}$ and $t_k = a + k\Delta t$. Applying the Mean Value Theorem to f and g, we find that for each k there are points t_k^* and t_k^{**} in $[t_{k-1}, t_k]$ such that

$$f'(t_k^*) = \frac{f(t_k) - f(t_{k-1})}{\Delta t} \text{ and } g'(t_k^{**}) = \frac{g(t_k) - g(t_{k-1})}{\Delta t}.$$

Thus the length of C is

$$\lim_{n\to\infty} \sum_{k=1}^{n} \sqrt{(f'(t_k^*))^2 + (g'(t_k^{**}))^2}\, \Delta t.$$

This last quantity is almost the limit of a Riemann sum. It is not the limit of a Riemann sum because t_k^* and t_k^{**} are, in general, different points. However, since f' and g' are both continuous, it can be shown that t_k^* and t_k^{**} get sufficiently close to each other as $n \to \infty$ for this not to matter. (The proof of this fact is outside of the scope of the text.) Applying the preceding final result, we find that the limit of our sum is the desired integral

$$\int_a^b \sqrt{(f'(t))^2 + (g'(t))^2}\, dt.$$

■

Examples and Explorations

EXAMPLE 1 **Parametrizing the unit circle**

Sketch the parametric curves determined by the following parametric equations:

(a) $x = \cos t,\ y = \sin t,\ t \in [0, \pi]$

(b) $x = \cos t,\ y = \sin t,\ t \in [\pi, 2\pi]$

(c) $x = \sin(3t),\ y = \cos(3t),\ t \in [0, 2\pi]$

(d) $x = \dfrac{1 - t^2}{1 + t^2},\ y = \dfrac{2t}{1 + t^2},\ t \in [0, \infty)$

SOLUTION

(a) We will start by eliminating the parameter, but also plot a few points along the way. If we square the equations for x and y, we obtain $x^2 = \cos^2 t$ and $y^2 = \sin^2 t$. Now adding these two equations and applying the Pythagorean identity, we have

$$x^2 + y^2 = \cos^2 t + \sin^2 t = 1.$$

Therefore $x^2 + y^2 = 1$, which means that the parametric curve is (at least a portion of) a circle of radius 1 and centered at the origin. We want to know where on this circle the parametric curve lies, and we also want to know the direction of motion as t increases. Consider the following three values of t: $t = 0$, $t = \frac{\pi}{2}$, and $t = \pi$. Corresponding to these values we have $(x(0), y(0)) = (\cos 0, \sin 0) = (1, 0)$,

$$\left(x\left(\frac{\pi}{2}\right), y\left(\frac{\pi}{2}\right)\right) = \left(\cos\frac{\pi}{2}, \sin\frac{\pi}{2}\right) = (0, 1), \text{ and } (x(\pi), y(\pi)) = (\cos\pi, \sin\pi) = (-1, 0).$$

In addition, as t increases (slightly) from 0, both x and y will be positive. Thus, we see that the parametric curve is the top half of the unit circle and the direction of motion is counterclockwise; see the first figure that follows at the left.

(b) We eliminated the parameter from these equations in part (a). A similar analysis of the behavior of the variables x and y as t increases from π to 2π shows that this parametrization gives us the bottom half of the unit circle, with the direction of motion still being counterclockwise; see the middle figure. The details are left to Exercise 35.

(c) We again eliminate the parameter by squaring each equation and adding. Once more, we obtain $x^2 + y^2 = 1$. As before, the curve is some portion of the unit circle centered at the origin. Now, however, when $t = 0$, we are at the point $(x(0), y(0)) = (\sin 0, \cos 0) = (0, 1)$. Furthermore, as t increases from 0, x increases and y decreases; therefore the direction of motion is clockwise. Finally, we note that for $t \in [0, 2\pi]$ both $x = \sin 3t$ and $y = \cos 3t$ go through three periods. Therefore, the unit circle is traversed three times in the clockwise direction with these equations; see the following figure at the right:

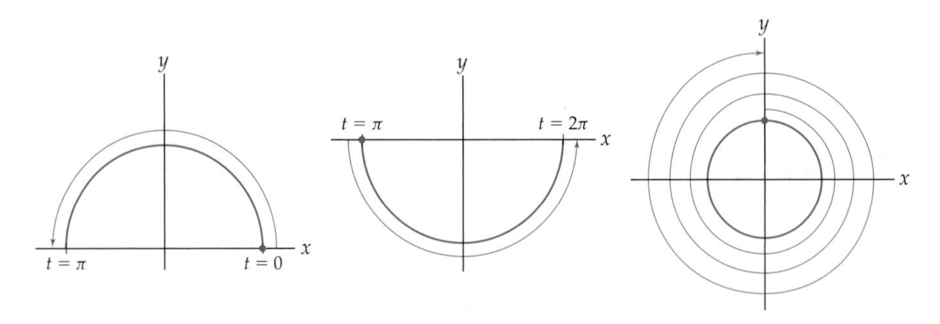

(d) We will see that this pair of equations is quite a different parametrization of the unit circle. Although it does not involve trigonometric functions, we may again eliminate the parameter by squaring each equation and adding the result:

$$\begin{aligned} x^2 + y^2 &= \left(\frac{1-t^2}{1+t^2}\right)^2 + \left(\frac{2t}{1+t^2}\right)^2 = \frac{1-2t^2+t^4+4t^2}{(1+t^2)^2} \\ &= \frac{1+2t^2+t^4}{(1+t^2)^2} = \frac{(1+t^2)^2}{(1+t^2)^2} = 1. \end{aligned}$$

Thus, the parametric curve is again at least a portion of the unit circle. To determine which portion (and in what direction and with what speed), we begin by evaluating x and y for two values of t:

t	0	1
$x = \dfrac{1-t^2}{1+t^2}$	1	0
$y = \dfrac{2t}{1+t^2}$	0	1

Since the parameter t can vary from 0 to ∞, we also evaluate the following limits:

$$\lim_{t\to\infty} x(t) = \lim_{t\to\infty} \frac{1-t^2}{1+t^2} = -1, \quad \lim_{t\to\infty} y(t) = \lim_{t\to\infty} \frac{2t}{1+t^2} = 0.$$

Putting all this information together, we arrive at the following parametric curve:

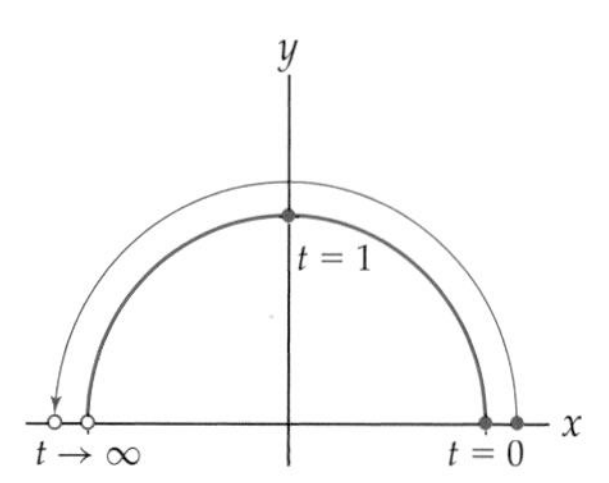

Note that every point on the top half of the unit circle is traversed exactly once, except for $(-1, 0)$, which is not part of the parametrization. In addition, unlike the motion in parts (a), (b), and (c), the motion here does *not* take place at a uniform speed. For example, as t increases from 0 to 1, the particle moves all the way from $(1, 0)$ to $(0, 1)$, but as t increases from 1, the particle moves from $(0, 1)$ toward, but never reaches $(-1, 0)$. □

EXAMPLE 2 **Graphing an astroid**

Sketch the graph of the curve described by the parametric equations

$$x = \cos^3\theta \quad \text{and} \quad y = \sin^3\theta.$$

SOLUTION

We begin by tabulating the coordinates of several points for some convenient values of θ:

θ	0	$\frac{\pi}{6}$	$\frac{\pi}{4}$	$\frac{\pi}{3}$	$\frac{\pi}{2}$	$\frac{2\pi}{3}$	$\frac{3\pi}{4}$	$\frac{5\pi}{6}$
$x = \cos^3\theta$	1	$\frac{3\sqrt{3}}{8}$	$\frac{\sqrt{2}}{4}$	$\frac{1}{8}$	0	$-\frac{1}{8}$	$-\frac{\sqrt{2}}{4}$	$-\frac{3\sqrt{3}}{8}$
$y = \sin^3\theta$	0	$\frac{1}{8}$	$\frac{\sqrt{2}}{4}$	$\frac{3\sqrt{3}}{8}$	1	$\frac{3\sqrt{3}}{8}$	$\frac{\sqrt{2}}{4}$	$\frac{1}{8}$
θ	π	$\frac{7\pi}{6}$	$\frac{5\pi}{4}$	$\frac{4\pi}{3}$	$\frac{3\pi}{2}$	$\frac{5\pi}{3}$	$\frac{7\pi}{4}$	$\frac{11\pi}{6}$
$x = \cos^3\theta$	-1	$-\frac{3\sqrt{3}}{8}$	$-\frac{\sqrt{2}}{4}$	$-\frac{1}{8}$	0	$\frac{1}{8}$	$\frac{\sqrt{2}}{4}$	$\frac{3\sqrt{3}}{8}$
$y = \sin^3\theta$	0	$-\frac{1}{8}$	$-\frac{\sqrt{2}}{4}$	$-\frac{3\sqrt{3}}{8}$	-1	$-\frac{3\sqrt{3}}{8}$	$-\frac{\sqrt{2}}{4}$	$-\frac{1}{8}$

We plot these points in the figure that follows on the left:

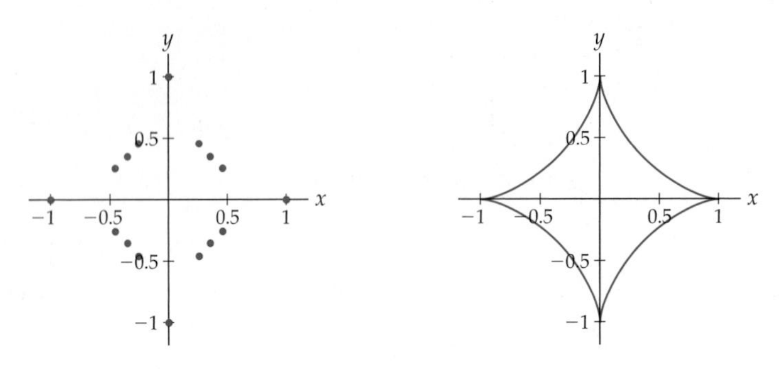

In addition, the slope of the curve at each point is a function of θ:

$$\frac{dy}{dx} = \frac{dy/d\theta}{dx/d\theta} = \frac{\sin^2\theta\cos\theta}{-\cos^2\theta\sin\theta} = -\tan\theta.$$

Thus, the slope of the tangent line to the curve when $\theta = 0$ is zero. We also see that the slope of the tangent line at every point in the first quadrant is negative, and since $\lim_{\theta\to\pi/2^-}(-\tan\theta) = -\infty$, the slopes are nearly vertical close to the point $(0, 1)$. When we analyze the tangents to the curve in the other three quadrants in a similar fashion, and connect the points appropriately, we obtain the graph on the right, known as an ***astroid***.

Finally, we note that we can also eliminate the parameter. Here we have $x^{2/3} = \cos^2\theta$ and $y^{2/3} = \sin^2\theta$. When we add these two equations together, we obtain $x^{2/3} + y^{2/3} = 1$, whose graph is the same astroid. □

EXAMPLE 3 **Finding parametric equations for the cycloid**

Find the parametric equations that describe the curve traced by a point on the circumference of a wheel as it rolls along a straight path.

SOLUTION

We see that this is a more realistic application of parametric equations. We may visualize the given curve as the path traced by a point on a bicycle tire as it rolls along a road. In this situation we wish to understand a curve, and parametric equations provide an effective tool to describe it.

We let the radius of the wheel be r and, for convenience, we place our wheel on a track formed by the x-axis, with the point tracing the curve starting at the origin. The gray circle in the figure that follows shows the initial position of the wheel, and the black circle shows the wheel after it has rolled a short distance. The blue curve is what we wish to describe with the use of parametric equations.

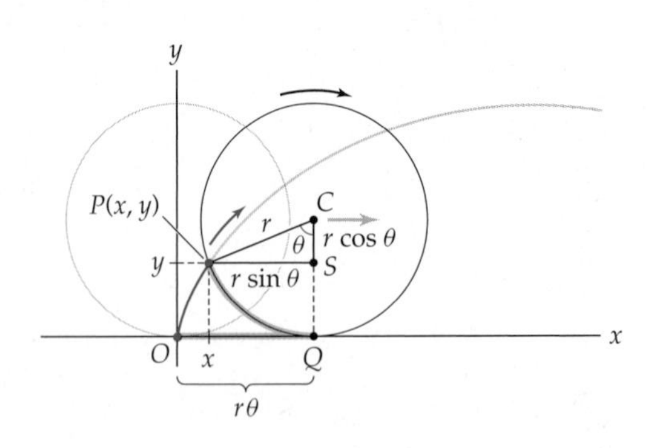

To parametrize the blue curve, we need to find a parameter that can be used to describe both the x- and y-coordinates of each point on the curve; one such parameter is the angle θ, measured in radians, through which the wheel has rotated. We need to express the coordinates of each point $P = P(x, y)$ on the curve in terms of the parameter θ. On the one hand, first note that the arc $\widehat{PQ}$ from P to Q subtended by θ has length $r\theta$ and that this length is precisely equal to the length of the horizontal segment $\overline{OQ}$. (To see this think of rolling the black circle back to its original position.) On the other hand, the length of $\overline{OQ}$ is also equal to x plus the length of segment $\overline{PS}$. Since $PS = r\sin\theta$ and $OQ = r\theta$, it follows that $x = r\theta - r\sin\theta$. Notice that we have now described the coordinate x in terms of the parameter θ (and the constant r).

Similarly, y plus the length of vertical segment $\overline{SC}$ is equal to the radius r. Since the length of SC is $r\cos\theta$, we have $y = r - r\cos\theta$. Therefore, we can express the cycloid with the parametric equations

$$x = r\theta - r\sin\theta, \quad y = r - r\cos\theta, \quad \theta \in \mathbb{R}.$$

The following graph shows the cycloid after two revolutions of the circle:

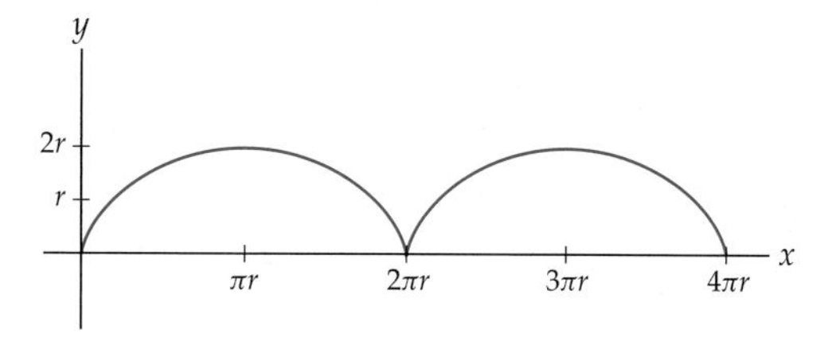

□

EXAMPLE 4

Finding tangent lines on the cycloid

Find the equations of the tangent lines to the cycloid $x = r\theta - r\sin\theta,\ y = r - r\cos\theta$ at $\theta = \frac{\pi}{2}$.

SOLUTION

To find the equations of the tangent lines, we must first determine the derivative $\frac{dy}{dx}$ in terms of the parameter θ. The parametric equations for the cycloid are

$$x = r\theta - r\sin\theta,\ y = r - r\cos\theta,\ \theta \in \mathbb{R}.$$

It is easy to find the derivatives of the coordinates x and y with respect to the parameter θ:

$$\frac{dx}{d\theta} = r - r\cos\theta, \quad \frac{dy}{d\theta} = r\sin\theta.$$

Now, using Definition 9.2, we have

$$\frac{dy}{dx} = \frac{dy/d\theta}{dx/d\theta} = \frac{r\sin\theta}{r - r\cos\theta} = \frac{\sin\theta}{1 - \cos\theta}.$$

This means that at $\theta = \frac{\pi}{2}$, the slope of the tangent line will be $\frac{\sin(\pi/2)}{1-\cos(\pi/2)} = 1$. The point on the cycloid corresponding to $\theta = \frac{\pi}{2}$ is

$$(x, y) = \left(r\left(\frac{\pi}{2} - \sin\frac{\pi}{2}\right), r - r\cos\frac{\pi}{2}\right) = \left(r\left(\frac{\pi}{2} - 1\right), r\right).$$

Therefore, the equation of the tangent line at $\theta = \frac{\pi}{2}$ will be

$$y - r = 1\left(x - r\left(\frac{\pi}{2} - 1\right)\right), \quad \text{or equivalently,} \quad y = x + r\left(2 - \frac{\pi}{2}\right).$$

The following figure shows the tangent line from this example as well as the horizontal tangent line from Example 5:

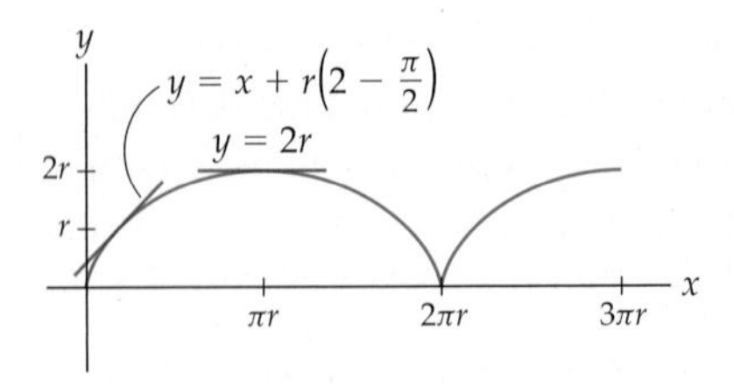

□

EXAMPLE 5

Horizontal and vertical tangent lines on the cycloid

Find all the points on the graph of the cycloid $x = r\theta - r\sin\theta$, $y = r - r\cos\theta$, $\theta \in \mathbb{R}$ at which the tangent line is either horizontal or vertical, and find the equations of the horizontal tangent lines.

SOLUTION

The graph of the cycloid will have a horizontal tangent line wherever the numerator of $\frac{dy}{dx} = \frac{\sin\theta}{1-\cos\theta}$ is zero and the denominator is nonzero. This occurs when θ is any odd multiple of π. Now we have to find the equations of these horizontal tangent lines. If θ is an odd multiple of π, then $\cos\theta = -1$ and therefore at such values of θ we are at the y-coordinate $y = r - r(-1) = 2r$. Thus the equation of the horizontal tangent line at this point is $y = 2r$, which should be obvious from looking at the graph of the cycloid. In addition, this horizontal line is tangent to the cycloid at every at point where the cycloid has a zero slope.

In the computations we just did, we found the locations and equations of the horizontal tangent lines on the graph of the cycloid. To find the vertical tangent lines, we must find the values of θ for which the denominator of $\frac{dy}{dx} = \frac{\sin\theta}{1-\cos\theta}$ is zero but the numerator is nonzero, and the values of α such that

$$\lim_{\theta\to\alpha} \frac{\sin\theta}{1-\cos\theta}$$

is undefined. In Exercise 73 you will show that the latter occurs at each even multiple of π.

□

EXAMPLE 6

Analyzing the concavity of a parametric curve

Let $x = x(t)$ and $y = y(t)$ be the parametric equations for a curve C. Show that if x and y are both twice-differentiable functions of t, then the concavity at a point on C is given by

$$\frac{d^2y}{dx^2} = \frac{\frac{dx}{dt}\frac{d^2y}{dt^2} - \frac{d^2x}{dt^2}\frac{dy}{dt}}{\left(\frac{dx}{dt}\right)^3}$$

when $\frac{dx}{dt} \neq 0$.

Use this result to confirm that the cycloid given by $x = r\theta - r\sin\theta$, $y = r - r\cos\theta$ is concave down everywhere except when θ is an even multiple of *pi*.

SOLUTION

We first note that by the chain rule we have

$$\frac{d^2x}{dt^2} = \frac{d}{dt}\left(\frac{dx}{dt}\right) = \frac{d}{dx}\left(\frac{dx}{dt}\right)\frac{dx}{dt}.$$

Therefore, when $\frac{dx}{dt} \neq 0$,

$$\frac{d}{dx}\left(\frac{dx}{dt}\right) = \frac{d^2x/dt^2}{dx/dt}.$$

Similarly, we may show that

$$\frac{d}{dx}\left(\frac{dy}{dt}\right) = \frac{d^2y/dt^2}{dx/dt}.$$

Now, since $\frac{dy}{dx} = \frac{dy/dt}{dx/dt}$, it follows that

$$\begin{aligned}
\frac{d^2y}{dx^2} &= \frac{d}{dx}\left(\frac{dy}{dx}\right) = \frac{d}{dx}\left(\frac{dy/dt}{dx/dt}\right) \\
&= \frac{\frac{dx}{dt}\frac{d}{dx}\left(\frac{dy}{dt}\right) - \frac{d}{dx}\left(\frac{dx}{dt}\right)\frac{dy}{dt}}{\left(\frac{dx}{dt}\right)^2} && \leftarrow \text{the quotient rule} \\
&= \frac{\frac{dx}{dt}\frac{d^2y/dt^2}{dx/dt} - \frac{d^2x/dt^2}{dx/dt}\frac{dy}{dt}}{\left(\frac{dx}{dt}\right)^2} && \leftarrow \text{the earlier equalities} \\
&= \frac{\frac{dx}{dt}\frac{d^2y}{dt^2} - \frac{d^2x}{dt^2}\frac{dy}{dt}}{\left(\frac{dx}{dt}\right)^3}. && \leftarrow \text{algebra}
\end{aligned}$$

For the cycloid, since $x = r\theta - r\sin\theta$ and $y = r - r\cos\theta$, we have

$$\frac{dx}{d\theta} = r - r\cos\theta, \quad \frac{d^2x}{d\theta^2} = r\sin\theta, \quad \frac{dy}{d\theta} = r\sin\theta, \quad \text{and} \quad \frac{d^2y}{d\theta^2} = r\cos\theta.$$

Combining these equations with the formula for $\frac{d^2y}{dx^2}$, we obtain

$$\begin{aligned}
\frac{d^2y}{dx^2} &= \frac{(r - r\cos\theta)(r\cos\theta) - (r\sin\theta)(r\sin\theta)}{(r - r\cos\theta)^3} \\
&= \frac{r^2(\cos\theta - 1)}{r^3(1 - \cos\theta)^3} && \leftarrow \text{algebra and } \sin^2\theta + \cos^2\theta = 1 \\
&= -\frac{1}{r(1 - \cos\theta)^2}. && \leftarrow \text{algebra}
\end{aligned}$$

Since r is a positive constant, the quotient $-\frac{1}{r(1-\cos\theta)^2} < 0$ when θ is not a multiple of 2π. Therefore, the cycloid is concave down when θ is not an even multiple of π. □

EXAMPLE 7 **Finding the arc length of a parametric curve**

Find the arc length of one arch of the cycloid traced by a point on the circumference of a wheel with radius r units (i.e., the portion of the graph of the cycloid that represents one full rotation of the wheel).

SOLUTION

One such arch is given by the portion of the cycloid corresponding to $\theta \in [0, 2\pi]$, since one full rotation of the wheel corresponds to 2π radians. For $\theta \in [0, 2\pi]$, the desired arc length is given by

$$\int_0^{2\pi} \sqrt{(f'(\theta))^2 + (g'(\theta))^2}\, d\theta,$$

where $f(\theta) = r\theta - r\sin\theta$ and $g(\theta) = r - r\cos\theta$. Thus, the arc length is

$$\int_0^{2\pi} \sqrt{(r - r\cos\theta)^2 + (r\sin\theta)^2}\, d\theta = r\int_0^{2\pi} \sqrt{2 - 2\cos\theta}\, d\theta = 8r.$$

The details of the last step (in which the definite integral is calculated to be equal to $8r$) are left for Exercise 40. □

? TEST YOUR UNDERSTANDING

- What is the definition of parametric equations?
- How do you graph parametric equations by plotting points? By eliminating the parameter?
- How do you find the derivative of a curve defined parametrically? How do you find the locations of any horizontal or vertical tangent lines on a parametric curve?
- The cycloid in Example 3 is the graph of some function of x—that is, $y = h(x)$—since the graph passes the vertical line test. Can you eliminate the parameter from the equations in Example 3 to find $h(x)$?
- How can a definite integral be used to calculate the arc length of a parametric curve? How does this definite integral arise from a limit of approximations arrived at with the use of line segments?

EXERCISES 9.1

Thinking Back

Eliminating a variable: Find a relationship between variables x and y by eliminating the variable t.

- $x = t^2,\ y = t^6$
- $x = \sin t,\ y = \sin t$
- $x = \sin t,\ y = \cos t$
- $x = \sinh t,\ y = \cosh t$

Arc length: Find the arc length of the following curves on the given intervals.

- $y = 3x - 4,\ x \in [0, 5]$
- $y = \sqrt{4 - x^2},\ x \in [0, 2]$
- $y = x^{3/2},\ x \in [0, 4]$
- $y = x^2,\ x \in [0, 1]$

Concepts

0. *Problem Zero:* Read the section and make your own summary of the material.

1. *True/False:* Determine whether each of the statements that follow is true or false. If a statement is true, explain why. If a statement is false, provide a counterexample.

(a) *True or False:* Every function $y = f(x)$ can be written in terms of parametric equations.

(b) *True or False:* Given parametric equations $x = x(t)$ and $y = y(t)$, the parameter can be eliminated to obtain the form $y = f(x)$.

(c) *True or False:* Every parametric curve passes the vertical line test.

(d) *True or False:* Every curve in the plane has a unique expression in terms of parametric equations.

(e) *True or False:* If the functions $x = f(t)$ and $y = g(t)$ are differentiable for every $t \in \mathbb{R}$, then the parametric curve defined by x and y is differentiable for every value of t.

(f) *True or False:* A curve parametrized by $x = x(t)$, $y = y(t)$ has a horizontal tangent line at $(x(t_0), y(t_0))$ if $y'(t) = 0$.

(g) *True or False:* A curve parametrized by $x = x(t)$, $y = y(t)$ has a horizontal tangent line at $(x(t_0), y(t_0))$ if $x'(t) \neq 0$ and $y'(t) = 0$.

(h) *True or False:* The cycloid curve associated with a circle of radius r is made up of a series of semicircles of radius r.

2. *Examples:* Construct examples of the thing(s) described in the following. Try to find examples that are different than any in the reading.

(a) Parametric equations $x = f(t),\ y = g(t)$ on the interval $[0, 1)$ that trace the unit circle exactly once clockwise, starting at the point $(1, 0)$.

(b) Parametric equations $x = f(t),\ y = g(t)$ on the interval $[0, 2\pi)$ that trace the circle centered at $(2, -3)$ with radius 5 exactly once counterclockwise, starting at the point $(7, -3)$.

(c) Parametric equations $x = f(t),\ y = g(t)$ whose graph is not the graph of a function $y = f(x)$.

3. If $x = x(t)$ and $y = y(t)$ are differentiable functions of t, determine the direction of motion along the curve when

(a) $x'(t) > 0$ and $y'(t) > 0$.
(b) $x'(t) > 0$ and $y'(t) < 0$.
(c) $x'(t) < 0$ and $y'(t) > 0$.
(d) $x'(t) < 0$ and $y'(t) < 0$.

4. Complete the following definition: ***Parametric equations*** are ______________.

Use the results of Exercise 3 to analyze the direction of motion for the parametric curves given by the equations in Exercises 5–8.

5. $x = t^2,\ y = t^3,\ t \in \mathbb{R}$
6. $x = \sin t,\ y = \cos t,\ t \in \mathbb{R}$
7. $x = e^t,\ y = \ln t,\ t > 0$
8. $x = t^3 - t,\ y = t^3 + t,\ t \in \mathbb{R}$

In Exercises 9–11 parametrizations are provided for portions of the same function. For each problem do the following:

(i) Eliminate the parameter to show that the curves are portions of the same function.
(ii) Describe the portion of the graph that each parametrization describes.
(iii) Discuss the direction of motion along the graph for each parametrization.

9. (a) $x = t,\ y = t^2 - 1,\ t \geq 0$
(b) $x = -t,\ y = t^2 - 1,\ t \geq 0$
10. (a) $x = t^2,\ y = t^3,\ t \geq 0$
(b) $x = t^2,\ y = t^3,\ t \leq 0$
11. (a) $x = t,\ y = \sin t,\ t \geq 0$
(b) $x = t - 1,\ y = \sin(t - 1),\ t \geq 1$
12. Suppose a parametric curve is given by parametric equations $x = x(t), y = y(t)$ for t in some interval I. How can we find the slope of the parametric curve at some point $(x(t_0), y(t_0))$? What is the equation of the tangent line to the parametric curve at the point t_0?
13. Explain how we can find the locations at which a parametric curve determined by $x = x(t)$ and $y = y(t)$ has horizontal or vertical tangent lines.
14. Show that the parametrization $x = 2t + 1,\ y = 4t^2 - 4$ for $t \in [-1, \infty)$ has the same graph as the one we plotted point by point in the reading.
15. Explain why the parametrization $x = \sin t + 1,\ y = \sin^2 t - 4$ for $t \in (-\infty, \infty)$ repeatedly traces the same small portion of the graph of the function $y = x^2 - 2x - 3$.

Skills

In Exercises 16–23 sketch the parametric curve by plotting points.

16. $x = t,\ y = t^2,\ t \in \mathbb{R}$
17. $x = 3t + 1,\ y = t,\ t \in [-2, 5]$
18. $x = 3t + 1,\ y = 2t,\ t \in [0, 8]$
19. $x = 1 - 2t,\ y = 5 - 3t,\ t \in \mathbb{R}$
20. $x = 2t - 1,\ y = 3t + 5,\ t \in \mathbb{R}$
21. $x = t^3 - t,\ y = t^3 + t,\ t \in \mathbb{R}$
22. $x = 2\sin^3 t,\ y = 2\cos^3 t,\ t \in [0, 2\pi]$
23. $x = \cos^5 t,\ y = \sin^5 t,\ t \in [0, 2\pi]$

In Exercises 24–34 sketch the parametric curve by eliminating the parameter.

24. $x = 2t - 1,\ y = 3t + 5,\ t \in \mathbb{R}$
25. $x = 2t - 1,\ y = 3t^2 + 5,\ t \in \mathbb{R}$
26. $x = t + 2,\ y = e^t,\ t \in \mathbb{R}$
27. $x = \tan t,\ y = \tan t,\ t \in \left(-\frac{\pi}{2}, \frac{\pi}{2}\right)$
28. $x = \cos 2t,\ y = -\sin 2t,\ t \in [0, 2\pi]$
29. $x = 3\cos t,\ y = 4\sin t,\ t \in [0, 2\pi]$
30. $x = \sin t,\ y = \cos 2t,\ t \in [0, 2\pi]$
31. $x = \cosh t,\ y = \sinh t,\ t \in \mathbb{R}$
32. $x = \sec t,\ y = \tan t,\ t \in \left(-\frac{\pi}{2}, \frac{\pi}{2}\right)$
33. $x = \csc t,\ y = \cot t,\ t \in (0, \pi)$
34. $x = \log_{10} t,\ y = \ln t,\ t \in (0, \infty)$
35. Finish Example 1 (b) by showing that the graph of the parametric equations $x = \cos t,\ y = \sin t,\ t \in [\pi, 2\pi]$ is the bottom half of the unit circle centered at the origin with a counterclockwise direction of motion.

In Exercises 36–39 provide a parametrization with the given properties

36. The curve is a circle centered at the origin. It is traced once, clockwise, starting at the point $(0, 1)$ with $t \in [0, 2\pi]$.
37. The curve is a circle centered at the origin. It is traced once, counterclockwise, starting at the point $(0, 3)$ with $t \in [0, 1]$.
38. The curve is a circle centered at the point (a, b). It is traced once, counterclockwise, starting at the point $(a+r, b)$ with $t \in [0, 2\pi]$.
39. The curve is a circle centered at the origin. It is traced once, counterclockwise, and contains all points of the unit circle except for $(0, -1)$ with $t \in \mathbb{R}$.
40. Complete the calculation in Example 7 by using the trigonometric identity $\sin^2\left(\frac{\theta}{2}\right) = \frac{1}{2}(1 - \cos\theta)$ to show that $\int_0^{2\pi} \sqrt{1 - \cos\theta}\, d\theta = 4\sqrt{2}$.

In Exercises 41–44 find an equation for the line tangent to the parametric curve at the given value ot t.

41. $x = 2t - 1,\ y = 3t + 5,\ t = -1$.
42. $x = t + 2,\ y = e^t,\ t = 0$.
43. $x = t^2,\ y = (2 - t)^2,\ t = \frac{1}{2}$.
44. $x = \cos^3 t,\ y = \sin^3 t,\ t = \frac{\pi}{4}$.

In Exercises 45–48 use Example 6 to find $\frac{d^2y}{dx^2}$ for the parametric curve at the given value of t. Note that these are the same parametric equations as in Exercises 41–44.

45. $x = 2t - 1,\ y = 3t + 5,\ t = -1$.
46. $x = t + 2,\ y = e^t,\ t = 0$.

47. $x = t^2,\ y = (2-t)^2,\ t = \frac{1}{2}$.

48. $x = \cos^3 t,\ y = \sin^3 t,\ t = \frac{\pi}{4}$.

In Exercises 49–53 sketch the parametric curve and find its length.

49. $x = 1 + t^2,\ y = 3 + 2t^3,\ t \in [0, 1]$

50. $x = \cos\theta + \theta\sin\theta,\ y = \sin\theta - \theta\cos\theta,\ \theta \in [0, 2\pi]$

51. $x = \sin^3\theta,\ y = \cos^3\theta,\ \theta \in [0, 2\pi]$

52. $x = e^t \cos t,\ y = e^t \sin t,\ t \in [0, 1]$

53. $x = 5 + 2t,\ y = e^t + e^{-t},\ t \in [0, 1]$

54. Show that the graph of the parametric equations

$$x = a + (c-a)t,\ y = b + (d-b)t,\ t \in [0, 1]$$

is a line segment from (a, b) to (c, d).

In Exercises 55–60 use the result of Exercise 54 to find parametric equations for the line segments connecting the given pairs of points in the direction indicated.

55. From $(1, -3)$ to $(6, 7)$

56. From $(6, 7)$ to $(1, -3)$

57. From $(1, 4)$ to $(-3, 5)$

58. From $(-3, 5)$ to $(1, 4)$

59. From $(\pi, 3)$ to $(\pi, 8)$

60. From $(0, e)$ to $(-6, e)$

61. Use the arc length formula for parametric equations and your answer to Exercise 55 to find the distance between the points $(1, -3)$ and $(6, 7)$. Verify your answer by using the distance formula to compute the distance.

62. A ***trochoid*** is a generalization of a cycloid in which the point tracing the path is on a spoke of the wheel, instead of on the circumference. Thus, if the radius of the wheel is r, the point is k units from the center of the wheel, such that either $k < r$ or $k > r$. (When $k > r$, you can think of the point being on a flange extending the radius of the wheel. This case occurs as train wheels roll, since there is an extension of each wheel beyond the portion of the wheel rolling on the track.) Find parametric equations for the trochoid.

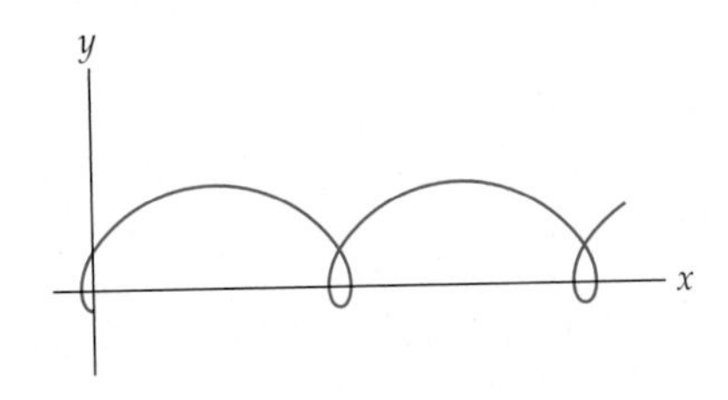

63. An ***epicycloid*** is another variation of a cycloid in which the point tracing the path is on the circumference of a wheel, but the wheel is rolling without slipping on the outside of another wheel, instead of along a horizontal track. If the radius of the rolling wheel is k and the radius of the fixed wheel is r, find parametric equations for the epicycloid.

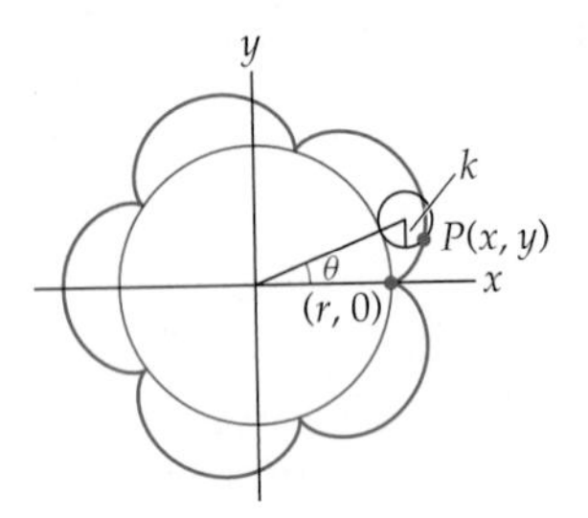

64. The ***involute*** of a circle is the curve described by the endpoint P of a thread as it unwinds from a fixed circular spool. For simplicity suppose that the radius of the circular spool is r and that when the spool is placed with its center at the origin, the point P starts at $(r, 0)$. Assume that the thread is unwinding counterclockwise. Find parametric equations for the point P. (*Hint: If the string is taut at all times as it unwinds, the length of segment PT is $r\theta$.*)

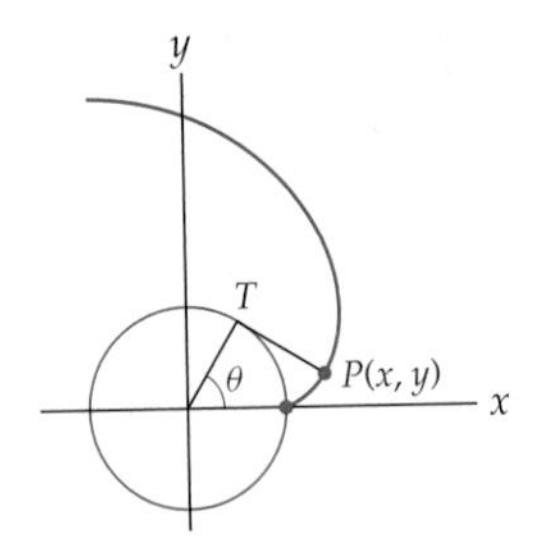

Applications

65. (a) Find an integral that represents the length of an elliptical track whose equations are given by the parametric equations $x = \sin\theta,\ y = 3\cos\theta,\ \theta \in [0, 2\pi]$, where x and y are in kilometers.

(b) Approximate the length of the track, using the midpoint method with 20 subintervals.

66. Annie needs to make a crossing in her kayak from an island to a north–south coastline 3 miles due east. It is foggy, so she cannot see any landmarks to steer by. Instead, she takes a compass heading due east and sticks to it all the way across. Tidal currents in the channel push her boat southward at a speed proportional to the distance from either shoreline. She paddles at 2 miles per hour from west to east. That is, her east–west position changes with time as $x'(t) = 2$. Her north–south position changes as $y'(t) = 1.778t^2 - 2.667t$. Her starting position was $(0, 0)$, at time $t = 0$.

(a) Solve the two given differential equations by integrating with respect to t to find a parametric description of Annie's path across the channel.

(b) How far south of her starting point does Annie make landfall?

(c) Use a numerical integration technique to determine the distance she paddles.

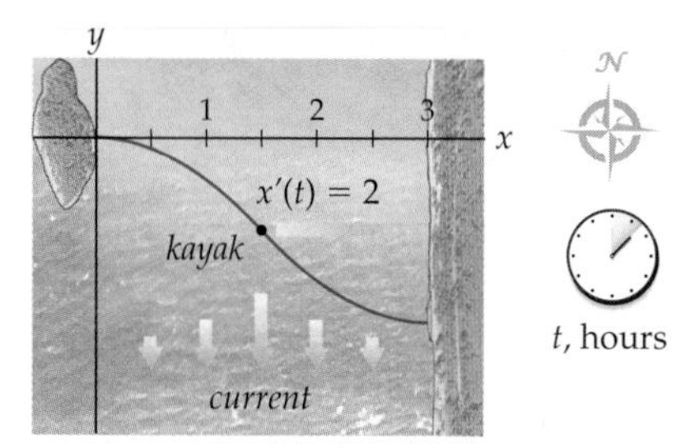

67. Annie has to make the same crossing as in Exercise 66 two weeks later, again in a fog. Remembering how far off course she was pushed previously, she takes a compass heading 15 degrees north of east this time. Thus, her eastward speed is $x'(t) = 2\cos\frac{\pi}{12}$ and her north–south velocity is $y'(t) = 0.444x(t)(x(t) - 3) + 2\sin\frac{\pi}{12}$.

(a) Solve the two given differential equations by integrating with respect to t to find a parametrization of Annie's path across the channel.

(b) How far south of her starting point does Annie make landfall this time?

(c) How far does she paddle? How long does she paddle?

Proofs

68. Show that every function $y = f(x)$ can be written as parametric equations.

69. Use your result from Exercise 68 to show that the arc length formula for a function $y = f(x)$ is a special case of the arc length formula for a parametric curve.

70. Let c and d be constants, and for $t \in [a, b]$ let $f(t)$ and $g(t)$ be differentiable functions with continuous first derivatives. Prove that the arc length of the parametric curve given by $x = f(t),\ y(t) = g(t)$ for $t \in [a, b]$ is equal to the arc length of the parametric curve defined by $x = c + f(t),\ y(t) = d + g(t)$ for $t \in [a, b]$ for every c and d in $\mathbb{R}$.

71. Let $k > 0$ be a constant, and let $f(t)$ and $g(t)$ be differentiable functions of t with continuous first derivatives for every $t \in [a, b]$. Prove that the arc length of the curve defined by the parametric equations

$$x = kf(t),\ y = kg(t),\ t \in [a, b]$$

is k times as long as the arc length of the curve defined by the parametric equations

$$x = f(t),\ y = g(t),\ t \in [a, b].$$

What is the arc length of the curve defined by the equations

$$x = f(kt),\ y = g(kt),\ t \in [a/k, b/k]?$$

72. Show that the parametric curve associated with the *linear* parametric equations

$$x = m_1 t + b_1,\ y = m_2 t + b_2,\ t \in \mathbb{R}$$

is a line if m_1 and m_2 are not both zero. For what values of m_1 and m_2 will y be a function of x? In this case, what is the slope of the line? What do the equations describe if m_1 and m_2 are both zero? Find parametric equations $x = f(t)$, $y = g(t)$ such that neither f nor g is a linear function, but the parametric curve associated with the equations is a line.

73. In Example 5 we saw that the cycloid

$$x = r\theta - r\sin\theta,\ y = r - r\cos\theta,\ \theta \in \mathbb{R}$$

has a horizontal tangent line at each odd multiple of π. Show that the cycloid has a vertical tangent at each even multiple of π by showing that $\lim_{\theta \to 2k\pi} \frac{dy}{dx}$ does not exist wherever k is an integer.

Thinking Forward

Parametric equations in three dimensions: Parametric equations can also be used to describe curves in three dimensions. Try to visualize the graphs described by the following parametric equations.

▶ A line through the origin:

$$x = t,\ y = -3t,\ z = 2t,\ t \in \mathbb{R}$$

▶ A helix centered on the z-axis:

$$x = \cos t,\ y = \sin t,\ z = t,\ t \in \mathbb{R}$$

▶ An expanding helix:

$$x = t\cos t,\ y = t\sin t,\ z = t,\ t \in [0, \infty)$$

9.2 POLAR COORDINATES

- Polar coordinates are defined
- Conversion formulas between rectangular and polar coordinates are derived
- Principles of graphing with polar coordinates are introduced

Plotting Points in Polar Coordinates

We begin by specifying a point called the ***pole*** and a ray emanating from the pole called the ***polar axis***, as follows:

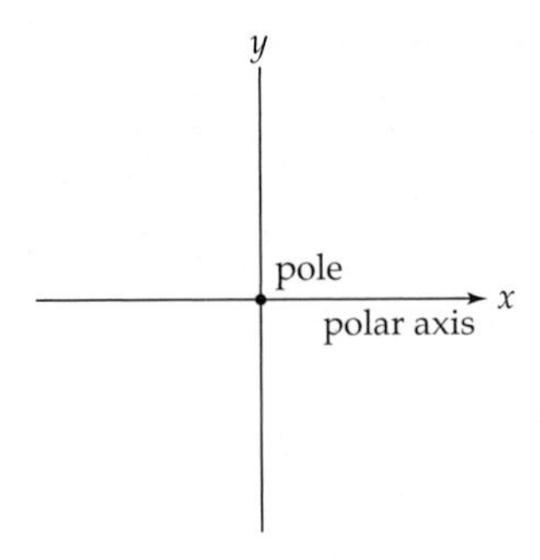

In the preceding figure you will notice that we've placed the pole at the origin of a rectangular coordinate system and the polar axis along the positive x-axis. This is the convention we always use with polar coordinates. As we will see later in the section, this convention will also allow us to derive formulas for converting between polar coordinates and rectangular coordinates.

In a polar system we use the coordinate pair (r, θ) to represent a point in the plane. The first coordinate r is the signed distance from the pole, while θ is an angular measure, in radians, from the polar axis. When $\theta > 0$, we rotate counterclockwise from the polar axis, and when $\theta < 0$, we rotate clockwise. When $r > 0$, we measure r units along the ray specified by the value of θ. When $r < 0$, we measure $|r|$ units in the opposite direction, which is equivalent to moving in the positive direction along the ray $\theta + \pi$, as shown here:

For $r > 0$ we move along the ray θ;
for $r < 0$ we move along $\theta + \pi$

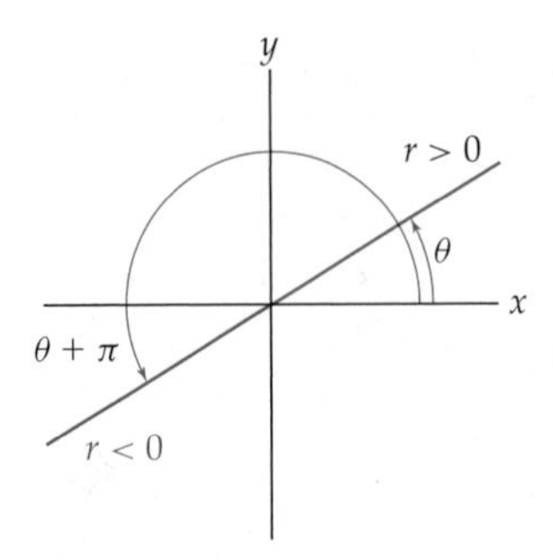

In the rectangular coordinate system, once scales on the axes are specified, there is a one-to-one correspondence between points in the plane and coordinates that name points. This relationship does not hold in the polar coordinate system. We have the following theorem:

THEOREM 9.5 **Equivalent Polar Coordinates**

(a) The polar coordinates $(r, \theta + 2\pi k)$ represent the same point for every integer k.

(b) The polar coordinates $(-r, \theta + \pi)$ represent the same point as (r, θ) for any value of θ.

(c) The polar coordinates $(0, \theta)$ represent the pole for any value of θ.

Converting Between Polar and Rectangular Coordinates

To convert from polar coordinates to rectangular coordinates, we will write the coordinates x and y as functions of the coordinates r and θ. Then, given r and θ, we will easily be able to find the corresponding x and y. The figure that follows shows a point in the first quadrant of the plane and its geometric representation in terms of both rectangular and polar coordinates. In the other three quadrants we can draw similar figures and reach the same conclusions as we do shortly; however, for simplicity we will use the first-quadrant picture as our general example.

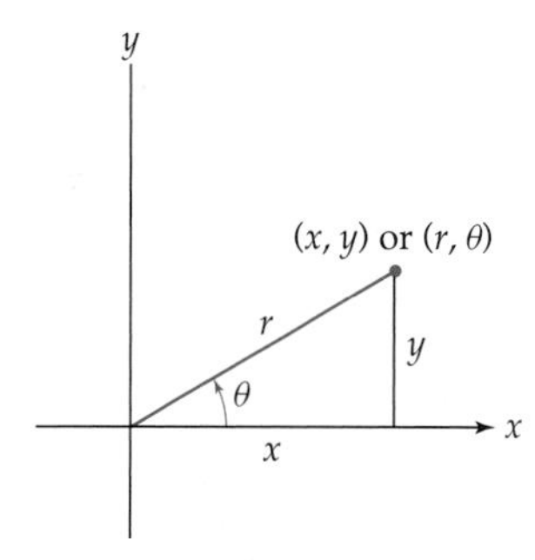

Using the triangle shown, we see that $\cos\theta = \frac{x}{r}$ and $\sin\theta = \frac{y}{r}$. This gives us the following theorem:

THEOREM 9.6 **Converting from Polar to Rectangular Coordinates**

If a point in the plane is represented by (r, θ) in polar coordinates, then the rectangular coordinates of the point are given by (x, y), where

$$x = r\cos\theta \quad \text{and} \quad y = r\sin\theta.$$

Note that given the polar coordinates (r, θ) of any point in the plane, there will be a unique pair of rectangular coordinates (x, y) representing the point.

To convert from rectangular coordinates to polar coordinates, we need to have a way of finding r and θ given any coordinates x and y. Applying the Pythagorean theorem and the definition of the tangent function to the triangle in the figure shown, we have the following relationships:

THEOREM 9.7 **Converting from Rectangular to Polar Coordinates**

If a point in the plane is represented by (x, y) in rectangular coordinates, then the polar coordinates (r, θ) of the point satisfy the following formulas:

$$r^2 = x^2 + y^2 \quad \text{and} \quad \tan\theta = \frac{y}{x}.$$

Notice that the formulas in Theorem 9.7 do not give unique values of r and θ for given values of x and y. This is to be expected, since every point in the plane has multiple polar coordinate representations. For example, the rectangular coordinates (1, 1) may be expressed as $\left(\sqrt{2}, \frac{\pi}{4} + 2k\pi\right)$ or $\left(-\sqrt{2}, \frac{3\pi}{4} + 2k\pi\right)$, for any positive integer k.

The Graphs of Some Simple Polar Coordinate Equations

In rectangular coordinates, the simplest possible equations are those of the form $x = a$ and $y = b$, where a and b are constants. Appropriately enough, the graphs of these equations are very simple as well: vertical and horizontal lines, respectively. In polar coordinates we first consider the equations $r = c$ and $\theta = \alpha$, where c and α are constants. For any real number c, the polar equation $r = c$ describes the set of points $|c|$ units from the pole. Therefore, the graph of $r = c$ is a circle with radius $|c|$ centered at the pole, as shown next at the left. For any real number α, the polar equation $\theta = \alpha$ describes the set of points with polar angle α. At first you might think that this would give a graph of the ray with polar angle θ. However, remember that when $\theta = \alpha$, the value of r can be either positive or negative. Therefore, the graph of $\theta = \alpha$ is a line through the pole, as shown here at the right:

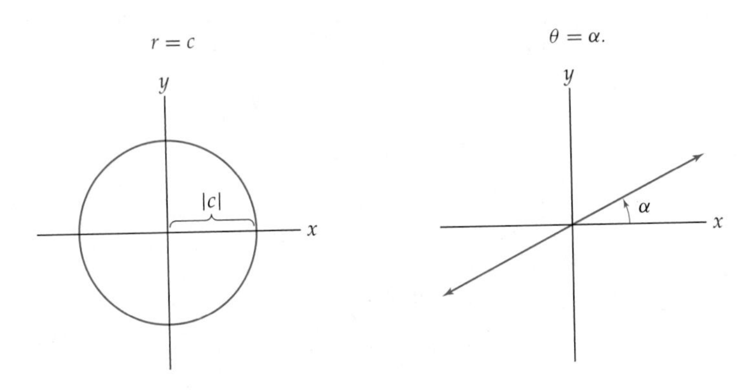

Two other simple categories of polar equations are those defined by

$$r = 2a\cos\theta \text{ and } r = 2a\sin\theta,$$

where $a \neq 0$. The following theorem tells us that these are the equations of circles passing through the pole and tangent to one of the coordinate axes.

THEOREM 9.8 **Circles Tangent to the Coordinate Axes at the Pole**

For $a \neq 0$, the graphs of the equations

$$r = 2a\cos\theta \text{ and } r = 2a\sin\theta$$

are the circles whose equations in rectangular coordinates are

$$(x-a)^2 + y^2 = a^2 \text{ and } x^2 + (y-a)^2 = a^2,$$

respectively.

Proof. We prove that $(x-a)^2 + y^2 = a^2$ is another equation for $r = 2a\cos\theta$ and leave the other case for Exercise 68. We multiply both sides of $r = 2a\cos\theta$ by r and use the facts that $r^2 = x^2 + y^2$ and $x = r\cos\theta$ to obtain

$$x^2 + y^2 = 2ax.$$

If we subtract $2ax$ from each side of the equation and then complete the square, we get

$$(x-a)^2+y^2=a^2,$$

as required. The graph of this equation is the circle with radius a and centered at the point $(a, 0)$. The circle is tangent to the y-axis and passes through the pole. ■

Examples and Explorations

EXAMPLE 1

Plotting points with polar coordinates

Plot the points that have polar coordinates $\left(2, \frac{\pi}{3}\right)$ and $\left(-1, -\frac{\pi}{2}\right)$.

SOLUTION

To plot $\left(2, \frac{\pi}{3}\right)$, we first locate the ray $\theta = \frac{\pi}{3}$ and move 2 units (in the positive direction) along this ray, as shown next at the left. Similarly, to plot the point $\left(-1, -\frac{\pi}{2}\right)$, we first find the ray $\theta = -\frac{\pi}{2}$ and move 1 unit in the negative direction along that ray. Note that this operation is equivalent to moving one unit in the positive direction along the ray $\theta = -\frac{\pi}{2} + \pi = \frac{\pi}{2}$, as shown here at the right:

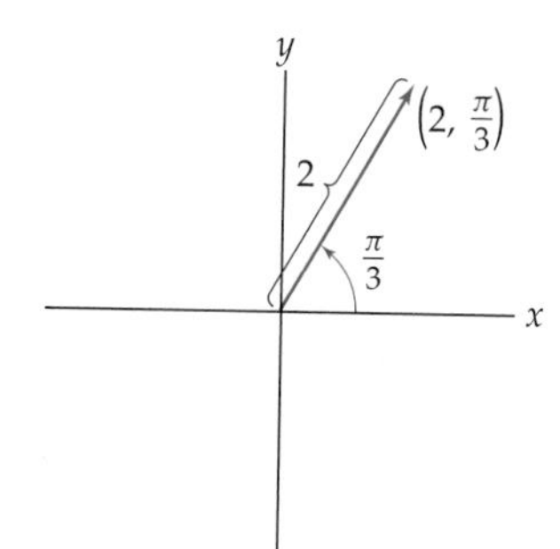

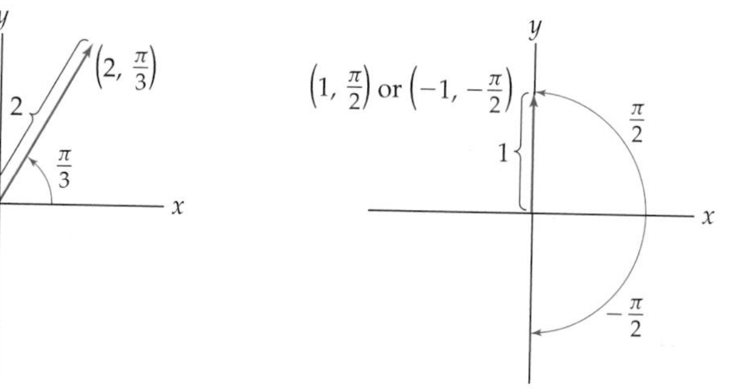

□

EXAMPLE 2

Converting from polar to rectangular coordinates

Find the rectangular coordinates of the points that have polar coordinates $\left(2, \frac{\pi}{3}\right)$ and $\left(-1, -\frac{\pi}{2}\right)$.

SOLUTION

We use the conversion formulas given in Theorem 9.6. For the first of the two pairs of coordinates we have

$$x = 2\cos\frac{\pi}{3} \text{ and } y = 2\sin\frac{\pi}{3}.$$

The rectangular coordinates are $(1, \sqrt{3})$.

The second coordinate pair converts to

$$x = -1\cos\left(-\frac{\pi}{2}\right) = 0 \text{ and } y = -1\sin\left(-\frac{\pi}{2}\right) = 1.$$

Compare these results with those you obtained in Example 1. □

EXAMPLE 3

Finding all the polar coordinate representations for a point

Find all of the pairs of polar coordinates for the point with rectangular coordinates $(1, \sqrt{3})$.

SOLUTION

We use the conversion formulas given in Theorem 9.7 along with the fact that the point $(1, \sqrt{3})$ lies in the first quadrant. We have $r^2 = 1^2 + (\sqrt{3})^2 = 4$ and $\tan\theta = \frac{\sqrt{3}}{1} = \sqrt{3}$. Thus, either $r = 2$ and $\theta = \frac{\pi}{3} + 2\pi k$, where $k \in \mathbb{Z}$, or $r = -2$ and $\theta = \frac{4\pi}{3} + 2\pi k$, where $k \in \mathbb{Z}$. Note that although the angles $\theta = \frac{4\pi}{3} + 2\pi k$ are in the third quadrant for each k, the fact that $r < 0$ still gives us a point in the first quadrant.

This example builds on Example 2. When we converted the point $\left(2, \frac{\pi}{3}\right)$ to rectangular coordinates, we got the unique answer of that example. However, since every point in the polar plane has infinitely many names, when we convert back to polar coordinates we expect infinitely many answers. □

EXAMPLE 4

Graphing polar equations by converting to rectangular coordinates

Graph the equations $r = 4\cos\theta$ and $r = 5\csc\theta$.

SOLUTION

We will begin with the polar equation $r = 4\cos\theta$. From Theorem 9.8 we know that this is the equation for the circle whose equation in rectangular coordinates is $(x - 2)^2 + y^2 = 4$, as shown next at the left.

Now consider the polar equation $r = 5\csc\theta$. We begin by multiplying both sides of the equation by $\sin\theta$ to obtain

$$r\sin\theta = 5.$$

By Theorem 9.6 we have $r\sin\theta = y$; combining this equation with the preceding one, we must have $y = 5$. Therefore, the graph of $r = 5\csc\theta$ is a horizontal line at height 5, as shown here at the right:

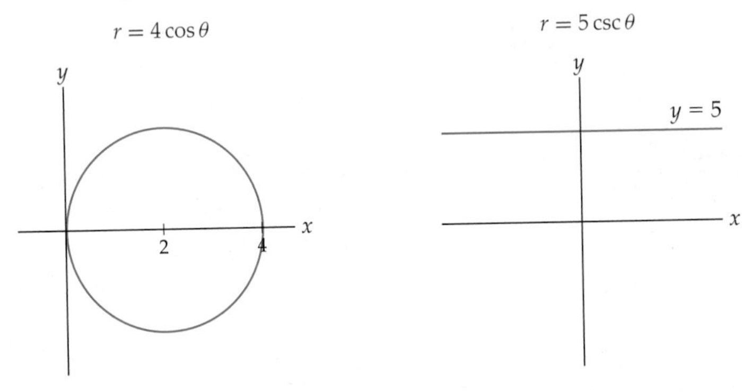

□

CAUTION

We graphed the polar equation $r = 5\csc\theta$ in Example 4 by transforming it to rectangular coordinates. Using Theorems 9.6 and 9.7, we can always transform an equation in polar coordinates to one in rectangular coordinates. However, as the next example demonstrates, we *cannot* expect that every polar equation will have a simple expression in rectangular coordinates. Therefore, we may not understand the polar equation any better by transforming it to rectangular coordinates.

EXAMPLE 5

Converting a polar equation to an equation in rectangular coordinates

Convert the polar equation $r = \cos 2\theta$ to rectangular coordinates.

SOLUTION

We begin by using the double-angle identity for cosine, $\cos 2\theta = \cos^2\theta - \sin^2\theta$. We now have $r = \cos^2\theta - \sin^2\theta$. Since $\cos\theta = \frac{x}{r}$ and $\sin\theta = \frac{y}{r}$, we have

$$r = \frac{x^2}{r^2} - \frac{y^2}{r^2},$$

or equivalently,

$$r^3 = x^2 - y^2.$$

Finally, because $r = \pm(x^2 + y^2)^{1/2}$, we obtain the equation

$$\pm(x^2 + y^2)^{3/2} = x^2 - y^2.$$

We could now use the techniques of Section 2.4 to graph $\pm(x^2 + y^2)^{3/2} = x^2 - y^2$. However, that would be making the problem of graphing the relatively simple polar equation $r = \cos 2\theta$ a complicated mess involving implicit functions. In the next section we will be discussing methods for graphing polar equations like $r = \cos 2\theta$ directly. □

? TEST YOUR UNDERSTANDING

- What are the formulas for converting between rectangular coordinates and polar coordinates? How are these formulas derived?
- Why does every point in the plane have a unique pair of rectangular coordinates that represents the point, but have infinitely many pairs of polar coordinates that represent it?
- Why are the polar coordinates $\left(8, -\frac{\pi}{6}\right)$ *not* a representation of the point with rectangular coordinates $(-4, 4\sqrt{3})$, even though these values of r, θ, x, and y satisfy the formulas $r^2 = x^2 + y^2$ and $\tan\theta = \frac{y}{x}$?
- Why is the graph of the equation $r = 2a\cos\theta$ a circle in the polar plane?
- Why is the graph of the equation $r = b\sec\theta$ a line in the polar plane?

EXERCISES 9.2

Thinking Back

Plotting in a rectangular coordinate system:

- Plot the points $(1, 1)$, $(1, 2)$, $(1, 3)$, and $(1, 4)$ in a rectangular coordinate system. Fill in the blank: These points all lie on the same ________ line.
- Plot the points $(-1, -2)$, $(0, -2)$, $(1, -2)$, and $(2, -2)$ in a rectangular coordinate system. Fill in the blank: These points all lie on the same ________ line.
- Plot the equations $x = -3$ and $y = 5$ in a rectangular coordinate system. Where do the graphs of these two equations intersect?

Completing the square:

- Consider the equation $x^2 + y^2 = 4x$. Use the method of completing the square to show that this equation is equivalent to $(x - 2)^2 + y^2 = 4$. (This calculation was used in Example 4.)
- Find the center and radius of the circle with equation

$$x^2 + y^2 + 6x - 8y - 44 = 0$$

by completing the square.

Concepts

0. *Problem Zero:* Read the section and make your own summary of the material.

1. *True/False:* Determine whether each of the statements that follow is true or false. If a statement is true, explain why. If a statement is false, provide a counterexample.

(a) *True or False:* Each point in the plane has a unique representation in rectangular coordinates.

(b) *True or False:* Each point in the plane has a unique representation in polar coordinates.

(c) *True or False:* If $b \neq c$, then the graphs of the polar equations $r = b$ and $r = c$ are different.

(d) *True or False:* If $\alpha \neq \beta$, then the graphs of the polar equations $\theta = \alpha$ and $\theta = \beta$ are different.

(e) *True or False:* The graph of $r = \csc\theta$ for $-\frac{\pi}{2} < \theta < \frac{\pi}{2}$ is a horizontal line in a polar coordinate system.

(f) *True or False:* In a polar coordinate system, the coordinates (r, θ) and $(r, \theta + \pi)$ represent the same point if and only if $r = 0$.

(g) *True or False:* When A and B are nonzero constants, the graph of $r = A\sin\theta + B\cos\theta$ is a circle in a polar coordinate system.

(h) *True or False:* Every function $r = f(\theta)$ in a polar coordinate system can be expressed in terms of rectangular coordinates x and y.

2. *Examples:* Construct examples of the thing(s) described in the following. Try to find examples that are different than any in the reading.

(a) Two pairs of polar coordinates for the point $(0, 3)$ given in rectangular coordinates.

(b) Two equations for the line $y = x$ in polar coordinates.

(c) The equations of two distinct circles with radius 2 tangent to the the x-axis at the pole in polar coordinates.

3. Explain why every point in the polar coordinate plane has infinitely many different polar coordinate representations.

4. Consider the point in the plane given by polar coordinates $(r, \theta) = \left(2, \frac{\pi}{3}\right)$.

(a) Express this point in polar coordinates where $r = 2$, but $\theta \neq \frac{\pi}{3}$, if possible.

(b) Express this point in polar coordinates where $\theta = \frac{\pi}{3}$, but $r \neq 2$, if possible.

(c) Express this point in polar coordinates where $r \neq 2$ and $\theta \neq \frac{\pi}{3}$, if possible.

5. Explain why the point $(r, \theta) = \left(8, -\frac{\pi}{3}\right)$ is not a polar representation of the point with rectangular coordinates $(x, y) = (-4, 4\sqrt{3})$, even though these values of r, θ, x, and y satisfy the formulas $r^2 = x^2 + y^2$ and $\tan\theta = \frac{y}{x}$. Include a picture with your explanation.

6. Explain why the graphs of $r = 3$ and $r = -3$ are identical in a polar coordinate system.

7. Find all values of c such that the graphs of $r = c$ and $r = -c$ are the same in a polar coordinate system.

8. Explain why the graphs of $\theta = \frac{\pi}{2}$ and $\theta = -\frac{\pi}{2}$ are identical in a polar coordinate system.

9. Find all values of α such that the graphs of $\theta = \alpha$ and $\theta = -\alpha$ are the same in a polar coordinate system.

10. In Example 5 we converted the simple polar equation $r = \cos 2\theta$ into the messy rectangular equation $\pm(x^2 + y^2)^{3/2} = x^2 - y^2$. Does this rectangular equation have a simpler rectangular coordinate form? If so, what is it? If not, why not?

11. Explain why the inequalities $r > 0$ and $0 < \theta < \frac{\pi}{2}$ together describe the points in the first quadrant. Use similar inequalities to describe the points in the third quadrant.

12. Explain why the inequality $0 \leq r \leq 2$ describes the points inside or on the circle with radius 2 and centered at the origin. Use a similar inequality to describe the points in the ***annulus*** shown here:

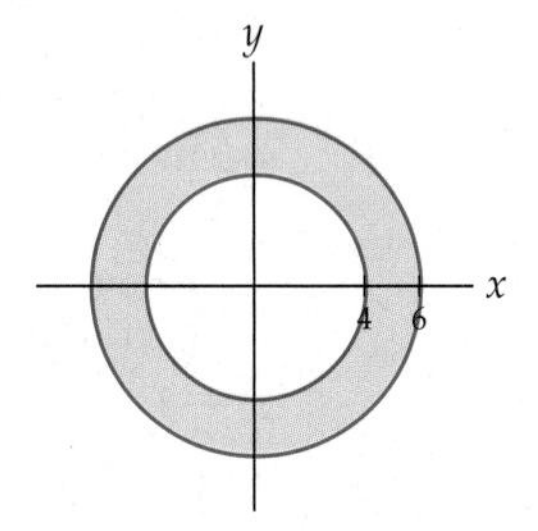

13. Find all polar coordinates that represent the point $(1, 0)$ given in rectangular coordinates.

14. Find all polar coordinates that represent the point $(0, 1)$ given in rectangular coordinates.

15. Find all polar coordinates that represent the point $(-1, 0)$ that is also given in polar coordinates.

16. Find all values of a and b such that (a, b) represents the same point whether it is given in rectangular coordinates or polar coordinates.

Skills

In Exercises 17–23 the polar coordinates for several sets of points are given. Find the rectangular coordinates for each of the points, and then plot and label the points in the same polar coordinate system.

17. $\left(3, \frac{\pi}{6}\right)$, $\left(-3, \frac{\pi}{6}\right)$, $\left(3, -\frac{\pi}{6}\right)$ and $\left(-3, -\frac{\pi}{6}\right)$

18. $(1, 0)$, $(2, 0)$, $(3, 0)$, and $(4, 0)$

19. $\left(5, \frac{\pi}{2}\right)$, $\left(-5, -\frac{3\pi}{2}\right)$, $\left(5, \frac{5\pi}{2}\right)$ and $\left(-5, -\frac{\pi}{2}\right)$

20. $\left(0, \frac{\pi}{6}\right)$, $\left(0, \frac{\pi}{3}\right)$, $(0, \pi)$ and $(0, -\pi)$

21. $\left(1, \frac{\pi}{4}\right)$, $\left(2, \frac{\pi}{4}\right)$, $\left(3, \frac{\pi}{4}\right)$ and $\left(4, \frac{\pi}{4}\right)$

22. $(2, 0)$, $\left(2, \frac{\pi}{4}\right)$, $\left(2, \frac{\pi}{2}\right)$, $(2, \pi)$ and $\left(2, \frac{3\pi}{2}\right)$

23. $\left(1, -\frac{\pi}{2}\right)$, $\left(2, -\frac{\pi}{2}\right)$, $\left(3, -\frac{\pi}{2}\right)$ and $\left(4, -\frac{\pi}{2}\right)$

In Exercises 24–31 find all polar coordinate representations for the point given in rectangular coordinates.

24. $(0, 0)$

25. $(1, 0)$

26. $(0, -3)$

27. $(0, 2)$

28. $(3, 4)$

29. $(6, 2\sqrt{3})$

30. $(-2, 0)$

31. $(3, -3)$

In Exercises 32–47 convert the equations given in polar coordinates to rectangular coordinates.

32. $\theta = \pi$

33. $\theta = \dfrac{\pi}{4}$

34. $\theta = -\dfrac{\pi}{6}$

35. $\theta = \dfrac{7\pi}{6}$

36. $r = 2\cos\theta$

37. $r = 5\sin\theta$

38. $r = -3\sec\theta$

39. $r = 6\csc\theta$

40. $r = \tan\theta$

41. $r = \sin 2\theta$

42. $r^2 = \sin\theta$

43. $r^2 = \cos\theta$

44. $r = \sin 4\theta$

45. $r = \theta$

46. $r = \cos 4\theta$

47. $r = \sin^3\theta$

In Exercises 48–55 convert the equations given in rectangular coordinates to equations in polar coordinates.

48. $x = 0$

49. $y = 0$

50. $x = 4$

51. $y = x$

52. $y = \sqrt{3}x$

53. $y = -3$

54. $y = x + 1$

55. $y = mx$

In Exercises 56–59 convert the equations given in polar coordinates to equations in rectangular coordinates.

56. $r = 1,\ r = -2,\ r = 3.$

57. $r = k$, for each positive integer k less than 10.

58. $\theta = 0,\ \theta = \dfrac{\pi}{4},\ \theta = -\dfrac{\pi}{2}.$

59. $\theta = \dfrac{k\pi}{6}$, for each positive integer k less than 12.

Applications

In Exercises 60 and 61 Ian is a rock climber. Rock protection these days relies on a process called camming. The idea for camming devices is that they push harder against the rock when they are loaded with the force of a fall.

60. Ian has one cam that is 1 inch wide, with a point where the rope attaches on its right corner, so that the device pivots on its left corner. In other words, the left corner of the cam becomes the center of a circle, and the point where the rope attaches follows the curve $r = 1$.

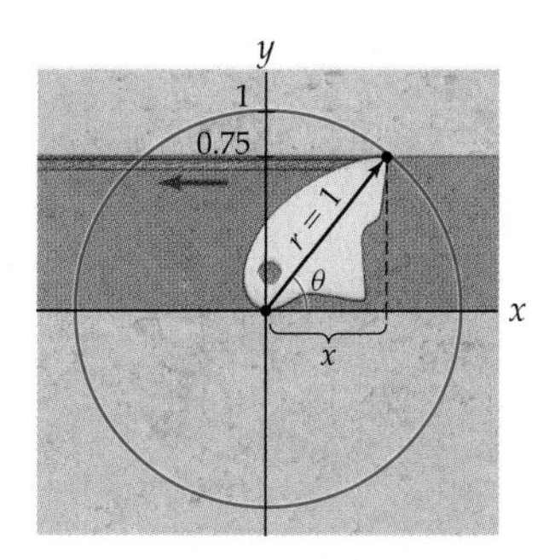

(a) If a crack is 0.75 inch wide, at what angle must Ian turn the device in order to put it into the crack? *(Hint: compute the x-coordinate of the right edge of the device.)*

(b) If Ian falls, the rope will pull on the right edge of the cam while the left edge remains wedged in a fixed position. What happens to the width of the cam in the crack?

61. Ian has another camming device, one that has two arms extending from a point where the rope attaches in the center.

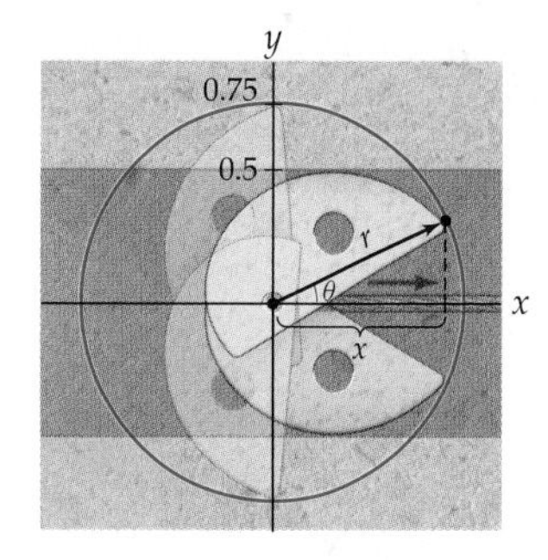

The outside edge of each arm follows the curve $r = 0.75$. The ends of the arms make an angle $\pm\theta$ with the rope, and the rope pulls along the ray $\theta = 0$. Ian can retract the arms so that θ is small in order to put the device into a crack, and then a spring pulls them back so that each arm wedges against the walls of the crack.

(a) Ian puts the device into a horizontal crack that is 1 inch tall. What angle do the arms make with the rope? *(Hint: Compute the y-coordinates of the two arm ends.)*

(b) What happens if Ian falls, causing the rope to pull outwards (rightwards) in the crack?

Proofs

62. Prove that the graph of the equation

$$r = k\sec\theta, \quad -\frac{\pi}{2} < \theta < \frac{\pi}{2},$$

is a vertical line for any value of $k \neq 0$.

63. Prove that the graph of the equation

$$r = k\csc\theta, \ 0 < \theta < \pi,$$

is a horizontal line for any value of $k \neq 0$.

64. Let $a \neq 0$. Prove that the graph of the equation

$$r = \frac{a}{1 - \cos\theta}$$

is a parabola in a polar coordinate system. When $a \neq 0$, what is the graph of the equation $r = \dfrac{a}{1-\sin\theta}$?

65. Let $a \neq 0$ and $0 < b < 1$. Prove that the graph of the equation

$$r = \frac{a}{1 - b\cos\theta}$$

is an ellipse in a polar coordinate system. When $a \neq 0$ and $b > 1$, what is the graph of the equation $r = \dfrac{a}{1-b\cos\theta}$?

66. In this problem you will prove the three parts of Theorem 9.5:

(a) Prove that the polar coordinates $(r, \theta + 2\pi k)$ represent the same point for every integer k.

(b) Prove that the point with polar coordinates $(-r, \theta + \pi)$ represents the same point as (r, θ) for any value of θ.

(c) Prove that the polar coordinates $(0, \theta)$ represent the pole for any value of θ.

67. Show that the graph of the equation $r = k\cos\theta$ is a circle tangent to the y-axis for any $k \neq 0$. What are the center and radius of the circle?

68. Finish the proof of Theorem 9.8 by showing that the graph of the equation $r = 2k\sin\theta$ is a circle with center $(0, k)$ and radius k tangent to the x-axis for any $k \neq 0$.

69. Modify the proof of Theorem 9.8 to show that the graph of the equation $r = k\sin\theta + l\cos\theta$ is a circle. Find the center and radius in terms of k and l.

70. Find and prove a formula for the distance between the points (r_1, θ_1) and (r_2, θ_2) when they are plotted in a polar coordinate system.

Thinking Forward

▶ *Understanding symmetry:*
When a function $y = f(x)$ has the property that $f(-x) = f(x)$ for every value of x in the domain of f, the function is said to be an *even* function and its graph in a rectangular coordinate system is symmetrical with respect to the y-axis. When a function $r = f(\theta)$ has the property that $f(-\theta) = f(\theta)$ for every value of θ in the domain of f, what geometrical property would the graph of $r = f(\theta)$ have when it is plotted in a polar coordinate system?

▶ *Understanding symmetry:*
When a function $y = f(x)$ has the property that $f(-x) = -f(x)$ for every value of x in the domain of f, the function is said to be an *odd* function and its graph in a rectangular coordinate system is symmetrical with respect to the origin. When a function $r = f(\theta)$ has the property that $f(-\theta) = -f(\theta)$ for every value of θ in the domain of f, what geometrical property would the graph of $r = f(\theta)$ have when it is plotted in a polar coordinate system?

9.3 GRAPHING POLAR EQUATIONS

- A general method for graphing polar curves is provided
- Symmetry tests for polar curves are discussed
- Several categories of polar curves are introduced

Using the θr-plane to Get Information About a Polar Graph

In Section 9.2 we introduced polar coordinates and studied how points and some simple equations are graphed. We can always transform an equation in polar coordinates to an equation in rectangular coordinates. Some polar equations have nice rectangular coordinate representations. Since we are now experts at graphing curves in a rectangular coordinate system, the transformed equation may be easier to graph. However, as we saw in Example 5 of Section 9.2, the transformation process to rectangular coordinates does not always result in a nice equation. To aid in our understanding of graphing polar equations in the ***polar plane*** (see the figure that follows at the left), we introduce the θr-plane (figure at right), which is simply a rectangular coordinate system with the variable θ plotted on the horizontal axis and the variable r plotted on the vertical axis.

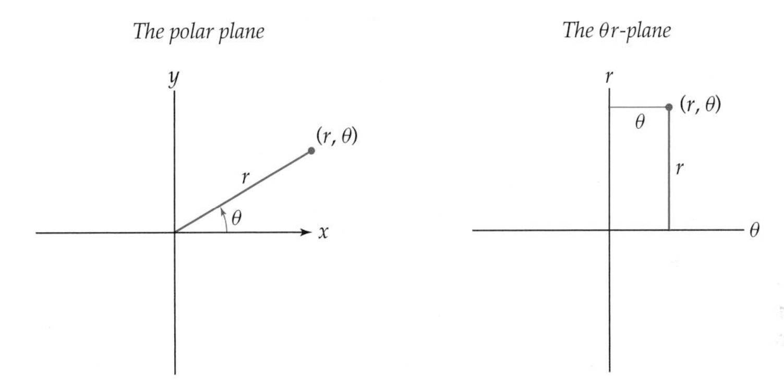

In Example 4 we graphed the equation $r = 4\cos\theta$ in the polar plane. We reproduce that graph next at the left. The figure at the right shows the graph of the same equation in the θr-plane.

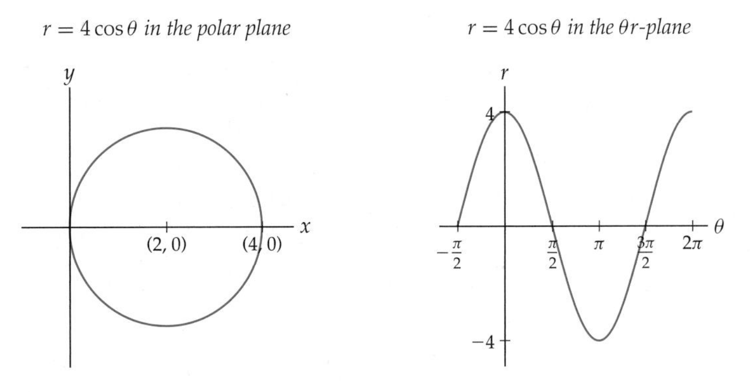

What is the connection between these two graphs? More generally, assuming that we are successful at graphing an equation $r = f(\theta)$ in the θr-plane, how can we use the properties of that graph to draw the desired curve in the polar plane? Our reason for doing this should be clear: We have a great deal of experience drawing curves in a rectangular coordinate system, and we wish to draw upon this experience to help us learn to understand curves in the polar plane.

First consider the inequalities that determine the four quadrants in the polar plane, as shown in the following table:

Quadrant	Inequality
I	$0 < \theta < \frac{\pi}{2}$
II	$\frac{\pi}{2} < \theta < \pi$
III	$\pi < \theta < \frac{3\pi}{2}$
IV	$\frac{3\pi}{2} < \theta < 2\pi$

The following diagrams show the same information in the polar plane (left) and in the θr-plane (right):

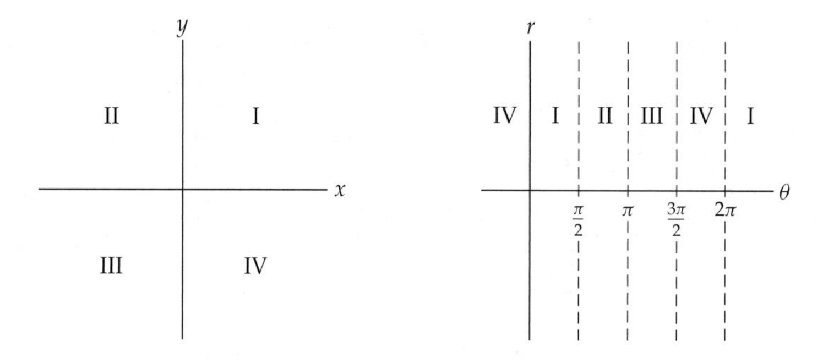

Notice that in the θr-plane the quadrant numbers start repeating; this is because, for example, quadrant I in the polar plane corresponds to angles $0 < \theta < \frac{\pi}{2}$ as well as angles $2\pi < \theta < \frac{5\pi}{2}$ and, in fact, to any angle in an interval of the form $\left(0 + 2\pi k, \frac{\pi}{2} + 2\pi k\right)$, where k is an integer.

The figure at the right illustrates the correspondence with the polar plane quadrants only when r is positive. Recall that when r is negative points are plotted in the diagonally opposite quadrant. The following figure shows this situation schematically:

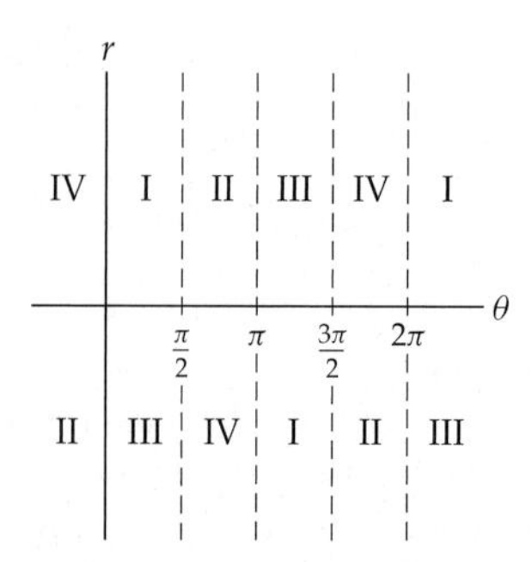

Given the graph of a polar equation in the θr-plane, what does this information tell us about the graph of the curve in the polar plane? Consider again the graph of $r = 4\cos\theta$. The following figure regraphs that equation with various pieces color-coded according to the sectors shown in the previous graph:

$r = 4\cos\theta$ in the θr-plane

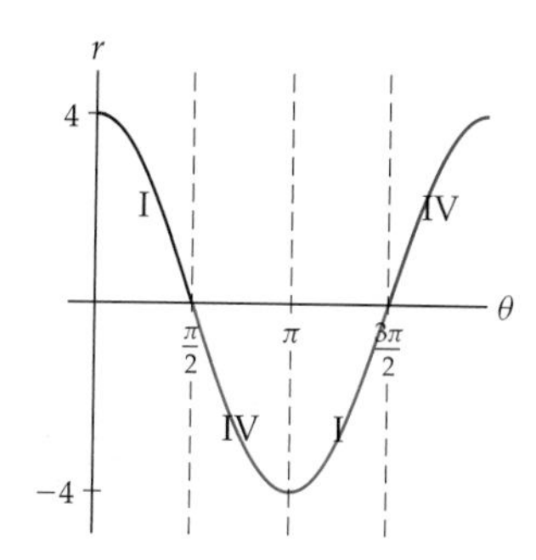

Notice that the black and blue portions of the graph will appear in the first quadrant when graphed in polar coordinates while the red and green portions will be in the fourth quadrant of the polar plane. This analysis tells us why the graph plotted appears only in the first and fourth quadrants; however, it does not tell us why the graph is a circle.

Certain points in the θr-plane can tell us a great deal about the corresponding graph in the polar plane. For example, consider those points in the θr-plane that lie *on* the θ-axis. The r-coordinate of each of these points is zero. Since every point in the polar form $(0, \theta)$ is drawn at the pole of the polar plane, each point on the θ-axis corresponds to the pole in the polar plane. In terms of the previous color-coded graph, this means that the points at $\theta = \frac{\pi}{2}$ and $\theta = \frac{3\pi}{2}$ (and, in fact, for θ any odd multiple of $\frac{\pi}{2}$) correspond to the pole in the polar plane.

The points in the θr-plane with θ-coordinate of the form πk and $\frac{\pi}{2} + \pi k$ for some integer k are also good reference points. Points with $\theta = \pi k$ will correspond to points in the polar plane that lie on the horizontal axis. Points with $\theta = \frac{\pi}{2} + \pi k$ will correspond to points on the vertical axis in the polar plane. In terms of the color-coded graph, this means that the only vertical intercept of $r = 4\cos\theta$ in the polar plane will be at the pole (since the points where $\theta = \frac{\pi}{2} + \pi k$ have $r = 0$). Moreover, there are only two horizontal intercepts of $r = 4\cos\theta$ in the polar plane. We have already identified the horizontal intercept at the pole, and the only other horizontal intercept appears at the single point with polar coordinates $(4, 0)$ or $(-4, \pi)$ or $(4, 2\pi)$, etc. Notice that by examining certain points on the graph in the θr-plane, we have obtained all of the quadrant and intercept information for the corresponding graph in the polar plane.

Symmetry in Polar Graphs

In Section 0.2 we discussed functions whose graphs are symmetrical with respect to the y-axis and functions whose graphs are symmetrical with respect to the origin. These functions are called *even* and *odd*, respectively. Graphs that do not represent functions can also show these symmetries. For example, the graph that follows at the left is symmetrical with respect to the vertical axis, but it is not the graph of a function in rectangular coordinates because it fails the vertical line test. Similarly, the graph at the right is symmetrical with respect to the origin, but it is not the graph of a function in rectangular coordinates.

Symmetry with respect to the vertical axis

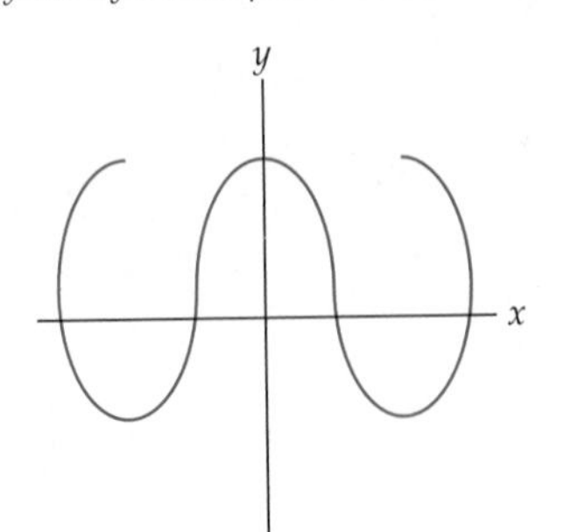

Symmetry with respect to the origin

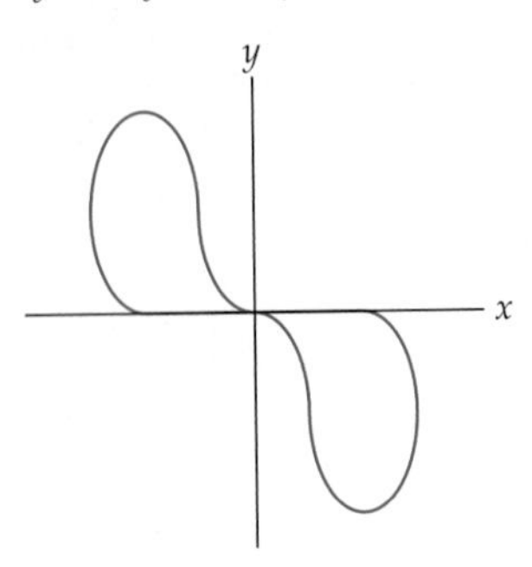

There is one more type of symmetry we'd like to consider: ***symmetry with respect to the x-axis***. Using rectangular coordinates, we have the following definition (compare this with the definition of y-axis symmetry given in Definition 0.9 of Section 0.2):

DEFINITION 9.9 **Symmetry with Respect to the x-axis**

A graph in the xy-plane is said to be ***symmetrical with respect to the x-axis*** if, for every point (x, y) on the graph, the point $(x, -y)$ is also a point on the graph.

Following is an example of a graph that is symmetrical with respect to the x-axis:

Symmetry with respect to the x-axis

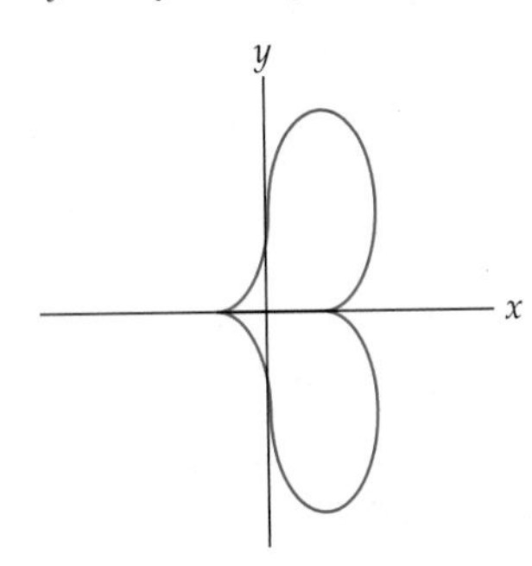

We have now considered three types of symmetry. It is an interesting fact that if a graph has any two of those types of symmetry, it also has the third! The proof is straightforward and is left for Exercise 56.

THEOREM 9.10 **Relationships Among Symmetries in the Plane**

Consider the following three types of symmetry:

(a) symmetry with respect to the vertical axis;

(b) symmetry with respect to the horizontal axis;

(c) symmetry with respect to the origin.

If a graph in the plane has any two of these symmetries, then it has the third as well.

We are particularly concerned about symmetries in the polar plane because many of the equations we will be graphing have one or all of these symmetries. Such symmetries may be recognized in the polar plane by understanding the relationships between r and θ. By writing the definitions of these symmetries in terms of r and θ, we arrive at the properties summarized in the following theorem:

THEOREM 9.11

Symmetries in the Polar Plane

(a) A graph is symmetrical with respect to the x-axis if, for every point (r, θ) on the graph, the point $(r, -\theta)$ is also on the graph.

(b) A graph is symmetrical with respect to the y-axis if, for every point (r, θ) on the graph, the point $(-r, -\theta)$ is also on the graph.

(c) A graph is symmetrical with respect to the origin if, for every point (r, θ) on the graph, the point $(-r, \theta)$ is also on the graph.

You can visualize these symmetry properties in the figure shown next at the left.

CAUTION Note that because every point in the polar plane has multiple representations in terms of its polar coordinates, the symmetries in Theorem 9.11 may also be expressed in terms of other relationships between coordinates on the graph. For example, if $(r, \pi - \theta)$ is on the graph whenever (r, θ) is on the graph, then the graph will be symmetrical with respect to the y-axis; see the following figure at the right:

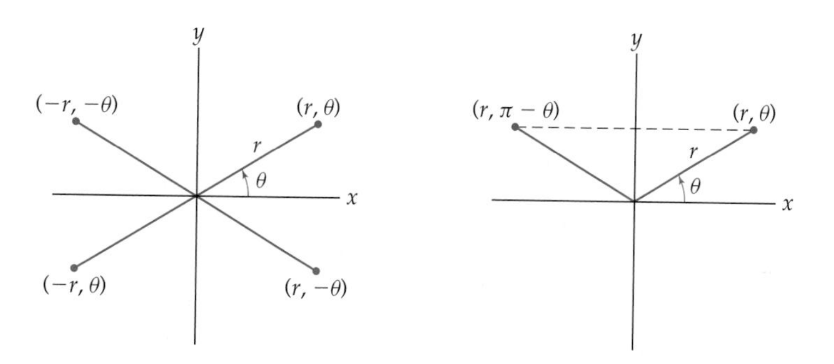

For example, we can see that the polar graph $r = 4\cos\theta$ is symmetrical with respect to the x-axis. Later we will examine a polar function that has all three types of symmetry mentioned in Theorem 9.11. Of course, polar graphs can lack all of these symmetries. For example, consider one of the simplest polar equations: $r = \theta,\ \theta \geq 0$. This simple equation has a quite beautiful graph, known as the ***spiral of Archimedes***.

The spiral of Archimedes

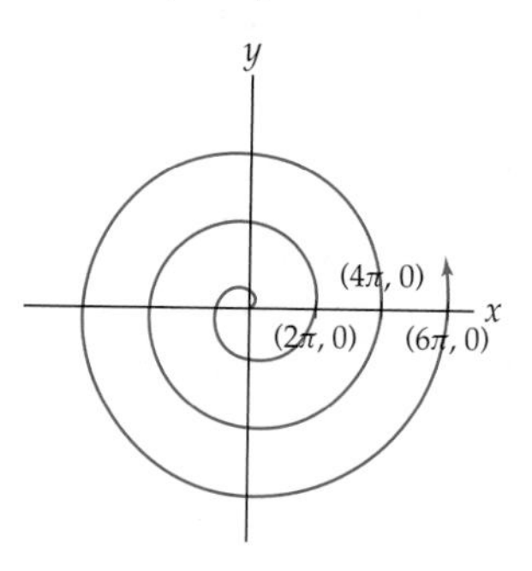

Before we proceed to our examples, we list several of the more common polar curves in the following compendium:

Equation	Name	Example
$r = a \pm b\cos\theta$ $r = a \pm b\sin\theta$, where $a,\ b \in \mathbb{R}$, $a \neq 0,\ b \neq 0$	Cardioid when $\left\lvert\frac{a}{b}\right\rvert = 1$	$r = 1 + \cos\theta$
	Limaçon when $\left\lvert\frac{a}{b}\right\rvert \neq 1$ ▶ with inner loop when $\left\lvert\frac{a}{b}\right\rvert < 1$	$r = 1 + 2\cos\theta$
	▶ dimpled when $1 < \left\lvert\frac{a}{b}\right\rvert < 2$	$r = 3 + 2\cos\theta$
	▶ convex when $\left\lvert\frac{a}{b}\right\rvert > 2$	$r = 5 + 2\cos\theta$
$r = a\sin k\theta$ $r = a\cos k\theta$, where $a \in \mathbb{R}$ and $k \in \mathbb{Z}$	Circle when $\lvert k\rvert = 1$	$r = \sin\theta$
	Rose curve with $2k$ petals when $k \neq 0$ is even	$r = \cos 2\theta$
	Rose curve with k petals when $k \neq \pm 1$ is odd	$r = \cos 3\theta$
$r^2 = \pm a\sin 2\theta$ $r^2 = \pm a\cos 2\theta$, where $a \in \mathbb{R}$ and $a \neq 0$	Lemniscate	$r^2 = \sin 2\theta$
		$r^2 = \cos 2\theta$

We will explore each of these curves shortly.

The names of several of the curves in the table are derived from the Greek and/or Latin. "Limaçon" comes from the Latin "limax," or snail. "Cardioid" is derived from the Greek word "kardia," or heart. "Lemniscate" has antecedents in both Greek and Latin; "lemnicos" and "lemniscus" both mean "ribbon."

Examples and Explorations

EXAMPLE 1

Showing that the graphs of polar equations have symmetry

Show that $r = 4\cos\theta$ is symmetrical with respect to the x-axis and that $r = \cos 2\theta$ is symmetrical with respect to the x-axis, the y-axis, and the origin.

SOLUTION

The cosine is an even function; therefore $4\cos(-\theta) = 4\cos\theta$ for every value of θ. Thus, if (r, θ) is on the graph of $r = 4\cos\theta$, then $(r, -\theta)$ will also be on the graph. Therefore, the graph of $r = 4\cos\theta$ is symmetrical with respect to the x-axis.

We next turn to the equation $r = \cos 2\theta$. Showing that the graph of this equation is symmetrical with respect to the x-axis is similar to our analysis of $r = 4\cos\theta$ and is left for Exercise 15. To show that the graph is symmetrical with respect to the y-axis we will show that $\cos(2(\pi - \theta)) = \cos 2\theta$ for every value of θ. We have

$$\begin{aligned}\cos(2(\pi - \theta)) &= \cos(2\pi - 2\theta) && \leftarrow \text{distributive property}\\ &= \cos(-2\theta) && \leftarrow \text{periodicity of cosine}\\ &= \cos 2\theta. && \leftarrow \text{symmetry of cosine}\end{aligned}$$

Therefore, $r = \cos 2\theta$ is symmetrical with respect to the y-axis. We are now done because, by Theorem 9.10, $r = \cos 2\theta$ must also be symmetrical with respect to the origin. □

EXAMPLE 2

Graphing a cardioid

Graph the polar equation $r = 1 + \cos\theta$.

SOLUTION

We take this opportunity to show how some of the properties of the graphs of the polar functions in our earlier compendium may be constructed. First, notice that since $1 + \cos\theta$ is periodic with period 2π, we need only consider values of θ in the interval $[0, 2\pi)$. Now consider the graph of $r = 1 + \cos\theta$ in the θr-plane. This is the graph of $\cos\theta$ shifted up one unit, as shown in the next figure. From certain points on this graph, we will be able to extract quadrant and intercept information about the polar graph.

$r = 1 + \cos\theta$ in the θr-plane

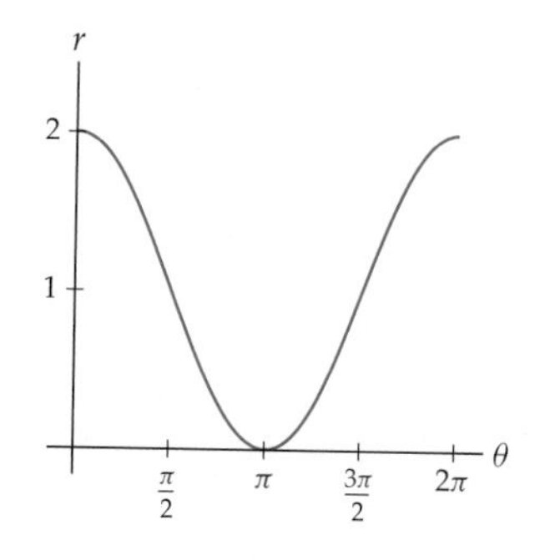

As we discussed earlier in this section, the points on the horizontal axis in the preceding graph correspond to the pole in the polar graph of the equation, so we know that the polar graph of $r = 1 + \cos\theta$ passes through the pole. Moreover, the points on the graph where θ is an integral multiple of π correspond to horizontal intercepts on the polar graph; in the given graph, the points at $\theta = 0$ and $\theta = 2\pi$ both correspond to the horizontal intercept $(r, \theta) = (2, 0)$ on the polar graph and the point at $\theta = \pi$ corresponds to the pole, as already mentioned. Finally, at odd multiples of $\theta = \frac{\pi}{2}$, the given graph indicates that the polar graph of $r = 1 + \cos\theta$ will have vertical intercepts at the polar coordinates $\left(1, \frac{\pi}{2}\right)$ and $\left(1, \frac{3\pi}{2}\right)$. This intercept information is recorded in the graph that follows at the left. We can also obtain quadrant information as shown in the following figure at the right:

Some points on the polar graph of the cardioid

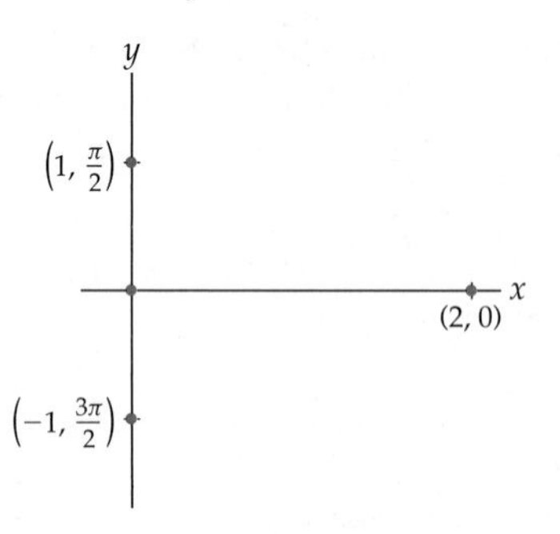

Sectors in the θr-plane indicate quadrants in the polar plane

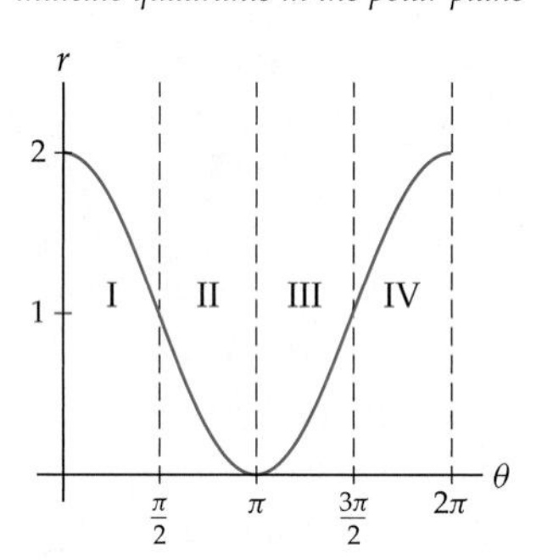

When we test for symmetry, we see that $1 + \cos(-\theta) = 1 + \cos\theta$; therefore the graph in the polar plane will be symmetrical with respect to the horizontal axis. This means that we need only consider values of θ in the interval $[0, \pi]$. The following table shows values of r for several angles $\theta \in [0, \pi]$:

θ	0	$\frac{\pi}{6}$	$\frac{\pi}{4}$	$\frac{\pi}{3}$	$\frac{\pi}{2}$	$\frac{2\pi}{3}$	$\frac{3\pi}{4}$	$\frac{5\pi}{6}$	π
r	2	$1 + \frac{\sqrt{3}}{2}$	$1 + \frac{\sqrt{2}}{2}$	$\frac{3}{2}$	1	$\frac{1}{2}$	$1 - \frac{\sqrt{2}}{2}$	$1 - \frac{\sqrt{3}}{2}$	0

Plotting the intercepts and points from the table gives us the figure that follows at the left. Finally, we connect the plotted points with a smooth curve and use the symmetry of the graph to obtain the graph on the right, called a cardioid.

Points on the graph of $r = 1 + \cos\theta$

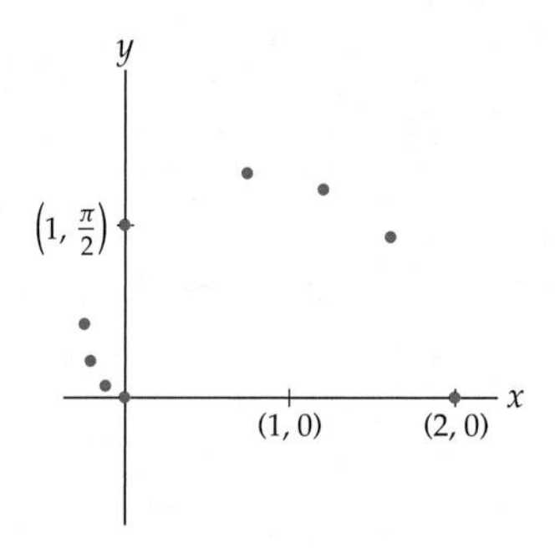

The cardioid $r = 1 + \cos\theta$

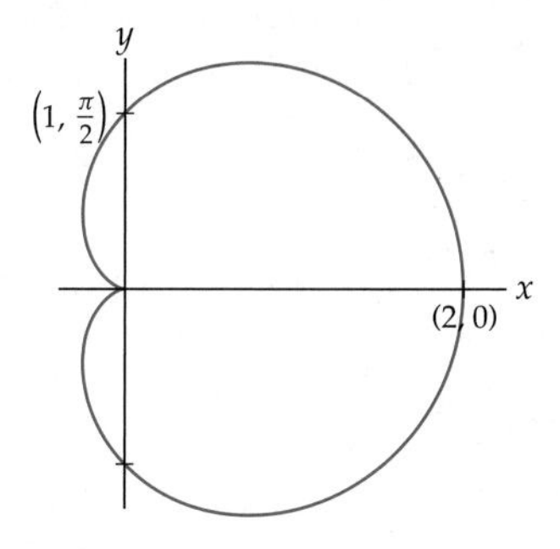

As mentioned in our chart before the examples, there are other equations whose graphs are cardioids in the polar plane. For example, the graph of the equation $r = 2(1 - \sin\theta)$ is another cardioid:

The cardioid $r = 2(1 - \sin\theta)$

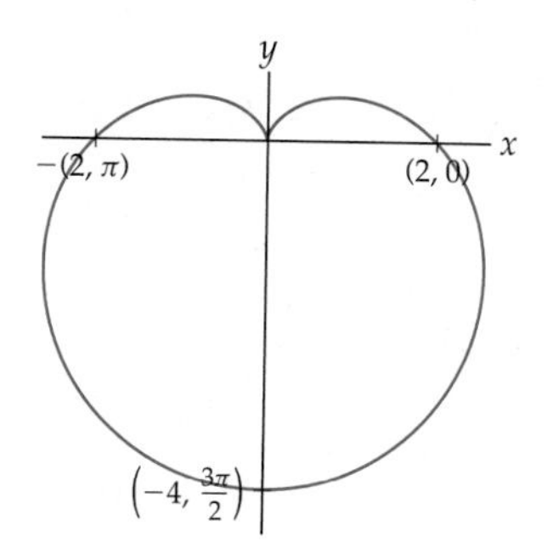

In the Exercise 39 we ask you to graph this curve. □

EXAMPLE 3

Graphing a limaçon

Graph the curve defined by the polar equation $r = \frac{1}{2} + \cos\theta$.

SOLUTION

The graph here is a member of the class of curves called the ***limaçons***. We again start by graphing the equation $r = \frac{1}{2} + \cos\theta$ in the θr-plane:

$r = \frac{1}{2} + \cos\theta$ *in the* θr*-plane*

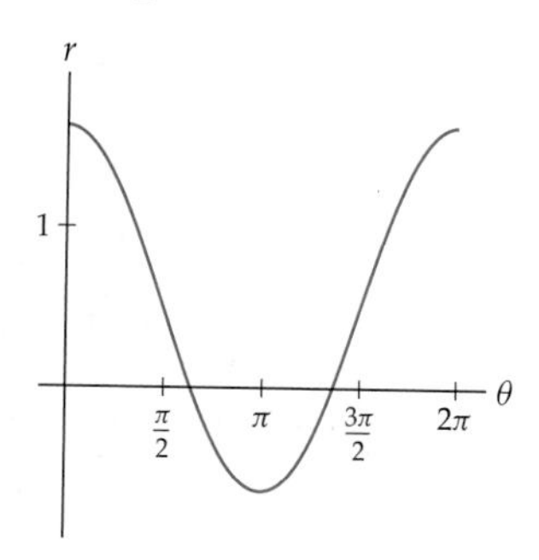

As in Example 2, $r = \frac{1}{2} + \cos\theta$ is periodic with period 2π and its graph in the polar plane will be symmetrical with respect to the x-axis. Therefore we need only consider values of θ in the interval $[0, \pi]$. Horizontal intercepts occur at integral multiples of π, resulting in the polar points $\left(\frac{3}{2}, 0\right)$ and $\left(-\frac{1}{2}, \pi\right)$. We will have a third horizontal intercept at the pole, corresponding to the value where $r = 0$. Here $0 = \frac{1}{2} + \cos\theta$ has the unique solution $\theta = \frac{2\pi}{3}$ on the interval $[0, \pi]$, and that solution corresponds to the polar point $\left(0, \frac{2\pi}{3}\right)$. The only additional vertical intercept occurs when $\theta = \frac{\pi}{2}$; here we have the polar point $\left(\frac{1}{2}, \frac{\pi}{2}\right)$.

Unlike the situation in Example 2, the sign of r varies when θ is in the interval $[0, \pi]$. We have $r > 0$ when $0 \le \theta < \frac{2\pi}{3}$ and $r < 0$ when $\frac{2\pi}{3} < \theta \le \pi$. Here all values of θ in the interval $\left(0, \frac{\pi}{2}\right)$ will be graphed in the first quadrant of the polar plane. Second-quadrant angles in the interval $\left(\frac{\pi}{2}, \frac{2\pi}{3}\right)$ will be graphed in the second quadrant of the polar plane, while second-quadrant angles in the interval $\left(\frac{2\pi}{3}, \pi\right)$ will be graphed in the fourth quadrant of the polar plane! The following table lists the values of r for several angles $\theta \in [0, \pi]$:

θ	0	$\frac{\pi}{6}$	$\frac{\pi}{4}$	$\frac{\pi}{3}$	$\frac{\pi}{2}$	$\frac{2\pi}{3}$	$\frac{3\pi}{4}$	$\frac{5\pi}{6}$	π
r	$\frac{3}{2}$	$\frac{1}{2}+\frac{\sqrt{3}}{2}$	$\frac{1}{2}+\frac{\sqrt{2}}{2}$	1	$\frac{1}{2}$	0	$\frac{1}{2}-\frac{\sqrt{2}}{2}$	$\frac{1}{2}-\frac{\sqrt{3}}{2}$	$-\frac{1}{2}$

The figure that follows at the left shows a plot of the intercepts and points given in the preceding table. By connecting these points with a smooth curve and using the symmetry of the graph, we obtain the limaçon shown at the right. As mentioned earlier, the term "limaçon" is derived from the Latin word for snail. The snail's "shell" is more obvious in the plotted points than in the final graph.

Points on the graph of $r = \frac{1}{2} + \cos\theta$

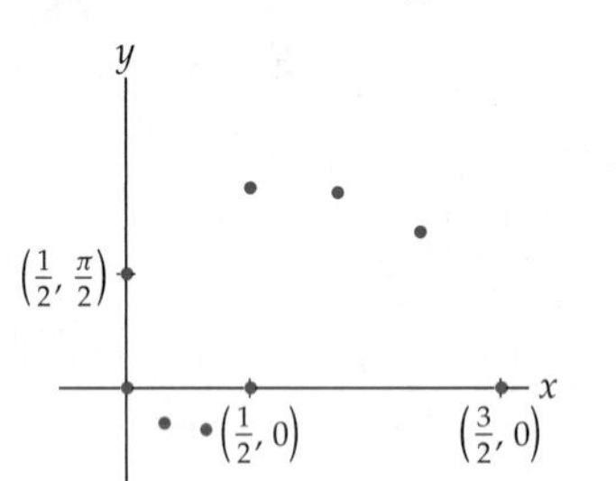

The limaçon $r = \frac{1}{2} + \cos\theta$

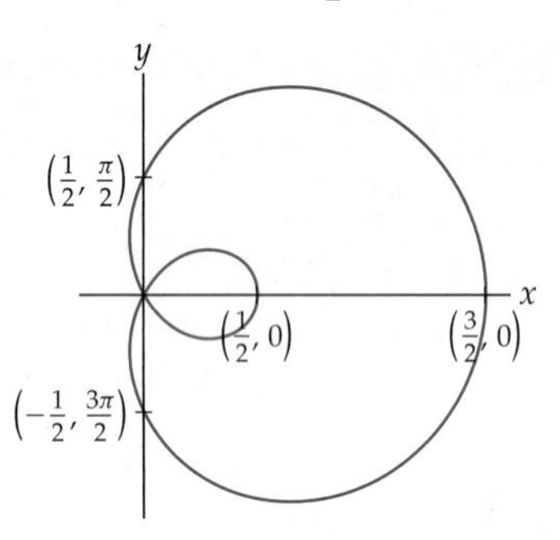

□

EXAMPLE 4

A rose with four petals

Graph the polar equation $r = \cos 2\theta$.

SOLUTION

In Example 1 we showed that the graph of $r = \cos 2\theta$ is symmetrical with respect to the x-axis, with respect to the y-axis, and with respect to the origin. These symmetries will save us a considerable amount of work as we graph the given curve. In particular, we need only graph the curve on the interval $\left[0, \frac{\pi}{2}\right]$ and then let the symmetry of the graph do the rest of the work.

We start our analysis by graphing the equation in the θr-plane:

$r = \cos 2\theta$ *in the* θr*-plane*

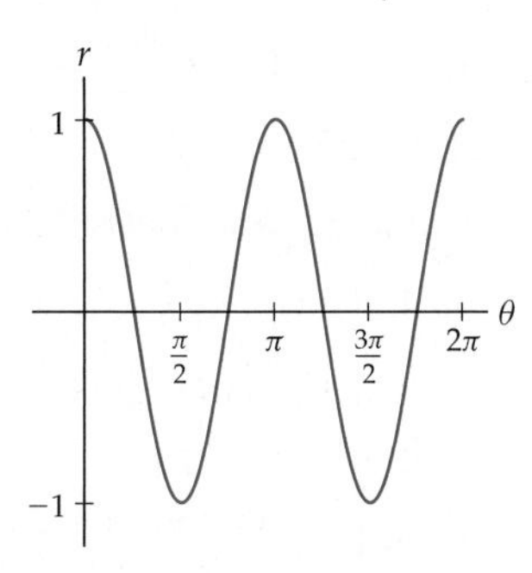

The only horizontal intercepts on the interval $\theta \in \left[0, \frac{\pi}{2}\right]$ will occur at the polar point $(1, 0)$ and when r is zero. We obtain the other horizontal intercept at the pole by finding the unique solution of the polar equation $0 = \cos 2\theta$ for $\theta \in \left[0, \frac{\pi}{2}\right]$. This occurs when $\theta = \frac{\pi}{4}$. On our interval there will be one vertical intercept, at the polar point $\left(-1, \frac{\pi}{2}\right)$.

From the graph, we can see that the sign of r varies when θ is in the interval $\left[0, \frac{\pi}{2}\right]$. We have $r > 0$ on the interval $\left[0, \frac{\pi}{4}\right)$ and $r < 0$ on $\left(\frac{\pi}{4}, \frac{\pi}{2}\right]$. Thus the values of θ in the interval $\left(0, \frac{\pi}{4}\right)$ will be graphed in the first quadrant while the values of θ in the interval $\left(\frac{\pi}{4}, \frac{\pi}{2}\right)$ will be graphed in the third quadrant. The following table lists values of r for several angles in the interval $\left[0, \frac{\pi}{2}\right]$:

θ	0	$\frac{\pi}{12}$	$\frac{\pi}{8}$	$\frac{\pi}{6}$	$\frac{\pi}{4}$	$\frac{\pi}{3}$	$\frac{3\pi}{8}$	$\frac{5\pi}{12}$	$\frac{\pi}{2}$
r	1	$\frac{\sqrt{3}}{2}$	$\frac{\sqrt{2}}{2}$	$\frac{1}{2}$	0	$-\frac{1}{2}$	$-\frac{\sqrt{2}}{2}$	$-\frac{\sqrt{3}}{2}$	-1

Plotting the intercepts and points from the table gives us the picture shown next at the left. We connect these points with a smooth curve and use the symmetry of the graph to obtain the four-petaled rose graphed at the right.

Points on the graph of $r = \cos 2\theta$

y x $(1, 0)$ $\left(-1, \frac{3\pi}{2}\right)$

The rose $r = \cos 2\theta$

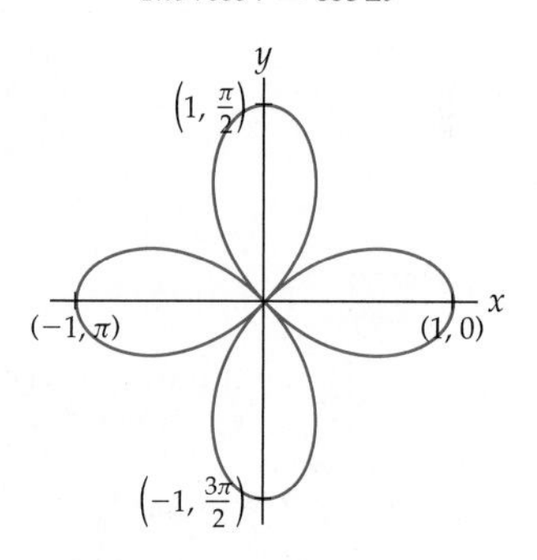

□

In Exercise 55 you are asked to show that when n is a positive odd integer, the polar rose $r = \cos n\theta$ or $r = \sin n\theta$ is traced twice on the interval $[0, 2\pi]$. This implies that when n is odd, we need only graph $r = \sin n\theta$ on the interval $[0, \pi]$. In addition, in Exercise 58 you are asked to show that for every integer n, the graph of $r = \sin n\theta$ is symmetrical with respect to the y-axis. These facts will prove useful in the next example.

EXAMPLE 5

A rose with three petals

Graph the polar equation $r = \sin 3\theta$.

SOLUTION

Because $r = \sin 3\theta$ is a rose defined with the sine function and the odd integer $n = 3$, we need only consider the graph of $r = \sin 3\theta$ for $\theta \in \left[0, \frac{\pi}{2}\right]$. The only horizontal intercepts of $r = \sin 3\theta$ on the interval $\left[0, \frac{\pi}{2}\right]$ will occur at the pole $(0, 0)$. However, the curve will pass through the pole twice, initially when θ is zero and again when θ is $\frac{\pi}{3}$. On our interval there will be one more vertical intercept at $\left(-1, \frac{\pi}{2}\right)$.

The sign of r varies when θ is in the interval $\left[0, \frac{\pi}{2}\right]$: $r > 0$ on $\left[0, \frac{\pi}{3}\right)$ and $r < 0$ on $\left(\frac{\pi}{3}, \frac{\pi}{2}\right]$. Thus the values of θ in the interval $\left(0, \frac{\pi}{3}\right)$ will be graphed in the first quadrant, while the values of θ in the interval $\left(\frac{\pi}{3}, \frac{\pi}{2}\right)$ will be graphed in the third quadrant. We again tabulate values of r for several angles in the interval $\left[0, \frac{\pi}{2}\right]$:

θ	0	$\frac{\pi}{18}$	$\frac{\pi}{12}$	$\frac{\pi}{9}$	$\frac{\pi}{6}$	$\frac{2\pi}{9}$	$\frac{\pi}{4}$	$\frac{5\pi}{18}$	$\frac{\pi}{3}$	$\frac{7\pi}{18}$	$\frac{5\pi}{12}$	$\frac{4\pi}{9}$	$\frac{\pi}{2}$
r	0	$\frac{1}{2}$	$\frac{\sqrt{2}}{2}$	$\frac{\sqrt{3}}{2}$	1	$\frac{\sqrt{3}}{2}$	$\frac{\sqrt{2}}{2}$	$\frac{1}{2}$	0	$-\frac{1}{2}$	$-\frac{\sqrt{2}}{2}$	$-\frac{\sqrt{3}}{2}$	-1

Plotting the intercepts and these points gives us the graph shown next at the left. Connecting the points with a smooth curve and using the symmetry of the graph, we obtain the three-petaled rose graphed at the right.

Points on the graph of $r = \sin 3\theta$

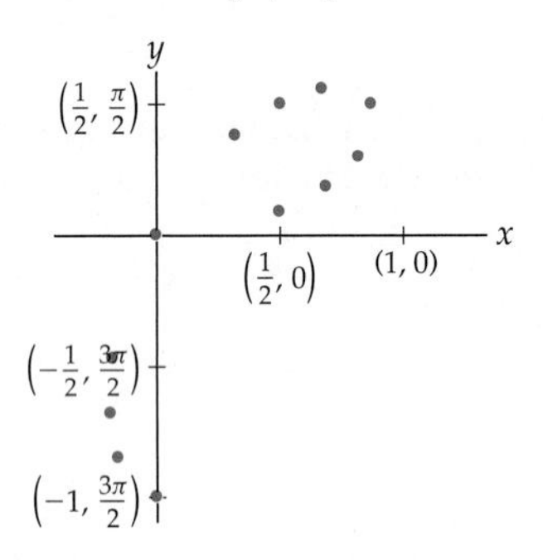

The rose $r = \sin 3\theta$

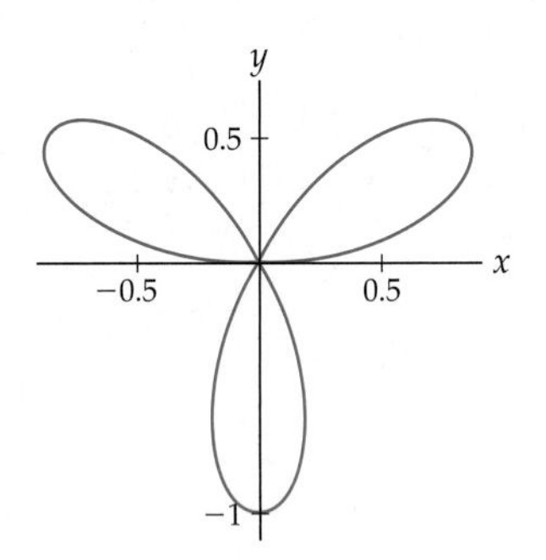

□

EXAMPLE 6

A lemniscate

Graph the polar equation $r^2 = \sin 2\theta$.

SOLUTION

First we note that θ cannot be a second- or fourth-quadrant angle, because if θ were in one of those quadrants, then $\sin 2\theta$ would be negative (since the sine is negative on the intervals $(\pi, 2\pi)$ and $(3\pi, 4\pi)$). In this case, because $r^2 \geq 0$, the equation $r^2 = \sin 2\theta$ could not hold. Therefore the graph of our equation will appear only in the first and third quadrants of the polar plane.

We next show that the graph of $r^2 = \sin 2\theta$ is symmetrical with respect to the origin. It suffices to show that for each point (r, θ) on the graph of $r^2 = \sin 2\theta$, the point $(r, \theta + \pi)$ is also on the graph. We have $\sin(2(\theta + \pi)) = \sin(2\theta + 2\pi) = \sin 2\theta = r^2$. Thus, $(r, \theta + \pi)$ satisfies the equation $r^2 = \sin 2(\theta + \pi)$, so $(r, \theta + \pi)$ is a point on the graph. Therefore the graph of $r^2 = \sin 2\theta$ is symmetrical with respect to the origin.

We again tabulate values of r for several angles in the interval $\left[0, \frac{\pi}{2}\right]$:

θ	0	$\frac{\pi}{12}$	$\frac{\pi}{8}$	$\frac{\pi}{6}$	$\frac{\pi}{4}$	$\frac{\pi}{3}$	$\frac{3\pi}{8}$	$\frac{5\pi}{12}$	$\frac{\pi}{2}$
r	0	$\frac{\sqrt{2}}{2}$	$\sqrt{\frac{\sqrt{2}}{2}}$	$\sqrt{\frac{\sqrt{3}}{2}}$	1	$\sqrt{\frac{\sqrt{3}}{2}}$	$\sqrt{\frac{\sqrt{2}}{2}}$	$\frac{\sqrt{2}}{2}$	0

Plotting the intercepts and points from the table gives us the graph shown next at the left. We connect these points with a smooth curve and use the symmetry of the graph to obtain the lemniscate graphed at the right.

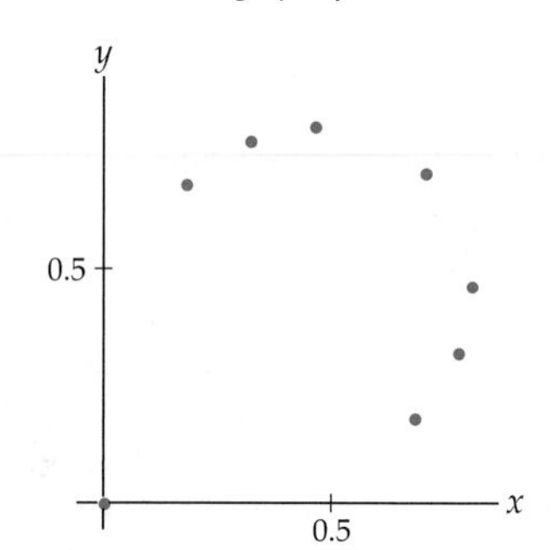

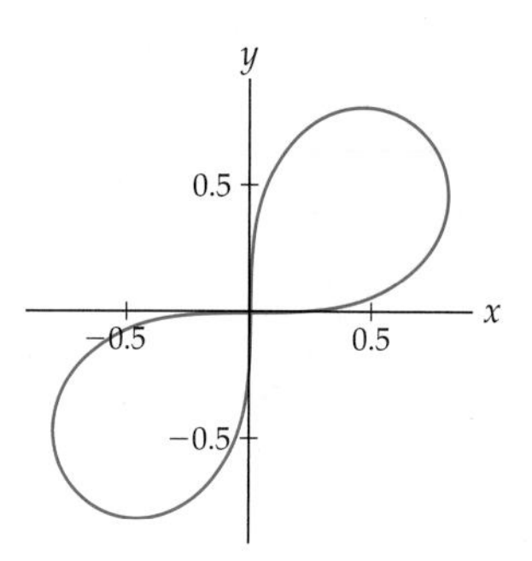

□

CHECKING THE ANSWER It is important to understand the basics of graphing polar curves by hand, but a graphing calculator or computer algebra system may be used to check your work. Calculators or software can also be used to graph many interesting examples that are too complicated to graph by hand. For example, the three figures that follow show some polar curves suggested by T. H. Fay (American Mathematical Monthly, 96 (1989), 442–443). The figures were graphed with the computer program *Mathematica*.

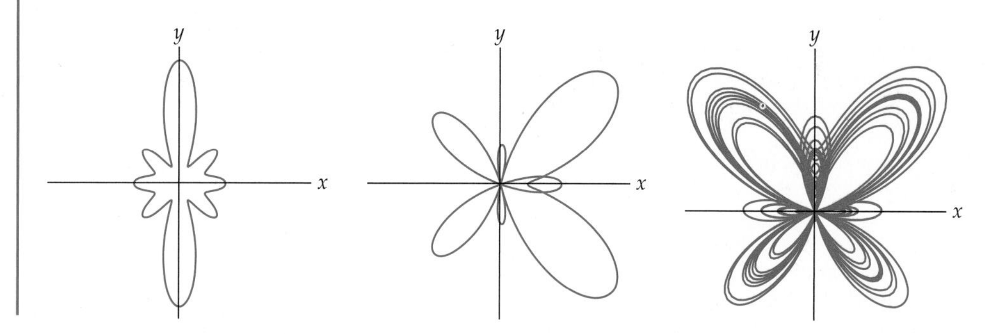

TEST YOUR UNDERSTANDING

- Given a polar equation, why is it helpful to graph the equation in the θr-plane first?
- When θ is a first-quadrant angle, in which quadrants might the polar point corresponding to θ on the graph of an equation $F(r, \theta) = 0$ appear? What about when θ is a second-, third-, or fourth-quadrant angle? Why does this happen?
- How are the symmetries of a graph in polar coordinates recognized?
- When a polar function is symmetrical about both the x- and y-axes, why is it also symmetric about the origin?
- Which polar equations have polar graphs that are circles, cardioids, limaçons, roses, etc.?

EXERCISES 9.3

Thinking Back

- *Finding points of intersection:* Graph the functions
$$y = 1 + \sin x \text{ and } y = \cos x,$$
and find all points of intersection.
- *Finding points of intersection:* Graph the functions
$$y = \frac{1}{2} + \cos x \text{ and } y = \frac{1}{2} - \cos x,$$
and find all points of intersection.
- *Translations of graphs:* What is the relationship between the graphs of the functions
$$y = f(x) \text{ and } y = f(x - k)$$
when $k > 0$?
- *Translations of graphs:* What is the relationship between the graphs of the functions
$$y = f(x) \text{ and } y = f(x) + k$$
when $k > 0$?

Concepts

0. *Problem Zero:* Read the section and make your own summary of the material.

1. *True/False:* Determine whether each of the statements that follow is true or false. If a statement is true, explain why. If a statement is false, provide a counterexample.

(a) *True or False:* If $\frac{\pi}{2} < \theta < \pi$, then the point (r, θ) is located in the second quadrant when it is plotted in a polar coordinate system.

(b) *True or False:* The graph of $r = \sin 5\theta$ is a five-petaled rose.

(c) *True or False:* The graph of $r = \cos 6\theta$ is a six-petaled rose.

(d) *True or False:* If a graph in the polar plane is symmetrical with respect to the origin, then for every polar point (r, θ) on the graph, the polar point $(-r, \theta + 2\pi)$ is also on the graph.

(e) *True or False:* The graph of a polar function $r = f(\theta)$ is symmetrical with respect to the y-axis if, for every point (r, θ) on the graph, the point $(r, -\theta)$ is also on the graph.

(f) *True or False:* When k is a positive integer, the polar roses $r = \sin k\theta$ and $r = \cos k\theta$ are symmetrical with respect to both the x-axis and y-axis if and only if k is even.

(g) *True or False:* In the rectangular coordinate system the graph of the equation $(x^2+y^2)^2 = k(x^2-y^2)$ is a lemniscate for every $k > 0$.

(h) *True or False:* In the rectangular coordinate system the only function $y = f(x)$ that is symmetrical with respect to both the y-axis and the origin is $y = 0$.

2. *Examples:* Construct examples of the thing(s) described in the following. Try to find examples that are different than any in the reading.

(a) An equation in polar coordinates whose graph is a cardioid.

(b) An equation in polar coordinates whose graph is a limaçon.

(c) An equation in polar coordinates whose graph is a lemniscate.

3. How can the graph of an equation $r = f(\theta)$ in the θr-plane be used to provide information about the graph of the same equation in the polar plane?

4. Explain the significance of the following figure:

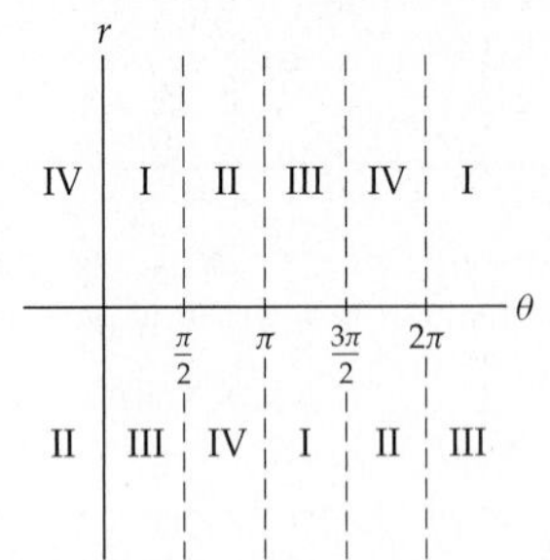

What is being shown in this figure?

5. This problem concerns the connections between the polar plane and the θr-plane.

(a) Which points in the θr-plane correspond to the pole in the polar plane, and why?

(b) Which points in the θr-plane correspond to the horizontal axis in the polar plane, and why?

(c) Which points in the θr-plane correspond to the vertical axis in the polar plane, and why?

6. What is the relationship between the graphs of $r = f(\theta)$ and $r = f(\theta - k)$ in the θr-plane for $k > 0$? What is the relationship between the two graphs in polar coordinates?

7. Fill in the blanks:

(a) On a graph that is symmetrical with respect to the x-axis in the polar plane, if the polar point $\left(2, \frac{\pi}{3}\right)$ is on the graph, then the polar point _______ is also on the graph.

(b) On a graph that is symmetrical with respect to the y-axis in the polar plane, if the polar point $\left(2, \frac{\pi}{3}\right)$ is on the graph, then the polar point _______ is also on the graph.

(c) On a graph that is symmetrical with respect to the y-axis in the Cartesian coordinate system, if the point $(3, 5)$ is on the graph, then the point _______ is also on the graph.

(d) On a graph that is symmetrical with respect to the origin in the polar plane, if the polar point $\left(2, \frac{\pi}{3}\right)$ is on the graph, then the polar point _______ is also on the graph.

(e) On a graph that is symmetrical with respect to the origin in the Cartesian coordinate system, if the point $(3, 5)$ is on the graph, then the point _______ is also on the graph.

8. Fill in the blanks:

(a) If the point $(r, \theta + \pi)$ is on the graph of a polar curve whenever the point (r, θ) is on the graph, then the curve is symmetrical _______.

(b) If the point _______ is on the graph of a polar curve whenever the point (r, θ) is on the graph, then the curve is symmetrical with respect to the origin. (There is more than one way to answer this question correctly!)

(c) If the point $(-r, -\theta)$ is on the graph of a polar curve whenever the point (r, θ) is on the graph, then the curve is symmetrical _______.

9. For each of the following, give an example of a polar equation whose graph has the type(s) of symmetry listed, if possible. If such an equation doesn't exist, explain why.

(a) symmetry about the x-axis

(b) symmetry about the y-axis

(c) symmetry about the origin

(d) symmetry about the origin but not about the x-axis

(e) symmetry about the x-axis, y-axis, and origin

(f) symmetry about the x-axis and y-axis but not about the origin

10. For each type of polar curve, give a general form of the polar equation for that curve:

(a) a line passing through the pole

(b) a circle centered at the pole

(c) a cardioid
(d) a limaçon
(e) a rose with an odd number of petals
(f) a rose with an even number of petals
(g) a lemniscate
(h) the spiral of Archimedes

11. What type of polar curve is named for a heart? For a snail? For a ribbon?

12. What is the difference between a cardioid and a limaçon?

13. Which kind(s) of symmetry does the rose $r = \cos 5\theta$ have? How many petals does this curve have? Which kind(s) of symmetry does the rose $r = \sin 8\theta$ have? How many petals does this curve have?

14. Using polar coordinates, the graphs of the equations $r = \cos k\theta$ and $r = \sin k\theta$ are roses with k petals when $k \geq 3$ is an odd integer. What are the polar graphs of these equations when $k = 1$? What are the graphs of these equations when k is a negative odd integer?

15. Finish Example 1 by showing that $r = \cos 2\theta$ is symmetrical with respect to the horizontal axis.

16. In this section we graphed many polar equations, and most of our examples followed similar procedures. Describe a general procedure for sketching the graph of a polar equation.

Skills

Graph the equations in Exercises 17–24 in the θr-plane. Label each arc of your curve with the quadrant in which the corresponding polar graph will occur.

17. $r = 1 - \cos\theta$

18. $r = \dfrac{1}{2} - \sin\theta$

19. $r = 2 + \sin\theta$

20. $r = \sin 2\theta$

21. $r = \cos^2\theta$

22. $r = \theta,\ r \leq 0$

23. $r = \sec\theta$

24. $r = 1 + \sin 3\theta$

Graph the equations in Exercises 25–32 in the polar plane. Compare your graphs with the corresponding graphs in Exercises 17–24.

25. $r = 1 - \cos\theta$

26. $r = \dfrac{1}{2} - \sin\theta$

27. $r = 2 + \sin\theta$

28. $r = \sin 2\theta$

29. $r = \cos^2\theta$

30. $r = \theta,\ r \leq 0$

31. $r = \sec\theta$

32. $r = 1 + \sin 3\theta$

Find the smallest interval necessary to draw a complete graph of the functions in Exercises 33–38, and then graph each function using a graphing calculator or computer algebra system.

33. $r = \cos(2\theta/5)$

34. $r = \dfrac{4\cos 3\theta + \cos 2\theta}{\cos\theta}$

35. $r = \dfrac{4\cos\theta + \cos 9\theta}{\cos\theta}$

36. $r = e^{\sin\theta} - 2\cos 4\theta$

37. $r = e^{\sin\theta} - 2\cos 4\theta + \sin^5(\theta/12)$

38. $r = e^{\cos 2\theta} - 1.5\cos 4\theta$

39. Graph the cardioid $r = 2(1 - \sin\theta)$.

For each pair of functions in Exercises 40–45,

(a) Algebraically find all values of θ where $f_1(\theta) = f_2(\theta)$.
(b) Sketch the two curves in the same polar coordinate system.
(c) Find all points of intersection between the two curves.

40. $f_1(\theta) = \sin 2\theta$ and $f_2(\theta) = -\sin 2\theta$

41. $f_1(\theta) = \sin\theta$ and $f_2(\theta) = \cos\theta$

42. $f_1(\theta) = 1$ and $f_2(\theta) = 2\sin 2\theta$

43. $f_1(\theta) = 1 + \sin\theta$ and $f_2(\theta) = 1 - \cos\theta$

44. $f_1(\theta) = \sin\theta$ and $f_2(\theta) = \sin 2\theta$

45. $f_1(\theta) = \sin^2\theta$ and $f_2(\theta) = \cos^2\theta$

46. Graph the cardioids $r = 1 + \cos\theta$, $r = 2(1 + \cos\theta)$, and $r = 3(1 + \cos\theta)$ in the same polar coordinate system. What are the relationships among these three graphs? How does changing the constant k in the graph of the cardioid $r = k(1 + \cos\theta)$ affect the graph? What happens when $k < 0$?

47. Graph the cardioids $r = 1 + \sin\theta$ and $r = 1 - \sin\theta$. What is the relationship between these two graphs?

48. Graph the limaçons $r = \dfrac{1}{2} + \sin\theta$, $r = 1 + \sin\theta$, $r = 2 + \sin\theta$, and $r = 3 + \sin\theta$. Make a conjecture about the behavior of graphs of limaçons of the form $r = a + \sin\theta$ for various values of a. In particular, try to understand which values of a will give a limaçon with an inner loop. Which values of a will give a limaçon with a dimple? Which values of a will give a convex limaçon?

49. Graph the limaçons $r = 3 + \cos\theta$, $r = 3 + 3\cos\theta$, $r = 3 + 4\cos\theta$, and $r = 3 + 6\cos\theta$. Make a conjecture about the behavior of graphs of limaçons of the form $r = 3 + b\cos\theta$ for various values of b. In particular, try to understand which values of b will give a limaçon with an inner loop. Which values of b will give a limaçon with a dimple? Which values of b will give result in a convex limaçon?

50. Consider the polar equation $r^2 = \cos 2\theta$.

(a) Show that the polar graph of this equation is symmetrical with respect to the x-axis.

(b) Show that the polar graph of this equation is symmetrical with respect to the origin.

(c) Explain why the polar graph of the equation is symmetrical with respect to the y-axis.

(d) Use the techniques of this section to graph the equation.

(e) How could you graph the equation with a calculator or computer algebra system?

51. Find an equation for the curve obtained as the set of midpoints of every chord with one endpoint at $(1, 0)$ on the unit circle $r = 1$. Express your answer (a) with parametric equations, (b) in rectangular coordinates, and (c) in polar coordinates.

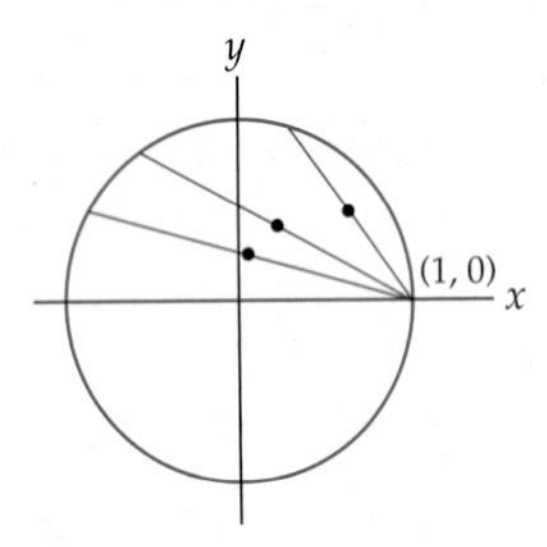

52. The graph of $r = \cos\left(\frac{5}{2}\theta\right)$ is the flower-like graph with 10 petals, shown next at the left:

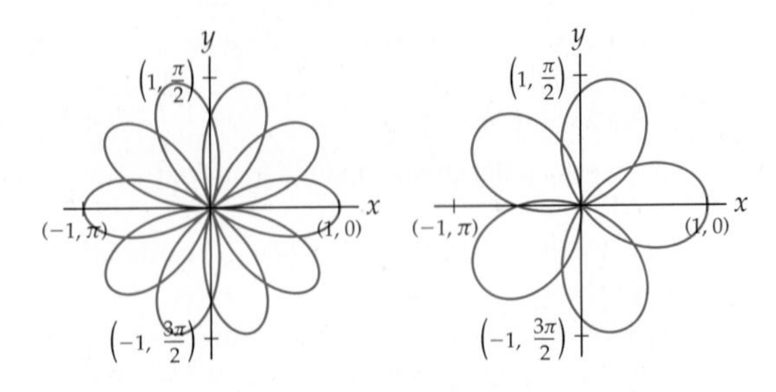

The graph of $r = \cos\left(\frac{5}{3}\theta\right)$ is also a flower-like graph, but with 5 petals, shown in the preceding figure at the right.

(a) Why does the graph of $r = \cos\left(\frac{5}{2}\theta\right)$ have 10 petals while the graph of $r = \cos\left(\frac{5}{3}\theta\right)$ has only 5?

(b) Let $\frac{p}{q}$ be a positive rational number reduced to lowest terms with $p > q$. How many petals will the curve $r = \cos\left(\frac{p}{q}\theta\right)$ have? What is the smallest interval of the form $[0, b]$ required to obtain the graphs of these curves? Explain how to calculate b in terms of p and q.

(c) What is the graph of $r = \cos k\theta$ when k is an irrational number?

Applications

53. Annie is using wood framing and a fabric shell to design a kayak. For simplicity, she considers making the cross section of the kayak follow the curve $r = A(1 - 0.5\sin\theta)$, where $A = 1$ at the middle of the kayak and becomes smaller as the cross section is taken farther from the middle of the boat. Sketch the cross section of the boat when $A = 1$.

54. Annie is concerned that her kayak will be too deep—that it will draw too much water or else have too much freeboard to be easy to paddle. She decides to investigate a cross-section function of the form

$$r = A(1 - 0.5\sin\theta)(1 + 0.2\cos^2\theta).$$

(a) How does Annie need to alter the choice of A so that the width of this boat is the same as that without the term $(1 + 0.2\cos^2\theta)$?

(b) Sketch this cross section with Annie's choice of A from part (a).

(c) Describe in words the effect that the multiplicative term $(1 + 0.2\cos^2\theta)$ has on the cross section of the kayak. How is that effect achieved?

Proofs

55. Prove that when n is a positive odd integer, the polar rose $r = \cos n\theta$ or $r = \sin n\theta$ is traced twice on the interval $[0, 2\pi]$ and thus has exactly n petals.

56. Prove Theorem 9.10. That is, show that if a graph in the plane has any two of three types of symmetry, namely, symmetry about the x-axis, symmetry about the y-axis, and symmetry about the origin, then it has the third type of symmetry as well.

57. Prove that, for every even integer n, the graph of $r = \sin n\theta$ is symmetrical with respect to the x-axis.

58. Prove that, for every integer n, the graph of $r = \sin n\theta$ is symmetrical with respect to the y-axis.

59. Prove that, for every even integer n, the graph of $r = \cos n\theta$ is symmetrical with respect to the y-axis.

60. Prove that, for every integer n, the graph of $r = \cos n\theta$ is symmetrical with respect to the x-axis.

Thinking Forward

- *Volume of a cylinder:* Find the volume of the right circular cylinder with height 2 units and cross-sectional circle given by the equation $r = 4\cos\theta$.
- *Volume of a sphere:* Find the volume of the sphere whose equator is given by the equation $r = 4\cos\theta$.

9.4 COMPUTING ARC LENGTH AND AREA WITH POLAR FUNCTIONS

- A formula for finding the area bounded by a polar curve is derived
- We explain how polar functions can be expressed with parametric equations
- A formula for finding the arc length of a polar curve is derived

Computing the Area Bounded by Polar Curves

We've seen that many interesting regions in the plane are bounded by polar functions of the form $r = f(\theta)$. We will start this section by discussing how to compute the area of such a region without transforming the equation into rectangular coordinates.

When working in rectangular coordinates, we used rectangles as our most "basic" shape—that is, as the building blocks for approximating areas of more complicated regions. When working in polar coordinates, it is more natural to approximate a region with sectors of circles. For example, consider a region R in the polar plane bounded by two rays $\theta = \alpha$ and $\theta = \beta$ and a polar function of the form $r = f(\theta)$, as shown next at the left. (Regions of this type will be the typical polar regions we wish to approximate; compare our work here with our work in rectangular coordinates when we considered the area under the graph of a function $y = f(x)$ from $x = a$ to $x = b$.) The following figure at the right shows the region R approximated with "wedges" (i.e., sectors of circles):

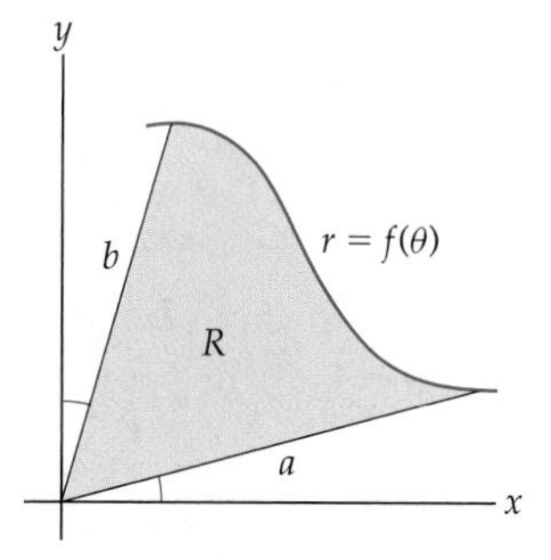

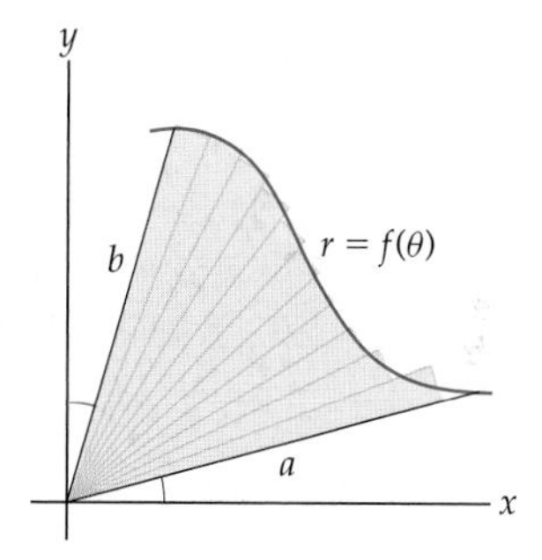

Of course, before we can use sectors of circles to approximate areas, we must determine how to find the area of a sector of a circle. Consider a sector of a circle with radius r and that has a central angle ϕ (measured in radians):

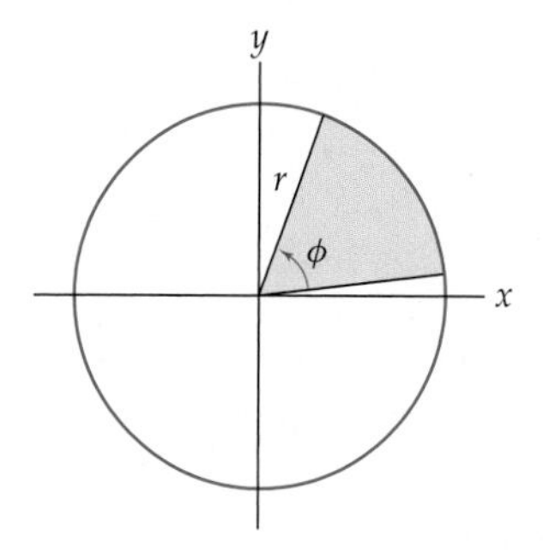

The area of the sector is a fraction of the area of the circle. Specifically, it is $\frac{\phi}{2\pi}$ times the area of the entire circle $\left(\text{i.e., } \frac{\phi}{2\pi}(\pi r^2) = \frac{1}{2}\phi r^2\right)$, as stated in the following theorem:

THEOREM 9.12 **The Area of a Sector of a Circle**

In a circle of radius r, the area of a sector with central angle ϕ, in radians, is given by

$$\text{Area} = \frac{1}{2}\phi r^2.$$

We will now return to the region R and its approximation with wedges. Although we could easily approximate the area of this region with a fixed number of wedges, we want to find the *exact* area; therefore we need to fix a general notation. The procedure is strikingly similar to our work with Riemann sums in Chapter 4, except that instead of dealing with approximating rectangles in rectangular coordinates, we will be dealing with approximating sectors in polar coordinates.

Let n be a positive integer, and define

$$\Delta\theta = \frac{\beta - \alpha}{n} \quad \text{and} \quad \theta_k = \alpha + k\Delta\theta.$$

We also choose some point θ_k^* in each interval $[\theta_{k-1}, \theta_k]$ and let S_k be the sector of the circle with radius $f(\theta_k^*)$ centered at the origin and extending from the radius with polar angle θ_{k-1} to the radius with polar angle θ_k. These are the sectors shown earlier as wedges.

By Theorem 9.12, the area of the kth sector S_k is $\frac{1}{2}(f(\theta_k^*))^2\Delta\theta$. The approximate area of the region R is the sum of the areas of the sectors:

$$\text{Area of region } R \approx \sum_{k=1}^{n} \frac{1}{2}(f(\theta_k^*))^2\Delta\theta.$$

Notice that this is a Riemann sum for the function $\frac{1}{2}(f(\theta))^2$ on the interval $[\alpha, \beta]$. Therefore, the area of R will be the limit of the Riemann sum as $n \to \infty$. That is,

$$\text{Area of region } R = \lim_{n\to\infty} \sum_{k=1}^{n} \frac{1}{2}(f(\theta_k^*))^2\Delta\theta.$$

We now have the following theorem:

THEOREM 9.13 **The Area of a Region in the Polar Plane Bounded by a Function $r = f(\theta)$**

Let α and β be real numbers such that $0 \le \beta - \alpha \le 2\pi$. Let R be the region in the polar plane bounded by the rays $\theta = \alpha$ and $\theta = \beta$ and a positive continuous function $r = f(\theta)$. Then the area of region R is

$$\frac{1}{2}\int_{\alpha}^{\beta} (f(\theta))^2\, d\theta.$$

Computing Arc Length in Polar Coordinates

Every polar function of the form $r = f(\theta)$ can be easily rewritten in terms of parametric equations of the form $x = x(\theta)$ and $y = y(\theta)$. Recall that to transform from polar coordinates to rectangular coordinates we use

$$x = r\cos\theta, \; y = r\sin\theta.$$

If $r = f(\theta)$, we immediately obtain

$$x = f(\theta)\cos\theta, \; y = f(\theta)\sin\theta.$$

These are the parametric equations that we sought.

For example, to express the cardioid with equation $r = 1 + \cos\theta$ in terms of parametric equations, we have

$$x = (1 + \cos\theta)\cos\theta, \; y = (1 + \cos\theta)\sin\theta.$$

By thinking of polar functions as parametric equations, we can use the work we just did to find the arc length of a polar curve. If we combine the parametric equations $x = f(\theta)\cos\theta$ and $y = f(\theta)\sin\theta$ with Theorem 9.4 of Section 9.1, which tells us how to compute the arc length of a curve defined by parametric equations, we arrive at the following theorem:

THEOREM 9.14

The Arc Length of a Polar Curve

Let $r = f(\theta)$ be a differentiable function of θ such that $f'(\theta)$ is continuous for all $\theta \in [\alpha, \beta]$. Furthermore, assume that $r = f(\theta)$ is a one-to-one function from $[\alpha, \beta]$ to the graph of the function. Then the length of the polar graph of $r = f(\theta)$ on the interval $[\alpha, \beta]$ is

$$\int_\alpha^\beta \sqrt{(f'(\theta))^2 + (f(\theta))^2}\, d\theta.$$

Proof. Theorem 9.4 tells us that when x and y are functions of the parameter θ, the arc length of the curve on the interval $[\alpha, \beta]$ is

$$\int_\alpha^\beta \sqrt{\left(\frac{dx}{d\theta}\right)^2 + \left(\frac{dy}{d\theta}\right)^2}\, d\theta.$$

Combining this integral with the parametric equations for the curve, $x = f(\theta)\cos\theta$ and $y = f(\theta)\sin\theta$, we have arc length

$$\int_\alpha^\beta \sqrt{\left(\frac{d}{d\theta}(f(\theta)\cos\theta)\right)^2 + \left(\frac{d}{d\theta}(f(\theta)\sin\theta)\right)^2}\, d\theta.$$

Taking the derivatives inside the radical gives

$$\int_\alpha^\beta \sqrt{(f'(\theta)\cos\theta - f(\theta)\sin\theta)^2 + (f'(\theta)\sin\theta + f(\theta)\cos\theta)^2}\, d\theta.$$

After expanding the terms under of the radical and simplifying (you will do this in Exercise 58), we obtain our result:

$$\int_\alpha^\beta \sqrt{(f'(\theta))^2 + (f(\theta))^2}\, d\theta.$$

■

Examples and Explorations

EXAMPLE 1

Finding the area bounded by a polar rose

Calculate the area bounded by the curve $r = \cos 2\theta$.

SOLUTION

This is the first of three examples in which we use Theorem 9.13 to find areas of polar regions. As with all area computations, we must understand the area we are trying to compute. When we can, we will use the symmetry of the region to simplify our work.

Recall that we graphed this polar rose in Example 4 in Section 9.3. We reproduce that graph here, with one portion of the graph shaded:

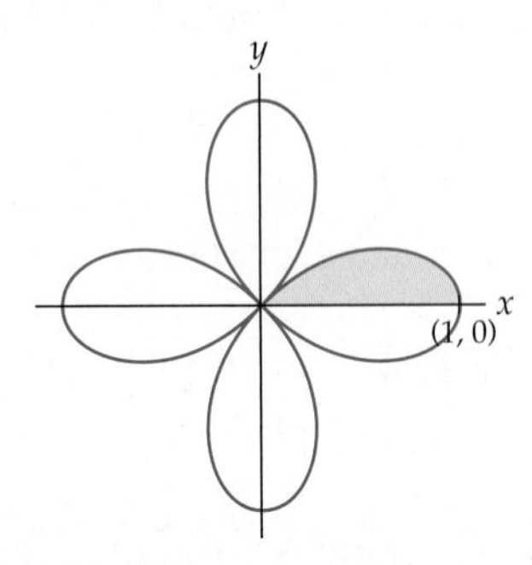

To compute the area of the curve we will use the symmetry of the rose. We will calculate only the area of the shaded region of the figure—that is one-half of one of the petals (and one-eighth of the total area). The shaded region corresponds to the values of θ in $\left[0, \frac{\pi}{4}\right]$. Therefore the area of the shaded region is

$$\frac{1}{2}\int_0^{\pi/4} \cos^2 2\theta \, d\theta.$$

The value of this integral is $\frac{\pi}{16}$. (You should be able to work out the details.) Since the area of the shaded region is one-eighth of the region bounded by the polar rose, the area bounded by the rose is $\frac{\pi}{2}$ square units. □

CAUTION In Example 1 we used the symmetry of the region to help us compute the area of a polar rose. Using such symmetries will often make area computations easier. We could also have computed the entire area from the integral

$$\frac{1}{2}\int_0^{2\pi} \cos^2 2\theta \, d\theta,$$

because the entire curve gets traced once on the interval $[0, 2\pi]$. However, it is not always correct to integrate over the interval $[0, 2\pi]$ to find the area of a polar rose. When $n \geq 3$ is odd, any polar rose of the form $r = \cos n\theta$ or $r = \sin n\theta$ is traced twice on the interval $[0, 2\pi]$. Therefore, to find the areas bounded by these roses, we could use integrals of the form

$$\frac{1}{2}\int_0^{\pi} \cos^2 n\theta \, d\theta \text{ or } \frac{1}{2}\int_0^{\pi} \sin^2 n\theta \, d\theta.$$

For these functions, integrating over the interval $[0, 2\pi]$ would give an answer that is twice the correct value.

EXAMPLE 2 **Finding the area between the loops of a limaçon**

Calculate the area between the interior and exterior loops of the limaçon $r = \frac{1}{2} + \cos\theta$.

SOLUTION

This is the limaçon we graphed in Example 3 in Section 9.3. We reproduce that graph here, again with one portion shaded:

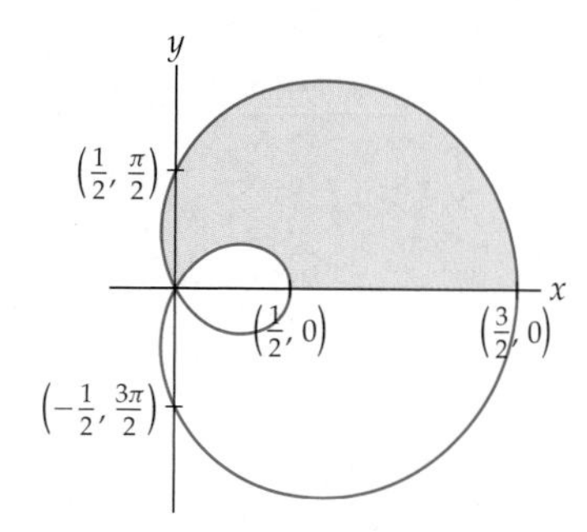

To compute the area between the loops of the limaçon, we will make use once more of the symmetry of the region. We will calculate the area of the shaded region of the figure. This area is one-half of our desired area. To do our calculation, we will first find the area between the x-axis and top half of the outer loop of the limaçon. Next, we will find the area between the x-axis and the top half of the inner loop. The difference of these areas will give us the area of the shaded region.

To compute the area between the x-axis and the top half of the outer loop of the limaçon, we use the integral

$$\frac{1}{2}\int_0^{2\pi/3}\left(\frac{1}{2}+\cos\theta\right)^2 d\theta.$$

The upper limit of this integral is $\frac{2\pi}{3}$, because that is the smallest positive value of θ at which $r = 0$ and the curve goes through the pole. The value of the integral is $\frac{\pi}{4} + \frac{3\sqrt{3}}{16}$. The area between the x-axis and the top half of the inner loop of the limaçon is given by the integral

$$\frac{1}{2}\int_{\pi}^{4\pi/3}\left(\frac{1}{2}+\cos\theta\right)^2 d\theta.$$

We integrate here over the interval $\left[\pi, \frac{4\pi}{3}\right]$ because this is the interval that traces the desired portion of the curve. Note that we could also have integrated over the interval $\left[\frac{2\pi}{3}, \pi\right]$. This would have given us the area between the x-axis and the bottom half of the inner loop of the limaçon. Since the two halves of the inner loop have equal area, the values of the two integrals would be equal. The value of either of these integrals is $\frac{\pi}{8} - \frac{3\sqrt{3}}{16}$. Therefore, the shaded region has area

$$\frac{\pi}{4} + \frac{3\sqrt{3}}{16} - \left(\frac{\pi}{8} - \frac{3\sqrt{3}}{16}\right) = \frac{\pi + 3\sqrt{3}}{8}.$$

The area between the two loops is twice this value, $\frac{\pi+3\sqrt{3}}{4}$. □

EXAMPLE 3

Finding the area between two polar curves

Calculate the area of the region in the polar plane that is inside the circle $r = 3\cos\theta$, but outside the cardioid $r = 1 + \cos\theta$.

SOLUTION

The region in question is shown as the shaded part of the following graph:

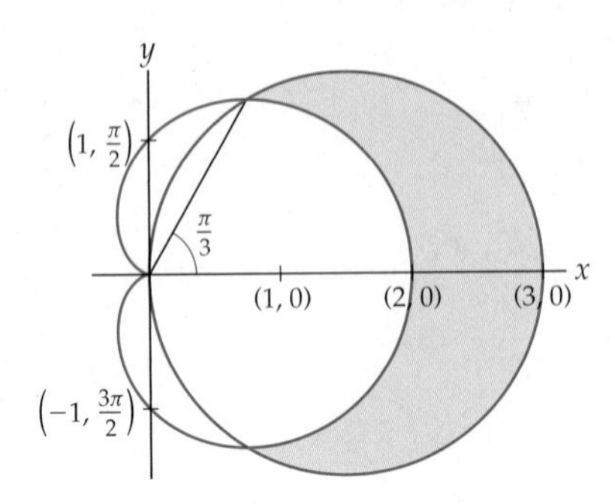

Using the symmetry of the region, we know that it is enough to compute the area of the top half of the region and then multiply by 2. Solving $3\cos\theta = 1 + \cos\theta$ for $\theta \in [0, \pi]$, we see that the two curves intersect when $\theta = \frac{\pi}{3}$. Therefore the area of the top half of the region is given by the definite integral

$$\frac{1}{2}\int_0^{\pi/3} ((3\cos\theta)^2 - (1+\cos\theta)^2)\,d\theta.$$

The value of this integral is $\frac{\pi}{2}$. Since the area of the shaded region is one-half of the original region, the desired area is π square units. □

EXAMPLE 4

Computing the arc length of a polar curve

Compute the arc length of the cardioid defined by the polar equation $r = 1 + \cos\theta$.

SOLUTION

By Theorem 9.14, we need only evaluate the integral

$$\int_0^{2\pi} \sqrt{\left(\frac{d}{d\theta}(1+\cos\theta)\right)^2 + (1+\cos\theta)^2}\,d\theta.$$

After differentiating and simplifying, we find that the integral is equal to

$$\int_0^{2\pi} \sqrt{\sin^2\theta + 1 + 2\cos\theta + \cos^2\theta}\,d\theta = \int_0^{2\pi} \sqrt{2 + 2\cos\theta}\,d\theta.$$

The value of this definite integral is 8 (as you will show in Exercise 16). Therefore the arc length of the cardioid $r = 1 + \cos\theta$ is 8 units. □

? TEST YOUR UNDERSTANDING

- Why is the integral $\int_\alpha^\beta \frac{1}{2}(f(\theta))^2\,d\theta$ used to compute the area of a region in the polar plane bounded by the function $r = f(\theta)$? Why not just use the integral $\int_\alpha^\beta f(\theta)\,d\theta$?
- Given two polar equations $r = f(\theta)$ and $r = g(\theta)$, why is it insufficient to solve the equation $f(\theta) = g(\theta)$ for θ in order to find the points of intersection of the two graphs? How do you find all points of intersection on the graphs of f and g?
- How can the symmetries of the graphs of polar functions be used to simplify area calculations? Why is it important to know where curves intersect themselves or each other when you are calculating an area?
- How do you transform a polar function of the form $r = f(\theta)$ into parametric equations $x = x(\theta)$ and $y = y(\theta)$?
- How do you compute the arc length of a polar function $r = f(\theta)$?

EXERCISES 9.4

Thinking Back

Calculate each of the following definite integrals (these are precisely the definite integrals that were encountered in the reading for this section):

- $\frac{1}{2}\int_0^{\pi/4} \cos^2 2\theta \, d\theta$
- $\frac{1}{2}\int_0^{\pi} \cos^2 3\theta \, d\theta$
- $\frac{1}{2}\int_0^{2\pi/3} \left(\frac{1}{2} + \cos\theta\right)^2 \, d\theta$
- $\frac{1}{2}\int_{\pi}^{4\pi/3} \left(\frac{1}{2} + \cos\theta\right)^2 \, d\theta$
- $\frac{1}{2}\int_0^{\pi/3} ((3\cos\theta)^2 - (1+\cos\theta)^2) \, d\theta$

Concepts

0. *Problem Zero:* Read the section and make your own summary of the material.

1. *True/False:* Determine whether each of the statements that follow is true or false. If a statement is true, explain why. If a statement is false, provide a counterexample.
   (a) *True or False:* To approximate the area of the region in the polar plane bounded by the function $r = f(\theta)$ and the rays $\theta = \alpha$ and $\theta = \beta$, we can use a sum of areas of sectors of circles.
   (b) *True or False:* Suppose we subdivide the interval of angles $\theta \in \left[\frac{\pi}{4}, \frac{\pi}{2}\right]$ into four equal subintervals. Then $\Delta\theta = \frac{\pi}{16}$.
   (c) *True or False:* The area of the region in the polar plane bounded by the function $r = f(\theta)$ and the rays $\theta = \alpha$ and $\theta = \beta$ is given by the definite integral $\int_\alpha^\beta f(\theta)\, d\theta$.
   (d) *True or False:* The area between the continuous function $r = f(\theta)$ and the θ-axis on the interval $[\alpha, \beta]$ in the θr-plane is given by the integral $\int_\alpha^\beta |f(\theta)|\, d\theta$.
   (e) *True or False:* Since the graph of $r = 2\cos\theta$ is a circle with radius 1, the value of the integral $\frac{1}{2}\int_0^{2\pi} (2\cos\theta)^2\, d\theta$ is π.
   (f) *True or False:* The polar equation $r = \cos 4\theta$ is traced twice as θ varies from 0 to 2π.
   (g) *True or False:* Every polar function of the form $r = f(\theta)$ can be written with parametric equations.
   (h) *True or False:* If $r = f(\theta)$ is a differentiable function for $0 \le \theta \le 2\pi$, then the integral $\int_0^{2\pi} \sqrt{(f(\theta))^2 + (f'(\theta))^2}\, d\theta$ represents the length of the curve $r = f(\theta)$ for $0 \le \theta \le 2\pi$.

2. *Examples:* Construct examples of the thing(s) described in the following. Try to find examples that are different than any in the reading.
   (a) An integral that represents the area of a circle with radius a in polar coordinates.
   (b) An integral that represents the circumference of a circle with radius a in polar coordinates.
   (c) An integral that represents the circumference of a circle with radius a in terms of parametric equations.

3. When we use rectangular coordinates to approximate the area of a region, we subdivide the region into vertical strips and use a sum of areas of rectangles to approximate the area. Explain why we use a "wedge" (i.e., a sector of a circle) and not a rectangle when we use polar coordinates to compute an area.

4. When we investigated area in rectangular coordinates in Chapter 4, we often tried to find the areas of regions under curves $y = f(x)$ from $x = a$ to $x = b$. In the polar plane, the typical region whose area we wish to find is a region R bounded by two rays $\theta = \alpha$ and $\theta = \beta$ and a polar function of the form $r = f(\theta)$. Why is this our basic type of region in the polar plane?

5. In this section we described a method for approximating the area of a polar region bounded by two rays $\theta = \alpha$ and $\theta = \beta$ and a polar function $r = f(\theta)$. The method involved a "subdivide, approximate, and add" strategy in which we fixed some general notation. Draw a carefully labeled picture that illustrates the roles of $\Delta\theta$, θ_k^*, and $f(\theta_k^*)$ for one approximating sector S_k.

6. Explain how we arrive at the definite integral formula $\frac{1}{2}\int_\alpha^\beta (f(\theta))^2 \, d\theta$ in Theorem 9.13 for computing the area bounded by a polar function $r = f(\theta)$ on an interval $[\alpha, \beta]$. (Your explanation should include a limit of Riemann sums.) What would the integral $\int_\alpha^\beta f(\theta) \, d\theta$ represent?

7. Why do we require that $0 \le \beta - \alpha \le 2\pi$ in the statement of Theorem 9.13?

8. Consider the three-petaled polar rose defined by $r = \cos 3\theta$. Explain why the definite integral $\frac{1}{2}\int_0^{2\pi} \cos^2 3\theta \, d\theta$ calculates *twice* the area bounded by the petals of this rose.

9. Explain how the symmetries of the graphs of polar functions can be used to simplify area calculations.

10. Explain how to use parametric equations to transform a polar function $r = f(\theta)$.

11. Explain how to use parametric equations to transform the function $y = f(x)$.

12. What is the formula for computing the arc length of a polar curve $r = f(\theta)$ where $\theta \in [\alpha, \beta]$? What conditions on the polar function $f(\theta)$ are necessary for this formula to hold?

13. Rick wants to show that the circumference of a unit circle is 2π. He decides to use the arc length formula given in Theorem 9.14 with the function $r = \sin\theta$ on the interval $[0, 2\pi]$ and obtains 2π as the length. Explain the two mistakes he made and how they cancelled to give the correct answer.

14. Give a geometric explanation of why
$$\frac{n}{2}\int_0^{2\pi/n} r^2\, d\theta = \pi r^2$$
for any positive real number r and any positive integer n. Would the equation also hold for non-integer values of n?

15. The following integral expression may be used to find the area of a region in the polar coordinate plane:
$$\frac{1}{2}\int_0^{\pi/4} \sin^2\theta\, d\theta + \frac{1}{2}\int_{\pi/4}^{\pi/2} \cos^2\theta\, d\theta.$$
Sketch the region and then compute its area. (If you prefer, you may use a simpler integral to compute the same area.)

16. Complete Example 4 by evaluating the integral $\int_0^{2\pi} \sqrt{2 + 2\cos\theta}\, d\theta$.

Skills

In Exercises 17–25 find a definite integral expression that represents the area of the given region in the polar plane, and then find the exact value of the expression.

17. The region enclosed by the spiral $r = \theta$ and the x-axis on the interval $0 \le \theta \le \pi$.

18. The region inside one loop of the lemniscate $r^2 = \sin 2\theta$.

19. The region between the two loops of the limaçon $r = 1 + \sqrt{2}\cos\theta$.

20. The region between the two loops of the limaçon $r = \sqrt{3} - 2\sin\theta$.

21. The region inside the cardioid $r = 3 - 3\sin\theta$ and outside the cardioid $r = 1 + \sin\theta$.

22. The region inside both of the cardioids $r = 3 - 3\sin\theta$ and $r = 1 + \sin\theta$.

23. The region inside the circle $x^2 + y^2 = 1$ to the right of the vertical line $x = \frac{1}{2}$.

24. The region bounded by the limaçon $r = 1 + k\sin\theta$, where $0 < k < 1$. Bonus: Explain why the area approaches π as $k \to 0$.

25. The graph of the polar equation $r = \sec\theta - 2\cos\theta$ for $\theta \in \left(-\frac{\pi}{2}, \frac{\pi}{2}\right)$ is called a ***strophoid***. Graph the strophoid and find the area bounded by the loop of the graph.

In Exercises 26–30 find a definite integral that represents the length of the specified polar curve, and then find the exact value of the integral.

26. The spiral $r = \theta$ for $0 \le \theta \le 2\pi$.

27. The spiral $r = e^{\theta}$ for $0 \le \theta \le 2\pi$.

28. The spiral $r = e^{\theta}$ for $2k\pi \le \theta \le 2(k+1)\pi$.

29. The spiral $r = e^{\alpha\theta}$ for $0 \le \theta \le 2\pi$, where α is a nonzero constant.

30. The cardioid $r = 2 - 2\sin\theta$, for $0 \le \theta \le 2\pi$.

In Exercises 31–36 find a definite integral that represents the length of the specified polar curve, and then use a graphing calculator or computer algebra system to approximate the value of the integral.

31. One petal of the polar rose $r = \cos 2\theta$.

32. One petal of the polar rose $r = \cos 3\theta$.

33. One petal of the polar rose $r = \cos 4\theta$.

34. The entire limaçon $r = 1 + 2\sin\theta$.

35. The inner loop of the limaçon $r = 1 + 2\sin\theta$.

36. The limaçon $r = 2 + \cos\theta$.

37. Complete Example 2 by evaluating the integral expression
$$\int_0^{2\pi/3}\left(\frac{1}{2} + \cos\theta\right)^2 d\theta - \int_{\pi}^{4\pi/3}\left(\frac{1}{2} + \cos\theta\right)^2 d\theta.$$

Each of the integrals or integral expressions in Exercises 38–44 represents the area of a region in the plane. Use polar coordinates to sketch the region and evaluate the expressions.

38. $\displaystyle \frac{1}{2}\int_0^{2\pi} (1 + \sin\theta)^2\, d\theta$

39. $\displaystyle \int_0^{\pi} (1 + \cos\theta)^2\, d\theta$

40. $\displaystyle \int_{-\pi/2}^{\pi/2} (2 - \sin\theta)^2\, d\theta$

41. $\displaystyle \int_0^{\pi/2} \sin 2\theta\, d\theta$

42. $\displaystyle 3\int_0^{\pi/2} \sin^2 3\theta\, d\theta$

43. $\displaystyle \frac{1}{2}\int_0^{2\pi} (2 + \sin 4\theta)^2\, d\theta$

44. $\displaystyle \int_{-\pi/4}^{\pi/2}\left(\frac{\sqrt{2}}{2} + \sin\theta\right)^2 d\theta - \int_{-\pi/2}^{-\pi/4}\left(\frac{\sqrt{2}}{2} + \sin\theta\right)^2 d\theta$

45. Use Theorem 9.13 to show that the area of the circle defined by the polar equation $r = a$ is πa^2.

46. Use Theorem 9.13 to show that the area of the circle defined by the polar equation $r = 2a\cos\theta$ is πa^2.

47. Use Theorem 9.14 to show that the circumference of the circle defined by the polar equation $r = a$ is $2\pi a$.

48. Use Theorem 9.14 to show that the area of the circle defined by the polar equation $r = 2a\sin\theta$ is $2\pi a$.

49. Find the area interior to two circles with the same radius if each circle passes through the center of the other. (*Hint: Consider the circles $r = a$ and $r = 2a\cos\theta$.*)

Applications

50. Annie is designing a kayak with central cross section given by $r = 1 - 0.5\sin\theta$, where r is measured in feet.

Kayak and central cross section

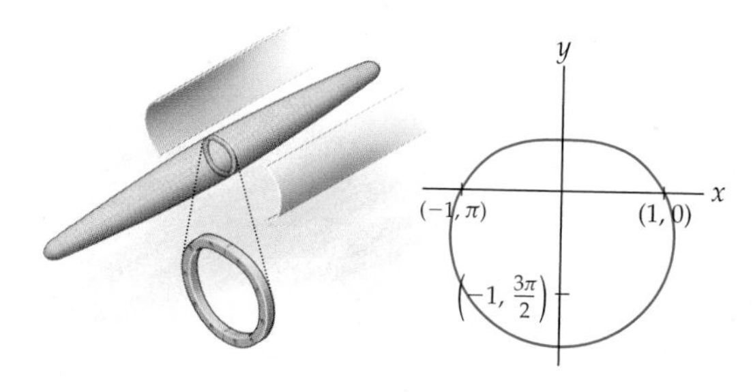

She will cover the kayak with an impermeable polyurethane fabric. The fabric comes in rolls 1 meter wide. Annie needs to use the length of the fabric to run the length of her boat, so she hopes to be able to use just two widths of the fabric to cover the boat. Will she be able to?

51. Annie's second potential design for the central cross section of her kayak is

$$r = 0.83(1 - 0.5\sin\theta)(1 + 0.2\cos^2\theta).$$

New kayak and its central cross section

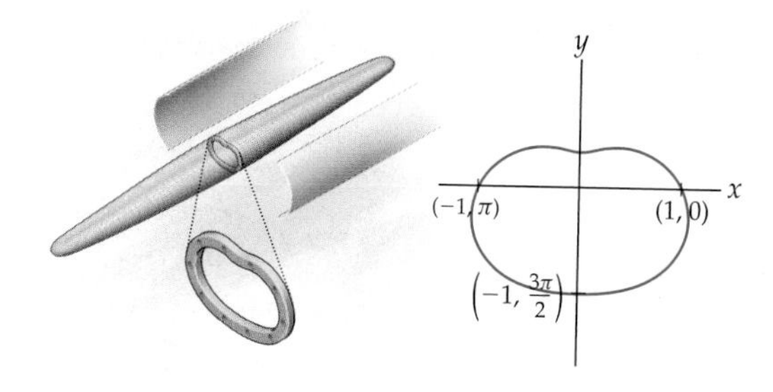

Will this design allow Annie to use just two widths of the fabric? *(Hint: Use a numerical integration technique to approximate the length of the cross section.)*

52. Ian sometimes sews his own outdoor gear. He wants to make a body-hugging climbing pack. The bottom of the pack is the area outside the circle $r = 14$, but inside $r = 14\sqrt{2}\sin\theta$, where r is measured in inches. Ian requires the pack to carry 2500 cubic inches of gear. If the pack has vertical sides, how tall does Ian need to make it?

Proofs

53. Prove that the area of a sector with central angle ϕ in a circle of radius r is given by $A = \dfrac{1}{2}\phi r^2$.

54. Prove that the area enclosed by one petal of the polar rose $r = \cos 3\theta$ is the same as the area enclosed by one petal of the polar rose $r = \sin 3\theta$.

55. Prove that the area enclosed by all of the petals of the polar rose $r = \cos 2n\theta$ is the same for every positive integer n.

56. Prove that the area enclosed by all of the petals of the polar rose $r = \sin(2n+1)\theta$ is the same for every positive integer n.

57. Prove that the part of the polar curve $r = \dfrac{1}{\theta}$ that lies inside the circle defined by the polar equation $r = 1$ has infinite length.

58. Complete the proof of Theorem 9.14 by verifying that

$$(f'(\theta)\cos\theta - f(\theta)\sin\theta)^2 + (f'(\theta)\sin\theta + f(\theta)\cos\theta)^2$$

is equal to

$$(f'(\theta))^2 + (f(\theta))^2.$$

Thinking Forward

The cardioid defined by $r = 1 + \cos\theta$ and its interior is translated one unit perpendicular to the xy-plane to define a "cylinder."

- *Volume:* Find the volume of the cylinder.
- *Surface area:* Find the surface area of the cylinder. Remember to include the top and bottom of the solid in your calculations.

9.5 CONIC SECTIONS*

- Circles, ellipses, parabolas, and hyperbolas
- Using rectangular coordinates to express conics
- Using polar coordinates to express conics

Circles and Ellipses

When the line $y = x$ is revolved about the y-axis, we obtain a double cone. The point where the two cones meet is called the ***vertex***. Any plane that does not pass through the vertex will intersect the cone in a curve called a ***conic section***. There are four basic curves that may be formed when a plane intersects this cone:

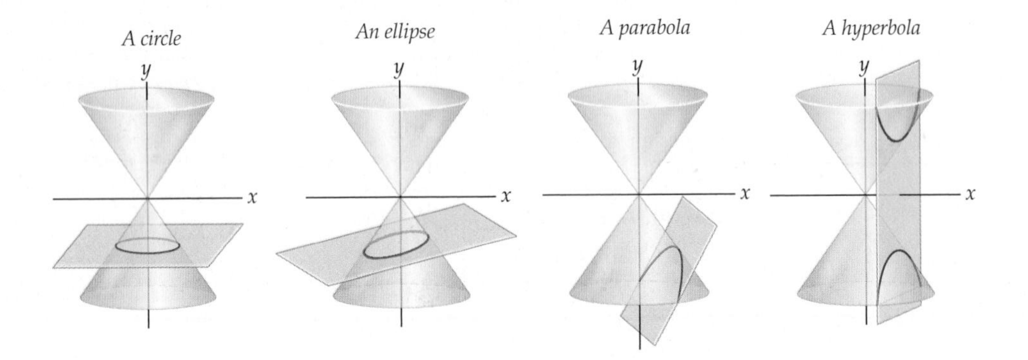

A plane through the vertex will intersect the cone in either a single point, a line, or a pair of intersecting lines. These intersections are considered to be ***degenerate*** conic sections. For the remainder of this section we will assume that our conic sections are not degenerate.

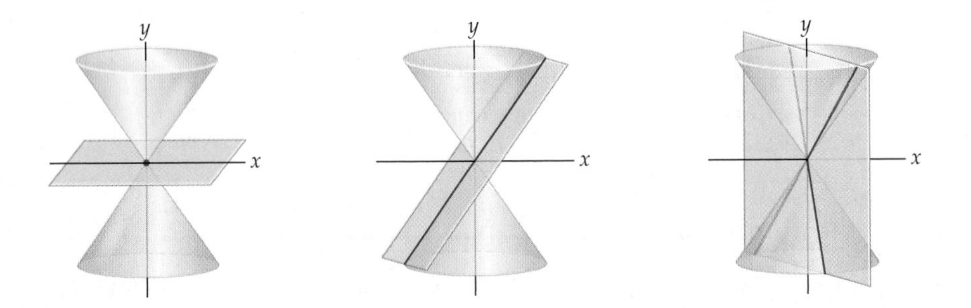

As shown, a plane perpendicular to the y-axis but that misses the vertex of the cone intersects the cone in a circle. If such a plane is tilted slightly, the curve of intersection will be an ellipse. As we know, the equation of the circle with center (x_0, y_0) and radius R can be written in the form $(x - x_0)^2 + (y - y_0)^2 = R^2$. When the circle is centered at the origin, its equation can be written in the form $x^2 + y^2 = R^2$. Such an equation can also be written in the form

$$\frac{x^2}{R^2} + \frac{y^2}{R^2} = 1.$$

If we allow the denominators of the two quotients in this equation to be different positive constants, we obtain the equation

$$\frac{x^2}{A^2} + \frac{y^2}{B^2} = 1.$$

You may know that the graph of such an equation is an ellipse.

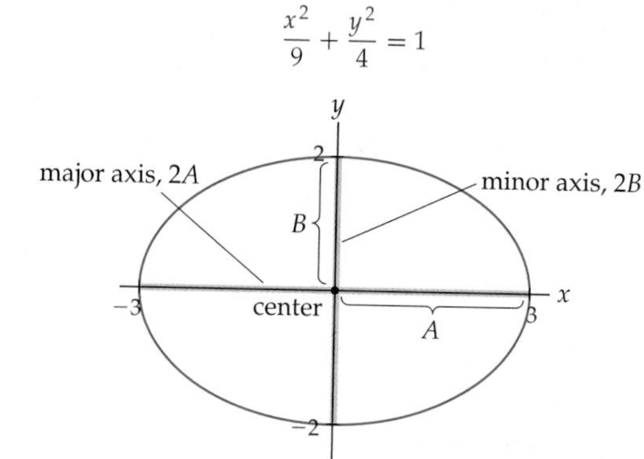

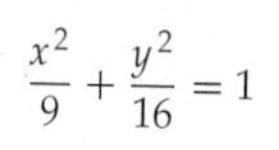

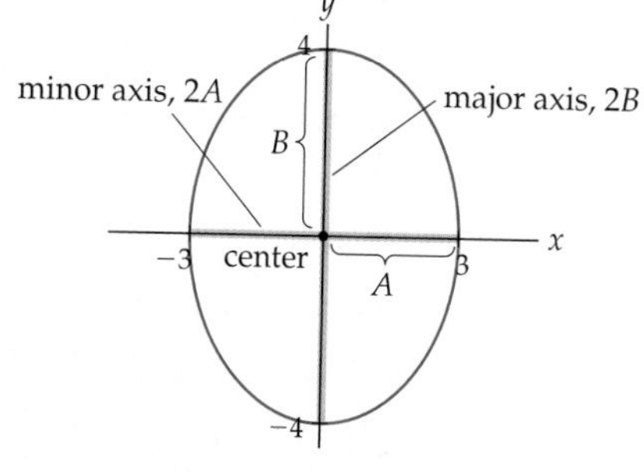

When $A > B$, the graph of the ellipse $\frac{x^2}{A^2} + \frac{y^2}{B^2} = 1$ is wider than it is tall. In this case, the segment of length $2A$ is called the ***major axis*** of the ellipse and the segment of length $2B$ is called the ***minor axis***. When $A < B$, the ellipse is taller than it is wide. In this case, $2B$ is the length of the major axis and $2A$ is the length of the minor axis. We may consider a circle to be an ellipse in which the major and minor axes are equal.

More classically, each conic section has a geometric definition. For example, the geometric definition for an ellipse is given in Definition 9.15:

DEFINITION 9.15

Ellipse

Given two points in a plane, called ***foci***, an ***ellipse*** is the set of points in the plane such that the sum of the distances to the two foci is a constant.

We will show that the curve with equation $\frac{x^2}{A^2} + \frac{y^2}{B^2} = 1$, where $A > B$, satisfies Definition 9.15. The argument when $A < B$ is similar.

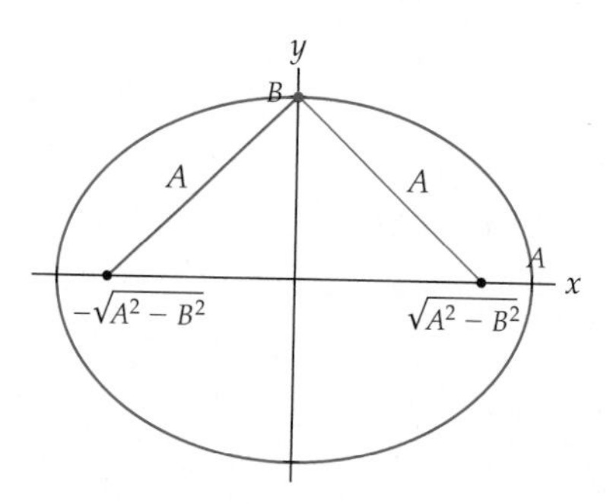

Let $\mathcal{V}$ represent the graph of the equation $\frac{x^2}{A^2} + \frac{y^2}{B^2} = 1$. From this equation, we may see that $\mathcal{V}$ is symmetric with respect to both the x- and y-axes. Furthermore, the points $(-A, 0)$, $(A, 0)$, $(0, -B)$, and $(0, B)$ are all on $\mathcal{V}$. We will show that the sum of the distances from every point on $\mathcal{V}$ to the points $(-\sqrt{A^2 - B^2}, 0)$ and $(\sqrt{A^2 - B^2}, 0)$ is the constant $2A$. This will demonstrate that $\mathcal{V}$ is an ellipse and that $(-\sqrt{A^2 - B^2}, 0)$ and $(\sqrt{A^2 - B^2}, 0)$ are the coordinates of the foci. We begin by letting $C = \sqrt{A^2 - B^2}$ and (x, y) be a point on $\mathcal{V}$. Also,

$$C^2 = A^2 - B^2 \quad \text{and} \quad y^2 = B^2 - \frac{B^2x^2}{A^2}.$$

Let D_1 represent the distance from (x, y) to the point $(C, 0)$. Thus,

$$\begin{aligned}
D_1^2 &= (x - C)^2 + y^2 \\
&= x^2 - 2Cx + C^2 + B^2 - \frac{B^2x^2}{A^2} && \leftarrow y^2 = B^2 - \frac{B^2x^2}{A^2} \\
&= A^2 - 2Cx + x^2 - \frac{B^2x^2}{A^2} && \leftarrow C^2 = A^2 - B^2 \\
&= \frac{A^4 - 2A^2Cx + (A^2 - B^2)x^2}{A^2} && \leftarrow \text{algebra} \\
&= \frac{A^4 - 2A^2Cx + C^2x^2}{A^2} && \leftarrow C^2 = A^2 - B^2
\end{aligned}$$

Since both the numerator and denominator of the last expression are perfect squares, $D_1 = \frac{A^2 - Cx}{A}$. In Exercise 55 you will be asked to perform a similar computation to show that the distance D_2 from the point (x, y) to the point $(-C, 0)$ is $D_2 = \frac{A^2 + Cx}{A}$. The sum of these distances is

$$D_1 + D_2 = \frac{A^2 - Cx}{A} + \frac{A^2 + Cx}{A} = 2A.$$

This gives us the following theorem (the proof of part (b) is similar to the preceding argument and is left for Exercise 56):

THEOREM 9.16 **The Equation and Foci of an Ellipse**

(a) If $A > B > 0$, the graph of the equation $\frac{x^2}{A^2} + \frac{y^2}{B^2} = 1$ is an ellipse with foci $(\pm\sqrt{A^2 - B^2}, 0)$.

(b) If $0 < A < B$, the graph of the equation $\frac{x^2}{A^2} + \frac{y^2}{B^2} = 1$ is an ellipse with foci $(0, \pm\sqrt{B^2 - A^2})$.

For example, the foci of the ellipse $\frac{x^2}{25} + \frac{y^2}{16} = 1$ are $(\pm\sqrt{25 - 16}, 0) = (\pm 3, 0)$. Furthermore, the sum of the distances from every point on this ellipse to the foci is 10 units, the length of the major axis.

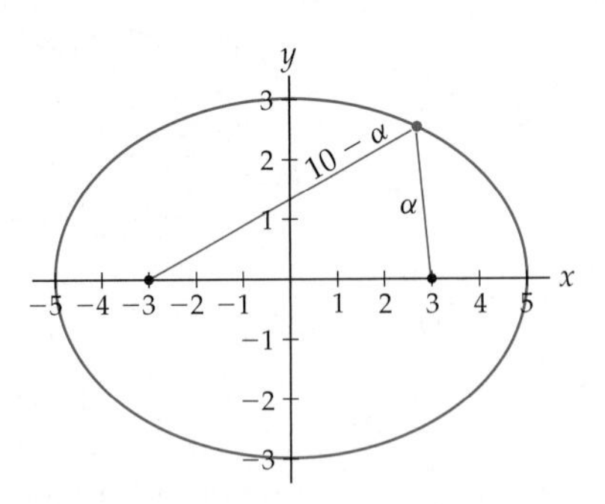

Parabolas

As a conic section, a parabola is obtained when the plane slicing a cone is parallel to a side of the cone. We studied parabolas extensively in Chapter 0. We know that the graph of the equation $y = x^2$ is an upwards-opening parabola with its vertex at the origin. We also saw that the graph of the equation $y = a(x - x_0)^2 + y_0$ is a parabola, as long as $a \neq 0$. The vertex of the new parabola is (x_0, y_0), while the constant a scales the graph of $y = x^2$ vertically. In addition, if $a < 0$, the new parabola opens downwards.

We may also define a parabola geometrically:

DEFINITION 9.17

Parabola

A ***parabola*** is the set of all points in the plane equidistant from a given fixed line in the plane, called the ***directrix*** of the parabola, and a given fixed point called the ***focus*** of the parabola.

We show that the curve $\mathcal{P}$ with equation $y = ax^2$, where $a > 0$, satisfies Definition 9.17. We will see that the focus is on the positive y-axis with coordinates $\left(0, \frac{1}{4a}\right)$ and the directrix is the horizontal line with equation $y = -\frac{1}{4a}$.

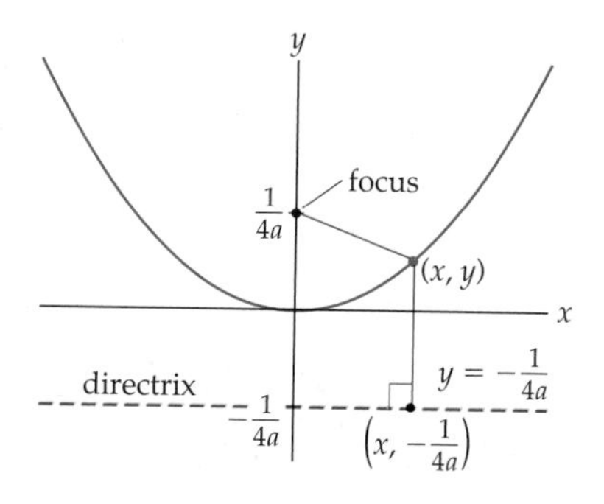

If we choose an arbitrary point (x, y) on the parabola, the distance from (x, y) to the line with equation $y = -\frac{1}{4a}$ is $y + \frac{1}{4a}$, while its distance from the point $\left(0, \frac{1}{4a}\right)$ is

$$
\begin{aligned}
\sqrt{x^2 + \left(y - \frac{1}{4a}\right)^2} &= \sqrt{x^2 + y^2 - \frac{y}{2a} + \frac{1}{16a^2}} && \leftarrow \text{algebra} \\
&= \sqrt{\frac{y}{a} + y^2 - \frac{y}{2a} + \frac{1}{16a^2}} && \leftarrow y = ax^2 \\
&= \sqrt{y^2 + \frac{y}{2a} + \frac{1}{16a^2}} && \leftarrow \text{algebra} \\
&= y + \frac{1}{4a}. && \leftarrow \text{the quantity is a perfect square}
\end{aligned}
$$

Since the distances from any point on $\mathcal{P}$ to the directrix and the focus are equal, $\mathcal{P}$ satisfies Definition 9.17. We generalize this result in the following theorem:

THEOREM 9.18

The Focus and Directrix of a Parabola

Let $y = a(x - x_0)^2 + y_0$ with $a \neq 0$ be the equation of a parabola. The coordinates of the focus of the parabola are $\left(x_0, y_0 + \frac{1}{4a}\right)$, and the equation of the directrix is $y = y_0 - \frac{1}{4a}$.

Hyperbolas

The final conic section we will discuss is the hyperbola. A hyperbola is formed when a plane intersects both halves of the double cone (in a nondegenerate way). The most familiar hyperbola is the graph of the function $y = \frac{1}{x}$, although the graphs of the equations $x^2 - y^2 = 1$ and $y^2 - x^2 = 1$ are also hyperbolas.

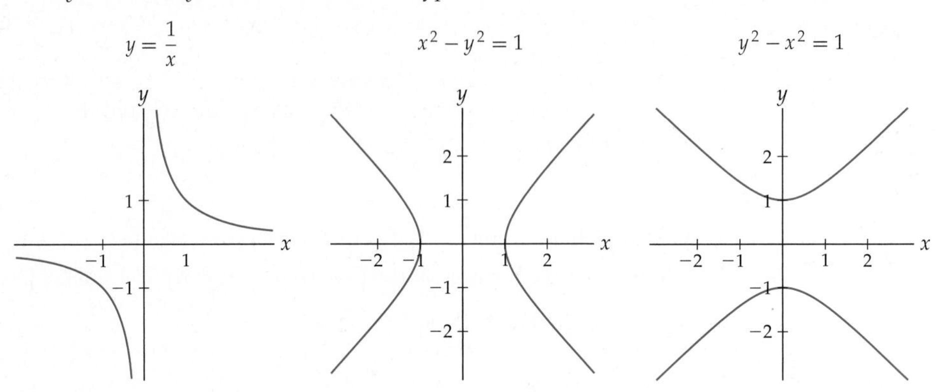

We will be discussing primarily hyperbolas with defining equations $\frac{x^2}{A^2} - \frac{y^2}{B^2} = 1$ and $\frac{y^2}{B^2} - \frac{x^2}{A^2} = 1$, where A and B are positive constants.

Hyperbolas may also be defined geometrically:

DEFINITION 9.19 **Hyperbola**

Given two points in a plane, called ***foci***, a ***hyperbola*** is the set of points in the plane for which the difference of the distances to the two foci is constant. The line containing the foci is called the ***focal axis***, the midpoint of the segment connecting the foci is called the ***center*** of the hyperbola, and the points where the focal axis intersects the hyperbola are called the ***vertices*** of the hyperbola.

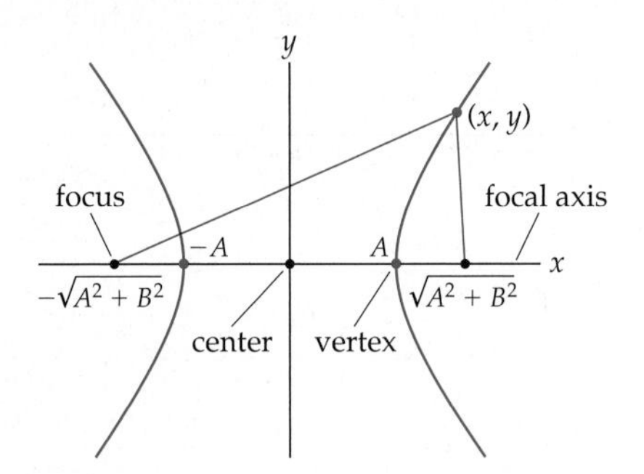

The proof of the following theorem is quite similar to the proof of Theorem 9.16 and is left for Exercises 57 and 58:

THEOREM 9.20 **The Equation, Foci, and Asymptotes of a Hyperbola**

Let A and B be positive.

(a) The graph of the equation $\frac{x^2}{A^2} - \frac{y^2}{B^2} = 1$ is a hyperbola with foci $(\pm\sqrt{A^2 + B^2}, 0)$.

(b) The graph of the equation $\frac{y^2}{B^2} - \frac{x^2}{A^2} = 1$ is a hyperbola with foci $(0, \pm\sqrt{A^2 + B^2})$.

The lines $y = \frac{B}{A}x$ and $y = -\frac{B}{A}x$ are asymptotes for these hyperbolas.

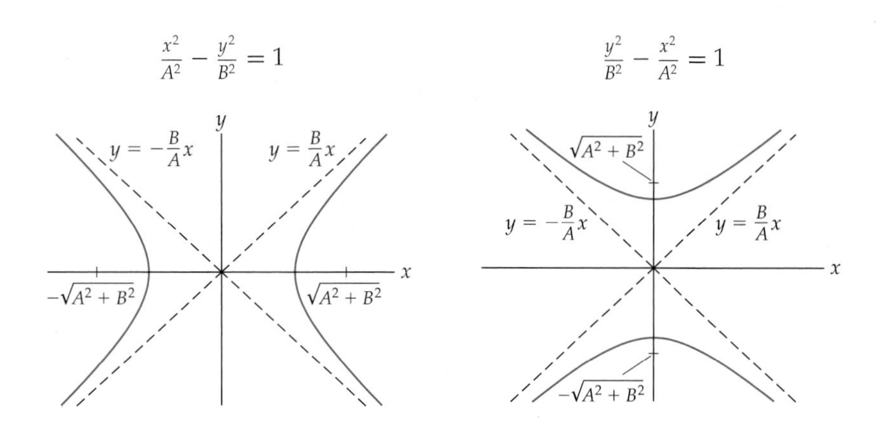

For example, since $A = 3$ and $B = 4$ in the equation $\frac{x^2}{9} - \frac{y^2}{16} = 1$, the foci of this hyperbola are $(\pm 5, 0)$ and the graph has asymptotes $y = \frac{4}{3}x$ and $y = -\frac{4}{3}x$.

Conic Sections in Polar Coordinates

In Definiton 9.17 we saw that a parabola is the set of all points in the plane equidistant from a given fixed line in the plane, where the line is called the directrix of the parabola and the fixed point is called the focus of the parabola. If we let F and l be the focus and directrix of the parabola, respectively, we may rephrase this definition by saying that a parabola is the set of all points P in the plane such that the ratio of the distance from F to P to the distance from l to P is 1. If we let FP represent the length of the segment from F to P, we have

$$\frac{FP}{\text{the distance from } l \text{ to } P} = 1.$$

We may generalize this idea to provide alternative geometric definitions for ellipses and hyperbolas. We have already seen that circles have very simple representations in polar coordinates. The alternative definitions will allow us to find simple polar coordinate representations for parabolas, ellipses, and hyperbolas. For consistency, we may orient any of these three conics as shown here:

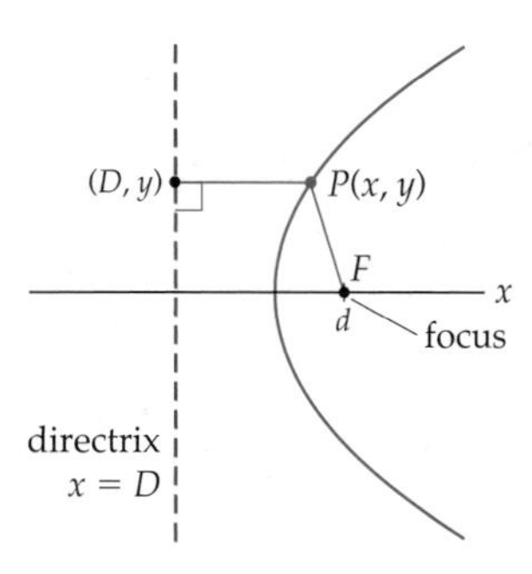

That is, we will use rightwards-opening parabolas with equations $x = c(y - y_0)^2$ for $c > 0$. For the parabola there is a single vertical directrix, l, with equation $x = D$. If we let $P = (x, y)$ be a point on the parabola, then the distance from P to l is DP and the distance from P to the focus, F, is FP. Here we have

$$\frac{FP}{DP} = 1.$$

The ellipses we consider will have equations $\frac{x^2}{A^2}+\frac{y^2}{B^2}=1$, where $A > B$ and, thus, where the major axis of the ellipse is on the x-axis. The hyperbolas we will consider are centered at the origin, open to the left and right and, therefore, have equations $\frac{x^2}{A^2}-\frac{y^2}{B^2}=1$. Ellipses and hyperbolas also have directrices. In fact, every ellipse and every hyperbola has two directrices.

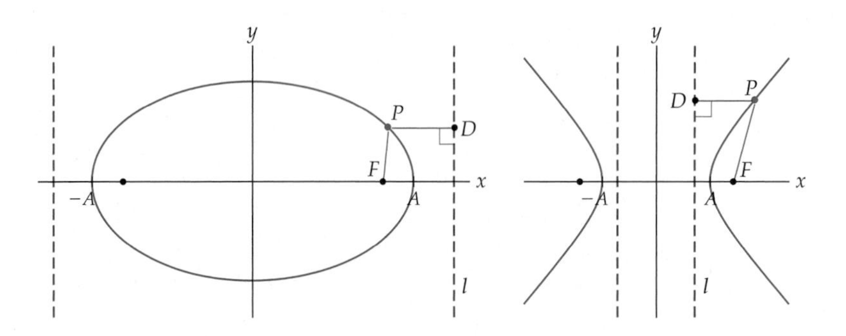

If F is the focus of an ellipse or a hyperbola $\mathcal{V}$, and D is a point on the directrix closest to that focus, then for every point P on $\mathcal{V}$, we have

$$\frac{FP}{DP}=e,$$

where e is a positive constant called the *eccentricity* of the curve. Recall that when $A > B$, the foci of the ellipse $\frac{x^2}{A^2}+\frac{y^2}{B^2}=1$ are $(\pm\sqrt{A^2-B^2},0)$ and the foci of the hyperbola $\frac{x^2}{A^2}-\frac{y^2}{B^2}=1$ are $(\pm\sqrt{A^2+B^2},0)$.

DEFINITION 9.21

The Eccentricity of Conic Sections

(a) The ***eccentricity*** e of a parabola is 1.

(b) When $A > B > 0$, the ***eccentricity*** of the ellipse $\frac{x^2}{A^2}+\frac{y^2}{B^2}=1$ is $e=\frac{\sqrt{A^2-B^2}}{A}$, and when $0 < A < B$, $e=\frac{\sqrt{B^2-A^2}}{B}$.

(c) The ***eccentricity*** of the hyperbola $\frac{x^2}{A^2}-\frac{y^2}{B^2}=1$ is $e=\frac{\sqrt{A^2+B^2}}{A}$, and for $\frac{y^2}{B^2}-\frac{x^2}{A^2}=1$, $e=\frac{\sqrt{A^2+B^2}}{B}$.

Note that from Definition 9.21 (b), $0 < e < 1$ for an ellipse, and from Definition 9.21 (c), $e > 1$ for a hyperbola. For example, the eccentricity of the ellipse $\frac{x^2}{25}+\frac{y^2}{16}=1$ is $e=\frac{\sqrt{25-16}}{5}=\frac{3}{5}$.

In Exercise 59 you will show that, for an ellipse or a hyperbola, the eccentricity is also

$$e=\frac{\text{the distance between the foci}}{\text{the distance between the vertices}}.$$

We use the eccentricity to define the directrices of ellipses and hyperbolas:

DEFINITION 9.22

Directrices of Ellipses and Hyperbolas

(a) When $A > B$, the ***directrices*** of the ellipse $\frac{x^2}{A^2} + \frac{y^2}{B^2} = 1$ are the vertical lines $x = \frac{A}{e}$ and $x = -\frac{A}{e}$, where e is the eccentricity of the ellipse. When $B > A$, the ***directrices*** of the ellipse are the horizontal lines $y = \frac{B}{e}$ and $y = -\frac{B}{e}$, where e is the eccentricity of the ellipse.

(b) The ***directrices*** of the hyperbola $\frac{x^2}{A^2} - \frac{y^2}{B^2} = 1$ are the vertical lines $x = \frac{A}{e}$ and $x = -\frac{A}{e}$, where e is the eccentricity of the hyperbola. The ***directrices*** of the hyperbola $\frac{y^2}{B^2} - \frac{x^2}{A^2} = 1$ are the horizontal lines $y = \frac{B}{e}$ and $y = -\frac{B}{e}$, where e is the eccentricity of the hyperbola.

Continuing with the brief example we started before, we see that the directrices of the ellipse $\frac{x^2}{25} + \frac{y^2}{16} = 1$ are the lines $x = \pm\frac{A}{e} = \pm\frac{5}{3/5} = \pm\frac{25}{3}$.

With these definitions for eccentricity and directrices established, we have the following theorem:

THEOREM 9.23

Eccentricity as the Ratio of Two Distances

Let F be the focus of a parabola, an ellipse, or a hyperbola, and let l be the directrix of the curve closest to F. If P is a point on the curve and D is the point on l closest to P, then

$$\frac{FP}{DP} = e.$$

The proof of Theorem 9.23 is mostly computational and is left for Exercises 60 and 61.

Note that the eccentricity of an ellipse is in the interval $(0, 1)$, the eccentricity of a parabola is 1 and the eccentricity of a hyperbola is greater than 1. We are now ready to express the equations for ellipses, parabolas, and hyperbolas in polar coordinates. We use a polar coordinate system in which the pole is at a focus, F, of the conic section. We position the parabola so that the directrix is perpendicular to the polar axis and the parabola opens to the right. For the ellipse and hyperbola, the focal axis should coincide with the polar axis. Let V represent any of these conics and P be a point on V. Let $x = u$ be the equation of the directrix closest to the pole. The following figure illustrates the situation for an ellipse:

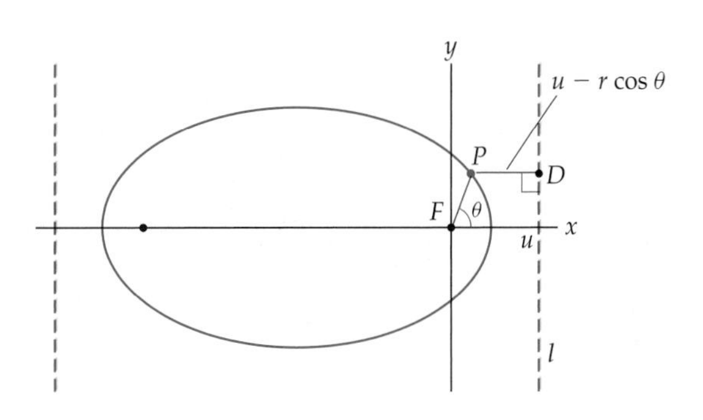

Now, the distance from the focus to the curve is $FP = r$. The distance from P to the directrix is $DP = u - r\cos\theta$. By Theorem 9.23 we have

$$\frac{FP}{DP} = \frac{r}{u - r\cos\theta} = e.$$

When we solve this equation for r, we get the following theorem:

THEOREM 9.24 **Conics in Polar Coordinates**

The graph of the polar equation

$$r = \frac{eu}{1 + e\cos\theta}$$

is a conic with eccentricity e and directrix $x = u$.

Examples and Explorations

EXAMPLE 1 **Finding the foci, eccentricity, and directrices for an ellipse**

The graph of each of the equations that follow is an ellipse. Find the foci, eccentricity, and directrices for each.

(a) $\dfrac{x^2}{16} + \dfrac{y^2}{4} = 1$ **(b)** $9x^2 + 25y^2 = 900$

(c) $\dfrac{x^2}{16} + \dfrac{y^2}{25} = 1$ **(d)** $x^2 + 4x + 9y^2 - 18y - 23 = 0$

SOLUTION

(a) Theorem 9.16 tells us that when $A > B$, the foci of the ellipse with equation $\frac{x^2}{A^2} + \frac{y^2}{B^2} = 1$ are $(\pm\sqrt{A^2 - B^2}, 0)$. Therefore, the foci of the ellipse with equation $\frac{x^2}{4^2} + \frac{y^2}{2^2} = 1$ have coordinates $(\pm\sqrt{4^2 - 2^2}, 0) = (\pm 2\sqrt{3}, 0)$.

From Definition 9.21, the eccentricity of the ellipse with equation $\frac{x^2}{A^2} + \frac{y^2}{B^2} = 1$ is $e = \frac{\sqrt{A^2 - B^2}}{A}$ when $A > B$. Here we have $e = \frac{2\sqrt{3}}{4} = \frac{\sqrt{3}}{2}$.

From Definition 9.22, the directrices of the ellipse with equation $\frac{x^2}{A^2} + \frac{y^2}{B^2} = 1$ are the vertical lines $x = \pm\frac{A}{e}$ when $A > B$. Here, the directrices have equations

$$x = \pm\frac{4}{\sqrt{3}/2} = \pm\frac{8\sqrt{3}}{3}.$$

(b) Before we use the theorem and definitions we mentioned in part (a), we need to rewrite the equation in the correct form. Here, $9x^2 + 25y^2 = 900$ is equivalent to

$$\frac{x^2}{10^2} + \frac{y^2}{6^2} = 1.$$

Therefore, the foci of this ellipse have coordinates $(\pm\sqrt{10^2 - 6^2}, 0) = (\pm 8, 0)$. The eccentricity of the ellipse is $e = \frac{8}{10} = \frac{4}{5}$, and the directrices have equations

$$x = \pm\frac{10}{4/5} = \pm\frac{25}{2}.$$

(c) Theorem 9.16 also tells us that when $A < B$, the foci of the ellipse with equation

$$\frac{x^2}{A^2} + \frac{y^2}{B^2} = 1$$

are $(0, \pm\sqrt{B^2 - A^2})$. Therefore, the foci of the ellipse with equation $\frac{x^2}{16} + \frac{y^2}{25} = 1$ have coordinates $(0, \pm\sqrt{5^2 - 4^2}) = (0, \pm 3)$.

We use Definitions 9.21 and 9.22 to find the eccentricity and directrices of this ellipse. The eccentricity will be

$$e = \frac{\sqrt{B^2 - A^2}}{B},$$

and the directrices will be horizontal lines with equations

$$y = \frac{B}{e} \quad \text{and} \quad y = -\frac{B}{e}.$$

Here, we have $e = \frac{\sqrt{5^2 - 4^2}}{5} = \frac{3}{5}$. The directrices have equations $y = \pm\frac{5}{3/5} = \pm\frac{25}{3}$.

(d) For our final equation in this group, we complete the square for each variable in the equation $x^2 + 4x + 9y^2 - 18y - 23 = 0$. After completing the squares, we have the equation

$$(x + 2)^2 + 9(y - 1)^2 = 36.$$

Dividing both sides of the equation by 36, we obtain

$$\frac{(x + 2)^2}{6^2} + \frac{(y - 1)^2}{2^2} = 1.$$

The graph of this equation is the graph of the ellipse $\frac{x^2}{6^2} + \frac{y^2}{2^2} = 1$ translated two units to the left and one unit up. The ellipse with equation $\frac{x^2}{6^2} + \frac{y^2}{2^2} = 1$ has foci $(\pm\sqrt{6^2 - 2^2}, 0) = (\pm 4\sqrt{2}, 0)$. The eccentricity of this ellipse is $e = \frac{4\sqrt{2}}{6} = \frac{2\sqrt{2}}{3}$, and the directrices have equations $x = \pm\frac{6}{2\sqrt{2}/3} = \pm\frac{9\sqrt{2}}{2}$. Therefore, the foci of the ellipse with equation $\frac{(x+2)^2}{6^2} + \frac{(y-1)^2}{2^2} = 1$ are $(-2 \pm 4\sqrt{2}, 1)$. Translating an ellipse does not change its eccentricity, but the directrices are $x = -2 \pm \frac{9\sqrt{2}}{2}$. □

EXAMPLE 2 **Finding the focus and directrix for a parabola**

Find the focus and directrix for the parabola with equation $y = 4x^2 + 16x + 65$.

SOLUTION

We begin by completing the square. Here, we have $y = 4(x + 4)^2 + 1$. By Theorem 9.18, the focus of the parabola with equation $y = a(x - x_0)^2 + y_0$ is $\left(x_0, y_0 + \frac{1}{4a}\right)$ and the equation of the directrix is $y = y_0 - \frac{1}{4a}$. Here $a = 4$, $x_0 = -4$, and $y_0 = 1$. Therefore, the focus is the point $\left(-4, 1 + \frac{1}{4\cdot 4}\right) = \left(-4, \frac{17}{16}\right)$ and the directrix has equation $y = 1 - \frac{1}{4\cdot 4} = \frac{15}{16}$. □

EXAMPLE 3 **Finding the foci, asymptotes, eccentricity, and directrices for a hyperbola**

The graph of each of the equations that follow is a hyperbola. Find the foci, asymptotes, eccentricity, and directrices for each.

(a) $\frac{x^2}{16} - \frac{y^2}{4} = 1$ **(b)** $9x^2 - 25y^2 = 900$

(c) $\frac{y^2}{25} - \frac{x^2}{16} = 1$ **(d)** $x^2 + 4x - 9y^2 - 18y - 41 = 0$

SOLUTION

(a) Theorem 9.20 tells us that the foci of the hyperbola with equation $\frac{x^2}{A^2} - \frac{y^2}{B^2} = 1$ are

$$(\sqrt{A^2+B^2}, 0) \quad \text{and} \quad (-\sqrt{A^2+B^2}, 0).$$

Therefore, the foci of the hyperbola with equation $\frac{x^2}{4^2} - \frac{y^2}{2^2} = 1$ have coordinates

$$(\pm\sqrt{4^2+2^2}, 0) = (\pm 2\sqrt{5}, 0).$$

The same theorem also tells us that the hyperbola will have asymptotes $y = \pm\frac{B}{A}x$. The asymptotes for this hyperbola have equations $y = \pm\frac{2}{4}x = \pm\frac{1}{2}x$.

From Definition 9.21, the eccentricity of the hyperbola with equation $\frac{x^2}{A^2} - \frac{y^2}{B^2} = 1$ is $e = \frac{\sqrt{A^2+B^2}}{A}$. Here we have $e = \frac{2\sqrt{5}}{4} = \frac{\sqrt{5}}{2}$.

From Definition 9.22, the directrices of the hyperbola with equation $\frac{x^2}{A^2} - \frac{y^2}{B^2} = 1$ are the vertical lines $x = \pm\frac{A}{e}$. Here, the directrices have equations $x = \pm\frac{4}{\sqrt{5}/2} = \pm\frac{8\sqrt{5}}{5}$.

(b) Before we use the theorem and definitions we mentioned in part (a), we need to rewrite the equation in the correct form. Here, $9x^2 - 25y^2 = 900$ is equivalent to $\frac{x^2}{10^2} - \frac{y^2}{6^2} = 1$. Therefore, the foci of this hyperbola have coordinates $(\pm\sqrt{10^2+6^2}, 0) = (\pm 2\sqrt{34}, 0)$. The asymptotes have equations $y = \pm\frac{6}{10}x = \pm\frac{3}{5}x$. The eccentricity of the hyperbola is $e = \frac{2\sqrt{34}}{10} = \frac{\sqrt{34}}{5}$, and the directrices have equations $x = \pm\frac{10}{\sqrt{34}/5} = \pm\frac{25\sqrt{34}}{17}$.

(c) Theorem 9.20 also tells us that the foci of the hyperbola with equation $\frac{y^2}{B^2} - \frac{x^2}{A^2} = 1$ are $(0, \pm\sqrt{A^2+B^2})$. Therefore, the foci of the hyperbola with equation $\frac{y^2}{25} - \frac{x^2}{16} = 1$ have coordinates $(0, \pm\sqrt{4^2+5^2}) = (0, \pm\sqrt{41})$. The asymptotes of a hyperbola with this type of equation are again $y = \pm\frac{B}{A}x$. Here, they have the equations $y = \pm\frac{5}{4}x$.

We use Definitions 9.21 and 9.22 to find the eccentricity and directrices of this hyperbola. The eccentricity will be

$$e = \frac{\sqrt{A^2+B^2}}{B},$$

and the directrices will be horizontal lines with equations

$$y = \frac{B}{e} \quad \text{and} \quad y = -\frac{B}{e}.$$

Thus, $e = \frac{\sqrt{4^2+5^2}}{5} = \frac{\sqrt{41}}{5}$ and the directrices have equations $y = \pm\frac{5}{\sqrt{41}/5} = \pm\frac{25\sqrt{41}}{41}$.

(d) For our final equation in this group, begin by completing the square for each variable in the equation $x^2 + 4x - 9y^2 - 18y - 41 = 0$. You should confirm that this results in the equation

$$(x+2)^2 - 9(y+1)^2 = 36.$$

Dividing both sides of the equation by 36, we obtain

$$\frac{(x+2)^2}{6^2} - \frac{(y+1)^2}{2^2} = 1,$$

whose graph is the hyperbola with equation $\frac{x^2}{6^2} - \frac{y^2}{2^2} = 1$, translated two units to the left and one unit down. The hyperbola with equation $\frac{x^2}{6^2} - \frac{y^2}{2^2} = 1$ has foci $(\pm\sqrt{6^2+2^2}, 0) = (\pm 2\sqrt{10}, 0)$. The asymptotes have equations $y = \frac{2}{6}x = \frac{1}{3}x$ and $y = -\frac{1}{3}x$. The eccentricity of this hyperbola is $e = \frac{2\sqrt{10}}{6} = \frac{\sqrt{10}}{3}$, and the directrices have equations $x = \pm\frac{6}{\sqrt{10}/3} = \pm\frac{9\sqrt{10}}{5}$.

Therefore, the foci of our hyperbola are $(-2 \pm 2\sqrt{10}, -1)$. The asymptotes are the lines $y + 1 = \frac{1}{3}(x+2)$ and $y + 1 = -\frac{1}{3}(x+2)$, or equivalently, $y = \frac{1}{3}x - \frac{1}{3}$ and $y = -\frac{1}{3}x - \frac{5}{3}$. Translating the hyperbola did not change its eccentricity, but the directrices are $x = -2 \pm \frac{9\sqrt{10}}{5}$. □

EXAMPLE 4

Using polar coordinates to graph conics

Use polar coordinates to graph each of the following equations:

(a) $r = \dfrac{1}{1+\cos\theta}$ **(b)** $r = \dfrac{2}{1+2\cos\theta}$ **(c)** $r = \dfrac{1}{2+\cos\theta}$

SOLUTION

By Theorem 9.24, the graph of the equation $r = \frac{eu}{1+e\cos\theta}$ is a conic with a focus at the pole and with eccentricity e and directrix $x = u$.

(a) For $r = \frac{1}{1+\cos\theta}$ the eccentricity is $e = 1$, so the graph is a parabola. The directrix will be the vertical line $x = 1$.

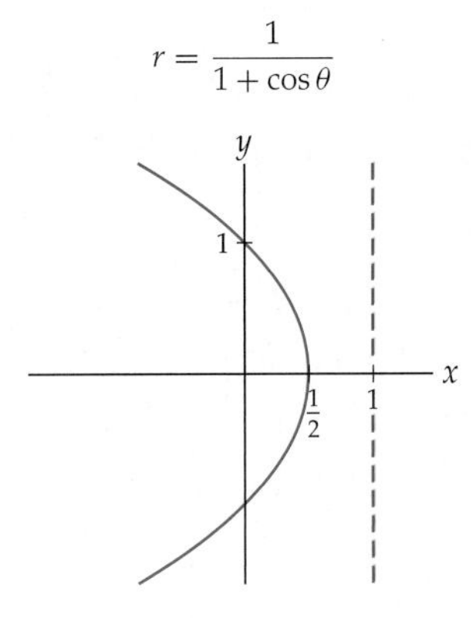

(b) For $r = \frac{2}{1+2\cos\theta}$ the eccentricity is $e = 2$, so the graph is a hyperbola. The directrix will be the vertical line $x = 1$.

$$r = \frac{2}{1 + 2\cos\theta}$$

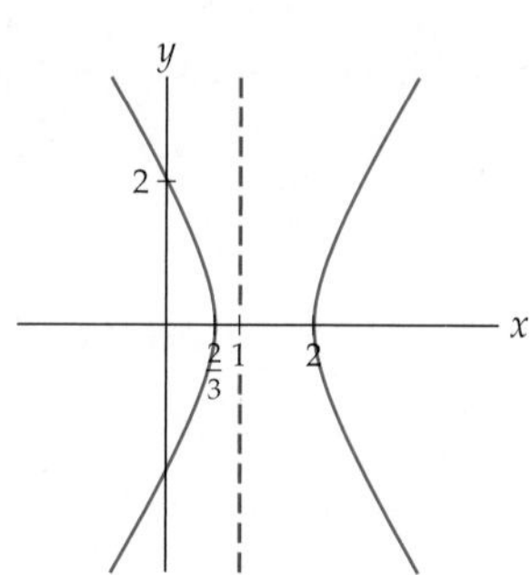

(c) Note that

$$r = \frac{1}{2 + \cos\theta} = \frac{1/2}{1 + 1/2\cos\theta}.$$

Here the eccentricity is $e = \frac{1}{2}$, so the graph is an ellipse. The directrix will be the vertical line $x = 1$.

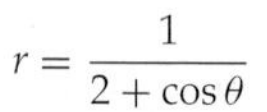

$$r = \frac{1}{2 + \cos\theta}$$

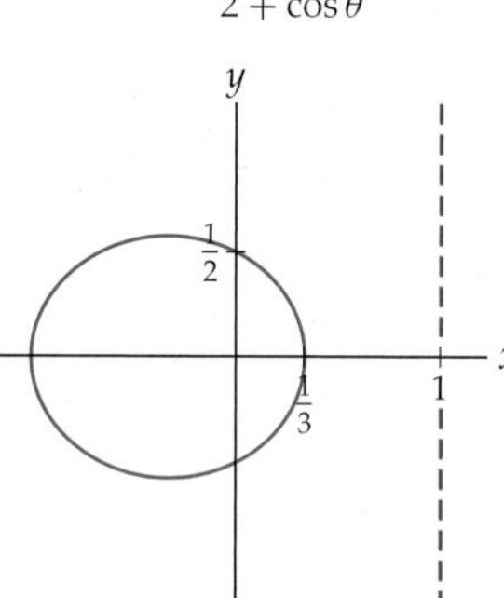

□

? TEST YOUR UNDERSTANDING

- What is a conic section? How many different types of conic sections are there? What are the degenerate conic sections? What are the nondegenerate conic sections?
- What is the definition of a circle? What is the definition of an ellipse? What are the foci of an ellipse? What are the major and minor axes of an ellipse? What are the directrices of an ellipse? How do you recognize the equation for an ellipse in rectangular coordinates?
- What is the definition of a parabola? What is the focus of a parabola? What is the directrix of a parabola? How do you recognize the equation for a parabola in rectangular coordinates?
- What is the definition of a hyperbola? What are the foci of a hyperbola? What are the directrices of a hyperbola? How do you recognize the equation for a hyperbola in rectangular coordinates?
- What are the eccentricities of the conic sections? Specifically, what is the eccentricity of a parabola? In what intervals are the eccentricities for a ellipse and a hyperbola? How do you recognize the equation for a parabola, an ellipse, or a hyperbola in polar coordinates?

EXERCISES 9.5

Thinking Back

Finding the center and radius for a circle by completing the square: Each of the equations that follow is an equation for a circle. Find the center and radius for each.

- ▶ $x^2 + y^2 - y - \frac{3}{4} = 0$
- ▶ $x^2 + 4x + y^2 - 21 = 0$
- ▶ $x^2 + 3x + y^2 + y + \frac{3}{2} = 0$

Completing the square: Find the sets of points satisfying each of the following equations.

- ▶ $x^2 - 6x + y^2 + 2y + 10 = 0$
- ▶ $x^2 - 8x + y^2 - 10y + 40 = 0$
- ▶ $x^2 - 6x + y + 40 = 0$
- ▶ $x + y^2 - 4y - 5 = 0$

Concepts

0. *Problem Zero:* Read the section and make your own summary of the material.

1. *True/False:* Determine whether each of the statements that follow is true or false. If a statement is true, explain why. If a statement is false, provide a counterexample.

(a) *True or False:* A conic section is the intersection of a double cone and a plane.

(b) *True or False:* A point is a conic section.

(c) *True or False:* An ellipse is the set of points in the plane equidistant from two distinct points.

(d) *True or False:* Given a line $\mathcal{L}$ and a point P not on $\mathcal{L}$, a parabola is the set of points in the plane equidistant from $\mathcal{L}$ and P.

(e) *True or False:* A parabola is also a hyperbola.

(f) *True or False:* If the focus of a parabola is on the directrix of the parabola, the parabola is degenerate.

(g) *True or False:* Given two distinct points in a plane, a hyperbola is the set of points in the plane for which the quotient of the distances to the two points is constant.

(h) *True or False:* The graph of the polar equation $r = \frac{1}{1+\sin\theta}$ is a parabola.

2. *Examples:* Construct examples of the thing(s) described in the following. Try to find examples that are different than any in the reading.

(a) A degenerate conic section.

(b) The equation for a hyperbola whose asymptotes are the coordinate axes.

(c) The equation for a parabola that opens to the left, in polar coordinates.

Complete the definitions in Exercises 3–5.

3. Given two points in a plane, called ________, an ***ellipse*** is the set of points in the plane for which ________.

4. Given a fixed line in the plane, called the ________, and a given fixed point called the ________, a ***parabola*** is the set of all points in the plane ________.

5. Given two points in a plane, called ________, a ***hyperbola*** is the set of points in the plane for which ________.

6. Explain why there are infinitely many different ellipses with the same foci.

7. Explain why there are infinitely many different hyperbolas with the same foci.

8. Three noncollinear points determine a unique circle. Do three noncollinear points determine a unique ellipse? If so, explain why. If not, provide three noncollinear points that are on two distinct ellipses.

9. The graph of the equation $\frac{x^2}{A^2} + \frac{y^2}{B^2} = 1$ is an ellipse for any nonzero constants A and B.

(a) If $A > B$, what is the eccentricity of the ellipse?

(b) If $A < B$, what is the eccentricity of the ellipse?

(c) Explain why the eccentricity, e, of an ellipse is always between 0 and 1.

(d) If $A > B$, what is $\lim_{A\to B} e$? What happens to the shape of the ellipse as $A \to B$?

(e) If $A > B$, what is $\lim_{A\to\infty} e$? What happens to the shape of the ellipse as $A \to \infty$?

10. The graph of the equation $\frac{x^2}{A^2} - \frac{y^2}{B^2} = 1$ is a hyperbola for any nonzero constants A and B.

(a) What is the eccentricity of the hyperbola?

(b) Explain why the eccentricity, e, of a hyperbola is always greater than 1.

(c) What is $\lim_{A\to 0} e$? What happens to the shape of the hyperbola as $A \to 0$?

(d) What is $\lim_{A\to\infty} e$? What happens to the shape of the hyperbola as $A \to \infty$?

11. When $A > B$, the directrices of the ellipse with equation $\frac{x^2}{A^2} + \frac{y^2}{B^2} = 1$ are the lines with equations $x = \pm\frac{A}{e}$, where e is the eccentricity of the ellipse. What are the equations for the directrices when $A < B$?

12. The directrices of the hyperbola with equation $\frac{x^2}{A^2} - \frac{y^2}{B^2} = 1$ are the lines with equations $x = \pm\frac{A}{e}$, where e is the eccentricity of the hyperbola. What are the equations for the directrices of the hyperbola with equation $\frac{y^2}{B^2} - \frac{x^2}{A^2} = 1$?

13. Sketch the graphs of the equations

$$r = \frac{1}{1+\cos\theta} \quad \text{and} \quad r = \frac{1}{1+\sin\theta}.$$

What is the relationship between these graphs? What is the eccentricity of each graph?

14. Sketch the graphs of the equations

$$r = \frac{2}{1+2\cos\theta} \quad \text{and} \quad r = \frac{2}{1+2\sin\theta}.$$

What is the relationship between these graphs? What is the eccentricity of each graph?

15. Sketch the graphs of the equations

$$r = \frac{2}{2+\cos\theta} \quad \text{and} \quad r = \frac{2}{2+\sin\theta}.$$

What is the relationship between these graphs? What is the eccentricity of each graph?

16. Let α, β, and γ be nonzero constants. Show that the graph of $r = \frac{\alpha}{\beta+\gamma\cos\theta}$ is a conic section with eccentricity $\left|\frac{\gamma}{\beta}\right|$ and directrix $x = \frac{\alpha}{\gamma}$.

17. Let α, β, and γ be nonzero constants. Show that the graph of $r = \frac{\alpha}{\beta+\gamma\sin\theta}$ is a conic section with eccentricity $\left|\frac{\gamma}{\beta}\right|$ and directrix $y = \frac{\alpha}{\gamma}$.

Skills

Complete the square to describe the conics in Exercises 18–21.

18. $2x^2 + 4x + y^2 - 6y - 3 = 0$

19. $2x^2 + 4x - y^2 - 6y - 23 = 0$

20. $4x + y^2 - 6y - 3 = 0$

21. $y^2 - 8y - 4x^2 - 8x - 13 = 0$

Use Cartesian coordinates to express the equations for the parabolas determined by the conditions specified in Exercises 22–31.

22. directrix $x = 3$, focus $(0, 1)$

23. directrix $y = 0$, focus $(0, 1)$

24. directrix $x = -1$, focus $(2, 5)$

25. directrix $y = -6$, focus $(2, -8)$

26. directrix $x = x_0$, focus (x_1, y_1), where $x_0 \neq x_1$

27. directrix $y = y_0$, focus (x_1, y_1), where $y_0 \neq y_1$

28. $r = \frac{3}{1+\cos\theta}$

29. $r = \frac{4}{1+\sin\theta}$

30. $r = \frac{\alpha}{1+\cos\theta}$

31. $r = \frac{\alpha}{1+\sin\theta}$

Use Cartesian coordinates to express the equations for the ellipses determined by the conditions specified in Exercises 32–37.

32. foci $(\pm 1, 0)$, major axis 4

33. foci $(\pm\sqrt{5}, 1)$, major axis 6

34. foci $(3, \pm 5)$, major axis 12

35. foci $\left(0, \pm\frac{3\sqrt{3}}{2}\right)$, minor axis 3

36. foci $(\pm\alpha, 0)$, major axis 4α

37. foci $(0, \pm\alpha)$, minor axis 2α

Use Cartesian coordinates to express the equations for the hyperbolas determined by the conditions specified in Exercises 38–43.

38. foci $(\pm 2, 0)$, directrices $x = \pm 1$

39. foci $(\pm 6, 0)$, directrices $x = \pm 1$

40. foci $(0, \pm 4)$, directrices $y = \pm 1$

41. foci $(0, \pm 4)$, directrices $y = \pm 2$

42. foci $(3, 1)$ and $(7, 1)$, directrix $x = 4$

43. foci $(3, 1)$ and $(3, 9)$, directrix $y = 4$

Use polar coordinates to graph the conics in Exercises 44–51.

44. $r = \frac{2}{1+3\cos\theta}$

45. $r = \frac{2}{1-3\cos\theta}$

46. $r = \frac{2}{1+\cos\theta}$

47. $r = \frac{5}{1+\cos\theta}$

48. $r = \frac{2}{4+\cos\theta}$

49. $r = \frac{2}{4+\sin\theta}$

50. $r = \frac{1}{1-\cos\theta}$

51. $r = \frac{2}{2+\sin\theta}$

Applications

In the late sixteenth century Johannes Kepler showed that the planets in our solar system revolve around the sun in elliptical orbits. Exercises 52–54 deal with the orbits of the Earth and Mars. These orbits are described by their eccentricities. If the coordinates for the orbit of a planet are oriented so that the major axis of the ellipse is aligned along the x-axis, then the Cartesian equation for the ellipse is $\frac{x^2}{A^2} + \frac{y^2}{B^2} = 1$, with $A > B$. Recall that in this case $2A$ is called the major axis of the ellipse. The ***semimajor*** axis has length A.

52. Show that the eccentricity satisfies the equation $B^2 = A^2(1 - e^2)$.

53. Measurements indicate that Earth's orbital eccentricity is 0.0167 and its semimajor axis is 1.00000011 astronomical units.

(a) Write a Cartesian equation for Earth's orbit.

(b) Give a polar coordinate equation for Earth's orbit, assuming that the sun is the focus of the elliptical orbit.

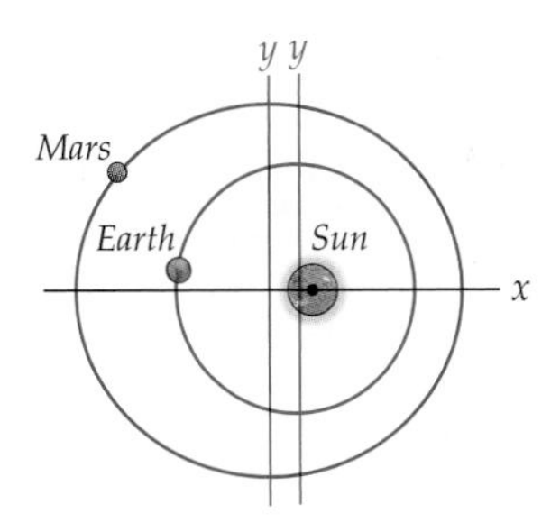

54. Measurements indicate that the orbital eccentricity of Mars is 0.0935 and its semimajor axis is 1.517323951 astronomical units.
    (a) Write a Cartesian equation for the orbit of Mars.
    (b) Do x and y have the same meaning as in Exercise 53?
    (c) Give a polar coordinate equation for the orbit of Mars, assuming that the sun is the focus of the elliptical orbit.

Proofs

55. Let $A > B > 0$. Show that the distance from any point on the graph of the curve with equation $\frac{x^2}{A^2} + \frac{y^2}{B^2} = 1$ to the point $(-C, 0)$ is $D_2 = \frac{A^2 + Cx}{A^2}$, where $C = \sqrt{A^2 - B^2}$.
56. Prove Theorem 9.16 (b). That is, show that if $0 < A < B$, the graph of the curve with equation $\frac{x^2}{A^2} + \frac{y^2}{B^2} = 1$ is an ellipse with foci $(0, \pm\sqrt{B^2 - A^2})$.
57. Prove Theorem 9.20 (a). That is, show that the graph of the equation $\frac{x^2}{A^2} - \frac{y^2}{B^2} = 1$ satisfies Definition 9.19, where the points with coordinates $(\pm\sqrt{A^2 + B^2}, 0)$ are the foci of the hyperbola.
58. Prove Theorem 9.20 (b). That is, show that the graph of the equation $\frac{y^2}{B^2} - \frac{x^2}{A^2} = 1$ satisfies Definition 9.19, where the points with coordinates $(0, \pm\sqrt{A^2 + B^2})$ are the foci of the hyperbola.
59. Prove that for an ellipse or a hyperbola the eccentricity is given by

$$e = \frac{\text{the distance between the foci}}{\text{the distance between the vertices}}.$$

In Exercises 60 and 61 we ask you to prove Theorem 9.23 for ellipses and hyperbolas.

60. Consider the ellipse with equation $\frac{x^2}{A^2} + \frac{y^2}{B^2} = 1$, where $A > B$. Let F be the focus with coordinates $(\sqrt{A^2 - B^2}, 0)$. Let $e = \frac{\sqrt{A^2 - B^2}}{A}$ and l be the vertical line with equation $x = \frac{A}{e}$. Show that for any point P on the ellipse, $\frac{FP}{DP} = e$, where D is the point on l closest to P.
61. Consider the hyperbola with equation $\frac{x^2}{A^2} - \frac{y^2}{B^2} = 1$. Let F be the focus with coordinates $(\sqrt{A^2 + B^2}, 0)$. Let $e = \frac{\sqrt{A^2 + B^2}}{A}$ and l be the vertical line with equation $x = \frac{A}{e}$. Show that for any point P on the hyperbola, $\frac{FP}{DP} = e$, where D is the point on l closest to P.

Thinking Forward

Analogs of conic sections in three dimensions: Sketch each of the following.

- Given two points in three-dimensional space, a certain type of ***ellipsoid*** may be described as the set of points for which the sum of the distances to the two points is a constant.
- A certain type of ***paraboloid*** is the set of all points in three-dimensional space equidistant from a given plane and a given fixed point not on the plane.
- Given two points in three-dimensional space, a certain type of ***hyperboloid*** may be described as the set of points for which the difference of the distances to the two points is a constant.

CHAPTER REVIEW, SELF-TEST, AND CAPSTONES

Before you progress to the next chapter, be sure you are familiar with the definitions, concepts, and basic skills outlined here. The capstone exercises at the end bring together ideas from this chapter and look forward to future chapters.

Definitions

Give precise mathematical definitions or descriptions of each of the concepts that follow. Then illustrate the definition or description with a graph or an algebraic example.

- *Parametric equations* in two variables
- The *slope* of a parametric curve $x = x(t)$, $y = y(t)$ at the point $(x(t_0), y(t_0))$
- *Symmetry with respect to the x-axis*

Theorems

Fill in the blanks to complete each of the following theorem statements:

- Let C be a curve in the plane with parametrization $x = f(t)$, $y = g(t)$ for $t \in [a, b]$ such that the parametrization is a ________ function from the interval $[a, b]$ to the curve C. If $x = f(t)$ and $y = g(t)$ are differentiable functions of t such that $f'(t)$ and $g'(t)$ are ________ on $[a, b]$, then the length of the curve C is given by ________.
- For $a \neq 0$, the graph of the equation $r = 2a\cos\theta$ is the ________ whose equation in rectangular coordinates is ________.
- For $a \neq 0$, the graph of the equation $r = 2a\sin\theta$ is the ________ whose equation in rectangular coordinates is ________.
- A graph is symmetrical with respect to the x-axis if, for every point (r, θ) on the graph, the point ________ is also on the graph.
- A graph is symmetrical with respect to the y-axis if, for every point (r, θ) on the graph, the point ________ is also on the graph.
- A graph is symmetrical with respect to the origin if, for every point (r, θ) on the graph, the point ________ is also on the graph.
- Let α and β be real numbers such that $0 \le \beta - \alpha \le 2\pi$. Let R be the region in the polar plane bounded by the rays $\theta = \alpha$ and $\theta = \beta$ and a positive continuous function $r = f(\theta)$. Then the area of region R is ________.
- Let $r = f(\theta)$ be a differentiable function of θ such that $f'(\theta)$ is continuous for all $\theta \in [\alpha, \beta]$. Furthermore, assume that $r = f(\theta)$ is a one-to-one function from $[\alpha, \beta]$ to the graph of the function. Then the length of the polar graph of $r = f(\theta)$ on the interval $[\alpha, \beta]$ is ________.

Algebraic Rules

The calculus of parametric equations: Let $x = f(t)$ and $y = g(t)$, where f and g are differentiable functions.

- $\dfrac{dy}{dx} =$ ________.
- The arc length of the curve defined by the parametric equations on the interval $[a, b]$ is________.

Conversion formulas: Fill in the blanks to convert between rectangular coordinates and polar coordinates.

- $r =$________.
- $\theta =$________.
- $x =$________.
- $y =$________.

The calculus of polar functions: Let $r = f(\theta)$, where f is a positive differentiable function.

- The area enclosed by the function f and the rays $\theta = \alpha$ and $\theta = \beta$ is ________, provided that ________.
- The arc length of the curve defined by f on the interval $[a, b]$ is ________.

Skill Certification: Parametric Equations and Polar Coordinates

Sketching parametric equations: Sketch the curves defined by the given sets of parametric equations. Indicate the direction of motion on each curve.

1. $x = t^2$, $y = t^3$, $t \in [-2, 2]$
2. $x = 3t + 1$, $y = -2t + 3$, $t \in [0, 1]$
3. $x = \sin t$, $y = \cos t$, $t \in [0, 4\pi]$
4. $x = \sin t$, $y = \cos^2 t$, $t \in [0, 4\pi]$
5. $x = \sec t$, $y = \tan t$, $t \in \left(-\frac{\pi}{2}, \frac{\pi}{2}\right)$
6. $x = \tan t$, $y = 1 + \sec^2 t$, $t \in \left(-\frac{\pi}{2}, \frac{\pi}{2}\right)$

7. $x = \sinh t$, $y = \cosh t$, $t \in \mathbb{R}$
8. $x = \sinh t$, $y = \cosh^2 t$, $t \in \mathbb{R}$

Lengths of parametric curves: Find the arc lengths of the curves defined by the parametric equations on the specified intervals.

9. $x = t^2$, $y = t^3$, $t \in [-2, 2]$
10. $x = 3t + 1$, $y = -2t + 3$, $t \in [0, 1]$
11. $x = \sin kt$, $y = \cos kt$, $t \in [0, 2\pi]$, where k is a constant
12. $x = \cosh t$, $y = t$, $t \in [0, 1]$

Graphs of polar functions: Use polar coordinates to graph each of the following functions.

13. $r = \sin\theta$
14. $r = 3\cos\theta$
15. $r = \sin 3\theta$
16. $r = 3\cos 4\theta$
17. $r = 2 - 2\sin\theta$
18. $r = 4 - 2\cos\theta$
19. $r = 2 - \sqrt{8}\sin\theta$
20. $r = 3 - 2\cos\theta$
21. $r^2 = -\sin 2\theta$
22. $r^2 = 3\cos 2\theta$

The arc length of polar functions: Find the arc lengths of the following polar functions.

23. $r = a$, where a is a positive constant, for $\theta \in [0, 2\pi]$
24. $r = a\cos\theta$, where a is a positive constant, for $\theta \in [0, \pi]$
25. $r = a\sin\theta$ for $\theta \in [0, \pi]$
26. $r = \sin\theta + \cos\theta$ for $\theta \in [0, \pi]$
27. $r = 1 - \sin\theta$ for $\theta \in [0, 2\pi]$
28. $r = a\sin\theta + b\cos\theta$, where a and b are positive constants for $\theta \in [0, \pi]$
29. $r = \sec\theta$ for $\theta \in \left[-\frac{\pi}{6}, \frac{\pi}{6}\right]$
30. $r = \csc\theta$ for $\theta \in \left[\frac{\pi}{4}, \frac{3\pi}{4}\right]$

Areas of regions bounded by polar functions: Find the areas of the following regions.

31. The area bounded by the function $r = a$, where a is a positive constant.
32. The area bounded by the function $r = a\cos\theta$, where a is a positive constant.
33. The area bounded by one petal of $r = \cos 2\theta$.
34. The area bounded by one petal of $r = \sin 3\theta$.
35. The area bounded by $r = 1 - \sin\theta$
36. The area between the inner and outer loops of $r = 1 - \sqrt{2}\cos\theta$.
37. The area bounded by one loop of $r^2 = -\sin 2\theta$.
38. The area bounded by one loop of $r^2 = 4\cos 2\theta$.
39. The area that is inside both lemniscates $r^2 = \cos 2\theta$ and $r^2 = \sin 2\theta$.
40. The area inside both polar roses $r = \sin 3\theta$ and $r = \cos 3\theta$.

Capstone Problems

A. A ***hypocycloid*** is another generalization of a cycloid in which the point tracing the path is on the circumference of a wheel, but the wheel is rolling without slipping on the inside of another wheel. If the radius of the rolling wheel is k and the radius of the fixed wheel is r, find parametric equations for the hypocycloid.

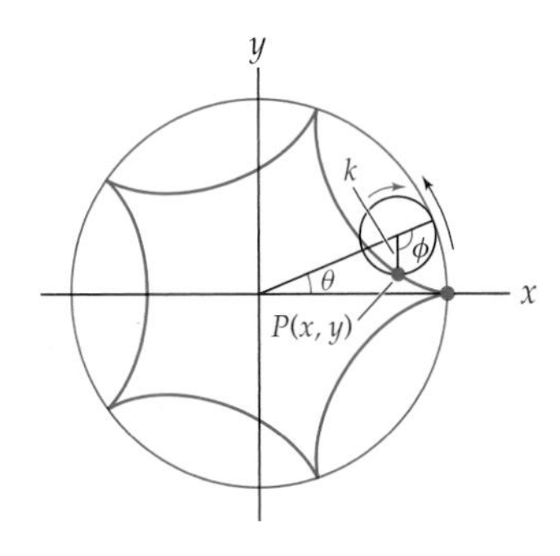

What is the path if the radius of the smaller wheel is exactly one-half the radius of the larger wheel?

B. A certain set of parametric equations has the properties that

$$x'(t) = x(t),\ y'(t) = x(t)y(t),\ x(0) = 1 \text{ and } y(0) = e.$$

(a) Solve the system of differential equations. *(Hint: Start by solving the initial-value problem $x'(t) = x(t)$ with $x(0) = 1$.)*

(b) Graph the parametric curve defined by $x = x(t)$ and $y = y(t)$ by eliminating the parameter.

C. When $k \geq 2$ is an integer, the polar graph of $r = \sin k\theta$ or $r = \cos k\theta$ is a rose.

(a) How many petals does the rose have when k is an integer?

(b) What can you say about the symmetries of either $r = \sin k\theta$ or $r = \cos k\theta$ when k is rational? Use a graphing calculator or a computer algebra system to graph several cases before answering the question.

(c) What can you say about the polar graph of $r = \sin k\theta$ or $r = \cos k\theta$ when k is irrational?

D. Let r_1 and r_2 be positive real numbers and $0 < \theta_2 - \theta_1 < \pi$. Prove that the area of the triangle with vertices $(0, 0)$, (r_1, θ_1), and (r_2, θ_2) in the polar plane is $\frac{r_1 r_2}{2}\sin(\theta_2 - \theta_1)$.

CHAPTER 10

Vectors

10.1 Cartesian Coordinates

Three-Dimensional Space in Rectangular Coordinates
An Introduction to Planes
Distances, Spheres, and Cylinders
Quadric Surfaces
Examples and Explorations

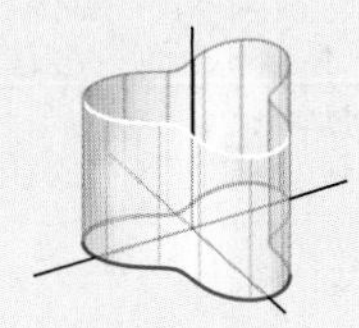

10.2 Vectors

The Algebra and Geometry of Vectors
Unit Vectors
Using Vectors to Analyze Forces
Examples and Explorations

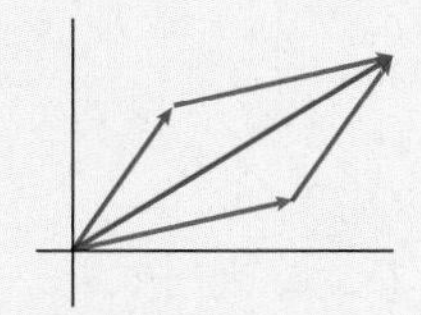

10.3 Dot Product

The Dot Product
Projections
The Triangle Inequality
Examples and Explorations

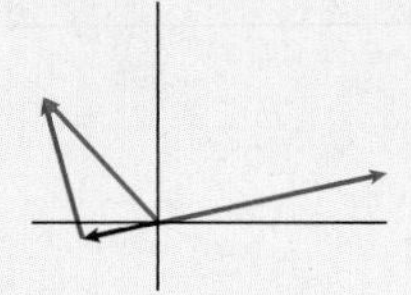

10.4 Cross Product

Determinants of 3×3 Matrices
The Cross Product
Algebraic Properties of the Cross Product
The Geometry of the Cross Product
Triple Scalar Product
Examples and Explorations

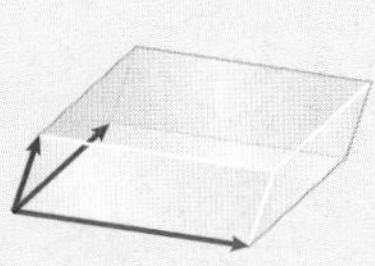

10.5 Lines in Three-Dimensional Space

Equations for Lines
Two Lines in $\mathbb{R}^3$
The Distance from a Point to a Line (Revisited)
Examples and Explorations

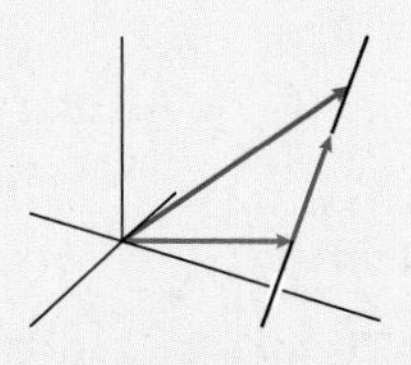

10.6 Planes

The Equation of a Plane
Geometric Conditions That Determine Planes
Distances
Intersecting Planes
Examples and Explorations

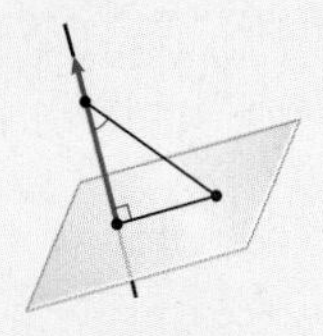

Chapter Review, Self-Test, and Capstones

10.1 CARTESIAN COORDINATES

- The three-dimensional rectangular coordinate system
- Computing distances between two points
- The equation of a sphere

Three-Dimensional Space in Rectangular Coordinates

In this chapter we will study vectors and see how they are used in the descriptions of lines and planes in three dimensions. We will see that vectors have interrelated geometric and analytic properties. By studying the connections between the geometric and algebraic properties of vectors we will be able to gain a fuller understanding of vectors, lines, and planes.

In previous chapters we used a two-dimensional coordinate system almost exclusively. Until we reached Chapter 9, we focused on rectangular or Cartesian coordinates to name the points in the plane. Beginning with this section we expand this concept to describe a three-dimensional Cartesian system. In Chapter 13 we will discuss other useful coordinate systems for three-dimensional space. To construct a three-dimensional rectangular system we choose an origin O and three mutually perpendicular coordinate axes intersecting at O. The coordinate axes are ordered and usually labeled x, y, and z. We will typically draw the system as shown here:

A three-dimensional Cartesian coordinate system

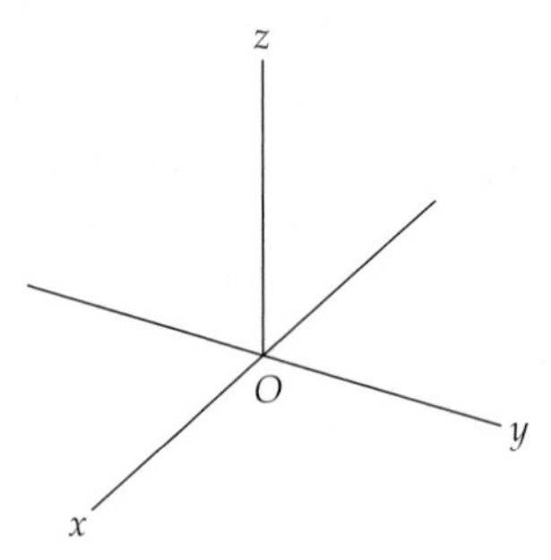

In this figure the z-axis is vertical, increasing as you go up, the y-axis increases from left to right, and the x-axis should be interpreted as pointing straight out from the page, increasing as it comes toward you. The figure shown is an example of a ***right-handed system***, because it obeys the following "right-hand rule": If the index finger of the right hand points in the positive x direction and the middle finger of the right hand points in the positive y direction, then the right thumb will naturally point in the positive z direction as the hand shown here illustrates:

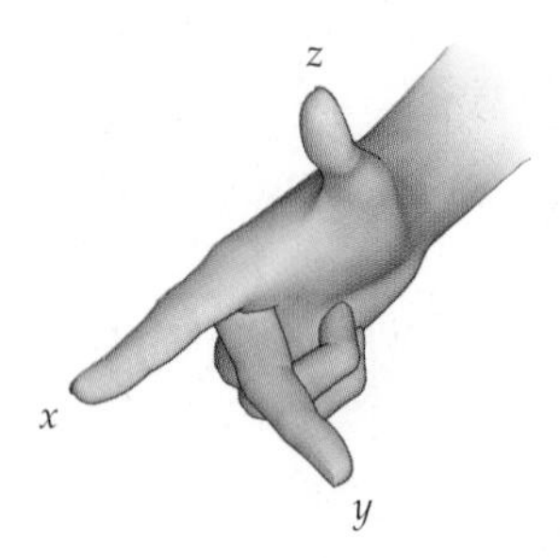

The coordinate system shown next at the left is also a right-handed system. However, there are labelings of the axes that give left-handed systems, as in the figure on the right.

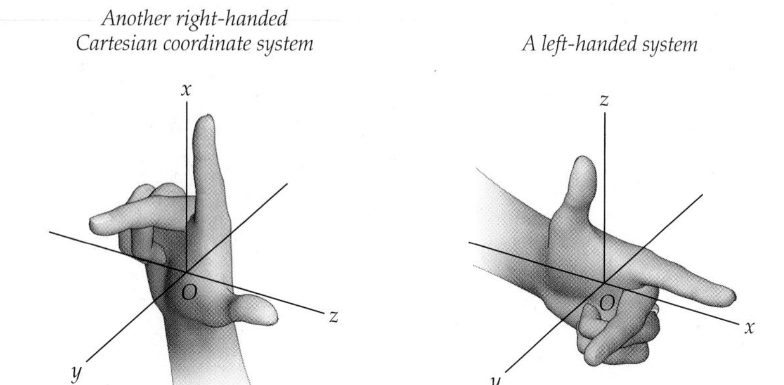

We used a different labeling of the axes when we discussed surfaces of revolution in Chapter 6. That labeling also gave us a right-handed system.

In Exercise 14 we will ask you about other possible labelings of the axes and about which labelings give right-handed systems. The notion of right-handedness will be crucial in Section 10.4, when we discuss the cross product. For now, it is just a convenient convention.

Once the axes are drawn, to complete the Cartesian system, a linear scale should be added to each axis, as is done with any two-dimensional rectangular system. Given the axes and their scales, there is a one-to-one correspondence between the points in the three-dimensional system and ordered triples of real numbers (a, b, c).

We will often use the abbreviations ***2-space*** or $\mathbb{R}^2$ (pronounced "R two") to denote two-dimensional space. Similarly, we will use ***3-space*** or $\mathbb{R}^3$ for three-dimensional space.

To locate a point (a, b, c) in a three-dimensional Cartesian coordinate system we move the appropriate number of units along each of the three coordinate axes. For example, the point $(2, 3, -5)$ is the unique point 2 units in the positive x direction from the ***yz-plane***, 3 units in the positive y direction from the ***xz-plane***, and 5 units in the negative z direction from the ***xy-plane***. The following figure below is illustrative:

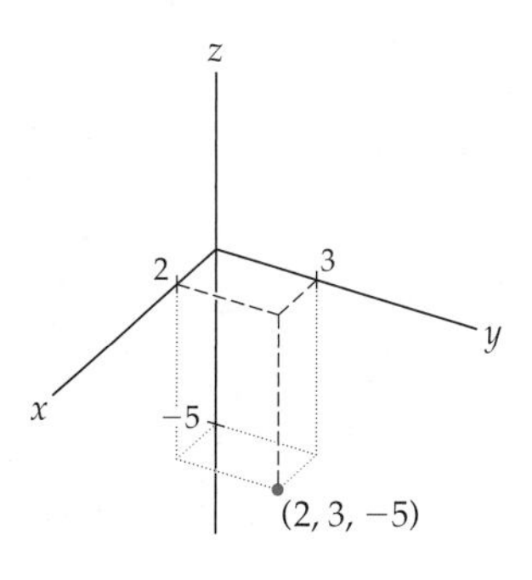

An Introduction to Planes

Two distinct intersecting lines in $\mathbb{R}^3$ determine a unique plane. In particular, when taken in pairs, the coordinate axes determine ***coordinate planes***. The x- and y-axes determine the xy-plane. Similarly, the x- and z-axes determine the xz-plane, and the y- and z-axes determine the yz-plane.

As you know, in a two-dimensional rectangular system the coordinate axes divide the plane into four parts known as quadrants. Similarly, in a three-dimensional system, the three coordinate planes divide the system into eight parts, known as ***octants***. The octant in which x, y, and z are all positive is called the ***first octant***. The other seven octants are usually not numbered; however, they can be distinguished by the signs of the coordinates.

For example, the octant determined by the inequalities $x < 0$, $y < 0$, and $z > 0$ lies at the top (since $z > 0$), left (since $y < 0$) and back (since $x < 0$) of the following system:

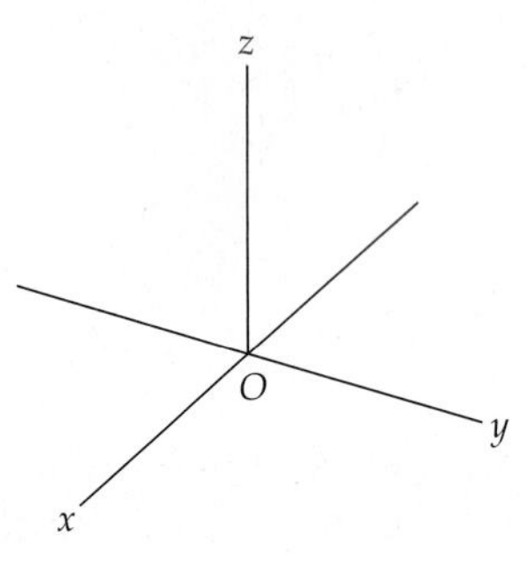

In mathematics, as in linguistics, it is important to realize that even simple statements are sensitive to the contexts in which they are made. For example, consider the mathematical sentence "$x = 3$." The three figures that follow illustrate different geometrical interpretations of this sentence. On a number line, $x = 3$ is a point, as we see on the left. In a two-dimensional system, $x = 3$ is the equation for a line parallel to the y-axis, as shown in the middle. In a three-dimensional system, the same equation $x = 3$ represents a plane parallel to the yz-plane. Every point on this plane has the property that its x-coordinate is 3.

Three interpretations of the equation $x = 3$

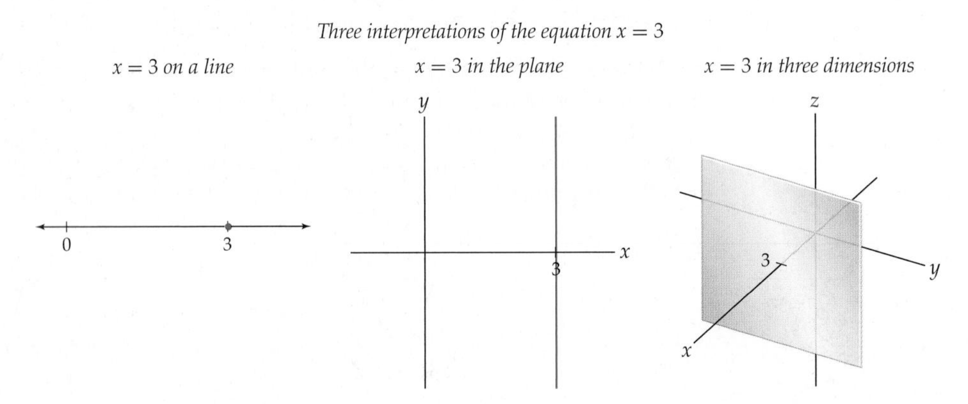

More generally, any equation of the form $x = a$, $y = b$, or $z = c$ represents a plane parallel to one of the coordinate planes. Of course there are planes in 3-space other than those parallel to the coordinate planes. We discuss planes fully in Section 10.6.

Distances, Spheres, and Cylinders

We know that in 2-space the distance between the points (x_1, y_1) and (x_2, y_2) can be computed with the Pythagorean theorem;

$$\text{distance} = \sqrt{(x_2 - x_1)^2 + (y_2 - y_1)^2}.$$

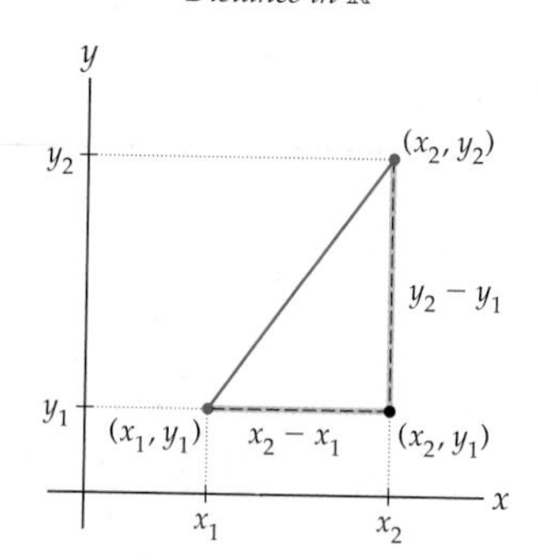

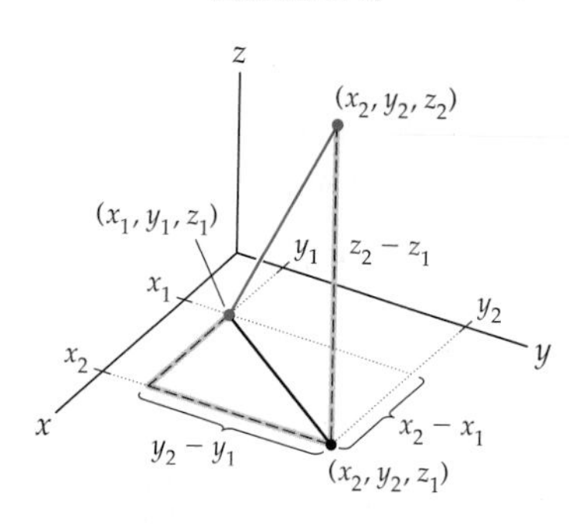

To find the distance between two points in $\mathbb{R}^3$ we need to extend this formula appropriately.

DEFINITION 10.1 The Distance Formula in $\mathbb{R}^3$

Let (x_1, y_1, z_1) and (x_2, y_2, z_2) be two points in $\mathbb{R}^3$. We define the ***distance*** between these points to be

$$\sqrt{(x_2 - x_1)^2 + (y_2 - y_1)^2 + (z_2 - z_1)^2}.$$

For example, to find the distance between the points $(2, 1, -4)$ and $(-3, 2, 1)$ we compute

$$\text{distance} = \sqrt{(2 - (-3))^2 + (1 - 2)^2 + (-4 - 1)^2} = \sqrt{25 + 1 + 25} = \sqrt{51}.$$

The distance between the points is $\sqrt{51} \approx 7.14$ units.

To find the ***midpoint*** of a line segment connecting points $P(x_1, y_1, z_1)$ and $Q(x_2, y_2, z_2)$ we may just average the corresponding coordinates. That is, the midpoint of the segment PQ is the point $\left(\frac{x_1+x_2}{2}, \frac{y_1+y_2}{2}, \frac{z_1+z_2}{2}\right)$. We ask you to prove this fact in Exercise 62. We may also use the idea of the midpoint of a segment to define what it means for two points to be symmetric about a third point. We say that $P(x_1, y_1, z_1)$ and $Q(x_2, y_2, z_2)$ are symmetric about a point $R(x_3, y_3, z_3)$ if R is the midpoint of the segment PQ. For example, to find the point Q symmetric to $P(-1, 2, 4)$ about the point $R(2, 7, -1)$, we see that Q must have coordinates $(5, 12, -6)$ so that R is the midpoint of PQ.

A circle can be defined to be the set of points in a plane at a fixed distance from a given central point. Using the distance formula, we know that the equation of a circle in the xy-plane can be written as

$$(x - x_0)^2 + (y - y_0)^2 = r^2,$$

where the center of the circle is (x_0, y_0) and the radius (distance) is r. We may use a similar equation to describe the points on a sphere.

DEFINITION 10.2 Spheres

A ***sphere*** is the set of points in three-dimensional space at a fixed distance from a given central point. The sphere with center (x_0, y_0, z_0) and radius $r > 0$ is given by the formula

$$(x - x_0)^2 + (y - y_0)^2 + (z - z_0)^2 = r^2.$$

For example, the equation of the sphere centered at $(-2, 3, 1)$ with radius 4 is

$$(x+2)^2 + (y-3)^2 + (z-1)^2 = 16.$$

In common usage a cylinder is a circle translated a finite distance in a direction out of the plane of the circle. We now define a cylinder in a more general way. The curves we translate can be any curves in the plane.

DEFINITION 10.3

Cylinders

Let C be a curve in some plane $\mathcal{P}$, and let l be a line that intersects $\mathcal{P}$, but does not lie in $\mathcal{P}$. A ***cylinder*** is the set of all points in $\mathbb{R}^3$ that are on lines parallel to l that intersect C. The curve C is called the ***directrix*** of the cylinder. The lines in the cylinder parallel to l are called ***rulings*** of the cylinder.

In the following examples each directrix is a curve in the xy-plane and the rulings are parallel to the z-axis:

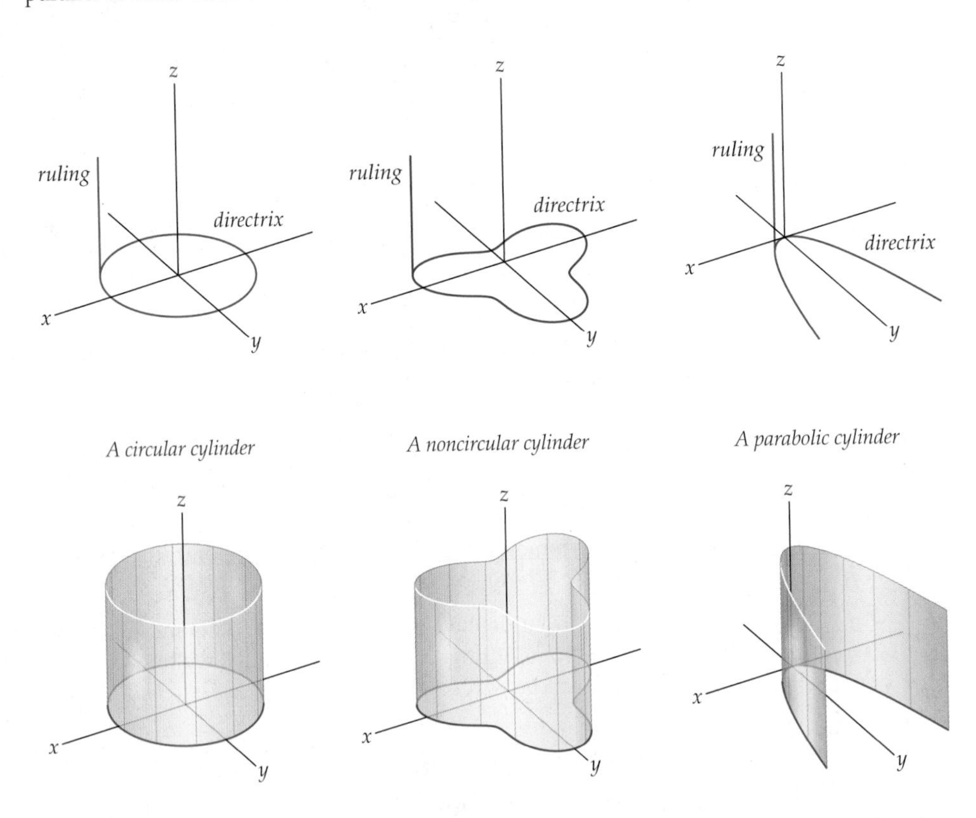

Most of the cylinders we will consider have rulings that are perpendicular to the plane containing the directrix. These cylinders are said to be ***right*** cylinders. The preceding examples are all right cylinders. As we mentioned earlier in this section, a given equation can have different graphs depending upon the context. When we consider the equation $y = x^2$ in $\mathbb{R}^2$, we see that the graph is the familiar parabola. But when we consider the graph of this equation in $\mathbb{R}^3$, we observe that it is the right parabolic cylinder shown in the preceding, rightmost figure. In fact, the graph of any curve given by a function $y = f(x)$ is a right cylinder when considered in $\mathbb{R}^3$. (Think of translating the graph in the direction of the z-axis.)

Quadric Surfaces

A conic section can be thought of as the graph of a second-degree polynomial in two variables,

$$Ax^2 + By^2 + Cxy + Dx + Ey + F = 0,$$

where $A, B, C, \ldots, F$ are constants. ***Quadric surfaces*** are the three-dimensional analogs of conic sections. A quadric surface is the graph of a second-degree polynomial in three variables; its equation is

$$Ax^2 + By^2 + Cz^2 + Dxy + Exz + Fyz + Gx + Hy + Iz + J = 0,$$

where $A, B, C, \ldots, J$ are constants. We will not need to consider this most general form for quadric surfaces because, under suitable translations and rotations, every quadric surface can be written in one of the following two forms:

$$Cz = Ax^2 + By^2 \quad \text{and} \quad Cz^2 = Ax^2 + By^2 + D.$$

If any of the constants in either of the preceding equations is zero, the corresponding quadric surface is said to be ***degenerate***. The only type of degenerate quadric surface we will mention in detail is the one we get when the constant $D = 0$ in the second equation. The graphs of these surfaces are the cones we discuss next. All other degenerate quadric surfaces are right cylinders of conic sections. For example, if we let $A = C = 1$ and $B = 0$ in the equation $Cz = Ax^2 + By^2$, we have $z = x^2$. The graph of this equation in $\mathbb{R}^3$ is the right parabolic cylinder with directrix $z = x^2$ shown here:

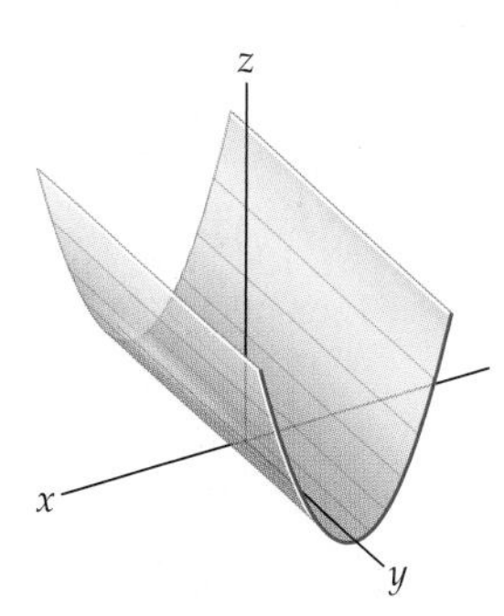

The cones and all nondegenerate quadric surfaces can be written in one of the following six forms for positive constants a, b, and c, possibly after another rotation:

1. Elliptic Cones: $z^2 = \frac{x^2}{a^2} + \frac{y^2}{b^2}$
2. Ellipsoids: $\frac{x^2}{a^2} + \frac{y^2}{b^2} + \frac{z^2}{c^2} = 1$
3. Hyperboloids of One Sheet: $\frac{x^2}{a^2} + \frac{y^2}{b^2} - \frac{z^2}{c^2} = 1$
4. Hyperboloids of Two Sheets: $\frac{x^2}{a^2} + \frac{y^2}{b^2} - \frac{z^2}{c^2} = -1$
5. Elliptic Paraboloids: $z = \frac{x^2}{a^2} + \frac{y^2}{b^2}$
6. Hyperbolic Paraboloids: $z = \frac{x^2}{a^2} - \frac{y^2}{b^2}$

To help us understand these surfaces we will look at:

- the ***intercepts***, the places where the graphs cross the coordinate axes,
- the ***traces***, the intersections with the coordinate planes, and
- the ***sections***, intersections with arbitrary planes parallel to the coordinate planes.

1. **Elliptic Cones**: $z^2 = \frac{x^2}{a^2} + \frac{y^2}{b^2}$.

   The only place where the graph intersects one of the coordinate axes is the origin. The horizontal sections are ellipses with equations $c^2 = \frac{x^2}{a^2} + \frac{y^2}{b^2}$. The trace in the xz-plane is the pair of intersecting lines $z = \pm\frac{x}{a}$. Similarly, the trace in the yz-plane is the pair of intersecting lines $z = \pm\frac{y}{b}$. The following graph is one such elliptic cone:

   The elliptic cone $z^2 = \frac{x^2}{4} + \frac{y^2}{9}$

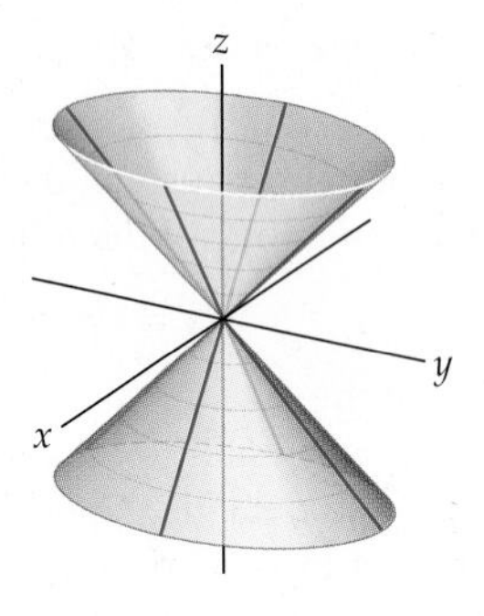

2. **Ellipsoids**: $\frac{x^2}{a^2} + \frac{y^2}{b^2} + \frac{z^2}{c^2} = 1$.

   An ellipsoid is a three-dimensional analog of an ellipse. The x-intercepts are $(\pm a, 0, 0)$, the y-intercepts are $(0, \pm b, 0)$, and the z-intercepts are $(0, 0, \pm c)$. To find the trace in the xy-plane we set $z = 0$ and see that we obtain the ellipse with equation $\frac{x^2}{a^2} + \frac{y^2}{b^2} = 1$. Similarly, the traces in the xz-plane and yz-plane are the ellipses $\frac{x^2}{a^2} + \frac{z^2}{c^2} = 1$ and $\frac{y^2}{b^2} + \frac{z^2}{c^2} = 1$, respectively.

   The ellipsoid $\frac{x^2}{9} + \frac{y^2}{16} + \frac{z^2}{4} = 1$

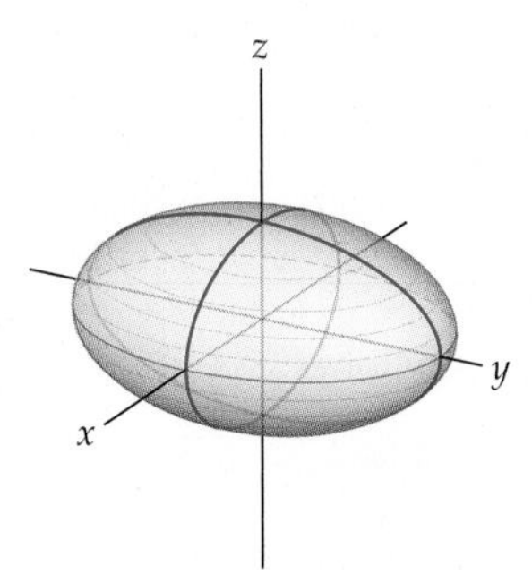

3. **Hyperboloids of One Sheet**: $\frac{x^2}{a^2} + \frac{y^2}{b^2} - \frac{z^2}{c^2} = 1$.

   The x-intercepts are $(\pm a, 0, 0)$ and the y-intercepts are $(0, \pm b, 0)$. There are no z-intercepts, since the equation $-\frac{z^2}{c^2} = 1$ has no solutions. The trace in the xy-plane is the ellipse with equation $\frac{x^2}{a^2} + \frac{y^2}{b^2} = 1$. Similarly, every section parallel to the xy-plane is an ellipse, and the size of these ellipses increases with distance from the xy-plane.

   When we set $y = 0$ to find the trace in the xz-plane, we see that we obtain a hyperbola with equation $\frac{x^2}{a^2} - \frac{z^2}{c^2} = 1$. Similarly, the trace in the yz-plane is the hyperbola with equation $\frac{y^2}{b^2} - \frac{z^2}{c^2} = 1$.

The hyperboloid $\frac{x^2}{9} + \frac{y^2}{16} - \frac{z^2}{4} = 1$

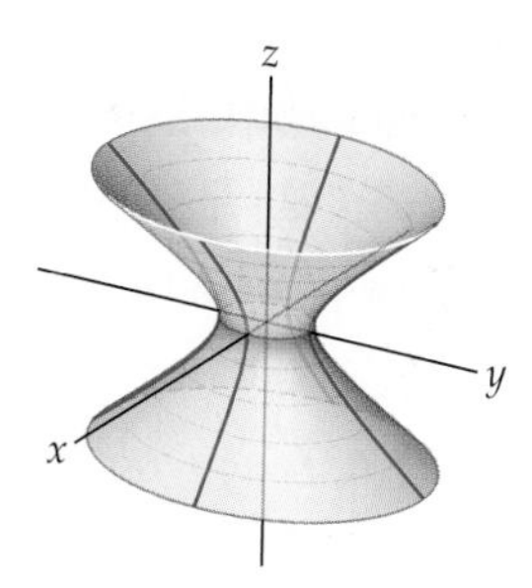

4. **Hyperboloids of Two Sheets**: $\frac{x^2}{a^2} + \frac{y^2}{b^2} - \frac{z^2}{c^2} = -1$.

   As the name implies, the graph of a hyperboloid of two sheets consists of two pieces. There are no x- or y-intercepts because, when $z = 0$, the equation $\frac{x^2}{a^2} + \frac{y^2}{b^2} = -1$ has no real solutions. We can see that there are z-intercepts at $(0, 0, \pm c)$ and that all points on the hyperboloid have the property that $z \geq c$ or $z \leq -c$. For these values of z, the sections parallel to the xy-plane are all ellipses. The trace in the xz-plane is the hyperbola with equation $\frac{x^2}{a^2} - \frac{z^2}{c^2} = -1$. Similarly, the trace in the yz-plane is the hyperbola with equation $\frac{y^2}{b^2} - \frac{z^2}{c^2} = -1$.

The hyperboloid $\frac{x^2}{9} + \frac{y^2}{16} - \frac{z^2}{4} = -1$

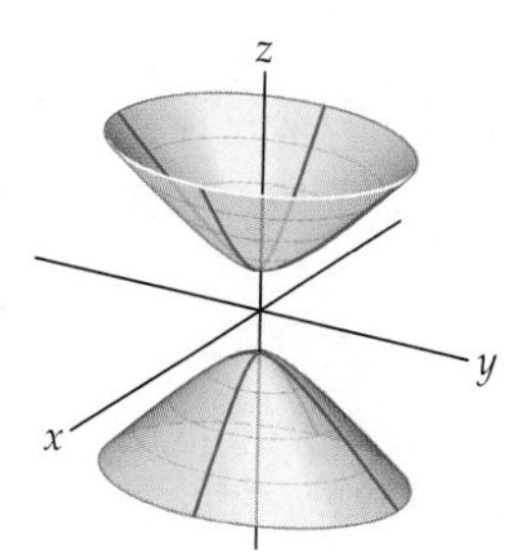

5. **Elliptic Paraboloids**: $z = \frac{x^2}{a^2} + \frac{y^2}{b^2}$.

   The only place where the graph of this elliptic paraboloid intersects any of the coordinate axes is the origin. The trace in the xz-plane is the parabola with equation $z = \frac{x^2}{a^2}$, and the trace in the yz-plane is the parabola with equation $z = \frac{y^2}{b^2}$. When $a = b$, the graph is the surface of revolution formed when the graph of $z = \frac{x^2}{a^2}$ (in the xz-plane) is revolved around the z-axis.

The elliptic paraboloid $z = \frac{x^2}{4} + \frac{y^2}{9}$

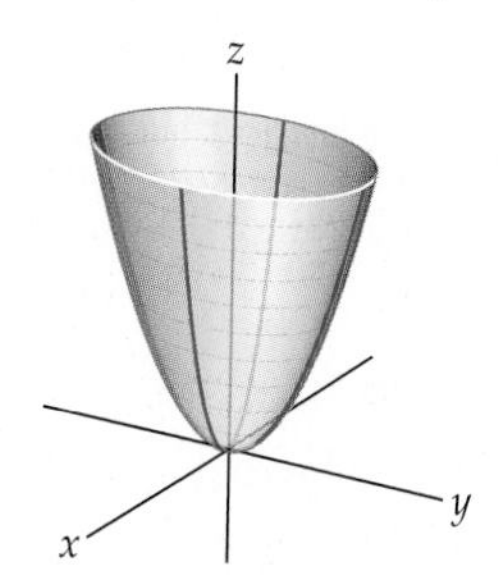

6. **Hyperbolic Paraboloids**: $z = \frac{x^2}{a^2} - \frac{y^2}{b^2}$.

   Again, the only place where this paraboloid intersects any of the coordinate axes is at the origin. The trace in the xy-plane is the pair of intersecting lines $y = \pm\frac{b}{a}x$. The trace in the xz-plane is the (upwards-opening) parabola with equation $z = \frac{x^2}{a^2}$, and the trace in the yz-plane is the (downwards-opening) parabola with equation $z = -\frac{y^2}{b^2}$.

The hyperbolic paraboloid $z = \frac{x^2}{4} - \frac{y^2}{9}$

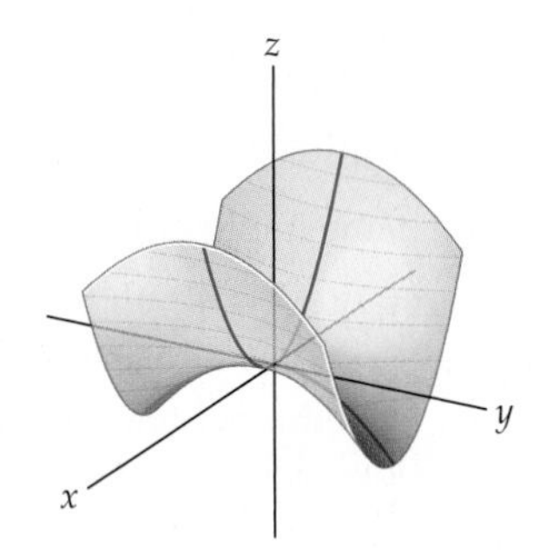

Examples and Explorations

EXAMPLE 1

Graphing a cylinder

Graph the cylinder in $\mathbb{R}^3$ defined by $z = \sin y,\ 0 \le y \le 2\pi$.

SOLUTION

We begin by graphing the sine curve in the yz-plane, as shown next on the left, and then translate this graph, using the x-axis as a ruling, to obtain the graph on the right.

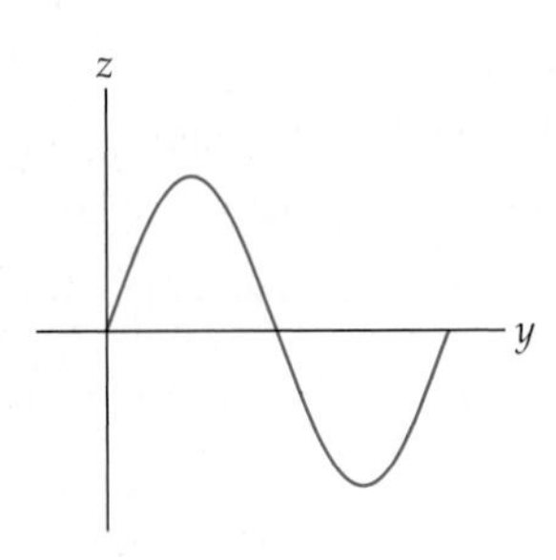

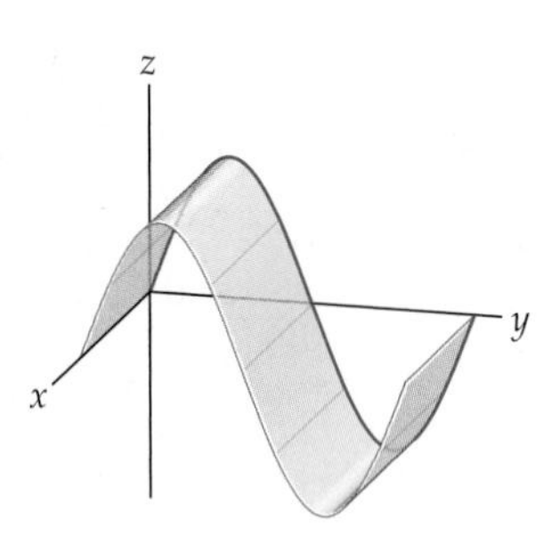

□

EXAMPLE 2

Inscribing a cube in a sphere

Find the side length of the largest cube that can be inscribed in a sphere of radius 1.

SOLUTION

The largest cube that can be inscribed in the sphere has a main diagonal that is a diameter of the sphere.

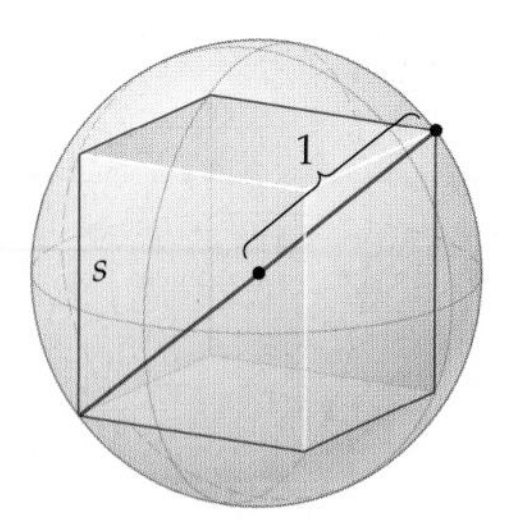

Therefore, the main diagonal of the cube will have length 2 units. If we let s be the length of each side of the cube, then, by the distance formula, we have $s^2 + s^2 + s^2 = 2^2$, or equivalently, $3s^2 = 4$. So, each side of the cube should measure $\frac{2}{3}\sqrt{3}$ units. □

EXAMPLE 3

Inscribing a sphere between a cube and a sphere

Find the radius of the largest sphere that can be inscribed between the cube and sphere from Example 2.

SOLUTION

The following figure shows the sphere whose diameter we wish to find, along with the radius of the larger sphere:

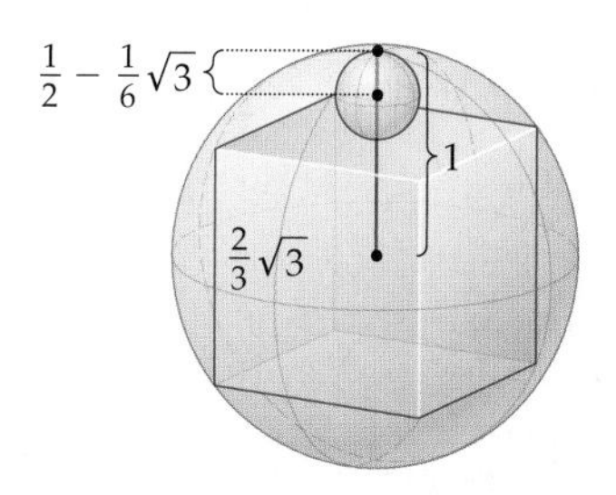

The large sphere has radius 1 and each side of the cube measures $\frac{2}{3}\sqrt{3}$ units. Therefore, the portion of this radius that extends above the cube has length $1 - \frac{1}{2} \cdot \frac{2}{3}\sqrt{3} = 1 - \frac{1}{3}\sqrt{3}$ units. This is the diameter of the smaller sphere. Therefore the radius of the smaller sphere is $\frac{1}{2} - \frac{1}{6}\sqrt{3}$ units. □

? TEST YOUR UNDERSTANDING

- What is the difference between a right-handed coordinate system and a left-handed coordinate system?
- What are the coordinate planes in three-dimensional space? How are they drawn?
- How are the distances between two points computed in two-dimensional space and three-dimensional space? How could the distance formula be generalized to find the distance between two points in n-dimensional space?
- What are the graphical representations of the equations $x = a$ and $y = b$ in $\mathbb{R}^2$? What are the graphical representations of the equations $x = a$, $y = b$, and $z = c$ in 3-space?
- How do you write an equation of a sphere with a given center and radius?

EXERCISES 10.1

Thinking Back

Lines parallel to the coordinate axes: Provide equations for the specified lines.

- The line parallel to the x-axis and containing the point (π, e).
- The line parallel to the y-axis and containing the point (π, e).
- The line parallel to the y-axis and containing the point $(0, 5)$.

Circles in $\mathbb{R}^2$: Provide equations for the specified circles.

- The circle in $\mathbb{R}^2$ with center $(3, 6)$ and tangent to the x-axis.
- The circle in $\mathbb{R}^2$ with center $(3, 6)$ and tangent to the y-axis.
- The circle in $\mathbb{R}^2$ with center $(-4, 3)$ and containing the point $(5, -2)$.

Concepts

0. *Problem Zero:* Read the section and make your own summary of the material.

1. *True/False:* Determine whether each of the statements that follow is true or false. If a statement is true, explain why. If a statement is false, provide a counterexample.

(a) *True or False:* The distance between two distinct points can be zero.

(b) *True or False:* In the Cartesian plane the equation $x = 5$ represents a line.

(c) *True or False:* In three-dimensional space the equation $x = 5$ represents a line.

(d) *True or False:* In four-dimensional space the equation $x = 5$ represents a plane.

(e) *True or False:* The equation

$$x^2 + y^2 + z^2 + 2x - 4y - 10z + 50 = 0$$

represents a sphere with center $(-1, 2, 5)$.

(f) *True or False:* If a sphere in $\mathbb{R}^3$ has its center in the first octant and is tangent to each of the coordinate planes, then its center is at the point (c, c, c) for some constant c.

(g) *True or False:* When two distinct spheres intersect, they intersect in either a point or a circle.

(h) *True or False:* Three noncollinear points in $\mathbb{R}^3$ determine a unique plane.

2. *Examples:* Construct examples of the thing(s) described in the following. Try to find examples that are different than any in the reading.

(a) A plane parallel to the yz-plane.

(b) A sphere tangent to the xy-plane.

(c) A sphere tangent to all three coordinate planes.

3. Consider the equations $y = 5$ and $x = -3$.

(a) What do these equations represent in a two-dimensional system?

(b) What do these equations represent in a three-dimensional system?

4. Consider the equation $x + 2y = 4$.

(a) What does this equation represent in a two-dimensional system?

(b) What does this equation represent in a three-dimensional system?

5. What are the coordinates of the vertices of a cube with side length 2, whose center is at the origin, and whose faces are parallel to the coordinate planes?

6. The sides of a $2 \times 3 \times 4$ rectangular solid are parallel to the coordinate planes. The coordinates of four of its vertices are $(1, -2, 3)$, $(-1, -2, -1)$, $(-1, 1, 3)$, and $(1, -2, 3)$. What are the coordinates of the other four vertices?

7. What is the definition of a sphere?

8. What is the definition of a cylinder? What is the directrix? What is a ruling?

9. Consider the equation $x^2 + y^2 = 4$.

(a) What does this equation represent in a two-dimensional system?

(b) What does this equation represent in a three-dimensional system?

10. Consider the equation $z = y^2$.

(a) What does this equation represent in the yz-plane?

(b) What does this equation represent in a three-dimensional system?

11. What point is symmetric about the origin to the point $(5, -6, 7)$?

12. What point is symmetric to the point $(-1, 3, 6)$ with respect to the xy-plane?

13. What point is symmetric to the point $(3, -7, -4)$ with respect to the plane $z = 1$?

14. Find all of the x, y, z labelings of the axes in the diagram that follows. Determine which of your labelings are right-handed systems and which are left-handed systems.

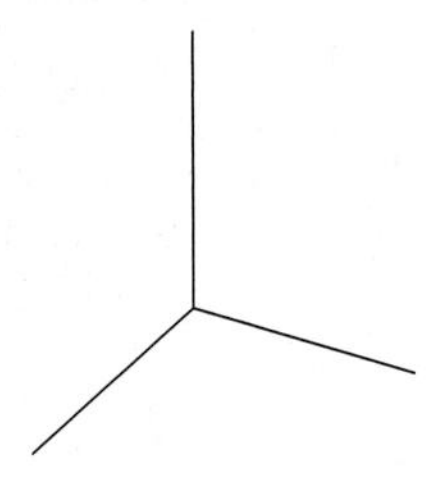

15. Show that exchanging two of the axes labels on a right-handed system creates a left-handed system. What does exchanging two pairs of axes labels do (for example, exchanging x and y and then exchanging the "new" y and z)?

16. Sketch the point $(2, 3, 4)$ in a three-dimensional Cartesian coordinate system, and then answer the following questions:
    (a) Do the coordinates $(2, 3, 4)$ represent a unique point in a three-dimensional Cartesian coordinate system?
    (b) Are there any other coordinates (a, b, c) that would have the same location as $(2, 3, 4)$ in your graph?
    (c) How far from the xy-plane is the point $(2, 3, 4)$?
    (d) How far from the xz-plane is the point $(2, 3, 4)$?
    (e) How far from the yz-plane is the point $(2, 3, 4)$?
    (f) How far from the x-axis is the point $(2, 3, 4)$?
    (g) How far from the origin is the point $(2, 3, 4)$?
    (h) How far from the point $(1, -2, 3)$ is the point $(2, 3, 4)$?
    (i) What is the equation of the sphere with center $(2, 3, 4)$ and passing through the point $(1, -2, 3)$?
    (j) What is the equation of the plane parallel to the xy-plane and that contains the point $(2, 3, 4)$?
    (k) What is the equation of the plane parallel to the xz-plane and that contains the point $(2, 3, 4)$?
    (l) What is the equation of the plane parallel to the yz-plane and that contains the point $(2, 3, 4)$?

17. How does the Pythagorean theorem generalize to higher dimensions? In particular, how would you compute the distance between two points in four-dimensional space? Five-dimensional space? n-dimensional space?

18. The points in the first octant satisfy the inequalities $x > 0$, $y > 0$, $z > 0$. For each of the other seven octants, find a set of inequalities that describes the points in the octant. Use the inequalities to label the octants of the following right-handed coordinate system:

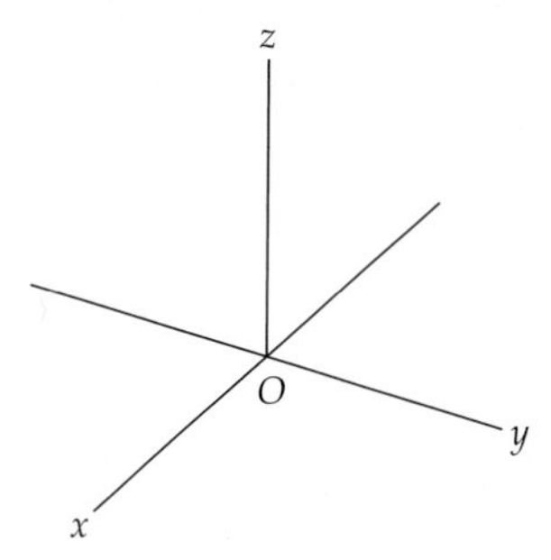

19. Use inequalities to describe each of the following sets, and sketch the regions:
    (a) The region in the first octant above the plane $z = 3$.
    (b) The region inside the sphere with center $(1, 2, 4)$ and radius 5.
    (c) The region inside the right circular cylinder with directrix $x^2 + y^2 - 4x + 6y - 12 = 0$.

20. Sketch the right circular cylinders:
$$x^2 + y^2 = 1,\ x^2 + z^2 = 4,\ \text{and } y^2 + z^2 = 9.$$

Skills

Find the distance between the given pair of points in Exercises 21–28.

21. $(-2, 3)$ and $(5, -6)$
22. $(4, 0)$ and $(-5, 12)$
23. $(1, 4, 7)$ and $(-2, 3, 5)$
24. $(3, 0, -1)$ and $(2, -8, 0)$
25. $(-1, 4, -3)$ and $(-4, 3, 1)$
26. $(4, 5, 8, -2)$ and $(-1, 3, -3, 6)$
27. $(-1, 3, 5, 2, 0)$ and $(0, 6, 1, -2, 3)$
28. $(0, 0, \ldots, 0)$ and $(1, 1, \ldots, 1)$ in n-space

In Exercises 29–35 find an equation of a sphere with the specified characteristics.

29. center $(3, -2, 5)$ and radius 5
30. center $(4, -2, -3)$ and radius 3
31. center $(2, 5, -7)$ and tangent to the xy-plane
32. center $(2, 5, -7)$ and tangent to the yz-plane
33. center $(2, 5, -7)$ and containing the origin
34. containing the point $(1, 4, 7)$ and whose center is $(-2, 3, 5)$
35. containing the point $(3, 0, -1)$ and whose center is $(2, -8, 0)$

Use the midpoint formula to find the equations of the spheres in Exercises 36 and 37.

36. the sphere in which the segment with endpoints $(3, -2, 6)$ and $(5, 6, 4)$ is a diameter
37. the sphere in which the segment with endpoints $(6, -1, 4)$ and $(-3, 3, 1)$ is a diameter

In Exercises 38 and 39 find the center and radius of the sphere with the given equation.

38. $x^2 + y^2 + z^2 - 2x + 6y - 8z + 17 = 0$
39. $x^2 + y^2 + z^2 + 3y - 5z + 3 = 0$
40. Find the equations of the intersections of the sphere
$$x^2 + y^2 + z^2 - 2x + 6y - 8z + 17 = 0$$
with each of the coordinate planes.

Graph the quadric surfaces given by the equations in Exercises 41–48.

41. $x^2 = \frac{y^2}{9} + \frac{z^2}{9}$

42. $z^2 = \frac{x^2}{9} + \frac{y^2}{25}$

43. $x^2 + y^2 + 1 = z^2$

44. $x^2 + y^2 - z^2 = 1$

45. $z = \frac{x^2}{9} + \frac{y^2}{25}$

46. $9x^2 + 16y^2 + 16z^2 = 144$

47. $z = x^2 - y^2$

48. $z = x^2 + y^2$

49. A circle is inscribed in a square so that each side of the square is tangent to the circle. A smaller circle is inscribed in the square so that this circle is tangent to two sides of the square and is tangent to the larger circle, as shown in the following figure:

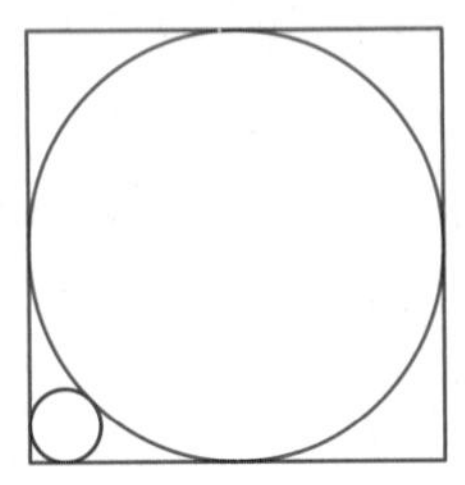

What is the ratio of the radius of the smaller circle to the radius of the larger circle?

50. A sphere is inscribed in a cube so that each face of the cube is tangent to the sphere. A smaller sphere is inscribed in the cube so that this sphere is tangent to three sides of the cube and is tangent to the larger sphere, as shown in the following figure:

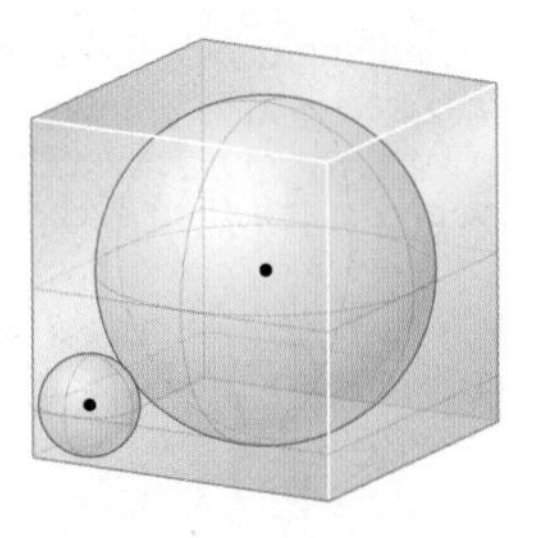

What is the ratio of the radius of the smaller sphere to the radius of the larger sphere? (*Hint: Try Exercise 49 first.*)

51. Show that the triangle with vertices $(5, 4, -1)$, $(3, 6, -1)$, and $(3, 4, 1)$ is equilateral.

52. Show that the triangle with vertices $(1, 2, -2)$, $(-3, 2, -6)$, and $(-3, 6, -2)$ is equilateral.

53. Show that the points $(1, 5, 0)$, $(3, 8, 6)$, and $(7, -7, 4)$ are the vertices of a right triangle and find its area.

A ***regular tetrahedron*** is a solid with four faces in which each face is an equilateral triangle of the same size. In Exercises 54 and 55 you are asked some basic questions about regular tetrahedra.

54. (a) Show that the three points $(1, 0, 0)$, $(0, 1, 0)$, and $(0, 0, 1)$ are the vertices of an equilateral triangle.

(b) Determine the two values of a so that the four points $(1, 0, 0)$, $(0, 1, 0)$, $(0, 0, 1)$, and (a, a, a) are the vertices of a regular tetrahedron.

55. Find the equations of the spheres that circumscribe the two tetrahedra you determined in Exercise 54 (b). (*Hint: The center of the sphere is the point (x_0, y_0, z_0), where x_0 is the mean of the x-coordinates of the four vertices of the tetrahedron, etc.*)

56. Show that the six points $(1, 0, 0)$, $(-1, 0, 0)$, $(0, 1, 0)$, $(0, -1, 0)$, $(0, 0, 1)$, and $(0, 0, -1)$ form the vertices of a regular octahedron. (A ***regular octahedron*** is an eight-sided solid in which each face is an equilateral triangle of the same size, and in which four triangles come together at each vertex.)

57. Find the volume of the octahedron given in Exercise 56. (*Hint: Recall that the volume of a pyramid is $\frac{1}{3}$ · height · area of base.*)

Applications

58. Ian is doing a high traverse. One morning he looks at the map and notes that if he considers his camp to be at the origin, then his objective is at $(5.9, 3.3, -0.37)$. All distances are in miles.

(a) How far away is his objective, as the crow flies?

(b) In order to reach his objective, Ian has to go over a high pass that lies at $(4.2, 4.4, 0.15)$ relative to his camp. Find a more realistic estimate of how far he has to go to his objective than that from part (a).

59. Annie is in a kayak in the middle of a channel between Orcas Island and Blakely Island, two of the San Juan Islands in Washington State. She knows from readings she has taken in the past that if she considers the town of Deer Harbor to be at the origin, then the summit of Constitution Peak is at $(7.9, 4.0)$, while the summit of Blakely Peak is at $(9.3, -3.4)$. All distances are given in miles. She pulls out her compass and finds that the summit of Constitution Peak is 15 degrees east of north from her, while the summit of Blakely Peak is at 100 degrees from north. How far is Annie from Deer Harbor?

60. The Subaru reflecting telescope on Mauna Kea, Hawaii, has a mirror in the shape of a circular paraboloid with a diameter of 8.3 meters and a focal length of 15 meters. While there are some tricks to how that focal length plays out in practice, if we put the center of the telescope at the origin, pointed straight up, then the effective focus would be at $(0, 0, p)$, where p satisfies $4pz = x^2 + y^2$, with all distances given in meters. How high above the xy-plane is the edge of the telescope?

A reflecting telescope

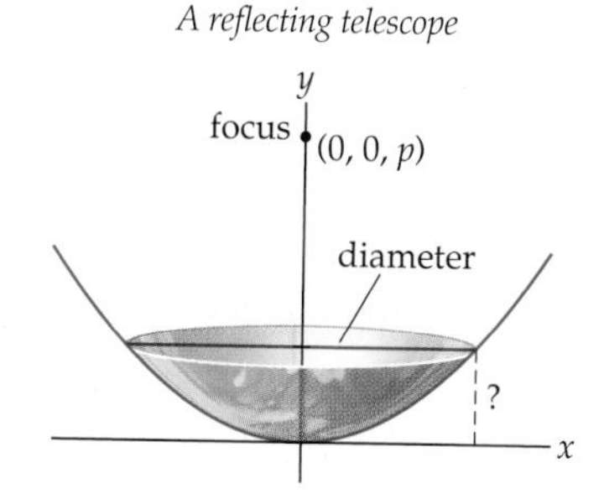

61. The Hyper Potato Chip Company makes potato chips in the shape of hyperbolic paraboloids. Each chip satisfies the equation $z = \frac{x^2}{a^2} - \frac{y^2}{a^2}$ when the center of the chip is placed at the origin. The height of each Hyper chip is 0.16 inch, the length is 2.6 inches, and the width is 1.6 inches. Find the parameter a and write the equation of a Hyper chip.

A Hyper chip

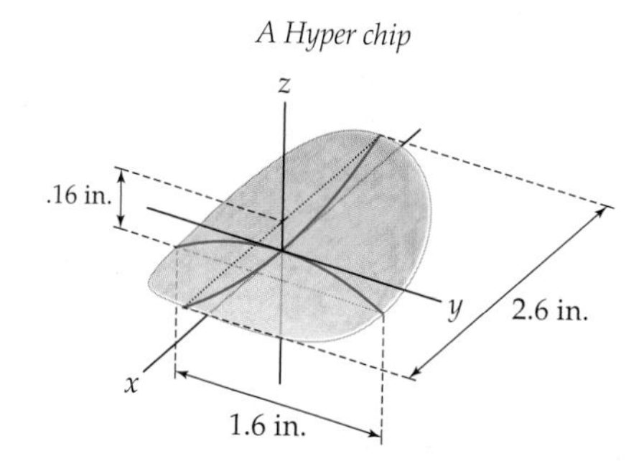

Proofs

62. Prove that the midpoint of the line segment connecting the point (x_1, y_1, z_1) to the point (x_2, y_2, z_2) is $\left(\frac{x_1+x_2}{2}, \frac{y_1+y_2}{2}, \frac{z_1+z_2}{2}\right)$.

63. Given any edge $\mathcal{E}$ of a tetrahedron, there is exactly one edge $\mathcal{E}'$ that does not share a face with $\mathcal{E}$. We will call $\mathcal{E}$ and $\mathcal{E}'$ ***opposite*** edges of the tetrahedron. Prove that the line segments connecting the midpoints of the opposite edges of a regular tetrahedron bisect each other.

64. Prove that the midpoint of the line segment connecting the points $(x_1, x_2, \ldots, x_n)$ and $(y_1, y_2, \ldots, y_n)$ in $\mathbb{R}^n$ is $\left(\frac{x_1+y_1}{2}, \frac{x_2+y_2}{2}, \ldots, \frac{x_n+y_n}{2}\right)$.

Thinking Forward

Hyperspheres: A four-dimensional ***hypersphere*** is the set of all points in $\mathbb{R}^4$ that are the same distance from a given central point.

- ▶ What is the equation of the hypersphere with center $(1, 2, 3, 4)$ and passing through the point $(2, 5, 3, -4)$?
- ▶ What is the equation of the hypersphere in which the segment with endpoints $(1, 2, 3, 4)$ and $(2, 5, 3, -4)$ is a diameter?
- ▶ Hyperspheres can be defined in $\mathbb{R}^n$ for any integer $n > 3$. How would you define a hypersphere in $\mathbb{R}^n$?

10.2 VECTORS

- Vectors in $\mathbb{R}^2$ as ordered pairs and in $\mathbb{R}^3$ as ordered triples
- The geometry of vectors
- Using vectors to analyze forces

The Algebra and Geometry of Vectors

The mass and temperature of an object are examples of physical quantities that can be represented adequately with a single numeric value, a magnitude known as a ***scalar***. Other quantities, such as displacements, velocities, and forces, need two measures—a magnitude *and* a direction—to represent them. Such quantities are represented by ***vectors***.

DEFINITION 10.4 Vectors in $\mathbb{R}^2$ and $\mathbb{R}^3$

A ***vector*** in $\mathbb{R}^2$ is an ordered pair $\langle x, y \rangle$ subject to the following operations of addition and scalar multiplication:

(a) The ***sum*** of vectors $\langle x_1, y_1 \rangle$ and $\langle x_2, y_2 \rangle$ is given by

$$\langle x_1, y_1 \rangle + \langle x_2, y_2 \rangle = \langle x_1 + x_2, y_1 + y_2 \rangle.$$

(b) The ***scalar multiple*** of vector $\langle x, y \rangle$ by a real number c, a ***scalar***, is given by

$$c\langle x, y \rangle = \langle cx, cy \rangle.$$

Similarly, a ***vector*** in $\mathbb{R}^3$ is an ordered triple $\langle x, y, z \rangle$ subject to the following operations of addition and scalar multiplication:

(c) The ***sum*** of vectors $\langle x_1, y_1, z_1 \rangle$ and $\langle x_2, y_2, z_2 \rangle$ is given by

$$\langle x_1, y_1, z_1 \rangle + \langle x_2, y_2, z_2 \rangle = \langle x_1 + x_2, y_1 + y_2, z_1 + z_2 \rangle.$$

(d) The ***scalar multiple*** of vector $\langle x, y, z \rangle$ by scalar c is given by

$$c\langle x, y, z \rangle = \langle cx, cy, cz \rangle.$$

We will often use boldface type to denote vectors; for example $\mathbf{u} = \langle 1, -5 \rangle$ is a vector in $\mathbb{R}^2$ and $\mathbf{v} = \langle \pi, e, -0.3 \rangle$ is a vector in $\mathbb{R}^3$. The entries of a vector are referred to as ***components***. Thus, for the preceding vector $\mathbf{v}$, the x-component is π, the y-component is e, and the z-component is -0.3.

We will (almost) exclusively deal with vectors in $\mathbb{R}^2$ and $\mathbb{R}^3$; however, it should be clear that vectors with n components may be defined in an analogous fashion, as a set of ordered n-tuples with an addition and a scalar multiplication.

There are two geometric models that we may use for vectors in $\mathbb{R}^2$ and $\mathbb{R}^3$. First, we may think of the vectors as ***position vectors.*** In this model we interpret the vector $\langle a, b \rangle$ from $\mathbb{R}^2$ as an arrow in the plane whose ***initial point*** is at the origin and whose ***terminal point*** is (a, b). Similarly, we interpret the vector $\langle a, b, c \rangle$ from $\mathbb{R}^3$ as an arrow in 3-space whose initial point is at the origin and whose terminal point is (a, b, c).

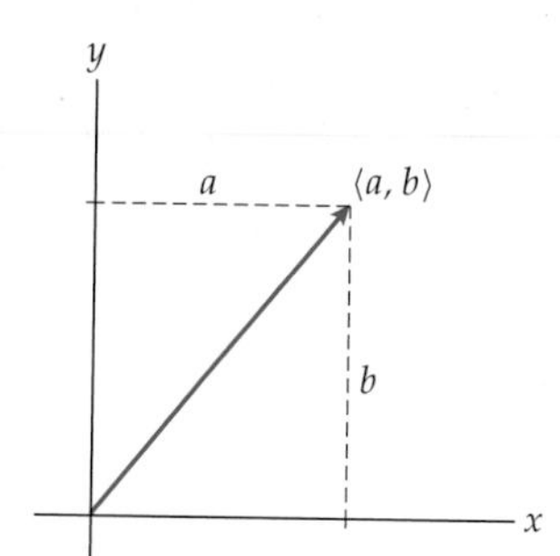

The position vector $\langle a, b \rangle$ *in* $\mathbb{R}^2$

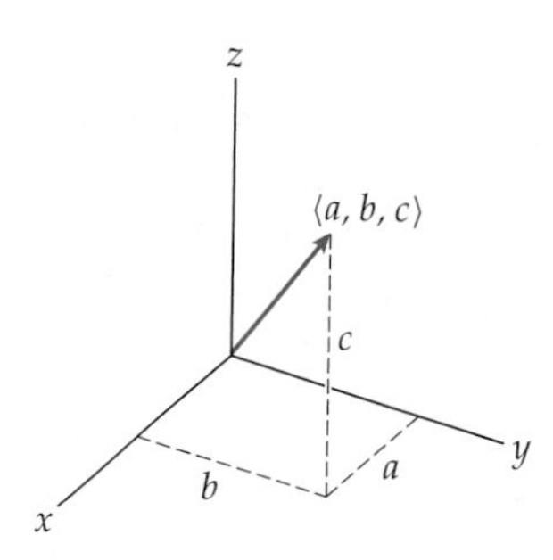

The position vector $\langle a, b, c \rangle$ *in* $\mathbb{R}^3$

Our second geometric model is a variation of the preceding one. We may think of a vector $\langle a, b \rangle$ as any translation of the the position vector from the previous model. Thus, any parallel arrow with the same length pointing in the same direction is another model for the vector $\langle a, b \rangle$.

Any parallel translation of a vector is the same vector

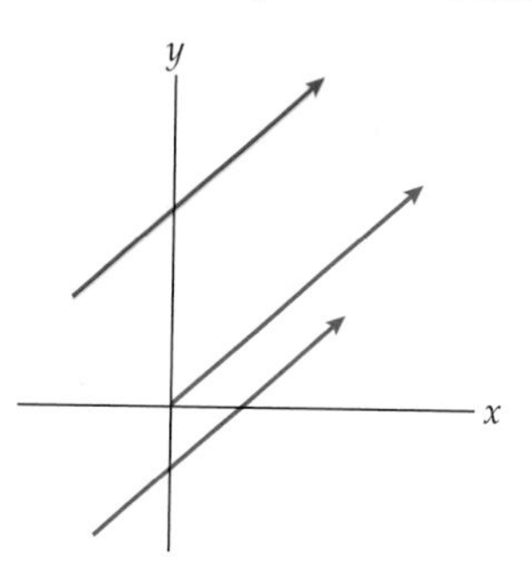

A similar model may be applied to vectors in $\mathbb{R}^3$.

Using the second model, we may geometrically add two vectors **u** and **v** as follows: As we show next, we place vector **u** in the diagram and then place the initial point of **v** at the terminal point of **u**. The vector $\mathbf{u} + \mathbf{v}$ is the vector that extends from the initial point of **u** to the terminal point of **v**.

Constructing $\mathbf{u} + \mathbf{v}$ *geometrically*

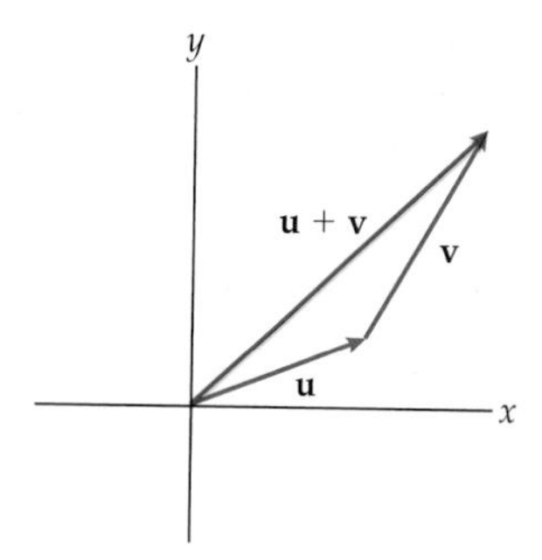

Another important concept is the magnitude of a vector, along with its synonyms "norm" and "length." We use the Pythagorean theorem in our definition.

DEFINITION 10.5 **The Magnitude, Norm, or Length of a Vector**

The ***magnitude, norm,*** or ***length*** of a vector $\mathbf{v}$ is denoted by $\|\mathbf{v}\|$.
In $\mathbb{R}^2$, when $\mathbf{v} = \langle a, b\rangle$,

$$\|\mathbf{v}\| = \sqrt{a^2 + b^2}.$$

In $\mathbb{R}^3$, when $\mathbf{v} = \langle a, b, c\rangle$,

$$\|\mathbf{v}\| = \sqrt{a^2 + b^2 + c^2}.$$

The magnitude of $\langle a, b\rangle$ *is* $\sqrt{a^2 + b^2}$

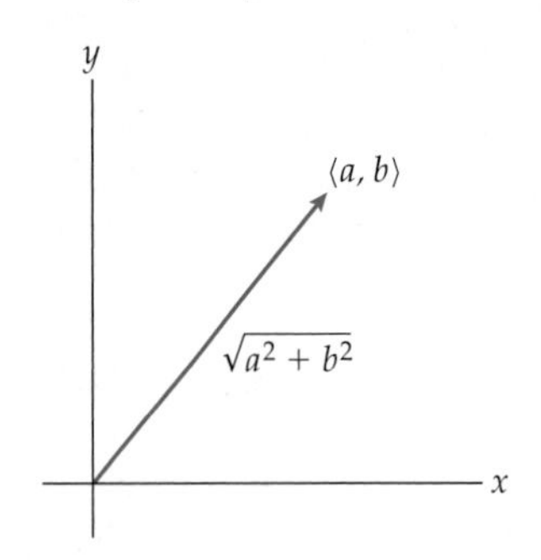

The magnitude of $\langle a, b, c\rangle$ *is* $\sqrt{a^2 + b^2 + c^2}$

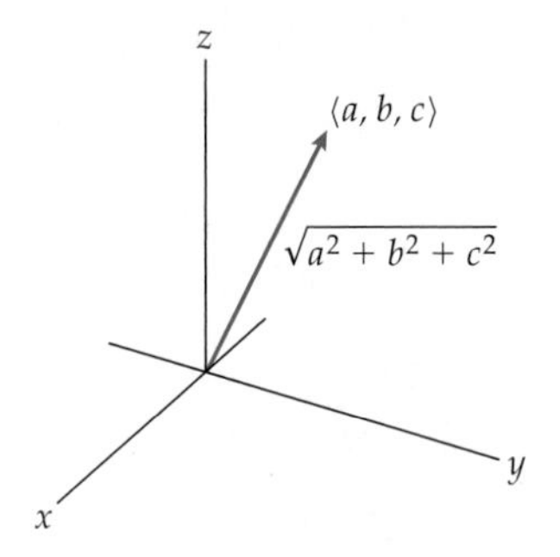

For example, the norm of the vector $\mathbf{v} = \langle 1, -2\rangle$ is $\|\mathbf{v}\| = \sqrt{1^2 + (-2)^2} = \sqrt{5}$.

$\|\mathbf{v}\| = \sqrt{5}$

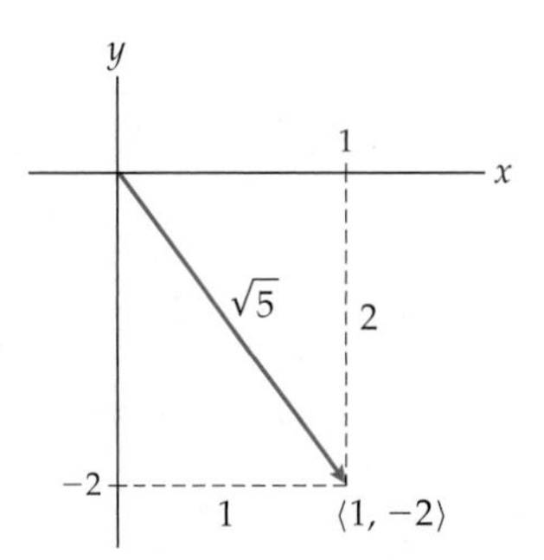

When we multiply a vector by a scalar $k > 0$, we obtain the vector $k\mathbf{v}$ that has the same direction as $\mathbf{v}$ but whose magnitude is equal to the magnitude of $\mathbf{v}$ multiplied by k. When $k < 0$, $k\mathbf{v}$ is the vector with direction opposite to the direction of the vector $\mathbf{v}$ that has magnitude equal to the magnitude of $\mathbf{v}$ multiplied by $|k|$. The following figure shows a vector $\mathbf{v}$ along with two scalar multiples of $\mathbf{v}$:

A vector $\mathbf{v}$ *and two scalar multiples of* $\mathbf{v}$

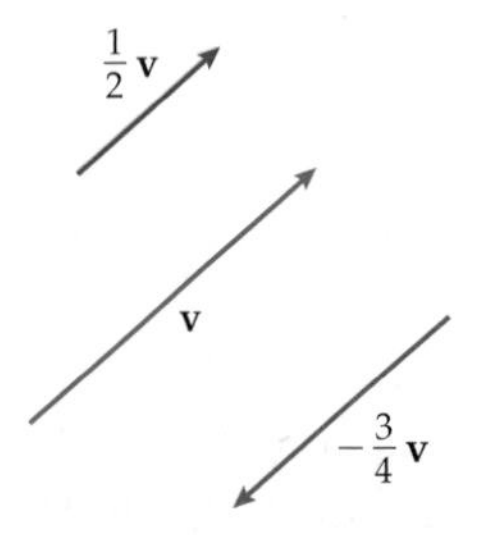

DEFINITION 10.6

Equal Vectors

Two vectors are said to be ***equal*** if and only if their corresponding components are equal.

In $\mathbb{R}^2$, $\langle x_1, y_1\rangle = \langle x_2, y_2\rangle$ if and only if $x_1 = x_2$ and $y_1 = y_2$.

In $\mathbb{R}^3$, $\langle x_1, y_1, z_1\rangle = \langle x_2, y_2, z_2\rangle$ if and only if $x_1 = x_2$, $y_1 = y_2$, and $z_1 = z_2$.

DEFINITION 10.7

The Zero Vector in $\mathbb{R}^2$ and $\mathbb{R}^3$

The ***zero vector*** in $\mathbb{R}^2$ is $\mathbf{0} = \langle 0, 0\rangle$ and the ***zero vector*** in $\mathbb{R}^3$ is $\mathbf{0} = \langle 0, 0, 0\rangle$.

Vectors in $\mathbb{R}^2$ and $\mathbb{R}^3$ obey many of the familiar algebraic properties that real numbers do.

THEOREM 10.8

Algebraic Properties of Vectors

(a) *Vector addition is commutative:*

For any two vectors $\mathbf{u}$ and $\mathbf{v}$ with the same number of components,

$$\mathbf{u} + \mathbf{v} = \mathbf{v} + \mathbf{u}.$$

(b) *Vector addition is associative:*

For any three vectors $\mathbf{u}$, $\mathbf{v}$, and $\mathbf{w}$, each with the same number of components,

$$(\mathbf{u} + \mathbf{v}) + \mathbf{w} = \mathbf{u} + (\mathbf{v} + \mathbf{w}).$$

(c) *Scalar multiplication distributes over vector addition:*

For any scalar c and any two vectors $\mathbf{u}$ and $\mathbf{v}$ with the same number of components,

$$c(\mathbf{u} + \mathbf{v}) = c\mathbf{u} + c\mathbf{v}.$$

In Exercises 57, 58, and 59 we ask you to use Definition 10.4 to prove the three parts of Theorem 10.8. The following figure illustrates geometrically why vector addition is commutative. Since we may get from point A to point B via either of the paths $\mathbf{u} + \mathbf{v}$ or $\mathbf{v} + \mathbf{u}$, and the figure formed is a parallelogram, addition is commutative.

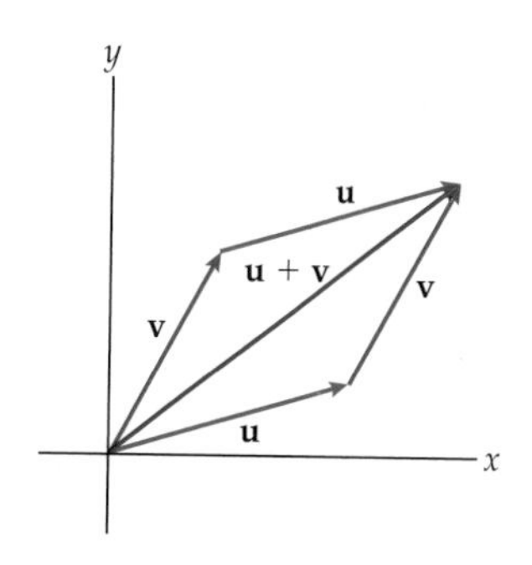

By simply relabeling two of the vectors from the preceding illustration of addition, we see how we may subtract vectors geometrically. If we let $\mathbf{w} = \mathbf{u} + \mathbf{v}$, then $\mathbf{v} = \mathbf{w} - \mathbf{u}$, as the following figure illustrates:

Constructing **w** − **u** *geometrically*

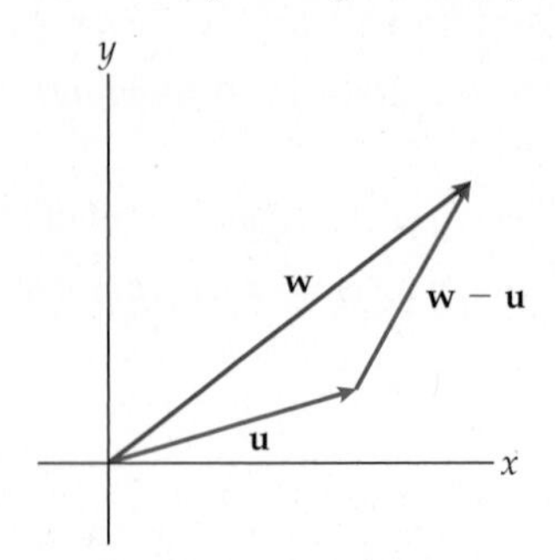

We see that when **u** and **w** are placed at the same initial point, **w** − **u** is the vector that extends from the terminal point of **u** to the terminal point of **w**.

We may use scalar multiplication to define what it means for two vectors to be parallel.

DEFINITION 10.9

Parallel Vectors

Two vectors are said to be ***parallel*** if and only if one of the vectors is a scalar multiple of the other.

As a consequence of this definition, the zero vector is parallel to every other vector, since, given any vector **v**, $0\mathbf{v} = \mathbf{0}$.

We also mention that there is a vector extending from a point $P(x_0, y_0, z_0)$ to a point $Q(x_1, y_1, z_1)$. In this case we use the notation

$$\overrightarrow{PQ} = \langle x_1 - x_0, y_1 - y_0, z_1 - z_0 \rangle.$$

That is, $\overrightarrow{PQ}$ is the vector whose initial point is P and whose terminal point is Q.

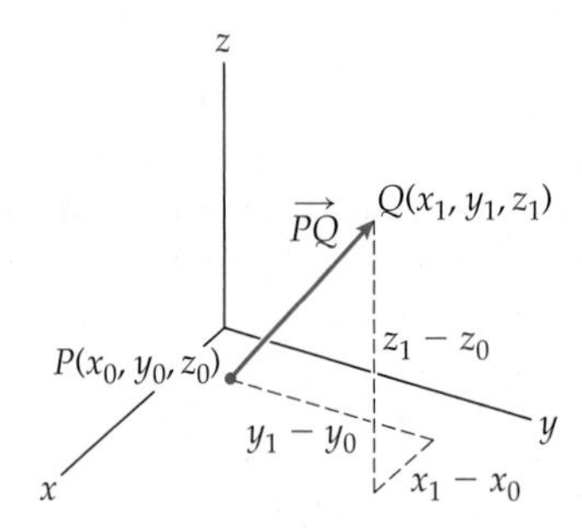

For example, if $P(8, 5, -3)$ and $Q(13, 5, -8)$, then $\overrightarrow{PQ} = \langle 13 - 8, 5 - 5, -8 - (-3) \rangle = \langle 5, 0, -5 \rangle$.

Unit Vectors

Often our primary interest is the direction of a nonzero vector **v**. At such times we may use a vector with length 1 and that is parallel to **v**. Vectors with length equal to 1 unit are called ***unit*** vectors.

DEFINITION 10.10 Unit Vectors

A vector $\mathbf{u}$ is said to be a ***unit vector*** if $\|\mathbf{u}\| = 1$.

To find a unit vector $\mathbf{u}$ with the same direction as a given vector $\mathbf{v} \neq \mathbf{0}$, we may multiply $\mathbf{v}$ by the reciprocal of its norm, as the following theorem states:

THEOREM 10.11 **Unit Vectors**

Given any nonzero vector $\mathbf{v}$, the vector $\dfrac{1}{\|\mathbf{v}\|}\mathbf{v}$ is a unit vector in the direction of $\mathbf{v}$.

We ask you to prove Theorem 10.11 in Exercise 60.

It is convenient to have a special notation for the unit vectors pointing in the positive direction of each of the coordinate axes. Hence, we make the following definition:

DEFINITION 10.12 Standard Basis Vectors in $\mathbb{R}^2$ and $\mathbb{R}^3$

The ***standard basis vectors*** in $\mathbb{R}^2$ are

$$\mathbf{i} = \langle 1, 0\rangle \text{ and } \mathbf{j} = \langle 0, 1\rangle.$$

The ***standard basis vectors*** in $\mathbb{R}^3$ are

$$\mathbf{i} = \langle 1, 0, 0\rangle, \ \mathbf{j} = \langle 0, 1, 0\rangle, \text{ and } \mathbf{k} = \langle 0, 0, 1\rangle.$$

The standard basis vectors in $\mathbb{R}^2$

The standard basis vectors in $\mathbb{R}^3$

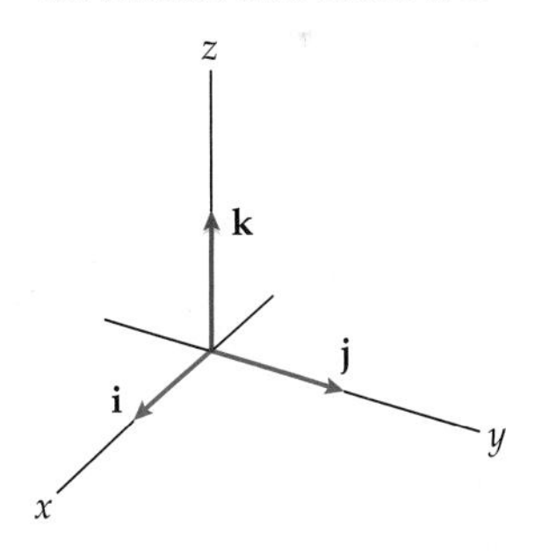

Any vector in $\mathbb{R}^2$ or $\mathbb{R}^3$ can be written in terms of the appropriate standard basis vectors. The vector $\mathbf{v} = \langle a, b\rangle$ can be written as $\mathbf{v} = a\mathbf{i} + b\mathbf{j}$. Thus, $\langle 8, -9\rangle = 8\mathbf{i} - 9\mathbf{j}$. Similarly, the vector $\mathbf{v} = \langle a, b, c\rangle$ can be written as $\mathbf{v} = a\mathbf{i} + b\mathbf{j} + c\mathbf{k}$. Thus, $\langle -7, 2, -3\rangle = -7\mathbf{i} + 2\mathbf{j} - 3\mathbf{k}$.

Using Vectors to Analyze Forces

There are many applications involving vectors. Any quantity with both a magnitude and a direction may be analyzed with the use of vectors. Two common vector applications are ***displacements*** and ***forces***. We discuss the analysis of forces here.

According to Newton's second law of motion, the vector sum of all of the forces acting on a body equals the mass m of the body times its acceleration: $\mathbf{F} = m\mathbf{a}$. Thus, if a body is not in motion, the sum of the forces acting upon it must be zero. As a simple example, the figure that follows shows a weight of 50 pounds suspended by a single rope. The rope exerts a force equal to 50 pounds, acting upwards, as the force of gravity acts downwards.

The vector forces in each direction must have the same magnitude

$-\mathbf{v}$

$\mathbf{v}$

This same principle applies in more complicated contexts, as we will see shortly. A mass m is suspended from a ceiling by two chains that form angles α and β. The sum of the three vectors $\mathbf{u}$, $\mathbf{v}$, and $\mathbf{w}$ must be zero as long as the object is not in motion.

The sum of the vectors $\mathbf{u}$, $\mathbf{v}$, *and* $\mathbf{w}$ *must be zero*

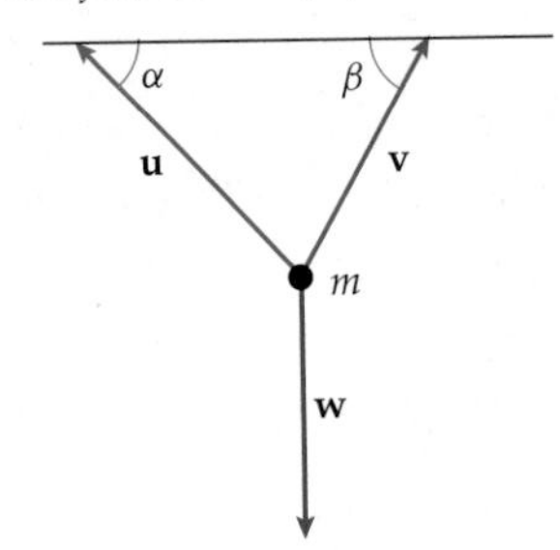

We explore a specific example of this type in Example 3.

Examples and Explorations

EXAMPLE 1

Adding and subtracting vectors

Let $\mathbf{u} = \langle 3, 8 \rangle$ and $\mathbf{v} = \langle 2, -4 \rangle$. Compute $\mathbf{u} + \mathbf{v}$ and $\mathbf{u} - \mathbf{v}$, and then sketch all four of these vectors.

SOLUTION

Using Definition 10.4, we have

$$\mathbf{u} + \mathbf{v} = \langle 3 + 2, 8 + (-4) \rangle = \langle 5, 4 \rangle \text{ and } \mathbf{u} - \mathbf{v} = \langle 3 - 2, 8 - (-4) \rangle = \langle 1, 12 \rangle.$$

To add the two vectors geometrically we place the initial point of one of the vectors at the terminal point of the other. In the figure that follows at the left we have placed $\mathbf{v}$ at the tip of $\mathbf{u}$. The sum is the vector extending from the initial point of $\mathbf{u}$ to the terminal point of $\mathbf{v}$.

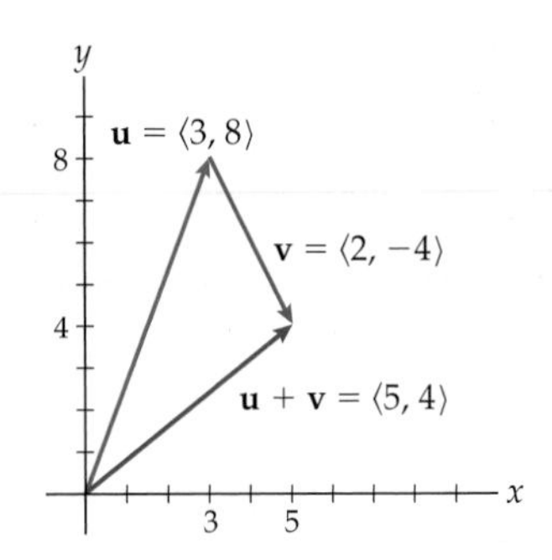

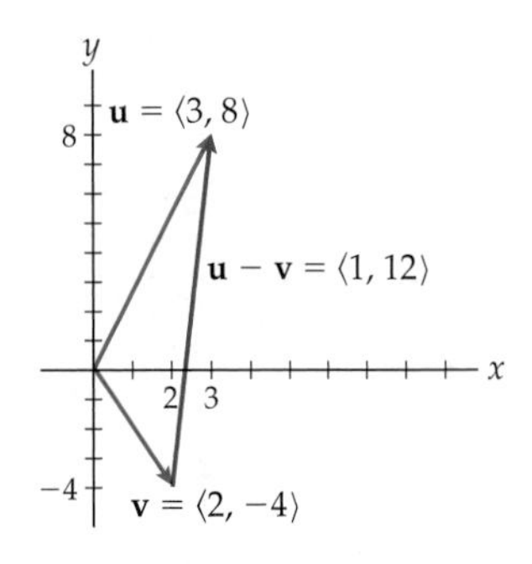

To subtract two vectors geometrically we position their initial points together, as we see in the right-hand figure. The vector $\mathbf{u} - \mathbf{v}$ extends from the terminal point of $\mathbf{v}$ to the terminal point of $\mathbf{u}$. □

EXAMPLE 2

Finding a unit vector in the same direction as a given nonzero vector

Find a unit vector with the same direction as the vector $\mathbf{v} = \langle 4, -3, 5 \rangle$.

SOLUTION

The vector $\mathbf{v}$ has norm $\|\mathbf{v}\| = \sqrt{4^2 + (-3)^2 + 5^2} = \sqrt{50} = 5\sqrt{2}$. The unit vector in the direction of $\mathbf{v}$ is the vector $\frac{1}{\|\mathbf{v}\|}\mathbf{v} = \frac{1}{5\sqrt{2}}\langle 4, -3, 5 \rangle$. If we wish, we may distribute the scalar and rationalize the denominators to obtain $\left\langle \frac{2}{5}\sqrt{2}, -\frac{3}{10}\sqrt{2}, \frac{1}{2}\sqrt{2} \right\rangle$. □

EXAMPLE 3

Finding the forces in cables suspending an object

The figure that follows shows a 50-pound weight suspended by two cables. Find the force in each of the cables attached to the ceiling.

The sum of the vector forces equals zero

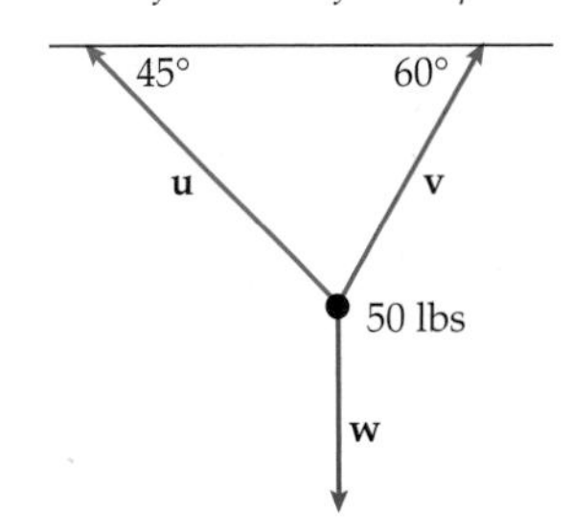

SOLUTION

Since the weight is stationary, the sum of the three vectors is $\mathbf{u}+\mathbf{v}+\mathbf{w} = \mathbf{0}$. We decompose each vector into its horizontal and vertical components, using the standard convention that motion to the right is positive on the x-axis and motion upward is positive on the y-axis. Therefore,

$$\mathbf{u} = -\cos 45^\circ \|\mathbf{u}\|\mathbf{i} + \sin 45^\circ \|\mathbf{u}\|\mathbf{j} = -\frac{\sqrt{2}}{2}\|\mathbf{u}\|\mathbf{i} + \frac{\sqrt{2}}{2}\|\mathbf{u}\|\mathbf{j},$$

$$\mathbf{v} = \cos 60^\circ \|\mathbf{v}\|\mathbf{i} + \sin 60^\circ \|\mathbf{v}\|\mathbf{j} = \frac{1}{2}\|\mathbf{v}\|\mathbf{i} + \frac{\sqrt{3}}{2}\|\mathbf{v}\|\mathbf{j}, \text{ and}$$

$$\mathbf{w} = -50\mathbf{j}.$$

In order to have $\mathbf{u}+\mathbf{v}+\mathbf{w}=\mathbf{0}$, the sum of the three x-components and the sum of the three y-components must both be zero. So,

$$\frac{\sqrt{2}}{2}\|\mathbf{u}\| = \frac{1}{2}\|\mathbf{v}\| \text{ and } \frac{\sqrt{2}}{2}\|\mathbf{u}\| + \frac{\sqrt{3}}{2}\|\mathbf{v}\| = 50 \text{ pounds.}$$

From the first of these equations, we have $\sqrt{2}\|\mathbf{u}\| = \|\mathbf{v}\|$, which we can substitute into the second equation to obtain $\|\mathbf{u}\| = \frac{100}{\sqrt{2}+\sqrt{6}} \approx 25.9$ pounds. Therefore $\|\mathbf{v}\| \approx 36.6$ pounds. □

? TEST YOUR UNDERSTANDING

- What are the geometric and analytic interpretations of a vector in $\mathbb{R}^2$? In $\mathbb{R}^3$?
- How are vectors added and subtracted geometrically? Algebraically?
- What does it mean for one vector to be a scalar multiple of another vector? How can you tell whether two vectors are parallel?
- What is a unit vector? How do you find a unit vector with the same direction as a given vector?
- How can you use vectors to analyze the forces acting on an object?

EXERCISES 10.2

Thinking Back

Properties of addition: Provide definitions for the following properties.

- the commutative property of addition for real numbers
- the associative property of addition for real numbers
- the distributive property of multiplication over addition for real numbers

Computing distances: Find the distances between the specified pairs of points.

- $(-3, 4)$ and $(6, -5)$
- $(11, 12, 13)$ and $(11, 12, -2)$
- $(5, -2, -1)$ and $(3, 0, -4)$

Concepts

0. *Problem Zero:* Read the section and make your own summary of the material.

1. *True/False:* Determine whether each of the statements that follow is true or false. If a statement is true, explain why. If a statement is false, provide a counterexample.

(a) *True or False:* Every vector has a norm.

(b) *True or False:* Given any vector $\mathbf{v}$, there is a unit vector in the direction of $\mathbf{v}$.

(c) *True or False:* The unit vector in the direction of a given nonzero vector $\mathbf{v}$ is always shorter than $\mathbf{v}$.

(d) *True or False:* Given two points A and B in 3-space, the vector from A to B is denoted by $\overrightarrow{BA}$.

(e) *True or False:* The vector $\langle a, b\rangle$ in $\mathbb{R}^2$ can be interpreted as the vector $\langle a, b, 0\rangle$ in $\mathbb{R}^3$.

(f) *True or False:* Standard basis vectors are unit vectors.

(g) *True or False:* For any nonzero vector $\mathbf{v}$, the vector $-\mathbf{v}$ has the same length as $\mathbf{v}$ but points in the opposite direction.

(h) *True or False:* If $\mathbf{v} = \overrightarrow{AB}$, then $-\mathbf{v} = \overrightarrow{BA}$.

2. *Examples:* Construct examples of the thing(s) described in the following. Try to find examples that are different than any in the reading.

(a) A unit vector in $\mathbb{R}^2$ that is not parallel to either $\mathbf{i}$ or $\mathbf{j}$.

(b) A vector in $\mathbb{R}^3$ with magnitude 5.

(c) Two distinct vectors parallel to $\langle 1, 2, -3\rangle$, each with norm 3.

In Exercises 3–8, let $A, B, C, D, \ldots, Z$ be points in $\mathbb{R}^3$. Simplify the given quantity.

3. $\overrightarrow{AB} + \overrightarrow{BC}$

4. $\overrightarrow{AB} + \overrightarrow{BC} + \overrightarrow{CD}$

5. $\overrightarrow{AB} + \overrightarrow{BA}$

6. $\overrightarrow{BD} - \overrightarrow{CD}$

7. $\overrightarrow{AB} + \overrightarrow{BC} + \overrightarrow{CA}$

8. $\overrightarrow{AB} + \overrightarrow{BC} + \cdots + \overrightarrow{YZ}$

9. How do you add two vectors algebraically? Geometrically?

10. How do you subtract two vectors algebraically? Geometrically?

11. Find a vector parallel to $\langle a, b, c\rangle$ but twice as long.

12. Find a vector parallel to $\langle a, b, c\rangle$ but half as long and pointing in the opposite direction.

13. If the initial point of the vector $\langle 2, 3, -5\rangle$ is the point $(-3, 2, 4)$, what is the terminal point of the vector?

14. If the terminal point of the vector $\langle 2, 3, -5\rangle$ is the point $(-3, 2, 4)$, what is the initial point of the vector?

15. Find the terminal point of a vector of magnitude 5 that is parallel to the vector $\langle 1, 2, 3\rangle$ and whose initial point is $(0, 3, -2)$.

16. Let $\mathbf{v}_0 = \langle a, b\rangle$ and let $\mathbf{v} = \langle x, y\rangle$. Describe the sets of points in $\mathbb{R}^2$ satisfying the following properties:

(a) $\|\mathbf{v}\| = 4$

(b) $\|\mathbf{v}\| \le 4$

(c) $\|\mathbf{v} - \mathbf{v}_0\| = 4$

17. Let $\mathbf{v}_0 = \langle a, b, c\rangle$ and let $\mathbf{v} = \langle x, y, z\rangle$. Describe the sets of points in $\mathbb{R}^3$ satisfying the following properties:

(a) $\|\mathbf{v}\| = 4$

(b) $\|\mathbf{v}\| \le 4$

(c) $\|\mathbf{v} - \mathbf{v}_0\| = 4$

18. Let $\mathbf{v} = \langle w, x, y, z\rangle$. Describe the sets of points in $\mathbb{R}^4$ satisfying $\|\mathbf{v}\| = 4$.

19. How do you generalize the ideas of this section to vectors with four components? To vectors with n components?

20. What is the set of all position vectors in $\mathbb{R}^2$ of magnitude 5?

21. What is the set of all position vectors in $\mathbb{R}^3$ of magnitude 5?

22. Consider the vector $\mathbf{v} = \langle 2, 3, 7\rangle$.

(a) Graph $\mathbf{v}$.

(b) What vector is symmetric about the origin to $\mathbf{v}$? Graph that vector.

(c) What vector is symmetric about the xy-plane to $\mathbf{v}$? Graph that vector.

(d) What vector is symmetric about the point $(0, 3, 0)$ to $\mathbf{v}$? Graph that vector.

(e) What vector is symmetric about the plane $z = 2$ to $\mathbf{v}$? Graph that vector.

Skills

In each of Exercises 23–28, find $\mathbf{u} + \mathbf{v}$ and $\mathbf{u} - \mathbf{v}$. Also, sketch $\mathbf{u}$, $\mathbf{v}$, $\mathbf{u} + \mathbf{v}$, and $\mathbf{u} - \mathbf{v}$.

23. $\mathbf{u} = \langle 2, -6\rangle,\ \mathbf{v} = \langle 6, 2\rangle$

24. $\mathbf{u} = \langle 3, -4\rangle,\ \mathbf{v} = \langle -1, 5\rangle$

25. $\mathbf{u} = \langle 1, -2\rangle,\ \mathbf{v} = \langle -3, 6\rangle$

26. $\mathbf{u} = \langle 4, 0\rangle,\ \mathbf{v} = \langle 0, 3\rangle$

27. $\mathbf{u} = \langle 1, -4, 6\rangle,\ \mathbf{v} = \langle 2, -4, 7\rangle$

28. $\mathbf{u} = \langle 3, 6, 11\rangle,\ \mathbf{v} = \langle 1, -2, 3\rangle$

In Exercises 29–32 find $\overrightarrow{PQ}$.

29. $P = (3, 6),\ Q = (-3, -2)$

30. $P = (5, -3),\ Q = (3, -6)$

31. $P = (1, -2, 5),\ Q = (1, -7, -2)$

32. $P = (-1, 5, 4),\ Q = (-1, 6, 4)$

In Exercises 33–36 find the norm of the vector.

33. $\mathbf{v} = \langle 3, -4\rangle$

34. $\mathbf{v} = \left\langle \frac{1}{2}, \frac{1}{3}, \frac{1}{4}\right\rangle$

35. $\mathbf{v} = \left\langle \frac{1}{3}, \frac{1}{3}, -\frac{1}{3}\right\rangle$

36. $\mathbf{v} = \langle 0, 0, 6\rangle$

In Exercises 37–42, find $\|\mathbf{v}\|$ and find the unit vector in the direction of $\mathbf{v}$.

37. $\mathbf{v} = \langle 3, -4\rangle$

38. $\mathbf{v} = \left\langle \sqrt{\frac{1}{3}}, -\sqrt{\frac{2}{3}}\right\rangle$

39. $\mathbf{v} = \left\langle \frac{1}{5}, \frac{1}{3}\right\rangle$

40. $\mathbf{v} = \langle 2, 1, -5\rangle$

41. $\mathbf{v} = \langle 1, 1, 1\rangle$

42. $\mathbf{v} = \left\langle \frac{\sqrt{2}}{2}, \frac{1}{2}, \frac{1}{2}\right\rangle$

In Exercises 43–51 find a vector with the given properties.

43. Find a vector in the direction of $\langle 3, 1, 2\rangle$ and with magnitude 5.

44. Find a vector in the direction of $\langle -1, 2, 3\rangle$ and with magnitude 3.

45. Find a vector in the direction of $\langle 8, -7, 2\rangle$ and with magnitude 2.

46. Find a vector in the direction of $\langle 9, 0, -6\rangle$ and with magnitude 7.

47. Find a vector in the direction opposite to $\langle -1, -4, -6\rangle$ and with magnitude 7.

48. Find a vector in the direction opposite to $\langle -4, 5, -1\rangle$ and with magnitude 3.

49. Find a vector in the direction opposite to $\langle 0, -3, 4\rangle$ and with magnitude 10.

50. Find a unit vector in the direction opposite to $\langle 3, 4, 5\rangle$.

51. Find a vector of length 3 that points in the direction opposite to $\langle 1, -2, 3\rangle$.

52. Let $P = (2, 3, 0, -1)$ and $Q = (3, -2, 1, 6)$ be points in four-dimensional space.

(a) Find $\overrightarrow{PQ}$.

(b) Find $\|\overrightarrow{PQ}\|$.

(c) Find the unit vector in the direction of $\overrightarrow{PQ}$.

Applications

53. A weight of 100 pounds is suspended by two ropes as shown in the accompanying figure. What are the magnitudes of the forces in each of the ropes?

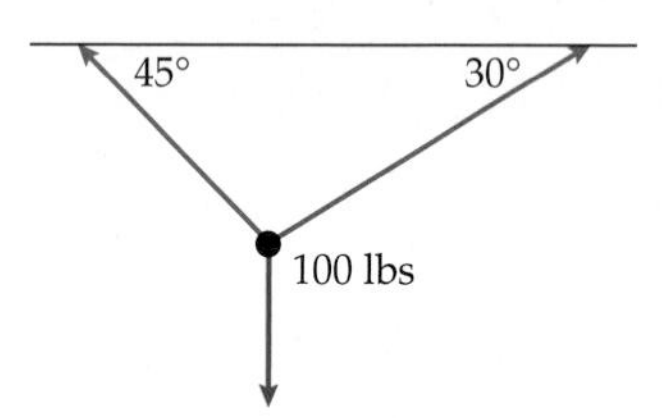

54. Each bolt attached in a certain ceiling is known to be able to withstand a force of 200 pounds before it pulls out. What is the maximum weight of an object that can be suspended in the following system?

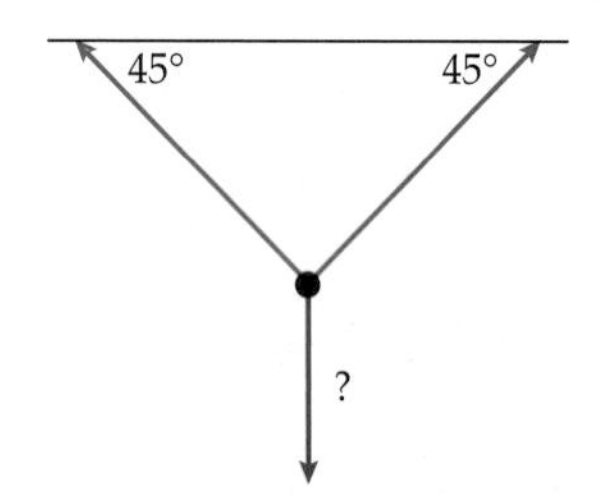

55. Annie has to make a crossing by kayak from Stuart Island to Speiden Island, two islands in the San Juan archipelago that are part of Washington State.

(a) If she crosses due south during a slack tide at a leisurely 2 mph, what is a vector describing her velocity?

(b) If she crosses in a perfectly southeastern direction at slack tide at 2 mph, what is a vector describing her velocity?

(c) The tides can pump through the Stuart–Speiden channel at a good clip. The outgoing tide can move due west at 2 mph. When faced with this situation, Annie turns on the jets and paddles southeast at 3 mph. What is a vector describing her velocity?

56. Ian is descending Middle Cascade Glacier in North Cascades National Park in Washington State.

(a) Near the top of the glacier, he is descending a 30-degree slope due northwards at 3 mph. Give a vector describing his velocity.

(b) Toward the middle of the glacier the slope steepens. Now Ian is descending directly northeast at a 45-degree angle. His speed is only 2 mph. What is a vector describing his velocity?

(c) Now Ian needs to exit the glacier before he runs into the icefall. He contours due east, so that he is neither ascending nor descending. He is moving only at 1.5 mph now. What is a vector describing his velocity?

Proofs

57. Prove part (a) of Theorem 10.8 for vectors in $\mathbb{R}^3$; that is, show that for $\mathbf{u} = \langle u_1, u_2, u_3 \rangle$ and $\mathbf{v} = \langle v_1, v_2, v_3 \rangle$,

$$\mathbf{u} + \mathbf{v} = \mathbf{v} + \mathbf{u}.$$

58. Prove part (b) of Theorem 10.8 for vectors in $\mathbb{R}^3$; that is, show that for $\mathbf{u} = \langle u_1, u_2, u_3 \rangle$, $\mathbf{v} = \langle v_1, v_2, v_3 \rangle$ and $\mathbf{w} = \langle w_1, w_2, w_3 \rangle$,

$$(\mathbf{u} + \mathbf{v}) + \mathbf{w} = \mathbf{u} + (\mathbf{v} + \mathbf{w}).$$

59. Prove part (c) of Theorem 10.8 for vectors in $\mathbb{R}^3$; that is, show that for $\mathbf{u} = \langle u_1, u_2, u_3 \rangle$, $\mathbf{v} = \langle v_1, v_2, v_3 \rangle$ and scalar c,

$$c(\mathbf{u} + \mathbf{v}) = c\mathbf{u} + c\mathbf{v}.$$

60. Prove Theorem 10.11; that is, show that when $\mathbf{v} \neq \mathbf{0}$, the scaled vector $\frac{1}{\|\mathbf{v}\|}\mathbf{v}$ is a unit vector with the same direction as $\mathbf{v}$.

61. Let c and d be a scalars and let $\mathbf{v}$ be a vector in $\mathbb{R}^3$. Show that the following distributive property holds:

$$(c + d)\mathbf{v} = c\mathbf{v} + d\mathbf{v}.$$

62. Use a vector argument to prove that the segment connecting the midpoints of two sides of a triangle is parallel to the third side of the triangle and half of its length.

63. Use vector methods to show that the diagonals of a parallelogram bisect each other.

64. Let Quad($PQRS$) denote the quadrilateral in the xy-plane with vertices P, Q, R, and S. If P' is the midpoint of side PQ, Q' is the midpoint of side QR, R' is the midpoint of side RS, and S' is the midpoint of side SP, prove that Quad($P'Q'R'S'$) is a parallelogram.

Thinking Forward

Perpendicular vectors in $\mathbb{R}^2$

- How many vectors are there in $\mathbb{R}^2$ that are perpendicular to a given nonzero vector $\langle a, b \rangle$?
- How many unit vectors are there in $\mathbb{R}^2$ that are perpendicular to a given nonzero vector $\langle a, b \rangle$?
- Find a vector in $\mathbb{R}^2$ that is perpendicular to the vector $\langle 1, 3 \rangle$.
- Find a unit vector in $\mathbb{R}^2$ that is perpendicular to the vector $\langle 1, 3 \rangle$.

Perpendicular vectors in $\mathbb{R}^3$

- How many vectors are there in $\mathbb{R}^3$ that are perpendicular to a given nonzero vector $\langle a, b, c \rangle$?
- How many unit vectors are there in $\mathbb{R}^3$ that are perpendicular to a given nonzero vector $\langle a, b, c \rangle$?
- Find a unit vector in $\mathbb{R}^3$ that is perpendicular to both vectors $\mathbf{i}$ and $\mathbf{k}$.

10.3 DOT PRODUCT

- The dot product
- The geometry of the dot product
- Projecting one vector onto another

The Dot Product

In Section 10.2 we discussed how to multiply a vector $\mathbf{v}$ by a scalar k. Recall that $k\mathbf{v}$ is a vector parallel to $\mathbf{v}$ with magnitude $|k|$ times the magnitude of $\mathbf{v}$ and with the same direction if $k > 0$ and the opposite direction if $k < 0$. Here we will define the dot product, which allows us to multiply two vectors with the same number of components. In Section 10.4 we will discuss the cross product, a different method for multiplying two vectors in $\mathbb{R}^3$.

DEFINITION 10.13

Dot Product

Let $\mathbf{u} = \langle u_1, u_2 \rangle$ and $\mathbf{v} = \langle v_1, v_2 \rangle$ be vectors in $\mathbb{R}^2$. We define the ***dot product,*** $\mathbf{u} \cdot \mathbf{v}$, to be

$$\mathbf{u} \cdot \mathbf{v} = u_1 v_1 + u_2 v_2.$$

Let $\mathbf{u} = \langle u_1, u_2, u_3 \rangle$ and $\mathbf{v} = \langle v_1, v_2, v_3 \rangle$ be vectors in $\mathbb{R}^3$. We define the ***dot product,*** $\mathbf{u} \cdot \mathbf{v}$, to be

$$\mathbf{u} \cdot \mathbf{v} = u_1 v_1 + u_2 v_2 + u_3 v_3.$$

For example, the dot product of the vectors $\langle 1, 4 \rangle$ and $\langle 7, -3 \rangle$ from $\mathbb{R}^2$ is

$$\langle 1, 4 \rangle \cdot \langle 7, -3 \rangle = 1 \cdot 7 + 4 \cdot (-3) = 7 - 12 = -5,$$

and the dot product of the vectors $\langle 2, 7, 6 \rangle$ and $\langle -1, 2, -2 \rangle$ from $\mathbb{R}^3$ is

$$\langle 2, 7, 6 \rangle \cdot \langle -1, 2, -2 \rangle = 2 \cdot (-1) + 7 \cdot 2 + 6 \cdot (-2) = -2 + 14 - 12 = 0.$$

Note that in both cases the dot product is a scalar, not a vector. For most of the rest of this section we will discuss the significance of the dot product and what this scalar tells us about the relationship between the two vectors. Note also that Definition 10.13 could easily be generalized to allow us to find the dot product of two vector in $\mathbb{R}^n$.

Theorem 10.14 collects several basic algebraic properties of the dot product. We prove part (b) for vectors in $\mathbb{R}^3$ and leave the proofs of the other parts of the theorem for Exercise 62.

THEOREM 10.14

Algebraic Properties of the Dot Product

For any vectors $\mathbf{u}$, $\mathbf{v}$, and $\mathbf{w}$, and any scalar k,

(a) $\mathbf{u} \cdot \mathbf{v} = \mathbf{v} \cdot \mathbf{u}$

(b) $\mathbf{u} \cdot (\mathbf{v} + \mathbf{w}) = \mathbf{u} \cdot \mathbf{v} + \mathbf{u} \cdot \mathbf{w}$

(c) $k(\mathbf{u} \cdot \mathbf{v}) = (k\mathbf{u}) \cdot \mathbf{v} = \mathbf{u} \cdot (k\mathbf{v})$

(d) $\mathbf{v} \cdot \mathbf{v} = \|\mathbf{v}\|^2$

Proof. Let $\mathbf{u} = \langle u_1, u_2, u_3 \rangle$, $\mathbf{v} = \langle v_1, v_2, v_3 \rangle$, and $\mathbf{w} = \langle w_1, w_2, w_3 \rangle$. Using the definitions of vector addition and the dot product, we have

$$\begin{aligned}
\mathbf{u} \cdot (\mathbf{v} + \mathbf{w}) &= \langle u_1, u_2, u_3 \rangle \cdot (\langle v_1, v_2, v_3 \rangle + \langle w_1, w_2, w_3 \rangle) \\
&= \langle u_1, u_2, u_3 \rangle \cdot \langle v_1 + w_1, v_2 + w_2, v_3 + w_3 \rangle \\
&= u_1(v_1 + w_1) + u_2(v_2 + w_2) + u_3(v_3 + w_3) \\
&= u_1 v_1 + u_1 w_1 + u_2 v_2 + u_2 w_2 + u_3 v_3 + u_3 w_3 \\
&= (u_1 v_1 + u_2 v_2 + u_3 v_3) + (u_1 w_1 + u_2 w_2 + u_3 w_3) \\
&= \langle u_1, u_2, u_3 \rangle \cdot \langle v_1, v_2, v_3 \rangle + \langle u_1, u_2, u_3 \rangle \cdot \langle w_1, w_2, w_3 \rangle \\
&= \mathbf{u} \cdot \mathbf{v} + \mathbf{u} \cdot \mathbf{w}.
\end{aligned}$$

■

Note that this proof may be generalized to vectors in $\mathbb{R}^n$. Similarly, the other parts of Theorem 10.14 may be generalized to vectors in $\mathbb{R}^n$.

We obtain a scalar as the dot product of the two vectors **u** and **v**. What does this number tell us about the geometric relationship between the two vectors? Before we answer that question, we need to mention that the angle between two nonzero vectors **u** and **v** is the angle $0 \le \theta \le \pi$ created when **u** and **v** are considered as position vectors. (If either $\mathbf{u} = \mathbf{0}$ or $\mathbf{v} = \mathbf{0}$, then the angle between **u** and **v** is undefined.)

The angle between two nonzero vectors is $0 \le \theta \le \pi$

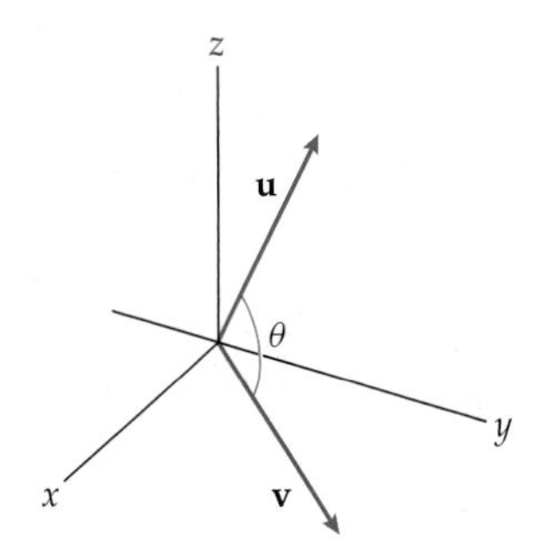

THEOREM 10.15

The Geometry of the Dot Product

Let **u** and **v** be vectors. Then the dot product

$$\mathbf{u} \cdot \mathbf{v} = \begin{cases} 0, & \text{if either } \mathbf{u} = \mathbf{0} \text{ or } \mathbf{v} = \mathbf{0} \\ \|\mathbf{u}\|\,\|\mathbf{v}\| \cos\theta, & \text{if } \mathbf{u} \ne \mathbf{0} \text{ and } \mathbf{v} \ne \mathbf{0}, \text{ where } \theta \text{ is the angle between } \mathbf{u} \text{ and } \mathbf{v}. \end{cases}$$

This theorem provides a relationship between the dot product of two vectors, their lengths, and the angle between them.

In Exercise 56 we ask you to prove Theorem 10.15 for vectors in $\mathbb{R}^2$. We will prove it for vectors in $\mathbb{R}^3$ in a moment. Before we do, let us remind you of the ***Law of Cosines***, which we will use in our proof.

THEOREM 10.16

Law of Cosines

In a triangle with side lengths a, b, and c, where θ is the angle between the sides of length a and b,

$$a^2 + b^2 - 2ab\cos\theta = c^2.$$

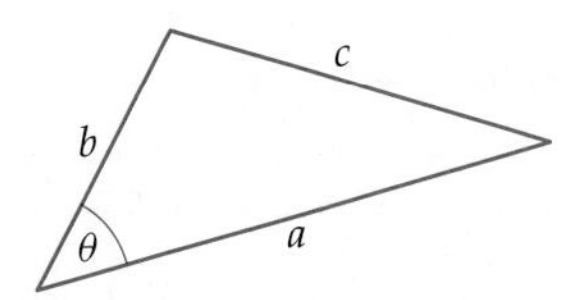

We are now ready to prove Theorem 10.15 for vectors in $\mathbb{R}^3$.

Proof. Let $\mathbf{u} = \langle u_1, u_2, u_3 \rangle$ and $\mathbf{v} = \langle v_1, v_2, v_3 \rangle$. If either $\mathbf{u}$ or $\mathbf{v}$ is $\mathbf{0}$, then

$$u_1 v_1 + u_2 v_2 + u_3 v_3 = 0 + 0 + 0 = 0,$$

and we are done.

If neither $\mathbf{u}$ nor $\mathbf{v}$ is $\mathbf{0}$ but they are parallel, then $\mathbf{u} = k\mathbf{v}$ for some scalar k, and θ, the angle between $\mathbf{u}$ and $\mathbf{v}$, is either 0 or π. If $\theta = 0$, then $\cos\theta = 1$ and $\mathbf{u} = k\mathbf{v}$ for some positive scalar k. Using the appropriate algebraic properties of Theorem 10.14, we have

$$\mathbf{u} \cdot \mathbf{v} = k\mathbf{v} \cdot \mathbf{v} = k\|\mathbf{v}\|^2 = \|\mathbf{u}\|\,\|\mathbf{v}\|.$$

In Exercise 58 we ask you to analyze the case where $\mathbf{u} = k\mathbf{v}$ and $k < 0$.

If neither $\mathbf{u}$ nor $\mathbf{v}$ is $\mathbf{0}$, and if $0 < \theta < \pi$, consider the following triangle formed from the vectors $\mathbf{u}$ and $\mathbf{v}$:

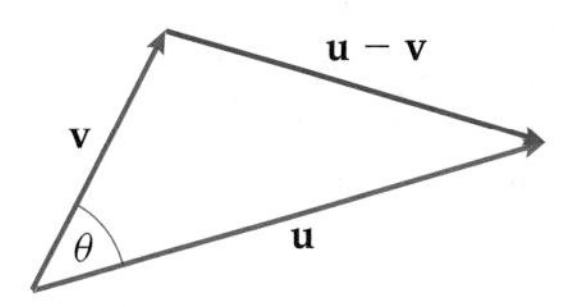

From the Law of Cosines we know that

$$\|\mathbf{u}\|^2 + \|\mathbf{v}\|^2 - 2\|\mathbf{u}\|\,\|\mathbf{v}\| \cos\theta = \|\mathbf{u} - \mathbf{v}\|^2.$$

Rearranging, we get

$$\|\mathbf{u}\|\,\|\mathbf{v}\| \cos\theta = \frac{1}{2}(\|\mathbf{u}\|^2 + \|\mathbf{v}\|^2 - \|\mathbf{u} - \mathbf{v}\|^2).$$

To finish our proof we will show that the right-hand side of this last equality is the dot product $\mathbf{u} \cdot \mathbf{v}$.

Since $\mathbf{u} = \langle u_1, u_2, u_3 \rangle$ and $\mathbf{v} = \langle v_1, v_2, v_3 \rangle$, we have $\mathbf{u} - \mathbf{v} = \langle u_1 - v_1, u_2 - v_2, u_3 - v_3 \rangle$. We also have the following norms:

$$\|\mathbf{u}\|^2 = u_1^2 + u_2^2 + u_3^2,$$
$$\|\mathbf{v}\|^2 = v_1^2 + v_2^2 + v_3^2, \quad \text{and}$$
$$\|\mathbf{u} - \mathbf{v}\|^2 = (u_1 - v_1)^2 + (u_2 - v_2)^2 + (u_3 - v_3)^2.$$

Therefore, $\frac{1}{2}(\|\mathbf{u}\|^2 + \|\mathbf{v}\|^2 - \|\mathbf{u} - \mathbf{v}\|^2)$ equals

$$\frac{1}{2}((u_1^2 + u_2^2 + u_3^2) + (v_1^2 + v_2^2 + v_3^2) - (u_1^2 - 2u_1v_1 + v_1^2 + u_2^2 - 2u_2v_2 + v_2^2 + u_3^2 - 2u_3v_3 + v_3^2))$$
$$= u_1 v_1 + u_2 v_2 + u_3 v_3 = \mathbf{u} \cdot \mathbf{v}.$$

We now have our result. ■

As a corollary to Theorem 10.15 we may immediately determine when the angle between two vectors is acute, right, or obtuse:

THEOREM 10.17

The Angle Between Two Vectors and the Dot Product

Let θ be the angle between nonzero vectors $\mathbf{u}$ and $\mathbf{v}$. Then

$$\theta \text{ is } \begin{cases} \text{acute,} & \text{if and only if } \mathbf{u}\cdot\mathbf{v} > 0 \\ \text{right,} & \text{if and only if } \mathbf{u}\cdot\mathbf{v} = 0 \\ \text{obtuse,} & \text{if and only if } \mathbf{u}\cdot\mathbf{v} < 0. \end{cases}$$

The details of the proof are left for Exercise 60.

DEFINITION 10.18

Orthogonal Curves and Vectors

(a) Two curves are said to be ***orthogonal*** at a point of intersection if the tangent lines to the curves at the point of intersection are perpendicular.

(b) Two nonzero vectors are ***orthogonal*** if the angle between them is a right angle.

(c) The zero vector is ***orthogonal*** to every vector.

Orthogonal curves

Orthogonal vectors

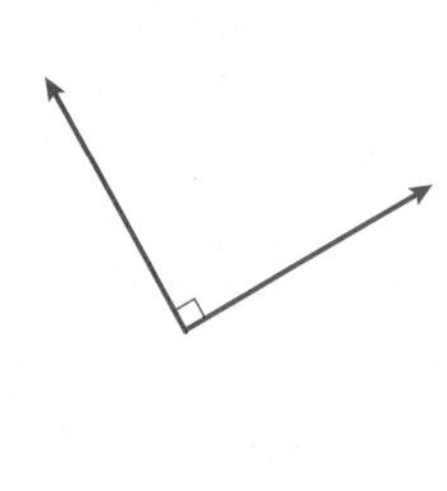

We have the following theorem, whose proof is left for Exercise 61:

THEOREM 10.19

The Dot Product Test for Orthogonality

Vectors $\mathbf{u}$ and $\mathbf{v}$ are orthogonal if and only if $\mathbf{u}\cdot\mathbf{v} = 0$.

Projections

Although our coordinate systems consist of mutually perpendicular axes, in certain applications one nonzero vector $\mathbf{u}$ is of particular significance. When this is the case, we may wish to write another vector $\mathbf{v}$ as a sum of two vectors, one of which is parallel to $\mathbf{u}$, the other orthogonal to $\mathbf{u}$.

DEFINITION 10.20

Vector Projections and Vector Components

Let $\mathbf{u}$ be a nonzero vector and let $\mathbf{v}$ be any vector. We define the ***vector projection of* v *onto* u**, denoted $\mathbf{v}_{\parallel}$, and the ***vector component of* v *orthogonal to* u**, denoted $\mathbf{v}_{\perp}$, as the pair of vectors having the following three properties:

(a) $\mathbf{v}_{\parallel}$ is parallel to $\mathbf{u}$.

(b) $\mathbf{v}_{\perp}$ is orthogonal to $\mathbf{u}$.

(c) $\mathbf{v}_{\parallel} + \mathbf{v}_{\perp} = \mathbf{v}$.

(Note that we will be using the notations $\mathbf{v}_{\parallel}$ and $\mathbf{v}_{\perp}$ only temporarily. The formal notations are given shortly.)

In Exercise 63 we ask you to prove that such a decomposition of $\mathbf{v}$ is unique.

Geometrically we have one of the situations shown in the following two figures:

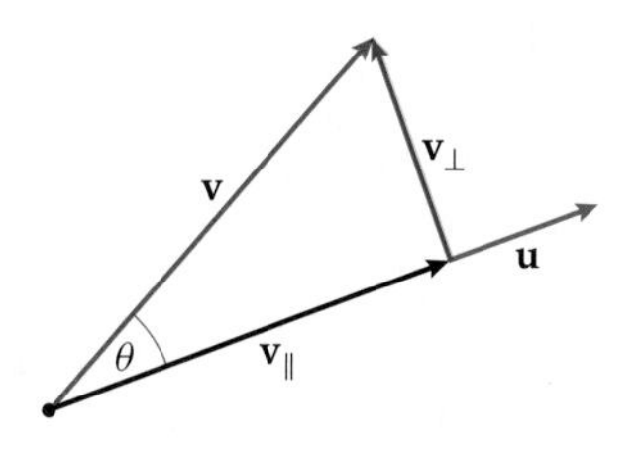

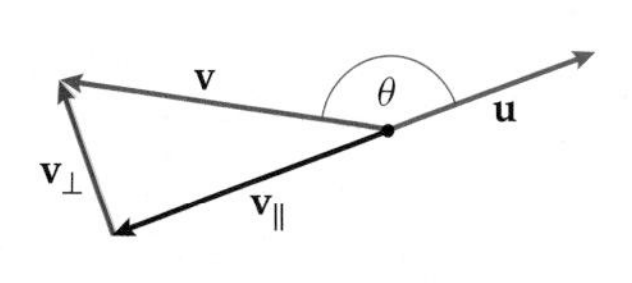

Since $\mathbf{u} \cdot \mathbf{v} = \|\mathbf{u}\| \|\mathbf{v}\| \cos\theta$, we have

$$\|\mathbf{v}\| \cos\theta = \frac{1}{\|\mathbf{u}\|}\mathbf{u} \cdot \mathbf{v} = \frac{\mathbf{u} \cdot \mathbf{v}}{\|\mathbf{u}\|}.$$

Observe that the signed quantity $\|\mathbf{v}\| \cos\theta$ is positive when $0 \le \theta < 90^\circ$ and negative when $90^\circ < \theta \le 180^\circ$. In either case, $\|\mathbf{v}_{\parallel}\| = \|\mathbf{v}\| |\cos\theta|$.

We use $\|\mathbf{v}\| \cos\theta = \frac{\mathbf{u} \cdot \mathbf{v}}{\|\mathbf{u}\|}$ in the following definition:

DEFINITION 10.21 The Component of a Projection

Let $\mathbf{u}$ be any nonzero vector. Then the ***component of the projection of* v *onto* u** is the scalar

$$\text{comp}_{\mathbf{u}}\mathbf{v} = \frac{\mathbf{u} \cdot \mathbf{v}}{\|\mathbf{u}\|}.$$

It bears repeating that $\text{comp}_{\mathbf{u}}\mathbf{v}$ is a signed quantity and its sign depends upon the size of the angle between $\mathbf{u}$ and $\mathbf{v}$.

From our earlier discussion, $\mathbf{v}_{\parallel}$ equals $\text{comp}_{\mathbf{u}}\mathbf{v}$ times the unit vector in the direction of $\mathbf{u}$, and we have

$$\mathbf{v}_{\parallel} = \frac{\mathbf{u} \cdot \mathbf{v}}{\|\mathbf{u}\|}\frac{\mathbf{u}}{\|\mathbf{u}\|} = \frac{\mathbf{u} \cdot \mathbf{v}}{\|\mathbf{u}\|^2}\mathbf{u}.$$

We now introduce the more formal notation for $\mathbf{v}_{\parallel}$.

DEFINITION 10.22 The Vector Projection

Let $\mathbf{u}$ be any nonzero vector. Then the ***vector projection of* v *onto* u** is

$$\text{proj}_{\mathbf{u}}\mathbf{v} := \frac{\mathbf{u} \cdot \mathbf{v}}{\|\mathbf{u}\|^2}\mathbf{u}.$$

Finally, since $\mathbf{v}_{\parallel} + \mathbf{v}_{\perp} = \mathbf{v}$, the vector component of $\mathbf{v}$ orthogonal to $\mathbf{u}$ is

$$\mathbf{v}_{\perp} = \mathbf{v} - \text{proj}_{\mathbf{u}}\mathbf{v}.$$

The Triangle Inequality

You are probably familiar with the statement: "The shortest distance between two points is a straight line." This is the famous *triangle inequality*. In the figure shown next, we have two paths from point A to point B. One is $\mathbf{u} + \mathbf{v}$, while the other is $\mathbf{u}$ followed by $\mathbf{v}$. Since

$\mathbf{u}+\mathbf{v}$ is the straight path between the two points, its length must be shorter than the length of the path given by $\mathbf{u}$ followed by $\mathbf{v}$. This inequality may be phrased in terms of vectors, as in the next theorem.

The shortest distance between points A and B is the straight line

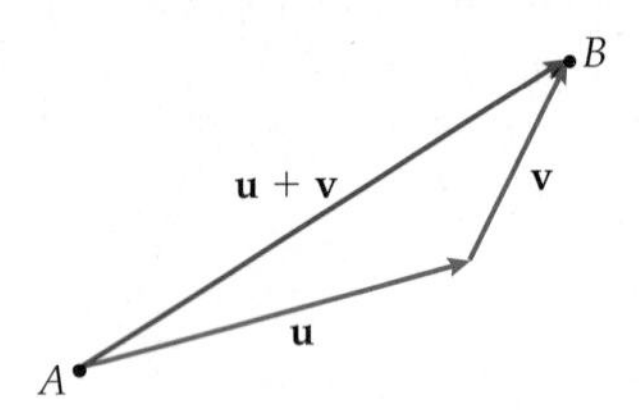

This figure illustrates why the statement about the shortest distance is a statement about triangles!

THEOREM 10.23 **The Triangle Inequality**

Given vectors $\mathbf{u}$ and $\mathbf{v}$, $\|\mathbf{u}+\mathbf{v}\| \le \|\mathbf{u}\| + \|\mathbf{v}\|$, and $\|\mathbf{u}+\mathbf{v}\| = \|\mathbf{u}\| + \|\mathbf{v}\|$ if and only if $\mathbf{u} = \mathbf{0}$ or $\mathbf{v} = \mathbf{0}$ or $\mathbf{u} = k\mathbf{v}$ for some positive scalar k.

Proof. First note that if $\mathbf{u} = \mathbf{0}$ or $\mathbf{v} = \mathbf{0}$, then $\|\mathbf{u}+\mathbf{v}\| = \|\mathbf{u}\| + \|\mathbf{v}\|$. If neither $\mathbf{u}$ nor $\mathbf{v}$ is $\mathbf{0}$, but $\mathbf{u} = k\mathbf{v}$ for some positive scalar k, then

$$\|\mathbf{u}+\mathbf{v}\| = \|k\mathbf{v}+\mathbf{v}\| = (k+1)\|\mathbf{v}\| = k\|\mathbf{v}\| + \|\mathbf{v}\| = \|\mathbf{u}\| + \|\mathbf{v}\|.$$

We now use the dot product to prove the inequality. We have

$$\begin{aligned}
\|\mathbf{u}+\mathbf{v}\|^2 &= (\mathbf{u}+\mathbf{v}) \cdot (\mathbf{u}+\mathbf{v}) \\
&= \mathbf{u}\cdot\mathbf{u} + \mathbf{u}\cdot\mathbf{v} + \mathbf{v}\cdot\mathbf{u} + \mathbf{v}\cdot\mathbf{v} \\
&= \|\mathbf{u}\|^2 + 2\mathbf{u}\cdot\mathbf{v} + \|\mathbf{v}\|^2 \\
&= \|\mathbf{u}\|^2 + 2\|\mathbf{u}\|\|\mathbf{v}\|\cos\theta + \|\mathbf{v}\|^2 \\
&\le \|\mathbf{u}\|^2 + 2\|\mathbf{u}\|\|\mathbf{v}\| + \|\mathbf{v}\|^2 \\
&= (\|\mathbf{u}\| + \|\mathbf{v}\|)^2.
\end{aligned}$$

So, $\|\mathbf{u}+\mathbf{v}\|^2 \le (\|\mathbf{u}\| + \|\mathbf{v}\|)^2$. Since magnitudes are nonnegative, we have our result,

$$\|\mathbf{u}+\mathbf{v}\| \le \|\mathbf{u}\| + \|\mathbf{v}\|.$$

■

Examples and Explorations

EXAMPLE 1 **Finding the angle between two vectors**

Find the angle between $\mathbf{u} = \langle 1, 2, 3\rangle$ and $\mathbf{v} = \langle 2, 1, -4\rangle$.

SOLUTION

Here,

$$\begin{aligned}
\mathbf{u}\cdot\mathbf{v} &= 2 + 2 - 12 = -8, \\
\|\mathbf{u}\| &= \sqrt{1^2 + 2^2 + 3^2} = \sqrt{14}, \quad \text{and} \\
\|\mathbf{v}\| &= \sqrt{2^2 + 1^2 + (-4)^2} = \sqrt{21}.
\end{aligned}$$

Therefore, if θ is the angle between $\mathbf{u}$ and $\mathbf{v}$, then $\cos\theta = \dfrac{\mathbf{u}\cdot\mathbf{v}}{\|\mathbf{u}\|\|\mathbf{v}\|} = \dfrac{-8}{\sqrt{14}\sqrt{21}} \approx -0.4666$. Thus, $\theta \approx 117.8°$.

□

EXAMPLE 2

Using the dot product to show orthogonality

Show that the triangle with vertices $P = (6, 2, 2)$, $Q = (2, 0, -1)$, and $R = (5, 1, -2)$ is a right triangle.

SOLUTION

We can form the vectors

$$\overrightarrow{PQ} = \langle -4, -2, -3 \rangle, \ \overrightarrow{PR} = \langle -1, -1, -4 \rangle, \text{ and } \overrightarrow{QR} = \langle 3, 1, -1 \rangle.$$

For $\triangle PQR$ to be a right triangle, we need one right angle. Using the dot product, we see that although $\overrightarrow{PQ}$ is orthogonal to neither $\overrightarrow{PR}$ nor $\overrightarrow{QR}$ (check this), we do have

$$\overrightarrow{PR} \cdot \overrightarrow{QR} = (-1)(3) + (-1)(1) + (-4)(-1) = 0.$$

Therefore the angle at vertex R is a right angle. Thus $\triangle PQR$ is a right triangle. □

EXAMPLE 3

Finding the projection of one vector onto another

Given vectors $\mathbf{u} = \langle 8, 2 \rangle$ and $\mathbf{v} = \langle -4, 5 \rangle$, find $\text{comp}_{\mathbf{u}}\mathbf{v}$, $\text{proj}_{\mathbf{u}}\mathbf{v}$ the vector projection of $\mathbf{v}$ onto $\mathbf{u}$, and the component of $\mathbf{v}$ orthogonal to $\mathbf{u}$.

SOLUTION

The component of the projection of $\mathbf{v}$ onto $\mathbf{u}$ is

$$\text{comp}_{\mathbf{u}}\mathbf{v} = \frac{\langle 8, 2 \rangle \cdot \langle -4, 5 \rangle}{\|\langle 8, 2 \rangle\|} = \frac{-22}{\sqrt{68}}.$$

The vector projection of $\mathbf{v}$ onto $\mathbf{u}$ is $\text{comp}_{\mathbf{u}}\mathbf{v}$ times the unit vector in the direction of $\mathbf{u}$. Thus,

$$\text{proj}_{\mathbf{u}}\mathbf{v} = \frac{\mathbf{u} \cdot \mathbf{v}}{\|\mathbf{u}\|^2}\mathbf{u} = \frac{\langle 8, 2 \rangle \cdot \langle -4, 5 \rangle}{\|\langle 8, 2 \rangle\|^2} \langle 8, 2 \rangle = -\frac{11}{34} \langle 8, 2 \rangle.$$

The vector component of $\mathbf{v}$ orthogonal to $\mathbf{u}$ is given by

$$\mathbf{v} - \text{proj}_{\mathbf{u}}\mathbf{v} = \langle -4, 5 \rangle - \left(-\frac{11}{34} \langle 8, 2 \rangle \right) = \left\langle -\frac{24}{17}, \frac{96}{17} \right\rangle.$$

The following figure shows these vectors:

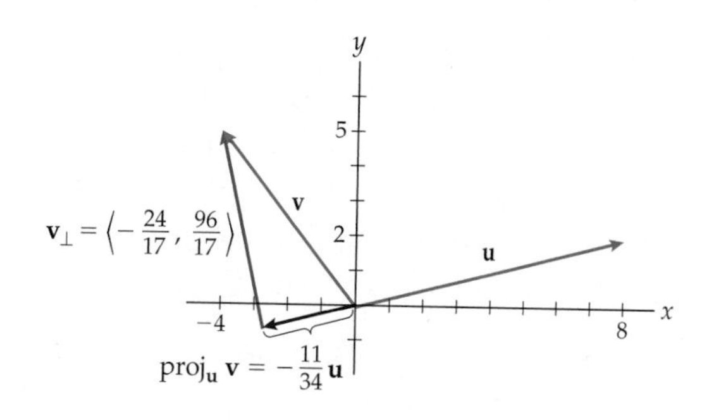

□

EXAMPLE 4 **Finding the distance from a point to a line**

Find the distance from the point $P = (1, 4, -2)$ to the line determined by the points $Q = (3, -2, 5)$ and $R = (0, 4, -3)$.

SOLUTION

The diagram that follows is a schematic illustrating the given situation. The distance we seek is merely the magnitude of the vector component of

$$\overrightarrow{QP} = \langle 1 - 3, 4 - (-2), -2 - 5 \rangle = \langle -2, 6, -7 \rangle$$

orthogonal to

$$\overrightarrow{QR} = \langle 0 - 3, 4 - (-2), -3 - 5 \rangle = \langle -3, 6, -8 \rangle.$$

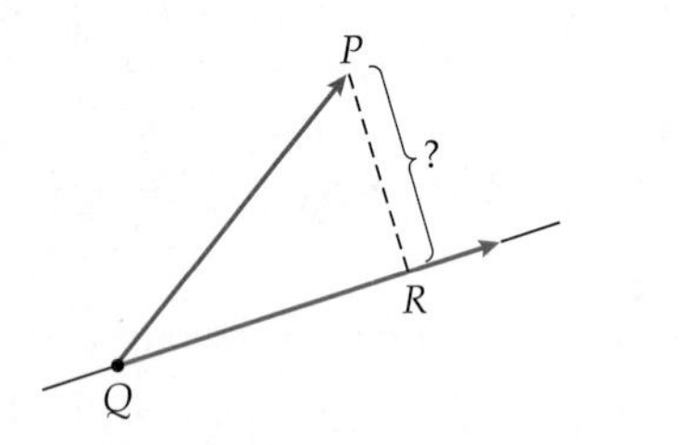

We may first find

$$\text{proj}_{\overrightarrow{QR}} \overrightarrow{QP} = \frac{\overrightarrow{QR} \cdot \overrightarrow{QP}}{\|\overrightarrow{QR}\|^2} \overrightarrow{QR} = \frac{\langle -3, 6, -8 \rangle \cdot \langle -2, 6, -7 \rangle}{\|\langle -3, 6, -8 \rangle\|^2} \langle -3, 6, -8 \rangle = \frac{98}{109} \langle -3, 6, -8 \rangle.$$

Thus the vector component of $\overrightarrow{QP}$ orthogonal to $\overrightarrow{QR}$ is

$$\overrightarrow{QP} - \text{proj}_{\overrightarrow{QR}} \overrightarrow{QP} = \langle -2, 6, -7 \rangle - \frac{98}{109} \langle -3, 6, -8 \rangle = \left\langle \frac{76}{109}, \frac{66}{109}, \frac{21}{109} \right\rangle.$$

The distance from P to the line determined by the points Q and R is the magnitude of this vector, or

$$\left\| \left\langle \frac{76}{109}, \frac{66}{109}, \frac{21}{109} \right\rangle \right\| = \sqrt{\left(\frac{76}{109}\right)^2 + \left(\frac{66}{109}\right)^2 + \left(\frac{21}{109}\right)^2} = \frac{\sqrt{10573}}{109} \approx 0.94. \qquad \square$$

? TEST YOUR UNDERSTANDING

- What is the definition of the dot product $\mathbf{u} \cdot \mathbf{v}$, and what is the geometric relationship among $\mathbf{u}$, $\mathbf{v}$, and $\mathbf{u} \cdot \mathbf{v}$?
- How is the angle between two vectors defined and how is it computed?
- What does it mean to project one vector onto another? When would you want to compute such a projection? How is the projection computed?
- How do you find the vector component of one vector orthogonal to another vector?
- What is the triangle inequality? How is it proved? Why is the inequality $|a+b| \leq |a|+|b|$, where a and b are real numbers, also called the triangle inequality?

EXERCISES 10.3

Thinking Back

- *Law of Cosines:* Use the Law of Cosines to find the measures of the angles of a triangle with side lengths 2, 3, and 4.
- *The distance between a point and a sphere:* Find the shortest distance from the point $(11, -3, 5)$ to the sphere $(x+2)^2 + (y-3)^2 + z^2 = 16$.
- *The Triangle Inequality:* Given $a = 5$ and $b = -3$, verify that the triangle inequality $|a+b| \leq |a| + |b|$ holds.

Concepts

0. *Problem Zero:* Read the section and make your own summary of the material.

1. *True/False:* Determine whether each of the statements that follow is true or false. If a statement is true, explain why. If a statement is false, provide a counterexample.

(a) *True or False:* For all vectors $\mathbf{u}$ and $\mathbf{v}$ in $\mathbb{R}^3$, $\mathbf{u}\cdot\mathbf{v} = \mathbf{v}\cdot\mathbf{u}$.

(b) *True or False:* The component of the projection of $\mathbf{v}$ onto $\mathbf{u}$, $\text{comp}_{\mathbf{u}}\mathbf{v}$, is a vector.

(c) *True or False:* The projection of $\mathbf{v}$ onto $\mathbf{u}$, $\text{proj}_{\mathbf{u}}\mathbf{v}$, is a vector in the direction of $\mathbf{v}$.

(d) *True or False:* Two curves are said to be *orthogonal* at a point if they intersect at the point.

(e) *True or False:* If θ is the angle between two nonzero vectors $\mathbf{u}$ and $\mathbf{v}$, then $\cos\theta = \dfrac{\mathbf{u}\cdot\mathbf{v}}{\|\mathbf{u}\|\|\mathbf{v}\|}$.

(f) *True or False:* If $\mathbf{u}$ and $\mathbf{v}$ are nonzero vectors such that $\text{proj}_{\mathbf{u}}\mathbf{v} = \text{proj}_{\mathbf{v}}\mathbf{u}$, then $\mathbf{u}$ and $\mathbf{v}$ are either equal or orthogonal.

(g) *True or False:* If $\mathbf{u}\cdot\mathbf{v} = \|\mathbf{u}\|\|\mathbf{v}\|$, where $\mathbf{u}$ and $\mathbf{v}$ are nonzero, then $\mathbf{u} = k\,\mathbf{v}$, where $k > 0$.

(h) *True or False:* The product $(\mathbf{u}\cdot\mathbf{v})\cdot\mathbf{w}$ is defined for vectors $\mathbf{u}$, $\mathbf{v}$, and $\mathbf{w}$.

2. *Examples:* Construct examples of the thing(s) described in the following. Try to find examples that are different than any in the reading.

(a) Two nonzero vectors in $\mathbb{R}^3$ whose dot product is zero.

(b) A function $y = f(x)$ that is orthogonal to the function $y = \sin x$ at $x = 0$.

(c) A vector $\mathbf{v} \neq \mathbf{i}$ such that $\text{proj}_{\mathbf{i}}\mathbf{v} = \mathbf{i}$.

3. State the definition of the dot product.

4. What does it mean geometrically for two vectors to be orthogonal at a point? What does it mean algebraically? What do we mean when we say that two curves are orthogonal at a point of intersection?

5. Why is the Law of Cosines a generalization of the Pythagorean theorem?

6. Let $\mathbf{v} = a\,\mathbf{i} + b\,\mathbf{j}$ and $\mathbf{w} = c\,\mathbf{i} + d\,\mathbf{j}$. Give conditions on the constants a, b, c, and d that guarantee that

(a) $\mathbf{v}$ is parallel to $\mathbf{w}$.

(b) $\mathbf{v}$ is perpendicular to $\mathbf{w}$.

7. Let $\mathbf{v} = a\mathbf{i} + b\mathbf{j} + c\mathbf{k}$ and $\mathbf{w} = \alpha\mathbf{i} + \beta\mathbf{j} + \gamma\mathbf{j}$. Give conditions on the constants a, b, c, α, β, and γ that guarantee that

(a) $\mathbf{v}$ is parallel to $\mathbf{w}$.

(b) $\mathbf{v}$ is perpendicular to $\mathbf{w}$.

8. What is the relationship between $\|\mathbf{v}\|$ and $\mathbf{v}\cdot\mathbf{v}$?

9. Let $\mathbf{u}$ be a nonzero vector.

(a) Show that $\mathbf{u}\cdot\mathbf{v} = \mathbf{u}\cdot\mathbf{w}$ does *not* necessarily imply that $\mathbf{v} = \mathbf{w}$.

(b) What geometric relationship must $\mathbf{u}$, $\mathbf{v}$, and $\mathbf{w}$ satisfy if $\mathbf{u}\cdot\mathbf{v} = \mathbf{u}\cdot\mathbf{w}$?

10. Let $\mathbf{u}$ and $\mathbf{v}$, be nonzero vectors.

(a) When does $\text{comp}_{\mathbf{u}}\mathbf{v} = \text{comp}_{\mathbf{v}}\mathbf{u}$?

(b) When does $\text{proj}_{\mathbf{u}}\mathbf{v} = \text{proj}_{\mathbf{v}}\mathbf{u}$?

11. What geometric relationship must two vectors have in order for $\|\mathbf{u}+\mathbf{v}\| = \|\mathbf{u}\| + \|\mathbf{v}\|$?

12. Consider the position vector $\mathbf{i} = \langle 1, 0\rangle$ in $\mathbb{R}^2$. Describe the set of position vectors $\mathbf{v}$ in $\mathbb{R}^2$ with the property that $\mathbf{v}\cdot\mathbf{i} = 0$.

13. Consider the position vector $\mathbf{i} = \langle 1, 0, 0\rangle$ in $\mathbb{R}^3$. Describe the set of position vectors $\mathbf{v}$ in $\mathbb{R}^3$ with the property that $\mathbf{v}\cdot\mathbf{i} = 0$.

14. Consider the position vector $\mathbf{i} = \langle 1, 0\rangle$ in $\mathbb{R}^2$. Describe the set of position vectors $\mathbf{v}$ in $\mathbb{R}^2$ with the property that $\text{proj}_{\mathbf{i}}\mathbf{v} = \mathbf{i}$.

15. Consider the position vector $\mathbf{i} = \langle 1, 0, 0\rangle$ in $\mathbb{R}^3$. Describe the set of position vectors $\mathbf{v}$ in $\mathbb{R}^3$ with the property that $\text{proj}_{\mathbf{i}}\mathbf{v} = \mathbf{i}$.

16. Let $\mathbf{v}_0 = \langle a, b\rangle$. Describe the set of points (x, y) such that, for $\mathbf{v} = \langle x, y\rangle$,

(a) $\mathbf{v}\cdot\mathbf{v}_0 = 0$.

(b) $(\mathbf{v}-\mathbf{v}_0)\cdot\mathbf{v}_0 = 0$.

(c) $(\mathbf{v}-\mathbf{v}_0)\cdot\mathbf{v} = 0$.

17. Recall that when two lines are perpendicular, their slopes are negative reciprocals.

(a) Find a vector parallel to the line $y = mx + b$.

(b) If $m \neq 0$, find the slope of any line perpendicular to $y = mx + b$ and find a vector parallel to that perpendicular line.

(c) Show that the dot product of the vectors in parts (a) and (b) is zero.

18. Let $\mathbf{v}_1$ and $\mathbf{v}_2$ be two nonzero position vectors in $\mathbb{R}^2$ that are not scalar multiples of each other. Explain why, given any vector $\mathbf{w}$ in $\mathbb{R}^2$, there are scalars c_1 and c_2 such that $\mathbf{w} = c_1\mathbf{v}_1 + c_2\mathbf{v}_2$.

19. To illustrate the concept in Exercise 18:
(a) Explain why the vectors $\mathbf{v}_1 = \langle 1, -2 \rangle$ and $\mathbf{v}_2 = \langle 3, 5 \rangle$ are not scalar multiples of each other.
(b) Find scalars c_1 and c_2 such that $\langle 5, 1 \rangle = c_1\mathbf{v}_1 + c_2\mathbf{v}_2$.

Skills

In Exercises 20-23, find the dot product of the given pairs of vectors and the angle between the two vectors.

20. $\mathbf{u} = \langle 1, 2 \rangle$, $\mathbf{v} = \langle 3, 5 \rangle$

21. $\mathbf{u} = \langle 2, 0, -5 \rangle$, $\mathbf{v} = \langle -3, 7, -1 \rangle$

22. $\mathbf{u} = \langle 3, -1, 2 \rangle$, $\mathbf{v} = \langle -4, -6, 3 \rangle$

23. $\mathbf{u} = \langle -5, 1, 3 \rangle$, $\mathbf{v} = \langle -3, 2, 7 \rangle$

In Exercises 24-27, find $\text{comp}_{\mathbf{u}}\mathbf{v}$, $\text{proj}_{\mathbf{u}}\mathbf{v}$, and the component of $\mathbf{v}$ orthogonal to $\mathbf{u}$.

24. $\mathbf{u} = \langle 1, 2 \rangle$, $\mathbf{v} = \langle 3, 5 \rangle$

25. $\mathbf{u} = \langle 3, -1, 2 \rangle$, $\mathbf{v} = \langle -4, -6, 3 \rangle$

26. $\mathbf{u} = \langle 2, 0, -5 \rangle$, $\mathbf{v} = \langle -3, 7, -1 \rangle$

27. $\mathbf{u} = \langle 3, 1, -2 \rangle$, $\mathbf{v} = \langle -6, -2, 4 \rangle$

In Exercises 28-31, find $\text{proj}_{\mathbf{u}}\mathbf{v}$ and $\text{proj}_{\mathbf{v}}\mathbf{u}$.

28. $\mathbf{u} = \langle 1, 4 \rangle$, $\mathbf{v} = \langle 2, -3 \rangle$

29. $\mathbf{u} = \langle 3, -4 \rangle$, $\mathbf{v} = \langle 16, 12 \rangle$

30. $\mathbf{u} = \langle 3, 0, 1 \rangle$, $\mathbf{v} = \langle 2, 2, -5 \rangle$

31. $\mathbf{u} = \langle 1, -5, -1 \rangle$, $\mathbf{v} = \langle 0, 1, 0 \rangle$

In Exercises 32–36, (a) compute $\mathbf{u} \cdot \mathbf{v}$, (b) find the angle between $\mathbf{u}$ and $\mathbf{v}$, and (c) find $\text{proj}_{\mathbf{u}}\mathbf{v}$.

32. Let $\mathbf{u} = \langle 1, 5 \rangle$ and $\mathbf{v} = \langle 2, 7 \rangle$.

33. Let $\mathbf{u} = \langle -2, 3, 5 \rangle$ and $\mathbf{v} = \langle 13, -5, 8 \rangle$.

34. Let $\mathbf{u} = \langle 0, 3, -4 \rangle$ and $\mathbf{v} = \langle -5, 6, 0 \rangle$.

35. Let $\mathbf{u} = \langle 2, 4, -1, 2 \rangle$ and $\mathbf{v} = \langle -1, 3, -2, 6 \rangle$.

36. Let $\mathbf{u} = \langle 5, -2, 3, 4 \rangle$ and $\mathbf{v} = \langle 0, -1, 1, 7 \rangle$.

For Exercises 37 and 38, let $P = (2, 5, 7)$, $Q = (-2, 1, -5)$, and $R = (-3, 0, 4)$.

37. Find the distance between the point P and the line determined by the points Q and R.

38. Find the altitude of triangle $\triangle PQR$ from vertex R to side $\overline{PQ}$.

39. Find the angle between the diagonal of a face of a cube and the adjoining edge of the cube that is *not* an edge of that face.

40. Find the angle between the diagonal of a cube and an adjoining edge of the cube.

41. Find the angle between the diagonal of a cube and an adjoining diagonal of one of the faces of the cube.

42. Find the angle between two distinct diagonals of a cube.

Exercises 43–53 deal with ***direction angles*** and ***direction cosines***. Let $\mathbf{v}$ be a nonzero vector in $\mathbb{R}^3$. The ***direction angles*** α, β, and γ of $\mathbf{v}$ are the angles that $\mathbf{v}$ makes with the positive x-, y- and z-axes, respectively. The ***direction cosines*** of $\mathbf{v}$ are $\cos\alpha$, $\cos\beta$, and $\cos\gamma$.

Find the direction angles and direction cosines for the vectors given in Exercises 43–46.

43. $\langle 1, 2, 3 \rangle$

44. $\langle -2, 0, 3 \rangle$

45. $\langle -1, 1, -4 \rangle$

46. $\langle -3, 4, 2 \rangle$

47. Show that for any vector $\mathbf{v}$ in $\mathbb{R}^3$,

$$\mathbf{v} = \|\mathbf{v}\| \left((\cos\alpha)\mathbf{i} + (\cos\beta)\mathbf{j} + (\cos\gamma)\mathbf{k} \right),$$

where α, β, and γ are the direction angles of $\mathbf{v}$.

48. Use Exercise 47 to show that if $\cos\alpha$, $\cos\beta$, and $\cos\gamma$ are the direction cosines of a vector $\mathbf{v}$, then

$$\cos^2\alpha + \cos^2\beta + \cos^2\gamma = 1.$$

In Exercises 49-51, two direction cosines are given. Use Exercise 48 to find the third direction cosine.

49. $\cos\alpha = \dfrac{1}{2}$, $\cos\beta = \dfrac{1}{2}$.

50. $\cos\beta = \dfrac{1}{4}$, $\cos\gamma = \dfrac{1}{3}$.

51. $\cos\alpha = \dfrac{1}{2}$, $\cos\gamma = \dfrac{\sqrt{3}}{4}$.

52. Let $\mathbf{v}$ be a vector in $\mathbb{R}^n$.
(a) How would you compute the direction angles and direction cosines for $\mathbf{v}$?
(b) Show that if $\alpha_1, \alpha_2, \ldots, \alpha_n$ are the direction angles for $\mathbf{v}$, then $\cos^2\alpha_1 + \cos^2\alpha_2 + \cdots + \cos^2\alpha_n = 1$.

53. Let $\mathbf{v} = \langle 1, 1, 1, \ldots, 1 \rangle$ be a vector in $\mathbb{R}^n$.
(a) Use Exercise 52 to find the direction angles $\alpha_1, \alpha_2, \ldots, \alpha_n$.
(b) Show that $\alpha_i \to \dfrac{\pi}{2}$ as $n \to \infty$.

Applications

54. Ian is climbing a glacier to a col, a gap in a ridge. The snow is steep, so he is zigzagging up, first northeast, then southeast, always 45 degrees away from due east, and always at 30 degrees from the horizontal. He travels at 0.5 mile per hour in whichever direction he heads.

(a) What is the component of his velocity horizontally, due east?

(b) The map shows that the col is $\frac{1}{4}$ mile east of him. How long will it take Ian to get there?

55. Annie is making a north-to-south crossing from one island to another, with a tidal current in the channel. The current in the channel is moving at 1 mph due west. Annie has pointed her kayak in the direction $\langle 1, -4 \rangle$ and is paddling at 2 mph.

(a) What angle does Annie's boat make with a direct southerly heading?

(b) What is the component of Annie's velocity toward the south?

(c) The crossing is 2 miles due south. How long will it take Annie?

Proofs

56. Let $\mathbf{u}$ and $\mathbf{v}$ be two nonzero vectors in $\mathbb{R}^2$. Prove that $\mathbf{u} \cdot \mathbf{v} = \|\mathbf{u}\|\|\mathbf{v}\| \cos\theta$, where θ is the angle between $\mathbf{u}$ and $\mathbf{v}$.

57. Show that for any vector $\mathbf{v}$ in $\mathbb{R}^3$,

$$\mathbf{v} = (\mathbf{v} \cdot \mathbf{i})\mathbf{i} + (\mathbf{v} \cdot \mathbf{j})\mathbf{j} + (\mathbf{v} \cdot \mathbf{k})\mathbf{k}.$$

58. Show that $\mathbf{u} \cdot \mathbf{v} = \|\mathbf{u}\|\|\mathbf{v}\| \cos\theta$ when $\mathbf{u}$ and $\mathbf{v}$ are nonzero vectors such that $\mathbf{u} = k\mathbf{v}$ with $k < 0$.

59. Use the fact that $\mathbf{u} \cdot \mathbf{v} = \|\mathbf{u}\|\|\mathbf{v}\| \cos\theta$ to prove the Cauchy–Schwarz inequality $|\mathbf{u} \cdot \mathbf{v}| \le \|\mathbf{u}\|\|\mathbf{v}\|$. What relationship must $\mathbf{u}$ and $\mathbf{v}$ have in order for $|\mathbf{u} \cdot \mathbf{v}| = \|\mathbf{u}\|\|\mathbf{v}\|$?

60. Let θ be the angle between nonzero vectors $\mathbf{u}$ and $\mathbf{v}$. Prove each of the following:

(a) θ is acute if and only if $\mathbf{u} \cdot \mathbf{v} > 0$

(b) θ is right if and only if $\mathbf{u} \cdot \mathbf{v} = 0$

(c) θ is obtuse if and only if $\mathbf{u} \cdot \mathbf{v} < 0$

61. Prove that vectors $\mathbf{u}$ and $\mathbf{v}$ are orthogonal if and only if $\mathbf{u} \cdot \mathbf{v} = 0$. (This is Theorem 10.19.)

62. Prove the following statements from Theorem 10.14 for vectors in $\mathbb{R}^3$:

(a) $\mathbf{u} \cdot \mathbf{v} = \mathbf{v} \cdot \mathbf{u}$

(b) $k(\mathbf{u} \cdot \mathbf{v}) = (k\,\mathbf{u}) \cdot \mathbf{v} = \mathbf{u} \cdot (k\,\mathbf{v})$

(c) $\mathbf{v} \cdot \mathbf{v} = \|\mathbf{v}\|^2$

63. Let $\mathbf{u}$ be a nonzero vector and let $\mathbf{v}$ be any vector. Show that the decomposition $\mathbf{v} = \mathbf{v}_{\|} + \mathbf{v}_{\perp}$, where $\mathbf{v}_{\|}$ is parallel to $\mathbf{u}$ and $\mathbf{v}_{\perp}$ is orthogonal to $\mathbf{u}$, is unique. (*Hint: Assume that there is another decomposition with these properties, and show that the two decompositions must be identical.*)

64. Use a vector argument to prove that a parallelogram is a rectangle if and only if the diagonals have the same length.

65. Use a vector argument to prove that a parallelogram is a rhombus if and only if the diagonals are perpendicular.

Thinking Forward

Lines and vectors: Find the specified unit vectors.

- Two unit vectors parallel to the line $y = \frac{3}{5}x - 7$.
- Two unit vectors parallel to the line $y = mx + b$.
- Two unit vectors perpendicular to the line $y = \frac{3}{5}x - 7$.
- Two unit vectors perpendicular to the line $y = mx + b$.

10.4 CROSS PRODUCT

- Multiplying two vectors in $\mathbb{R}^3$ using the cross product
- The geometry of the cross product
- The relationship between the algebra and the geometry of the cross product

Determinants of 3×3 Matrices

Before we get to the definition of the cross product, it will be convenient to discuss how to compute the determinant of a 3-by-3 matrix, or, more briefly, a 3×3 matrix. First, a ***matrix*** is a rectangular array of entries. Here we are interested primarily in 3×3 matrices (i.e., arrays with 3 rows and 3 columns):

$$\begin{bmatrix} a_1 & a_2 & a_3 \\ b_1 & b_2 & b_3 \\ c_1 & c_2 & c_3 \end{bmatrix}.$$

The ***determinant*** of a square matrix is a value derived from the entries of the matrix. In particular, for a 3×3 matrix we may use the following definition:

DEFINITION 10.24 The Determinant of a 3 x 3 Matrix

The ***determinant*** of the 3×3 matrix $A = \begin{bmatrix} a_1 & a_2 & a_3 \\ b_1 & b_2 & b_3 \\ c_1 & c_2 & c_3 \end{bmatrix}$, denoted by $\det A$, is the sum

$$\det \begin{bmatrix} a_1 & a_2 & a_3 \\ b_1 & b_2 & b_3 \\ c_1 & c_2 & c_3 \end{bmatrix} = a_1b_2c_3 - a_1b_3c_2 + a_2b_3c_1 - a_2b_1c_3 + a_3b_1c_2 - a_3b_2c_1.$$

At first it might seem difficult to remember how to compute this sum. Fortunately, there is a nice visual trick that can be used to help. Compare the products along the six colored "diagonals" shown in the following matrix with the final sum in Definition 10.24. Note that four of the six diagonals wrap around the matrix as they descend.

Every 3×3 matrix has six "diagonals"

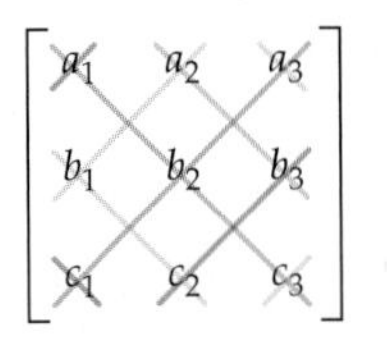

If the diagonal goes down from left to right, add the product. If the diagonal goes down from right to left, subtract the product. The resulting sum is the determinant of the 3×3 matrix.

For example, to compute the determinant of the matrix

$$M = \begin{bmatrix} 1 & -2 & 5 \\ 0 & 3 & -1 \\ -3 & 2 & 4 \end{bmatrix},$$

we add the products along the six diagonals, making sure to use the appropriate signs:

$$\begin{aligned}\det M &= (1)(3)(4) + (-2)(-1)(-3) + (5)(0)(2) - (1)(-1)(2) - (-2)(0)(4) - (5)(3)(-3) \\ &= 12 - 6 + 0 + 2 + 0 + 45 = 53.\end{aligned}$$

The determinant is defined for all square matrices, but for larger square matrices a recursive definition is used. For example, the determinant of a 4×4 matrix is a modified sum of the determinants of four 3×3 matrices, the determinant of a 5×5 matrix is a modified sum of the determinants of five 4×4 matrices, etc. As a result, these determinants have more summands than those that lie along diagonals. In general, the determinant of an $n \times n$ matrix involves the sum of $n!$ products, with each product containing n factors. For example, computing the determinant of a 5×5 matrix involves the sum of $5! = 120$ products, each with 5 factors. In this text we will be using only the determinants of 3×3 matrices to compute cross products. The study of matrices and determinants belongs to a branch of mathematics called *linear algebra*. Here we are introducing the minimum necessary to help us define the cross product.

The Cross Product

Our final vector product, the cross product, will be defined in terms of the determinant of a matrix containing the components of two vectors. Later in this section we will examine geometric properties of this product. We will also see that the cross product can be used to determine areas of parallelograms and, along with the dot product, to compute the volumes of solids known as parallelepipeds. In Section 10.6 we will see how the cross product is used to find equations of planes in $\mathbb{R}^3$.

The cross product differs from our two previous vector products in that it is defined only for vectors in $\mathbb{R}^3$. (Recall that when we multiply a vector by a scalar, the vector can have any number of components and that we may take the dot product on any two vectors with the same number of components.)

DEFINITION 10.25 The Cross Product

Let $\mathbf{u} = \langle u_1, u_2, u_3 \rangle = u_1\mathbf{i} + u_2\mathbf{j} + u_3\mathbf{k}$ and $\mathbf{v} = \langle v_1, v_2, v_3 \rangle = v_1\mathbf{i} + v_2\mathbf{j} + v_3\mathbf{k}$. Then the ***cross product*** of $\mathbf{u}$ and $\mathbf{v}$, denoted by $\mathbf{u} \times \mathbf{v}$, is

$$\mathbf{u} \times \mathbf{v} = \det \begin{bmatrix} \mathbf{i} & \mathbf{j} & \mathbf{k} \\ u_1 & u_2 & u_3 \\ v_1 & v_2 & v_3 \end{bmatrix} = (u_2v_3 - u_3v_2)\mathbf{i} + (u_3v_1 - u_1v_3)\mathbf{j} + (u_1v_2 - u_2v_1)\mathbf{k}.$$

For example, we'll compute the cross product of the vectors $\mathbf{u} = \langle -1, 3, 4 \rangle$ and $\mathbf{v} = \langle 2, 0, -5 \rangle$:

$$\begin{aligned}\mathbf{u} \times \mathbf{v} &= \det \begin{bmatrix} \mathbf{i} & \mathbf{j} & \mathbf{k} \\ -1 & 3 & 4 \\ 2 & 0 & -5 \end{bmatrix} \\ &= (3 \cdot (-5) - 4 \cdot 0)\mathbf{i} + (4 \cdot 2 - (-1) \cdot (-5))\mathbf{j} + (-1 \cdot 0 - 3 \cdot 2)\mathbf{k} \\ &= -15\mathbf{i} + 3\mathbf{j} - 6\mathbf{k} = \langle -15, 3, -6 \rangle.\end{aligned}$$

Shortly we will discuss the geometric relationship between $\mathbf{u}$, $\mathbf{v}$, and $\mathbf{u} \times \mathbf{v}$, but right now at least we know how to compute a cross product.

One simple geometric consequence of the definition is that the cross product of two parallel vectors is $\mathbf{0}$.

THEOREM 10.26 **The Cross Product of Parallel Vectors**

The cross product of two parallel vectors $\mathbf{u}$ and $\mathbf{v}$ in $\mathbb{R}^3$ is $\mathbf{u} \times \mathbf{v} = \mathbf{0}$.

The proof of Theorem 10.26 follows from the definition of the cross product and is left for Exercise 65.

Algebraic Properties of the Cross Product

The cross product may be the first product you've encountered that is *not* commutative. However, it is ***anticommutative***. That is, for every two vectors $\mathbf{u}$ and $\mathbf{v}$ in $\mathbb{R}^3$, we have the following:

THEOREM 10.27 **The Cross Product Is Anticommutative**

For any vectors $\mathbf{u}$ and $\mathbf{v}$ in $\mathbb{R}^3$,

$$\mathbf{v} \times \mathbf{u} = -(\mathbf{u} \times \mathbf{v}).$$

The proof of Theorem 10.27 follows from the definition of the cross product and is left for Exercise 66.

THEOREM 10.28 **Multiplication by a Scalar and the Cross Product**

For any vectors $\mathbf{u}$ and $\mathbf{v}$ in $\mathbb{R}^3$ and any scalar c,

$$c(\mathbf{u} \times \mathbf{v}) = (c\mathbf{u}) \times \mathbf{v} = \mathbf{u} \times (c\mathbf{v}).$$

Again the proof follows from the definition of the cross product. This proof is left for Exercise 67.

As with multiplication and addition of scalars, the cross product is distributive over addition. Because the cross product is not commutative, we need to state *two* distributive properties. Again, their proofs follow from the definition of the cross product; they are left for Exercise 68.

THEOREM 10.29 **Distributive Properties of the Cross Product**

Let $\mathbf{u}$, $\mathbf{v}$, and $\mathbf{w}$ be vectors in $\mathbb{R}^3$. Then

(a) $\mathbf{u} \times (\mathbf{v} + \mathbf{w}) = \mathbf{u} \times \mathbf{v} + \mathbf{u} \times \mathbf{w}$.

(b) $(\mathbf{u} + \mathbf{v}) \times \mathbf{w} = \mathbf{u} \times \mathbf{w} + \mathbf{v} \times \mathbf{w}$.

The final algebraic property we discuss here provides a relationship between the dot product and cross product and will be used shortly to help us understand the geometry of the cross product. It is known as Lagrange's identity.

THEOREM 10.30 **Lagrange's Identity**

Let $\mathbf{u}$ and $\mathbf{v}$ be vectors in $\mathbb{R}^3$. Then

$$\|\mathbf{u} \times \mathbf{v}\|^2 = \|\mathbf{u}\|^2\|\mathbf{v}\|^2 - (\mathbf{u} \cdot \mathbf{v})^2.$$

The proof of Theorem 10.30 entails expanding the quantities on both sides of Lagrange's identity. This task is left for Exercise 70.

The Geometry of the Cross Product

We begin by showing that the cross product $\mathbf{u} \times \mathbf{v}$ is orthogonal to both $\mathbf{u}$ and $\mathbf{v}$. As an immediate consequence the cross product is orthogonal to any plane containing both $\mathbf{u}$ and $\mathbf{v}$.

THEOREM 10.31 **The Cross Product u x v is Orthogonal to Both u and v**

Let $\mathbf{u}$ and $\mathbf{v}$ be vectors in $\mathbb{R}^3$. Then

(a) $\mathbf{u} \cdot (\mathbf{u} \times \mathbf{v}) = 0$.

(b) $\mathbf{v} \cdot (\mathbf{u} \times \mathbf{v}) = 0$.

Proof. We prove part (a) and leave part (b) for Exercise 69. Let $\mathbf{u} = \langle u_1, u_2, u_3 \rangle$ and $\mathbf{v} = \langle v_1, v_2, v_3 \rangle$. Then, by the definition of the cross product,

$$\mathbf{u} \times \mathbf{v} = \langle u_2 v_3 - u_3 v_2, u_3 v_1 - u_1 v_3, u_1 v_2 - u_2 v_1 \rangle$$

and

$$\begin{aligned} \mathbf{u} \cdot (\mathbf{u} \times \mathbf{v}) &= \langle u_1, u_2, u_3 \rangle \cdot \langle u_2 v_3 - u_3 v_2, u_3 v_1 - u_1 v_3, u_1 v_2 - u_2 v_1 \rangle \\ &= u_1(u_2 v_3 - u_3 v_2) + u_2(u_3 v_1 - u_1 v_3) + u_3(u_1 v_2 - u_2 v_1) \\ &= 0. \end{aligned}$$

■

Recall that for vectors $\mathbf{u}$ and $\mathbf{v}$, we have $\mathbf{u} \cdot \mathbf{v} = \|\mathbf{u}\| \|\mathbf{v}\| \cos\theta$, where θ is the angle between the vectors. The magnitude of the cross product is related to the magnitudes of the vectors and the angle between them.

THEOREM 10.32 **The Cross Product u x v and the Angle Between u and v**

Let $\mathbf{u}$ and $\mathbf{v}$ be nonzero vectors in $\mathbb{R}^3$ with the same initial point. Then

$$\|\mathbf{u} \times \mathbf{v}\| = \|\mathbf{u}\| \|\mathbf{v}\| \sin\theta,$$

where θ is the angle between $\mathbf{u}$ and $\mathbf{v}$.

Proof. By Lagrange's identity

$$\|\mathbf{u} \times \mathbf{v}\|^2 = \|\mathbf{u}\|^2 \|\mathbf{v}\|^2 - (\mathbf{u} \cdot \mathbf{v})^2.$$

But since $\mathbf{u} \cdot \mathbf{v} = \|\mathbf{u}\| \|\mathbf{v}\| \cos\theta$, we have

$$\|\mathbf{u} \times \mathbf{v}\|^2 = \|\mathbf{u}\|^2 \|\mathbf{v}\|^2 - (\|\mathbf{u}\| \|\mathbf{v}\| \cos\theta)^2 = \|\mathbf{u}\|^2 \|\mathbf{v}\|^2 (1 - \cos^2\theta) = \|\mathbf{u}\|^2 \|\mathbf{v}\|^2 \sin^2\theta.$$

Taking the square root of the leftmost and rightmost quantities produces our desired result. ■

Two nonparallel vectors determine a parallelogram. The cross product can help us find the area of that parallelogram.

The area of the parallelogram is $\|\mathbf{u}\| \|\mathbf{v}\| \sin\theta$

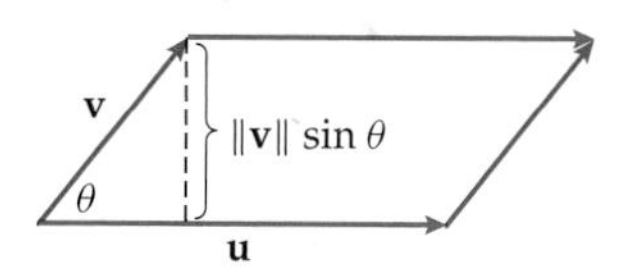

THEOREM 10.33

The Area of a Parallelogram

Let $\mathbf{u}$ and $\mathbf{v}$ be vectors in $\mathbb{R}^3$. Then the area of the parallelogram determined by $\mathbf{u}$ and $\mathbf{v}$ is given by $\|\mathbf{u} \times \mathbf{v}\|$.

Proof. We first consider the case where $\mathbf{u}$ and $\mathbf{v}$ are parallel vectors. Here the parallelogram determined by $\mathbf{u}$ and $\mathbf{v}$ is degenerate; that is, it is collapsed into a line segment and has zero area. By Theorem 10.26 $\mathbf{u} \times \mathbf{v} = \mathbf{0}$, so $\|\mathbf{u} \times \mathbf{v}\| = \|\mathbf{0}\| = 0$.

If $\mathbf{u}$ and $\mathbf{v}$ are not parallel, then the area of the parallelogram determined by $\mathbf{u}$ and $\mathbf{v}$ is the product of the length of one of the sides and the distance between that side and the opposite side.

In our parallelogram, $\|\mathbf{u}\|$ is the length of one side. In the parallelogram shown in the previous figure, we see that the distance from that side to the opposite side is $\|\mathbf{v}\| \sin\theta$. Therefore the area of the parallelogram is $\|\mathbf{u}\|\|\mathbf{v}\| \sin\theta = \|\mathbf{u} \times \mathbf{v}\|$. ■

We have already discussed what it means for a three-dimensional coordinate system to be left- or right-handed. Similarly, three vectors $\mathbf{u}$, $\mathbf{v}$, and $\mathbf{w}$ that cannot be translated into the same plane are said to form a ***right-handed triple*** if, when the index finger of the right hand points in the direction of $\mathbf{u}$ and the middle finger of the right hand points in the direction of $\mathbf{v}$, then the right thumb will naturally point in the direction of $\mathbf{w}$. In particular, for any two nonparallel vectors $\mathbf{u}$ and $\mathbf{v}$ in $\mathbb{R}^3$, the vectors $\mathbf{u}$, $\mathbf{v}$, and $\mathbf{u} \times \mathbf{v}$ always form a right-handed triple.

The vectors $\mathbf{u}$, $\mathbf{v}$, and $\mathbf{u} \times \mathbf{v}$ form a right-handed triple

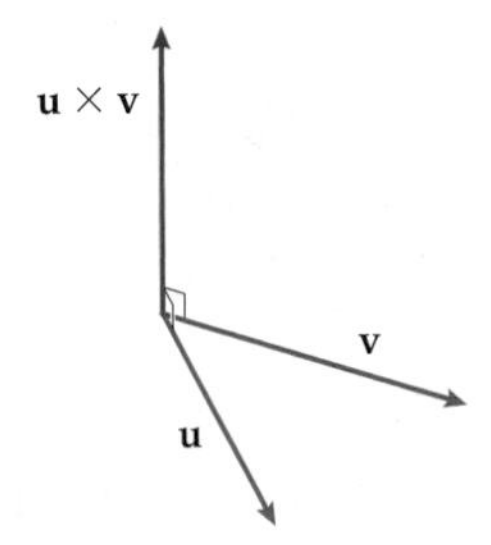

THEOREM 10.34

Two Nonparallel Vectors and Their Cross Product Form a Right-Handed Triple

Let $\mathbf{u}$ and $\mathbf{v}$ be nonparallel vectors in $\mathbb{R}^3$. Then the vectors $\mathbf{u}$, $\mathbf{v}$, and $\mathbf{u} \times \mathbf{v}$ form a right-handed triple.

Proof. Let $\mathbf{u}$ and $\mathbf{v}$ be nonparallel position vectors in $\mathbb{R}^3$. We may position our coordinate axes so that $\mathbf{u}$ lies along the positive x-axis and $\mathbf{v}$ lies in the xy-plane and has a positive y-coordinate. That is, $\mathbf{u} = \langle u_1, 0, 0\rangle$ and $\mathbf{v} = \langle v_1, v_2, 0\rangle$ with $u_1 > 0$ and $v_2 > 0$. The sign of v_1 may be positive, zero, or negative. Two of these possibilities are illustrated here:

Vectors $\mathbf{u}$ and $\mathbf{v}$ with $v_1 > 0$ *Vectors $\mathbf{u}$ and $\mathbf{v}$ with $v_1 < 0$*

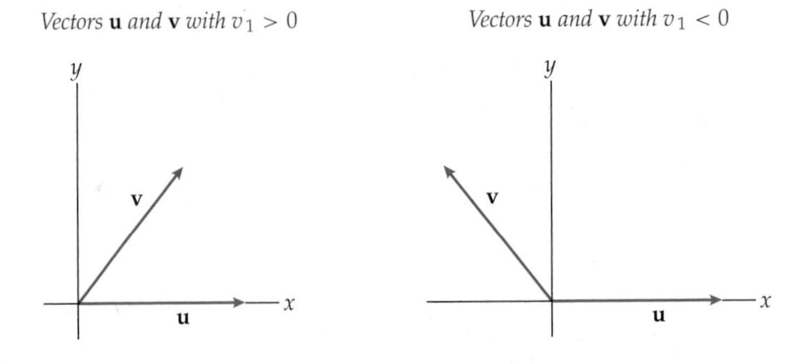

In any of the three cases, when we take the cross product we obtain

$$\mathbf{u} \times \mathbf{v} = \det \begin{bmatrix} \mathbf{i} & \mathbf{j} & \mathbf{k} \\ u_1 & 0 & 0 \\ v_1 & v_2 & 0 \end{bmatrix} = \langle 0, 0, u_1 v_2 \rangle.$$

Since $\mathbf{u} \times \mathbf{v}$ has a positive z-coordinate, we have our desired result. ■

The next theorem reiterates several of the basic geometric properties of the cross product that we have already proved. We summarize these properties in Theorem 10.35 because they may be used as an alternative (geometric) definition of the cross product. When they are used as such, the algebraic properties follow as a consequence.

THEOREM 10.35

The Geometry of the Cross Product

Let $\mathbf{u}$ and $\mathbf{v}$ be vectors in $\mathbb{R}^3$.

If $\mathbf{u}$ and $\mathbf{v}$ are parallel then $\mathbf{u} \times \mathbf{v} = \mathbf{0}$.

If $\mathbf{u}$ and $\mathbf{v}$ are not parallel, then $\mathbf{u} \times \mathbf{v}$ has the following properties:

(a) $\|\mathbf{u} \times \mathbf{v}\| = \|\mathbf{u}\| \|\mathbf{v}\| \sin \theta$.

(b) $\mathbf{u} \times \mathbf{v}$ is perpendicular to any plane containing both $\mathbf{u}$ and $\mathbf{v}$.

(c) $\mathbf{u}$, $\mathbf{v}$, and $\mathbf{u} \times \mathbf{v}$ form a right-handed triple.

Triple Scalar Product

We cannot randomly combine three vectors $\mathbf{u}$, $\mathbf{v}$, and $\mathbf{w}$ from $\mathbb{R}^3$ with dot and cross products. For example, the combination $(\mathbf{u} \cdot \mathbf{v}) \cdot \mathbf{w}$ is *not* defined, since $\mathbf{u} \cdot \mathbf{v}$ is a scalar and we need two vectors to form a dot product. Similarly, $(\mathbf{u} \cdot \mathbf{v}) \times \mathbf{w}$ is not defined. (Why?) However, the product $\mathbf{u} \cdot (\mathbf{v} \times \mathbf{w})$ is defined and has an important geometrical interpretation. In addition, we will see that $\mathbf{u} \cdot (\mathbf{v} \times \mathbf{w}) = (\mathbf{u} \times \mathbf{v}) \cdot \mathbf{w}$. Both of these are examples of ***triple scalar products***. In Exercise 80 we discuss the ***vector triple product*** $\mathbf{u} \times (\mathbf{v} \times \mathbf{w})$.

A ***parallelepiped*** is a three-dimensional analog of a parallelogram, in much the same way that a cube is a three-dimensional analog of a square. Specifically, a parallelepiped is a six-sided solid whose surface consists of three pairs of parallel faces, each of which is a parallelogram. Any three vectors $\mathbf{u}$, $\mathbf{v}$, and $\mathbf{w}$ in $\mathbb{R}^3$ that do not lie in the same plane will determine a parallelepiped.

The parallelepiped determined by $\mathbf{u}$, $\mathbf{v}$, *and* $\mathbf{w}$

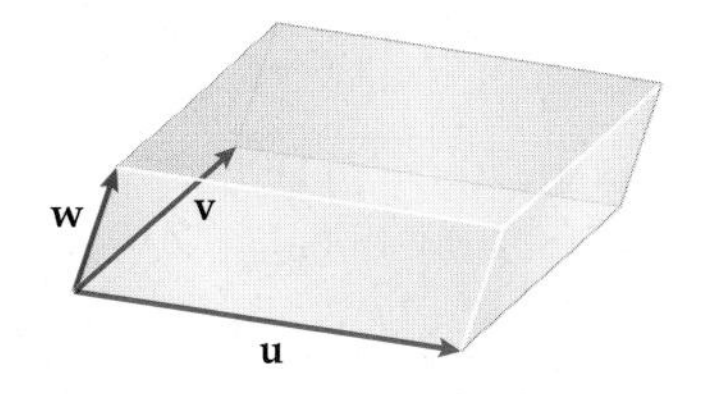

In Theorem 10.33 we saw that the area of a parallelogram involves a cross product. Similarly, the volume of the parallelepiped involves a triple scalar product.

THEOREM 10.36

The Triple Scalar Product and the Volume of Parallelepipeds

Let $\mathbf{u}$, $\mathbf{v}$, and $\mathbf{w}$ be vectors in $\mathbb{R}^3$. Then $|\mathbf{u} \cdot (\mathbf{v} \times \mathbf{w})|$ is the volume of the parallelepiped determined by $\mathbf{u}$, $\mathbf{v}$, and $\mathbf{w}$. Furthermore, the volume of the parallelepiped is $\mathbf{u} \cdot (\mathbf{v} \times \mathbf{w})$ if and only if $\mathbf{u}$, $\mathbf{v}$, and $\mathbf{w}$ form a right-handed triple.

Proof. Let $\mathcal{F}$ denote one of the sides of the parallelepiped determined by vectors $\mathbf{v}$ and $\mathbf{w}$. From Theorem 10.33 we know that the area of $\mathcal{F}$ is $\|\mathbf{v} \times \mathbf{w}\|$. The distance between $\mathcal{F}$ and the opposite face is the distance from the terminal end of $\mathbf{u}$ to $\mathcal{F}$. There are two cases to consider: $\mathbf{u}$ and $\mathbf{v} \times \mathbf{w}$ are either on the same side of $\mathcal{F}$ or on opposite sides of $\mathcal{F}$.

Suppose $\mathbf{u}$ and $\mathbf{v} \times \mathbf{w}$ are on the same side of $\mathcal{F}$. Then $\mathbf{u}$, $\mathbf{v}$, and $\mathbf{w}$ must form a right-handed triple. Since $\mathbf{v} \times \mathbf{w}$ is perpendicular to $\mathcal{F}$, the distance from the terminal end of $\mathbf{u}$ to $\mathcal{F}$ is the component, $\text{comp}_{\mathbf{v}\times\mathbf{w}}\mathbf{u} = \|\mathbf{u}\| \cos\phi$, of the projection of $\mathbf{u}$ onto $\mathbf{v}\times\mathbf{w}$, where ϕ is the angle between $\mathbf{u}$ and $\mathbf{v} \times \mathbf{w}$. The following figure shows an example of this case:

A parallelepiped determined by the right-handed triple $\mathbf{u}$, $\mathbf{v}$, *and* $\mathbf{w}$

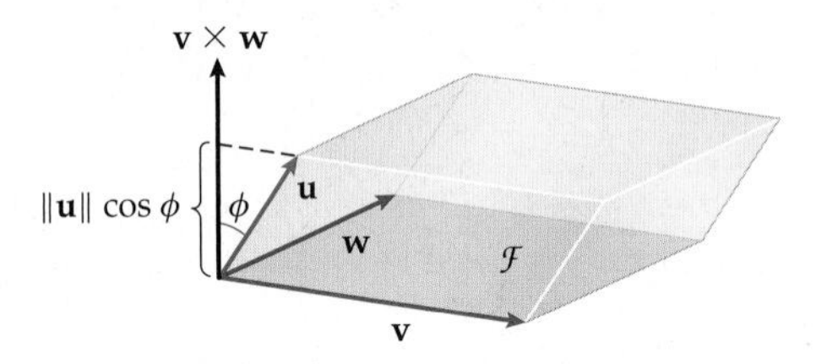

Here, since $\mathbf{u}$ and $\mathbf{v} \times \mathbf{w}$ are on the same side of $\mathcal{F}$, $\phi < 90^\circ$ and $\cos\phi > 0$. Thus the volume of the parallelepiped is $\|\mathbf{u}\|\,\|\mathbf{v} \times \mathbf{w}\| \cos\phi = \mathbf{u} \cdot (\mathbf{v} \times \mathbf{w})$.

In the case where $\mathbf{u}$ and $\mathbf{v} \times \mathbf{w}$ are on opposite sides of $\mathcal{F}$, $\phi > 90^\circ$ and $\cos\phi < 0$. Thus the volume of the parallelepiped is $-\|\mathbf{u}\|\,\|\mathbf{v} \times \mathbf{w}\| \cos\phi = -\mathbf{u} \cdot (\mathbf{v} \times \mathbf{w})$.

We now have our desired result: The volume of the parallelepiped is $\mathbf{u} \cdot (\mathbf{v} \times \mathbf{w})$ if and only if $\mathbf{u}$, $\mathbf{v}$, and $\mathbf{w}$ form a right-handed triple, and in any case the volume is $|\mathbf{u} \cdot (\mathbf{v} \times \mathbf{w})|$. ■

Theorem 10.36 also gives us a simple computational criterion for determining when three vectors lie in the same plane: The triple scalar product $\mathbf{u} \cdot (\mathbf{v} \times \mathbf{w})$ is equal to zero if and only if $\mathbf{u}$, $\mathbf{v}$, and $\mathbf{w}$ are coplanar.

The following theorem describes relationships between dot and cross products:

THEOREM 10.37

Properties of the Triple Scalar Product

Let $\mathbf{u}$, $\mathbf{v}$, and $\mathbf{w}$ be vectors in $\mathbb{R}^3$. Then

(a) $\mathbf{u} \cdot (\mathbf{v} \times \mathbf{w}) = \mathbf{v} \cdot (\mathbf{w} \times \mathbf{u}) = \mathbf{w} \cdot (\mathbf{u} \times \mathbf{v})$

(b) $\mathbf{u} \cdot (\mathbf{v} \times \mathbf{w}) = (\mathbf{u} \times \mathbf{v}) \cdot \mathbf{w}$

Part (b) of Theorem 10.37 says, roughly, that in a triple scalar product the dot and cross products can be "exchanged." We prove part (a) next and leave part (b) for Exercise 76.

Proof. First suppose that $\mathbf{u}$, $\mathbf{v}$, $\mathbf{w}$ form a right-handed triple. Then $\mathbf{v}$, $\mathbf{w}$, $\mathbf{u}$, and $\mathbf{w}$, $\mathbf{u}$, $\mathbf{v}$ do also. (Why?) Thus, from Theorem 10.36, the three triple scalar products $\mathbf{u} \cdot (\mathbf{v} \times \mathbf{w})$, $\mathbf{v} \cdot (\mathbf{w} \times \mathbf{u})$, and $\mathbf{w} \cdot (\mathbf{u} \times \mathbf{v})$ all give the volume of the parallelepiped determined by $\mathbf{u}$, $\mathbf{v}$ and $\mathbf{w}$. Therefore they are all equal.

For the converse, if **u**, **v**, **w** do not form a right-handed triple, then neither **v**, **w**, **u** nor **w**, **u**, **v** do. Here the volume of the parallelepiped determined by **u**, **v**, and **w** is

$$-\mathbf{u} \cdot (\mathbf{v} \times \mathbf{w}) = -\mathbf{v} \cdot (\mathbf{w} \times \mathbf{u}) = -\mathbf{w} \cdot (\mathbf{u} \times \mathbf{v}).$$

Therefore in this case we also have $\mathbf{u} \cdot (\mathbf{v} \times \mathbf{w}) = \mathbf{v} \cdot (\mathbf{w} \times \mathbf{u}) = \mathbf{w} \cdot (\mathbf{u} \times \mathbf{v})$. ■

Examples and Explorations

EXAMPLE 1

Using a cross product to find the area of a parallelogram

(a) Use Definition 10.25 to compute $\mathbf{u} \times \mathbf{v}$ for vectors $\mathbf{u} = \langle 2, 0, -3 \rangle$ and $\mathbf{v} = \langle -1, 4, 2 \rangle$.

(b) Find the area of the parallelogram determined by **u** and **v**.

SOLUTION

(a) We use Definition 10.25 to compute the cross product:

$$\begin{aligned} \mathbf{u} \times \mathbf{v} &= \det \begin{bmatrix} \mathbf{i} & \mathbf{j} & \mathbf{k} \\ 2 & 0 & -3 \\ -1 & 4 & 2 \end{bmatrix} \\ &= ((0)(2) - (-3)(4))\mathbf{i} + ((-3)(-1) - (2)(2))\mathbf{j} + ((2)(4) - (0)(-1))\mathbf{k} \\ &= 12\mathbf{i} - \mathbf{j} + 8\mathbf{k}. \end{aligned}$$

(b) The area of the parallelogram determined by **u** and **v** is

$$\|\mathbf{u} \times \mathbf{v}\| = \|12\mathbf{i} - \mathbf{j} + 8\mathbf{k}\| = \sqrt{12^2 + (-1)^2 + 8^2} = \sqrt{209} \approx 14.5 \text{ square units.}$$ □

CHECKING THE ANSWER

We may check the plausibility of our cross product $\mathbf{u} \times \mathbf{v}$ with the dot products:

$$\mathbf{u} \cdot (\mathbf{u} \times \mathbf{v}) = \langle 2, 0, -3 \rangle \cdot \langle 12, -1, 8 \rangle = 2 \cdot 12 + 0(-1) - 3 \cdot 8 = 0 \quad \text{and}$$

$$\mathbf{v} \cdot (\mathbf{u} \times \mathbf{v}) = \langle -1, 4, 2 \rangle \cdot \langle 12, -1, 8 \rangle = -1 \cdot 12 + 4(-1) + 2 \cdot 8 = 0.$$

These calculations show that the cross product we obtained is orthogonal to both of the vectors **u** and **v**.

EXAMPLE 2

Cross products involving the standard basis vectors

Find the cross products of the standard basis vectors:

$$\begin{matrix} \mathbf{i} \times \mathbf{i} & \mathbf{i} \times \mathbf{j} & \mathbf{i} \times \mathbf{k} \\ \mathbf{j} \times \mathbf{i} & \mathbf{j} \times \mathbf{j} & \mathbf{j} \times \mathbf{k} \\ \mathbf{k} \times \mathbf{i} & \mathbf{k} \times \mathbf{j} & \mathbf{k} \times \mathbf{k} \end{matrix}$$

SOLUTION

The cross product of two parallel vectors is **0**. Since every vector is parallel to itself, we immediately have

$$\mathbf{i} \times \mathbf{i} = \mathbf{j} \times \mathbf{j} = \mathbf{k} \times \mathbf{k} = \mathbf{0}.$$

Each of the basis vectors is a unit vector, and the angle between two *distinct* basis vectors is 90°. Therefore the norm of the cross product of two distinct basis vectors is 1 by property (a) of Theorem 10.35.

By property (b) of the same theorem, we know that the cross product of any two distinct basis vectors must be (plus or minus) the remaining basis vector. For example, $\mathbf{i} \times \mathbf{j} = \pm\mathbf{k}$. We need only determine which of $\mathbf{i}$, $\mathbf{j}$, and $\pm\mathbf{k}$ forms a right-handed triple. Here, $\mathbf{i}$, $\mathbf{j}$, and $\mathbf{k}$ form such a triple (check this), so $\mathbf{i} \times \mathbf{j} = \mathbf{k}$. Also, because the cross product is anticommutative, if we reverse the order within $\mathbf{i} \times \mathbf{j} = \mathbf{k}$, we will have $\mathbf{j} \times \mathbf{i} = -\mathbf{k}$.

The complete list of the cross products using the basis vectors is

$$\begin{array}{lll} \mathbf{i} \times \mathbf{i} = \mathbf{0} & \mathbf{i} \times \mathbf{j} = \mathbf{k} & \mathbf{i} \times \mathbf{k} = -\mathbf{j} \\ \mathbf{j} \times \mathbf{i} = -\mathbf{k} & \mathbf{j} \times \mathbf{j} = \mathbf{0} & \mathbf{j} \times \mathbf{k} = \mathbf{i} \\ \mathbf{k} \times \mathbf{i} = \mathbf{j} & \mathbf{k} \times \mathbf{j} = -\mathbf{i} & \mathbf{k} \times \mathbf{k} = \mathbf{0}. \end{array}$$

□

EXAMPLE 3 **Finding the area of a triangle determined by three points**

Find the area of the triangle with vertices $A = (4, -2)$, $B = (7, 3)$, and $C = (-1, 3)$.

SOLUTION

We first find the vectors

$$\begin{aligned} \overrightarrow{AB} &= \langle 7-4, 3-(-2) \rangle = \langle 3, 5 \rangle \text{ and} \\ \overrightarrow{AC} &= \langle -1-4, 3-(-2) \rangle = \langle -5, 5 \rangle. \end{aligned}$$

The area of the triangle with vertices at A, B, and C is one-half of the area of the parallelogram determined by $\overrightarrow{AB}$ and $\overrightarrow{AC}$. We would like to use the technique of Example 1 to find the area of the parallelogram, but in order to take the cross product of two vectors, they must have three components. Here we can treat the xy-plane as part of 3-space by thinking of the xy-plane as the plane in $\mathbb{R}^3$ with $z = 0$. That is, we let $\overrightarrow{AB} = \langle 3, 5, 0 \rangle$ and $\overrightarrow{AC} = \langle -5, 5, 0 \rangle$.

Now,

$$\begin{aligned} \text{Area } \Delta ABC &= \frac{1}{2}\|\overrightarrow{AB} \times \overrightarrow{AC}\| = \frac{1}{2}\left\| \det \begin{bmatrix} \mathbf{i} & \mathbf{j} & \mathbf{k} \\ 3 & 5 & 0 \\ -5 & 5 & 0 \end{bmatrix} \right\| \\ &= \frac{1}{2}\|(0-0)\mathbf{i} + (0-0)\mathbf{j} + (15-(-25)\mathbf{k}\| = \frac{1}{2}\|40\mathbf{k}\| \\ &= 20 \text{ square units.} \end{aligned}$$

□

EXAMPLE 4 **Finding the volume of a parallelepiped**

(a) Find the volume of the parallelepiped determined by $\mathbf{u} = \langle 0, 3, -2 \rangle$, $\mathbf{v} = \langle 5, 3, -1 \rangle$, and $\mathbf{w} = \langle -3, 2, 7 \rangle$.

(b) Do the vectors $\mathbf{u}$, $\mathbf{v}$, and $\mathbf{w}$ form a right-handed triple or a left-handed triple?

SOLUTION

(a) The volume of the parallelepiped determined by $\mathbf{u}$, $\mathbf{v}$, and $\mathbf{w}$ is the absolute value of the triple scalar product $\mathbf{u} \cdot (\mathbf{v} \times \mathbf{w})$. Although we could first evaluate the cross product $\mathbf{v} \times \mathbf{w}$ and then take the dot product of the resulting vector with $\mathbf{u}$, it is slightly more efficient to just take the absolute value of the determinant of the 3×3 matrix formed from the components of $\mathbf{u}$, $\mathbf{v}$, and $\mathbf{w}$ as the rows. (In Exercise 78, we ask you to explain why this always works.)

Thus, the required volume is

$$|\mathbf{u}\cdot(\mathbf{v}\times\mathbf{w})| = \left|\det\begin{bmatrix} 0 & 3 & -2 \\ 5 & 3 & -1 \\ -3 & 2 & 7 \end{bmatrix}\right| = |0+9-20+0-105-18| = |-134|$$

$$= 134 \text{ cubic units.}$$

(b) By Theorem 10.36, the vectors $\mathbf{u}$, $\mathbf{v}$, and $\mathbf{w}$ form a left-handed triple, since the triple scalar product $\mathbf{u}\cdot(\mathbf{v}\times\mathbf{w}) = -134 < 0$. □

? TEST YOUR UNDERSTANDING

- How do you find the determinant of a 3×3 matrix?
- What is the definition of the cross product?
- What are the geometric properties of the cross product?
- How do you find the area of a parallelogram determined by two vectors? How do you find the volume of a parallelepiped determined by three vectors?
- How do you find the area of a triangle determined by three points?

EXERCISES 10.4

Thinking Back

- *Coordinate system:* Explain what it means for a coordinate system to be right-handed and what it means for a coordinate system to be left-handed.
- *Orthogonal vectors:* Let $\mathbf{u}$ and $\mathbf{v}$ be two nonparallel position vectors in $\mathbb{R}^3$ lying in the xy-plane. Find two unit vectors orthogonal to both $\mathbf{u}$ and $\mathbf{v}$ *without* using the cross product.

Concepts

0. *Problem Zero:* Read the section and make your own summary of the material.

1. *True/False:* Determine whether each of the statements that follow is true or false. If a statement is true, explain why. If a statement is false, provide a counterexample.

(a) *True or False:* For any two vectors $\mathbf{u}$ and $\mathbf{v}$ in $\mathbb{R}^3$, $\mathbf{u}\times\mathbf{v} = \mathbf{v}\times\mathbf{u}$.

(b) *True or False:* If $\mathbf{u}$ and $\mathbf{v}$ are two vectors in $\mathbb{R}^3$, then $\mathbf{u}\times\mathbf{v} = \mathbf{u}\cdot\mathbf{v}$.

(c) *True or False:* If $\mathbf{u}\times\mathbf{v} = \mathbf{v}\times\mathbf{u}$, then $\mathbf{u}$ and $\mathbf{v}$ are parallel.

(d) *True or False:* If $\mathbf{u}$, $\mathbf{v}$, and $\mathbf{w}$ are vectors in $\mathbb{R}^3$, then $(\mathbf{u}\times\mathbf{v})\times\mathbf{w} = \mathbf{u}\times(\mathbf{v}\times\mathbf{w})$.

(e) *True or False:* The triple scalar product can be used to find the volume of a parallelepiped.

(f) *True or False:* If $\mathbf{u}$, $\mathbf{v}$, and $\mathbf{w}$ are vectors in $\mathbb{R}^3$, then $\mathbf{u}\cdot(\mathbf{v}\times\mathbf{w}) = -\mathbf{v}\cdot(\mathbf{u}\times\mathbf{w})$.

(g) *True or False:* If $\mathbf{u}$ and $\mathbf{v}$ are nonparallel vectors in $\mathbb{R}^3$, then $\dfrac{\mathbf{u}\cdot\mathbf{v}}{\|\mathbf{u}\times\mathbf{v}\|} = \cot\theta$, where θ is the angle between $\mathbf{u}$ and $\mathbf{v}$.

(h) *True or False:* If $\mathbf{u}$ and $\mathbf{v}$ are unit vectors in $\mathbb{R}^3$, then $\mathbf{u}\times\mathbf{v}$ is also a unit vector.

2. *Examples:* Construct examples of the thing(s) described in the following. Try to find examples that are different than any in the reading.

(a) Two nonzero vectors $\mathbf{u}$ and $\mathbf{v}$ in $\mathbb{R}^3$ such that $\mathbf{u}\times\mathbf{v} = \mathbf{v}\times\mathbf{u}$.

(b) Three vectors $\mathbf{u}$, $\mathbf{v}$, and $\mathbf{w}$ in $\mathbb{R}^3$ such that $(\mathbf{u}\times\mathbf{v})\times\mathbf{w} \neq \mathbf{u}\times(\mathbf{v}\times\mathbf{w})$.

(c) Three vectors $\mathbf{u}$, $\mathbf{v}$, and $\mathbf{w}$ in $\mathbb{R}^3$ such that $(\mathbf{u}\times\mathbf{v})\times\mathbf{w} = \mathbf{u}\times(\mathbf{v}\times\mathbf{w})$.

3. What is the definition of the cross product?

4. How is the determinant of a 3×3 matrix used in the computation of the determinant of two vectors $\mathbf{u} = \langle u_1, u_2, u_3\rangle$ and $\mathbf{v} = \langle v_1, v_2, v_3\rangle$?

5. If $\mathbf{u}$ and $\mathbf{v}$ are nonzero vectors in $\mathbb{R}^3$, what is the geometric relationship between $\mathbf{u}$, $\mathbf{v}$, and $\mathbf{u}\times\mathbf{v}$?

6. What is Lagrange's identity? How is it used to understand the geometry of the cross product?

7. If $\mathbf{u}$ and $\mathbf{v}$ are nonzero vectors in $\mathbb{R}^3$, why do the equations $\mathbf{u}\cdot(\mathbf{u}\times\mathbf{v}) = 0$ and $\mathbf{v}\cdot(\mathbf{u}\times\mathbf{v}) = 0$ tell us that the cross product is orthogonal to both $\mathbf{u}$ and $\mathbf{v}$?

8. What is meant by the parallelogram determined by vectors $\mathbf{u}$ and $\mathbf{v}$ in $\mathbb{R}^3$? How do you find the area of this parallelogram?

9. Sketch the parallelogram determined by the two vectors $\langle 1, 2\rangle$ and $\langle 3, -1\rangle$. How can you use the cross product to find the area of this parallelogram?

10. What is meant by the triangle determined by vectors $\mathbf{u}$ and $\mathbf{v}$ in $\mathbb{R}^3$? How do you find the area of this triangle?

11. If $\|\mathbf{u}\times\mathbf{v}\| = \|\mathbf{u}\|\|\mathbf{v}\|$, what is the geometric relationship between $\mathbf{u}$ and $\mathbf{v}$?

12. Give an example of three vectors in $\mathbb{R}^3$ that form a right-handed triple. Explain how you can use the same three vectors to form a left-handed triple.

13. Give an example of three nonzero vectors $\mathbf{u}$, $\mathbf{v}$, and $\mathbf{w}$ in $\mathbb{R}^3$ such that $\mathbf{u} \times \mathbf{v} = \mathbf{u} \times \mathbf{w}$ but $\mathbf{v} \neq \mathbf{w}$. What geometric relationship must the three vectors have for this to happen?

14. What is the definition of the triple scalar product for vectors $\mathbf{u}$, $\mathbf{v}$, and $\mathbf{w}$ in $\mathbb{R}^3$?

15. If the triple scalar product $\mathbf{u} \cdot (\mathbf{v} \times \mathbf{w})$ is equal to zero, what geometric relationship do the vectors $\mathbf{u}$, $\mathbf{v}$, and $\mathbf{w}$ have?

16. What is a parallelepiped? What is meant by the parallelepiped determined by the vectors $\mathbf{u}$, $\mathbf{v}$, and $\mathbf{w}$? How do you find the volume of the parallelepiped determined by $\mathbf{u}$, $\mathbf{v}$, and $\mathbf{w}$?

17. If $\mathbf{u}$, $\mathbf{v}$, and $\mathbf{w}$ are three vectors in $\mathbb{R}^3$, what is wrong with the expression $\mathbf{u} \times \mathbf{v} \times \mathbf{w}$?

18. If $\mathbf{u}$, $\mathbf{v}$, and $\mathbf{w}$ are three vectors in $\mathbb{R}^3$, which of the following products make sense and which do not?
    (a) $\mathbf{u} \cdot (\mathbf{v} \cdot \mathbf{w})$
    (b) $\mathbf{u} \cdot (\mathbf{v} \times \mathbf{w})$
    (c) $\mathbf{u} \times (\mathbf{v} \cdot \mathbf{w})$
    (d) $\mathbf{u} \times (\mathbf{v} \times \mathbf{w})$

19. If $\mathbf{u}$ and $\mathbf{v}$ are vectors in $\mathbb{R}^3$ such that $\mathbf{u} \cdot \mathbf{v} = 0$ and $\mathbf{u} \times \mathbf{v} = \mathbf{0}$, what can we conclude about $\mathbf{u}$ and $\mathbf{v}$?

20. If $\mathbf{u}$, $\mathbf{v}$, and $\mathbf{w}$ are three mutually orthogonal vectors in $\mathbb{R}^3$, explain why $\mathbf{u} \times (\mathbf{v} \times \mathbf{w}) = \mathbf{0}$.

21. If $\mathbf{u}$ and $\mathbf{v}$ are vectors in $\mathbb{R}^3$ such that $\mathbf{u} \times \mathbf{v} = \mathbf{v} \times \mathbf{u}$, what can we conclude about $\mathbf{u}$ and $\mathbf{v}$?

Skills

In Exercises 22–29 compute the indicated quantities when $\mathbf{u} = \langle 2, 1, -3 \rangle$, $\mathbf{v} = \langle 4, 0, 1 \rangle$, and $\mathbf{w} = \langle -2, 6, 5 \rangle$.

22. $\mathbf{u} \times \mathbf{v}$ and $\mathbf{v} \times \mathbf{u}$

23. $\mathbf{u} \times \mathbf{w}$ and $\mathbf{w} \times \mathbf{u}$

24. $\mathbf{v} \times \mathbf{w}$ and $\mathbf{w} \times \mathbf{v}$

25. $(\mathbf{u} \times \mathbf{v}) \times \mathbf{w}$ and $\mathbf{u} \times (\mathbf{v} \times \mathbf{w})$

26. $(\mathbf{u} \times \mathbf{v}) \cdot \mathbf{w}$ and $\mathbf{w} \cdot (\mathbf{v} \times \mathbf{u})$

27. Find the area of the parallelogram determined by the vectors $\mathbf{u}$ and $\mathbf{v}$.

28. Find the area of the parallelogram determined by the vectors $\mathbf{v}$ and $\mathbf{w}$.

29. Find the volume of the parallelepiped determined by vectors $\mathbf{u}$, $\mathbf{v}$, and $\mathbf{w}$. Do $\mathbf{u}$, $\mathbf{v}$, and $\mathbf{w}$ form a right-handed triple?

In Exercises 30–35 compute the indicated quantities when $\mathbf{u} = \langle -3, 1, -4 \rangle$, $\mathbf{v} = \langle 2, 0, 5 \rangle$, and $\mathbf{w} = \langle 1, 3, 13 \rangle$.

30. $\mathbf{u} \times \mathbf{v}$ and $\mathbf{v} \times \mathbf{u}$

31. $\mathbf{u} \times \mathbf{w}$ and $\mathbf{w} \times \mathbf{u}$

32. $\mathbf{v} \times \mathbf{w}$ and $\mathbf{w} \times \mathbf{v}$

33. $(\mathbf{u} \times \mathbf{v}) \cdot \mathbf{w}$ and $\mathbf{u} \cdot (\mathbf{v} \times \mathbf{w})$

34. Find the area of the parallelogram determined by the vectors $\mathbf{u}$ and $\mathbf{v}$.

35. Find the volume of the parallelepiped determined by the vectors $\mathbf{u}$, $\mathbf{v}$, and $\mathbf{w}$.

In Exercises 36–41 use the given sets of points to find:

(a) A nonzero vector $\mathbf{N}$ perpendicular to the plane determined by the points.
(b) Two unit vectors perpendicular to the plane determined by the points.
(c) The area of the triangle determined by the points.

36. $P(1, 4, 6)$, $Q(-3, 5, 0)$, $R(3, 2, -1)$

37. $P(4, 3, -2)$, $Q(0, 0, 3)$, $R(-1, 3, 6)$

38. $P(3, 1, 8)$, $Q(0, 6, -1)$, $R(-3, 5, -3)$

39. $P(2, -5, 1)$, $Q(-4, 5, 8)$, $R(-1, -5, 3)$

40. $P(4, -2)$, $Q(-2, 0)$, $R(1, -5)$
    (*Hint: Think of the xy-plane as part of* $\mathbb{R}^3$.)

41. $P(1, 6)$, $Q(0, -3)$, $R(-5, 4)$
    (*Hint: Think of the xy-plane as part of* $\mathbb{R}^3$.)

Suppose a triangle has side lengths a, b, and c. The ***semiperimeter*** of the triangle is defined to be $s = \frac{1}{2}(a+b+c)$. We can use the side lengths of the triangle to calculate its area by applying ***Heron's formula***:

$$\text{Area} = \sqrt{s(s-a)(s-b)(s-c)}.$$

Use Heron's formula to compute the areas of the triangles determined by the points P, Q, and R in Exercises 42–44:

42. P, Q, and R from Exercise 36

43. P, Q, and R from Exercise 37

44. P, Q, and R from Exercise 41

45. Heron's formula allows us to compute the area of a triangle from its side lengths.
    (a) Explain why knowing the side lengths of a quadrilateral is *not* sufficient to compute the area of the quadrilateral.
    (b) Explain why knowing the side lengths of a quadrilateral and the length of one diagonal is sufficient to compute the area of the quadrilateral.
    (c) Explain why knowing the side lengths of a quadrilateral and the measure of the angle at one vertex is also sufficient to compute the area of the quadrilateral.

Use the results of Exercise 45 to find the areas of the quadrilaterals $PQRS$ specified in Exercises 46–49.

46. $PQ = 6$, $QR = 7$, $RS = 8$, $SP = 9$, $PR = 10$

47. $PQ = 6$, $QR = 7$, $RS = 8$, $SP = 9$, $QS = 10$

48. $PQ = 6$, $QR = 7$, $RS = 8$, $SP = 6$, $\angle P = 60°$

49. $PQ = 7$, $QR = 8$, $RS = 8$, $SP = 9$, $\angle R = 60°$

In Exercises 50–53 the coordinates of points P, Q, R, and S are given. (a) Determine whether quadrilateral $PQRS$ is a parallelogram. (b) Find the area of quadrilateral $PQRS$.

50. $P(0,0)$, $Q(-1,3)$, $R(-4,-1)$, $S(-3,-2)$

51. $P(-1,3)$, $Q(2,5)$, $R(4,1)$, $S(1,-1)$

52. $P(2,7)$, $Q(3,-1)$, $R(1,-10)$, $S(-1,-2)$

53. $P(-1,3)$, $Q(2,5)$, $R(6,3)$, $S(4,-2)$

In Exercises 54–57 the coordinates of points P, Q, R, and S are given. (a) Show that the four points are coplanar. (b) Determine whether quadrilateral $PQRS$ is a parallelogram. (c) Find the area of quadrilateral $PQRS$.

54. $P(0,0,0)$, $Q(1,-2,5)$, $R(-1,2,11)$, $S(-2,4,6)$

55. $P(2,-3,8)$, $Q(-2,4,6)$, $R(7,18,-7)$, $S(15,4,-3)$

56. $P(1,2,6)$, $Q(4,1,-5)$, $R(3,6,8)$, $S(0,4,13)$

57. $P(3,4,-2)$, $Q(7,0,6)$, $R(2,1,7)$, $S(5,-2,13)$

Applications

Some crystals have rhombohedral structures. A rhombohedron is a parallelepiped in which all of the edge lengths are equal and each of the six faces is a congruent rhombus. Find the volumes of the rhombohedral crystals described in Exercises 58 and 59.

A rhombohedral crystal

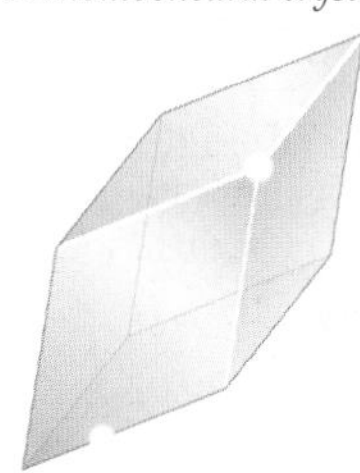

58. Each side length is 1 cm, and the acute angles in each face measure 60°. (*Hint: Let vector* **i** *form one of the edges of the rhombohedron, and let a second nonparallel edge be in the xy-plane.*)

59. Each side length is 2 cm, and the acute angles in each face measure 45°. (*See the hint in the previous exercise.*)

Turning a bolt with a wrench produces a ***torque*** vector that drives the bolt forward. The magnitude of the torque vector is $\|\mathbf{r}\|\|\mathbf{F}\|\sin\theta$, where $\mathbf{r}$ is the vector along the handle of the wrench, $\mathbf{F}$ is the force vector applied to the handle of the wrench, and θ is the angle between these two vectors. Therefore, the magnitude of the torque is $\|\mathbf{r}\times\mathbf{F}\|$. In Exercises 60 and 61, find the magnitude of the torque. Express each answer in foot-pounds.

The torque vector

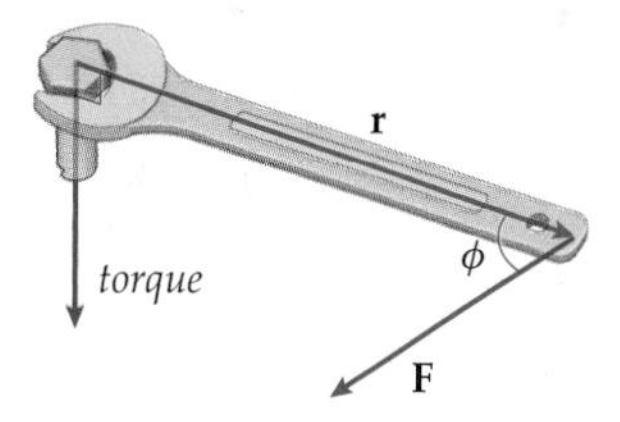

60. A force of 20 lb is applied to a wrench with a 6-inch handle at an angle of 60°.

61. A force of 40 lb is applied to a wrench with a 9-inch handle at an angle of 90°.

Proofs

62. Prove that the determinant of a 3×3 matrix with integer entries is an integer.

63. Let A be a 3×3 matrix with determinant D, and let A' be a 3×3 matrix obtained from A by exchanging two rows. Prove that $\det A' = -D$.

64. Let B be a 3×3 matrix with determinant d, and let B' be a 3×3 matrix obtained from B by exchanging two columns. Prove that $\det B' = -d$.

65. Use the definition of the cross product to prove that the cross product of two parallel vectors is $\mathbf{0}$. (This is Theorem 10.26.)

66. Use the definition of the cross product to prove that the cross product of two vectors $\mathbf{u}$ and $\mathbf{v}$ is anticommutative; that is, prove that $\mathbf{u}\times\mathbf{v} = -\mathbf{v}\times\mathbf{u}$. (This is Theorem 10.27.)

67. Let $\mathbf{u}$ and $\mathbf{v}$ be vectors in $\mathbb{R}^3$ and let c be a scalar. Prove that $c(\mathbf{u}\times\mathbf{v}) = (c\mathbf{u})\times\mathbf{v} = \mathbf{u}\times(c\mathbf{v})$. (This is Theorem 10.28).

68. Let $\mathbf{u}$, $\mathbf{v}$, and $\mathbf{w}$ be vectors in $\mathbb{R}^3$. Prove:

$$\mathbf{u}\times(\mathbf{v}+\mathbf{w}) = \mathbf{u}\times\mathbf{v}+\mathbf{u}\times\mathbf{w} \text{ and}$$

$$(\mathbf{u}+\mathbf{v})\times\mathbf{w} = \mathbf{u}\times\mathbf{w}+\mathbf{v}\times\mathbf{w}.$$

(This is Theorem 10.29.)

69. Let $\mathbf{u}$ and $\mathbf{v}$ be vectors in $\mathbb{R}^3$. Prove that $\mathbf{v}\cdot(\mathbf{u}\times\mathbf{v}) = 0$. (This is Theorem 10.31(b).)

70. Let $\mathbf{u}$ and $\mathbf{v}$ be vectors in $\mathbb{R}^3$. Prove Lagrange's identity, Theorem 10.30:

$$\|\mathbf{u}\times\mathbf{v}\|^2 = \|\mathbf{u}\|^2\|\mathbf{v}\|^2 - (\mathbf{u}\cdot\mathbf{v})^2.$$

71. Let $\mathbf{u}$, $\mathbf{v}$, and $\mathbf{w}$ be three vectors in $\mathbb{R}^3$ in which the components of each vector are integers.

(a) Prove that the volume of the parallelepiped determined by $\mathbf{u}$, $\mathbf{v}$ and $\mathbf{w}$ is an integer.

(b) Find examples of vectors $\mathbf{u}$ and $\mathbf{v}$ with integer components that show that the area of the parallelogram determined by $\mathbf{u}$ and $\mathbf{v}$ can be either an integer or an irrational number.

72. Let $\mathbf{u}$, $\mathbf{v}$, and $\mathbf{w}$ be three mutually perpendicular vectors in $\mathbb{R}^3$.

(a) Prove that $\mathbf{u} \times (\mathbf{v} \times \mathbf{w}) = \mathbf{0}$.

(b) Show that $|\mathbf{u} \cdot (\mathbf{v} \times \mathbf{w})| = \|\mathbf{u}\|\|\mathbf{v}\|\|\mathbf{w}\|$.

73. Let $\mathbf{u}$ and $\mathbf{v}$ be vectors in $\mathbb{R}^3$ such that $\mathbf{u} \cdot \mathbf{v} \neq 0$. Prove that if θ is the angle between $\mathbf{u}$ and $\mathbf{v}$, then $\tan\theta = \dfrac{\|\mathbf{u} \times \mathbf{v}\|}{\mathbf{u} \cdot \mathbf{v}}$.

74. Let $\mathbf{u}$, $\mathbf{v}$, and $\mathbf{w}$ be vectors in $\mathbb{R}^3$. Prove that $\mathbf{u} \times \mathbf{v} = \mathbf{u} \times \mathbf{w}$ if and only if $\mathbf{u}$ is parallel to $\mathbf{v} - \mathbf{w}$.

75. Let $\mathbf{u}$, $\mathbf{v}$, and $\mathbf{w}$ be vectors in $\mathbb{R}^3$ with $\mathbf{u} \neq \mathbf{0}$. Show that if $\mathbf{u} \times \mathbf{v} = \mathbf{u} \times \mathbf{w}$ and $\mathbf{u} \cdot \mathbf{v} = \mathbf{u} \cdot \mathbf{w}$, then $\mathbf{v} = \mathbf{w}$.

76. Let $\mathbf{u}$, $\mathbf{v}$, and $\mathbf{w}$ be vectors in $\mathbb{R}^3$. Prove that

$$\mathbf{u} \cdot (\mathbf{v} \times \mathbf{w}) = (\mathbf{u} \times \mathbf{v}) \cdot \mathbf{w}.$$

(This is part (b) of Theorem 10.37.)

77. Prove that, for vectors $\mathbf{r}$, $\mathbf{s}$, $\mathbf{u}$, and $\mathbf{v}$ in $\mathbb{R}^3$,

$$(\mathbf{r} \times \mathbf{s}) \cdot (\mathbf{u} \times \mathbf{v}) = (\mathbf{r} \cdot \mathbf{u})(\mathbf{s} \cdot \mathbf{v}) - (\mathbf{r} \cdot \mathbf{v})(\mathbf{s} \cdot \mathbf{u}).$$

78. Let $\mathbf{u} = \langle u_1, u_2, u_3 \rangle$, $\mathbf{v} = \langle v_1, v_2, v_3 \rangle$, and $\mathbf{w} = \langle w_1, w_2, w_3 \rangle$. Show that

$$\mathbf{u} \cdot (\mathbf{v} \times \mathbf{w}) = \det \begin{bmatrix} u_1 & u_2 & u_3 \\ v_1 & v_2 & v_3 \\ w_1 & w_2 & w_3 \end{bmatrix}.$$

79. Prove that if vectors $\mathbf{r}$, $\mathbf{s}$, $\mathbf{u}$, and $\mathbf{v}$ in $\mathbb{R}^3$ can all be translated to the same plane, then

$$(\mathbf{r} \times \mathbf{s}) \times (\mathbf{u} \times \mathbf{v}) = \mathbf{0}.$$

80. The product $\mathbf{u} \times (\mathbf{v} \times \mathbf{w})$ is an example of a vector called a ***vector triple product.***

(a) Show that if $\mathbf{v} = \langle v_1, v_2, v_3 \rangle$ and $\mathbf{w} = \langle w_1, w_2, w_3 \rangle$, then $\mathbf{i} \times (\mathbf{v} \times \mathbf{w}) = w_1\mathbf{v} - v_1\mathbf{w}$.

(b) Derive similar expressions from $\mathbf{j} \times (\mathbf{v} \times \mathbf{w})$ and $\mathbf{k} \times (\mathbf{v} \times \mathbf{w})$.

(c) Use your results from parts (a) and (b) to show that $\mathbf{u} \times (\mathbf{v} \times \mathbf{w}) = (\mathbf{u} \cdot \mathbf{w})\mathbf{v} - (\mathbf{u} \cdot \mathbf{v})\mathbf{w}$.

(d) Use your results from part (c) and the anticommutativity of the cross product to derive a similar expression for the vector triple product $(\mathbf{u} \times \mathbf{v}) \times \mathbf{w}$.

(e) Use your results from parts (c) and (d) to show that the cross product is *not* associative.

(f) Under what conditions is

$$\mathbf{u} \times (\mathbf{v} \times \mathbf{w}) = (\mathbf{u} \times \mathbf{v}) \times \mathbf{w}?$$

Thinking Forward

Planes in $\mathbb{R}^3$: Different geometric conditions can be used to specify a plane. The following questions ask you to explain why the specified conditions uniquely determine a plane.

- ▶ Explain why two nonparallel vectors and a point uniquely determine a plane containing both vectors and the point.
- ▶ Explain why a single nonzero vector and a point uniquely determine a plane containing the point. (*Hint: Think of the collection of vectors orthogonal to the given vector with the given point as the initial point of all of the vectors.*)
- ▶ Give two other sets of geometric conditions that would uniquely determine a plane in $\mathbb{R}^3$.

10.5 LINES IN THREE-DIMENSIONAL SPACE

- Using vectors to construct equations for lines in $\mathbb{R}^3$
- Parallel, intersecting, and skew lines
- Computing the distance from a point to a line

Equations for Lines

Recall that the general form for the equation of a line in the xy-plane is $ax+by=d$, where a, b, and d are scalars. An equation of this form is called ***a linear equation in two variables.*** In this section we will discuss how to write equations of lines in three-dimensional space.

It seems natural to think that the graph of a ***linear equation in three variables***,

$$ax+by+cz=d, \text{ where } a,\ b,\ c, \text{ and } d \text{ are scalars,}$$

would be a line in 3-space. However, as we will see in Section 10.6, the graphs of such equations are planes, not lines! (Recall our discussion of these equations in Section 10.2. There we examined the linear equation $x=3$ and saw that in 3-space the graph of the equation was a plane parallel to the yz-plane. The equation $x=3$ is a simple example of an equation of the form $ax+by+cz=d$, with $a=1$, $b=c=0$, and $d=3$.) So, if the graph of a linear equation in three variables is not a line, what types of equations *do* give lines?

As you know, a line in the plane can be determined by two distinct points or by a point and a direction, specified by a slope. In 3-space the situation is analogous: A line can be determined by two distinct points or by a single point and a direction, specified by a vector. To determine an equation for a line, $\mathcal{L}$, in 3-space we will use a point, P_0, on $\mathcal{L}$ and a *direction vector*, $\mathbf{d}$, parallel to $\mathcal{L}$. If we are given two points, we can immediately compute a direction vector and start by finding a parametrization for $\mathcal{L}$. If $P_0=(x_0,y_0,z_0)$ and $P_1=(x_1,y_1,z_1)$ are two distinct points on $\mathcal{L}$, then a direction vector for $\mathcal{L}$ is

$$\mathbf{d}=\overrightarrow{P_0P_1}=\langle x_1-x_0, y_1-y_0, z_1-z_0\rangle.$$

We now let P_0 be the point (x_0,y_0,z_0) and $\mathbf{d}$ be the vector $\langle a,b,c\rangle$. Point P_0 corresponds to the position vector $\mathbf{P_0}=\langle x_0,y_0,z_0\rangle$. Every point on $\mathcal{L}$ can be obtained by adding a suitable multiple of $\mathbf{d}$ to the position vector $\mathbf{P_0}$, as in the following figure:

A line in a three-dimensional Cartesian coordinate system

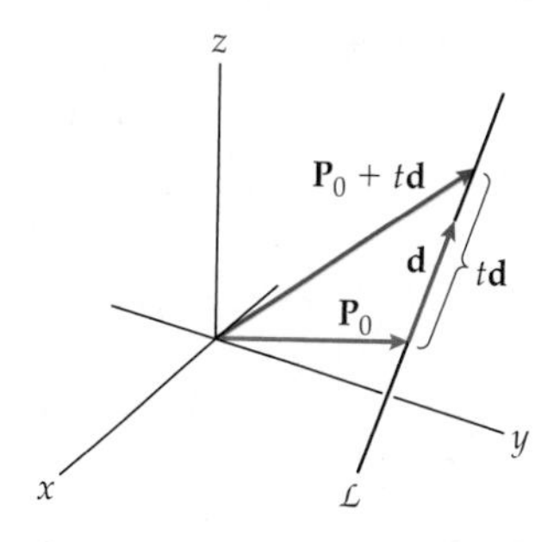

That is, every point on $\mathcal{L}$ is of the form $\mathbf{P_0}+t\mathbf{d}$, for some value of our parameter t. This result immediately gives us a ***vector function***, $\mathbf{r}(t)$, for the line $\mathcal{L}$. That is,

$$\mathbf{r}(t)=\mathbf{P_0}+t\mathbf{d}, \quad \text{or}$$
$$\mathbf{r}(t)=\langle x_0,y_0,z_0\rangle+t\,\langle a,b,c\rangle, \quad \text{or}$$
$$\mathbf{r}(t)=\langle x_0+at, y_0+bt, z_0+ct\rangle,$$

where $-\infty < t < \infty$. Any of these three equations is a parametrization for $\mathcal{L}$. From the last of the equations, we may immediately find parametric equations for the line $\mathcal{L}$:

$$x(t) = x_0 + at, \quad y(t) = y_0 + bt, \quad z(t) = z_0 + ct, \quad \text{where } -\infty < t < \infty.$$

Assuming that none of the components of the vector $\mathbf{d}$ is zero, we can solve each of the preceding equations for the parameter t to obtain

$$t = \frac{x - x_0}{a}, \ t = \frac{y - y_0}{b}, \ t = \frac{z - z_0}{c}.$$

We may use these equations to eliminate the parameter t to find a ***symmetric form*** for the line $\mathcal{L}$:

$$\frac{x - x_0}{a} = \frac{y - y_0}{b} = \frac{z - z_0}{c}.$$

Thus, we may use a vector equation, parametric equations, or the symmetric form to specify a line, and given any one of these, we may quickly obtain the others.

For example, to find an equation of the line $\mathcal{L}$ containing the points $P = (3, 2, -3)$ and $Q = (-1, 4, 2)$, we immediately have a direction vector

$$\mathbf{d} = \overrightarrow{PQ} = \langle -1 - 3, 4 - 2, 2 - (-3) \rangle = \langle -4, 2, 5 \rangle.$$

Using P for our point on $\mathcal{L}$, we obtain

$$\begin{aligned} \mathbf{r}(t) &= \langle 3, 2, -3 \rangle + t \langle -4, 2, 5 \rangle, -\infty < t < \infty, \quad \text{and} \\ \mathbf{r}(t) &= \langle 3 - 4t, 2 + 2t, -3 + 5t \rangle, -\infty < t < \infty. \end{aligned}$$

Either of these two equations is a vector function whose graph is $\mathcal{L}$. If we prefer to express $\mathcal{L}$ by means of parametric equations, we have

$$x(t) = 3 - 4t, \ y(t) = 2 + 2t, \ z(t) = -3 + 5t, -\infty < t < \infty.$$

Finally, if we want to use the symmetric form, we have

$$\frac{x - 3}{-4} = \frac{y - 2}{2} = \frac{z + 3}{5}.$$

Two Lines in $\mathbb{R}^3$

Given two lines in the plane, the lines either intersect, are parallel, or are identical. In three-dimensional space there is a fourth possibility: The lines may be ***skew***; that is, they neither are parallel nor intersect. For example, on a cube, some edges intersect and some are parallel, but the lines containing the highlighted edges in the following cube are skew:

Skew lines

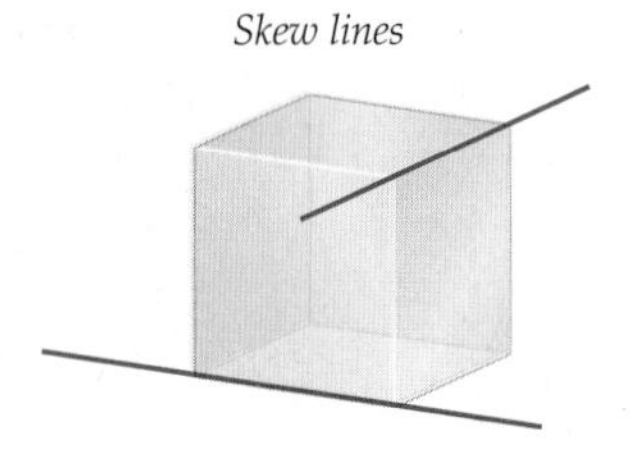

Given the equations of two lines, we can use the direction vectors of the lines to start our analysis of the relationship between the lines. If the direction vectors of the two lines are scalar multiples of each other, then the lines are either parallel or identical. If the direction vectors of the two lines are not multiples of each other, the lines either intersect or are skew. To complete the analysis, we determine whether or not the lines share a point. Parallel lines that share a point are identical. Nonparallel lines that share a point (obviously) intersect.

The Distance from a Point to a Line (Revisited)

The distance from a point, P, to a line, $\mathcal{L}$, is the distance from P to the closest point on $\mathcal{L}$. This is the distance along a line through P and perpendicular to $\mathcal{L}$. That distance can be computed with the technique of Example 4 from Section 10.3. Here, we provide an alternative method.

Let $\mathbf{r}(t) = \mathbf{P}_0 + t\mathbf{d}$, $-\infty < t < \infty$, be a parametrization for $\mathcal{L}$. Thus, P_0 is on $\mathcal{L}$ and $\mathbf{d}$ is parallel to $\mathcal{L}$, as in the following figure:

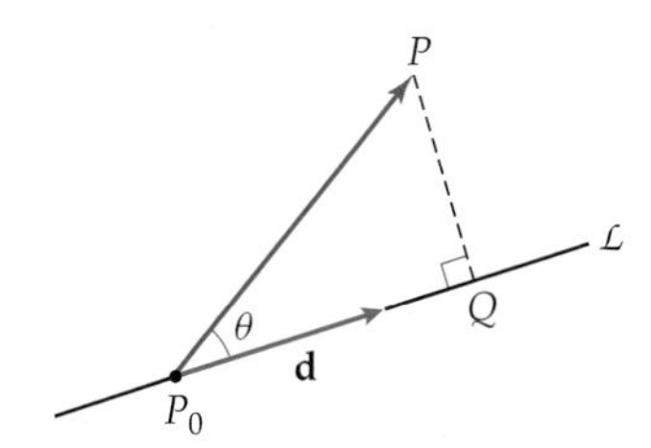

Now, suppose Q is the point on the line $\mathcal{L}$ that is closest to the point P, as in the figure shown. If we knew the location of the point Q, then we could immediately use the distance formula to calculate the length of $\overrightarrow{PQ}$. Typically, though, we do not know the location of Q. However, we do see that $\|\overrightarrow{PQ}\| = \|\overrightarrow{P_0P}\| \sin\theta$, where θ is the angle between $\overrightarrow{P_0P}$ and $\mathbf{d}$. Now, by Theorem 10.32,

$$\|\mathbf{d} \times \overrightarrow{P_0P}\| = \|\mathbf{d}\|\,\|\overrightarrow{P_0P}\| \sin\theta.$$

Thus, we have

$$\|\overrightarrow{PQ}\| = \|\overrightarrow{P_0P}\| \sin\theta = \frac{\|\mathbf{d} \times \overrightarrow{P_0P}\|}{\|\mathbf{d}\|}.$$

To summarize, we have the following theorem:

THEOREM 10.38 **The Distance from a Point to a Line**

Given a point P and a line $\mathcal{L}$ parameterized by $\mathbf{r}(t) = \mathbf{P}_0 + t\mathbf{d}$, the distance from P to $\mathcal{L}$ is

$$\frac{\|\mathbf{d} \times \overrightarrow{P_0P}\|}{\|\mathbf{d}\|}.$$

Examples and Explorations

EXAMPLE 1 **Expressing the equation of a line in symmetric form when one component of the direction vector is zero**

Find an equation of the line $\mathcal{L}$ containing the points $P = (-5, 6, -1)$ and $Q = (4, 6, -7)$. Write the answer as a parametrization, in terms of parametric equations, and in symmetric form.

SOLUTION

A direction vector for the line is

$$\overrightarrow{PQ} = \langle 4 - (-5), 6 - 6, -7 - (-1) \rangle = \langle 9, 0, -6 \rangle.$$

However, any nonzero multiple of $\overrightarrow{PQ}$ may be used instead. Here we will take $\mathbf{d} = \frac{1}{3}\overrightarrow{PQ} = \langle 3, 0, -2\rangle$ as our direction vector. Now, using P for our point on $\mathcal{L}$, we have

$$\mathbf{r}(t) = \langle -5, 6, -1\rangle + t\langle 3, 0, -2\rangle, \quad -\infty < t < \infty,$$
$$\mathbf{r}(t) = \langle -5 + 3t, 6, -1 - 2t\rangle, \quad -\infty < t < \infty.$$

Either of these two equations is a parametrization.

Using the second of the two forms, we have the parametric equations

$$x(t) = -5 + 3t,\ y(t) = 6,\ z(t) = -1 - 2t, \quad -\infty < t < \infty.$$

There is also a symmetric form for a line $\mathcal{L}$ if one of the three components of the direction vector $\mathbf{d}$ is zero. When $\mathbf{P_0} = \langle x_0, y_0, z_0\rangle$ and $\mathbf{d} = \langle a, 0, c\rangle$, then, as before, we have the parametrization

$$\begin{aligned}\mathbf{r}(t) &= \mathbf{P_0} + t\mathbf{d}, -\infty < t < \infty\\ &= \langle x_0 + at, y_0, z_0 + ct\rangle, -\infty < t < \infty.\end{aligned}$$

We have the following parametric equations for $\mathcal{L}$:

$$x(t) = x_0 + at,\ y(t) = y_0,\ z(t) = z_0 + ct, -\infty < t < \infty.$$

The symmetric form for $\mathcal{L}$ is

$$\frac{x - x_0}{a} = \frac{z - z_0}{c},\ y = y_0.$$

Thus, for our line $\mathcal{L}$, we have

$$\frac{x + 5}{3} = \frac{z + 1}{-2},\ y = 6.$$

There are analogous expressions when one of the other components of the direction vector is zero. □

EXAMPLE 2

Converting from symmetric form to parametric equations

Given the symmetric equations

$$\frac{x - 4}{5} = \frac{y + 2}{-3} = \frac{z - 1}{7}$$

for a line $\mathcal{L}$, find parametric equations for $\mathcal{L}$.

SOLUTION

We introduce the parameter, t, as follows:

$$t = \frac{x - 4}{5} = \frac{y + 2}{-3} = \frac{z - 1}{7}.$$

So,

$$\begin{aligned}x - 4 &= 5t\\ y + 2 &= -3t\\ z - 1 &= 7t,\end{aligned}$$

where $-\infty < t < \infty$. These equations are parametric equations for $\mathcal{L}$. We may also write the system in the form

$$x(t) = 4 + 5t,\ y(t) = -2 - 3t,\ z(t) = 1 + 7t, -\infty < t < \infty.$$ □

EXAMPLE 3 **Determining when two lines are parallel**

Consider the lines $\mathcal{L}_1$ and $\mathcal{L}_2$ respectively given by the parametrizations

$$\mathbf{r}_1(t) = \langle 3+5t, -1-6t, 4t\rangle, -\infty < t < \infty, \text{ and}$$

$$\mathbf{r}_2(t) = \langle 4-10t, 7+12t, -1-8t\rangle, -\infty < t < \infty.$$

Are lines $\mathcal{L}_1$ and $\mathcal{L}_2$ parallel? If they are, find the distance between them.

SOLUTION

Two distinct lines $\mathcal{L}_1$ and $\mathcal{L}_2$ are parallel if they have (nonzero) parallel direction vectors, $\mathbf{d}_1$ and $\mathbf{d}_2$, respectively. That is, the lines are parallel if there is a scalar, k, such that $\mathbf{d}_2$ is equal to $k\mathbf{d}_1$.

Lines $\mathcal{L}_1$ and $\mathcal{L}_2$ have direction vectors $\mathbf{d}_1 = \langle 5, -6, 4\rangle$ and $\mathbf{d}_2 = \langle -10, 12, -8\rangle$, respectively. Since $\mathbf{d}_2 = -2\mathbf{d}_1$, the lines are either parallel or identical. Two parallel lines are identical if they share a point. We may use any point on either of the lines and check whether that point is on the other line. One convenient point on $\mathcal{L}_1$ is $\mathbf{r}_1(0) = \langle 3, -1, 0\rangle$. Is $(3, -1, 0)$ also on $\mathcal{L}_2$? It is if there is a solution of the system of equations

$$\begin{aligned} 4 - 10t &= 3 \\ 7 + 12t &= -1 \\ -1 - 8t &= 0. \end{aligned}$$

Note that the unique solution of the last of these equations is $t = -\frac{1}{8}$, but this value does not satisfy either of the other equations. Therefore, there is no value of t that simultaneously satisfies all three of the preceding equations. This fact tells us that the lines $\mathcal{L}_1$ and $\mathcal{L}_2$ are parallel, but not identical.

To find the distance between the two lines, we choose any point on one of the lines and find the distance from that point to the other line. We already know that $(3, -1, 0)$ is on $\mathcal{L}_1$. We will therefore use this point as P. We need a point on $\mathcal{L}_2$, so we will use $\mathbf{r}_2(0) = \langle 4, 7, -1\rangle$. That is, we will let $P_0 = (4, 7, -1)$. Thus $\overrightarrow{P_0P} = \langle -1, -8, 1\rangle$. Finally, the direction vector $\mathbf{d}_2 = \langle -10, 12, -8\rangle$. We are now ready to use the distance formula given in Theorem 10.38. The distance from the point P to the line $\mathcal{L}_2$, and therefore the distance between $\mathcal{L}_1$ and $\mathcal{L}_2$, is

$$\frac{\|\langle -10, 12, -8\rangle \times \langle -1, -8, 1\rangle\|}{\|\langle -10, 12, -8\rangle\|} = \frac{\|\langle -52, 18, 92\rangle\|}{\sqrt{308}} = \sqrt{\frac{2873}{77}} \approx 6.1 \text{ units.}$$

As an extension of this example, consider the parametrization

$$\mathbf{r}_3(t) = \langle -12+15t, 17-18t, -12+12t\rangle, -\infty < t < \infty.$$

A direction vector for this parametrization is $\mathbf{d}_3 = \langle 15, -18, 12\rangle$, which is a scalar multiple of $\mathbf{d}_1$. Furthermore, the point $(3, -1, 0)$ is on the line determined by $\mathbf{r}_3(t)$, since the system of equations

$$\begin{aligned} -12 + 15t &= 3 \\ 17 - 18t &= -1 \\ -12 + 12t &= 0 \end{aligned}$$

has the solution $t = 1$. Thus, $\mathbf{r}_3$ is a different parametrization for $\mathcal{L}_1$.

Note that the two parameters are *not* equal at the point $(3, -1, 0)$; that is,

$$\mathbf{r}_1(0) = \langle 3, -1, 0\rangle = \mathbf{r}_3(1).$$

This is irrelevant! The significant concepts here are that the lines are parallel and they share a point; therefore they must represent the same line. □

EXAMPLE 4

Determining when two lines intersect

Do the lines $\mathcal{L}_1$ and $\mathcal{L}_2$, respectively given by

$\mathbf{r}_1(t) = \langle -8 - 5t, 3 - t, 4 \rangle, -\infty < t < \infty,$ and $\mathbf{r}_2(t) = \langle 6 + t, 4 + 2t, 1 + 3t \rangle, -\infty < t < \infty,$

intersect?

SOLUTION

We see immediately that the direction vectors for these lines are $\mathbf{d}_1 = \langle -5, -1, 0 \rangle$ and $\mathbf{d}_2 = \langle 1, 2, 3 \rangle$, which are *not* parallel. (Why?) Do the lines share a point? If so, the lines intersect. If not, they are skew.

Finding a candidate for a point of intersection here is more delicate than in Example 3, where we could have used *any* point on either of the lines. We *cannot* just equate the corresponding components, as written, and try to solve for t. If you try this, you will see that the system

$$-8 - 5t = 6 + t, \quad 3 - t = 4 + 2t, \quad 4 = 1 + 3t$$

has no solution. However, as we will see in a moment, the lines *do* intersect! The difficulty here is with the parameters. Although we've used the letter t as the parameter for each line, there are *two* distinct parameters, one for $\mathcal{L}_1$ and a second for $\mathcal{L}_2$. To proceed, we change the name of the variable for one of the parameters. Let

$$\mathbf{r}_2(u) = \langle 6 + u, 4 + 2u, 1 + 3u \rangle, -\infty < u < \infty.$$

Now we equate the corresponding components of $\mathbf{r}_1$ and $\mathbf{r}_2$ to obtain the system

$$-8 - 5t = 6 + u, \quad 3 - t = 4 + 2u, \quad 4 = 1 + 3u.$$

This system has the unique solution $t = -3$, $u = 1$. Thus, the lines do intersect at

$$\mathbf{r}_1(-3) = \mathbf{r}_2(1).$$

□

CHECKING THE ANSWER

To verify our conclusion, we evaluate

$$\mathbf{r}_1(-3) = \big\langle -8 - 5(-3), 3 - (-3), 4 \big\rangle = \langle 7, 6, 4 \rangle$$

and

$$\mathbf{r}_2(1) = \langle 6 + 1, 4 + 2 \cdot 1, 1 + 3 \cdot 1 \rangle = \langle 7, 6, 4 \rangle.$$

These calculations confirm that the point $(7, 6, 4)$ is on both lines.

EXAMPLE 5

Finding the distance from a point to a line

Find the distance from the point $P = (2, 0, 1)$ to the line given by

$$\mathbf{r}(t) = \langle 3 - t, -5 + 4t, 6 \rangle, -\infty < t < \infty.$$

SOLUTION

Here, $P_0 = (3, -5, 6)$ corresponds to $\mathbf{r}(0)$ and $\mathbf{d} = \langle -1, 4, 0 \rangle$ is the direction vector for the line. Thus, $\|\mathbf{d}\| = \sqrt{17}$ and $\overrightarrow{P_0P} = \langle -1, 5, -5 \rangle$. The norm of $\mathbf{d} \times \overrightarrow{P_0P}$ is computed as follows:

$$\|\mathbf{d} \times \overrightarrow{P_0P}\| = \left\| \det \begin{bmatrix} \mathbf{i} & \mathbf{j} & \mathbf{k} \\ -1 & 4 & 0 \\ -1 & 5 & -5 \end{bmatrix} \right\| = \|-20\mathbf{i} - 5\mathbf{j} - \mathbf{k}\| = \sqrt{426}.$$

By Theorem 10.38, the distance from P to the line is $\frac{\|\mathbf{d} \times \overrightarrow{P_0P}\|}{\|\mathbf{d}\|} = \sqrt{\frac{426}{17}} \approx 5.0$ units. □

? TEST YOUR UNDERSTANDING

- How do you find an equation of a line determined by two points?
- How do you find an equation of a line parallel to a given vector and passing through a given point?
- Given a parametrization of a line, parametric equations of a line, or the equation of the line in symmetric form, how do you find the other forms?
- What are parallel lines, intersecting lines, and skew lines in 3-space?
- How do you find the distance from a point to a line?

EXERCISES 10.5

Thinking Back

Lines in the plane: Write the equation of the specified lines in $\mathbb{R}^2$ as vector parametrizations and in symmetric form.

- $y = 2x + 5$
- The line perpendicular to the line $y = 2x + 5$ and containing the point $(2, -1)$.
- $y = mx + b$
- The line perpendicular to the line $y = mx + b$ and containing the point $(0, b)$.
- *The distance between two points in the plane:* What is the formula for computing the distance between points (x_1, y_1) and (x_2, y_2)?
- *The distance between a point and a line in the plane:* Describe a method for computing the distance between the point (x_0, y_0) and the line $y = mx + b$.

Concepts

0. *Problem Zero:* Read the section and make your own summary of the material.

1. *True/False:* Determine whether each of the statements that follow is true or false. If a statement is true, explain why. If a statement is false, provide a counterexample.

(a) *True or False:* Two parallel lines that share a point are identical.

(b) *True or False:* If the direction vector for a line is not a multiple of one of the standard basis vectors, then the line will intersect all three coordinate planes.

(c) *True or False:* If a line $\mathbf{r}(t) = \mathbf{P}_1 + t\mathbf{d}$, $-\infty < t < \infty$, is parallel to the xz-plane then $\mathbf{d}$ must have the form $\langle a, 0, c\rangle$ for some real numbers a and c.

(d) *True or False:* If $\mathbf{r}_1(t) = \mathbf{P}_1 + t\mathbf{d}_1$, $-\infty < t < \infty$, and $\mathbf{r}_2(t) = \mathbf{P}_2 + t\mathbf{d}_2$, $-\infty < t < \infty$, are vector parametrizations for lines in 3-space with $\mathbf{d}_1 \neq \mathbf{d}_2$, then the lines do not intersect.

(e) *True or False:* If $\mathbf{r}_1(t) = \mathbf{P}_1 + t\mathbf{d}_1$, $-\infty < t < \infty$, and $\mathbf{r}_2(t) = \mathbf{P}_2 + t\mathbf{d}_2$, $-\infty < t < \infty$, are vector parametrizations for lines in 3-space with $\mathbf{P}_1 = \mathbf{P}_2$, then the lines intersect.

(f) *True or False:* Let $\mathbf{r}_1(t) = \mathbf{P}_1 + t\mathbf{d}_1$, $-\infty < t < \infty$, and $\mathbf{r}_2(t) = \mathbf{P}_2 + t\mathbf{d}_2$, $-\infty < t < \infty$, be vector parametrizations for lines in 3-space. If $\mathbf{P}_1 \neq \mathbf{P}_2$ and $\mathbf{d}_1 \neq \mathbf{d}_2$, then the lines cannot be identical.

(g) *True or False:* Let $\mathbf{r}_1(t) = \mathbf{P}_1 + t\mathbf{d}_1$, $-\infty < t < \infty$, and $\mathbf{r}_2(t) = \mathbf{P}_2 + t\mathbf{d}_2$, $-\infty < t < \infty$, be vector parametrizations for lines in 3-space. If $\mathbf{P}_1 \neq \mathbf{P}_2$ and $\mathbf{d}_1 \neq \mathbf{d}_2$, then the lines cannot be parallel.

(h) *True or False:* If $\mathbf{r}_1(t) = \mathbf{P}_1 + t\mathbf{d}_1$, $-\infty < t < \infty$, and $\mathbf{r}_2(t) = \mathbf{P}_2 + t\mathbf{d}_2$, $-\infty < t < \infty$, are vector parametrizations for lines in 3-space with $\mathbf{d}_1 \neq \mathbf{d}_2$, then the lines are skew.

2. *Examples:* Construct examples of the thing(s) described in the following. Try to find examples that are different than any in the reading.

(a) A line in $\mathbb{R}^3$ that is parallel to the xz-plane.

(b) A line in $\mathbb{R}^3$ through the origin.

(c) A line parallel to the z-axis.

3. What is a linear equation in three variables? Give an example. What is the graph of a linear equation in three variables?

4. Provide two sets of geometric conditions that can be used to determine a line.

5. Let P and Q be distinct points in $\mathbb{R}^3$. Provide a step-by-step procedure for finding the equation of the line containing P and Q.

6. Let $P = (a, b, c)$ and $Q = (\alpha, \beta, \gamma)$ be distinct points in $\mathbb{R}^3$. Explain why the parametrization

$$x = a + (\alpha - a)t,\ y = b + (\beta - b)t,\ z = c + (\gamma - c)t,$$

for $0 \le t \le 1$, describes the line *segment* connecting P and Q. What parametrization would describe the segment from Q to P?

7. Explain how the slopes of two lines in $\mathbb{R}^2$ can be used to determine whether the lines are parallel or identical. If the two lines are not parallel or identical, what must be true about the lines?

8. Explain how the direction vectors of two lines in $\mathbb{R}^3$ can be used to determine whether the two lines are parallel or identical. If the two lines are not parallel or identical, what are the possibilities for them?

9. Find an equation of the line containing the point $(-1, 3, 7)$ with direction vector $\langle 2, -4, 9\rangle$.
   (a) Use a different direction vector to find another equation for the same line.
   (b) Find an equation for the same line with the form
$$x = at + 5,\ y = bt + y_0,\ z = ct + z_0.$$
10. Find an equation of the line containing the point (x_0, y_0, z_0) with direction vector $\langle a, b, c\rangle$.
   (a) Use a different direction vector to find another equation for the same line.
   (b) Assuming that $a \neq 0$, find an equation for the same line with the form
$$x = at + 5,\ y = bt + y_0,\ z = ct + z_0.$$
11. Find the points where the line
$$\mathbf{r}(t) = \langle 4 - 3t, 8 + 7t, 5 + t\rangle,\ -\infty < t < \infty,$$
intersects each of the coordinate planes.
12. If a, b, and c are nonzero, find the points where the line
$$\mathbf{r}(t) = \langle x_0 + at, y_0 + bt, z_0 + ct\rangle$$
intersects each of the coordinate planes. If $a = 0$, explain why the line is parallel to the yz-plane.
13. Find the points where the line
$$x = t + 2,\ y = 2t - 5,\ z = -4t - 7$$
intersects each of the coordinate planes.
14. Find the points where the line
$$x = -3t + 5,\ y = 4,\ z = 2t + 11$$
intersects the xy-plane and yz-plane. Explain why the line does *not* intersect the xz-plane.
15. Find the point where the line
$$x = -7,\ y = t,\ z = 5$$
intersects the xz-plane. Explain why the line does *not* intersect the other two coordinate planes.
16. Let
$$x = at + x_0,\ y = bt + y_0,\ z = ct + z_0$$
be parametric equations for a line $\mathcal{L}$ in $\mathbb{R}^3$. If x_0, y_0, and z_0 are all nonzero, give conditions on a, b, and c so that
   (a) $\mathcal{L}$ intersects all three coordinate planes.
   (b) $\mathcal{L}$ intersects the xy- and yz-planes, but not the xz-plane.
   (c) $\mathcal{L}$ intersects exactly one of three coordinate planes.
17. Let $\mathcal{L}$ be the line determined by the equation
$$\mathbf{r}(t) = \langle 2 + 7t, 3 - 5t, 2t\rangle,\ -\infty < t < \infty.$$
   (a) Give parametric equations for $\mathcal{L}$.
   (b) Write an equation for $\mathcal{L}$ in symmetric form.
18. Let $\mathcal{L}$ be the line determined by the equation
$$\frac{x}{4} = \frac{y+2}{5} = -\frac{z-8}{3}.$$
   (a) Provide a vector parametrization for $\mathcal{L}$.
   (b) Give parametric equations for $\mathcal{L}$.
19. Let $\mathcal{L}$ be the line determined by the system of equations
$$x(t) = 4,\ y(t) = 3 - 5t,\ z(t) = t,\ -\infty < t < \infty.$$
   (a) Provide a vector parametrization for $\mathcal{L}$.
   (b) Write an equation for $\mathcal{L}$ in symmetric form.
20. We wish to find the distance from the point P to the line $\mathcal{L}$ as shown in the figure that follows. We know the coordinates of points P and P_0, but we do not know the coordinates of point Q.
   (a) If you knew the measure of angle θ, explain how you would find the distance from point P to line $\mathcal{L}$.

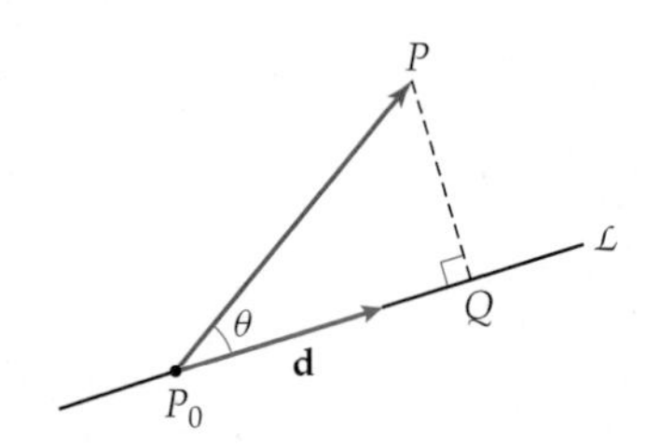

   (b) Using a cross product, explain how you can find the distance from point P to line $\mathcal{L}$ even if you do not know the measure of angle θ.

Skills

In Exercises 21 and 22, find the equation of the line containing the given points in slope–intercept form. Then, use the technique of this section to find a vector parametrization for the same line. Finally, show that your equations are equivalent.

21. $P(0, 5)$, $Q(2, -1)$
22. $P(3, -2)$, $Q(6, 4)$

In Exercises 23–28, find an equation of the line containing the given point and parallel to the given vector. Express your answer (a) as a vector parametrization, (b) in terms of parametric equations, and (c) in symmetric form.

23. $P(0, 0, 0)$, $\mathbf{d} = \langle 1, 2, -4\rangle$
24. $P(2, 3, 5)$, $\mathbf{d} = \langle 2, 3, 5\rangle$
25. $P(-1, 3, 7)$, $\mathbf{d} = \langle 2, 0, 4\rangle$
26. $P(x_0, y_0, z_0)$, $\mathbf{d} = \langle a, b, c\rangle$
27. $P(3, 1)$, $\mathbf{d} = \langle 2, 5\rangle$
28. $P(1, 3, -2, 4)$, $\mathbf{d} = \langle 4, -1, 5, 8\rangle$

In Exercises 29–34, find an equation of the line containing the given pair of points. Express your answer (a) as a vector parametrization, (b) in terms of parametric equations, and (c) in symmetric form.

29. $P(0, 0, 0)$, $Q(4, -1, 6)$
30. $P(3, -1, 7)$, $Q(5, 8, -2)$

31. $P(-4, 11, 0),\ Q(4, 11, 2)$

32. $P(x_0, y_0, z_0),\ Q(x_1, y_1, z_1)$

33. $P(1, 6),\ Q(4, 5)$

34. $P(3, -1, 2, 6),\ Q(1, 4, 5, -2)$

35. (a) Find a vector parametrization for the line containing the points $P(x_0, y_0, z_0)$ and $Q(x_1, y_1, z_1)$.

(b) Apply a restriction to your parameter from part (a) so that the result parametrizes the *segment* from P to Q.

In Exercises 36–39, use the result of Exercise 35 to find parametric equations for the line segment connecting point P to point Q.

36. $P(0, 2, 3),\ Q(4, 5, -1)$

37. $P(1, 7, 3),\ Q(-1, -2, 5)$

38. $P(0, 0, 0),\ Q(1, 2, 3)$

39. $P(3, -1, 4),\ Q(-1, 5, 9)$

In Exercises 40–45, determine whether the given pairs of lines are parallel, identical, intersecting, or skew. If the lines are parallel, compute the distance between them. If the lines intersect, find the point of intersection and the angle at which the lines intersect.

40. $\mathbf{r}_1(t) = \langle 2 + t, 5 - 3t, 6 + 7t\rangle$,
$\mathbf{r}_2(u) = \langle 3 - u, 4 + 3u, 5 - 2u\rangle$

41. $\mathbf{r}_1(t) = \langle 2t + 6, -t + 1, 3t\rangle$,
$\mathbf{r}_2(t) = \langle -4t + 3, 2t - 1, -6t + 2\rangle$

42. $\mathbf{r}_1(t) = \langle 3 - t, 7 + t, 4 + 5t\rangle$,
$\mathbf{r}_2(t) = \langle 3 - t, 4 + 3t, 5 - 2t\rangle$

43. $\mathbf{r}_1(t) = \langle 5t + 2, -4t, t - 7\rangle$,
$\mathbf{r}_2(t) = \langle -3t + 4, -t + 12, -2t - 1\rangle$

44. $\mathbf{r}_1(t) = \langle 4 + 5t, 6, 7 - 2t\rangle$,
$\mathbf{r}_2(t) = \langle 6 - 4t, -3 + 3t, -1 + 4t\rangle$

45. $\mathbf{r}_1(t) = \langle 2 + 5t, 6 - 4t, 8 + t\rangle$,
$\mathbf{r}_2(u) = \langle 2 - 10u, 3 + 8u, -2u\rangle$

In Exercises 46–49, calculate the distance from the given point to the given line in the following two ways: (a) using the method of Example 4 from Section 10.3 and (b) using Theorem 10.38 from this section.

46. Point $P(-6, 3, 0)$ to the line determined by the points $Q(3, -1, 5)$ and $R(4, 5, -2)$.

47. Point $P(0, 1, -2)$ to the line given by

$$\mathbf{r}(t) = \langle 1 + 5t, -6 + t, -4t\rangle.$$

48. Point $P(-6, 3, 0)$ to the line given by

$$\frac{x-3}{4} = \frac{y+2}{6} = z.$$

49. Point $P(2, 5)$ to the line $y = 2x - 3$.

50. Find the distance from the point $P(2, 3, -1, 4)$ to the line determined by the points $Q(1, 0, 4, 8)$ and $R(3, 4, -1, 6)$.

51. Find values for α such that the lines determined by

$$\mathbf{r}_1(t) = \langle 4 - t, -6 + 2t, 6 + 5t\rangle \quad \text{and}$$
$$\mathbf{r}_2(t) = \langle \alpha t, 2 - 2\alpha t, 3 + 15t\rangle$$

are (a) parallel and (b) orthogonal.

Applications

52. Emmy is a civil engineer at the Hanford Nuclear Reservation in Washington State. She has discovered a leak of toxic wastes in one of the tank farms of the facility. The tank farm is huge, and she does not know which tank is leaking. Worse yet, the tanks are all underground. Inspecting the tanks would require digging up the entire tank farm, an operation that is considered too expensive. Instead, Emmy has wells dug in several locations around the tank farm, to try to trace the leak back to its source. If the earth's surface is considered to be the xy-plane, then the bottom of the tank farm is the plane $z = -40$, where the distance is given in feet. Emmy's wells find contaminated groundwater at the points $(758, 60, -49)$ and $(1033, 247, -55)$. If the waste is leaking from the bottom of one of the tanks and is moving along a straight path, what is the location of the tank she should check first for the leak?

53. Ian is climbing Mount Logan in Canada's Icefield Range. He has made a second camp high on a ridge of the mountain. He does not know how high his camp is, but he notices that a summit far away lines up perfectly with a nearer minor point whose height he can read on his map. When Ian considers the summit of Mount Logan to be $(0, 0, 5.96)$, he finds that the far summit is at $(11.2, -5.6, 4.2)$ and the nearer point is at $(5.4, -2.5, 4.5)$. All the coordinates are given in kilometers. Ian's camp is due east of the summit of Mount Logan. How high is Ian on the mountain? How far is he from the summit?

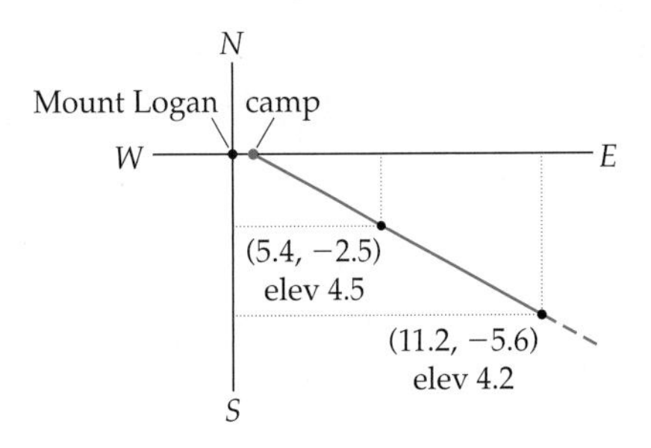

Proofs

54. Let $\mathcal{L}_1$ and $\mathcal{L}_2$ be lines in $\mathbb{R}^3$, with P_1 and Q_1 points on $\mathcal{L}_1$ and P_2 and Q_2 points on $\mathcal{L}_2$. Show that $\mathcal{L}_1$ is parallel to $\mathcal{L}_2$ if and only if $\overrightarrow{P_1Q_1}$ is parallel to $\overrightarrow{P_2Q_2}$.

55. Prove that the distance from the point P to the line given by the equation $\mathbf{r}(t) = \mathbf{P}_0 + t\mathbf{d}$ is given by $\dfrac{\|\mathbf{d} \times \overrightarrow{P_0P}\|}{\|\mathbf{d}\|}$.

Thinking Forward

- *Lines tangent to spheres*: Let $\mathcal{S}$ denote the unit sphere centered at the origin. How many lines are there that are tangent to $\mathcal{S}$? How many lines are there that are tangent to $\mathcal{S}$ at each point on the sphere? How many lines are there that are tangent to $\mathcal{S}$ and parallel to one of the coordinate planes? How many lines are there that are tangent to $\mathcal{S}$ and parallel to one of the coordinate axes?
- *Planes tangent to spheres*: Let $\mathcal{S}$ denote the unit sphere centered at the origin. How many planes are there that are tangent to $\mathcal{S}$? How many planes are there that are tangent to $\mathcal{S}$ at each point on the sphere? How many planes are there that are tangent to $\mathcal{S}$ and parallel to one of the coordinate planes? How many planes are there that are tangent to $\mathcal{S}$ and parallel to one of the coordinate axes?

10.6 PLANES

- Using a normal vector and a point to construct the equation for a plane
- Computing the distance from a point to a plane
- Determining whether two planes are parallel or intersect

The Equation of a Plane

In our final section of the chapter we discuss planes in three-dimensional space. Just as curves in two dimensions have tangent lines, surfaces have tangent planes, and those planes can be used as approximations for the surface. We will visit these topics in Chapter 12.

There are several geometric conditions that determine a unique plane. For example, each of the following sets of information uniquely determines one plane in $\mathbb{R}^3$:

- three noncollinear points;
- a line and a point not on the line;
- two distinct intersecting lines;
- two (distinct) parallel lines.

Our path to the equation of a plane, however, involves a point on the plane and a single vector, $\mathbf{N}$, called a ***normal vector***, orthogonal to the plane. We will also see, in examples and exercises, how we can find an equation for the plane given any of the conditions just listed.

Let $P_0 = (x_0, y_0, z_0)$ be a given point on a plane $\mathcal{P}$ and let $\mathbf{N} = \langle a, b, c\rangle$ be a vector orthogonal to $\mathcal{P}$. Place $\mathbf{N}$ so that its initial point is at P_0, and consider the vector $\overrightarrow{P_0P}$, where P is an arbitrary point on $\mathcal{P}$.

A plane is determined by a point on the plane and a vector orthogonal to the plane

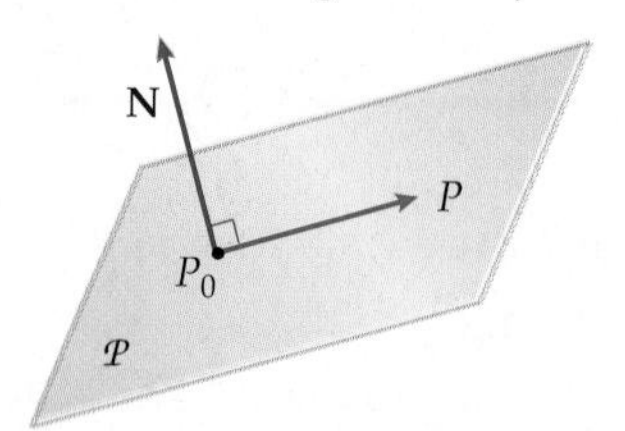

Note that since $\mathbf{N}$ is orthogonal to $\mathcal{P}$, it is orthogonal to every vector lying in $\mathcal{P}$. In particular, $\mathbf{N}$ is orthogonal to the vector $\overrightarrow{P_0P}$, and thus $\mathbf{N} \cdot \overrightarrow{P_0P} = 0$. If $P = (x, y, z)$, then

$$\mathbf{N} \cdot \overrightarrow{P_0P} = \langle a, b, c\rangle \cdot \langle x - x_0, y - y_0, z - z_0\rangle = 0.$$

Carrying out the dot product, we obtain the following equation for a plane:

$$a(x - x_0) + b(y - y_0) + c(z - z_0) = 0.$$

This equation is analogous to the point–slope form for the equation of a line that we studied in Chapter 0.

We may rearrange the last equation to obtain

$$ax + by + cz = ax_0 + by_0 + cz_0.$$

Since a, b, c, x_0, y_0, and z_0 are all constants, if we let $d = ax_0 + by_0 + cz_0$, then

$$ax + by + cz = d,$$

the ***general form for the equation of a plane***. Recall from the last section that equations of this form are called linear equations and the graph of a linear equation in three variables is a plane.

For example, to find the equation of the plane that has the normal vector $\mathbf{N} = \langle 5, -3, 8\rangle$ and contains the point $(2, -7, 1)$, we set up the equation

$$\langle 5, -3, 8\rangle \cdot \langle x - 2, y - (-7), z - 1\rangle = 0.$$

We can then write the equation in either of the forms

$$5(x - 2) - 3(y + 7) + 8(z - 1) = 0 \qquad \text{and} \qquad 5x - 3y + 8z = 39.$$

Geometric Conditions That Determine Planes

We now know how to find the equation of a plane given a normal vector to the plane and a point on the plane. When we are given different geometric conditions that determine a plane, we will follow this same procedure to find the equation of the plane.

Suppose we know three noncollinear points on a plane and we wish to find the equation for the plane. We can use the three points to find a normal vector for the plane and then use this normal vector and any one of the three points to find the equation of the plane, as we did at the beginning of this section. One normal vector to the plane containing P, Q, and R is the cross product $\mathbf{N} = \overrightarrow{PQ} \times \overrightarrow{PR}$, as shown in the following figure:

A plane is determined by three noncollinear points

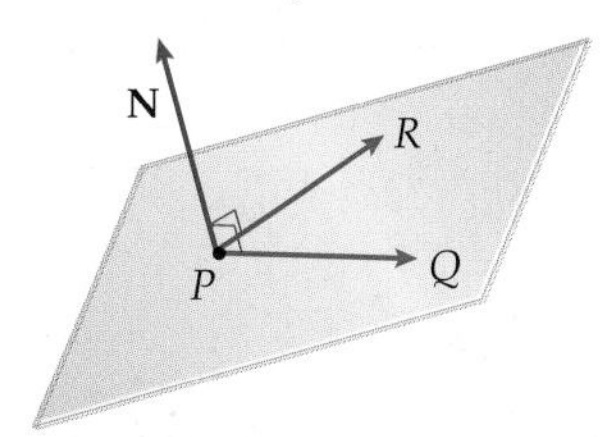

We will discuss how to find the equation of a plane by using one more of the sets of geometric conditions that determine the plane. You should then consider how to find equations for planes by using the other sets of geometric conditions.

A given line and a point not on the line define a plane uniquely. When you have the equation for a line, $\mathbf{r}(t) = \mathbf{P}_0 + t\mathbf{d}$, together with a point, P, not on the line, you immediately know that P and P_0 are two distinct points on some plane $\mathcal{P}$. In addition, the direction vector **d** for the line gives you one vector parallel to the plane. We already know that we can find a vector normal to $\mathcal{P}$ if we have two nonparallel vectors that are parallel to the plane (since we can then use their cross product). Vectors **d** and $\overrightarrow{P_0P}$ meet our requirements. Thus, we will use $\mathbf{N} = \mathbf{d} \times \overrightarrow{P_0P}$ and point P_0 to find the equation of the plane.

A plane is determined by a line and a point not on the line

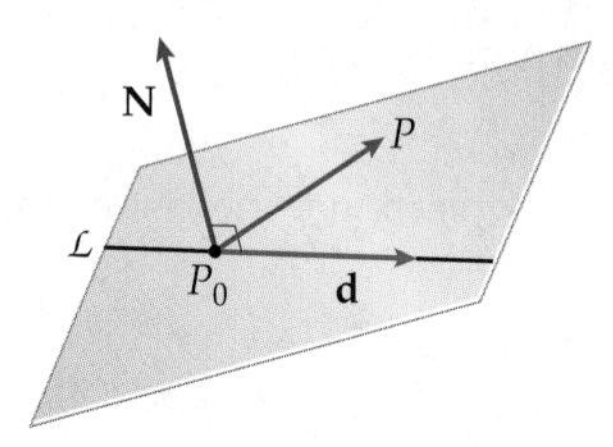

Distances

The distance from a point P, to a plane, $\mathcal{P}$, is the distance from P to the closest point on $\mathcal{P}$. This is the distance along a line through P and orthogonal to $\mathcal{P}$.

When we have the equation of a plane, we can immediately find a vector, **N**, normal to the plane. One way to find the distance from point P to plane $\mathcal{P}$ would be to find the equation of the line $\mathcal{L}$ containing P with direction vector **N**. We could then locate the point Q of intersection of $\mathcal{L}$ and $\mathcal{P}$. Finally we could use the distance formula to calculate the distance from P to Q.

Finding the distance from a point to a plane

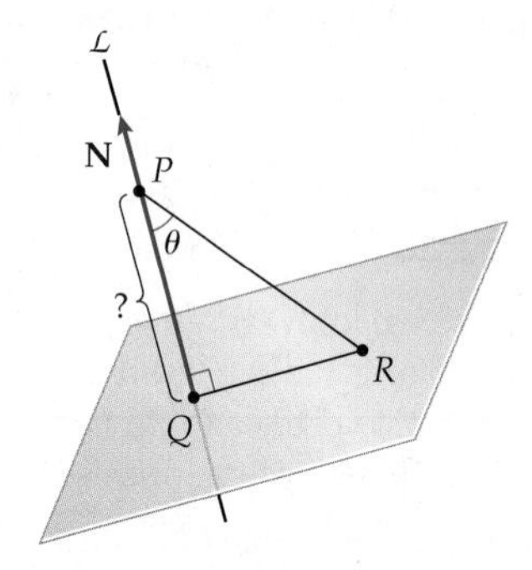

The method just outlined would work, but is more than we need to just find the distance from P to the plane. The extra work comes when we perform several steps to find the point of intersection, Q. It turns out that we do not need to find Q. We do need to find $\|\overrightarrow{QP}\|$, and this magnitude is $|\text{comp}_{\mathbf{N}}\, \overrightarrow{RP}\,|$, where R can be any point on the plane $\mathcal{P}$.

To summarize, we have the following theorem:

THEOREM 10.39 **The Distance from a Point to a Plane**

Given a point P and a plane $\mathcal{P}$ containing a point R with normal vector **N**, the distance from P to $\mathcal{P}$ is

$$|\text{comp}_{\mathbf{N}}\, \overrightarrow{RP}\,| = \frac{|\mathbf{N} \cdot \overrightarrow{RP}\,|}{\|\mathbf{N}\|}.$$

The following theorem is a direct consequence of Theorem 10.39:

THEOREM 10.40

The Distance Between a Plane and the Origin

Let $P = (x_0, y_0, z_0)$ be any point on the plane $\mathcal{P}$. The distance between the origin and $\mathcal{P}$ is given by $|\langle x_0, y_0, z_0\rangle \cdot \mathbf{n}|$ where $\mathbf{n}$ is a unit normal vector for $\mathcal{P}$.

The proof of Theorem 10.40 is left for Exercise 62. We may also adapt Theorem 10.39 to find the distance between parallel planes $\mathcal{P}_1$ and $\mathcal{P}_2$.

Finding the distance between two parallel planes

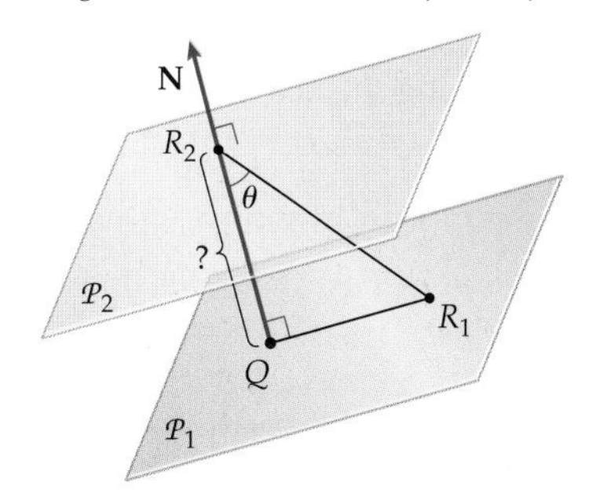

Since $\mathcal{P}_1$ and $\mathcal{P}_2$ are parallel, they have a common normal vector $\mathbf{N}$. We need a single point on each of the planes. Suppose that $R_1 = (x_1, y_1, z_1)$ is on $\mathcal{P}_1$ and $R_2 = (x_2, y_2, z_2)$ is on $\mathcal{P}_2$. We may now apply Theorem 10.39 to compute the distance to obtain the following:

THEOREM 10.41

The Distance Between Parallel Planes

Given a point R_1 on a plane $\mathcal{P}_1$ and a point R_2 on a parallel plane $\mathcal{P}_2$ with common normal vector $\mathbf{N}$, the distance from $\mathcal{P}_1$ to $\mathcal{P}_2$ is

$$\frac{|\mathbf{N} \cdot \overrightarrow{R_1R_2}\,|}{\|\mathbf{N}\|}.$$

For our final distance computation we compute the distance between two skew lines. When two lines $\mathcal{L}_1$ and $\mathcal{L}_2$ are skew, we can find a unique pair of parallel planes such that each of the parallel planes contains one of the skew lines. To visualize this situation, think of a translation that takes one of the lines onto the x-axis. The same translation moves the other line to another line somewhere in the coordinate system that is skew in relation to the x-axis. Now, rotate this new system around the x-axis until the skew line lies in a plane parallel to the xy-plane, say, $z = c$. This rotation leaves the x-axis fixed. The planes $z = 0$ and $z = c$ are the unique parallel planes that contain these transformed lines. Since our translation and rotation did not change the relative orientations of the lines, if you rotate the lines back and then translate the lines back to their original positions, these two operations together will take the planes $z = 0$ and $z = c$ to the unique pair of parallel planes containing the original skew lines. Furthermore, since the distance between the planes $z = 0$ and $z = c$ is $|c|$, that is also the distance between the original skew lines. This *will not* be our method for computing the distance between skew lines, but we hope that it makes the geometry easier to understand.

Our method computes the distance between the unique parallel planes that contain the skew lines. If we had the normal vector to these parallel planes, we could use Theorem 10.41

to compute the distance. The normal vector $\mathbf{N}$ may be computed from the direction vectors $\mathbf{d}_1$ and $\mathbf{d}_2$ for $\mathcal{L}_1$ and $\mathcal{L}_2$. As the following figure shows, the normal vector $\mathbf{N} = \mathbf{d}_1 \times \mathbf{d}_2$:

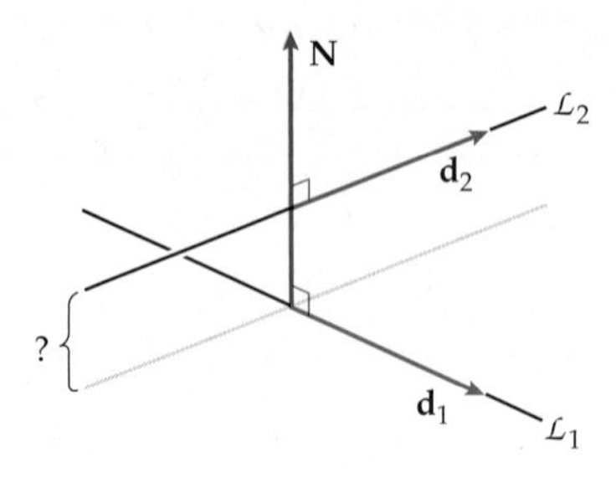

Combining this reasoning with Theorem 10.41, we have

THEOREM 10.42 **The Distance Between Skew Lines**

Let $\mathcal{L}_1$ and $\mathcal{L}_2$ be nonparallel lines with equations $\mathbf{r}_1(t) = P_1 + t\mathbf{d}_1$ and $\mathbf{r}_2(t) = P_2 + t\mathbf{d}_2$, respectively. Then the distance between $\mathcal{L}_1$ and $\mathcal{L}_2$ is given by

$$\frac{|(\mathbf{d}_1 \times \mathbf{d}_2) \cdot \overrightarrow{P_1P_2}|}{\|\mathbf{d}_1 \times \mathbf{d}_2\|}.$$

Intersecting Planes

We now turn our attention to intersections; two planes may intersect, and a line may intersect a plane. Two distinct planes in 3-space either intersect or are parallel. Given the equations of two planes, we can immediately tell whether or not they intersect by examining the normal vectors of the planes. If the normal vectors are parallel, then the planes are parallel and possibly even identical. The planes are identical if and only if the original equations are equivalent. As the following figure illustrates, when the normal vectors are not parallel, the planes will intersect in a line. To find the equation of the line of intersection, we will need a direction vector for the line and a point on the line. The direction vector can be found with the use of the cross product.

The line of intersection of two planes is orthogonal to each of the normal vectors

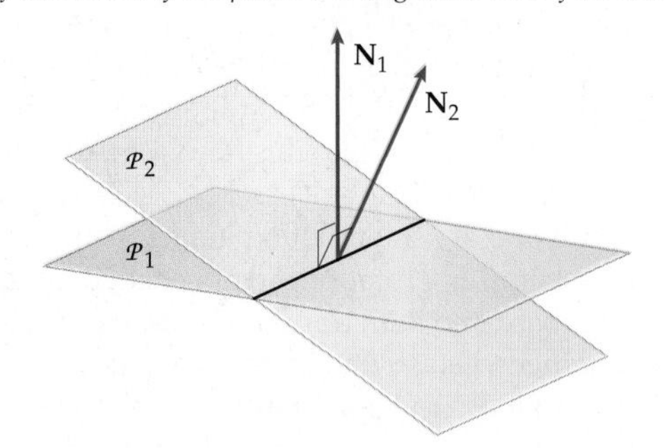

Thus, to find the equation of the line of intersection of two intersecting planes $\mathcal{P}_1$ and $\mathcal{P}_2$, we find their respective normal vectors $\mathbf{N}_1$ and $\mathbf{N}_2$. We also determine a single point on the line of intersection. We use the cross product $\mathbf{N}_1 \times \mathbf{N}_2$ as the direction vector for the line of intersection in order to determine the equation of the line.

Given a line and a plane in 3-space, one of three things must occur: The line and plane can intersect in a unique point, the line and plane can be parallel, or the line may lie *in* the plane. We can determine which of these situations occurs by looking at the dot product of the direction vector for the line and the normal vector for the plane. If $\mathbf{d}$ is the direction vector for the line and $\mathbf{N}$ is the normal vector for the plane, we consider $\mathbf{d} \cdot \mathbf{N}$. If $\mathbf{d} \cdot \mathbf{N} = 0$, the two vectors are orthogonal, so either the line and plane are parallel or the line is in the plane. If $\mathbf{d} \cdot \mathbf{N} \neq 0$, the line and plane intersect in a unique point.

Examples and Explorations

EXAMPLE 1

Finding the equation of a plane determined by three points

Show that the three points $P = (2, -1, 5)$, $Q = (9, 3, 7)$, and $R = (4, 7, -5)$ are noncollinear, and find the equation of the plane they determine.

SOLUTION

We form the vectors

$$\overrightarrow{PQ} = \langle 9 - 2, 3 - (-1), 7 - 5 \rangle = \langle 7, 4, 2 \rangle \text{ and } \overrightarrow{PR} = \langle 4 - 2, 7 - (-1), -5 - 5 \rangle = \langle 2, 8, -10 \rangle$$

and note that they are not parallel, since neither of them is a scalar multiple of the other. This implies that the three points are noncollinear. Next, we compute our normal vector. To find a normal vector we may take the cross product of the two vectors we just found. We let $\mathbf{N} = \overrightarrow{PQ} \times \overrightarrow{PR} = \langle 7, 4, 2 \rangle \times \langle 2, 8, -10 \rangle = \langle -56, 74, 48 \rangle$. Finally, since P, Q, and R are all in the plane, we are free to choose any of them as our designated point. We will use P. We now have an equation for the plane:

$$-56(x - 2) + 74(y + 1) + 48(z - 5) = 0.$$

Note that the components of $\mathbf{N}$ are all multiples of 2. This is because the components of both vectors are integers and all components of $\overrightarrow{PR}$ are multiples of 2. To make our calculation (slightly) simpler, we could have scaled $\overrightarrow{PR}$ by $\frac{1}{2}$ before we took the cross product. Scaling either $\overrightarrow{PQ}$ or $\overrightarrow{PR}$ by a nonzero constant leaves the relative geometry of the vectors unchanged. Certainly, if $\overrightarrow{PQ}$ and $\overrightarrow{PR}$ are not parallel, then $\overrightarrow{PQ}$ and $k(\overrightarrow{PR})$ are not parallel for any nonzero scalar k. Had we scaled $\overrightarrow{PR}$ by $\frac{1}{2}$, we would have obtained the normal vector

$$\tilde{\mathbf{N}} = \overrightarrow{PQ} \times \left(\frac{1}{2}\,\overrightarrow{PR}\right) = \langle -28, 37, 24 \rangle.$$

Using this as our normal vector, we would have gotten

$$-28(x - 2) + 37(y + 1) + 24(z - 5) = 0$$

as the equation for our plane. This equation is equivalent to the one we found before. □

EXAMPLE 2

Finding the equation of a plane determined by a line and a point not on the line

Show that the point $Q(3, -4, -1)$ is not on the line, $\mathcal{L}$, determined by

$$\mathbf{r}(t) = \langle 8 - 3t, -2 + 5t, 6 + 2t \rangle.$$

Give a general form for the equation of the plane $\mathcal{P}$ determined by Q and $\mathcal{L}$.

SOLUTION

If P were on $\mathcal{L}$, there would be a value of t that simultaneously satisfied the system of equations

$$8 - 3t = 3, \quad -2 + 5t = -4, \quad 6 + 2t = -1.$$

Since the solutions of each of the three equations are different, Q is not on $\mathcal{L}$.

We quickly note that the vector $\mathbf{d} = \langle -3, 5, 2\rangle$ is a direction vector for $\mathcal{L}$ and point $Q_0(8, -2, 6)$ is on $\mathcal{L}$. Another vector parallel to the plane of interest is $\overrightarrow{Q_0Q}$. Here,

$$\overrightarrow{Q_0Q} = \langle 3 - 8, -4 - (-2), -1 - 6\rangle = \langle -5, -2, -7\rangle.$$

To find a normal vector to the plane, we use the cross product of $\mathbf{d}$ and $\overrightarrow{Q_0Q}$:

$$\mathbf{d} \times \overrightarrow{Q_0Q} = \langle -3, 5, 2\rangle \times \langle -5, -2, -7\rangle = \langle -31, -31, 31\rangle.$$

Since we are interested primarily in the direction of this vector, we may scale it by any nonzero value to obtain simpler coefficients. We will scale the cross product by $-\frac{1}{31}$ and let $\mathbf{N} = \langle 1, 1, -1\rangle$.

We can use either Q or Q_0 as our point on the plane. If we use Q to obtain the equation for the plane, we arrive at the equation

$$1(x - 3) + 1(y + 4) - 1(z + 1) = 0.$$

A general form for the equation of the plane is $x + y - z = 0$. □

EXAMPLE 3 **Finding the distance from a point to a plane**

Find the distance from the point $P = (6, 3, -2)$ to the plane given by $3x - y + 2z = 4$.

SOLUTION

We need a point on the plane. Verify that $R = (0, 0, 2)$ is on the plane. We can immediately find the vector $\overrightarrow{RP}$ and a normal vector $\mathbf{N}$ to the plane. In this example we have $\overrightarrow{RP} = \langle 6, 3, -4\rangle$ and $\mathbf{N} = \langle 3, -1, 2\rangle$. Thus, the distance from P to the plane can be computed as follows:

$$|\text{comp}_{\mathbf{N}}\, \overrightarrow{RP}\,| = \frac{|\mathbf{N}\cdot \overrightarrow{RP}\,|}{\|\mathbf{N}\|} = \frac{|\langle 3, -1, 2\rangle \cdot \langle 6, 3, -4\rangle|}{\|\langle 3, -1, 2\rangle\|} = \frac{|18 - 3 - 8|}{\sqrt{3^2 + 1^2 + 2^2}} = \frac{7}{\sqrt{14}} = \frac{\sqrt{14}}{2}.$$ □

EXAMPLE 4 **Finding the distance between two parallel planes**

Show that the plane $\mathcal{P}_1$ with equation $-4x + 6y + 2z = 1$ and the plane $\mathcal{P}_2$ given by $2x - 3y - z = 5$ are parallel. Then compute the distance between $\mathcal{P}_1$ and $\mathcal{P}_2$.

SOLUTION

The plane $\mathcal{P}_1$ has the normal vector $\mathbf{N}_1 = \langle -4, 6, 2\rangle$ and the plane $\mathcal{P}_2$ has the normal vector $\mathbf{N}_2 = \langle 2, -3, -1\rangle$. Notice that the vectors $\mathbf{N}_1$ and $\mathbf{N}_2$ are scalar multiples of each other. This means that the planes $\mathcal{P}_1$ and $\mathcal{P}_2$ are parallel.

Now, the point $R_1 = (0, 0, 1/2)$ is on plane $\mathcal{P}_1$ and the point $R_2 = (0, 0, -5)$ is on plane $\mathcal{P}_2$. (Check these.) We use Theorem 10.41 to compute the distance between the planes:

$$\frac{|\langle 2, -3, -1\rangle \cdot \overrightarrow{R_1R_2}\,|}{\|\langle 2, -3, -1\rangle\|} = \frac{|\langle 2, -3, -1\rangle \cdot \langle 0, 0, -11/2\rangle|}{\|\langle 2, -3, -1\rangle\|} = \frac{11\sqrt{14}}{28}.$$ □

EXAMPLE 5

Finding the distance between two skew lines

Find the distance between the skew lines $\mathcal{L}_1$ and $\mathcal{L}_2$, respectively given by the vector functions

$$\mathbf{r}_1(t) = \langle 5+3t, -7+t, 2-5t\rangle \text{ and } \mathbf{r}_2(t) = \langle -2t, 7, 1-3t\rangle.$$

SOLUTION

Before we begin to compute the distance, we should really think about the relationship of the lines $\mathcal{L}_1$ and $\mathcal{L}_2$. Are they really skew? Might they intersect, be parallel, or even be identical?

Direction vectors for $\mathcal{L}_1$ and $\mathcal{L}_2$ are $\mathbf{d}_1 = \langle 3, 1, -5\rangle$ and $\mathbf{d}_2 = \langle -2, 0, -3\rangle$, respectively. Since these direction vectors are not scalar multiples of each other, the lines either intersect or are skew. Now, at this point we could look for a point of intersection. However, let us think about the process we are about to start. We are about to compute the distance between $\mathcal{L}_1$ and $\mathcal{L}_2$. If this distance is zero, the lines intersect. If the distance is positive, the lines are skew.

We first find the normal vector $\mathbf{N}$:

$$\mathbf{N} = \mathbf{d}_1 \times \mathbf{d}_2 = \langle 3, 1, -5\rangle \times \langle -2, 0, -3\rangle = \langle -3, 19, 2\rangle.$$

(Check this.) We will use $P_1 = \mathbf{r}_1(0) = (5, -7, 2)$ and $P_2 = \mathbf{r}_2(0) = (0, 7, 1)$ as our points on $\mathcal{L}_1$ and $\mathcal{L}_2$, respectively. Thus, $\overrightarrow{P_1P_2} = \langle -5, 14, -1\rangle$. Finally, we use Theorem 10.42 to find the distance between $\mathcal{L}_1$ and $\mathcal{L}_2$. This distance is

$$\frac{|\mathbf{N}\cdot \overrightarrow{P_1P_2}\,|}{\|\mathbf{N}\|} = \frac{|\langle -3, 19, 2\rangle \cdot \langle -5, 14, -1\rangle|}{\|\langle -3, 19, 2\rangle\|} = \frac{279}{374}\sqrt{374}.$$

Since the distance is nonzero, the lines are skew. □

EXAMPLE 6

Finding the line of intersection of two planes

Show that the planes whose equations are $-x+2y-4z=6$ and $3x+5y-z=4$ intersect, and find parametric equations for the line of intersection.

SOLUTION

The normal vector to $-x+2y-4z=6$ is $\mathbf{N}_1 = \langle -1, 2, -4\rangle$, while the normal vector to $3x+5y-z=4$ is $\mathbf{N}_2 = \langle 3, 5, -1\rangle$. These normal vectors are not parallel; therefore the planes must intersect. The line of intersection must be orthogonal to each of the normal vectors. To find the direction vector, $\mathbf{d}$, for the line of intersection, we use the cross product. Here,

$$\mathbf{d} = \mathbf{N}_1 \times \mathbf{N}_2 = \langle -1, 2, -4\rangle \times \langle 3, 5, -1\rangle = \langle 18, -13, -11\rangle.$$

To find the equation for the line of intersection, we now need only a point on the line, which will, of course, lie on each of the planes. Unless the line is parallel to one of the coordinate planes, it will intersect all three coordinate planes. In this example, our line will intersect all three coordinate planes, because all three components of $\mathbf{d}$ are nonzero. To simplify our search for a point of intersection, we can decide to find such a point in one of the coordinate planes. Here, we will look for the point of intersection in the xy-plane. Thus, we will set $z=0$ in each of the equations $-x+2y-4z=6$ and $3x+5y-z=4$, to obtain the system

$$-x+2y=6 \text{ and } 3x+5y=4,$$

which has the solution $x = -2$ and $y = 2$. Thus, the point $(-2, 2, 0)$ is in each of the planes and therefore on our line of intersection. Parametric equations for the line of intersection are

$$x(t) = -2 + 18t,\ y(t) = 2 - 13t,\ z(t) = -11t,\ -\infty < t < \infty.$$ □

EXAMPLE 7

Understanding the geometry of a line and a plane

Let $\mathcal{L}$ be the line given by $\mathbf{r}(t) = \langle 4 + t, -3 + 5t, 2 - 3t \rangle$ and $\mathcal{P}$ be the plane with equation $-x + 5y + 6z = 5$. Show that $\mathcal{L}$ and $\mathcal{P}$ intersect and find the point of intersection.

SOLUTION

The vector $\mathbf{d} = \langle 1, 5, -3 \rangle$ is a direction vector for $\mathcal{L}$, while the vector $\mathbf{N} = \langle -1, 5, 6 \rangle$ is normal to $\mathcal{P}$. We compute the dot product:

$$\mathbf{d} \cdot \mathbf{N} = \langle 1, 5, -3 \rangle \cdot \langle -1, 5, 6 \rangle = -1 + 25 - 18 = 6 \neq 0.$$

Since the dot product is *not* zero, the line and plane intersect. Because $\mathcal{L}$ and $\mathcal{P}$ intersect, there will be a value of the parameter $t = t_0$ such that $\mathbf{r}(t_0)$ is on the plane. That is, the coordinates $(x(t_0), y(t_0), z(t_0))$ given by

$$x(t_0) = 4 + t_0,\ y(t_0) = -3 + 5t_0,\ z(t_0) = 2 - 3t_0$$

satisfy the equation of the plane. Since $-x + 5y + 6z = 5$, we have

$$-x(t_0) + 5y(t_0) + 6z(t_0) = 5.$$

Equivalently,

$$-(4 + t_0) + 5(-3 + 5t_0) + 6(2 - 3t_0) = 5.$$

Solving this equation for t_0, we obtain $t_0 = 2$. We now have the point of intersection,

$$(x(2), y(2), z(2)) = (6, 7, -4).$$ □

? TEST YOUR UNDERSTANDING

- How do you use a point on a plane and a vector normal to the plane to find the equation of the plane?
- How do you use other geometric conditions to find the equation of a plane? For example, how do you find the equation of a plane containing three noncollinear points or a plane containing two intersecting lines?
- How do you find the distance from a point to a plane? How do you find the distance between two parallel planes? How do you find the distance between two skew lines?
- How do you determine when two planes intersect? How do you find the line of intersection when they do?
- How do you determine when a line intersects a plane? How do you find the point of intersection when it does?

EXERCISES 10.6

Thinking Back

- *Linear equations:* Explain why the equation

$$2x - 3y = 5$$

represents a line in the xy-plane, but represents a plane in a three-dimensional coordinate system.

- *Orthogonal vectors:* Show that $\mathbf{u} = \langle 1, 2, -3 \rangle$ is orthogonal to $a\mathbf{v} + b\mathbf{w}$, where $\mathbf{v} = \langle 1, 1, 1 \rangle$, $\mathbf{w} = \langle -1, 2, 1 \rangle$, and a and b are any real numbers. Interpret this statement geometrically.

Concepts

0. *Problem Zero:* Read the section and make your own summary of the material.

1. *True/False:* Determine whether each of the statements that follow is true or false. If a statement is true, explain why. If a statement is false, provide a counterexample.
   (a) *True or False:* The graph of every linear equation in $\mathbb{R}^3$ is a line.
   (b) *True or False:* Three distinct points determine a plane.
   (c) *True or False:* Two distinct lines in $\mathbb{R}^3$ always determine a plane.
   (d) *True or False:* Three concurrent lines determine a plane. (***Concurrent*** means that the three lines intersect in a common point.)
   (e) *True or False:* If two distinct lines in $\mathbb{R}^3$ are not skew, they determine a unique plane.
   (f) *True or False:* If two distinct planes in $\mathbb{R}^3$ intersect, they intersect in a unique line.
   (g) *True or False:* If the direction vector of a line is parallel to the normal vector to a plane, the line and plane are parallel.
   (h) *True or False:* Let $\mathcal{P}_1$ and $\mathcal{P}_2$ be distinct planes with normal vectors $\mathbf{N}_1$ and $\mathbf{N}_2$, respectively. If $\mathbf{N}_1 \times \mathbf{N}_2 = \mathbf{0}$, the planes are parallel.

2. *Examples:* Construct examples of the thing(s) described in the following. Try to find examples that are different than any in the reading.
   (a) A plane parallel to $x + 3y - 4z = 7$.
   (b) A line orthogonal to the plane $x + 3y - 4z = 7$.
   (c) A plane orthogonal to the line

$$x = 3t - 5,\ y = -2t + 7,\ z = -4.$$

3. Let $ax + by + cz = d$ be the equation of a plane with a, b, c, and d all nonzero. What are the coordinates of the intersection of the plane and the x-, y-, and z-axes? Explain how to use these points to sketch the plane.

4. Explain why two planes orthogonal to the same vector are either parallel or identical.

5. Let $\mathbf{v} = \langle a, b, c\rangle$ and $\mathbf{w} = \langle \alpha, \beta, \gamma\rangle$, where $\mathbf{v}$ and $\mathbf{w}$ are *not* parallel vectors. Explain why the planes $ax + by + cz = d$ and $\alpha x + \beta y + \gamma z = \delta$ intersect.

6. Explain why there are infinitely many different planes containing any given line in $\mathbb{R}^3$. What form does the equation of a plane containing the x-axis have?

7. What does it mean for three points to be collinear? How do you determine that three given points are collinear? What does it mean for three points to be noncollinear?

8. Explain why three noncollinear points determine a unique plane. Explain how you would use the coordinates of the points to find the equation of the plane. Explain why three collinear points do not determine a unique plane.

9. Explain why a line $\mathcal{L}$ and a point P not on $\mathcal{L}$ determine a unique plane. Explain how you would use the equation of $\mathcal{L}$ and the coordinates P to find the equation of the plane. Explain why P and $\mathcal{L}$ do *not* determine a unique plane if P is on $\mathcal{L}$.

10. Explain why two intersecting lines determine a unique plane. Explain how you would use the equations of the lines to find the equation of the plane.

11. Explain why two distinct parallel lines determine a unique plane. Explain how you would use the equations of the lines to find the equation of the plane.

12. Explain why two skew lines do *not* determine a plane.

13. Explain why any two skew lines lie on a unique pair of parallel planes.

14. The angle θ between two intersecting planes, called the ***dihedral angle***, is defined to be the angle between the two normal vectors to the planes, where

$$\theta = \cos^{-1}\frac{\mathbf{N}_1 \cdot \mathbf{N}_2}{\|\mathbf{N}_1\|\|\mathbf{N}_2\|}.$$

Draw a figure that illustrates the dihedral angle and explain why the definition given is a reasonable definition.

15. Given the equations

$$x = at + x_0,\ y = bt + y_0,\ z = ct + z_0$$

for a line $\mathcal{L}$ and

$$\alpha x + \beta y + \gamma z = \delta$$

for a plane $\mathcal{P}$, explain how to determine whether $\mathcal{L}$ is orthogonal to $\mathcal{P}$.

16. Explain how to tell when two planes are perpendicular.

17. When a line $\mathcal{L}$ intersects a plane $\mathcal{P}$ the angle between them is defined to be the complement of the acute angle between the direction vector for the line and the normal vector to the plane. Draw a figure that illustrates this angle, and explain why the definition given is a reasonable definition.

18. Explain the similarities in the derivations of the formulas for the distances from a point to a plane and from a point to a line.

19. Explain the derivation of the formula for finding the distance between two skew lines $\mathcal{L}_1$ and $\mathcal{L}_2$. Why does this formula work?

20. Two distinct nonparallel planes intersect in a line. Outline a procedure for finding the equation of the line of intersection.

Skills

In Exercises 21–30, find the equations of the planes determined by the given conditions.

21. The plane contains the origin and is normal to the vector $\langle 4, -1, 5\rangle$.

22. The plane contains the point $(-2, 3, -1)$ and is normal to the vector $3\mathbf{i} - 2\mathbf{j}$.

23. The plane contains the point $(2, -1, 6)$ and is normal to the vector $\langle 2, -1, 6\rangle$.

24. The plane contains the points $(1,0,0)$, $(0,1,0)$, and $(0,0,1)$.
25. The plane contains the points $(2,4,3)$, $(3,-5,0)$, and $(-4,1,6)$.
26. The plane contains the points $(-4,0,0)$, $(0,3,0)$, and $(0,0,5)$.
27. The plane contains the points $(x_0,0,0)$, $(0,y_0,0)$, and $(0,0,z_0)$.
28. The plane contains the point $(1,2,-5)$ and the line determined by $\mathbf{r}(t)=\langle -4+t, 3+5t, 2-3t\rangle$.
29. The plane contains the point $(-4,1,3)$ and the line determined by $\mathbf{r}(t)=\langle 3+7t, -2+t, 3+4t\rangle$.
30. The plane contains the point $(-4,1,3)$ and is normal to the line determined by $\mathbf{r}(t)=\langle -4+t, 3+5t, 2-3t\rangle$.
31. Show that the lines determined by

$$\mathbf{r}_1(t)=\langle -2-5t, 3+2t, 4t\rangle \text{ and}$$
$$\mathbf{r}_2(t)=\langle 8+15t, 1-6t, 3-12t\rangle$$

are parallel, and then find an equation of the plane containing both lines.
32. Show that the lines determined by

$$\mathbf{r}_1(t)=\langle 3-5t, -2+t, 6\rangle \text{ and}$$
$$\mathbf{r}_2(t)=\langle 4+15t, 5-3t, 4\rangle$$

are parallel, and then find an equation of the plane containing both lines.
33. Show that the lines determined by

$$\mathbf{r}_1(t)=\langle 7, 3-4t, 2+6t\rangle \text{ and}$$
$$\mathbf{r}_2(t)=\langle 6-t, 3+8t, 9-5t\rangle$$

intersect, and then find an equation of the plane containing the two lines.
34. Show that the lines determined by

$$\mathbf{r}_1(t)=\langle 3-t, 4-4t, -3+4t\rangle \text{ and}$$
$$\mathbf{r}_2(t)=\langle 5-t, -6+2t, -2+t\rangle$$

intersect, and then find an equation of the plane containing both lines.

In Exercises 35–38, find an equation of the line of intersection of the two given planes.

35. $x+2y+3z=4$ and $-2x+y-4z=6$
36. $y-5z=3$ and $6x-7y=5$
37. $x=4$ and $3x-5y+2z=-3$
38. $x-2z=7$ and $x-3y-4z=0$
39. Find the distance from the point $(2,0,-3)$ to the plane $3x-4y+5z=1$.
40. Show that the planes given by $2x-4y-3z=5$ and $-4x+8y+6z=1$ are parallel, and find the distance between the planes.
41. Show that the planes given by $2x-3y+5z=7$ and $-6x+9y-15z=8$ are parallel, and find the distance between the planes.
42. Show that the planes given by $y-7z=16$ and $2y-14z=5$ are parallel, and find the distance between the planes.
43. Show that the lines with the equations

$$\frac{x+1}{2}=\frac{y-3}{-4}=\frac{z-2}{5} \text{ and } \frac{x-4}{3}=\frac{y+1}{2}=\frac{z}{3}$$

are skew, find the equations of the parallel planes containing the lines, and find the distance between the lines.

Use your answers from Exercise 14 to find the angle between the indicated planes in Exercises 44 and 45.

44. $7x-3y+5z=6$ and $2x+3y-z=1$
45. $-x+7y-2z=5$ and $3x+5y-4z=2$

Use Exercise 17 to find the angle between the indicated lines and planes in Exercises 46 and 47.

46. $\mathbf{r}(t)=\langle 5+2t, 6-t, 4+5t\rangle$ and $3y+5z=-4$
47. $\mathbf{r}(t)=\langle 3+5t, 2-t, 4-3t\rangle$ and $-10x+2y+6z=7$

In Exercises 48–51, determine whether the given line is parallel to, intersects, or lies in the given plane. If the line is parallel to the plane, calculate its distance from the plane. If the line intersects the plane, find the point and angle at which they intersect.

48. $\mathbf{r}(t)=\langle t, -5-2t, -1+3t\rangle$ and $4x-5y+2z=5$
49. $\mathbf{r}(t)=\langle 3-2t, -4, 5-4t\rangle$ and $2x+5y-z=7$
50. $\mathbf{r}(t)=\langle 4+t, 6-5t, -3t\rangle$ and $3x+3y-4z=30$
51. $\mathbf{r}(t)=\langle 5+6t, 3-t, 0\rangle$ and $7y-5z=3$
52. At every point on a sphere $(x-a)^2+(y-b)^2+(z-c)^2=r^2$, there is some plane tangent to the sphere. Explain how to find the equation of the tangent plane at any given point.
53. Use your answer in Exercise 52 to find the equations of the planes tangent to the given spheres at the specified points.
    (a) $x^2+y^2+z^2=1$ at $\left(\frac{1}{2},\frac{1}{4},\frac{\sqrt{11}}{4}\right)$.
    (b) $(x-1)^2+(y+2)^2+z^2=9$ at $(3,0,1)$.

Applications

54. Emmy is trying to get information about the water table below the Hanford reservation. She has drilled wells that show that the water table can be found at $(0,0,-35)$ and $(300,0,-38)$. She drills one more well and finds the water table at $(0,300,-37)$.
    (a) Find a plane that approximates the water table.
    (b) If she drills another hole at $x=300$, $y=300$, how deep does she expect to find the water table?
55. Annie is sitting on a beach in the evening, looking out at a mooring buoy. She wonders how deep the water is at the buoy. She assumes that the beach slopes out as a plane. She is sitting at a point $(150,30,5)$ relative to the buoy, where a z-coordinate of zero represents sea level and the coordinates are given in feet. The point on the shore that is closest to the buoy looks as if it is around $(120,40,0)$ relative to the buoy. There is a piece of driftwood down the beach that seems to be at about $(140,60,5)$. How deep is the water at the buoy?

Proofs

56. Let $\mathcal{L}_1$ and $\mathcal{L}_2$ be two skew lines. Prove that no plane contains both $\mathcal{L}_1$ and $\mathcal{L}_2$.

57. Prove that the planes determined by the equations $ax + by + cz = d$ and $\alpha x + \beta y + \gamma z = \delta$ are perpendicular if and only if $a\alpha + b\beta + c\gamma = 0$.

58. Let **a**, **b**, and **c** be position vectors terminating in some plane $\mathcal{P}$. Show that $(\mathbf{a} \times \mathbf{b}) + (\mathbf{b} \times \mathbf{c}) + (\mathbf{c} \times \mathbf{a})$ is normal to $\mathcal{P}$.

59. (a) Show that the distance from the point $P(x_0, y_0, z_0)$ to the plane with equation $ax + by + cz + d = 0$ is
$$\frac{|ax_0 + by_0 + cz_0 + d|}{\sqrt{a^2 + b^2 + c^2}}.$$
(b) Use the result of part (a) to recalculate the distance in Exercise 39.

60. Let $\mathbf{r}_1(t) = \mathbf{P}_1 + t\mathbf{d}_1$ and $\mathbf{r}_2(t) = \mathbf{P}_2 + t\mathbf{d}_2$ be parametrizations for two nonparallel lines. Prove that the lines intersect if and only if the three vectors $\mathbf{P}_1 - \mathbf{P}_2$, $\mathbf{d}_1$, and $\mathbf{d}_2$ are coplanar.

61. Let $\mathbf{r}_1(t) = \mathbf{P}_0 + t\mathbf{d}_1$ and $\mathbf{r}_2(u) = \mathbf{Q}_0 + u\mathbf{d}_2$ respectively be the equations of lines $\mathcal{L}_1$ and $\mathcal{L}_2$. Show that $(\mathbf{P}_0 - \mathbf{Q}_0) \cdot (\mathbf{d}_1 \times \mathbf{d}_2) = 0$ if and only if $\mathcal{L}_1$ and $\mathcal{L}_2$ lie in the same plane.

62. Prove Theorem 10.40. That is, show that if $P = (x_0, y_0, z_0)$ is a point on plane $\mathcal{P}$, then the distance between the origin and $\mathcal{P}$ is given by $|\langle x_0, y_0, z_0\rangle \cdot \mathbf{n}|$, where **n** is a unit normal vector to $\mathcal{P}$.

Thinking Forward

- *A plane tangent to a surface:* A particular smooth surface has tangent vectors $\mathbf{v}_x = \mathbf{i} - 3\mathbf{j}$ and $\mathbf{v}_y = \mathbf{i} + 4\mathbf{k}$ at the point $P(2, -3, 4)$. Find the equation of the tangent plane to the surface by finding the normal vector to the plane $\mathbf{N} = \mathbf{v}_x \times \mathbf{v}_y$ containing the point P.
- *A plane tangent to a surface:* A particular smooth surface has tangent vectors $\mathbf{v}_x = \mathbf{i} + \alpha\mathbf{j}$ and $\mathbf{v}_y = \mathbf{i} + \beta\mathbf{k}$ at the point $P(x_0, y_0, z_0)$. Find the equation of the tangent plane to the surface by finding the normal vector to the plane $\mathbf{N} = \mathbf{v}_x \times \mathbf{v}_y$ containing the point P.

CHAPTER REVIEW, SELF-TEST, AND CAPSTONES

Before you progress to the next chapter, be sure you are familiar with the definitions, concepts, and basic skills outlined here. The capstone exercises at the end bring together ideas from this chapter and look forward to future chapters.

Definitions

Give precise mathematical definitions or descriptions of each of the concepts that follow. Then illustrate the definition with a sketch or an algebraic example.

- a *sphere* in $\mathbb{R}^3$
- a *cylinder* in $\mathbb{R}^3$, along with the *directrix* and the *rulings* of the cylinder
- *vectors* in $\mathbb{R}^2$ and $\mathbb{R}^3$
- the *scalar* multiple of a vector
- the *magnitude, norm,* or *length* of a vector
- the *dot product* of two vectors
- the *standard basis vectors* in $\mathbb{R}^2$ and $\mathbb{R}^3$
- *orthogonal* curves and vectors
- the *projection* of one vector onto another
- the *determinant* of a 3×3 matrix
- the *cross product* of two vectors from $\mathbb{R}^3$

Theorems

Fill in the blanks to complete each of the following theorem statements:

- Given any nonzero vector **v**, the vector ________ is a unit vector in the direction of **v**.
- For any vectors $\mathbf{u} = \langle u_1, u_2, u_3\rangle$ and $\mathbf{v} = \langle v_1, v_2, v_3\rangle$, $\mathbf{u} \cdot \mathbf{v} =$ ________.
- If **u** and **v** are two nonzero vectors, then $\mathbf{u} \cdot \mathbf{v} =$ ________ $\cos\theta$, where θ is ________.
- *Law of Cosines*: In a triangle with side lengths a, b, and c, where θ is the angle between the sides of length a and b, $a^2 + b^2 -$ ________ $= c^2$.
- Let θ be the angle between nonzero vectors **u** and **v**. Then

θ is
________ if and only if $\mathbf{u} \cdot \mathbf{v} > 0$
________ if and only if $\mathbf{u} \cdot \mathbf{v} = 0$
________ if and only if $\mathbf{u} \cdot \mathbf{v} < 0$.

- *The Triangle Inequality*: Given any vectors $\mathbf{u}$ and $\mathbf{v}$, $\|\mathbf{u}+\mathbf{v}\|$ ____ $\|\mathbf{u}\| + \|\mathbf{v}\|$.
- The cross product of two parallel vectors $\mathbf{u}$ and $\mathbf{v}$ in $\mathbb{R}^3$ is ________.
- For any vectors $\mathbf{u}$ and $\mathbf{v}$ in $\mathbb{R}^3$, $\mathbf{v} \times \mathbf{u} =$ ____ $(\mathbf{u} \times \mathbf{v})$.
- For any vectors $\mathbf{u}$ and $\mathbf{v}$ in $\mathbb{R}^3$ and any scalar c, $c(\mathbf{u} \times \mathbf{v}) =$ ________ $=$ ________.
- For vectors $\mathbf{u}$ and $\mathbf{v}$ in $\mathbb{R}^3$, $\mathbf{u} \cdot (\mathbf{u} \times \mathbf{v}) =$ ________ and $\mathbf{v} \cdot (\mathbf{u} \times \mathbf{v}) =$ ________.
- Let $\mathbf{u}$ and $\mathbf{v}$ be nonzero vectors in $\mathbb{R}^3$ with the same initial point. Then $\|\mathbf{u} \times \mathbf{v}\| =$ ________ $\sin\theta$, where θ is ________.
- If $\mathbf{u}$ and $\mathbf{v}$ are nonparallel vectors in $\mathbb{R}^3$, then $\mathbf{u}$, $\mathbf{v}$, and $\mathbf{u} \times \mathbf{v}$ form a ________ triple.
- If $\mathbf{u}$, $\mathbf{v}$, and $\mathbf{w}$ are vectors in $\mathbb{R}^3$, then ________ is the volume of the parallelepiped determined by $\mathbf{u}$, $\mathbf{v}$, and $\mathbf{w}$.
- The distance from a point P to a line $\mathcal{L}$ parametrized by $\mathbf{r}(t) = \mathbf{P}_0 + t\mathbf{d}$ is ________.
- The distance from a point P to a plane $\mathcal{P}$ containing a point R with normal vector $\mathbf{N}$ is ________.

Notation and Algebraic Rules for Vectors

Notation: Describe the meanings of each of the following mathematical expressions:

- $\mathbb{R}^2$
- $\mathbb{R}^3$
- $\langle a, b\rangle$
- $\langle a, b, c\rangle$
- $\overrightarrow{PQ}$
- $\|\mathbf{v}\|$
- $\mathbf{i}$
- $\mathbf{j}$
- $\mathbf{k}$
- $\mathbf{u} \cdot \mathbf{v}$
- $\mathbf{u} \times \mathbf{v}$
- $\text{comp}_{\mathbf{u}}\mathbf{v}$
- $\text{proj}_{\mathbf{u}}\mathbf{v}$

Algebraic Properties of Vector Arithmetic: Each of the statements that follow demonstrates a commutative rule, an associative rule, or a distributive rule of vector arithmetic. Fill in the blanks and give the name of the relevant property.

- For any two vectors $\mathbf{u}$ and $\mathbf{v}$ with the same number of components, $\mathbf{u} + \mathbf{v} =$ ________.
- For any three vectors $\mathbf{u}, \mathbf{v}$ and $\mathbf{w}$, each with the same number of components, $(\mathbf{u} + \mathbf{v}) + \mathbf{w} =$ ________.
- For any scalar c and any two vectors $\mathbf{u}$ and $\mathbf{v}$ with the same number of components, $c(\mathbf{u} + \mathbf{v}) =$ ________.
- For any vectors $\mathbf{u}$, $\mathbf{v}$, and $\mathbf{w}$, $\mathbf{u} \cdot (\mathbf{v} + \mathbf{w}) =$ ________.
- For any vectors $\mathbf{u}$ and $\mathbf{v}$, and any scalar k, $k(\mathbf{u} \cdot \mathbf{v}) =$ ________ $=$ ________.
- Let $\mathbf{u}$, $\mathbf{v}$, and $\mathbf{w}$ be vectors in $\mathbb{R}^3$. Then $\mathbf{u} \times (\mathbf{v} + \mathbf{w}) =$ ________.
- $\mathbf{u}$, $\mathbf{v}$, and $\mathbf{w}$ be vectors in $\mathbb{R}^3$. Then $(\mathbf{u}+\mathbf{v}) \times \mathbf{w} =$ ________.

Skill Certification: Working in $\mathbb{R}^3$

Distances between points: Find the distance between each pair of points.

1. $(1, 2, -3)$ and $(4, 7, -3)$
2. $(-5, 7, 0)$ and $(0, 6, -3)$
3. $(1, 5, -2)$ and $(3, 9, -1)$
4. $(4, 6, 2)$ and $(1, 3, -5)$

Equations of Spheres: Find the equation of the specified sphere.

5. center $(2, -3, 4)$, radius 6
6. center $(2, -3, 4)$, tangent to the xz-plane
7. the segment with endpoints $(1, 5, -2)$ and $(3, 9, -1)$ is a diameter
8. center $(4, 6, 2)$, $(1, 3, -5)$ is a point on the sphere

Products and Norms: In Exercises 9–24 let $\mathbf{u} = \mathbf{i}$, $\mathbf{v} = 2\mathbf{j}$, and $\mathbf{w} = \mathbf{i} + 2\mathbf{j} + \frac{1}{2}\mathbf{k}$.

9. Sketch the position vectors $\mathbf{u}$, $\mathbf{v}$, and $\mathbf{w}$.
10. Sketch the parallelogram determined by $\mathbf{u}$ and $\mathbf{v}$.
11. Sketch the parallelepiped determined by $\mathbf{u}$, $\mathbf{v}$, and $\mathbf{w}$.
12. Compute $\|\mathbf{u}\|$, and use the result to label your sketches in Exercises 9–11.
13. Compute $\|\mathbf{v}\|$, and use the result to label your sketches in Exercises 9–11.
14. Compute $\|\mathbf{w}\|$, and use the result to label your sketches in Exercises 9–11.
15. Compute $\mathbf{u} \cdot \mathbf{v}$.
16. Compute $\mathbf{u} \times \mathbf{v}$.
17. Compute the angle between $\mathbf{u}$ and $\mathbf{v}$.
18. Compute the area of the parallelogram determined by $\mathbf{u}$ and $\mathbf{v}$.
19. Compute the lengths of the diagonals of the parallelogram determined by $\mathbf{u}$ and $\mathbf{v}$.
20. Find a vector orthogonal to both $\mathbf{u}$ and $\mathbf{v}$.
21. Find a unit vector orthogonal to both $\mathbf{u}$ and $\mathbf{v}$.
22. Compute the volume of the parallelepiped determined by $\mathbf{u}$, $\mathbf{v}$, and $\mathbf{w}$.
23. Compute the areas of the six faces of the parallelepiped determined by $\mathbf{u}$, $\mathbf{v}$, and $\mathbf{w}$.
24. Compute the lengths of the four diagonals of the parallelepiped determined by $\mathbf{u}$, $\mathbf{v}$, and $\mathbf{w}$. (*Hint:* $\mathbf{u} + \mathbf{v} + \mathbf{w}$ *is one of the diagonals.*)

Products and Norms: Let $\mathbf{u} = \langle 2, 4, -1 \rangle$, $\mathbf{v} = \langle 0, -3, 2 \rangle$, and $\mathbf{w} = \langle -1, 1, 5 \rangle$. Use these vectors to find the specified quantities in Exercises 25–36.

25. $\|\mathbf{u}\|$

26. $\|\mathbf{v}\|$

27. $\mathbf{u} \cdot \mathbf{v}$

28. $\mathbf{u} \times \mathbf{v}$

29. the angle between $\mathbf{u}$ and $\mathbf{v}$

30. the area of the parallelogram determined by $\mathbf{u}$ and $\mathbf{v}$

31. the lengths of the diagonals of the parallelogram determined by $\mathbf{u}$ and $\mathbf{v}$

32. a vector orthogonal to both $\mathbf{u}$ and $\mathbf{v}$

33. a unit vector orthogonal to both $\mathbf{u}$ and $\mathbf{v}$

34. the volume of the parallelepiped determined by $\mathbf{u}$, $\mathbf{v}$, and $\mathbf{w}$

35. the areas of the six faces of the parallelepiped determined by $\mathbf{u}$, $\mathbf{v}$, and $\mathbf{w}$

36. the lengths of the four diagonals of the parallelepiped determined by $\mathbf{u}$, $\mathbf{v}$, and $\mathbf{w}$

Lines and Planes Determined by Points: Consider the three points $P(2, -3, 5)$, $Q(3, 1, 0)$, and $R(-1, 0, 7)$. Find the following.

37. an equation for the line containing P and Q

38. an equation for the line containing P and R

39. an equation for the plane containing P, Q, and R

40. the area of the triangle determined by P, Q, and R

Intersections and Distances: Let P be the point with coordinates $(1, 2, -3)$, $\mathcal{L}$ be the line with equation $\mathbf{r}(t) = \langle 2 + 5t, -3 + t, -4t \rangle$, and $\mathcal{P}$ be the plane with equation $x - 3y + 4z = 5$. Find the following.

41. the distance from P to $\mathcal{L}$

42. the distance from P to $\mathcal{P}$

43. the point at which $\mathcal{L}$ intersects $\mathcal{P}$

44. the angle at which $\mathcal{L}$ intersects $\mathcal{P}$

Capstone Problems

A. A weight of p pounds is suspended by two ropes as shown in the figure that follows. What are the magnitudes of the forces in each of the ropes?

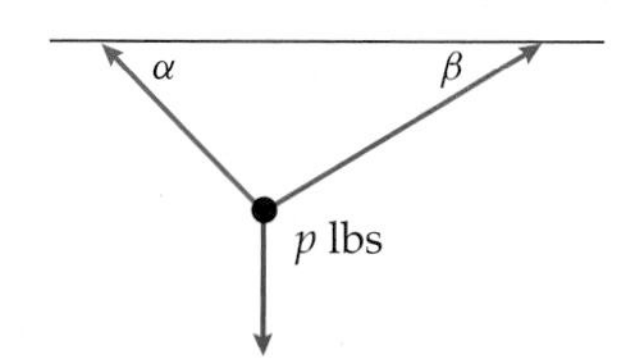

B. Explain why the solid with vertices $(1, 0, 0)$, $(-1, 0, 0)$, $(0, 1, 0)$, $(0, -1, 0)$, $(0, 0, 1)$, and $(0, 0, -1)$ is a regular octahedron.

(a) Find the area of each face.

(b) Find the equations of the three lines that form the edges of the face of the octahedron with vertices $(1, 0, 0)$, $(0, 1, 0)$, and $(0, 0, 1)$.

(c) Find the equation of each face of the octahedron.

(d) Find the volume of the octahedron.

C. Find the equation of the sphere with center $(1, -3, 5)$ and tangent to the plane with equation $3x - 5y - 2z = 6$. Find the equation of the sphere with center (α, β, γ) and tangent to the plane with equation $ax + by + cz = d$.

D. One molecule of methane is composed of one carbon atom (chemical symbol C) and four hydrogen atoms (chemical symbol H). Therefore the chemical formula for methane is CH_4. Geometrically, if we picture the carbon atom at the center of the molecule, the four hydrogen atoms would lie at the four vertices of a regular tetrahedron. (Recall that a regular tetrahedron is a solid with four congruent faces, each of which is an equilateral triangle.) Each of the hydrogen atoms is bonded to the carbon atom, but the hydrogens are not bonded to each other. Following is a schematic of methane inside a regular tetrahedron:

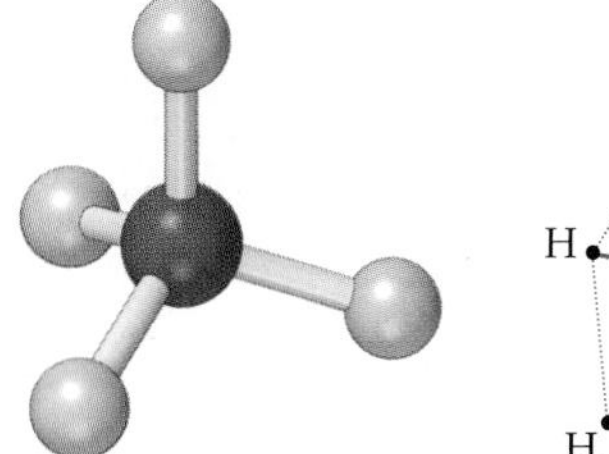

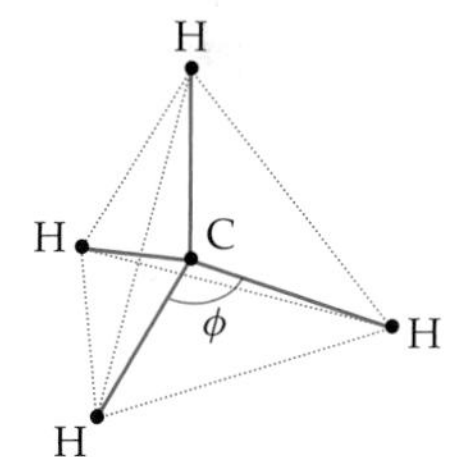

To aid you in visualizing the molecule, it might help to place the carbon atom at the origin of a three-dimensional coordinate system with two of the hydrogen atoms in the xz-plane at $(a, 0, c)$ and $(-a, 0, c)$. Then the other two hydrogen atoms must lie in the yz-plane at $(0, a, -c)$ and $(0, -a, -c)$, as shown here:

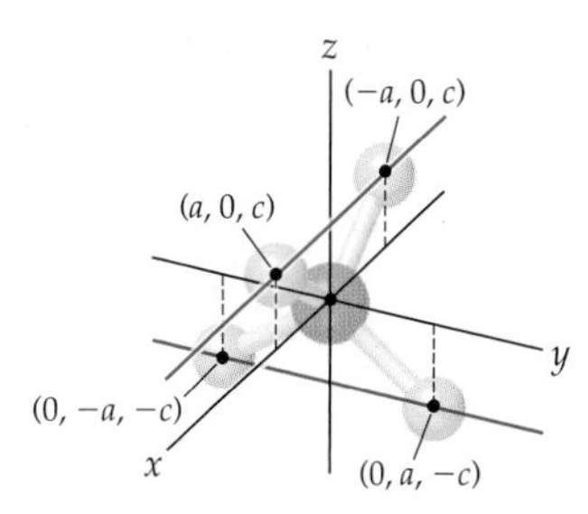

Since the carbon is at the origin, the coordinates of these four points also give the components of the vectors from the carbon atom to the hydrogen atoms.

(a) In methane the H–C–H bond angles are all equal. Explain why this means that

$$\langle a, 0, c \rangle \cdot \langle -a, 0, c \rangle = \langle a, 0, c \rangle \cdot \langle 0, a, -c \rangle.$$

(b) What are the H–C–H bond angles in methane?

CHAPTER 11

Vector Functions

11.1 Vector-Valued Functions

Parametric Equations in $\mathbb{R}^3$
Vector-Valued Functions
Parametrized Curves
Limits and Continuity of Vector Functions
Examples and Explorations

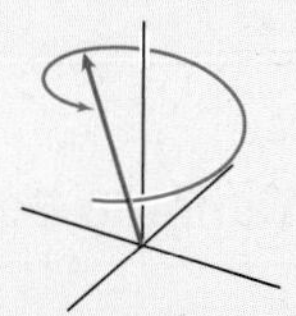

11.2 The Calculus of Vector Functions

The Derivative of a Vector Function
The Geometry of the Derivative
Velocity and Acceleration
Derivatives of Vector Products
Integration of a Vector Function
Examples and Explorations

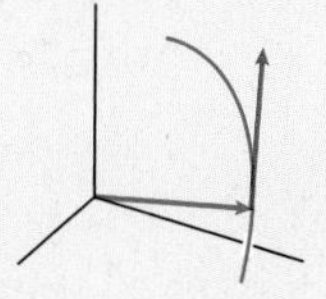

11.3 Unit Tangent and Unit Normal Vectors

Unit Tangent and Unit Normal Vectors
The Osculating Plane and the Binormal Vector
Examples and Explorations

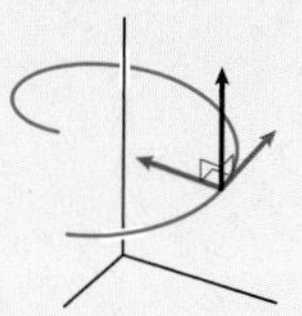

11.4 Arc Length Parametrizations and Curvature

The Arc Length of Space Curves
Arc Length Parametrizations
Examples and Explorations

11.5 Motion

Motion Under the Force of Gravity
Displacement and Distance Travelled
Tangential and Normal Components of Acceleration
Examples and Explorations

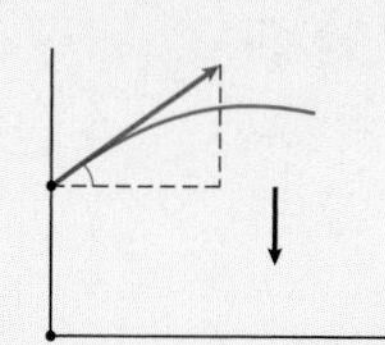

Chapter Review, Self-Test, and Capstones

11.1 VECTOR-VALUED FUNCTIONS

- Defining curves in $\mathbb{R}^3$ with parametric equations in three variables
- Vector functions in $\mathbb{R}^2$ and $\mathbb{R}^3$
- Curves defined with parametric equations in $\mathbb{R}^2$ and $\mathbb{R}^3$

Parametric Equations in $\mathbb{R}^3$

For most of this book we have studied functions $f : X \to Y$ whose domains and codomains are subsets of the real numbers. In this chapter we broaden our study to include functions whose codomains are subsets of $\mathbb{R}^2$ or $\mathbb{R}^3$. Such functions are called vector-valued functions or vector functions. The graph of such a function is a curve in the plane, with codomain $\mathbb{R}^2$, or a curve in 3-space, with codomain $\mathbb{R}^3$. We will still use the basic concepts of calculus to analyze these functions, their graphs, and their applications.

Recall that in Chapter 9 we defined parametric equations in $\mathbb{R}^2$ to be a pair of functions

$$x = x(t) \quad \text{and} \quad y = y(t),$$

where t is a *parameter* defined on some interval I of real numbers. At the time we also remarked that the definition could be extended to three-dimensional space, which we do here:

DEFINITION 11.1 Parametric Equations in $\mathbb{R}^3$

Parametric equations in $\mathbb{R}^3$ are triples of functions

$$x = x(t), y = y(t), \text{ and } z = z(t),$$

where the ***parameter*** t is defined on some interval I of real numbers.

The ***parametric curve***, or ***space curve***, associated with the equations is the set of points

$$\{(x(t), y(t), z(t)) \in \mathbb{R}^3 \mid t \in I\}.$$

For example, the equations

$$x = \cos t, \; y = \sin t, \; z = t, \; t \in [0, 4\pi]$$

are parametric equations in three variables. (See Example 1.)

Vector-Valued Functions

Another way to express a curve defined by parametric equations, using slightly different notation, is with a vector-valued function. Our definition follows:

DEFINITION 11.2

Vector Functions with Three Components

Let

$$x = x(t), y = y(t), \text{ and } z = z(t)$$

be three real-valued functions, each of which is defined on some interval $I \subseteq \mathbb{R}$. A ***vector function***, or ***vector-valued function***, $\mathbf{r}(t)$, with three components is a function of the form

$$\mathbf{r}(t) = \langle x(t), y(t), z(t) \rangle = x(t)\mathbf{i} + y(t)\mathbf{j} + z(t)\mathbf{k}.$$

The variable t in the vector function is called the ***parameter***. The functions $x = x(t)$, $y = y(t)$, and $z = z(t)$ are called the ***components*** of $\mathbf{r}(t)$. The ***vector curve***, or ***space curve***, associated with the vector function is the set of points

$$\{(x(t), y(t), z(t)) \mid t \in I\}$$

in $\mathbb{R}^3$.

Note that for every value of $t \in I$, the value of the function is a vector. When we consider a vector-valued function that represents the position of a particle, we always interpret the vector as a position vector and the vector curve as the set of terminal points of all the position vectors of the particle. That is, the collection of points

$$(x(t), y(t), z(t)) \text{ for } t \in I,$$

is the vector curve for $\mathbf{r}(t) = \langle x(t), y(t), z(t) \rangle$ since these are the terminal points of the vectors

$$\langle (x(t), y(t), z(t) \rangle \text{ for } t \in I.$$

A vector curve is illustrated in the following figure:

The collection of terminal points of the position vectors forms the graph of a vector function

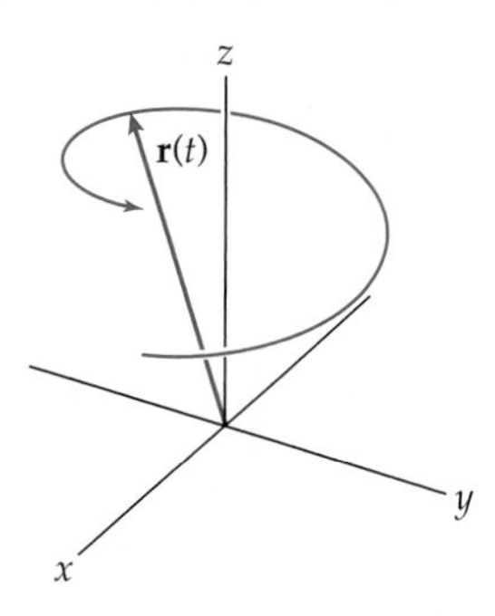

We could similarly define a vector function with two components to be a function of the form $\mathbf{r}(t) = \langle x(t), y(t) \rangle = x(t)\mathbf{i} + y(t)\mathbf{j}$, where $x(t)$ and $y(t)$ are functions of t on some interval $I \subseteq \mathbb{R}$. This is an alternative way of expressing a function defined by two parametric equations.

Note that we will sometimes consider a vector function and a real-valued function of a single variable, $f(t)$, simultaneously. When we do, we will refer to $f(t)$ as a ***scalar function***.

Parametrized Curves

The space curve associated with the vector function $\mathbf{r}(t) = \langle x(t), y(t), z(t) \rangle$ is the set of points $\{(x(t), y(t), z(t)) \mid t \in I \subseteq \mathbb{R}\}$. Along with the curve itself, the parameter, t, imposes directionality on the curve, pointing in the direction in which t is increasing. This directionality is indicated with an arrow.

To understand a space curve, it can be helpful to analyze its projection onto one of the coordinate planes. For example, rather than immediately drawing $\mathbf{r}(t) = \langle x(t), y(t), z(t)\rangle$ in 3-space, we may consider the image of the parametric curve defined by $x = x(t)$ and $y = y(t)$ in the xy-plane. We could similarly consider the projections onto the yz- and xz-planes. If we judiciously select which two components to consider, we will obtain the best information.

For example, consider the function $\mathbf{r}(t) = \langle \cos t, \sin t, t\rangle$ for $t \in [0, 4\pi]$. Temporarily ignoring the z-component of this function, we have the parametric equations $x = \cos t$ and $y = \sin t$. From our work in Chapter 9, we recognize that these equations describe the unit circle centered at the origin. In addition, the curve is traced twice, counterclockwise, starting at the point $(1, 0)$ as t increases on the interval $[0, 4\pi]$, as in the following figure:

The projection of the vector function onto the xy-plane

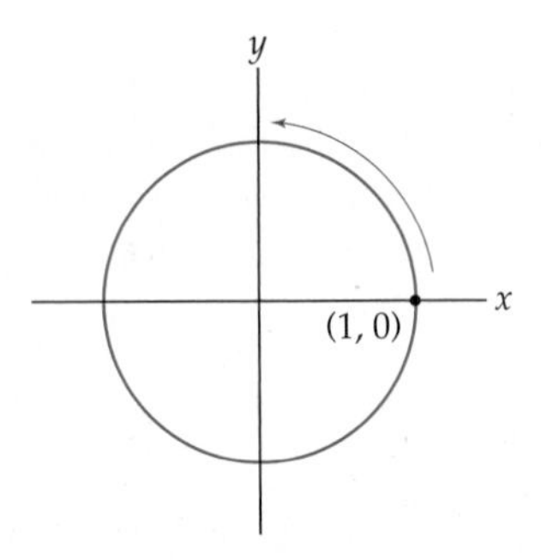

To incorporate the z-coordinate, we note that z increases as t increases. The complete vector curve will spiral around the z-axis as t increases, and we obtain the following circular helix:

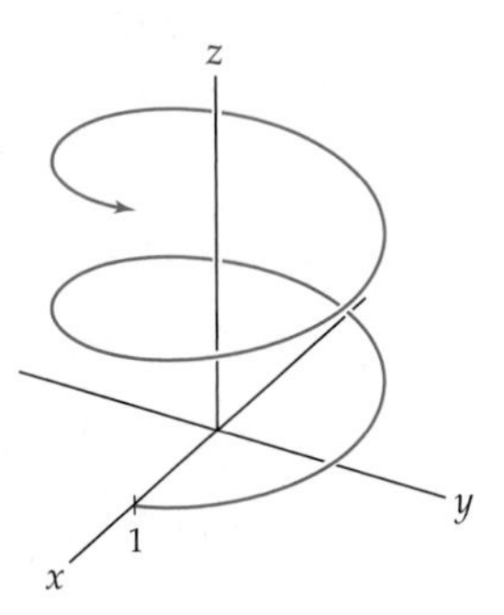

To visualize a vector curve, it may help to tabulate the points in 3-space that correspond to a few values of the parameter t. We do this in the examples that follow.

Scalars, scalar functions, and vector-valued functions may be combined algebraically in various ways. For example, the product of the scalar k and the vector function $\mathbf{r}(t) = \langle x(t), y(t), z(t)\rangle$ is

$$k\,\mathbf{r}(t) = k\langle x(t), y(t), z(t)\rangle = \langle k\,x(t), k\,y(t), k\,z(t)\rangle.$$

The product of a scalar function $f(t)$ and the vector function $\mathbf{r}(t) = \langle x(t), y(t), z(t)\rangle$ is

$$f(t)\,\mathbf{r}(t) = f(t)\,\langle x(t), y(t), z(t)\rangle = \langle f(t)\,x(t), f(t)\,y(t), f(t)\,z(t)\rangle.$$

The dot product and cross product of the vector functions $\mathbf{r}_1(t) = \langle x_1(t), y_1(t), z_1(t)\rangle$ and $\mathbf{r}_2(t) = \langle x_2(t), y_2(t), z_2(t)\rangle$ are, respectively,

$$\begin{aligned}\mathbf{r}_1(t)\cdot\mathbf{r}_2(t) &= \langle x_1(t), y_1(t), z_1(t)\rangle\cdot\langle x_2(t), y_2(t), z_2(t)\rangle\\ &= x_1(t)\,x_2(t) + y_1(t)\,y_2(t) + z_1(t)\,z_2(t) \quad \text{and}\end{aligned}$$

$$\begin{aligned}\mathbf{r}_1(t)\times\mathbf{r}_2(t) &= \langle x_1(t), y_1(t), z_1(t)\rangle\times\langle x_2(t), y_2(t), z_2(t)\rangle\\ &= \big\langle y_1(t)\,z_2(t) - y_2(t)\,z_1(t), x_2(t)\,z_1(t) - x_1(t)\,z_2(t), x_1(t)\,y_2(t) - x_2(t)\,y_1(t)\big\rangle.\end{aligned}$$

Note that the dot product is a scalar function and the cross product is a vector function. The structures of the various products formed with two component vector functions are similar.

Limits and Continuity of Vector Functions

The limit of a vector-valued function may be defined in terms of the limits of the components.

DEFINITION 11.3 The Limit of a Vector Function

Let $\mathbf{r}(t) = \langle x(t), y(t), z(t)\rangle$ be a vector-valued function defined on a punctured interval around t_0. The ***limit*** of $\mathbf{r}(t)$ as t approaches t_0, denoted by $\lim_{t\to t_0}\mathbf{r}(t)$, is defined by

$$\lim_{t\to t_0}\mathbf{r}(t) = \lim_{t\to t_0}\langle x(t), y(t), z(t)\rangle = \lim_{t\to t_0} x(t)\,\mathbf{i} + \lim_{t\to t_0} y(t)\,\mathbf{j} + \lim_{t\to t_0} z(t)\,\mathbf{k},$$

provided that each of the limits, $\lim_{t\to t_0} x(t)$, $\lim_{t\to t_0} y(t)$, and $\lim_{t\to t_0} z(t)$, exists.

Similarly, the ***limit*** of the vector function $\mathbf{r}(t) = \langle x(t), y(t)\rangle$ is

$$\lim_{t\to t_0}\mathbf{r}(t) = \lim_{t\to t_0}\langle x(t), y(t)\rangle = \lim_{t\to t_0} x(t)\mathbf{i} + \lim_{t\to t_0} y(t)\mathbf{j},$$

provided that each of the limits, $\lim_{t\to t_0} x(t)$ and $\lim_{t\to t_0} y(t)$, exists.

As in Definition 11.3, our definition of the continuity of a vector-valued function relies on the continuity of the component functions.

DEFINITION 11.4 The Continuity of a Vector Function

Let $\mathbf{r}(t)$ be a vector-valued function defined on an open interval $I \subseteq \mathbb{R}$, and let $t_0 \in I$. The function $\mathbf{r}(t)$ is said to be ***continuous*** at t_0 if

$$\lim_{t\to t_0}\mathbf{r}(t) = \mathbf{r}(t_0).$$

For a vector function $\mathbf{r}(t) = \langle x(t), y(t), z(t)\rangle$ this definition means that $\mathbf{r}$ is continuous at t_0 if and only if $\lim_{t\to t_0}\mathbf{r}(t) = \langle x(t_0), y(t_0), z(t_0)\rangle$. The analogous statement holds for a vector function with two components. We also define continuity on an interval, using the ideas of Chapter 1.

DEFINITION 11.5 The Continuity of a Vector Function on an Interval

A vector function $\mathbf{r}(t)$ is ***continuous on an interval*** I if it is continuous at every interior point of I, right continuous at any closed left endpoint, and left continuous at any closed right endpoint.

Examples and Explorations

EXAMPLE 1 Graphing a circular helix by plotting points

Graph the circular helix defined by $\mathbf{r}(t) = \langle \cos t, \sin t, t \rangle$ for $t \in [0, 4\pi]$ by plotting points on the curve.

SOLUTION

We've tabulated the coordinates of several points on the curve for values of t in the interval $[0, 4\pi]$:

t	0	$\pi/2$	π	$3\pi/2$	2π
(x, y, z)	$(1, 0, 0)$	$(0, 1, \pi/2)$	$(-1, 0, \pi)$	$(0, -1, 3\pi/2)$	$(1, 0, 2\pi)$

t		$5\pi/2$	3π	$7\pi/2$	4π
(x, y, z)		$(0, 1, 5\pi/2)$	$(-1, 0, 3\pi)$	$(0, -1, 7\pi/2)$	$(1, 0, 4\pi)$

These points and the helix are shown in the following figure:

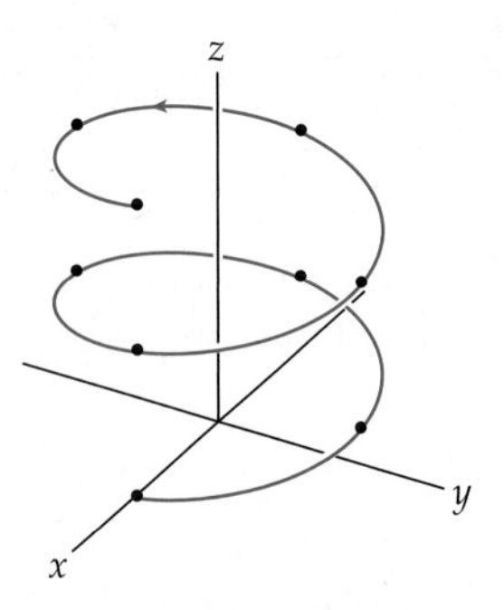

□

EXAMPLE 2 Graphing a spiral

Graph the vector curve defined by $\mathbf{r}(t) = \langle t \sin t, t \cos t \rangle$ for $t \in [0, 4\pi]$.

SOLUTION

We have already seen that the graph of the parametric equations $x = \sin t$, $y = \cos t$ for $t \in [0, 4\pi]$ is a circle with radius 1 and centered at the origin. The graph of this parametrization starts at $(0, 1)$ when $t = 0$, and the circle is traced twice, in a clockwise direction for

$t \in [0, 4\pi]$. The graph of $\mathbf{r}(t)$ is related. If we square $x = t \sin t$ and $y = t \cos t$ and add, we obtain

$$x^2 + y^2 = t^2 \sin^2 t + t^2 \cos^2 t = t^2.$$

This is not the equation of a circle, but we know that $\sqrt{x^2 + y^2} = t$ is the distance from the origin. That is, the distance from the origin increases with t as the particle revolves around the origin. When $t = 0$, we have $(x, y) = (0, 0)$, and the graph will spiral out clockwise as t increases. To obtain a more precise graph, we will plot a few values of (x, y) corresponding to values of the parameter in the table.

We also evaluate a few reference points for the space curve in the following table:

t	0	$\pi/2$	π	$3\pi/2$	2π
(x, y)	$(0, 0)$	$(\pi/2, 0)$	$(0, -\pi)$	$(-3\pi/2, 0)$	$(0, 2\pi)$
t		$5\pi/2$	3π	$7\pi/2$	4π
(x, y)		$(5\pi/2, 0)$	$(0, -3\pi)$	$(-7\pi/2, 0)$	$(0, 4\pi)$

We plot these points along with the spiral in the following figure:

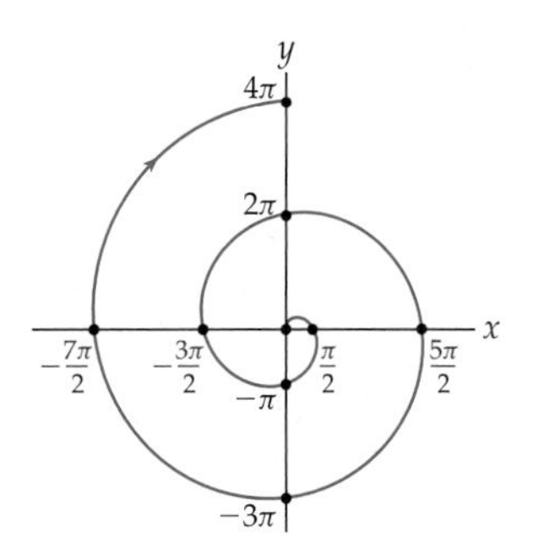

□

EXAMPLE 3 Graphing a conical helix

Graph the vector curve defined by $\mathbf{r}(t) = \langle t \sin t, t, t \cos t \rangle$ for $t \in [0, 4\pi]$.

SOLUTION

We begin by considering a projection of this curve onto the xz-plane. The graph in the xz-plane will be defined by the parametric equations $x = t \sin t,\ z = t \cos t$. The graph of these equations is the spiral we graphed in Example 2, except in the xz-plane rather than in the xy-plane.

Note that when we include the y-coordinate, we obtain a helix that winds around the y-axis. We evaluate a few reference points for the space curve in the following table:

t	0	π	2π	3π	4π
(x, y, z)	$(0, 0, 0)$	$(0, \pi, -\pi)$	$(0, 2\pi, 2\pi)$	$(0, 3\pi, -3\pi)$	$(0, 4\pi, 4\pi)$

These points and the curve are shown next. This curve is an example of a *conical helix* because the graph can be drawn on the surface of a cone.

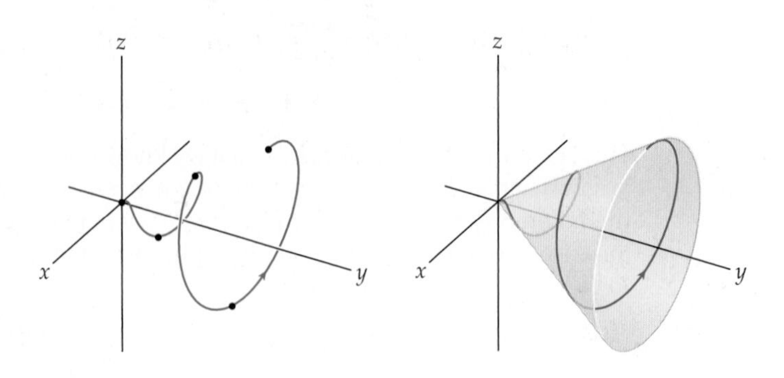

□

CHECKING THE ANSWER Elegant graphs such as the ones we have shown may be produced with a computer algebra system like Maple, Mathematica, or Matlab. However, many simple curves may be drawn by hand using the techniques introduced in this section.

EXAMPLE 4 **Evaluating the limit of a vector function**

Evaluate the limits of the following vector functions if they exist:

$$\lim_{t\to 0}\left\langle t\sin\frac{1}{t}, \frac{t}{e^t-1}\right\rangle \quad \text{and} \quad \lim_{t\to 2}\left\langle \frac{t-2}{t^2-4}, \frac{2^t-4}{t-2}, \frac{t}{t-2}\right\rangle$$

SOLUTION

The limit of the first vector function exists if and only if the two limits

$$\lim_{t\to 0} t\sin\frac{1}{t} \quad \text{and} \quad \lim_{t\to 0}\frac{t}{e^t-1}$$

exist. We may use the Squeeze Theorem to show that $\lim_{t\to 0} t\sin\frac{1}{t} = 0$ and L'Hôpital's Rule to show that $\lim_{t\to 0}\frac{t}{e^t-1} = 1$. Therefore, $\lim_{t\to 0}\left\langle t\sin\frac{1}{t}, \frac{t}{e^t-1}\right\rangle = \langle 0, 1\rangle$.

For the other vector function, the limits of the first two components exist, since

$$\lim_{t\to 2}\frac{t-2}{t^2-4} = \lim_{t\to 2}\frac{1}{t+2} = \frac{1}{4} \quad \text{and} \quad \lim_{t\to 2}\frac{2^t-4}{t-2} = \lim_{t\to 2}(\ln 2)2^t = 4\ln 2.$$

However, the limit of the third component, $\lim_{t\to 2}\frac{t}{t-2}$, does not exist. Therefore,

$$\lim_{t\to 2}\left\langle \frac{t-2}{t^2-4}, \frac{2^t-4}{t-2}, \frac{t}{t-2}\right\rangle$$

does not exist. □

TEST YOUR UNDERSTANDING

- What are parametric equations in two variables? How does the definition generalize to parametric equations in three variables?
- What techniques can you use to graph parametric equations in three variables?
- What is a vector-valued function? How is the definition related to the definition of parametric equations?
- What techniques can you use to graph a vector function in three variables?
- If one or more of the components of a vector function is discontinuous, how does that affect the graph of the function?

EXERCISES 11.1

Thinking Back

Parametric equations for the unit circle: Find parametric equations for the unit circle centered at the origin of the xy-plane that satisfy the given conditions.

- The graph is traced counterclockwise once on the interval $[0, 2\pi]$ starting at the point $(1, 0)$.
- The graph is traced clockwise once on the interval $[0, 2\pi]$ starting at the point $(1, 0)$.
- The graph is traced counterclockwise twice on the interval $[0, 2\pi]$ starting at the point $(1, 0)$.
- *Parametric equations for a circle:* Find parametric equations whose graph is the circle with radius ρ centered at the point (a, b) in the xy-plane such that the graph is traced counterclockwise $k > 0$ times on the interval $[0, 2\pi]$ starting at the point $(a + \rho, b)$.

Concepts

0. *Problem Zero:* Read the section and make your own summary of the material.

1. *True/False:* Determine whether each of the statements that follow is true or false. If a statement is true, explain why. If a statement is false, provide a counterexample.

(a) *True or False:* Every curve in the plane has a unique expression in terms of a vector-valued function with two components.

(b) *True or False:* Every space curve has a unique expression in terms of a vector-valued function with three components.

(c) *True or False:* The graph of two parametric equations in two variables is also the graph of a vector-valued function with two components.

(d) *True or False:* Every vector-valued function in three variables can be expressed in terms of three parametric equations.

(e) *True or False:* The graph of three parametric equations is also the graph of a vector-valued function with three components.

(f) *True or False:* If $\mathbf{r}(t)$ is a vector-valued function with domain $\mathbb{R}$, $\lim_{t\to c} \mathbf{r}(t)$ exists for every $c \in \mathbb{R}$.

(g) *True or False:* If a point t_0 is in the domain of a vector function $\mathbf{r}(t)$, then $\lim_{t\to t_0} \mathbf{r}(t) = \mathbf{r}(t_0)$.

(h) *True or False:* Every vector function is continuous at every point in its domain.

2. *Examples:* Construct examples of the thing(s) described in the following. Try to find examples that are different than any in the reading.

(a) Parametric equations with three components for a circular helix winding around the x-axis

(b) A vector-valued function with two components whose graph is a spiral starting at the origin

(c) A vector-valued function with three components whose graph is a circle contained in the plane 5 units from the yz-plane.

3. Let $y = f(x)$. What is the definition of $\lim_{x\to c} f(x) = L$?

4. Let $\mathbf{r}(t) = \langle x(t), y(t), z(t)\rangle$. What is the definition of $\lim_{t\to c} \mathbf{r}(t)$?

5. Explain why we do *not* need an "epsilon–delta" definition for the limit of a vector-valued function.

6. Let $y = f(x)$. State the definition for the continuity of the function f at a point c in the domain of f.

7. Let $\mathbf{r}(t) = \langle x(t), y(t), z(t)\rangle$. Provide a definition for the continuity of the vector function $\mathbf{r}$ at a point c in the domain of $\mathbf{r}$.

8. Most of the parametric equations and vector-valued functions we have studied have component functions that are continuous. What happens when one of the component functions is discontinuous at a point? For example, the "floor" function $z(t) = \lfloor t \rfloor$ has a jump discontinuity for every integer t. What is the graph of the equations

$$x = \cos 2\pi t,\ y = \sin 2\pi t, z = \lfloor t \rfloor,\ t \in \mathbb{R}?$$

9. Let $\mathbf{r}(t) = \langle x(t), y(t)\rangle,\ t \in [a, b]$, be a vector-valued function, where $a < b$ are real numbers and the functions $x(t)$ and $y(t)$ are continuous. Explain why the graph of $\mathbf{r}$ is contained in some circle centered at the origin. (*Hint: Think about the Extreme Value Theorem.*)

10. Let $\mathbf{r}(t) = \langle x(t), y(t)\rangle,\ t \in [a, \infty)$, be a vector-valued function, where a is a real number. Explain why the graph of $\mathbf{r}$ may or may not be contained in a circle centered at the origin. (*Hint: Graph the functions* $\mathbf{r}_1(t) = \left\langle \frac{1}{t}, \frac{1}{t}\right\rangle$ *and* $\mathbf{r}_2(t) = \langle t, t\rangle$, *both with domain* $[1, \infty)$.)

11. Let $\mathbf{r}(t) = \langle x(t), y(t), z(t)\rangle,\ t \in [a, b]$, be a vector-valued function, where $a < b$ are real numbers and the functions $x(t)$, $y(t)$, and $z(t)$ are continuous. Explain why the graph of $\mathbf{r}$ is contained in some sphere centered at the origin.

12. Let $\mathbf{r}(t) = \langle x(t), y(t), z(t)\rangle,\ t \in [a, \infty)$, be a vector-valued function, where a is a real number. Explain why the graph of $\mathbf{r}$ may or may not be contained in some sphere centered at the origin. (*Hint: Consider the functions* $\mathbf{r}_1(t) = \langle \cos t, \sin t, 1/t\rangle$ *and* $\mathbf{r}_2(t) = \langle \cos t, \sin t, t\rangle$, *both with domain* $[1, \infty)$.)

13. As we saw in Example 1, the graph of the vector-valued function $\mathbf{r}(t) = \langle \cos t, \sin t, t\rangle$, for $t \in [0, 2\pi]$ is a circular helix that spirals counterclockwise around the z-axis and climbs as t increases. Find another parametrization for this helix so that the motion along the helix is faster for a given change in the parameter.

14. As we saw in Example 1, the graph of the vector-valued function $\mathbf{r}(t) = \langle \cos t, \sin t, t\rangle$, for $t \in [0, 2\pi]$ is a circular helix that spirals counterclockwise around the z-axis and climbs as t increases. Find another parametrization for this helix so that the motion is downwards.

15. Let $\mathbf{r}(t) = \langle x(t), y(t) \rangle$, $t \in [a, \infty)$, be a vector-valued function, where a is a real number. Under what conditions would the graph of $\mathbf{r}$ have a horizontal asymptote as $t \to \infty$? Provide an example illustrating your answer.

16. Let $\mathbf{r}(t) = \langle x(t), y(t) \rangle$, $t \in [a, \infty)$, be a vector-valued function, where a is a real number. Under what conditions would the graph of $\mathbf{r}$ have a vertical asymptote as $t \to \infty$? Provide an example illustrating your answer.

17. What is the dot product of the vector functions $\mathbf{r}_1(t) = \langle x_1(t), y_1(t) \rangle$ and $\mathbf{r}_2(t) = \langle x_2(t), y_2(t) \rangle$?

18. Compute the cross product of the vector functions $\mathbf{r}_1(t) = \langle x_1(t), y_1(t) \rangle$ and $\mathbf{r}_2(t) = \langle x_2(t), y_2(t) \rangle$ by thinking of $\mathbb{R}^2$ as the xy-plane in $\mathbb{R}^3$. That is, let $\mathbf{r}_1(t) = \langle x_1(t), y_1(t), 0 \rangle$ and $\mathbf{r}_2(t) = \langle x_2(t), y_2(t), 0 \rangle$, and take the cross product of these vector functions.

In Exercises 19–21 sketch the graph of a vector-valued function $\mathbf{r}(t) = \langle x(t), y(t) \rangle$ with the specified properties. Be sure to indicate the direction of increasing values of t.

19. Domain $t \geq 0$, $\mathbf{r}(0) = \langle 0, 3 \rangle$, $\mathbf{r}(1) = \langle -2, 1 \rangle$, $\lim_{t \to \infty} x(t) = -5$, and $\lim_{t \to \infty} y(t) = -\infty$.

20. Domain $t \geq 0$, $\mathbf{r}(0) = \langle 1, 0 \rangle$, $\mathbf{r}(2) = \langle 0, -1 \rangle$, $\lim_{t \to \infty} x(t) = -2$, and $\lim_{t \to \infty} y(t) = -3$.

21. Domain $t \in \mathbb{R}$, $\mathbf{r}(2) = \langle 1, 1 \rangle$, $\mathbf{r}(0) = \langle -3, 3 \rangle$, $\mathbf{r}(-2) = \langle -5, -5 \rangle$, $\lim_{t \to -\infty} x(t) = \infty$, $\lim_{t \to -\infty} y(t) = -\infty$, and $\lim_{t \to \infty} \mathbf{r}(t) = \langle 0, 0 \rangle$.

22. Given a vector-valued function $\mathbf{r}(t)$ with domain $\mathbb{R}$, what is the relationship between the graph of $\mathbf{r}(t)$ and the graph of $k\mathbf{r}(t)$, where $k > 1$ is a scalar?

23. Given a vector-valued function $\mathbf{r}(t)$ with domain $\mathbb{R}$, what is the relationship between the graph of $\mathbf{r}(t)$ and the graph of $\mathbf{r}(kt)$, where $k > 1$ is a scalar?

24. Explain why the graph of every vector-valued function $\mathbf{r}(t) = \langle \cos t, \sin t, f(t) \rangle$ lies on the surface of the cylinder $x^2 + y^2 = 1$ for every continuous function f.

25. Explain why the graph of every vector-valued function $\mathbf{r}(t) = \langle \cos t, \sin t, \cos t \rangle$ lies on the intersection of the two cylinders $x^2 + y^2 = 1$ and $y^2 + z^2 = 1$.

Skills

Find parametric equations for each of the vector-valued functions in Exercises 26–34, and sketch the graphs of the functions, indicating the direction for increasing values of t.

26. $\mathbf{r}(t) = \langle \sin 3t, \cos 3t \rangle$ for $t \in [0, 2\pi]$

27. $\mathbf{r}(t) = \langle 2 - \sin t, 4 + \cos t \rangle$ for $t \in [0, 2\pi]$

28. $\mathbf{r}(t) = \langle \sin t, \cos 2t \rangle$ for $t \in [0, 2\pi]$

29. $\mathbf{r}(t) = \langle 1 + \sin t, 3 - \cos 2t \rangle$ for $t \in [0, 2\pi]$

30. $\mathbf{r}(t) = (3 + t)\mathbf{i} + \left(3 - \frac{1}{t}\right)\mathbf{j}$ for $t > 0$

31. $\mathbf{r}(t) = \langle t, t^2, t^3 \rangle$ for $t \in [0, 2]$

32. $\mathbf{r}(t) = \langle \cos^2 t, 4 \sin t, t \rangle$ for $t \in [0, 2\pi]$

33. $\mathbf{r}(t) = \langle \cos^2 t, \sin 2t \rangle$ for $t \in [0, 2\pi]$

34. $\mathbf{r}(t) = \langle t \sin t, t \cos t, t \rangle$ for $t \in [0, 4\pi]$

Evaluate and simplify the indicated quantities in Exercises 35–41.

35. $\langle 1, 3t, t^3 \rangle + \langle t, t^2, t^3 \rangle$

36. $\langle 1, t, t^2 \rangle - \langle t, t^2, t^3 \rangle$

37. $5 \langle \cos t, \sin t \rangle$

38. $t \langle \sin t, \cos t \rangle$

39. $\langle \sin t, \cos t \rangle \cdot \langle \cos t, -\sin t \rangle$

40. $((\sin t)\,\mathbf{i} + (\cos t)\,\mathbf{j} + t\mathbf{k}) \times ((\cos t)\,\mathbf{i} + (\sin t)\,\mathbf{j})$

41. $\langle 1, t, t^2 \rangle \times \langle t, t^2, t^3 \rangle$

Evaluate the limits in Exercises 42–45.

42. $\lim_{t \to 0} \langle \sin 3t, \cos 3t \rangle$

43. $\lim_{t \to \pi} \langle \sin t, \cos t, \sec t \rangle$

44. $\lim_{t \to 1^-} \left\langle \ln t, \frac{e^t - 1}{t - 1}, e^t \right\rangle$

45. $\lim_{t \to 0^+} \left\langle \frac{\sin t}{t}, \frac{1 - \cos t}{t}, \left(1 + \frac{1}{t}\right)^t \right\rangle$

Find and graph the vector function $\mathbf{r}(t) = \langle x(t), y(t) \rangle$ determined by the differential equations in Exercises 46–48.

46. $x'(t) = x$, $y'(t) = x^2$, $x(0) = 1$, $y(0) = 2$. (*Hint: Start by solving the initial-value problem* $x'(t) = x$, $x(0) = 1$.)

47. $x'(t) = 1 + x^2$, $y'(t) = x^2$, $x(0) = 0$, $y(0) = 1$ (*Hint: Start by solving the initial-value problem* $x'(t) = 1 + x^2$, $x(0) = 0$.)

48. $x'(t) = -y$, $y'(t) = x$, $x(0) = 1$, $y(0) = 0$ (*Hint: What familiar pair of functions have the given properties?*)

49. Show that the graph of the vector function $\mathbf{r}(t) = \langle 3 \sin t, 5 \cos t, 4 \sin t \rangle$ is a circle. (*Hint: Show that the graph lies on a sphere and in a plane.*)

Applications

50. Annie is conscious of tidal currents when she is sea kayaking. This activity can be tricky in an area south-southwest of Cattle Point on San Juan Island in Washington State. Annie is planning a trip through that area and finds that the velocity of the current changes with time and can be expressed by the vector function

$$\left\langle 0.4 \cos\left(\frac{\pi(t - 8)}{6}\right), -1.1 \cos\left(\frac{\pi(t - 11)}{6}\right) \right\rangle,$$

where t is measured in hours after midnight, speeds are given in knots, and $\langle 0, 1\rangle$ points due north.

Cattle Point on San Juan Island

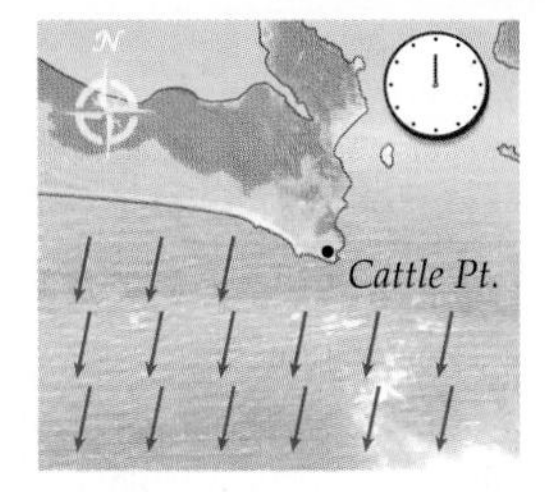

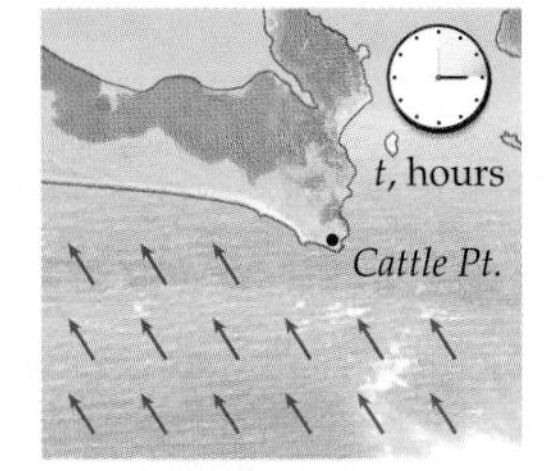

(a) What is the velocity of the current at 8:00 A.M.?

(b) What is the velocity of the current at 11:00 A.M.?

(c) Annie needs to paddle through here heading southeast, 135 degrees from north. She wants the current to push her. What is the best time for her to pass this point? (*Hint: Find the dot product of the given vector function with a vector in the direction of Annie's travel, and determine when the result is maximized.*)

51. Arne is a wingsuit base jumper in Norway. He is working out a jump from a cliff high above a fjord. After a couple of seconds, he will be 1500 meters above the bottom of the fjord, will reach his terminal velocity of 14 meters per second towards the ground, and will travel 30 meters per second horizontally. He calls the time when this happens $t = 0$. Below the cliff from which he jumped, the ground slopes to a second cliff, 300 meters above the water of the fjord. Arne must clear that second cliff, 2000 meters due south from his point at $t = 0$. For $t \geq 0$, his position function is given by

$$\left\langle A\sin\left(\frac{\pi t}{21}\right), \frac{A}{2}t, 1200 - 14t\right\rangle.$$

Approximately how large can A be while still allowing Arne to clear the second cliff?

The cliffs above the fjord

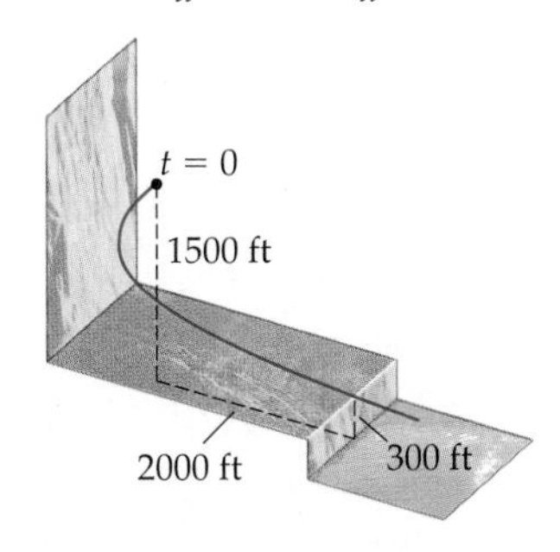

52. The DNA molecule takes the shape of a double helix—two helices that stay a roughly uniform distance apart.

(a) Neglecting actual dimensions, we can model one strand of DNA using the vector function

$$h_1(t) = \langle\cos(t), \sin(t), \alpha t\rangle.$$

Sketch the graph of h_1. What is the effect of the parameter α?

(b) A second strand of DNA can be constructed by shifting the first. Does the graph of

$$h_2(t) = \langle\cos(t), \sin(t), \alpha t + \beta\rangle$$

ever intersect that of h_1?

(c) The distance between two curves is the minimum distance between any two points on the curves. What is the distance between h_1 and h_2 if $\alpha = 1$ and $\beta = \pi$? (*Hint: Write two points on the curves using parameters t_1 and t_2, expand the formula for the distance between them, and then use a trigonometric identity for addition. Then let $s = t_1 - t_2$ and minimize.*)

A DNA molecule

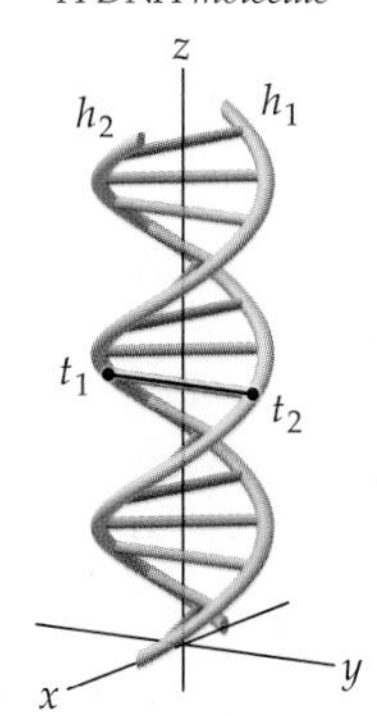

53. Every description of the DNA molecule says that the strands of the helices run in opposite directions. This is meant as a statement about chemistry, not about the geometric shape of the double helix. Consider two helices

$$h_1(t) = \langle\cos t, \sin t, \alpha t\rangle, \text{ and}$$

$$h_2(t) = \langle\sin t, \cos t, \alpha t\rangle.$$

(a) Sketch these two helices in the same coordinate system, and show that they run geometrically in different directions.

(b) Explain why it is impossible for these two helices to fail to intersect, and hence why they could not form a configuration for DNA.

Proofs

54. Let $\mathbf{r}(t) = \langle x(t), y(t) \rangle$ be a vector-valued function defined on an open interval containing the point t_0. Prove that $\mathbf{r}(t)$ is continuous at t_0 if and only if $x(t)$ and $y(t)$ are both continuous at t_0.

55. Prove that the graph of the vector function $\mathbf{r}(t) = \langle t \sin t, t \cos t, t \rangle$, where $t \geq 0$, is a conical helix by showing that it lies on the graph of the cone described by $z = \sqrt{x^2 + y^2}$.

56. Let c_1 and c_2 be scalars, $\mathbf{r}_1(t)$ and $\mathbf{r}_2(t)$ be continuous vector functions with two components, and t_0 be a point in the domains of both $\mathbf{r}_1$ and $\mathbf{r}_2$. Prove that
$$\lim_{t \to t_0} (c_1\mathbf{r}_1(t) + c_2\mathbf{r}_2(t)) = c_1\mathbf{r}_1(t_0) + c_2\mathbf{r}_2(t_0).$$

57. Let $\mathbf{r}_1(t)$ and $\mathbf{r}_2(t)$ be continuous vector functions with two components, and let t_0 be a point in the domains of both $\mathbf{r}_1$ and $\mathbf{r}_2$. Prove that
$$\lim_{t \to t_0} (\mathbf{r}_1(t) \cdot \mathbf{r}_2(t)) = \mathbf{r}_1(t_0) \cdot \mathbf{r}_2(t_0).$$

58. Let $\mathbf{r}_1(t)$ and $\mathbf{r}_2(t)$ be continuous vector functions with three components, and let t_0 be a point in the domains of both $\mathbf{r}_1$ and $\mathbf{r}_2$. Prove that
$$\lim_{t \to t_0} (\mathbf{r}_1(t) \times \mathbf{r}_2(t)) = \mathbf{r}_1(t_0) \times \mathbf{r}_2(t_0).$$

59. Prove that the dot product of the continuous vector-valued functions $\mathbf{r}_1(t) = \langle x_1(t), y_1(t) \rangle$ and $\mathbf{r}_2(t) = \langle x_2(t), y_2(t) \rangle$ is a continuous scalar function.

60. If α, β, and γ are nonzero constants, the graph of a vector function of the form $\mathbf{r}(t) = \langle \alpha t, \beta t^2, \gamma t^3 \rangle$ is called a *twisted cubic*. Prove that a twisted cubic intersects any plane in at most three points.

61. Let $x_1(t)$, $x_2(t)$, $y_1(t)$, and $y_2(t)$ be differentiable scalar functions. Prove that the dot product of the vector-valued functions $\mathbf{r}_1(t) = \langle x_1(t), y_1(t) \rangle$ and $\mathbf{r}_2(t) = \langle x_2(t), y_2(t) \rangle$ is a differentiable scalar function.

Thinking Forward

- *The derivative of a vector function:* Give a definition for the differentiability of a vector-valued function.
- *The integral of a vector function:* Give a definition for the definite integral of a vector-valued function.

11.2 THE CALCULUS OF VECTOR FUNCTIONS

- Differentiation and integration of vector functions
- Velocity and acceleration for vector-valued functions
- The geometry of the derivative of a vector-valued function

The Derivative of a Vector Function

In order for the vector function $\mathbf{r}(t) = \langle x(t), y(t), z(t) \rangle$ to be differentiable, we need each of the component functions x, y, and z to be differentiable.

DEFINITION 11.6 The Derivative of a Vector Function

Let
$$x = x(t), y = y(t), \text{ and } z = z(t)$$
be three real-valued functions, each of which is differentiable at every point in some interval $I \subseteq \mathbb{R}$. The ***derivative*** of the vector function $\mathbf{r}(t) = \langle x(t), y(t), z(t) \rangle$ is
$$\mathbf{r}'(t) = \langle x'(t), y'(t), z'(t) \rangle = x'(t)\mathbf{i} + y'(t)\mathbf{j} + z'(t)\mathbf{k}.$$
Similarly, the ***derivative*** of the vector function $\mathbf{r}(t) = \langle x(t), y(t) \rangle$ is
$$\mathbf{r}'(t) = \langle x'(t), y'(t) \rangle = x'(t)\mathbf{i} + y'(t)\mathbf{j}.$$
In these cases we say that the function $\mathbf{r}$ is ***differentiable***.

For example, if $\mathbf{r}(t) = \langle \sin t, \cos t, t^2 \rangle$ for $t \in \mathbb{R}$, then $\mathbf{r}'(t) = \langle \cos t, -\sin t, 2t \rangle$.

Recall that in Chapter 2 the derivative of a (scalar) function was defined in terms of a limit. Thus, the derivatives of the component functions, $x'(t)$, $y'(t)$, and $z'(t)$, are defined in terms of a limit, and we have the following theorem, the proof of which is left for Exercise 62:

THEOREM 11.7

The Derivative of a Vector-Valued Function

Let $\mathbf{r}(t)$ be a differentiable vector function with either two or three components. The derivative of $\mathbf{r}(t)$ is given by

$$\mathbf{r}'(t) = \lim_{h \to 0} \frac{\mathbf{r}(t+h) - \mathbf{r}(t)}{h}.$$

When you prove Theorem 11.7, be sure to interpret $\mathbf{r}(t+h)$ correctly. For a vector function in $\mathbb{R}^3$,

$$\mathbf{r}(t+h) = \langle x(t+h), y(t+h), z(t+h) \rangle.$$

In $\mathbb{R}^2$, $\mathbf{r}(t+h)$ has the analogous meaning.

THEOREM 11.8

The Chain Rule for Vector-Valued Functions

Let $t = f(\tau)$ be a differentiable real-valued function of τ, and let $\mathbf{r}(t)$ be a differentiable vector function with either two or three components such that $f(\tau)$ is in the domain of $\mathbf{r}$ for every value of τ on some interval I. Then

$$\frac{d\mathbf{r}}{d\tau} = \frac{d\mathbf{r}}{dt}\frac{dt}{d\tau}.$$

Proof. We will prove the Chain Rule for the case where $\mathbf{r}(t) = \langle x(t), y(t) \rangle$ and leave the case where $\mathbf{r}(t)$ has three components to Exercise 67. In either case, the hard work was actually done in Chapter 2 when we proved the Chain Rule for the composition of two scalar functions. By the definition of the derivative of a vector-valued function,

$$\begin{aligned}
\frac{d}{d\tau}(\mathbf{r}(t)) &= \frac{d}{d\tau}(\langle x(t), y(t) \rangle) && \leftarrow \text{the definition of } \mathbf{r} \\
&= \left\langle \frac{d}{d\tau}(x(t)), \frac{d}{d\tau}(y(t)) \right\rangle && \leftarrow \text{the definition of the derivative of a vector function} \\
&= \left\langle \frac{dx}{dt}\frac{dt}{d\tau}, \frac{dy}{dt}\frac{dt}{d\tau} \right\rangle && \leftarrow \text{the Chain Rule for scalar functions} \\
&= \left\langle \frac{dx}{dt}, \frac{dy}{dt} \right\rangle \frac{dt}{d\tau} && \leftarrow \text{multiplication of a vector function by a scalar function} \\
&= \frac{d\mathbf{r}}{dt}\frac{dt}{d\tau} && \leftarrow \text{the definition of the derivative of a vector function}
\end{aligned}$$

■

Returning to the example we used earlier, let $\mathbf{r}(t) = \langle \sin t, \cos t, t^2 \rangle$ for $t \in \mathbb{R}$, and let $t = e^{3\tau}$. Then

$$\begin{aligned}
\frac{d\mathbf{r}}{d\tau} &= \langle \cos t, -\sin t, 2t \rangle (3e^{3\tau}) = \langle \cos e^{3\tau}, -\sin e^{3\tau}, 2e^{3\tau} \rangle (3e^{3\tau}) \\
&= \langle 3e^{3\tau} \cos e^{3\tau}, -3e^{3\tau} \sin e^{3\tau}, 6e^{6\tau} \rangle.
\end{aligned}$$

The Geometry of the Derivative

We use Theorem 11.7 to interpret the geometry of the derivative. Recall from Section 10.2 that the difference $\mathbf{v} - \mathbf{w}$ of two vectors with the same initial point may be interpreted as the vector that extends from the terminal point of $\mathbf{w}$ to the terminal point of $\mathbf{v}$. Thus, for position vectors $\mathbf{r}(t+h)$ and $\mathbf{r}(t)$, the numerator of the quotient $\frac{\mathbf{r}(t+h)-\mathbf{r}(t)}{h}$ is the vector that points from the terminal point of $\mathbf{r}(t)$ to the terminal point of $\mathbf{r}(t+h)$, as shown here at the left:

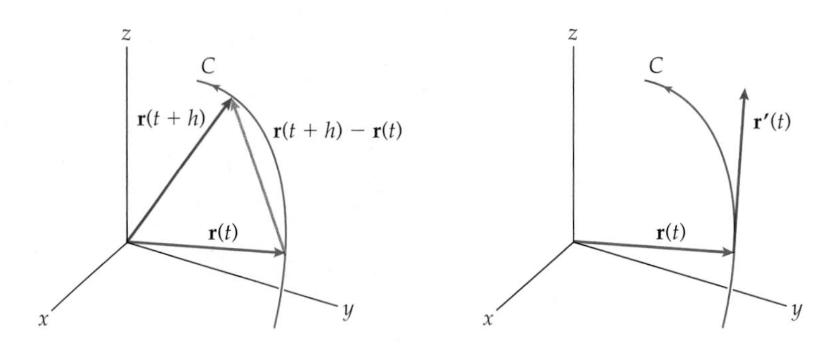

If the scalar, h, in the denominator of the quotient is positive, the direction of $\frac{\mathbf{r}(t+h)-\mathbf{r}(t)}{h}$ is the same as the direction of $\mathbf{r}(t+h) - \mathbf{r}(t)$, although their magnitudes may differ. (A similar argument may be made when $h < 0$.) As with the derivative of a function of a single variable, the derivative is tangent to the curve at $\mathbf{r}(t)$. As shown in the graph on the right, we will always assume that the initial point of the derivative $\mathbf{r}'(t)$ is positioned at the terminal point of the vector $\mathbf{r}(t)$. This ensures that the derivative is tangent to the curve C at $\mathbf{r}(t)$.

We may also use the derivative of the vector function to construct a tangent line to the vector curve.

DEFINITION 11.9 **The Tangent Line to a Vector Curve**

Let $\mathbf{r}(t) = \langle x(t), y(t), z(t)\rangle$ be a differentiable vector function on some interval $I \subseteq \mathbb{R}$, and let t_0 be a point in I such that $\mathbf{r}'(t_0) \neq \mathbf{0}$. The ***tangent line*** to the vector curve defined by $\mathbf{r}(t)$ at $\mathbf{r}(t_0)$ is given by

$$\mathcal{L}(t) = \mathbf{r}(t_0) + t\,\mathbf{r}'(t_0).$$

From Definition 11.9 we see that the tangent line is the line containing the point $\mathbf{r}(t_0)$ whose direction vector is $\mathbf{r}'(t_0)$.

Velocity and Acceleration

Recall that when t represents time and $y = s(t)$ represents the position function of a particle moving along a straight path, $s'(t)$ and $s''(t)$ represent the velocity and acceleration of the particle, respectively. We define the velocity and acceleration of a particle moving along a space curve determined by a vector function in an analogous fashion.

DEFINITION 11.10 **Velocity, Speed, and Acceleration along a Space Curve**

Let $\mathbf{r}(t) = \langle x(t), y(t), z(t) \rangle$ be a differentiable vector-valued function defined at every point in some time interval $I \subseteq \mathbb{R}$, and let C be the space curve defined by $\mathbf{r}(t)$.

(a) The ***velocity***, $\mathbf{v}$, of the particle as it moves along C is given by

$$\mathbf{v}(t) = \mathbf{r}'(t) = \langle x'(t), y'(t), z'(t) \rangle.$$

(b) The ***speed*** of the particle as it moves along C is given by

$$\|\mathbf{v}(t)\| = \|\langle x'(t), y'(t), z'(t) \rangle\|.$$

(c) In addition, if $\mathbf{r}(t)$ is twice differentiable, the ***acceleration***, $\mathbf{a}$, of the particle as it moves along C is given by

$$\mathbf{a}(t) = \mathbf{v}'(t) = \mathbf{r}''(t) = \langle x''(t), y''(t), z''(t) \rangle.$$

Note that velocity and acceleration are vectors, but speed is a scalar, since it is the norm of the velocity.

Derivatives of Vector Products

Recall that there are three products that apply to vectors: Vectors may be multiplied by scalars, two vectors with the same number of components may be multiplied by means of the dot product, and two three-component vectors may be multiplied via the cross product. The same products apply to vector functions. Fortunately, the derivatives of these product functions follow the rules we would predict.

THEOREM 11.11 **Derivatives of Products of Vector Functions**

Let k be a scalar, $f(t)$ be a differentiable scalar function, and $\mathbf{r}(t)$ be a differentiable vector function. Then

(a) $\frac{d}{dt}(k\mathbf{r}(t)) = k\mathbf{r}'(t)$.

(b) $\frac{d}{dt}(f(t)\mathbf{r}(t)) = f'(t)\mathbf{r}(t) + f(t)\mathbf{r}'(t)$.

Furthermore, if $\mathbf{r}_1(t)$ and $\mathbf{r}_2(t)$ are differentiable vector functions with both having either two or three components, then

(c) $\frac{d}{dt}(\mathbf{r}_1(t) \cdot \mathbf{r}_2(t)) = \mathbf{r}_1'(t) \cdot \mathbf{r}_2(t) + \mathbf{r}_1(t) \cdot \mathbf{r}_2'(t)$.

Finally, if $\mathbf{r}_1(t)$ and $\mathbf{r}_2(t)$ are both differentiable three-component vector functions, then

(d) $\frac{d}{dt}(\mathbf{r}_1(t) \times \mathbf{r}_2(t)) = \mathbf{r}_1'(t) \times \mathbf{r}_2(t) + \mathbf{r}_1(t) \times \mathbf{r}_2'(t)$.

Recall that the dot product is commutative, so the order of the products in Theorem 11.11 (c) is not significant, but since the cross product is not commutative, the order of the products in Theorem 11.11 (d) is significant.

The proof of each part of Theorem 11.11 follows directly from the definitions of the derivative and the relevant product. We will prove Theorem 11.11(c) when $\mathbf{r}_1(t)$ and $\mathbf{r}_2(t)$

are differentiable vector functions with two components. The proofs of all other parts of Theorem 11.11 are left for Exercises 63–66.

Proof. Let $\mathbf{r}_1(t) = \langle x_1(t), y_1(t)\rangle$ and $\mathbf{r}_2(t) = \langle x_2(t), y_2(t)\rangle$. Then

$$\mathbf{r}_1(t)\cdot\mathbf{r}_2(t) = \langle x_1(t), y_1(t)\rangle\cdot\langle x_2(t), y_2(t)\rangle = x_1(t)\,x_2(t) + y_1(t)\,y_2(t).$$

Thus, when we take the derivative, we have

$$\begin{aligned}\frac{d}{dt}(\mathbf{r}_1(t)\cdot\mathbf{r}_2(t)) &= \frac{d}{dt}(x_1(t)\,x_2(t) + y_1(t)\,y_2(t))\\ &= x_1'(t)\,x_2(t) + x_1(t)\,x_2'(t) + y_1'(t)\,y_2(t) + y_1(t)\,y_2'(t)\\ &= (x_1'(t)\,x_2(t) + y_1'(t)\,y_2(t)) + (x_1(t)\,x_2'(t) + y_1(t)\,y_2'(t))\\ &= \langle x'_1(t), y'_1(t)\rangle\cdot\langle x_2(t), y_2(t)\rangle + \langle x_1(t), y_1(t)\rangle\cdot\langle x_2'(t), y_2'(t)\rangle\\ &= \mathbf{r}_1'(t)\cdot\mathbf{r}_2(t) + \mathbf{r}_1(t)\cdot\mathbf{r}_2'(t).\end{aligned}$$

■

When the magnitude of a vector function is constant, the function and its derivative are orthogonal, as stated in the next theorem. This result is not terribly surprising: In $\mathbb{R}^2$, any vector function with constant magnitude k has a graph that is at least a portion of a circle with radius k centered at the origin, and every tangent to a circle is orthogonal to the radius at the point of tangency. The three-dimensional case is similar: Every vector function with constant magnitude k has a graph that lies on the boundary of a sphere of radius k and that is centered at the origin. The following figures illustrate the two- and three-dimensional cases (left and right, respectively).

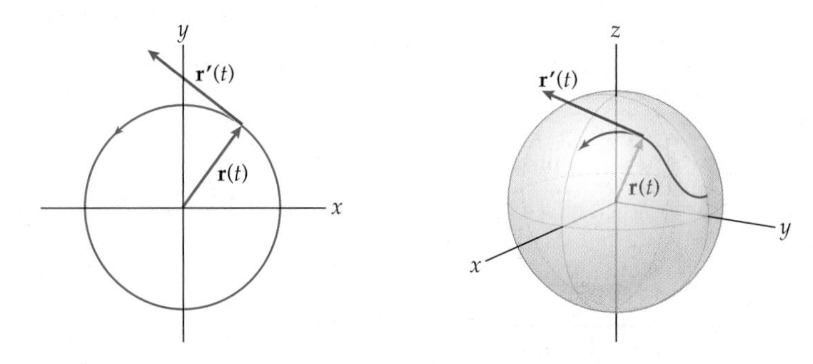

THEOREM 11.12

A Vector Function with a Constant Magnitude Is Orthogonal to Its Derivative

Let $\mathbf{r}(t)$ be a differentiable vector function such that $\|\mathbf{r}(t)\| = k$ for some constant k. Then

$$\mathbf{r}(t)\cdot\mathbf{r}'(t) = 0.$$

Proof. Since the magnitude of $\mathbf{r}(t)$ is constant, so is its square. That is,

$$\|\mathbf{r}(t)\|^2 = k^2,$$

where k^2 is a constant. However, recall that

$$\|\mathbf{r}(t)\|^2 = \mathbf{r}(t)\cdot\mathbf{r}(t).$$

So we have

$$\mathbf{r}(t)\cdot\mathbf{r}(t) = k^2.$$

We now take the derivative of each side of the latter equation, using Theorem 11.11 (c) to take the derivative of the left side, and obtain

$$\mathbf{r}(t) \cdot \mathbf{r}'(t) + \mathbf{r}'(t) \cdot \mathbf{r}(t) = 2\mathbf{r}(t) \cdot \mathbf{r}'(t) = 0.$$

Since their dot product is zero, the vectors $\mathbf{r}(t)$ and $\mathbf{r}'(t)$ are orthogonal. ■

Integration of a Vector Function

Recall from Chapter 4 that every function that is continuous on an interval $[a, b]$ is integrable on $[a, b]$. We use this property in the following definition:

DEFINITION 11.13

The Integral of a Vector Function

Let $x = x(t)$, $y = y(t)$, and $z = z(t)$ be three real-valued functions with antiderivatives

$$\int x(t)\,dt, \quad \int y(t)\,dt, \quad \text{and} \quad \int z(t)\,dt,$$

respectively, on some interval $I \subseteq \mathbb{R}$. Then the vector function

$$\int \mathbf{r}(t)\,dt = \int (x(t)\mathbf{i} + y(t)\mathbf{j} + z(t)\mathbf{k})\,dt = \mathbf{i}\int x(t)\,dt + \mathbf{j}\int y(t)\,dt + \mathbf{k}\int z(t)\,dt$$

is an ***antiderivative*** of the vector function $\mathbf{r}(t) = \langle x(t), y(t), z(t)\rangle$.

Similarly, if $x(t)$, $y(t)$, and $z(t)$ are all integrable on the interval $[a, b]$, then the ***definite integral*** of the vector function $\mathbf{r}(t)$ from a to b is the vector

$$\int_a^b \mathbf{r}(t)\,dt = \int_a^b (x(t)\mathbf{i} + y(t)\mathbf{j} + z(t)\mathbf{k})\,dt = \mathbf{i}\int_a^b x(t)\,dt + \mathbf{j}\int_a^b y(t)\,dt + \mathbf{k}\int_a^b z(t)\,dt.$$

Note that every antiderivative of a vector function is another vector function and any two antiderivatives of the same function differ by a constant (vector). However, the definite integral of a vector function is unique and is a constant (vector). In order for a vector function $\mathbf{r}(t) = \langle x(t), y(t), z(t)\rangle$ to be integrable, each of its component functions must be integrable. This condition tells us that we could also find $\int_a^b \mathbf{r}(t)\,dt$ as the limit of a Riemann sum if necessary. We explore antiderivatives and definite integrals in the examples and exercises.

Examples and Explorations

EXAMPLE 1

The derivative of a vector function

Find the derivative of the vector function $\mathbf{r}(t) = \langle \cos t, \sin t, t\rangle$, and find the tangent lines to the curve defined by $\mathbf{r}$ at $t = \frac{\pi}{4}$ and $t = \pi$.

SOLUTION

To find the derivative of $\mathbf{r}(t)$, we take the derivative of each component function to obtain $\mathbf{r}'(t) = \langle -\sin t, \cos t, 1\rangle$.

We now find the tangent line when $t = \frac{\pi}{4}$. Since

$$\mathbf{r}\left(\frac{\pi}{4}\right) = \left\langle \frac{\sqrt{2}}{2}, \frac{\sqrt{2}}{2}, \frac{\pi}{4} \right\rangle \quad \text{and} \quad \mathbf{r}'\left(\frac{\pi}{4}\right) = \left\langle -\frac{\sqrt{2}}{2}, \frac{\sqrt{2}}{2}, 1 \right\rangle,$$

the equation of the tangent line at $t = \frac{\pi}{4}$ is

$$\mathcal{L}_1(\tau) = \left\langle \frac{\sqrt{2}}{2}, \frac{\sqrt{2}}{2}, \frac{\pi}{4} \right\rangle + \tau \left\langle -\frac{\sqrt{2}}{2}, \frac{\sqrt{2}}{2}, 1 \right\rangle.$$

Because we may use any nonzero multiple of the direction vector as the direction vector for our line, we may also express the line as

$$\mathcal{L}_1(\tau) = \left\langle \frac{\sqrt{2}}{2}, \frac{\sqrt{2}}{2}, \frac{\pi}{4} \right\rangle + \tau \langle -1, 1, \sqrt{2} \rangle.$$

Similarly, when $t = \pi$, the tangent line will pass through the terminal point of the position vector $\mathbf{r}(\pi) = \langle -1, 0, \pi \rangle$ and will have the direction vector $\mathbf{r}'(\pi) = \langle 0, -1, 1 \rangle$. The equation of this line is

$$\mathcal{L}_2(\tau) = \langle -1, 0, \pi \rangle + \tau \langle 0, -1, 1 \rangle.$$

□

EXAMPLE 2 **Using derivatives to understand the graph of a vector function**

Graph the vector function $\mathbf{r}(t) = t^2\mathbf{i} + (3t - 2t^3)\mathbf{j}$.

SOLUTION

Let $\mathcal{C}$ represent the graph of the function. We begin by noting that $\mathcal{C}$ will lie entirely in the first and fourth quadrants, since $x = t^2$. We also note that because $x = t^2$ is an even function and $y = 3t - 2t^3$ is an odd function, $\mathcal{C}$ will be symmetric with respect to the x-axis. To understand the complete graph we need only analyze the vector function for positive values of t. We first note that the only time $\mathcal{C}$ will intersect the y-axis is when $x = 0$, but $\mathcal{C}$ will intersect the *x-axis* for the three values of t when $3t - 2t^3 = 0$, namely, $t = 0$ and $t = \pm\sqrt{\frac{3}{2}}$.

We take the derivative and obtain

$$\mathbf{r}'(t) = 2t\mathbf{i} + (3 - 6t^2)\mathbf{j}.$$

We can find where the curve $\mathcal{C}$ has horizontal tangents by finding the values where $\frac{dy}{dx} = 0$. By the chain rule we know that $\frac{dy}{dx} = \frac{dy/dt}{dx/dt}$. Here we have

$$\frac{dy}{dx} = \frac{dy/dt}{dx/dt} = \frac{3 - 6t^2}{2t}.$$

We see that the slope of the tangent lines to $\mathcal{C}$ will be zero when $t = \pm\frac{1}{\sqrt{2}}$ and will be undefined when $t = 0$. For $t > 0$, we note that $\frac{dx}{dt}$ is positive, $\frac{dy}{dt} > 0$ when $t \in (0, 1/\sqrt{2})$, and $\frac{dy}{dt} < 0$ when $t > \frac{1}{\sqrt{2}}$. Therefore, the graph will have a relative maximum when $t = \frac{1}{\sqrt{2}}$. We also note that $y > 0$ for $t \in (0, \sqrt{3/2})$ and $y < 0$ for $t > \sqrt{3/2}$. Incorporating this information into a graph, we obtain the figure that follows on the left. We then use symmetry to obtain the complete graph on the right.

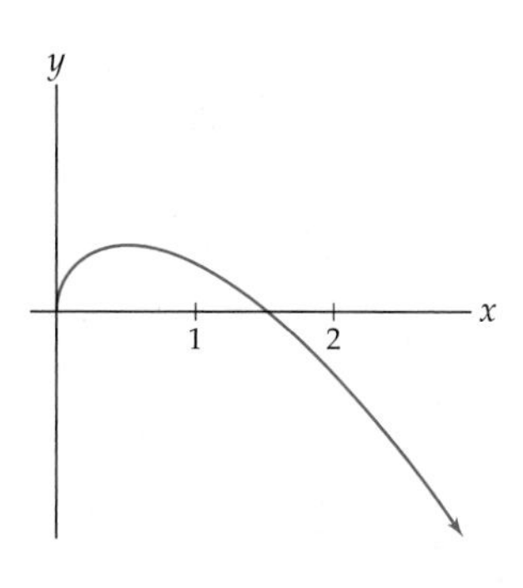

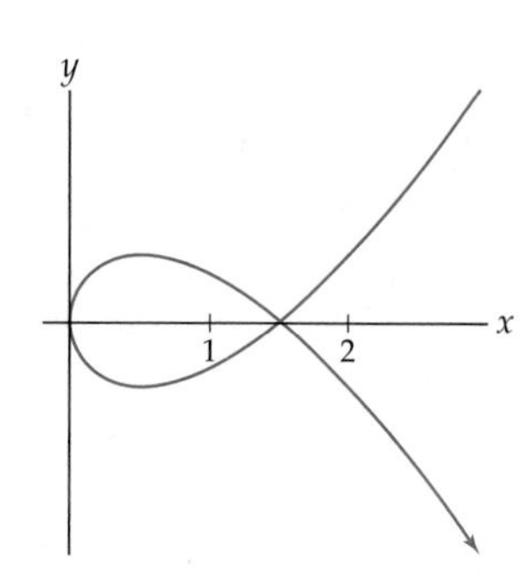

□

EXAMPLE 3 **The integral of a vector function**

Find the antiderivative of the vector function $\mathbf{r}(t) = \langle \cos t, \sin t, t \rangle$, and compute the definite integral of $\mathbf{r}$ from 0 to π.

SOLUTION

We integrate

$$\int \mathbf{r}(t)\, dt = \int \langle \cos t, \sin t, t \rangle\, dt = \mathbf{i}\int \cos t\, dt + \mathbf{j}\int \sin t\, dt + \mathbf{k}\int t\, dt = \left\langle \sin t, -\cos t, \frac{t^2}{2} \right\rangle + \mathbf{C},$$

where $\mathbf{C}$ is a vector constant and has the form $\mathbf{C} = \langle c_1, c_2, c_3 \rangle$ for scalar constants c_1, c_2, and c_3.

We use the Fundamental Theorem of Calculus to evaluate the definite integral

$$\int_0^\pi \mathbf{r}(t)\, dt.$$

We already have an antiderivative $\left\langle \sin t, -\cos t, \frac{t^2}{2} \right\rangle$ for $\mathbf{r}(t)$. Thus, to compute the definite integral, we evaluate this antiderivative at 0 and π and take the difference. That is,

$$\int_0^\pi \mathbf{r}(t)\, dt = \left\langle \sin t, -\cos t, \frac{t^2}{2} \right\rangle \Big|_0^\pi = \left\langle 0, 1, \frac{\pi^2}{2} \right\rangle - \langle 0, -1, 0 \rangle = \left\langle 0, 2, \frac{\pi^2}{2} \right\rangle.$$

□

Example 3 was a rather mechanical and artificial use of the integral. In a slightly more satisfying example, we are given a formula for a tangent vector and are asked to find a vector function that contains a particular point.

EXAMPLE 4 **Finding a position vector given its velocity vector**

A particle is moving along a space curve. The velocity vector for the curve is given by $\mathbf{v}(t) = \langle \sin t, t^2, t^3 \rangle$. When $t = 0$, the position of the particle is $(3, -2, 5)$. Find the position function for the curve.

SOLUTION

We begin by integrating $\mathbf{v}(t) = \mathbf{r}'(t) = \langle \sin t, t^2, t^3 \rangle$:

$$\mathbf{r}(t) = \int \mathbf{v}(t)\, dt = \int \langle \sin t, t^2, t^3 \rangle\, dt = (-\cos t + c_1)\,\mathbf{i} + \left(\frac{t^3}{3} + c_2\right)\mathbf{j} + \left(\frac{t^4}{4} + c_3\right)\mathbf{k}.$$

We now use the initial position $\mathbf{r}(0) = 3\mathbf{i} - 2\mathbf{j} + 5\mathbf{k} = (-\cos 0 + c_1)\,\mathbf{i} + \left(\frac{0^3}{3} + c_2\right)\mathbf{j} + \left(\frac{0^4}{4} + c_3\right)\mathbf{k}$. Solving the system

$$3 = -1 + c_1,\ -2 = 0 + c_2,\ 5 = 0 + c_3,$$

we obtain $c_1 = 4$, $c_2 = -2$, and $c_3 = 5$. Thus,

$$\mathbf{r}(t) = (-\cos t + 4)\,\mathbf{i} + \left(\frac{t^3}{3} - 2\right)\mathbf{j} + \left(\frac{t^4}{4} + 5\right)\mathbf{k}. \qquad \square$$

We may check our answer to Exercise 4 in two steps. First we take the derivative of $\mathbf{r}(t) = (-\cos t + 4)\,\mathbf{i} + \left(\frac{t^3}{3} - 2\right)\mathbf{j} + \left(\frac{t^4}{4} + 5\right)\mathbf{k}$ and obtain

$$\mathbf{r}'(t) = (\sin t)\mathbf{i} + t^2\mathbf{j} + t^3\mathbf{k},$$

giving us the correct tangent vector function.

Next, we evaluate

$$\mathbf{r}(0) = (-\cos 0 + 4)\,\mathbf{i} + \left(\frac{0^3}{3} - 2\right)\mathbf{j} + \left(\frac{0^4}{4} + 5\right)\mathbf{k} = \langle 3, -2, 5\rangle.$$

Thus, we have the correct solution.

EXAMPLE 5

Finding the angle between two intersecting vector functions

Show that the conical helices defined by the vector functions $\mathbf{r}_1(t) = \langle t\cos t, t\sin t, t\rangle$ and $\mathbf{r}_2(t) = \langle t\sin t, t, t\cos t\rangle$ intersect at the origin, and then find the angle of intersection between the curves.

SOLUTION

Since both $\mathbf{r}_1(0) = (0, 0, 0)$ and $\mathbf{r}_2(0) = (0, 0, 0)$, the two helices intersect at the origin. We now find the tangent vectors to both graphs. The derivatives are

$$\mathbf{r}_1'(t) = \langle \cos t - t\sin t, \sin t + t\cos t, 1\rangle \quad \text{and} \quad \mathbf{r}_2'(t) = \langle \sin t + t\cos t, 1, \cos t - t\sin t\rangle.$$

The tangent vectors when $t = 0$ are $\mathbf{r}_1'(0) = \langle 1, 0, 1\rangle$ and $\mathbf{r}_2'(0) = \langle 0, 1, 1\rangle$. Recall that the cosine of the angle θ between two vectors $\mathbf{u}$ and $\mathbf{v}$ is given by $\cos\theta = \frac{\mathbf{u}\cdot\mathbf{v}}{\|\mathbf{u}\|\|\mathbf{v}\|}$. Thus, the cosine of the angle between the helices at the origin is

$$\cos\theta = \frac{\langle 1, 0, 1\rangle \cdot \langle 0, 1, 1\rangle}{\|\langle 1, 0, 1\rangle\|\|\langle 0, 1, 1\rangle\|} = \frac{1}{\sqrt{2}\sqrt{2}} = \frac{1}{2}.$$

Therefore, the (acute) angle of intersection between the two curves is $60°$. $\square$

? TEST YOUR UNDERSTANDING

- We defined the derivative of a vector-valued function in terms of the derivatives of its component functions. Why do we bother stating and proving Theorem 11.7, which says that we can find a derivative by using the limit of a difference quotient?
- Given a vector-valued function $\mathbf{r}$ and its graph C, what geometric relationship does the derivative function have to C?
- Under what conditions are a vector-valued function and its derivative orthogonal?
- For a vector-valued function $\mathbf{r}(t)$, when would the velocity vector $\mathbf{v}(t)$ and acceleration vector $\mathbf{a}(t)$ be orthogonal?
- Given an acceleration vector $\mathbf{a}(t)$, how would you find the position vector $\mathbf{r}(t)$?

EXERCISES 11.2

Thinking Back

Differentiability: Use the definition of the derivative to find the derivatives of the following functions.

- $f(x) = x^2$
- $f(x) = x^3 - 1$
- $f(x) = \sqrt{x+1}$

Integrability: Use the definition of the definite integral to evaluate the following integral.

- $\int_1^4 (x+3)\,dx$
- $\int_0^2 x^2\,dx$
- $\int_1^3 (x^2-3)\,dx$

Concepts

0. *Problem Zero:* Read the section and make your own summary of the material.

1. *True/False:* Determine whether each of the statements that follow is true or false. If a statement is true, explain why. If a statement is false, provide a counterexample.

(a) *True or False:* If $\mathbf{r}(t)$ is a vector-valued function with domain $\mathbb{R}$, then $\lim_{h\to 0} \mathbf{r}(t+h)$ exists.

(b) *True or False:* If $\mathbf{r}(t)$ is a vector-valued function with domain $\mathbb{R}$, then $\lim_{h\to 0} \frac{\mathbf{r}(t+h)-\mathbf{r}(t)}{h}$ exists.

(c) *True or False:* Given a differentiable vector-valued function $\mathbf{r}(t)$, if $\|\mathbf{r}(t)\|$ is constant, then $\mathbf{r}(t)\cdot\mathbf{r}'(t) = 0$.

(d) *True or False:* Given a differentiable vector-valued function $\mathbf{r}(t)$, if $\mathbf{r}(t)\cdot\mathbf{r}'(t) = 0$, then $\|\mathbf{r}(t)\|$ is constant.

(e) *True or False:* Given a twice-differentiable vector-valued function $\mathbf{r}(t)$, if $\|\mathbf{r}(t)\|$ is constant, then $\mathbf{r}'(t)\cdot\mathbf{r}''(t) = 0$.

(f) *True or False:* Given a differentiable vector-valued function $\mathbf{r}(t)$, if $\|\mathbf{r}(t)\|$ is constant, then $\mathbf{r}(t)\times\mathbf{r}'(t) \neq 0$.

(g) *True or False:* Given a twice-differentiable vector-valued function, the velocity and acceleration vectors are never equal.

(h) *True or False:* If $\mathbf{r}(t)$ is a vector-valued function defined on the interval $[a,b]$, then $\int_a^b \mathbf{r}(t)\,dt$ exists.

2. *Examples:* Construct examples of the thing(s) described in the following. Try to find examples that are different than any in the reading.

(a) A differentiable vector function.

(b) A differentiable vector function $\mathbf{r}(t) = \langle x(t), y(t), z(t)\rangle$ in which none of the component functions is a constant but $\|\mathbf{r}(t)\|$ is constant.

(c) A twice-differentiable vector function $\mathbf{r}(t) = \langle x(t), y(t), z(t)\rangle$ in which none of the component functions is a constant but $\mathbf{r}(t) = \mathbf{r}'(t)$ for every t.

3. State what it means for a scalar function $y = f(x)$ to be differentiable at a point c.

4. State what it means for a vector function $\mathbf{r}(t) = \langle x(t), y(t), z(t)\rangle$ to be differentiable.

5. Explain why we do not need to explicitly use a limit in the definition for the derivative of a vector function $\mathbf{r}(t) = \langle x(t), y(t), z(t)\rangle$.

6. Given a differentiable vector-valued function $\mathbf{r}(t)$, explain why $\mathbf{r}'(t_0)$ is tangent to the curve defined by $\mathbf{r}(t)$ when the initial point of $\mathbf{r}'(t_0)$ is placed at the terminal point of $\mathbf{r}(t_0)$.

7. Give an example of a vector-valued function $\mathbf{r}(t)$ that is *not* differentiable at at least one point in its domain. Explain why your example is not differentiable at that point. Graph the function, and discuss the problem there is with constructing a tangent vector at the point of nondifferentiability.

8. In our discussion of the geometry of the derivative we graphed a differentiable vector function $\mathbf{r}(t)$ and the difference quotient $\frac{\mathbf{r}(t+h)-\mathbf{r}(t)}{h}$ for positive values of h. Draw the corresponding picture for negative values of h, and explain why $\lim_{h\to 0^-} \frac{\mathbf{r}(t+h)-\mathbf{r}(t)}{h}$ will result in the same tangent vector.

9. State what it means for a scalar function $y = f(x)$ to be integrable on an interval $[a,b]$.

10. State what it means for a vector function $\mathbf{r}(t) = \langle x(t), y(t), z(t)\rangle$ to be integrable.

11. Explain why we do not need to explicitly use a limit in the definition for the definite integral of a vector function $\mathbf{r}(t) = \langle x(t), y(t), z(t)\rangle$.

12. This exercise has to do with the integrability of vector functions.

(a) Explain why every vector function $\mathbf{r}(t) = \langle x(t) y(t), z(t)\rangle$ defined on $[a,b]$ is integrable if each of the component functions $x(t)$, $y(t)$, and $z(t)$ is continuous.

(b) When a vector function $\mathbf{r}(t)$ defined on $[a,b]$ is integrable, $\mathbf{r}(t) = \int_a^b \mathbf{r}(t)\,dt$ may be difficult to evaluate. Explain why.

13. Complete the following definition: If $\mathbf{r}(t) = \langle x(t), y(t), z(t)\rangle$ is a differentiable position function, then the ***velocity*** vector $\mathbf{v}(t)$ is ....

14. Complete the following definition: If $\mathbf{r}(t) = \langle x(t), y(t), z(t)\rangle$ is a twice-differentiable position function, then the ***acceleration*** vector $\mathbf{a}(t)$ is ....

15. Let $t = f(\tau)$ be a differentiable real-valued function of τ, and let $\mathbf{r}(t)$ be a differentiable vector function with either two or three components such that $f(\tau)$ is in the domain of $\mathbf{r}$ for every value of τ on some interval I. Find $\frac{d}{d\tau}(\mathbf{r}(f(\tau)))$.

16. Find the derivative of the vector-valued function $\mathbf{r}(t) = t^2\mathbf{i} - 6t^2\mathbf{j},\ t > 0$, and show that the derivative at any point t_0 is a scalar multiple of the derivative at any other point. What does this property say about the graph of $\mathbf{r}$? Use that information to sketch the graph of $\mathbf{r}$.

17. Find the derivative of the vector-valued function $\mathbf{r}(t) = \langle t^3, 5t^3, -2t^3\rangle,\ t > 0$, and show that the derivative at any point t_0 is a scalar multiple of the derivative at any other point. What does this property say about the graph of $\mathbf{r}$? Sketch the graph of $\mathbf{r}$.

18. The graph of every vector-valued function $\mathbf{r}(t)$ is a curve in the xy-plane. If $\mathbf{r}$ is twice differentiable, explain how information about $\mathbf{r}'(t)$ and $\mathbf{r}''(t)$ can be used to graph $\mathbf{r}$.

19. What is the relationship between the graph of a differentiable vector function $\mathbf{r}(t)$ and the graph of $\int \mathbf{r}'(t)\,dt$, one of the antiderivatives of $\mathbf{r}'(t)$?

20. Given a differentiable vector function $\mathbf{r}(t)$ defined on $[a, b]$, explain why the integral $\int_a^b \|\mathbf{r}'(t)\|\ dt$ would be a scalar, not a vector.

Skills

In Exercises 21–23 you are given a vector function $\mathbf{r}$ and a scalar function $t = f(\tau)$. Compute $\frac{d\mathbf{r}}{d\tau}$ in the following two ways:

(a) By using the chain rule $\frac{d\mathbf{r}}{d\tau} = \frac{d\mathbf{r}}{dt}\frac{dt}{d\tau}$.

(b) By substituting $t = f(\tau)$ into the formula for $\mathbf{r}$.

Ensure that your two answers are consistent.

21. $\mathbf{r}(t) = \langle t, t^2, t^3\rangle, t = \sin\tau$

22. $\mathbf{r}(t) = \langle \cos t, \sin t, t\rangle, t = \tau^3$

23. $\mathbf{r}(t) = \langle t^2\sin t^2, t, t^2\cos t^2\rangle, t = \sqrt{\tau}$

In Exercises 24–29 a vector function and a point on the graph of the function are given. Find an equation for the line tangent to the curve at the specified point, and then find an equation for the plane orthogonal to the tangent line containing the given point.

24. $\mathbf{r}(t) = \langle t, t^2, t^3\rangle,\ (2, 4, 8)$

25. $\mathbf{r}(t) = te^t\mathbf{i} + t\ln t\,\mathbf{j},\ (e, 0, 0)$

26. $\mathbf{r}(t) = \left\langle \sec t, \frac{1}{t+1}, e^t\ln(t+1)\right\rangle,\ (1, 1, 0)$

27. $\mathbf{r}(t) = \langle \sin t, \cos t, 2\sin 2t\rangle,\ (1, 0, 0)$

28. $\mathbf{r}(t) = \langle t, t, t^{3/2}\rangle,\ (4, 4, 8)$

29. $\mathbf{r}(t) = \cos 3t\,\mathbf{i} + \sin 4t\,\mathbf{j} + t\,\mathbf{k},\ \left(0, 0, \frac{\pi}{2}\right)$

Find the velocity and acceleration vectors for the position vectors given in Exercises 30–34.

30. $\mathbf{r}(t) = \langle t, t^2, t^3\rangle$

31. $\mathbf{r}(t) = te^t\mathbf{i} + t\ln t\,\mathbf{j}$

32. $\mathbf{r}(t) = \left\langle \sec t, \frac{1}{t}, e^t\ln t\right\rangle$

33. $\mathbf{r}(t) = \langle \sin t, \cos t, 2\sin 2t\rangle$

34. $\mathbf{r}(t) = \cos 3t\,\mathbf{i} + \sin 4t\,\mathbf{j} + t\,\mathbf{k}$

In Exercises 35–39 a vector function $\mathbf{r}(t)$ and scalar function $t = f(\tau)$ are given. Find $\frac{d\mathbf{r}}{d\tau}$.

35. $\mathbf{r}(t) = \langle t, t^2, t^3\rangle,\ t = \tau^3 + 1$

36. $\mathbf{r}(t) = te^t\mathbf{i} + t\ln t\,\mathbf{j},\ t = e^\tau$

37. $\mathbf{r}(t) = \langle \sec t, 1/t, e^t\ln t\rangle,\ t = \tau^{-1}$

38. $\mathbf{r}(t) = \langle \sin t, \cos t, 2\sin 2t\rangle,\ t = \sqrt{\tau^2 + 1}$

39. $\mathbf{r}(t) = \cos 3t\,\mathbf{i} + \sin 4t\,\mathbf{j} + t\,\mathbf{k},\ t = 5\tau - 2$

Evaluate the integrals in Exercises 40–44.

40. $\int \langle t, t^2, t^3\rangle\,dt$

41. $\int \langle \sin t, \cos t, \tan t\rangle\,dt$

42. $\int_1^2 \langle te^t, \ln t, \tan^{-1} t\rangle\,dt$

43. $\int_0^{2\pi} \langle \sin t, \cos t, t\rangle\,dt$

44. $\int_0^{2\pi} \|\langle \sin t, \cos t, t\rangle\|\,dt$

Use the given velocity vectors $\mathbf{v}(t) = \mathbf{r}'(t)$ and initial positions in Exercises 45–48 to find the position function $\mathbf{r}(t)$.

45. $\mathbf{v}(t) = \langle t, t^2\rangle,\ \mathbf{r}(0) = \langle 3, -4\rangle$

46. $\mathbf{v}(t) = \langle 0, \sec t\tan t, \sec^2 t\rangle,\ \mathbf{r}(\pi) = \langle -2, 4, 3\rangle$

47. $\mathbf{v}(t) = e^t\,\mathbf{i} + \ln t\,\mathbf{j},\ \mathbf{r}(1) = \mathbf{i} - 6\mathbf{j}$

48. $\mathbf{v}(t) = \cos t\,\mathbf{i} + \sin t\,\mathbf{j} + t\,\mathbf{k},\ \mathbf{r}(0) = -\mathbf{i} + 5\mathbf{j}$

Use the given acceleration vectors $\mathbf{a}(t) = \mathbf{v}'(t) = \mathbf{r}''(t)$ and initial conditions in Exercises 49–52 to find the position function $\mathbf{r}(t)$.

49. $\mathbf{a}(t) = \langle 2t, 3t^2\rangle,\ \mathbf{v}(0) = \langle 1, -2\rangle,\ \mathbf{r}(0) = \langle 2, -3\rangle$

50. $\mathbf{a}(t) = \langle 1, \sec t\tan t, \sec^2 t\rangle$, $\mathbf{v}(\pi) = \langle 0, 3, -1\rangle,\ \mathbf{r}(\pi) = \langle 4, 5, -2\rangle$

51. $\mathbf{a}(t) = -32\mathbf{j},\ \mathbf{v}(0) = 5\,\mathbf{i} + 5\,\mathbf{j},\ \mathbf{r}(0) = 26\,\mathbf{j}$

52. $\mathbf{a}(t) = \langle \sin 3t, t^2, \cos 3t\rangle,\ \mathbf{v}(0) = \langle 0, 0, 0\rangle,\ \mathbf{r}(0) = \langle 2, -5, 3\rangle$

53. Find all points of intersection between the vector function $\mathbf{r}(t) = \langle t, t^2, t^3\rangle$ and the plane defined by $3x - 3y + z = 1$.

Find all points of intersection between the graphs of the vector functions in Exercises 54–56, and find the acute angle of intersection of the curves at those points.

54. $\mathbf{r}_1(t) = \langle t, t^2 \rangle$ and $\mathbf{r}_2(t) = \langle t^2, t \rangle$

55. $\mathbf{r}_1(t) = \langle 3\cos t, 3\sin t \rangle$ and $\mathbf{r}_2(t) = \langle 2 + 2\sin t, 2\cos t \rangle$

56. $\mathbf{r}_1(t) = \langle \cos t, \sin t, t \rangle$ and $\mathbf{r}_2(t) = \langle \sin t, \cos t, t \rangle$

57. A certain vector function has the properties that its graph is a space curve passing through the point (0, 1, 2) and that its tangent vector at every point on the curve is equal to the position vector. Find the position function and sketch its graph.

Applications

58. Annie is making a crossing from south to north between two islands. The distance between the islands is 2 miles. A current in the channel pushes her off her line and slows her progress towards the opposite island. Her position t hours after leaving the southern island is

$$\langle 0.255t^3 - 0.479t^2, 2t - 0.958t^2 + 0.511t^3 \rangle.$$

Space coordinates are measured in miles from her starting position; time is measured in hours.

(a) What is Annie's actual velocity vector at any time t?

(b) Where is Annie's speed a minimum? What is her speed at that point?

59. Arne is a wingsuit base jumper in Norway. He wants to jump off a cliff above a fjord, 900 meters atop the valley floor. He knows that when he jumps, his downward acceleration due to gravity will be constant while there will be an upward acceleration depending on the air catching his wingsuit. He will convert about half of that upward acceleration into forward motion, which will cause drag. We ignore that, since he is still basically falling. Thus, his acceleration will be approximately

$$\langle -0.5dy'(t), -9.8 - dy'(t) \rangle,$$

where $y(t)$ denotes his vertical distance (in meters) from the jump point. The constant $d \approx 0.36$ is a parameter describing air resistance. Note that after he jumps, y will always be negative.

(a) What is Arne's velocity at time t in terms of his position relative to the start?

(b) Observe that the second coordinate of his velocity is his vertical speed, which is actually $y'(t)$. Use this fact to write an expression for $y(t)$ in terms of $y'(t)$.

(c) Arne's horizontal speed away from his jump point is the first coordinate of his velocity; call it $x'(t)$. Write an expression for $x'(t)$ in terms of $y'(t)$.

(d) Use parts (b) and (c) and integrate one more time to find a vector in terms of $y(t)$ describing Arne's position at any time t.

(e) The cliff Arne jumps from is vertical for the first 100 meters down, but it quickly slopes out below that. To survive the jump, Arne must clear a point 200 meters below him and 80 meters horizontally away. He quickly makes a calculation showing that it will take him more than 7.4 seconds to fall 200 meters. Can he make it?

Proofs

60. Prove that the tangent vector is always orthogonal to the position vector for the vector-valued function $\mathbf{r}(t) = \langle \sin t, \sin t \cos t, \cos^2 t \rangle$.

61. Let $\mathbf{r}(t)$ be a vector-valued function whose graph is a curve C, and let $\mathbf{a}(t)$ be the acceleration vector. Prove that if $\mathbf{a}(t)$ is always zero, then C is a straight line.

62. Prove Theorem 11.7 for vectors in $\mathbb{R}^2$. That is, let $x(t)$ and $y(t)$ be two scalar functions, each of which is differentiable on an interval $I \subseteq \mathbb{R}$, and let $\mathbf{r}(t) = \langle x(t), y(t) \rangle$ be a vector function. Prove that

$$\mathbf{r}'(t) = \lim_{h \to 0} \frac{\mathbf{r}(t+h) - \mathbf{r}(t)}{h}.$$

63. Let k be a scalar and $\mathbf{r}(t)$ be a differentiable vector function. Prove that $\frac{d}{dt}(k\mathbf{r}(t)) = k\mathbf{r}'(t)$. (This is Theorem 11.11 (a).)

64. Let $f(t)$ be a differentiable scalar function and $\mathbf{r}(t)$ be a differentiable vector function. Prove that

$$\frac{d}{dt}(f(t)\mathbf{r}(t)) = f'(t)\mathbf{r}(t) + f(t)\mathbf{r}'(t).$$

(This is Theorem 11.11 (b).)

65. Let $\mathbf{r}_1(t)$ and $\mathbf{r}_2(t)$ be differentiable vector functions with three components each. Prove that

$$\frac{d}{dt}(\mathbf{r}_1(t) \cdot \mathbf{r}_2(t)) = \mathbf{r}_1'(t) \cdot \mathbf{r}_2(t) + \mathbf{r}_1(t) \cdot \mathbf{r}_2'(t).$$

(This is Theorem 11.11 (c).)

66. Let $\mathbf{r}_1(t)$ and $\mathbf{r}_2(t)$ both be differentiable three-component vector functions. Prove that

$$\frac{d}{dt}(\mathbf{r}_1(t) \times \mathbf{r}_2(t)) = \mathbf{r}_1'(t) \times \mathbf{r}_2(t) + \mathbf{r}_1(t) \times \mathbf{r}_2'(t).$$

(This is Theorem 11.11 (d).)

67. Let $t = f(\tau)$ be a differentiable real-valued function of τ, and let $\mathbf{r}(t)$ be a differentiable vector function with three components such that $f(\tau)$ is in the domain of $\mathbf{r}$ for every value of τ on some interval I. Prove that $\frac{d\mathbf{r}}{d\tau} = \frac{d\mathbf{r}}{dt}\frac{dt}{d\tau}$. (This is Theorem 11.8.)

68. Let $\mathbf{r}(t)$ be a differentiable vector function. Prove that $\frac{d}{dt}(\|\mathbf{r}\|) = \frac{1}{\|\mathbf{r}\|}\mathbf{r} \cdot \mathbf{r}'$. (*Hint:* $\|\mathbf{r}\|^2 = \mathbf{r} \cdot \mathbf{r}$.)

69. Let $\mathbf{r}(t)$ be a differentiable vector function such that $\mathbf{r}(t) \cdot \mathbf{r}'(t) = 0$ for every value of t. Prove that $\|\mathbf{r}(t)\|$ is a constant.

70. For constants α, β, and γ, the graph of a vector-valued function of the form

$$\mathbf{r}(t) = \alpha e^{\gamma t}(\cos t)\mathbf{i} + \alpha e^{\gamma t}(\sin t)\mathbf{j} + \beta e^{\gamma t}\mathbf{k},\ t \geq 0$$

is called a ***concho-spiral***. Prove that the angle between the tangent vector to a concho-spiral and the vector $\mathbf{k}$ is constant.

71. Prove that if a particle moves along a curve at a constant speed, then the velocity and acceleration vectors are orthogonal.

Thinking Forward

- *Unit Tangent Vectors in* $\mathbb{R}^2$*:* Given the vector function $\mathbf{r}(t) = \langle t^2, t^3 \rangle$, find a vector of length 1 tangent to the curve at the point $\mathbf{r}(1)$. How many vectors in $\mathbb{R}^2$ are there of length 1 that are orthogonal to this tangent vector?
- *Unit Tangent Vectors in* $\mathbb{R}^3$*:* Given the vector function $\mathbf{r}(t) = \langle t, t^2, t^3 \rangle$, find a vector of length 1 tangent to the curve at the point $\mathbf{r}(1)$. How many vectors in $\mathbb{R}^3$ are there of length 1 that are orthogonal to this tangent vector?
- *Orthogonal Vectors:* For any constants a, b, and c, show that the velocity and acceleration vectors for
$$\mathbf{r}(t) = \langle a \sin bt, a \cos bt, ct \rangle$$
are orthogonal.
- *Horizontal Vectors:* For any constants a, b, and c, show that the acceleration vector for
$$\mathbf{r}(t) = \langle a \sin bt, a \cos bt, ct \rangle$$
is parallel to the xy-plane.

11.3 UNIT TANGENT AND UNIT NORMAL VECTORS

- The unit tangent and principal unit normal vectors to a space curve
- The osculating plane, the plane of "best fit" to a space curve
- The binormal vector to a space curve

Unit Tangent and Unit Normal Vectors

Our goal for the remainder of this chapter is to understand the graphs of vector functions and the motion of particles whose trajectories can be specified by a vector function. No matter what parametrization, $\mathbf{r}(t)$, we use for a curve C, the derivative vector $\mathbf{r}'(t)$ will be tangent to C at the terminal point of $\mathbf{r}(t)$, assuming that $\mathbf{r}(t)$ is differentiable and $\mathbf{r}'(t) \neq \mathbf{0}$. We will be interested primarily in the direction of the tangent vector, not its magnitude, so we make the following definition:

DEFINITION 11.14 **The Unit Tangent Vector**

Let $\mathbf{r}(t)$ be a differentiable vector function on some interval $I \subseteq \mathbb{R}$ such that $\mathbf{r}'(t) \neq \mathbf{0}$ on I. The ***unit tangent function*** is defined to be

$$\mathbf{T}(t) = \frac{\mathbf{r}'(t)}{\|\mathbf{r}'(t)\|}.$$

As we do with the tangent vector $\mathbf{r}'(t)$, we will always position the initial point of the unit tangent vector $\mathbf{T}(t)$ at the terminal point of $\mathbf{r}(t)$ to ensure that $\mathbf{T}(t)$ is tangent to the curve C, as shown in the following figure:

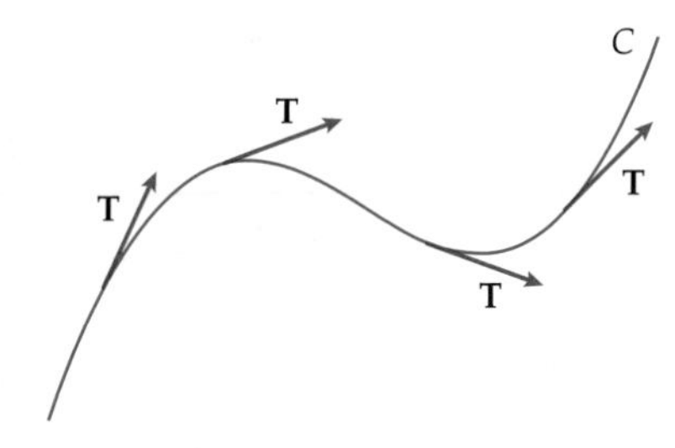

Since the magnitude of the unit tangent vector is constant, Theorem 11.12 tells us that $\mathbf{T}(t)$ and its derivative are always orthogonal. We use this property in the following definition:

DEFINITION 11.15

The Principal Unit Normal Vector

Let $\mathbf{r}(t)$ be a differentiable vector function on some interval $I \subseteq \mathbb{R}$ such that the derivative of the unit tangent vector $\mathbf{T}'(t) \neq \mathbf{0}$ on I. The ***principal unit normal vector*** at $\mathbf{r}(t)$ is defined to be

$$\mathbf{N}(t) = \frac{\mathbf{T}'(t)}{\|\mathbf{T}'(t)\|}.$$

Definition 11.15 ensures that the principal unit normal vector is a unit vector, that it is orthogonal to the tangent vectors, and that it points in the direction in which C is bending. The following figure illustrates these properties:

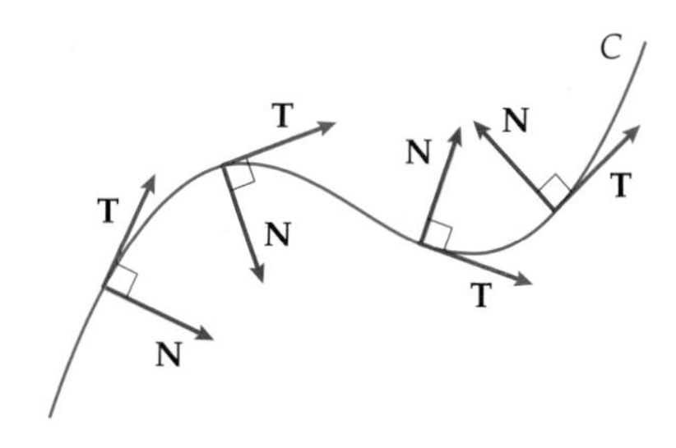

The Osculating Plane and the Binormal Vector

We do not expect curves, even curves in the plane, to contain straight segments. For example, the parabola defined by the equation $y = x^2$ does not contain any linear pieces. However, we have mentioned that the tangent line is the best linear approximation to a differentiable function at the point of tangency. Analogously, we do not expect portions of space curves to be planar. That is, there may not be a plane that contains even a small segment of a space curve, but we may use the unit tangent and principal unit normal vectors to define a plane, the ***osculating plane***, in which the curve fits "best."

DEFINITION 11.16

The Osculating Plane and the Binormal Vector

Let $\mathbf{r}(t) = \langle x(t), y(t), z(t)\rangle$ be a differentiable vector function on some interval $I \subseteq \mathbb{R}$ such that the derivative of the unit tangent vector $\mathbf{T}'(t_0) \neq 0$ where $t_0 \in I$. The ***binormal vector* B** at $\mathbf{r}(t_0)$ is defined to be

$$\mathbf{B}(t_0) = \mathbf{T}(t_0) \times \mathbf{N}(t_0),$$

and the ***osculating plane*** at $\mathbf{r}(t_0)$ is defined by

$$\mathbf{B}(t_0) \cdot \langle x - x(t_0), y - y(t_0), z - z(t_0)\rangle = 0.$$

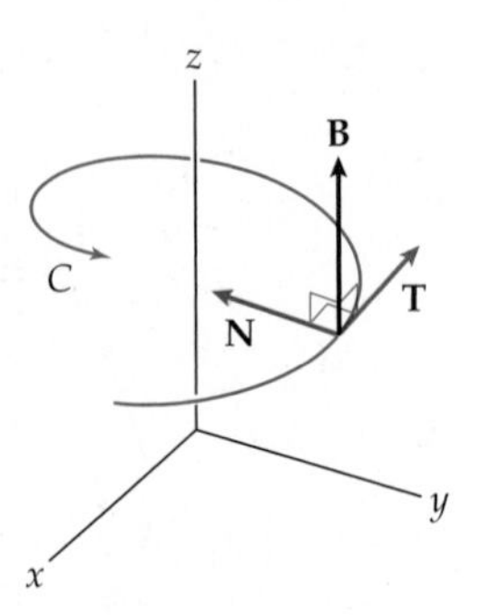

The preceding figure illustrates Definition 11.16. At a point on the curve C the vectors **T** and **N** determine a unique plane, the osculating plane. (*Osculating* comes from the Latin "osculum" meaning "kiss.") From its definition, the binormal vector **B** is orthogonal to both **T** and **N**, and therefore to the osculating plane. Vectors **T**, **N**, and **B** are three mutually perpendicular unit vectors. (We ask you to prove this in Exercise 47.) In the order **T**, **N**, **B** they also form a right-hand coordinate system, called the **TNB**-frame, ***moving frame***, or ***Frenet frame***. This frame allows for a more in-depth understanding of the motion of a particle moving along a space curve.

When we are working with a curve in the xy-plane, since both **T** and **N** lie in the plane, the xy-plane *is* the osculating plane. In this case, we may think of the xy-plane as part of a three-dimensional coordinate system, with every binormal vector parallel to the z-axis. Thus, for a planar curve, at every point where the binormal vector exists, $\mathbf{B}(t_0) = \pm\mathbf{k}$.

Examples and Explorations

EXAMPLE 1

Finding a unit tangent vector and a unit normal vector in the plane

Find the unit tangent and principal unit normal vectors to the graph of the sine function $y = \sin x$ at $x = \frac{\pi}{2}$ and at $x = \frac{5\pi}{4}$.

SOLUTION

Any function of the form $y = f(x)$ has the parametrization $x = t,\ y = f(t)$. Thus, we may use the vector function $\mathbf{r}(t) = \langle t, \sin t\rangle$ to compute the desired vectors. The unit tangent vector

$$\mathbf{T}(t) = \frac{\mathbf{r}'(t)}{\|\mathbf{r}'(t)\|} = \frac{\langle 1, \cos t\rangle}{\|\langle 1, \cos t\rangle\|} = \frac{\langle 1, \cos t\rangle}{\sqrt{1+\cos^2 t}}.$$

To compute the principal unit normal vector $\mathbf{N}(t) = \frac{\mathbf{T}'(t)}{\|\mathbf{T}'(t)\|}$, we first use the quotient rule to find $\mathbf{T}(t)$. You may check that:

$$\mathbf{T}'(t) = \frac{\sin t\,\langle \cos t, -1\rangle}{(1+\cos^2 t)^{3/2}}.$$

Dividing $\mathbf{T}'(t)$ by its magnitude and simplifying, we obtain

$$\mathbf{N}(t) = \frac{\sin t\,\langle \cos t, -1\rangle}{|\sin t|\sqrt{1+\cos^2 t}}.$$

Evaluating the unit tangent vector and principal unit normal vector at $t = \frac{\pi}{2}$, we get

$$\mathbf{T}\left(\frac{\pi}{2}\right) = \langle 1, 0\rangle \text{ and } \mathbf{N}\left(\frac{\pi}{2}\right) = \langle 0, -1\rangle.$$

This result should not be a surprise. At $\frac{\pi}{2}$ we expect there to be a horizontal tangent and a vertical normal vector pointing into the curve. In addition, since we chose a parametrization that traced the curve from left to right, the tangent vector should (and does) point to the right.

At $t = \frac{5\pi}{4}$ we obtain

$$\mathbf{T}\left(\frac{5\pi}{4}\right) = \frac{\langle 1, \cos(5\pi/4)\rangle}{\sqrt{1+\cos^2(5\pi/4)}} = \frac{\langle 1, -\sqrt{2}/2\rangle}{\sqrt{1+1/2}} = \frac{1}{\sqrt{3}}\langle\sqrt{2}, -1\rangle \text{ and } \mathbf{N}\left(\frac{5\pi}{4}\right) = \frac{1}{\sqrt{3}}\langle 1, \sqrt{2}\rangle. \quad \square$$

CHECKING THE ANSWER

Now, given that we are dealing with the function $y = \sin x$, perhaps it seemed that the preceding computations were (unnecessarily) complicated. If they did, you were right! For a differentiable function of the form $y = f(x)$, we have other techniques for finding the desired unit vectors. Since $f'(x_0)$ is the slope of f at x_0, we can immediately obtain the tangent vector, $\langle f'(x), 1\rangle$. If we divide this vector by its magnitude, we will obtain the equation of the unit tangent vector at x_0. Recall that two nonvertical lines in the plane are perpendicular if and only if their slopes are negative reciprocals. Therefore, a normal vector at x_0 will be a multiple of $\pm\langle 1, -f'(x_0)\rangle$. To obtain the principal unit normal vector at x_0, we again divide by the magnitude and choose the correct sign, the one that points *into* the curve. Example 2 repeats these computations with a space curve, for which the additional complexity *is* necessary. Before continuing to that example, repeat the calculations in Example 1, using the analysis just outlined to find the unit tangent and principal unit normal vectors to $y = \sin x$ at $x = \frac{\pi}{2}$ and at $x = \frac{5\pi}{4}$. Make sure your results agree with the answers we found in Example 1.

EXAMPLE 2

Finding the equation of an osculating plane

Find the unit tangent and principal unit normal vectors to the graph of the helix defined by the vector function $\mathbf{r}(t) = \langle \cos t, \sin t, t\rangle$ at $t = \frac{\pi}{2}$ and at $t = \frac{5\pi}{4}$. Then, use these vectors to construct the equations of the osculating planes at those points on the curve.

SOLUTION

We start by computing

$$\mathbf{r}'(t) = \langle -\sin t, \cos t, 1\rangle.$$

Thus,

$$\mathbf{T}(t) = \frac{\mathbf{r}'(t)}{\|\mathbf{r}'(t)\|} = \frac{\langle -\sin t, \cos t, 1\rangle}{\|\langle -\sin t, \cos t, 1\rangle\|} = \frac{\langle -\sin t, \cos t, 1\rangle}{\sqrt{2}}.$$

To compute $\mathbf{N}(t)$ we will divide $\mathbf{T}'(t)$ by its magnitude. In this example, since the radical in the denominator of the final quotient in the preceding equation is a constant, it will cancel when we divide. Thus, to make our computation slightly simpler, we will divide the derivative of $\langle -\sin t, \cos t, 1\rangle$ by its magnitude. We obtain

$$\mathbf{N}(t) = \frac{\langle -\cos t, -\sin t, 0\rangle}{\|\langle -\cos t, -\sin t, 0\rangle\|} = \frac{\langle -\cos t, -\sin t, 0\rangle}{\sqrt{1}} = \langle -\cos t, -\sin t, 0\rangle.$$

We now evaluate these functions to obtain the required vectors at $t = \frac{\pi}{2}$ and at $t = \frac{5\pi}{4}$. At $t = \frac{\pi}{2}$ we have

$$\mathbf{T}\left(\frac{\pi}{2}\right) = \frac{\langle -\sin(\pi/2), \cos(\pi/2), 1\rangle}{\sqrt{2}} = \frac{\langle -1, 0, 1\rangle}{\sqrt{2}} \text{ and}$$

$$\mathbf{N}\left(\frac{\pi}{2}\right) = \langle -\cos(\pi/2), -\sin(\pi/2), 0\rangle = \langle 0, -1, 0\rangle.$$

At $t = \frac{5\pi}{4}$ we have

$$\mathbf{T}\left(\frac{5\pi}{4}\right) = \frac{\langle -\sin(5\pi/4), \cos(5\pi/4), 1\rangle}{\sqrt{2}} = \frac{\langle \sqrt{2}/2, -\sqrt{2}/2, 1\rangle}{\sqrt{2}} = \left\langle \frac{1}{2}, -\frac{1}{2}, \frac{\sqrt{2}}{2}\right\rangle \quad \text{and}$$

$$\mathbf{N}\left(\frac{5\pi}{4}\right) = \left\langle -\cos\left(\frac{5\pi}{4}\right), -\sin\left(\frac{5\pi}{4}\right), 0\right\rangle = \left\langle \frac{\sqrt{2}}{2}, \frac{\sqrt{2}}{2}, 0\right\rangle.$$

Finally, we construct the equations of the osculating planes. For the equation of a plane, we need a point on the plane and a vector orthogonal to the plane. For an osculating plane at a point t_0, we will use the terminal point of the vector $\mathbf{r}(t_0)$ and the binormal vector $\mathbf{B}(t_0) = \mathbf{T}(t_0) \times \mathbf{N}(t_0)$, respectively.

At $t = \frac{\pi}{2}$ we have $\mathbf{r}\left(\frac{\pi}{2}\right) = \left\langle 0, 1, \frac{\pi}{2}\right\rangle$ and

$$\mathbf{B}\left(\frac{\pi}{2}\right) = \mathbf{T}\left(\frac{\pi}{2}\right) \times \mathbf{N}\left(\frac{\pi}{2}\right) = \frac{\langle -1, 0, 1\rangle}{\sqrt{2}} \times \langle 0, -1, 0\rangle = \frac{\langle 1, 0, 1\rangle}{\sqrt{2}}.$$

For simplicity we may use $\langle 1, 0, 1\rangle$ as the vector orthogonal to the osculating plane and obtain the equation $x + z = \frac{\pi}{2}$ for the osculating plane.

Similarly, at $t = \frac{5\pi}{4}$ we have $\mathbf{r}\left(\frac{5\pi}{4}\right) = \left\langle -\frac{\sqrt{2}}{2}, -\frac{\sqrt{2}}{2}, \frac{5\pi}{4}\right\rangle$ and

$$\mathbf{B}\left(\frac{5\pi}{4}\right) = \left\langle \frac{1}{2}, -\frac{1}{2}, \frac{\sqrt{2}}{2}\right\rangle \times \left\langle \frac{\sqrt{2}}{2}, \frac{\sqrt{2}}{2}, 0\right\rangle = \left\langle -\frac{1}{2}, \frac{1}{2}, \frac{\sqrt{2}}{2}\right\rangle.$$

Again, for simplicity we may use $\left\langle -1, 1, \sqrt{2}\right\rangle$ as the vector orthogonal to the osculating plane. Therefore, when $t = \frac{5\pi}{4}$, we obtain the equation $-x + y + \sqrt{2}z = \frac{5\pi\sqrt{2}}{4}$ for the osculating plane. □

CHECKING THE ANSWER

There are several things we can check in our answers to Example 2.

- As the parameter t increases, the helix is spiraling upwards. Therefore, each tangent vector should have a positive z-component. Note that $\mathbf{T}(t) = \frac{\langle -\sin t, \cos t, 1\rangle}{\sqrt{2}}$. The z-component is always $\frac{1}{\sqrt{2}}$.
- Each of the answers should be a unit vector. (They are, as you can check.)
- At each of the points, the tangent and normal vectors should be orthogonal. Here we can use the dot product. For example, at $t = \frac{5\pi}{4}$,

$$\mathbf{T}\left(\frac{5\pi}{4}\right) \cdot \mathbf{N}\left(\frac{5\pi}{4}\right) = \left\langle \frac{1}{2}, -\frac{1}{2}, \frac{\sqrt{2}}{2}\right\rangle \cdot \left\langle \frac{\sqrt{2}}{2}, \frac{\sqrt{2}}{2}, 0\right\rangle = 0.$$

? TEST YOUR UNDERSTANDING

- How is the unit tangent vector defined? Why is the unit tangent vector useful?
- How is the principal unit normal vector defined? Why are the unit tangent vector at a point P and the principal unit normal vector at P always orthogonal?
- Why does the principal unit normal vector always point "into" the curve?
- How is the osculating plane defined? What is its geometric relationship to a space curve at a point?
- What is the binormal vector?

EXERCISES 11.3

Thinking Back

- *Unit vectors:* If $\mathbf{v}$ is a nonzero vector, explain why $\dfrac{\mathbf{v}}{\|\mathbf{v}\|}$ is a unit vector.
- *Equation of a plane:* Find the equation of the plane determined by the vectors $\langle 1, -3, 5\rangle$ and $\langle 2, 4, -1\rangle$ and containing the point $(0, 3, -2)$.

Concepts

0. *Problem Zero:* Read the section and make your own summary of the material.

1. *True/False:* Determine whether each of the statements that follow is true or false. If a statement is true, explain why. If a statement is false, provide a counterexample.

(a) *True or False:* For every position vector, $\mathbf{r}(t)$, $\mathbf{r}'(t) \cdot \mathbf{r}''(t) = 0$.

(b) *True or False:* If $\|\mathbf{T}'(t)\| = 0$, then $\mathbf{N}(t)$ is not defined.

(c) *True or False:* The cross product of two unit vectors is another unit vector.

(d) *True or False:* If C is a curve in the xy-plane and the principal unit normal vector, $\mathbf{N}$, is defined at the point P_0 on C, then the xy-plane is the osculating plane for C at P_0.

(e) *True or False:* If a vector function has a unit tangent vector and a principal unit normal vector at a point, then it has a binormal vector at that point also.

(f) *True or False:* If the unit tangent vector, $\mathbf{T}(t)$, the principal unit normal vector, $\mathbf{N}(t)$, and the binormal vector $\mathbf{B}(t)$, are all defined for some value of t, then $\mathbf{B}(t) \times \mathbf{N}(t) = -\mathbf{T}(t)$.

(g) *True or False:* If the unit tangent vector, $\mathbf{T}(t)$, the principal unit normal vector, $\mathbf{N}(t)$, and the binormal vector $\mathbf{B}(t)$, are all defined at some point t, then $\mathbf{T}(t) \cdot (\mathbf{N}(t) \times \mathbf{B}(t)) = 1$.

(h) *True or False:* A space curve can have infinitely many different osculating planes.

2. *Examples:* Construct examples of the thing(s) described in the following. Try to find examples that are different than any in the reading.

(a) A differentiable vector-valued function $\mathbf{r}(t)$ that does not have a unit tangent vector at any point.

(b) A twice-differentiable vector-valued function with a unit tangent vector at every point but that does not have a unit normal vector at any point.

(c) A nonconstant differentiable vector function $\mathbf{r}(t) = \langle x(t), y(t)\rangle$ for which the xy-plane is *not* the osculating plane.

3. Imagine that you are driving on a twisting mountain road. Describe the unit tangent vector, principal unit normal vector, and binormal vector as you ascend, descend, twist right, and twist left.

4. Given a differentiable vector-valued function $\mathbf{r}(t)$, what is the definition of the unit tangent vector $\mathbf{T}(t)$?

5. Under what conditions does a differentiable vector-valued function $\mathbf{r}(t)$ *not* have a unit tangent vector at a point in the domain of $\mathbf{r}(t)$?

6. Let $\mathcal{L}$ be a straight line in $\mathbb{R}^3$. Find vector functions with the following properties:

(a) The graph of $\mathbf{r}_1(t)$ is $\mathcal{L}$, and $\mathbf{r}_1(t)$ has a unit tangent vector for every value of t.

(b) The graph of $\mathbf{r}_2(t)$ is $\mathcal{L}$, and there is a least one value of t at which $\mathbf{r}_2(t)$ does not have a unit tangent vector.

7. Given a differentiable vector-valued function $\mathbf{r}(t)$, what is the relationship between $\mathbf{r}'(t_0)$ and $\mathbf{T}(t_0)$ at a point t_0 in the domain of $\mathbf{r}(t)$?

8. Given a differentiable vector-valued function $\mathbf{r}(t)$, what are the advantages and disadvantages in using $\mathbf{r}'$ or $\mathbf{T}$ to analyze the behavior of $\mathbf{r}(t)$?

9. Given a twice-differentiable vector-valued function $\mathbf{r}(t)$, what is the definition of the principal unit normal vector $\mathbf{N}(t)$?

10. Given a twice-differentiable vector-valued function $\mathbf{r}(t)$, why does the principal unit normal vector $\mathbf{N}(t)$ point into the curve? (*Hint: Use the definition!*)

11. Given a twice-differentiable vector-valued function $\mathbf{r}(t)$ and a point t_0 in its domain, what are the geometric relationships between the unit tangent vector $\mathbf{T}(t_0)$, the principal unit normal vector $\mathbf{N}(t_0)$, and $\left.\dfrac{d\mathbf{N}}{dt}\right|_{t_0}$?

12. Under what conditions does a twice-differentiable vector-valued function $\mathbf{r}(t)$ *not* have a principal unit normal vector at a point in the domain of $\mathbf{r}(t)$?

13. Given a twice-differentiable vector-valued function $\mathbf{r}(t)$, what is the definition of the binormal vector $\mathbf{B}(t)$?

14. We've seen that the graph of a continuous function $y = f(x)$ is the same as the graph of the vector-valued function $\mathbf{r}(t) = \langle t, f(t)\rangle$. What is the direction of the binormal vector $\mathbf{B}(t)$ when the graph of f is concave up? When the graph is concave down? What happens to $\mathbf{B}(t)$ at a point of inflection?

15. Under what conditions does a twice-differentiable vector-valued function $\mathbf{r}(t)$ not have a binormal vector at a point in the domain of $\mathbf{r}(t)$?

16. Given a twice-differentiable vector-valued function $\mathbf{r}(t)$ and a point t_0 in its domain, what is the osculating plane at $\mathbf{r}(t_0)$?

17. Carefully outline the steps you would use to find the equation of the osculating plane at a point $\mathbf{r}(t_0)$ on the graph of a vector function.

18. If the osculating plane is the same at every point on a space curve, what must be true about the curve?

19. Given the three mutually perpendicular vectors $\mathbf{T}(t_0)$, $\mathbf{N}(t_0)$, and $\mathbf{B}(t_0)$ at a point $\mathbf{r}(t_0)$, how many distinct planes contain the point and two of the vectors?

20. Let C be the graph of a vector-valued function $\mathbf{r}$. The plane determined by the vectors $\mathbf{N}(t_0)$ and $\mathbf{B}(t_0)$ and containing the point $\mathbf{r}(t_0)$ is called the ***normal*** plane for C at $\mathbf{r}(t_0)$. Find the equation of the normal plane to the helix determined by $\mathbf{r}(t) = \langle \cos t, \sin t, t\rangle$ for $t = \pi$.

21. Let C be the graph of a vector-valued function $\mathbf{r}$. The plane determined by the vectors $\mathbf{T}(t_0)$ and $\mathbf{B}(t_0)$ and containing the point $\mathbf{r}(t_0)$ is called the ***rectifying*** plane for C at $\mathbf{r}(t_0)$. Find the equation of the rectifying plane to the helix determined by $\mathbf{r}(t) = \langle \cos t, \sin t, t\rangle$ when $t = \pi$.

Skills

For each of the vector-valued functions in Exercises 22–28, find the unit tangent vector.

22. $\mathbf{r}(t) = \langle t, t^2\rangle$

23. $\mathbf{r}(t) = \langle t^2 + 5, 5t, 4t^3\rangle$

24. $\mathbf{r}(t) = \langle \cos \alpha t, \sin \alpha t\rangle$

25. $\mathbf{r}(t) = \langle \cos^3 t, \sin^3 t\rangle$

26. $\mathbf{r}(t) = \langle t, t^2, t^3\rangle$

27. $\mathbf{r}(t) = \langle 3\sin t, 5\cos t, 4\sin t\rangle$

28. $\mathbf{r}(t) = \langle \sin 2t, \cos 2t, t\rangle$

For each of the vector-valued functions in Exercises 29–34, find the unit tangent vector and the principal unit normal vector at the specified value of t.

29. $\mathbf{r}(t) = \langle t, t^2\rangle,\ t = 1$

30. $\mathbf{r}(t) = \langle a \sinh t, b \cosh t\rangle$, where a and b are positive, $t = 0$

31. $\mathbf{r}(t) = \langle \cos \alpha t, \sin \alpha t\rangle$, where $\alpha > 0,\ t = \pi$

32. $\mathbf{r}(t) = \langle \cos^3 t, \sin^3 t\rangle,\ t = \dfrac{\pi}{4}$

33. $\mathbf{r}(t) = \langle 3\sin t, 5\cos t, 4\sin t\rangle,\ t = \pi$

34. $\mathbf{r}(t) = \langle \sin 2t, \cos 2t, t\rangle,\ t = \dfrac{\pi}{4}$

For each of the vector-valued functions in Exercises 35–39, find the unit tangent vector, the principal unit normal vector, the binormal vector, and the equation of the osculating plane at the specified value of t.

35. $\mathbf{r}(t) = \left\langle t, t^2, \dfrac{2}{3}t^3\right\rangle$ at $t = 1$

36. $\mathbf{r}(t) = \left\langle \dfrac{t}{\sqrt{2}}, \sqrt{t}, \dfrac{1}{3}t^{3/2}\right\rangle$ at $t = 1$

37. $\mathbf{r}(t) = \langle e^t, e^{-t}, \sqrt{2}t\rangle$ at $t = 0$

38. $\mathbf{r}(t) = \langle 3\sin t, 5\cos t, 4\sin t\rangle$ at $t = \dfrac{\pi}{4}$

39. $\mathbf{r}(t) = \langle \sin 2t, \cos 2t, t\rangle$ at $t = \dfrac{\pi}{2}$

Using the definitions of the normal plane and rectifying plane in Exercises 20 and 21, respectively, find the equations of these planes at the specified points for the vector functions in Exercises 40–42. Note: These are the same functions as in Exercises 35, 37, and 39.

40. $\mathbf{r}(t) = \left\langle t, t^2, \dfrac{2}{3}t^3\right\rangle$ at $t = 1$

41. $\mathbf{r}(t) = \langle e^t, e^{-t}, \sqrt{2}t\rangle$ at $t = 0$

42. $\mathbf{r}(t) = \langle \sin 2t, \cos 2t, t\rangle$ at $t = \dfrac{\pi}{2}$

Applications

43. Annie is kayaking with her friend Jim. Jim is inexperienced, and soon his boat is upside down. Being inexperienced, he also fails to execute a roll, so he has to swim. Since they are both in single kayaks, they cannot get him back in the boat, so Jim grabs his kayak and Annie's kayak, and Annie tows the whole lot towards shore. There is a current, so if they did not paddle, they would follow a path defined by the vector function $\mathbf{r}(t) = \langle t^{1.5}, t(1.7 - t)\rangle$, where t represents time in hours.
   (a) For $0 \le t \le 1.7$, where would they move fastest if they did not paddle at all?
   (b) Their best strategy to get to shore is to paddle exactly orthogonally to the current, towards the land that is on the inside of the path of the current. If they get things in order and start paddling at $t = 0.1$, in what direction are they paddling?

44. Ian is planning a return from a mountain in the Cascades. He will descend to around 6000 feet and then follow a contour, staying high and out of the trees to make travel easy, until he is at a point directly above a trail. He will then descend directly to the trail. The contour at 6000 feet follows the curve $\mathbf{r}(x) = \langle x, 0.45x^2 + x\rangle$. Ian uses his map to determine that he will be at the point where he needs to descend to the trail when he can first see a mountain peak that is on a compass heading of 30 degrees; that is, when the tangent to the level curve is 30 degrees east of north.
   (a) Where does that happen? What is the unit tangent vector at that point?
   (b) When he reaches that point, he will descend directly down the slope, orthogonally to the contour. Downhill is generally north. What compass heading will that be? What is the unit normal vector there?

Proofs

45. Prove that the cross product of two orthogonal unit vectors is a unit vector.

46. Let $y = f(x)$ be a function with domain $\mathbb{R}$ and that is differentiable at every point in its domain, and let $\mathbf{r}(t) = \langle t, f(t) \rangle$.

(a) Explain why the graph of $\mathbf{r}$ is the same as the graph of $y = f(x)$.

(b) Prove that the unit tangent vector at $\mathbf{r}(\alpha)$ is parallel to the tangent line to f at α.

(c) Prove that the principal unit normal vector for $\mathbf{r}$ is undefined at every inflection point of f.

47. Let $\mathbf{r}(t) = \langle x(t), y(t), z(t) \rangle$ be a differentiable vector function on some interval $I \subseteq \mathbb{R}$ such that the derivative of the unit tangent vector $\mathbf{T}'(t_0) \neq 0$, where $t_0 \in I$. Prove that the binormal vector

$$\mathbf{B}(t_0) = \mathbf{T}(t_0) \times \mathbf{N}(t_0)$$

(a) is a unit vector;

(b) is orthogonal to both $\mathbf{T}(t_0)$ and $\mathbf{N}(t_0)$.

Also, prove that $\mathbf{T}(t_0)$, $\mathbf{N}(t_0)$, and $\mathbf{B}(t_0)$ form a right-handed coordinate system.

Thinking Forward

- *Equation of a circle:* Find an equation for the circle of radius ρ whose center is ρ units from the point (a, b) in the direction of the vector $\langle \alpha, \beta \rangle$.
- *Equation of a sphere:* Find an equation for the sphere of radius ρ whose center is ρ units from the point (a, b, c) in the direction of the vector $\langle \alpha, \beta, \gamma \rangle$.

11.4 ARC LENGTH PARAMETRIZATIONS AND CURVATURE

- The arc length of a curve in $\mathbb{R}^3$
- Arc length parametrizations
- The curvature of planar curves and space curves

The Arc Length of Space Curves

We begin this section with a reminder of some of the work we did in Section 9.1. Let $x = x(t)$ and $y = y(t)$ be differentiable functions of t. In Definition 9.3 we said that the arc length of the parametric curve determined by x and y on an interval $[a, b]$ was

$$\lim_{n \to \infty} \sum_{k=1}^{n} \sqrt{(x(t_k) - x(t_{k-1}))^2 + (y(t_k) - y(t_{k-1}))^2},$$

where $\Delta t = \frac{b-a}{n}$ and $t_k = a + k\Delta t$. Then, in Theorem 9.4 we argued that the arc length could be evaluated with the use of $\int_a^b \sqrt{(x'(t))^2 + (y'(t))^2}\, dt$.

We now express Definition 9.3 and Theorem 9.4 in the language of vector functions. The vector function $\mathbf{r}(t) = x(t)\mathbf{i} + y(t)\mathbf{j}$ has the same graph as the parametric equations. Definition 9.3 would say that the arc length of the vector curve on the interval $[a, b]$ would be $\lim_{n \to \infty} \sum_{k=1}^{n} \|\mathbf{r}(t_k) - \mathbf{r}(t_{k-1})\|$, where $\Delta t = \frac{b-a}{n}$ and $t_k = a + k\Delta t$. Similarly, $\int_a^b \|\mathbf{r}'(t)\|\, dt$ represents the length of the vector curve for $t \in [a, b]$. These formulas are valid, since

$$\|\mathbf{r}(t_k) - \mathbf{r}(t_{k-1})\| = \sqrt{(x(t_k) - x(t_{k-1}))^2 + (y(t_k) - y(t_{k-1}))^2} \text{ and}$$

$$\|\mathbf{r}'(t)\| = \sqrt{(x'(t))^2 + (y'(t))^2},$$

by the Pythagorean theorem and the definition of the norm of a vector, respectively.

The great thing about these forms is that they carry over to space curves (or even to curves in higher dimensions)! Thus, we have the following definition:

DEFINITION 11.17 **The Arc Length of a Vector Curve**

Let C be a curve in $\mathbb{R}^3$ given by the vector function $\mathbf{r}(t) = \langle x(t), y(t), z(t)\rangle$ for $t \in [a, b]$, where x, y, and z are differentiable functions of t and such that the function is one-to-one from the interval $[a, b]$ to the curve C. The ***length of the curve*** C from a to b, denoted by $l(a, b)$, is

$$l(a, b) = \lim_{n\to\infty} \sum_{k=1}^{n} \|\mathbf{r}(t_k) - \mathbf{r}(t_{k-1})\|,$$

where $\Delta t = \frac{b-a}{n}$ and $t_k = a + k\Delta t$.

The following figure illustrates the idea behind the definition of arc length:

The arc length of a vector curve may be approximated by using segments

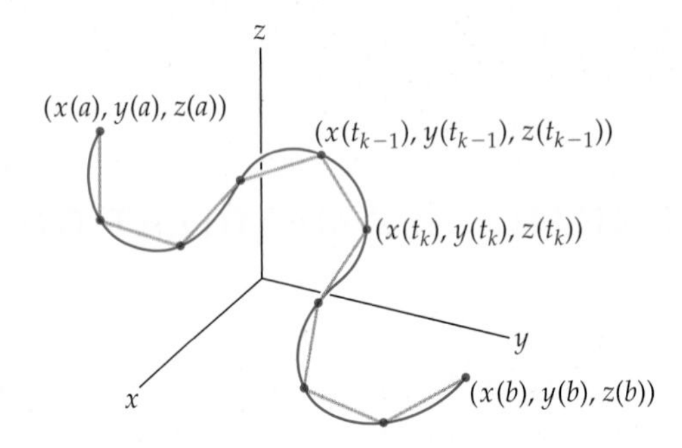

We also have the following theorem:

THEOREM 11.18 **The Arc Length of a Vector Curve**

Let C be a curve in $\mathbb{R}^3$ given by the vector function $\mathbf{r}(t) = \langle x(t), y(t), z(t)\rangle$ such that the function is one-to-one from the interval $[a, b]$ to the curve C. If x, y, and z are differentiable functions of t such that x', y', and z' are continuous on $[a, b]$, then the length of the curve C from $\mathbf{r}(a)$ to $\mathbf{r}(b)$ is given by

$$l(a, b) = \int_a^b \|\mathbf{r}'(t)\| dt.$$

A proof of Theorem 11.18 may be adapted from the proof of Theorem 9.4 in Section 9.1 and is left to you.

Arc Length Parametrizations

We have already discussed the geometry of the relationship between a differentiable vector function and its derivative. The derivative gives us information about the rate and direction of motion of a particle moving along a vector curve. Recall that for a real-valued function of the form $y = f(x)$, the second derivative gives us more subtle information about the shape of the curve and the sign of f'' tells us whether a curve is concave up or concave down. Space curves are even more complicated than planar curves. A single number cannot tell us everything about how a space curve is bending. However, we do want to know about the "curvature" of a space curve. We will get to that definition later in this section.

To help us define curvature in a relatively easy way, we need to settle on a particular type of parametrization for a given vector curve. You should have already realized that a given vector curve has many different parametrizations. For example, in Chapter 9 we examined several parametrizations for the unit circle. In fact, every vector curve has infinitely many different parametrizations. For example, for every nonzero value of k, the vector function $\mathbf{r}(t) = \langle \cos kt, \sin kt \rangle$ for $t \geq 0$ provides a different parametrization of the unit circle. The differences between two of these parametrizations are the rate at which the particle traces the circle and the direction of motion. For every value of k, the particle starts at the point $(1, 0)$ when $t = 0$. The motion is clockwise when $k < 0$ and counterclockwise when $k > 0$. The rate at which the particle moves increases linearly with k. For example, when $k = 1$, the particle traces the unit circle once on the interval $[0, 2\pi)$, but if $k = 5$, the particles traces the circle 5 times on the same interval.

Similarly, every vector curve in $\mathbb{R}^3$ has infinitely many parametrizations. The particular type of parametrization that we need for our upcoming definition of curvature is called an "arc length parametrization."

DEFINITION 11.19 **Arc Length Parametrization**

Let C be the graph of a differentiable vector function $\mathbf{r}(t)$ defined on an interval I. The function $\mathbf{r}(t)$ is said to be an ***arc length parametrization*** for C if

$$\int_c^d \|\mathbf{r}'(t)\|\, dt = d - c$$

for every c and $d \in I$. Also, C is said to be ***parametrized by arc length.***

When a curve is parametrized by arc length, the length of every segment of the curve is equal to the change in the value of the parameter. For example, $\mathbf{r}_1(t) = \langle \cos t, \sin t \rangle$ and $\mathbf{r}_2(t) = \langle \cos 2t, \sin 2t \rangle$ are both parametrizations of the unit circle. The function $\mathbf{r}_1(t)$ is an arc length parametrization of the unit circle, since

$$\int_c^d \|\mathbf{r}_1'(t)\| dt = \int_c^d \|\langle -\sin t, \cos t \rangle\| dt = \int_c^d \sqrt{\sin^2 t + \cos^2 t}\, dt = \int_c^d dt = d - c.$$

However, the function $\mathbf{r}_2(t)$ is not an arc length parametrization of the unit circle, since

$$\begin{aligned}\int_c^d \|\mathbf{r}_2'(t)\|\, dt &= \int_c^d \|\langle -2\sin 2t, 2\cos 2t \rangle\| dt \\ &= \int_c^d \sqrt{4\sin^2 2t + 4\cos^2 2t}\, dt = \int_c^d 2\, dt = 2(d - c).\end{aligned}$$

In the next theorem we will see that when we use an arc length parametrization, the derivative of $\mathbf{r}$ with respect to arc length is particularly simple.

THEOREM 11.20 **The Derivative of a Vector Function with Respect to Arc Length**

Let C be a curve in $\mathbb{R}^3$ given by the differentiable vector function $\mathbf{r}(t) = \langle x(t), y(t), z(t) \rangle$ such that $\mathbf{r}(t)$ is one-to-one from the interval $[a, b]$ to the curve C and has a nonzero derivative. If we define the real-valued function $s(t)$ by

$$s(t) = \int_a^t \|\mathbf{r}'(\tau)\| d\tau,$$

then

$$\frac{d\mathbf{r}}{ds} = \mathbf{T}(t)$$

is the unit tangent vector function for C.

Proof. We first note that

$$\int_a^t \|\mathbf{r}'(\tau)\| d\tau = l(a, t) = s(t).$$

That is, $s(t)$ represents the arc length of C between the terminal points of $\mathbf{r}(a)$ and $\mathbf{r}(t)$.

Now, by the Fundamental Theorem of Calculus

$$s'(t) = \frac{ds}{dt} = \|\mathbf{r}'(t)\|,$$

and by the chain rule,

$$\frac{d\mathbf{r}}{dt} = \frac{d\mathbf{r}}{ds}\frac{ds}{dt}.$$

Solving for $\frac{d\mathbf{r}}{ds}$ and using the fact that $\frac{ds}{dt} = \|\mathbf{r}'(t)\|$, we have

$$\frac{d\mathbf{r}}{ds} = \frac{\mathbf{r}'(t)}{\|\mathbf{r}'(t)\|}.$$

We see that $\frac{d\mathbf{r}}{ds}$ has the same direction as $\mathbf{r}'(t)$ and remind you that any nonzero vector divided by its magnitude is a unit vector. Thus, $\frac{d\mathbf{r}}{ds} = \mathbf{T}(t)$, the unit tangent vector to C. ■

With most parametrizations of a curve, both the direction *and* magnitude of the tangent vectors change as a particle traverses the curve. However, Theorem 11.20 tells us that when a curve is parametrized by arc length, since the derivative is a unit vector, only the directions of the tangent vectors are changing. This change in the unit tangent vectors is the basis for the following definition:

DEFINITION 11.21 Curvature

Let C be the graph of a vector function $\mathbf{r}(s)$ defined on an interval I, parametrized by arc length s, and with unit tangent vector $\mathbf{T}$. The ***curvature*** κ of C at a point on the curve is the scalar given by

$$\kappa = \left\|\frac{d\mathbf{T}}{ds}\right\|.$$

The magnitude of the rate of change of the unit tangent vector defines curvature

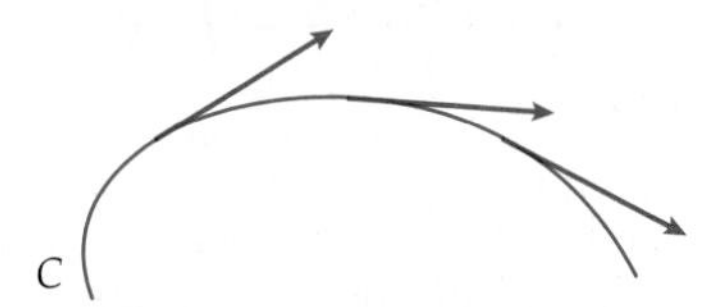

The following theorem tells us the relationship between the derivative of the unit tangent vector, curvature, and the principal unit normal vector:

THEOREM 11.22 The Relationship Between $\frac{d\mathbf{T}}{ds}$, κ, and $\mathbf{N}$

Let C be the graph of a vector function $\mathbf{r}(s)$ defined on an interval I and parametrized by arc length s. If the unit tangent vector $\mathbf{T}$ has a nonzero derivative, then

$$\frac{d\mathbf{T}}{ds} = \kappa\mathbf{N}.$$

The proof of Theorem 11.22 follows from the facts that the vector functions $\frac{d\mathbf{T}}{ds}$, $\frac{d\mathbf{T}}{dt}$, and $\mathbf{N}$ are scalar multiples of each other and that $\|\mathbf{N}\| = 1$. The details are left for Exercise 64.

Although Definition 11.21 provides an elegant way to express and understand curvature, it is usually impractical in computations, because, unfortunately, arc length parametrizations are frequently difficult to construct.

The next two theorems provide more computationally efficient alternatives for computing the curvature.

THEOREM 11.23 **Formulas for the Curvature of a Space Curve**

Let C be the graph of a twice-differentiable vector function $\mathbf{r}(t)$ defined on an interval I with unit tangent vector $\mathbf{T}(t)$. Then the curvature κ of C at a point on the curve is given by

(a) $\kappa = \dfrac{\|\mathbf{T}'(t)\|}{\|\mathbf{r}'(t)\|}$.

(b) $\kappa = \dfrac{\|\mathbf{r}'(t) \times \mathbf{r}''(t)\|}{\|\mathbf{r}'(t)\|^3}$.

Proof. Let s be an arc length parameter for C.

(a) By the chain rule,

$$\kappa = \left\|\frac{d\mathbf{T}}{ds}\right\| = \left\|\frac{d\mathbf{T}/dt}{ds/dt}\right\| = \frac{\|\mathbf{T}'(t)\|}{\|\mathbf{r}'(t)\|}.$$

(b) By the chain rule and Theorem 11.20,

$$\mathbf{r}'(t) = \frac{d\mathbf{r}}{dt} = \frac{d\mathbf{r}}{ds}\frac{ds}{dt} = \mathbf{T}\frac{ds}{dt}.$$

Differentiating with respect to t by the product rule, we obtain

$$\mathbf{r}''(t) = \frac{d\mathbf{T}}{dt}\frac{ds}{dt} + \mathbf{T}\frac{d^2s}{dt^2}.$$

Now, first by the chain rule and then by Theorem 11.22, we have

$$\mathbf{r}''(t) = \frac{d\mathbf{T}}{ds}\frac{ds}{dt}\frac{ds}{dt} + \mathbf{T}\frac{d^2s}{dt^2} = \kappa\mathbf{N}\left(\frac{ds}{dt}\right)^2 + \mathbf{T}\frac{d^2s}{dt^2}.$$

Using the preceding equations to form the cross product $\mathbf{r}'(t) \times \mathbf{r}''(t)$, we obtain

$$\begin{aligned}\mathbf{r}'(t) \times \mathbf{r}''(t) &= \mathbf{T}\frac{ds}{dt} \times \left(\kappa\mathbf{N}\left(\frac{ds}{dt}\right)^2 + \mathbf{T}\frac{d^2s}{dt^2}\right)\\ &= \kappa\,(\mathbf{T} \times \mathbf{N})\left(\frac{ds}{dt}\right)^3 + (\mathbf{T} \times \mathbf{T})\,\frac{ds}{dt}\frac{d^2s}{dt^2}\\ &= \kappa\mathbf{B}\left(\frac{ds}{dt}\right)^3 = \kappa\mathbf{B}\|\mathbf{r}'(t)\|^3,\end{aligned}$$

since $\mathbf{T} \times \mathbf{N} = \mathbf{B}$ and the cross product of any three-component vector with itself is $\mathbf{0}$. Finally, if we take the norms of the first and last parts in the preceding chain of equalities and divide by $\|\mathbf{r}'(t)\|^3$, we have our result. ■

Even with Theorem 11.23, the computation of the curvature can be quite laborious. We provide a complete calculation in Example 3.

It is also convenient to have formulas for the curvature of planar curves. The following theorem gives us a simpler method for computing the curvature of a function of the form $y = f(x)$ and for finding the curvature of a curve defined by a vector function with two components:

THEOREM 11.24 **Formulas for Curvature in the Plane**

(a) Let $y = f(x)$ be a twice-differentiable function. Then the curvature of the graph of f is given by

$$\kappa = \frac{|f''(x)|}{(1 + (f'(x))^2)^{3/2}}.$$

(b) Let C be the graph of a vector function $\mathbf{r}(t) = \langle x(t), y(t) \rangle$ in the xy-plane, where x and y are twice-differentiable functions of t such that $x'(t)$ and $y'(t)$ are not simultaneously zero. Then the curvature κ of C at a point on the curve is given by

$$\kappa = \frac{|x'(t)y''(t) - x''(t)y'(t)|}{((x'(t))^2 + (y'(t))^2)^{3/2}}.$$

We ask you to prove Theorem 11.24 in Exercises 65 and 66.

We make two final definitions to aid our understanding of curves. For a vector curve C with a positive curvature at a point, we define the radius of curvature and osculating circle.

DEFINITION 11.25 **Radius of Curvature and Osculating Circle**

Let C be the graph of a vector function $\mathbf{r}$, and let $\mathbf{r}(t_0)$ be the position vector for a point on C at which the curvature $\kappa > 0$.

(a) The ***radius of curvature*** ρ of C at $\mathbf{r}(t_0)$ is given by

$$\rho = \frac{1}{\kappa}.$$

(b) The ***osculating circle*** to the curve C at $\mathbf{r}(t_0)$ is the circle in the osculating plane, with radius ρ, and whose center is the terminal point of the position vector $\mathbf{r}(t_0) + \rho\mathbf{N}(t_0)$.

As we mentioned in Section 11.3, when a curve C is planar, the osculating plane *is* the plane. In this case, the osculating circle also lies in the plane. The figure appearing next at the left shows a portion of the sine curve and the osculating circles when $x = \frac{\pi}{2}$ and when $x = \frac{5\pi}{4}$. In Example 2 we discuss how to find the equations of these circles. The figure at the right is a space curve, also with two osculating circles.

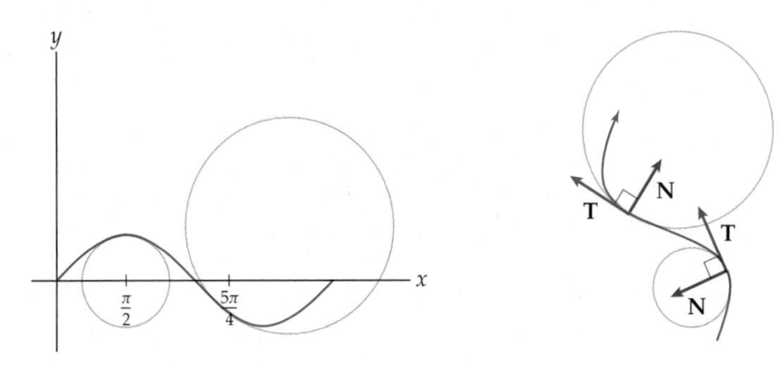

In two or three dimensions, an osculating circle is the circle that fits the shape of the curve "the best," in that the curvatures of the curve and the osculating circle are the same at the point of tangency.

Examples and Explorations

EXAMPLE 1 **Finding the arc length of a space curve**

Find the arc length of the helix defined by $\mathbf{r}(t) = \langle \cos \alpha t, \sin \alpha t, \beta t \rangle$ on the interval $[a, b]$, where a, b, α, and β are constants.

SOLUTION

The integral

$$\int_a^b \|\mathbf{r}'(t)\| \, dt = \int_a^b \|\langle -\alpha \sin \alpha t, \alpha \cos \alpha t, \beta \rangle\| \, dt$$

represents the specified arc length and equals

$$\int_a^b \sqrt{\alpha^2 \sin^2 \alpha t + \alpha^2 \cos^2 \alpha t + \beta^2} \, dt = \int_a^b \sqrt{\alpha^2 + \beta^2} \, dt = (b - a)\sqrt{\alpha^2 + \beta^2}.$$

□

EXAMPLE 2 **Finding osculating circles in the plane**

Find the equations of the osculating circles to the graph of the sine function $y = \sin x$ at $x = \frac{\pi}{2}$ and at $x = \frac{5\pi}{4}$.

SOLUTION

This is a continuation of Example 1 in Section 11.3. In that example, we found the unit tangent and principal unit normal vectors to the curve at $x = \frac{\pi}{2}$ and at $x = \frac{5\pi}{4}$. Next, we need the curvature. Using Theorem 11.24(a), we have

$$\kappa = \frac{|f''(x)|}{(1 + (f'(x))^2)^{3/2}} = \frac{|-\sin x|}{(1 + \cos^2 x)^{3/2}}.$$

At $x = \frac{\pi}{2}$, we have $\kappa = 1$. The radius of curvature $\rho = \frac{1}{\kappa} = 1$. Therefore, the center of the osculating circle will be one unit from the point $\left(\frac{\pi}{2}, \sin\left(\frac{\pi}{2}\right)\right) = \left(\frac{\pi}{2}, 1\right)$ in the direction of the principal unit normal vector $\mathbf{N}\left(\frac{\pi}{2}\right) = \langle 0, -1 \rangle$. Thus, the osculating circle will have its center at $\left(\frac{\pi}{2}, 0\right)$ and have radius 1. The equation of this circle is $\left(x - \frac{\pi}{2}\right)^2 + y^2 = 1$.

At $x = \frac{5\pi}{4}$,

$$\kappa = \frac{|-\sin(5\pi/4)|}{\left(1 + \cos^2(5\pi/4)\right)^{3/2}} = \frac{\sqrt{2}/2}{(3/2)^{3/2}} = \frac{2}{3\sqrt{3}}.$$

The radius of curvature is the reciprocal of κ, so $\rho = \frac{3\sqrt{3}}{2}$. The principal unit normal vector $\mathbf{N}\left(\frac{5\pi}{4}\right) = \frac{1}{\sqrt{3}}\langle 1, \sqrt{2} \rangle$. The osculating circle will have its center at the terminal point of the position vector $\left\langle \frac{5\pi}{4}, \sin\left(\frac{5\pi}{4}\right)\right\rangle + \rho\mathbf{N}\left(\frac{5\pi}{4}\right) = \left\langle \frac{5\pi}{4} + \frac{3}{2}, \sqrt{2} \right\rangle$ and have radius $\frac{3\sqrt{3}}{2}$. The equation of the circle is $\left(x - \frac{5\pi}{4} - \frac{3}{2}\right)^2 + (y - \sqrt{2})^2 = \frac{27}{4}$. The graph of the function, along with these two osculating circles, appears on the left preceding Example 1. □

EXAMPLE 3 **Finding the center and radius of an osculating circle to a space curve**

Find the center and radius of the osculating circle to the graph of the helix defined by the vector function $\mathbf{r}(t) = \langle \cos t, \sin t, t \rangle$ at $t = \frac{\pi}{2}$ and at $t = \frac{5\pi}{4}$.

SOLUTION

This is also a continuation of an example from Section 11.3. In Example 2 of that section, we found the unit tangent and principal unit normal vectors to the helix at $t = \frac{\pi}{2}$ and at $t = \frac{5\pi}{4}$. Now we need the curvature. Using Theorem 11.23(b), we have

$$\kappa = \frac{\|\mathbf{r}'(t) \times \mathbf{r}''(t)\|}{\|\mathbf{r}'(t)\|^3} = \frac{\|\langle -\sin t, \cos t, 1\rangle \times \langle -\cos t, -\sin t, 0\rangle\|}{\|\langle -\sin t, \cos t, 1\rangle\|^3} = \frac{\|\langle \sin t, -\cos t, 1\rangle\|}{\sqrt{2}^3} = \frac{1}{2}.$$

The curvature at every point on the helix is $\frac{1}{2}$. (Why does it make sense that the curvature on this helix is constant?) The radius of curvature is constantly 2. Finally, we only need the center points for the two circles. At $t = \frac{\pi}{2}$, the center of the osculating circle will be

$$\mathbf{r}\left(\frac{\pi}{2}\right) + 2\mathbf{N}\left(\frac{\pi}{2}\right) = \left\langle 0, 1, \frac{\pi}{2}\right\rangle + 2\langle 0, -1, 0\rangle = \left\langle 0, -1, \frac{\pi}{2}\right\rangle.$$

The osculating circle will therefore have its center at $\left(0, -1, \frac{\pi}{2}\right)$ and have radius 2.

At $t = \frac{5\pi}{4}$, the center of the osculating circle will be

$$\mathbf{r}\left(\frac{5\pi}{4}\right) + 2\mathbf{N}\left(\frac{5\pi}{4}\right) = \left\langle -\frac{\sqrt{2}}{2}, -\frac{\sqrt{2}}{2}, \frac{5\pi}{4}\right\rangle + 2\left\langle \frac{\sqrt{2}}{2}, \frac{\sqrt{2}}{2}, 0\right\rangle = \left\langle \frac{\sqrt{2}}{2}, \frac{\sqrt{2}}{2}, \frac{5\pi}{4}\right\rangle.$$

The osculating circle will therefore have its center at $\left(\frac{\sqrt{2}}{2}, \frac{\sqrt{2}}{2}, \frac{5\pi}{4}\right)$ and have radius 2. □

? TEST YOUR UNDERSTANDING

- How is the arc length of a vector-valued function defined? Why is this definition consistent with the definitions of arc length you learned for differentiable functions of the form $y = f(x)$ and for parametric equations $x = x(t)$ and $y = y(t)$?
- What does it mean for a vector function to be parametrized by arc length?
- How is curvature defined? Why is the definition of curvature difficult to use?
- For a twice-differentiable function $y = f(x)$, how is the curvature at a point x_0 related to the concavity of the graph at x_0?
- What is the radius of curvature at a point P on a curve? How is the osculating circle at P defined?

EXERCISES 11.4

Thinking Back

- *The arc length of the sine function:* Find the arc length of the sine function $y = e^{x/2} + e^{-x/2}$ for $0 \le x \le 1$.
- *The arc length of the cycloid:* Find the arc length of the cycloid $x = \theta - \sin\theta,\ y = 1 - \cos\theta$, for $0 \le \theta \le 2\pi$.

Concepts

0. *Problem Zero:* Read the section and make your own summary of the material.

1. *True/False:* Determine whether each of the statements that follow is true or false. If a statement is true, explain why. If a statement is false, provide a counterexample.

(a) *True or False:* For every differentiable vector-valued function $\mathbf{r}$, $\int_a^b \|\mathbf{r}'(t)\|\, dt \ge (b - a)$.

(b) *True or False:* Every vector-valued function has an arc length parametrization.

(c) *True or False:* Every straight line has constant curvature.

(d) *True or False:* The radius of curvature is the reciprocal of the curvature.

(e) *True or False:* The radius of curvature is zero at every point on a straight line.

(f) *True or False:* If a vector-valued function $\mathbf{r}$ is differentiable for some value of t, its curvature at that point is positive.

(g) *True or False:* If a vector-valued function $\mathbf{r}$ is twice differentiable at some value t_0, then the osculating circle is defined at $\mathbf{r}(t_0)$.

(h) *True or False:* If a vector-valued function $\mathbf{r}$ has a principal unit normal vector for some value t_0, then the osculating circle at $\mathbf{r}(t_0)$ is defined.

2. *Examples:* Construct examples of the thing(s) described in the following. Try to find examples that are different than any in the reading.
   (a) An arc length parametrization for a straight line in the plane.
   (b) A space curve with constant curvature.
   (c) A space curve with varying curvature.
3. Define what it means for a curve to be parametrized by arc length.
4. Explain how a ruler made of string could be used to understand an arc length parametrization.
5. Why are the concepts of "concave up" and "concave down" insufficient for quantifying how a space curve bends?
6. Let C be the graph of the vector-valued function $\mathbf{r}(t)$. Define the curvature at a point on C.
7. What makes Definition 11.21 for curvature easy to understand? What makes it difficult to use?
8. Let P_0 be a point on a curve C with positive curvature κ. Define the radius of curvature at P_0.
9. Compare the definition of curvature with the four formulas given in Theorems 11.23 and 11.24. Discuss when each formula would be the easiest to use.
10. Sam and Ben are arguing about the curvature of a function $y = f(x)$. Sam claims that the maximum curvature of every such function occurs at a relative extremum. Ben disagrees. Who is right? If Sam is right, prove the result. If Ben is right, provide a counterexample that disproves Sam's claim.
11. Recall that the graph of the vector function $\mathbf{r}(t) = \langle a + \alpha t, b + \beta t, c + \gamma t\rangle$ is a straight line in $\mathbb{R}^3$ if constants α, β, and γ are not all zero.
    (a) Show that the curvature at every point on this line is zero and the radius of curvature at every point is undefined.
    (b) Explain why what you showed in part (a) makes sense.
12. Let $\alpha > 0$.
    (a) Explain why the graph of the vector function $\mathbf{r}(t) = \langle a + \alpha\cos t, b + \alpha\sin t\rangle$ is a circle C with center (a, b) and radius α.
    (b) Show that the curvature at every point on C is $\dfrac{1}{\alpha}$ and the radius of curvature at every point is α.
    (c) Explain why the osculating circle at every point on C *is* the circle C.
    (d) Explain why the answers to parts (a), (b), and (c) make sense.
13. Give a step-by-step procedure for finding the equation of the osculating circle at a point on a planar curve where the curvature is nonzero.
14. Give a step-by-step procedure for finding the center and radius of the osculating circle at a point on a space curve where the curvature is nonzero.
15. Show that the second derivative of the function of $y = x^2$ is constant, but its curvature varies with x.
16. Show that the curvature of the function of $y = \sqrt{1 - x^2}$, $x \in (-1, 1)$, is constant, but its second derivative varies with x.

In Exercises 17–21 sketch the graph of a vector-valued function $\mathbf{r}(t) = x(t)\mathbf{i} + y(t)\mathbf{j}$ with the specified properties.

17. $\mathbf{r}(1) = \langle 1, 2\rangle,\ \kappa(1) = 3,\ \mathbf{N}(1) = \left\langle \frac{3}{5}, \frac{4}{5}\right\rangle$
    $\mathbf{r}(3) = \langle -3, 5\rangle,\ \kappa(3) = 3,\ \mathbf{N}(3) = \left\langle \frac{1}{\sqrt{2}}, \frac{1}{\sqrt{2}}\right\rangle$
18. $\mathbf{r}(0) = \langle 0, 0\rangle,\ \kappa(0) = 0,\ \mathbf{N}(0)$ is undefined
    $\mathbf{r}(2) = \langle 3, 7\rangle,\ \kappa(2) = 1,\ \mathbf{N}(2) = \langle 1, 0\rangle$
19. $\mathbf{r}(1) = \langle 3, 5\rangle,\ \rho(1) = 3,\ \mathbf{N}(1) = \langle 1, 0\rangle$
    $\mathbf{r}(4) = \langle -2, 4\rangle,\ \rho(4) = 2,\ \mathbf{N}(4) = \left\langle \frac{\sqrt{3}}{3}, \frac{\sqrt{6}}{3}\right\rangle$
20. κ is always 0, $\mathbf{r}(0) = \langle 2, 0\rangle,\ \mathbf{r}(1) = \langle 0, 2\rangle$
21. κ is always $\frac{1}{2}$, $\mathbf{r}(0) = \langle 2, 0\rangle,\ \mathbf{r}(1) = \langle 0, 2\rangle$

Skills

Find the arc length of the curves defined by the vector-valued functions on the specified intervals in Exercises 22–27.

22. $\mathbf{r}(t) = \langle 3t - 4, -2t + 5, t + 3\rangle,\ [1, 5]$
23. $\mathbf{r}(t) = \langle 3\cos 4t, 3\sin 4t\rangle,\ \left[0, \frac{\pi}{2}\right]$
24. $\mathbf{r}(t) = \langle t - \sin t, 1 - \cos t\rangle,\ [0, 2\pi]$
25. $\mathbf{r}(t) = \left\langle 4\sin t, t^{3/2}, -4\cos t\right\rangle,\ [0, 4]$
26. $\mathbf{r}(t) = \left\langle e^t, \sqrt{2}t, e^{-t}\right\rangle,\ [0, 1]$
27. $\mathbf{r}(t) = \left\langle e^t\sin t, e^t\cos t, e^t\right\rangle,\ [0, \pi]$

In Exercises 28–30 show that the given vector-valued functions are not arc length parametrizations. Then find an arc length parametrization for the curves defined by those functions.

28. $\mathbf{r}(t) = \langle 3\cos t, 3\sin t\rangle$
29. $\mathbf{r}(t) = \langle 3 + 2t, 4 - t, -1 + 5t\rangle$
30. $\mathbf{r}(t) = \langle \cos t, \sin t, t\rangle$

In Exercises 31–35 find the curvature of the given function at the indicated value of x. Then sketch the curve and the osculating circle at the indicated point.

31. $y = e^x,\ x = 0$
32. $y = \sin x,\ x = \frac{\pi}{2}$
33. $y = \csc x,\ x = \frac{\pi}{2}$

34. $y = \sqrt{x}$, $x = 2$

35. $y = \frac{1}{3}(x^2 + 2)^{3/2}$, $x = 1$

Find the curvature of each of the functions defined by the parametric equations in Exercises 36–38.

36. $x = k\cos t$, $y = k\sin t$

37. $x = a\cos t$, $y = b\sin t$

38. $x = \cos t + t\sin t$, $y = \sin t - t\cos t$

Find the curvature of each of the vector-valued functions defined in Exercises 39–44.

39. $\mathbf{r}(t) = \langle t - \sin t, 1 - \cos t\rangle$

40. $\mathbf{r}(t) = \langle t, t^2, t^3\rangle$

41. $\mathbf{r}(t) = \langle t\sin t, t\cos t, t\rangle$

42. $\mathbf{r}(t) = \langle t\sin t, t\cos t, 2t\rangle$

43. $\mathbf{r}(t) = \langle t\sin t, t\cos t, \alpha t\rangle$, where α is a constant

44. $\mathbf{r}(t) = \langle \sin t, \cos t, \cosh t\rangle$

45. Find the value(s) of x where the curvature is greatest on the graph of $y = x^3$.

46. Show that the curvature on the parabola defined by $y = x^2$ is greatest at the origin.

47. Show that the curvature is constant at every point on the circular helix defined by $\mathbf{r}(t) = \langle a\cos t, a\sin t, bt\rangle$, where a and b are positive constants.

48. Find the points on the graphs of $y = e^x$ and $y = \ln x$ where the curvature is maximal. Explain how the two answers are related.

49. Let $\mathbf{r}(t) = \left\langle \cos t, \sin t, \frac{1}{t}\right\rangle$, $t > 0$. Show that $\lim_{t\to\infty} \kappa = 1$ and $\lim_{t\to 0^+} \kappa = 0$.

50. Find the curvature on the graph of the elliptical helix defined by $\mathbf{r}(t) = \langle a\cos t, b\sin t, ct\rangle$, where a, b, and c are positive constants.

Exercises 51 and 52 derive the equations necessary to define the ***torsion*** τ of a space curve. (Torsion measures the rate at which a space curve twists away from the osculating plane.) In each of these exercises, let C be a space curve and let $\mathbf{r}(s)$ be an arc length parametrization for C such that $\frac{d\mathbf{T}}{ds}$ and $\frac{d\mathbf{N}}{ds}$ exist at every point on C.

51. Explain why $\frac{d\mathbf{B}}{ds}$ exists at every point on C. (*Hint: Differentiate* $\mathbf{B} = \mathbf{T} \times \mathbf{N}$ *with respect to arc length.*)

52. Use Exercise 51 and Theorem 11.12 to show that $\frac{d\mathbf{B}}{ds}$ is parallel to $\mathbf{N}$. (*Hint: Recall that* $\frac{d\mathbf{T}}{ds} = \kappa\mathbf{N}$.) Since $\frac{d\mathbf{B}}{ds}$ and $\mathbf{N}$ are parallel, we define the torsion τ to be the scalar such that $\frac{d\mathbf{B}}{ds} = -\tau\mathbf{N}$.

53. Use Exercise 52 to show that $\frac{d\mathbf{N}}{ds} = -\kappa\mathbf{T} + \tau\mathbf{B}$, where κ is the curvature. (*Hint: Differentiate* $\mathbf{N} = \mathbf{B} \times \mathbf{T}$ *with respect to arc length.*)

Use the definition of torsion in Exercise 52 to compute the torsion of the vector functions in Exercises 54–56.

54. $\mathbf{r}(t) = \langle \cos t, \sin t, t\rangle$

55. $\mathbf{r}(t) = \langle \cosh t, \sinh t, t\rangle$

56. $\mathbf{r}(t) = \langle 3\sin t, 5\cos t, 4\sin t\rangle$

57. Use the results of Exercise 52 to show that the torsion of a planar curve C is zero at every point on C.

Applications

58. Ian will walk along a contour on the side of a mountain to get from a point he calls the origin to the point $(2, 0)$, with distances measured in miles. Unfortunately, the mountainside is riven by gullies and ridges. Thus, his route will follow the curve

$$\mathbf{r}(t) = \left\langle 1.2t - 0.5t^2 + 0.1t^4, 0.15\sin(6\pi t)\right\rangle.$$

How far does Ian actually have to walk?

59. Travelling along a contour on a mountainside is usually most difficult at the points of greatest curvature of the route, since these indicate deep gullies or sharp, rocky promontories that must be passed.

(a) Sketch the graph of Ian's route from the previous problem.

(b) Sketch each component of the curve describing his route. What information does that give you about the curvature?

(c) Roughly where does Ian expect the greatest difficulties?

Proofs

60. Use Theorem 11.24 to prove that the curvature of a linear function $y = mx + b$ is zero for every value of x.

61. Use Theorem 11.24 to prove that the curvature is zero at a point of inflection of a twice-differentiable function $y = f(x)$.

62. Prove that a planar curve with constant curvature is either a line or a circle. Give an example of a space curve with constant curvature that is neither a line nor a circle.

63. Prove that the magnitude of the acceleration of a particle moving along a curve, $\mathcal{C}$, with a constant speed is proportional to the curvature of $\mathcal{C}$.

64. Let C be the graph of a vector function $\mathbf{r}(s)$ defined on an interval I and parametrized by arc length s.

(a) If the unit tangent vector $\mathbf{T}$ has a nonzero derivative, use the chain rule and the definitions of the unit tangent vector $\mathbf{T}$ and principal unit normal vector $\mathbf{N}$ to show that $\frac{d\mathbf{T}}{ds}$, $\frac{d\mathbf{T}}{dt}$, and $\mathbf{N}$ are scalar multiples of each other. In particular, argue that $\frac{d\mathbf{T}}{ds} = \alpha\mathbf{N}$ for some scalar α.

(b) Use the fact that $\mathbf{N}$ is a unit vector to show that the constant α in part (a) must equal κ.

65. Let $y = f(x)$ be a twice-differentiable function. Show that the curvature of the graph of f is given by

$$\kappa = \frac{|f''(x)|}{(1 + (f'(x))^2)^{3/2}}.$$

This is part (a) of Theorem 11.24. (*Hint: Use the parametrization* $x = t,\ y = f(t)$.)

66. Let C be the graph of a vector function $\mathbf{r}(t) = \langle x(t), y(t)\rangle$ in the xy-plane, where $x(t)$ and $y(t)$ are twice-differentiable functions of t. Show that the curvature κ of C at a point on the curve is given by

$$\kappa = \frac{|x'(t)y''(t) - x''(t)y'(t)|}{((x'(t))^2 + (y'(t))^2)^{3/2}}.$$

This is part (b) of Theorem 11.24.

Thinking Forward

A decomposition of the acceleration vector: Find $\text{comp}_{\mathbf{v}(t)}\mathbf{a}(t)$, where $\mathbf{v}$ and $\mathbf{a}$ are the velocity and acceleration vectors, respectively, of the following functions.

- $\mathbf{r}(t) = \langle t, t^2, t^3\rangle$
- $\mathbf{r}(t) = \langle \cos t, \sin t, t\rangle$
- $\mathbf{r}(t) = \langle e^t \sin t, e^t \cos t, e^t\rangle$

11.5 MOTION

- The behavior of objects subject to the force of gravity
- The displacement and distance of projectiles
- The tangential and normal components of the acceleration vector

Motion Under the Force of Gravity

In Chapter 2 we discussed motion along a straight path. The motions we examined were of two types: straight up and down, and straight back and forth. Now that we have studied vector functions, we are ready to widen the discussion to motion along curves in a plane and motion along a space curve. We begin by discussing motion under the force of gravity.

Imagine a catapult that can propel a rock with an initial velocity of 30 miles per hour. A number of questions arise:

- If the catapult always releases the rock at the same angle to the horizontal, how far will the rock travel?
- Along what path would the rock travel?
- What is the maximum height reached by the rock?
- How long will it take until the rock hits the ground?
- How fast will the rock be travelling upon impact?
- If the catapult can be moved to the top of a hill, would the rock travel farther?
- If the angle at which the catapult releases the rock can be adjusted, what angle would maximize the distance the rock travels?
- Will the motion be planar?

We will answer some of these questions in this section and leave others to the exercises.

To begin our analysis we will assume that wind resistance and friction are negligible and that our object is a point mass. Recall that (on Earth) the gravitational constant g is 32 feet per second per second in the English system, and 9.8 meters per second per second in the metric system. We know from experience that a thrown object travels in a curved path. When we ignore wind resistance, we may assume that the motion is planar and use a two-dimensional coordinate system that places the initial position of the object at $(0, h)$, where h is the starting height of the projectile (as it is released). The following diagram illustrates the starting point of our analysis:

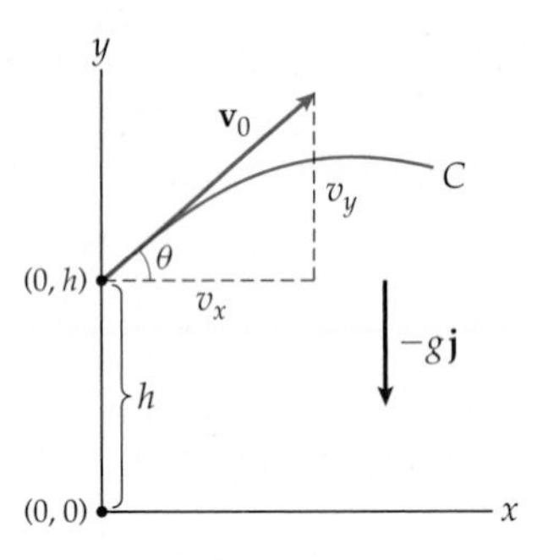

We assume that the projectile moves along a planar curve C given by the vector function $\mathbf{r}(t) = \langle x(t), y(t)\rangle$. The initial velocity of the vector function is

$$\mathbf{v}_0 = \mathbf{v}(0) = \mathbf{r}'(0) = v_x\mathbf{i} + v_y\mathbf{j} = \|\mathbf{v}_0\| \cos\theta\mathbf{i} + \|\mathbf{v}_0\| \sin\theta\mathbf{j}.$$

We also assume that the acceleration of the projectile after it is released is due solely to gravity, which is constant and acts downwards. Therefore, $\mathbf{r}''(t) = \mathbf{a}(t) = -g\mathbf{j}$. We will now be able to find the vector function $\mathbf{v}$ by integrating $\mathbf{a}$. That is,

$$\mathbf{v}(t) = \int \mathbf{a}(t)\,dt = \int \langle 0, -g\rangle\,dt = \langle c_1, -gt + c_2\rangle,$$

where c_1 and c_2 are constants of integration. We may evaluate these constants because we know the initial velocity $\mathbf{v}(0) = \langle \|\mathbf{v}_0\| \cos\theta, \|\mathbf{v}_0\| \sin\theta\rangle = \langle c_1, c_2\rangle$. Consequently,

$$c_1 = \|\mathbf{v}_0\| \cos\theta \text{ and } c_2 = \|\mathbf{v}_0\| \sin\theta.$$

We have the velocity function

$$\mathbf{v}(t) = \langle \|\mathbf{v}_0\| \cos\theta, -gt + \|\mathbf{v}_0\| \sin\theta\rangle.$$

Now, we integrate $\mathbf{v}$ to find $\mathbf{r}$:

$$\begin{aligned}\mathbf{r}(t) &= \int \mathbf{v}(t)\,dt = \int \langle \|\mathbf{v}_0\| \cos\theta, -gt + \|\mathbf{v}_0\| \sin\theta\rangle\,dt \\ &= \left\langle (\|\mathbf{v}_0\| \cos\theta)\,t + C_1, -\frac{1}{2}gt^2 + (\|\mathbf{v}_0\| \sin\theta)\,t + C_2\right\rangle.\end{aligned}$$

Here C_1 and C_2 are new constants of integration that may be evaluated by using the initial position of the projectile. We positioned our coordinate system so that $\mathbf{r}(0) = \langle 0, h\rangle$. So, we obtain $C_1 = 0$ and $C_2 = h$; thus, the parametrization for C is

$$\mathbf{r}(t) = \left\langle (\|\mathbf{v}_0\| \cos\theta)\,t, -\frac{1}{2}gt^2 + (\|\mathbf{v}_0\| \sin\theta)\,t + h\right\rangle.$$

It follows that the path of the projectile is a parabola, because when we eliminate the parameter t from the equations

$$x = (\|\mathbf{v}_0\| \cos\theta)\,t, \quad y = -\frac{1}{2}gt^2 + (\|\mathbf{v}_0\| \sin\theta)\,t + h,$$

we obtain the quadratic function

$$y = -\frac{g}{2\|\mathbf{v}_0\|^2 \cos^2\theta}x^2 + (\tan\theta)x + h.$$

(Recall that $\mathbf{v}_0$, θ and h are all constants.)

We will consider extended examples next.

Displacement and Distance Travelled

We are all familiar with the sentence:

The shortest distance between two points is a straight line.

This shortest distance is the displacement, while the length of the curve traversed by an object is the distance travelled. In terms of vector functions we have the following definition:

DEFINITION 11.26

Displacement Versus Distance Travelled

Let the vector-valued function $\mathbf{r}(t)$ represent the position function for a particle travelling on a curve C. If the domain of $\mathbf{r}$ is an interval I containing points a and b:

(a) The ***displacement*** of the the particle from a to b is given by the vector $\mathbf{r}(b) - \mathbf{r}(a)$.

(b) The ***distance travelled*** by the particle from a to b is given by the scalar $\int_a^b \|\mathbf{r}'(t)\|dt$.

The following figure illustrates these definitions:

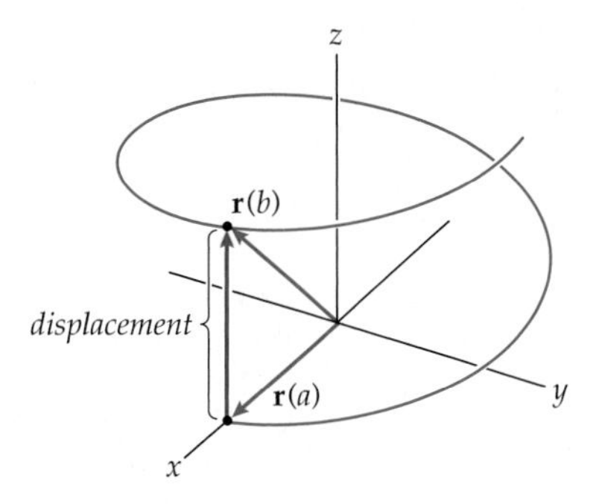

The displacement vector is the vector with initial point $\mathbf{r}(a)$ and terminal point $\mathbf{r}(b)$. The distance travelled is this length of the curve connecting $\mathbf{r}(a)$ and $\mathbf{r}(b)$. As a quick example, for the helix defined by the vector function $\mathbf{r}(t) = \langle \cos t, \sin t, t\rangle$, the displacement vector from 0 to 2π is

$$\mathbf{r}(2\pi) - \mathbf{r}(0) = \langle \cos 2\pi, \sin 2\pi, 2\pi\rangle - \langle \cos 0, \sin 0, 0\rangle = \langle 1, 0, 2\pi\rangle - \langle 1, 0, 0\rangle = \langle 0, 0, 2\pi\rangle,$$

while the distance travelled by a particle on the helix from 0 to 2π is

$$\begin{aligned}\int_0^{2\pi} \|\mathbf{r}'(t)\|dt &= \int_0^{2\pi} \|\langle -\sin t, \cos t, 1\rangle\|\,dt = \int_0^{2\pi} \sqrt{\sin^2 t + \cos^2 t + 1}\,dt \\ &= \int_0^{2\pi} \sqrt{2}\,dt = 2\sqrt{2}\pi.\end{aligned}$$

Tangential and Normal Components of Acceleration

We have already seen that every vector in $\mathbb{R}^3$ can be decomposed into a sum of three vectors that are directed along the x-, y-, and z-axes. Doing this requires nothing more than writing a vector $\mathbf{v}$ in the form $\mathbf{v} = a\mathbf{i} + b\mathbf{j} + c\mathbf{k}$. Rather than using this type of decomposition, however, when we are working with curves resulting from the motion of objects in $\mathbb{R}^3$, it is often convenient to decompose the acceleration vector into a sum of two vectors, one of which is in the direction of the unit tangent vector $\mathbf{T}$ and the other is in the direction of the principal unit normal vector **N**. These two components are known as the ***tangential component of acceleration*** and the ***normal component of acceleration***, respectively, and are illustrated as follows:

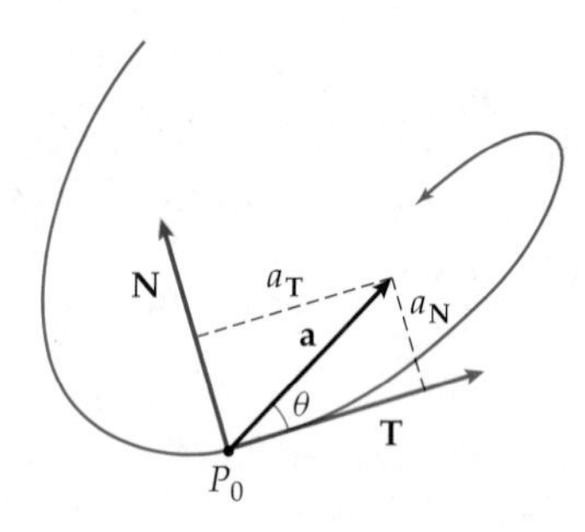

Even the graph of a very complicated vector function in $\mathbb{R}^3$ is locally planar, as long as the function is twice differentiable. As we mentioned in Section 11.3, the osculating plane fits a space curve "best" at a point P_0 on the curve, and the osculating plane is determined by the point P_0 and the unit tangent and principal unit normal vectors at P_0.

THEOREM 11.27

The Tangential and Normal Components of Acceleration

Let $\mathbf{r}(t)$ be a twice-differentiable vector function with either two or three components and with derivatives $\mathbf{r}'(t) = \mathbf{v}(t)$ and $\mathbf{r}''(t) = \mathbf{a}(t)$. Then the tangential component of acceleration a_T is given by

$$a_{\mathbf{T}} = \frac{\mathbf{v} \cdot \mathbf{a}}{\|\mathbf{v}\|}$$

and the normal component of acceleration a_N is given by

$$a_{\mathbf{N}} = \frac{\|\mathbf{v} \times \mathbf{a}\|}{\|\mathbf{v}\|}.$$

Proof. Since vectors $\mathbf{v}$ and $\mathbf{T}$ are scalar multiples, $\text{comp}_{\mathbf{T}}\mathbf{a} = \text{comp}_{\mathbf{v}}\mathbf{a}$. From the figure shown, if θ is the angle between vectors $\mathbf{a}$ and $\mathbf{v}$, then

$$a_{\mathbf{T}} = \|\mathbf{a}\| \cos\theta = \frac{\|\mathbf{v}\|\|\mathbf{a}\| \cos\theta}{\|\mathbf{v}\|} = \frac{\mathbf{v} \cdot \mathbf{a}}{\|\mathbf{v}\|}.$$

Similarly,

$$a_{\mathbf{N}} = \|\mathbf{a}\| \sin\theta = \frac{\|\mathbf{v}\|\|\mathbf{a}\| \sin\theta}{\|\mathbf{v}\|} = \frac{\|\mathbf{v} \times \mathbf{a}\|}{\|\mathbf{v}\|}.$$

■

Note that if $\mathbf{r}(t) = \langle x(t), y(t)\rangle$, then to take the cross product in Theorem 11.27, we would let $\mathbf{v}(t) = \langle x'(t), y'(t), 0\rangle$, and let $\mathbf{a}(t) = \langle x''(t), y''(t), 0\rangle$. The normal component of acceleration is also called ***centripetal acceleration***.

Examples and Explorations

EXAMPLE 1

The Human Cannonball

Chuckles the Human Cannonball is to be shot from the circus cannon. The muzzle of the cannon is 10 feet above ground level and is initially at an angle of 30° with the horizontal. The muzzle velocity of the cannon is 30 mph. The square net to catch Chuckles measures 20 feet on each side and is placed at a height of 6 feet, 30 feet directly in front of the muzzle of the cannon. Should Chuckles call an ambulance?

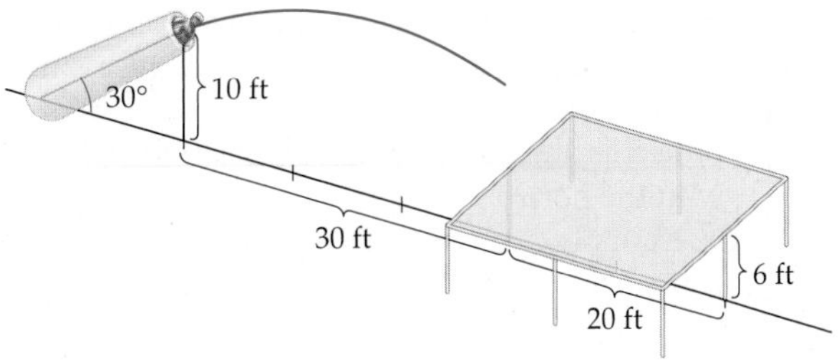

SOLUTION

We place our coordinate system so that the muzzle of the cannon is at $(0, 10)$. The leading edge of the net will then be at $(30, 6)$. The muzzle speed of 30 mph is equivalent to 44 feet/sec. At the angle of 30°, Chuckles' initial velocity function will be

$$\mathbf{v}_0 = 44\cos 30°\mathbf{i} + 44\sin 30°\mathbf{j} = 22\sqrt{3}\mathbf{i} + 22\mathbf{j}.$$

Since, in general, $\mathbf{r}(t) = \left\langle (\|\mathbf{v}_0\|\cos\theta)\,t, -\frac{1}{2}gt^2 + (\|\mathbf{v}_0\|\sin\theta)\,t + h\right\rangle$, here we have

$$\mathbf{r}(t) = \left\langle (\|\mathbf{v}_0\|\cos\theta)\,t, -\frac{1}{2}gt^2 + (\|\mathbf{v}_0\|\sin\theta)\,t + h\right\rangle = \left\langle 22\sqrt{3}t, -16t^2 + 22t + 10\right\rangle,$$

where we have used the gravitational constant $g = 32$ feet/sec^2. We next find the time it takes until Chuckles would reach the net's height of 6 feet. To do this we set the y-component of $\mathbf{r}(t)$ equal to 6 and solve for t.

That is, we solve $-16t^2 + 22t + 10 = 6$ and obtain the roots $t = \frac{11 \pm \sqrt{185}}{16}$. In this context, only the positive root $t = \frac{11 + \sqrt{185}}{16}$ is relevant. It would take Chuckles approximately 1.53 seconds before he reached a height of 6 feet. We now use the x-component of $\mathbf{r}(t)$ to find the horizontal distance that Chuckles would travel in 1.53 seconds and obtain the distance $22\sqrt{3}(1.53) \approx 58.3$ feet. Since the net extends only from 30 feet to 50 feet from the cannon muzzle, Chuckles should prepare for a bad landing. □

EXAMPLE 2

The Human Cannonball, part 2

Chuckles the Human Cannonball is unhappy with the planned scenario of Example 1. If the angle of the cannon can be adjusted, at what angle should Chuckles ask to be shot so that he hits the middle of the net? You may assume that all other constants given in Example 1 remain the same.

SOLUTION

We now have the vector function

$$\mathbf{r}(t) = \langle (44\cos\theta)\,t, -16t^2 + (44\sin\theta)\,t + 10\rangle.$$

We need to find the values of t and θ such that the x-component is 40 and the y-component is 6. That is, we need to solve the system of equations

$$(44\cos\theta)\,t = 40 \text{ and } -16t^2 + (44\sin\theta)\,t + 10 = 6$$

simultaneously for t and θ. There are four pairs of values for t and θ that satisfy this system of equations, but only the two pairs in which t is positive make sense contextually. One pair is $t \approx 0.94$ second and $\theta \approx 14°$. The other pair is $t \approx 2.68$ seconds and $\theta \approx 70°$. (We ask you to solve the same system of equations in Exercise 25.) □

In Example 1 Chuckles would have gone over 55 feet if he had been shot at an angle of 30°. It makes sense that if he is shot at a smaller angle of elevation, he wouldn't travel as far. In Exercise 33 we will ask you to prove that in a ground-to-ground trajectory, a projectile always travels the greatest horizontal distance when it is shot at an angle of 45° with the horizontal. Therefore, it also makes sense that there would be a second solution in Example 2 in which the angle of elevation would be greater than 45°. This doesn't, of course, tell us that the numbers we found are correct, only that they are reasonable.

EXAMPLE 3

The Human Cannonball, part 3

Chuckles the Human Cannonball is happier with the scenario of Example 2, but he needs to decide which of the two angles would be preferable. The big top tent he is under is 50 feet tall.

SOLUTION

If Chuckles is shot at the smaller angle, he would hit the net at a more shallow angle, which might lead to his skidding off the net. For this reason the steeper angle might be preferable. However, we should also consider his speed upon impact and the maximum height he would reach if he were shot at the two angles. We consider his speed upon impact and the maximum height he would reach if he were shot at an angle of 70°. We leave the analysis at the initial angle of 14° for Exercise 26.

Since the velocity function under the force of gravity is

$$\mathbf{v}(t) = \left\langle \|\mathbf{v}_0\| \cos\theta, -gt + \|\mathbf{v}_0\| \sin\theta \right\rangle,$$

we have

$$\begin{aligned}\mathbf{v}(t) &= \left\langle \|\mathbf{v}_0\| \cos\theta, -gt + \|\mathbf{v}_0\| \sin\theta \right\rangle = \langle 44\cos 70°, -32t + 44\sin 70° \rangle \\ &= \langle 15.0, -32t + 41.3 \rangle.\end{aligned}$$

To find Chuckles' speed upon impact, recall that speed is the magnitude of velocity and that, at an angle of 70°, he will hit the net after about 2.7 seconds. Therefore, his speed upon impact will be

$$\|\mathbf{v}(2.7)\| = \left\| \langle 15.0, -32(2.7) + 41.3 \rangle \right\| = \|\langle 15.0, -45 \rangle\| \approx 47 \text{ feet per second.}$$

To find the maximum height we consider the y-component of the position function $\mathbf{r}(t)$. This is

$$y(t) = -16t^2 + (44\sin 70°)t + 10 = -16t^2 + 41.3t + 10.$$

The maximum height will be attained when $y'(t) = -32t + 41.3 = 0$. (Why?) Thus, the maximum height occurs at about 1.3 seconds. The corresponding height is

$$y(1.3) = -16(1.3)^2 + 41.3(1.3) + 10 \approx 37 \text{ feet.}$$

He has plenty of room to spare! □

EXAMPLE 4

Finding centripetal acceleration

A girl is swinging a small plane attached to a string 2 feet long in a circle over her head at a rate of 20 revolutions per minute. If the circle is horizontal at a height of 6 feet above the ground, find a vector function that models the path of the plane and find the centripetal acceleration of the plane.

SOLUTION

The vector function

$$\mathbf{r}(t) = \left\langle 2\cos\left(\frac{2\pi}{3}t\right), 2\sin\left(\frac{2\pi}{3}t\right), 6\right\rangle$$

has the motion described, where t is measured in seconds. Note that every 3 seconds the plane would complete exactly one revolution, as specified. The velocity and acceleration vectors are, respectively,

$$\mathbf{v}(t) = \left\langle -\frac{4\pi}{3}\sin\left(\frac{2\pi}{3}t\right), \frac{4\pi}{3}\cos\left(\frac{2\pi}{3}t\right), 0\right\rangle \quad \text{and}$$

$$\mathbf{a}(t) = \left\langle -\frac{8\pi^2}{9}\cos\left(\frac{2\pi}{3}t\right), -\frac{8\pi^2}{9}\sin\left(\frac{2\pi}{3}t\right), 0\right\rangle.$$

The centripetal acceleration, or equivalently, the normal component of acceleration, $a_{\mathbf{N}}$, is given by

$$a_{\mathbf{N}} = \frac{\|\mathbf{v}\times\mathbf{a}\|}{\|\mathbf{v}\|} = \frac{\left\|\left\langle -\frac{4\pi}{3}\sin\left(\frac{2\pi}{3}t\right), \frac{4\pi}{3}\cos\left(\frac{2\pi}{3}t\right), 0\right\rangle \times \left\langle -\frac{8\pi^2}{9}\cos\left(\frac{2\pi}{3}t\right), -\frac{8\pi^2}{9}\sin\left(\frac{2\pi}{3}t\right), 0\right\rangle\right\|}{\left\|\left\langle -\frac{4\pi}{3}\sin\left(\frac{2\pi}{3}t\right), \frac{4\pi}{3}\cos\left(\frac{2\pi}{3}t\right), 0\right\rangle\right\|}$$

$$= \frac{\|\langle 0, 0, 32\pi^3/27\rangle\|}{4\pi/3} = \frac{32\pi^3/27}{4\pi/3} = \frac{8\pi^2}{9}.$$

Note that in this example the velocity and acceleration vectors are orthogonal. When this occurs, the normal component of acceleration will be the magnitude of the acceleration vector, as we find here. □

? TEST YOUR UNDERSTANDING

- ▶ How are the vector functions giving the velocity and position of an object moving under the acceleration due to gravity derived from the gravitational constant?
- ▶ Why does a projectile subject to gravity move in a parabolic arch?
- ▶ How are the component functions for the position and velocity of a projectile used to determine the height reached by the object, the distance travelled by the object, and the travel time of the projectile?
- ▶ What are the normal and tangential components of acceleration? How is the osculating plane related to the components of acceleration?
- ▶ What is centripetal acceleration?

EXERCISES 11.5

Thinking Back

- ▶ *Projecting one vector onto another:* Show that the formula for the projection of a vector $\mathbf{v}$ onto a nonzero vector $\mathbf{u}$ is given by $\text{proj}_{\mathbf{u}}\mathbf{v} = \frac{\mathbf{u}\cdot\mathbf{v}}{\|\mathbf{u}\|}$, where $\mathbf{u} \neq \mathbf{0}$.
- ▶ *The difference between a vector and its projection onto another vector:* If $\mathbf{u} \neq \mathbf{0}$ and $\mathbf{v}$ is an arbitrary vector, what is the geometric relationship between $\mathbf{v}$, $\text{proj}_{\mathbf{u}}\mathbf{v}$, and $\mathbf{v} - \text{proj}_{\mathbf{u}}\mathbf{v}$?

Concepts

0. *Problem Zero:* Read the section and make your own summary of the material.

1. *True/False:* Determine whether each of the statements that follow is true or false. If a statement is true, explain why. If a statement is false, provide a counterexample.

(a) *True or False:* When the wind resistance is negligible, a projectile travels on a parabolic path if its initial velocity has a nonzero horizontal component.

(b) *True or False:* The velocity of a projectile is zero when it has reached its maximum height.

(c) *True or False:* If the wind resistance is negligible, the horizontal component of velocity is constant.

(d) *True or False:* The acceleration of a projectile is $-g\mathbf{k}$, where g is the gravitational constant.

(e) *True or False:* The velocity of a moving particle is defined to be the absolute value of its speed.

(f) *True or False:* Centripetal acceleration and the normal component of acceleration are identical.

(g) *True or False:* The sum of the tangential component of acceleration and the normal component of acceleration is the acceleration.

(h) *True or False:* If $\|\mathbf{r}(b) - \mathbf{r}(a)\| = \int_a^b \|\mathbf{r}'(t)\|\, dt$ for every a and b in the domain of $\mathbf{r}$, then the curve defined by $\mathbf{r}$ is parametrized by arc length.

2. *Examples:* Construct examples of the thing(s) described in the following. Try to find examples that are different than any in the reading.

(a) A nonconstant vector function $\mathbf{r}(t)$ defined on an interval $[a, b]$ for which

$$\|\mathbf{r}(b) - \mathbf{r}(a)\| = \int_a^b \|\mathbf{r}'(t)\| dt.$$

(b) A vector function with a nonzero acceleration vector such that the normal component of acceleration is always zero.

(c) A vector function with a nonzero acceleration vector such that the tangential component of acceleration is always zero.

3. Explain why an object thrown horizontally on a planet with a smaller gravitational force than Earth's would travel farther.

In our derivation of the position function $\mathbf{r}(t) = \left\langle (\|\mathbf{v}_0\| \cos\theta)\, t, -\frac{1}{2} g t^2 + (\|\mathbf{v}_0\| \sin\theta)\, t + h \right\rangle$ for the motion of a projectile, we ignored air resistance and wind effects. In Exercises 4 and 5 we ask you to think about how these effects would change the model.

4. Air resistance is a type of friction. It is a vector that is proportional to, and acting in the opposite direction from, the velocity. How would incorporating air resistance change the development of our model of the motion of a projectile?

5. We may model the wind as a vector with a direction. How would including the wind change the development of our model of the motion of a projectile?

6. A projectile reaches its maximum height H at time $t_0 > 0$. What fraction of H does it reach at time $\frac{1}{2} t_0$?

7. Given that an object is moving along the graph of a vector function $\mathbf{r}(t)$ defined on an interval $[a, b]$, what is meant by the displacement of the object from $t = a$ to $t = b$?

8. If an object is moving along the graph of a vector function $\mathbf{r}(t)$ defined on an interval $[a, b]$, how can the distance the object travels from $t = a$ to $t = b$ be calculated?

9. Define each of the following: the tangential component of acceleration, the normal component of acceleration, and centripetal acceleration. How are each of these quantities computed?

10. In the graph that follows, a portion of a curve is drawn, along with the unit tangent vector, $\mathbf{T}$, principal unit normal vector, $\mathbf{N}$, and acceleration vector, $\mathbf{a}$, at point P. Add the vectors $\text{proj}_{\mathbf{T}}\mathbf{a}$ and $\text{proj}_{\mathbf{N}}\mathbf{a}$ to the graph, and label the magnitudes of these vectors appropriately.

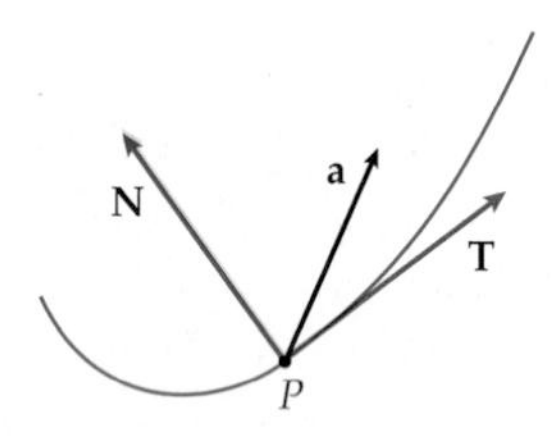

11. Imagine that you are driving on a twisting mountain road. Describe the tangential and normal components of acceleration as you ascend, descend, twist right, and twist left.

12. Two marbles roll off a table at the same moment. One of the marbles has a horizontal speed of 1 inch per second, and the other has a horizontal speed of 4 inches per second.

(a) Which marble hits the ground first?

(b) If the surface of the table is 3 feet from the ground, what are the speeds of the marbles as they hit?

13. Let $f(t)$ be a nonconstant continuous function.

(a) Explain why a particle moving along the vector function $\mathbf{r}(t) = \langle f(t), (f(t))^2 \rangle$ is on a parabolic path.

(b) What condition must $f(t)$ satisfy so that $\mathbf{r}(t) \cdot \mathbf{a}(t) = 0$?

Skills

In Exercises 14–17, find (a) the displacement vectors from $\mathbf{r}(a)$ to $\mathbf{r}(b)$, (b) the magnitude of the displacement vector, and (c) the distance travelled by a particle on the curve from a to b.

14. $\mathbf{r}(t) = \langle t, t^2 \rangle$, $a = 2$, $b = 3$

15. $\mathbf{r}(t) = \langle t \sin t, t \cos t \rangle$, $a = 0$, $b = \pi$

16. $\mathbf{r}(t) = \langle e^t \sin t, e^t \cos t, e^t \rangle$, $a = 0$, $b = 1$

17. $\mathbf{r}(t) = \langle \alpha \sin\beta t, \alpha \cos\beta t, \gamma t \rangle$, $a = 0$, $b = 1$

Find the tangential and normal components of acceleration for the position functions in Exercises 18–22

18. $\mathbf{r}(t) = \langle t, t^2 \rangle$

19. $\mathbf{r}(t) = \langle \sin 3t, \cos 4t \rangle$

20. $\mathbf{r}(t) = \langle t \sin t, t \cos t \rangle$

21. $\mathbf{r}(t) = \langle e^t \sin t, e^t \cos t, e^t \rangle$

22. $\mathbf{r}(t) = \langle 2 \sin t, 3 \cos t, \cos 2t \rangle$

23. Find the tangential and normal components of acceleration for a particle moving along the circular helix defined by $\mathbf{r}(t) = \langle \cos t, \sin t, t \rangle$.

24. Find the tangential and normal components of acceleration for a particle moving along the conical helix defined by $\mathbf{r}(t) = \langle t \cos t, t \sin t, t \rangle$.

Applications

25. Finish Example 2 by solving the system of equations

$$(44\cos\theta)\,t = 40 \text{ and } -16t^2 + (44\sin\theta)\,t + 10 = 6$$

for θ and t. (*Hint: Solve the first equation for* $\cos\theta$ *and the second equation for* $\sin\theta$*, and use a Pythagorean identity.*)

26. Finish Example 3 by calculating the maximum height Chuckles the Human Cannonball will reach and the speed at which he will hit the net if he is shot from the cannon at an elevation of 14°.

Complete Exercises 27–31 using the appropriate gravitational constants from the table that follows. In each exercise you may ignore wind resistance.

	g in m/sec^2	g in ft/sec^2
Earth	9.8	32
Moon	1.6	5.2
Mars	3.7	12.1
Jupiter	24.8	81

27. In 1968 Bob Beamon set an Olympic record in the long jump in the Mexico City Olympic Games with a jump of 8.90 meters.

(a) Assuming that his angle of elevation was 30°, how fast was he running as he jumped?

(b) Assuming that he could have performed the long jump on the moon with the same initial velocity and same angle of elevation, how far would Beamon have gone?

28. There is considerable controversy surrounding who hit the longest home run in professional baseball. However, any ball that travels 500 feet in the air is considered "monumental." Assuming that a hitter contacts a ball 4 feet from the ground and that the ball leaves the bat with a 45° angle of elevation and subsequently travels a horizontal distance of 500 feet, answer the following questions:

(a) How fast is the speed of the ball right after contact?

(b) What is the maximum height reached by the ball?

(c) How long does the ball spend in the air after being hit?

(d) On May 22, 1963, Mickey Mantle, playing for the New York Yankees, hit a home run that struck a point on the façade of the right-field roof of the old Yankee Stadium approximately 115 feet high and 370 feet from home plate. Explain why this is not enough information to determine how far the ball would have travelled if it had not hit the façade.

29. In 1974 Nolan Ryan, then playing for the California Angels, was credited with throwing one of the fastest pitches in professional baseball history. The pitch reached a speed of 100.9 mph.

(a) Assume that the horizontal distance from the pitcher's mound to the catcher's glove was 60 feet, that the ball left Ryan's hand at a height of 6 feet, and that it was caught by the catcher with his hand at a height of 3 feet. What was the angle of elevation of the pitch as it left Ryan's hand?

(b) If Ryan had thrown the ball at the same speed and from the same height at an angle of elevation of 45°, what horizontal distance would the ball have travelled before it hit the ground?

30. A certain type of tank gun can be shot with a muzzle velocity of 1500 meters per second.

(a) At what angle should the cannon be shot to hit a target 1 kilometer away? You may assume that the target and the muzzle of the cannon are at the same height.

(b) What is the range of the cannon? Assume that the muzzle of the cannon is 5 meters high.

(c) If the tank is repositioned atop a small hill so that the muzzle of the cannon is now 20 meters above the surrounding area, what is the new range of the cannon?

(d) If the cannon were on Jupiter with its muzzle 5 meters high, what would its range be?

31. An aircraft is flying toward the tank gun discussed in Exercise 30. Assume that the muzzle of the gun is 5 meters high and the gun can fire with a maximum angle of elevation of 60°. At what altitude should the pilot of the plane fly to keep out of the gun's range?

32. In 1971, during NASA's Apollo 14 mission to the moon, astronaut Alan Shepard hit a golf ball that is estimated to have travelled 2400 feet on the lunar surface, almost a half mile. Assuming a similar terrain, and ignoring wind resistance, how far would his shot have travelled on Earth?

Proofs

33. Show that a ball thrown with an angle of elevation of 45° will travel farther than if it is thrown at any other angle. (*Hint: Assume that the initial and final heights of the ball are equal, that the initial velocity is fixed, and that friction and wind resistance are negligible.*)

34. Prove that an object thrown at an acute angle of elevation travels along a parabolic path, if wind resistance is ignored.

35. Let $\mathbf{r}(t)$ be a vector function whose graph is a space curve containing distinct points P and Q. Prove that if the acceleration is always $\mathbf{0}$, then the graph of $\mathbf{r}$ is a straight line.

36. Assuming that the initial height of a projectile is zero, prove that doubling the initial velocity of the projectile has the effect of multiplying the maximum height of the projectile and the horizontal distance travelled by the projectile by a factor of 4.

37. Prove that the normal component of acceleration is 0 at a point of inflection on the graph of a twice-differentiable function $y = f(x)$. (*Hint: The graph of* $y = f(x)$ *is also the graph of the vector function* $\mathbf{r}(t) = \langle t, f(t)\rangle$*.*)

38. Prove that if a particle moves along a curve at a constant speed, then $a_{\mathbf{N}} = \|\mathbf{a}\|$ and $a_{\mathbf{T}} = 0$. Is the converse true? That is, if $a_{\mathbf{N}} = \|\mathbf{a}\|$ and $a_{\mathbf{T}} = 0$, must the particle be moving at a constant speed?

Thinking Forward

- *The graph of a function $f : \mathbb{R}^2 \to \mathbb{R}$*

  A vector-valued function with two components can be thought of as a function with a domain that is a subset of $\mathbb{R}$ and a codomain $\mathbb{R}^2$. We've seen that the graph of such a function is a curve in the plane. We may also define functions that have a subset of $\mathbb{R}^2$ as the domain and the codomain $\mathbb{R}$. In general terms, what is the graph of such a function? (One such function is $f(x, y) = x^2 - 3x \sin y^2$.)

- *The graph of a function $f : \mathbb{R}^3 \to \mathbb{R}$*

  A vector-valued function with three components can be thought of as a function with a domain that is a subset of $\mathbb{R}$ and a codomain $\mathbb{R}^3$. The graph of such a function is a space curve. We may also define functions that have a subset of $\mathbb{R}^3$ as the domain and the codomain $\mathbb{R}$. In general terms, what is the graph of such a function? (One such function is $f(x, y, z) = x + 2y - 5xyz$.)

CHAPTER REVIEW, SELF-TEST, AND CAPSTONES

Before you progress to the next chapter, be sure you are familiar with the definitions, concepts, and basic skills outlined here. The capstone exercises at the end bring together ideas from this chapter and look forward to future chapters.

Definitions

Give precise mathematical definitions or descriptions of each of the concepts that follow. Then illustrate the definition with a graph or an algebraic example.

- *parametric equations* in $\mathbb{R}^2$ and $\mathbb{R}^3$
- a *vector function* or *vector-valued function* in $\mathbb{R}^2$ and $\mathbb{R}^3$
- a *space curve*
- the *limit* of a vector function $\mathbf{r}(t)$ as t approaches t_0
- the *derivative* of a vector function $\mathbf{r}(t)$
- an *antiderivative* of a vector function $\mathbf{r}(t)$
- the *definite integral* of a vector function $\mathbf{r}(t)$ on an interval $[a, b]$
- the *unit tangent function* for a differentiable vector function $\mathbf{r}(t)$
- the *principal unit normal vector* at $\mathbf{r}(t)$
- the *binormal vector* at $\mathbf{r}(t_0)$
- the *osculating plane* at $\mathbf{r}(t_0)$
- an *arc length parametrization* for the graph of a differentiable function $\mathbf{r}(t)$
- the *curvature* at a point on the graph of a vector function $\mathbf{r}(s)$
- the *radius of curvature* at a point on the graph of a vector function $\mathbf{r}(t)$
- the *osculating circle* at a point on the graph of a vector function $\mathbf{r}(t)$

Theorems

Fill in the blanks to complete each of the following theorem statements:

- The derivative of a vector function $\mathbf{r}(t)$ is given by $\mathbf{r}'(t) = \lim_{h\to 0}$ ________.
- *The Chain Rule for Vector-Valued Functions*: Let $t = f(\tau)$ be a differentiable real-valued function of τ, and let $\mathbf{r}(t)$ be a differentiable vector function with either two or three components such that $f(\tau)$ is in the domain of $\mathbf{r}$ for every value of τ on some interval I. Then $\dfrac{d\mathbf{r}}{d\tau} =$ ________.
- Let C be the graph of a twice-differentiable vector function $\mathbf{r}(t)$ defined on an interval I and with unit tangent vector $\mathbf{T}(t)$. Then the curvature κ of C at a point on the curve is given by $\kappa = \dfrac{\|\quad\|}{\|\quad\|}$ and $\kappa = \dfrac{\|\quad\times\quad\|}{\|\quad\|^3}$.
- Let $y = f(x)$ be a twice-differentiable function. Then the curvature of the graph of f is given by $\kappa = \dfrac{|\quad|}{(\quad)^{3/2}}$.
- Let C be the graph of a vector function $\mathbf{r}(t) = \langle x(t), y(t)\rangle$ in the xy-plane, where $x(t)$ and $y(t)$ are twice-differentiable functions of t such that $x'(t)$ and $y'(t)$ are not simultaneously zero. Then the curvature κ of C at a point on the curve is given by $\kappa = \dfrac{|\quad|}{(\quad)^{3/2}}$.

Notation and Rules

Notation: Describe the meanings of each of the following mathematical expressions.

- $\mathbf{r}(t)$
- $\mathbf{v}(t)$
- $\mathbf{a}(t)$
- $\mathbf{T}(t)$
- $\mathbf{N}(t)$
- $\mathbf{B}(t)$
- κ
- ρ
- a_{T}
- a_{N}

Derivatives of Products: Fill in the following blanks to complete a differentiation rule for vectors.

- Let k be a scalar and $\mathbf{r}(t)$ be a differentiable vector function. Then $\frac{d}{dt}(k\mathbf{r}(t)) =$ ______.
- Let $f(t)$ be a differentiable scalar function and $\mathbf{r}(t)$ be a differentiable vector function. Then $\frac{d}{dt}(f(t)\mathbf{r}(t)) =$ ______.
- If $\mathbf{r}_1(t)$, $\mathbf{r}_2(t)$ are differentiable vector functions, both having either two or three components, then $\frac{d}{dt}(\mathbf{r}_1(t) \cdot \mathbf{r}_2(t)) =$ ______.
- If $\mathbf{r}_1(t)$ and $\mathbf{r}_2(t)$ are both differentiable three-component vector functions, then $\frac{d}{dt}(\mathbf{r}_1(t) \times \mathbf{r}_2(t)) =$ ______.
- Let $\mathbf{r}(t)$ be a differentiable vector function such that $\|\mathbf{r}(t)\| = k$ for some constant k. Then $\mathbf{r}(t) \cdot \mathbf{r}'(t) =$ ______.

Skill Certification: The Calculus of Vector Functions

Sketching vector functions: Sketch the following vector functions.

1. $\mathbf{r}(t) = \langle t, t^3 \rangle,\ t \in \mathbb{R}$
2. $\mathbf{r}(t) = \langle \cos t, t \rangle,\ t \in \mathbb{R}$
3. $\mathbf{r}(t) = \langle t, \sin 2t, \cos 2t \rangle,\ t \in [0, 4\pi]$
4. $\mathbf{r}(t) = \langle t, t\cos t, t\sin t \rangle,\ t \in [0, 4\pi]$

Finding limits: Find the given limits if they exist. If a limit does not exist, explain why.

5. $\lim\limits_{t\to 0} \left\langle \dfrac{1/(3+t) - 1/3}{t}, \dfrac{(3+t)^2 - 9}{t} \right\rangle$
6. $\lim\limits_{t\to -3} \left\langle \dfrac{|t+3|}{t+3}, \dfrac{|t+3|}{t-3} \right\rangle$
7. $\lim\limits_{t\to 0} \left\langle \dfrac{\sin t}{t}, \dfrac{1-\cos t}{t}, (1+t)^{1/t} \right\rangle$
8. $\lim\limits_{t\to 1} \left\langle (\ln t)^{t-1}, t^{(1/t)-1}, \dfrac{\sin(\ln t)}{t-1} \right\rangle$

Velocity and acceleration vectors: Find the velocity and acceleration vectors for the given vector functions.

9. $\mathbf{r}(t) = \langle t, 2t-3, 3t+5 \rangle$
10. $\mathbf{r}(t) = \langle t, 2t^2, 3t^3 \rangle$
11. $\mathbf{r}(t) = \langle \cos 2t, \sin 3t \rangle$
12. $\mathbf{r}(t) = \langle e^t, t, e^{-t} \rangle$

Unit tangent vectors: Find the unit tangent vector for the given function at the specified value of t.

13. $\mathbf{r}(t) = \langle t, t^3 \rangle,\ t = 2$
14. $\mathbf{r}(t) = \langle 3\sin 2t, 3\cos 2t \rangle,\ t = \frac{\pi}{3}$
15. $\mathbf{r}(t) = \langle \alpha \sin \beta t, \alpha \cos \beta t \rangle,\ t = 0$
16. $\mathbf{r}(t) = \langle t, 5\sin 3t, 5\cos 3t \rangle,\ t = \frac{\pi}{6}$
17. $\mathbf{r}(t) = \langle t, 2t\sin t, 2t\cos t \rangle,\ t = \pi$
18. $\mathbf{r}(t) = \langle t, \alpha \sin \beta t, \alpha \cos \beta t \rangle,\ t = 0$

Principal unit normal vectors: Find the principal unit normal vector for the given function at the specified value of t.

19. $\mathbf{r}(t) = \langle t, t^3 \rangle,\ t = 2$
20. $\mathbf{r}(t) = \langle 3\sin 2t, 3\cos 2t \rangle,\ t = \frac{\pi}{3}$
21. $\mathbf{r}(t) = \langle \alpha \sin \beta t, \alpha \cos \beta t \rangle$, where α and β are positive, $t = 0$
22. $\mathbf{r}(t) = \langle t, 5\sin 3t, 5\cos 3t \rangle,\ t = \frac{\pi}{6}$
23. $\mathbf{r}(t) = \langle t, 2t\sin t, 2t\cos t \rangle,\ t = \pi$
24. $\mathbf{r}(t) = \langle t, \alpha \sin \beta t, \alpha \cos \beta t \rangle,\ t = 0$

Binormal vectors and osculating planes: Find the binormal vector and equation of the osculating plane for the given function at the specified value of t.

25. $\mathbf{r}(t) = \langle t, t^3 \rangle,\ t = 2$
26. $\mathbf{r}(t) = \langle 3\sin 2t, 3\cos 2t \rangle,\ t = \frac{\pi}{3}$
27. $\mathbf{r}(t) = \langle \alpha \sin \beta t, \alpha \cos \beta t \rangle,\ t = 0$
28. $\mathbf{r}(t) = \langle t, 5\sin 3t, 5\cos 3t \rangle,\ t = \frac{\pi}{6}$
29. $\mathbf{r}(t) = \langle t, 2t\sin t, 2t\cos t \rangle,\ t = \pi$
30. $\mathbf{r}(t) = \langle t, \alpha \sin \beta t, \alpha \cos \beta t \rangle,\ t = 0$

Osculating circles: Find the equation of the osculating circle to the given function at the specified value of t.

31. $\mathbf{r}(t) = \langle t, t^3 \rangle,\ t = 2$
32. $\mathbf{r}(t) = \langle 3\sin 2t, 3\cos 2t \rangle,\ t = \frac{\pi}{3}$
33. $\mathbf{r}(t) = \langle \alpha \sin \beta t, \alpha \cos \beta t \rangle,\ t = 0$

Osculating circles: Find the center and radius of the osculating circle to the given vector function at the specified value of t.

34. $\mathbf{r}(t) = \langle t, 5\sin 3t, 5\cos 3t \rangle,\ t = \frac{\pi}{6}$
35. $\mathbf{r}(t) = \langle t, 2t\sin t, 2t\cos t \rangle,\ t = \pi$
36. $\mathbf{r}(t) = \langle t, \alpha \sin \beta t, \alpha \cos \beta t \rangle,\ t = 0$

Osculating circles: Find the equation of the osculating circle to the given scalar function at the specified point.

37. $f(x) = x^2$, $(0, 0)$

38. $f(x) = \cos x$, $(0, 1)$

39. $f(x) = e^x$, $(0, 1)$

40. $f(x) = \ln x$, $(1, 0)$

Capstone Problems

A. A certain vector function $\mathbf{r}(t)$ has the properties that its graph is a space curve passing through the point $(4, -3, 6)$ and that its derivative is $\mathbf{r}'(t) = -2\mathbf{r}(t)$. Find $\mathbf{r}(t)$.

B. Let $y = f(x)$ be a twice-differentiable function.

(a) Carefully outline the steps required to find the equation of the osculating circle at an arbitrary point x_0 in the domain of f.

(b) Write a program, using a computer algebra system such as Maple, Mathematica, or Matlab, to animate the osculating circle moving along the graph of the function $y = \sin x$.

C. Let $\mathbf{r}(t) = \langle x(t), y(t), z(t) \rangle$ be a twice-differentiable function.

(a) Carefully outline the steps required to find the unit tangent vector, principal unit normal vector, and binormal vector at an arbitrary point t_0 in the domain of $\mathbf{r}$.

(b) Write a program, using a computer algebra system such as Maple, Mathematica, or Matlab, to animate the Frenet frame moving along the graph of the function $\mathbf{r}(t) = \langle \cos t, \sin t, t \rangle$.

D. Suppose a small planet is orbiting the sun. Its position relative to the center of the sun is given by some vector function $\mathbf{r} = \mathbf{r}(t)$. When we put Newton's second law of motion together with his law of gravitation, with the sun at the origin, we find that the position of the planet at time t satisfies $\frac{d^2\mathbf{r}}{dt^2} = -\frac{GM\mathbf{r}}{r^3}$, where G is the gravitational constant, M is the mass of the sun, and $r = \|\mathbf{r}\|$. Define $\mathbf{u} = \mathbf{r} \times \frac{d\mathbf{r}}{dt}$. It turns out that $\mathbf{u}$ is a constant vector.

(a) Use the relation $\mathbf{v}_1 \times (\mathbf{v}_1 \times \mathbf{v}_2) = -\|\mathbf{v}_1\|^2\mathbf{v}_2$ for orthogonal vectors $\mathbf{v}_1$ and $\mathbf{v}_2$ to show that

$$\frac{d^2\mathbf{r}}{dt^2} \times \mathbf{u} = \frac{GM}{r}\frac{d\mathbf{r}}{dt}.$$

(b) Antidifferentiate the result of part (a), using the fact that $\mathbf{u}$ is constant to show that

$$\frac{d\mathbf{r}}{dt} \times \mathbf{u} = GM\frac{\mathbf{r}}{r} + \mathbf{w}$$

for some constant vector $\mathbf{w}$.

(c) Since $\frac{d\mathbf{r}}{dt} \times \mathbf{u}$ is orthogonal to $\mathbf{u}$ and $\mathbf{r}$ is orthogonal to $\mathbf{u}$, it follows that $\mathbf{w}$ is orthogonal to $\mathbf{u}$. If we let θ denote the angle from $\mathbf{w}$ to $\mathbf{r}$, then (r, θ) represents polar coordinates for the position of the planet in the $\mathbf{rw}$-plane. Use this fact and the relation

$$(\mathbf{v}_1 \times \mathbf{v}_2) \cdot \mathbf{v}_3 = \mathbf{v}_1 \cdot (\mathbf{v}_2 \times \mathbf{v}_3)$$

to show that

$$\|\mathbf{u}\|^2 = GMr + \|\mathbf{w}\|r\cos\theta.$$

(d) Convert the last expression in part (c) to Cartesian coordinates to conclude that the satellite's orbit around the planet is an ellipse (or circle) whenever $\|\mathbf{w}\| < GM$. This is Kepler's first law of planetary motion: Planets follow an elliptical path around the sun. The law can be generalized to describe the orbit of any relatively small body around a much larger body.

CHAPTER 12

Multivariable Functions

12.1 Functions of Two and Three Variables

Functions of Two Variables
Graphing a Function of Two Variables
Functions of Three Variables
Level Curves and Level Surfaces
Examples and Explorations

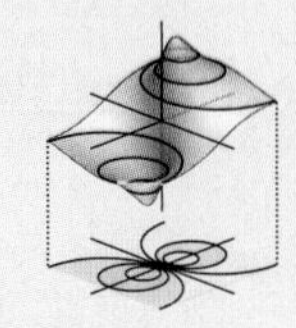

12.2 Open Sets, Closed Sets, Limits, and Continuity

Open Sets and Closed Sets
The Limit of a Function of Two or More Variables
The Continuity of a Function of Two or More Variables
Examples and Explorations

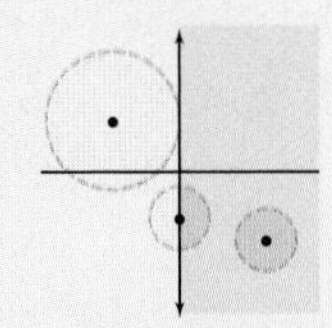

12.3 Partial Derivatives

Partial Derivatives of Functions of Two and Three Variables
Higher Order Partial Derivatives
Finding a Function When the Partial Derivatives Are Given
Examples and Explorations

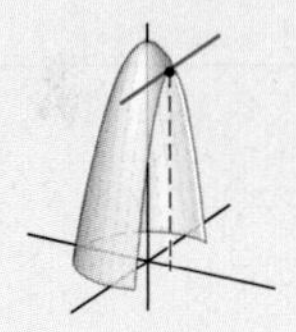

12.4 Directional Derivatives and Differentiability

The Directional Derivative
Differentiability
Examples and Explorations

$D_{\mathbf{u}} f(x_0, y_0)$

12.5 The Chain Rule and the Gradient

The Chain Rule
The Gradient
Using the Gradient to Compute the Directional Derivative
Examples and Explorations

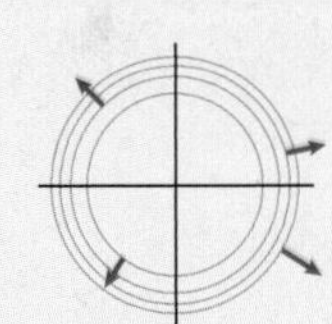

12.6 Extreme Values

The Gradient at a Local Extremum
The Second-Derivative Test for Classifying Local Extrema
Examples and Explorations

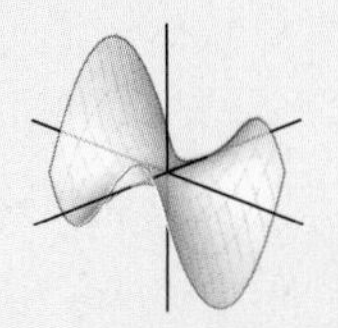

12.7 Lagrange Multipliers

Constrained Extrema
Optimizing a Function on a Closed and Bounded Set
Optimizing a Function with Two Constraints
Examples and Explorations

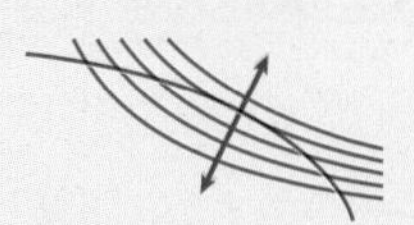

Chapter Review, Self-Test, and Capstones

12.1 FUNCTIONS OF TWO AND THREE VARIABLES

- Functions of two and three variables
- The graphs of linear functions of two variables and surfaces of revolution
- Level curves and level surfaces

Functions of Two Variables

Recall that a function $f : A \to B$ is an assignment that associates to each element x of the domain set A exactly one element $f(x)$ of the codomain set B. Calculus is a tool that allows us to analyze and understand functions. Thus far we have focused primarily on the following types of functions:

- For the first third of this text, we studied primarily the calculus of functions for which the sets A and B are subsets of the real numbers.
- When we studied sequences in Chapter 7, we insisted that the domain be the set of natural numbers while the target set was usually a subset of the real numbers.
- The vector-valued functions we studied in Chapter 11 had domains that were subsets of the real numbers and target sets that were subsets of $\mathbb{R}^n$ for a fixed value of $n \geq 2$.

We now turn our attention to functions whose domains are subsets of either $\mathbb{R}^2$ or $\mathbb{R}^3$ but that have a target set that is a subset of the real numbers. We will first consider functions whose domains are subsets of $\mathbb{R}^2$ and soon discuss functions whose domains are subsets of $\mathbb{R}^3$. We refer to these as functions of two variables and functions of three variables, respectively.

For example, consider the addition of two real numbers x and y. We may take addition to be a function $f : \mathbb{R}^2 \to \mathbb{R}$ that uses the rule

$$f(x, y) = x + y.$$

Note that we already have a slight, but common, abuse of notation. Since we denote elements of $\mathbb{R}^2$ in one of the notations $\langle x, y\rangle$ or (x, y), perhaps we should write $f(\langle x, y\rangle)$ or $f((x, y))$; however, we use the simpler and more common notation $f(x, y)$.

The order of the input variables x and y is significant in a function of two variables. Consider the function $g : R^2 \to \mathbb{R}$ defined by the rule $g(x, y) = 5x - y$. For this function we do not generally have $g(a, b) = g(b, a)$. For example, $g(1, 0) = 5 \cdot 1 - 0 = 5$ and $g(0, 1) = 5 \cdot 0 - 1 = -1$. (When would we have $g(a, b) = g(b, a)$?)

As with functions whose domains are subsets of $\mathbb{R}$, we often do not specify the domain of a function when it is a subset of $\mathbb{R}^2$. If the function is defined by an equation, we assume that the domain is the largest subset of $\mathbb{R}^2$ for which the equation is defined. The range of a function of two variables is the set of possible outputs.

DEFINITION 12.1 The Domain and Range of a Function of Two Variables

If f is a function that takes an element from an unspecified subset of $\mathbb{R}^2$ into $\mathbb{R}$, then we will take the ***domain*** of f to be the largest subset of $\mathbb{R}^2$ for which f is defined:

$$\text{Domain}(f) = \{(x, y) \in \mathbb{R}^2 | f(x, y) \text{ is defined}\}.$$

The ***range*** of f is the set of all possible outputs of f. That is,

$$\text{Range}(f) = \{z \in \mathbb{R} \mid \text{there exists } (x, y) \in \text{Domain}(f) \text{ with } f(x, y) = z\}.$$

For example, the domain of the function $h(x, y) = \sqrt{xy}$ consists of all points $(x, y) \in \mathbb{R}^2$ such that the product $xy \geq 0$. The latter is the case on the coordinate axes and when the signs of x and y are the same (i.e., in the first and third quadrants). The range of h consists of all nonnegative real numbers.

The domain of $h(x, y) = \sqrt{xy}$

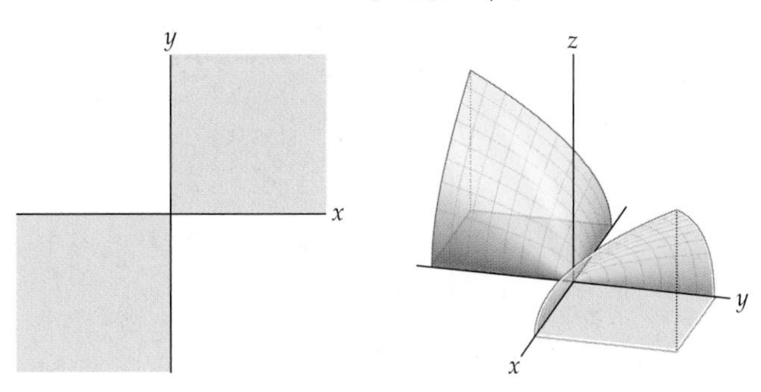

Graphing a Function of Two Variables

When we wish to analyze the behavior of a function, it is often helpful to understand the graph of the function.

DEFINITION 12.2

The Graph of a Function of Two Variables

The ***graph*** of a function of two variables, f, is the collection of ordered triples whose first two coordinates are in the domain of f. That is,

$$\text{Graph}(f) = \{(x, y, f(x, y)) \in \mathbb{R}^3 \mid (x, y) \in \text{Domain}(f)\}.$$

In other words the graph of a function of two variables is a particular subset of a three-dimensional Cartesian coordinate system. Consider the first function we mentioned in this chapter: $f(x, y) = x + y$. The graph of f is the subset of $\mathbb{R}^3$ given by

$$\text{Graph}(f) = \{(x, y, x + y) \mid (x, y) \in \mathbb{R}^2\}.$$

If we let $z = f(x, y) = x + y$, we see that the graph of f consists of exactly those points in the plane $z = x + y$.

The graph of $f(x, y) = x + y$

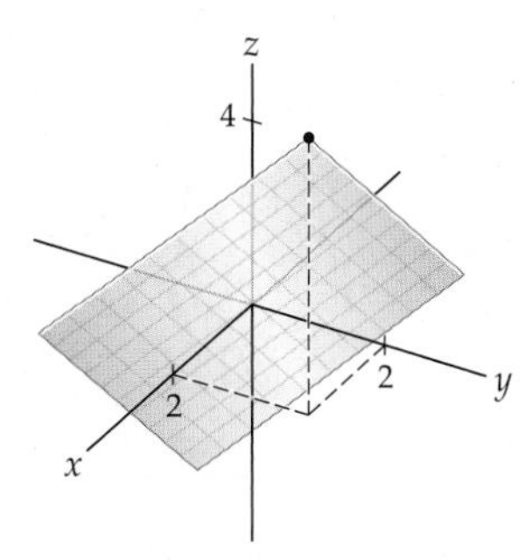

In Exercise 69 we ask you to prove that when a, b, and c are constants, the graph of the function $f(x, y) = ax + by + c$ is a plane.

In Chapter 6 we discussed another type of surface that may be readily expressed by a function of two variables. Recall that when we revolve the graph of a continuous function f that exists on an interval $[a, b]$, where $0 \leq a < b$, around the y-axis, we obtain a surface of revolution. As a reminder, consider the surface we obtain when we revolve the graph of the function $f(x) = x^2$, on the interval $[0, 2]$, around the z-axis. Following are the graphs of the function and the surface:

The function $z = x^2$

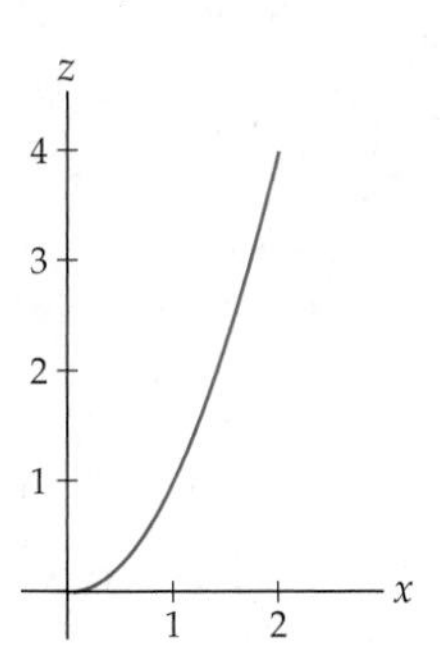

The surface of revolution $z = x^2 + y^2$

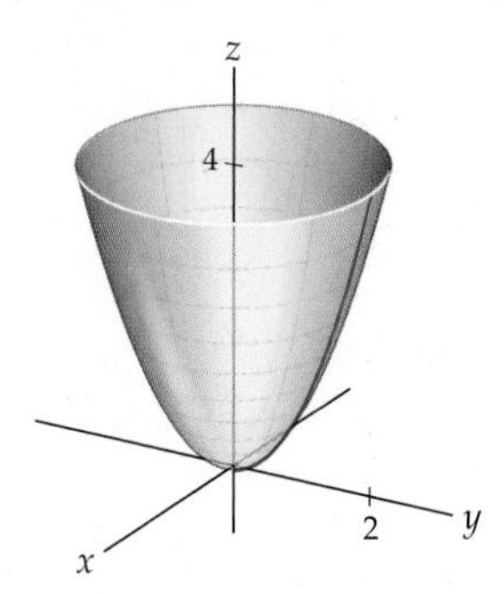

It is quite easy to express this surface as a function of two variables. When the interval we are using is $[a, b]$ with $0 \leq a < b$, we replace every occurrence of the variable x by the quantity $\sqrt{x^2 + y^2}$ in the formula for the function and evaluate $z = f(\sqrt{x^2 + y^2})$. In our simple example here we have $z = (\sqrt{x^2 + y^2})^2 = x^2 + y^2$. We provide other examples in the subsections that follow.

We will see that graphs of functions of two variables can be considerably more complicated than graphs of functions of a single variable. As we progress through this chapter, we will discuss limits, continuity, and differentiability for functions of two and three variables. Understanding these concepts will enable us to analyze the functions more thoroughly. In general, if a function of two variables is sufficiently well behaved, its graph will be a surface in three-dimensional space $\mathbb{R}^3$. The graph of the plane $z = x + y$ is one such example. But graphs of functions of two variables can be much more complicated. Here we see a graph of a portion of the surface defined by the function $f(x, y) = \sin(xy)$:

The graph of $f(x, y) = \sin(xy)$

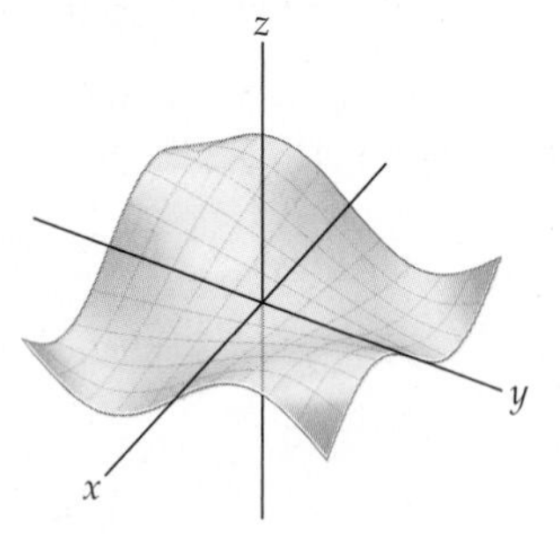

Functions of Three Variables

We now turn our attention to functions of three variables (i.e., those functions whose domains are subsets of $\mathbb{R}^3$ and whose codomains are subsets of $\mathbb{R}$). An example of a function of three variables is given by the formula $f(x, y, z) = xy + z - 7$. As before, we usually do not specify the domain of a function when it is a subset of $\mathbb{R}^3$. Rather, we assume that the domain is the largest subset of $\mathbb{R}^3$ for which the rule is defined. The range of a function of three variables is the set of possible outputs of the function.

DEFINITION 12.3

The Domain and Range of a Function of Three Variables

If f is a function that takes an element from an unspecified subset of $\mathbb{R}^3$ into $\mathbb{R}$, then we will take the ***domain*** of f to be the largest subset of $\mathbb{R}^3$ for which f is defined:

$$\text{Domain}(f) = \{(x, y, z) \in \mathbb{R}^3 \mid f(x, y, z) \text{ is defined}\}.$$

The ***range*** of f is the set of all possible outputs of f. That is,

$$\text{Range}(f) = \{w \in \mathbb{R} \mid \text{ there exists } (x, y, z) \in \text{ Domain}(f) \text{ with } f(x, y, z) = w\}.$$

For example, the domain of the function $f(x, y, z) = \sqrt{xyz}$ consists of all points $(x, y, z) \in \mathbb{R}^3$ such that the product $xyz \geq 0$. The latter is the case on the coordinate planes, in the first octant (when x, y, and z are all positive), and in those octants where exactly one of the signs of x, y, and z is the positive. The range of this function consists of all nonnegative real numbers.

DEFINITION 12.4

The Graph of a Function of Three Variables

The ***graph*** of a function of three variables, f, is the collection of ordered quadruples whose first three coordinates are in the domain of f. That is,

$$\text{Graph}(f) = \{(x, y, z, f(x, y, z)) \in \mathbb{R}^4 \mid (x, y, z) \in \text{Domain}(f)\}.$$

In other words the graph of a function of three variables is a particular subset of a four-dimensional Cartesian coordinate system! Note that we are not providing even a single example of the graph of a function of three variables here. We have mentioned that the graph of a sufficiently well-defined function of two variables is a surface (basically a two-dimensional object) in three-dimensional space $\mathbb{R}^3$. The graph of even the simplest function of three variables is a so-called hypersurface (three-dimensional object) existing in four-dimensional space, $\mathbb{R}^4$. Later in this section we will discuss how to use level surfaces to help visualize such objects, but it is often more efficient to try to understand them by using the techniques of calculus and *not* try to sketch them.

Level Curves and Level Surfaces

A topographic map renders a portion of the surface of the Earth in a two-dimensional form:

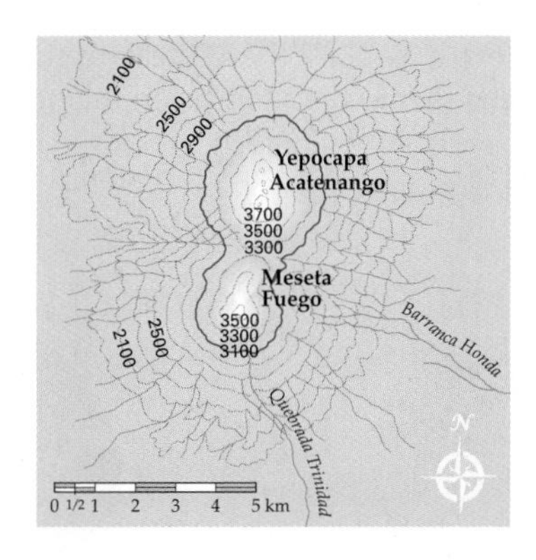

The map shows the Fuego and Acatenango volcanic complexes in Guatemala. Each curve on the map displays those points on the surface of these complexes that are at the same elevation. For example, the curve marked 3100 displays the points on the surface at 3100 meters. Thus, if you walked along that curve, you would neither ascend nor descend as you circumnavigated the volcanos. This topographic map gives us a way to understand the surface of the indicated region in Guatemala, although the map itself is flat. We will use this idea to help us understand functions of two variables. We make the following definition:

DEFINITION 12.5 **Level Curves**

Let f be a function of two variables and let c be a point in the range of f. The ***level curve*** for f at height c is the curve in the plane with equation $f(x, y) = c$.

Thus, each point on the level curve is at the same height on the surface determined by the function. The collection of level curves for a function of two variables, f, in effect, gives us a topographic map for the surface of the graph of f. For example, consider $f(x, y) = \sqrt{9 - (x^2 + y^2)}$, a function of two variables. The range of f is the interval $[0, 3]$. If we choose the values 0, 1, 2, and 3 for c, we have the equations

$$\sqrt{9 - (x^2 + y^2)} = 0,\ \sqrt{9 - (x^2 + y^2)} = 1,\ \sqrt{9 - (x^2 + y^2)} = 2 \text{ and } \sqrt{9 - (x^2 + y^2)} = 3.$$

The graphs of these four equations are the following level curves, although note that the only point that satisfies the equation $\sqrt{9 - (x^2 + y^2)} = 3$ is the origin $(0, 0)$:

Level curves for $f(x, y) = \sqrt{9 - (x^2 + y^2)}$

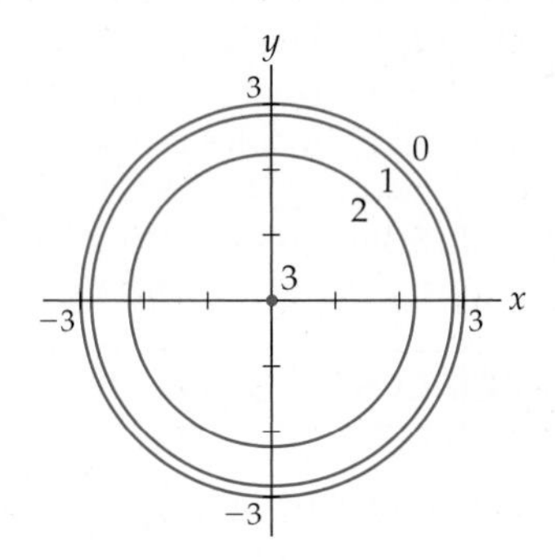

Now, if we stack these circles at the appropriate heights and interpolate similar circles between them, we can envision the hemisphere that is the graph of f:

The graph of $f(x, y) = \sqrt{9 - (x^2 + y^2)}$

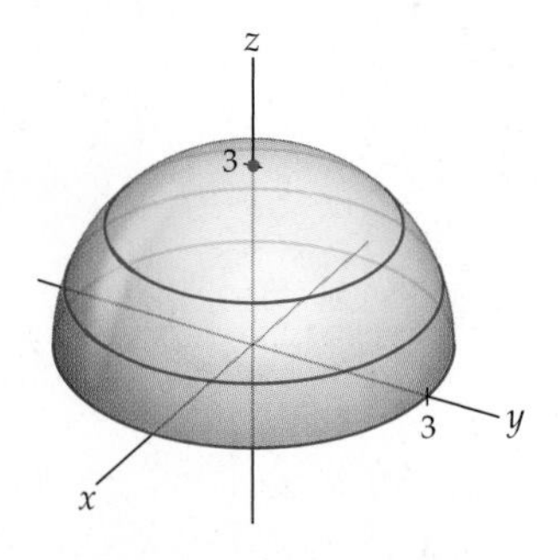

As we mentioned earlier, the graph of a function of three variables is a four-dimensional object. To aid our understanding of these functions, we can sometimes use level surfaces analogous to the level curves we have defined.

DEFINITION 12.6

Level Surfaces

Let f be a function of three variables and let c be a point in the range of f. The ***level surface*** for f at height c is the surface in three-dimensional space with equation $f(x, y, z) = c$.

We now consider an example quite similar to the one we just discussed. Let the function $g(x, y, z) = \sqrt{9 - (x^2 + y^2 + z^2)}$. The range of g is the interval $[0, 3]$. We will find the level surfaces for $c = 0$, 1, 2, and 3. When $c = 0$, we have $0 = \sqrt{9 - (x^2 + y^2 + z^2)}$, or equivalently, $x^2 + y^2 + z^2 = 9$. Thus, the level surface for $c = 0$ is a sphere of radius 3 and centered at the origin. You may check that the level surface for $c = 1$ is the sphere of radius $\sqrt{8}$ and centered at the origin; the level surface for $c = 2$ is the sphere of radius $\sqrt{5}$, also centered at the origin; and the level "surface" for $c = 3$ is the single point $(0, 0, 0)$. We see that each level surface of the graph of g is a sphere and that the radii of the level spheres get smaller as c increases from 0 to 3. The level surfaces are concentric spheres centered at the origin.

Three level surfaces for $g(x, y, z) = \sqrt{9 - (x^2 + y^2 + z^2)}$

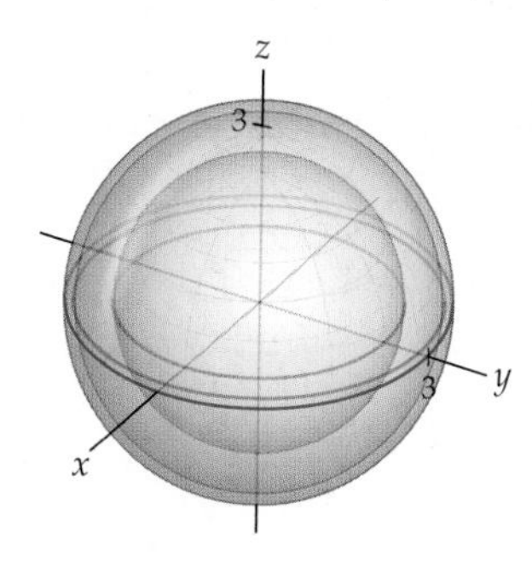

In fact, the graph of g is the "top" half of the hypersphere defined by the equation $w^2 + x^2 + y^2 + z^2 = 9$. The spheres that are the level surfaces for g are analogous to the circles that are the level curves for the function f we discussed before. Understanding the geometry of a function with one fewer variable can help us understand the graph of a figure that exists in a higher dimension.

Examples and Explorations

EXAMPLE 1

Understanding the basic properties of functions of two variables

For each of the following functions of two variables, evaluate the function at the given values and find the domain and range of the function:

(a) $f(x, y) = 3x - 4y + 5;\ (3, 4),\ (4, 3)$

(b) $g(x, y) = \dfrac{x^2 - y^2}{x};\ (2, 5),\ (3, 0)$

(c) $h(r, s) = \dfrac{\ln(2r + s)}{2r + s};\ (1, 0),\ (0, e)$

SOLUTION

(a) We first evaluate the function for the specified pairs of coordinates:

$$f(3, 4) = 3 \cdot 3 - 4 \cdot 4 + 5 = 9 - 16 + 5 = -2 \quad \text{and}$$
$$f(4, 3) = 3 \cdot 4 - 4 \cdot 3 + 5 = 12 - 12 + 5 = 5.$$

In particular, we note that the order of the coordinates is significant when we evaluate this function. The domain of the function $f(x, y) = 3x - 4y + 5$ is $\mathbb{R}^2$, since the function is defined for every ordered pair in $\mathbb{R}^2$. Finally, the range of f is $\mathbb{R}$, because every real number can be obtained (in infinitely many different ways). For example, if we let r be an arbitrary nonzero real number, the ordered pair $\left(\frac{r-5}{3}, 0\right)$ has the property that $f\left(\frac{r-5}{3}, 0\right) = r$. Thus, the function f maps $\mathbb{R}^2$ *onto* $\mathbb{R}$.

(b) We evaluate the function g for the specified pairs of coordinates:

$$g(2, 5) = \frac{2^2 - 5^2}{2} = \frac{4 - 25}{2} = -\frac{21}{2} \quad \text{and} \quad g(3, 0) = \frac{3^2 - 0^2}{3} = 3.$$

The domain of the function g is the set of all ordered pairs $\{(x, y) \in \mathbb{R}^2 \mid x \neq 0\}$, since g is defined for all ordered pairs except when $x = 0$. As with f, the range of g is $\mathbb{R}$, because every real number can be obtained as an output from g. For example, $g(r, 0) = r$ and $g(r, r) = 0$ for every nonzero real number r.

(c) Before we evaluate the function h for the specified pairs of coordinates, note that we are using different letters to represent our variables, but h is just another function of two variables. When we evaluate h for a particular coordinate pair, we assume that the first value in the pair is the r-value and the second is the s-value, since we wrote the function as $h(r, s)$. Thus,

$$h(1, 0) = \frac{\ln(2 \cdot 1 + 0)}{2 \cdot 1 + 0} = \frac{\ln 2}{2} \quad \text{and} \quad h(0, e) = \frac{\ln(2 \cdot 0 + e)}{2 \cdot 0 + e} = \frac{1}{e}.$$

Because the domain of the natural logarithm is all positive real numbers, the domain of the function h is the set of all ordered pairs $\{(r, s) \in \mathbb{R}^2 \mid 2r + s > 0\}$, shown here:

The domain of $h(r, s) = \dfrac{\ln(2r + s)}{2r + s}$

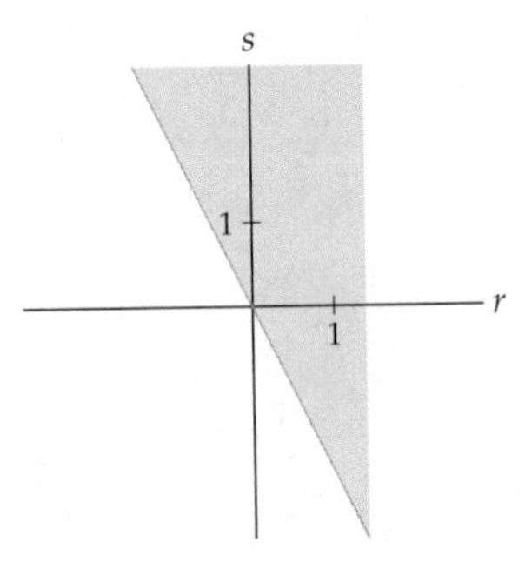

Finding the range of h is slightly more difficult. As an aid, we note that if we let $w = 2r+s$, we may represent the function in the form $h(w) = \frac{\ln w}{w}$. Using the techniques of single-variable calculus, you may show that this function has the following graph:

The graph of $h(w) = \dfrac{\ln w}{w}$

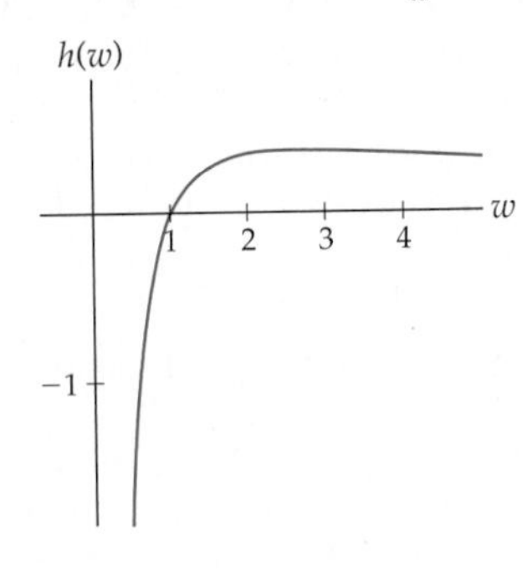

It can be shown that the maximum of this function, $1/e$, occurs when $w = e$ and that $\lim_{w \to 0^+} \frac{\ln w}{w} = -\infty$. Therefore the range of the function $\frac{\ln w}{w}$ is $(-\infty, 1/e]$. Since the quantity $2r + s$ can take on any real value, this interval will also be the range of $h(r, s) = \frac{\ln(2r+s)}{2r+s}$. □

EXAMPLE 2 **Composing functions of two variables**

Consider the functions

$$g(t) = t^3,\ h(t) = t^2 - 7,\ f(x, y) = x^2 - y,\ w(x, y) = \frac{x}{y},\ \text{ and } \mathbf{r}(t) = \langle \sin t,\ \cos t \rangle$$

For each of the following, simplify the expression or explain why it is not possible to evaluate that particular expression:

(a) $f(g(t))$ **(b)** $g(f(x, y))$ **(c)** $w(g(t), h(t))$ **(d)** $w(\mathbf{r}(t))$ **(e)** $f(w(x, y))$

SOLUTION

(a) This composition is not defined. The range of the function g is $\mathbb{R}$, but the domain of f is $\mathbb{R}^2$.

(b) This composition is defined, and

$$g(f(x, y)) = g(x^2 - y) = (x^2 - y)^3.$$

(c) This composition is also defined, and

$$w(g(t), h(t)) = \frac{g(t)}{h(t)} = \frac{t^3}{t^2 - 7}.$$

(d) The output of the vector function $\mathbf{r}$ is the ordered pair $\langle \sin t, \cos t \rangle$, so this composition is defined and

$$w(\mathbf{r}(t)) = \frac{\sin t}{\cos t} = \tan t.$$

(e) This composition is not defined. The range of the function w is a subset of $\mathbb{R}$, but the domain of f is $\mathbb{R}^2$. □

EXAMPLE 3 **Graphing surfaces of revolution**

Sketch the surfaces formed when the graphs of the following functions on the specified intervals are revolved around the z-axis:

(a) $f(x) = x - 1$ on the interval $[0, 2]$ **(b)** $g(x) = x^3 - x$ on the interval $[-1, 0]$

Also, find the equation for each surface as a function of two variables.

SOLUTION

We begin by noting that, for each function, we are considering the dependent variable to be z. Thus the graph of the function f on the interval $[0, 2]$ is a line segment, and the graph of g on the interval $[-1, 0]$ is a portion of the cubic function, both in the xz-plane.

(a) For our first function, since our interval is $[0, 2]$, all we need to do to obtain the equation of the first surface of revolution as a function of two variables is to replace x with $\sqrt{x^2 + y^2}$ in the formula for the function f and obtain

$$z = f\left(\sqrt{x^2 + y^2}\right) = \sqrt{x^2 + y^2} - 1.$$

The graphs of the function and the surface are as follows:

The function $f(x) = x - 1$

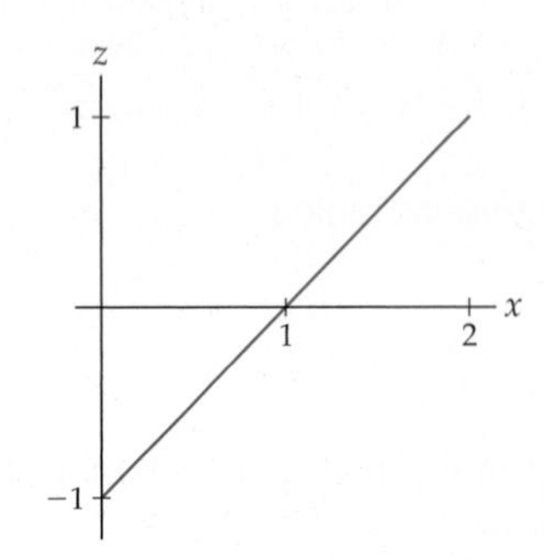

The surface of revolution $z = \sqrt{x^2 + y^2} - 1$

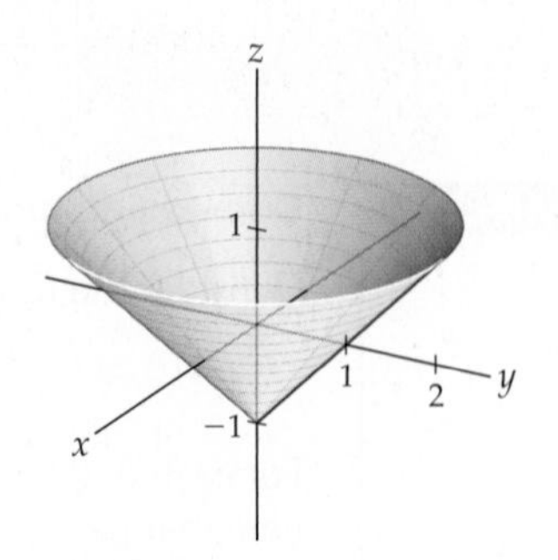

(b) For our second function, our interval is $[-1, 0]$. When we have an interval $[a, b]$ with $a < b \leq 0$, we consider the function $g(-x)$ on the interval $[-b, -a]$ instead. Here we have

$$g(-x) = ((-x)^3 - (-x)) = x - x^3 \text{ on } [0, 1],$$

and we proceed as we did in part (a). To obtain the equation of the second surface of revolution as a function of two variables, we replace the occurrences of x with $\sqrt{x^2 + y^2}$ in the formula for $g(-x)$ and obtain

$$z = g\left(-\sqrt{x^2 + y^2}\right) = (x^2 + y^2)^{1/2} - (x^2 + y^2)^{3/2}.$$

Here are the graphs of the functions and the surface:

The functions
$g(x)$ on $[-1, 0]$ and $g(-x)$ on $[0, 1]$

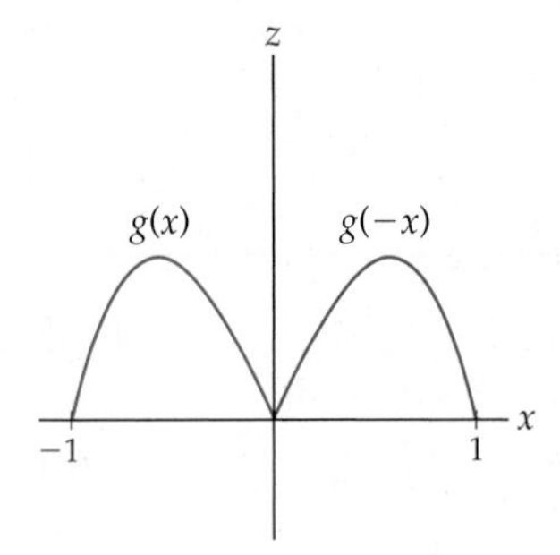

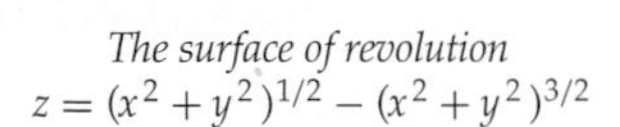

The surface of revolution
$z = (x^2 + y^2)^{1/2} - (x^2 + y^2)^{3/2}$

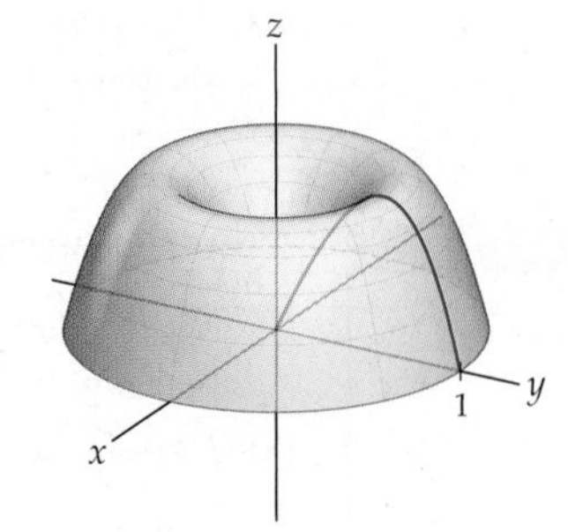

□

EXAMPLE 4

Drawing level curves

Let $f(x, y) = -\dfrac{8x}{x^2 + y^2 + 1}$. Sketch level curves $f(x, y) = c$ for $c = -3, -2, -1, 0, 1, 2,$ and 3.

SOLUTION

For each value of c, we will graph the set of points that are solutions of the equation $-\dfrac{8x}{x^2 + y^2 + 1} = c$. We begin with $c = 0$. The equation $-\dfrac{8x}{x^2 + y^2 + 1} = 0$ holds exactly when $x = 0$. Thus, the level curve here is the y-axis.

Next, when $c = 1$, we have the equation

$$-\frac{8x}{x^2 + y^2 + 1} = 1.$$

The denominator of the function is never zero. Therefore, we obtain the equivalent equation $-8x = x^2 + y^2 + 1$ when we multiply both sides of the given equation by $x^2 + y^2 + 1$. Then,

$$\begin{aligned} -8x &= x^2 + y^2 + 1 \\ x^2 + 8x + y^2 &= -1 && \leftarrow \text{arithmetic} \\ (x+4)^2 + y^2 &= 15 && \leftarrow \text{completing the square} \end{aligned}$$

At this point we recognize that the level curve when $c = 1$ is the circle with radius $\sqrt{15}$ and centered at $(-4, 0)$. Nearly identical calculations reveal that the level curve when $c = -1$ is given by the equation $(x - 4)^2 + y^2 = 15$, whose graph is the circle with radius $\sqrt{15}$ and centered at $(4, 0)$.

The other level curves we wish to find are also circles. Similar calculations provide the information in the following table:

c	equation	radius	center
2	$(x+2)^2 + y^2 = 3$	$\sqrt{3}$	$(-2, 0)$
-2	$(x-2)^2 + y^2 = 3$	$\sqrt{3}$	$(2, 0)$
3	$\left(x + \frac{4}{3}\right)^2 + y^2 = \frac{7}{9}$	$\frac{\sqrt{7}}{3}$	$\left(-\frac{4}{3}, 0\right)$
-3	$\left(x - \frac{4}{3}\right)^2 + y^2 = \frac{7}{9}$	$\frac{\sqrt{7}}{3}$	$\left(\frac{4}{3}, 0\right)$

Therefore, we have the following graph of the level curves:

Level curves for $f(x, y) = -\dfrac{8x}{x^2 + y^2 + 1}$

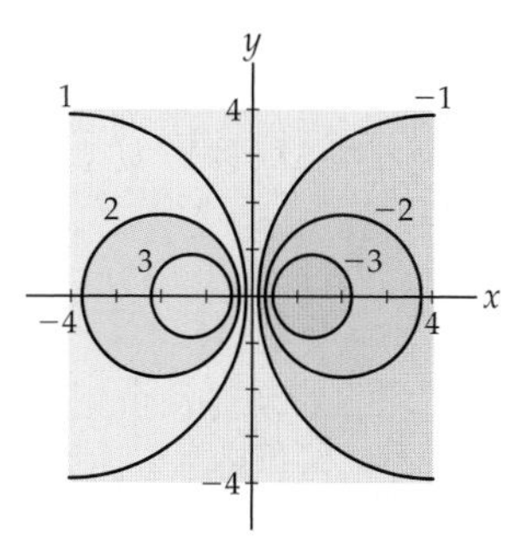

We may extend the analysis we used to obtain these level curves to find the range of the function f. (See Exercise 8.)

Finally, the graph of f shown next was generated by a computer algebra system. The breaks between the colors correspond to the level curves we just found. If you have access to such a program, it can greatly aid your understanding of the graphs of functions of two variables.

$$f(x, y) = -\frac{8x}{x^2 + y^2 + 1}$$

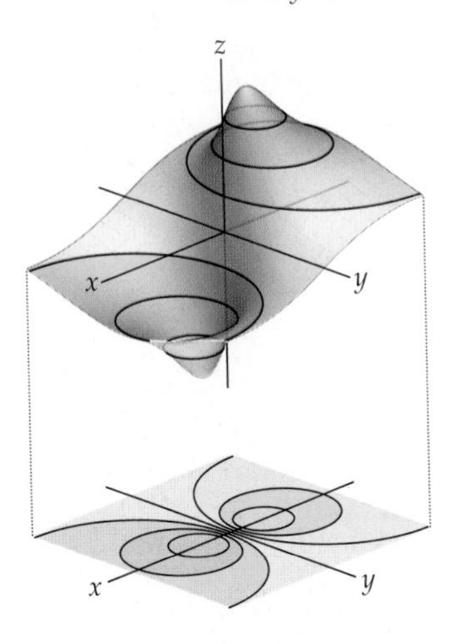

□

EXAMPLE 5

Understanding the basic properties of functions of three variables

For each of the following functions of three variables, evaluate the function at the given values and find the domain and range of the function:

(a) $f(x, y, z) = xyz + xy + z;\ (1, 2, 3)$ **(b)** $g(x, y, z) = \frac{x^2 - y^2}{x + z};\ (-3, 2, 4)$

(c) $h(r, s, t) = \frac{\ln(r + s + t)}{r + s + t};\ (e, e^2, -e)$

SOLUTION

(a) We first evaluate the function for the specified triple of coordinates:

$$f(1, 2, 3) = 1 \cdot 2 \cdot 3 + 1 \cdot 2 + 3 = 6 + 2 + 3 = 11.$$

The domain of the function $f(x, y, z) = xyz + xy + z$ is $\mathbb{R}^3$, since the function is defined for every ordered triple in $\mathbb{R}^3$. Finally, the range of f is $\mathbb{R}$, because every real number can be obtained (in infinitely many different ways). For example, let r be an arbitrary real number, then the ordered triple $(0, 0, r)$ has the property that $f(0, 0, r) = r$. Thus, the function f maps $\mathbb{R}^3$ *onto* $\mathbb{R}$.

(b) We first evaluate the function g for the specified triple of coordinates:

$$g(-3, 2, 4) = \frac{(-3)^2 - 2^2}{-3 + 4} = \frac{9 - 4}{1} = 5.$$

The domain of the function g is the set of all ordered triples $\{(x, y, z) \in \mathbb{R}^3 \mid x + z \neq 0\}$, since g is defined for all triples except when $x + z = 0$. That is, the domain of g consists of all points in $\mathbb{R}^3$ except those points on the plane defined by the equation $x + z = 0$. The range of g is $\mathbb{R}$, because every real number can be obtained as an output from g. For example, if we let r be an arbitrary nonzero real number, the ordered pair $(r, 0, 0)$ has the property that $g(r, 0, 0) = r$.

(c) This function is quite similar to the function found in part (c) of Example 1. Thus, our analysis will be quite similar. We have

$$h(e, e^2, -e) = \frac{\ln(e + e^2 - e)}{e + e^2 - e} = \frac{2}{e^2}.$$

Since the domain of the natural logarithm is all positive real numbers, the domain of the function h is the set of all ordered triples $\{(r, s, t) \in \mathbb{R}^3 \mid r + s + t > 0\}$. That is, the domain consists of all points in $\mathbb{R}^3$ above the plane defined by the equation $r + s + t = 0$.

As in part (c) of Example 1, the range of $h(r, s, t) = \frac{\ln(r+s+t)}{r+s+t}$ is the interval $(-\infty, 1/e]$. □

? TEST YOUR UNDERSTANDING

- What is a function of two variables? What is a function of three variables? What is a function of n variables?
- How do you determine the domain of a function of two or three variables? What is the range of a function?
- Why is the graph of a linear function of two variables a plane?
- How many dimensions are required to sketch the graph of a function of two variables? How many dimensions are required to sketch the graph of a function of three variables?
- What is a level curve? What is a level surface? How can level curves be used to understand a function of two variables? How can level surfaces be used to understand a function of three variables?

EXERCISES 12.1

Thinking Back

- *The definition of a function:* Explain why Definition 0.1 is general enough to include functions of two and three variables.
- *The domain and range of a function of two variables:* Explain why Definition 0.2 is not general enough to define the domain or range of a function of two or three variables.

Concepts

0. *Problem Zero:* Read the section and make your own summary of the material.

1. *True/False:* Determine whether each of the statements that follow is true or false. If a statement is true, explain why. If a statement is false, provide a counterexample.

(a) *True or False:* The domain of a function of two variables is a subset of $\mathbb{R}^2$.

(b) *True or False:* The range of a function of two variables is a subset of $\mathbb{R}^2$.

(c) *True or False:* The graph of a function of two variables is a subset of $\mathbb{R}^3$.

(d) *True or False:* The domain of a function of three variables is a subset of $\mathbb{R}$.

(e) *True or False:* The range of a function of three variables is a subset of $\mathbb{R}$.

(f) *True or False:* The graph of a function of three variables is a subset of $\mathbb{R}^4$.

(g) *True or False:* The graph of a linear function of two variables is a plane.

(h) *True or False:* If a function $f : \mathbb{R} \to \mathbb{R}$ is continuous on an interval $[0, p]$ then the surface formed when the graph of f is rotated about the y-axis may be expressed as a function of two variables.

2. *Examples:* Construct examples of the thing(s) described in the following. Try to find examples that are different than any in the reading.

(a) A function of two variables whose graph is a plane.

(b) A function of two variables whose graph is the surface of revolution formed when the graph of a function of a single variable is revolved around the y-axis.

(c) A function of two variables for which each level curve is a parabola.

3. Let $f : \mathbb{R} \to \mathbb{R}$ be a function of a single variable. Explain why the graph of f is a subset of $\mathbb{R}^2$.

4. Let $f : \mathbb{R}^2 \to \mathbb{R}$ be a function of two variables. Explain why the graph of f is a subset of $\mathbb{R}^3$.

5. Let $f : \mathbb{R}^3 \to \mathbb{R}$ be a function of three variables. Explain why the graph of f is a subset of $\mathbb{R}^4$.

6. (a) Graph $f(x) = x$.

(b) Graph the function $f(x, y) = x + y$. Explain why the graph of $f(x, y) = x + y$ contains the origin, $(0, 0, 0)$.

(c) The graph of the function

$$f(x, y, z) = x + y + z$$

is a ***hyperplane*** of dimension 3 in $\mathbb{R}^4$. Explain why "hyperplane" is a good name for the graph. ("Hyper" is from a Greek word meaning *over* or *beyond*. The Latin equivalent is super, meaning essentially the same thing.) Explain why the graph of f contains the origin, $(0, 0, 0, 0)$.

(d) Fill in the blanks: The graph of the function $f(x_1, x_2, \ldots, x_n) = x_1 + x_2 + \cdots + x_n$ is a hyperplane of dimension ______ in ______.

7. (a) Graph the function $f(x) = \sqrt{1 - x^2}$.

(b) Graph the function $f(x, y) = \sqrt{1 - x^2 - y^2}$. (*Hint: Let $z = \sqrt{1 - x^2 - y^2}$, and then square both sides of the equation.*)

(c) The graph of the function

$$f(x, y, z) = \sqrt{1 - x^2 - y^2 - z^2}$$

is half of a ***hypersphere*** of dimension 3. Explain why "hypersphere" is a good name for the graph and why the graph is only half of the hypersphere. What equation defines the entire hypersphere?

(d) Fill in the blanks: The graph of the function $f(x_1, x_2, \ldots, x_n) = \sqrt{1 - x_1^2 - x_2^2 - \cdots - x_n^2}$ is half of a ______ of dimension ______ in ______.

8. Find the range of the function $f(x, y) = -\dfrac{8x}{x^2 + y^2 + 1}$ from Example 4 by finding the largest value, M, and smallest value, m, of c for which the equation

$$-\frac{8x}{x^2 + y^2 + 1} = c$$

has a solution. (*Hint: Find the two values for c such that the level "curves" are a single point. Then show that the given equation has a solution for every value of c between those two values.*)

9. Let $z = f(x, y)$ be a function of two variables. Explain why the two sets $\{(x, y) \mid (x, y, z) \in \text{Graph}(f)\}$ and $\text{Domain}(f)$ are identical.

10. Let $z = f(x, y)$ be a function of two variables. Explain why the two sets $\{z \mid (x, y, z) \in \text{Graph}(f)\}$ and $\text{Range}(f)$ are identical.

11. Let $w = f(x, y, z)$ be a function of three variables. Explain why the two sets $\{(x, y, z) \mid (x, y, z, w) \in \text{Graph}(f)\}$ and $\text{Domain}(f)$ are identical.

12. Let $w = f(x, y, z)$ be a function of three variables. Explain why the two sets $\{w \mid (x, y, z, w) \in \text{Graph}(f)\}$ and $\text{Range}(f)$ are identical.

In Exercises 13–21, provide a rough sketch of the graph of a function of two variables with the specified level "curve(s)." (There are many possible correct answers to each question.)

13. One level curve consists of a single point.

14. One level curve consists of exactly two points.

15. One level curve is a circle together with the point that is the center of the circle.

16. One level curve consists of all the points (m, n) where m and n are both integers.

17. All of the level curves are circles except one, which is a point.

18. All of the level curves are circles.

19. Some level curves consist of two concentric circles.

20. Some level curves consist of infinitely many concentric circles.

21. All of the level curves are squares.

Skills

In Exercises 22–28, evaluate the given function at the specified points in the domain, and then find the domain and range of the function.

22. $f(x, y) = x^2 y^3$, $\left(\frac{1}{2}, \frac{4}{3}\right)$, $(0, \pi)$

23. $f(x, y) = x^2 - y^2$, $(1, 5)$, $(-3, -2)$

24. $g(x, y) = \dfrac{x^2 + y^2}{x + y}$, $(\pi, 1)$, $(4, 5)$

25. $g(x, y) = \dfrac{\ln(xy)}{\sqrt{1 - x}}$, $(-e, -1)$, $\left(\frac{1}{2}, 2\right)$

26. $f(x, y) = \dfrac{\sin(x - y)}{x - y}$, $(0, \pi)$, $\left(\frac{\pi}{2}, \frac{\pi}{3}\right)$

27. $f(x, y, z) = x^2 + y^2 + z^2$, $(1, 0, -5)$, $\left(\frac{1}{2}, -1, \frac{1}{3}\right)$

28. $f(x, y, z) = \dfrac{x + y + z}{xyz}$, $(1, -2, 6)$, $(-3, -2, 4)$

In Exercises 29–36, let

$$g_1(t) = \sin t,\ g_2(t) = \cos t,\ g_3(t) = 1 - t,$$

$$f_1(x, y) = x^2 + y^2,\ f_2(x, y) = \frac{x^2}{y^2},\ f_3(x, y, z) = \frac{x + y}{y + z},$$

$$\mathbf{r}_1(t) = \langle 1 + t, t - 1 \rangle, \mathbf{r}_2(t) = \langle t, t^2, t^3 \rangle.$$

Either simplify the specified composition or explain why the composition cannot be formed.

29. $f_1(g_1(t), g_2(t))$

30. $g_1(f_1(x, y), f_2(x, y))$

31. $g_1(g_2(t))$

32. $f_1(\mathbf{r}_1(t))$

33. $f_3(g_2(t), g_1(t), g_3(t))$

34. $f_1(f_2(x, y), f_3(x, y, z))$

35. $\mathbf{r}_2(g_3(t))$

36. $\mathbf{r}_2(f_3(x, y, z))$

In Exercises 37–42, sketch the surface of revolution formed when the given function on the specified interval is revolved around the z-axis and find a function of two variables with the surface as its graph.

37. $f(x) = x,\ [0, 3]$

38. $f(x) = \sqrt{x},\ [0, 4]$

39. $f(x) = x^2,\ [0, 2]$

40. $f(x) = x^{3/2},\ [0, 1]$

41. $f(x) = \sin x,\ \left[0, \frac{\pi}{2}\right]$

42. $f(x) = \cos x,\ \left[0, \frac{\pi}{2}\right]$

In Exercises 43–52, sketch the level curves $c = -3, -2, -1, 0, 1, 2, 3$ if they exist for the specified function.

43. $f(x, y) = x + y$

44. $f(x, y) = \dfrac{3x}{y}$

45. $f(x, y) = \dfrac{3x}{y^2}$

46. $f(x, y) = \sqrt{xy}$

47. $f(x, y) = 4 - (x^2 + y^2)$

48. $f(x, y) = x^2 - y^2$

49. $f(x, y) = \sin(x + y)$

50. $f(x, y) = \cos(xy)$

51. $f(x, y) = y \csc x$

52. $f(x, y) = y \sec x$

In Exercises 53–58, determine the level surfaces $c = -3, -2, -1, 0, 1, 2, 3$ if they exist for the specified function.

53. $f(x, y, z) = x + 2y + 3z$

54. $f(x, y, z) = -2x - y + 4z$

55. $f(x, y, z) = \dfrac{x}{y - z}$

56. $f(x, y, z) = \dfrac{x^2 + y^2}{z}$

57. $f(x, y, z) = x^2 + y^2 + z^2$

58. $f(x, y, z) = \sqrt{1 - (x^2 + y^2 + z^2)}$

In Exercises 59–62, match the given function of two variables with the surfaces in Figures I–IV and with the level curves in Figures A–D. Make sure you explain your reasoning.

59. $f(x, y) = x^2 - y^2$

60. $f(x, y) = e^{-(x+1)^2-(y+1)^2} + e^{-(x-1)^2-(y-1)^2}$

61. $f(x, y) = e^{-x^2} + e^{-y^2}$

62. $f(x, y) = \sin x + \sin y$

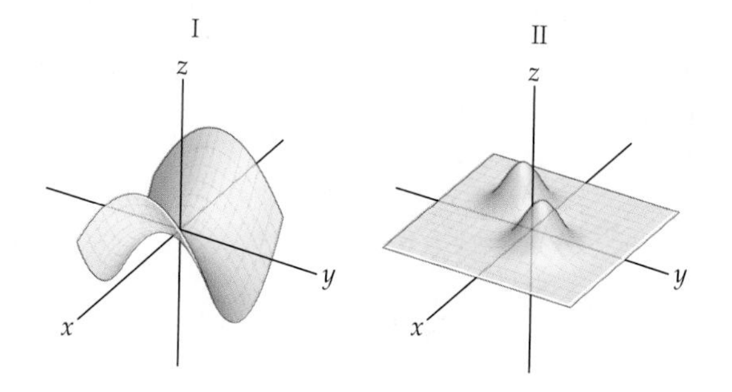

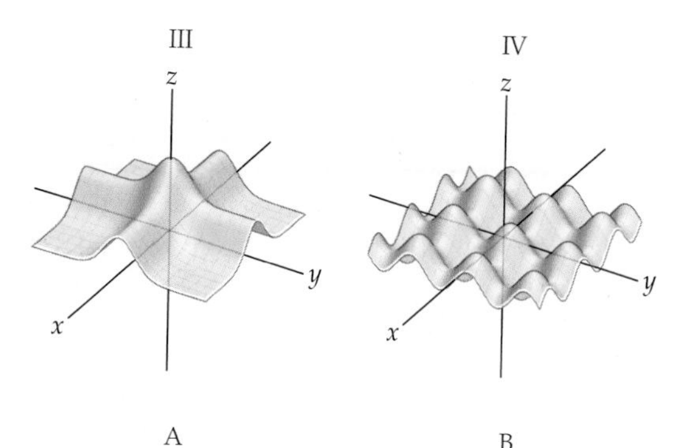

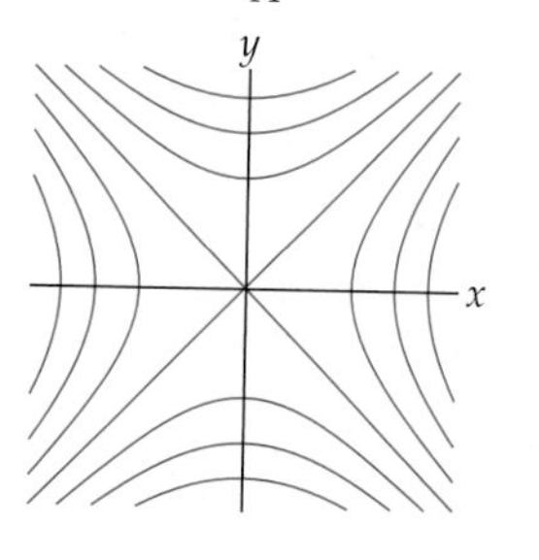

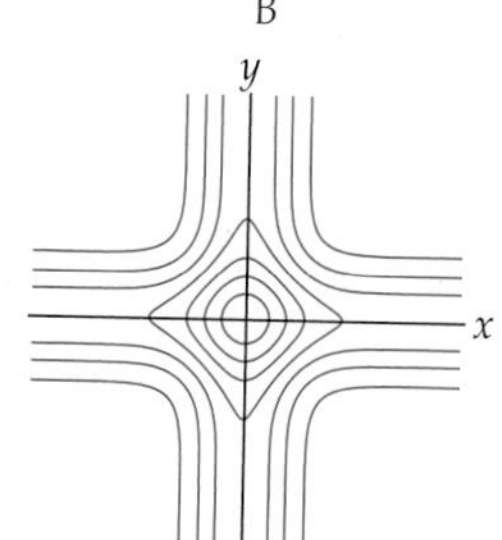

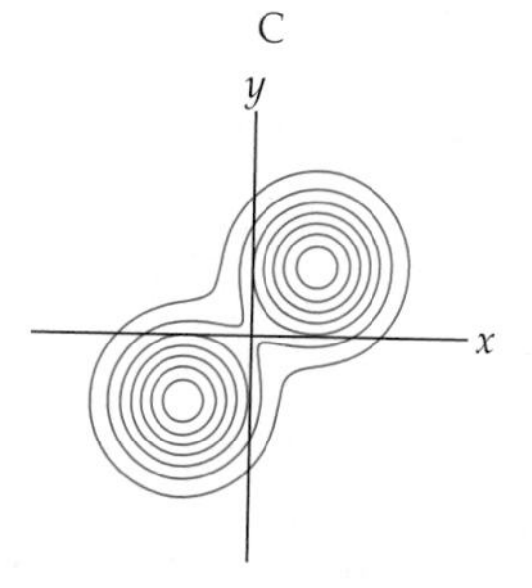

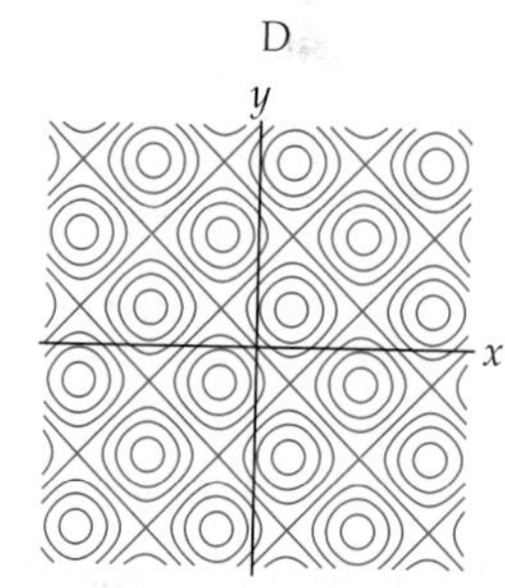

63. We may express the volume of a right circular cylinder by using the function of two variables, $V(r, h) = \pi r^2 h$, where r is the radius of either end of the cylinder and h is the height of the cylinder. What is the domain of the function $V(r, h)$? Express the surface area, S, of a cylinder as a function of variables r and h.

64. Express the volume, V, and surface area, S, of a right circular cone with radius r and height h as functions of two variables. What is the domain of each function?

65. Express the volume, V, and surface area, S, of a rectangular parallelepiped (i.e., a box) with side lengths x, y, and z as functions of three variables. What is the domain of each function?

66. Express the formulas for converting from polar coordinates to rectangular coordinates found in Section 9.2 as functions of two variables. What is the domain of each function?

Applications

67. Leila has been gathering data on the population density of caribou in a valley of the Selkirk Range in British Columbia, Canada. In winter, the caribou stay close to the bottom of the valley. Leila models the population for February with the function

$$p_F(x, y) = 13e^{-(y+0.5x-2)^2}e^{-0.2(x-1)^2},$$

where x and y are measured in miles from the center of the valley.

(a) Where are the caribou most likely to be found?

(b) During the summer, there is a stream in the valley. What is a vector in the direction of the stream?

68. Emmy is still trying to track down a leak in her Hanford tank farm. She has modeled the depth of a layer of basalt under the farm as a plane, but she has come to realize that there is a joint in the basalt that drops the level by a foot north of the curve with equation $x = 50 + 0.3y^{1.25}$. Thus, her model for the basalt layer is now

$$b(x, y) = -40 - \frac{1}{350}x + \frac{1}{725}y - u(x - 50 + 0.3y^{1.25}),$$

where u is the unit step function. The toxic solution must flow along the impenetrable layer of basalt. What is the route of its flow, and what is a vector giving the direction of its flow at any point?

Proofs

69. For constants a, b, and c, a function of two variables of the form $f(x, y) = ax + by + c$ is called a ***linear function of two variables***. Show that the graph of the linear function $f(x, y) = ax + by + c$ is a plane with normal vector $\langle a, b, -1 \rangle$ containing the point $(0, 0, c)$.

70. Let $z = f(x, y)$ be a function of two variables. Prove that when $c_1 \neq c_2$, the level curves defined by the equations $f(x, y) = c_1$ and $f(x, y) = c_2$ do not intersect.

71. Let $z = f(x, y)$ be a function of two variables. Prove that if the level curves defined by the equations $f(x, y) = c_1$ and $f(x, y) = c_2$ intersect, then the curves are identical. (*Hint: See Exercise 70.*)

72. Let $w = f(x, y, z)$ be a function of three variables. Prove that when $c_1 \neq c_2$, the level surfaces defined by the equations $f(x, y, z) = c_1$ and $f(x, y, z) = c_2$ do not intersect.

73. Let $w = f(x, y, z)$ be a function of three variables. Prove that if the level surfaces defined by the equations $f(x, y, z) = c_1$ and $f(x, y, z) = c_2$ intersect, then the surfaces are identical. (*Hint: See Exercise 72.*)

Thinking Forward

▶ *A kind of derivative for a function of two variables:* Explain why the derivative of the function $\frac{3x}{y^4}$ is $\frac{3}{y^4}$ if x is the variable and y is a constant and is $-\frac{12x}{y^5}$ if y is the variable and x is a constant. What is the derivative if both x and y are constants?

▶ *A kind of derivative for a function of three variables:* Explain why the derivative of the function $xe^{-4z}\sin y$ is $e^{-4z}\sin y$ if x is the variable and y and z are constants, and the derivative is $xe^{-4z}\cos y$ if y is the variable and x and z are constants, and the derivative is $-4xe^{-4z}\sin y$ if z is the variable and x and y are constants. What is the derivative if x, y, and z are all constants?

12.2 OPEN SETS, CLOSED SETS, LIMITS, AND CONTINUITY

- ▶ Open sets and closed sets
- ▶ The limit for a function of two or more variables
- ▶ The continuity of functions of two or more variables

Open Sets and Closed Sets

The tools of calculus introduced in Chapters 1 and 2 allow us to analyze functions of a single variable. The definitions of the limit, continuity, and differentiability require that our functions be defined on an open interval. Some theorems, such as the Extreme Value Theorem, require that our functions be defined on a closed interval. To understand these concepts for functions of two and three variables, we will spend the next few pages defining terms like *open sets, closed sets,* and *bounded sets*. It may be helpful to review the basic calculus concepts and theorems you studied in those earlier chapters before proceeding through this section.

We first define the following:

DEFINITION 12.7 **Open Disks and Open Balls**

Let $\epsilon > 0$.

(a) Let $(x_0, y_0) \in \mathbb{R}^2$. A subset of $\mathbb{R}^2$ of the form

$$\{(x, y) \mid (x - x_0)^2 + (y - y_0)^2 < \epsilon\}$$

is said to be an ***open disk*** in $\mathbb{R}^2$.

(b) Let $(x_0, y_0, z_0) \in \mathbb{R}^3$. A subset of $\mathbb{R}^3$ of the form

$$\{(x, y, z) \mid (x - x_0)^2 + (y - y_0)^2 + (z - z_0)^2 < \epsilon\}$$

is said to be an ***open ball*** in $\mathbb{R}^3$.

Thus, an open disk is a subset of $\mathbb{R}^2$ contained within a circle, not including the points on the circumference of the circle. Similarly, an open ball is a subset of $\mathbb{R}^3$ contained within a sphere, not including the points on the surface of the sphere.

An open disk in $\mathbb{R}^2$

An open ball in $\mathbb{R}^3$

These basic objects allow us to define open sets in $\mathbb{R}^2$ and $\mathbb{R}^3$.

DEFINITION 12.8 **Open Sets in $\mathbb{R}^2$ and $\mathbb{R}^3$**

(a) A subset S of $\mathbb{R}^2$ is said to be ***open*** if, for every point $(x, y) \in S$, there is an open disk D such that $(x, y) \in D \subseteq S$.

(b) A subset S of $\mathbb{R}^3$ is said to be ***open*** if, for every point $(x, y, z) \in S$, there is an open ball B such that $(x, y, z) \in B \subseteq S$.

A set S in the plane is open if, for every point in S, there is an open disk that both contains the point and is a subset of the set S

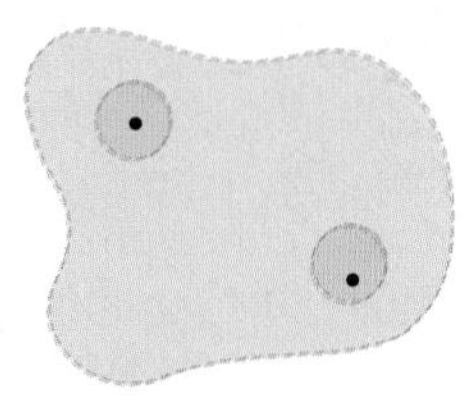

The following four examples illustrate the definition of an open set in $\mathbb{R}^2$ (consider the analogous examples in $\mathbb{R}^3$):

1. Every open disk is itself an open set. For example, the unit open disk
$$U = \{(x, y) \mid x^2 + y^2 < 1\}$$
is an open set because, for every point (x, y) in U, there is another smaller, open disk D such that $(x, y) \in D \subseteq U$.

The unit disk $U = \{(x, y) \mid x^2 + y^2 < 1\}$ is an open set

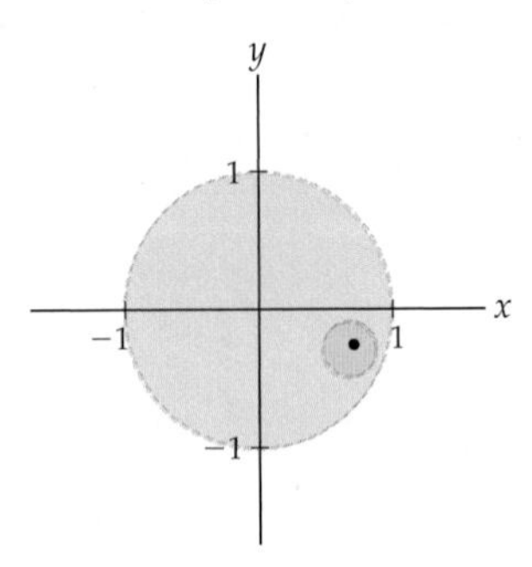

2. If we let W be the unit disk together with any point on the circumference of the circle, say, $(1, 0)$, then the set W will not be an open set. That is,
$$W = \{(x, y) \mid x^2 + y^2 < 1\} \cup \{(1, 0)\}.$$
The point $(1, 0)$ is in W, so this set is not open, because there is no open disk containing $(1, 0)$ that is a subset of W.

The set $W = \{(x, y) \mid x^2 + y^2 < 1\} \cup \{(1, 0)\}$ is not an open set

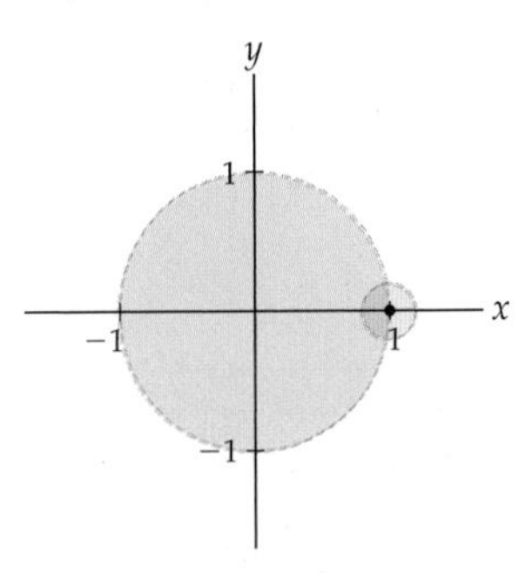

3. Similarly, if we let $\overline{U} = \{(x, y) \mid x^2 + y^2 \leq 1\}$, our set includes all points inside and on the circumference of the unit circle. Like the set W, this set is also not open.
4. The ***empty set***, denoted by $\emptyset$, contains no elements and is a subset of every set. As a subset of $\mathbb{R}^2$, the empty set is open, since it vacuously satisfies Definition 12.8 (a).

DEFINITION 12.9 **The Complement of a Set in $\mathbb{R}^2$ and $\mathbb{R}^3$**

(a) Let A be a subset of $\mathbb{R}^2$. The ***complement*** of A, denoted A^c, is the set

$$A^c = \{(x, y) \in \mathbb{R}^2 \mid (x, y) \notin A\}.$$

(b) Let A be a subset of $\mathbb{R}^3$. The ***complement*** of A, denoted A^c, is the set

$$A^c = \{(x, y, z) \in \mathbb{R}^3 \mid (x, y, z) \notin A\}.$$

That is, the complement of a set A is the set of all points *not in* A. For example, the complement of the unit disk $U = \{(x, y) \mid x^2 + y^2 < 1\}$ is the set $U^c = \{(x, y) \mid x^2 + y^2 \geq 1\}$.

The complement of the unit disk in $\mathbb{R}^2$

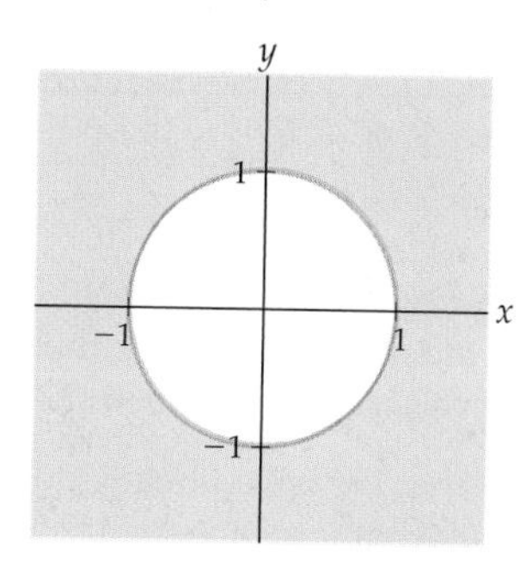

We ask you to prove the following theorem in Exercise 63:

THEOREM 12.10 **The Complement of the Complement of a Set Is the Original Set**

If S is a subset of $\mathbb{R}^2$ or $\mathbb{R}^3$, then $(S^c)^c = S$.

We use the idea of the complement of a set to define a closed set.

DEFINITION 12.11 **Closed Sets in $\mathbb{R}^2$ and $\mathbb{R}^3$**

A subset S of $\mathbb{R}^2$ or $\mathbb{R}^3$ is said to be ***closed*** if its complement, S^c, is open.

For example, the set $U^c = \{(x, y) \mid x^2 + y^2 \geq 1\}$ is a closed set because it is the complement of the open set $U = \{(x, y) \mid x^2 + y^2 < 1\}$. Note that a set is not closed just because it is not open. Later, in Example 1, we will show that the set $\{(x, y) \mid x \geq 0 \text{ and } y > 0\}$ is neither open nor closed. In that respect, this set is similar to a half-open, half-closed interval like $[0, 1)$.

We ask you to prove the following theorem in Exercise 64:

THEOREM 12.12 **The Complement of a Closed Set Is an Open Set**

If S is a closed subset of $\mathbb{R}^2$ or $\mathbb{R}^3$, then S^c is an open set.

We now define the boundary of a set.

DEFINITION 12.13

The Boundary of a Subset of $\mathbb{R}^2$ or $\mathbb{R}^3$

(a) Let A be a subset of $\mathbb{R}^2$. A point (x, y) is said to be a ***boundary point*** of A if every open disk containing (x, y) in $\mathbb{R}^2$ intersects both A and A^c.

(b) Let A be a subset of $\mathbb{R}^3$. A point (x, y, z) is said to be a ***boundary point*** of A if every open ball containing (x, y, z) in $\mathbb{R}^3$ intersects both A and A^c.

The set of all boundary points of a set A is called the ***boundary*** of A and is denoted by ∂A.

The boundary of the unit disk $U = \{(x, y) \mid x^2 + y^2 < 1\}$ consists exactly of the points on the unit circle. That is, $\partial U = \{(x, y) \mid x^2 + y^2 = 1\}$.

Every open disk containing a point on the unit circle intersects both U and U^c

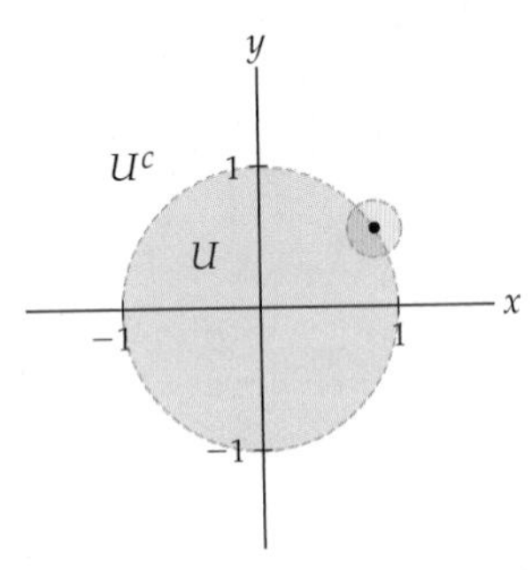

There are many interesting theorems about open and closed sets in $\mathbb{R}^2$ and $\mathbb{R}^3$. We ask you to prove several of these in the exercises.

Recall that we have already defined what it means for a function to be bounded. We now define what it means for a set to be bounded.

DEFINITION 12.14

Bounded Subsets of $\mathbb{R}^2$ and $\mathbb{R}^3$

(a) A subset A of $\mathbb{R}^2$ is said to be ***bounded*** if A is a subset of some open disk D in $\mathbb{R}^2$.

(b) A subset A of $\mathbb{R}^3$ is said to be ***bounded*** if A is a subset of some open ball B in $\mathbb{R}^3$.

A subset of $\mathbb{R}^2$ or $\mathbb{R}^3$ that is not bounded is said to be ***unbounded***.

For example, the set A in the following figure is bounded, while the set consisting of the coordinate axes in $\mathbb{R}^2$ or $\mathbb{R}^3$ is unbounded.

A is a bounded set *The points on the coordinate axes form an unbounded set*

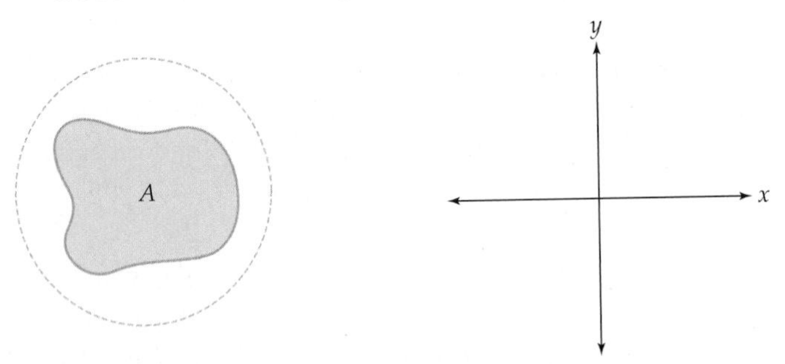

Now that we have set forth these preliminary definitions, after a brief review of the definition of the limit of a function of a single variable, we will be ready to define the limit of a function of two or three variables.

The Limit of a Function of Two or More Variables

Recall that for a function, f, of a single variable, we defined $\lim_{x\to a} f(x) = L$ if, for every $\epsilon > 0$, there is a $\delta > 0$ such that $|f(x) - L| < \epsilon$ whenever $0 < |x - a| < \delta$. The definition makes precise the informal idea that a function f has a limit L at a point a if the output values of f are "close to" L for values of the input variable x that are "close to" a. We will use this definition as a model to define the limit for functions of two and three variables. Before we get to the new definition, however, note that the definition of the limit implies that the function f be defined on a punctured open interval. In Exercise 7 we ask how this is accomplished.

There is very little that needs to be changed in the definition of a limit for a function of a single variable to form a definition for the limit of a function of two or three variables. Certainly the types of domains are different: The domain of a function of two variables consists of a set of ordered pairs, and the domain of a function of three variables consists of a set of ordered triples.

DEFINITION 12.15

The Limit of a Function of Two or More Variables

Let f be a function of two or more variables. The limit of f at $\mathbf{a}$ is L if, for every $\epsilon > 0$, there is a $\delta > 0$ such that $|f(\mathbf{x}) - L| < \epsilon$ whenever $0 < \|\mathbf{x} - \mathbf{a}\| < \delta$. In this case we write $\lim_{\mathbf{x}\to\mathbf{a}} f(\mathbf{x}) = L$.

In the case where f is a function of two variables, we may let $\mathbf{x} = \langle x, y\rangle$ and $\mathbf{a} = \langle a, b\rangle$, therefore $\lim_{\mathbf{x}\to\mathbf{a}} f(\mathbf{x}) = L$ means that for every $\epsilon > 0$, there is a $\delta > 0$ such that

$$|f(x, y) - L| < \epsilon \text{ whenever } 0 < \sqrt{(x - a)^2 + (y - b)^2} < \delta.$$

Similarly, if f is a function of three variables, we may let $\mathbf{x} = \langle x, y, z\rangle$ and $\mathbf{a} = \langle a, b, c\rangle$. Here, $\lim_{\mathbf{x}\to\mathbf{a}} f(\mathbf{x}) = L$ means that for every $\epsilon > 0$, there is a $\delta > 0$ such that

$$|f(x, y, z) - L| < \epsilon \text{ whenever } 0 < \sqrt{(x - a)^2 + (y - b)^2 + (z - c)^2} < \delta.$$

The vector notation of Definition 12.15 can be used in place of the preceding two expanded statements. We will use the more succinct notation when appropriate.

Limits of combinations of functions of two or three variables obey the same types of algebraic properties we saw for limits of functions of a single variable in Chapter 1.

THEOREM 12.16

Rules for Calculating Limits of Combinations

If $\lim_{\mathbf{x}\to\mathbf{a}} f(\mathbf{x})$ and $\lim_{\mathbf{x}\to\mathbf{a}} g(\mathbf{x})$ exist, then the following rules hold for their combinations:

Constant Multiple Rule: $\lim_{\mathbf{x}\to\mathbf{a}} kf(\mathbf{x}) = k \lim_{\mathbf{x}\to\mathbf{a}} f(\mathbf{x})$ for any real number k.

Sum Rule: $\lim_{\mathbf{x}\to\mathbf{a}} (f(\mathbf{x}) + g(\mathbf{x})) = \lim_{\mathbf{x}\to\mathbf{a}} f(\mathbf{x}) + \lim_{\mathbf{x}\to\mathbf{a}} g(\mathbf{x})$.

Difference Rule: $\lim_{\mathbf{x}\to\mathbf{a}} (f(\mathbf{x}) - g(\mathbf{x})) = \lim_{\mathbf{x}\to\mathbf{a}} f(\mathbf{x}) - \lim_{\mathbf{x}\to\mathbf{a}} g(\mathbf{x})$.

Product Rule: $\lim_{\mathbf{x}\to\mathbf{a}} (f(\mathbf{x})g(\mathbf{x})) = \left(\lim_{\mathbf{x}\to\mathbf{a}} f(\mathbf{x})\right)\left(\lim_{\mathbf{x}\to\mathbf{a}} g(\mathbf{x})\right)$.

Quotient Rule: $\lim_{\mathbf{x}\to\mathbf{a}} \dfrac{f(\mathbf{x})}{g(\mathbf{x})} = \dfrac{\lim_{\mathbf{x}\to\mathbf{a}} f(\mathbf{x})}{\lim_{\mathbf{x}\to\mathbf{a}} g(\mathbf{x})}$, if $\lim_{\mathbf{x}\to\mathbf{a}} g(\mathbf{x}) \neq 0$.

Another important theorem concerning limits of functions of a single variable has an analog for functions of two or more variables. Theorem 1.8 tells us that the function $f : \mathbb{R} \to \mathbb{R}$ has the two-sided limit $\lim_{x\to a} f(x) = L$ if and only if the two one-sided limits $\lim_{x\to a^+} f(x)$ and $\lim_{x\to a^-} f(x)$ are both equal to L. Before we state the analogous theorem for functions of two or more variables, we define what we mean by the limit of a function of two or more variables *along a path*.

DEFINITION 12.17

The Limit of a Function of Two or Three Variables Along a Path

(a) Let f be a function of two variables defined on an open set $S \subseteq \mathbb{R}^2$. If the graph of the continuous vector function $\langle x(t), y(t)\rangle$ is a curve $C \subset S$ such that $\lim_{t\to t_0} \langle x(t), y(t)\rangle = \langle a, b\rangle$, then we define the ***limit of*** f ***as*** (x, y) ***approaches*** (a, b) ***along*** C, denoted by $\lim_{\substack{(x,y)\to(a,b)\\ C}} f(x, y)$, to be $\lim_{t\to t_0} f(x(t), y(t))$.

(b) Let f be a function of three variables defined on an open set $S \subseteq \mathbb{R}^3$. If the graph of the continuous vector function $\langle x(t), y(t), z(t)\rangle$ is a curve $C \subset S$ such that $\lim_{t\to t_0} \langle x(t), y(t), z(t)\rangle = \langle a, b, c\rangle$, then we define the ***limit of*** f ***as*** (x, y, z) ***approaches*** (a, b, c) ***along*** C, denoted by $\lim_{\substack{(x,y,z)\to(a,b,c)\\ C}} f(x, y, z)$, to be $\lim_{t\to t_0} f(x(t), y(t), z(t))$.

The following schematic illustrates three such curves containing a point (a, b) in an open subset of $\mathbb{R}^2$.

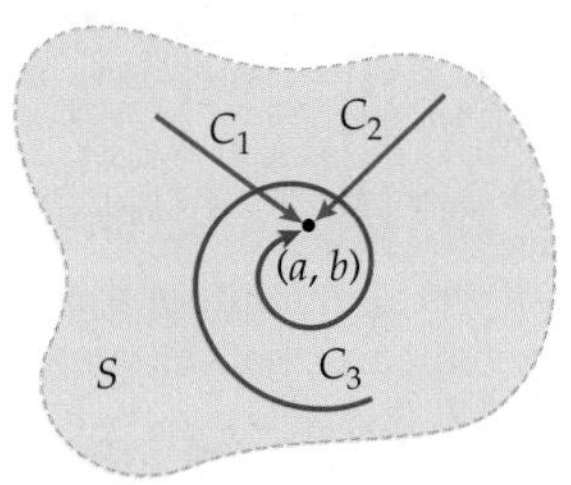

THEOREM 12.18

The Limit of a Function of Two or More Variables Along Distinct Paths

Let f be a function of two or more variables defined on an open set S. Then the limit $\lim_{\mathbf{x}\to\mathbf{a}} f(\mathbf{x}) = L$ exists if and only if $\lim_{\substack{\mathbf{x}\to\mathbf{a}\\ C}} f(\mathbf{x}) = L$ for every path C containing the point (a, b).

Given an open set S containing a point $\mathbf{a}$, there are infinitely many paths containing $\mathbf{a}$ that are also contained in S. Thus, it is impractical to prove that a limit of two or more variables exists by showing that the limit of the function along each path is the same. Theorem 12.18 is usually applied when the limit does not exist and we can find two distinct paths C_1 and C_2 such that

$$\lim_{\substack{\mathbf{x}\to\mathbf{a}\\ C_1}} f(\mathbf{x}) \neq \lim_{\substack{\mathbf{x}\to\mathbf{a}\\ C_2}} f(\mathbf{x}).$$

For example, consider the function $f(x, y) = \frac{xy}{x^2+y^2}$. We will show that $\lim_{(x,y)\to(0,0)} f(x, y)$ does not exist by evaluating the limit along two paths containing the origin and getting different values. First consider the limit along the x-axis (i.e., when $y = 0$):

$$\lim_{\substack{(x,y)\to(0,0)\\ x\text{-axis}}} f(x, y) = \lim_{\substack{(x,y)\to(0,0)\\ x\text{-axis}}} \frac{xy}{x^2+y^2} = \lim_{x\to 0} \frac{x\cdot 0}{x^2+0} = \lim_{x\to 0} 0 = 0.$$

However, along the line $y = x$,

$$\lim_{\substack{(x,y)\to(0,0)\\ y=x}} f(x,y) = \lim_{\substack{(x,y)\to(0,0)\\ y=x}} \frac{xy}{x^2+y^2} = \lim_{x\to 0}\frac{x\cdot x}{x^2+x^2} = \lim_{x\to 0}\frac{x^2}{2x^2} = \lim_{x\to 0}\frac{1}{2} = \frac{1}{2}.$$

Since the two limits do not agree, it follows from Theorem 12.18 that $\lim_{(x,y)\to(0,0)} f(x,y)$ does not exist.

The Continuity of a Function of Two or More Variables

Recall that if f is a function of a single variable, f is said to be continuous at a point a in the domain of f if $\lim_{x\to a} f(x) = f(a)$. We will define what it means for a function of two or more variables to be continuous at a point in its domain with an analogous definition.

DEFINITION 12.19 The Continuity of Functions of Two or Three Variables

Let f be a function of two or three variables defined on an open set S, and let $\mathbf{a}$ be a point in S. Then f is ***continuous*** at $\mathbf{a}$ if $\lim_{\mathbf{x}\to\mathbf{a}} f(\mathbf{x}) = f(\mathbf{a})$.

Also, f is ***continuous on*** S if f is continuous at every point $\mathbf{a} \in S$.

If the domain of f is $\mathbb{R}^2$ or $\mathbb{R}^3$ and f is continuous at every point in its domain, then f is ***continuous everywhere***.

Since continuity is defined in terms of a limit, it follows that sums, differences, and products of continuous functions will also be continuous. In addition, quotients will be continuous as long as we do not divide by zero. Most of the functions we work with on a regular basis are continuous at every point in their domains. For example, the function $f(x,y) = \frac{xy}{x^2+y^2}$ we discussed just after Theorem 12.18 is continuous at every point in its domain. This domain is the set of all points in $\mathbb{R}^2$ except the origin, so if a and b are not both zero, we will have

$$\lim_{(x,y)\to(a,b)} \frac{xy}{x^2+y^2} = \frac{ab}{a^2+b^2}.$$

This function is discontinuous at the origin for two reasons: The function is not defined at the origin, and $\lim_{(x,y)\to(0,0)} \frac{xy}{x^2+y^2}$ does not exist.

Examples and Explorations

EXAMPLE 1 Understanding open sets, closed sets, and boundaries

For each of the following subsets of $\mathbb{R}^2$, determine whether the set is open, closed, or neither open nor closed. Find the boundary of each set as well.

(a) $S_1 = \{(x,y) \in \mathbb{R}^2 \mid x \ge 0\}$ **(b)** $S_2 = \{(x,y) \in \mathbb{R}^2 \mid y \ne 0\}$

(c) $S_3 = \{(x,y) \in \mathbb{R}^2 \mid x \ge 0 \text{ and } y > 0\}$

SOLUTION

(a) The set S_1 is the right half-plane. The complement of S_1 is $S_1^c = \{(x,y) \in \mathbb{R}^2 \mid x < 0\}$. Any point (a,b) with $a < 0$ is in S_1^c. The open ball $B_1 = \{(x,y) \mid (x-a)^2 + (y-b)^2 < a^2\}$ contains (a,b) and is a subset of S_1^c. Therefore, S_1^c is an open set. Its complement $(S_1^c)^c = S_1$ is a closed set. The boundary of S_1 is the y-axis, since every open ball containing a point on the y-axis contains points of both S_1 and S_1^c but points off the y-axis do not share this property. The following figure is illustrative:

The set S_1 is closed because S_1^c is open

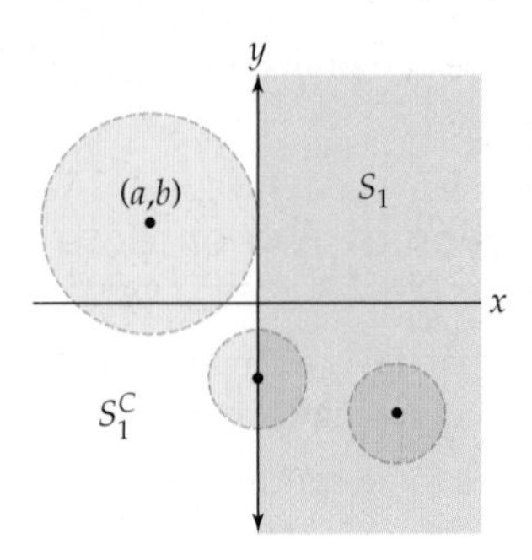

(b) The set S_2 consists of all points in the plane, except for those on the x-axis. We will show that this set is open. If we choose the point $(a, b) \in S_2$, then $b \neq 0$. The open ball $B_2 = \{(x, y) \mid (x-a)^2 + (y-b)^2 < b^2\}$ contains (a, b) and is a subset of S_2 whether $b > 0$ or $b < 0$. (In the illustration that follows, $b < 0$.) Therefore, S_2 is an open set. The boundary of S_2 is the x-axis, since every open ball containing a point on the x-axis contains points of both S_2 and S_2^c but points not on the x-axis do not share this property.

The set S_2 is open

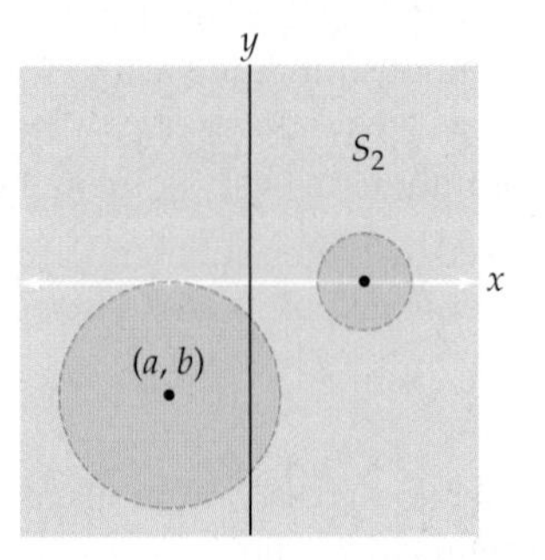

(c) The set S_3 consists of all points in the first quadrant, along with those points on the positive y-axis. We will show that S_3 is neither open nor closed. S_3 is not open because, although $(0, 1) \in S_3$, we cannot find an open ball containing $(0, 1)$ that is also a subset of S_3, since every open ball containing $(0, 1)$ contains points of both S_3 and S_3^c. To show that S_3 is not closed, we will show that S_3^c is not open. Although $(1, 0) \in S_3^c$, every open ball that contains $(1, 0)$ intersects both S_3 and S_3^c. Therefore, S_3 is neither open nor closed. The boundary of S_3 is $\partial S_3 = \{(x, 0) \mid x \geq 0\} \cup \{(0, y) \mid y \geq 0\}$.

The set S_3 is neither open nor closed

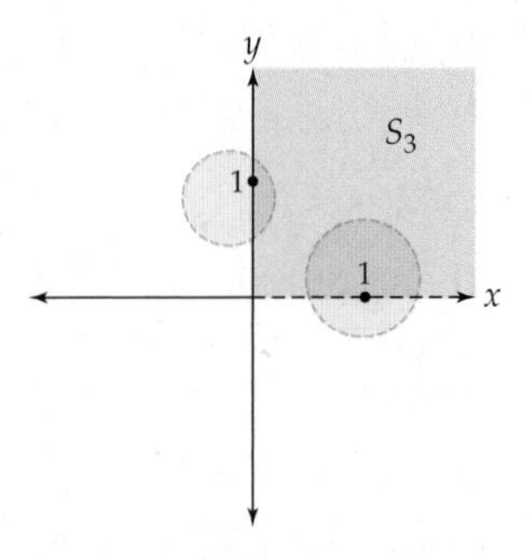

□

EXAMPLE 2

Evaluating limits of functions of two and three variables

Evaluate the limits:

(a) $\lim_{(x,y,z)\to(2,-2,3)} (xy^2 + 4xyz)$ **(b)** $\lim_{(x,y)\to(0,0)} \frac{x^2y^3 + 5xy - 3}{x + y - 1}$ **(c)** $\lim_{(x,y)\to(0,3)} \frac{(\sin x)(\cos y)}{e^{xy} - 2}$

SOLUTION

(a) The function in the first limit is a polynomial function of three variables. Just as polynomials of a single variable are continuous for every real number, polynomials of two variables are continuous for every point in $\mathbb{R}^2$ and polynomials of three variables are continuous for every point in $\mathbb{R}^3$. Therefore, to determine the limit, we may evaluate the function $f(x, y, z) = xy^2 + 4xyz$ at $(2, -2, 3)$. We obtain

$$\lim_{(x,y,z)\to(2,-2,3)} (xy^2 + 4xyz) = 2(-2)^2 + 4 \cdot 2(-2)3 = -40.$$

(b) Similarly, rational functions of two or three variables are also continuous where they are defined. The values where the second function $g(x, y) = \frac{x^2y^3+5xy-3}{x+y-1}$ is discontinuous are those points where the denominator is zero, namely, all points along the line defined by the equation $x + y = 1$. Since the origin does not satisfy this equation, the function g is continuous at $(0, 0)$. Therefore,

$$\lim_{(x,y)\to(0,0)} \frac{x^2y^3 + 5xy - 3}{x + y - 1} = \frac{0^20^3 + 5 \cdot 0 \cdot 0 - 3}{0 + 0 - 1} = 3.$$

(c) Since the function in our final limit is an algebraic combination of the sine, cosine, and exponential functions, all of which are transcendental functions, it will also be continuous where it is defined. The points where the function $h(x, y) = \frac{(\sin x)(\cos y)}{e^{xy}-2}$ will be discontinuous will be the solutions of the equation $e^{xy} = 2$, or equivalently, points on the hyperbola defined by the equation $y = \frac{\ln 2}{x}$. The point $(0, 3)$ does not satisfy this equation, so the function h is continuous at $(0, 3)$. Therefore,

$$\lim_{(x,y)\to(0,3)} \frac{(\sin x)(\cos y)}{e^{xy} - 2} = \frac{(\sin 0)(\cos 3)}{e^{0\cdot 3} - 2} = 0.$$

□

EXAMPLE 3

Showing that a limit does not exist

Explain why $\lim_{(x,y)\to(6,3)} \frac{x^2+xy-6y^2}{x^2-4y^2}$ does not exist.

SOLUTION

The function $f(x, y) = \frac{x^2+xy-6y^2}{x^2-4y^2}$ is a rational function of two variables, but the point $(6, 3)$ is not in the domain of f because the polynomial that forms the denominator of f evaluates to 0 at $(6, 3)$. However, the numerator also evaluates to 0 at that point. Thus, this limit is an indeterminate form of the type $\frac{0}{0}$. As we did with rational functions of a single variable, we may try to simplify the quotient, so that we may use a multivariable version of the Cancellation Theorem to evaluate the limit. Since $\frac{x^2+xy-6y^2}{x^2-4y^2} = \frac{(x-2y)(x+3y)}{(x-2y)(x+2y)}$, we have

$$\lim_{(x,y)\to(6,3)} \frac{x^2 + xy - 6y^2}{x^2 - 4y^2} = \lim_{(x,y)\to(6,3)} \frac{(x - 2y)(x + 3y)}{(x - 2y)(x + 2y)}.$$

Now, we are tempted to cancel the common factor that occurs in the numerator and denominator and say that

$$\lim_{(x,y)\to(6,3)} \frac{(x - 2y)(x + 3y)}{(x - 2y)(x + 2y)} = \lim_{(x,y)\to(6,3)} \frac{x + 3y}{x + 2y}.$$

However, the equality between the two limits is not valid. There is an analog to the Cancellation Theorem for limits of functions of more than one variable, but the analog would apply only when the cancellation of factors introduces a single point into the domain of the reduced function. In our example, although the functions $f(x, y) = \frac{x^2+xy-6y^2}{x^2-4y^2}$ and $g(x, y) = \frac{x+3y}{x+2y}$ are equivalent when $x \neq 2y$, they are not equivalent for points on the line $x = 2y$. However, Definition 12.15 requires that, for $\lim_{(x,y)\to(a,b)} f(x, y)$ to exist, the function f must be defined on an open set containing the point (a, b), with the possible exception that f does not have to be defined at the point (a, b) itself. Thus, although $\lim_{(x,y)\to(6,3)} \frac{x+3y}{x+2y} = \frac{5}{4}$, we cannot conclude that our original limit has the same value. The function $f(x, y) = \frac{x^2+xy-6y^2}{x^2-4y^2}$ is undefined along the line defined by $x = 2y$, and thus f is undefined for infinitely many points in every open set containing the point $(6, 3)$. Therefore $\lim_{(x,y)\to(6,3)} \frac{x^2+xy-6y^2}{x^2-4y^2}$ does not exist.

However, if C is any curve in the plane that intersects the line $x = 2y$ only at the point $(6, 3)$, then

$$\lim_{\substack{(x,y)\to(6,3)\\ C}} \frac{x^2+xy-6y^2}{x^2-4y^2} = \lim_{(x,y)\to(6,3)} \frac{x+3y}{x+2y} = \frac{5}{4}.$$ □

EXAMPLE 4 **Evaluating limits along paths**

Show that $\lim_{(x,y)\to(0,0)} \frac{x^2y}{x^4+y^2}$ does not exist by evaluating the limit along the curves defined by the parabolas $y = mx^2$.

SOLUTION

In Exercise 13 you will show that

$$\lim_{\substack{(x,y)\to(0,0)\\ C}} \frac{x^2y}{x^4+y^2} = 0$$

when C is either the x- or y-axis. However, the fact that these limits exist and agree is not enough to conclude that the limit is zero. When we evaluate the limit of the function along a parabola of the form $y = mx^2$, we will see that the result depends upon the value of m. From this observation, we conclude that the limit does not exist. We have

$$\lim_{\substack{(x,y)\to(0,0)\\ y=mx^2}} \frac{x^2y}{x^4+y^2} = \lim_{x\to 0} \frac{x^2\cdot mx^2}{x^4+(mx^2)^2} = \lim_{x\to 0} \frac{mx^4}{x^4(1+m^2)} = \lim_{x\to 0} \frac{m}{1+m^2} = \frac{m}{1+m^2}.$$

Since this limit is a nonconstant function of m, $\lim_{(x,y)\to(0,0)} \frac{x^2y}{x^4+y^2}$ does not exist. □

EXAMPLE 5 **Using polar coordinates to evaluate limits**

Evaluate the limits

(a) $\lim_{(x,y)\to(0,0)} \frac{x^2y}{x^2+y^2}$ **(b)** $\lim_{(x,y)\to(0,0)} \frac{xy}{x^2+y^2}$

if they exist.

SOLUTION

In Chapter 9 we saw that we could translate between rectangular coordinates and polar coordinates by using the equations $r^2 = x^2 + y^2$, $\tan\theta = \frac{y}{x}$, $x = r\cos\theta$, and $y = r\sin\theta$.

Changing to polar coordinates can be useful when a function contains the factor $x^2 + y^2$, which we find in both of the given limits.

(a) First,

$$\lim_{(x,y)\to(0,0)} \frac{x^2 y}{x^2+y^2} = \lim_{r\to 0} \frac{(r\cos\theta)^2(r\sin\theta)}{r^2} = \lim_{r\to 0} \frac{r^3 \sin\theta\cos^2\theta}{r^2} = \lim_{r\to 0} r\sin\theta\cos^2\theta.$$

Since we are taking the limit as $r \to 0$, and the limit contains a factor of r, its value is zero for every value of θ.

(b) We also convert the second limit to polar coordinates and obtain

$$\lim_{(x,y)\to(0,0)} \frac{xy}{x^2+y^2} = \lim_{r\to 0} \frac{(r\cos\theta)(r\sin\theta)}{r^2} = \lim_{r\to 0} \frac{r^2 \sin\theta\cos\theta}{r^2} = \lim_{r\to 0} \sin\theta\cos\theta.$$

Here, the value of the limit depends upon the angle θ. Therefore, the limit can be different along different paths approaching the origin. For example,

$$\lim_{\substack{(x,y)\to(0,0)\\ \theta=0}} \frac{xy}{x^2+y^2} = \lim_{r\to 0} 0 = 0 \quad \text{and} \quad \lim_{\substack{(x,y)\to(0,0)\\ \theta=\pi/4}} \frac{xy}{x^2+y^2} = \lim_{r\to 0} \frac{1}{2} = \frac{1}{2}.$$

Thus $\lim_{(x,y)\to(0,0)} \frac{xy}{x^2+y^2}$ does not exist. □

EXAMPLE 6

Examining the continuity of a piecewise-defined function

Determine where the functions

(a) $f(x,y) = \begin{cases} \dfrac{x^2 y}{x^2+y^2}, & \text{if } (x,y) \neq (0,0) \\ 0, & \text{if } (x,y) = (0,0) \end{cases}$ **(b)** $g(x,y) = \begin{cases} \dfrac{xy}{x^2+y^2}, & \text{if } (x,y) \neq (0,0) \\ 0, & \text{if } (x,y) = (0,0) \end{cases}$

are continuous.

SOLUTION

These are the same functions we discussed in Example 5, except that their values have been defined at the origin. Since rational functions are continuous everywhere in their domains, both of the given functions are continuous everywhere, with the possible exception of the origin.

(a) Because we already showed that

$$\lim_{(x,y)\to(0,0)} \frac{x^2 y}{x^2+y^2} = \lim_{(x,y)\to(0,0)} f(x,y) = 0 = f(0,0),$$

the function f is also continuous at the origin. Therefore, f is continuous everywhere.

(b) Although g is continuous everywhere except the origin, there is no way to define the function at $(0,0)$ to make g continuous, since we have already seen that $\lim_{(x,y)\to(0,0)} g(x,y)$ does not exist. □

? TEST YOUR UNDERSTANDING

- What are open and closed sets in $\mathbb{R}$? In $\mathbb{R}^2$? In $\mathbb{R}^3$?
- What is a bounded set in $\mathbb{R}$? In $\mathbb{R}^2$? In $\mathbb{R}^3$?
- What is the definition of the limit of a function of one variable? Of two variables? Of three variables? What does it mean intuitively for these limits to exist?
- What do we mean by the limit of a function of two or three variables along a path? What is the analogous concept for a function of a single variable?
- What is the definition of the continuity of a function of one variable at a point? Of two variables? Of three variables?

EXERCISES 12.2

Thinking Back

- *Intervals:* What is meant by an open interval in $\mathbb{R}$? What is meant by a closed interval in $\mathbb{R}$? Is every interval either open or closed?
- *Limits:* If f is a function of a single variable, what is the intuitive interpretation of $\lim_{x\to a} f(x)$? What is the formal definition?

Concepts

0. *Problem Zero:* Read the section and make your own summary of the material.

1. *True/False:* Determine whether each of the statements that follow is true or false. If a statement is true, explain why. If a statement is false, provide a counterexample.
   (a) *True or False:* The complement of a closed subset of $\mathbb{R}^2$ is an open subset of $\mathbb{R}^2$.
   (b) *True or False:* Every subset of $\mathbb{R}^2$ is either open or closed.
   (c) *True or False:* There are subsets of $\mathbb{R}^2$ that are both open and closed.
   (d) *True or False:* If a subset S of $\mathbb{R}^3$ is neither open nor closed, then S^c is neither open nor closed.
   (e) *True or False:* If $\lim_{(x,y)\to(0,0)} f(x,y) = L$, then $f(0,0) = L$.
   (f) *True or False:* If
   $$\lim_{(x,y)\to(a,b)} f(x,y) = L, \quad \text{then} \quad \lim_{\substack{(x,y)\to(a,b)\\ C}} f(x,y) = L$$
   for every path C containing (a,b).
   (g) *True or False:* If
   $$\lim_{\substack{(x,y)\to(a,b)\\ C}} f(x,y) = L$$
   for every path C containing (a,b), then $f(x,y)$ is continuous at (a,b).
   (h) *True or False:* If $f(x,y)$ is continuous everywhere and $\lim_{(x,y)\to(0,0)} f(x,y) = L$, then $f(0,0) = L$.

2. *Examples:* Construct examples of the thing(s) described in the following. Try to find examples that are different than any in the reading.
   (a) An open subset of $\mathbb{R}^3$.
   (b) A closed subset of $\mathbb{R}^3$.
   (c) A subset of $\mathbb{R}^3$ that is both open and closed.

3. If
   $$\lim_{\substack{(x,y)\to(1,-3)\\ C_1}} f(x,y) = 5 \quad \text{and} \quad \lim_{\substack{(x,y)\to(1,-3)\\ C_2}} f(x,y) = 5,$$
   where C_1 and C_2 are two distinct curves in $\mathbb{R}^2$ containing the point $(1,-3)$, what can you say about $\lim_{(x,y)\to(1,-3)} f(x,y)$?

4. If
   $$\lim_{\substack{(x,y)\to(-2,0)\\ C_1}} g(x,y) = 5 \quad \text{and} \quad \lim_{\substack{(x,y)\to(-2,0)\\ C_2}} g(x,y) = 8,$$
   where C_1 and C_2 are two distinct curves in $\mathbb{R}^2$ containing the point $(-2,0)$, what can you say about $\lim_{(x,y)\to(-2,0)} g(x,y)$?

5. If
   $$\lim_{\substack{(x,y)\to(3,-7)\\ C}} f(x,y) = 5$$
   for every curve C in $\mathbb{R}^2$ containing the point $(3,-7)$, what can you say about $\lim_{(x,y)\to(3,-7)} f(x,y)$? What can you say about $f(3,-7)$?

6. If $g : \mathbb{R}^3 \to \mathbb{R}$ is continuous at the point (a,b,c) and C is a curve in $\mathbb{R}^3$ containing the point (a,b,c), what can we say about
   $$\lim_{\substack{(x,y,z)\to(a,b,c)\\ C}} g(x,y,z)?$$

7. How does the definition of the limit of a function of a single variable, f, imply that f is defined on the union of two open intervals?

8. How does the definition of the limit of a function of two variables, f, imply that f is defined on an open subset of $\mathbb{R}^2$?

9. How does the definition of the limit of a function of three variables, f, imply that f is defined on an open subset of $\mathbb{R}^3$?

10. Explain how the definition of the limit of a function of two or three variables along a path simplifies to the limit of a function of a single variable.

11. Review the definition of continuity of a function of a single variable, f, at a point. Why is it necessary for f to be defined on the union of two open intervals?

12. Review the definition of continuity of a function of two or three variables, f, at a point. Why is it necessary for f to be defined on an open subset of $\mathbb{R}^2$ or $\mathbb{R}^3$?

13. Show that when C is either the x- or y-axis, we have
   $$\lim_{\substack{(x,y)\to(0,0)\\ C}} \frac{x^2y}{x^4+y^2} = 0.$$

14. Copy the figure that follows onto a sheet of paper. Now cut a slit along the dashed line, leave the left side of the paper on the table, and gently raise the right side of the paper along the slit.

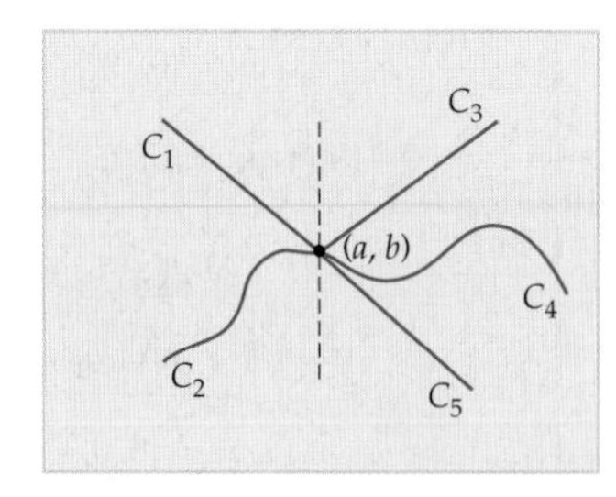

(a) Explain how the paper may be interpreted as the graph of a function of two variables, $f(x, y)$.

(b) If the tabletop is the xy-plane, explain why

$$\lim_{\substack{(x,y)\to(a,b) \\ C_1}} f(x, y) = \lim_{\substack{(x,y)\to(a,b) \\ C_2}} f(x, y) = 0.$$

(c) Explain why

$$\lim_{\substack{(x,y)\to(a,b) \\ C_3}} f(x, y) = \lim_{\substack{(x,y)\to(a,b) \\ C_4}} f(x, y) = \lim_{\substack{(x,y)\to(a,b) \\ C_5}} f(x, y) > 0.$$

(d) Explain why $\lim_{(x,y)\to(a,b)} f(x, y)$ does not exist.

15. Provide a definition for $\lim_{(x,y)\to(a,b)} f(x, y) = \infty$. Model your definition on Definitions 1.9 and 12.15.

16. Provide a definition for $\lim_{(x,y,z)\to(a,b,c)} f(x, y, z) = \infty$. Model your definition on Definitions 1.9 and 12.15.

17. Find functions $f(x, y)$ and $g(x, y)$ and a point $(a, b) \in \mathbb{R}^2$ such that

$$\lim_{(x,y)\to(a,b)} f(x, y) + \lim_{(x,y)\to(a,b)} g(x, y) \neq \lim_{(x,y)\to(a,b)} (f(x, y)+g(x, y)).$$

Does this example contradict the sum rule for limits of a function of two variables?

18. Find functions $f(x, y)$ and $g(x, y)$ and a point $(a, b) \in \mathbb{R}^2$ such that

$$\left(\lim_{(x,y)\to(a,b)} f(x, y)\right)\left(\lim_{(x,y)\to(a,b)} g(x, y)\right) \neq \lim_{(x,y)\to(a,b)} (f(x, y)g(x)).$$

Does this example contradict the product rule for limits of a function of two variables?

19. Let f be a function of two variables that is continuous everywhere.

(a) Explain why the function $\frac{f(x,y)}{x-y}$ is continuous if and only if $x \neq y$.

(b) Use Definition 12.15 to explain why $\lim_{(x,y)\to(a,a)} \frac{f(x,y)}{x-y}$ does not exist for any real number a.

20. Let f be a function of three variables that is continuous everywhere.

(a) Explain why the function $\frac{f(x,y,z)}{x+y+z}$ is continuous if and only if $x + y + z \neq 0$.

(b) Use Definition 12.15 to explain why

$$\lim_{(x,y,z)\to(a,b,-(a+b))} \frac{f(x,y,z)}{x+y+z}$$

does not exist for any pair (a, b) of real numbers.

Skills

In Exercises 21–26, (a) determine whether the given subset of $\mathbb{R}^2$ is open, closed, both open and closed, or neither open nor closed, (b) find the complement of the set, and (c) find the boundary of the given set.

21. all points (x, y) such that $x > 0$ and $y > 0$

22. all points on the coordinate axes

23. all points satisfying the inequality $|x| + |y| \leq 1$

24. all points (x, y) such that $|x| < 1$ and $|y| \leq 2$

25. the empty set

26. $\mathbb{R}^2$

In Exercises 27–32, (a) determine whether the given subset of $\mathbb{R}^3$ is open, closed, both open and closed, or neither open nor closed, (b) find the complement of the set, and (c) find the boundary of the given set.

27. all points (x, y, z) such that $x > 0,\ y < 0$ and $z < 0$

28. all points on the coordinate axes

29. all points on the xy-plane

30. all points (x, y, z) such that $|x| < 1$, $|y| < 2$, and $|z| > 3$

31. the empty set

32. $\mathbb{R}^3$

Evaluate the limits in Exercises 33–40 if they exist.

33. $\lim_{(x,y)\to(-2,\pi)} x^2y^3 \sin y$

34. $\lim_{(x,y,z)\to(3,-4,\pi/4)} \frac{x^2y}{\tan z}$

35. $\lim_{(x,y)\to(1,2)} \frac{x^2+y^2}{x^2-y^2}$

36. $\lim_{(x,y)\to(-2,1)} \frac{x^3-y^3}{x^2-y^2}$

37. $\lim_{\substack{(x,y)\to(3,3) \\ x=3}} \frac{x^3-y^3}{x^2-y^2}$

38. $\lim_{\substack{(x,y)\to(3,3) \\ y=3}} \frac{x^3-y^3}{x^2-y^2}$

39. $\lim_{\substack{(x,y)\to(3,3) \\ y=x}} \frac{x^3-y^3}{x^2-y^2}$

40. $\lim_{(x,y)\to(3,3)} \frac{x^3-y^3}{x^2-y^2}$

In Exercises 41–46, use polar coordinates to analyze the given limits.

41. $\lim_{(x,y)\to(0,0)} \frac{x^2}{x^2+y^2}$

42. $\lim_{(x,y)\to(0,0)} \frac{y^2}{x^2+y^2}$

43. $\lim_{(x,y)\to(0,0)} \frac{x^2y^2}{x^2+y^2}$

44. $\lim_{(x,y)\to(0,0)} \frac{x^2y^3}{x^4+2x^2y^2+y^4}$

45. $\lim_{(x,y)\to(0,0)} \frac{xy}{\sqrt{x^2+y^2}}$

46. $\lim_{(x,y)\to(0,0)} \frac{\sin(x^2+y^2)}{x^2+y^2}$

Determine the domains of the functions in Exercises 47–56, and find where the functions are continuous.

47. $f(x, y) = \dfrac{x^2}{x^2 - y^2}$

48. $f(x, y, z) = \dfrac{xy^2}{x + y - z}$

49. $f(x, y) = \sqrt{x^2 + y}$

50. $f(x, y, z) = \ln(x^2 + y^2 + z^2)$

51. $f(x, y, z) = \dfrac{\sin(x + y + z)}{\sqrt{x + y + z}}$

52. $f(x, y, z) = e^{-xyz}$

53. $f(x, y) = \begin{cases} \dfrac{xy}{\sqrt{x^2 + y^2}}, & \text{if } (x, y) \neq (0, 0) \\ 0, & \text{if } (x, y) = (0, 0) \end{cases}$

54. $f(x, y) = \begin{cases} \dfrac{xy}{x^2 + y^2}, & \text{if } (x, y) \neq (0, 0) \\ 0, & \text{if } (x, y) = (0, 0) \end{cases}$

55. $f(x, y) = \begin{cases} \dfrac{\sin(x^2 + y^2)}{x^2 + y^2}, & \text{if } (x, y) \neq (0, 0) \\ 1, & \text{if } (x, y) = (0, 0) \end{cases}$

56. $f(x, y) = \begin{cases} e^{-1/(x^2+y^2)}, & \text{if } (x, y) \neq (0, 0) \\ 0, & \text{if } (x, y) = (0, 0) \end{cases}$

Applications

The *ideal gas law* states that the pressure, P, volume, V, temperature in degrees Kelvin, T, and amount, n, of a gas are related by the equation $P = \frac{nRT}{V}$, where R is the *universal gas constant*. In Exercises 57–60, assume that the amount of the gas is constant and evaluate the specified limits.

57. If the volume is held fixed, what is $\lim_{T \to 0} P$? Explain why this makes sense.

58. If the volume is held fixed, what is $\lim_{T \to \infty} P$? Explain why this makes sense.

59. If the temperature is held fixed, what is $\lim_{V \to 0^+} P$? Explain why this makes sense.

60. If the temperature is held fixed, what is $\lim_{V \to \infty} P$? Explain why this makes sense.

61. Emmy is charting a layer of basalt beneath a Hanford tank farm. She has determined that on the south end of the tank farm the basalt lies at

$$b_1(x, y) = -40 - \frac{1}{350}x + \frac{1}{725}y,$$

while on the north the plane of the top of the basalt seems to be at

$$b_2(x, y) = -39.5 - \frac{1}{150}x + \frac{1}{300}y.$$

(a) If the surface of the basalt layer is continuous, what is the line of intersection of these planes on the Earth's surface?

(b) If the surface of the basalt layer is continuous, where should it be when $x = 150$ at the intersection of the planes?

(c) Emmy drills a test hole at that point and finds that the basalt lies at -41.2 feet. What conclusions might she draw from this result?

62. Annie is intrigued by the currents off the western point of Patos Island, one of the San Juan Islands that are part of Washington State. The currents coming off the north side of the island can be faster than those coming off the south side. She models the speed of the current as a function of x and y, using $s(x, y) = 1 + \dfrac{0.25y}{\sqrt{x^2 + y^2}}$, where distances are given in miles, the origin is a point in the water just west of the point of the island, and Patos Island can be considered to lie to the right of the origin. For simplicity Annie just lets the current speed function go right over the island, although obviously there is no water there.

(a) Show that the speed of the current far north of the island is approximately 1.25 mph.

(b) Show that the speed of the current far south of the island is approximately 0.75 mph.

(c) Show that if Annie paddles through the origin along a line $x = my$ with m constant, then her speed is constant.

(d) The speed of the current is evidently not defined at the origin, but does $\lim_{(x,y) \to (0,0)} s(x, y)$ exist? Explain. Can you imagine what the water looks like at the origin?

Proofs

63. Prove Theorem 12.10. That is, show that $(S^c)^c = S$ when S is a subset of $\mathbb{R}^2$ or $\mathbb{R}^3$.

64. Prove that if S is a closed subset of $\mathbb{R}^2$ or $\mathbb{R}^3$, then S^c is an open set. This is Theorem 12.12.

65. Let S be a subset of $\mathbb{R}^2$ or $\mathbb{R}^3$. Prove that a set S is open if and only if $\partial S \cap S = \emptyset$.

66. Let S be a subset of $\mathbb{R}^2$ or $\mathbb{R}^3$. Prove that a set S is closed if and only if $\partial S \subseteq S$.

67. Let S be a subset of $\mathbb{R}^2$ or $\mathbb{R}^3$. Prove that $\partial S = \partial (S^c)$.

68. Let S be a subset of $\mathbb{R}^2$ or $\mathbb{R}^3$. Prove that ∂S is a closed set.

69. Let S be a subset of $\mathbb{R}^2$ or $\mathbb{R}^3$. Prove that $\partial (\partial S) \subseteq \partial S$.

70. Prove that the empty set is both an open subset and a closed subset of $\mathbb{R}^2$.

71. Prove that $\mathbb{R}^2$ is both an open subset and a closed subset of $\mathbb{R}^2$.

Thinking Forward

- *The derivative along a cut edge:* Let $f(x, y) = x^2y^3$. Find the rate of change of f in the (positive) y direction when the surface is cut by the plane with equation $x = 2$. Find the rate of change of f in the (positive) x direction when the surface is cut by the plane with equation $y = 1$.
- *The derivative when two variables are held fixed:* Let $f(x, y, z) = x^2y^3\sqrt{z}$. Find the rate of change of f in the (positive) z-direction when the values of x and y are constant. Find the rate of change of f in the (positive) y direction when the values of x and z are constant. Find the rate of change of f in the (positive) x direction when the values of y and z are constant.

12.3 PARTIAL DERIVATIVES

- Partial derivatives of functions of two and three variables
- The geometry of partial derivatives
- Higher order partial derivatives

Partial Derivatives of Functions of Two and Three Variables

Recall that a function f of a single variable is differentiable at a point c in its domain if $\lim_{h\to 0} \frac{f(c+h)-f(c)}{h}$ exists. When this limit exists, the derivative $f'(c)$ gives us the slope of the tangent line to the graph of the function at the point $(c, f(c))$. In the next section we will discuss what it means for a function of two variables, $f(x, y)$, to be differentiable at a point in its domain. However, right now we remark that the graph of a function of two variables is a surface and the analogous tangent "object" to a function of two variables is a plane, not a line. In fact, a function of two variables whose graph is sufficiently smooth at a point (x_0, y_0) in the domain of f has infinitely many tangent lines at the point $(x_0, y_0, f(x_0, y_0))$. Each of these tangent lines lies in the plane tangent to the surface at $(x_0, y_0, f(x_0, y_0))$. In order to understand the plane tangent to a surface, we begin by defining the partial derivatives of a function of two variables.

DEFINITION 12.20 **Partial Derivatives for a Function of Two Variables**

Let f be a function of two variables defined on an open set S containing the point (x_0, y_0).

(a) The ***partial derivative with respect to*** x ***at*** (x_0, y_0), denoted by $f_x(x_0, y_0)$, is the limit

$$\lim_{h\to 0} \frac{f(x_0 + h, y_0) - f(x_0, y_0)}{h},$$

provided that this limit exists.

(b) Similarly, the ***partial derivative with respect to*** y ***at*** (x_0, y_0), denoted by $f_y(x_0, y_0)$, is the limit

$$\lim_{h\to 0} \frac{f(x_0, y_0 + h) - f(x_0, y_0)}{h},$$

provided that this limit exists.

We may also define the partial-derivative functions $f_x(x, y)$ and $f_y(x, y)$. The domains of these functions are the sets of all points where the respective limits mentioned in Definition 12.22 exist. The partial derivatives measure the rates of change of the function f in the x and y directions. For example, when $f(x, y) = 9 - x^2 - y^2$,

$$f_x(2, 1) = \lim_{h\to 0} \frac{f(2 + h, 1) - f(2, 1)}{h} = \lim_{h\to 0} \frac{(9 - (2 + h)^2 - 1^2) - 4}{h}$$

$$= \lim_{h\to 0} \frac{-4h - h^2}{h} = \lim_{h\to 0}(-4 - h) = -4.$$

The following figure shows the graph of f:

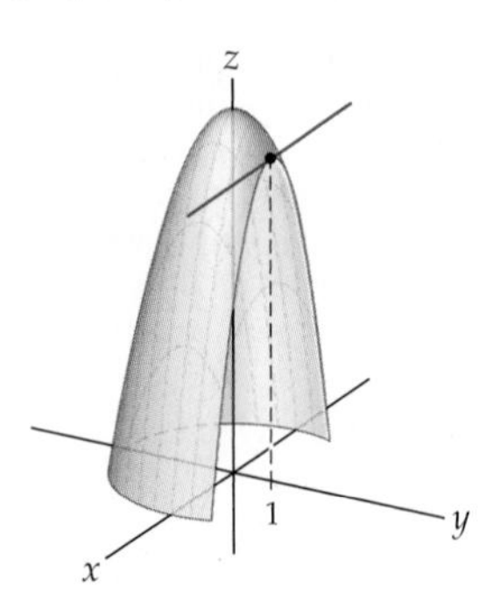

The cut edge toward the right is the intersection of the surface with the plane $y = 1$. When we compute $f_x(2, 1)$, we are finding the rate of change of this curve in the (positive) x direction.

In Exercise 23 we ask you to use the definition to show that $f_y(2, 1) = -2$. This partial derivative represents the rate of change of the surface in the positive y direction at the point $(2, 1)$. We may visualize this partial derivative as the rate of change of the curve created when the surface is cut by the plane $x = 2$.

More generally, a partial derivative with respect to x provides the rate of change of the function in the x direction. That is, if we fix the y-variable at y_0, we may think of the function $f(x, y_0)$ as a function of a single variable x, and the partial derivative tells us the rate of change of that function. To visualize this, we may think of the curve created when the surface $f(x, y)$ is cut by the vertical plane $y = y_0$. The partial derivative $f_x(x_0, y_0)$ is the slope of the tangent line of this function at (x_0, y_0). Similarly, when we take the partial derivative of the function with respect to y, we fix x_0 and the result is a function of the single variable y.

We may also define partial derivatives for functions of three variables.

DEFINITION 12.21 Partial Derivatives for a Function of Three Variables

Let f be a function of three variables defined on an open set S containing the point (x_0, y_0, z_0).

(a) The ***partial derivative with respect to*** x ***at*** (x_0, y_0, z_0), denoted by $f_x(x_0, y_0, z_0)$, is the limit

$$\lim_{h\to 0} \frac{f(x_0 + h, y_0, z_0) - f(x_0, y_0, z_0)}{h},$$

provided that this limit exists.

(b) The ***partial derivative with respect to*** y ***at*** (x_0, y_0, z_0), denoted by $f_y(x_0, y_0, z_0)$, is the limit

$$\lim_{h\to 0} \frac{f(x_0, y_0 + h, z_0) - f(x_0, y_0, z_0)}{h},$$

provided that this limit exists.

(c) The ***partial derivative with respect to*** z ***at*** (x_0, y_0, z_0), denoted by $f_z(x_0, y_0, z_0)$, is the limit

$$\lim_{h\to 0} \frac{f(x_0, y_0, z_0 + h) - f(x_0, y_0, z_0)}{h},$$

provided that this limit exists.

From the preceding definitions, we see that to compute the partial derivative f_x for a function of two variables, we treat the variable y like a constant and take the derivative of the function as if it were a function of the single variable x. Similarly, to compute f_y, we treat the variable x like a constant and take the derivative of the function as if it were a function of the single variable y. When f is a function of three variables, to find f_x, we treat *both* y and z like constants and take the derivative of the function as though it were a function of the single variable x. Analogous statements hold for finding partial derivatives with respect to y and z. This discussion leads to the following shortcuts:

- Let $f(x, y)$ be a function of two variables. The partial derivative $f_x(x, y)$ may be computed by treating y like a constant and taking the derivative of the resulting function of the single variable x, using any appropriate rules for finding derivatives. Similarly, the partial derivative $f_y(x, y)$ may be computed by treating x like a constant and taking the derivative of the resulting function of the single variable y, using any appropriate rules for finding derivatives.
- Let $f(x, y, z)$ be a function of three variables. The partial derivative $f_x(x, y, z)$ may be computed by treating y and z like constants and taking the derivative of the resulting function of the single variable x, using any appropriate rules for finding derivatives. The partial derivative $f_y(x, y, z)$ may be computed by treating x and z like constants and taking the derivative of the resulting function of the single variable y, using any appropriate rules for finding derivatives. Finally, the partial derivative $f_z(x, y, z)$ may be computed by treating x and y like constants and taking the derivative of the resulting function of the single variable z, using any appropriate rules for finding derivatives.

For example, if $f(x, y, z) = x\sin(yz) + y^2$, we have the following partial derivatives:

$$f_x(x, y, z) = \sin(yz), \quad f_y(x, y, z) = xz\cos(yz) + 2y, \quad \text{and } f_z(x, y, z) = xy\cos(yz).$$

There are alternative notations for partial derivatives. For example, when $w = f(x, y, z)$, we may use the following notation:

$$\begin{aligned}
\frac{\partial w}{\partial x} &= \frac{\partial}{\partial x}(f) = \frac{\partial f}{\partial x} = f_x(x, y, z),\\
\frac{\partial w}{\partial y} &= \frac{\partial}{\partial y}(f) = \frac{\partial f}{\partial y} = f_y(x, y, z),\\
\frac{\partial w}{\partial z} &= \frac{\partial}{\partial z}(f) = \frac{\partial f}{\partial z} = f_z(x, y, z).
\end{aligned}$$

Note that we are using the notation $\frac{\partial}{\partial x}$ as an operator meaning "take the partial derivative with respect to x" of the function, and $\frac{\partial w}{\partial x}$ is the partial derivative.

Higher Order Partial Derivatives

In Section 12.6 we will see how partial derivatives are used to locate points in the domain where the relative extremes of a function of two variables occur. We will also see how higher order partial derivatives may be used to classify those points as maxima, minima, or neither maxima nor minima.

DEFINITION 12.22 Second-Order Partial Derivatives

Let $f(x, y)$ be a function of two variables. We define the following four second-order partial derivatives for f:

(a) $\dfrac{\partial^2 f}{\partial x^2} = \dfrac{\partial}{\partial x}\left(\dfrac{\partial f}{\partial x}\right) = f_{xx}(x, y)$.

(b) $\dfrac{\partial^2 f}{\partial x \partial y} = \dfrac{\partial}{\partial x}\left(\dfrac{\partial f}{\partial y}\right) = f_{yx}(x, y)$.

(c) $\dfrac{\partial^2 f}{\partial y \partial x} = \dfrac{\partial}{\partial y}\left(\dfrac{\partial f}{\partial x}\right) = f_{xy}(x, y)$.

(d) $\dfrac{\partial^2 f}{\partial y^2} = \dfrac{\partial}{\partial y}\left(\dfrac{\partial f}{\partial y}\right) = f_{yy}(x, y)$.

Similarly, there are nine different second-order partial derivatives for a function of three variables. Note that in parts (b) and (c) of the definition above we have $\dfrac{\partial^2 f}{\partial x \partial y} = f_{yx}(x, y)$ and $\dfrac{\partial^2 f}{\partial y \partial x} = f_{xy}(x, y)$, respectively. These are *not* typographical errors. When we write $\dfrac{\partial^2 f}{\partial x \partial y}$, we mean take the partial derivative of the function f first with respect to variable y and then take the partial derivative of the result with respect to x. The notation f_{yx} has an analogous meaning. The order in which the partial derivatives is taken in part (c) is reversed. The partial derivatives in parts (b) and (c) are referred to as ***mixed partial derivatives.***

When $f(x, y) = x \cos y + x^2$, we have the following partial derivatives:

$$f_x(x, y) = \cos y + 2x \text{ and } f_y(x, y) = -x \sin y.$$

The second-order partial derivatives are:

$$f_{xx}(x, y) = 2,\ f_{xy}(x, y) = -\sin y = f_{yx}(x, y),\ \text{and } f_{yy}(x, y) = -x \cos y.$$

Note that the mixed partial derivatives $f_{xy}(x, y)$ and $f_{yx}(x, y)$ are equal. This does not always happen, but the mixed partial derivatives will be equal everywhere in an open set S if the mixed second-order partial derivatives are continuous on S.

THEOREM 12.23 Clairaut's Theorem: The Equality of the Mixed Second-Order Partial Derivatives

Let $f(x, y)$ be a function defined on an open subset S of $\mathbb{R}^2$. If the second-order partial derivatives of f are continuous everywhere in S, then $f_{xy}(x, y) = f_{yx}(x, y)$ at every point in S.

We will omit the proof of Clairaut's theorem.

In our earlier example, the function and all of its partial derivatives were continuous everywhere in $\mathbb{R}^2$. Therefore, by Clairaut's theorem the mixed partial derivatives f_{xy} and f_{yx} are guaranteed to be equal, as we computed.

Although we will rarely have the need, we may similarly find third- or even higher-order partial derivatives for functions of two or more variables. We explore a few examples shortly.

Finding a Function When the Partial Derivatives Are Given

Earlier in this section we saw that the function $f(x, y) = x \cos y + x^2$ has the partial derivatives

$$f_x(x, y) = \cos y + 2x \quad \text{and} \quad f_y(x, y) = -x \sin y.$$

Is there a way to reverse this process? That is, given two functions $g(x, y)$ and $h(x, y)$, is it possible to find a function $F(x, y)$ such that $g(x, y) = F_x(x, y)$ and $h(x, y) = F_y(x, y)$? Although we cannot do this for every pair of functions g and h, we can do it when the functions satisfy the following theorem:

THEOREM 12.24

The Existence of a Function with Specified Partial Derivatives

Let $g(x, y)$ and $h(x, y)$ be functions with continuous partial derivatives $g_y(x, y)$ and $h_x(x, y)$, respectively. Then there exists a function $F(x, y)$ such that $F_x(x, y) = g(x, y)$ and $F_y(x, y) = h(x, y)$ if and only if $g_y(x, y) = h_x(x, y)$.

For example, consider the pair of functions $g(x, y) = y^2e^{xy^2}$ and $h(x, y) = 2xye^{xy^2} - \sin y$. There is a function $F(x, y)$ such that

$$F_x(x, y) = g(x, y) = y^2e^{xy^2} \text{ and } F_y(x, y) = h(x, y) = 2xye^{xy^2} - \sin y$$

since $g_y(x, y) = (2y + 2xy^3)e^{xy^2} = h_x(x, y)$. In a moment we will outline a procedure for finding F. We first note, however, that for the pair $\hat{g}(x, y) = e^{xy}$ and $\hat{h}(x, y) = e^{xy}$, such a function does *not* exist, since $\hat{g}_y(x, y) = xe^{xy} \neq ye^{xy} = \hat{h}_x(x, y)$. Theorem 12.24 tells us that a function F exists only when functions g and h have the specified relationship.

Given that the functions g and h satisfy the condition in Theorem 12.24, we now outline the steps required to find a function $F(x, y)$. We will use the given pair of functions $g(x, y) = y^2e^{xy^2}$ and $h(x, y) = 2xye^{xy^2} - \sin y$. To find F, we start by considering the form of a function F whose partial derivative with respect to x is $g(x, y) = F_x(x, y) = y^2e^{xy^2}$. We integrate $g(x, y)$ with respect to x, treating y like a constant:

$$\int y^2e^{xy^2}\,dx = e^{xy^2} + q(y) = F(x, y).$$

Note:

- Instead of having a simple "constant of integration," we have an unknown function $q(y)$, because when we take the partial derivative of any function of y with respect to x, the derivative is zero.
- We can check our work in two ways. First, we may compute $\frac{\partial}{\partial x}(F(x, y))$ to ensure that it would result in the function g that we were given. Second, we will take $\frac{\partial}{\partial y}(F(x, y))$ to ensure that this partial derivative is consistent with the function $h(x, y)$ that we were given. We carry out this step next.

We compute

$$\frac{\partial}{\partial y}(F(x, y)) = \frac{\partial}{\partial y}(e^{xy^2} + q(y)) = 2xye^{xy^2} + q'(y).$$

In order for this partial derivative, $2xye^{xy^2} + q'(y)$, to equal $h(x, y) = 2xye^{xy^2} - \sin y$, we must have $q'(y) = -\sin y$. This happens when $q(y) = \cos y + C$, where C is any constant. Therefore, $F(x, y) = e^{xy^2} + \cos y + C$ is the function we sought. We may check this result by showing that $F_x(x, y) = g(x, y) = y^2e^{xy^2}$.

To summarize, given functions $g(x, y)$ and $h(x, y)$, there is a function $F(x, y)$ such that $g(x, y) = F_x(x, y)$ and $h(x, y) = F_y(x, y)$ if and only if $g_y(x, y) = h_x(x, y)$. When this is the case, you may find F as follows:

- Evaluate $\int g(x, y)\,dx = F(x, y)$. Be sure to include a "function of integration," $q(y)$, as part of the antiderivative.
- Compute $\frac{\partial}{\partial y}(F(x, y))$.

- Set $\frac{\partial}{\partial y}(F(x, y)) = h(x, y)$, and solve the resulting equation for $q'(y)$.
- Finish constructing $F(x, y)$ by finding an antiderivative for $q'(y)$.
- Check your work by showing that $F_x(x, y) = g(x, y)$.

Theorem 12.24 may be rephrased in terms of a first-order differential equation. A differential equation of the form

$$g(x, y) + h(x, y)\frac{dy}{dx} = 0$$

is said to be ***exact*** if $g_y(x, y) = h_x(x, y)$. In this case, the implicitly defined function $F(x, y) + C = 0$, where C is a constant, is the solution of the differential equation. Using the same functions as before, we see that the differential equation

$$y^2e^{xy^2} + (2xye^{xy^2} - \sin y)\frac{dy}{dx} = 0$$

is exact, since if $g(x, y) = y^2e^{xy^2}$ and $h(x, y) = 2xye^{xy^2} - \sin y$, then

$$g_y(x, y) = (2y + 2xy^3)e^{xy^2} = h_x(x, y).$$

The solution of this differential equation is the function $F(x, y) = e^{xy^2} + \cos y + C = 0$.

Examples and Explorations

EXAMPLE 1

Using the definition to compute partial derivatives

Use the definition to find the indicated partial derivatives for the following functions:

(a) f_y when $f(x, y) = \frac{x}{y}$ **(b)** $\frac{\partial g}{\partial z}$ when $g(x, y, z) = \frac{z^2}{xy^3}$

SOLUTION

(a)

$$f_y(x, y) = \lim_{h\to 0}\frac{f(x, y+h) - f(x, y)}{h} \quad \leftarrow \text{the definition of } f_y$$

$$= \lim_{h\to 0}\frac{\frac{x}{y+h} - \frac{x}{y}}{h} \quad \leftarrow \text{the function is } \frac{x}{y}$$

$$= \lim_{h\to 0} -\frac{hx}{hy(y+h)} \quad \leftarrow \text{algebra}$$

$$= \lim_{h\to 0} -\frac{x}{y(y+h)} \quad \leftarrow \text{Cancellation Theorem}$$

$$= -\frac{x}{y^2} \quad \leftarrow \text{evaluation of the limit}$$

(b)

$$\frac{\partial g}{\partial z} = \lim_{h\to 0}\frac{g(x, y, z+h) - g(x, y, z)}{h} \quad \leftarrow \text{the definition of } \frac{\partial g}{\partial z}$$

$$= \lim_{h\to 0}\frac{\frac{(z+h)^2}{xy^3} - \frac{z^2}{xy^3}}{h} \quad \leftarrow \text{the function is } \frac{z^2}{xy^3}$$

$$= \lim_{h\to 0}\frac{2zh + h^2}{hxy^3} \quad \leftarrow \text{algebra}$$

$$= \lim_{h\to 0}\frac{2z + h}{xy^3} \quad \leftarrow \text{Cancellation Theorem}$$

$$= \frac{2z}{xy^3} \quad \leftarrow \text{evaluation of the limit}$$

□

We can check our answers by using "shortcuts" for finding derivatives. Since $f(x, y) = xy^{-1}$, to find f_y we treat x as though it were a constant and take the derivative of the resulting function of y. We obtain $f_y(x, y) = -xy^{-2}$, which is equivalent to the derivative we found in part (a). Similarly, if we treat both x and y as constants in the function g and take the derivative of the resulting function of z, we obtain the result we found in part (b).

EXAMPLE 2 **Finding the partial derivatives of a function of two variables at a point and interpreting them graphically**

Let $f(x, y) = \frac{x}{y^2} - \frac{y}{x}$.

(a) Find the first-order partial derivatives f_x and f_y.

(b) Evaluate $f_x(-1, 2)$ and $f_y(-1, 2)$, and find equations for the lines tangent to the surface in the x and y directions at the point $(-1, 2, f(-1, 2))$.

(c) Find an equation for the plane containing the point $(-1, 2, f(-1, 2))$ and the lines from part (b).

SOLUTION

(a) The first-order partial derivatives are

$$f_x(x, y) = \frac{1}{y^2} + \frac{y}{x^2} \text{ and } f_y(x, y) = -\frac{2x}{y^3} - \frac{1}{x}.$$

(b) At $(-1, 2)$, we have $f(-1, 2) = \frac{7}{4}$, $f_x(-1, 2) = \frac{9}{4}$, and $f_y(-1, 2) = \frac{5}{4}$. These partial derivatives represent the rates of change of the function f in the x direction and y direction, respectively, at the point $(-1, 2)$. Therefore, direction vectors of the lines tangent to the surface at $(-1, 2)$ are $\mathbf{i} + \frac{9}{4}\mathbf{k}$ and $\mathbf{j} + \frac{5}{4}\mathbf{k}$. Thus, the line tangent to the surface at $\left(-1, 2, \frac{7}{4}\right)$ in the x direction is given by the parametric equations

$$x = -1 + t, \ y = 2, z = \frac{7}{4} + \frac{9}{4}t.$$

Similarly, the line tangent to the surface at $\left(-1, 2, \frac{7}{4}\right)$ in the y direction is given by the parametric equations

$$x = -1, \ y = 2 + t, z = \frac{7}{4} + \frac{5}{4}t.$$

(c) To obtain the normal vector for the plane containing the two intersecting lines we found in part (b), we take the cross product of the direction vectors for the lines. That is,

$$\mathbf{N} = \left\langle 0, 1, \frac{5}{4} \right\rangle \times \left\langle 1, 0, \frac{9}{4} \right\rangle = \left\langle \frac{9}{4}, \frac{5}{4}, -1 \right\rangle.$$

The equation for the plane orthogonal to this vector and containing the point $\left(-1, 2, \frac{7}{4}\right)$ is

$$\left\langle \frac{9}{4}, \frac{5}{4}, -1 \right\rangle \cdot \left\langle x + 1, y - 2, z - \frac{7}{4} \right\rangle = 0,$$

or equivalently, $\frac{9}{4}x + \frac{5}{4}y - z = -\frac{3}{2}$. □

EXAMPLE 3 **Finding the equation of a plane containing the tangent lines in the *x* and *y* directions**

Let $f(x, y)$ be a function with first-order partial derivatives $f_x(a, b)$ and $f_y(a, b)$ at a point (a, b) in the domain of f.

(a) Find equations for the lines tangent to the surface in the x and y directions at the point $(a, b, f(a, b))$.

(b) Find an equation for the plane containing the point $(a, b, f(a, b))$ and the lines from part (a).

SOLUTION

(a) Our analysis follows the work we did in Example 2. The partial derivatives represent the rates of change of the function f in the x and y directions, respectively, at the point (a, b). Therefore, direction vectors of the lines tangent to the surface at (a, b) are $\mathbf{i}+f_x(a, b)\mathbf{k}$ and $\mathbf{j}+f_y(a, b)\mathbf{k}$. Thus, the line tangent to the surface at $(a, b, f(a, b))$ in the x direction is given by the parametric equations

$$x = a + t,\ y = b,\ z = f(a, b) + f_x(a, b)t.$$

Similarly, the line tangent to the surface at $(a, b, f(a, b))$ in the y direction is given by the parametric equations

$$x = a,\ y = b + t,\ z = f(a, b) + f_y(a, b)t.$$

(b) To obtain the normal vector for the plane containing the two intersecting lines we found in part (b), we take the cross product of the direction vectors for the lines. That is,

$$\mathbf{N} = \langle 0, 1, f_y(a, b)\rangle \times \langle 1, 0, f_x(a, b)\rangle = \langle f_x(a, b), f_y(a, b), -1\rangle.$$

The equation for the plane orthogonal to this vector and containing the point $(a, b, f(a, b))$ is

$$\langle f_x(a, b), f_y(a, b), -1\rangle \cdot \langle x - a, y - b, z - f(a, b)\rangle = 0,$$

or equivalently, $f_x(a, b)(x - a) + f_y(a, b)(y - b) = z - f(a, b)$. □

CHECKING THE ANSWER

We may use the general form for the equation of the plane to check our result from Example 2. We had the point $(-1, 2, f(-1, 2)) = \left(-1, 2, \frac{7}{4}\right)$ and partial derivatives $f_x(-1, 2) = \frac{9}{4}$ and $f_y(-1, 2) = \frac{5}{4}$. Therefore, the equation of the plane is

$$\frac{9}{4}(x + 1) + \frac{5}{4}(y - 2) = z - \frac{7}{4},$$

which we see is equivalent to our final answer in Example 2.

EXAMPLE 4 **Computing higher order partial derivatives**

Let $f(x, y) = \frac{x}{y^2} - \frac{y}{x}$.

(a) Find all second-order partial derivatives for f, and show that the mixed second-order partial derivatives are equal.

(b) Show that the mixed third-order partial derivatives f_{xxy}, f_{xyx}, and f_{yxx} are all equal.

SOLUTION

(a) We found the first-order partial derivatives $\frac{\partial f}{\partial x} = \frac{1}{y^2} + \frac{y}{x^2}$ and $\frac{\partial f}{\partial y} = -\frac{2x}{y^3} - \frac{1}{x}$ in Example 2. We have

$$\frac{\partial^2 f}{\partial x^2} = \frac{\partial}{\partial x}\left(\frac{1}{y^2} + \frac{y}{x^2}\right) = -\frac{2y}{x^3},$$

$$\frac{\partial^2 f}{\partial y \partial x} = \frac{\partial}{\partial y}\left(\frac{1}{y^2} + \frac{y}{x^2}\right) = -\frac{2}{y^3} + \frac{1}{x^2},$$

$$\frac{\partial^2 f}{\partial x \partial y} = \frac{\partial}{\partial x}\left(-\frac{2x}{y^3} - \frac{1}{x}\right) = -\frac{2}{y^3} + \frac{1}{x^2},$$

$$\frac{\partial^2 f}{\partial y^2} = \frac{\partial}{\partial y}\left(-\frac{2x}{y^3} - \frac{1}{x}\right) = \frac{6x}{y^4}.$$

Note that the partial derivatives $\frac{\partial^2 f}{\partial y \partial x}$ and $\frac{\partial^2 f}{\partial x \partial y}$ are equal.

(b) We may compute the required third-order partial derivatives by taking the appropriate partial derivatives of the results from part (a).

$$f_{xxy} = \frac{\partial}{\partial y}\left(\frac{\partial^2 f}{\partial x^2}\right) = \frac{\partial}{\partial y}\left(-\frac{2y}{x^3}\right) = -\frac{2}{x^3},$$

$$f_{xyx} = \frac{\partial}{\partial x}\left(\frac{\partial^2 f}{\partial y \partial x}\right) = \frac{\partial}{\partial x}\left(-\frac{2}{y^3} + \frac{1}{x^2}\right) = -\frac{2}{x^3}.$$

Since $f_{yxx} = \frac{\partial}{\partial x}\left(\frac{\partial^2 f}{\partial x \partial y}\right)$ and we already know that $\frac{\partial^2 f}{\partial x \partial y} = \frac{\partial^2 f}{\partial y \partial x}$, we may conclude that $f_{yxx} = -\frac{2}{x^3}$ as well. □

EXAMPLE 5

Finding a function when the partial derivatives are given

Show that there is a function $F(x, y)$ such that $F_x(x, y) = g(x, y)$ and $F_y(x, y) = h(x, y)$ when

$$g(x, y) = y\cos(xy) \text{ and } h(x, y) = x\cos(xy) + 3y^2.$$

Then find F.

SOLUTION

We use Theorem 12.24 to show that there is such a function. Since the partial derivatives

$$g_y(x, y) = \cos(xy) - xy\sin(xy) = h_x(x, y),$$

Theorem 12.24 guarantees that a function $F(x, y)$ exists.

Now, to find F, we integrate $g(x, y)$ with respect to x:

$$\int y\cos(xy)\,dx = \sin(xy) + q(y) = F(x, y).$$

As we remarked in our discussion earlier in this section, we may check our work by finding both first-order partial derivatives of this function to be sure that $\frac{\partial}{\partial x}(F(x, y)) = g(x, y)$ and $\frac{\partial}{\partial y}(F(x, y))$ is consistent with $h(x, y)$. You should check that $\frac{\partial}{\partial x}(F(x, y)) = g(x, y)$. Checking the consistency of the other partial derivative is part of our procedure. Here,

$$\frac{\partial}{\partial y}(F(x, y)) = \frac{\partial}{\partial y}(\sin(xy) + q(y)) = x\cos(xy) + q'(y).$$

In order for this partial derivative, $x\cos(xy) + q'(y)$, to equal $h(x, y) = x\cos(xy) + 3y^2$, we must have $q'(y) = 3y^2$. This happens when $q(y) = y^3 + C$, where C is any constant. Therefore, $F(x, y) = \sin(xy) + y^3 + C$ is the function we sought.

If we are given a point on the graph of F, we may use that information to evaluate the constant C. For example, here, if we want the graph of F to contain the point $(0, 2, 5)$, then we must have

$$F(0, 2) = \sin(0 \cdot 2) + 2^3 + C = 8 + C = 5.$$

Therefore, $C = -3$, and our function is $F(x, y) = \sin(xy) + y^3 - 3$. □

We may check our answer by finding the partial derivatives of our function

$$F(x, y) = \sin(xy) + y^3 + C$$

Since $F_x(x, y) = y\cos(xy) = g(x, y)$ and $F_y(x, y) = x\cos(xy) + 3y^2 = h(x, y)$, our work was correct.

EXAMPLE 6

Solving an exact differential equation

Show that the differential equation

$$y\cos(xy) + (x\cos(xy) + 3y^2)\frac{dy}{dx} = 0$$

is exact. Solve the initial-value problem $F\left(\frac{\pi}{3}, 6\right) = 200$.

SOLUTION

Showing that the differential equation is exact requires the same steps that we performed in Example 5. If we let $g(x, y) = y\cos(xy)$ and $h(x, y) = x\cos(xy) + 3y^2$, then $g_y(x, y) = \cos(xy) - xy\sin(xy) = h_x(x, y)$, so the differential equation is exact. Furthermore, the general solution of the differential equation is $F(x, y) + C = 0$, where $F(x, y)$ is the same function we found in Example 5. That is,

$$F(x, y) = \sin(xy) + y^3 + C = 0.$$

To solve the initial-value problem, we determine the value of C that satisfies $F\left(\frac{\pi}{3}, 6\right) = 200$. When $x = \frac{\pi}{3}$ and $y = 6$, we have $\sin(2\pi) + 6^3 + C = 200$. The unique solution of this equation is $C = -16$. Therefore, the solution to the initial-value problem is

$$\sin(xy) + y^3 - 16 = 0$$

□

? TEST YOUR UNDERSTANDING

- What are the definitions of the partial derivatives with respect to x and y for a function $f(x, y)$? How are these definitions similar to the definition for the derivative of a function of a single variable? How are they different?
- What are the geometrical interpretations for $f_x(a, b)$ and $f_y(a, b)$ at a point (a, b) in the domain of a function f? When these partial derivatives exist, how can we find the equations of lines tangent to the surface defined by f in the x and y directions? How can we find the equation of the plane containing these lines?
- How may the shortcuts for derivatives of a function of a single variable be applied to find partial derivatives of functions of two or three variables?
- How are higher order partial derivatives computed? What is meant by mixed second-order partial derivatives? What conditions are sufficient to guarantee that the mixed second-order partial derivatives are equal?
- What condition(s) must $g(x, y)$ and $h(x, y)$ satisfy for these functions to equal the first-order partial derivatives of a function $F(x, y)$? That is, when will $g(x, y) = F_x(x, y)$ and $h(x, y) = F_y(x, y)$? What is the procedure for finding F?

EXERCISES 12.3

Thinking Back

- *Finding a direction vector for a tangent line:* Find a direction vector for the line tangent to the curve $y = x^3$ when $x = 2$.
- *Finding the equation of the plane containing two intersecting lines:* Show that the lines given by $\mathbf{r}_1(t) = \langle 3t - 4, -4t + 1, t\rangle$ and $\mathbf{r}_2(t) = \langle -t + 2, 2t - 9, -2t + 7\rangle$ intersect, and find the equation of the plane containing the lines.

Concepts

0. *Problem Zero:* Read the section and make your own summary of the material.
1. *True/False:* Determine whether each of the statements that follow is true or false. If a statement is true, explain why. If a statement is false, provide a counterexample.
   (a) *True or False:* The partial derivative of a function $f(x, y)$ with respect to x is defined by $\lim_{h\to 0} \frac{f(x+h, y) - f(x, y)}{h}$ if the limit exists.
   (b) *True or False:* The partial derivative of a function $g(r, s)$ with respect to s is defined by $\lim_{h\to 0} \frac{g(r, s+h) - g(r, s)}{h}$ if the limit exists.
   (c) *True or False:* The second-order partial derivative of a function $f(x, y)$ with respect to x and y is defined by $\lim_{h\to 0} \frac{f(x+h, y+h) - f(x, y)}{h}$ if the limit exists.

(d) *True or False:* If (a, b) is a point in the domain of a function $f(x, y)$ with continuous partial derivatives, then $f_x(a, b) = f_y(a, b)$.

(e) *True or False:* If (a, b) is a point in the domain of a function $f(x, y)$ with continuous second-order partial derivatives, then $f_{xy}(a, b) = f_{yx}(a, b)$.

(f) *True or False:* If (a, b) is a point in the domain of a function $f(x, y)$ at which the second-order partial derivatives exist but are not continuous, then $f_{xy}(a, b) \neq \frac{\partial^2 f}{\partial y \partial x}(a, b)$.

(g) *True or False:* If $f(x, y) = |x|y$, then $f_x(x, y)$ at $(0, 0)$ does not exist.

(h) *True or False:* If $f(x, y) = |x|y$, then $f_y(x, y)$ at $(0, 0)$ does not exist.

2. *Examples:* Construct examples of the thing(s) described in the following. Try to find examples that are different than any in the reading.

(a) A function $f(x, y)$ such that $f_x(x, y) = y^2 e^{xy^2}$.

(b) A function $g(x, y)$ such that $\frac{\partial g}{\partial y} = xy + e^{4x} \sin 3y$.

(c) A function $F(x, y)$ such that $F_x(x, y) = \sec^2 x + \frac{1}{2}y^2 + \frac{4}{3}e^{4x} \cos 3y$ and $F_y(x, y) = xy - e^{4x} \sin 3y$.

3. Draw a line on a piece of paper. If the paper is on a horizontal surface, how does the pitch or slope of the line change as you rotate the paper on the tabletop?

4. Draw a line on a piece of paper. If the paper is on an angled surface, how does the pitch or slope of the line change as you rotate the paper on the surface? Why would it *not* make sense to talk about the slope of a plane?

5. Let (x_0, y_0) be a point in the domain of the function $f(x, y)$ at which $f_x(x_0, y_0)$ and $f_y(x_0, y_0)$ both exist. Explain how these partial derivatives may be interpreted as the slopes of lines tangent to the surface defined by $f(x, y)$ at the point (x_0, y_0).

6. Let $f(x)$ be a differentiable function of a single variable x.

(a) What is the relationship between the graph of $f(x)$ and the graph of the function of two variables, $g(x, y) = f(x)$?

(b) For what values of x and y do the first-order partial derivatives of g exist?

(c) What are $\frac{\partial g}{\partial x}$ and $\frac{\partial g}{\partial y}$? Why do these partial derivatives make sense?

7. Let $f(y)$ be a differentiable function of a single variable y.

(a) What is the relationship between the graph of $f(y)$ and the graph of the function of two variables, $g(x, y) = f(y)$?

(b) For what values of x and y do the first-order partial derivatives of g exist?

(c) What are $\frac{\partial g}{\partial x}$ and $\frac{\partial g}{\partial y}$? Why do these partial derivatives make sense?

8. The function $f(x, y) = |xy|$ is graphed in the figure that follows. Use the definition of the partial derivatives to show that $f_x(0, 0) = f_y(0, 0) = 0$. What are the equations of the lines tangent to the surface in the x and y directions at $(0, 0, 0)$? What is the equation of the plane containing these two lines?

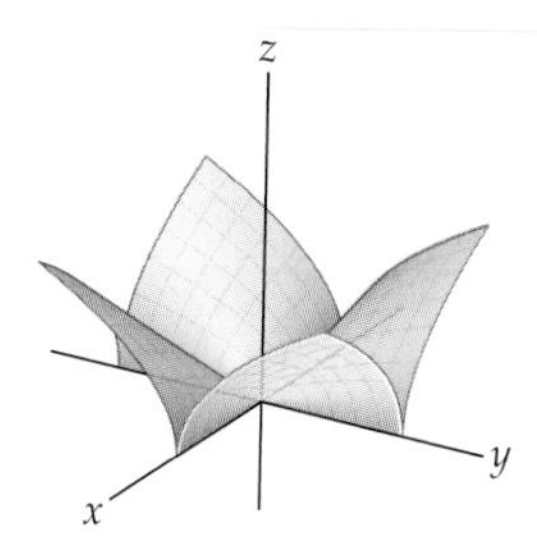

9. The function $g(x, y) = x^2 - y^2$ is graphed in the figure that follows. Use the definition of the partial derivatives to show that $g_x(0, 0) = g_y(0, 0) = 0$. What are the equations of the lines tangent to the surface in the x and y directions at $(0, 0, 0)$? What is the equation of the plane containing these two lines?

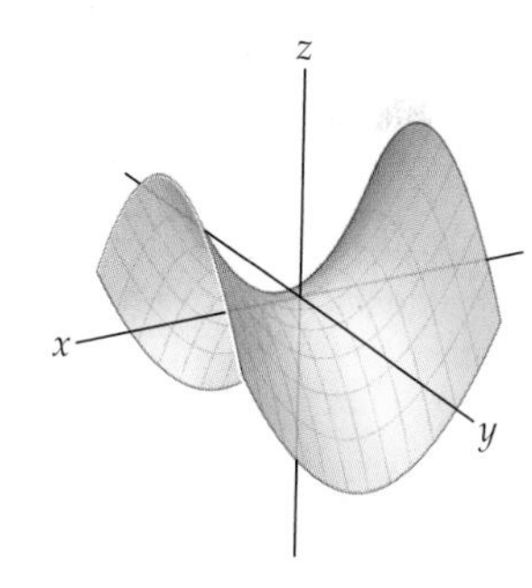

10. Compare the graphs in Exercises 8 and 9. Which of these surfaces do you think has a well-defined tangent plane at $(0, 0, 0)$?

11. Let $f(x, y) = \begin{cases} 0, & \text{if } xy = 0 \\ 1, & \text{if } xy \neq 0. \end{cases}$ Use the definition of the partial derivatives to show that $f_x(0, 0) = f_y(0, 0) = 0$. Explain why this example shows that the existence of the partial derivatives at a point (x_0, y_0) for a function $f(x, y)$ does *not* guarantee that f is continuous at (x_0, y_0).

12. Consider the function f defined in Exercise 11.

(a) Use the definition of the partial derivatives to show that $f_x(x_0, 0)$ exists for every value of x_0 but $f_y(x_0, 0)$ exists only when $x_0 = 0$.

(b) Use the definition of the partial derivatives to show that $f_y(0, y_0)$ exists for every value of y_0 but $f_x(0, y_0)$ exists only when $y_0 = 0$.

13. Let $f(x, y, z) = \begin{cases} 0, & \text{if } xyz = 0 \\ 1, & \text{if } xyz \neq 0. \end{cases}$ Use the definition of the partial derivatives to show that $f_x(0, 0, 0) = f_y(0, 0, 0) = f_z(0, 0, 0) = 0$. Explain why this example shows that the existence of the partial derivatives at a point (x_0, y_0, z_0) for a function $f(x, y, z)$ does *not* guarantee that f is continuous at (x_0, y_0, z_0).

14. Consider the function f defined in Exercise 13. Use the definition of the partial derivatives to show that $f_x(x_0, y_0, 0)$ and $f_y(x_0, y_0, 0)$ exist for every value of x_0 and y_0 but $f_z(x_0, y_0, 0)$ exists only when at least one of x_0 and y_0 is zero.

15. Assume that $f(x, y)$ is a function of two variables with partial derivatives of every order. Assume also that the order in which the partial derivatives are taken is significant.

(a) How many different second-order partial derivatives does f have?

(b) How many different third-order partial derivatives does f have?

(c) How many different nth-order partial derivatives does f have?

16. Assume that $g(x, y, z)$ is a function of three variables with partial derivatives of every order. Assume also the order in which the partial derivatives are taken is significant.

(a) How many different second-order partial derivatives does g have?

(b) How many different third-order partial derivatives does g have?

(c) How many different nth-order partial derivatives does g have?

17. Assume that $f(x, y)$ is a function of two variables with partial derivatives of every order. Assume also that the order in which the partial derivatives are taken is *not* significant.

(a) How many different second-order partial derivatives does f have?

(b) How many different third-order partial derivatives does f have?

(c) How many different nth-order partial derivatives does f have?

18. Assume that $g(x, y, z)$ is a function of three variables with partial derivatives of every order. Assume also that the order in which the partial derivatives are taken is *not* significant.

(a) How many different second-order partial derivatives does g have?

(b) How many different third-order partial derivatives does g have?

(c) How many different nth-order partial derivatives does g have?

19. Let $f(x, y)$ and $g(x, y)$ be functions of two variables with the property that $\dfrac{\partial f}{\partial x} = \dfrac{\partial g}{\partial x}$ for every point $(x, y) \in \mathbb{R}^2$. What is the relationship between f and g?

20. Let $f(x, y)$ and $g(x, y)$ be functions of two variables with the property that $\dfrac{\partial f}{\partial y} = \dfrac{\partial g}{\partial y}$ for every point $(x, y) \in \mathbb{R}^2$. What is the relationship between f and g?

21. Let $f(x, y)$ and $g(x, y)$ be functions of two variables with the property that $\dfrac{\partial f}{\partial x} = \dfrac{\partial g}{\partial x}$ and $\dfrac{\partial f}{\partial y} = \dfrac{\partial g}{\partial y}$ for every point $(x, y) \in \mathbb{R}^2$. What is the relationship between f and g?

22. Let $f(x, y, z)$ and $g(x, y, z)$ be functions of three variables with the property that $\dfrac{\partial f}{\partial y} = \dfrac{\partial g}{\partial y}$ for every point $(x, y, z) \in \mathbb{R}^3$. What is the relationship between f and g?

Skills

Use the definition of the partial derivative to find the partial derivatives specified in Exercises 23–26.

23. $f_y(2, 1)$ when $f(x, y) = 9 - x^2 - y^2$

24. $f_x(0, -3)$ and $f_y(0, -3)$ when $f(x, y) = \dfrac{3x}{y^2}$

25. $\dfrac{\partial f}{\partial x}, \dfrac{\partial f}{\partial y}$ and $\dfrac{\partial f}{\partial z}$ when $f(x, y, z) = \dfrac{xy^2}{z}$

26. $\dfrac{\partial g}{\partial x}$ and $\dfrac{\partial g}{\partial y}$ when $g(x, y) = \dfrac{\sqrt{y}}{x+1}$

Find the first-order partial derivatives for the functions in Exercises 27–36.

27. $f(x, y) = e^x \sin(xy)$

28. $g(x, y) = \tan^{-1}(xy^2)$

29. $f(x, y) = x^y$

30. $g(x, y) = \dfrac{\ln(y^2+1)}{x}$

31. $f(x, y) = x \sin y$

32. $g(x, y) = x \cos y$

33. $f(r, \theta) = r \sin \theta$

34. $g(r, \theta) = r \cos \theta$

35. $f(x, y, z) = \dfrac{xy^2}{x+z}$

36. $g(x, y, z) = 2xy + 2xz + 2yz$

In Exercises 37–42, use the partial derivatives you found in Exercises 27–32 and the point (x_0, y_0) specified to (a) find the equation of the line tangent to the surface defined by the function in the x direction, (b) find the equation of the line tangent to the surface defined by the function in the y direction, and (c) find the equation of the plane containing the lines you found in parts (a) and (b).

37. Exercise 27, $\left(0, \dfrac{\pi}{2}\right)$

38. Exercise 28, $(1, 0)$

39. Exercise 29, $(e, 3)$

40. Exercise 30, $(3, 0)$

41. Exercise 31, $\left(2, \dfrac{\pi}{3}\right)$

42. Exercise 32, $(1, \pi)$

In Exercises 43–50, compute all of the second-order partial derivatives for the functions in Exercises 27–34 and show that the mixed partial derivatives are equal.

43. Exercise 27

44. Exercise 28

45. Exercise 29

46. Exercise 30

47. Exercise 31

48. Exercise 32

49. Exercise 33

50. Exercise 34

For the partial derivatives given in Exercises 51–54, find the most general form for a function of two variables, $f(x, y)$, with the given partial derivative.

51. $\dfrac{\partial f}{\partial x} = 0$

52. $\dfrac{\partial f}{\partial y} = 0$

53. $\dfrac{\partial f}{\partial x^2} = 0$

54. $\dfrac{\partial^2 f}{\partial y \partial x} = 0$

For the partial derivatives given in Exercises 55–58, find the most general form for a function of three variables, $f(x, y, z)$, with the given partial derivative.

55. $\dfrac{\partial f}{\partial x} = 0$

56. $\dfrac{\partial f}{\partial y} = 0$

57. $\dfrac{\partial f}{\partial x^2} = 0$

58. $\dfrac{\partial^2 f}{\partial y \partial x} = 0$

For each pair of functions in Exercises 59–62, use Theorem 12.24 to show that there is a function of two variables, $F(x, y)$, such that $\dfrac{\partial F}{\partial x} = g(x, y)$ and $\dfrac{\partial F}{\partial y} = h(x, y)$. Then find F.

59. $g(x, y) = e^x \cos y,\ h(x, y) = -e^x \sin y + 2y$

60. $g(x, y) = \dfrac{1}{xy} - \sin x,\ h(x, y) = -\dfrac{\ln x}{y^2}$

61. $g(x, y) = \dfrac{y}{1+x^2y^2},\ h(x, y) = \dfrac{x}{1+x^2y^2}$

62. $g(x, y) = \dfrac{2x}{y},\ h(x, y) = -\dfrac{x^2}{y^2}$

Solve the exact differential equations in Exercises 63–66.

63. $e^y + (xe^y - 7)\dfrac{dy}{dx} = 0$

64. $y\cos(xy) - 3 + (x\cos(xy) + 2)\dfrac{dy}{dx} = 0$

65. $e^x \ln y + x^3 + \dfrac{e^x}{y}\dfrac{dy}{dx} = 0$

66. $\dfrac{12x^3}{y^5} - \dfrac{15x^4}{y^6}\dfrac{dy}{dx} = 0$

Applications

67. Recall that the volume, V, of a cylinder is given by $V(r, h) = \pi r^2 h$. Find $\dfrac{\partial V}{\partial r}$ and $\dfrac{\partial V}{\partial h}$. What are the physical interpretations of these partial derivatives?

68. The pressure P of an ideal gas is given by $P(n, T, V) = \dfrac{nRT}{V}$, where n is the amount of the gas, R is a constant, T is the temperature of the gas in degrees Kelvin, and V is the volume of the gas. Find $\dfrac{\partial P}{\partial n}$, $\dfrac{\partial P}{\partial V}$, and $\dfrac{\partial P}{\partial T}$. What are the physical interpretations of these partial derivatives?

69. Annie is covering the kayak she is building with fabric. She knows that the shape of the fabric, when stretched across the ribs of her kayak, satisfies the differential equation $f_{xx} + f_{yy} = 0$.

(a) Annie has no interest in solving that equation; she is just building a kayak. You, however, should show that $\alpha e^{nx} \sin ny$ and $\beta e^{ny} \sin nx$ are solutions for any real numbers n, α, and β.

(b) Assuming that she needed something to do besides apply the fabric to her kayak, Annie measured and normalized the function describing the ribs near the bow to find that they have the shape given by

$$f(x, 0) = 0.04 \sin\left(\frac{\pi x}{2}\right),$$
$$f(x, 1) = 0.04\, e^{\pi/2} \sin\left(\frac{\pi x}{2}\right),$$
$$f(0, y) = 0,$$
$$f(1, y) = 0.04\, e^{\pi y/2}.$$

What solution $f(x, y)$ from part (a) satisfies these conditions? Sketch the solution.

70. Alex is modeling traffic patterns at a bottleneck on a freeway as it leaves Denver. He uses a well-known equation $u_x + u_t = 0$, where $u = u(x, t)$ is a scaled traffic density at a point x on the highway at time t.

(a) Show that $u_n(x, t) = \sin(n(x - t))$ is a solution of the equation for any integer n.

(b) Alex finds that in heavy traffic at the bottleneck the solution of his equation can look like

$$24 + \sum_{n=1}^{N} \frac{\sin(n(x-t))}{n} \quad \text{for some integer } N.$$

Show that this is a solution, and plot it for $N = 8$. Can you tell what happens as N gets large?

Proofs

71. Let $z = (f(x, y))^n$. Show that

$$\frac{\partial z}{\partial x} = n(f(x, y))^{n-1}\frac{\partial f}{\partial x} \text{ and } \frac{\partial z}{\partial y} = n(f(x, y))^{n-1}\frac{\partial f}{\partial y}.$$

72. Let $z = f(x, y)g(x, y)$. Show that

$$\frac{\partial z}{\partial x} = \frac{\partial f}{\partial x}g(x, y) + f(x, y)\frac{\partial g}{\partial x} \quad \text{and}$$
$$\frac{\partial z}{\partial y} = \frac{\partial f}{\partial y}g(x, y) + f(x, y)\frac{\partial g}{\partial y}.$$

73. Let $z = \dfrac{f(x, y)}{g(x, y)}$. Show that

$$\frac{\partial z}{\partial x} = \frac{f_x(x, y)g(x, y) - f(x, y)g_x(x, y)}{(g(x, y))^2} \quad \text{and}$$
$$\frac{\partial z}{\partial y} = \frac{f_y(x, y)g(x, y) - f(x, y)g_y(x, y)}{(g(x, y))^2}.$$

74. Let $p(x)$ be a polynomial of degree n, $q(y)$ be any function of y, and $f(x, y) = p(x)q(y)$. Prove that

$$\frac{\partial^{n+1} f}{\partial x^{n+1}} = 0.$$

75. Let $f(x)$ be a differentiable function of x, $g(y)$ be a differentiable function of y, and $h(x, y) = f(x) + g(y)$. Prove that

$$\frac{\partial^2 h}{\partial x \partial y} = \frac{\partial^2 h}{\partial y \partial x}.$$

76. Let $f(x)$ be a differentiable function of x, $g(y)$ be a differentiable function of y, and $h(x, y) = f(x)g(y)$. Prove that

$$\frac{\partial^2 h}{\partial x \partial y} = \frac{\partial^2 h}{\partial y \partial x}.$$

Thinking Forward

A chain rule for functions of two variables: Let $f(x, y) = x^2y^3$, $x = s\cos t$, and $y = s\sin t$.

- Find $\frac{\partial f}{\partial x}$ and $\frac{\partial f}{\partial y}$.
- Find $\frac{\partial x}{\partial s}$ and $\frac{\partial x}{\partial t}$.
- Find $\frac{\partial y}{\partial s}$ and $\frac{\partial y}{\partial t}$.
- Substitute the expressions for x and y into $f(x, y)$.
- Find $\frac{\partial f}{\partial s}$ and $\frac{\partial f}{\partial t}$ using the preceding results.
- Make a conjecture about the relationship between $\frac{\partial f}{\partial s}$, $\frac{\partial f}{\partial x}$, $\frac{\partial f}{\partial y}$, $\frac{\partial x}{\partial s}$, and $\frac{\partial y}{\partial s}$.

12.4 DIRECTIONAL DERIVATIVES AND DIFFERENTIABILITY

- Directional derivatives
- Differentiability of functions of two and three variables
- The tangent plane to a function of two variables

The Directional Derivative

In Section 12.3 we defined the partial derivatives for a function of two or three variables. Recall that when $f(x, y)$ is a function of two variables, the partial derivatives $\frac{\partial f}{\partial x}$ and $\frac{\partial f}{\partial y}$ tell us the rate of change of the function f in the (positive) x and y directions, respectively, when these derivatives exist. How can we compute the rate of change of f in a different direction? To do this we define the ***directional derivative*** of f in the direction of a specified unit vector $\mathbf{u}$.

DEFINITION 12.25 The Directional Derivative of a Function of Two Variables

Let $f(x, y)$ be a function of two variables defined on an open set containing the point (x_0, y_0), and let $\mathbf{u} = \langle \alpha, \beta \rangle$ be a unit vector. The ***directional derivative*** of f at (x_0, y_0) in the direction of $\mathbf{u}$, denoted by $D_{\mathbf{u}} f(x_0, y_0)$, is given by

$$\lim_{h \to 0} \frac{f(x_0 + \alpha h, y_0 + \beta h) - f(x_0, y_0)}{h},$$

provided that this limit exists.

Note that when $\mathbf{u} = \mathbf{i}$ or $\mathbf{u} = \mathbf{j}$, the directional derivative is $\frac{\partial f}{\partial x}$ or $\frac{\partial f}{\partial y}$, respectively. That is, for a point (x_0, y_0) in the domain of f at which the partial derivatives exist,

$$D_{\mathbf{i}} f(x_0, y_0) = f_x(x_0, y_0) \text{ and } D_{\mathbf{j}} f(x_0, y_0) = f_y(x_0, y_0).$$

Thus, the directional derivative generalizes the partial derivatives. The figure shown next indicates the geometry of the directional derivative. We may visualize $D_{\mathbf{u}}f(x_0, y_0)$ as the slope of the cut edge at the point $(x_0, y_0, f(x_0, y_0))$ when we slice the surface $z = f(x, y)$ with a vertical plane parallel to the vector $\mathbf{u}$ containing the point.

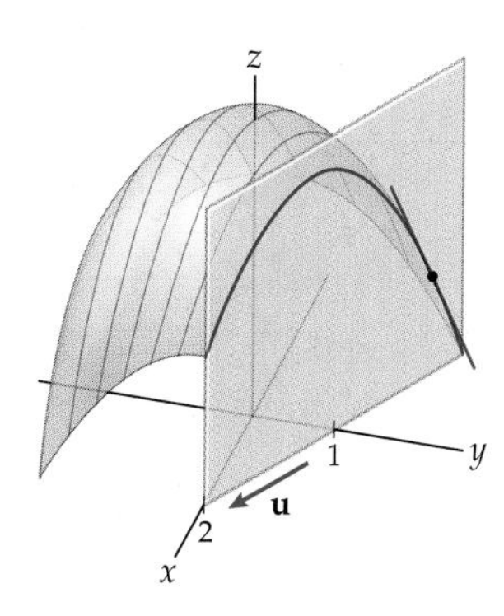

To compute the directional derivative of the function $f(x, y) = \frac{x^2}{y}$ at the point $(-1, 2)$ in the direction of the vector $\mathbf{v} = \langle 3, -4\rangle$, we first need to find a unit vector $\mathbf{u}$ with the same direction as $\mathbf{v}$. We have

$$\mathbf{u} = \frac{1}{\|\mathbf{v}\|}\mathbf{v} = \frac{1}{\sqrt{3^2 + (-4)^2}}\langle 3, -4\rangle = \left\langle \frac{3}{5}, -\frac{4}{5}\right\rangle.$$

Thus, the directional derivative

$$D_{\mathbf{u}}f(-1, 2) = \lim_{h\to 0} \frac{\dfrac{(-1+3h/5)^2}{2-4h/5} - \dfrac{(-1)^2}{2}}{h} = \lim_{h\to 0} \frac{-20h + 9h^2}{h(50-20h)} = \lim_{h\to 0} \frac{-20+9h}{50-20h} = -\frac{2}{5}.$$

Geometrically, the number $-\frac{2}{5}$ describes the slope of the line tangent to the cut edge of the graph when the surface defined by $f(x, y) = \frac{x^2}{y}$ is sliced by the vertical plane parallel to the vector $\mathbf{u} = \left\langle \frac{3}{5}, -\frac{4}{5}\right\rangle$ and containing the point $(-1, 2)$. We may use all of this information to find the equation of the line tangent to the surface at the point $(-1, 2, f(-1, 2)) = \left(-1, 2, \frac{1}{2}\right)$. Here, that equation is

$$x = -1 + \frac{3}{5}t,\ y = 2 - \frac{4}{5}t,\ z = \frac{1}{2} - \frac{2}{5}t.$$

More generally, for a point (x_0, y_0) in the domain of a function $f(x, y)$ and a unit vector $\mathbf{u} = \langle\alpha, \beta\rangle$ for which the directional derivative $D_{\mathbf{u}}f(x_0, y_0)$ exists, the equation of the line tangent to the surface defined by f at the point $(x_0, y_0, f(x_0, y_0))$ will be

$$x = x_0 + \alpha t,\ y = y_0 + \beta t,\ z = f(x_0, y_0) + D_{\mathbf{u}}f(x_0, y_0)t.$$

We may also define the directional derivative for a function of three variables.

DEFINITION 12.26 **The Directional Derivative of a Function of Three Variables**

Let $f(x, y, z)$ be a function of three variables defined on an open set containing the point (x_0, y_0, z_0), and let $\mathbf{u} = \langle\alpha, \beta, \gamma\rangle$ be a unit vector. The ***directional derivative*** of f at (x_0, y_0, z_0) in the direction of $\mathbf{u}$, denoted by $D_{\mathbf{u}}f(x_0, y_0, z_0)$, is given by

$$\lim_{h\to 0} \frac{f(x_0 + \alpha h, y_0 + \beta h, z_0 + \gamma h) - f(x_0, y_0, z_0)}{h},$$

provided that this limit exists.

As with a directional derivative for a function of two variables, when $\mathbf{u}$ is one of the standard basis vectors,

$$D_{\mathbf{i}} f(x_0, y_0, z_0) = f_x(x_0, y_0, z_0),$$
$$D_{\mathbf{j}} f(x_0, y_0, z_0) = f_y(x_0, y_0, z_0),$$
$$D_{\mathbf{k}} f(x_0, y_0, z_0) = f_z(x_0, y_0, z_0).$$

Differentiability

Recall that for a function $y = f(x)$ defined on an open interval containing the point x_0, we defined the derivative of f in terms of the limit of the difference quotient; that is,

$$f'(x_0) = \lim_{h \to 0} \frac{f(x_0 + h) - f(x_0)}{h},$$

provided that the limit exists. Unfortunately, we cannot use a definition analogous to this to define the differentiability of a function of two or more variables. In Exercise 98 of Section 2.2 we asked you to show that if a function $y = f(x)$ is differentiable at x_0 and $\Delta y = f(x_0 + \Delta x) - f(x_0)$, then

$$\Delta y = f'(x_0)\Delta x + \epsilon \Delta x,$$

where ϵ is a function satisfying $\lim_{\Delta x \to 0} \epsilon = 0$. This property tells us that the function f is "locally linear" near x_0 and, furthermore, that we may use the tangent line at $(x_0, f(x_0))$ to approximate the function f on some interval containing x_0. We will use an analogous property to define differentiability for a function of two or three variables.

DEFINITION 12.27 **Differentiability for Functions of Two Variables**

Let $z = f(x, y)$ be a function of two variables defined on an open set containing the point (x_0, y_0), and let $\Delta z = f(x_0 + \Delta x, y_0 + \Delta y) - f(x_0, y_0)$. The function f is said to be ***differentiable*** at (x_0, y_0) if the partial derivatives $f_x(x_0, y_0)$ and $f_y(x_0, y_0)$ both exist and

$$\Delta z = f_x(x_0, y_0)\Delta x + f_y(x_0, y_0)\Delta y + \epsilon_1 \Delta x + \epsilon_2 \Delta y,$$

where ϵ_1 and ϵ_2 are functions of Δx and Δy, and both go to zero as $(\Delta x, \Delta y) \to (0, 0)$.

In general, it is cumbersome to use Definition 12.27 to show that a function of two variables is differentiable. Fortunately, when the the partial derivatives are continuous at (x_0, y_0), we may use the following theorem:

THEOREM 12.28 **The Continuity of the Partial Derivatives Guarantees Differentiability**

If $f(x, y)$ is a function of two variables with partial derivatives $\frac{\partial f}{\partial x}$ and $\frac{\partial f}{\partial y}$ that are continuous on some open set containing the point (x_0, y_0), then f is differentiable at (x_0, y_0).

Theorem 12.28 allows us to verify much more readily where a function is differentiable, for most of the common functions we study. For example, consider the function $f(x, y) = \frac{x^2}{y}$

that we discussed earlier in the section. The partial derivatives $f_x(x, y) = \frac{2x}{y}$ and $f_y(x, y) = -\frac{x^2}{y^2}$ of this function are continuous at every point in the domain of f. That is, they are continuous on the set $S = \{(x, y) \mid y \neq 0\}$. Therefore, f is differentiable on S. As this example illustrates, every rational function of two variables will be differentiable everywhere the function is defined, because the only places where the partial derivatives will be discontinuous are the points where the function itself is discontinuous.

It bears repeating that a function with partial derivatives at a point (x_0, y_0) may or may not be differentiable at (x_0, y_0). For example, the function

$$f(x, y) = \begin{cases} 0, & \text{if } x = 0 \text{ or } y = 0 \\ 1, & \text{otherwise} \end{cases}$$

has the property that $f_x(0, 0) = 0 = f_y(0, 0)$. Although these partial derivatives exist, f is not differentiable at the origin, because f does not satisfy Definition 12.27. In fact, the function is not even continuous at the origin.

When a function of two variables, $f(x, y)$, is differentiable at a point (x_0, y_0), we may modify the equality $\Delta z = f_x(x_0, y_0)\Delta x + f_y(x_0, y_0)\Delta y + \epsilon_1 \Delta x + \epsilon_2 \Delta y$ from Definition 12.27 to find the equation of the plane tangent to the surface defined by f at (x_0, y_0). If we let $\Delta x = x - x_0$, $\Delta y = y - y_0$, and $\Delta z = z - f(x_0, y_0)$, we immediately have the equation of the tangent plane when we drop the terms $\epsilon_1 \Delta x$ and $\epsilon_2 \Delta y$.

THEOREM 12.29

Using the Partial Derivatives to Find the Equation of the Tangent Plane

Let $f(x, y)$ be a function of two variables that is differentiable at the point (x_0, y_0). Then the equation of the plane tangent to the surface defined by $f(x, y)$ at (x_0, y_0) is

$$f_x(x_0, y_0)(x - x_0) + f_y(x_0, y_0)(y - y_0) = z - f(x_0, y_0).$$

Before we give the definition of what it means for a function of three variables to be differentiable, recall that the graph of a function of three variables is a three-dimensional object existing in four-dimensional space, $\mathbb{R}^4$. If this graph is sufficiently well behaved, the natural tangent object to the surface is a ***hyperplane*** in $\mathbb{R}^4$. Perhaps the easiest way to think about a hyperplane is by considering the general equation that defines one. Just as the equation $a_1 x_1 + a_2 x_2 = b$ defines a line in $\mathbb{R}^2$ and the equation $a_1 x_1 + a_2 x_2 + a_3 x_3 = b$ defines a plane in $\mathbb{R}^3$, the equation $a_1 x_1 + a_2 x_2 + a_3 x_3 + a_4 x_4 = b$ defines a hyperplane in $\mathbb{R}^4$.

DEFINITION 12.30

Differentiability for Functions of Three Variables

Let $w = f(x, y, z)$ be a function of three variables defined on an open set containing the point (x_0, y_0, z_0), and let $\Delta w = f(x_0 + \Delta x, y_0 + \Delta y, z_0 + \Delta z) - f(x_0, y_0, z_0)$. The function f is said to be ***differentiable*** at (x_0, y_0, z_0) if the partial derivatives $f_x(x_0, y_0, z_0)$, $f_y(x_0, y_0, z_0)$, and $f_z(x_0, y_0, z_0)$ all exist and

$$\Delta w = f_x(x_0, y_0, z_0)\Delta x + f_y(x_0, y_0, z_0)\Delta y + f_z(x_0, y_0, z_0)\Delta z + \epsilon_1 \Delta x + \epsilon_2 \Delta y + \epsilon_3 \Delta z,$$

where ϵ_1, ϵ_2, and ϵ_3 all go to zero as $(\Delta x, \Delta y, \Delta z) \to (0, 0, 0)$.

The following theorem for functions of three variables is an analog of Theorem 12.28 for functions of two variables:

THEOREM 12.31

The Continuity of the Partial Derivatives Guarantees Differentiability

If $f(x, y, z)$ is a function of three variables with partial derivatives $\frac{\partial f}{\partial x}$, $\frac{\partial f}{\partial y}$, and $\frac{\partial f}{\partial z}$ that are continuous on some open set containing the point (x_0, y_0, z_0), then f is differentiable at (x_0, y_0, z_0).

The hyperplane that may be used to approximate a differentiable function of three variables, $w = f(x, y, z)$, "close to" a point of differentiability (x_0, y_0, z_0) is given by the equation

$$f_x(x_0, y_0, z_0)(x - x_0) + f_y(x_0, y_0, z_0)(y - y_0) + f_z(x_0, y_0, z_0)(z - z_0) = w - f(x_0, y_0, z_0).$$

This equation may be generalized to even higher dimensions.

Examples and Explorations

EXAMPLE 1

Computing directional derivatives

Find the directional derivative of $f(x, y) = \frac{x}{y^2}$ in the direction of the vector $3\mathbf{i} - 2\mathbf{j}$ at the point $(5, -3)$.

SOLUTION

Before we compute any directional derivative, we need to have a unit vector in the correct direction. We divide the vector $3\mathbf{i} - 2\mathbf{j}$ by its magnitude $\sqrt{3^2 + (-2)^2} = \sqrt{13}$ and obtain the unit vector $\mathbf{u} = \frac{3}{\sqrt{13}}\mathbf{i} - \frac{2}{\sqrt{13}}\mathbf{j}$. Now,

$$D_{\mathbf{u}} f(5, -3) = \lim_{h \to 0} \frac{\dfrac{5 + (3/\sqrt{13})h}{(-3 - (2/\sqrt{13})h)^2} - \dfrac{5}{(-3)^2}}{h}$$

$$= \lim_{h \to 0} \frac{1}{h} \cdot \frac{9(5 + (3/\sqrt{13})h) - 5(-3 - (2/\sqrt{13})h)^2}{9(-3 - (2/\sqrt{13})h)^2}$$

$$= \lim_{h \to 0} \frac{1}{h} \cdot \frac{45 + \frac{27}{\sqrt{13}}h - 45 - \frac{60}{\sqrt{13}}h - \frac{20}{13}h^2}{9\left(-3 - \frac{2}{\sqrt{13}}h\right)^2} = \lim_{h \to 0} \frac{-\frac{33}{\sqrt{13}} - \frac{20}{13}h}{9\left(-3 - \frac{2}{\sqrt{13}}h\right)^2}$$

$$= -\frac{11}{27\sqrt{13}}.$$

□

EXAMPLE 2

Using partial derivatives to determine where a function is differentiable

Find the first-order partial derivatives for the functions

(a) $f(x, y) = \dfrac{x^2\sqrt{y}}{x - 1}$ **(b)** $g(x, y) = \dfrac{\ln x}{\tan^{-1} y}$ **(c)** $P(r, \theta, \phi) = \dfrac{r\theta}{\phi}$

Use the partial derivatives to determine the sets on which the functions are differentiable.

SOLUTION

(a) The partial derivatives of f are

$$\frac{\partial f}{\partial x} = \frac{x(x - 2)\sqrt{y}}{(x - 1)^2} \quad \text{and} \quad \frac{\partial f}{\partial y} = \frac{x^2}{2(x - 1)\sqrt{y}}.$$

These partial derivatives are continuous wherever the denominators are nonzero and $y > 0$. They are continuous on the set $S_1 = \{(x, y) \mid x \neq 1 \text{ and } y > 0\}$, and f is differentiable at every point in S_1.

(b) The partial derivatives of g are

$$\frac{\partial g}{\partial x} = \frac{1}{x \tan^{-1} y} \quad \text{and} \quad \frac{\partial g}{\partial y} = -\frac{\ln x}{(1+y^2)(\tan^{-1} y)^2}.$$

Since the domain of g consists of just the points in $\mathbb{R}^2$ for which $x > 0$ and $y \neq 0$, both of these partial derivatives are continuous only on the set $S_2 = \{(x, y) \mid x > 0 \text{ and } y \neq 0\}$. Our function g is differentiable at every point in S_2.

(c) Finally, for P we have

$$\frac{\partial P}{\partial r} = \frac{\theta}{\phi}, \quad \frac{\partial P}{\partial \theta} = \frac{r}{\phi}, \quad \text{and} \quad \frac{\partial P}{\partial \phi} = -\frac{r\theta}{\phi^2}.$$

These partial derivatives are continuous everywhere in the domain of P. That is, they are continuous on the set $S_3 = \{(r, \theta, \phi) \mid \phi \neq 0\}$. The function P is differentiable at every point in its domain. □

EXAMPLE 3

Finding the equation of the tangent plane

Find the equation of the tangent plane to the function

(a) $f(x, y) = \dfrac{x^2\sqrt{y}}{x-1}$ at $(3, 4)$.

(b) $g(x, y) = \dfrac{\ln x}{\tan^{-1} y}$ at $(e, 1)$.

SOLUTION

These are the two functions we discussed in Example 2.

(a) From the work we have already done, we know that the partial derivatives of f are continuous on an open set containing the point $(3, 4)$. At $(3, 4)$, we have $f(3, 4) = 9$, $f_x(3, 4) = \frac{3(3-2)\sqrt{4}}{(3-1)^2} = \frac{3}{2}$, and $f_y(3, 4) = \frac{3^2}{2(3-1)\sqrt{4}} = \frac{9}{8}$. Thus, the equation of the tangent plane is

$$\frac{3}{2}(x-3) + \frac{9}{8}(y-4) = z - 9, \quad \text{or equivalently,} \quad 12x + 9y - 8z = 0.$$

(b) At $(e, 1)$, we have $g(e, 1) = \frac{4}{\pi}$, $g_x(e, 1) = \frac{4}{\pi e}$, and $g_y(e, 1) = -\frac{8}{\pi^2}$. Thus, the equation of the tangent plane is

$$\frac{4}{\pi e}(x - e) - \frac{8}{\pi^2}(y - 1) = z - \frac{4}{\pi}.$$

□

? TEST YOUR UNDERSTANDING

- What is a directional derivative? What is the geometric significance of a directional derivative for a function of two variables? What is the relationship between a directional derivative and a partial derivative?
- Why do we need to use a unit vector when we are computing a directional derivative? What happens if you do not use a unit vector when you compute a directional derivative? How do you compute a directional derivative for a function of two variables? For a function of three variables?
- What is the definition of differentiability for a function of two variables? How is the definition of differentiability for a function of two variables related to the definition of differentiability for a function of a single variable?

- What is the definition of differentiability for a function of three variables? How could you generalize this definition to a function of four (or more) variables?
- When $f(x, y)$ is a differentiable function of two variables at (x_0, y_0), why is
$$f_x(x_0, y_0)(x - x_0) + f_y(x_0, y_0)(y - y_0) = z - f(x_0, y_0)$$
the equation of the tangent plane at (x_0, y_0)?

EXERCISES 12.4

Thinking Back

- *Collinear tangent lines:* Show that the left- and right-hand derivatives exist for the function $f(x) = x^3$ at $x = 2$. Show also that the lines with slopes $f'_-(2)$ and $f'_+(2)$ at the point $(2, 8)$ are collinear.
- *Noncollinear tangent lines:* Show that the left- and right-hand derivatives exist for the function $f(x) = |x| + 3$ at $x = 0$. Show also that the lines with slopes $f'_-(0)$ and $f'_+(0)$ at the point $(0, 3)$ are not collinear.

Concepts

0. *Problem Zero:* Read the section and make your own summary of the material.

1. *True/False:* Determine whether each of the statements that follow is true or false. If a statement is true, explain why. If a statement is false, provide a counterexample.

(a) *True or False:* If $f(x, y)$ is a function of two variables, then $D_{\mathbf{j}} f(a, b) = f_y(a, b)$, provided that these derivatives exist at (a, b).

(b) *True or False:* If $\mathbf{u}$ is a unit vector and the directional derivative $D_{\mathbf{u}} f(a, b)$ exists for a function of two variables, f, then $D_{\mathbf{u}} f(a, b) = -D_{-\mathbf{u}} f(a, b)$.

(c) *True or False:* If $f(x, y)$ is a function of two variables and $D_{\mathbf{u}} f(a, b)$ exists for some unit vector $\mathbf{u}$, then $D_{\mathbf{v}} f(a, b)$ exists for every unit vector $\mathbf{v}$.

(d) *True or False:* If $f(x, y, z)$ is a function of three variables and $\mathbf{v} \in \mathbb{R}^3$ is a nonzero vector, then $D_{\mathbf{v}/\|\mathbf{v}\|} f(a, b, c) = \frac{1}{\|\mathbf{v}\|} D_{\mathbf{v}} f(a, b, c)$.

(e) *True or False:* If the partial derivatives $f_x(a, b)$ and $f_y(a, b)$ both exist for a function of two variables, $f(x, y)$, then f is differentiable at the point (a, b).

(f) *True or False:* A function of two variables, $f(s, t)$, is differentiable at the point (a, b) if the partial derivatives f_s and f_t are continuous on an open set containing the point (a, b).

(g) *True or False:* If a function of two variables, $f(x, y)$, is differentiable at the point (a, b), then $D_{\mathbf{u}} f(a, b) = D_{\mathbf{v}} f(a, b)$ for all unit vectors $\mathbf{u}$ and $\mathbf{v}$.

(h) *True or False:* If $f(x, y)$ is differentiable at the point (a, b), then there is a unit vector $\mathbf{u}$ such that $D_{\mathbf{u}} f(a, b) \neq 1$.

2. *Examples:* Construct examples of the thing(s) described in the following. Try to find examples that are different than any in the reading.

(a) A function of two variables, $f(x, y)$, such that $D_{\mathbf{u}} f(0, 0) = 0$ for every unit vector $\mathbf{u} \in \mathbb{R}^2$.

(b) A function of three variables, $f(x, y, z)$, such that $f_x(0, 0, 0) = f_y(0, 0, 0) = f_z(0, 0, 0) = 0$ but f is not differentiable at $(0, 0, 0)$.

(c) A function of three variables, $f(x, y, z)$, such that $D_{\mathbf{u}} f(0, 0, 0) = 0$ for every unit vector $\mathbf{u} \in \mathbb{R}^3$.

3. How many unit vectors are there in $\mathbb{R}^1$? How many unit vectors are there in $\mathbb{R}^n$ for $n > 1$?

4. Let $\mathbf{u}$ be a unit vector in $\mathbb{R}^2$.

(a) Explain why $-\mathbf{u}$ is a unit vector.

(b) If (a, b) is a point in the domain of the function of two variables, $f(x, y)$, at which $D_{\mathbf{u}} f(a, b)$ exists, what is the relationship between $D_{\mathbf{u}} f(a, b)$ and $D_{-\mathbf{u}} f(a, b)$?

5. What is the definition of the directional derivative for a function of two variables, $f(x, y)$? Be sure to include the words "unit vector" in your definition.

6. What is the definition of the directional derivative for a function of three variables, $f(x, y, z)$? Be sure to include the words "unit vector" in your definition.

7. Let $\mathbf{v}$ be a vector in $\mathbb{R}^n$ and let f be a function of n variables. How would we define the directional derivative of f in the direction of a unit vector $\mathbf{u} \in \mathbb{R}^n$ at $\mathbf{v}$?

8. Let $f(x, y) = x + y$ and $\mathbf{u} = \langle \alpha, \beta \rangle$ be a unit vector.

(a) Use the definition of the directional derivative to find $D_{\mathbf{u}} f(1, 2)$.

(b) Explain why $\langle k\alpha, k\beta \rangle$ is a unit vector only when $|k| = 1$.

(c) Assume that $|k| \neq 1$ and evaluate the limit
$$\lim_{h \to 0} \frac{(1 + (k\alpha)h) + (2 + (k\beta)h) - (1 + 2)}{h}.$$

(d) Use your results from parts (a) and (c) to explain why it is necessary to use a unit vector in the definition of the directional derivative.

9. Let $f(x, y) = xy$ and $\mathbf{u} = \langle \alpha, \beta \rangle$ be a unit vector.

(a) Use the definition of the directional derivative to find $D_{\mathbf{u}} f(-1, 3)$.

(b) Assume that $|k| \neq 1$ and evaluate
$$\lim_{h \to 0} \frac{(-1 + (k\alpha)h)(3 + (k\beta)h) - (-1 \cdot 3)}{h}.$$

(c) Use your results from parts (a) and (b) to explain why it is necessary to use a unit vector in the definition of the directional derivative.

10. Let $f(x, y) = e^x \sin y$ and $\mathbf{u} = \langle \alpha, \beta \rangle$ be a unit vector.

(a) Use the definition of the directional derivative to find $D_{\mathbf{u}} f(0, \pi)$.

(b) Assume that $|k| \neq 1$ and evaluate

$$\lim_{h \to 0} \frac{e^{k\alpha h} \sin(\pi + k\beta h)}{h}.$$

(*Hint:* $\sin(A + B) = \sin A \cos B + \sin B \cos A$.)

(c) Use your results from parts (a) and (b) to explain why it is necessary to use a unit vector in the definition of the directional derivative.

11. Let $f(x)$ be a function of a single variable. Define the directional derivative of f in the direction of the unit vector $\mathbf{u} = \langle \alpha \rangle$ at a point c. What are the only possible values for α?

12. Use your definition from Exercise 11 to show that the directional derivative of a function of a single variable $f(x)$ at a point c in the direction of $\mathbf{i}$ is $f'(c)$ and that the directional derivative of f at c in the direction of $-\mathbf{i}$ is $-f'(c)$.

13. What does it mean for a function of two variables, $f(x, y)$, to be differentiable at a point (a, b)?

14. If the function $f(x, y)$ is differentiable at a point (a, b), explain why the tangent lines to the graph of f at (a, b) in the x and y directions are sufficient to determine the tangent plane to the surface.

15. Let $\mathbf{u}_1$ and $\mathbf{u}_2$ be two nonparallel unit vectors in $\mathbb{R}^2$. If the function $f(x, y)$ is differentiable at a point (a, b), explain why the tangent lines to the graph of f at (a, b) in the $\mathbf{u}_1$ and $\mathbf{u}_2$ directions are sufficient to determine the tangent plane to the surface.

16. What does it mean for a function of three variables, $f(x, y, z)$, to be differentiable at a point (a, b, c)?

17. If the function $f(x, y, z)$ is differentiable at a point (a, b, c), explain why the tangent lines to the graph of f at (a, b, c) in the x, y, and z directions are sufficient to determine the tangent hyperplane to the surface.

18. Let $\mathbf{u}_1$, $\mathbf{u}_2$, and $\mathbf{u}_3$ be three unit vectors in $\mathbb{R}^3$ that cannot be put into the same plane. If the function $f(x, y, z)$ is differentiable at a point (a, b, c), explain why the tangent lines to the graph of f at (a, b, c) in the $\mathbf{u}_1$, $\mathbf{u}_2$, and $\mathbf{u}_3$ directions are sufficient to determine the tangent hyperplane to the surface.

19. Using Definition 12.30 as a model, provide a definition of differentiability for a function of n variables.

20. Using Theorem 12.31 as a model, provide a conjecture you think would be sufficient to guarantee that a function of n variables is differentiable at a point in its domain.

Skills

In Exercises 21–28, find the directional derivative of the given function at the specified point P and in the direction of the given unit vector $\mathbf{u}$.

21. $f(x, y) = x^2 - y^2$ at $P = (2, 3)$, $\mathbf{u} = \left\langle \frac{\sqrt{2}}{2}, \frac{\sqrt{2}}{2} \right\rangle$

22. $f(x, y) = x^2 - y^2$ at $P = (2, 3)$, $\mathbf{u} = \frac{3}{5}\mathbf{i} - \frac{4}{5}\mathbf{j}$

23. $f(x, y) = \frac{x}{y^2}$ at $P = (-2, 1)$, $\mathbf{u} = \left\langle \frac{\sqrt{10}}{10}, -\frac{3\sqrt{10}}{10} \right\rangle$

24. $f(x, y) = \frac{x}{y^2}$ at $P = (-2, 1)$, $\mathbf{u} = -\frac{5}{13}\mathbf{i} + \frac{12}{13}\mathbf{j}$

25. $f(x, y) = \sqrt{\frac{y}{x}}$ at $P = (4, 9)$, $\mathbf{u} = \left\langle -\frac{\sqrt{17}}{17}, -\frac{4\sqrt{17}}{17} \right\rangle$

26. $f(x, y) = \sqrt{\frac{y}{x}}$ at $P = (4, 9)$, $\mathbf{u} = \frac{15}{17}\mathbf{i} + \frac{8}{17}\mathbf{j}$

27. $f(x, y, z) = x^2 + y^2 - z^3$ at $P = (2, -2, 2)$, $\mathbf{u} = \left\langle \frac{3}{5}, 0, \frac{4}{5} \right\rangle$

28. $f(x, y, z) = x^2 + y^2 - z^3$ at $P = (2, -2, 2)$,
$\mathbf{u} = \frac{2}{3}\mathbf{i} - \frac{2}{3}\mathbf{j} + \frac{1}{3}\mathbf{k}$

In Exercises 29–34, find the equation of the line tangent to the surface at the given point P and in the direction of the given unit vector $\mathbf{u}$. Note that these are the same functions, points, and vectors as in Exercises 21–26.

29. $f(x, y) = x^2 - y^2$ at $P = (2, 3)$, $\mathbf{u} = \left\langle \frac{\sqrt{2}}{2}, \frac{\sqrt{2}}{2} \right\rangle$

30. $f(x, y) = x^2 - y^2$ at $P = (2, 3)$, $\mathbf{u} = \frac{3}{5}\mathbf{i} - \frac{4}{5}\mathbf{j}$

31. $f(x, y) = \frac{x}{y^2}$ at $P = (-2, 1)$, $\mathbf{u} = \left\langle \frac{\sqrt{10}}{10}, -\frac{3\sqrt{10}}{10} \right\rangle$

32. $f(x, y) = \frac{x}{y^2}$ at $P = (-2, 1)$, $\mathbf{u} = -\frac{5}{13}\mathbf{i} + \frac{12}{13}\mathbf{j}$

33. $f(x, y) = \sqrt{\frac{y}{x}}$ at $P = (4, 9)$, $\mathbf{u} = \left\langle -\frac{\sqrt{17}}{17}, -\frac{4\sqrt{17}}{17} \right\rangle$

34. $f(x, y) = \sqrt{\frac{y}{x}}$ at $P = (4, 9)$, $\mathbf{u} = \frac{15}{17}\mathbf{i} + \frac{8}{17}\mathbf{j}$

In Exercises 35–38, find the directional derivative of the given function at the specified point P and in the direction of the given vector $\mathbf{v}$.

35. $f(x, y) = x^2 - y^2$ at $P = (3, 3)$, $\mathbf{v} = \langle -1, 5 \rangle$

36. $f(x, y) = \frac{x}{y^2}$ at $P = (9, -3)$, $\mathbf{v} = 2\mathbf{i} + 7\mathbf{j}$

37. $f(x, y) = \sqrt{\frac{y}{x}}$ at $P = (1, 16)$, $\mathbf{v} = \langle 2, -1 \rangle$

38. $f(x, y, z) = x^2 + y^2 - z^3$ at $P = (0, -2, 5)$,
$\mathbf{v} = -\mathbf{i} + 3\mathbf{j} + 5\mathbf{k}$

In Exercises 39–42, show that the directional derivative of the given function at the specified point P is zero for every unit vector $\mathbf{u}$.

39. $f(x, y) = xy + 2x - y$ at $P = (1, -2)$

40. $f(x, y) = 3x^2 - 4xy + 2y^2$ at $P = (0, 0)$

41. $f(x, y) = (x + 1)y^2$ at $P = (-1, 0)$

42. $f(x, y, z) = x^2 + y^2 - z^3$ at $P = (0, 0, 0)$

Find all points where the first-order partial derivatives of the functions in Exercises 43–54 are continuous. Then use Theorems 12.28 and 12.31 to determine the sets in which the functions are differentiable.

43. $f(x, y) = x^2 - y^2$

44. $f(x, y) = \dfrac{x}{y^2}$

45. $f(x, y) = \dfrac{x}{x^2 + y^2 - 1}$

46. $f(x, y) = \dfrac{x}{x^2 + y^2}$

47. $f(x, y) = \sin(xy)$

48. $f(x, y) = \cos(xy)$

49. $f(x, y) = \tan(xy)$

50. $f(x, y) = \tan(x + y)$

51. $f(x, y) = \sqrt{\dfrac{y}{x}},\ x > 0,\ y \geq 0$

52. $f(x, y) = \ln(xy^2)$

53. $f(x, y, z) = x^2 + y^2 - z^3$

54. $f(x, y, z) = \dfrac{x}{y^2 + z^2}$

Use the first-order partial derivatives of the functions in Exercises 55–64 to find the equation of the plane tangent to the graph of the function at the indicated point P. Note that these are the same functions as in Exercises 43–52.

55. $f(x, y) = x^2 - y^2$, $P = (1, -3)$

56. $f(x, y) = \dfrac{x}{y^2}$, $P = (-4, 7)$

57. $f(x, y) = \dfrac{x}{x^2 + y^2}$, $P = (-3, 0)$

58. $f(x, y) = \dfrac{x}{x^2 + y^2 - 1}$, $P = (1, -3)$

59. $f(x, y) = \sin(xy)$, $P = \left(2, \dfrac{\pi}{2}\right)$

60. $f(x, y) = \cos(xy)$, $P = (\pi, -3)$

61. $f(x, y) = \tan(xy)$, $P = \left(1, -\dfrac{\pi}{4}\right)$

62. $f(x, y) = \tan(x + y)$, $P = (0, \pi)$

63. $f(x, y) = \sqrt{\dfrac{y}{x}}$, $P = (1, 9)$

64. $f(x, y) = \ln(xy^2)$, $P = (1, -3)$

Use the first-order partial derivatives of the functions in Exercises 65 and 66 to find the equation of the hyperplane tangent to the graph of the function at the indicated point P. Note that these are the same functions as in Exercises 53 and 54.

65. $f(x, y, z) = x^2 + y^2 - z^3$, $P = (1, -5, 3)$

66. $f(x, y, z) = \dfrac{x}{y^2 + z^2}$, $P = (4, 0, 2)$

Applications

67. Ian is travelling along a glacier on a line directly northeast. The elevation of the glacier in that area is described by the function

$$f(x, y) = 1.2 - 0.2x^2 - 0.3y^2 + 0.1xy - 0.25x,$$

where x, y, and f are given in miles.

(a) If Ian is at the point $(0, 0)$, how steeply is he descending?

(b) In what direction would Ian have to turn in order to contour across (i.e., neither ascend nor descend) the glacier?

68. Alex is laying out a new road that will descend from the mountains near Boulder, Colorado. The local topography is described by the function

$$f(x, y) = 1.1 - 0.2x + 0.05y^2 + 0.1xy.$$

All distances are given in miles. The technical requirements for the road stipulate that the road must be built at a 6% grade. That is, the road will descend 6 feet for every 100 feet it travels horizontally. If Alex's current position is $(-0.5, 0)$ and the road is going generally southeast, in which direction does he need to look to survey the next section of road?

Proofs

69. Let (a, b) be a point in the domain of the function of two variables, $f(x, y)$, and $\mathbf{u}$ be a unit vector for which $D_{\mathbf{u}} f(a, b)$ exists. Prove that $D_{-\mathbf{u}} f(a, b) = -D_{\mathbf{u}} f(a, b)$.

70. Let $f(x, y)$ be a function of two variables and k be a constant. Prove that if f is differentiable at the point (a, b) and $D_{\mathbf{u}} f(a, b) = k$ for every unit vector $\mathbf{u} \in \mathbb{R}^2$, then $k = 0$.

71. Let $f(x, y)$ be a function of two variables. Prove that if f is differentiable at the point (a, b), then there is a unit vector, $\mathbf{u}$, such that $D_{\mathbf{u}} f(a, b) \neq 1$. *(Hint: See Exercise 70.)*

Thinking Forward

- *Paraboloid:* Sketch the paraboloid that is the graph of the function $f(x, y) = 4 - x^2 - y^2$. What is the point (a, b) at which the function attains its maximum value? What is the directional derivative $D_{\mathbf{u}} f(a, b)$ for any unit vector $\mathbf{u}$?
- *Hyperboloid:* Sketch the hyperboloid that is the graph of the function $g(x, y) = x^2 - y^2$. Does the function g have any extreme values? What is the directional derivative $D_{\mathbf{u}} g(0, 0)$ for any unit vector $\mathbf{u}$?

12.5 THE CHAIN RULE AND THE GRADIENT

- The chain rule for functions of two or more variables
- The gradient
- The gradient and directional derivatives

The Chain Rule

In Theorem 12.29 we saw that if a function of two variables, $f(x, y)$, is differentiable at a point (x_0, y_0) in its domain, then the equation of the plane tangent to the surface is given by

$$f_x(x_0, y_0)(x - x_0) + f_y(x_0, y_0)(y - y_0) = z - f(x_0, y_0).$$

Just as a line tangent to the graph of a function of a single variable may be used to approximate the function close to the point of tangency, we may use the tangent plane to approximate a differentiable function of two variables. If we let $\Delta x = x - x_0$, $\Delta y = y - y_0$, and $\Delta z = z - f(x_0, y_0)$, the previous equation becomes

$$\Delta z = f_x(x_0, y_0)\Delta x + f_y(x_0, y_0)\Delta y.$$

This equation may be interpreted to say that, close to the point of tangency, the difference between the value of the function f and the corresponding value on the tangent plane is approximately $f_x(x_0, y_0)\Delta x + f_y(x_0, y_0)\Delta y$. We will use this property to prove the chain rule for functions of two or more variables.

Recall that when $y = f(x)$ is a function of the single variable x, and x is a function of another variable t, we may find the rate of change of f with respect to t with the chain rule, $\frac{dy}{dt} = \frac{dy}{dx}\frac{dx}{dt}$. If any of our functions is a function of two or more variables, we need to adapt the chain rule to fit the particular functions we have.

THEOREM 12.32 **Chain Rule (Version I)**

Given functions $z = f(x, y)$, $x = u(t)$, and $y = v(t)$, for all values of t at which u and v are differentiable, and if f is differentiable at $(u(t), v(t))$, then

$$\frac{dz}{dt} = \frac{\partial z}{\partial x}\frac{dx}{dt} + \frac{\partial z}{\partial y}\frac{dy}{dt}.$$

For example, if $z = x^2 \sin y$, $x = t^3$, and $y = e^{5t}$, then

$$\frac{dz}{dt} = \frac{\partial z}{\partial x}\frac{dx}{dt} + \frac{\partial z}{\partial y}\frac{dy}{dt} = (2x \sin y)(3t^2) + (x^2 \cos y)(5e^{5t}).$$

Although this equation is correct, it is better to replace x and y with their respective functions of t and simplify the result:

$$\frac{dz}{dt} = (2(t^3)\sin(e^{5t}))(3t^2) + ((t^3)^2\cos(e^{5t}))(5e^{5t}) = 6t^5\sin(e^{5t}) + 5t^6 e^{5t}\cos(e^{5t}).$$

In Example 1 we verify this derivative by first replacing x and y with their respective functions of t and then taking the derivative of the resulting function.

THEOREM 12.33

Chain Rule (Version II)

Given functions $z = f(x, y)$, $x = u(s, t)$, and $y = v(s, t)$, for all values of s and t at which u and v are differentiable, and if f is differentiable at $(u(s, t), v(s, t))$, then

$$\frac{\partial z}{\partial s} = \frac{\partial z}{\partial x}\frac{\partial x}{\partial s} + \frac{\partial z}{\partial y}\frac{\partial y}{\partial s} \quad \text{and} \quad \frac{\partial z}{\partial t} = \frac{\partial z}{\partial x}\frac{\partial x}{\partial t} + \frac{\partial z}{\partial y}\frac{\partial y}{\partial t}.$$

For example, if $z = e^x \cos y$, $x = s^2t^3$, and $y = \frac{s}{t}$, then

$$\frac{\partial z}{\partial s} = \frac{\partial z}{\partial x}\frac{\partial x}{\partial s} + \frac{\partial z}{\partial y}\frac{\partial y}{\partial s} = (e^x \cos y)(2st^3) + (-e^x \sin y)\left(\frac{1}{t}\right).$$

Again, although this equation is correct, it is better to replace x and y with their respective functions of s and t and simplify the result:

$$\frac{\partial z}{\partial s} = \left(e^{s^2t^3} \cos\left(\frac{s}{t}\right)\right)(2st^3) + \left(-e^{s^2t^3} \sin\left(\frac{s}{t}\right)\right)\left(\frac{1}{t}\right) = e^{s^2t^3}\left(2st^3 \cos\left(\frac{s}{t}\right) - \frac{1}{t}\sin\left(\frac{s}{t}\right)\right).$$

In Example 2 we continue this example by finding $\frac{\partial z}{\partial t}$ and verifying that both of these partial derivatives are correct by replacing x and y with their respective functions of s and t and then taking the partial derivatives of the resulting function.

We now prove Theorem 12.32. The proof of Theorem 12.33 is similar and is left for Exercise 65.

Proof. Let $x = u(t)$, $y = v(t)$, and t_0 be a point in the domains of both u and v, where the functions are differentiable and where f is differentiable at $(x_0, y_0) = (u(t_0), v(t_0))$. By Definition 12.27, at (x_0, y_0) we have

$$\Delta z = \frac{\partial f}{\partial x}\Delta x + \frac{\partial f}{\partial y}\Delta y + \epsilon_1 \Delta x + \epsilon_2 \Delta y,$$

where $\epsilon_1 \to 0$ and $\epsilon_2 \to 0$ as $(\Delta x, \Delta y) \to (0, 0)$. If we divide both sides of this equation by Δt, we obtain

$$\frac{\Delta z}{\Delta t} = \frac{\partial f}{\partial x}\frac{\Delta x}{\Delta t} + \frac{\partial f}{\partial y}\frac{\Delta y}{\Delta t} + \epsilon_1 \frac{\Delta x}{\Delta t} + \epsilon_2 \frac{\Delta y}{\Delta t}.$$

If we let $\Delta x = u(t_0 + \Delta t) - u(t_0)$ and $\Delta y = v(t_0 + \Delta t) - v(t_0)$, then $\Delta x \to 0$ and $\Delta y \to 0$ as $\Delta t \to 0$. Consequently, $\epsilon_1 \to 0$ and $\epsilon_2 \to 0$ as $\Delta t \to 0$. Taking the limit of the quotients, we get:

$$\begin{aligned} \frac{dz}{dt} &= \lim_{\Delta t \to 0} \frac{\Delta z}{\Delta t} = \lim_{\Delta t \to 0}\left(\frac{\partial f}{\partial x}\frac{\Delta x}{\Delta t} + \frac{\partial f}{\partial y}\frac{\Delta y}{\Delta t} + \epsilon_1 \frac{\Delta x}{\Delta t} + \epsilon_2 \frac{\Delta y}{\Delta t}\right) \\ &= \frac{\partial f}{\partial x}\frac{dx}{dt} + \frac{\partial f}{\partial y}\frac{dy}{dt}. \end{aligned}$$

■

After examining Theorems 12.32 and 12.33, you may be able to guess that there is a more comprehensive version of the chain rule, which we state in the following theorem:

THEOREM 12.34

Chain Rule (Complete Version)

Given functions $z = f(x_1, x_2, \ldots, x_n)$ and $x_i = u_i(t_1, t_2, \ldots, t_m)$ for $1 \le i \le n$, for all values of $t_1, t_2, \ldots, t_m$ at which each u_i is differentiable, and if f is differentiable at $(u_1(t_1, t_2, \ldots, t_m), u_2(t_1, t_2, \ldots, t_m), \ldots, u_n(t_1, t_2, \ldots, t_m))$, then

$$\frac{\partial z}{\partial t_j} = \frac{\partial z}{\partial x_1}\frac{\partial x_1}{\partial t_j} + \frac{\partial z}{\partial x_2}\frac{\partial x_2}{\partial t_j} + \cdots + \frac{\partial z}{\partial x_n}\frac{\partial x_n}{\partial t_j}, \text{ for } 1 \le j \le m.$$

The Gradient

In Section 12.6 we will be finding extrema for functions of two variables. Recall that the extreme values of a function of a single variable occur at the function's critical points (i.e., those points at which the derivative either is zero or does not exist). For functions of two variables, we will use the *gradient* of the function to find extrema.

DEFINITION 12.35 **The Gradient**

Let $z = f(x, y)$ be a function of two variables. The ***gradient*** of f is the vector function defined by

$$\nabla f(x, y) = \frac{\partial f}{\partial x}\mathbf{i} + \frac{\partial f}{\partial y}\mathbf{j} = \langle f_x(x, y), f_y(x, y) \rangle .$$

Similarly, if $w = f(x, y, z)$ is a function of three variables, then the ***gradient*** of f is the vector function defined by

$$\nabla f(x, y, z) = \frac{\partial f}{\partial x}\mathbf{i} + \frac{\partial f}{\partial y}\mathbf{j} + \frac{\partial f}{\partial z}\mathbf{k} = \langle f_x(x, y, z), f_y(x, y, z), f_z(x, y, z) \rangle .$$

The domain of the gradient is the set of all points in the domain of f at which the partial derivatives exist.

The symbol ∇f is read "the gradient of f," "grad f," or "del f."

Recall that when the first-order partial derivatives of f are continuous in a neighborhood of a point P, the function is differentiable at P. Therefore, if the gradient of f is continuous in a neighborhood of a point P, the function is differentiable at P. In particular, a differentiable function of two variables, $f(x, y)$, can have an extreme value only at a point at which the gradient is the zero vector, since such points are the only places where the tangent plane might be horizontal (a necessary, but insufficient, condition for having an extreme). Therefore, our first step for locating extreme values will be to find those places where the gradient is zero. We see that for a function of two or three variables, the gradient plays the role that the derivative plays for functions of a single variable.

Using the Gradient to Compute the Directional Derivative

In Section 12.4 we introduced the directional derivative and defined it in terms of a limit. As we will see in the following theorem, the gradient provides a shortcut for finding the directional derivative when the first-order partial derivatives of the function exist:

THEOREM 12.36 **The Gradient and the Directional Derivative**

Let $f(x, y)$ be a function of two variables and (x_0, y_0) be a point in the domain of f at which f is differentiable. Then, for every unit vector $\mathbf{u} \in \mathbb{R}^2$,

$$D_{\mathbf{u}} f(x_0, y_0) = \nabla f(x_0, y_0) \cdot \mathbf{u}.$$

Similarly, if $\mathbf{u} \in \mathbb{R}^3$ is a unit vector, $f(x, y, z)$ is a function of three variables, and (x_0, y_0, z_0) is a point in the domain of f at which f is differentiable, then

$$D_{\mathbf{u}} f(x_0, y_0, z_0) = \nabla f(x_0, y_0, z_0) \cdot \mathbf{u}.$$

Recall that in Section 12.4 we used Definition 12.26 to find the directional derivative of the function $f(x, y) = \frac{x^2}{y}$ at the point $(-1, 2)$ in the direction of the unit vector $\mathbf{u} = \left\langle \frac{3}{5}, -\frac{4}{5} \right\rangle$

and saw that $D_{\mathbf{u}}f(-1,2) = -\frac{2}{5}$. We now compute the same derivative, but using Theorem 12.36. First we need the gradient of f:

$$\nabla f(x,y) = \frac{\partial f}{\partial x}\mathbf{i} + \frac{\partial f}{\partial y}\mathbf{j} = \frac{2x}{y}\mathbf{i} - \frac{x^2}{y^2}\mathbf{j}.$$

At $(-1,2)$, $\nabla f(-1,2) = -\mathbf{i} - \frac{1}{4}\mathbf{j}$. Now, using the theorem, we have

$$D_{\mathbf{u}}f(-1,2) = \nabla f(-1,2)\cdot\mathbf{u} = \left(-\mathbf{i} - \frac{1}{4}\mathbf{j}\right)\cdot\left(\frac{3}{5}\mathbf{i} - \frac{4}{5}\mathbf{j}\right) = -\frac{2}{5}.$$

Although the computation using the definition was not terrible, Theorem 12.36 certainly saves time, particularly when the limit in the definition is difficult to work with.

We now prove Theorem 12.36.

Proof. We prove the theorem when f is a function of two variables and leave the case when f is a function of three variables for Exercise 66. By Definition 12.26 we have

$$D_{\mathbf{u}}f(x_0,y_0) = \lim_{h\to 0}\frac{f(x_0+\alpha h, y_0+\beta h) - f(x_0,y_0)}{h},$$

where $\mathbf{u} = \langle\alpha,\beta\rangle$. We now define the function $F(h) = f(x_0+\alpha h, y_0+\beta h)$ and note that F is a function of the single variable h, because α, β, x_0, and y_0 are all constants. Thus,

$$D_{\mathbf{u}}f(x_0,y_0) = \lim_{h\to 0}\frac{F(h)-F(0)}{h} = F'(0),$$

by the definition of the derivative. If we let $x = x_0 + \alpha\cdot h$ and $y = y_0 + \beta\cdot h$, then, by the chain rule, Theorem 12.32, it follows that

$$F'(h) = \frac{\partial f}{\partial x}\frac{\partial x}{\partial h} + \frac{\partial f}{\partial x}\frac{\partial x}{\partial h} = f_x(x,y)\alpha + f_y(x,y)\beta.$$

Now, when $h = 0$, $x = x_0$ and $y = y_0$. Thus, if we evaluate the equation for $F'(h)$ at $h = 0$, we have

$$F'(0) = D_{\mathbf{u}}f(x_0,y_0) = f_x(x_0,y_0)\alpha + f_y(x_0,y_0)\beta = \nabla f(x_0,y_0)\cdot\mathbf{u},$$

the result we seek. ■

The gradient of a function $f(x,y)$ gives us information about two important geometric properties of the function. Theorem 12.37 tells us that the gradient ∇f points in the direction in which f increases most rapidly, and Theorem 12.38 says that the gradient at (x_0,y_0) is orthogonal to the level curve $f(x,y) = c_0$, where $c_0 = f(x_0,y_0)$. In the figure that follows, the green gradient vectors are orthogonal to the level curves and point in the direction in which the function increases most rapidly. Generally, that means that they point in the direction in which the level curves are most tightly spaced. If you are a hiker familiar with topographic maps, this is saying that the shortest path to the peak crosses the level curves quickly. It may be a difficult trek, but it does minimize the distance you have to travel.

The gradient vectors are orthogonal to the level curves

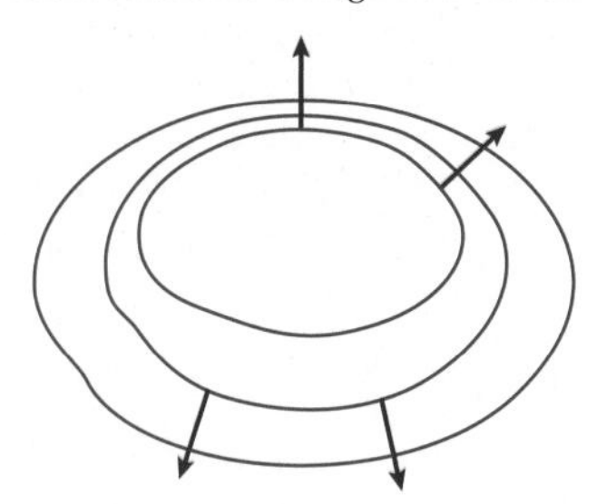

THEOREM 12.37

The Gradient Points in the Direction of Greatest Increase

Let f be a function of two or three variables, and let P be a point in the domain of f at which f is differentiable. Then the gradient of f at P points in the direction in which f increases most rapidly.

Proof. We prove the theorem when f is a function of two variables and leave the case when f is a function of three variables for Exercise 74. By Theorem 12.36,

$$D_{\mathbf{u}}f(x_0, y_0) = \nabla f(x_0, y_0) \cdot \mathbf{u} = \|\nabla f(x_0, y_0)\| \|\mathbf{u}\| \cos\theta,$$

where θ is the angle between the vectors $\nabla f(x_0, y_0)$ and $\mathbf{u}$. Now, since $\mathbf{u}$ is a unit vector, $\|\mathbf{u}\| = 1$. Thus,

$$D_{\mathbf{u}}f(x_0, y_0) = \|\nabla f(x_0, y_0)\| \cos\theta.$$

The quantity on the right is greatest when $\theta = 0$. Therefore, the directional derivative is greatest when $\mathbf{u}$ is parallel to the gradient. ■

Going back to the function $f(x, y) = \frac{x^2}{y}$, we have $\nabla f(-1, 2) = -\mathbf{i} - \frac{1}{4}\mathbf{j}$. This vector points in the direction in which f increases most quickly at the point $(-1, 2)$, and the rate of change of f in that direction is the magnitude

$$\left\|\nabla f(-1, 2)\right\| = \left\|-\mathbf{i} - \frac{1}{4}\mathbf{j}\right\| = \sqrt{(-1)^2 + \left(-\frac{1}{4}\right)^2} = \frac{\sqrt{17}}{4}.$$

The function decreases most rapidly in the opposite direction, $-\nabla f(-1, 2) = \mathbf{i} + \frac{1}{4}\mathbf{j}$, and the rate of change of f in this direction is $-\frac{\sqrt{17}}{4}$.

THEOREM 12.38

Gradient Vectors are Orthogonal to Level Curves

Let f be a function of two variables, and let (x_0, y_0) be a point in the domain of f at which f is differentiable. If C is the level curve containing the point $c_0 = f(x_0, y_0)$, then $\nabla f(x_0, y_0)$ and C are orthogonal at $f(x_0, y_0)$.

Proof. Let $x = g(t), y = h(t)$ be a parametrization for the level curve C at height $f(x_0, y_0) = c_0$ such that $g(t_0) = x_0$ and $h(t_0) = y_0$. The curve C has equation $f(x, y) = c_0$, or in terms of the parametrization, $F(t) = f(g(t), h(t)) = c_0$. Now, by Theorem 12.32 and the fact that $F(t)$ is a constant function, we have

$$F'(t) = \frac{\partial f}{\partial x}\frac{\partial x}{\partial t} + \frac{\partial f}{\partial y}\frac{\partial y}{\partial t} = 0.$$

In particular, at $t = t_0$

$$F'(t_0) = \nabla f(x_0, y_0) \cdot \langle g'(t_0), h'(t_0) \rangle = 0.$$

But $\langle g'(t_0), h'(t_0) \rangle$ is the tangent vector to the level curve C at $t = t_0$. Therefore, the equation tells us that the gradient at (x_0, y_0) and the level curve are orthogonal at $f(x_0, y_0)$. ■

Examples and Explorations

EXAMPLE 1

Verifying a derivative (chain rule, version I)

Let $z = x^2 \sin y$, $x = t^3$, and $y = e^{5t}$. Verify that

$$\frac{dz}{dt} = 6t^5 \sin(e^{5t}) + 5t^6 e^{5t} \cos(e^{5t})$$

by replacing x and y with their respective functions of t and then taking the derivative of the resulting function.

SOLUTION

Earlier in the section we used Theorem 12.32 to obtain the derivative shown. Now we express z as a function of t:

$$z = (t^3)^2 \sin(e^{5t}) = t^6 \sin(e^{5t}).$$

Taking the derivative with respect to t, we have

$$\frac{dz}{dt} = 6t^5 \sin(e^{5t}) + 5t^6 e^{5t} \cos(e^{5t}).$$

□

EXAMPLE 2 **Using the chain rule (version II) to find a partial derivative**

Let $z = e^x \cos y$, $x = s^2t^3$, and $y = \frac{s}{t}$. Earlier in the section we found $\frac{\partial z}{\partial s}$. Now we use Theorem 12.33 to find $\frac{\partial z}{\partial t}$, and then we verify that both of these partial derivatives are correct by replacing x and y with their respective functions of s and t and taking the appropriate partial derivatives of the resulting function.

SOLUTION

By Theorem 12.33,

$$\frac{\partial z}{\partial t} = \frac{\partial z}{\partial x}\frac{\partial x}{\partial t} + \frac{\partial z}{\partial y}\frac{\partial y}{\partial t} = (e^x \cos y)(3s^2t^2) + (-e^x \sin y)\left(-\frac{s}{t^2}\right).$$

This result is correct, but it is preferable to write the function as a function of just s and t. We use $x = s^2t^3$ and $y = \frac{s}{t}$ to do so:

$$\frac{\partial z}{\partial t} = \left(e^{s^2t^3} \cos\left(\frac{s}{t}\right)\right)(3s^2t^2) + \left(-e^{s^2t^3} \sin\left(\frac{s}{t}\right)\right)\left(-\frac{s}{t^2}\right) = e^{s^2t^3}\left(3s^2t^2 \cos\left(\frac{s}{t}\right) + \frac{s}{t^2} \sin\left(\frac{s}{t}\right)\right).$$

To verify these partial derivatives, we first write z as a function of s and t; that is,

$$z = e^{s^2t^3} \cos\left(\frac{s}{t}\right).$$

We now find $\frac{\partial z}{\partial s}$ and $\frac{\partial z}{\partial t}$, using the appropriate derivative rules:

$$\frac{\partial z}{\partial s} = 2st^3 e^{s^2t^3} \cos\left(\frac{s}{t}\right) - \frac{1}{t} e^{s^2t^3} \sin\left(\frac{s}{t}\right).$$

If we factor this equation, we obtain what we had earlier in the section, namely,

$$\frac{\partial z}{\partial s} = e^{s^2t^3}\left(2st^3 \cos\left(\frac{s}{t}\right) - \frac{1}{t} \sin\left(\frac{s}{t}\right)\right).$$

Finally,

$$\frac{\partial z}{\partial t} = 3s^2t^2 e^{s^2t^3} \cos\left(\frac{s}{t}\right) + \frac{s}{t^2} e^{s^2t^3} \sin\left(\frac{s}{t}\right).$$

This is equivalent to the result we obtained before. □

EXAMPLE 3 **Using the chain rule**

If $w = f(x, y, z)$, $x = u(s, t)$, $y = v(s, t)$, and $z = w(s, t)$ are differentiable functions at every point in their domains, what are $\frac{\partial w}{\partial s}$ and $\frac{\partial w}{\partial t}$?

SOLUTION

We have

$$\frac{\partial w}{\partial s} = \frac{\partial w}{\partial x}\frac{\partial x}{\partial s} + \frac{\partial w}{\partial y}\frac{\partial y}{\partial s} + \frac{\partial w}{\partial z}\frac{\partial z}{\partial s} \quad \text{and} \quad \frac{\partial w}{\partial t} = \frac{\partial w}{\partial x}\frac{\partial x}{\partial t} + \frac{\partial w}{\partial y}\frac{\partial y}{\partial t} + \frac{\partial w}{\partial z}\frac{\partial z}{\partial t}.$$

□

EXAMPLE 4 **Finding and interpreting the gradient**

Find the gradient of each of the following functions:

(a) $f(x,y) = e^y \sin\left(\frac{x}{y}\right)$ **(b)** $g(x,y,z) = xy^2 - \frac{z}{xy^3}$

Then find the direction in which the function f increases most rapidly at the point $(\pi, 1)$ and the rate of change of f in that direction. Finally, find the direction in which g decreases most rapidly at the point $(2, -1, 4)$ and the rate of change of g in that direction.

SOLUTION

(a) The gradient of f is

$$\nabla f(x,y) = \frac{e^y}{y}\cos\left(\frac{x}{y}\right)\mathbf{i} + e^y\left(\sin\left(\frac{x}{y}\right) - \frac{x}{y^2}\cos\left(\frac{x}{y}\right)\right)\mathbf{j}.$$

At $(\pi, 1)$, $\nabla f(\pi, 1) = e\cos(\pi)\,\mathbf{i} + e(\sin(\pi) - \pi\cos(\pi))\,\mathbf{j} = -e\mathbf{i} + \pi e\mathbf{j}$. This is the direction in which f increases most rapidly at $(\pi, 1)$, and the magnitude

$$\|\nabla f(\pi, 1)\| = \|-e\mathbf{i} + \pi e\mathbf{j}\| = \sqrt{e^2 + \pi^2 e^2} = e\sqrt{1+\pi^2}$$

is the rate of change of f in that direction.

(b) The gradient of g is

$$\nabla g(x,y,z) = \left(y^2 + \frac{z}{x^2y^3}\right)\mathbf{i} + \left(2xy + \frac{3z}{xy^4}\right)\mathbf{j} - \frac{1}{xy^3}\mathbf{k}.$$

At $(2, -1, 4)$,

$$\begin{aligned}\nabla g(2,-1,4) &= \left((-1)^2 + \frac{4}{2^2(-1)^3}\right)\mathbf{i} + \left(2\cdot 2\cdot(-1) + \frac{3\cdot 4}{2\cdot(-1)^4}\right)\mathbf{j} - \frac{1}{2\cdot(-1)^3}\mathbf{k} \\ &= 2\mathbf{j} + \frac{1}{2}\mathbf{k}.\end{aligned}$$

This is the direction in which g increases most rapidly at $(2, -1, 4)$, so g decreases most rapidly in the opposite direction, $-2\mathbf{j} - \frac{1}{2}\mathbf{k}$. The rate of decrease in this direction is the negative of the magnitude

$$\|\nabla g(2,-1,4)\| = \sqrt{2^2 + (1/2)^2} = \frac{1}{2}\sqrt{17}.$$

□

EXAMPLE 5 **Finding a function given its gradient**

Find the function $f(x,y)$ whose gradient is $\nabla f(x,y) = (2x + x^2y^2)e^{xy^2}\mathbf{i} + 2x^3ye^{xy^2}\mathbf{j}$.

SOLUTION

This is really the same type of problem we saw in Example 5 in Section 12.3. In that example we were given functions $g(x,y)$ and $h(x,y)$ and we used them to find a function f such that $f_x(x,y) = g(x,y)$ and $f_y(x,y) = h(x,y)$. Now we have $g(x,y) = f_x(x,y) = (2x + x^2y^2)e^{xy^2}$ and $h(x,y) = f_y(x,y) = 2x^3ye^{xy^2}$. Recall that, to ensure that such a function f existed, we showed that $\frac{\partial g}{\partial y} = \frac{\partial h}{\partial x}$. Here,

$$\frac{\partial}{\partial y}((2x + x^2y^2)e^{xy^2}) = (6x^2y + 2x^3y^3)e^{xy^2} = \frac{\partial}{\partial x}(2x^3ye^{xy^2}),$$

so we should be able to find a function f with the given gradient.

Since the function $2x^3ye^{xy^2}$ is slightly simpler than the other function, we choose to integrate this function with respect to y as the first step in finding f:

$$f(x, y) = \int 2x^3ye^{xy^2}\,dy = x^2e^{xy^2} + q(x).$$

(If you are wondering how we integrated this function, try a u-substitution with $u = xy^2$.) Next, to find $q(x)$, we differentiate $f(x, y)$ with respect to x and ensure that the resulting derivative equals $(2x + x^2y^2)e^{xy^2}$. That is,

$$\frac{\partial}{\partial x}(x^2e^{xy^2} + q(x)) = (2x + x^2y^2)e^{xy^2} + q'(x).$$

Thus, we must have $q'(x) = 0$ and $q(x) = C$, where C is a constant. We now have $f(x, y) = x^2e^{xy^2} + C$. □

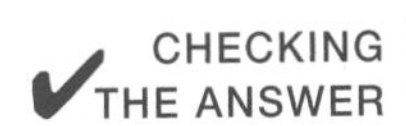
CHECKING THE ANSWER

We may check our answer by showing that for

$$f(x, y) = x^2e^{xy^2} + C, \quad \text{we have} \quad \nabla f(x, y) = (2x + x^2y^2)e^{xy^2}\mathbf{i} + 2x^3ye^{xy^2}\mathbf{j}.$$

EXAMPLE 6 **Showing that gradient vectors are orthogonal to level curves**

Show that the gradient vectors are orthogonal to the level curves for the function $f(x, y) = x^2 + y^2$.

SOLUTION

The graph of $f(x, y) = x^2 + y^2$ is a paraboloid, each of whose level curves is a circle centered at the origin. The gradient is $\nabla f(x, y) = 2x\mathbf{i} + 2y\mathbf{j}$; thus, every gradient vector points directly outward from the origin. For the point (x_0, y_0) the gradient is $\nabla f(x_0, y_0) = 2x_0\mathbf{i} + 2y_0\mathbf{j}$, and a tangent vector to the level curve containing the point (x_0, y_0) is $y_0\mathbf{i} - x_0\mathbf{j}$. The following figure shows the level curves $x^2 + y^2 = c$ for $c = 1, 2, 3$, and 4.

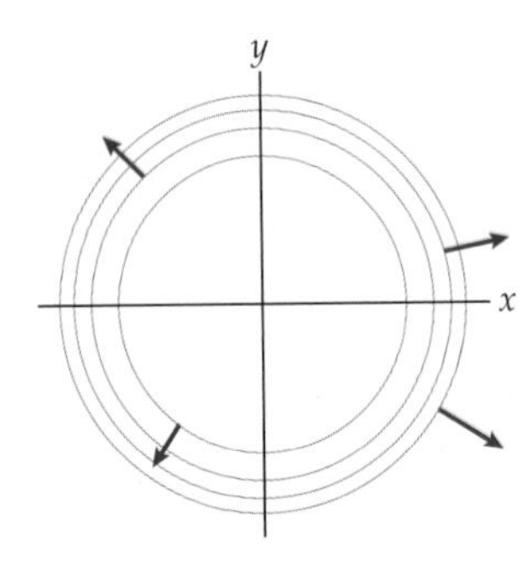

The gradient vector and tangent vector are orthogonal. □

? TEST YOUR UNDERSTANDING

- If f is a function of variables $x_1, x_2, \ldots, x_n$, and each of these variables is a function of $t_1, t_2, \ldots, t_m$, how would we find $\frac{\partial f}{\partial t_1}$? How many summands would there be in this partial derivative?
- What is the gradient? What does the gradient tell us about a function?
- If f is a function of two variables, what is the relationship between the gradient of f and the level curves of f?
- How can the gradient be used to compute the directional derivative of a function?
- Given the gradient of a function of two variables, how would we find the function?

EXERCISES 12.5

Thinking Back

- *Chain rule:* If f is a function of x and x is a function of t, how is the chain rule used to find the rate of change of f with respect to t?
- *Critical points:* What is the definition of a critical point for a function of a single variable? How do we use critical points to locate the extrema of the function?

Concepts

0. *Problem Zero:* Read the section and make your own summary of the material.

1. *True/False:* Determine whether each of the statements that follow is true or false. If a statement is true, explain why. If a statement is false, provide a counterexample.

(a) *True or False:* If $z=f(x,y)$ and $x=u(s,t)$, then $\frac{\partial z}{\partial s} = \frac{\partial z}{\partial x}\frac{\partial x}{\partial s}$.

(b) *True or False:* If $z = f(x,y)$ and $x = u(t)$, then $\frac{dz}{dt} = \frac{\partial z}{\partial x}\frac{dx}{dt}$.

(c) *True or False:* If $z = f(x,y)$, then $\nabla f(x,y) = \frac{\partial z}{\partial x} + \frac{\partial z}{\partial y}$.

(d) *True or False:* If $z = f(x,y)$ and $\nabla f(1,3) = \mathbf{0}$, then the graph of f is differentiable at $(1,3)$.

(e) *True or False:* If $z = f(x,y)$ and $\mathbf{u}$ is a unit vector, then $D_{\mathbf{u}} f(a,b) = \nabla f(a,b) \cdot \mathbf{u}$.

(f) *True or False:* If $f(x,y,z)$ is differentiable at (a,b,c) and $\mathbf{u} \in \mathbb{R}^3$ is a unit vector, then $D_{\mathbf{u}} f(a,b,c) = \nabla f(a,b,c) \cdot \mathbf{u}$.

(g) *True or False:* If $w = f(x,y,z)$ is differentiable at (a,b,c), then $-\nabla f(a,b,c)$ points in the direction in which f is decreasing most rapidly at (a,b,c).

(h) *True or False:* If $z = f(x,y)$ is differentiable at $(-1,0)$, and if $f(-1,0) = 4$, then $\nabla f(-1,0)$ is orthogonal to the level curve $f(x,y) = 4$ at $(-1,0)$.

2. *Examples:* Construct examples of the thing(s) described in the following. Try to find examples that are different than any in the reading.

(a) A function $z = f(x,y)$ for which $\nabla f(0,0) = \mathbf{0}$ but f is not differentiable at $(0,0)$.

(b) A function $z = f(x,y)$ for which $\nabla f(0,0) = \mathbf{0}$ for every point in $\mathbb{R}^2$.

(c) A function $z = f(x,y)$ and a unit vector $\mathbf{u}$ such that $D_{\mathbf{u}} f(0,0) \neq \nabla f(0,0) \cdot \mathbf{u}$.

3. Let $z = e^{-x}(3xy - 4x + y^2)$, $x = \sin t$, and $y = \cos t$.

(a) Find $\frac{dz}{dt}$ by using the Chain Rule, Theorem 12.32.

(b) Find $\frac{dz}{dt}$ by evaluating $f(x(t), y(t)) = f(\sin t, \cos t)$ and taking the derivative of the resulting function.

(c) Show that your answers from parts (a) and (b) are the same. Which method was easier?

4. Let $z = e^{-xy^2}$, $x = s\sin t$, and $y = s^2\cos t$.

(a) Find $\frac{\partial z}{\partial s}$ by using the Chain Rule, Theorem 12.33.

(b) Find $\frac{\partial z}{\partial s}$ by evaluating $f(x(s,t), y(s,t)) = f(s\sin t, s^2\cos t)$ and taking the partial derivative with respect to s of the resulting function.

(c) Show that your answers from parts (a) and (b) are the same. Which method was easier?

5. Explain why the chain rule from Chapter 2 is a special case of Theorem 12.34 with $n = 1$ and $m = 1$.

6. Explain why Theorem 12.32 is a special case of Theorem 12.34 with $n = 2$ and $m = 1$.

7. Explain why Theorem 12.33 is a special case of Theorem 12.34 with $n = 2$ and $m = 2$.

8. Consider the function $f(x,y) = 2x + 3y$.

(a) Why is the graph of f a plane?

(b) In what direction is f increasing most rapidly at the point $(-1,4)$?

(c) In what direction is f increasing most rapidly at the point (x_0, y_0)?

(d) Why are your answers to parts (b) and (c) the same?

9. Continue with the function $f(x,y) = 2x + 3y$ from Exercise 8.

(a) What are the level curves of f?

(b) Show that every gradient vector, $\nabla f(x,y)$, is orthogonal to every level curve of f.

10. Consider the function $f(x,y) = ax + by$, where neither a nor b is zero.

(a) Why is the graph of f a plane?

(b) In what direction is f increasing most rapidly at the point $(2,-3)$?

(c) In what direction is f increasing most rapidly at the point (x_0, y_0)?

(d) Why are your answers to parts (b) and (c) the same?

11. Continue with the function $f(x,y) = ax + by$ from Exercise 10.

(a) What are the level curves of f?

(b) Show that every gradient vector, $\nabla f(x,y)$, is orthogonal to every level curve of f.

12. If a function $f(x,y)$ is differentiable at (a,b), explain how to use the gradient $\nabla f(a,b)$ to find the equation of the plane tangent to the graph of f at (a,b).

13. If a function $f(x,y,z)$ is differentiable at (a,b,c), explain how to use the gradient $\nabla f(a,b,c)$ to find the equation of the hyperplane tangent to the graph of f at (a,b,c).

14. Sketch level curves $z = 1, 4, 9$, and 16 for the function $z = x^2 + y^2$. Include the graphs of three gradient vectors on each level curve. What do you observe?

15. Sketch level curves $z = 9, 16, 21$, and 24 for the function $z = 25 - x^2 - y^2$. Include the graphs of three gradient vectors on each level curve. What do you observe?

16. Sketch level curves $z = 1, 4, 9$, and 16 for the function $z = \frac{x^2}{4} + \frac{y^2}{9}$. Include the graphs of three gradient vectors on each level curve. What do you observe?

17. Sketch level curves $z = -1, 0$, and 1 for the function $z = x^2 - y^2$. Include the graphs of three gradient vectors on each level curve. What do you observe?

18. When would you have to use the definition of the directional derivative rather than the shortcut $D_{\mathbf{u}}f(x_0, y_0) = \nabla f(x_0, y_0) \cdot \mathbf{u}$?

19. Imagine using a topographic map to plan a hike on a mountain. If your hike stays on a single contour line, what does that mean about the difficulty of the hike? If you intentionally hike perpendicularly to the contour lines, what does *that* mean about the difficulty of the hike?

20. An isotherm is a curve on a weather map connecting points on the map that have the same temperature. From a point on an isotherm, which direction would result in the greatest temperature change?

Skills

Use Theorem 12.32 to find the indicated derivatives in Exercises 21–26. Express your answers as functions of a single variable.

21. $\frac{dz}{dt}$ when $z = \sin x \cos y$, $x = e^t$, and $y = t^3$.

22. $\frac{dz}{dt}$ when $z = x^3 e^y$, $x = \sin t$, and $y = \cos t$.

23. $\frac{dx}{dt}$ when $x = r\cos\theta$, $r = t^2 - 5$, and $\theta = t^3 + 1$.

24. $\frac{dy}{dt}$ when $y = r\sin\theta$, $r = t^3$, and $\theta = \sqrt{t}$.

25. $\frac{dr}{dt}$ when $r = \sqrt{x^2 + y^2}$, $x = \sqrt{t}$, and $y = t^2$.

26. $\frac{d\theta}{dt}$ when $\theta = \tan^{-1}\left(\frac{y}{x}\right)$, $x = e^t$, and $y = e^{2t}$.

Use Theorem 12.33 to find the indicated derivatives in Exercises 27–30. Express your answers as functions of two variables.

27. $\frac{\partial z}{\partial s}$ when $z = x^2y^3$, $x = t\sin s$, and $y = s\cos t$.

28. $\frac{\partial z}{\partial t}$ when $z = x^2y^3$, $x = t\sin s$, and $y = s\cos t$.

29. $\frac{\partial z}{\partial r}$ when $z = (x^2 + xy)e^y$, $x = r\cos\theta$, and $y = r\sin\theta$.

30. $\frac{\partial z}{\partial \theta}$ when $z = (x^2 + xy)e^y$, $x = r\cos\theta$, and $y = r\sin\theta$.

Use Theorem 12.34 to find the indicated derivatives in Exercises 31–36. Be sure to simplify your answers.

31. $\frac{dx}{dt}$ when $x = \rho\sin\phi\cos\theta$, $\rho = t^2$, $\phi = t^3$, and $\theta = t^4$.

32. $\frac{dx}{dt}$ when $x = \rho\sin\phi\sin\theta$, $\rho = \sqrt{t}$, $\phi = \sqrt[3]{t}$, and $\theta = \sqrt[4]{t}$.

33. $\frac{\partial w}{\partial \rho}$ when $w = (x^2 + z)e^y$, $x = \rho\sin\phi\cos\theta$, $y = \rho\sin\phi\sin\theta$, $z = \rho\cos\phi$.

34. $\frac{\partial w}{\partial \theta}$ when $w = (x^2 + z)e^y$, $x = \rho\sin\phi\cos\theta$, $y = \rho\sin\phi\sin\theta$, $z = \rho\cos\phi$.

35. $\frac{d\rho}{dt}$ when $\rho = \sqrt{x^2 + y^2 + z^2}$, $x = \sqrt{t}$, $y = t^2$, $z = t^3$.

36. $\frac{d\theta}{dt}$ when $\theta = \tan^{-1}\left(\frac{yz}{x}\right)$, $x = e^t$, $y = e^{2t}$, $z = e^{3t}$.

Find the gradient of the given functions in Exercises 37–42.

37. $z = x^2\sin y + y\sin x$

38. $z = \tan^{-1}\left(\frac{y}{x}\right)$

39. $f(x, y) = \sqrt{x^2 + y^2}$

40. $f(x, y) = y\sec^{-1}\left(\frac{1}{x}\right)$

41. $f(x, y, z) = \sqrt{x^2 + y^2 + z^2}$

42. $f(x, y, z) = \cos^{-1}\left(\frac{z}{\sqrt{x^2 + y^2 + z^2}}\right)$

In Exercises 43–48:

(a) Find the direction in which the given function increases most rapidly at the specified point.

(b) Find the rate of change of the function in the direction you found in part (a).

(c) Find the direction in which the given function decreases most rapidly at the specified point.

Note: These are the same functions as in Exercises 37–42.

43. $z = x^2\sin y + y\sin x$ at $\left(\pi, \frac{\pi}{2}\right)$

44. $z = \tan^{-1}\left(\frac{y}{x}\right)$ at $(1, \sqrt{3})$

45. $f(x, y) = \sqrt{x^2 + y^2}$ at $(2, -3)$

46. $f(x, y) = y\sec^{-1}x$ at $(2, 5)$

47. $f(x, y, z) = \sqrt{x^2 + y^2 + z^2}$ at $(2, -1, -2)$

48. $f(x, y, z) = \cos^{-1}\left(\frac{z}{\sqrt{x^2 + y^2 + z^2}}\right)$ at $(\sqrt{10}, -1, 5)$

In Exercises 49–54, find the directional derivative of the given function at the specified point P and in the specified direction $\mathbf{v}$. Note that some of the direction vectors are *not* unit vectors.

49. $z = e^x\tan y$, $P = \left(0, \frac{\pi}{4}\right)$, $\mathbf{v} = 3\mathbf{i} - \mathbf{j}$

50. $z = 3x^4 - 5y^2$, $P = (1, -2)$, $\mathbf{v} = \left\langle \frac{5}{13}, -\frac{12}{13} \right\rangle$

51. $z = \ln\left(\frac{x}{y^2}\right)$, $P = (7, 1)$, $\mathbf{v} = -\mathbf{i} - 4\mathbf{j}$

52. $z = \sqrt{x^2 + y^2}$, $P = (-3, 4)$, $\mathbf{v} = \langle 4, -3 \rangle$

53. $w = x \sin y \cos z$, $P = \left(3, \frac{\pi}{4}, -\frac{\pi}{2}\right)$, $\mathbf{v} = \mathbf{i} - 2\mathbf{j} + 3\mathbf{k}$

54. $w = \ln\left(\frac{x^2}{yz}\right) + \ln\left(\frac{z}{xy}\right) - \ln\left(\frac{x}{y^2}\right)$, $P = (3, 5, 8)$, $\mathbf{v} = 13\mathbf{i} + 21\mathbf{j} + 34\mathbf{k}$

In Exercises 55–60, find a function of two variables with the given gradient.

55. $\nabla f(x, y) = \left\langle -\frac{3}{x}, \frac{1}{y}\right\rangle$

56. $\nabla f(x, y) = (2x + \cos x \cos y)\,\mathbf{i} - \sin x \sin y\,\mathbf{j}$

57. $\nabla f(x, y) = -\frac{y}{x^2 + y^2}\mathbf{i} + \frac{x}{x^2 + y^2}\mathbf{j}$

58. $\nabla f(x, y) = \langle y^2 e^{xy^2}, 2xye^{xy^2}\rangle$

59. $\nabla f(x, y) = \left\langle \frac{y}{(x + y)^2}, -\frac{x}{(x + y)^2}\right\rangle$

60. $\nabla f(x, y) = \left(\frac{1}{y} - \frac{y}{x^2}\right)\mathbf{i} + \left(\frac{1}{x} - \frac{x}{y^2}\right)\mathbf{j}$

Applications

61. Ian is travelling along a glacier. The elevation of the glacier in his area is described by the function

$$f(x, y) = 1.2 - 0.2x^2 - 0.3y^2 + 0.1xy - 0.25x,$$

where x, y, and f are given in miles.

(a) What is the direction that Ian would need to go to descend most steeply if he is at the point $(0.5, -0.5)$?

(b) In what direction would Ian have to turn in order to contour (i.e., , neither ascend nor descend) across the glacier?

62. Emmy is tracking down another source of contamination at Hanford, this time in a warehouse containing numerous 55-gallon drums of waste. Detectors placed throughout the facility have measured a certain amount of radiation at different points. From these measurements, Emmy has constructed a polynomial approximation to the radiation given by

$$f(x, y) = -0.0156x^2 - 0.0392y^2 + 0.0268xy + 1.7x + 20,$$

where the radiation is measured in microrems per hour and the distances x and y are measured in feet from a corner of the warehouse. At what location in the warehouse should Emmy start her search for the contaminated drum?

Proofs

63. Prove that $\|\nabla f(x, y, z)\| = 1$ for every point in the domain of the function $f(x, y, z) = \sqrt{x^2 + y^2 + z^2}$ except the origin.

64. Use at least two methods to prove that $\frac{dr}{dt} = 0$ when $r = \sqrt{x^2 + y^2}$, $x = \alpha \cos t$, and $y = \alpha \sin t$ if α is a constant.

65. Prove Theorem 12.33. That is, show that if $z = f(x, y)$, $x = u(s, t)$, and $y = v(s, t)$, then, for all values of s and t at which u and v are differentiable, and if f is differentiable at $(u(s, t), v(s, t))$, it follows that

$$\frac{\partial z}{\partial s} = \frac{\partial z}{\partial x}\frac{\partial x}{\partial s} + \frac{\partial z}{\partial y}\frac{\partial y}{\partial s} \quad \text{and} \quad \frac{\partial z}{\partial t} = \frac{\partial z}{\partial x}\frac{\partial x}{\partial t} + \frac{\partial z}{\partial y}\frac{\partial y}{\partial t}.$$

66. Prove Theorem 12.36 when f is a function of three variables. That is, show that if (x_0, y_0, z_0) is a point in the domain of $f(x, y, z)$ at which the first-order partial derivatives of f exist, and if $\mathbf{u} \in \mathbb{R}^3$ is a unit vector for which the directional derivative $D_{\mathbf{u}} f(x_0, y_0, z_0)$ also exists, then $D_{\mathbf{u}} f(x_0, y_0, z_0) = \nabla f(x_0, y_0, z_0) \cdot \mathbf{u}$.

In Exercises 67–72 you will prove several basic properties of the gradient for functions of two variables. In each exercise, assume that f and/or g is differentiable.

67. Prove that $\nabla f = 0$ if f is the constant function $f(x, y) = c$.

68. Prove that $\nabla(\alpha f)(x, y) = \alpha \nabla f(x, y)$, where α is a constant.

69. Prove that

$$\nabla(f(x, y) + g(x, y)) = \nabla f(x, y) + \nabla g(x, y).$$

70. Prove that

$$\nabla(\alpha f(x, y) + \beta g(x, y)) = \alpha \nabla f(x, y) + \beta \nabla g(x, y),$$

where α and β are constants.

71. Prove that

$$\nabla(f(x, y)g(x, y)) = f(x, y)\nabla g(x, y) + g(x, y)\nabla f(x, y).$$

72. Prove that

$$\nabla\left(\frac{f(x, y)}{g(x, y)}\right) = \frac{g(x, y)\nabla f(x, y) - f(x, y)\nabla g(x, y)}{(g(x, y))^2},$$

where $g(x, y) \neq 0$.

73. Analogous properties hold for functions of three variables. What would you have to change in the proofs in Exercises 67–72 to make them work for functions of three variables?

74. Let $f(x, y, z)$ be a function of three variables, and let P be a point in the domain of f at which f is differentiable. Prove that the gradient of f at P points in the direction in which f increases most rapidly. (This is Theorem 12.37.)

Thinking Forward

- *The gradient at a maximum:* If a function of two variables, $f(x, y)$, is differentiable at a point (x_0, y_0) where the function has a maximum, what is $\nabla f(x_0, y_0)$?
- *The gradient at a minimum:* If a function of three variables, $f(x, y, z)$, is differentiable at a point (x_0, y_0, z_0) where the function has a minimum, what is $\nabla f(x_0, y_0, z_0)$?

12.6 EXTREME VALUES

- Local and global extrema of functions of two variables
- Critical points of functions of two variables
- Using critical points to find the local extrema of functions of two variables

The Gradient at a Local Extremum

Many of the ideas in this section parallel the analogous concepts for functions of a single variable. It would be worthwhile to review the definitions of local extrema and critical points for functions of a single variable. Before proceeding in this section, we also suggest that you think about how the first- and second-derivative tests can be adapted for functions of two variables, if at all. Further, we suggest that when you are done with this section, think about how the concepts presented here might be adapted to analyze functions of three or more variables.

Suppose a function $f(x, y)$ has a local maximum at some point (x_0, y_0). This means that the value of $f(x_0, y_0)$ is greater than or equal to all other "nearby" values of the function. More precisely, we have the following definition:

DEFINITION 12.39 Local and Global Extrema of a Function of Two Variables

(a) A function $f(x, y)$ has a ***local*** or ***relative maximum*** at (x_0, y_0) if $f(x_0, y_0) \geq f(x, y)$ for every point (x, y) in some open disk containing (x_0, y_0).

(b) A function $f(x, y)$ has a ***local*** or ***relative minimum*** at (x_0, y_0) if $f(x_0, y_0) \leq f(x, y)$ for every point (x, y) in some open disk containing (x_0, y_0).

(c) A function $f(x, y)$ has a ***local*** or ***relative extremum*** at (x_0, y_0) if f has either a local maximum or a local minimum at (x_0, y_0).

(d) A function $f(x, y)$ has a ***global*** or ***absolute maximum*** at (x_0, y_0) if $f(x_0, y_0) \geq f(x, y)$ for every point (x, y) in the domain of f.

(e) A function $f(x, y)$ has a ***global*** or ***absolute minimum*** at (x_0, y_0) if $f(x_0, y_0) \leq f(x, y)$ for every point (x, y) in the domain of f.

(f) A function $f(x, y)$ has a ***global*** or ***absolute extremum*** at (x_0, y_0) if f has either a global maximum or a global minimum at (x_0, y_0).

Clearly, every global extremum is also a local extremum. At a local extremum, the plane tangent to the function must be either horizontal or undefined. Consider the following three graphs:

Local maximum with horizontal tangent plane

Local minimum with horizontal tangent plane

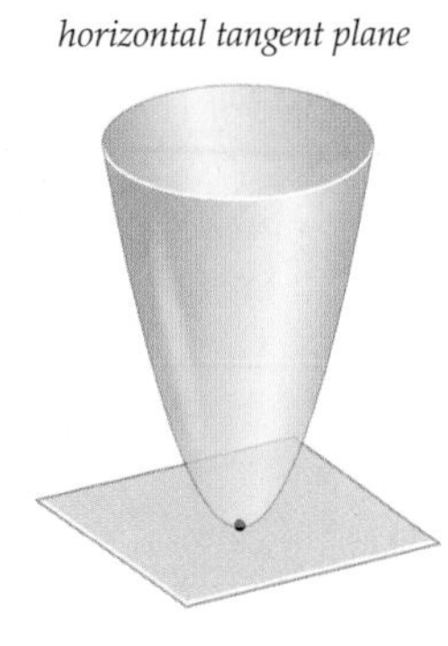

Local maximum with no tangent plane

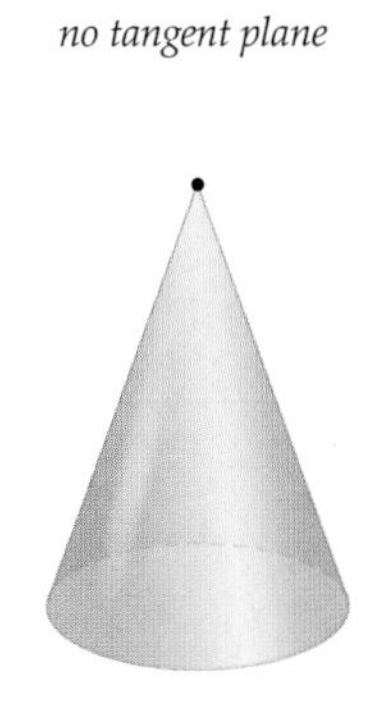

When a function has a horizontal tangent plane at a point P, its gradient at P is zero. Such a point is called a ***stationary point***. When a function is not differentiable at a point, its gradient there is either zero or undefined. Points at which either of these things occur are called ***critical points***.

DEFINITION 12.40 Stationary Points of a Function of Two Variables

A point (x_0, y_0) in the domain of a function $f(x, y)$ is called a ***stationary point*** of f if f is differentiable at (x_0, y_0) and $\nabla f(x_0, y_0) = \mathbf{0}$.

In the first two of the preceding three graphs, the functions have stationary points where the graphs have the horizontal tangent planes. The condition that $f(x, y)$ needs to be differentiable at (x_0, y_0) is necessary, since $\nabla f(x_0, y_0) = \mathbf{0}$ does not ensure that f is differentiable at (x_0, y_0). Consider, for example, the function

$$f(x, y) = \begin{cases} 0, & \text{if } x = 0 \text{ or } y = 0 \\ 1, & \text{otherwise.} \end{cases}$$

In Section 12.4, we saw that $\nabla f(0, 0) = \mathbf{0}$ but that f is *not* differentiable at the origin. Therefore, $(0, 0)$ is not a stationary point for f.

DEFINITION 12.41 Critical Points of a Function of Two Variables

A point (x_0, y_0) in the domain of a function $f(x, y)$ is called a ***critical point*** of f if (x_0, y_0) is a stationary point of f or if f is not differentiable at (x_0, y_0).

Therefore, each of the three graphs shown before has a critical point at its local extremum. In fact, a local extremum can occur only at a critical point.

THEOREM 12.42 **Local Extrema Occur at Critical Points**

If $f(x, y)$ has a local extremum at (x_0, y_0), then (x_0, y_0) is a critical point of f.

The proof of Theorem 12.42 is similar to the proof of Theorem 3.3 in Section 3.1 and is left for Exercise 56.

The converse of Theorem 12.42 is not true. For example, consider the function $g(x, y) = x^2 - y^2$:

$g(x, y) = x^2 - y^2$

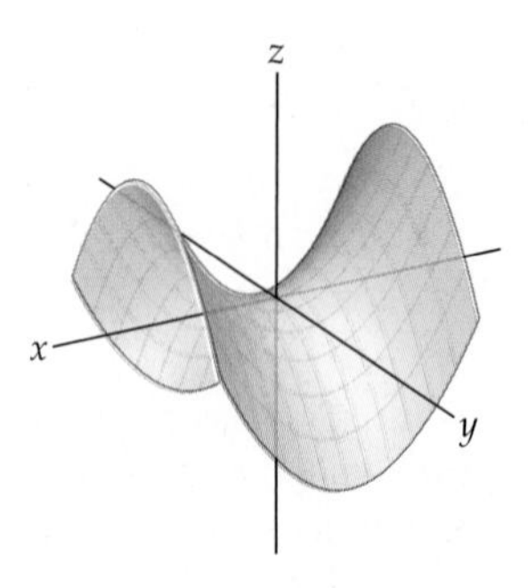

Here we have $\nabla g(0, 0) = \mathbf{0}$, but the function has neither a maximum nor a minimum at $(0, 0)$. This example leads us to the definition of a ***saddle point***.

DEFINITION 12.43 **Saddle Points of a Function of Two Variables**

A point (x_0, y_0) in the domain of a function $f(x, y)$ is called a ***saddle point*** of f if (x_0, y_0) is a stationary point of f at which there is neither a maximum nor a minimum.

The term "saddle point" comes from the fact that the graphs of the surfaces near the simplest saddle points look like saddles, as we see in the graph of $g(x, y) = x^2 - y^2$. However, the graphs near other saddle points can look far more complicated. For example, the graph of $f(x, y) = x^2y - y^3$, which also has a saddle point at the origin, is as follows:

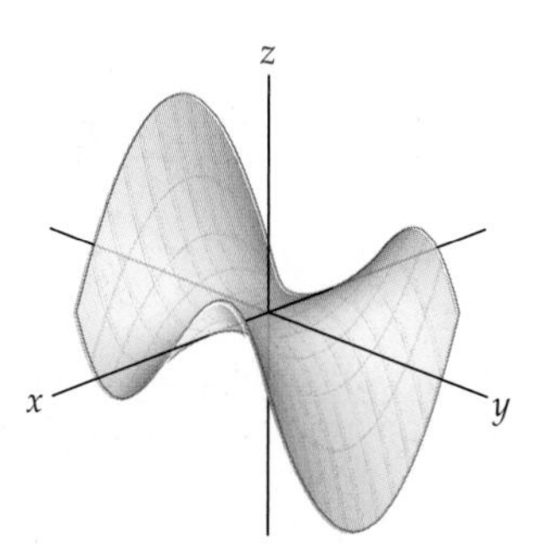

The Second-Derivative Test for Classifying Local Extrema

We will now discuss a method for determining whether a function has a local maximum, local minimum, or saddle point at each stationary point. This test uses the second-order partial derivatives of the function. You may be wondering why we are not starting by discussing a test that uses the first-order partial derivatives. The short answer is that such a test is somewhat impractical. You will explore the reason for that in Exercise 7.

Before we discuss the second-order partial derivative test, we define the ***discriminant*** of a function of two variables.

DEFINITION 12.44 **The Hessian and the Discriminant of a Function of Two Variables**

Let $f(x, y)$ be a function with continuous second-order partial derivatives on some open set S.

(a) The ***Hessian*** of f is the 2×2 matrix of second-order partial derivatives:

$$H_f = \begin{bmatrix} \dfrac{\partial^2 f}{\partial x^2} & \dfrac{\partial^2 f}{\partial y \partial x} \\ \dfrac{\partial^2 f}{\partial x \partial y} & \dfrac{\partial^2 f}{\partial y^2} \end{bmatrix}.$$

(b) The ***discriminant*** of f is the determinant of the Hessian. That is,

$$\det(H_f) = \frac{\partial^2 f}{\partial x^2}\frac{\partial^2 f}{\partial y^2} - \left(\frac{\partial^2 f}{\partial x \partial y}\right)^2.$$

Note that since we require that the second-order partial derivatives be continuous to form either the Hessian or the discriminant, the determinant of the Hessian matrix is equivalent to the discriminant because Theorem 12.23 guarantees the equality of the mixed second-order partial derivatives.

The discriminant of a function of two variables is a related function of two variables. For example, when $f(x, y) = x^3 e^{2y}$, we have the first-order partial derivatives

$$\frac{\partial f}{\partial x} = 3x^2 e^{2y} \quad \text{and} \quad \frac{\partial f}{\partial y} = 2x^3 e^{2y}$$

and the second-order partial derivatives

$$\frac{\partial^2 f}{\partial x^2} = 6xe^{2y}, \quad \frac{\partial^2 f}{\partial y^2} = 4x^3 e^{2y}, \quad \text{and} \quad \frac{\partial^2 f}{\partial x \partial y} = 6x^2 e^{2y}.$$

Therefore, the discriminant is

$$\det(H_f(x, y)) = (6xe^{2y})(4x^3 e^{2y}) - (6x^2 e^{2y})^2 = -12x^4 e^{4y}.$$

We finally note that the Hessian may be generalized to functions of three or more variables, although we will not be using the Hessian in these contexts. We will be using the Hessian and the discriminant to analyze the stationary points of a function of two variables. The Hessian of a function of three variables may be used in a similar manner.

THEOREM 12.45 **The Second-Order Partial-Derivative Test for Classifying Stationary Points**

Let $f(x, y)$ be a function with continuous second-order partial derivatives on some open disk containing the point at (x_0, y_0). If f has a stationary point at (x_0, y_0), then

(a) f has a relative maximum at (x_0, y_0) if $\det(H_f(x_0, y_0)) > 0$ with $f_{xx}(x_0, y_0) < 0$ or $f_{yy}(x_0, y_0) < 0$.

(b) f has a relative minimum at (x_0, y_0) if $\det(H_f(x_0, y_0)) > 0$ with $f_{xx}(x_0, y_0) > 0$ or $f_{yy}(x_0, y_0) > 0$.

(c) f has a saddle point at (x_0, y_0) if $\det(H_f(x_0, y_0)) < 0$.

(d) If $\det(H_f(x_0, y_0)) = 0$, no conclusion may be drawn about the behavior of f at (x_0, y_0).

Before we prove Theorem 12.45, we consider two examples. First, for the function

$$f(x, y) = x^2 + xy + y^2 + 3x - 6y,$$

we have $f_x(x, y) = 2x + y + 3$ and $f_y(x, y) = x + 2y - 6$. When we solve the system of equations

$$2x + y + 3 = 0 \quad \text{and} \quad x + 2y - 6 = 0,$$

we find the only stationary point of f, namely, $(-4, 5)$. The second-order partial derivatives are $f_{xx}(x, y) = 2$, $f_{yy}(x, y) = 2$, and $f_{xy}(x, y) = 1$. Therefore, $\det(H_f(-4, 5)) = 2 \cdot 2 - 1^2 = 3$. Since $\det(H_f(-4, 5)) > 0$, $f_{xx}(-4, 5) > 0$, and $f_{yy}(-4, 5) > 0$, it follows by Theorem 12.45 that f has a local minimum at $(-4, 5)$.

Next, if we let $g(x, y) = x^6 + y^6$, then $\nabla g(x, y) = 6x^5\mathbf{i} + 6y^5\mathbf{j}$. The point $(0, 0)$ is the only stationary point for g. The second-order partial derivatives are $g_{xx}(x, y) = 30x^4$, $f_{yy}(x, y) = 30y^4$, and $f_{xy}(x, y) = 0$. Therefore, $\det(H_g(0, 0)) = 0$. Thus, no conclusion can be drawn from Theorem 12.45 about the behavior of g at the origin. However, since $g(0, 0) = 0$ and $g(x, y) > 0$ for every point in $\mathbb{R}^2$ except the origin, we may deduce that g has an absolute minimum of 0 at the origin, without using the theorem.

We will prove Theorem 12.45 in a moment, but perhaps the following remarks will help explain the reasoning in the proof: Recall that the sign of the second derivative of a function of a single variable determines whether a function is concave up or concave down. Similarly, the sign of the second-order directional derivative $D_{\mathbf{u}}(D_{\mathbf{u}}f(x, y)) = D^2_{\mathbf{u}}f(x, y)$ may be used to determine whether the function is concave up or concave down in the direction of the unit vector $\mathbf{u}$. If $D^2_{\mathbf{u}}f(x_0, y_0) > 0$ for every unit vector at a stationary point (x_0, y_0), then the function has a local minimum at the stationary point. Similarly, if $D^2_{\mathbf{u}}f(x_0, y_0) < 0$ for every unit vector at the stationary point, then the function has a local maximum there.

One final note before we turn to the proof: Recall that the discriminant of a quadratic polynomial $y = ax^2 + bx + c$ is the quantity $b^2 - 4ac$ and the discriminant of a function of two variables is given in Definition 12.44. In the proof of Theorem 12.45, we use this term in both contexts.

Proof. In Exercise 55 you will prove that for a function $f(x, y)$ with continuous second-order partial derivatives, the second directional derivative in the direction of the unit vector $\mathbf{u} = \langle \alpha, \beta \rangle$ is given by

$$D^2_{\mathbf{u}}f(x, y) = \alpha^2 f_{xx}(x, y) + 2\alpha\beta f_{xy}(x, y) + \beta^2 f_{yy}(x, y).$$

Now let (x_0, y_0) be a stationary point of f, and let

$$A = f_{xx}(x_0, y_0), \quad B = f_{xy}(x_0, y_0), \quad \text{and} \quad C = f_{yy}(x_0, y_0).$$

Before proceeding further, note that $AC - B^2$ is the discriminant, $\det(H_f(x_0, y_0))$. We have

$$D^2_{\mathbf{u}}f(x_0, y_0) = \alpha^2 A + 2\alpha\beta B + \beta^2 C.$$

Now, if $\alpha \neq 0$ and we let $m = \dfrac{\beta}{\alpha}$, then

$$D^2_{\mathbf{u}}f(x_0, y_0) = \alpha^2\left(A + 2\frac{\beta}{\alpha}B + \left(\frac{\beta}{\alpha}\right)^2 C\right) = \alpha^2\left(A + 2mB + m^2C\right).$$

The polynomial $q(m) = A + 2Bm + Cm^2$ is a quadratic in m and has the same roots as $D^2_{\mathbf{u}}f(x_0, y_0)$. Now, if the discriminant, $4B^2 - 4AC$, of this *polynomial* is negative, then $q(m)$ has no real roots. Thus, if $AC - B^2 > 0$, then $q(m)$ has no real roots, and the sign of $q(m)$ never changes. In order for $AC - B^2$ to be positive, the signs of A and C must be the same. So, if $A = f_{xx}(x_0, y_0)$, $C = f_{yy}(x_0, y_0)$, and $AC - B^2 = \det(H_f(x_0, y_0))$ are all positive, the function is concave up in every direction at the stationary point (x_0, y_0), and it follows that f has a local minimum at (x_0, y_0). If A and C are both negative with $AC - B^2 > 0$, the function is concave down in every direction at (x_0, y_0), and it follows that f has a local maximum at (x_0, y_0). If $AC - B^2 < 0$, the polynomial $q(m)$

has two real roots, and then the function f is concave up in some directions and concave down in other directions. Therefore, f has a saddle at the stationary point. If $AC - B^2 = 0$, we cannot draw any conclusions about the function. In Exercises 10-12 we ask you to show that a function can have a local maximum, local minimum, or saddle point when $AC - B^2 = 0$. Finally, if $\alpha = 0$, then $\beta \neq 0$, and we may modify the argument by factoring β^2 instead of α^2 out of the equation for $D_{\mathbf{u}}^2 f(x, y)$. ■

Examples and Explorations

EXAMPLE 1

Finding and analyzing stationary points

Find the local extrema of the function $f(x, y) = x^3 - 3xy - y^3$. Classify each extreme point as a local or global minimum or maximum.

SOLUTION

Our function is a polynomial. Every polynomial in two variables is differentiable at every point in $\mathbb{R}^2$, so the only critical points of f will be stationary points. The gradient of f is

$$\nabla f(x, y) = (3x^2 - 3y)\mathbf{i} + (-3x - 3y^2)\mathbf{j}.$$

To find the stationary points, we look for the points that make the gradient $\mathbf{0}$. We look for solutions of the system of equations

$$3x^2 - 3y = 0 \quad \text{and} \quad -3x - 3y^2 = 0.$$

Using the first equation, we see that $y = x^2$. Substituting x^2 for y in the second equation and simplifying, we obtain $x(x^3 + 1) = 0$. This equation has solutions $x = 0$ and $x = -1$. Since $x^2 = y$, we have the two stationary points $(0, 0)$ and $(-1, 1)$.

Now, the second-order partial derivatives of f are

$$f_{xx}(x, y) = 6x, \quad f_{yy}(x, y) = -6y, \quad \text{and} \quad f_{xy}(x, y) = -3.$$

The discriminant is $\det(H_f(x, y)) = (6x)(-6y) - (-3)^2 = -36xy - 9$. At the stationary points, we have $\det(H_f(0, 0)) = -9$ and $\det(H_f(-1, 1)) = 27$. By Theorem 12.45, since $\det(H_f(0, 0)) < 0$, there is a saddle point at $(0, 0)$, and since $\det(H_f(-1, 1)) > 0$ and $f_{xx}(-1, 1) = -6 < 0$, f has a local maximum at $(-1, 1)$.

Finally, it is good to know if there is an absolute maximum at $(-1, 1)$ or just a relative maximum. We have $f(-1, 1) = 1$, but when $x = 0$, $\lim_{y \to -\infty} f(0, y) = \lim_{y \to -\infty} (-y^3) = \infty$. Therefore, f has just a relative maximum at $(-1, 1)$. □

EXAMPLE 2

Tabulating information about stationary points

Find the local extrema of the function $g(x, y) = 3x^2 y - 3y + y^3$. Classify each extreme point as a local or global minimum or maximum.

SOLUTION

Again, the function is a polynomial, so the only critical points of g will be stationary points. The gradient of g is

$$\nabla g(x, y) = g_x(x, y)\mathbf{i} + g_y(x, y)\mathbf{j} = 6xy\mathbf{i} + (3x^2 - 3 + 3y^2)\mathbf{j}.$$

To find the stationary points, we look for solutions of the system of equations

$$6xy = 0 \quad \text{and} \quad 3x^2 - 3 + 3y^2 = 0.$$

From the first equation, we see that one of x and y must be zero. Substituting $x = 0$ into the second equation and solving for y, we have $y = \pm 1$. Similarly, if $y = 0$, solving for x yields $x = \pm 1$. Thus, we have exactly four stationary points: $(0, 1)$, $(0, -1)$, $(1, 0)$, and $(-1, 0)$.

The second-order partial derivatives of g are

$$g_{xx}(x, y) = 6y, \quad g_{yy}(x, y) = 6y, \quad \text{and} \quad g_{xy}(x, y) = 6x.$$

The discriminant is $\det(H_g(x, y)) = (6y)(6y) - (6x)^2 = 36(y^2 - x^2)$. When there are several stationary points, it is convenient to tabulate the relevant information about the discriminant and second-order partial derivatives in order to classify the stationary points. Accordingly, we construct the following table:

(x_0, y_0)	$\det(H_g(x_0, y_0)) = 36(y_0^2 - x_0^2)$	$g_{xx}(x_0, y_0) = 6y_0$	$g(x_0, y_0)$	conclusion
$(0, 1)$	36	6	-2	minimum
$(0, -1)$	36	-6	2	maximum
$(1, 0)$	-36		0	saddle point
$(-1, 0)$	-36		0	saddle point

It remains to determine whether the function has an absolute minimum at $(0, 1)$ or just a relative minimum, and an absolute maximum at $(0, -1)$ or just a relative maximum. Note that when $x = 0$, we have the limits

$$\lim_{y\to\infty} g(0, y) = \lim_{y\to\infty}(-3y + y^3) = \infty \quad \text{and} \quad \lim_{y\to-\infty} g(0, y) = \lim_{y\to-\infty}(-3y + y^3) = -\infty.$$

Therefore, g has a relative minimum at $(0, 1)$ and a relative maximum at $(0, -1)$. □

EXAMPLE 3

Optimizing a function of two variables

Max is planning to construct a box with five wooden sides and a glass front. If the wood costs \$5 per square foot and the glass costs \$10 per square foot, what dimensions should Max use for the box to minimize the cost of the materials if the box needs to have a volume of 12 cubic feet?

SOLUTION

Following is a schematic of the box:

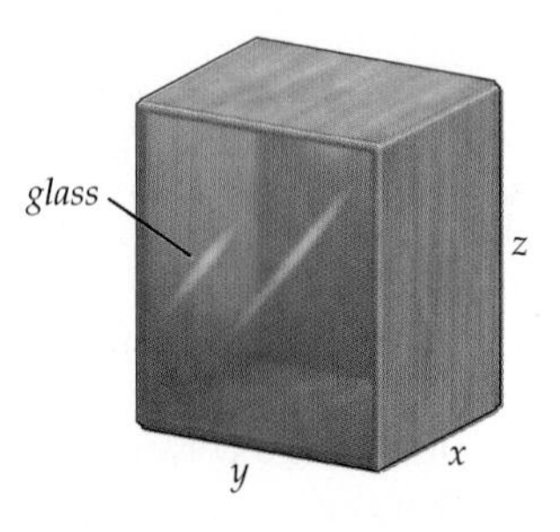

Using the dimensions in the figure and the fact that the volume of the box is to be 12 cubic feet, we have the constraint $V = xyz = 12$ between the variables. The glass front has area yz. Therefore, the cost of the glass will be $\$10yz$. The back wooden face will cost $\$5yz$, the left and right sides will each cost $\$5xz$, and the top and bottom will each cost $\$5xy$. Thus, in dollars, the cost of the box will be $10xy + 10xz + 15yz$. Now, from our constraint on the volume, we have $z = \frac{12}{xy}$, which we may use in our equation for the cost to rewrite our function with just two variables:

$$C(x, y) = 10xy + \frac{120}{y} + \frac{180}{x}, \quad \text{where } x > 0 \quad \text{and} \quad y > 0.$$

To find the stationary points, we find the first-order partial derivatives

$$C_x = 10y - \frac{180}{x^2} \quad \text{and} \quad C_y = 10x - \frac{120}{y^2}.$$

Setting these partial derivatives equal to zero and solving simultaneously, we obtain $x = 3$ and $y = 2$. We also have the second-order partial derivatives

$$C_{xx} = \frac{360}{x^3}, \quad C_{yy} = \frac{240}{y^3}, \quad \text{and} \quad C_{xy} = 10.$$

Thus, the discriminant at (3, 2) is $\det(H_C(3, 2)) = \left(\frac{360}{27}\right)\left(\frac{240}{8}\right) - 10^2 = 300 > 0$, and since $C_{xx}(3, 2) > 0$ as well, this stationary point gives us a minimum cost for the box. These values tell us that the glass should be a square 2 feet on each side and the other dimension of the box should be 3 feet. □

EXAMPLE 4

Using the derivative to find the distance from a point to a plane

In Theorem 10.39 in Chapter 10 we showed that the distance between a point P and a plane $\mathcal{P}$ is given by $\frac{|\mathbf{N} \cdot \overrightarrow{RP}|}{\|\mathbf{N}\|}$, where R is any point in the plane and $\mathbf{N}$ is a normal vector for $\mathcal{P}$. Use the minimization techniques discussed in this section to show that the distance from the point $P(x_0, y_0, z_0)$ to the plane $\mathcal{P}$ with equation $ax + by + cz = d$ is also given by

$$\frac{|ax_0 + by_0 + cz_0 - d|}{\sqrt{a^2 + b^2 + c^2}}.$$

SOLUTION

Let (x, y, z) be an arbitrary point in the plane $\mathcal{P}$. We wish to minimize the distance from P to (x, y, z) given by

$$\text{distance} = \sqrt{(x - x_0)^2 + (y - y_0)^2 + (z - z_0)^2}.$$

However, as we have previously mentioned, if we minimize the square of the distance, we accomplish the same goal. We will minimize

$$\text{distance}^2 = (x - x_0)^2 + (y - y_0)^2 + (z - z_0)^2$$

subject to the constraint that $ax+by+cz = d$. Now, at least one of the coefficients a, b, and c is nonzero. Without loss of generality, we assume that $c \neq 0$, so we have $z = \frac{d-ax-by}{c}$. Thus, we will minimize the function

$$D(x, y) = (x - x_0)^2 + (y - y_0)^2 + \left(\frac{d - ax - by}{c} - z_0\right)^2.$$

Here,

$$\nabla D = \left(2(x - x_0) - \frac{2a}{c}\left(\frac{d - ax - by}{c} - z_0\right)\right)\mathbf{i} + \left(2(y - y_0) - \frac{2b}{c}\left(\frac{d - ax - by}{c} - z_0\right)\right)\mathbf{j}.$$

Setting $\nabla D = \mathbf{0}$ and solving for x and y is cumbersome, but the unique stationary point for D is

$$\left(\frac{ad - aby_0 - acz_0 + b^2x_0 + c^2x_0}{a^2 + b^2 + c^2}, \frac{bd - abx_0 - bcz_0 + a^2y_0 + c^2y_0}{a^2 + b^2 + c^2}\right).$$

In Exercise 13, you are asked to provide the details of this result. In Exercise 14, you are also asked to show that this point provides an absolute minimum for the function D. In Exercise 15, you are asked to show that this minimal value is $\frac{(ax_0+by_0+cz_0-d)^2}{a^2+b^2+c^2}$, and we have the desired result. □

? TEST YOUR UNDERSTANDING

- What are stationary points? What are critical points? What are saddle points?
- Where do the local extrema of a function of two variables occur? How do we use the first-order partial derivatives to find local extrema?
- How do we use the second-order partial derivatives to classify the stationary points of a function of two variables?
- What is the main idea in the proof of Theorem 12.45 that is used to classify the stationary points of a function of two variables?
- How could the ideas from this section be adapted to find the extrema of functions of three variables? How would the definitions have to be modified? Could Theorem 12.45 be generalized to handle three variables?

EXERCISES 12.6

Thinking Back

- *First-Derivative Test:* Review the first-derivative test for functions of a single variable. Explain how the test works, what conditions a function must satisfy to make the test useful, and when, if ever, the test might fail.
- *Second-Derivative Test:* Review the second-derivative test for functions of a single variable. Explain how the test works, what conditions a function must satisfy to make the test useful, and when, if ever, the test might fail.

Concepts

0. *Problem Zero:* Read the section and make your own summary of the material.

1. *True/False:* Determine whether each of the statements that follow is true or false. If a statement is true, explain why. If a statement is false, provide a counterexample.

(a) *True or False:* If $\nabla f(a, b) = \mathbf{0}$, then f has a stationary point at (a, b).

(b) *True or False:* If (a, b) is a stationary point of a function $f(x, y)$, then (a, b) is a critical point of f.

(c) *True or False:* If (a, b) is a critical point of a function $f(x, y)$, then (a, b) is a stationary point of f.

(d) *True or False:* If (a, b) is a saddle point of a function $f(x, y)$, then (a, b) is a critical point of f.

(e) *True or False:* If $f(x, y)$ has a local maximum at (a, b), then $f(a, b) > f(x, y)$ for every point in some open disk containing (a, b).

(f) *True or False:* The function $f(x, y) = \pi$ has an absolute minimum at every point in $\mathbb{R}^2$.

(g) *True or False:* A function $f(x, y)$ can have an absolute maximum and an absolute minimum at every point in $\mathbb{R}^2$.

(h) *True or False:* If the graph of $f(x, y)$ is a plane and $D_{\mathbf{u}} f(a, b) \neq 0$ for some point (a, b), then f has no extrema.

2. *Examples:* Construct examples of the thing(s) described in the following. Try to find examples that are different than any in the reading.

(a) A function of two variables with a local minimum at $(1, -3)$.

(b) A function of two variables with a point of nondifferentiability at $(4, -5)$.

(c) A function of two variables with a saddle point at $(0, 0)$.

3. How do you find the critical points of a function of two variables, $f(x, y)$? What is the significance of the critical points?

4. What is a stationary point of a function of two variables, $f(x, y)$? What, if anything, is the difference between a critical point and a stationary point of f?

5. What is a saddle point of a function of two variables, $f(x, y)$?

6. Explain why every saddle point is both a stationary point and a critical point.

7. Review the first-derivative test for a function of a single variable, set forth in Theorem 3.8 from Section 3.2. Explain why it might be difficult to design or implement an analogous test for a function of two or more variables.

8. In the proof of Theorem 12.45 we argued that, for a unit vector $\langle \alpha, \beta \rangle$, if $\alpha = 0$, then $\beta \neq 0$. Explain why by finding the only possible values of β.

9. In the proof of Theorem 12.45 we mentioned that if the quantity $AC - B^2 > 0$, then the signs of A and C are the same. Explain why.

Theorem 12.45 is inconclusive when the discriminant, $\det(H_f)$, is zero at a stationary point. In Exercises 10–12 we ask you to illustrate this fact by analyzing three functions of two variables with stationary points at the origin.

10. Show that the function $f(x, y) = x^4 + y^4$ has a stationary point at the origin. Show that the discriminant $\det(H_f(0, 0)) = 0$. Explain why f has an absolute minimum at the origin.

11. Show that the function $g(x, y) = -(x^4 + y^4)$ has a stationary point at the origin. Show that the discriminant $\det(H_g(0, 0)) = 0$. Explain why g has an absolute maximum at the origin.

12. Show that the function $h(x, y) = x^3 + y^3$ has a stationary point at the origin. Show that the discriminant $\det(H_h(0, 0)) = 0$. Show that there are points arbitrarily close to the origin such that $h(x, y) > 0$. Show that there are points arbitrarily close to the origin such that $h(x, y) < 0$. Explain why all this shows that h has a saddle at the origin.

In Exercises 13–16 we ask you to complete some of the details we omitted in Example 4.

13. Show that the unique solution of the system of equations

$$2(x - x_0) - \frac{2a}{c}\left(\frac{d - ax - by}{c} - z_0\right) = 0 \quad \text{and}$$

$$2(y - y_0) - \frac{2b}{c}\left(\frac{d - ax - by}{c} - z_0\right) = 0$$

is

$$\left(\frac{ad - aby_0 - acz_0 + b^2x_0 + c^2x_0}{a^2 + b^2 + c^2}, \frac{bd - abx_0 - bcz_0 + a^2y_0 + c^2y_0}{a^2 + b^2 + c^2}\right).$$

14. Use Theorem 12.45 to show that the point

$$\left(\frac{ad - aby_0 - acz_0 + b^2x_0 + c^2x_0}{a^2 + b^2 + c^2}, \frac{bd - abx_0 - bcz_0 + a^2y_0 + c^2y_0}{a^2 + b^2 + c^2}\right)$$

provides an absolute minimum for the function

$$D(x, y) = (x - x_0)^2 + (y - y_0)^2 + \left(\frac{d - ax - by}{c} - z_0\right)^2.$$

15. Show that the minimal value of

$$D(x, y) = (x - x_0)^2 + (y - y_0)^2 + \left(\frac{d - ax - by}{c} - z_0\right)^2,$$

is $\frac{(ax_0 + by_0 + cz_0 - d)^2}{a^2 + b^2 + c^2}$ by evaluating

$$D\left(\frac{ad - aby_0 - acz_0 + b^2x_0 + c^2x_0}{a^2 + b^2 + c^2}, \frac{bd - abx_0 - bcz_0 + a^2y_0 + c^2y_0}{a^2 + b^2 + c^2}\right).$$

16. Let $\mathcal{P}$ be the plane $ax + by + cz = d$, $\mathbf{N} = \langle a, b, c \rangle$ be the normal vector to $\mathcal{P}$, R be a point on $\mathcal{P}$, and P be the point (x_0, y_0, z_0). Show that the distance formula we derived for computing the distance from point P to plane $\mathcal{P}$ in Chapter 10, $\frac{|\mathbf{N} \cdot \overrightarrow{RP}|}{\|\mathbf{N}\|}$, is equivalent to the distance formula we derived in Example 4. That is, show that

$$\frac{|\mathbf{N} \cdot \overrightarrow{RP}\,|}{\|\mathbf{N}\|} = \frac{|ax_0 + by_0 + cz_0 - d|}{\sqrt{a^2 + b^2 + c^2}}.$$

17. Let P be the point (x_1, y_1, z_1), P_0 be the point (x_0, y_0, z_0), $\mathbf{d}$ be the vector $\langle a, b, c \rangle$, and $\mathcal{L}$ be the line parametrized by $\mathbf{r}(t) = \mathbf{P}_0 + t\mathbf{d}$. In Chapter 10 we showed that the distance from P to $\mathcal{L}$ is $\frac{\|\mathbf{d} \times \overrightarrow{P_0P}\|}{\|\mathbf{d}\|}$. Explain how to derive this distance formula by minimizing a function of a single variable.

18. Show that $f(x, y) = \sqrt{x^2 + y^2}$ has a critical point at $(0, 0)$. Explain why f has an absolute minimum at $(0, 0)$ and why you cannot use Theorem 12.45 to show this.

19. Every function of a single variable, $f(x) = x^n$, where n is a positive integer greater than 1 will have a critical point at $x = 0$. For which values of n will there be a relative minimum at $x = 0$? For which values of n will there be an inflection point at $x = 0$? Are there any other possibilities for the behavior of the function at $x = 0$?

20. Consider a function of two variables, $f(x, y) = x^m y^n$, where m and n are positive integers. What conditions do m and n have to satisfy in order for f to have a relative minimum at the origin? What conditions do m and n have to satisfy in order for f to have a saddle point at the origin? Are there any other possibilities for the behavior of the function at the origin? It may be helpful to refer to your answers to Exercise 19 before you answer this question.

Skills

In Exercises 21–26, find the discriminant of the given function.

21. $f(x,y) = e^{2x}\cos y$

22. $g(x,y) = e^{-3x}\sin(2y)$

23. $f(s,t) = s^3 e^{t/2}$

24. $g(s,t) = s^2 t e^{-st^2}$

25. $f(\theta,\phi) = \cos\theta\sin\phi$

26. $g(\theta,\phi) = \cos\theta\cos\phi$

In Exercises 27–30, use the result from Example 4 to find the distance from the point P to the given plane.

27. $P = (3,5,-2)$, $x - 3y + 4z = 8$

28. $P = (-1,3,5)$, $5x - 2y - 7z + 2 = 0$

29. $P = (4,0,-3)$, $12y - 5z = 7$

30. $P = (-3,2,6)$, $x = 3y + 5z$

In Exercises 31–52, find the relative maxima, relative minima, and saddle points for the given functions. Determine whether the function has an absolute maximum or absolute minimum as well.

31. $f(x,y) = 3x^2 + 3x + 6y^2 - 7$

32. $g(x,y) = 4x^2 - 8xy + y^2 + 3y + 5$

33. $f(x,y) = x^3 + y^3 - 12x - 3y + 15$

34. $g(x,y) = x^3 + 6x^2 + 6y^2 - 4$

35. $f(x,y) = x^3 - 12xy + y^3$

36. $g(x,y) = x^4 - 8x^2y + y^4$

37. $f(x,y) = 2x^2 - 2xy + 4y^2 + 3x + 9y - 5$

38. $g(x,y) = 8x^3 - 3xy^2 + 2y^3 - 4x - 2$

39. $f(x,y) = 8xy + \dfrac{1}{x} + \dfrac{1}{y}$

40. $g(x,y) = e^x\cos y$

41. $f(x,y) = x\sin y$

42. $g(x,y) = x^2 + \sin y - 3$

43. $f(x,y) = e^{x^2}\cos y$

44. $g(x,y) = 5x - 4y^2 + x\ln y,\ y > 0$

45. $f(x,y) = e^{xy}$

46. $g(x,y) = \dfrac{1}{x^2+y^2}$

47. $f(x,y) = \dfrac{1}{x^2+y^2-1}$

48. $g(x,y) = \dfrac{1}{x^2+y^2+1}$

49. $f(x,y) = x^2y^2$

50. $g(x,y) = xy^2$

51. $f(x,y) = x^3y^2$

52. $g(x,y) = x^3y$

Applications

53. Bob plans to build a rectangular wooden box. The plywood he will use for the bottom costs \$2 per square foot. The pine for the sides costs \$5 per square foot, and the oak for the top costs \$7 per square foot. What should the dimensions of the box be to minimize the cost of a box with a volume of 20 cubic feet?

54. Bob plans to build a rectangular wooden box. The wood he will use for the bottom costs \$$A$ per square foot. The wood for the sides costs \$$B$ per square foot, and the wood for the top costs \$$C$ per square foot. What should the dimensions of the box be to minimize the cost of a box with a volume of V cubic feet?

Proofs

55. Prove that for a function $f(x,y)$ with continuous second-order partial derivatives, the second directional derivative in the direction of the unit vector $\mathbf{u} = \langle a,b\rangle$ is given by

$$D^2_{\mathbf{u}} f(x,y) = a^2 f_{xx}(x,y) + 2abf_{xx}(x,y) + b^2 f_{yy}(x,y).$$

56. Prove Theorem 12.42. That is, show that if $f(x,y)$ has a local extremum at (x_0,y_0), then (x_0,y_0) is a critical point of f. (*Hint: Consider the functions of a single variable, $g(x) = f(x,y_0)$ and $h(y) = f(x_0,y)$.*)

57. Prove that if the square of the distance from the point (x_0,y_0) to the line with equation $\alpha x + \beta y = \gamma$ is minimized, then the distance from (x_0,y_0) to the line is also minimized.

58. Prove that if the square of the distance from the point (x_0,y_0) to the curve with equation $g(x,y) = 0$ is minimized, then the distance from (x_0,y_0) to the curve is also minimized.

Thinking Forward

- *A function without extrema:* What is the domain of the function $f(x,y) = 3x - 4y$? Explain why $f(x,y)$ has no local extrema on its domain.
- *Extrema on a closed and bounded set:* How could we find the maximum and minimum values of the function $f(x,y) = 3x - 4y$ if we restrict the domain to the disk $x^2 + y^2 \le 1$?

12.7 LAGRANGE MULTIPLIERS

- Constrained extrema
- The method of Lagrange multipliers
- The Extreme Value Theorem for functions of two or more variables

Constrained Extrema

In Example 3 of the previous section, we discussed how to construct a box with five sides made of wood and one side made of glass. We saw that if the wood costs \$5 per square foot and the glass costs \$10 per square foot, the cost of constructing the box is $C(x, y) = 10xy + 10xz + 15yz$. In Section 12.6, we discussed how to optimize functions of two independent variables. Fortunately, we were also told that the volume of the box was to be 12 cubic feet, giving the ***constraint*** equation $12 = xyz$. This constraint equation allowed us to eliminate one of the variables. In the example, we showed that the values $x = 3$ and $y = z = 2$ together minimize the function subject to the constraint. Now, consider the gradient of the function $C(x, y, z) = 10xy + 10xz + 15yz$ at the point $(3, 2, 2)$. Since

$$\nabla C(x, y, z) = (10y + 10z)\mathbf{i} + (10x + 15z)\mathbf{j} + (10x + 15y)\mathbf{k},$$

we have $\nabla C(3, 2, 2) = 40\mathbf{i} + 60\mathbf{j} + 60\mathbf{k}$. In addition, if we write the constraint equation in the form $g(x, y, z) = xyz - 12 = 0$, we may also compute the gradient of this function at the point $(3, 2, 2)$. When we do, we obtain

$$\nabla g(x, y, z) = yz\mathbf{i} + xz\mathbf{j} + xy\mathbf{k} \quad \text{and} \quad \nabla g(3, 2, 2) = 4\mathbf{i} + 6\mathbf{j} + 6\mathbf{k}.$$

Note that $\nabla C(3, 2, 2)$ is a multiple of $\nabla g(3, 2, 2)$. This is not a coincidence; it is a consequence of the main theorem of this section, presented next. The technique we discuss here is called the ***method of Lagrange multipliers***, or, more simply, ***Lagrange's method***. This technique tells us that the gradient of a function of several variables is always a scalar multiple of the gradient of the constraint equation at a point that optimizes the function.

THEOREM 12.46 **The Method of Lagrange Multipliers**

Let f and g be functions with continuous first-order partial derivatives. If $f(x, y)$ has a relative extremum at a point (x_0, y_0) subject to the constraint $g(x, y) = 0$, then $\nabla f(x_0, y_0)$ and $\nabla g(x_0, y_0)$ are parallel. Thus, if $\nabla g(x_0, y_0) \neq \mathbf{0}$,

$$\nabla f(x_0, y_0) = \lambda \nabla g(x_0, y_0)$$

for some scalar λ.

Although we have stated Theorem 12.46 in terms of functions of two variables, it is also true if f and g are both functions of three or more variables.

DEFINITION 12.47 **Lagrange Multiplier**

The scalar λ mentioned in Theorem 12.46 is called a ***Lagrange multiplier***.

The following figure shows a schematic illustrating Theorem 12.46.

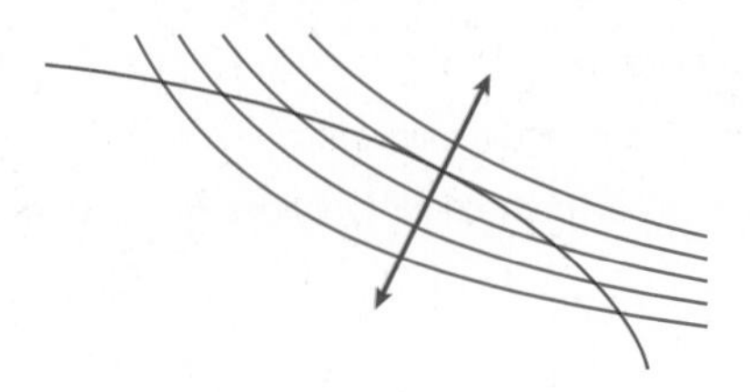

The blue curves are level curves for a function $f(x, y)$. The red curve is the graph of a constraint equation $g(x, y) = 0$. At an extreme of f subject to the constraint, the gradients of both f and g are parallel. These are the green vectors shown in the figure.

Before we prove Theorem 12.46, we again consider our earlier example, to see how Lagrange's method works. We optimize the function $C(x, y, z) = 10xy+10xz+15yz$ subject to the constraint $g(x, y, z) = xyz - 12 = 0$. We've seen that

$$\nabla C(x, y, z) = (10y+10z)\mathbf{i}+(10x+15z)\mathbf{j}+(10x+15y)\mathbf{k} \text{ and } \nabla g(x, y, z) = yz\mathbf{i}+xz\mathbf{j}+xy\mathbf{k}.$$

By Theorem 12.46, we must then have

$$\nabla C(x, y, z) = \lambda \nabla g(x, y, z)$$

for some constant λ. This equation gives rise to the following system of three equations with four unknowns:

$$\begin{aligned} 10y + 10z &= \lambda yz, \\ 10x + 15z &= \lambda xz, \quad \text{and} \\ 10x + 15y &= \lambda xy. \end{aligned}$$

We are unable find a unique solution for such a system. We need a fourth equation relating the variables. Fortunately we have the constraint equation $xyz = 12$. The resulting system of four equations in four unknowns is nonlinear, since each of the original three equations has a term that is the product of three variables. Solving a nonlinear system can require numerical tricks. Indeed, many nonlinear systems of equations are impossible to solve exactly. Here, a little perseverance does pay off. It is often helpful to solve each of the equations involving λ for λ and equate the results. When we do this for the preceding system, we obtain

$$\lambda = \frac{10y + 10z}{yz} = \frac{10x + 15z}{xz} = \frac{10x + 15y}{xy},$$

or equivalently,

$$\frac{10}{z} + \frac{10}{y} = \frac{10}{z} + \frac{15}{x} = \frac{10}{y} + \frac{15}{x}.$$

Since $\frac{10}{z} + \frac{15}{x} = \frac{10}{y} + \frac{15}{x}$, we must have $y = z$, and since $\frac{10}{z} + \frac{10}{y} = \frac{10}{z} + \frac{15}{x}$, we must also have $y = z = \frac{2}{3}x$. Now, because $xyz = 12$, it follows that $x = 3$ and $y = z = 2$, the same result we obtained in Section 12.6. Note that we did not solve the system for λ: The actual value of λ is not significant.

We are now ready to prove Theorem 12.46.

Proof. Let $z = f(x, y)$ be the function of two variables that we wish to optimize subject to the constraint $g(x, y) = 0$. Assume that $\nabla g(x, y) \neq \mathbf{0}$ for any point in the domain of g, let C be the curve determined by $g(x, y) = 0$, and let $\mathbf{r}(t) = \langle x(t), y(t) \rangle$ be a parametrization for C. Along the curve C, z may be expressed as a function of the single variable t: $z = f(x(t), y(t))$. By the chain rule, the derivative $\frac{dz}{dt}$ is given by

$$\frac{dz}{dt} = \frac{\partial f}{\partial x}\frac{dx}{dt} + \frac{\partial f}{\partial y}\frac{dy}{dt},$$

or equivalently,

$$\frac{dz}{dt} = f_x(x, y)x'(t) + f_y(x, y)y'(t).$$

But note that $\nabla f(x, y) = f_x(x, y)\mathbf{i} + f_y(x, y)\mathbf{j}$ and $\mathbf{r}'(t) = x'(t)\mathbf{i} + y'(t)\mathbf{j}$ is the tangent function to the curve C. Now, if t_0 is a point that optimizes f subject to the constraint $g(x, y) = 0$, then the derivative $\left.\frac{dz}{dt}\right|_{t=t_0} = 0$. So, at t_0, the dot product $\nabla f(x(t_0), y(t_0)) \cdot \mathbf{r}'(t_0) = 0$. Thus, the vectors $\nabla f(x(t_0), y(t_0))$ and $\mathbf{r}'(t_0)$ are orthogonal. But by Theorem 12.38, $\nabla g(x(t_0), y(t_0))$ is also orthogonal to C at $(x(t_0), y(t_0))$. Therefore, $\nabla f(x(t_0), y(t_0))$ and $\nabla g(x(t_0), y(t_0))$ are parallel, since they are orthogonal to the same vector. ■

Optimizing a Function on a Closed and Bounded Set

Recall that the Extreme Value Theorem tells us that every continuous function on a closed interval will have both a maximum value and a minimum value on the interval. We have the following analogous theorem for functions of two variables:

THEOREM 12.48 **The Extreme Value Theorem for a Function of Two Variables**

Let f be a continuous function of two variables defined on the closed and bounded set S. Then there exist points (x_M, y_M) and (x_m, y_m) in S such that $f(x_M, y_M)$ is the maximum value of f on S and $f(x_m, y_m)$ is the minimum value of f on S.

The proof of Theorem 12.48 lies outside the scope of this text. An analogous theorem also holds when f is a function of more than two variables, and we will use that theorem when appropriate.

Given a continuous function f defined on a closed and bounded set S, we may use the following outline to find those points (x_M, y_M) and (x_m, y_m) which maximize and minimize f on S:

1. Find the stationary points and other critical points of f.
2. Select only those critical points that lie in S.
3. Evaluate the function at each of the critical points found in step 2.
4. Use the method of Lagrange multipliers to locate the points on the boundary of S that maximize and minimize f.
5. Evaluate the function at each of the critical points on the boundary of S.
6. Use the extrema from steps 3 and 5 to find the maximum and minimum values of the function f on S.

Optimizing a Function with Two Constraints

We may extend the ideas of this section to optimize a function of three variables, $w = f(x, y, z)$, subject to two constraints. In this context, the constraints, $g(x, y, z) = 0$ and $h(x, y, z) = 0$, define surfaces in $\mathbb{R}^3$. Assuming that these surfaces intersect in a curve, we would be attempting to find the maximum and minimum values of f on this curve of intersection. To do this, we attempt to solve the system given by

$$\nabla f(x_0, y_0, z_0) = \lambda \nabla h(x_0, y_0, z_0) + \mu \nabla h(x_0, y_0, z_0),$$

where λ and μ are both Lagrange multipliers. (See Example 5.)

Examples and Explorations

EXAMPLE 1 **Optimizing a function subject to a constraint**

Find the maximum and minimum of the function $f(x, y) = xy$ subject to the constraint $x^2 + 4y^2 = 16$.

SOLUTION

We could solve the equation $x^2 + 4y^2 = 16$ for either x or y and use the resulting equation to rewrite the function f in terms of a single variable. However, we will use Lagrange's method to optimize the function. We start by writing the constraint in the form $g(x, y) = x^2 + 4y^2 - 16 = 0$ and find

$$\nabla f(x, y) = y\mathbf{i} + x\mathbf{j} \quad \text{and} \quad \nabla g(x, y) = 2x\mathbf{i} + 8y\mathbf{j}.$$

We now set up the system of equations determined by $\nabla f(x, y) = \lambda \nabla g(x, y)$, where λ is a constant. Here we have

$$y = 2\lambda x \quad \text{and} \quad x = 8\lambda y.$$

Our third equation relating the variables is the constraint equation $x^2 + 4y^2 = 16$. We solve each of the two equations for λ and equate the results:

$$\frac{y}{2x} = \lambda = \frac{x}{8y}.$$

We may now deduce that $x^2 = 4y^2$. Using this relationship in the constraint equation, we find that $2x^2 = 16$. Therefore, $x = \pm 2\sqrt{2}$, and from this we have $y = \pm\sqrt{2}$. Thus, we have the four pairs of coordinates

$$(2\sqrt{2}, \sqrt{2}),\ (2\sqrt{2}, -\sqrt{2}),\ (-2\sqrt{2}, \sqrt{2}), \quad \text{and} \quad (-2\sqrt{2}, -\sqrt{2}).$$

Finally, we evaluate the function $f(x, y)$ at each of these points and obtain

$$f(2\sqrt{2}, \sqrt{2}) = 4,\ f(2\sqrt{2}, -\sqrt{2}) = -4,\ f(-2\sqrt{2}, \sqrt{2}) = -4, \quad \text{and} \quad f(-2\sqrt{2}, -\sqrt{2}) = 4.$$

The maximum value of f subject to the constraint is 4. This maximum occurs at both $(2\sqrt{2}, \sqrt{2})$ and $(-2\sqrt{2}, -\sqrt{2})$. The minimum value of the function is -4. The minimum occurs at both $(-2\sqrt{2}, \sqrt{2})$ and $(2\sqrt{2}, -\sqrt{2})$. □

We may check our answers by evaluating the gradients of f and g at the points we just found and showing that they are indeed parallel. For example, since

$$\nabla f(x, y) = y\mathbf{i} + x\mathbf{j} \quad \text{and} \quad \nabla g(x, y) = 2x\mathbf{i} + 8y\mathbf{j},$$

we have

$$\nabla f(2\sqrt{2}, \sqrt{2}) = \sqrt{2}\mathbf{i} + 2\sqrt{2}\mathbf{j} \quad \text{and} \quad \nabla g(2\sqrt{2}, \sqrt{2}) = 4\sqrt{2}\mathbf{i} + 8\sqrt{2}\mathbf{j}.$$

Note that $\nabla f(2\sqrt{2}, \sqrt{2}) = \frac{1}{4}\nabla g(2\sqrt{2}, \sqrt{2})$. The gradients are indeed parallel when $(x, y) = (2\sqrt{2}, \sqrt{2})$. Similarly, at each of the other three points we found in Example 1, we have $\nabla f(x_0, y_0) = \frac{1}{4}\nabla g(x_0, y_0)$.

EXAMPLE 2

Analyzing the result of an optimization problem

Analyze the result of Example 1 by graphing the level curves of the function $f(x, y) = xy$ corresponding to the constrained extrema, along with the constraint equation $x^2 + 4y^2 = 16$.

SOLUTION

In Example 1 we found that the maximum and minimum of $f(x, y) = xy$ subject to the constraint $x^2 + 4y^2 = 16$ were 4 and -4, respectively. We graph the level curves $xy = 4$ and $xy = -4$ along with the following ellipse, which is the graph of the equation $x^2 + 4y^2 = 16$:

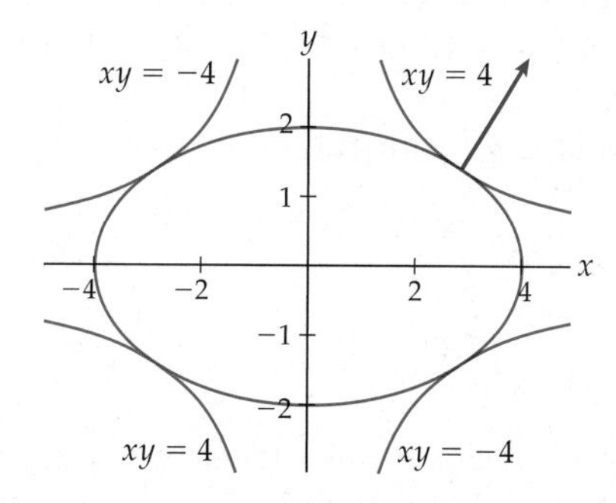

Note that the level curves and the ellipse are tangent at the maxima and minima. In the first and third quadrants, the level curve is the graph of $xy = 4$. This graph intersects the ellipse at $(2\sqrt{2}, \sqrt{2})$ and $(-2\sqrt{2}, -\sqrt{2})$. Since the curves are tangent at these two points, the gradients of the two curves would be orthogonal to the curves at those points. Therefore, the gradients would be parallel at those points. We show only one of these gradients in the preceding figure, because they overlap.

The figure that follows at the left shows the ellipse, together with four additional level curves of the function f. Note that the ellipse is tangent only to the level curves $f(x, y) = 4$ and $f(x, y) = -4$.

$x^2 + 4y^2 = 16$, along with the level curves $f(x, y) = \pm 2, \pm 4, \pm 6$

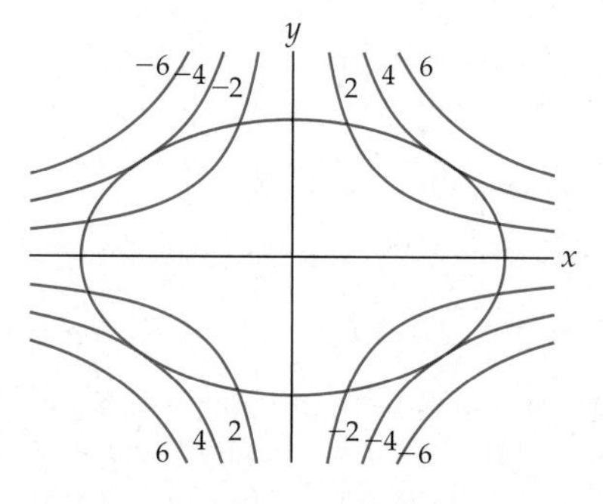

The graph of $f(x, y) = xy$ restricted to $x^2 + 4y^2 = 16$

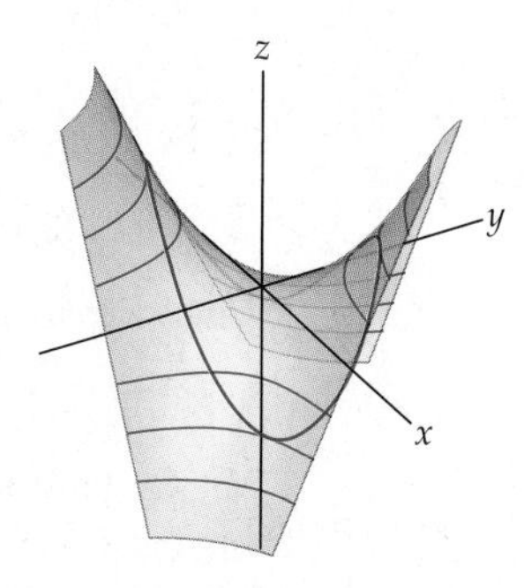

The figure at the right shows the graph of the surface defined by $f(x, y) = xy$, together with the curve on that surface resulting from the constraint $x^2 + 4y^2 = 16$. On this graph, you see the two maxima and two minima that we found. □

EXAMPLE 3

Understanding the difference between a function and a constraint

Find the maximum and minimum of the function $f(x, y) = x^2 + 4y^2 - 16$ subject to the constraint $xy = 0$.

SOLUTION

Here we have reversed the roles of the function and the constraint given in Example 1. We could proceed as outlined before, but note that since we now have the constraint $xy = 0$, either x or y must be zero. If $x = 0$, then $f(0, y) = 4y^2 - 16$, which has minimum value -16 when $y = 0$, but no maximum value, since the function increases without bound as $y \to \pm\infty$. Similarly, if $y = 0$, then $f(x, 0) = x^2 - 16$. Again, there is a minimum of -16 when $x = 0$ and no maximum value. Therefore, subject to the constraint, the function $f(x, y) = x^2 + 4y^2 - 16$ has a minimum of $-16 = f(0, 0)$ and no maximum value.

Our point here is that if you confuse the function and the constraint, you may obtain vastly different results. Make sure that you understand which function you are trying to optimize and which function is your constraint. □

EXAMPLE 4

Optimizing a function on a closed and bounded domain

Find the maximum and minimum of the function $f(x, y) = x^3 + 9x^2 + 6y^2$ on the region $x^2 + y^2 \leq 9$.

SOLUTION

Since f is a polynomial, the only critical points of f are stationary points. We have $\nabla f(x, y) = (3x^2 + 18x)\mathbf{i} + 12y\mathbf{j}$. The stationary points of f will satisfy the equations

$$3x(x + 6) = 0 \quad \text{and} \quad 12y = 0.$$

The only points satisfying these two equations are $(0, 0)$ and $(-6, 0)$. Of these points, only $(0, 0)$ satisfies the inequality $x^2 + y^2 \leq 9$, so it is a point where the maximum or minimum may occur. We ignore the other point, since it is outside the domain.

We now consider points on the boundary of the region. These points are given by the constraint $x^2 + y^2 = 9$. We let $g(x, y) = x^2 + y^2 - 9 = 0$. Now, $\nabla g(x, y) = 2x\mathbf{i} + 2y\mathbf{j}$. By Lagrange's method, we have the system of equations $3x^2 + 18x = 2\lambda x$ and $12y = 2\lambda y$. From the second of these equations, we see that either $y = 0$ or $\lambda = 6$. If $y = 0$, then by the constraint equation, $x = \pm 3$. Using $\lambda = 6$ in the first of the equations, we have $3x^2 + 6x = 0$, so either $x = -2$ or $x = 0$. Again using the constraint when $x = -2$, we obtain $y = \pm\sqrt{5}$, and when $x = 0$, we have $y = \pm 3$.

We now have a list of seven points where the maximum and minimum of the function on the given domain may occur:

$$(0, 0),\ (3, 0),\ (-3, 0),\ (-2, \sqrt{5}),\ (-2, -\sqrt{5}),\ (0, 3), \quad \text{and} \quad (0, -3).$$

When we evaluate the function at these seven points, we see that

$$f(0, 0) = 0,\ f(3, 0) = 108,\ f(-3, 0) = 54,\ f(-2, \pm\sqrt{5}) = 58, \quad \text{and} \quad f(0, \pm 3) = 54.$$

We see that the maximum of f on the region is 108 and the minimum of the function on the region is 0. Note that it was *not* necessary to use the second-derivative test to analyze f at $(0, 0)$, since we were looking for the maximum and minimum of the function on the given region and we had to analyze the behavior of f only at the finite number of points we obtained in the previous steps. When we evaluated the function at the seven points we found, the minimum value of the function and the maximum value of the function at those points had to give the two values we wanted.

Although we successfully answered the question posed in this example, in Exercise 13 you will show that the function $f(x, y) = x^3 + 9x^2 + 6y^2$ with domain $\mathbb{R}^2$ has only a relative minimum at $(0, 0)$, not an absolute minimum, and that f has a saddle point at the other stationary point, $(-6, 0)$. □

EXAMPLE 5 **Optimizing a function of three variables with two constraints**

Find the points in $\mathbb{R}^3$ that are farthest from the origin and that lie on both the cylinder $y^2 + z^2 = 1$ and the plane $x + 2y + 3z = 0$.

SOLUTION

As usual, it is sufficient (and easier) to maximize the square of the distance from the origin, $D(x, y, z) = x^2 + y^2 + z^2$, subject to the constraints $g(x, y, z) = y^2 + z^2 - 1 = 0$ and $h(x, y, z) = x + 2y + 3z = 0$. As the following figure illustrates, geometrically we are looking for the point(s) on the ellipse formed by the intersection of the cylinder and plane farthest from the origin.

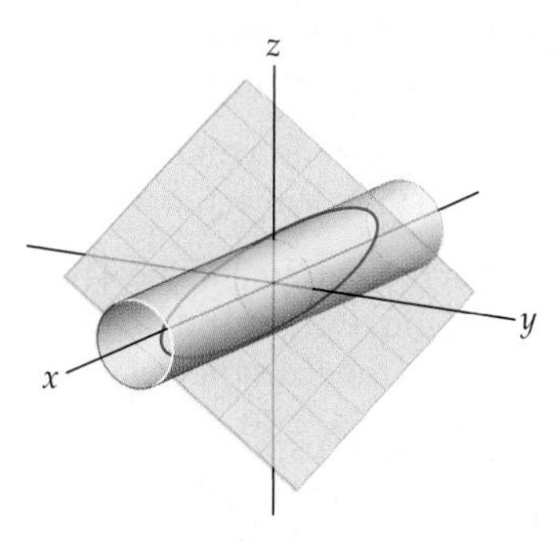

The gradients of the three functions we are interested in are

$$\begin{aligned} \nabla D(x, y, z) &= 2x\mathbf{i} + 2y\mathbf{j} + 2z\mathbf{k}, \\ \nabla g(x, y, z) &= 2y\mathbf{j} + 2z\mathbf{k}, \quad \text{and} \\ \nabla h(x, y, z) &= \mathbf{i} + 2\mathbf{j} + 3\mathbf{k}. \end{aligned}$$

We wish to solve the system $\nabla D(x, y, z) = \lambda \nabla g(x, y, z) + \mu \nabla h(x, y, z)$ subject to $g(x, y, z) = 0$ and $h(x, y, z) = 0$. This leads to the following system of five equations:

$$\begin{aligned} 2x &= 0 + \mu, \\ 2y &= \lambda \cdot 2y + \mu \cdot 2, \\ 2z &= \lambda \cdot 2z + \mu \cdot 3, \\ y^2 + z^2 &= 1, \text{ and} \\ x + 2y + 3z &= 0. \end{aligned}$$

Since $\mu = 2x$, we may eliminate μ to obtain a system with only four equations in four variables:

$$y = \lambda y + 2x, \quad z = \lambda z + 3x, \quad y^2 + z^2 = 1, \quad x + 2y + 3z = 0.$$

A little thought tells us that if either $y = 0$ or $z = 0$, the system does not have a solution. So we may solve the first two of these equations for λ and equate the results. We obtain

$$\frac{y - 2x}{y} = \frac{z - 3x}{z},$$

or equivalently, $x(2z - 3y) = 0$. So, either $x = 0$ or $z = \frac{3}{2}y$.

If $x = 0$, then when we solve the equations $y^2 + z^2 = 1$ and $2y + 3z = 0$ for y and z, we have $(x, y, z) = \left(0, \frac{3}{\sqrt{13}}, -\frac{2}{\sqrt{13}}\right)$ or $(x, y, z) = \left(0, -\frac{3}{\sqrt{13}}, \frac{2}{\sqrt{13}}\right)$. These two points are both 1 unit from the origin.

Now, using $z = \frac{3}{2}y$ we solve the system $y^2 + \left(\frac{3}{2}y\right)^2 = 1$ and $x + 2y + 3 \cdot \frac{3}{2}y = 0$ and obtain the points $\left(-\sqrt{13}, \frac{2}{\sqrt{13}}, \frac{3}{\sqrt{13}}\right)$ and $\left(\sqrt{13}, -\frac{2}{\sqrt{13}}, -\frac{3}{\sqrt{13}}\right)$. These two points are each $\sqrt{14}$ units from the origin. They are the two points on the cylinder and plane that are farthest from the origin. The other two points are the points on the cylinder and plane that are closest to the origin. □

? TEST YOUR UNDERSTANDING

- What is a constraint equation? Does every continuous function have a maximum value and minimum value subject to a constraint?
- What is a Lagrange multiplier? What is the method of Lagrange multipliers? If we try to optimize the function of three variables, $w = f(x, y, z)$, subject to the constraint $g(x, y, z) = 0$, how many equations do we obtain when we use the method of Lagrange multipliers? How many variables are in the system?
- What methods do you know to find the maxima and minima of a function of two or more variables subject to a constraint?
- What is the Extreme Value Theorem for a function of two variables? Three variables? What is the relationship between the Extreme Value Theorem for a function of a single variable to the Extreme Value Theorem for a function of two or more variables?
- Outline the steps you should take to find the maximum value and minimum value of a function of two variables on a closed and bounded set. Would the steps change if you had a function of three variables on a closed and bounded set?

EXERCISES 12.7

Thinking Back

- *Optimizing a function of two variables subject to a constraint:* If you wish to find the maximum and minimum of the function $f(x, y) = xy$ subject to the constraint $x^2 + 4y^2 = 16$, you may eliminate one variable of f by solving the constraint equation for either x or y and rewriting f in terms of a single variable. Do this and then use the techniques of Chapter 3 to find the maximum and minimum of the resulting function.
- *Optimizing a function of three variables subject to a constraint:* If you wish to find the maximum and minimum of the function $f(x, y, z) = xyz$ subject to the constraint $x^2 + 4y^2 + 9z^2 = 16$, you may eliminate one variable of f by solving the constraint equation for x, y, or z and rewriting f in terms of two variables. Do this and then use the techniques of Section 12.6 to find the maximum and minimum of the resulting function.

Concepts

0. *Problem Zero:* Read the section and make your own summary of the material.

1. *True/False:* Determine whether each of the statements that follow is true or false. If a statement is true, explain why. If a statement is false, provide a counterexample.

(a) *True or False:* Every function of two variables $f(x, y)$ subject to a constraint $g(x, y) = 0$ has both a maximum and a minimum value.

(b) *True or False:* The function $f(x, y, z) = \frac{xy}{y+z}$ has both a minimum and a maximum on the region defined by $x^2 + y^2 + z^2 \leq 1$.

(c) *True or False:* The function $f(x, y, z) = e^{\sin xy}\cos(xy+z)$ has both a minimum and a maximum on the region defined by $x^2 + y^2 + z^2 \leq 1$.

(d) *True or False:* The function $f(x, y) = xy + 2y - 5xy^2$ has both a minimum and a maximum on the region defined by $x^2 + y^2 \geq 1$.

(e) *True or False:* The function $y = x^2$ has both a maximum and a minimum for $x \in (-2, 2)$.

(f) *True or False:* If $f(x_0, y_0)$ is the maximum value of f on a region $\mathcal{R}$, then $f(x_0, y_0)$ is the maximum value of f.

(g) *True or False:* If $f(x_0, y_0)$ is the maximum value of f on its domain, then $f(x_0, y_0)$ is the maximum value of f on every closed and bounded region $\mathcal{R}$.

(h) *True or False:* If $f(x, y)$ is a continuous function and if m and M are the minimum and maximum values of f on its domain, then $m \leq f(x, y) \leq M$ on every closed and bounded region $\mathcal{R}$.

2. *Examples:* Construct examples of the thing(s) described in the following. Try to find examples that are different than any in the reading.
   (a) A function of two variables $f(x, y)$, along with a constraint $g(x, y) = 0$, that could be optimized with the method of Lagrange multipliers.
   (b) A function of two variables $f(x, y)$, along with a constraint $g(x, y) = 0$, that could be optimized by eliminating one of the variables.
   (c) A function of two variables $f(x, y)$ that has neither a maximum nor a minimum value on the set $x^2 + y^2 \leq 4$.
3. What is the distinction between the function $z = f(x, y)$ and the constraint equation $g(x, y) = 0$ when we use the method of Lagrange multipliers?
4. Outline a method for optimizing a function of two variables, $z = f(x, y)$.
5. Describe two methods for optimizing a function $z = f(x, y)$ subject to a constraint.
6. What is meant by a "closed" subset of $\mathbb{R}^2$? What is meant by a "bounded" subset of $\mathbb{R}^2$?
7. Give an example of each of the following:
   (a) A subset of $\mathbb{R}^2$ that is neither closed nor bounded.
   (b) A subset of $\mathbb{R}^2$ that is closed but not bounded.
   (c) A subset of $\mathbb{R}^2$ that is not closed but is bounded.
   (d) A subset of $\mathbb{R}^2$ that is closed and bounded.
8. What is meant by a "closed" subset of $\mathbb{R}^3$? What is meant by a "bounded" subset of $\mathbb{R}^3$?
9. Give an example of each of the following:
   (a) A subset of $\mathbb{R}^3$ that is neither closed nor bounded.
   (b) A subset of $\mathbb{R}^3$ that is closed but not bounded.
   (c) A subset of $\mathbb{R}^3$ that is not closed but is bounded.
   (d) A subset of $\mathbb{R}^3$ that is closed and bounded.
10. Outline the steps required to find the minimum and maximum of a function of two variables on a closed and bounded set $\mathcal{R}$.
11. Given a function of three variables, $w = f(x, y, z)$, and a constraint equation $g(x, y, z) = 0$, how many equations would we obtain if we tried to optimize f by the method of Lagrange multipliers?
12. Given a function of n variables, $z = f(x_1, x_2, \ldots, x_n)$, and a constraint equation, $g(x_1, x_2, \ldots, x_n) = 0$, how many equations would we obtain if we tried to optimize f by the method of Lagrange multipliers?
13. In Example 4 we found that the function $f(x, y) = x^3 + 9x^2 + 6y^2$ has stationary points at $(0, 0)$ and $(-6, 0)$.
   (a) Use the second-derivative test to show that f has a saddle point at $(-6, 0)$.
   (b) Use the second-derivative test to show that f has a relative minimum at $(0, 0)$.
   (c) Use the value of $f(-10, 0)$ to argue that f has a relative minimum at $(0, 0)$, and not an absolute minimum, without using the second-derivative test.

In Exercises 14–17, by considering the function $f(x, y) = x^2y$ subject to the constraint $x+y = 0$, you will explore a situation in which the method of Lagrange multipliers does not provide an extremum of a function.

14. Show that the only point given by the method of Lagrange multipliers for the function $f(x, y) = x^2y$ subject to the constraint $x + y = 0$ is $(0, 0)$.
15. Explain why $(0, 0)$ is *not* an extremum of $f(x, y) = x^2y$ subject to the constraint $x + y = 0$.
16. Why does the method of Lagrange multipliers fail with this function?
17. Optimize $f(x, y) = x^2y$ subject to the constraint $ax + by = 0$ for nonzero constants a and b. Are there any nonzero values of a and b for which the method of Lagrange multipliers succeeds?
18. Explain how you could use the method of Lagrange multipliers to find the extrema of a function of two variables, $f(x, y)$, subject to the constraint that (x, y) is on the boundary of the rectangle $\mathcal{R}$ defined by $a \leq x \leq b$ and $c \leq y \leq d$.
19. Explain the steps you would take to find the extrema of a function of two variables, $f(x, y)$, if (x, y) is a point in the rectangle $\mathcal{R}$ defined by $a \leq x \leq b$ and $c \leq y \leq d$.
20. Explain how you could use the method of Lagrange multipliers to find the extrema of a function of two variables, $f(x, y)$, subject to the constraint that (x, y) is a point on the boundary of a triangle $\mathcal{T}$ in the xy-plane.
21. Explain the steps you would take to find the extrema of a function of two variables, $f(x, y)$, if (x, y) is a point in a triangle $\mathcal{T}$ in the xy-plane.
22. When you use the method of Lagrange multipliers to find the maximum and minimum of $f(x, y) = x + y$ subject to the constraint $xy = 1$, you obtain two points. Is there a relative maximum at one of the points and a relative minimum at the other? Which is which?
23. Let $f(x, y)$ be a differentiable function such that $\nabla f(x, y) \neq \mathbf{0}$ for every point in the domain of f, and let $\mathcal{R}$ be a closed, bounded subset of $\mathbb{R}^2$. Explain why the maximum and minimum of f restricted to $\mathcal{R}$ occur on the boundary of $\mathcal{R}$.

Skills

In Exercises 24–32, find the maximum and minimum of the function f subject to the given constraint. In each case explain why the maximum and minimum must both exist.

24. $f(x, y) = x + y$ when $x^2 + y^2 = 16$
25. $f(x, y) = x + y$ when $x^2 + 4y^2 = 16$
26. $f(x, y) = xy$ when $x^2 + y^2 = 16$
27. $f(x, y) = xy$ when $x^2 + 4y^2 = 16$
28. $f(x, y, z) = x + y + z$ when $x^2 + y^2 + z^2 = 9$
29. $f(x, y, z) = x + y + z$ when $x^2 + 4y^2 + 16z^2 = 64$
30. $f(x, y, z) = xyz$ when $x^2 + y^2 + z^2 = 9$

31. $f(x, y, z) = xyz$ when $x^2 + 4y^2 + 16z^2 = 64$

32. $f(x, y, z) = xy$ when $x^2 + y^2 + z^2 = 9$

In Exercises 33–38, find the point on the given curve closest to the specified point. Recall that if you minimize the square of the distance, you have minimized the distance as well.

33. $(0, 0)$ and $x^3 + y^3 = 1$

34. $(0, 0)$ and $x^2 + y^3 = 1$

35. $(0, 0)$ and $\sqrt{x} + \sqrt{y} = 1$

36. $(1, 1)$ and $\sqrt{x} + \sqrt{y} = 2$

37. $(0, 0)$ and $x^n + y^n = 1$ for n a positive, odd integer

38. $(0, 0)$ and $x^n + y^n = 1$ where $n > 2$ is a positive, even integer

In Exercises 39–43, find the point on the given surface closest to the specified point.

39. $xyz = 1$ and $(0, 0, 0)$

40. $x + 2y - 3z = 7$ and $(0, 0, 0)$

41. $ax + by + cz = d$ and $(0, 0, 0)$

42. $ax + by + cz = d$ and (α, β, γ)

43. $x + 2y - 3z = 4$ and $(-1, 5, 3)$

In Exercises 44–49, find the maximum and minimum of the given function on the specified region. Also, give the points where the maximum and minimum occur.

44. $f(x, y) = 3x^2 - 5y^2$ on the circular region $x^2 + y^2 \leq 1$

45. $f(x, y) = 3x^2 + 5y^2$ on the square with vertices $(1, 1)$, $(-1, 1)$, $(-1, -1)$, and $(1, -1)$

46. $f(x, y) = x^2 + y$ on the circular region $x^2 + y^2 \leq 4$

47. $f(x, y) = x^2 + y$ on the square with vertices $(1, 0)$, $(0, 1)$, $(-1, 0)$, and $(0, -1)$

48. $f(x, y) = \frac{x}{y}$ on the circular region $x^2 + (y - 2)^2 \leq 1$

49. $f(x, y) = \frac{x}{y}$ on the rectangle given by $-1 \leq x \leq 1$, $1 \leq y \leq 4$

50. Find the dimensions of the rectangular solid with maximum volume such that all sides are parallel to the coordinate planes; one vertex is at the origin; and the vertex diagonally opposite is in the first octant and on the plane with equation $2x + 5y + 6z = 30$.

51. Find the dimensions of the rectangular solid with maximum volume such that all sides are parallel to the coordinate planes; one vertex is at the origin; and the vertex diagonally opposite is in the first octant and on the plane with equation $ax + by + cz = d$, where a, b, c, and d are positive real numbers.

52. Find the points in $\mathbb{R}^3$ closest to the origin and that lie on the cylinder $x^2 + y^2 = 4$ and the plane $x - 3y + 2z = 6$.

53. Find the points in $\mathbb{R}^3$ closest to the origin and that lie on the cone $z^2 = x^2 + y^2$ and the plane $x + 2y = 6$.

Applications

54. Julia plans to make a cylindrical vase in which the bottom of the vase is to be 0.3 cm thick and the curved, lateral part of the vase is to be 0.2 cm thick. If the vase needs to have a volume of 1 liter, what should its dimensions be to minimize its weight?

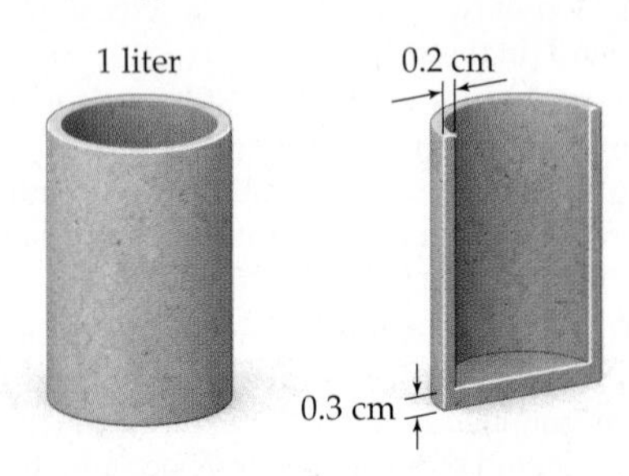

55. Bob is building a toy chest for his son that will be a rectangular box with an open top. The base will be made from plywood that costs \$2 per square foot, and the sides will be made from oak that costs \$5 per square foot. What dimensions should Bob make the toy chest if he wants it to have a capacity of 10 cubic feet but wishes to minimize the cost of the wood?

Proofs

56. Let $\mathcal{T}$ be a triangle with side lengths a, b, and c. The ***semiperimeter*** of $\mathcal{T}$ is defined to be $s = \frac{1}{2}(a + b + c)$. Heron's formula for the area A of a triangle is

$$A = \sqrt{s(s-a)(s-b)(s-c)}.$$

Use Heron's formula and the method of Lagrange multipliers to prove that, for a triangle with perimeter P, the equilateral triangle maximizes the area.

57. Prove that a square maximizes the area of all rectangles with perimeter P.

58. Prove that if you minimize the square of the distance from the origin to a point (x, y) subject to the constraint $g(x, y) = 0$, you have minimized the distance from the origin to (x, y) subject to the same constraint.

59. Prove that if you minimize the square of the distance from the origin to a point (x, y, z) subject to the constraint

$g(x, y, z) = 0$, then you have minimized the distance from the origin to (x, y, z) subject to the same constraint.

60. Let a and b be nonzero real numbers and let $r > 0$. Prove that the point closest to the origin on the circle $(x-a)^2 + (y-b)^2 = r^2$ is on the line that contains the points $(0, 0)$ and (a, b).

61. Let (α, β) and (a, b) be distinct points in $\mathbb{R}^2$, and let r be a positive real number. Use the result of Exercise 60 to prove that the point closest to (α, β) on the circle $(x-a)^2 + (y-b)^2 = r^2$ is on the line that contains the points (α, β) and (a, b).

62. Let a, b, and c be nonzero real numbers and let $r > 0$. Prove that the point closest to the origin on the sphere $(x-a)^2 + (y-b)^2 + (z-c)^2 = r^2$ is on the line that contains the points $(0, 0, 0)$ and (a, b, c).

63. Let (α, β, γ) and (a, b, c) be distinct points in $\mathbb{R}^3$, and let r be a positive real number. Use the result of Exercise 62 to prove that the point closest to (α, β, γ) on the sphere $(x-a)^2+(y-b)^2+(z-c)^2 = r^2$ is on the line that contains the points (α, β, γ) and (a, b, c).

The ***arithmetic mean*** of the real numbers $a_1, a_2, \ldots, a_n$ is $\frac{1}{n}(a_1 + a_2 + \cdots + a_n)$. If $a_i > 0$ for $1 \le i \le n$, then the ***geometric mean*** of $a_1, a_2, \ldots, a_n$ is $(a_1 a_2 \cdots a_n)^{1/n}$. In Exercises 64–66 we ask you to prove that the geometric mean is always less than the arithmetic mean for a set of positive numbers.

64. Use the method of Lagrange multipliers to show that $\sqrt{xy} \le \frac{1}{2}(x + y)$ when x and y are both positive.

65. Use the method of Lagrange multipliers to show that $\sqrt[3]{xyz} \le \frac{1}{3}(x + y + z)$ when x, y, and z are all positive.

66. Use the method of Lagrange multipliers to show that $(a_1 a_2 \cdots a_n)^{1/n} \le \frac{1}{n}(a_1 + a_2 + \cdots + a_n)$ when $a_1, a_2, \ldots, a_n$ are all positive.

Thinking Forward

- *A Double Summation:* Let

$$\sum_{i=1}^{m}\sum_{j=1}^{n} f(i,j) = \sum_{i=1}^{m}\left(\sum_{j=1}^{n} f(i,j)\right).$$

Evaluate $\displaystyle\sum_{i=1}^{15}\sum_{j=1}^{10} ij^2$.

- *Reordering a Double Summation:* Explain why

$$\sum_{i=1}^{m}\sum_{j=1}^{n} f(i,j) = \sum_{j=1}^{n}\sum_{i=1}^{m} f(i,j).$$

CHAPTER REVIEW, SELF-TEST, AND CAPSTONES

Before you progress to the next chapter, be sure you are familiar with the definitions, concepts, and basic skills outlined here. The capstone exercises at the end bring together ideas from this chapter and look forward to future chapters.

Definitions

Give precise mathematical definitions or descriptions of each of the following concepts that follow. Then illustrate the definition with a graph or an algebraic example.

- a *level curve* for $f(x, y)$
- a *level surface* for $f(x, y, z)$
- an *open disk* in $\mathbb{R}^2$
- an *open ball* in $\mathbb{R}^3$
- an *open subset* of $\mathbb{R}^2$ or $\mathbb{R}^3$
- the *complement* of a subset $\mathbb{R}^2$ or $\mathbb{R}^3$
- a *closed subset* of $\mathbb{R}^2$ or $\mathbb{R}^3$
- the *boundary* of a subset of $\mathbb{R}^2$ or $\mathbb{R}^3$
- a *bounded subset* of $\mathbb{R}^2$ or $\mathbb{R}^3$
- the ϵ–δ definition of the statement $\lim_{\mathbf{x}\to\mathbf{a}} f(\mathbf{x}) = L$, where f is a function of two or more variables
- the limit of a function of two or three variables *along a path*
- what it means in terms of a limit for a function f of two or three variables to be *continuous* at a point in the domain of f
- the limit definition of the *partial derivatives* $f_x(x_0, y_0)$ and $f_y(x_0, y_0)$
- the limit definition of the *partial derivatives* $f_x(x_0, y_0, z_0)$, $f_y(x_0, y_0, z_0)$, and $f_z(x_0, y_0, z_0)$
- the limit definition of the *directional derivative* of a function of two or three variables, f, at a point P in the direction of a unit vector $\mathbf{u}$
- the *gradient* of a function f of two or three variables
- a function f of two or three variables has a *local minimum* or a *local maximum* at a point P
- a function f of two or three variables has a *global minimum* or a *global maximum* at a point P

- a function f of two or three variables has a *stationary point* at a point P
- a function f of two or three variables has a *critical point* at a point P
- a function $f(x, y)$ has a *saddle point* at a point (x_0, y_0)
- the *Hessian* and *discriminant* of a function $f(x, y)$

Theorems

Fill in the blanks to complete each of the following theorem statements:

- If S is a closed subset of $\mathbb{R}^3$, then S^c is ________ subset of $\mathbb{R}^3$.
- Let $f(x, y)$ be a function defined on an open subset S of $\mathbb{R}^2$. If the second-order partial derivatives of a function $f(x, y)$ are ________ in a (an) _____ subset S of $\mathbb{R}^2$, then $f_{xy}(x, y) = f_{yx}(x, y)$ at every point in S.
- If $f(x, y)$ is a function of two variables with the partial derivatives $\frac{\partial f}{\partial x}$ and $\frac{\partial f}{\partial y}$ that are ________ on ________ containing the point (x_0, y_0), then f is differentiable at (x_0, y_0).
- Let $f(x, y)$ be a function of two variables that is differentiable at the point (x_0, y_0). The equation of the plane tangent to the surface defined by $f(x, y)$ at (x_0, y_0) is ________.
- *The chain rule*: Given functions $z = f(x, y)$, $x = u(s, t)$, and $y = v(s, t)$, for all values of s and t at which u and v are differentiable, and if f is differentiable at $(u(s, t), v(s, t))$, $\frac{\partial z}{\partial s} =$ ________ and $\frac{\partial z}{\partial t} =$ ________.
- Let $f(x, y)$ be a function of two variables and (x_0, y_0) be a point in the domain of f at which the first-order partial derivatives of f exist. If $\mathbf{u} \in \mathbb{R}^2$ is a ________ for which the directional derivative $D_{\mathbf{u}} f(x_0, y_0)$ also exists, then $D_{\mathbf{u}} f(x_0, y_0) =$ ________.
- If $f(x, y)$ is a function of two or three variables and P is a point in the domain of f at which f is differentiable, then the ________ of f at P points in the direction in which f ________.
- If $f(x, y)$ has a local extremum at (x_0, y_0), then (x_0, y_0) is a ________ of f.

Notation, Algebraic Rules, and Optimization

Notation: Describe the meanings of each of the following mathematical expressions:

For $A \subset \mathbb{R}^2$,

- A^c
- ∂A
- $f_x(x_0, y_0)$
- $f_y(x_0, y_0)$
- $f_z(x_0, y_0, z_0)$
- $\frac{\partial w}{\partial x}$
- $\frac{\partial w}{\partial y}$
- $\frac{\partial w}{\partial z}$
- $D_{\mathbf{u}} f(x_0, y_0)$
- $\nabla f(x, y)$
- $\nabla f(x, y, z)$

Limits of combinations: Fill in the blanks to complete the limit rules. You may assume that $\lim_{\mathbf{x}\to\mathbf{a}} f(\mathbf{x})$ and $\lim_{\mathbf{x}\to\mathbf{a}} g(\mathbf{x})$ exist and that k is a scalar.

- $\lim_{\mathbf{x}\to\mathbf{a}} kf(\mathbf{x}) =$ ________.
- $\lim_{\mathbf{x}\to\mathbf{a}} (f(\mathbf{x}) + g(\mathbf{x})) =$ ________.
- $\lim_{\mathbf{x}\to\mathbf{a}} (f(\mathbf{x}) - g(\mathbf{x})) =$ ________.
- $\lim_{\mathbf{x}\to\mathbf{a}} (f(\mathbf{x})g(\mathbf{x})) =$ ________.
- $\lim_{\mathbf{x}\to\mathbf{a}} (f(\mathbf{x})g(\mathbf{x})) =$ ________, provided that ________.

Function optimization: Fill in the blanks to complete optimization facts.

- Let $f(x, y)$ be a function with continuous second-order partial derivatives on some open disk containing the point (x_0, y_0). If (x_0, y_0) is a stationary point in the domain of f, then
  (a) f has a relative maximum at (x_0, y_0) if $\det(H_f(x_0, y_0))$________ with $f_{xx}(x_0, y_0)$________ or $f_{yy}(x_0, y_0)$________.
  (b) f has a relative minimum at (x_0, y_0) if $\det(H_f(x_0, y_0))$________ with $f_{xx}(x_0, y_0)$________ or $f_{yy}(x_0, y_0)$________.
  (c) f has a saddle point at (x_0, y_0) if $\det(H_f(x_0, y_0))$ ________.
  (d) No conclusion may be drawn about the behavior of f at (x_0, y_0) if $\det(H_f(x_0, y_0))$________.
- Let f and g be functions with continuous first-order partial derivatives. If $f(x, y)$ has a relative extremum at a point (x_0, y_0) subject to the constraint $g(x, y) = 0$, then ________ and ________ are parallel.

Skill Certification: The Calculus of Multivariable Functions

Level curves: Sketch the level curves $f(x, y) = c$ of the following functions for $c = -3, -2, -1, 0, 1, 2$, and 3:

1. $f(x, y) = -\dfrac{2y}{x}$
2. $f(x, y) = y^2 - x^2$
3. $f(x, y) = x - \sin^{-1} y$
4. $f(x, y) = x \sec y$

Evaluating limits: Evaluate the following limits, or explain why the limit does not exist.

5. $\displaystyle\lim_{(x,y)\to(1,2)} \frac{x^2+y^2}{x^2-y^2}$
6. $\displaystyle\lim_{(x,y)\to(4,\pi)} \tan^{-1}\left(\frac{y}{x}\right)$
7. $\displaystyle\lim_{(x,y)\to(0,0)} \frac{x^3+y^3}{x^2+y^2}$
8. $\displaystyle\lim_{(x,y)\to(0,0)} \frac{x+y}{x^2+y^2}$
9. $\displaystyle\lim_{(x,y,z)\to(1,0,-1)} \frac{\sin(xy)}{x^2-y^2+z^2}$
10. $\displaystyle\lim_{(x,y,z)\to(0,0,0)} \frac{x^2+y^2}{x^2+y^2+z^2}$
11. $\displaystyle\lim_{\substack{(x,y)\to(0,0)\\ x=0}} \frac{x^2+y^2}{x^2-y^2}$
12. $\displaystyle\lim_{\substack{(x,y)\to(0,0)\\ y=0}} \frac{x^2+y^2}{x^2-y^2}$

Continuity: Find the set of points where the function is continuous.

13. $f(x, y) = \dfrac{x+y}{x-y}$
14. $f(x, y) = \dfrac{x+y}{x^2+y^2}$
15. $f(x, y) = \ln xy$
16. $f(x, y) = \cos\left(\dfrac{1}{x-y}\right)$
17. $f(x, y, z) = \dfrac{x+y+z}{x-2y+3z}$
18. $f(x, y, z) = \dfrac{x+y+z}{x^2+y^2+z^2}$

Partial derivatives: Find all first- and second-order partial derivatives for the following functions:

19. $f(x, y) = \dfrac{x+y}{x^2+y^2}$
20. $f(x, y) = \dfrac{x^2-y^2}{x^2+y^2}$
21. $f(x, y) = xye^{x^2}$
22. $f(x, y) = \tan^{-1}\left(\dfrac{y}{x}\right)$
23. $f(x, y, z) = xz^2e^y$
24. $f(x, y, z) = \ln(x+y+z)$

Directional derivatives: Find the directional derivative of the given function at the specified point P in the direction of the given vector. Note: The given vectors may not be unit vectors.

25. $f(x, y) = \dfrac{y}{x}$, $P = (4, 3)$, $\mathbf{v} = \langle 2, -3\rangle$
26. $f(x, y) = \dfrac{y}{x}$, $P = (4, 3)$, $\mathbf{v} = \langle 3, -2\rangle$
27. $f(x, y) = \dfrac{x+y}{x^2+y^2}$, $P = (1, 2)$, $\mathbf{v} = \langle 3, -2\rangle$
28. $f(x, y) = \dfrac{xy}{x^2+y^2}$, $P = (-2, 1)$, $\mathbf{v} = \left\langle \dfrac{3}{5}, -\dfrac{4}{5}\right\rangle$
29. $f(x, y, z) = xy^2z^3$, $P = (0, 0, 0)$, $\mathbf{v} = \langle 1, -2, -1\rangle$
30. $f(x, y, z) = \sqrt{\dfrac{xy}{z}}$, $P = (2, 3, 1)$, $\mathbf{v} = \langle 2, 1, -2\rangle$

Gradients: Find the gradient of the given function, and find the direction in which the function increases most rapidly at the specified point P.

31. $f(x, y) = \dfrac{y}{x}$, $P = (4, 3)$
32. $f(x, y) = \dfrac{x}{y}$, $P = (4, 3)$
33. $f(x, y) = x \sin y$, $P = \left(3, \dfrac{\pi}{2}\right)$
34. $f(x, y) = \tan^{-1}\left(\dfrac{y}{x}\right)$, $P = (-2, 2)$
35. $f(x, y, z) = \dfrac{x^2y}{z}$, $P = (0, 3, -1)$
36. $f(x, y, z) = \ln(x+y+z)$, $P = (e, 0, -1)$

Extrema: Find the local maxima, local minima, and saddle points of the given functions.

37. $f(x, y) = 2x^2 + y^2 + y + 5$
38. $f(x, y) = x^3 - 12xy - y^3$
39. $f(x, y) = x^3 + y^3 - 6x^2 + 3y^2 - 4$
40. $f(x, y) = x^4 + y^4 + 4xy$

Capstone Problems

A. *Chain Rule:* Let $z = f(x, y)$, $x = x(s, t)$, and $y = y(t)$.

(a) Find $\dfrac{\partial z}{\partial s}$ and $\dfrac{\partial z}{\partial t}$.

(b) Use the results of part (a) to find $\dfrac{\partial z}{\partial s}$ and $\dfrac{\partial z}{\partial t}$ when $z = x^2 \sin y$, $x = e^{s/t}$, and $y = t^3$. Express your answer in terms of s and t.

B. *Laplace's equation:* A function $f(x, y)$ is said to satisfy ***Laplace's equation*** if $\dfrac{\partial^2 f}{\partial x^2} + \dfrac{\partial^2 f}{\partial y^2} = 0$.

(a) Show that the function $f(x, y) = \tan^{-1}\left(\dfrac{y}{x}\right)$ satisfies Laplace's equation.

(b) Show that the function $f(x, y) = \ln\left(\dfrac{y}{x}\right)$ does not satisfy Laplace's equation.

C. *Extrema on a closed and bounded set:* Find and classify the maxima, minima, and saddle points of the function $f(x, y) = y^3 - 3x^2y$ on the square region with vertices $(\pm 3, \pm 3)$.

D. *The Method of Least Squares:* The ***method of least squares*** is one way to fit a line to a collection of data points. The object of this method is to find the slope m and y-intercept b for a ***regression line*** $y = mx + b$ that provides the minimal sum for the squares of the vertical distances from the data points to the line, as indicated in the following figure:

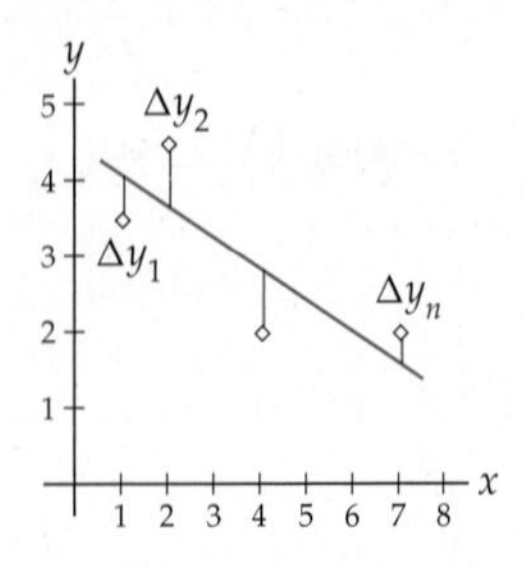

Most scientific calculators have built-in programs to find the equation of the regression line for a collection of data points (x_1, y_1), $(x_2, y_2), \ldots, (x_n, y_n)$, but here we ask you to first prove that:

(a) $$m = \frac{(\sum x_k)(\sum y_k) - n(\sum x_k y_k)}{(\sum x_k)^2 - n(\sum x_k^2)} \quad \text{and}$$

$$b = \frac{1}{n}\left(\sum y_k - m \sum x_k\right),$$

where each of the summations is evaluated from $k = 1$ to $k = n$.

Use the results of part (a) to find the equations of the regression lines for the following collections of data:

(b) $(1, 2)$, $(3, 4)$, $(5, 5)$, $(6, 8)$

(c) $(1, 2)$, $(1, 4)$, $(5, 2)$, $(5, 4)$

CHAPTER 13

Double and Triple Integrals

13.1 Double Integrals over Rectangular Regions
- Volumes
- Double and Triple Summations
- Double Integrals over Rectangular Regions
- Iterated Integrals and Fubini's Theorem
- Examples and Explorations

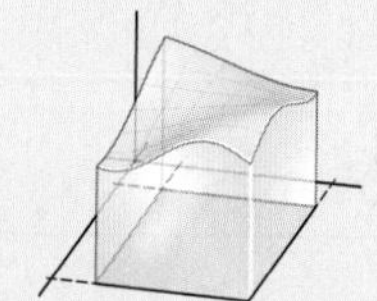

13.2 Double Integrals over General Regions
- General Regions in the Plane
- Double Integrals over General Regions
- Algebraic Properties of Double Integrals
- Examples and Explorations

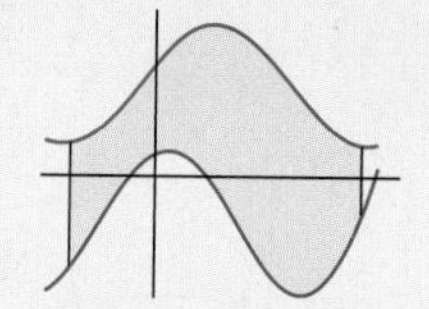

13.3 Double Integrals in Polar Coordinates
- Polar Coordinates and Double Integrals
- Double Integrals in Polar Coordinates over General Regions
- Examples and Explorations

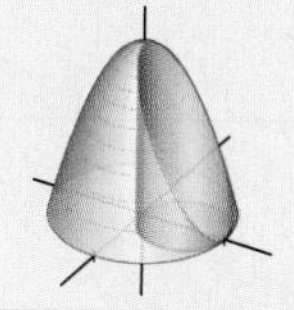

13.4 Applications of Double Integrals
- The Mass of a Planar Region
- The Center of Mass and First Moments
- Moments of Inertia
- Probability Distributions
- Examples and Explorations

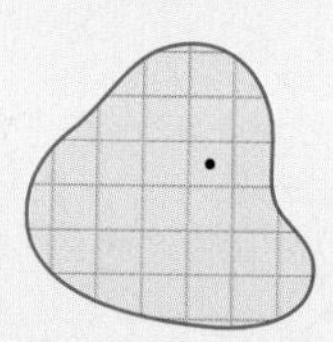

13.5 Triple Integrals
- Triple Integrals over Rectangular Solids
- Iterated Integrals and Fubini's Theorem
- Triple Integrals over General Regions
- Applications of Triple Integration
- Examples and Explorations

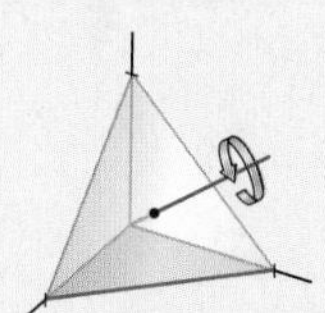

13.6 Integration with Cylindrical and Spherical Coordinates
- Cylindrical Coordinates
- Spherical Coordinates
- Integration with Spherical Coordinates
- Examples and Explorations

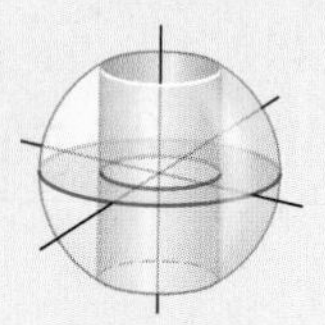

13.7 Jacobians and Change of Variables
- Change of Variables in $\mathbb{R}^2$
- Change of Variables in $\mathbb{R}^3$
- Examples and Explorations

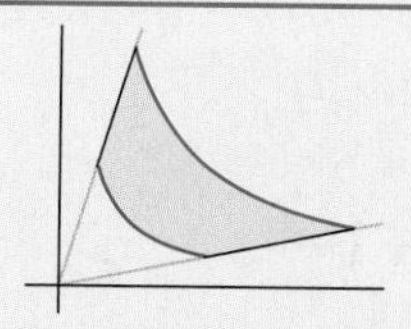

Chapter Review, Self-Test, and Capstones

13.1 DOUBLE INTEGRALS OVER RECTANGULAR REGIONS

- Computing volumes bounded by functions of two variables
- Defining double integrals as the limit of a Riemann sum
- Using Fubini's theorem to evaluate double integrals as iterated integrals

Volumes

Consider the following two questions:

- How can we compute the volume of a solid?
- How can we compute the mass of a solid whose density varies from point to point?

The way in which we will answer these two questions is quite similar to the methods we used in single-variable calculus. For example, to find the volume of the solid bounded above by the surface and below by the xy-plane, in the left-hand figure that follows, we subdivide the volume into vertical slices whose volumes we can reasonably approximate, as we see in the middle figure, and then, take a limit as the size of the pieces goes to zero and, simultaneously, the number of pieces goes to infinity. This brief outline summarizes much of what we will do in this chapter.

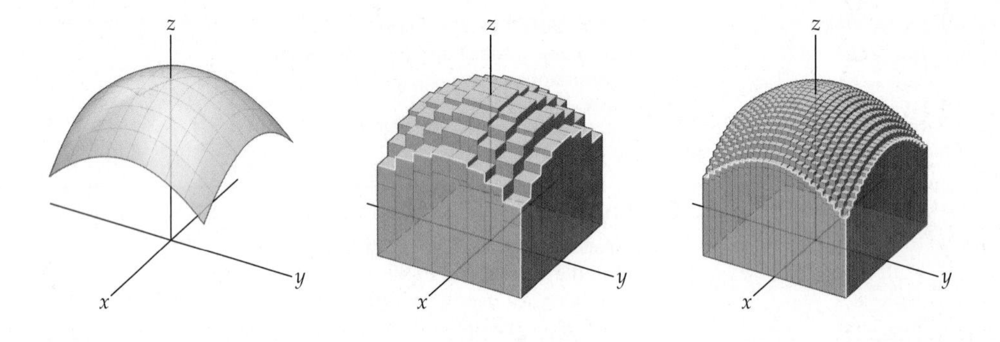

We will begin by reviewing and extending the summation notation we introduced in Chapter 4.

Double and Triple Summations

Recall that

$$\sum_{k=1}^{n} a_k = a_1 + a_2 + a_3 + \cdots + a_{n-1} + a_n.$$

We take this idea a step or two further by introducing double and triple sums.

DEFINITION 13.1 **Double Sums and Triple Sums**

(a) The ***double summation*** $\sum_{j=1}^{m}\sum_{k=1}^{n} a_{jk}$ is given by

$$\begin{aligned}\sum_{j=1}^{m}\sum_{k=1}^{n} a_{jk} &= \sum_{j=1}^{m}\left(\sum_{k=1}^{n} a_{jk}\right)\\ &= \sum_{j=1}^{m}(a_{j1}+a_{j2}+a_{j3}+\cdots+a_{jn})\\ &= a_{11}+a_{12}+a_{13}+\cdots+a_{1n}\\ &\quad + a_{21}+a_{22}+a_{23}+\cdots+a_{2n}\\ &\quad + \vdots \quad \vdots \quad \vdots \quad \cdots \quad \vdots\\ &\quad + a_{m1}+a_{m2}+a_{m3}+\cdots+a_{mn}.\end{aligned}$$

(b) Similarly, the ***triple summation*** $\sum_{i=1}^{l}\sum_{j=1}^{m}\sum_{k=1}^{n} a_{ijk}$ is given by

$$\sum_{i=1}^{l}\sum_{j=1}^{m}\sum_{k=1}^{n} a_{ijk} = \sum_{i=1}^{l}\left(\sum_{j=1}^{m}\left(\sum_{k=1}^{n} a_{ijk}\right)\right).$$

For example, consider the following double sum:

$$\sum_{j=1}^{2}\sum_{k=1}^{3} j^2k = (1^2\cdot1+1^2\cdot2+1^2\cdot3)+(2^2\cdot1+2^2\cdot2+2^2\cdot3) = 1+2+3+4+8+12 = 30.$$

We know that, for finite sums, addition obeys both the commutative and associative rules. In Exercises 71 and 72 you will use these properties to prove the following theorem:

THEOREM 13.2 **Changing the Order of a Finite Double or Triple Sum**

For positive integers l, m, and n,

(a) $\sum_{j=1}^{m}\sum_{k=1}^{n} a_{jk} = \sum_{k=1}^{n}\sum_{j=1}^{m} a_{jk}$

(b) $\sum_{i=1}^{l}\sum_{j=1}^{m}\sum_{k=1}^{n} a_{ijk} = \sum_{j=1}^{m}\sum_{k=1}^{n}\sum_{i=1}^{l} a_{ijk} = \sum_{k=1}^{n}\sum_{i=1}^{l}\sum_{j=1}^{m} a_{ijk}$

Note that there are three additional permutations of the summations in Theorem 13.2(b) equal to those listed.

Double Integrals over Rectangular Regions

Starting with Section 13.2, we will allow more complicated domains, but we begin by considering functions defined on rectangular regions. Let $a < b$ and $c < d$ be real numbers and

$\mathcal{R}$ be the rectangle in the xy-plane defined by

$$\mathcal{R} = \{(x, y) \mid a \leq x \leq b \text{ and } c \leq y \leq d\},$$

as shown in the following figure at the left:

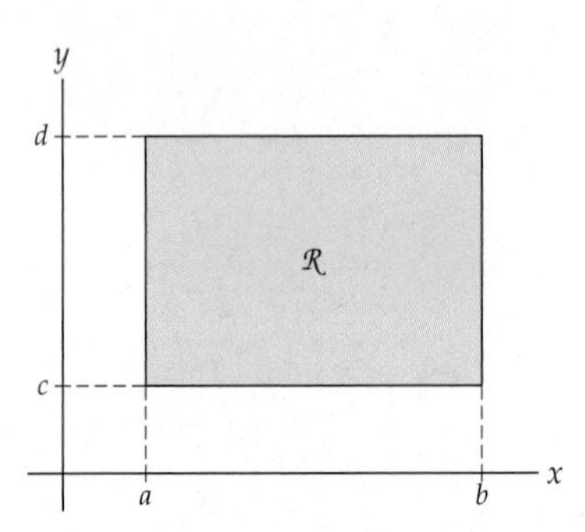

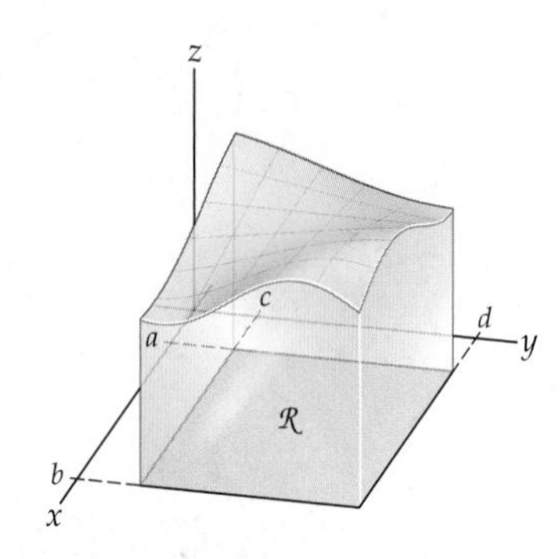

We also assume that $f(x, y) > 0$ is a function defined at every point of $\mathcal{R}$, as we see in the figure on the right. Here is our procedure for finding the volume V of the solid bounded below by $\mathcal{R}$ and bounded above by the graph of f:

- We subdivide the interval $[a, b]$ into m equal subintervals, each of width $\Delta x = \frac{b-a}{m}$, and also let $x_j = a + j\Delta x$ for $0 \leq j \leq m$.
- Similarly, we subdivide the interval $[c, d]$ into n equal subintervals, each of width $\Delta y = \frac{d-c}{n}$, and let $y_k = a + k\Delta y$ for $0 \leq k \leq n$.
- The preceding subdivisions partition the rectangle into $m \times n$ congruent rectangles as follows:

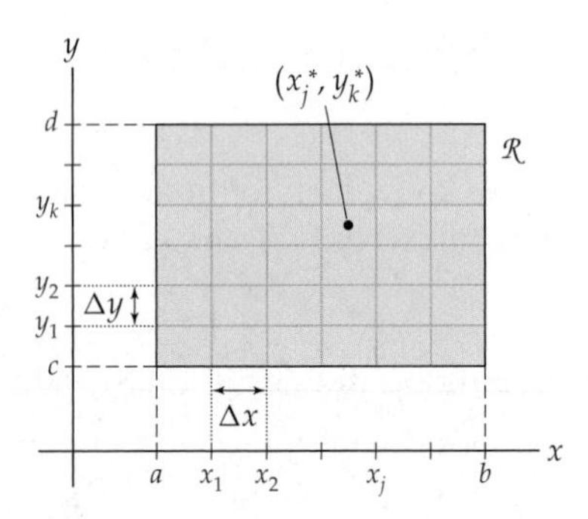

- For each $j = 1, 2, \ldots, m$ and $k = 1, 2, \ldots, n$, we select a point (x_j^*, y_k^*) in the subrectangle $\mathcal{R}_{jk} = \{(x, y) \mid x_{j-1} \leq x \leq x_j \text{ and } y_{k-1} \leq y \leq y_k\}$. One such point is shown in the previous figure.
- We let ΔA be the common area of each subrectangle. That is, $\Delta A = \Delta x \Delta y$.
- The product $f(x_j^*, y_k^*)\Delta A$ approximates the volume of the solid bounded below by the rectangle $\mathcal{R}_{jk}$ and above by the graph of f.
- When we sum these approximate volumes over all of the subrectangles, we obtain an approximation for the volume of V.

The summations

$$\sum_{j=1}^{m}\sum_{k=1}^{n} f(x_j^*, y_k^*)\Delta A = \sum_{k=1}^{n}\sum_{j=1}^{m} f(x_j^*, y_k^*)\Delta A$$

are both ***Riemann sums*** for f on $\mathcal{R}$.

We now define Δ to be the length of the diagonal of each subrectangle. That is,

$$\Delta = \sqrt{(\Delta x)^2 + (\Delta y)^2}.$$

Taking the limit as $\Delta \to 0$ ensures that $m \to \infty$ and $n \to \infty$. We use the limit of the Riemann sum as $\Delta \to 0$ to define the volume of the solid V, provided that the limit exists. The following definitions formalize these ideas:

DEFINITION 13.3

Riemann Sums for Functions of Two Variables

Let $a < b$ and $c < d$ be real numbers, let $\mathcal{R}$ be the rectangle defined by

$$\mathcal{R} = \{(x, y) \mid a \le x \le b \text{ and } c \le y \le d\},$$

and let $f(x, y)$ be a function defined on $\mathcal{R}$. The sums

$$\sum_{j=1}^{m} \sum_{k=1}^{n} f(x_j^*, y_k^*)\Delta A = \sum_{k=1}^{n} \sum_{j=1}^{m} f(x_j^*, y_k^*)\Delta A$$

are ***Riemann sums*** for f on $\mathcal{R}$, where $x_j,\ x_j^*,\ y_k,\ y_k^*,\ \Delta x,\ \Delta y$, and ΔA are defined as outlined before.

DEFINITION 13.4

Double Integrals

Let $a < b$ and $c < d$ be real numbers, let $\mathcal{R}$ be the rectangle defined by

$$\mathcal{R} = \{(x, y) \mid a \le x \le b \text{ and } c \le y \le d\},$$

and let $f(x, y)$ be a function defined on $\mathcal{R}$. Provided that the limits exist, the ***double integral*** of f over $\mathcal{R}$ is

$$\iint_{\mathcal{R}} f(x, y)\, dA = \lim_{\Delta \to 0} \sum_{j=1}^{m} \sum_{k=1}^{n} f(x_j^*, y_k^*)\Delta A = \lim_{\Delta \to 0} \sum_{k=1}^{n} \sum_{j=1}^{m} f(x_j^*, y_k^*)\Delta A,$$

where the double sums are Riemann sums, as outlined in Definition 13.3, and where $\Delta = \sqrt{(\Delta x)^2 + (\Delta y)^2}$. When the limits exist, the function f is said to be ***integrable*** on $\mathcal{R}$.

Note that as $\Delta \to 0$, the increment of area $\Delta A \to 0$. For every function that is continuous on a rectangular region, the limit will exist and the function will be integrable.

DEFINITION 13.5

The Volume of a Solid and the Signed Volume

Let $a < b$ and $c < d$ be real numbers, let $\mathcal{R}$ be the rectangle defined by

$$\mathcal{R} = \{(x, y) \mid a \le x \le b \text{ and } c \le y \le d\},$$

and let $f(x, y)$ be an integrable function defined on $\mathcal{R}$.

(a) The ***signed volume*** between the graph of f and the rectangle $\mathcal{R}$ is defined to be the double integral $\iint_{\mathcal{R}} f(x, y)\, dA$.

(b) The ***(absolute) volume*** between the graph of f and the rectangle $\mathcal{R}$ is defined to be the double integral $\iint_{\mathcal{R}} |f(x, y)|\, dA$.

As a consequence of Definition 13.5, if $f(x, y) \geq 0$ on $\mathcal{R}$, then the double integral $\iint_{\mathcal{R}} f(x, y)\, dA$ represents the volume of the solid bounded above by the graph of f and below by the rectangle $\mathcal{R}$.

In Example 1, we compute the volume of the solid bounded above by the graph of the function $f(x, y) = x^2 y$ on the rectangle

$$\mathcal{R} = \{(x, y) \mid 1 \leq x \leq 3 \text{ and } 2 \leq y \leq 5\}$$

by using Definition 13.4 to evaluate the double integral $\iint_{\mathcal{R}} x^2 y\, dA$. We will see that such computations are rather lengthy and time consuming. Fortunately, we have an alternative.

Iterated Integrals and Fubini's Theorem

As we mentioned, using the definition to evaluate a double integral is a lengthy process. When we can, we will use iterated integrals to evaluate double integrals.

DEFINITION 13.6 Iterated Integrals

Let a, b, c, and d be real numbers. We define the following ***iterated integrals***:

(a) $$\int_a^b \int_c^d f(x, y)\, dy\, dx = \int_a^b \left(\int_c^d f(x, y)\, dy \right) dx$$

(b) $$\int_c^d \int_a^b f(x, y)\, dx\, dy = \int_c^d \left(\int_a^b f(x, y)\, dx \right) dy.$$

The general idea here is that the subdivisions that are used to construct a Riemann sum for a double integral over a rectangle $\mathcal{R}$ may be added most easily in one of two orders:

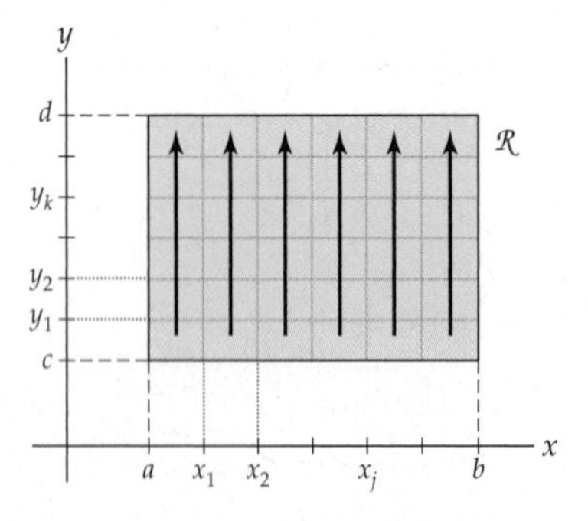

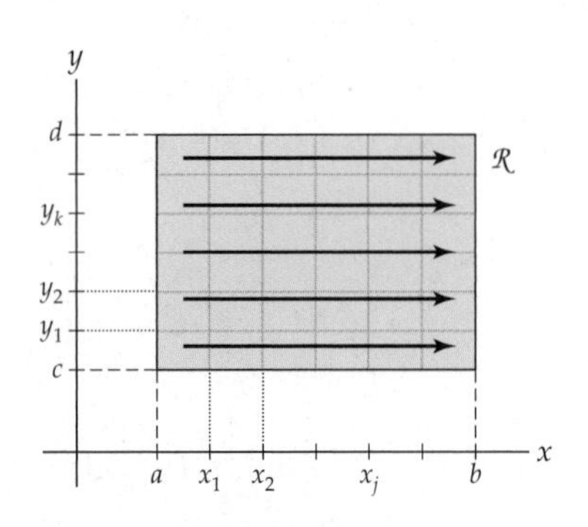

As the figures show, either we sum over the subrectangles, first moving up each column and then moving to the right, or we sum along the rows from left to right and then move up. The first order of summation gives us part (a) of Definition 13.6. The other order gives us the iterated integral in part (b).

We will evaluate an iterated integral after we introduce Fubini's theorem, which tells us that we may use an iterated integral, rather than the definition of the double integral, to evaluate a double integral.

THEOREM 13.7

Fubini's Theorem

Let $a < b$ and $c < d$ be real numbers, let $\mathcal{R}$ be the rectangle defined by

$$\mathcal{R} = \{(x, y) \mid a \leq x \leq b \text{ and } c \leq y \leq d\},$$

and let $f(x, y)$ be continuous on $\mathcal{R}$. Then

$$\iint_{\mathcal{R}} f(x, y)\, dA = \int_c^d \int_a^b f(x, y)\, dx\, dy = \int_a^b \int_c^d f(x, y)\, dy\, dx.$$

Proof. A rigorous proof of Fubini's theorem is beyond the scope of this text, but we provide a less formal argument. From the definition of the integral as a limit of a Riemann sum, we have

$$\int_c^d \int_a^b f(x, y)\, dx\, dy = \int_c^d \left(\int_a^b f(x, y)\, dx \right) dy = \lim_{n \to \infty} \sum_{k=1}^{n} \left(\lim_{m \to \infty} \sum_{j=1}^{m} f(x_j^*, y_k^*) \Delta x \right) \Delta y,$$

where $\Delta x = \frac{b-a}{m}$, $\Delta y = \frac{d-c}{n}$, and (x_j^*, y_k^*) is a sample point in the jkth subrectangle. We know that for the finite summations

$$\sum_{k=1}^{n} \sum_{j=1}^{m} f(x_j^*, y_k^*) \Delta x \Delta y = \sum_{j=1}^{m} \sum_{k=1}^{n} f(x_j^*, y_k^*) \Delta y \Delta x.$$

Fubini's theorem follows from the fact that when f is continuous, the orders of both the limits and the sums may be interchanged:

$$\begin{aligned} \int_c^d \int_a^b f(x, y)\, dx\, dy &= \lim_{n \to \infty} \sum_{k=1}^{n} \left(\lim_{m \to \infty} \sum_{j=1}^{m} f(x_j^*, y_k^*) \Delta x \right) \Delta y \\ &= \lim_{m \to \infty} \sum_{j=1}^{m} \left(\lim_{n \to \infty} \sum_{k=1}^{n} f(x_j^*, y_k^*) \Delta y \right) \Delta x \\ &= \int_a^b \int_c^d f(x, y)\, dy\, dx. \end{aligned}$$

■

To illustrate Fubini's theorem, let

$$\mathcal{R} = \{(x, y) \mid 1 \leq x \leq 3 \text{ and } 2 \leq y \leq 5\}.$$

Then

$$\iint_{\mathcal{R}} x^2 y\, dA = \int_2^5 \int_1^3 x^2 y\, dx\, dy = \int_1^3 \int_2^5 x^2 y\, dy\, dx.$$

For rectangular regions, the level of difficulty of the two iterated integrals we may use to evaluate the double integral is the same. In Section 13.2 we will discuss how to evaluate double integrals over more complicated regions. At that time we will see that, for a more complicated region, one of the two orderings of the iterated integrals can make the evaluation process much simpler. Here we evaluate

$$\iint_{\mathcal{R}} x^2y\,dA = \int_2^5\int_1^3 x^2y\,dx\,dy \qquad \leftarrow \text{Fubini's theorem}$$

$$= \int_2^5\left(\int_1^3 x^2y\,dx\right)dy \qquad \leftarrow \text{evaluation procedure for the iterated integral}$$

$$= \int_2^5 \left[\frac{1}{3}x^3y\right]_{x=1}^{x=3} dy \qquad \leftarrow \text{the Fundamental Theorem of Calculus}$$

$$= \int_2^5 \left(\frac{27}{3}y - \frac{1}{3}y\right)dy = \int_2^5 \frac{26}{3}y\,dy \qquad \leftarrow \text{evaluation of the inner antiderivative}$$

$$= \left[\frac{13}{3}y^2\right]_{y=2}^{y=5} \qquad \leftarrow \text{the Fundamental Theorem of Calculus}$$

$$= \frac{13}{3}(25-4) = 91. \qquad \leftarrow \text{evaluation of the outer antiderivative}$$

Note that when we evaluate the inner integral with respect to x, we treat the other variable, y, as a constant. Not until we evaluate the outer integral with respect to y do we consider y as a variable. In Exercise 17 you will use the other order of integration to evaluate this double integral.

Examples and Explorations

EXAMPLE 1 Using the definition to evaluate a double integral

Let

$$\mathcal{R} = \{(x, y) \mid 1 \le x \le 3 \text{ and } 2 \le y \le 5\}.$$

Use the definition of the double integral to evaluate $\iint_{\mathcal{R}} x^2y\,dA$.

SOLUTION

Before we evaluate the integral, we note that it represents the volume of the solid bounded above by the graph of $f(x, y) = x^2y$ and below by the xy-plane on the rectangular region $\mathcal{R}$, since $x^2y \ge 0$ on $\mathcal{R}$. This solid is shown here:

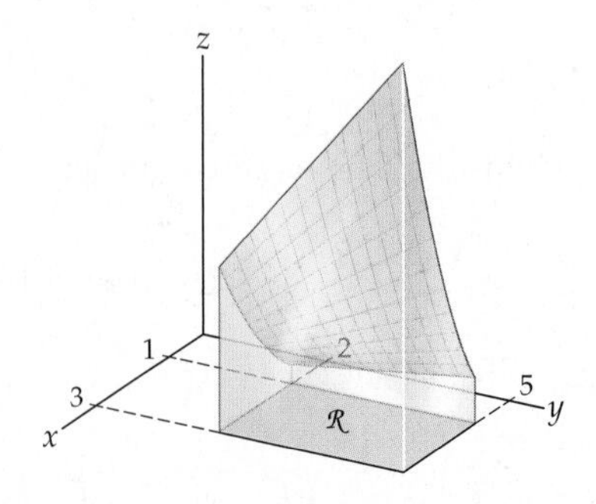

To use Definition 13.5 to evaluate the integral, for our starred points (x_j^*, y_k^*) we will choose $(x_j, y_k) = (1 + j\Delta x, 2 + k\Delta y)$ for each j and k. Thus, the double integral of x^2y on $\mathcal{R}$ is given by

$$\iint_{\mathcal{R}} x^2y\,dA = \lim_{\Delta\to 0}\sum_{j=1}^{m}\sum_{k=1}^{n}(1+j\Delta x)^2(2+k\Delta y)\Delta A.$$

Recall that using the definition to compute a definite integral of a function of a single variable can be quite time consuming. Still, we will use the definition to evaluate this double integral, but will use Fubini's theorem to evaluate other double integrals.

We start working on the Riemann sum. Since the index of the "interior" summation is k,

$$\sum_{j=1}^{m}\sum_{k=1}^{n}(1+j\Delta x)^2(2+k\Delta y)\Delta A=\sum_{j=1}^{m}(1+j\Delta x)^2\,\Delta A\Big(\sum_{k=1}^{n}(2+k\Delta y)\Big).$$

Now, using properties of summations from Chapter 4, we have

$$\sum_{k=1}^{n}(2+k\Delta y)=2n+\frac{1}{2}n(n+1)\Delta y.$$

Combining the preceding two equations gives

$$\sum_{j=1}^{m}(1+j\Delta x)^2\Delta A\Big(\sum_{k=1}^{n}(2+k\Delta y)\Big)=\sum_{j=1}^{m}(1+j\Delta x)^2\Delta A\left(2n+\frac{1}{2}n(n+1)\Delta y\right).$$

Again using properties of summations from Chapter 4, we obtain

$$\sum_{j=1}^{m}(1+j\Delta x)^2=\sum_{j=1}^{m}(1+2j\Delta x+j^2(\Delta x)^2)=m+m(m+1)\Delta x+\frac{1}{6}m(m+1)(2m+1)(\Delta x)^2.$$

Incorporating this result into the original Riemann sum, we have

$$\sum_{j=1}^{m}\sum_{k=1}^{n}(1+j\Delta x)^2(2+k\Delta y)\Delta A=$$
$$\left(m+m(m+1)\Delta x+\frac{1}{6}m(m+1)(2m+1)(\Delta x)^2\right)\left(2n+\frac{1}{2}n(n+1)\Delta y\right)\Delta A.$$

But recall that $\Delta x=\frac{b-a}{m}=\frac{2}{m}$, $\Delta y=\frac{d-c}{n}=\frac{3}{n}$ and $\Delta A=\Delta x\Delta y=\frac{6}{mn}$, so the Riemann sum equals

$$\left(m+m(m+1)\frac{2}{m}+\frac{1}{6}m(m+1)(2m+1)\left(\frac{2}{m}\right)^2\right)\left(2n+\frac{1}{2}n(n+1)\frac{3}{n}\right)\frac{6}{mn}$$
$$=6\left(1+\frac{2(m+1)}{m}+\frac{2(m+1)(2m+1)}{3m^2}\right)\left(2+\frac{3(n+1)}{2n}\right).$$

We are finally ready to evaluate the limit of the Riemann sum! We get

$$\iint_{\mathcal{R}}x^2y\,dA=\lim_{\Delta\to 0}6\left(1+\frac{2(m+1)}{m}+\frac{2(m+1)(2m+1)}{3m^2}\right)\left(2+\frac{3(n+1)}{2n}\right)$$
$$=\lim_{m\to\infty}\left(\lim_{n\to\infty}6\left(1+\frac{2(m+1)}{m}+\frac{2(m+1)(2m+1)}{3m^2}\right)\left(2+\frac{3(n+1)}{2n}\right)\right)$$
$$=6\left(1+2+\frac{4}{3}\right)\left(2+\frac{3}{2}\right)=91.$$

□

Note that we evaluated this double integral earlier in the section by using Fubini's theorem. Of course, we obtained the same result.

EXAMPLE 2 **Using Fubini's theorem**

Use Fubini's theorem to evaluate the following double integrals:

(a) $\iint_{\mathcal{R}_1}xy\sin(x^2y)\,dA$, where $\mathcal{R}_1=\{(x,y)\mid 0\le x\le 1 \text{ and } 0\le y\le\pi\}$.

(b) $\iint_{\mathcal{R}_2}(\sin x+e^{-2y})\,dA$, where $\mathcal{R}_2=\{(x,y)\mid 0\le x\le\pi \text{ and } 0\le y\le 1\}$.

(c) $\iint_{\mathcal{R}_3}\ln y\,dA$, where $\mathcal{R}_3=\left\{(x,y)\mid 0\le x\le 5 \text{ and } \frac{1}{e}\le y\le e\right\}$.

SOLUTION

(a) Our first integral represents the volume of the solid bounded above by the graph of $xy\sin(x^2y)$ and below by the xy-plane on the rectangular region $\mathcal{R}_1$, since $xy\sin(x^2y) \geq 0$ on $\mathcal{R}_1$. Following is a graph of this solid:

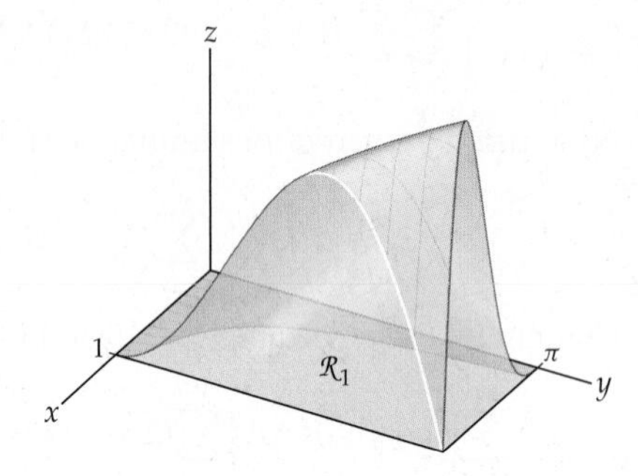

By Fubini's theorem,

$$\iint_{\mathcal{R}_1} xy\sin(x^2y)\,dA = \int_0^1\int_0^\pi xy\sin(x^2y)\,dy\,dx = \int_0^\pi\int_0^1 xy\sin(x^2y)\,dx\,dy.$$

As we mentioned earlier, for rectangular regions the levels of difficulty of the two iterated integrals given by Fubini's theorem should be quite similar. We choose the second integral here. We begin by evaluating $\int_0^1 xy\sin(x^2y)\,dx$. Note that when we integrate the "interior" integral with respect to x, we treat y as a constant. Therefore, we may use substitution to find an antiderivative for the function $xy\sin(x^2y)$ with respect to x. The function $-\frac{1}{2}\cos(x^2y)$ is such an antiderivative. We may check this by taking the partial derivative $\frac{\partial}{\partial x}\left(-\frac{1}{2}\cos(x^2y)\right)$ and observing that it is, indeed, $xy\sin(x^2y)$. Therefore,

$$\int_0^1 xy\sin(x^2y)\,dx = \left[-\frac{1}{2}\cos(x^2y)\right]_{x=0}^{x=1} = -\frac{1}{2}(\cos(1^2y) - \cos(0^2y)) = \frac{1}{2}(1-\cos y).$$

We now have

$$\iint_{\mathcal{R}_1} xy\sin(x^2y)\,dA = \int_0^\pi\int_0^1 xy\sin(x^2y)\,dx\,dy = \int_0^\pi \frac{1}{2}(1-\cos y)\,dy.$$

The integral on the right is a definite integral like those we saw in Chapter 4. We have

$$\int_0^\pi \frac{1}{2}(1-\cos y)\,dy = \left[\frac{1}{2}(y-\sin y)\right]_{y=0}^{y=\pi} = \frac{\pi}{2}.$$

Note that the answer here is a constant. Recall that when $f(x)$ is an integrable function on an interval $[a,b]$, the value of the definite integral $\int_a^b f(x)\,dx$ is a constant. Similarly, when $f(x,y)$ is an integrable function on a rectangle $\mathcal{R}$, the value of the double integral $\iint_{\mathcal{R}} f(x,y)\,dA$ is a constant.

Before we proceed to the other double integrals, we summarize the steps we just used:

- Use Fubini's theorem to express the double integral as an iterated integral.
- Choose an order of integration. The levels of difficulty of the two iterated integrals given by Fubini's theorem may differ. Choose the order that is easier to evaluate.
- Whichever of the two iterated integrals you choose, work on the "interior" integral first, treating the other variable as a constant. (In the example we just

completed, the interior variable was x and the exterior variable was y. You may use any of the integration techniques from Chapter 5 in the process.)

- When you are done with this interior integration, you should have a new definite integral that involves just the exterior variable. (In the preceding example, the last step involved evaluating a definite integral with respect to y.)
- Evaluate the exterior integral.
- The result should be a constant.

(b) Our second integral represents the volume of the solid bounded above by the graph of $\sin x + e^{-2y}$ and below by the xy-plane on the rectangular region $\mathcal{R}_2$, since $\sin x + e^{-2y} \geq 0$ on $\mathcal{R}_2$. Following is a graph of this solid:

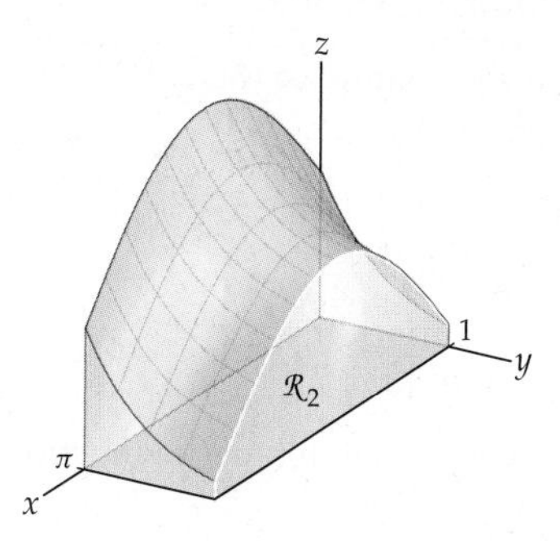

Again by Fubini's theorem,

$$\iint_{\mathcal{R}_2} (\sin x + e^{-2y})\, dA = \int_0^{\pi} \int_0^1 (\sin x + e^{-2y})\, dy\, dx = \int_0^1 \int_0^{\pi} (\sin x + e^{-2y})\, dx\, dy.$$

We evaluate the first iterated integral here. We begin by evaluating $\int_0^1 (\sin x + e^{-2y})\, dy$. We integrate with respect to y, treating x as a constant. Therefore,

$$\int_0^1 (\sin x + e^{-2y})\, dy = \left[(\sin x)y - \frac{1}{2}e^{-2y}\right]_{y=0}^{y=1} = \sin x + \frac{1 - e^{-2}}{2}.$$

We now integrate this function with respect to x on the interval $[0, \pi]$:

$$\int_0^{\pi} \left(\sin x + \frac{1 - e^{-2}}{2}\right) dx = \left[-\cos x + \frac{1 - e^{-2}}{2}x\right]_{x=0}^{x=\pi} = 2 + \frac{\pi(1 - e^{-2})}{2}.$$

(c) Our third integral represents a signed volume, since the graph of $\ln y$ takes on both positive and negative values on the rectangular region $\mathcal{R}_3$, as shown in the following figure:

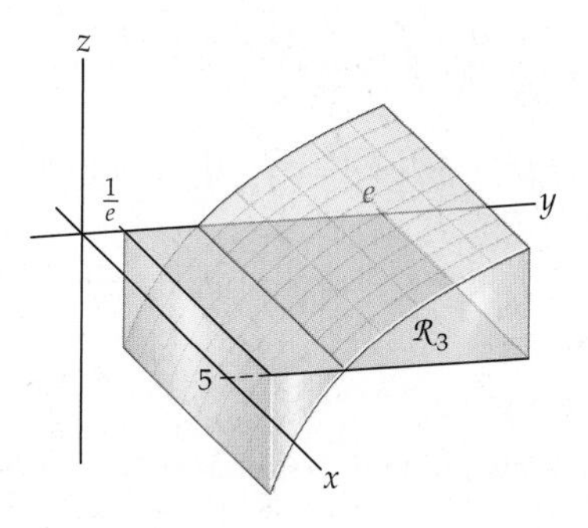

Here we have

$$\iint_{\mathcal{R}_3} \ln y\, dA = \int_0^5 \int_{1/e}^{e} \ln y\, dy\, dx = \int_{1/e}^{e} \int_0^5 \ln y\, dx\, dy.$$

We now evaluate the second iterated integral. Note that the function $\ln y$ is constant with respect to x; therefore,

$$\int_0^5 \ln y \, dx = \left[(\ln y)x\right]_{x=0}^{x=5} = 5 \ln y.$$

We integrate this function with respect to y on the interval $\left[\frac{1}{e}, e\right]$:

$$\int_{1/e}^{e} 5 \ln y \, dy = 5\left[y \ln y - y\right]_{y=1/e}^{y=e} = 5\left((e \ln e - e) - \left(\frac{1}{e} \ln \frac{1}{e} - \frac{1}{e}\right)\right) = 10e^{-1}. \quad \square$$

EXAMPLE 3

Computing volume

Find the volume of the solid between the graph of the function $f(x, y) = x \sin y$ and the rectangle $\mathcal{R}$,

$$\mathcal{R} = \{(x, y, 0) \mid -1 \le x \le 2 \text{ and } 0 \le y \le \pi\}$$

in the xy-plane.

SOLUTION

Following is a graph of the function f on the rectangular region $\mathcal{R}$:

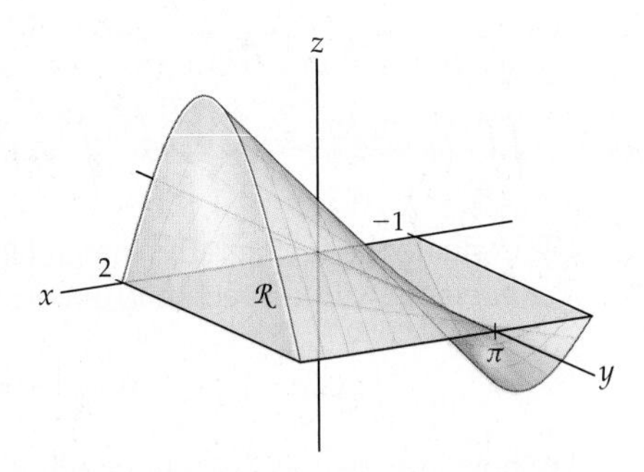

Note that $f(x, y) \ge 0$ on the rectangle $\mathcal{R}_1 = \{(x, y) \mid 0 \le x \le 2 \text{ and } 0 \le y \le \pi\}$ and $f(x, y) \le 0$ on the rectangle $\mathcal{R}_2 = \{(x, y) \mid -1 \le x \le 0 \text{ and } 0 \le y \le \pi\}$. Therefore the volume we want is given by the difference

$$\iint_{\mathcal{R}} |x \sin y| \, dA = \iint_{\mathcal{R}_1} x \sin y \, dA - \iint_{\mathcal{R}_2} x \sin y \, dA.$$

We will use Fubini's theorem to compute both of these double integrals:

$$\iint_{\mathcal{R}_1} x \sin y \, dA = \int_0^2 \int_0^{\pi} x \sin y \, dy \, dx \quad \text{and} \quad \iint_{\mathcal{R}_2} x \sin y \, dA = \int_{-1}^{0} \int_0^{\pi} x \sin y \, dy \, dx.$$

Both iterated integrals require us to compute the integral

$$\int_0^{\pi} x \sin y \, dy = \left[-x \cos y\right]_{y=0}^{y=\pi} = 2x.$$

We now finish the evaluation of the two iterated integrals:

$$\iint_{\mathcal{R}_1} x \sin y \, dA = \int_0^2 2x \, dx = \left[x^2\right]_{x=0}^{x=2} = 4 \text{ and } \iint_{\mathcal{R}_2} x \sin y \, dA = \int_{-1}^{0} 2x \, dx = \left[x^2\right]_{x=-1}^{x=0} = -1.$$

Therefore, the volume $\iint_{\mathcal{R}} |x \sin y| \, dA = 4 - (-1) = 5$. $\square$

? TEST YOUR UNDERSTANDING

- What are double sums and triple sums?
- What is a Riemann sum for a function of two variables over a rectangular region?
- What is the definition of a double integral?
- What is an iterated integral? How are iterated integrals defined? What is the difference between a double integral and an iterated integral?
- How is the volume between the graph of a function of two variables $f(x, y)$ and the xy-plane defined? Not every double integral gives a volume. Why? Which double integrals represent volumes? Which do not?

EXERCISES 13.1

Thinking Back

- *Using summation formulas:* Verify that the formulas $\sum_{k=1}^{n} k = \frac{n(n+1)}{2}$ and $\sum_{k=1}^{n} k^2 = \frac{n(n+1)(2n+1)}{6}$ hold when $n = 4$ and $n = 7$.
- *Using the definition to compute a definite integral:* Evaluate the definite integral $\int_1^4 x^2\,dx$ as the limit of a right Riemann sum.

Concepts

0. *Problem Zero:* Read the section and make your own summary of the material.

1. *True/False:* Determine whether each of the statements that follow is true or false. If a statement is true, explain why. If a statement is false, provide a counterexample.

(a) *True or False:*

$$\sum_{j=1}^{10}\sum_{k=1}^{15} jk^2 = \sum_{k=1}^{10}\sum_{j=1}^{15} j^2k$$

(b) *True or False:*

$$\sum_{j=1}^{m}\sum_{k=1}^{n} j^2k^3 = \left(\sum_{j=1}^{m} j^2\right)\left(\sum_{k=1}^{n} k^3\right)$$

(c) *True or False:*

$$\sum_{j=1}^{m}\sum_{k=1}^{n} e^{jk^2} = \left(\sum_{j=1}^{m} e^{j}\right)\left(\sum_{k=1}^{n} e^{k^2}\right)$$

(d) *True or False:* If $f(x, y)$ is continuous on the rectangle $\mathcal{R} = \{(x, y) \mid a \le x \le b \text{ and } c \le y \le d\}$, then $\iint_{\mathcal{R}} f(x, y)\,dA$ exists.

(e) *True or False:* If $f(x, y)$ is continuous on the rectangle $\mathcal{R} = \{(x, y) \mid a \le x \le b \text{ and } c \le y \le d\}$, then $\iint_{\mathcal{R}} f(x, y)\,dA = \int_a^b \int_c^d f(x, y)\,dx\,dy$.

(f) *True or False:* If $f(x, y)$ is continuous on the rectangle $\mathcal{R} = \{(x, y) \mid a \le x \le b \text{ and } c \le y \le d\}$, then $\iint_{\mathcal{R}} f(x, y)\,dA = \int_a^b \int_c^d f(x, y)\,dy\,dx$.

(g) *True or False:* If $f(x, y)$ is continuous on the rectangle $\mathcal{R} = \{(x, y) \mid a \le x \le b \text{ and } c \le y \le d\}$, then $\int_b^a \int_d^c f(x, y)\,dy\,dx = \int_a^b \int_c^d f(x, y)\,dy\,dx$.

(h) *True or False:* If $f(x, y)$ is continuous on the rectangle $\mathcal{R} = \{(x, y) \mid a \le x \le b \text{ and } c \le y \le d\}$, then $\int_c^d \int_a^b f(x, y)\,dx\,dy = \int_a^b \int_c^d f(x, y)\,dy\,dx$.

2. *Examples:* Construct examples of the thing(s) described in the following. Try to find examples that are different than any in the reading.

(a) Two different sigma notation expressions of the same double sum.

(b) A double summation whose value is zero although none of the summands is zero.

(c) A nonzero function $f(x, y)$ defined on the rectangle $\mathcal{R} = \{(x, y) \mid -2 \le x \le 2 \text{ and } -3 \le y \le 3\}$ such that $\iint_{\mathcal{R}} f(x, y)\,dA = 0$.

3. Express the sum $3e^4 + 3e^9 + 3e^{16} + 4e^4 + 4e^9 + 4e^{16}$ using double-summation notation.

4. Express the sum $\frac{2}{5} + \frac{2}{6} + \frac{2}{7} + \frac{2}{8} + \frac{4}{5} + \frac{4}{6} + \frac{4}{7} + \frac{4}{8}$ using double-summation notation.

5. How many summands are in $\sum_{j=3}^{13}\sum_{k=5}^{20} \frac{j^2}{e^{jk}}$?

6. How many summands are in $\sum_{j=j_0}^{m}\sum_{k=k_0}^{n} j^k$?

7. How many summands are in $\sum_{i=2}^{15}\sum_{j=3}^{17}\sum_{k=4}^{19} \frac{i}{j+k}$?

8. How many summands are in $\sum_{i=i_0}^{l}\sum_{j=j_0}^{m}\sum_{k=k_0}^{n} k^i j^k$?

9. Discuss the similarities and differences between the definition of the definite integral found in Chapter 4 and the definition of the double integral found in this section.

10. Explain how to construct a Riemann sum for a function of two variables over a rectangular region.

11. Explain how to construct a *midpoint* Riemann sum for a function of two variables over a rectangular region for which each (x_j^*, y_k^*) is the midpoint of the subrectangle

$$\mathcal{R}_{jk} = \{(x, y) \mid x_{j-1} \le x_j^* \le x_j \text{ and } y_{k-1} \le y_k^* \le y_k\}.$$

Refer to your answer to Exercise 10 or to Definition 13.3.

12. What is the difference between a double integral and an iterated integral?

13. State Fubini's theorem.

14. Explain why using an iterated integral to evaluate a double integral is often easier than using the definition of the double integral to evaluate the integral.

15. Explain how the Fundamental Theorem of Calculus is used in evaluating the iterated integral $\int_a^b \int_c^d f(x, y)\, dy\, dx$.

16. Explain how the Fundamental Theorem of Calculus is used in evaluating the iterated integral $\int_c^d \int_a^b f(x, y)\, dx\, dy$.

17. Earlier in this section, we showed that we could use Fubini's theorem to evaluate the integral $\iint_{\mathcal{R}} x^2 y\, dA$ and we showed that $\int_2^5 \int_1^3 x^2 y\ \, dx\, dy = 91$. Now evaluate the double integral by evaluating the iterated integral $\int_1^3 \int_2^5 x^2 y\, dy\, dx$.

Explain why it would be difficult to evaluate the double integrals in Exercises 18 and 19 as iterated integrals.

18. $\iint_{\mathcal{R}} e^{xy}\, dA$,
where $\mathcal{R} = \{(x, y) \mid 1 \le x \le 2 \text{ and } 1 \le y \le 3\}$

19. $\iint_{\mathcal{R}} \cos(xy)\, dA$,
where $\mathcal{R} = \left\{(x, y) \mid \frac{\pi}{4} \le x \le \frac{\pi}{2} \text{ and } \frac{\pi}{2} \le y \le \pi\right\}$

20. Use the results of Exercises 18 and 19 to explain why it may not be possible to evaluate a double integral by using an iterated integral.

21. Show that $\int_a^b \int_c^d f(x, y)\, dy\, dx$ does not always equal $\int_c^d \int_a^b f(x, y)\, dx\, dy$ by evaluating these two iterated integrals for $f(x, y) = \frac{x-y}{(x+y)^3}$ when $a = c = 0$ and $b = d = 1$. Why does this result not violate Fubini's theorem?

22. Outline the steps required to find the volume of the solid bounded by the graph of a function $f(x, y)$ and the xy-plane for $a \le x \le b$ and $c \le y \le d$.

Skills

Evaluate the sums in Exercises 23–28.

23. $\displaystyle\sum_{j=1}^{3}\sum_{k=1}^{2} j^k$

24. $\displaystyle\sum_{j=1}^{3}\sum_{k=1}^{2} k^j$

25. $\displaystyle\sum_{j=1}^{3}\sum_{k=1}^{4} (3j - 4k)$

26. $\displaystyle\sum_{j=1}^{m}\sum_{k=1}^{n} (j - k)$

27. $\displaystyle\sum_{i=1}^{4}\sum_{j=1}^{3}\sum_{k=1}^{2} ij^2k^3$

28. $\displaystyle\sum_{i=1}^{l}\sum_{j=1}^{m}\sum_{k=1}^{n} ij^2k^3$

Use Definition 13.4 to evaluate the double integrals in Exercises 29–32.

29. $\iint_{\mathcal{R}} xy\, dA$
where $\mathcal{R} = \{(x, y) \mid 0 \le x \le 2 \text{ and } 1 \le y \le 4\}$

30. $\iint_{\mathcal{R}} x^2 y\, dA$
where $\mathcal{R} = \{(x, y) \mid -1 \le x \le 0 \text{ and } 0 \le y \le 2\}$

31. $\iint_{\mathcal{R}} xy^3\, dA$
where $\mathcal{R} = \{(x, y) \mid -2 \le x \le 2 \text{ and } -1 \le y \le 1\}$

32. $\iint_{\mathcal{R}} x^2 y^3\, dA$
where $\mathcal{R} = \{(x, y) \mid 1 \le x \le 3 \text{ and } 0 \le y \le 2\}$

Evaluate each of the integrals in Exercises 33–36 as an iterated integral, and then compare your answers with those you found in Exercises 29–32.

33. The integral in Exercise 29.

34. The integral in Exercise 30.

35. The integral in Exercise 31.

36. The integral in Exercise 32.

Evaluate each of the double integrals in Exercises 37–54 as iterated integrals.

37. $\iint_{\mathcal{R}} (3 - x + 4y)\, dA$,
where $\mathcal{R} = \{(x, y) \mid 0 \le x \le 1 \text{ and } -1 \le y \le 3\}$

38. $\iint_{\mathcal{R}} y^2\, dA$,
where $\mathcal{R} = \{(x, y) \mid -3 \le x \le 2 \text{ and } -2 \le y \le 2\}$

39. $\iint_{\mathcal{R}} (2 - 3x^2 + y^2)\, dA$,
where $\mathcal{R} = \{(x, y) \mid -3 \le x \le 2 \text{ and } 3 \le y \le 5\}$

40. $\iint_{\mathcal{R}} (x - e^y)\, dA$,
where $\mathcal{R} = \{(x, y) \mid -3 \le x \le 2 \text{ and } -2 \le y \le 2\}$

41. $\iint_{\mathcal{R}} \sin(x + 2y)\, dA$,
where $\mathcal{R} = \left\{(x, y) \mid 0 \le x \le \pi \text{ and } 0 \le y \le \frac{\pi}{2}\right\}$

42. $\iint_{\mathcal{R}} x \sin x \cos y\, dA$,
where $\mathcal{R} = \{(x, y) \mid -3 \le x \le 2 \text{ and } -2 \le y \le 2\}$

43. $\iint_{\mathcal{R}} xe^{xy}\, dA$,
where $\mathcal{R} = \{(x, y) \mid 0 \le x \le 1 \text{ and } 0 \le y \le \ln 5\}$

44. $\iint_{\mathcal{R}} x^2 \cos(xy)\, dA$,
where $\mathcal{R} = \{(x, y) \mid 0 \le x \le \pi \text{ and } 0 \le y \le 1\}$

45. $\iint_{\mathcal{R}} \frac{x}{x+y}\, dA$,
where $\mathcal{R} = \{(x, y) \mid 1 \le x \le 4 \text{ and } 0 \le y \le 3\}$

46. $\iint_{\mathcal{R}} x^3 e^{x^2 y}\, dA$,
where $\mathcal{R} = \{(x, y) \mid 0 \le x \le 4 \text{ and } -1 \le y \le 1\}$

47. $\iint_{\mathcal{R}} y \sin x\, dA$,
where $\mathcal{R} = \left\{(x, y) \mid 0 \le x \le \frac{\pi}{2} \text{ and } 0 \le y \le 1\right\}$

48. $\iint_{\mathcal{R}} \sin(2x + y)\, dA$,
where $\mathcal{R} = \left\{(x, y) \mid 0 \le x \le \frac{\pi}{2} \text{ and } 0 \le y \le \frac{\pi}{2}\right\}$

49. $\iint_{\mathcal{R}} y^2 \sin x\, dA$,
where $\mathcal{R} = \{(x, y) \mid 0 \le x \le \pi \text{ and } 0 \le y \le 3\}$

50. $\iint_{\mathcal{R}} xy\sin(x^2)\,dA$,
where $\mathcal{R} = \{(x, y) \mid 0 \le x \le \sqrt{\pi} \text{ and } 0 \le y \le 1\}$

51. $\iint_{\mathcal{R}} y\cos(xy)\,dA$,
where $\mathcal{R} = \left\{(x, y) \mid 0 \le x \le \frac{\pi}{2} \text{ and } 0 \le y \le 1\right\}$

52. $\iint_{\mathcal{R}} e^{x+y}\,dA$,
where $\mathcal{R} = \{(x, y) \mid 0 \le x \le 1 \text{ and } 0 \le y \le 1\}$

53. $\iint_{\mathcal{R}} x^2 e^{xy}\,dA$,
where $\mathcal{R} = \{(x, y) \mid 0 \le x \le 1 \text{ and } 0 \le y \le 1\}$

54. $\iint_{\mathcal{R}} \frac{dA}{x^2+2xy+y^2}$,
where $\mathcal{R} = \{(x, y) \mid 1 \le x \le 2 \text{ and } 0 \le y \le 1\}$

In Exercises 55–58, find the signed volume between the graph of the given function and the xy-plane over the specified rectangle in the xy-plane.

55. $f(x, y) = 3x - 2y^5 + 1$,
where $\mathcal{R} = \{(x, y) \mid -4 \le x \le 6 \text{ and } 0 \le y \le 7\}$

56. $f(x, y) = x^2 - y^3 + 4$,
where $\mathcal{R} = \{(x, y) \mid 0 \le x \le 3 \text{ and } -2 \le y \le 3\}$

57. $f(x, y) = y^3 e^{xy^2}$,
where $\mathcal{R} = \{(x, y) \mid 0 \le x \le 2 \text{ and } -2 \le y \le 3\}$

58. $f(x, y) = xy^3 e^{x^2y^2}$,
where $\mathcal{R} = \{(x, y) \mid -2 \le x \le -1 \text{ and } -2 \le y \le 0\}$

In Exercises 59–64, find the volume between the graph of the given function and the xy-plane over the specified rectangle.

59. $f(x, y) = xy$,
where $\mathcal{R} = \{(x, y) \mid -2 \le x \le 3 \text{ and } -1 \le y \le 5\}$

60. $f(x, y) = -2x^2y^3$,
where $\mathcal{R} = \{(x, y) \mid -2 \le x \le 3 \text{ and } -1 \le y \le 5\}$

61. $f(x, y) = \sin x \cos y$,
where $\mathcal{R} = \{(x, y) \mid 0 \le x \le \pi \text{ and } 0 \le y \le \pi\}$

62. $f(x, y) = \frac{y}{x} e^y$,
where $\mathcal{R} = \{(x, y) \mid 1 \le x \le e \text{ and } 0 \le y \le 2\}$

63. $f(x, y) = \frac{x}{y} + \frac{y}{x}$,
where $\mathcal{R} = \{(x, y) \mid 1 \le x \le 3 \text{ and } 1 \le y \le 5\}$

64. $f(x, y) = x^2 y e^{xy}$,
where $\mathcal{R} = \{(x, y) \mid 0 \le x \le 1 \text{ and } 0 \le y \le 2\}$

Use midpoint Riemann sums with the specified numbers of subintervals to approximate the iterated integrals in Exercises 65–68. (See Exercise 11 for the definition of a midpoint Riemann sum for a double integral.)

65. $\int_0^1 \int_0^{3/2} e^{xy}\,dx\,dy$. Let each subrectangle be a square with side length $\frac{1}{2}$ unit.

66. $\int_0^{\pi} \int_0^1 \sin(xy)\,dy\,dx$. Use four subrectangles by dividing each of the intervals $[0, \pi]$ and $[0, 1]$ into two equal pieces.

67. $\int_0^1 \int_0^{\pi/2} \cos(xy)\,dx\,dy$. Use six subrectangles by dividing the interval $\left[0, \frac{\pi}{2}\right]$ into three equal pieces and the interval $[0, 1]$ into two equal pieces.

68. $\int_0^1 \int_0^1 y\sin(x^2)\,dx\,dy$. Let each subrectangle be a square with side length $\frac{1}{3}$ unit.

Applications

69. Emmy oversees the operations of a number of sedimentation tanks, into which a toxic solution is dumped so that the toxic materials will settle out. The base of each tank is a square with side lengths of 80 feet, but she does not know the volume of any of the tanks. She defines a function

$$r(t) = \begin{cases} \frac{-(t-10)^2}{100} + 1, & \text{if } t < 10 \\ 1, & \text{if } t \ge 10. \end{cases}$$

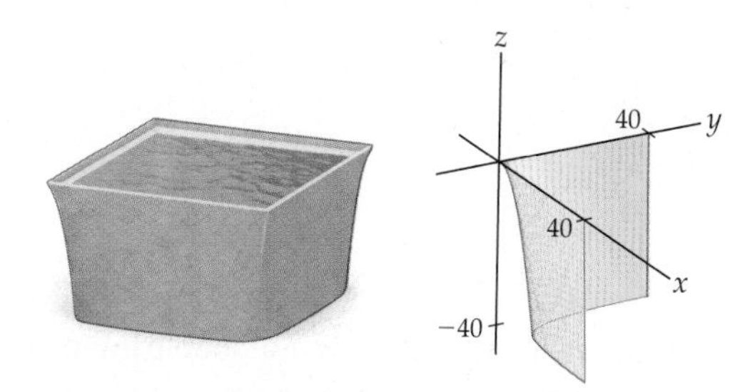

With this function, one-fourth of a tank is described by the surface $-12r(x)r(y)$ for $x \in [0, 40)$, $y \in [0, 40)$. The rest of the tank is the same function reflected about the lines $x = 40$ and $y = 40$ and the point $(40, 40)$. What is the volume of each tank?

70. Leila has been assigned the task of estimating the number of caribou in a rectangular management unit (an area of land) during the summer. She imposes coordinates on the unit, which is 4 miles wide by 5 miles long (i.e., $[0, 4] \times [0, 5]$). She does not have time to commission a count of population, but she knows from a past study that the density of caribou in this region in July is approximated by $d(x, y) = 0.08x^2y^2 - 0.456x^2y - 0.08x^2 - 0.328xy^2 + 1.87xy + 0.328x - 0.061y^2 + 0.347y + 0.061$. Roughly how many caribou can be found in the management unit?

Proofs

71. Prove Theorem 13.2(a). That is, prove that $\sum_{j=1}^{m}\sum_{k=1}^{n} a_{jk} = \sum_{k=1}^{n}\sum_{j=1}^{m} a_{jk}$.
72. List the other five orders for the summations in $\sum_{i=1}^{l}\sum_{j=1}^{m}\sum_{k=1}^{n} a_{ijk}$, and prove that all six orders for the summations are equal.
73. Let $a < b$ and $c < d$ be real numbers, and let $\mathcal{R}$ be the rectangle defined by
$$\mathcal{R} = \{(x, y) \mid a \le x \le b \text{ and } c \le y \le d\}$$
in the xy-plane. If $g(x)$ is continuous on the interval $[a, b]$ and $h(y)$ is continuous on $[c, d]$, use Fubini's theorem to prove that
$$\iint_{\mathcal{R}} g(x)h(y)\, dA = \left(\int_a^b g(x)\, dx\right)\left(\int_c^d h(y)\, dy\right).$$
74. Let a and b be positive real numbers, and let
$$\mathcal{R} = \{(x, y) \mid -a \le x \le a \text{ and } -b \le y \le b\}.$$
Assuming that g and h are continuous on their domains, prove that $\iint_{\mathcal{R}} g(x)h(y)\, dA = 0$ if either g or h is an odd function.
75. Let $a < b$ and $c < d$ be real numbers, and let $\mathcal{R}$ be the rectangle in the xy-plane defined by
$$\mathcal{R} = \{(x, y) \mid a \le x \le b \text{ and } c \le y \le d\}.$$
Prove that $\iint_{\mathcal{R}} dA = (b-a)(d-c)$. What is the relationship between $\mathcal{R}$ and the product $(b-a)(d-c)$?

Thinking Forward

- *Riemann sums for functions of three variables:* Provide a definition of a Riemann sum for a function of three variables. Model your definition on Definition 13.3.
- *Triple integrals:* Provide a definition for a triple integral $\iiint_{\mathcal{S}} f(x, y, z)\, dV$, where $\mathcal{S}$ is a rectangular box in $\mathbb{R}^3$ given by $\mathcal{S} = \{(x, y, z) \mid a_1 \le x \le a_2,\ b_1 \le y \le b_2 \text{ and } c_1 \le z \le c_2\}$. Model your definition on Definition 13.4.

13.2 DOUBLE INTEGRALS OVER GENERAL REGIONS

- Analyzing bounded regions in the plane
- Defining and computing double integrals on general regions
- The algebraic properties of double integrals

General Regions in the Plane

In Section 13.1 we defined the double integral of a function of two variables over a rectangular region. We now wish to generalize this concept to define a double integral over a general region Ω (omega) bounded by a simple closed curve—that is, a two-dimensional region in which the boundary may be represented by a finite collection of curves with differentiable parametrizations. Following is an example of such a region:

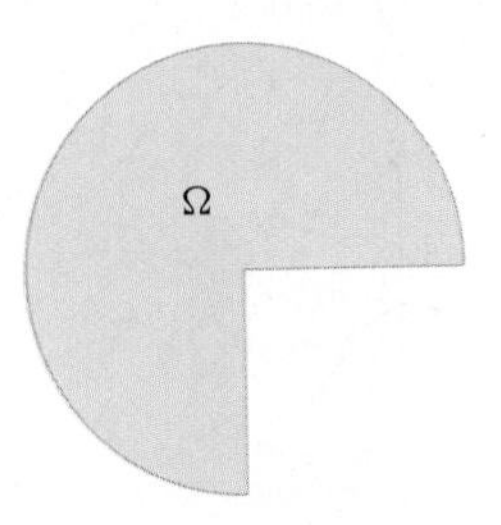

Unlike definite integrals, which are defined on finite intervals, for double intervals over regions more complicated than a rectangle we have to make an effort to understand the region Ω over which the double integral is defined.

In particular, the regions we use must be subdivided into subregions that fall into one of the following two types:

DEFINITION 13.8

Type I and Type II Regions

(a) Let $y = g_1(x)$ and $y = g_2(x)$ be two functions defined on the interval $[a, b]$ such that $g_1(x) \le g_2(x)$ for every $x \in [a, b]$. The region Ω bounded above by $g_2(x)$, below by $g_1(x)$, on the left by the line $x = a$, and on the right by the line $x = b$ is said to be a ***type I region***.

(b) Let $x = h_1(y)$ and $x = h_2(y)$ be two functions defined on the interval $[c, d]$ such that $h_1(y) \le h_2(y)$ for every $y \in [c, d]$. The region Ω bounded on the left by $h_1(x)$, on the right by $h_2(y)$, below by the line $y = c$, and above by the line $y = d$ is said to be a ***type II region***.

A type I region is shown next at the left, and a type II region next on the right.

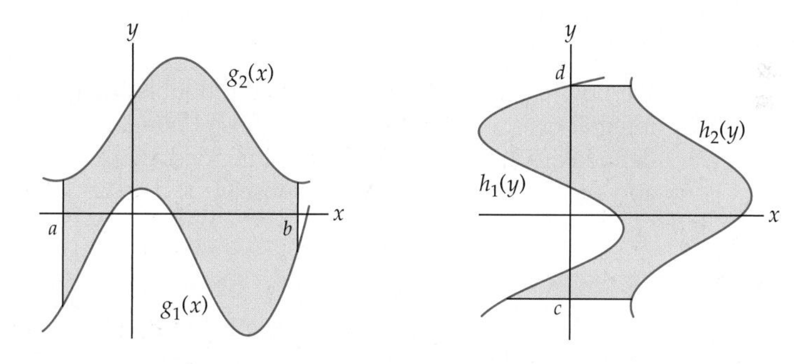

Many regions naturally fall into one or the other of these two categories. Some regions, however, may be considered to be either a type I or a type II region. For example, consider the area in the first quadrant between the functions $y = \sqrt{x}$ and $y = x^3$:

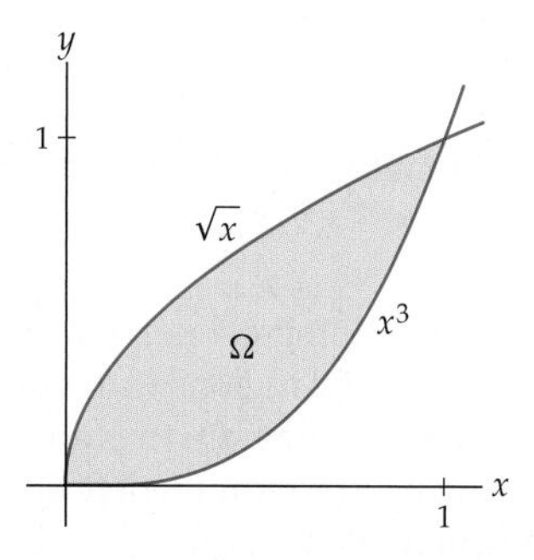

As a type I region, Ω is bounded above by $y = \sqrt{x}$ and below by $y = x^3$ for $x \in [0, 1]$. As a type II region, Ω is bounded on the left by $x = y^2$ and on the right by $x = \sqrt[3]{y}$ for $y \in [0, 1]$.

Many regions need to be subdivided so that their subregions fall more simply into one of the two categories. For example, the region Ω bounded above by the semicircle with equation $y = \sqrt{4 - x^2}$ and below by the quarter circle $y = -\sqrt{4 - x^2}$ on the interval $[-2, 0]$ and by the x-axis on the interval $[0, 2]$ may be subdivided into the two subregions Ω_1 and Ω_2, as pictured next. We see that Ω_1 and Ω_2 are type I regions.

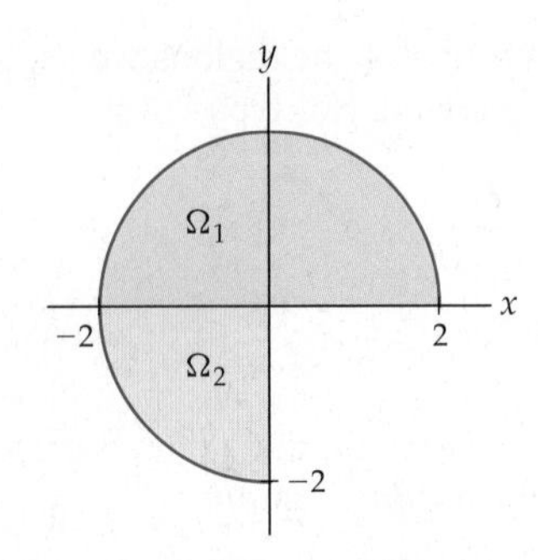

Note, however, that there are always many ways to subdivide such a region. If we wish, we may divide this region into three quarter circles, each of which may be considered to be a type I or a type II region. For example, Ω_2, shown in the preceding figure, is also a type II region, bounded on the left by $x = -\sqrt{4 - y^2}$ and on the right by the y-axis on the interval $-2 \le y \le 0$.

Double Integrals over General Regions

We now expand our definition of the double integral to allow us to compute a double integral over a general region Ω whose boundary is a simple closed curve. Let $f(x, y)$ be defined on Ω. Note that since Ω is bounded, there exists a rectangle $\mathcal{R} = \{(x, y) \mid a \le x \le b$ and $c \le y \le d\}$ such that Ω is a subset of $\mathcal{R}$ (i.e., $\Omega \subseteq \mathcal{R}$).

DEFINITION 13.9 **Double Integrals over General Regions**

Let Ω be a region in the xy-plane bounded by a simple closed curve, let $f(x, y)$ be a function defined on Ω, and let $\mathcal{R} = \{(x, y) \mid a \le x \le b$ and $c \le y \le d\}$ be a rectangle containing Ω. We define the ***double integral*** of f over Ω to be

$$\iint_{\Omega} f(x, y)\, dA = \iint_{\mathcal{R}} F(x, y)\, dA,$$

where

$$F(x, y) = \begin{cases} f(x, y), & \text{if } (x, y) \in \Omega \\ 0, & \text{if } (x, y) \notin \Omega, \end{cases}$$

provided that the double integral $\iint_{\mathcal{R}} F(x, y)\, dA$ exists.

As with double integrals defined on rectangular regions, we will try to evaluate double integrals on general regions with Fubini's theorem. To use Fubini's theorem in this context, when we wish to integrate the function $f(x, y)$ over the type I region Ω bounded above by $g_2(x)$, below by $g_1(x)$, on the left by the line $x = a$, and on the right by the line $x = b$, we set up the iterated integral

$$\iint_{\Omega} f(x, y)\, dA = \int_a^b \int_{g_1(x)}^{g_2(x)} f(x, y)\, dy\, dx.$$

Similarly, to integrate $f(x, y)$ over the type II region bounded on the left by $h_1(x)$, on the right by $h_2(y)$, below by the line $y = c$, and above by the line $y = d$, we set up the iterated integral

$$\iint_{\Omega} f(x, y)\, dA = \int_c^d \int_{h_1(y)}^{h_2(y)} f(x, y)\, dx\, dy.$$

Note that in each case the limits of the outer integration are constants, and the limits of the inner integration are functions of the outer variable of integration.

For example, let Ω be the following triangular region in the first quadrant:

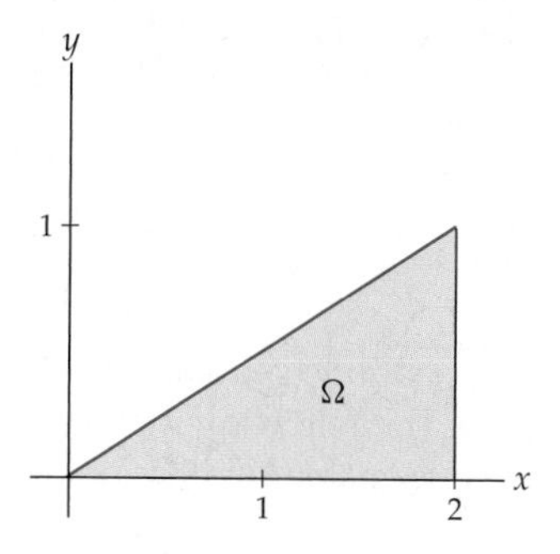

To evaluate the double integral of a function $f(x, y)$ on this region, we may treat Ω either as the type I region bounded below by $y = 0$ and above by $y = \frac{1}{2}x$ for $x \in [0, 2]$ or as the type II region bounded on the left by the function $x = 2y$ and on the right by the function $x = 2$ for $y \in [0, 1]$. Therefore,

$$\iint_{\Omega} f(x, y)\, dA = \int_0^2 \int_0^{x/2} f(x, y)\, dy\, dx = \int_0^1 \int_{2y}^2 f(x, y)\, dx\, dy.$$

Depending upon the function $f(x, y)$, one of these two iterated integrals may be significantly easier to evaluate. When $f(x, y) = \cos(x^2)$, treating Ω as a type I region results in a significantly simpler integral, since the inner integral may be easily evaluated with the Fundamental Theorem of Calculus:

$$\int_0^{x/2} \cos(x^2)\, dy = \left[y \cos(x^2)\right]_{y=0}^{y=x/2} = \frac{1}{2}x\cos(x^2).$$

To finish evaluating the iterated integral, we now use substitution to find an antiderivative for the resulting function. We have

$$\int_0^2 \frac{1}{2}x\cos(x^2)\, dx = \left[\frac{1}{4}\sin(x^2)\right]_{x=0}^{x=2} = \frac{1}{4}\sin 4.$$

By contrast, if we try to evaluate the double integral by using the iterated integral

$$\int_0^1 \int_{2y}^2 \cos(x^2)\, dx\, dy,$$

we are immediately stymied by the fact that the function $\cos(x^2)$ does *not* have a simple antiderivative when we are integrating first with respect to the variable x.

This example illustrates how crucial it is to be able to analyze a given planar region as both a type I region and a type II region. When we write a double integral as an iterated integral, one of the two possible orders of integration can result in a computation that is significantly easier than the other. In such a case, the relatively simple chore of selecting the appropriate order of integration saves a great deal of work integrating.

Algebraic Properties of Double Integrals

The algebraic properties set forth in the next theorem are similar to the algebraic properties of definite integrals we studied in Chapter 4. The proofs of these properties follow from the definition of the double integral and are left for Exercises 67 and 68.

THEOREM 13.10

Algebraic Properties of the Double Integral

Let $f(x, y)$ and $g(x, y)$ be integrable functions on the general region Ω, and let $c \in \mathbb{R}$. Then

(a) $\displaystyle\iint_{\Omega} cf(x, y)\, dA = c\iint_{\Omega} f(x, y)\, dA.$

(b) $\displaystyle\iint_{\Omega} (f(x, y) + g(x, y))\, dA = \iint_{\Omega} f(x, y)\, dA + \iint_{\Omega} g(x, y)\, dA.$

As we mentioned earlier in the section, we may also need to decompose a region into subregions. The following theorem tells us that the double integral is well behaved with respect to reasonable subdivisions:

THEOREM 13.11

Subdividing the Region on Which a Double Integral Is Defined

Let $f(x, y)$ be an integrable function on the general region Ω. If Ω_1 and Ω_2 are general regions that are subsets of Ω that do not overlap, except possibly on their boundaries, and if $\Omega = \Omega_1 \cup \Omega_2$, then

$$\iint_{\Omega} f(x, y)\, dA = \iint_{\Omega_1} f(x, y)\, dA + \iint_{\Omega_2} f(x, y)\, dA.$$

Example 1 provides an application of Theorem 13.11 that illustrates how to split a region into subregions for integration.

Examples and Explorations

EXAMPLE 1

Expressing a planar region as a type I region and a type II region

Express the double integral $\iint_{\Omega} f(x, y)\, dA$ as an iterated integral over the following region Ω, itself expressed as both a type I region and a type II region:

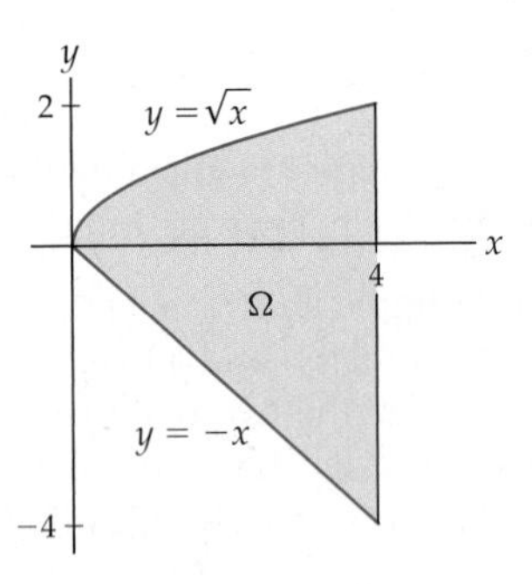

SOLUTION

It is slightly easier to express Ω as type I region. To envision the boundary components of a type I region, it often helps to draw a "typical" vertical strip, as shown in the figure at the left.

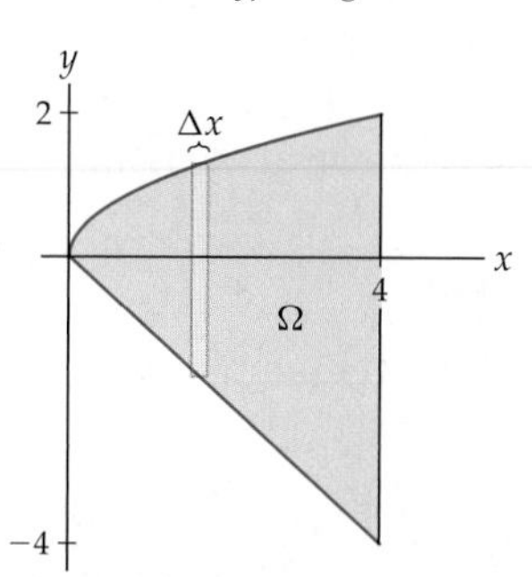

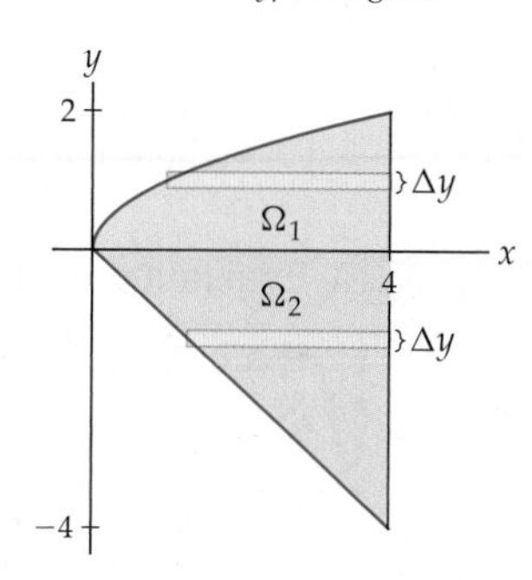

Each vertical strip that crosses the region is bounded below by the function $y = -x$ and above by the function $y = \sqrt{x}$. Such vertical strips would cover Ω as x varies over the interval $[0, 4]$. Thus, to integrate $f(x, y)$ over Ω as a type I region, we have

$$\iint_{\Omega} f(x, y)\, dA = \int_0^4 \int_{-x}^{\sqrt{x}} f(x, y)\, dy\, dx.$$

To understand the region as a type II region, we draw two horizontal strips. Above the x-axis, these strips are bounded on the left by the function $x = y^2$ and on the right by the line $x = 4$. Such strips would cover Ω_1 as y varies over the interval $[0, 2]$, as pictured at the right. Similarly, below the x-axis such strips are bounded on the left by the function $x = -y$ and on the right by the line $x = 4$. They would cover Ω_2 as y varies over the interval $[-4, 0]$. Thus, to integrate a function $f(x, y)$ over Ω as a type II region, we would have

$$\iint_{\Omega} f(x, y)\, dA = \int_{-4}^{0} \int_{-y}^{4} f(x, y)\, dx\, dy + \int_0^2 \int_{y^2}^{4} f(x, y)\, dx\, dy.$$

Note that we have used Theorem 13.11 to break the integral over Ω into the integrals over Ω_1 and Ω_2. Also, observe that for all three iterated integrals shown, the limits of the outer integrals are constants. The limits of the inner integrals are functions of the outer variable, although they may be constant functions. □

EXAMPLE 2

Drawing a region determined by an iterated integral

Sketch the regions determined by the iterated integrals:

(a) $\displaystyle\int_{-3\pi/4}^{\pi/4} \int_{\cos x}^{\sin x} xy\, dy\, dx$ **(b)** $\displaystyle\int_0^2 \int_0^{\sqrt{2-y}} xe^{x^2}\, dx\, dy$

SOLUTION

(a) The region described by the first integral is bounded below by the function $y = \cos x$ and above by $y = \sin x$ for values of x in the interval $[-3\pi/4, \pi/4]$. This region is shown here at the left:

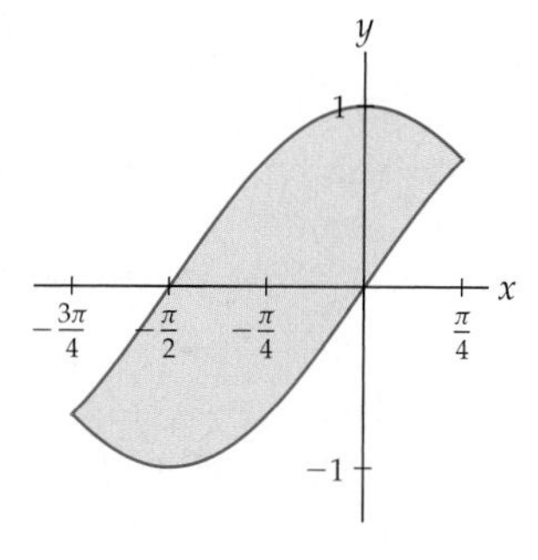

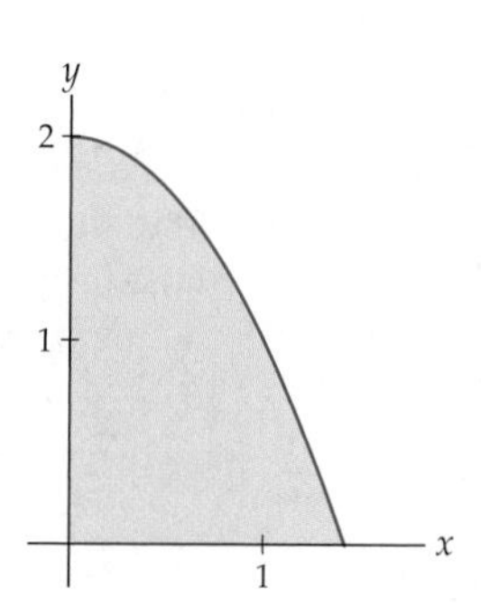

(b) The region described by the second integral is bounded on the left by the line $x = 0$ and on the right by the function $x = \sqrt{2-y}$ for values of y in the interval $[0, 2]$. For convenience, to sketch the function $x = \sqrt{2-y}$, we may rewrite it as $y = 2 - x^2$, where $x \geq 0$. This region is shown at the right. □

EXAMPLE 3

Evaluating an iterated integral

Evaluate the iterated integrals from Example 2.

SOLUTION

To evaluate an iterated integral, we begin by evaluating the inner integral.

(a) Thus, when we evaluate $\int_{-3\pi/4}^{\pi/4} \int_{\cos x}^{\sin x} xy\,dy\,dx$, we start with

$$\int_{\cos x}^{\sin x} xy\,dy = \left[\frac{1}{2}xy^2\right]_{y=\cos x}^{y=\sin x} = \frac{1}{2}x(\sin^2 x - \cos^2 x) = -\frac{1}{2}x\cos(2x).$$

For the rightmost equality, we used the double-angle identity for the cosine. We now integrate this function on the interval $\left[-\frac{3\pi}{4}, \frac{\pi}{4}\right]$ to complete the evaluation of the iterated integral:

$$\int_{-3\pi/4}^{\pi/4} \left(-\frac{1}{2}x\cos(2x)\right) dx = \left[-\left(\frac{1}{4}x\sin(2x) + \frac{1}{8}\cos(2x)\right)\right]_{-3\pi/4}^{\pi/4} = -\frac{\pi}{4}.$$

Note that we used integration by parts to find the antiderivative of the function.

(b) For the second iterated integral, $\int_0^2 \int_0^{\sqrt{2-y}} xe^{x^2}\,dx\,dy$, we again work first on the inner integral. Here we use substitution to find the antiderivative:

$$\int_0^{\sqrt{2-y}} xe^{x^2}\,dx = \left[\frac{1}{2}e^{x^2}\right]_0^{\sqrt{2-y}} = \frac{1}{2}(e^{2-y} - 1).$$

We now evaluate the outer integral:

$$\int_0^2 \frac{1}{2}(e^{2-y} - 1)\,dy = \left[-\frac{1}{2}(e^{2-y} + y)\right]_0^2 = \frac{1}{2}(e^2 - 3).$$ □

EXAMPLE 4

Reversing the order of integration in an iterated integral

Reverse the order of integration in the iterated integral $\int_0^2 \int_0^{\sqrt{2-y}} xe^{x^2}\,dx\,dy$. Then evaluate the new iterated integral and show that the result is the same as the value obtained in Example 3.

SOLUTION

This is one of the integrals we have been discussing since Example 2. In that example, we drew the region Ω, expressed as a type II region. To reverse the order of integration we must first express Ω as a type I region. Using the work we did in Example 2, we see that the region is bounded above by $y = 2 - x^2$ and below by $y = 0$, where x is in the interval $\left[0, \sqrt{2}\right]$. Thus,

$$\int_0^2 \int_0^{\sqrt{2-y}} xe^{x^2}\,dx\,dy = \int_0^{\sqrt{2}} \int_0^{2-x^2} xe^{x^2}\,dy\,dx.$$

Again, working on the inner integral, we have

$$\int_0^{2-x^2} xe^{x^2}\,dy = \left[xe^{x^2}y\right]_0^{2-x^2} = (2x - x^3)e^{x^2}.$$

To finish this example we evaluate the outer integral. Using the techniques of integration by substitution and integration by parts, we obtain

$$\int_0^{\sqrt{2}} (2x - x^3)e^{x^2}\,dx = \left[\frac{1}{2}(3 - x^2)e^{x^2}\right]_0^{\sqrt{2}} = \frac{1}{2}(e^2 - 3).$$

As we have previously mentioned, when we evaluate an iterated integral, sometimes one order of integration leads to a computation that is significantly simpler than the other order of integration. Which order of integration did you find easier for this integral, the one we did in Example 3, or the one we just completed? □

? TEST YOUR UNDERSTANDING

- What is a general region in the xy-plane? What is a type I region? What is a type II region? Is every general region either a type I region or a type II region?
- How is the definition of the double integral over a rectangular region used to define the double integral of a function over a general region? Why does this definition make sense?
- How is a region that is neither a type I region nor a type II region decomposed into subregions that are type I or type II?
- Which order of integration in an iterated integral is associated with a type I region, and which order of integration in an iterated integral is associated with a type II region? Why?
- How can the shape of a region Ω help or hinder in the evaluation of the double integral $\iint_\Omega f(x, y)\,dA$? Does the interval $[a, b]$ ever make a definite integral $\int_a^b g(x)\,dx$ easier or harder to evaluate?

EXERCISES 13.2

Thinking Back

- *Finding the area between two curves:* If $g_1(x)$ and $g_2(x)$ are two continuous functions such that $g_1(x) \le g_2(x)$ on the interval $[a, b]$, find a definite integral representing the area between the graphs of the two functions on $[a, b]$.
- *Finding the area between two curves:* If $h_1(y)$ and $h_2(y)$ are two continuous functions such that $h_1(y) \le h_2(y)$ on the interval $[c, d]$, find a definite integral representing the area between the graphs of the two functions on $[c, d]$.

Concepts

0. *Problem Zero:* Read the section and make your own summary of the material.

1. *True/False:* Determine whether each of the statements that follow is true or false. If a statement is true, explain why. If a statement is false, provide a counterexample.

(a) *True or False:* Every rectangular region in the plane, with its sides parallel to the coordinate axes, may be considered to be either a type I region or a type II region.

(b) *True or False:* Every rectangular region in the plane may be considered to be a type I region or a type II region.

(c) *True or False:* Every region in the plane can be expressed as either a single type I region or a single type II region.

(d) *True or False:* A general region in the plane with a polygonal boundary can be decomposed into finitely many type I and type II regions.

(e) *True or False:* To evaluate the double integral $\iint_\Omega f(x,y)\,dA$ where Ω is a type I region, you integrate first with respect to y.

(f) *True or False:* If Ω is the set of all points satisfying the inequality $x^2+y^2 \le 4$, then $\iint_\Omega f(x,y)\,dA = \int_{-2}^{2}\int_{-\sqrt{4-y^2}}^{\sqrt{4-y^2}} f(x,y)\,dx\,dy$.

(g) *True or False:* If f is a continuous function and $\Omega = \Omega_1 \cup \Omega_2$, then $\iint_\Omega f(x,y)\,dA = \iint_{\Omega_1} f(x,y)\,dA + \iint_{\Omega_2} f(x,y)\,dA$.

(h) *True or False:* If f is a positive continuous function defined on a region Ω, and if $\Gamma \subset \Omega$, then $\iint_\Gamma f(x,y)\,dA \le \iint_\Omega f(x,y)\,dA$.

2. *Examples:* Construct examples of the thing(s) described in the following. Try to find examples that are different than any in the reading.

(a) A non-rectangular region that can be expressed as a single type I region or as a single type II region.

(b) A rectangular region Ω that can be expressed neither as a single type I region nor as a single type II region.

(c) A region Ω that can be expressed as a single type II region, but that requires two type I regions intersecting only on their boundaries, in order to express Ω simply.

3. Explain the difference between a type I region and a type II region.

4. Let $\mathcal{R} = \{(x,y) \mid a \le x \le b \text{ and } c \le y \le d\}$ be a rectangular region. Explain why $\mathcal{R}$ is both a type I region and a type II region.

Which of the iterated integrals in Exercises 5–8 could correctly be used to evaluate the double integral $\iint_{\mathcal{R}} f(x,y)\,dA$, where $f(x,y)$ is a continuous function and $\mathcal{R}$ is the rectangular region bounded by the lines $x=1$, $x=4$, $y=2$, and $y=6$? For each incorrect integral, how could it be changed to give the correct value?

5. $\displaystyle\int_1^4\int_2^6 f(x,y)\,dy\,dx$

6. $\displaystyle\int_4^1\int_6^2 f(x,y)\,dy\,dx$

7. $\displaystyle\int_1^4\int_2^6 f(x,y)\,dx\,dy$

8. $\displaystyle\int_2^6\int_1^4 f(x,y)\,dx\,dy$

Which of the iterated integrals in Exercises 9–12 could correctly be used to evaluate the double integral $\iint_\Omega f(x,y)\,dA$, where $f(x,y)$ is a continuous function and Ω is the right triangular region bounded below by the x-axis, on the left by the y-axis, and along the hypotenuse by the line $y=-x+2$? For each incorrect integral, how could it be changed to give the correct value?

9. $\displaystyle\int_0^2\int_0^{-x+2} f(x,y)\,dy\,dx$

10. $\displaystyle\int_0^{-x+2}\int_0^2 f(x,y)\,dx\,dy$

11. $\displaystyle\int_0^2\int_0^{-y+2} f(x,y)\,dx\,dy$

12. $\displaystyle\int_2^0\int_{-y+2}^0 f(x,y)\,dx\,dy$

13. The following region Ω is bounded by the functions $y=\frac{1}{2}x$ and $y=\sqrt{x}$:

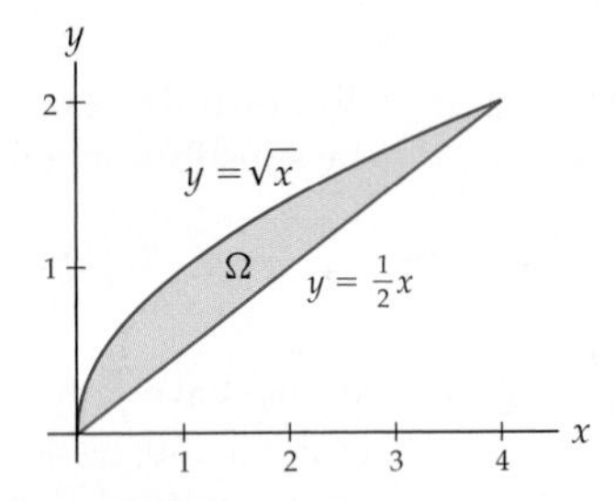

Express Ω as a type I region and as a type II region. Referring to Definition 13.8, if Ω is a type I region, what are a and b? If Ω is a type II region, what are c and d?

14. Explain why the double integral $\iint_\Omega dA$ gives the area of the region Ω. Illustrate your explanation with an example.

15. Let $g_1(x)$ and $g_2(x)$ be two continuous functions such that $g_1(x) \le g_2(x)$ on the interval $[a,b]$, and let Ω be the region in the xy-plane bounded by g_1 and g_2 on $[a,b]$. Use your answer to Exercise 14 to set up an iterated integral whose value is the area of Ω. How is this iterated integral related to the definite integral you would have used to compute the area of Ω in Chapter 4?

16. Let $h_1(y)$ and $h_2(y)$ be two continuous functions such that $h_1(x) \le h_2(x)$ on the interval $[c,d]$, and let Ω be the region in the xy-plane bounded by h_1 and h_2 on $[c,d]$. Use your answer to Exercise 14 to set up an iterated integral whose value is the area of Ω. How is this iterated integral related to the definite integral you would have used to compute the area of Ω in Chapter 4?

17. Use the results of Exercises 15 and 16 to find the area of the region Ω shown in Exercise 13.

18. Express the area of the region Ω between the function $f(x)=x^2$ and the x-axis on the interval $[-3,3]$ as an iterated integral, integrating first with respect to x. Express the area of Ω as a sum of two iterated integrals, integrating first with respect to y in each. Now evaluate your integrals.

19. Express the area of the region Ω between the function $f(x)=x^3$ and the x-axis on the interval $[-3,3]$ as a sum of two iterated integrals, integrating first with respect to x in each. Express the area of Ω as a sum of two different iterated integrals, integrating first with respect to y. Now evaluate your integrals.

20. When you wish to evaluate the definite integral $\int_a^b f(x)\,dx$ of a continuous function f, the interval $[a,b]$ is never an impediment to using the Fundamental Theorem of Calculus. However, when you wish to evaluate the double integral $\iint_\Omega g(x,y)\,dA$ of a continuous function g over a region Ω, the region can make the evaluation process easier or harder. Why?

Skills

Let $f(x, y)$ be a continuous function. For each region Ω shown in Exercises 21–24, set up one or more (if necessary) iterated integrals to compute $\iint_\Omega f(x, y)\, dA$, (a) where you integrate first with respect to y and (b) where you integrate first with respect to x.

21.

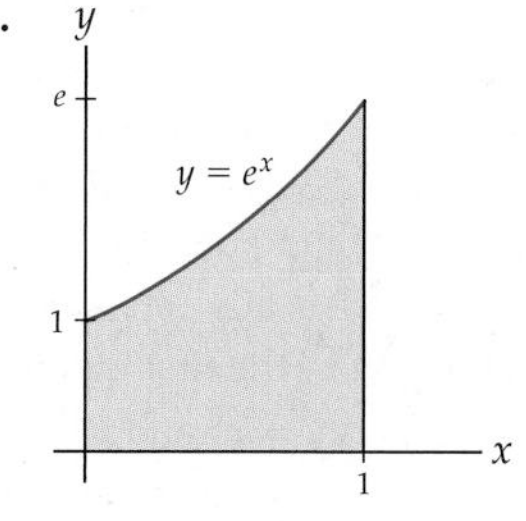

22.

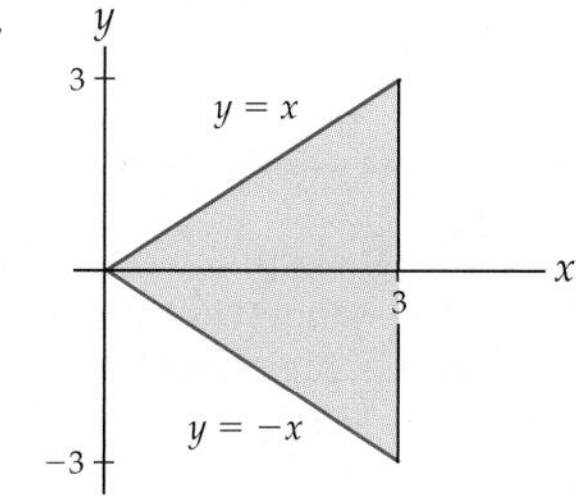

23.

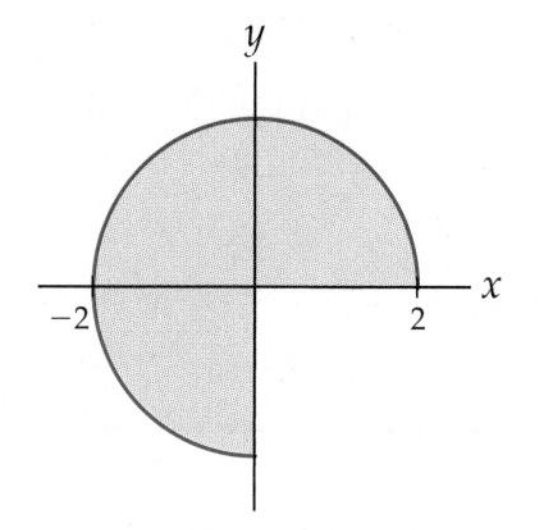

24.

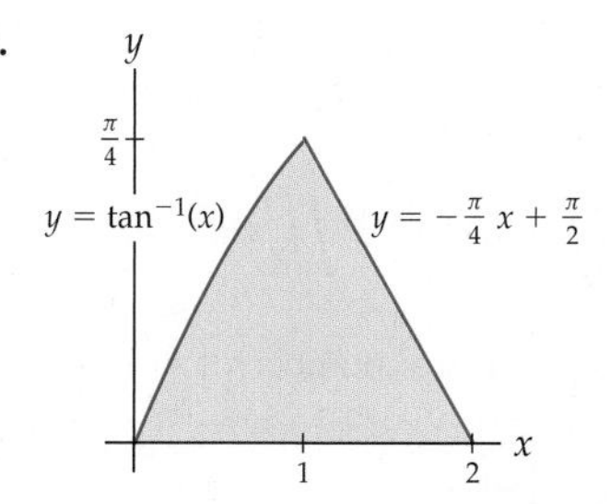

Let $f(x, y)$ be a continuous function. Sketch each region Ω described in Exercises 25–28. Then set up one or more (if necessary) iterated integrals to compute $\iint_\Omega f(x, y)\, dA$, (a) where you integrate first with respect to y and (b) where you integrate first with respect to x.

25. $\Omega = \left\{(x, y) \mid 0 \le x \le \frac{\pi}{4} \text{ and } \sin x \le y \le \cos x\right\}$

26. $\Omega = \{(x, y) \mid -2 \le x \le 2 \text{ and } |x| \le y \le 2\}$

27. $\Omega = \{(x, y) \mid |x| + |y| \le 1\}$

28. $\Omega = \{(x, y) \mid x^2 + y^2 \le 9\}$

In Exercises 29–34, sketch the region determined by the limits of the iterated integrals and then give another iterated integral (or a sum of iterated integrals if necessary) using the opposite order of integration.

29. $\displaystyle\int_1^2 \int_{\ln x}^{e^x} f(x, y)\, dy\, dx$

30. $\displaystyle\int_0^2 \int_x^{-x+4} f(x, y)\, dy\, dx$

31. $\displaystyle\int_0^{\pi/2} \int_0^{\sin y} f(x, y)\, dx\, dy$

32. $\displaystyle\int_0^8 \int_0^{\sqrt[3]{y}} f(x, y)\, dx\, dy$

33. $\displaystyle\int_0^{\pi/2} \int_{\sin y}^1 f(x, y)\, dx\, dy$

34. $\displaystyle\int_0^8 \int_{\sqrt[3]{y}}^2 f(x, y)\, dx\, dy$

In Exercises 35–40, find the volume of the solid bounded above by the given function over the specified region Ω.

35. $f(x, y) = 10 - 2x + y$, with Ω the region from Exercise 21.

36. $f(x, y) = 10 - 2x + y$, with Ω the region from Exercise 22.

37. $f(x, y) = \sqrt{4 - x^2 - y^2}$, with Ω the region from Exercise 23.

38. $f(x, y) = x^2 y$, with Ω the region from Exercise 25.

39. $f(x, y) = \sin x \cos y$, with Ω the region from Exercise 26.

40. $f(x, y) = 1 - |x| - |y|$, with Ω the region from Exercise 27.

Find the volumes of the solids described in Exercises 41–44.

41. The portion of the first octant bounded by the coordinate planes and the plane $3x + 4y + 6z = 12$.

42. The solid bounded above by the plane with equation $2x + 3y - z = 2$ and bounded below by the triangle with vertices $(1, 0, 0)$, $(4, 0, 0)$, and $(0, 2, 0)$.

43. The solid bounded above by the paraboloid with equation $z = 8 - x^2 - y^2$ and bounded below by the rectangle $\mathcal{R} = \{(x, y) \mid 1 \le x \le 2 \text{ and } 0 \le y \le 2\}$ in the xy-plane.

44. The solid bounded above by the hyperboloid with equation $z = x^2 - y^2$ and bounded below by the square with vertices $(2, 2, -4)$, $(2, -2, -4)$, $(-2, -2, -4)$, and $(-2, 2, -4)$.

Evaluate the iterated integrals in Exercises 45–48 by reversing the order of integration. Explain why it is easier to reverse the order of integration than evaluate the given iterated integral.

45. $\displaystyle\int_0^9 \int_{\sqrt{y}}^3 \sqrt{1 + x^3}\, dx\, dy$

46. $\displaystyle\int_0^{\sqrt{\pi}} \int_x^{\sqrt{\pi}} \cos(y^2)\, dy\, dx$

47. $\displaystyle\int_0^{\sqrt{3}} \int_{\tan^{-1} x}^{\pi/3} \sec y\, dy\, dx$

48. $\displaystyle\int_0^1 \int_{\pi/4}^{\cot^{-1} y} \csc x\, dx\, dy$

In Exercises 49–58, sketch the region determined by the iterated integral and then evaluate the integral. For some of these integrals, it may be helpful to reverse the order of integration.

49. $\int_0^3 \int_{x+2}^{2x-3} (x^2 + 3xy)\, dy\, dx$

50. $\int_0^3 \int_{y+2}^{2y-3} (x^2 + 3xy)\, dx\, dy$

51. $\int_4^9 \int_2^{\sqrt{x}} (x^3 + y^2)\, dy\, dx$

52. $\int_4^9 \int_2^{\sqrt{y}} (x^2 + 3xy)\, dx\, dy$

53. $\int_0^{\pi/4} \int_{\tan x}^{\sec x} y\, dy\, dx$

54. $\int_0^1 \int_{-e^x}^{e^x} \sin(e^x)\, dy\, dx$

55. $\int_0^1 \int_y^1 e^{x^2}\, dx\, dy$

56. $\int_0^{\pi/4} \int_0^{\sec x} \sec x\, dy\, dx$

57. $\int_{\pi/4}^{3\pi/4} \int_0^{\csc y} \csc y\, dx\, dy$

58. $\int_0^3 \int_0^{\sqrt{9-y^2}} \sqrt{9 - x^2}\, dx\, dy$

In Exercises 59–62, evaluate the double integral over the specified region.

59. $\iint_\Omega x e^{x^3}\, dA$, where Ω is the triangular region with vertices $(0, 0)$, $(2, 0)$, and $(2, 2)$.

60. $\iint_\Omega x e^{x^3}\, dA$, where Ω is the triangular region in the first quadrant bounded below by the x-axis, bounded above by the line $y = mx$, where $m > 0$, and bounded on the right by the line with equation $x = 1$.

61. $\iint_\Omega x^2 y^3\, dA$, where Ω is the region in the first quadrant bounded by the graphs of the curves $y = x^2$ and $x = y^2$.

62. $\iint_\Omega dA$, where Ω is the region in the first quadrant bounded by the graphs of the curves $y = x^m$ and $y = x^n$, where m and n are distinct positive integers.

Applications

63. Emmy oversees the operation of a sedimentation lagoon that was built and lined using the natural contours of the terrain. The bottom of the lagoon is the part of the surface

$$z = \frac{1}{10}|x| + \frac{1}{10}|y| - 3$$

that lies below the $z = 0$ plane, where all units are in meters and $z = 0$ represents the water level. What is the volume of the lagoon?

64. Leila is designing a new summer range management unit for caribou in the Selkirk Mountains in the Idaho panhandle. The old unit was laid out as a rectangle, which had nothing to do with the behavior of the caribou. The new one is supposed to resemble the actual area in which the caribou live. Leila has used a study which indicates that the density of caribou in this region in July is approximated by $d(x, y) = 0.08x^2y^2 - 0.456x^2y - 0.08x^2 - 0.328xy^2 + 1.87xy + 0.328x - 0.061y^2 + 0.347y + 0.061$. Her proposed southern boundary for the management unit is a mountain ridge that roughly follows the curve $0.0195x^4$, while the northern border is a political boundary at $\frac{x}{4} + 4$. The western boundary is a state line on which she places the y-axis. Roughly how many caribou can be found in the management unit?

Proofs

65. Let $f(x, y)$ be an integrable function on the rectangle $\mathcal{R} = \{(x, y) \mid a \le x \le b \text{ and } c \le y \le d\}$, and let $\alpha \in \mathbb{R}$. Use the definition of the double integral to prove that

$$\iint_\mathcal{R} \alpha f(x, y)\, dA = \alpha \iint_\mathcal{R} f(x, y)\, dA.$$

66. Let $f(x, y)$ and $g(x, y)$ be integrable functions on the rectangle $\mathcal{R} = \{(x, y) \mid a \le x \le b \text{ and } c \le y \le d\}$. Use the definition of the double integral to prove that

$$\iint_\mathcal{R} (f(x, y) + g(x, y))\, dA = \iint_\mathcal{R} f(x, y)\, dA + \iint_\mathcal{R} g(x, y)\, dA.$$

67. Prove Theorem 13.10 (a). That is, show that if $f(x, y)$ is an integrable function on the general region Ω and $c \in \mathbb{R}$, then

$$\iint_\Omega \alpha f(x, y)\, dA = \alpha \iint_\Omega f(x, y)\, dA.$$

68. Prove Theorem 13.10 (b). That is, show that if $f(x, y)$ and $g(x, y)$ are integrable functions on the general region Ω, then

$$\iint_\Omega (f(x, y) + g(x, y))\, dA = \iint_\Omega f(x, y)\, dA + \iint_\Omega g(x, y)\, dA.$$

69. Let a, b, and c be positive real numbers. Prove that the volume of the pyramid with vertices $(0, 0, 0)$, $(a, 0, 0)$, $(0, b, 0)$, and $(0, 0, c)$ is $\frac{1}{6}abc$.

70. Let a and c be positive real numbers. Prove that the volume of the right-square pyramid with vertices $(a, a, 0)$, $(-a, a, 0)$, $(a, -a, 0)$, $(-a, -a, 0)$, and $(0, 0, c)$ is $\frac{4}{3}a^2c$. *(Hint: Use the result of Exercise 69.)*

Thinking Forward

- *Three iterated integrals:* Let $[a_1, a_2]$, $[b_1, b_2]$, and $[c_1, c_2]$ be three closed intervals. Explain why the triple integral
$$\int_{a_1}^{a_2}\int_{b_1}^{b_2}\int_{c_1}^{c_2} dz\,dy\,dx$$
computes the volume of the rectangular solid with length $a_2 - a_1$, width $b_2 - b_1$, and height $c_2 - c_1$.

- *Three more iterated integrals:* Evaluate the triple integral
$$\int_0^2\int_0^{-(3/2)x+3}\int_0^{4-2x-(4/3)y} dz\,dy\,dx,$$
and give a physical interpretation to the integral.

13.3 DOUBLE INTEGRALS IN POLAR COORDINATES

- Expressing a double integral with polar coordinates
- Finding areas of regions bounded by functions expressed with polar coordinates
- Finding volumes of solids bounded by functions expressed with polar coordinates

Polar Coordinates and Double Integrals

In Chapter 9 we saw that every point in the coordinate plane can be expressed with polar coordinates (r, θ), where θ, in radians, measures the counterclockwise rotation from the positive x-axis and r measures the signed distance that the point is from the origin on the line determined by θ. In Chapter 9 we allowed r and θ to be any real numbers. In this section, where we discuss how to use polar coordinates to evaluate double integrals, we will insist that $r \geq 0$ and that θ be a real number in an interval of width 2π, typically $\theta \in [0, 2\pi]$ or $\theta \in [-\pi, \pi]$.

In the first two sections of this chapter we discussed how to use rectangular coordinates to integrate functions of two variables. Here, we extend this idea to functions and regions that are more naturally expressed with polar coordinates. That is, we wish to find
$$\int_{\mathcal{R}} f(r, \theta)\,dA,$$
where f is a function of r and θ and the region $\mathcal{R}$ is also expressed in terms of r and θ. In Section 13.1 we began with a basic rectangular region, which we then partitioned into subrectangles. Here we start with a polar "rectangle" defined by the inequalities
$$0 \leq a \leq r \leq b \quad \text{and} \quad \alpha \leq \theta \leq \beta,$$
as shown in the following figure at the left:

A polar "rectangle"

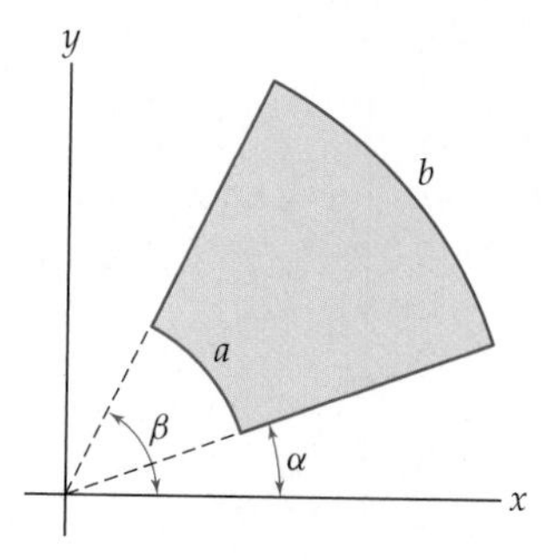

The subdivided rectangle

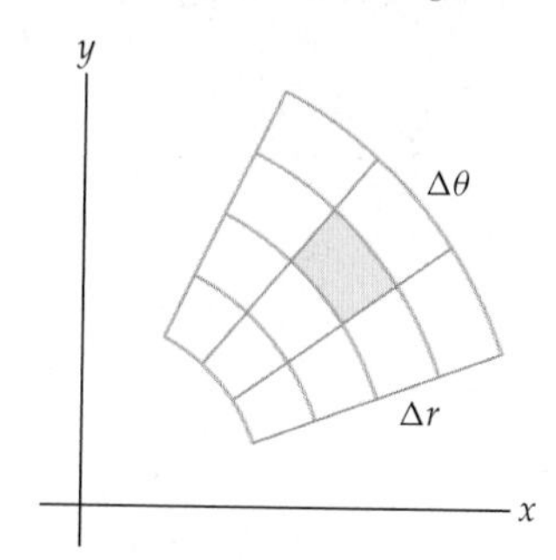

In the right-hand figure, we see the same region divided into "subrectangles." We follow a procedure analogous to that we used in Section 13.1. We assume that $f(r, \theta)$ is a function defined at every point of $\mathcal{R}$. We will outline the steps required to find the volume V of the solid bounded below by $\mathcal{R}$ and bounded above by the graph of f in a moment, but first we need to understand how to compute the area of one of the subregions, shown in the right-hand figure.

The areas of the "subrectangles" depend upon the values of Δr and $\Delta\theta$. Consider the ***annulus*** with inner radius r_{j-1} and outer radius of r_j, as shown in the following figure at the left:

The annulus with inner and outer radii r_{j-1} and r_j

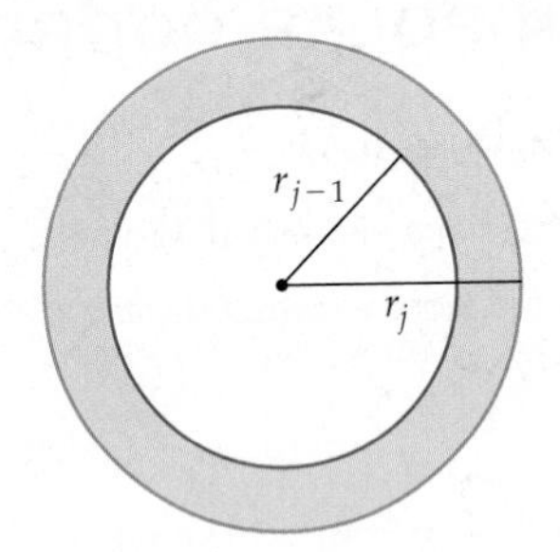

A slice of the annulus

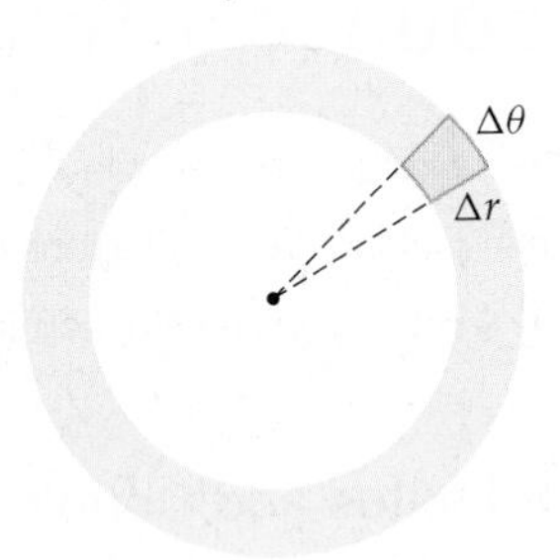

The area of the annulus is given by

$$\pi(r_j^2 - r_{j-1}^2) = \pi(r_j + r_{j-1})(r_j - r_{j-1}) = 2\pi\frac{r_j + r_{j-1}}{2}\Delta r.$$

Note that if we let $\hat{r}_j = \frac{r_j + r_{j-1}}{2}$, then $\hat{r}_j \in [r_{j-1}, r_j]$. Thus, the area of the annulus is $2\pi\hat{r}_j\Delta r$. The area of a "slice" of the annulus corresponding to an angular rotation of $\Delta\theta$ will be

$$(2\pi\hat{r}_j\Delta r)\frac{\Delta\theta}{2\pi} = \hat{r}_j\Delta r\Delta\theta.$$

We are now ready to approximate the volume V of the solid bounded below by $\mathcal{R}$ and above by the graph of f:

- We subdivide the interval $[a, b]$ into m equal subintervals, each of width $\Delta r = \frac{b-a}{m}$, and we also let $r_j = a + j\Delta r$ for $0 \le j \le m$.
- Similarly, we subdivide the interval $[\alpha, \beta]$ into n equal subintervals, each of width $\Delta\theta = \frac{\beta-\alpha}{n}$, and we let $\theta_k = \alpha + k\Delta\theta$ for $0 \le k \le n$.
- The subdivisions we just created partition the polar rectangle into $m \times n$ polar rectangles, like the one shown earlier at the right.
- For each $j = 1, 2, \ldots, m$ and $k = 1, 2, \ldots, n$, we select a point (r_j^*, θ_k^*) in the subrectangle $\mathcal{R}_{jk} = \{(r, \theta) \mid r_{j-1} \le r_j^* \le r_j \text{ and } \theta_{k-1} \le \theta_k^* \le \theta_k\}$.

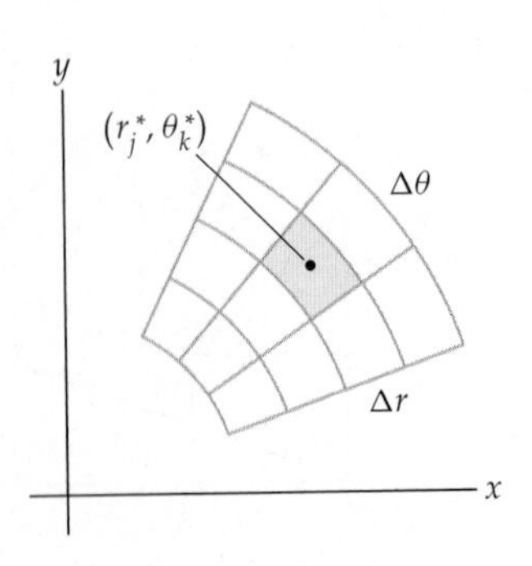

One such point is shown in the preceding figure.

- The area of a subregion is given by $\hat{r}_j \Delta r \Delta\theta$, where $\hat{r}_j \in [r_{j-1}, r_j]$.
- The product $f(r_j^*, \theta_k^*)\hat{r}_j \Delta r \Delta\theta$ approximates the volume of the solid bounded below by the rectangle $\mathcal{R}_{jk}$ and above by the graph of f.
- When we sum these approximate volumes over all of the subregions, we obtain an approximation for the volume of V:

$$\sum_{j=1}^{m}\sum_{k=1}^{n} f(r_j^*, \theta_k^*)\,\hat{r}_j \Delta r \Delta\theta.$$

The limit of this summation provides the iterated integral of the function on the polar region $\mathcal{R}$, namely,

$$\int_\alpha^\beta \int_a^b f(r,\theta) r\,dr\,d\theta = \lim_{\Delta\to 0} \sum_{j=1}^{m}\sum_{k=1}^{n} f(r_j^*, \theta_k^*)\,\hat{r}_j \Delta r \Delta\theta,$$

where $\Delta = \sqrt{(\Delta r)^2 + (\Delta\theta)^2}$.

We evaluate this type of iterated integral just as we evaluated iterated integrals expressed with rectangular coordinates. For example, we may prove that the volume of a sphere with radius R is $\frac{4}{3}\pi R^3$ by integrating the function $f(x, y) = \sqrt{R^2 - (x^2 + y^2)}$ over the polar rectangle $\mathcal{R} = \{(r,\theta) \mid 0 \le r \le R \text{ and } 0 \le \theta \le 2\pi\}$. Before we integrate, however, we must express f in polar coordinates. Recall that $r^2 = x^2 + y^2$. Therefore $z = \sqrt{R^2 - r^2}$. The volume of the top hemisphere is given by the integral

$$\int_0^{2\pi} \int_0^R \sqrt{R^2 - r^2}\; r\,dr\,d\theta.$$

The inner integral is evaluated as

$$\int_0^R \sqrt{R^2 - r^2}\; r\,dr = \left[-\frac{1}{3}(R^2 - r^2)^{3/2}\right]_0^R = \frac{1}{3}R^3.$$

We now evaluate the outer integral:

$$\int_0^{2\pi} \frac{1}{3}R^3\,d\theta = \left[\frac{1}{3}R^3\theta\right]_0^{2\pi} = \frac{2}{3}\pi R^3.$$

Therefore the volume of the entire sphere is $\frac{4}{3}\pi R^3$.

Double Integrals in Polar Coordinates over General Regions

We may also use polar coordinates to evaluate an iterated integral over a more general region

$$\Omega = \{(r,\theta) \mid f_1(\theta) \le r \le f_2(\theta) \text{ and } \alpha \le \theta \le \beta\}.$$

For a polar function $g(r,\theta)$ defined on the region Ω, we have

$$\int_\alpha^\beta \int_{f_1(\theta)}^{f_2(\theta)} g(r,\theta)\, r\,dr\,d\theta.$$

If $g(r,\theta) \ge 0$ on Ω, then the iterated integral represents the volume of the solid bounded above by the function $g(r,\theta)$ over the region Ω. If $g(r,\theta)$ takes on both positive and negative values on Ω, then the double integral represents the signed volume of the solid between the graph of the function $g(r, \theta)$ and the coordinate plane over the region Ω.

For example, consider the following figures:

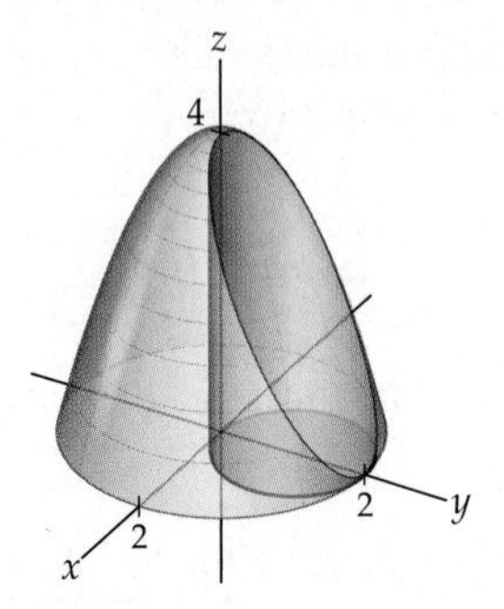

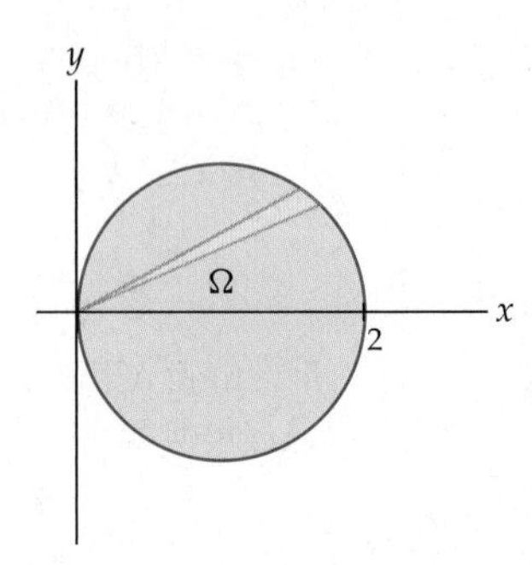

To find the volume of the solid bounded below by the disk whose boundary is the circle with equation $r = 2\cos\theta$ in the coordinate plane and bounded above by the paraboloid $z = 4 - x^2 - y^2$, we evaluate the integral

$$\int_0^{\pi} \int_0^{2\cos\theta} (4 - r^2)\, r\, dr\, d\theta.$$

Note that:

- Our limits of integration for θ are 0 and π, because the entire circle is traced once over this interval.
- Our limits of integration for r are 0 and $2\cos\theta$, since every "slice" emanating from the origin has those boundaries, as we see in the right-hand figure.
- We have replaced $x^2 + y^2$ with r^2 to express the function with polar coordinates.
- If we had preferred, we could have used the expression

  $$2\int_0^{\pi/2} \int_0^{2\cos\theta} (4 - r^2)\, r\, dr\, d\theta$$

  to evaluate the volume, since both the circular region and the surface are symmetric with respect to the xz-plane.

To evaluate either of the preceding integrals, we begin with the inner integration:

$$\int_0^{2\cos\theta} (4 - r^2)\, r\, dr = \left[2r^2 - \frac{1}{4}r^4\right]_0^{2\cos\theta} = 8\cos^2\theta - 4\cos^4\theta.$$

To finish the computation of the volume, we have

$$\int_0^{\pi} (8\cos^2\theta - 4\cos^4\theta)\, d\theta = \left[\frac{5}{2}\theta + \frac{5}{2}\sin\theta\cos\theta - \sin\theta\cos^3\theta\right]_0^{\pi} = \frac{5}{2}\pi.$$

Every double integral expressed with rectangular coordinates may also be expressed with polar coordinates. Recall that we may use the equations

$$x = r\cos\theta \quad \text{and} \quad y = r\sin\theta$$

to change from rectangular coordinates to polar coordinates. Therefore, the double integral

$$\iint_{\Omega} f(x, y)\, dA$$

may also be expressed as

$$\iint_{\Omega} f(r\cos\theta, r\sin\theta)\, r\, dr\, d\theta.$$

The value of making this transformation depends upon the region Ω and the particular function f we are trying to integrate. If Ω has a more "natural" expression in polar coordinates, or if f has a simpler antiderivative in polar coordinates, the change to polar coordinates may make the evaluation of the integral considerably easier. We look at such an integral in Example 4.

Examples and Explorations

EXAMPLE 1 **Finding the area bounded by a polar rose**

Use a double integral to calculate the area bounded by the curve $r = \cos 4\theta$.

SOLUTION

As with all area computations, we must understand the region whose area we are trying to compute. When we can, we will use the symmetry of the region to simplify our work. In Chapter 9 we saw that the graph of an equation of the form $r = \cos n\theta$ or $r = \sin n\theta$ is a polar rose when $n > 1$ is an integer. In addition, if n is even, the figure has $2n$ petals; if n is odd, the figure has n petals. The graph of this curve is as follows:

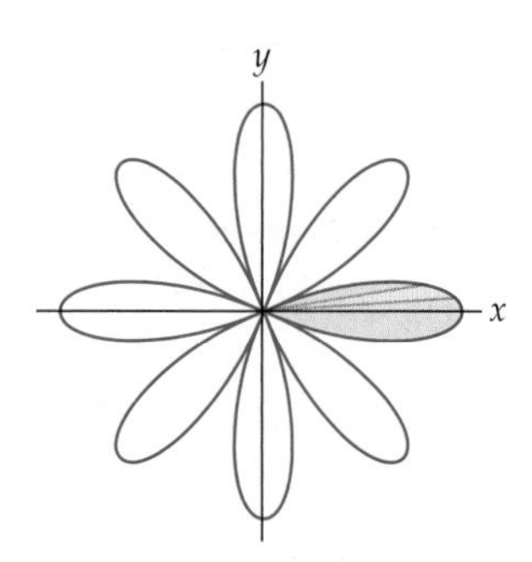

To compute the area bounded by the curve, we will use the symmetry of the rose. Note that we have shaded one of the eight petals of the rose. The values of θ for which the graph passes through the origin correspond to the roots of $\cos 4\theta$ (i.e., odd multiples of $\pi/8$). A sample radial slice of each petal starts at the origin and ends on the curve. The shaded region corresponds to the values of θ in the interval $[-\pi/8, \pi/8]$. The double integral whose value is the area of the shaded region is

$$\int_{-\pi/8}^{\pi/8} \int_0^{\cos 4\theta} r\,dr\,d\theta.$$

Evaluating the inner integral first, we obtain

$$\int_0^{\cos 4\theta} r\,dr = \left[\frac{1}{2}r^2\right]_0^{\cos 4\theta} = \frac{1}{2}\cos^2 4\theta.$$

Now, evaluating the outer integral, we have

$$\int_{-\pi/8}^{\pi/8} \frac{1}{2}\cos^2 4\theta\,d\theta = \left[\frac{1}{4}\theta + \frac{1}{16}\sin 4\theta\cos 4\theta\right]_{-\pi/8}^{\pi/8} = \frac{\pi}{16}.$$

Since the area of the shaded region is one-eighth of the region bounded by the polar rose, the area bounded by the rose is $\frac{\pi}{2}$ square units. □

EXAMPLE 2 **Finding the area between two polar curves**

Use a double integral to calculate the area of the region in the polar plane that is inside both the circle $r = 3\cos\theta$ and the cardioid $r = 1 + \cos\theta$.

SOLUTION

By the symmetry of the graphs, half of the region whose area we wish to calculate is the shaded portion of the following figure:

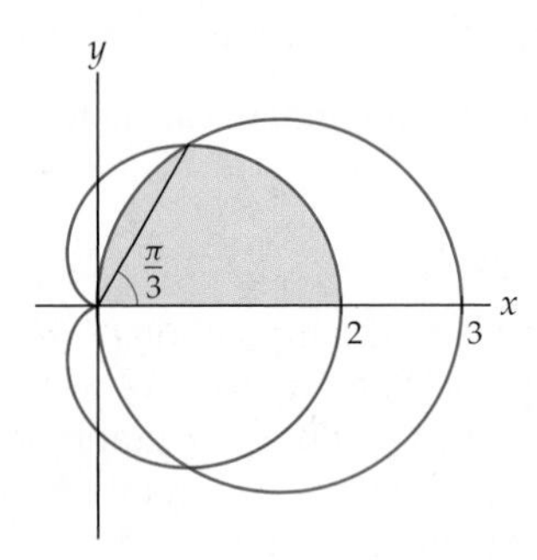

We will compute the area of the top half of the region and then multiply by 2. Solving the equation $3\cos\theta = 1 + \cos\theta$, we see that the two curves intersect when $\theta = \frac{\pi}{3}$. In the figure, we have drawn the ray $\theta = \frac{\pi}{3}$. The iterated integral that represents the area of the portion of the shaded region for values of $\theta \le \frac{\pi}{3}$ is

$$\int_0^{\pi/3} \int_0^{1+\cos\theta} r\,dr\,d\theta,$$

and the iterated integral that represents the area of the portion of the shaded region for $\frac{\pi}{3} \le \theta \le \frac{\pi}{2}$ is

$$\int_{\pi/3}^{\pi/2} \int_0^{3\cos\theta} r\,dr\,d\theta.$$

In Exercise 12 you will show that the values of these two integrals are $\frac{1}{4}\pi + \frac{9}{16}\sqrt{3}$ and $\frac{3}{8}\pi - \frac{9}{16}\sqrt{3}$, respectively. Therefore, the area of the shaded region is $\frac{5}{8}\pi$ and the value of the area we wish to compute is $\frac{5}{4}\pi$ square units. □

EXAMPLE 3 **Using a double integral to prove the area formula for the circle**

Use a double integral to prove that the area of a circle with radius R is πR^2.

SOLUTION

We will compute the area of the circle whose equation is $r = R$. The area is given by the iterated integral

$$\int_0^{2\pi} \int_0^R r\,dr\,d\theta.$$

The inner integral is

$$\int_0^R r\,dr = \left[\frac{1}{2}r^2\right]_0^R = \frac{1}{2}R^2.$$

We now evaluate the outer integral:

$$\int_0^{2\pi} \frac{1}{2}R^2\, d\theta = \left[\frac{1}{2}R^2\theta\right]_0^{2\pi} = \pi R^2.$$

In Exercises 63 and 64 you will prove this result twice more by finding the area of the circles with equations $r = 2R\cos\theta$ and $r = 2R\sin\theta$. □

EXAMPLE 4 **Changing an integral from rectangular to polar coordinates**

Evaluate the integral $\displaystyle\int_0^2 \int_{\sqrt{3}x}^{\sqrt{16-x^2}} (x^2 + y^2)\, dy\, dx$.

SOLUTION

We may try to evaluate the integral as it is written. To do this, we evaluate the inner integral first:

$$\begin{aligned}\int_{\sqrt{3}x}^{\sqrt{16-x^2}} (x^2 + y^2)\, dy &= \left[x^2 y + \frac{1}{3}y^3\right]_{\sqrt{3}x}^{\sqrt{16-x^2}} \\ &= \left(x^2\sqrt{16 - x^2} + \frac{1}{3}(16 - x^2)^{3/2}\right) - \left(\sqrt{3}x^3 + \frac{1}{3}3\sqrt{3}x^3\right) \\ &= x^2\sqrt{16 - x^2} + \frac{1}{3}(16 - x^2)^{3/2} - 2\sqrt{3}x^3.\end{aligned}$$

Now, to finish the problem, we would need to evaluate the integral

$$\int_0^2 \left(x^2\sqrt{16 - x^2} + \frac{1}{3}(16 - x^2)^{3/2} - 2\sqrt{3}x^3\right) dx.$$

With diligence, we would use trigonometric substitution to find an antiderivative for the integrand, but we present an alternative.

We may replace the original integrand $x^2 + y^2$ with its polar coordinate equivalent r^2. We shall similarly see that the region described by the limits of the integral may be expressed with polar coordinates. These are two essential things to consider when you are deciding between using rectangular and polar coordinates. The region in question is bounded above by the graph of the function $y = \sqrt{16 - x^2}$ and bounded below by the line whose equation is $y = \sqrt{3}x$, where $x \in [0, 2]$. Here is the region:

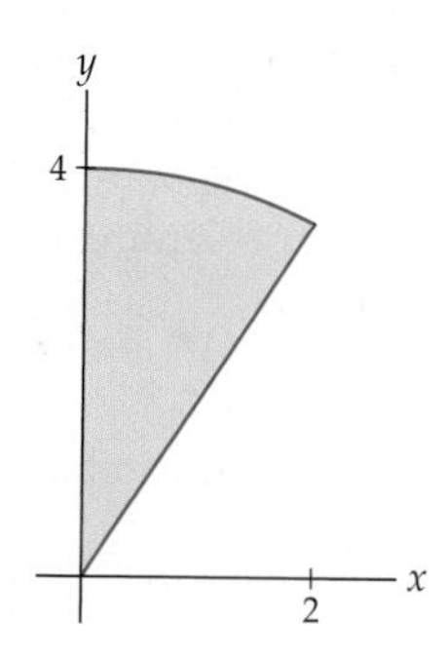

This region is a sector of the circle $r = 4$, where $\frac{\pi}{3} \le \theta \le \frac{\pi}{2}$. We may now rewrite the integral in terms of polar coordinates. Whenever we convert from rectangular to polar coordinates, we must replace either $dy\,dx$ or $dx\,dy$ by $r\,dr\,d\theta$. As we already mentioned, in this example we replace $x^2 + y^2$ with r^2. The limits for the integration with respect to r extend from 0 to 4, and the limits for θ extend from $\frac{\pi}{3}$ to $\frac{\pi}{2}$. Therefore we have

$$\int_0^2 \int_{\sqrt{3}x}^{\sqrt{16-x^2}} (x^2 + y^2)\,dy\,dx = \int_{\pi/3}^{\pi/2} \int_0^4 r^3\,dr\,d\theta.$$

We first evaluate the inner integral:

$$\int_0^4 r^3\,dr = \left[\frac{1}{4}r^4\right]_0^4 = 64.$$

We now finish by evaluating the outer integral:

$$\int_{\pi/3}^{\pi/2} 64\,d\theta = [64\theta]_{\pi/3}^{\pi/2} = 64\left(\frac{\pi}{2} - \frac{\pi}{3}\right) = \frac{32}{3}\pi.$$

□

EXAMPLE 5

Using polar coordinates to compute a volume

Find the volume of the solid bounded above by the sphere $x^2 + y^2 + z^2 = 8$ and bounded below by the cone with equation $z = \sqrt{x^2 + y^2}$.

SOLUTION

The sphere and cone intersect along a circle. If we rewrite the equations in the forms $z^2 = 8 - x^2 - y^2$ and $z^2 = x^2 + y^2$ and equate the results, we see that $x^2 + y^2 = 4$. The disk bounded by the circle defined by this equation is the region over which we need to integrate the difference of the two functions. The graph that follows at the left shows the entire hemisphere and cone, along with the disk in the xy-plane, over which we will integrate. The open figure depicted at the right shows a quarter of the left-hand figure.

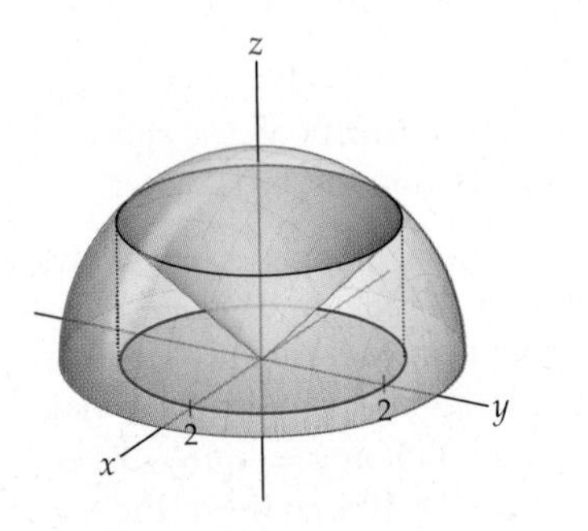

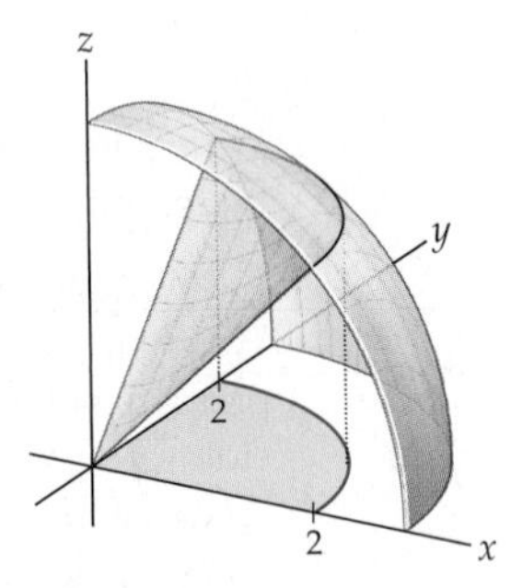

To find the volume of the solid, we will integrate the difference of the two functions over the indicated circle in the xy-plane. In polar coordinates, the equations of the top hemisphere and the cone are $z = \sqrt{8 - r^2}$ and $z = r$, respectively. The equation of the circle of intersection of the sphere and cone projected onto the xy-plane is $r = 2$. Therefore, the iterated integral representing the volume is

$$\int_0^{2\pi} \int_0^2 \left(\sqrt{8 - r^2} - r\right) r\,dr\,d\theta.$$

We now evaluate the iterated integral by first working on the inner integral:

$$\int_0^2 \left(r\sqrt{8 - r^2} - r^2\right) dr = \left[-\frac{1}{3}(8 - r^2)^{3/2} - \frac{1}{3}r^3\right]_0^2 = \frac{16}{3}(\sqrt{2} - 1).$$

We next integrate the rightmost quantity with respect to θ over the interval $[0, 2\pi]$:

$$\int_0^{2\pi} \frac{16}{3}(\sqrt{2} - 1)\, d\theta = \frac{32}{3}\pi(\sqrt{2} - 1).$$

This is the volume of the region we wished to find. □

CHECKING THE ANSWER

The volume we were asked to find is the volume of the solid of revolution created when the region in the first quadrant bounded above by the graph of $y = \sqrt{8 - x^2}$ and below by the line $y = x$ is rotated about the y-axis, as we illustrate in the following figures:

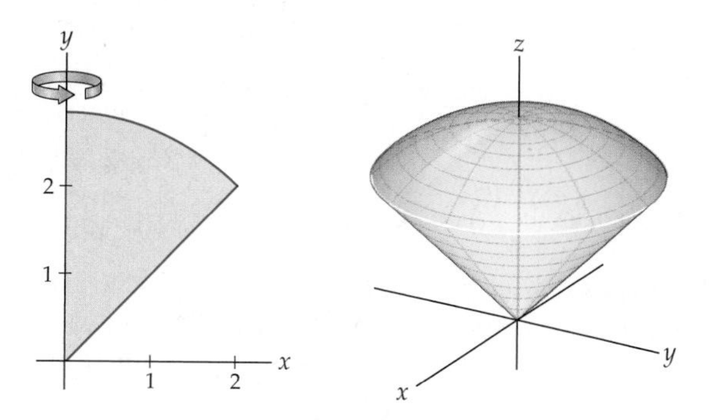

We may evaluate the volume of this region using either the washer method or the shell method, both of which we discussed in Chapter 6. Using the shell method, we have

$$2\pi \int_0^2 x\left(\sqrt{8 - x^2} - x\right) dx.$$

You should check that the value of this integral is also $\frac{32}{3}\pi(\sqrt{2} - 1)$.

TEST YOUR UNDERSTANDING

- How are polar coordinates used to express double integrals?
- What are the formulas needed to convert from rectangular to polar coordinates? What are the formulas needed to convert from polar to rectangular coordinates?
- How are iterated integrals used to calculate areas between functions that are expressed in polar coordinates? How can a single integral be used to express such areas?
- How are iterated integrals used to calculate the volumes bounded by functions that are expressed in polar coordinates?
- How is an integral that is expressed in rectangular coordinates rewritten in terms of polar coordinates? When would such rewriting be advantageous? When would it be disadvantageous? How would an integral that is expressed in polar coordinates be expressed in terms of rectangular coordinates?

EXERCISES 13.3

Thinking Back

Each of the integral expressions that follow represents the area of a region in the plane bounded by a function expressed in polar coordinates. Use the ideas from this section and from Chapter 9 to sketch the regions, and then evaluate each integral.

- $\displaystyle \frac{1}{2}\int_0^{\pi} \cos^2 3\theta\, d\theta$
- $\displaystyle 4\int_0^{\pi/4} \cos^2 2\theta\, d\theta$
- $\displaystyle \int_0^{2\pi/3} \left(\frac{1}{2} + \cos\theta\right)^2 d\theta - \int_{\pi}^{4\pi/3} \left(\frac{1}{2} + \cos\theta\right)^2 d\theta$
- $\displaystyle \frac{1}{2}\int_{\pi/6}^{5\pi/6} ((3\sin\theta)^2 - (1 + \sin\theta)^2)\, d\theta$

Concepts

0. *Problem Zero:* Read the section and make your own summary of the material.

1. *True/False:* Determine whether each of the statements that follow is true or false. If a statement is true, explain why. If a statement is false, provide a counterexample.

(a) *True or False:* To approximate the area of the region in the polar plane bounded by the function $r = f(\theta)$ and the rays $\theta = \alpha$ and $\theta = \beta$, we can use a sum of areas of sectors of circles.

(b) *True or False:* Suppose we subdivide the interval of angles $\theta \in \left[\frac{\pi}{4}, \frac{\pi}{2}\right]$ into four equal subintervals. Then $\Delta\theta = \frac{\pi}{16}$.

(c) *True or False:* If $f(\theta) \ge 0$ on the interval $[\alpha, \beta]$, then the area of the region in the polar plane bounded by the function $r = f(\theta)$ and the rays $\theta = \alpha$ and $\theta = \beta$ is given by the iterated integral $\int_\alpha^\beta \int_0^{f(\theta)} dr\, d\theta$.

(d) *True or False:* If $f(\theta) \ge 0$ on the interval $[\alpha, \beta]$, then the volume of the solid bounded above by the function $g(r, \theta) = r^2$, over the region in the polar plane bounded by the function $r = f(\theta)$ and the rays $\theta = \alpha$ and $\theta = \beta$, is given by the iterated integral $\int_\alpha^\beta \int_0^{f(\theta)} r^2\, dr\, d\theta$.

(e) *True or False:* Since the graph of $r = 2\cos 2\theta$ is a circle with radius 1, the value of the integral $\int_0^{2\pi} \int_0^{2\cos\theta} r\, dr\, d\theta$ is π.

(f) *True or False:* The polar equation $r = \sin 4\theta$ is traced twice as θ varies from 0 to 2π.

(g) *True or False:*

$$\int_0^\pi \int_1^5 (x^3 + y^3)^2\, dy\, dx = \int_0^\pi \int_1^5 (\theta^3 + r^3)^2\, dr\, d\theta.$$

(h) *True or False:*

$$\int_0^2 \int_0^{\sqrt{4-x^2}} (x^2 + y^2)^3\, dy\, dx = \int_0^{\pi/2} \int_0^2 r^7\, dr\, d\theta.$$

2. *Examples:* Construct examples of the thing(s) described in the following. Try to find examples that are different than any in the reading.

(a) An iterated integral that represents the area of a circle with radius R express with polar coordinates.

(b) An iterated integral using polar coordinates that represents the volume of a sphere with radius R.

(c) An iterated integral in rectangular coordinates that would be easier to evaluate by using polar coordinates.

3. Let $\alpha < \beta$ and $a < b$. In polar coordinates, a polar rectangle is bounded by the two rays $\theta = \alpha$ and $\theta = \beta$ and the two circles $r = a$ and $r = b$. Sketch a polar rectangle and explain why this is the basic region for integration in the polar coordinate plane.

4. When we use rectangular coordinates to approximate the area of a region, we subdivide the region into vertical strips and approximate the area by using a sum of areas of rectangles. Explain why we use a wedge (i.e., a sector of a circle), and not a rectangle, when we employ polar coordinates to compute an area.

5. In this section we described a method for approximating the volume bounded by a function f above a polar region bounded by two rays $\theta = \alpha$ and $\theta = \beta$ and between two circles $r = a$ and $r = b$. The method employed a "subdivide, approximate, and add" strategy that involved using some general notation. Draw a carefully labeled picture that illustrates the roles of $\Delta\theta$, θ_k^*, Δr, and r_j^* for one approximating region.

6. In Section 9.4 we showed that the area in the polar coordinate plane bounded by the function $r = f(\theta)$ on the interval $[\alpha, \beta]$ is given by the integral $\frac{1}{2}\int_\alpha^\beta (f(\theta))^2\, d\theta$. In this section we discussed how to use the iterated integral $\int_\alpha^\beta \int_0^{f(\theta)} r\, dr\, d\theta$ to compute the same area. Explain why the value of these two integrals is the same.

7. Let $0 < f_1(\theta) < f_2(\theta)$ on the interval $[\alpha, \beta]$. What does the integral $\int_\alpha^\beta \int_{f_1(\theta)}^{f_2(\theta)} dr\, d\theta$ represent in a rectangular θr-coordinate system? What does the integral represent in a polar coordinate system?

8. Why do we require that $0 \le \beta - \alpha \le 2\pi$ when we are trying to find the volume of a solid bounded above by the graph of a function $z = g(r, \theta)$, over a region bounded by the polar functions $r = f_1(\theta)$ and $r = f_2(\theta)$, where $f_1(\theta) \le f_2(\theta)$ on the interval $[\alpha, \beta]$? If $0 \le \beta - \alpha \le 2\pi$, does that condition ensure that the integral

$$\int_\alpha^\beta \int_{f_1(\theta)}^{f_2(\theta)} g(r, \theta)\, r\, dr\, d\theta$$

will represent the volume we want?

9. Consider the three-petaled polar rose defined by $r = \cos 3\theta$. Explain why the iterated integral $\int_0^{2\pi} \int_0^{\cos 3\theta} r\, dr\, d\theta$ calculates *twice* the area bounded by the petals of this rose.

10. Explain how the symmetries of the graphs of polar functions can be used to simplify area calculations.

11. Give a geometric explanation why

$$n \int_0^{2\pi/n} \int_0^R r\, dr\, d\theta = \pi R^2$$

for any positive real number R and any positive integer n. Would the equation also hold for nonintegral values of n?

12. Complete Example 2 by showing that

$$\int_0^{\pi/3} \int_0^{1+\cos\theta} r\, dr\, d\theta = \frac{1}{4}\pi + \frac{9}{16}\sqrt{3}$$

and

$$\int_{\pi/3}^{\pi/2} \int_0^{3\cos\theta} r\, dr\, d\theta = \frac{3}{8}\pi - \frac{9}{16}\sqrt{3}.$$

In Exercises 13–20, we explore the relationship between the shell method for finding volumes of solids of revolution discussed in Chapter 6 and the method of double integrals using polar coordinates.

13. Sketch a function $z = f(x)$ in the xz-plane such that $f(x) \ge 0$ on the interval $[0, b]$. Use the shell method to find an integral that represents the volume of the solid of revolution obtained when the region bounded above by the graph of f and bounded below by the x-axis on the interval $[0, b]$ is rotated about the z-axis.

14. Explain why the function $z = g(x, y) = f(\sqrt{x^2 + y^2})$ is the equation of the surface obtained when the graph of f is rotated about the z-axis. Sketch the surface obtained when your function from Exercise 13 is rotated about the z-axis.

15. Use the techniques of Section 13.2 to obtain an iterated integral that employs rectangular coordinates to represent the volume of the solid that is bounded above by the graph of the function g from Exercise 14 and below by the xy-plane over the circular disk $x^2 + y^2 \le b^2$.

16. Use the techniques of this section to obtain an iterated integral that employs polar coordinates to represent the volume of the solid discussed in Exercise 15.

17. Show that the integrals from Exercises 13 and 16 evaluate to the same quantity.

18. Let $0 < a < b$. Use the shell method to find an integral that represents the volume of the solid of revolution obtained when the region bounded above by the graph of f and bounded below by the x-axis on the interval $[a, b]$ is rotated about the z-axis.

19. Use the techniques of this section to obtain an iterated integral that employs polar coordinates to represent the volume of the solid bounded above by the graph of the function g from Exercise 14 and below by the xy-plane over the annulus $a^2 \le x^2 + y^2 \le b^2$.

20. Show that the integrals from Exercises 18 and 19 evaluate to the same quantity.

Skills

Each of the integrals or integral expressions in Exercises 21–28 represents the area of a region in the plane. Use polar coordinates to sketch the region and evaluate the expressions.

21. $\int_0^{2\pi}\int_0^{1+\sin\theta} r\,dr\,d\theta$

22. $2\int_0^{\pi}\int_0^{1+\cos\theta} r\,dr\,d\theta$

23. $2\int_{-\pi/2}^{\pi/2}\int_0^{2-\sin\theta} r\,dr\,d\theta$

24. $2\int_0^{\pi/2}\int_0^{\sin\theta} r\,dr\,d\theta$

25. $2\int_0^{\pi/2}\int_0^{\sin 3\theta} r\,dr\,d\theta$

26. $\int_0^{2\pi}\int_0^{2+\sin 4\theta} r\,dr\,d\theta$

27. $2\int_{-\pi/4}^{\pi/2}\int_0^{(\sqrt{2}/2)+\sin\theta} r\,dr\,d\theta - 2\int_{-\pi/2}^{-\pi/4}\int_0^{(\sqrt{2}/2)+\sin\theta} r\,dr\,d\theta$

28. $2\int_0^{2\pi/3}\int_0^{(1/2)+\cos\theta} r\,dr\,d\theta - 2\int_{\pi}^{4\pi/3}\int_0^{(1/2)+\cos\theta} r\,dr\,d\theta$

In Exercises 29–38, find an iterated integral in polar coordinates that represents the area of the given region in the polar plane and then evaluate the integral.

29. The region enclosed by the spiral $r = \theta$ and the x-axis on the interval $0 \le \theta \le \pi$.

30. The region inside one loop of the lemniscate $r^2 = \sin 2\theta$.

31. The region between the two loops of the limaçon $r = 1 + \sqrt{2}\cos\theta$.

32. The region between the two loops of the limaçon $r = \sqrt{3} - 2\sin\theta$.

33. The region inside the cardioid $r = 3 - 3\sin\theta$ and outside the cardioid $r = 1 + \sin\theta$.

34. The region where the two cardioids $r = 3 - 3\sin\theta$ and $r = 1 + \sin\theta$ overlap.

35. The region inside the circle $x^2 + y^2 = 1$ and to the right of the vertical line $x = \dfrac{1}{2}$.

36. One loop of the curve $r = 4\sin 3\theta$.

37. The region bounded by the limaçon $r = 1 + k\sin\theta$, where $0 < k < 1$. Explain why it makes sense for the area to approach π as $k \to 0$.

38. The graph of the polar equation $r = \sec\theta - 2\cos\theta$ is called a ***strophoid***. Graph the strophoid and find the area bounded by the loop of the graph.

Each of the integrals or integral expressions in Exercises 39–46 represents the volume of a solid in $\mathbb{R}^3$. Use polar coordinates to describe the solid, and evaluate the expressions.

39. $2\int_0^{2\pi}\int_0^{4} r\sqrt{16 - r^2}\,dr\,d\theta$

40. $2\int_0^{2\pi}\int_0^{R} r\sqrt{R^2 - r^2}\,dr\,d\theta$

41. $\int_0^{2\pi}\int_0^{2} (4r - r^3)\,dr\,d\theta$

42. $\int_0^{\pi}\int_0^{R} (R^2 r - r^3)\,dr\,d\theta$

43. $\int_0^{2\pi}\int_0^{3} (6r - 2r^2)\,dr\,d\theta$

44. $h\int_0^{2\pi}\int_0^{R}\left(r - \frac{r^2}{R}\right)dr\,d\theta$

45. $\int_{-\pi/4}^{5\pi/4}\int_0^{(\sqrt{2}/2)+\sin\theta} r\,dr\,d\theta$

46. $2\int_{-\pi/4}^{\pi/2}\int_0^{(\sqrt{2}/2)+\sin\theta} r\,dr\,d\theta - 2\int_{-\pi/2}^{-\pi/4}\int_0^{(\sqrt{2}/2)+\sin\theta} r\,dr\,d\theta$

In Exercises 47–56, use polar coordinates to find an iterated integral that represents the volume of the solid described and then find the volume of the solid.

47. The region enclosed by the paraboloids $z = x^2 + y^2$ and $z = 16 - x^2 - y^2$.

48. The region enclosed by the paraboloids $z = x^2 + y^2$ and $z = R^2 - x^2 - y^2$.

49. The region between the cone with equation $z = \sqrt{x^2 + y^2}$ and the unit sphere centered at the origin.

50. The region between the cone with equation $z = \sqrt{x^2 + y^2}$ and the sphere centered at the origin and with radius R.

51. The region bounded above by the unit sphere centered at the origin and bounded below by the plane $z = \frac{3}{5}$.

52. The region bounded above by the unit sphere centered at the origin and bounded below by the plane $z = h$ where $0 \le h \le 1$.

53. The region between two spheres with radius 1 if each passes through the center of the other.

54. The region between two spheres with radius R if each passes through the center of the other.

55. The region bounded below by the graph of the cone with equation $z = \sqrt{x^2 + y^2}$ and bounded above by the plane $z = h$, where $h > 0$.

56. The region bounded below by the graph of the cone with equation $z = \sqrt{x^2 + y^2}$ and bounded above by the plane $z = 8 - \frac{x}{2}$.

Sketch the region of integration for each of integrals in Exercises 57–60, and then evaluate the integral by converting to polar coordinates.

57. $\displaystyle\int_0^{3\sqrt{2}/2} \int_x^{\sqrt{9-x^2}} dy\,dx$

58. $\displaystyle\int_0^1 \int_{(\sqrt{3}/3)y}^{\sqrt{4-y^2}} \sqrt{x^2 + y^2}\,dx\,dy$

59. $\displaystyle\int_0^4 \int_0^{\sqrt{16-x^2}} e^{x^2} e^{y^2}\,dy\,dx$

60. $\displaystyle\int_{1/2}^1 \int_{-\sqrt{1-x^2}}^{\sqrt{1-x^2}} \left(2\ln x + \ln\left(1 + \left(\frac{y}{x}\right)^2\right)\right) dy\,dx$

Applications

61. Leila has been assigned the task of determining the risk to elk herds from the wolf population in a certain region of Idaho. She needs to find out the number of wolves near two distinct elk herds, one centered at the origin and the other 12 miles due north. She estimates the density of wolves in the region as 0.08 wolf per square mile. How many wolves would she expect to find within 12 miles of both herds?

62. Emmy needs to determine the volume of a sedimentation tank. The tank is circular with a radius of 75 feet as viewed from above, with a small concrete island that contains circulation and monitoring equipment in the center. The island has a radius of 7 feet. The depth of the tank at any distance r from the center is $d(r) = 0.15 \times 10^{-13} r^8 - 15$, where the surface of the solution in the tank is at depth zero. What is the volume of the tank?

Proofs

63. Use a double integral to prove that the area of the circle with radius R and equation $r = 2R\cos\theta$ is πR^2.

64. Use a double integral to prove that the area of the circle with radius R and equation $r = 2R\sin\theta$ is πR^2.

65. Use a double integral in polar coordinates to prove that the volume of a sphere with radius R is $\frac{4}{3}\pi R^3$.

66. Let h and R be positive real numbers. Explain why the region bounded above by the graph of the function $f(x, y) = h - \frac{h}{R}\sqrt{x^2 + y^2}$ and below by the xy-plane is a cone with height h and radius R. Use a double integral with polar coordinates to prove that the volume of this cone is $\frac{1}{3}\pi R^2 h$.

67. Use a double integral with polar coordinates to prove that the area of a sector with central angle ϕ in a circle of radius R is given by $A = \frac{1}{2}\phi R^2$.

68. Use a double integral with polar coordinates to prove that the area enclosed by one petal of the polar rose $r = \cos 3\theta$ is the same as the area enclosed by one petal of the polar rose $r = \sin 3\theta$.

69. Use a double integral with polar coordinates to prove that the combined area enclosed by all of the petals of the polar rose $r = \cos 2n\theta$ is the same for every positive integer n.

70. Use a double integral with polar coordinates to prove that the combined area enclosed by all of the petals of the polar rose $r = \sin(2n + 1)\theta$ is the same for every positive integer n.

Thinking Forward

▶ *A triple integral using cylindrical coordinates:* Show that the triple integral

$$\int_0^{2\pi} \int_0^R \int_0^h r\,dz\,dr\,d\theta = \pi R^2 h.$$

Explain why this integral gives the volume of a right circular cylinder with radius R and height h.

▶ *A triple integral using spherical coordinates:* Evaluate the triple integral

$$\int_0^{\pi} \int_0^{2\pi} \int_0^R \rho^2 \sin\phi\,d\rho\,d\theta\,d\phi.$$

Note that this integral gives the volume of a sphere with radius R.

13.4 APPLICATIONS OF DOUBLE INTEGRALS

- Using an iterated integral to find the mass of an object in the plane
- Finding the center of mass and centroid of a planar region
- Using double integrals to define the moment of inertia and the radius of gyration

The Mass of a Planar Region

In Section 6.4 we saw that the mass of a linear rod parallel to the x-axis with varying density $\rho(x) > 0$ is given by the integral

$$\int_a^b \rho(x)\,dx.$$

In the current section we first discuss how to find the mass of a region lying in the xy-plane when the density is a function of both variables x and y. Recall that in Sections 13.1 and 13.2 we discussed a strategy for computing the volume of a solid bounded above by a positive function $f(x, y)$ over a rectangular region in the plane and over a more general region in $\mathbb{R}^2$, respectively. We employ the same strategy here to find the mass of an object in the plane when the density of the object is given by the function $\rho(x, y)$, where $\rho(x, y) > 0$ for every point in the domain, as indicated in the following figure.

Region in the plane with variable density $\rho(x, y)$

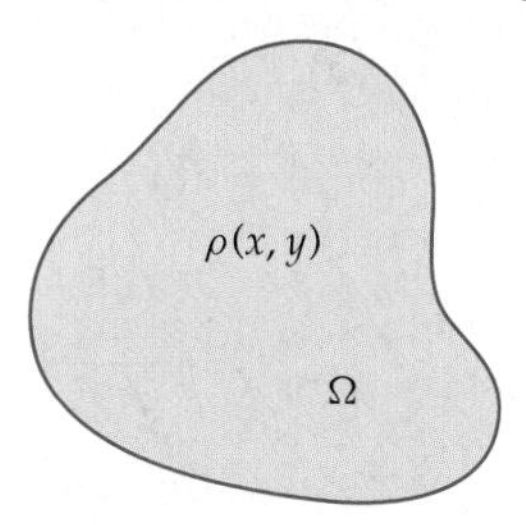

The technique of subdividing, approximating, and adding is precisely what is required here to approximate the mass of a planar region whose density is given by $\rho(x, y)$. Therefore, in the limit, to find the exact value of the mass of such a region in the plane, we may use the double integral

$$\text{Mass} = \iint_\Omega \rho(x, y)\,dA.$$

Depending upon the density function ρ and the region Ω, we may choose to evaluate the double integral as either a type I or type II region, using rectangular coordinates, or we may choose to use polar coordinates. For example, consider the triangular region Ω with vertices $(1, 1)$, $(2, 0)$, and $(2, 3)$. If the density at every point in Ω is proportional to the distance the point is from the y-axis, then $\rho(x, y) = kx$, where $k > 0$ is a constant of proportionality. It is somewhat easier to treat Ω as a type I region. The equations of the boundary segments are shown in the following figure:

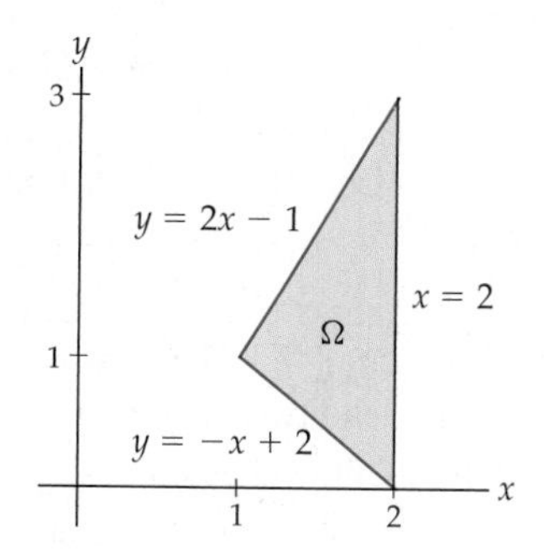

Therefore, the integral

$$\int_1^2 \int_{-x+2}^{2x-1} kx\, dy\, dx$$

represents the mass of the region. In Exercise 15 you will show that the value of this integral is $\frac{5}{2}k$. In Example 2, we discuss finding the mass of a region when the use of polar coordinates provides a simpler computation.

The Center of Mass and First Moments

Consider a system of two point masses on the x-axis: mass m_1 at point x_1 and mass m_2 at point x_2. The ***center of mass***, $\overline{x}$, of this simple system may be found with the formula:

$$\overline{x} = \frac{m_1 x_1 + m_2 x_2}{m_1 + m_2}.$$

For example, if a mass of 3 grams is located at $x = 2$ and a mass of 6 grams is located at $x = 5$, the center of mass of this system is found at $\overline{x} = \frac{3 \cdot 2 + 6 \cdot 5}{3+6} = 4$, as shown in the following figure:

3 grams
6 grams
x
2
$\overline{x} = 4$
5

This value conforms to our experience that the center of mass of the system should be closer to the larger mass.

Similarly, for n point masses $m_1, m_2, m_3, \ldots, m_n$, at the x-coordinates $x_1, x_2, x_3, \cdots, x_n$, respectively, the center of mass of this system is given by

$$\overline{x} = \frac{m_1 x_1 + m_2 x_2 + \cdots + m_n x_n}{m_1 + m_2 + \cdots + m_n} = \frac{\sum_{k=1}^{n} m_k x_k}{\sum_{k=1}^{n} m_k}.$$

We may generalize this idea to point masses in two (or three) dimensions:

m_1
(x_1, y_1)
m_n
(x_n, y_n)
m_3
(x_3, y_3)
m_2
(x_2, y_2)

Thus, for point masses, $m_1, m_2, m_3, \ldots, m_n$, at the coordinate pairs (x_1, y_1), (x_2, y_2), $(x_3, y_3), \ldots, (x_n, y_n)$, the center of mass of this system is given by the point $(\overline{x}, \overline{y})$, where

$$\overline{x} = \frac{\sum_{k=1}^{n} m_k x_k}{\sum_{k=1}^{n} m_k} \quad \text{and} \quad \overline{y} = \frac{\sum_{k=1}^{n} m_k y_k}{\sum_{k=1}^{n} m_k}.$$

Note that the computations for $\overline{x}$ and $\overline{y}$ are independent of each other; $\overline{x}$ depends only on the masses and their x-coordinates and $\overline{y}$ depends only on the masses and their y-coordinates.

We now turn our attention to finding the center of mass of a planar region with a density function $\rho(x, y)$. Such a region is called a ***lamina*** (*plural* laminæ). We employ our usual strategy of subdividing the region into small pieces, approximating the mass of each small subregion by a single value of the density function over the subregion and then adding all those values. In this context, we may approximate the x- and y-coordinates of the center of mass independently. The following figure provides a schematic:

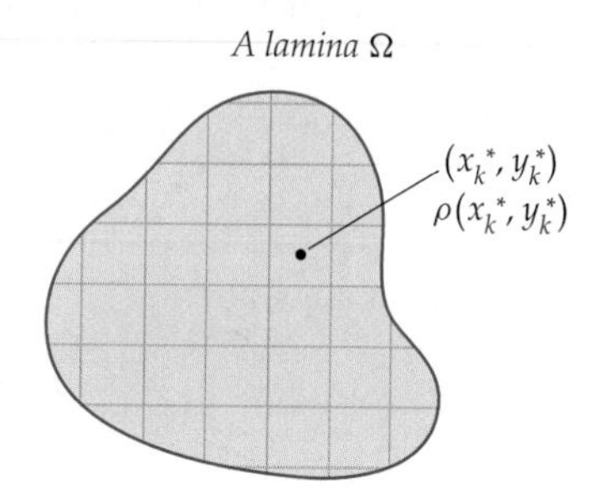

The approximate mass of each subregion is given by its (approximate) density $\rho(x_k^*, y_k^*)$ times its area ΔA. In the denominators that follow, we sum over all of the subregions. In the numerators, we need the extra factors x_k^* and y_k^*, as we are trying to approximate $\overline{x}$ and $\overline{y}$, respectively. Thus, $\overline{x}$ and $\overline{y}$ may be approximated by

$$\overline{x} \approx \frac{\sum_{k=1}^{n} x_k^* \rho(x_k^*, y_k^*) \Delta A}{\sum_{k=1}^{n} \rho(x_k^*, y_k^*) \Delta A} \quad \text{and} \quad \overline{y} \approx \frac{\sum_{k=1}^{n} y_k^* \rho(x_k^*, y_k^*) \Delta A}{\sum_{k=1}^{n} \rho(x_k^*, y_k^*) \Delta A}.$$

In the limit, as the mesh of our grid goes to zero, we obtain the exact coordinates of $\overline{x}$ and $\overline{y}$:

$$\overline{x} = \frac{\iint_\Omega x\, \rho(x, y)\, dA}{\iint_\Omega \rho(x, y)\, dA} \quad \text{and} \quad \overline{y} = \frac{\iint_\Omega y\, \rho(x, y)\, dA}{\iint_\Omega \rho(x, y)\, dA}.$$

We use the integrals in the numerators in the following definition:

DEFINITION 13.12 First Moments About the x- and y-axes

Let Ω be a lamina in the xy-plane in which the density at each point is given by the continuous function $\rho(x, y)$.

(a) The ***first moment*** of the mass in Ω about the y-axis is

$$M_y = \iint_\Omega x\, \rho(x, y)\, dA.$$

(b) The ***first moment*** of the mass in Ω about the x-axis is

$$M_x = \iint_\Omega y\, \rho(x, y)\, dA.$$

Thus, if we let $m = \iint_\Omega \rho(x, y)\, dA$, then m is the mass of the lamina Ω and the coordinates of the center of mass are

$$\overline{x} = \frac{M_y}{m} \quad \text{and} \quad \overline{y} = \frac{M_x}{m}.$$

The first moments are also known as the ***linear moments***. Upon first glance, you might think that there are errors in the definitions of the first moment. We see, for example, that $M_y = \iint_\Omega x\, \rho(x, y)\, dA$. To help remember this notation, recall that the factor x in the integrand measures the distance from the y-axis—hence the designation M_y. The first moments M_y and M_x provide measures of the distribution of the mass with respect to the y- and x-axes, respectively. It is possible to generalize these concepts to measure the distributions about other vertical and horizontal lines. For example, to measure the distribution of the mass about the vertical line $x = x_0$, we would evaluate the integral $\iint_\Omega (x - x_0)\, \rho(x, y)\, dA$. We will not be using these more general moments in this text.

Returning to our earlier example, we again consider the triangular region Ω with vertices $(1, 1)$, $(2, 0)$, and $(2, 3)$. Recall that the density at every point in Ω was proportional to the point's distance from the y-axis. At that time, we computed the mass of the lamina as

$$m = \int_1^2 \int_{-x+2}^{2x-1} kx \, dy \, dx = \frac{5}{2}k,$$

where k is a constant of proportionality. The integrals we need for the first moments require only minor adjustments:

$$M_y = \int_1^2 \int_{-x+2}^{2x-1} kx^2 \, dy \, dx = \frac{17}{4}k,$$

and

$$M_x = \int_1^2 \int_{-x+2}^{2x-1} kxy \, dy \, dx = \frac{27}{8}k.$$

Therefore,

$$\bar{x} = \frac{M_y}{m} = \frac{(17/4)k}{(5/2)k} = \frac{17}{10} \quad \text{and} \quad \bar{y} = \frac{M_x}{m} = \frac{(27/8)k}{(5/2)k} = \frac{27}{20}.$$

Note that the center of mass, $(17/20, 27/20)$, is located within the region Ω. When a region is convex, the center of mass lies within the region. Physically, when the center of mass lies within the region, the lamina may be balanced at that point.

When the density of a two-dimensional region Ω is constant, the center of mass is also called the ***centroid*** of Ω. In Exercise 20, you will explain why the centroid relates only to the geometry of the region and not to its mass.

Moments of Inertia

The moment of inertia of an object measures how difficult it is to change the angular momentum of the object. The moment of inertia of a spinning wheel is what makes it difficult for the wheel to stop when it is rotating. The integrals required to calculate the moments of inertia about the x- and y-axes are quite similar to those needed for the first moments.

DEFINITION 13.13 Moments of Inertia

Let Ω be a lamina in the xy-plane for which the density at each point is given by the continuous function $\rho(x, y)$.

(a) The ***moment of inertia about the*** y***-axis*** of the lamina is

$$I_y = \iint_\Omega x^2 \, \rho(x, y) \, dA.$$

(b) The ***moment of inertia about the*** x***-axis*** of the lamina is

$$I_x = \iint_\Omega y^2 \, \rho(x, y) \, dA.$$

(c) The ***moment of inertia about the origin*** of the lamina, also known as the ***polar moment***, is

$$I_o = \iint_\Omega (x^2 + y^2) \, \rho(x, y) \, dA.$$

The moments of inertia are also known as the ***second moments***. We may use these moments of inertia to compute the ***radius of gyration*** about the y-axis, the x-axis, and the origin. If the lamina has mass m, these are

$$R_y = \sqrt{\frac{I_y}{m}}, \quad R_x = \sqrt{\frac{I_x}{m}}, \quad \text{and} \quad R_o = \sqrt{\frac{I_o}{m}},$$

respectively. The radii of gyration are the radial distances at which the mass of the lamina could be concentrated without changing its rotational inertia. We compute the moments of inertia and radii of gyration for a region in Examples 4 and 5.

Probability Distributions

We've seen that when we integrate a density function ρ over a laminar region Ω, we obtain the mass of the lamina. Similarly, if we know a population density over a planar region, we may compute the population. We provide such a computation in Example 6. We may also use an iterated integral to find the probability of an event when we know a ***joint probability distribution function***. Recall that a continuous probability distribution function on an interval $[a, b]$ is a positive-valued function f defined on $[a, b]$ such that $\int_a^b f(x)\,dx = 1$. A joint probability distribution function of two variables on a subset $X \subset \mathbb{R}^2$ is a positive-valued function $f : X \to [0, \infty)$ such that $\iint_X f(x, y)\,dA = 1$. To compute the probability of an event $Z \subset X$, we evaluate the integral $\iint_Z f(x, y)\,dA$. Joint probability distribution functions of more than two variables are defined in an analogous manner, but we shall not be using them in this text.

Examples and Explorations

EXAMPLE 1

Finding the centroid of a triangular region

Find the centroid of the triangular region with vertices $(1, 1)$, $(2, 0)$, and $(2, 3)$.

SOLUTION

This is the same triangular region we used earlier in the section, but here we assume that the density ρ is constant:

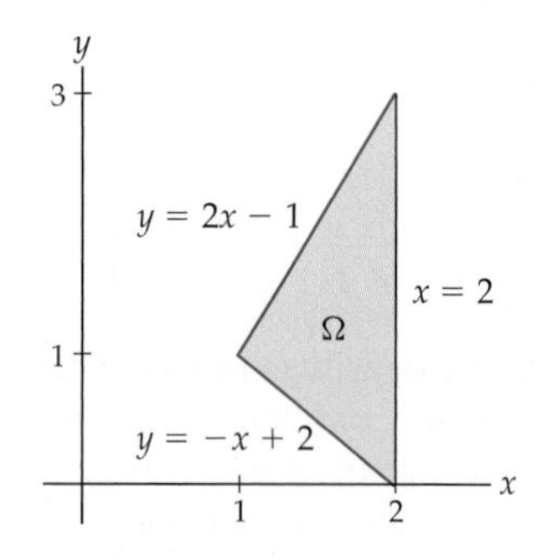

Since ρ is constant, we may factor it out of the integral. In our computation for $\bar{x}$ we have

$$\bar{x} = \frac{\iint_\Omega x\,\rho\,dA}{\iint_\Omega \rho\,dA} = \frac{\rho\iint_\Omega x\,dA}{\rho\iint_\Omega dA} = \frac{\iint_\Omega x\,dA}{\iint_\Omega dA}.$$

Similarly, we would have

$$\overline{y} = \frac{\iint_\Omega y\,dA}{\iint_\Omega dA}.$$

The integrals in the rightmost denominators each evaluate to the area of Ω. For this example, we may use either the indicated double integral or, since Ω is a triangle, the area formula for triangles to compute the area of Ω. We cannot avoid setting up an iterated integral, however, because we need to evaluate the integrals in the numerators. Here we need to evaluate the following three related integrals:

$$\text{Area} = \int_1^2 \int_{-x+2}^{2x-1} dy\,dx, \quad M_y = \int_1^2 \int_{-x+2}^{2x-1} x\,dy\,dx, \quad \text{and} \quad M_x = \int_1^2 \int_{-x+2}^{2x-1} y\,dy\,dx.$$

These integrals evaluate to

$$\text{Area} = \frac{3}{2}, \quad M_y = \frac{5}{2}, \quad \text{and} \quad M_x = 2.$$

Therefore, we have

$$\overline{x} = \frac{M_y}{\text{Area of } \Omega} = \frac{5/2}{3/2} = \frac{5}{3} \quad \text{and} \quad \overline{y} = \frac{M_x}{\text{Area of } \Omega} = \frac{2}{3/2} = \frac{4}{3}.$$

Thus, the centroid of the triangle is $(5/3, 4/3)$. Note that this point is inside the triangle. As we mentioned when we computed the center of mass in our earlier example, if a region Ω is convex, the centroid must be located within Ω. □

EXAMPLE 2 **Using polar coordinates to find a mass**

Find the mass of the semicircular lamina whose boundary on the left is given by the line $x = 1$ and on the right is given by $(x-1)^2 + y^2 = 1$ if the density at every point is proportional to its distance from the origin.

SOLUTION

The lamina we have described is shown here:

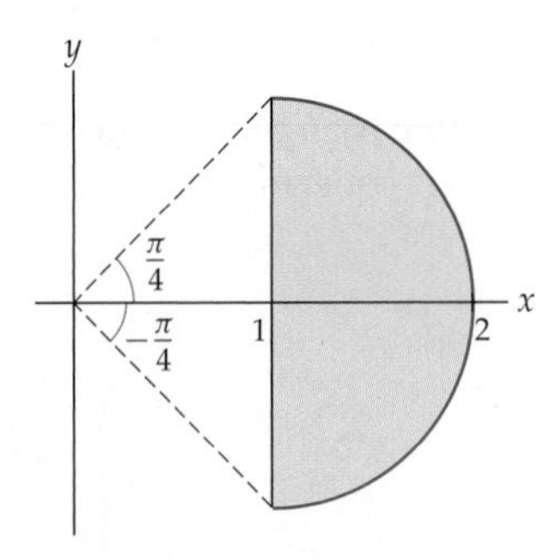

Since the density is proportional to the distance from the origin, it may be expressed easily in polar coordinates as a function of the radial distance r. That is, $\rho(r) = kr$, where $k > 0$ is a constant of proportionality. The equation of the circle in polar coordinates is $r = 2\cos\theta$, and the vertical line $x = 1$ in polar coordinates is given by $r = \sec\theta$. Finally, we use $r\,dr\,d\theta$ for our increment of area, dA. Therefore, the mass of the plate is given by

$$\int_{-\pi/4}^{\pi/4} \int_{\sec\theta}^{2\cos\theta} kr^2\,dr\,d\theta.$$

To evaluate this iterated integral, we first evaluate the inner integral:

$$\int_{\sec\theta}^{2\cos\theta} kr^2\,dr = \left[\frac{1}{3}kr^3\right]_{\sec\theta}^{2\cos\theta} = \frac{1}{3}k(8\cos^3\theta - \sec^3\theta).$$

We may now use the techniques of Chapter 5 to evaluate the outer integral:

$$\int_{-\pi/4}^{\pi/4} \frac{1}{3}k(8\cos^3\theta - \sec^3\theta)\,d\theta = \frac{1}{9}k(17\sqrt{2} + 3\ln(\sqrt{2}-1)).$$

You will show the details of this integration in Exercise 22. □

EXAMPLE 3 **Using polar coordinates to find a center of mass**

Find the center of mass of the semicircular lamina from Example 2.

SOLUTION

To start, we note that the y-coordinate of the center of mass is $\bar{y} = 0$, since the region Ω and the density function are symmetric with respect to the x-axis. Now, we have already computed the mass of the plate, using the integral

$$\int_{-\pi/4}^{\pi/4}\int_{\sec\theta}^{2\cos\theta} kr^2\,dr\,d\theta.$$

We still need to calculate M_y, the first moment of the mass about the y-axis. For M_y, all we have to do is introduce x into the preceding integrand. Since we are using polar coordinates, we will use the fact that $x = r\cos\theta$ instead. Therefore,

$$M_y = \int_{-\pi/4}^{\pi/4}\int_{\sec\theta}^{2\cos\theta} kr^3\cos\theta\,dr\,d\theta = \frac{1}{60}k(157\sqrt{2} + 15\ln(\sqrt{2}-1)).$$

You will show the details of this integration in Exercise 23. Thus, the x-coordinate of the center of mass is given by

$$\frac{M_y}{\text{mass}} = \bar{x} = \frac{\frac{k}{60}(157\sqrt{2} + 15\ln(\sqrt{2}-1))}{\frac{k}{9}(17\sqrt{2} + 3\ln(\sqrt{2}-1))} = \frac{471\sqrt{2} + 45\ln(\sqrt{2}-1)}{340\sqrt{2} + 60\ln(\sqrt{2}-1)} \approx 1.4638.$$

Again, as a quick check on the reasonableness of the answer, the point $(\bar{x}, \bar{y}) \approx (1.4638, 0)$ is within the region Ω. □

In Example 3, we used the symmetry of the region Ω and the density function to claim that $\bar{y} = 0$. If we wish, we may evaluate an iterated integral to compute M_x. Here, since the variable of integration on the inner integral is r, we may factor the integral as follows:

$$M_x = \int_{-\pi/4}^{\pi/4}\int_{\sec\theta}^{2\cos\theta} kr^3\sin\theta\,dr\,d\theta = \left(\int_{-\pi/4}^{\pi/4}\sin\theta\,d\theta\right)\left(\int_{\sec\theta}^{2\cos\theta} kr^3\,dr\right).$$

Now, since sine is an odd function and the interval $[-\pi/4, \pi/4]$ is symmetric with respect to the origin, it follows that $\int_{-\pi/4}^{\pi/4}\sin\theta\,d\theta = 0$, and therefore M_x and $\bar{y}$ are both zero, as we reasoned before.

EXAMPLE 4 **Finding the moments of inertia for a triangular region**

Find the moments of inertia for the triangular lamina Ω in the xy-plane and with vertices $(1, 1)$, $(2, 0)$, and $(2, 3)$ if the density at every point in Ω is proportional to the distance the point is from the y-axis.

SOLUTION

This is the same region we used in previous examples. We have $\rho(x, y) = kx$, where $k > 0$ is a constant of proportionality. Earlier in the section we saw that

$$M_y = \int_1^2 \int_{-x+2}^{2x-1} kx^2 \, dy \, dx \quad \text{and} \quad M_x = \int_1^2 \int_{-x+2}^{2x-1} kxy \, dy \, dx.$$

The integrals we need for I_y and I_x require the extra factors x and y, respectively. In Exercises 18 and 19 you will show that the values of these integrals are

$$I_y = \int_1^2 \int_{-x+2}^{2x-1} kx^3 \, dy \, dx = \frac{147}{20}k \quad \text{and} \quad I_x = \int_1^2 \int_{-x+2}^{2x-1} kxy^2 \, dy \, dx = \frac{28}{5}k.$$

From Definition 13.13, we see that $I_o = I_y + I_x$. Here we have $I_o = \frac{147}{20}k + \frac{28}{5}k = \frac{259}{20}k.$ □

EXAMPLE 5

Finding the radii of gyration for a triangular region

Find the radii of gyration about the x-axis, y-axis, and origin for the triangular lamina Ω with vertices $(1, 1)$, $(2, 0)$, and $(2, 3)$ if the density at every point in Ω is proportional to the the distance the point is from the y-axis.

SOLUTION

In Example 4 we computed the moments of inertia for this region, and earlier in this section we saw that the mass of Ω is $m = \frac{5}{2}k$. The radii of gyration with respect to the y-axis, x-axis, and origin are, respectively,

$$R_y = \sqrt{\frac{I_y}{m}} = \sqrt{\frac{(147/20)k}{(5/2)k}} = \sqrt{\frac{147}{50}}, \quad R_x = \sqrt{\frac{I_x}{m}} = \sqrt{\frac{(28/5)k}{(5/2)k}} = \frac{\sqrt{56}}{5}, \quad \text{and}$$

$$R_o = \sqrt{\frac{I_o}{m}} = \sqrt{\frac{(259/20)k}{(5/2)k}} = \sqrt{\frac{259}{50}}.$$

□

EXAMPLE 6

Determining a population from a population density

A biologist is culturing a population of bacteria in a circular petri dish with a radius of 5 centimeters. She introduces bacteria into the dish and incubates the dish for 24 hours. At the end of that time she estimates that the colony of bacteria in the dish contains 100 bacteria per square centimeter at the center of the dish, with the population decreasing linearly to the edge of the dish. She estimates the population density to be 10 bacteria per square centimeter at the edge. Approximately how many bacteria are in the dish?

SOLUTION

This problem is nearly identical to those in which we were given a (mass) density function ρ for a laminar region Ω. To find the total mass, we used an iterated integral to integrate the density function ρ over the region Ω. We will do the same here to find the population of bacteria. We start by imposing a polar coordinate system on the circular petri dish, with the center of the dish at the origin. Since there are 100 bacteria per square centimeter at the origin and the density decreases linearly to 10 bacteria per square centimeter at the edge of the dish, the function $\rho(r, \theta) = 100 - 18r$ gives us the population density at every point in the petri dish for $r \in [0, 5]$. Therefore, the population of bacteria in the dish is given by the iterated integral

$$\int_0^{2\pi} \int_0^5 (100 - 18r)\, r\, dr\, d\theta.$$

The value of this integral is $1000\pi \approx 3100$ bacteria. Note that we have rounded our answer, since the biologist clearly approximated her values for the two population densities. □

? TEST YOUR UNDERSTANDING

- Given a system of point masses in the plane, how is the center of mass of the system computed? How is this idea used to find the center of mass of a lamina Ω whose density function is $\rho(x, y)$?
- What is the centroid of a region Ω? What is the relationship between the centroid of a region and its center of mass? Why is the centroid of Ω independent of the mass of Ω?
- How is the moment of inertia of a region Ω with respect to the x- and y-axes defined? How are these definitions related to the definitions of the first moments of Ω with respect to the x- and y-axes? What is meant by the polar moment?
- How are the radii of gyration with respect to the x- and y-axes, R_x, and R_y defined for a region Ω? What integrals need to be computed to find R_x and R_y?
- How are the integrals for masses, first moments, and moments of inertia related? Which of the three integrals is easiest to step up? Given one of these integrals, how would you modify it to set up the other integrals? When would rectangular coordinates be easier to use for these integrals? When would polar coordinates be easier to use?

EXERCISES 13.4

Thinking Back

- *Mass of a rod:* Suppose a thin rod with a radius of 1 centimeter and a length of 20 centimeters has a varying density such that the density of the rod x centimeters from the left end is given by the function $\rho(x) = 10 + 0.01x^2$ grams per cubic centimeter. Find the mass of the rod.
- *More questions about the rod:* How far from the left end of the rod is the center of mass in the previous problem? How far from the left end of the rod is the moment of inertia? How far from the left end of the rod is the radius of gyration?

Concepts

0. *Problem Zero:* Read the section and make your own summary of the material.

1. *True/False:* Determine whether each of the statements that follow is true or false. If a statement is true, explain why. If a statement is false, provide a counterexample.

(a) *True or False:* The center of mass of a lamina, Ω, in the xy-plane is a point in Ω.

(b) *True or False:* The centroid of a region, Ω, in the xy-plane is a point in Ω.

(c) *True or False:* The center of mass of a circular region is the center of the circle for every density function $\rho(x, y)$.

(d) *True or False:* The centroid of a circular region is the center of the circle.

(e) *True or False:* The first moment of the mass in a lamina Ω about the x-axis is given by $M_x = \iint_\Omega x\,\rho(x, y)\, dA$, where $\rho(x, y)$ is the lamina's density function.

(f) *True or False:* If M_x and I_x are the first and second moments, respectively, for a lamina in the xy-plane, then $xM_x = I_x$.

(g) *True or False:* If I_x, I_y, and I_o are the moments of inertia of a lamina about the x-axis, y-axis, and origin, respectively, then $I_o = I_x + I_y$.

(h) *True or False:* If m is the mass of a lamina, I_y is the moment of inertia of the lamina about the y-axis, and R_y is the radius of gyration about the y-axis, then $R_y^2 = I_y/m$.

2. *Examples:* Construct examples of the thing(s) described in the following. Try to find examples that are different than any in the reading.
   (a) A lamina Ω in the xy-plane and a density function $\rho(x, y)$ such that the center of mass of Ω and the centroid of Ω are the same.
   (b) A lamina Ω in the xy-plane and a non-constant density function $\rho(x, y)$ such that the center of mass of Ω and the centroid of Ω are the same.
   (c) A lamina Ω in the xy-plane such that the center of mass is *not* in Ω.

Identify the quantities determined by the integral expressions in Exercises 3–11. If x and y are both measured in centimeters and $\rho(x, y)$ is a density function in grams per square centimeter, give the units of the expression.

3. $\iint_\Omega dA$

4. $\iint_\Omega x\,dA$ and $\iint_\Omega y\,dA$

5. $\dfrac{\iint_\Omega x\,dA}{\iint_\Omega dA}$ and $\dfrac{\iint_\Omega y\,dA}{\iint_\Omega dA}$

6. $\iint_\Omega \rho(x, y)\,dA$

7. $\iint_\Omega x\,\rho(x, y)\,dA$ and $\iint_\Omega y\,\rho(x, y)\,dA$

8. $\dfrac{\iint_\Omega x\,\rho(x, y)\,dA}{\iint_\Omega \rho(x, y)\,dA}$ and $\dfrac{\iint_\Omega y\,\rho(x, y)\,dA}{\iint_\Omega \rho(x, y)\,dA}$

9. $\iint_\Omega x^2\rho(x, y)\,dA$ and $\iint_\Omega y^2\rho(x, y)\,dA$

10. $\iint_\Omega (x^2 + y^2)\rho(x, y)\,dA$

11. $\sqrt{\dfrac{\iint_\Omega x^2\,\rho(x, y)\,dA}{\iint_\Omega \rho(x, y)\,dA}}$ and $\sqrt{\dfrac{\iint_\Omega y^2\,\rho(x, y)\,dA}{\iint_\Omega \rho(x, y)\,dA}}$

Throughout this section we computed several integrals relating to the triangular region Ω with vertices $(1, 1)$, $(2, 0)$, and $(2, 3)$. In Exercises 12–19 you are asked to provide the details of those computations.

12. Show that the area of Ω is $\frac{3}{2}$ by using the area formula for triangles and by evaluating the integral

$$\int_1^2 \int_{-x+2}^{2x-1} dy\,dx.$$

13. Show that when the density of the region is constant, the first moment about the y-axis is

$$M_y = \int_1^2 \int_{-x+2}^{2x-1} x\,dy\,dx = \frac{5}{2}.$$

14. Show that when the density of the region is constant, the first moment about the x-axis is

$$M_x = \int_1^2 \int_{-x+2}^{2x-1} y\,dy\,dx = 2.$$

15. Show that when the density of the region is proportional to the distance from the y-axis, the mass of Ω is given by

$$\int_1^2 \int_{-x+2}^{2x-1} kx\,dy\,dx = \frac{5}{2}k.$$

16. Show that when the density of the region is proportional to the distance from the y-axis, the first moment about the y-axis is

$$M_y = \int_1^2 \int_{-x+2}^{2x-1} kx^2\,dy\,dx = \frac{17}{4}k.$$

17. Show that when the density of the region is proportional to the distance from the y-axis, the first moment about the x-axis is

$$M_x = \int_1^2 \int_{-x+2}^{2x-1} kxy\,dy\,dx = \frac{27}{8}k.$$

18. Show that when the density of the region is proportional to the distance from the y-axis, the moment of inertia about the y-axis is

$$I_y = \int_1^2 \int_{-x+2}^{2x-1} kx^3\,dy\,dx = \frac{147}{20}k.$$

19. Show that when the density of the region is proportional to the distance from the y-axis, the first moment about the x-axis is

$$I_x = \int_1^2 \int_{-x+2}^{2x-1} kxy^2\,dy\,dx = \frac{28}{5}k.$$

20. Explain why the location of the centroid relates only to the geometry of the region and not its mass.

21. Find the moments of inertia about the x- and y-axes for the semicircular lamina described in Example 2. Assume that the density at every point is proportional to the distance of the point from the origin.

22. Complete Example 2 by showing that

$$\int_{-\pi/4}^{\pi/4} \frac{k}{3}(8\cos^3\theta - \sec^3\theta)\,d\theta = \frac{k}{9}(17\sqrt{2} + 3\ln(\sqrt{2} - 1)).$$

23. Complete Example 3 by showing that

$$\int_{-\pi/4}^{\pi/4} \int_{\sec\theta}^{2\cos\theta} kr^3\cos\theta\,dr\,d\theta = \frac{k}{60}(157\sqrt{2} + 15\ln(\sqrt{2} - 1)).$$

Skills

In Exercises 24–30, let $\mathcal{T}$ be the triangular region with vertices $(0, 0)$, $(1, 1)$, and $(1, -1)$.

24. Find the centroid of $\mathcal{T}$.

25. If the density at each point in $\mathcal{T}$ is proportional to the point's distance from the y-axis, find the mass of $\mathcal{T}$.

26. If the density at each point in $\mathcal{T}$ is proportional to the point's distance from the y-axis, find the center of mass of $\mathcal{T}$.

27. If the density at each point in $\mathcal{T}$ is proportional to the point's distance from the y-axis, find the moments of inertia about the x- and y-axes. Use these answers to find the radii of gyration of $\mathcal{T}$ about the x- and y-axes.

28. If the density at each point in $\mathcal{T}$ is proportional to the point's distance from the x-axis, find the mass of $\mathcal{T}$.

29. If the density at each point in $\mathcal{T}$ is proportional to the point's distance from the x-axis, find the center of mass of $\mathcal{T}$.

30. If the density at each point in $\mathcal{T}$ is proportional to the point's distance from the x-axis, find the moments of inertia about the x- and y-axes. Use these answers to find the radii of gyration of $\mathcal{T}$ about the x- and y-axes.

In Exercises 31–37, let $\mathcal{T}_2$ be the triangular region with vertices $(1, 0)$, $(2, 1)$, and $(2, -1)$. Note that $\mathcal{T}_2$ is a translation of the triangle we used in Exercises 24–30.

31. Find the centroid of $\mathcal{T}_2$.

32. If the density at each point in $\mathcal{T}_2$ is proportional to the point's distance from the y-axis, find the mass of $\mathcal{T}_2$.

33. If the density at each point in $\mathcal{T}_2$ is proportional to the point's distance from the y-axis, find the center of mass of $\mathcal{T}_2$.

34. If the density at each point in $\mathcal{T}_2$ is proportional to the point's distance from the y-axis, find the moments of inertia about the x- and y-axes. Use these answers to find the radii of gyration of $\mathcal{T}_2$ about the x- and y-axes.

35. If the density at each point in $\mathcal{T}_2$ is proportional to the square of the point's distance from the y-axis, find the mass of $\mathcal{T}_2$.

36. If the density at each point in $\mathcal{T}_2$ is proportional to the square of the point's distance from the y-axis, find the center of mass of $\mathcal{T}_2$.

37. If the density at each point in $\mathcal{T}_2$ is proportional to the square of the point's distance from the y-axis, find the moments of inertia about the x- and y-axes. Use these answers to find the radii of gyration of $\mathcal{T}_2$ about the x- and y-axes.

In Exercises 38–44, let $\mathcal{R}$ be the rectangular region with vertices $(0, 0)$, $(b, 0)$, $(0, h)$, and (b, h).

38. Find the centroid of $\mathcal{R}$.

39. If the density at each point in $\mathcal{R}$ is proportional to the point's distance from the y-axis, find the mass of $\mathcal{R}$.

40. If the density at each point in $\mathcal{R}$ is proportional to the point's distance from the y-axis, find the center of mass of $\mathcal{R}$.

41. If the density at each point in $\mathcal{R}$ is proportional to the point's distance from the y-axis, find the moments of inertia about the x- and y-axes. Use these answers to find the radii of gyration of $\mathcal{R}$ about the x- and y-axes.

42. If the density at each point in $\mathcal{R}$ is proportional to the square of the point's distance from the y-axis, find the mass of $\mathcal{R}$.

43. If the density at each point in $\mathcal{R}$ is proportional to the square of the point's distance from the y-axis, find the center of mass of $\mathcal{R}$.

44. If the density at each point in $\mathcal{R}$ is proportional to the square of the point's distance from the y-axis, find the moments of inertia about the x- and y-axes. Use these answers to find the radii of gyration of $\mathcal{R}$ about the x- and y-axes.

In Exercises 45–51, let $\mathcal{C} = \{(x, y) \mid x^2 + y^2 \le 1\}$.

45. Find the centroid of $\mathcal{C}$.

46. If the density at each point in $\mathcal{C}$ is proportional to the point's distance from the y-axis, find the mass of $\mathcal{C}$.

47. If the density at each point in $\mathcal{C}$ is proportional to the point's distance from the y-axis, find the center of mass of $\mathcal{C}$.

48. If the density at each point in $\mathcal{C}$ is proportional to the point's distance from the y-axis, find the moments of inertia about the x- and y-axes. Use these answers to find the radii of gyration of $\mathcal{C}$ about the x- and y-axes.

49. If the density at each point in $\mathcal{C}$ is proportional to the point's distance from the origin, find the mass of $\mathcal{C}$.

50. If the density at each point in $\mathcal{C}$ is proportional to the point's distance from the origin, find the center of mass of $\mathcal{C}$.

51. If the density at each point in $\mathcal{C}$ is proportional to the point's distance from the origin, find the moments of inertia about the x-axis, the y-axis, and the origin. Use these answers to find the radii of gyration of $\mathcal{C}$ about the x-axis, the y-axis, and the origin.

In Exercises 52–58, let $\mathcal{S} = \{(x, y) \mid x^2 + y^2 \le 1 \text{ and } x \ge 0\}$.

52. Find the centroid of $\mathcal{S}$.

53. If the density at each point in $\mathcal{S}$ is proportional to the point's distance from the y-axis, find the mass of $\mathcal{S}$.

54. If the density at each point in $\mathcal{S}$ is proportional to the point's distance from the y-axis, find the center of mass of $\mathcal{S}$.

55. If the density at each point in $\mathcal{S}$ is proportional to the point's distance from the y-axis, find the moments of inertia about the x- and y-axes. Use these answers to find the radii of gyration of $\mathcal{S}$ about the x- and y-axes.

56. If the density at each point in $\mathcal{S}$ is proportional to the point's distance from the origin, find the mass of $\mathcal{S}$.

57. If the density at each point in $\mathcal{S}$ is proportional to the point's distance from the origin, find the center of mass of $\mathcal{S}$.

58. If the density at each point in $\mathcal{S}$ is proportional to the point's distance from the origin, find the moments of inertia about the x-axis, the y-axis, and the origin. Use these answers to find the radii of gyration of $\mathcal{S}$ about the x-axis, the y-axis, and the origin.

59. Let Ω be a lamina in the xy-plane. Suppose Ω is composed of two non-overlapping laminæ Ω_1 and Ω_2, as follows:

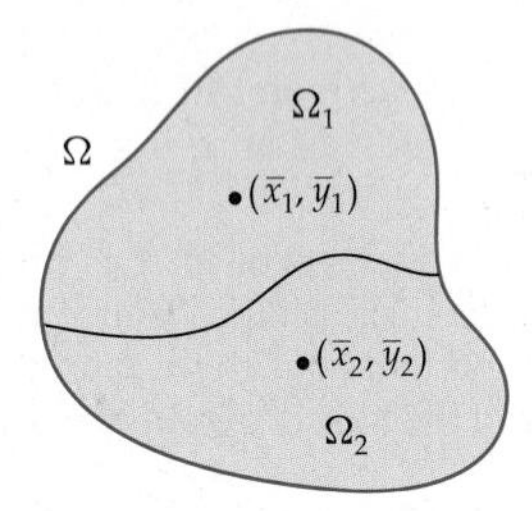

Show that if the masses and centers of masses of Ω_1 and Ω_2 are m_1 and m_2, and $(\overline{x}_1, \overline{y}_1)$ and $(\overline{x}_2, \overline{y}_2)$, respectively, then the center of mass of Ω is $(\overline{x}, \overline{y})$, where

$$\overline{x} = \frac{m_1\overline{x}_1 + m_2\overline{x}_2}{m_1 + m_2} \quad \text{and} \quad \overline{y} = \frac{m_1\overline{y}_1 + m_2\overline{y}_2}{m_1 + m_2}.$$

60. Let Ω be a lamina in the xy-plane. Suppose Ω is composed of n non-overlapping laminæ Ω_1, $\Omega_2, \ldots, \Omega_n$. Show that if the masses of these laminæ are $m_1, m_2, \ldots,$ m_n and the centers of masses are $(\overline{x}_1, \overline{y}_1), (\overline{x}_2, \overline{y}_2), \ldots,$ $(\overline{x}_n, \overline{y}_n)$, then the center of mass of Ω is $(\overline{x}, \overline{y})$, where

$$\overline{x} = \frac{\sum_{k=1}^{n} m_k\overline{x}_k}{\sum_{k=1}^{n} m_k} \quad \text{and} \quad \overline{y} = \frac{\sum_{k=1}^{n} m_k\overline{y}_k}{\sum_{k=1}^{n} m_k}.$$

Use the results of Exercises 59 and 60 to find the centers of masses of the laminæ in Exercises 61–67.

61. In the following lamina, all angles are right angles and the density is constant:

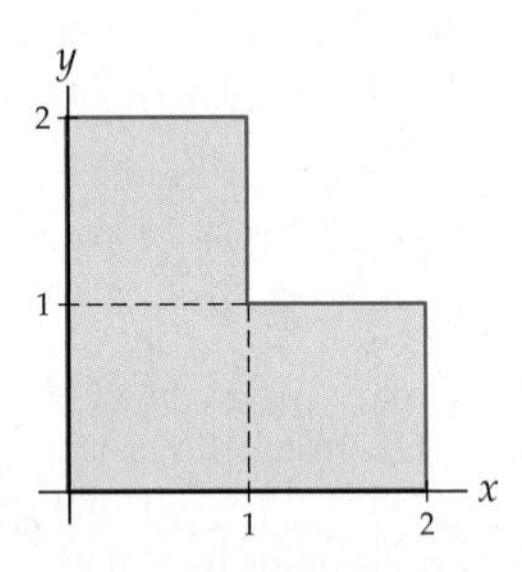

62. Use the lamina from Exercise 61, but assume that the density is proportional to the distance from the x-axis.

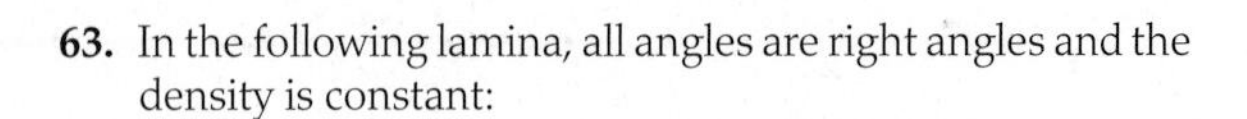

63. In the following lamina, all angles are right angles and the density is constant:

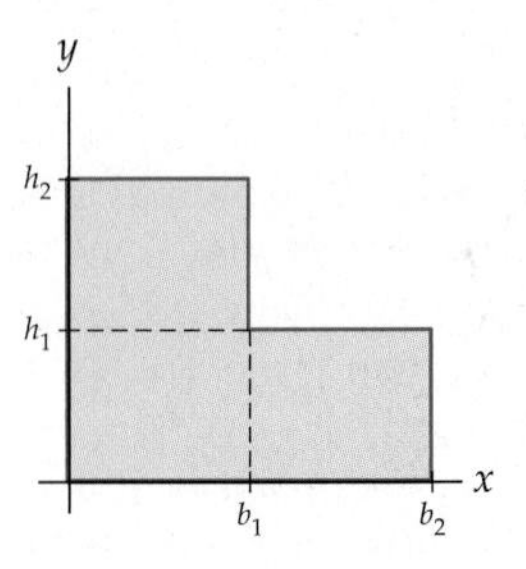

64. In the following lamina, all angles are right angles and the density is constant:

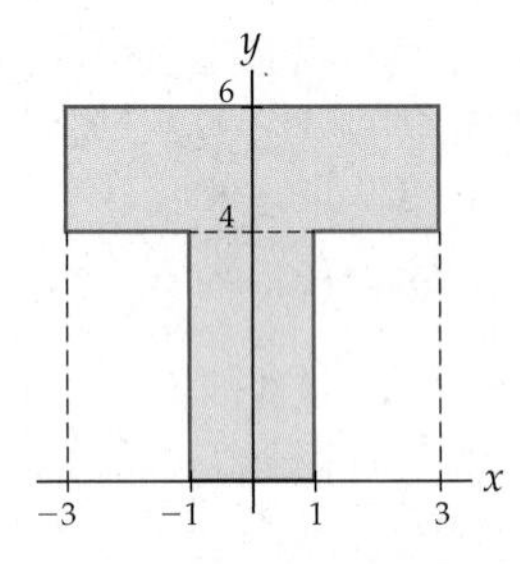

65. Use the lamina from Exercise 64, but assume that the density is proportional to the distance from the x-axis.

66. The lamina in the figure that follows is bounded above by the lines with equations $y = x + 2a$ and $y = -x + 2a$ and below by the x-axis on the interval $-a \leq x \leq a$. The density of the lamina is constant.

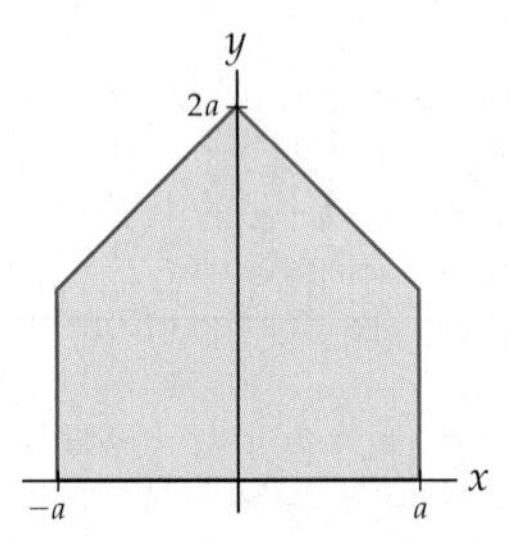

67. Use the lamina from Exercise 66, but assume that the density is proportional to the distance from the x-axis.

Applications

68. Leila has been required to track down a grizzly that attacked a person. The bear once had a radio collar, but has lost it. From the old data, Leila knows that the bear's range is roughly a region bounded by the x-axis, the y-axis, and the line $y = 4.3 - 0.8x$. The bear is equally likely to be found in any part of that region. The probability density function for the bear's location is $f(x, y) =$ 0.0873. There is a trail that lies roughly along the y-axis. How far must Leila's team bushwhack from the trail in order for her team to have a 50% chance of finding the bear on the first day of the search?

69. In Exercise 68, what is the expected distance from the road at which the bear is likely to be found? (*Hint: Compute the first moment in the x-direction for the distribution.*)

Proofs

Recall that a ***median*** of a triangle is a segment connecting a vertex of a triangle to the midpoint of the opposite side. Let $\mathcal{T}$ be the triangle with vertices $(0,0)$, $(a,0)$, and (c,d). In Exercises 70–72, prove the given statements.

70. The medians of triangle $\mathcal{T}$ are ***concurrent***; that is, all three medians intersect at the same point, P.

71. Use the integral definition for the centroid to show that the centroid of $\mathcal{T}$ is point P from Exercise 70.

72. Prove that the centroid of triangle $\mathcal{T}$ is two-thirds of the way from each vertex to the opposite side.

73. Prove that the centroid of a circle is the center of the circle.

74. Recall that an annulus is the region between two concentric circles. Prove that the centroid of an annulus is the common center of the two circles.

Thinking Forward

▶ *A triple iterated integral:* Let α, β, γ, δ, ϵ, and ζ be real numbers. Evaluate the triple iterated integral

$$\int_{\alpha}^{\beta}\int_{\gamma}^{\delta}\int_{\epsilon}^{\zeta} dz\,dy\,dx.$$

What does this integral represent?

▶ *A triple iterated integral of a density function:* Let α, β, γ, δ, ϵ, and ζ be real numbers, and let $\rho(x,y,z)$ be a function giving the density at each point of a three-dimensional rectangular solid. What does the triple integral

$$\int_{\alpha}^{\beta}\int_{\gamma}^{\delta}\int_{\epsilon}^{\zeta} \rho(x,y,z)\,dz\,dy\,dx.$$

represent?

13.5 TRIPLE INTEGRALS

▶ Generalizing integration to functions of three variables

▶ Computing triple integrals defined on rectangular solids and on more general regions

▶ Computing mass, center of mass, and moments of inertia of three-dimensional solids

Triple Integrals over Rectangular Solids

In Section 13.1 we defined the double integral of a function of two variables over a rectangular region in the plane. We will be building upon Definition 13.4 to define the triple integral of a function of three variables over a region in $\mathbb{R}^3$ that is a rectangular solid, such as the following one:

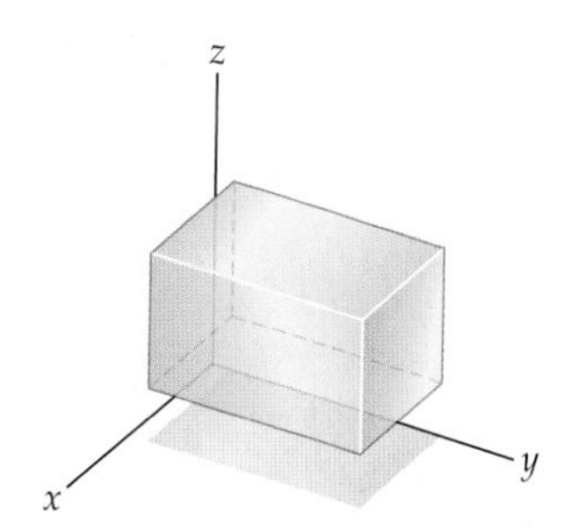

Later in this section we will extend our definition to more general regions in $\mathbb{R}^3$. We will define the triple integral in terms of the limit of a Riemann sum, just as we did for the double integral. Rather than belabor the details, we quickly outline the steps needed to

define a triple integral. If you have questions, then you should reread Section 13.1, which discusses the steps more completely for double integrals. We start with Riemann sums for a function of three variables.

DEFINITION 13.14

Riemann Sums for a Function of Three Variables

Let $a_1 < a_2$, $b_1 < b_2$, and $c_1 < c_2$ be real numbers, let $\mathcal{R}$ be the rectangular solid defined by

$$\mathcal{R} = \{(x, y, z) \mid a_1 \le x \le a_2,\ b_1 \le y \le b_2, \text{ and } c_1 \le z \le c_2\},$$

and let $f(x, y, z)$ be a function defined on $\mathcal{R}$. The sum

$$\sum_{i=1}^{l}\sum_{j=1}^{m}\sum_{k=1}^{n} f(x_i^*, y_j^*, z_k^*)\Delta V$$

is a ***Riemann Sum*** for f on $\mathcal{R}$, where

$$\Delta x = \frac{a_2 - a_1}{l}, \quad \Delta y = \frac{b_2 - b_1}{m}, \quad \Delta z = \frac{c_2 - c_1}{n}, \quad \Delta V = \Delta x \Delta y \Delta z,$$

$$x_i = a_1 + i\Delta x \text{ for } 0 \le i \le l,\ y_j = b_1 + j\Delta y \text{ for } 0 \le j \le m,\ z_k = c_1 + k\Delta z \text{ for } 0 \le k \le n,$$

$$\text{and } x_i^* \in [x_{i-1}, x_i],\ y_j^* \in [y_{j-1}, y_j],\ z_k^* \in [z_{k-1}, z_k].$$

Note that the nested summations in Definition 13.14 may be done in any of $3! = 6$ orders. The order we give in the definition is just one of those six, and we could have used any of the other five instead.

We now define Δ to be the length of the space diagonal of each rectangular solid. That is,

$$\Delta = \sqrt{(\Delta x)^2 + (\Delta y)^2 + (\Delta z)^2}.$$

When we take the limit $\Delta \to 0$, this ensures that $l \to \infty$, $m \to \infty$, and $n \to \infty$ simultaneously. We use the limit of the Riemann sum as $\Delta \to 0$ to define the triple integral, provided that the limit exists.

DEFINITION 13.15

Triple Integrals

Let $a_1 < a_2$, $b_1 < b_2$, and $c_1 < c_2$ be real numbers, let $\mathcal{R}$ be the rectangular solid defined by

$$\mathcal{R} = \{(x, y, z) \mid a_1 \le x \le a_2,\ b_1 \le y \le b_2, \text{ and } c_1 \le z \le c_2\},$$

and let $f(x, y, z)$ be a function defined on $\mathcal{R}$. Provided that the limit exists, the ***triple integral*** of f over $\mathcal{R}$ is

$$\iiint_{\mathcal{R}} f(x, y, z)\, dV = \lim_{\Delta \to 0} \sum_{i=1}^{l}\sum_{j=1}^{m}\sum_{k=1}^{n} f(x_i^*, y_j^*, z_k^*)\Delta V,$$

where the triple sum in the equation is a Riemann sum as outlined in Definition 13.14, and where $\Delta = \sqrt{(\Delta x)^2 + (\Delta y)^2 + (\Delta z)^2}$. When the triple integral exists on $\mathcal{R}$, the function f is said to be ***integrable*** on $\mathcal{R}$.

Note that as $\Delta \to 0$, the increment of volume $\Delta V \to 0$ as well. The summations in Definition 13.15 can be ordered in five other ways to obtain the same value for the triple integral.

At this point you may be wondering what the value of a particular triple integral represents about the function $f(x, y, z)$. As we have seen before, the answer depends upon the context. For example, if f is the constant function $f(x, y, z) = 1$, the triple integral $\iiint_{\mathcal{R}} dV$ would be a complicated way to compute the volume of the rectangular solid $\mathcal{R}$. If this were the only application of triple integration, we would not be bothering with the concept. However, recall that in Section 13.4 we showed that the double integral $\int_a^b \int_c^d \rho(x, y)\, dy\, dx$ gives the mass of the laminar rectangle $\mathcal{R} = \{(x, y) \mid a \leq x \leq b \text{ and } c \leq y \leq d\}$ if $\rho(x, y)$ is the density function for the lamina. Similarly, if $\rho(x, y, z) \geq 0$ is a density function of a rectangular solid $\mathcal{R}$, the triple integral $\iiint_{\mathcal{R}} \rho(x, y, z)\, dV$ gives the mass of the solid. Later in this section we will see how to use triple integrals to find the center of mass and moments of inertia for three-dimensional regions also. These will be our primary applications of triple integration.

Iterated Integrals and Fubini's Theorem

As we saw in Section 13.1, it is usually simpler to evaluate a double integral by using an iterated integral rather than the definition. The situation here is similar.

DEFINITION 13.16 Iterated Triple Integrals

Let a_1, a_2, b_1, b_2, c_1, and c_2 be real numbers. The ***iterated triple integral*** is

$$\int_{a_1}^{a_2} \int_{b_1}^{b_2} \int_{c_1}^{c_2} f(x, y, z)\, dz\, dy\, dx = \int_{a_1}^{a_2} \left(\int_{b_1}^{b_2} \left(\int_{c_1}^{c_2} f(x, y, z)\, dz \right) dy \right) dx.$$

Again, there are five other orderings we could use to construct an iterated triple integral. For example,

$$\int_{b_1}^{b_2} \int_{c_1}^{c_2} \int_{a_1}^{a_2} f(x, y, z)\, dx\, dz\, dy = \int_{b_1}^{b_2} \left(\int_{c_1}^{c_2} \left(\int_{a_1}^{a_2} f(x, y, z)\, dx \right) dz \right) dy.$$

For a region that is a rectangular solid, the different orderings for the integrals usually do not significantly change the level of difficulty in a problem. The ordering shown in Definition 13.16 is the one we will try first in most problems.

We have a version of Fubini's theorem that tells us that we may use an iterated triple integral, rather than the definition of the triple integral, to evaluate a triple integral.

THEOREM 13.17 **Fubini's Theorem for Triple Integrals**

Let a_1, a_2, b_1, b_2, c_1, and c_2 be real numbers, and let $\mathcal{R}$ be the rectangular solid defined by

$$\mathcal{R} = \{(x, y, z) \mid a_1 \leq x \leq a_2,\ b_1 \leq y \leq b_2, \text{ and } c_1 \leq z \leq c_2\}.$$

If $f(x, y, z)$ is continuous on $\mathcal{R}$, then

$$\iiint_{\mathcal{R}} f(x, y, z)\, dV = \int_{a_1}^{a_2} \int_{b_1}^{b_2} \int_{c_1}^{c_2} f(x, y, z)\, dz\, dy\, dx$$

and is also equal to any of the other five possible orderings for the iterated triple integral.

The proof of this version of Fubini's theorem is again beyond the scope of the text, but here is an example illustrating how it is used: Let

$$\mathcal{R} = \{(x, y, z) \mid 1 \le x \le 3,\ 2 \le y \le 6,\ \text{and } 4 \le z \le 7\}.$$

Assume that the density at each point of $\mathcal{R}$ is proportional to the distance of the point from the xy-plane. That is, $\rho(x, y, z) = kz$, where k is a constant of proportionality. We may use Theorem 13.17 to find the mass of $\mathcal{R}$. The mass of the solid is given by

$$\iiint_{\mathcal{R}} kz\, dV = \int_1^3 \int_2^6 \int_4^7 kz\, dz\, dy\, dx.$$

Here we have

$$\begin{aligned}
\iiint_{\mathcal{R}} kz\, dV &= \int_1^3 \int_2^6 \int_4^7 kz\, dz\, dy\, dx && \leftarrow \text{Fubini's theorem}\\
&= \int_1^3 \left(\int_2^6 \left(\int_4^7 kz\, dz \right) dy \right) dx && \leftarrow \text{evaluation procedure for the iterated integral}\\
&= \int_1^3 \left(\int_2^6 \left[\frac{1}{2} kz^2 \right]_{z=4}^{z=7} dy \right) dx && \leftarrow \text{the Fundamental Theorem of Calculus}\\
&= \int_1^3 \left(\int_2^6 \left(\frac{1}{2} k(49 - 16) \right) dy \right) dx && \leftarrow \text{evaluation of the innermost antiderivative}\\
&= \int_1^3 \left[\frac{33}{2} ky \right]_{y=2}^{y=6} dx = \int_1^3 66k\, dx && \leftarrow \text{evaluation of the next integral}\\
&= \left[66kx \right]_{x=1}^{x=3} = 132k. && \leftarrow \text{evaluation of the outer integral}
\end{aligned}$$

If we had used any of the five other orders of integration to evaluate the triple integral, we would have obtained the same result.

Triple Integrals over General Regions

We now expand our definition of the triple integral to allow us to compute a triple integral over more general regions. The general regions we will use are of three types. First, consider a region Ω_{xy} in the xy-plane and two functions $z = g_1(x, y)$ and $z = g_2(x, y)$ defined on Ω_{xy} such that $g_1(x, y) \le g_2(x, y)$ for every point in Ω_{xy}. We define the three-dimensional region Ω to be the set

$$\Omega = \{(x, y, z) \mid g_1(x, y) \le z \le g_2(x, y) \text{ for } (x, y) \in \Omega_{xy}\}.$$

For example, in the following figure, Ω_{xy} is the circle in the xy-plane, with radius 2, and centered at the origin, and the solid shown is bounded below by the plane with equation $g_1(x, y) = \frac{1}{2}x - \frac{1}{2}y - 2$ and is bounded above by the paraboloid with equation $g_2(x, y) = 4 - x^2 - y^2$:

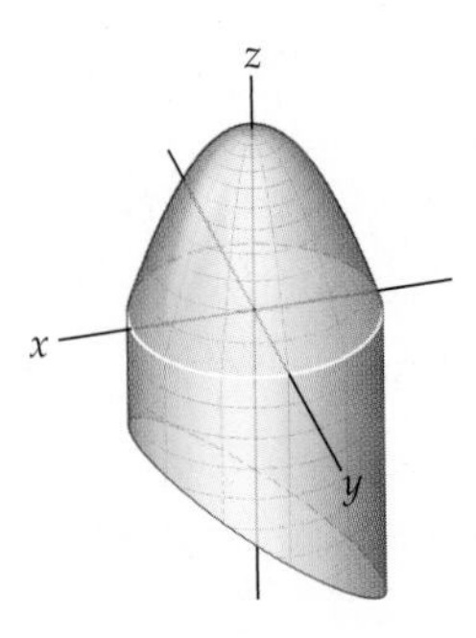

The other two types of regions we will be using are defined in a similar manner. In one, we start with a region Ω_{xz} in the xz-plane and consider two functions $y = h_1(x,z)$ and $y = h_2(x,z)$ such that $h_1(x,z) \le h_2(x,z)$ for every $(x,z) \in \Omega_{xz}$. In the other, we start with a region Ω_{yz} in the yz-plane and consider two functions $x = k_1(y,z)$ and $x = k_2(y,z)$ such that $k_1(y,z) \le k_2(y,z)$ for every $(y,z) \in \Omega_{yz}$.

DEFINITION 13.18 **Triple Integrals over General Regions**

Let Ω be a general region in $\mathbb{R}^3$ of one of the three types just described, and let

$$\mathcal{R} = \{(x,y,z) \mid a_1 \le x \le a_2,\ b_1 \le y \le b_2, \text{ and } c_1 \le z \le c_2\}$$

be a rectangular solid containing Ω. We define the ***triple integral*** of f over Ω to be

$$\iiint_\Omega f(x,y,z)\,dV = \iiint_{\mathcal{R}} F(x,y,z)\,dV,$$

where

$$F(x,y,z) = \begin{cases} f(x,y,z), & \text{if } (x,y,z) \in \Omega \\ 0, & \text{if } (x,y,z) \notin \Omega \end{cases}$$

provided that the triple integral $\iiint_{\mathcal{R}} F(x,y,z)\,dV$ exists.

We will use Fubini's theorem to evaluate triple integrals on such general regions. To use Fubini's theorem in this context, when we wish to integrate the function $f(x,y,z)$ over a region Ω bounded below by $g_1(x,y)$ and above by $g_2(x,y)$ over a region Ω_{xy} in the xy-plane, we evaluate

$$\iiint_\Omega f(x,y,z)\,dV = \iint_{\Omega_{xy}} \int_{g_1(x,y)}^{g_2(x,y)} f(x,y,z)\,dz\,dA.$$

Once we have evaluated the inner integral $\int_{g_1(x,y)}^{g_2(x,y)} f(x,y,z)\,dz$, we evaluate the remaining double integral, using the techniques we discussed in Sections 13.2 and 13.3, depending upon the particular integral we are facing. We evaluate integrals over the other general regions in an analogous manner.

For example, to evaluate the triple integral of the function $f(x,y,z) = 5x - 3y$ over the tetrahedral region Ω bounded by the coordinate planes and the plane containing the points $(2,0,0)$, $(0,3,0)$, and $(0,0,6)$, we first try to visualize Ω. The region, which lies in the first octant, is shown here at the left:

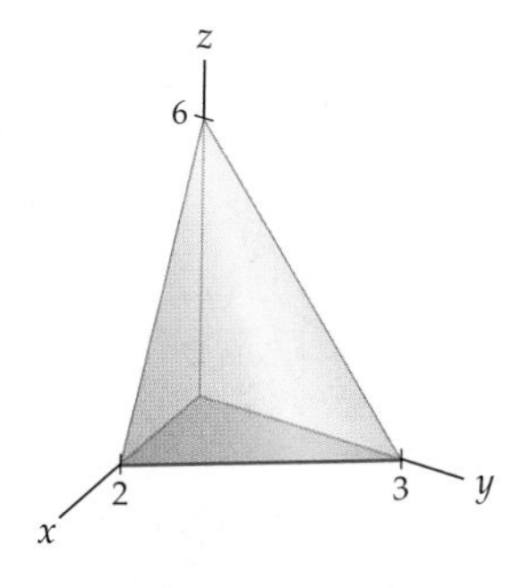

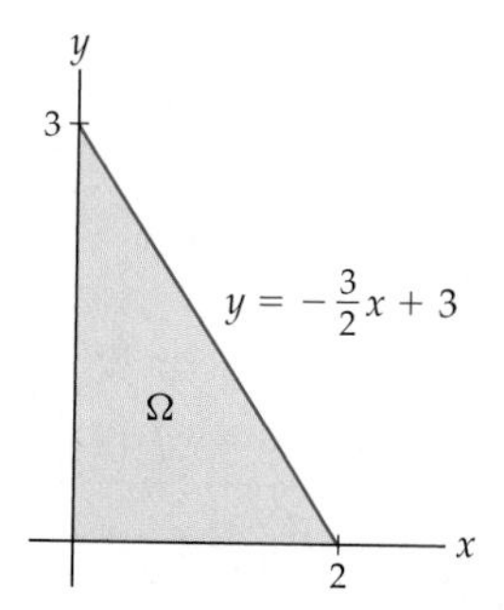

Because this region is relatively simple, we may consider it as belonging to *any* of the three general types we outlined earlier. We will consider it to be of the first type. The equation of the plane containing the three specified points as a function of x and y is

$g_2(x, y) = -3x - 2y + 6$. Since Ω is bounded below by the xy-plane, we use $g_1(x, y) = 0$. Therefore,

$$\iiint_{\Omega} (5x - 3y)\, dV = \iint_{\Omega_{xy}} \int_0^{-3x-2y+6} (5x - 3y)\, dz\, dA,$$

where Ω_{xy} is the triangular region shown at the right. Here it is particularly easy to find Ω_{xy}, since it is the bottom face of the tetrahedron. More generally, however, Ω_{xy} is the projection of Ω onto the xy-plane. To evaluate the double integral over Ω_{xy} we may treat this region as either a type I region or a type II region, as described in Section 13.2. For this example, we will treat Ω_{xy} as a type I region. Thus we have

$$\iint_{\Omega_{xy}} \int_0^{-3x-2y+6} (5x - 3y)\, dz\, dA = \int_0^2 \int_0^{-(3/2)x+3} \int_0^{-3x-2y+6} (5x - 3y)\, dz\, dy\, dx.$$

Before we evaluate this triple integral, notice that the limits of the innermost integral are functions of the variables of the outer two integrals (although one of them is a constant function), the limits of the middle integral are functions of the variable of the outermost integral (although, again, one of them is a constant function), and the limits of the outermost integral are constants. When we complete the evaluation of the iterated integral, we will see that our answer is also a constant. These are properties you should *always* have when you set up and evaluate an iterated triple integral. Now,

$$\int_0^2 \int_0^{-(3/2)x+3} \int_0^{-3x-2y+6} (5x - 3y)\, dz\, dy\, dx$$

$$= \int_0^2 \int_0^{-(3/2)x+3} \left[(5x - 3y)z\right]_{z=0}^{z=-3x-2y+6} dy\, dx \qquad \leftarrow \text{the Fundamental Theorem of Calculus}$$

$$= \int_0^2 \int_0^{-(3/2)x+3} (-15x^2 - xy + 6y^2 + 30x - 18y)\, dy\, dx \qquad \leftarrow \text{evaluation and multiplication}$$

$$= \int_0^2 \left[-15x^2y - \frac{1}{2}xy^2 + 2y^3 + 30xy - 9y^2\right]_{y=0}^{y=-(3/2)x+3} dx \qquad \leftarrow \text{the Fundamental Theorem of Calculus}$$

$$= \int_0^2 \left(\frac{117}{8}x^3 - \frac{261}{4}x^2 + \frac{171}{2}x - 27\right) dx \qquad \leftarrow \text{evaluation and simplification}$$

$$= \left[\frac{117}{32}x^4 - \frac{87}{4}x^3 + \frac{171}{4}x^2 - 27x\right]_{x=0}^{x=2} = \frac{3}{2}. \qquad \leftarrow \text{the Fundamental Theorem of Calculus}$$

In this example, we see that the evaluation of even a relatively simple triple integral over a simple region can necessitate a lengthy computation. None of the steps in this computation are terribly difficult, but there are certainly many points at which small mistakes can lead to an incorrect result. Care must be taken in setting up and evaluating triple integrals, as we will see in the examples that follow.

Applications of Triple Integration

The applications we discussed in Section 13.4 generalize to three-dimensional space in a natural way. We have already mentioned that if we know the density function $\rho(x, y, z)$ for a region $\Omega \subset \mathbb{R}^3$, then $\iiint_{\Omega} \rho(x, y, z)\, dV$ represents the mass of Ω. The definitions for the first and second moments and for the center of mass of a three-dimensional region are closely related to the definitions we made for laminæ in $\mathbb{R}^2$.

DEFINITION 13.19

First Moments About the Coordinate Planes

Let Ω be a region in $\mathbb{R}^3$ in which the density at each point is given by the continuous function $\rho(x, y, z)$.

(a) The ***first moment*** of the mass in Ω with respect to the yz-plane is

$$M_{yz} = \iiint_\Omega x\,\rho(x, y, z)\,dV.$$

(b) The ***first moment*** of the mass in Ω with respect to the xz-plane is

$$M_{xz} = \iiint_\Omega y\,\rho(x, y, z)\,dV.$$

(c) The ***first moment*** of the mass in Ω with respect to the xy-plane is

$$M_{xy} = \iiint_\Omega z\,\rho(x, y, z)\,dV.$$

Just as we did with laminæ, the center of mass of a region $\Omega \subset \mathbb{R}^3$ is computed from the first moments and the mass of Ω. If the mass of Ω is $m = \iiint_\Omega \rho(x, y, z)\,dV$, then the center of mass of Ω is $(\overline{x}, \overline{y}, \overline{z})$, where

$$\overline{x} = \frac{M_{yz}}{m}, \quad \overline{y} = \frac{M_{xz}}{m}, \quad \text{and} \quad \overline{z} = \frac{M_{xy}}{m}.$$

We may also define moments of inertia about the three coordinate axes and use them to compute the radius of gyration about each axis.

DEFINITION 13.20

Moments of Inertia About the Coordinate Axes

Let Ω be a region in $\mathbb{R}^3$ in which the density at each point is given by the continuous function $\rho(x, y, z)$.

(a) The ***moment of inertia of Ω about the x-axis*** is

$$I_x = \iiint_\Omega (y^2 + z^2)\,\rho(x, y, z)\,dV.$$

(b) The ***moment of inertia of Ω about the y-axis*** is

$$I_y = \iiint_\Omega (x^2 + z^2)\,\rho(x, y, z)\,dV.$$

(c) The ***moment of inertia of Ω about the z-axis*** is

$$I_z = \iiint_\Omega (x^2 + y^2)\,\rho(x, y, z)\,dV.$$

To compute the ***radii of gyration*** about the x-, y-, and z-axes, we use

$$R_x = \sqrt{I_x/m}, \quad R_y = \sqrt{I_y/m}, \quad \text{and} \quad R_z = \sqrt{I_z/m},$$

respectively. These are the radial distances at which the mass of the solid could be concentrated without changing its rotational inertia about the specified axis. We will provide examples of all of these computations in the examples that follow.

Examples and Explorations

EXAMPLE 1 **Finding the volume of a rectangular parallelepiped**

Set up the six iterated triple integrals that could be used to find the volume of the rectangular parallelepiped defined by

$$\mathcal{R} = \{(x, y, z) \mid -1 \le x \le 3,\ 2 \le y \le 4, \text{ and } 1 \le z \le 6\}.$$

Evaluate one of the triple integrals to show that the volume of the parallelepiped is the product of its length, width, and height.

SOLUTION

The following rectangular solid measures 4 units in the x direction, 2 units in the y direction, and 5 units in the z direction:

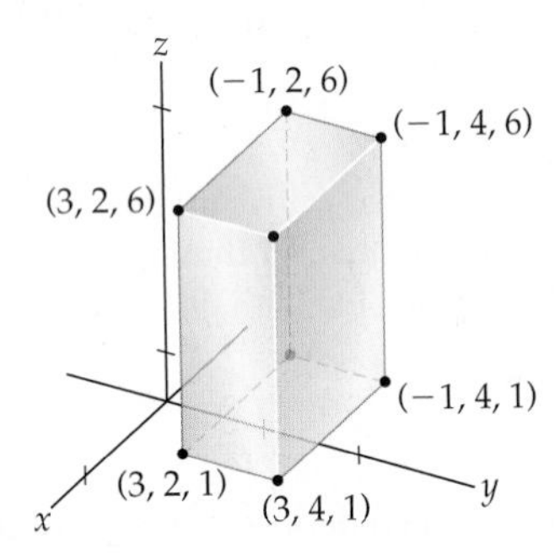

The volume of this solid is $4 \times 2 \times 5 = 40$ cubic units.

The first iterated integral we could use to compute the volume is

$$\text{Volume} = \iiint_{\mathcal{R}} dV = \int_{-1}^{3}\int_{2}^{4}\int_{1}^{6} dz\,dy\,dx.$$

This order of integration is typically the first one we try. We will discuss the reasons for this in Example 3. Here we note that the innermost integration is with respect to z, the middle integration is with respect to y, and the outer integration is with respect to x. Therefore, the limits on the inner, middle, and outer integrals are the values that correspond to those variables from the definition of the parallelepiped. To obtain the other five possible iterated integrals, we permute the differentials dx, dy, and dz, moving the limits of integration in the corresponding way. Thus, we also have

$$\begin{aligned}\text{Volume} &= \int_{2}^{4}\int_{-1}^{3}\int_{1}^{6} dz\,dx\,dy = \int_{-1}^{3}\int_{1}^{6}\int_{2}^{4} dy\,dz\,dx \\ &= \int_{1}^{6}\int_{-1}^{3}\int_{2}^{4} dy\,dx\,dz = \int_{2}^{4}\int_{1}^{6}\int_{-1}^{3} dx\,dz\,dy = \int_{1}^{6}\int_{2}^{4}\int_{-1}^{3} dx\,dy\,dz.\end{aligned}$$

We now evaluate the triple integral, using the first ordering we provided. Note that we work from the innermost integral to the outermost integral:

$$\begin{aligned}\int_{-1}^{3}\int_{2}^{4}\int_{1}^{6} dz\,dy\,dx &= \int_{-1}^{3}\int_{2}^{4} [z]_1^6\,dy\,dx = \int_{-1}^{3}\int_{2}^{4} 5\,dy\,dx \\ &= \int_{-1}^{3} [5y]_2^4\,dx = \int_{-1}^{3} 10\,dx \\ &= [10x]_{-1}^{3} = 40.\end{aligned}$$

Our answers agree: The volume of $\mathcal{R}$ is 40 cubic units. As we mentioned earlier in the section, we would not have bothered to define triple integration if this were the only application. For example, in Example 2 we find the mass of the solid $\mathcal{R}$, assuming that the density is a nonconstant function. □

EXAMPLE 2

Finding the mass of a rectangular parallelepiped

Find the mass of the solid $\mathcal{R}$ from Example 1, assuming that the density at every point of $\mathcal{R}$ is proportional to the distance of the point from the xz-plane.

SOLUTION

Instead of integrating the constant function $f(x, y, z) = 1$ to find the volume of $\mathcal{R}$, here we integrate the density function $\rho(x, y, z) = ky$, where k is a constant of proportionality, to find the mass of the solid. Almost all of the work involved in setting up the iterated triple integral was done in Example 1; the only change is the integrand. We choose the same order of variables that we used before:

$$\begin{aligned} \text{Mass} &= \iiint_{\mathcal{R}} ky\, dV = \int_{-1}^{3}\int_{2}^{4}\int_{1}^{6} ky\, dz\, dy\, dx \\ &= \int_{-1}^{3}\int_{2}^{4} \left[kyz\right]_1^6 dy\, dx = \int_{-1}^{3}\int_{2}^{4} 5ky\, dy\, dx \\ &= \int_{-1}^{3} \left[\tfrac{5}{2}ky^2\right]_2^4 dx = \int_{-1}^{3} 30k\, dx \\ &= \left[30kx\right]_{-1}^{3} = 120k. \end{aligned}$$

Note again that there are five other orders we could have used for the iterated integral. For this particular problem, the levels of difficulty for the integrations would be similar. □

EXAMPLE 3

Finding the volume of a pyramid

Use a triple integral to find the volume of the pyramid $\mathcal{P}$ whose base is the square with vertices $(1, 0, 0)$, $(0, 1, 0)$, $(-1, 0, 0)$, and $(0, -1, 0)$ and whose top vertex is $(0, 0, 1)$.

SOLUTION

The pyramid $\mathcal{P}$ is shown in the following figure at the left:

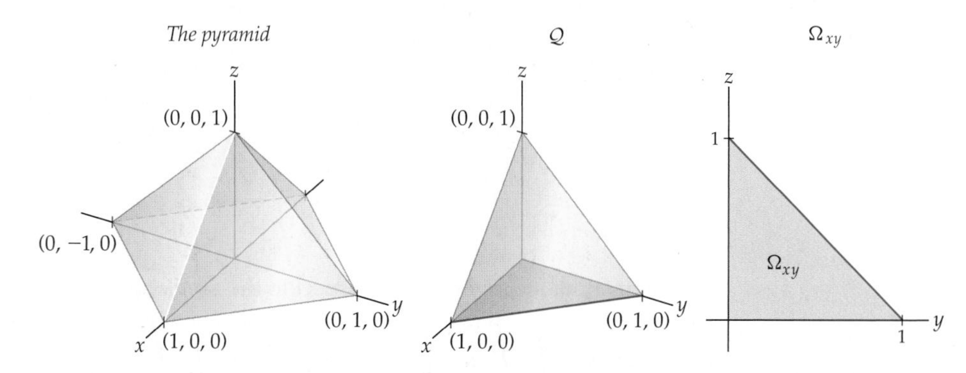

By the symmetry of $\mathcal{P}$, the pyramid $\mathcal{Q}$ in the middle figure has a volume that is one-quarter of the volume of $\mathcal{P}$. We will find the volume of $\mathcal{Q}$ and then multiply by 4 to obtain the volume of $\mathcal{P}$. The oblique planar face of $\mathcal{Q}$ has equation $x + y + z = 1$. We mentioned in Example 1 that we would explain the reason for typically choosing the order of integration specified by $dz\, dy\, dx$. When we have a function of two variables, we tend to think of z as a function of x and y. Similarly, when we have a function of a single variable, we tend to think of y as a function x. This is the convention we have grown accustomed to and that may cause us to prefer it. Notice that in the iterated integral we evaluated in Example 2, the upper limit on the innermost integral is a function of two variables. We preferred to express the equation of the oblique planar face as a function of x and y. Similarly, when we provide the limits on the middle integral, we need to express one or both boundaries as functions of a single variable. When we do so, we prefer to express the limits on this integral as functions of x. The projection of $\mathcal{Q}$ onto the xy-plane gives the region Ω_{xy} shown at the right. The right-hand boundary of Ω_{xy} may be expressed either as $y = -x + 1$ or as $x = -y + 1$. We prefer the former, which is the upper limit that we use on the middle integral. In Example 6 we will use one of the other orders of integration to construct an iterated triple integral, but for now the volume of $\mathcal{Q}$ is given by the following iterated triple integral:

$$\begin{aligned}\text{Volume of } \mathcal{Q} &= \int_0^1 \int_0^{-x+1} \int_0^{-x-y+1} dz\,dy\,dx = \int_0^1 \int_0^{-x+1} [z]_0^{-x-y+1}\,dy\,dx \\ &= \int_0^1 \int_0^{-x+1} (-x - y + 1)dy\,dx = \int_0^1 \left[-xy - \frac{1}{2}y^2 + y\right]_0^{-x+1} dx \\ &= \int_0^1 \left(\frac{1}{2}x^2 - x + \frac{1}{2}\right) dx = \left[\frac{1}{6}x^3 - \frac{1}{2}x^2 + \frac{1}{2}x\right]_0^1 = \frac{1}{6} \text{ cubic units.}\end{aligned}$$

Multiplying this answer by 4, we find that the volume of $\mathcal{P}$ is $\frac{2}{3}$ cubic unit. □

Recall that the volume of a pyramid is $\frac{1}{3}$(area of the base)(height). Each side of the square base of pyramid $\mathcal{P}$ has length $\sqrt{2}$. The height of $\mathcal{P}$ is 1. Therefore, the volume of $\mathcal{P}$ is $\frac{1}{3}\sqrt{2} \cdot \sqrt{2} \cdot 1 = \frac{2}{3}$ cubic unit, as we just found.

EXAMPLE 4 Finding the center of mass of a pyramid

Find the center of mass of the pyramid $\mathcal{P}$ from Example 3, assuming that the density of the pyramid is uniform.

SOLUTION

By the symmetries of $\mathcal{P}$ and its density function, the first moments of the mass with respect to the yz- and xz-planes are both zero. Therefore, if we let $(\bar{x}, \bar{y}, \bar{z})$ be the center of mass of $\mathcal{P}$, then both $\bar{x}$ and $\bar{y}$ are zero. To find $\bar{z}$, we could compute both the mass of $\mathcal{P}$ and the first moment M_{xy} of $\mathcal{P}$ when the density function $\rho(x, y, z) = k$, where k is a constant. However, note that the z-coordinates for the centers of mass of the pyramids $\mathcal{P}$ and $\mathcal{Q}$ are equal. Although the x- and y-coordinates of the centers of mass of $\mathcal{P}$ and $\mathcal{Q}$ would be different, we already know that $\bar{x} = \bar{y} = 0$ for $\mathcal{P}$. Therefore, using our work from Example 3, we find that the mass of $\mathcal{Q}$ is given by

$$m = \int_0^1 \int_0^{-x+1} \int_0^{-x-y+1} k\,dz\,dy\,dx = \frac{1}{6}k.$$

The moment of inertia of $\mathcal{Q}$ with respect to the xy-plane is

$$\begin{aligned}M_{xy} &= \int_0^1 \int_0^{-x+1} \int_0^{-x-y+1} kz\,dz\,dy\,dx = \int_0^1 \int_0^{-x+1} \left[\frac{1}{2}kz^2\right]_0^{-x-y+1} dy\,dx \\ &= \int_0^1 \int_0^{-x+1} \frac{1}{2}k(-x - y + 1)^2 dy\,dx = \int_0^1 \left[-\frac{1}{6}k(-x - y + 1)^3\right]_0^{-x+1} dx \\ &= \int_0^1 \frac{1}{6}k(-x + 1)^3\,dx = \left[-\frac{1}{24}k(-x + 1)^4\right]_0^1 = \frac{1}{24}k.\end{aligned}$$

Therefore, the z-coordinate of the center of mass of both $\mathcal{P}$ and $\mathcal{Q}$ is

$$\bar{z} = \frac{M_{xy}}{m} = \frac{k/24}{k/6} = \frac{1}{4},$$

and the center of mass of $\mathcal{P}$ is $(0, 0, 1/4)$. Note that the point $(0, 0, 1/4)$ lies inside the pyramid. This shows that our result is plausible. □

EXAMPLE 5 **Finding the moments of inertia and radii of gyration for a pyramid**

Find the moments of inertia about the coordinate axes for the pyramid $\mathcal{Q}$ from Example 3, assuming that its density is uniform. Use those moments to find the radius of gyration about each coordinate axis.

SOLUTION

In Example 4 we saw that the mass of $\mathcal{Q}$ is

$$m = \int_0^1 \int_0^{-x+1} \int_0^{-x-y+1} k\, dz\, dy\, dx = \frac{1}{6}k.$$

To find the moment of inertia about the x-axis, we modify this integral by including the factor $(y^2 + z^2)$ in the integrand. Thus, we have

$$I_x = \int_0^1 \int_0^{-x+1} \int_0^{-x-y+1} k(y^2 + z^2)\, dz\, dy\, dx.$$

The evaluation procedure for this integral is quite similar to those in Examples 3 and 4. In Exercise 15 you will show that $I_x = \frac{1}{30}k$.

To find the radius of gyration about the x-axis, we compute $R_x = \sqrt{\frac{I_x}{m}}$, where m is the mass of $\mathcal{Q}$. We get

$$R_x = \sqrt{\frac{I_x}{m}} = \sqrt{\frac{(1/30)k}{(1/6)k}} = \frac{\sqrt{5}}{5}.$$

Now, to find the moments of inertia about the y- and z-axes, and the corresponding radii of gyration, we could evaluate the integrals

$$I_y = \int_0^1 \int_0^{-x+1} \int_0^{-x-y+1} k(x^2 + z^2)\, dz\, dy\, dx, I_z = \int_0^1 \int_0^{-x+1} \int_0^{-x-y+1} k(x^2 + y^2)\, dz\, dy\, dx,$$

and follow that by calculating $R_y = \sqrt{I_y/m}$ and $R_z = \sqrt{I_z/m}$. You will do exactly that in Exercises 16 and 17. Here we ask you to convince yourself that pyramid $\mathcal{Q}$ has rotational symmetry about the line given by the parametric equations

$$x = t,\ y = t,\ z = t$$

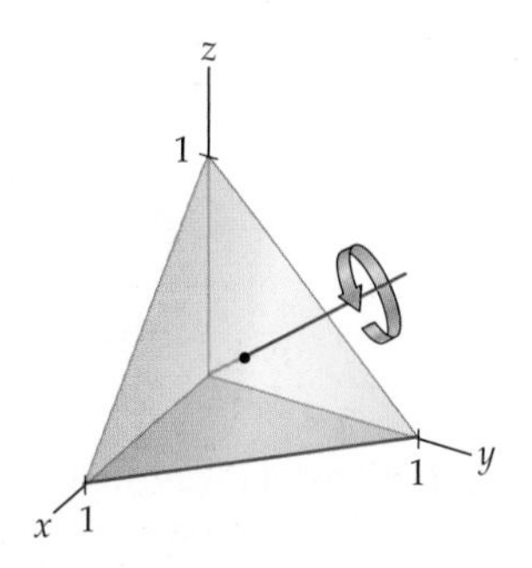

A rotation of $\frac{\pi}{3}$ about this line takes each coordinate axis to one of the others, depending upon the direction of the rotation. In the rotation shown in the preceding figure, the x-axis would rotate onto the y-axis, etc. This means that the moments of inertia and the radii of gyration for all three coordinate axes must be equal. That is,

$$I_x = I_y = I_z = \frac{1}{30}k \quad \text{and} \quad R_x = R_y = R_z = \frac{\sqrt{5}}{5}.$$

□

EXAMPLE 6 **Finding the mass of a slice of a cylinder**

Find the mass of the slice of the right circular cylinder $x^2 + z^2 = 4$ bounded on the left by the xz-plane and on the right by the plane with equation $x - y + z = -4$ if the density at each point in the cylinder is proportional to the distance of the point from the xz-plane.

SOLUTION

The region we have described is shown here:

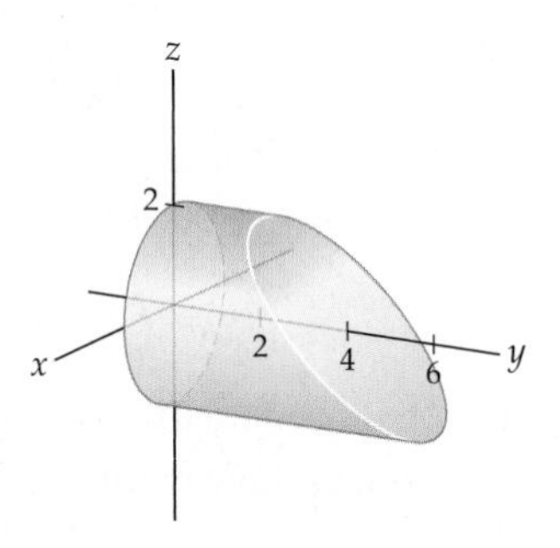

The density function we need to integrate is $\rho(x, y, z) = ky$, where k is a positive constant of proportionality. Let $\mathcal{C}$ represent the slice of the cylinder under consideration. We need to evaluate the triple integral

$$\iiint_{\mathcal{C}} ky\, dV.$$

No matter which order of integration we settle upon, the function ρ will be relatively simple to integrate. The order we will use depends upon the characteristics of the region. In particular, we will consider the projection of $\mathcal{C}$ onto the coordinate planes and decide which is simplest. Since the cylinder is symmetric with respect to the y-axis, the projection of $\mathcal{C}$ onto the xz-plane will be the circle with radius 2, centered at the origin of that plane. Let $\mathcal{C}_{xz}$ be this projection.

$\mathcal{C}_{xz}$

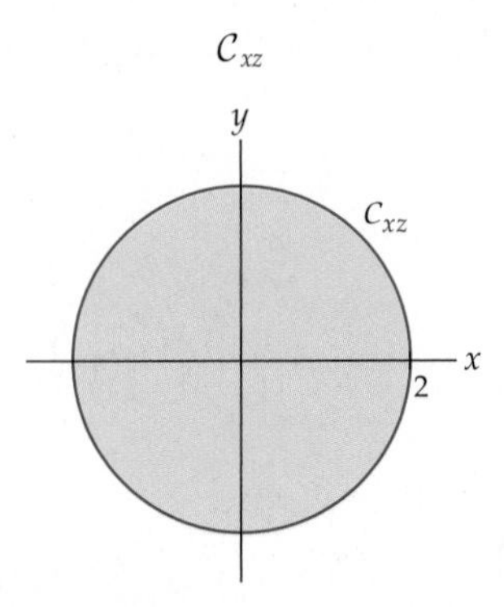

The projections of $\mathcal{C}$ onto the other coordinate planes would be more difficult to analyze; thus we will integrate *first* with respect to y. The limits on this integral are $y = 0$ and

$y = x + z + 4$, the equations of the planes that form the boundaries of $\mathcal{C}$. We now have

$$\iiint_{\mathcal{C}} ky\, dV = \iint_{\mathcal{C}_{xz}} \int_0^{x+z+4} ky\, dy\, dA.$$

We next turn our attention to setting up iterated integrals to evaluate the double integral over $\mathcal{C}_{xz}$. The region $\mathcal{C}_{xz}$ is bounded above by the semicircle with equation $z = \sqrt{4 - x^2}$ and below by the semicircle with equation $z = -\sqrt{4 - x^2}$. Thus, we have

$$\iiint_{\mathcal{C}} ky\, dV = \int_{-2}^{2} \int_{-\sqrt{4-x^2}}^{\sqrt{4-x^2}} \int_0^{x+z+4} ky\, dy\, dz\, dx.$$

The evaluation of this integral is left for you in Exercise 18. As we see in this example, the main difficulty in evaluating a triple integral often lies in analyzing the region of integration. When you evaluate the preceding integral, you will see that it is a chore, but you will still be using only basic integration techniques that were first introduced in Chapter 5. □

? TEST YOUR UNDERSTANDING

- What is a Riemann sum for a function of three variables? How many different orderings are possible for the nested summations? Does the order of the summations matter?
- How is the triple integral for a function f over a rectangular solid defined? How is this definition similar to the definition of the double integral of a function over a rectangular region? How is it different?
- What is an iterated triple integral? What is Fubini's theorem? How are iterated integrals used to compute triple integrals? How many choices are possible when you order the variables of integration in an iterated triple integral?
- What are the three types of three-dimensional regions we discussed in this section? What are the orders of integration associated with each of these different regions?
- What applications of triple integration did we discuss in this section? How are integrals for masses, first moments of mass, and moments of inertia related to each other?

EXERCISES 13.5

Thinking Back

- *Using iterated double integrals to compute area:* Let Ω be the region in the first quadrant bounded by the x-axis and the lines with equations $x = 1$ and $y = x$. Set up iterated double integrals that represent the area of Ω in which Ω is treated first as a type I region and then as a type II region.
- *Using iterated double integrals to compute mass:* Let Ω be the first-quadrant region bounded by the y-axis and the lines with equations $y = x$ and $y = 2-x$. If the density at each point of Ω is given by the function $\rho(x, y)$, set up iterated double-integral expressions that represent the mass of Ω in which Ω is treated first as a type I region and then as a type II region.

Concepts

0. *Problem Zero:* Read the section and make your own summary of the material.

1. *True/False:* Determine whether each of the statements that follow is true or false. If a statement is true, explain why. If a statement is false, provide a counterexample.

(a) *True or False:* If $f(x, y, z)$ is a continuous function on the rectangular solid defined by $\mathcal{R} = \{(x, y, z) \mid a_1 \le x \le a_2,\ b_1 \le y \le b_2,\ \text{and}\ c_1 \le z \le c_2\}$, then $\iiint_{\mathcal{R}} f(x, y, z)\, dV = \int_{a_1}^{a_2} \int_{b_1}^{b_2} \int_{c_1}^{c_2} f(x, y, z)\, dx\, dy\, dz$.

(b) *True or False:* If $f(x, y, z)$ is a continuous function on the rectangular solid defined by $\mathcal{R} = \{(x, y, z) \mid a_1 \le x \le a_2,\ b_1 \le y \le b_2,\ \text{and}\ c_1 \le z \le c_2\}$, then $\iiint_{\mathcal{R}} f(x, y, z)\, dV = \int_{a_1}^{a_2} \int_{c_1}^{c_2} \int_{b_1}^{b_2} f(x, y, z)\, dy\, dz\, dx$.

(c) *True or False:* If $f(x, y, z)$ is a continuous function of three variables and $\Omega = \Omega_1 \cup \Omega_2$ is a subset of $\mathbb{R}^3$, then $\iiint_{\Omega} f(x, y, z)\, dV = \iiint_{\Omega_1} f(x, y, z)\, dV + \iiint_{\Omega_2} f(x, y, z)\, dV$.

(d) *True or False:* If Ω is a bounded subset of $\mathbb{R}^3$, then there is a rectangular solid $\mathcal{R}$ with its sides parallel to the coordinate planes such that $\Omega \subseteq \mathcal{R}$.

(e) *True or False:* If f is a positive continuous function defined on a region Ω and $\Gamma \subseteq \Omega$, then $\iiint_\Gamma f(x, y, z)\, dV \le \iiint_\Omega f(x, y, z)\, dV$.

(f) *True or False:* If $\rho(x, y, z)$ gives the density at every point of a region Ω, then the first moment of the mass in Ω with respect to the xy-plane is $M_{xy} = \iiint_\Omega xy\, \rho(x, y, z)\, dV$.

(g) *True or False:* If $\rho(x, y, z)$ gives the density at every point of a region Ω, then the moment of inertia of Ω about the x-axis is $I_x = \iiint_\Omega (y^2 + z^2)\, \rho(x, y, z)\, dV$.

(h) *True or False:* If $g_1(x, y) \le g_2(x, y)$ on the square region $\mathcal{R} = \{(x, y) \mid 0 \le x \le 2 \text{ and } 0 \le y \le 2\}$, then the iterated integral $\int_{g_1(x,y)}^{g_2(x,y)} \int_0^2 \int_0^2 dV$ represents the volume of the solid bounded below by $g_1(x, y)$ and above by $g_2(x, y)$ on $\mathcal{R}$.

2. *Examples:* Construct examples of the thing(s) described in the following. Try to find examples that are different than any in the reading.

(a) An iterated triple integral over a rectangular solid such that the integral gives the mass of the solid.

(b) An iterated triple integral over a non-rectangular solid such that the integral gives the volume of the solid.

(c) An iterated triple integral over a solid such that the integral gives the moment of inertia of the solid about the y-axis.

3. Explain why

$$\sum_{i=1}^{l}\sum_{j=1}^{m}\sum_{k=1}^{n} ij^2k^3 = \sum_{j=1}^{m}\sum_{k=1}^{n}\sum_{i=1}^{l} ij^2k^3.$$

4. Explain how to construct a Riemann sum for a function of three variables over a rectangular solid.

5. Explain how to construct a *midpoint* Riemann sum for a function of three variables over a rectangular solid for which each (x_i^*, y_j^*, z_k^*) is the midpoint of the subsolid $\mathcal{R}_{ijk} = \{(x, y, z) \mid x_{i-1} \le x_i^* \le x_i, y_{j-1} \le y_j^* \le y_j,$ and $z_{k-1} \le z_k^* \le z_k\}$. Refer either to your answer to Exercise 4 or to Definition 13.14.

6. Discuss the similarities and differences between the definition of the double integral found in Section 13.1 and the definition of the triple integral found in this section.

7. What is the difference between a triple integral and an *iterated* triple integral?

8. Let $f(x, y, z)$ be a continuous function of three variables, let

$$\Omega_{xy} = \{(x, y) \mid a \le x \le b \text{ and } h_1(x) \le y \le h_2(x)\}$$

be a set of points in the xy-plane, and let

$$\Omega = \{(x, y, z) \mid (x, y) \in \Omega_{xy} \text{ and } g_1(x, y) \le z \le g_2(x, y)\}$$

be a set of points in 3-space. Find an iterated triple integral equal to the the triple integral $\iiint_\Omega f(x, y, z)\, dV$. How would your answer change if

$$\Omega_{xy} = \{(x, y) \mid a \le y \le b \text{ and } h_1(y) \le x \le h_2(y)\}?$$

9. Let $f(x, y, z)$ be a continuous function of three variables, let

$$\Omega_{yz} = \{(y, z) \mid a \le y \le b \text{ and } h_1(y) \le z \le h_2(y)\}$$

be a set of points in the yz-plane, and let

$$\Omega = \{(x, y, z) \mid (y, z) \in \Omega_{yz} \text{ and } g_1(y, z) \le x \le g_2(y, z)\}$$

be a set of points in 3-space. Find an iterated triple integral equal to the triple integral $\iiint_\Omega f(x, y, z)\, dV$. How would your answer change if

$$\Omega_{yz} = \{(y, z) \mid a \le z \le b \text{ and } h_1(z) \le y \le h_2(z)\}?$$

10. Let $f(x, y, z)$ be a continuous function of three variables, let

$$\Omega_{xz} = \{(x, z) \mid a \le x \le b \text{ and } h_1(x) \le z \le h_2(x)\}$$

be a set of points in the xz-plane, and let

$$\Omega = \{(x, y, z) \mid (x, z) \in \Omega_{xz} \text{ and } g_1(x, z) \le y \le g_2(x, z)\}$$

be a set of points in 3-space. Find an iterated triple integral equal to the triple integral $\iiint_\Omega f(x, y, z)\, dV$. How would your answer change if

$$\Omega_{xz} = \{(x, z) \mid a \le z \le b \text{ and } h_1(z) \le x \le h_2(z)\}?$$

11. Let $\rho(x, y, z)$ be a density function defined on the rectangular solid $\mathcal{R}$ where

$$\mathcal{R} = \{(x, y, z) \mid -1 \le x \le 3,\ 0 \le y \le 2, \text{ and } 2 \le z \le 7\}.$$

Set up iterated integrals representing the mass of $\mathcal{R}$, using all six distinct orders of integration.

12. Let $\rho(x, y, z)$ be a density function defined on the rectangular solid $\mathcal{R}$ where

$$\mathcal{R} = \{(x, y, z) \mid a_1 \le x \le a_2, b_1 \le y \le b_2, \text{ and } c_1 \le z \le c_2\}.$$

Set up iterated integrals representing the mass of $\mathcal{R}$, using all six distinct orders of integration.

13. Let $\rho(x, y, z)$ be a density function defined on the tetrahedron Ω with vertices $(0, 0, 0)$, $(2, 0, 0)$, $(0, 4, 0)$, and $(0, 0, 3)$. Set up iterated integrals representing the mass of Ω, using all six distinct orders of integration.

14. Let $\rho(x, y, z)$ be a density function defined on the tetrahedron Ω with vertices $(0, 0, 0)$, $(a, 0, 0)$, $(0, b, 0)$, and $(0, 0, c)$, where a, b, and c are positive real numbers. Set up iterated integrals representing the mass of Ω, using all six distinct orders of integration.

In Exercises 15–17, evaluate the three integrals that can be used to find the moments of inertia for the pyramid $\mathcal{Q}$ described in Example 5 and then use those values to find the radii of gyration about the coordinate axes. Recall that the mass of $\mathcal{Q}$ is $\frac{1}{6}k$.

15. Show that

$$I_x = \int_0^1 \int_0^{-x+1} \int_0^{-x-y+1} k(y^2 + z^2)\, dz\, dy\, dx = \frac{1}{30}k.$$

16. Show that

$$I_y = \int_0^1 \int_0^{-x+1} \int_0^{-x-y+1} k(x^2+z^2)\,dz\,dy\,dx = \frac{1}{30}k.$$

Use your answer to show that $R_y = \frac{\sqrt{5}}{5}$.

17. Show that

$$I_z = \int_0^1 \int_0^{-x+1} \int_0^{-x-y+1} k(x^2+y^2)\,dz\,dy\,dx = \frac{1}{30}k.$$

Use your answer to show that $R_z = \frac{\sqrt{5}}{5}$.

18. Complete Example 6 by evaluating the iterated integral

$$\int_{-2}^{2} \int_{-\sqrt{4-x^2}}^{\sqrt{4-x^2}} \int_0^{x+z+4} ky\,dy\,dz\,dx.$$

Identify the quantities determined by the integral expressions in Exercises 19–24. If x, y, and z are all measured in centimeters and $\rho(x,y,z)$ is a density function in grams per cubic centimeter on the three-dimensional region Ω, give the units of the expression.

19. $\iiint_\Omega dV$

20. $\iiint_\Omega \rho(x,y,z)\,dV$

21. $\iiint_\Omega x\,\rho(x,y,z)\,dV$

22. $\dfrac{\iiint_\Omega z\,\rho(x,y,z)\,dV}{\iiint_\Omega \rho(x,y,z)\,dV}$

23. $\iiint_\Omega (x^2+z^2)\rho(x,y,z)dV$

24. $\sqrt{\dfrac{\iiint_\Omega (x^2+z^2)\,\rho(x,y,z)\,dV}{\iiint_\Omega \rho(x,y,z)\,dV}}$

Skills

Evaluate the iterated integrals in Exercises 25–30.

25. $\int_0^1 \int_2^4 \int_{-1}^5 (x+yz^2)\,dx\,dy\,dz$

26. $\int_1^3 \int_{-1}^5 \int_0^3 xy^2z\,dz\,dy\,dx$

27. $\int_0^1 \int_0^{6-y} \int_0^{3-3y-(1/2)z} (y-z)\,dx\,dz\,dy$

28. $\int_0^1 \int_0^{3-3y} \int_0^{6y} \frac{x}{y}\,dz\,dx\,dy$

29. $\int_0^{\pi/2} \int_0^{\cos y} \int_0^{\sin y} (2x+y)\,dz\,dx\,dy$

30. $\int_0^1 \int_0^3 \int_0^{\sin^{-1} x} (x+y)\,dz\,dy\,dx$

Evaluate the triple integrals in Exercises 31–34 over the specified rectangular solid regions.

31. $\iiint_{\mathcal{R}} x^2yz\,dV$, where

$\mathcal{R} = \{(x,y,z) \mid -2 \le x \le 1,\ 0 \le y \le 3, \text{ and } 1 \le z \le 5\}$

32. $\iiint_{\mathcal{R}} (x+2y+3z)\,dV$, where

$\mathcal{R} = \{(x,y,z) \mid 0 \le x \le 4,\ 1 \le y \le 5, \text{ and } 2 \le z \le 7\}$

33. $\iiint_{\mathcal{R}} z\sin x\cos y\,dV$, where

$\mathcal{R} = \left\{(x,y,z) \mid 0 \le x \le \pi,\ \frac{3\pi}{2} \le y \le 2\pi, \text{ and } 1 \le z \le 3\right\}$

34. $\iiint_{\mathcal{R}} \ln(xyz^2)\,dV$, where

$\mathcal{R} = \{(x,y,z) \mid 1 \le x \le 3,\ 1 \le y \le e, \text{ and } 1 \le z \le 2\}$

Describe the three-dimensional region expressed in each iterated integral in Exercises 35–44.

35. $\int_{-2}^4 \int_2^6 \int_0^5 f(x,y,z)\,dy\,dx\,dz$

36. $\int_{-1}^5 \int_{-3}^2 \int_4^8 f(x,y,z)\,dz\,dx\,dy$

37. $\int_0^3 \int_0^3 \int_0^{3-y} f(x,y,z)\,dz\,dy\,dx$

38. $\int_0^2 \int_0^3 \int_{4x/3}^4 f(x,y,z)\,dz\,dx\,dy$

39. $\int_0^3 \int_0^{1-y/3} \int_0^{2-(2/3)y-2z} f(x,y,z)\,dx\,dz\,dy$

40. $\int_0^2 \int_0^{1-3x/2} \int_{2x+(4/3)y-4}^0 f(x,y,z)\,dz\,dy\,dx$

41. $\int_{-3}^3 \int_{-\sqrt{9-x^2}}^{\sqrt{9-x^2}} \int_{-\sqrt{9-x^2-y^2}}^{\sqrt{9-x^2-y^2}} f(x,y,z)\,dz\,dy\,dx$

42. $\int_{-3}^3 \int_{-\sqrt{9-x^2}}^{\sqrt{9-x^2}} \int_{-3}^3 f(x,y,z)\,dz\,dy\,dx$

43. $\int_{-3}^3 \int_{-\sqrt{9-y^2}}^{\sqrt{9-y^2}} \int_0^{9-y^2-z^2} f(x,y,z)\,dx\,dz\,dy$

44. $\int_{-3}^3 \int_{-\sqrt{9-z^2}}^{\sqrt{9-z^2}} \int_0^{3-\sqrt{x^2+z^2}} f(x,y,z)\,dy\,dx\,dz$

In Exercises 45–52, rewrite the indicated integral with the specified order of integration.

45. Exercise 35 with the order $dx\,dy\,dz$.

46. Exercise 36 with the order $dy\,dz\,dx$.

47. Exercise 37 with the order $dx\,dy\,dz$.

48. Exercise 38 with the order $dx\,dy\,dz$.

49. Exercise 39 with the order $dz\,dy\,dx$.

50. Exercise 40 with the order $dx\,dy\,dz$.

51. Exercise 41 with the order $dy\,dx\,dz$.

52. Exercise 42 with the order $dy\,dx\,dz$.

Find the masses of the solids described in Exercises 53–56.

53. The first-octant solid bounded by the coordinate planes and the plane $3x + 4y + 6z = 12$ if the density at each point is proportional to the distance of the point from the xz-plane.

54. The solid bounded above by the plane with equation $2x + 3y - z = 2$ and bounded below by the triangle with vertices $(1, 0, 0)$, $(4, 0, 0)$, and $(0, 2, 0)$ if the density at each point is proportional to the distance of the point from the xy-plane.

55. The solid bounded above by the paraboloid with equation $z = 8 - x^2 - y^2$ and bounded below by the rectangle $\mathcal{R} = \{(x, y, 0) \mid 1 \le x \le 2 \text{ and } 0 \le y \le 2\}$ in the xy-plane if the density at each point is proportional to the square of the distance of the point from the origin.

56. The solid bounded above by the hyperboloid with equation $z = x^2 - y^2$ and bounded below by the square with vertices $(2, 2, -4)$, $(2, -2, -4)$, $(-2, -2, -4)$, and $(-2, 2, -4)$ if the density at each point is proportional to the distance of the point from the plane with equation $z = -4$.

Applications

In Exercises 57–60, let $\mathcal{R}$ be the rectangular solid defined by $\mathcal{R} = \{(x, y, z) \mid 0 \le x \le 4,\ 0 \le y \le 3,\ 0 \le z \le 2\}$.

57. Assume that the density of $\mathcal{R}$ is uniform throughout.

(a) Without using calculus, explain why the center of mass is $(2, 3/2, 1)$.

(b) Verify that the center of mass is $(2, 3/2, 1)$, using the appropriate integral expressions.

58. Assume that the density of $\mathcal{R}$ is uniform throughout, and find the moment of inertia about the x-axis and the radius of gyration about the x-axis.

59. Assume that the density at each point in $\mathcal{R}$ is proportional to the distance of the point from the xy-plane.

(a) Without using calculus, explain why the x- and y-coordinates of the center of mass are $\bar{x} = 2$ and $\bar{y} = \frac{3}{2}$, respectively.

(b) Use an appropriate integral expression to find the z-coordinate of the center of mass.

60. Assuming that the density at each point in $\mathcal{R}$ is proportional to the distance of the point from the xy-plane, find the moment of inertia about the x-axis and the radius of gyration about the x-axis.

In Exercises 61–64, let $\mathcal{R}$ be the rectangular solid defined by $\mathcal{R} = \{(x, y, z) \mid 0 \le a_1 \le x \le a_2,\ 0 \le b_1 \le y \le b_2,\ 0 \le c_2 \le z \le c_2\}$.

61. Assume that the density of $\mathcal{R}$ is uniform throughout.

(a) Without using calculus, explain why the center of mass is $\left(\frac{a_1+a_2}{2}, \frac{b_1+b_2}{2}, \frac{c_1+c_2}{2}\right)$.

(b) Verify that $\left(\frac{a_1+a_2}{2}, \frac{b_1+b_2}{2}, \frac{c_1+c_2}{2}\right)$ is the center of mass by using the appropriate integral expressions.

62. Assume that the density of $\mathcal{R}$ is uniform throughout, and find the moment of inertia about the x-axis and the radius of gyration about the x-axis.

63. Assume that the density at each point in $\mathcal{R}$ is proportional to the distance of the point from the yz-plane.

(a) Without using calculus, explain why the y- and z-coordinates of the center of mass are $\bar{y} = \frac{b_1+b_2}{2}$ and $\bar{z} = \frac{c_1+c_2}{2}$, respectively.

(b) Use an appropriate integral expression to find the x-coordinate of the center of mass.

64. Assuming that the density at each point in $\mathcal{R}$ is proportional to the distance of the point from the yz-plane, find the first moment of inertia about the x-axis and the radius of gyration about the x-axis.

Let a, b, and c be positive real numbers. In Exercises 65–68, let $\mathcal{T}$ be the tetrahedron with vertices $(0, 0, 0)$, $(a, 0, 0)$, $(0, b, 0)$, and $(0, 0, c)$.

65. Assume that the density at each point in $\mathcal{T}$ is uniform throughout.

(a) Find the x-coordinate of the center of mass of $\mathcal{T}$.

(b) Explain how to use your answer from part (a) to find the y- and z-coordinates of the center of mass without doing any other computations.

66. Assume that the density of $\mathcal{T}$ is uniform throughout. Set up the integrals required to find the moment of inertia about the x-axis and the radius of gyration about the x-axis.

67. Assume that the density at each point in $\mathcal{T}$ is proportional to the distance of the point from the xz-plane. Set up the integral expressions required to find the center of mass of $\mathcal{T}$.

68. Assuming that the density at each point in $\mathcal{T}$ is proportional to the distance of the point from the yz-plane, set up the integrals required to find the first moment of inertia about the x-axis and the radius of gyration about the x-axis.

Proofs

69. Let $\mathcal{R} = \{(x, y, z) \mid a_1 \le x \le a_2,\ b_1 \le y \le b_2, \text{ and } c_1 \le z \le c_2\}$. If $\alpha(x)$, $\beta(y)$, and $\gamma(z)$ are integrable on the intervals $[a_1, a_2]$, $[b_1, b_2]$, and $[c_1, c_2]$, respectively, use Fubini's theorem to prove that

$$\iiint_{\mathcal{R}} \alpha(x)\beta(y)\gamma(z)\, dV = \left(\int_{a_1}^{a_2} \alpha(x)\, dx\right)\left(\int_{b_1}^{b_2} \beta(y)\, dy\right)\left(\int_{c_1}^{c_2} \gamma(z)\, dz\right).$$

70. Let a, b, and c be positive real numbers, and let $\mathcal{R} = \{(x, y, z) \mid -a \le x \le a,\ -b \le y \le b, \text{ and } -c \le z \le c\}$. Prove that $\iiint_{\mathcal{R}} \alpha(x)\beta(y)\gamma(z)\, dV = 0$ if any of α, β, and γ is an odd function.

71. Let $\mathcal{R} = \{(x, y, z) \mid a_1 \le x \le a_2,\ b_1 \le y \le b_2, \text{ and } c_1 \le z \le c_2\}$. Prove that

$$\iiint_{\mathcal{R}} dV = (a_2 - a_1)(b_2 - b_1)(c_2 - c_1).$$

What is the relationship between $\mathcal{R}$ and the product $(a_2 - a_1)(b_2 - b_1)(c_2 - c_1)$?

72. Let $f(x, y, z)$ be an integrable function on the rectangular solid $\mathcal{R} = \{(x, y, z) \mid a_1 \le x \le a_2,\ b_1 \le y \le b_2, \text{ and } c_1 \le z \le c_2\}$, and let $\kappa \in \mathbb{R}$. Use the definition of the triple integral to prove that

$$\iiint_{\mathcal{R}} \kappa f(x, y, z)\, dV = \kappa \iiint_{\mathcal{R}} f(x, y, z)\, dV.$$

73. Let $f(x, y, z)$ and $g(x, y, z)$ be integrable functions on the rectangular solid $\mathcal{R} = \{(x, y, z) \mid a_1 \le x \le a_2,\ b_1 \le y \le b_2, \text{ and } c_1 \le z \le c_2\}$. Use the definition of the triple integral to prove that

$$\iiint_{\mathcal{R}} (f(x, y, z) + g(x, y, z))\, dV = \iiint_{\mathcal{R}} f(x, y, z)\, dV + \iiint_{\mathcal{R}} g(x, y, z)\, dV.$$

Thinking Forward

- *Cylindrical coordinates:* When we use polar coordinates in the xy-plane and the usual z-coordinate, we are using *cylindrical coordinates*. Let Ω_{xy} be a region in the xy-plane such that for each point (r, θ) in Ω_{xy}, $h_1(\theta) \le r \le h_2(\theta)$ and $\alpha \le \theta \le \beta$. Then, if $g_1(x, y) \le z \le g_2(x, y)$ for each (x, y) in Ω_{xy}, explain why

$$\iiint_{\Omega} f(x, y, z)\, dV = \iint_{\Omega_{xy}} \int_{g_1(x,y)}^{g_2(x,y)} f(x, y, z)\, dz\, dA$$
$$= \int_{\alpha}^{\beta} \int_{h_1(\theta)}^{h_2(\theta)} \int_{g_1(r\cos\theta,\, r\sin\theta)}^{g_2(r\cos\theta,\, r\sin\theta)} f(r\cos\theta, r\sin\theta, z)\, r\, dz\, dr\, d\theta.$$

- *Cylindrical coordinates:* Use the results of the previous exercise to explain why the triple integral

$$\int_0^{2\pi} \int_0^3 \int_0^{9-r^2} kr^2\, dz\, dr\, d\theta$$

represents the mass of a solid bounded below by the xy-plane and bounded above by the paraboloid with equation $z = 9 - x^2 - y^2$ if the density at every point in the solid is proportional to the distance of the point from the origin. Then determine the mass of the solid.

13.6 INTEGRATION WITH CYLINDRICAL AND SPHERICAL COORDINATES

- ▶ Generalizing polar coordinates to three dimensions
- ▶ Converting between rectangular coordinates, cylindrical coordinates, and spherical coordinates
- ▶ Constructing triple integrals with cylindrical and spherical coordinates

Cylindrical Coordinates

In this section we will discuss two generalizations of polar coordinates to three-dimensional space: *cylindrical coordinates* and *spherical coordinates*. We will also discuss how these coordinate systems may be used in triple integrals. We will see that the cylindrical coordinate system uses two linear measures and one angular measure to locate each point in $\mathbb{R}^3$ and the spherical coordinate system uses one linear measure and two angular measures to locate points.

In the cylindrical coordinate system, we use polar coordinates in the xy-plane and the usual z-coordinate to locate positions off of the xy-plane. As we see in the following figure, given a point P with rectangular coordinates (x, y, z), r measures the distance from the projection of P in the xy-plane to the origin and θ measures the angular rotation from the x-axis to the ray containing the origin and the point $(x, y, 0)$:

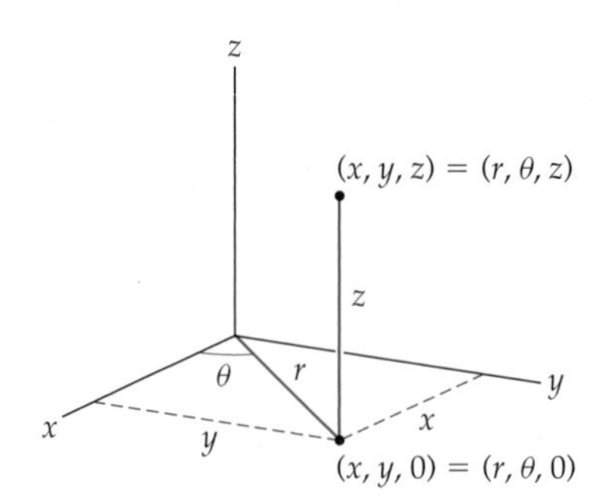

That is, if (x, y, z) is a point in $\mathbb{R}^3$ given in rectangular coordinates, then, in cylindrical coordinates we have (r, θ, z), where

$$r^2 = x^2 + y^2, \quad \tan\theta = \frac{y}{x}, \quad \text{and} \quad z = z.$$

Note that the values for r and θ are computed just as they were when we were discussing polar coordinates in the plane. Since our primary motivation in this chapter is integration, we will typically choose $r \geq 0$ and restrict values of θ to an interval of length 2π, usually either $[0, 2\pi]$ or $[-\pi, \pi]$.

When we have the coordinates of a point (r, θ, z) in cylindrical coordinates, the rectangular coordinates are (x, y, z), where

$$x = r\cos\theta, \quad y = r\sin\theta, \quad \text{and} \quad z = z.$$

From Section 13.5, we know that, for a three-dimensional region, Ω, that is bounded below by the function $g_1(x, y)$ and above by $g_2(x, y)$ on the domain Ω_{xy}, the triple integral of the function $f(x, y, z)$ is

$$\iiint_{\Omega} f(x, y, z)\, dV = \iint_{\Omega_{xy}} \int_{g_1(x,y)}^{g_2(x,y)} f(x, y, z)\, dz\, dA.$$

In Section 13.5 we discussed how to use rectangular coordinates to evaluate the exterior double integral over the region Ω_{xy}. However, in Section 13.3 we saw that the increment of area dA is given by $r\,dr\,d\theta$ in polar coordinates. Therefore, to use cylindrical coordinates to evaluate the triple integral of f, we replace each occurrence of x in f, g_1, and g_2 with $r\cos\theta$ and each occurrence of y with $r\sin\theta$ and we use $r\,dz\,dr\,d\theta$ in place of the increment of volume dV. We also need to express the region Ω_{xy} in terms of *polar* coordinates, as we did in Section 13.3.

For example, we will compute the mass of the region Ω bounded above by the paraboloid with equation $g(x, y) = 4 - x^2 - y^2$ and bounded below by the xy-plane, assuming that the density at each point is proportional to the distance of the point from the xy-plane. The triple integral we wish to evaluate is $\iiint_\Omega kz\,dV$. Using rectangular coordinates, we could evaluate

$$\int_{-2}^{2}\int_{-\sqrt{4-x^2}}^{\sqrt{4-x^2}}\int_{0}^{4-x^2-y^2} kz\,dz\,dy\,dx.$$

However, using cylindrical coordinates, we will have a simpler iterated integral. With cylindrical coordinates, the equation of the paraboloid is $z = 4 - r^2$. The projection of the solid onto the xy-plane is a circle with radius 2 and centered at the origin. The equation of this circle is $r = 2$ for $\theta \in [0, 2\pi]$. Finally, our increment of volume is $dV = r\,dz\,dr\,d\theta$. Therefore, the mass may be computed by

$$\begin{aligned}
\int_0^{2\pi}\int_0^2\int_0^{4-r^2} kzr\,dz\,dr\,d\theta &= \int_0^{2\pi}\int_0^2 \left[\frac{1}{2}krz^2\right]_0^{4-r^2} dr\,d\theta \\
&= \int_0^{2\pi}\int_0^2 \frac{1}{2}kr(4-r^2)^2 dr\,d\theta \\
&= -\int_0^{2\pi} \left[\frac{1}{12}k(4-r^2)^3\right]_0^2 d\theta \\
&= \int_0^{2\pi} \frac{16}{3}k\,d\theta = \left[\frac{16}{3}k\theta\right]_0^{2\pi} = \frac{32\pi}{3}k.
\end{aligned}$$

Every iterated triple integral requires three integrations, so we expect the computation to require at least that much work, but as we see in this example, using the cylindrical coordinate system here provides a much simpler path toward the solution. Evaluating the equivalent iterated triple integral in rectangular coordinates would be even more laborious.

Consider using cylindrical coordinates to evaluate triple integrals when the projection of the region of integration Ω onto the xy-plane, Ω_{xy}, has a "natural" expression in polar coordinates. This will always occur when Ω displays rotational symmetry about the z-axis.

Spherical Coordinates

Let (x, y, z) be the rectangular coordinates of a point P in $\mathbb{R}^3$. As we just discussed, we may also express the coordinates of P as cylindrical coordinates (r, θ, z), where

$$r^2 = x^2 + y^2, \quad \tan\theta = \frac{y}{x}, \quad \text{and} \quad z = z.$$

In the spherical system, the coordinates of a point P in $\mathbb{R}^3$ are given by (ρ, θ, ϕ), where ρ is the distance that P is from the origin, θ is the same angle we used in the cylindrical system, and $\phi \in [0, \pi]$ is the angle that the ray from the origin through P makes with the positive z-axis, as we see in the figure that follows at the left. The figure at the right shows the relationships between the spherical, cylindrical, and rectangular coordinates.

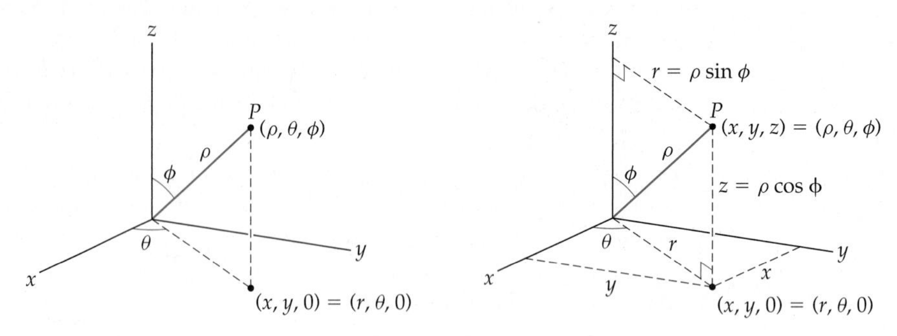

The angles θ and ϕ will be measured in radians. We will always assume that $0 \le \phi \le \pi$ and that θ lies in some interval of length 2π, typically $\theta \in [0, 2\pi]$ or $\theta \in [-\pi, \pi]$. In the figures that follow, we show a sphere, a plane, and a cone. These are the graphs obtained when ρ, θ, and ϕ, respectively, are constant.

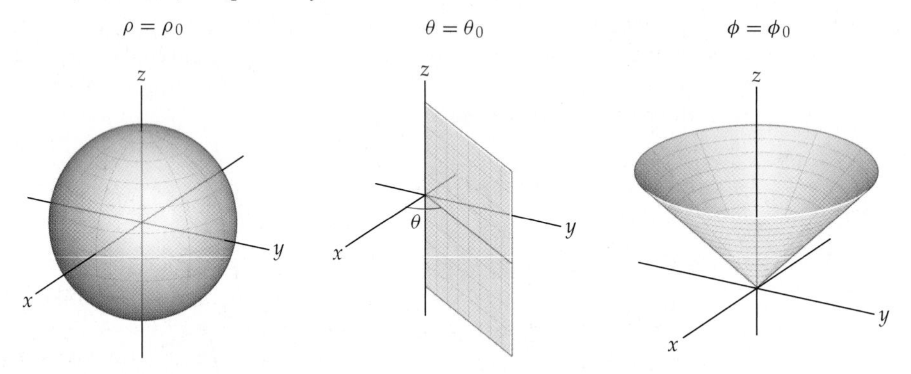

We summarize the relationships between the variables of our three-dimensional coordinate systems in the following theorem:

THEOREM 13.21

Converting Between the Three-Dimensional Coordinate Systems

Let P be a point in $\mathbb{R}^3$ with coordinates (x, y, z) in the rectangular coordinate system, (r, θ, z) in the cylindrical coordinate system, and (ρ, θ, ϕ) in the spherical coordinate system.

(a) The cylindrical coordinates and rectangular coordinates for P are related by the following equations:

$$r = \sqrt{x^2 + y^2}, \quad \tan\theta = \frac{y}{x}, \quad \text{and} \quad z = z$$
$$x = r\cos\theta, \quad y = r\sin\theta, \quad \text{and} \quad z = z.$$

(b) The cylindrical coordinates and spherical coordinates for P are related by the following equations:

$$\rho = \sqrt{r^2 + z^2}, \quad \theta = \theta, \quad \text{and} \quad \tan\phi = \frac{r}{z}$$
$$r = \rho\sin\phi, \quad \theta = \theta, \quad \text{and} \quad z = \rho\cos\phi.$$

(c) The rectangular coordinates and spherical coordinates for P are related by the following equations:

$$\rho = \sqrt{x^2 + y^2 + z^2}, \quad \tan\theta = \frac{y}{x}, \quad \text{and} \quad \cos\phi = \frac{z}{\sqrt{x^2 + y^2 + z^2}}$$
$$x = \rho\sin\phi\cos\theta, \quad y = \rho\sin\phi\sin\theta, \quad \text{and} \quad z = \rho\cos\phi.$$

For example, if the rectangular coordinates of a point P are $(1, \sqrt{3}, 2\sqrt{3})$, then

$$r = \sqrt{x^2 + y^2} = \sqrt{1^2 + (\sqrt{3})^2} = 2, \ z = 2\sqrt{3}, \text{ and } \tan\theta = \frac{y}{x} = \frac{\sqrt{3}}{1} = \sqrt{3}.$$

Therefore, $\theta = \frac{\pi}{3}$ and the cylindrical coordinates for P are $\left(2, \frac{\pi}{3}, 2\sqrt{3}\right)$. Now that we have the cylindrical coordinates for P, we may use either those or the rectangular coordinates for P to find the spherical coordinates for P. We'll use the cylindrical coordinates. We have

$$\rho = \sqrt{r^2 + z^2} = \sqrt{2^2 + (2\sqrt{3})^2} = 4, \ \theta = \frac{\pi}{3}, \quad \text{and} \quad \tan\phi = \frac{r}{z} = \frac{2}{2\sqrt{3}} = \frac{1}{\sqrt{3}}.$$

Therefore, since $\phi \in [0, \pi]$, we have $\phi = \frac{\pi}{6}$. So the spherical coordinates for P are $\left(4, \frac{\pi}{3}, \frac{\pi}{6}\right)$.

Integration with Spherical Coordinates

Recall that to integrate a function $f(r, \theta)$ over a region $\Omega \subset \mathbb{R}^2$, we use $dA = r\,dr\,d\theta$ for our increment of area. That is,

$$\iint_\Omega f(r, \theta)\,dA = \iint_\Omega f(r, \theta)\,r\,dr\,d\theta.$$

Similarly, as we saw earlier in this section, when we integrate a function $f(r, \theta, z)$ over a region $\Omega \subset \mathbb{R}^3$, we use $dV = r\,dz\,dr\,d\theta$ for our increment of volume and obtain

$$\iiint_\Omega f(r, \theta, z)\,dV = \iiint_\Omega f(r, \theta, z)\,r\,dz\,dr\,d\theta.$$

When we use spherical coordinates, we must also adapt our increment of volume. Here, we will have $dV = \rho^2 \sin\phi\,d\rho\,d\theta\,d\phi$. To see this, start by considering an increment of volume in spherical coordinates determined by $\Delta\rho = \rho_i - \rho_{i-1}$, $\Delta\theta = \theta_j - \theta_{j-1}$, and $\Delta\phi = \phi_k - \phi_{k-1}$. When $\Delta\rho$, $\Delta\theta$, and $\Delta\phi$ are small, this region is nearly a rectangular solid, as shown next. Thus, its volume may be approximated by the product of its three dimensions.

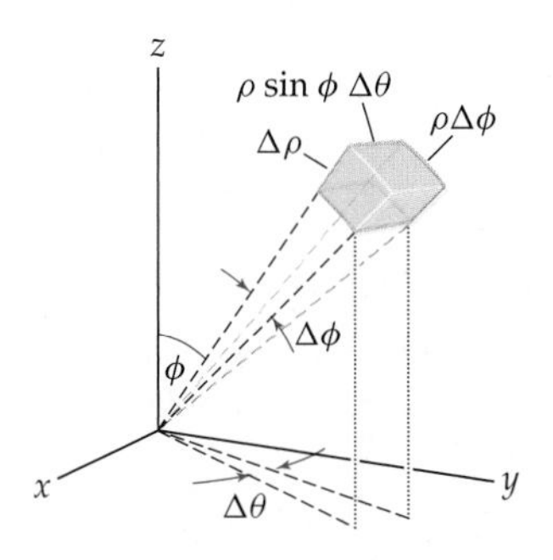

The measure of the side of this solid determined by $\Delta\rho$ is just that linear measure. The approximate measure of the side determined by $\Delta\phi$ is $\rho\Delta\phi$, since it is the arc of the circle of radius ρ subtended by the angle $\Delta\phi$. The projection of the third side into the xy-plane is shown in the figure. This projection is an arc of a circle with radius r (in cylindrical coordinates) subtended by the angle $\Delta\theta$. Therefore, the measure of the projection is $r\Delta\theta$. However, since $r = \rho\sin\phi$, the approximate measure of the third side is $r\Delta\theta = \rho\sin\phi\Delta\theta$. Multiplying these three factors together, we have $\Delta V = \rho^2 \sin\phi\,\Delta\rho\,\Delta\theta\,\Delta\phi$. When we take the limit as $\Delta\rho$, $\Delta\theta$, and $\Delta\phi$ all tend to zero, we have the volume differential $dV = \rho^2 \sin\phi\,d\rho\,d\theta\,d\phi$. Therefore, when we have a function $f(\rho, \theta, \phi)$ given in spherical coordinates and defined on a three-dimensional region Ω, the integral of f on Ω is

$$\iiint_\Omega f(\rho, \theta, \phi)\,\rho^2 \sin\phi\,d\rho\,d\theta\,d\phi.$$

Typically, the spherical coordinate system is used when the boundaries of a solid are spheres or parts of spheres. For example, let $\mathcal{E}$ be the portion of the unit sphere centered at the origin in the first octant, as shown in the following figure:

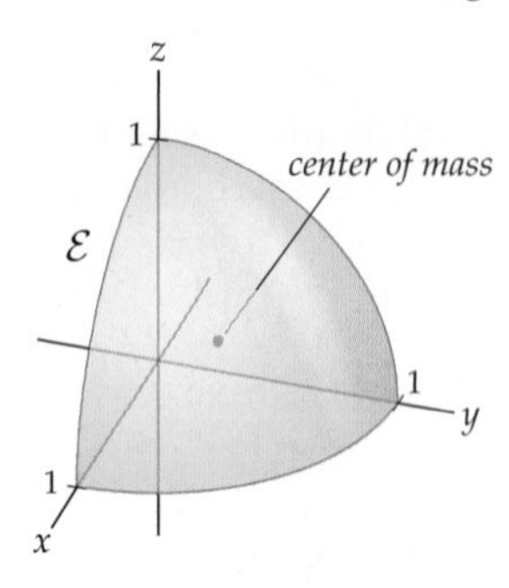

We will find the center of mass of $\mathcal{E}$, assuming that it has uniform density k. To find the x-coordinate of the center of mass, $\bar{x}$, we start by computing the first moment with respect to the yz-plane. We have

$$M_{yz} = \iiint_{\mathcal{E}} kx\,dV = \int_0^{\pi/2}\int_0^{\pi/2}\int_0^1 k\rho^3\sin^2\phi\cos\theta\,d\rho\,d\theta\,d\phi,$$

since $x = \rho\sin\phi\cos\theta$ and $dV = \rho^2\sin\phi\,d\rho\,d\theta\,d\phi$. In Exercise 18 you will show that $M_{yz} = \frac{1}{16}\pi k$. Similarly, the mass, m, of $\mathcal{E}$ is given by the integral

$$m = \iiint_{\mathcal{E}} k\,dV = \int_0^{\pi/2}\int_0^{\pi/2}\int_0^1 k\rho^2\sin\phi\,d\rho\,d\theta\,d\phi.$$

However, we may circumvent this computation because the value of that integral has to be one-eighth of the volume of a sphere times the (uniform) density k. That is, $m = \frac{1}{8}\left(\frac{4}{3}\pi\right)k = \frac{1}{6}\pi k$. In Exercise 19 we ask that you evaluate the integral to obtain this result. Therefore, $\bar{x} = \frac{M_{yz}}{m} = \frac{3}{8}$. Because of the symmetry of $\mathcal{E}$, the y- and z-coordinates of the center of mass must also be $\frac{3}{8}$. So the center of mass of $\mathcal{E}$ is $\left(\frac{3}{8}, \frac{3}{8}, \frac{3}{8}\right)$. (In Exercises 20 and 21, you are asked to show that M_{xy} and M_{xz} both equal $\frac{1}{16}\pi k$ to confirm this result.)

Examples and Explorations

EXAMPLE 1

Finding the volume of a region between a cylinder and a sphere

Use rectangular, cylindrical, and spherical coordinates to set up triple integrals representing the volume inside the sphere with equation $x^2+y^2+z^2 = 4$ but outside the cylinder with equation $x^2+y^2 = 1$.

SOLUTION

The sphere is centered at the origin and has radius 2. The z-axis is the axis of symmetry for the cylinder with radius 1. The graph is as follows:

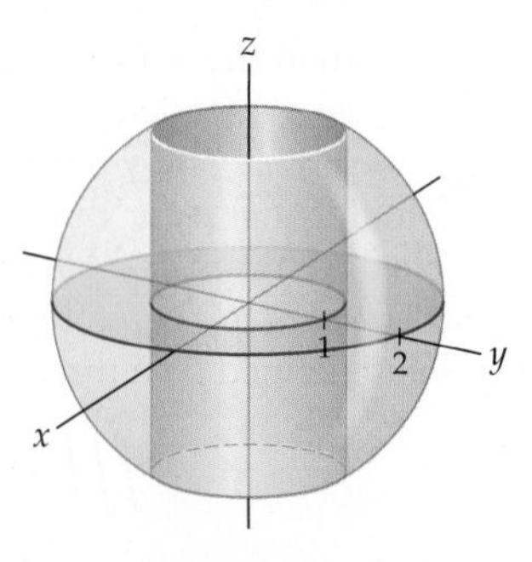

Let Ω represent the region. Since Ω is symmetric with respect to the z-axis, when we use rectangular coordinates we will integrate first with respect to z. The projection of Ω onto the xy-plane is the annulus Ω_{xy}, shown here:

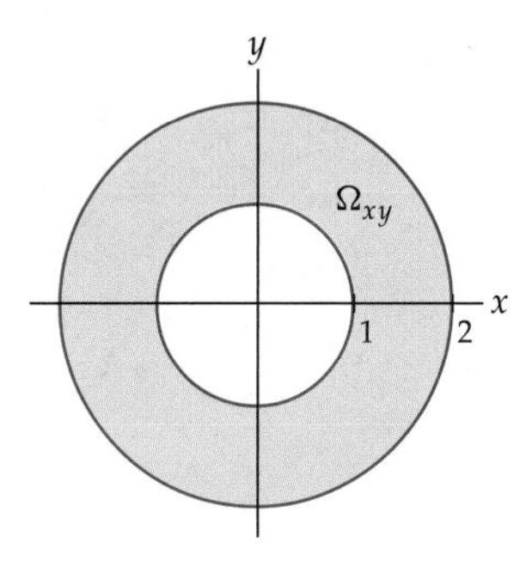

It would require several subdivisions to represent Ω_{xy} as a sum of type I and type II regions. However, we can use rectangular coordinates to represent the volume of Ω by setting up a triple integral giving the volume of the entire sphere and then subtracting a triple integral that represents the portion of the volume of the sphere inside the cylinder. That is, the volume of Ω is

$$\int_{-2}^{2}\int_{-\sqrt{4-x^2}}^{\sqrt{4-x^2}}\int_{-\sqrt{4-x^2-y^2}}^{\sqrt{4-x^2-y^2}} dz\,dy\,dx - \int_{-1}^{1}\int_{-\sqrt{1-x^2}}^{\sqrt{1-x^2}}\int_{-\sqrt{4-x^2-y^2}}^{\sqrt{4-x^2-y^2}} dz\,dy\,dx.$$

Note that it is more natural to represent Ω_{xy} with polar coordinates than with rectangular coordinates. This is why it will be easier to represent the volume of Ω with cylindrical coordinates than it was with rectangular coordinates. In cylindrical coordinates, the equations of the upper and lower hemispheres are $z = \sqrt{4-r^2}$ and $z = -\sqrt{4-r^2}$, respectively. We must also remember to include the factor r in the integrand when we use cylindrical coordinates. Therefore, the triple integral representing the volume of Ω is

$$\int_{0}^{2\pi}\int_{1}^{2}\int_{-\sqrt{4-r^2}}^{\sqrt{4-r^2}} r\,dz\,dr\,d\theta.$$

Lastly, we will use spherical coordinates to set up a triple integral for the volume. In spherical coordinates, the equation of the sphere is $\rho = 2$. Since $x = \rho\sin\phi\cos\theta$ and $y = \rho\sin\phi\sin\theta$, the equation of the cylinder $x^2 + y^2 = 1$ in spherical coordinates is

$$(\rho\sin\phi\cos\theta)^2 + (\rho\sin\phi\sin\theta)^2 = 1.$$

We ask you to show in Exercise 22 that this equation simplifies to $\rho = \csc\phi$. To set up the iterated integral, we also need the correct interval of values for ϕ. The sphere and cylinder intersect when $\csc\phi = 2$. The values of ϕ satisfying this relationship within the interval $[0, \pi]$ are $\phi = \frac{\pi}{6}$ and $\phi = \frac{5\pi}{6}$. Recall that when we integrate with spherical coordinates, we must include the factor $\rho^2\sin\phi$ in the integrand. Therefore, the volume of Ω is given by

$$\int_{\pi/6}^{5\pi/6}\int_{0}^{2\pi}\int_{\csc\phi}^{2} \rho^2\sin\phi\,d\rho\,d\theta\,d\phi.$$

□

EXAMPLE 2

Finding the volume of a region between a cone and a sphere

Use rectangular, cylindrical, and spherical coordinates to set up triple integrals representing the volume bounded below by the cone with equation $z = \sqrt{x^2+y^2}$ and bounded above by the unit sphere centered at the origin.

SOLUTION

The z-axis is the axis of symmetry for the cone. Its vertex is at the origin and it opens upward. Here is the graph:

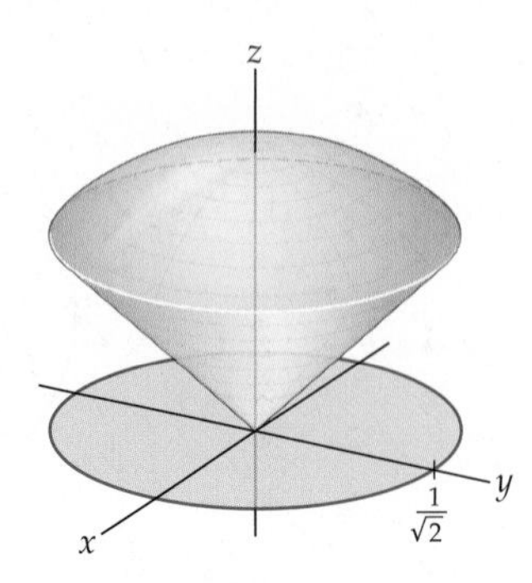

Let Ω represent the region. Since Ω is symmetric with respect to the z-axis, when we use rectangular coordinates we will integrate first with respect to z. The cone and sphere intersect at height $z = \frac{1}{\sqrt{2}}$ in a circle with radius $\frac{1}{\sqrt{2}}$. Therefore, the projection of Ω onto the xy-plane is a circle Ω_{xy} with radius $\frac{1}{\sqrt{2}}$ and centered at the origin. The equation for this circle is $x^2 + y^2 = \frac{1}{2}$.

The region Ω_{xy} may be treated as either a type I or type II region. We will treat it as a type I region. Using rectangular coordinates, we find that the volume of Ω is

$$\int_{-1/\sqrt{2}}^{1/\sqrt{2}} \int_{-\sqrt{(1/2)-x^2}}^{\sqrt{(1/2)-x^2}} \int_{\sqrt{x^2-y^2}}^{\sqrt{1-x^2-y^2}} dz\,dy\,dx.$$

We will now use an iterated integral with cylindrical coordinates to represent the volume of Ω. In cylindrical coordinates, the equation of the upper hemisphere of the unit sphere is $z = \sqrt{1-r^2}$ and the equation of the cone is $z = r$. We include the factor r in the integrand when we use cylindrical coordinates, and we obtain the integral

$$\int_0^{2\pi} \int_0^{1/\sqrt{2}} \int_r^{\sqrt{1-r^2}} r\,dz\,dr\,d\theta$$

for the volume of Ω.

Finally, we will use spherical coordinates to set up a triple integral for the volume. In spherical coordinates, the equation of the sphere is $\rho = 1$ and the equation of the cone is $\phi = \frac{\pi}{4}$. Recall that when we integrate with spherical coordinates, we must include the factor $\rho^2 \sin\phi$ in the integrand. Therefore, the volume of Ω is given by

$$\int_0^{\pi/4} \int_0^{2\pi} \int_0^1 \rho^2 \sin\phi\,d\rho\,d\theta\,d\phi.$$

□

? TEST YOUR UNDERSTANDING

- How do you convert between rectangular and polar coordinates in $\mathbb{R}^2$? How do you convert between rectangular and cylindrical coordinates in $\mathbb{R}^3$? How do you convert between rectangular and spherical coordinates? How do you convert between cylindrical and spherical coordinates?
- What do x, y, and z represent in rectangular coordinates? What range of values is used for x, y, and z in rectangular coordinates? If α, β, and γ are constants, what are the graphs of $x = \alpha$, $y = \beta$, and $z = \gamma$ in rectangular coordinates?

- What do r, θ, and z represent in cylindrical coordinates? What range of values is used for r in cylindrical coordinates? What range of values is used for θ in cylindrical coordinates? What range of values is used for z in cylindrical coordinates? If α, β, and γ are constants, what are the graphs of $r = \alpha$, $\theta = \beta$, and $z = \gamma$ in cylindrical coordinates?
- What do ρ, θ, and ϕ represent in spherical coordinates? What range of values is used for ρ in spherical coordinates? What range of values is used for θ in spherical coordinates? What range of values is used for ϕ in spherical coordinates? If α, β, and γ are constants, what are the graphs of $\rho = \alpha$, $\theta = \beta$, and $\phi = \gamma$ in spherical coordinates?
- Under what conditions should you use rectangular, cylindrical, or spherical coordinates in evaluating triple integrals?

EXERCISES 13.6

Thinking Back

- *Graphing with polar coordinates:* Use polar coordinates in $\mathbb{R}^2$ to graph each of the following equations:
  - $r = 3$
  - $\theta = \dfrac{\pi}{3}$
  - $r = 2\cos\theta$
  - $r = 1 + 2\sin\theta$
- *Rectangular versus polar coordinates:* When is it easier to use the rectangular coordinate system in $\mathbb{R}^2$? When is it easier to use the polar coordinate system?
- *Integrating with polar coordinates:* Let Ω be a region in $\mathbb{R}^2$. Provide a double integral that represents the area of Ω when you integrate with polar coordinates.

Concepts

0. *Problem Zero:* Read the section and make your own summary of the material.

1. *True/False:* Determine whether each of the statements that follow is true or false. If a statement is true, explain why. If a statement is false, provide a counterexample.

(a) *True or False:* With cylindrical coordinates, the graph of the function $z = \sqrt{2r\cos\theta - r^2}$ is a hemisphere.

(b) *True or False:* With spherical coordinates, the graph of $\phi = \dfrac{\pi}{2}$ is the xy-plane.

(c) *True or False:* With spherical coordinates, for every value of $a \in (0, \pi)$ the intersection of the graphs of $\phi = a$ and $\rho = 1$ is a circle with radius 1.

(d) *True or False:* The graph of $z = r$ in cylindrical coordinates is the same as the graph of $\phi = \dfrac{\pi}{4}$ in spherical coordinates.

(e) *True or False:* If f is an integrable function of three variables on a region Ω, then

$$\iiint_\Omega f(x, y, z)\, dz\, dy\, dx = \iiint_\Omega f(r, \theta, z)\, dz\, dr\, d\theta.$$

(f) *True or False:* If f is an integrable function of three variables on a region Ω, then

$$\iiint_\Omega f(x, y, z)\, dz\, dy\, dx = \iiint_\Omega f(x(r,\theta), y(r,\theta), z)\, r\, dz\, dr\, d\theta.$$

(g) *True or False:* The integral

$$\int_0^{2\pi}\int_0^{2\pi}\int_0^R \rho^2 \sin\phi\, d\rho\, d\theta\, d\phi = \frac{4}{3}\pi R^3.$$

(h) *True or False:* The rectangular, cylindrical, and spherical coordinate systems are the only coordinate systems in $\mathbb{R}^3$.

2. *Examples:* Construct examples of the thing(s) described in the following. Try to find examples that are different than any in the reading.

(a) A region in $\mathbb{R}^3$ that is most easily expressed with rectangular coordinates.

(b) A region in $\mathbb{R}^3$ that is most easily expressed with cylindrical coordinates.

(c) A region in $\mathbb{R}^3$ that is most easily expressed with spherical coordinates.

3. What are the graphs of the constant functions $x = x_0$, $y = y_0$, and $z = z_0$ in the rectangular coordinate system?

4. What are the graphs of the constant functions $r = r_0$, $\theta = \theta_0$, and $z = z_0$ in the cylindrical coordinate system?

5. What are the graphs of the constant functions $\rho = \rho_0$, $\theta = \theta_0$, and $\phi = \phi_0$ in the spherical coordinate system?

6. For what values of $\alpha \in [0, \pi]$ in the spherical coordinate system is the graph $\phi = \alpha$ *not* a cone?

Fill in the blanks for the conversion formulas in Exercises 7–12.

7. To convert from rectangular to cylindrical coordinates:
$r =$ ________, $\theta =$ ________, $z =$ ________.

8. To convert from cylindrical to rectangular coordinates: $x =$ ________, $y =$ ________, $z =$ ________.
9. To convert from rectangular to spherical coordinates: $\rho =$ ________, $\theta =$ ________, $\phi =$ ________.
10. To convert from spherical to rectangular coordinates: $x =$ ________, $y =$ ________, $z =$ ________.
11. To convert from cylindrical to spherical coordinates: $\rho =$ ________, $\theta =$ ________, $\phi =$ ________.
12. To convert from spherical to cylindrical coordinates: $r =$ ________, $\theta =$ ________, $z =$ ________.
13. What are the six forms used to express the volume increment dV when you use rectangular coordinates to evaluate a triple integral? How do you decide which order to use?
14. The volume increment $dV =$ ________ when you use cylindrical coordinates to evaluate a triple integral. Why is this the standard order of integration for cylindrical coordinates?
15. The volume increment $dV =$ ________ when you use spherical coordinates to evaluate a triple integral. Why is this the standard order of integration for spherical coordinates?
16. What geometric conditions do you look for when you are deciding which coordinate system to use in $\mathbb{R}^3$?
17. What geometric conditions do you look for when you are deciding which coordinate system to use when you are evaluating a triple integral?

In Exercises 18–21, we ask you to confirm a result from earlier in the section. Let $\mathcal{E}$ be the portion of the unit sphere centered at the origin in the first octant. Assume that $\mathcal{E}$ has uniform density k.

18. Show that the first moment of $\mathcal{E}$ is $M_{yz} = \frac{1}{16}\pi k$.
19. Show that the mass of $\mathcal{E}$ is $\frac{1}{6}\pi k$ by evaluating the integral
$$\iiint_{\mathcal{E}} k\,dV = \int_0^{\pi/2}\int_0^{\pi/2}\int_0^1 k\rho^2 \sin\phi\,d\rho\,d\theta\,d\phi.$$
20. Set up the appropriate triple integral with spherical coordinates to show that $M_{xy} = \frac{1}{16}\pi k$.
21. Set up the appropriate triple integral with spherical coordinates to show that $M_{xz} = \frac{1}{16}\pi k$.
22. From Example 1, recall that $x^2+y^2 = 1$ is the equation of the cylinder with radius 1, whose axis of symmetry is the z-axis. Show that the equation of this cylinder in spherical coordinates is $\rho = \csc\phi$.

Skills

Find the coordinates specified in Exercises 23–28.

23. Give the cylindrical and spherical coordinates for the point with rectangular coordinates $(1, 0, 0)$.
24. Give the cylindrical and spherical coordinates for the point with rectangular coordinates $(-6, 6, 6)$.
25. Give the rectangular and spherical coordinates for the point with cylindrical coordinates $(\sqrt{48}, \pi/3, 4)$.
26. Give the rectangular and spherical coordinates for the point with cylindrical coordinates $(4, \pi/3, 6)$.
27. Give the rectangular and cylindrical coordinates for the point with spherical coordinates $(8, \pi/2, \pi)$.
28. Give the rectangular and cylindrical coordinates for the point with spherical coordinates $(6, \pi/4, \pi/4)$.

Describe the graphs of the equations in Exercises 29–38 in $\mathbb{R}^3$, and provide alternative equations in the specified coordinate systems.

29. Change $x = 4$ to the cylindrical and spherical systems.
30. Change $z = x+y$ to the cylindrical and spherical systems.
31. Change $r = 2$ to the rectangular and spherical systems.
32. Change $r = 4\sin\theta$ to the rectangular and spherical systems.
33. Change $\theta = \frac{\pi}{2}$ to the rectangular system.
34. Change $\theta = \alpha$, where α is a constant, to the rectangular system.
35. Change $\rho = 2$ to the rectangular and cylindrical systems.
36. Change $\phi = \frac{\pi}{4}$ to the rectangular and spherical systems.
37. Change $\phi = \frac{\pi}{2}$ to the rectangular and cylindrical systems.
38. Change $\phi = \frac{3\pi}{4}$ to the rectangular and cylindrical systems.

The iterated integrals in Exercises 39–42 use cylindrical coordinates. Describe the solids determined by the limits of integration.

39. $\displaystyle\int_0^{2\pi}\int_0^3\int_0^r f(r,\theta,z)\,r\,dz\,dr\,d\theta$
40. $\displaystyle\int_0^{\pi}\int_1^2\int_0^{r^2} f(r,\theta,z)\,r\,dz\,dr\,d\theta$
41. $\displaystyle\int_0^{\pi}\int_0^{2\sin\theta}\int_0^{\sqrt{16-r^2}} f(r,\theta,z)\,r\,dz\,dr\,d\theta$
42. $\displaystyle\int_0^{\pi/2}\int_0^1\int_0^{\sqrt{1-r^2}} f(r,\theta,z)\,r\,dz\,dr\,d\theta$

The iterated integrals in Exercises 43–46 use spherical coordinates. Describe the solids determined by the limits of integration.

43. $\displaystyle\int_{\pi/2}^{\pi}\int_0^{2\pi}\int_0^2 f(\rho,\theta,\phi)\rho^2\sin\phi\,d\rho\,d\theta\,d\phi$

44. $\displaystyle\int_0^{\pi/2}\int_0^{\pi/2}\int_0^1 f(\rho,\theta,\phi)\rho^2\sin\phi\,d\rho\,d\theta\,d\phi$

45. $\displaystyle\int_0^{\pi/4}\int_0^{2\pi}\int_0^{3\sec\theta} f(\rho,\theta,\phi)\rho^2\sin\phi\,d\rho\,d\theta\,d\phi$

46. $\displaystyle\int_{\pi/4}^{\pi/2}\int_0^{2\pi}\int_0^{3\csc\theta} f(\rho,\theta,\phi)\rho^2\sin\phi\,d\rho\,d\theta\,d\phi$

Use a triple integral with either cylindrical or spherical coordinates to find the volumes of the solids described in Exercises 47–56.

47. The region inside both the sphere with equation $x^2+y^2+z^2=4$ and the cylinder with equation $x^2+(y-1)^2=1$.

48. The region inside the cylinder with equation $x^2+(y-1)^2=1$, bounded below by the xy-plane and bounded above by the cone with equation $z=\sqrt{x^2+y^2}$.

49. The region bounded below by the xy-plane, bounded above by the sphere with radius 2 and centered at the origin, and outside the cylinder with equation $x^2+y^2=1$.

50. The region bounded below by the plane with equation $z=c$ and bounded above by the sphere with equation $x^2+y^2+z^2=R^2$, where c and R are constants such that $0<c<R$.

51. The region bounded above by the plane with equation $z=x$ and bounded below by the paraboloid with equation $z=x^2+y^2$.

52. The region bounded above by the sphere with equation $\rho=2$ and bounded below by the cone with equation $\phi=\frac{\pi}{3}$.

53. The region in the next figure which is bounded below by the xy-plane, bounded above by the hyperboloid with equation $x^2+y^2-z^2=1$, and inside the cylinder with equation $x^2+y^2=5$.

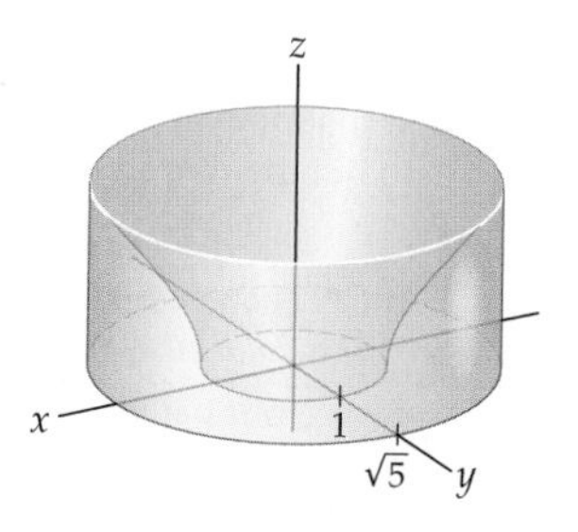

54. The first-octant region bounded above by the sphere with equation $\rho=R$ and bounded below by the plane with equation $z=x+y$.

55. The region bounded above by the sphere with equation $\rho=R$ and bounded below by the cone with equation $\phi=\alpha$. Explain why the volume should be zero if $\alpha=0$ and $\frac{4}{3}\pi R^3$ if $\alpha=\pi$.

56. The region bounded above by the sphere with equation $\rho=R$ and bounded below by the plane with equation $z=b$, where $0\le b\le R$. Explain why the volume should be $\frac{2}{3}\pi R^3$ if $b=0$ and zero if $b=R$. Use that result to find the volume of the region when $-R\le b<0$.

Find the specified quantities for the solids described in Exercises 57–66.

57. The mass of the region from Exercise 47, assuming that the density at every point is proportional to the square of the point's distance from the z-axis.

58. The mass of the region from Exercise 48, assuming that the density at every point is proportional to the square of the point's distance from the xy-plane.

59. The center of mass of the region from Exercise 49, assuming that the density at every point is proportional to the point's distance from the z-axis.

60. The center of mass of the region from Exercise 50, assuming that the density at every point is proportional to the point's distance from the xy-plane.

61. The mass of the region from Exercise 51, assuming that the density at every point is proportional to the square of the point's distance from the xy-plane.

62. The moment of inertia about the z-axis of the region from Exercise 52, assuming that the density at every point is inversely proportional to the point's distance from the z-axis.

63. The moment of inertia about the z-axis of the region from Exercise 53, assuming that the density at every point is inversely proportional to the point's distance from the z-axis.

64. The mass of the region from Exercise 54, assuming that the density at every point is proportional to the point's distance from the xy-plane.

65. The center of mass of the region from Exercise 55, assuming that the density of the region is constant.

66. The moment of inertia about the xy-plane of the region from Exercise 56, assuming that the density of the region is constant.

Applications

67. Emmy is responsible for a tank that is circular when viewed from above. The tank has a radius of 75 feet. An island with a radius of 8 feet lies at the center of the tank and contains monitoring equipment. Emmy knows that the sides of the tank are vertical but the bottom of the tank is the surface of a sphere of radius 150 feet, forming a bowl at the bottom. At the edge of the tank, the depth is 8 feet. What is the volume of the tank?

68. Annie is a sea-kayaking guide who finds herself with time on her hands in the winter. She is building a wood-and-fabric kayak for herself, but is concerned about its capacity for cargo. The outside surface of the kayak satisfies the equation

$$r = \frac{(49 - l^2)(2 - \sin\theta)}{147},$$

where the coordinates are cylindrical, r is the distance from a line joining the tips of the boat, and l is the horizontal distance from the center.

(a) What is the volume of the kayak?

(b) When Annie sits in a kayak, the seat and her body fill about 5 cubic feet of the boat. The stringers, which are supports inside that run the length of the kayak, take up space, so the interior has a radius that is 1.5 inches smaller than what you calculated in part (a). Approximately how much cargo space will she actually have in the kayak?

Proofs

69. Let a be a constant. Prove that the equation of the plane $x = a$ is $r = a\sec\theta$ in cylindrical coordinates.

70. Let b be a constant. Prove that the equation of the plane $y = b$ is $r = b\csc\theta$ in cylindrical coordinates.

71. Let a be a constant. Prove that the equation of the plane $x = a$ is $\rho = a\csc\phi\sec\theta$ in spherical coordinates.

72. Let b be a constant. Prove that the equation of the plane $y = b$ is $r = b\csc\phi\csc\theta$ in spherical coordinates.

73. Let R be the radius of the base of a cone and h be the height of the cone. Use cylindrical coordinates to set up and evaluate a triple integral proving that the volume of the cone is $\frac{1}{3}\pi R^2 h$.

74. Let R be the radius of a sphere. Use cylindrical coordinates to set up and evaluate a triple integral proving that the volume of the sphere is $\frac{4}{3}\pi R^3$.

75. Repeat Exercise 73, but using spherical coordinates.

76. Repeat Exercise 74, but using spherical coordinates.

Thinking Forward

A non-standard coordinate system in $\mathbb{R}^2$: Imagine a coordinate system in $\mathbb{R}^2$ in which the coordinate axes u and v are *not* perpendicular, as shown in the figure.

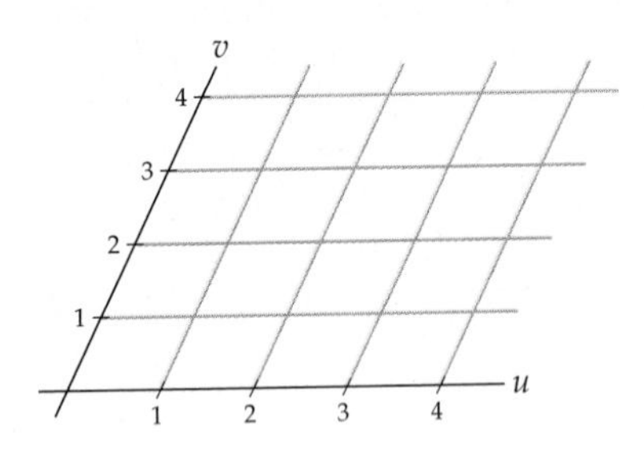

- Explain why every point in $\mathbb{R}^2$ has unique coordinates in this coordinate system.
- What is the area of each of the small parallelograms in the figure?
- If $\mathbf{a}$ is a vector parallel to the u-axis and $\mathbf{b}$ is a vector parallel to the v-axis, what is the area of the parallelogram determined by $\mathbf{a}$ and $\mathbf{b}$?

A non-standard coordinate system in $\mathbb{R}^3$: Imagine a three-dimensional analog to the coordinate system described in the previous problem, except that now there are three coordinate axes meeting at an origin but the three axes are not mutually perpendicular.

- What is the volume of each parallelepiped determined by a unit change in the u, v, and w directions?
- If $\mathbf{a}$, $\mathbf{b}$, and $\mathbf{c}$ are vectors parallel to the u-, v-, and w-axes, respectively, what is the volume of the parallelepiped determined by $\mathbf{a}$, $\mathbf{b}$, and $\mathbf{c}$?

13.7 JACOBIANS AND CHANGE OF VARIABLES

- Using nonstandard coordinate systems in $\mathbb{R}^2$
- Using nonstandard coordinate systems in $\mathbb{R}^3$
- Using Jacobians to construct double and triple integrals

Change of Variables in $\mathbb{R}^2$

The first integration technique that we discussed in Chapter 5 was integration by substitution. At that time we saw that

$$\int_a^b f(g(x))g'(x)\,dx = \int_c^d f(u)\,du,$$

where $u = g(x)$, $c = g(a)$, and $d = g(b)$. The point of changing the variable in that context was to obtain an integral $\int_c^d f(u)\,du$ that was simpler to evaluate than the original integral. We studied related phenomena when we introduced polar coordinates, cylindrical coordinates, and spherical coordinates. Some integrals are easier to evaluate when we represent them in those coordinate systems. No matter how comfortable we may feel with the rectangular coordinate system, analyzing an integral with an alternative coordinate system often makes the evaluation of a double or triple integral simpler. For example, we discussed how to evaluate a double integral $\iint_\Omega g(x, y)\,dA$ when we analyzed Ω as a type I region or a type II region in Section 13.2, but then, in Section 13.3, we extended our discussion by analyzing Ω in terms of polar coordinates. In Section 13.3 we saw that

$$\iint_\Omega g(x, y)\,dA = \int_\alpha^\beta \int_{f_1(\theta)}^{f_2(\theta)} g(r\cos\theta, r\sin\theta)\, r\,dr\,d\theta$$

when the boundary components of Ω are determined by the polar functions $r = f_1(\theta)$ and $r = f_2(\theta)$ for $\theta \in [\alpha, \beta]$. In the current section we will learn how to generalize this idea so that we can use other coordinate systems to analyze double and triple integrals.

Typically we start with some region Ω in $\mathbb{R}^2$. We wish to find a function $T : \Omega \to \Omega'$ such that Ω' is a simpler subset of $\mathbb{R}^2$. Such functions are called ***transformations***. In particular, we will require the transformations that we use in this section to be both one-to-one and differentiable at every point in the interior of Ω. As the following schematic illustrates, we consider the points in Ω to be given by the rectangular coordinates (x, y) and the points in our target to be given by coordinates (u, v):

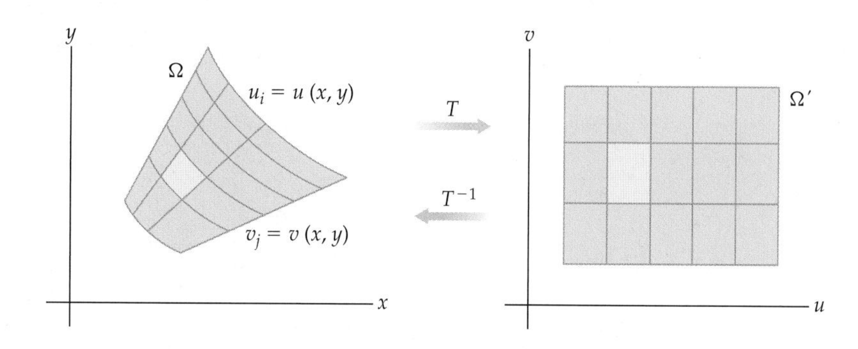

The goal is to determine a transformation T such that $T(\Omega) = \Omega'$ may be more simply analyzed in the new coordinate system. In our schematic, the image set Ω' is a rectangle, but the image set does not need to be that simple.

The coordinates u and v are both functions of x and y. That is,

$$u = u(x, y) \quad \text{and} \quad v = v(x, y).$$

Since we require T to be one-to-one on the interior of Ω, it is also invertible. Therefore, the x- and y-coordinates in the interior of Ω are functions of u and v; that is,

$$x = x(u, v) \quad \text{and} \quad y = y(u, v).$$

The grid lines in Ω in the previous figure at the left are level curves of the form $u_i = u(x, y)$ for $1 \leq i \leq m$ and $v_j = v(x, y)$ for $1 \leq j \leq n$. Optimally, the boundaries of Ω are also level curves. If we let $\Delta u = u_{i+1} - u_i$ and $\Delta v = v_{j+1} - v_j$, then when the mesh of the grid is relatively small, each subregion, such as the highlighted region in the left-hand figure, may be approximated by a parallelogram. We expand this subregion in the following figure:

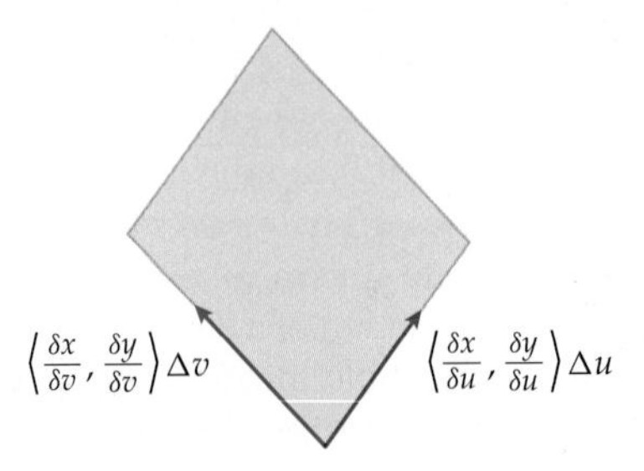

We may normalize the vectors $\left\langle \frac{\partial x}{\partial u}, \frac{\partial y}{\partial u} \right\rangle$ and $\left\langle \frac{\partial x}{\partial v}, \frac{\partial y}{\partial v} \right\rangle$ at a typical point (x, y) so that they both have unit length. In this case, the sides of the parallelogram have length $\left\langle \frac{\partial x}{\partial u}, \frac{\partial y}{\partial u} \right\rangle \Delta u$ and $\left\langle \frac{\partial x}{\partial v}, \frac{\partial y}{\partial v} \right\rangle \Delta v$. As we know from Chapter 10, the area of this parallelogram is given by the magnitude of the cross product

$$\left(\left\langle \frac{\partial x}{\partial u}, \frac{\partial y}{\partial u}, 0 \right\rangle \Delta u\right) \times \left(\left\langle \frac{\partial x}{\partial v}, \frac{\partial y}{\partial v}, 0 \right\rangle \Delta v\right) = \left(\left\langle \frac{\partial x}{\partial u}, \frac{\partial y}{\partial u}, 0 \right\rangle \times \left\langle \frac{\partial x}{\partial v}, \frac{\partial y}{\partial v}, 0 \right\rangle\right) \Delta u \Delta v.$$

Recall that the cross product

$$\left\langle \frac{\partial x}{\partial u}, \frac{\partial y}{\partial u}, 0 \right\rangle \times \left\langle \frac{\partial x}{\partial v}, \frac{\partial y}{\partial v}, 0 \right\rangle = \det \begin{bmatrix} \mathbf{i} & \mathbf{j} & \mathbf{k} \\ \frac{\partial x}{\partial u} & \frac{\partial y}{\partial u} & 0 \\ \frac{\partial x}{\partial v} & \frac{\partial y}{\partial v} & 0 \end{bmatrix} = \left(\frac{\partial x}{\partial u}\frac{\partial y}{\partial v} - \frac{\partial x}{\partial v}\frac{\partial y}{\partial u}\right)\mathbf{k}.$$

Furthermore, the quantity $\frac{\partial x}{\partial u}\frac{\partial y}{\partial v} - \frac{\partial x}{\partial v}\frac{\partial y}{\partial u}$ is the determinant of the 2×2 matrix:

$$\det \begin{bmatrix} \frac{\partial x}{\partial u} & \frac{\partial y}{\partial u} \\ \frac{\partial x}{\partial v} & \frac{\partial y}{\partial v} \end{bmatrix} = \frac{\partial x}{\partial u}\frac{\partial y}{\partial v} - \frac{\partial x}{\partial v}\frac{\partial y}{\partial u}.$$

We use this determinant in the next definition.

DEFINITION 13.22 The Jacobian for a Transformation in $\mathbb{R}^2$

Let Ω and Ω' be subsets of $\mathbb{R}^2$. If the transformation $T : \Omega \to \Omega'$ has a differentiable inverse with $x = x(u, v)$ and $y = y(u, v)$, then we define the ***Jacobian*** of the transformation T, denoted by $\frac{\partial(x,y)}{\partial(u,v)}$, to be the determinant of the matrix of partial derivatives; that is

$$\frac{\partial(x, y)}{\partial(u, v)} = \det \begin{bmatrix} \frac{\partial x}{\partial u} & \frac{\partial y}{\partial u} \\ \frac{\partial x}{\partial v} & \frac{\partial y}{\partial v} \end{bmatrix} = \frac{\partial x}{\partial u}\frac{\partial y}{\partial v} - \frac{\partial x}{\partial v}\frac{\partial y}{\partial u}.$$

Using Definition 13.22, we see that the area of the parallelogram we were discussing is

$$\left|\frac{\partial(x, y)}{\partial(u, v)}\right| \Delta u \Delta v.$$

Now, to approximate the integral of a function $g(x, y)$ over a region Ω, we may use the Riemann sum

$$\sum_{i=1}^{m}\sum_{j=1}^{n} g(x_i, y_j)\Delta A = \sum_{i=1}^{m}\sum_{j=1}^{n} g(x(u_i, v_j), y(u_i, v_j)) \left|\frac{\partial(x, y)}{\partial(u, v)}\right| \Delta u \Delta v.$$

When we take the limit of this Riemann sum as $m \to \infty$ and $n \to \infty$, we have

$$\iint_{\Omega} g(x, y)\, dA = \iint_{\Omega'} g(x(u, v), y(u, v)) \left|\frac{\partial(x, y)}{\partial(u, v)}\right| du\, dv.$$

If the preceding iterated integral is simple enough, we may use it to evaluate the double integral.

Our formulas for converting between rectangular and polar coordinates define a transformation if we omit the origin and insist that $\Omega \in [0, 2\pi)$. Recall that to convert from rectangular to polar coordinates, we use

$$r = \sqrt{x^2 + y^2} \quad \text{and} \quad \tan\theta = \frac{y}{x}.$$

To convert from polar coordinates to rectangular coordinates, we use

$$x = r\cos\theta \quad \text{and} \quad y = r\sin\theta.$$

Thus, subject to the restrictions we mentioned, the Jacobian of the transformation is

$$\frac{\partial(x, y)}{\partial(r, \theta)} = \det \begin{bmatrix} \frac{\partial x}{\partial r} & \frac{\partial y}{\partial r} \\ \frac{\partial x}{\partial \theta} & \frac{\partial y}{\partial \theta} \end{bmatrix} = \det \begin{bmatrix} \cos\theta & \sin\theta \\ -r\sin\theta & r\cos\theta \end{bmatrix} = r(\cos^2\theta + \sin^2\theta) = r.$$

This is the result we expect, since we know that

$$\iint_{\Omega} g(x, y)\, dA = \iint_{\Omega'} g(r\cos\theta, r\sin\theta) \left|\frac{\partial(x, y)}{\partial(r, \theta)}\right| dr\, d\theta = \iint_{\Omega'} g(r\cos\theta, r\sin\theta)\, r\, dr\, d\theta.$$

That is, the factor r we introduced in order to use polar coordinates to evaluate a double integral is just the Jacobian of the transformation.

To find a change of variable in order to simplify an integral $\iint_{\Omega} f(x, y)\, dA$, we determine invertible and differentiable functions $u = u(x, y)$ and $v = v(x, y)$ that allow us to analyze Ω more easily. To use Definition 13.22 to find the Jacobian we must also find the inverse of the transformation, which requires us to determine the functions $x = x(u, v)$ and $y = y(u, v)$.

Rather than go through this step, we may use the fact that $\frac{\partial(x,y)}{\partial(u,v)}\frac{\partial(u,v)}{\partial(x,y)} = 1$. The proof of this equation is left for Exercise 60.

Change of Variables in $\mathbb{R}^3$

The situation is quite similar in three dimensions. As we have already seen, a three-dimensional region may be easier to analyze with a coordinate system other than the rectangular system in $\mathbb{R}^3$. Rather than derive the analogous iterated triple integral in as great a detail as we did for the two-dimensional case, we will summarize the changes. We start with some region Ω initially expressed with rectangular coordinates in $\mathbb{R}^3$. A function $T : \Omega \to \Omega'$ is a transformation in $\mathbb{R}^3$. We require our transformations in $\mathbb{R}^3$ to be both invertible and differentiable in the interior of Ω. We consider the points in Ω to be given by the rectangular coordinates (x, y, z) and the points in our target to be given by coordinates (u, v, w).

The coordinates u, v, and w are all functions of x, y, and z. That is,

$$u = u(x, y, z), \quad v = v(x, y, z), \quad \text{and} \quad w = w(x, y, z).$$

Since T is invertible, it follows that x, y, and z are each functions of u, v, and w:

$$x = x(u, v, w), \quad y = y(u, v, w), \quad \text{and} \quad z = z(u, v, w).$$

We use level surfaces to subdivide Ω:

$$u_i = u(x, y, z) \text{ for } 1 \le i \le l, v_j = v(x, y, z) \text{ for } 1 \le j \le m, \text{ and } w_k = w(x, y, z) \text{ for } 1 \le k \le n.$$

Optimally, the boundaries of Ω are level surfaces.

We subdivide Ω into small pieces, each of which may be approximated with a small parallelepiped, as follows:

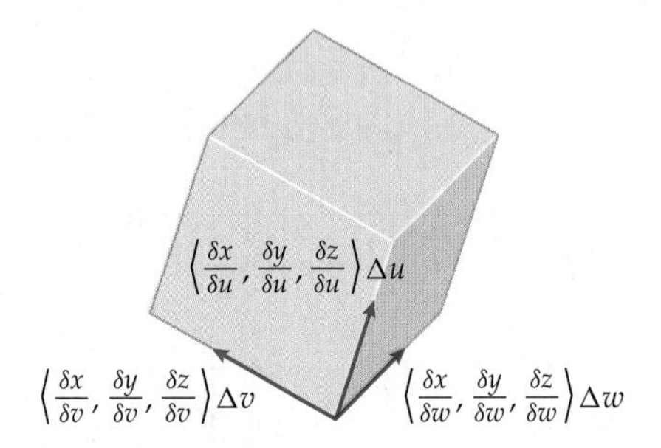

The volume of the parallelepiped is given by the triple scalar product

$$\left(\left(\left\langle \frac{\partial x}{\partial u}, \frac{\partial y}{\partial u}, \frac{\partial z}{\partial u} \right\rangle \Delta u\right) \times \left(\left\langle \frac{\partial x}{\partial v}, \frac{\partial y}{\partial v}, \frac{\partial z}{\partial v} \right\rangle \Delta v\right)\right) \cdot \left(\left\langle \frac{\partial x}{\partial w}, \frac{\partial y}{\partial w}, \frac{\partial z}{\partial w} \right\rangle \Delta w\right) =$$

$$\left(\left\langle \frac{\partial x}{\partial u}, \frac{\partial y}{\partial u}, \frac{\partial z}{\partial u} \right\rangle \times \left\langle \frac{\partial x}{\partial v}, \frac{\partial y}{\partial v}, \frac{\partial z}{\partial v} \right\rangle\right) \cdot \left\langle \frac{\partial x}{\partial w}, \frac{\partial y}{\partial w}, \frac{\partial z}{\partial w} \right\rangle \Delta u \Delta v \Delta w.$$

Recall that the triple scalar product is equal to the determinant of an appropriate 3×3 matrix. We use this matrix to define the Jacobian of a transformation in $\mathbb{R}^3$.

DEFINITION 13.23

The Jacobian for a Transformation in $\mathbb{R}^3$

Let Ω and Ω' be subsets of $\mathbb{R}^3$. If the transformation $T : \Omega \to \Omega'$ has a differentiable inverse with $x = x(u, v, w)$, $y = y(u, v, w)$, and $z = z(u, v, w)$, then we define the ***Jacobian*** of the transformation T, denoted by $\frac{\partial(x,y,z)}{\partial(u,v,w)}$, to be the determinant of the matrix of partial derivatives; that is,

$$\frac{\partial(x, y, z)}{\partial(u, v, w)} = \det \begin{bmatrix} \frac{\partial x}{\partial u} & \frac{\partial y}{\partial u} & \frac{\partial z}{\partial u} \\ \frac{\partial x}{\partial v} & \frac{\partial y}{\partial v} & \frac{\partial z}{\partial v} \\ \frac{\partial x}{\partial w} & \frac{\partial y}{\partial w} & \frac{\partial z}{\partial w} \end{bmatrix}.$$

Thus, we may express the volume of the parallelepiped as $\left|\frac{\partial(x,y,z)}{\partial(u,v,w)}\right| \Delta u \Delta v \Delta w$. Using this expression in a Riemann sum and taking the appropriate limit, we see that

$$\iiint_{\Omega} g(x, y, z)\, dV = \iiint_{\Omega'} g(x(u, v, w), y(u, v, w), z(u, v, w)) \left|\frac{\partial(x, y, z)}{\partial(u, v, w)}\right| du\, dv\, dw.$$

If the preceding iterated integral is simple enough, we may use it to evaluate the triple integral.

The procedure just given is consistent with the evaluation procedure we established when we introduced spherical coordinates in Section 13.6. From Theorem 13.21, we know that, given the spherical coordinates (ρ, θ, ϕ) of a point, we may obtain the rectangular coordinates of the point with

$$x = \rho \sin\phi \cos\theta, \quad y = \rho \sin\phi \sin\theta, \quad \text{and} \quad z = \rho \cos\phi.$$

In Exercise 52 you are asked to compute the Jacobian of this transformation and show that

$$\frac{\partial(x, y, z)}{\partial(\rho, \theta, \phi)} = \det \begin{bmatrix} \frac{\partial x}{\partial \rho} & \frac{\partial y}{\partial \rho} & \frac{\partial z}{\partial \rho} \\ \frac{\partial x}{\partial \theta} & \frac{\partial y}{\partial \theta} & \frac{\partial z}{\partial \theta} \\ \frac{\partial x}{\partial \phi} & \frac{\partial y}{\partial \phi} & \frac{\partial z}{\partial \phi} \end{bmatrix} = -\rho^2 \sin\phi.$$

Therefore, we again find that

$$\begin{aligned} \iiint_{\Omega} g(x, y, z)\, dV &= \iiint_{\Omega'} g(x(\rho, \theta, \phi), y(\rho, \theta, \phi), z(\rho, \theta, \phi)) \left|\frac{\partial(x, y, z)}{\partial(\rho, \theta, \phi)}\right| d\rho\, d\theta\, d\phi \\ &= \iiint_{\Omega'} g(x(\rho, \theta, \phi), y(\rho, \theta, \phi), z(\rho, \theta, \phi))\, \rho^2 \sin\phi\, d\rho\, d\theta\, d\phi. \end{aligned}$$

Examples and Explorations

EXAMPLE 1

Evaluating a double integral by changing the variable

Evaluate the double integral $\iint_{\Omega} \frac{1}{(x+y)^2}\, dA$, where Ω is the trapezoidal region in the first quadrant bounded by the lines $x + y = 1$ and $x + y = 4$ and the x- and y-axes.

SOLUTION

Following is the region Ω:

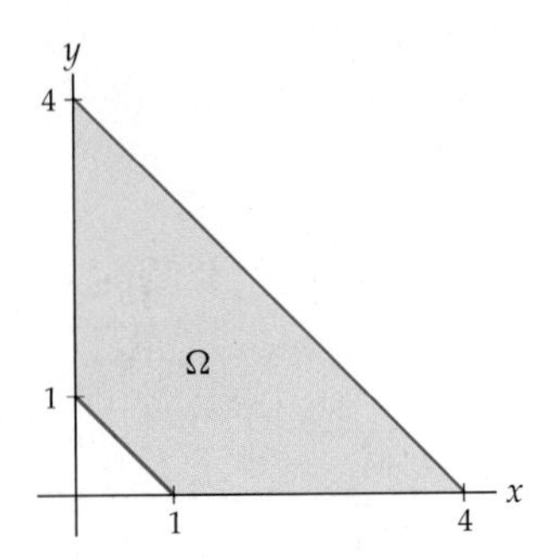

It would be possible to decompose Ω into two type I regions or two type II regions, but we will use a change of variable that allows us to evaluate the integral with a single iterated integral. The difficulty here is that, relative to the x- and y-axes, the region is mildly complicated. If one of our coordinate axes were parallel to the parallel sides of the trapezoid and the other were perpendicular to those sides, we would be able to treat Ω more simply. Fortunately, we may accomplish this aim with the change of variable

$$u = x + y \quad \text{and} \quad v = y - x.$$

Note that the level curves for $u = x + y$ are lines with slope -1 and the level curves for v are lines with slope 1, as we suggested. With this transformation, the equation of the lines $x+y = 1$ and $x+y = 4$ are expressed more simply as $u = 1$ and $u = 4$, respectively. Solving the system $u = x+y$ and $v = y-x$ simultaneously, we may express x and y as the following functions of u and v:

$$x = \frac{1}{2}(u - v) \quad \text{and} \quad y = \frac{1}{2}(u + v).$$

Using these functions, we see that the x-axis ($y = 0$) has equation $v = -u$ and the y-axis ($x = 0$) has equation $v = u$. Therefore, under the transformation, the graph of Ω' in the uv-plane is as follows:

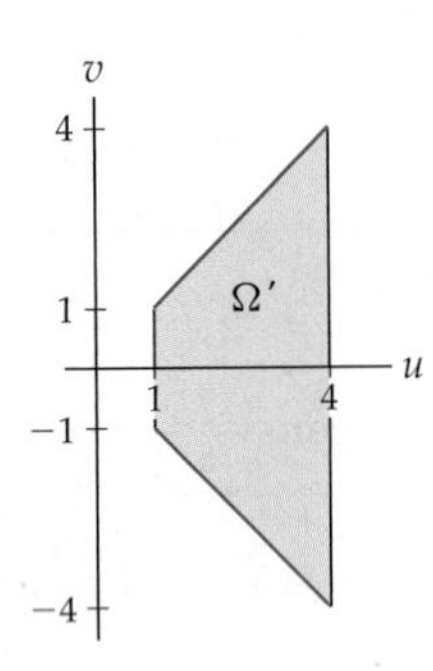

Our new variables make it easier to treat Ω' as a type I region. Before we set up the appropriate double integral, we need the Jacobian of the transformation. It is

$$\frac{\partial(x, y)}{\partial(u, v)} = \det\begin{bmatrix} \frac{\partial x}{\partial u} & \frac{\partial y}{\partial u} \\ \frac{\partial x}{\partial v} & \frac{\partial y}{\partial v} \end{bmatrix} = \det\begin{bmatrix} \frac{1}{2} & \frac{1}{2} \\ -\frac{1}{2} & \frac{1}{2} \end{bmatrix} = \frac{1}{2}.$$

Now, the double integral will be

$$\iint_{\Omega} \frac{1}{(x+y)^2}\, dA = \frac{1}{2}\int_1^4 \int_{-u}^{u} \frac{1}{u^2}\, dv\, du.$$

Note that because the Jacobian is a constant for this transformation, we were able to factor it out of the iterated integral. The most difficult part of the process is finding the appropriate change of variable to simplify Ω and, therefore, the double integral. The transformation we used was *not* the only one we could have used, but it sufficed to simplify the integral. We leave the evaluation of the iterated integral to Exercise 22. □

EXAMPLE 2 **Evaluating another double integral by changing the variable**

Evaluate the double integral $\iint_{\Omega_2} \frac{x}{y}\, dA$, where Ω_2 is the first-quadrant region bounded by the lines $y = \frac{1}{4}x$ and $y = 4x$ and the curves $y = \frac{1}{x}$ and $y = \frac{4}{x}$.

SOLUTION

Following is the region:

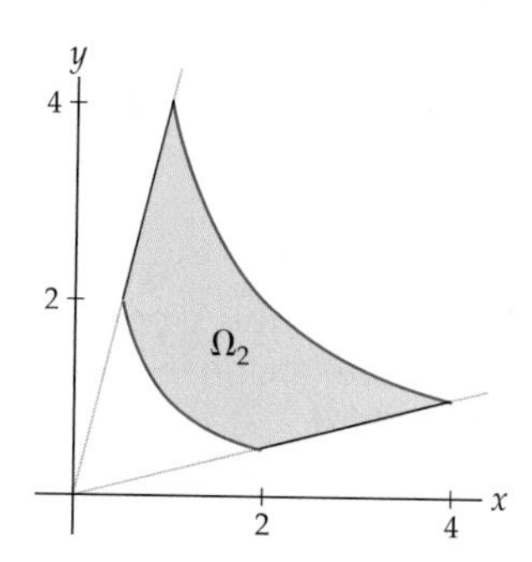

It would be possible to decompose Ω_2 into three type I regions or three type II regions, but we will use a change of variable that allows us to evaluate the integral with a single iterated integral. Note that the curves $y = \frac{1}{x}$ and $y = \frac{4}{x}$ may also be expressed as $xy = 1$ and $xy = 4$, respectively. If we let $u = xy$, then we may express these two curves as $u = 1$ and $u = 4$. The lines $y = \frac{1}{4}x$ and $y = 4x$ may be expressed as $\frac{y}{x} = \frac{1}{4}$ and $\frac{y}{x} = 4$, respectively. If we let $v = \frac{y}{x}$, we may express these lines as $v = \frac{1}{4}$ and $v = 4$. Thus, in the uv-plane, the graph of Ω_2 is transformed to the rectangle Ω'_2, as shown in the following figure:

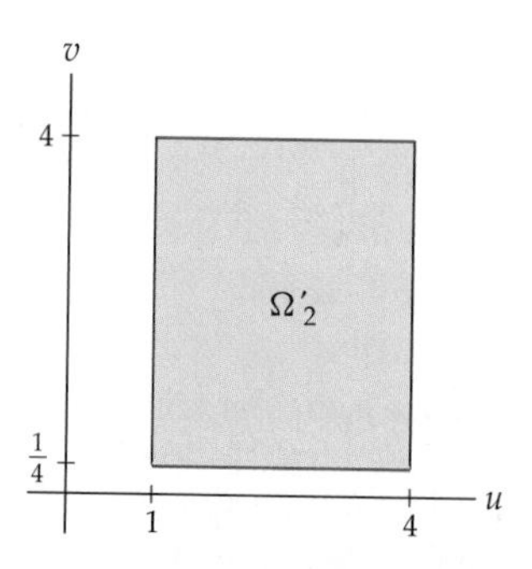

We will treat Ω'_2 as a type I region, although we could just as easily treat it as a type II region.

Before we set up the appropriate double integral, we need the Jacobian of the transformation. Here we will use the fact that $\frac{\partial(x,y)}{\partial(u,v)}\frac{\partial(u,v)}{\partial(x,y)} = 1$. We have

$$\frac{\partial(u,v)}{\partial(x,y)} = \det\begin{bmatrix} \frac{\partial u}{\partial x} & \frac{\partial v}{\partial x} \\ \frac{\partial u}{\partial y} & \frac{\partial v}{\partial y} \end{bmatrix} = \det\begin{bmatrix} y & -\frac{y}{x^2} \\ x & \frac{1}{x} \end{bmatrix} = \frac{2y}{x}.$$

Now, since $\frac{y}{x} = v$, it follows that $\frac{\partial(u,v)}{\partial(x,y)} = 2v$. Thus, $\frac{\partial(x,y)}{\partial(u,v)} = \frac{1}{2v}$.

Alternatively, we could solve the system

$$u = xy \quad \text{and} \quad v = \frac{y}{x}$$

for x and y. In Exercise 25, you will show that

$$x = \sqrt{\frac{u}{v}} = u^{1/2}v^{-1/2} \quad \text{and} \quad y = \sqrt{uv} = u^{1/2}v^{1/2}$$

and to use these equations to show that $\frac{\partial(x,y)}{\partial(u,v)} = \frac{1}{2v}$. Either way, we need to express the Jacobian in terms of the new variables, here u and v.

We are now ready to rewrite the double integral in terms of the new variables. Since $v = \frac{y}{x}$, we have $\frac{x}{y} = \frac{1}{v}$; thus, the double integral will be

$$\iint_{\Omega_2} \frac{x}{y}\,dA = \frac{1}{2}\int_1^4 \int_{1/4}^4 \frac{1}{v^2}\,dv\,du.$$

The factor $\frac{1}{2}$ in front of the integral was originally a factor of the Jacobian. Again, the most difficult part of this process is finding the appropriate change of variable to simplify Ω_2 and, therefore, the double integral. We leave the evaluation of the iterated integral to Exercise 26. □

EXAMPLE 3 **Finding the area of an elliptical annulus**

Find the area between the ellipses with equations $4x^2 + 9y^2 = 36$ and $4x^2 + 9y^2 = 144$.

SOLUTION

Let $\mathcal{S}$ represent this annulus, shown in the following figure:

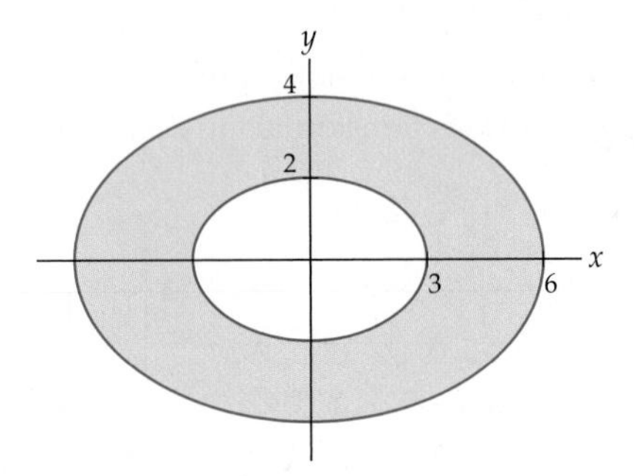

If this were a circular annulus, it would be quite easy to find its area. Fortunately, using a suitably chosen transformation we can obtain a circular annulus. Let $u = 2x$ and $v = 3y$. The equations of the ellipses are $4x^2 + 9y^2 = 36$ and $4x^2 + 9y^2 = 144$, or equivalently, $(2x)^2 + (3y)^2 = 36$ and $(2x)^2 + (3y)^2 = 144$. Now, using the transformation, we have

$$u^2 + v^2 = 36 \quad \text{and} \quad u^2 + v^2 = 144.$$

That is, $\mathcal{S}'$ is a circular ellipse with inner radius 6 and outer radius 12. In order to find the Jacobian of the transformation, we solve for x and y in terms of u and v. Here we have $x = \frac{1}{2}u$ and $y = \frac{1}{3}v$. Thus, the Jacobian of the transformation is

$$\frac{\partial(x,y)}{\partial(u,v)} = \det\begin{bmatrix} \frac{\partial x}{\partial u} & \frac{\partial y}{\partial u} \\ \frac{\partial x}{\partial v} & \frac{\partial y}{\partial v} \end{bmatrix} = \det\begin{bmatrix} \frac{1}{2} & 0 \\ 0 & \frac{1}{3} \end{bmatrix} = \frac{1}{6}.$$

The area of the ellipse is given by the double integral

$$\iint_S dA = \iint_{S'} \left| \frac{\partial(x, y)}{\partial(u, v)} \right| dA = \iint_{S'} \frac{1}{6}\, dA.$$

In our previous examples, we used rectangular coordinates to evaluate the transformed integrals, but here it is simpler to use polar coordinates. We obtain

$$\iint_S dA = \frac{1}{6} \int_0^{2\pi} \int_6^{12} r\, dr\, d\theta = 18\pi. \qquad \square$$

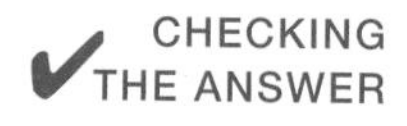

CHECKING THE ANSWER

We could also have computed the area of this annulus by means of the area formula for an ellipse, or

$$\text{Area} = \pi ab,$$

where a and b are the lengths of the semimajor and semiminor axes of the ellipse. In this example, the area of the larger ellipse is $4 \cdot 6\pi = 24\pi$ and the area of the smaller ellipse is $2 \cdot 3\pi = 6\pi$. Therefore, the area of the annulus is 18π, as we saw before.

? TEST YOUR UNDERSTANDING

- What is the integration-by-substitution formula for evaluating a definite integral? Why is this formula consistent with the change-of-variable formula that uses Jacobians for evaluating a double integral? What is the "Jacobian" in the integration-by-substitution formula?
- How is a Jacobian related to an area in $\mathbb{R}^2$? How is it related to a volume in $\mathbb{R}^3$?
- What is the Jacobian when you convert from rectangular coordinates to polar coordinates in $\mathbb{R}^2$? The Jacobians for the transformations from rectangular coordinates to polar coordinates in $\mathbb{R}^2$ and from rectangular coordinates to cylindrical coordinates in $\mathbb{R}^3$ are equal. Why does this make sense?
- What is the Jacobian for the transformation from rectangular coordinates to spherical coordinates in $\mathbb{R}^3$?
- What do we look for when we are considering a coordinate system to help us evaluate a double or triple integral?

EXERCISES 13.7

Thinking Back

- *Integration by substitution:* What is the integration-by-substitution formula for definite integrals? How is it derived?
- *Integrating with polar coordinates:* Let Ω be a region in $\mathbb{R}^2$. Give a double integral that represents the area of Ω when you integrate with polar coordinates.
- *Integrating with cylindrical coordinates:* Let Ω be a region in $\mathbb{R}^3$. Give a triple integral that represents the volume of Ω when you integrate with cylindrical coordinates.
- *Integrating with spherical coordinates:* Let Ω be a region in $\mathbb{R}^3$. Give a triple integral that represents the volume of Ω when you integrate with spherical coordinates.

Concepts

0. *Problem Zero:* Read the section and make your own summary of the material.

1. *True/False:* Determine whether each of the statements that follow is true or false. If a statement is true, explain why. If a statement is false, provide a counterexample.

(a) *True or False:* There is a transformation in $\mathbb{R}^2$ that takes a circle to a rectangle.

(b) *True or False:* There is a transformation in $\mathbb{R}^2$ that takes a parallelogram to a square.

(c) *True or False:* There is a transformation in $\mathbb{R}^3$ that takes a cylinder to a rectangular solid.

(d) *True or False:* There is a transformation in $\mathbb{R}^3$ that takes a sphere to a rectangular solid.

(e) *True or False:* If $u = 2x + y$ and $v = x - 2y$, then $x = \frac{2}{5}u + \frac{1}{5}v$ and $y = \frac{1}{5}u - \frac{2}{5}v$.

(f) *True or False:* If $u = 2x + y$ and $v = x - 2y$, then $\frac{\partial(x,y)}{\partial(u,v)} = \frac{1}{5}$.

(g) *True or False:* Given a rectangular region $\mathcal{R}$ in the xy-plane and a square region $\mathcal{R}'$ in the uv-plane, there is a transformation taking $\mathcal{R}$ to $\mathcal{R}'$.

(h) *True or False:* Given a rectangular region $\mathcal{R}$ in the xy-plane and a square region $\mathcal{R}'$ in the uv-plane, there is a unique transformation taking $\mathcal{R}$ to $\mathcal{R}'$.

2. *Examples:* Construct examples of the thing(s) described in the following. Try to find examples that are different than any in the reading.
   (a) A transformation $T : \mathbb{R}^2 \to \mathbb{R}^2$ whose Jacobian is a positive constant.
   (b) A transformation $T : \mathbb{R}^2 \to \mathbb{R}^2$ whose Jacobian varies.
   (c) A transformation $T : \mathbb{R}^3 \to \mathbb{R}^3$ whose Jacobian is a negative constant.

3. Let T be a transformation in $\mathbb{R}^2$. How is the Jacobian of the transformation related to the change in the increment of area, ΔA?

Let $\mathcal{R}$ be the rectangle in the Cartesian coordinate system with vertices $(1, 0)$, $(3, 2)$, $(2, 3)$, and $(0, 1)$. Use $\mathcal{R}$ to answer Exercises 4–9.

4. Set up and evaluate iterated double integrals equal to $\iint_{\mathcal{R}} x^2y\,dA$, treating $\mathcal{R}$ as a union of type I regions.

5. Set up and evaluate iterated double integrals equal to $\iint_{\mathcal{R}} x^2y\,dA$, treating $\mathcal{R}$ as a union of type II regions.

6. Show that the boundaries of $\mathcal{R}$ are the lines with equations $x + y = 1$, $x + y = 5$, $x - y = 1$, and $x - y = -1$.
   (a) Use these equations to explain why the change of variables $u = x+y$ and $v = x-y$ results in the simpler system of equations in the uv-coordinate system.
   (b) Sketch the graph of the transformed rectangle $\mathcal{R}'$ in the uv-coordinate system.

7. Use your transformation from Exercise 6 to express x and y as functions of u and v, and find the Jacobian of the transformation.

8. Find the Jacobian of the transformation from Exercise 6, using the fact that $\frac{\partial(x,y)}{\partial(u,v)}\frac{\partial(u,v)}{\partial(x,y)} = 1$.

9. Set up and evaluate a single iterated double integral equal to $\iint_{\mathcal{R}} x^2y\,dA$ in which you integrate with respect to u and v rather than with respect to x and y.

Let a and b be constants. Use the transformation $T : \mathbb{R}^2 \to \mathbb{R}^2$ defined by $u = x + a$ and $v = y + b$ to answer Exercises 10–13.

10. What does this transformation do to the point with coordinates (x_0, y_0)? What does the transformation do to a vertical line $x = x_0$? What does the transformation do to a horizontal line $y = y_0$?

11. What does this transformation do to a region Ω? In particular, if $a = 3$, $b = 4$, and Ω is the rectangle defined by the inequalities $0 \le x \le 1$ and $0 \le y \le 2$, what is the image of Ω under the given transformation?

12. What are x and y as functions of u and v? What is the Jacobian of this transformation? How does your answer conform to the answer to Exercise 3?

13. Explain why this transformation would or would not be useful for evaluating a double integral.

Let a and b be positive constants. Use the transformation $T : \mathbb{R}^2 \to \mathbb{R}^2$ defined by $u = ax$ and $v = by$ to answer Exercises 14–17.

14. What does this transformation do to the point with coordinates (x_0, y_0)? What does the transformation do to a vertical line $x = x_0$? What does the transformation do to a horizontal line $y = y_0$?

15. What does this transformation do to a region Ω? In particular, if $a = 3$, $b = \frac{1}{2}$, and Ω is the square defined by the inequalities $0 \le x \le 1$ and $0 \le y \le 1$, what is the image of Ω under the given transformation?

16. What are x and y as functions of u and v? What is the Jacobian of this transformation? How does your answer conform to the answer to Exercise 3?

17. If we allow either a or b to be negative, how would that change your answers to Exercises 14 and 16?

Let $\theta \in [0, 2\pi)$. Use the transformation $T : \mathbb{R}^2 \to \mathbb{R}^2$ defined by $u = (\sin\theta)x - (\cos\theta)y$ and $v = (\cos\theta)x + (\sin\theta)y$ to answer Exercises 18 and 19.

18. What does this transformation do to a region Ω? In particular, if $\theta = \frac{\pi}{6}$ and Ω is the rectangle defined by the inequalities $0 \le x \le 1$ and $0 \le y \le 2$, what is the image of Ω under the given transformation?

19. What are x and y as functions of u and v? What is the Jacobian of this transformation? How does your answer conform to the answer to Exercise 3?

20. Let $T : \mathbb{R}^3 \to \mathbb{R}^3$ be a transformation. How is the Jacobian of the transformation related to the change of the increment of volume, ΔV?

Exercises 21–23 continue the work started in Example 1.

21. Evaluate the double integral $\iint_{\Omega} \frac{1}{(x+y)^2}\,dA$ without using a change of variables. Note that you will need to treat Ω as the union of two type I regions or two type II regions.

22. Complete Example 1 by evaluating the iterated integral $\frac{1}{2}\int_1^4 \int_{-u}^{u} \frac{1}{u^2}\,dv\,du$.

23. Find the areas of the trapezoid Ω and the transformed trapezoid Ω' in Example 1. What is the relationship between these two areas and the Jacobian of the transformation?

Exercises 24–26 continue the work started in Example 2.

24. We mentioned that to evaluate the double integral $\iint_{\Omega_2} \frac{x}{y}\,dA$ without using a change of variables, you would need to decompose Ω_2 into a union of three type I regions or three type II regions. What are those decompositions?

25. Show that if $u = xy$ and $v = \frac{y}{x}$, then

$$x = \sqrt{\frac{u}{v}} = u^{1/2}v^{-1/2} \quad \text{and} \quad y = \sqrt{uv} = u^{1/2}v^{1/2}.$$

Use these formulas to show that $\frac{\partial(x,y)}{\partial(u,v)} = \frac{1}{2v}$.

26. Complete Example 2 by evaluating the iterated integral

$$\iint_{\Omega_2} \frac{x}{y}\, dA = \frac{1}{2}\int_1^4 \int_{1/4}^4 \frac{1}{v^2}\, dv\, du.$$

Skills

In Exercises 27–32, functions $x = x(u, v)$ and $y = y(u, v)$ are given that determine transformations from an xy-coordinate system to a uv-coordinate system in $\mathbb{R}^2$. Use these functions to determine a region in the xy-plane that has the image specified for the given values of u and v, and find the Jacobian of the transformation.

27. $x = u - v$ and $y = u + v$ for $0 \le u \le 2$ and $0 \le v \le 1$

28. $x = 3u + 4v$ and $y = 4u - 3v$ for $1 \le u \le 3$ and $1 \le v \le 5$

29. $x = \frac{u}{v}$ and $y = uv$ for $1 \le u \le 2$ and $1 \le v \le 3$

30. $x = u^2 + v^2$ and $y = u^2 - v^2$ for $0 \le u \le 4$ and $0 \le v \le 4$

31. $x = u \sin v$ and $y = u \cos v$ for $0 \le u \le 2$ and $0 \le v \le \pi$

32. $x = u \sec v$ and $y = u \tan v$ for $0 \le u \le 2$ and $0 \le v \le \frac{\pi}{4}$

For each double integral in Exercises 33–38, (a) sketch the region Ω, (b) use the specified transformation to sketch the transformed region, and (c) use the transformation to evaluate the integral.

33. $\iint_\Omega (x + y)^2\, dA$, where Ω is the trapezoid with vertices $(1, 3)$, $(3, 1)$, $(9, 3)$, and $(3, 9)$. Use the transformation given by $u = x + y$ and $v = x - y$.

34. Evaluate the integral from Exercise 33, but use the transformation given by $u = x + y$ and $v = 3x - y$.

35. Evaluate the integral from Exercise 33, but use the transformation given by $u = x + y$ and $v = \frac{x}{y}$.

36. $\iint_\Omega xy\, dA$, where Ω is the trapezoid with vertices $(2, 3)$, $(3, 2)$, $(5, 3)$, and $(3, 5)$. Use the transformation from Exercise 33.

37. Evaluate the double integral from Exercise 36, using the transformation given by $u = x + y$ and $v = 2y - x$.

38. $\iint_\Omega \left(\frac{y^3}{x} - xy\right) dA$, where Ω is the region in the first and second quadrants that is bounded above by the hyperbola $y^2 - x^2 = 12$, bounded below by the hyperbola $y^2 - x^2 = 3$, and bounded on the left and right by the lines $y = -2x$ and $y = 2x$, respectively. Use the transformation given by $u = \frac{y}{x}$ and $v = y^2 - x^2$.

Evaluate the double integrals in Exercises 39–48. Use suitable transformations as necessary.

39. $\iint_\Omega (x - y)^3\, dA$, where Ω is the parallelogram with vertices $(0, 0)$, $(3, 0)$, $(5, 2)$, and $(2, 2)$.

40. $\iint_\Omega xy\, dA$, where Ω is the parallelogram from Exercise 39.

41. $\iint_\Omega \frac{2y - x}{3x + y + 1}\, dA$, where Ω is the parallelogram with vertices $(0, 0)$, $(2, 1)$, $(1, 4)$, and $(-1, 3)$.

42. $\iint_\Omega (3x^2 - 5xy - 2y^2)\, dA$, where Ω is the parallelogram from Exercise 41.

43. $\iint_\Omega \left(\frac{x^2}{y^2} + x^2y^2\right) dA$, where Ω is the following region:

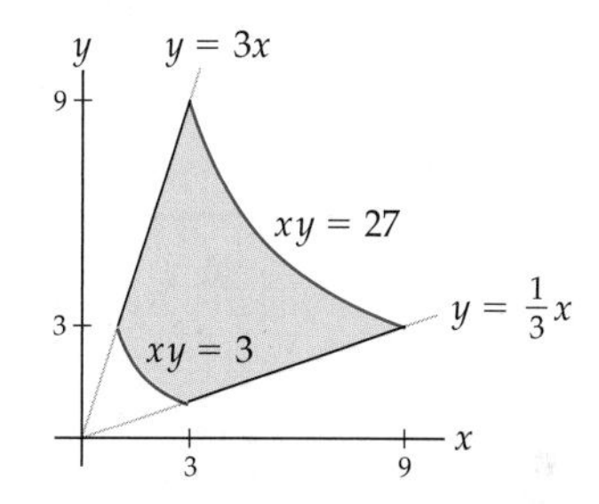

44. $\iint_\Omega xy^3\, dA$, where Ω is the region from Exercise 43.

45. $\iint_\Omega \frac{y^2}{x^3}\, dA$, where Ω is the following region:

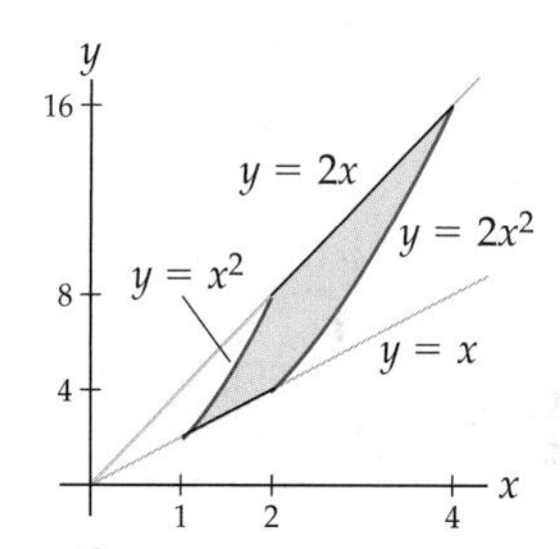

46. $\iint_\Omega \frac{y + xy}{x^2}\, dA$, where Ω is the region from Exercise 45.

47. $\iint_\Omega \frac{x^3}{y^3}\, dA$, where Ω is the following region:

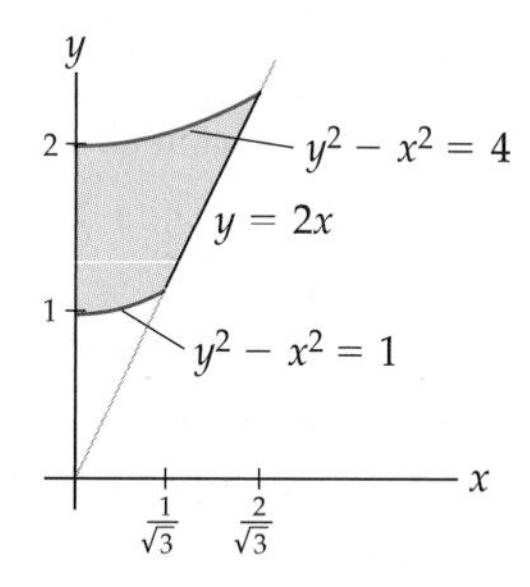

48. $\iint_\Omega x^2\left(1 - \frac{1}{y^2}\right) dA$, where Ω is the region from Exercise 47.

Applications

49. Joe plans to build a swimming pool in his backyard in the shape of parallelogram to fit the contours of his property in the dimensions shown in the following figure:

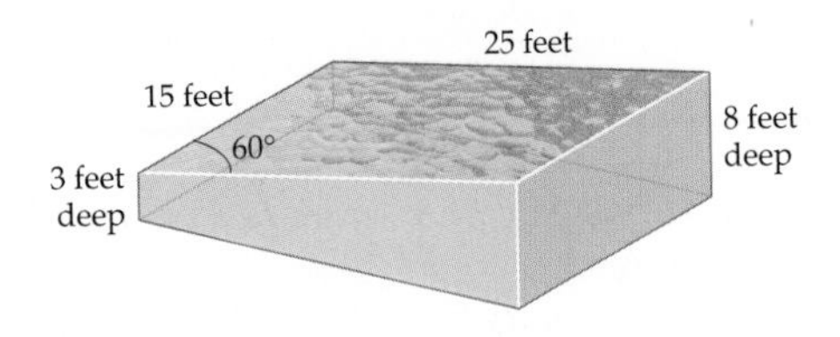

The shallow end of the pool will start at a depth of 3 feet and will increase linearly to a depth of 8 feet on the opposite side of the pool. How much water will the pool hold?

50. Joe also plans to build a goldfish pond in the front of his house in the shape of an ellipse with a 5-foot major axis and a 3-foot minor axis. The water in the pond will be deepest at the center of the pond. The depth of the water, in feet, at any point in the pond will be given by the function $d(x, y) = 2\sqrt{\frac{x^2}{25} + \frac{y^2}{9}} - 2$, where $(x, y) = (0, 0)$ represents the center of the surface of the pond. How much water will the pond hold?

Proofs

51. The formulas for converting from cylindrical coordinates to rectangular coordinates are

$$x = r\cos\theta, \quad y = r\sin\theta, \quad \text{and} \quad z = z.$$

Prove that the Jacobian $\frac{\partial(x,y,z)}{\partial(r,\theta,z)} = r$.

52. The formulas for converting from spherical coordinates to rectangular coordinates are

$$x = \rho\sin\phi\cos\theta, \quad y = \rho\sin\phi\sin\theta, \quad \text{and} \quad z = \rho\cos\phi.$$

Prove that the Jacobian $\frac{\partial(x,y,z)}{\partial(\rho,\theta,\phi)} = -\rho^2\sin\phi$.

Let α, β, γ, and δ be constants. A transformation $T : \mathbb{R}^2 \to \mathbb{R}^2$, where

$$x = \alpha u + \beta v \quad \text{and} \quad y = \gamma u + \delta v,$$

is called a ***linear transformation*** of $\mathbb{R}^2$. Use this transformation to answer Exercises 53–55.

53. Prove that a linear transformation takes a line $ax + by = c$ in the xy-plane to a line in the uv-plane if the Jacobian of the transformation is nonzero.

54. Prove that there is a linear transformation that takes a line in the xy-plane to a point in the uv-plane if the Jacobian of the transformation is zero.

55. Assuming that the Jacobian is nonzero, find expressions for u and v as functions of x and y.

56. Let α_i, β_i, and γ_i be constants for $i = 1, 2,$ and 3. A transformation $T : \mathbb{R}^3 \to \mathbb{R}^3$ defined by

$$x = \alpha_1 u + \beta_1 v + \gamma_1 w,$$
$$y = \alpha_2 u + \beta_2 v + \gamma_2 w, \quad \text{and}$$
$$z = \alpha_3 u + \beta_3 v + \gamma_3 w.$$

is called a ***linear transformation*** of $\mathbb{R}^3$. Prove that this transformation takes a plane $ax + by + cz = d$ in the xyz-coordinate system to a plane in the uvw-coordinate system if the Jacobian of the transformation is nonzero.

57. Use a change of variables to prove that the area of the ellipse with equation $\left(\frac{x}{a}\right)^2 + \left(\frac{y}{b}\right)^2 = 1$ is πab.

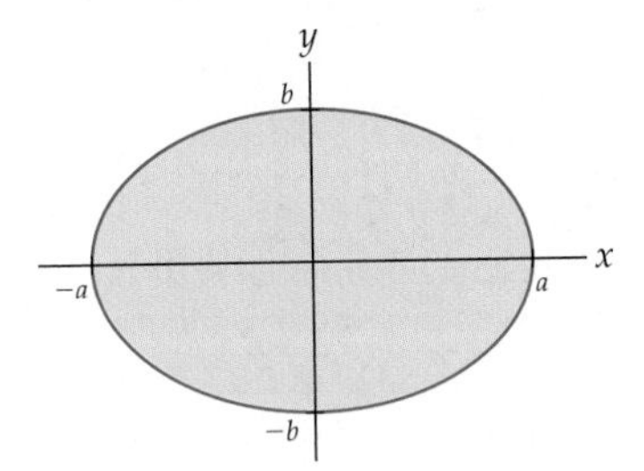

58. Use a change of variables to prove that the volume of the ellipsoid with equation $\left(\frac{x}{a}\right)^2 + \left(\frac{y}{b}\right)^2 + \left(\frac{z}{c}\right)^2 = 1$ is $\frac{4}{3}\pi abc$.

59. Prove the following chain rule for Jacobians: If x and y are differentiable functions of u and v, and if u and v are differentiable functions of s and t, then

$$\frac{\partial(x, y)}{\partial(s, t)} = \frac{\partial(x, y)}{\partial(u, v)}\frac{\partial(u, v)}{\partial(s, t)}.$$

60. Let Ω and Ω' be subsets of $\mathbb{R}^2$. Use the results of Exercise 59 to prove that if a transformation $T : \Omega \to \Omega'$ is invertible, and if both T and T^{-1} are differentiable, then

$$\frac{\partial(x, y)}{\partial(u, v)}\frac{\partial(u, v)}{\partial(x, y)} = 1.$$

Thinking Forward

Determinants of 3×3 *matrices:* Let

$$A = \begin{bmatrix} a_{11} & a_{12} & a_{13} \\ a_{21} & a_{22} & a_{23} \\ a_{31} & a_{32} & a_{33} \end{bmatrix}.$$

The determinant of A, $\det(A)$, may be defined in terms of determinants of 2×2 matrices as follows:

$$\det(A) = a_{11} \det \begin{bmatrix} a_{22} & a_{23} \\ a_{32} & a_{33} \end{bmatrix} - a_{12} \det \begin{bmatrix} a_{21} & a_{23} \\ a_{31} & a_{33} \end{bmatrix} + a_{13} \det \begin{bmatrix} a_{21} & a_{22} \\ a_{31} & a_{32} \end{bmatrix}.$$

- Show that the previous definition of $\det(A)$ provides the same result as the procedure we used to find the determinant of a 3×3 matrix earlier in this section.
- Use the preceding definition to find

$$\det \begin{bmatrix} 1 & 3 & 4 \\ -2 & 5 & -1 \\ -4 & 2 & -3 \end{bmatrix}.$$

- In an analogous fashion, find a recursion formula for computing the determinant of a 4×4 matrix in terms of determinants of 3×3 matrices.

CHAPTER REVIEW, SELF-TEST, AND CAPSTONES

Before you progress to the next chapter, be sure you are familiar with the definitions, concepts, and basic skills outlined here. The capstone exercises at the end bring together ideas from this chapter and look forward to the next chapter.

Definitions

Give precise mathematical definitions or descriptions of each of the concepts that follow. Then illustrate the definition with a graph or an algebraic example.

- a *Riemann sum* for a function $z = f(x, y)$ defined on a rectangle $\mathcal{R} = \{(x, y) \mid a \le x \le b \text{ and } c \le y \le d\}$
- the *double integral* of f over

$$\mathcal{R} = \{(x, y) \mid a \le x \le b \text{ and } c \le y \le d\}$$

- the *double integral* of f over a general region

$$\Omega \subseteq \{(x, y) \mid a \le x \le b \text{ and } c \le y \le d\}$$

- the *Jacobian* of a transformation $T : \Omega \to \Omega'$

Theorems

Fill in the blanks to complete each of the following theorem statements:

- *Fubini's theorem:* Let $a < b$ and $c < d$ be real numbers, and let $\mathcal{R}$ be the rectangle defined by

$$\mathcal{R} = \{(x, y) \mid a \le x \le b \text{ and } c \le y \le d\}.$$

If $f(x, y)$ is continuous on $\mathcal{R}$, then $\iint_{\mathcal{R}} f(x, y)\, dA =$ ________.

- If $f(x, y)$ is an integrable function and $c \in \mathbb{R}$, then $\iint_\Omega cf(x, y)\, dA =$ ________.
- If $f(x, y)$ and $g(x, y)$ are integrable functions, then $\iint_\Omega (f(x, y) + g(x, y))\, dA =$ ________.
- Let $f(x, y)$ be an integrable function on the general region Ω. If Ω_1 and Ω_2 are general regions that are subsets of ________, and if $\Omega = \Omega_1 \cup \Omega_2$, then $\iint_\Omega f(x, y)\, dA =$ ________.

Notation and Algebraic Rules

Notation: Give the meanings of each of the following mathematical expressions:

- $\displaystyle\sum_{j=1}^{m} \sum_{k=1}^{n} a_{jk}$
- $\displaystyle\sum_{i=1}^{l} \sum_{j=1}^{m} \sum_{k=1}^{n} a_{ijk}$
- $\displaystyle\iint_{\mathcal{R}} f(x, y)\, dA$
- $\displaystyle\int_a^b \int_{g_1(x)}^{g_2(x)} f(x, y)\, dy\, dx$
- $\displaystyle\int_c^d \int_{h_1(y)}^{h_2(y)} f(x, y)\, dx\, dy$
- $\displaystyle\iiint_{\mathcal{R}} f(x, y, z)\, dV$
- $\displaystyle\int_a^b \int_{h_1(x)}^{h_2(x)} \int_{g_1(x,y)}^{g_2(x,y)} f(x, y, z)\, dz\, dy\, dx.$
- $\dfrac{\partial(x, y)}{\partial(u, v)}$
- $\dfrac{\partial(x, y, z)}{\partial(u, v, w)}$

Converting between coordinate systems: Provide the conversion formulas between each of the coordinate systems that follow. Include a sketch indicating the reason for each formula.

- Given rectangular coordinates x and y, the polar coordinates are $r =$ ________ and $\tan\theta =$ ________.

- Given polar coordinates r and θ, the rectangular coordinates are $x =$ ________ and $y =$ ________.
- Given rectangular coordinates x, y, and z, the cylindrical coordinates are $r =$ ________, $\tan\theta =$ ________, and $z =$ ________.
- Given cylindrical coordinates r, θ, and z, the rectangular coordinates are $x =$ ________, $y =$ ________, and $z =$ ________.
- Given rectangular coordinates x, y, and z, the spherical coordinates are $\rho =$ ________, $\theta =$ ________, and $\phi =$ ________.
- Given spherical coordinates ρ, $\tan\theta$, and $\cos\phi$, the spherical coordinates are $x =$ ________, $y =$ ________, and $z =$ ________.
- Given cylindrical coordinates r, θ, and z, the spherical coordinates are $\rho =$ ________, $\theta =$ ________, and $\tan\phi =$ ________.
- Given spherical coordinates ρ, θ, and ϕ, the cylindrical coordinates are $r =$ ________, $\theta =$ ________, and $z =$ ________.

First and Second Moments: Let $\rho(x, y)$ be a continuous density function for a lamina Ω in the xy-plane. Provide iterated integrals or iterated integral expressions that could be used to compute each of the following quantities:

- the mass of Ω
- the first moment of the mass in Ω about the y-axis
- the first moment of the mass in Ω about the x-axis
- the center of mass of Ω
- the moment of inertia of the mass in Ω about the y-axis
- the moment of inertia of the mass in Ω about the x-axis
- the moment of inertia of the mass in Ω about the origin
- the radius of gyration of the mass in Ω about the y-axis
- the radius of gyration of the mass in Ω about the x-axis
- the radius of gyration of the mass in Ω about the origin

Skill Certification: Integrating Functions of Two and Three Variables

Using the definition to evaluate a double integral: Evaluate the given double integrals as a limit of a Riemann sum. For each integral, let $\mathcal{R} = \{(x, y) \mid 0 \le x \le 2 \text{ and } 1 \le y \le 4\}$.

1. $\displaystyle\iint_{\mathcal{R}} (x + 2y)\, dA$

2. $\displaystyle\iint_{\mathcal{R}} (xy^2)\, dA$

Evaluating a double integral as an iterated integral: Use Fubini's theorem to evaluate the given double integrals. For each integral, show that you obtain the same result when you integrate using both possible orders of integration when $\mathcal{R} = \{(x, y) \mid 0 \le x \le 2 \text{ and } 1 \le y \le 4\}$.

3. $\displaystyle\iint_{\mathcal{R}} (x + 2y)\, dA$

4. $\displaystyle\iint_{\mathcal{R}} (xy^2)\, dA$

Evaluating iterated integrals: Sketch the region determined by the limits of the given iterated integrals, and then evaluate the integrals.

5. $\displaystyle\int_0^1 \int_{x^2}^{\sqrt{x}} x^2y^3\, dy\, dx$

6. $\displaystyle\int_0^2 \int_0^{\sqrt{4-x^2}} y^3\, dy\, dx$

7. $\displaystyle\int_0^1 \int_{-\sqrt{1-y^2}}^{\sqrt{1-y^2}} \frac{x}{y+1}\, dx\, dy$

8. $\displaystyle\int_1^2 \int_0^{1/x} \sqrt{xy}\, dy\, dx$

9. $\displaystyle\int_1^4 \int_0^{1/y} \frac{x}{y}\, dx\, dy$

10. $\displaystyle\int_0^2 \int_0^y x\sqrt{y^2 - x^2}\, dx\, dy$

Reversing the order of integration: Sketch the region determined by the limits of the given iterated integrals, and then evaluate the integrals by reversing the order of integration.

11. $\displaystyle\int_0^4 \int_{\sqrt{x}}^2 y\cos x\, dy\, dx$

12. $\displaystyle\int_0^{\sqrt{\pi}} \int_x^{\sqrt{\pi}} \sin y^2\, dy\, dx$

13. $\displaystyle\int_0^9 \int_{\sqrt{y}}^3 \frac{1}{1+x^3}\, dx\, dy$

14. $\displaystyle\int_0^{16} \int_{\sqrt[4]{x}}^2 \frac{1}{1+y^5}\, dy\, dx$

15. $\displaystyle\int_0^4 \int_y^4 e^{y/x}\, dx\, dy$

16. $\displaystyle\int_0^1 \int_{\sqrt{x}}^1 e^{x/y^2}\, dy\, dx$

Using polar coordinates to evaluate iterated integrals: Sketch the region determined by the limits of the given iterated integrals, and then evaluate the integrals.

17. $\displaystyle\int_0^{\pi/2} \int_0^3 r^2\, dr\, d\theta$

18. $\displaystyle\int_{\pi/2}^{3\pi/2} \int_1^4 r^{3/2}\, dr\, d\theta$

19. $\displaystyle\int_0^{\pi/2} \int_0^{2\sin\theta} r^3\, dr\, d\theta$

20. $\displaystyle\int_0^{2\pi} \int_0^{1+\sin\theta} r\, dr\, d\theta$

Using polar coordinates to evaluate iterated integrals: Evaluate the given iterated integrals by converting them to polar coordinates. Include a sketch of the region.

21. $\displaystyle\int_0^2 \int_0^{\sqrt{4-y^2}} e^{x^2+y^2}\, dx\, dy$

22. $\displaystyle\int_0^4 \int_0^{\sqrt{4y-y^2}} \frac{1}{\sqrt{x^2+y^2}}\, dx\, dy$

23. $\displaystyle\int_{-3}^3 \int_{-\sqrt{9-x^2}}^{\sqrt{9-x^2}} \frac{x+2y}{x^2+y^2}\, dy\, dx$

24. $\displaystyle\int_{-5}^0 \int_{-\sqrt{25-x^2}}^0 \frac{3}{(4+x^2+y^2)^3}\, dy\, dx$

Evaluating triple integrals: Each of the triple integrals that follows represents the volume of a solid. Sketch the solid and evaluate the integral.

25. $\displaystyle\int_0^2 \int_1^4 \int_{-2}^3 dx\, dy\, dz$

26. $\displaystyle\int_1^5 \int_{-2}^0 \int_{-1}^3 dz\, dy\, dx$

27. $\displaystyle\int_0^4 \int_0^{3-(3/4)x} \int_0^{2-(1/2)x-(2/3)y} dz\, dy\, dx$

28. $\displaystyle\int_0^3 \int_0^{2-2y/3} \int_0^{4-(4/3)y-2z} dx\, dz\, dy$

29. $\displaystyle\int_{-2}^{2}\int_{0}^{\sqrt{4-x^2}}\int_{0}^{y} dz\,dy\,dx$

30. $\displaystyle\int_{0}^{2\pi}\int_{0}^{2}\int_{0}^{5} r\,dz\,dr\,d\theta$

31. $\displaystyle\int_{0}^{2\pi}\int_{0}^{3}\int_{0}^{\sqrt{9-r^2}} r\,dz\,dr\,d\theta$

32. $\displaystyle\int_{0}^{2\pi}\int_{0}^{4}\int_{r}^{4} r\,dz\,dr\,d\theta$

33. $\displaystyle\int_{0}^{2\pi}\int_{0}^{4}\int_{0}^{4-r} r\,dz\,dr\,d\theta$

34. $\displaystyle\int_{0}^{\pi}\int_{0}^{2}\int_{0}^{r\sin\theta} r\,dz\,dr\,d\theta$

35. $\displaystyle\int_{0}^{\pi}\int_{0}^{2\pi}\int_{0}^{5} \rho^2\sin\phi\,d\rho\,d\theta\,d\phi$

36. $\displaystyle\int_{0}^{\pi/4}\int_{0}^{2\pi}\int_{0}^{4\sec\theta} \rho^2\sin\phi\,d\rho\,d\theta\,d\phi$

Capstone Problems

A. Show that

$$\int_0^1\int_0^1 \frac{x-y}{(x+y)^3}\,dy\,dx \neq \int_0^1\int_0^1 \frac{x-y}{(x+y)^3}\,dx\,dy.$$

Explain why this does not contradict Fubini's theorem.

B. A hole with radius $\frac{R}{2}$ is drilled though the center of a sphere with radius R, as shown in the following figure. Find the volume of the resulting solid. *(Hint: You may assume that the equation of the sphere is $x^2+y^2+z^2=R^2$ and the equation of the cylinder through the sphere is $x^2+y^2=(R/2)^2$.)*

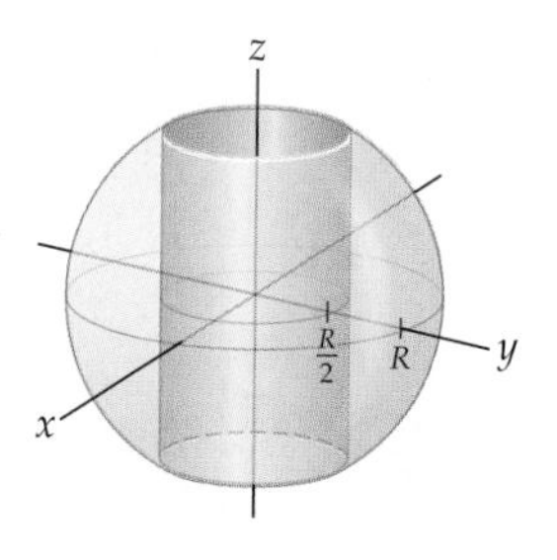

C. Find the center of mass and moment of inertia about the z-axis for the solid from Problem B, assuming that it has a uniform density.

D. Evaluate the integral

$$\iint_T \ln\left(\frac{y-x}{x+y}\right)\,dA,$$

where T is the first-quadrant triangular region with vertices $(2, 3)$, $(2, 4)$, and $(3, 4)$.

CHAPTER 14

Vector Analysis

14.1 Vector Fields

Calculus and Vector Analysis
Vector Fields
Conservative Vector Fields
Examples and Explorations

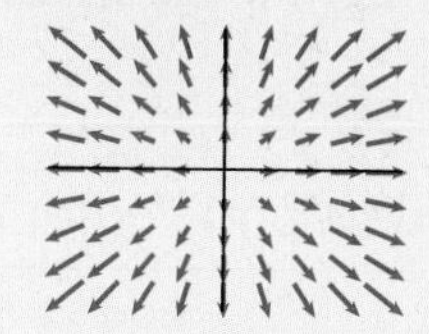

14.2 Line Integrals

Integrals Along Curves in Space
Line Integrals of Multivariate Functions
Line Integrals of Vector Fields
The Fundamental Theorem of Line Integrals
Examples and Explorations

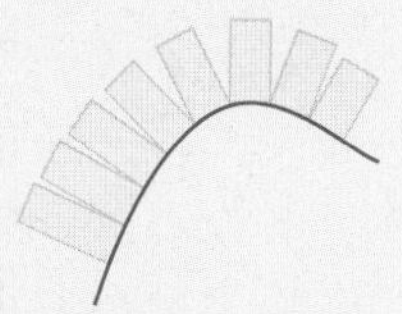

14.3 Surfaces and Surface Integrals

Parametrized Surfaces
Surface Area and Surface Integrals of Multivariate Functions
The Flux of a Vector Field Across a Surface
Examples and Explorations

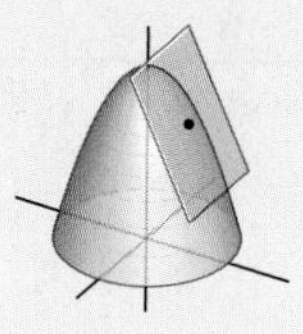

14.4 Green's Theorem

The Fundamental Theorem of Calculus and Green's Theorem
The Del Operator, the Divergence, and the Curl
Green's Theorem
Examples and Explorations

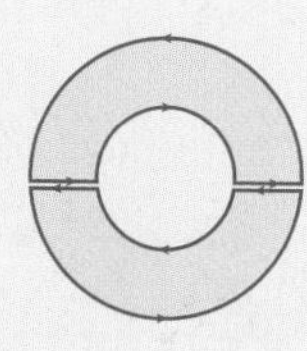

14.5 Stokes' Theorem

A Generalization of Green's Theorem
Stokes' Theorem
Examples and Explorations

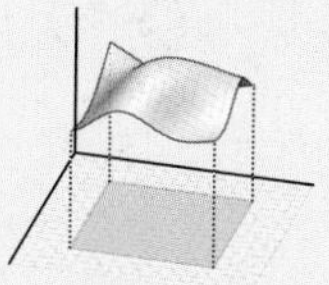

14.6 The Divergence Theorem

The Divergence Theorem
A Summary of Theorems About Vector Analysis
Examples and Explorations

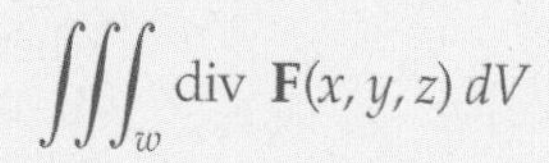

Chapter Review, Self-Test, and Capstones

14.1 VECTOR FIELDS

- An overview of vector analysis
- Assigning a vector to each point in Euclidean space
- Conservative vector fields and their potential functions

Calculus and Vector Analysis

In this final chapter, we unify and generalize our understanding of integration and the Fundamental Theorem of Calculus. We will generalize the original notion of integration that was introduced in Chapter 4 to enable us to integrate over curves other than the x-axis, extend the integration results of Chapter 13 to allow us to perform double integration over regions other than regions in the xy-plane, and see new multivariate extensions of the Fundamental Theorem of Calculus. The basic ideas will be the same, but the context in which we implement them will be more subtle and more varied. Now would be an excellent time to review what you have studied about the integration of functions of one variable, vector-valued functions, and multivariate functions.

Here are two examples of the types of questions we will address in this chapter. First, consider a fence whose base is a straight line, like the x-axis, and another fence whose base is along an undulating hill:

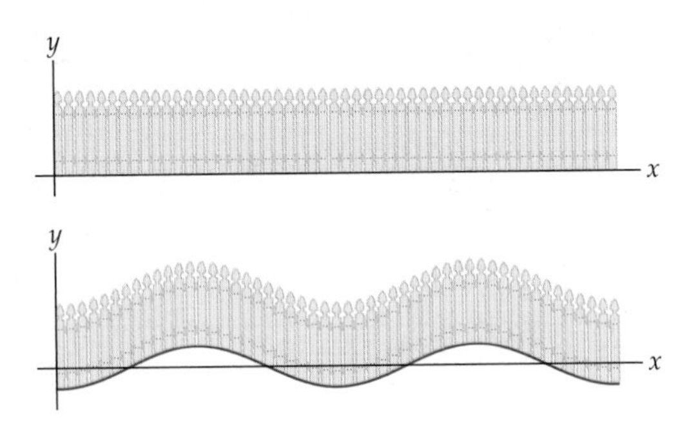

Both fences clearly have well-defined areas. To find the area of the first, we integrate

$$A = \int h(x)\, dx,$$

where $h(x)$ is a function that tells us the height of the fence above the x-axis. The area of the second fence is not quite this integral, since the curve of its base will distort the fence. Still, the fence height and its base appear smooth, so its area ought to be expressible as a one-variable integration of something, where we integrate along the green curve in the figure. The resulting integral will combine ideas we used for computing arc lengths and areas in previous chapters. The resulting integral will have the form

$$A = \int h(t)\varphi(t)\, dt,$$

where $\varphi(t)$ is a factor that accounts for the bending of the base.

Similarly, in Section 6.4 we discussed work W as the product of force F and distance d, or $W = Fd$. From this equation, we saw how to choose an appropriate integral to calculate the work done in pumping water out the top of a tank, opposite the downward force of gravity. If we wanted to know the work required to move an object along a more complicated path, and opposed to a variable force, we would expect that this, too, could be expressed as an integral, perhaps now with multiple inputs.

In Section 14.2 we will see how to solve both of these problems. For now, we introduce a new mathematical object that is necessary for the solution.

Vector Fields

One immediate question that confronts us when considering how to generalize the Fundamental Theorem of Calculus to several dimensions is the issue of what, exactly, we are trying to integrate. As we saw in the preceding example we might be interested in finding the area of a fence whose base is curved; we might also be interested in the work done by that fence as it stays upright in a windstorm, when whirling wind exerts forces of different strengths and in different directions on the fence at different points. This idea of varying forces associated with points in space or points in the plane motivates the notion of a *vector field.*

A vector field is a function that assigns a vector to each point in its domain. Thus, a vector field in the plane has inputs that are points in the xy-plane and outputs that are vectors in the xy-plane; a vector field in $\mathbb{R}^3$ has inputs that are points in $\mathbb{R}^3$ and outputs that are vectors in $\mathbb{R}^3$.

DEFINITION 14.1

Vector Field

A ***vector field in*** $\mathbb{R}^2$ is a function $\mathbf{F}(x, y)$ with domain $D \subseteq \mathbb{R}^2$ and whose outputs are vectors in $\mathbb{R}^2$ of the form

$$\mathbf{F}(x, y) = \langle F_1(x, y), F_2(x, y)\rangle$$

for each point (x, y) in D.

Similarly, a ***vector field in*** $\mathbb{R}^3$ is a function $\mathbf{G}(x, y, z)$ with domain $D \subseteq \mathbb{R}^3$ and whose outputs are vectors in $\mathbb{R}^3$ of the form

$$\mathbf{G}(x, y, z) = \langle G_1(x, y, z), G_2(x, y, z), G_3(x, y, z)\rangle$$

for each point (x, y, z) in D.

In this chapter, the only vector fields we are interested in are those whose domains are well-behaved enough to support vector analysis. Thus, for the remainder of our discussion of vector fields, we will always suppose that every vector field has a domain D that is open, connected, and simply connected. This means that D is open in the sense of Chapter 12; that, for any two points P and Q in D, there is a path from P to Q that lies in D; and that any loop in D can be smoothly contracted to a point in D. In general, a two-dimensional region is simply connected if it does not contain any holes.

A region in $\mathbb{R}^2$ *that is simply connected*

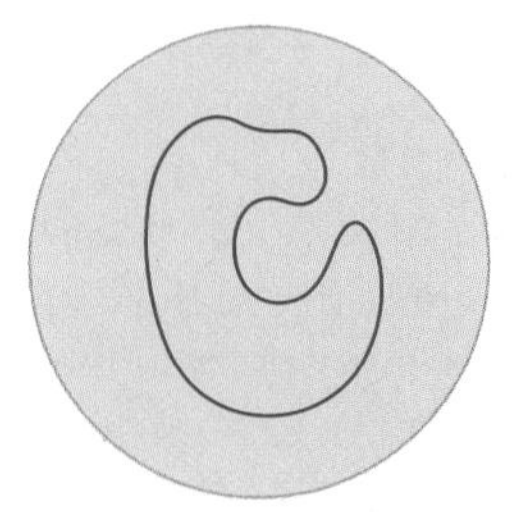

A region in $\mathbb{R}^2$ *that is not simply connected*

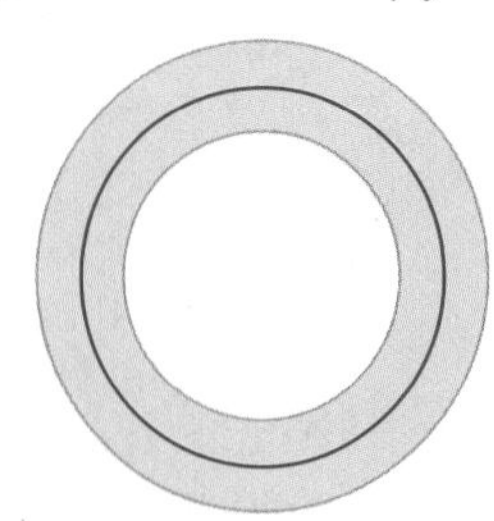

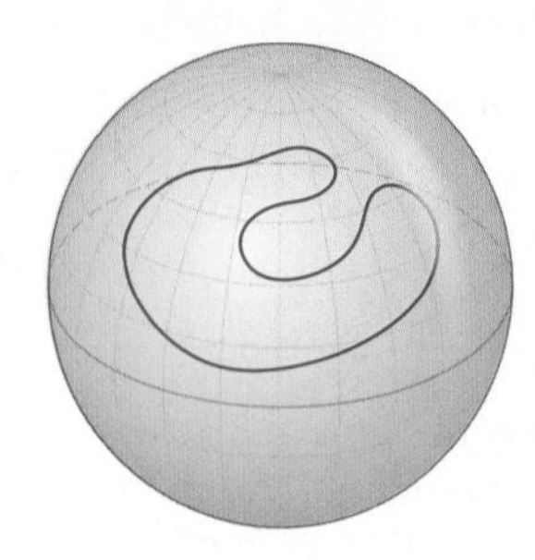

A region in $\mathbb{R}^3$ that is simply connected

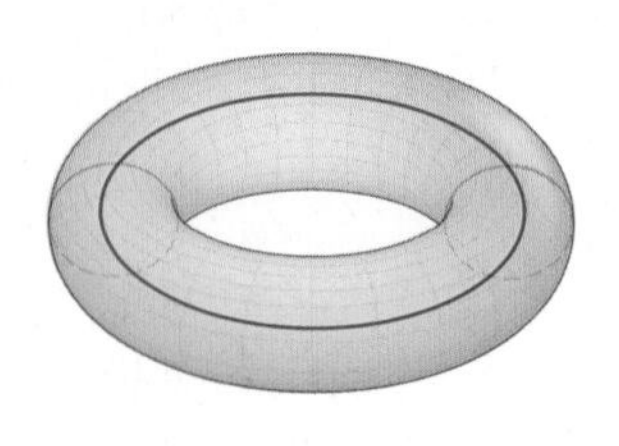

A region in $\mathbb{R}^3$ that is not simply connected

A three-dimensional region may have holes and yet still be simply connected, as shown in the following sequence of figures:

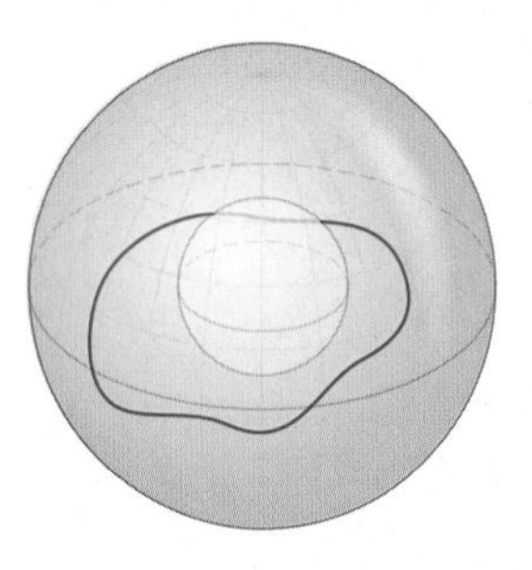

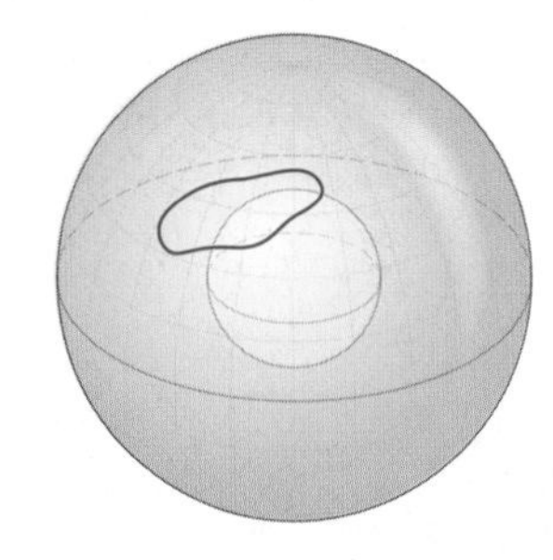

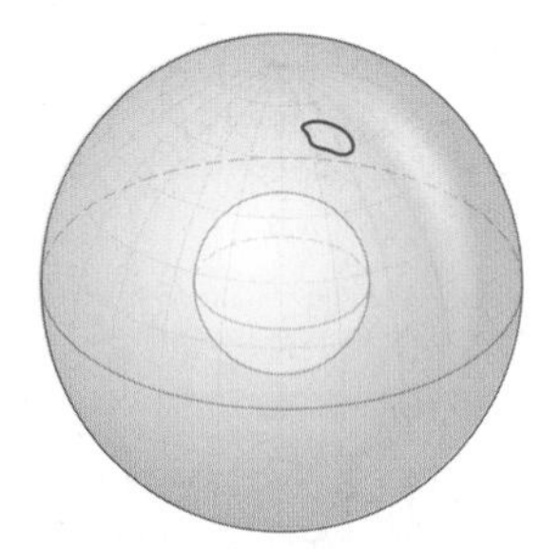

We will also usually assume that the vector field in question is sufficiently smooth so that the component functions are continuous and have continuous first partial derivatives.

The notion of a vector field in $\mathbb{R}^n$ may of course be defined analogously for any natural number n, but we will not need such vector fields in this text.

Following are pictures of the vector fields $\mathbf{F}(x, y) = 2\mathbf{i} + 0\mathbf{j}$ and $\mathbf{G}(x, y) = \mathbf{i} - \mathbf{j}$, where at each point (x, y) we draw the vector $\langle F_1(x, y), F_2(x, y)\rangle$:

$\mathbf{F}(x, y) = 2\mathbf{i} + 0\mathbf{j}$

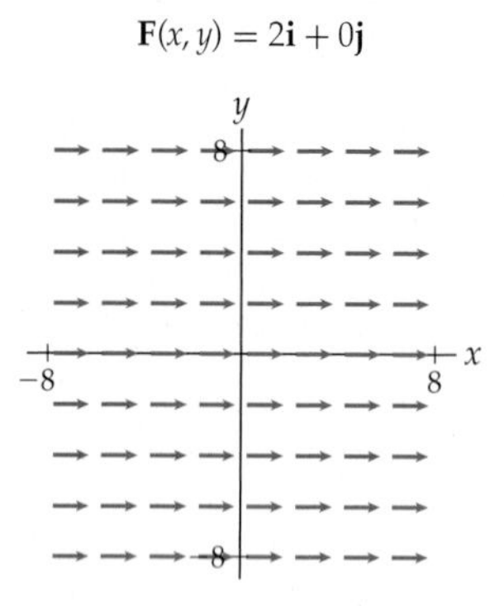

$\mathbf{G}(x, y) = \mathbf{i} - \mathbf{j}$

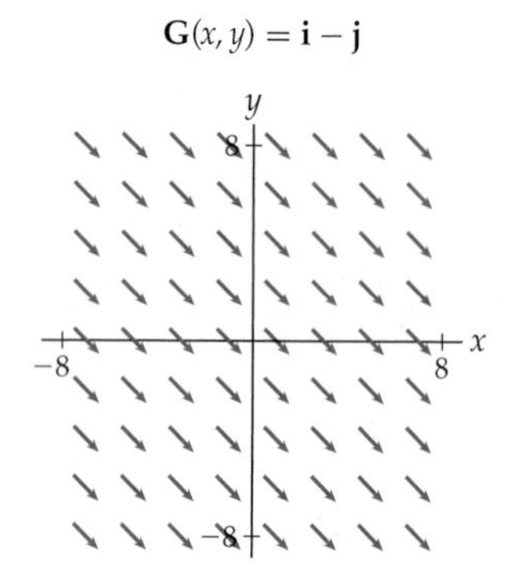

Note that these diagrams show only some of the vectors. A true representation would look like a solid block of color, because every point in the plane would have a vector associated with it. The following vector fields $\mathbf{F}(x, y) = \langle x, y\rangle$ and $\mathbf{G}(x, y) = \langle \sin x, \sin y\rangle$ are slightly more interesting:

$\mathbf{F}(x, y) = \langle x, y \rangle$

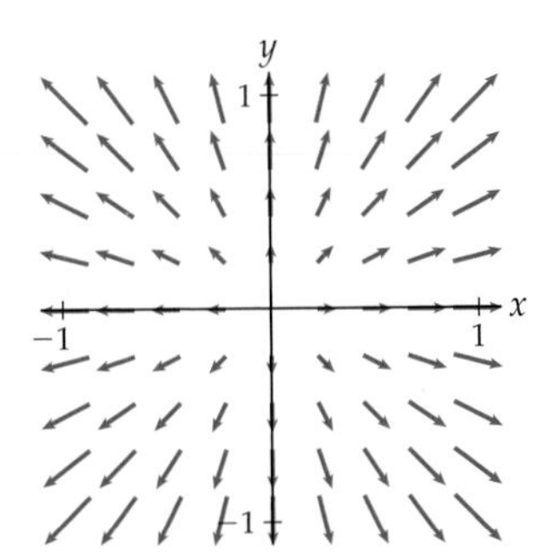

$\mathbf{G}(x, y) = \langle \sin x, \sin y \rangle$

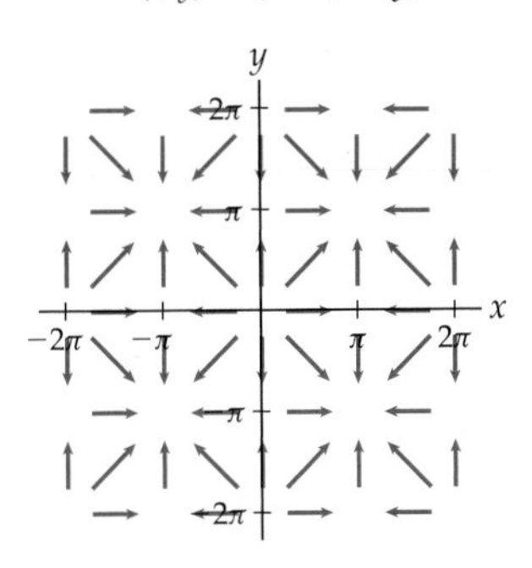

Note that in the vector field $\mathbf{F}$ the magnitudes of the vectors in the codomain increase as we move away from the origin and have constant magnitude on concentric circles about the origin. For example, $\|\mathbf{F}(x, y)\| = 1$ if and only if (x, y) lies on the unit circle. In addition, all vectors emanating from points near the origin point away from the origin. In such a situation the origin is called a ***source***. In all of our vector fields, the vectors are drawn so that their lengths are representative of their magnitudes.

In the vector field $\mathbf{G}$, all vectors emanating from points near the origin again point away from the origin, so the origin is a source. In addition because of the periodicity of the sine function, all coordinate pairs of the form $(2k\pi, 2n\pi)$, where k and n are both integers, are also sources. Note, however, that all of the vectors that emanate from points surrounding (π, π) point towards (π, π). Such a point is called a ***sink***. Again by the periodicity of the sine, all points of the form $((2k+1)\pi, (2n+1)\pi)$, where k and n are both integers, are also sinks.

Conservative Vector Fields

Some vector fields, such as $\mathbf{F}(x, y) = \langle x, y \rangle$, can be written as the gradient of a function. For example, we might have

$$\mathbf{F}(x, y) = \nabla f(x, y), \quad \text{where} \quad f(x, y) = \frac{x^2}{2} + \frac{y^2}{2} \quad \text{and} \quad \nabla f = \frac{\partial f}{\partial x}\mathbf{i} + \frac{\partial f}{\partial y}\mathbf{j}$$

This is one of the infinitely many possible choices for the function $f(x, y)$. It is also true that

$$\mathbf{F}(x, y) = \nabla g(x, y), \quad \text{where} \quad g(x, y) = \frac{x^2}{2} + \frac{y^2}{2} + \alpha,$$

for any choice of the constant α.

If a vector field can be expressed as a gradient, then it has many mathematically desirable attributes, including being easy to integrate along curves, a property we will see in the next section. Such fields are known as *conservative vector fields*.

DEFINITION 14.2 Conservative Vector Field

A ***conservative vector field*** $\mathbf{F}$ is a vector field that can be written as the gradient of some function f. That is,

$$\mathbf{F}(x, y) = \nabla f(x, y) = \frac{\partial f}{\partial x}\mathbf{i} + \frac{\partial f}{\partial y}\mathbf{j}$$

if $f(x, y)$ is a function of two variables, or

$$\mathbf{F}(x, y, z) = \nabla f(x, y, z) = \frac{\partial f}{\partial x}\mathbf{i} + \frac{\partial f}{\partial y}\mathbf{j} + \frac{\partial f}{\partial z}\mathbf{k}$$

if $f(x, y, z)$ is a function of three variables.

In either case, any function f whose gradient is equal to $\mathbf{F}$ is called a ***potential function*** for $\mathbf{F}$.

Recall that if a function $g(x, y)$ has continuous first and second partial derivatives, then

$$\frac{\partial^2 g}{\partial y \partial x} = \frac{\partial^2 g}{\partial x \partial y}.$$

So, to verify that $\mathbf{F}(x, y) = \langle F_1(x, y), F_2(x, y) \rangle$ is a conservative vector field with a potential function $f(x, y)$ (with continuous second partial derivatives), it is sufficient to check that

$$\frac{\partial F_1}{\partial y} = \frac{\partial F_2}{\partial x},$$

since equality of these first partial derivatives of the components of $\mathbf{F}$ means that the mixed second-order partial derivatives of f are equal. In fact, a vector field

$$\mathbf{F}(x, y) = \langle F_1(x, y), F_2(x, y) \rangle,$$

whose first partial derivatives $\frac{\partial F_1}{\partial y}$ and $\frac{\partial F_2}{\partial x}$ are both continuous on an open subset of $\mathbb{R}^2$, is conservative if and only if $\frac{\partial F_1}{\partial y} = \frac{\partial F_2}{\partial x}$.

The situation is similar for vector fields in $\mathbb{R}^3$, except there are more mixed partial derivatives to check. Note that potential functions are not unique. As we saw earlier, if $f(x, y)$ is a potential function for a conservative vector field $\mathbf{F}(x, y)$, then so is $f(x, y) + \alpha$ for any constant α.

In some cases it will be enough to know that a vector field is conservative; in others, it will be important to find a potential function for the field. By thinking through the definitions of conservative vector fields and potential functions, we can see that the question is one of recovering a function from its partial derivatives. That is easy to do by integrating, but there is a subtle point to consider.

For example, let

$$\mathbf{G}(x, y, z) = 6xz\mathbf{i} + \cos y\mathbf{j} + 3x^2\mathbf{k}.$$

By checking all six mixed second partial derivatives, we find that $\mathbf{G}$ is indeed conservative. Now, we want a potential function $g(x, y, z)$ whose gradient is $\mathbf{G}$. Any g that satisfies

$$\frac{\partial g}{\partial x} = 6xz, \qquad \frac{\partial g}{\partial y} = \cos y, \qquad \frac{\partial g}{\partial z} = 3x^2$$

will work. We let $G_1(x, y, z) = 6xz$, $G_2(x, y, z) = \cos y$, and $G_3(x, y, z) = 3x^2$ and begin by integrating G_1 with respect to x:

$$\int G_1(x, y, z)\, dx = \int 6xz\, dx = 3x^2 z + \alpha = g(x, y, z).$$

But the gradient of this function is not $\mathbf{G}(x, y, z)$. Here we have

$$\nabla g(x, y, z) = 6xz\mathbf{i} + 0\mathbf{j} + 3x^2\mathbf{k}.$$

We are missing a term that will give the correct partial derivative for y. Since the terms of g that do not involve x are constants with respect to differentiation by x, we can fix the problem by integrating all the terms of G_2 that do not involve x, integrating the terms of G_3 that involve neither x nor y, and adding the results to the function g that we have already found. This gives

$$\int G_2(x, y, z)\, dy = \int \cos y\, dy = \sin y + \alpha.$$

Since none of the terms of G_3 involve only z, when we add $\sin y + \alpha$ to our previous function and set the constant term to zero, we obtain

$$g(x, y, z) = 3x^2 z + \sin y.$$

You may check that

$$\nabla g(x, y, z) = \langle 6xz, \cos y, 3x^2 \rangle = \mathbf{G}(x, y, z).$$

More generally, to find the potential functions $f(x, y)$ and $g(x, y, z)$ for the respective conservative vector fields

$$\mathbf{F}(x, y) = \langle F_1(x, y), F_2(x, y) \rangle \quad \text{and} \quad \mathbf{G}(x, y, z) = \langle G_1(x, y, z), G_2(x, y, z), G_3(x, y, z) \rangle,$$

we construct

$$f(x, y) = \int F_1(x, y)\, dx + B \quad \text{and} \quad g(x, y, z) = \int G_1(x, y, z)\, dx + B + C,$$

where B is the integral with respect to y of the terms in $F_2(x, y)$ or $G_2(x, y, z)$ that have no x factor and C is the integral with respect to z of the terms in $G_3(x, y, z)$ that have no x or y factor.

It is also possible, and sometimes more convenient, to perform these integrations in other orders. Consider the complexities of the component functions when you are deciding.

Examples and Explorations

EXAMPLE 1

Drawing vector fields in the plane

Sketch or use graphing software to create plots of the following vector fields in $\mathbb{R}^2$:

(a) $\mathbf{F}(x, y) = \langle x, -y \rangle$

(b) $\mathbf{G}(x, y) = \dfrac{y}{x^2 + y^2}\mathbf{i} + \dfrac{x}{x^2 + y^2}\mathbf{j}$

SOLUTION

Following are the two vector fields:

$\mathbf{F}(x, y) = \langle x, -y \rangle$

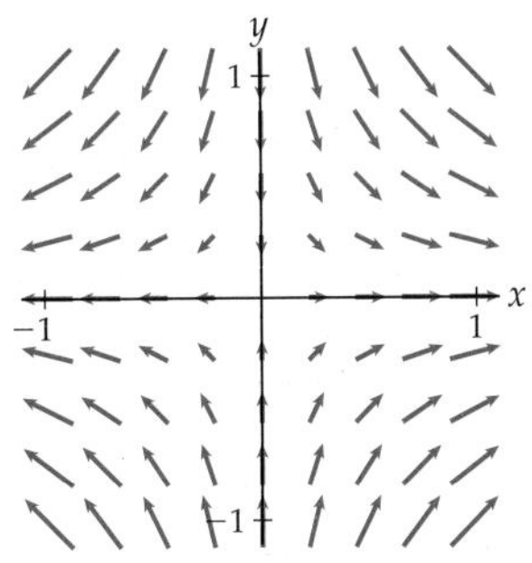

$\mathbf{G}(x, y) = \dfrac{y}{x^2 + y^2}\mathbf{i} + \dfrac{x}{x^2 + y^2}\mathbf{j}$

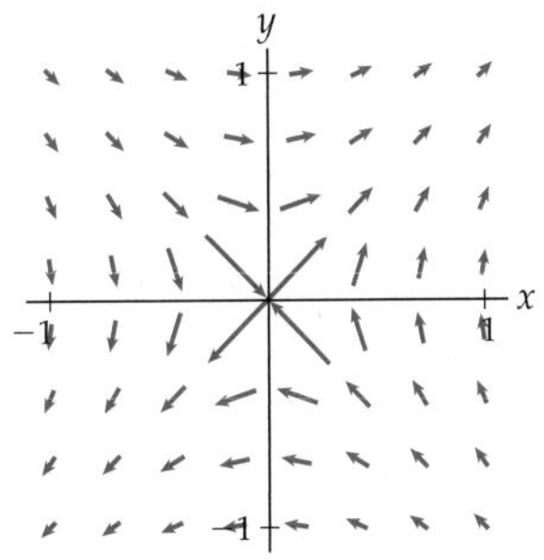

Each vector in the field is computed using the component functions. For example, $\mathbf{F}(1, 1/2) = \langle 1, -1/2 \rangle$. Our graphs are drawn with software, but they may be drawn by hand. □

EXAMPLE 2 **Drawing vector fields in $\mathbb{R}^3$**

Sketch or use graphing software to create plots of the given vector fields in $\mathbb{R}^3$. Identify any sources or sinks in each vector field.

(a) $\mathbf{F}(x, y, z) = \langle x, y, z \rangle$ **(b)** $\mathbf{G}(x, y, z) = y\mathbf{i} - x\mathbf{j} + z\mathbf{k}$

SOLUTION

Three-dimensional vector fields can be a little difficult to understand, even when we use graphing software. As an aid, it may be useful to visualize the analogous vector field one dimension lower.

(a) The vector field $\mathbf{F}(x, y) = \langle x, y \rangle$ that we drew earlier in the section is a two-dimensional analog for the field shown next. Each vector points away from the origin, their magnitudes increase as you move away from the origin, and the magnitudes are constant on spheres centered at the origin. Note that to determine any particular vector in the field, we evaluate the function at a point. For example, $\mathbf{F}(1, 2, 3) = \langle 1, 2, 3 \rangle$.

$\mathbf{F}(x, y, z) = \langle x, y, z \rangle$

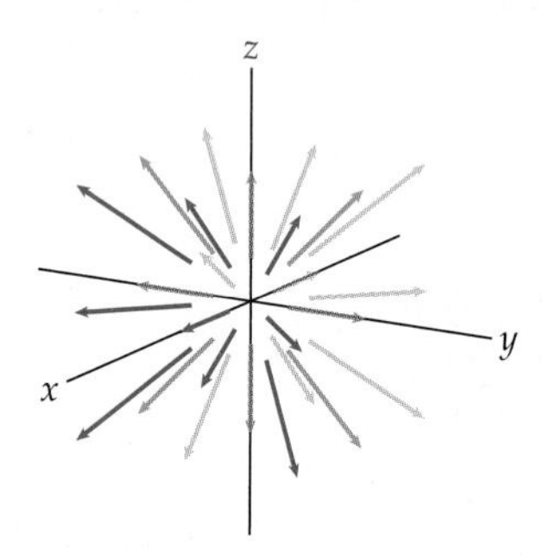

Note that the origin is a source in this vector field, since all vectors emanating from points near the origin point away from it.

(b) We first sketch the two-dimensional vector field $\mathbf{G}(x, y) = y\mathbf{i} - x\mathbf{j}$ shown next at the left. We see that the vectors are tangent to concentric circles centered at the origin and that their magnitudes increase as you move away from the origin. In the three-dimensional vector field $\mathbf{G}(x, y, z) = y\,\mathbf{i} - x\,\mathbf{j} + z\,\mathbf{k}$ shown at the right, the $\mathbf{k}$-components of the vectors increase as you move away from the xy-plane as well.

$\mathbf{G}(x, y) = y\mathbf{i} - x\mathbf{j}$

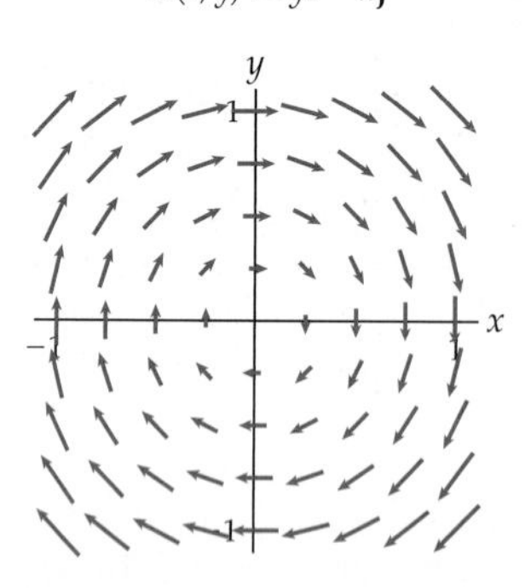

$\mathbf{G}(x, y, z) = y\mathbf{i} - x\mathbf{j} + z\mathbf{k}$

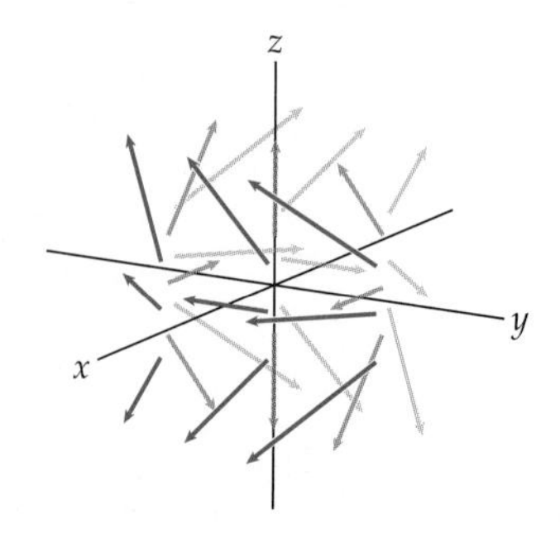

This vector field has no sources or sinks. □

EXAMPLE 3

Determining whether a vector field is conservative

Determine whether the vector fields from Example 1 are conservative.

(a) $\mathbf{F}(x, y) = \langle x, -y \rangle$ **(b)** $\mathbf{G}(x, y) = \frac{y}{x^2+y^2}\mathbf{i} + \frac{x}{x^2+y^2}\mathbf{j}$

SOLUTION

A vector field $\mathbf{F}(x, y) = \langle F_1(x, y), F_2(x, y) \rangle$ is conservative if and only if

$$\frac{\partial F_1}{\partial y} = \frac{\partial F_2}{\partial x}$$

when $\frac{\partial F_1}{\partial y}$ and $\frac{\partial F_2}{\partial x}$ are both continuous on some open subset of $\mathbb{R}^2$.

(a) For the vector field $\mathbf{F}(x, y) = \langle x, -y \rangle$,

$$\frac{\partial F_1}{\partial y} = \frac{\partial}{\partial y}(x) = 0 \quad \text{and} \quad \frac{\partial F_2}{\partial x} = \frac{\partial}{\partial x}(-y) = 0.$$

Since these partial derivatives are the same, $\mathbf{F}(x, y)$ is indeed conservative.

(b) For the vector field $\mathbf{G}(x, y) = \frac{y}{x^2+y^2}\mathbf{i} + \frac{x}{x^2+y^2}\mathbf{j}$,

$$\frac{\partial G_1}{\partial y} = \frac{\partial}{\partial y}\left(\frac{y}{x^2+y^2}\right) = \frac{x^2-y^2}{(x^2+y^2)^2} \quad \text{and} \quad \frac{\partial G_2}{\partial x} = \frac{\partial}{\partial x}\left(\frac{x}{x^2+y^2}\right) = \frac{y^2-x^2}{(x^2+y^2)^2}.$$

Since $\frac{\partial G_1}{\partial y} \neq \frac{\partial G_2}{\partial x}$ and these partial derivatives are continuous on every open subset of $\mathbb{R}^2$ not containing the origin, $\mathbf{G}(x, y)$ is not conservative. □

EXAMPLE 4

Finding a potential function for a vector field in $\mathbb{R}^2$

Find a potential function for the vector field $\mathbf{F}(x, y) = \langle x, -y \rangle$ from Example 1.

SOLUTION

Since

$$\mathbf{F}(x, y) = \langle x, -y \rangle,$$

$$f(x, y) = \int x\, dx + B = \frac{x^2}{2} + \alpha + B,$$

where α is an arbitrary constant and B is the integral with respect to y of the terms in $F_2(x, y)$ in which the factor x does not appear. In this case, that is all of $F_2(x, y)$, so

$$B = \int (-y)\, dy = \frac{-y^2}{2} + \beta,$$

where β is an arbitrary constant. Setting the constants equal to zero since they do not affect the gradient of $f(x, y)$, we have

$$f(x, y) = \frac{x^2}{2} - \frac{y^2}{2}.$$

□

We can verify this solution by computing ∇f directly:

$$\nabla f(x, y) = f_x(x, y)\mathbf{i} + f_y(x, y)\mathbf{j} = x\mathbf{i} - y\mathbf{j} = \mathbf{F}(x, y).$$

EXAMPLE 5 **Finding a potential function for a vector field in $\mathbb{R}^3$**

Find a potential function for

$$\mathbf{G}(x, y, z) = \left(\frac{y}{x^2+1} + z\right)\mathbf{i} + \tan^{-1} x\mathbf{j} + (x + \cos z)\,\mathbf{k}.$$

SOLUTION

We start by integrating $G_1(x, y, z)$. We have

$$g(x, y, z) = \int \left(\frac{y}{x^2+1} + z\right) dx + B + C = y\tan^{-1} x + xz + \alpha + B + C,$$

where α is an arbitrary constant, B is the integral with respect to y of the terms in $G_2(x, y, z)$ in which the factor x does not appear, and C is the integral with respect to z of the terms in $G_3(x, y, z)$ in which neither the factor x nor the factor y appears. In this case $B = 0$, because the only term in $G_2(x, y, z)$ is $\tan^{-1} x$ and this part of the function $g(x, y, z)$ is already recovered by the integral of $G_1(x, y, z)$. For C, however, there is a term with no x- or y-component, so

$$C = \int \cos z\, dz = \sin z + \beta,$$

where β is an arbitrary constant. Combining these results and setting the arbitrary constants to zero, we see that

$$g(x, y, z) = y\tan^{-1} x + xz + \sin z$$

is a potential function for $\mathbf{G}(x, y, z)$. □

CHECKING THE ANSWER

We can verify this solution by computing $\nabla g(x, y, z)$ directly:

$$\nabla g(x, y, z) = \langle g_x, g_y, g_z\rangle = \left(\frac{y}{x^2+1} + z\right)\mathbf{i} + \tan^{-1} x\,\mathbf{j} + (x + \cos z)\,\mathbf{k} = \mathbf{G}(x, y, z).$$

? TEST YOUR UNDERSTANDING

- Which curve is longer, the curve $y = 0$ for $0 \le x \le \pi$ or the curve $y = \sin x$ for $0 \le x \le \pi$? Why? With the answer to this question in mind, which of the fences pictured in the illustration on page 1086 has a greater area? Why?
- Which of the vector fields pictured in this section would carry an object caught within towards the origin? Why?
- If $f(x, y, z)$ is a potential function for $\mathbf{F}(x, y, z)$, then what is $\nabla f(x, y, z)$?
- What does it mean to say that a vector field $\mathbf{F}(x, y)$ is conservative? How would you show that a vector field failed to be conservative?
- Given a conservative vector field, how do we find its potential function? How does the procedure for finding a potential function for a conservative vector field in the plane differ from the procedure for finding a potential function for a conservative vector field in $\mathbb{R}^3$?

EXERCISES 14.1

Thinking Back

- *Work as an integral of force and distance:* Find the work done in moving an object along the x-axis from the origin to $x = \frac{\pi}{2}$ if the force acting on the object at a given value of x is $F(x) = x \sin x$.

Vector geometry: Use properties of vectors to obtain each of the specified quantities.

- Find a unit vector that points in the direction of $\mathbf{F}(3, 3, \sqrt{13})$ for the vector field

$$\mathbf{F} = z(x - y)^2\mathbf{i} + \sqrt{y}\,\mathbf{j} + \mathbf{k}.$$

- Find the equation of a line through the origin and parallel to $\mathbf{F}(3, 3, \sqrt{13})$.

Calculus of vector-valued functions: Calculate each of the following.

- $\frac{d}{dt}(\mathbf{r}(t))$, where $\mathbf{r}(t) = 3\cos^2 t\mathbf{i} + 5t\mathbf{j} + \frac{t}{t^2+1}\mathbf{k}$.
- $\int \mathbf{r}(t)\,dt$, where $\mathbf{r}(t) = e^t\mathbf{i} + t^3\mathbf{j} - 4\mathbf{k}$.

Concepts

0. *Problem Zero:* Read the section and make your own summary of the material.

1. *True/False:* Determine whether each of the statements that follow is true or false. If a statement is true, explain why. If a statement is false, provide a counterexample.

(a) *True or False:* A vector field is a function whose outputs are scalars.

(b) *True or False:* A vector field is a function whose outputs are vectors.

(c) *True or False:* A vector field is a function whose inputs are scalars.

(d) *True or False:* A vector field in $\mathbb{R}^3$ is a function whose inputs are points in $\mathbb{R}^3$.

(e) *True or False:* A conservative vector field has infinitely many potential functions.

(f) *True or False:* Every vector field $\mathbf{F}(x, y)$ is the gradient of some function $f(x, y)$.

(g) *True or False:* If two functions have the same gradient, they are the same function.

(h) *True or False:* Work is the integral of force times distance.

2. *Examples:* Construct examples of the thing(s) described in the following. Try to find examples that are different than any in the reading.

(a) A vector field in $\mathbb{R}^2$ and another in $\mathbb{R}^3$.

(b) A conservative vector field.

(c) A vector field that is not conservative.

3. What are the inputs of a vector field in the Cartesian plane?

4. What are the inputs of a vector field in $\mathbb{R}^3$?

5. What are the outputs of a vector field in the Cartesian plane?

6. What are the outputs of a vector field in $\mathbb{R}^3$?

7. What does it mean to say that a vector field is conservative?

8. Do the vectors in the range of $\mathbf{F}(x, y) = x\mathbf{i} + y\mathbf{j}$ point towards or away from the origin?

9. What is the difference between the graphs of

$$\mathbf{G}(x, y) = \mathbf{i} + \mathbf{j} \text{ and } \mathbf{F}(x, y) = 2\mathbf{i} + 2\mathbf{j}?$$

10. What is the difference between the graphs of

$$\mathbf{G}(x, y, z) = -\mathbf{i} - \mathbf{j} - \mathbf{k} \text{ and } \mathbf{F}(x, y, z) = \mathbf{i} + \mathbf{j} + \mathbf{k}?$$

11. What is the difference between the graphs of

$$\mathbf{G}(x, y) = x\mathbf{i} + y\mathbf{j} \text{ and } \mathbf{F}(x, y) = -x\mathbf{i} + -y\mathbf{j}?$$

12. What is the difference between the graphs of

$$\mathbf{G}(x, y, z) = 2\mathbf{i} - 3\mathbf{j} + z\mathbf{k} \text{ and } \mathbf{F}(x, y, z) = -2\mathbf{i} + 3\mathbf{j} - z\mathbf{k}?$$

13. Consider the vector field $\mathbf{F}(x, y, z) = \langle yz, xz, xy\rangle$. Find a vector field $\mathbf{G}(x, y, z)$ with the property that, for all points in $\mathbb{R}^3$, $\mathbf{G}(x, y, z) = 2\mathbf{F}(x, y, z)$.

14. What is the difference between vector fields

$$\mathbf{F}(x, y) = \langle x^2, y^2\rangle \text{ and } \mathbf{G}(x, y) = \langle x^2, y^2, 0\rangle?$$

15. How would you show that a given vector field in $\mathbf{R}^2$ is not conservative?

16. How would you show that a given vector field in $\mathbf{R}^3$ is not conservative?

Skills

In Exercises 17–24, find a potential function for the given vector field.

17. $\mathbf{F}(x, y) = \langle 3x^2 \cos y, -x^3 \sin y\rangle$

18. $\mathbf{F}(x, y) = \langle e^y \sec^2 x, e^y \tan x\rangle$

19. $\mathbf{G}(x, y) = \langle 5x^4 + y, x - 12y^3\rangle$

20. $\mathbf{G}(x, y) = \mathbf{i} - \mathbf{j}$

21. $\mathbf{F}(x, y, z) = yz\mathbf{i} + xz\mathbf{j} + xy\mathbf{k}$

22. $\mathbf{F}(x, y, z) = e^{y^2}\mathbf{i} + (2xye^{y^2} + \sin z)\mathbf{j} + y\cos z\mathbf{k}$

23. $\mathbf{G}(x, y, z) = \cos y\mathbf{i} + (\sin z - x\sin y)\mathbf{j} + y\cos z\mathbf{k}$

24. $\mathbf{G}(x, y, z) = \langle ye^{xy+z}, xe^{xy+z}, e^{xy+z}\rangle$

Sketch the vector fields in Exercises 25–32.

25. $\mathbf{F}(x, y) = 2\mathbf{i} + 0\mathbf{j}$

26. $\mathbf{F}(x, y) = 0\mathbf{i} + 2\mathbf{j}$

27. $\mathbf{F}(x, y) = \mathbf{i} + \mathbf{j}$

28. $\mathbf{F}(x, y) = 2\mathbf{i} + 2\mathbf{j}$

29. $\mathbf{F}(x, y) = \mathbf{i} - \mathbf{j}$

30. $\mathbf{F}(x, y) = -\mathbf{i} + \mathbf{j}$

31. $\mathbf{F}(x, y) = x\mathbf{i} + 2y\mathbf{j}$

32. $\mathbf{F}(x, y) = -2x\mathbf{i} - 3y\mathbf{j}$

Show that the vector fields in Exercises 33–40 are not conservative.

33. $\mathbf{F}(x, y) = \langle xy, -y\rangle$

34. $\mathbf{F}(x, y) = \langle x^2 + y^2, \cos y\rangle$

35. $\mathbf{G}(x, y) = \dfrac{1}{x^2 + y}\mathbf{i} + \dfrac{y}{x}\mathbf{j}$

36. $\mathbf{G}(x, y) = y\mathbf{i} - x\mathbf{j}$

37. $\mathbf{F}(x, y, z) = 2\mathbf{i} - z\mathbf{j} + e^{yz}\mathbf{k}$

38. $\mathbf{F}(x, y, z) = \tan(yz)\mathbf{i} + (xz\sec^2(yz) - 2)\mathbf{j} + 4z^3\mathbf{k}$

39. $\mathbf{G}(x, y, z) = \langle 3, yz, z + 12\rangle$

40. $\mathbf{G}(x, y, z) = \langle e^y + z, xe^y + z, x + y\rangle$

Determine whether or not each of the vector fields in Exercises 41–48 is conservative. If the vector field is conservative, find a potential function for the field.

41. $\mathbf{F}(x, y) = e^y\mathbf{i} + \sin y\mathbf{j}$

42. $\mathbf{F}(x, y) = \left\langle \tan^{-1} y, \dfrac{x}{1 + y^2}\right\rangle$

43. $\mathbf{G}(x, y) = \langle 2x + y\cos(xy), x\cos(xy) - 1\rangle$

44. $\mathbf{G}(x, y) = yx^2\mathbf{i} + e^y\mathbf{j}$

45. $\mathbf{F}(x, y, z) = \langle ye^{2z} + 1, xe^{2z}, 2xye^{2z}\rangle$

46. $\mathbf{F}(x, y, z) = \mathbf{i} + 2\mathbf{j} - 3\mathbf{k}$

47. $\mathbf{G}(x, y, z) = (z - y)\mathbf{i} - xy\mathbf{j} + (xz + y)\mathbf{k}$

48. $\mathbf{G}(x, y, z) = \langle \sin(yz), \cos(yz), x^2\rangle$

Applications

49. Construct a vector field to describe each of the situations that follow. There may be more than one choice of constant that gives an accurate answer. *(Hint: Examples from the section and earlier exercises may be useful, as well as plotting a few vectors from a given field.)*

(a) A thin film of water flowing across a flat plate from right to left at a constant rate.

(b) A thin film of liquid flowing across a flat plate diagonally from top left to bottom right at a constant rate.

(c) A thin film of liquid flowing on a flat plate rotating counterclockwise about the origin.

50. Write a vector field to describe each of the flows that follow. As in Exercise 49, there may be more than one choice of constant that gives an accurate answer.

(a) A thin film of liquid flowing across a flat plate from top to bottom at a constant rate.

(b) A thin film of liquid flowing across a flat plate in the direction of $3\mathbf{i} + 2\mathbf{j}$ at a constant rate.

(c) Fluid rotating clockwise around the origin.

51. Annie uses a vector field to model the current in a channel in the San Juan Islands in Washington State. She has imposed a coordinate system centered at the origin of the area she is interested in, and she is looking at the region $[-1, 1] \times [-1, 1]$. The velocity of the current is given by $\langle 0.9 - x^2 + 0.5y^2, xy\rangle$.

(a) Sketch the vector field.

(b) Is the vector field conservative? If so, what is an equation of a potential function? Explain the relationship between the potential function and the current.

(c) Can you tell where the land that influences the current lies?

Proofs

52. Prove that a conservative vector field has infinitely many potential functions.

53. Prove that if two functions $f(x, y)$ and $g(x, y)$ have the same gradient, then they differ by at most a constant.

Thinking Forward

Integration of vector fields: In order to answer the questions posed at the beginning of this section, we ought to be able to use the "subdivide, approximate, and add" strategy from Chapter 6. The results—area and work—should be scalars.

- ▶ How might the integral to find the arc length of $\mathbf{r}(t)$ from $t = a$ to $t = b$ in Section 11.4 be modified to produce the area of a fence whose base is $\mathbf{r}(t)$ and whose height is given by $h(t) = h(x(t), y(t))$?

- ▶ Write down some possible interpretations of what integrating a vector field along a curve to find work could mean, bearing in mind that the result should be a scalar.

14.2 LINE INTEGRALS

- ▶ Defining the integral of a multivariate function along a curve
- ▶ Defining the integral of a vector field along a curve
- ▶ Using the Fundamental Theorem of Line Integrals

Integrals Along Curves in Space

In this section, we address both of the problems posed at the beginning of the chapter: how to find the area of a fence whose base is not the x-axis and how to compute work along a curve in space. We will also see an extension of the Fundamental Theorem of Calculus.

Since both of our motivating questions have to do with curves in space, we recall the discussion in Section 11.4 of parametrized curves in $\mathbb{R}^2$ and $\mathbb{R}^3$. In general, in order for us to be able to compute the arc length of a curve C, we need the curve to be described by a vector function $\mathbf{r}(t) = \langle x(t), y(t), z(t)\rangle$ such that $\mathbf{r} : [a, b] \to C$ is one-to-one, $\|\mathbf{r}'(t)\| \neq 0$ for all $t \in [a, b]$, and each of the component functions $x(t)$, $y(t)$, and $z(t)$ is differentiable and has a continuous first derivative. Throughout the rest of the chapter, we will call such curves and parametrizations *smooth*. For a smooth curve C with smooth parametrization $\mathbf{r}(t)$ and endpoints $\mathbf{r}(a)$, $\mathbf{r}(b)$, Theorem 11.18 gives the arc length $\mathcal{L}$ of C as

$$\mathcal{L}(a, b) = \int_a^b \|\mathbf{r}'(t)\|\, dt.$$

Line Integrals of Multivariate Functions

Our process for computing the area of a fence whose base is a smooth curve C uses the same ideas we found in Definition 4.9, when we defined the definite integral. When we computed the area under a curve on an interval, we used the "subdivide, approximate, and add" strategy. We saw that the area is the limit of the sum of areas of approximating subrectangles. More formally,

$$A = \lim_{n\to\infty} \sum_{k=1}^{n} (\text{height of } k\text{th rectangle} \times \text{width of } k\text{th rectangle}).$$

In Chapter 4 this equation became

$$A = \lim_{n\to\infty} \sum_{k=1}^{n} f(x_k^*)\Delta x = \int_a^b f(x)\, dx.$$

Here we want to say the same thing, adjusting for the fact that our curve C is smoothly parametrized by $\mathbf{r}(t)$ for $a \le t \le b$. So now the width of an approximating subrectangle is given by

$$\|\mathbf{r}'(t)\| = \sqrt{\left(\frac{dx}{dt}\right)^2 + \left(\frac{dy}{dt}\right)^2 + \left(\frac{dz}{dt}\right)^2},$$

and the height is given by $f(t_k^*)$, where, as before, t_k^* is an arbitrary value of t in the kth subdivision of $[a, b]$, as seen in the following figure:

Approximating the area of a function defined on a curve

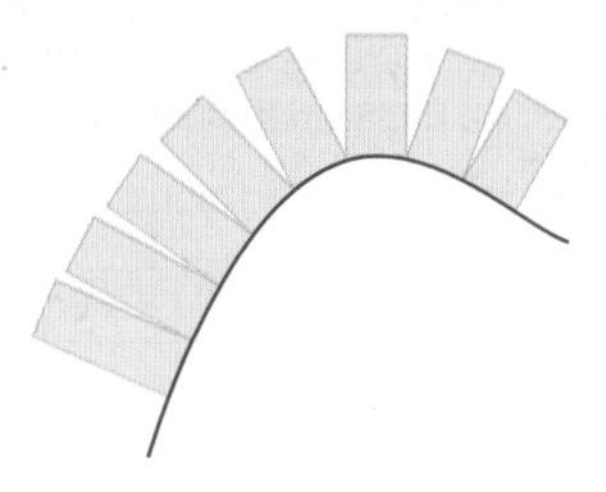

Combining these insights, we define the integral of a multivariate function along a curve in space.

DEFINITION 14.3

Line Integral of a Multivariate Function

Let C be a curve in $\mathbb{R}^3$ with a smooth parametrization $\mathbf{r}(t) = \langle x(t), y(t), z(t)\rangle$ for $t \in [a, b]$. Then the ***integral of*** $f(x, y, z)$ ***along*** C is

$$\int_C f(x, y, z)ds = \int_a^b f(x(t), y(t), z(t))\|\mathbf{r}'(t)\|\, dt.$$

The definition for curves in $\mathbb{R}^2$ is analogous. They are the same as curves in $\mathbb{R}^3$, except that all z-components are zero. Also, note the new notation:

$$ds = \|\mathbf{r}'(t)\|\, dt.$$

The integral along C is sometimes referred to as the ***integral with respect to arc length***.

Line integrals of multivariate functions act like the single-variable integrals of Chapter 4 because, after the substitutions for $f(x, y, z)$ and $\|\mathbf{r}'(t)\|$ in terms of t, they *are* single-variable integrals. Note that if $f(x, y, z)$ is constantly equal to 1, the line integral returns just the arc length of C. This situation is parallel to that for single-variable integrals: The interval $[a, b]$ has length $b - a$, and

$$\int_a^b 1\, dx = b - a.$$

Line integrals are additive, like single-variable integrals. That is, if $C = C_1 \cup C_2$, where C_1 and C_2 overlap in a single point, then

$$\int_C f(x, y, z)\, ds = \int_{C_1} f(x, y, z)\, ds + \int_{C_2} f(x, y, z)\, ds.$$

This property is particularly useful when a natural parametrization of C by $\mathbf{r}(t)$ has $\mathbf{r}'(t) = \mathbf{0}$ at a point. We can then rewrite the integral along C as a sum of two integrals that do not have this problem. If a curve C is not itself smooth, but can be written as a finite sum of smooth curves, we say that C is ***piecewise smooth***.

Another point of similarity between line integrals and single-variable integrals is that the direction of integration matters. Recall from Chapter 4 that

$$\int_a^b f(x)\, dx = -\int_b^a f(x)\, dx.$$

The same is true for the integral along a curve C with endpoints P and Q. If $\mathbf{r}_1(t)$ is a smooth parametrization of C that starts at P and ends at Q, and if $\mathbf{r}_2(t)$ is a smooth parametrization of C that starts at Q and ends at P, then the integral along C using $\mathbf{r}_1(t)$ will be the negative of the integral along C using $\mathbf{r}_2(t)$.

Line Integrals of Vector Fields

In Chapter 6 we discussed the concept of *work*. We now generalize this concept to include work done by a variable force acting along a curve in space. We represent the force acting in each of the **i**, **j**, and **k** directions as a vector field whose component functions vary with location in space:

$$\mathbf{F}(x, y, z) = F_1(x, y, z)\mathbf{i} + F_2(x, y, z)\mathbf{j} + F_3(x, y, z)\mathbf{k}.$$

Here $F_1(x, y, z)$ measures the action of $\mathbf{F}(x, y, z)$ in the **i** direction, $F_2(x, y, z)$ measures the action of $\mathbf{F}(x, y, z)$ in the **j** direction, and $F_3(x, y, z)$ measures the action of $\mathbf{F}(x, y, z)$ in the **k** direction.

To compute work done by a vector field along a curve, we follow our usual strategy of subdividing, approximating, and adding, followed by taking a limit as the size of the subdivisions approaches zero. In this case, the limit will be the integral of the vector field along the curve. Subdividing the curve uses the position vector $\langle x_n - x_{n-1}, y_n - y_{n-1}, z_n - z_{n-1}\rangle$ to approximate the curve between points (x_n, y_n, z_n) and $(x_{n-1}, y_{n-1}, z_{n-1})$. To compute the action of the vector field $\mathbf{F}(x, y, z)$ on this small piece of the curve, $F_1(x_n, y_n, z_n)$ is used to approximate F_1, the action of $\mathbf{F}(x, y, z)$ in the **i** direction for that subdivision. The components $F_2(x_n, y_n, z_n)$ and $F_3(x_n, y_n, z_n)$ are similarly used in the **j** and **k** directions. This approximation is a constant force in a fixed direction, so we multiply $F_1(x_n, y_n, z_n) \cdot (x_n - x_{n-1})$ to get the action of $\mathbf{F}(x, y, z)$ in the **i** direction. Also, summing in the **j** and **k** directions, we have

$$W \approx F_1(x_n, y_n, z_n) \cdot (x_n - x_{n-1}) + F_2(x_n, y_n, z_n) \cdot (y_n - y_{n-1}) + F_3(x_n, y_n, z_n) \cdot (z_n - z_{n-1})$$

as the approximate work done by $\mathbf{F}(x, y, z)$ on the nth subdivided piece of the curve. Adding the action along the n pieces of the subdivided curve, we obtain

$$\sum_{n=1}^{M}(F_1(x_n, y_n, z_n) \cdot (x_n - x_{n-1}) + F_2(x_n, y_n, z_n) \cdot (y_n - y_{n-1}) + F_3(x_n, y_n, z_n) \cdot (z_n - z_{n-1})).$$

As M approaches infinity, the change in x, y, and z becomes $\mathbf{r}'(t)$ and $\mathbf{F}(x_n, y_n, z_n)$ becomes the value of the field at a point; thus, the limit of the summation becomes

$$\int_a^b \left(F_1(x, y, z) \cdot \frac{\partial}{\partial x}\mathbf{r}(t) + F_2(x, y, z) \cdot \frac{\partial}{\partial y}\mathbf{r}(t) + F_3(x, y, z) \cdot \frac{\partial}{\partial z}\mathbf{r}(t)\right) dt$$

$$= \int_a^b \mathbf{F}(x, y, z) \cdot \mathbf{r}'(t)dt = \int_C \mathbf{F}(x, y, z) \cdot d\mathbf{r}.$$

This result motivates our definition of the line integral of a vector field.

DEFINITION 14.4 Line Integral of a Vector Field

Suppose that C is a smooth curve in $\mathbb{R}^3$ with a smooth parametrization $\mathbf{r}(t)$ for $t \in [a, b]$ and with a vector field $\mathbf{F}(x, y, z) = F_1\mathbf{i} + F_2\mathbf{j} + F_3\mathbf{k}$ whose component functions are each continuous on C and whose domain is open, connected, and simply connected. Then the ***line integral of*** $\mathbf{F}(x, y, z)$ ***along*** C is

$$\int_C \mathbf{F}(x, y, z) \cdot d\mathbf{r} = \int_a^b (F_1(x, y, z)x'(t) + F_2(x, y, z)y'(t) + F_3(x, y, z)z'(t))\, dt.$$

Two alternative ways of writing this integral are

$$\int_C \mathbf{F}(x, y, z) \cdot d\mathbf{r} = \int_C (F_1\, dx + F_2\, dy + F_3\, dz) = \int_C \mathbf{F} \cdot \mathbf{T}\, ds,$$

where $\mathbf{T}$ is the unit tangent vector. You are encouraged to check that all of the representations really are equivalent by carrying out the relevant substitutions and simplifications. For example, since the unit tangent vector is $\mathbf{T} = \frac{\mathbf{r}'(t)}{\|\mathbf{r}'(t)\|}$ and the differential is $ds = \|\mathbf{r}'(t)\|\,dt$, upon cancellation we see that $\int_C \mathbf{F}\cdot\mathbf{T}\,ds$ is equivalent to the line integral given in Definition 14.4.

As with the line integrals of multivariate functions discussed earlier in the section, changing the direction of travel along C changes the sign. If $\mathbf{r}(t)$ and $\mathbf{q}(t)$ are smooth parametrizations that traverse the curve C in opposite directions, then

$$\int_C \mathbf{F}(x,y,z)\cdot d\mathbf{r} = -\int_C \mathbf{F}(x,y,z)\cdot d\mathbf{q}.$$

The Fundamental Theorem of Line Integrals

Line integrals of vector fields can be complicated to compute in practice, depending on the curve C and the choice of $\mathbf{r}(t)$ to describe it. In the case of conservative vector fields, however, the fact that $\mathbf{F}(x,y,z)$ is a gradient allows us to reduce the line integral of the vector field to subtraction.

THEOREM 14.5 **The Fundamental Theorem of Line Integrals**

Let C be a smooth curve that is the graph of the vector function $\mathbf{r}(t)$ defined on the interval $[a,b]$ with $P = \mathbf{r}(a)$ and $Q = \mathbf{r}(b)$. If $\mathbf{F}(x,y,z)$ is a conservative vector field with $\mathbf{F}(x,y,z) = \nabla f(x,y,z)$ on an open, connected, and simply connected domain containing the curve C, then

$$\int_C \mathbf{F}(x,y,z)\cdot d\mathbf{r} = f(Q) - f(P).$$

Before we prove Theorem 14.5, note that $P = \mathbf{r}(a)$ and $Q = \mathbf{r}(b)$. These equations represent a common abuse of notation, because P and Q are points, but $\mathbf{r}(a)$ and $\mathbf{r}(b)$ are vectors. By this we mean that if P is the point (x_0, y_0, z_0), then $\mathbf{r}(a) = \langle x_0, y_0, z_0\rangle$ and $f(P) = f(x_0, y_0, z_0)$.

Proof. If $\mathbf{F}(x,y,z) = \nabla f(x,y,z)$ and $\mathbf{r}(t)$ is a smooth parametrization of C with $\mathbf{r}(a) = P$ and $\mathbf{r}(b) = Q$, and if $\|\mathbf{r}'(t)\| \neq 0$ for all $t \in [a,b]$, then

$$\mathbf{F}(x,y,z) = \frac{\partial f}{\partial x}\mathbf{i} + \frac{\partial f}{\partial y}\mathbf{j} + \frac{\partial f}{\partial z}\mathbf{k} \quad \text{and} \quad \mathbf{r}'(t) = \frac{dx}{dt}\mathbf{i} + \frac{dy}{dt}\mathbf{j} + \frac{dz}{dt}\mathbf{k}.$$

Therefore,

$$\mathbf{F}\cdot d\mathbf{r} = \left(\frac{\partial f}{\partial x}\frac{dx}{dt} + \frac{\partial f}{\partial y}\frac{dy}{dt} + \frac{\partial f}{\partial z}\frac{dz}{dt}\right) dt.$$

By the chain rule, we have

$$\mathbf{F}\cdot d\mathbf{r} = \frac{d}{dt}(f(x(t), y(t), z(t)))\,dt = \frac{d}{dt}(f(t))\,dt.$$

Therefore,

$$\begin{aligned}\int_C \mathbf{F}\cdot d\mathbf{r} &= \int_a^b \frac{d}{dt}(f(t))\,dt \\ &= f(x(b), y(b), z(b)) - f(x(a), y(a), z(a)) \\ &= f(Q) - f(P).\end{aligned}$$

■

Theorem 14.5 is very useful, since it allows us to avoid finding an explicit parametrization of C and many important vector fields in the sciences are conservative. In Exercise 65, you will show that a consequence of this theorem is that line integrals of conservative vector fields depend only on the endpoints of curves, not on the curves themselves. Such integrals are therefore said to be ***independent of path***.

Examples and Explorations

EXAMPLE 1

Area of a fence whose base is a curve in the plane

Find the area of a wall whose base is the part of the circle of radius 2 centered at the origin, lying in the first quadrant, and whose height at point (x, y) is given by $f(x, y) = 2x + y$.

SOLUTION

Following is an illustration of the wall, along with the plane defined by $z = 2x + y$:

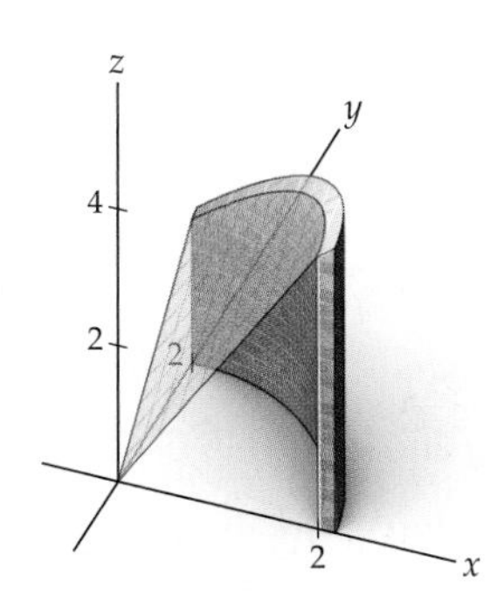

The quarter-circle base of the wall may be parametrized by $\mathbf{r}(t) = \langle 2\cos t, 2\sin t\rangle$, where $0 \le t \le \frac{\pi}{2}$, so

$$\|\mathbf{r}'(t)\| = \sqrt{4\sin^2 t + 4\cos^2 t} = 2.$$

We then substitute into the two-variable version of Definition 14.3:

$$\begin{aligned}\int_C f(x, y)\, ds &= \int_a^b f(x(t), y(t))\|\mathbf{r}'(t)\|\, dt \\ &= \int_0^{\pi/2} (2(2\cos t) + 2\sin t)\cdot 2\, dt \\ &= \big[8\sin t - 4\cos t\big]_0^{\pi/2} = 12.\end{aligned}$$

As usual, when we are computing an area, it is important to remember that integration gives the *signed* area; thus, if $f(x, y)$ were negative at any point on C, we would need to compensate by integrating $|f(x, y)|$. □

EXAMPLE 2

Integrating a multivariate function along a curve in $\mathbb{R}^3$

Find

$$\int_C f(x, y, z)\, ds,$$

where C is the line segment from $(1, 4, 2\sqrt{3})$ to $(3, 7, 4\sqrt{3})$ and

$$f(x, y, z) = x + y + z.$$

SOLUTION

Recall that the line segment from the point (a, b, c) to the point (d, e, f) may be parametrized by

$$\mathbf{r}(t) = \big\langle a + t(d-a), b + t(e-b), c + t(f-c)\big\rangle, \quad \text{for } 0 \le t \le 1.$$

In this case we have

$$\mathbf{r}(t) = \langle 1+2t, 4+3t, 2\sqrt{3}+2\sqrt{3}\,t\rangle \text{ for } t \in [0,1].$$

So,

$$\mathbf{r}'(t) = \langle 2, 3, 2\sqrt{3}\rangle, \qquad \|\mathbf{r}'(t)\| = 5,$$

and on the curve,

$$f(x(t), y(t), z(t)) = 5 + 2\sqrt{3} + (5+2\sqrt{3})t.$$

Substituting gives

$$\begin{aligned}\int_C f(x,y,z)\,ds &= \int_0^1 (5+2\sqrt{3}+(5+2\sqrt{3})t)5\,dt \\ &= 5\left[(5+2\sqrt{3})t + \frac{1}{2}(5+2\sqrt{3})t^2\right]_0^1 \\ &= \frac{75}{2} + 15\sqrt{3}.\end{aligned}$$

□

EXAMPLE 3

Computing line integrals of vector fields

Find the line integral $\int_C \mathbf{F}\cdot d\mathbf{r}$ for each of the following vector fields along the given curve:

(a) $\mathbf{F}(x,y) = \langle -y, x\rangle$, where C_1 is the straight segment from $(3,0)$ to $(8,\pi)$.

(b) $\mathbf{F}(x,y,z) = \langle e^y, xe^z, y^2\rangle$, where C_2 is the curve given by $\mathbf{r}(t) = \langle t, 2, t^2\rangle$, for $-2 \le t \le 2$.

SOLUTION

(a) The curve C_1 is smoothly parametrized by

$$\mathbf{r}(t) = \langle 3+5t, \pi t\rangle, \qquad 0 \le t \le 1.$$

Thus, $\mathbf{r}'(t) = \langle 5, \pi\rangle$. So,

$$\begin{aligned}\int_{C_1} \mathbf{F}(x,y)\cdot d\mathbf{r} &= \int_0^1 \langle -\pi t, 3+5t\rangle \cdot \langle 5, \pi\rangle dt \\ &= \int_0^1 3\pi\,dt = 3\pi.\end{aligned}$$

(b) For C_2 and $\mathbf{F}(x,y,z)$,

$$\mathbf{r}(t) = \langle t, 2, t^2\rangle, \qquad \text{for} \qquad -2 \le t \le 2.$$

Therefore, $\mathbf{r}'(t) = \langle 1, 0, 2t\rangle$. So,

$$\begin{aligned}\int_{C_2} \mathbf{F}(x,y,z)\cdot d\mathbf{r} &= \int_{-2}^{2} \langle e^2, te^{t^2}, 4\rangle \cdot \langle 1, 0, 2t\rangle\,dt \\ &= \int_{-2}^{2} (e^2 + 8t)\,dt \\ &= [e^2 t + 4t^2]_{-2}^{2} \\ &= (2e^2+16) - (-2e^2+16) = 4e^2.\end{aligned}$$

□

EXAMPLE 4

Applying the Fundamental Theorem of Line Integrals

Use the Fundamental Theorem of Line Integrals to evaluate the following:

(a) The line integral

$$\int_{C_1} \mathbf{F}(x, y) \cdot d\mathbf{r},$$

where $\mathbf{F}(x, y) = \left\langle 6x^2 \ln(y+1), \frac{2x^3}{y+1} \right\rangle$ and C_1 is the straight segment from $(1, 1)$ to $(3, 7)$.

(b) The line integral

$$\int_{C_2} \mathbf{G}(x, y, z) \cdot d\mathbf{r},$$

where $\mathbf{G}(x, y, z) = \langle yze^{xyz}, xze^{xyz}, xye^{xyz} \rangle$ and C_2 is the circular helix about the z-axis given by

$$\mathbf{r}(t) = \langle \cos t, \sin t, t \rangle, \text{ for } t \in \left[\frac{\pi}{4}, \frac{3\pi}{4}\right].$$

SOLUTION

(a) The Fundamental Theorem of Line Integrals makes it quite easy to carry out these integrations; it is enough to verify that the vector field in question is conservative, compute the potential function, substitute according to the theorem, and subtract. In the case of the first field,

$$\frac{\partial}{\partial y}(6x^2 \ln(y+1)) = \frac{6x^2}{y+1} = \frac{\partial}{\partial x}\left(\frac{2x^3}{y+1}\right),$$

so $\mathbf{F}$ is conservative. Integrating, we find a potential function for $\mathbf{F}$:

$$f(x, y) = 2x^3 \ln(y+1).$$

By the Fundamental Theorem,

$$\int_{C_1} \mathbf{F}(x, y) \cdot d\mathbf{r} = f(3, 7) - f(1, 1) = 54 \ln 8 - 2 \ln 2 = 162 \ln 2 - 2 \ln 2 = 160 \ln 2.$$

(b) In the case of the second field, it is easier to integrate $g_1(x, y, z) = yze^{xyz}$ with respect to x and check that the resulting function can be adjusted as necessary to produce a potential function for $\mathbf{G}$. Integration yields

$$g(x, y, z) = e^{xyz}; \qquad \mathbf{G}(x, y, z) = \nabla g(x, y, z).$$

Therefore, $g(x(t), y(t), z(t)) = e^{t \cos t \sin t}$. Evaluating $\mathbf{r}(\pi/4)$ and $\mathbf{r}(3\pi/4)$ gives us $(\sqrt{2}/2, \sqrt{2}/2, \pi/4)$ as the starting point of C_2 and $(-\sqrt{2}/2, \sqrt{2}/2, 3\pi/4)$ as the endpoint of C_2. Then

$$\int_{C_2} \mathbf{G}(x, y, z) \cdot d\mathbf{r} = g\left(-\frac{\sqrt{2}}{2}, \frac{\sqrt{2}}{2}, \frac{3\pi}{4}\right) - g\left(\frac{\sqrt{2}}{2}, \frac{\sqrt{2}}{2}, \frac{\pi}{4}\right) = e^{-3\pi/8} - e^{\pi/8}.$$

□

EXAMPLE 5

Evaluating line integrals of vector functions along closed curves

Integrate each of the following vector fields along the specified closed curve C:

(a) $\mathbf{F}(x, y) = (y+2)\mathbf{i} + (x+3)\mathbf{j}$, where C is the unit circle in the plane, traversed counterclockwise starting and ending at $(1, 0)$.

(b) $\mathbf{G}(x, y) = y\mathbf{i} - y^2\mathbf{j}$, where C is the square with vertices $(1, 0)$, $(0, 1)$, $(-1, 0)$, and $(0, -1)$, traversed counterclockwise starting and ending at $(1, 0)$.

SOLUTION

(a) The vector field $\mathbf{F}(x, y)$ is conservative, since $\frac{\partial}{\partial y}(y+2) = 1 = \frac{\partial}{\partial x}(x+3)$, so we can use the Fundamental Theorem of Line Integrals. A potential function for $\mathbf{F}(x, y)$ is

$$f(x, y) = 2x + xy + 3y.$$

The unit circle is smoothly parametrized by $\mathbf{r}(t) = (\cos t)\mathbf{i} + (\sin t)\mathbf{j}$ for $t \in [0, 2\pi]$. So, $\mathbf{r}'(t) = \langle -\sin t, \cos t\rangle$. Since we are integrating over a closed curve, we expect to end where we started. As we know, when we evaluate $\mathbf{r}(t)$ at $t = 0$ and $t = 2\pi$, we start and stop at $(1, 0)$. Applying the Fundamental Theorem gives

$$\int_C \mathbf{F}(x, y) \cdot d\mathbf{r} = f(1, 0) - f(1, 0) = 0.$$

The zero value we obtain here is not a coincidence: In Exercise 64 you will prove that the integral of *any* conservative vector field along any closed curve is equal to zero.

(b) The vector field $\mathbf{G}(x, y)$ is not conservative, since $\frac{\partial}{\partial y}(y) = 1$ but $\frac{\partial}{\partial x}(y^2) = 0$. Because the vector field is not conservative, we may not use the Fundamental Theorem of Line Integrals. For this, piecewise-smooth, simple closed curve, we will perform a separate integration for each of the sides of the square, as indicated in the following figure:

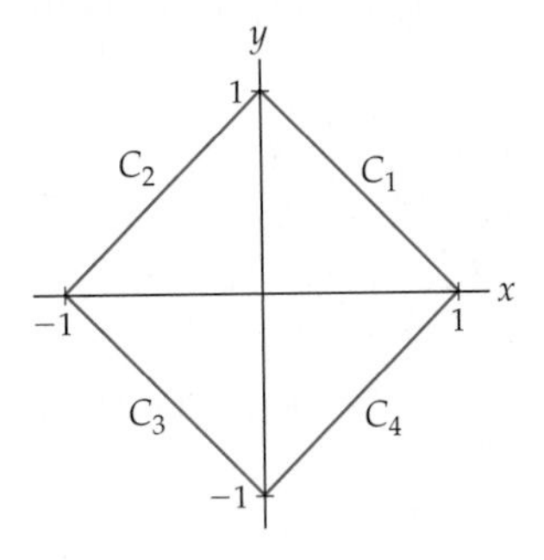

We parametrize each side of the square as follows:

- For side C_1 we use $x = 1 - t$ and $y = t$, where $0 \le t \le 1$.
- For C_2 we use $x = 1 - t$ and $y = 2 - t$, where $1 \le t \le 2$.
- For C_3 we use $x = -3 + t$ and $y = 2 - t$, where $2 \le t \le 3$.
- Finally, for C_4 we use $x = -3 + t$ and $y = -4 + t$, where $3 \le t \le 4$.

To evaluate the line integral, we have

$$\int_C \mathbf{G}(x, y) \cdot d\mathbf{r} = \int_{C_1} \mathbf{G}(x, y) \cdot d\mathbf{r} + \int_{C_2} \mathbf{G}(x, y) \cdot d\mathbf{r} + \int_{C_3} \mathbf{G}(x, y) \cdot d\mathbf{r} + \int_{C_4} \mathbf{G}(x, y) \cdot d\mathbf{r}.$$

On C_1 we have $\mathbf{G}(x(t), y(t)) = \langle t, -t^2\rangle$ and $d\mathbf{r} = \langle -1, 1\rangle dt$. Therefore,

$$\begin{aligned}\int_{C_1} \mathbf{G}(x, y) \cdot d\mathbf{r} &= \int_0^1 \langle t, -t^2\rangle \cdot \langle -1, 1\rangle\, dt \\ &= \int_0^1 (-t - t^2)\, dt \\ &= \left[-\frac{1}{2}t^2 - \frac{1}{3}t^3\right]_0^1 = -\frac{5}{6}.\end{aligned}$$

On C_2 we have $\mathbf{G}(x(t), y(t)) = \langle 2-t, -(2-t)^2\rangle$ and $d\mathbf{r} = \langle -1, -1\rangle\, dt$. Hence,

$$\int_{C_2} \mathbf{G}(x,y)\cdot d\mathbf{r} = \int_1^2 \langle 2-t, -(2-t)^2\rangle \cdot \langle -1,-1\rangle\, dt$$
$$= \int_1^2 (t-2+(2-t)^2)dt$$
$$= \left[\frac{1}{2}t^2 - 2t - \frac{1}{3}(2-t)^3\right]_1^2 = -\frac{1}{6}.$$

The integrals for sides C_3 and C_4 are similar. You should check that the values of these integrals are $-\frac{1}{6}$ and $-\frac{5}{6}$, respectively. Therefore,

$$\int_C \mathbf{G}(x,y)\cdot d\mathbf{r} = -\frac{5}{6} - \frac{1}{6} - \frac{1}{6} - \frac{5}{6} = -2$$

□

EXAMPLE 6

Computing the work done along a curve

Find the work done by the vector field $\mathbf{F}(x,y) = y^2\mathbf{i} - x^2\mathbf{j}$ along the curve parametrized by $x = t^2$, $y = t^3$, for $0 \le t \le 2$.

SOLUTION

To find the work, we compute the line integral $\int_C \mathbf{F}(x,y)\cdot d\mathbf{r}$, where $\mathbf{r}(t) = \langle t^2, t^3\rangle$. Here we have $\mathbf{F}(x(t), y(t)) = \langle t^6, -t^4\rangle$ and $d\mathbf{r} = \langle 2t, 3t^2\rangle\, dt$. Thus,

$$\int_C \mathbf{F}(x,y)\cdot d\mathbf{r} = \int_0^2 \langle t^6, -t^4\rangle \cdot \langle 2t, 3t^2\rangle\, dt$$
$$= \int_0^2 (2t^7 - 3t^6)\, dt = \left[\frac{1}{4}t^8 - \frac{3}{7}t^7\right]_0^2 = \frac{64}{7}.$$

□

? TEST YOUR UNDERSTANDING

- How does the integral of a multivariate function along a smooth curve resemble a single-variable integral from Chapter 4? Why does it do so?
- What does ds stand for in terms of $\mathbf{r}(t)$?
- What does $d\mathbf{r}$ stand for in terms of $\mathbf{r}(t)$?
- Why does the integral of any conservative vector field along any smooth or piecewise-smooth curve depend only on the endpoints and not on the curve?
- Why does the integral of any conservative vector field along any smooth or piecewise-smooth closed curve always equal zero?

EXERCISES 14.2

Thinking Back

Average value: Review the average value formula from Section 4.6. Use the formula to compute the average value of the following functions.

- $g(x) = x^2 + x + 2$ on $[1, 5]$
- $f(x) = \cos x$ on $[\pi, 2\pi]$

Arc length: Review the discussion of arc length in Sections 6.3 and 11.4. Compute the length of the following curves.

- The plane curve $\langle t, t^2\rangle$ from $t = 4$ to $t = 6$
- The circular helix given by $\mathbf{r}(t) = \cos z\mathbf{i} + \sin z\mathbf{j} + z\mathbf{k}$ from $z = 0$ to $z = 2\pi$

Concepts

0. *Problem Zero:* Read the section and make your own summary of the material.

1. *True/False:* Determine whether each of the statements that follow is true or false. If a statement is true, explain why. If a statement is false, provide a counterexample.
   (a) *True or False:* The integral of a multivariate function along a smooth curve C produces a scalar.
   (b) *True or False:* The integral of a vector field $\mathbf{F}(x, y)$ along a smooth curve C produces a vector.
   (c) *True or False:* A smooth parametrization $\mathbf{r}(t)$ must have $\mathbf{r}'(t) \neq 0$ for all points on the curve.
   (d) *True or False:* If a curve is not itself smooth, but can be written as a finite sum of smooth curves, it is possible to integrate either a multivariate function or a vector field along the curve by integrating over each smooth piece separately.
   (e) *True or False:* If a vector field is conservative, it cannot be integrated along any curve.
   (f) *True or False:* The Fundamental Theorem of Line Integrals applies to the integral of a conservative vector field along a smooth curve.
   (g) *True or False:* Every integral of a vector field along a smooth curve can be evaluated with the Fundamental Theorem of Line Integrals.
   (h) *True or False:* The integral of a vector field along a smooth closed curve is always zero.

2. *Examples:* Construct examples of the thing(s) described in the following. Try to find examples that are different than any in the reading.
   (a) A smooth parametrization of the straight line between two points in $\mathbb{R}^3$.
   (b) A smooth curve C that is not a straight line, and a smooth parametrization of C.
   (c) A vector field $\mathbf{F}(x, y)$ such that the integral of $\mathbf{F}(x, y)$ around the unit circle is zero.

3. Write, but do not evaluate, the line integral $\int_C f(x, y, z)\, ds$ as an integral explicitly in terms of t if $f(x, y, z) = e^{xyz}$ and $\mathbf{r}(t) = t\mathbf{i} + t^2\mathbf{j} + t^4\mathbf{k}$ for $a \le t \le b$.

4. Make a chart of all the new notation in this section, including what each item means in terms you already understand.

For each integral in Exercises 5–8, give the vector field that is being integrated.

5. $\displaystyle\int_C (x + y)\, dx + xy\, dy$

6. $\displaystyle\int_C \cos(x^2 y)\, dx - 3e^y\, dy$

7. $\displaystyle\int_C xy^2\, dx + (xy - z)\, dy + \cos y\, dz$

8. $\displaystyle\int_C xyz\, dx + \sin(zy)\, dy + (z - x)\, dz$

9. Write the first alternative form of $\int_C \mathbf{F}(x, y) \cdot d\mathbf{r}$ for $\mathbf{F}(x, y)$ and C_1 from Example 3. (This alternative form is described immediately after Definition 14.4 and is used in Exercises 5–8.)

10. Write the first alternative form of $\int_C \mathbf{G}(x, y, z) \cdot d\mathbf{r}$ for $\mathbf{G}(x, y, z)$ and C_2 from Example 3. (This alternative form is described immediately after Definition 14.4 and is used in Exercises 5–8.)

11. Translate ds into terms of $\mathbf{r}(t)$.

12. Translate $d\mathbf{r}$ into terms of $\mathbf{r}(t)$.

13. Write $\int_C \mathbf{F}(x, y) \cdot d\mathbf{r}$ explicitly as an integral of t, where $\mathbf{F}(x, y) = \langle x^2 y, x - y\rangle$ and $\mathbf{r}(t) = \langle 2t, e^t\rangle$ for $a \le t \le b$.

14. Write $\int_C \mathbf{F}(x, y.z) \cdot d\mathbf{r}$ explicitly as an integral of t, where $\mathbf{F}(x, y, z) = \langle zy - x, xz - y, xy - z\rangle$ and $\mathbf{r}(t) = \langle t, \cos t, \sin t\rangle$ for $a \le t \le b$.

15. Give a smooth parametrization $\mathbf{r}(t)$ for the unit circle, starting and ending at $(1, 0)$ and travelling in the counterclockwise direction.

16. Give a smooth parametrization $\mathbf{r}(t)$ for the unit circle, starting and ending at $(1, 0)$ and travelling in the clockwise direction.

17. Give a smooth parametrization $\mathbf{r}(t)$ for the circular helix with radius 2 and centered about the y-axis, starting at $(2, 0, 0)$ and completing two full revolutions.

18. Give a smooth parametrization $\mathbf{r}(t)$ for the straight line between the points $(\pi, e, 1)$ and $(3, 7, 13)$.

19. In Chapter 4 we defined the ***average value*** of an integrable function f on the interval $[a, b]$ to be $\frac{1}{b-a}\int_a^b f(x)\, dx$. Generalize this expression to a formula that will give the average value of $f(x, y, z)$ along a smooth curve C parametrized by $\mathbf{r}(t)$ for $a \le t \le b$.

20. Use your intuition from Exercise 19 to propose a formula for the average value of $\mathbf{F}(x, y, z)$ along a smooth curve C parametrized by $\mathbf{r}(t)$ for $a \le t \le b$.

Skills

In Exercises 21–28, evaluate the multivariate line integral of the given function over the specified curve.

21. $g(x, y) = x$, with C the graph of $y = x^2$ in the xy-plane from the origin to $(2, 4)$.

22. $f(x, y) = e^{xy}$, with C the line with equation $x = 3y$ for $-1 \le y \le 1$.

23. $f(x, y) = x^2 + y^2$, with C the unit circle traversed counterclockwise.

24. $f(x, y) = x^2 + y^2$, with C the unit circle traversed *clockwise*.

25. $f(x, y, z) = e^{x+y+z}$, with C the straight line segment from the origin to $(1, 2, 3)$.

26. $g(x, y, z) = xyz$, with C the curve parametrized by $\mathbf{r}(t) = \left\langle \frac{\sqrt{2}}{3}t^3, t^2, t \right\rangle$ for $1 \le t \le 4$.

27. $f(x, y, z) = e^{x^2+y+z^2}$, with C the circular helix of radius 1, centered about the z-axis, and parametrized by $\mathbf{r}(t) = \langle \cos t, \sin t, t \rangle$ from height 0 to π.

28. $f(x, y, z) = e^{x^2+y^2+z^2}$, with C the circular helix of radius 1, centered about the y-axis, and parametrized by $\mathbf{r}(t) = \langle \cos t, t, \sin t \rangle$ for $\pi \le y \le 3\pi$.

Evaluate each of the vector field line integrals in Exercises 29–36 over the indicated curves.

29. $\mathbf{F}(x, y) = \mathbf{i} - \mathbf{j}$, with C the curve with equation $y - x^2 = 1$ for $5 \le x \le 10$.

30. $\mathbf{F}(x, y) = y\,\mathbf{i} - x\,\mathbf{j}$, with C the circle of radius 2, centered at the origin, and traversed counterclockwise starting at $(2, 0)$.

31. $\mathbf{F}(x, y)$ and C are as in Exercise 29, but C is traversed in the reverse direction, from $x = 10$ to $x = 5$.

32. $\mathbf{F}(x, y, z) = 2^z\,\mathbf{i} + \ln x\,\mathbf{j} + xz\,\mathbf{k}$, with C the curve parametrized by $\langle t, \ln t, t \rangle$ for $1 \le t \le e^2$.

33. $\mathbf{F}(x, y) = y \sin xy\,\mathbf{i} + x \cos xy\,\mathbf{j}$, with C the cardioid given by $r = 1 + 2\cos\theta$ from $\theta = 0$ to $\theta = 2\pi$.

34. $\mathbf{F}(x, y, z) = yz\,\mathbf{i} + x\,\mathbf{j} + z^2\,\mathbf{k}$, with C the straight line segment from the origin to $(1, 0, 4)$.

35. $\mathbf{F}(x, y, z) = -x^2 y\,\mathbf{i} + x^3\,\mathbf{j} + y^2\,\mathbf{k}$, with C the circular helix parametrized by $x = \cos t$, $y = \sin t$, $z = t$, for $2\pi \le t \le 3\pi$.

36. $\mathbf{F}(x, y, z) = \frac{2}{1-z}\,\mathbf{i} + \mathbf{j} - \frac{1}{z^2+1}\,\mathbf{k}$, with C the curve parametrized by $x = \ln(1+t)$, $y = \ln(1-t^2)$, $z = t$, for $3 \le t \le 7$.

Use the Fundamental Theorem of Line Integrals, if applicable, to evaluate the integrals in Exercises 37–44. Otherwise, show that the vector field is not conservative.

37. $\mathbf{F}(x, y) = \left\langle \frac{1}{x} + \ln y, \frac{1}{y} + \frac{x}{y} \right\rangle$, with C the straight line segment from (π, e) to $(1, \pi)$.

38. $\mathbf{F}(x, y) = \langle 2^x \ln 2, 2y \rangle$, with C the straight line segment from $(3, 7)$ to $(0, 1)$.

39. $\mathbf{F}(x, y, z) = \langle -z, 1, x \rangle$, with C the circular helix given by $x = \cos t$, $y = t$, $z = \sin t$, for $0 \le t \le 2\pi$.

40. $\mathbf{F}(x, y) = \left\langle \frac{xe^x}{(x+y+4)^2}, \frac{-e^x}{(x+y+4)^2} \right\rangle$, with C the cardioid given by $r = (1 + 2\sin\theta)$ from $\theta = 0$ to $\theta = \pi$.

41. $\mathbf{F}(x, y, z) = yz^{xy} \ln z\,\mathbf{i} + xz^{xy} \ln z\,\mathbf{j} + \frac{z^{xy}}{z}\,\mathbf{k}$, with C any curve from $(0, 0, 1)$ to $(2, 16, 3)$.

42. $\mathbf{F}(x, y, z) = (\cos(zy) - \sin(zy))\,\mathbf{i} - (xz\sin(zy) + xz\cos(zy))\,\mathbf{j} - (xy\sin(zy) + xy\cos(zy))\,\mathbf{k}$, with C any curve from $\left(5, \frac{\pi}{2}, \frac{\pi}{3}\right)$ to $(2, 0, 2\pi)$.

43. $\mathbf{F}(x, y, z) = z \sin x\,\mathbf{i} + x \ln(y+4)\,\mathbf{j} + y\,\mathbf{k}$, with C the unit circle traversed counterclockwise starting at the point $(1, 0)$.

44. $\mathbf{F}(x, y, z) = (z^3 + 1)\,\mathbf{i} + x \cos z\,\mathbf{j} + xye^{xyz}\,\mathbf{k}$, with C the curve parametrized by $\mathbf{r}(t) = t^2\mathbf{i} + t^3\mathbf{j} - t\mathbf{k}$ for $0 \le t \le 4$.

Evaluate the line integrals in Exercises 45–50.

45. $\int_C \mathbf{F}(x, y) \cdot d\mathbf{r}$, where $\mathbf{F}(x, y) = -y\,\mathbf{i} + x\,\mathbf{j}$ and C is the spiral $x = t\cos t$, $y = t\sin t$, for $\pi \le t \le 2\pi$.

46. $\int_C \mathbf{F}(x, y, z) \cdot d\mathbf{r}$, where

$$\mathbf{F}(x, y, z) = \frac{y^2}{z}\,\mathbf{i} + \frac{2xy}{z}\,\mathbf{j} - \frac{xy^2}{z^2}\,\mathbf{k}$$

and C is the portion of the conical helix $x = t\cos t$, $y = t\sin t$, $z = t$, for $t = \pi$ to $t = \frac{3\pi}{2}$.

47. $\int_C f(x, y)\, ds$, where $f(x, y) = (x^2 + y^2)^{1/2}$ and C is the curve parametrized by $x = t\sin t$, $y = t\cos t$, for $0 \le t \le 4\pi$.

48. $\int_C \mathbf{F}(x, y, z) \cdot d\mathbf{r}$, where

$$\mathbf{F}(x, y, z) = \frac{\ln(z+4)}{x}\,\mathbf{i} + xyz\,\mathbf{j} + y\,\mathbf{k}$$

and C is the curve parametrized by $x = 4t$, $y = t^2$, $z = 1 - t$, for $0 \le t \le 1$.

49. $\int_C \mathbf{F}(x, y, z) \cdot d\mathbf{r}$, where

$$\mathbf{F}(x, y, z) = (yze^{xyz} + 2)\,\mathbf{i} + (xze^{xyz} - 1)\,\mathbf{j} + xye^{xyz}\,\mathbf{k}$$

and C is the curve of intersection of the surface $z = \sqrt{x^2 + y^2}$ and the plane $z - x + y = 10$.

50. $\int_C f(x, y, z)\, ds$, where $f(x, y, z) = (x+z)3^y$ and C is the curve parametrized by $x = 2 - t$, $y = 4t$, $z = t + 5$, for $t = 1$ to $t = 4$.

Applications

In Exercises 51–54, assume that a thin wire in space follows the given curve C and that the density of the wire at any point on C is given by $\rho(x, y, z)$. Find the mass of the wire by computing the line integral of the density function along the specified curve.

51. C is the portion of the unit circle that lies in the first quadrant, and $\rho(x, y) = \ln(e^2\sqrt{x^2 + y^2})$.

52. C is the straight line through $(5, -10)$ and $(0, 2)$, and $\rho(x, y) = e^{x+y}$.

53. C is parametrized by $\mathbf{r}(t) = \left\langle 2t + 1, 10 - t, \frac{t^2}{2} \right\rangle$ from $t = 0$ to $t = 1$, and $\rho(x, y, z) = (x + 2y)\sqrt{2z + 5}$.

54. C is the conical helix parametrized by $\mathbf{r}(t) = \langle t\cos t, t\sin t, t \rangle$ from $t = 0$ to $t = 3\pi$, and $\rho(x, y, z) = \sqrt{x^2 + y^2 + z^2}$.

In Exercises 55–58, find the work done by the given vector field **F** along the specified curve C.

55. $\mathbf{F}(x, y) = y\mathbf{i} - x\mathbf{j}$, where C is the portion of the circle of radius 2, centered at the origin, and in the first quadrant, traversed counterclockwise from (2, 0) to (0, 2).

56. $\mathbf{F}(x, y) = xe^{y}\mathbf{i} + 2y\mathbf{j}$, where C is the curve $y = \ln x$ from $(e, 1)$ to $(e^2, 2)$.

57. $\mathbf{F}(x, y, z) = x \ln y\mathbf{i} + z\mathbf{j} + z^2\mathbf{k}$, where C is the curve parametrized by $\langle t, e^t, e^{-t}\rangle$ from $(1, e, 1/e)$ to $(\ln 2, 2, 1/2)$.

58. $\mathbf{F}(x, y, z) = yz\mathbf{i} + xz\mathbf{j} + xy\mathbf{k}$, where C is the curve that is on the hyperbolic saddle and is parametrized by $\langle \cos t, \sin t, \cos^2 t - \sin^2 t\rangle$ for $0 \le t \le \frac{\pi}{4}$.

For a function $f(x, y)$ defined along a curve C in the plane, we may define the ***average value*** of f on C to be the integral of f along C, divided by the length of C:

$$f\text{avg}(C) = \frac{\int_C f(x, y)\, ds}{\int_C ds}.$$

A similar definition applies to functions of three variables. (See also Exercise 19.) In Exercises 59 and 60, use these definitions to find the average value of the specified function along the given curve.

59. $f(x, y) = \sin^{-1} y \cos^{-1} x$, with C the portion of the unit circle that lies in the first quadrant.

60. $g(x, y, z) = xze^{xyz}$, with C the straight line segment from the origin to the point (1, 1, 1).

61. If C is a smooth closed curve in the plane, the ***flux*** of a vector field $\mathbf{F}(x, y)$ across C measures the net flow out of the region enclosed by C and is defined to be

$$\int_C \mathbf{F}(x, y) \cdot \mathbf{n}\, ds,$$

where $\mathbf{n}$ is a unit vector that is perpendicular to C and points in the "outward" direction. Find the flux of $\mathbf{F}(x, y) = xe^{x^2+y^2}\mathbf{i} + ye^{x^2+y^2}\mathbf{j}$ across the boundary of the unit circle. (On the unit circle, $\mathbf{n} = \langle x, y\rangle$.)

62. Ian is planning a trip into the Wind River Range of mountains in Wyoming. He will carry ice gear, rock gear, and backpacking gear to a total of 80 pounds. He will follow a trail so that his position is given by

$$\langle 0.7t^2 + 1, 1.5t(2 - t), 0.1t^3\rangle \text{ for } t \in [0, 1].$$

Distances are given in miles. How much work does Ian have to do to carry his pack on this trail?

63. Annie is making a crossing from one island to another, with a strong current in the channel. Her path is given by $\mathbf{r}(t) = \langle t, 0.2t(1.2 - t)\rangle$ for $t \in [0, 1.2]$, while the velocity of the current is $V = \langle 0, 0.9x(1.2 - x)\rangle$, with the times in hours and the speeds in miles per hour.

(a) Is the velocity of the current a conservative vector field?

(b) Evaluate the line integral of the velocity of the current along the curve of Annie's paddling path.

(c) What is the physical significance of the line integral?

Proofs

64. Prove that if **F** is a conservative vector field, then the line integral of **F** along any smooth closed curve C is zero.

65. Prove that if $\mathbf{F}(x, y, z)$ is a conservative vector field, then line integrals of $\mathbf{F}(x, y, z)$ are independent of path.

Thinking Forward

Integration across a surface: In the next section, we will use the results of this section to enable us to deal with integrals on surfaces. We already know some of the basics, developed in Chapter 11.

- Our original computation of the arc length of a space curve in Chapter 11 used the "subdivide, approximate, and add" strategy and then took a limit, to end up with $\int \|\mathbf{r}(t)\|\, dt = \int ds$ as the integral used to compute arc length. How could this construction be modified to compute the area of a surface S that is not part of the xy-plane?
- What should be added to the answer to the previous problem to compute the average value of a function $f(x(t), y(t), z(t))$ across a smooth surface S?
- How could the integral of a vector field along a curve that we developed in this section be generalized to integrate a vector field across a surface?

14.3 SURFACES AND SURFACE INTEGRALS

- Using parametric equations to define surfaces
- Computing the integral of a multivariate function on a surface
- Using an integral to compute the flow through a surface

Parametrized Surfaces

In this section we will address the question of how to compute double integrals over surfaces that do not lie in a coordinate plane, and we will see how to integrate a vector field across a surface. In essence, this section generalizes double integrals in the same way that the previous section generalized integrals of functions of one variable. Surface area in the context of integrating over surfaces plays a role analogous to that of arc length in the context of line integrals: In both cases, the basic size of the object is calculated.

The surfaces we are most familiar with are those that are the graphs of functions of two variables, $z = f(x, y)$. One such surface, given by $z = x^2 - y^2$ for $x \in [-1, 3]$ and $y \in [-1, 3]$, is the following:

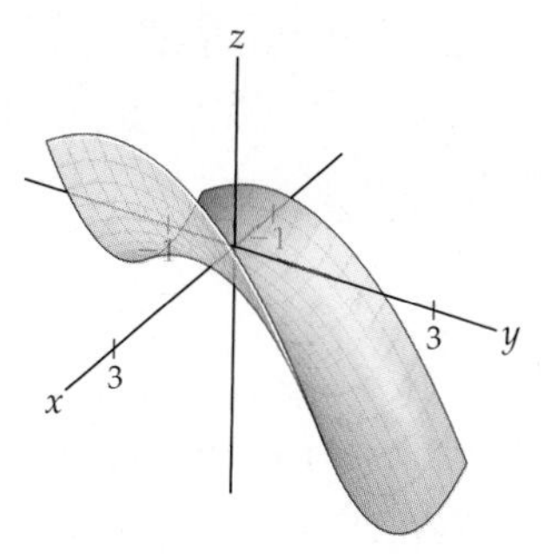

As we have done in other contexts, we will move back and forth between expressing a surface S in terms of points and in terms of position vectors.

Some surfaces cannot be expressed in the form $z = f(x, y)$ but may still be obtained by distorting a portion of a plane while preserving its two-dimensionality. In general, a surface is the image of a region of the uv-plane under a change of variables like those we studied in Section 13.7. Just as a curve C is called a parametrized curve when described by $\mathbf{r}(t) = x(t)\mathbf{i} + y(t)\mathbf{j} + z(t)\mathbf{k}$, a surface S given by

$$\mathbf{r}(u, v) = x(u, v)\mathbf{i} + y(u, v)\mathbf{j} + z(u, v)\mathbf{k},$$

for points $(u, v) \in D$, where D is a region in the uv-plane, is known as a ***parametrized surface***. Typically D is a relatively simple subset of the uv-plane, such as a rectangle or a circle. For instance, consider the half of the unit sphere in $\mathbb{R}^3$ that lies on the nonnegative side of the xz-plane:

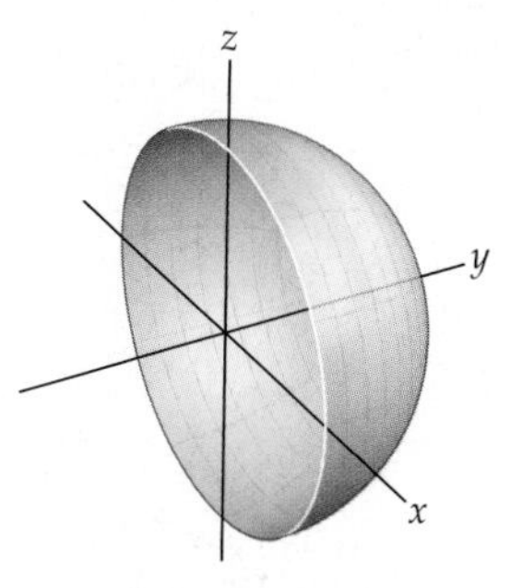

This surface can be parametrized by

$$\mathbf{r}(u,v) = \langle \sin u \sin v, \cos u, \sin u \cos v \rangle, \text{ where } u \in \left[0, \frac{\pi}{2}\right] \text{ and } v \in [0, 2\pi].$$

Since we are interested in doing calculus, we want to deal only with smooth surfaces and parametrizations. A parametrization

$$\mathbf{r}(u,v) = x(u,v)\mathbf{i} + y(u,v)\mathbf{j} + z(u,v)\mathbf{k} \text{ for } (u,v) \in D$$

is ***smooth*** if it is differentiable, if it has a continuous first derivative, and if $\nabla\mathbf{r}(u,v) \neq \mathbf{0}$ for all $(u,v) \in D$. A surface $\mathcal{S}$ is smooth if it has a smooth parametrization. We will also work only with surfaces that are ***orientable***; that is, surfaces for which there is a consistent choice of normal vector $\mathbf{n}$. Planes and spheres have this property. The Möbius band is an example of a ***nonorientable*** surface, since there is no consistent choice of normal vector.

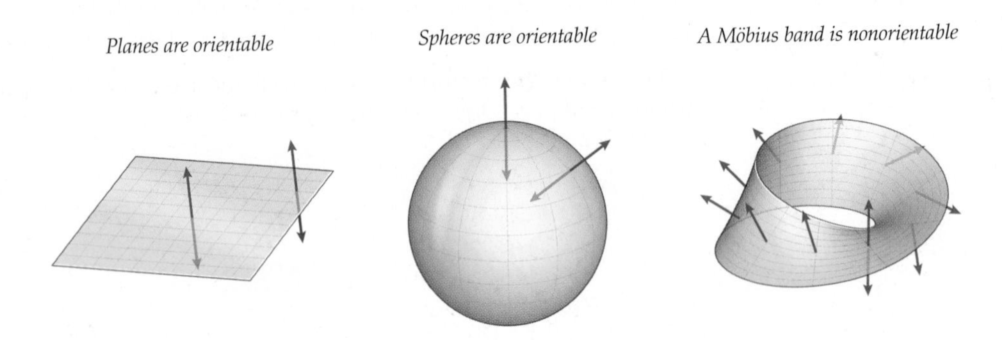

Nonorientable surfaces will not be discussed in this book, although they are interesting in their own right.

Surface Area and Surface Integrals of Multivariate Functions

If we restrict our attention to smooth surfaces, we can use the "subdivide, approximate, and add" strategy first to approximate surface area and then to take the limit of increasingly fine approximations. This limit will result in an integration that yields the exact surface area.

Just as we approximated curves by their tangent vectors, we approximate surfaces by their tangent planes. Recall from Section 12.3 that the vectors $\langle 1, 0, f_x(x_0, y_0)\rangle$ and $\langle 0, 1, f_y(x_0, y_0)\rangle$ both lie in the plane tangent to the graph of $z = f(x,y)$ at the point $(x_0, y_0, f(x_0, y_0))$.

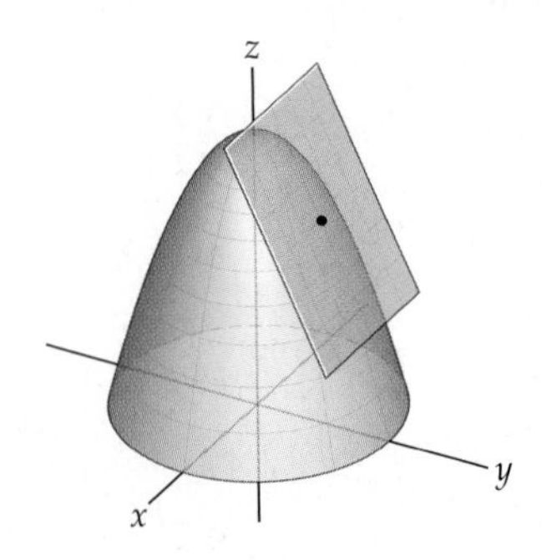

Since these vectors lie in the tangent plane, the parallelogram they determine is a piece of the tangent plane. We will use such parallelograms to approximate the area of the surface.

Starting with the familiar case, suppose we are interested in the area of the surface S described by

$$x\mathbf{i} + y\mathbf{j} + z(x, y)\mathbf{k}$$

for $(x, y) \in D \subseteq \mathbb{R}^2$, with $z(x, y)$ smooth on D. By partitioning D into n subregions and choosing (x_n, y_n) in each region, we can use the area of the parallelogram with sides $\langle 1, 0, z_x(x_n, y_n)\rangle\Delta x$ and $\langle 0, 1, z_y(x_n, y_n)\rangle\Delta y$,

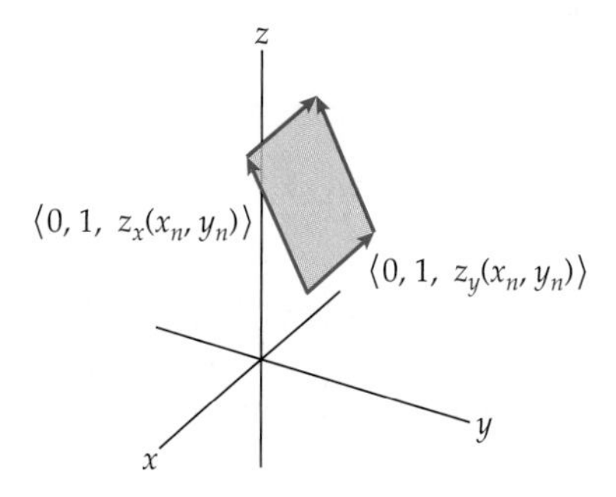

as shown in the preceding figure, to approximate the area of the part of the surface that lies over the nth subregion:

$$\begin{aligned} A_n &\approx \left\|\langle 1, 0, z_x(x_n, y_n)\rangle\Delta x \times \langle 0, 1, z_y(x_n, y_n)\rangle\Delta y\right\| \\ &\approx \sqrt{(z_x(x_n, y_n))^2 + (z_y(x_n, y_n))^2 + 1}\Delta x\Delta y. \end{aligned}$$

Adding approximations over the subregions, we have

$$\text{Surface area} \approx \sum_{n=1}^{M} \sqrt{(z_x(x_n, y_n))^2 + (z_y(x_n, y_n))^2 + 1}\Delta x\Delta y.$$

The limit as the sizes of the subregions approach zero gives the exact surface area:

$$\begin{aligned} \text{Surface area} &= \int_D \sqrt{(z_x)^2 + (z_y)^2 + 1}\, dA \\ &= \iint_S \sqrt{(z_x)^2 + (z_y)^2 + 1}\, dy\, dx. \end{aligned}$$

If instead S is a general surface parametrized by $\mathbf{r}(u, v) = \langle x(u, v), y(u, v), z(u, v)\rangle$, then the approximation is done with the formulas

$$\mathbf{r}_u(u, v) = x_u\mathbf{i} + y_u\mathbf{j} + z_u\mathbf{k} \quad \text{and} \quad \mathbf{r}_v(u, v) = x_v\mathbf{i} + y_v\mathbf{j} + z_v\mathbf{k},$$

both of which lie in the plane tangent to S at the point $(x(u, v), y(u, v), z(u, v))$. Substituting $\|\mathbf{r}_u \times \mathbf{r}_v\|$ for $\|\langle 1, 0, z_x(x_n, y_n)\rangle \times \langle 0, 1, z_y(x_n, y_n)\rangle\|$ in the approximation for A_n, and dA for $\Delta x\, \Delta y$ yields

$$\text{Surface area} = \int_D \|\mathbf{r}_u(x, y, z) \times \mathbf{r}_v(x, y, z)\|\, dA.$$

This equation motivates our definition of the area of a surface and simplifies some notation as well.

DEFINITION 14.6 Surface Area

The ***surface area*** of a smooth surface S is

$$\int_S 1\, dS.$$

(a) If S is given by $z = f(x, y)$ for $(x, y) \in D \subseteq \mathbb{R}^2$, then

$$dS = \sqrt{\left(\frac{\partial z}{\partial x}\right)^2 + \left(\frac{\partial z}{\partial y}\right)^2 + 1}\; dA.$$

(b) If S is parametrized by $\mathbf{r}(u, v)$ for $(u, v) \in D$, then

$$\begin{aligned} dS &= \|\mathbf{r}_u \times \mathbf{r}_v\|\, dA \\ &= \|\mathbf{r}_u \times \mathbf{r}_v\|\, du\, dv. \end{aligned}$$

Once we know how to find the area of a surface, it is easy to see what we ought to do to integrate a function $f(x, y, z)$ on that surface: multiply the area integrand dS by $f(x, y, z)$, and integrate. Then the integral of a function $f(x, y, z)$ over a surface S is

$$\int_S f(x, y, z)\, dS$$

The motivation for this definition of the integral of a function over a surface is the same as our motivation for the integral of a function along a curve. In each case, the integrand becomes

$$(\text{value of function }) \times (\text{size of region}),$$

where the size of the region is a length or an area, depending on the dimension. Thus, our definition of the integral of a multivariate function over a surface is as follows:

DEFINITION 14.7 Surface Integral of a Multivariate Function

The ***integral of*** $f(x, y, z)$ ***over a smooth surface*** S is

$$\begin{aligned} \int_S f(x, y, z)\, dS &= \iint_D f(x(u, v), y(u, v), z(u, v))\|\mathbf{r}_u \times \mathbf{r}_v\|\, dA \\ &= \iint_D f(x(u, v), y(u, v), z(u, v))\|\mathbf{r}_u \times \mathbf{r}_v\|\, du\, dv, \end{aligned}$$

where D is the domain of the smooth parametrization of S by $\mathbf{r}(u, v)$.

This definition is more tricky in practice than it appears, since S may be quite complex. One of the attractive features of the Divergence Theorem, which appears later in this chapter, is that it provides an alternative way of evaluating some surface integrals.

The Flux of a Vector Field Across a Surface

Integrating a vector field over a surface is similar to integrating a vector field along a curve, although in the case of a surface there is a conceptual change beyond the increase in dimension. The interaction of vector field and surface of most general interest is the *flux* through the surface; that is, the measurement of what is transferred through the surface by the action of the vector field. As an example, consider the passage of fluid through a membrane. If we know the rates and directions of flow at every point on the surface, we should be able to find the total flow through the surface by integrating the rates of flow at every point on the surface. We are thus summing the action of vector field $\mathbf{F}(x, y, z)$ through the surface, in the direction normal to the surface at that point. Because we are interested only in the direction that is perpendicular, and not a magnitude, we will represent the normal direction by a unit normal vector $\mathbf{n}$. Because we have restricted our attention to surfaces that are orientable, any surface we consider has a continuous and well-defined unit normal vector. We need only to decide in which direction we want to travel across the surface. For instance, the vectors $\mathbf{k}$ and $-\mathbf{k}$ are both unit vectors perpendicular to the xy-plane; to evaluate the flux of a vector field through the xy-plane, as shown in the figure that follows, we would need to know which direction we were interested in, $\mathbf{k}$ or $-\mathbf{k}$.

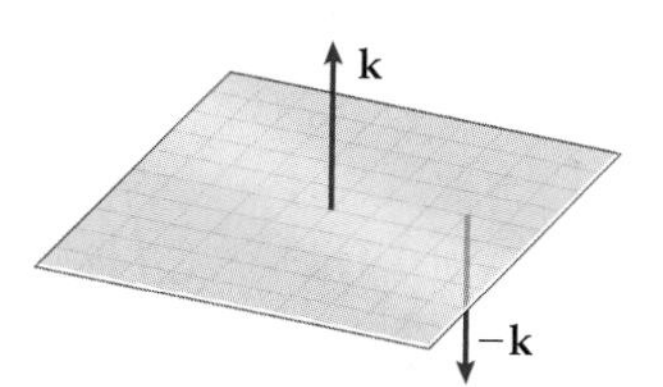

Now that we have resolved to count the activity of $\mathbf{F}(x, y, z)$ in the direction normal to S, our usual strategy of "subdivide, approximate, and add" will involve subdividing the surface into small subregions, then choosing an arbitrary point (x_n, y_n, z_n) in each subregion, and approximating the total flow through that region by computing $\mathbf{F}(x_n, y_n, z_n) \cdot \mathbf{n}$, where $\mathbf{n}$ is the unit normal vector at (x_n, y_n, z_n).

DEFINITION 14.8

Flux of a Vector Field Across a Surface

If $\mathbf{F}(x, y, z)$ is a continuous vector field defined on an oriented surface S given by $\mathbf{r}(u, v)$ for $(u, v) \in D$ with unit normal vector $\mathbf{n}$, then the ***flux*** of $\mathbf{F}(x, y, z)$ through S is

$$\int_S \mathbf{F}(x, y, z) \cdot \mathbf{n}\, dS.$$

(a) If S is the graph of $z = z(x, y)$, then

$$\int_S \mathbf{F}(x, y, z) \cdot \mathbf{n}\, dS = \iint_D (\mathbf{F}(x, y, z) \cdot \mathbf{n}) \|\mathbf{z}_x \times \mathbf{z}_y\|\, dy\, dx.$$

(b) If S is parametrized by u and v, then

$$\int_S \mathbf{F}(x, y, z) \cdot \mathbf{n}\, dS = \iint_D (\mathbf{F}(x, y, z) \cdot \mathbf{n}) \|\mathbf{r}_u \times \mathbf{r}_v\|\, du\, dv.$$

Note that $\mathbf{n}$, being a unit normal vector to S, satisfies

$$\mathbf{n} = \pm \frac{\mathbf{r}_u \times \mathbf{r}_v}{\|\mathbf{r}_u \times \mathbf{r}_v\|}.$$

We might be tempted to cancel the magnitudes and simply integrate $\mathbf{F} \cdot (\mathbf{r}_u \times \mathbf{r}_v)$, but this is correct only if $\mathbf{r}_u \times \mathbf{r}_v$ happens to point in the desired direction.

Examples and Explorations

EXAMPLE 1

Computing surface areas

Find the areas of the following surfaces:

(a) The portion of the plane $z = 3x + 5y$ that lies above the rectangle $[0, 1] \times [1, 5]$ in the xy-plane.

(b) The unit sphere.

SOLUTION

(a) The portion of the plane we seek is the graph of the function $f(x, y) = 3x + 5y$ with $f_x = 3$ and $f_y = 5$. Then

$$dS = \sqrt{9 + 25 + 1}\, dA = \sqrt{35}\, dA.$$

So,

$$\int_S dS = \int_1^5 \int_0^1 \sqrt{35}\, dx\, dy = 4\sqrt{35}.$$

(b) The unit sphere is parametrized by

$$\mathbf{r}(u, v) = \langle \cos u \sin v, \sin u \sin v, \cos v \rangle \text{ for } u \in [0, 2\pi] \text{ and } v \in [0, \pi].$$

Thus,

$$\mathbf{r}_u = \langle -\sin u \sin v, \cos u \sin v, 0 \rangle \quad \text{and} \quad \mathbf{r}_v = \langle \cos u \cos v, \sin u \cos v, -\sin v \rangle.$$

You should verify that

$$\|\mathbf{r}_u \times \mathbf{r}_v\| = \sin v.$$

So,

$$\int_S dS = \int_0^{2\pi} \int_0^{\pi} \sin v\, dv\, du = 4\pi.$$

□

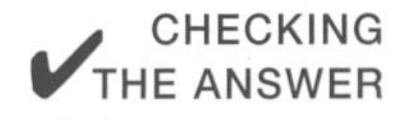
CHECKING THE ANSWER

(a) The surface here is a parallelogram whose four vertices are $(0, 1, f(0, 1)) = (0, 1, 5)$, $(1, 1, f(1, 1)) = (1, 1, 8)$, $(1, 5, f(1, 5)) = (1, 5, 28)$, and $(0, 5, f(0, 5)) = (0, 5, 25)$. This parallelogram is congruent to the parallelogram determined by the vectors $\langle 1, 0, 3 \rangle$ and $\langle 0, 4, 20 \rangle$. As we saw in Section 10.4, the area of the latter parallelogram is $\|\langle 1, 0, 3 \rangle \times \langle 0, 4, 20 \rangle\|$. You may check that the value of this norm is also $4\sqrt{35}$.

(b) The surface area of a sphere with radius R is given by $4\pi R^2$. Since the radius was 1 in our example, we obtained the correct result.

EXAMPLE 2

Integrating a multivariate function on a surface

Integrate the indicated functions over the accompanying surfaces:

(a) The function $g(x, y, z) = 4x^2 + 4y^2 + 1$ over the portion of the surface $\mathcal{S}$ defined by $z = 1 - x^2 - y^2$ and that lies above the unit disk in the xy-plane.

(b) The function $h(x, y, z) = e^{x-2z}$, where $\mathcal{S}$ is the surface parametrized by

$$\mathbf{r}(u, v) = \langle 2u + 3v, 7v, u + 5v \rangle \text{ for } 0 \le u \le \pi \text{ and } 2 \le v \le 4.$$

SOLUTION

(a) This surface is the graph of z as a function of x and y, where $\frac{\partial z}{\partial x} = -2x$ and $\frac{\partial z}{\partial y} = -2y$, so we have

$$dS = \sqrt{4x^2 + 4y^2 + 1}\, dA$$

and

$$\int_S g(x, y, z)\, dS = \iint_{\text{unit disk}} (4x^2 + 4y^2 + 1)\sqrt{4x^2 + 4y^2 + 1}\, dA.$$

It is easier to evaluate this integral with polar coordinates. Recall that $x^2 + y^2 = r^2$ and $dA = r\,dr\,d\theta$. Since we are integrating above the unit disk, we have

$$\begin{aligned}\int_0^{2\pi}\int_0^1 (4r^2 + 1)\sqrt{4r^2 + 1}\, r\,dr\,d\theta &= \int_0^{2\pi}\int_0^1 r(4r^2 + 1)^{3/2}\, dr\,d\theta \\ &= \int_0^{2\pi} \frac{1}{20}\left[(4r^2 + 1)^{5/2}\right]_{r=0}^{r=1} d\theta \\ &= \int_0^{2\pi} \frac{5^{5/2} - 1}{20}\, d\theta = \frac{\pi(25\sqrt{5} - 1)}{10}.\end{aligned}$$

(b) This surface is parametrized by $\mathbf{r}(u, v) = \langle 2u + 3v, 7v, u + 5v\rangle$. We have

$$\mathbf{r}_u = \langle 2, 0, 1\rangle \quad \text{and} \quad \mathbf{r}_v = \langle 3, 7, 5\rangle.$$

Then

$$dS = \|\mathbf{r}_u \times \mathbf{r}_v\| dA = \|\langle -7, -7, 14\rangle\|\, dA = 7\sqrt{6}\, dA.$$

We rewrite h in terms of the parameters u and v, giving

$$h(x(u, v), y(u, v), z(u, v)) = e^{(2u+3v)-2(u+5v)} = e^{-7v}.$$

So,

$$\begin{aligned}\int_S h(x, y, z)\, dS &= 7\sqrt{6}\int_D e^{-7v} dA \\ &= 7\sqrt{6}\int_0^{\pi}\int_2^4 e^{-7v}\, dv\,du \\ &= 7\sqrt{6}\int_0^{\pi}\left[-\frac{1}{7}e^{-7v}\right]_2^4 du \\ &= \sqrt{6}\int_0^{\pi}(e^{-14} - e^{-28})\, du = \frac{\sqrt{6}\pi(e^{14} - 1)}{e^{28}}.\end{aligned}$$

□

EXAMPLE 3 **The mass of a lamina**

Recall that a lamina is a two-dimensional object with a density function ρ and that the mass is the integral of the density function over the region occupied by the lamina.

Find the mass of each lamina:

(a) S is the portion of the cone $z = \sqrt{x^2 + y^2}$ that lies above the disk of radius 3 in the xy-plane and centered at the origin. S has the constant density function $\rho(x, y, z) = k$.

(b) S is the same surface, but with density function $\rho(x, y, z) = x^2 + y^2 + z^2$.

SOLUTION

(a) This surface, shown next, is a graph of z in terms of x and y, with

$$\frac{\partial z}{\partial x} = \frac{x}{\sqrt{x^2+y^2}} \quad \text{and} \quad \frac{\partial z}{\partial y} = \frac{y}{\sqrt{x^2+y^2}}.$$

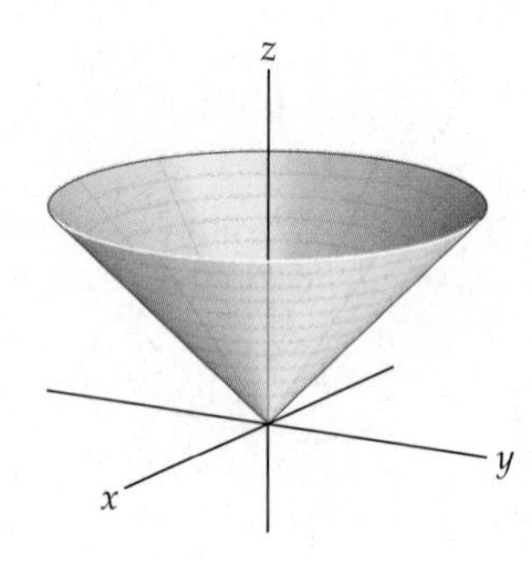

So,

$$dS = \sqrt{\frac{x^2}{x^2+y^2} + \frac{y^2}{x^2+y^2} + 1}\, dA = \sqrt{2}\, dA.$$

The mass of this lamina is given by

$$\int_S k\, dS = \int_D k\sqrt{2}\, dA,$$

where D is the circle with radius 3 and centered at the origin. Because of the symmetry of the lamina, it is convenient to integrate with polar coordinates. We have

$$\int_0^{2\pi}\int_0^3 k\sqrt{2}\, r\, dr\, d\theta = \int_0^{2\pi}\left[\frac{k\sqrt{2}r^2}{2}\right]_0^3 d\theta$$
$$= \int_0^{2\pi} \frac{9k\sqrt{2}}{2}\, d\theta = 9k\sqrt{2}\pi.$$

(b) In this case the differential, dS, is the same as before, but

$$\rho(x,y,z) = x^2+y^2+z^2 = x^2+y^2+\left(\sqrt{x^2+y^2}\right)^2 = 2x^2+2y^2,$$

since $z = \sqrt{x^2+y^2}$. So,

$$\int_S \rho(x,y,z)\, dS = \int_D (2x^2+2y^2)\sqrt{2}\, dA$$
$$= \int_0^{2\pi}\int_0^3 2r^2\sqrt{2}\, r\, dr\, d\theta$$
$$= \int_0^{2\pi}\int_0^3 2\sqrt{2}r^3\, dr\, d\theta$$
$$= \int_0^{2\pi}\left[\frac{\sqrt{2}}{2}r^4\right]_0^3 d\theta$$
$$= \int_0^{2\pi} \frac{81\sqrt{2}}{2}\, d\theta = 81\sqrt{2}\pi.$$

□

EXAMPLE 4 **Finding the flux through a surface**

Find the fluxes of

(a) $\mathbf{F}(x, y, z) = yz\mathbf{i} + 3x\cos z\mathbf{j} - 4y^2x\mathbf{k}$ through the rectangle $1 \le y \le 10$ and $2 \le z \le 4$ in the yz-plane in the positive x direction.

(b) $\mathbf{F}(x, y, z) = x\mathbf{i} + y\mathbf{j} + z\mathbf{k}$ through the upper half of the unit sphere in the direction of the unit normal vector with positive z-component.

SOLUTION

(a) This surface is just a piece of the yz-plane. It is parametrized by

$$y\mathbf{j} + z\mathbf{k} \qquad 1 \le y \le 10,\ 2 \le z \le 4.$$

Because of the orientation, $\mathbf{n} = \mathbf{i}$, so

$$\mathbf{F}(x, y, z) \cdot \mathbf{n} = (yz\mathbf{i} + 3x\cos z\mathbf{j} - 4y^2x\mathbf{k}) \cdot \mathbf{i} = yz.$$

The flux of $\mathbf{F}(x, y, z)$ through S is

$$\int_S \mathbf{F} \cdot \mathbf{n}\, dS = \int_1^{10} \int_2^4 yz\, dz\, dy = 297.$$

(b) Here the surface is given by

$$z = \sqrt{1 - x^2 - y^2}.$$

The choice of $\mathbf{n}$ should have a positive z-component. We recall from Chapter 12 that if $z = f(x, y)$, then the vector

$$\left\langle 1, 0, \frac{\partial z}{\partial x} \right\rangle \times \left\langle 0, 1, \frac{\partial z}{\partial y} \right\rangle = \left\langle -\frac{\partial z}{\partial x}, -\frac{\partial z}{\partial y}, 1 \right\rangle$$

is normal to the surface. This vector already has a positive z-component, so all we have to do is scale it to form a unit vector. We have

$$\mathbf{v} = \left\langle -\frac{\partial z}{\partial x}, -\frac{\partial z}{\partial y}, 1 \right\rangle = \left\langle \frac{x}{\sqrt{1 - x^2 - y^2}}, \frac{y}{\sqrt{1 - x^2 - y^2}}, 1 \right\rangle$$

perpendicular to the surface. This equation yields

$$\frac{\mathbf{v}}{\|\mathbf{v}\|} = \mathbf{n} = \left\langle x, y, \sqrt{1 - x^2 - y^2} \right\rangle$$

as the desired unit normal. We can also use that information to compute

$$dS = \sqrt{\frac{x^2}{1 - x^2 - y^2} + \frac{y^2}{1 - x^2 - y^2} + 1}\, dA = \frac{1}{\sqrt{1 - x^2 - y^2}}\, dA.$$

If we let D be the unit disk in the xy-plane and centered at the origin, then the flux through S in the positive z direction is

$$\begin{aligned}
\int_S \mathbf{F}(x, y, z) \cdot \mathbf{n}\, dS &= \int_D \left(\langle x, y, z \rangle \cdot \left\langle x, y, \sqrt{1 - x^2 - y^2} \right\rangle \right) \frac{1}{\sqrt{1 - x^2 - y^2}}\, dA \\
&= \int_D \left(\left\langle x, y, \sqrt{1 - x^2 - y^2} \right\rangle \cdot \left\langle x, y, \sqrt{1 - x^2 - y^2} \right\rangle \right) \frac{1}{\sqrt{1 - x^2 - y^2}}\, dA \\
&= \int_D \frac{1}{\sqrt{1 - x^2 - y^2}}\, dA \\
&= \int_0^{2\pi} \int_0^1 \frac{r}{\sqrt{1 - r^2}}\, dr\, d\theta \\
&= \int_0^{2\pi} \left[-(1 - r^2)^{1/2} \right]_0^1 d\theta \\
&= \int_0^{2\pi} 1\, d\theta = 2\pi.
\end{aligned}$$

□

? TEST YOUR UNDERSTANDING

- How was the "subdivide, approximate, and add" approach used in motivating the definition of surface area?
- What is measured by $\int_S dS$?
- What is measured by $\int_S f(x, y, z)\, dS$?
- Why does this section refer only to vector fields in $\mathbb{R}^3$?
- What is measured by $\int_S \mathbf{F} \cdot \mathbf{n}\, dS$?

EXERCISES 14.3

Thinking Back

- *Area:* Finding the area of a region in the xy-plane is one of the motivating applications of integration. It is also a special case of the surface area calculation developed in this section. Find the area of the region in the xy-plane bounded by the curves $y = x^2$ and $x = \sqrt{y}$.
- *Double integrals:* The double integrals of Chapter 13 were our first exposure to surface integrals, on the very well-behaved surface of the xy-plane. Evaluate $f(x, y) = x^2 - y^2$ on the region bounded by the x-axis and the curve defined by $y = 4 - x^2$.
- *Average value:* Recall that the average value of a function of two variables $f(x, y)$ on a region Ω is given by the quotient
$$\frac{\iint_\Omega f(x, y)\, dA}{\iint_\Omega dA}.$$
Compute the average value of $g(x, y) = xy$ on the first-quadrant region bounded by $y = x$ and $y = x^2$.

Concepts

0. *Problem Zero:* Read the section and make your own summary of the material.

1. *True/False:* Determine whether each of the statements that follow is true or false. If a statement is true, explain why. If a statement is false, provide a counterexample.

(a) *True or False:* The result of integrating a vector field over a surface is a vector.

(b) *True or False:* The result of integrating a function over a surface is a scalar.

(c) *True or False:* For a region R in the xy-plane, $dS = dA$.

(d) *True or False:* In computing $\int_S f(x, y, z)\, dS$, the direction of the normal vector is irrelevant.

(e) *True or False:* If $f(x, y, z)$ is defined on an open region containing a smooth surface S, then $\int_S f(x, y, z)\, dS$ measures the flow through S in the positive z direction determined by $f(x, y, z)$.

(f) *True or False:* If $\mathbf{F}(x, y, z)$ is defined on an open region containing a smooth surface S, then $\int_S \mathbf{F}(x, y, z) \cdot \mathbf{n}\, dS$ measures the flow through S in the direction of $\mathbf{n}$ determined by the field $\mathbf{F}(x, y, z)$.

(g) *True or False:* In computing
$$\int_S \mathbf{F}(x, y, z) \cdot \mathbf{n}\, dS,$$
the direction of the normal vector is irrelevant.

(h) *True or False:* In computing
$$\int_S \mathbf{F}(x, y, z) \cdot \mathbf{n}\, dS$$
with $\mathbf{n}$ pointing in the correct direction, we could use a scalar multiple of $\mathbf{n}$, since the length will cancel in the dS term.

2. *Examples:* Construct examples of the thing(s) described in the following. Try to find examples that are different than any in the reading.

(a) Two different surfaces with the same area. (Try to make these very different, not just shifted copies of each other.)

(b) Let S be the surface parametrized by
$$\mathbf{r}(u, v) = x(u, v)\mathbf{i} + y(u, v)\mathbf{j} + z(u, v)\mathbf{k}.$$
Give two different unit normal vectors to S at the point $r(u_0, v_0)$.

(c) A smooth surface that can be smoothly parametrized as $\mathbf{r}(x, z) = \langle x, f(x), z\rangle$.

3. Why do surface integrals of multivariate functions *not* include an $\mathbf{n}$ term, whereas surface integrals of vector fields *do* include this term?

4. Make a chart of all the new notation, definitions, and theorems in this section, including what each new item means in terms you already understand.

5. Compute a general formula for dS for any plane $ax+by+cz=k$ if $c \neq 0$.

6. Relate the integrand $\mathbf{F}\cdot\mathbf{n}\,dS$ to the discussions of work in Sections 6.4 and 14.2.

7. Give a smooth parametrization of the upper half of the unit sphere in terms of x and y.

8. Compute dS for your parametrization in Exercise 7.

9. Give a smooth parametrization, in terms of u and v, of the sphere of radius k and centered at the origin.

10. Compute dS for your parametrization in Exercise 9.

11. Given a smooth surface S described as a function $z=f(x,y)$, calculate the upwards-pointing normal vector for S.

12. Give a smooth parametrization for a "generalized cylinder" S, given by extending the curve $y=x^2$ upwards and downwards from $z=-2$ to $z=3$.

13. Compute $\mathbf{n}$ for the surface S in Exercise 12.

14. Generalize your answer to Exercise 12 to give a parametrization and a normal vector for the extension of any differentiable plane curve $y=f(x)$ through $a \le z \le b$.

15. Use what you know about average value from previous sections to propose a formula for the average value of a multivariate function $f(x,y,z)$ on a smooth surface S.

16. Use what you know about averages to propose a formula for the average rate of flux of a vector field $\mathbf{F}(x,y,z)$ through a smooth surface S in the direction of $\mathbf{n}$.

17. If S is parametrized by $\mathbf{r}(u,v)$, why is $dS = \|\mathbf{r}_u \times \mathbf{r}_v\|\,du\,dv$ the correct factor to use to account for distortion of area?

18. Examine the parametrization of surfaces and the computation of dS in light of the discussion of change of variables and the Jacobian in Section 13.7. How are these ideas related?

19. Give a formula for a normal vector to the surface S determined by $x=f(y,z)$, where $f(y,z)$ is a function with continuous partial derivatives.

20. Give a formula for a normal vector to the surface S determined by $y=g(x,z)$, where $g(x,z)$ is a function with continuous partial derivatives.

Skills

Find the areas of the given surfaces in Exercises 21–26.

21. S is the portion of the plane with equation $y-z=\frac{\pi}{2}$ that lies above the rectangle determined by $0 \le x \le 4$ and $3 \le y \le 6$.

22. S is the portion of the plane with equation $x=y+z$ that lies above the region in the xy-plane that is bounded by $y=x$, $y=5$, $y=10$, and the y-axis.

23. S is the portion of the saddle surface determined by $z=x^2-y^2$ that lies above and/or below the annulus in the xy-plane determined by the circles with radii $\frac{\sqrt{3}}{2}$ and $\sqrt{2}$ and centered at the origin.

24. S is the portion of the surface determined by $x=9-y^2-z^2$ that lies on the positive side of the yz-plane (i.e., where $x \ge 0$).

25. S is the portion of the surface parametrized by $\mathbf{r}(u,v) = \langle 3u-v, v+u, v-u\rangle$ whose preimage (the domain in the uv-plane) is the unit square $[0,1]\times[0,1]$.

26. S is the lower branch of the hyperboloid of two sheets $z^2=x^2+y^2+1$ that lies below the annulus determined by $1 \le r \le 2$ in the xy-plane.

Integrate the given function over the accompanying surface in Exercises 27–34.

27. $f(x,z)=e^{-(x^2+z^2)}$, where S is the unit disk centered at the point $(0,2,0)$ and in the plane $y=2$.

28. $f(x,y,z)=x^2-y+3z$, where S is the portion of the plane with equation $2x-6y+3z=1$ whose preimage in the xz-plane is the region bounded by the coordinate axes and the lines with equations $z=4$ and $x=z$.

29. $f(x,y,z) = xyz^2$, where S is the portion of the cone $\sqrt{3}z=\sqrt{x^2+y^2}$ that lies within the sphere of radius 4 and centered at the origin.

30. $f(x,y,z)=e^z$, where S is the portion of the unit sphere in the first octant.

31. $f(x,y,z)=\frac{y}{x}\sqrt{4z^2+1}$, where S is the portion of the paraboloid $z=x^2+y^2$ that lies above the rectangle determined by $1 \le x \le e$ and $0 \le y \le 2$ in the xy-plane.

32. $f(x,y,z) = e^{\sqrt{8y^2-4z+1}}(4z+8x^2+1)^{-1/2}$, where S is the portion of the saddle surface with equation $z=y^2-x^2$ that lies above the region in the xy-plane bounded by the line with equation $y=x$ in the first quadrant, the line with equation $y=-x$ in the second quadrant, and the unit circle.

33. $f(x,y,z)=x-z+y^2$, where S is given by

$$\mathbf{r}(u,v) = (u+v)\,\mathbf{i} + 2\sqrt{u^2+v^2}\,\mathbf{j} + (u-v)\,\mathbf{k}$$

on the region in the uv-plane bounded by the graphs of $v=u$ and $v=u^2$.

34. $f(x,y,z) = \frac{1}{x\ln(z+2y)}$, where S is the surface given by $\mathbf{r}(u,v) = 2u\,\mathbf{i} + (u+v)\,\mathbf{j} + (2u-2v)\,\mathbf{k}$ for $3 \le u \le 7$ and $2 \le v \le 10$.

Find the flux of the given vector field through the given surface in Exercises 35–42.

35. $\mathbf{F}(x,y,z) = \cos(xyz)\,\mathbf{i} + \mathbf{j} - yz\mathbf{k}$, where S is the portion of the surface with equation $z=y^3-y^2$ that lies above and/or below the rectangle determined by $-3 \le x \le 2$ and $-1 \le y \le 1$ in the xy-plane, with $\mathbf{n}$ pointing in the positive z direction.

36. $\mathbf{F} = \left\langle x\ln(xz), 5z, \frac{1}{y^2+1}\right\rangle$, where S is the region of the plane with equation $12x-9y+3z=10$, where $2 \le x \le 3$ and $5 \le y \le 10$, with $\mathbf{n}$ pointing upwards.

37. $\mathbf{F}(x, y, z) = -xz\,\mathbf{i} - yz\,\mathbf{j} + z^2\,\mathbf{k}$, where S is the cone with equation $z = \sqrt{x^2 + y^2}$ between $z = 2$ and $z = 4$, with $\mathbf{n}$ pointing outwards.

38. $\mathbf{F}(x, y, z) = \langle yz, xz, xy \rangle$, where S is the portion of the saddle determined by $z = x^2 - y^2$ that lies above the region in the xy-plane bounded by the x-axis and the parabola with equation $y = 1 - x^2$.

39. $\mathbf{F}(x, y, z) = z \cos yz\,\mathbf{j} + z \sin yz\,\mathbf{k}$, where S is the portion of the plane with equation $2x - 8y - 10z = 42$ that lies on the positive side of the rectangle with corners $(0, -\pi, 0)$, $(0, \pi, 0)$, $(0, \pi, \pi)$, and $(0, -\pi, \pi)$ in the yz-plane.

40. $\mathbf{F}(x, y, z) = \mathbf{i} + \mathbf{j} + \mathbf{k}$, where S is the lower half of the unit sphere, with $\mathbf{n}$ pointing outwards.

41. $\mathbf{F}(x, y, z) = 5\,\mathbf{i} + 13\,\mathbf{j} + 2\,\mathbf{k}$, where S is the unit sphere, with $\mathbf{n}$ pointing outwards.

42. $\mathbf{F}(x, y, z) = -y\,\mathbf{i} + x\,\mathbf{j} - e^{yz}\,\mathbf{k}$, where S is the cylinder with equation $x^2 + y^2 = 9$ from $z = 2$ to $z = 4$, with $\mathbf{n}$ pointing outwards.

Evaluate the integrals in Exercises 43–48.

43. Find $\int_S 1\, dS$, where S is the portion of the surface determined by $z = x^2 - \sqrt{3}y$ that lies above the region in the xy-plane bounded by the x-axis and the lines with equations $y = 2x$ and $x = 3$.

44. Find $\int_S \mathbf{F}(x, y, z) \cdot \mathbf{n}\, dS$ if

$$\mathbf{F}(x, y, z) = \frac{\ln(x^2 + y^2 + 1)}{z + 3}\mathbf{i} + \frac{y}{y + 1}\mathbf{j} + e^{z^2}\mathbf{k},$$

where S is the portion of the sphere with radius 2, centered at the origin, and that lies below the plane with equation $z = -\sqrt{2}$, with $\mathbf{n}$ pointing outwards.

45. Find the integral of $f(x, y, z) = z^3 + z(x^2 + 2^y)$ on the portion of the unit sphere that lies in the first octant, above the rectangle $\left[0, \frac{1}{2}\right] \times \left[0, \frac{1}{3}\right]$ in the xy-plane.

46. Find $\int_S 1\, dS$, where S is the portion of the surface with equation $x = e^{yz} - e^{-yz}$ that lies on the positive side of the circle of radius 3 and centered at the origin in the yz-plane.

47. Find $\int_S \mathbf{F}(x, y, z) \cdot \mathbf{n}\, dS$ if

$$\mathbf{F}(x, y, z) = 2xz\mathbf{i} + 2yz\mathbf{j} - 18\mathbf{k}$$

and S is the portion of the hyperboloid $x^2 + y^2 - 9 = z^2$ that lies between the planes $z = -4$ and $z = 0$, with $\mathbf{n}$ pointing outwards.

48. Find the integral of $f(x, y, z) = x - y - z$ on the portion of the plane with equation $10x - \sqrt{33}y + 36z = 30$ with $2 \le x \le 7$ and $1 \le z \le 2$.

Applications

Find the masses of the laminæ in Exercises 49 and 50.

49. The lamina occupies the region of the hyperbolic saddle with equation $z = x^2 - y^2$ that lies above and/or below the disk of radius 2 about the origin in the xy-plane where the density is uniform.

50. The lamina occupies the region of the hyperboloid with equation $z^2 + 4 = x^2 + y^2$ that lies above and/or below the disk of radius 5 about the origin in the xy-plane, and the density function, $\rho(x, y, z)$, is proportional to distance from the origin.

Find the flux of the given vector field through a permeable membrane described by surface S in Exercises 51 and 52 .

51. $\mathbf{F}(x, y, z) = y\,\mathbf{i} + x\,\mathbf{j} + \mathbf{k}$, where S is the paraboloid with equation $z = 4x^2 + y^2 + 1$ that lies above the annulus determined by $1 \le x^2 + y^2 \le 4$ in the xy-plane.

52. $\mathbf{F}(x, y, z) = -z\,\mathbf{i} + y\,\mathbf{j} + x\,\mathbf{k}$, where S is the surface with equation $y = \cos z$ for $1 \le x \le 5$ and $0 \le z \le \frac{\pi}{2}$.

53. Suppose that an electric field is given by

$$\mathbf{E} = 2y\mathbf{i} + 2xy\mathbf{j} + yz\mathbf{k}.$$

Compute the flux $\int_S \mathbf{E} \cdot \mathbf{n}\, dA$ of the field through the unit cube $[0, 1] \times [0, 1] \times [0, 1]$.

54. Suppose that an electric field is given by

$$\mathbf{E} = \frac{K}{(x^2 + y^2 + z^2)^{3/2}}(x\mathbf{i} + y\mathbf{j} + z\mathbf{k}),$$

where K is a constant. Show that the flux of the field through any sphere centered at the origin is constant.

Mac owns a farm in the Palouse, a large agricultural region in the northwestern United States. One field on his farm can be modeled as the surface $s(x, y) = 0.24\sqrt{x^2 + y^2}$ on the square $[0, 0.25] \times [0, 0.25]$, where all distances are given in miles. Use this information to answer Exercises 55 and 56.

55. What is the actual area of Mac's field?

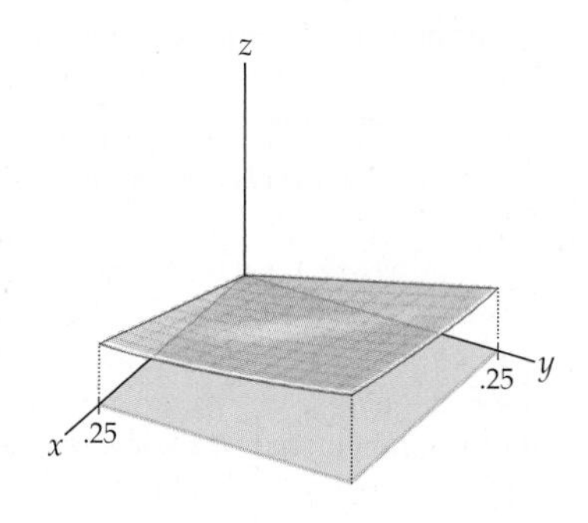

56. Mac is concerned with evapotranspiration from his land: the process wherein moisture from the soil evaporates, or is taken up by plants and then transpired into the air. The problem is that certain parts of his field tend to be drier than others, and since the whole field is tilted so as to face south, the entire field dries out more quickly than if it were flat. The rate at which the field dries out depends on how much solar radiation hits Mac's land (i.e., a flux of radiation into the field). On a hot day in July at noon, the rays of the sun hit the earth parallel to the line $\langle 0, t, -3t \rangle$ for $t \in \mathbb{R}$. Compare the solar flux through this field with that through the horizontal rectangle $[0, 0.25] \times [0, 0.25]$ in the xy-plane that underlies it.

Proofs

57. Show that reversing the orientation of a surface S reverses the sign of $\int_S \mathbf{F}(x, y, z) \cdot \mathbf{n}\, dS$.

58. Show that the two definitions of dS in Definition 14.6 are equivalent, by showing that if S is a surface described by $\mathbf{r}(x, y) = (x, y, z(x, y))$ for $(x, y) \in D$, then

$$\|\mathbf{r}_x(x, y) \times \mathbf{r}_y(x, y)\| = \sqrt{\left(\frac{\partial z}{\partial x}\right)^2 + \left(\frac{\partial z}{\partial y}\right)^2 + 1}.$$

59. Let R be a simply connected region in the xy-plane. Show that the portion of the paraboloid with equation $z = x^2 + y^2$ determined by R has the same area as the portion of the saddle with equation $z = x^2 - y^2$ determined by R.

60. Let a, b, and c be nonzero constants. Find a general formula for the area of the portion of the plane with equation $ax + by + cz = k$ that lies above a rectangle $[\alpha, \beta] \times [\gamma, \delta]$ in the xy-plane.

Thinking Forward

- *Integrating vector fields over three-dimensional regions:* Let W be a three-dimensional region in $\mathbb{R}^3$. If a vector field measures the movement of a gas or a fluid in space, what should the integral of a field over W measure?
- *Comparing double integrals and surface integrals:* Think about how surface integrals compared and contrasted with double integrals in the plane. What changes and what is the same in these two cases?
- *Computations in $\mathbb{R}^4$:* In $\mathbb{R}^3$, a surface is determined by (1) its normal vector at every point and (2) one point on the surface. An intuitive way to think of this is that, in $\mathbb{R}^3$, $\mathbf{n}$ tells us which direction is not occupied by the surface because it is perpendicular to the surface. It can be difficult to visualize, but the same procedure can lifted to solids in $\mathbb{R}^4$. How might we write an integral to determine the volume of a three-dimensional region W in $\mathbb{R}^4$? How might we integrate over surfaces in $\mathbb{R}^4$?

14.4 GREEN'S THEOREM

- Defining the del operator, ∇
- Understanding the relationships between the divergence and the curl
- Extending the Fundamental Theorem of Calculus into the plane

The Fundamental Theorem of Calculus and Green's Theorem

In this section, and in the two that follow, we generalize the spirit of the Fundamental Theorem of Calculus in a new way. Recall that the Fundamental Theorem says that, under appropriate conditions,

$$\int_a^b f(x)\,dx = F(b) - F(a),$$

where $F(x)$ is an antiderivative of $f(x)$. One way of interpreting this theorem is that it relates the behavior of the function $F(x)$ on the boundary of a set in $\mathbb{R}$ to the behavior of the derivative of $F(x)$ in the interior of the region. The region in question is the interval $[a, b]$; its boundary is the set containing points a and b, and its interior is the open interval (a, b).

Given a function $g(x, y)$ defined on a bounded region $\mathcal{R}$ of the plane, it is reasonable, therefore, to ask if there is some relationship between $\int_{\mathcal{R}} g(x, y)\,dA$ and an operation on the curve C that is the boundary of $\mathcal{R}$. Since the boundary curve is more complicated than the two endpoints of an interval in $\mathbb{R}$, we expect that this operation will be more complicated; we shall see that it is an integration. Similarly, we can hope that there is a relationship between $f(x, y, z)$ defined on a bounded region $\mathcal{R}$ in $\mathbb{R}^3$ and an operation on the surface $\mathcal{S}$ that forms the boundary of $\mathcal{R}$. As in the case of line and surface integrals, we will sometimes find that vector fields are relevant to the discussion.

The Del Operator, the Divergence, and the Curl

The theorems of the next three subsections are improved by compact notation involving two quantities associated with vector fields: the *divergence* and the *curl*. To state these quantities, it is convenient and traditional to define a vector operator: the *del operator*, ∇.

DEFINITION 14.9 **The Del Operator, ∇**

The ***del operator***, ∇, is the vector of operations

$$\nabla = \frac{\partial}{\partial x}\mathbf{i} + \frac{\partial}{\partial y}\mathbf{j} + \frac{\partial}{\partial z}\mathbf{k}$$

or, in $\mathbb{R}^2$,

$$\nabla = \frac{\partial}{\partial x}\mathbf{i} + \frac{\partial}{\partial y}\mathbf{j}.$$

This operator is used in dot or cross products with vector fields. For example, if

$$\mathbf{F}(x, y, z) = (x + y)\mathbf{i} + e^{xyz}\mathbf{j} + (x - y + z^2)\mathbf{k},$$

then

$$\begin{aligned}\nabla \cdot \mathbf{F}(x, y, z) &= \frac{\partial}{\partial x}(x + y) + \frac{\partial}{\partial y}e^{xyz} + \frac{\partial}{\partial z}(x - y + z^2)\\ &= 1 + xze^{xyz} + 2z\end{aligned}$$

and

$$\begin{aligned}\nabla \times \mathbf{F}(x,y,z) &= \left(\frac{\partial}{\partial x}\mathbf{i} + \frac{\partial}{\partial y}\mathbf{j} + \frac{\partial}{\partial z}\mathbf{k}\right) \times ((x+y)\mathbf{i} + e^{xyz}\mathbf{j} + (x-y+z^2)\mathbf{k})\\ &= \left(\frac{\partial}{\partial y}(x-y+z^2) - \frac{\partial}{\partial z}(e^{xyz})\right)\mathbf{i} + \left(\frac{\partial}{\partial z}(x+y) - \frac{\partial}{\partial x}(x-y+z^2)\right)\mathbf{j}\\ &\quad + \left(\frac{\partial}{\partial x}(e^{xyz}) - \frac{\partial}{\partial y}(x+y)\right)\mathbf{k}\\ &= (-1 - xye^{xyz})\mathbf{i} - \mathbf{j} + (yze^{xyz} - 1)\mathbf{k}.\end{aligned}$$

For a vector field in $\mathbb{R}^3$, the operation $\nabla \cdot \mathbf{F}(x,y,z)$ produces a scalar that is the sum of $\mathbf{F}$'s directional derivatives in each of the $\mathbf{i}$, $\mathbf{j}$, and $\mathbf{k}$ directions. If $\mathbf{F}(x,y,z)$ represents the flow of a gas or fluid, then the sign of this scalar corresponds to whether the gas or fluid is compressing ($\nabla \cdot \mathbf{F}(x,y,z)$ is negative) or expanding ($\nabla \cdot \mathbf{F}(x,y,z)$ is positive). The scalar is known as the *divergence* of the vector field $\mathbf{F}(x,y,z)$.

DEFINITION 14.10

Divergence of a Vector Field

The ***divergence*** of vector field $\mathbf{F}$ is the dot product of ∇ and $\mathbf{F}$.

(a) In $\mathbb{R}^2$, if $\mathbf{F}(x,y) = F_1(x,y)\mathbf{i} + F_2(x,y)\mathbf{j}$, then

$$\text{div } \mathbf{F}(x,y) = \nabla \cdot \mathbf{F}(x,y) = \frac{\partial F_1}{\partial x} + \frac{\partial F_2}{\partial y}.$$

(b) In $\mathbb{R}^3$, if $\mathbf{F}(x,y,z) = F_1(x,y,z)\mathbf{i} + F_2(x,y,z)\mathbf{j} + F_3(x,y,z)\mathbf{k}$, then

$$\text{div } \mathbf{F}(x,y,z) = \nabla \cdot \mathbf{F}(x,y,z) = \frac{\partial F_1}{\partial x} + \frac{\partial F_2}{\partial y} + \frac{\partial F_3}{\partial z}.$$

The divergence at a point of a vector field measures the magnitude of the rate of change of the vector field away from a source or towards a sink.

In a similar vein, the cross product $\nabla \times \mathbf{F}$ is called the *curl* of the field. At a point in the vector field, the curl corresponds physically to the circulation or rotation of the field at that point. This interpretation of the curl is motivated by applications in engineering and physics.

DEFINITION 14.11

Curl of a Vector Field

The ***curl*** of a vector field $\mathbf{F}(x,y,z) = F_1(x,y,z)\mathbf{i} + F_2(x,y,z)\mathbf{j} + F_3(x,y,z)\mathbf{k}$ is the cross product of ∇ with $\mathbf{F}(x,y,z)$:

$$\text{curl } \mathbf{F} = \nabla \times \mathbf{F}(x,y,z) = \left(\frac{\partial F_3}{\partial y} - \frac{\partial F_2}{\partial z}\right)\mathbf{i} + \left(\frac{\partial F_1}{\partial z} - \frac{\partial F_3}{\partial x}\right)\mathbf{j} + \left(\frac{\partial F_2}{\partial x} - \frac{\partial F_1}{\partial y}\right)\mathbf{k}.$$

If $\mathbf{F}(x,y) = (F_1(x,y), F_2(x,y))$ is a vector field in $\mathbb{R}^2$, we define the curl of $\mathbf{F}$ to be curl $(F_1(x,y), F_2(x,y), 0)$. So,

$$\begin{aligned}\text{curl } \mathbf{F} &= \text{curl } (F_1(x,y), F_2(x,y), 0) = \nabla \times \langle F_1(x,y), F_2(x,y), 0\rangle\\ &= \left(\frac{\partial F_2}{\partial x} - \frac{\partial F_1}{\partial y}\right)\mathbf{k}.\end{aligned}$$

Using the same field as before, namely, $\mathbf{F}(x,y,z) = (x+y)\mathbf{i} + e^{xyz}\mathbf{j} + (x-y+z^2)\mathbf{k}$, we have

$$\text{div } \mathbf{F} = 1 + xze^{xyz} + 2z,$$

which is a scalar because it is the result of a dot product, and

$$\text{curl } \mathbf{F} = (-1 - xye^{xyz})\mathbf{i} - \mathbf{j} + (yze^{xyz} - 1)\mathbf{k},$$

which is a vector because it is the result of a cross product.

Recall that the cross product is not commutative: $\nabla \times \mathbf{F} \neq \mathbf{F} \times \nabla$. In Chapter 10, we saw that the cross product may be computed as the determinant of a 3×3 matrix. We may use that approach to find the curl as well:

$$\begin{aligned} \nabla \times \mathbf{F}(x, y, z) &= \nabla \times \langle F_1(x, y, z), F_2(x, y, z), F_3(x, y, z) \rangle \\ &= \det \begin{bmatrix} \mathbf{i} & \mathbf{j} & \mathbf{k} \\ \frac{\partial}{\partial x} & \frac{\partial}{\partial y} & \frac{\partial}{\partial z} \\ F_1(x, y, z) & F_2(x, y, z) & F_3(x, y, z) \end{bmatrix}. \end{aligned}$$

There is a special relationship between the divergence and curl of a field, on the one hand, and the curl of a conservative vector field, on the other:

THEOREM 14.12 **Divergence of Curl and Curl of a Gradient**

(a) If $\mathbf{F} = \langle F_1(x, y, z), F_2(x, y, z) \rangle$ is a vector field in $\mathbb{R}^2$ or $\mathbf{F} = \langle F_1(x, y, z), F_2(x, y, z), F_3(x, y, z) \rangle$ is a vector field in $\mathbb{R}^3$, for which F_1, F_2, and F_3 have continuous second-order partial derivatives, then

$$\text{div}(\text{curl } \mathbf{F}) = \nabla \cdot (\nabla \times \mathbf{F}) = 0.$$

(b) For a multivariate function f in $\mathbb{R}^2$ or $\mathbb{R}^3$ with continuous second partial derivatives,

$$\text{curl } \nabla f = \nabla \times (\nabla f) = \mathbf{0}.$$

In each case proofs may be obtained by using the equality of mixed partial derivatives. We ask you to prove these identities in Exercises 55 and 56.

Green's Theorem

Our next extension of the Fundamental Theorem of Calculus is in the plane. Consider a region $\mathcal{R}$ of the plane whose boundary is a smooth, simple closed curve C, oriented counterclockwise.

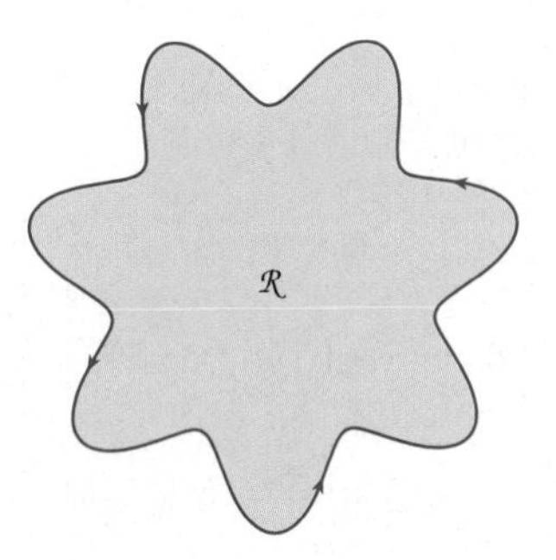

Green's Theorem asserts that integrating along the boundary curve C is equal to integrating a double integral over the interior of $\mathcal{R}$. Recall the notation from Section 14.2: For a vector field $\mathbf{F}(x, y) = \langle F_1(x, y), F_2(x, y) \rangle$ and a curve $\mathbf{r}(t) = \langle x(t), y(t) \rangle$, for $a \leq t \leq b$,

$$\int_C \mathbf{F}(x, y) \cdot d\mathbf{r} = \int_C F_1(x, y)\, dx + F_2(x, y)\, dy = \int_a^b \left(F_1(x, y)\frac{dx}{dt} + F_2(x, y)\frac{dy}{dt} \right) dt.$$

THEOREM 14.13 **Green's Theorem**

Let $\mathbf{F}(x, y) = \langle F_1(x, y), F_2(x, y)\rangle$ be a vector field defined on a region $\mathcal{R}$ in the plane whose boundary is a smooth or piecewise-smooth, simple closed curve C. If $\mathbf{r}(t)$ is a parametrization of C in the counterclockwise direction (as viewed from the positive z-axis), then

$$\int_C \mathbf{F}(x, y) \cdot d\mathbf{r} = \int_C F_1(x, y)\, dx + F_2(x, y)\, dy = \iint_R \left(\frac{\partial F_2}{\partial x} - \frac{\partial F_1}{\partial y}\right) dA.$$

Proof. A thorough proof of Green's Theorem is beyond the scope of this text. However, we will sketch the proof for a region that can be described both as the set of points lying between smooth functions $g_1(x)$ and $g_2(x)$, for $a \le x \le b$, with $g_1(a) = g_2(a)$ and $g_1(b) = g_2(b)$, and also as the set of points lying between smooth functions $h_1(y)$ and $h_2(y)$, for $c \le y \le d$, with $h_1(c) = h_2(c)$ and $h_1(d) = h_2(d)$.

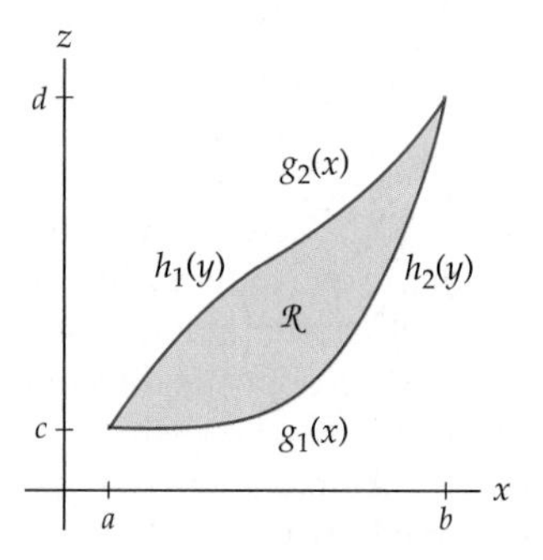

To prove Green's Theorem for this region, let C be the boundary curve traversed in the counterclockwise direction. Then $C = C_1 \cup C_2$, where $C_1 = \mathbf{r}_1(t) = (t, g_1(t))$ for $a \le t \le b$ and $C_2 = \mathbf{r}_2(t) = (t, g_2(t))$ for $a \le t \le b$. (Since we are going around the boundary of R counterclockwise, our path along C_2 will be in the negative direction.)

Fix a vector field $\mathbf{F}(x, y) = F_1(x, y)\mathbf{i} + F_2(x, y)\mathbf{j}$. Then, for a given $x \in [a, b]$,

$$\int_{g_1(x)}^{g_2(x)} \frac{\partial F_1}{\partial y}\, dy = \left[F_1(x, y)\right]_{g_1(x)}^{g_2(x)} = F_1(x, g_2(x)) - F_1(x, g_1(x)).$$

So,

$$\int_a^b \int_{g_1(x)}^{g_2(x)} \frac{\partial F_1}{\partial y}\, dy\, dx = \int_a^b F_1(x, g_2(x)) - F_1(x, g_1(x))\, dx$$

$$= -\int_b^a F_1(x, g_2(x))\, dx - \int_a^b F_1(x, g_1(x))\, dx = -\int_C F_1(x, y)\, dx.$$

A parallel argument shows that

$$\int_c^d \int_{h_1(y)}^{h_2(y)} \frac{\partial F_2}{\partial x}\, dx\, dy = \int_c^d F_2(h_2(y), y) - F_2(h_1(y), y)\, dy$$

$$= \int_c^d F_2(h_2(y), y)\, dy + \int_d^c F_2(h_1(y), y)\, dy = \int_C F_2(x, y)\, dy.$$

Combining the two results gives the desired equality:

$$\int_C \mathbf{F}(x, y) \cdot d\mathbf{r} = \iint_R \left(\frac{\partial F_2}{\partial x} - \frac{\partial F_1}{\partial y}\right) dA.$$ ■

Green's Theorem can be used to evaluate line integrals over simple closed curves with reverse (clockwise) parametrizations. Since reversing the direction of travel in a parametrization of a curve changes the sign of the resulting line integral, we can multiply both sides of the preceding equation by -1 to obtain a formula for clockwise-traversed curves. (See Example 5.)

The relationship between interior double integrals and boundary line integrals given by Green's Theorem also applies to regions $\mathcal{S}$ that are not themselves bounded by smooth or piecewise-smooth simple closed curves, provided that $\mathcal{S}$ can be written as a sum of such regions. For instance, if Q is the annular region $Q = \{(x, y) \mid 4 \le x^2 + y^2 \le 16\}$ shown next at the left, we may represent this region as the union of $\mathcal{R}_1$ and $\mathcal{R}_2$, where each of the subregions has a piecewise-smooth simple closed boundary, as in the figure at the right:

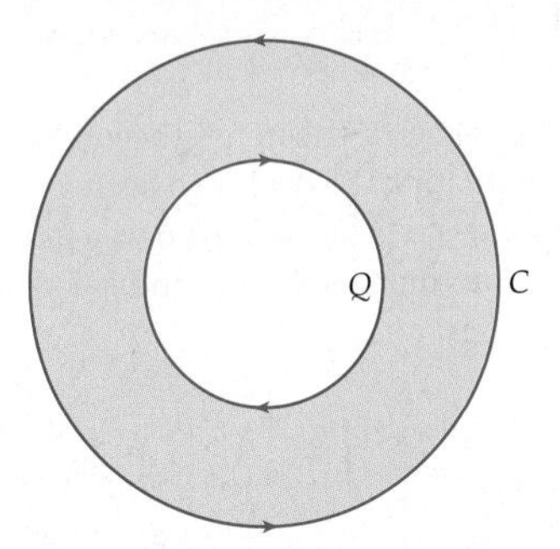

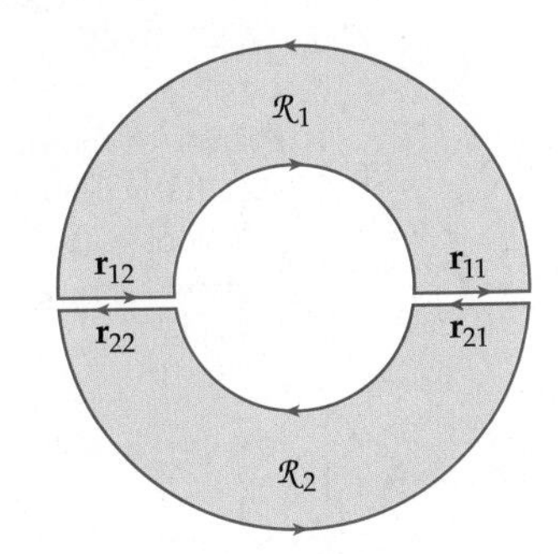

Moreover, if both $\mathcal{R}_1$ and $\mathcal{R}_2$ have parametrizations with counterclockwise orientations, as indicated in the two figures, then the line integrals along the pairs $\mathbf{r}_{11}$, $\mathbf{r}_{21}$ and $\mathbf{r}_{12}$, $\mathbf{r}_{22}$ will cancel in the total summation, since we are integrating the same vector field $\mathbf{F}(x, y)$ along the same curve in opposite directions. So these curves can be disregarded in our treatment of Q. Looking at the curves that form the boundary of Q, we see that it is important that the inner curve D be parametrized in the *clockwise* direction, because of how its components must be parametrized as boundaries of $\mathcal{R}_1$ and $\mathcal{R}_2$. Taking all this into account, we may conclude that

$$\int_C \mathbf{F}(x, y) \cdot d\mathbf{r} + \int_D \mathbf{F}(x, y) \cdot d\mathbf{r} = \iint_Q \left(\frac{\partial F_2}{\partial y} - \frac{\partial F_1}{\partial x} \right) dA,$$

where C is the outer edge traversed counterclockwise and D is the inner edge traversed clockwise.

Green's Theorem can be restated in terms of both the divergence and the curl.

THEOREM 14.14

Green's Theorem: Curl and Divergence Expressions

Let $\mathcal{R}$ be a region in the plane to which Green's Theorem applies, with smooth boundary curve C oriented in the counterclockwise direction by $\mathbf{r}(t) = \langle (x(t), y(t)) \rangle$ for $a \le t \le b$, with vector field $\mathbf{F}(x, y) = \langle (F_1(x, y), F_2(x, y)) \rangle$ defined on $\mathcal{R}$.

(a) Green's Theorem, curl form:
A unit vector perpendicular to the xy-plane and thus to the region R in the positive direction is just $\mathbf{n} = \mathbf{k}$. So we can rewrite Green's Theorem as

$$\int_C \mathbf{F}(x, y) \cdot d\mathbf{r} = \iint_{\mathcal{R}} \left(\frac{\partial F_2}{\partial x} - \frac{\partial F_1}{\partial y} \right) dA = \iint_{\mathcal{R}} \operatorname{curl} \mathbf{F} \cdot \mathbf{k}\, dA.$$

(b) Green's Theorem, divergence form:
If we restrict our attention to the plane, we see that a unit vector that lies in the xy-plane and is perpendicular to the curve C is given by

$$\mathbf{n} = \frac{y'(t)}{\sqrt{(x'(t))^2 + (y'(t))^2}}\mathbf{i} + \frac{-x'(t)}{\sqrt{(x'(t))^2 + (y'(t))^2}}\mathbf{j}.$$

Then Green's Theorem is equivalent to the statement

$$\int_C \mathbf{F}(x, y) \cdot \mathbf{n}\, ds = \iint_{\mathcal{R}} \operatorname{div} \mathbf{F}\, dA.$$

Note that in Theorem 14.14 (a) we treat **F** as a vector field in $\mathbb{R}^3$ whose **k**-component is zero, since the integrand on the right-hand side of the equation is the **k**-component of the curl of a vector field in $\mathbb{R}^3$.

Proof.

(a) The curl form of Green's Theorem is the direct result of labeling the terms in the original statement of the theorem.

(b) The divergence form requires a little manipulation:

$$\int_C \mathbf{F}(x,y)\cdot\mathbf{n}\,ds$$

$$= \int_a^b \left(F_1(x,y)\frac{y'(t)}{\sqrt{(x'(t))^2+(y'(t))^2}} + F_2(x,y)\frac{-x'(t)}{\sqrt{(x'(t))^2+(y'(t))^2}} \right)\sqrt{(x'(t))^2+(y'(t))^2}\,dt$$

$$= \int_C F_1(x,y)\,dy - F_2(x,y)\,dx = \int_C -F_2(x,y)\,dx + F_1(x,y)\,dy.$$

Applying Green's Theorem to this last line integral gives

$$\int_C -F_2(x,y)\,dx + F_1(x,y)\,dy = \iint_{\mathcal{R}} \left(\frac{\partial F_1}{\partial x} + \frac{\partial F_2}{\partial y}\right) dA$$

$$= \iint_{\mathcal{R}} \operatorname{div} \mathbf{F}\, dA.$$

■

These equivalent formulations do not affect our application of Green's Theorem; rather, they foreshadow the generalization of the theorem to regions that are surfaces other than the xy-plane, to Stokes' Theorem, and to a higher dimensional analog called the Divergence Theorem. There are various statements of Green's Theorem, including the divergence and curl forms, throughout mathematical literature.

Examples and Explorations

EXAMPLE 1

Computing the divergence and the curl

(a) Compute the divergence of $\mathbf{F}(x,y) = x^2\mathbf{i} + xy\mathbf{j}$.

(b) Compute the divergence and curl of $\mathbf{G}(x,y,z) = \dfrac{1}{xyz}\mathbf{i} + e^{xyz}\mathbf{j} + \dfrac{1}{z^2+1}\mathbf{k}$.

SOLUTION

(a) We use Definition 14.10 to find the divergence of **F**:

$$\operatorname{div} \mathbf{F}(x,y) = \nabla\cdot\mathbf{F}(x,y)$$

$$= \frac{\partial F_1}{\partial x} + \frac{\partial F_2}{\partial y}$$

$$= 2x + x = 3x.$$

(b) We again use Definition 14.10 to find the divergence of **G**, or

$$\mathbf{G}(x,y,z) = G_1(x,y,z)\mathbf{i} + G_2(x,y,z)\mathbf{j} + G_3(x,y,z)\mathbf{k} = \frac{1}{xyz}\mathbf{i} + e^{xyz}\mathbf{j} + \frac{1}{z^2+1}\mathbf{k}.$$

We have

$$\operatorname{div} \mathbf{G}(x,y,z) = \nabla\cdot\mathbf{G}(x,y,z)$$

$$= \frac{\partial G_1}{\partial x} + \frac{\partial G_2}{\partial y} + \frac{\partial G_3}{\partial z}$$

$$= -\frac{yz}{(xyz)^2} + xze^{xyz} - \frac{2z}{(z^2+1)^2}.$$

We now use Definition 14.11 to compute the curl:

$$\begin{aligned}\text{curl}\,\mathbf{G}(x,y,z) &= \nabla\times\mathbf{G}(x,y,z)\\ &= \left(\frac{\partial G_3}{\partial y}-\frac{\partial G_2}{\partial z}\right)\mathbf{i}+\left(\frac{\partial G_1}{\partial z}-\frac{\partial G_3}{\partial x}\right)\mathbf{j}+\left(\frac{\partial G_2}{\partial x}-\frac{\partial G_1}{\partial y}\right)\mathbf{k}\\ &= (0-xye^{xyz})\mathbf{i}+\left(-\frac{xy}{(xyz)^2}-0\right)\mathbf{j}+\left(yze^{xyz}+\frac{xz}{(xyz)^2}\right)\mathbf{k}\\ &= -xye^{xyz}\mathbf{i}-\frac{xy}{(xyz)^2}\mathbf{j}+\left(yze^{xyz}+\frac{xz}{(xyz)^2}\right)\mathbf{k}.\end{aligned}$$

□

EXAMPLE 2

Verifying Green's Theorem

Verify Green's Theorem for the unit disk in the xy-plane with boundary curve C, traversed counterclockwise, where $\mathbf{F}(x,y)=x\mathbf{i}+2x\mathbf{j}$.

SOLUTION

Let R be the unit disk and C be the boundary curve parametrized by

$$\mathbf{r}(t)=\langle\cos t,\sin t\rangle,\quad\text{for}\quad 0\le t\le 2\pi.$$

So, $\frac{d\mathbf{r}}{dt}=\langle-\sin t,\cos t\rangle$, for $0\le t\le 2\pi$. We also express the vector field $\mathbf{F}$ in terms of the parameter t:

$$\mathbf{F}(x(t),y(t))=\langle x(t),2x(t)\rangle=\langle\cos t,2\cos t\rangle.$$

Then

$$\int_C\mathbf{F}(x,y)\cdot d\mathbf{r}=\int_0^{2\pi}(-\cos t\sin t+2\cos^2 t)\,dt.$$

You should verify that $\frac{1}{2}\cos^2 t+\sin t\cos t+t$ is an antiderivative of $-\cos t\sin t+2\cos^2 t$. Therefore,

$$\int_0^{2\pi}(-\cos t\sin t+2\cos^2 t)\,dt=\left[\frac{1}{2}\cos^2 t+\sin t\cos t+t\right]_0^{2\pi}=2\pi.$$

We will now perform the double integration given by Green's Theorem. Since $\mathbf{F}(x,y)=x\mathbf{i}+2x\mathbf{j}$, we have $\frac{\partial F_2}{\partial x}=2$ and $\frac{\partial F_1}{\partial y}=0$. Therefore,

$$\begin{aligned}\iint_R\left(\frac{\partial F_2}{\partial x}-\frac{\partial F_1}{\partial y}\right)dA &= \iint_R 2\,dA\\ &= \int_0^{2\pi}\int_0^1 2\,r\,dr\,d\theta\\ &= \int_0^{2\pi}\left[r^2\right]_0^1 d\theta=2\pi.\end{aligned}$$

As expected, our results agree. □

EXAMPLE 3

Computing with Green's Theorem

Evaluate $\int_C\mathbf{F}(x,y)\cdot d\mathbf{r}$, where $\mathbf{F}(x,y)=xy\mathbf{i}+xy^3\mathbf{j}$ and C is the boundary of the rectangle in the xy-plane and whose sides are the coordinate axes and the lines $y=2$ and $x=3$.

SOLUTION

This problem shows some of the strength of Green's Theorem. Parametrizing the four distinct pieces of C would not be difficult in this case, but it would be tedious. Meanwhile, the region $\mathcal{R}$ that C encloses is very easy to describe. Applying Green's Theorem, we have

$$\begin{aligned}\int_C \mathbf{F}(x,y)\cdot d\mathbf{r} &= \iint_R \left(\frac{\partial F_2}{\partial x} - \frac{\partial F_1}{\partial y}\right) dA \\ &= \int_0^3 \int_0^2 (y^3 - x)dy\,dx \\ &= \int_0^3 \left[\frac{1}{4}y^4 - xy\right]_0^2 dx \\ &= \int_0^3 (4-2x)\,dx \\ &= \left[4x - x^2\right]_0^3 = 3.\end{aligned}$$

□

EXAMPLE 4 **Using Green's Theorem with a conservative vector field**

Use Green's Theorem to evaluate

$$\int_C \mathbf{F}(x,y)\cdot d\mathbf{r},$$

where C is the boundary of the first-quadrant region $\mathcal{R}$ that lies between the curves $y = x^2$ and $y = x$ and where the conservative vector field $\mathbf{F}(x,y) = \nabla f(x,y)$, in which $f(x,y) = e^{xy}$.

SOLUTION

Computing the partial derivatives of $f(x,y)$, we have

$$\mathbf{F}(x,y) = ye^{xy}\mathbf{i} + xe^{xy}\mathbf{j}.$$

Then,

$$\begin{aligned}\int_C \mathbf{F}(x,y)\cdot d\mathbf{r} &= \int_R \left(\frac{\partial F_2}{\partial x} - \frac{\partial F_1}{\partial y}\right) dA \\ &= \int_R ((e^{xy} + xye^{xy}) - (e^{xy} + xye^{xy}))\,dA = 0.\end{aligned}$$

Green's Theorem, together with the equality of the second-order mixed partial derivatives, gives us another proof of the Fundamental Theorem of Line Integrals for smooth or piecewise-smooth simple closed curves. If $\mathbf{F}(x,y) = F_1(x,y)\mathbf{i} + F_2(x,y)\mathbf{j}$ is conservative, then, by the equality of the mixed partial derivatives,

$$\frac{\partial F_2}{\partial x} = \frac{\partial F_1}{\partial y}.$$

So, for a conservative vector field and a smooth or piecewise-smooth simple closed curve C that is the boundary of a region $\mathcal{R}$,

$$\int_C \mathbf{F}(x,y)\cdot d\mathbf{r} = \int_R \left(\frac{\partial F_2}{\partial x} - \frac{\partial F_1}{\partial y}\right) dA = \int_{\mathcal{R}} 0\,dA = 0.$$

□

Applying the Fundamental Theorem of Line Integrals to

$$\int_C \mathbf{F}(x,y)\cdot d\mathbf{r}$$

directly, where C is parametrized by $\mathbf{r}(t)$ for $a \le t \le b$, gives

$$\int_C \mathbf{F}(x,y)\cdot d\mathbf{r} = f(x(b),y(b)) - f(x(a),y(a)) = 0,$$

since the curve's being closed entails that $(x(b),y(b)) = (x(a),y(a))$.

EXAMPLE 5 **Using Green's Theorem for curves with reverse orientation**

Use Green's Theorem to evaluate $\int_C \mathbf{F}(x, y) \cdot d\mathbf{r}$, where $\mathbf{F}(x, y) = y\mathbf{i} + 2x\mathbf{j}$ and C is the boundary of the region bounded by the x-axis and the curve $y = 1 - x^2$, traversed in the clockwise direction.

SOLUTION

From Section 14.2, we know that reversing the direction of travel along a curve changes the sign of the line integrals along that curve. Here, since $\mathbf{F}(x, y) = y\mathbf{i} + 2x\mathbf{j}$, we have

$$\frac{\partial}{\partial x}(2x) - \frac{\partial}{\partial y}(y) = 2 - 1 = 1.$$

So, in this case, Green's Theorem implies that

$$\begin{aligned}\int_C \mathbf{F}(x, y) \cdot d\mathbf{r} &= -\iint_R 1\, dA \\ &= -\int_{-1}^{1}\int_{0}^{1-x^2} 1\, dy\, dx \\ &= \int_{-1}^{1} x^2 - 1\, dx \\ &= \left[\frac{1}{3}x^3 - x\right]_{-1}^{1} = -\frac{4}{3}.\end{aligned}$$

□

EXAMPLE 6 **Analyzing water flow in a river channel**

The Los Angeles River is channelized for much of its length. The channelized portion of the river flows in concrete banks with uniform width and slope and runs straight from north to south. Therefore, the velocity vector of the flow is $\mathbf{F} = 0\mathbf{i} + f(y)\mathbf{j}$. Suppose we measure the rate of flow at two points that are close to one another.

(a) One day the flow is found to have speed 0.4 feet per second at the top and bottom edges of a rectangle $\mathcal{R}$ with boundary C enclosing the river. Find $\int_C \mathbf{F} \cdot d\mathbf{r}$.

(b) Suppose that another day we find that the rate of flow at the lower edge of the rectangle is slightly higher than the flow at the top. Use the divergence form of Green's Theorem to explain what unusual phenomenon is occurring in Los Angeles that day.

SOLUTION

(a) Since the flow at the top and bottom is the same, the edge vectors have opposite signs, and the x-component of the velocity field is zero, it follows that the integral is zero.

(b) We have

$$0 < \int_C \mathbf{F} \cdot \mathbf{n}\, ds = \iint_{\mathcal{R}} \frac{\partial f}{\partial y}\, dA.$$

But since the river is channelized, $\frac{\partial f}{\partial y} > 0$ at some points in $\mathcal{R}$. This means that the flow is increasing, so it must be raining. □

? TEST YOUR UNDERSTANDING

- What is the definition of the divergence of a vector field? What does the divergence measure?
- What is the definition of the curl of a vector field? What does the curl measure?
- How is Green's Theorem similar to the Fundamental Theorem of Calculus from Chapter 4?

- How can Green's Theorem be used to compute line integrals around smooth or piecewise-smooth simple closed curves that are parametrized in the *clockwise* direction?
- In our examples, we used Green's Theorem to reduce line integrals to double integrals (though the reverse direction is also sometimes useful, as we will see in the exercises). In what way is this use different from our use of the Fundamental Theorem of Calculus from Chapter 4?

EXERCISES 14.4

Thinking Back

Remembering the Fundamental Theorem of Calculus: Answer the following questions about the Fundamental Theorem of Calculus.

- Consider the "region" in $\mathbb{R}^1$ given by the interval $[a, b]$. What is the boundary of this region? What is its interior?
- Suppose that $g(x)$ is a differentiable function on $[a, b]$. Express $g(b) - g(a)$ in terms of a function on the interior of $[a, b]$.
- Express $\int_a^b f(x)\,dx$ in terms of something on the boundary of $[a, b]$.

Intractable Integrals: One of the virtues of Green's Theorem—in both directions—is that it gives us a way to rewrite difficult integrals in a form we hope will be easier to manipulate.

- Give an example of a differentiable function $f(x, y)$ that is not obviously integrable.

Concepts

0. *Problem Zero:* Read the section and make your own summary of the material.

1. *True/False:* Determine whether each of the statements that follow is true or false. If a statement is true, explain why. If a statement is false, provide a counterexample.

(a) *True or False:* The del operator, ∇, converts vectors into scalars.

(b) *True or False:* The del operator, ∇, measures the rotation of a vector field.

(c) *True or False:* The divergence of a vector field is a scalar.

(d) *True or False:* The curl of a vector field is a vector.

(e) *True or False:* The curl of a gradient vector field is $\mathbf{0}$.

(f) *True or False:* Both the Fundamental Theorem of Calculus (Theorem 4.24) and Green's Theorem relate the integral of a function on a (mathematically well-behaved) region to a quantity measured on the boundary of that region.

(g) *True or False:* The curl of a vector field measures how much the field is compressing or expanding.

(h) *True or False:* The conclusion of Green's Theorem does not depend on the direction of parametrization of the boundary curve in question.

2. *Examples:* Construct examples of the thing(s) described in the following. Try to find examples that are different than any in the reading.

(a) A smooth simple closed curve parametrized in the counterclockwise direction.

(b) A smooth simple closed curve parametrized in the clockwise direction.

(c) A simple closed curve that is not smooth, but is piecewise smooth, parametrized in the counterclockwise direction.

3. In what sense is the integrand of the double integral in Green's Theorem the antiderivative of the vector field?

4. Make a chart of all the new notation, definitions, and theorems in this section, including what each new item means in terms you already understand.

5. Give two examples of quantities that may be computed by $\int_C \mathbf{F} \cdot d\mathbf{r}$.

6. Give three examples of quantities that may be computed by $\iint_{\mathcal{R}} G(x, y)\,dA$.

7. Explain how your answer to Exercise 6 is relevant to the discussion of Green's Theorem.

8. Draw the annular region bounded by $r = 1$ and $r = 2$. Divide it into two regions, each of which is suitable for the application of Green's Theorem.

9. If the velocity of a flow of a gas at a point (x, y, z) is represented by $\mathbf{F}$ and the gas is compressing at that point, what does this imply about the divergence of $\mathbf{F}$ at the point?

10. Give an example of a field with negative divergence at the origin.

11. If the velocity of a flow of a gas at a point (x, y, z) is represented by $\mathbf{F}$ and the gas is expanding at that point, what does this imply about the divergence of $\mathbf{F}$ at the point?

12. Give an example of a field with positive divergence at $(1, 0, \pi)$.

13. Use the vector field $\mathbf{F}(x, y) = x^2 e^y \mathbf{i} + \cos x \sin y\,\mathbf{j}$ and Green's Theorem to write the line integral of $\mathbf{F}(x, y)$ about the unit circle, traversed counterclockwise, as a double integral. Do not evaluate the integral.

14. Use the same vector field as in Exercise 13 together with the divergence form of Green's Theorem to write the line integral of $\mathbf{F}(x, y)$ about the unit circle as a double integral. Do not evaluate the integral.

15. Use the same vector field as in Exercise 13, and compute the $\mathbf{k}$-component of the curl of $\mathbf{F}(x, y)$.

16. Use the curl form of Green's Theorem to write the line integral of $\mathbf{F}(x, y)$ about the unit circle as a double integral. Do not evaluate the integral.

Skills

Compute the divergence of the vector fields in Exercises 17–22.

17. $\mathbf{F}(x, y) = yx^2\mathbf{i} - x\cos y\,\mathbf{j}$

18. $\mathbf{G}(x, y) = \langle x\cos(xy), y\cos(xy)\rangle$

19. $\mathbf{G}(x, y, z) = yz^2\mathbf{i} - x\sin z\,\mathbf{j} + x^2e^{xy}\mathbf{k}$

20. $\mathbf{G}(x, y, z) = (x^2 + y - z)\mathbf{i} - 2y\cos z\,\mathbf{j} + e^{x^2+y^2+z^2}\mathbf{k}$

21. $\mathbf{F}(x, y, z) = \sin^{-1}(xy)\mathbf{i} + \ln(y + z)\mathbf{j} + \dfrac{1}{(2x + 3y + 5z + 1)}\mathbf{k}$

22. $\mathbf{F}(x, y, z) = xe^{yz}\mathbf{i} + ye^{xz}\mathbf{j} + ze^{xy}\mathbf{k}$

Compute the curl of the vector fields in Exercises 23–28.

23. $\mathbf{F}(x, y, z) = \sin^{-1}(xy)\mathbf{i} + \ln(y + z)\mathbf{j} + \dfrac{1}{(2x + 3y + 5z + 1)}\mathbf{k}$

24. $\mathbf{G}(x, y, z) = (x^2 + y - z)\mathbf{i} - 2y\cos z\,\mathbf{j} + e^{x^2+y^2+z^2}\mathbf{k}$

25. $\mathbf{F}(x, y, z) = xe^{yz}\mathbf{i} + ye^{xz}\mathbf{j} + ze^{xy}\mathbf{k}$

26. $\mathbf{F}(x, y) = -4x^2y\mathbf{i} + 4xy^2\mathbf{j}$

27. $\mathbf{F}(x, y) = \cos(x + y)\mathbf{i} + \sin(x - y)\mathbf{j}$

28. $\mathbf{G}(x, y, z) = (2y + 3z)\mathbf{i} + 2xy^2\mathbf{j} + (zx - y)\mathbf{k}$

Use Green's Theorem to evaluate the integrals in Exercises 29–34.

29. Find $\int_C \mathbf{F}\cdot d\mathbf{r}$, where $\mathbf{F}(x, y) = -4x^2y\mathbf{i} + 4xy^2\mathbf{j}$ and C is the unit circle traversed counterclockwise.

30. Find $\int_C \mathbf{F}\cdot d\mathbf{r}$, where $\mathbf{F}(x, y) = (y^3 + 3x^2y)\mathbf{i} + 2x\mathbf{j}$ and C is the boundary of the half of the unit circle that lies above the x-axis, traversed counterclockwise.

31. Find $\int_C \mathbf{F}\cdot d\mathbf{r}$, where $\mathbf{F}(x, y) = (x + 2y)\mathbf{i} + (x - 2y)\mathbf{j}$ and C is the boundary of the region bounded by the curves $x = y^2, x = 4$, traversed counterclockwise.

32. Find $\int_C \mathbf{F}\cdot d\mathbf{r}$, where $\mathbf{F}(x, y) = y\ln(x+1)^2\mathbf{i} + \mathbf{j}$ and C is the boundary of the square with vertices at $(0, 0)$, $(0, 2)$, $(2, 0)$, and $(2, 2)$, traversed counterclockwise.

33. Find $\int_C \mathbf{F}\cdot d\mathbf{r}$, where

$$\mathbf{F}(x, y) = xe^{x^2+y^2}\mathbf{i} + (10\sin y + x)\mathbf{j}$$

and C is the boundary of the region in the third quadrant bounded by $y = x$ and the x-axis, for $x \in [-1, 0]$, traversed counterclockwise.

34. Find $\int_C \mathbf{F}\cdot d\mathbf{r}$, where $\mathbf{F}(x, y) = xy\mathbf{i} + 4(x - y)\mathbf{j}$ and C is the boundary of the region bounded by the curves $x = 1, x = e, y = e^x$, and $y = \ln x$, traversed counterclockwise.

Verify Green's Theorem by working Exercises 35–42 in pairs. In each pair, evaluate the desired integrals first directly and then using the theorem.

35. Directly compute (i.e., without using Green's Theorem) $\int_C \mathbf{F}(x, y)\cdot d\mathbf{r}$, where $\mathbf{F}(x, y) = y^2\mathbf{i} - xy\mathbf{j}$ and C is the unit circle traversed counterclockwise.

36. Use Green's Theorem to evaluate the line integral in Exercise 35.

37. Directly compute (i.e., without using Green's Theorem) $\int_C \mathbf{F}(x, y)\cdot d\mathbf{r}$, where

$$\mathbf{F}(x, y) = (2^x + y)\mathbf{i} + (2x - y)\mathbf{j}$$

and C is the triangle described by $x = 0, y = 0$, and $y = 1 - x$, traversed counterclockwise.

38. Use Green's Theorem to evaluate the line integral in Exercise 37.

39. Directly compute (i.e., without using Green's Theorem) $\iint_{\mathcal{R}} (3y - 3x)\,dA$, where $\mathcal{R}$ is the portion of the disk of radius 2, centered at the origin, and lying above the x-axis.

40. Use Green's Theorem to evaluate the double integral in Exercise 39.

41. Directly compute (i.e., without using Green's Theorem) $\iint_{\mathcal{R}} (e^x + e^y)\,dA$, where $\mathcal{R}$ is the region bounded by the lines $x = 1$, $x = \ln 2$, $y = 0$, and $y = 2$.

42. Use Green's Theorem to evaluate the double integral in Exercise 41.

Evaluate the integrals in Exercises 43–46 directly or using Green's Theorem.

43. $\int_C \mathbf{F}\cdot d\mathbf{r}$, where $\mathbf{F}(x, y) = y2^x\mathbf{i} + xy^e\mathbf{j}$ and C is the boundary of the square with vertices at $(0, 0)$, $(0, 1)$, $(1, 0)$, and $(1, 1)$, traversed clockwise.

44. $\int_C \mathbf{F}(x, y)\cdot d\mathbf{r}$, where $\mathbf{F}(x, y) = \frac{1}{2}x^2y^3\mathbf{i} + xy\mathbf{j}$ and C is the circle of radius 3, centered at the origin, traversed clockwise.

45. $\iint_{\mathcal{R}} (2xe^{x^2+y^2} + 2ye^{-(x^2+y^2)})\,dA$, where $\mathcal{R}$ is the annulus bounded by $r = 1$ and $r = 2$.

46. $\iint_{\mathcal{R}} (3xy - 4x^2y)\,dA$, where $\mathcal{R}$ is the unit disk.

Applications

47. Find the work done by the vector field

$$\mathbf{F}(x, y) = (\cos(x^2) + 4xy^2)\mathbf{i} + (2^y - 4x^2y)\mathbf{j}$$

in moving an object around the unit circle, starting and ending at $(1, 0)$.

48. Find the work done by the vector field

$$\mathbf{F}(x, y) = x^3y^2\mathbf{i} + (y - x)\mathbf{j}$$

in moving an object around the triangle with vertices $(1, 1)$, $(2, 2)$, and $(3, 1)$, starting and ending at $(2, 2)$.

49. Find the work done by the vector field

$$F(x, y) = \frac{xe^{\sqrt{x^2+y^2}}}{\sqrt{x^2+y^2}}\mathbf{i} + \frac{ye^{\sqrt{x^2+y^2}}}{\sqrt{x^2+y^2}}\mathbf{j}$$

in moving an object around the unit circle, starting and ending at $(1, 0)$.

50. Find the work done by the vector field

$$\mathbf{F}(x, y) = (\cos x - 3ye^x)\mathbf{i} + \sin x \sin y\mathbf{j}$$

in moving an object around the periphery of the rectangle with vertices $(0, 0)$, $(2, 0)$, $(2, \pi)$, and $(0, \pi)$, starting and ending at $(2, \pi)$.

51. The current through a certain passage of the San Juan Islands in Washington State is given by

$$F = \langle F_1(x, y), F_2(x, y)\rangle = \langle 0, 1.152 - 0.8x^2\rangle.$$

Consider a disk $\mathcal{R}$ of radius 1 mile and centered on this region. Denote the boundary of the disk by $\partial\mathcal{R}$.

(a) Compute $\iint_{\mathcal{R}} \left(\frac{\partial F_2}{\partial x} - \frac{\partial F_1}{\partial y}\right) dA$.

(b) Show that

$$\int_{\partial\mathcal{R}} F \cdot n\, ds = \int_0^{2\pi} (1.152 - 0.8\cos^2\theta)\sin\theta\, d\theta.$$

Conclude that Green's Theorem is valid for the current in this area of the San Juan Islands.

(c) What do the integrals from Green's Theorem tell us about this region of the San Juan Islands?

52. Emmy has to examine how well a waste tank with radius 50 feet is stirred. The current at the edge of the tank moves at $\frac{\pi}{4}$ radians per minute in a counterclockwise direction. Emmy knows that if the vector field of the current velocity is $\mathbf{v} = \langle v_1(x, y), v_2(x, y)\rangle$, then the average value of $\frac{\partial v_2}{\partial x} - \frac{\partial v_1}{\partial y}$ in the tank must be at least 1.5 in order to get adequate mixing in the tank. Does she need to speed up stirring for this tank?

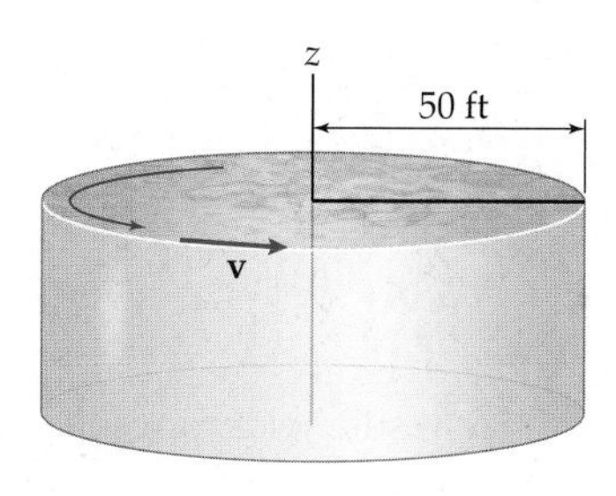

53. Traffic can be described by fluid properties such as density and momentum; hence it is usually modeled as a continuum. Consider a straight road with multiple lanes going due north. Fast traffic uses the lanes farther left, and slower traffic uses the ones to the right. Denote the velocity of traffic at any point of the road by $\mathbf{v}(x, y) = \langle 0, v_2(x, y)\rangle$. Also; that is, assume that the speed in any particular lane is constant; that is, $\frac{\partial v_2}{\partial y} = 0$.

(a) What can we say about the curl $\frac{\partial v_2}{\partial x} - \frac{\partial v_1}{\partial y}$ at any point on the road? What does it mean for that quantity to be small? What if it is large? Does this quantity have anything to do with traffic safety?

(b) Suppose that $v_2(x, y) = 75 - \alpha x$ in miles per hour, where x is measured in miles. Compute both sides of Green's Theorem for the traffic flow over a segment of highway described by $S = [0, 0.0113] \times [0, 0.5]$ with boundary C. In other words, verify that

$$\int_C \mathbf{v}(x, y) \cdot d\mathbf{r} = \iint_S \left(\frac{\partial v_2}{\partial x} - \frac{\partial v_1}{\partial y}\right) dx\, dy.$$

(c) What is the significance of either of the integrals from part (b) for the traffic? Comment on whether the integral on the left or the integral on the right would be easier for traffic engineers to calculate from measurements.

54. Traffic engineers use road tubes to measure the number and speed of cars passing a line in the road. Suppose traffic engineers set road tubes on a freeway whose inner edge lies along the circle $x^2 + y^2 = 0.2$ and whose outer edge is $x^2 + y^2 = 0.2113$. The traffic is flowing counterclockwise in the first quadrant of the circles. The road tubes lie along the lines $y = 0$ and $x = 0$, and the measurements show that the speed of cars along the inner circle is 69 mph while along the outside circle it is 61 mph. Denote the stretch of road in the first quadrant by $\mathcal{S}$ with oriented boundary C, and denote the velocity of the cars at any point by $\mathbf{v}(x, y)$.

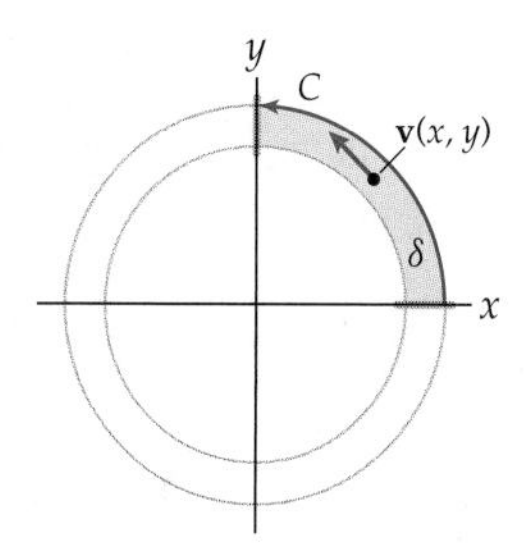

(a) Compute $\int_C \mathbf{v}(x, y) \cdot d\mathbf{r}$. What is the significance of the sign of this quantity?

(b) Use Green's Theorem to compute the average curl of the traffic on this stretch of road:

$$\frac{1}{A}\iint_S \left(\frac{\partial v_2}{\partial x} - \frac{\partial v_1}{\partial y}\right) dx\, dy.$$

Here, A is the area of the roadway.

Proofs

55. Let $\mathbf{F}$ be a vector field in $\mathbb{R}^2$ or $\mathbb{R}^3$. Prove that div curl $\mathbf{F} = 0$. (This is Theorem 14.12, part (a).)

56. Let f be a function of two or three variables. Prove that curl $\nabla f = \mathbf{0}$. (This is Theorem 14.12, part (b).)

57. Prove that, for any conservative vector field $\mathbf{F}(x, y)$,

$$\iint_R \left(\frac{\partial F_2}{\partial x} - \frac{\partial F_1}{\partial y} \right) dA = 0$$

for any simply connected region $R \subset \mathbb{R}^2$ whose boundary is smooth or piecewise smooth.

58. Give an example of a vector field defined on a simply connected region $\mathcal{R} \subset \mathbb{R}^2$ whose boundary is smooth or piecewise smooth, that is not a conservative field, but whose integral over the unit disk does have the property that

$$\iint_{\mathcal{R}} \left(\frac{\partial F_2}{\partial x} - \frac{\partial F_1}{\partial y} \right) dA = 0.$$

Your example shows that the converse of Exercise 57 fails.

Thinking Forward

- *Generalizing Green's Theorem:* How might we generalize Green's Theorem to two-dimensional regions that are surfaces in $\mathbb{R}^3$, rather than patches in the xy-plane? What sort of statement do you expect?
- *Another generalization of Green's Theorem:* How might we generalize Green's Theorem to higher dimensions? What sort of relationship might we hope to find between a triple integral over a (reasonably well-behaved) three-dimensional region of space and a double integral over the surface that is the boundary of this region?

14.5 STOKES' THEOREM

- Generalizing Green's Theorem to smooth surfaces other than the plane
- Expressing Stokes' Theorem in terms of the curl
- Computations using Stokes' Theorem

A Generalization of Green's Theorem

The goal of this section is to generalize Green's Theorem to regions that are two dimensional but that do not lie in the xy-plane. For this to work, we will need our surfaces to be well behaved enough that they act like a small portion of the xy-plane that floated off into space and were deformed only in a smooth (or piecewise-smooth) way, as shown in the following figure:

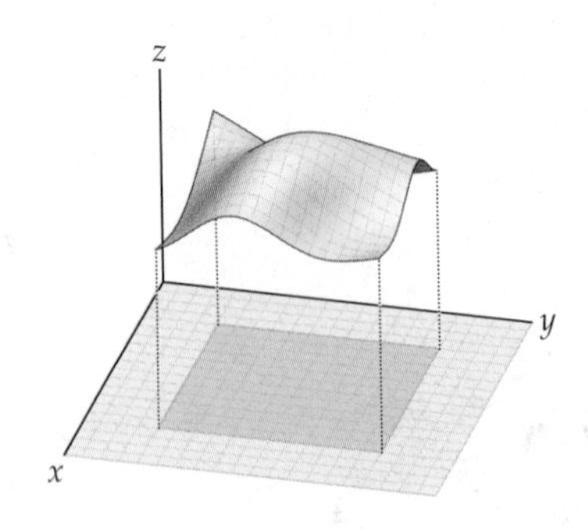

Recall from Section 14.3 that we restrict our attention to surfaces that are smooth or piecewise smooth. In these circumstances, Green's Theorem still holds, provided that we take account of the distortion of the surface.

Stokes' Theorem

The generalized form of Green's Theorem known as *Stokes' Theorem* is traditionally written in terms of the curl of a vector field. Recall that the curl form of Green's Theorem states that, under appropriate restrictions on C and $\mathcal{R}$,

$$\int_C \mathbf{F}(x, y) \cdot d\mathbf{r} = \iint_{\mathcal{R}} \text{curl } \mathbf{F}(x, y) \cdot \mathbf{k}\, dA.$$

In the context of a general surface, what matters is that $\mathbf{k}$ is a unit vector normal to $\mathcal{R}$ and pointing in the direction from which C was parametrized counterclockwise. Generalizing gives Stokes' Theorem:

THEOREM 14.15

Stokes' Theorem

Let $\mathcal{S}$ be a smooth or piecewise-smooth oriented surface with a smooth or piecewise-smooth boundary curve C. Suppose that $\mathcal{S}$ has an (oriented) unit normal vector $\mathbf{n}$ and that C has a parametrization that traverses C in the counterclockwise direction with respect to $\mathbf{n}$. If $\mathbf{F}(x, y, z) = F_1(x, y, z)\mathbf{i} + F_2(x, y, z)\mathbf{j} + F_3(x, y, z)\mathbf{k}$ is a vector field on an open region containing $\mathcal{S}$, then

$$\int_C \mathbf{F}(x, y, z) \cdot d\mathbf{r} = \iint_{\mathcal{S}} \text{curl } \mathbf{F}(x, y, z) \cdot \mathbf{n}\, d\mathcal{S}.$$

If $\mathcal{S}$ is a region of the xy-plane, then the unit normal is $\mathbf{k}$ and the right-hand side of the preceding equation is

$$\iint_{\mathcal{S}} \left(\frac{\partial F_2}{dx} - \frac{\partial F_1}{dy} \right) dA,$$

which is Green's Theorem. Another way to think of this is that Green's Theorem is a special case of Stokes' Theorem for regions in the xy-plane.

Proof. A thorough proof of Stokes' Theorem is beyond the scope of this course; however, we do provide a sketch.

To sketch the proof, first consider a plane P in space, other than the xy-plane. Since P is a plane, we could relabel points and treat the new plane as the xy-plane. This would involve a change of variables using the techniques of Chapter 13. Green's Theorem will hold here, since the relabeled plane behaves exactly like the old xy-plane in every respect. (For an example, consider substituting the xz-plane for the xy-plane. It is not too difficult to figure out what Green's Theorem should be in the xz-plane.)

Now consider an oriented smooth surface $\mathcal{S}$ in space. Since $\mathcal{S}$ is smooth, we can seek to understand it by subdividing it into many small subregions, each of which is closely approximated by the tangent plane at a point in the subregion. On this tiny piece of tangent plane, Green's Theorem applies. With a careful trace through definitions and notation, we can show that when we modify Green's Theorem to account for the plane no longer being the xy-plane, the new integrand will be curl $\mathbf{F}(x, y, z) \cdot \mathbf{n}\, d\mathcal{S}$, as is required by Stokes' Theorem.

Next we recall the discussion of Green's Theorem for regions that are sums of regions bounded by simple closed curves, like the annular region discussed in the previous section. As the following figure indicates, we approximate the surface with small portions of approximating tangent planes. It takes a bit more work to show that the line integrals around adjoining pieces of these approximating tangent planes cancel as they do for the annular region. (The subtlety in this case is that the *planes* are different.)

A smooth surface approximated with portions of tangent planes

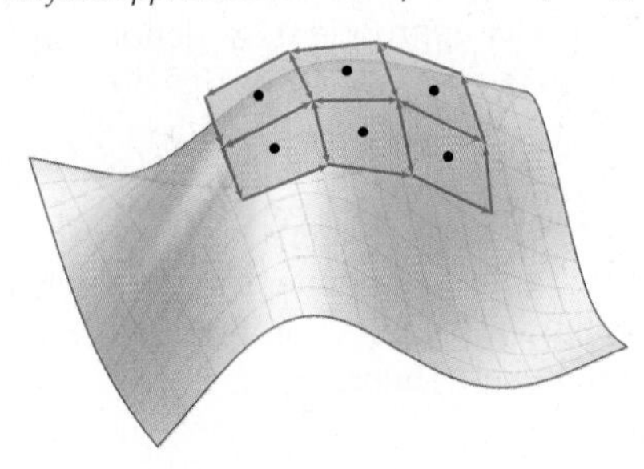

Once this is done, we resort to our usual plan of taking a limit of approximations to find the exact value. The outline of the proof of Stokes' Theorem follows the same "subdivide, approximate, and add" strategy that we have used in previous integration results. The major steps are as follows:

- Show that Green's Theorem can be applied in any plane and that when it is, the integrand is as desired.
- Subdivide S into subregions, and argue that S is well approximated by gluing lots of subdivision tangent planes together.
- Argue that applying Green's Theorem to all these tiny tangent planes cancels all shared borders, leaving only the perimeter.
- Approximate the true integral by calculating the sum of Green's Theorem on tangent planes.
- Add up the subregion approximations, and take a limit as the subregions become arbitrarily small. Addition becomes integration in the limit. ■

Examples and Explorations

EXAMPLE 1

Verifying Stokes' Theorem

Verify Stokes' Theorem for the vector fields

(a) $\mathbf{F}(x, y, z) = yz\mathbf{i} + xz\mathbf{j} + xy\mathbf{k}$ and **(b)** $\mathbf{G}(x, y, z) = -y\mathbf{i} + x\mathbf{j} + e^z\mathbf{k}$

on the surface defined by $S = \{(x, y, z) \mid z = 1 - x^2 - y^2,\ x^2 + y^2 \le 1\}$, with outward unit normal vector.

SOLUTION

The graph of S is the part of the downwards-opening paraboloid $z = 1 - x^2 - y^2$ that lies above the xy-plane, with boundary curve, C, equal to the unit circle in the xy-plane.

To verify Stokes' Theorem, we separately compute

$$\iint_S \text{curl}\,\mathbf{F}(x, y, z) \cdot \mathbf{n}\, dS \quad \text{and} \quad \int_C \mathbf{F}(x, y, z) \cdot d\mathbf{r}$$

and compare the results.

(a) For $\mathbf{F}(x, y, z) = yz\mathbf{i} + xz\mathbf{j} + xy\mathbf{k}$,

$$\text{curl}\,\mathbf{F}(x, y, z) = \nabla \times \mathbf{F}(x, y, z) = \mathbf{0}.$$

This equation implies that

$$\iint_S \text{curl } \mathbf{F}(x,y,z) \cdot \mathbf{n}\, dS = 0.$$

Turning to the line integral of $\mathbf{F}(x,y,z)$ about the boundary of S, we observe that $\mathbf{F}(x,y,z)$ is conservative:

$$\mathbf{F}(x,y,z) = \nabla f, \quad \text{where} \quad f(x,y,z) = xyz.$$

By the Fundamental Theorem of Line Integrals,

$$\int_C \mathbf{F}(x,y,z) \cdot d\mathbf{r} = 0,$$

because C is closed. (Note that Theorem 14.12 also explains why curl $\mathbf{F}(x,y,z) = \mathbf{0}$, since $\mathbf{F}(x,y,z)$ is conservative.)

(b) Recall that, for a surface S given by $z = f(x,y)$, a normal vector is $\mathbf{N} = \left\langle -\frac{\partial z}{\partial x}, -\frac{\partial z}{\partial y}, 1 \right\rangle$. This vector in general needs to be scaled before it becomes a unit normal vector, and it may point in the opposite of the desired direction, depending on orientation of S. In this case,

$$\mathbf{N} = \langle 2x, 2y, 1 \rangle,$$

which does point in the desired direction, away from the z-axis. The scaling will cancel in the integration.

Combining the above with the fact that curl $\mathbf{G}(x,y,z) = \text{curl}\,(-y\mathbf{i} + x\mathbf{j} + e^z\mathbf{k}) = 2\mathbf{k}$, we have

$$\begin{aligned}
\iint_S \text{curl } \mathbf{G}(x,y,z) \cdot \mathbf{N}\, dS &= \iint 2\, dA \\
&= \int_0^{2\pi} \int_0^1 2\, r\, dr\, d\theta \\
&= \int_0^{2\pi} \left[r^2\right]_0^1 \, d\theta \\
&= \int_0^{2\pi} 1\, d\theta = 2\pi.
\end{aligned}$$

To compute the line integral $\int_C \mathbf{G}(x,y,z) \cdot d\mathbf{r}$ directly, we first note that C is parametrized by

$$\mathbf{r}(t) = \cos t\,\mathbf{i} + \sin t\,\mathbf{j} + 0\mathbf{k} \quad \text{for} \quad 0 \le t \le 2\pi$$

with

$$d\mathbf{r} = \langle -\sin t, \cos t, 0 \rangle\, dt \quad \text{for} \quad 0 \le t \le 2\pi.$$

On the curve C,

$$\mathbf{G}(x,y,z) = -y\mathbf{i} + x\mathbf{j} + e^z\mathbf{k} = -\sin t\,\mathbf{i} + \cos t\,\mathbf{j} + 1\mathbf{k}.$$

Substituting gives

$$\int_C \mathbf{G}(x,y,z) \cdot d\mathbf{r} = \int_0^{2\pi} (\sin^2 t + \cos^2 t)\, dt = \int_0^{2\pi} 1\, dt = 2\pi.$$

Again, we have verified Stokes' Theorem. □

EXAMPLE 2 **Using Stokes' Theorem for a graph $y = f(x, z)$**

Use Stokes' Theorem to compute $\int_C \mathbf{F}(x, y, z) \cdot d\mathbf{r}$, where

$$\mathbf{F}(x, y, z) = xyz\mathbf{i} + (y-2)\mathbf{j} + yz\mathbf{k}$$

and C is the boundary of the region of the plane

$$x + y + z = 4$$

that lies to the right of the triangular region $0 \le z \le 4$, $0 \le x \le z$ in the xz-plane, oriented in the positive y direction.

SOLUTION

The surface shown next is the graph of $y = 4 - x - z$.

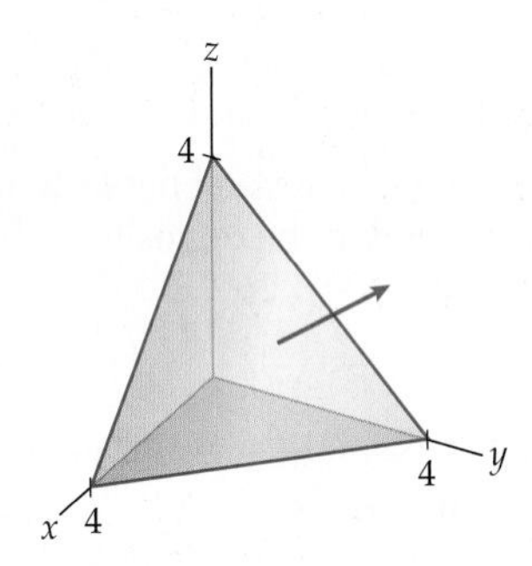

Modifying our previous calculation of $\mathbf{N}$ for surfaces $z = f(x, y)$, we have the normal vector

$$\mathbf{N} = \left\langle -\frac{\partial y}{\partial x}, 1, -\frac{\partial y}{\partial z} \right\rangle = \langle 1, 1, 1 \rangle.$$

Computing curl $\mathbf{F}(x, y, z)$ yields:

$$\text{curl } \mathbf{F}(x, y, z) = z\mathbf{i} + xy\mathbf{j} - xz\mathbf{k}.$$

Then,

$$\begin{aligned}
\iint_S \text{curl } \mathbf{F}(x, y, z) \cdot \mathbf{N}\, dS &= \int_0^2 \int_0^z (z + xy - xz)\, dx\, dz \\
&= \int_0^2 \int_0^z (z + x(4 - x - z) - xz)\, dx\, dz \\
&= \int_0^2 \left[xz + \frac{1}{2}(4 - 2z)x^2 - \frac{1}{3}x^3 \right]_0^z dz \\
&= \int_0^2 \left(3z^2 - \frac{4}{3}z^3 \right) dz \\
&= \left[z^3 - \frac{1}{3}z^4 \right]_0^2 = \frac{8}{3}.
\end{aligned}$$

□

EXAMPLE 3 **Using Stokes' Theorem for an arbitrary surface**

Use Stokes' Theorem to evaluate the integral of the vector field

$$\mathbf{F}(x, y, z) = \langle e^{xyz}, -xy^2z, xyz^2 \rangle$$

around the curve C given by $y^2 + z^2 = 9$ in the plane $x = 5$ and traversed in the counterclockwise direction when viewed from the right (i.e., where $x > 5$).

SOLUTION

Here any smooth orientable surface whose boundary is C will do. We make the simple choice of a disk of radius 3 in the plane $x = 5$ and with center $(5, 0, 0)$. This plane is parallel to the yz-plane. Hence, a unit normal vector is $\mathbf{i}$.

Evaluating $\mathbf{F}(x, y, z)$, we have

$$\begin{aligned}\text{curl }\mathbf{F} &= \nabla \times \left\langle e^{xyz}, -xy^2z, xyz^2\right\rangle \\ &= \left\langle xz^2 + xy^2, xye^{xyz} - yz^2, -y^2z - xze^{xyz}\right\rangle.\end{aligned}$$

If we let D be the disk with radius 3, centered on the x-axis, and in the plane $x = 5$, substituting according to Stokes' Theorem gives

$$\begin{aligned}\int_C \mathbf{F}(x, y, z)\cdot d\mathbf{r} &= \iint_S \text{curl }\mathbf{F}(x, y, z)\cdot \mathbf{n}\, dS = \iint_D (xz^2 + xy^2)\, dA \\ &= \iint_D x(z^2 + y^2)\, dA.\end{aligned}$$

To finish the integration, we will use polar coordinates. Here, $r^2 = z^2 + y^2$ and $dA = r\,dr\,d\theta$. Furthermore, since $x = 5$, we have

$$\begin{aligned}\iint_D x(z^2 + y^2)\, dA &= \int_0^{2\pi}\int_0^3 5r^2\, r\, dr\, d\theta = \int_0^{2\pi}\left[\frac{5}{4}r^4\right]_0^3 d\theta \\ &= \int_0^{2\pi}\frac{405}{4}\, d\theta = \frac{405\pi}{2}.\end{aligned}$$

□

? TEST YOUR UNDERSTANDING

- What is the relationship between Green's Theorem and Stokes' Theorem?
- What is the relationship between Stokes' Theorem and the Fundamental Theorem of Calculus?
- Why are both sides of the equation in Stokes' Theorem potentially more complicated than they are in either the Fundamental Theorem of Calculus or Green's Theorem?
- Suppose that S_1 and S_2 are two different smooth oriented surfaces with a common boundary curve C and a common orientation relative to C. (That is, travelling counterclockwise along C with respect to S_1 is the same direction as travelling counterclockwise with respect to S_2.) Let $\mathbf{F}(x, y, z)$ be a vector field that is defined on an open set containing both surfaces. What does Stokes' Theorem imply about

$$\iint_{S_1} \mathbf{F}(x, y, z)\cdot \mathbf{n_1}\, dS \quad \text{and} \quad \iint_{S_2} \mathbf{F}(x, y, z)\cdot \mathbf{n_2}\, dS$$

if, for example, S_1 is the hemisphere that is the graph of $z = \sqrt{1 - x^2 - y^2}$ and S_2 is the paraboloid that is the graph of $z = 4(x^2 + y^2 - 1)$, as shown in the figures that follow? Both of these graphs are bounded by the unit circle in the xy-plane.

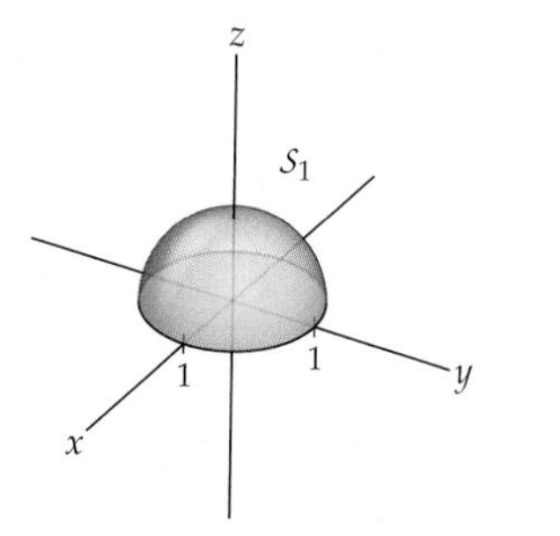

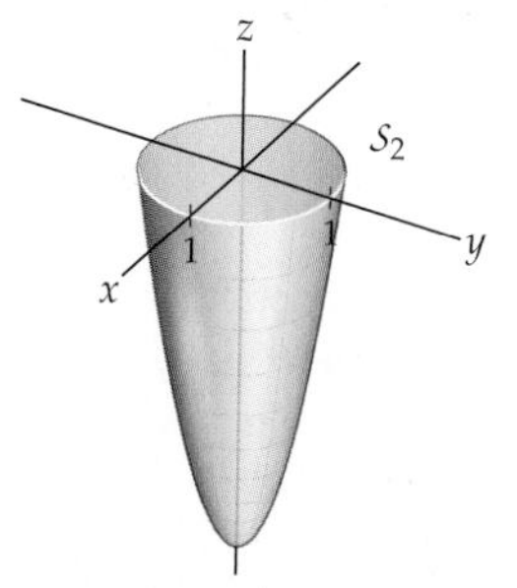

- In the examples, why are $\mathbf{n}$ and dS never explicitly separated? (It may be useful to look at Section 14.3.)

EXERCISES 14.5

Thinking Back

Antecedents: Stokes' Theorem is an immediate generalization of Green's Theorem, but it is related to other theorems as well. For instance, it can also be viewed as a two-dimensional version of the Fundamental Theorem of Line Integrals.

- ▶ Review the Fundamental Theorem of Line Integrals in Section 14.2. For conservative vector fields, what does the theorem imply about $\int_{C_1} \mathbf{F} \cdot d\mathbf{r}$ and $\int_{C_2} \mathbf{F} \cdot d\mathbf{r}$ if C_1 and C_2 are two smooth curves with common initial and terminal points and $\mathbf{F}$ is smooth?
- ▶ Review the Fundamental Theorem of Calculus in Chapter 4. What parts of the Stokes' Theorem equation correspond to the interval of integration, the integrand, and the antiderivative?

Concepts

0. *Problem Zero:* Read the section and make your own summary of the material.
1. *True/False:* Determine whether each of the statements that follow is true or false. If a statement is true, explain why. If a statement is false, provide a counterexample.
   (a) *True or False:* Stokes' Theorem asserts that the flux of a vector field through a smooth surface with a smooth boundary is equal to the line integral of this field about the boundary of the surface.
   (b) *True or False:* Stokes' Theorem can be interpreted as a generalization of Green's Theorem.
   (c) *True or False:* Stokes' Theorem applies only to conservative vector fields.
   (d) *True or False:* Stokes' Theorem is always used as a way to evaluate difficult surface integrals.
   (e) *True or False:* Stokes' Theorem can be interpreted as a generalization of the Fundamental Theorem of Line Integrals.
   (f) *True or False:* If $\mathbf{F}(x, y, z)$ is a conservative vector field, then Stokes' Theorem and Theorem 14.12 together give an alternative proof of the Fundamental Theorem of Line Integrals for simple closed curves.
   (g) *True or False:* Stokes' Theorem can be interpreted as a generalization of the Fundamental Theorem of Calculus.
   (h) *True or False:* Stokes' Theorem can be used to evaluate surface area.
2. *Examples:* Construct examples of the thing(s) described in the following. Try to find examples that are different than any in the reading.
   (a) A smooth surface with a smooth boundary.
   (b) A surface that is not smooth, but that has a smooth boundary.
   (c) A surface that is smooth, but does not have a smooth boundary.
3. Write two different normal vectors for a smooth surface $\mathcal{S}$ given by $(x, y, g(x, y))$ at the point $(x_0, y_0, g(x_0, y_0))$.
4. Make a chart of all the new notation, definitions, and theorems in this section, including what each new thing means in terms you already understand.
5. Suppose that $\mathcal{S}_1$ is the upper half of the unit sphere, with outwards-pointing normal $\mathbf{n}_1$, and $\mathcal{S}_2$ is a balloon-shaped surface whose boundary is the unit circle and whose orientation leads to counterclockwise parametrization of the unit circle. If $\mathbf{F}(x, y, z)$ is a smooth vector field defined on a region large enough to include both surfaces, what is the relationship between $\iint_{\mathcal{S}_1} \text{curl}\,\mathbf{F} \cdot \mathbf{n}_1\, d\mathcal{S}$ and $\iint_{\mathcal{S}_2} \text{curl}\,\mathbf{F} \cdot \mathbf{n}_2\, d\mathcal{S}$?
6. In what way is Green's Theorem a special case of Stokes' Theorem?
7. If $\text{curl}\,\mathbf{F}(x, y, z) \cdot \mathbf{n}$ is constantly equal to 1 on a smooth surface $\mathcal{S}$ with a smooth boundary curve C, then Stokes' Theorem can reduce the integral for the surface area to a line integral. State this integral.
8. In what way is Stokes' Theorem a generalization of the Fundamental Theorem of Line Integrals?
9. Why is the orientation of $\mathcal{S}$ important to the statement of Stokes' Theorem? What will change if the orientation is reversed?
10. Draw a picture of a smooth surface with subdivisions and locally approximating tangent planes.
11. Give an example of a vector field whose orientation does not affect the outcome of Stokes' Theorem.
12. In your own words, explain how the "subdivide, approximate, and add" strategy from Chapter 4 is applied in the sketch of the proof of Stokes' Theorem.
13. Why is dA in Green's Theorem replaced by dS in Stokes' Theorem?
14. Given a smooth surface $\mathcal{S}$ with boundary curve the unit circle traversed counterclockwise, rewrite
$$\iint_{\mathcal{S}} \text{curl}\,\mathbf{F}(x, y, z) \cdot \mathbf{n}\, dS$$
as a double integral in the plane.
15. Given an integral of the form $\int_C \mathbf{F} \cdot d\mathbf{r}$, what considerations would lead you to evaluate the integral with Stokes' Theorem?
16. Given an integral of the form $\iint_{\mathcal{S}} \text{curl}\,\mathbf{F}(x, y, z) \cdot \mathbf{n}\, dS$, what considerations would lead you to evaluate the integral with Stokes' Theorem?
17. Why does the statement of Stokes' Theorem require that the surface $\mathcal{S}$ be smooth or piecewise smooth? What, if anything, goes wrong if this condition is not met?
18. Why does the statement of Stokes' Theorem require that the boundary curve C be smooth or piecewise smooth? What, if anything, goes wrong if this condition is not met?

Skills

Evaluate the integrals in Exercises 19–32. In some cases, Stokes' Theorem will help; in other cases, it may be preferable to evaluate the integrals directly.

19. $\int_C \mathbf{F}(x,y,z) \cdot d\mathbf{r}$, where C is the boundary of the triangle in the plane $y = 2$ and with vertices $(1,2,0)$, $(0,2,0)$, and $(0,2,1)$, with the normal vector pointing in the positive y direction and
$$\mathbf{F}(x,y,z) = 3yz\mathbf{i} + e^x\mathbf{j} + x^2z\mathbf{k}.$$

20. $\int_C \mathbf{F}(x,y,z) \cdot d\mathbf{r}$, where C is in the plane $z = 12 - x - y$, with upwards-pointing normal vector, and is the boundary of the region that lies above the square in the xy-plane and with vertices $(3,5,0)$, $(3,7,0)$, $(4,5,0)$, and $(4,7,0)$, and where
$$\mathbf{F}(x,y,z) = (y - z)\mathbf{i} + 2x\mathbf{j} + 5xz\mathbf{k}.$$

21. $\iint_S \operatorname{curl} \mathbf{F}(x,y,z) \cdot \mathbf{n}\, dS$, where S is the portion of the plane $x + y - z = 0$ with upwards-pointing normal vector and
$$\mathbf{F}(x,y,z) = yze^{xyz}\mathbf{i} + xze^{xyz}\mathbf{j} + xye^{xyz}\mathbf{k}.$$

22. $\iint_S \operatorname{curl} \mathbf{F}(x,y,z) \cdot \mathbf{n}\, dS$, where S is the portion of the paraboloid $z = x^2 + y^2$ that lies above the square $-1 \le x \le 1$, $0 \le y \le 2$ in the xy-plane with upwards-pointing normal vector and
$$\mathbf{F}(x,y,z) = x^3yz\mathbf{i} + xy^2z\mathbf{j} + yz\mathbf{k}.$$

23. $\int_C \mathbf{F}(x,y,z) \cdot d\mathbf{r}$, where C is the closed curve in the plane $y = x$ and formed by the curves $x = z$ and $x^2 = z$, traversed counterclockwise with respect to normal vector $\mathbf{n} = \langle 1, -1, 0 \rangle$, and where
$$\mathbf{F}(x,y,z) = 7xy\mathbf{i} - z\mathbf{j} + 3xyz\mathbf{k}.$$

24. $\int_C \mathbf{F}(x,y,z) \cdot d\mathbf{r}$, where C is the boundary of the region in the plane $z = 2x - y + 10$ and that lies above the curves $x = 1$, $x = 2$, and $y = e^x$, and where
$$\mathbf{F}(x,y,z) = (3x + y)\mathbf{i} + (y - 2z)\mathbf{j} + (2 + 3z)\mathbf{k}.$$

25. $\iint_S \operatorname{curl} \mathbf{F}(x,y,z) \cdot \mathbf{n}\, dS$, where S is the cap of the unit sphere that lies below the xy-plane and inside the cylinder $x^2 + y^2 = \frac{1}{9}$ with outwards-pointing normal vector and where
$$\mathbf{F}(x,y,z) = -yz^2\mathbf{i} + xz^2\mathbf{j} + 3^{-xyz}\mathbf{k}.$$

26. $\iint_S \operatorname{curl} \mathbf{F}(x,y,z) \cdot \mathbf{n}\, dS$, where S is the portion of the hyperbolic paraboloid $z = x^2 - y^2$ that lies inside the elliptical cylinder $4x^2 + 9y^2 = 36$ with upwards-pointing normal vector and $\mathbf{F}(x,y,z) = (1 - yz\sin(xyz))\mathbf{i} - (1 + xz\sin(xyz))\mathbf{j} + (1 - xy\sin(xyz))\mathbf{k}$.

27. $\int_C \mathbf{F}(x,y,z) \cdot d\mathbf{r}$, where C is the curve in the plane $x - y + z = 20$ and that lies above the curves $y = 4$ and $y = x^2$ in the xy-plane, traversed counterclockwise with respect to $\mathbf{n} = \langle 1, -1, 1 \rangle$, and where $\mathbf{F}(x,y,z) = (2x - 3y + 4z)\mathbf{i} + (5x + y - z)\mathbf{j} + (x + 4y + 2z)\mathbf{k}$.

28. $\int_C \mathbf{F}(x,y,z) \cdot d\mathbf{r}$, where C is the curve on the paraboloid $z = x^2 + y^2$ that lies above the unit circle, traversed counterclockwise with respect to the outwards-pointing normal vector, and where
$$\mathbf{F}(x,y,z) = (3x + y - z)\mathbf{i} + (4y - 2z)\mathbf{j} + (x - 3z)\mathbf{k}.$$

29. $\iint_S \operatorname{curl} \mathbf{F}(x,y,z) \cdot \mathbf{n}\, dS$, where S is the portion of the surface $y = \sqrt{4 - x^2 - z^2}$ that lies between $y = 4$ and $y = \sqrt{3}$ with normal vector pointing in the positive y direction and where
$$\mathbf{F}(x,y,z) = (-4z - xz^2)\mathbf{i} + \sin(xyz)\mathbf{j} + (4x + x^2z)\mathbf{k}.$$

30. $\iint_S \operatorname{curl} \mathbf{F}(x,y,z) \cdot \mathbf{n}\, dS$, where S is the portion of the surface $z = e^{x^2+y^2}$ that lies above the plane $z = \frac{1}{4}$ with upwards-pointing normal vector and where
$$\mathbf{F}(x,y,z) = \ln(2x + 1)\mathbf{i} + 2^{y+1}\mathbf{j} + xe^y\mathbf{k}.$$

31. $\int_C \mathbf{F} \cdot d\mathbf{r}$, where C is the intersection of the surface $z = e^{-(x^2+y^2)}$ and the cylinder $x^2 + y^2 = 9$ and where
$$\mathbf{F}(x,y,z) = \left\langle 3x + 3, 4x + \ln(y^2 + 1) - z, 2x + y \right\rangle.$$

32. $\iint_S \operatorname{curl} \mathbf{F}(x,y,z) \cdot \mathbf{n}\, dS$, where S is the bottom sheet of the hyperboloid $z^2 = x^2 + y^2 + 1$ that lies above the plane $z = 2\sqrt{2}$ with outwards-pointing normal vector.

The hypothesis of Stokes' Theorem requires that the vector field $\mathbf{F}$ be defined and continuously differentiable on an open set containing the surface S bounded by C, that C be simple, smooth, and closed, and that S be oriented and smooth. For each of the situations in Exercises 33–36, show that Stokes' Theorem does not apply.

33. $\mathbf{F}(x,y,z) = \left\langle \frac{x}{x^2+y^2}, -\frac{y}{x^2+y^2}, z^2 \right\rangle$, and C is the unit circle.

34. $\mathbf{F}(x,y,z) = \langle \ln(xy + 1) + 5^x3^y2^z, 4xz^2 \rangle$, and C is the boundary of the square in the plane $z = 6$ and with vertices at $(2,0,6)$, $(-2,0,6)$, $(2,4,6)$, and $(-2,4,6)$.

35. S is the portion of the cone $z = \sqrt{x^2 + y^2}$ inside the cylinder $x^2 + y^2 = k^2$ for some $k \ge 0$.

36. S is the pyramid with vertices at $(0,0,6)$, $(2,0,0)$, $(-2,0,0)$, $(0,3,0)$, and $(0,-3,0)$.

Applications

For a given vector field $\mathbf{F}(x,y,z)$ and simple closed curve C, traversed counterclockwise to a chosen normal vector $\mathbf{n}$, the ***circulation*** of $\mathbf{F}(x,y,z)$ around C measures the rotation of the fluid about C in the direction counterclockwise to the aforementioned chosen normal vector and is defined to be $\int_C \mathbf{F}(x,y,z) \cdot d\mathbf{r}$. Find the circulation of the given vector field around C in Exercises 37 and 38.

37. $\mathbf{F}(x,y,z) = \mathbf{i} + \mathbf{j} + \mathbf{k}$, and C is the curve of intersection of the plane $\theta = \frac{\pi}{4}$ or $\theta = \frac{5\pi}{4}$ and the unit sphere.

38. $\mathbf{F}(x,y,z) = \langle 3y, -x, e^{x+y} \rangle$, and C is the intersection of the cone $z = \sqrt{x^2 + y^2}$ and the unit sphere.

Consider once again the notion of the rotation of a vector field. If a vector field $\mathbf{F}(x, y, z)$ has curl $\mathbf{F} = \mathbf{0}$ at a point P, then the field is said to be ***irrotational*** at that point. Show that the fields in Exercises 39–42 are irrotational at the given points.

39. $\mathbf{F}(x, y, z) = \langle -\sin x, 3y^3, 4z + 12\rangle$, $P = (2, 3, 4)$.

40. $\mathbf{F}(x, y, z) = \langle y + z, x + z, x + y\rangle$, $P = (1, -1, 1)$.

41. $\mathbf{F}(x, y, z) = \langle 2xyz, x^2z, x^2y\rangle$, $P = (1, 1, 1)$.

42. $\mathbf{F}(x, y, z) = \nabla f(x, y, z)$, where $f(x, y, z)$ is a function defined on $\mathbb{R}^3$ and has continuous first and second partial derivatives, and where $P = (x_0, y_0, z_0)$.

Find the work done by the given vector field moving around the curve in the indicated direction in Exercises 43 and 44.

43. $\mathbf{F}(x, y, z) = yz\mathbf{i} + 2xz\mathbf{j} + xy\mathbf{k}$, and C is the curve formed by the intersection of the plane $12x + 2y - z = 15$ and the cylinder $y^2 + z^2 = 4$, traversed counterclockwise with respect to the normal vector $\mathbf{n} = \langle 12, 2, -1\rangle$.

44. $\mathbf{F}(x, y, z) = \left\langle \frac{2x}{x^2+y^2+z^2+1}, \frac{2y}{x^2+y^2+z^2+1}, \frac{2z}{x^2+y^2+z^2+1} \right\rangle$, and C is the curve created by the intersection of the plane $z = 4$ with the surface $x^2 + 2y - z = 0$, when $y \geq 0$, together with the line segment connecting the points $(2, 0, 4)$ and $(-2, 0, 4)$, traversed counterclockwise with respect to the normal vector $\mathbf{k}$.

45. The current through a certain region of the San Juan Islands in Washington State is given by $\mathbf{F} = \langle 0, 1.152 - 0.8x^2\rangle$. Consider a disk R of radius 1 mile centered on this region. Denote the boundary of the disk by ∂R.

(a) Compute $\iint_R \nabla \times \mathbf{F} \cdot \mathbf{n}\, dA$.

(b) Show that

$$\iint_{\partial R} \mathbf{F} \cdot d\mathbf{r} = \int_0^{2\pi} (1.152 - 0.8\cos^2\theta)\sin\theta\, d\theta = 0.$$

Conclude that Stokes' Theorem is valid for the current in this region of the San Juan Islands.

(c) What do the integrals from Stokes' Theorem tell us about this region of the San Juan Islands?

Proofs

46. (a) Use Stokes' Theorem and the Fundamental Theorem of Line Integrals to show that

$$\iint_S \text{curl } \mathbf{F}(x, y, z) \cdot \mathbf{n}\, dS = 0$$

for any conservative vector field $\mathbf{F}$.

(b) Without using Stokes' Theorem, show that

$$\iint_S \text{curl } \mathbf{F}(x, y, z) \cdot \mathbf{n}\, dS = 0$$

for any conservative vector field $\mathbf{F}$.

47. Show that if C is a smooth curve in a plane $z = r$, where r is an arbitrary constant, that is parallel to the xy-plane, and if $\mathbf{F}(x, y, z) = \langle F_1(x, y, z), F_2(x, y, z), F_3(x, y, z)\rangle$, then $\int_C \mathbf{F} \cdot d\mathbf{r}$ does not depend on F_3.

48. State and prove a version of Exercise 47 for smooth curves in planes of the form $x = r$ and of the form $y = r$.

49. Let C be a smooth simple closed curve in the plane $ax + by + cz = k$ with nonzero constants a, b, c, and k, and let $\mathbf{F}(x, y, z) = \langle F_1(x, y, z), F_2(x, y, z), F_3(x, y, z)\rangle$ be a vector field whose component functions are linear. Show that $\int_C \mathbf{F} \cdot d\mathbf{r}$ depends only on the area enclosed by C.

50. Show that $\int_C \mathbf{F} \cdot d\mathbf{r} = 0$ does not imply that $\mathbf{F}$ is conservative, by (1) computing $\int_C \mathbf{F} \cdot d\mathbf{r}$ for C the unit circle and $\mathbf{F}(x, y, z) = x\mathbf{i} - y\mathbf{j} + 2^z \ln(z + 1)\mathbf{k}$ and (2) showing that $\mathbf{F}$ is not conservative but has zero curl.

51. Suppose S_1 and S_2 are two smooth surfaces with the same smooth boundary curve C and that traversing the boundary in the counterclockwise direction determined by the orientation of the surfaces describes the same direction of travel along C. Moreover, let $\mathbf{F}(x, y, z)$ be a smooth vector field defined on a region containing both surfaces. Show that

$$\iint_{S_1} \text{curl } \mathbf{F}(x, y, z) \cdot \mathbf{n}\, dS = \iint_{S_2} \text{curl } \mathbf{F}(x, y, z) \cdot \mathbf{n}\, dS$$

In this case, the integral is ***surface independent***, a property analogous to the path independence of line integrals for conservative vector fields.

Thinking Forward

Generalizing to Higher Dimensions: Stokes' Theorem is about a surface inside of $\mathbb{R}^3$. But the essence of the theorem has to do with integrating over a surface, and does not use any properties that $\mathbb{R}^3$ has which $\mathbb{R}^4$ does not.

- ▶ How would a version of Stokes' Theorem for $\mathbb{R}^4$ need to change to accommodate the extra dimension?
- ▶ If we take the perspective that Stokes' Theorem is a two-dimensional version of the Fundamental Theorem of Line Integrals, it is natural to ask what a three-dimensional version of Stokes' Theorem would look like. Make a conjecture about a three-dimensional version of Stokes' Theorem. (Note that the previous question asks about a surface in $\mathbb{R}^4$, while this question is about a three-dimensional region in $\mathbb{R}^4$.)

14.6 THE DIVERGENCE THEOREM

- Extending the Fundamental Theorem of Calculus to three-dimensional regions
- The Divergence Theorem
- Justifying the Divergence Theorem

The Divergence Theorem

Our final theorem of vector analysis presents the Divergence Theorem, another generalization of Green's Theorem and, in turn, of the Fundamental Theorem of Calculus. Here the essential boundary-to-interior relationship is extended to three-dimensional solids, whose boundaries are surfaces.

To set up the Divergence Theorem, we will need to discuss bounded three-dimensional regions in space. The sort of solids we are interested in are regions $W \subseteq \mathbb{R}^3$ whose boundaries are smooth (or piecewise-smooth), simple closed oriented surfaces. Intuitively, we want the boundary of W to be a reasonably well-behaved surface with a choice of inward and outward normal vector. Recall our requirement that all vector fields we use in this chapter be smooth; that is, that their component functions be continuous and have continuous first derivatives.

To see how Green's Theorem can be scaled up in dimension, let W be a reasonably well-behaved region in $\mathbb{R}^3$. For example, W might be described as

$$\{(x, y, z) \mid a \leq x \leq b,\ \ g_1(x) \leq y \leq g_2(x),\ \ f_1(x, y) \leq z \leq f_2(x, y)\}$$

or analogous solids for each of the other five permutations of x, y, and z.

Recall the divergence form of Green's Theorem, which reads, under the usual assumptions about $\mathbf{F}(x, y, z)$, C, and R,

$$\int_C \mathbf{F}(x, y) \cdot \mathbf{n}\, ds = \iint_R \operatorname{div} \mathbf{F}\, dA.$$

If W is as described, then Green's Theorem applies to every slice of W that is parallel to the xy-plane. On the basis of past experience, it is reasonable to suppose that subdividing the height range into small subintervals, using a choice from within each subinterval to approximate, and adding the approximations will, in the limit, yield the correct value.

On the left-hand side of the preceding equation, we have an integral of line integrals, which should result in a two-dimensional, or surface, integral. On the right-hand side of the equation, we have a double integral, which should result in a volume.

The Divergence Theorem says that this is in essence what happens:

THEOREM 14.16 **Divergence Theorem**

Suppose W is a region in $\mathbb{R}^3$ bounded by a smooth or piecewise-smooth closed oriented surface $\mathcal{S}$. If $\mathbf{F}(x, y, z)$ is defined on an open region containing W, then

$$\iiint_W \operatorname{div} \mathbf{F}(x, y, z)\, dV = \iint_{\mathcal{S}} \mathbf{F}(x, y, z) \cdot \mathbf{n}\, d\mathcal{S}$$

where $\mathbf{n}$ is the outwards unit normal vector.

Proof. Again, a thorough proof of this theorem is beyond the scope of this book. We sketch a direct proof of the Divergence Theorem for very well-behaved regions in $\mathbb{R}^3$.

To prove the Divergence Theorem for the type of region W that we have described, it suffices to prove that

$$\iiint_W \frac{\partial F_1}{\partial x}\,dV = \iint_S F_1(x,y,z)\mathbf{i}\cdot\mathbf{n}\,dS,$$

$$\iiint_W \frac{\partial F_2}{\partial y}\,dV = \iint_S F_2(x,y,z)\mathbf{j}\cdot\mathbf{n}\,dS, \quad \text{and}$$

$$\iiint_W \frac{\partial F_3}{\partial z}\,dV = \iint_S F_3(x,y,z)\mathbf{k}\cdot\mathbf{n}\,dS$$

where $\mathbf{n} = \langle n_1, n_2, n_3\rangle$ and $\mathbf{F} = \langle F_1, F_2, F_3\rangle$. We prove the last equation for regions W as we have described them. Such regions can be understood as the regions between $S_1 = \{(x,y,z) \mid z = f_1(x,y)\}$ and $S_2 = \{(x,y,z) \mid z = f_2(x,y)\}$ for points (x,y) in the region D of the xy-plane described by $a \le x \le b, g_1(x) \le y \le g_2(x)$. Then

$$\iint_S F_3(x,y,z)\mathbf{k}\cdot\mathbf{n}\,dS = \iint_{S_2} F_3(x,y,z)\mathbf{k}\cdot\mathbf{n}\,dS + \iint_{S_1} F_3(x,y,z)\mathbf{k}\cdot\mathbf{n}\,dS.$$

We let $M = \left\|\left\langle \frac{\partial F_2}{\partial x}, \frac{\partial F_2}{\partial y}, -1\right\rangle\right\|$ On S_2, $\mathbf{n}$ has a positive $\mathbf{k}$-component, so $\mathbf{n} = \frac{1}{M}\left\langle -\frac{\partial F_2}{\partial x}, -\frac{\partial F_2}{\partial y}, 1\right\rangle$, and on S_1, $\mathbf{n}_1$ has a negative $\mathbf{k}$-component, so $\mathbf{n}_1 = \frac{1}{M}\left\langle \frac{\partial F_2}{\partial x}, \frac{\partial F_2}{\partial y}, -1\right\rangle$, and $\mathbf{n}_1 = -\mathbf{n}$. Substituting, we have

$$\begin{aligned}\iint_S F_3(x,y,z)\,\mathbf{k}\cdot\mathbf{n}\,dS &= \iint_{S_2} F_3(x,y,z)\,\mathbf{k}\cdot\mathbf{n}\,dS - \iint_{S_1} F_3(x,y,z)\,\mathbf{k}\cdot\mathbf{n}\,dS \\ &= \iint_S\left(\int_{f_1(x,y)}^{f_2(x,y)} \frac{\partial F_3}{\partial z}\,dz\right)dS = \iiint_W \frac{\partial F_3}{\partial z}\,dV.\end{aligned}$$

■

One way to give a physical intuition for the Divergence Theorem is to recall that the divergence of a vector field in $\mathbb{R}^3$ measures whether or not a gas represented by the field is expanding or compressing. If a gas occupies a region of space, and it is expanding, then the left-hand side of the Divergence Theorem—the integral of div $\mathbf{F}$—will be positive. At the same time, if a gas in a region is expanding, we would expect that it will have a positive flux flowing out through its boundary. This is part of what the Divergence Theorem asserts: Since both sides of the equation give the same value, in particular they have the same sign. The finer detail in the theorem keeps track of the exact relationship.

The Divergence Theorem can be extended to relate surface and volume integrals for solids bounded by more than one closed surface. This procedure is similar to that for Green's Theorem and, as in that case, requires some bookkeeping to make sure that the theorem is being applied correctly. An example is given in Example 3.

A Summary of Theorems About Vector Analysis

Having arrived at the end of a chapter on vector analysis, which is in turn at the end of a three-semester calculus sequence, it is appropriate to examine some of the most salient results side by side, to appreciate their similarities as well as their differences arising from context. A summary of the chief results of this chapter and their precursors follows. You may find it a useful exercise to reconsider the notation, background assumptions, and context pertaining to each theorem.

Theorem Name	Integral Statement	Reference
Fundamental Theorem of Calculus	$\int_a^b f(x)\,dx = F(b) - F(a)$	Theorem 4.24
Fundamental Theorem of Line Integrals	$\int_C \nabla f(x,y,z) \cdot d\mathbf{r} = f(Q) - f(P)$	Theorem 14.5
Green's Theorem	$\iint_R \left(\frac{\partial F_2}{\partial x} - \frac{\partial F_1}{\partial y}\right) dA = \int_C \mathbf{F}(x,y) \cdot d\mathbf{r}$	Theorem 14.13
Stokes' Theorem	$\iint_S \operatorname{curl} \mathbf{F}(x,y,z) \cdot \mathbf{n}\,dS = \int_C \mathbf{F}(x,y,z) \cdot d\mathbf{r}$	Theorem 14.15
Divergence Theorem	$\iiint_W \operatorname{div} \mathbf{F}(x,y,z)\,dV = \iint_S \mathbf{F}(x,y,z) \cdot \mathbf{n}\,dS$	Theorem 14.16

Examples and Explorations

EXAMPLE 1

Verifying the Divergence Theorem

Verify the conclusion of the Divergence Theorem for the vector field

$$\mathbf{F}(x,y,z) = x^2\mathbf{i} + y^2\mathbf{j} + z^2\mathbf{j}$$

with region $\mathcal{R}$, the unit ball centered at the origin.

SOLUTION

To compute

$$\iint_S \mathbf{F}(x,y,z) \cdot \mathbf{n}\,dS,$$

recall that S, the unit sphere that is the boundary of $\mathcal{R}$, is parametrized by

$$\mathbf{r}(\phi,\theta) = (\cos\theta\sin\phi, \sin\theta\sin\phi, \cos\phi), \quad \text{for } 0 \le \phi \le \pi \text{ and } 0 \le \theta \le 2\pi.$$

Then

$$\mathbf{r}_\phi(\phi,\theta) = \langle\cos\theta\cos\phi, \sin\theta\cos\phi, -\sin\phi\rangle \text{ and } \mathbf{r}_\theta(\phi,\theta) = \langle-\sin\theta\sin\phi, \cos\theta\sin\phi, 0\rangle.$$

The cross product,

$$\mathbf{r}_\phi \times \mathbf{r}_\theta = \left\langle\cos\theta\sin^2\phi, \sin\theta\sin^2\phi, \cos\phi\sin\phi\right\rangle,$$

and $\|\mathbf{r}_\phi \times \mathbf{r}_\theta\| = \sin\phi$. For all values of ϕ and θ, the vector $\mathbf{r}_\phi \times \mathbf{r}_\theta$ points away from the origin and thus in the direction of the outwards unit normal. Therefore, we let

$$\mathbf{n} = \frac{\mathbf{r}_\phi \times \mathbf{r}_\theta}{\sin\phi} = \langle\cos\theta\sin\phi, \sin\theta\sin\phi, \cos\phi\rangle.$$

On S,

$$\mathbf{F}(x,y,z) = \left\langle\cos^2\theta\sin^2\phi, \sin^2\theta\sin^2\phi, \cos^2\phi\right\rangle.$$

So,

$$\begin{aligned}\mathbf{F}(x,y,z) \cdot \mathbf{n} &= \left\langle\cos^2\theta\sin^2\phi, \sin^2\theta\sin^2\phi, \cos^2\phi\right\rangle \cdot \langle\cos\theta\sin\phi, \sin\theta\sin\phi, \cos\phi\rangle \\ &= \cos^3\theta\sin^3\phi + \sin^3\theta\sin^3\phi + \cos^3\phi.\end{aligned}$$

Substituting, we have

$$\iint_S \mathbf{F}(x,y,z)\cdot\mathbf{n}\,dS = \int_0^{2\pi}\int_0^{\pi}(\cos^3\theta\sin^3\phi+\sin^3\theta\sin^3\phi+\cos^3\phi)\,d\phi\,d\theta$$
$$=\int_0^{2\pi}\left[(\cos^3\theta+\sin^3\theta)\left(-\frac{1}{3}\sin^2\phi\cos\phi-\frac{2}{3}\cos\phi\right)-\frac{1}{3}\cos^2\phi\sin\phi-\frac{2}{3}\phi\right]_0^{\pi}d\theta$$
$$=\int_0^{2\pi}\frac{4}{3}(\cos^3\theta+\sin^3\theta)\,d\theta$$
$$=\frac{4}{3}\left[\sin\theta-\frac{1}{3}\sin^3\theta-\cos\theta+\frac{1}{3}\cos^3\theta\right]_0^{2\pi}=0.$$

To compute the triple integral given by the Divergence Theorem, we find

$$\operatorname{div}\mathbf{F}(x,y,z)=2x+2y+2z.$$

Since $\mathcal{R}$ is the unit sphere, this integral is most convenient to evaluate in spherical coordinates. Converting div $\mathbf{F}(x,y,z)$ into spherical coordinates gives

$$\iiint_R \operatorname{div}\mathbf{F}(x,y,z)\,dV$$
$$=2\int_0^{2\pi}\int_0^{\pi}\int_0^1(\rho\cos\theta\sin\phi+\rho\sin\theta\sin\phi+\rho\cos\phi)\,\rho^2\sin\phi\,d\rho\,d\phi\,d\theta$$
$$=2\int_0^{2\pi}\int_0^{\pi}\int_0^1\rho^3(\cos\theta\sin^2\phi+\sin\theta\sin^2\phi+\cos\phi\sin\phi)\,d\rho\,d\phi\,d\theta$$
$$=2\int_0^{2\pi}\int_0^{\pi}\left[\frac{1}{4}\rho^4(\cos\theta\sin^2\phi+\sin\theta\sin^2\phi+\cos\phi\sin\phi)\right]_0^1 d\phi\,d\theta$$
$$=\frac{1}{2}\int_0^{2\pi}\int_0^{\pi}(\cos\theta\sin^2\phi+\sin\theta\sin^2\phi+\cos\phi\sin\phi)\,d\phi\,d\theta$$
$$=\frac{1}{2}\int_0^{2\pi}\left[(\cos\theta+\sin\theta)\left(\frac{1}{2}\phi-\frac{1}{4}\sin 2\phi\right)+\frac{1}{2}\sin^2\phi\right]_0^{\pi}d\theta$$
$$=\frac{\pi}{4}\int_0^{2\pi}(\cos\theta+\sin\theta)\,d\theta$$
$$=\frac{\pi}{4}\left[\sin\theta-\cos\theta\right]_0^{2\pi}=0.$$

□

EXAMPLE 2 **Using the Divergence Theorem**

Use the Divergence Theorem to evaluate the following flux integrals:

(a)
$$\iint_S \mathbf{F}(x,y,z)\cdot\mathbf{n}\,dS,$$

where

$$\mathbf{F}(x,y,z)=x\mathbf{i}+ze^{xz}\mathbf{j}+xyz\mathbf{k}$$

and $\mathcal{S}$ is the surface of the rectangular solid bounded by the planes $x=0$, $x=5$, $y=0$, $y=3$, $z=0$, and $z=7$.

(b)
$$\iint_S \mathbf{G}(x,y,z)\cdot\mathbf{n}\,dS$$

where

$$\mathbf{G}(x,y,z)=(e^{yz}-x^2yz)\mathbf{i}+xy^2z\mathbf{j}+2z\mathbf{k}$$

and $\mathcal{S}$ is the surface of the ellipsoid $x^2+y^2+4z^2=16$.

SOLUTION

(a) Computing the divergence of $\mathbf{F}(x, y, z)$, we find that

$$\text{div } \mathbf{F} = 1 + xy.$$

Applying the Divergence Theorem allows us to rewrite the integral:

$$\begin{aligned}
\iint_S \mathbf{F}(x, y, z) \cdot \mathbf{n}\, dS &= \iiint_R (1 + xy)\, dz\, dy\, dx \\
&= \int_0^5 \int_0^3 \int_0^7 (1 + xy)\, dz\, dy\, dx \\
&= \int_0^5 \int_0^3 (7 + 7xy)\, dy\, dx \\
&= \int_0^5 \left[7y + \frac{7}{2}xy^2\right]_0^3 \\
&= \int_0^5 \left(21 + \frac{63}{2}x\right) dx \\
&= \left[21x + \frac{63}{4}x^2\right]_0^5 = \frac{1995}{4}.
\end{aligned}$$

(b) Computing the divergence of $\mathbf{G}(x, y, z)$, we find that

$$\text{div } \mathbf{G} = -2xyz + 2xyz + 2 = 2.$$

Applying the Divergence Theorem allows us to rewrite the surface integral as a volume integral, which we evaluate by means of cylindrical coordinates.

$$\begin{aligned}
\iint_S \mathbf{G} \cdot \mathbf{n}\, dS &= \iiint_R \text{div } \mathbf{G}\, dV \\
&= \int_0^{2\pi} \int_0^4 \int_{-\sqrt{4-(r^2/4)}}^{\sqrt{4-(r^2/4)}} 2\, r\, dz\, dr\, d\theta \\
&= \int_0^{2\pi} \int_0^4 \left[2rz\right]_{-\sqrt{4-(r^2/4)}}^{\sqrt{4-(r^2/4)}} dr\, d\theta \\
&= \int_0^{2\pi} \int_0^4 4r\sqrt{4 - \frac{1}{4}r^2}\, dr\, d\theta \\
&= \int_0^{2\pi} \left[-\frac{16}{3}\left(4 - \frac{1}{4}r^2\right)^{3/2}\right]_0^4 d\theta \\
&= \int_0^{2\pi} \frac{128}{3}\, d\theta \\
&= \frac{256\pi}{3}.
\end{aligned}$$

□

EXAMPLE 3 **Using the Divergence Theorem for a region bounded by two surfaces**

Use the Divergence Theorem to evaluate

$$\iint_{S_1} \mathbf{F}(x, y, z) \cdot \mathbf{n}\, dS - \iint_{S_2} \mathbf{F}(x, y, z) \cdot \mathbf{n}\, dS,$$

where

$$\mathbf{F}(x, y, z) = x\mathbf{i} + y\mathbf{j} + z\mathbf{k},$$

S_1 is the surface of the sphere of radius 4 centered at the origin, and S_2 is the unit sphere.

SOLUTION

The region is bounded by two concentric spheres:

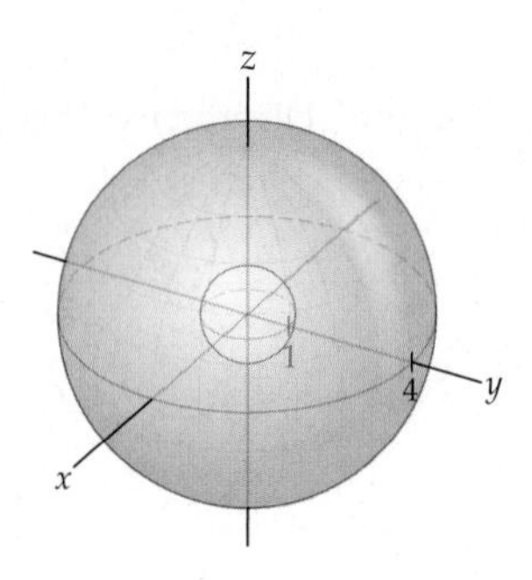

Computing div **F**, we find that div $\mathbf{F} = 1 + 1 + 1 = 3$. Now, applying the Divergence Theorem gives

$$\iint_{S_1} \mathbf{F}(x, y, z) \cdot \mathbf{n}\, dS - \iint_{S_2} \mathbf{F}(x, y, z) \cdot \mathbf{n}\, dS = \int_0^{2\pi} \int_0^{\pi} \int_1^4 3\rho^2 \sin\phi \, d\rho\, d\phi\, d\theta$$
$$= \int_0^{2\pi} \int_0^{\pi} \left[\rho^3 \sin\phi\right]_1^4 d\phi\, d\theta$$
$$= \int_0^{2\pi} \int_0^{\pi} 63 \sin\phi \, d\phi\, d\theta$$
$$= \int_0^{2\pi} \left[-63\cos\phi\right]_0^{\pi} d\theta$$
$$= \int_0^{2\pi} 126\, d\theta = 252\pi.$$

□

EXAMPLE 4 **Calculating an electric field intensity by using Gauss' Law**

Gauss' Law is a method for calculating an electric field intensity **E** generated by simple charge distributions. Given some charge density $\rho(\mathbf{x})$ in a region Ω enclosed by a surface S, Gauss' Law states that the electric field at any point $\mathbf{x} \in \Omega$ satisfies

$$\nabla \cdot \mathbf{E}(\mathbf{x}) = \frac{\rho(\mathbf{x})}{\epsilon_0},$$

where ϵ_0 is a constant called the ***permittivity of free space***. Integrate both sides of this equation and apply the Divergence Theorem to establish the following integral form of Gauss' Law:

$$\iint_S \operatorname{div} \mathbf{E}\, dA = \frac{1}{\epsilon_0} \iiint_\Omega \rho\, dV.$$

SOLUTION

Integrating both sides of the the equation, we see that

$$\iiint_\Omega \nabla \cdot \mathbf{E}\, dV = \frac{1}{\epsilon_0} \iiint_\Omega \rho\, dV.$$

Applying the Divergence Theorem to the left side and noting that the boundary of the region Ω is S gives

$$\iint_S \operatorname{div} \mathbf{E}\, dA = \frac{1}{\epsilon_0} \iiint_\Omega \rho\, dV.$$

□

? TEST YOUR UNDERSTANDING

- What is the relationship between the Divergence Theorem and the Fundamental Theorem of Calculus?
- What relationship exists among the Divergence Theorem, Green's Theorem, and Stokes' Theorem?
- If we wish to use the Divergence Theorem to compute the flux of a vector field flowing into a closed bounded region $W \subseteq \mathbb{R}^3$, how does the theorem need to be adapted?
- In Example 1, what properties of the vector field caused the outcome to be zero?
- How do the change-of-variables factors arise in each of the integrals in Example 1?

EXERCISES 14.6

Thinking Back

A Long Look Back: Here at end of the chapter, which is in turn the end of the course, and moreover the end of a multisemester calculus odyssey, it is appropriate to reflect on the unifying questions, insights, and results that characterize the subject.

- How do the definitions of differentiation and integration that we have made throughout this text rely on the earlier idea of a limit that we defined in Chapter 1?
- How does the strategy of subdivision, approximation, and addition interact with limits in the justifications of the theorems given in Chapter 14?
- Are there any questions about integration that you had before reading this chapter that are now resolved? What are they?
- Are there any questions about integration that you had before reading this chapter that you still have? What are they?
- Do you have any new questions about integration now that had not occurred to you before reading this chapter? What are they?

Concepts

0. *Problem Zero:* Read the section and make your own summary of the material.

1. *True/False:* Determine whether each of the statements that follow is true or false. If a statement is true, explain why. If a statement is false, provide a counterexample.

(a) *True or False:* The Fundamental Theorem of Line Integrals, Green's Theorem, Stokes' Theorem, and the Divergence Theorem bear a family resemblance to the Fundamental Theorem of Calculus and can in some way be interpreted as generalizations of that theorem.

(b) *True or False:* The vector field **F** in the statement of the Divergence Theorem must satisfy the condition we have required throughout this chapter that the component functions of **F** be continuous with continuous derivatives on an open region that contains the region in which we wish to apply the theorem.

(c) *True or False:* The Divergence Theorem is interesting only as a tool for evaluating difficult flux integrals.

(d) *True or False:* It is often, though not always, easier to integrate over a region in $\mathbb{R}^3$ than to evaluate a surface integral.

(e) *True or False:* The Divergence Theorem is a generalization of the technique of integration by parts.

(f) *True or False:* The Divergence Theorem is a consequence of Stokes' Theorem.

(g) *True or False:* The Divergence Theorem can be used to calculate the flux through a region that is bounded by two distinct smooth simple closed surfaces.

(h) *True or False:* The Divergence Theorem is a generalization of Green's Theorem.

2. *Examples:* Construct examples of the thing(s) described in the following. Try to find examples that are different than any in the reading.

(a) A simple closed surface that is smooth.

(b) A simple closed surface that is not smooth, but is piecewise smooth.

(c) A simple closed surface that is neither smooth nor piecewise smooth.

3. Does the Divergence Theorem apply to surfaces that are not closed?

4. Make a chart of all the new notation, definitions, and theorems in this section, including what each new item means in terms you already understand.

5. Give an intuitive explanation of why the paraboloid $z = x^2 + y^2$ is smooth but the surface of the rectangular solid bounded by the coordinate planes and the planes $x = 14$, $y = 5$, and $z = 32$ is only piecewise smooth.

6. Give an intuitive explanation of why the surface of a tetrahedron is piecewise smooth but the surface of the cone $z = \sqrt{x^2 + y^2}$ is not.

7. (a) Let $\mathbf{F}(x, y, z) = x\mathbf{i} + y\mathbf{j} + z\mathbf{k}$, and let S be the unit sphere. Make a sketch of the field. Without evaluating the integral, do you expect that $\iint_S \mathbf{F} \cdot \mathbf{n}\, dS$ will be positive, negative, or zero? Why?
   (b) Let $\mathbf{F}(x, y, z) = 3\mathbf{i} + 7\mathbf{j} + 9\mathbf{k}$, and let S be the unit sphere. Make a sketch of the field. Without evaluating the integral, do you expect that $\iint_S \mathbf{F} \cdot \mathbf{n}\, dS$ will be positive, negative, or zero? Why?
   (c) Use the Divergence Theorem to check your guesses in parts (a) and (b).

8. Recall from Section 14.4 that the divergence of a vector field is related to the expansion or compression of a gas whose motion is represented by that field. If a gas whose motion is represented by a vector field is expanding (or compressing) in a region of space, what effect should that have on the flux of the vector field out of (or into) a closed region W? Making some additional assumptions about the direction of the vectors in $\mathbf{F}$ may help you to think about this situation.

9. If $\mathbf{F}(x, y, z) = \nabla f(x, y, z)$ is a conservative vector field, what will div $\mathbf{F}$ be?

10. Give an example of a conservative vector field whose divergence is not uniformly equal to zero in $\mathbb{R}^3$.

11. Give an example of a conservative vector field whose divergence is uniformly equal to zero in $\mathbb{R}^3$.

12. Give an example of a non-conservative vector field whose divergence is uniformly equal to zero in $\mathbb{R}^3$.

13. Give an example of a non-conservative vector field whose divergence is never equal to zero in $\mathbb{R}^3$.

14. Considering your answers to Exercises 10–13, what do you conjecture is true or not true about conservative vector fields in the context of the Divergence Theorem?

Decide whether or not each of the integrals in Exercises 15–20 can be evaluated by means of the Divergence Theorem. (You do not need to evaluate the integrals.)

15. $\iint_S \mathbf{F}(x, y, z) \cdot \mathbf{n}\, dS$, where S is the cone $y = \sqrt{x^2 + z^2}$ between $y = 0$ and $y = 4$ and where
$$\mathbf{F}(x, y, z) = xyz\mathbf{i} + 2^{xyz}\mathbf{j} + \frac{x}{y^2 + 1}\mathbf{k}.$$

16. $\iint_S \mathbf{F}(x, y, z) \cdot \mathbf{n}\, dS$, where S is the surface of the tetrahedron bounded by the three coordinate planes and the plane $3x + 2y + 6z = 12$ and where
$$\mathbf{F}(x, y, z) = z \sin y\mathbf{i} - x \cos z\mathbf{j} + z \cos y\mathbf{k}.$$

17. $\iint_S \mathbf{F}(x, y, z) \cdot \mathbf{n}\, dS$, where S is the unit sphere and $\mathbf{F}(x, y, z) = \langle e^{\sqrt{x^2+y^2+z^2}}, \ln(x^2y^2z^2), 5x^3z \rangle$.

18. $\iint_S \mathbf{F}(x, y, z) \cdot \mathbf{n}\, dS$, where S is the portion of the surface $x = z^3$ that lies between the planes $y = 13$ and $y = 42$ and where
$$\mathbf{F}(x, y, z) = \left\langle \cos(xy), \frac{1}{y^2z^2 + 1}, x + y \right\rangle.$$

19. $\iint_S \mathbf{F}(x, y, z) \cdot \mathbf{n}\, dS$, where S is the portion of the hyperboloid of two sheets $x^2 + y^2 + 1 = z^2$ that lies between the planes $z = 5$ and $z = 10$ and where
$$\mathbf{F}(x, y, z) = \langle x^2 - 2y + z^2, 4x + z - 2, 2^{xyz} \rangle.$$

20. $\iint_S \mathbf{F}(x, y, z) \cdot \mathbf{n}\, dS$, where S is the surface of the torus parametrized by
$$x = (3 + \cos v)\cos u, \ \ y = (3 + \cos v)\sin u, \ \ z = \sin v$$
and where $\mathbf{F}(x, y, z) = (3x - z)\mathbf{i} - x^2z\mathbf{j} + xy^7\mathbf{k}$.

Skills

In Exercises 21–24, compute the divergence of the given vector field.

21. $\mathbf{F}(x, y, z) = x\mathbf{i} + y\mathbf{j} + z\mathbf{k}$

22. $\mathbf{F}(x, y, z) = yz \cos z\mathbf{i} + (z - x)\mathbf{j} + e^x y\mathbf{k}$

23. $\mathbf{F}(x, y, z) = xe^{xyz}\mathbf{i} + ye^{xyz}\mathbf{j} + ze^{xyz}\mathbf{k}$

24. $\mathbf{F}(x, y, z) = \cos x\mathbf{i} - y \sin zy\mathbf{j} + \cos z\mathbf{k}$

In Exercises 25–40, evaluate the integral
$$\iint_S \mathbf{F}(x, y, z) \cdot \mathbf{n}\, dS$$
for the specified function $\mathbf{F}(x, y, z)$ and the given surface S. In each integral, $\mathbf{n}$ is the outwards-pointing normal vector.

25. $\mathbf{F}(x, y, z) = 4x^3yz\mathbf{i} + 6x^2y^2z\mathbf{j} + 6x^2yz^2\mathbf{k}$, and S is the surface of the first-octant cube with side length π and with one vertex at the origin.

26. $\mathbf{F}(x, y, z) = xy^2\mathbf{i} + y(z - 3x)\mathbf{j} + 4xyz\mathbf{k}$, and S is the surface of the region W bounded by the planes $y = 0$, $y = z$, $z = 3$, $x = 0$, and $x = 4$.

27. $\mathbf{F}(x, y, z) = e^z x \sin y\mathbf{i} + e^z \cos y\mathbf{j} + e^x \tan^{-1} y\mathbf{k}$, and S is the surface of the region W that lies within the unit sphere and above the plane $z = \frac{\sqrt{2}}{2}$.

28. $\mathbf{F}(x, y, z) = x^3\mathbf{i} + y^3\mathbf{j} + z^3\mathbf{k}$, and S is the sphere of radius 3 and centered at the origin.

29. $\mathbf{F}(x, y, z) = 15xz^2\mathbf{i} + 15yx^2\mathbf{j} + 15y^2z\mathbf{k}$, and S is the surface of the lower half of the unit sphere, along with the unit circle in the plane.

30. $\mathbf{F}(x, y, z) = 4xy\mathbf{i} - yz\mathbf{j} + z^3x\mathbf{k}$, and S is the surface of the region W that lies within the cone $z = \sqrt{x^2 + y^2}$ and between the planes $z = 1$ and $z = 4$.

31. $\mathbf{F}(x, y, z) = \langle xz, yz, xyz \rangle$, and S is the surface of the cylinder with equation $x^2 + y^2 = 9$ for $-2 \le z \le 2$.

32. $\mathbf{F}(x, y, z) = \langle e^{x^2+y^2+z^2}, 3y - z, ye^x \rangle$, and S is the unit sphere.

33. $\mathbf{F}(x, y, z) = \sin y \cos z\,\mathbf{i} + yz^2\mathbf{j} + zx^2\mathbf{k}$, and S is the surface of the region W bounded by the paraboloid $y = x^2 + z^2$ and the planes $y = 1$ and $y = 4$.

34. $\mathbf{F}(x, y, z) = \langle 2x - y, yz - ye^{2x}, x^2 + yz \rangle$, and S is the surface of the region that lies within $x = y^2 + z^2$ and $x = 2z + 1$.

35. $\mathbf{F}(x, y, z) = \left\langle x^2y^2, 2xy, \frac{2xz^3}{3} \right\rangle$, and S is the surface of the region that lies within the hyperboloid with equation $x^2 = z^2 + y^2 + 1$ and between the planes $x = 1$ and $x = 4$.

36. $\mathbf{F}(x, y, z) = x2^z \ln 2\mathbf{i} + (x - y + z)\mathbf{j} + (2y + 4z)\mathbf{k}$, and S is the surface of the first-octant region W bounded by the coordinate planes and the plane $x + y + 2z = 2$.

37. $\mathbf{F}(x, y, z) = \left\langle \frac{xe}{2}, zx - ye, \frac{3ez}{2} + \ln(x^2y^2z^2 + 1) \right\rangle$, and S is the surface of the sphere of radius 3 and centered at $(2, 1, 3)$.

38. $\mathbf{F}(x, y, z) = \langle e^{x^2+z^2}, 4y, e^{x^2+z^2} \rangle$, and S is the surface of the right circular cylinder of radius 1 whose axis is the line $\langle 2, t, 2 \rangle$ that runs parallel to the y-axis. The left and right sides of the cylinder lie in the planes $y = 3$ and $y = 0$, respectively.

39. $\mathbf{F}(x, y, z) = 8x\mathbf{i} - 13y\mathbf{j} + (13z - 12e^y)\mathbf{k}$, and S is the surface of the region W that lies above the region $|y| \le |x|, y \ge 0$ in the xy-plane, within the unit circle, and below the saddle $z = x^2 - y^2$.

40. $\mathbf{F}(x, y, z) = 4\cos z + 2x\mathbf{i} + (3y - z)\mathbf{j} + 12z\mathbf{k}$, and S is the surface of the region W bounded below by $z = x^2 + y^2$ and above by the sphere $x^2 + y^2 + z^2 = 2$ centered at the origin.

In Exercises 41–44, find the fluxes of the vector fields through the given surfaces in the direction of the outwards-pointing normal vector.

41. $\mathbf{F}(x, y, z) = \langle xe^y, \ln(xyz), xyz^2 \rangle$, and S is the surface of the rectangular solid with vertices $(1, 1, 1)$, $(1, 5, 1)$, $(1, 1, 4)$, $(1, 5, 4)$, $(7, 1, 1)$, $(7, 5, 1)$, $(7, 1, 4)$, and $(7, 5, 4)$.

42. $\mathbf{F}(x, y, z) = 2xz\,\mathbf{i} + 4xy\,\mathbf{j} - 8zy\,\mathbf{k}$, and S is the surface determined by $y = 4 - x^2, y \ge 0, 0 \le z \le 5$.

43. $\mathbf{F}(x, y, z) = \langle y\cos z, 3y, \sin(xy) \rangle$, and S is the surface of the pyramid with the square base in the xy-plane and with vertices $(1, 1, 0)$, $(-1, 1, 0)$, $(1, -1, 0)$, $(-1, -1, 0)$ and apex $(0, 0, 4)$.

44. $\mathbf{F}(x, y, z) = \left\langle x^2y^2z^2, z - x - y, \frac{y}{z+1} \right\rangle$, and S is the surface of the region bounded by $z = x^2$, $x = 0$, and $4x + 2z = 4$.

Applications

45. The current through a certain region of the San Juan Islands in Washington State is given by $\mathbf{F} = \langle 0, 1.152 - 0.8x^2 \rangle$. Consider a disk R of radius 1 centered on this region. Denote the circle that comprises the boundary of the disk by ∂R.

(a) Compute $\iint_R \nabla \cdot \mathbf{F}\, dA$.

(b) Show that $\int_{\partial R} \mathbf{F} \cdot \mathbf{n}\, ds = 0$. Conclude that the Divergence Theorem is valid for the current in this region of the San Juan Islands.

(c) What do the values of the integrals from the Divergence Theorem tell us about this region of the San Juan Islands?

46. Suppose that an electric field is given by $\mathbf{E} = \langle 2y, 2xy, yz \rangle$. Use the Divergence Theorem to compute the flux $\iint_S \mathbf{E}\, d\mathbf{A}$ of the field through the surface of the unit cube $[0, 1] \times [0, 1] \times [0, 1]$.

47. Consider a straight road with multiple lanes going due north. The road is 0.0113 mile wide. Denote the velocity of traffic at any point of the road by $\mathbf{v}(x, y) = \langle 0, v_2(x, y) \rangle$. Traffic engineers are using road tubes along the lines $y = 0$ and $y = 1$ to monitor the speed of traffic.

(a) Suppose that the road tubes indicate that $v_2(x, 1) = v_2(x, 0) - 5$. Compute

$$\int_C \mathbf{v} \cdot \mathbf{n}\, ds < 0,$$

and discuss what this inequality means for traffic in this stretch of road.

(b) Compute the average divergence

$$\frac{1}{A} \iint_S \nabla \cdot \mathbf{v}\, dx\, dy$$

of $\mathbf{v}$ on this stretch, where A is the area of the roadway. What is the significance of the average divergence as regards the traffic?

Proofs

48. Show that if $\mathbf{F}(x, y, z) = \operatorname{curl} \mathbf{G}(x, y, z)$ for some smooth vector field $\mathbf{G}$, and if S is a smooth or piecewise-smooth simple closed surface, then

$$\iint_S \mathbf{F} \cdot \mathbf{n}\, dS = 0.$$

49. Show that if $\mathbf{F}$ is conservative, it does not follow that $\iint_S \mathbf{F}(x, y, z) \cdot \mathbf{n}\, dS = 0$, by using the Divergence Theorem to evaluate $\iint_S \mathbf{F}(x, y, z) \cdot \mathbf{n}\, dS$, where $\mathbf{F}(x, y, z) = \nabla f(x, y, z)$, $f(x, y, z) = x^2 + y^2 + z^2$, and S is the unit sphere with outwards-pointing normal vector.

50. Suppose that $\mathbf{F}(x, y, z) = \operatorname{curl} \mathbf{G}(x, y, z)$ for some smooth vector field $\mathbf{G}$ and W is a region in $\mathbb{R}^3$ that is bounded by a smooth simple closed surface. Use the table of integral theorems on page 1145 to equate $\iiint_W \operatorname{div} \mathbf{F}(x, y, z)\, dV$ with a line integral.

51. Suppose S is any smooth or piecewise-smooth simple closed surface enclosing a region W in $\mathbb{R}^3$, and let

$$\mathbf{F}(x, y, z) = \langle ax + b, cy + d, ez + f \rangle,$$

where a, b, c, d, e, and f are scalars and a, b, and c are not all zero. Find a condition for the relationship of a, b, and c that will guarantee that $\iint_S \mathbf{F} \cdot \mathbf{n}\, dS$ is positive for any choice of smooth or piecewise-smooth simple closed surface S in $\mathbb{R}^3$.

52. Let $\mathbf{F}(x, y, z)$ and S be as in Exercise 51. Find a condition for the relationship of a, b, and c which will guarantee that $\iint_S \mathbf{F} \cdot \mathbf{n}\, dS = 0$ for any choice of smooth or piecewise-smooth simple closed surface S in $\mathbb{R}^3$.

53. Show that if $\iint_S \mathbf{F}(x, y, z) \cdot \mathbf{n}\, dS = 0$, it does not follow that $\mathbf{F}$ is conservative.

Thinking Forward

Calculus in $\mathbb{R}^4$ and beyond: The results of this chapter have illustrated several ways that the Fundamental Theorem of Calculus generalizes to two and three dimensions. Earlier chapters have shown us how to integrate and find partial derivatives of functions taking any finite number of inputs. It is reasonable to ask, then, if the Fundamental Theorem has analogs in $\mathbb{R}^n$, $n \geq 4$.

- What do you think the Divergence Theorem for $\mathbb{R}^4$ might say? What is the guiding intuition for a Divergence Theorem in $\mathbb{R}^n$?
- As in the case of generalizing Green's Theorem to $\mathbb{R}^3$, there is more than one way to try to generalize the Divergence Theorem to $\mathbb{R}^4$ and beyond. What is a way to generalize the Divergence Theorem that is different from the preceding answer you gave.

Calculus for functions from $\mathbb{R}^n$ to $\mathbb{R}^m$: Vector fields can be construed as functions from, say, $\mathbb{R}^3$ to $\mathbb{R}^3$, both multivariate and vector-valued. There are many applications for functions of the general form

$$\mathbf{f}(x_1, \ldots, x_n) = \langle f_1(x_1, \ldots, x_n), \ldots f_m(x_1, \ldots, x_n) \rangle,$$

where n and m are natural numbers. Consider $f : \mathbb{R}^4 \to \mathbb{R}^5$. There are $4 \times 5 = 20$ partial derivatives of f, but if f has smooth component functions, we can still expect to do calculus.

- What might an integral of f represent? How might we handle the partial derivatives of f in a matrix?

Analysis: This book is an introduction to, among other things, the theory of integration. But there are more questions to ask.

- Is there a way to integrate a function with dense discontinuities—for example, the function whose output is 1 if the input is irrational and zero otherwise? This function cannot be integrated by the methods of Chapter 4, but there is a way to expand the notion of integration that will allow us to integrate the function on finite intervals. This and other questions are addressed in analysis courses.

CHAPTER REVIEW, SELF-TEST, AND CAPSTONES

Be sure you are familiar with the definitions, concepts, and basic skills outlined here. The capstone exercises at the end bring together ideas from this chapter.

Definitions

Give precise mathematical definitions or descriptions of each of the concepts that follow. Then illustrate the definition with a graph or an algebraic example.

- a *vector field*
- a *conservative vector field*
- a *potential function*
- a *smooth parametrization*
- the *integral* of a function of two or three variables along a curve C
- the *line integral* of a vector field $\mathbf{F}(x, y, z)$ along a curve C
- the *surface area* of a smooth surface S
- The *integral* of $f(x, y, z)$ over a smooth surface S
- the *flux* of a vector field across a surface
- the *divergence* of a vector field
- the *curl* of a vector field

Theorems

Fill in the blanks to complete each of the following theorem statements:

- *The Fundamental Theorem of Line Integrals*: If $\mathbf{F}(x, y, z)$ is a ________ vector field with $\mathbf{F}(x, y, z) = \nabla f(x, y, z)$ on an ________, ________, and ________ domain containing the curve C with endpoints P and Q, then

$$\int_C \mathbf{F}(x, y, z) \cdot d\mathbf{r} = \_\_\_\_\_\_\_\_.$$

- For any vector field $\mathbf{F}$ in $\mathbb{R}^2$ or $\mathbb{R}^3$,

$$\text{div (curl } \mathbf{F}) = \_\_\_\_\_\_\_\_ = \_\_\_\_\_\_\_\_.$$

- For multivariate functions f in $\mathbb{R}^2$ or $\mathbb{R}^3$ with continuous second partial derivatives,

$$\text{curl } \nabla f = \_\_\_\_\_\_\_\_ = \_\_\_\_\_\_\_\_.$$

- *Green's Theorem*: Let $\mathbf{F}(x, y) = \langle F_1(x, y), F_2(x, y)\rangle$ be a vector field defined on a region R in the xy-plane whose boundary is a ________ curve C. If $\mathbf{r}(t)$ is a parametrization of C in the counterclockwise direction (as viewed from the positive z-axis), then

$$\int_C \mathbf{F}(x, y) \cdot d\mathbf{r} = \int_C \_\_\_\_\_ \, dx + \_\_\_\_\_ \, dy$$
$$= \iint_R (\_\_\_\_\_ - \_\_\_\_\_) \, dA.$$

- *Stokes' Theorem*: Let $\mathcal{S}$ be a _____ or _____ surface with a _____ or _____ boundary curve C. Suppose that $\mathcal{S}$ has an (oriented) unit normal vector $\mathbf{n}$ and that C has a parametrization which traverses C in the _____ direction with respect to $\mathbf{n}$. If $\mathbf{F}(x, y, z) = F_1(x, y, z)\mathbf{i} + F_2(x, y, z)\mathbf{j} + F_3(x, y, z)\mathbf{k}$ is a vector field on an open region containing $\mathcal{S}$, then

$$\int_C \mathbf{F}(x, y, z) \cdot d\mathbf{r} = \iint_{\mathcal{S}} \_\_\_\_\_ \, d\mathcal{S}.$$

- *Divergence Theorem*: Suppose W is a region in $\mathbb{R}^3$ bounded by a ________ or ________ surface $\mathcal{S}$. If $\mathbf{F}(x, yz)$ is defined on an open region containing W, then

$$\iiint_W \text{div } \mathbf{F}(x, y, z) \, dV = \iint_{\mathcal{S}} \_\_\_\_\_\_\_\_ \, d\mathcal{S}$$

where $\mathbf{n}$ is the outwards ________.

Notation and Integration Rules

Notation: Describe the meanings of each of the following mathematical expressions.

- ∇
- div $\mathbf{F}(x, y, z)$
- curl $\mathbf{F}(x, y, z)$

Integration Theorems: Assuming that the necessary hypotheses are present for each of the following integration theorems, provide an alternative method for evaluating the given integral.

- *The Fundamental Theorem of Calculus*:

$$\int_a^b f(x) \, dx = \_\_\_\_\_\_\_\_\_\_\_\_$$

- *The Fundamental Theorem of of Line Integrals*:

$$\int_C \nabla f(x, y, z) \cdot d\mathbf{r} = \_\_\_\_\_\_\_\_\_\_\_\_$$

- *Green's Theorem*:

$$\iint_R \left(\frac{\partial f_2}{\partial x} - \frac{\partial f_1}{\partial y}\right) dA = \_\_\_\_\_\_\_\_\_\_\_\_$$

- *Stokes' Theorem*:

$$\iint_{\mathcal{S}} \text{curl } \mathbf{F}(x, y, z) \cdot \mathbf{n} \, d\mathcal{S} = \_\_\_\_\_\_\_\_\_\_\_\_$$

- *Divergence Theorem*:

$$\iiint_W \text{div } \mathbf{F}(x, y, z) \, dV = \_\_\_\_\_\_\_\_\_\_\_\_$$

Skill Certification: Vector Calculus

Potential Functions: Find a potential function for each vector field.

1. $\mathbf{F}(x, y) = \langle 3x^2y^2, 2x^3y\rangle$

2. $\mathbf{F}(x, y) = \dfrac{y^2}{x}\mathbf{i} + 2y \ln x \mathbf{j}$

3. $\mathbf{G}(x, y, z) = xze^{y^2}\mathbf{i} + 2xyze^{y^2}\mathbf{j} + xe^{y^2}\mathbf{k}$

4. $\mathbf{G}(x, y, z) = \langle 2xy^2ze^{x^2yz}, (1 + x^2yz)e^{x^2yz}, x^2y^2e^{x^2yz}\rangle$

Conservative Vector Fields: Determine whether the vector fields that follow are conservative. If the field is conservative, find a potential function for it.

5. $\mathbf{F}(x, y) = \dfrac{2x}{y}\mathbf{i} + \dfrac{x^2}{y^2}\mathbf{j}$

6. $\mathbf{F}(x, y) = \langle (1 + xy)e^{xy}, e^{xy}\rangle$

7. $\mathbf{F}(x, y) = \langle y^3e^{xy^2}, (1 + 2xy^2)e^{xy^2}\rangle$

8. $\mathbf{F}(x, y, z) = \left\langle \frac{y}{z}, \frac{x}{z}, -\frac{xy}{z^2} \right\rangle$
9. $\mathbf{G}(x, y, z) = \cos y \sin z\mathbf{i} - x \sin y \sin z\mathbf{j} - x \cos y \cos z\mathbf{k}$
10. $\mathbf{G}(x, y, z) = y^2 z\mathbf{i} + 2xyz\mathbf{j} + xy^2\mathbf{k}$

Line Integrals: Evaluate the line integral of the given function over the specified curve.

11. $\int_C \mathbf{F}(x, y) \cdot d\mathbf{r}$, where $\mathbf{F}(x, y) = 7y^2\mathbf{i} - 3xy\mathbf{j}$ and C is the line segment $x = t$, $y = 3t$ for $0 \le t \le 1$.
12. $\int_C \mathbf{F}(x, y) \cdot d\mathbf{r}$, where $\mathbf{F}(x, y) = \langle 8x^2 y, -9xy^2 \rangle$ and C is the curve parametrized by $x = t^2$, $y = t^3$ for $0 \le t \le 1$.
13. $\int_C \mathbf{F}(x, y) \cdot d\mathbf{r}$, where $\mathbf{F}(x, y) = \frac{x}{x^2+y^2}\mathbf{i} - \frac{y}{x^2+y^2}\mathbf{j}$ and C is the unit circle centered at the origin.
14. $\int_C \mathbf{F}(x, y, z) \cdot d\mathbf{r}$, where $\mathbf{F}(x, y, z) = yz\mathbf{i} + xz\mathbf{j} + xy\mathbf{k}$ and C is the curve parametrized by $x = t^2$, $y = t^{-2}$, $z = t$ for $1 \le t \le 3$.
15. $\int_C \mathbf{F}(x, y, z) \cdot d\mathbf{r}$, where $\mathbf{F}(x, y, z) = \langle e^x, e^y, e^z \rangle$ and C is the line segment from $(0, 0, 0)$ to $(1, 1, 1)$.
16. $\int_C \mathbf{F}(x, y, z) \cdot d\mathbf{r}$, where $\mathbf{F}(x, y, z) = \langle e^x, e^y, e^z \rangle$ and C is the curve parametrized by $x = t$, $y = t^2$, $z = t^3$ for $0 \le t \le 1$. Compare your answer with that of Exercise 15.

Surface Area: Find the area of each of the following surfaces.

17. The portion of the plane $2x + 3y + 4z = 12$ that lies in the first octant.
18. The portion of the plane $2x + 3y + 4z = 12$ that lies inside the cylinder with equation $x^2 + y^2 = 9$.
19. The portion of the sphere with equation $x^2 + y^2 + z^2 = 16$ that lies inside the cylinder with equation $x^2 + y^2 = 9$.
20. The portion of the paraboloid with equation $z = 9 - x^2 - y^2$ that lies above the xy-plane.

Flux: Find the outward flux of the given vector field through the specified surface.

21. $\mathbf{F}(x, y, z) = \langle x, y, z \rangle$ and $\mathcal{S}$ is the portion of the cone with equation $z = \sqrt{x^2 + y^2}$ for $z \le 3$.
22. $\mathbf{F}(x, y, z) = \langle x, y, z \rangle$ and $\mathcal{S}$ is the sphere with equation $x^2 + y^2 + z^2 = 9$.
23. $\mathbf{F}(x, y, z) = xyz\mathbf{i}$ and $\mathcal{S}$ is the portion of the plane with equation $x + y + z = 4$ in the first octant.
24. $\mathbf{F}(x, y, z) = x^2\mathbf{i} + y^2\mathbf{j} + z^2\mathbf{k}$ and $\mathcal{S}$ is the top half of the unit sphere centered at the origin.

Divergence and Curl: Find the divergence and curl of the following vector fields.

25. $\mathbf{F}(x, y) = (3x + 4y)\mathbf{i} + (x - 5y)\mathbf{j}$
26. $\mathbf{F}(x, y) = x^2 y\mathbf{i} + xy^3\mathbf{j}$
27. $\mathbf{F}(x, y, z) = (x - y)\mathbf{i} + (y - z)\mathbf{j} + (z - x)\mathbf{k}$
28. $\mathbf{F}(x, y, z) = (y^2 - z^2)\mathbf{i} + (z^2 - x^2)\mathbf{j} + (x^2 - y^2)\mathbf{k}$

Green's Theorem: Use Green's Theorem to evaluate the integral $\int_C \mathbf{F} \cdot d\mathbf{r}$ for the given vector field and curve.

29. $\mathbf{F}(x, y) = (y^2 + 1)\mathbf{i} + 2xy\mathbf{j}$ and C is the circle with equation $x^2 + y^2 = 9$, traversed counterclockwise.
30. $\mathbf{F}(x, y) = \langle e^x \cos y, e^x \sin y \rangle$ and C is the ellipse with equation $3x^2 + 4y^2 = 25$, traversed counterclockwise.
31. $\mathbf{F}(x, y) = xy\mathbf{j}$ and C is the square with vertices $(\pm 3, \pm 3)$, traversed counterclockwise.
32. $\mathbf{F}(x, y) = y\mathbf{i} - x\mathbf{j}$ and C is the square with vertices $(\pm 3, \pm 3)$, traversed counterclockwise.

Stokes' Theorem: Evaluate the integrals that follow. In some cases Stokes' Theorem will help; in other cases it may be preferable to evaluate the integrals directly.

33. $\int_C \mathbf{F}(x, y, z) \cdot d\mathbf{r}$, where

$$\mathbf{F}(x, y, z) = (y^2 - x^2)\mathbf{i} + (z^2 - y^2)\mathbf{j} + (x^2 - y^2)\mathbf{k},$$

C is the triangle with vertices $(2, 0, 0)$, $(0, 6, 0)$, and $(0, 0, 3)$, and $\mathbf{n}$ points upwards.
34. $\int_C \mathbf{F}(x, y, z) \cdot d\mathbf{r}$, where $\mathbf{F}(x, y, z) = \langle y^3, -4z, 4x^2 \rangle$, C is the circle in the xy-plane parametrized by $x = 2\cos t$, $y = 2\sin t$, and $\mathbf{n}$ points upwards.
35. $\iint_{\mathcal{S}} \operatorname{curl} \mathbf{F}(x, y, z) \cdot \mathbf{n}\, d\mathcal{S}$, where

$$\mathbf{F}(x, y, z) = \langle 5y, 5x, z^2 \rangle,$$

$\mathcal{S}$ is the portion of the unit sphere $x^2 + y^2 + z^2 = 1$ above the plane $z = \frac{1}{2}$, and $\mathbf{n}$ is the upwards-pointing normal vector.
36. $\iint_{\mathcal{S}} \operatorname{curl} \mathbf{F}(x, y, z) \cdot \mathbf{n}\, d\mathcal{S}$, where

$$\mathbf{F}(x, y, z) = \langle x^3 yz, xy^2 z, yz \rangle,$$

$\mathcal{S}$ is the portion of the hyperboloid $z = x^2 - y^2 + 1$ that lies above the square with vertices $(\pm 1, \pm 1)$, and $\mathbf{n}$ is the upwards-pointing normal vector.

Divergence Theorem: Evaluate the integral

$$\iint_{\mathcal{S}} \mathbf{F}(x, y, z) \cdot \mathbf{n}\, d\mathcal{S}$$

for the specified vector field $\mathbf{F}$ and surface $\mathcal{S}$. In each case, $\mathbf{n}$ is the outwards-pointing normal vector.

37. $\mathbf{F}(x, y, z) = x^2 y\mathbf{i} + y^2 z\mathbf{j} + xz^2\mathbf{k}$ and $\mathcal{S}$ is the hemisphere bounded above by $x^2 + y^2 + z^2 = 1$ and bounded below by the xy-plane.
38. $\mathbf{F}(x, y, z) = x^2\mathbf{i} + y^2\mathbf{j} + z^2\mathbf{k}$ and $\mathcal{S}$ is the surface of the rectangular solid described by the inequalities $0 \le x \le 3$, $0 \le y \le 2$, and $0 \le z \le 1$.
39. $\mathbf{F}(x, y, z) = x^3 z\mathbf{i} + xy\mathbf{j} + 4yz\mathbf{k}$ and $\mathcal{S}$ is the boundary of the frustum of the cone with equation $x = \sqrt{y^2 + z^2}$ between the planes $x = 2$ and $x = 5$.
40. $\mathbf{F}(x, y, z) = xy\mathbf{i} + xyz\mathbf{j} + yz\mathbf{k}$ and $\mathcal{S}$ is the boundary of the cylinder with equation $x^2 + z^2 = 4$ for $0 \le y \le 5$.

Capstone Problems

A. Let $\mathbf{F} = \mathbf{F}(x, y, z)$ and $\mathbf{G} = \mathbf{G}(x, y, z)$. Prove that

$$\nabla(\mathbf{F} \cdot \mathbf{G}) = (\mathbf{F} \cdot \nabla)\mathbf{G} + (\mathbf{G} \cdot \nabla)\mathbf{F} + \mathbf{F} \times (\nabla \times \mathbf{G}) + \mathbf{G} \times (\nabla \times \mathbf{F}).$$

B. The ***Laplacian*** of a function f of two or three variables is the divergence of the gradient. That is,

$$\nabla^2 f = \nabla \cdot (\nabla f).$$

The ***vector Laplacian*** of a vector field

$$\mathbf{F}(x, y, z) = \langle F_1(x, y, z), F_2(x, y, z), F_3(x, y, z)\rangle$$

is

$$\nabla^2 \mathbf{F} = \langle \nabla^2 F_1, \nabla^2 F_2, \nabla^2 F_3 \rangle.$$

Prove that, for a vector field $\mathbf{F} = \mathbf{F}(x, y, z)$,

$$\nabla \times (\nabla \times \mathbf{F}) = \nabla(\nabla \cdot \mathbf{F}) - \nabla^2 \mathbf{F}.$$

C. Show that $\int_C \mathbf{F} \cdot d\mathbf{r} = \pi\rho^2$ when $\mathbf{F}(x, y) = -y\mathbf{i} + x\mathbf{j}$ and C is the circle with radius ρ centered at the origin and traversed counterclockwise.

(a) Formulate a conjecture for the value of the integral $\int_C \mathbf{F} \cdot d\mathbf{r}$, where $\mathbf{F}(x, y) = -y\mathbf{i} + x\mathbf{j}$ and C is any simple closed curve.

(b) Prove your conjecture.

(c) Use your conjecture to find the area of the ellipse with parametrization

$$x = a \cos t \quad \text{and} \quad y = b \sin t.$$

D. Let $\mathbf{F} = \mathbf{F}(x, y, z)$ and $\mathbf{G} = \mathbf{G}(x, y, z)$ be vector fields defined on a region R enclosed by an orientable surface S with outwards unit normal vector $\mathbf{n}$. Prove that if

$$\nabla \cdot \mathbf{F} = \nabla \cdot \mathbf{G} \quad \text{and} \quad \nabla \times \mathbf{F} = \nabla \times \mathbf{G}$$

on R and $\mathbf{F} \cdot \mathbf{n} = \mathbf{G} \cdot \mathbf{n}$ on S, then $\mathbf{F} = \mathbf{G}$ on S.

ANSWERS TO ODD PROBLEMS

Chapter 0

Section 0.1

1. F, F, F, T, F, F, T, T.
3. See the reading at the beginning of this section.
5. Domain(f) = $\{x \in \mathbb{R} \mid f(x) \text{ is defined}\}$; for $f(x) = \sqrt{x}$ the domain is Domain(f) = $\{x \in \mathbb{R} \mid x \geq 0\}$.
7. (a) Yes: $\sqrt{3+1} = 2$; (b) No: $\sqrt{1+1} \neq 1$; (c) No: $\sqrt{-5+1}$ is undefined.
9. (a) $f(2) = 5$; (b) There is no x with $f(x) = 0$; (c) You can solve $y = x^2 + 1$ for x if and only if $y \in [1, \infty)$.
11. One example is $f(2) = 3$, $f(4) = 1$, $f(6) = 3$, $f(8) = 2$, $f(10) = 1$.
13. Yes, it could be a function; however, it would not be one-to-one.
15. (a) This is well-defined as the function $f(x) = x$, for $x \in \mathbb{R}$. (b) Not well-defined, since we would need $f(0) = 0$, and $f(3) = 3$, but 0 and 3 are not in the target set $(0, 3)$. (c) This is well-defined as the multivariable function $f(x, y, z) = (x, y, z)$, for $(x, y, z) \in \mathbb{R}^3$.
17. See the reading near Theorem 0.27.
19. $f(2) = f(-2)$
21. (a) $f(x) < 0$ for all $x \in I$; (b) $f(b) < f(a)$ for all $b > a$ in I
23. Average rate of change of f on $[a, b]$ is given by $\frac{f(b)-f(a)}{b-a}$. Here $f(b) - f(a)$ is the difference Δy of y-values, or the "rise." Similarly, $b - a$ is the difference Δx of x-values, or the "run."
25. The largest possible δ for the local maximum at $x = 2$ is approximately $\delta = 1.5$.
27. Domain $[1, \infty)$, range $[0, \infty)$
29. Domain $x \neq -2$, range $y \neq 0$
31. Domain $\mathbb{R}$, range $(0, 1]$
33. $(-\infty, 0] \cup [2, \infty)$
35. $(-\infty, -3] \cup [1, \infty)$
37. $(-\infty, -3) \cup (1, \infty)$
39. $(-\infty, -3) \cup (-3, -1] \cup [1, 3) \cup (3, \infty)$
41. $(1, 2) \cup (2, \infty)$
43. (a) 17; (b) $a^6 + 1$; (c) $(x^2 + 1)^2 + 1 = x^4 + 2x^2 + 2$.
45. (a) $\sqrt{25 + 9 + 4} = \sqrt{38}$; (b) $\sqrt{9 + 0 + 16} = 5$; (c) $\sqrt{x^2 + y^2 + z^2}$
47. (a) $(9, -3, 13)$; (b) $(17, 2, 8)$; (c) $(3a + b, a, b)$.
49. $f(-1) = -2$, $f(0) = 1$, $f(1) = 4$, $f(2) = 8$
51. One good window for this graph is $x \in [-1, 1]$, $y \in [-0.3, 1]$.
53.

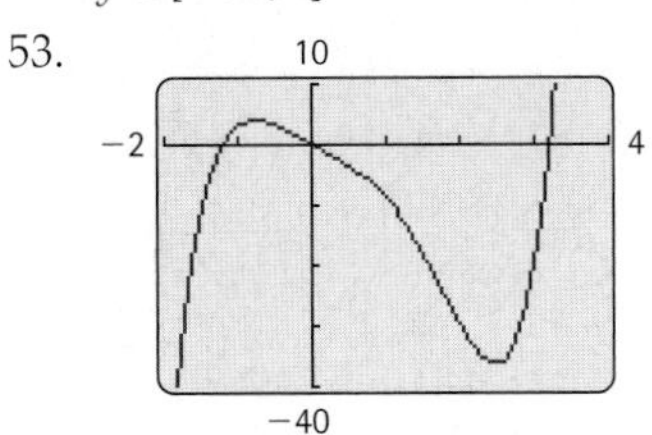

55.

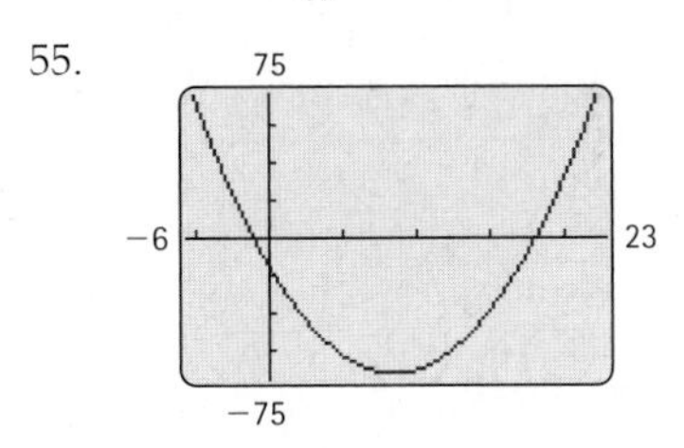

57. (a) Domain $\mathbb{R}$, range $[-1, \infty)$; (b) roots at $x = 0$ and $x = 2$, y-intercept at $y = 0$, local and global minimum at $x = 1$ (with a value of $f(1) = -1$), no local or global maxima, no inflection points; (c) positive on $(-\infty, 0) \cup (2, \infty)$, negative on $(0, 2)$, decreasing on $(-\infty, 1)$, increasing on $(1, \infty)$, concave up on $\mathbb{R}$, concave down nowhere; (d) no asymptotes.
59. (a) Domain $\mathbb{R}$, range $[-4.75, \infty)$; (b) roots at $x \approx -1.6$ and $x = 0$, y-intercept at $y = 0$, local and global minimum at $x = -1$ (with a value of $f(-1) = -4.75$), local minimum at $x = 2$ (with value $f(2) = 2$), local maximum at $x = 1$ (with value $f(1) = 3.25$), inflection points at $x \approx 0$ and $x \approx 1.5$; (c) positive on $(-\infty, 1.6) \cup (0, \infty)$, negative on $(-1.6, 0)$, decreasing on $(-\infty, -1) \cup (1, 2)$, increasing on $(-1, 1) \cup (2, \infty)$, concave up on $(-\infty, 0) \cup (1.5, \infty)$, concave down on $(0, 1.5)$; (d) no asymptotes.
61. (a) Domain $\mathbb{R}$, range $(1, 2]$; (b) no roots, y-intercept at $y = 2$, local and global maximum at $x = 0$ (with a value of $f(0) = 2$), no local or global minima, inflection points at $x = -1$ and $x = 1$; (c) positive on $(-\infty, \infty)$, negative nowhere, increasing on $(-\infty, 0)$, decreasing on $(0, \infty)$, concave up on $(-\infty, -1) \cup (1, \infty)$, concave down on $(-1, 1)$;
(d) horizontal asymptote at $y = 1$.
63. One example is the graph of $f(x) = 2^{-x}$.
65. Not possible.

67.

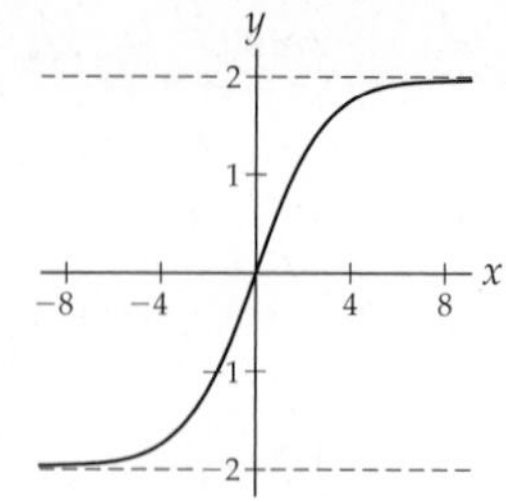

69. One example is the graph of $f(x) = \frac{(x^2-1)(x^2-4)}{x}$.

71.

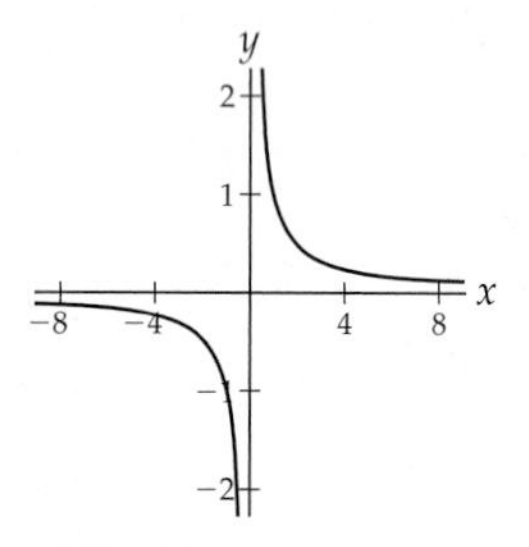

73. 4.2

75. $\frac{\sqrt{10}-\sqrt{2}}{8}$

77. $-\frac{100}{99}$

79. y, 512, x

81. y, x

83. (a) Independent variables a and b, dependent variable H, function equation $H(a,b) = \sqrt{a^2+b^2}$; (b) independent variables x, y, and z, dependent variable V, function equation $V(x,y,z) = xyz$.

85. 356 groundhogs

87. (a) $I(t) = \begin{cases} 36{,}000, & \text{if } 0 \le t < 4 \\ 38{,}500, & \text{if } 4 \le t < 6 \\ 49{,}000, & \text{if } t \ge 6. \end{cases}$

(b) $M(t) = \begin{cases} 36{,}000t, & \text{if } 0 \le t < 4 \\ 144{,}000 + 38{,}500(t-4), & \text{if } 4 < t \le 6 \\ 221{,}000 + 49{,}000(t-6), & \text{if } t \ge 6. \end{cases}$

(c) You will have earned a total of one million dollars after approximately 21.9 years.

89. The range of f is the set of real numbers y such that there is some x with $3x - 1 = y$. This range is $\mathbb{R}$, since for any real number y, the number $x = \frac{y+1}{3}$ has the property that $f(x) = 3\left(\frac{y+1}{3}\right) - 1 = y$.

91. It suffices to show that f is not one-to-one if and only if it fails the Horizontal Line Test. Not one-to-one means there exists some $a \ne b$ in the domain of f such that $f(a) = f(b)$. This is precisely what it means for f to intersect the horizontal line $y = f(a)$ in two places, namely, $(a, f(a))$ and $(b, f(b)) = (b, f(a))$.

93. For all a and b in $(-\infty, \infty)$, we have $a < b \Rightarrow -3a > -3b \Rightarrow 1 - 3a > 1 - 3b \Rightarrow f(a) > f(b)$.

95. For all real numbers a and b, $\frac{f(b)-f(a)}{b-a} = \frac{(-2b+4)-(-2a+4)}{b-a} = \frac{-2(b-a)}{b-a} = -2$.

Section 0.2

1. F, T, T, F, F, F, F, F.

3. $(f-g)(x) = (f+(-g))(x) = f(x)+(-g)(x) = f(x)-g(x)$.

5. (a) One answer is $(1, 2)$; (b) one answer is $f\left(\frac{1}{2}x\right)$; (c) $(-1, 4)$; (d) $(2, -5)$.

7. (a) The domain is the set of x such that $x \in [2, \infty)$ and $f(x) \in [-10, \infty)$. Since the range of f, $[-3, 3]$, always satisfies the second condition, the domain of $g \circ f$ is $[2, \infty)$. (b) We would need to know which x-values guarantee that $g(x) \in [2, \infty)$. (c) Not enough information for $f \circ f$; the domain of $g \circ g$ is $[-10, \infty)$.

9. $f(2x)$

11. If f is odd then $f(0)$ is either 0 or undefined.

13. (a) $f(-2) = 1$, $f(-1) = -2$, $f(3) = 4$ ($f(0)$ could be anything); (b) $f(-2) = -1$, $f(-1) = 2$, $f(0) = 0$, and $f(3) = -4$.

15. $(f^{-1})^{-1} = f$.

17. f^{-1} has domain $(-\infty, 3]$ and range $[-1, 1)$.

19. The missing entries in the first column are $-3, -2, 0, 0$. The missing entries in the second column are $-2, 3, 3, -6$. The missing entries in the third column are $-1, 2, 2, -7$.

21. Domain $x \ne 2$; $(f+g)(x) = x^2 + 1 + \frac{1}{x-2}$; $(f+g)(1) = 1$.

23. Domain $(0, 2) \cup (2, \infty)$; $\left(\frac{g}{h}\right)(x) = \left(\frac{1}{x-2}\right)/\sqrt{x}$; $\left(\frac{g}{h}\right)(1) = -1$.

25. Domain $x \ne 2, x \ne \frac{5}{2}$; $(g \circ g)(x) = \frac{1}{\frac{1}{x-2} - 2}$; $(g \circ g)(1) = -\frac{1}{3}$.

27. Domain $x \ne 7$; $g(x-5) = \frac{1}{(x-5)-2}$; $g(1-5) = -\frac{1}{6}$.

29. Domain $x \ge -\frac{1}{3}$; $h(3x+1) = \sqrt{3x+1}$; $h(3(1)+1) = 2$

31. The new row is 3, 5, 9, 7, 9, 3, 7.

33. The new row is 2, 3, 2, 2, 3, 2, 3.

35. The new row is 0, 1, 3, 2, 3, 0, 2.

37. For $x = 1, 2, 3, 4, 5, 6, 7$ we have $g(x-1) = 1, 0, 1, 1, 0, 1, 0$, respectively.

39. 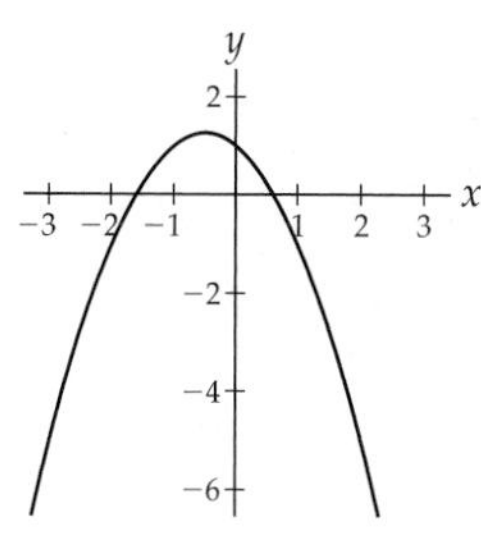

41. 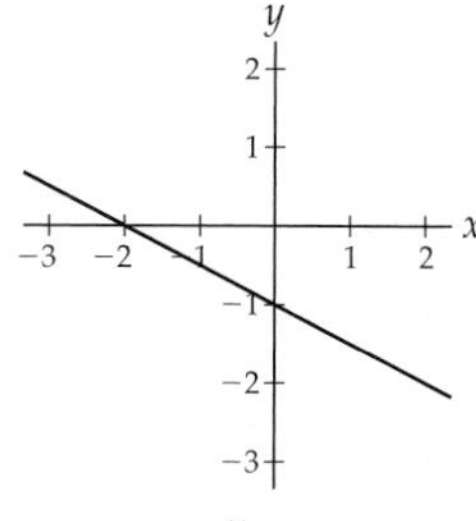

43. 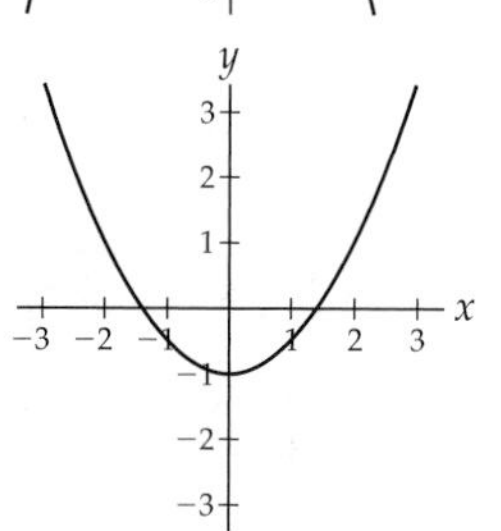

45. 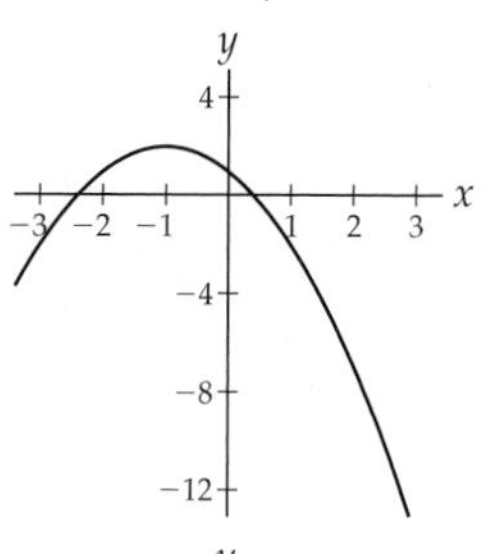

47. 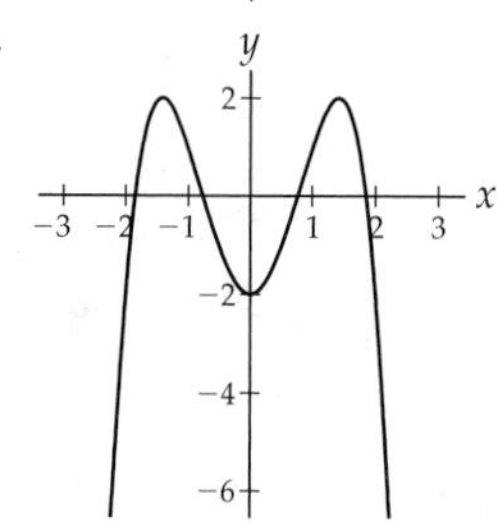

49. 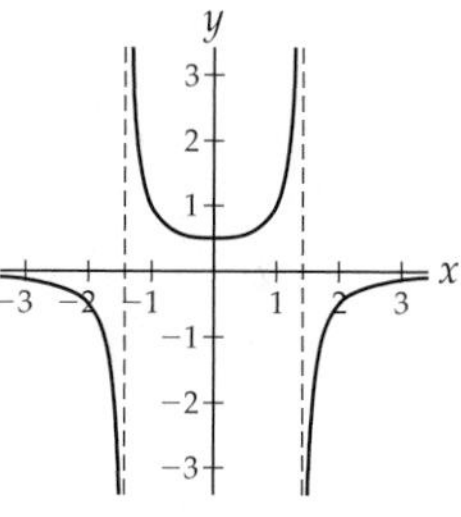

51. $(f+g)(-1) = 0$, $(f+g)(0) = 1$, $(f+g)(1) = 0$, $(f+g)(2) = 9$, $(f+g)(3) = 14$;
$$(f+g)(x) = \begin{cases} x+1, & \text{if } x \le 0 \\ x^2 - x, & \text{if } 0 < x < 2 \\ x^2+5, & \text{if } x \ge 2. \end{cases}$$

53. $(g \circ f)(-1) = 1$, $(g \circ f)(0) = -1$, $(g \circ f)(1) = -1$, $(g \circ f)(2) = 5$, $(g \circ f)(3) = 5$;
$$(g \circ f)(x) = \begin{cases} -(2x+1), & \text{if } x \le 0 \\ -x^2, & \text{if } 0 < x < \sqrt{2} \\ 5, & \text{if } x \ge \sqrt{2}. \end{cases}$$

55. $f(-1-1) = -3$, $f(0-1) = -1$, $f(1-1) = 1$, $f(2-1) = 1$, $f(3-1) = 4$;
$$f(x-1) = \begin{cases} 2x-1, & \text{if } x \le 1 \\ (x-1)^2, & \text{if } x > 1. \end{cases}$$

57. One way: $g(x) = 3x+5$, $h(x) = x^2$; another way: $g(x) = x+5$, $h(x) = 3x^2$.

59. One way: $g(x) = \frac{6}{x}$, $h(x) = x+1$; another way: $g(x) = 6x$, $h(x) = \frac{1}{x+1}$.

61. Even, since $f(-x) = (-x)^4 + 1 = f(x)$.

63. Neither, since $f(-x) \neq f(x)$ and $f(-x) \neq -f(x)$.

65. Odd, since $f(-x) = -f(x)$.

67. $f(g(x)) = 2 - 3\left(-\frac{1}{3}x + \frac{2}{3}\right) = x$ and $g(f(x)) = -\frac{1}{3}(2-3x) + \frac{2}{3} = x$.

69. $f(g(x)) = f\left(\frac{x}{1+x}\right) = \dfrac{x/(1+x)}{1-(x/(1+x))} = x$ (for $x \neq -1$); similarly, $g(f(x)) = x$ for $x \neq 1$.

71. 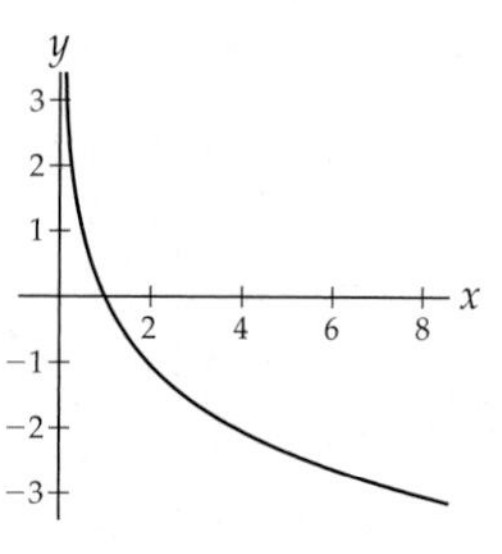

73. f is invertible on the restricted domain $[0, 2]$, and its inverse has the following graph:

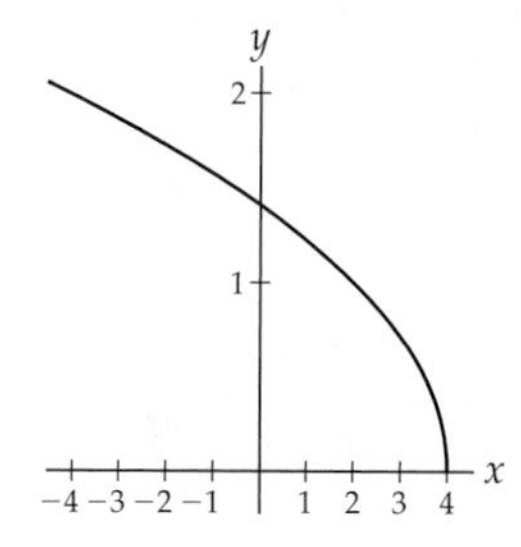

75. $f^{-1}(x) = \frac{1-2x}{5}$

77. $f^{-1}(x) = \frac{1}{x-1}$

79. $f^{-1}(x) = \frac{-x-1}{x-1}$

81. (a) $S(x) = x^2$; (b) $C(S) = 4.25S + 200$; (c) $C(x) = 4.25x^2 + 200$; $C(x)$ is the composition $C(S(x))$.

83. (a) $C(n) = 20 + 12n$. (b) Suppose there were a and b in the domain $(-\infty, \infty)$ of $C(n)$ with $C(a) = C(b)$. In this case $20 + 12a = 20 + 12b$, so $12a = 12b$, and therefore we must have $a = b$. (c) $n(C) = \frac{C-20}{12}$. This function describes the number n of shirts that you can have made for C dollars. (d) $n(150) \approx 10.83$, so you can make 10 full shirts.

85. $f(2x) = (2x)^k = 2^k x^k = 2^k f(x)$.

87. (a) If (x, y) is on the graph of f, then $y = f(x)$, so $kf(x) = ky$; thus (x, ky) is a point on the graph of $kf(x)$. (b) If (x, y) is on the graph of f, then $y = f(x)$, so $f\left(k\left(\frac{1}{k}x\right)\right) = f(x) = y$, so $\left(\frac{1}{k}x, y\right)$ is on the graph of $f(kx)$.

89. If f is odd then for all x in the domain of f we have $f(-x) = -f(x)$. In particular, $f(-0) = -f(0)$, so $f(0) = -f(0)$. This means we must have $f(0) = 0$.

91. Since $f^{-1}(f(x)) = x$ for all x in the domain of f, we know that f^{-1} is defined for all values $f(x)$ in the range of f. Therefore the domain of f^{-1} is the range of f. Similarly, since $f(f^{-1}(x)) = x$, the range of f^{-1} is in the domain of f.

93. If $f(a) = b$ then $f^{-1}(f(a)) = f^{-1}(b)$ and therefore $a = f^{-1}(b)$.

Section 0.3

1. T, F, T, F, F, T, T, F.

3. The graphs look like:

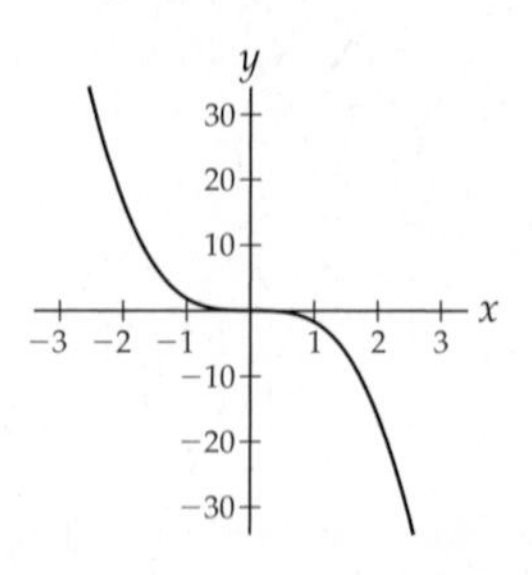

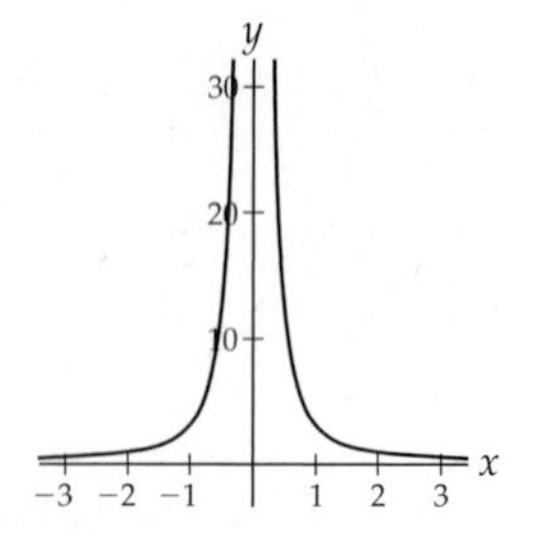

5. 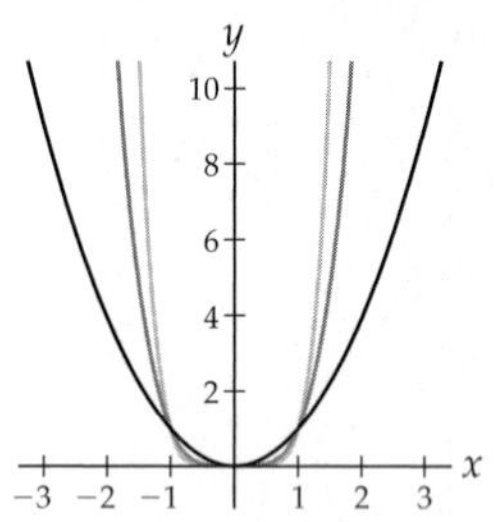

7. Yes; $\frac{1}{f(x)} = \frac{1}{Ax^k} = \frac{1}{A}x^{-k}$, so $C = \frac{1}{A}$ and $r = -k$.

9. Leading coefficient 3, leading term $3x^4$, degree 4, constant term 0, coefficients $a_1 = 4$, $a_3 = -11$.

11. If $b^2 - 4ac < 0$, then $\sqrt{b^2 - 4ac}$ is not a real number, so by the quadratic formula the quadratic has no real roots. If $f(x) = 3x^2 + 2x + 6$, then $b^2 - 4ac = 2^2 - 4(3)(6) = -68$ is negative.

13. After reducing, $f(x)$ becomes $\frac{x-1}{x+2}$, which has a vertical asymptote at $x = -2$.

15. One answer is $f(x) = \frac{3x^2+1}{(x+2)(x-2)}$.

17. (a) Graph the functions given in part (b) to see three possible answers; (b) $f(x) = \frac{-(x-1)(x-3)(x+1)}{(x-2)^2(x+1)}$; $f(x) = \frac{-(x-1)(x-3)(x+1)^2}{(x-2)^2(x+1)^2}$; $f(x) = \frac{-(x-1)^2(x-3)(x+1)}{(x-2)(x+1)(x^2+1)}$.

19. 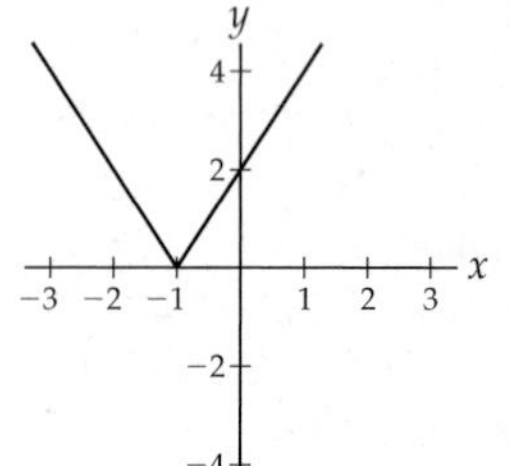

21. 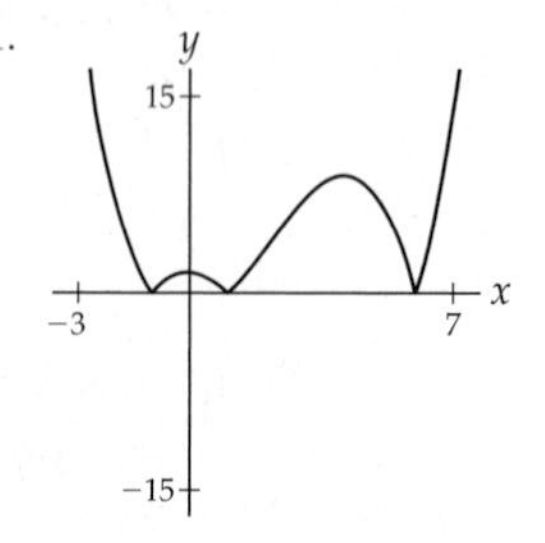

23. If $x \geq 0$ then $\sqrt{x^2} = x$, and if $x < 0$ then $\sqrt{x^2} = -x$.

25. Domain $\mathbb{R}$, zero at $x = 0$

27. Domain $(0, \infty)$, never zero

29. Domain $(-\infty, -1) \cup (1, \infty)$, never zero

31. Domain $(-\infty, -1) \cup (-1, 3]$, zero at $x = 3$

33. Domain $(-\infty, -1] \cup [6, \infty)$, zero at $x = -1$

35. $y + 2 = -(x - 3)$

37. $y - 4 = 2(x + 1)$

39. One possible answer is $f(x) = 2x^2$.

41. One possible answer is $f(x) = x(x - 2)(x - 4)$.

43. 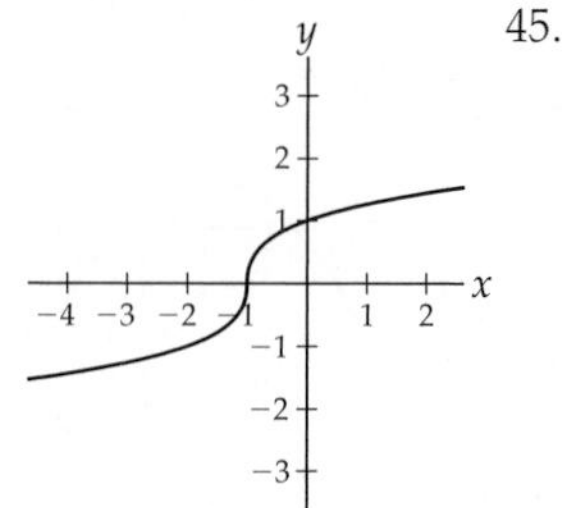

45. 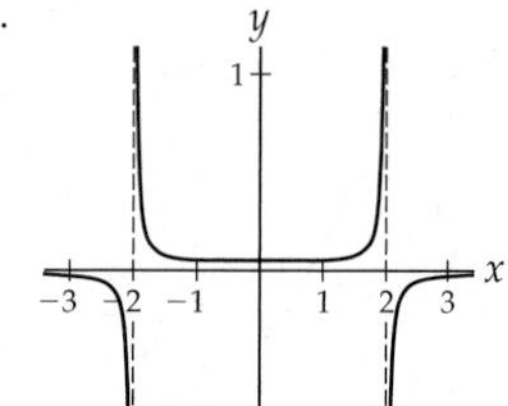

47. 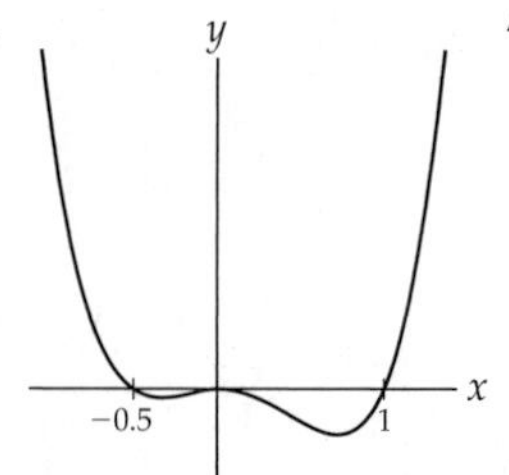

49. 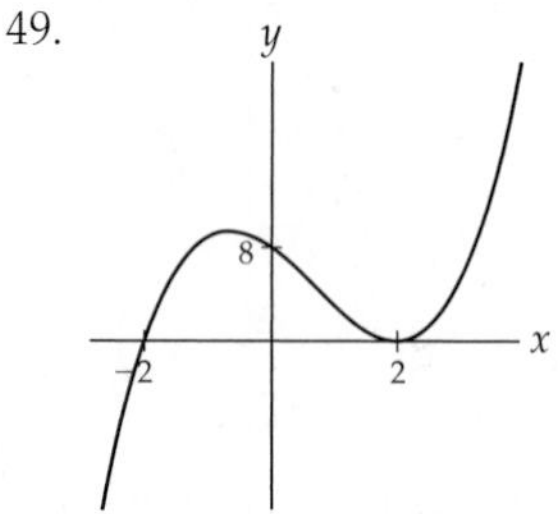

51. The graph of f has roots at $x = 0$, $x = -3$, and $x = 1$, vertical asymptotes at $x = \pm 2$, and no horizontal asymptotes.

53. 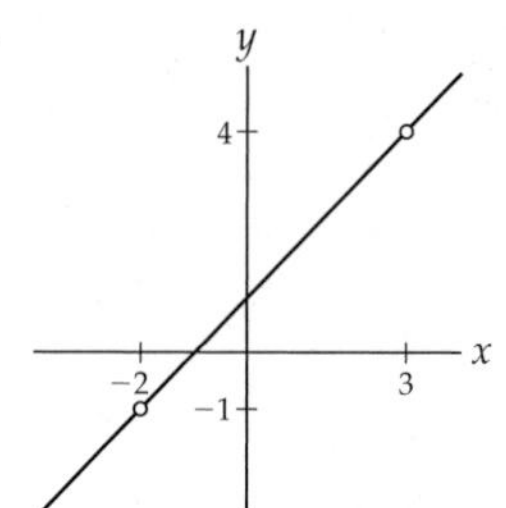

55. The graph of f has a hole at $x = -1$, roots at $x = -\frac{3}{2}$ and $x = 1$, a vertical asymptote at $x = 3$, and no horizontal asymptotes.

57. The parts of the graph of $y = \sin x$ that are below the x-axis should be reflected across the x-axis to obtain the graph of $g(x)$.

59. The graph should look like the one you found in Exercise 57, but shifted down two units.

61. $f(x) = \frac{2}{x}$

63. $f(x) = -(x + 2)(x - 1)^3$

65. $f(x) = (x - 2)^2 + 2$

67. $f(x) = \frac{-4(x^2+1)(x-4)}{(x+2)(x-2)(x-4)}$

69. $f(x) = \frac{-4(x-1)(x+1)}{(x^2+1)^2}$

71. $f(x) = |2(x + 1)(x - 3)|$

73. $|5-3x| = \begin{cases} 5-3x, & \text{if } x \le \frac{5}{3} \\ 3x-5, & \text{if } x > \frac{5}{3} \end{cases}$

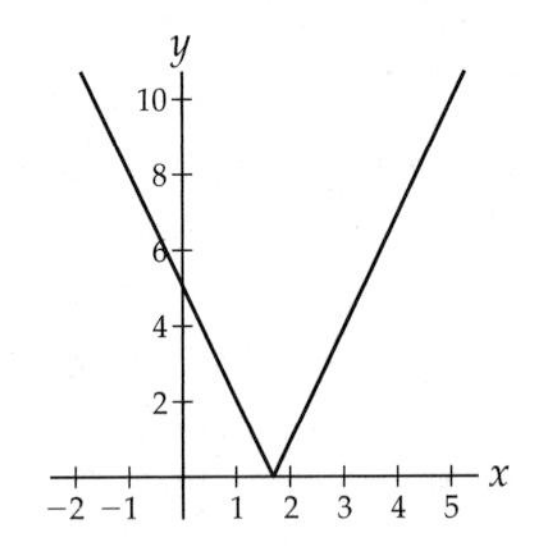

75. $|x^2+1| = x^2+1$ for all real numbers x

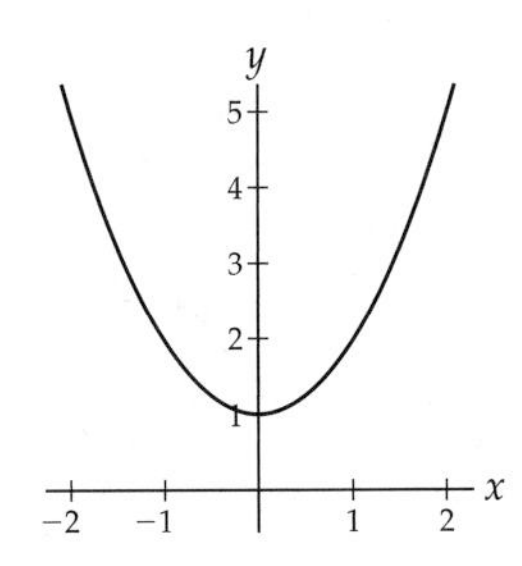

77. $|9-x^2| = \begin{cases} 9-x^2, & \text{if } -3 \le x \le 3 \\ -(9-x^2), & \text{if } x < -3 \text{ or } x > 3. \end{cases}$

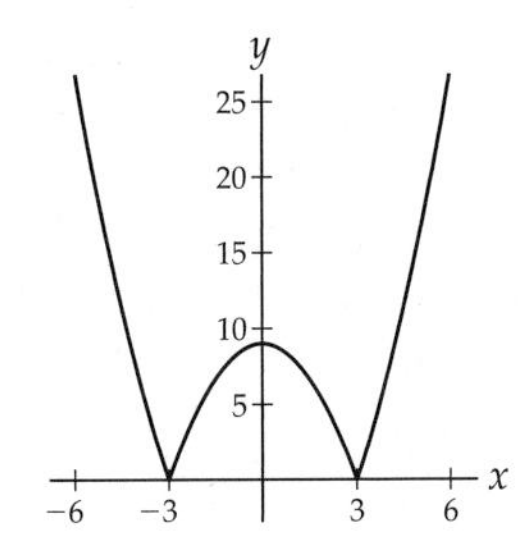

79. $|x^2-3x-4| = \begin{cases} x^2-3x-4, & \text{if } x \le -1 \\ -(x^2-3x-4), & \text{if } -1 < x < 4 \\ x^2-3x-4, & \text{if } x \ge 4. \end{cases}$

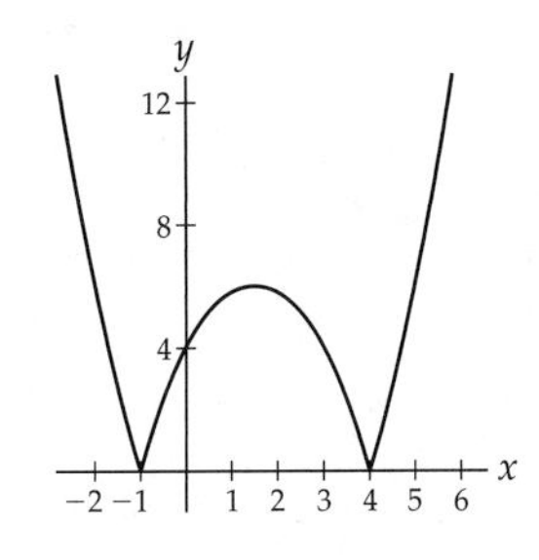

81. The minimum possible degree is 3.

83. No, since polynomials increase or decrease without bound and do not have asymptotes. A rational function might be a better choice.

85. If $f(x) = Ax^k$ and $g(x) = Bx^l$ are power functions, then $(f \circ g)(x) = f(Bx^l) = A(Bx^l)^k = (AB^k)x^{lk}$ is also a power function.

87. If $f(x) = mx + b$ and $g(x) = nx + c$ then $f(g(x)) = f(nx+c) = m(nx+c) + b = (mn)x + (mc+b)$.

89. If $f(x) = Ax^3 +$ lower-degree terms and $g(x) = Bx^3 +$ lower-degree terms, then $f(x)g(x) = ABx^6 +$ lower-degree terms. Since f and g are cubic we know that A and B are nonzero. Thus AB must also be nonzero, and therefore fg is of degree 6.

91. (a) If $f(x) = k$ for all k, then $f(x) = k = 0x + k$ is also a linear function. (b) If $f(x) = mx + b$ is a linear function with $m \neq 0$, then f is a polynomial of degree 1 with coefficients $a_1 = m$ and $a_0 = b$. If $f(x) = mx + b$ with $m = 0$ then f is a polynomial of degree zero with sole coefficient $a_0 = b$.

93. A number of the form $f(c) = \frac{p(c)}{q(c)}$ is equal to zero when its numerator $p(c)$ is zero but its denominator $q(c)$ is not.

Section 0.4

1. F, T, T, T, T, F, T, F.

3. A function is exponential if it can be written in the form $f(x) = Ab^x$; the variable is in the exponent and a constant is in the base. For a power function, the situation is reversed. x^x is neither a power nor an exponential function because a variable appears in both the base and the exponent.

5. $\sqrt{3} \approx 1.73205$. We have $2^{1.7} \approx 3.2490$, $2^{1.73} \approx 3.3173$, $2^{1.732} \approx 3.3219$, $2^{1.7320} \approx 3.3219$, $2^{1.73205} \approx 3.3220$, and so on. Each of these approximations gets closer to the value of $2^{\sqrt{3}}$.

7. $f(x) = 3(e^{-2})^x \approx 3(0.135)^x$, $g(x) = -2e^{(\ln 3)x} \approx -2e^{1.0986x}$.

9. We must have $b > 0$ and $b \neq 1$, since those conditions are necessary for b^x to be an exponential function.

11. The graph that passes through (2, 1) is $\log_2 x$; the graph that passes through $\left(2, \frac{1}{2}\right)$ is $\log_4 x$.

13. $f(x) = 2\ln x$ is the graph of $\ln x$ stretched vertically by a factor of 2; $g(x) = -\ln x$ is the graph of $\ln x$ reflected over the x-axis.

15. If θ is an angle in standard position, then $\sin\theta$ is the vertical coordinate y of the point (x, y) where the terminal edge of θ intersects the unit circle.

17. The terminal edges of the angles $\frac{\pi}{4}$, $\frac{9\pi}{4}$, and $-\frac{7\pi}{4}$ all meet the unit circle at the same point (and in particular, at the same y-coordinate).

19. $\cos\theta = x$ is $-\frac{\sqrt{8}}{3}$ if θ is in the third quadrant; $\cos\theta = \frac{\sqrt{8}}{3}$ if θ is in the fourth quadrant; θ cannot be in the first or second quadrant.

21. See Definition 0.27 for the restricted domains of the trigonometric functions, and thus the ranges of the inverse trigonometric functions. The domain of $\arcsin x$ is $[-1, 1]$, the domain of $\arctan x$ is all of $\mathbb{R}$, and the domain of $\operatorname{arcsec} x$ is $(-\infty, -1] \cup [1, \infty)$. Their ranges are the restricted domains of $\sin x$, $\tan x$, and $\sec x$, respectively.

23. Only (a) and (c) are defined.

25. (a) Negative; (b) Negative; (c) Positive; (d) Positive

27. $(2, 3) \cup (3, \infty)$

29. $(2, \infty)$

31. $\ldots \cup \left(-\frac{5\pi}{2}, -\frac{3\pi}{2}\right) \cup \left(-\frac{\pi}{2}, \frac{\pi}{2}\right) \cup \left(\frac{5\pi}{2}, \frac{7\pi}{2}\right) \cup \ldots$

33. -2

35. 4

37. -1

39. $\sqrt{2}$

41. π

43. $\frac{3\pi}{4}$

45. $x = \frac{\ln 3}{\ln(3/2)} \approx 2.70951$

47. $x = -\frac{17}{15}$

49. $x = \pi k$, where k is any integer

51. $\frac{\sqrt{35}}{6}$

53. $-\frac{17}{18}$

55. Negative

57. $\sqrt{1 - x^2}$

59. $x^2 + 1$

61. $\sqrt{1 - \left(\frac{3}{x}\right)^2}$

63. $1 - 2(5x)^2$

65. Start with the graph of $y = \left(\frac{1}{2}\right)^x$, then reflect over the x-axis and shift up by 10 units.

67.

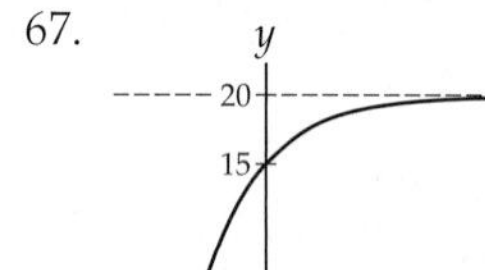

69.

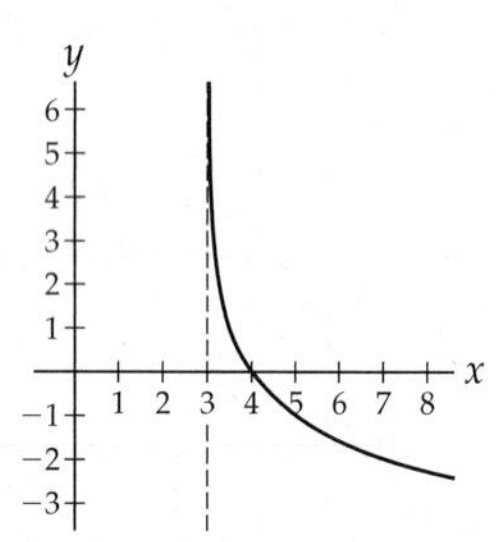

71.

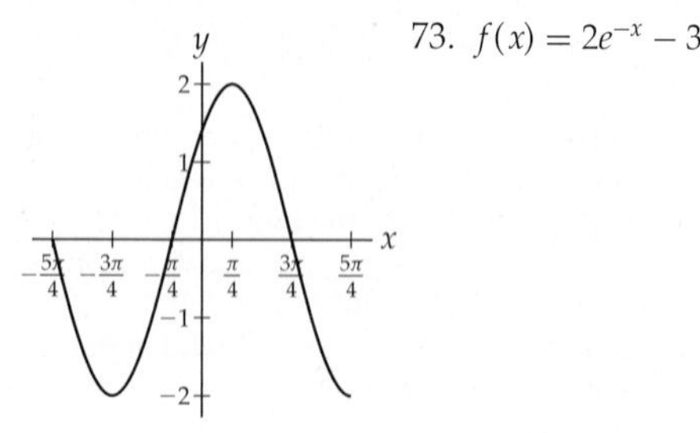

73. $f(x) = 2e^{-x} - 3$

75. $f(x) = -5e^{-x} + 10$

77. $f(x) = -\cos 2x$

79. (a) $I(t) \approx 10,000e^{0.08t}$; (b) $I(t) = 10,000(1.085)^t$.

81. (a) $S(t) = 250e^{-\ln 2/29t} \approx 250e^{-0.0239t}$, or equivalently, $S(t) = 250(0.97638)^t$; (b) 2.39%; (c) 156 years

83. 211.97 feet

85. Seeking a contradiction, suppose that A and b are nonzero real numbers with $Ab^x = 0$ for some real number x. Since $A \neq 0$ we know $b^x = 0$, and therefore that $x(\ln b) = \ln(b^x) = \ln 0$. But this is a contradiction because $\ln 0$ is undefined, so the product $x(\ln b)$ of real numbers cannot equal $\ln 0$.

87. $\log_2 x = \log_3 x \Leftrightarrow \frac{\ln x}{\ln 2} = \frac{\ln x}{\ln 3} \Leftrightarrow (\ln 3)(\ln x) = (\ln 2)(\ln x) \Leftrightarrow (\ln x)(\ln 3 - \ln 2) = 0 \Leftrightarrow \ln x = 0 \Leftrightarrow x = 1$.

89. $y = \log_b x$ if and only if $b^y = x$. Since the only solution of $b^y = 1$ is $y = 0$ (if $b \neq 0$), we know that $\log_b 1 = 0$.

91. Since $x = b^{\log_b x}$, we have $\log_b(x^a) = \log_b((b^{\log_b x})^a) = \log_b(b^{(\log_b x)a} = (\log_b x)a = a \log_b x$.

93. (a) $\log_b\left(\frac{1}{x}\right) = \log_b(x^{-1}) = -\log_b x$;
(b) $\log_b\left(\frac{x}{y}\right) = \log_b(xy^{-1}) = \log_b x - \log_b y$.

95. (a) For any angle θ, $(\cos\theta, \sin\theta)$ are the coordinates of the point where the terminal edge of θ meets the unit circle. Since the equation of the unit circle is $x^2 + y^2 = 1$ and we have $x = \cos\theta$ and $y = \sin\theta$, we must have $\sin^2\theta + \cos^2\theta = 1$.

97. For any angle θ, $\sin\theta$ is the y-coordinate where the terminal edge of θ meets the unit circle. The angle $-\theta$ is the angle of the same magnitude as θ but opening in the clockwise direction from the x-axis, and therefore its terminal edge will be the same as the terminal edge of θ except reflected across the x-axis. Therefore the y-coordinates of these two terminal edges have the same magnitude but opposite signs; in other words, $\sin(-\theta) = -\sin\theta$. The remaining even/odd identities can be proved in a similar fashion.

99. $\sin 2\theta = \sin(\theta + \theta) = \sin\theta\cos\theta + \sin\theta\cos\theta = 2\sin\theta\cos\theta$. The identity for $\cos 2\theta$ is proved similarly, and the alternative forms follow from the first two forms and the Pythagorean identity.

Section 0.5

1. F, T, T, F, T, T, F, F.

3. If C is true, then D must be true. If C is false, then D may or may not be true.

5. "For all $x > 0$, we have $x > -2$." and "If $x > 0$, then $x > -2$."

7. The original statement is true. The converse is "Every rectangle is a square," which is false. The contrapositive is "Everything that is not a rectangle is not a square," which is true.

9. The contrapositive is "Not(Q)$\Rightarrow$Not(P)," which is logically equivalent to $P \Rightarrow Q$.

11. False.

13. True.

15. True.

17. True. The negation is "For all real numbers x, $x \leq 2$ and $x \geq 3$."

19. True. The negation is "There exists a real number that is both rational and irrational."

21. True. The negation is "There exists x such that, for all y, $y \neq x^2$." (In other words, "There exists x for which there is no y with $y = x^2$.)"

23. True. The negation is "There is some integer x greater than 1 for which $x < 2$."

25. False. One counterexample is $x = 1.35$.

27. False.

29. False. One counterexample is $x = -1$.

31. True. One example is $x = 3$.

33. True. The negation is "There exist real numbers a and b such that $a < b$ but $3a + 1 \geq 3b + 1$."

35. True.

37. False.

39. True.

41. True. One example is $x = -1$, since for all y we have $|y| > -1$.

43. True.

45. False. The only counterexample where the two sides of the double implication are not equivalent is $x = 0$, $y = 0$.

47. (a) $B \Rightarrow$ (Not A); (b) (Not B) $\Rightarrow A$

49. (a) (Not A) $\Rightarrow$ (Not B); (b) $A \Rightarrow B$

51. (a) $C \Rightarrow$ (A and B); (b) Not(C) $\Rightarrow$ (Not A) or (Not B)

53. (a) (B and C) $\Rightarrow A$; (b) ((Not B) or (Not C)) $\Rightarrow$ (Not A)

55. (a) The converse is "If x is rational, then x is a real number." (b) The contrapositive is "If x is irrational, then x is not a real number." (c) $x = \pi$ is a counterexample to the original and the contrapositive.

57. (a) The converse is "If $x \geq 3$, then $x > 2$." (b) The contrapositive is "If $x < 3$, then $x \leq 2$." which is false. (c) $x = 2.5$ is a counterexample to both the original and the contrapositive.

59. (a) The converse is "If $\sqrt{x}$ is not a real number, then x is negative." (b) The contrapositive is "If $\sqrt{x}$ is a real number, then x is nonnegative. (c) No possible counterexamples for any of the statements.

61. (a) The converse is "If $|x| = -x$, then $x \leq 0$." (b) The contrapositive is "If $|x| \neq -x$, then $x > 0$." (c) No possible counterexamples for any of the statements.

63. (a) The converse is "If $x^2 > x$, then x is not zero." (b) The contrapositive is "If $x^2 \leq x$, then $x = 0$." (c) $x = 0.5$ is a counterexample to the original statement and its contrapositive; the converse happens to be true.

65. (a) The converse is "If there is some integer n such that $x = 2n + 1$, then x is odd." (b) The contrapositive is "If there is no integer n such that $x = 2n + 1$, then x is not odd." (c) No possible counterexamples for any of the statements.

67. Linda wears a red hat, Alina wears a blue hat, Phil wears a green hat, Stuart wears a yellow hat.

69. Liz and Rein are liars, and Zubin tells the truth. Here is an argument that proves this solution is correct: If Liz told the truth then everyone would, which would contradict Rein's and Zubin's statements. Therefore Liz must be a liar. Now if Rein is telling the truth, then by Rein's statement, Zubin must tell the truth also, but Zubin's statement would be false; another contradiction. Therefore Rein must be lying. This makes Zubin's statement true, so Zubin is telling the truth.

71. Kate and Hyun are liars and Jaan tells the truth. Here's why: Seeking a contradiction, suppose Kate tells the truth. Then by her statement both Hyun and Jaan lie. If Hyun lies then her statement means that Kate also lies, which is a contradiction. Therefore we can assume that Kate lies. Thus Hyun's statement is false, so Hyun is a liar. But by Kate's false statement, at least one of Hyun or Jaan tells the truth. Therefore Jaan tells the truth.

73. Suppose x is irrational and r is rational. Seeking a contradiction, suppose $x - r$ is rational. Since the sum of two rational numbers is rational, we know that $(x - r) + r = x$ must be rational. But this contradicts the hypothesis that x is irrational; therefore $x - r$ must be an irrational number.

75. If n is even and m is odd then there are integers k and l such that $n = 2k$ and $m = 2l + 1$. Then $n + m = (2k) + (2l + 1) = 2(k + l) + 1$. Since $k + l$ is an integer, this means that $n + m$ is odd.

77. Let $n = 2k$ and $m = 2l + 1$ and compute n times m.

79. If $M = \left(\frac{x_1+x_2}{2}, \frac{y_1+y_2}{2}\right)$, the distance formula easily shows that dist$(M, P) =$ dist(M, Q).

81. If a, b, and c are any real numbers with $b \neq 0$ and $c \neq 0$, then $\frac{a/b}{c} = \frac{a/b}{c/1} = \frac{a(1)}{bc} = \frac{a}{bc}$.

83. Follow the outline given in the problem.

85. $|a - b| + |b| \geq |(a - b) + b| = |a|$ so $|a - b| \geq |a| - |b|$.

87. $|x - c| > \delta$ if and only if $x - c > \delta$ or $x - c < -\delta$, which is equivalent to having $x > c + \delta$ or $x < c - \delta$, meaning that $x \in (c + \delta, \infty) \cup x \in (-\infty, c - \delta)$.

For answers to odd-numbered Skill Certification exercises in the Chapter Review, please visit the Book Companion Web Site at www.whfreeman.com/tkcalculus.

Chapter 1

Section 1.1

1. T, F, T, F, F, T, F, T.

3. $\lim_{x \to 1} f(x) = 5$; cannot say anything about $f(1)$

5. $\lim_{x \to 2^-} f(x)$ is not equal to 8.

7. f has a horizontal asymptote at $y = 3$ and a vertical asymptote at $x = 1$.

9. (a) The terms get smaller; $\lim_{k \to \infty} \frac{1}{3^k} = 0$. (b) For every $k > 8$ the terms will be less than 0.0001.

11. (a) The terms get larger and larger as more numbers are added. (b) For every $k > 44$ the terms will be greater than 1000.

13. (a) By solving $s(t) = 0$ we see that the orange hits the surface of the moon at about $t = 2.75$ seconds. The average rate of change on $[1.75, 2.75]$ is -11.925 feet per second; on $[2.25, 2.75]$ it is -13.25, on $[2.5, 2.75]$ it is -13.9125, and on $[2.625, 2.75]$ it is -14.2438. (b) We might estimate -14.5 feet per second on impact.

15. (a) 6.25 for four rectangles, and 5.8125 for eight. (b) If we did this for more and more rectangles, the area approximation would decrease to become closer and closer to the actual area under the curve. We might estimate 5.5 square units from our earlier work. (The actual answer turns out to be about 5.33.)

17. Let the graph of the function escape out to $\pm\infty$ at $x = 2$ with a vertical asymptote at the last minute.

19. (a) $g(1)$ is of the form $\frac{0}{0}$ and therefore is not defined. A calculator graph is shown here. It is not immediately clear that $g(x)$ is not defined at $x = 1$. Tracing along the graph near $x = 1$ might reveal a hole in the graph, depending on your calculator. (b) Even though $g(1)$ is undefined, the function values do approach 0 as $x \to 1$, so the limit exists and is equal to 0.

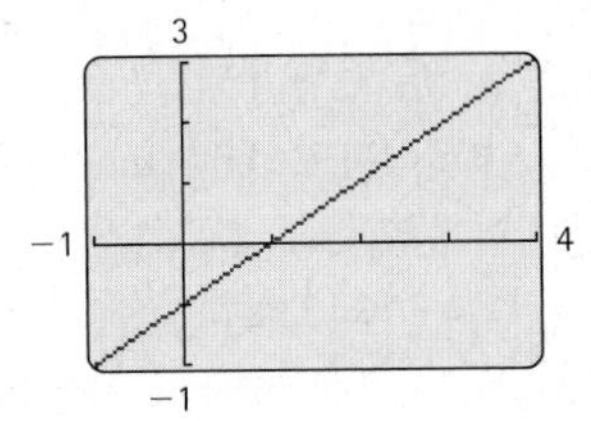

21.

23.

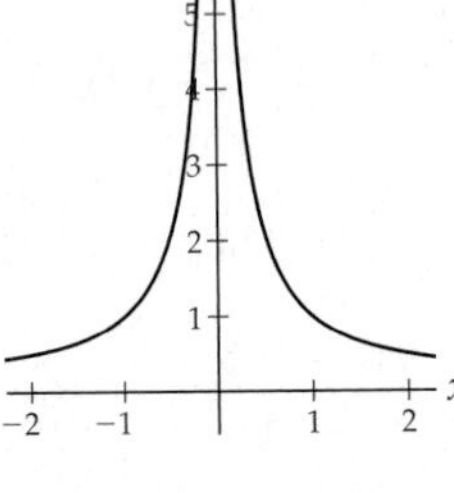

25.

27.

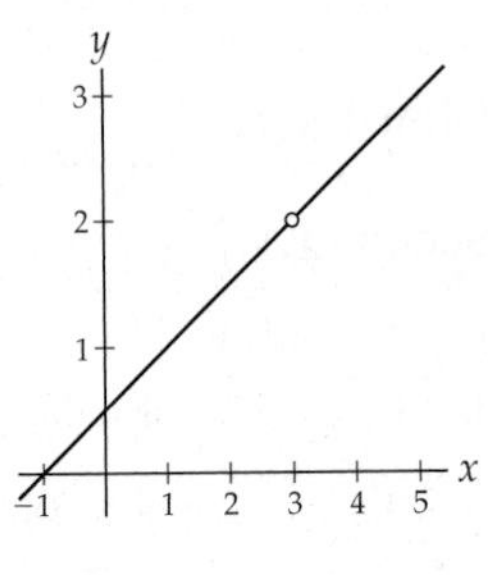

29.

31. $-2, -2, -2, -2$

33. $-2, 2$, DNE, 2

35. $\infty, -\infty$, DNE, undefined

37. $-1, -1, -1, -2$

39. 7

41. DNE: $-\infty$ from the left and ∞ from the right.

43. $\frac{1}{7}$

45. DNE: ∞ from the left and $-\infty$ from the right.

47. -3

49. 0

51. DNE: $\sin x$ oscillates between -1 and 1 as $x \to \infty$.

53. DNE: $-\infty$ from the left and ∞ from the right.

55. 2

57. $\frac{1}{2}$

59. 1

61. DNE: ∞ from the left and $-\infty$ from the right.

63. DNE: 4 from the left and -5 from the right.

65. 2

67. -93

69. DNE: $-\infty$ from the left and ∞ from the right.

71. 0

73. 1

75. (a) $S(0) = \frac{12}{3} = 4$; (b) 17, 19, 21; (c) 22; (d) 22.

77. (a) Your y-range will have to be quite large; (b) ∞; (c) population explosion past what any finite-sized planet can handle; (d) in this model no world would be left after 2027.

79. For $k = 101$ we have $\frac{101(101+1)}{2} = 5151$, and for larger k the quantity $\frac{k(k+1)}{2}$ will be still larger.

81. By the previous problem, if x is within 0.01 of $x = 1$ then $(x-1)^2$ is in the interval $(0, 0.0001)$, and thus $0 < (x-1)^2 < 0.0001$. Therefore $\frac{1}{(x-1)^2} > \frac{1}{0.0001} = 10,000$.

Section 1.2

1. F, F, F, F, F, F, T, T.

3. See the reading at the beginning of the section.

5. (a) $(1, 1.5) \cup (1.5, 2)$; (b) $(0, 0.25) \cup (0.25, 0.5)$; (c) $(0, 1) \cup (1, 2)$

7. If $x \in (2-\delta, 2) \cup (2, 2+\delta)$, then $f(x) \in (5-\varepsilon, 5+\varepsilon)$.

9. If $x \in (N, \infty)$, then $f(x) \in (2-\varepsilon, 2+\varepsilon)$.

11. If $x \in (1, 1+\delta)$, then $f(x) \in (M, \infty)$.

13. Combine the two figures above Theorem 1.8. The two blue one-sided δ-bars combine to make a punctured δ-interval around $x = c$.

15. $(1.75, 6.25)$

17. There is some $M > 0$ for which there is no $N > 0$ that would guarantee that if $x > N$, then $\frac{1000}{x} > M$.

19. For all $\epsilon > 0$, there exists a $\delta > 0$ such that if $x \in (-3-\delta, -3) \cup (-3, -3+\delta)$, then $\sqrt{x+7} \in (2-\epsilon, 2+\epsilon)$.

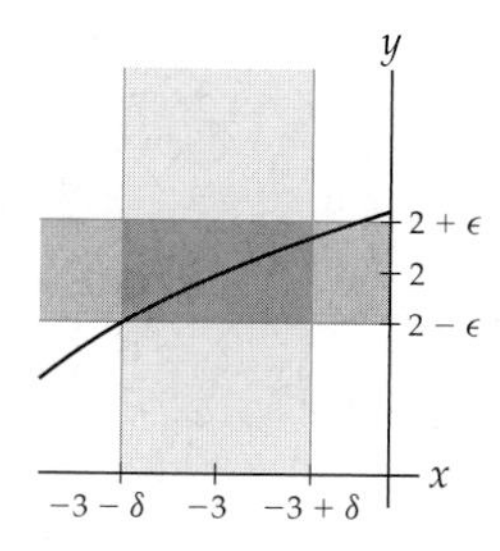

21. For all $\epsilon > 0$, there exists a $\delta > 0$ such that if $x \in (-1-\delta, -1) \cup (-1, -1+\delta)$, then $x^3 - 2 \in (-3-\epsilon, -3+\epsilon)$.

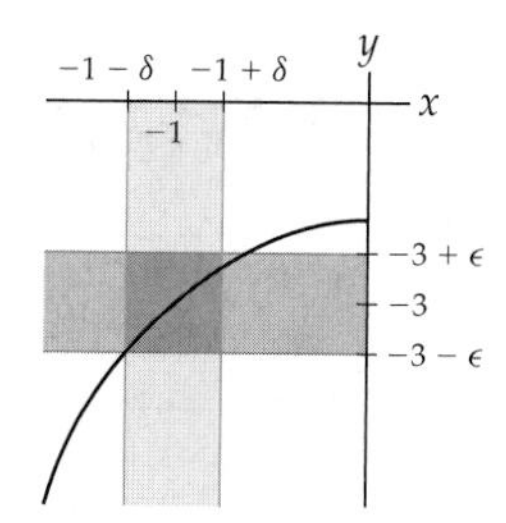

23. For all $\epsilon > 0$, there exists a $\delta > 0$ such that if $x \in (2-\delta, 2) \cup (2, 2+\delta)$, then $\frac{x^2-4}{x-2} \in (4-\epsilon, 4+\epsilon)$.

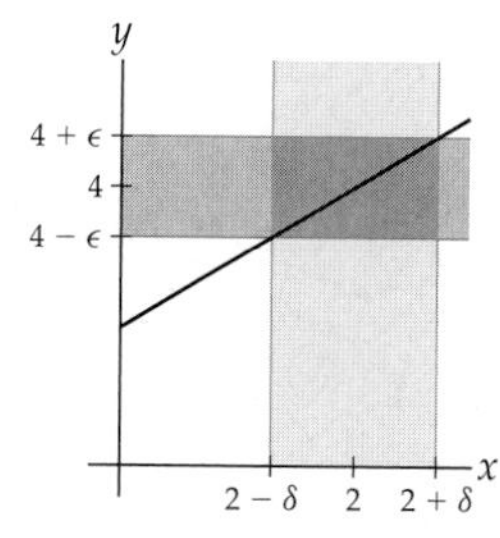

25. For all $\epsilon > 0$, there exists a $\delta > 0$ such that if $x \in (0, \delta)$, then $\sqrt{x} \in (-\epsilon, \epsilon)$.

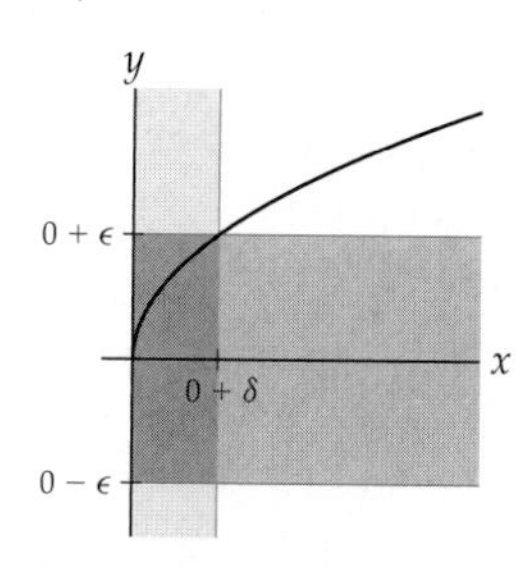

27. For all $M > 0$, there exists a $\delta > 0$ such that if $x \in (-2, -2+\delta)$, then $\frac{1}{x+2} \in (M, \infty)$.

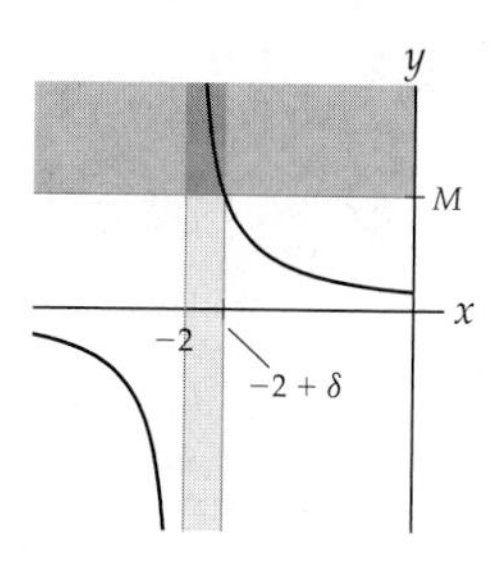

29. For all $M > 0$, there exists a $\delta > 0$ such that if $x \in (2, 2+\delta)$, then $\frac{1}{2^x-4} \in (M, \infty)$.

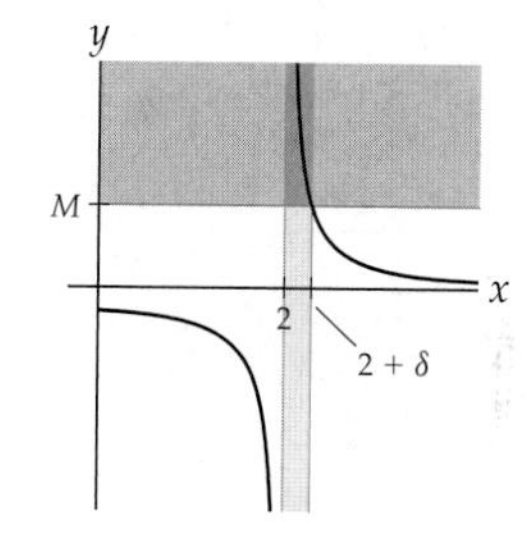

31. For all $\epsilon > 0$, there is some $N > 0$ such that if $x \in (N, \infty)$, then $\frac{x}{1-2x} \in (-0.5-\epsilon, -0.5+\epsilon)$.

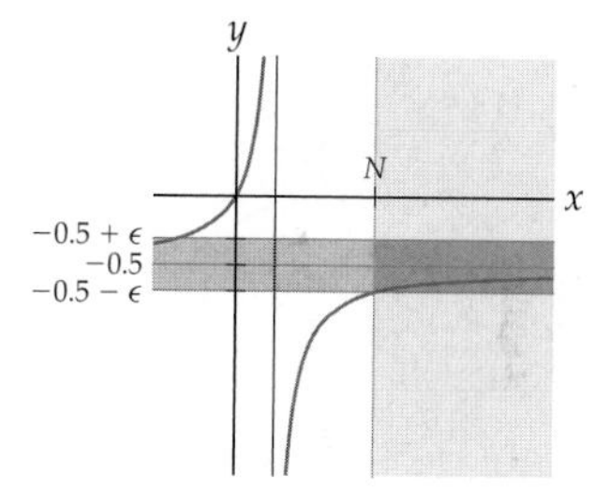

33. For all $M > 0$, there is some $N > 0$ such that if $x \in (N, \infty)$, then $x^3 + x + 1 \in (M, \infty)$.

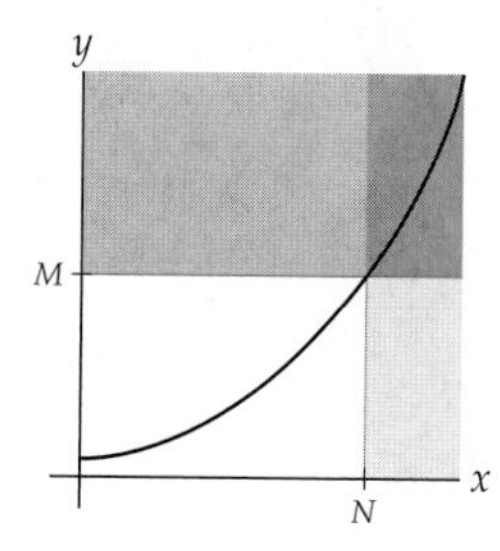

35. For all $\epsilon > 0$, there exists a $\delta > 0$ such that if $h \in (-\delta, 0) \cup (0, \delta)$, then $\frac{(2+h)^2-4}{h} \in (4-\epsilon, 4+\epsilon)$.

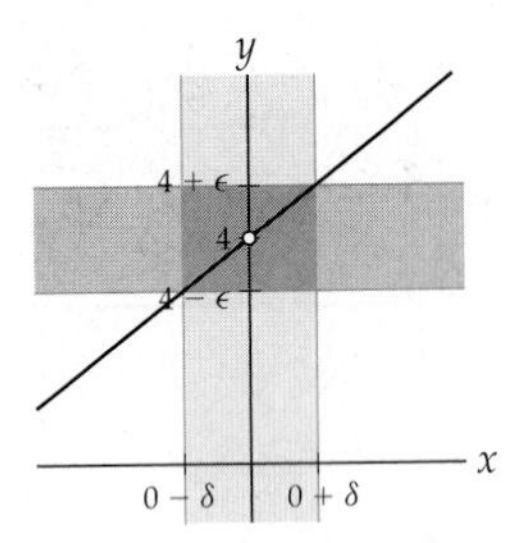

37. For all $M < 0$, there exists some $\delta > 0$ such that if $x \in (c, c+\delta)$, then $f(x) \in (-\infty, M)$.

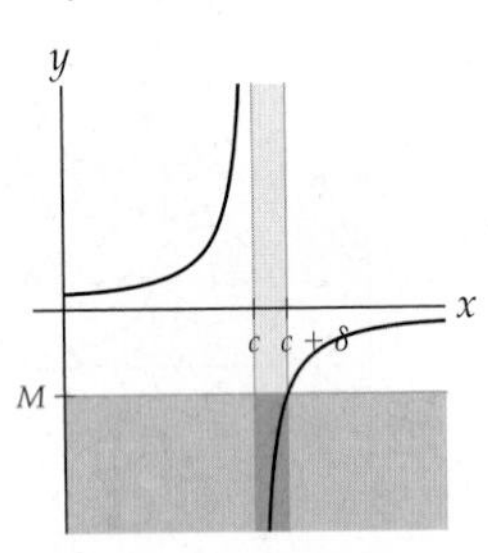

39. For all $\epsilon > 0$ there is some $N < 0$ such that if $x \in (-\infty, N)$, then $f(x) \in (L-\epsilon, L+\epsilon)$.

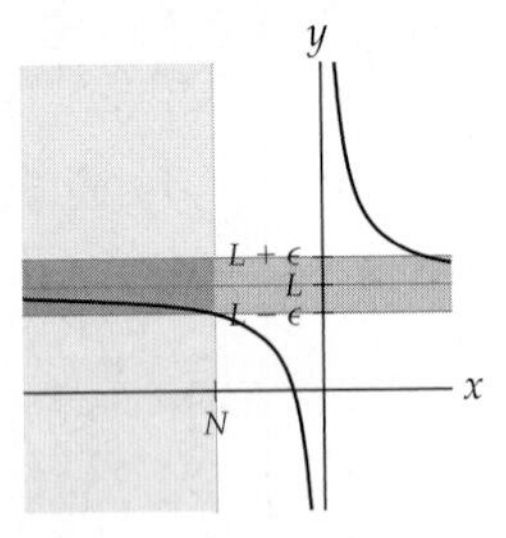

41. For all $M > 0$, there is some $N < 0$ such that if $x \in (-\infty, N)$, then $f(x) \in (M, \infty)$.

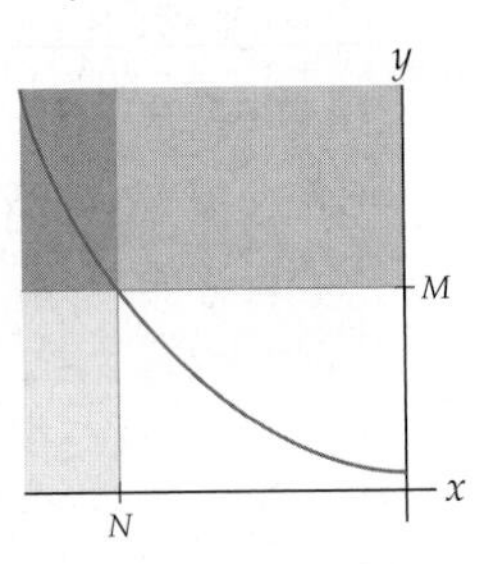

43. $\delta = (8.5)^{1/3} - 2 \approx 0.0408$ 45. $\delta = 3$

47. $\delta = \frac{\pi}{4}$

49. $\delta \approx 1.11$, using a calculator approximation to solve $\frac{1-\cos x}{x} = \frac{1}{2}$.

51. $\delta \approx 0.016662$ 53. $\delta = 1$

55. $\delta \approx 0.0005$ 57. $N = 5$

59. $N = e^{100}$ 61. $N = \dfrac{\ln\left(\frac{1}{2}\right)}{\ln 3} \approx -0.63093$

63. $N = \sqrt{104} \approx 10.198$

65. (a) 149.3 months; (b) yes, since $\lim_{t\to\infty} 50t = \infty$.

67. (a) \$18.68; (b) between \$16.20 and \$21.50; (c) $L = 6000$, $c = 18.68$, $\epsilon = 1000$, $\delta = 2.48$.

69. (a) For all $\epsilon > 0$, there exists $\delta > 0$ such that if $x \in (2-\delta, 2) \cup (2, 2+\delta)$, then $7 - x \in (5-\epsilon, 5+\epsilon)$. (b) $2-\delta < x < 2+\delta$ and $x \neq 2$ means that $-\delta < x-2 < \delta$ and $x \neq 2$. This means that $|x-2| < \delta$ and $x - 2 \neq 0$, and thus $0 < |x-2| < \delta$. (c) $5-\epsilon < 7-x < 5+\epsilon$ means that $-\epsilon < 2-x < \epsilon$, and thus that $|2-x| < \epsilon$, which is equivalent to $|x-2| < \epsilon$. (d) For a given $\epsilon > 0$, choose $\delta = \epsilon$.
(e) Given any $\epsilon > 0$, let $\delta = \epsilon$. If $x \in (2-\delta, 2) \cup (2, 2+\delta$, then $0 < |x-2| < \delta$ by part (b). Thus by part (d), $|x-2| < \epsilon$, and therefore $7 - x \in (5-\epsilon, 5+\epsilon)$ by part (c).

71. (a) For all $\epsilon > 0$, there exists $N > 0$ such that if $x \in (N, \infty)$, then $\frac{1}{x} \in (0-\epsilon, 0+\epsilon)$. (b) $x \in (N, \infty)$ is equivalent to $x > N$. (c) $\frac{1}{x} \in (-\epsilon, \epsilon)$ means that $-\epsilon < \frac{1}{x} < \epsilon$. For the limit we are considering, $\frac{1}{x}$ approaches $y = 0$ from above, so we actually wish to show that we can guarantee $0 < \frac{1}{x} < \epsilon$. (d) For a given $\epsilon > 0$, choose $N = \frac{1}{\epsilon}$. (e) Given any $\epsilon > 0$, let $N = \frac{1}{\epsilon}$. If $x \in (N, \infty)$ then $x > N > 0$ by part (b). Thus by our choice from part (d), $0 < \frac{1}{\epsilon} < x$, and thus $0 < \frac{1}{x} < \epsilon$. By part (c) this means $-\epsilon < \frac{1}{x} < \epsilon$ and thus $\frac{1}{x} \in (0-\epsilon, 0+\epsilon)$.

Section 1.3

1. T, T, F, T, T, F, F, F.

3. For all $\epsilon > 0$, there exists a $\delta > 0$ such that if $0 < |x+2| < \delta$, then $\left|\frac{3}{x+1} + 3\right| < \epsilon$.

5. $0 < |x-5| < 0.01$; $|f(x)+2| < 0.5$

7. $x \in (1.9, 2) \cup (2, 2.1)$

9. $x^2 - 1 \in (-3.5, -2.5)$ (note that there are in fact no x-values with this property)

11. $f(x) \in (L-\varepsilon, L+\varepsilon)$

13. The statement is false. One counterexample is $x = 1.5$, since $x = 1.5$ satisfies $0 < |x-2| < 1$, but not $|x^2 - 4| < 0.5$, since $(1.5)^2 = 2.25$.

15. The statement is false. One counterexample is $x = 0.74$; this value satisfies $0 < |x-0| < 0.75$ but not $|x^2 - 0| < 0.5$, since $(0.74)^2 = 0.5476$.

17. The statement is true. Sketch the graph of $y = x^2$ and show an ϵ interval around $y = 4$ with $\epsilon = 0.4$ (i.e., show the horizontal "bar" between $y = 3.6$ and $y = 4.4$). Then draw the δ interval around $x = -2$ with $\delta = 0.075$, excluding $x = -2$ (i.e., show the vertical "bar" between $x = -2.075$ and $x = -1.925$, excluding $x = -2$). Indicate on your graph that every x-value in $(-2.075, -2) \cup (-2, -1.925)$ has an $f(x)$-value in $(3.6, 4.4)$.

19. (a) $\delta = \frac{5}{325}$; (b) $\delta = \frac{.01}{325}$; (c) $\delta = 1$

21. See the proof of Theorem 1.11.

23. $\delta = \frac{0.25}{3}$

25. $\delta = 0.25$

27. $N = \sqrt{1000} \approx 31.62$

29. $\delta = \epsilon$

31. $\delta = \frac{\epsilon}{4}$

33. $\delta = \sqrt{\frac{\epsilon}{5}}$

35. $\delta = \sqrt{\epsilon}$

37. $\delta = \min(1, 2\epsilon)$

39. $\delta = \min(1, \epsilon/5)$

41. $\delta = \varepsilon^2$

43. $\delta = \frac{1}{M}$

45. With $C(r) = 0.25(2\pi r(5)) + 0.5(2\pi r^2) + 0.1(2(2\pi r) + 5)$, we have (a) $r \approx 1.94$ inches; (b) $|C(r) - 30| < 10$; (c) $r \in (1.43, 2.38)$, so $|r - 1.94| < 0.44$.

47. Given $\epsilon > 0$, choose $\delta = \epsilon/2$. Then if $0 < |x-1| < \delta$, we have $|(2x+4) - 6| = 2|x-1| < 2\delta = \epsilon$.

49. Given $\epsilon > 0$, choose $\delta = \epsilon$. Then if $0 < |x+6| < \delta$, we have $|(x+2) - (-4)| = |x+6| < \delta = \epsilon$.

51. Given $\epsilon > 0$, choose $\delta = \frac{\epsilon}{6}$. Then if $0 < |x-4| < \delta$, we have $|(6x-1) - (23)| = |6x - 24| = 6|x-4| < 6\delta = \epsilon$.

53. Given $\epsilon > 0$, choose $\delta = \sqrt{\epsilon/3}$. Then if $0 < |x| < \delta$, we have $|(3x^2+1) - 1| = |3x^2| = 3|x|^2 < 3\delta^2 = 3(\sqrt{\epsilon/3})^2 = \epsilon$.

55. Given $\epsilon > 0$, choose $\delta = \sqrt{\frac{\epsilon}{2}}$. Then if $0 < |x-1| < \delta$, we have $|(2x^2 - 4x + 3) - 1| = |2x^2 - 4x + 2| = 2|x^2 - 2x + 1| = 2|x-1|^2 < 2\delta^2 = \epsilon$.

57. Given $\epsilon > 0$, choose $\delta = \epsilon$. Then if $0 < |x-1| < \delta$, we have $\left|\frac{x^2-1}{x-1} - 2\right| = \left|\frac{x^2-2x+1}{x-1}\right| = |x-1| < \delta = \epsilon$.

59. Given $\epsilon > 0$, choose $\delta = \epsilon^2$. Then if $5 < x < 5 + \delta$, we have $0 < x - 5 < \delta$ and thus $|\sqrt{x-5} - 0| = \sqrt{x-5} < \sqrt{\delta} = \epsilon$.

61. Given $M > 0$, choose $\delta = 1/M$. Then if $x \in (-2, -2+\delta)$, we have $0 < x + 2 < \delta$, and thus $\frac{1}{x+2} > \frac{1}{\delta} = \frac{1}{1/M} = M$.

63. Given $\epsilon > 0$, choose $N = \frac{1}{\epsilon}$. Then if $x > N$, we have $\left|\frac{2x-1}{x} - 2\right| = \left|\frac{2x-1-2x}{x}\right| = \frac{1}{|x|} < \frac{1}{N} = \epsilon$.

65. Given $M > 0$, choose $N = \frac{M+5}{3}$. Then if $x > N$, we have $3x - 5 > 3N - 5 = M$.

67. Given $\epsilon > 0$, choose $\delta = \min(1, \epsilon/5)$. Then if $0 < |x-3| < \delta$, we have $|(x^2 - 2x - 3) - 0| = |x^2 - 2x - 3| = |x-3||x+1| < \delta|x+1| \le \delta(5) \le \epsilon$.

69. Given $\epsilon > 0$, choose $\delta = \min(1, \epsilon/5)$. Then if $0 < |x-5| < \delta$, we have $|(x^2 - 6x + 7) - 2| = |x^2 - 6x + 5| = |x-5||x-1| < \delta|x-1| \le \delta(5) = \epsilon$.

71. Given $\epsilon > 0$, choose $\delta = \min(1, \epsilon/3)$. Then if $0 < |x-2| < \delta$, we have $\left|\frac{4}{x^2} - 1\right| = \frac{|2-x||2+x|}{x^2} < \delta\frac{|2+x|}{x^2} \le \delta(3) = \epsilon$.

Section 1.4

1. T, T, F, F, F, F, T, F.

3. $\lim_{x\to 1} f(x) = f(1)$

5. See the reading.

7. Define $f(1)$ to be 0.

9. Removable discontinuity, neither left nor right continuous. One possible graph is:

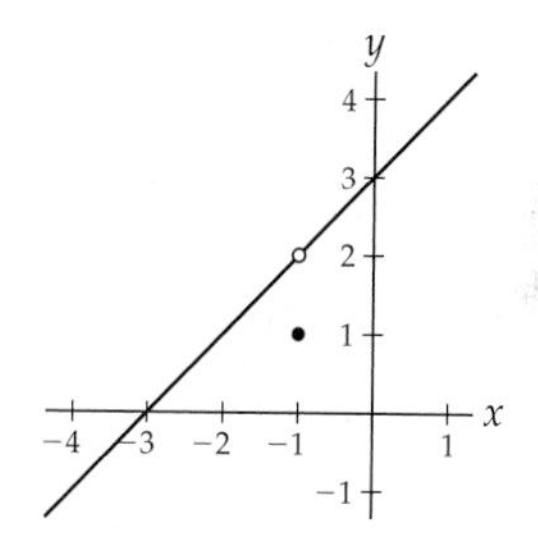

11. Jump discontinuity, neither left nor right continuous. One possible graph is:

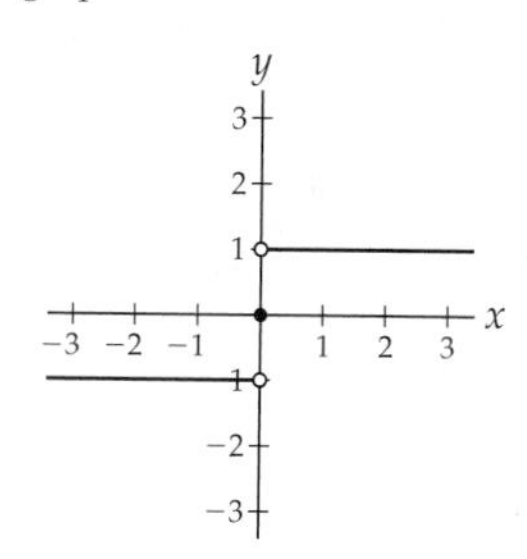

13. For all $\epsilon > 0$, there exists $\delta > 0$ such that whenever $x \in (c - \delta, c + \delta)$, we also have $f(x) \in (f(c) - \epsilon, f(c) + \epsilon)$.

15. For all $\epsilon > 0$, there exists $\delta > 0$ such that whenever $x \in (c, c + \delta)$, we also have $f(x) \in (f(c) - \epsilon, f(c) + \epsilon)$.

17. See the reading.

19. The graph below is not continuous on $[1, 3]$, and there is no minimum value on that interval.

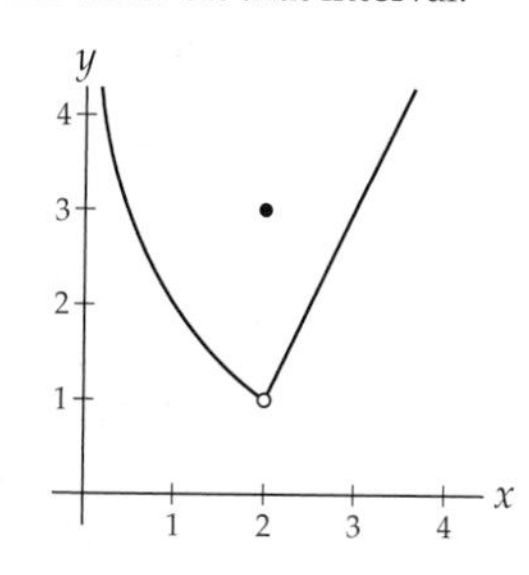

21.

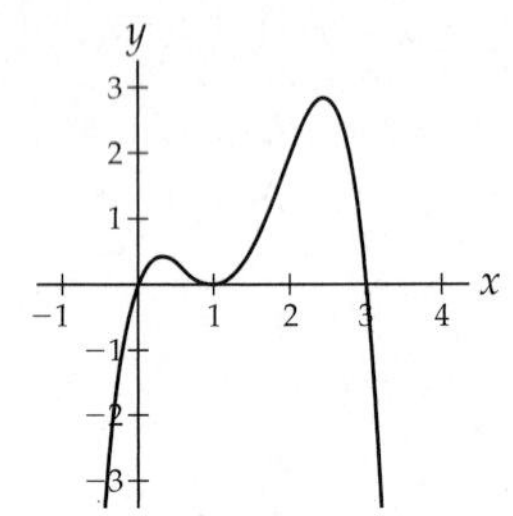

23. Continuous on $(-\infty, 2) \cup (2, \infty)$, removable discontinuity at $x = 2$, neither left- nor right continuous at $x = 2$.

25. Continuous on $(-\infty, -1) \cup (-1, 1] \cup (1, \infty)$, removable discontinuity at $x = -1$, infinite discontinuity at $x = 1$, left continuous at $x = 1$.

27. Not possible.

29.

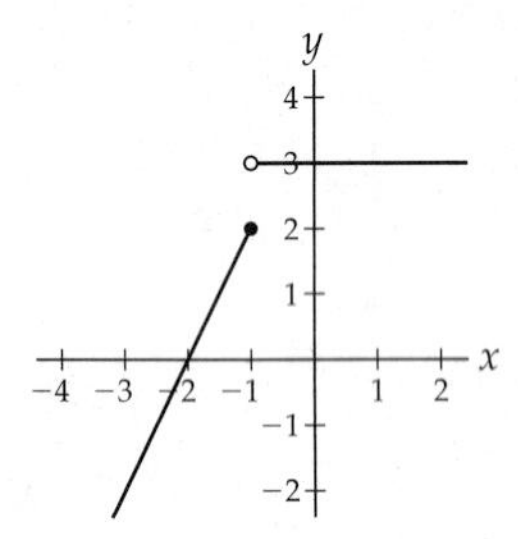

31. Not possible.

33. 6

35. -17

37. Can't do this yet, since 0 is not in the domain of x^{-3}.

39. Continuous at $x = 3$.

41. Jump discontinuity and left continuous at $x = 2$.

43. Continuous at $x = 1$, jump discontinuity and right continuous at $x = 2$.

45. Not continuous at $x = 2$.

47. Continuous at $x = 1$.

49. (a) Since f is continuous on $[-2, 2]$, the Extreme Value Theorem guarantees that f will attain global maximum and minimum values on $[-2, 2]$. (b) The maximum value on that interval occurs twice, at $x = -2$ and at $x = 2$, and the minimum value occurs both at $x \approx -1.22$ and at $x \approx 1.22$.

51. (a) Since f is continuous on $[-1, 1]$, the Extreme Value Theorem guarantees that f will attain global maximum and minimum values on $[-1, 1]$. (b) The maximum value on that interval occurs at $x = 0$, and the minimum value occurs twice, at $x = -1$ and $x = 1$.

53. (a) Since f is continuous on $[0, 2]$, the Extreme Value Theorem guarantees that f will attain global maximum and minimum values on $[0, 2]$. (b) The minimum value on that interval occurs at $x = \frac{4}{3}$, and the maximum value occurs twice, at $x = 0$ and $x = 2$.

55. (a) Since f is continuous on $[0, 2]$ and $K = 0$ is between $f(0) = 5$ and $f(2) = -11$, the Intermediate Value Theorem guarantees that there is some $c \in (0, 2)$ such that $f(c) = 0$. (b) Using a graph we can approximate that in $(0, 2)$ there is one such value of c, namely, $c \approx 1.5$.

57. (a) Since f is continuous on $[-2, 4]$ and $K = -4$ is between $f(-2) = -22$ and $f(4) = 14$, the Intermediate Value Theorem guarantees that there is some $c \in (-2, 4)$ such that $f(c) = -4$. (b) Using a graph we can approximate that in $(-2, 4)$ there are three such values of c, namely, $c = 1$, $c \approx -0.73$, and $c \approx 2.73$.

59. (a) Since f is continuous on $[2, 4]$ and $K = -4$ is between $f(2) = -6$ and $f(4) = 14$, the Intermediate Value Theorem guarantees that there is some $c \in (2, 4)$ such that $f(c) = -4$. (b) Using a graph we can approximate that in $(2, 4)$ there is one such value of c, namely, $c \approx 2.73$.

61. (a) Since f is continuous everywhere, the Intermediate Value Theorem applies on any interval $[a, b]$. For example, since $f(-3) = -25 < -15$ and $f(0) = 2 > -15$, the Intermediate Value Theorem guarantees that there is some $c \in (-3, 0)$ for which $f(c) = -15$. (b) Using a graph we can approximate that $c \approx -2.57$ is such a value.

63. (a) Since f is continuous everywhere, the Intermediate Value Theorem applies on any interval $[a, b]$. For example, since $f(0) = 0 < \frac{1}{2}$ and $f\left(\frac{\pi}{2}\right) = 1 > \frac{1}{2}$, the Intermediate Value Theorem guarantees that there is some $c \in \left(0, \frac{\pi}{2}\right)$ for which $f(c) = \frac{1}{2}$. (b) $c = \frac{\pi}{6}$ is such a value.

65. (a) Since f is continuous everywhere, the Intermediate Value Theorem applies on any interval $[a, b]$. For example, since $f\left(-\frac{1}{3}\right) = 0 < 1$ and $f(2) = 7 > 1$, the Intermediate Value Theorem guarantees that there is some $c \in (0, 7)$ for which $f(c) = 1$. (b) $c = 0$ is such a value.

67. Positive on $(-\infty, -2) \cup \left(-\frac{1}{2}, \infty\right)$, negative on $\left(-2, -\frac{1}{2}\right)$

69. Positive on $(-\infty, -2) \cup (-1, 1) \cup (2, \infty)$, negative on $(-2, -1) \cup (1, 2)$

71. Positive on $(2, \infty)$, negative on $(-\infty, 2)$

73. Positive on $(0, 2]$, negative on $(-\infty, 0) \cup (2, \infty)$

75. The length $H(t)$ of Alina's hair is a continuous function because her hair can't suddenly get longer or shorter without going through a continuous change, so the theorems apply. We are given that $H(0) = 2$ and $H(6) = 42$. The Extreme Value Theorem says that Alina's hair had a maximum and a minimum length sometime in the last 6 years. The Intermediate Value Theorem says that Alina's hair has been every length between 2 and 42 inches at some point during the last 6 years.

77. The Extreme Value Theorem says that at some point in the past year, there was a time when Phil's wagon contained the most gas and a time when it contained the least. The Intermediate Value Theorem says that for every amount between 0 and 19 gallons of gas, there was at least one time in the past year that Phil's wagon contained that amount of gas.

79. (a) $M(v) = \begin{cases} 8{,}500, & \text{if } 0 \le v < 30 \\ 10{,}000, & \text{if } 30 \le v < 60 \\ 11{,}500, & \text{if } 60 \le v < 90 \\ 13{,}000, & \text{if } v = 90. \end{cases}$
(b) $M(0) = 8,500$, $M(30) = 10,000$, $M(59) = 10,000$, $M(61) = 11,500$, and $M(90) = 13,000$; (c) $M(v)$ fails to be continuous at $v = 30$, $v = 60$, and $v = 90$.

81. Given $\epsilon > 0$, let $\delta = \epsilon/3$. If $0 < |x-2| < \delta$, then $|(3x-5)-(3(2)-5)| = |3x-6| = 3|x-2| < 3\delta = \epsilon$.

83. Given $\epsilon > 0$, choose $\delta = \epsilon$. If $0 < |x-c| < \delta$, then $||x|-|c|| < |x-c| < \delta = \epsilon$.

85. f is not continuous at $c = 2$. For example, consider $\epsilon = 1$. No matter how small we take $\delta > 0$, there are irrational values of x with $0 < |x-2| < \delta$ for which $|f(x)-f(c)| = |x^2-0| = x^2$ is outside of the interval $(0-\epsilon, 0+\epsilon) = (-1, 1)$.

87. If $A > 0$ then for large magnitude N we will have $f(-N) < 0 < f(N)$, and if $A < 0$ then for large magnitude N we will have $f(-N) > 0 > f(N)$. In either case the Intermediate Value Theorem applies to the continuous cubic function $f(x)$ on $[-N, N]$ and therefore $f(x)$ has at least one real root on $[-N, N]$.

89. Suppose $c \ge 0$; the proof for $c \le 0$ is similar. Given $\epsilon > 0$, let $\delta = \min\left(1, \frac{\epsilon}{2c+1}\right)$. If $0 < |x-c| < \delta$, then since $\delta \le 1$ we have $c-1 < x < c+1$. Then $|x^2-c^2| = |x-c|\cdot|x+c| < \delta\cdot|x+c| \le \delta|(c+1)+c| = \delta(2c+1) = \frac{\epsilon}{2c+1}(2c+1) = \epsilon$.

91. Suppose $c > 0$; the proof for $c < 0$ is similar. Given $\epsilon > 0$, let $\delta = \min\left(1, \epsilon\cdot\frac{(c-1)^2c^2}{2c+1}\right)$. If $0 < |x-c| < \delta$, then since $\delta \le 1$ we have $c-1 < x < c+1$. Then $\left|\frac{1}{x^2}-\frac{1}{c^2}\right| = \frac{|c^2-x^2|}{|x^2c^2|} = \frac{|c-x|\cdot|c+x|}{x^2c^2} < \frac{\delta|c+x|}{x^2c^2} \le \frac{\delta|c+(c+1)|}{(c-1)^2c^2} = \epsilon$.

93. Suppose $c > 0$. Given $\epsilon > 0$, let $\delta = \min\left(1, \epsilon\cdot\frac{\sqrt{c-1}+\sqrt{c}}{2c+1}\right)$. If $0 < |x-c| < \delta$, then since $\delta \le 1$ we have $c-1 < x < c+1$. Then $|\sqrt{x}-\sqrt{c}| = \frac{|x^2-c^2|}{\sqrt{x}+\sqrt{c}} = \frac{|x-c|\cdot|x+c|}{\sqrt{x}+\sqrt{c}} < \frac{\delta|x+c|}{\sqrt{x}+\sqrt{c}} \le \frac{\delta|(c+1)+c|}{\sqrt{c-1}+\sqrt{c}} = \epsilon$.

Section 1.5

1. T, T, F, F, T, F, F, T.

3. See Theorem 1.20.

5. Consider the limit $\lim_{x\to 1} \frac{x-1}{x^2-1}$.

7. $f(x) = x$, $g(x) = \frac{1}{x}$, $c = 0$ is one example. This does not contradict the product rule for limits because one of the limits does not exist as $x \to 0$.

9. For all $\epsilon > 0$, there exists $\delta > 0$ such that if $0 < |x-c| < \delta$, then $|(f(x)-g(x))-(L-M)| < \epsilon$, where $L = \lim_{x\to c} f(x)$ and $M = \lim_{x\to c} g(x)$.

11. To get this function we add the power function $\sqrt{x}$ and the constant function 1, and then compose the result with the power function x^3. Thus by the sum and composition rules for limits, and the fact that power and constant functions are continuous, we know that this function f is continuous and thus we can calculate its limits at domain points by evaluation.

13. -2

15. Not enough information.

17. $\frac{1}{2}$

19. The graph of g looks just like the graph of f but with a hole at $x = 1$. $f(1) = 2$ but $g(1)$ is undefined, while both functions approach 2 as x approaches 1.

21. Limits as $x \to c$ are not determined by function values at the point $x = c$; they are determined by the behavior of the function in question as x approaches c.

23. Consider a picture like the ones that follow the Squeeze Theorem for limits in the reading, but think of them upside-down.

25. Using the constant multiple rule followed by the continuity of linear functions, we have $\lim_{x\to 1} 15(3-2x) = 15\lim_{x\to 1}(3-2x) = 15(3-2(1)) = 15$.

27. Applying the sum rule, the product rule, the constant multiple rule, and then the continuity of power and linear functions, in that order, we have $\lim_{x\to 3}(3x+x^2(2x+1)) = \lim_{x\to 3}(3x) + \lim_{x\to 3}(x^2(2x+1)) = \lim_{x\to 3}(3x) + \lim_{x\to 3} x^2 \cdot \lim_{x\to 3}(2x+1) = 3(3)+(3^2)(2(3)+1) = 72$.

29. -1

31. 4

33. 2

35. 2

37. $-\frac{1}{10}$

39. $\frac{4}{7}$

41. 0

43. 3

45. -1

47. 0

49. $3e^{1.7(4)}+1$

51. $\frac{1}{4}$

53. 2

55. 0

57. 1

59. $\frac{\pi}{6}$

61. $\frac{2}{\pi}$

63. 6

65. 3

67. -1

69. -1

71. Continuous on $\mathbb{R}$.

73. Continuous on $(-\infty, -2)$ and on $(-2, \infty)$; neither left nor right continuous at $x = -2$.

75. Continuous on $(-\infty, 1)$ and on $(1, \infty)$; neither left nor right continuous at $x = 1$.

77. Continuous on $(-\infty, \pi)$ and (π, ∞); right but not left continuous at $x = \pi$.

79. 0

81. 0

83. 0

85. 0

87. (a) $T(63{,}550) = \$14381.50$; (b) this value for $T(63,550)$ matches up with the value of the next piece $14,381.5 + 0.31(m - 63,550)$ when $m = 63,550$. This means that when you change tax brackets, the higher tax rate only applies to the amount over 63, 550.

89. If $f(x) = a_n x^n + a_{n-1}x^{n-1} + \cdots + a_1 x + a_0$ is a polynomial function and c is a real number, then by the sum rule, constant multiple rule, and limit of a constant, we have $\lim_{x \to c} f(x) = a_n \lim_{x \to c} x^n + a_{n-1} \lim_{x \to c} x^{n-1} + \cdots + a_1 \lim_{x \to c} x + a_0$. Since power functions with positive integer powers are continuous everywhere, we can solve each of the component limits by evaluation, which gives us $\lim_{x \to c} f(x) = a_n c^n + a_{n-1}c^{n-1} + \cdots + a_1 x + a_0 = f(c)$. Therefore the polynomial function f is continuous at $x = c$.

91. Given $\epsilon > 0$, we can choose $\delta > 0$ so that if $x \in (c - \delta, c) \cup (c, c + \delta)$,then $f(x)$ is within $\frac{\epsilon}{|k|}$ of L. Then $|kf(x) - kL| = |k||f(x) - L| < |k|(\epsilon/|k|) = \epsilon$.

93. If $\lim_{x \to c} f(x)$ exists and $\lim_{x \to c} g(x)$ exists and is nonzero, then by the product and reciprocal rules for limits we have $\lim_{x \to c} \frac{f(x)}{g(x)} = \lim_{x \to c} f(x) \frac{1}{g(x)} = \lim_{x \to c} f(x) \lim_{x \to c} \frac{1}{g(x)} = \lim_{x \to c} f(x) \frac{1}{\lim_{x \to c} g(x)} = \frac{\lim_{x \to c} f(x)}{\lim_{x \to c} g(x)}$.

95. Given $c \in \mathbb{R}$, $\lim_{x \to c} f(x) = \lim_{x \to c} Ab^x = A \lim_{x \to c} (e^{\ln b})^x = A(\lim_{x \to c} e^x)^{\ln b} = A(e^c)^{\ln b} = A(e^{\ln b})^c = Ab^c = f(c)$.

97. Mimic the proof in the reading that $\sin x$ is continuous everywhere.

99. If $x = c$ is in the domain of $\sec x$, then c is not a multiple of $\frac{\pi}{2}$, and $\cos c \neq 0$. Therefore $\lim_{x \to c} \sec x = \lim_{x \to c} \frac{1}{\cos x}$ is the quotient of $\lim_{x \to c} 1$ by $\lim_{x \to c} \cos x = \cos c$, or $\frac{1}{\cos c} = \sec c$.

Section 1.6

1. T, T, F, T, T, T, F, F.

3. $\lim_{x \to c} \frac{f(x)}{g(x)} = \frac{L}{M}$

5. $\lim_{x \to c} \frac{f(x)}{g(x)}$ is infinite

7. $\infty + \infty \to \infty$, $\infty - \infty$ is indeterminate, $\infty + 1 \to \infty$, and $0 + \infty \to \infty$.

9. See Theorems 1.27 and 1.28.

11. (a) The numerator approaches 1 while the denominator gets larger and larger, which makes the quotient get smaller and smaller; (b) see the reading; (c) the denominator approaches 1 while the numerator gets smaller and smaller, which makes the quotient get smaller and smaller.

13. (a) The numerator approaches 1 while the denominator gets smaller and smaller, which makes the quotient get larger and larger; (b) the numerator gets larger and larger, making the quotient get larger and larger, while at the same time the denominator gets smaller and smaller, making the quotient get larger and larger still; (c) the denominator approaches 1 while the numerator gets larger and larger, which makes the quotient get larger and larger.

15. (a) $\lim_{x \to 2} \frac{(x-2)^2}{x-2}$; (b) $\lim_{x \to 2} \frac{2x-2}{x-2}$; (c) $\lim_{x \to 2} \frac{x-2}{(x-2)^3}$

17. (a) $\lim_{x \to \infty} x\left(\frac{1}{x^2}\right)$; (b) $\lim_{x \to \infty} x\left(\frac{1}{x}\right)$; (c) $\lim_{x \to \infty} x^2\left(\frac{1}{x}\right)$

19. (a) $\lim_{x \to 0+} x^x$; (b) $\lim_{x \to \infty} (e^{-x^2})^{1/x}$; (c) $\lim_{x \to \infty} (e^{-x^2})^{-\frac{1}{x}}$.

21. (a); $\lim_{x \to \infty} x^{1/x}$ (b) $\lim_{x \to \infty} (2^x)^{1/x}$; (c) $\lim_{x \to \infty} x^{1/\ln x}$.

23. Root at $x = -1$, hole at $x = 3$, no asymptotes.

25. Root at $x = -1$, hole at $x = 2$, vertical asymptote at $x = -2$, horizontal asymptote at $y = 1$.

27. Root at $x = -1$, no holes, vertical asymptotes at $x = \pm 2$, horizontal asymptote at $y = 0$.

29. No roots, no holes, no vertical asymptotes, horizontal asymptotes at $y = 0$ on the left and $y = \frac{1}{2}$ on the right.

31. Root at $x = 0$, no holes, no asymptotes.

33. Root at $-\frac{\tan 1}{3} \approx -0.519$, no holes, no vertical asymptotes, horizontal asymptotes at $y = -\frac{\pi}{2} + 1$ on the left and $\frac{\pi}{2} + 1$ on the right.

35. $\lim_{x \to 0^+} -4x^{-3} = -\infty$, $\lim_{x \to 0^-} -4x^{-3} = \infty$

37. 0

39. $-\infty$

41. $-\infty$

43. $-\frac{1}{3}$

45. DNE; $\lim_{x \to 0^-} \frac{x^2+1}{x(x-1)} = \infty$, $\lim_{x \to 0^+} \frac{x^2+1}{x(x-1)} = -\infty$

47. -1

49. -9

51. $-\infty$

53. 0

55. 0

57. 0

59. ∞

61. 0

63. $-\infty$

65. ∞

67. 0

69. $-\infty$ from the left, ∞ from the right

71. 0

73. ∞

75. 1

77. e^2

79. e^3

81. ∞; population explosion impossible for any finite-sized planet to contain.

83. (a) $\lim_{t \to \infty} s(t)$ does not exist, since both sine and cosine oscillate between -1 and 1 as $x \to \infty$. (b) There is no friction to damp the oscillations of the spring. (c) $s(t) = \sqrt{2}\sin\left(\sqrt{\frac{9}{2}}t\right) + 2\cos\left(\sqrt{\frac{9}{2}}t\right)$

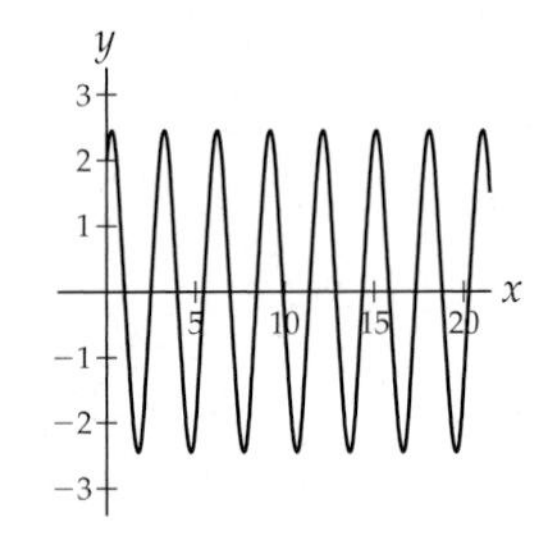

85. $\lim_{x\to\infty}(a_nx^n + a_{n-1}x^{n-1} + \cdots + a_1x + a_0) =$
$\lim_{x\to\infty}\left(a_nx^n\left(1 + \frac{a_{n-1}}{a_nx} + \cdots + \frac{a_1}{a_nx^{n-1}} + \frac{a_0}{a_nx^n}\right)\right) =$
$\left(\lim_{x\to\infty} a_nx^n\right)(1 + 0 + \cdots + 0 + 0) = \lim_{x\to\infty} a_nx^n$.

87. Mimic the proof of the first part of Theorem 1.24, but using $-M < 0$.

89. Given $\epsilon > 0$, choose $N_1 > 0$ to get $f(x)$ within $\frac{\epsilon}{2}$ of L and choose $N_2 > 0$ to get $g(x)$ within $\frac{\epsilon}{2}$ of M. Then for $N = \max(N_1, N_2)$ and $x > N$ we have $(L + M) - \epsilon < f(x) + g(x) < (L + M) + \epsilon$.

91. By Theorem 1.25 and the previous exercise, $\lim_{x\to\infty} x^{-k} = \lim_{x\to\infty}\frac{1}{x^k} \to \frac{1}{\infty} \to 0$.

93. By Theorem 1.25 and the previous exercise, $\lim_{x\to\infty} e^{-x} = \lim_{x\to\infty}\frac{1}{e^x} \to \frac{1}{\infty} \to 0$.

For answers to odd-numbered Skill Certification exercises in the Chapter Review, please visit the Book Companion Web Site at www.whfreeman.com/tkcalculus.

Chapter 2

Section 2.1

1. T, T, F, F, T, T, T, T.

3. We only know one point on the line.

5. d is the difference $s(b) - s(a)$ in position, t is the difference $b - a$ in time, and $r = \frac{d}{t}$.

7. They are all the same.

9. The secant lines have slopes 0.995, 0.9995, and 0.99995, so again 1 is a good estimate for the slope of the tangent line.

11. (a) In order, the x-coordinate of the first dot, the x-coordinate of the second dot, the y-coordinate of the first dot, and the y-coordinate of the second dot. (b) The horizontal and vertical distances, respectively, between the first and second dot. (c) The slope of the line connecting the two dots, and the slope of the line tangent to the graph at the first dot.

13. The answers are the same as those in Exercise 11.

15. (c) < (d) < (a) < (b)

17. $x = 0.5$; $x = -1$ (or $x = 2$); $x = -0.4$ and $x = 1.3$

19.

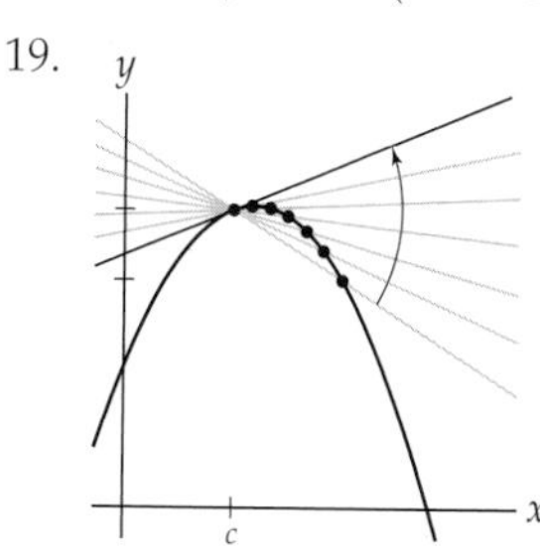

21. At $x = 1$ your graph should have a height of 2 and a horizontal tangent line. At $x = 3$ your graph should be drawn so that its tangent line has a slope of 2.

23. Your graph should go through the points $(-1, 2)$ and $(1, -2)$, and the tangent line at each of those points should have slope 3.

25.

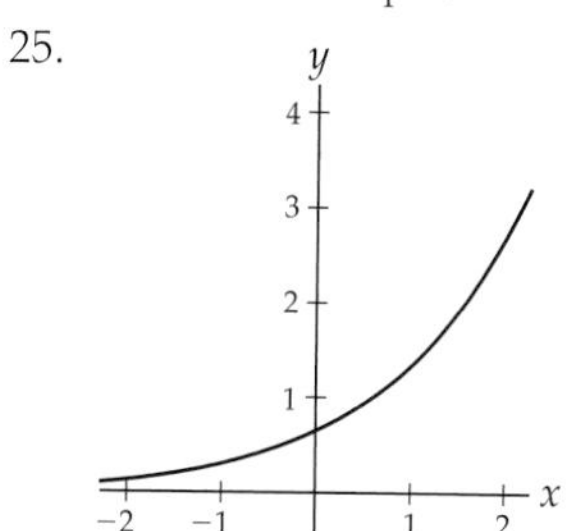

27. Your graph of $y = f'(x)$ should have zeroes at $x = -1.5$, $x = 0$, and $x = 1.5$, should be positive on $(-\infty, -1.5)$ and $(0, 1.5)$, and should be negative on $(-1.5, 0)$ and $(1.5, \infty)$.

29. Clearly the graph of f' should have a root at $x = 0$ and should be positive on $(-\infty, 0)$ and negative on $(0, \infty)$. There is one more thing to notice here, however; the graph of f gets very "flat" at its ends, so its tangent lines have very shallow slopes as $x \to \infty$ and as $x \to -\infty$. Therefore, the graph of the derivative should be approaching zero as $x \to \infty$ and as $x \to -\infty$. Thus your graph of f' should also have a two-sided horizontal asymptote at $y = 0$

31. Your graph should have horizontal tangent lines at $x = -2$ and at $x = 0$. Moreover, the tangent lines to the graph of your function should have a negative slope everywhere on $(-\infty, -2)$, a positive slope everywhere on $(-2, 0)$, and a negative slope everywhere on $(0, \infty)$.

33.

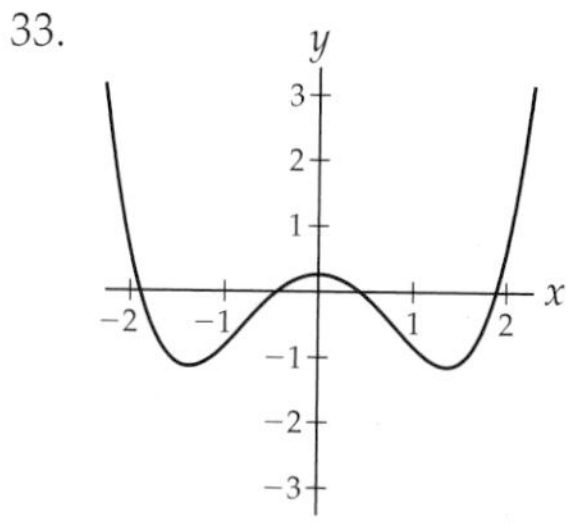

35. For $h = 1$, $h = 0.5$, $h = 0.25$, and $h = 0.1$, respectively, we have secant lines with slopes -3, -2.5, -2.25, and -2.1. One estimate of $f'(1)$ is -2.

37. For $h = 1$, $h = 0.5$, $h = 0.25$, and $h = 0.1$, respectively, we have secant lines with slopes 2, 1.25, 1.0625, and 1.01. One estimate of $f'(0)$ is 1.

39. For $h = 1$, $h = 0.5$, $h = 0.25$, and $h = 0.1$, respectively, we have secant lines with slopes 0.6931, 0.4463, 0.2425, and 0.0995. One estimate of $f'(0)$ is 0.

41. For $h = 1$, $h = 0.5$, $h = 0.25$, and $h = 0.1$, respectively, we have secant lines with slopes -0.4597, -0.2448, -0.1244, and -0.04996. One estimate of $f'\left(\frac{\pi}{2}\right)$ is 0.

43. For $h = 1$, $h = 0.5$, $h = 0.25$, and $h = 0.1$, respectively, we have secant lines with slopes 1, 1, 1, and 1. One estimate of $f'(3)$ is 1.

45. Average velocities on $[0, 0.5]$, $[0, 0.25]$, $[0.0.1]$ are -8, -4, and -1.6 feet per second, respectively. We would expect the initial velocity to be 0.

47. Average velocities on $[2, 2.1]$ and $[2, 2.01]$ are -65.6 and -64.16 feet per second. Average velocities on $[1.9, 2]$ and $[1.99, 2]$ are -62.4 and -63.84 feet per second, respectively. We might expect the velocity at time $t = 2$ to be about 64 feet per second.

49. Although this book does know what you did today, we will not repeat it here.

51. (a) Her average rate of change is zero on $[0, 30]$; this means that after 30 minutes she is exactly as far away from the oak tree as she was at time $t = 0$. (b) The second 10 minutes. (c) One approximation might be $\frac{140-280}{10-5} = -28$ feet per minute. The fact that the sign is negative indicates that she is moving *toward* the oak tree. (d) $t \approx 4$ minutes, $t \approx 12$ minutes, $t \approx 25$ minutes. These are the times she changes direction. (e) $(4, 12)$ and $(25, 30)$. These are the times she is moving toward the oak tree.

53. (a) $h(12)$ is the average height of a 12-year-old, in feet. $h'(12)$ is measured in feet per year and represents the instantaneous rate of change of the height of a 12-year-old person; that is, $h'(12)$ is the rate at which an average 12-year-old is growing (in feet per year). (b) $h(12)$ is positive; $h'(12)$ is positive. (c) $h(t)$ might have a maximum at $t \approx 60$ years (people may tend to slouch or get shorter as they get older); $h'(t)$ might have a maximum at $t \approx 14$ years (growth spurt).

55. 5 miles per hour

57. (a) About 139.163 billion dollars; (b) about 3.445 billion dollars per year; (c) about 3.479 billion dollars per year; (d) something a little bit less than the previous answer (the actual answer turns out to be about 3.436 billion dollars a year).

59. If $f(x) = mx + c$ with $m > 0$, then on any interval $[a, b]$ the average rate of change of f is $\frac{f(b)-f(a)}{b-a} = \frac{(mb+c)-(ma+c)}{b-a} = m$, which is positive by assumption. This means that for all $b > a$ we have $f(b) = f(a) + m(b - a)$, and thus $f(b) > f(a)$; this the definition of what it means for f to be increasing.

Section 2.2

1. F, F, F, F, F, F, F, T.

3. (a) $f'(5) = \lim_{h\to 0} \frac{f(5+h)-f(5)}{h} = \lim_{z\to 5} \frac{f(z)-f(5)}{z-5}$; (b) $f'(x) = \lim_{h\to 0} \frac{f(x+h)-f(x)}{h} = \lim_{z\to x} \frac{f(z)-f(x)}{z-x}$; (c) $f'_+(-2) = \lim_{h\to 0^+} \frac{f(-2+h)-f(-2)}{h} = \lim_{z\to -2^+} \frac{f(z)-f(-2)}{z+2}$.

5. Let $z = x + h$. Then $h \to 0$ is equivalent to $z \to x$, and $z - x = h$.

7. $\lim_{x\to 2}(4x^3 - 5x + 1) = 23$ and $\lim_{h\to 0} \frac{(4(2+h)^3-5(2+h)+1)-23}{h}$ exists.

9. f is differentiable, and thus continuous, at $x = c$.

11. Continuous yes, differentiable no, left and right differentiable yes.

13. See the graphs in the reading in the discussion about differentiability.

15. (a) If $f(x) = 3x^2 - 1$ then $f'(-4) = -24$. (b) If $f(x) = 3x^2 - 1$ then $\left.\frac{df}{dx}\right|_{-4} = -24$. (c) $\left.\frac{d}{dx}\right|_{-4}(3x^2 - 1) = -24$.

17. $3x^2 - 2$, $3x^2 - 2$, $3x^2 - 2$, and $(3x^2 - 2)(x^3 - 2x + 1)$; the last notation represents a product of the derivative and the function.

19. For all $\epsilon > 0$, there exists $\delta > 0$ such that if $h \in (-\delta, 0) \cup (0, \delta)$, then $\frac{(3+h)^2-9}{h} \in (f'(3) - \epsilon, f'(3) + \epsilon)$.

21. Since $f'(4) = \frac{1}{2\sqrt{4}} = \frac{1}{4}$, the tangent line to f at $x = 4$ has equation $y = f(4) + f'(4)(x - 4) = 2 + \frac{1}{4}(x - 4)$. The value of $f(x)$ at $x = 4.1$ should be close to the value of this linear function at $x = 4.1$, which is $2 + \frac{1}{4}(4.1 - 4) = 2.025$. This is rather close to the calculator approximation of $\sqrt{4.1} = 2.02485$.

23. $\lim_{h\to 0} \frac{(-3+h)^2-(-3)^2}{h} = -6$; $\lim_{z\to -3} \frac{z^2-9}{z-3} = -6$

25. $\lim_{h\to 0} \frac{\frac{1}{-1+h}+1}{h} = -1$; $\lim_{z\to -1} \frac{\frac{1}{z}+1}{z+1} = -1$

27. $\lim_{h\to 0} \frac{(1-(-1+h)^3)-(1-(-1)^3)}{h} = -3$; $\lim_{z\to -1} \frac{(1-z^3)-2}{z+1} = -3$

29. $\lim_{h\to 0} \frac{\sqrt{9+h}-3}{h} = \frac{1}{6}$; $\lim_{z\to 9} \frac{\sqrt{z}-3}{z-9} = \frac{1}{6}$

31. $\lim_{h\to 0} \frac{\frac{(2+h)-1}{(2+h)+3} - \frac{2-1}{2+3}}{h} = \frac{4}{25}$; $\lim_{z\to 2} \frac{\frac{z-1}{z+3} - \frac{1}{5}}{z-2} = \frac{4}{25}$

33. $\lim_{h\to 0} \frac{e^h-1}{h} = 1$; $\lim_{z\to 0} \frac{e^{z-1}}{z-0} = 1$

35. $\lim_{h\to 0} \frac{\sin h-0}{h} = 1$; $\lim_{z\to 0} \frac{\sin z-0}{z-0} = 1$

37. $\lim_{h\to 0} \frac{\tan h-0}{h} = 1$; $\lim_{z\to 0} \frac{\tan z-0}{z-0} = 1$

39. $\lim_{h\to 0} \frac{(-2(x+h)^2)-(-2x^2)}{h} = -4x$

41. $\lim_{h\to 0} \frac{((x+h)^3+2)-(x^3+2)}{h} = 3x^2$

43. $\lim_{z\to x} \frac{\frac{2}{z+1} - \frac{2}{x+1}}{z-x} = -2\frac{1}{(x+1)^2}$

45. $\lim_{z\to x} \frac{\frac{1}{z^2} - \frac{1}{x^2}}{z-x} = \frac{-2}{x^3}$

47. $\lim_{h\to 0} \frac{3\sqrt{x+h}-3\sqrt{x}}{h} = \frac{3}{2\sqrt{x}}$

49. $\lim_{z\to x} \frac{\sqrt{2z+1}-\sqrt{2x+1}}{z-x} = \frac{1}{\sqrt{1+2x}}$

51. $\lim_{h\to 0} \frac{\frac{(x+h)-1}{(x+h)+3} - \frac{x-1}{x+3}}{h} = \frac{4}{(x+3)^2}$

53. $\lim_{h\to 0} \frac{\frac{(x+h)^3}{x+h+1} - \frac{x^3}{x+1}}{h} = \frac{2x^3+3x^2}{(x+1)^2}$

55. $6x^2$, $12x$, and 12.

57. $6x$ and 12

59. $y = -6x - 9$

61. $y + 1 = -3(x - 1)$

63. $y - 2 = -(x - 3)$

65. Continuous everywhere except at $x = 2$, and neither left nor right continuous at that point. Not differentiable, but left and right differentiable, at $x = -1$. Not differentiable, and not left or right differentiable, at $x = 2$.

67. Continuous everywhere except at $x = -1$, where it is only left continuous. Not differentiable at $x = -1$. Left but not right differentiable at $x = -1$.

69. $\lim_{x\to 0} \frac{1}{x}$ does not exist, and neither does $\lim_{h\to 0} \frac{\frac{1}{0+h} - \frac{1}{0}}{h}$ since $\frac{1}{0}$ is undefined. Therefore f is neither continuous nor differentiable at $x = 0$, from either side.

71. $\lim_{h\to 0^-} \frac{|(2+h)^2-4|-0}{h} = -4$, $\lim_{h\to 0^+} \frac{|(2+h)^2-4|-0}{h} = 4$, so f is both left and right differentiable at $x = 2$, but not differentiable at $x = 2$.

73. $f'_-(2) = \lim_{h\to 0^-} \frac{((2+h)+4)-3(2)}{h} = 1$;
$f'_+(2) = \lim_{h\to 0^+} \frac{3(2+h)-3(2)}{h} = 3$; $f'(2)$ does not exist.

75. $f'_-(1) = \lim_{h\to 0^-} \frac{(1+h)^2-(1)^2}{h} = 2$;
$f'_+(1) = \lim_{h\to 0^+} \frac{(2(1+h)-1)-(1)^2}{h} = 2$; $f'(1) = 2$.

77. Continuous but not differentiable at $x = 0$.

79. Not continuous and not differentiable at $x = 1$.

81. $f(1) = -4$ is less than zero and $f(3) = 4$ is greater than zero. Since f is continuous on $[1, 3]$, the Intermediate Value Theorem tells us that f must have a root between $x = 1$ and $x = 3$.

83. $f(0) = 1$ is greater than zero and $f(1) = -1$ is less than zero. Since f is continuous on $[0, 1]$, the Intermediate Value Theorem tells us that f must have a root between $x = 0$ and $x = 1$.

85. $f(-2) = -7$ is less than zero and $f(1) = 2$ is greater than zero. Since f is continuous on $[-2, 1]$, the Intermediate Value Theorem tells us that f must have a root between $x = -2$ and $x = 1$.

87. $\lim_{h\to 0} \frac{s(1+h)-s(1)}{h} = -32$.

89. (a) The graph has a bump/hill shape with roots at $x = 0$ and at $x = 2$. The physical interpretation is that your velocity is always positive and that you start out with a zero velocity, speed up, and then eventually slow back down to a zero velocity. (b) The graph has a root at about $x = 1$ and is positive on $[0, 1)$ and negative on $(1, 2]$ (for example, a line with negative slope passing through $(1, 0)$). This represents the fact that you were speeding up, and then slowing down, during your trip.

91. (a) $S(3) = 200 + 8(10)(3) = \$440$, $S(6) = 200 + 8(10)(6) = \$680$, and $S(8) = 200 + 8(10)(6) + 11.5(10)(2) = \$910$. $S'(3) = 80$ dollars per week, $S'(8) = 11.5(10) = \$115$ dollars per week. Cannot compute $S'(6)$ because the rate at which you are paid changes at that point.
(b) $S(t) = 200 + 8(10)t$, if $t \leq 6$ and $S(t) = 680 + 11.5(10)(t - 6)$, if $t > 6$. (c) Continuous but not differentiable. (d) $\lim_{t\to 6} s(t) = s(6) = 680$, but $\lim_{h\to 0+} \frac{s(6+h)-s(6)}{h} = 115$ while $\lim_{h\to 0^-} \frac{s(6+h)-s(6)}{h} = 80$.

93. $\lim_{h\to 0} \frac{(m(x+h)+b)-(mx+b)}{h} = \lim_{h\to 0} \frac{mh}{h} = \lim_{h\to 0} m = m$.

95. The tangent line to f at $x = c$ has slope $f'(c)$ and passes through the point $(c, f(c))$. Putting this slope and point into the point-slope form of a line gives the desired answer.

97. $f'(c)$ exists when $\lim_{h\to 0} \frac{f(c+h)-f(c)}{h}$ exists, which is true precisely when $f'_+(c) = \lim_{h\to 0^+} \frac{f(c+h)-f(c)}{h}$ and $f'_-(c) = \lim_{h\to 0^-} \frac{f(c+h)-f(c)}{h}$ exist and are equal.

Section 2.3

1. T, F, T, F, F, T, T, T.

3. $\frac{d}{dx}(kf(x)) = k\frac{df}{dx}$; $\frac{d}{dx}(f(x)+g(x)) = \frac{df}{dx} + \frac{dg}{dx}$; $\frac{d}{dx}(f(x)-g(x)) = \frac{df}{dx} - \frac{dg}{dx}$

5. See the first paragraph in the reading.

7. The power rule only applies to functions of the form x^k, and 3^x cannot be written in this form because there is a variable in the exponent.

9. The rule is $f'(x)g(x)h(x) + f(x)g'(x)h(x) + f(x)g(x)h'(x)$. Then $y' = (2)(x^2+x+1)(1-3x^4) + (2x-1)(2x+1)(1-3x^4) + (2x-1)(x^2+x+1)(-12x^3)$.

11. 5

13. 8

15. f has negative slope for $x < 0$ and positive slope for $x > 0$, so $f'(x)$ is negative on $x < 0$ and positive on $x > 0$. Then f' has positive slope everywhere, so $f''(x)$ is positive everywhere. Continue this type of argument for the remaining graphs.

17. As you can see, this algebra is no fun. If you remember to simplify functions before differentiating then you will not have this problem.

19. No; we can say f is continuous at $x=2$, but it might approach that breakpoint $x=2$ with different slopes from the left and the right.

21. $a=-2, b=6$

23. 4

25. 1

27. $\frac{27}{2}$

29. (a) $f'(x)=(2x)(x+1)+(x^2)(1)$. (b) Multiply out first to get $f(x)=x^3+x^2$; then $f'(x)=3x^2+2x$. (c) Multiply out the answer in part (a) to get the answer in part (b).

31. (a) $f'(x)=\frac{7}{2}x^{5/2}(2-5x^3)+x^{7/2}(-15x^2)$. (b) Multiply out first to get $f(x)=2x^{7/2}-5x^{13/2}$; then $f'(x)=7x^{5/2}-\frac{65}{2}x^{11/2}$. (c) Multiply out the answer in part (a) to get the answer in part (b).

33. (a) $f'(x)=\frac{(2x-3x^2)\sqrt{x}-(x^2-x^3)(1/2)x^{-1/2}}{x}$. (b) Expand first to get $f(x)=x^{1.5}-x^{2.5}$; then $f'(x)=1.5x^{0.5}-2.5x^{1.5}$. (c) Multiply out the answer in part (a) to get the answer in part (b).

35. $f'(x)=-3(7x^6)$

37. $f'(x)=12x$

39. $f'(x)=2(3)-4(5x^4)$

41. $f'(x)=2x+1$

43. $f'(x)=81x^2+108x+36$

45. $f'(x)=-6x^2$

47. $f'(x)=\frac{d}{dx}(x-1)=1$, for $x\neq -1$

49. $f'(x)=0$

51. $f'(x)=\frac{8}{5}x^{3/5}$

53. $f'(x)=-\frac{8}{5}x^{-13/5}-2$

55. $f'(x)=\frac{(7x^6-15x^4)(1-3x^4)-(x^7-3x^5+4)(-12x^3)}{(1-3x^4)^2}$

57. $f'(x)=-3x^{-5/2}$

59. $f'(x)=\frac{23}{(4+5x)^2}$

61. It is easier to differentiate after simplifying to $f(x)=\frac{1}{x^2-5x+6}$. We have $f'(x)=\frac{0(x^2-5x+6)-(1)(2x-5)}{(x^2-5x+6)^2}=\frac{-2x+5}{(x^2-5x+6)^2}$.

63. $f'(x)=\frac{2x(x^3+5x^2-3x)-x^2(3x^2+10x-3)}{(x^3+5x^2-3x)^2}$

65. $f'(x)=\begin{cases}-1, & \text{if } x<0\\ \text{DNE}, & \text{if } x=0\\ 1, & \text{if } x>0.\end{cases}$

67. $f'(x)=\begin{cases}-2, & \text{if } x<\frac{1}{2}\\ \text{DNE}, & \text{if } x=\frac{1}{2}\\ 2, & \text{if } x>\frac{1}{2}.\end{cases}$

69. $f'(x)=\begin{cases}3x^2, & \text{if } x<1\\ \text{DNE}, & \text{if } x=1\\ 1, & \text{if } x>1.\end{cases}$

71. $f'(x)=\begin{cases}-2x, & \text{if } x\leq 0\\ 2x, & \text{if } x>0.\end{cases}$

73. $f(x)=\frac{1}{2}x^6-\frac{2}{3}x^3+4x+1$

75. $f(x)=-\frac{4}{7}x^7+x+\frac{18}{7}$

77. $f(x)=-\frac{3}{10}x^{10}+4x^6+\frac{1}{5}x^5-8x+2$

79. (a) $a(t)=0.024t$, $s(t)=0.004t^3+400t$ (assuming initial position is $s_0=0$ at Venus). $a_0=0$ thousands of miles per hour per hour, $v_0=400$ miles per hour, $s_0=0$ miles. (b) Yes, since the velocity is always positive. (c) No, the acceleration is increasing (since $a(t)=0.024t$ grows larger as t increases). (d) Approximately 140 hours, at which time it will be going about 635.2 thousand miles per hour.

81. If $a(t)=-32$ then (by antidifferentiating) $v(t)=-32t+C$, and $v(0)=v_0$ implies that $C=v_0$. Now (by antidifferentiating) $s(t)=-16t^2+v_0t+K$, and $s(0)=s_0$ implies that $K=s_0$.

83. (a) $s(t)=-\frac{2}{5}t^3-17t^2+575$; (b) $s(5)=100$, $s'(5)=-200$, $s''(5)=-46$, $s'(0)=0$; (c) 575 feet; (d) $a(t)=-\frac{12}{5}t-34$ is *not* constant.

85. Suppose k is a positive integer. $\frac{d}{dx}(x^k)=\lim_{h\to 0}\frac{(x+h)^k-x^k}{h}$

$$=\lim_{h\to 0}\frac{(x^k+khx^{k-1}+h^2\cdot(\text{lower-order terms}))-x^k}{h}$$
$$=\lim_{h\to 0}\frac{khx^{k-1}+h^2\cdot(\text{lower-order terms})}{h}$$
$$=\lim_{h\to 0}(kx^{k-1}+h\cdot(\text{lower-order terms}))=kx^{k-1}.$$

87. $(f-g)'(x)=\frac{d}{dx}(f+(-g))$
$$=\frac{d}{dx}(f)+\frac{d}{dx}(-g)=\frac{d}{dx}(f)+(-1)\frac{d}{dx}(g)f'(x)-g'(x).$$

89. $\left(\frac{f}{g}\right)'(x)=\lim_{h\to 0}\frac{\frac{f(x+h)}{g(x+h)}-\frac{f(x)}{g(x)}}{h}$

$$=\lim_{h\to 0}\frac{\left(\frac{f(x+h)g(x)-f(x)g(x+h)}{g(x+h)g(x)}\right)}{h}$$
$$=\lim_{h\to 0}\frac{f(x+h)g(x)-f(x)g(x+h)}{hg(x+h)g(x)}$$
$$=\lim_{h\to 0}\frac{f(x+h)g(x)-f(x)g(x)+f(x)g(x)-f(x)g(x+h)}{hg(x+h)g(x)}$$
$$=\lim_{h\to 0}\frac{g(x)(f(x+h)-f(x))-f(x)(g(x+h)-g(x))}{hg(x+h)g(x)}$$
$$=\lim_{h\to 0}\frac{\left(g(x)\frac{f(x+h)-f(x)}{h}-f(x)\frac{g(x+h)-g(x)}{h}\right)}{g(x+h)g(x)}$$
$$=\frac{g(x)\left(\lim_{h\to 0}\frac{f(x+h)-f(x)}{h}\right)-f(x)\left(\lim_{h\to 0}\frac{g(x+h)-g(x)}{h}\right)}{\lim_{h\to 0}(g(x+h)g(x))}$$
$$=\frac{g(x)f'(x)-f(x)g'(x)}{\lim_{h\to 0}(g(x+h)g(x))}=\frac{f'(x)g(x)-f(x)g'(x)}{(g(x))^2}.$$

91. (a) Since any two functions with the same derivative differ by a constant, it suffices to prove that the derivative of $\frac{1}{k+1}x^{k+1}$ is x^k. By the power rule and the constant multiple rule we have $\frac{d}{dx}\left(\frac{1}{k+1}x^{k+1}\right) = \frac{1}{k+1}\frac{d}{dx}(x^{k+1}) = \frac{1}{k+1}(k+1)x^{(k+1)-1} = x^k$. (b) If $k=-1$ then there will be division by zero in the calculation from part (a). Therefore, this proof does not work in the case where $f(x) = x^{-1}$.

Section 2.4

1. T, F, F, F, T, T, T, F.

3. (a) $(g(h(x)))' = g'(h(x))h'(x)$ and (b) $\frac{dg}{dx} = \frac{dg}{dh}\frac{dh}{dx}$.

5. (a) $(f(u(v(w(x)))))' = f'(u(v(w(x))))u'(v(w(x)))v'(w(x))w'(x)$; (b) $\frac{df}{dx} = \frac{df}{du}\frac{du}{dv}\frac{dv}{dw}\frac{dw}{dx}$

7. (a) $\frac{d}{dx}((3x+\sqrt{x})^2) = 2(3x+\sqrt{x})\left(3+\frac{1}{2}x^{-1/2}\right)$. (b) $\frac{d}{dx}((3x+\sqrt{x})(3x+\sqrt{x})) = \left(3+\frac{1}{2}x^{-1/2}\right)(3x+\sqrt{x}) + (3x+\sqrt{x})\left(3+\frac{1}{2}x^{-1/2}\right)$. (c) $\frac{d}{dx}(9x^2+6x^{3/2}+x) = 18x+9x^{1/2}+1$.

9. -2

11. 6

13. 4

15. -12

17. If $y=y(x)$ then $\frac{d}{dx}(y^3) = \frac{d}{dx}((y(x))^3) = 3(y(x))^2y'(x)$.

19. $(x+1)(y^2+y-1) = 1$ is shown in the right graph, and $xy^2+y=1$ is shown in the left graph.

21. $f'(x) = -(x^3+1)^{-2}(3x^2)$

23. $f'(x) = (3x^2+1)^9 + 54x^2(3x^2+1)^8$

25. $f'(x) = -\frac{1}{2}(x^2+1)^{-3/2}(2x)$

27. $f'(x) = -2(x\sqrt{x+1})^{-3}\left(\sqrt{x+1}+x\left(\frac{1}{2}(x+1)^{-1/2}\right)\right)$

29. $f'(x) = \frac{(-x^{-2}-6x)(x^5-x^{-1/2})-(x^{-1}-3x^2)(5x^4+(1/2)x^{-3/2})}{(x^5-x^{-1/2})^2}$

31. $-1(x^{1/3}-2x)^{-2}\left(\frac{1}{3}x^{-2/3}-2\right)$

33. $-\frac{1}{2}x^{-3/2}(x^2-1)^3 + x^{-1/2}3(x^2-1)^2(2x)$

35. $f'(x) = \frac{1}{2}(3x-4(2x+1)^6)^{-1/2}(3-24(2x+1)^5(2))$

37. $f'(x) = 100(5(3x^4-1)^3+3x-1)^{99}(15(3x^4-1)^2(12x^3)+3)$

39. $f'(x) = 3\left(-\frac{2}{3}\right)((x^2+1)^8-7x)^{-5/3}(8(x^2+1)^7(2x)-7)$

41. $f'(x) = 7(5x^4-3x^2)^6(20x^3-6x)(2x^3+1) + (5x^4-3x^2)^7(6x^2)$

43. $-9((2x+1)^{-10}-1)^{-10}(-10(2x+1)^{-6})$

45. $f'(x) = 8(x^4-\sqrt{3-4x})^7\left(4x^3-\frac{1}{2}(3-4x)^{-1/2}(-4)\right)+5$

47. $\frac{(-3)(x^5+2x^4+x^3)-(-2-3x)(5x^4+8x^3+3x^2)}{(x^5+2x^4+x^3)^2}$

49. 0

51. 162

53. $3s^2s'$

55. 0

57. s^2+2rss'

59. Combine the graphs of $y=\sqrt{9-\frac{1}{4}x^2}$ and $y=-\sqrt{9-\frac{1}{4}x^2}$.

61. Combine the graphs of $y=\sqrt{\frac{1}{3}x^2-\frac{16}{3}}$ and $y=-\sqrt{\frac{1}{3}x^2-\frac{16}{3}}$.

63. $f(x) = (x^2+1)^5$

65. $f(x) = \frac{1}{10}(x^2+1)^5 - \frac{1}{10}$

67. $f(x) = \frac{2}{9}(3x+1)^{3/2} + \frac{7}{9}$

69. $\frac{dy}{dx} = \frac{4x}{y}$

71. $\frac{dy}{dx} = \frac{-6x-y^2}{2xy}$

73. $\frac{dy}{dx} = \frac{-3(y^2-y+6)}{(3x+1)(2y-1)}$

75. $\frac{dy}{dx} = \frac{5y}{\frac{3}{2}(3y-1)^{-1/2}-5x}$

77. $\frac{dy}{dx} = \frac{(3y-1)^2}{2y(3y-1)-3(y^2+1)}$

79. $\frac{dy}{dx} = \frac{2x^3y^2(y+1)-y^2(y+1)^2}{x^4y^2-x^2(y+1)^2}$

81. (a) Slope is $-\frac{1}{\sqrt{3}}$ at the point $\left(\frac{1}{2},\frac{\sqrt{3}}{2}\right)$ and $\frac{1}{\sqrt{3}}$ at the point $\left(\frac{1}{2},-\frac{\sqrt{3}}{2}\right)$. (b) Slope is -1 at the point $\left(\frac{\sqrt{2}}{2},\frac{\sqrt{2}}{2}\right)$, and 1 at the point $\left(-\frac{\sqrt{2}}{2},\frac{\sqrt{2}}{2}\right)$. (c) $(1,0)$ and $(-1,0)$. (d) $\left(\frac{\sqrt{2}}{2},\frac{\sqrt{2}}{2}\right)$ and $\left(-\frac{\sqrt{2}}{2},-\frac{\sqrt{2}}{2}\right)$

83. (a) $(1,-1)$; (b) $(-3,1)$; (c) none; (d) $(-3,1)$

85. If Linda sells magazines at a rate of $\frac{dm}{dt} = 12$ magazine subscriptions per week, and she makes $\frac{dD}{dm} = 4$ dollars per magazine, then by the chain rule, the amount of money she makes each week is $\frac{dD}{dt} = \frac{dD}{dm}\frac{dm}{dt} = 4(12) = 48$ dollars per week.

87. (a) $\frac{dA}{dr} = 2\pi r$. (b) No; Yes. (c) $\frac{dA}{dt} = \frac{d}{dt}(\pi(r(t))^2) = 2\pi r(t)r'(t) = 2\pi r\frac{dr}{dt}$. (d) Yes; Yes. (e) $\left.\frac{dA}{dt}\right|_{r=24} = 2\pi(24)(2) = 96\pi$.

89.
$$\frac{d}{dx}\left(\frac{f(x)}{g(x)}\right) = \frac{d}{dx}(f(x)(g(x))^{-1})$$
$$= \frac{d}{dx}(f(x))\cdot(g(x))^{-1} + f(x)\cdot\frac{d}{dx}((g(x))^{-1})$$
$$= f'(x)\cdot(g(x))^{-1} + f(x)\cdot(-g(x))^{-2}g'(x)$$
$$= \frac{f'(x)}{g(x)} - \frac{f(x)g'(x)}{(g(x))^2} = \frac{f'(x)g(x)-f(x)g'(x)}{(g(x))^2}.$$

91. Mimic the proof in Example 5(a).

93. Let $y = x^{-k}$, where $-k$ is a negative integer. Then $x^k y = 1$, so by implicit differentiation and the product rule, we have $kx^{k-1}y + x^k y' = 0$, and therefore $y' = \frac{-kx^{k-1}x^{-k}}{x^k} = -kx^{-k-1}$.

Section 2.5

1. T, T, F, F, T, T, F, T.

3. No, since the base is a variable. No, since the exponent is a variable.

5. When $k = 1$ we have $\frac{d}{dx}(e^x) = \frac{d}{dx}(e^{1x}) = 1e^{1x} = e^x$. When $b = e$ we have $\frac{d}{dx}(b^x) = \frac{d}{dx}(e^x) = (\ln e)e^x = (1)e^x = e^x$.

7. $\ln x$ is of the form $\log_b x$ with $b = e$.

9. The graph passing through $(0, 2)$ is $2(2^x)$; the graph passing through $(0, 1)$ and $(1, 4)$ is 4^x; the graph passing through $(2, 4)$ is 2^x.

11. $f(3) = 2^9 = 512$ and $g(3) = 8^2 = 64$. The solutions of $2^{(x^2)} = (2^x)^2$ are $x = 0$ and $x = 2$.

13. Consider that logarithmic functions are the inverses of exponential functions, and use the definition of one-to-one.

15. Logarithmic differentiation is the process of applying $\ln|x|$ to both sides of an equation $y = f(x)$ and then differentiating both sides in order to solve for $f'(x)$. It is useful for finding derivatives of functions that involve multiple products or quotients, and functions that have variables in both a base and an exponent.

17. $f'(x) = -(2 - e^{5x})^{-2}(-5e^{5x})$

19. $f'(x) = 6xe^{-4x} - 12x^2e^{-4x}$

21. $f'(x) = \frac{(-1)e^x - (1-x)e^x}{e^{2x}}$

23. $e^x(x^2 + 3x - 1) + e^x(2x + 3)$

25. $f'(x) = 3x^2$, with $x > 0$

27. $f'(x) = e^{(e^x)}e^x$

29. $f'(x) = e^{ex+1}$

31. $f'(x) = 0$

33. $f'(x) = -3x^{-4}e^{2x} + x^{-3}(2e^{2x})$

35. $f'(x) = 2x\log_2 x + \frac{x}{\ln 2} + 3x^2$

37. $f'(x) = \frac{1}{x^2+e^{\sqrt{x}}}\left(2x + e^{\sqrt{x}}\left(\frac{1}{2}x^{-1/2}\right)\right)$

39. $f'(x) = \frac{1}{2}(\log_2(3x-5))^{-1/2}\left(\frac{1}{\ln 2}\frac{1}{3^x-5}(\ln 3)3^x\right)$

41. $f'(x) = 2x\ln(\ln x) + x^2\frac{1}{\ln x}\cdot\frac{1}{x}$

43. $f'(x) = (\ln 2)2^{x^2}(2x)$

45. $f'(x) = \begin{cases} (\ln 2)2^x, & \text{if } x < -2 \\ \text{DNE}, & \text{if } x = -2 \\ -\frac{2}{x^3}, & \text{if } x > -2. \end{cases}$

47. $f'(x) = \begin{cases} 2x, & \text{if } x < 1 \\ \text{DNE}, & \text{if } x = 1 \\ \frac{1}{x}, & \text{if } x \geq 1. \end{cases}$

49. $f'(x) = \frac{1}{2}(x\ln|2^x+1|)^{-1/2}\left(\ln|2^x+1| + \frac{(\ln 2)x2^x}{2^x+1}\right)$

51. $f'(x) = \frac{2^x\sqrt{x^3-1}}{\sqrt{x}(2x-1)}\left(\ln 2 + \frac{3x^2}{2(x^3-1)} - \frac{1}{2x} - \frac{2}{2x-1}\right)$

53. $f'(x) = x^{\ln x}(2)(\ln x)\left(\frac{1}{x}\right)$

55. $f'(x) = \left(\frac{x}{x-1}\right)^x\left(\ln x - \ln(x-1) + 1 - \frac{x}{x-1}\right)$

57. $f'(x) = (\ln x)^{\ln x}\left(\frac{\ln(\ln x)}{x} + \frac{1}{x}\right)$

59. $f(x) = \frac{e^{4x}}{3x^5-1}$

61. $f(x) = \frac{1}{2}\ln|x^2+3| + C$

63. $f(x) = \ln(1 + e^x) + C$

65. (a) $A(t) = 1000e^{0.077t}$; (b) $A(30) \approx \$10{,}074.42$; (c) $t \approx 18$ years.

67. (a) 1250 people; (b) 45,000 people; (c) about 119 days; (d) $P'(t) = \frac{45{,}000(0.12)(35)e^{-0.12t}}{(1+35e^{-0.12t})^2}$, which is always positive; this means that the graph of $P(t)$ is always moving up as we move from left to right.

69. $\frac{d}{dx}(e^{kx}) = e^{kx}\cdot\frac{d}{dx}(kx) = e^{kx}\cdot k = ke^{kx}$

71. If $f(x) = Ae^{kx}$ is exponential, then $f'(x) = Ake^{kx} = k(Ae^{kx}) = kf(x)$, so $f'(x)$ is proportional to $f(x)$.

73. For $x > 0$ we have $|x| = x$ and therefore $\frac{d}{dx}(\ln|x|) = \frac{d}{dx}(\ln x) = \frac{1}{x}$. For $x < 0$ we have $|x| = -x$ and therefore by the chain rule we have $\frac{d}{dx}(\ln|x|) = \frac{d}{dx}(\ln(-x)) = \frac{1}{-x}(-1) = \frac{1}{x}$.

Section 2.6

1. T, F, F, T, F, T, F, T.

3. See the proof in the reading.

5. (a) If x is in degrees, then the slope of the graph of $\sin x$ at $x = 0$ is very small, and in particular not equal to $\cos 0 = 1$. To convince yourself of this, graph $\sin x$ (in degrees) together with the line $y = x$ (which has slope 1 at $x = 0$).

7. (a) $\cos(3x^2)$ is a composition, not a product, but the product rule was applied; (b) the chain rule was applied incorrectly, with the derivative of $3x^2$ written on the inside instead of the outside.

9. No, because to differentiate $\sin^{-1}(\sin x) = x$ we would first need to know how to differentiate $\sin^{-1} x$, which is exactly what we would be trying to prove.

11. The expressions are equal; the algebraic expression is clearly easier to evaluate at a given number.

13. (a) Use the fact that $\frac{1}{2}e^{-x} \geq 0$ for all x, and split the expressions in the definitions of $\cosh x$ and $\sinh x$ into sums. (b) Calculate the two limits by dividing top and bottom by e^x and show they are both equal to 1.

15. (a) Think about the graph of the sum of $\frac{1}{2}e^x$ and $-\frac{1}{2}e^{-x}$.
(b) Calculate $\lim_{x\to-\infty} \dfrac{\frac{1}{2}(e^x - e^{-x})}{\frac{1}{2}e^{-x}}$ by dividing top and bottom by e^{-x}, and Show that this limit is equal to 1.

17. $f'(x) = \dfrac{2x\cos x + (x^2+1)\sin x}{\cos^2 x}$

19. $f'(x) = -\csc^2 x + \csc x \cot x$

21. $f'(x) = 0$

23. $f'(x) = 3\sec x \sec^2 x + 3\sec x \tan^2 x$

25. $f'(x) = \cos(\cos(\sec x))(-\sin(\sec x))(\sec x \tan x)$

27. $f'(x) = e^{\csc^2 x}(2\csc x)(-\csc x \cot x)$

29. $f'(x) = \dfrac{-(\ln 2)2^x(5x\sin x) + 2^x(5\sin x + 5x\cos x)}{25x^2\sin^2 x}$

31. $f'(x) = \sqrt{\sin x \cos x} + \frac{1}{2}x(\sin x \cos x)^{-1/2}(\cos^2 x - \sin^2 x)$

33. $f'(x) = \dfrac{(6x\ln x + 3x)\tan x - 3x^2\ln x \sec^2 x}{\tan^2 x}$

35. $f'(x) = \cos(\ln x)\left(\frac{1}{x}\right)$

37. $f'(x) = \dfrac{6x}{\sqrt{1-9x^4}}$

39. $f'(x) = 2x\arctan x^2 + x^2\left(\dfrac{2x}{1+x^4}\right)$

41. $f'(x) = \dfrac{2x}{|x^2|\sqrt{x^4-1}}$

43. $f'(x) = \dfrac{1}{\sqrt{1-\sec^4 x}}(2\sec x)(\sec x \tan x)$

45. $\dfrac{\frac{\tan^{-1}x}{\sqrt{1-x^2}} - \frac{\sin^{-1}x}{1+x^2}}{(\tan^{-1}x)^2}$

47. $f'(x) = \dfrac{1}{\operatorname{arcsec}(\sin^2 x)}\dfrac{1}{\sin^2 x\sqrt{\sin^4 x - 1}}(2\sin x \cos x)$

49. $f'(x) = \sec(1+\tan^{-1}x)\tan(1+\tan^{-1}x) \times \left(\dfrac{1}{1+x^2}\right)$

51. $f'(x) = \sinh x^3 + 3x^3\cosh x^3$

53. $f'(x) = \sinh(\ln(x^2+1))\left(\dfrac{1}{x^2+1}\right)(2x)$

55. $f'(x) = \frac{1}{2}(\cosh^2 x + 1)^{-1/2}(2\cosh x \sinh x)$

57. $f'(x) = \dfrac{3x^2}{\sqrt{x^6+1}}$

59. $f'(x) = \dfrac{\frac{1}{\sqrt{x^2+1}}(\cosh^{-1}x) - \sinh^{-1}x\left(\frac{1}{\sqrt{x^2-1}}\right)}{(\cosh^{-1}x)^2}$

61. $f'(x) = \cos(e^{\sinh^{-1}x})e^{\sinh^{-1}x}\dfrac{1}{x^2+1}$

63. $f'(x) = (\sin x)^x\left(\ln(\sin x) + \dfrac{x\cos x}{\sin x}\right)$

65. $f'(x) = (\sin x)^{\cos x}\left(-\sin x \ln(\sin x) + \dfrac{\cos^2 x}{\sin x}\right)$

67. $f(x) = \sin^{-1} 2x$

69. $f(x) = \frac{1}{6}\ln(1+9x^2)$

71. $f(x) = \sin^{-1}\left(\frac{3x}{2}\right)$

73. $f(x) = \sinh^{-1} 2x$

75. $f(x) = -\frac{1}{6}\ln(1-9x^2)$

77. $f(x) = \sinh^{-1}\left(\frac{3x}{2}\right)$

79. (a) To simplify things, let $C = \dfrac{\sqrt{4km - f^2}}{2m}$; since k, m, and f are all constants, so is C. Then follow the given hint. (b) Set $s(0) = s_0$ and $s'(0) = v_0$ and solve for A and B.

81. (a) 625 feet, according to the model (the official number is 630). (b) About 82 degrees from horizontal. (c) About 13 degrees from horizontal.

83. Mimic the proof in the reading for $\frac{d}{dx}(\sin x)$, except using a sum identity for cosine in the second step.

85. $\frac{d}{dx}(\csc x) = \frac{d}{dx}\left(\frac{1}{\sin x}\right) = \frac{(0)(\sin x) - (1)(\cos x)}{(\sin x)^2} = \frac{-\cos x}{\sin^2 x} = -\left(\frac{1}{\sin x}\right)\left(\frac{\cos x}{\sin x}\right) = -\csc x \cot x$.

87. Differentiating both sides of $\tan(\tan^{-1}x) = x$ gives us $\sec^2(\tan^{-1}x)\frac{d}{dx}(\tan^{-1}x) = 1$, and therefore $\frac{d}{dx}(\tan^{-1}x) = \frac{1}{\sec^2(\tan^{-1}x)}$. By one of the Thinking Back problems in this section, we have $\sec^2(\tan^{-1}x) = 1 + x^2$, and therefore $\frac{d}{dx}(\tan^{-1}x) = \frac{1}{1+x^2}$.

89. $x^2 - y^2 = \cosh^2 t - \sinh^2 t = \left(\frac{e^t - e^{-t}}{2}\right)^2 - \left(\frac{e^t + e^{-t}}{2}\right)^2 = \frac{e^{2t} - 2e^te^{-t} + e^{-2t} - e^{2t} - 2e^te^{-t} + e^{-2t}}{2} = 1$. For the bonus question, consider whether $\cosh t$ can ever be negative.

91. Expand the right-hand side using the definitions and simplify to get the left-hand side.

93. Mimic the proof in the reading that was given for the derivative of $\sinh x$.

95. Mimic the proof in the reading that was given for the derivative of $\sinh^{-1} x$.

97. $\frac{e^y - e^{-y}}{2} = x \Longrightarrow e^y - \frac{1}{e^y} = 2x \Longrightarrow \frac{e^{2y}-1}{e^y} = 2x \Longrightarrow e^{2y} - 2xe^y - 1 = 0$. This is quadratic in e^y, and the quadratic formula gives $e^y = \frac{2x \pm \sqrt{4x^2+4}}{2}$, and thus $y = \ln(x \pm \sqrt{x^2+1})$. Since logarithms are only defined for positive numbers we have $y = \ln(x + \sqrt{x^2+1})$.

99. Start with $y = \tanh x = \dfrac{\sinh x}{\cosh x}$ and use the exercises above.

For answers to odd-numbered Skill Certification exercises in the Chapter Review, please visit the Book Companion Web Site at www.whfreeman.com/tkcalculus.

Chapter 3

Section 3.1

1. T, T, T, F, F, T, F, F.

3. $f'(1)$ is either 0 or does not exist. If f is differentiable at $x = 1$ then $f'(1)$ must equal 0.

5. $x = -2, x = 0, x = 4, x = 5$.

7. f' has at least two zeroes in the interval $[-4, 2]$.

9. There is some $c \in (-2, 4)$ with $f'(c) = -\frac{1}{3}$.

11. If f is continuous on $[a, b]$, differentiable on (a, b), and if $f(a) = f(b) = 0$, then there exists at least one value $c \in (a, b)$ for which the tangent line to f at $x = c$ is horizontal.

13. The graph of $f(x) = (x - 2)(x - 6)$ is one example.

15. Your graph should have roots at $x = -2$ and $x = 2$ and horizontal tangent lines at three places between these roots.

17. One example is an "upside-down V" with roots at $x = -2$ and $x = 2$ where the top point of the V occurs at $x = -1$.

19. One example is the function f that is equal to $x + 3$ for $-3 \le x < -1$ and equal to 0 for $x = -1$.

21. Draw a graph that happens to have a slight "cusp" just at the place where its tangent line would have been equal to the average rate of change.

23. $f'(x) = 0$ at $x = \frac{3}{2}$ and $f'(x)$ does not exist at $x = 3$. f has a local maximum at $x = \frac{3}{2}$ and a local minimum at $x = 3$.

25. $f'(x) = 0$ at $x \approx 0.5$, $x \approx 2$, and $x \approx 3.5$. f has a local minimum at $x \approx 0.5$ and a local minimum at $x = 3.5$. There is neither a maximum nor a minimum at $x \approx 2$.

27. One critical point at $x = -0.65$, a local minimum.

29. Critical points $x = -3, x = 0, x = 1$, a local minimum, maximum, and minimum, respectively.

31. One critical point at $x = \ln\left(\frac{3}{2}\right)$, a local maximum.

33. Undefined at $x = 0$. One critical point at $x = \frac{e}{2}$, a local maximum.

35. Critical points at points of the form $x = \pi k$ where k is an integer; local minima at the odd multiples $x = \pi(2k + 1)$, local maxima at the even multiples $x = \pi(2k)$.

37. From the graph, f appears to be continuous on $[-3, 1]$ and differentiable on $(-3, 1)$, and moreover $f(-3) = f(1) = 0$, so Rolle's Theorem applies. Therefore there is some $c \in (-3, 1)$ such that $f'(c) = 0$. In this example there are three such values of c, namely $c \approx -2.3$, $c = -1$, and $c = 0.3$.

39. From the graph, f appears to be continuous on $[0, 4]$ and differentiable on $(0, 4)$, and moreover $f(0) = f(4) = 0$, so Rolle's Theorem applies. Therefore there is some $c \in (0, 4)$ such that $f'(c) = 0$. In this example there are three such values of c, namely $c \approx 0.5$, $c \approx 2$, and $c \approx 3.5$.

41. f is continuous and differentiable everywhere, and $f(0) = f(3) = 0$, so Rolle's Theorem applies; $c = \frac{1}{3}(4 - \sqrt{7})$ and $c = \frac{1}{3}(4 + \sqrt{7})$.

43. f is continuous and differentiable everywhere, and $f(-2) = f(2) = 0$, so Rolle's Theorem applies; $c \approx -1.27279$, $c \approx 0$, and $c \approx 1.27279$.

45. f is continuous and differentiable everywhere, and $\cos\left(-\frac{\pi}{2}\right) = \cos\left(\frac{3\pi}{2}\right) = 0$, so Rolle's Theorem applies; $c = 0, c = \pi$.

47. f is continuous and differentiable everywhere, and $f(0) = f(2) = 0$, so Rolle's Theorem applies; $c = \sqrt{2}$.

49. f appears continuous on $[0, 2]$ and differentiable on $(0, 2)$; there is one value $x = c$ that satisfies the conclusion of the Mean Value Theorem, roughly at $x \approx 1.2$.

51. f appears continuous on $[-3, 0]$ and differentiable on $(-3, 0)$; there are two values $x = c$ that satisfy the conclusion of the Mean Value Theorem, at $c \approx -2.8$ and $c \approx -0.9$.

53. f is not continuous or differentiable on $[-3, 2]$, so the Mean Value Theorem does not apply.

55. f is continuous and differentiable on $[-2, 3]$; $c \approx -0.5275$ and $c \approx 2.5275$.

57. f is continuous on $[0, 1]$ and differentiable on $(0, 1)$, so the Mean Value Theorem applies; $c \approx 0.4028$.

59. f is continuous and differentiable everywhere, so the Mean Value Theorem applies; $c = cos^{-1}\left(\frac{2}{\pi}\right) \approx 0.88$.

61. $C(h)$ is a differentiable function, and $C'(4) = 0.6 \neq 0$, so $C(h)$ cannot have a local minimum at $h = 4$.

63. (a) Let $s(t)$ be Alina's distance from the grocery store (in miles) at time t. Then $s(0) = 20$ and $s(0.5) = 0$; since $s(t)$ is continuous and differentiable, the MVT applies, and tells you that there is some time $c \in (0, 0.5)$ where $s'(c) = \frac{0-20}{0.5-0} = -40$ miles per hour (the negative sign means she is travelling towards the store).
(b) Questions to think about: Could Alina possibly have been travelling under 40 miles per hour for her whole trip? Or over 40 miles an hour for her whole trip? If her velocity was less than 40 miles per hour at some point, and more than 40 miles per hour at some other point, can she have avoided going *exactly* 40 miles per hour at some time during her trip? Why or why not?

65. See the proof in the reading.

67. Suppose r_1, r_2, and r_3 are the roots of f. Since f is continuous and differentiable everywhere, Rolle's Theorem guarantees that f' will have at least one root on $[r_1, r_2]$, and at least one root on $[r_2, r_3]$; therefore f' has at least two roots.

69. $g(x)$ is continuous on $[a, b]$ and differentiable on (a, b) because f is. Moreover, $g(a) = f(a) - f(a) = 0$ and $g(b) = f(b) - f(a) = f(a) - f(a) = 0$, so Rolle's Theorem applies to $g(x)$. Therefore we can conclude that there is some $c \in (a, b)$ such that $g'(c) = 0$. Since $g'(x) = f'(x) - 0 = f'(x)$, this also means that $f'(c) = 0$, as desired.

Section 3.2

1. T, T, T, F, F, F, F, F.

3. See the first paragraph in the reading.

5. Yes; yes; no.

7. A function can only change sign at roots and discontinuities or non-domain points, and the critical points of f are exactly these types of points for f'.

9. 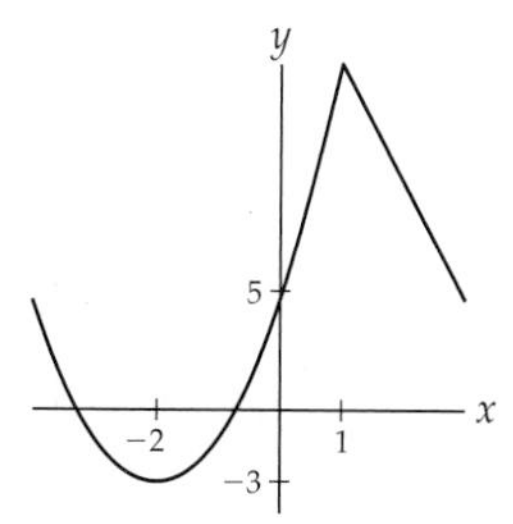

11. If $a < b \le 0$ then $a^4 > b^4$, and if $0 < a < b$ then $a^4 < b^4$. With derivatives, we have $f'(x) = 4x^3$, which is negative for $x < 0$ and positive for $x > 0$.

13. $x = \pm\sqrt{15}$;

15. There is some constant C such that $g(x) - h(x) = C$ for all x in their domains.

17. f is shown in Graph II and f' is shown in Graph I.

19. 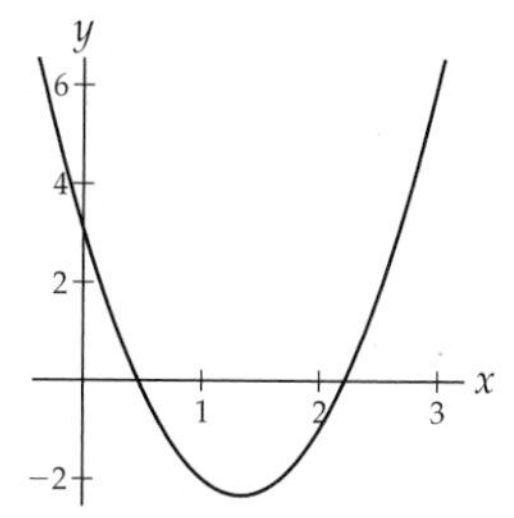

21. Your graph of f' should have roots at $x = 0$ and $x = 2$, and should be positive on $(0, \infty)$ and negative on $(-\infty, 0)$. In particular this means that your graph of f' should "bounce" off the x-axis at $x = 2$.

23. 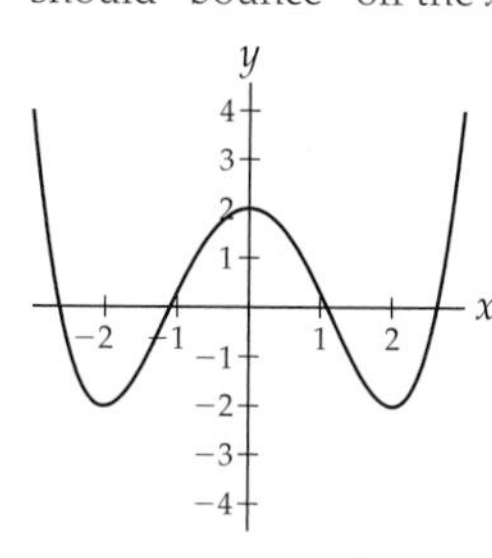

25. Your graph of f should have horizontal tangent lines at $x = 1$ at $x = 3$, and should be decreasing on $(-\infty, 1)$ and $(1, 3)$ and increasing on $(3, \infty)$. The graph of f has an inflection point, but not a local extremum, at $x = 1$, and a local minimum at $x = 3$.

27. Increasing on $(-\infty, 0)$ and $(3, \infty)$; decreasing on $(0, 3)$.

29. Decreasing on $(-\infty, -2)$ and $(2, \infty)$; increasing on $(-2, 2)$.

31. Decreasing on $(-\infty, 1)$ and increasing on $(1, \infty)$.

33. Decreasing on $(-\infty, 0)$ and increasing on $(0, \infty)$.

35. Increasing on intervals of the form $[2k + 1, 2k + 3]$ where k is an integer, and decreasing elsewhere.

37. Increasing on intervals of the form $\left[-\frac{\pi}{4} + \pi k, \frac{\pi}{4} + \pi k\right]$, decreasing elsewhere.

39. Local maximum at $x = 0$, local minimum at $x = 2$

41. Local maximum at $x = -\frac{1}{2}$

43. No local extrema

45. Local maxima at all odd integers, local minima at all even integers

47. No local extrema

49. Local maximum at $x = 0$

51. Possible pictures of f and f', respectively:

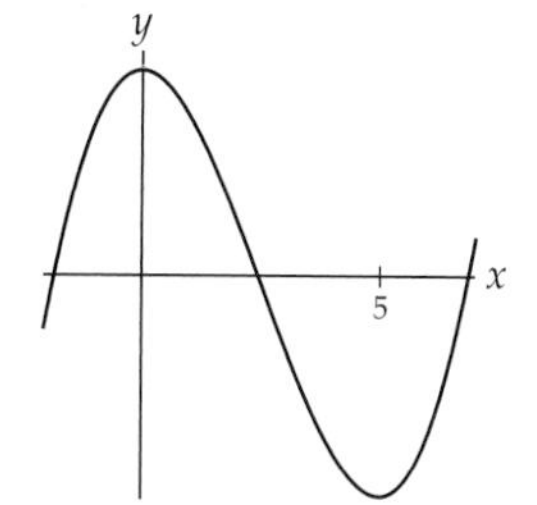

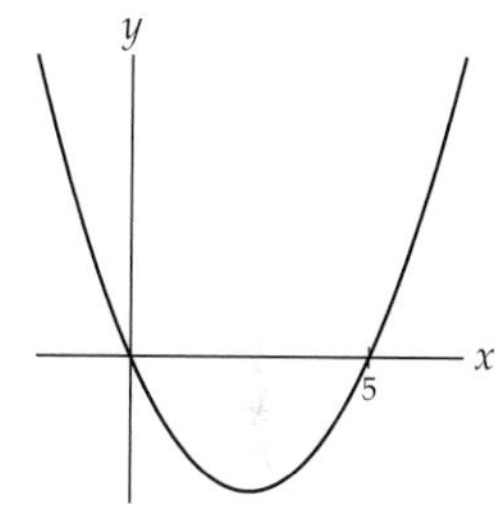

53. Possible pictures of f and f', respectively:

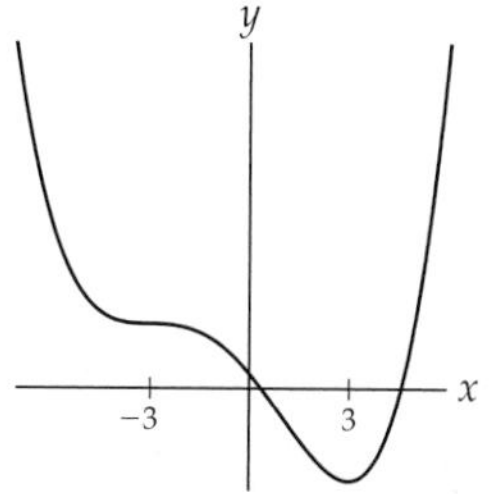

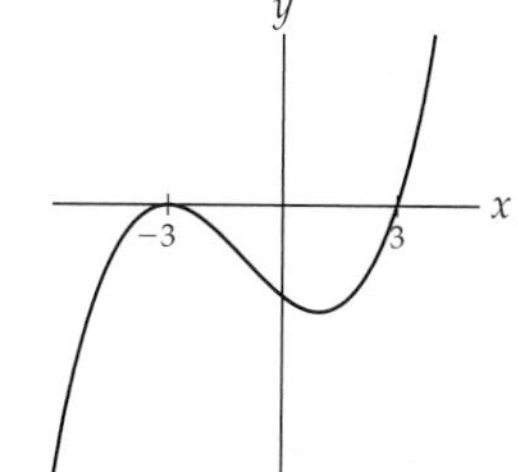

55. Your graph of f should be increasing on $(-\infty, 1)$, $(3, 8)$, and $(8, \infty)$ and decreasing on $(1, 3)$. It should also have horizontal tangent lines at $x = 1$ (a local maximum) and $x = 8$ (not an extremum), and be nondifferentiable at $x = 3$ (a local minimum). Your graph of f' should be positive on $(-\infty, 1)$, $(3, 8)$, and $(8, \infty)$ and negative on $(1, 3)$. It should also have roots at $x = 1$ and $x = 8$ and be undefined at $x = 3$.

57. Defined everywhere, roots at $x = -\frac{1}{3}$ and $x = 2$. Positive on $\left(-\infty, -\frac{1}{3}\right) \cup (2, \infty)$ and negative elsewhere. Local minimum at $x = \frac{5}{6}$. Increasing on $\left(\frac{5}{6}, \infty\right)$ and decreasing elsewhere. $\lim_{x\to-\infty} f(x) = \infty$ and $\lim_{x\to\infty} f(x) = \infty$.

59. Defined everywhere, roots at $x = -1$ and $x = 1$. Positive on $(-1, 1)$ and negative elsewhere. Local minimum at $x = 1$. Increasing on $\left(-\infty, -\frac{1}{3}\right)$ and $(1, \infty)$ and decreasing elsewhere. $\lim_{x\to\infty} f(x) = \infty$ and $\lim_{x\to-\infty} f(x) = -\infty$.

61. Defined everywhere, root at $x = 0$. Positive on $(0, \infty)$ and negative elsewhere. Critical point at $x = 2$ that is *not* a local extremum. Increasing everywhere (yes, even including $x = 2$ – although answering $(-\infty, 2) \cup (2, \infty)$ is sufficient for our purposes here). $\lim_{x\to-\infty} f(x) = -\infty$ and $\lim_{x\to\infty} f(x) = \infty$.

63. Defined everywhere, root at $x = -\frac{11}{2}$. Positive on $\left(-\frac{11}{2}, \infty\right)$ and negative elsewhere. Local maximum at $x = -2$, local minimum at $x = -\frac{5}{3}$. Increasing on $(-\infty, -2) \cup \left(-2, -\frac{5}{3}\right)$ and decreasing elsewhere. $\lim_{x\to-\infty} f(x) = -\infty$ and $\lim_{x\to\infty} f(x) = \infty$.

65. Defined everywhere, roots at $x = \pm 1$. Positive on $(-1, 1)$ and negative elsewhere. Local maxmimum at $x = 0$. Increasing on $(-\infty, 0)$ and decreasing elsewhere. $\lim_{x\to-\infty} f(x) = -\infty$ and $\lim_{x\to\infty} f(x) = -\infty$.

67. Defined everywhere except at $x = 1$, where there is a vertical asymptote; root at $x = -1$. Positive on $(-\infty, -1) \cup (1, \infty)$ and negative elsewhere it is defined. No local extrema. Always decreasing. $\lim_{x\to-\infty} f(x) = 0$ and $\lim_{x\to\infty} f(x) = 0$, so there is a horizontal asymptote at $y = 0$.

69. Defined everywhere; no roots. Positive everywhere. Local minimum at $x = 0$. Increasing on $(0, \infty)$ and decreasing elsewhere. $\lim_{x\to\infty} f(x) = \infty$ and $\lim_{x\to-\infty} f(x) = \infty$.

71. Defined except at $x = -3$ and $x = 2$ where there are vertical asymptotes; root at $x = 1$. Positive on $(-\infty, -3) \cup (2, \infty)$ and negative elsewhere. Local maximum at $x = 1$, local minimum at $x = \frac{11}{3}$. Horizontal asymptote at $y = 1$.

73. Defined except at $x = 2$ where there is a vertical asymptote; roots at $x = 0$ and $x = 1$. Positive on $(1, 2) \cup (2, \infty)$ and negative elsewhere. Local maximum at $x = 0$, local minima at $x = 3 - \sqrt{5}$ and $x = 3 + \sqrt{5}$. Increasing on $(-\infty, 0) \cup (3 - \sqrt{5}, 2) \cup (3 + \sqrt{5}, \infty)$ and decreasing elsewhere. No horizontal asymptote.

75. Defined on $(0, \infty)$; root at $x = 1$. Positive on $(1, \infty)$ and negative elsewhere it is defined. Local minimum at $x = \frac{1}{e}$. Increasing on $\left(\frac{1}{e}, \infty\right)$ and decreasing elsewhere it is defined. $\lim_{x\to 0^+} f(x) = \infty$, so there is a vertical asymptote at $x = 0$. $\lim_{x\to\infty} f(x) = \infty$.

77. Defined everywhere, root at $x = 0$. Positive everywhere except $x = 0$. Local minimum at $x = 0$, local maximum at $x = \frac{-2}{\ln 3}$. Increasing on $\left(-\infty, \frac{-2}{\ln 3}\right) \cup (0, \infty)$ and negative elsewhere. $\lim_{x\to-\infty} f(x) = 0$, so there is a horizontal asymptote on the left at $y = 0$. $\lim_{x\to\infty} f(x) = \infty$, so there is no horizontal asymptote on the right.

79. Defined on $(0, \infty)$, root at $x = 1$. Positive on $(1, \infty)$ and negative elsewhere it is defined. Local maximum at $x = e$. Increasing on $(0, e)$ and decreasing elsewhere it is defined. $\lim_{x\to 0^+} f(x) = -\infty$, so there is a vertical asymptote at $x = 0$. $\lim_{x\to\infty} f(x) = 0$, so there is a horizontal asymptote at $y = 0$.

81. Defined everywhere, no roots. Positive everywhere. Local maximum at $x = 1 - \frac{1}{\sqrt{3}}$, local minimum at $x = 1 + \frac{1}{\sqrt{3}}$. Increasing on $\left(-\infty, 1 - \frac{1}{\sqrt{3}}\right) \cup \left(1 + \frac{1}{\sqrt{3}}, \infty\right)$ and decreasing elsewhere. $\lim_{x\to-\infty} f(x) = 0$, so there is a horizontal asymptote on the left at $y = 0$. $\lim_{x\to\infty} f(x) = \infty$, so there is no horizontal asymptote on the right.

83. (a) $(0, 2.5)$; (b) $t = 2.5$; (c) $(0, 1)$ and $(3.3, 4)$; (d) $(3.5, 3.8)$ is such an interval; Bubbles is moving towards the left side of the tunnel and is slowing down while doing so.

85. (a) When $t = (1/0.51)((2n + 1)(\pi/2) - 2.04)$; at these times the tide is going neither in nor out, and thus these are times of low and high tides. (b) The high tides occur when $n = 1, 5, 9, \ldots$ in the expression from part (a); the low tides occur when $n = 3, 7, 11, \ldots$. (c) Just before low tide, as the current is moving south.

87. Given that $f(x) = mx + b$ where $m \neq 0$ (so f is nonconstant), show that f' is either always positive or always negative. Since $f'(x) = m$ is a (nonzero) constant, this is clearly the case.

89. Mimic the proof from the reading of part (a) of the theorem.

91. Mimic the proof from the reading of part (a) of the theorem.

Section 3.3

1. F, F, F, T, T, F, F, F.

3. See the discussion following Definition 3.9.

5. The converse is "If f is concave up on I, then f'' is positive on I," which is *false* ($f''(x)$ might be zero at some point).

7. The graph of $f(x) = -x^3$ has this property at $x = 0$.

9. $f''(x) = 30x^4$, so $f''(0) = 0$. However, $f''(-1) > 0$ and $f''(1) > 0$, so f'' does not change sign at $x = 0$; therefore $x = 0$ is not an inflection point of f.

11. Linear

13. 5, 4.
15. Draw an arc connecting $(0, 5)$ and $(3, 1)$ that is concave up.
17. Draw an arc connecting $(0, -5)$ and $(3, -1)$ that is concave up.
19. The function $f(x) = -e^x$ has these properties on all of $\mathbb{R}$.
21. Possible graph of f' followed by possible graph of f'':

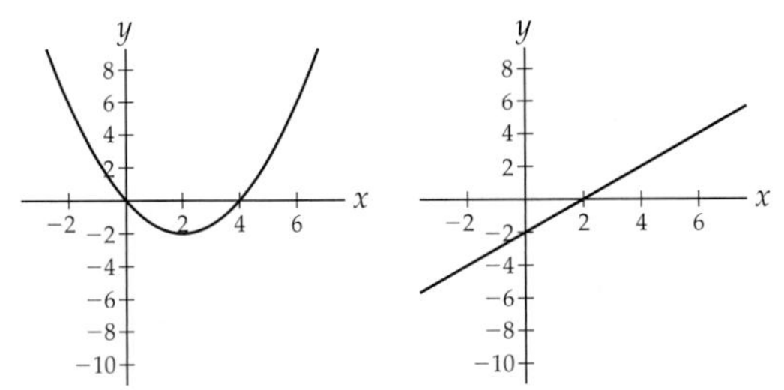

23. Your graph of f' should have roots at $x = -3$ and at $x = 0$ and should be positive on $(-\infty, -3)$ and negative on $(-3, 0)$ and $(0, \infty)$. Your graph of f'' should have roots at $x \approx -2$ and $x = 0$ and should be negative on $(-\infty, -2) \cup (0, \infty)$ and positive on $(-2, 0)$.
25. Possible graph of f followed by possible graph of f''

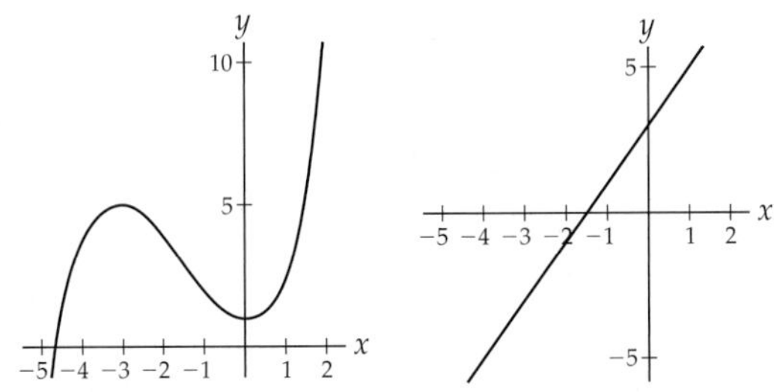

27. Your graph of f' should be decreasing on $(-\infty, 2)$ and increasing on $(2, \infty)$, with a local minimum at $x = 2$. Your graph of f should be concave down on $(-\infty, 2)$ and concave up on $(2, \infty)$, with an inflection point at $x = 2$.
29. Local maximum at $x = 0$, local minimum at $x = 2$
31. Local maximum at $x = -\frac{1}{2}$
33. No local extrema
35. Local maxima at all odd integers, local minima at all even integers
37. No local extrema
39. Local maximum at $x = 0$
41. Always concave up; no inflection points.
43. Concave up on $(-\infty, 0) \cup (1, \infty)$, concave down on $(0, 1)$; inflection points at $x = 0$ and $x = 1$.
45. Concave up on $(-\infty, -1) \cup (0, \infty)$, concave down on $(-1, 0)$; inflection points at $x = -1$ and $x = 0$.
47. Concave up on $\left(-\infty, \ln\left(\frac{9}{16}\right)\right)$, concave down on $\left(\ln\left(\frac{9}{16}\right), \infty\right)$; inflection point at $\ln\left(\frac{9}{16}\right)$.
49. Concave up on $\left(-\infty, 1 - \frac{\sqrt{2}}{2}\right) \cup \left(1 + \frac{\sqrt{2}}{2}, \infty\right)$, concave down on $\left(1 - \frac{\sqrt{2}}{2}, 1 + \frac{\sqrt{2}}{2}\right)$; inflection points at $x = 1 \pm \frac{\sqrt{2}}{2}$.
51. Concave down on intervals of the form $\left(\frac{\pi}{4} + 2\pi k, \frac{5\pi}{4} + 2\pi k\right)$ for integers k, concave up elsewhere; inflection points at points of the form $\frac{\pi}{4} + \pi k$ for integers k.
53. 55.

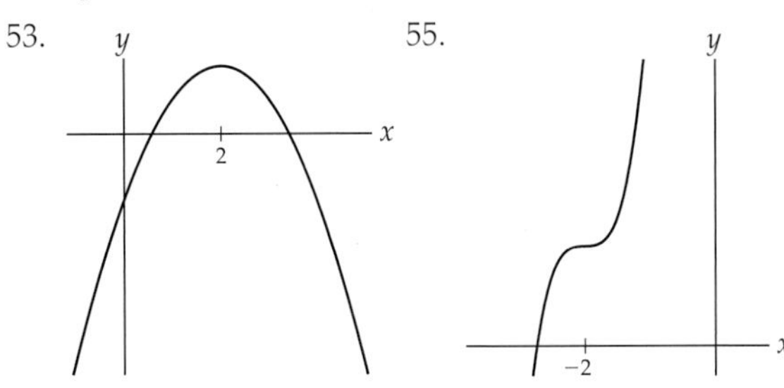

57.

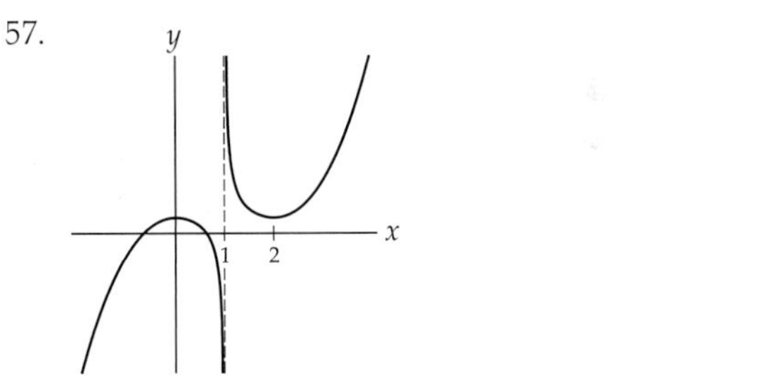

59. Among other things, your graph should have roots at $x = -3$ and $x = -1$, a local minimum at $x = -2$, and inflection points at $x = -1$ and $x = 0$.
61. Among other things, your graph should have roots at $x = -3$, $x = 1$, and $x = 4$, a local maximum at $x = -2$, a local minimum at $x = 3$, and inflection points at $x = -1$, $x = 1$, and $x = 2$.
63. Defined everywhere, roots at $x = -3$ and $x = 0$. Positive on $(-\infty, -3) \cup (0, \infty)$ and negative elsewhere. Local minimum at $x = -\frac{3}{2}$. Increasing on $\left(-\frac{3}{2}, \infty\right)$ and decreasing elsewhere. No inflection points. Concave up everywhere. $\lim_{x \to \infty} f(x) = \infty$ and $\lim_{x \to -\infty} f(x) = \infty$.
65. Defined everywhere, roots at $x = -3$ and $x = 0$. Positive on $(-3, 0) \cup (0, \infty)$ and negative elsewhere. Local maximum at $x = -2$, local minimum at $x = 0$. Increasing on $(-\infty, -3) \cup (0, \infty)$ and decreasing elsewhere. Inflection point at $x = -1$. Concave up on $(-1, \infty)$ and concave down elsewhere. $\lim_{x \to \infty} f(x) = \infty$ and $\lim_{x \to -\infty} f(x) = -\infty$.
67. Defined everywhere, roots at $x = -1$ and $x = 0$. Positive on $(0, \infty)$ and negative elsewhere. Local maximum at $x = \frac{1}{3}$, local minimum at $x = 1$. Increasing on $(-\infty, -1) \cup \left(-\frac{1}{3}, \infty\right)$ and decreasing elsewhere. Inflection point at $x = -\frac{2}{3}$. Concave up on $\left(-\frac{2}{3}, \infty\right)$ and concave down elsewhere. $\lim_{x \to \infty} f(x) = \infty$ and $\lim_{x \to -\infty} f(x) = -\infty$.

69. Defined everywhere, roots at $x = -2$ and $x = 0$. Positive on $(-\infty, -2) \cup (0, \infty)$ and negative elsewhere. Local minimum at $x = -\frac{3}{2}$. Increasing on $\left(-\frac{3}{2}, \infty\right)$ and decreasing elsewhere. Inflection points at $x = -1$ and $x = 0$. Concave up on $(-\infty, -1) \cup (0, \infty)$ and concave down elsewhere.

71. Defined for $x \geq 0$, roots at $x = 0$ and $x = 4$. Positive on $(0, 4)$ and negative elsewhere it is defined. Local maximum at $x = \frac{4}{3}$. Increasing on $\left(0, \frac{4}{3}\right)$ and decreasing elsewhere it is defined. No inflection points. Always concave down. $\lim_{x\to\infty} f(x) = -\infty$.

73. Defined everywhere except at $x = 1$ and $x = 4$; root at $x = -1$. Positive on $(4, \infty)$ and negative elsewhere it is defined. No local extrema. Decreasing everywhere. No inflection points. Concave up on $(4, \infty)$ and concave down elsewhere it is defined. $\lim_{x\to 1} f(x) = -\frac{2}{3}$ but $f(1)$ is not defined, so f has a hole at $x = 1$. $\lim_{x\to 4} f(x)$ is infinite (∞ from the right and $-\infty$ from the left), so f has a vertical asymptote at $x = 4$. $\lim_{x\to\pm\infty} f(x) = 1$, so f has a horizontal asymptote at $x = 1$.

75. Defined everywhere, root at $x = 1$. Positive on $(-\infty, 1)$ and negative elsewhere. Local minimum at $x = 2$. Increasing on $(2, \infty)$ and decreasing elsewhere. Inflection point at $x = 3$. Concave up on $(-\infty, 3)$ and concave down elsewhere. $\lim_{x\to-\infty} f(x) = \infty$ and $\lim_{x\to\infty} f(x) = 0$, so there is a horizontal asymptote on the right – but not the left – at $y = 0$.

77. Local minimum at $x = 1$.

79. Defined everywhere, roots at $x = 0$ and $x = 1$. Never positive, and negative everywhere except at the roots. Local minimum at $x = \frac{1}{8}$, local maximum at $x = 0$. Increasing on $(-\infty, 0) \cup (1, \infty)$ and decreasing elsewhere. Inflection point at $x = 1$. Concave up on $(-\infty, 0) \cup (0, 1)$ and concave down elsewhere. $\lim_{x\to\infty} f(x) = \infty$ and $\lim_{x\to-\infty} f(x) = -\infty$.

81. Defined for $x > 0$, no roots. Always positive. Local minimum at $x = 1$. Increasing on $(1, \infty)$ and decreasing elsewhere it is defined. Inflection point at $x = e$. Concave up on $(0, e)$ and concave down elsewhere it is defined. $\lim_{x\to 0^+} f(x) = \infty$ so there is a vertical asymptote at $x = 0$. $\lim_{x\to\infty} f(x) = \infty$ so there is no horizontal asymptote.

83. f has a local minimum at $x = 0$, no local maxima, and no inflection points.

85. f has no local extrema and no inflection points.

87. (a) $[0, 15]$; he is north of the corner, walking north and slowing down. (b) $[60, 80]$; he is south of the corner, walking south and speeding up. (c) $[15, 40]$; he is walking south and speeding up (first north and then south of the corner). (d) At $t = 40$; he is south of the corner, walking south; he just finished speeding up (in the southward direction) and is now just starting to slow down (but still walking in the southward direction).

89. The glacier is concave down for $x \in (2.532, 5.294)$.

91. Mimic the proof of part (a) of the same theorem.

93. If $f(x) = ax^2 + bx + c$, then $f''(x) = 2a$. If $a > 0$ then $f''(x) > 0$ for all x, so f is always concave up. If $a < 0$ then $f''(x) < 0$ for all x, so f is always concave down.

95. If $f''(x) = 0$ on an interval I, then $f'(x) = k$ must be constant, and therefore $f(x) = kx + c$ must be linear.

Section 3.4

1. F, T, T, F, F, F, F, F.

3. The derivative may or may not be zero (or fail to exist) at the endpoints of the interval.

5. There will not be any endpoint extrema; there may or may not be a global maximum (or minimum) on the interval.

7. For example, f might have a vertical asymptote on I.

9. (a) Global minimum at $x = 0$, no global maximum.
(b) Global minimum at $x = 5$, global maximum at $x = 2$.
(c) Global minimum at $x = 0$, no global maximum.
(d) Global minimum at $x = 0$, no global maximum.
(e) Global minimum at $x = 0$, no global maximum.
(f) Global minimum at $x = 0$, no global maximum.

11. (a) Global maximum at $x = -1$ with value $f(-1) = 7$, global minimum at $x = -3$ with value $f(-3) = -45$. (b) Global maximum at $x = 0$ with value $f(0) = 0$, global minimum at $x = 2$ with value $f(2) = -20$. (c) No global maximum, global minimum at $x = 2$ with value $f(2) = -20$. (d) Global maximum at $x = -1$ with value $f(-1) = 7$, no global minimum.

13. (a) Global maximum at $x = -1$ with value $f(-1) = 11$, global minimum at $x = 1$ with value $f(1) = -5$. (b) No global maximum or global minimum. (c) Global maximum at $x = -1$ with value $f(-1) = 11$, no global minimum. (d) Global maximum at $x = 0$ with value $f(0) = 0$, global minimum at $x = 3$ with value $f(3) = -117$.

15. (a) Global maximum at $x = 0$ with value $f(0) = 1$, global minimum at $x = 3$ with value $f(3) = \frac{1}{1+\sqrt{3}}$. (b) No global maximum or global minimum. (c) Global maximum at $x = 1$ with value $f(1) = \frac{1}{2}$, global minimum at $x = 2$ with value $f(2) = \frac{1}{1+\sqrt{2}}$. (d) Global maximum at $x = 0$ with value $f(0) = 1$, no global minimum.

17. (a) Global maximum at $x = 4$ with value $f[4] = 56$, global minimum at $x = 1$ with value $f(1) = -2$. (b) No global maximum, global minimum at $x = 1$ with value $f(1) = -2$. (c) Global maximum at $x = 0$ with value $f(0) = 0$, no global minimum. (d) No global maximum, no global minimum.

19. (a) Global minimum at $x = 0$ and $x = 4$ with value $f(0) = f(4) = \frac{1}{\sqrt{3}}$; no global maximum. (b) Global minimum at $x = 10$ with value $f(10) = \frac{1}{3\sqrt{7}}$, no global maximum. (c) Global minimum at $x = 0$ with value $f(0) = \frac{1}{\sqrt{3}}$, no global maximum. (d) No global minimum, no global maximum.

21. The most optimal solution is to let the horizontal straight edges have length zero, resulting in a circle of radius $\frac{60}{\pi}$.

23. The horizontal top and bottom edges should have length ≈ 19.6903 units, and the vertical edges should have length 13.1268 units.

25. $x = 18$ and $y = 18$

27. $a = 50$, $b = 50$

29. 200

31. $\left(-\frac{1}{5}, \frac{2}{5}\right)$

33. $\left(\frac{2-\sqrt{6}}{2}, 1.75\right)$ or $\left(\frac{2+\sqrt{6}}{2}, 1.75\right)$

35. A square pen with sides of length $\frac{175}{2}$ feet will produce the maximum area of 7656.25 square feet.

37. Four parallel north-south fences of length 125 feet, and two east-west fences of length 250 feet. The resulting pen will have an area of 31, 250 square feet.

39. 180 feet of border fencing

41. About 143 feet of border fencing

43. The base of the box should have sides of length 6.26 inches, and the height of the box should be approximately 10.65 inches. The resulting largest possible volume is approximately 417.48 cubic inches.

45. The largest possible volume is 11, 664 cubic inches; the largest possible surface area is 3, 332.6 square inches.

47. The largest possible volume is 14, 851 cubic inches; the largest possible surface area is 462.6 square inches.

49. At $t = 0$ minutes; at the right endpoint of the model (one such endpoint could be $t = 5$).

51. The minimum area occurs when you cut the wire so that 4.399 inches of wire are used to make the circle. The maximum area occurs when you don't cut the wire at all, and use all 10 inches to make the circle.

53. The minimum area occurs when you cut the wire so that 0.577 inches of wire are used to make the circle. The maximum area occurs when you don't cut the wire at all, and use all 10 inches to make the circle.

55. To get the minimum surface area, the oil drums should be constructed so they are $\frac{40}{\pi 20^{2/3}} \approx 3.0767$ feet high with a radius of $\frac{20^{1/3}}{\pi} \approx 1.8533$ feet. There is no global maximum.

57. To minimize the cost, the cans should be made with a radius of $2\left(\frac{5}{\pi}\right)^{1/3} \approx 2.335$ inches, and a height of ≈ 11.6754 inches. It is not possible to maximize the cost.

59. The steam pipe should be buried along the 800-foot side of the parking lot for 425 feet, and then diagonally under the parking lot to the opposite corner.

61. (a) You would sell the most (975 paintings) at \$5.00 apiece. You would sell the least (15 paintings) at \$45.00 apiece. (b): $R(c) = c(0.6c^2 - 54c + 1230)$.
(c) You would earn the most money (\$8327.10) by selling the paintings for \$15.28 apiece. You would earn the least money (\$672.90) by selling the paintings for \$44.72 apiece. (d) Although you sell the most paintings when charging \$5.00 apiece, you only earn \$5.00 for each of those paintings.

63. If the sides of the rectangle have length x and y, then $P = 2x + 2y$, so $y = \frac{1}{2}(P - 2x)$. Therefore the area is $A = xy = x\left(\frac{1}{2}(P - 2x)\right)$, which has derivative $A'(x) = \frac{1}{2}P - 2x$. Thus the only critical point of $A(x)$ is $x = \frac{P}{4}$. Since $A'\left(\frac{P}{4} - 1\right) > 0$ and $A'\left(\frac{P}{4} + 1\right) < 0$, the first-derivative test tells us that $A(x)$ has a local maximum at $x = \frac{P}{4}$; in fact it is the global maximum of $A(x)$ on $\left[0, \frac{P}{2}\right]$. Therefore A is maximized when $x = \frac{P}{4}$ $\left(\text{and thus } y = \frac{P}{4}\right)$, i.e., when the rectangle is a square.

Section 3.5

1. F, T, T, F, T, T, T, F.

3. $V = \pi y^2 s$, $SA = 2\pi ys + 2\pi y^2$

5. $V = \frac{\pi}{4}h^3$, $SA = \frac{3}{2}\pi h^2$

7. See Theorem 3.13(a) for the statement of the theorem. A triangle with legs of lengths 3 and 4 will have a hypotenuse of length $\sqrt{3^2 + 4^2} = \sqrt{25} = 5$.

9. $\frac{dV}{dt} = 4\pi r^2 \frac{dr}{dt}$

11. $\frac{dV}{dt} = \frac{E^2}{2\pi^2}\frac{dE}{dt}$, or equivalently, $\frac{dV}{dt} = 2r^2\frac{dE}{dt}$.

13. (a) $\frac{dV}{dt} = 4\pi r^2 \frac{dr}{dt}$; (b) $\frac{dS}{dt} = 8\pi r \frac{dr}{dt}$.

15. $\frac{dV}{dt} = 2\pi rh \frac{dr}{dt}$

17. $\frac{dV}{dt} = \frac{3\pi}{4}h^2 \frac{dh}{dt}$

19. $A = \frac{p^2}{16}$

21. $SA = 4\pi h + 8\pi$

23. $SA = \pi r\sqrt{r^2 + 25} + \pi r^2$

25. $A = \frac{c^2}{4}$

27. $f' = 2uu' + kv'$

29. $f' = v + tv' + kv'$

31. $f' = 2v'\sqrt{u + w} + \frac{v(u' + w')}{\sqrt{u + w}}$

33. $f' = w'(u + t)^2 + 2w(u + t)(u' + 1)$

35. $f' = \frac{1}{k}(u't + u + w')$

37. $\left.\frac{dA}{dt}\right|_{r=12} = 96\pi \frac{\text{in}^2}{\text{sec}}$; $\left.\frac{dA}{dt}\right|_{r=24} = 192\pi \frac{\text{in}^2}{\text{sec}}$; $\left.\frac{dA}{dt}\right|_{r=100} = 800\pi \frac{\text{in}^2}{\text{sec}}$.

39. $\left.\frac{dV}{dt}\right|_{s=8} = 6(8)^2 = 384$ cubic inches per minute.

41. $\left.\frac{dV}{dt}\right|_{x=\sqrt[3]{55}} = 6(\sqrt[3]{55})^2 \approx 86.77$ cubic inches per minute.

43. $\left.\frac{dr}{dt}\right|_{r=12} = \frac{5}{24\pi}\frac{\text{in}}{\text{sec}}$

45. $16\frac{\text{in}^2}{\text{sec}}$

47. Suppose x is Stuart's distance from the streetlight, l is the length of his shadow, and $y = x + l$ is the distance from the tip of his shadow to the streetlight. Then $\left.\frac{dy}{dt}\right|_{x=10} = -\frac{40}{7}\frac{\text{ft}}{\text{sec}}$; this does not depend on x.

49. $\frac{2}{\sqrt{128}} \approx 0.177$ feet per second downwards.

51. The area is growing by 6 square feet per second at that particular moment.

53. $\left.\frac{dr}{dt}\right|_{h=4} = \frac{1}{8\pi}\frac{\text{in}}{\text{sec}}$

55. $\left.\frac{dh}{dt}\right|_{h=4} = \frac{1}{12\pi}\frac{\text{in}}{\text{sec}}$

57. $\left.\frac{dh}{dt}\right|_{h=3} = -\frac{25}{72\pi}\frac{\text{in}}{\text{min}}$

59. (a) $\frac{dA}{dt} = 2b - a$ for all times t. (b) $1 \le a < 21$ and $0 < b \le 10$, with $0 \le t < 5$. (c) The area is increasing when $2b > a$ and decreasing when $2b < a$. In terms of time, the area is increasing from time $t = 0$ to time $t = 2.375$ and then decreasing from time $t = 2.375$ to time $t = 5$.

61. (a) We see that $x(t)/(13200 - 11710) = \tan(\pi/2 - \theta(t))$. It follows that $x' = -1490\sec^2(\pi/2 - \theta)\theta'$. (b) Now $\theta' = -(11\pi/180)/60$ radians per minute (negative because the sun is going down), and when it hits Ian the shadow makes $\tan\theta = 1490/3100$. It follows that $x' = 25.4$ feet per minute.

63. $\frac{d}{dr}\left(\frac{4}{3}\pi r^3\right) = 4\pi r^2$

65. Mimic the work in Example (c).

Section 3.6

1. F, F, F, F, T, F, T, F.

3. See the discussion at the beginning of the section.

5. $y = 2(x-1)$ is the line tangent to f at $x = 1$, and $y = x - 1$ is the line tangent to $g(x)$ at $x = 1$. Very close to $x = 1$ the graphs of f and g are very close to the graphs of these tangent lines.

7. $\lim_{x\to\infty}\frac{x}{2^x} \to \frac{\infty}{\infty}$ and $\lim_{x\to\infty}\frac{1/2^x}{1/x} \to \frac{0}{0}$; the first form works better.

9. $\lim_{x\to\infty}\frac{\ln x}{x^2} \to \frac{\infty}{\infty}$ and $\lim_{x\to\infty}\frac{x^{-2}}{\frac{1}{\ln x}} \to \frac{0}{0}$; the first form works better.

11. $\lim_{x\to 0}\frac{x}{\sin x} \to \frac{0}{0}$ and $\lim_{x\to 0}\frac{\csc x}{x^{-1}} \to \frac{\infty}{\infty}$; the first form works better.

13. The only error is in the second application of L'Hôpital's Rule (it does not apply).

15. (a) $\lim_{x\to 1}\frac{2x+1}{1} = 3$; (b) $\lim_{x\to 1}(x+2) = 3$

17. (a) $\lim_{x\to\infty}\frac{1}{-6x} = 0$; (b) $\lim_{x\to\infty}\frac{\frac{1}{x}-\frac{1}{x^2}}{\frac{2}{x^2}-3} = \frac{0-0}{0-3} = 0$

19. (a) $\lim_{x\to\infty}\frac{3e^{3x}}{-2e^{2x}} = \lim_{x\to\infty}\frac{3e^x}{-2} = -\infty$; (b) $\lim_{x\to\infty}\frac{1}{\frac{1}{e^{3x}}-\frac{1}{e^x}} = -\infty$

21. $-8\ln 2$

23. $\frac{1}{2}$

25. 0

27. 0

29. ∞

31. 0

33. 0

35. 0

37. 1

39. 1

41. 0

43. 0

45. -1

47. 1

49. ∞

51. 1

53. e

55. 1

57. 0

59. 1

61. 1

63. 1

65. $\lim_{x\to\infty}\frac{x+100}{x} = 1$, so neither dominates.

67. $\lim_{x\to\infty}\frac{2x^{100}}{100x^2} = \infty$, so $v(x)$ dominates $u(x)$.

69. $\lim_{x\to\infty}\frac{e^x}{2^x} = \infty$, so $v(x)$ dominates $u(x)$.

71. Neither function dominates.

73. $\lim_{x\to\infty}\frac{0.001e^{0.001x}}{100x^{100}} = \infty$, so $u(x)$ dominates $v(x)$.

75. 0

77. ∞

79. ∞

81. On I, f has a global maximum at $x = 1$ and a global minimum at $x = \frac{1}{e}$, since $\lim_{x\to 0^+} f(x) = 0$. On J, f has no global maximum and a global minimum at $x = \frac{1}{3}$, since $\lim_{x\to\infty} f(x) = \infty$.

83. On I, f has a global maximum at $x = 3$ and a global minimum at $x = 0$. On J, f has a global maximum at $x = 3$ but no global minimum.

85. On I, f has no global maximum and a global minimum at $x = \pi$, since $\lim_{x\to 0^+} f(x) = \infty$. On J, f has no global minimum and no global maximum, since $\lim_{x\to 0^+} f(x) = \infty$ and $\lim_{x\to 2\pi^-} f(x) = -\infty$.

87. $\frac{\Delta w}{\Delta t} = \frac{w(t) - w(0)}{t - 0} = \frac{w(t)}{t}$.

89. Split into a sum of two limits, the first of which is $\lim_{t\to\infty} E(t) = \frac{E_0}{1+0.2W_0}$ and the second of which is of the form $\frac{\infty}{\infty}$ and which by L'Hopital's Rule is equal to 0.

91. If $f(x) = Ae^{kx}$ is exponential and $g(x) = x^2$, then by applying L'Hôpital's rule twice, we have $\lim_{x\to\infty}\frac{Ae^{kx}}{x^2} = \lim_{x\to\infty}\frac{Ake^{kx}}{2x} = \lim_{x\to\infty}\frac{Ak^2e^{kx}}{2}$, which in the limit is equal to ∞.

93. If $f(x) = Ae^kx$ is exponential and $g(x) = Bx^r$ is a power function, then by applying L'Hôpital's rule r times, we have $\lim_{x\to\infty} \frac{Ae^{kx}}{Bx^r} = \lim_{x\to\infty} \frac{Ake^{kx}}{Brx^{r-1}} = \cdots = \lim_{x\to\infty} \frac{Ak^r e^{kx}}{Br(r-1)(r-2)\cdots(1)}$, which is of the form $\frac{\infty}{\text{constant}} \to \infty$.

95. $\lim_{x\to\infty} \ln\left(\left(1+\frac{r}{x}\right)^x\right) = \lim_{x\to\infty} x\ln\left(1+\frac{r}{x}\right) =$ $\lim_{x\to\infty} \frac{\ln\left(1+\frac{r}{x}\right)}{\frac{1}{x}} = \lim_{x\to\infty} \frac{\frac{1}{1+\frac{r}{x}}\left(\frac{-r}{x^2}\right)}{\frac{-1}{x^2}} = \lim_{x\to\infty} \frac{r}{1+\frac{r}{x}} = r$, and thus $\lim_{x\to\infty} \left(1+\frac{r}{x}\right)^x = e^r$.

For answers to odd-numbered Skill Certification exercises in the Chapter Review, please visit the Book Companion Web Site at www.whfreeman.com/tkcalculus.

Chapter 4

Section 4.1

1. T, T, T, F, F, F, F, F.

3. The terms in the sum do not appear to have a consistent pattern.

5. The sum of the squares of the integers greater than or equal to 3 and less than or equal to 87.

7. (a) $b_p + b_{p+1} + \cdots + b_{q-1} + b_q$. (b) Index is i, starting value is p, ending value is q, b_i describes each term in the sum. (c) p and q must be nonnegative integers; b_i can be any real number; also, we must have $p \le q$.

9. One possible answer is $a_k = k^2$, $m = 3$, $n = 7$.

11. The first sum is $1 + \frac{1}{2} + \frac{1}{5} + \frac{1}{10} + \cdots + \frac{1}{65}$; the second sum is $2\left(\frac{1}{2} + \frac{1}{4} + \frac{1}{10} + \frac{1}{20} + \cdots + \frac{1}{130}\right)$.

13. (a) $\sum_{k=1}^{5} \frac{2}{k^2}$; (b) $\sum_{k=2}^{6} \frac{2}{(k-1)^2}$; (c) $\sum_{k=0}^{4} \frac{2}{(k+1)^2}$.

15. (a) $\sum_{k=1}^{2} k = 1 + 2 = 3$, and $\frac{2(2+1)}{2} = \frac{6}{2} = 3$; (b) $\sum_{k=1}^{8} k = 1+2+3+4+5+6+7+8 = 36$, and $\frac{8(8+1)}{2} = \frac{72}{2} = 36$; (c) $\sum_{k=1}^{9} k = 1+2+3+4+5+6+7+8+9 = 45$, and $\frac{9(9+1)}{2} = \frac{90}{2} = 45$.

17. See Theorem 4.4.

19. The area under f' should be related to the function f, since in the example the area under $v(t) = s'(t)$ is related to the distance travelled $s(t)$.

21. $\sum_{k=1}^{8} 3$; $m = 1$, $n = 8$, $a_k = 3$

23. $\sum_{k=1}^{5} \frac{k+2}{k^3}$ and $\sum_{k=3}^{7} \frac{k}{(k-2)^3}$ are the most obvious ways.

25. $\sum_{k=2}^{10} (k^2 + 1)$

27. $\sum_{k=1}^{n} \frac{k}{n}$

29. $16 + 25 + 36 + 49 + 64 + 81 = 271$

31. $0 + \frac{1}{8} + \frac{1}{2} + \frac{9}{8} + 2 + \frac{25}{8} = \frac{55}{8}$

33. Approximately $0.17321 + 0.17607 + 0.17889 + 0.18166 + 0.18439 + 0.18708 + 0.18973 + 0.19235 + 0.19494 + 0.19748 \approx 1.8558$

35. $3n - \frac{n(n+1)}{2}$; -4750; $-123{,}750$; $-497, 500$.

37. $\frac{n(n+1)(2n+1)}{6} + 2\frac{n(n+1)}{2} + n - 13$; 348, 537; 42, 042, 737; 334, 835, 487.

39. $\frac{1}{n^4}\left(\frac{n^2(n+1)^2}{4} - n\right)$; 0.255024; 0.251001; 0.2505

41. $2a_0 + 2a_1 + 2a_2 - a_{101} + \sum_{k=3}^{100} a_k$

43. $\sum_{k=2}^{25} (3k^2 + 2k - 1) - 52$

45. 48

47. $\lim_{n\to\infty} \frac{2n^3+6n^2+10n}{6n^3} = \frac{1}{3}$

49. $\lim_{n\to\infty} \frac{n+n(n+1)+\frac{1}{6}n(n+1)(2n+1)}{n^3-1} = \frac{1}{3}$

51. $\lim_{n\to\infty} \left(1 + \frac{n^2+n}{n^2} + \frac{2n^3+3n^2+n}{6n^3}\right) = 1 + 1 + \frac{2}{6} = \frac{7}{3}$

53. Answers can vary depending on what you choose to use as your velocity in each subinterval. Using the initial velocity of each 5-second chunk we approximate that the distance travelled is $d \approx 5v(0) + 5v(5) + 5v(10) + 5v(15) + 5v(20) + 5v(25) + 5v(30) + 5v(35) = 2310$ feet.

55. (a) In any year t, $B(t)$ is the sum of the increases $I(1), I(2), \ldots I(t)$. (b) The graph of $I(t)$ is a series of rectangles with heights $I(1)$, $I(2)$, and so on. For any year t, $B(t)$ is the sum of the areas of the first through tth rectangles. (c) In any given year, if I is positive then B will increase that year; if I is negative then B will decrease that year. It seems like I could be related to the derivative of B.

57. (a) $A(n) = \sum_{k=1}^{n} 10(0.7)^k$; (b) $A(3) = 15.33$ billion, $A(4) = 17.73$ billion, and $A(5) = 19.41$ billion; (c) $A(10) = 22.67$ billon, so the total might be about 23 billion dollars.

59. $\sum_{k=0}^{n} 3a_k = 3a_0 + 3a_1 + 3a_2 + \cdots + 3a_n = 3(a_0 + a_1 + a_2 + \cdots + a_n) = 3\sum_{k=0}^{n} a_k$.

61. Pairing the first and last terms in the sum $1 + 2 + 3 + \ldots + n$ gives a sum of $n + 1$. Pairing the second and second-to-last terms also gives a sum of $n + 1$. Continuing in this fashion we obtain $\frac{n}{2}$ pairs whose sum is $n + 1$.

Section 4.2

1. F, T, F, F, F, T, T, F.

3. (a) Counting partial squares as half-squares, the blob has an approximate area of $46\left(\frac{1}{4}\right) + 36\left(\frac{1}{8}\right) = 16$ square units. (b) Two possible ways to improve the approximation are to count partial squares more accurately depending on their sizes, or to use a finer grid.

5. (a) The right sum is an over-approximation, and the left sum an under-approximation, in this example. Both the midpoint sum and the trapezoid sum look like very close approximations. (b) The midpoint sum was off by the smallest amount, with an error of only about 0.0417.
7. (a) 7.85; (b) No; (c) upper sum is ≥ 8.2, lower sum is ≤ 7.5; (d) No; (e) No.
9. The heights of the rectangles used with the upper sum are by definition always greater than or equal to the heights of the rectangles used for any other Riemann sum.
11. (a) $M_1 = 0$, $M_2 = 2$; (b) $M_1 = 0$, $M_2 = \frac{2}{3}$ $\left(\text{or } \frac{4}{3}\right)$, $M_3 = 2$; (c) $M_1 = 0$, $M_2 = \frac{1}{2}$, $M_3 = \frac{3}{2}$, $M_4 = 2$.
13. (a) Meters; (b) the area of a rectangle; (c) looks like distance should be related to the area under the velocity curve.
15. Part of the sum indicates that we have $\Delta x = \frac{1}{2}$, whereas another part tells us that $\Delta x = 0.25$.
17. Right sum with $f(x) = x^2$, $N = 4$, $\Delta x = \frac{1}{2}$, and $x_k = 1 + \frac{k}{2}$; thus $a = 1$ and $b = 3$.
19. Trapezoid sum with $f(x) = \sin x$, $N = 100$, $\Delta x = 0.05$, and $x_k = 0 + 0.05k$ (so $x_{k-1} = 0 + 0.05(k-1)$); thus $a = 0$ and $b = 0 + 0.05(100) = 5$.
21. 1.219
23. 51.88
25. $\frac{8}{3}$
27. (a) 5; (b) 6.875
29. (a) 1.16641; (b) 1.26997
31. (a) 51.3434; (b) 52.9562
33. (a) 2; (b) 2.148; (c) 2.25
35. Exact = 26; LS = 24.5; RS = 27.5; midpoint = 26; upper = 27.5; lower = 24.5; trapezoid = 26
37. Exact $= \frac{\pi}{2}$; LHS ≈ 1.366; RHS ≈ 1.366; midpoint ≈ 1.630; upper ≈ 1.866; lower ≈ 0.866; trapezoid ≈ 1.366.
39. $\sum_{k=1}^{4} \frac{1}{4}\sqrt{\frac{k}{4} + 1}$
41. $\sum_{k=1}^{6} e^{(2k+3)/4} \left(\frac{1}{2}\right)$
43. $\sum_{k=1}^{4} \frac{\sin(\frac{\pi}{4}(k-1)) + \sin(\frac{\pi}{4}k)}{2} (\frac{\pi}{4})$
45. (a) 1320 feet; (b) 3080 feet
47. With a right sum we approximate that the change in temperature is -24.2385 degrees.
49. (a) $c_1 \approx 6114.15$, $c_2 \approx 8722.35$; (b) plotting $g(t) = 6114.15 + 8722.35(\sin((t-90)\pi/105) - 2/\pi)$ with the four data points between 90 and 195 days gives a relatively good fit.
51. Since f is increasing, we have $x_{k-1} \leq x_k$; now apply Definition 4.6.
53. Draw a picture of a concave up function with a trapezoid sum approximation. Compare the trapezoid to the area under the curve. What do you notice?

Section 4.3

1. F, T, F, F, F, F, T, F.
3. $\int_a^b f(x)dx$; the definite integral of f on $[a, b]$
5. $\int_a^b f(x)dx = \lim_{n\to\infty} \sum_{k=1}^{n} f(x_k^*)\Delta x$, where $\Delta x = \frac{b-a}{n}$, $x_k = a + k\Delta x$,a nd x_k^* is a point in the kth subinterval $[x_{k-1}, x_k]$.
7. The region between the graph of f and the x-axis from $x = a$ to $x = a$ has a width of zero.
9. A larger portion of the region between the graph of $f(x) = x^2 - 4$ and the x-axis on $[-3, 3]$ is negative, so the definite integral is negative.
11. 0
13. (a) Negative; (b) negative
15. (a) $\int_a^b (f(x) + g(x))dx = \int_a^b f(x)dx + \int_a^b g(x)dx$, but $\int_a^b f(x) \cdot g(x)dx \neq \int_a^b f(x)dx \cdot \int_a^b g(x)dx$. (b) With $f(x) = x$ and $g(x) = 1$, $\int_0^1 f(x)dx = \frac{1}{2}$, $\int_0^1 g(x)dx = 1$, and $\int_0^1 (f(x) + g(x))dx = \frac{3}{2}$. (c) With $f(x) = x$ and $g(x) = x$, $\int_0^1 f(x)dx = \frac{1}{2}$ and $\int_0^1 g(x)dx = \frac{1}{2}$, but $\int_0^1 f(x)g(x) = \frac{1}{3}$.
17. (a) $\sum_{k=1}^{n} \left(\frac{3k}{n}\right)^2 \left(\frac{3}{n}\right) = \frac{27}{n^3}\left(\frac{n(n+1)(2n+1)}{6}\right)$ (b) $\sum_{k=1}^{n} \left(\frac{3}{n}(k-1)\right)^2 \left(\frac{3}{n}\right) = \frac{27}{n^3}\left(\frac{n(n+1)(2n+1)}{6} - 2\frac{n(n+1)}{2} + n\right)$ (c) For $n = 100$, RHS ≈ 9.13545 and LHS ≈ 8.86545; for $n = 1000$, RHS ≈ 9.0135 and LHS ≈ 8.9865. (d) Both expressions approach 9 as $n \to \infty$.
19. (a) Infinite discontinuity (vertical asymptote) at $x = 0$; (b) the area accumulated would be infinite; (c) a sequence of Riemann sum approximations would not converge to a real number as $n \to \infty$; (d) compare the two graphs and speculate.
21. 2
23. 264
25. $\frac{\pi}{2}$
27. $\frac{1}{2}$
29. 5
31. 14
33. Not enough information.
35. Not enough information.
37. -8
39. $-\frac{52}{3}$
41. (a) $\sum_{k=1}^{n} \left(5 - \left(2 + \frac{3k}{n}\right)\right)\left(\frac{3}{n}\right) = \frac{3}{n}(3n) - \frac{9}{n^2}\left(\frac{n(n+1)}{2}\right)$. (b) 4.455, 4.4955; (c) $\frac{9}{2}$.
43. (a) $\sum_{k=1}^{n} 2\left(\frac{k}{n}\right)^2 \left(\frac{1}{n}\right) = \frac{2}{n^3}\left(\frac{n(n+1)(2n+1)}{6}\right)$. (b) 0.6767, 0.667667; (d) $\frac{2}{3}$.

45. (a) $\sum_{k=1}^{n}\left(\left(2+\frac{k}{n}\right)+1\right)^2\left(\frac{1}{n}\right)=\frac{1}{n}(9n)+\frac{6}{n^2}\left(\frac{n(n+1)}{2}\right)$ $+\frac{1}{n^3}\left(\frac{n(n+1)(2n+1)}{6}\right)$. (b) 12.3684, 12.3368; (c) $\frac{37}{3}$.

47. $\frac{62}{3}$

49. -93

51. $\frac{136}{3}$

53. (a) $\int_0^{40}(-0.22t^2+8.8t)dt$; (b) 2346.67 feet.

55. $\int_a^b cf(x)dx = \lim_{n\to\infty}\sum_{k=1}^{n} cf(x_k^*)\Delta x =$ $c\lim_{n\to\infty}\sum_{k=1}^{n} f(x_k^*)\Delta x = c\int_a^b f(x)dx$.

57. If $\int_a^b f(x)dx = \lim_{n\to\infty}\sum_{k=1}^{n} f(x_k)\Delta x$, then $\int_b^a f(x)dx = \lim_{n\to\infty}\sum_{k=1}^{n} f(x_{n-k})(-\Delta x) =$ $-\lim_{n\to\infty}\sum_{k=1}^{n} f(x_k)\Delta x = -\int_a^b f(x)dx$.

59. The area in question is a trapezoid with heights a and b and width $b-a$, therefore area $\frac{a+b}{2}(b-a)=\frac{1}{2}(b^2-a^2)$.

61. Simplify $\sum_{k=1}^{n}\left(a+\frac{k(b-a)}{n}\right)^2\left(\frac{b-a}{n}\right)$, apply sum formulas, and then take the limit as $n\to\infty$. Remember that a and b are constants.

Section 4.4

1. T, F, F, T, F, F, F, F.

3. An antiderivative of f is just one function whose derivative is f, while the indefinite integral $\int f(x)dx$ is the family of all antiderivatives of f.

5. See Definitions 4.9 and 4.15. The definite integral is a number while the indefinite integral is a family of functions; the definite integral concerns area while the indefinite integral concerns antiderivatives; we calculate definite integrals using Riemann sums, but we calculate indefinite integrals by antidifferentiating.

7. All blanks should contain $\frac{1}{7}x^7$.

9. At this point we don't know of a function whose derivative is $\sec x$, but we do know of a function whose derivative is $\sin x$ (it is $-\cos x$).

11. See the theorems in this section.

13. $\frac{d}{dx}\left(\frac{1}{2}e^{(x^2)}(x^2-1)\right)=x^3e^{(x^2)}$

15. $\frac{d}{dx}(x(\ln x-1))=1(\ln x-1)+x\left(\frac{1}{x}-0\right)=\ln x$

17. One easy example is $f(x)=x$, $g(x)=x^2$.

19. $F'(x)=\csc^2 x$ for both $x<0$ and $x>0$. The piecewise antiderivative $F(x)$ here differs from the one in Theorem 4.18 by being shifted up 100 units on the right of the y-axis. Notice from the graphs that this does transformation does not affect the slope of the graph of $F(x)$.

21. $\frac{1}{3}x^3-\frac{1}{2}x^6-7x+C$

23. $\frac{3}{4}x^4+\frac{5}{3}x^3-\frac{3}{2}x^2-5x+C$

25. $\frac{1}{3}x^3+\frac{1}{\ln 2}(2^x)+4x+C$

27. $\frac{2}{3}x^{3/2}+2x^{1/2}+C$

29. $\frac{3.2}{\ln(1.43)}(1.43)^x-50x+C$

31. $2e^{2x-6}+C$

33. $\frac{1}{3}\sec(3x)+C$

35. $3(\tan x-x)+C$

37. $\pi x+C$

39. $\frac{7}{2}\sin^{-1}(2x)+C$

41. $\frac{1}{3}\sec^{-1}(3x)+C$

43. $\ln|1+x^2|+C$

45. x^2e^x+C

47. $\ln|3x^2+1|+C$

49. $e^x-e^{2x}+C$

51. $\frac{1}{18}(x^3+1)^6+C$

53. $\frac{x^2}{\ln x}+C$

55. $e^x\sqrt{x}+C$

57. $\ln x\cos x+C$

59. $\frac{3}{2}\sinh 2x+C$

61. $x\tanh x+C$

63. $\frac{a}{b}e^{bx+c}+C$

65. $\frac{1}{ab}\sec(ax)+C$

67. $\frac{a}{b}\tan^{-1}(bx)+C$

69. (a) $v(t)=-32t+42$ feet per second; (b) $s(t)=-16t^2+42t+4$ feet.

71. See the proof in the reading.

73. $\frac{d}{dx}\left(\frac{1}{k}e^{kx}\right)=\left(\frac{1}{k}\right)(k)e^{kx}=e^{kx}$, and $\frac{d}{dx}\left(\frac{1}{\ln b}b^x\right)=\left(\frac{1}{\ln b}\right)(\ln b)b^x=b^x$.

75. Each of the integrands is the derivative of the corresponding inverse trigonometric function in its solution.

77. See the proof in the reading.

79. By the sum rule for differentiation, and the fact that the derivative of a constant is zero, $\frac{d}{dx}(G(x))=\frac{d}{dx}(F(x)+C)$ implies that $G'(x)=F'(x)+0$.

Section 4.5

1. F, F, T, F, F, T, T, F.

3. The Fundamental Theorem of Calculus relates the concept of signed area under a curve to antiderivatives, which enables us to quickly find the exact areas under curves, and connects the important concepts of definite integrals and indefinite integrals.

5. The Mean Value Theorem

7. $\frac{1}{b-a}\int_a^b v(t)dt$ is equal to $\frac{s(b)-s(a)}{b-a}$, so $\int_a^b v(t)dx=s(b)-s(a)$. Since $s(t)$ is an antiderivative of $v(t)$, this is an example of the equation in the Fundamental Theorem.

9. $f(4)\approx 4.125$

11. $f(-2)\approx 2.5$

13. (a) $\lim_{n\to\infty}\left(\frac{3}{n}(n)+\frac{6}{n^2}\frac{n(n+1)}{2}+\frac{3}{n^3}\frac{n(n+1)(2n+1)}{6}\right)=7$; (b) $3\left(\frac{1}{3}\right)(2^3-1^3)=7$; (c) $\left[3\left(\frac{1}{3}\right)x^3\right]_1^2=7$

15. (a) $\frac{100}{101}$; (b) 5000

17. Equivalent to FTC: (a), (d), and (e).

19. $\frac{12}{5}$

21. $\frac{42}{\ln 2}$

23. $-\frac{2}{3}(5)^{-3/2}+\frac{2}{3}(2)^{-3/2}$

25. $4-2\sqrt{3}$

27. $\sqrt{2}-1$

29. $\frac{\pi}{8}$

31. 2π

33. $-\frac{1}{2e}+\frac{1}{2}$

35. $\frac{\pi}{4}$

37. $\frac{3}{\ln 4}$

39. $\ln 7-\ln 2$

41. $\ln 9-\ln 14$

43. $\ln(32)-\ln(8)$

45. $\frac{\pi}{6}$

47. $\frac{3\pi}{2}-\pi\ln 8$

49. $-\frac{3}{20}$

51. $\frac{3}{2}$

53. $\frac{1}{e}+e$

55. $\frac{1}{2}\cot\left(\frac{\pi^2}{16}\right)-\frac{1}{2}\cot\left(\frac{\pi^2}{4}\right)$

57. $-\frac{3}{2}\ln 5$

59. $\frac{25}{2}$

61. $\frac{343}{12}$

63. $3-\frac{1}{\sqrt{2}}$

65. $\int_0^{24} 0.05t\,dt=\left[\frac{0.05}{2}t^2\right]_0^{24}=14.4$ gallons.

67. (a) Supposing that the "positive" direction is to the right, the figurine moves right along the track for the first five seconds, then left for the next five seconds, then right again, then left again. (b) approximately 5.27 inches, 3.125 inches, and 0 inches. (c) approximately 14.84 inches.

69. $\frac{9}{14}$

71. (a) 80, and there is a line of about 5 people at noon; (b) 50 customers per hour; (c) at 3:00 P.M. there are about 40 people in the line; (d) every customer has been served.

73. $\int_a^b x\,dx=\left[\frac{1}{2}x^2\right]_a^b=\frac{1}{2}b^2-\frac{1}{2}a^2=\frac{1}{2}(b^2-a^2)$.

75. If F is an antiderivative of f, then kF is an antiderivative of kf. Therefore $\int_a^b kf(x)dx=kF(b)-kF(a)=k(F(b)-F(a))=k[F(x)]_a^b=k\int_a^b f(x)dx$.

77. See the proof in the reading.

79. If $G(x)=F(x)+C$ then $[G(x)]_a^b=[F(x)+C]_a^b=(F(b)+C)-(F(a)+C)=F(b)-F(a)=[F(x)]_a^b$.

Section 4.6

1. F, F, T, F, F, F, F, T.

3. Three (although two of them will be equal); positive; negative

5. (a) $\int_{-3}^{4} f(x)dx$; (b) $\int_{-3}^{-1} f(x)dx-\int_{-1}^{2} f(x)dx+\int_2^4 f(x)dx$

7. negative; positive

9. One possible example is $f(x)=x$ on $[-2,2]$.

11. (a) See graph that follows; (b) $\int_{-2}^{4} f(x)dx$; (c) $\int_{-2}^{-1} f(x)dx-\int_{-1}^{3} f(x)dx+\int_3^4 f(x)dx$.

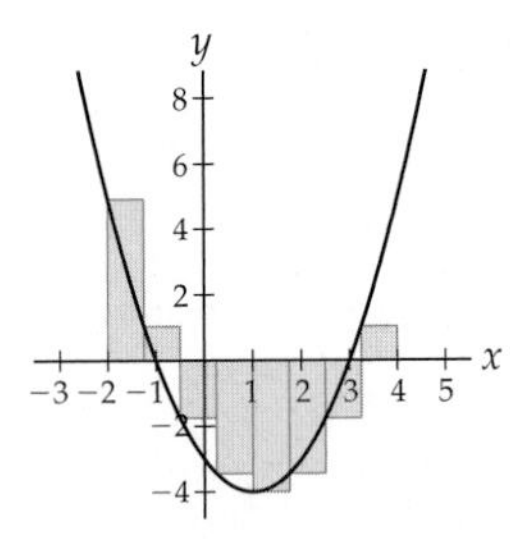

13. (a) See graph that follows; (b) $\int_0^1 (g(x)-f(x))dx+\int_1^3 (f(x)-g(x))dx+\int_3^4 (g(x)-f(x))dx$.

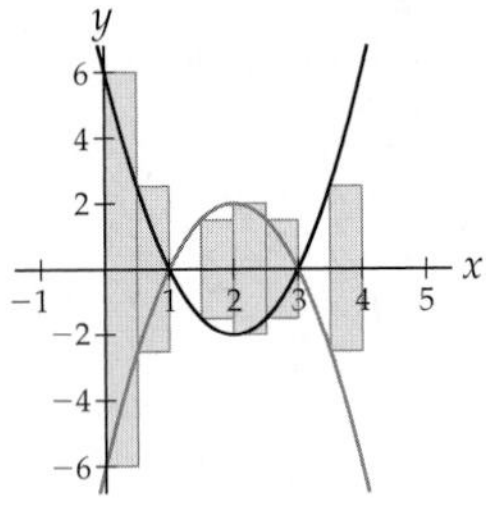

15. Approximately $-\frac{10}{3}$

17. Draw any graph where the average height of the graph on $[-2,5]$ is 10, and the slope from $(-2,f(-2))$ to $(5,f(5))$ on your graph is -3.

19. See the reading.

21. f is continuous on $[1,5]$, so the Mean Value Theorem for Integrals applies and says that there is some $c\in(1,5)$ for which $f(c)=c(c-6)$ is equal to the average value $\frac{1}{5-1}\int_1^5 x(x-6)dx$ of the function f. You can compute this average value; it is equal to $-\frac{23}{3}$. Therefore, there is some $c\in(1,5)$ such that $f(c)=-\frac{23}{3}$. We can solve the equation $f(c)=-\frac{23}{3}$ to find such a value of c. Actually there are two such values: $c\approx 1.8453$ and $c\approx 4.1547$.

23. Using a left sum with $n=10$, we have (a) 43.333; (b) 80.

25. Using a left sum with $n=8$, we have (a) -11.1973; (b) 12.2229.

27. (a) $\frac{5}{2}$; (b) $\frac{13}{2}$

29. (a) $-\frac{4}{3}$; (b) $\frac{109}{12}$

31. (a) $\frac{25}{2}$; (b) 25.2

33. (a) 54, (b) $\frac{194}{3}$

35. (a) $\frac{2}{9}$, (b) $\frac{2}{9}$

37. (a) $\frac{\pi}{2}$, (b) $\frac{\pi}{2}$

39. Using a left sum with $n=8$, we have 22.

41. $\frac{13}{2}$

43. $\frac{38}{6}$

45. 15

47. $\frac{19}{3}$

49. $\frac{145}{12}$

51. $-2+\sqrt{3}+\frac{\pi}{3}$

53. $(f(-1)+f(-0.75)+f(-0.5)+f(-0.25)+f(0)+f(0.25)+f(0.5)+f(0.75))\left(\frac{1}{4}\right)\approx 1.981$

55. One possible approximation is $\left(f\left(\frac{\pi}{4}\right)+f\left(\frac{\pi}{2}\right)+\cdots+f(2\pi)\right)\left(\frac{\pi}{4}\right)=0$

57. 7

59. 4

61. -1

63. $\frac{1}{2}(e^2-e^{-2})$

65. $\frac{11}{28}$

67. 0

69. $c=1$

71. $c=\frac{-3+\sqrt{201}}{12}$

73. $-1+\sqrt{3}$

75. (a) There is some time $c\in(0,4)$ when $f'(c)=\frac{f(4)-f(0)}{4-0}=1.44$; therefore there is some time c in the first four days when the rate of growth of the plant is 1.44 centimeters per day. (b) There is some time c during the first four days when the height of the plant is equal to the average height of the plant over those four days. Since this average height is 1.92 feet, we know that there is some $c\in(0,4)$ such that $f(c)=1.92$. We can solve for $c=2.3094$.

77. (a) $\int_{1.51}^{5.29}(55-f(x))dx+\int_{5.29}^{18}(r(x)-f(x))dx+\int_{18}^{30.71}(s(x)-g(x))dx+\int_{30.71}^{34.49}(55-g(x))dx$; (b) 561.9 square feet (too big!)

79. Write out the Riemann sum for each side; the absolute values on the right-hand side factor out of both the sum and the limit.

81. See the reading concerning the development of average value in the plant height example.

Section 4.7

1. F, F, F, T, F, T, T, F.

3. $\int_2^x t^2 dt$

5. x is the independent variable; $A(x)$ is the dependent variable, which represents the signed area under the graph of $f(t)$ from $t=0$ to $t=x$; t does not affect the dependent or independent variable.

7. (a) As x increases, the signed area $A(x)$ accumulated also increases, but at a rate that decreases. Thus the rate of change of the rate of change of $A(x)$ is negative, so $A(x)$ is concave down). (b) $A''(x)=\frac{d}{dx}\left(\frac{d}{dx}\int_0^x f(t)dt\right)=\frac{d}{dx}(f(x))=f'(x)$; therefore, f decreasing $\Rightarrow f'$ negative $\Rightarrow A''$ negative $\Rightarrow A$ concave down.

9. (a) $A'(x)$ and $B'(x)$ are both equal to x^2, so $A(x)$ and $B(x)$ differ by a constant. (b) $A(x)-B(x)=\int_0^x t^2\,dt-\int_3^x t^2\,dt=\int_0^3 t^2\,dt$, which is a constant. (c) $\int_0^3 t^2\,dt$ is the signed area under the graph of $y=x^2$ from $x=0$ to $x=3$.

11. $F(x)=A(g(x))$, where $A(x)=\int_3^x \sin t\,dt$ is outside and $g(x)=x^2$ is inside.

13. $A(5)<A(0)<A(-1)<A(-2)$

15. $A(x)$ is positive on approximately $[0,2]$ and negative on $[2,6]$. $A(x)$ increases on $[0,1]$, then decreases on $[1,5]$, then increases again on $[5,6]$. The graph of A has the following shape:

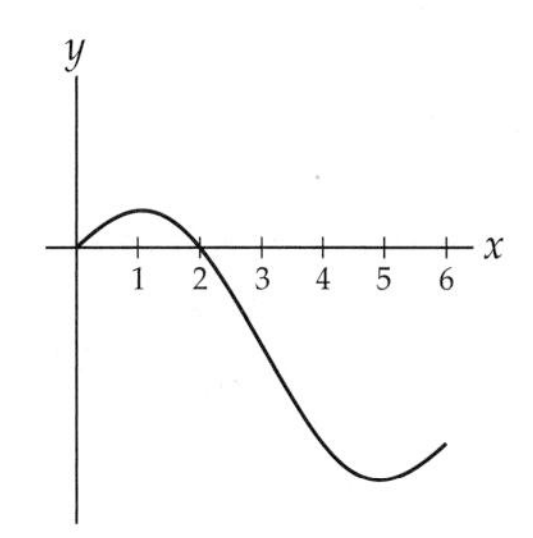

17. See the proof of Theorem 4.35.

19. (a) $f(b)-f(a)$; (b) 0.

21. (a) $\ln x=\int_1^x \frac{1}{t}\,dt$ is defined on $(0,\infty)$ because $\frac{1}{x}$ is continuous on $(0,\infty)$. Moreover, $\ln x=\int_1^x \frac{1}{t}\,dt$ is zero for $x=1$, increases without bound as $x\to\infty$, and decreases without bound as $x\to 0^+$; therefore $\ln x$ has range $(-\infty,\infty)$. (b) Follows from the fact that we define e^x to be the inverse of $\ln x$.

23. $\ln 10-1$

25. $\log_b x=\frac{1}{\ln b}\int_1^x \frac{1}{t}dt$, or $\log_b x=\int_1^x \frac{1}{(\ln b)t}\,dt$

27. (a) $A(2)$ is the signed area under the graph of t^2+1 from $t=1$ to $t=2$; (b) $A(2)=\frac{10}{3}$, $A(5)=\frac{136}{3}$; (c) $A(x)=\frac{1}{3}x^3+x-\frac{4}{3}$

29. (a) $A(x)$ is the signed area under the graph of $\frac{1}{t}$ from $t=3$ to $t=x$; (b) $A(2)=\ln 2-\ln 3$, $A(5)=\ln 5-\ln 3$; (c) $A(x)=\ln x-\ln 3$

31. $\int_0^x \sin^2(3^t)\,dt$, $\int_2^x \sin^2(3^t)\,dt$, $\int_{-1}^x \sin^2(3^t)\,dt$

33. $\int_0^x e^{(t^2)}\,dt$, $\int_{-2}^x e^{(t^2)}\,dt$, $\int_{10}^x e^{(t^2)}\,dt$

35. $-e^{x^2+1}$

37. $\cos(x^2)(2x)$

39. $\int_1^{\sqrt{x}} \ln t\,dt+\frac{1}{2}\sqrt{x}\ln\sqrt{x}$

41. 0

43. $-2\ln|x^2|-4$

45. $54x$

47. $\sin((x+2)^2)-\sin(x^2)$

49. $f(x)=\frac{1}{2}\ln|2x-1|+3$

51. $f(x)=\int_2^x \frac{1}{t^3+1}dt$

53. $f(x)=\int_1^x e^{-t^2}\,dt$

55. Using left and right sums with 9 rectangles, $1.92897<\ln 10<2.82897$.

57. Using left and right sums with four rectangles, $0.877269<\ln e<1.14881$.

59. (a) $s(t) = \int_0^t \sin(0.1w^2)dw$; (b) about 2.41143 feet to the right of the starting position; (c) each rectangle has a height measured in feet per second and a width measured in seconds, thus an area that is measured in feet; (d) by looking at the graph of $v(t) = \sin(0.1t^2)$, the velocity oscillates faster and faster between 1 and -1 feet per second.

61. (a) $F_{t_0}(t) = \int_{t_0}^t 86400f(x)dx = 86400[-1.033x^3 + 441.75x^2 - 53605x]_{t_0}^t$; (b) $F_{90}(195) \approx 58.9$ billion cubic feet; (c) about 17.971 billion cubic feet, about 76% of which flows through between days 90 and 195.

63. If G is an antiderivative of f, then by the Fundamental Theorem of Calculus we have $\int_a^b f(x)\,dx = G(b) - G(a)$. Therefore, $A(x) = \int_0^x f(t)\,dt = G(x) - G(0)$. Since A and G differ by a constant, they have the same derivative, namely, f; thus A is an antiderivative of f.

65. See the proof given in the reading.

67. This is a just a rewording of the Second Fundamental Theorem, since $F(x) = \int_a^x f(t)\,dt$ and the Second Fundamental Theorem tells us that the derivative of F is f.

69. (a) $F' = f$, F is continuous on $[a, b]$, and F is differentiable on (a, b). (b) There exists a number $c \in (a, b)$ such that $F'(c) = \frac{F(b)-F(a)}{b-a}$. (c) $f(c) = F'(c) = \frac{F(b)-F(a)}{b-a} = \frac{\int_a^b f(t)\,dt - \int_a^a f(t)\,dt}{b-a} = \frac{\int_a^b f(t)\,dt}{b-a}$.

71. Since $\frac{1}{t}$ is continuous for $t > 0$, the first part of the Second Fundamental Theorem guarantees that if $\ln x = \int_1^x \frac{1}{t}\,dt$, then $\ln x$ is continuous and differentiable on $(0, \infty)$.

73. $\ln 1 = \int_1^1 \frac{1}{t}\,dt = 0$. Since the signed area under $\frac{1}{t}$ is positive over any interval in $(0, \infty)$, if $0 < x < 1$, then $\ln x = \int_1^x \frac{1}{t}\,dt = -\int_x^1 \frac{1}{t}\,dt$ is negative. Similarly, if $x > 1$, then $\ln x = \int_1^x \frac{1}{t}\,dt$ is positive.

75. (a) $\frac{d}{dx}(\ln ax) = \frac{d}{dx}\int_1^{ax} \frac{1}{t}\,dt = \frac{1}{ax}(a) = \frac{1}{x}$. (b) $\ln ax$ and $\ln x$ have the same derivative, so they must differ by a constant, that is, $\ln ax = \ln x + C$. (c) $\ln a = \ln 1 + C$, so $\ln a = 0 + C$, so $C = \ln a$. Then just rename x to be b to complete the proof.

77. $\ln \frac{a}{b} = \ln(ab^{-1}) = \ln a + \ln(b^{-1}) = \ln a - \ln b$

For answers to odd-numbered Skill Certification exercises in the Chapter Review, please visit the Book Companion Web Site at www.whfreeman.com/tkcalculus.

Chapter 5

Section 5.1

1. T, F, F, F, T, F, T, F.

3. Both integrals turn into $\int \frac{1}{u}\,du$ after a change of variables; $u = x^2 + 1$ in the first case, $u = \ln x$ in the second.

5. Three possible answers are $\int 3x^2 \sin(x^3)\,dx$, $\int \frac{\sin(\ln x)}{x}\,dx$, and $\int \frac{\sin(\sqrt{x}+2)}{2\sqrt{x}}\,dx$.

7. Three possible answers are $\int \frac{3}{\sqrt{3x+1}}\,dx$, $\int \frac{\cos x}{\sqrt{\sin x}}\,dx$, and $\int \frac{1}{x\sqrt{\ln x}}\,dx$.

9. $du = (2x + 1)\,dx$

11. $du = \cos x\,dx$

13. $\int_{-1}^5 u^2\,du = \frac{126}{3} = 42$; $\int_{x=-1}^{x=5} u^2\,du = \frac{1}{3}(5)^6 - \frac{1}{3}(-1)^6 = 5208$; $\int_{u(-1)}^{u(5)} u^2\,du = \int_1^{25} u^2\,du = \frac{1}{3}(25)^3 - \frac{1}{3}(1)^3 = 5208$

15. Three of the many such integrals are in Exercises 34, 37, and 51.

17. (a) $\frac{1}{2}\int u^2\,du = \frac{1}{6}(x^2 - 1)^3 + C$; (b) $\int (x^5 - 2x^3 + x)\,dx = \frac{1}{6}x^6 - \frac{1}{2}x^4 + \frac{1}{2}x^2 + C$; (c) the answers differ by a constant: $\frac{1}{6}(x^2 - 1)^3 = \frac{1}{6}x^6 - \frac{1}{2}x^4 + \frac{1}{2}x^2 - \frac{1}{6}$.

19. (a) $-\frac{1}{3}\int_{x=-2}^{x=2} e^u\,du = -\frac{1}{3}[e^{5-3x}]_{-2}^2 = -\frac{1}{3}(e^{-1} - e^{11})$; (b) $-\frac{1}{3}\int_{11}^{-1} e^u\,du = -\frac{1}{3}[e^u]_{11}^{-1} = -\frac{1}{3}(e^{-1} - e^{11})$

21. $\frac{1}{9}(3x + 1)^3 + C$

23. $4\ln(x^2 + 1) + C$

25. $\frac{1}{2}(\sqrt{x} + 3)^2 + C$

27. $-\frac{1}{2}\cot x^2 + C$

29. $\frac{1}{3}\ln|3x + 1| + C$

31. $\frac{1}{2\pi}\sin^2 \pi x + C$

33. $\frac{1}{2}\sec 2x + C$

35. $-\frac{1}{6}\cot^6 x + C$

37. $\frac{1}{11}x^{11} + \frac{1}{4}x^8 + \frac{1}{5}x^5 + C$

39. $-\frac{4}{5}\cos(x^{5/4}) + C$

41. $\frac{1}{2\ln 2}2^{x^2+1} + C$

43. $\frac{2}{3}(\ln x)^{3/2} + C$

45. $\frac{1}{4}(\ln x)^2 + C$

47. $\frac{3}{2}\ln(x^2 + 1) + \tan^{-1} x + C$

49. $-\ln|2 - e^x| + C$

51. $-2e^{-x} - x + C$

53. $-\frac{1}{8}\cos^8 x + C$

55. $(\sin x^2)^{1/2} + C$

57. $\cos\left(\frac{1}{x}\right) + C$

59. $\frac{1}{3}(x^2 + 1)^{3/2} + C$

61. $\frac{2}{5}(x + 1)^{5/2} - \frac{2}{3}(x + 1)^{3/2} + C$

63. $\frac{2}{27}(3x+1)^{3/2} - \frac{2}{9}(3x+1)^{1/2} + C$

65. $-\cos(e^x) + C$

67. $\frac{1}{103}(x-1)^{103} + \frac{2}{102}(x-1)^{102} + \frac{1}{101}(x-1)^{101} + C$

69. $\frac{1}{2}\cosh x^2 + C$

71. $\frac{3}{8}(10^{4/3} - 1)$

73. 0

75. $2\sqrt{\ln 3} - 2\sqrt{\ln 2}$

77. $\frac{\pi}{8}$

79. $1/4 + (5/2)\ln 2 - (1/2)\ln 10$

81. (a) 0; (b) $c = 1$.

83. 4.27569 inches

85. 33.2128 cubic inches

87. (a) $\int_0^{k\pi} \sin\left(2\left(x - \frac{\pi}{4}\right)\right) dx = -\frac{1}{2}\sin(2k\pi) = 0$ for any integer k; (b) f is a general sine function with period π whose graph is half above and half below the x-axis.

89. Since $\frac{d}{dx}(f(u(x))) = f'(u(x))u'(x)$, we have $\int f'(u(x))u'(x)\,dx = f(u(x)) + C$.

91. See the proof of Theorem 5.3.

Section 5.2

1. F, T, F, F, T, F, T, F.

3. $\int u\,dv = uv - \int v\,du$; (b) $\int u(x)v'(x)\,dx = u(x)v(x) - \int v(x)u'(x)\,dx$

5. $\int x\sin x\,dx$; $\int x\ln x\,dx$

7. $\int \ln x\,dx$

9. If $u = 1$, then integrating dv is equivalent to solving the original integral.

11. Exercises 28, 27, and 30 are three such examples.

13. $du = 3\cos 3x\,dx$ and $dv = dx$; $\int u\,dv = \int \sin 3x\,dx$, and $\int v\,du = \int 3x\cos 3x\,dx$. The first integral is clearly easier, and the second integral can be rewritten in terms of the first using integration by parts.

15. $[g(x)]_a^b - [h(x)]_a^b = (g(b) - g(a)) - (h(b) - h(a)) = (g(b) - h(b)) - (g(a) - h(a)) = [g(x) - h(x)]_a^b$.

17. (a) $\frac{d}{dx}(x\ln x) = \ln x + 1$; (b) $\int(\ln x + 1)\,dx = x\ln x + C$; (c) $\int \ln x\,dx = x\ln x - \int 1\,dx$

19. $u = x$, $v = \frac{1}{\ln 2}2^x$

21. $u = \ln x$, $v = -x^{-2}$

23. $u = \tan^{-1}x$, $dv = dx$

25. (a) $x^3\ln x - \frac{1}{3}x^3 + C$; (b) $\frac{1}{3}(x^3\ln x^3 - x^3) + C$; (c) Your answers should differ by a constant, which in this case is $C = 0$.

27. $xe^x - e^x + C$

29. $\frac{1}{2}x^2\ln x - \frac{1}{4}x^2 + C$

31. $-\frac{1}{2}\cos x^2 + C$

33. $\frac{1}{3}x^2e^{3x} - \frac{2}{9}xe^{3x} + \frac{2}{27}e^{3x} + C$

35. $-xe^x - e^x + C$

37. $\frac{3}{2}e^{x^2} + C$

39. $3(x\ln x - x)$

41. $-e^{-x}(x^2+1) - 2xe^{-x} - 2e^{-x} + C$

43. $3x + \ln x + C$

45. $\frac{1}{3}x^3 - 2(xe^x - e^x) + \frac{1}{2}e^{2x} + C$

47. $\frac{2}{3}x^{3/2}\ln x - \frac{4}{9}x^{3/2} + C$

49. $\frac{1}{2}x^2e^{x^2} - \frac{1}{2}e^{x^2} + C$

51. $-x\cot x + \ln|\sin x| + C$

53. $x^3\sin x + 3x^2\cos x - 6x\sin x - 6\cos x + C$

55. $-\frac{1}{2}x^2\cos x^2 + \frac{1}{2}\sin x^2 + C$

57. $x\tan^{-1}3x - \frac{1}{6}\ln|1 + 9x^2| + C$

59. $-\frac{1}{5}e^{2x}\cos x + \frac{2}{5}e^{2x}\sin x + C$

61. $\frac{1}{\ln(2/3)}(2/3)^x + C$

63. $\frac{1}{2}e^{-x}\sin x - \frac{1}{2}e^{-x}\cos x + C$

65. $x\tan x - \ln|\sec x| + C$

67. $\frac{1}{2}x^2\tan^{-1}x + \frac{1}{2}\tan^{-1}x - \frac{1}{2}x + C$

69. $\frac{1}{2}x\sqrt{x^2+1} - \frac{1}{2}\sinh^{-1}x + C$

71. $2\ln 2 - 1$

73. $-\frac{2}{e}$

75. $\frac{1}{2}e^{\pi} + \frac{1}{2}$

77. $\frac{1}{4}(\pi + 2\ln 2)$

79. (a) $-\frac{2}{e}$; (b) $1 + \frac{-2+e}{e}$

81. (a) $-\frac{1}{4} - \frac{3\pi}{8}$; (b) $\frac{5\pi}{8}(2 + \pi)$

83. 4.27569 inches

85. About 15.47 cubic inches

87. (a) $-\frac{5}{\pi}\cos(\pi t) + \frac{5}{\pi^2}\sin(\pi t)$; (b) the particle moves right, then left, then right again, a greater distance each time – see the picture of $s(t)$ that follows; (c) at $t = 3$ seconds it is more than 4 feet to the right, and it is just turning around to come back to the starting position; (d) after a long time the particle oscillates back and forth by greater and greater distances, moving faster and faster to cover more distance in the same time period.

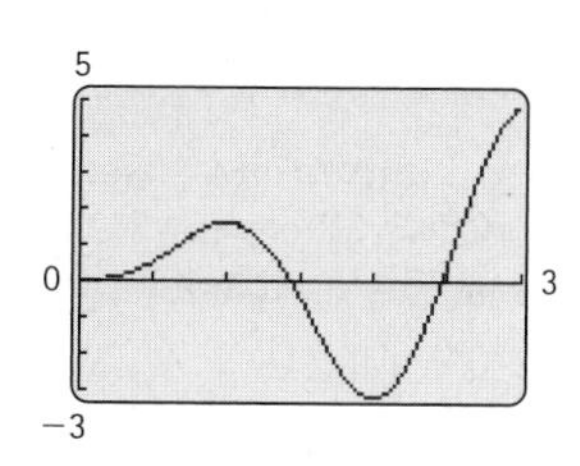

89. (a) See the proof of Theorem 5.8;
(b) $\frac{d}{dx}(x\ln x - x) = \ln x + x\left(\frac{1}{x}\right) - 1 = \ln x$.

91. (a) Let $u = \tan^{-1} x$ and $dv = dx$ to get $x\tan^{-1} x - \int \frac{x}{x^2+1}\,dx$, and then use substitution with $w = x^2+1$ to turn this into $x\tan^{-1} x - \frac{1}{2}\int \frac{1}{w}dw = x\tan^{-1} x - \frac{1}{2}\ln|w| + C = x\tan^{-1} x - \frac{1}{2}\ln(x^2+1) + C$.
(b) $\frac{d}{dx}(x\tan^{-1} x - \frac{1}{2}\ln(x^2+1)) = \tan^{-1} x + x\left(\frac{1}{x^2+1}\right) - \frac{1}{2}\left(\frac{1}{x^2+1}\right)(2x) = \tan^{-1} x$.

Section 5.3

1. F, T, T, T, T, F, F, T.

3. A rational function is a function that can be expressed as a quotient of polynomials. A rational function is proper if the degree of its numerator is strictly less than the degree of its denominator, and improper otherwise.

5. $p(x) = q(x)(x^2 - x + 3) + (3x + 1)$; $\frac{p(x)}{q(x)} = (x^2 - x + 3) + \frac{3x-1}{q(x)}$

7. Divide both sides by $q(x)$ and simplify.

9. $m(x)$ will have degree 2 and the degree of $R(x)$ will be strictly less than the degree of $q(x)$.

11. Hint: Try an example, and then think about it in general.

13. If $q(x)$ is an irreducible quadratic then $\frac{p(x)}{q(x)}$ is its own partial fractions decomposition; there is nothing further we can decompose. If $q(x)$ is a reducible quadratic then $q(x)$ is a product $l_1(x)l_2(x)$ of two linear functions, and we can obtain a partial fractions decomposition of the form $\frac{A_1}{l_1(x)} + \frac{A_2}{l_2(x)}$; in the example given, we have $l_1(x) = x - 1$ and $l_2(x) = x - 2$.

15. $y = x^2 - 3x + 5 = \left(x - \frac{3}{2}\right)^2 + \frac{11}{4}$ can be obtained from the graph of $y = x^2$ by translating $\frac{3}{2}$ units to the right and $\frac{11}{4}$ units up.

17. $\ln|x-2| - \ln|x+1| + C$

19. $\ln|x-1| - 2(x-1)^{-1} + C$

21. $x + \ln|x| + \frac{6}{x} + C$

23. $\frac{5}{2}x^2 - 5x + 8\ln|x-2| + 27\ln|x+3| + C$

25. $\frac{1}{7}\left(2\ln|x-2| - \ln|x^2+3| + \sqrt{3}\tan^{-1}\left(\frac{1}{\sqrt{3}}x\right)\right) + C$

27. $2\ln|x-1| - \ln|x^2+x+1| + C$

29. $\ln|x^2-4x+5| + 2\tan^{-1}(x-2) + C$

31. $3\ln|x-2| + (x-2)^{-1} + \frac{1}{2}(x-2)^{-2} + C$.

33. $\frac{1}{3}\ln|3x+1| - 2\ln|x-3| + C$

35. $\frac{1}{2}\ln|x^2+1| + \tan^{-1} x - \frac{1}{2}\ln|x^2+3| + C$

37. $-x^{-1} - 2\ln|x| + \ln|x^2+1| - \tan^{-1} x + C$

39. $-\ln|x+1| + \frac{3}{2}\ln(x^2+1) + \tan^{-1} x + C$

41. $x^2 + 3\ln|x^2-4| + C$

43. $\frac{1}{3}x^3 - \frac{5}{2}x^2 + 6x + C$

45. $\frac{2}{3}x^3 - 2x + \frac{1}{2}\ln|x^2-4x+1| + C$

47. $\ln 2 - \ln 6 + \ln 5$

49. $\ln 3 - \ln 4 + \ln 2$

51. $2 + \ln 4 - \ln 2$

53. $-\ln|\cos x| + \frac{1}{1+\ln x} + C$

55. $\ln|1+\ln x| + \frac{1}{1+\ln x} + C$

57. (a) $-\frac{\ln 3}{2}$; (b) $\frac{1}{4}$

59. (a) $\frac{1}{2}\left(\ln 5 - \frac{2}{5}\right) \approx 0.605$; (b) $x \approx 0.669$; note that $f(0.668) \approx 0.605$.

61. $r(t) = \ln\left(\frac{p(t)(K-p_0)}{p_0(K-p(t))}\right)$, and thus $p(t) = \frac{p_0Ke^{r(t)}}{K-p_0+p_0e^{r(t)}}$.

63. Either mimic the example after Theorem 5.14 by adding and subtracting $\left(\frac{b}{2}\right)^2$, or else just simply multiply out $f(x) = \left(x + \frac{b}{2}\right)^2 + \left(c - \frac{b^2}{4}\right)$ and show that it is equal to $x^2 + bx + c$.

Section 5.4

1. F, F, F, F, F, F, T, T.

3. $\frac{d}{dx}\left(-\frac{1}{5}\cos^5 x + \frac{1}{7}\cos^7 x\right) = \cos^4 x\sin x - \cos^6 x\sin x = \cos^4 x\sin x(1-\cos^2 x) = \cos^4 x\sin x\sin^2 x = \sin^3 x\cos^4 x$.

5. Multiply the integrand by $\frac{\csc x+\cot x}{\csc x+\cot x}$.

7. Three possible examples are the integrals in Exercises 21, 22, and 41.

9. Reduce powers with double-angle identities

11. Apply the identity $\cot^2 x = \csc^2 x - 1$ repeatedly until the integrand is converted into a sum of expressions that we can integrate. The power of cotangent will eventually decrease to zero.

13. Use parts with $u = \csc^{k-2} x$ and $dv = \csc^2 x$.

15. Reduce powers with double-angle identities.

17. Use $\tan^2 x = \sec^2 x - 1$ to rewrite the integrand entirely in terms of secant, saving one copy of $\sec x\tan x$ for the differential of the substitution $u = \sec x$.

19. (a) Rewrite as $\int(1-\cos^2 x)\cos^2 x\,dx = \int \cos^2 x\,dx - \int \cos^4 x\,dx$ and then apply double-angle identities to reduce powers;
(b) $\int\left(\frac{1}{2}\sin 2x\right)^2 dx = \frac{1}{8}\left(x - \frac{1}{4}\sin 4x\right) + C$.

21. $\frac{1}{2}x - \frac{1}{12}\sin(6x) + C$

23. $\sin x - \frac{2}{3}\sin^3 x + \frac{1}{5}\sin^5 x + C$

25. $-\frac{1}{2}\ln|\cos 2x| + C$

27. $\frac{1}{4}\ln|\sec 4x + \tan 4x| + C$

29. $-\frac{1}{4}\cot^4 x + \frac{1}{2}\cot^2 x + \ln|\sin x| + C$

31. $\frac{1}{4}\sec 2x \tan 2x + \frac{1}{4}\ln|\sec 2x + \tan 2x| + C$

33. $\frac{3}{8}x + \frac{1}{4}\sin 2x + \frac{1}{32}\sin 4x + C$

35. One way is to use parts with $u = \csc^2 x$ and $dv = \csc^2 x\,dx$ to get $-\csc^2 x \cot x - 2\int \csc^2 x \cot^2 x\,dx$. Then use substitution with $u = \cot x$ to get $-\csc^2 x \cot x + \frac{2}{3}\cot^3 x + C$.

37. $\frac{1}{5}\sin^5 x - \frac{1}{7}\sin^7 x + C$

39. $\frac{1}{3}\sec^3 x - \sec x + C$

41. $-\frac{1}{8}x - \frac{1}{96}\sin(12x) + C$

43. $\frac{1}{8}\sec^8 x + C$

45. $\frac{1}{7}\sec^7 x - \frac{2}{5}\sec^5 x + \frac{1}{3}\sec^3 x + C$

47. $\frac{1}{14}\sin^{14} x - \frac{1}{8}\sin^{16} x + \frac{1}{18}\sin^{18} x + C$

49. $-\frac{1}{2}\csc x \cot x + \frac{1}{2}\ln|\csc x + \cot x| + C$

51. $\frac{1}{3}\sec^3 x + C$

53. $-\ln|\cos x| + C$

55. $\sin x + C$

57. $\frac{1}{\cos x} + C$

59. $-\frac{1}{2}\cos^2 x \ln(\cos x) + \frac{1}{4}\cos^2 x + C$

61. $\frac{2}{9}\cos^{9/2} x - \frac{2}{5}\cos^{5/2} x + C$

63. $\frac{1}{3}\cosh^3 x + C$

65. $\frac{1}{7}\cosh^7 x + C$

67. $\frac{16}{15}$

69. $\frac{4}{3}$

71. $\frac{4}{15}$

73. $\frac{8}{15}$

75. $\frac{1}{3}$

77. (a) 0; (b) $\frac{32}{15}$; (c) $\frac{16}{15\pi}$

79. (a) 1.5; (b) $\frac{1}{2} + \frac{3\sqrt{3}}{\pi}$

81. $\ln(7 + 4\sqrt{3})$ inches

83. See the proof of Theorem 5.17 in the reading.

85. (a) $\int \sin^4 x\,dx = -\frac{1}{4}\sin^3 x \cos x + \frac{3}{8}x - \frac{3}{16}\sin 2x + C$, and $\int \sin^8 x\,dx = -\frac{1}{8}\sin^7 x \cos x - \frac{7}{48}\sin^5 x \cos x - \frac{35}{192}\sin^3 x \cos x + \frac{35}{128}x - \frac{35}{256}\sin 2x$. (b) The formula reduces the power of $\sin x$ in the integrand by two every time we apply it. (c) Apply integration by parts with $u = \sin^{k-1} x$ and $dv = \sin x\,dx$, so that $v = -\cos x$ and $du = (k-1)\sin^{k-2} x \cos x\,dx$. After applying the integration by parts formula, multiply out the integrand to obtain a term of the form $(k-1)\int \sin^k x\,dx$. Then solve for $\int \sin^k x\,dx$.

Section 5.5

1. T, T, T, F, F, T, F, T.

3. (a) $1 = \sec^2 u \frac{du}{dx}$, so $dx = \sec^2 u\,du$; (b) $\frac{dx}{du} = \sec^2 u$, so $dx = \sec^2 u\,du$

5. See the reading.

7. In terms of the Pythagorean identity, sine and cosine behave the same way; we can always use $x = \sin u$ instead of $x = \cos u$.

9. $x = a\sin u$ makes sense only for $x \in [-1, 1]$, but the domain of $a^2 - x^2$ is $(-\infty, \infty)$. On the other hand, $x = a\tan u$ makes sense for $x \in (-\infty, \infty)$, i.e., everywhere that $x^2 + a^2$ is defined.

11. (a) Use algebra; (b) use conventional substitution with $u = 4 + x^2$; (c) use trigonometric substitution with $x = 2\tan u$; (d) use algebra after writing $16 - x^4 = (4 - x^2)(4 + x^2)$.

13. See the proof of Theorem 5.18.

15. See the proof of Theorem 5.18.

17. $\sqrt{1 - x^2}$ is not defined on [2, 3]; its domain is [−1, 1].

19. Three of the many such integrals are in Exercises 41, 42, and 46.

21. $\sqrt{1 - x^2}$

23. $\frac{x}{\sqrt{x^2+4}}$

25. $\tan^2 u = \frac{(x-5)^2}{9-(x-5)^2}$

27. $\sin(2\cos^{-1} x) = 2x\sqrt{1 - x^2}$

29. $(x + 3)^2 - 11$; $x + 3 = \sqrt{11}\sec u$

31. $\left(x - \frac{5}{2}\right)^2 - \frac{21}{4}$; $x - \frac{5}{2} = \frac{\sqrt{21}}{2}\sec u$

33. $12 - 2(x + 2)^2$; $x + 2 = \sqrt{6}\sin u$

35. (a) $\int \frac{1}{9\tan^2 u+9}(3\sec^2 u)\,du = \frac{1}{3}\tan^{-1}\frac{x}{3} + C$;
(b) $\frac{1}{9}\int \frac{1}{\left(\frac{x}{3}\right)^2+1}\,dx = \frac{1}{3}\tan^{-1}\frac{x}{3} + C$.

37. (a) $\frac{1}{2}\int \frac{1}{u}\,du = \frac{1}{2}\ln|4 + x^2| + C$
(b) $\int \frac{2\tan u}{4+4\tan^2 u}(2\sec^2 u)\,du = -\ln\left|\frac{2}{\sqrt{x^2+4}}\right| + C$

39. $-\frac{\sqrt{4-x^2}}{x} - \sin^{-1}\frac{x}{2} + C$

41. $(x^2 - 1)^{1/2} + C$

43. $\sin^{-1}\frac{x}{\sqrt{3}} + C$

45. $\sqrt{9 - x^2} - 3\ln\left(\frac{\sqrt{9-x^2}+3}{x}\right) + C$

47. $\frac{1}{2}(x - 2)\sqrt{1 - (x - 2)^2} + \frac{1}{2}\sin^{-1}(x - 2) + C$

49. $\frac{\sqrt{x^2-9}}{9x} + C$

51. $\frac{x}{\sqrt{1-x^2}} + C$

53. $\frac{1}{3}\tan^{-1}\left(\frac{x-2}{3}\right) + C$

55. $-\frac{1}{4}\sqrt{9-4x^2}+\frac{3}{2}\sin^{-1}\frac{2x}{3}+C$

57. $\frac{1}{2}(x-4)\sqrt{(x-4)^2+9}+\frac{9}{2}\ln\left|\frac{1}{3}\sqrt{(x-4)^2+9}+\frac{1}{3}(x-4)\right|+C$

59. $\sqrt{3-x^2}\left(\frac{15}{8}x-\frac{1}{4}x^3\right)+\frac{27}{8}\sin^{-1}\left(\frac{x}{\sqrt{3}}\right)+C$

61. $\frac{1}{2}x\sqrt{2-x^2}+\sin^{-1}\left(\frac{x}{\sqrt{2}}\right)+C$

63. If $x>1$, we have $\sqrt{x^2-1}-\sec^{-1}x+C$; if $x<-1$, we have $\sqrt{x^2-1}+\sec^{-1}x+C$.

65. $\ln(2(\sqrt{x^2-2}+x))+C$

67. $\frac{1}{5}(x^2+1)^{5/2}-\frac{1}{3}(x^2+1)^{3/2}$

69. $x\ln(1+x^2)-2x+2\tan^{-1}x+C$

71. $\frac{1}{3}(e^{2x}+1)^{3/2}+C$

73. $-\frac{1}{3}\frac{\sqrt{e^{2x}+3}}{e^x}+C$

75. $\frac{1}{2}x^2-2\ln(x^2+4)+C$

77. $-\frac{2}{9}x+\frac{1}{9}x^3-\frac{23}{9\sqrt{6}}\tan^{-1}\left(\sqrt{\frac{3}{2}}x\right)+C$

79. $\frac{1}{3}(20)^{3/2}-\frac{1}{3}(4)^{3/2}$

81. $\frac{1}{3}\ln 10-\frac{1}{3}\ln(3+\sqrt{34})$

83. $-\sqrt{3}+\sqrt{15}$

85. $\frac{2}{3}-\frac{1}{3\sqrt{7}}$

87. $\sinh^{-1}\frac{x}{2}+C$

89. $\sinh^{-1}x-\tanh(\sinh^{-1}x)+C$

91. (a) $4\pi\int_{-\sqrt{42}}^{\sqrt{42}}\sqrt{42-x^2}\,dx$; (b) using trigonometric substitution with $x=\sqrt{42}\sin u$, we obtain a volume of approximately 829 cubic millimeters.

93. (a) $2\int_{-r}^{r}\sqrt{r^2-x^2}\,dx$

(b) $2\int_{-\pi/2}^{\pi/2}\sqrt{r^2-r^2\sin^2 u}(r\cos u)\,du=$

$2r^2\int_{-\pi/2}^{\pi/2}\cos^2 u\,du=2r^2\left(\frac{1}{2}\right)\int_{-\pi/2}^{\pi/2}(1+\cos 2u)\,du=$

$r^2\left[u+\frac{1}{2}\sin 2u\right]_{-\pi/2}^{\pi/2}=\pi r^2$

Section 5.6

1. F, F, F, F, T, T, F, T.

3. The interval over which we are integrating could be infinite, or the function to be integrated could have a discontinuity or a vertical asymptote on the interval.

5. The integrand $\frac{1}{e^x}$ is continuous on $[0,1]$, in particular with no vertical asymptotes; this integral is not improper.

7. The Fundamental Theorem of Calculus does not apply here because $(x-3)^{-4/3}$ is not continuous at $x=3$, inside the interval $[0,5]$. Applying it anyway would give the incorrect answer $\frac{-3}{2^{1/3}}+\frac{3}{(-3)^{1/3}}$, when the correct answer is that the improper integral diverges.

9. If $p\ge 1$ and $x\in[0,1]$ then $\frac{1}{x^p}$ is greater than $\frac{1}{x}$, whose improper integral on $[0,1]$ is known to diverge.

11. Since $\frac{1}{x+1}<\frac{1}{x}$ for $x\ge 1$ and $\int_1^\infty\frac{1}{x}\,dx$ diverges, this comparison does not tell us anything.

13. $\frac{1}{x+1}\le\frac{1}{x}$, but knowing that an improper integral is less than a divergent one does not give us any information. Instead, note that the graph of $y=\frac{1}{x+1}$ is just a horizontal translation of the graph of $y=\frac{1}{x}$. This means their improper integrals are related; how?

15. We can split at any point in $[1,\infty)$, for example at $x=2$ to get: $\int_1^2\frac{1}{x-1}\,dx+\int_2^\infty\frac{1}{x-1}\,dx$.

17. $\int_0^2\frac{1}{2x^2-10x+12}\,dx+\int_2^{2.5}\frac{1}{2x^2-10x+12}\,dx+\int_{2.5}^3\frac{1}{2x^2-10x+12}\,dx+\int_3^5\frac{1}{2x^2-10x+12}\,dx$

19. Choosing $x=\pm 1$ for the splits, we have

$$\int_{-\infty}^{-1}\frac{e^{-x^2}}{x^2}\,dx+\int_{-1}^{0}\frac{e^{-x^2}}{x^2}\,dx+\int_0^1\frac{e^{-x^2}}{x^2}\,dx+\int_1^\infty\frac{e^{-x^2}}{x^2}\,dx.$$

21. 3

23. Diverges

25. 100

27. Diverges

29. $2^{1/3}+1$

31. Diverges

33. Diverges

35. $\frac{1}{2}$

37. 2

39. 2

41. 0

43. Diverges

45. Diverges

47. Diverges

49. 1

51. 2

53. Diverges

55. Diverges

57. Diverges by comparison with $\frac{1}{x}$

59. Converges by comparison with $\frac{1}{x^2}$

61. Diverges by comparison with $\frac{1}{x}$

63. Converges by comparison with $\frac{1}{x^2}$

65. Converges by limit comparison with $\frac{1}{x^2}$

67. Converges by limit comparison with $\frac{1}{x^2}$

69. Diverges by comparison with $\frac{1}{x^{1/3}}$

71. About 1.25 pounds.

73. About 3.23 years.

75. If $p > 1$, then $\int_1^\infty \frac{1}{x^p}\,dx = \lim_{B\to\infty}\int_1^B x^{-p}\,dx = \lim_{B\to\infty}\left[\frac{1}{-p+1}x^{-p+1}\right]_1^B = \lim_{B\to\infty}\left(\frac{1}{1-p}B^{1-p} - \frac{1}{1-p}\right) = \frac{1}{p-1}$. Note that we used the fact that $p > 1$ to take the limit, since in that case we have $1-p < 0$ and $B^{1-p} \to 0$ as $B \to \infty$.

77. If $p \geq 1$, then $\int_0^1 \frac{1}{x^p}\,dx = \lim_{A\to 0^+}\int_A^1 x^{-p}\,dx = \lim_{A\to 0^+}\left[\frac{1}{-p+1}x^{-p+1}\right]_A^1 = \lim_{A\to 0^+}\left(\frac{1}{1-p} - \frac{1}{1-p}A^{1-p}\right)$ diverges.

79. Mimic the proof of the case of Theorem 5.23 in the reading.

81. We will suppose that $d > c$; the case where $d < c$ is similar. We have $\int_{-\infty}^d f(x)\,dx + \int_d^\infty f(x)\,dx = (\int_{-\infty}^c f(x)\,dx + \int_c^d f(x)\,dx) + \int_d^\infty f(x)\,dx = \int_{-\infty}^c f(x)\,dx + (\int_c^d f(x)\,dx + \int_d^\infty f(x)\,dx) = \int_{-\infty}^c f(x)\,dx + \int_c^\infty f(x)\,dx$.

Section 5.7

1. T, T, F, T, T, T, T, F.

3. Use pictures similar to the three immediately before Theorem 5.24, but with an increasing function.

5. One example is $f(x) = -x^2 + 2x + 3$ on $[0, 3]$, with $n = 3$ left and right sums.

7. The same example from Problem 5 works here as well.

9. Midpoint sum

11. Do the approximation separately on $[a, c]$ and on $[c, b]$ and calculate bounds for the error on each piece using Theorem 5.27. The total error will be less than or equal to the sum of the absolute values of the two errors you calculated.

13. See the discussion in the reading prior to Theorem 5.27.

15. (a) LEFT(n) < MID(n) < TRAP(n) < RIGHT(n); (b) between TRAP(n) and MID(n).

17. Using the weighted average from Exericse 61, the $n = 1000$ Simpson's Rule approximation is 1.09861229, which we would expect to be an even better approximation.

19. By the old formula we have $\int_1^5 (3x^2 - x + 4)\,dx = 3\int_1^5 x^2\,dx - \int_1^5 x\,dx + \int_1^5 4\,dx = 3\left(\frac{1}{3}\right)(5^3 - 1^3) - \left(\frac{1}{2}\right)(5^2 - 1^2) + 4(5-1) = 128$ and by the new formula we also have $\int_1^5 (3x^2 - x + 4)\,dx = \frac{5-1}{6}((3(1)^2 - 1 + 4) + 4(3(3)^2 - 3 + 4) + (3(5^2) - 5 + 4)) = 128$.

21. Mimic Example 3 but with $n = 4$ instead of $n = 6$.

23. RIGHT(4) $= 40$. Since $f'(x) = 2x + 1$ is positive on $[0, 4]$ we know that $f(x) = x^2 + x$ is monotonically increasing on $[0, 4]$. Therefore we can guarantee that $|E_{\text{RIGHT}(4)}| \leq |f(4) - f(0)|\left(\frac{4-0}{4}\right) = 20$. The actual area is about 29.33, which is within the error bounds of our approximation.

25. LEFT(6) $= 0.37404$. Since $f'(x) = -e^{-x}$ is negative on $[1, 3]$ we know that $f(x) = e^{-x}$ is monotonically decreasing on $[1, 3]$. Therefore we can guarantee that $|E_{\text{LEFT}(6)}| \leq |f(3) - f(1)|\left(\frac{3-1}{6}\right) = |e^{-3} - e^{-1}|\left(\frac{1}{3}\right) = 0.106031$. The actual area is approximately 0.318092, which is within the error bounds for our approximation.

27. TRAP(6) $= 0.321033$. Since $f''(x) = e^{-x}$ is positive on $[1, 3]$ we know that $f(x) = e^{-x}$ is concave up on $[1, 3]$. Moreover, $|f''(x)| = |e^{-x}| \leq e^{-1}$ for all $x \in [1, 3]$. Therefore we can guarantee that $|E_{\text{TRAP}(6)}| \leq \frac{e^{-1}(3-1)^3}{12(6)^2} = 0.00681258$.

29. MID(12) $= 7.63016$. Since $f''(x) = -\frac{1}{x^2}$ is negative on $[1, 7]$ we know that $f(x) = \ln x$ is concave down on $[1, 7]$. Moreover $|f''(x)| = |-\frac{1}{x^2}| \leq \frac{1}{1^2} = 1$ for all $x \in [1, 7]$. Therefore $|E_{\text{MID}(12)}| \leq \frac{1(7-1)^3}{24(12)^2} = 0.0625$.

31. TRAP(8) $= 3.25174$; using the fact that $|f''(x)| \leq 2$ for all $x \in [0, 2]$ (this is not the best possible bound), we can guarantee that $|E_{\text{TRAP}(8)}| \leq 0.0208333$.

33. LEFT(9) ≈ 0.996; $|E_{\text{LEFT}(9)}| \leq 0.484$.

35. SIMP(12) $= 24$. Since $f^{[4]} = 0$ for all $x \in [-2, 4]$ we can guarantee that $E_{\text{SIMP}(12)} \leq 0$, i.e., that the estimate is in fact exact.

37. SIMP(4) $= 3.7266$. Since $f'[4] = 24$ for all $x \in [-1, 2]$ we can guarantee that $E_{\text{SIMP}(4)} \leq 0.1266$.

39. SIMP(8) $= 0$. Since $f'[4] \leq 1$ for all $x \in [-\pi, \pi]$ we can guarantee that $E_{SIMP(8)} \leq 0.013$.

41. Since $f(x) = xe^x$ is monotonically increasing on $[-1, 1]$, Theorem 5.25 guarantees we can have $|E_{\text{LEFT}(n)}| \leq 0.5$ if we choose $n = 13$ or higher. We have LEFT(13) $= 0.509077$.

43. Since $f(x) = \frac{1}{x}$ is monotonically decreasing on $[2, 4]$, Theorem 5.25 guarantees we can have $|E_{\text{LEFT}(n)}| \leq 0.05$ if we choose $n = 10$ or higher. We have RIGHT(10) $= 0.668771$.

45. Since $f(x) = x^2$ is concave up on $[0, 3]$ with $f''(x) = 2$ for all x, Theorem 5.27 guarantees we can have $|E_{\text{MID}(n)}| \leq 0.1$ if we choose $n = 5$ or higher. We have MID(5) $= 8.91$.

47. Since $f(x) = x^{2/3}$ is concave down on $[1, 4]$ with $|f''(x)| \leq \frac{2}{9}$ for all $x \in [1, 4]$, Theorem 5.27 guarantees we can have $|E_{\text{MID}(n)}| \leq 0.005$ if we choose $n = 10$ or higher. We have TRAP(10) $= 5.44578$.

49. Since $f(x) = 10e^{-x}\,dx$ has $|f^{[4]}(x)| \le 10$ for all $x \in [0, 2]$, Theorem 5.30 guarantees we can have $|E_{SIMP(n)}| \le 0.01$ if we choose $n = 4$ or higher, meaning that we should use 2 or more parabola-topped rectangles. We have SIMP(4) = 8.64956.

51. Since $f(x) = \sin x$ has $|f^{[4]}(x)| \le 1$ for all x, Theorem 5.30 guarantees we can have $|E_{SIMP(n)}| \le 0.0005$ if we choose $n = 8$ or higher, meaning that we should use 4 or more parabola-topped rectangles. We have SIMP(8) = 2.00027.

53. (a) LEFT(4) = 32.96 degrees, RIGHT(4) = 19.99 degrees, MID(4) = 25.6717 degrees, TRAP(4) = 26.48 degrees, and SIMP(4) = 25.94 degrees. (b) $|E_{LEFT(4)}| < 12.97$, $|E_{RIGHT(4)}| < 12.97$, $|E_{MID(4)}| < 0.625$, $|E_{TRAP(4)}| < 1.25$, and $|E_{SIMP(4)}| < 0.208$.

55. Hints: For (a) you need to use four parabolas, over the intervals [1993, 1995], [1995, 1997], [1997, 1999], and [1999, 2001]. In part (b) note that the area under the graph of r on an interval is the change in the GDP over that interval. For part (c) you can make a table of difference quotients to estimate r', and then repeat that process to estimate higher derivatives.

57. Remember that the period of time must be in seconds. A discrete integral for the total flow is $\phi \approx (1/3)(60 \times 60 \times 60 \times 24(700 + 4 \times 1000 + 2 \times 6300 + 4 \times 4000 + 2 \times 500 + 4 \times 650 + 700)) + 60(24(700 \times 5) \times 60) = 65{,}275{,}200{,}000$. The answer is in cubic feet of water. Inasmuch as the last flow is 700 cfs, as is the first, and the flow is periodic, it seems obvious that we should take the flow to be a constant 700 cfs for the last five days of the year, which is where that last term comes from.

59. Mimic the proof of part (a) of the theorem given in the reading.

61. Use the expanded sum expressions for the trapezoid sum with n subdivision points called $x_0, x_2, x_4, \dots x_{2n}$, the midpoint sum with the same n subdivision points and thus midpoints $x_1, x_3, x_5, \dots, x_{2n-1}$, and compare to Simpson's Rule with all $2n$ subdivision points $x_0, x_1, x_2, \dots x_{2n}$.

63. Generalize the method shown for $n = 6$ in Example 3 to arbitrary n.

For answers to odd-numbered Skill Certification exercises in the Chapter Review, please visit the Book Companion Web Site at www.whfreeman.com/tkcalculus.

Chapter 6

Section 6.1

1. F, F, F, T, T, T, T, T.

3. (a) 2.25π; (b) 5.25π

5. Mimic the example and illustrations in the reading at the beginning of the section.

7. (a) $\Delta x = \frac{1}{2}$, subdivision points $x_0 = 0$, $x_1 = 0.5$, $x_2 = 1$, $x_3 = 1.5$, and $x_4 = 2$; (b) using midpoints, $x_1^* = 0.25$, $x_2^* = 0.75$, $x_3^* = 1.25$, and $x_4^* = 1.75$; (c) corresponding values $f(x_1^*) = 0.0625$, $f(x_2^*) = 0.5625$, $f(x_3^*) = 1.5625$, and $f(x_4^*) = 3.0625$;
(d) $(16\pi - \pi((0.75)^2)^2)\left(\frac{1}{2}\right) \approx 24.6357$.

9. $\pi \int_2^3 (1+x)^2\,dx$; the solid obtained by rotating the region between the graph of $f(x) = 1 + x$ and the x-axis on [2, 3] around the x-axis.

11. $\pi \int_0^2 (4 - x^2)\,dx$; the solid obtained by rotating the region between the graph of $f(x) = \sqrt{4 - x^2}$ and the x-axis on [0, 2] around the x-axis.

13. They are equal.

15. The second one is larger.

17. The second one is larger.

19. The region between $f(x) = x + 1$ and the line $y = 4$ from $x = 1$ to $x = 3$, rotated around the x-axis.

21. The region between $f(x) = x - 1$ and the x-axis on [1, 3], rotated around the x-axis.

23. The region between the graph of $f(x) = 2x + 1$ and the line $y = 5$ on the x-interval [0, 2], rotated about the y-axis.

25. $\pi \int_{-3}^0 2^2\,dy + \pi \int_0^3 (2^2 - (f^{-1}(y))^2)\,dy$

27. $\pi + 2\pi + 3\pi + 4\pi = 10\pi$

29. $3\pi + (2 + 2\sqrt{2})\pi + (3 + 2\sqrt{3})\pi + 8\pi \approx 70.03$

31. $\pi \displaystyle\int_1^3 \left(\frac{3}{x}\right)^2 dx = 6\pi$

33. $\pi \int_0^1 (3^2 - 1^2)\,dy + \pi \displaystyle\int_1^3 \left(\left(\frac{3}{y}\right)^2 - 1^2\right) dy = 8\pi + 4\pi = 12\pi$

35. $\pi \int_0^3 (5^2 - (y+2)^2)\,dy = 36\pi$

37. $\pi \int_2^5 (3^2 - (3 - (x-2))^2)\,dx = 18\pi$

39. $\pi \int_0^3 (3^2 - ((y+2) - 2)^2)\,dy = 18\pi$

41. $\pi \int_0^4 (2^2 - (\sqrt{4-y})^2)\,dy + \pi \int_4^5 (2^2)\,dy = 8\pi + 4\pi = 12\pi$

43. $\pi \int_0^4 ((3 - (\sqrt{4-y}))^2 - 1^2)\,dy + \pi \int_4^5 (3^2 - 1^2)\,dy = 8\pi + 8\pi = 16\pi$

45. $\pi \int_1^{5/3} ((5-x)^2 - (2x)^2)\,dx + \pi \int_{5/3}^4 ((2x)^2 - (5-x)^2)\,dx = \frac{112\pi}{27} = \frac{1813\pi}{27} = \frac{1925\pi}{27}$

47. $\pi \int_0^4 \left((\sqrt{y})^2 - \left(\frac{y}{2}\right)^2\right) dy = \frac{8\pi}{3}$

49. $\pi \int_0^2 ((2x+1)^2 - (x^2+1)^2)\,dx = \frac{104\pi}{15}$

51. $\pi \int_1^4 (x^2 - ((x-2)^2)^2)\,dx = \frac{72\pi}{5}$

53. $\pi \int_1^4 ((x+1)^2 - ((x-2)^2 + 1)^2)\,dx = \frac{117\pi}{5}$

55. $\pi \int_0^{\pi} (1 - \cos x)^2\,dx \approx 14.80$

57. $\pi \int_0^2 (\pi - \cos^{-1}(1-y))^2\,dy \approx 18.4399$

59. (a) 481 feet; (b) 91,636,272 cubic feet;
(c) $\int_0^{481} \left(756 - \frac{756}{481}x\right)^2 dx = 91,636,272.$

61. The cone can be obtained by rotating the region under $y = \frac{3}{5}x$ around the x-axis; this solid has volume
$\pi \int_0^5 \left(\frac{3}{5}x\right)^2 dx = 15\pi.$

63. A cone of radius r can be obtained by rotating the region under the graph of $y = \frac{r}{h}x$ around the x-axis; this solid has volume $\pi \int_0^h \left(\frac{r}{h}x\right)^2 dx = \frac{1}{3}\pi r^2 h.$

Section 6.2

1. T, T, F, F, F, T, F, T.

3. (a) Washer of volume $\frac{15\pi}{2}$; (b) shell of volume $\frac{15\pi}{4}$; (c) cylinder of volume $\frac{3\pi}{4}$; (d) disk of volume $\frac{9\pi}{2}$; (e) shell of volume $\frac{9\pi}{4}$; (f) washer of volume $\frac{21\pi}{2}$.

5. Shells have the same types of dimensions as washers, but what we think of as a height on a shell is considered a thickness on a washer, and vice-versa.

7. (a) $\Delta x = 0.5$, subdivision points $x_0 = 0$, $x_1 = 0.5$, $x_2 = 1$, $x_3 = 1.5$, and $x_4 = 2$; (b) midpoints $x_1^* = 0.25$, $x_2^* = 0.75$, $x_3^* = 1.25$, and $x_4^* = 1.75$; (c) corresponding values $f(x_1^*) = 0.0625$, $f(x_2^*) = 0.5625$, $f(x_3^*) = 1.5625$, and $f(x_4^*) = 3.0625$; (d) $2\pi rh\Delta x = 2\pi \cdot \frac{3}{4} \cdot \frac{9}{16} \cdot \frac{1}{2} = \frac{27}{64}\pi$

9. $\int_0^2 2\pi y^3\, dy = 2\pi \int_0^2 y(y^2)\, dy$; the solid obtained by rotating the region between the graph of $f(x) = \sqrt{x}$, the y-axis, and the line $y = 2$ around the x-axis

11. The region between the graph of $f(x) = \cos x$ and the x-axis on the interval $\left[0, \frac{\pi}{4}\right]$, rotated around the y-axis

13. The region between $y = x^2 + 4$ and $y = 5$ from $x = 0$ to $x = 1$, rotated around the x-axis

15. The region between $y = e^x$ and $y = e$ from $x = 0$ to $x = 1$, rotated around the x-axis

17. $2\pi \int_0^2 y(3 - f^{-1}(y))\, dy$

19. $2\pi \int_0^2 x(f(x) + 3)\, dx$

21. With shells: $2(2\pi \int_0^2 y(2 - y)\, dy)$; with disks: $2(\pi \int_0^2 x^2\, dx)$; disks are easier in this case.

23. With shells: $2\pi \int_0^1 (3 - x)(x^2 + 1)\, dx$; with washers: $\pi \int_0^1 (3^2 - 2^2)\, dy + \pi \int_1^2 ((3 - \sqrt{y - 1})^2 - 2^2)\, dy$; shells are easier.

25. $3.093 + 8.099 + 9.572 + 5.154 = 25.92$

27. $10.953 + 14.082 + 16.671 + 18.426 = 60.132$

29. $2\pi \int_0^2 x(4 - x^2)\, dx = 8\pi$

31. $2\pi \int_0^2 (2 - x)(4 - x^2)\, dx = \frac{40\pi}{3}$

33. $2\pi \int_2^5 x(x - 2)\, dx = 36\pi$

35. $2\pi \int_0^3 (3 - y)(5 - (y + 2))\, dy = 18\pi$

37. $2\pi \int_2^5 (x - 2)(x - 2)\, dx = 18\pi$

39. $2\pi \int_0^2 x(5 - (4 - x^2))\, dx = 12\pi$

41. $2\pi \int_0^2 (3 - x)(5 - (4 - x^2))\, dx = 16\pi$

43. $2\pi \int_1^5 (5 - x)((x + 1) + 2)\, dx = \frac{256\pi}{3}$

45. $2\pi \int_0^1 y((2 + \sqrt{y}) - (2 - \sqrt{y}))\, dy$
$+ 2\pi \int_1^4 y((2 + \sqrt{y}) - y)\, dy = \frac{8\pi}{5} + \frac{64\pi}{5} = \frac{72\pi}{5}$

47. $2\pi \int_0^1 (y + 1)((2 + \sqrt{y}) - (2 - \sqrt{y}))\, dy$
$+ 2\pi \int_1^4 (y + 1)((2 + \sqrt{y}) - y)\, dy = \frac{64\pi}{15} + \frac{287\pi}{15} = \frac{117\pi}{5}$

49. $2\pi \int_0^2 x(4 - (4 - x^2))\, dx = 2\pi \int_0^2 x^3\, dx = 8\pi$

51. $2\pi \int_0^6 x((x - 3)^2 + 2)\, dx = 180\pi$

53. $2\pi \int_0^1 y(\sqrt{y} - y^2)\, dy = \frac{3\pi}{10}$

55. With disks: $\pi \int_0^3 (9 - x^2)^2\, dx = \frac{648\pi}{5}$

57. With shells: $2\pi \int_0^2 x(x^2 - 4x + 4)\, dx = \frac{8\pi}{3}$

59. With disks: $\pi \int_0^{2/3} 9\, dx + \pi \int_{2/3}^2 \left(\frac{2}{x}\right)^2 dx = 10\pi$; with shells: $2\pi \int_0^1 2y\, dy + 2\pi \int_1^3 \left(\frac{2}{y}\right) y\, dy = 10\pi$

61. With shells: $2\pi \int_{-2}^0 (3 - x)(2 + x)\, dx$
$+2\pi \int_0^2 (3 - x)(2 - x)\, dx = \frac{44\pi}{3} + \frac{28\pi}{3}$

63. $2\pi \int_{0.6}^3 y \left(2\sqrt{\frac{1}{0.15}(3 - y)}\right) dy \approx 125.463$ cubic centimeters.

65. See the proof of Theorem 9.4.

67. The sphere can be obtained by rotating the region under $y = \sqrt{9 - x^2}$ around the x-axis; with the shell method, this solid has volume $2\pi \int_0^3 x(2\sqrt{9 - x^2})\, dx$. We can solve this integral with the substitution $u = 9 - x^2$, $du = -2x\, dx$ to obtain a volume of 36π, just as predicted by the volume formula.

69. A sphere of radius r can be obtained by rotating the region between the graph of $y = \sqrt{r^2 - x^2}$ and the x-axis on $[-r, r]$ around the x-axis. Using integration by substitution with $u = r^2 - x^2$, $du = -2x\, dx$, the volume of this solid is $2\pi \int_0^r x(\sqrt{r^2 - x^2} - (-\sqrt{r^2 - x^2}))\, dx = 2\pi \int_0^r x(2\sqrt{r^2 - x^2}) = \frac{4}{3}\pi r^3$. Technical note: Usually we can't do the shell method along the x-axis when we've rotated a region around the x-axis; it works here only because the solid is a perfect sphere, so it is equivalent to rotating the region between the graph of $x = \sqrt{r^2 - y^2}$ and the y-axis from $y = -r$ to $y = r$ around the y-axis.

Section 6.3

1. T, F, T, F, T, F, F, T.
3. See Definition 9.6.
5. See the proof of Theorem 9.7.
7. $y = e^x$
9. No; $f(x)$ has a vertical asymptote at $x = \frac{\pi}{2}$.
11. Both arc lengths are given by the definite integral $\int_0^2 \sqrt{1+4x^2}\,dx$, so they are equal. This makes sense because $f(x) = x^2 - 3$ and $g(x) = 5 - x^2$ differ by a vertical flip and shift only; these transformations should not affect the arc length.
13. $2\int_{-5}^{5} \sqrt{1+\left(\frac{-x}{\sqrt{25-x^2}}\right)^2}\,dx$
15. (a) 3.79009; (b) 3.81994; (c) it is unclear which approximation might be the best.
17. See the reading before Definition 9.9.
19. 16.97
21. 2.3
23. $2 + 4\sqrt{3} \approx 8.9282$
25. $\Delta y_1 = 7$, $\Delta y_2 = 1$, $\Delta y_3 = 1$, $\Delta y_4 = 7$; 16.97
27. $\Delta y_1 \approx 0.405$, $\Delta y_2 \approx 0.288$, $\Delta y_3 \approx 0.223$, $\Delta y_4 \approx 0.182$; approximately 2.3
29. $\Delta y_1 = 2\sqrt{2}$, $\Delta y_2 = 0$, $\Delta y_3 = -2\sqrt{2}$; $2 + 4\sqrt{3} \approx 8.9282$
31. $\int_{-1}^{4} \sqrt{10}\,dx = 5\sqrt{10}$
33. $\int_1^3 \sqrt{1+9x}\,dx = \frac{2}{27}(28)^{3/2} - \frac{2}{27}(10)^{3/2}$
35. $\int_{-1}^{1} \sqrt{28+18x}\,dx = \frac{1}{27}(46)^{3/2} - \frac{1}{27}(10)^{3/2}$
37. $\int_{-3}^{3} \frac{3}{\sqrt{9-x^2}}\,dx = 3\pi$
39. $\frac{1}{2}\int_0^1 \frac{1+x}{\sqrt{x}}\,dx = \frac{4}{3}$
41. $2\int_0^2 x^{-1/3}\,dx = 3(2^{2/3})$
43. $\int_1^2 \frac{1+16x^2}{8x}\,dx = 3 + \frac{\ln 2}{8} \approx 3.087$
45. $\int_{\pi/4}^{3\pi/4} \csc x\,dx = \ln(2\sqrt{2}+3)$
47. $f(x) = e^{3x}$, $[a,b] = [-2,5]$
49. $f(x) = \frac{1}{x}$, $[a,b] = [2,3]$
51. $f(x) = \ln|\cos x|$, $[a,b] = [0, \frac{\pi}{4}]$
53. One approximation is 3.09573
55. One approximation is 9.05755
57. $8\pi\sqrt{5} + 40\pi\sqrt{37}$
59. $\frac{\pi}{64}(35\sqrt{377} + 91\sqrt{1385} + 189\sqrt{3737} + 341\sqrt{8297})$
61. $\pi\sqrt{\pi^2+4}$
63. $4\sqrt{2}\pi$
65. $\frac{\pi}{6}(17\sqrt{17} - 1)$
67. $\frac{\pi}{18}(343 - 13^{3/2}$
69. $\frac{\pi}{16}(4e^4\sqrt{16e^8+1} + \ln(4e^4 + \sqrt{16e^8+1}) - 4e^{-4}\sqrt{16e^{-8}+1} - \ln(4e^{-4} + \sqrt{16e^{-8}+1})$
71. $2\pi\sqrt{2} + \pi\ln(\sqrt{2}+1) - \pi\ln(\sqrt{2}-1)$
73. $\int_0^{\sqrt{6}} \sqrt{1+x^2}\,dx = \sqrt{\frac{21}{2}} + \frac{1}{2}\ln(\sqrt{6}+\sqrt{7} \approx 4.0545$ inches
75. See the discussion in the reading preceding Definition 9.6.
77. (a) The distance from $(a, f(a)) = (a, ma+c)$ to $(b, f(b)) = (b, mb+c)$ is given by $\sqrt{(b-a)^2 + ((mb+c)-(ma+c))^2} = \sqrt{(b-a)^2 + m^2(b-a)^2} = (b-a)\sqrt{1+m^2}$. Since the graph of $f(x) = mx + c$ is a line, this is the exact arc length. (b) If $f(x) = mx + c$, then $f'(x) = m$, which is constant, so the arc length is given by $\int_a^b \sqrt{1+m^2}\,dx = [\sqrt{1+m^2}]_a^b = \sqrt{1+m^2}(b-a)$.
79. If $f(x) = \sqrt{r^2 - x^2}$ (the top half of the circle), then $f'(x) = -\frac{x}{\sqrt{r^2-x^2}}$. Thus the circumference of the circle is twice the arc length of the top half: $2\int_{-r}^{r} \sqrt{1+\left(-\frac{x}{\sqrt{r^2-x^2}}\right)^2}\,dx = 2\int_{-r}^{r} \frac{r}{\sqrt{r^2-x^2}}\,dx$. Using trigonometric substitution with $x = r\sin u$ (and changing the limits of integration accordingly), this is equal to $2r\int_{-\pi/2}^{\pi/2} 1\,du = 2r[u]_{-\pi/2}^{\pi/2} = 2r\left(\frac{\pi}{2} - \left(-\frac{\pi}{2}\right)\right) = 2\pi r$.
81. (a) Use the fact that $\frac{s+t}{q} = \frac{t}{p}$. (b) Use the fact that $C = \sqrt{A^2+B^2}$. (c) Start with the fact that the surface area of the frustum is $S = \pi q(s+t) - \pi pt$. (d) Combine the given relationship with the equation $S = \pi(p+q)s$.
83. Apply the definite integral formula in Theorem 9.10 to the function $f(x) = \sqrt{r^2 - x^2}$ on the interval $[-r, r]$, and then solve the integral. Simplify the integrand as much as possible before attempting to integrate!

Section 6.4

1. F, T, F, T, F, T, F, T.
3. See the reading at the beginning of the section.
5. The depth at the bottom of the pool is constant, but the depth along the side walls varies at different parts of the wall.
7. If you know x_k, Δx, and N, then you can solve for a and b, since $x_k = a + k\,\Delta x$ and $\Delta x = \frac{b-a}{N}$.

9. It would require 38,422.9 foot-pounds of work to lift the entire full hot tub; it would take significantly less work to pump the water out of the tub, since only the water at the bottom of the tub has to move 4 feet (the rest of the water has a shorter distance to travel).

11. Hint: Think about the definitions of Δy and y_k.

13. Hydrostatic force is calculated using the *area* of a slice of the pool wall, not the volume of a slice of the entire pool.

15. See the paragraphs in the reading just before the Examples and Explorations subsection. The extra dimension Δx becomes the differential dx when we pass from a Riemann sum to a definite integral.

17. In the numerators each individual centroid coordinate is multiplied by the area of the corresponding rectangle, which makes centroid coordinates from larger rectangles count more than those from smaller rectangles.

19. 1920π grams

21. 2000 foot-pounds

23. 1497.6 pounds

25. $\Delta x = 6, x_0 = 0, x_1 = 6, x_2 = 12, x_3 = 18, x_4 = 24$. If we choose $x_1^* = 0, x_2^* = 6, x_3^* = 12$, and $x_4^* = 18$, then the mass is approximately 3746.95 grams.

27. $\Delta y = 1, y_0 = 0, y_1 = 1, y_2 = 2, y_3 = 3, y_4 = 4$. If we choose $y_1^* = 0.5, y_2^* = 1.5, y_3^* = 2.5$, and $y_4^* = 3.5$, then the work is approximately 3301.97 foot-pounds.

29. $\Delta y = 1, y_0 = 0, y_1 = 1, y_2 = 2, y_3 = 3, y_4 = 4$. If we choose $y_1^* = 0.5, y_2^* = 1.5, y_3^* = 2.5$, and $y_4^* = 3.5$, then the force is 7987.2 pounds.

31. $(\bar{x}, \bar{y}) \approx (5.417, 1.142)$

33. $(\bar{x}, \bar{y}) \approx (0.819, 4.604)$

35. $(\bar{x}, \bar{y}) = \left(\frac{363}{65}, \frac{15}{13}\right)$

37. $(\bar{x}, \bar{y}) = \left(\frac{4}{5}, \frac{32}{7}\right)$

39. $(\bar{x}, \bar{y}) \approx \left(\frac{\pi}{2}, 0.6\right)$

41. 46,436 grams

43. $m = (1.5)^2 \int_0^{12} (4.2 + 0.4x - 0.03x^2)\, dx = 139.32$ grams

45. 125 grams

47. (a) The solid of revolution obtained by revolving the region between the graph of $y = 8 - \frac{2}{25}x^2$ and the x-axis on $[0, 10]$ around the y-axis; (b) $\rho(y) = -\frac{0.97}{8}y + 1.12$; (c) $m = \pi \int_0^8 \left(-\frac{0.97}{8}y + 1.12\right)\left(100 - \frac{25}{2}y\right) dy$ $= 1001.12$ ounces

49. 6.126 foot-pounds

51. $W = (62.4)(40) \int_0^3 (3 - y)\, dy = 11{,}232$ foot-pounds

53. $W = (62.4)\left(\frac{4\pi}{25}\right) \int_0^{10} y^2(10 - y)\, dy = 26{,}138.1$ foot-pounds

55. $W = (62.4)(16\pi) \int_0^{13} (16 - y)\, dy = 387{,}366$ foot-pounds

57. $W = (42.3)(100)(2) \int_0^8 \sqrt{64 - (8 - y)^2}(16 - y)\, dy + (42.3)(100)(2) \int_0^{16} \sqrt{64 - (16 - y)^2}(16 - y)\, dy \approx 6{,}289{,}650$ foot-pounds.

59. $W = (20)(18) + (.25) \int_0^{18} (18 - y)\, dy = 400.5$ foot-pounds.

61. 3.1765 pounds

63. $F = (62.4)(8) \int_0^3 (3 - y)\, dy = 2246.4$ pounds

65. $F = (62.4)(0.8\pi) \int_0^1 (1 - y)\, dy = 78{,}4142$ pounds

67. $F = 62.4 \int_0^{200} \left(\frac{150}{200}y + 250\right)(200 - y)\, dy = 3.744 \times 10^8$ pounds

69. $F = 62.4 \int_0^8 (6 + y)(12 - y)\, dy + 62.4 \int_8^{12} (20)(12 - y)\, dy \approx$ 47,257 pounds.

71. Leila should put the repeater approximately at coordinates $(4.85, 4.26)$.

73. $m = \rho V = \rho A \Delta x$. Adding up the mass on each cross section gives a Riemann sum of the form $\sum_{k=1}^{n} \rho(x_k^*) A(x_k^*) \Delta x$. Since ρ and A are continuous functions, this Riemann sum converges to a definite integral.

75. $F = \omega A d = \omega l d \Delta y$. Adding up the hydrostatic force on each cross section gives a Riemann sum of the form $\sum_{k=1}^{n} \omega(x_k^*) l(x_k^*) d(x_k^*) \Delta y$. Since ω, l, and A are continuous functions, this Riemann sum converges to a definite integral.

Section 6.5

1. T, F, T, F, T, T, T, T.

3. A solution of a differential equation is one of possibly many functions that makes the differential equation true, while a solution of an initial-value problem is the unique function that makes both the differential equation and the initial value true.

5. If $y(x) = \sqrt{x + C}$ then $\frac{dy}{dx} = \frac{1}{2}(x + C)^{-1/2} = \frac{1}{2\sqrt{x+C}} = \frac{1}{2y}$.

7. (a) The initial-value problem is $\frac{dv}{dt} = -9.8$ with $v(0) = 0$. The solution is $v(t) = -9.8t$. (b) The initial-value problem is $\frac{ds}{dt} = -9.8t$ with $s(0) = 100$. The solution is $s(t) = 100 - 4.9t^2$.

9. The x and y parts of the differential equation have been "separated" into factors. A differential equation in separable form can be solved by dividing both sides by $q(y)$, integrating both sides, and then solving for y, if each of these steps are possible.

11. $A = e^{-C}$

13. The solutions of the differential equation flow through the slope field; this means that we can trace out approximate solution curves in the slope field. For an initial-value problem we can begin at the initial value and then trace out a curve by flowing along with the slope field to obtain a sketch of the unique solution.

15. When using Euler's method to approximate a solution of $\frac{dy}{dx} = g(x, y)$ we have at each step a vertical change of $\Delta y_k = g(x_k, y_k)\Delta x$. The value $g(x_k, y_k)$ is the same as $\left.\frac{dy}{dx}\right|_{(x_k, y_k)}$.

17. (a) 500. (b) Since $0 < P < 500$ we have $0 < 1 - \frac{P}{500} < 1$; since P is growing we have $\frac{dP}{dt} > 0$. Therefore k must be positive. (c) When P is small, the quantity $\frac{P}{500}$ is also small, so $\frac{dP}{dt} \approx kP(1-0) = kP$. (d) When $P \approx 500$ we have $\frac{P}{500} \approx 1$, and therefore $\frac{dP}{dt} \approx kP(1-1) = 0$.

19. $y(x) = \frac{1}{7}x^7 + 2x^4 + 16x + C$

21. $y(x) = Ae^{x^2}$

23. $y(x) = Ae^{1/3x^3}$

25. $y = \pm\sqrt{6x + 2\ln|x| + C}$

27. $y(x) = \ln\left(\frac{1}{2}x^2 + C\right)$

29. $y(x) = 4e^{3x}$

31. $y(x) = -2x^3 + 5$

33. $y(x) = 3e^{-x} + 1$

35. $y(x) = \sqrt{6x + 4}$

37. $y(x) = -e^{1/2x^2}$

39. $y(x) = -\ln(-e^x + 1 + e^{-2})$

41. $y(x) = \frac{1}{2}\ln|1 + x^2| + 4$

43. $y = 4900 - 3900e^{-x/500}$

45. $y = \frac{1}{3}(1 + 5e^{6x})$

47. $y = \frac{2e^{3x}}{1+2e^{3x}}$

49. $y(x) = \frac{1}{1-\sin x}$

51. $y(x) = \sqrt[3]{x^3 - 6x + 512}$

53. $(x_1, y_1) = (0.25, 1.5)$, $(x_2, y_2) = (0.5, 0.75)$, $(x_3, y_3) = (0.75, 0.375)$, $(x_4, y_4) = (1, 0.1875)$. A plot of these points gives:

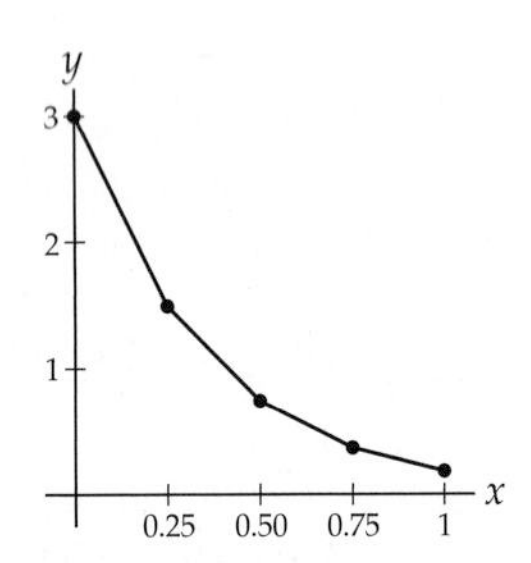

55. $(x_1, y_1) = (0.5, 0)$, $(x_2, y_2) = (1, 0.25)$, $(x_3, y_3) = (1.5, 0.875)$, $(x_4, y_4) = (2, 2.0625)$. A plot of these points gives:

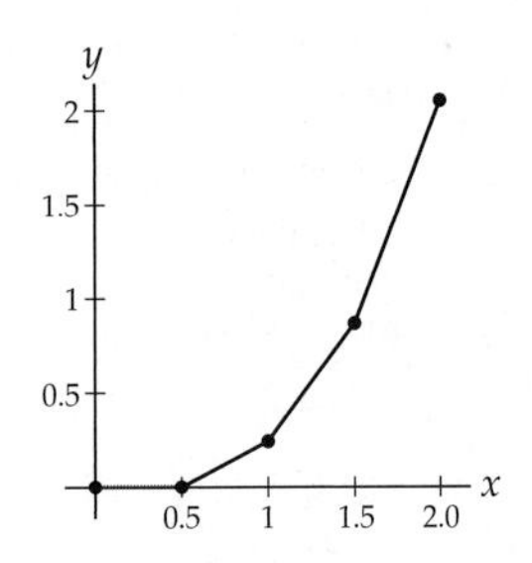

57. $(x_1, y_1) = (0.5, 2)$, $(x_2, y_2) = (1, 2.125)$, $(x_3, y_3) = (1.5, 2.36)$, $(x_4, y_4) = (2, 2.68)$. A plot of these points gives:

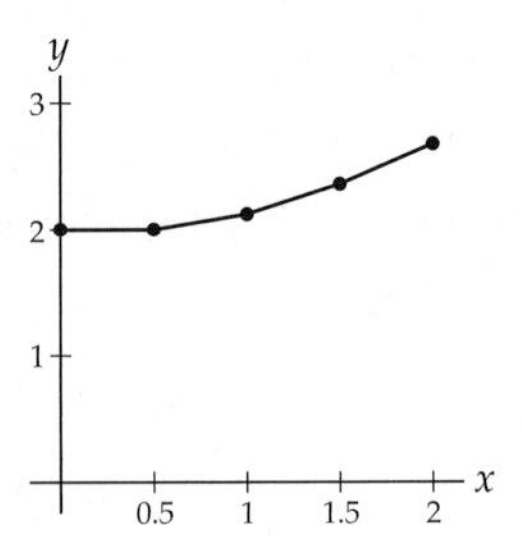

59.

61.

63.

65. (a) $\frac{dP}{dt} = 0.0139P$; with initial value $P(0) = 1.08$ million the solution is $P(t) = 1.08e^{0.0139t}$. (b) Between 55 and 56 years ago. (c) Just under 50 years.

67. (a) $\frac{dC}{dt} = -kC$ for some proportionality constant k. The solution is $C(t) = Ae^{-kt}$ for some constant A that represents the initial amount $C(0)$ of carbon when the organism died. (b) $k = \frac{\ln 2}{5730} \approx 0.00012$. (c) About 19,188 years old.

69. (a) $\frac{dP}{dt} = 0.1P\left(1 - \frac{P}{1000}\right)$, and by separation of variables followed by partial fractions (or the formula in Theorem 9.21) we obtain $P(t) = \frac{1000(100)e^{0.1t}}{1000+100(e^{0.1t}-1))}$. (b) The graph is shown next. (c) According to the logistic model, the population will be growing the fastest in about 21.97 years.

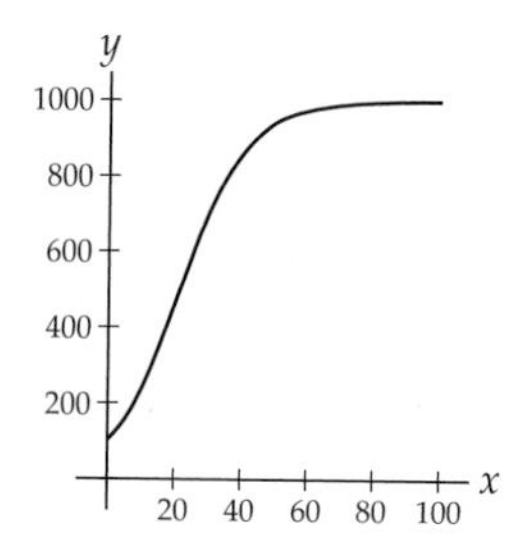

71. (a) We'll set $t = 0$ at 8 pm and measure time in hours. The differential equation is $\frac{dT}{dt} = k(70 - T)$, and its solution is $T(t) = 70 + 18.8e^{kt}$. (b) $k \approx 0.1433$. (c) Just after 5 pm.

73. (a) The differential equation is $\frac{dP}{dt} = kP^{1.1}$, and its solutions are of the form $P(t) = (-0.1kt + C)^{-10}$. (b) With $t = 0$ set at 2005 we have $P(t) = (-0.1(0.008)t + 0.83)^{-10}$. (c) In mid-2013.

75. By separation of variables we have $\int \frac{1}{Q} dQ = \int k\,dt$, which gives $\ln|Q| = kt + C$. Thus $|Q| = e^{kt+C}$ so $Q = Ae^{kt}$. Since $Q(0) = Q_0$ we have $Q_0 = Ae^{k(0)}$ and thus $A = Q_0$.

77. $\lim_{t\to\infty} P(t) = \lim_{t\to\infty} \frac{LP_0e^{kt}}{L+P_0(e^{kt}-1)} = \lim_{t\to\infty} \frac{LP_0}{Le^{-kt}+P_0-P_0e^{-kt}} = \frac{LP_0}{P_0} = L$.

79. $\lim_{t\to\infty}(A - (A - T_0)e^{-kt} = A - 0 = A$. If the model represents cooling, then $T > A$ and thus $A - T < 0$; in this case we wish the rate of change $\frac{dT}{dt} = k(A - T)$ to be negative, which it will be if $k > 0$. For a model of heating we have $T < A$ and thus $A - T > 0$ and we want a positive rate of change, which again requires $k > 0$.

For answers to odd-numbered Skill Certification exercises in the Chapter Review, please visit the Book Companion Web Site at www.whfreeman.com/tkcalculus.

Chapter 7

Section 7.1

1. T, T, F, F, F, F, F, T.

3. A function whose domain is $\mathbb{Z}^+$

5. The subscript k for the term a_k

7. $1, 3, 5, 7, 9$, $a_k = 2k - 1$

9. $a_k = a_{k-1} + 1$, for $k \geq 2$.

11. There exists an $N > 0$ such that for all $k > N$, $a_{k+1} > a_k$.

13. There exists an $N > 0$ such that for all $k > N$, $a_{k+1} < a_k$.

15. Bounded above: there exists a real number M such that $M > a_k$ for every k. Bounded below: there exists a real number m such that $m < a_k$ for every k. Bounded: both bounded above and bounded below.

17. $\left\{(-1)^k k\right\}$

19. The necessary inequalities will not hold if some or all of the terms are negative.

21. $f_1 = f_2 = 1$, $f_3 = 2$, $f_4 = 3$ and $f_5 = 5$

23. Every nonempty subset S of the real numbers that is bounded below has a greatest lower bound.

25. $a_k = (1/2)(1 + (-1)^k)$

27. $a_k = k\left(\frac{1}{3}\right)^k$

29. $a_k = \frac{k+1}{2k+1}$

31. $2, 0, \frac{2}{3}, 0, \frac{2}{5}$

33. $\frac{\cos(x)}{x+1}, \frac{\cos(2x)}{x^2+4}, \frac{\cos(3x)}{x^3+9}, \frac{\cos(4x)}{x^4+16}, \frac{\cos(5x)}{x^5+25}$

35. $1, \frac{\sqrt{2}}{2}, \frac{\sqrt{3}}{3}, \frac{1}{2}, \frac{\sqrt{5}}{5}$

37. 2

39. -1

41. 1

43. $3, 3\left(\frac{1}{2}\right), 3\left(\frac{1}{2}\right)^2, 3\left(\frac{1}{2}\right)^3, 3\left(\frac{1}{2}\right)^4$

45. $-2, -2(-3), -2(-3)^2, -2(-3)^3, -2(-3)^4$

47. $\{-3 + 10k\}_{k=0}^{\infty}$

49. $\{1 - 2k\}_{k=0}^{\infty}$

51. $((k+1)^2 - 5(k+1)) - (k^2 - 5k) = 2k - 4 > 0$ for $k \geq 3$, so the sequence is eventually strictly increasing.

53. $\frac{k+1}{k+3} - \frac{k}{k+2} = \frac{2}{(k+2)(k+3)} > 0$ for $k \geq 0$, so the sequence is strictly increasing.

55. $\frac{(k+1)^2/(k+1)!}{k^2/k!} = \frac{k+1}{k^2} < 1$ for $k \geq 2$, so the sequence is eventually strictly decreasing.

57. $\frac{3^{k+1}/(2\cdot4\cdot6\cdots(2k)(2k+2))}{3^k/(2\cdot4\cdot6\cdots(2k))} = \frac{3}{2k+2} < 1$ for $k \geq 1$, so the sequence is strictly decreasing.

59. When $f(x) = \frac{\sqrt{x+1}}{x}$, $f'(x) = -\frac{x+2}{2x^2\sqrt{x+1}}$, which is negative for $x > 0$, so the sequence is strictly decreasing.

61. When $f(x) = \frac{\sin x}{x}$, $f'(x) = \frac{x\cos x - \sin x}{x^2}$, which changes sign infinitely often for $x > 0$. The test fails, but the sequence is neither increasing nor decreasing.

63. Decreasing and unbounded

65. Not monotonic, bounded

67. Not monotonic, bounded

69. Not monotonic, bounded

71. Eventually decreasing, bounded

73. Not monotonic, unbounded

75. $x_5 \approx 1.26097$

77. $x_7 \approx -25.1327$

79. $x_{k+1} = x_k - \frac{x_k^2 - a}{2x_k} = \frac{1}{2}\left(x_k + \frac{a}{x_k}\right)$

81. $x_3 \approx 1.7321$

83. $x_6 \approx 10.0499$

85. (a) The signs of x_k alternate, and starting with x_2 the magnitudes of the terms increase. Therefore, two successive terms will never be very close. (b) $f(2) = 0$.

87. To the nearest milligram the levels are: 100; 175; 231; 273; 305; 329; 347.

89. 100; 105; 110.25; 115.76; 121.55. $n = 0$ represents the initial investment. The sequence is increasing without bound.

91. Since $\{a_k\}$ is increasing $a_{k+1} \geq a_k$ for every k. Since $\{a_k\}$ is decreasing $a_{k+1} \leq a_k$ for every k. Therefore $a_{k+1} = a_k$ for every k.

93. Use the difference test $S_{n+1} - S_n = a_{n+1} > 0$, therefore $\{S_n\}$ is an increasing sequence.

95. If each case, we may multiply both sides of the given inequality by a_k, maintaining the direction of the inequality since each term is positive.

Section 7.2

1. T, F, F, T, F, T, T, F.

3. By the definition of convergence, only the terms after N that depend on the given ϵ determine convergence. These are precisely the terms that define the tail of the sequence.

5. If the sequences $\{a_k\}$ and $\{b_k\}$ both converge, the sequence $\{a_k - b_k\}$ converges by Theorem 7.11.

7. Let $a_k = b_k = k$

9. Let $a_k = k^2$ and $b_k = k$

11. Let $a_k = \frac{1}{k}$ and $b_k = \frac{1}{k^2}$

13. $\left\{\frac{(-1)^k}{k}\right\}$

15. Such a sequence is not possible. The first term of the sequence is an upper bound for the sequence.

17. $a_k = (k!)^2$

19. The limit $\lim_{x\to\infty} \frac{x^3}{e^x}$ is an indeterminate form of the type $\frac{\infty}{\infty}$. We may use L'Hôpital's Rule to show that $\lim_{x\to\infty} \frac{x^3}{e^x} = 0$.

21. For $0 < r < 1$ the sequence is bounded, monotonically decreasing and converges to zero. For $r = 1$ the sequence is constant, and converges to c. For $r > 1$ the sequence is unbounded, monotonically increasing and diverges.

23. Constant, therefore monotonic and bounded, converges to 11

25. Not eventually monotonic, bounded, no limit

27. Not eventually monotonic, bounded, converges to zero

29. Strictly decreasing, bounded, converges to -2

31. Eventually strictly decreasing, bounded, converges to zero

33. Strictly increasing, bounded, converges to 1

35. Strictly decreasing, bounded, converges to 1

37. Eventually strictly decreasing, bounded, converges to zero

39. Strictly decreasing, bounded, converges to 1

41. Not monotonic, bounded, does not converge

43. Strictly increasing, bounded, converges to 8

45. Eventually strictly increasing, bounded, converges to 42

47. Eventually strictly increasing, bounded below, but not bounded above, diverges to ∞

49. Not monotonic, bounded, converges to zero

51. Monotonically decreasing, bounded, converges to zero

53. Monotonically increasing, unbounded, diverges

55. The sine function is bounded between -1 and 1. So $-1 \leq \sin(2^k) \leq 1$, and thus $-\frac{1}{2^k} \leq \frac{\sin(2^k)}{2^k} \leq \frac{1}{2^k}$. Both $\left\{-\frac{1}{2^k}\right\}$ and $\left\{\frac{1}{2^k}\right\}$ converge to zero by Theorem 7.15. By Theorem 7.19 the limit is therefore zero.

57. $\lim_{k\to\infty} \frac{k-7}{3k+5} = \lim_{k\to\infty} \frac{1-(7/k)}{3+(5/k)} = \frac{1}{3}$

59. Multiply the numerator and denominator by $\sqrt{k^2+k}+k$, then divide the numerator and denominator of the new expression by k. The limit is $\frac{1}{2}$.

61. To the nearest cent the answers are 106.00, 106.09, 106.15, 106.17, 106.18, and 106.18. The sequence is bounded by $100e^{0.06}$

63. Assume that the sequence converges to L for $r < -1$, then by Theorem 7.12 the sequence $\{|r^k|\}_{k=0}^{\infty}$ converges to $|L|$. However, this sequence is also geometric and its convergence contradicts the fact that geometric sequences diverge when the base is greater than 1.

65. The limit is indeterminate of the type ∞^0. By L'Hôpital's Rule $\lim_{x\to\infty} \frac{\ln x}{x} = 0$. Exponentiating we have $\lim_{x\to\infty} x^{1/x} = e^0 = 1$.

67. If $c = 0$, the sequence is constant and converges to zero. So assume $c \neq 0$. Let $\epsilon > 0$ be given. Since the sequence $\{a_k\} \to L$, there is an $N > 0$ such that for every $k > N$, $|a_k - L| < \frac{\epsilon}{|c|}$. Thus, for $k > N$, $|ca_k - cL| < \epsilon$. Therefore, $\{ca_k\} \to cL$.

69. Let $\epsilon > 0$ be given. Since $\{b_k\} \to M$, there is an $N_1 > 0$ such that $b_k - M| < \frac{\epsilon}{2(1+|L|)}$ and an $N_2 > 0$ such that $b_k - M| < 1$ for $k > N_1$ and $k > N_2$, respectively. For $k > N_2$ we also have $|b_k| < 1 + |M|$. Similarly, since $\{a_k\} \to L$, there is an $N_3 > 0$ such that for every $k > N_3$, $|a_k - L| < \frac{\epsilon}{2(1+|M|)}$. Let $N = \max(N_1, N_2, N_3)$. Now, $|a_kb_k - LM| = |a_kb_k - Lb_k + Lb_k - LM| \leq |a_k - L||b_k| + |b_k - M||L|$. From our analysis above, for $k > N$ this last quantity is less than or equal to $\frac{\epsilon}{2(1+|M|)}(1+|M|) + \frac{\epsilon}{2(1+|L|)}|L| < \frac{\epsilon}{2} + \frac{\epsilon}{2} = \epsilon$. Therefore, $a_kb_k \to LM$.

71. See the proof of Theorem 1.6.

73. See the proof of Theorem 1.29.

75. If $\lim_{x\to\infty} a(x) = L$, then there is an $N > 0$ such that for every $x > N$, $|a(x) - L| < \epsilon$. Since $a_k = a(k)$, for every $k > N$ we have $|a_k - L| < \epsilon$. Therefore, $a_k \to L$.

77. Let $\epsilon > 0$ be given. Since $a_k \to \infty$, there is an $N > 0$ such that for every $k > N$ we have $a_k > \frac{1}{\epsilon}$. For $k > N$, we therefore have $\frac{1}{a_k} < \epsilon$. So, $\frac{1}{a_k} \to 0$.

79. The terms of the sequence $\left\{\left(\frac{1}{k}\right)^p\right\}$ are the reciprocals of the terms of the sequence $\{k^p\}$, so you may use your answer from Exercise 78.

81. If $|r| > 1$, then the subsequence $\{r^{2k}\}$ is monotonically increasing and unbounded, therefore the sequence r^n diverges.

83. Use L'Hôpital's Rule to show that $\lim_{x\to\infty} \frac{x^r}{(1+p)^x} = 0$.

85. If the sequence $\{a_k\}$ converges to L, then for $\epsilon = 1$ there is an $N > 0$ such that $|a_k - L| < 1$ for $k > N$. Thus, for $n > N$ we have $|a_k| < \max(|L-1|, |L+1|)$. Let $M = \max(|a_1|, |a_2|, \dots, |a_N|, |L-1|, |L+1|)$. For every k, $|a_k| < M$.

87. Let $\epsilon > 0$. There is an $N_1 > 0$ such that for $k > N_1$, $|a_{2k} - L| < \epsilon$. Similarly, there is an $N_2 > 0$ such that for $k > N_2$, $|a_{2k+1} - L| < \epsilon$. Let $N = \max(N_1, N_2)$. For every $k > N$ we have $|a_k - L| < \epsilon$. Thus, $a_k \to L$.

Section 7.3

1. F, F, F, T, F, T, T, F.

3. The sequence of partial sums is the sequence a_n where $S_n = \sum_{k=1}^{n} a_k$. When we add any list of numbers, we always add one new summand at a time. This is precisely what we do when we form the sequence of partial sums.

5. $\sum_{k=1}^{\infty} a_k = \sum_{k=1}^{9} a_k + \sum_{k=10}^{\infty} a_k$. To find the sum of $\sum_{k=10}^{\infty} a_k$, you would need to know $\sum_{k=1}^{9} a_k$.

7. $\sum_{k=1}^{\infty} a_k = \sum_{j=s}^{\infty} a_{j-s+1}$

9. A series that can be written in the form $\sum_{k=0}^{\infty} cr^k$. A geometric series will converge if $|r| < 1$.

11. Since a geometric series has the form $\sum_{k=0}^{\infty} cr^k$, the only way for a geometric series to contain a zero term is if one or both of c or r is zero. When this occurs the series converges to c.

13. $\sum_{k=1}^{\infty} \frac{1}{2^k}$

15. Use $\sum_{k=1}^{\infty} 1$ and $\sum_{k=1}^{\infty}(-1)$. Then each term of the series $\sum_{k=1}^{\infty}(1-1)$ is zero.

17. Use $\sum_{k=0}^{\infty} \left(\frac{1}{2}\right)^k$ for both series. This series converges to 2, but the product series $\sum_{k=0}^{\infty} \left(\frac{1}{4}\right)^k$ converges to $\frac{4}{3}$.

19. The series $\sum_{k=0}^{\infty} 2^k$ and $\sum_{k=0}^{\infty} 4^k$ both diverge, but the series. $\sum_{k=0}^{\infty} \frac{2^k}{4^k}$ converges to 2.

21. $\frac{1}{3}, \frac{1}{6}, \frac{1}{11}, \frac{1}{18}, \frac{1}{27}$

23. $-1, 1, -1, 1, -1$

25. $1, \frac{1}{2}, \frac{1}{3}, \frac{1}{4}, \frac{1}{5}$

27. $1, -\frac{1}{2!}, \frac{1}{4!}, -\frac{1}{6!}, \frac{1}{8!}$

29. $\frac{1}{3}, \frac{1}{2}, \frac{13}{22}, \frac{64}{99}, \frac{203}{297}$

31. $-1, 0, -1, 0, -1$

33. $1, \frac{3}{2}, \frac{11}{6}, \frac{25}{12}, \frac{137}{60}$

35. $1, \frac{1}{2!}, \frac{13}{24}, \frac{389}{720}, \frac{4357}{8064}$

37. $2 - \frac{1}{2^{100}}$

39. $2^{1001} - 1$

41. $3\left(\left(\frac{2}{3}\right)^6 - \left(\frac{2}{3}\right)^{151}\right)$

43. 9

45. 11

47. The series converges to 6.

49. The series diverges.

51. The series converges to $\frac{3\pi}{3-e}$.

53. The series converges to $\frac{9}{40}$.

55. The series converges to $\frac{100}{3}$.

57. The series converges to $-\frac{192}{7}$.

59. The series converges to $\frac{64}{3}$.

61. $S_n = \left(1 - \frac{1}{3}\right) + \left(\frac{1}{2} - \frac{1}{4}\right) + \left(\frac{1}{3} - \frac{1}{5}\right) + \cdots + \left(\frac{1}{n-1} - \frac{1}{n+1}\right) + \left(\frac{1}{n} - \frac{1}{n+2}\right) = 1 + \frac{1}{2} - \frac{1}{n+1} - \frac{1}{n+2}$. The series converges to $\frac{3}{2}$.

63. $S_n = \left(1 - \frac{1}{3}\right) + \left(\frac{1}{3} - \frac{1}{3^2}\right) + \left(\frac{1}{3^2} - \frac{1}{3^3}\right) + \cdots + \left(\frac{1}{3^n} - \frac{1}{3^{n+1}}\right) = 1 - \frac{1}{3^{n+1}}$. The series converges to 1.

65. The series diverges.

67. $x \in (-1, 1)$

69. The series converges for every value of x except odd multiples of $\frac{\pi}{2}$.

71. $237\sum_{k=1}^{\infty}(0.001)^k = \frac{79}{333}$

73. $3051\sum_{k=1}^{\infty}(0.0001)^k = \frac{339}{1111}$

75. $1 + 27\sum_{k=1}^{\infty}(0.01)^k = \frac{14}{11}$

77. $\frac{1}{10}\left(1 + 9\sum_{k=1}^{\infty}(0.1)^k\right) = \frac{1}{5}$

79. 4 meters

81. $\frac{2}{3}$

83. $\frac{\sqrt{3}}{4}$

85. Let $\{S_n\}$ be the sequence of partial sums for the first series and $\{T_n\}$ be the sequence of partial sums for the second series. We know that these sequences converge to L and M, respectively. By Theorem 7.11 this sequence, $\{a_n + b_n\}$, that is the sequence of partial sums for the series $\sum_{k=1}^{\infty}(a_k + b_k)$, converges to $L + M$.

87. The series $\sum_{k=0}^{\infty}(cr^k \cdot bv^k)$ is a convergent geometric series since $|rv| < 1$. The series of the products *cannot* converge to the product of the sums of the two series.

89. The series $\sum_{k=0}^{\infty} \frac{cr^k}{bv^k} = \left(\frac{c}{b}\right)\sum_{k=0}^{\infty}\left(\frac{r}{v}\right)^k$, which is a geometric series. The series will converge if $\left|\frac{r}{v}\right| < 1$.

Section 7.4

1. F, T, F, F, T, F, F, T.

3. $\sum_{k=1}^{\infty} \frac{1}{k^p}$

5. For a geometric series $\sum_{k=1}^{\infty} cr^k$, the limit $\lim_{k\to\infty} cr^k = 0$ if and only if $|r| < 1$. The geometric series converges if and only if $|r| < 1$.

7. The series diverges.

9. If $f : [1, \infty) \to \mathbb{R}$ is a continuous function that is eventually positive and decreasing, then the improper integral $\int_1^{\infty} f(x)dx$ and the series $\sum_{k=1}^{\infty} f(k)$ either both converge or both diverge. The statement is valid because the "tail" of a series determines its convergence or divergence.

11. R_n is an upper bound on the error in using S_n to approximate the sum of the series. If we can ensure that R_n is small, we can ensure the quality of the approximation is good.

13. If $a(x)$ is not continuous, then the improper integral $\int_1^{\infty} a(x)dx$ may not be defined.

15. $\sin(\pi x)$

17. The divergence test fails since $\lim_{k\to\infty} \frac{1}{k^2} = 0$.

19. The series diverges since $\lim_{k\to\infty} \frac{k}{3k+100} = \frac{1}{3}$.

21. The divergence test fails since $\lim_{k\to\infty} \frac{k!}{k^k} = 0$.

23. The divergence test fails since $\lim_{k\to\infty} \frac{k^2+1}{k!} = 0$.

25. The function $f(x) = e^{-x}$ is continuous, positive, and decreasing on the interval $[0, \infty)$. The series converges because the improper integral $\int_0^{\infty} e^{-x}dx = 1$.

27. The function $f(x) = \frac{1}{x\sqrt{\ln x}}$ is continuous, positive, and decreasing on the interval $[2, \infty)$. The series diverges because the improper integral $\int_2^{\infty} \frac{dx}{x\sqrt{\ln x}}$ diverges.

29. The function $f(x) = \frac{2x-4}{x^2-4x+3}$ is continuous, positive, and decreasing on the interval $[4, \infty)$. The series diverges because the improper integral $\int_4^{\infty} \frac{2x-4}{x^2-4x+3}dx$ diverges.

31. The function $f(x) = \frac{1}{(x-2)^2}$ is continuous, positive, and decreasing on the interval $[3, \infty)$. The series converges because the improper integral $\int_3^{\infty} \frac{dx}{(x-2)^2}$ converges.

33. A convergent p-series, $p = 3/2$

35. The function $f(x) = \frac{k}{2+k^2}$ satisfies the conditions of the integral test and the integral $\int_1^{\infty} \frac{x}{2+x^2}dx$ diverges.

37. The function $f(x) = \frac{\ln x}{x}$ is continuous, positive and eventually decreasing. Since the integral $\int_1^{\infty} \frac{\ln x}{x}dx$ diverges, the series diverges.

39. The series converges. Use the integral test.

41. This telescoping series converges to 1. You can also use the integral test to show that the series converges.

43. The series converges. Use the integral test.

45. (b) $S_{10} \approx 1.581950$, (c) $R_{10} = e^{-10} \approx 0.000045$, (d) $(1.581950, 1.581995)$, (e) 14

47. (b) $S_{10} \approx 1.5$, (c) $R_{10} = 0.1$, (d) $(1.5, 1.6)$, (e) 1000000

49. $p > 1$

51. $p > 1$

53. (a) We have $q_2 = (0.22)(q_1 + h)$, $q_4 = (0.11)^2 q_0 + 0.33h(0.11)$, and more generally $q_{2N} = (0.11)^N q_0 + 0.33h\sum_{k=0}^{N-1}(0.11)^k$. (b) $q_3 = 0.11q_1 + 0.61h$, $q_5 = (0.11)^2 q_1 + 0.61h(0.11)$, and more generally $q_{2N+1} = (0.11)^N q_1 + 0.61h\sum_{k=0}^{N-1}(0.11)^k$. (c) The sum during the even-numbered years is $0.33h\frac{1}{1-0.11} \approx 0.371h$. So Leila needs $0.371h = P$. That is, $h \approx 2.697P$.

55. Since f is continuous, positive, and decreasing, each summand $f(k) \le \int_{k-1}^{k} f(x)dx$. Therefore, $\sum_{k=n+1}^{\infty} f(k) \le \int_n^{\infty} f(x)dx$.

57. Let $c > 0$. There are several cases to consider. When $r \le -1$, $\lim_{k\to\infty} cr^k$ does not exist. When $r = 1$, $\lim_{k\to\infty} cr^k = c$. When $r > 1$, $\lim_{k\to\infty} cr^k = \infty$. The situations when $c < 0$ are similar.

Section 7.5

1. T, F, F, F, F, F, F, F.

3. Use the comparison test to analyze the series $\sum_{k=1}^{\infty}(-a_k)$.

5. Let $\sum_{k=1}^{\infty} a_k$ and $\sum_{k=1}^{\infty} b_k$ be two series with nonnegative terms such that $0 \le a_k \le b_k$ for every positive integer k larger than some positive integer K. If the series $\sum_{k=1}^{\infty} b_k$ converges, then the series $\sum_{k=1}^{\infty} a_k$ converges. This is valid because the convergence or divergence of a series only depends upon the "tail" of the series.

7. Let $\sum_{k=1}^{\infty} a_k$ and $\sum_{k=1}^{\infty} b_k$ be two series whose terms are eventually positive. The three parts of the theorem remain unchanged. This is valid because the convergence or divergence of a series only depends upon the "tail" of the series.

9. If the series converges and you compare it to a divergent series you will not gain any useful information.

11. The improper integral $\int_2^\infty \frac{dx}{x\ln x}$ diverges.

13. The value of the limit is zero. Since p-series diverge for $0 < p < 1$, we don't get any useful information from the limit comparison test.

15. b_k, a_k, converges

17. b_k, a_k, diverges

19. The p-series $\sum_{k=1}^\infty \frac{1}{k^{3/2}}$ converges and $\lim_{k\to\infty} \frac{(k^{3/2}-k-1)/(5k^3+3)}{1/k^{3/2}} = \frac{1}{5}$. Therefore the given series converges.

21. Compare the series to the harmonic series using the limit comparison test to show that the series diverges.

23. The terms of this series are positive starting with $k = 2$. Compare the series to the convergent p-series $\sum_{k=1}^\infty \frac{1}{k^{3/2}}$ using the limit comparison test to show that the given series converges.

25. The terms of this series are positive. Compare the series to the harmonic series using either comparison test to show that the given series diverges.

27. The terms of this series are positive. Compare the series to the convergent p-series with $p = 3/2$. Since $\lim_{k\to\infty} \frac{(1+\ln k)/k^2}{1/k^{3/2}} = 0$, by the limit comparison test the given series converges.

29. The terms of this series are positive. Compare the series to the convergent p-series with $p = 2$. Since $\lim_{k\to\infty} \frac{(1+\ln k)/k^2}{1/k^2} = 0$, by the limit comparison test the given series converges.

31. A divergent p-series, $p = \frac{1}{2}$

33. Use either comparison test with the p-series $\sum_{k=1}^\infty \frac{1}{k^{7/4}}$ to show that the series converges.

35. Use the limit comparison test with the harmonic series to show that the series diverges.

37. Use the comparison test with the harmonic series to show that the series diverges.

39. Use the limit comparison test with the p-series $\sum_{k=1}^\infty \frac{1}{k^{3/2}}$ to show that the series converges.

41. The series converges. Use the integral test.

43. The series converges. Use the integral test.

45. The series converges. Use the integral test.

47. Use the limit comparison test with the harmonic series to show that the series diverges.

49. $p \in \mathbb{R}$

51. It converges. Write $\alpha(n) = a_1 n + a_2 n^2 + \cdots + a_N n^N$. Since all the coefficients are positive, then $\alpha(n) > a_1 n$. It follows that $V = \sum_{n=0}^\infty w_n < W \sum_{n=0}^\infty (r^{a_1})^n = \frac{W}{1-r^{a_1}}$.

53. Since $\sum_{k=1}^\infty a_k$ converges and all terms are nonnegative, there is an $N > 0$ such that for all $n > N$ we have $a_k < 1$. Therefore, when $n > N$ we also have $a_k^2 < a_k$. Therefore, by the comparison test the series $\sum_{k=1}^\infty a_k^2$ converges.

55. We compare the series $\sum_{k=1}^\infty (a_k \cdot b_k)$ to the series using the limit comparison test. We have $\lim_{k\to\infty} \frac{a_k \cdot b_k}{a_k} = \lim_{k\to\infty} b_k$. By the divergence test, the value of this limit is zero. Therefore, since the series $\sum_{k=1}^\infty a_k$ converges, by the limit comparison test the series $\sum_{k=1}^\infty (a_k \cdot b_k)$ converges.

57. Since $\lim_{k\to\infty} \frac{a_k}{b_k} = 0$, there is an $N > 0$ such that for every $k > N$ $0 < a_k < b_k$. Therefore, by the comparison test the series $\sum_{k=1}^\infty a_k$ converges.

Section 7.6

1. T, T, T, F, T, F, F, F.

3. 1

5. 1

7. The ratio test could be used on the series $\sum_{k=1}^\infty (-a_k)$.

9. The series satisfies the hypotheses of the ratio test. Since the terms are positive and $\lim_{k\to\infty} \frac{a_{k+1}}{a_k} > 1$, $\lim_{k\to\infty} a_k \neq 0$, therefore the series diverges by the divergence test.

11. 1

13. The root test could be used on the series $\sum_{k=1}^\infty (-a_k)$.

15. It is not immediately apparent how to evaluate $\lim_{k\to\infty} \left(\frac{1}{k!}\right)^{1/k}$. It is more practical to use the ratio test.

17. It is a convergent geometric series. You could also show this using the integral, comparison, limit comparison, ratio, or root tests.

19. $\lim_{k\to\infty} \frac{a_{k+1}}{a_k}$ does not exist, but the series is the sum of two convergent geometric series; $\sum_{k=0}^\infty \frac{1}{10^k} = \frac{10}{9}$ and $\frac{1}{5}\sum_{k=0}^\infty \frac{1}{10^k} = \frac{2}{9}$. So the sum is $\frac{4}{3}$.

21. $\frac{1}{90}$

23. $\frac{1}{(k+2)(k+1)}$

25. 12

27. $n^3 - n$

29. The series converges by the ratio test.

31. The ratio test is inconclusive, but the series converges.

33. The ratio test is inconclusive, but the series diverges by the divergence test.

35. The series diverges.

37. The root test is inconclusive, but the series converges.

39. The series converges.

41. Each term of the series is positive, and it converges by the ratio test.

43. Compare the series to $\sum_{k=1}^{\infty} \frac{1}{k}$ using the limit comparison test to show that the series diverges.
45. The series converges. You may show this by comparing it to the convergent p-series, $\sum_{k=1}^{\infty} \frac{1}{k^3}$ using the limit comparison test.
47. Each term of the series is positive, and it converges by the limit comparison test.
49. Each term of the series is positive, and it converges by the root test.
51. Each term of the series is positive, and it converges by the ratio test.
53. Each term of the series is positive, and it diverges by the ratio test.
55. Each term of the series is positive, and it converges by the root test.
57. Each term of the series is positive, and it converges by the ratio test.
59. Each term of the series is greater than the corresponding term of the divergent series $\sum_{k=1}^{\infty} \frac{1}{3k}$, thus the series diverges by the comparison test.
61. (a) Use the ratio test. (b) Use the divergence test. (c) By the definition of dominance.
63. (a) Use the ratio test. (b) Use the divergence test. (c) By the definition of dominance.
65. (a) $420\frac{1-r_{11}}{1-r} \approx 1920$. (b) $f_{10} + \int_{10}^{\infty} 0.95^x dx \approx 1931$.
67. See Exercise 5.
69. Use L'Hôpital's Rule to show that $\lim_{x\to\infty} x^{1/x} = 1$. From this we may show that $\lim_{k\to\infty} (ck^m)^{1/k} = 1$. Every p-series has this form.
71. (a) For large values of k, each term of the series is less than $(c_1 + c_2 + \cdots + c_n)k^{m_n}$. (b) Part (a) shows that the root test is inconclusive for the series.

Section 7.7

1. F, T, T, F, T, F, T, T.
3. A series of the form $\sum_{k=1}^{\infty}(-1)^{k+1}a_k$ or $\sum_{k=1}^{\infty}(-1)^k a_k$ where $\{a_k\}$ is a sequence of positive numbers.
5. If $\lim_{k\to\infty} a_k \neq 0$ the series will diverge by the divergence test.
7. Since the signs of the terms of the series are alternating and their magnitudes are monotonically decreasing, the terms of the sequence of partial sums have the stated property.
9. The series $\sum_{k=1}^{\infty} |a_k|$ must converge.
11. Look at the definitions. For absolute convergence you must only show that the series $\sum_{k=1}^{\infty} |a_k|$ converges; to show conditional convergence, you must show that the series $\sum_{k=1}^{\infty} |a_k|$ diverges and the series $\sum_{k=1}^{\infty} a_k$ converges.
13. If all of the terms of the series $\sum_{k=1}^{\infty} a_k$ are positive, then $\sum_{k=1}^{\infty} a_k = \sum_{k=1}^{\infty} |a_k|$, so if the series converges it is absolutely convergent.
15. Montonically decreasing, $\lim_{x\to\infty} f(x) = 0$, $\int_1^{\infty} f(x)\, dx$, $\sum_{k=1}^{\infty} f(k)$
17. By Theorem 7.24
19. Let $\sum_{k=1}^{\infty} a_k = \sum_{k=1}^{\infty} b_k = \sum_{k=1}^{\infty} \frac{1}{k}$
21. The series diverges.
23. The series converges.
25. The series converges absolutely.
27. The series converges.
29. The series diverges by the divergence test.
31. The series converges absolutely.
33. The series converges absolutely.
35. The series converges absolutely.
37. The series is a convergent telescoping series. It converges absolutely.
39. The series diverges.
41. The series converges conditionally.
43. The series converges absolutely.
45. The series converges conditionally.
47. The series converges absolutely.
49. The series converges absolutely.
51. The series converges absolutely.
53. (b) $S_{10} = 0.817$, $L \in (0.817, 0.827)$, (c) $n = 999$
55. (b) $S_{10} = 0.5403023058681397174018$, $L \in (0.5403023058681397174009, 0.5403023058681397174018)$, (c) $n = 5$
57. (b) $S_{10} = 0.36787946$, $L \in (0.36787943, 0.36787946)$, (c) $n = 11$
59. When $p > 1$ the series converges absolutely, when $0 < p \leq 1$ the series converges conditionally and when $p \leq 0$ the series diverges.
61. The series converges absolutely for every value of p.
63. (a) As long as the denominator isn't zero we have

$$\frac{A_n - A_{n-1}}{A_{n-1} - A_{n-2}} = \frac{8069 - A_{n-1}(1+1/7)}{8069 - A_{n-2}(1+1/7)}$$
$$= \frac{8069 - (8069 - A_{n-2}(1/7))(1+1/7)}{8069 - A_{n-2}(1+1/7)}$$
$$= \frac{8069 - 8069}{8069 - A_{n-2}(1+1/7)} - \frac{(8069 - A_{n-2}(1+1/7))(1/7)}{8069 - A_{n-2}(1+1/7)} = -\frac{1}{7}.$$

(b) It is a telescoping sum. (c) Since it is an alternating series and the nth term goes to zero as $n \to \infty$. The alternating series test applies and shows that it converges. It is also a geometric series, whose sum is 7060.375.

65. If $\sum_{k=1}^{\infty} a_k$ converges, then for some $N > 0$ every $a_k < 1$ when $k > N$. Thus, for $k > N$, $a_k^2 < a_k$. By the comparison test, the series $\sum_{k=N+1}^{\infty} a_k^2$ converges and thus the series $\sum_{k=1}^{\infty} a_k^2$ converges.

67. Model your proof on the argument found in the proof of the alternating series test.

69. This is the contrapositive of Theorem 7.41.

71. Compare the series to the harmonic series using the limit comparison test.

73. Use the hint!

For answers to odd-numbered Skill Certification exercises in the Chapter Review, please visit the Book Companion Web Site at www.whfreeman.com/tkcalculus.

Chapter 8

Section 8.1

1. F, T, T, F, T, T, T, F.

3. A series of the form $\sum_{k=0}^{\infty} a_k x^k$.

5. The bases of the factors $(x-k)^k$ vary with the index.

7. The interval of values of x for which the series converges. It is often best to use the ratio test for absolute convergence to find those values for which $\lim_{k\to\infty} \left|\frac{a_{k+1}}{a_k}\right| |x - x_0| < 1$. The interval of convergence can be an open interval centered at x_0, a half-open interval centered at x_0, a closed interval centered at x_0, or $\mathbb{R}$.

9. When $x = 2$ we have the series $\sum_{k=0}^{\infty}(-1)^k \frac{k}{3k+1}$. Since $\lim_{k\to\infty}(-1)^k \frac{k}{3k+1} \neq 0$, the power series diverges when $x = 2$.

11. When $x = 1$ we have the alternating harmonic series, which converges conditionally. When $x = -1$ we have the harmonic series, which diverges. These two pieces of information, together with Theorem 8.3, tell us the series has the interval of convergence $(-1, 1]$.

13. 6

15. 2

17. If the radius of convergence is ρ the limit $\lim_{k\to\infty}\left|\frac{a_{k+1}}{a_k}\rho\right| = 1$, which means that the ratio test for absolute convergence fails at the endpoints of the interval of convergence.

19. $\frac{1}{\rho}$

21. $\mathbb{R}$

23. $\mathbb{R}$

25. $(-1, 1]$

27. $[-3, -1)$

29. $[\pi - 1, \pi + 1]$

31. $\left[\frac{1}{2}, \frac{5}{2}\right]$

33. $\left(-\frac{5}{2}, \frac{7}{2}\right)$

35. $(-6, 2)$

37. 0

39. $\left[-\frac{8}{3}, -\frac{2}{3}\right]$

41. $(1, 4)$

43. $\mathbb{R}$

45. $\mathbb{R}$

47. $\mathbb{R}$

49. 1

51. $\mathbb{R}$

53. The powers of $x + 1$ are negative; $x < -2$ and $x > 0$

55. Powers of $\sin x$; $\mathbb{R}$

57. Powers of $\frac{x+2}{x-3}$; $x \le \frac{1}{2}$

59. (a) On $[0, \infty)$. (b) There are many ways to answer this question. Use enough terms that the error is not visible on the graph. Since the range of the function is approximately $[0, 2500]$ on the interval, to ensure that the error is smaller than, e.g. 2%, we should require $39500^{2/3}(500/19200)^n/n! < 50$, which occurs for $n = 2$. It follows that one or two terms suffice to get a good picture.

61. The statement follows from the contrapositive of part (a) of Theorem 8.2.

63. At $x_0 + \rho$ consider the series $\sum_{k=0}^{\infty} \left|a_k((x+\rho) - x_0)^k\right| = \sum_{k=0}^{\infty}\left|a_k\rho^k\right|$. This is precisely the series of absolute values when the original series is evaluated at $x_0 - \rho$. Therefore, if the original series converges absolutely at one endpoint of the interval of convergence, it converges absolutely at the other endpoint as well.

65. If the series converges absolutely at $x_0 + \rho$, then by Exercise 63 it converges absolutely at $x_0 - \rho$ as well.

67. By Exercise 66, if the radius of convergence, ρ, of both series is finite, then $\rho = \sqrt{\rho}$. The only solutions to this equation are 0 and 1.

69. Use the ratio test for absolute convergence.

71. (a) Since $a_{k+2} = a_k$ for every $k > 0$, the series $\sum_{k=0}^{\infty} a_k x^k$ may be rewritten as $a_0 \sum_{k=0}^{\infty} x^{2k} + a_1 \sum_{k=0}^{\infty} x^{2k+1}$. These two series both have radius of convergence 1, so the series has radius of convergence 1. (b) The two series mentioned above diverge at ± 1. (c) $\frac{a_0 + a_1 x}{1 - x^2}$

Section 8.2

1. T, T, F, T, F, F, T, F.

3. At $x = 0$, we have $\ln 2 - \sum_{k=1}^{\infty} \frac{1}{k}$ which diverges, since the harmonic series diverges. At $x = 4$, we have $\ln 2 + \sum_{k=1}^{\infty} \frac{(-1)^{k-1}}{k}$ that converges conditionally.

5. The second Taylor polynomial has the same value, the same slope for its tangent line, and the same concavity as the function f at the point x_0.

7. A Maclaurin polynomial is an approximation to the function f at zero. A Taylor polynomial at x_0 is an approximation to f at x_0.

9. A Taylor polynomial of degree n is an approximation to the function f at x_0. It is possible for a Taylor series for a function f to converge to f on the interval of convergence for the series. In this case the series is an alternate way to represent the function on the interval of convergence.

11. The single point 0; an interval with one of the forms $(-\rho, \rho)$, $(-\rho, \rho]$, $[-\rho, \rho)$, $[-\rho, \rho]$ where $\rho > 0$, or $\mathbb{R}$.

13. $P_1(x) = 5 - 2x$, $P_2(x) = P_3(x) = 5 - 2x + 3x^2$. For any polynomial function f of degree n, the mth Maclaurin polynomial for f, $P_m(x) = f(x)$ when $m \geq n$.

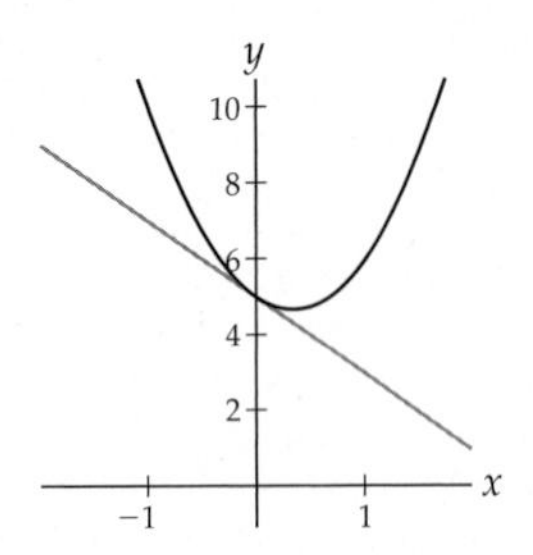

15. $P_1(x) = 6 + 4(x-1)$,
$P_2(x) = P_3(x) = 6 + 4(x-1) + 3(x-1)^2$. For any polynomial function f of degree n, the mth Taylor polynomial for f, $P_m(x) = f(x)$ when $m \geq n$.

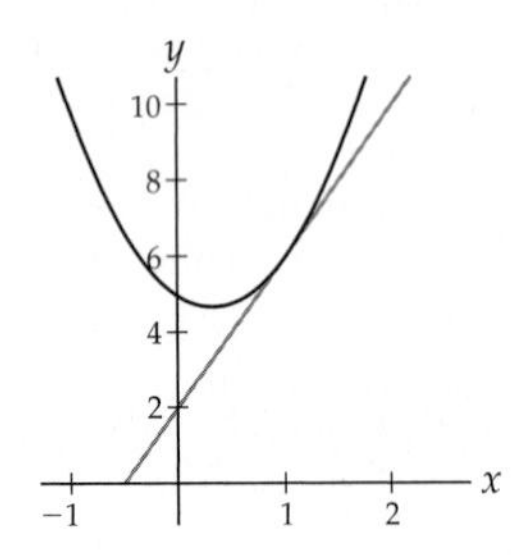

17. $P_1(x) = (ax_0^3 + bx_0^2 + cx_0 + d) + (3ax_0^2 + 2bx_0 + c)(x - x_0)$,
$P_2(x) = (ax_0^3 + bx_0^2 + cx_0 + d) + (3ax_0^2 + 2bx_0 + c)(x - x_0) + (3ax_0 + b)(x - x_0)^2$, $P_3(x) = P_4(x) = (ax_0^3 + bx_0^2 + cx_0 + d) + (3ax_0^2 + 2bx_0 + c)(x - x_0) + (3ax_0 + b)(x - x_0)^2 + a(x - x_0)^3$.
For any polynomial function f of degree n, the mth Taylor polynomial for f, $P_m(x) = f(x)$ when $m \geq n$.

19. (a) $P_1(x) = P_2(x) = 0$. (b) The third derivative of f is undefined at 0.

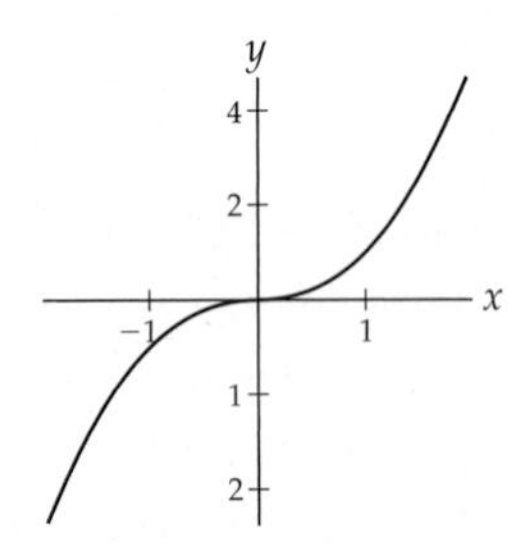

21. $1 - \frac{1}{2}x^2 + \frac{1}{24}x^4$

23. $x - \frac{1}{6}x^3$

25. $1 - 2x^2 + \frac{2}{3}x^4$

27. $1 - \frac{1}{2}x - \frac{1}{8}x^2 - \frac{1}{16}x^3 - \frac{5}{128}x^4$

29. $x^2 - \frac{1}{6}x^4$

31. $\sum_{k=0}^{\infty} \frac{(-1)^k}{(2k)!} x^{2k}$

33. $\sum_{k=0}^{\infty} \frac{(-1)^k}{(2k+1)!} x^{2k+1}$

35. $\sum_{k=0}^{\infty} \frac{(-4)^k x^{2k}}{(2k)!}$

37. $1 - \frac{1}{2}x - \sum_{k=2}^{\infty} \frac{1 \cdot 3 \cdot 5 \cdots (2k-3)}{2^k k!} x^k$

39. $\sum_{k=0}^{\infty} \frac{(-1)^k}{(2k+1)!} x^{2k+2}$

41. $-\left(x - \frac{\pi}{2}\right) + \frac{1}{6}\left(x - \frac{\pi}{2}\right)^3$

43. $-(x - \pi) + \frac{1}{6}(x - \pi)^3$

45. $\ln 3 + \frac{1}{3}(x-3) - \frac{1}{18}(x-3)^2 + \frac{1}{81}(x-3)^3 - \frac{1}{324}(x-3)^4$

47. $-2\left(x - \frac{\pi}{4}\right) + \frac{4}{3}\left(x - \frac{\pi}{4}\right)^3$

49. $\sum_{k=0}^{\infty} \frac{(-1)^{k+1}}{2k+1}\left(x - \frac{\pi}{2}\right)^{2k+1}$

51. $\sum_{k=0}^{\infty} \frac{(-1)^{k+1}}{2k+1}(x - \pi)^{2k+1}$

53. $\ln 3 + \sum_{k=1}^{\infty} \frac{(-1)^{k+1}}{k3^k}(x-3)^k$

55. $\sum_{k=0}^{\infty} \frac{(-1)^{k+1} 2^{2k+1}}{k!}\left(x - \frac{\pi}{4}\right)^{2k+1}$

57. When p is a positive integer, the coefficients defined by $\binom{p}{k} = \frac{p(p-1)(p-2) \cdot (p-k+1)}{k!}$ are all 0 for integers $k > p$.

59. $\sum_{k=0}^{\infty} \binom{1/3}{k} x^k$

61. $\sum_{k=0}^{\infty} \binom{-1/2}{k} x^k$

63. (a) and (b) $f'(0) = \lim_{h \to 0} \frac{e^{-1/h^2}}{h} = 0$. Using induction, show that $f^{(k)}(0) = 0$ for every $k \in \mathbb{Z}^+$. (c) The Maclaurin series is the constant function 0. (d) The series made of (infinitely many) zeroes converges to zero for every value of x, but the function $f(x) = 0$ if and only if $x = 0$.

65. $1, 1 - \frac{1}{4}x^2, 1 - \frac{1}{4}x^2 + \frac{1}{64}x^4, 1 - \frac{1}{4}x^2 + \frac{1}{64}x^4 - \frac{1}{2304}x^6$

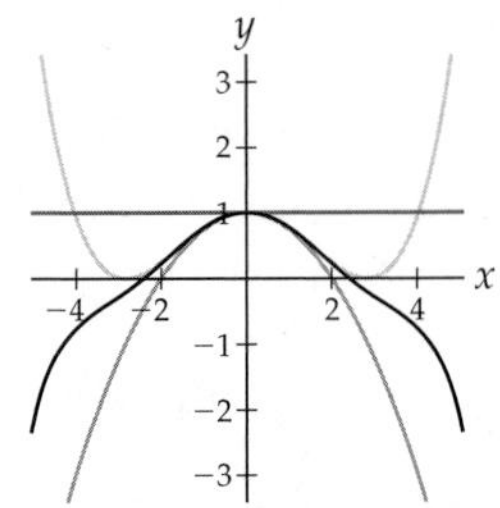

67. $\frac{1}{2}x, \frac{1}{2}x - \frac{1}{16}x^3, \frac{1}{2}x - \frac{1}{16}x^3 + \frac{1}{384}x^5, \frac{1}{2}x - \frac{1}{16}x^3 + \frac{1}{384}x^5 - \frac{1}{18{,}432}x^7$

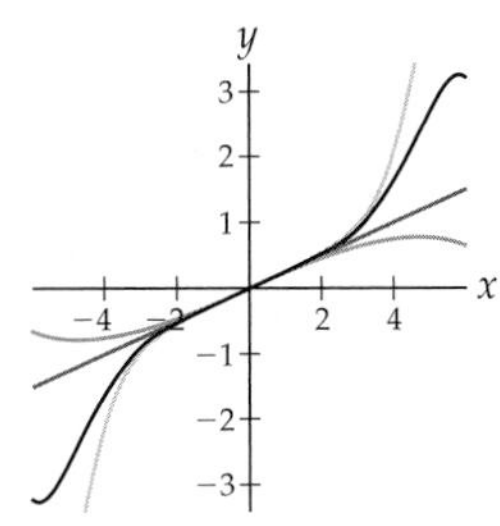

69. When $k = 1$ $\frac{d}{dx}(\ln x) = \frac{1}{x} = (-1)^{1-1}\frac{(1-1)!}{x^1}$. Next assume the statement holds for k, $\frac{d^{k+1}}{dx^{k+1}}(\ln x) = -(k+1)(-1)^{k-1}\frac{(k-1)!}{x^{k+1}} = (-1)^k\frac{k!}{x^{k+1}}$. Therefore, the statement holds for $k \geq 1$.

Section 8.3

1. F, T, F, T, T, T, F, T.

3. $a_k = \frac{f^{(k)}(0)}{k!}$

5. $c_k = \frac{h^{(k)}(x_0)}{k!}$

7. $\frac{1}{b}f\left(-\frac{a}{b}x\right) = \frac{1}{b}\frac{1}{1+(a/b)x} = \frac{1}{ax+b}$

9. $R_3(x) = R_4(x) = 0$.

11. Taylor's Theorem gives us the formula $R_n(x) = \int_{x_0}^{x}\frac{(x-t)^n}{n!}f^{(n+1)}(t)dt$ for computing the nth remainder in terms of an integral.

13. When you know a few such series you can use them to quickly find the Maclaurin series for many related functions.

15. If the Maclaurin series for $f(x)$ is $\sum_{k=0}^{\infty} a_k x^k$, the Maclaurin series for $cx^m f(x)$ is $\sum_{k=0}^{\infty} ca_k x^{k+m}$. The intervals of convergence for the two series are the same.

17. $e^{-\pi}$

19. $\frac{\pi}{6}$

21. 0

23. $-\frac{\sin c}{120}x^5$

25. $\frac{\cos c}{120}x^5$

27. $\frac{120c^4 - 240c^2 + 24}{5!(1+c^2)^5}$

29. $-\frac{7}{256(1-c)^{9/2}}x^5$

31. $\frac{5\sin c + c\cos c}{120}x^5$

33. $R_n(x) = \frac{f^{(n+1)}(c)x^{n+1}}{(n+1)!}$. Since $f(x) = \cos x$, the derivatives cycle through $-\sin x$, $-\cos x$, $\sin x$ and $\cos x$. $\lim_{n\to\infty}|R_n(x)| \leq \frac{|x|^{n+1}}{(n+1)!} = 0$

35. $R_n(x) = \frac{f^{(n+1)}(c)x^{n+1}}{(n+1)!}$. Since $f(x) = \sin x$, the derivatives cycle through $\cos x$, $-\sin x$, $-\cos x$ and $\sin x$. $\lim_{n\to\infty}|R_n(x)| \leq \frac{|x|^{n+1}}{(n+1)!} = 0$

37. $-\frac{\sin c}{120}\left(x - \frac{\pi}{2}\right)^5$

39. $\frac{\cos c}{120}(x - \pi)^5$

41. $\frac{1}{5c^5}(x-3)^5$

43. $\left(\frac{88(1+\tan^2 c)^2\tan^2 c + 16(1+\tan^2 c)^3}{120} + \frac{16\tan^4 c(1+\tan^2 c)}{120}\right)\left(x - \frac{\pi}{4}\right)^5$

45. $|R_n x| \leq \frac{\left|x - \frac{\pi}{2}\right|^{n+1}}{(n+1)!}$

47. $|R_n x| \leq \frac{|x-\pi|^{n+1}}{(n+1)!}$

49. $\frac{1}{(n+1)c^{n+1}}(x-3)^{n+1}$

51. $\sum_{k=0}^{\infty}\frac{x^{3k}}{k!}$, $\mathbb{R}$

53. $\sum_{k=0}^{\infty}(-1)^{k+1}\frac{5^{2k+1}x^{4k+2}}{(2k+1)!}$, $\mathbb{R}$

55. $\frac{1}{8}\sum_{k=0}^{\infty}\left(-\frac{1}{8}\right)^k x^{3k}$, $(-2, 2)$

57. $\sum_{k=0}^{\infty}\frac{x^{2k}}{(2k)!}$, $\mathbb{R}$

59. $1 + \sum_{k=1}^{\infty}(-1)^k\frac{2^{2k-1}x^{2k}}{(2k)!}$, $\mathbb{R}$

61. $x + x^2 + \frac{1}{3}x^3 - \frac{1}{30}x^5$ $\mathbb{R}$

63. $x^3 + x^4 + \frac{1}{2}x^5 - \frac{1}{3}x^6$ $\mathbb{R}$

65. $x + \frac{5}{6}x^3 + \frac{101}{120}x^5 + \frac{4241}{5040}x^7 (-1, 1)$

67. $e^{-0.3} = \sum_{k=0}^{\infty}\frac{(-0.3)^k}{k!} \approx \sum_{k=0}^{3}\frac{(-0.3)^k}{k!} = 0.7405$. Use $\sum_{k=0}^{6}\frac{(-0.3)^k}{k!}$.

69. $\ln 1.5 = \sum_{k=1}^{\infty}\frac{(-1)^{k+1}0.5^k}{k} \approx \sum_{k=1}^{7}\frac{(-1)^{k+1}0.5^k}{k} \approx 0.4058$. Use $\sum_{k=1}^{15}\frac{(-1)^{k+1}0.5^k}{k}$.

71. $\tan^{-1} 0.4 = \sum_{k=0}^{\infty}\frac{(-1)^k 0.4^{2k+1}}{2k+1} \approx \sum_{k=0}^{2}\frac{(-1)^k 0.4^{2k+1}}{2k+1} \approx 0.3807$. Use $\sum_{k=0}^{5}\frac{(-1)^k 0.4^{2k+1}}{2k+1}$.

73. $\sin\frac{\pi}{90} = \sum_{k=0}^{\infty}\frac{(-1)^k(\pi/90)^{2k+1}}{(2k+1)!} \approx \frac{\pi}{90} \approx 0.0349$. Use $\sum_{k=0}^{1}\frac{(-1)^k(\pi/90)^{2k+1}}{(2k+1)!}$.

75. $\cos\frac{\pi}{36} = \sum_{k=0}^{\infty}\frac{(-1)^k(\pi/36)^{2k}}{(2k)!} \approx \sum_{k=0}^{1}\frac{(-1)^k(\pi/36)^{2k}}{(2k)!} \approx 0.9962$. Use $\sum_{k=0}^{2}\frac{(-1)^k(\pi/36)^{2k}}{(2k)!}$.

77. (a) $e^{0.3} = \sum_{k=0}^{\infty}\frac{(0.3)^k}{k!}$, (b) $|R_5(x)| \leq \frac{e^{0.3}(0.3)^6}{720}$, (c) 7

79. (a) $\ln 0.5 = -\sum_{k=1}^{\infty}\frac{(0.5)^k}{k}$, (b) $\frac{1}{6}$, (c) 999, 999

81. (a) $\sin 1 = \sum_{k=0}^{9}\frac{(-1)^k}{(2k+1)!}$, (b) $\frac{1}{720}$, (c) 11

83. (a) $\lim_{h\to 0}\frac{((\sin h)/h) - 1}{h} = 0$. (b) $\sum_{k=0}^{\infty}\frac{(-1)^k}{(2k+1)!}x^{2k}$

85. (a) $y = \frac{1}{100}x + 35$. (b) From Theorem 8.10, the error term is of form $\frac{w''(c)x^2}{2}$ for some $c \in [0, 300]$.
(d) Assuming that the quadratic is somewhat close to $w(t)$, the error term is approximated by $w_2''(x)x^2/2$. This is bounded above by $\max_{0\leq x\leq 300} w_2''(x)300^2/2 = \frac{1}{90000}90000/2 = 1/2$. We estimate that the line is at most half a foot off of the water table over the entire 300 foot interval.

87. The partial proof of Theorem 8.9 in the text shows that the theorem holds for $k=1$ and $k=2$. Assume the theorem is valid for an arbitrary value of k. Applying integration by parts to the integral with $u=f^{(n+1)}(t)$ and $dv=\frac{(x-t)^n}{n!}dt$, we see that the integral in the statement equals $\frac{f^{(n+1)}(x_0)}{(n+1)!}(x-x_0)^{n+1}+\int_{x_0}^{x}\frac{(x-t)^{n+1}}{n!}f^{(n+2)}(t)dt$. When we combine this with the summation $\sum_{k=0}^{n}\frac{f^{(k)}(x_0)}{(k)!}(x-x_0)^k$ we have the result.

89. Model your proof on the argument found after Theorem 8.10, showing that the remainder for $\sin x$ goes to 0 as $n\to\infty$.

91. Model your proof on the argument found Example 2.

Section 8.4

1. T, T, T, T, F, F, F, T.

3. $f(x_0)=a_0, f'(x_0)=a_1, f''(x_0)=2a_2, f^{(k)}(x_0)=k!a_k$

5. $G(x)=7+\sum_{k=0}^{\infty}\frac{a_k}{k}(x-x_0)^k$

7. Since $\frac{d^{k-1}}{dx^{k-1}}\left(\frac{1}{1-x}\right)=\frac{(k-1)!}{(1-x)^k}$ to find the Maclaurin series for $\frac{1}{(1-x)^k}$ we could take the $(k-1)^{\text{st}}$ derivative of the series for $\frac{1}{1-x}$, term by term, and divide each term by $k!$.

9. $2\sum_{k=0}^{\infty}\frac{(-1)^k3^k}{k!}x^k$

11. (a) $\left(-\frac{1}{2},\frac{1}{2}\right)$, (b) $\sum_{k=0}^{\infty}(k+1)2^{k+1}x^k$, (c) $\sum_{k=1}^{\infty}\frac{2^{k-1}}{k}x^k$

13. (a) $(-\infty,\infty)$, (b) $\sum_{k=0}^{\infty}\frac{(-1)^{k+1}(k+1)}{(2k+2)!}x^k$,
(c) $\sum_{k=1}^{\infty}\frac{(-1)^{k+1}}{k(2k-2)!}x^k$

15. (a) $(-6,-4]$, (b) $\sum_{k=0}^{\infty}(-1)^{k+1}(x+5)^k$,
(c) $\sum_{k=2}^{\infty}\frac{(-1)^{k+1}}{(k-1)k}(x+5)^k$

17. The function $\frac{1}{1+x^3}$ is continuous on the interval $[0,0.5]$. (a) Use Theorem 5.30 to find the number of terms required to give an approximation accurate to within 0.001 of its actual value. Then form the Simpson's approximation with the appropriate number of terms. (b) $\sum_{k=0}^{\infty}(-1)^kx^{3k}$ (c) Evaluate the definite integral of the series from part (b) term by term from 0 to 0.5. Sum all of the terms that are greater than 0.001. This is the approximation.

19. The function $\sin(x^2)$ is continuous on the interval $[0,0.5]$. (a) Use Theorem 5.30 to find the number of terms required to give an approximation accurate to within 0.001 of its actual value. Then form the Simpson's approximation with the appropriate number of terms. (b) $\sum_{k=0}^{\infty}\frac{(-1)^k}{(2k+1)!}x^{4k+2}$ (c) Evaluate the definite integral of the series from part (b) term by term from 0 to 0.5. Sum all of the terms that are greater than 0.001. This is the approximation.

21. $\frac{d}{dx}\left(\sum_{k=0}^{\infty}\frac{(-1)^k}{(2k+1)!}x^{2k+1}\right)=\sum_{k=0}^{\infty}\frac{(-1)^k}{(2k+1)!}\frac{d}{dx}\left(x^{2k+1}\right)=$
$\sum_{k=0}^{\infty}\frac{(-1)^k}{(2k)!}x^{2k}$

23. $\frac{d}{dx}\left(\sum_{k=0}^{\infty}\frac{(-1)^k}{k!}x^k\right)=\sum_{k=0}^{\infty}\frac{(-1)^k}{k!}\frac{d}{dx}\left(x^k\right)=$
$\sum_{k=0}^{\infty}\frac{(-1)^k}{k!}x^k$

25. (a) $\sum_{k=0}^{\infty}(-1)^kx^{2k}, (-1,1)$,
(b) $\sum_{k=1}^{\infty}(-1)^{k+1}kx^{2k-1}, (-1,1)$

27. (a) $\sum_{k=0}^{\infty}(-2)^kx^{3k}, \left(-\frac{1}{2},\frac{1}{2}\right)$,
(b) $\sum_{k=1}^{\infty}(-2)^{k-3}kx^{3k-1}, \left(-\frac{1}{2},\frac{1}{2}\right)$

29. (a) $\sum_{k=0}^{\infty}\frac{1}{2^{2k+2}}x^{2k}, (-2,2)$, (b) $\sum_{k=1}^{\infty}\frac{k}{2^{2k+2}}x^{2k-1}, (-2,2)$

31. (a) $\sum_{k=0}^{\infty}\frac{(-1)^k}{(2k+1)!}x^{4k+2}$, (b) $\sum_{k=0}^{\infty}\frac{(-1)^k}{(2k)!}x^{4k+1}$

33. (a) $\sum_{k=0}^{\infty}\frac{(-1)^k}{k!}x^{2k}$, (b) $\sum_{k=0}^{\infty}\frac{(-1)^k}{k!}x^{2k+1}$

35. (a) $\sum_{k=0}^{\infty}\frac{1}{(-2)^{k+1}}(x-3)^k, (1,5)$,
(b) $\sum_{k=0}^{\infty}\frac{k+1}{(-2)^{k+2}}(x-3)^k, (1,5)$

37. (a) $\sum_{k=0}^{\infty}(-1)^{k+1}(x-3)^k, (2,4)$,
(b) $\sum_{k=0}^{\infty}(-1)^k(k+1)(x-3)^k, (2,4)$

39. (a) $-\frac{1}{8}\sum_{k=0}^{\infty}\left(-\frac{3}{8}\right)^k(x-3)^k, \left(\frac{1}{3},\frac{17}{3}\right)$,
(b) $-\frac{1}{24}\sum_{k=0}^{\infty}\left(-\frac{3}{8}\right)^{k+1}(k+1)(x-3)^k, \left(\frac{1}{3},\frac{17}{3}\right)$

41. (a) $\sum_{k=0}^{\infty}\frac{(-1)^k}{(2k+1)!}x^{6k+3}$, (b) $2+\sum_{k=0}^{\infty}\frac{(-1)^k}{(6k+4)(2k+1)!}x^{6k+4}$

43. (a) $\sum_{k=0}^{\infty}\left(\frac{-1}{3}\right)^k\frac{x^{2k}}{k!}$, (b) $\sum_{k=0}^{\infty}\left(\frac{-1}{3}\right)^k\frac{x^{2k+1}}{(2k+1)(k!)}$

45. (a) $\sum_{k=0}^{\infty}(-1)^kx^{3k}$, (b) $-5+\sum_{k=0}^{\infty}\frac{(-1)^k}{3k+1}x^{3k+1}$

47. (a) $\sum_{k=0}^{\infty}(-1)^k\frac{5^{2k+1}}{(2k+1)!}x^{4k+4}$,
(b) $1+\sum_{k=0}^{\infty}(-1)^k\frac{5^{2k+1}}{(4(k+1)(2k+1)!}x^{4k+5}$

49. (a) $\sum_{k=0}^{\infty}\frac{(-3)^kx^{2k+2}}{k!}$, (b) $1+\sum_{k=0}^{\infty}\frac{(-3)^kx^{2k+3}}{(2k+3)k!}$

51. $\sum_{k=0}^{\infty}\frac{(-1)^k}{(3k+2)(2k+1)!}2^{6k+3}$

53. $2\sum_{k=0}^{\infty}\frac{(-1)^k}{(2k+1)3^kk!}$

55. $\sum_{k=0}^{\infty}\frac{(-1)^k}{3k+1}(0.3)^{3k+1}$

57. $\sum_{k=0}^{\infty}\frac{(-1)^k5^{2k+1}}{(4k+5)(2k+1)!}(2^{4k+5}-1)$

59. $\sum_{k=0}^{\infty}\frac{(-3)^k}{(2k+3)k!}-\sum_{k=0}^{\infty}\frac{(-3)^k}{2^{2k+3}(2k+3)k!}$

61. $\sum_{k=0}^{10}\frac{(-1)^k2^{6k+3}}{(3k+2)(2k+1)!}\approx 0.45254$

63. $\sum_{k=0}^{3}\frac{2(-1)^k}{(2k+1)3^kk!}\approx 1.79824$

65. $\sum_{k=0}^{1} \frac{(-1)^k}{(3k+1)} 0.3^{3k+1} \approx 0.29798$

67. $\sum_{k=0}^{27} \frac{(-1)^k 5^{2k+1}}{(4k+5)(2k+1)!}(2^{4k+5} - 1) \approx -0.039348$

69. $\sum_{k=0}^{9} \frac{(-1)^k 3^k}{(2k+3)k!}(1 - 0.5^{2k+3} \approx 0.48101$

71. $y_0''(x) = \sum_{k=1}^{\infty} \frac{(3k)(3k-1)x^{3k-2}}{(2\cdot 3)(5\cdot 6)\cdots((3k-1)\cdot 3k)} =$
$x + \sum_{k=2}^{\infty} \frac{x^{3k-2}}{(2\cdot 3)(5\cdot 6)\cdots((3k-4)\cdot(3k-3))}$. On the other side of the equation we see that
$xy_0(x) = x + x\sum_{k=1}^{\infty} \frac{x^{3k}}{(2\cdot 3)(5\cdot 6)\cdots((3k-1)\cdot 3k)} =$
$x + \sum_{k=1}^{\infty} \frac{x^{3k+1}}{(2\cdot 3)(5\cdot 6)\cdots((3k-1)\cdot 3k)} =$
$x + \sum_{k=2}^{\infty} \frac{x^{3k-2}}{(2\cdot 3)(5\cdot 6)\cdots((3(k-1)-1)\cdot 3(k-1))}$. These are the same, so y_0 is a solution to Airy's equation.

73. The two series are the Maclaurin series for some function $f(x)$. Thus, the series $\sum_{k=0}^{\infty}(a_k - b_k)x^k$ is a Maclaurin series for the zero function. Thus, each coefficient $a_k - b_k = 0$.

75. Let f be an odd function with $f(x) = \sum_{k=0}^{\infty} a_k x^k$. Since f is odd, we have $f(x) = -f(-x)$ for every x in the domain. This means that $f(x) = \sum_{k=0}^{\infty}(-1)^{k+1}a_k x^k$. From Exercise 73 we have that $a_k = (-1)^{k+1}a_k$ for every k. This is only possible if $a_{2k} = 0$ for every k.

For answers to odd-numbered Skill Certification exercises in the Chapter Review, please visit the Book Companion Web Site at www.whfreeman.com/tkcalculus.

Chapter 9

Section 9.1

1. T, F, F, F, F, F, T, F.

3. (a) To the right and up. (b) To the right and down. (c) To the left and up. (d) To the left and down.

5. Up and to the left when $t < 0$, up and to the right when $t > 0$.

7. Up and to the right

9. (i) $y = x^2 - 1$, (ii) (a) the right half of the parabola, (b) the left half of the parabola; (iii) motion to the right as t increases for (a), and motion to the left as t increases for (b).

11. The graphs are the portions of $y = \sin x$ to the right of the y-axis. The direction of motion for both is to the right.

13. There is a horizontal tangent line at a value, t_0, where $\frac{dy}{dt} = 0$ and $\frac{dx}{dt} \neq 0$ or when $\lim_{t\to t_0} \frac{dy/dt}{dx/dt} = 0$. There is a vertical tangent line at value t_0 where $\frac{dy}{dt} \neq 0$ and $\frac{dx}{dt} = 0$ or when $\lim_{t\to t_0} \frac{dy/dt}{dx/dt} = \pm\infty$.

15. Since the sine function is periodic and $-1 \le \sin t \le 1$ for all values of t, the values of x and y will also be bounded; $0 \le x \le 2$ and $-4 \le y \le -3$.

17.

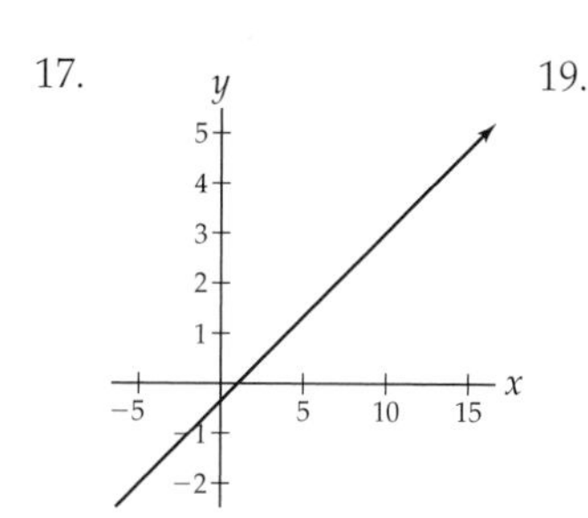

19.

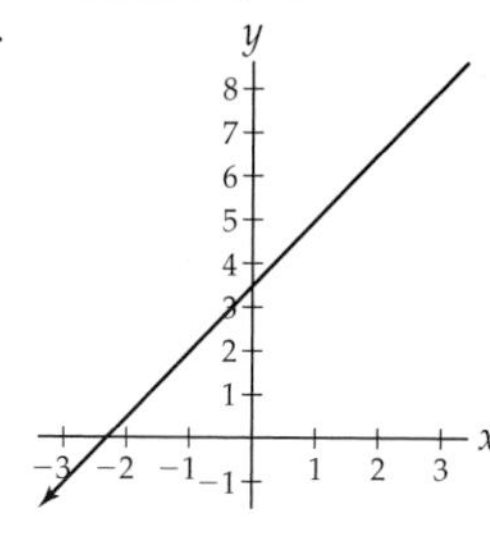

21.

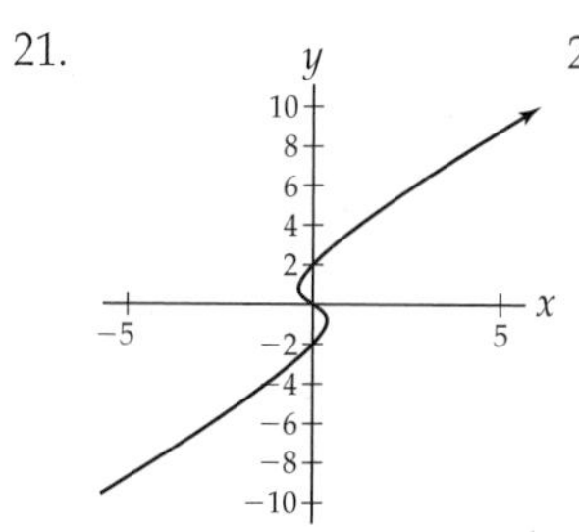

23.

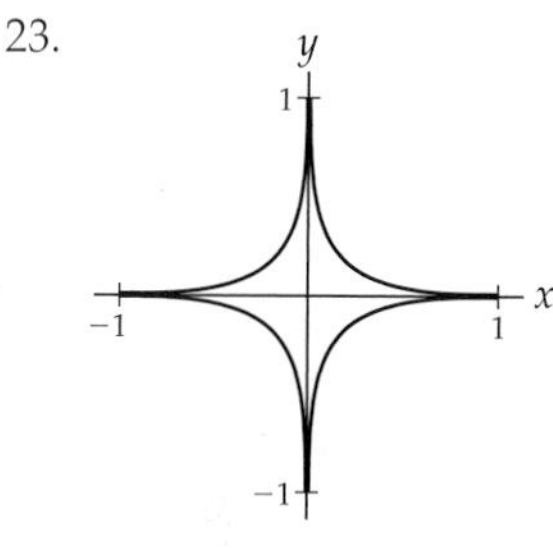

25.

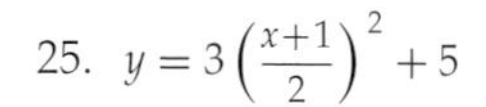
$y = 3\left(\frac{x+1}{2}\right)^2 + 5$

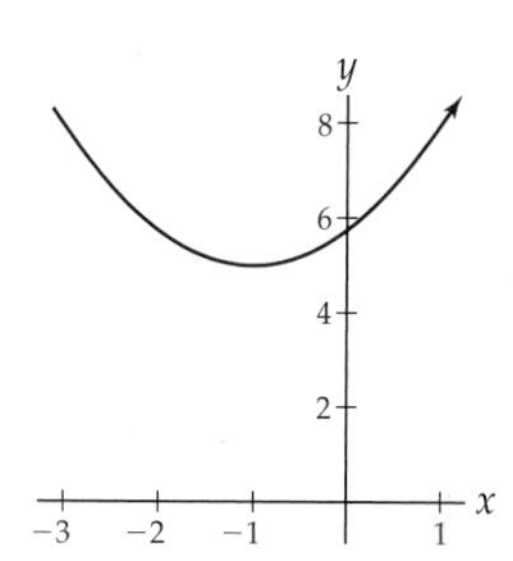

27. $y = x$

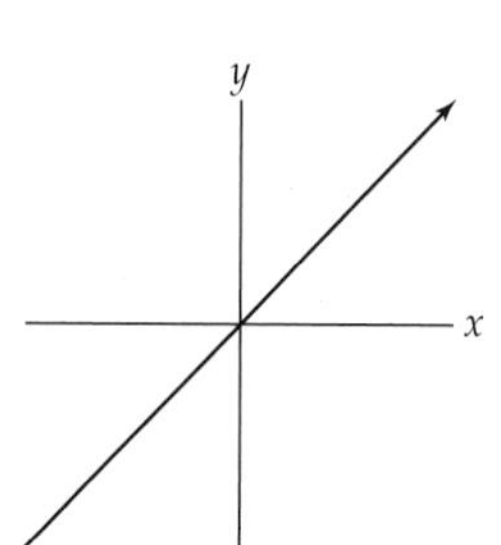

29.

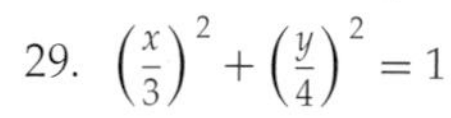
$\left(\frac{x}{3}\right)^2 + \left(\frac{y}{4}\right)^2 = 1$

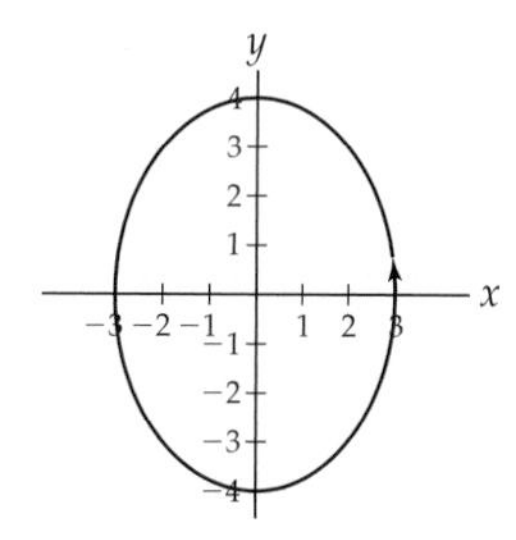

31. $x^2 - y^2 = 1$

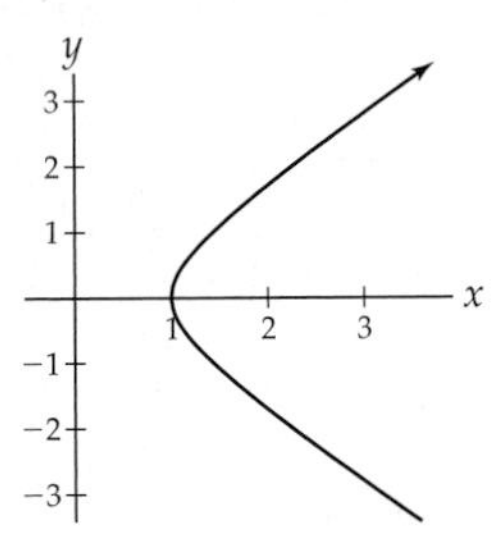

33. $x^2 - y^2 = 1$, same graph as in Exercise 31

35. $x^2 + y^2 = 1$, when $t = \pi$, $(x, y) = (-1, 0)$, $t = \frac{3\pi}{2}$, $(x, y) = (0, -1)$ and $t = 2\pi$, $(x, y) = (1, 0)$

37. $x = -3\sin(2\pi t)$, $y = 3\cos(2\pi t)$

39. $x = \frac{2t}{1+t^2}$, $x = \frac{1-t^2}{1+t^2}$

41. $y - 2 = \frac{3}{2}(x + 3)$

43. $y - \frac{9}{4} = -3\left(x - \frac{1}{4}\right)$

45. 0

47. 8

49. $\frac{2}{27}\left(10\sqrt{10} - 1\right)$

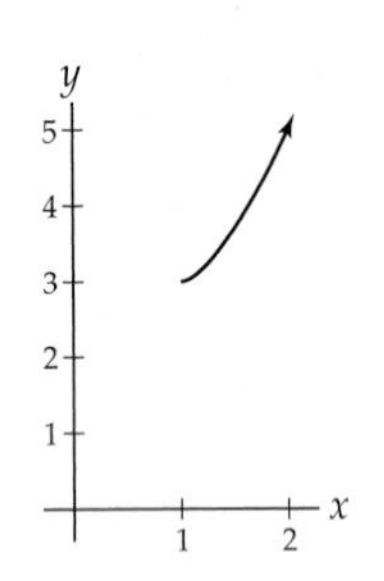

51. 6

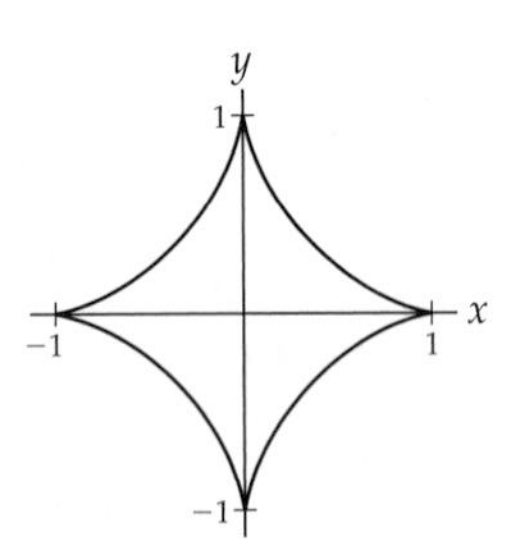

53. $e - \frac{1}{e}$

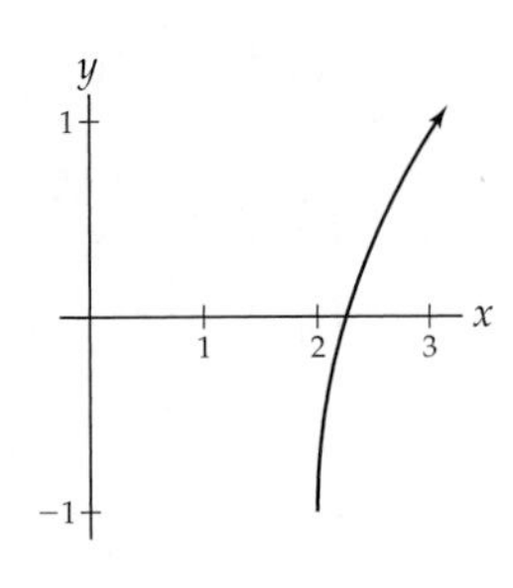

55. $x = 1 + 5t$, $y = -3 + 10t$, $t \in [0, 1]$

57. $x = 1 - 4t$, $y = 4 + t$, $t \in [0, 1]$

59. $x = \pi$, $y = 3 + 5t$, $t \in [0, 1]$

61. $5\sqrt{5}$

63. $x = (k + r)\cos\theta - k\cos\left(\frac{(k+r)\theta}{k}\right)$,
$y = (k + r)\sin\theta - r\sin\left(\frac{(k+r)\theta}{k}\right)$

65. $\int_0^{2\pi} \sqrt{\cos^2 t + 9\sin^2 t}\,dt \approx 13.4$

67. (a) $x(t) = 1.932t$, $y(t) = 0.552t^3 - 1.287t^2 + 0.518t$, (b) 0.232 miles, (c) 3.044 miles

69. When $x = t$ and $y = f(t)$, we have $\int_a^b \sqrt{(x'(t))^2 + (y'(t))^2}dt = \int_a^b \sqrt{1 + (f'(t))^2}dt$.

71. $\int_a^b \sqrt{((kf)'(t))^2 + ((kg)'(t))^2}dt =$ $k\int_a^b \sqrt{(f'(t))^2 + (g'(t))^2}dt = k$ times the arc length of the second set of parametric equations. The arc length of the final set of parametric equations is the same as the length of the second set of parametric equations.

73. $\frac{dy}{dx} = \frac{\sin\theta}{1-\cos\theta}$. By L'Hôpital's rule $\lim_{\theta\to 2k\pi^+} \frac{\sin\theta}{1-\cos\theta} = \lim_{\theta\to 2k\pi^+} \frac{\cos\theta}{\sin\theta} = \infty$ and $\lim_{\theta\to 2k\pi^-} \frac{\sin\theta}{1-\cos\theta} = \lim_{\theta\to 2k\pi^-} \frac{\cos\theta}{\sin\theta} = -\infty$. Thus, there is a vertical tangent line at each even multiple of π.

Section 9.2

1. T, F, F, F, T, T, T, T.

3. The value of θ in the polar coordinates for a point P measures the angular rotation of P from the x-axis. Adding an integer rotation of 2π doesn't change the location of the point.

5. The point $\left(8, -\frac{\pi}{3}\right)$ is in the fourth quadrant. The point $(-4, 4\sqrt{3})$ is in the second quadrant.

7. The graphs of $r = c$ and $r = -c$ are the same for every real number c.

9. $\alpha = k\frac{\pi}{2}$ for any integer k.

11. Every point, (r, θ), is on a ray in the first quadrant when $r > 0$ and $0 < \theta < \frac{\pi}{2}$. The points in the third quadrant are given by the inequalities $r > 0$ and $\pi < \theta < \frac{3\pi}{2}$.

13. $(1, 2k\pi)$ and $(-1, (2k + 1)\pi)$ for every integer k.

15. $(1, (2k + 1)\pi)$ and $(-1, 2k\pi)$ for every integer k.

17. $\left(\frac{3\sqrt{3}}{2}, \frac{3}{2}\right), \left(-\frac{3\sqrt{3}}{2}, -\frac{3}{2}\right), \left(\frac{3\sqrt{3}}{2}, -\frac{3}{2}\right), \left(-\frac{3\sqrt{3}}{2}, \frac{3}{2}\right)$

19. $(0, 5)$, $(0, -5)$, $(0, 5)$, $(0, 5)$

21. $\left(\frac{\sqrt{2}}{2}, \frac{\sqrt{2}}{2}\right), \left(\sqrt{2}, \sqrt{2}\right), \left(\frac{3\sqrt{2}}{2}, \frac{3\sqrt{2}}{2}\right), \left(2\sqrt{2}, 2\sqrt{2}\right)$

23. $(0, -1)$, $(0, -2)$, $(0, -3)$, $(0, -4)$

25. $(1, 2k\pi)$ and $(-1, (2k + 1)\pi)$ for any integer k.

27. $\left(2, \frac{\pi}{2} + 2\pi k\right)$ and $\left(-2, -\frac{\pi}{2} + 2\pi k\right)$ for any integer k

29. $\left(4\sqrt{3}, \frac{\pi}{6} + 2\pi k\right)$ and $\left(-4\sqrt{3}, \frac{7\pi}{6} + 2\pi k\right)$ for any integer k.

31. $\left(3\sqrt{2}, -\frac{\pi}{4} + 2\pi k\right)$ and $\left(-3\sqrt{2}, \frac{3\pi}{4} + 2\pi k\right)$ for any integer k.

33. $y = x$

35. $y = \frac{\sqrt{3}}{3}x$

37. $x^2 + \left(y - \frac{5}{2}\right)^2 = \left(\frac{5}{2}\right)^2$

39. $y = 6$

41. $(x^2 + y^2)^{3/2} = 2xy$

43. $\pm(x^2 + y^2)^{3/2} = x.$

45. $x \tan\sqrt{x^2 + y^2} = y$

47. $(x^2 + y^2)^2 = y^3$

49. $\theta = 0.$

51. $\theta = \frac{\pi}{4}.$

53. $r = -3\csc\theta.$

55. $\tan\theta = m$

57. $x^2 + y^2 = k^2$

59. For $k = 1$ and $k = 7$, $y = \frac{\sqrt{3}}{3}x$. For $k = 2$ and $k = 8$, $y = \sqrt{3}x$. For $k = 3$ and $k = 9$, $x = 0$. For $k = 4$ and $k = 10$, $y = -\sqrt{3}x$. For $k = 5$ and $k = 11$, $y = -\frac{\sqrt{3}}{3}x$. For $k = 6$ and $k = 12$, $y = 0$.

61. (a) Assuming the arms of the cam remain symmetrically placed, each $\frac{3}{4}$ inch arm must fit in a half inch of the crack. Thus, we require $0.75\sin\theta = \frac{1}{2}$, so $\theta = \sin^{-1}\frac{2}{3}$. (b) This causes the cam to push even harder against the walls of the crack.

63. Since $\csc\theta = \frac{r}{y}$, we have, $r = k\frac{r}{y}$, or equivalently $y = k$.

65. We have $a = r - br\cos\theta$. In rectangular coordinates this can be written $\sqrt{x^2 + y^2} = a + bx$. Squaring both sides, completing the square, and simplifying, we may obtain $\left(x - \frac{ab}{1-b^2}\right)^2 + \frac{y^2}{1-b^2} = \left(\frac{a}{1-b^2}\right)^2$ which is the equation of an ellipse if $0 < b < 1$. If $b > 1$ the equation gives a hyperbola.

67. Since $\cos\theta = \frac{x}{r}$, we have $r = a\frac{x}{r}$, or equivalently $r^2 = x^2 + y^2 = ax$. When you complete the square for this equation you obtain $\left(x - \frac{a}{2}\right)^2 + y^2 = \left(\frac{a}{2}\right)^2$. So the graph of the equation is a circle with center $\left(\frac{a}{2}, 0\right)$ and radius $\frac{a}{2}$.

69. Since $\cos\theta = \frac{x}{r}$ and $\sin\theta = \frac{y}{r}$, we have $r = k\frac{y}{r} + l\frac{x}{r}$, or equivalently $r^2 = x^2 + y^2 = lx + ky$. When you complete the square for this equation you obtain $\left(x - \frac{l}{2}\right)^2 + \left(y - \frac{k}{2}\right)^2 = \frac{k^2+l^2}{4}$. So the graph of the equation is a circle with center $\left(\frac{l}{2}, \frac{k}{2}\right)$ and radius $\frac{\sqrt{k^2+l^2}}{2}$.

Section 9.3

1. F, T, F, T, F, T, T, T.

3. The θ-intercepts of the graph in the θr-plane correspond to the places where the polar graph passes through the pole. The location of the graph in the θr-plane corresponds to the quadrant in which the polar graph is drawn.

5. (a) All θ-intercepts of the graph in the θr-plane correspond to the places where the polar graph passes through the pole. (b) The points on the θ-axis correspond to the points on the horizontal axis in the polar plane. (c) All points on the vertical lines $\theta = (2k + 1)\frac{\pi}{2}$ correspond to the points on the vertical axis in the polar plane.

7. (a) $\left(2, -\frac{\pi}{3}\right)$ (b) $\left(2, \frac{2\pi}{3}\right)$ (c) $(-3, 5)$ (d) $\left(2, \frac{4\pi}{3}\right)$ (e) $(-3, -5)$

9. (a) $r = 2\cos\theta$ (b) $r = 2\sin\theta$ (c) $r = \sin 2\theta$ (d) $r^2 = \sin 2\theta$ (e) $r = \sin 2\theta$ (f) not possible

11. A cardioid, a limaçon, and a lemniscate.

13. $r = \cos 5\theta$: symmetry about the horizontal axis and rotational symmetry, five petals. $r = \sin 8\theta$: symmetry about the x- and y-axes, symmetry about the origin, rotational symmetry, sixteen petals.

15. Since $\cos(-2\theta) = \cos 2\theta$ the graph will be symmetrical with respect to the horizontal axis.

17.

19.

21.

23.

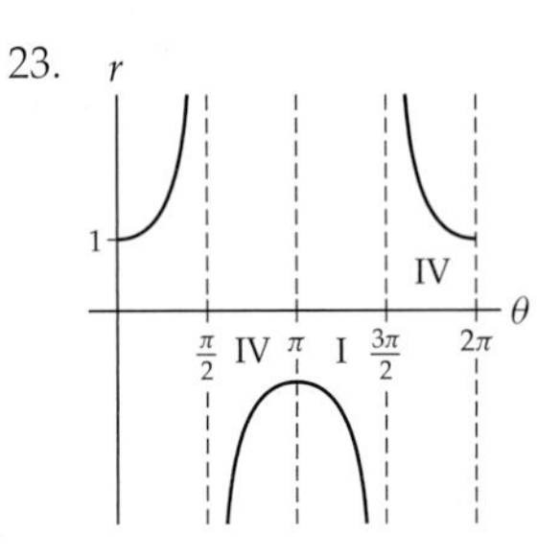

25.

27.

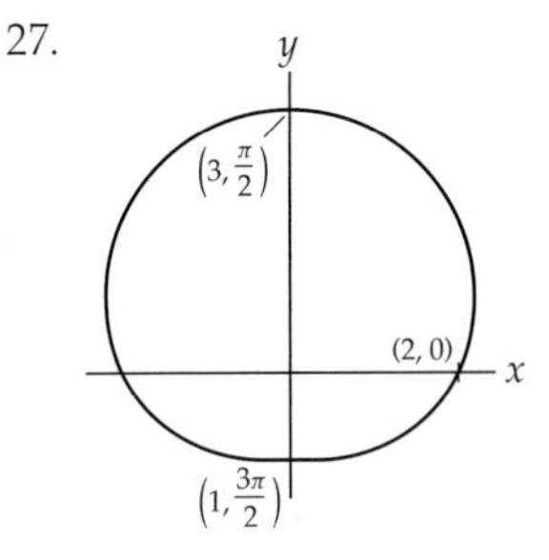

29. 31.

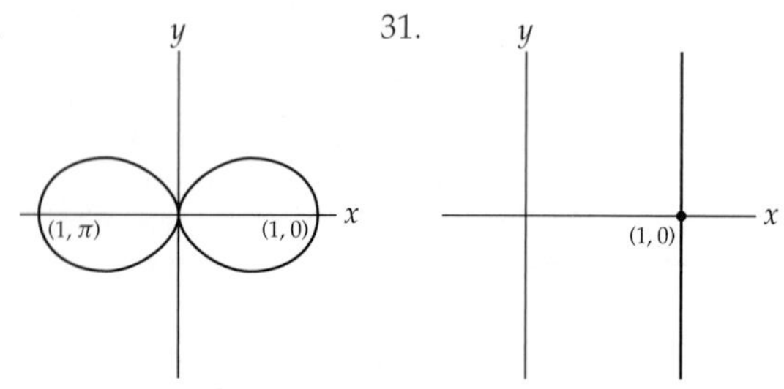

33. Use the interval $[0, 10\pi]$.

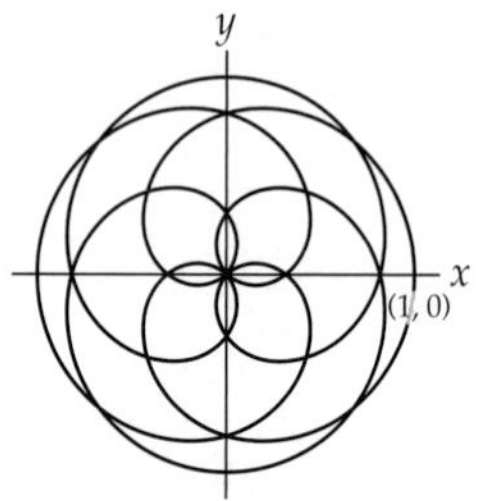

35. Use the intervals $\left[0, \frac{\pi}{2}\right) \cup \left(\frac{\pi}{2}, \frac{3\pi}{2}\right) \cup \left(\frac{3\pi}{2}, 2\pi\right]$.

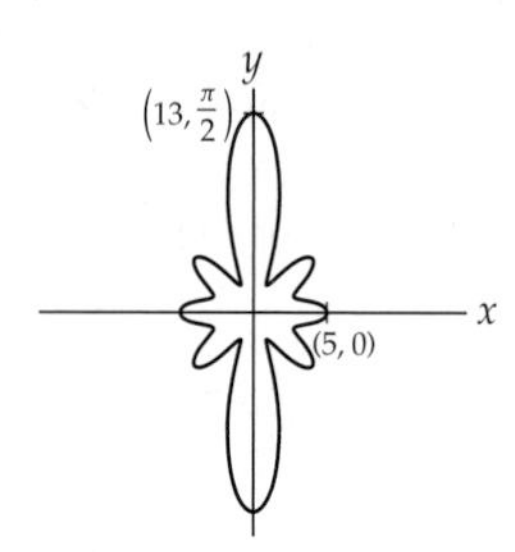

37. Use the interval $[0, 24\pi]$.

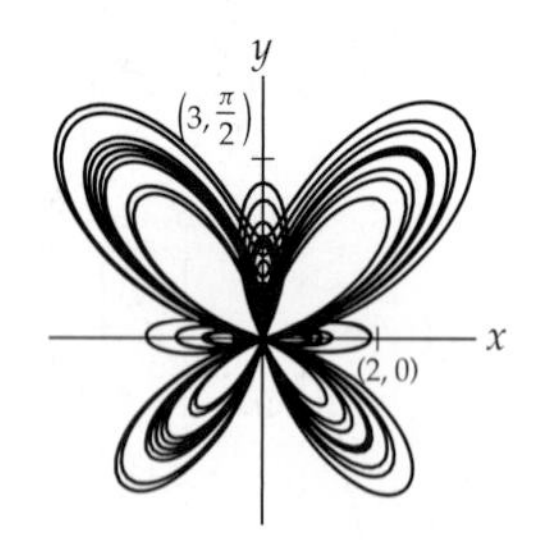

39.

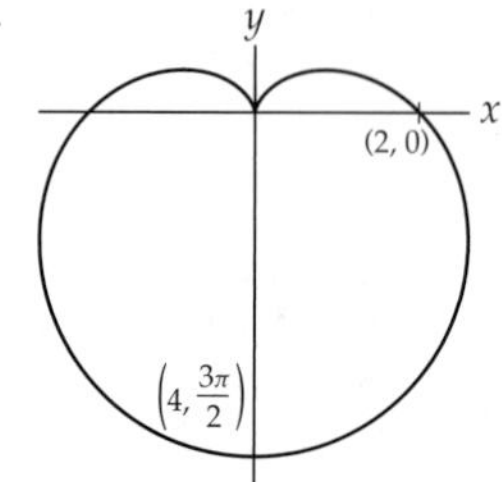

41. (a) $\frac{\pi}{4} + k\pi$ for every integer k. (c) The curves intersect at $(0, 0)$ and $\left(\frac{\sqrt{2}}{2}, \frac{\pi}{4}\right)$.

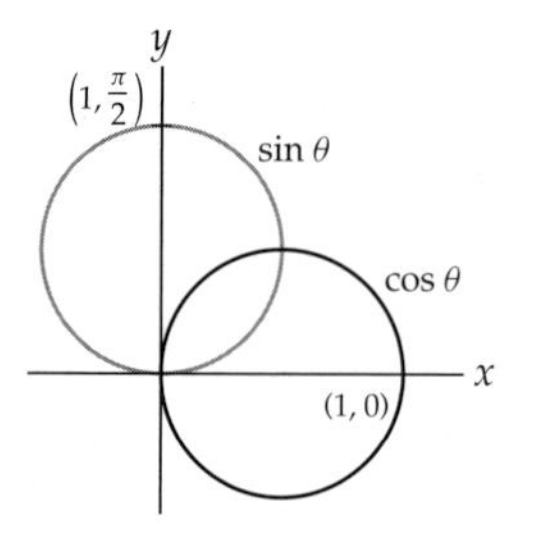

43. (a) $\frac{3\pi}{4} + k\pi$ for every integer k. (b) Graph is shown below. (c) The curves intersect at $(0, 0)$, $\left(1 + \frac{\sqrt{2}}{2}, \frac{3\pi}{4}\right)$ and $\left(1 - \frac{\sqrt{2}}{2}, \frac{7\pi}{4}\right)$.

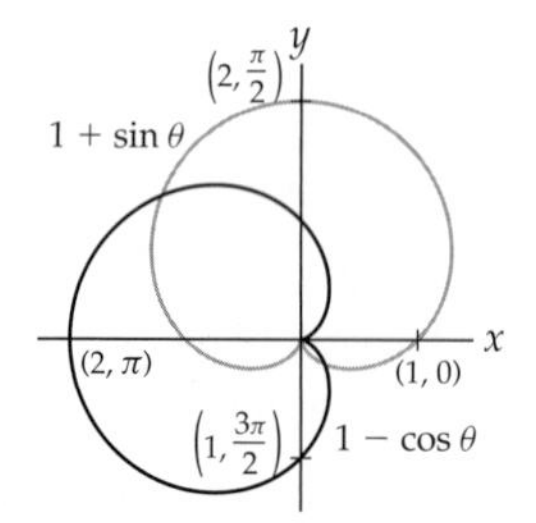

45. (a) $\frac{\pi}{4} + \frac{k\pi}{2}$ for $k = 0, 1, 2$ and 3. (b) Graph is shown below. (c) The curves intersect at $(0, 0)$, $\left(\frac{1}{2}, \frac{\pi}{4} + \frac{k\pi}{2}\right)$ for every integer k.

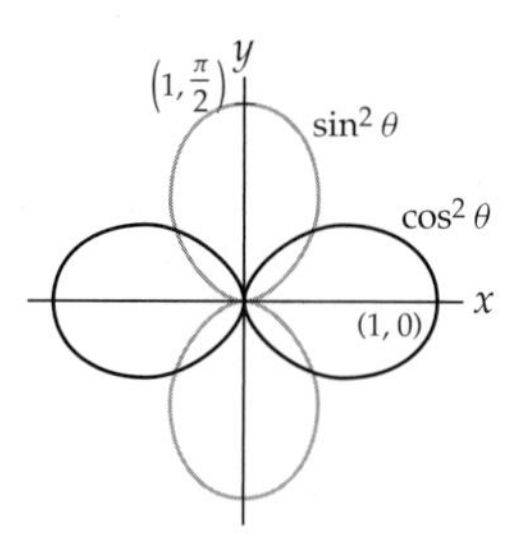

47. They are the mirror images of each other.

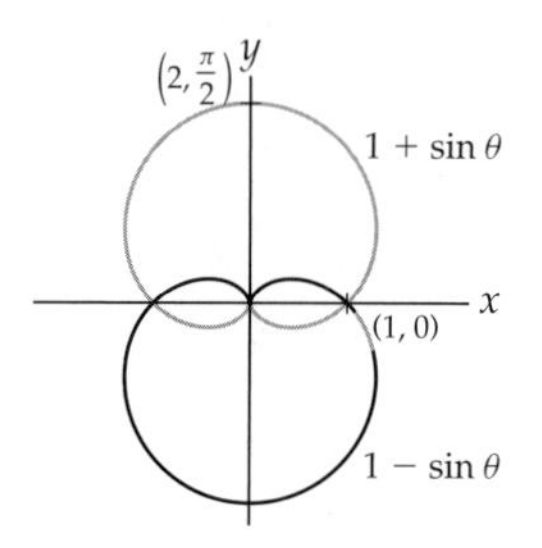

49. When $0 \le b \le \frac{3}{2}$ the limaçon will be convex. When $\frac{3}{2} < b \le 3$ there will be a dimple. When $b = 3$ the curve is a cardioid.

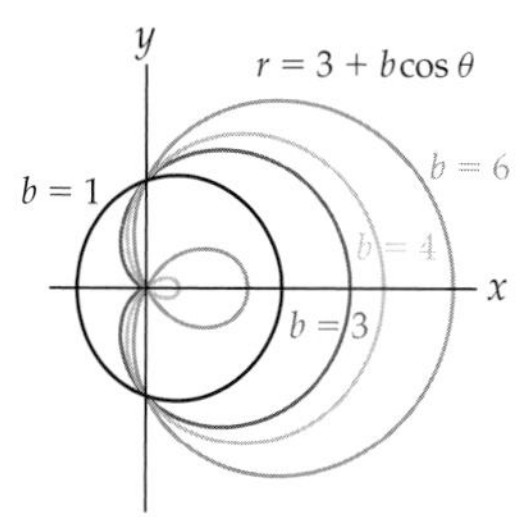

51. (a) $x = \frac{1}{2} + \frac{1}{2}\cos t,\ y = \frac{1}{2}\sin t,\ t \in \mathbb{R}$,
(b) $\left(x - \frac{1}{2}\right)^2 + y^2 = \frac{1}{4}$, (c) $r = \cos\theta$

53.

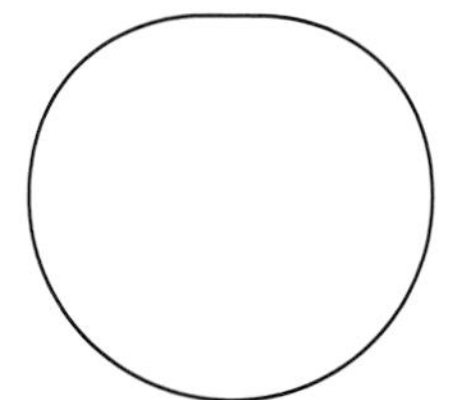

55. We will prove the statement for $f(\theta) = \sin n\theta$. The argument for $f(\theta) = \cos n\theta$ is similar. Let (r, θ) be a point on the graph of $f(\theta) = \sin n\theta$. That is, $r = \sin n\theta$. If n is a positive odd integer, then by the angle sum identity for sine we have $\sin(n(\theta + \pi)) = -\sin(n\theta)$. Therefore the point $(-r, \sin(n(\theta + \pi)) = (-r, -\sin(n\theta)$ has the same location as (r, θ). Thus, the graph traced on the interval $[\pi, 2\pi]$is the same as the graph traced on the interval $[0, \pi]$.

57. Let (r, θ) be a point on the graph of $f(\theta) = \sin n\theta$ where n is even. We have $f(-\theta) = \sin(-n\theta) = \sin(2\pi - n\theta) = \sin(n\theta)$. Therefore, the point $(r, -\theta)$ is also on the graph. Thus, the graph is symmetrical with respect to the x-axis.

59. Let (r, θ) be a point on the graph of $f(\theta) = \cos n\theta$, where n is even. It suffices to show that $(r, \pi - \theta)$ is also on the graph. We have $f(\pi - \theta) = \cos(n(\pi - \theta)) = \cos(n\theta)$. Therefore, the point $(r, \pi - \theta)$ is also on the graph.

Section 9.4

1. T, T, F, T, F, F, T, F.

3. For a function $r = f(\theta)$ in polar coordinates, θ is an angular rotation. A small change $\Delta\theta$ determines a wedge-shaped slice of the region.

5.

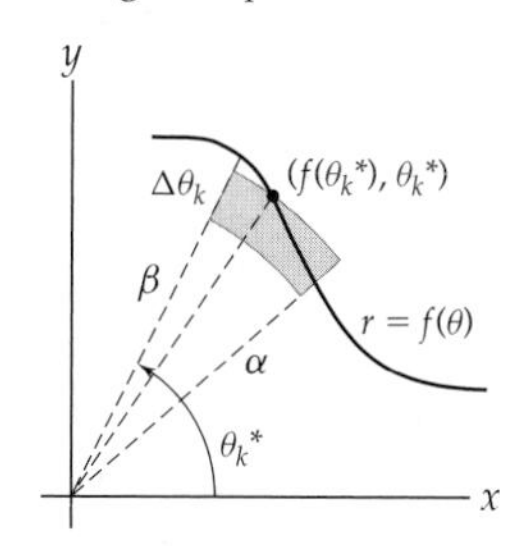

7. To compute an area we need $\beta > \alpha$ for the limits of integration in the integral. So that we don't compute a portion of the area twice, we need $\beta - \alpha < 2\pi$.

9. It is often simpler to compute a known fraction of the total area and then finish by multiplying by the appropriate factor. Review Example 2 to see how this may be done.

11. $x = t,\ y = f(t)$

13. The graph of the equation $r = \sin\theta$ is a circle with radius $\frac{1}{2}$. Since he integrated on the interval $[0, 2\pi]$ the circle is traced twice, so he got the correct answer.

15. The expression computes the area inside the two circles below. The area is $\frac{\pi - 2}{8}$.

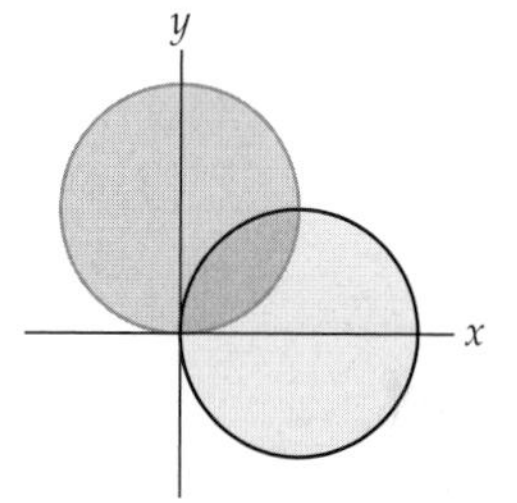

17. $\frac{1}{2}\int_0^{\pi} \theta^2 d\theta = \frac{\pi^3}{6}$

19. $\int_0^{3\pi/4} (1+\sqrt{2}\cos\theta)^2 d\theta - \int_{3\pi/4}^{\pi} (1+\sqrt{2}\cos\theta)^2 d\theta = 3+\pi$

21. $\int_{-\pi/2}^{\pi/6} ((3 - 3\sin\theta)^2 - (1 + \sin\theta)^2) d\theta = 8\pi + 9\sqrt{3}$

23. $\int_0^{\pi/3} (1 - \sec^2\theta) d\theta = \frac{\pi}{3} - \frac{\sqrt{3}}{4}$

25. $\int_0^{\pi/4} (\sec\theta - 2\cos\theta)^2\, d\theta = 2 - \frac{\pi}{2}$

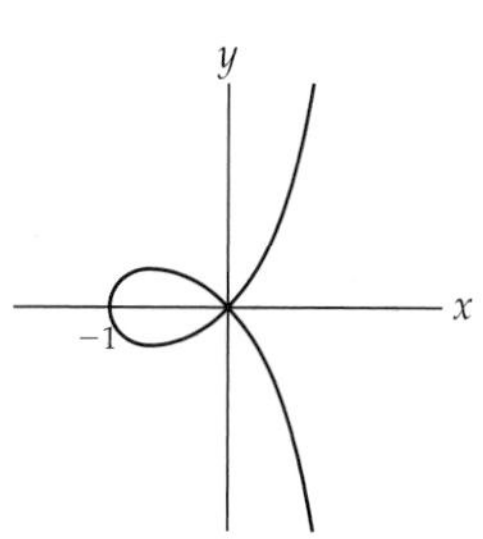

27. $\int_0^{2\pi} \sqrt{2} e^{\theta} d\theta = \sqrt{2}(e^{2\pi} - 1)$

29. $\int_0^{2\pi} \sqrt{1 + \alpha^2}\, e^{\alpha\theta} d\theta = \frac{\sqrt{1+\alpha^2}}{\alpha}(e^{2\pi\alpha} - 1)$

31. $2\int_0^{\pi/4} \sqrt{4\sin^2 2\theta + \cos^2 2\theta}\, d\theta \approx 2.422$

33. $2\int_0^{\pi/8} \sqrt{16\sin^2 4\theta + \cos^2 4\theta}\, d\theta \approx 2.145$

35. $\int_{7\pi/6}^{11\pi/6} \sqrt{(1+2\sin\theta)^2 + 4\cos^2\theta}\, d\theta \approx 2.682$

37. $\frac{3\sqrt{3}+\pi}{4}$

39. $\frac{3\pi}{2}$

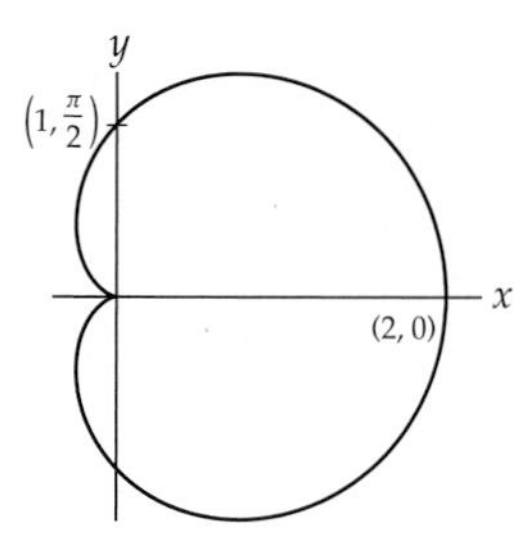

41. 1

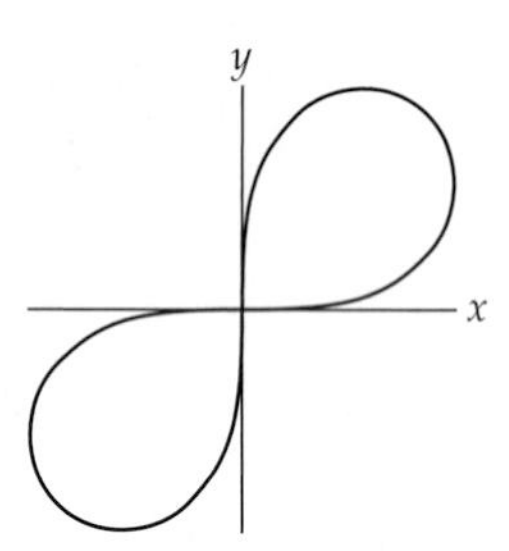

43. $\frac{9\pi}{2}$

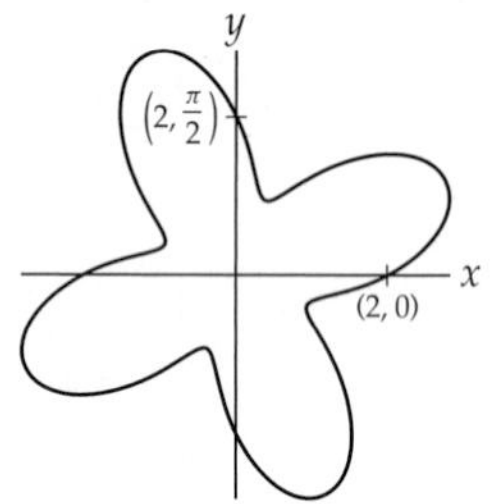

45. The area is $\frac{1}{2}\int_0^{2\pi} a^2\, d\theta = \pi a^2$.

47. The circumference is $\int_0^{2\pi} a\, d\theta = 2\pi a$.

49. $a^2\left(\frac{2\pi}{3} - \frac{\sqrt{3}}{2}\right)$

51. The length of the central cross section of the kayak is about 6.15 feet or about 1.87 meters. Annie should be able to use just two widths of the fabric.

53. $A = \frac{1}{2}\int_{\theta_0}^{\theta_0+\phi} r^2 d\theta = \frac{1}{2}\phi r^2$.

55. For any positive integer n, the area is given by $n\int_0^{2\pi/n} \cos^2(2n\theta)\, d\theta = \frac{\pi}{2}$.

57. The length of the curve is given by the improper integral $\int_1^{\infty} \sqrt{\frac{1}{\theta^2} + \frac{1}{\theta^4}}\, d\theta$ which diverges to ∞.

Section 9.5

1. T, T, F, T, F, T, F, T.

3. See Definition 9.15

5. See Definition 9.19

7. Since a hyperbola is the set of points in the plane for which the difference between the distances of the foci is a constant, and there are infinitely many differences which can be obtained in this way, there are infinitely many different hyperbolas with the same foci.

9. (a) $\frac{\sqrt{A^2-B^2}}{A}$, (b) $\frac{\sqrt{B^2-A^2}}{B}$, (c) Since $\sqrt{A^2 - B^2} < A$ and $\sqrt{B^2 - A^2} < B$, (d) It becomes more circular. (e) It elongates and flattens.

11. $y = \pm\frac{B}{e}$

13.

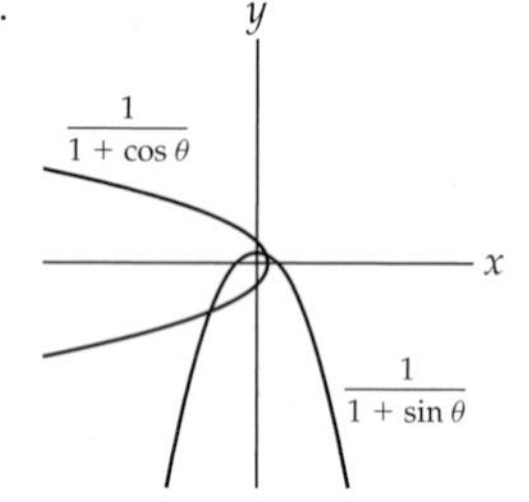

Both are parabolas with a focus at the origin; $r = \frac{1}{1+\cos\theta}$ opens to the left, the other opens down. All parabolas have eccentricity one.

15.

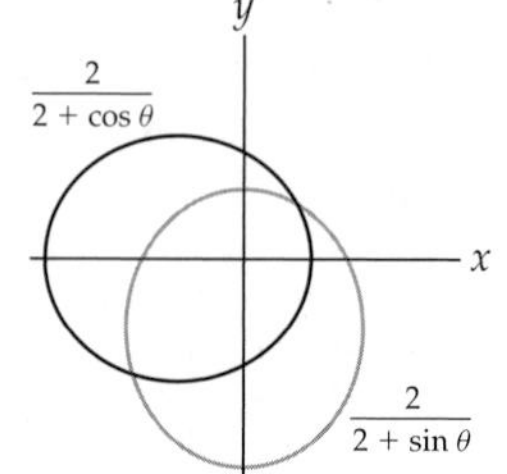

Both are ellipses with a focus at the origin; $r = \frac{2}{2+\cos\theta}$ has its major axis on the x-axis, the other has its major axis on the y-axis. Each has eccentricity $\frac{1}{2}$.

17. $\frac{\alpha}{\beta+\gamma\sin\theta} = \frac{\frac{\gamma}{\beta}\frac{\alpha}{\gamma}}{1+\frac{\gamma}{\beta}\sin\theta}$

19. The equation is equivalent to $\frac{(x+1)^2}{8} - \frac{(y+3)^2}{16} = 1$. Its graph is a hyperbola with center $(-1, -3)$ opening to the left and right.

21. The equation is equivalent to $\frac{(y-4)^2}{25} - \frac{4(x+1)^2}{25} = 1$. Its graph is a hyperbola with center $(-1, 4)$ opening up and down.

23. $y = \frac{1}{2}x^2 + \frac{1}{2}$

25. $y = -\frac{1}{4}x^2 - 7$

27. $y = \frac{1}{2(y_1 - y_0)}(x - x_1)^2 + \frac{y_0 + y_1}{2}$

29. $y = -\frac{1}{8}x^2 + 2$

31. $y = -\frac{1}{2\alpha}x^2 + \frac{\alpha}{2}$

33. $\frac{x^2}{9} + \frac{(y-1)^2}{4} = 1$

35. $\frac{4x^2}{9} + \frac{y^2}{9} = 1$

37. $\frac{x^2}{\alpha^2} + \frac{y^2}{2\alpha^2} = 1$

39. $\frac{x^2}{6} - \frac{y^2}{30} = 1$

41. $\frac{y^2}{8} - \frac{x^2}{8} = 1$

43. $\frac{(y-5)^2}{4} - \frac{(x-3)^2}{12} = 1$

45.

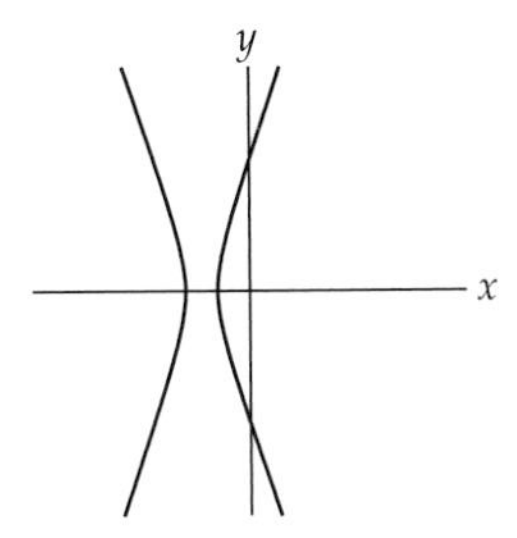

47.

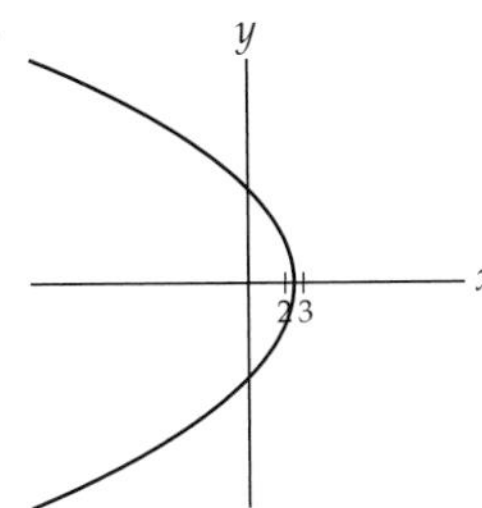

49.

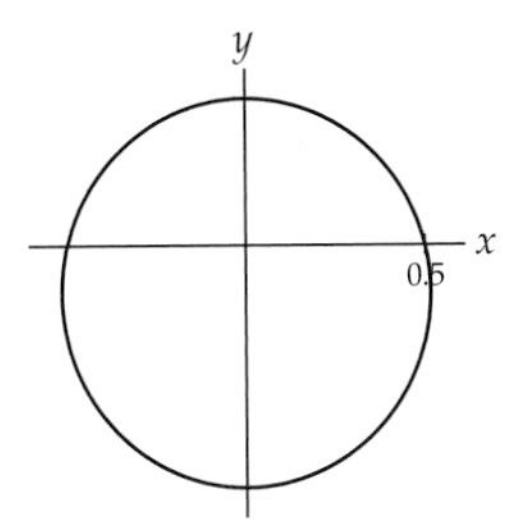

51.

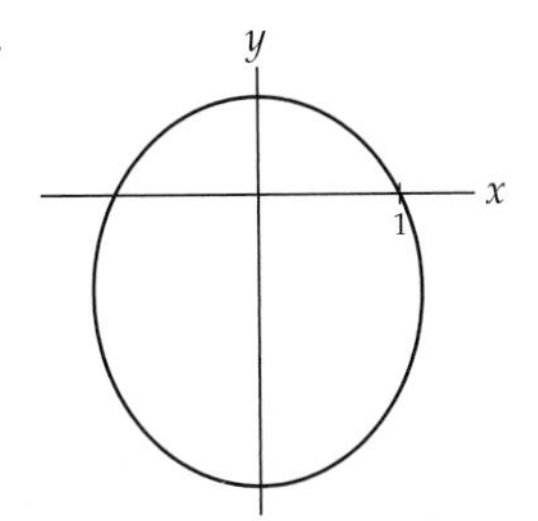

53. (a) $\frac{x^2}{1.00000011^2} + \frac{y^2}{0.9998606552^2} = 1,$

(b) $r = \frac{1.00000011}{1 + 0.0167\cos\theta}$

55. Model your argument on the proof that the distance from any point on the graph of the curve to the point $(0, \sqrt{A^2 - B^2})$ is $D_1 = \frac{A^2 - Cx}{A^2}$.

57. Model your proof on the analogous argument for an ellipse by showing that the distance from any point (x, y) on the curve in the first quadrant to the focus $(\sqrt{A^2 + B^2}, 0)$ is $D_1 = \frac{Cx - A^2}{A}$ and the distance from (x, y) to the focus $(-\sqrt{A^2 + B^2}, 0)$ is $D_2 = \frac{Cx + A^2}{A}$, where $C = \sqrt{A^2 + B^2}$. Thus, the difference of the distances is $2A$ for every point on the curve.

59. For an ellipse with equation $\frac{x^2}{A^2} + \frac{y^2}{B^2} = 1$ where $A > B > 0$, the distance between the foci is $2\sqrt{A^2 - B^2}$ and the distance between the vertices is $2A$. We have $\frac{2\sqrt{A^2 - B^2}}{2A} = e$. The situation when $0 < A < B$ is similar, as is the computation for a hyperbola.

61. Let $P = (x, y)$ be a point in the first quadrant on the hyperbola defined by $\frac{x^2}{A^2} - \frac{y^2}{B^2} = 1$. The coordinates of the focus on the positive x-axis is $F = (\sqrt{A^2 + B^2}, 0)$ and the equation of the directrix which intersects the positive x-axis is $x = \frac{A}{e}$. Let D be the point on the directrix closest to P. Then $DP = x - \frac{A}{e}$ and $FP = \sqrt{(x - \sqrt{A^2 + B^2})^2 + y^2}$. Use the fact that $y^2 = \frac{B^2}{A^2}x^2 - B^2$ to show that $\frac{FP}{DP} = e$.

For answers to odd-numbered Skill Certification exercises in the Chapter Review, please visit the Book Companion Web Site at www.whfreeman.com/tkcalculus.

Chapter 10

Section 10.1

1. F, T, F, F, F, T, T, T.

3. (a) $y = 5$ is a line parallel to the x-axis; $x = -3$ is a line parallel to the y-axis (b) $y = 5$ is a plane parallel to the xz-plane; $x = -3$ is a plane parallel to the yz-plane

5. $(1, 1, 1)$, $(1, 1, -1)$, $(1, -1, 1)$, $(1, -1, -1)$, $(-1, 1, 1)$, $(-1, 1, -1)$, $(-1, -1, 1)$, and $(-1, -1, -1)$

7. The set of all points in $\mathbb{R}^3$ equidistant from a given point (x_0, y_0, z_0).

9. (a) A circle with radius 2 centered at the origin, (b) a right circular cylinder with radius 2 centered on the z-axis.

11. $(-5, 6, -7)$

13. $(3, -7, 6)$

15. There are three possible exchanges: x with y, x with z and y with z. We will discuss the first of these. The others are similar. If the placement of x, y, and z form a right-hand coordinate system, then the axes can be oriented to be superimposed on the graph below left. Switching the labels on the x and y axes we obtain the left-hand system below right. Exchanging two pairs of the labels will give another right-handed system.

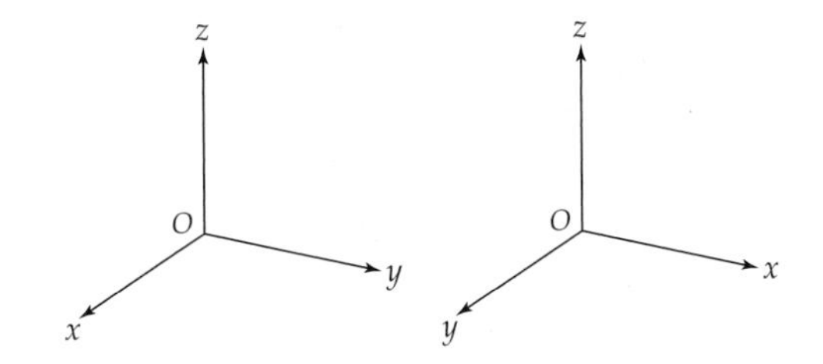

17. The distance between the points $(a_1, a_2, \ldots, a_n)$ and $(b_1, b_2, \ldots, b_n)$ is $\sqrt{(a_1 - b_1)^2 + (a_2 - b_2)^2 + \cdots + (a_n - b_n)^2}$.

19. (a) $x > 0, y > 0, z > 3$,
(b) $(x-1)^2 + (y-2)^2 + (z-4)^2 < 25$,
(c) $(x-2)^2 + (y+3)^2 < 25$

21. $\sqrt{130}$

23. $\sqrt{14}$

25. $\sqrt{26}$

27. $\sqrt{51}$

29. $(x-3)^2 + (y+2)^2 + (z-5)^2 = 25$

31. $(x-2)^2 + (y-5)^2 + (z+7)^2 = 49$

33. $(x-2)^2 + (y-5)^2 + (z+7)^2 = 78$

35. $(x-2)^2 + (y+8)^2 + z^2 = 66$

37. $\left(x - \frac{3}{2}\right)^2 + (y-1)^2 + \left(z - \frac{5}{2}\right)^2 = \frac{53}{2}$

39. Center $\left(0, -\frac{3}{2}, \frac{5}{2}\right)$, radius $\frac{\sqrt{22}}{2}$

41.

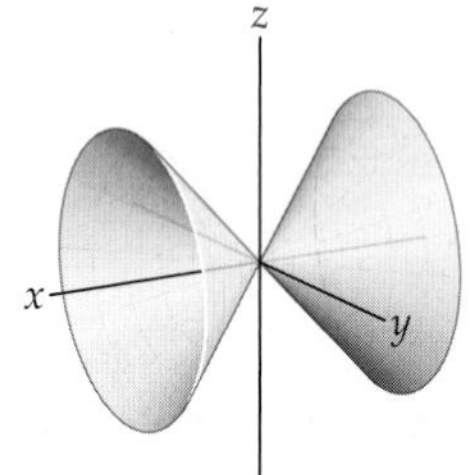

43.

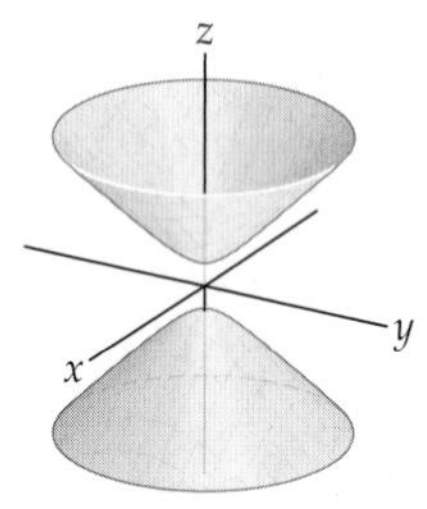

45.

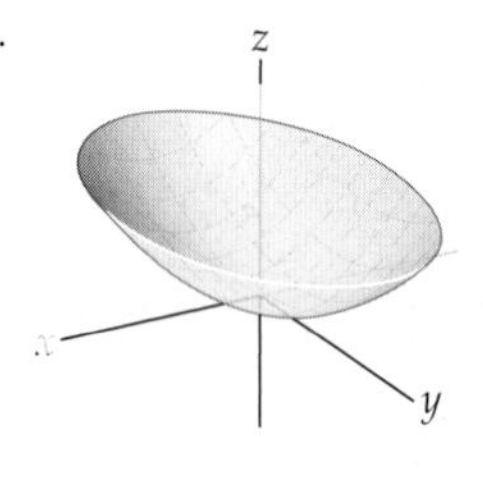

47.

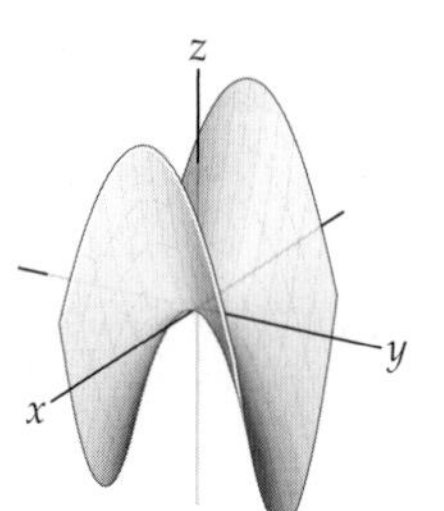

49. $3 - 2\sqrt{2}$

51. Use the distance formula to show that the length of each side of the triangle is $\sqrt{8}$.

53. Use the distance formula to show that the lengths of the sides of the triangle are 7, 14 and $\sqrt{245}$. Since $7^2 + 14^2 = (\sqrt{245})^2$, the triangle is right. The area is 49 square units.

55. $\left(x - \frac{1}{2}\right)^2 + \left(y - \frac{1}{2}\right)^2 + \left(z - \frac{1}{2}\right)^2 = \frac{3}{4}$ and $\left(x - \frac{1}{6}\right)^2 + \left(y - \frac{1}{6}\right)^2 + \left(z - \frac{1}{6}\right)^2 = \frac{3}{4}$

57. $\frac{4}{3}$

59. The compass readings give the slopes of lines from her boat to the respective peaks. Thus her boat is on the line $(y - 4.0) = \tan(75\pi/180)(x - 7.9)$ from Constitution Peak, and it is on the line $(y + 3.4) = \tan\left(\frac{-10\pi}{180}\right)(x - 9.3)$. Those lines intersect at $(6.07, -2.83)$. Thus, she is 6.7 miles from Deer Harbor.

61. $a = \pm 3.81$

63. Consider the regular tetrahedron with vertices $(1, 0, 0)$, $(0, 1, 0)$, $(0, 0, 1)$ and $(1, 1, 1)$ from Exercise 54. The midpoint of the segment connecting the midpoints of the opposite edges is the point $\left(\frac{1}{2}, \frac{1}{2}, \frac{1}{2}\right)$.

Section 10.2

1. T, F, F, F, T, T, T, T.

3. $\overrightarrow{AC}$

5. $\mathbf{0}$

7. $\mathbf{0}$

9. Algebraically, you add them "componentwise." Geometrically, you place the initial point of a vector **w** at the terminal point of a vector **v**. The vector **v** + **w** is the vector that has the same initial point as **v** and the same terminal point as **w**.

11. $\langle 2a, 2b, 2c \rangle$

13. $(-1, 5, -1)$

15. $\left(\frac{5\sqrt{14}}{14}, 3 + \frac{10\sqrt{14}}{14}, -2 + \frac{15\sqrt{14}}{14}\right)$

17. (a) The set of all position vectors whose terminal points are on the sphere of radius 4 centered at the origin.
(b) The set of all position vectors whose terminal points are in or on the sphere of radius 4 centered at the origin.
(c) The set of all vectors whose initial point is (a, b, c) and whose terminal point is on the sphere of radius 4 centered at (a, b, c).

19. Use vectors with 4 (or n) components. Algebraically, you still add vectors component by component and when you multiply a vector by a scalar you multiply each component by that scalar.

21. The set of all position vectors whose terminal points are on the sphere of radius 5 centered at the origin.

23. $\mathbf{u} + \mathbf{v} = \langle 8, -4 \rangle$; $\mathbf{u} - \mathbf{v} = \langle -4, -8 \rangle$

25. $\mathbf{u} + \mathbf{v} = \langle -2, 4 \rangle$; $\mathbf{u} - \mathbf{v} = \langle 4, -8 \rangle$

27. $\mathbf{u} + \mathbf{v} = \langle 3, -8, 13 \rangle$; $\mathbf{u} - \mathbf{v} = \langle -1, 0, -1 \rangle$

29. $\langle -6, -8 \rangle$

31. $\langle 0, -5, -7 \rangle$

33. 5

35. $\frac{\sqrt{3}}{3}$

37. $\|\mathbf{v}\| = 5, \left\langle \frac{3}{5}, -\frac{4}{5} \right\rangle$

39. $\|\mathbf{v}\| = \frac{\sqrt{34}}{15}, \frac{15}{\sqrt{34}} \left\langle \frac{1}{5}, \frac{1}{3} \right\rangle$

41. $\|\mathbf{v}\| = \sqrt{3}, \frac{1}{\sqrt{3}} \langle 1, 1, 1 \rangle$

43. $\frac{5}{\sqrt{14}} \langle 3, 1, 2 \rangle$

45. $\frac{2}{\sqrt{117}}\langle 8, -7, 2\rangle$

47. $\frac{7}{\sqrt{53}}\langle 1, 4, 6\rangle$

49. $\frac{7}{\sqrt{53}}\langle 1, 4, 6\rangle$

51. $-\frac{3}{\sqrt{14}}\langle 1, -2, 3\rangle$

53. The magnitude of the force in the rope on the left is $50(3\sqrt{2}-\sqrt{6})$ pounds and the magnitude of the force in the rope on the right is $100(\sqrt{3}-1)$ pounds.

55. (a) $\langle 0, -2\rangle$, (b) $\langle\sqrt{2}, -\sqrt{2}\rangle$, (c) $\langle\sqrt{3}-2, -\sqrt{3}\rangle$

57. $\mathbf{u}+\mathbf{v} = \langle u_1, u_2, u_3\rangle + \langle v_1, v_2, v_3\rangle = \langle u_1+v_1, u_2+v_2, u_3+v_3\rangle = \langle v_1+u_1, v_2+u_2, v_3+u_3\rangle = \langle v_1, v_2, v_3\rangle + \langle u_1, u_2, u_3\rangle = \mathbf{v}+\mathbf{u}$.

59. $c(\mathbf{u}+\mathbf{v}) = c\left(\langle u_1, u_2, u_3\rangle + \langle v_1, v_2, v_3\rangle\right) = c\left(\langle u_1+v_1, u_2+v_2, u_3+v_3\rangle\right) = \big\langle c(u_1+v_1), c(u_2+v_2), c(u_3+v_3)\big\rangle = \langle cu_1+cv_1, cu_2+cv_2, cu_3+cv_3\rangle = \langle cu_1, cu_2, cu_3\rangle + \langle cv_1, cv_2, cv_3\rangle = c\,\langle u_1, u_2, u_3\rangle + c\,\langle v_1, v_2, v_3\rangle = c\mathbf{u}+c\mathbf{v}$.

61. $(c+d)\mathbf{v} = (c+d)\langle v_1, v_2, v_3\rangle = \langle (c+d)v_1, (c+d)v_2, (c+d)v_3\rangle = \langle cv_1+dv_1, cv_2+dv_2, cv_3+dv_3\rangle = \langle cv_1, cv_2, cv_3\rangle + \langle dv_1, dv_2, dv_3\rangle = c\langle v_1, v_2, v_3\rangle + d\langle v_1, v_2, v_3\rangle = c\mathbf{v}+d\mathbf{v}$.

63. Let $\mathbf{u} = \langle u_1, u_2\rangle$ and $\mathbf{v} = \langle v_1, v_2\rangle$ be two position vectors. The vertices of the parallelogram determined by $\mathbf{u}$ and $\mathbf{v}$ are $(0,0)$, (u_1, u_2), (v_1, v_2) and (u_1+v_1, u_2+v_2). The midpoint of both opposite pairs of vertices is $\left(\frac{u_1+v_1}{2}, \frac{u_2+v_2}{2}\right)$.

Section 10.3

1. T, F, F, F, T, T, T, F.

3. See Definition 10.13.

5. When the angle between the sides with lengths a and b is right, the Law of Cosines reduces to the Pythagorean theorem since $\cos 90° = 0$.

7. (a) Assuming none of the constants are zero, we need $\frac{\alpha}{a} = \frac{\beta}{b} = \frac{\gamma}{c}$; (b) $a\alpha + b\beta + c\gamma = 0$.

9. (a) Let $\mathbf{u} = \mathbf{v} = \mathbf{i}$ and $\mathbf{w} = \mathbf{i}+\mathbf{j}$; (b) $\mathbf{u}$ must be orthogonal to $\mathbf{v}-\mathbf{w}$

11. The two vectors must be parallel and point in the same direction.

13. Any position vector that lies in the yz-plane.

15. Any vector of the form $\langle 1, v_2, v_3\rangle$.

17. (a) $\langle 1, m\rangle$, (b) slope $-\frac{1}{m}$, $\langle -m, 1\rangle$, (c) $\langle 1, m\rangle\cdot\langle -m, 1\rangle = -m+m = 0$.

19. (a) If there were a scalar c such that $c\mathbf{v}_1 = \mathbf{v}_2$, in order for the first components to be equal we would need $c = 3$, but $3\mathbf{v}_1 = \langle 3, -6\rangle \neq \mathbf{v}_2$. (b) $2\mathbf{v}_1 + 1\mathbf{v}_2 = \langle 5, 1\rangle$.

21. -1, $\cos^{-1}\left(-\frac{1}{\sqrt{1711}}\right)$

23. 38, $\cos^{-1}\left(\frac{38}{\sqrt{2170}}\right)$

25. $\text{proj}_{\mathbf{u}}\mathbf{v} = \mathbf{0}$, $\text{comp}_{\mathbf{u}}\mathbf{v} = 0$, and the component of $\mathbf{v}$ orthogonal to $\mathbf{u}$ is $\mathbf{v}$

27. $\text{comp}_{\mathbf{u}}\mathbf{v} = -2\sqrt{14}$; $\text{proj}_{\mathbf{u}}\mathbf{v} = \mathbf{v}$; and the component of $\mathbf{v}$ orthogonal to $\mathbf{u}$ is $\mathbf{0}$

29. $\text{proj}_{\mathbf{u}}\mathbf{v} = \mathbf{0}$, $\text{proj}_{\mathbf{v}}\mathbf{u} = \mathbf{0}$

31. $\text{proj}_{\mathbf{u}}\mathbf{v} = -\frac{5}{27}\langle 1, -5, -1\rangle$, $\text{proj}_{\mathbf{v}}\mathbf{u} = \langle 0, -5, 0\rangle$

33. (a) -1; (b) $\cos^{-1}\left(-\frac{1}{2\sqrt{2451}}\right)$; (c) $-\frac{1}{38}\mathbf{u}$

35. (a) 24; (b) $\cos^{-1}\left(\frac{24\sqrt{2}}{50}\right)$; (c) $\frac{24}{25}\mathbf{u}$

37. $\frac{48}{83}\sqrt{166}$

39. $\frac{\pi}{2}$

41. $\cos^{-1}\left(\frac{\sqrt{6}}{3}\right)$

43. $\cos\alpha = \frac{1}{\sqrt{14}}$, $\alpha = \cos^{-1}\left(\frac{1}{\sqrt{14}}\right)$, $\cos\beta = \frac{2}{\sqrt{14}}$, $\beta = \cos^{-1}\left(\frac{2}{\sqrt{14}}\right)$, $\cos\gamma = \frac{3}{\sqrt{14}}$, $\gamma = \cos^{-1}\left(\frac{3}{\sqrt{14}}\right)$

45. $\cos\alpha = -\frac{1}{\sqrt{18}}$, $\alpha = \cos^{-1}\left(-\frac{1}{\sqrt{18}}\right)$, $\cos\beta = \frac{1}{\sqrt{18}}$, $\beta = \cos^{-1}\left(\frac{1}{\sqrt{18}}\right)$, $\cos\gamma = -\frac{4}{\sqrt{18}}$, $\gamma = \arccos\left(-\frac{4}{\sqrt{18}}\right)$

47. Let $\mathbf{v} = \langle a, b, c\rangle$. From their definitions, $\cos\alpha = \frac{a}{\|\mathbf{v}\|}$, $\cos\beta = \frac{b}{\|\mathbf{v}\|}$ and $\cos\gamma = \frac{c}{\|\mathbf{v}\|}$. Thus, $\|\mathbf{v}\|\left((\cos\alpha)\mathbf{i} + (\cos\beta)\mathbf{j} + (\cos\gamma)\mathbf{k}\right) = a\mathbf{i} + b\mathbf{j} + c\mathbf{k} = \mathbf{v}$.

49. $\cos\gamma = \pm\frac{\sqrt{2}}{2}$

51. $\cos\beta = \pm\frac{3}{4}$

53. (a)$\alpha_i = \cos^{-1}\left(\frac{1}{\sqrt{n}}\right)$; (b) $\lim_{n\to\infty}\cos^{-1}\left(\frac{1}{\sqrt{n}}\right) = \frac{\pi}{2}$

55. (a) $\tan^{-1}\left(\frac{1}{4}\right)$, (b) $4\sqrt{4/17} \approx 1.9$, (c) slightly more than 1 hour.

57. If $\mathbf{v} = \langle a, b, c\rangle$, then $\mathbf{v}\cdot\mathbf{i} = a$, $\mathbf{v}\cdot\mathbf{j} = b$ and $\mathbf{v}\cdot\mathbf{k} = c$. Thus, $(\mathbf{v}\cdot\mathbf{i})\mathbf{i} + (\mathbf{v}\cdot\mathbf{j})\mathbf{j} + (\mathbf{v}\cdot\mathbf{k})\mathbf{k} = a\mathbf{i} + b\mathbf{j} + c\mathbf{k} = \mathbf{v}$.

59. Taking the absolute value of each side of $\mathbf{u}\cdot\mathbf{v} = \|\mathbf{u}\|\|\mathbf{v}\|\cos\theta$, we see that $|\mathbf{u}\cdot\mathbf{v}| = \|\mathbf{u}\|\|\mathbf{v}\|\,|\cos\theta|$. Since $|\cos\theta| \le 1$ we have our result. In order to have the equality, the two vectors must be parallel.

61. If either $\mathbf{u}$ or $\mathbf{v}$ is $\mathbf{0}$, then $\mathbf{u}\cdot\mathbf{v} = 0$ and $\mathbf{u}$ and $\mathbf{v}$ are orthogonal, since $\mathbf{0}$ is orthogonal to every vector. If neither $\mathbf{u}$ nor $\mathbf{v}$ is $\mathbf{0}$, then $\mathbf{u}\cdot\mathbf{v} = \|\mathbf{u}\|\|\mathbf{v}\|\cos\theta$ by Theorem 10.15, where θ is the angle between the position vectors $\mathbf{u}$ and $\mathbf{v}$. Since neither $\mathbf{u}$ nor $\mathbf{v}$ is $\mathbf{0}$, the dot product is only zero if $\cos\theta = 0$, that is, if $\mathbf{u}$ and $\mathbf{v}$ are orthogonal.

63. Assume that $\mathbf{v} = \mathbf{v}_\parallel + \mathbf{v}_\perp$ and $\mathbf{v} = \mathbf{v}'_\parallel + \mathbf{v}'_\perp$ are two sums such that $\mathbf{v}_\parallel$ and $\mathbf{v}'_\parallel$ are both parallel to $\mathbf{u}$ and $\mathbf{v}_\perp$ and $\mathbf{v}'_\perp$ are both orthogonal to $\mathbf{u}$. Then $\mathbf{v}_\parallel + \mathbf{v}_\perp = \mathbf{v}'_\parallel + \mathbf{v}'_\perp$, so $\mathbf{v}_\parallel - \mathbf{v}'_\parallel = \mathbf{v}'_\perp - \mathbf{v}_\perp$. The difference on the left will also be parallel to $\mathbf{u}$ and the difference on the right will also be orthogonal to $\mathbf{u}$. Thus, they are orthogonal to each other. The only way for this to happen is if they are both $\mathbf{0}$. Therefore, the two decompositions are the same.

65. Let **u** and **v** be two nonparallel position vectors. The diagonals of the parallelogram determined by **u** and **v** are the vectors $\mathbf{u}+\mathbf{v}$ and $\mathbf{u}-\mathbf{v}$. The dot product $(\mathbf{u}+\mathbf{v})\cdot(\mathbf{u}-\mathbf{v}) = \mathbf{u}\cdot\mathbf{u}-\mathbf{u}\cdot\mathbf{v}+\mathbf{v}\cdot\mathbf{u}+\mathbf{v}\cdot\mathbf{v} = \|\mathbf{u}\|^2-\|\mathbf{v}\|^2$, since the dot product is commutative. If the parallelogram is a rhombus, then $\|\mathbf{u}\| = \|\mathbf{v}\|$. Thus, the dot product of the vectors forming the diagonals is zero, so they are perpendicular. Conversely, if the diagonals of the parallelogram are perpendicular, we must have $\|\mathbf{u}\|^2 - \|\mathbf{v}\|^2 = 0$ and $\|\mathbf{u}\| = \|\mathbf{v}\|$, so the parallelogram is a rhombus.

Section 10.4

1. F, F, T, F, T, T, T, F.
3. See Definition 10.25.
5. The cross product, $\mathbf{u}\times\mathbf{v}$, is orthogonal to both **u** and **v**. The magnitude of $\mathbf{u}\times\mathbf{v}$ is $\|\mathbf{u}\|\|\mathbf{v}\|\sin\theta$ where θ is the angle between **u** and **v**. The vectors **u**, **v** and $\mathbf{u}\times\mathbf{v}$ form a right-hand triple.
7. The dot product is zero if and only if the vectors are orthogonal.
9. Embed the vectors in $\mathbb{R}^3$ as $\langle 1,2,0\rangle$ and $\langle 3,-1,0\rangle$. The cross product $\langle 1,2,0\rangle\times\langle 3,-1,0\rangle = \langle 0,0,-7\rangle$. The norm of this vector is 7, which is the area of the parallelogram in square units.
11. They are orthogonal.
13. Let $\mathbf{u} = \langle 1,0,0\rangle$, $\mathbf{v} = \langle 2,1,1\rangle$ and $\mathbf{w} = \langle 4,1,1\rangle$. If $\mathbf{u}\times\mathbf{v} = \mathbf{u}\times\mathbf{w}$, then **u** is parallel to $\mathbf{v}-\mathbf{w}$.
15. As position vectors, the three vectors lie in the same plane.
17. The cross product is not associative, so we must specify whether we mean $(\mathbf{u}\times\mathbf{v})\times\mathbf{w}$ or $\mathbf{u}\times(\mathbf{v}\times\mathbf{w})$.
19. At least one of them is **0**.
21. They are parallel.
23. $\langle 23,-4,14\rangle$, $\langle -23,4,-14\rangle$
25. $\langle -46,3,-22\rangle$, $\langle -42,-30,-38\rangle$
27. $\sqrt{213}$
29. 106 cubic units. No, they form a left-handed triple.
31. $\langle 25,35,-10\rangle$, $\langle -25,-35,10\rangle$
33. 0, 0
35. 0
37. (a) $\langle -24,7,15\rangle$ (b) $\pm\dfrac{1}{5\sqrt{34}}\langle -24,7,15\rangle$, (c) $\frac{5}{2}\sqrt{34}$.
39. (a) $\langle 20,-9,30\rangle$; (b) $\pm\dfrac{1}{\sqrt{1381}}\langle 20,-9,30\rangle$; (c) $\dfrac{\sqrt{1381}}{2}$
41. (a) **k**; (b) $\pm$**k**; (c) 26
43. $\dfrac{\sqrt{850}}{2}$
45. (a) The lengths of the sides of a quadrilateral do not determine the quadrilateral's area. For example, a square with side lengths 1 unit has an area of 1 square unit, but a rhombus with side lengths 1 and opposite interior angles of 30° has area $\frac{\sqrt{3}}{2}$. (b) If you know the side lengths and the length of one diagonal you can decompose the quadrilateral into two triangles. (c) The side lengths adjacent to the angle whose measure is known determine a unique triangle. You may find the length of the third side of the triangle, which is a diagonal of the quadrilateral and use the reasoning from (b).
47. $\frac{3}{4}\left(\sqrt{759}+\sqrt{2079}\right)$
49. $16\sqrt{759}+6\sqrt{70}$
51. (a) Parallelogram, (b) 16 square units
53. (a) Not a parallelogram, (b) $\frac{49}{2}$ square units
55. (a) $\overrightarrow{PQ} = \overrightarrow{SR}$ but P, Q, R and S are not collinear, (b) parallelogram, (b) $7\sqrt{254}$ square units
57. (a) $\overrightarrow{PQ}$ and $\overrightarrow{SR}$ are parallel, (b) not a parallelogram, (b) $\frac{7}{2}\sqrt{146}$ square units
59. $4\sqrt{2\sqrt{2}-2}\approx 3.64$ cubic centimeters
61. 30 foot-pounds
63. If $A = \begin{bmatrix} a & b & c \\ d & e & f \\ g & h & i \end{bmatrix}$, then $\det A = aei+bfg+cdh-afh-bdi-ceg$. Show that when you exchange any two rows of A, you obtain the negative of this sum.
65. If $\mathbf{v} = \langle a,b,c\rangle$ and α is a scalar, then $\mathbf{v}\times(\alpha\mathbf{v}) = (\alpha bc-\alpha bc)\mathbf{i}+(\alpha ac-\alpha ac)\mathbf{j}+(\alpha ab-\alpha ab)\mathbf{k} = \mathbf{0}$.
67. Let $\mathbf{u} = \langle u_1,u_2,u_3\rangle$ and $\mathbf{v} = \langle v_1,v_2,v_3\rangle$. Show that $c(\mathbf{u}\times\mathbf{v})$, $(c\mathbf{u})\times\mathbf{v}$, and $\mathbf{u}\times(c\mathbf{v})$ are all equal $\langle c(u_2v_3-u_3v_2), c(u_3v_1-u_1v_3), c(u_1v_2-u_2v_1)\rangle$.
69. Let $\mathbf{u} = \langle u_1,u_2,u_3\rangle$ and $\mathbf{v} = \langle v_1,v_2,v_3\rangle$, the triple scalar product $\mathbf{v}\cdot(\mathbf{u}\times\mathbf{v}) = \det\begin{bmatrix} v_1 & v_2 & v_3 \\ u_1 & u_2 & u_3 \\ v_1 & v_2 & v_3 \end{bmatrix}$. Show that this determinant is zero.
71. (a) The volume of the parallelepiped determined by **u**, **v** and **w** is the absolute value of the determinant of the matrix in which the rows are the components of **u**, **v** and **w**. Since each component is an integer, the determinant and its absolute value are integers. (b) The area of the parallelogram determined by **i** and **j** is 1. The area of the parallelogram determined by **i** and $\mathbf{j}+\mathbf{k}$ is $\sqrt{2}$.
73. Divide the left and right sides of the equations $\|\mathbf{u}\times\mathbf{v}\| = \|\mathbf{u}\|\|\mathbf{v}\|\sin\theta$ and $\mathbf{u}\cdot\mathbf{v} = \|\mathbf{u}\|\|\mathbf{v}\|\cos\theta$ to obtain the result.

75. If $\mathbf{u} \times \mathbf{v} = \mathbf{u} \times \mathbf{w}$, then $\mathbf{u}$ is parallel to $\mathbf{v} - \mathbf{w}$ and if $\mathbf{u} \cdot \mathbf{v} = \mathbf{u} \cdot \mathbf{w}$, then $\mathbf{u}$ is orthogonal to $\mathbf{v} - \mathbf{w}$. Thus, $\mathbf{v} - \mathbf{w}$ is orthogonal to itself. This can only happen if $\mathbf{v} - \mathbf{w} = \mathbf{0}$.

77. Let $\mathbf{r} = \langle r_1, r_2, r_3 \rangle$, $\mathbf{s} = \langle s_1, s_2, s_3 \rangle$, $\mathbf{u} = \langle u_1, u_2, u_3 \rangle$ and $\mathbf{v} = \langle v_1, v_2, v_3 \rangle$. Show that both sides of the equality are equal to $r_1s_2u_1v_2 + r_1s_3u_1v_3 + r_2s_1u_2v_1 + r_2s_3u_2v_3 + r_3s_2u_3v_2 + r_3s_2u_3v_2 - (r_1s_2u_2v_1 + r_1s_3u_3v_1 + r_2s_1u_1v_2 + r_2s_3u_3v_2 + r_3s_1u_1v_3 + r_3s_2u_2v_3)$.

79. If $\mathbf{r}$, $\mathbf{s}$, $\mathbf{u}$, and $\mathbf{v}$ all lie in some plane $\mathcal{P}$, then the cross products $\mathbf{r} \times \mathbf{s}$ and $\mathbf{u} \times \mathbf{v}$ are both orthogonal to $\mathcal{P}$. Therefore, these two vectors are parallel and the cross product of two parallel vectors is $\mathbf{0}$.

Section 10.5

1. T, F, T, F, T, F, F, F.

3. A linear equation in three variables is an equation that can be written in the form $Ax + By + Cz = D$, where A, B, C, and D are real numbers. An example is $2x - 3y + \pi z = 12$. The graph of every linear equation in three variables is a plane.

5. Let $P = (a, b, c)$ and $Q = (\alpha, \beta, \gamma)$. The direction vector for the line is $\pm \langle a - \alpha, b - \beta, c - \gamma \rangle$. We will choose the positive sign for this vector, and we may choose either point P or point Q; we will choose Q. The equation of the line as a vector function is $\mathbf{r}(t) = \langle \alpha + (a - \alpha)t, \beta + (b - \beta)t, \gamma + (c - \gamma)t \rangle$.

7. If the slopes of the two lines are equal, the lines are either parallel or identical. Two lines in $\mathbb{R}^2$ that aren't parallel or identical intersect in a unique point.

9. (a) $\mathbf{r}_1(t) = \langle -1 + 2t, 3 - 4t, 7 + 9t \rangle$
(b) $\mathbf{r}_2(t) = \langle -1 + 4t, 3 - 8t, 7 + 18t \rangle$
(c) $x = 2t + 5, y = -4t - 9, z = 9t + 34$

11. $\left(0, \frac{52}{3}, \frac{19}{3}\right); \left(\frac{52}{7}, 0, \frac{27}{7}\right); (19, -27, 0)$

13. $(0, -9, 1); \left(\frac{9}{2}, 0, -17\right); \left(\frac{1}{4}, -\frac{17}{2}, 0\right)$

15. $(-7, 0, 5)$. The line is parallel to the y-axis.

17. (a) $x = 2 + 7t, y = 3 - 5t, z = 2t$. (b) $\frac{x-2}{7} = \frac{y-3}{-5} = \frac{z}{2}$.

19. (a) $\mathbf{r}(t) = \langle 4, 3 - 5t, t \rangle$; (b) $x = 4, \frac{y-3}{-5} = z$

21. In slope-intercept form, $y = -3x + 5$. Next, $x = 2t$ and $y = 5 - 6t$. Eliminate the parameter t to obtain the previous equation.

23. (a) $\mathbf{r}(t) = \langle t, 2t, -4t \rangle$; (b) $x = t, y = 2t, z = -4t$; (c) $x = \frac{y}{2} = -\frac{z}{4}$

25. (a) $\mathbf{r}(t) = \langle 2t - 1, 3, 4t + 7 \rangle$; (b) $x = 2t - 1, y = 3, z = 4t + 7$; (c) $y = 3, \frac{x+1}{2} = \frac{z-7}{4}$

27. (a) $\mathbf{r}(t) = \langle 2t + 3, 5t + 1 \rangle$; (b) $x = 2t + 3, y = 5t + 1$; (c) $\frac{x-3}{2} = \frac{y-1}{5}$

29. (a) $\mathbf{r}(t) = \langle 4t, -t, 6t \rangle$; (b) $x = 4t, y = -t, z = -6t$; (c) $\frac{x}{4} = -y = \frac{z}{6}$

31. (a) $\mathbf{r}(t) = \langle 8t - 4, 11, 2t \rangle$; (b) $x = 8t - 4, y = 11, z = 2t$; (c) $y = 11, \frac{x+4}{8} = \frac{z}{2}$

33. (a) $\mathbf{r}(t) = \langle 3t + 1, -t + 6 \rangle$; (b) $x = 3t + 1, y = -t + 6$; (c) $\frac{x-1}{3} = -y + 6$

35. (a) $\mathbf{r}(t) = \langle (x_1 - x_0)t + x_0, (y_1 - y_0)t + y_0, (z_1 - z_0)t + z_0 \rangle$; (b) $\mathbf{r}(t) = \langle (x_1 - x_0)t + x_0, (y_1 - y_0)t + y_0, (z_1 - z_0)t + z_0 \rangle$, $0 \le t \le 1$

37. $\mathbf{r}(t) = \langle -2t + 1, -9t + 7, 2t + 3 \rangle$, $0 \le t \le 1$

39. $\mathbf{r}(t) = \langle -4t + 3, 6t - 1, 5t + 4 \rangle$, $0 \le t \le 1$

41. The lines are parallel. They are $\sqrt{\frac{117}{7}}$ units apart.

43. The lines intersect at $(-8, 8, -9)$.

45. The lines are parallel. They are $5\sqrt{\frac{61}{21}}$ units apart.

47. $\sqrt{\frac{1084}{21}}$

49. $\frac{4\sqrt{5}}{5}$

51. (a) $\alpha = -3$, (b) $\alpha = 15$

53. To the nearest tenth of a kilometer, he is at an elevation of 4.7 km and is 0.7 km east of the summit.

55. See the proof preceding Theorem 10.38.

Section 10.6

1. F, F, F, F, T, T, F, T.

3. $\left(\frac{d}{a}, 0, 0\right), \left(0, \frac{d}{b}, 0\right), \left(0, 0, \frac{d}{c}\right)$. You can plot these three points and construct the triangle with these points as vertices to understand the plane that contains them.

5. If the vectors normal to the two planes are not parallel, the two planes intersect in a line.

7. It means that the three points lie on a single line. Let P, Q, and R be the points. If the two vectors $\overrightarrow{PQ}$ and $\overrightarrow{PR}$ are parallel, the points are collinear. If the three points don't lie on the same line, by definition they are noncollinear.

9. Three noncollinear points determine a unique plane. Any two distinct points on $\mathcal{L}$ along with P determine the plane. Choose any point Q on $\mathcal{L}$. Construct the vector $\overrightarrow{PQ}$. A normal vector to the plane is given by $\mathbf{N} = \overrightarrow{PQ} \times \mathbf{d}$, where $\mathbf{d}$ is the direction vector for $\mathcal{L}$. Use P and $\mathbf{N}$ to find the equation of the plane. In $\mathbb{R}^3$ there are infinitely many different planes containing every line.

11. Choose two distinct points on one of the lines and one point on the other. These three points determine a unique plane. (See the answer to Exercise 9.) Every point on both lines lies in the plane determined by the three chosen points.

13. Let $\mathbf{d}$ and $\mathbf{d}'$ be the direction vectors for the lines $\mathcal{L}$ and $\mathcal{L}'$, respectively. Since the lines are skew, $\mathbf{N} = \mathbf{d} \times \mathbf{d}' \neq \mathbf{0}$. Any point from $\mathcal{L}$ along with $\mathbf{N}$ and any point from $\mathcal{L}'$ along with $\mathbf{N}$ determine the unique pair of parallel planes.

15. The direction vector for $\mathcal{L}$ is $\mathbf{d} = \langle a, b, c\rangle$ and the normal vector for $\mathcal{P}$ is $\mathbf{N} = \langle \alpha, \beta, \gamma\rangle$. The line and plane are orthogonal if and only if $\mathbf{d}$ and $\mathbf{N}$ are scalar multiples.

17. Let Q be the point of intersection between $\mathcal{L}$ and $\mathcal{P}$. If $\mathcal{L}$ is orthogonal to $\mathcal{P}$, then every line in $\mathcal{P}$ passing through Q is orthogonal to $\mathcal{L}$, and the angle between $\mathcal{L}$ and $\mathcal{P}$ is $\frac{\pi}{2}$. Otherwise, the angle described is the minimum angle that $\mathcal{L}$ can form with any line in $\mathcal{P}$ through Q.

19. The formula builds upon the computation for the component of a vector in the direction of a given vector. We choose arbitrary points P_1 and P_2 on lines $\mathcal{L}_1$ and $\mathcal{L}_2$, respectively, and $|\text{comp}_{\mathbf{N}}\overrightarrow{P_1P_2}|$, where $\mathbf{N}$ is the cross product of the direction vectors of $\mathcal{L}_1$ and $\mathcal{L}_2$. The formula works because the skew lines lie on a unique pair of parallel planes.

21. $4x - y + 5z = 0$

23. $2x - y + 6z = 41$

25. $12x - 5y + 19z = 61$

27. $y_0z_0x + x_0z_0y + x_0y_0z = x_0y_0z_0$

29. $3x + 7y - 7z = -26$

31. The direction vectors for the lines $\langle -5, 2, 4\rangle$ and $\langle 15, -6, -12\rangle$ are scalar multiples of each other, so the lines are parallel; $14x + 55y - 10z = 137$.

33. The point $(7, -5, 14)$ is on both lines; $14x + 3y + 2z = 111$.

35. $\mathbf{r}(t) = \left\langle -\frac{8}{5} - 11t, \frac{14}{5} - 2t, 5t\right\rangle$

37. $\mathbf{r}(t) = \langle 4, 3 + 2t, 5t\rangle$

39. $\sqrt{2}$

41. The normal vectors are parallel, so the planes are parallel. They are $\frac{29}{3\sqrt{38}}$ units apart.

43. The direction vectors of the lines are $\langle 2, -4, 5\rangle$ and $\langle 3, 2, 3\rangle$. Since these vectors aren't scalar multiples of each other, the lines either intersect, or are skew. The planes containing the lines are $-22x + 9y + 16z = 81$ and $-22x + 9y + 16z = -97$, respectively, so the lines must be skew. The distance between the lines is $\frac{178}{\sqrt{821}}$.

45. $\cos^{-1}\left(\frac{4\sqrt{3}}{9}\right)$

47. $\frac{\pi}{2}$

49. The line and plane are parallel. They are $\frac{13}{15}\sqrt{30}$ units apart.

51. They intersect at $\left(\frac{143}{7}, \frac{3}{7}, 0\right)$ at an angle of $\frac{\pi}{2} - \cos^{-1}\left(\frac{7}{\sqrt{2738}}\right)$.

53. (a) $\frac{1}{2}x + \frac{1}{4}y + \frac{\sqrt{11}}{4}z = 1$, (b) $2x + 2y + z = 7$

55. The plane of the beach is $3x + y - 16z = 400$. The water is about 25 feet deep at the buoy.

57. The planes are perpendicular if and only if their normal vectors are orthogonal, that is if and only if $\langle a, b, c\rangle \cdot \langle \alpha, \beta, \gamma\rangle = a\alpha + b\beta + c\gamma = 0$.

59. (a) Let $\mathcal{P}$ be the plane. By Theorem 10.39 the distance from P to $\mathcal{P}$ is $\frac{|\mathbf{N}\cdot\overrightarrow{RP}|}{\|\mathbf{N}\|}$, where $\mathbf{N} = \langle a, b, c\rangle$ is the normal vector to $\mathcal{P}$ and R is a point on $\mathcal{P}$. At least one of the constants, a, b, and c is nonzero. Assume $c \neq 0$, the other cases are similar. Since R is on $\mathcal{P}$, it has coordinates $\left(x_1, y_1, -\frac{d+ax_1+by_1}{c}\right)$. Thus, $\overrightarrow{RP} = \left\langle x_0 - x_1, y_0 - y_1, z_0 + \frac{d+ax_1+by_1}{c}\right\rangle$. So, $\frac{|\mathbf{N}\cdot\overrightarrow{RP}|}{\|\mathbf{N}\|} = \frac{|ax_0-ax_1+by_0-by_1+cz_0+d+ax_1+by_1|}{\sqrt{a^2+b^2+c^2}} = \frac{|ax_0+by_0+cz_0+d|}{\sqrt{a^2+b^2+c^2}}$.
(b) $\sqrt{2}$

61. If $\mathcal{L}_1$ and $\mathcal{L}_2$ lie in the same plane, then they intersect, are parallel, or are identical. If they are parallel or identical then $\mathbf{d}_1 \times \mathbf{d}_2 = \mathbf{0}$, so the result holds. If $\mathcal{L}_1$ and $\mathcal{L}_2$ intersect, then $\mathbf{d}_1 \times \mathbf{d}_2$ is orthogonal to every vector in their plane, so again the result holds. Conversely, if $(\mathbf{P}_0 - \mathbf{Q}_0)\cdot(\mathbf{d}_1 \times \mathbf{d}_2) = 0$, the vector $\mathbf{d}_1 \times \mathbf{d}_2$ is orthogonal to every vector connecting a point on line $\mathcal{L}_1$ and a point on $\mathcal{L}_2$, so $\mathbf{d}_1 \times \mathbf{d}_2$ is the normal vector to the plane containing the two lines.

For answers to odd-numbered Skill Certification exercises in the Chapter Review, please visit the Book Companion Web Site at www.whfreeman.com/tkcalculus.

Chapter 11

Section 11.1

1. F, F, T, T, T, F, F, F.

3. The limit $\lim_{x\to c} f(x) = L$ means that for all $\epsilon > 0$, there exists $\delta > 0$ such that if $x \in (c - \delta, c) \cup (c, c + \delta)$, then $f(x) \in (L - \epsilon, L + \epsilon)$.

5. The limit of vector-valued function is defined in terms of the limits of the components of the functions. Each component is a function of a single variable. These limits are defined using an "epsilon-delta" definition.

7. The function $\mathbf{r}$ is said to be ***continuous*** at a point c in the domain of $\mathbf{r}$ if $\lim_{t\to c}\mathbf{r}(t) = \mathbf{r}(c)$.

9. Since the functions $x(t)$ and $y(t)$ are both continuous functions, by the Extreme Value Theorem each of these functions has a minimum value and a maximum value on the interval $[a, b]$. Let M be the largest of the absolute values of these four numbers. The graph of $\mathbf{r}(t)$ on $[a, b]$ is contained within the circle $x^2 + y^2 = 2M^2$.

11. Since the functions $x(t)$, $y(t)$ and $z(t)$ are all continuous functions, by the Extreme Value Theorem each of these functions has a maximum value on the interval $[a, b]$. Let M be the maximum of these three maxima. The graph of $\mathbf{r}(t)$ on $[a, b]$ is contained within the sphere $x^2 + y^2 + z^2 = 3M^2$.

13. $\mathbf{r}(t) = \langle \cos 2t, \sin 2t, 2t \rangle$, $[0, \pi]$

15. We need $\lim_{t\to\infty} y(t) = L$, where L is the value of the horizontal asymptote and one of the limits; $\lim_{t\to\infty} x(t) = \infty$ or $\lim_{t\to\infty} x(t) = -\infty$. For example, the function $\mathbf{r}(t) = \left\langle t, 5 - \frac{1}{t} \right\rangle$.

17. $x_1(t)x_2(t) + y_1(t)y_2(t)$

19. 21.

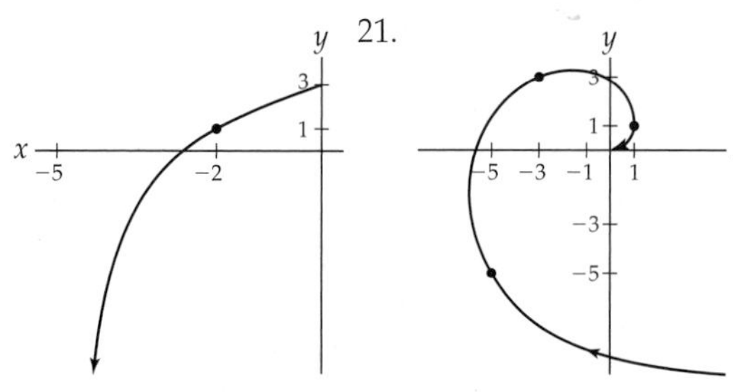

23. The graphs are the same. The rate at which the curve is drawn is faster when k is large.

25. Every point on the graph of $\mathbf{r}(t)$ satisfies the equations of both cylinders.

27. $x(t) = 2 - \sin t$, $y(t) = 4 + \cos t$.

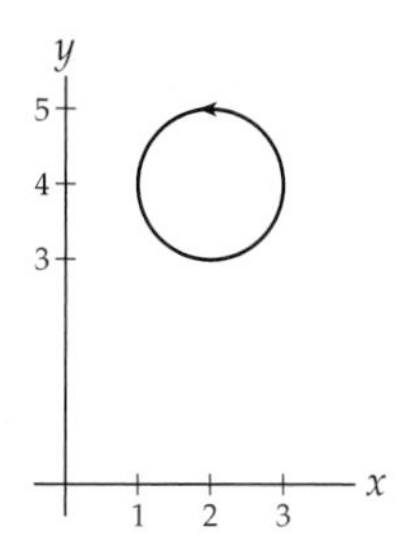

29. $x(t) = 1 + \sin t$, $y(t) = 3 - \cos 2t$. The graph is the portion of the parabola shown below. The motion starts at (1, 2) and the particle moves to the right for $t \in \left[0, \frac{\pi}{2}\right]$ and $t \in \left[\frac{3\pi}{2}, 2\pi\right]$ and to the left for $t \in \left[\frac{\pi}{2}, \frac{3\pi}{2}\right]$

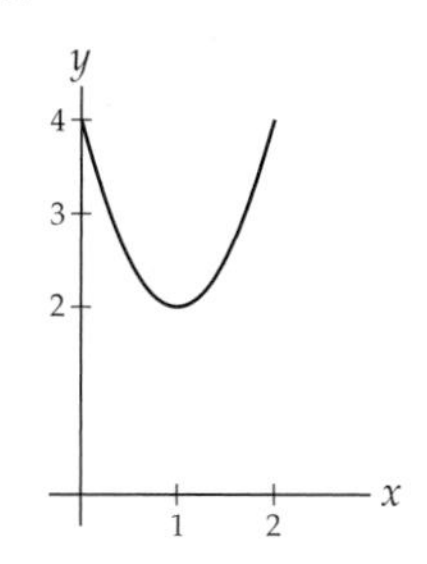

31. $x(t) = t$, $y(t) = t^2$, $z(t) = t^3$

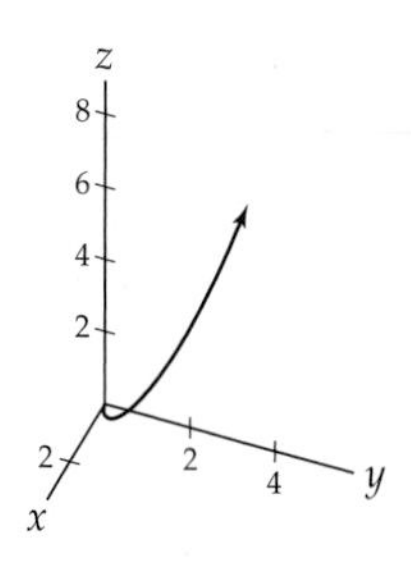

33. $x(t) = \cos^2 t$, $y(t) = \sin 2t$

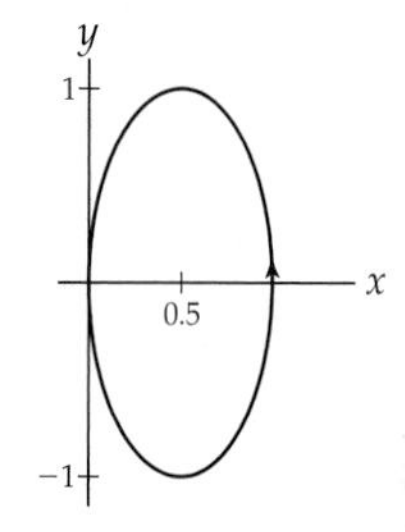

35. $\langle 1 + t, 3t + t^2, 2t^3 \rangle$

37. $\langle 5\cos t, 5\sin t \rangle$

39. 0

41. **0**

43. $\langle 0, -1, -1 \rangle$

45. $\langle 1, 0, 1e \rangle$

47. $x(t) = \tan t$, $y(t) = \tan t - t + 1$

49. Every point on the graph of $\mathbf{r}(t)$ satisfies both the equation of the sphere with equation $x^2 + y^2 + z^2 = 25$ and the plane with equation $4x - 3z = 0$.

51. $A \approx 24.7$

53. These curves both lie on the cylinder with equation $x^2 + y^2 = 1$. One progresses clockwise around it, the other counterclockwise. Therefore, they must meet.

55. The parametric equations $x(t) = t\sin t$, $y(t) = t\cos t$ and $z(t) = t$ satisfy the equation $z = \sqrt{x^2 + y^2}$. Therefore, the graph of $\mathbf{r}(t)$ lies on the graph of the cone.

57. Let $\mathbf{r}_1(t) = \langle x_1(t), y_1(t) \rangle$ and $\mathbf{r}_2(t) = \langle x_2(t), y_2(t) \rangle$. The components $\mathbf{r}_1(t)$ and $\mathbf{r}_2(t)$ are continuous. We know that products and sums of continuous functions are continuous. Therefore, the function $\mathbf{r}_1(t) \cdot \mathbf{r}_2(t) = x_1(t)x_2(t) + y_1(t)y_2(t)$ is continuous. The result follows from this.

59. See the proof in Exercise 57.

61. We know that products and sums of differentiable functions are differentiable. Therefore, the function $\mathbf{r}_1(t) \cdot \mathbf{r}_2(t) = x_1(t)x_2(t) + y_1(t)y_2(t)$ is differentiable.

Section 11.2

1. F, F ,T, T, F, F, F, F.

3. The function is differentiable at c if $\lim_{h\to 0} \frac{f(c+h)-f(c)}{h}$ exists.

5. A vector function is differentiable at a point c if each of its component functions is differentiable at c. The differentiability of these scalar functions at a point has already been defined.

7. Let $\mathbf{r}(t) = \langle t, |t|\rangle$. The graph of this function is the same as the graph of $y = |x|$. The corner at the origin means that the function doesn't have a well-defined tangent vector at the origin.

9. Let $f(x)$ be a function defined on the interval $[a, b]$. The function is integrable if $\lim_{n\to\infty} \sum_{k=1}^{n} f(x_k^*)\Delta x$ exists, where $\Delta x = \frac{b-a}{n}$, $x_k = a + k\Delta x$, and $x_k^* \in [x_{k-1}, x_k]$.

11. A vector function is integrable on an interval $[a, b]$ if each of its component functions is integrable on $[a, b]$. The integrability of these scalar functions on $[a, b]$ has already been defined.

13. $\langle x'(t), y'(t), z'(t)\rangle$

15. $\frac{d}{d\tau}(\mathbf{r}(f(\tau))) = \mathbf{r}'(f(\tau)f'(\tau)$.

17. $\mathbf{r}'(t) = \langle 3t^2, 15t^2, -6t^2\rangle$. If t_0 and t_1 are positive real numbers, then $\mathbf{r}'(t_0) = \langle 3t_0^2, 15t_0^2, -6t_0^2\rangle = 3t_0^2 1, 5, -2$ and $\mathbf{r}'(t_1) = \langle 3t_1^2, 15t_1^2, -6t_1^2\rangle = 3t_1^2 1, 5, -2$. Since $\mathbf{r}'(t_0)$ and $\mathbf{r}'(t_1)$ are both multiples of the same vector, they are scalar multiples of each other. The graph of $\mathbf{r}(t)$ is a portion of a straight line.

19. They differ by a constant vector.

21. $\langle \cos\tau, 2\sin\tau\cos\tau, 3\sin^2\tau\cos\tau\rangle$

23. $\left\langle \sin\tau + \tau\cos\tau, \frac{1}{2\sqrt{\tau}}, \cos\tau - \tau\sin\tau\right\rangle$

25. $x = e + 2et$, $y = t$, $z = 0$; $2ex + y = 1 + 2e^2$

27. $x = 1 + t$, $y = 0$, $z = -4t$; $x - 4z = 1$

29. $x = 3t, y = 4t, z = t + \pi/2$; $3x + 4y + z = \pi/2$

31. $\mathbf{v}(t) = \langle (t+1)e^t, 1 + \ln t\rangle$, $\mathbf{a}(t) = \langle (t+2)e^t, \frac{1}{t}\rangle$

33. $\mathbf{v}(t) = \langle \cos t, -\sin t, 4\cos 2t\rangle$, $\mathbf{a}(t) = \langle -\sin t. -\cos t, -8\sin 2t\rangle$

35. $\langle 3\tau^2, 6\tau^2(\tau^3 + 1), 9\tau^2(\tau^3 + 1)^2\rangle$

37. $\left\langle -\frac{1}{\tau^2}\sec\frac{1}{\tau}\tan\frac{1}{\tau}, 1, \frac{1}{\tau^2}e^{1/\tau}(\ln\tau - \tau)\right\rangle$

39. $\langle -15\sin(15\tau - 6), 20\cos(20\tau - 8), 5\rangle$

41. $\langle -\cos t + c_1, \sin t + c_2, -\ln|\cos t| + c_3\rangle$

43. $\langle 0, 0, 2\pi^2\rangle$

45. $\langle \frac{1}{2}t^2 + 3, \frac{1}{3}t^3 - 4\rangle$

47. $\langle e^t + 1 - e, t\ln t - t - 5\rangle$

49. $\left\langle \frac{1}{3}t^3 + t + 2, \frac{1}{4}t^4 - 2t - 3\right\rangle$

51. $\langle 5t, -16t^2 + 5t + 26\rangle$

53. $(1, 1, 1)$

55. The circles intersect at $\left(\frac{9}{4}, \pm\frac{\sqrt{63}}{4}\right)$. The angle of intersection at both points is $\cos^{-1}\frac{3}{4}$ radians.

57. The function is $\mathbf{r}(t) = \langle 0, e^t, 2e^t\rangle$. The graph is the ray with parametric equations $x = 0$, $y = s$, $z = 2s$ for $s > 0$.

59. (a) $\langle -0.5dy(t), -9.8t - dy(t)\rangle$, (b) $-\frac{9.8t + y'(t)}{d}$, (c) $x'(t) = 4.9t + 0.5y'(t)$, (d) $\langle 2.45t^2 + 0.5y(t), y(t)\rangle$, (e) No, he should not jump.

61. If $\mathbf{a}(t) = \langle 0, 0, 0\rangle$, then $\mathbf{v}(t) = \langle \alpha, \beta, \gamma\rangle$, where α, β, and γ are constants. We also have $\mathbf{r}(t) = \langle \alpha t + \delta, \beta t + \epsilon, \gamma t + \lambda\rangle$, where δ, ϵ, and λ are also constants. The graph of $\mathbf{r}(t)$ is a straight line.

63. We will prove the case where $\mathbf{r}(t) = \langle x(t), y(t)\rangle$. The case where $\mathbf{r}(t)$ has three components is similar. We have $k\mathbf{r}(t) = \langle kx(t), ky(t)\rangle$ and $\frac{d}{dt}(k\mathbf{r}(t)) = \left\langle \frac{d}{dt}(kx(t)), \frac{d}{dt}(ky(t))\right\rangle = \langle kx'(t), ky'(t)\rangle = k\langle x'(t), y'(t)\rangle = k\mathbf{r}'(t)$.

65. Let $\mathbf{r}_1(t) = \langle x_1(t), y_1(t), z_1(t)\rangle$ and $\mathbf{r}_2(t) = \langle x_2(t), y_2(t), z_2(t)\rangle$. Then
$$\frac{d}{dt}(\mathbf{r}_1(t)\cdot\mathbf{r}_2(t)) = \frac{d}{dt}(x_1x_2 + y_1y_2 + z_1 + z_2)$$
$$= x_1'x_2 + x_1x_2' + y_1'y_2 + y_1y_2' + z_1'z_2 + z_1z_2'$$
$$= \frac{d}{dt}(\langle x_1, y_1, z_1\rangle)\cdot\langle x_2, y_2, z_2\rangle + \langle x_1, y_1, z_2\rangle\cdot\frac{d}{dt}(\langle x_2, y_2, z_2\rangle)$$
$$= \mathbf{r}_1'(t)\cdot\mathbf{r}_2(t) + \mathbf{r}_1(t)\cdot\mathbf{r}_2'(t).$$

67. Let $\mathbf{r}(t) = \langle x(t), y(t), z(t)\rangle$. Then $\frac{d}{d\tau}(\mathbf{r}(t)) = \frac{d}{d\tau}(\langle x, y, z\rangle)$
$$= \left\langle \frac{dx}{dt}\frac{dt}{d\tau}, \frac{dy}{dt}\frac{dt}{d\tau}, \frac{dz}{dt}\frac{dt}{d\tau}\right\rangle = \left\langle \frac{dx}{dt}, \frac{dy}{dt}, \frac{dz}{dt}\right\rangle\frac{dt}{d\tau} = \frac{d\mathbf{r}}{dt}\frac{dt}{d\tau}.$$

69. We will show that $\|\mathbf{r}(t)\|^2$ is a constant. Since $\mathbf{r}$ is continuous, this is equivalent. Taking the derivative we have $\frac{d}{dt}\left(\|\mathbf{r}(t)\|^2\right) = \frac{d}{dt}(\mathbf{r}(t)\cdot\mathbf{r}(t)) = 2\mathbf{r}(t)\cdot\mathbf{r}'(t) = 0$. Since the derivative of $\|\mathbf{r}(t)\|^2$ is a zero, we must have $\|\mathbf{r}(t)\|$ is a constant.

71. Apply the result of Exercise 69.

Section 11.3

1. F, T, F, T, T, T, T, T.

3. All of these vectors have magnitude one. The unit tangent vector points straight ahead. The principal unit normal vector is orthogonal to the unit tangent vector and points "into" the curve. Together these two vectors determine the osculating plane that is the plane in which the curve fits "best" at the point of tangency. The binormal vector is normal to the osculating plane. The tangent vector, normal vector and binormal vector form a right-hand triple, so when the curve is bending to the left, it points up, and when the road bends to the right, it points down.

5. The unit tangent vector won't exist at any point t_0 at which $\mathbf{r}'(t_0) = \mathbf{0}$.

7. If $\mathbf{r}'(t_0) \neq \mathbf{0}$, then they point in the same direction, but $\mathbf{T}(t_0)$ is a unit vector, and $\mathbf{r}'(t_0)$ can have a different length.

9. $\mathbf{N}(t) = \dfrac{\mathbf{T}'(t)}{\|\mathbf{T}'(t)\|}$

11. All of these vectors have magnitude one. The unit tangent vector points straight in the direction of motion of the curve. The principal unit normal vector is orthogonal to the unit tangent vector and points "into" the curve. Together these two vectors determine the osculating plane that is the plane in which the curve fits "best" at the point of tangency. The derivative $\left.\dfrac{d\mathbf{N}}{dt}\right|_{t_0}$ is orthogonal to both the tangent vector and normal vector.

13. At each point t_0 at which $\mathbf{r}(t_0) \neq \mathbf{0}$, $\mathbf{B}(t_0) = \mathbf{T}(t_0) \times \mathbf{N}(t_0)$, where $\mathbf{T}$ and $\mathbf{N}$ are the unit tangent vector and principal unit normal vector, respectively.

15. It isn't defined when $\mathbf{r}'(t) = \mathbf{0}$.

17. Find $\mathbf{r}'(t_0)$. If $\mathbf{r}'(t_0) \neq \mathbf{0}$ continue. Find the unit tangent vector at t_0, $\mathbf{T}(t_0)$. If $\mathbf{T}(t_0) \neq \mathbf{0}$, continue. Find the principal unit normal vector at t_0, $\mathbf{N}(t_0)$. Find the binormal vector at t_0, $\mathbf{B}(t_0) = \mathbf{T}(t_0) \times \mathbf{N}(t_0)$. The osculating plane contains the point $\mathbf{r}(t_0)$ and has $\mathbf{B}(t_0)$ as its normal vector.

19. 3

21. $x = 1$

23. $\dfrac{1}{\sqrt{144t^4+4t^2+25}}\langle 2t, 5, 12t^2\rangle$

25. $\dfrac{1}{|\sin(t)\cos(t)|}\langle -\cos^2 t \sin t, \sin^2 t \cos t\rangle$

27. $\dfrac{1}{5}\langle 3\cos t, -5\sin t, 4\cos t\rangle$

29. $\mathbf{T}(1) = \left\langle \frac{\sqrt{5}}{5}, \frac{2\sqrt{5}}{5}\right\rangle$, $\mathbf{N}(1) = \left\langle -\frac{2\sqrt{5}}{5}, -\frac{\sqrt{5}}{5}\right\rangle$

31. $\mathbf{T}(\pi) = \langle -\sin\alpha\pi, \cos\alpha\pi\rangle$, $\mathbf{N}(\pi) = \langle -\cos\alpha\pi, -\sin\alpha\pi\rangle$

33. $\mathbf{T}(\pi) = \left\langle -\frac{3}{5}, 0, -\frac{4}{5}\right\rangle$, $\mathbf{N}(\pi) = \langle 0, 1, 0\rangle$

35. $\mathbf{T}(1) = \left\langle \frac{\sqrt{14}}{14}, \frac{\sqrt{14}}{7}, \frac{3\sqrt{14}}{14}\right\rangle$, $\mathbf{N}(1) = \frac{1}{\sqrt{266}}\langle -11, -8, 9\rangle$, $\mathbf{B}(1) = \frac{1}{\sqrt{123}}\langle 7, -7, 5\rangle$, $7x - 7y + 5z = 5$

37. $\mathbf{T}(0) = \left\langle \frac{1}{2}, -\frac{1}{2}, \frac{\sqrt{2}}{2}\right\rangle$, $\mathbf{N}(0) = \left\langle \frac{\sqrt{2}}{2}, \frac{\sqrt{2}}{2}, 0\right\rangle$, $\mathbf{B}(0) = \left\langle -\frac{1}{2}, \frac{1}{2}, \frac{\sqrt{2}}{2}\right\rangle$, $-x + y + \sqrt{2}z = 0$

39. $\mathbf{T}\left(\frac{\pi}{2}\right) = \left\langle -\frac{2\sqrt{5}}{5}, 0, \frac{\sqrt{5}}{5}\right\rangle$, $\mathbf{N}\left(\frac{\pi}{2}\right) = \left\langle \frac{\sqrt{2}}{2}, \frac{\sqrt{2}}{2}, 0\right\rangle$, $\mathbf{B}\left(\frac{\pi}{2}\right) = \left\langle -\frac{\sqrt{5}}{5}, 0, -\frac{2\sqrt{5}}{5}\right\rangle$, $x + 2z = \pi$

41. Normal plane: $x - y + \sqrt{2}z = 0$. Rectifying plane: $x + y = 2$.

43. (a) Their tangent vector is $\langle 1.5t^{0.5}, 1.7 - 2t\rangle$, so that their speed is $4t^2 - 4.45t + 2.89$. This is concave up, so their maximum speed will occur when t is a maximum on the interval at $t = 1.7$. (b) Approximately $\langle 0.95, -0.30\rangle$.

45. For any two vectors $\mathbf{a}$ and $\mathbf{b}$ in $\mathbb{R}^3$ $\|\mathbf{a} \times \mathbf{b}\| = \|\mathbf{a}\|\|\mathbf{b}\| \sin\theta$. If $\mathbf{a}$ and $\mathbf{b}$ are orthogonal unit vectors, then $\|\mathbf{a}\|$, $\|\mathbf{b}\|$ and $\sin\theta$ are all one. Thus, $\mathbf{a} \times \mathbf{b}$ is a unit vector.

47. The three properties follow from the definition of the three vectors and the geometric properties of the cross product. Since $\mathbf{T}(t_0)$ and $\mathbf{N}(t_0)$ are orthogonal unit vectors, their cross product $\mathbf{B}(t_0)$ is another unit vector and is orthogonal to them both. Furthermore, we know that for any two nonparallel vectors $\mathbf{a}$ and $\mathbf{b}$, $\mathbf{a}$, $\mathbf{b}$, and $\mathbf{a} \times \mathbf{b}$ form a right-hand triple.

Section 11.4

1. F, F, T, T, F, F, F, T.

3. Informally, it means that for every unit change in the parameter, a unit change occurs along the curve. For the formal definition see Defintion 11.19.

5. There are more ways for a curve to bend. Think of a spring. The concepts of "concave up" and "concave down" are insufficient to describe how it is bending.

7. So that curvature has a consistent definition, the rate of change of the tangent vector with respect to arc length provides a relatively simple definition. Unfortunately, it is often difficult to find an arc length parametrization for a curve.

9. When you have an arc length parametrization for the curve, use the definition. Use Theorem 11.23(a) when you have a space curve and $\mathbf{T}'(t)$ is easy to compute. Use Theorem 11.23(b) when you have a twice-differentiable parametrization for the space curve. Use Theorem 11.24(a) for a planar curve defined by a twice-differentiable function $y = f(x)$. Use Theorem 11.24(b) for a planar curve defined by a vector function.

11. (a) By Theorem 11.23, $\kappa = \dfrac{\|\mathbf{r}'(t) \times \mathbf{r}''(t)\|}{\|\mathbf{r}'(t)\|^3}$. So, if $\mathbf{r}(t) = \langle a + \alpha t, b + \beta t, c + \gamma t\rangle$, then $\mathbf{r}'(t) = \langle \alpha, \beta, \gamma\rangle$ and $\mathbf{r}''(t) = \langle 0, 0, 0\rangle$, so $\kappa = 0$. Since $\kappa = 0$, the radius of curvature is undefined. (b) A straight line doesn't curve, so this is consistent with our concept of curvature.

13. Let (x_0, y_0) be a point on the curve. Find the curvature, κ, at (x_0, y_0) using the appropriate part of Theorem 11.24. If $\kappa \neq 0$, the radius of curvature $\rho = \dfrac{1}{\kappa}$. Find the normal vector, $\mathbf{N}$ at (x_0, y_0). The radius of the osculating circle is ρ. The center of the osculating circle is the terminal point of the position vector $\langle x_0, y_0\rangle + \rho\mathbf{N}$.

15. $y'' = 2$ and by Theorem 11.24, $\kappa = \dfrac{2}{(1+4x^2)^{3/2}}$.

17. 19.

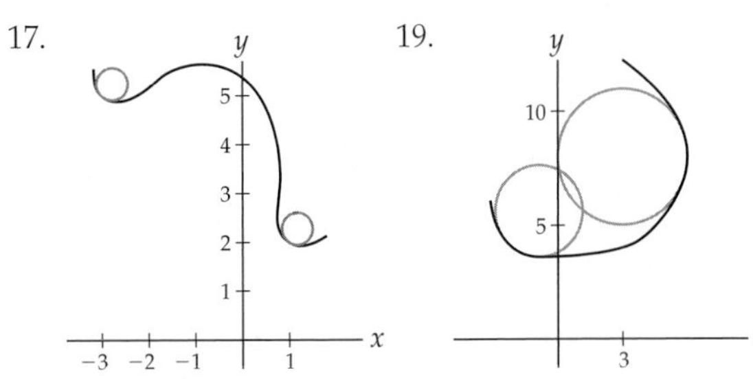

21.

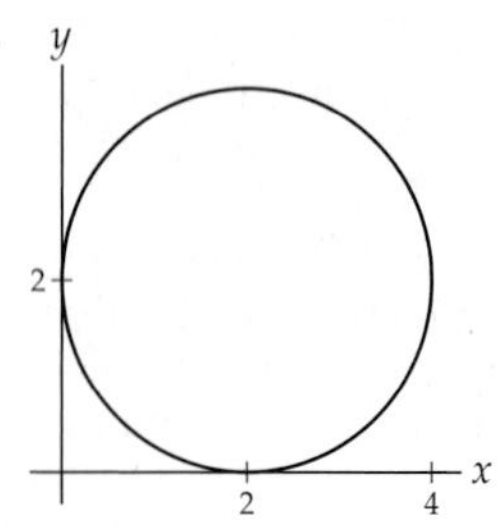

23. 6π

25. $\frac{488}{27}$

27. $\sqrt{3}\,(e^{\pi}-1)$

29. The arc length of the curve on the interval [0, 1] is $\sqrt{30}$; $\mathbf{r}(s) = \left\langle 3 + \frac{\sqrt{30}}{15}s, 4 - \frac{\sqrt{30}}{30}s, -1 + \frac{\sqrt{30}}{6}s \right\rangle$

31. $\kappa = \frac{\sqrt{2}}{4}$

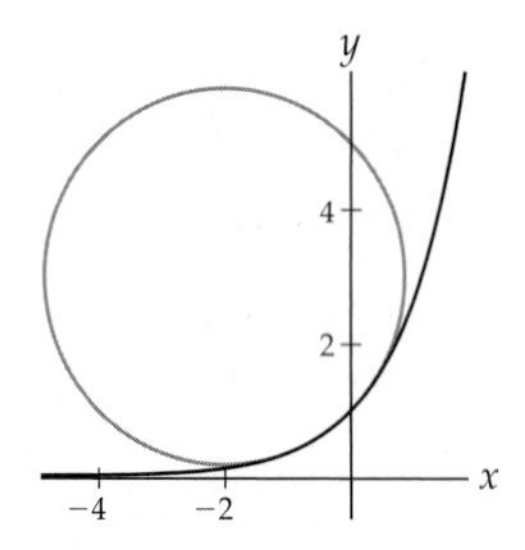

33. $\kappa = 1$

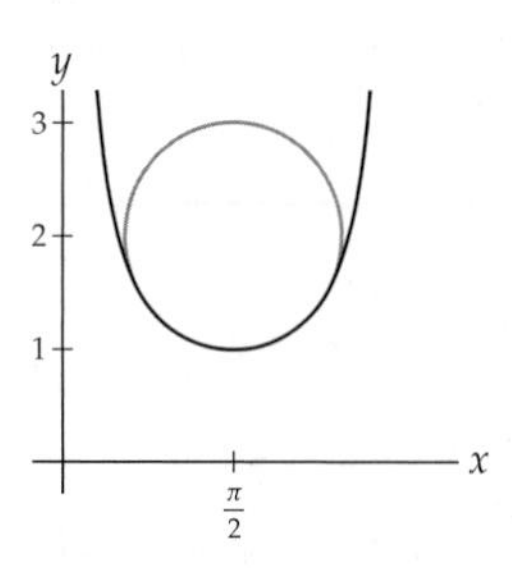

35. $\frac{\sqrt{3}}{6}$

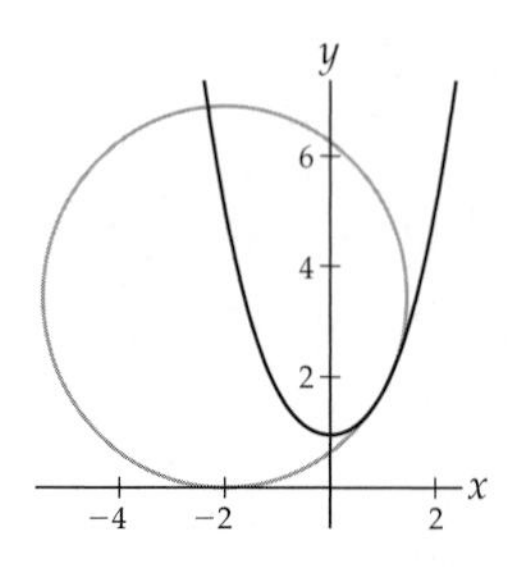

37. $\frac{ab}{\left(a^2\sin^2 t + b^2\cos^2 t\right)^{3/2}}$

39. $\frac{1}{2\sqrt{2(1-\cos t)}}$

41. $\sqrt{\frac{t^4+5t^2+8}{(t^2+2)^3}}$

43. $\sqrt{\frac{t^4+(4+\alpha^2)t^2+4+4\alpha^2}{(t^2+1+\alpha^2)^3}}$

45. $\pm\frac{1}{45^{1/4}}$

47. $\kappa = \frac{a}{a^2+b^2}$

49. $\kappa = t^3\sqrt{\frac{t^6+t^2+4}{(t^4+1)^3}}$ has the stated limits.

51. We differentiate: $\frac{d}{ds}(\mathbf{B}) = \frac{d}{ds}(\mathbf{T}\times\mathbf{N}) = \frac{dT}{ds}\times\mathbf{N} + \mathbf{T}\times\frac{dN}{ds}$. Since the quantities on the right exist at every point on C, the derivative exists.

53. We differentiate: $\frac{d}{ds}(\mathbf{N}) = \frac{d}{ds}(\mathbf{B}\times\mathbf{T}) = \frac{dB}{ds}\times\mathbf{T} + \mathbf{B}\times\frac{dT}{ds} = -\tau\mathbf{N}\times\mathbf{T} + \mathbf{B}\times(\kappa\mathbf{N}) = \tau\mathbf{B} - \kappa\mathbf{T}$

55. $\frac{6+t^2}{4+t^2+t^4}$

57. By choosing the coordinate axes appropriately we may assume that the curve C lies in the xy-plane. Thus, the binormal vector $\mathbf{B}(t) = \pm\mathbf{k}$, and in either case is a constant. Therefore, $\frac{d\mathbf{B}}{ds} = 0$. Thus, the torsion is zero.

59. $\mathbf{r}(t)$ $x(t)$

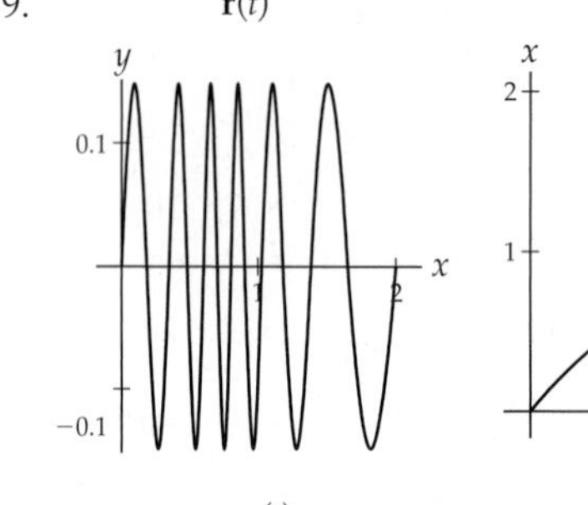

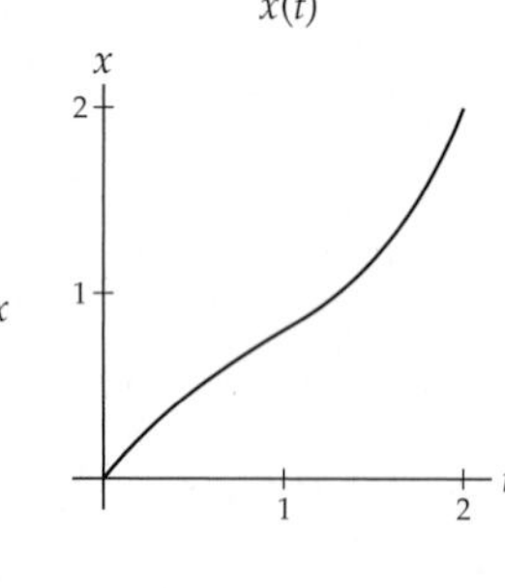

$y(t)$

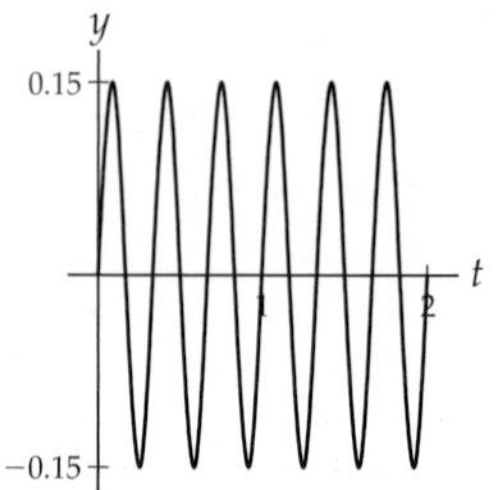

(b) When the x-coordinate changes slowly, that means that the y-coordinate is changing quickly relative to it. The greatest curvature should occur near the point where the slope of the x-coordinate is smallest, while the y-coordinate changes quickly. (c) The slope of the first component of the curve is a minimum at about $t = 0.91$, and the second component turns quickly at about $t = 0.92$. We expect the curvature to be greatest at about that point.

61. By Theorem 11.24 $\kappa = \frac{|f''(x)|}{\left(1+(f'(x))^2\right)^{3/2}}$. Let x be a point of inflection of a twice-differentiable function, f. Since $f''(x) = 0$ at each inflection point, we have $\kappa = 0$.

63. Let $\mathbf{r}(t)$ be a twice differentiable vector-valued function with graph C. By Theorem 11.23 the curvature at each point on C is $\kappa = \frac{\|\mathbf{r}'(t)\times\mathbf{r}''(t)\|}{\|\mathbf{r}'(t)\|^3}$. But recall that $\|\mathbf{r}'(t)\times\mathbf{r}''(t)\| = \|\mathbf{r}'(t)\|\,\|\mathbf{r}''(t)\|\sin\theta$ where θ is the angle between $\mathbf{r}'(t)$ and $\mathbf{r}''(t)$. Since the speed of the particle is constant, by Theorem 11.12 $\mathbf{r}'(t)$ and $\mathbf{r}''(t)$ are orthogonal. Therefore, $\kappa = \frac{\|\mathbf{r}''(t)\|}{\|\mathbf{r}'(t)\|^2}$. But the denominator of this expression is the square of the speed, which we are assuming is constant. Therefore, the curvature is proportional to the magnitude of the acceleration.

65. Let $\mathbf{r}(t) = \langle t, f(t), 0\rangle$. So, $\mathbf{r}'(t) = \langle 1, f'(t), 0\rangle$ and $\mathbf{r}''(t) = \langle 0, f''(t), 0\rangle$. By Theorem 11.23 $\kappa = \frac{\|\mathbf{r}'(t)\times\mathbf{r}''(t)\|}{\|\mathbf{r}'(t)\|^3} = \frac{\|\langle 0,0,f''(t)\rangle\|}{\|\langle 1,f'(t),0\rangle\|} = \frac{|f''(t)|}{(1+(f'(t))^2)^{3/2}}$.

Section 11.5

1. T, F, T, T, F, T, F, F.

3. When the downward acceleration due to the force of gravity is smaller, an object takes longer to fall to the ground. When the object is thrown horizontally, it still takes longer to hit the ground. Therefore, it travels farther.

5. You can include the wind velocity as a vector added to the velocity vector of the projectile.

7. $\mathbf{r}(b) - \mathbf{r}(a)$

9. These are the magnitudes of the acceleration vector, parallel to the tangent vector and parallel to the principal normal unit vector, respectively. Their values are given by $a_{\mathrm{T}} = \frac{\mathbf{v}\cdot\mathbf{a}}{\|\mathbf{v}\|}$ and $a_{\mathrm{N}} = \frac{\|\mathbf{v}\times\mathbf{a}\|}{\|\mathbf{v}\|}$.

11. The tangential component of acceleration always points in the instantaneous direction of your motion, tangent to the curve defined by your path. The normal component of acceleration is orthogonal to your direction of motion and points "into" the curve, so if you are curving to the left, it points to your left; if your are ascending, it points upward, etc.

13. (a) If you eliminate the parameter, the function may be expressed in the form $y = x^2$. (b) This will occur when $\|\mathbf{r}(t)\|$ is a constant and more generally when $f'(t)f''(t) + 4f(t)(f'(t))^3 + 4(f(t))^2 f'(t)f''(t) = 0$.

15. (a) $\langle 0, -\pi\rangle$, (b) π, (c) $\frac{1}{2}(\pi\sqrt{\pi^2+1} + \ln(\pi + \sqrt{\pi^2+1}))$

17. (a) $\langle \alpha\sin\beta, \alpha(\cos\beta - 1), \gamma\rangle$, (b) $\sqrt{\alpha^2 - 2\alpha\cos\beta + 1 + \gamma^2}$, (c) $\sqrt{\alpha^2\beta^2 + \gamma^2}$

19. $a_{\mathrm{T}} = \frac{-27\sin 3t\cos 3t + 64\sin 4t\cos 4t}{\sqrt{9\cos^2 3t + 16\sin^2 4t}}$, $a_{\mathrm{N}} = \frac{|48\cos 3t\cos 4t + 36\sin 3t\sin 4t|}{\sqrt{9\cos^2 3t + 16\sin^2 4t}}$

21. $a_{\mathrm{T}} = \sqrt{3}e^t$, $a_{\mathrm{N}} = \sqrt{2}e^t$

23. $a_{\mathrm{T}} = 0$, $a_{\mathrm{N}} = 1$

25. The values of t are the roots of $16t^4 - 129t^2 + 101 = 0$. These are approximately ± 0.94 seconds and ± 2.68 seconds. Only the positive values make sense in the context of the problem. When $t = 0.94$, $\theta \approx 14^\circ$ and when $t = 2.68$, $\theta \approx 70^\circ$.

27. (a) 10 meters per second, (b) 54 meters

29. (a) 0°, (b) 292 feet

31. 86,000 meters

33. The function $\mathbf{r}(t) = \langle x(t), y(t)\rangle = \left\langle (\|\mathbf{v}_0\|\cos\theta)\,t, -\frac{1}{2}gt^2 + (\|\mathbf{v}_0\|\sin\theta)\,t\right\rangle$ models the motion of the ball. The ball will hit the ground when the y-component is zero. This will occur when $t = \frac{\|\mathbf{v}_0\|\sin\theta}{g}$. At this time it will have travelled $\frac{\|\mathbf{v}_0\|\sin\theta\cos\theta}{g}$ units horizontally. Thus, the x-component will be maximized when $\theta = 45^\circ$.

35. If $\mathbf{a}(t) = \langle 0, 0, 0\rangle$, then the velocity vector is $\mathbf{v}(t) = \langle \alpha, \beta, \gamma\rangle$ for constants α, β and γ. The position function will be $\mathbf{r}(t) = \langle a + \alpha t, b + \beta t, c + \gamma t\rangle$, where a, b, and c are three more constants. The graph of $\mathbf{r}(t)$ is a straight line.

37. Let $\mathbf{r}(t) = \langle t, f(t), 0\rangle$. Then $\mathbf{v}(t) = \langle 1, f'(t), 0\rangle$ and $\mathbf{a}(t) = \langle 0, f''(t), 0\rangle$. We have $\mathbf{v}(t)\times\mathbf{a}(t) = \langle 0, 0, f''(t)\rangle$. At a point of inflection we have $a_{\mathrm{T}} = \frac{\|\mathbf{v}(t)\times\mathbf{a}(t)\|}{\|\mathbf{v}(t)\|} = 0$.

For answers to odd-numbered Skill Certification exercises in the Chapter Review, please visit the Book Companion Web Site at www.whfreeman.com/tkcalculus.

Chapter 12

Section 12.1

1. T, F, T, F, T, T, T, T.

3. The graph of f is the set $\{(x, f(x)) | x \in \mathbb{R}\}$.

5. The graph of f is the set $\{(x, y, z, f(x, y, z)) | (x, y, z) \in \mathbb{R}^3\}$.

7.

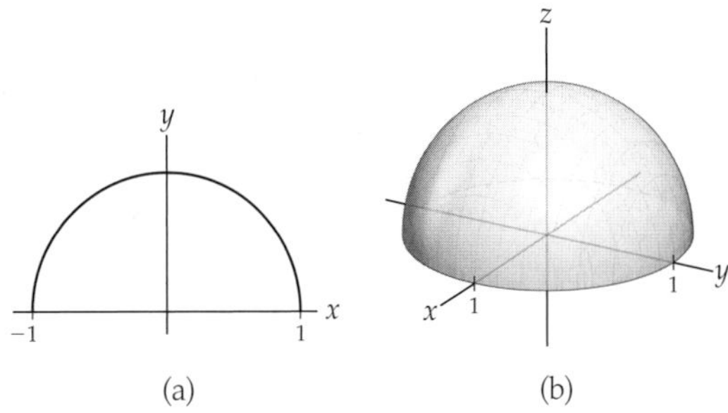

(c) The prefix *hyper* means *above*. The range of the square root function is nonnegative real numbers. $x^2 + y^2 + z^2 + w^2 = 1$. (d) Hypersphere, n, $\mathbb{R}^{n+1}$.

9. A point (x, y, z) is on the graph of f if and only if (x, y) is in the domain of the function f.

11. A point (x, y, z, w) is on the graph of f if and only if (x, y, z) is in the domain of the function f.

13.

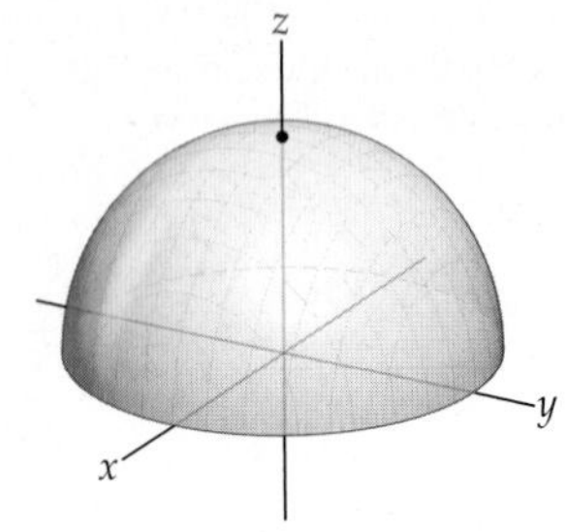

15.

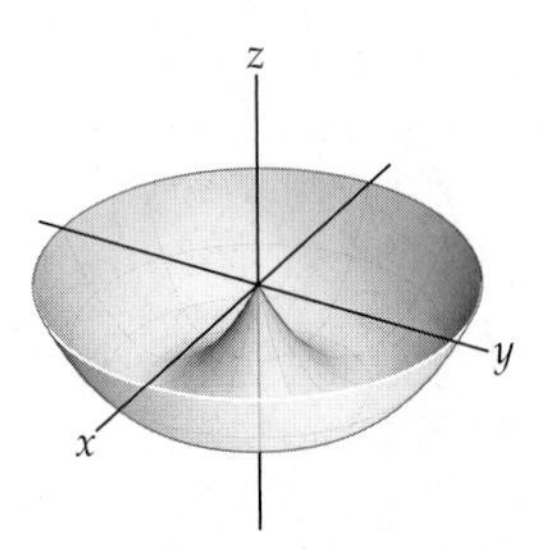

17. See the answer to Exercise 13

19. See the answer to Exercise 15

21.

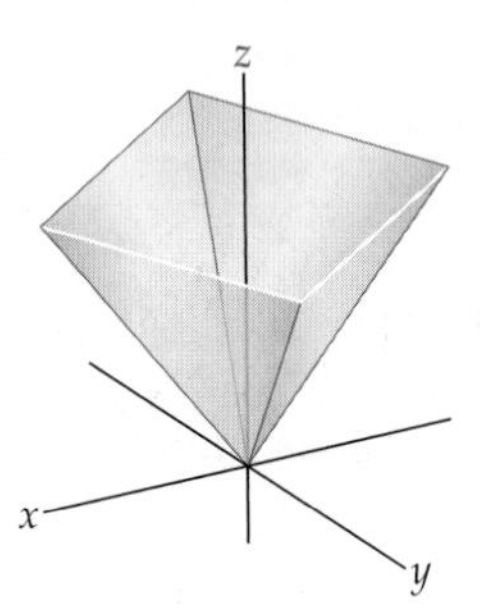

23. $-24, 5$, Domain$(f) = \mathbb{R}^2$, Range$(f) = \mathbb{R}$

25. $(1+e)^{-1/2}, 0$, Domain$(f) = \{(x, y) | 0 < x < 1 \text{ and } y > 0\} \cup \{(x, y) | x < 0 \text{ and } y < 0\}$, Range$(f) = \mathbb{R}$

27. $26, \frac{49}{36}$, Domain$(f) = \mathbb{R}^3$, Range$(f) = \{w \in \mathbb{R} | w \geq 0\}$

29. 1

31. $\sin(\cos t)$

33. $\frac{\cos t + \sin t}{\sin t + 1 - t}$

35. $\langle 1 - t, (1-t)^2, (1-t)^3 \rangle$

37. $F(x, y) = \sqrt{x^2 + y^2}$

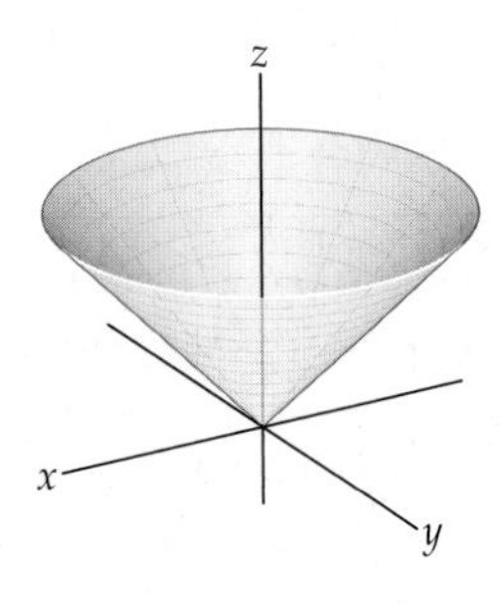

39. $F(x, y) = x^2 + y^2$

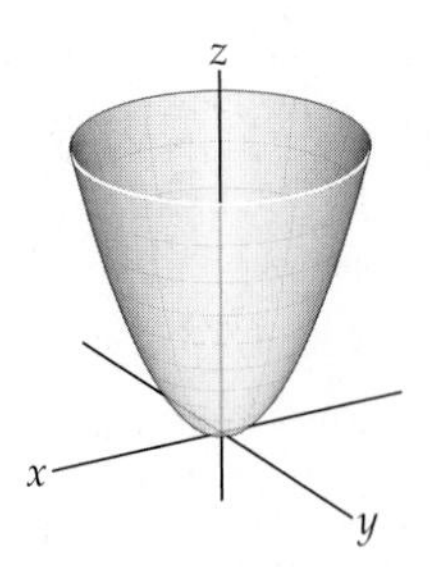

41. $F(x, y) = \sin\sqrt{x^2 + y^2}$

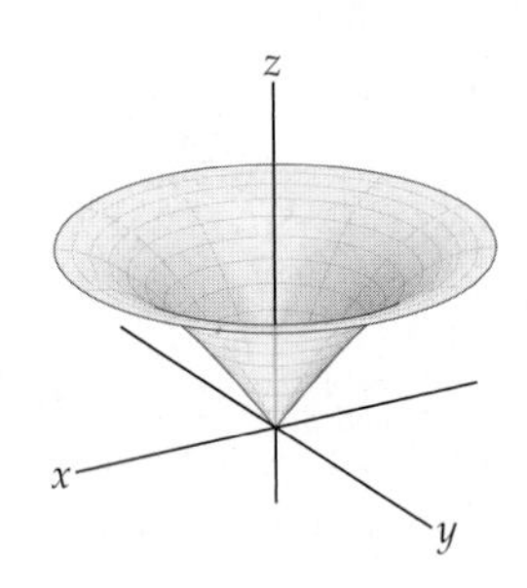

43.

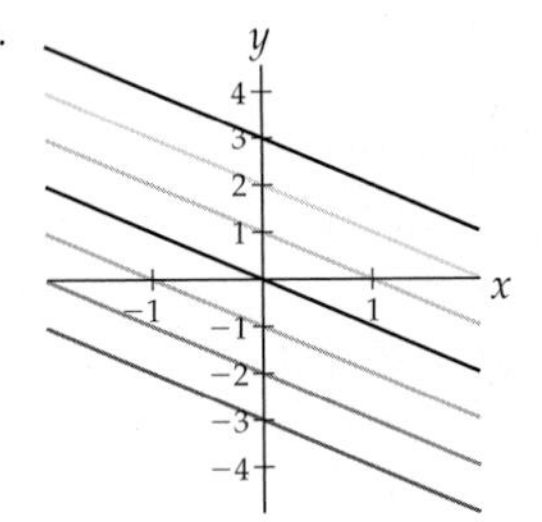

45.

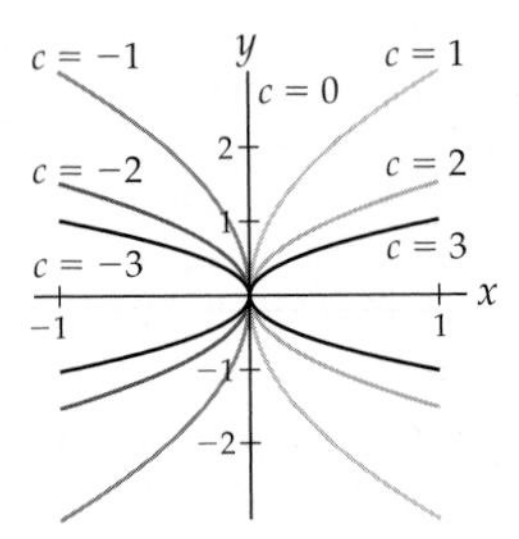

47.

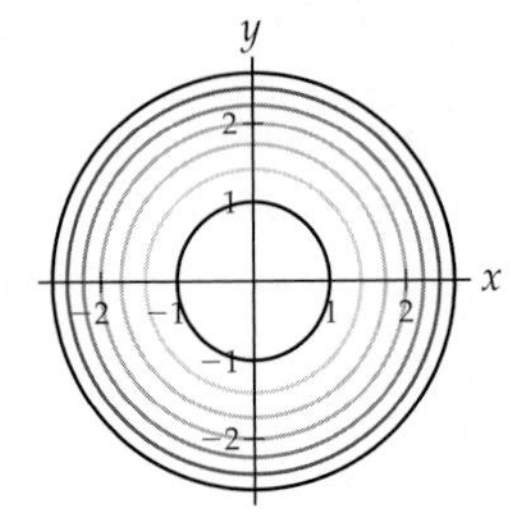

49. Only the level curves for $c = -1, 0$ and 1 are defined. The level curves for $c = -1$ are the parallel lines defined by $x + y = -\frac{\pi}{2} + 2k\pi$ where $k \in \mathbb{Z}$. The level curves for $c = 0$ are the parallel lines defined by $x + y = k\pi$ where $k \in \mathbb{Z}$. The level curves for $c = 1$ are the parallel lines defined by $x + y = \frac{\pi}{2} + 2k\pi$ where $k \in \mathbb{Z}$.

51.

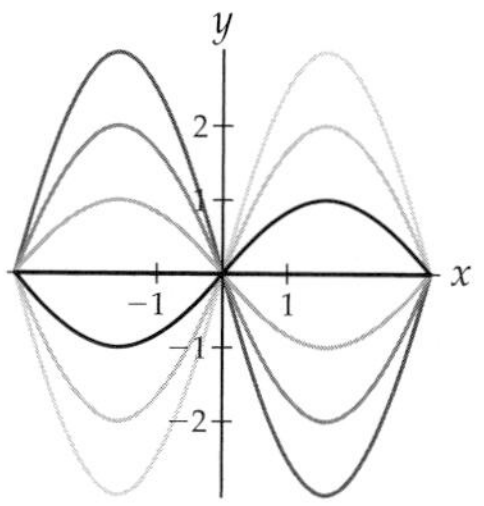

53. The level surfaces are the planes with equations $x + 2y + 3z = c$ for $c = -3, -2, -1, 0, 1, 2, 3$.

55. The level surface $\frac{x}{y-z} = c$ consists of every point on the plane $x - cy + cz = 0$ except those points where $y = z$.

57. The level surfaces for $c = -3, -2$ and -1 are undefined. The level surface $x^2 + y^2 + z^2 = 0$ is the origin. The level surfaces $x^2 + y^2 + z^2 = c$ for $c = 1, 2$, and 3 are the spheres centered at the origin with radii 1, $\sqrt{2}$, and $\sqrt{3}$, respectively.

59. I A

61. III B

63. Domain(V)$= \{(r, h) | r > 0 \text{ and } h > 0\}$, $S(r, h) = 2\pi r^2 + 2\pi rh$.

65. $V(x, y, z) = xyz$, $S(x, y, z) = 2xy + 2xz + 2yz$, Domain($V$)=Domain($S$)$= \{(x, y, z) | x > 0, y > 0 \text{ and } z > 0\}$.

67. (a) From examining a plot, we can see that the caribou are distributed up and down the valley along the line $y = -0.5x + 2$. (b) From examining a plot, we can see that the caribou are distributed up and down the valley along the line $y = -0.5x + 2$. Thus the stream roughly follows the bottom of the valley, $\langle 4, -2 \rangle$.

69. The graph of the equation $z = ax + by + c$ is a plane. The normal vector to the plane $ax + by - z + c = 0$ is $\langle a, b, -1 \rangle$. When $x = 0$ and $y = 0$ we have $z = c$.

71. If c_1 is not equal to c_2, then since f is a function, $f(x, y)$ cannot equal both c_1 and c_2 for the same input (x, y). Thus, c_1 must equal c_2 in order for the level curves to intersect. But this implies that the level curves are identical.

73. You may model your proof on the solution of Exercise 71.

Section 12.2

1. T, F, T, T, F, T, F, T.

3. Nothing, the limit may or may not exist.

5. $\lim_{(x,y)\to(3,-7)} f(x, y) = 5$, $f(3, -7)$ could be any real number, or could even be undefined.

7. When $\lim_{x\to c} f(x) = L$, the function f must be defined on $(c - \delta, c) \cup (c, c + \delta)$ for some $\delta > 0$.

9. When $\lim_{(x,y,z)\to(a,b,c)} f(x, y, z) = L$, the function f must be defined on a punctured ball defined by the inequality $0 < \sqrt{(x-a)^2 + (y-b)^2 + (z-c)^2} < \delta$ for some $\delta > 0$.

11. For f to be continuous at c, we must have $\lim_{x\to c} f(x) = f(c)$. For the limit to exist, f must be defined on $(c - \delta, c) \cup (c, c + \delta)$ for some $\delta > 0$.

13. Along the x-axis $\lim_{\substack{(x,y)\to(0,0)\\ C}} \frac{x^2 y}{x^4 + y^2} = \lim_{x\to 0} \frac{x^2 \cdot 0}{x^4 + 0} = 0$. Along the y-axis $\lim_{\substack{(x,y)\to(0,0)\\ C}} \frac{x^2 y}{x^4 + y^2} = \lim_{y\to 0} \frac{0 \cdot y}{0 + y^2} = 0$.

15. We say that $\lim_{(x,y)\to(a,b)} f(x, y) = \infty$ if for every $M > 0$ there is a $\delta > 0$ such that $f(x, y) > M$ whenever $\sqrt{(x-a)^2 + (y-b)^2} < \delta$.

17. Let $f(x, y) = \frac{x^2}{x^2 + y^2}$ and $g(x, y) = \frac{y^2}{x^2 + y^2}$ then $\lim_{(x,y)\to(0,0)} (f(x, y) + g(x, y)) = 1$; but neither of the individual limits exists as $(x, y) \to (0, 0)$. This does not contradict the sum rule for limits of a function of two variables since neither of the individual limits exists.

19. (a) The quotient of two continuous functions is continuous, as long as the denominator is not zero. (b) For every real number a, every open disk containing the point (a, a) contains infinitely many points of the line $y = x$. Therefore, the quotient is not even defined for infinitely many points in any open ball containing (a, a).

21. (a) Open, (b) $\{(x, y) | x \le 0 \text{ or } y \le 0\}$, (c) $\{(x, 0) | x \ge 0\} \cup \{(0, y) | y \ge 0\}$

23. (a) Closed, (b) $\{(x, y) | |x| + |y| > 1\}$, (c) $\{(x, y) | |x| + |y| = 1\}$

25. (a) Both open and closed, (b) $\mathbb{R}^2$, (c) the empty set

27. (a) Open, (b) $\{(x, y, z) | x \le 0, y \ge 0 \text{ or } z \ge 0\}$, (c) $\{(x, y, 0) \mid x \ge 0,\ y \le 0\} \cup \{(x, 0, z) \mid x \ge 0,\ z \le 0\} \cup \{(0, y, z) \mid y \le 0,\ z \le 0\}$

29. (a) Closed, (b) $\{(x, y, z) | z \ne 0\}$ (c) the xy-plane

31. (a) Both open and closed, (b) $\mathbb{R}^3$, (c) the empty set

33. 0

35. $-\frac{5}{3}$

37. $\frac{9}{2}$

39. does not exist

41. $\lim_{r\to 0} \frac{r^2 \cos^2\theta}{r^2} = \lim_{r\to 0} \cos^2\theta$. Since the limit is a nonconstant function of θ, the limit does not exist.

43. $\lim_{r\to 0} \frac{(r^2\cos^2\theta)(r^2\sin^2\theta)}{r^2} = \lim_{r\to 0} r^2 \cos^2\theta \sin^2\theta = 0$.

45. $\lim_{r\to 0} \frac{(r\cos\theta)(r\sin\theta)}{r} = \lim_{r\to 0} r\cos\theta\sin\theta = 0$.

47. Domain(f)$= \{(x, y) | x^2 \ne y^2\}$. f is continuous on its domain.

49. Domain(f)$= \{(x, y) | y \ge -x^2\}$. f is continuous on the set $\{(x, y) | y > -x^2\}$.

51. Domain(f)$= \{(x,y,z)|x+y+z>0\}$. f is continuous on its domain.

53. Domain(f)$= \mathbb{R}^2$. f is continuous everywhere.

55. Domain(f)$= \mathbb{R}^2$. f is continuous everywhere.

57. The limit is zero. The pressure would decrease as the gas cools off.

59. The limit is infinite. The pressure increases without bound as the gas is compressed.

61. (a) $y = \frac{232}{119}x - \frac{4350}{17}$, (b) -40.38 feet, (c) There are many possibilities. The surface might not be as planar as her previous results led her to believe. She might have been unlucky and hit a pocket or hole in the basalt. Or again, there might be a joint in the basalt layer that would cause a discontinuity in its depth.

63. $x \in S$ if and only if $x \notin S^c$. $x \notin S^c$ if and only if $x \in (S^c)^c$. Therefore, $S = (S^c)^c$.

65. We prove the statement when $S \subseteq \mathbb{R}^2$. The proof for $\mathbb{R}^3$ is similar. First, assume S is open. Thus, for every $x \in S$ there is an open disk D such that $x \in D \subseteq S$. This implies that $x \notin \partial S$. Therefore, $\partial S \cap S = \emptyset$. Now, assume S is not open. There exists an $x \in S$ such that for every open disk D containing x, $D \cap S^c$ is nonempty. Therefore, $x \in \partial S \cap S$.

67. We prove the statement when $S \subseteq \mathbb{R}^2$. The proof for $\mathbb{R}^3$ is similar. $x \in \partial S$ if and only if for every open disk D containing x, D intersects both S and S^c. Since $S = (S^c)^c$, this occurs if and only if $x \in \partial(S^c)$.

69. By Exercise 68, ∂S is a closed set, and by Exercise 66 the boundary of a closed set is a subset of the set. Therefore, $\partial(\partial S) \subseteq \partial S$.

71. The complement of $\emptyset$ in $\mathbb{R}^2$ is $\mathbb{R}^2$. By Exercise 70, $\emptyset$ is both open and closed. Since the complement of an open set is closed and the complement of a closed set is open, $\mathbb{R}^2$ is both open and closed.

Section 12.3

1. T, T, F, F, T, F, F, F.

3. The slope remains zero.

5. $f_x(x_0, y_0)$ and $f_y(x_0, y_0)$ represent the slopes of the curves formed by the intersection of the surface and the planes $y = y_0$ and $x = x_0$, respectively. The lines tangent to these curves are also tangent to the surface.

7. (a) The graph of g is the "cylinder" created when the graph of f is translated in the x direction. (b) $\frac{\partial g}{\partial x}$ exists for every value of x and y if y is in the domain of f. $\frac{\partial g}{\partial y}$ exists for all values of x and y where $f'(y)$ exists. (c) $\frac{\partial g}{\partial x} = 0$ and $\frac{\partial g}{\partial y} = f'(y)$

9. $g_x(0,0) = \lim_{h\to 0} \frac{h^2-0}{h} = 0 = \lim_{h\to 0} \frac{-h^2-0}{h} = g_y(0,0)$. The tangent line in the x direction is the x-axis and the tangent line in the y direction is the y-axis. The plane containing these two lines is $z = 0$.

11. $f_x(0,0) = \lim_{h\to 0} \frac{f(h,0)-f(0,0)}{h} = \lim_{h\to 0} \frac{0-0}{h} = 0$. The computation for $f_y(0,0)$ is similar. The function is discontinuous at $(0,0)$.

13. $f_x(0,0,0) = \lim_{h\to 0} \frac{f(h,0,0)-f(0,0,0)}{h} = \lim_{h\to 0} \frac{0-0}{h} = 0$. The computations for $f_y(0,0,0)$ and $f_z(0,0,0)$ are similar. The function is discontinuous at $(0,0,0)$.

15. (a) 2^2, (b) 2^3, (c) 2^n

17. (a) 3, (b) 4, (c) $n+1$

19. $f(x,y) - g(x,y) = h(y)$, where $h(y)$ is a function of y

21. $f(x,y) - g(x,y) = C$, where C is a constant

23. -2

25. $\frac{\partial f}{\partial x} = \frac{y^2}{z}$, $\frac{\partial f}{\partial y} = \frac{2xy}{z}$, $\frac{\partial f}{\partial z} = -\frac{xy^2}{z^2}$

27. $\frac{\partial f}{\partial x} = e^x(\sin(xy) + y\cos(xy))$, $\frac{\partial f}{\partial y} = xe^x\cos(xy)$

29. $\frac{\partial f}{\partial x} = yx^{y-1}$, $\frac{\partial f}{\partial y} = (\ln x)x^y$

31. $\frac{\partial f}{\partial x} = \sin y$, $\frac{\partial f}{\partial y} = x\cos y$

33. $\frac{\partial f}{\partial r} = \sin\theta$, $\frac{\partial f}{\partial \theta} = r\cos\theta$

35. $\frac{\partial f}{\partial x} = \frac{y^2 z}{(x+z)^2}$, $\frac{\partial f}{\partial y} = \frac{2xy}{x+z}$, $\frac{\partial f}{\partial z} = -\frac{xy^2}{(x+z)^2}$

37. (a) $x = t, y = \frac{\pi}{2}, z = \frac{\pi}{2}t$, (b) $x = 0, y = \frac{\pi}{2} + t, z = 0$, (c) $\frac{\pi}{2}x = z$

39. (a) $x = e + t, y = 3, z = e^3 + 3e^2 t$, (b) $x = e, y = 3 + t, z = e^3 + e^3 t$, (c) $3e^2 x + e^3 y - z = 5e^3$

41. (a) $x = 2 + t, y = \frac{\pi}{3}, z = \sqrt{3} + \frac{\sqrt{3}}{2}t$, (b) $x = 2, y = \frac{\pi}{3} + t, z = \sqrt{3} + t$, (c) $\sqrt{3}x + 2y - 2z = \frac{2\pi}{3}$

43. $\frac{\partial^2 f}{\partial x^2} = e^x(\sin(xy) + 2y\cos(xy) - y^2\sin(xy))$, $\frac{\partial^2 f}{\partial x \partial y} = \frac{\partial^2 f}{\partial y \partial x} = e^x(\cos(xy) + x\cos(xy) - xy\sin(xy))$, $\frac{\partial^2 f}{\partial y^2} = -x^2 e^x \sin(xy)$

45. $\frac{\partial^2 f}{\partial x^2} = y(y-1)x^{y-2}$, $\frac{\partial^2 f}{\partial x \partial y} = \frac{\partial^2 f}{\partial y \partial x} = x^{y-1}(1 + y\ln x)$, $\frac{\partial^2 f}{\partial y^2} = (\ln x)^2 x^y$

47. $\frac{\partial^2 f}{\partial x^2} = 0$, $\frac{\partial^2 f}{\partial x \partial y} = \frac{\partial^2 f}{\partial y \partial x} = \cos y$, $\frac{\partial^2 f}{\partial y^2} = -x\sin y$

49. $\frac{\partial^2 f}{\partial r^2} = 0$, $\frac{\partial^2 f}{\partial r \partial \theta} = \frac{\partial^2 f}{\partial \theta \partial r} = \cos\theta$, $\frac{\partial^2 f}{\partial \theta^2} = -r\sin\theta$

51. $f(x,y) = h(y)$

53. $f(x,y) = xh_1(y) + h_2(y)$

55. $f(x,y,z) = h(y,z)$

57. $f(x,y,z) = xh_1(y,z) + h_2(y,z)$

59. $\frac{\partial g}{\partial y} = \frac{\partial h}{\partial x}$, $F(x,y) = e^x\cos y + y^2 + C$

61. $\frac{\partial g}{\partial y} = \frac{\partial h}{\partial x}$, $F(x, y) = tan^{-1}(xy) + C$

63. $xe^y - 7y + C = 0$

65. $e^x \ln y + \frac{1}{4}x^4 + C = 0$

67. $\frac{\partial V}{\partial r} = 2\pi rh$. This tells us how fast the volume changes with a unit change in the radius. $\frac{\partial V}{\partial h} = \pi r^2$. This tells us how fast the volume changes with a unit change in the height.

69. (b) $0.04e^{\pi y/2} \sin \frac{\pi x}{2}$

71. Both results follow from the Chain Rule (Theorem 2.12), the power rule, and the definition of the partial derivative.

73. Both results follow from the Quotient Rule (Theorem 2.11) and the definition of the partial derivative.

75. $\frac{\partial h}{\partial y} = g'(y)$, so $\frac{\partial^2 h}{\partial x \partial y} = 0$. Similarly, $\frac{\partial h}{\partial x} = f'(x)$, so $\frac{\partial^2 h}{\partial y \partial x} = 0$.

Section 12.4

1. T, T, F, F, F, T, F, T.

3. There are two unit vectors in $\mathbb{R}^1$, $\mathbf{i}$ and $-\mathbf{i}$. There are infinitely many unit vectors in $\mathbb{R}^n$ when $n > 1$.

5. The directional derivative of $f(x, y)$ at (x_0, y_0) in the direction of the unit vector $\mathbf{u} = \langle a, b \rangle$ is the limit $\lim_{h \to 0} \frac{f(x_0 + a \cdot h, y_0 + b \cdot h) - f(x_0, y_0)}{h}$, provided that this limit exists.

7. $D_{\mathbf{u}}f(\mathbf{v}) = \lim_{h \to 0} \frac{f(\mathbf{v} + h\mathbf{u}) - f(\mathbf{v})}{h}$, provided that this limit exists.

9. (a) $D_{\mathbf{u}} = 3\alpha - \beta$, (b) $3k\alpha - k\beta$, (c) The limit from part (b) is different for each value of k. We want the value of the directional derivative to depend upon the direction of $\mathbf{u}$, not its magnitude.

11. $\lim_{h \to 0} \frac{f(c + \alpha h) - f(c)}{h}$, $\alpha = \pm 1$

13. See Definition 12.27.

15. If $f(x, y)$ is differentiable at the point (a, b), all of the lines tangent to the surface defined by f at (a, b) lie in the same plane, so any two distinct lines in that plane may be used to determine the equation of the plane.

17. If $f(x, y, z)$ is differentiable at the point (a, b, c), all of the lines tangent to the graph of f at (a, b, c) lie in the same hyperplane, so any three non-coplanar lines in that hyperplane may be used to determine the equation of the hyperplane.

19. Let $y = f(\mathbf{x})$ be a function of n variables defined on an open set containing the point $\mathbf{x_0}$ and let $\Delta y = f(\mathbf{x_0} + \Delta \mathbf{x}) - f(\mathbf{x_0})$. The function f is said to be ***differentiable*** at $\mathbf{x_0}$ if the partial derivatives $f_{x_i}(\mathbf{x_0})$ exist for each $1 \leq i \leq n$ and $\Delta y = \langle f_{x_1}(\mathbf{x_0}), f_{x_2}(\mathbf{x_0}), \ldots, f_{x_n}(\mathbf{x_0}) \rangle \cdot \Delta \mathbf{x} + \epsilon \cdot \Delta \mathbf{x}$, where $\epsilon \to \mathbf{0}$ as $\Delta x \to \mathbf{0}$.

21. $-\sqrt{2}$

23. $-11\frac{\sqrt{10}}{10}$

25. $-7\frac{\sqrt{17}}{816}$

27. $-\frac{36}{5}$

29. $x = 2 + \frac{\sqrt{2}}{2}t, y = 3 + \frac{\sqrt{2}}{2}t, z = -5 - \sqrt{2}t$

31. $x = -2 + \frac{\sqrt{10}}{10}t, y = 1 - \frac{3\sqrt{10}}{10}t, z = -2 - 11\frac{\sqrt{10}}{10}t$

33. $x = 4 - \frac{\sqrt{17}}{17}t, y = 9 - \frac{4\sqrt{17}}{17}t, z = \frac{3}{2} - 7\frac{\sqrt{17}}{816}$

35. $-\frac{18}{13}\sqrt{26}$

37. $-\frac{33}{40}\sqrt{5}$

39. For every unit vector $\mathbf{u} = \langle \alpha, \beta \rangle$, $\lim_{h \to 0} \frac{(1+\alpha h)(-2+\beta h)+2(1+\alpha h)-(-2+\beta h)-2}{h} = 0.$

41. For every unit vector $\mathbf{u} = \langle \alpha, \beta \rangle$, $\lim_{h \to 0} \frac{\alpha h(\beta h)^2}{h} = 0.$

43. f has continuous first-order partial derivatives and is differentiable at every point in $\mathbb{R}^2$.

45. The function has continuous first-order partial derivatives and is differentiable at every point in $\mathbb{R}^2$ except when $x^2 + y^2 = 1$.

47. The function has continuous first-order partial derivatives and is differentiable at every point in $\mathbb{R}^2$.

49. The function has continuous first-order partial derivatives and is differentiable at every point in $\mathbb{R}^2$ at which the product xy is not an odd multiple of $\frac{\pi}{2}$

51. The function has continuous first-order partial derivatives and is differentiable at every point such that $x > 0$ and $y > 0$.

53. The function has continuous first-order partial derivatives and is differentiable at every point in $\mathbb{R}^3$.

55. $2(x - 1) + 6(y + 3) = z + 8$

57. $x + 9z = -6$

59. $\pi x + 4y + 2z = 4\pi$

61. $\pi x - 4y + 2z = 2\pi - 2$

63. $9x - y + 6z = 18$

65. $2(x - 1) - 10(y + 5) - 27(z - 3) = w + 1$

67. (a) At about a 10 degree slope. (b) He must move in the direction $(0, 1)$ or $(0, -1)$, due south or due north.

69. Let $\mathbf{u} = \langle \alpha, \beta \rangle$. By Definition 12.26, $D_{-\mathbf{u}}f(a, b) = \lim_{\eta \to 0} \frac{f(a - \alpha\eta, b - \beta\eta) - f(a, b)}{\eta}$. If we let $h = -\eta$, we have $D_{-\mathbf{u}}f(a, b) = \lim_{h \to 0} \frac{f(a + \alpha h, b + \beta h) - f(a, b)}{-h} = -D_{\mathbf{u}}f(a, b).$

71. By Exercise 70, f cannot be differentiable at (a, b) if $D_{\mathbf{u}}f(a, b) = 1$ for every unit vector $\mathbf{u}$.

Section 12.5

1. F, F, F, F, F, T, T, T.

3. $e^{-\sin t}(2\sin t\cos t - 3\sin t\cos^2 t - \cos^3 t + 3\cos^2 t$ $-3\sin^2 t - 4\cos t)$.

5. When $n = m = 1$, $\frac{\partial z}{\partial t_j} = \frac{\partial z}{\partial x_1}\frac{\partial x_1}{\partial t_j} + \frac{\partial z}{\partial x_2}\frac{\partial x_2}{\partial t_j} + \cdots + \frac{\partial z}{\partial x_n}\frac{\partial x_n}{\partial t_j}$ simplifies to $\frac{dz}{dt_1} = \frac{dz}{dx_1}\frac{dx_1}{dt_1}$.

7. When $n = m = 2$, $\frac{\partial z}{\partial t_j} = \frac{\partial z}{\partial x_1}\frac{\partial x_1}{\partial t_j} + \frac{\partial z}{\partial x_2}\frac{\partial x_2}{\partial t_j} + \cdots + \frac{\partial z}{\partial x_n}\frac{\partial x_n}{\partial t_j}$ simplifies to $\frac{\partial z}{\partial t_1} = \frac{\partial z}{\partial x_1}\frac{\partial x_1}{\partial t_1} + \frac{\partial z}{\partial x_2}\frac{\partial x_2}{\partial t_1}$ and $\frac{\partial z}{\partial t_2} = \frac{\partial z}{\partial x_1}\frac{\partial x_1}{\partial t_2} + \frac{\partial z}{\partial x_2}\frac{\partial x_2}{\partial t_2}$.

9. (a) The level curves of f are the lines with equations $y = -\frac{2}{3}x + C$ where $C \in \mathbb{R}$. (b) The vector $\langle 3, -2\rangle$ may be used as a direction vector for every level curve, $\nabla f(x, y) = \langle 2, 3\rangle$, and $\langle 3, -2\rangle \cdot \langle 2, 3\rangle = 0$.

11. (a) The level curves of f are the lines with equations $y = -\frac{a}{b}x + C$ where $C \in \mathbb{R}$. (b) The vector $\langle b, -a\rangle$ may be used as a direction vector for every level curve, $\nabla f(x, y) = \langle a, b\rangle$, and $\langle b, -a\rangle \cdot \langle a, b\rangle = 0$.

13. $\nabla f(a, b, c) \cdot (\langle x, y, z\rangle - \langle a, b, c\rangle) = w - f(a, b, c)$.

15. The level curves are concentric circles centered at the origin. The gradient vectors are orthogonal to the level curves and point toward the origin. The gradient vectors increase in magnitude as you get further from the origin.

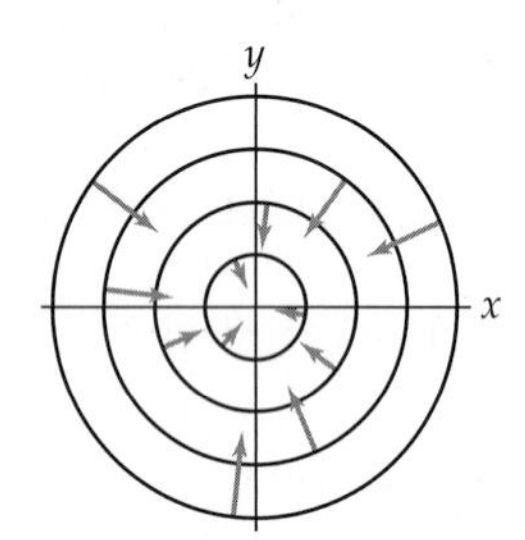

17. The level curves are hyperbolas. The gradient vectors are orthogonal to the level curves and point toward the x-axis and away from the y-axis. The gradient vectors increase in magnitude as you get further from the origin.

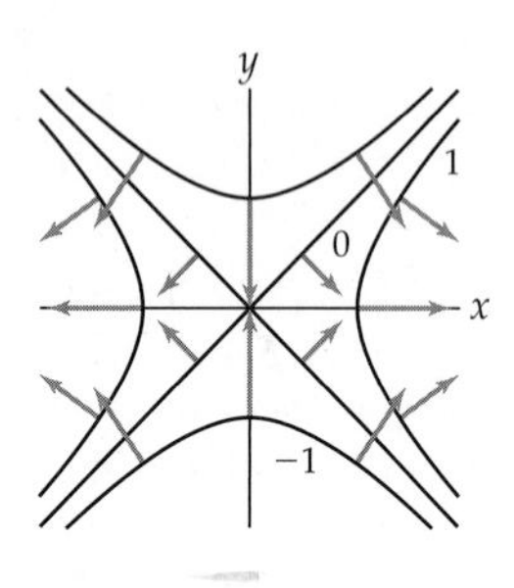

19. Staying on a contour line means that you are always staying at the same elevation, so the hike is relatively easy. Walking perpendicularly to the contour lines means you are always either ascending or descending, making the hike more rigorous.

21. $e^t\cos(e^t)\cos(t^3) - 3t^2\sin(e^t)\sin(t^3)$

23. $2t\cos(t^3 + 1) + (15t^2 - 3t^4)\sin(t^3 + 1)$

25. $\frac{1+4t^3}{2\sqrt{t+t^4}}$

27. $2s^3t^2\cos s\cos^3 t\sin s + 3s^2t^2\cos^3 t\sin^2 s$

29. $e^{r\sin\theta}(2r\cos^2\theta + 2r\sin\theta\cos\theta + r^2\sin\theta\cos^2\theta$ $+r^2\sin^2\theta\cos\theta)$

31. $2t\sin(t^3)\cos(t^4) + 3t^4\cos(t^3)\cos(t^4) - 4t^5\sin(t^3)\sin(t^4)$

33. $e^{\rho\sin\phi\sin\theta}(2\rho\sin^2\phi\cos^2\theta + \rho^2\sin^3\phi\sin\theta\cos^2\theta$ $+\rho\sin\phi\cos\phi\sin\theta + \cos\phi)$

35. $\frac{1}{2}(1 + 4t^3 + 6t^5)(t + t^4 + t^6)^{-1/2}$

37. $\langle 2x\sin y + y\cos x, x^2\cos y + \sin x\rangle$

39. $\frac{1}{\sqrt{x^2+y^2}}\langle x, y\rangle$

41. $\frac{1}{\sqrt{x^2+y^2+z^2}}\langle x, y, z\rangle$

43. (a) $\left\langle \frac{3\pi}{2}, 0\right\rangle$, (b) $\frac{3\pi}{2}$, (c) $-\left\langle \frac{3\pi}{2}, 0\right\rangle$

45. (a) $\left\langle \frac{2}{\sqrt{13}}, -\frac{3}{\sqrt{13}}\right\rangle$, (b) 1, (c) $-\left\langle \frac{2}{\sqrt{13}}, -\frac{3}{\sqrt{13}}\right\rangle$

47. (a) $\left\langle \frac{2}{3}, -\frac{1}{3}, -\frac{2}{3}\right\rangle$, (b) 1, (c) $-\left\langle \frac{2}{3}, -\frac{1}{3}, -\frac{2}{3}\right\rangle$

49. $\frac{\sqrt{10}}{10}$

51. $\frac{55\sqrt{17}}{119}$

53. $\frac{9\sqrt{7}}{14}$

55. $f(x, y) = \ln\left|\frac{y}{x^3}\right| + C$

57. $f(x, y) = \tan^{-1}\left(\frac{y}{x}\right) + C$

59. $f(x, y) = \frac{x}{x+y} + C$

61. (a) $\langle 0.5, -0.35\rangle$, (b) $\pm\langle 0.35, 0.5\rangle$.

63. $\nabla f(x, y, z) = \left\langle \frac{x}{\sqrt{x^2+y^2+z^2}}, \frac{y}{\sqrt{x^2+y^2+z^2}}, \frac{z}{\sqrt{x^2+y^2+z^2}}\right\rangle$, so

$$\|\nabla f(x, y, z)\| = \left(\frac{x}{\sqrt{x^2+y^2+z^2}}\right)^2 + \left(\frac{y}{\sqrt{x^2+y^2+z^2}}\right)^2 + \left(\frac{z}{\sqrt{x^2+y^2+z^2}}\right)^2 = 1$$

65. We have $z = f(x, y)$, $x = u(s, t)$ and $y = v(s, t)$. Assume that u, v and $z = f(u(s, t), v(s, t)$ are differentiable at a point (s_0, t_0) and that $x_0 = u(s_0, t_0)$ and $y_0 = v(s_0, t_0)$. We will show that $\frac{\partial z}{\partial s} = \frac{\partial z}{\partial x}\frac{\partial x}{\partial s} + \frac{\partial z}{\partial y}\frac{\partial y}{\partial s}$. The remaining equality is quite similar. By the definition of the partial derivative at s_0:
$\frac{\partial z}{\partial s} = \lim_{\Delta s \to 0} \frac{f(u(s_0+\Delta s, t_0), v(s_0+\Delta s, t_0)) - f(u(s_0, t_0), v(s_0, t_0))}{\Delta s} =$
$\lim_{\Delta s \to 0} \frac{f(u(s_0+\Delta s, t_0), v(s_0+\Delta s, t_0)) - f(x_0, y_0)}{\Delta s} = \lim_{\Delta s \to 0} \frac{\Delta z}{\Delta s}$. Now, when Δs is sufficiently small, we may replace the numerator in the equation above by $f_x(x_0, y_0)\Delta x + f_y(x_0, y_0)\Delta y$. We now have
$\frac{\partial z}{\partial s} = \lim_{\Delta s \to 0} \frac{f_x(x_0, y_0)\Delta x + f_y(x_0, y_0)\Delta y}{\Delta s} =$
$\lim_{\Delta s \to 0} \left(f_x(x_0, y_0)\frac{\Delta x}{\Delta s} + f_y(x_0, y_0)\frac{\Delta y}{\Delta s}\right) = \frac{\partial z}{\partial x}\frac{\partial x}{\partial s} + \frac{\partial z}{\partial y}\frac{\partial y}{\partial s}$.

67. $\frac{\partial f}{\partial x} = 0 = \frac{\partial f}{\partial y}$, so $\nabla f = 0$.

69. $\nabla(f(x, y) + g(x, y))$
$= \langle f_x(x, y) + g_x(x, y), f_y(x, y) + g_y(x, y)\rangle$
$= \langle f_x(x, y), f_y(x, y)\rangle + \langle g_x(x, y), g_y(x, y)\rangle$
$= \nabla f(x, y) + \nabla g(x, y)$.

71. $\nabla(f(x, y)g(x, y)) = \langle f_x(x, y)g(x, y)$
$+ f(x, y)g_x(x, y), f_y(x, y)g(x, y) + f(x, y)g_y(x, y)\rangle$
$= g(x, y)\langle f_x(x, y), f_y(x, y)\rangle + f(x, y)\langle g_x(x, y), g_y(x, y)\rangle$
$= f(x, y)\nabla g(x, y) + g(x, y)\nabla f(x, y)$.

73. Use three component vectors rather than two component vectors.

Section 12.6

1. F, T, F, T, F, T, T, T.

3. A critical point of f is a point at which either $\nabla f = \mathbf{0}$ or ∇f does not exist. The only place a function can have an extreme value at an interior point of its domain is at a critical point.

5. A saddle point is a stationary point at which there is neither a maximum nor a minimum.

7. The first derivative test for a function of a single variable requires that you check the sign of the first derivative to the left and right of a critical point. To use an analogous test for a function of two variables, we would have to check the sign of the directional derivative at infinitely many points encircling each critical point.

9. If A and C have opposite signs then both terms AC and $-B^2$ are negative, so their sum is negative.

11. $g(x, y) = -f(x, y)$ from Exercise 10, so since $f(x, y)$ has a minimum at the origin, g has a maximum at the origin.

13. Plow through the algebra!

15. Plow through the algebra!

17. The distance from P to an arbitrary point on $\mathcal{L}$ is given by
$$d(t) = \sqrt{(x_0 - x_1 + at)^2 + (y_0 - y_1 + bt)^2 + (z_0 - z_1 + ct)^2}.$$
To find the distance from P to $\mathcal{L}$ we may minimize the function $d(t)$.

19. f will have an absolute minimum at $x = 0$ if n is even and an inflection point if n is odd. There are no other possibilities.

21. $-4e^{4x}$

23. $-\frac{3}{4}s^4e^t$

25. $\cos^2\theta\sin^2\phi - \sin^2\theta\cos^2\phi$

27. $\frac{14}{13}\sqrt{26}$

29. $\frac{8}{13}$

31. $f\left(-\frac{1}{2}, 0\right) = -\frac{31}{4}$ is the absolute minimum

33. $f(-2, -1) = 33$, local maximum; $f(-2, 1) = 29$, saddle; $f(2, -1) = 1$, saddle; $f(2, 1) = -3$, local minimum

35. $f(0, 0) = 0$, saddle; $f(4, 4) = -64$, local minimum

37. $f\left(-\frac{3}{2}, -\frac{3}{2}\right) = -14$, absolute minimum

39. $f\left(\frac{1}{2}, \frac{1}{2}\right) = 6$, local minimum

41. Every point of the form $(0, k\pi)$ where $k \in \mathbb{Z}$ is a stationary point. They are all saddle points.

43. There is a saddle point at every point of the form $(0, k\pi)$ where $k \in \mathbb{Z}$.

45. $f(0, 0) = 1$, saddle

47. $f(0, 0) = -1$, local maximum

49. f has an absolute minimum at every point of the form $(0, y)$ or $(x, 0)$.

51. f has saddle point at every point of the form $(0, y)$, a local minimum at every point of the form $(x, 0)$ for $x > 0$, and a local maximum at every point of the form $(x, 0)$ for $x < 0$.

53. The box should have a square base measuring $\frac{10}{9}\sqrt[3]{\frac{81}{5}}$ feet on each side and a height of $\sqrt[3]{\frac{81}{5}}$ feet.

55. By Theorem 12.36, $D_{\mathbf{u}}f(x, y) = af_x(x, y) + bf_y(x, y)$. Taking the directional derivative, again, we obtain $D_{\mathbf{u}}(D_{\mathbf{u}}f(x, y)) = D_{\mathbf{u}}^2 f(x, y) = D_{\mathbf{u}}(af_x(x, y) + bf_y(x, y)) = a^2f_{xx}(x, y) + abf_{yx}(x, y) + baf_{xy}(x, y) + b^2f_{yy}(x, y)$. Since the function has continuous second-order partial derivatives, the mixed second-order partial derivatives are equal and we have the result.

57. Let (x, y) be a point on the line. The distance from (x_0, y_0) to (x, y) is given by $d = \sqrt{(x - x_0)^2 + (y - y_0)^2}$ and the square of the distance from (x_0, y_0) to (x, y) is given by $D = (x - x_0)^2 + (y - y_0)^2$. The gradients of these two functions are
$\nabla d = \frac{1}{2\sqrt{(x-x_0)^2 + (y-y_0)^2}}(2(x - x_0)\mathbf{i} + 2(y - y_0)\mathbf{j})$ and $\nabla D = 2(x - x_0)\mathbf{i} + 2(y - y_0)\mathbf{j}$. The critical points of these two functions are the same and would have the same minima.

Section 12.7

1. F, F, T, F, F, F, F, T.

3. The function f represents the quantity we wish to maximize or minimize. The constraint equation provides an interrelationship between the variables of f.

5. We can either eliminate one of the variables by solving the constraint equation for one of the variables, or use the method of Lagrange multipliers.

7. (a) $\{(x, y)|x > 0\}$, (b) $\{(x, y)|x \geq 0\}$, (c) $\{(x, y)|x^2 + y^2 < 1\}$, (d) $\{(x, y)|x^2 + y^2 \leq 1\}$

9. (a) $\{(x, y, z)|x > 0\}$, (b) $\{(x, y, z)|x \geq 0\}$, (c) $\{(x, y, z)|x^2 + y^2 + z^2 < 1\}$, (d) $\{(x, y, z)|x^2 + y^2 + z^2 \leq 1\}$

11. 4

13. $f_{yy}(x, y) = 12$ and the discriminant of f is $\det(H_f(x, y)) = 72(x + 3)$. (a) $\det(H_f(-6, 0)) < 0$ so there is a saddle point at $(-6, 0)$. (b) $\det(H_f(0, 0)) > 0$ and $f_{yy}(0, 0) > 0$ so there is a relative minimum at $(0, 0)$. (c) From Example 4 we already know (i) that $f(0, 0) = 0$ is the absolute minimum of f on the region defined by $x^2 + y^2 \leq 9$ and (ii) the point $(0, 0)$ is an interior point to that region, so there is, at least, a relative minimum of f at the origin. Since $f(-10, 0) = -100 < 0 = f(0, 0)$, the minimum is only relative, not absolute.

15. When $y = -x$ we are looking for the critical points of the function $g(x) = -x^3$. The only critical point of g is its inflection point at $x = 0$.

17. When $y = -\frac{a}{b}x$ we are looking for the critical points of the function $h(x) = -\frac{a}{b}x^3$. The only critical point of h is its inflection point at $x = 0$.

19. Find the extrema of the function f using the techniques of Section 12.6. Choose only the critical points in $\mathcal{R}$. Evaluate f at the critical points. Use the results of Exercise 18 to find the extrema on the boundary of $\mathcal{R}$. Choose the largest and smallest values of the function in the interior of $\mathcal{R}$ and on the boundary of $\mathcal{R}$.

21. Find the extrema of the function f using the techniques of Section 12.6. Choose only the critical points in $\mathcal{T}$. Evaluate f at the critical points. Use the results of Exercise 20 to find the extrema on the boundary of $\mathcal{T}$. Choose the largest and smallest values of the function in the interior of $\mathcal{T}$ and on the boundary of $\mathcal{T}$.

23. The Extreme Value Theorem guarantees that a maximum and minimum of f occurs on $\mathcal{R}$. If the gradient exists everywhere and is never zero, f does not have any critical points, so the maximum and minimum must occur on the boundary of $\mathcal{R}$.

25. $f\left(\frac{8\sqrt{5}}{5}, \frac{2\sqrt{5}}{5}\right) = 2\sqrt{5}$ is the maximum and $f\left(-\frac{8\sqrt{5}}{5}, -\frac{2\sqrt{5}}{5}\right) = -2\sqrt{5}$ is the minimum. Both exist because the ellipse is a closed and bounded set.

27. $f(2\sqrt{2}, \sqrt{2}) = f(-2\sqrt{2}, -\sqrt{2}) = 4$ is the maximum and $f(2\sqrt{2}, -\sqrt{2}) = f(-2\sqrt{2}, \sqrt{2}) = -4$ is the minimum. Both exist because the ellipse is a closed and bounded set.

29. $f\left(\frac{32}{\sqrt{21}}, \frac{8}{\sqrt{21}}, \frac{2}{\sqrt{21}}\right) = 2\sqrt{21}$ is the maximum and $f\left(-\frac{32}{\sqrt{21}}, -\frac{8}{\sqrt{21}}, -\frac{2}{\sqrt{21}}\right) = -2\sqrt{21}$ is the minimum. Both exist because the ellipsoid is a closed and bounded set.

31. The maxima and minima occur at the eight points $\left(\pm\frac{8}{\sqrt{3}}, \pm\frac{4}{\sqrt{3}}, \pm\frac{2}{\sqrt{3}}\right)$. When an even number of the signs are negative the function has a maximum of $\frac{64}{3\sqrt{3}}$. When an odd number of the signs are negative the function has a minimum of $-\frac{64}{3\sqrt{3}}$. Both exist because the ellipsoid is a closed and bounded set.

33. $(1, 0)$ and $(0, 1)$

35. $\left(\frac{1}{4}, \frac{1}{4}\right)$

37. $(1, 0)$ and $(0, 1)$

39. $(1, 1, 1), (-1, -1, 1), (-1, 1, -1), (1, -1, -1)$

41. $\left(\frac{ad}{a^2+b^2+c^2}, \frac{bd}{a^2+b^2+c^2}, \frac{cd}{a^2+b^2+c^2}\right)$

43. $\left(-\frac{5}{7}, \frac{39}{7}, \frac{15}{7}\right)$

45. The maximum of 8 occurs at each corner of the square and the minimum is $0 = f(0, 0)$.

47. The maximum is $1 = f(1, 0) = f(0, 1) = f(-1, 0)$. The minimum is $-1 = f(0, -1)$.

49. The maximum is $1 = f(1, 1)$. The minimum is $-1 = f(-1, 1)$.

51. $x = \frac{d}{3a}, y = \frac{d}{3b}, z = \frac{d}{3c}$

53. $\left(\frac{6}{5}, \frac{12}{5}, \frac{6\sqrt{5}}{5}\right)$

55. The base should be a square $5\sqrt[3]{2/5}$ feet on each side. The height should be $\sqrt[3]{2/5}$ feet.

57. Maximize $f(x, y) = xy$ subject to the constraint that $g(x, y) = 2x + 2y - P = 0$.

59. The distance is $d = \sqrt{x^2 + y^2 + z^2}$ and the square of the distance is $D = x^2 + y^2 + z^2$, $\nabla d = \frac{x}{\sqrt{x^2+y^2+z^2}}\mathbf{i} + \frac{y}{\sqrt{x^2+y^2+z^2}}\mathbf{j} + \frac{z}{\sqrt{x^2+y^2+z^2}}\mathbf{k}$ and $\nabla D = 2x\mathbf{i} + 2y\mathbf{j} + 2z\mathbf{k}$. The system of equations $\nabla d = \lambda\nabla g$ and $\nabla D = \lambda\nabla g$ have the same solutions.

61. Rotating, scaling or translating the point and the circle does not change the essential geometry of the system, if $P = (\alpha, \beta)$ is not the center of the circle (a, b), the point closest to P on the circle will be on the line containing both P and (a, b).

63. Rotating, scaling or translating the point and the sphere does not change the essential geometry of the system, if $P = (\alpha, \beta, \gamma)$ is not the center of the sphere (a, b, c), the point closest to P on the sphere will be on the line containing both P and (a, b, c).

65. Let $A > 0$, use the method of Lagrange multipliers to show that the function $f(x, y, z) = \sqrt[3]{xyz}$ has its maximum value when $x = y = z = A$, subject to the constraint equation $g(x, y, z) = \frac{1}{3}(x + y + z) - A = 0$. Thus, $\sqrt[3]{xyz} \leq \frac{1}{3}(x + y + z)$ when x, y and z are all positive.

For answers to odd-numbered Skill Certification exercises in the Chapter Review, please visit the Book Companion Web Site at www.whfreeman.com/tkcalculus.

Chapter 13

Section 13.1

1. T, T, F, T, F, T, T, T.
3. $\sum_{j=3}^{4}\sum_{k=2}^{4} je^{k^2}$.
5. 176
7. 3360
9. A definite integral is the limit of a Riemann sum of function of a single variable $f(x)$ over a closed interval $[a, b]$ as the number of subintervals goes to ∞. A double integral is the limit of a double Riemann sum of function of a two variables $f(x, y)$ over a rectangular region that has be subdivided into smaller subrectangles in both the x and y directions as the numbers of pieces in those two directions goes to ∞.
11. For each subrectangle $\mathcal{R}_{jk}$ choose $(x_j^*, y_k^*) = \left(\frac{x_{j-1}+x_j}{2}, \frac{y_{k-1}+y_k}{2}\right)$.
13. See Theorem 13.7.
15. To evaluate the integral first find an antiderivative for f with respect to y. You use the FTC to evaluate the inner integral by evaluating this function at d and c and evaluating the difference. The result will be the definite integral of a function of the single variable x. Use the FTC to evaluate this integral, if possible.
17. 91
19. The first step in the integration will be fine, but the resulting integral will be one without a simple antiderivative. Say you integrate first with respect to y. You will obtain $\int_{\pi/4}^{\pi/2} \frac{1}{x}\left(\sin(\pi x) - \sin\left(\frac{\pi}{2}x\right)\right) dx$. The function $\frac{1}{x}\left(\sin(\pi x) - \sin\left(\frac{\pi}{2}x\right)\right)$ does *not* have a simple antiderivative.
21. $\int_0^1\int_0^1 \frac{x-y}{(x+y)^3} dy\,dx = \frac{\pi}{2}$ and $\int_0^1\int_0^1 \frac{x-y}{(x+y)^3} dx\,dy = -\frac{\pi}{2}$. This does not violate Fubini's Theorem because the function is not continuous everywhere on the rectangle $\mathcal{R} = \{(x, y)|0 \leq x \leq 1 \text{ and } 0 \leq y1\}$.
23. 20
25. −48
27. 1260
29. 15
31. 0
33. 15
35. 0
37. 26
39. $\frac{340}{3}$
41. 0
43. $\frac{4-\ln 5}{\ln 5}$
45. $\frac{9}{2} - 8\ln 2 + \frac{7}{2}\ln 7$
47. $\frac{1}{2}$
49. 18
51. $\frac{2}{\pi}$
53. $\frac{1}{2}$
55. $-\frac{1175650}{3}$
57. $\frac{e^{18}-e^{8}-10}{4}$
59. $\frac{169}{2}$
61. 4
63. $4\ln 5 + 12\ln 3$
65. 2.288
67. 1.386
69. Approximately 64533.3 cubic feet
71. The result follows from the commutative property of addition.
73. By Fubini's Theorem
$$\iint_{\mathcal{R}} g(x)h(y)dA = \int_a^b\int_c^d g(x)h(y)dy\,dx = \int_a^b g(x)\int_c^d h(y)dy\,dx = \left(\int_a^b g(x)dx\right)\left(\int_c^d h(y)dy\right).$$
75. Use the result of Exercise 73. The product $(b - a)(d - c)$ is the area of rectangle $\mathcal{R}$.

Section 13.2

1. T, F, F, T, T, T, F, T.
3. A type I region is bounded below by a function $g_1(x)$, above by a function $g_2(x)$, on the left by $x = a$ and on the right by $x = b$, for constants $a < b$. A type II region is bounded on the left by a function $h_1(y)$, on the right by a function $h_2(y)$, above by $y = d$ and below by $y = c$, for constants $c < d$.
5. Correct
7. Reverse dx and dy.
9. Correct
11. Correct
13. As a type I region, Ω is bounded below by the function $y = \frac{1}{2}x$ and above by the function $y = \sqrt{x}$ on the interval $[a, b] = [0, 4]$. As a type II region, Ω is bounded below on the left by the function $x = y^2$ and on the right by the function $x = 2y$ on the interval $[c, d] = [0, 2]$.
15. The area of $\Omega = \int_a^b\int_{g_1(x)}^{g_2(x)} dy\,dx = \int_a^b [y]_{g_1(x)}^{g_2(x)} dx = \int_a^b (g_2(x) - g_2(x))\,dx$.
17. $\frac{4}{3}$
19. $\int_{-3}^{0}\int_{-x^3}^{0} dy\,dx + \int_0^3\int_0^{x^3} dy\,dx = \int_{-27}^{0}\int_{-3}^{\sqrt[3]{y}} dx\,dy + \int_0^{27}\int_{\sqrt[3]{y}}^{3} dx\,dy = 0$

21. (a) $\int_0^1 \int_0^{e^x} f(x, y)\,dy\,dx$,
(b) $\int_0^1 \int_0^1 f(x, y)\,dx\,dy + \int_1^e \int_{\ln y}^1 f(x, y)\,dx\,dy$

23. (a) $\int_{-2}^2 \int_0^{\sqrt{4-x^2}} f(x, y)\,dy\,dx + \int_{-2}^0 \int_{-\sqrt{4-x^2}}^0 f(x, y)\,dy\,dx$,
(b) $\int_{-2}^2 \int_{-\sqrt{4-y^2}}^0 f(x, y)\,dx\,dy + \int_0^2 \int_0^{\sqrt{4-y^2}} f(x, y)\,dx\,dy$

25. (a) $\int_0^{\pi/4} \int_{\sin x}^{\cos x} f(x, y)\,dy\,dx$,
(b)
$\int_0^{\sqrt{2}/2} \int_0^{\sin^{-1} y} f(x, y)\,dx\,dy + \int_{\sqrt{2}/2}^1 \int_0^{\cos^{-1} y} f(x, y)\,dx\,dy$

27. (a) $\int_{-1}^0 \int_{-x-1}^{x+1} f(x, y)\,dy\,dx + \int_0^1 \int_{x-1}^{-x+1} f(x, y)\,dy\,dx$,
(b) $\int_{-1}^0 \int_{-y-1}^{y+1} f(x, y)\,dx\,dy + \int_0^1 \int_{y-1}^{-y+1} f(x, y)\,dx\,dy$

29. $\int_0^1 \int_1^{e^y} f(x, y)\,dx\,dy + \int_1^e \int_1^2 f(x, y)\,dx\,dy +$
$\int_e^{e^2} \int_{\ln y}^2 f(x, y)\,dx\,dy$

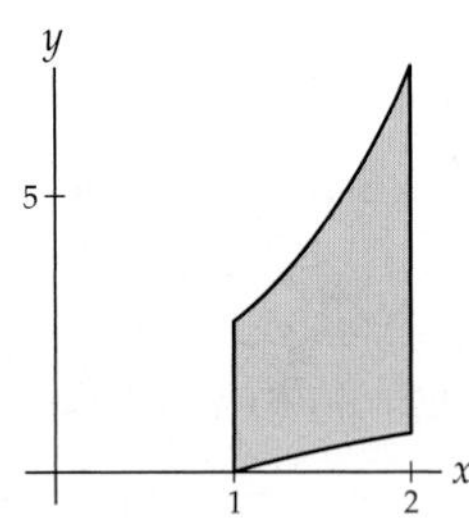

31. $\int_0^1 \int_{\sin^{-1} x}^{\pi/2} f(x, y)\,dy\,dx$

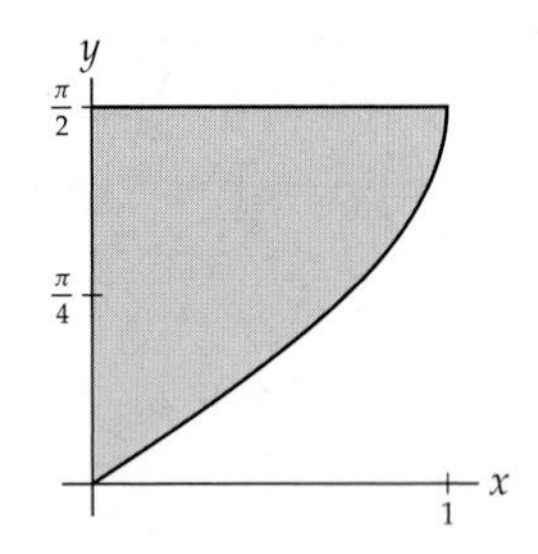

33. $\int_0^1 \int_0^{\sin^{-1} x} f(x, y)\,dy\,dx$

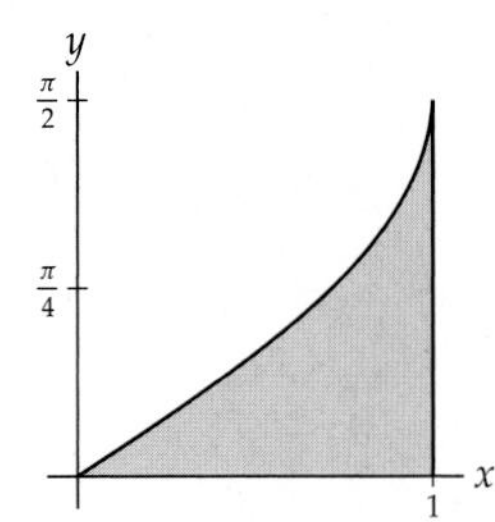

35. $\frac{e^2}{4} + 10e - \frac{49}{4}$

37. 4π

39. 0

41. 4

43. $\frac{26}{3}$

45. $\frac{2}{9}(56\sqrt{7} - 1)$. The function $\sqrt{1 + x^3}$ does not have a simple antiderivative.

47. 1. The function $\sec y$ has a simpler antiderivative when integrated with respect to x than it does when integrated with respect to y.

49. $-\frac{927}{8}$

51. $\frac{101027}{90}$

53. $\frac{\pi}{8}$

55. $\frac{1}{2}(e - 1)$

57. 2

59. $\frac{1}{3}(e^8 - 1)$

61. $\frac{3}{110}$

63. 200 cubic meters

65. Using the notation of Definition 13.4, the equality holds since $\iint_{\mathcal{R}} \alpha f(x, y)\,dA = \lim_{\Delta \to 0} \sum_{j=1}^{m} \sum_{k=1}^{n} \alpha f(x_j^*, y_k^*)\Delta A =$
$\alpha \lim_{\Delta \to 0} \sum_{k=1}^{n} \sum_{j=1}^{m} f(x_j^*, y_k^*)\Delta A = \alpha \iint_{\mathcal{R}} f(x, y)dA$.

67. Let $\mathcal{R} = \{(x, y) | a \le x \le b \text{ and } c \le y \le d\}$ be a rectangle containing Ω. From Exercise 65, $\iint_{\mathcal{R}} \alpha F(x, y)\,dA = \alpha \iint_{\mathcal{R}} F(x, y)dA$, where $F(x, y) = f(x, y)$ on Ω and is zero everywhere else in $\mathcal{R}$. Thus, $\alpha F(x, y) = \alpha f(x, y)$ for every point in Ω and is zero everywhere else in $\mathcal{R}$, also, so the statement holds.

69. The volume is given by the integral
$\int_0^a \int_0^{-b/ay+b} \left(c - \frac{c}{a}x - \frac{c}{b}y\right)\,dy\,dx = \frac{1}{6}abc$.

Section 13.3

1. T, T, F, F, F, F, T, T.

3. The values $\theta = \alpha$, $\theta = \beta$, $r = a$ and $r = b$ provide constant boundaries for the region.

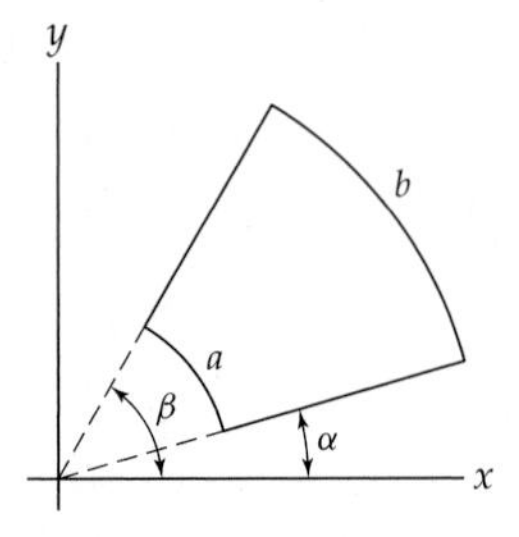

5.

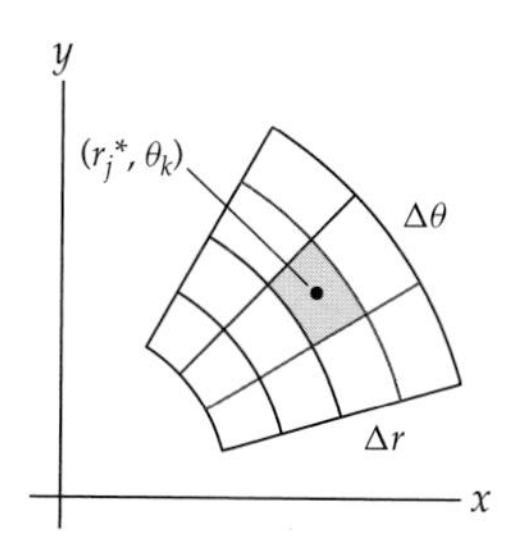

7. In the rectangular coordinate system the integral represents the area between the graphs of $f_1(\theta)$ and $f_2(\theta)$ on the interval $[\alpha, \beta]$. In a polar system, the integral represents the volume of the solid bounded above by the function $f(r, \theta) = \frac{1}{r}$ on the polar region bounded by the two functions between the rays $\theta = \alpha$ and $\theta = \beta$.

9. Because the curve is traced twice on the interval $[0, 2\pi]$.

11. For any positive integer n, the integral represents the area of a sector of a circle whose central angle is $\frac{2\pi}{n}$. This area would be $\frac{\pi R^2}{n}$. The equation holds for every nonzero value of n.

13. $2\pi \int_0^b x f(x)\, dx$

15. $\int_{-b}^{b} \int_{-\sqrt{b^2-x^2}}^{\sqrt{b^2-x^2}} f(\sqrt{x^2+y^2})\, dy\, dx$

17. From Exercise 16 we have the integral $\int_0^{2\pi} \int_0^b r f(r)\, dr\, d\theta = \left(\int_0^{2\pi} d\theta\right)\left(\int_0^b r f(r)\, dr\right) =$ $2\pi \int_0^b r f(r)\, dr$, which equals the answer to Exercise 14.

19. $\int_0^{2\pi} \int_a^b r f(r) dr d\theta$

21. $\frac{3\pi}{2}$

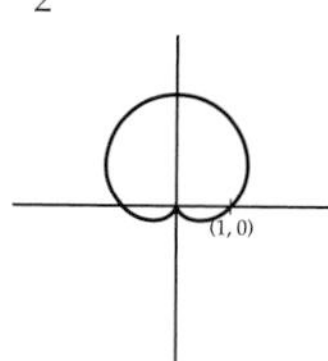

23. $\frac{9\pi}{2}$

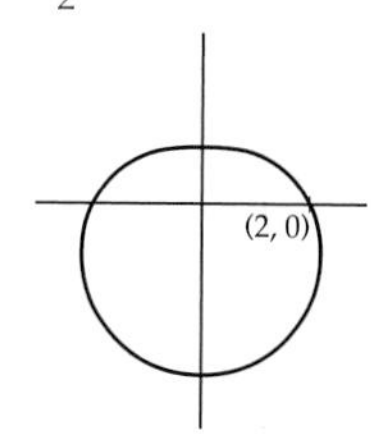

25. $\frac{\pi}{4}$

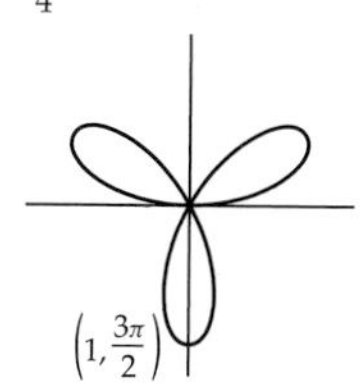

27. The area between the inner and outer loops of the limaçon is $\frac{3+\pi}{2}$

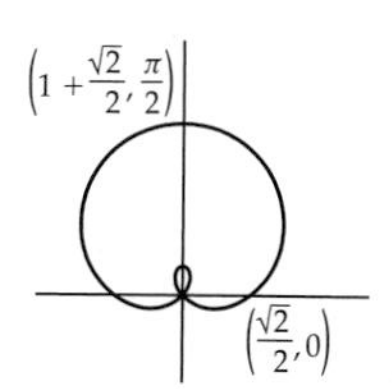

29. $\int_0^{\pi} \int_0^{\theta} r\, dr\, d\theta = \frac{\pi^3}{6}$

31. $2\int_0^{3\pi/4} \int_0^{1+\sqrt{2}\cos\theta} r\, dr\, d\theta - 2\int_{\pi}^{5\pi/4} \int_0^{1+\sqrt{2}\cos\theta} r\, dr\, d\theta =$ $3 + \pi$

33. $2\int_{-\pi/2}^{\pi/6} \int_{1+\sin\theta}^{3-3\sin\theta} r\, dr\, d\theta = 8\pi + 9\sqrt{3}$

35. $2\int_0^{\pi/3} \int_{\sec\theta/2}^{1} r\, dr\, d\theta = \frac{\pi}{3} - \frac{\sqrt{3}}{4}$

37. $\int_0^{2\pi} \int_0^{1+k\sin\theta} r\, dr\, d\theta = \pi\left(1 + \frac{k^2}{2}\right)$. When $k = 0$, "the limaçon" is a circle with radius one.

39. A sphere of radius 4 centered at the origin, $\frac{256}{3}\pi$.

41. The portion of the hyperboloid $z = 4 - r^2$ above the xy-plane, 8π.

43. The portion of the cone $z = 6 - 2r$ above the xy-plane, 18π.

45. The cylinder formed by the outer loop of the limaçon $r = \frac{\sqrt{2}}{2} + \sin\theta$ with height 1, $\frac{3}{4}(1+\pi)$.

47. $\int_0^{2\pi} \int_0^{2\sqrt{2}} (16 - 2r^2) r\, dr\, d\theta = 64\pi$

49. $\int_0^{2\pi} \int_0^{\sqrt{2}/2R} (\sqrt{R^2 - r^2} - r) r\, dr\, d\theta = \frac{\pi R^3}{3}(2 - \sqrt{2})$

51. $\int_0^{2\pi} \int_0^{4/5} \left(\sqrt{1 - r^2} - \frac{3}{5}\right) r\, dr\, d\theta = \frac{52}{375}\pi$

53. $\int_0^{2\pi} \int_0^{\sqrt{3}/2} (2\sqrt{1 - r^2} - 1) r\, dr\, d\theta = \frac{5}{12}\pi$

55. $\int_0^{2\pi} \int_0^{h} (h - r)\, r\, dr\, d\theta = \frac{1}{3}\pi h^3$

57. $\frac{9\pi}{8}$

59. $\frac{\pi}{4} e^{16}$

61. Approximately 14 wolves.

63. $\int_0^{\pi} \int_0^{2R\cos\theta} r\, dr\, d\theta = \pi R^2$

65. $2\int_0^{2\pi} \int_0^{R} \sqrt{R^2 - r^2} r\, dr\, d\theta = \frac{4}{3}\pi R^3$

67. $\int_0^{\phi} \int_0^{R} r\, dr\, d\theta = \frac{1}{2}\phi R^2$

69. $\int_0^{2\pi} \int_0^{\cos 2n\theta} r\, dr\, d\theta = \frac{\pi}{2}$ for every positive integer n

Section 13.4

1. F, F, F, T, F, F, T, T.

3. The area of Ω. The units are square centimeters.

5. The coordinates of the centroid of Ω. The units are centimeters.

7. The first moments M_y and M_x. The units are grams·centimeters.

9. The second moments I_y and I_x. The units are grams·square centimeters.

11. The radii of gyration R_y and R_x. The units are centimeters.

13. $\int_1^2 \int_{-x+2}^{2x-1} x\,dy\,dx = \int_1^2 \left[xy\right]_{-x+2}^{2x-1} dx =$
$\int_1^2 (3x^2 - 3x)\,dx = \left[x^3 - \frac{3}{2}x^2\right]_1^2 = (8-6) - \left(1 - \frac{3}{2}\right) = \frac{5}{2}.$

15. This is the same integral as in Exercise 13, with an extra factor of k.

17. $\int_1^2 \int_{-x+2}^{2x-1} kxy\,dy\,dx = \frac{k}{2}\int_1^2 x\left[y^2\right]_{-x+2}^{2x-1} dx$
$= \frac{k}{2}\int_1^2 (3x^3 - 3x)\,dx = \frac{k}{2}\left[\frac{3}{4}x^4 - \frac{3}{2}x^2\right]_1^2$
$= \frac{k}{2}\left((12-6) - \left(\frac{3}{4} - \frac{3}{2}\right)\right) = \frac{27}{8}k.$

19. $\int_1^2 \int_{-x+2}^{2x-1} kxy^2\,dy\,dx = \frac{k}{3}\int_1^2 x\left[y^3\right]_{-x+2}^{2x-1} dx$
$= 3k\int_1^2 (x^4 - 2x^3 + 2x^2 - x)\,dx$
$= 3k\left[\frac{1}{5}x^5 - \frac{1}{2}x^4 + \frac{2}{3}x^3 - \frac{1}{2}x^2\right]_1^2$
$= 3k\left(\left(\frac{32}{5} - 8 + \frac{16}{3} - 2\right) - \left(\frac{1}{5} - \frac{1}{2} + \frac{2}{3} - \frac{1}{2}\right)\right) = \frac{28}{5}k.$

21. $I_y = \left(\frac{673}{175}\sqrt{2} - \frac{1}{5}\ln(\sqrt{2}+1)\right)k,$
$I_x = \left(\frac{821}{2100}\sqrt{2} + \frac{1}{20}\ln(\sqrt{2}+1)\right)k$

23. $\int_{\sec\theta}^{2\cos\theta} kr^3\cos\theta\,dr = \frac{k}{4}\left(16\cos^5\theta - \sec^3\theta\right),$
$\int_{-\pi/4}^{\pi/4} \frac{k}{4}\left(16\cos^5\theta - \sec^3\theta\right) d\theta$
$= \frac{k}{60}(157\sqrt{2} + 15\ln(\sqrt{2}-1))$

25. $\frac{2}{3}k$, where k is the constant of proportionality

27. $I_x = \frac{2}{15}k,\ I_y = \frac{2}{5}k,\ R_x = \frac{\sqrt{5}}{5},\ R_y = \frac{\sqrt{15}}{5}$, where k is the constant of proportionality

29. $\left(\frac{3}{4}, 0\right)$

31. $\left(\frac{5}{3}, 0\right)$

33. $\left(\frac{17}{10}, 0\right)$

35. $\frac{17}{6}k$, where k is the constant of proportionality

37. $I_x = \frac{49}{90}k,\ I_y = \frac{43}{5}k,\ R_x = 7\frac{\sqrt{255}}{255},\ R_y = \frac{\sqrt{21930}}{85}$, where k is the constant of proportionality

39. $\frac{1}{2}kb^2h$, where k is the constant of proportionality

41. $I_y = \frac{1}{4}kb^4h,\ I_x = \frac{1}{6}kb^2h^3,\ R_y = \frac{\sqrt{2}}{2}b,\ R_x = \frac{\sqrt{3}}{3}h$, where k is the constant of proportionality

43. $\left(\frac{3}{4}b, \frac{1}{2}h\right)$

45. $(0, 0)$

47. $(0, 0)$

49. $\frac{2}{3}\pi k$, where k is the constant of proportionality

51. $I_x = I_y = \frac{1}{5}k\pi,\ I_o = \frac{2}{5}k\pi;\ R_x = R_y = \frac{\sqrt{30}}{10};\ R_o = \frac{\sqrt{15}}{5}.$

53. $\frac{2}{3}k$, where k is the constant of proportionality

55. $I_y = \frac{4}{15}k,\ I_x = \frac{2}{15}k,\ R_y = \frac{\sqrt{10}}{5},\ R_x = \frac{\sqrt{5}}{5}$, where k is the constant of proportionality

57. $\left(\frac{3}{2\pi}, 0\right)$

59. The masses of Ω_1 and Ω_2 may be treated as point masses located at $(\bar{x}_1, \bar{y}_1)$ and $(\bar{x}_2, \bar{y}_2)$, respectively. The center of mass of this two-point system is given by $\bar{x} = \frac{m_1\bar{x}_1 + m_2\bar{x}_2}{m_1 + m_2}$ and $\bar{y} = \frac{m_1\bar{y}_1 + m_2\bar{y}_2}{m_1 + m_2}$.

61. $\left(\frac{5}{6}, \frac{5}{6}\right)$

63. $\left(\frac{b_1^2h_2 + b_2^2h_1 - b_1b_2h_1}{2(b_1h_2 + b_2h_1 - b_1h_1)}, \frac{b_1h_2^2 + b_2h_1^2 - b_1h_1^2}{2(b_1h_2 + b_2h_1 - b_1h_1)}\right)$

65. $\left(0, \frac{260}{57}\right)$

67. $\left(0, \frac{15}{14}a\right)$

69. Approximately 1.79

71. We assume that $0 < c < a$ and leave the case where $0 < a \le c$ to you. Using the result of Exercise 70 we have

$$\bar{x} = \frac{\int_0^c \int_0^{d/cx} x\,dy\,dx + \int_0^c \int_0^{d/c-a(x-a)} x\,dy\,dx}{ad/2} = \frac{a+c}{3}$$

and

$$\bar{x} = \frac{\int_0^c \int_0^{d/cx} y\,dy\,dx + \int_0^c \int_0^{d/c-a(x-a)} y\,dy\,dx}{ad/2} = \frac{d}{3}.$$

73. Use polar coordinates and the circle $r = a$, where a is a positive constant. The first moments are $M_x = \int_0^{2\pi}\int_0^a r^2\cos\theta\,dr\,d\theta = 0$ and $M_y = \int_0^{2\pi}\int_0^a r^2\sin\theta\,dr\,d\theta = 0$. Therefore, the origin is the centroid of the circle.

Section 13.5

1. F, T, F, T, T, F, T, F.

3. The two summations contain the same summands, although the summands are rearranged. By the commutative and associative properties of addition, they are equal.

5. For each subsolid $\mathcal{R}_{i,j,k}$ choose
$(x_i^*, y_j^*, z_k^*) = \left(\frac{x_{i-1}+x_i}{2}, \frac{y_{j-1}+y_j}{2}, \frac{z_{k-1}+z_k}{2}\right)$.

7. A triple integral is the limit of a Riemann sum over a rectangular solid $\mathcal{R}$ as the limit of the mesh of subdivisions of $\mathcal{R}$ goes to zero. An iterated triple integral is the result of three consecutive integrations in the order specified by the placement of the variables of integration.

9. $\int_a^b \int_{h_1(y)}^{h_2(y)} \int_{g_1(y,z)}^{g_2(y,z)} f(x,y,z)\,dx\,dz\,dy$,
$\int_a^b \int_{h_1(z)}^{h_2(z)} \int_{g_1(y,z)}^{g_2(y,z)} f(x,y,z)\,dx\,dy\,dz$

11. $\int_{-1}^3 \int_0^2 \int_2^7 \rho(x,y,z)dz\,dy\,dx$,
$\int_0^2 \int_{-1}^3 \int_2^7 \rho(x,y,z)dz\,dx\,dy$,
$\int_{-1}^3 \int_2^7 \int_0^2 \rho(x,y,z)\,dy\,dz\,dx$,
$\int_2^7 \int_{-1}^3 \int_0^2 \rho(x,y,z)\,dy\,dx\,dz$,
$\int_0^2 \int_2^7 \int_{-1}^3 \rho(x,y,z)\,dx\,dz\,dy$,
$\int_2^7 \int_0^2 \int_{-1}^3 \rho(x,y,z)\,dx\,dy\,dz$

13. $\int_0^2 \int_0^{4-2x} \int_0^{3-(3/2)x-(3/4)y} \rho(x,y,z)\,dz\,dy\,dx$,
$\int_0^4 \int_0^{2-(1/2)y} \int_0^{3-(3/2)x-(3/4)y} \rho(x,y,z)\,dz\,dx\,dy$,
$\int_0^2 \int_0^{3-(3/2)x} \int_0^{4-2x-(4/3)z} \rho(x,y,z)\,dy\,dz\,dx$,
$\int_0^3 \int_0^{2-(2/3)z} \int_0^{4-2x-(4/3)z} \rho(x,y,z)\,dy\,dx\,dz$,
$\int_0^4 \int_0^{4-(4/3)z} \int_0^{2-(1/2)y-(2/3)z} \rho(x,y,z)\,dx\,dz\,dy$,
$\int_0^3 \int_0^{3-(3/4)y} \int_0^{2-(1/2)y-(2/3)z} \rho(x,y,z)\,dx\,dy\,dz$

15. $\int_0^{-x-y+1} k(y^2+z^2)dz$
$= k\left(y^2(-x-y+1) + \frac{1}{3}(-x-y+1)^3\right)$,
$\int_0^{-x+1} k\left(y^2(-x-y+1) + \frac{1}{3}(-x-y+1)^3\right)dy$
$= \frac{k}{6}(x-1)^4$, $\int_0^1 \frac{k}{6}(x-1)^4\,dx = \frac{k}{30}$

17. $\int_0^{-x-y+1} k(x^2+y^2)dz = k(-x-+1)(x^2+y^2)$,
$\int_0^{-x+1} k(-x-+1)(x^2+y^2)\,dy =$
$\frac{k}{12}(7x^4 - 12x^3 + 12x^2 - 4x + 1)$,
$\int_0^1 \frac{k}{12}(7x^4 - 12x^3 + 12x^2 - 4x + 1)\,dx = \frac{k}{30}$

19. The volume of Ω. The units are cubic centimeters.

21. The first moment M_{yz}. The units are grams·centimeters.

23. The moment of inertia about the y-axis, I_y. The units are grams·square centimeters.

25. 36

27. $\frac{145}{48}$

29. $\frac{1}{3} + \frac{\pi}{8}$

31. 162

33. 8

35. The rectangular solid given by
$\mathcal{R} = \{(x,y,z) \mid 2 \le x \le 6, 0 \le y \le 5 \text{ and } -2 \le z \le 4\}$.

37. The "half-cube":

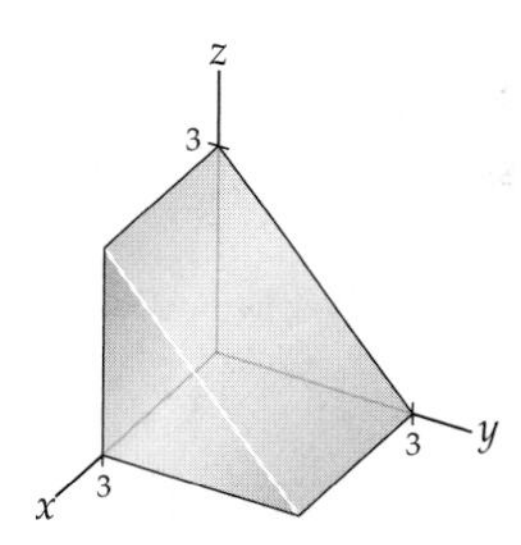

39. The tetrahedron with vertices (0, 0, 0), (2, 0, 0), (0, 3, 0) and (0, 0, 1).

41. A sphere with radius 3 centered at the origin.

43. The region bounded on the right by the paraboloid $x = 9 - y^2 - z^2$ and bounded on the left by the yz-plane.

45. $\int_{-2}^4 \int_0^5 \int_2^6 f(x,y,z)\,dy\,dx\,dz$

47. $\int_0^3 \int_0^{3-z} \int_0^3 f(x,y,z)\,dx\,dy\,dz$

49. $\int_0^2 \int_0^{3-(3/2)x} \int_0^{1-(1/2)x-(1/3)y} f(x,y,z)\,dz\,dy\,dx$

51. $\int_{-3}^{3}\int_{-\sqrt{9-z^2}}^{\sqrt{9-z^2}}\int_{-\sqrt{9-x^2-z^2}}^{\sqrt{9-x^2-z^2}} f(x,y,z)\,dy\,dx\,dz$

53. $3k$, where k is the constant of proportionality

55. $\frac{31412}{315}k$, where k is the constant of proportionality

57. (a) The centroid of a rectangular box is its midpoint.
(b) For example, $\bar{x} = \frac{\int_0^4\int_0^3\int_0^2 x\,dz\,dy\,dx}{\int_0^4\int_0^3\int_0^2 dz\,dy\,dx} = 2.$

59. (a) Since the density is proportional to the distance from the xy-plane, the location of the x- and y-coordinates will be the same as in Exercise 57. (b) $\bar{z} = \frac{4}{3}$

61. (a) The centroid of a rectangular box is its midpoint.
(b) For example, $\bar{x} = \frac{\int_{a_1}^{a_2}\int_{b_1}^{b_2}\int_{c_1}^{c_2} x\,dz\,dy\,dx}{\int_{a_1}^{a_2}\int_{b_1}^{b_2}\int_{c_1}^{c_2} dz\,dy\,dx} = \frac{a_1+a_2}{2}.$

63. (a) Since the density is proportional to the distance from the yz-plane, the location of the y- and z-coordinates will be the same as in Exercise 61. (b) $\bar{x} = \frac{2(a_2^3-a_1^3)}{3(a_2^2-a_1^2)}$

65. (a) $\bar{x} = \frac{a}{4}$. (b) The y-coordinate of the center of mass will also be one-fourth of the distance from the xz-plane to the point $(0, b, 0)$, i.e., $\bar{y} = \frac{b}{4}$. Similarly, $\bar{z} = \frac{c}{4}$.

67. The mass of the tetrahedron is
$m = \int_0^a\int_0^{b-(b/a)x}\int_0^{c(1-(x/a)-(y/b))} ky\,dz\,dy\,dx$, where k is a constant of proportionality. The center of mass is given by $(\bar{x}, \bar{y}, \bar{z})$, where
$\bar{x} = \frac{1}{m}\int_0^a\int_0^{b-(b/a)x}\int_0^{c(1-(x/a)-(y/b))} kxy\,dz\,dy\,dx,$
$\bar{y} = \frac{1}{m}\int_0^a\int_0^{b-(b/a)x}\int_0^{c(1-(x/a)-(y/b))} ky^2\,dz\,dy\,dx,$ and
$\bar{z} = \frac{1}{m}\int_0^a\int_0^{b-(b/a)x}\int_0^{c(1-(x/a)-(y/b))} kyz\,dz\,dy\,dx.$

69. By Fubini's Theorem $\iiint_{\mathcal{R}} \alpha(x)\beta(y)\gamma(z)\,dV$
$= \int_{a_1}^{a_2}\int_{b_1}^{b_2}\int_{c_1}^{c_2} \alpha(x)\beta(y)\gamma(z)dz\,dy\,dx$
$= \int_{a_1}^{a_2}\alpha(x)\int_{b_1}^{b_2}\beta(y)\int_{c_1}^{c_2}\gamma(z)dz\,dy\,dx$
$= \left(\int_{a_1}^{a_2}\alpha(x)\,dx\right)\left(\int_{b_1}^{b_2}\beta(y)\,dy\right)\left(\int_{c_1}^{c_2}\gamma(z)dz\right).$

71. Use the result of Exercise 69. The product $(a_2-a_1)(b_2-b_1)(c_2-c_1)$ is the volume of rectangular solid $\mathcal{R}$.

73. Using the notation of Definition 13.15, the equality holds since $\iiint_{\mathcal{R}} (f(x,y,z)+g(x,y,z))\,dV$
$= \lim_{\Delta\to 0}\sum_{i=1}^{l}\sum_{j=1}^{m}\sum_{k=1}^{n}\left(f\left(x_i^*, y_j^* .z_k^*\right)+g\left(x_i^*, y_j^* .z_k^*\right)\right)\Delta V$
$= \lim_{\Delta\to 0}\sum_{i=1}^{l}\sum_{j=1}^{m}\sum_{k=1}^{n} f(x_i^*, y_j^* .z_k^*)\Delta V +$
$\lim_{\Delta\to 0}\sum_{i=1}^{l}\sum_{j=1}^{m}\sum_{k=1}^{n} g(x_i^*, y_j^* .z_k^*)\Delta V$
$= \iiint_{\mathcal{R}} f(x,y,z)\,dV + \iiint_{\mathcal{R}} g(x,y,z)\,dV.$

Section 13.6

1. T, T, F, T, T, T, F, F.

3. $x = x_0$ is a plane parallel to the yz-plane, $y = y_0$ is a plane parallel to the xz-plane and $z = z_0$ is a plane parallel to the xy-plane.

5. $\rho = \rho_0$ is a sphere centered at the origin, $\theta = \theta_0$ is a vertical half-plane starting at the z-axis, and $\phi = \phi_0$ is a cone whose axis of symmetry is the z-axis.

7. $r = \sqrt{x^2+y^2}$, $\theta = \tan^{-1}\left(\frac{y}{x}\right)$, $z = z$.

9. $\rho = \sqrt{x^2+y^2+z^2}$, $\theta = \tan^{-1}\left(\frac{y}{x}\right)$,
$\phi = \cos^{-1}\left(\frac{z}{\sqrt{x^2+y^2+z^2}}\right).$

11. $\rho = \sqrt{r^2+z^2}$, $\theta = \theta$, $\phi = \tan^{-1}\left(\frac{r}{z}\right)$.

13. $dx\,dy\,dz$, $dx\,dz\,dy$, $dy\,dx\,dz$, $dy\,dz\,dx$, $dz\,dx\,dy$, $dz\,dy\,dx$. You make your choice based on the shape of the region of integration.

15. $\rho^2\sin\phi\,d\rho\,d\theta\,d\phi$. Since ρ is usually expressed as a function of θ and ϕ, this usually provides the simplest order of integration.

17. The region of integration Ω usually determines the coordinate system you choose. You usually start with the coordinate system that provides the simplest expression for Ω.

19. $\int_0^{\pi/2}\int_0^{\pi/2}\int_0^1 k\rho^2\sin\phi\,d\rho\,d\theta\,d\phi$
$= \frac{k}{3}\int_0^{\pi/2}\int_0^{\pi/2}\sin\phi\,d\theta\,d\phi = \frac{\pi k}{6}\int_0^{\pi/2}\sin\phi\,d\phi = \frac{\pi k}{6}.$

21. $M_{xz} = \int_0^{\pi/2}\int_0^{\pi/2}\int_0^1 k\rho^3\sin^2\phi\sin\theta\,d\rho\,d\theta\,d\phi$

23. Cylindrical $(1, 0, 0)$, spherical $\left(1, 0, \frac{\pi}{2}\right)$

25. Rectangular $(2\sqrt{3}, 6, 4)$, spherical $\left(8, \frac{\pi}{3}, \frac{\pi}{3}\right)$

27. Rectangular $(0, 0, -8)$, cylindrical $(0, 0, -8)$

29. A plane parallel to the yz-plane, cylindrical $r = 4\sec\theta$, spherical $\rho = 4\csc\phi\sec\theta$.

31. A right circular cylinder centered on the z-axis, rectangular $x^2+y^2 = 4$, spherical $\rho = 2\csc\phi$.

33. The half of the yz-plane in which $y \geq 0$.

35. A sphere centered at the origin with radius 2, rectangular $x^2+y^2+z^2 = 4$, cylindrical $r^2+z^2 = 4$.

37. The xy-plane, rectangular $z = 0$, cylindrical $z = 0$.

39. The region bounded below by the $r\theta$-plane and bounded above by the cone $z = r$ on the circle with radius 3 centered at the origin.

41. The region inside the right circular cylinder with equation $r = 2\sin\theta$, bounded below by the $r\theta$-plane and bounded above by the sphere with radius 4 centered at the origin.

43. The hemisphere below the xy-plane of the sphere with radius 2 centered at the origin.

45. The region inside the upward opening right circular cone that has the z-axis as its axis of symmetry, height 3, and base with radius 3.

47. $\dfrac{48\pi-64}{9}$

49. $2\sqrt{3}\pi$

51. $\dfrac{\pi}{32}$

53. $\dfrac{16}{3}\pi$

55. $\frac{2}{3}\pi R^3(1-\cos\alpha)$. If $\alpha=0$ the solid is "empty," so the volume should be zero. If $\alpha=\pi$, the solid is a sphere, so the volume should be $\frac{4}{3}\pi R^3$.

57. $\dfrac{1920\pi-3328}{225}k$, where k is the constant of proportionality.

59. $(x,y,z)=\left(0,0,\dfrac{94}{15\sqrt{3}+40\pi}\right)$.

61. $\dfrac{7\pi}{1024}k$, where k is the constant of proportionality.

63. $\dfrac{16\pi}{3}k$, where k is the constant of proportionality.

65. $x=y=0,\ z=\frac{3}{8}R(\cos\alpha+1)$.

67. Approximately 31800 cubic feet

69. Since $x=r\cos\theta=a$, when we solve for r we have $r=a\sec\theta$.

71. Since $x=\rho\sin\phi\cos\theta=a$, when we solve for ρ we have $\rho=a\csc\phi\sec\theta$.

73. $$\int_0^{2\pi}\int_0^R\int_{hr/R}^h r\,dz\,dr\,d\theta=\int_0^{2\pi}\int_0^R\left(hr-\frac{h}{R}r^2\right)dr\,d\theta$$
$$=\int_0^{2\pi}\frac{1}{6}R^2h\,d\theta=\frac{1}{3}\pi R^2h.$$

75. $$\int_0^{\tan^{-1}(R/h)}\int_0^{2\pi}\int_0^{h\sec\phi}\rho^2\sin\phi\,d\rho\,d\theta\,d\phi$$
$$=\int_0^{\tan^{-1}(R/h)}\int_0^{2\pi}\frac{1}{3}h^3\frac{\sin\phi}{\cos^3\phi}d\theta\,d\phi$$
$$=\int_0^{\tan^{-1}(R/h)}\frac{2}{3}\pi h^3\frac{\sin\phi}{\cos^3\phi}d\phi=\frac{1}{3}\pi R^2h.$$

Section 13.7

1. T, T, T, T, T, F, T, F.

3. It is a scaling factor. For example, in the simplest case when the Jacobian is a constant, k, the area of the domain set Ω is k times the area of the target set Ω'.

5. $$\int_0^1\int_{-y+1}^{y+1}x^2y\,dx\,dy+\int_1^2\int_{y-1}^{y+1}x^2y\,dx\,dy+\int_2^3\int_{y-1}^{5-y}x^2y\,dx\,dy=\frac{4}{5}+\frac{17}{2}+\frac{97}{10}=19.$$

7. $x=\dfrac{u+v}{2},\ y=\dfrac{u-v}{2},\ \dfrac{\partial(x,y)}{\partial(u,v)}=-\dfrac{1}{2}$

9. $$\frac{1}{16}\int_1^5\int_{-1}^1(u+v)^2(u-v)\,dv\,du=19$$

11. It translates the region a units horizontally and b units vertically. The rectangle is translated to the rectangle in the uv-plane where $3\le u\le 4$ and $4\le v\le 6$.

13. It would not be particularly useful because it only translates the region and would not provide a simpler expression for the limits of integration.

15. It scales the region by a factor of a units horizontally and b units vertically. The rectangle is scaled to the rectangle in the uv-plane where $0\le u\le 3$ and $0\le v\le\frac{1}{2}$.

17. If $a<0$, in addition to a scaling factor, there is a reflection about the vertical axis, and if $b<0$ there is a reflection about the horizontal axis. The sign of the Jacobian changes if either a or b is negative, but it does not change if both are negative.

19. $x=(\sin\theta)u+(\cos\theta)v$ and $y=-(\cos\theta)u+(\sin\theta)v$, $\dfrac{\partial(x,y)}{\partial(u,v)}=1$. The transformation rotates points about the origin. It does not change the area of the region, therefore the Jacobian should be, and is, 1.

21. $$\iint_\Omega\frac{1}{(x+y)^2}dA=\int_0^1\int_{1-x}^{4-x}\frac{1}{(x+y)^2}dy\,dx+\int_1^4\int_0^{4-x}\frac{1}{(x+y)^2}dy\,dx=\ln 4$$

23. Area$(\Omega)=\dfrac{15}{2}$ square units and Area$(\Omega')=15$ square units, Area$(\Omega)=\dfrac{\partial(x,y)}{\partial(u,v)}\cdot$Area$(\Omega')$

25. The quotient $\frac{u}{v}=xy\cdot\frac{x}{y}=x^2$. Therefore, $x=\sqrt{\frac{u}{v}}$. Similarly the product $uv=xy\cdot\frac{y}{x}=y^2$. Therefore, $y=\sqrt{uv}$. So, $\dfrac{\partial(x,y)}{\partial(u,v)}=\begin{bmatrix}\frac{1}{2\sqrt{uv}} & \frac{1}{2}\sqrt{\frac{v}{u}}\\ -\frac{1}{2}\sqrt{\frac{u}{v^3}} & \frac{1}{2}\sqrt{\frac{u}{v}}\end{bmatrix}=\dfrac{1}{2v}$.

27. The rectangle with vertices $(0,0)$, $(2,2)$, $(1,3)$ and $(-1,1)$; $\dfrac{\partial(x,y)}{\partial(u,v)}=2$.

29. The region in the first quadrant bounded by the hyperbolas with equations $y=\frac{1}{x}$ and $y=\frac{4}{x}$ and lines $y=x$ and $y=9x$, $\dfrac{\partial(x,y)}{\partial(u,v)}=\dfrac{2u}{v}$.

31. The right half of the circle with radius two centered at the origin, $\dfrac{\partial(x,y)}{\partial(u,v)}=-u$.

33. 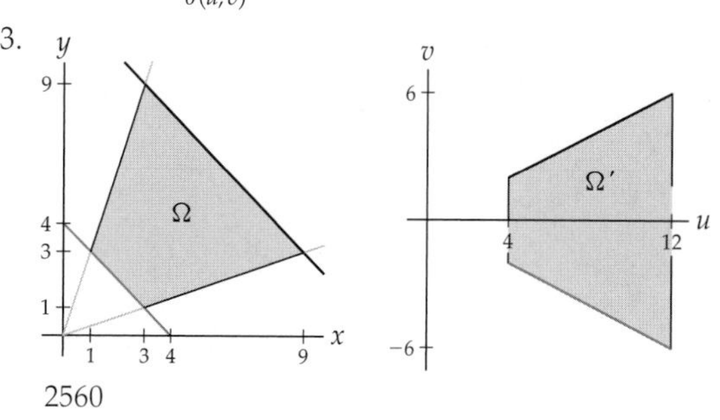

2560

35. 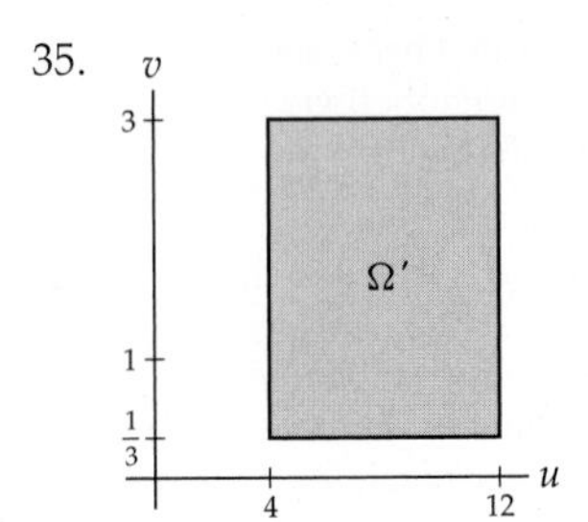

2560

37. 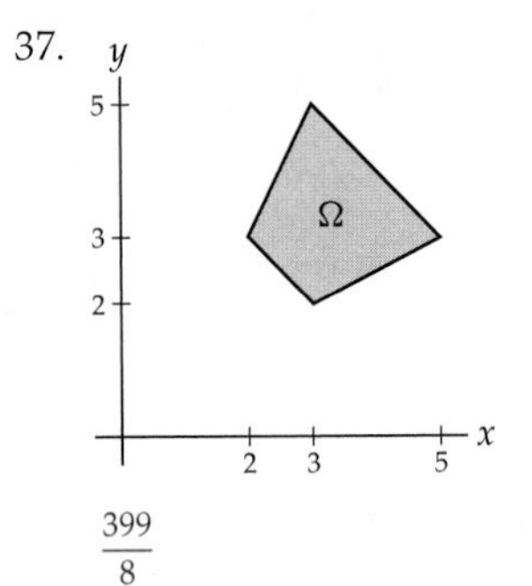

$\frac{399}{8}$

39. $\frac{81}{2}$

41. $\frac{21}{2}\ln 2$

43. $\frac{160}{3} + 6552\ln 3$

45. $\frac{15}{8}$

47. $\frac{5}{64}$

49. 1100 cubic feet.

51. $\frac{\partial(x,y,z)}{\partial(r,\theta,z)} = \det\begin{bmatrix} \cos\theta & \sin\theta & 0 \\ -r\sin\theta & r\cos\theta & 0 \\ 0 & 0 & 1 \end{bmatrix} = r.$

53. Using the transformation in the equation $ax + by = c$ we obtain $(a\alpha + b\gamma)u + (a\beta + b\delta)v = c$, which is a linear equation, if the coefficients of u and v are not both zero. This is the case when the Jacobian is nonzero.

55. $u = \frac{1}{\alpha\delta - \beta\gamma}(\delta x - \beta y)$ and $v = \frac{1}{\alpha\delta - \beta\gamma}(-\gamma x + \alpha y)$

57. Use the transformation given by $x = au$ and $y = bv$. You will transform the equation to the equation of the unit circle centered at the origin.

59. Show that $\det\begin{bmatrix} \frac{\partial x}{\partial u} & \frac{\partial y}{\partial u} \\ \frac{\partial x}{\partial v} & \frac{\partial y}{\partial v} \end{bmatrix} \det\begin{bmatrix} \frac{\partial u}{\partial s} & \frac{\partial v}{\partial s} \\ \frac{\partial u}{\partial t} & \frac{\partial v}{\partial t} \end{bmatrix} =$

$$\det\begin{bmatrix} \frac{\partial x}{\partial u}\frac{\partial u}{\partial s} + \frac{\partial x}{\partial v}\frac{\partial v}{\partial s} & \frac{\partial y}{\partial u}\frac{\partial u}{\partial s} + \frac{\partial y}{\partial v}\frac{\partial v}{\partial s} \\ \frac{\partial x}{\partial u}\frac{\partial u}{\partial t} + \frac{\partial x}{\partial v}\frac{\partial v}{\partial t} & \frac{\partial y}{\partial u}\frac{\partial u}{\partial t} + \frac{\partial y}{\partial v}\frac{\partial v}{\partial t} \end{bmatrix}.$$

For answers to odd-numbered Skill Certification exercises in the Chapter Review, please visit the Book Companion Web Site at www.whfreeman.com/tkcalculus.

Chapter 14

Section 14.1

1. F, T, F, T, T, F, F, T

3. A vector field in the Cartesian plane, or $\mathbb{R}^2$, has domain $D \subseteq \mathbb{R}^2$.

5. Vectors in the Cartesian plane.

7. A vector field $\mathbf{F}$ is conservative if it is the gradient of some function f.

9. For every point $(x, y) \in \mathbb{R}^2$, $\mathbf{F}(x, y)$ and $\mathbf{G}(x, y)$ are parallel; only for $(0, 0)$ are they the same vector.

11. For every point $(x, y) \in \mathbb{R}^2$, $\mathbf{F}(x, y) = -\mathbf{G}(x, y)$; at $(0, 0)$ they the same vector.

13. $\mathbf{G}(x, y, z) = \langle 2yz, 2xz, 2xy \rangle$

15. Let $\mathbf{F}(x, y) = \langle f_1(x, y), f_2(x, y) \rangle$. If $f_{1,y} \neq f_{2,x}$, then $\mathbf{F}$ is not conservative.

17. $f(x, y) = x^3 \cos y$

19. $g(x, y) = x^5 + xy - 3y^4$

21. $f(x, y, z) = xyz$

23. $g(x, y, z) = x\cos y + y\sin z$

25. 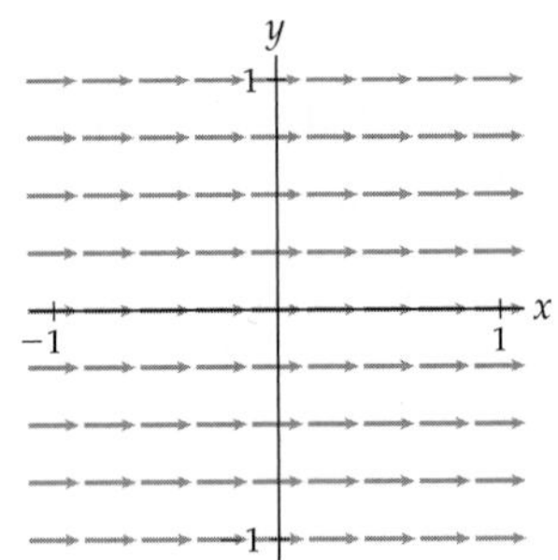

27. 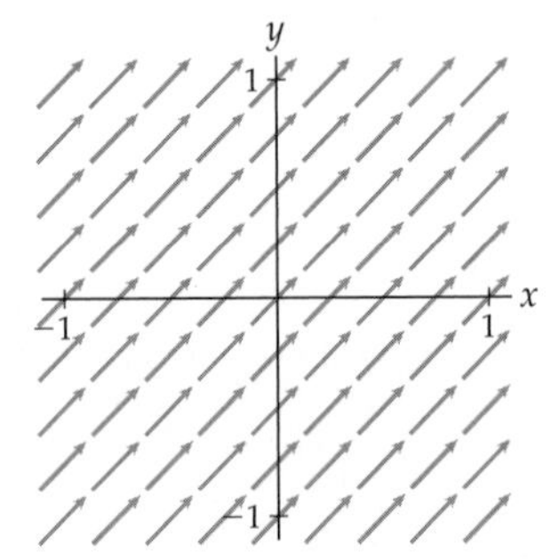

29. 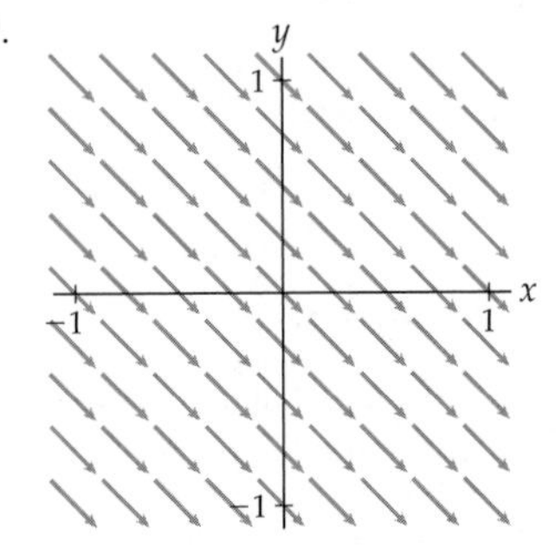

31.

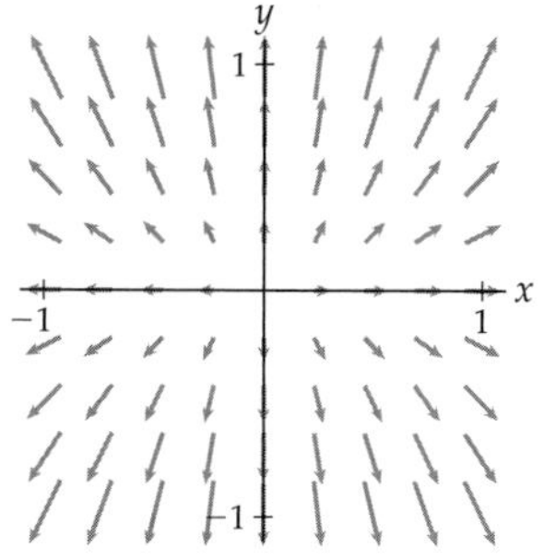

33. $\frac{\partial}{\partial y}(xy) = y \neq \frac{\partial}{\partial x}(-y) = 0$

35. $\frac{\partial}{\partial y}\left(\frac{1}{x^2+y}\right) = -(x^2+y)^{-2} \neq \frac{\partial}{\partial x}\left(\frac{y}{x}\right) = \frac{-y}{x^2}$

37. $\frac{\partial}{\partial z}(-z) = -1 \neq \frac{\partial}{\partial y}(e^{yz}) = ze^{yz}$

39. $\frac{\partial}{\partial z}(yz) = y \neq \frac{\partial}{\partial y}(z+12)$

41. Not conservative

43. $g(x, y) = x^2 + \sin(xy) - y$

45. $f(x, y, z) = xye^{2z} + x$

47. Not conservative

49. (a) Since the fluid is moving at a constant rate, $\mathbf{F}(x, y)$ should have a constant magnitude. Since the motion is horizontal only, there is no $\mathbf{j}$ component to the field. Since the motion is in the negative $\mathbf{i}$ direction, a reasonable field is $\mathbf{F}(x, y) = -\alpha\mathbf{i}$, for any positive constant α. (b) $\mathbf{F}(x, y) = \langle -\alpha, -\alpha \rangle$, for any real number $\alpha > 0$. (c) $\mathbf{G}(x, y) = -y\mathbf{i} + x\mathbf{j}$

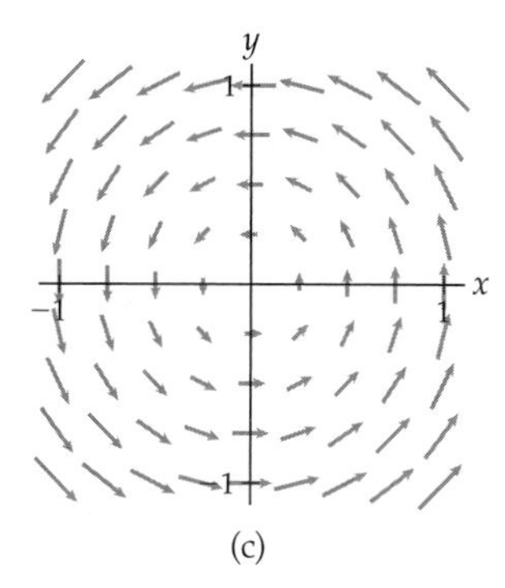

(c)

51. (b) It is conservative. Potential functions have the form $Q(x, y) = 0.9x - \frac{x^3}{3} + \frac{1}{2}xy^2 + k$. This function represents, in some sense, the amount of water that needs to get through a passage, or if you prefer, the pressure of the water that must pass through. (c) There are points of land due north and due south of this channel, forcing the water through a relatively narrower passage.

53. Suppose that $f(x, y)$ and $g(x, y)$ satisfy $\nabla f(x, y) = \nabla g(x, y)$. Then

$$f(x, y) = \int f_x(x, y)\,dx + B_f + \alpha_f$$

and $g(x, y) = \int g_x(x, y)\,dx + B_g + \alpha_g$, where B_f is the integral with respect to y of all the terms in $f_y(x, y)$ that do not involve an x term, B_g is the integral with respect to y of all the terms in $g_y(x, y)$ that do not involve an x term, and α_f, α_g are arbitrary constants. Since $g_y(x, y) = f_y(x, y)$, $B_f = B_g$, and $f(x, y) = g(x, y) + \alpha$.

Section 14.2

1. T, F, T, T, F, T, F, F

3. $\int_a^b e^{t^7}\sqrt{1 + 4t^2 + 16t^6}dt$

5. $\mathbf{F}(x, y) = (x + y)\mathbf{i} + xy\mathbf{j}$

7. $\mathbf{F}(x, y, z) = \langle xy^2, xy - z, \cos y \rangle$

9. $\int_C \mathbf{F}(x, y) \cdot d\mathbf{r} = \int_C (-y)\,dx + x\,dy$

11. $ds = \|\mathbf{r}'(t)\|\,dt$

13. $\int_a^b (4t^2e^t + 2te^t - e^{2t})\,dt$

15. $\mathbf{r}(t) = \cos t\mathbf{i} + \sin t\mathbf{j},\ 0 \le t \le 2\pi$

17. $\mathbf{r}(t) = \langle 2\cos t, t, 2\sin t \rangle,\ 0 \le t \le 4\pi$

19. $f_{avg}(C) = \frac{\int_C f(x,y,z)ds}{\int_C ds}$

21. $\frac{1}{12}(17^{3/2} - 1)$

23. $\mathbf{r}(t) = \langle \cos t, \sin t \rangle, 0 \le t \le 2\pi$; $\|\mathbf{r}(t)\| = 1$, $\int_C f(x, y)ds = 2\pi$.

25. $\mathbf{r}(t) = \langle t, 2t, 3t \rangle, 0 \le t \le 1$. Then $\int_C f\,ds = \sqrt{14}\int_0^1 e^{6t}dt = \frac{\sqrt{14}(e^6-1)}{6}$.

27. $\sqrt{2}e\,(e^{\pi} - 1)$

29. $\mathbf{r}(t) = \langle t, t^2 + 1 \rangle$, $\mathbf{r}'(t) = \langle 1, 2t \rangle$, $\int_C \mathbf{F}(x, y) \cdot d\mathbf{r} = \int_5^{10}(1 - 2t)\,dt = -70$

31. 70

33. 0 since $\mathbf{F}(x, y)$ is conservative and C is closed.

35. π

37. $\ln\pi - \pi - 1$

39. $\mathbf{F}$ is not conservative: $\frac{\partial f_1}{\partial z} = -1 \neq 1 = \frac{\partial f_3}{\partial x}$

41. $3^{32} - 1$, $\mathbf{F}(x, y, z) = \nabla(z^{xy})$; the integral is path independent.

43. $\mathbf{F}$ is not conservative: $\frac{\partial f_2}{\partial z} = 0 \neq 1 = \frac{\partial f_3}{\partial y}$

45. $\frac{7}{3}\pi^3$

47. $\|\mathbf{r}'(t)\| = \|\langle \sin t + t\cos t, \cos t - t\sin t \rangle\| = \sqrt{1 + t^2}$, $\int_C f(x, y)ds = \frac{1}{3}((16\pi)^{3/2} - 1)$

49. $\mathbf{F}(x, y, z) = \nabla f(x, y, z)$, where $f(x, y, z) = e^{xyz} + 2x - y$. The intersection of the surfaces is an ellipse, in particular a closed curve. So the integral is zero.

51. 4π

53. 112

55. -2π

57. $\frac{1}{3}(\ln 2)^3 + \ln 2 - \frac{31}{24}e^{-3}$

59. $\frac{\pi^2}{12}$

61. $2\pi e$

63. (a) No; (b) 0; (c) It is a measure of how much the current helps or hurts Annie's progress.

65. Suppose that $\mathbf{F}(x, y, z)$ is a conservative vector field. Then $\mathbf{F}(x, y, z) = \nabla f(x, y, z)$ for some $f(x, y, z)$. Let $\mathbf{r}(t)$, $a \le t \le b$ be a smooth parametrization of C. By the Fundamental Theorem of Line Integrals,

$$\int_C \mathbf{F}(x, y, z) \cdot d\mathbf{r}$$

$= f\left(x(b), y(b), z(b)\right) - f\left(x(a), y(a), z(a)\right)$, which depends only on the endpoints of C and not on C itself.

Section 14.3

1. F, T, T, T, F, T, F, T.

3. Surface integrals of multivariate functions only need to account for the distortion of the surface's area compared to a comparable region in the parameter plane (xy or uv); surface integrals of vector fields need to account for the action of $\mathbf{F}(x, y, z)$ through the surface, that is, in the normal direction.

5. $dS = \sqrt{\frac{a^2+b^2+c^2}{c^2}}$

7. $\mathbf{r}(x, y) = \langle x, y, \sqrt{1 - x^2 - y^2} \rangle$.

9. $\mathbf{r}(u, v) = \langle k \cos u \cos v, k \sin u \cos v, k \sin v \rangle, 0 \le u \le 2\pi, 0 \le v \le \pi$.

11. $\mathbf{n} = -f_x \mathbf{i} - f_y \mathbf{j} + \mathbf{k}$

13. $\mathbf{n} = 2t\mathbf{i} - \mathbf{j}$

15. $f_{avg}(S) = \frac{\int_S f(x,y,z)dS}{\int_S 1 dS}$

17. $\|\mathbf{r}_u \times \mathbf{r}_v\|$ is the area of the parallelogram with sides $\mathbf{r}_u$ and $\mathbf{r}_v$; the area of this parallelogram approximates the area of the portion of the tangent plane used to approximate the area of $\mathcal{S}$; in the limit, the analogy is exact.

19. $\mathbf{n} = \mathbf{i} - f_y \mathbf{j} - f_z \mathbf{k}$

21. $12\sqrt{2}$

23. $\frac{19\pi}{6}$

25. $2\sqrt{6}$

27. $\pi - \frac{\pi}{e}$

29. 0

31. $4e^2 + 14$

33. $\frac{20}{21}\sqrt{3}$

35. $-\frac{12}{5}$

37. 240π

39. $\mathbf{n} = \langle 2, -8, -10 \rangle$; $\int_S \mathbf{F} \cdot \mathbf{n}\, dS = -2 \int_\pi^\pi \int_0^1 4z \cos yz + 5z \sin yz\, dA = -20\pi$.

41. 0

43. $\int_S 1\, dS = \int_0^3 \int_0^{2x} 2\sqrt{x^2 + 1}\, dy\, dx = \frac{4}{3}(10^{3/2} - 1)$.

45. $\frac{61}{324} + \frac{2^{-2/3}}{\ln 2} - \frac{1}{2 \ln 2}$

47. -832π

49. $\frac{k\pi}{6}(17^{3/2} - 1)$

51. 3π

53. $\frac{596(37^{3/2}+48)}{5(37^{3/2}-5)}$

55. 0.0643 square miles

57. Reversing the orientation of $\mathcal{S}$ means substituting $-\mathbf{n}$ for $\mathbf{n}$. Factoring -1 out of the integral gives the desired result.

59. In both cases, the surface area is $\int 1 dS$ with limits determined by R. Computing dS for each surface, we see that they are equal. ($dS = \sqrt{4x^2 + 4y^2 + 1}$)

Section 14.4

1. F, F, T, T, T, T, F, F.

3. The terms in the integrand are the mixed partial derivatives of the component functions of the vector field.

5. Work done by $\mathbf{F}$ moving a particle along C, flux perpendicular to C.

7. Setting $G(x, y) = \frac{\partial f_2}{\partial x} - \frac{\partial f_1}{\partial y}$, allows the use of Green's Theorem to evaluate the integral.

9. At the point (x, y, z), div $\mathbf{F}$ will be negative.

11. At the point (x, y, z), div $\mathbf{F}$ will be positive.

13. $-\iint_R (\sin x \sin y + x^2 e^y) dA$, where R is the unit circle

15. $-\sin x \sin y - x^2 e^y$

17. $2xy + x \sin y$

19. 0

21. $\frac{y}{\sqrt{1-(xy)^2}} + \frac{1}{y+z} - \frac{5}{(2x+3y+5z+1)^2}$

23. $-\left(\frac{3}{(2x+3y+5z+1)^2} + \frac{1}{y+z}\right)\mathbf{i} + \left(\frac{2}{(2x+3y+5z+1)^2}\right)\mathbf{j} - \frac{x}{\sqrt{1-x^2y^2}}\mathbf{k}$

25. $\langle xze^{xy} - xye^{xz}, xye^{yz} - yze^{xy}, yze^{xz} - xze^{yz} \rangle$

27. $(\cos(x - y) + \sin(x + y))\mathbf{k}$

29. Using Green's Theorem we have $\int_0^{2\pi} \int_0^1 4r^3\, dr\, d\theta = 2\pi$.

31. $-\frac{32}{3}$

33. $\frac{1}{4}(e^2 - 2e + 3)$

35. 0. Note that $\mathbf{F}$ is not conservative, so this is not the result of applying the Fundamental Theorem of Line Integrals.

37. $\frac{1}{2}$

39. 16

41. $4 - 2e + (e^2 - 1)(\ln 2 - 1)$

43. $\frac{1}{\ln 2} - \frac{1}{e+1}$

45. 0

47. Work is given by $\int_C \mathbf{F} \cdot d\mathbf{r}$. Using Green's Theorem to evaluate the integral, we have $W = -2\pi$.

49. $\mathbf{F}$ is conservative, so $W = 0$.

51. (a) 0; (c) They tell us that there is no creation of water inside the passage. The water flowing in and out is balanced.

53. (a) Since speed increases as we move left among the lanes, we can say that $\frac{\partial v_2}{\partial x} < 0$, while the fact that $v_1 \equiv 0$ means that $\frac{\partial v_1}{\partial y} = 0$. This means that the curl is always negative – the traffic tends to have a rotation to it. If the curl is low, it means all lanes are running nearly the same speed, but if it is high, then the variation in speeds among the lanes is high. Generally, low curl is safer. (b) The curl in this case is $-\alpha$, so the area integral on the right is approximately -0.00568α. The curve integral on the left divides into four segments. Since $d\mathbf{r}$ goes in opposite directions on the top and bottom and the speeds are the same there, those segments cancel. The integral on the left segment is $\int_{0.5}^{0} 75\,dy$, while that on the right is $\int_0^{0.5}(75 - 0.0113\alpha)\,dy$. Thus the integral on the left is $\int_0^{0.5} -0.0113\alpha$, which ends up the same as the other side. (c) These describe the average curl for the traffic flow on this stretch of road, i.e., the circulation of traffic on this stretch. The higher the number, the less safe the road. The curve integral is what engineers would be able to compute using road tubes they put down to measure the number and speed of passing cars. In other words, they could put road tubes along $y = 0$ and $y = 0.5$ to compute the curve integral.

55. Directly computing div curl $\mathbf{F}$, we find that div curl $\mathbf{F}$ is

$$\frac{\partial^2 F_3}{\partial x \partial y} - \frac{\partial^2 F_2}{\partial x \partial z} + \frac{\partial^2 F_1}{\partial y \partial z} - \frac{\partial^2 F_3}{\partial y \partial x} + \frac{\partial^2 F_2}{\partial z \partial x} - \frac{\partial^2 F_1}{\partial z \partial y} = 0.$$

57. If $\mathbf{F}$ is a conservative vector field, then by the Fundamental Theorem of Line Integrals, $\int_C \mathbf{F} \cdot d\mathbf{r} = 0$ for any closed simple smooth or piecewise-smooth curve. By Green's Theorem, we have

$$\iint_R \left(\frac{\partial F_2}{\partial x} - \frac{\partial F_1}{\partial y}\right) dA = \int_C \mathbf{F} \cdot d\mathbf{r} = 0$$

Section 14.5

1. F, T, F, F, T, T, T, T.

3. $\left\langle -\frac{\partial g}{\partial x}, -\frac{\partial g}{\partial y}, 1\right\rangle$ and $\left\langle \frac{\partial g}{\partial x}, \frac{\partial g}{\partial y}, -1\right\rangle$

5. By Stokes' Theorem, both integrals are equivalent to $\int_C \mathbf{F}(x, y, z) \cdot d\mathbf{r}$. So they are equal to each other.

7. $SA = \int_C \mathbf{F}(x, y, z) \cdot d\mathbf{r}$

9. The orientation of $\mathcal{S}$ determines the direction of travel along C; in Stokes' Theorem, C is traversed counterclockwise with respect to n. If the orientation is reversed, the integrals will change sign.

11. Any conservative vector field will have $\int_C \mathbf{F} \cdot d\mathbf{r} = 0$, so any conservative vector field is an example of this.

13. Both notations refer to area. In the case of Green's Theorem, dA refers to ordinary area. In the case of Stokes' Theorem, dS keeps track of any distortion to the area of the surface $\mathcal{S}$ caused by transforming a patch of the uv-plane to parametrize $\mathcal{S}$.

15. Answers will vary. When the line integral has a piecewise-continuous boundary that requires the evaluation of several smooth pieces, but by using the right-hand side of Stokes' Theorem, we can obtain the same result with a single area integral.

17. Answers will vary. If a surface for an intended application of Stokes' Theorem is not smooth or piecewise smooth, there is no guarantee that the incremental application of Green's Theorem in the tangent planes is accurate, or even well defined.

19. $\frac{35}{12}$

21. 0

23. $\frac{1}{12}$

25. $\frac{16\pi}{81}$

27. $\frac{320}{3}$.

29. 8π

31. 36π

33. $\mathbf{F}$ is not defined on the z-axis.

35. $\mathcal{S}$ is not smooth; in particular, $\mathbf{n}$ is not well-defined at the cone point, $(0, 0, 0)$.

37. This vector field is conservative, so $\int_C \mathbf{F} \cdot d\mathbf{r} = 0$.

39. curl $\mathbf{F} = \mathbf{0}$

41. curl $\mathbf{F} = \mathbf{0}$

43. -60π

45. (a) 0; (b) $r = r(\theta) = \langle x(\theta), y(\theta)\rangle = \langle \cos\theta, \sin\theta\rangle$. Thus on ∂R we have $\mathbf{F} = \langle 0, 1.152 - 0.8\cos^2\theta\rangle$ and $d\mathbf{r} = \frac{d}{d\theta}\langle \cos\theta, \sin\theta\rangle d\theta = \langle -\sin\theta, \cos\theta\rangle d\theta$, hence $\int_{\partial R} \mathbf{F} d\mathbf{r} = \int_0^{2\pi}(1.152 - 0.9\cos^2\theta)\cos\theta\, d\theta = 0.$; (c) They tell us that there is no net circulation of water in this zone.

47. Using Stokes' Theorem to compute the integral gives $\mathbf{n} = \pm\mathbf{k}$, depending on the direction in which C is traversed. The $\mathbf{k}$ component of curl $\mathbf{F}$ does not depend on F_3, so $\int_C \mathbf{F} \cdot d\mathbf{r}$ does not depend on F_3.

49. Since $\mathbf{n} = \pm\langle a, b, c\rangle$ on all of the surface, and the partial derivatives of F_1, F_2, F_3 are all constants, curl $\mathbf{F} \cdot \mathbf{n}\, dS$ will be some constant r. Then $\int_C \mathbf{F} \cdot d\mathbf{r} = r \iint_S dA$, which depends only on area.

51. This is a direct consequence of Stokes' Theorem. By the theorem, both sides of the equation are equal to $\int_C \mathbf{F}(x, y, z) \cdot d\mathbf{r}$.

Section 14.6

1. T, T, F, T, F, F, T, T.

3. No. Such surfaces have no interior (or infinite interior).

5. On the surface of the paraboloid, there is always a well-defined choice of normal vector; intuitively speaking, the graph has no sharp corners. The rectangular solid, however, is (piecewise) comprised of smooth planes, but the intersections of these planes form sharp corners on the surface, which do not have well-defined normal vectors.

7. (a) positive (b) zero

9. $f_{xx} + f_{yy} + f_{zz}$

11. $G(x, y, z) = 3\mathbf{i} - 4\mathbf{j} + 57\mathbf{k}$.

13. $\langle x + yz, y + 2xz, z + xy\rangle$

15. Not possible: S is neither smooth nor piecewise-smooth.
17. Not possible: $\mathbf{F}(x, y, z)$is not defined throughout the region enclosed by $\mathcal{S}$.
19. Possible: this integral satisfies the conditions of the Divergence Theorem.
21. div $\mathbf{F} = 3$
23. div $\mathbf{F} = 3e^{xyz} + 3xyze^{xyz}$
25. $3\pi^{7}$
27. 0
29. 12π
31. 0
33. $\frac{21\pi}{2}$
35. 675π
37. $36\pi e$
39. $\frac{8}{3}$
41. $18(e^5 + \ln 5 - e) + 4320$
43. 16
45. (a) 0; (b) $\int_{\partial R} F \cdot n \, dS = \int_0^{2\pi} (1.152 - 0.9\cos^2\theta)\sin\theta \, d\theta = 0$; (c) They tell us that the same amount of water flows into the region as flows out. The fact that the divergence vanishes everywhere in the region shows that there are no sources or sinks of water in the region.
47. (a) No traffic can leave the roadway to either side, so this means that there is a net flux of cars into this stretch of road: traffic is slowing down. This condition could not persist indefinitely, since the road will fill up. (b) By the Divergence Theorem, it is the -0.0565 computed in part (a) divided by the area 0.0113 of the road; i.e., -5. The divergence here is $\frac{\partial v_2}{\partial y}$, which is the acceleration along the road. Since this is negative, it confirms that traffic is slowing down.
49. 12π
51. $a + b + c > 0$
53. Let $\mathcal{S}$ be the unit sphere and let $\mathbf{F}(x, y, z) = \langle -2xz, 2yz, xy \rangle$. Then $\mathbf{F}$ is not conservative, since $\frac{\partial f_1}{\partial z} = -2x \neq y = \frac{\partial f_3}{\partial x}$. However, div $\mathbf{F} = 0$, so by the Divergence Theorem, $\iint_{\mathcal{S}} \mathbf{F}(x, y, z) \cdot \mathbf{n} \, dS = 0$.

For answers to odd-numbered Skill Certification exercises in the Chapter Review, please visit the Book Companion Web Site at www.whfreeman.com/tkcalculus.

INDEX

absolute area, 375–376, 380–381
absolute convergence, 646
  defined, 645
  ratio test for, 647–648
absolute value, 39–40
  function, 39
absolute volume, 995
abuse of notation, 904, 1100
acceleration, 157, 864–865
  centripetal, 894, 896–897
  defined, 865
  normal components of, 893–894
  tangential components of, 893–894
accumulation function, 317–318
  area, 388–390
algebraic function, 35–43. *see also* absolute value function; polynomial function; power function; rational function
  defined, 35
algebraic rules
  of exponents, 49
  for logarithms, 49
alternating harmonic series, 642, 645
alternating series, 641–645
  defined, 641
  remainder, 644
alternating series test, 642, 645–646
annulus, 1018
anticommutative, 816
antiderivative, 252, 358, 559
  defined, 195, 354
  elementary, 358
  of exponential function, 356
  by guess-and-check, 358
  and indefinite integrals, 354–355
  of inverse trigonometric functions, 53
  of logarithm function, 422
  of power function, 355
  of trignometric functions, 356, 446
  vector function, 867
antidifferentiation, 355–358, 559
  approximating
  area under a curve, 329–330
  centroid, 546
  convergent series, 620–621
  length of curve, 526
  limits with tables, 133–134
  with rectangles, 329–330
  roots, 181–183
  square roots, 181
  surface area, 530
  volume, 500–501, 514–515
  volume of solid, 514–515
  volume with rectangular solid, 1061
arc length
  approximating, 526–527, 532–534
  defined, 573
  definite integral for, 527–528, 534–535
  of parametric curve, 712–713, 719–720
  parametrization, 882–886
  of polar curve, 750–751
  of vector curve, 882
arch, 719
Archimedes, spiral of, 737
arcsec*x*, *see* inverse secant function
arcsin *x*, *see* inverse sine function
arctan *x*, *see* inverse tangent function
area, 334, 378
  absolute, 376
  between curves, 377–378
  of circle, 328, 1022–1023
  elliptical annulus, 1076–1077
  function, 394
  line integral, 1101
  polar coordinates, 749
  polar curves, 749–750, 753–754
  sector, 749
  signed, 376
  triangle, 822
  between two graphs, 377
  under a curve, 342–344
area accumulation function, 388–390
arithmetic function operations, 19
arithmetic mean, 987
arithmetic sequence, 581
associated slope function, 156, 159–160, 170
astroid, 715, 716
asymptote, 7, 39, 81–82, 141, 273
average, 243
  radius, 515, 529
  rate of change, 7, 12–13, 382
  value, 378–379, 380
  velocity, 159, 162–163, 364

b^x, *see* exponential function
back-substitution, 414
ball
  open, 919
base, 49
base conversion formula, 49
bell curve, 353
Bessel function, 681
binomial coefficients, 678
binomial series, 677–678
binomial theorem, 151
binormal vector, 875
bisection method, 122, 182
bob, 150, 234
boundary
  defined, 922
  point, 922
  of set, 922
bounded
  sequence, 583–585, 600
  sets, 922
boundedness, 583, 601

calculating
  definite integrals, 368
  derivatives, 175–177, 187–196
  indefinite integrals, 355, 357, 359
  limits, 116, 132, 143, 144
Cancellation Theorem for limits, 125
cardioid, 739, 740
  length, 757
carrying capacity, 564
Cartesian coordinate system, 778–787
Cartesian plane, 4
catenary, 228
Cauchy Mean Value Theorem, 304
center of mass, 545, 1035, 1047
  and first moments, 1030–1032
  pyramid, 1050–1051
centripetal acceleration, 894, 896–897
centroid, 432, 545–547, 552, 553, 558, 1032, 1033–1034
chain rule, 201–204, 205, 230, 358, 359, 391, 395, 408–409, 863, 868, 955–956, 960
change of variables, 408
  Jacobian, 1069–1077
checking
  accuracy of an approximation, 483, 486–487
  binormal vector, 878
  center of mass, 1035
  continuity, 192
  cross products, 821
  definite integrals, 411
  differentiability, 192
  equations, 12
  graphs of polar curves, 745
  indefinite integral, 404
  intersection points, 758
  solution to exact differential equation, 941–942
  volume of pyramid, 1050
  volume of solid of revolution, 502
circular functions, 227
circular reference, *see* reference, circular
circulation, 1123
circumference of a circle, 535–536
Clairaut's theorem, 936

closed formula, 480, 580–581, 585
closed set, 919, 921, 925
codomain, 2
coefficient, 35, 36, 200
combining sums, 322
commutative, 20
comparison test, 471
 for indefinite integrals, 475
 series, 626–627, 629
complement, of set, 921
completing the square, 435, 462
components, 792, 829–830
compositions, 19–20, 24, 54, 57, 201, 911
 limit of, 123
compound interest, 604
compression, 22
concave down, 264
concave up, 6, 264
concavity, 6, 264, 268–269, 483
 parametric curve, 718
 and second derivative, 266
concho-spiral, 874
conclusion, 64
concurrent, 845, 1041
conditional convergence, 645, 650
cone, 293
 elliptic, 783, 784
 surface area of, 293
 volume of, 293
conic sections
 defined, 758
 degenerate, 758
 eccentricity of, 764
 focus-directrix definition of, 761
 nondegenerate, 783
 polar equations of, 763
cosh x, 227
$\cosh^{-1} x$, 229, 357
conic section, 758–770
 defined, 758
 in polar coordinates, 763–766
conjugate, 137, 178
conservative vector field, 1089–1091, 1092
constant function, 15, 36, 188
constant multiple of functions, 19
constant multiple rule
 for definite integrals, 345, 347
 for derivatives, 189
 for indefinite integrals, 357, 359
 for limits, 123, 923
 for sums, 319
 for series, 608
 for sequences, 596
constant term, 36
constraint, 281, 977, 979–980
continuous, 100, 109, 110, 111, 117, 123, 124, 131, 929
 of area function, 389
 everywhere, 925
 function of two or three variables, 925
 functions, 123
 growth rate, 571
 on an interval, 110
 one-sided, 110
 at a point, 109
 on a set, 925
 vector function, 855, 856
contrapositive 5, 65
convergence, 89, 312, 328, 468, 469, 470–471
 absolute, 645, 646
 conditional, 645
 limit, 100
 Maclaurin series, 684–686
 power series, 681–691
 sequences, 594–600
 series, 607–610
 tests, 617
converse, 64
converting between coordinate systems, 725–726, 1060, 1081
coordinate planes, 779
coordinate systems
 conversion, 725–726, 1060, 1081
coplanar, 820
cosecant function, 51, 446
cosine function, 51
cost function, 282
cotangent function, 51
counterexample, 65–66
critical point, 200, 212, 240, 245, 253, 266
 and local extrema, 241, 244, 967
cross product, 815
 area, 821
cross-sectional area, 501–502
cube
 sphere, 786–787
cubic polynomial, 36
curl, 1123, 1127–1128
 of gradient, 1124
 of vector field, 1123
curvature, 882, 884, 885–886
 radius of, 886
curve sketching, 270–273
 first derivative and, 256
 strategies, 267–268
cycloid, 716
 arc length, 719–720
 tangent lines, 717–718
cylinder, 293, 782
 graphing, 786
 right, 780
 surface area of, 293
 volume of, 293, 500, 505
cylindrical coordinates, 1058–1059

decay, exponential, 48
decreasing
 function, 7, 250–251, 254–255, 481
 sequence, 582
definite integral, 329, 343, 346, 349, 350–351, 354, 364, 366, 367, 375, 376, 378, 379, 380–382, 402, 414–415, 426, 544–545
 for arc length, 528–529
 calculating, 411, 423
 of cross-sectional area, 501–502
 formulas, 346–348, 349, 485
 properties of, 344–346
 vector function, 867
 for volume by shells, 516
degenerate
 conic section, 758
 quadric surface, 783
degree, 11
del operator, 1122
delta (δ), 90
delta-epsilon
 and limit, 91
 proofs, 102, 104
Δx, 343
density
 mass and, 542, 548–550
 population, 1036–1037
dependent variable, 2
derivative, 156
 of area function, 392, 393
 chain rule, 201
 of constant, 187
 constant multiple rule, 189
 defined, 79
 directional, 946–948, 957–959
 of exponential function, 212–214
 exponential rule, 213
 as function, 170
 of hyperbolic functions, 227–228
 of identity function, 187
 of inverse functions, 216–217
 of inverse trigonometric function, 225–226
 limits for, 137
 of linear function, 187
 at local extrema, 240–241
 logarithmic differentiation, 219–220
 of logarithmic function, 214–216
 partial, 933–935
 of piecewise function, 179
 at point, 169
 of power function, 188
 power rule, 188
 product rule, 191
 quotient rule, 191
 of rational function, 191
 reciprocal rule, 200
 second, 175, 263, 264, 265, 266–267
 second-order partial, 936
 sum rule, 189
 of trigonometric function, 224–225
 units, 159
 vector function, 862–863
determinant, 814–815
differ by a constant, 354
difference of two squares, 12
difference quotient, 159
difference rule
 for derivatives, 189
 for limits, 123
differentiability, 170–172, 948–950
 of area function, 390
 of exponential functions, 213
 on an interval, 171
 one-sided, 171
 at a point, 170

and partial derivatives, 948, 950
vector function, 862
differential, defined, 174
differential equations, 223, 402, 443, 559, 575
exact, 938, 941–942
separable, 560–561, 565–566
slope field, 561–562, 566
differentiating, 194, 217–218, 230
compositions of functions, 201–202
defined, 159
piecewise-defined function, 194–195, 218–219
power functions, 187
rules, 192, 215, 359
dihedral angle, 845
direct proof, 66, 68
direction angles, 812
direction cosines, 812
direction vector, 827
directional derivative, 946–948
gradient and, 957–959
directrix, 765
cylinder, 782
ellipse, 764, 770
hyperbola, 764, 770
parabola, 761, 763, 767
discontinuities, 117–118
defined, 110
improper integrals and, 469
infinite, 110, 111
jump, 110, 111
at a point, 110
removable, 110, 111
discontinuity
at a point, 110
discontinuous, 111
discriminant, 37, 969
disk, 502–503, 504–506, 514
defined, 500
open, 919
displacement, 797
vs. distance travelled, 893
distance, 364, 368, 543, 765, 766, 780–781
approximations, 318
defined, 503–504
formula, 159, 532, 781
origin to plane, 839
parallel planes, 839–840, 842
point to line, 810, 829, 832
point to plane, 838–839, 842, 973
skew lines, 839, 840, 843
travelled, displacement vs., 893
divergence
of curl, 1123, 1124
of harmonic series, 620–621
of improper integral, 468, 469, 470–471, 475–476
of p-series, 620–621
of sequence, 89, 312, 594–596, 598, 611
of series, 89, 607–608, 609
test, 617–618, 623
Theorem, 1143–1144, 1145–1148
of vector field, 1123
divisible, 63
DNA, 861
does not exist, 80, 81
domain, 2, 3, 8, 11, 25, 40, 53, 56
function of three variables, 907
function of two variables, 904
dominance, 311, 628
sequences, 598–599
doomsday model, 89, 150
dot product, 803–806
orthogonality, 806, 809
double helix, 861
double integral, 995
algebraic properties of, 1009–1010
applications of, 1029–1037
circle, 1022–1023
evaluation, 998–999
Fubinis theorem and, 997–998
general region, 1008–1009
iterated integrals and, 996–997
and moments of inertia, 1031–1032
polar coordinates, 1017–1021, 1023–1024
rectangular regions, 993–996
double-angle identities, 52, 446, 448
dS, 1112
du in terms of dx, 417
dummy variable, 389
dx, 343

e, 126–127
eccentricity, 764, 765
e^{kx}, *see* exponential function
electric field intensity, 1148
elementary antiderivative, 358
ellipse, 759–760
directrix, 765
eccentricity, 764
focus, 759, 760, 764
ellipsoid, 773, 783, 784
elliptic cone, 783, 784
elliptic paraboloids, 785
Elvis, 99, 290
empty set, 921
epicycloid, 722
epsilon (ϵ), 91
epsilon-delta definition of limit, 91
equal vectors, 795
equation, 559
error, 481, 482, 484
from Simpson's Rule , 486
escape velocity, 575
estimating instantaneous velocity, 162–163
estimating the slope of a tangent line, 161–162
Euler's formula, 702
Euler's method, 562–563, 567
evaluation notation, 365, 366
even
function, 22, 27, 52, 735
integer, 63
symmetry, 22
exact area, 349
exact differential equation, 941–942
exists, 80, 91
expanded sum, 321
exponent, 35
exponential
decay, 48
function, 47–48, 127–128, 212–214, 218–219, 356
growth, 48, 563
rule, 212, 356
extrema, 6
absolute, 966
global, 278, 280–281, 966
local, 240–241, 245, 252, 267, 966
relative, 966
Extreme Value Theorem, 113, 116, 242, 979

factorial, 151, 580
factorial-like, 634
sequence, 587–588
factoring, 11
family of antiderivatives, 354
Fermat's theorem for local extrema, 241
Fibonacci sequence, 580
first moment, 1031, 1047
first-derivative test, 252–254, 266, 269–270
flaming tent, 284
floor function, 388
flux, 1108, 1117
of a vector field across a surface, 1112–1113
focus
ellipse, 759, 760, 764
hyperbola, 762
parabola, 761
foot-pounds, 543
for all, 63
force, 543
in cables, 799–800
fractals, 616
Frenet frame, 876
frustums, 529, 536–537
Fubini's Theorem, 996–998, 1043–1044
function, 2-3, 8-9
absolute value, 39–40
algebraic, 35–43
area accumulation, 388–390
behavior, 10–11
composition, 19, 201–202, 911
concavity, 6
constant, 15
decreasing, 15
defined by integral, 390
differentiable, 170, 171, 172
domain, 3, 904, 907
even, 22, 27, 52, 735
exotic, 115
exponential, 47–48, 127–128, 212–214, 218–219, 356
graph of, 4
hyperbolic, 227–228
identity, 15
implicit, 203
increasing, 6

function (*Cont.*)
inverse, 23–24
inverse hyperbolic, 229
inverse trigonometric, 53, 128, 225–226, 231
logarithmic, 49, 127, 214–216, 217–218
multivariable, 14
natural logarithm, 49
negative, 15
odd, 22
one-to-one, 5
periodic, 224
piecewise-defined, 13, 114–115, 131, 179, 194–195, 218
polynomial, 36–37
positive, 6
power, 35–36
quadratic, 433, 435, 485
range, 3, 904, 907
rational, 37–39
with same derivative, 251–252
scalar, 853
three variable, 906–907
transcendental, 47
trigonometric, 50–52
two-variable, 904–906
vector, 853
Fundamental Theorem of Algebra, 37
Fundamental Theorem of Calculus, 364–366, 367, 368–370
Fundamental Theorem of Line Integrals, 1100–1101, 1103

Gabriel's Horn, 480
gas
ideal law, 932
constant, 932
Gauss' Law, 1148
geometric mean, 987
geometric sequences, 581, 598, 602, 609
geometric series, 609–610
repeating decimals, 612–613
global behavior of a polynomial, 147
global extrema, 6, 114, 282
global maximum, 280
global minimum, 15
gradient, 961–962
defined, 957
and directional derivative, 957
and level curves, 959, 962
graph
defined, 4
function of three variables, 907
function of two variables, 905–906
of inverse function, 24, 29
level curves, 908
level surface, 909
polar curves, 733–735
of polynomial functions, 36, 37, 40–41
of position, velocity and, 160–161
of power functions, 35, 36
properties of, 5-7
of rational functions, 38, 41–42
space curves, 853, 857
window, 9-10
graphing calculator, 348, 534
gravity
Earth, 899
force, and motion, 891–892
Jupiter, 899
Mars, 899
moon, 899
greatest lower bound, 584
Green's Theorem, 1125–1126
curl form, 1123, 1126–1127
divergence form, 1123, 1126–1127
Fundamental Theorem of Calculus and, 1122
Green's Theorem, curl form, 1140
Green's Theorem, divergence form, 1140
growth, 48
guess-and-check method for integrals, 360
gyration, radius of, 1033, 1036, 1047, 1051

half-life, 571
harmonic series, 620
alternating, 642
headlights, 156
helix, 723, 854
conical, 857
graphing, 856
hemisphere, 908
Heron's formula, 824, 986
Hessian, 969
higher order partial derivative, 935–936
hole, 38
horizontal
compression, 22
reflection, 22
stretch, 22
translation, 22
horizontal asymptote, 39, 82, 144
Horizontal Asymptote Theorem, 141
horizontal line test, 5
Human Cannonball, 894–896
hydrostatic force, 543–544, 550–552
hyperbola, 762–763
asymptotes, 762
directrix, 764
eccentricity, 764
focus, 762, 764
hyperbolic cosine, 227
Maclaurin series, 686–687
hyperbolic derivatives, 228
hyperbolic functions, 227–228, 357
derivatives of, 228
inverse, 229, 357
hyperbolic integrals, 450
hyperbolic sine, 229
hyperbolic tangent, 227
hyperboloid, 773, 783, 784–785
hyperplane, 916, 949
hypersphere, 791, 936
hypocycloid, 775
hypothesis, 64, 248

ideal gas law, 932
identities, 51, 52, 444–445, 447–448
identity function, 15, 187
if and only if, 65
if...then, 64
i, *j*, and *k* components, 822, 1099, 1123
implication, 64–65, 67–68
implicit differentiation, 203–204, 205–208
implicit function, 203, 204, 206–207
improper integrals, 467–468, 474–475
comparison Test for, 471
defined, 467
as limits of definite integrals, 472–473
over discontinuities, 469
over unbounded intervals, 468
of power functions, 470, 473–474
improper rational functions, 432, 433, 436
increasing function, 6, 7, 250–251, 254–255
sequence, 582
indefinite integral, 354–355
constant multiple rules for, 357–358, 359
sum rules for, 357–358, 359
independent of path, 1101
independent variable, 2
indeterminate form, 125, 139, 141, 146, 153, 306
L'Hôpital's rule for, 303–304
logarithms for, 304–305
index, 318, 579
changing, 585, 613
induction, 321
inertia, moments of, 1032–1033, 1035–1036, 1047, 1051–1052
infinite discontinuity, 110
infinite limit, 80, 94, 138–139
at infinity, 80, 94, 139–141
inflating balloon, 296–297
inflection point, 6, 265–266, 268–269
initial edge, 50
initial-value problem applications of, 563–565
defined, 402
differential equations and, 559–560
instantaneous rate of change, 159, 169
instantaneous velocity, 79, 162–163
integer, 63
integrable function, 344, 995, 1042
integral, 355
definite, *see* definite integral
double, *see* double integral
of hyperbolic functions, 357
indefinite, *see* indefinite integral
improper, *see* improper integrals
iterated, 996–997, 1011–1013
iterated triple, 1043
as limit of Riemann sums, 344
numerical approximation of, 480–481
with respect to arc length, 1098
test, 618–619
triple, *see* triple integral
integrand, 355
integration
changing the variable, 1069, 1072
formula, 355–357, 404
limits of, 343, 346

with spherical coordinates, 1061–1062
vector function, 867
and volume, 1001
integration by parts
for definite integrals, 423
formula, 420, 421
implementation strategies, 420–422
definite integral, 429
definite integrals, 426
formula, 424
more than once, 428
integration by substitution, 409, 412
and back-substitution, 414
for definite integrals, 411, 414–415
formula for, 408
selection, 413–414
Intermediate Value Theorem, 113, 114, 116, 117
intersecting lines, 758
and plane, 779
interval of convergence, 661–662, 664–667
inverse function, 23–24, 29
derivatives of, 216–217
hyperbolic, 229, 231, 357
secant, 53, 226
sine, 53, 226, 423
tangent, 53, 423
trigonometric, 53–54, 225–226, 356
involute, 722
irrational numbers, 115–116
irreducible, 37
irrotational, 1142
iterated integral, 996–997, 1011–1013
triple, 1043

Jacobian, 1071, 1073
jello, 549
joules, 543
jump discontinuity, 110

Kepler, Johannes, 772
first law of planetarymotion, 902

L'Hôpital's rule, 302–304, 305–308
Lagrange multiplier, 977
Lagrange's form for remainder, 683
Lagrange's Identity, 816
lamina, 1030, 1031, 1115–1116
Laplace's equation, 989
Laplacian, 1155
lateral surface area, 293
Law of Cosines, 804, 805
law of similar triangles, 294, 297
leading coefficient, 36
leading term, 36
of polynomial, 147
least squares, 989
least upper bound, 584, 590
Least Upper Bound Axiom, 113, 585
left continuous, 109
left derivative, 171
left differentiability, 171
left-handed systems, 779
left limit, 92
left sum, 332, 333, 336–337, 482
Leibniz notation, 173–175, 201, 202
lemniscate, 744–745
length
arc, *see* arc length
of curve, 531–532
parametric curve, 719–720
of vector, 794
level curves, 908, 912–914
and gradient, 959, 962
level surfaces, 909
limaçon, 738, 741–742
area, 752–753
limit, 78, 80, 84–85
of algebraic functions, 124
of average rate of change, 78–79
calculating, 108, 116, 123, 130–131, 132, 143–145, 308–309
cancellation theorem for, 125
and continuity, 109–110, 114–115, 131
of combinations of functions, 123–124
converges to, 100, 607
definition of
algebraic, 100
formal, 90–91
for derivatives, 137
evaluation, 928
examples of, 78–79
of functions, 79–81, 923, 924, 927
identifying, 82–83
indeterminate, 141–142, 146, 302, 303
infinite, 80, 94, 104–105, 138–139, 148
at infinity, 80, 94, 95–96, 139–140, 148
of integration, 343, 346, 415
non-indeterminate, 142
notation, 85
one-sided, 80, 92–93, 104–105
of partial sums, 606, 608
along paths, 928
of Riemann sums, 344, 345
rules, 129–130, 596
of sequences, 78, 100, 595, 600–601
squeeze theorem for, 125, 132–133
of sum, 323–324, 528
with tables of values, 82, 133–134
trigonometric, 142–143
two-sided, 81, 93
uniqueness of, 91–92
of vector function, 855, 858
limit comparison test, 478, 627–630
line integral
along closed curves, 1103–1104
Fundamental Theorem of, 1100–1101, 1103, 1129
of multivariate function, 1097–1098
of vector field, 1099–1100, 1102
line of intersection, 840
line segments, 526
linear approximation, 173, 181
linear equation, 827
linear function, 36, 187, 918
linear moments, 1031
linear transformation, 1080
ln x, *see* natural logarithm function
local extrema, 6, 240, 244
and critical points, 968
gradient at, 966–968
second-derivative test, 968–971
local linearity, 173, 181–182
local maximum, 6, 240
local minimum, 240
logarithmic differentiation, 216, 219–220
logarithmic functions, 49–50
algebraic rules for, 49
continuity of exponential and, 127
derivatives of, 214–216
graphs of, 50
natural, 49, 392
logarithmic growth, 50
logarithms
approximation, with Riemann sums, 397
to calculate limit, 308–309
for indeterminate forms, 304–305
$\log_b x$, *see* logarithmic function
logistic growth model, 442, 564
logistic model, 564
lower sum, 333

Maclaurin polynomials, 672–673, 675
Maclaurin series, 674, 675–676, 684
approximation, 690
convergence, 687
evaluation, 687
hyperbolic cosine, 686–687
integration, 696–697
manipulation, 684–686, 688
multiplication, 689–690
for sine function, 674, 683
Taylor series derivation from, 688–689
magnitude, 794
major axis, 759
mass, 150, 548–550, 1034–1035
center of, 545, 1030
defined, 542
first moment of, 1031, 1047
of lamina, 1030
of planar region, 1029–1030
of rectangular parallelepiped, 1049
of slice of cylinder, 1052
parallelepiped, 1063
mass of a lamina, 1048
matrix, 814
maximum,
absolute, 966
global, 6, 278, 966
local, 6, 240, 253, 267, 278, 966
relative, 966
mean
arithmetic, 987
geometric, 987
mean value, see average value
Mean Value Theorem, 242–243, 246
Cauchy, 304
for Integrals, 379–380, 382–383
median, of triangle, 1041
method of least squares, 989

midpoint, 775
  sum, 332, 333, 337, 483, 484, 487–488
  trapezoid, 484
min, 105
minimum,
  absolute, 966
  global, 6, 278, 966
  local, 6, 240, 253, 267, 278, 966
  relative, 966
minor axis, 759
mixed partial derivative, 936
modelling, with polynomial functions, 40–41
moment
  first, 1030–1032, 1047
  linear, 1031
  polar, 1032
  second, 1033
moment of inertia, 1032–1033, 1035–1036, 1047, 1051–1052
monic, 441
monotonic, 481, 587, 601–602
  function, 482
  sequence, 582, 599–600
  tests, 583
moving frame, 876
multiplicity, 38
multipliers, Lagrange, 977
multivariable function, 14
Möbius band, 1110

natural exponential function, 48
natural growth rate, 564, 571
natural logarithm function, 49, 391–392, 422
negative angle, 50
negative function, 15
net area, 332
net change, 366, 371
Net Change Theorem, 366–367
Newton's Law of Cooling and Heating, 564, 567–568
Newton's method, 181–182, 589–590
newton-meters, 543
newtons, 543
non-indeterminate form, 142
nonlinear equations, 978
nonorientable surface, 1110
norm, 794
normal component of acceleration, 894
normal distribution, 353
normal plane, 880
normal vector, 836
n^{th} derivative, 175

octahedron, 790
octants, 780
odd
  function, 22, 27–28, 735
  integer, 63
  symmetry, 22
one-sided continuity, 110
one-sided limits, 80, 92–93
open ball, 919
open disk, 919
open set, 920, 924
operator notation, 174
optimization, 278–286, 981
orientable surface, 1139
orthogonal curves, 806
orthogonal vectors, 806
osculating circle, 886, 887
osculating plane, 875, 877–878

p-series, 470, 619–620
Pappus' Centroid Theorem, 558
parabola, 773, 774
  directrix, 774, 781
  focus, 774, 781
paraboloid, 760–761
  elliptic, 761
  hyperbolic, 761
parallel lines, 828
parallel vectors, 796, 816
parallelepiped, 815
  volume, 819, 820, 822–823, 1048–1049
parallelogram, area of, 817
parameter, 708, 852
  elimination, 710
parameterization, 227
parametric curve, 708, 711, 712, 718–720, 852
parametric equations, 707, 709–710, 716–717, 830, 852
parametrization, 712
  arc length, 882–883
parametrized surface, 1109–1110
parity, 445
partial derivative, 936–938, 939
  and differentiability, 948, 949, 950–951
  of functions of two and three variables, 933–935
  graphical interpretation, 934
  higher order, 935–936, 940–941
  mixed, 936
partial fractions, 433–434
partial sums, 328
  sequence of, 606
parts, *see* integration by parts
path independent, 1101
periodic function, 224
permittivity, 1148
piecewise, 42
piecewise function, 13, 39, 42, 114–115, 131, 179, 194–195, 218
piecewise smooth, 1098
plane
  coordinate, 780
  general form, 837
  point and line, 841–842
  three points, 841
planetary orbits, 902
polar axis, 724
polar coordinates
  circles, 726
  equivalent, 725
  plotting, 724–725, 727
  representations, 728
polar curve
  arc length, 751
polar equations
  symmetry, 739
polar moment, 1032
polar plane, 733
pole, 724
polynomial function, 36–37, 147
polynomial long division, 433, 435, 436
polynomials
  Maclaurin, 673
  Taylor, 673
population, 218–219
  density, 1036–1037
  growth, 572
position function, 157, 159
position vectors, 792, 793
positive angle, 50
positive function, 6
potential function, 1089, 1090, 1093–1094
power function, 35, 40, 111, 355, 358, 470, 473–474
power rule, 188–189
  for antiderivatives, 355
  integer powers, 189
  rational powers, 207–208
power series, 660–661, 662–664, 667–668
  differentiation, 694–695
  integration, 695–696
  manipulation, 684–686
principal unit normal vector, 875
probability density function, 1040
probability distribution function, 1033
product notation, 640
product, of functions, 19
product rule, 359
  for derivatives, 191
    reversing, 420
  for limits, 123
    reversing, 420
projection
  component, 807
  vector, 806
proof, 66, 68
  by contradiction, 67, 69
  direct, 66, 68
proper rational function, 432, 434
properties of integrals, 346
proportional, 214
punctured interval, 90
Pythagorean identities, 52, 444, 447, 449–450
Pythagorean Theorem, 294

QED, 64
■, *see* QED
quadratic, 485
  formula, 12
  function, 433, 435, 485
  polynomial, 36
quadric surface, 783
quantifiers, 63, 64
quartic polynomial, 36
quintic polynomial, 36
quotient of functions, 19

quotient rule,
for derivatives, 191
for limits, 123
reversing, 358, 359

radian, 50
radius of convergence, 662, 664–665, 666–667
radius of curvature, 886, 887
radius of gyration, 1033, 1036, 1047, 1051–1052
range, 3, 8
function of three variables, 907
function of two variables, 904
rate of change, 7, 214
ratio test, 583
absolute convergence, 647–648
series, 633, 635, 636
rational function, 37–39, 41
Horizontal Asymptote Theorem for, 141
improper, 432–433, 436
proper, 432–433
rational number, 37, 63
reciprocal rule, for derivatives, 200
rectifying plane, 880
recursion formula, 580, 586
recursive sequences, 580
recursively defined sequences, 580–581
reducing powers, 446, 448
reduction formula, 453
reference angle, 54
reference, circular, *see* circular reference
reflection, 22, 25
region between two curves, 547
regression line, 989
related rates, 291–298
remainder,
alternating series, 644
of convergent series, 621, 622
of function, 681
Lagrange's form, 683
polynomial, 435
removable discontinuity, 111
repeated root, 41
repeating decimals as geometric series, 612–613
representative
disk, 505
slice, 506, 548, 549
restricted domain, 24, 28
reversing differentiation rules, 358
rhombohedron, 825
Riemann sum, 330–331, 553–554, 750, 994, 999
error bounds, 483–484
for functions of two variables, 1042
for functions of two variables, 995
left sum, 332, 333, 336–337
limit of, 995, 997
lower sum, 333, 334
midpoint sum, 332, 333, 336–337, 483–484
notation, 335
right sum, 332, 333, 335, 336
trapezoid sum, 334, 336–337, 483–484
upper sum, 333, 334
for volumes by shells, 515–516
right continuous, 109
right cylinder, 782
right derivative, 171
right differentiability, 171
right limit, 80, 92, 93
right sum, 332, 333, 335, 336, 344, 481, 482, 486–487
right-handed system, 778, 779
right-handed triple, 818
ripples in a pond, 295
Rolle's Theorem, 241–242, 243, 244–246
root, 6
root test, 634–636, 637–638
modifying, 651
series, 634
rose, 742–744, 1021
area, 751–752
rotational symmetry, 22
$\mathbb{R}^3$, 781
$\mathbb{R}^2$, 781
ruling, 782

saddle point, 968
same derivative, 251–252, 354
scalar
defined, 792
function, 853
multiple, 792
$\sec^{-1} x$, *see* inverse secant function
secant
function, 51
inverse function, 53
line, 158, 161–162
second derivative, 175, 264, 268–269
test, 266–267, 269–270
Second Fundamental Theorem of Calculus, 390–391
proof of, 392–393
second moment, 1047
second-order partial derivatives, 936
section, 783
sector, area of, 750
semimajor axis, 772
semiperimeter, 824, 986
separable differential equation, 443, 560–561
separation of variables, 561, 565
sequence, 78, 161–162, 585, 586
analyzing, 587–588
arithmetic, 581
of average velocities, 162–163
bounded, 583–585, 587, 599–600, 601–602
convergence of, 89, 312, 594–599
decreasing, 582
divergence of, 89, 312, 594–599
dominance relationships for, 598
factorial, 580
Fibonacci, 580
geometric, 581, 598, 602
increasing, 582
indexing of, 585
limit, 78, 100
monotonic, 582–583, 587, 599–600, 601–602
notation, 581
of numbers, 579–580
of partial sums, 606, 607
patterns, 86
recursive, 580–581
subsequence, 597
tail of, 595
unbounded, 584
series, 89, 606
analyzing, 610–611, 636–637
algebra of, 608
alternating, 641–645
alternating harmonic, 642
approximating, 620–621
binomial, 677–678
comparison test, 626–627, 629
convergence, 89, 607–608, 617
convergent, 620–621
divergence, 89, 607–608, 617–618
evaluations for, 18
geometric, 609–610, 612–613
harmonic, 619–620
index of, 613
*k*th term of, 606
limit comparison test, 627–630
Maclaurin, 674
power, 493
of real numbers, 606
tail, 610
Taylor, 151, 674
telescoping, 610, 611
set
boundary, 922
bounded, 922
closed, 921
complement, 921
open, 920
set notation, 3
shadow from a streetlight, 297
shells, 514–520
Sierpinski triangle, 616
$\sum$, *see* sigma notation
sigma notation, 318–320, 321–322
sign changes, 114, 117
sign chart, 11, 256, 271
signed area, 332, 344, 350, 364, 369–370, 376, 380–381
signed volume, 995
similar triangles, 293–294
simplifying before differentiating, 194
simply connected, 1087
Simpson's Rule, 484–486, 488–489
$\sin^{-1} x$, *see* inverse sine function
sine function, 51
$\sinh x$, 228
$\sinh^{-1} x$, 229, 357
sink, 1092
skew lines, 828
slant length, 529

slice, 506
slope, 156
  associated function, 170
  field, 561, 562, 566
  function, 156, 159–160
  of parametric curve, 711
  of tangent line, 158–159, 161–162
slope-intercept form, 173
smooth, 156, 1097
  piecewise, 1098
  curve, 1097
  parametrization, 1097, 1110
  surface, 1110
solid of revolution, 466, 502
solution, 559, 566
solving equations, 14
source, 1089
space curve, 852, 853–854, 865, 881–882, 887–888
speed, 865
sphere, 293, 781
  cube, 786–787
  surface area of, 536–537
spherical coordinates, 1059–1062
spiral, 856–857
  of Archimedes, 737
splitting a sum, 320
spring, 234
Squeeze Theorem, 125–126, 132–133, 393
standard basis vector, 797, 821–822
standard normal distribution, 353
standard position, 50
stationary point, 967, 969, 971–972
Stokes' Theorem, 1135–1139, 1145
strange gravity, 199
stretch, 22
strophoid, 756, 1027
subsequence, 597
substitution, 408, 412–413
  definite integrals and, 411
  selection of, 409–411, 413–414
sum
  formulas, 320, 322–324, 347–348
  of functions, 19
  identities, 52
  Riemann, *see* Riemann sum
  of series, 607
  telescoping, 367
sum rule, 923
  for definite integrals, 345, 349
  for derivatives, 189
  for differentiation, 359
  for double integrals, 995
  for indefinite integrals, 357, 359
  for limits, 123
  for sequences, 596
  for series, 607
  for sums, 319
  for triple integrals, 1042
summation
  double, 993
  triple, 993
supergrowth, 572
surface area, 466, 1110–1112
  approximation, 529–530
  definite integral for, 530–531, 537
  formulas, 293
  of sphere, 536–537
surface independent, 1156
surface integral, 1114–1115
  of multivariate function, 1110–1112
surface of revolution, 529, 530, 912
symmetric difference quotient, 165
symmetric form, 828, 829–830
symmetry, 745
  even, 22
  odd, 22
  in polar graphs, 735–739
  rotational, 22
  x-axis, 726
  y-axis, 22
  z-axis, 1062, 1064
synthetic division, 206

tail of a sequence, 595
$\tan^{-1} x$, *see* inverse tangent function
tangent function, 51
  inverse, 53
tangent line, 18, 156, 158–159, 161–162, 171, 172–173, 177, 206–207, 864
  vertical, 711
tangent plane, 949, 951, 967
tangent vector, 870
  unit, 874
tangential component of acceleration, 894
$\tanh x$, 227
$\tanh^{-1} x$, 229, 357
target, 2
Taylor polynomials, 673
Taylor series, 151, 674, 676–677
  remainder, 681
Taylor's theorem, 682
telescoping series, 610
telescoping sum, 367
term-by-term differentiation and integration, 694–696
terminal edge, 50
tests for convergence and divergence of series, 617–618
tetrahedron, 790
there exists, 63
thickness of shell, 515
3-space, 779
TNB frame, 876
torque, 825
torsion, 890
torus, 558
trace, 783
transcendental function, 47
transformation, 20, 22, 26–27, 1069
  Jacobian for, 1071
  linear, 1080
translations, 20, 25, 279
trapezoid sum, 334, 336–337, 484, 488
triangle inequality, 8, 72
trigonometric
  antiderivatives, 356
  derivatives, 224
  fucntions, 50, 51, 57
  identities, 52, 445, 447–448
  limits, 142
trigonometric substitution, 454–463
triple integral, 1041–1053
triple scalar product, 820, 1072
trochoid, 722
twisted cubic, 862
2-space, 779
type I region, 1007
type II region, 1007

unbounded sets, 922
uniqueness of limits, 91–92
unit circle, 50
  parametrization, 713–715
unit normal vector, 875, 876–877
unit tangent vector, 874, 876–877
unit vector, 796–797, 799
units of the derivative, 159
universal gas constant, 932
upper bound, least, 113, 584
upper sum, 333, 334
using algebra first, 358
u-substitution, 408

variable, 2
vector, 792
  adding, 798–799
  angle between, 808
  components, 792, 806
  curve, 853
  initial point, 792
  Laplacian, 1155
  normal, 836
  position, 792
  projection, 806, 809
  subtracting, 798–799
  sum, 792
  terminal point, 792
  unit, 796–797
vector curl, 1123
vector field, 1087–1089, 1091–1092
  conservative, 1089–1091, 1093
vector function, 827
  continuity of, 855, 856
  derivative, 862–863
  integral, 867
  limit, 855, 858
  with three components, 853
vector triple product, 819, 826
vector-valued function, 852–853
velocity, 157, 160–161, 865
  average, 159
  curve, 318
  definite integral of, 368
  function, 157
  instantaneous, 79, 162–163
  vector, 869–870
vertex, 758
  conic section, 758
vertical
  asymptote, 39, 144
  compression, 22
  line test, 4

reflection, 22
stretch, 22
tangent line, 171
translation, 22
volume, 296–297, 507–508, 1062–1064
absolute, 995
approximation, 500–501, 504–505
of cone, 293, 506–507
of cylinder, 293, 505
as definite integral, 501–502, 503
using polar coordinates, 1024–1025
pyramid, 1049–1050
of shell, 514–520
signed, 995
solid of revolution, 502
of sphere, 293, 500
and integration, 1002

washers, 502, 503, 506
water flow, 1130
weight, 542
weight of water, 543
weight-density, 543
wombat, 218–219
work, 513, 543, 544, 547–548, 1099
line integral, 1105

x-axis, 375–376
x-coordinate, 27, 1035, 1062
xy-plane, 779
xz-plane, 779

y-axis, 506-507, 519–520
symmetry, 22
y-intercept, 6
yz-plane, 779

z-axis, 1047, 1062, 1064
zero vector, 795

Definition of the Limit

$\lim_{x \to c} f(x) = L$ means that:

For all $\epsilon > 0$, there exists a $\delta > 0$ such that if $x \in (c - \delta, c) \cup (c, c + \delta)$, then $f(x) \in (L - \epsilon, L + \epsilon)$.

Equivalently, in terms of absolute value inequalities, $\lim_{x \to c} f(x) = L$ means that:

For all $\epsilon > 0$, there exists a $\delta > 0$ such that if $0 < |x - c| < \delta$, then $|f(x) - L| < \epsilon$.

The Extreme Value Theorem

If f is continuous on a closed interval $[a, b]$, then there exist values M and m in the interval $[a, b]$ such that $f(M)$ is the maximum value of $f(x)$ on $[a, b]$ and $f(m)$ is the minimum value of $f(x)$ on $[a, b]$.

The Intermediate Value Theorem

If f is continuous on a closed interval $[a, b]$, then for any K strictly between $f(a)$ and $f(b)$, there exists at least one $c \in (a, b)$ such that $f(c) = K$.

Definition of the Derivative

The derivative of a function $f(x)$ is defined to be the function:

$$f'(x) = \lim_{h \to 0} \frac{f(x+h) - f(x)}{h}$$

or, equivalently:

$$f'(x) = \lim_{z \to x} \frac{f(z) - f(x)}{z - x}.$$

Derivative Rules

$$\frac{d}{dx}(kf(x)) = kf'(x)$$

$$\frac{d}{dx}(f(x) + g(x)) = f'(x) + g'(x)$$

$$\frac{d}{dx}(f(x)g(x)) = f'(x)g(x) + f(x)g'(x)$$

$$\frac{d}{dx}\left(\frac{f(x)}{g(x)}\right) = \frac{f'(x)g(x) - f(x)g'(x)}{(g(x))^2}$$

$$\frac{d}{dx}(f(g(x))) = f'(g(x))g'(x)$$

Rolle's Theorem

If f is continuous on $[a, b]$ and differentiable on (a, b), and if $f(a) = f(b) = 0$, then there exists at least one value $c \in (a, b)$ for which $f'(c) = 0$.

The Mean Value Theorem

If f is continuous on $[a, b]$ and differentiable on (a, b), then there exists at least one value $c \in (a, b)$ such that

$$f'(c) = \frac{f(b) - f(a)}{b - a}.$$

Derivatives of Basic Functions

$$\frac{d}{dx}(x^k) = kx^{k-1}$$

$$\frac{d}{dx}(e^{kx}) = ke^{kx}$$

If $b > 0$ and $b \neq 1$, then $\frac{d}{dx}(b^x) = (\ln b)b^x$

$$\frac{d}{dx}(\ln |x|) = \frac{1}{x}$$

If $b > 0$ and $b \neq 1$, then $\frac{d}{dx}(\log_b |x|) = \frac{1}{(\ln b)x}$

$$\frac{d}{dx}(\sin x) = \cos x$$

$$\frac{d}{dx}(\cos x) = -\sin x$$

$$\frac{d}{dx}(\tan x) = \sec^2 x$$

$$\frac{d}{dx}(\cot x) = -\csc^2 x$$

$$\frac{d}{dx}(\sec x) = \sec x \tan x$$

$$\frac{d}{dx}(\csc x) = -\csc x \cot x$$

$$\frac{d}{dx}(\sin^{-1} x) = \frac{1}{\sqrt{1 - x^2}}$$

$$\frac{d}{dx}(\tan^{-1} x) = \frac{1}{1 + x^2}$$

$$\frac{d}{dx}(\sec^{-1} x) = \frac{1}{|x|\sqrt{x^2 - 1}}$$

$$\frac{d}{dx}(\sinh x) = \cosh x$$

$$\frac{d}{dx}\cosh x = \sinh x$$

$$\frac{d}{dx}\tanh x = \operatorname{sech}^2 x$$